ARX-3681
UCF

D. Demus, J. Goodby, G. W. Gray,
H.-W. Spiess, V. Vill

Handbook
of Liquid Crystals

Handbook of Liquid Crystals

D. Demus, J. Goodby,
G. W. Gray,
H.-W. Spiess, V. Vill

Vol. 1:
Fundamentals

Vol. 2 A:
Low Molecular Weight Liquid Crystals I

Vol. 2 B:
Low Molecular Weight Liquid Crystals II

Vol. 3:
High Molecular Weight Liquid Crystals

Further title of interest:

J. L. Serrano:
Metallomesogens

ISBN 3-527-29296-9

D. Demus, J. Goodby, G. W. Gray,
H.-W. Spiess, V. Vill

Handbook of Liquid Crystals

Vol. 1:
Fundamentals

 WILEY-VCH

Weinheim • New York • Chichester
Brisbane • Singapore • Toronto

Prof. Dietrich Demus
Veilchenweg 23
06118 Halle
Germany

Prof. John W. Goodby
School of Chemistry
University of Hull
Hull, HU6 7RX
U. K.

Prof. George W. Gray
Merck Ltd.
Liquid Crystals
Merck House
Poole BH15 1TD
U.K.

Prof. Hans-Wolfgang Spiess
Max-Planck-Institut für
Polymerforschung
Ackermannweg 10
55128 Mainz
Germany

Dr. Volkmar Vill
Institut für Organische Chemie
Universität Hamburg
Martin-Luther-King-Platz 6
20146 Hamburg
Germany

This book was carefully produced. Nevertheless, authors, editors and publisher do not warrant the information contained therein to be free of errors. Readers are advised to keep in mind that statements, data, illustrations, procedural details or other items may inadvertently be inaccurate.

Library of Congress Card No. applied for.

A catalogue record for this book is available from the British Library.

Deutsche Bibliothek Cataloguing-in-Publication Data:

Handbook of liquid crystals / D. Demus ... – Weinheim ; New York
; Chichester ; Brisbane ; Singapore ; Toronto : Wiley-VCH
   ISBN 3-527-29502-X

    Vol. 1. Fundamentals. – 1998
   ISBN 3-527-29270-5

© WILEY-VCH Verlag GmbH. D-69469 Weinheim (Federal Republic of Germany), 1998

Printed on acid-free and chlorine-free paper.

All rights reserved (including those of translation in other languages). No part of this book may be reproduced in any form – by photoprinting, microfilm, or any other means – nor transmitted or translated into machine language without written permission from the publishers. Registered names, trademarks, etc. used in this book, even when not specifically marked as such, are not to be considered unprotected by law.
Composition and Printing: Fa. Konrad Triltsch Druck- und Verlagsanstalt GmbH, D-97070 Würzburg.
Bookbinding: Wilhelm Osswald & Co., D-67433 Neustadt
Printed in the Federal Republic of Germany.

The Editors

D. Demus
studied chemistry at the Martin-Luther-University, Halle, Germany, where he was also awarded his Ph. D. In 1981 he became Professor, and in 1991 Deputy Vice-Chancellor of Halle University. From 1992–1994 he worked as a Special Technical Advisor for the Chisso Petrochemical Corporation in Japan. Throughout the period 1984–1991 he was a member of the International Planning and Steering Committee of the International Liquid Crystal Conferences, and was a non-executive director of the International Liquid Crystal Society. Since 1994 he is active as an Scientific Consultant in Halle. He has published over 310 scientific papers and 7 books and he holds 170 patients.

J. W. Goodby
studied for his Ph. D. in chemistry under the guidance of G. W. Gray at the University of Hull, UK. After his post-doctoral research he became supervisor of the Liquid Crystal Device Materials Research Group at AT&T Bell Laboratories. In 1988 he returned to the UK to become the Thorn-EMI/STC Reader in Industrial Chemistry and in 1990 he was appointed Professor of Organic Chemistry and Head of the Liquid Crystal Group at the University of Hull. In 1996 he was the first winner of the G. W. Gray Medal of the British Liquid Crystal Society.

G. W. Gray
studied chemistry at the University of Glasgow, UK, and received his Ph. D. from the University of London before moving to the University of Hull. His contributions have been recognised by many awards and distinctions, including the Leverhulme Gold Medal of the Royal Society (1987), Commander of the Most Excellent Order of the British Empire (1991), and Gold Medallist and Kyoto Prize Laureate in Advanced Technology (1995). His work on structure/property relationships has had far reaching influences on the understanding of liquid crystals and on their commercial applications in the field of electro-optical displays. In 1990 he became Research Coordinator for Merck (UK) Ltd, the company which, as BDH Ltd, did so much to commercialise and market the electro-optic materials which he invented at Hull University. He is now active as a Consultant, as Editor of the journal "Liquid Crystals" and as author/editor for a number of texts on Liquid Crystals.

H. W. Spiess
studied chemistry at the University of Frankfurt/Main, Germany, and obtained his Ph. D. in physical chemistry for work on transition metal complexes in the group of H. Hartmann. After professorships at the University of Mainz, Münster and Bayreuth he was appointed Director of the newly founded Max-Planck-Institute for Polymer Research in Mainz in 1984. His main research focuses on the structure and dynamics of synthetic polymers and liquid crystalline polymers by advanced NMR and other spectroscopic techniques.

V. Vill
studied chemistry and physics at the University of Münster, Germany, and acquired his Ph. D. in carbohydrate chemistry in the gorup of J. Thiem in 1990. He is currently appointed at the University of Hamburg, where he focuses his research on the synthesis of chiral liquid crystals from carbohydrates and the phase behavior of glycolipids. He is the founder of the LiqCryst database and author of the Landolt-Börnstein series *Liquid Crystals*.

List of Contributors

Volume 1, Fundamentals

Barois, P. (**VII**:6.1)
Centre de Recherche P. Pascal
Ave. A. Schweitzer
33600 Pessac Cedex
France

Blinov, L. M. (**VII**:9)
Institute of Crystallography
Russian Academy of Sciences
Leninsky prosp. 59
117333 Moscow
Russia

Bouligand, Y. (**VII**:7)
Histophysique et Cytophysique
Ecole Pratique des Hautes Etudes
10, rue A. Bocquel
49100 Angers
France

Cladis, P. E. (**VII**:6.3 and **VII**:6.4)
Lucent Technologies
Bell Laboratories Innov.
600 Mountain Ave.
MH1E437
Murray Hill, NJ 07974
USA

Crossland, W. A.; Wilkinson, T. D. (**IX**:2)
University of Cambridge
Dept. of Engineering
Trumpington Street
Cambridge CB2 1PZ
U.K.

Demus, D. (**VI**)
Veilchenweg 23
06118 Halle
Germany

Dunmur, D. A.; Toriyama, K. (**VII**:1–5)
Department of Chemistry
University of Southampton
Southampton SO17 1BJ
U.K.

Gleeson, H. F. (**VIII**:6–8 and **IX**:3)
Department of Physics
Schuster Laboratory
University of Manchester
Manchester, MJ3 9PL
U.K.

Goodby, J. W. (**II** and **V**)
School of Chemistry
University of Hull
Hull, HU6 7RX
U. K.

Gray, G. W. (**I** and **II**)
Merck Ltd.
Liquid Crystals
Merck House
Poole BH15 1TD
U.K.

Jérôme, B. (**VII**:10)
FOM-Institute for Atomic and Molecular
Physics
Kruislaan 407
1098 SJ Amsterdam
The Netherlands

Kapustina, O. (**VII**:11)
Shvernik St. 4
117036 Moscow
Russia

Leslie, F. M. (**III**:1)
University of Strathclyde
Dept. of Mathematics
Livingstone Tower
26 Richmond Street
Glasgow G1 1XH
U.K.

Leigh, W. J. (**IX**:4)
Department of Chemistry
McMaster University
1280 Main Street West
Hamilton, ON L8S 4M1
Canada

Noack, F.[†] (**VII**:13)
IV. Physikalisches Institut
Universität Stuttgart
Pfaffenwaldring 57
70550 Stuttgart
Germany

Osipov, M. (**III**:2)
Institute of Crystallography
Academy of Science of Russia
Leninski pr. 59
Moscow 117333
Russia

Palffy-Muhoray, P. (**VII**:12)
Kent State University
Liquid Crystal Institute
Kent, OH 44242
USA

Pollmann, P. (**VII**:6.2.4)
Universität GH Paderborn
FB 13, Chemie u. Chemietechnik
FG Physikal. Chemie
33095 Paderborn
Germany

Richardson, R. M. (**VIII**:4)
School of Chemistry
University of Bristol
Cantock's Close
Bristol BS8 1TS
U.K.

Sage, I. (**IX**:1)
DERA
Electronics Sector
St. Andrews Road
Malvern, WR14 3PS
U.K.

Schmidt, C. (**VIII**:1)
Institut für Makromolekulare Chemie
Universität Freiburg
Sonnenstraße 5
79104 Freiburg
Germany

Schneider, F.; Kneppe, H. (**VII**:8)
Universität GH Siegen
Physikal. Chemie
57068 Siegen
Germany

Seddon, J. M. (**VIII**:3)
Liquid Crystals Group
Dept. of Chemistry
Imperial College
London SW7 2AY
U.K.

Spiess, H.-W. (**VIII**:1)
Max-Planck-Institut für
Polymerforschung
Ackermannweg 10
55128 Mainz
Germany

Templer, R. H. (**VIII**:2)
Imperial College of Science
Technology and Medicine
Dept. of Chemistry
Exhibition Road
London SW7 2AY
U.K.

Thiemann, T. (**IV**)
Institute of Advanced Materials Study
Kyushu University
6-1, Kasuga-koh-en, Kasuga-shi
Fukuoka 816
Japan

Thoen, J. (**VII**:6.2.1)
Katholieke Universiteit Leuven
Departement Natuurkunde
Laboratorium voor Akoestiek en
Thermische Fysica
Celstijnenlaan 200 D
3001 Leuven (Heverlee)
Belgium

Vill, V. (**IV**)
Institut für Organische Chemie
Universität Hamburg
Martin-Luther-King-Platz 6
20146 Hamburg
Germany

Wedler, W. (**VII**:6.2.2 and **VII**:6.2.3)
Tektronix, Inc.; CPID
M/S 63-424
26600 SW Parkway
Wilsonville, OR 97070-1000
USA

Wilson, M. R. (**III**:3)
Department of Chemistry
University of Durham
South Road
Durham, DH1 3LE
U.K.

Workentin, M. S. (**IX**:4)
Department of Chemistry
University of Western Ontario
London, ON N6A 5B7
Canada

Outline

Volume 1

Chapter I: Introduction and Historical Development 1
George W. Gray

Chapter II: Guide to the Nomenclature and Classification of Liquid Crystals 17
John W. Goodby and George W. Gray

Chapter III: Theory of the Liquid Crystalline State 25
   1   Continuum Theory for Liquid Crystals 25
        Frank M. Leslie
   2   Molecular Theories of Liquid Crystals 40
        M. A. Osipov
   3   Molecular Modelling 72
        Mark R. Wilson

Chapter IV: General Synthetic Strategies 87
Thies Thiemann and Volkmar Vill

Chapter V: Symmetry and Chirality in Liquid Crystals 115
John W. Goodby

Chapter VI: Chemical Structure and Mesogenic Properties 133
Dietrich Demus

Chapter VII: Physical Properties 189
   1   Tensor Properties of Anisotropic Materials 189
        David Dunmur and Kazuhisa Toriyama
   2   Magnetic Properties of Liquid Crystals 204
        David Dunmur and Kazuhisa Toriyama
   3   Optical Properties 215
        David Dunmur and Kazuhisa Toriyama
   4   Dielectric Properties 231
        David Dunmur and Kazuhisa Toriyama
   5   Elastic Properties 253
        David Dunmur and Kazuhisa Toriyama
   6   Phase Transitions 281

6.1	Phase Transitions Theories.	281
	Philippe Barois	
6.2	Experimental Methods and Typical Results	310
6.2.1	Thermal Methods	310
	Jan Thoen	
6.2.2	Density.	334
	Wolfgang Wedler	
6.2.3	Metabolemeter.	350
	Wolfgang Wedler	
6.2.4	High Pressure Investigations.	355
	P. Pollmann	
6.3	Fluctuations and Liquid Crystal Phase Transitions	379
	P. E. Cladis	
6.4	Re-entrant Phase Transitions in Liquid Crystals.	391
	P. E. Cladis	
7	Defects and Textures.	406
	Y. Bouligand	
8	Flow Phenomena and Viscosity	454
	Frank Schneider and Herbert Kneppe	
9	Behavior of Liquid Crystals in Electric and Magnetic Fields	477
	Lev M. Blinov	
10	Surface Alignment.	535
	Blandine Jérôme	
11	Ultrasonic Properties.	549
	Olga A. Kapustina	
12	Nonlinear Optical Properties of Liquid Crystals.	569
	P. Palffy-Muhoray	
13	Diffusion in Liquid Crystals	582
	F. Noack †	

Chapter VIII: Characterization Methods		595
1	Magnetic Resonance.	595
	Claudia Schmidt and Hans Wolfgang Spiess	
2	X-Ray Characterization of Liquid Crystals: Instrumentation	619
	Richard H. Templer	
3	Structural Studies of Liquid Crystals by X-ray Diffraction	635
	John M. Seddon	
4	Neutron Scattering.	680
	Robert M. Richardson	
5	Light Scattering from Liquid Crystals.	699
	Helen F. Gleeson	
6	Brillouin Scattering from Liquid Crystals	719
	Helen F. Gleeson	
7	Mössbauer Studies of Liquid Crystals	727
	Helen F. Gleeson	

Chapter IX: Applications 731
   1   Displays 731
          Ian C. Sage
   2   Nondisplay Applications of Liquid Crystals. 763
          William A. Crossland and Timothy D. Wilkinson
   3   Thermography Using Liquid Crystals 823
          Helen F. Gleeson
   4   Liquid Crystals as Solvents for Spectroscopic, Chemical
          Reaction, and Gas Chromatographic Applications 839
          William J. Leigh and Mark S. Workentin

Index Volume 1 897

Volume 2 A

Part I: Calamitic Liquid Crystals 1

Chapter I: Phase Structures of Calamitic Liquid Crystals 3
John W. Goodby

Chapter II: Phase Transitions in Rod-Like Liquid Crystals 23
Daniel Guillon

Chapter III: Nematic Liquid Crystals 47
   1   Synthesis of Nematic Liquid Crystals. 47
          Kenneth J. Toyne
   2   Physical Properties 60
   2.1   Elastic Properties of Nematic Liquid Crystals 60
          Ralf Stannarius
   2.2   Dielectric Properties of Nematic Liquid Crystals 91
          Horst Kresse
   2.3   Diamagnetic Properties of Nematic Liquid Crystals. 113
          Ralf Stannarius
   2.4   Optical Properties of Nematic Liquid Crystals 128
          Gerhard Pelzl
   2.5   Viscosity 142
          Herbert Kneppe and Frank Schneider
   2.6   Dynamic Properties of Nematic Liquid Crystals 170
          R. Blinc and I. Muševič
   3   Applications 199
   3.1   TN, STN Displays 199
          Harald Hirschmann and Volker Reiffenrath
   3.2   Active Matrix Addressed Displays. 230
          Eiji Kaneko

3.3   Dynamic Scattering 243
        Birendra Bahadur
3.4   Guest-Host Effect 257
        Birendra Bahadur

Chapter IV: Chiral Nematic Liquid Crystals 303
1   The Synthesis of Chiral Nematic Liquid Crystals 303
        Christopher J. Booth
2   Chiral Nematics: Physical Properties and Applications 335
        Harry Coles

Chapter V: Non-Chiral Smectic Liquid Crystals 411
1   Synthesis of Non-Chiral Smectic Liquid Crystals 411
        John W. Goodby
2   Physical Properties of Non-Chiral Smectic Liquid Crystals 441
        C. C. Huang
3   Nonchiral Smectic Liquid Crystals – Applications 470
        David Coates

Volume 2 B

Part 2: Discotic Liquid Crystals 491

Chapter VI: Chiral Smectic Liquid Crystals 493
1   Synthesis of Chiral Smectic Liquid Crystals 493
        Stephen M. Kelly
2   Ferroelectric Liquid Crystals 515
        Sven T. Lagerwall
3   Antiferroelectric Liquid Crystals 665
        Kouichi Miyachi and Atsuo Fukuda

Chapter VII: Synthesis and Structural Features 693
Andrew N. Cammidge and Richard J. Bushby

Chapter VIII: Discotic Liquid Crystals: Their Structures and Physical Properties . . . 749
S. Chandrasekhar

Chapter IX: Applicable Properties of Columnar Discotic Liquid Crystals 781
Neville Boden and Bijou Movaghar

Part 3: Non-Conventional Liquid-Crystalline Materials 799

Chapter X: Liquid Crystal Dimers and Oligomers 801
Corrie T. Imrie and Geoffrey R. Luckhurst

Chapter XI: Laterally Substituted and Swallow-Tailed Liquid Crystals 835
Wolfgang Weissflog

Chapter XII: Phasmids and Polycatenar Mesogens 865
Huu-Tinh Nguyen, Christian Destrade, and Jacques Malthête

Chapter XIII: Thermotropic Cubic Phases 887
Siegmar Diele and Petra Göring

Chapter XIV: Metal-containing Liquid Crystals 901
Anne Marie Giroud-Godquin

Chapter XV: Biaxial Nematic Liquid Crystals 933
B. K. Sadashiva

Chapter XVI: Charge-Transfer Systems 945
Klaus Praefcke and D. Singer

Chapter XVII: Hydrogen-Bonded Systems 969
Takashi Kato

Chapter XVIII: Chromonics 981
John Lydon

Index Volumes 2A and 2B 1009

Volume 3

Part 1: Main-Chain Thermotropic Liquid-Crystalline Polymers 1

Chapter I: Synthesis, Structure and Properties 3
    1   Aromatic Main Chain Liquid Crystalline Polymers 3
         Andreas Greiner and Hans-Werner Schmidt
    2   Main Chain Liquid Crystalline Semiflexible Polymers 26
         Emo Chiellini and Michele Laus
    3   Combined Liquid Crystalline Main-Chain/Side-Chain Polymers 52
         Rudolf Zentel
    4   Block Copolymers Containing Liquid Crystalline Segments 66
         Guoping Mao and Christopher K. Ober

Chapter II: Defects and Textures in Nematic Main-Chain Liquid Crystalline Polymers 93
Claudine Noel

Part 2: Side-Group Thermotropic Liquid-Crystalline Polymers 121

Chapter III: Molecular Engineering of Side Chain Liquid Crystalline Polymers by Living Polymerizations 123
Coleen Pugh and Alan L. Kiste

Chapter IV: Behavior and Properties of Side Group Thermotropic Liquid Crystal Polymers 207
Jean-Claude Dubois, Pierre Le Barny, Monique Mauzac, and Claudine Noel

Chapter V: Physical Properties of Liquid Crystalline Elastomers 277
Helmut R. Brand and Heino Finkelmann

Part 3: Amphiphilic Liquid Crystals 303

Chapter VI: Amphotropic Liquid Crystals 305
Dieter Blunk, Klaus Praefcke and Volkmar Vill

Chapter VII: Lyotropic Surfactant Liquid Crystals 341
C. Fairhurst, S. Fuller, J. Gray, M. C. Holmes, G. J. T. Tiddy

Chapter VIII: Living Systems 393
Siegfried Hoffmann

Chapter IX: Cellulosic Liquid Crystals 451
Peter Zugenmaier

Index Volumes 1 – 3 483

Contents

Chapter I: Introduction and Historical Development 1
George W. Gray

1	Introduction	1
2	The Early Years up to About 1925	2
3	The Second Phase from 1925 to 1959	5
4	The Third Phase from 1960 to the Present Time	8
4.1	Lyotropic Liquid Crystals	11
4.2	Theory	12
4.3	Polymer Dispersed Liquid Crystals (PDLCs) and Anchoring	12
4.4	Materials and New Phases	13
5	Conclusions	14
6	References	14

Chapter II: Guide to the Nomenclature and Classification of Liquid Crystals . . . 17
John W. Goodby and George W. Gray

1	Introduction	17
2	General Definitions	18
3	Structural Features	19
4	Polymeric Liquid Crystals	20
5	Notation of Thermotropic Liquid Crystalline Properties	20
5.1	Description of the Solid State	20
5.1.1	Description of Soft Crystals	20
5.2	Description of the Liquid Crystalline Phases	21
5.2.1	Nematic and Chiral Nematic Phases	21
5.2.2	Smectic Liquid Crystals	21
5.2.3	Chiral Smectic Liquid Crystals	21
5.2.4	Columnar Phases	22
5.2.5	Plastic Crystals	23

5.2.6	Condis Crystals	23
5.2.7	Cubic	23
5.2.8	Re-entrants	23
5.3	Description of the Clearing Parameters	23
6	Stereochemistry	23
7	References	23

Chapter III: Theory of the Liquid Crystalline State 25

1	Continuum Theory for Liquid Crystals	25
	Frank M. Leslie	
1.1	Introduction	25
1.2	Equilibrium Theory for Nematics	26
1.2.1	The Frank–Oseen Energy	26
1.2.2	A Virtual Work Formulation	27
1.2.3	Body Forces and Moments	28
1.2.4	The Equilibrium Equations	29
1.2.5	Boundary Conditions	30
1.2.6	Proposed Extensions	31
1.3	Equilibrium Theory for Smectic Liquid Crystals	32
1.3.1	An Energy Function for SmC Liquid Crystals	32
1.3.2	Equilibrium Equations	33
1.4	Dynamic Theory for Nematics	35
1.4.1	Balance Laws	35
1.4.2	A Rate of Work Hypothesis	36
1.4.3	The Viscous Stress	36
1.4.4	Equations of Motion	37
1.5	References	38
2	Molecular Theories of Liquid Crystals	40
	M. A. Osipov	
2.1	Introduction	40
2.2	Microscopic Definition of the Order Parameters for Nematic and Smectic Phases	41
2.2.1	Uniaxial Nematic Phase	41
2.2.2	Biaxial Nematic Phase	42
2.2.3	Smectic A and C Phases	43
2.3	Anisotropic Intermolecular Interactions in Liquid Crystals	44
2.3.1	Hard-core Repulsion	44
2.3.2	Electrostatic and Dispersion Interactions	44
2.3.3	Model Potentials	46
2.4	Molecular Theory of the Nematic Phase	47
2.4.1	Mean-field Approximation and the Maier–Saupe Theory	47
2.4.2	Short-range Orientational Correlations	51

2.4.3	Excluded Volume Effects and the Onsager Theory	52
2.4.4	Packing Effects in Thermotropic Nematics	54
2.4.5	The Role of Molecular Biaxiality	56
2.4.6	Density Functional Approach to the Statistical Theory of Liquid Crystals	58
2.5	Molecular Models for Simple Smectic Phases	60
2.5.1	Mean-field Theory of the Nematic–Smectic A Transition	60
2.5.2	Phase Diagram of a Hard-rod Fluid	63
2.5.3	The Role of Intermolecular Attraction	65
2.5.4	Smectic A–Smectic C Transition	65
2.6	Conclusions	69
2.7	References	69
3	Molecular Modelling	72

Mark R. Wilson

3.1	Techniques of Molecular Modelling	72
3.1.1	Molecular Mechanics	72
3.1.2	Molecular Dynamics and Monte Carlo Simulation	74
3.1.3	Quantum Mechanical Techniques	78
3.2	Applications of Molecular Modelling	78
3.2.1	Determination of Molecular Structure	78
3.2.2	Determination of Molecular Properties	80
3.2.3	Determination of Intermolecular Potentials	81
3.2.4	Large-Scale Simulation of Liquid Crystals	82
3.3	References	85

Chapter IV: General Synthetic Strategies ... 87
Thies Thiemann and Volkmar Vill

1	Introduction	87
2	General Guidelines	87
2.1	Thermotropic Calamitic Liquid Crystals	87
2.2	Columnar Liquid Crystals	88
2.3	Concept of Leading Structures	89
3	Brief Survey of the History of Liquid Crystal Synthesis	89
4	Common Reactions in Liquid Crystal Synthesis	93
4.1	Methods of Aryl–Aryl Bond Formation	93
4.2	Alkyl-Functionalization of Arenes	95
4.3	Aryl–Cycloalkyl Linkages	95
4.3.1	4-E-Substituted Cyclohexylarenes	96
4.3.2	3-Substituted Arylcyclopentanes	96
4.4	Linking Blocks via Simultaneous Construction of Ethenyl and Ethynyl Bridges: Tolanes and Ethenes	97
5	Building Blocks and Their Precursors	98

6	Chirality: the Preparation and Use of Enantiomers	105
6.1	Synthesis	105
6.2	Esterification: Chiral Alkanoic Acids and Chiral Alkanols	106
6.3	Etherification	107

| 7 | Liquid Crystal Synthesis in Education and a Note on the Purification of Liquid Crystals | 107 |

| 8 | References | 107 |

Chapter V: Symmetry and Chirality in Liquid Crystals ... 115
John W. Goodby

1	Symmetry Operations in Liquid Crystals	115
2	Molecular Asymmetry	115
2.1	Group Priority	117
2.2	Diastereoisomers	117
2.3	Dissymmetry	117
3	Space Symmetry in Liquid Crystals	118
4	Mesophase Phase Symmetry	122
4.1	The Chiral Nematic Phase	123
4.2	Helical Smectic Phases	126
5	Frustrated Phases	129
5.1	Double Twist and Blue Phase Helical Structures	129
5.2	Twist Grain Boundary Phases	130
6	References	132

Chapter VI: Chemical Structure and Mesogenic Properties ... 133
Dietrich Demus

1	Introduction	133
2	Rod-Like (Calamitic) Liquid Crystalline Compounds	135
2.1	General	135
2.2	Ring Systems	136
2.2.1	Six-Membered Rings	136
2.2.2	Ring Systems with More than Six Atoms	139
2.2.3	Rings with Three to Five Atoms	142
2.3	Linking Groups	144
2.4	Terminal Substituents	146
2.5	Lateral Substituents	151
3	Liquid Crystals with Unconventional Molecular Shapes	153
3.1	Acyclic Compounds	154

3.2	Flexible Cyclic Compounds and Cyclophanes	154
3.3	Compounds with Large Lateral Substituents	154
3.3.1	Acyclic Lateral Substituents	154
3.3.2	Lateral Ring-Containing Substituents	155
3.3.3	Lateral Two-Ring-Containing Substituents	157
3.4	Swallow-Tailed Compounds	158
3.5	Polycatenar Compounds	160
3.5.1	Bicatenar Compounds	160
3.5.2	Tricatenar Compounds	160
3.5.3	Tetracatenar Compounds	160
3.5.4	Pentacatenar and Hexacatenar (Phasmidic) Compounds	161
3.5.5	Summary of the Phase Behaviour of Compounds with More than Two Flexible Chains	162
3.6	Twins and Oligomers	162
3.6.1	Fused Twins	163
3.6.2	Ligated Twins	163
3.6.3	Twins with Lateral–Terminal and Lateral–Lateral Linking	164
3.6.4	Twins winth Tail-to-Tail (Terminal–Terminal) Linking	164
3.6.5	Cyclic Dimers and Oligomers	166
3.6.6	Calamitic–Discotic Dimers	166
3.6.7	Star-Like Compounds	167
3.7	Epitaxygens	168
3.8	Associated Liquid Crystals	169
3.9	Salt-Like Compounds and Metal Complexes	170
3.9.1	Salts	170
3.9.2	Inverse Salts	170
3.9.3	Metallomesogens	171
4	Discotics	171
4.1	Derivatives of Benzene and Cyclohexane	172
4.2	Large Ring Systems	172
4.3	Complex-Forming Salts and Related Compounds	174
4.4	Pyramidal (Bowlic), Tubular and Related Compounds	176
4.5	Substituted Sugars	177
5	Conclusion	177
6	References	179

Chapter VII: Physical Properties 189

1	Tensor Properties of Anisotropic Materials	189
	David Dunmur and Kazuhisa Toriyama	
1.1	Macroscopic and Microscopic Properties	194
1.2	References	202

2	Magnetic Properties of Liquid Crystals	204
	David Dunmur and Kazuhisa Toriyama	
2.1	Magnetic Anisotropy	204
2.2	Types of Magnetic Polarization	206
2.2.1	Diamagnetism	206
2.2.2	Paramagnetism	206
2.2.3	Ferromagnetism	208
2.3	Diamagnetic Liquid Crystals	208
2.4	Paramagnetic Liquid Crystals	210
2.5	Ferromagnetic Liquid Crystals	212
2.6	Applications of Magnetic Properties	212
2.7	References	214
3	Optical Properties	215
	David Dunmur and Kazuhisa Toriyama	
3.1	Symmetry of Liquid Crystal Phases and the Optical Indicatrix	217
3.2	Molecular Theory of Refractive Indices	217
3.3	Optical Absorption and Linear Dichroism	221
3.4	Refractive Indices and Liquid Crystal Phase Structure	223
3.5	Optics of Helicoidal Liquid Crystal Structures	226
3.6	References	229
4	Dielectric Properties	231
	David Dunmur and Kazuhisa Toriyama	
4.1	Dielectric Response of Isotropic Fluids	231
4.2	Dielectric Properties of Anisotropic Fluids	235
4.2.1	The Electric Permittivity at Low Frequencies: The Static Case	235
4.2.1.1	Nematic Phase	235
4.2.2	Frequency Dependence of the Electric Permittivity: Dielectric Relaxation	245
4.3	References	251
5	Elastic Properties	253
	David Dunmur and Kazuhisa Toriyama	
5.1	Introduction to Torsional Elasticity	253
5.2	Director Distribution Defects and Fluctuations	259
5.2.1	Defects in Liquid Crystals	260
5.2.2	Fluctuations	262
5.3	Curvature Elasticity of Liquid Crystals in Three Dimensions	263
5.4	Electric and Magnetic Field-induced Deformations	265
5.4.1	Director Distribution in Magnetic Fields	266
5.4.2	Director Distribution in Electric Fields	269
5.4.3	Fréedericksz Transitions as a Method for Measuring Elastic Constants	270
5.4.3.1	Capacitance Method	270
5.4.3.2	Birefringence Method	271
5.4.4	Fréedericksz Transition for Chiral Nematics	272
5.4.5	Fréedericksz Transitions for Smectic Phases	273

5.5	Molecular Aspects of Torsional Elasticity.	274
5.5.1	van der Waals Theory	274
5.5.2	Results from Lattice Models.	277
5.5.3	Mean Field and Hard Particle Theories	277
5.5.4	Computer Simulations.	278
5.6	Experimental Aspects of Elastic Constants and Comparison with Theoretical Predictions.	279
5.7	References.	280
6	Phase Transitions	281
6.1	Phase Transitions Theories	281
	Philippe Barois	
6.1.1	Introduction	281
6.1.2	The Isotropic–Nematic Transition.	281
6.1.2.1	Mean Field Approach (Landau–de Gennes).	281
6.1.2.2	Fluctuations	284
6.1.2.3	Isotropic–Nematic Transition in Restricted Geometries.	284
6.1.3	The Nematic–Smectic A Transition.	285
6.1.3.1	Definition of an Order Parameter	285
6.1.3.2	Mean Field Description (McMillan, de Gennes)	286
6.1.3.3	Analogy with Superconductors	286
6.1.3.4	Critical Exponents and Scaling	289
6.1.3.5	Renormalization Group Procedures.	288
6.1.3.6	Dislocation Loops Theory.	288
6.1.4	The Smectic A–Smectic C Transition.	291
6.1.4.1	The Superfluid Helium Analogy.	291
6.1.4.2	The N–SmA–SmC Point.	291
6.1.5	The Smectic A–Hexatic Smectic B Transition	293
6.1.6	Phase Transitions in Chiral Liquid Crystals.	294
6.1.6.1	Chirality in Nematic and Smectic Liquid Crystals	294
6.1.6.2	Mean Field Chiral N–SmA–SmC model	295
6.1.6.3	Twist Grain Boundary Phases.	296
6.1.7	Frustrated Smectics	298
6.1.7.1	Polar Smectics.	298
6.1.7.2	The Model of Frustrated Smectics (Prost).	299
6.1.7.3	The Mean Field Model	301
6.1.7.4	Critical Properties of the Isolated SmA–SmA Critical Point	304
6.1.7.5	The Re-entrant Phenomenon	306
6.1.8	Conclusions	307
6.1.9	References.	308
6.2	Experimental Methods and Typical Results.	310
6.2.1	Thermal Methods	310
	Jan Thoen	
6.2.1.1	Introduction	310
6.2.1.2	Theoretical Background.	310

6.2.1.3	Experimental Methods.	314
6.2.1.4	Calorimetric Results.	323
6.2.1.5	Photoacoustic and Photopyroelectric Results	331
6.2.1.6	Acknowledgements	332
6.2.1.7	References.	332
6.2.2	Density.	334
	Wolfgang Wedler	
6.2.2.1	Instrumentation	334
6.2.2.2	General Conclusions from Density Studies on Liquid Crystals.	334
6.2.2.3	Studies of Calamitic Compounds	338
6.2.2.4	Discotics.	343
6.2.2.5	Lyotropics and Anhydrous Soaps	343
6.2.2.6	Polymeric Liquid Crystals.	343
6.2.2.7	Further Studies.	343
6.2.3	Metabolemeter.	350
	Wolfgang Wedler	
6.2.3.1	Thermobarograms.	352
6.2.3.2	References.	355
6.2.4	High Pressure Investigations	355
	P. Pollmann	
6.2.4.1	Introduction.	355
6.2.4.2	Phase transitions at High Pressures	356
6.2.4.3	Critical Phase Transitions Under Pressure.	368
6.2.4.4	Pressure-Volume-Temperature Behavior	371
6.2.4.5	Appendix	375
6.2.4.6	References.	376
6.3	Fluctuations and Liquid Crystal Phase Transitions	379
	P. E. Cladis	
6.3.1	Introduction	379
6.3.2	The Nematic–Isotropic Transition.	380
6.3.3	The Uniaxial–Biaxial Nematic Transition.	381
6.3.4	Type I Smectic A's: Nematic–Smectic A Transition	381
6.3.5	Type II Smectic A's: Cholesteric–Smectic A Transition	383
6.3.6	Transitions between Tilted and Orthogonal Smectic Phases	384
6.3.7	B–SmB–SmA Transitions.	385
6.3.8	Fluctuations at Critical Points.	386
6.3.8.1	BP_{III}–Isotropic.	386
6.3.8.2	SmA_d and SmA_1.	386
6.3.8.3	NAC Multicritical Point.	387
6.3.9	Conclusion.	387
6.3.10	References and Notes.	387
6.4	Re-entrant Phase Transitions in Liquid Crystals	391
	P. E. Cladis	
6.4.1	Introduction.	391
6.4.2	R1: Re-entrance from Frustration.	393

6.4.3	Re-entrance from Geometric Complexity	400
6.4.4	R3: Re-entrance from Competing Fluctuations	402
6.4.5	Conclusions	403
6.4.6	References	403
7	**Defects and Textures**	406
	Yves Bouligand	
7.1	Definitions, Conventions and Methods	406
7.1.1	Local Molecular Alignment	706
7.1.2	Microscopic Preparations of Liquid Crystals	406
7.1.2.1	Thermotropic Textures	407
7.1.2.2	Lyotropic Textures	409
7.1.2.3	Liquid Monocrystals	410
7.1.3	Images of Liquid Crystals in Polarizing Microscopy	410
7.1.4	Other Microscopic Methods	413
7.2	Strong Deformations Resulting in Defects	413
7.2.1	Singular Points	413
7.2.2	Singular Lines	414
7.2.3	Walls	415
7.2.4	Interface Defects (Points and Lines)	415
7.2.5	The Nature of Thick and Thin Threads	418
7.3	Topological Structure of Defects	418
7.3.1	The Volterra Process	418
7.3.2	The Volterra Process in Nematic, Smectic A and Cholesteric Phases	418
7.3.3	A Different Version of the Volterra Process	420
7.3.4	Continuous and Discontinuous Defects in Cholesteric Phases	423
7.3.5	Core Disjunction into Pairs of Disclinations	427
7.3.6	Optical Contrast	428
7.3.7	Classification of Defects	429
7.4	Geometrical Constraints and Textures	430
7.4.1	Parallel Surfaces and Caustics in Liquid Crystals	430
7.4.2	Dupin's Cyclides and Focal Conics	430
7.4.3	Slight Deformations of Dupin's Cyclides	433
7.4.4	Textures Produced by Parallel Fluid Layers	434
7.4.4.1	Planar Textures	434
7.4.4.2	Focal Conics and Polygonal Textures	435
7.4.4.3	Fan Textures	436
7.4.4.4	Texture Distribution in Lamellar Mesophases	437
7.4.4.5	Illusory Conics	439
7.4.4.6	Walls, Pseudowalls and Broken Aspects	439
7.4.5	Origin of Spirals in Chiral Liquid Crystals	442
7.4.6	Defects and Mesophase Growth	443
7.4.7.2	Disclination Lines	445
7.4.7.3	Focal Curves	446
7.4.7.4	Translation Dislocations	447

7.4.7.5	Simulations of Defects and Textures	447
7.4.7.6	Defect Nucleation	448
7.4.7.7	Textures and Defect Associations	448
7.4.7.8	Crystals of Defects	450
7.5	References	450
8	Flow Phenomena and Viscosity	454
	F. Schneider and H. Kneppe	
8.1	Nematic Liquid Crystals	455
8.1.1	Shear Viscosity Coefficients η_1, η_2, η_3, and η_{12}	455
8.1.2	Rotational Viscosity	456
8.1.3	Flow Alignment	458
8.1.4	Viscous Flow under the Influence of Elastic Torques	460
8.1.5	Transverse Pressure	462
8.1.6	Backflow	463
8.1.7	Discotic Liquid Crystals	464
8.1.8	Influence of Temperature and Order Parameter on the Viscosity Coefficients	464
8.1.9	Concluding Remarks	465
8.2	Cholesteric Liquid Crystals	466
8.2.1	Helical Axis Parallel to the Shear Gradient (Case I)	466
8.2.2	Helical Axis Parallel to the Flow Direction (Case II)	467
8.2.3	Helical Axis Normal to v and grad v (Case III)	467
8.2.4	Torque Generation Under Flow	468
8.3	Biaxial Nematic Liquid Crystals	468
8.3.1	Shear Viscosity Coefficients	469
8.3.2	Rotational Viscosity Coefficients	470
8.3.3	Flow Alignment	470
8.4	SmC Phase	470
8.4.1	Shear Flow with a Fixed Director Orientation	471
8.4.2	Rotational Viscosity	472
8.4.3	Flow Alignment	473
8.4.4	SmC* Phase	473
8.5	SmA Phase	474
8.6	References	475
9	Behavior of Liquid Crystals in Electric and Magnetic Fields	477
	Lev M. Blinov	
9.1	Introduction	477
9.2	Direct Influence of an Electric or Magnetic Field on Liquid Crystal Structure	478
9.2.1	Shift of Phase Transition Temperatures	478
9.2.1.1	Second-Order Transitions	478
9.2.1.2	Strong First-Order Transitions	479
9.2.1.3	Weak First-Order Transitions	481
9.2.2	Influence of the Field on Order Parameters	482
9.2.3	Changes in Symmetry	483
9.2.3.1	Induced Biaxiality in Nematics	483

9.2.3.2	The Kerr Effect	484
9.2.4	Specific Features of Twisted Phases and Polymers	485
9.2.4.1	Blue Phases	485
9.2.4.2	Twist Grain Boundary Phases	487
9.2.4.3	Comment on Polymer Liquid Crystals	488
9.3	Distortions due to Direct Interaction of a Field with the Director	488
9.3.1	Nematics	488
9.3.1.1	Classical Frederiks Transition	488
9.3.1.2	Field-Induced Periodic Structures	492
9.3.1.3	Flexo-Electric Phenomena	493
9.3.2	Twisted Nematics and Cholesterics	497
9.3.2.1	Twist and Supertwist Structures	497
9.3.2.2	Instability of the Planar Cholesteric Texture	499
9.3.2.3	Field Untwisting of the Cholesteric Helix	501
9.3.2.4	Flexoelectric Effects	503
9.3.3	Smectics and Discotics	505
9.3.3.1	Field Behavior of Achiral Smectics	505
9.3.3.2	Chiral Ferroelectrics and Antiferroelectrics	508
9.4	Electrohydrodynamic Instabilities	515
9.4.1	Nematics	515
9.4.1.1	Classification of Instabilities	515
9.4.1.2	Isotropic Modes	516
9.4.1.3	Anisotropic Modes	521
9.4.2	Cholesterics and Smectics	526
9.4.2.1	Cholesterics	526
9.4.2.2	Smectics A	527
9.4.2.3	Smectics C	528
9.5	Conclusion	529
9.6	References	529
10	Surface Alignment	535
	Blandine Jérôme	
10.1	Introduction	535
10.2	Macroscopic Alignment of Nematic Liquid Crystals	536
10.2.1	Definitions	536
10.2.2	Anchoring Directions	537
10.2.3	Anchoring Energy	528
10.3	Microscopic Surface Order of Nematic Liquid Crystals	540
10.3.1	Surface Orientational Order	540
10.3.2	Microscopic Anchoring Mechanisms at Solid Substrates	541
10.3.3	The Nematic/Isotropic and Nematic/Vapor Interfaces	543
10.4	Orientation of Other Liquid Crystals	544
10.4.1	Smectic and Chiral Liquid Crystals	544
10.4.2	Polymer Liquid Crystals	546
10.4.3	Lyotropic Liquid Crystals	546
10.5	References	547

11	Ultrasonic Properties	549
	Olga A. Kapustina	
11.1	Structural Transformation in Liquid Crystals	549
11.1.1	Orientation Phenomena in Nematics	550
11.1.1.1	Homogeneous Distortion Stage	550
11.1.1.2	Spatially Periodic Distortion Stage	554
11.1.1.3	Inhomogeneous Distortion Stage	555
11.1.2	Cholesterics in an Ultrasonic Field	557
11.1.2.1	Periodic Distortion	557
11.1.2.2	Storage Mode	557
11.1.2.3	Focal-Conic to Planar Texture Transition	558
11.1.2.4	Bubble Domain Texture	558
11.1.2.5	Fingerprint Texture Transition	559
11.1.3	Smectic Phase in an Ultrasound Field	559
11.2	Wave Interactions in Nematics	559
11.3	Acousto-electrical Interactions in Nematics	562
11.4	Ultrasound Studies of Liquid Crystals	563
11.4.1	Ultrasonic Spectroscopy	563
11.4.2	Photoacoustic Spectroscopy	565
11.4.3	Acoustic Emission	565
11.4.4	Monitoring Boundary Effects	565
11.4.5	Acoustic Microscopy	566
11.5	References	566
12	Nonlinear Optical Properties of Liquid Crystals	569
	P. Palffy-Muhoray	
12.1	Introduction	569
12.2	Interaction between Electromagnetic Radiation and Liquid Crystals	569
12.2.1	Maxwell's Equations	569
12.2.2	Nonlinear Susceptibility and Hyperpolarizability	570
12.3	Nonlinearities Originating in Director Reorientation	571
12.3.1	D. C. Kerr Effect	571
12.4	Optical Field-Induced Reorientation	572
12.5	Nonlinearities without Director Reorientation	574
12.5.1	Optical-Field-Induced Orientational Order	574
12.5.2	Thermal Effects	575
12.5.3	Conformational Effects	575
12.5.4	Electronic Response	576
12.6	Optical Harmonic Generation	577
12.6.1	Bulk Second Harmonic Generation	577
12.6.2	Surface Second Harmonic Generation	578
12.6.3	Third Harmonic Generation	578
12.7	Materials and Potential Applications	578
12.8	References	579

| 13 | Diffusion in Liquid Crystals | 582 |

F. Noack†

13.1	Introduction	582
13.2	Theoretical Concepts	582
13.2.1	The Diffusion Tensor	582
13.2.2	Basic Models	583
13.2.3	Model Refinements	585
13.3	Experimental Techniques	585
13.3.1	Tracer Techniques	585
13.3.2	Quasielastic Neutron Scattering	586
13.3.3	Magnetic Resonance	586
13.4	Selected Results	587
13.4.1	The Experimental Dilemma	588
13.4.2	Nematic Mesophase	588
13.4.3	Nematic Homologues	589
13.4.4	Smectic Mesophases	590
13.4.5	Solute Diffusion	590
13.4.6	Lyotropic Mesophases	591
13.4.7	Selected Diffusion Constants	592
13.5	References	592

Chapter VIII: Characterization Methods . . . 595

| 1 | Magnetic Resonance | 595 |

Claudia Schmidt and Hans Wolfgang Spiess

1.1	Introduction	595
1.2	Basic Concepts of NMR (Nuclear Magnetic Resonance)	596
1.2.1	Anisotropy of Spin Interactions	595
1.2.2	Exchange and Motional Narrowing	598
1.2.3	Spin Relaxation	598
1.2.4	Advanced Techniques	599
1.2.5	Multidimensional Spectroscopy	600
1.3	Applications of NMR	602
1.3.1	Phase Behavior	602
1.3.2	Molecular Orientation and Conformation	603
1.3.3	Molecular Dynamics	607
1.3.4	Liquid-Crystalline Polymers	608
1.3.5	Liquid Crystals in Microconfined Environments	610
1.3.6	Viscoelastic Properties	611
1.4	ESR (Electron Spin Resonance) of Liquid Crystals	613
1.5	Summary	615
1.6	References	616

| 2 | X-Ray Characterization of Liquid Crystals: Instrumentation | 619 |

Richard H. Templer

2.1	Origins of X-Rays	619
2.2	Generation of X-Rays in the Laboratory	620
2.3	X-Ray Cameras	621
2.3.1	The Debye–Scherrer and Flat Film Cameras	622
2.3.2	The Guinier Camera	623
2.3.3	The Franks Camera	625
2.3.4	The Huxley–Holmes Mirror–Monochromator Camera	626
2.3.5	The Elliott Toroid Camera and Others	626
2.3.6	What Camera Should be Used	627
2.4	The Recording of X-Ray Diffraction Patterns	627
2.4.1	X-Ray Film	627
2.4.2	Image Plates	628
2.4.3	Multi-Wire Proportional Counters	629
2.4.4	Opto-Electronic X-Ray Imaging Devices	630
2.4.5	What Detector Should be Used	631
2.5	Holding a Liquid Crystal in the X-Ray Beam	631
2.6	Controlling the Sample Environment	632
2.7	References	634
3	Structural Studies of Liquid Crystals by X-Ray Diffraction	635
	John M. Seddon	
3.1	Introduction	635
3.2	Bragg's Law and Powder Diffraction	636
3.3	Diffraction Patterns: Unaligned and Aligned Samples	638
3.3.1	Unaligned Samples	638
3.3.2	Thermotropic Cubic Phases	641
3.3.3	Aligned Samples	641
3.4	Diffracted Intensity: Molecular Transforms and Reciprocal Lattices	641
3.5	Translational Order	644
3.5.1	Long-range Order	644
3.5.2	Smectic Order Parameters	646
3.5.3	Quasi-long-range Order	647
3.5.4	Short-range Order	648
3.5.5	Lattice Orientational Order	648
3.5.6	Diffuse Scattering	649
3.6	Nematic Phase	649
3.6.1	Biaxial Nematics	650
3.6.2	Critical Behavior of the N–SmA Transition	650
3.6.3	Orientational Order Parameters	651
3.7	Smectic A and Smectic C Phases	652
3.7.1	Smectic A polymorphism	653
3.7.2	Smectic C and C* Phases	654
3.7.3	Ferroelectric Smectic C Phases	655
3.8	Modulated and Incommensurate Fluid Smectic Phases	656
3.8.1	Modulated Phases	656

3.8.2	Incommensurate Phases	658
3.8.3	Re-entrant Phases	658
3.8.4	Liquid Crystal Surfaces: X-ray Reflectivity	659
3.9	Twist Grain Boundary Phases	659
3.10	Hexatic Phases	661
3.11	Ordered Smectic Phases	664
3.12	Crystalline Packing and Conformation of Mesogens	667
3.13	Columnar Liquid Crystals	668
3.14	Liquid Crystal Polymers	669
3.14.1	Block Copolymers	671
3.15	References	671
4	Neutron Scattering	680
	Robert M. Richardson	
4.1	Introduction	680
4.2	Neutron Scattering Experiments	681
4.3	Neutron Diffraction from Isotopically Labelled Samples	682
4.3.1	Orientational Order in Low Molar Mass Materials	682
4.3.2	The Background to Small Angle Neutron Scattering from Polymers	684
4.3.3	Small Angle Neutron Scattering (SANS) Results from Side Chain Liquid Crystal Polymers	685
4.3.4	SANS Results from Main Chain Liquid Crystal Polymers	686
4.4	Dynamics of Liquid Crystals	687
4.4.1	Time Resolved Diffraction	687
4.4.2	Coherent Inelastic Neutron Scattering	687
4.4.3	Background to Incoherent Quasi-Elastic Neutron Scattering (IQENS)	687
4.4.4	Instruments for High Resolution IQENS	688
4.4.5	IQENS Measurements of Translational Diffusion	689
4.4.6	IQENS from the Localized Motion of Calamitic Molecules	690
4.4.7	IQENS from Other Types of Mesophase	695
4.4.8	Medium Resolution IQENS Studies	695
4.5	Conclusions	696
4.6	References	697
5	Light Scattering from Liquid Crystals	699
	Helen F. Gleeson	
5.1	Light Scattering from Nematic Liquid Crystals	699
5.1.1	Static Light Scattering from Nematic Liquid Crystals	699
5.1.2	Dynamic Light Scattering from Nematic Liquid Crystals	702
5.1.3	Forced Rayleigh Scattering	708
5.2	Cholesteric and Blue Phases	708
5.3	Light Scattering from the Smectic A Phase	709
5.4	Light Scattering from Achiral and Chiral Smectic C Phases	709
5.5	Pretransitional Light Scattering Studies	713
5.5.1	The Isotropic to Nematic Phase Transition	713
5.5.2	The Isotropic to Cholesteric or Blue Phase Transition	713

5.5.3	Smectic Phase Transitions	714
5.6	References	716
6	Brillouin Scattering from Liquid Crystals	719
	Helen F. Gleeson	
6.1	Brillouin Scattering in the Isotropic and Nematic Phases	720
6.2	Brillouin Scattering in the Cholesteric and Isotropic Phases	722
6.3	Brillouin Scattering in the Smectic A Phase	723
6.4	Brillouin Scattering in the Smectic C Phase	725
6.5	References	725
7	Mössbauer Studies of Liquid Crystals	727
	Helen F. Gleeson	
	References	730

Chapter IX: Applications . . . 731

1	Displays	731
	Ian C. Sage	
1.1	Introduction	731
1.2	Display Construction	731
1.3	Quasi-Fréedericksz Effect Displays in Nematic Liquid Crystals	733
1.3.1	The Fréedericksz Transition	733
1.3.2	Transitions in Twisted Nematic Layers	735
1.3.3	Optical Properties of Nematic Layers	739
1.3.4	Nematic Devices with Wider Viewing Angle	744
1.3.5	Dichroic Dyed Displays	745
1.3.6	Materials for Twisted and Supertwisted Nematic Displays	746
1.4	Scattering Mode Liquid Crystal Devices	748
1.5	Addressing Nematic Liquid Crystal Displays	751
1.6	Ferroelectric Liquid Crystal Displays	754
1.7	References	760
2	Nondisplay Applications of Liquid Crystals	763
	William A. Crossland and Timothy D. Wilkinson	
2.1	Liquid Crystal Spatial Light Modulation	763
2.1.1	Polarized Light and Birefingence	763
2.1.1.1	The Half Wave plate	765
2.1.1.2	The Quarter Wave plate	766
2.1.1.3	Linear Polarizers	766
2.1.2	Electro-Optic Effects in Chiral Smectic C Phases	766
2.1.3	The FLC Spatial Light Modulator	768
2.1.4	Binary Intensity Modulation	768
2.1.5	Binary Phase Modulation	770
2.1.6	The FLC Optically Addressed Spatial Light Modulator (OASLM)	771

2.2	Optical Correlation	773
2.2.1	The Positive Lens and the Fourier Transform	773
2.2.2	Correlation by Fourier Transform	775
2.2.3	The Matched Filter	777
2.2.4	The Joint Transform Correlator	783
2.3	Optical Interconnects	785
2.3.1	Computer-Generated Holograms	786
2.3.2	Polarization Insensitive Holographic Replay	790
2.3.3	Holographic Interconnects	792
2.3.3.1	The One to n Holographic Switch	793
2.3.3.2	The n to n Holographic Switch	796
2.3.4	Shadow Routed Crossbars	798
2.3.4.1	The 1 to n Shadow Logic Switch	799
2.3.4.2	The n by n Crossbar Switch	800
2.3.4.3	The OCPM (Optically Connected Parallel Machine) Optical Switch	800
2.3.4.4	The ATM (Asynchronous Transfer Mode) Switch	802
2.4	Wavelength Tuneable Filters and Lasers	803
2.4.1	The Digitally Tuneable Wavelength Filter	803
2.4.2	Digitally Tuneable Fiber Laser	806
2.4.3	Liquid Crystal Birefringent Wavelength Filters	807
2.4.4	Fabry Perot Based Wavelength Filters	809
2.5	Optical Neural Networks and Smart Pixels	810
2.5.1	The Optical Vector Processor	811
2.5.2	Computer-Generated Holograms as Synapses	812
2.5.3	Hybrid Opto-Electronic Neural Networks	813
2.5.4	Smart Pixels and SLMs	814
2.6	Other Applications	817
2.6.1	Three-Dimensional Video	817
2.6.2	α-Si/Novelty Filter	818
2.6.3	In-fiber Liquid Crystal Devices	818
2.6.4	Liquid Crystal Lenses	819
2.6.5	Optical Aberration Correction	819
2.6.6	Switchable Phase Delays for Phase Array Antennae	820
2.7	References	820
3	Thermography Using Liquid Crystals *Helen F. Gleeson*	823
3.1	Introduction	823
3.2	Device Structures	825
3.3	Engineering and Aerodynamic Research	827
3.4	Medical Thermography	830
3.5	Thermal Mapping and Nondestructive Testing	831
3.6	Radiation Detection	833
3.7	References	834

4	Liquid Crystals as Solvents for Spectroscopic, Chemical Reaction, and Gas Chromatographic Applications	839
	William J. Leigh and Mark S. Workentin	
4.1	Introduction	839
4.2	Liquid Crystals as Solvents in Spectroscopy	840
4.2.1	Nuclear Magnetic Resonance Spectroscopy	840
4.2.1.1	Solute Structure Determination [4, 10–17]	841
4.2.1.2	Chemical Shift and Indirect Coupling Anisotropies [237]	842
4.2.1.3	Quadrupolar Coupling Constants [10, 11, 261, 262]	843
4.2.1.4	Dynamics of Intramolecular Motions [8, 270–272]	844
4.2.1.5	Enantiomeric Purity of Optically Active Solutes	845
4.2.2	Electron Paramagnetic Resonance Spectroscopy [2, 3, 9, 299]	845
4.2.3	Polarized Optical Absorption and Emission Spectroscopy [1, 2, 304, 305]	845
4.2.4	Enantiomeric Purity and Structure of Optically Active Solutes [338]	847
4.3	Liquid Crystals as Solvents in Chemical Reactions [344–348, 350, 440]	848
4.3.1	Potential Effects of Anisotropic Solvents on Solute Chemical Reactivity	848
4.3.2	Solubility Factors and Phase Separation	849
4.3.3	Selection of a Liquid Crystal as a Solvent	851
4.3.4	Selected Highlights: Reactions in Liquid Crystalline Solvents	852
4.3.5	Reactions in Liquid Crystals: 1981–1996	856
4.4	Liquid Crystals in Gas Chromatographic Applications [444–449]	857
4.4.1	Liquid Crystals as Anisotropic Stationary Phases	857
4.4.2	Application of Liquid Crystal Stationary Phases for the Determination of Thermodynamic Data for Nonmesogenic Solutes	859
4.5	References	885

Index Volume 1 897

General Introduction

Liquid crystals are now well established in basic research as well as in development for applications and commercial use. Because they represent a state intermediate between ordinary liquids and three-dimensional solids, the investigation of their physical properties is very complex and makes use of many different tools and techniques. Liquid crystals play an important role in materials science, they are model materials for the organic chemist in order to investigate the connection between chemical structure and physical properties, and they provide insight into certain phenomena of biological systems. Since their main application is in displays, some knowledge of the particulars of display technology is necessary for a complete understanding of the matter.

In 1980 VCH published the *Handbook of Liquid Crystals*, written by H. Kelker and R. Hatz, with a contribution by C. Schumann, which had a total of about 900 pages. Even in 1980 it was no easy task for this small number of authors to put together the *Handbook*, which comprised so many specialities; the *Handbook* took about 12 years to complete. In the meantime the amount of information about liquid crystals has grown nearly exponentially. This is reflected in the number of known liquid-crystalline compounds: in 1974 about 5000 (D. Demus, H. Demus, H. Zaschke, *Flüssige Kristalle in Tabellen*) and in 1997 about 70000 (V. Vill, electronic data base LIQCRYST). According to a recent estimate by V. Vill, the current number of publications is about 65000 papers and patents. This development shows that, for a single author or a small group of authors, it may be impossible to produce a representative review of all the topics that are relevant to liquid crystals – on the one hand because of the necessarily high degree of specialization, and on the other because of the factor of time.

Owing to the regrettable early decease of H. Kelker and the poor health of R. Hatz, neither of the former main authors was able to continue their work and to participate in a new edition of the *Handbook*. Therefore, it was decided to appoint five new editors to be responsible for the structure of the book and for the selection of specialized authors for the individual chapters. We are now happy to be able to present the result of the work of more than 80 experienced authors from the international scientific community.

The idea behind the structure of the *Handbook* is to provide in Volume 1 a basic overview of the fundamentals of the science and applications of the entire field of liquid crystals. This volume should be suitable as an introduction to liquid crystals for the non-specialist, as well as a source of current knowledge about the state-of-the-art for the specialist. It contains chapters about the historical development, theory, synthesis and chemical structure, physical properties, characterization methods, and applications of all kinds of liquid crystals. Two subsequent volumes provide more specialized information.

The two volumes on *Low Molecular Weight Liquid Crystals* are divided into parts dealing with calamitic liquid crystals (containing chapters about phase structures, nematics, cholesterics, and smectics), discotic liquid crystals, and non-conventional liquid crystals.

The last volume is devoted to polymeric liquid crystals (with chapters about main-chain and side-group thermotropic liquid crystal polymers), amphiphilic liquid crystals, and natural polymers with liquid-crystalline properties.

The various chapters of the *Handbook* have been written by single authors, sometimes with one or more coauthors. This provides the advantage that most of the chapters can be read alone, without necessarily having read the preceding chapters. On the other hand, despite great efforts on the part of the editors, the chapters are different in style, and some overlap of several chapters could not be avoided. This sometimes results in the discussion of the same topic from quite different viewpoints by authors who use quite different methods in their research.

The editors express their gratitude to the authors for their efforts to produce, in a relatively short time, overviews of the topics, limited in the number of pages, but representative in the selection of the material and up to date in the cited references.

The editors are indebted to the editorial and production staff of WILEY-VCH for their constantly good and fruitful cooperation, beginning with the idea of producing a completely new edition of the *Handbook of Liquid Crystals* continuing with support for the editors in collecting the manuscripts of so many authors, and finally in transforming a large number of individual chapters into well-presented volumes.

In particular we thank Dr. P. Gregory, Dr. U. Anton, and Dr. J. Ritterbusch of the Materials Science Editorial Department of WILEY-VCH for their advice and support in overcoming all difficulties arising in the partnership between the authors, the editors, and the publishers.

The Editors

Chapter I
Introduction and Historical Development

G. W. Gray

1 Introduction

It is with a sense of responsibility that I begin this summary of the historical development of liquid crystals, because one of the two authors of the original *Handbook of Liquid Crystals* of 1980 [1] was Professor Hans Kelker, a friend and a very well informed authority on the history of the subject and the personalities involved in the earlier stages of its emergence. Those who attended the Twelfth International Liquid Crystal Conference in Freiburg in 1988, which marked the centenary of the discovery of liquid crystals, and heard Professor Kelker's plenary lecture – *Some Pictures of the History of Liquid Crystals* [2] – which was part of a conference session devoted to a historical review of the field, will know this. Here he demonstrated that he was in possession of a very wonderful collection of manuscripts and photographs relating to the scientists who, in the latter part of the nineteenth century and the early part of the twentieth century, laid the foundations of our present-day knowledge of liquid crystals.

I am not in that privileged situation, but I have worked in the field for 50 years, beginning my first experiments on aromatic carboxylic acids in October 1947. I have therefore worked through approaching half of the historical span of the subject, including the most recent years during which the subject has expanded and deepened so markedly. I hope this first-hand experience will counterbalance my lack of detailed historical knowledge of the earlier years, as possessed by Professor Kelker. Were he alive today, I hope he would not disapprove of what I write in this chapter.

The history of the development of liquid crystals may be divided into three phases:
1. The period from their discovery in the latter part of the nineteenth century through to about 1925, the years during which the initial scepticism by some that a state of matter was possible in which the properties of anisotropy and fluidity were combined, through to a general acceptance that this was indeed true, and publication of a first classification of liquid crystals into different types.
2. The period from 1925 to about 1960, during which general interest in liquid crystals was at a fairly low level. It was a niche area of academic research, and only relatively few, but very active, scientists were devoted to extending knowledge of liquid crystals. Two world wars and their aftermaths of course contributed greatly to the retardation of this field during this period. Taking the aftermaths of the wars into account, probably at least 15 years were effectively lost to progress during this second phase.
3. The period from 1960 until today is by contrast marked by a very rapid development in activity in the field, triggered of course by the first indications that technological applications could be found for liquid crystals. These early indications

were justified, and led to today's strong electro-optical display industry. The quest for new applications stimulated research and the flow of financial support into the involved areas of chemistry, physics, electrical and electronic engineering, biology, etc. As a marker of this activity, the numbers of papers and patents published in 1968 was about 2000 and this had risen to 6500 in 1995.

2 The Early Years up to About 1925

The question as to when liquid crystals were discovered must now be addressed. In pinpointing a discovery, it is necessary to distinguish simple observations of an unusual phenomenon or effect from observations that develop into an understanding of the meaning and significance of that phenomenon or effect. If we accept that the latter criteria must be met to justify the word discovery, then the credit for the discovery of liquid crystals must go to Friederich Reinitzer, a botanist of the Institute for Plant Physiology of the German University of Prague, who in a paper submitted on May 3, 1888 [3], described his observations of the colored phenomena occurring in melts of cholesteryl acetate and cholesteryl benzoate. In addition, he noted the "double melting" behavior in the case of cholesteryl benzoate, whereby the crystals transformed at 145.5 °C into a cloudy fluid, which suddenly clarified only on heating to 178.5 °C. Subsequent cooling gave similar color effects (but see later) to those observed on cooling the melt of cholesteryl acetate. Today of course we know that the colored phenomena reported by Reinitzer are characteristic of many cholesteric or chiral nematic (N*) liquid crystal phases.

In his article [3], Reinitzer acknowledges that other workers before him had observed curious color behavior in melts of cholesteryl systems. He mentions that Planar in Russia and Raymann in Paris had noted violet colors reflected from cholesteryl chloride and that Lobisch in Germany had observed a bluish-violet flourescence in the case of cholesteryl amine and cholesteryl chloride. Two things distinguish these earlier observations from those of Reinitzer. These are Reinitzer's recording of the "double melting" property of cholesteryl benzoate, and the fact that Reinitzer carried out preliminary studies on thin films of cholesteryl benzoate and noted the range of spectral colors reflected as the temperature decreased until crystallization occurred and the complementary nature of the colored light when the sample was viewed in transmission. Moreover, Reinitzer knew of the excellent work of the German physicist Professor Otto Lehmann, then at the Polytechnical School at Aachen, in designing and developing polarization microscopes, and recognized that Lehmann could advise on the optical behavior of his cholesteryl esters.

The approach to Lehmann was made in March 1888 and the correspondence is excellently documented in Kelker and Knoll's article [2]. This interaction led to agreement that Reinitzer's materials were homogeneous systems of which Lehmann wrote in August 1889: "It is of high interest for the physicist that crystals can exist with a softness, being so considerable that one could call them nearly liquid." This led quickly to the submission by Lehmann, by then at the University of Karlsruhe, of his paper *Über fliessende Kristalle* to the *Zeitschrift für Physikalische Chemie* [4].

Significantly, this uses for the first time the term liquid crystal. As a consequence of the above events and the development of our understanding of liquid crystals which stemmed from them, we must clearly acknowledge Reinitzer as the true discoverer of liquid crystals and the date of the event as March 14, 1888.

It should be noted that the discovery related exclusively to materials we now class as thermotropic liquid crystals, wherein the liquid crystal phases form either on heating crystals or on cooling isotropic liquids, that is, as a consequence of thermal effects. In addition to thermotropic liquid crystals, a second class of fluid anisotropic materials is known, namely, lyotropic liquid crystals where the disruptive effect on the crystal lattice involves a solvent (often water), coupled where necessary with thermal change. Here, the order of the crystal is broken down by the solvent and the molecules form micelles which then arrange themselves in an ordered way, while allowing fluidity. Excess of solvent completes the decrease in order and an isotropic solution is formed. Observations of anisotropy and optical birefringence in such systems were indeed made well before Reinitzer's discovery, but like the observations of Planar, Raymann, and Lobisch, there was no followthrough to a realization of the full significance of what was being seen. These observations were made by Mettenheimer [5], Valentin [6], and Virchow [7] in the period 1834–1861, and involved studies of biological samples derived from nerve tissue, for example, myelin, a complex lipoprotein which can be separated into fractions and which forms a sheath round nerve cells. In water-containing sodium oleate, these sheaths develop what have been called myelinic forms visible microscopically, especially in polarized light, as fluid, birefringent entities. Progress on these anisotropic systems was however impeded by the complexity and lack of reproducibility of the biological systems involved, and whilst predating the studies of Reinitzer and Lehmann are not generally regarded as marking the discovery of liquid crystals.

Following publication of his paper in 1889 [4], Lehmann continued work with liquid crystals and indeed dominated the scene in the late 1800s and the early part of the twentieth century, continuing to publish on liquid crystals until the year of his death in 1922.

Turning to purely synthetic materials, unlike the cholesteryl esters which were of natural origin, examples of liquid crystal behavior were found in these by Lehmann in 1890. The materials were azoxy ethers prepared by Gattermann and Ritschke [8]. The next ten years or so saw studies of p-methoxycinnamic acid and in 1902 the synthesis by Meyer and Dahlem [9] of the first smectogen, ethyl p-azoxybenzoate, although not recognized structurally for what it was at that time. Through studying such materials, Lehmann did however recognize that all liquid crystals are not the same, and indeed in 1907 he examined the first liquid crystal material exhibiting two liquid crystal phases. This material had what was later shown to be a smectic A (SmA) and a cholesteric (N*) phase. Significantly in the context of much later work in the field of applications, he also reported on the aligning effects of surfaces on liquid crystals.

Despite the growing number of compounds shown to exhibit liquid crystal phases (and in a short number of years Vorländer contributed about 250), the acceptance of liquid crystals as a novel state of matter was not universal. Tammann in particular [10] persisted in the view that liquid crystals were colloidal suspensions, and was in bitter argument with Lehmann and Schenk who upheld the view that they were homogeneous systems existing in a new state distinct from the crystalline solid and isotropic liquid states. Nernst [11] too did not subscribe to the latter view and believed that liquid crystals were mixtures of tautomers.

There was however a steadily growing body of evidence supporting the view that liquid crystals represent a true state of matter and acceptance of this slowly grew, aided by the excellent reviews of 1905 by Schenk (*Kristalline Flüssigkeiten und flüssige Kristalle*) [12] and Vorländer (*Kristallinisch-flüssige Substanzen*) [13]. There then followed the important review of optical effects by Stumpf [14] and, much later, an important paper was that by Oseen [15] on a kinetic theory of liquid crystals. The real seal of acceptance of liquid crystals for what they are, i.e., a fascinating and distinct state of matter, was however given in 1922 in the famous publication by G. Friedel [16] in the *Annales de Physique*, entitled *Les États Mesomorphes de la Matière*.

Here, in connection with Friedel's article and on a personal note, I well remember my research supervisor, Professor and later Sir Brynmor Jones, sending me to the library to find the appropriate journal, requiring that I produce a complete translation from French of all 273 pages in order to be "fully familiar with all that had been written". This I dutifully did in the fullness of time, and on taking my translation to show my supervisor, he then reached up to a shelf and withdrew a black notebook saying "now you can compare the quality of your translation with mine!" I learned much from that exercise, as will anyone who repeats it today.

In addition to containing a wealth of information on microscopic techniques and materials, Friedel's article represented in 1922 the first classification of liquid crystals into types, i.e., nematic, smectic and cholesteric. Today, of course, cholesterics are known simply as chiral nematics with no need that they be derived from cholesterol, and we recognize the existence of several polymorphic smectic forms, whereas Friedel allowed for only one (today's smectic A; SmA).

Friedel did however understand the layered nature of smectics, firstly through the stepped edges possessed by smectic droplets with a free surface, and secondly through his detailed studies of the optical microscopic textures of thin films of smectic phases. He understood the optical discontinuities, i.e., the defects, of the smectic focal-conic texture and saw the relationship of the black lines delineating ellipses of different eccentricities and their associated hyperbolae in terms of focal-conic "domains" which may be divided into a series of parallel, curved surfaces known as Dupin cyclides. He also understood that the optically extinct homeotropic textures of smectics of the type he studied gave positive uniaxial interference figures consistent with systems of layers lying flat to the surface. His microscopic studies demonstrated the immense value of the optical microscope as a precise scientific instrument in studies of all types of liquid crystal phases.

Friedel's article, coupled with the publications on synthesis and studies of new liquid crystal materials by organic chemists in Germany, notably Vorländer (see, for example, his monograph *Chemische Kristallographie der Flüssigkeiten* of 1924 [17]), firmly cemented in place all the earlier observations, providing a firm basis on which to build the future structure of the subject.

Before moving on to phase two of the history, we might just return to Reinitzer, the discoverer of liquid crystals, and recognize the quality of his powers of observation, for not only did he focus on the color effects and double melting, but also he noted the blue color appearing in the isotropic melt just before the sample turned into the cloudy cholesteric phase. About this, he said the following: "there appeared (in the clear melt) at a certain point a deep blue colour which spread rapidly through the whole mass and almost as quickly disappeared, again leaving in its place a uniform turbidity. On further cooling, a similar colour effect appeared for the second time to be followed by crystallisation of the mass and a simultaneous disappearance of the colour effect." The turbid state and the second color effect were of course due to the cholesteric phase, but the first transient blue color we now know was associated with the optically isotropic 'blue phases' we are familiar with today. Although Lehmann believed that this transient effect represented a different state, the full significance of Reinitzer's observations had to wait until the 1980s when these isotropic cubic phases became a focus of attention in condensed matter physics.

A further point concerning the first phase of our history of liquid crystals is about nomenclature, a matter about which scientists of today still love to argue. In the early years, however, the debate was sparked by Friedel who strongly objected to Lehmann's term liquid crystal, on the basis that liquid crystals were neither true liquids nor true crystals. The term does of course remain in widespread use today, simply because the juxtaposition of two contradictory terms carries an element of mystery and attraction. Friedel preferred the term mesomorphic to describe the liquid crystal state, and the associated term mesophase, reflecting the intermediate nature of these phases between the crystalline and isotropic liquid states. These terms are again widely used today and coexist happily with the Lehmann terminology. A useful term springing from Friedel's nomenclature is the word mesogen (and also nematogen and smectogen), used to describe a material that is able to produce mesophases. The associated term mesogenic is used by some to indicate that a material does form liquid crystal phases and by others to indicate that a compound is structurally suited to give mesophases, but may not, if, for example, the melting point of the crystalline state is too high. Then the isotropic liquid is produced directly from the crystal, and, on cooling, crystallization may occur too quickly for even a monotropic liquid crystal phase to form. Yet this compound may show strong tendencies to be mesomorphic if binary phase diagrams of state are examined using a standard material as the second component. My view is that the term mesogenic should be used to describe a structural compatibility with mesophase formation, without the requirement that a phase is actually formed. After all, if the compound does really form a mesophase, the description of it as mesomorphic is perfectly adequate.

Finally, on the subject of nomenclature, Friedel of course gave us today's terms smectic and nematic with their well-known Greek derivations.

3 The Second Phase from 1925 to 1959

In the first part of this period, Vorländer and his group in Halle contributed strongly to the growing number of compounds known to form liquid crystal phases, some showing up to three different mesophases. Based upon his work came the recognition that elongated molecular structures (lath- or rod-like molecules) were particularly suited to mesophase formation. His work also showed that if the major axis of a molecule were long enough, protrusions could be tolerated without sacrifice of the liquid crystal properties. Thus 1,4-disubstituted naphthalenes with a strong extension of the major axis through the 1,4-substituents were liquid crystalline, despite the protruding second ring of the naphthalene core. It is interesting that Vorländer records that the materials behaved as liquid crystallline resins or lacquers (an early thought perhaps about the potential of liquid crystals for applications).

In his book *Nature's Delicate Phase of Matter*, Collings [18] remarks that over 80 doctoral theses stemmed from Vorländer's group in the period 1901–1934. Further evidence of Vorländer's productivity is found in the fact that five of the 24 papers presented at the very important and first ever symposium on liquid crystals held in 1933 under the auspices of the Faraday Society in London, Liquid Crystals and Anisotropic Fluids – A General Discussion [19], were his. Perhaps the most important consequence of Vorländer's studies was that in laying down the foundations of the relationship between molecular structure and liquid crystal properties, attention was focused upon the molecules as the fundamental structural units of the partially ordered phases. Up to then, even Lehmann had been uncertain about the units involved in the ordering and what occurred at the actual transitions.

The Faraday Meeting of 1933 was of great importance in bringing together the small number of active, but often isolated, scientists involved at that time in liquid crystal research. This propagated knowledge and understanding, but, as we shall see, it also generated some dispute.

As early as 1923, de Broglie and E. Friedel (the son of G. Friedel) had shown [20] that X-ray reflections could be obtained from a system of sodium oleate containing water, and that the results were consistent with a lamellar or layered structure. This X-ray work was extended [21] in 1925 to Vorländer's thermotropic ethyl *p*-azoxybenzoate, confirming G. Friedel's conclusions of a layered structure stemming from his microscopic studies of smectic defect structures. Further, in the period 1932–1935, Herrmann [22], who also contributed to the 1933 Faraday Discussion, was decisive in confirming the lamellar nature of smectics by X-ray studies which included Vorländer's material exhibiting more than one smectic phase. The latter work substantiated a change from a statistical order in the layers of one smectic to a hexagonal ordering in the lower temperature phase. A tilted lamellar structure was also found by Herrmann for some thallium soaps [23].

Amongst other names of historical interest featured on the Faraday Discussion program were, for example, Fréedericksz and Zolina (forces causing orientation of an anisotropic liquid), Zocher (magnetic field effects on nematics), Ostwald, Lawrence (lyotropic liquid crystals), Bernal, Sir W. H. Bragg (developing the concept of Dupin cyclides in relation to Friedel's earlier studies of focal-conics), and also Ornstein and Kast who presented new arguments in favor of the swarm theory, which was first put forward in 1907–1909 by Bose [24]. This theory had proposed that the nematic phase consisted of elongated swarms of some 10^6 molecules, and in the 1930s much effort was given to proving the existence of these swarms, which were used to explain some, but not all, of the physical properties of nematics. However, at the 1933 Faraday Meeting, the presentation of Oseen [25] and the strong reservations expressed by Zocher during the discussions were already casting shadows of doubt on the swarm theory. Today of course we accept that definitive proof of the existence of swarms was never obtained, and by 1938 Zocher was expressing further strong reservations about the theory [26], proposing alternatively that the nematic phase is a continuum, such that the molecular orientation changes in a continuous manner throughout the bulk of the mesophase. This was called the distortion hypothesis and together with Oseen's work marked the beginning of the modern continuum theory of liquid crystals. However, developments here had to wait until after the second world war when Frank [27], at a further Faraday discussion in 1958, and consequent upon his re-examination of Oseen's treatment, presented it as a theory of curvature elasticity, to be advanced in the next historical phase by names such as Ericksen, Leslie, de Gennes, and the Orsay Group in France.

The period following the war up until the 1950s is also significant for the work of Chatelain [28], in collaboration with Falgueirettes. Using surface alignment techniques, they measured the refractive indices of different nematics, and Chatelain produced his theoretical treatment of the values of the ordinary and extraordinary indices of refraction of an oriented nematic melt.

Following Vorländer, other chemists were now becoming interested in new liquid crystal materials, and in the early 1940s we find publications on structure/property relations by Weygand and Gabler [29]. Later, in the 1950s, Wiegand [30] in Germany and the author in the UK were also making systematic changes in the structures of mesogens to establish the effects on liquid crystal behavior. The author's work included not only systematic modifications to aromatic core structures, but also studies of many homologous series, establishing clearly that within series systematic changes in transition temperature always occur, within the framework of a limited number of patterns of behavior. In the period 1951–1959, the author published some 20 papers on structure/property relations in liquid crystals. These are rather numerous to reference here, but in the account of the third historical phase from 1960 until today, reference to relevant reviews and books is given.

Lyotropic liquid crystals also progressed during this second phase. Lawrence's paper at the 1933 Faraday meeting discussed the phase diagrams for different compositions of fatty acid salts, recognizing the different phase types involved and the transitions undergone with change of temperature and/or water content. Examples of aromatic materials, including dyes, giving lyotropic phases were also found, and solvents other than water as the lyophase were explored.

The early work of Robinson et al. [31] was also done in this period. This involved solutions of poly-γ-benzyl-L-glutamate in organic solvents. These solutions exhibited the selective light reflecting properties of thermotropic cholesteric liquid crystals.

This period of history also saw the publication of work by Eaborn and Hartshorne [32] on di-isobutylsilandiol, which generated a mesophase. This was a puzzling result at the time, as the molecular shape was inconsistent with views of the time that liquid crystal formation required rod-shaped molecules. Light would be shed on this only after the discovery of liquid crystal phases formed by disc-shaped molecules in the early 1970s.

Finally, it should be noted that in this period, in 1957, a very important review on liquid crystals was published by Brown and Shaw [33]. This did much to focus the attention of other scientists on the subject and certainly contributed to the increase in liquid crystal research, which was to herald the strong developments in the early 1970s.

The period 1925–1959 may be usefully summarized now. Although the level of activity in the field was limited, important developments did occur in relation to:

- the influence of external fields (electric and magnetic) on liquid crystals;
- the orienting influences of surfaces;
- measurements of the anisotropic physical properties of aligned liquid crystals;
- the range of new liquid crystal materials and structure/property relationships;

- the development of theories of the liquid crystal state ranging from the swarm theory to the emerging continuum theory;
- increased awareness of the value of polarizing optical microscopy for the identification of mesophases, the determination of transition temperatures, and reaching a fuller understanding of defect textures.

4 The Third Phase from 1960 to the Present Time

The first ten years of this period saw several important developments which escalated interest and research in liquid crystals. Among these, there was the publication by Maier and Saupe [34] of their papers on a mean field theory of the nematic state, focusing attention on London dispersion forces as the attractive interaction amongst molecules and upon the order parameter. This theory must be regarded as the essential starting point for the advances in theoretical treatments of the liquid crystal state which followed over the years.

There was also much activity in the field of new liquid crystal materials, notably by Demus et al. [35] in Germany and by the author who, in 1962, produced his monograph *Molecular Structure and the Properties of Liquid Crystals* [36], published by Academic Press.

Also, further X-ray studies began to advance knowledge of the structure of liquid crystal phases, particularly smectics. The work of de Vries and Diele should be mentioned, and later on, notably that by Levelut and co-workers in France and Leadbetter in England (see, for example, [37, 38]), work which culminated in the 1980s in a clear structural elucidation and classification of smetic liquid crystals. This distinguished the true lamellar smectics with little or no correlation between layers from lamellar systems, previously regarded as smectics, which possess three-dimensional order and are really soft crystals. Today, the true smectics are labeled SmA, for example, and the crystal phases are referred to simply by a letter such as K, or by CrK. The phase once known as SmD and first observed by the author and co-workers in laterally nitro substituted biphenyl carboxylic acids [39] is now recognized [40] as a cubic liquid crystal phase. Several other examples of cubic thermotropic liquid crystal phases are now known [41, 42].

Such studies focused attention on the microscopic textures of liquid crystal phases. The defects characterizing these textures are now well understood through rather beautiful studies by workers such as Kléman [43], and from textures it is now possible to go a long way towards characterizing the phase behavior of new materials. A great deal of work on phase characterization has been done, and two reference sources are important [44, 45]. Through such detailed studies of phase behavior, new phenomena were often recognized and explained, for example, the re-entrance phenomenon through the work of Cladis [46], and the existence of the blue phases (BPI, BPII, and BPIII) through the work of several groups [47]. We should remember of course that Reinitzer did observe blue phases many years earlier and knew, without understanding the situation, that something occurred between the isotropic liquid and the N^* phase on cooling many chiral materials. Reflectance microscopy played a big part in the eventual elucidation of these phases as cubic phases involving double twist cylinders.

The widened interest in liquid crystals exemplified above had its origins in a number of events, such as the publication of the Brown and Shaw review [33] and the author's monograph [36], but a very big part was played by the launch of the now regular International Liquid Crystal Conferences (ILCCs). We owe these now biennial conferences to Glenn Brown, the first one being organised by him in 1965 in Kent State University in Ohio. Attended by some 90 delegates from different countries, this meeting was most important in providing that small group of research workers with an identity and the opportunity to meet and discuss problems and new ideas. Glenn Brown was indeed successful in obtaining funding for the second ILCC, held again in Kent State University, in 1968. The liquid crystal community owes a great deal to Glenn Brown for his vision in making these meetings possible. He created for liquid crystals a community of scientists within which the vigorous development of new research results and the technological innovation of the 1970s was to generate and continue unabated to the present time.

At the 1965 ILCC Meeting, the focus on applications was on Fergason's presentations on thermography using cholesteric liquid crystals, but in 1968, the meeting was attended by a group of researchers from Radio Corporation of America (RCA) in Princeton, where work under Heilmeier, Castellano, Goldmacher, and Williams was being done on display devices based on liquid crystals. The liquid crystal community was having its eyes opened to the potential of liquid crystals for application in electro-optical displays. The seminal work of the RCA group, initially on dynamic scattering displays, cannot be overstressed for its importance to the field. Two years later, at the 1970 ILCC in Berlin, display applications of liquid crystals were being discussed freely, and later when the patents of Schadt and Helfrich and of Fergason on the twisted nematic liquid crystal electro-optical display mode came into the public domain, activity intensified. At the fourth ILCC in 1972, again at Kent State, display applications dominated the meeting, and two years later at the 1974 ILCC in Stockholm, the author's presentation was on materials for use in *five* different display types, i.e., dynamic scattering, Fréedericksz, twisted nematic, cholesteric memory, and cholesteric-nematic phase change display devices.

Electro-optical liquid crystal display devices were now well established, and the twisted nematic device was obviously the superior one, based as it was upon a field effect in a pure nematic of positive dielectric anisotropy rather than upon the conductivity anisotropy, generated by ionic dopants in nematics of negative dielectric anisotropy, as in dynamic scattering displays.

In 1970, the author and co-workers obtained a research grant from the UK Ministry of Defence for work on room temperature liquid crystal materials that would function well in electro-optic displays. When our attention was directed to liquid crystal materials of positive dielectric anisotropy rather than negative dielectric anisotropy, we were able to make rapid progress, drawing on the store of fundamental knowledge relating molecular structure to liquid crystal properties, and, as a consequence, following patenting, the synthesis and behavior of the 4-alkyl- and 4-alkoxy-4'-cyanobiphenyls, designed for twisted nematic displays, were published in 1973 [48]. The history of the events leading up to the discovery of the cyanobiphenyls has been nicely documented by Hilsum [49], the originator of the research program under the author at the University of Hull and the coordinator of the associated programs on physics/devices at DRA, Malvern, and eventually on commercial production at BDH Ltd (now Merck UK Ltd).

The advent of the cyanobiphenyls made available the materials for the manufacture of high quality, reliable liquid crystal displays, and in so doing provided the secure basis upon which today's burgeoning electro-optical liquid crystal display device industry rests. The availability of these materials also spawned intense interest in other related families of materials, and during the later 1970s, the cyclohexane analogs of the biphenyls [50] and the pyrimidine analogs [51] became available, widening the choice of physical properties available to device engineers.

The market place welcomed the first simple, direct drive twisted nematic liquid crystal (LC) displays, but in so doing, created a demand for devices capable of portraying more complex data, particularly important for the display of Chinese/Japanese characters. Multiplex driven liquid crystal displays with some capability in this direction required the exploration of more complex mixtures incorporating ester components. However, the limitations of multiplex addressing were quickly exposed, encouraging interest in new device forms and addressing techniques. One development was the discovery of the supertwisted nematic display [52, 53] and of addressing methods for twisted nematic displays using thin film transistors. These two possibilities have progressed forwards successfully, each having to overcome its own particular problems. For a survey, see the review by Schadt [54]. As a result, in the late 1980s and early 1990s really excellent full color LC displays for direct view and for projection, involving where necessary high definition resolution, have come to the fore and dominate today's marketplace. The devices have steadily improved in viewing angle, brightness, definition, and color quality, and the most up to date displays (supertwisted nematic and active matrix twisted nematic) are technological products of great quality which lead the liquid crystal display device industry into the new millennium in a most confident mood.

Work in other display areas has of course occurred. Through the seminal work of R. B. Meyer and the research of Clark and Lagerwall [55] on surface stabilized ferroelectric liquid crystal devices based on chiral smectic C liquid crystal materials, the potential for ferroelectric devices has been fully explored in recent years. With their faster switching capability, they are attractive, and the difficulties over addressing schemes and the manufacture of ferroelectric displays will perhaps soon be overcome to give the marketplace a further liquid crystal device.

In the search for novel materials, particularly new ferroelectric materials, new phase types were also discovered, notably the antiferroelectric phase [56] which, with tristable switching characteristics, also has potential for display use, possibly overcoming some of the difficulties with ferroelectric systems and providing a further display device of high quality.

A further development in the display area concerns liquid crystal devices using in-plane-switching techniques, giving much improved viewing angle of the display [57, 58]. Here the molecules switch across the surface of the display cell, and this technology is now being adopted by three companies. Plasma switching of other types of liquid crystal display is another interesting technology awaiting further development [59].

Going back to the earlier 1970s, with the advent of the cyanobiphenyls and later the cyclohexane and pyrimidine analogs, not only was the display device industry provided with a wealth of novel, useful materials, but also those in fundamental research were given a range of stable, room temperature liquid crystal materials for study by a growing panoply of experimental and theoretical techniques. Fundamental research therefore moved forward rapidly with very good readily available materials and the funding for such work that was released by the potential for technological applications. Publications of research papers and patents escalated, as mentioned earlier, and in the journals *Molecular Crystals and Liquid Crystals* and *Liquid Crystals* the field now has its own dedicated literature shop windows.

The knowledge base in fundamental science was also extended greatly by the discovery in Chandrasekhar's group [60] of liquid crystals formed by disc-shaped molecules. Capable of forming discotic nematic phases and a range of columnar phases, these materials currently attract much interest and technological applications for them are possible.

This period also saw the growth of work on liquid crystal polymers both of the main chain and the side group varieties, and amongst others, the names of Blumstein and Finkelmann are associated with the first advances in this field which attracts many workers today (see [61, 62]). Main chain liquid crystal polymers became known for their ability to form high tensile strength fibers and moldings, and side group liquid crystal polymers also have applications in optical components such as brightness enhancement films for display devices. Anisotropic networks, elastomers, and gels are also intriguing systems that have been reviewed by Kelly [63].

Fundamental studies of liquid crystals have also opened up much interest in metallomesogens, i.e., materials incorporating metal centers and capable of giving liquid crystal phases. Many materials ranging from the calamitic (rod-like) type to the disc-like metallophthalocyanines are now known and with appropriate metal centers have interesting magnetic characteristics. Indeed, such have been the developments in this area that a text devoted to the subject has been published [64].

If this section of this chapter hardly reads as a chronological history, this is because the last 15 years have seen so many developments in different directions that a simple pattern of evolution does not exist. Instead, developments have occurred in an explosive way, emanating outward from the core of fundamental knowledge acquired up to the end of the 1970s. We have already looked briefly at display applications, liquid crystals from disc-shaped molecules, liquid crystal polymers, and metallomesogens, but to follow all the developments of recent years radiating out from the central core is hardly possible in a short chapter like this.

Only a few topics can be selected for brief mention, and if some areas of development are excluded through shortness of space, the author can at least feel confident that justice is done to them in the later pages of this four volume *Handbook of Liquid Crystals*.

4.1 Lyotropic Liquid Crystals

Activity here may have been less intense than in the thermotropic liquid crystal area. Important developments began, however, through the work of the Swedish group under Ekwall, Fontell, Lawson, and Flautt, and the 1970s saw the publication by Winsor of his R-theory of fused micellar phases [65] and of Friberg's valuable book *Lyotropic Liquid Crystals* [66].

The importance of lyotropic liquid crystals in the oil industry, the food industry, and the detergent industry is high because of the need to know the exact behavior of amphiphile/water/oil systems and the role played by the micellar phases in the context of efficient extraction of oil from natural rocks and whether a system will flow, the processes of baking and the uses of emulsifiers, and the general efficiency of soaps and detergents. Especial mention should be made of the work of Tiddy and his studies of the complex phase relations involving lamellar, hexagonal, and viscous isotropic (cubic) micellar phases and his extension of the field to include zwitterionic systems (see, for example, [67]), as well as cationic and anionic amphiphiles. The importance of lyotropic liquid crystal concepts in relation to biological systems and the role played therein by liquid crystals must also be noted. A recent review of lyotropic liquid crystals by Hiltrop is to be found in [68].

4.2 Theory

Theoretical treatments of liquid crystals such as nematics have proved a great challenge since the early models by Onsager and the influential theory of Maier and Saupe [34] mentioned before. Many people have worked on the problems involved and on the development of the continuum theory, the statistical mechanical approaches of the mean field theory and the role of repulsive, as well as attractive forces. The contributions of many theoreticians, physical scientists, and mathematicians over the years has been great – notably of de Gennes (for example, the Landau–de Gennes theory of phase transitions), McMillan (the nematic–smectic A transition), Leslie (viscosity coefficients, flow, and elasticity), Cotter (hard rod models), Luckhurst (extensions of the Maier–Saupe theory and the role of flexibility in real molecules), and Chandrasekhar, Madhusudana, and Shashidhar (pretransitional effects and near-neighbor correlations), to mention but some. The development of these theories and their significance are fully documented in the second edition of Chandrasekhar's excellent monograph [69] entitled simply *Liquid Crystals*, and supported by earlier reviews in [70].

In many of the above studies, the Gay–Berne potential describing the interactions between anisotropic particles has been used, and this can be separated into repulsive and attractive parts, enabling studies of the roles played by each in mesophase formation. Computer simulation has been used to investigate Gay–Berne fluids, and phase diagrams giving isotropic, nematic, smectic, and crystalline phases have been produced (see, for example, Hashim et al. [71]). Simulations aimed ultimately at the prediction of the phase behavior of compounds of given molecular structure (the molecular dynamics approach), avoiding synthesis, is another area of growth, supported by the increasing power of computers, and results from studies of molecular mechanics, used in the determination of molecular structure and lowest energy conformations, are proving to be increasingly useful (see Chap. III, Sec. 3 of this volume [72]).

4.3 Polymer Dispersed Liquid Crystals (PDLCs) and Anchoring

The behavior of liquid crystals at surfaces is of course of great importance in normal flat panel electro-optical displays, and the subject of anchoring is an area of strong research activity, where the work of Barbero and Durand is particularly noteworthy [73].

In recent years, PDLCs have attracted much attention because sheets consisting of droplets of nematic liquid crystal in an amorphous polymer matrix can be made by a number of techniques. The orientation in the nematic droplets can be manipulated by electric fields, changing the appearance of the sheet from cloudy to clear, and opening up possibilities for electrically switchable windows and panels, and for large area signs and advertising boards. The high voltages initially required for the operation of such systems have been reduced considerably and problems of off-axis haze in the viewing panels have been diminished. The full impact of such switchable panels has not yet been realized, and some of the problems have focused attention on the fact that in such systems the liquid crystal is constrained in a confined geometry, and the ratio of surface contact area to bulk volume is high. This has led to important research on liquid crystals in confined geometries and a text on this subject by Crawford and Zumer was published in 1996 [74]. This valuable book embraces other important aspects of confinement than that in PDLC systems, and includes porous polymer network assemblies in nematic liquid crystals, polymer stabilized cholesterics with their implications for reflective cholesteric displays, liquid crystal gel dispersions, filled nematics, and anisotropic gels.

4.4 Materials and New Phases

The applications of liquid crystals have unquestionably added incentive to the quest for new liquid crystal materials with superior properties such as viscosity, elastic constants, transition temperatures, and stability. In recent years this has catalyzed work on chiral materials as dopants for ferroelectric displays and for antiferroelectric materials with structures avoiding the number of potentially labile ester groups that were present in the original materials in which antiferroelectric properties were discovered.

The quest for new materials, whether driven by their potential for applications or simply by natural scientific curiosity about structure/property relations (i.e., as part of fundamental research programs) has always been and is today a vital part of the liquid crystal scenario. Indeed, it is often the case that the free-thinking fundamental research on new materials opens the door to new applications or improvements in existing device performance. The fascination of the liquid crystal field in fact derives from this continuing materials–knowledge–applications knock-on effect. Importantly, it can occur in both directions.

Thus studies of the significance of the position of the location of the double bond in a terminal alkyl chain led to the alkenylbicyclohexane systems [75], which today provide excellent materials for supertwisted nematic devices, and the work on ferroelectric materials led to the discovery of the antiferroelectric phase.

A notable example of phase discovery was that of the twist grain boundary smectic A* phase (TGBA*) by Goodby et al. in 1989 [76]. This new liquid crystal phase is a frustrated smectic in which the opposing tendencies to twist and be lamellar are accommodated through a regular array of dislocations, the possibility of which was predicted by de Gennes [77] in 1972 when he saw the analogous roles played by the director in an SmA phase and the magnetic vector potential in a superconductor. The analogous TGBC* phase is also now known.

In the field of natural products, there is growing interest in the increasing range of carbohydrates that are being found to be liquid crystalline [78–80], swinging interest back to the role of liquid crystals in biological systems, where recent studies of lyotropic mesomorphism in deoxyguanosine cyclic monophosphates [81] and of the significance of hydrogen bonding [82] and charge transfer [83] in low molar mass liquid crystals may be of further relevance.

Studies of the liquid crystal properties of terminally [84] and laterally [85] connected di-, tri- and tetra-mesogens are probing the behavior in the area between monomeric liquid crystals and liquid crystal polymers. The intercalated smectic phase (SmA$_{cal}$) has been found [84] and the somewhat related organosiloxanes [86] are proving to be most interesting materials with properties intermediate between low molar mass and polymeric liquid crystals, developed apparently through a microsegregation of the siloxane parts of the molecules into layers, leaving the organic moieties as appendages. Materials with both ferroelectric and antiferroelectric properties are provided and, significantly for applications, have almost temperature insensitive tilt angles and polarizations.

Finally, a breakdown of the division between calamitic liquid crystals and the columnar phases formed by disc-shaped molecules has occurred through the discovery of materials that give both types of mesophase in single compounds. For example, some six-ring, double swallow-tailed mesogens reported by Weissflog et al. [87] exhibit the very interesting phase sequence SmC–oblique columnar–SmC–nematic, combining not only calamitic and columnar phases, but also a re-entrant SmC phase.

5 Conclusions

Thus in the above selected areas and in many others that cannot be mentioned specifically here, the development of knowledge and understanding of liquid crystal systems goes on in a manner that is quite prolific.

It remains to be said that much has happened in the short history of liquid crystals and that the field is vigorous in research today, both in the areas of fundamental science and application driven investigations. Saying that the position is healthy today may lead to the question: "What of the future?" Fortunately, the author is acting here in the capacity of historian, and it is not in the historian's job description to predict the future. It will only be said that the future prospects for liquid crystals look healthy, but they will only be maintained so if fundamental research by scientists of imagination is adequately funded to enable the exploration of new ideas, new aspects, and new possibilities, because history does demonstrate that many of the discoveries significant for applications and technology derive from sound basic science or a sound knowledge of established basic science.

6 References

[1] H. Kelker, W. Hatz, *Handbook of Liquid Crystals*, VCH, Deerfield Beach, FL **1980**.
[2] H. Kelker, P. M. Knoll, *Liq. Cryst.* **1989**, *5*, 19.
[3] F. Reinitzer, *Monatsh. Chem.* **1888**, *9*, 421; for an English translation see *Liq. Cryst.* **1989**, *5*, 7.
[4] O. Lehmann, *Z. Phys. Chem.* **1889**, *4*, 462.
[5] C. Mettenheimer, *Corr. Blatt Verein gem. Arb. Förderung wissenschaftl. Heilkunde* **1857**, *24*, 331.
[6] G. Valentin, *Die Untersuchung der Pflanzen- und der Tiergewebe im polarisierten Licht*, Engelmann, Leipzig **1861**.
[7] R. Virchow, *Virchows Arch. path. Anat. Physiol.* **1853**, *6*, 571.
[8] L. Gattermann, A. Ritschke, *Ber. Dtsch. Chem. Ges.* **1890**, *23*, 1738.
[9] F. Meyer, K. Dahlem, *Liebigs Ann.* **1902**, *320*, 122.
[10] G. Tammann, *Ann. Physik.* **1901**, *4*, 524; **1902**, *8*, 103; **1906**, *19*, 421.
[11] W. Nernst, *Z. Elektrochem.* **1906**, *12*, 431.
[12] R. Schenk, *Kristalline Flüssigkeiten und flüssige Kristalle*, Engelmann, Leipzig **1905**.
[13] D. Vorländer, *Kristallinisch-flüssige Substanzen*, Enke, Stuttgart **1905**.

[14] F. Stumpf, *Radioakt. Elektron.* **1918**, *XV*, 1.
[15] C. W. Oseen, *Die anisotropen Flüssigkeiten*, Borntrager, Berlin **1929**.
[16] G. Friedel, *Ann. Physique* **1922**, *18*, 273.
[17] D. Vorländer, *Chemische Kristallographie der Flüssigkeiten*, Akadem. Verlagsanstalt, Leipzig **1924**.
[18] P. J. Collings, *Nature's Delicate Phase of Matter*, Princeton U. P., **1990**.
[19] General Discussion Meeting held by the Faraday Society, April **1933**.
[20] M. de Broglie, E. Friedel, *C. R.* **1923**, *176*, 738.
[21] E. Friedel, *C. R.* **1925**, *180*, 269.
[22] K. Herrmann, A. H. Krummacher, *Z. Kristallogr.* **1932**, *81*, 31; **1935**, *92*, 49.
[23] K. Herrmann, *Trans. Faraday Soc.* **1933**, *29*, 27.
[24] E. Bose, *Phys. Z.* **1907**, *8*, 513; **1908**, *9*, 708; **1909**, *10*, 230; E. Bose, F. Conrat, *Phys. Z.* **1908**, *9*, 916.
[25] C. W. Oseen, *Trans. Faraday Soc.* **1933**, *29*, 883.
[26] H. Zocher, *Trans. Faraday Soc.* **1933**, *29*, 1062; H. Zocher, G. Ungar, *Z. Phys. Chem.* **1938**, *110*, 529; see also H. Zocher in *Liquid Crystals and Plastic Crystals*, Vol. 1 (Eds.: G. W. Gray, P. A. Winsor), Ellis Horwood, Chichester **1974**, Chap. 3.1, p. 64.
[27] F. C. Frank, *Discuss. Faraday Soc.* **1958**, *59*, 958.
[28] P. Chatelain, *C. R.* **1948**, *227*, 136; *Bull. Soc. Franc. Mineral. Crist.* **1954**, *77*, 353; **1955**, *78*, 262; P. Chatelain, O. Pellet, *Bull. Soc. Franc. Mineral. Crist.* **1950**, *73*, 154; J. Falgueirettes, *C. R.* **1952**, *234*, 2619.
[29] C. Weygand, R. Gabler, *J. Prakt. Chem.* **1940**, *155*, 332; C. Weygand, R. Gabler, J. Hoffmann, *Z. Phys. Chem.* **1941**, *50B*, 124; C. Weygand, R. Gabler, N. Bircon, *J. Prakt. Chem.* **1941**, *158*, 26.
[30] C. Wiegand, *Z. Naturforsch.* **1954**, *3b*, 313; **1955**, *9b*, 516.
[31] C. Robinson, *Trans. Faraday Soc.* **1956**, *52*, 571; C. Robinson, J. C. Ward, R. B. Beevers, *Discuss. Faraday Soc.* **1958**, *25*, 29.
[32] C. Eaborn, N. H. Hartshorne, *J. Chem. Soc.* **1955**, 549.
[33] G. H. Brown, W. G. Shaw, *Chem. Rev.* **1957**, *57*, 1097.
[34] W. Maier, A. Saupe, *Z. Naturforsch.* **1958**, *13a*, 564; **1959**, *14a*, 882; **1960**, *15a*, 287.
[35] D. Demus, L. Richter, C. E. Rurup, H. Sackmann, H. Schubert, *J. Phys. (Paris)* **1975**, *36*, C1-349.
[36] G. W. Gray, *Molecular Structure and the Properties of Liquid Crystals*, Academic, London **1962**.
[37] J. Doucet, A. M. Levelut, M. Lambert, *Phys. Rev. Lett.* **1974**, *32*, 301; J. Doucet, A. M. Levelut, M. Lambert, M. Liebert, L. Strzelecki, *J. Phys. (Paris)* **1975**, *36*, C1-13; J. J. Benattar, A. M. Levelut, L. Liebert, F. Moussa, *J. Phys. (Paris)* **1979**, *40*, C3-115.
[38] A. J. Leadbetter in *The Molecular Physics of Liquid Crystals* (Eds.: G. W. Gray, G. R. Luckhurst), Academic, London **1979**, Chap. 13 and references therein and later in *Thermotropic Liquid Crystals* (Ed.: G. W. Gray), Wiley and Sons, New York **1987**, Chap. 1 and references therein.
[39] G. W. Gray, B. Jones, F. Marson, *J. Chem. Soc.* **1957**, 393; D. Demus, G. Kunicke, J. Neelson, H. Sackmann, *Z. Naturforsch.* **1968**, *23a*, 84; S. Diele, P. Brand, H. Sackmann, *Mol. Cryst. Liq. Cryst.* **1972**, *17*, 84 and 163.
[40] G. E. Etherington, A. J. Leadbetter, X. J. Wang, G. W. Gray, A. R. Tajbakhsh, *Liq. Cryst.* **1986**, *1*, 209.
[41] D. Demus, A. Gloza, H. Hartung, I. Rapthel, A. Wiegeleben, *Cryst. Res. Technol.* **1981**, *16*, 1445.
[42] A. M. Levelut, B. Donnio, D. W. Bruce, *Liq. Cryst.* **1997**, *22*, 753.
[43] M. Kléman, *Liquid Crystals and Plastic Crystals*, Vol. 1 (Eds.: G. W. Gray, P. A. Winsor), Ellis Horwood, Chichester **1974**, Chap. 3.3, p. 76.
[44] D. Demus, L. Richter, *Textures of Liquid Crystals*, VEB Deutscher Verlag für Grundstoffindustrie, **1978**.
[45] G. W. Gray, J. W. Goodby, *Smectic Liquid Crystals, Textures and Structures*, Leonard Hill, Glasgow **1984**.
[46] P. E. Cladis, *Phys. Rev. Lett.* **1975**, *35*, 48; P. E. Cladis, R. K. Bogardus, W. B. Daniels, G. N. Taylor, *Phys. Rev. Lett.* **1977**, *39*, 720.
[47] D. Coates, G. W. Gray, *Phys. Lett.* **1973**, *45A*, 115; D. Armitage, F. P. Price, *J. Appl. Phys.* **1976**, *47*, 2735; H. Stegemeyer, K. Bergmann in *Liquid Crystals of One- and Two-Dimensional Order* (Eds.: W. Helfrich, A. Heppke), Springer, Berlin **1980**, p. 161; H. Stegemeyer, T. H. Blumel, K. Hiltrop, H. Onusseit, F. Porsch, *Liq. Cryst.* **1986**, *1*, 3; P. P. Crooker, *Liq. Cryst.* **1989**, *5*, 751.
[48] G. W. Gray, K. J. Harrison, J. A. Nash, *Electron. Lett.* **1973**, *9*, 130; G. W. Gray, K. J. Harrison, J. A. Nash, J. Constant, J. S. Hulme, J. Kirton, E. P. Raynes, *Liquid Crystals and Ordered Fluids*, Vol. 2 (Eds.: R. S. Porter, J. F. Johnson), Plenum, New York **1973**, p. 617.
[49] C. Hilsum in *Technology of Chemicals and Materials for Electronics* (Ed.: E. R. Howells), Ellis Horwood, Chichester **1991**, Chap. 3.
[50] R. Eidenschink, D. Erdmann, J. Krause, L. Pohl, *Angew. Chem., Int. Ed. Engl.* **1977**, *16*, 100.
[51] A. Boller, M. Cereghetti, M. Schadt, H. Scherrer, *Mol. Cryst. Liq. Cryst.* **1977**, *42*, 215.
[52] C. M. Waters, E. P. Raynes, V. Brimmel, *S. I. D. Proc. Jpn. Display* **1983**, 396.

[53] T. J. Scheffer, J. Nehring, *Appl. Phys. Lett.* **1984**, *45*, 1021.
[54] M. Schadt in *Liquid Crystals* (Eds.: H. Baumgartel, E. U. Frank, W. Grunbein; Guest Ed.: H. Stegemeyer), Springer, N. Y. **1994**, Chap. 6, p. 195.
[55] N. A. Clark, S. T. Lagerwall, *Appl. Phys. Lett.* **1980**, *36*, 899.
[56] A. D. L. Chandani, E. Gorecka, Y. Ouchi, H. Takezoe, A. Fukuda, *Jpn. J. Appl. Phys.* **1988**, *27*, L729.
[57] Society for Information Display, *Information Display* **1996**, *12* (12), 13.
[58] *Eurodisplay '96* **1997**, *16* (1), 16.
[59] S. Kataoka, *International Lecture*, Institute of Electrical and Electronic Engineers, Savoy Place, London, March **1997**.
[60] S. Chandrasekhar, B. K. Sadashiva, K. A. Suresh, *Pramana* **1977**, *9*, 471.
[61] *Side Chain Liquid Crystal Polymers* (Ed.: C. B. McArdle), Blackie, Glasgow **1989**.
[62] *Liquid Crystalline Polymers* (Ed.: C. Carfagna), Pergamon, Oxford **1994**.
[63] S. M. Kelly, *Liq. Cryst. Today* **1996**, 6 (4), 1.
[64] *Metallomesogens* (Ed.: J. L. Serrano), VCH, Weinheim **1996**.
[65] P. A. Winsor in *Liquid Crystals and Plastic Crystals*, Vol. 1 (Eds.: G. W. Gray, P. A. Winsor), Ellis Horwood, Chichester **1974**, Chap. 5, p. 199.
[66] S. Friberg, *Lyotropic Liquid Crystals* (Ed.: R. F. Gould), Advances in Chemistry Series, American Chemical Society, Washington, DC **1976**.
[67] H. Morgans, G. Williams, G. J. T. Tiddy, A. R. Katritzky, G. P. Savage, *Liq. Cryst.* **1993**, *15*, 899.
[68] K. Hiltrop in *Liquid Crystals* (Eds.: H. Baumgartel, E. U. Frank, W. Grunbein; Guest Ed.: H. Stegemeyer), Springer, N. Y. **1994**, Chap. 4, p. 143.
[69] S. Chandrasekhar, *Liquid Crystals*, 2nd ed., Cambridge U. P., Cambridge **1992**.
[70] *The Molecular Physics of Liquid Crystals* (Eds.: G. R. Luckhurst, G. W. Gray), Academic, London **1979**.
[71] R. Hashim, G. R. Luckhurst, S. Romano, *J. Chem. Soc., Faraday Trans.* **1995**, *21*, 2141; G. R. Luckhurst, R. A. Stephens, R. W. Phippen, *Liq. Cryst.* **1990**, *8*, 451.
[72] M. R. Wilson in *Handbook of Liquid Crystals*, Vol. 1 (Eds.: D. Demus, J. W. Goodby, G. W. Gray, H. W. Spiess and V. Vill), Wiley-VCH, Weinheim **1998**, p. 53.

[73] G. Barbero, G. Durand in *Liquid Crystals in Complex Geometries* (Eds.: G. P. Crawford, S. Zumer), Taylor and Francis, London **1996**, Chap. 2, p. 21.
[74] *Liquid Crystals in Complex Geometries* (Eds.: G. P. Crawford, S. Zumer), Taylor and Francis, London **1996**.
[75] M. Schadt, R. Buchecker, K. Muller, *Liq. Cryst.* **1989**, *5*, 293.
[76] J. W. Goodby, M. A. Waugh, S. M. Stein, E. Chin, R. Pindak, J. S. Patel, *Nature, Lond.* **1989**, *337*, 449; *J. Am. Chem. Soc.* **1989**, *111*, 8119; G. Strajer, R. Pindak, M. A. Waugh, J. W. Goodby, J. S. Patel, *Phys. Rev. Lett.* **1990**, *64*, 1545.
[77] P. G. de Gennes, *Solid State Commun.* **1972**, *10*, 753.
[78] J. W. Goodby, *Mol. Cryst. Liq. Cryst.* **1984**, *110*, 205; J. W. Goodby, J. A. Haley, G. Mackenzie, M. J. Watson, D. Plusquellec, V. Ferrieres, *J. Mater. Chem.* **1995**, *5*, 2209.
[79] H. Prade, R. Miethchen, V. Vill, *J. Prakt. Chem.* **1995**, *337*, 427; J. W. Goodby, J. A. Haley, M. J. Watson, G. Mackenzie, S. M. Kelly, P. Letellier, P. Gode, P. Goethals, G. Ronco, B. Harmouch, P. Martin, P. Villa, *Liq. Cryst.* **1997**, *22*, 497.
[80] M. A. Marcus, P. L. Finn, *Liq. Cryst.* **1988**, *3*, 38.
[81] G. P. Spada, S. Bonazzi, A. Gabriel, S. Zanella, F. Ciuchi and P. Mariani, *Liq. Cryst.* **1997**, *22*, 341.
[82] K. Willis, J. E. Luckhurst, D. J. Price, J. M. Frechet, H. Kihara, T. Kato, G. Ungar, D. W. Bruce, *Liq. Cryst.* **1996**, *21*, 585 and references therein.
[83] K. Praefcke, D. Singer, A. Eckert, *Liq. Cryst.* **1994**, *16*, 53; B. Neumann, D. Joachimi, C. Tschierske, *Liq. Cryst.* **1997**, *22*, 509.
[84] A. E. Blatch, I. D. Fletcher, G. R. Luckhurst, *Liq. Cryst.* **1995**, *18*, 801.
[85] J. Anderesch, C. Tschierske, *Liq. Cryst.* **1996**, *21*, 51; J. Anderesch, S. Diele, D. Lose, C. Tschierske, *Liq. Cryst.* **1996**, *21*, 103.
[86] M. Ibn-Elhaj, A. Skoulios, D. Guillon, J. Newton, P. Hodge, H. J. Coles, *Liq. Cryst.* **1995**, *19*, 373.
[87] W. Weissflog, M. Rogunova, I. Letkos, S. Diele, G. Pelzl, *Liq. Cryst.* **1995**, *19*, 541.

Chapter II
Guide to the Nomenclature and Classification of Liquid Crystals

John W. Goodby and George W. Gray

1 Introduction

Nomenclature in liquid crystal systems is a nonsystematic language that is still, like any modern language, very much alive. Thus, many changes to currently acceptable terms, introductions of new notations, and deletions of out-moded notation have been made since the conception of the currently used nomenclature system. As the nomenclature system is in somewhat of a fluxional state it is not wise to assume that all definitions and accompanying notations are sacrosanct. Nevertheless, in some areas the topic of nomenclature has settled down into an internationally accepted, but unrecognized (by Scientific Societies) notation system, while in other areas, where research is still very active, changes to notation are still common. Members of the International Liquid Crystal Society (ILCS) and the International Union of Pure and Applied Chemists (IUPAC) are, however, attempting to create the first widely accepted naming system for liquid crystals. The descriptions and notations that follow are in agreement with the current proposals of the ILCS and IUPAC.

Notation for liquid crystals really started with the naming of the nematic and smectic phases in the early 1920s by Friedel [1]. However, it was the discovery of the existence of a variety of smectic phases in the 1950s–60s which lead Sackmann and Demus to propose the current lettering scheme for smectic liquid crystals [2]. Originally only three smectic phases were defined, SmA, SmB and SmC, but more followed rapidly as new phases were discovered. The notation introduced by Sackmann and Demus was dependent on the thermodynamic properties of mesophases and their ability to mix with one another, thus the miscibility of a material of undefined phase type with a standard material of known/defined mesophase morphology became the criterion for phase classification. Immiscibility, on the other hand has no special significance. Consequently all materials should have become standardized with those labelled by Sackmann and Demus.

Shortly after the introduction of the notation system, confusion set in with the notations for the phases G and H becoming interchanged (which was later resolved by agreement between the Hull and Halle Research Groups [3]. In addition, the D phase had been introduced as a smectic phase but later it turned out to be cubic; the B phase was split into two, the tilted B and orthogonal B phases, which were later to be redefined as the B and G phases; two E phases were thought to exist, one being uniaxial and the other biaxial, that were later defined as all being biaxial; and of course there was the perennial problem as to whether or not a phase was a soft crystal or a real smectic phase. This latter debate finally gave rise to a change in notation for soft crystal phases, the Sm notation being dropped, but with the hangover of the B notation being used in both smectic and soft crystal phases.

As our understanding of smectic phases increased, and structural studies using X-ray diffraction became more prevalent, there was an attempt to use crystallographic notation to describe the structures of smectic phases and, in addition, subscripts and superscripts were introduced to describe certain structural features, for example the subscript 2 was introduced to describe bilayer structures. By and large, however, there has been a general resistance to moving over to a full blown crystallographic notation system simply because there is a general feeling that a small change in structure within a miscibility class would lead to an unnecessary change in notation, consequently leading to complications and confusion.

To some degree problems of notation did arise with the naming of columnar mesophases. Originally they were called discotic liquid crystals, and indeed they also acquired a crystallographic notation. Both of these notations have, however, fallen out of favor and the naming of the state has been redefined. As research in disc-like systems remains relatively active, it is to be expected that further phases will be discovered, and as our understanding of the structures of these phases increases changes may be made to our current notation.

Notation in chiral phases is in flux basically because many new phases have been recently discovered, for example, antiferroelectric phases, blue phases and twist grain boundary phases. Even the use of an asterisk to indicate the presence of chirality is a hotly debated topic because chiral systems have broken symmetries and sometimes helical structures. Thus the debate is an issue over which aspect of chirality does the asterisk represent – broken symmetry or helicity. Thus further developments and changes in notation may be expected to occur in this area in the future.

Another problem that exists in notation is the relationship between thermotropic and lyotropic nomenclature. In some cases continuous behavior has been seen between thermotropic and lyotropic mesophases suggesting that they should share the same notation; however, at this point in time this has not occurred.

The notation scheme given below will be used wherever possible in the *Handbook of Liquid Crystals*; however for the reasons given above, readers should take care in its implementation, and they should also remember that the literature has suffered many changes over the years and so nomenclature used years ago may not tally with today's notation.

2 General Definitions

Liquid crystal state – recommended symbol LC – a mesomorphic state having long-range orientational order and either partial positional order or complete positional disorder.

Mesomorphic state – a state of matter in which the degree of molecular order is intermediate between the perfect three-dimensional, long-range positional and orientational order found in solid crystals and the absence of long-range order found in isotropic liquids, gases and amorphous solids.

Liquid crystal – a substance in the liquid crystal state.

Crystal phase – phase with a long-range periodic positional/translational order.

Liquid phase – phase with no long-range periodic or orientational order.

Mesophase or *liquid crystal phase* – phase that does not possess long-range positional ordering, but does have long-range orientational order. A phase occurring over a defined range of temperature or pressure or concentration within the mesomorphic state.

Thermotropic mesophase – a mesophase formed by heating a solid or cooling an isotropic liquid, or by heating or cooling a thermodynamically stable mesophase.

Lyotropic mesophase – a mesophase formed by dissolving an amphiphilic mesogen in suitable solvents, under appropriate conditions of concentration and temperature.

Calamitic mesophase – a mesophase formed by molecules or macromolecules with rod or lath-like molecular structures.

Columnar phase – phase that is formed by stacking of molecules in columns. Note that sugars, etc. are not necessarily discotic; discotic reflects a disc-like molecular shape. Also phasmids are columnar, but not necessarily discotic.

Mesogen (mesomorphic compound) – a compound that under suitable conditions of temperature, pressure and concentration can exist as a mesophase.

Calamitic mesogen – a mesogen composed of molecules or macromolecules with rod or lath-like molecular structures.

Discotic mesogen – a mesogen composed of relatively flat, disc- or sheet-shaped molecules.

Pyramidal or bowlic mesogen – a mesogen composed of molecules derived from a semirigid conical core.

Polycatenary mesogen – a mesogen composed of molecules having an elongated rigid core with several flexible chains attached to the end(s).

Swallow-tailed mesogen – a mesogen composed of molecules with an elongated rigid core with a flexible chain attached at one end and a branched flexible chain, with branches of about the same length at the other.

Mesogenic dimers, trimers etc. – a mesogen consisting of molecules with two, three, or more linked mesogenic units usually of identical structure.

Sanidic mesogen – a mesogen composed of molecules with board-like shapes.

Amphiphilic mesogen – a compound composed of molecules consisting of two parts with contrasting character, which may be hydrophilic and hydrophobic, that is lipophobic and lipophilic.

Amphotropic material – a compound which can exhibit thermotropic as well as lyotropic mesophases.

3 Structural Features

Molecules of liquid crystalline compounds are subdivided into the *central core* (mesogenic group), the *linking groups*, and *lateral groups* as well as *terminal groups*, depending on whether or not the groups lie along the long axis of the molecule. In relation to disc-like molecules, the central rigid region is called the core and the outer region the periphery; linking groups have the same definition as for calamitic rod-like systems.

The term *mesogenic* means that the structure is generally compatible with mesophase formation. A compound that forms real mesophases is however *mesomorphic*.

4 Polymeric Liquid Crystals

Liquid crystalline polymers are classified either as *main-chain* liquid crystalline polymers (MCLCP), as polymers with mesogenic *side groups* (SGLCP), or as liquid crystalline elastomers.

5 Notation of Thermotropic Liquid Crystalline Properties

To denote one phase transition, the abbreviation T together with an index is used (for example, T_{N-I} for a nematic–isotropic transition; T_{C-A} for a smectic C–smectic A transition).

The complete transition sequence is characterized by the description of the solid state (Sec. 5.1), the liquid crystalline transitions (Sec. 5.2) and the clearing parameter (Sec. 5.3).

The phase symbol is followed by the upper temperature limit as measured during the heating and not during the cooling cycle.

Crystal types should be arranged in an ascending order of transition temperature.

If a transition temperature of a liquid crystalline phase is lower than the melting point, this phase only occurs monotropically.

Monotropic transitions appear in round brackets.

In this context, square brackets mean virtual transitions.

Examples:

Cr 34 N 56 I designates a compound melting at 34 °C into the nematic phase; at 56 °C it changes into the isotropic phase; normal behavior.

Cr 56.5 (SmA 45) I designates a compound melting at 56.5 °C into the isotropic phase. Below 45 °C, a monotropic A phase exists.

Cr 120 B 134 I [N 56 I_e] designates a compound melting at 120 °C into the crystal B phase. At 134 °C the isotropic phase is formed. The virtual nematic clearing point of 56 °C is one obtained by extrapolation from mixtures.

Cr_1 78 Cr_2 212 N ? I_{decomp} designates a compound with a crystal to crystal transition at 78 °C and a melting point of 212 °C to a nematic phase. The clearing point is unknown because decomposition takes place.

Cr_1 112 (Cr_2 89) I designates a material with a metastable crystal phase formed on cooling the isotropic melt slowly and having a lower melting point than the stable crystal phase.

5.1 Description of the Solid State

Cr   crystalline phase
Cr_2 second crystalline phase
g   glassy state
T_g glass-transition temperature

5.1.1 Description of Soft Crystals

The following phases should just be called B, E, etc. (i.e. retaining their historic classification, but losing their smectic code letter Sm) because they are no longer regarded as true smectics as they have long range positional order. In fact they are soft or disordered crystals. The single letter notation (B, E, etc.) is preferred to B_{Cr}, E_{Cr} etc.

B	crystal B
E	crystal E
E_{mod}	crystal E phase with modulated in-plane structure
G	crystal G
H	crystal H
J	crystal J
K	crystal K

Tilted Chiral Soft Crystal Phases

A superscript asterisk (*) is used throughout to denote the presence of chirality.

J*	tilted chiral crystal J
H*	tilted chiral crystal H
G*	tilted chiral crystal G
K*	tilted chiral crystal K

Chiral Orthogonal Soft Crystal Phases

A superscript asterisk (*) is used throughout to denote the presence of chirality.

B*	chiral orthogonal crystal B phase
E*	chiral orthogonal crystal E phase

5.2 Description of the Liquid Crystalline Phases

5.2.1 Nematic and Chiral Nematic Phases

n	director
N	nematic
N*	chiral nematic (cholesteric)
N_u	uniaxial nematic phase
N_b	biaxial nematic phase (sanidic phase)

also

N^*_∞	infinite pitch cholesteric, i.e. at a helix inversion or compensation point

BP	blue phases. These are designated BP_I, BP_{II}, BP_{III} or BP_{fog}, and BPS.

5.2.2 Smectic Liquid Crystals

Use Sm for smectic instead of S (unless spelt out), this avoids subscripts and double subscripts

SmA	Smectic A
	SmA_1 monolayer
	SmA_2 bilayer
	SmA_d interdigitated bilayer
	$Sm\tilde{A}$ modulated bilayer
SmB or SmB_{hex}	Smectic B; hexatic B (SmB preferred)
SmC	Smectic C
	SmC_1 monolayer
	SmC_2 bilayer
	SmC_d interdigitated bilayer
	$Sm\tilde{C}$ modulated bilayer
	SmC_{alt} alternating tilt phase – see smectic O
SmF	Smectic F
SmI	Smectic I
SmM	Smectic M – not found common use so far
SmO	Smectic O – not found common use so far. Use carefully in relation to smectic C_{alt} and antiferroelectric smectic C* phase
	SmO and SmM should be defined in the text

Intercalated mesophases take the subscript c, e.g., SmA_c.

Biaxial variants of uniaxial smectic phases take the subscript b, e.g., SmA_b.

5.2.3 Chiral Smectic Liquid Crystals

Usually given the notation * to indicate the presence of chirality, e.g., SmA*, SmC*.

Twist Grain Boundary Phases (TGB)

Orthogonal TGB Phases

TGBA* structure based on the smectic A phase. The phase can be either commensurate or incommensurate depending on the commensurability of the helical structure with respect to the rotation of the smectic A blocks in the phase.

TGBB* proposed structure based on a helical hexatic B phase.

Tilted TGB Phases

TGBC phase poses the following problems for nomenclature:

TGBC where the normal smectic C* helix is expelled to the screw dislocations.

TGBC* phase where the blocks have a local helix associated with the out of plane structure.

Possibly too early to define notation yet; therefore spell out notation in text.

Tilted Chiral Phases

SmC* chiral C phase
SmC^*_∞ infinite pitch smectic C* – ferroelectric
SmC^*_α unwound antiferroelectric phase occurring at higher temperature, and above a ferroelectric SmC* phase in a phase sequence
SmC^*_γ ferrielectric phases that occur on cooling ferroelectric C* phases
SmC^*_A antiferroelectric C* phase

Try not to use α, β and γ notations, simply spell out the phase type, e.g. ferrielectric SmC* phase, or SmC* (ferri).

The above notations may also be applied to SmI* and SmF* phases as and when required – for example, SmI^*_A is an antiferroelectric I* phase.

Orthogonal Phases

SmA* chiral orthogonal smectic A
SmB*/ chiral orthogonal hexatic smectic
SmB^*_{hex} B phase (SmB* preferred)

Analogous Achiral Systems

Here, antiferroelectric-like structures are often observed, i.e. zigzag layer ordering. The following designation should be used and reference to antiferroelectric-like ordering should be suppressed:

SmC_{alt} alternating tilt smectic C phase – see also smectic O for cross-referencing

Unknown Phases

Label as SmX_1, SmX_2, etc.

5.2.4 Columnar Phases

N_D nematic discotic phase
Col_h hexagonal discotic
Col_{ho} ordered hexagonal columnar phase
Col_{hd} disordered hexagonal columnar phase
Col_{ro} ordered rectangular columnar phase
Col_{rd} disordered rectangular columnar phase
Col_t tilted columnar phase
ϕ phasmidic phase, but Col is preferred

For nematic phases – spell out positive and negative birefringent situations in the text. In addition spell out if the nematic phase is composed of single molecular entities or short columns which exist in a disordered nematic state.

Care should be taken when using the term discotic; columnar is preferred.

Unknown Discotic Phases

Col_1, Col_2, etc.

5.2.5 Plastic Crystals

Rotationally (3-D) disordered crystals that may be derived from globular molecules leading to isotropic phases. This classification does not apply to crystal smectic phases composed of elongated molecules, although these could be described as anisotropic plastic crystals.

5.2.6 Condis Crystals

Crystals in which the positional and conformational order in the packing of molecules arranged in parallel is lost to some degree.

5.2.7 Cubic

Cubic thermotropic liquid crystalline phases are designated Cub.

CubD   Cubic D phase

5.2.8 Re-entrants

Use the same notation as for nematic or other appropriate phase with subscript re-entrant, i.e., N_{re} or SmC_{re} or Col_{hre}.

5.3 Description of the Clearing Parameters

I   isotropic, standard case
I_{decomp} decomposition at clearing temperature
I_e extrapolated temperature

6 Stereochemistry

(#)   unknown chirality

One Chiral Centre (R or S)

(S)   chiral (S)
(R)   chiral (R)
(S)/(R)   racemate

Two Chiral Centres

This situation is more complex and depends upon whether the two centres are the same or different. If they are the same the notation should be: chiral (*S,S* or *R,R*); or for racemic materials (*S,S* or *R,R*); meso-compounds (optically inactive by internal compensation should be denoted as (*S,R* or *R,S*).

7 References

[1] G. Friedel, *Ann. Physique* **1922**, *18*, 272.
[2] H. Sackmann, D. Demus, *Mol. Cryst. Liq. Cryst.* **1966**, *2*, 81.
[3] D. Demus, J. W. Goodby, G. W. Gray, H. Sackmann, *Mol. Cryst. Liq. Cryst. Lett.* **1980**, *56*, 311.

Chapter III
Theory of the Liquid Crystalline State

1 Continuum Theory for Liquid Crystals

Frank M. Leslie

1.1 Introduction

Continuum theory for liquid crystals has its origins in the work of Oseen [1] and Zocher [2] in the 1920s. The former essentially derived the static version of the theory for nematics that has been used extensively in device modelling, while the latter successfully applied it to Fréedericksz transitions [3]. Later Frank [4] gave a more direct formulation of the energy function employed in this theory, and stimulated interest in the subject after a period of relative dormancy. Soon thereafter Ericksen [5] set the theory within a mechanical framework, and generalized his interpretation of static theory to propose balance laws for dynamical behaviour [6]. Drawing upon Ericksen's work, Leslie [7] used ideas prevalent in continuum mechanics to formulate constitutive equations, thus completing dynamic theory.

This continuum theory models many static and dynamic phenomena in nematic liquid crystals rather well, and various accounts of both the theory and its applications are available in the books by de Gennes and Prost [8], Chandrasekhar [9], Blinov [10] and Virga [11], and also in the reviews by Stephen and Straley [12], Ericksen [13], Jenkins [14] and Leslie [15]. Given this success of continuum theory for nematics, much current interest in continuum modelling of liquid crystals now centres upon appropriate models for smectic liquid crystals, liquid crystalline polymers and (to a lesser extent) lyotropics. Otherwise, interest in nematics is largely confined to studies of behaviour at solid interfaces, this including discussions as to whether or not one should include an additional surface term in the Frank–Oseen energy as proposed by Nehring and Saupe [16], although there has also been some activity into the modelling of defects, particularly using a modified theory proposed recently for this purpose by Ericksen [17].

In this section I aim to describe in some detail continuum theory for nematics, and also to draw attention to some points of current interest, particularly surface conditions and surface terms. At this juncture it does seem premature to discuss new developments concerning smectics, polymers and lyotropics, although a brief discussion of an equilibrium theory for certain smectics seems appropriate, given that it relates to earlier work on this topic. Throughout, to encourage a wider readership, we endeavour to employ vector and matrix notation, avoiding use of Cartesian tensor notation.

1.2 Equilibrium Theory for Nematics

1.2.1 The Frank–Oseen Energy

Continuum theory generally employs a unit vector field $\mathbf{n}(\mathbf{x})$ to describe the alignment of the anisotropic axis in nematic liquid crystals, this essentially ignoring variations in degrees of alignment which appear to be unimportant in many macroscopic effects. This unit vector field is frequently referred to as a director. In addition, following Oseen [1] and Frank [4], it commonly assumes the existence of a stored energy density W such that at any point

$$W = W(\mathbf{n}, \nabla \mathbf{n}) \qquad (1)$$

the energy is therefore a function of the director and its gradients at that point. Since nematic liquid crystals lack polarity, $\mathbf{n}$ and $-\mathbf{n}$ are physically indistinguishable and therefore one imposes the condition that

$$W(\mathbf{n}, \nabla \mathbf{n}) = W(-\mathbf{n}, -\nabla \mathbf{n}) \qquad (2)$$

and invariance to rigid rotations requires that

$$W(\mathbf{n}, \nabla \mathbf{n}) = W(\underline{P}\mathbf{n}, \underline{P}\nabla \mathbf{n}\underline{P}^T) \qquad (3)$$

where $\underline{P}$ is any proper orthogonal matrix, the superscript denoting the transpose of the matrix. While the above suffices for chiral nematics or cholesterics, for non-chiral nematics, invariably referred to simply as nematics, material symmetry requires that (Eq. 3) be extended to

$$W(\mathbf{n}, \nabla \mathbf{n}) = W(\underline{Q}\mathbf{n}, \underline{Q}\nabla \mathbf{n}\underline{Q}^T) \qquad (4)$$

where the matrix $\underline{Q}$ belongs to the full orthogonal group, the function is therefore isotropic rather than hemitropic.

Oseen and Frank both consider an energy function that is quadratic in the gradients of the director $\mathbf{n}$, in which case the conditions (Eq. 2) and (Eq. 3) lead to

$$\begin{aligned}2W = &\, K_1(\text{div }\mathbf{n})^2 + K_2(\mathbf{n}\cdot\text{curl }\mathbf{n} + q)^2 \\ &+ K_3|\mathbf{n}\times\text{curl }\mathbf{n}|^2 + (K_2 + K_4) \\ &\cdot \text{div}[(\mathbf{n}\cdot\text{grad})\mathbf{n} - (\text{div }\mathbf{n})\mathbf{n}]\end{aligned} \qquad (5)$$

where the Ks and q are constants. The above is the form appropriate to cholesterics or chiral nematics, q being related to the natural pitch of the characteristic helical configurations found in these materials, through

$$p = 2\pi/q. \qquad (6)$$

For ordinary nematics, however, the conditions (Eq. 2) and (Eq. 4) yield

$$\begin{aligned}2W = &\, K_1(\text{div }\mathbf{n})^2 + K_2(\mathbf{n}\cdot\text{curl }\mathbf{n})^2 \\ &+ K_3|\mathbf{n}\times\text{curl }\mathbf{n}|^2 + (K_2 + K_4) \\ &\cdot \text{div}[(\mathbf{n}\cdot\text{grad})\mathbf{n} - (\text{div }\mathbf{n})\mathbf{n}]\end{aligned} \qquad (7)$$

the coefficient q necessarily zero for such materials. This latter energy can alternatively be expressed as

$$\begin{aligned}2W = &\, (K_1 - K_2 - K_4)(\text{div }\mathbf{n})^2 \\ &+ K_2 tr(\nabla\mathbf{n}\nabla\mathbf{n}^T) + K_4 tr(\nabla\mathbf{n})^2 \\ &+ (K_3 - K_2)(\mathbf{n}\cdot\text{grad})\mathbf{n}\cdot(\mathbf{n}\cdot\text{grad})\mathbf{n}\end{aligned} \qquad (8)$$

$tr\underline{A}$ and $\underline{A}^T$ denoting the trace and transpose of the matrix $\underline{A}$, respectively. This latter form can be more convenient for some purposes. It is common to refer to the constants K_1, K_2 and K_3 as the splay, twist and bend elastic constants, respectively, while the K_4 term is sometimes omitted since the last term in the form (Eq. 7) can clearly be expressed as a surface integral.

Given that nematics tend to align uniformly with the anisotropic axis everywhere parallel, Ericksen [18] argues that this must represent a state of minimum energy, and thus assumes that

$$W(\mathbf{n}, \nabla\mathbf{n}) > W(\mathbf{n}, 0), \quad \nabla\mathbf{n}\neq 0 \qquad (9)$$

and as a consequence the coefficients in (Eq. 7) or (Eq. 8) must satisfy

$$K_1 > 0, \quad K_2 > 0, \quad K_3 > 0, \\ 2K_1 > K_2 + K_4 > 0, \quad K_2 > K_4 \qquad (10)$$

Jenkins [19] discusses the corresponding restrictions placed upon the energy (Eq. 5) by assuming that the characteristic twisted helical configuration represents a minimum of energy.

To conclude this section we note an identity derived by Ericksen [6] from the condition (Eq. 3) by selecting

$$\underline{P} = \underline{I} + \varepsilon \underline{R}, \quad \underline{R}^T = -\underline{R} \qquad (11)$$

$\underline{I}$ being the unit matrix and ε a small parameter. With this choice one can quickly show that

$$\mathbf{n} * \frac{\partial W}{\partial \mathbf{n}} + \nabla \mathbf{n} \left(\frac{\partial W}{\partial \nabla \mathbf{n}} \right)^T + \nabla \mathbf{n}^T \frac{\partial W}{\partial \nabla \mathbf{n}}$$
$$= \frac{\partial W}{\partial \mathbf{n}} * \mathbf{n} + \frac{\partial W}{\partial \nabla \mathbf{n}} \nabla \mathbf{n}^T + \left(\frac{\partial W}{\partial \nabla \mathbf{n}} \right)^T \nabla \mathbf{n} \quad (12)$$

a result required below. In the above, the notation $\mathbf{a} * \mathbf{b}$ represents the 3×3 matrix with $(i, j)^{th}$ element $a_i b_j$.

1.2.2 A Virtual Work Formulation

The approach adopted by Ericksen [5] to equilibrium theory for both nematic and cholesteric liquid crystals appeals to a principle of virtual work, which for any volume V of material bounded by surface S takes the form

$$\delta \int_V W \, dv = \int_V (\mathbf{F} \cdot \delta \mathbf{x} + \mathbf{G} \cdot \Delta \mathbf{n}) \, dv$$
$$+ \int_S (\mathbf{t} \cdot \delta \mathbf{x} + \mathbf{s} \cdot \Delta \mathbf{n}) \, ds \qquad (13)$$

where

$$\Delta \mathbf{n} = \delta \mathbf{n} + (\delta \mathbf{x} \cdot \text{grad}) \mathbf{n} \qquad (14)$$

$\mathbf{F}$ denotes body force per unit volume, $\mathbf{t}$ surface force per unit area, and $\mathbf{G}$ and $\mathbf{s}$ are generalized body and surface forces, respectively. With the common assumption of incompressibility the virtual displacement $\delta \mathbf{x}$ is not arbitrary, but is subject to the constraint

$$\text{div } \delta \mathbf{x} = 0 \qquad (15)$$

and of course the variations $\delta \mathbf{n}$ and $\Delta \mathbf{n}$ are constrained by

$$\delta \mathbf{n} \cdot \mathbf{n} = \Delta \mathbf{n} \cdot \mathbf{n} = 0 \qquad (16)$$

due to $\mathbf{n}$ being of fixed magnitude.

Through consideration of an arbitrary, infinitesimal, rigid displacement in which $\Delta \mathbf{n}$ is zero, it quickly follows from (Eq. 13) that

$$\int_V \mathbf{F} \, dv + \int_S \mathbf{t} \, ds = 0 \qquad (17)$$

which of course expresses the fact that the resultant force is zero in equilibrium. Similarly, consideration of an arbitrary, infinitesimal, rigid rotation $\boldsymbol{\omega}$, in which

$$\delta \mathbf{x} = \boldsymbol{\omega} \times \mathbf{x}, \quad \Delta \mathbf{n} = \boldsymbol{\omega} \times \mathbf{n} \qquad (18)$$

yields from (Eq. 13) following rearrangement of the triple scalar products

$$\int_V (\mathbf{x} \times \mathbf{F} + \mathbf{n} \times \mathbf{G}) \, dv + \int_S (\mathbf{x} \times \mathbf{t} + \mathbf{n} \times \mathbf{s}) \, ds = 0$$
$$(19)$$

which one interprets as a statement that the resulting moment is zero in equilibrium. Hence (Eq. 19) relates the generalized forces $\mathbf{G}$ and $\mathbf{s}$ to the body and surface moments $\mathbf{K}$ and $\boldsymbol{\ell}$, respectively, through

$$\mathbf{K} = \mathbf{n} \times \mathbf{G}, \quad \boldsymbol{\ell} = \mathbf{n} \times \mathbf{s} \qquad (20)$$

which allows the determination of the generalized body force.

By first expressing the left hand side of the statement of virtual work (Eq. 13) as using (Eq. 15)

$$\delta \int_V W \, dv = \int_V (\delta W + (\delta \mathbf{x} \cdot \text{grad}) W) \, dv \qquad (21)$$

this taking account of the change of volume in the virtual displacement, and then reorganizing the resultant volume integral so that it has a similar format to the right hand side of (Eq. 13), it is possible to obtain expressions for the surface force $\mathbf{t}$ and the generalized surface force **s**, and also two equations required to hold in equilibrium [5]. Denoting by $\boldsymbol{\nu}$ the unit outwards normal to the surface S and bearing in mind the constraints (Eq. 15) and (Eq. 16), one finds from the surface integrals

$$\mathbf{t} = -p\boldsymbol{\nu} + \underline{T}^s \boldsymbol{\nu}, \quad \underline{T}^s = -\nabla \mathbf{n}^T \frac{\partial W}{\partial \nabla \mathbf{n}}$$

$$\mathbf{s} = \beta \mathbf{n} + \underline{S}^s \boldsymbol{\nu}, \quad \underline{S}^s = \frac{\partial W}{\partial \nabla \mathbf{n}} \quad (22)$$

and also from the volume integrals

$$\mathbf{F} - \text{grad } p + \text{div } \underline{T}^s = 0,$$

$$\mathbf{G} - \frac{\partial W}{\partial \mathbf{n}} + \text{div } \underline{S}^s = \gamma \mathbf{n}, \quad (23)$$

with p an arbitrary pressure due to incompressibility, and β and γ arbitrary scalars due to the director having fixed magnitude. In equations (Eq. 23) the divergence applies to the second of the indices of the matrices $\underline{T}^s$ and $\underline{S}^s$. The former of (Eq. 23) clearly represents the point form of the balance of forces (Eq. 17), while the latter can be shown to be the point form of the balance of moments (Eq. 19), this requiring some manipulation involving the identity (Eq. 12).

1.2.3 Body Forces and Moments

While the action of gravity upon a liquid crystal is identical to that on other materials, external magnetic and electric fields have a rather different effect upon these anisotropic liquids than they do on isotropic materials. Both can give rise to body forces and moments as is to be expected from rather simple arguments common in magnetostatics and electrostatics. To fix ideas, consider a magnetic field **H** which induces a magnetization **M** in the material, and this in turn gives rise to a body force **F** and a body moment **K**, given by

$$\mathbf{F} = (\mathbf{M} \cdot \text{grad})\mathbf{H}, \quad \mathbf{K} = \mathbf{M} \times \mathbf{H} \quad (24)$$

In an isotropic material the induced magnetization is necessarily parallel to the field and the couple is zero, but for a nematic or cholesteric liquid crystal the magnetization can have an anisotropic contribution of the form

$$\mathbf{M} = \chi_\perp \mathbf{H} + \chi_a \mathbf{n} \cdot \mathbf{H} \mathbf{n}, \quad \chi_a = \chi_\| - \chi_\perp \quad (25)$$

$\chi_\|$ and $\chi_\perp$ denoting the diamagnetic susceptibilities when **n** and **H** are parallel and perpendicular, respectively. As a consequence a body moment can occur given by

$$\mathbf{K} = \chi_a \mathbf{n} \cdot \mathbf{H} \mathbf{n} \times \mathbf{H} \quad (26)$$

and it immediately follows from the first of equations (Eq. 20) that

$$\mathbf{G} = \chi_a \mathbf{n} \cdot \mathbf{H} \mathbf{H} \quad (27)$$

any contribution parallel to the director being simply absorbed in the scalar γ in equations (Eq. 23). In general the anisotropy of the diamagnetic susceptibility χ_a is positive, but it is also rather small. Consequently one can ignore the influence of the liquid crystal upon the applied field.

Similar expressions arise for an electric field **E**, this creating an electric displacement **D** of the form

$$\mathbf{D} = \varepsilon_\perp \mathbf{E} + \varepsilon_a \mathbf{n} \cdot \mathbf{E} \mathbf{n}, \quad \varepsilon_a = \varepsilon_\| - \varepsilon_\perp \quad (28)$$

$\varepsilon_\|$ and $\varepsilon_\perp$ denoting the corresponding dielectric permittivities, which gives rise to similar body forces and moments. However, as Deuling [20] points out, there can be one important difference between the effects of magnetic and electric fields upon liquid crystals, in that an electric field can give rise to significant permittivities, and thus one must allow for the influence of the liquid crystal upon the applied field by employing the appropriate reduced version of Maxwell's equations.

Associated with a magnetic field there is an energy

$$\psi = \frac{1}{2} \mathbf{M} \cdot \mathbf{H} \quad (29)$$

and if one regards the energy ψ as simply a function of the director **n** and position **x**, then one can write

$$\mathbf{F} = \frac{\partial \psi}{\partial \mathbf{x}}, \quad \mathbf{G} = \frac{\partial \psi}{\partial \mathbf{n}} \qquad (30)$$

the former using the fact that in equilibrium the magnetic field is irrotational. However, if the field applied is dependent upon the director, as can occur for an electric field, then (Eq. 30) are not valid.

1.2.4 The Equilibrium Equations

The equations (Eq. 23) representing balance of forces and moments constitute six equations for four unknowns, two components of the unit vector **n** and the scalars p and γ. However, this apparent overdeterminacy does not materialize if the external body force and moment meet a certain requirement [5]. To see this combine the two equations as follows

$$\mathbf{F} + \nabla \mathbf{n}^T \mathbf{G} - \operatorname{grad} p + \operatorname{div} \underline{T}^s$$
$$+ \nabla \mathbf{n}^T \operatorname{div} \underline{S}^s - \nabla \mathbf{n}^T \frac{\partial W}{\partial \mathbf{n}} = 0 \qquad (31)$$

and by appeal to equations (Eq. 22) this reduces to

$$\mathbf{F} + \nabla \mathbf{n}^T \mathbf{G} = \operatorname{grad}(p + W) \qquad (32)$$

clearly limiting the body force and moment. However, when (Eq. 30) applies, the above at once yields

$$p + W - \psi = p_o \qquad (33)$$

where p_o is an arbitrary constant pressure, and thus the equation for the balance of forces integrates to give the pressure, removing the potential overdeterminacy. Also the balance of moments becomes

$$\operatorname{div} \underline{S}^s - \frac{\partial W}{\partial \mathbf{n}} + \frac{\partial \psi}{\partial \mathbf{n}} = \gamma \mathbf{n} \qquad (34)$$

which can be written in forms more convenient for particular problems as discussed by Ericksen [21].

Frequently one selects a particular form for the director **n** so that it is immediately a unit vector, this representation invariably involving two angles θ and ϕ so that

$$\mathbf{n} = \mathbf{f}(\theta, \phi), \quad \mathbf{n} \cdot \frac{\partial \mathbf{f}}{\partial \theta} = \mathbf{n} \cdot \frac{\partial \mathbf{f}}{\partial \phi} = 0 \qquad (35)$$

Initially for purposes of illustration it is convenient to restrict θ and ϕ to be functions of a single Cartesian coordinate, say z, and consider the application of an external magnetic field **H**, so that (Eq. 30) holds. In this event

$$W = W(\theta, \phi, \theta', \phi'), \quad \psi = \chi(\mathbf{x}, \theta, \phi) \qquad (36)$$

where the prime denotes differentiations with respect to z. Employing the chain rule, one obtains

$$\mathbf{n}' = \frac{\partial \mathbf{f}}{\partial \theta} \theta' + \frac{\partial \mathbf{f}}{\partial \phi} \phi' \qquad (37)$$

and thus

$$\frac{\partial W}{\partial \theta} = \frac{\partial W}{\partial \mathbf{n}} \cdot \frac{\partial \mathbf{f}}{\partial \theta} + \frac{\partial W}{\partial \mathbf{n}'} \cdot \left(\frac{\partial^2 \mathbf{f}}{\partial \theta^2} \theta' + \frac{\partial^2 \mathbf{f}}{\partial \theta \partial \phi} \phi' \right)$$

$$\frac{\partial W}{\partial \theta'} = \frac{\partial W}{\partial \mathbf{n}'} \cdot \frac{\partial \mathbf{f}}{\partial \theta}, \quad \frac{\partial \chi}{\partial \theta} = \frac{\partial \psi}{\partial \mathbf{n}} \cdot \frac{\partial \mathbf{f}}{\partial \theta} \qquad (38)$$

with similar expressions for the derivatives with respect to ϕ and ϕ'. Combining the above it follows that

$$\left(\frac{\partial W}{\partial \theta'} \right)' - \frac{\partial W}{\partial \theta} + \frac{\partial \chi}{\partial \theta}$$
$$= \left[\left(\frac{\partial W}{\partial \mathbf{n}'} \right)' - \frac{\partial W}{\partial \mathbf{n}} + \frac{\partial \psi}{\partial \mathbf{n}} \right] \cdot \frac{\partial \mathbf{f}}{\partial \theta} \qquad (39)$$

and similarly

$$\left(\frac{\partial W}{\partial \phi'} \right)' - \frac{\partial W}{\partial \phi} + \frac{\partial \chi}{\partial \phi}$$
$$= \left[\left(\frac{\partial W}{\partial \mathbf{n}'} \right)' - \frac{\partial W}{\partial \mathbf{n}} + \frac{\partial \psi}{\partial \mathbf{n}} \right] \cdot \frac{\partial \mathbf{f}}{\partial \phi} \qquad (40)$$

Hence, employing Eqs. (34) and (35) the former can be rewritten as

$$\left(\frac{\partial W}{\partial \theta'}\right)' - \frac{\partial W}{\partial \theta} + \frac{\partial \chi}{\partial \theta} = 0,$$

$$\left(\frac{\partial W}{\partial \phi'}\right)' - \frac{\partial W}{\partial \phi} + \frac{\partial \chi}{\partial \phi} = 0 \qquad (41)$$

For the general case when θ and ϕ are arbitrary functions of Cartesian coordinates, essentially a repetition of the above leads to

$$\mathrm{div}\left(\frac{\partial W}{\partial \nabla \theta}\right) - \frac{\partial W}{\partial \theta} + \frac{\partial \chi}{\partial \theta} = 0,$$

$$\mathrm{div}\left(\frac{\partial W}{\partial \nabla \phi}\right) - \frac{\partial W}{\partial \phi} + \frac{\partial \chi}{\partial \phi} = 0 \qquad (42)$$

which are clearly more convenient to use than (Eq. 34).

In addition, as Ericksen [21] also shows, the above reformulation can be extended to include curvilinear coordinate systems (y_1, y_2, y_3) introduced by

$$\mathbf{x} = \mathbf{x}(y_1, y_2, y_3). \qquad (43)$$

Denoting the Jacobian of this transformation by

$$J = \frac{\partial(x_1, x_2, x_3)}{\partial(y_1, y_2, y_3)} \qquad (44)$$

and introducing the notation

$$\overline{W} = JW = \overline{W}\left(y_i, \theta, \phi, \frac{\partial \theta}{\partial y_i}, \frac{\partial \phi}{\partial y_i}\right)$$

$$\overline{\psi} = J\psi = \overline{\chi}(y_i, \theta, \phi) \qquad (45)$$

where y_i is short for y_1, y_2, and y_3, and similarly for the partial derivatives, one can show that (Eq. 34) can be recast as

$$\left(\frac{\partial \overline{W}}{\partial \theta_{,i}}\right)_{,i} - \frac{\partial \overline{W}}{\partial \theta} + \frac{\partial \overline{\chi}}{\partial \theta} = 0,$$

$$\left(\frac{\partial \overline{W}}{\partial \phi_{,i}}\right)_{,i} - \frac{\partial \overline{W}}{\partial \phi} + \frac{\partial \overline{\chi}}{\partial \phi} = 0 \qquad (46)$$

where

$$\theta_{,i} = \frac{\partial \theta}{\partial y_i}, \quad \phi_{,i} = \frac{\partial \phi}{\partial y_i} \qquad (47)$$

and the repeated index is summed over the values 1, 2 and 3.

Rather clearly Eqs. (41), (42) and (46) are appropriate forms of the Euler–Lagrange equations for the integral

$$\int_V (W - \psi) \mathrm{d}v \qquad (48)$$

this the formulation of equilibrium theory initially adopted by, for example, Oseen [1] and Frank [4].

For electric fields, certainly for the special cases generally considered, the outcome is rather similar in that the equation for the balance of forces integrates to give the pressure, and one can recast that for balance of moments in the same way as above. However, a general treatment does not appear to be available.

1.2.5 Boundary Conditions

In general the choice of boundary conditions for the alignment at a liquid crystal–solid interface is one of two options, either strong or weak anchoring [8]. Strong anchoring as the term suggests implies that the alignment is prescribed at the boundary by a suitable prior treatment of the solid surface and remains fixed in the presence of competing agencies to realign it. Most commonly this fixed direction is in the plane of the surface (planar) or it is perpendicular (homeotropic), but it need not be so. Weak anchoring, first proposed by Papoular and Rapini [22] assigns an energy to the liquid crystal–solid interface, and assumes a balance between the moment or torque in the liquid crystal from the Frank–Oseen energy and that arising from the interfacial energy. Denoting by w this latter energy per unit area, the simplest assumption is a dependence upon the director **n** and a fixed direction $\mathbf{n}_o$ at the interface, so that

$$w = w(\mathbf{n}, \mathbf{n}_o) \qquad (49)$$

but equally one can have

$$w = w(\mathbf{n}, \mathbf{\nu}, \mathbf{\tau}) \qquad (50)$$

where $\mathbf{\nu}$ is again the unit normal and $\mathbf{\tau}$ a fixed unit vector on the surface. By essentially a repetition of the approach of section 2, Jenkins and Barratt [23] show that this leads to the boundary condition

$$\frac{\partial W}{\partial \nabla \mathbf{n}} \mathbf{\nu} + \frac{\partial w}{\partial \mathbf{n}} = \lambda \mathbf{n} \qquad (51)$$

W again denoting the Frank–Oseen energy and λ an arbitrary scalar.

If one introduces the representation (Eq. 35), it follows that

$$W = W(\theta, \phi, \nabla\theta, \nabla\phi), \quad w = \omega(\theta, \phi) \qquad (52)$$

and using the methods of the previous section the boundary condition (Eq. 51) becomes

$$\frac{\partial W}{\partial \nabla \theta} \cdot \mathbf{\nu} + \frac{\partial \omega}{\partial \theta} = 0, \quad \frac{\partial W}{\partial \nabla \phi} \cdot \mathbf{\nu} + \frac{\partial \omega}{\partial \phi} = 0 \qquad (53)$$

Frequently these boundary conditions can be reduced to equations for θ and ϕ at the interface by eliminating the derivatives of θ and ϕ, after they have been obtained from the equilibrium equations.

More recently, however, it has become apparent that the situation can be more complex with the realization that surface anchoring in nematics can be bistable, as found by Jerome, Pieranski and Boix [24] and Monkade, Boix and Durand [25]. Shortly thereafter, Barberi, Boix and Durand [26] showed that one can switch the surface alignment from one anchoring to the other using an electric field. As Nobili and Durand [27] discuss, one must consider rather more complex forms for the surface energy than previously in order to model these effects adequately, and they measure some relevant parameters, Sergan and Durand [28] describing further measurements.

1.2.6 Proposed Extensions

Some 25 years ago, Nehring and Saupe [16] proposed that one should add terms linear in second gradients of the director to the Frank–Oseen energy, this ultimately entailing the inclusion of a single additional term, namely

$$K_{13} \, \text{div}((\text{div } \mathbf{n})\mathbf{n}) \qquad (54)$$

which clearly proves to be a surface term in the sense that the volume integral integrates to give a surface integral over the boundary. On the grounds of a microscopic calculation they argue that the coefficient in (Eq. 54) is comparable in magnitude to the other coefficients in the Frank–Oseen energy, and so the term should be added.

More recently Oldano and Barbero [29] include such a term and consider the variational problem for static solutions to conclude that in general there are no continuous solutions to this problem, since it is not possible to satisfy all of the boundary conditions that arise in the variational formulation. These are of two types, one corresponding to weak anchoring as described in the previous section, but the other lacking a physical interpretation. This has given rise to some controversy, initially with Hinov [30], but later with contributions from Barbero and Strigazzi [31], Barbero, Sparavigna and Strigazzi [32], Barbero [33], Pergamenshchik [34] and Faetti [35,36]. In general the arguments rest solely upon a resolution of the variational problem, and mostly favour the inclusion of higher order derivatives in the bulk elastic energy in different ways to overcome this difficulty. More recently Barbero and Durand [37] relate this additional term to temperature-induced surface transitions in the alignment, and Barberi, Barbero, Giocondo and Moldovan [38] attempt to measure the corresponding coefficient. Also Lavrentovich and Pergamenshchik [39] argue that such a term explains their observations of stripe domains in experiments with very thin films, and also provide a measurement of the coefficient. While these latter developments relating to experimental observations are of interest, it is rather early to draw any conclusions.

A further recent innovation is due to Ericksen [17] who proposes an extension to the Frank–Oseen theory in order to improve solutions modelling defects. To this end he incorporates some variation in the degree of alignment or the order parameter, and therefore proposes an energy of the form

$$W = W(s, \mathbf{n}, \nabla s, \nabla \mathbf{n}) \tag{55}$$

where $\mathbf{n}$ is again a unit vector describing alignment and s is a scalar representing the degree of order or alignment. By allowing this scalar to tend to zero as one approaches point or line defects, it is possible to avoid the infinite energies that can occur with the Frank–Oseen energy. Some account of analyses based on this development are to be found in the book by Virga [11].

1.3 Equilibrium Theory for Smectic Liquid Crystals

1.3.1 An Energy Function for SmC Liquid Crystals

The first to give serious thought to continuum theory for smectics appears to have been de Gennes' group at Orsay [8]. Amongst other things they present an energy function for SmC liquid crystals [40], albeit restricted to small perturbations of planar layers. More recently, however, Leslie, Stewart, and Nakagawa [41] derive an energy for such smectics which is not limited to small perturbations, but which is identical to the Orsay energy when so restricted [42]. Below our aim is to present this energy, and show its relationship to that proposed by the Orsay group.

One can conveniently describe the layering in smectic liquid crystals by employing a density wave vector $\mathbf{a}$, which following Oseen [43] and de Gennes [8] is subject to the constraint

$$\text{curl } \mathbf{a} = 0 \tag{56}$$

provided that defects or dislocations in the layering are absent. To describe SmC configurations de Gennes adds a second vector $\mathbf{c}$ perpendicular to $\mathbf{a}$ and in the direction of inclination of the tilt of alignment with respect to $\mathbf{a}$. Leslie, Stewart and Nakagawa [41] invoke two simplifications to reduce mathematical complexity that clearly restrict the range of applicability of their energy, but equally appear very reasonable in many cases. They firstly assume that the layers, although deformable, remain of constant thickness, and also that the angle of tilt with respect to the layer normal remains constant, the latter excluding thermal and pretransitional effects. With these assumptions there is no loss of generality in choosing both $\mathbf{a}$ and $\mathbf{c}$ to be unit vectors, identifying $\mathbf{a}$ with the layer normal, the constraint (Eq. 56) still applying.

As in nematic theory one assumes the existence of an elastic energy W which here is a function of both $\mathbf{a}$ and $\mathbf{c}$ and their gradients, so that

$$W = W(\mathbf{a}, \mathbf{c}, \nabla \mathbf{a}, \nabla \mathbf{c}) \tag{57}$$

The absence of polarity in the alignment implies that the energy is independent of a simultaneous change of sign in **a** and **c**, and thus

$$W(\mathbf{a}, \mathbf{c}, \nabla\mathbf{a}, \nabla\mathbf{c}) = W(-\mathbf{a}, -\mathbf{c}, -\nabla\mathbf{a}, -\nabla\mathbf{c}) \tag{58}$$

Invariance to rigid rotations adds the requirement that

$$W(\mathbf{a}, \mathbf{c}, \nabla\mathbf{a}, \nabla\mathbf{c}) = W(\underline{P}\mathbf{a}, \underline{P}\mathbf{c}, \underline{P}\nabla\mathbf{a}\underline{P}^T, \underline{P}\nabla\mathbf{c}\underline{P}^T) \tag{59}$$

for all proper orthogonal matrices $\underline{P}$. While these conditions suffice for chiral materials, for non-chiral materials symmetry requires that (Eq. 59) is extended to include all orthogonal matrices.

Assuming a quadratic dependence upon gradients one finds for non-chiral materials that [44]

$$\begin{aligned}2W = &K_1^a(tr\nabla\mathbf{a})^2 + K_2^a(\mathbf{c}\cdot\nabla\mathbf{ac})^2 \\&+ 2K_3^a(tr\nabla\mathbf{a})(\mathbf{c}\cdot\nabla\mathbf{ac}) \\&+ K_1^c(tr\nabla\mathbf{c})^2 + K_2^c tr(\nabla\mathbf{c}\nabla\mathbf{c}^T) \\&+ K_3^c(\nabla\mathbf{cc}\cdot\nabla\mathbf{cc}) \\&+ 2K_4^c(\nabla\mathbf{cc}\cdot\nabla\mathbf{ca}) \\&+ 2K_1^{ac}(tr\nabla\mathbf{c})(\mathbf{c}\cdot\nabla\mathbf{ac}) \\&+ 2K_2^{ac}(tr\nabla\mathbf{c})(tr\nabla\mathbf{a})\end{aligned} \tag{60}$$

the coefficients being constants. This expression omits three surface terms associated with

$$tr(\nabla\mathbf{c})^2 - (tr\nabla\mathbf{c})^2, \quad tr(\nabla\mathbf{a})^2 - (tr\nabla\mathbf{a})^2,$$
$$tr(\nabla\mathbf{c}\nabla\mathbf{a}) - (tr\nabla\mathbf{a})(tr\nabla\mathbf{c}), \tag{61}$$

partly on the grounds that the above suffices for present purposes.

For chiral materials Carlsson, Stewart and Leslie [45] show that one must add two terms

$$2K_4^a \mathbf{b}\cdot\nabla\mathbf{ac}, \quad 2K_5^c \mathbf{b}\cdot\nabla\mathbf{ca} \tag{62}$$

plus an additional surface term associated with div **b**, where **b** completes the orthonormal triad with the vectors **a** and **c**, so that

$$\mathbf{b} = \mathbf{a}\times\mathbf{c} \tag{63}$$

The latter of (Eq. 62) describes the characteristic twist of the **c** director about the layer normal, while the former implies a non-planar equilibrium configuration, and consequently is generally omitted.

One can of course employ any two members of the orthonormal triad **a**, **b** and **c** to describe smectic configurations, and employing **b** and **c** the energy (Eq. 60) becomes [44]

$$\begin{aligned}2W = &A_{12}(\mathbf{b}\cdot\text{curl }\mathbf{c})^2 + A_{21}(\mathbf{c}\cdot\text{curl }\mathbf{b})^2 \\&+ 2A_{11}(\mathbf{b}\cdot\text{curl }\mathbf{c})(\mathbf{c}\cdot\text{curl }\mathbf{b}) \\&+ B_1(\text{div }\mathbf{b})^2 + B_2(\text{div }\mathbf{c})^2 \\&+ B_3\left[\frac{1}{2}(\mathbf{b}\cdot\text{curl }\mathbf{b}+\mathbf{c}\cdot\text{curl }\mathbf{c})\right]^2 \\&+ 2B_{13}(\text{div }\mathbf{b})\left[\frac{1}{2}(\mathbf{b}\cdot\text{curl }\mathbf{b}+\mathbf{c}\cdot\text{curl }\mathbf{c})\right] \\&+ 2C_1(\text{div }\mathbf{c})(\mathbf{b}\cdot\text{curl }\mathbf{c}) \\&+ 2C_2(\text{div }\mathbf{c})(\mathbf{c}\cdot\text{curl }\mathbf{b})\end{aligned} \tag{64}$$

As Carlsson, Stewart and Leslie [42] discuss, the terms associated with the coefficients A_{12}, A_{21}, B_1, B_2 and B_3 represent independent deformations of the uniformly aligned planar layers, the remaining terms being coupling terms. Also, as the notation for the coefficients anticipates, this expression reduces to the Orsay energy [40] when one considers small perturbations of uniformly aligned planar layers. The coefficients in the two energies (Eq. 60) and (Eq. 64) are related by

$$\begin{aligned}&K_1^a = A_{21}, \quad K_2^a = 2A_{11} + A_{12} + A_{21} + B_3 - B_1, \\&\quad 2K_3^a = -(2A_{11} + 2A_{21} + B_3) \\&K_1^c = B_2 - B_3, \quad K_2^c = B_3, \quad K_3^c = B_1 - B_3, \\&K_4^c = B_{13} \\&K_1^{ac} = C_1 + C_2, \quad K_2^{ac} = -C_2\end{aligned} \tag{65}$$

as Leslie, Stewart, Carlsson and Nakagawa [44] demonstrate.

1.3.2 Equilibrium Equations

Adopting a similar approach to that by Ericksen for nematics, Leslie, Stewart and Nakagawa [41] assume a principle of virtual work for a volume V of smectic liquid crystal bounded by a surface S of the form

$$\delta \int_V W \, dv = \int_V (\mathbf{F} \cdot \delta \mathbf{x} + \mathbf{G}^a \cdot \Delta \mathbf{a} + \mathbf{G}^c \cdot \Delta \mathbf{c}) \, dv$$
$$+ \int_S (\mathbf{t} \cdot \delta \mathbf{x} + \mathbf{s}^a \cdot \Delta \mathbf{a} + \mathbf{s}^c \cdot \Delta \mathbf{c}) \, ds \quad (66)$$

where

$$\Delta \mathbf{a} = \delta \mathbf{a} + (\delta \mathbf{x} \cdot \text{grad}) \mathbf{a},$$
$$\Delta \mathbf{c} = \delta \mathbf{c} + (\delta \mathbf{x} \cdot \text{grad}) \mathbf{c} \quad (67)$$

$\mathbf{F}$ denotes external body force per unit volume, $\mathbf{G}^a$ and $\mathbf{G}^c$ generalized external body forces per unit volume, $\mathbf{t}$ surface force per unit area, and $\mathbf{s}^a$ and $\mathbf{s}^c$ generalized surface forces per unit area. As before, on account of the assumed incompressibility, the virtual displacement is not arbitrary, but is subject to

$$\text{div} \, \delta \mathbf{x} = 0 \quad (68)$$

and the variation $\delta \mathbf{a}$ must satisfy

$$\text{curl} \, \delta \mathbf{a} = \text{curl}(\Delta \mathbf{a} - \nabla \mathbf{a} \delta \mathbf{x}) = 0 \quad (69)$$

on account of (Eq. 56). In addition of course

$$\mathbf{a} \cdot \delta \mathbf{a} = \mathbf{c} \cdot \delta \mathbf{c} = \mathbf{a} \cdot \Delta \mathbf{a} = \mathbf{c} \cdot \Delta \mathbf{c} = 0$$
$$\mathbf{a} \cdot \delta \mathbf{c} + \mathbf{c} \cdot \delta \mathbf{a} = \mathbf{a} \cdot \Delta \mathbf{c} + \mathbf{c} \cdot \Delta \mathbf{a} = 0 \quad (70)$$

given that $\mathbf{a}$ and $\mathbf{c}$ are mutually orthogonal unit vectors.

By a repetition of the arguments for a nematic regarding an infinitesimal, rigid displacement and rotation, one deduces that

$$\int_V \mathbf{F} \, dv + \int_S \mathbf{t} \, ds = 0$$
$$\int_V (\mathbf{x} \times \mathbf{F} + \mathbf{a} \times \mathbf{G}^a + \mathbf{c} \times \mathbf{G}^c) \, dv$$
$$= \int_S (\mathbf{x} \times \mathbf{t} + \mathbf{a} \times \mathbf{s}^a + \mathbf{c} \times \mathbf{s}^c) \, ds = 0 \quad (71)$$

representing balance of forces and moments, respectively. Hence, as above, it is possible to relate the generalized forces to a body moment $\mathbf{K}$ and surface moment $\boldsymbol{\ell}$ as follows

$$\mathbf{K} = \mathbf{a} \times \mathbf{G}^a + \mathbf{c} \times \mathbf{G}^c, \quad \boldsymbol{\ell} = \mathbf{a} \times \mathbf{s}^a + \mathbf{c} \times \mathbf{s}^c \quad (72)$$

essentially allowing the determination of the generalized body forces.

By rewriting the left hand side of (Eq. 66) just as before, it follows that the surface force and generalized surface forces are given respectively by [41]

$$\mathbf{t} = -p\boldsymbol{\nu} + \nabla \mathbf{a}^T (\boldsymbol{\beta} \times \boldsymbol{\nu}) + \underset{\sim}{T^s} \boldsymbol{\nu},$$

$$\underset{\sim}{T^s} = -\nabla \mathbf{a}^T \frac{\partial W}{\partial \nabla \mathbf{a}} - \nabla \mathbf{c}^T \frac{\partial W}{\partial \nabla \mathbf{c}}$$

$$\mathbf{s}^a = \alpha \mathbf{a} + \mu \mathbf{c} + \boldsymbol{\nu} \times \boldsymbol{\beta} + \underset{\sim}{S^a} \boldsymbol{\nu}, \quad \underset{\sim}{S^a} = \frac{\partial W}{\partial \nabla \mathbf{a}}$$

$$\mathbf{s}^c = \lambda \mathbf{c} + \mu \mathbf{a} + \underset{\sim}{S^c} \boldsymbol{\nu}, \quad \underset{\sim}{S^c} = \frac{\partial W}{\partial \nabla \mathbf{c}} \quad (73)$$

where $\boldsymbol{\nu}$ denotes the unit surface normal, p an arbitrary pressure due to the assumed incompressibility, $\boldsymbol{\beta}$ an arbitrary vector arising from the constraint (Eq. 69), and α, λ and μ arbitrary scalars stemming from the constraints (Eq. 70). In addition one obtains the balance laws

$$\mathbf{F} - \text{grad} \, p - \nabla \mathbf{a}^T (\text{curl} \, \boldsymbol{\beta}) + \text{div} \, \underset{\sim}{T^s} = 0$$

$$\text{div} \, \underset{\sim}{S^a} - \frac{\partial W}{\partial \mathbf{a}} + \mathbf{G}^a + \gamma \mathbf{a} + \kappa \mathbf{c} + \text{curl} \, \boldsymbol{\beta} = 0$$

$$\text{div} \, \underset{\sim}{S^c} - \frac{\partial W}{\partial \mathbf{c}} + \mathbf{G}^c + \tau \mathbf{c} + \kappa \mathbf{a} = 0 \quad (74)$$

γ, κ and τ being arbitrary scalars again arising from the constrains (Eq. 70). The first of equations (Eq. 74) is clearly the point form of the balance of forces, the first of equations (Eq. 71), and the remaining two are equivalent to the second of (Eq. 71) representing balance of moments, this requiring the generalization of the identity (Eq. 12). Also one can show that the surface moment $\boldsymbol{\ell}$ is given by

$$\boldsymbol{\ell} = (\boldsymbol{\beta} \cdot \mathbf{a}) \boldsymbol{\nu} - (\mathbf{a} \cdot \boldsymbol{\nu}) \boldsymbol{\beta} + \mathbf{a} \times \frac{\partial W}{\partial \nabla \mathbf{a}} \boldsymbol{\nu}$$
$$+ \mathbf{c} \times \frac{\partial W}{\partial \nabla \mathbf{c}} \boldsymbol{\nu} \quad (75)$$

this combining Eqs. (72) and (73).

If external forces and moments are absent, it is possible to combine equations (Eq. 74) to obtain the integral

$$p + W = p_o \qquad (76)$$

where p_o is an arbitrary constant pressure. With certain restrictions on the external body force and moment, a similar result follows when these terms are present. Hence as for a nematic, the balance of forces need not concern us except possibly to compute surface forces, and we therefore have two Euler–Lagrange type equations representing balance of moments.

Boundary conditions for the above theory tend to be very similar to those employed in nematics, with little so far emerging by way of genuine smectic boundary conditions.

1.4 Dynamic Theory for Nematics

1.4.1 Balance Laws

To derive a dynamic theory one can of course extend the above formulation of equilibrium theory employing generalized body and surface forces as in the initial derivation [7,15]. Here, however, we prefer a different approach [46], which, besides providing an alternative, is more direct in that it follows traditional continuum mechanics more closely, although introducing body and surface moments usually excluded, as well as a new kinematic variable to describe alignment of the anisotropic axis.

As above we assume that the liquid crystal is incompressible, and thus the velocity vector $\mathbf{v}$ is subject to the constraint

$$\text{div } \mathbf{v} = 0 \qquad (77)$$

with the result that conservation of mass reduces to the statement that the density ρ is constant in a homogeneous material. As before the director $\mathbf{n}$ is constrained to be of unit magnitude. Since thermal effects are excluded, our two balance laws are those representing conservation of linear and angular momentum, which for a volume V of liquid crystal bounded by surface S take the forms

$$\frac{\mathrm{d}}{\mathrm{d}t}\int_V \rho \mathbf{v}\,\mathrm{d}v = \int_V \mathbf{F}\,\mathrm{d}v + \int_S \mathbf{t}\,\mathrm{d}s$$

$$\frac{\mathrm{d}}{\mathrm{d}t}\int_V \rho \mathbf{x}\times\mathbf{v}\,\mathrm{d}v$$

$$= \int_V (\mathbf{x}\times\mathbf{F}+\mathbf{K})\,\mathrm{d}v + \int_S (\mathbf{x}\times\mathbf{t}+\boldsymbol{\ell})\,\mathrm{d}s \qquad (78)$$

where $\mathbf{F}$ and $\mathbf{K}$ are body force and moment per unit volume, $\mathbf{t}$ and $\boldsymbol{\ell}$ surface force and moment per unit area, respectively, $\mathbf{x}$ the position vector, and the time derivative the material time derivative. The inertial term associated with the director in the latter of equations (Eq. 78) is omitted on the grounds that it is negligible.

The force and moment on a surface with unit normal $\boldsymbol{\nu}$ are given by

$$\mathbf{t} = -p\boldsymbol{\nu} + \underline{T}\boldsymbol{\nu}, \quad \boldsymbol{\ell} = \underline{L}\boldsymbol{\nu} \qquad (79)$$

where p is an arbitrary pressure arising from the assumed incompressibility, and $\underline{T}$ and $\underline{L}$ are the stress and couple stress matrices or tensors, respectively. As a consequence one can express (Eq. 78) in point form

$$\rho\frac{\mathrm{d}\mathbf{v}}{\mathrm{d}t} = \mathbf{F} - \text{grad } p + \text{div } \underline{T},$$

$$0 = \mathbf{K} + \hat{\mathbf{T}} + \text{div } \underline{L} \qquad (80)$$

where $\hat{\mathbf{T}}$ is the axial vector associated with the asymmetric matrix $\underline{T}$, so that

$$\hat{T}_x = T_{zy} - T_{yz}, \quad \hat{T}_y = T_{xz} - T_{zx},$$

$$\hat{T}_z = T_{yx} - T_{xy}, \qquad (81)$$

and the divergence is with respect to the second index of both $\underline{T}$ and $\underline{L}$. Finally we remark that for isotropic liquids $\mathbf{K}$ and $\underline{L}$ are generally assumed to be absent, so that angular momentum reduces to $\hat{\mathbf{T}}$ being zero, or equivalently that the stress matrix $\underline{T}$ is symmetric.

1.4.2 A Rate of Work Hypothesis

Somewhat analogous to our earlier principle of virtual work, we here assume that the rate of working of external forces and moments either goes into increasing the kinetic and elastic energies, or is dissipated as viscous dissipation. Thus for a volume V of liquid crystal bounded by surface S

$$\int_V (\mathbf{F} \cdot \mathbf{v} + \mathbf{K} \cdot \mathbf{w}) \, dv + \int_S (\mathbf{t} \cdot \mathbf{v} + \boldsymbol{\ell} \, \mathbf{w}) \, ds$$

$$= \frac{d}{dt} \int_V \left(\frac{1}{2} \rho \mathbf{v} \cdot \mathbf{v} + W \right) dv + \int_V D \, dv \quad (82)$$

in which $\mathbf{w}$ denotes the local angular velocity, W is the Frank–Oseen energy, and D represents the rate of viscous dissipation. With the aid of equations (Eq. 79) the above can be expressed in point form, and following some simplification through use of equations (Eq. 80) one obtains

$$tr(\underline{T}\underline{V}^T) + tr(\underline{L}\underline{W}^T) - \mathbf{w} \cdot \hat{\mathbf{T}} = \frac{dW}{dt} + D \quad (83)$$

where the velocity and angular velocity gradient matrices $\underline{V}$ and $\underline{W}$ take the forms

$$\underline{V} = \left[\frac{\partial v_i}{\partial x_j} \right], \quad \underline{W} = \left[\frac{\partial w_i}{\partial x_j} \right] \quad (84)$$

respectively.

Noting that the material time derivative of the director $\mathbf{n}$ is given by

$$\frac{d}{dt} \mathbf{n} = \mathbf{w} \times \mathbf{n} \quad (85)$$

and also that one can show that

$$\frac{d}{dt}(\nabla \mathbf{n}) = \nabla \left(\frac{d\mathbf{n}}{dt} \right) - \nabla \mathbf{n} \underline{V} \quad (86)$$

it is possible by using the chain rule to express the material derivative of the energy W in a form linear in the velocity gradients, angular velocity gradients and the angular velocity. Also by appeal to the identity (Eq. 12), the resultant expression cancels the contributions from the static terms on the left hand side of (Eq. 83), given by Eqs. (20) and (22), and thus the rate of work postulate reduces to

$$tr(\underline{T}^d \underline{V}^T) + tr(\underline{L}^d \underline{W}^T) - \mathbf{w} \cdot \hat{\mathbf{T}}^d = D > 0 \quad (87)$$

the superscript d denoting dynamic contributions, and noting that the rate of viscous dissipation is necessarily positive. Alternatively of course one can argue from the linearity of these terms in the velocity and angular velocity gradients that the static contributions are indeed given by our earlier expressions.

To complete our dynamic theory, it is necessary to prescribe forms for the dynamic stress and couple stress. The simplest choice consistent with known effects appears to be that at a material point

$$\underline{T}^d \text{ and } \underline{L}^d \text{ are functions of } \underline{V}, \mathbf{n} \text{ and } \mathbf{w} \quad (88)$$

all evaluated at that point at the given instant. However, since dependence upon the gradients of the angular velocity is excluded, it follows at once from the inequality (Eq. 87), given that these gradients occur linearly, that

$$\underline{L}^d = 0 \quad (89)$$

in agreement with the earlier formulation [7]. Hence the rate of viscous dissipation inequality reduces to

$$D = tr(\underline{T}^d \underline{V}^T) - \mathbf{w} \cdot \hat{\mathbf{T}} > 0 \quad (90)$$

which we employ below to limit viscous coefficients.

1.4.3 The Viscous Stress

Invariance at once requires that the assumption (Eq. 88) for the dynamic part of the stress matrix be replaced by

$\underset{\sim}{T}^d$ is a function of $\underset{\sim}{A}$, **n** and ω (91)

all evaluated at that point at that instant, where

$$2\underset{\sim}{A} = \underset{\sim}{V} + \underset{\sim}{V}^T, \quad \omega = \mathbf{w} - \frac{1}{2}\text{curl } \mathbf{v} \quad (92)$$

$\underset{\sim}{A}$ being the familiar rate of strain tensor or matrix, and ω the angular velocity relative to the background rotation of the continuum. While it is possible to derive the viscous stress from the above assumption [46], we opt here for a more direct approach, and replace (Eq. 91) by

$\underset{\sim}{T}^d$ is a function of $\underset{\sim}{A}$, **n** and **N** (93)

all evaluated at that point at that instant, where using (Eq. 85)

$$\mathbf{N} = \omega \times \mathbf{n} = \frac{d}{dt}\mathbf{n} - \frac{1}{2}(\underset{\sim}{V} - \underset{\sim}{V}^T)\mathbf{n} \quad (94)$$

this restricting the dependence upon the relative rotation, essentially discounting the component parallel to the director **n**, which can be shown to be zero in any event [46].

Nematic symmetry requires that (Eq. 93) be an isotropic function and independent of a change of sign in the director **n**. Thus assuming a linear dependence upon $\underset{\sim}{A}$ and **N** one finds that

$$\underset{\sim}{T}^d = \alpha_1 \mathbf{n} \cdot \underset{\sim}{A}\mathbf{nn} * \mathbf{n} + \alpha_2 \mathbf{N} * \mathbf{n} + \alpha_3 \mathbf{n} * \mathbf{N}$$
$$+ \alpha_4 \underset{\sim}{A} + \alpha_5 \underset{\sim}{A}\mathbf{n} * \mathbf{n} + \alpha_6 \mathbf{n} * \underset{\sim}{A}\mathbf{n}, \quad (95)$$

where again $\mathbf{a} * \mathbf{b}$ denotes the matrix with $(i, j)^{\text{th}}$ element $a_i b_j$, and the αs are constants. Ignoring thermal effects, the above is also the form that one obtains for cholesterics or chiral nematics.

Somewhat straightforwardly it follows from the above that

$$\hat{\mathbf{T}}^d = \mathbf{n} \times \mathbf{g} \quad (96)$$

where the vector **g** takes the form

$$\mathbf{g} = -\gamma_1 \mathbf{N} - \gamma_2 \underset{\sim}{A}\mathbf{n},$$
$$\gamma_1 = \alpha_3 - \alpha_2, \quad \gamma_2 = \alpha_6 - \alpha_5 \quad (97)$$

Also employing Eqs. (92), (94) and (96) the viscous dissipation inequality (Eq. 90) becomes

$$D = tr(\underset{\sim}{T}^d \underset{\sim}{A}) - \omega \cdot \hat{\mathbf{T}}^d = tr(\underset{\sim}{T}^d \underset{\sim}{A}) - \mathbf{g} \cdot \mathbf{N} > 0 \quad (98)$$

The above of course restricts possible values for the viscous coefficients and one can readily deduce from it that [47]

$$\begin{aligned}\alpha_4 > 0, \quad &2\alpha_4 + \alpha_5 + \alpha_6 > 0, \\ &3\alpha_4 + 2\alpha_5 + 2\alpha_6 + 2\alpha_1 > 0 \\ \gamma_1 > 0, \quad &(\alpha_2 + \alpha_3 + \gamma_2)^2 < 4\gamma_1(2\alpha_4 + \alpha_5 + \alpha_6)\end{aligned} \quad (99)$$

the calculation aided by choosing axes parallel to **n** and **N**. However, in many cases it proves simpler to deduce consequences of (Eq. 98) by writing down the dissipation function for the particular flow under consideration, rather than try to derive results from the above inequalities.

By invoking Onsager relations, Parodi [48] argues that one restrict the viscous coefficients to satisfy the relationship

$$\gamma_2 = \alpha_3 + \alpha_2 \quad (100)$$

but subsequently Currie [49] shows that this relationship also follows as a result of a stability argument. As a consequence this condition between the viscous coefficients is now generally accepted, and leads to some simplification in the use of the theory. For example, when (Eq. 100) holds, Ericksen [21] shows that the viscous stress and the vector **g** follow directly from the dissipation function D through

$$\underset{\sim}{T}^d = \frac{1}{2}\frac{\partial D}{\partial \underset{\sim}{V}}, \quad \mathbf{g} = -\frac{1}{2}\frac{\partial D}{\partial \dot{\mathbf{n}}}, \quad \dot{\mathbf{n}} = \frac{d}{dt}\mathbf{n} \quad (101)$$

results that we require below.

1.4.4 Equations of Motion

If one combines the Eqs. (20), (22), (79), (89) and (96), it is possible to express the balance law for angular moment, the second of equations (Eq. 80) in the form

$$\mathbf{n} \times \left(\operatorname{div} \underline{S}^s - \frac{\partial W}{\partial \mathbf{n}} + \mathbf{G} + \mathbf{g} \right) = 0 \qquad (102)$$

this involving some manipulation and the use of the identity (Eq. 12), a result perhaps more easily derived employing Cartesian tensor notation. Equivalently (Eq. 102) becomes

$$\operatorname{div} \underline{S}^s - \frac{\partial W}{\partial \mathbf{n}} + \mathbf{G} + \mathbf{g} = \gamma \mathbf{n} \qquad (103)$$

a rather natural extension of the second of equations (Eq. 23) given the result (Eq. 89). With angular momentum in this form one can simplify the balance law for linear momentum, the first of equations (Eq. 80), in a manner rather similar to the derivation of (Eq. 32) to give

$$\rho \frac{d\mathbf{v}}{dt} = \mathbf{F} + \nabla \mathbf{n}^T \mathbf{G} - \operatorname{grad}(p + W)$$
$$+ \nabla \mathbf{n}^T \mathbf{g} + \operatorname{div} \underline{T}^d \qquad (104)$$

Further, if the body force and moment satisfy (Eq. 30), this last equation reduces to

$$\rho \frac{d\mathbf{v}}{dt} = -\operatorname{grad} \tilde{p} + \nabla \mathbf{n}^T \mathbf{g} + \operatorname{div} \underline{T}^d,$$
$$\tilde{p} = p + W - \psi, \qquad (105)$$

which is clearly simpler than the original.

As in Section 1.2.4, if one introduces a representation for the director $\mathbf{n}$ referred to Cartesian axes that trivially satisfies the constraint upon it, say

$$\mathbf{n} = \mathbf{f}(\theta, \phi), \quad \mathbf{n} \cdot \frac{\partial \mathbf{f}}{\partial \theta} = \mathbf{n} \cdot \frac{\partial \mathbf{f}}{\partial \phi} = 0 \qquad (106)$$

it also follows that

$$\nabla \mathbf{n} = \frac{\partial \mathbf{f}}{\partial \theta} * \nabla \theta + \frac{\partial \mathbf{f}}{\partial \phi} * \nabla \phi, \quad \dot{\mathbf{n}} = \frac{\partial \mathbf{f}}{\partial \theta} \dot{\theta} + \frac{\partial \mathbf{f}}{\partial \phi} \dot{\phi} \qquad (107)$$

As before, if an external field satisfying (Eq. 30) is present, one has

$$W = W(\theta, \phi, \nabla\theta, \nabla\phi), \quad \psi = \chi(\mathbf{x}, \theta, \phi) \qquad (108)$$
$$D = 2\Delta(\theta, \phi, \dot{\theta}, \dot{\phi}, \nabla \mathbf{v}), \quad \dot{\theta} = \frac{d\theta}{dt}, \quad \dot{\phi} = \frac{d\phi}{dt}$$

and proceeding as earlier and noting that

$$\frac{\partial \Delta}{\partial \dot{\theta}} = \frac{1}{2} \frac{\partial D}{\partial \dot{\mathbf{n}}} \cdot \frac{\partial \mathbf{f}}{\partial \theta}, \quad \frac{\partial \Delta}{\partial \dot{\phi}} = \frac{1}{2} \frac{\partial D}{\partial \dot{\mathbf{n}}} \cdot \frac{\partial \mathbf{f}}{\partial \phi} \qquad (109)$$

one can show that Eqs. (103) and (105) can be recast as

$$\operatorname{div}\left(\frac{\partial W}{\partial \nabla \theta}\right) - \frac{\partial W}{\partial \theta} + \frac{\partial \chi}{\partial \theta} - \frac{\partial \Delta}{\partial \dot{\theta}} = 0$$

$$\operatorname{div}\left(\frac{\partial W}{\partial \nabla \phi}\right) - \frac{\partial W}{\partial \phi} + \frac{\partial \chi}{\partial \phi} - \frac{\partial \Delta}{\partial \dot{\phi}} = 0 \qquad (110)$$

and

$$\rho \dot{\mathbf{v}} = -\operatorname{grad} \tilde{p} + \operatorname{div}\left(\frac{\partial \Delta}{\partial \underline{V}}\right) - \frac{\partial \Delta}{\partial \dot{\theta}} \nabla \theta - \frac{\partial \Delta}{\partial \dot{\phi}} \nabla \phi$$
$$(111)$$

respectively, the divergence in the latter with respect to the second index. As Ericksen [21] shows, it is possible to present the above reformulation of the equations in terms of curvilinear coordinates. However, given our rather restrictive notation, an attempt to summarise this here is more likely to confuse than to enlighten, and therefore we refer the interested reader to the original paper.

Boundary conditions for dynamic theory simply add the customary non-slip hypothesis upon the velocity vector to the conditions described above for static theory, with only a very occasional reference to the inclusion of a surface viscosity for motions of the director at a solid interface.

1.5 References

[1] C. W. Oseen, *Ark. Mat. Astron. Fys.* **1925**, *19A*, 1.
[2] H. Zocher, *Phys. Z.* **1927**, *28*, 790.
[3] H. Zocher, *Trans. Faraday Soc.* **1933**, *29*, 945.
[4] F. C. Frank, *Disc. Faraday Soc.* **1958**, *25*, 19.
[5] J. L. Ericksen, *Arch. Rat. Mech. Anal.* **1962**, *9*, 371.
[6] J. L. Ericksen, *Trans. Soc. Rheol.* **1961**, *5*, 23.

[7] F. M. Leslie, *Arch. Rat. Mech. Anal.* **1968**, *28*, 265.
[8] P. G. de Gennes, J. Prost, *The Physics of Liquid Crystals*, 2nd edn, Oxford University Press **1993**.
[9] S. Chandrasekhar, *Liquid Crystals*, 2nd edn. Cambridge University Press **1992**.
[10] L. M. Blinov, *Electro-optical and Magneto-optical Properties of Liquid Crystals*, J. Wiley and Sons, New York **1983**.
[11] E. G. Virga, *Variational Theories for Liquid Crystals*, Chapman and Hall, London **1994**.
[12] M. J. Stephen, J. P. Straley, *Rev. Mod. Phys.* **1974**, *46*, 617.
[13] J. L. Ericksen, *Adv. Liq. Cryst.* **1976**, *2*, 233.
[14] J. T. Jenkins, *Ann. Rev. Fluid. Mech.* **1978**, *10*, 197.
[15] F. M. Leslie, *Adv. Liq. Cryst.* **1979**, *4*, 1.
[16] J. Nehring, A. Saupe, *J. Chem. Phys.* **1971**, *54*, 337.
[17] J. L. Ericksen, *Arch. Rat. Mech. Anal.* **1991**, *113*, 97.
[18] J. L. Ericksen, *Phys. Fluids* **1966**, *9*, 1205.
[19] J. T. Jenkins, *J. Fluid. Mech.* **1971**, *45*, 465.
[20] H. J. Deuling, *Mol. Cryst. Liq. Cryst.* **1972**, *19*, 123.
[21] J. L. Ericksen, *Q. J. Mech. Appl. Math.* **1976**, *29*, 203.
[22] A. Rapini, M. Papoular, *J. Phys. (Paris) Colloq.* **1969**, *30*, C4, 54.
[23] J. T. Jenkins, P. J. Barratt, *Q. J. Mech. Appl. Math.* **1974**, *27*, 111.
[24] B. Jerome, P. Pieranski, M. Boix, *Europhys. Lett.* **1988**, *5*, 693.
[25] M. Monkade, M. Boix, G. Durand, *Europhys. Lett.* **1988**, *5*, 697.
[26] R. Barberi, M. Boix, G. Durand, *Appl. Phys. Lett.* **1989**, *55*, 2506.
[27] M. Nobili, G. Durand, *Europhys. Lett.* **1994**, *25*, 527.
[28] V. Sergan, G. Durand, *Liq. Cryst.* **1995**, *18*, 171.
[29] C. Oldano, G. Barbero, *J. Phys. (Paris) Lett.* **1985**, *46*, 451.
[30] H. P. Hinov, *Mol. Cryst. Liq. Cryst.* **1987**, *148*, 157.
[31] G. Barbero, A. Strigazzi, *Liq. Cryst.* **1989**, *5*, 693.
[32] G. Barbero, A. Sparavigna, A. Strigazzi, *Nuovo Cim. D* **1990**, *12*, 1259.
[33] G. Barbero, *Mol. Cryst. Liq. Cryst.* **1991**, *195*, 199.
[34] V. M. Pergamenshchik, *Phys. Rev. E* **1993**, *48*, 1254.
[35] S. Faetti, *Phys. Rev. E* **1994**, *49*, 4192.
[36] S. Faetti, *Phys. Rev. E* **1994**, *49*, 5332.
[37] G. Barbero, G. Durand, *Phys. Rev. E* **1993**, *48*, 1942.
[38] R. Barberi, G. Barbero, M. Giocondo, R. Moldovan, *Phys. Rev. E* **1994**, *50*, 2093.
[39] O. D. Lavrentovich, V. M. Pergamenshchik, *Phys. Rev. Lett.* **1994**, *73*, 979.
[40] Orsay Group on Liquid Crystals, *Solid State Commun.* **1971**, *9*, 653.
[41] F. M. Leslie, I. W. Stewart, M. Nakagawa, *Mol. Cryst. Liq. Cryst.* **1991**, *198*, 443.
[42] T. Carlsson, I. W. Stewart, F. M. Leslie, *Liq. Cryst.* **1991**, *9*, 661.
[43] C. W. Oseen, *Trans. Faraday Soc.* **1933**, *29*, 883.
[44] F. M. Leslie, I. W. Stewart, T. Carlsson, M. Nakagawa, *Cont. Mech. Thermodyn.* **1991**, *3*, 237.
[45] T. Carlsson, I. W. Stewart, F. M. Leslie, *J. Phys. A* **1992**, *25*, 2371.
[46] F. M. Leslie, *Cont. Mech. Thermodyn.* **1992**, *4*, 167.
[47] F. M. Leslie, *Q. J. Mech. Appl. Math.* **1966**, *19*, 357.
[48] P. Parodi, *J. Phys. (Paris)* **1970**, *31*, 581.
[49] P. K. Currie, *Mol. Cryst. Liq. Cryst.* **1974**, *28*, 335.

2 Molecular Theories of Liquid Crystals

Mikhail A. Osipov

2.1 Introduction

Molecular theories can provide important additional information about the properties and structure of liquid crystals because they enable one to understand (at least partially and in a simplified way) the onset of liquid crystalline order at the microscopic level. However, the development of a realistic molecular theory for liquid crystals appears to be a challenging problem for two main reasons. First, the intermolecular interaction potentials are not known exactly. These potentials are generally expected to be rather complex reflecting the relatively complex structure of typical mesogenic molecules. Second, it is hardly possible to do good statistical mechanics with such complex potentials even if they are known. At present there exists no regular method to calculate even the pair correlation function for a simple anisotropic fluid and one has to rely on rather crude approximations. The latter difficulty can, in principle, be overcome by means of computer simulations. Such simulations, however, appear to be very time consuming if realistic molecular models are employed. Even in the case of simple potentials one sometimes needs very large systems (for example, up to 8000 particles to simulate the polar smectic phase with long-range dipole–dipole interactions [1]). On the other hand, such simulations with realistic potentials usually require extensive interpretation, as in the case of a real experiment. In fact, one would have to make too many simulations to trace a relation between the features of the molecular structure and the macroscopic parameters of liquid crystal phases in an empirical way. Thus, in general, computer simulations are not a substitution for a molecular theory, but merely an independent source of information. In particular, computer simulations provide a unique tool for testing and generating various approximations that are inevitably used in any molecular theory.

Thus, the primary goal of a molecular theory is to obtain a qualitative insight into the molecular origin of various effects in liquid crystals. First of all, it is important to understand which properties of liquid crystals are actually determined by some basic and simple characteristics of molecular structure (like, for example, the elongated or disc-like molecular shape or the polarizability anisotropy), and which properties are particularly sensitive to the details of molecular structure, including flexibility, biaxiality or even the location of particular elements of structure. In this short chapter we will not be able to address all of these questions. However, we make an attempt to present the basics of the molecular theory of the nematic and simple smectic phases. We will also provide some additional references for a reader with more specific interests.

We start with the microscopic definitions and discussion of the nematic and smectic order parameters and then proceed with some elementary information about anisotropic intermolecular interactions in liquid crystals. Then we discuss in more detail the main molecular theories of the nematic–isotropic phase transition and conclude with a consideration of molecular models for smectic A and smectic C phases.

2.2 Microscopic Definition of the Order Parameters for Nematic and Smectic Phases

2.2.1 Uniaxial Nematic Phase

The uniaxial nematic phase possesses a quadrupole-type symmetry and is characterized by the order parameter $Q_{\alpha\beta}$ which is a symmetric traceless second-rank tensor:

$$Q_{\alpha\beta} = S\left(n_\alpha n_\beta - \frac{1}{3}\delta_{\alpha\beta}\right) \quad (1)$$

where the unit vector n is the director that specifies the preferred orientation of the primary molecular axes. We note that the corresponding macroscopic axis is nonpolar and therefore it is better represented by the quadratic combination $n_\alpha n_\beta$. The quantity S is the scalar order parameter that characterizes the degree of nematic ordering.

From the microscopic point of view the orientation of a rigid molecule i can be specified by the unit vectors a_i and b_i in the direction of long and short molecular axes, respectively; $(a_i \cdot b_i) = 0$. It should be noted that the definition of the primary molecular axis is somewhat arbitrary for a molecule without symmetry elements. Sometimes the axis a_i is taken along the main axis of the molecular inertia tensor. However, in many cases it is quite difficult to know beforehand which molecular axis will actually be ordered along the director. This can be a particularly difficult problem for a guest molecule in a nematic host [2].

Here we assume for simplicity that the primary axis is well defined. Then the scalar order parameter is given by the average

$$S = \langle P_2((a_i \cdot b_i))\rangle \quad (2)$$

where $P_2(x)$ is the second Legendre polynomial and the brackets $\langle ... \rangle$ denote the statistical averaging.

We note that the averaging in Eq. (2) is performed with the one-particle distribution function $f_1((a \cdot n))$ that determines the probability of finding a molecule with a given orientation of the long axis at a given point in space. In the uniaxial nematic phase the distribution function depends only on the angle ω between the long axis and the director. Then Eq. (2) can be rewritten as

$$S = \frac{1}{2}\int P_2(\cos\omega) f_1(\cos\omega) \, d\cos\omega \quad (3)$$

In mixtures of liquid crystals, the molecules of different components may possess different degrees of nematic ordering. In this case the nematic order parameters, S_α, for different components α are calculated separately using different distribution functions $f_\alpha(\cos\omega)$.

The definition of the order parameter (Eq. 2) is not entirely complete because one has to know the orientation of the director. In the general case the director appears self-consistently as a result of the breakdown of symmetry during the phase transition and sometimes its orientation is unknown beforehand (as, for example, in simulations). Thus, from the theoretical point of view it is more consistent to define directly the tensor order parameter as a true thermodynamic average:

$$Q_{\alpha\beta} = \left\langle a_\alpha a_\beta - \frac{1}{3}\delta_{\alpha\beta} \right\rangle = S\left(n_\alpha n_\beta - \frac{1}{3}\delta_{\alpha\beta}\right) \quad (4)$$

where the order parameter S is given by Eq. (2).

The molecular statistical definition of the order parameter can be clarified if one considers, for example, the magnetic susceptibility of the nematic phase. The anisotropic part of the susceptibility tensor $\chi_{\alpha\beta}$ can be considered as an order parameter [3] because it vanishes in the isotropic phase and is nonzero in the nematic phase. In addition, the macroscopic magnetic susceptibility can be written as a sum of contributions from individual molecules:

$$\chi_{\alpha\beta} = \rho \sum_i \langle \chi^M_{\alpha\beta} \rangle \quad (5)$$

where $\chi^M_{\alpha\beta}$ is the molecular susceptibility tensor and ρ is the number density.

Equation (4) is valid with high accuracy because induced magnetic dipoles interact only weakly. By contrast, interaction between the induced electric dipoles is strong and produces substantial local field effects which do not allow one to express the dielectric permittivity of the nematic phase in terms of the molecular parameters in a simple way (see, for example, [4]).

The molecular susceptibility $\chi^M_{\alpha\beta}$ can be diagonalized:

$$\chi^M_{\alpha\beta} = \chi_{11} b_\alpha b_\beta + \chi_{22} c_\alpha c_\beta + \chi_{33} a_\alpha a_\beta \quad (6)$$

where the orthogonal unit vectors a, b and c are the principal axes.

Now we have to average Eq. (5) to get the expression for the order parameter. In the nematic phase the orientational distribution function depends only on the primary molecular axis a. Thus the two short axes b and c are completely equivalent in the statistical sense and one obtains

$$\ldots \langle b_\alpha b_\beta \rangle = \langle c_\alpha c_\beta \rangle = \frac{1}{2}\left(\delta_{\alpha\beta} - \langle a_\alpha a_\beta \rangle\right) \quad (7)$$

where we have used the general relation

$$\delta_{\alpha\beta} = a_\alpha a_\beta + b_\alpha b_\beta c_\alpha c_\beta$$

Finally the macroscopic magnetic susceptibility can be expressed as

$$\chi_{\alpha\beta} = \bar{\chi}\delta_{\alpha\beta} + \Delta\chi\left\langle a_\alpha a_\beta - \frac{1}{3}\delta_{\alpha\beta}\right\rangle \quad (8)$$

where $\bar{\chi} = \rho(\chi_{11}+\chi_{22}+\chi_{33})$ is the average susceptibility and $\Delta\chi = \rho(\chi_{33}-(\chi_{11}+\chi_{22})/2)$ is the anisotropy of the susceptibility.

One can readily see from Eq. (7) that the traceless part of the average magnetic susceptibility is proportional to the nematic tensor order parameter $Q_{\alpha\beta}$ given by Eq. (3).

2.2.2 Biaxial Nematic Phase

Biaxial nematic ordering has been observed so far only in lyotropic systems. At the same time tilted smectic phases are also biaxial and thus the expressions presented in this section can be of more general use.

In the biaxial nematic phase it is possible to define two orthogonal directors n and m. In this case the magnetic susceptibility tensor can be rewritten as

$$\chi_{\alpha\beta} = \bar{\chi}\delta_{\alpha\beta} + \Delta\chi Q_{\alpha\beta} + \Delta\chi_\perp B_{\alpha\beta} \quad (9)$$

where $\Delta\chi_\perp = \rho(\chi_{11}-\chi_{22})$ is the transverse anisotropy of the molecular susceptibility that represents biaxiality, and where

$$B_{\alpha\beta} = \langle(b_\alpha b_\beta - c_\alpha c_\beta)\rangle \quad (10)$$

Now the traceless part of the magnetic susceptibility is a sum of two terms proportional to two tensor order parameters of the biaxial nematic phase: $Q_{\alpha\beta}$ and $B_{\alpha\beta}$.

In the uniaxial nematic phase $B_{\alpha\beta}=0$ and the tensor order parameter $Q_{\alpha\beta}$ is uniaxial. By contrast, in the biaxial phase the order parameter $Q_{\alpha\beta}$ can be written as a sum of a uniaxial and a biaxial part:

$$Q_{\alpha\beta} = S\left(n_\alpha n_\beta - \frac{1}{3}\delta_{\alpha\beta}\right)$$
$$+\Delta Q\left(m_\alpha m_\beta - l_\alpha l_\beta\right) \quad (11)$$

where the unit vector $\boldsymbol{l}=[\boldsymbol{n}\times\boldsymbol{m}]$ and where S is the largest eigenvalue of the tensor $Q_{\alpha\beta}$ (for prolate molecules).

The parameters S and ΔQ can be expressed as [8]

$$S = \langle P_2((\boldsymbol{a}\cdot\boldsymbol{n}))\rangle \quad (12)$$

$$\Delta Q = \frac{1}{3}\left(\langle P_2((\boldsymbol{a}\cdot\boldsymbol{m}))\rangle - \langle P_2((\boldsymbol{a}\cdot\boldsymbol{l}))\rangle\right) \quad (13)$$

The tensor order parameter $B_{\alpha\beta}$ can be expressed in the same general form as $Q_{\alpha\beta}$:

$$B_{\alpha\beta} = S'\left(n_\alpha n_\beta - \frac{1}{3}\delta_{\alpha\beta}\right)$$
$$+D\left(m_\alpha m_\beta - l_\alpha l_\beta\right) \quad (14)$$

where

$$S' = \langle P_2((\boldsymbol{b}\cdot\boldsymbol{n}))\rangle - \langle P_2((\boldsymbol{c}\cdot\boldsymbol{n}))\rangle \quad (15)$$

$$D = \frac{1}{3}\left(\langle P_2((\boldsymbol{b}\cdot\boldsymbol{m}))\rangle - \langle P_2((\boldsymbol{b}\cdot\boldsymbol{l}))\rangle \right.$$
$$\left. + \langle P_2((\boldsymbol{c}\cdot\boldsymbol{m}))\rangle - \langle P_2((\boldsymbol{c}\cdot\boldsymbol{l}))\rangle\right) \quad (16)$$

Here the parameter S' characterizes the tendency of the molecular short axes to be ordered along the main director $\boldsymbol{n}$ and the parameter ΔQ describes the ordering of the primary axis $\boldsymbol{a}$ along the director $\boldsymbol{m}$. In the case of perfect ordering of the primary molecular axes ($S=1$) the parameter $S'=0$, $\Delta Q = 0$ and the biaxial ordering in the phase is described by the parameter $B_{\alpha\beta}=D\left(m_\alpha m_\beta - l_\alpha l_\beta\right)$.

2.2.3 Smectic A and C Phases

Smectic ordering in liquid crystals is usually characterized by the complex order parameter $\rho_\alpha e^{i\psi}$ introduced by de Gennes [3]. Here $\rho_\alpha=\langle\cos(\boldsymbol{q}\cdot\boldsymbol{r})\rangle$ is the amplitude of the density wave, ψ is the phase and $\boldsymbol{q}$ is the wave vector. This order parameter appears naturally in the Fourier expansion of the one-particle density $\rho(\boldsymbol{r})$.

The order parameter of the SmC phase appears to be more complex because in this phase the director is not parallel to the wave vector $\boldsymbol{q}$. In a simple case, it is just possible to use the tilt angle Θ as an order parameter. However, this parameter does not specify the direction of the tilt and thus it is analogous to the scalar nematic order parameter S. The full tensor order parameter of the SmC phase can be constructed in several different ways. One is to define the pseudovector $\boldsymbol{w}$ [8, 9] that describes the rotation of the director with respect to the smectic plane normal:

$$w_\alpha = \delta_{\alpha\beta\gamma} Q_{\beta\nu} \hat{q}_\gamma \hat{q}_\nu \quad (17)$$

where $\delta_{\alpha\beta\gamma}$ is the Levi–Civita antisymmetric tensor, $\hat{\boldsymbol{q}}$ is the unit vector along $\boldsymbol{q}$ and $Q_{\alpha\beta}$ is the (biaxial) nematic order parameter.

The pseudo vector order parameter $\boldsymbol{w}$ vanishes in the SmA phase. If we neglect the biaxiality of the SmC phase, the order parameter $Q_{\alpha\beta}$ is expressed in terms of the director $\boldsymbol{n}$ and the vector order parameter (Eq. 16) is simplified:

$$\boldsymbol{w} = (\boldsymbol{n}\cdot\hat{\boldsymbol{q}})[\boldsymbol{n}\times\hat{\boldsymbol{q}}] \quad (18)$$

We note that the vector $\boldsymbol{w}$ is perpendicular to the tilt plane in the SmC phase and the absolute value of w is related to the tilt angle, $w=\sin 2\Theta/2 \approx \Theta$ at small $\Theta^2 \ll 1$. This pseudo vector order parameter can be used both in the theory of nonchiral and chiral SmC phases [8, 9].

2.3 Anisotropic Intermolecular Interactions in Liquid Crystals

2.3.1 Hard-core Repulsion

Liquid crystals are composed of relatively large molecules with strongly anisotropic shapes. In general, there exist a number of anisotropic interactions between such molecules that can be responsible for the nematic ordering. Historically the first consistent molecular theory of the nematic phase was proposed by Onsager [10]. Onsager showed that the nematic ordering can be stabilized by the hard-core repulsion between rigid rod-like molecules without any attraction forces. The steric repulsion between rigid particles is a limiting case of a strong short-range repulsion interaction that does not allow molecules to penetrate each other. The corresponding model interaction potential is discontinuous and can be written as

$$V_s(1,2) = \Omega(r_{12} - \xi_{12}) \quad (19)$$

where $\Omega(x)$ is a step function, $\Omega(r_{12} - \xi_{12}) = \infty$ if $r_{12} < \xi_{12}$ and thus the molecules penetrate each other; otherwise $\Omega(r_{12} - \xi_{12}) = 0$. Here ξ_{12} is the minimum distance of approach between the centers of the molecules 1 and 2 for a given relative orientation. The function ξ_{12} is determined by the molecular shape and depends on the relative orientation of the two molecules. For hard spheres $\xi_{12} = D$ where D is the diameter of the sphere. For any two prolate molecules, ξ_{12} varies between the diameter D and the length L.

2.3.2 Electrostatic and Dispersion Interactions

Hard-core repulsion between anisotropic molecules, discussed in the previous subsection, can be the driving force of the I–N transition in lyotropic systems. In contrast, in thermotropic liquid crystals the transition occurs at some particular temperature and therefore some attraction interaction must be involved. The corresponding molecular theory, based on anisotropic dispersion interactions, was proposed by Maier and Saupe [11, 12].

The dispersion (or Van der Waals) interaction appears in the second-order perturbation theory. The initial interaction potential is the electrostatic one. For the two molecules i and j the electrostatic interaction energy can be written as

$$V_{el}(1,2) = \iint \frac{e(\mathbf{r}_i)e(\mathbf{r}_j)}{|\mathbf{R}_i - \mathbf{R}_j + \mathbf{r}_i - \mathbf{r}_j|} d\mathbf{r}_i d\mathbf{r}_j \quad (20)$$

where $e(\mathbf{r}_i)$ is the charge distribution of the molecule i and $\mathbf{R}_i$ is the position of the center of mass.

The electrostatic interaction can be expanded in terms of molecular multipoles. For neutral molecules one obtains

$$V_{el}(i,j) = \frac{1}{R_{ij}^3} U_{dd}(i,j) + \frac{1}{R_{ij}^4} U_{dq}(i,j) + \ldots \quad (21)$$

where the first term is the dipole–dipole interaction potential

$$U_{dd}(i,j) = (\vec{\mu}_i \cdot \vec{\mu}_j) - 3(\vec{\mu}_i \cdot \mathbf{u}_{ij})(\vec{\mu}_j \cdot \mathbf{u}_{ij}) \quad (22)$$

and the second term is the dipole–quadrupole interaction energy

$$U_{dq}(i,j) = U_{dq}^{ij} - U_{dq}^{ji} \quad (23)$$

where

$$U_{\text{dq}}^{ij} = \frac{3}{2} Tr \, \mathbf{Q}_j \left(\vec{\mu}_i \cdot \mathbf{u}_{ij} \right) + 3 \left(\vec{\mu}_i \cdot \mathbf{Q}_j \cdot \mathbf{u}_{ij} \right)$$
$$- \frac{15}{2} \left(\vec{\mu}_i \cdot \mathbf{Q}_j \cdot \vec{\mu}_i \right) \left(\vec{\mu}_i \cdot \mathbf{u}_{ij} \right) \quad (24)$$

and where U_{dq}^{ij} is obtained by permutation of the indices i and j in Eq. (23). Here $\mathbf{R}_{ij} = \mathbf{R}_i - \mathbf{R}_j$ is the intermolecular vector, $\vec{\mu}_i$ is the molecular dipole and $\mathbf{Q}_i$ is the molecular quadrupole tensor, $\mathbf{u}_{ij} = \mathbf{R}_{ij}/|\mathbf{R}_{ij}|$.

The dispersion interaction energy is obtained in the second-order perturbation theory:

$$V_{\text{disp}}(i,j) = \sum_{ni,nj}{}' \quad (25)$$
$$\cdot \frac{\langle o_i o_j | V_{\text{el}}(i,j) | n_i n_j \rangle \langle n_i n_j | V_{\text{el}}(i,j) | o_i o_j \rangle}{E_{oioj} - E_{ninj}}$$

where $|o_i\rangle$ and $|n_i\rangle$ represent the ground state and the excited state of the molecule i, respectively and $E_{oioj} - E_{ninj}$ is the excitation energy of the system.

Substituting the multipole expansion (Eq. 20) into Eq. (24) and taking into account the leading term (that contains r_{ij}^{-6}) one obtains

$$V_{\text{disp}}(i,j) \approx \frac{1}{r_{ij}^6} \sum_{ni,nj}{}' \quad (26)$$
$$\cdot \frac{\langle o_i o_j | U_{\text{dd}}(i,j) | n_i n_j \rangle \langle n_i n_j | U_{\text{dd}}(i,j) | o_i o_j \rangle}{E_{oioj} - E_{ninj}}$$

where $U_{\text{dd}}(i,j)$ is given by Eq. (21).

We note that the full dispersion interaction energy (Eq. 24) can be approximated to by the dipole–dipole term (Eq. 25) only if the molecules are sufficiently far apart. However, Eq. (25) is often used in molecular theories of liquid crystals to draw qualitative conclusions. For example, the potential (Eq. 25) has been used in the Maier–Saupe theory.

The dipole–dipole dispersion interaction can be simplified if we assume that the molecular short axes are oriented randomly around the long axes $\mathbf{a}_i$ and $\mathbf{a}_j$. Then it is valid to average the potential (Eq. 25) over all $\mathbf{b}_i$ and $\mathbf{b}_j$ with the constraints $(\mathbf{b}_i \cdot \mathbf{a}_i) = 0$, $(\mathbf{b}_j \cdot \mathbf{a}_j) = 0$. It can be shown that the averaging results in the following expression for the effective uniaxial potential [13, 14]:

$$V_{\text{eff}}\left(\mathbf{a}_i, \mathbf{u}_{ij}; \mathbf{a}_j\right)$$
$$= \text{const} - J_{ij}^2 \left(\mathbf{a}_i \cdot \mathbf{u}_{ij}\right)^2 - J_{ji}^2 \left(\mathbf{a}_j \cdot \mathbf{u}_{ij}\right)^2$$
$$- J_{ij} \left[\left(\mathbf{a}_i \cdot \mathbf{a}_j\right) - \left(\mathbf{a}_i \cdot \mathbf{u}_{ij}\right)\left(\mathbf{a}_j \cdot \mathbf{u}_{ij}\right)\right]^2 \quad (27)$$

Here the coefficients can be expressed in terms of the electric dipole and quadrupole matrix elements [13, 14].

In the nematic phase there is no positional order and the molecular centers are distributed randomly. If one neglects the positional correlations, the interaction potential (Eq. 26) can be further simplified by averaging over all directions of the intermolecular vector. The resulting effective potential appears to be very simple:

$$V_{\text{eff}}\left(\left(\mathbf{a}_i \cdot \mathbf{a}_j\right)\right) = -\frac{1}{60} E r_{ij}^{-6} (\Delta \alpha)^2 P_2 \left(\left(\mathbf{a}_i \cdot \mathbf{a}_j\right)\right) \quad (28)$$

where $\Delta \alpha$ is the anisotropy of the molecular polarizability and E is the average excitation energy.

Equation (27) presents a simple anisotropic attraction potential that favors nematic ordering. This potential has been used in the original Maier–Saupe theory [11, 12]. We note that the interaction energy (Eq. 27) is proportional to the anisotropy of the molecular polarizability $\Delta \alpha$. Thus, this anisotropic interaction is expected to be very weak for molecules with low dielectric anisotropy. Such molecules, therefore, are not supposed to form the nematic phase. This conclusion, however, is in conflict with experimental results. Indeed, there exist a number of materials (for example, cyclohexylcyclohexanes [15]) which form the nematic phase but exhibit very small anisotropy of the polarizability. This is an indication that anisotropic dispersion forces do not make the major contribution to the stabilization of the nematic phase. On the other hand, the well known success of the Maier–Saupe theory (as discussed below) is mainly determined by its mathematical form and not by the particular intermolecular interaction that has been taken into account.

As shown by Gelbart and Gelbart [16], the predominant orientational interaction in nematics must be the isotropic dispersion attraction modulated by the anisotropic molecular hard core. The isotropic part of the dispersion interaction is generally larger than the anisotropic part because it is proportional to the average molecular polarizability $\bar{\alpha}$. And the anisotropy of this effective potential comes from that of the asymmetric molecular shape. Thus this effective potential is a combination of intermolecular attraction and repulsion. It can be written as

$$V_{\text{eff}}(1,2) = J_{\text{att}}(r_{12})\Theta(r_{12} - \xi_{12}) \quad (29)$$

where the step-function $\Theta(r_{12} - \xi_{12})$ determines the steric cut-off. $\Theta(r_{12} - \xi_{12}) = 0$ if the molecules penetrate each other (i.e. if $r_{12} < \xi_{12}$) and $\Theta(r_{12} - \xi_{12}) = 1$ otherwise.

We note that Eq. (24) contains also the induction interaction, that is the interaction between the permanent multipoles of the molecule i and the polarizability of the molecule j. This interaction corresponds to the terms with $n_i = 0$ or $n_j = 0$. The induction interaction can play an important role if the molecular hard core is strongly polar [18].

Electrostatic interactions between molecules with permanent electric multipoles are also strongly anisotropic. The corresponding interaction potentials, however, vanish after the integration over the intermolecular vector r_{ij} and thus they do not contribute in the mean-field approximation. The dipole–dipole and dipole–quadrupole potentials vanish also after an orientational averaging because they are polar. At the same time, the electrostatic interactions can be very important if the molecules possess large permanent dipoles. In this case the dipole–dipole interaction gives rise to strong short-range dipolar correlations including the formation of dimers with antiparallel dipoles. At present the statistical theory of strongly polar nematics is in its early stage (see, however, [6, 7]).

2.3.3 Model Potentials

Realistic intermolecular interaction potentials for mesogenic molecules can be very complex and are generally unknown. At the same time molecular theories are often based on simple model potentials. This is justified when the theory is used to describe some general properties of liquid crystal phases that are not sensitive to the details on the interaction. Model potentials are constructed in order to represent only the qualitative mathematical form of the actual interaction energy in the simplest possible way. It is interesting to note that most of the popular model potentials correspond to the first terms in various expansion series. For example, the well known Maier–Saupe potential $JP_2((a_i \cdot n))$ is just the first nonpolar term in the Legendre polynomial expansion of an arbitrary interaction potential between two uniaxial molecules, averaged over the intermolecular vector r_{ij}:

$$\bar{V}((a_i \cdot a_j)) = \int dr_{ij}\, V(a_i, r_{ij}, a_j)$$
$$= J_0 + J_2 P_2((a_i \cdot a_j)) + \ldots \quad (30)$$

where we have taken into account only nonpolar terms. Here $V(a_i, r_{ij}, a_j)$ is an arbitrary interaction potential between two uniaxial rigid molecules. It depends only on the long axes a_i, a_j and on the intermolecular vector r_{ij}.

The partially averaged potential (Eq. 29) can be used in the molecular theory of the nematic–isotropic transition (also being supplemented by the P_4 term [19]). However, several other properties of nematics cannot be described in this way. For example, the full anisotropy of the Frank elastic constants can be accounted for only taking into account the explicit dependence of the interaction potential on the intermolecular vector [20]. In this case appropriate model potentials can be obtained using some more general expansion of the full potential $V(a_i, r_{ij}, a_j)$. This potential can be expanded in terms of the spherical invariants

$$V(a_i, r_{ij}, a_j) = \sum_{l,m,k} J_{lmk}(r_{ij}) T^{lmk}(a_i, u_{ij}, a_j) \quad (31)$$

The set $(T^{lmk}(a_i, u_{ij}, a_j))$ is a complete orthogonal set of basis functions [86] that contain the vector a_i to the power l, the vector u_{ij} to the power m and the vector a_j to the power k. The explicit expressions for the lower order invariants have been given, for example, by Van der Meer [14]. The invariants with one zero index are just Legendre polynomials. For example $T^{202}(a_i, u_{ij}, a_j) = P_2((a_i \cdot a_j))$.

The invariants with $l+m+k$ odd are pseudoscalars and therefore the corresponding coupling constants $J^{lmk}(r_{ij})$ are pseudoscalars as well. These terms can appear only in the interaction potential between chiral molecules. The first nonpolar chiral term of the general expansion (Eq. 30) reads:

$$V*(a_i, r_{ij}, a_j) = J*(r_{ij})([a_i \times a_j] \cdot r_{ij}) \quad (32)$$

The potential (Eq. 31) promotes the twist of the long axes of neighboring molecules and is widely used in the statistical theory of cholesteric ordering [13].

Finally we note that there exist some special model potentials that combine an attraction at large separation and repulsion at short distances. The most popular potential of this kind is the Gay–Berne potential [22] which is a generalization of the Lennard–Jones potential for anisotropic particles. The Gay–Berne potential is very often used in computer simulations but not in the molecular theory because it is rather complex.

2.4 Molecular Theory of the Nematic Phase

2.4.1 Mean-field Approximation and the Maier–Saupe Theory

The simplest molecular theory of the nematic–isotropic (N–I) transition can be developed in the mean-field approximation. According to the general definition, in the mean-field approximation one neglects all correlations between different molecules. This is obviously a crude and unrealistic approximation but, on the other hand, it enables one to obtain very simple and useful expressions for the free energy. This approximation also appears to be sufficient for a qualitative description of the N–I transition. More precise and detailed theories of the nematic state are based on more elaborate statistical models that will be discussed briefly in Sec. 2.4.3.

In the language of statistical mechanics the mean-field approximation is equivalent to the assumption that the pair distribution function $f_2(1, 2)$ can be represented as a product of the two one-particle distributions, that is $f_2 = f_1(1) f_2(2)$. The same representation is applied also to all n-particle distribution functions. This definition can be used to derive the mean-field expression for the free energy.

We note that the general Gibbs expression for the free energy F can be written in terms of the N-particle distribution function $f_N = Z^{-1} \exp(-H/kT)$ in the following way:

$$F = \int H f_N \, d\Gamma - kT \int f_N \ln f_N \, d\Gamma \qquad (33)$$

where H is the Hamiltonian of the system and $d\Gamma$ denotes the integration over all microscopic variables. On the level of pair interactions the Hamiltonian H can be represented as a sum of the kinetic energy plus the sum of interaction potentials for all molecular pairs

$$H = \sum_i E_k^i + \sum_{i,j} V(i,j) \qquad (34)$$

In the mean-field approximation the N-particle distribution is factorized as $f_N = \Pi_i f_1(i)$ and the general expression (Eq. 32) is reduced to the following mean-field free energy (in the absence of the external field):

$$F = \text{const} \qquad (35)$$
$$+ \frac{1}{2}\rho^2 \int V(\omega_i, \omega_j, r_{ij}) f_1(\omega_i, r_i)$$
$$\cdot f_1(\omega_j, r_j) \, d\omega_i \, d\omega_j \, dr_i \, dr_j$$
$$+ \rho kT \int f_1(\omega_i, r_i) \cdot \ln f_1(\omega_i, r_i) \, d\omega_i \, dr_i$$

where ω_i specifies the orientation of the molecule i. In Eq. (34) the free energy is represented as a functional of the one particle distribution function $f_1(i)$. The equilibrium distribution is determined by minimization of the free energy (Eq. 34):

$$f_i(\omega_i, r_i) = \frac{1}{Z} \exp\Big[-\beta\rho \int V(\omega_i, r_{ij}, \omega_j)$$
$$\cdot f_1(\omega_j, r_j) \, d\omega_j \, dr_j\Big] \qquad (36)$$

where

$$Z = \int \exp\Big[-\beta\rho \int V(\omega_i, r_{ij}, \omega_j)$$
$$\cdot f_1(\omega_j, r_j) \, d\omega_j \, dr_j\Big] d\omega_i \, dr_i \qquad (37)$$

It should be noted that Eqs. (35) and (36) are rather general and not restricted to the nematic phase. The distribution function $f_1(i)$ depends on the position r_i; therefore it can describe the molecular distribution in smectic phases as well.

In the nematic phase there is no positional order and the distribution function $f_1(i)$ depends only on the molecular orientation. In this case Eq. (35) can be simplified:

$$f_1(\omega_1) = \frac{1}{Z_0} \exp[-\beta U_{MF}(\omega_i)] \qquad (38)$$

where the mean-field potential $U_{MF}(\omega_1)$ is given by

$$U_{MF}(\omega_i) = \int \tilde{V}(\omega_1, \omega_2) f_1(\omega_2) \, d\omega_2 \qquad (39)$$

Here the effective pair potential $\tilde{V}(\omega_1, \omega_2)$ reads:

$$\tilde{V}(\omega_1, \omega_2) = \int U(\omega_1, r_{12}, \omega_2) \, dr_{12} \qquad (40)$$

One can readily see from Eq. (37) that in the mean field approximation each molecule feels some average mean-field potential produced by other molecules. This mean-field potential is just the pair interaction energy averaged over the position and orientation of the second molecule.

It is important to note that the mathematical form of Eqs. (37) and (38) appears to be rather general and goes far beyond the mean-field approximation. In fact, the one-particle distribution can always be written in the form of Eq. (37) with some unknown one-particle potential. In several advanced statistical theories this effective potential can be explicitly expressed in terms of the correlation functions. For example, such an expression can be obtained in the density functional theory discussed in Sec. 2.4.4.

The mean-field approximation was originally developed to describe lattice systems like ferromagnetics or ferroelectric crystals. In the case of liquids, however, some formal problems arise. In Eq. (39) the integral over r_{12} diverges because any attraction interaction diverges if the molecules are allowed to penetrate each other. The obvious solution to this problem is to take into account the hard-core repulsion that restricts the minimum distance between attraction centers. This can be done by introducing a steric cut-off into the integral in Eq. (39). Then the attraction potential is substituted by the effective potential $V_{\text{eff}}(1, 2) = V(1, 2)\,\Theta(r_{12}-\xi_{12})$ where $\Theta(r_{12}-\xi_{12})$ is a step-function (see Eq. 29). We note however, that the introduction of a steric cut-off is equivalent to taking account of simple short-range steric correlations. These correlations also give rise to the additional contribution to the free energy that is called packing entropy. This entropy is discussed in detail in Sec. 2.4.2.

If one neglects the asymmetry of the molecular shape (i.e. puts the function $\xi_{12}=D=$ const in the effective potential) and uses the dipole–dipole dispersion interaction potential (Eq. 27), one arrives at the Maier–Saupe theory. In this theory the interaction potential contains only the $P_2((\boldsymbol{a}_1 \cdot \boldsymbol{a}_2))$ term and as a result it is possible to obtain the closed equation for the nematic order parameter S. Substituting the potential $V(1, 2) = -J(r_{12}) P_2((\boldsymbol{a}_1 \cdot \boldsymbol{a}_2))\,\Theta(r_{12}-D)$ into Eqs. (27–29), multiplying both sides of Eq. (27) by $P_2((\boldsymbol{a}_1 \cdot \boldsymbol{a}_2))$ and integrating over $\boldsymbol{a}_2$, we obtain the equation

$$S = \frac{1}{Z_0} \int P_2(\cos\Theta) \cdot \exp\left[\frac{\rho}{\tilde{T}} S P_2(\cos\Theta)\right] \frac{d\cos\Theta}{2} \quad (41)$$

where $\cos\Theta = (\boldsymbol{a} \cdot \boldsymbol{n})$ and the dimensionless temperature $\tilde{T}=kT/J_0$ where $J_0 = \int_D^\infty J(r) r^2 dr$.

The free energy of the nematic phase can be written in the form

$$F = \frac{1}{2}\rho^2 J_0 S^2 - kT \cdot \ln \int \exp\left[\frac{\rho}{\tilde{T}} S P_2(\cos\Theta)\right] \frac{d\cos\Theta}{2} \quad (42)$$

Equations (40) and (41) describe the first order nematic–isotropic transition. At high temperatures Eq. (40) has only the isotropic solution $S=0$. At $\tilde{T}\approx 0.223$ two other solutions appear. One of them is always unstable but the other one does correspond to the minimum of the free energy F and characterizes the nematic phase. The actual nematic–isotropic phase transition takes place when the free energy of the nematic phase becomes equal to that of the isotropic phase. This happens at $\tilde{T}=\tilde{T}_{N-I}\approx 0.220$. At the transition temperature the order parameter $S\approx 0.44$.

Finally, the isotropic solution loses its stability at $\tilde{T}=0.2$ and below this temperature there exists only the nematic stable solution. Thus, $\tilde{T}\approx 0.223$ is the upper limit of metastability of the nematic phase and $\tilde{T}=0.2$ is the lower limit of metastability of the isotropic phase.

The remarkable feature of the Maier–Saupe theory lies in its simplicity and universality. In particular, the temperature variation of the order parameter and its value at the transition point are predicted to be universal, that is independent of intermolecular forces and the molecular structure. This prediction appears to be supported by experiments. In Fig. 1 the results of the theory are compared with some experimental data for the parameter S presented by Luckhurst et al. [23]. One can see that the agreement is surprisingly good taking into account the number of approximations and

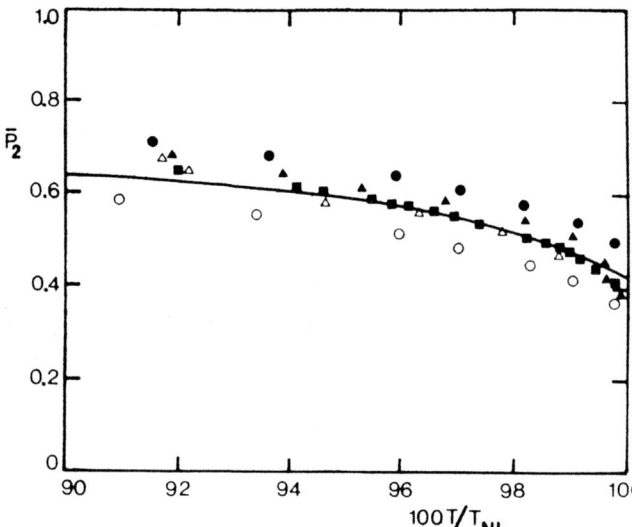

Figure 1. The temperature dependence of the order parameter S, at constant pressure, for 4,4'-dimethoxyazobenzene (●), 4,4'-diethoxyazobenzene (△), anisaldazine (▲), 2,4-nonadienoic acid (■) and 2,4-undecadienoic acid (○). The curve is predicted by the Maier–Saupe theory (after Luckhurst [23]).

simplifications involved in the Maier–Saupe theory.

It should be noted, however, that the agreement between the Maier–Saupe theory and experiment is not so good if one considers some other parameters of the N–I transition. For example, the discontinuity in entropy is overestimated several times. In particular, the difference between the transition temperature T_{N-I} and the lower limit of stability of the isotropic phase T^* is strongly overestimated. In the Maier–Saupe theory the parameter $\gamma = (T_{N-I} - T^*)/T_{N-I}$ is about 0.1 while experimentally $T_{N-I} - T^* \approx 1-2\,C$ and therefore $\gamma \sim (3-6)10^{-3}$. This means that the mean-field theory overestimates the difference between the isotropic and the nematic phase. The discrepancy is clearly related to the neglect of short-range orientational correlations between anisotropic molecules that make the local structure rather similar in both phases.

The more serious problem in the original Maier–Saupe theory is related to the choice of the anisotropic dispersion interaction potential (Eq. 27) (see the detailed discussion in [24]). An estimate of the transition temperature T_{N-I} yields the value that is an order of magnitude too small [24, 25]. Moreover, there exist materials (for example, cyclohexylcyclohexanes [15] or alkylbicyclooctanes [24]) which possess very low anisotropy of the molecular polarizability but nevertheless form stable nematic phases. These examples enable us to conclude that anisotropic dispersion forces certainly cannot be a dominant general mechanism of the stabilization of the nematic phase.

It was first noticed by Gelbart and Gelbart [16] that the predominant anisotropic interaction in nematics results from a coupling between the isotropic attraction and the anisotropic hard-core repulsion. This coupling is represented by the effective potential $V_{eff}(1,2) = V(1,2)\,\Theta(r_{12} - \xi_{12})$. This potential can be averaged over all orientations of the intermolecular vector and then can be expanded in Legendre polynomials. The first term of the expansion has the same structure as the Maier–Saupe potential $J(r_{12})P_2((a_1 \cdot a_2))$ but with the coupling constant J determined by the anisotropy of the molecular shape and the *average* molecular polarizability rather than by the polarizability *anisotropy*. We note that the particular interpretation of the coupling constant J does not influence any of the results of the Maier–Saupe theory except for the absolute value of the transition temperature. Thus, the success of the Maier–Saupe theory is mainly determined by its general mathematical structure and by the form of the model potential.

2.4.2 Short-range Orientational Correlations

There are several ways to improve the Maier–Saupe theory. One way is to take into account the asymmetry of the molecular shape and to account for the excluded volume effects. This approach will be discussed in the next subsection. The second possible way is to improve the statistical part of the theory by taking into account some intermolecular correlations. Finally, the third way is to improve the model potential.

Strong short-range orientational correlations can be conveniently taken into account in the cluster approximation. The simplest version of the cluster approximation in the theory of liquid crystals was proposed by Ypma et al. [26] who used the general approach of Callen and Strieb developed in the theory of ferromagnetism.

In the two-particle cluster approximation the interaction between two neighboring molecules is taken into account exactly while the interaction with the rest of the nearest neighbors is treated in the mean-field approximation. The two-particle Hamiltonian is written as

$$H_2(1,2) = -JP_2((\boldsymbol{a}_1 \cdot \boldsymbol{a}_2)) \\ + \left(1 - \frac{1}{\sigma}\phi\left[P_2((\boldsymbol{a}_1 \cdot \boldsymbol{n})) + P_2((\boldsymbol{a}_2 \cdot \boldsymbol{n}))\right]\right) \quad (43)$$

where σ is the number of nearest neighbors and the parameter ϕ is the strength of the mean-field which is to be determined in a self-consistent way. The pair distribution function is assumed to have the Boltzmann form with the Hamiltonian (Eq. 42):

$$f_2(1,2) = \frac{1}{Z_2}\exp[-\beta H_2(1,2)] \quad (44)$$

The one-particle distribution function is given by the same type of expression as in the mean-field theory:

$$f_1(1,2) = \frac{1}{Z_1}\exp[-\beta\phi P_2((\boldsymbol{a}_1 \cdot \boldsymbol{n}))] \quad (45)$$

The parameter ϕ is determined by the self-consistency relation that states that the order parameter S, calculated with the one-particle distribution function (Eq. 44), must be equal to the one calculated with the pair distribution function (Eq. 43):

$$\int P_2(\boldsymbol{a}_1 \cdot \boldsymbol{n}) f_1(1) d\boldsymbol{a}_1 = \\ \int P_2((\boldsymbol{a}_1 \cdot \boldsymbol{n})) f_2(1,2) d\boldsymbol{a}_1 d\boldsymbol{a}_2 \quad (46)$$

Finally from Eqs. (42)–(45) one can obtain the following simple expression for the free energy

$$\beta F = -\frac{1}{2}N\sigma \ln Z_2 + N(\sigma-1)\ln Z_1 \quad (47)$$

where

$$Z_1 = \int \exp[-\beta\phi P_2((\boldsymbol{a}_1 \cdot \boldsymbol{n}))] d\boldsymbol{a}_1 \\ Z_2 = \int \exp[-\beta H_2(1,2)] d\boldsymbol{a}_1 d\boldsymbol{a}_2 \quad (48)$$

The self-consistency relation (Eq. 45) is obtained by minimization of the free energy (Eq. 46) with respect to ϕ.

The use of a cluster approximation reduces the discrepancy between theory and experiment. For example, the difference $T_{N-I} - T^*$ is reduced several times compared to the Maier–Saupe theory. A detailed comparison of different versions of the cluster approximation with experimental data has been given by Chandrasekhar [61].

Finally we note that it is possible to include the $P_4(\cos\Theta)$ term in the model potential of the Maier–Saupe theory, as has been done by Luckhurst et al. [19] and Chandrasekhar and Madhusudana [28]. This procedure also leads to some quantitative improvement.

2.4.3 Excluded Volume Effects and the Onsager Theory

As discussed in the previous subsection, it is important to take into account the excluded volume effects even in a simple mean-field theory based on an anisotropic attraction interaction. The excluded volume effects are determined by hard-core repulsion that does not allow molecules to penetrate each other. It is interesting to note that by doing so we already go beyond the formal mean-field approximation. Indeed, with excluded volume effects the internal energy of the nematic phase can be written as

$$U = \frac{1}{2}\rho^2 \int V_s(1,2) h_2^0(1,2)$$
$$\cdot f_1(1) f_1(2) \mathrm{d}(1) \mathrm{d}(2) \qquad (49)$$

where $h_2^0(1,2) = \exp[-\beta V_s(1,2)]$ is, in fact, a simple correlation function between the rigid molecules 1 and 2. Here $V_s(1,2)$ is the steric repulsion potential (Eq. 19).

It is obvious that in this approximation we do take into account some short-range steric correlations between rigid molecules. We note that these correlations contribute not only to the internal energy but also to the entropy of the nematic. Excluded volume effects restrict the molecular motion and therefore the total entropy of the fluid is reduced. This additional contribution to the entropy is called the packing entropy.

A simple expression for the packing entropy at low densities was first derived by Onsager [10] who considered nematic ordering in a system of long rigid rods. In this system the rods interact only sterically and are supposed to be very long, $L/D \gg 1$, where L is the length and D is the diameter of the rod. At low densities it is possible to express the free energy of such a system in the form of the virial expansion:

$$\beta F = \rho \ln\rho + \rho$$
$$\cdot \int f_1(\omega_1)[\ln f_1(\omega_1) - 1]\mathrm{d}\omega_1$$
$$+ \frac{1}{2}\rho^2 \int f_1(\omega_1) f_1(\omega_2) \qquad (50)$$
$$\cdot B(\omega_1, \omega_2)\mathrm{d}\omega_1\, \mathrm{d}\omega_2 + \ldots$$

where $B(\omega_1, \omega_2)$ is the excluded volume for the two rods:

$$B(\omega_1, \omega_2) = \int \mathrm{d}\mathbf{r}_{12}\left(\exp[-\beta V_s(1,2)] - 1\right) \quad (51)$$

For two spherocylinders the excluded volume is expressed as

$$B(1,2) = 2L^2 D |\sin\gamma_{12}| + 2\pi L D^2 + \frac{4}{3}\pi d^3 \qquad (52)$$

where γ_{12} is the angle between the long axes of the two spherocylinders.

All terms in Eq. (49) are purely entropical in nature because the system is athermal. The second term is the orientational entropy and the third one is the packing entropy that is related to the second virial coefficient for two rigid rods. The expansion in Eq. (49) is actually performed in powers of the packing fraction $\eta = \rho v_0 \approx pi\, \rho D^2 L \ll 1$ if $L/D \gg 1$, where v_0 is the volume of a spherocylinder. At a very low volume fraction of rods the higher order terms in the expansion can be neglected [29].

We note that the free energy (Eq. 49) has the same general mathematical form as the mean-field free energy (Eq. 34). In the gas consisting of long rods, the role of the effective anisotropic potential is played by the excluded volume $B(a_1, a_2)$ multiplied by kT. Thus it is not surprising that the minimization of the free energy (Eq. 49) yields practically the same equation for the orientational distribution function as in the Maier–Saupe theory:

$$f_1((a_1 \cdot n)) = \frac{1}{Z_0}$$
$$\cdot \exp\left[-I\int |\sin\gamma_{12}| f_1((a_2 \cdot n)) \frac{da_2}{4\pi}\right] \quad (53)$$

where $I = 2\rho L^2 D \approx \eta L/\pi D$.

From the mathematical point of view the only difference between Eq. (52) and the corresponding equation in the Maier–Saupe theory is in the form of the effective interaction potential. In the Maier–Saupe theory the potential is $-JP_2(\cos\gamma_{12})$ while in the Onsager theory the potential has a different form $kTI|\sin\gamma_{12}|$ that also contains a substantial contribution from higher-order Legendre polynomials. However, from the physical standpoint the most important difference between the Maier–Saupe and the Onsager theories is in the nature of the transition. Maier–Saupe theory describes the N–I transition in thermotropic liquid crystals where the ordering appears at some particular temperature. By contrast, in the Onsager theory the transition occurs when the volume fraction of rods is increased. In Eq. (52) the bifurcation point (pseudo-second-order transition) corresponds to $I = 1$ and thus the critical packing fraction appears to be of the order D/L, $\eta = \pi D/L \ll 1$ if $D/L \ll 1$. This is a crude estimate of the actual transition density. Therefore, in the case of very long rods the transition takes place at very low density. In the limiting case of $L/D \to \infty$ the Onsager theory, which is based on the virial expansion, appears to be asymptotically exact [29].

The actual N–I transition for a gas consisting of long rods is more complex because the system separates into a dilute isotropic phase and the more concentrated nematic phase that already possesses a high degree of orientational order. This is related to the fact that the homogeneous system of long rods appears to be mechanically unstable (with respect to density fluctuations) within some density interval around $\eta_{cr} = \pi D/L$. The two coexisting densities ρ_1 and ρ_2 can be determined in the usual way by equating the chemical potentials and the pressures of the two phases:

$$\mu_1(\rho_1) = \mu_2(\rho_2)$$
$$P_1(\rho_1) = P_2(\rho_2) \quad (54)$$

where the pressure P and the chemical potential μ can be expressed in terms of the free energy in the following way:

$$P = -\left(\frac{\partial F}{\partial V}\right)_{T,N} = \rho \frac{\partial (F/V)}{\partial \rho} - \frac{F}{V} \quad (55)$$

$$\mu = -\left(\frac{\partial F}{\partial N}\right)_{T,V} = \rho \frac{\partial (F/V)}{\partial \rho} \quad (56)$$

Equations (53) and (54), supplemented by Eq. (52) for the orientational distribution function and Eq. (49) for the free energy, can be used to determine the coexisting densities and the value of the order parameter at the transition. Eq. (52) was first solved approximately by Onsager who used the trial function

$$f_1(\cos\Theta) = \left(\frac{\alpha}{4\pi \sinh\alpha}\right) \cosh(\alpha\cos\Theta) \quad (57)$$

and thus reduced the integral Eq. (52) to the equation for the single parameter α. The nematic order parameter S can be expressed in terms of α as

$$S = 1 - \frac{3}{\alpha}\cosh\alpha + \frac{3}{\alpha^2} \tag{58}$$

Eq. (52) has been solved numerically by Lasher [30] and Lekkerkerker et al. [31] by employing the Legendre polynomial expansion and by Lee and Meyer [32] by a direct numerical method. The results do not differ significantly from those obtained with the Onsager trial function (Eq. 55) and therefore the approximation (Eq. 55) appears to be sufficient for most practical purposes. A more detailed description of the Onsager theory and its generalizations can be found in reviews [33–36]. Here we do not consider it any more because this chapter is focused on the theory of thermotropic liquid crystals. In this context the main consequence of the Onsager approach is the conclusion that the excluded volume effects can be very important in stabilizing the nematic phase. This is expected to be true also in the case of thermotropic nematics because they are also composed of molecules with relatively rigid cores. Therefore, the packing entropy must be taken into account in any consistent theory of the N–I transition. We discuss this contribution in more detail in the following subsection.

2.4.4 Packing Effects in Thermotropic Nematics

The Onsager expression for the packing entropy is valid at very low densities and the theory can be applied directly to dilute solutions of rigid particles like tobacco mosaic virus or helical synthetic polypeptides. At the same time typical thermotropic nematics are composed of molecules with an axial ratio of the order of 3 or 4 and the packing fraction is of the order of 1. Thus, the direct use of the Onsager theory for such systems can result in large errors. The Onsager expression for the packing entropy has been generalized to the case of condensed nematic phases by several authors [37–40] using different approximations. The corresponding expression for the packing entropy can be written in the following general way:

$$S_p = -\frac{1}{2}\lambda(\eta)\rho^2 k_B$$
$$\cdot \int \Theta(r_{12} - \xi_{12}) f_1((a_1 \cdot n))$$
$$\cdot f_1((a_2 \cdot n)) da_1 da_2 dr_{12} \tag{59}$$

Here the coefficient $\lambda(\eta)$ depends on the packing fraction η. In the Onsager theory, which corresponds to the limit of small η, the factor $\lambda = 1$. The frequently used approximations for the packing entropy have been derived from the scaled particle theory [37] and were proposed by Parsons and Lee [38, 41]. The equations of state derived from various molecular theories for a condensed solution of hard rods are given in the review of Sato and Teramoto [36]. One can readily see that even for long rods with $L/D = 50$ the Onsager equation of state deviates significantly from the results of the Parsons–Lee or scaled particle theory already at relatively small packing fractions $\eta \approx 0.3$. We note that recently the Parsons–Lee approximation has been tested by Jackson et al. [5] for the system of relatively short spherocylinders with axial ratios $L/D = 3$ and $L/D = 5$. It has been shown that the results agree very well with Parsons–Lee theory up to packing fractions of $\eta = 0.5$. In the Parsons–Lee approximation the function $\lambda(\eta)$ is written in a simple form

$$\lambda_{PL} = \frac{1}{4} \frac{4 - 3\eta}{(1-\eta)^3} \tag{60}$$

Taking into account the packing entropy (Eq. 57), one can write the model free energy for the nematic phase

$$F/V = kT\rho \int f_1((\boldsymbol{a}_1 \cdot \boldsymbol{n}))$$
$$\cdot \ln f_1((\boldsymbol{a}_1 \cdot \boldsymbol{n})) d\boldsymbol{a}_1 + \frac{1}{2} kT\rho^2 \lambda(\eta)$$
$$\cdot \int f_1((\boldsymbol{a}_1 \cdot \boldsymbol{n})) f_1((\boldsymbol{a}_2 \cdot \boldsymbol{n}))$$
$$\cdot V_{\text{excl}}((\boldsymbol{a}_1 \cdot \boldsymbol{a}_2)) d\boldsymbol{a}_1 d\boldsymbol{a}_2 + \frac{1}{2} \rho^2$$
$$\cdot \int f_1((\boldsymbol{a}_1 \cdot \boldsymbol{n})) f_1((\boldsymbol{a}_2 \cdot \boldsymbol{n})) V_{\text{att}}((\boldsymbol{a}_1, \boldsymbol{a}_2, \boldsymbol{r}_{12}))$$
$$\cdot g_{\text{HC}}(\boldsymbol{a}_1, \boldsymbol{a}_2, \boldsymbol{r}_{12}) d\boldsymbol{a}_1 d\boldsymbol{a}_2 d\boldsymbol{r}_{12} \quad (61)$$

where $V_{\text{excl}}(1, 2)$ is the excluded volume for the two rigid particles, $V_{\text{att}}(1, 2)$ is the attraction interaction potential and $g_{\text{HC}}(1, 2)$ is the pair correlation function for the reference hard-core fluid.

For simplicity one can approximate the pair correlation function by the steric cut-off $\Theta(r_{12} - \xi_{12}) = \exp(-\beta V_s(1, 2))$. In this case the free energy (Eq. 59) corresponds to the so-called generalized Van der Waals theory considered in detail by Gelbart and Barboy [43, 44].

We note that in Eq. (59) the attraction interaction is taken into account as a perturbation. This means that one neglects a part of the orientational correlations determined by attraction. A simple way to improve the theory is to combine Eq. (59) with the two-particle cluster approach [40].

The first two terms in Eq. (59) represent the free energy of the reference hard-core fluid. The general structure of this reference free energy is similar in different approaches but the particular dependence on the packing fraction η can be quite different. This depends on the particular approach (used in the theory of hard-sphere fluids) that has been generalized to the nematic state. For example, the Parsons–Lee approximation is based on the Carnahan–Starling equation of state, the approach of Ypma and Vertogen is a generalization of the Percus–Yevick approximation and the approach of Cotter is based on the scaled particle theory. The alternative way is to use the so-called y-expansion of the hard-core free energy proposed by Gelbart and Barboy [44]. This is an expansion in powers of $\eta/(1-\eta)$ which is much more reliable at high densities compared with the usual virial expansion in powers of η. The y-expansion has also been used by Mulder and Frenkel [51] in the interpretation of the results of computer simulations for a system of hard ellipsoids.

The free energy (Eq. 59) with some modifications has been used in the detailed description of the N–I transition by Gelbart et al. [42, 43], Cotter [37] and Ypma and Vertogen [40]. The work of Ypma and Vertogen also contains a critical comparison with other approaches. One interesting conclusion of this analysis is related to the effective axial ratio of a mesogenic molecule. It has been shown that a good agreement with the experiment for the majority of the parameters of the N–I transition can be achieved only if one assumes that the effective geometrical anisotropy of the mesogenic molecule is much smaller than its actual value. This result has been interpreted in terms of molecular clusters that are supposed to be the building units of the nematic phase. The anisotropy of such a cluster is assumed to be much smaller than that of a single molecule. On the other hand, this result can be attributed to the quantitative inaccuracy of the free energy (Eq. 59). It is not excluded that the theory based on Eq. (59) overestimates a contribution from the excluded volume effects in thermotropic nematics.

The same problem can be viewed in a different way. According to the results of computer simulations [45] the nematic ordering in an athermal system of elongated rigid particles is formed only if the axial ratio is more than three. At the same time, the effective value of L/D for typical mesogenic molecules is usually assumed to be more than three. Thus, any dense fluid, composed of such molecules, must be in the nematic phase at all temperatures. In reality, however, some additional contribution from attractive forces is required to stabilize the nematic phase. As a result, we conclude that there exists some very delicate balance between repulsion and attraction in real thermotropic liquid crystals. The discrepancy from experiment can also be related to molecular flexibility; the anisotropy of the hard core may not be sufficiently large to stabilize the nematic phase. Thus, a fully consistent molecular theory of nematic liquid crystals must take into account the molecular flexibility in some way. This is a particularly difficult problem and so far the flexibility has been accounted for only in the context of the generalized mean-field theory applied separately to each small molecular fragment (see the review of Luckhurst and references therein [46]).

2.4.5 The Role of Molecular Biaxiality

The majority of the existing molecular theories of nematic liquid crystals are based on simple uniaxial molecular models like spherocylinders. At the same time typical mesogenic molecules are obviously biaxial. (For example, the biaxiality of the phenyl ring is determined by its breadth-to-thickness ratio which is of the order of two.) If this biaxiality is important, even a very good statistical theory may result in a poor agreement with experiment when the biaxiality is ignored. Several authors have suggested that even a small deviation from uniaxial symmetry can account for important features of the N–I transition [29, 42, 47, 48].

In the uniaxial nematic phase composed of biaxial molecules the orientational distribution function depends on the orientation of both the molecular long axis a and the short axis b, i.e. $f_1(1) = f_1((a \cdot n), (b \cdot n))$. The influence of the biaxiality on the distribution function is suitably described by the order parameter D:

$$D = \langle P_2((b \cdot n)) \rangle - \langle P_2((c \cdot n)) \rangle \quad (62)$$

which characterizes the difference in the tendencies of the two short axes b and c to orient along the director. The order parameter D appears to be rather small (roughly of the order 0.1 [47]) and it is often neglected in simple molecular theories. However, it has some influence on the N–I transition as discussed in ref. [47, 49]. At the same time, except for the parameter D, the molecular biaxiality does not directly manifest itself in the Maier–Saupe theory. If one neglects the parameter D, the equations of the theory will depend on some effective uniaxial potential which is equal to the true potential between biaxial molecules averaged over independent rotations of the two molecules about their long axes.

However, the contribution of the biaxiality is nontrivial if the hard-core repulsion between biaxial molecules is taken into account. The hard-core repulsion is described by the Maier function $\exp[-\beta V_s(1, 2)]$ that depends nonlinearly on the repulsive potential $V_s(1, 2)$. In this case the free energy depends on the effective uniaxial potential that is an average of the Maier function. This theory accounts for some biaxial steric correlations and the result of such averaging cannot be interpreted as a hard-core repulsion between some uniaxial particles. Thus, it is the biaxiality of the molecular shape that seems to be of primary importance.

The biaxiality of molecular shape can be directly taken into account in the context of the Onsager theory. The first attempt to do this has been made by Straley [49]. How-

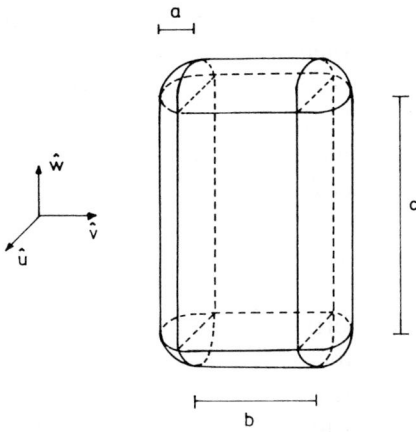

Figure 2. Spheroplatelet as a simple model for a biaxial particle [50].

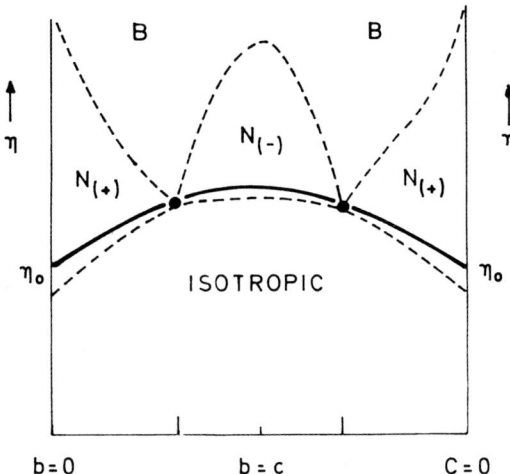

Figure 3. Phase diagram of the hard spheroplatelet fluid in the Onsager approximation for a family of particles with constant volume and different axial ratio. The points $b=0$ and $c=0$ correspond to the spherocylinders (after Mulder [50]).

ever, the full analyses can be performed only if the explicit expression for the excluded volume of the two biaxial particles is available. This expression has been derived by Mulder [50] for two spheroplatelets (see Fig. 2) with an arbitrary relative orientation. The excluded volume for such biaxial particles can be written in the analytical form

$$V_{\text{excl}}(\omega_1, \omega_2)$$
$$= \frac{32\pi a^3}{3} + 8\pi a^2 (b+c) + 8abc$$
$$+ 4abc \left[|v_1 \times w_2| + |v_2 \times w_1| \right]$$
$$+ 4ab^2 |v_1 \times v_2| + 4ac^2 |w_1 \times w_2|$$
$$+ b^2 c \left[|u_1 \times v_2| + |v_1 \times u_2| \right]$$
$$+ bc^2 \left[|u_1 \times w_2| + |w_1 \times u_2| \right] \quad (63)$$

where the unit vectors w, u and v are in the direction of the main axes of the spheroplatelet and the parameters a, b and c are shown on Fig. 2. The plausible constant volume section of the phase diagram of a fluid composed of such biaxial particles, obtained in the Onsager approximation, is shown in Fig. 3. The limiting points $c=0$ and $b=0$ correspond to the same spherocylinder with diameter α. In this limiting case the theory is reduced to the usual Onsager theory. The midpoint $b=c$ corresponds to the plate-like particle with the C_4 symmetry axis. Such particles form the uniaxial discotic nematic phase. For $0<c<b$ and for $0<b<c$ the particles are biaxial. However, for $c \gg b > c$ and for $b \gg c > 1$ they are rod-like while for $b \approx c$ they are plate-like. Thus, somewhere between these two domains one should find the crossover shape that corresponds to a boundary between the transitions into two different nematic phases N_+ and N_-. In the N_- phase the molecular planes (and the long axes) are oriented approximately along the director, while in the N_+ phase they are oriented perpendicular to the director. The crossover shape corresponds also to the transition from the isotropic to the biaxial nematic phase. For very long spheroplatelets the crossover shape is characterized by the relation $b \approx c^{1/2}$ in dimensionless units.

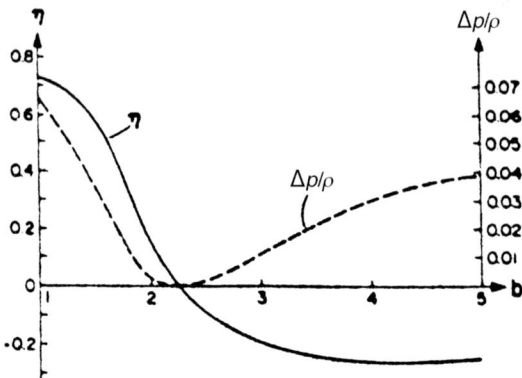

Figure 4. The dependence of the order parameter and the discontinuity in density at the N–I phase transition. The liquid is composed of hard rectangular parallelepipeds with dimensions $a=1$, b and $c=5$ (after Gelbart and Barboy [42]).

The phase diagram, shown in Fig. 3, was obtained from the Onsager approximation and, therefore, it is expected to be correct only at low densities. In the case of liquid densities the role of shape biaxiality has been analyzed by Gelbart [43]. Gelbart has considered the nematic ordering of rectangular parallelepipeds having dimensions $a < b < c$ (with a and c fixed, $c > a$) with the help of y-expansion. In Fig. 4, taken from [43], the nematic order parameter η and the discontinuity in density at the transition point are presented as a function of the breadth of the parallelepiped b that varies between $a=1$ and $c=5$. One can readily see that both the order parameter and the density gap decrease strongly with the increasing molecular biaxiality b/a. For the limiting case of $b=a=1$, the rod limit, the order parameter at the transition is very large. However, for values of b/a slightly less than 2 both the order parameter and the density gap are rather close to the typical experimental data. Finally, the value $b/a \approx 2.25$ corresponds to the crossover shape and at this point the order parameter and the density gap vanish identically.

Thus we see that the effect of shape biaxiality can indeed be very strong. It seems that this can be the main reason why the molecular theories, based on rod-like molecular models, overestimate the first orderness of the nematic–isotropic transition.

2.4.6 Density Functional Approach to the Statistical Theory of Liquid Crystals

One can find in the literature a large number of molecular theories of liquid crystals proposed by different authors using different approximations. However, the majority of these theories and the corresponding expressions for the free energy can be derived in a systematic way with the help of some general approaches. A very fruitful approach of this kind is the density functional theory that has been first applied to liquid crystals by Sluckin and Shukla [52] and by Singh [53].

In the density functional approach the free energy of a liquid crystal is represented as a functional of the one-particle density $\rho(\mathbf{r}, \omega) = \rho(\mathbf{r}) f_1(\omega)$ where $\rho(\mathbf{r})$ is the number density and $f_1(\mathbf{r}, \omega)$ is the one-particle distribution function. The equilibrium distribution function is determined, as usual, by minimization of the free energy functional $F[\rho]$. The general structure of this functional is unknown, of course, but the functional derivatives are known and can be expressed in terms of the correlation functions.

In the general case the free energy can be represented as a sum of two terms, $F = \Phi + H$ where Φ is the free energy of the system of noninteracting particles

$$\Phi = kT \int \rho(\mathbf{x}) [\ln \rho(\mathbf{x}) \Lambda - 1] d\mathbf{x}$$
$$+ \int \rho(\mathbf{x}) U_e(\mathbf{x}) d\mathbf{x} \qquad (64)$$

where $x = (r, \omega)$, $U_e(x)$ is the external potential and Λ is a constant. The potential H is determined by the intermolecular interactions and its functional derivatives are related to the direct correlation functions. For example, the second derivative can be written as

$$\delta^2 H / \delta\rho(x_1)\delta\rho(x_2) = -kT C_2(x_1, x_2) \quad (65)$$

where $C_2(x_1, x_2)$ is the direct pair correlation function related to the full correlation function by the Ornstein–Zernike equation. Now Eq. (63) can be used to develop a theory of the N–I transition. For this purpose one can perform the functional Taylor expansion of the free energy of the nematic phase around its value in the isotropic phase. The expansion is performed in powers of $\Delta\rho$ where $\Delta\rho = \rho_N - \rho_I$ is the difference between the one-particle densities in the nematic and isotropic phases. Now the free energy of the nematic phase up to the second order in $\Delta\rho$ reads:

$$F_N = F_I + kT \quad (66)$$
$$\cdot \int \rho_N(x_1)[\ln \rho_N(x_1) - 1] dx_1$$
$$+ \int \rho_N(x_1) U_e / x_1 \, dx_1$$
$$- \int C_1(x_1) \delta\rho(x_1) dx_1$$
$$- \frac{1}{2} \int C_2(x_1, x_2) \delta\rho(x_1) \delta\rho(x_2) dx_1 dx_2$$

where $C_1 = \delta H / \delta\rho$ and where the direct correlation functions $C_1(1)$ and $C_2(1, 2)$ are calculated for the isotropic phase. The higher order terms in the expansion (Eq. 64) depend on the higher order direct correlation functions.

Taking into account the equilibrium condition for the isotropic distribution function ρ_I:

$$\ln \rho_I + C_1 - \rho_I U_e(1) = \text{const}$$

we arrive at the following self-consistent equation for the one-particle distribution function

$$\rho_N(r_1, \omega_1)$$
$$= \rho_I \exp[-\beta U_e(r_1, \omega_1)$$
$$+ \int C_2(r_{12}, \omega_1, \omega_2) \delta\rho(r_1, \omega_1)$$
$$\cdot \delta\rho(r_2, \omega_2) dr_2 \, d\omega_2] \quad (67)$$

This general equation applies both to nematic and smectic phases because the one-particle density may depend on position. In the case of a uniform nematic phase without an external potential we obtain

$$f_1(\omega_1) =$$
$$Z^{-1} \exp\left[\int \tilde{C}_2(\omega_1, \omega_2) \delta\rho(\omega_2) d\omega_2\right] \quad (68)$$

We note that Eq. (66) has practically the same mathematical form as the mean-field equation (Eq. 39) for the orientational distribution function. In the density functional approach the role of the effective pair potential is played by the direct correlation function $\tilde{C}_2(\omega_1, \omega_2)$. In the case of uniaxial molecules the function $\tilde{C}_2(1, 2) = \tilde{C}_2((a_1 \cdot a_2))$. Expanding this function in Legendre polynomials, truncation after the P_2 term and substituting into Eq. (66), we arrive exactly at the Maier–Saupe equation for the orientational distribution function. This means that the general mathematical structure of the Maier–Saupe theory is not restricted to the mean-field approximation. The same equation has been derived in the context of a very general density functional approach. The only approximation has been related to the neglect of many-body direct correlation functions. Thus, the main equations of the Maier–Saupe theory remain valid generally on the level of pair correlations. This seems to be the main reason why this approach appears to be so successful in spite of its simplicity.

In the density functional approach the parameters of the N–I transition can be expressed in terms of the direct correlation function. Unfortunately, this function cannot be calculated exactly even for simple models and thus the general theory has to be accompanied by some practical approximations. We note that in this way it is possible to derive various approximate expressions for the free energy that have been obtained in various molecular theories. In other words, in the context of the density functional approach, various molecular theories usually correspond to some approximations for the direct correlation functions. We can illustrate this idea by the following examples.

The extended mean-field theory corresponds to the following approximate direct correlation function:

$$C_{MF}(1,2) \approx -V_{att}(1,2)\Theta(r_{12}-\xi_{12})/kT \quad (69)$$

The free energy of the Onsager theory can be obtained substituting the direct-correlation function by its first virial term

$$C_s(1,2) \approx \exp[-\beta V_s(1,2)] - 1 \quad (70)$$

Several more elaborate molecular theories correspond to the perturbative approximation

$$C_2(1,2) \approx C_{HC}(1,2) + g_{HC}(1,2)V_{att}(1,2) \quad (71)$$

where the functions $C_{HC}(1,2)$ and $g_{HC}(1,2)$ are calculated for the reference hard-core system.

Further approximations, including the ones discussed in Sec. 2.4.3, can be obtained by substituting $g_{HC}(1,2)$ with the steric cut-off function $\Theta(r_{12}-\xi_{12})$ and by using the Parsons approximation for the $C_{HC}(1,2)$:

$$C_{HC}(1,2) = C_{hs}(r_{12}/\xi_{12}) \quad (72)$$

where $C_{hs}(1,2)$ is the direct correlation function for a hard-sphere fluid. The more detailed discussion of various approximations for the direct correlation function can be found in the paper of Sluckin [54]. Recently some of these approximations have been tested against computer simulations [55].

The brief discussion of the density functional theory, presented above, enables one to conclude that this is a very powerful approach. It appears to be particularly helpful in the derivation of general expressions for various elasticity coefficients of the nematic phase. Indeed, it is also possible to expand the free energy of the distorted nematic state with respect to the homogeneous state in the same way as in Eq. (64). In this case the difference in the one-particle densities of the two states $\delta \rho$ is proportional to the gradients of the director and can be arbitrarily small. Then the functional expansion appears to be quantitatively correct. In this way it is possible to derive formally exact expressions for the Frank elastic constants [56], helical twisting power [57] and flexoelectric coefficients [58]. It should be noted, however, that so far the density functional theory has been formulated only for a system of rigid molecules. Thus, the corresponding general expressions are restricted to this simple class of molecular model.

2.5 Molecular Models for Simple Smectic Phases

2.5.1 Mean-field Theory of the Nematic–Smectic A Transition

Smectic phases are characterized by some positional order and therefore the one-particle distribution function $f_1(r, \omega)$ depends both on position r and the orientation ω. In the simplest smectic A phase there exists only one macroscopic direction that is parallel to the wave vector $\mathbf{k}$ of the periodic smectic structure. By contrast, in the smectic C phase there are two different macroscopic axes because the director is tilted with respect to $\mathbf{k}$ and thus the phase appears to be biaxial. As a result the distribution function of the smectic C phase generally depends on the orientation of both molecular long and short axes, if the molecules are biaxial.

For the smectic A phase the one-particle distribution function can be expanded in a complete set of basic functions:

$$f_1(\mathbf{r},\mathbf{a}) = f_1(\cos\omega, z)$$
$$= \sum_{l,m} f_{l,m} P_l(\cos\omega)\cos(kmz) \quad (73)$$

where $\cos\omega = (\mathbf{a}\cdot\hat{\mathbf{k}})$ and where $\hat{\mathbf{k}} = \mathbf{k}/k$.

From Eq. (71) one can readily see that the dominant (i.e. lowest order) order parameters for the nematic–smectic A transition are:

$$S\langle P_2(\cos\omega)\rangle$$
$$\sigma = \langle P_2(\cos\omega)\cos(kz)\rangle$$
$$\tau = \langle \cos(kz)\rangle \quad (74)$$

Here S is the usual nematic order parameter, τ is the purely translational order parameter and the parameter σ characterizes a coupling between orientational and translational ordering.

The simple theory of the nematic–smectic A transition has been proposed by McMillan [59] (and independently by Kobayashi [60]) by extending the Maier–Saupe approach to include the possibility of translational ordering. The McMillan theory is a classical mean-field theory and therefore the free energy is given by the general Eq. (34). For the smectic A phase it can be rewritten as

$$F = \text{const} + \frac{1}{2}\rho^2 \int V(\mathbf{a}_i,\mathbf{a}_j,\mathbf{r}_{ij})$$
$$\cdot f_1(\mathbf{a}_i,\mathbf{r}_i)\,f_1(\mathbf{a}_j,\mathbf{r}_j)\,d\mathbf{a}_i\,d\mathbf{a}_j\,d\mathbf{r}_i\,d\mathbf{r}_j$$
$$+ \rho kT \int f_1(\mathbf{a}_i,\mathbf{r}_i) \ln f_i(\mathbf{a}_i,\mathbf{r}_i)\,d\mathbf{a}_i\,d\mathbf{r}_i \quad (75)$$

where ρ is the mean density.

In the McMillan model the pair interaction potential is specified as

$$V(\mathbf{a}_i,\mathbf{a}_j,\mathbf{r}_{ij}) = -J_2(r_{12})\left(\delta + P_2((\mathbf{a}_1\cdot\mathbf{a}_2))\right) \quad (76)$$

where δ is a dimensionless constant.

McMillan used a particular form for the coupling constant $J_2(r_{12})$:

$$J_2(r_{12}) = \frac{V}{Nr_0^3\pi^{3/2}} \exp\left[-(r_{12}/r_0)^2\right] \quad (77)$$

where r_0 is some length of the order of the length of the rigid molecular core.

We note that the model potential (Eq. 74) is strongly simplified because in Eq. (74) the positional and orientational degrees of molecular freedom are decoupled. It has been pointed out by many authors [74–76] that in the general case the interaction potential must depend on the coupling between the intermolecular vector $\mathbf{r}_{12}$ and the molecular primary axes $\mathbf{a}_1$ and $\mathbf{a}_2$. We discuss the role of these terms in the next subsection.

The coupling constant (Eq. 75) can be expanded in Fourier series retaining only the leading term:

$$J_2(r_{12}) \approx -V[1 + \alpha \cos(kz)] \quad (78)$$

where

$$\alpha = 2\exp\left[-(kr_0/2)\right] \quad (79)$$

In the McMillan model the parameter α characterizes the strength of the interaction that induces the smectic ordering. The parameter α decreases with the increasing smectic period $d = 2\pi/k$ which is of the order of molecular length. Thus α is supposed to increase with increasing chain length.

The mean-field equilibrium distribution function is given by the general Eq. (35). Substituting Eq. (76) into Eq. (75) and then into Eq. (35) we obtain the distribution function of the smectic A phase

$$f_1(\cos\omega, z)$$
$$= Z^{-1} \exp\left[\frac{V_0}{kt}(\delta\alpha\tau\cos(kz) + \alpha\sigma\cos(kz)\right.$$
$$\left. \cdot P_2(\cos\omega) + SP_2(\cos\omega)\right] \quad (80)$$

Multiplying both sides of Eq. (77) with the functions $P_2(\cos\omega)$, $\cos(kz)$ and $\cos(kz)P_2(\cos\omega)$ and integrating over ω and z, one obtains the equations for the three order parameters

$$S = (d/Z) \int P_2(\cos\omega)$$
$$\cdot \exp\left[(V_0/kT)SP_2(\cos\omega)\right]$$
$$\cdot I_0(\kappa) \, d\cos\omega$$

$$\sigma = (d/Z) \int P_2(\cos\omega)$$
$$\cdot \exp\left[(V_0/kT)SP_2(\cos\omega)\right]$$
$$\cdot I_1(\kappa) \, d\cos\omega$$

$$\tau = (d/Z) \int \exp\left[(V_0/kT)SP_2(\cos\omega)\right]$$
$$\cdot I_0(\kappa) \, d\cos\omega \quad (81)$$

where $\kappa = (V_0/kT)[\alpha\sigma P_2(\cos\omega) + \delta\alpha\tau]$ and the function $I_n(\kappa)$ is the n-th order modified Bessel function that appears after the integration over z.

Equation (78) together with the expressions (Eqs. 73 and 74) for the free energy can be used to calculate numerically the parameters of the nematic–smectic A phase transition. The corresponding phase diagram, taken from the original paper by McMillan [59], that includes the isotropic, nematic and smectic A phases is shown in Fig. 5. The inset in Fig. 5 also presents a typical phase diagram of a homologous series of compounds showing the transition temperature versus alkyl chain length.

In general the McMillan theory provides a good qualitative and sometimes even quantitative description of the nematic–smectic A phase transition. The theory accounts successfully for the decrease in the transition entropy with the breadth of the nematic phase and even enables one to locate the tricritical point in a reasonable way. A more detailed discussion of the McMillan theory can be found, for example, in the book of Chandrasekhar [61].

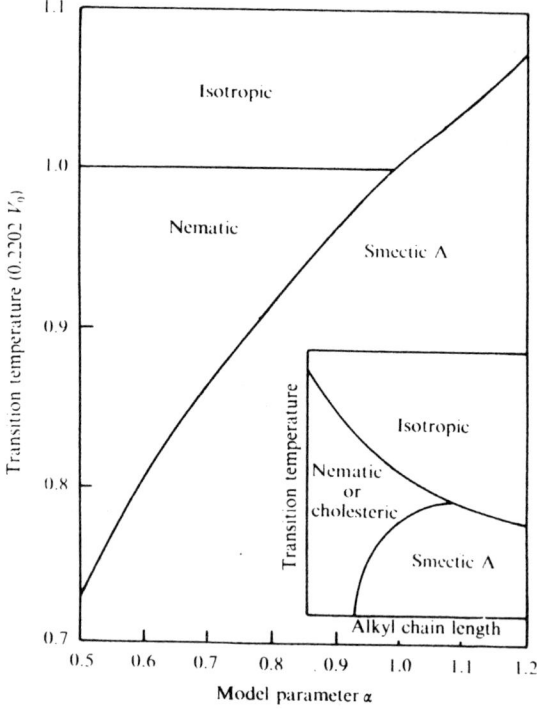

Figure 5. Phase diagram of a liquid crystal system according to the McMillan theory. Inset: typical phase diagram for a homologous series of compounds (after McMillan [59]).

The McMillan theory has been further refined by several authors [62–64] to improve the quantitative agreement with experiment. However, the basic structure of the theory remains the same. This theory presents another example of a successful application of a simple mean-field approach. On the other hand, there are several limitations of the McMillan theory that cannot be ignored. Firstly, the theory is based on the semi phenomenological potential that does not allow determination of the smectic period in a self-consistent way. Secondly, the model potential (Eq. 74) is of the attractive type and therefore the McMillan theory does not consider the relative role of intermolecular attraction and hard-core repulsion in the stabilization of the smectic A phase. This problem appears to be of particular importance after it was shown by computer simulations that the smectic A phase can be formed in a system of hard spherocylinders without any attraction interaction. Finally, the potential (Eq. 74) does not depend on a coupling between orientational and translational degrees of freedom. This means that McMillan theory does not account for an interaction that forces the director to be normal to the smectic plane. The corresponding free energy is then unstable with respect to smectic C fluctuations.

Several authors have used different approaches to overcome these limitations in order to develop a more sophisticated statistical theory of the smectic A phase. We will discuss briefly some of the recent theories in the following two subsections.

2.5.2 Phase Diagram of a Hard-rod Fluid

The computer simulation studies of Frenkel et al. [45] indicate that the excluded volume effects for molecular hard cores must play an important role in the stabilization of the smectic phase. In particular, it has been shown that hard spherocylinders, interacting only via hard-core repulsion, can form nematic, smectic A, and columnar phases.

In the system of very long spherocylinders the nematic–isotropic transition can be quantitatively described by the Onsager theory discussed in Sec. 2.3.2. At the same time, this approximation is expected to provide only a qualitative description of the nematic–smectic A transition in the same system. The reason is that long spherocylinders undergo a transition into the nematic phase at very small packing fractions $\eta \sim D/L \ll 1$. By contrast, the nematic–smectic A transition is expected to occur at large $\eta \sim 1$. The corresponding critical packing fractions can be estimated in the following way [65]. The Onsager theory is based on the virial expansion in powers of the number density $\rho = \eta/v_0$ where v_0 is the molecular volume. The transition to the nematic phase is determined by a balance between the orientational entropy $\rho \int f_1(1) \ln f_1(1) \, d(1)$ which is a maximum in the disordered state and the packing entropy $\sim \rho^2 \langle V_{\text{excl}}(1,2) \rangle$ which is a maximum in the orientationally ordered state. Thus the critical packing fraction is estimated as $\eta_{\text{N-I}} \sim v_0/\langle V_{\text{excl}} \rangle$. For long rods one finds $\langle V_{\text{excl}} \rangle \sim L^2 D$ and $v_0 \sim D^2 L$. Thus $\eta_{\text{N-I}} \sim D/L \ll 1$.

We note that in the system of long rods the nematic phase is strongly ordered. Then the transition into the smectic A phase is expected to take place in the nearly perfectly aligned system of rods. For parallel rods, however, the excluded volume $\langle V_{\text{excl}} \rangle \sim D^2 L$ and thus $\eta_{\text{N-A}} \sim 1$. This means that one cannot rely on the virial expansion even in the case of very long rods.

One possibility for improving the theory is to take into account higher order terms in the virial expansion. This has been done by Mulder for an aligned hard-rod fluid [66]. Mulder has taken into account the third- and fourth-order terms and has been able to obtain the numerical values of the transition density and the smectic period in very good agreement with the results of computer simulations [67]. The critical packing fraction and the dimensionless smectic wavelength observed are $\eta_{\text{N-A}} = 0.36$ and $\lambda \approx 1.27$ while the theoretical results are $\eta_{\text{N-A}} \approx 0.37$ and $\lambda \approx 1.34$ [66]. Recently Poniwierski performed an asymptotic analysis of the nematic–smectic A transition in the system of rods with orientational freedom and in the limit $L/D \to \infty$ [65]. He has shown that orientational fluctuations do not destroy the smectic A phase but the transition is shifted towards higher densities.

A different and more sophisticated approach to the theory of the nematic–smectic A transition is based on the nonlocal density functional theory developed for inhomogeneous hard-core fluids [68]. The nonlocal free energy functional is defined in the following way [69–71]:

$$F[\rho(r)] = F_{id}[\rho(r)] + H[\rho(r)] \quad (82)$$

where $F_{id}[\rho]$ is the ideal gas contribution (see Sec. 2.4.4).

The excess of free energy $H[\rho]$ is assumed to have a form resembling the local density approximation:

$$H[\rho(r)] = \int \rho(r)\, \Delta\psi(\bar{\rho}(r))\, dr \quad (83)$$

where $\Delta\psi$ is the excess of free energy per particle and $\bar{\rho}(r)$ is some auxiliary density that depends on $\rho(r)$. For a homogeneous isotropic fluid $\bar{\rho} = \rho$. In the inhomogeneous state $\bar{\rho}(r)$ is related to $\rho(r)$ in a nonlocal way:

$$\bar{\rho}(r) = \int w_{\text{eff}}(r - r')\, \rho(r')\, dr' \quad (84)$$

where $\omega_{\text{eff}}(r - r')$ is some weighting function.

The form of the function $\omega_{\text{eff}}(r)$ is different in different versions of the smoothed-density approximation proposed by Somoza and Tarazona [71, 72] and by Poniwierski and Sluckin [69, 73]. The density functional model of Somoza and Tarazona is based on the reference system of parallel hard ellipsoids that can be mapped into hard spheres. In the Poniwierski and Sluckin theory the effective weight function is determined by the Maier function for hard spherocylinders and the expression for $\Delta\psi(\rho)$ is obtained from the Carnahan–Starling ex-

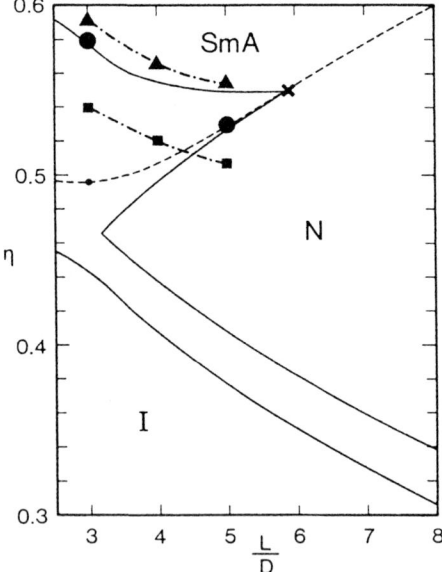

Figure 6. Phase diagram of a fluid of hard spherocylinders in the (axial ratio/order parameter) plane. The circles are the simulation results for the smectic A transition [45]. The N–SmA transition obtained in [45] is denoted by squares N and triangles SmA (after Poniwierski and Sluckin [69]).

cess of free energy for hard spheres with the same packing fraction.

Both groups have obtained phase diagrams for a fluid of hard spherocylinders that are in qualitative agreement with the results of computer simulations [45]. Figure 6 shows the phase diagram in the (L/D), η plane, obtained by Poniwierski and Sluckin [69], because it gives a more reasonable value for the tricritical point. One can see from Fig. 6 that the nematic–smectic A transition is first order for $L/D < 5.9$ and it is second order for $L/D > 5.9$. The location of the I–N–SmA triple point in Fig. 6 is in good agreement with simulations. The same good agreement has also been obtained by Somoza and Tarazona [71].

2.5.3 The Role of Intermolecular Attraction

The results of several molecular theories that describe the smectic ordering in a system of hard spherocylinders enable us to conclude that the contribution from hard-core repulsion can be described by the smoothed-density approximation. On the other hand, a realistic theory of thermotropic smectics can only be developed if the intermolecular attraction is taken into account. The interplay between hard-core repulsion and attraction in smectic A liquid crystals has been considered by Kloczkowski and Stecki [17] using a very simple model of hard spherocylinders with an additional attractive r^{-6} potential. Using the Onsager approximation, the authors have obtained equations for the order parameters that are very similar to the ones found in the McMillan theory but with explicit expressions for the model parameters. The more general analysis has been performed by Mederos and Sullivan [76] who have treated the anisotropic attraction interaction by the mean-field approximation while the hard-core repulsion has been taken into account using the nonlocal density functional approach proposed by Somoza and Tarazona.

In [76] the intermolecular attraction potential has been taken in the form

$$V_{att}(1,2) = V_1(r_{12}) + V_2(r_{12}) P_2((a_1 \cdot a_2)) + V_3(r_{12}) \left[P_2((a_1 \cdot u_{12})) + P_2((a_2 \cdot u_{12})) \right] \quad (85)$$

where $u_{12} = r_{12}/r_{12}$.

We note that the last term in Eq. (82) has been omitted from the McMillan–Kobayashi theory. This term explicitly describes a coupling between the molecular long axis and the intermolecular vector. The effect of such coupling seems to be very important in smectic liquid crystals because this energy is obviously minimized when the molecules are packed in layers with their long axes parallel to the layer normal. (That is if $r_{12} \perp a_1 \| a_2$). The magnitude of the coupling constant V_3 in the last term is comparable to that of V_2 for rod-like molecules with an elongation typical of that for mesogens [77] and can even be predominant for weakly anisotropic molecules [78]. It should be noted also that the last term in Eq. (82) does not contribute to the free energy of the nematic phase in the mean-field approximation as it vanishes after averaging over all orientations of the intermolecular vector.

Using the specific expressions for the coupling constants in Eq. (82), Mederos and Sullivan obtained the temperature–density phase diagrams shown in Figs. 7a and b. These two diagrams have been obtained for the same value of the geometrical anisotropy $\sigma_\| / \sigma_\perp$ and for different values of the reduced strength of the symmetry breaking potential given by the last term in Eq. (82). We see that the smectic phase is stabilized with the increasing strength of the symmetry breaking potential. By contrast, the nematic phase tends to disappear. Thus the coupling between orientational and translational degrees of freedom is important indeed and it should also be taken into account in the description of the hard-core repulsion in smectic phases. It is not excluded, however, that the role of such interaction is overestimated in the Mederos–Sullivan theory because in this treatment the hard-core repulsion alone does not lead to the smectic A phase.

2.5.4 Smectic A–Smectic C Transition

The transition from the smectic A phase into the smectic C one is accompanied by the tilt of the molecular long axes with respect to

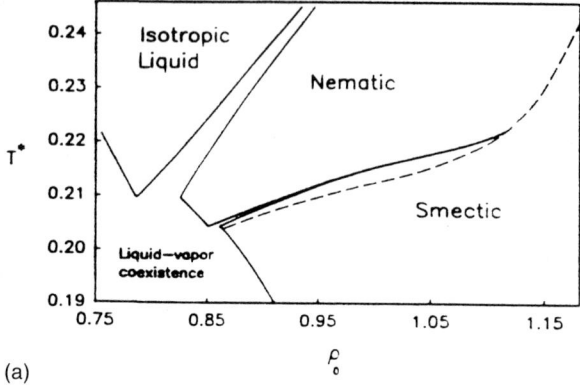

(a)

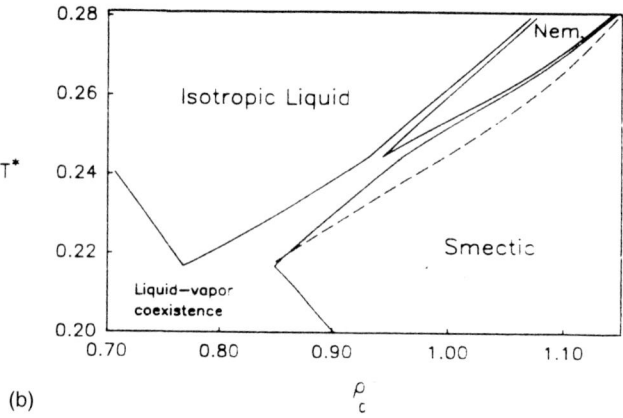

(b)

Figure 7. Temperature-density phase diagrams for $\sigma_\parallel/\sigma_\perp = 1.8$ and at $\varepsilon_3/\varepsilon_1 = 0.28$ (a) and $\varepsilon_3/\varepsilon_1 = 0.34$ (b) for a liquid crystal with both steric repulsion and attraction interactions between molecules. The parameter ε_3 measures the strength of the attraction potential that depends on the coupling between orientation and translation (after Mederos and Sullivan [76]).

the smectic plane normal e. We note that the resulting structure is unfavorable from the packing point of view [79]. Tilted molecules occupy more area in the smectic plane and therefore in the smectic C phase there is more excluded area in the layer than in the smectic A phase. As a result the packing entropy is decreased. This means that the smectic C phase is not expected to be formed in the system of hard rods, and, indeed, it has not been found in computer simulations of Veerman and Frenkel [45].

Thus, the smectic C phase can be formed only if there exists some specific intermolecular interaction that favors the tilt. Different interactions of this kind have been considered in the literature [79, 80–84] and some early theories have been analyzed in detail by Van der Meer [14].

Different molecular models for the SmC phase can be separated into two main classes that actually correspond to different molecular mechanisms of the smectic A–smectic C transition. Some models (for example, those of McMillan and Meyer [81] and Wulf [80]) imply that the molecular rotation about the long axis is frozen out in the smectic C phase. It seems to be even more important that in these models the smectic A–smectic C transition is governed by the ordering of molecular short axes while the tilt of the long axes occurs as a consequence. By contrast, in other models [82] the transition is directly related to the tilt of the long axes and the biaxiality of the smectic C phase is neglected.

We note that the more recent molecular models for the smectic C phase [79, 83, 84] also fall into one of these classes. For example, the model of Somoza and Tarazona [83] is based on steric interactions between molecules with biaxial shape. This interaction is assumed to be the driving force of the transition into the biaxial smectic C phase. At the same time, in the theory of Van der Meer and Vertogen [79] the molecular tilt is caused by the induction interaction between the off-center transverse dipoles and the polarizable core of neighboring molecules. This induction interaction is quadratic in dipole and therefore the free rotation around the molecular long axis does not destroy the smectic C phase. The recent theory of Poniwierski and Sluckin is based on the uniaxial molecular model in which hard cylinders carry axial quadrupoles.

It should be noted that the assumption of a strongly asymmetric orientational distribution of molecular short axes in the smectic C phase seems to be in contradiction with experiments [85]. Some other models also do not have any experimental support so far. Goodby et al. [87] and de Jeu [88] have studied the influence of electric dipole and molecular shape on the stability of the smectic C phase. The results do not support the models of Wulf [80] and Cabib and Benguigui [82] but reveal the importance of transverse dipoles. Thus there is some experimental evidence in favor of the model proposed by Van der Meer and Vertogen.

In this model the molecular tilt is determined by induction interaction between the off-center dipole and the polarizable core of the neighboring molecules. After averaging over the rotation around the molecular long axes the corresponding interaction potential reads [79]:

$$V_{ind}(1,2) = -J_{ind}\left(r_{12}(a_1 \cdot u_{12})^2\right) \quad (86)$$

where the coupling constant $J_{ind} \propto \bar{\alpha}\mu^2$ and where μ is the molecular dipole and $\bar{\alpha}$ is the average molecular polarizability.

The interaction energy (Eq. 83) promotes the tilt of the director in the smectic C phase. This can be seen in the following way. Let us consider the case of perfect nematic ordering. Then the potential (Eq. 83) is reduced to $-J_{ind}(n \cdot u_{12})^2$. Now let us average this potential over all orientations of the intermolecular unit vector u_{12} within the smectic plane. For any two molecules within one plane the vector $u_{12} \perp e$ and one obtains

$$\langle V_{ind}(1,2)\rangle = -J_{ind}\cos^2\Theta \quad (87)$$

where Θ is the tilt angle, $\cos\Theta = (n \cdot e)$.

In the model of Van der Meer and Vertogen the induction interaction energy (Eq. 83) is counterbalanced by the hard-core repulsion coupled with isotropic attraction between molecular hard cores. The resulting interaction potential is presented in the form of an expansion:

$$\begin{aligned}V_{eff}(1,2) = &(v_1 - J_{ind})\\ &\cdot [(a_1 \cdot u_{12})^2 + (a_2 \cdot u_{12})^2\\ &+ v_2(a_1 \cdot a_2)(a_1 \cdot u_{12})(a_2 \cdot u_{12})\\ &+ v_3(a_1 \cdot u_{12})^2(a_2 \cdot u_{12})^2]\end{aligned} \quad (88)$$

where the coefficients v_1, v_2 and v_3 are expressed in terms of the shape anisotropy and the attraction interaction strength.

In Eq. (85) the constant v_1 is positive and thus, without the induction interaction, the potential (Eq. 85) stabilizes the smectic A phase. Taking into account the packing entropy, Van der Meer and Vertogen have obtained the following simple free energy in the case of perfect nematic ordering:

$$\Delta F = \frac{1}{2}D_0\langle\cos\phi\rangle^2 - kT\ln\frac{1}{2\pi}$$
$$\cdot \int d\phi\, \exp[\beta D_0\langle\cos\phi\rangle\cos\phi] \quad (89)$$

where

$$D_0 = 2\left[(1+T/T_p)B_2 - C_2\right] \cdot P_2(\cos\Theta) - (1-T/T_p) \cdot B_4 P_4(\cos\Theta) \quad (90)$$

Here $\langle \cos\phi \rangle$ is the smectic order parameter and the coefficients B_2, C_2 and B_4 are presented in [79]. The induction interaction strength is adsorbed in C_2.

The simple free energy (Eq. 86) can be used to describe the transitions between nematic, smectic A and smectic C phases. The second order smectic A–smectic C transition temperature is given by

$$T_{AC} = T_p\left(\frac{3C_2}{3B_2 - 5B_4} - 1\right) \quad (91)$$

In this model the temperature variation of the tilt angle is the reduced temperature scale and does not depend on any molecular parameters. The corresponding temperature variation for the smectic A–smectic C and nematic–smectic C transitions is shown

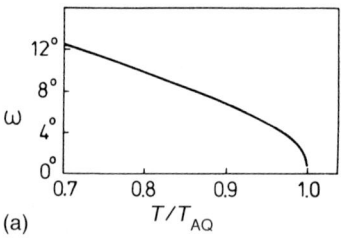

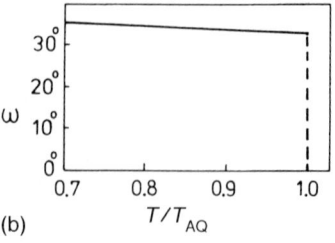

Figure 8. Temperature dependence of the tilt angle in the SmC phase near the second-order A–C transition (a) and near the first-order N–C transition (b) (after Van der Meer and Vertogen [79]).

in Fig. 8 a and b, respectively, taken from [79].

The molecular theory of Van der Meer and Vertogen is based on a specific molecular model that is not in contradiction with experiment. At the same time Barbero and Durand [89] have shown that the molecular tilt is an intrinsic property of any layered quadrupolar structure. This idea has been used by Poniwierski and Sluckin in their model [84] that presents a rather general mechanism for the stabilization of the smectic C phase. It is interesting to note that the mathematical form of the interaction potential in the Poniwierski–Sluckin theory is similar to the potential (Eq. 85). The energy of electrostatic interaction between two axial quadrupoles, employed in [84], can be written as

$$U_{qq}(1,2) = \frac{3}{4}\frac{q^2}{r^5}$$
$$\cdot [1 + 2(\boldsymbol{a}_1\cdot\boldsymbol{a}_2)^2 - 5(\boldsymbol{a}_1\cdot\boldsymbol{u}_{12})^2 - 5(\boldsymbol{a}_2\cdot\boldsymbol{u}_{12})^2$$
$$- 20(\boldsymbol{a}_1\cdot\boldsymbol{a}_2)(\boldsymbol{a}_1\cdot\boldsymbol{u}_{12})(\boldsymbol{a}_2\cdot\boldsymbol{u}_{12})$$
$$+ 35(\boldsymbol{a}_1\cdot\boldsymbol{u}_{12})^2(\boldsymbol{a}_2\cdot\boldsymbol{u}_{12})^2] \quad (92)$$

The last three terms of the potential (Eq. 90) have the same mathematical form as the corresponding terms in Eq. (85). This means that the quadrupolar-type potential appears to be a good model potential for the theory of smectic A–smectic C transition.

The interaction potentials (Eqs. 85 and 90) essentially depend on a coupling between the molecular orientation and the intermolecular vector. We note that this coupling could be neglected in the first approximation in the theory of the nematic–smectic A transition, as it is done, for example, in the McMillan theory. At the same time this coupling just determines the effect in the theory of transition into the smectic C phase.

Finally we note that both theories, discussed above, neglect the biaxiality of the

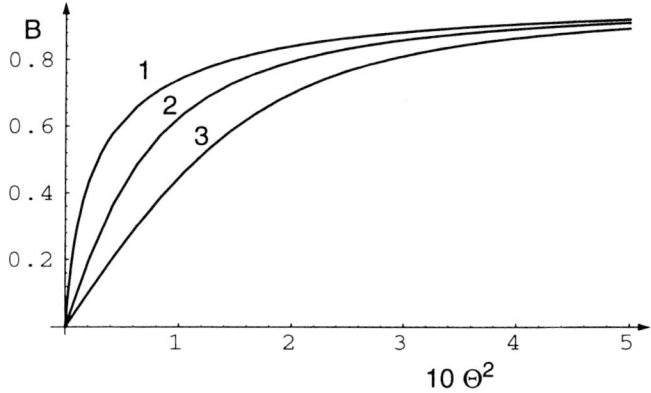

Figure 9. Biaxiality order parameter B as a function of the tilt angle in the SmC phase. The curves (3)–(1) correspond to the increasing strength of the biaxial part of the interaction potential [8].

smectic C phase. In real systems some biaxiality is always induced by the tilt and it is only a question of how large the corresponding contribution is to the free energy. In Fig. 9 we present the value of the biaxiality order parameter B (that determines the nonpolar ordering of molecular short axes in the smectic C phase) as a function of the tilt angle. This dependence has been calculated theoretically in [8] using a simple model. The value of the biaxiality order parameter seems to be overestimated. However, even if the actual value is several times smaller, the problem of interaction between biaxiality and the tilt in the smectic C phase deserves further attention.

2.6 Conclusions

The literature on the molecular theory of liquid crystals is enormous and in this chapter we have been able to cover only a small part of it. We have mainly been interested in the models for the nematic–isotropic, nematic–smectic A and smectic A–smectic C phase transitions. The existing theory includes also extensive calculations of the various parameters of the liquid crystal phases: Frank elastic constants, dielectric susceptibility, viscosity, flexoelectric coefficients and so on. This part of the molecular theory of liquid crystals remains completely beyond the scope of this review. We also did not consider the theory of more complex liquid crystalline phases including the cholesteric phase, the ferroelectric smectic C* phase, re-entrant phases, bilayer and incommensurate smectic phases and phases with hexatic ordering. The majority of these theories, however, employ the same general ideas and approximations that have been discussed above. Thus the present review presents some basic information necessary for the understanding of the existing molecular theory of liquid crystals.

Acknowledgements

The author would like to acknowledge interesting and stimulating discussions on the molecular theory of liquid crystals with T. J. Sluckin, D. A. Dunmur, G. R. Luckhurst, S. Hess, E. M. Terentjev, B. Mulder, A. Somoza and P. I. C. Teixeira.

2.7 References

[1] B. Berardi, S. Orlandi, C. Zannoni, *Chem. Phys. Lett.* **1996**, *261*, 363.
[2] A. Ferrarini, G. J. Moro, P. L. Nordio, G. R. Luckhurst, *Mol. Phys.* **1992**, *77*, 1.
[3] P. G. de Gennes, *The Physics of Liquid Crystals*, Oxford University Press, Oxford, **1974**.
[4] E. M. Averjanov, M. A. Osipov, *Sov. Phys. Uspihi* **1990**, *161*, 93.

[5] S. C. McGrother, D. C. Williamson, G. Jackson, *J. Chem. Phys.* **1996**, *104*, 6755.
[6] D. A. Dunmur, P. Palffy-Muhoray, *Mol. Phys.* **1992**, *76*, 1015; A. G. Vanakaras, D. J. Photinos, *Mol. Phys.* **1995**, *85*, 1089.
[7] A. Perera, G. N. Patey, *J. Chem. Phys.* **1989**, *91*, 3045.
[8] M. A. Osipov, S. A. Pikin, *J. Phys. (France)* **1995**, *II/5*, 1223.
[9] S. A. Pikin, M. A. Osipov in *Ferroelectric Liquid Crystals* (Ed.: G. W. Taylor), Gordon and Breach, New York, **1992**.
[10] L. Onsager, *Ann. N. Y. Acad. Sci.* **1949**, *51*, 627.
[11] W. Maier, A. Saupe, *Z. Naturforsch.* **1959**, *14a*, 882.
[12] W. Maier, A. Saupe, *Z. Naturforsch.* **1960**, *15a*, 287.
[13] B. W. Van der Meer, G. Vertogen in *Molecular Physics of Liquid Crystals* (Eds.: G. R. Luckhurst, G. W. Gray), Academic Press, New York, **1979**.
[14] B. W. Van der Meer, Thesis, Groningen, **1979**.
[15] L. Pohl, R. Eidenschink, J. Krause, G. Weber, *Phys. Lett. A* **1978**, *65*, 169.
[16] W. M. Gelbart, A. Gelbart, *Mol. Phys.* **1977**, *33*, 1387.
[17] A. Kloczkowski, J. Stecki, *Mol. Phys.* **1985**, *55*, 689.
[18] M. A. Osipov, H. Stegemeyer, A. Sprick, *Phys. Rev. E* **1996**, *54*, 6387.
[19] R. L. Humphries, P. G. James, G. R. Luckhurst, *J. Chem. Soc., Faraday Trans. II* **1972**, *68*, 1031.
[20] R. G. Priest, *J. Phys. (France)* **1975**, *36*, 437; *J. Chem. Phys.* **1976**, *65*, 408.
[21] B. W. Van der Meer, G. Vertogen, *Phys. Lett. A* **1979**, *71*, 488.
[22] J. G. Gray, B. J. Berne, *J. Chem. Phys.* **1981**, *74*, 3116.
[23] G. R. Luckhurst in *The Molecular Physics of Liquid Crystals* (Eds.: G. R. Luckhurst, G. W. Gray), Academic Press, London, **1979**.
[24] H. Toriumi, E. T. Samulski, *Mol. Cryst. Liq. Cryst.* **1983**, *101*, 163.
[25] W. H. de Jeu, J. van der Veen, *Mol. Cryst. Liq. Cryst.* **1977**, *40*, 1.
[26] J. G. T. Ypma, G. Vertogen, H. T. Koster, *Mol. Cryst. Liq. Cryst.* **1976**, *37*, 57.
[27] N. V. Madhusudana, S. Chandrasekhar, *Solid State Commun.* **1973**, *13*, 377.
[28] S. Chandrasekhar, N. V. Madhusudana, *Acta Crystallogr.* **1971**, *27*, 303.
[29] J. P. Straley, *Mol. Cryst. Liq. Cryst.* **1973**, *24*, 7.
[30] G. Lasher, *J. Chem. Phys.* **1970**, *53*, 4141.
[31] H. N. W. Lekkerkerker, Ph. Coulon, R. van der Haegen, R. Deblieck, *J. Chem. Phys.* **1984**, *80*, 3247.
[32] Sin-Doo Lee, R. B. Meyer, *J. Chem. Phys.* **1986**, *84*, 3443.
[33] G. J. Vroege, H. N. W. Lekkerkerker, *Rep. Progr. Phys.* **1992**, *55*, 1241.
[34] A. N. Semenov, A. R. Khokhlov, *Sov. Phys. Usp.* **1988**, *31*, 988.
[35] T. Odjik, *Macromolecules* **1986**, *19*, 2313.
[36] T. Sato, A. Teramoto in *Advances in Polymer Science*, Vol. 126, Springer-Verlag, Berlin, Heidelberg, **1996**.
[37] M. A. Cotter in *Molecular Physics of Liquid Crystals* (Eds.: G. R. Luckhurst, G. W. Gray), Academic Press, New York, **1979**.
[38] Sin-Doo Lee, *J. Chem. Phys.* **1987**, *87*, 4972.
[39] A. R. Khokhlov, A. N. Semenov, *J. Stat. Phys.* **1985**, *38*, 161.
[40] J. G. J. Ypma, G. Vertogen, *Phys. Rev. A* **1978**, *17*, 1490.
[41] J. D. Parsons, *Phys. Rev. A* **1979**, *19*, 1225.
[42] W. M. Gelbart, B. Barboy, *Acc. Chem. Res.* **1980**, *13*, 290.
[43] W. M. Gelbart, *J. Phys. Chem.* **1982**, *86*, 4298.
[44] B. Barboy, W. M. Gelbart, *J. Stat. Phys.* **1980**, *22*, 709.
[45] J. A. C. Veerman, D. Frenkel, *Phys. Rev. A* **1990**, *41*, 3237; D. Frenkel, *J. Phys. Chem.* **1988**, *92*, 3280.
[46] G. R. Luckhurst in *Recent Advances in Liquid Crystalline Polymers* (Ed.: L. L. Chapoy), Elsevier, London, **1986**, p. 105.
[47] G. R. Luckhurst, C. Zannoni, P. L. Nordio, U. Segre, *Mol. Phys.* **1975**, *330*, 1345.
[48] C. S. Shih, R. Alben, *J. Chem. Phys.* **1972**, *57*, 3055.
[49] J. P. Straley, *Phys. Rev. A* **1974**, *10*, 1881.
[50] B. Mulder, *Phys. Rev. A* **1989**, *39*, 360.
[51] B. M. Mulder, D. Frenkel, *Mol. Phys.* **1985**, *55*, 1171.
[52] T. J. Sluckin, P. Shukla, *J. Phys. A: Math. Gen.* **1983**, *16*, 1539.
[53] Y. Singh, *Phys. Rev. A* **1984**, *30*, 583.
[54] T. J. Sluckin, *Mol. Phys.* **1983**, *49*, 221.
[55] M. P. Allen et al., *Phys. Rev. E* **1995**, *52*, R95.
[56] A. Poniwierski, J. Stecki, *Mol. Phys.* **1979**, *38*, 1931.
[57] M. A. Osipov, Chap. 1 in *Liquid Crystalline and Mesomorphic Polymers* (Eds.: V. P. Shibaev, L. Lam), Springer-Verlag, New York, **1994**.
[58] M. A. Osipov, S. A. Pikin, E. M. Terentjev, *Sov. Scientific Rev.*, harwood acad. publ. **1989**, *11*, 191.
[59] W. L. McMillan, *Phys. Rev. A* **1971**, *4*, 1238; **1972**, *6*, 936.
[60] C. Kobayashi, *Mol. Cryst. Liq. Cryst.* **1971**, *13*, 137.
[61] S. Chandrasekhar, *Liquid Crystals*, 2nd edn., Cambridge University Press, **1992**.
[62] L. Shen, H. K. Sim, Y. M. Shin, C. W. Woo, *Mol. Cryst. Liq. Cryst.* **1977**, *39*, 299.
[63] H. T. Tan, *Phys. Rev. A* **1977**, *16*, 1715.
[64] P. J. Photinos, A. Saupe, *Phys. Rev. A* **1975**, *13*, 1926.

[65] A. Poniwierski, *Phys. Rev. A* **1992**, *45*, 5605.
[66] B. Mulder, *Phys. Rev. A* **1987**, *35*, 3095.
[67] A. Stroobants, H. N. W. Lekkerkerker, D. Frenkel, *Phys. Rev. Lett.* **1986**, *57*, 1452; *Phys. Rev. A* **1987**, *36*, 2929.
[68] P. Tarazona, *Phys. Rev. A* **1985**, *41*, 2672; W. A. Curtin, N. W. Ashcroft, *Phys. Rev. A* **1985**, *32*, 2909; T. F. Meister, D. M. Kroll, *Phys. Rev. A* **1985**, *31*, 4055.
[69] A. Poniwierski, T. J. Sluckin, *Phys. Rev. A* **1991**, *43*, 6837.
[70] A. Poniwierski, T. J. Sluckin, *Mol. Phys.* **1991**, *73*, 199.
[71] A. M. Somoza, P. Tarazona, *Phys. Rev. A* **1990**, *41*, 965.
[72] A. M. Somoza, P. Tarazona, *Phys. Rev. Lett.* **1988**, *61*, 2566; *J. Chem. Phys.* **1989**, *91*, 517.
[73] A. Poniwierski, R. Holyst, *Phys. Rev. Lett.* **1988**, *61*, 2461.
[74] L. Senbetu, C.-W. Woo, *Phys. Rev. A* **1978**, *17*, 1529.
[75] M. D. Lipkin, D. W. Oxtoby, *J. Chem. Phys.* **1983**, *79*, 1939.
[76] L. Mederos, D. E. Sullivan, *Phys. Rev. A* **1989**, *39*, 854.
[77] B. Tjipto-Margo, D. E. Sullivan, *J. Chem. Phys.* **1988**, *88*, 6620.
[78] C. G. Gray, K. E. Gubbins, *Theory of Molecular Fluids*, Clarendon Press, Oxford, **1984**.
[79] B. W. Van der Meer, G. Vertogen, *J. Phys. Colloq. (France)* **1979**, *40*, C3–222.
[80] A. Wulf, *Phys. Rev. A* **1975**, *11*, 365.
[81] A. Meyer, W. L. McMillan, *Phys. Rev. A* **1974**, *9*, 899.
[82] D. Cabib, L. Benguigui, *J. Phys. (France)* **1977**, *38*, 419.
[83] A. M. Somoza, P. Tarazona, *Phys. Rev. Lett.* **1988**, *61*, 2566.
[84] A. Poniwierski, T. J. Sluckin, *Mol. Phys.* **1991**, *73*, 199.
[85] H. Hervet, F. Volino et al., *J. Phys. (Lett.) France* **1974**, *35*, L-151; *Phys. Rev. (Lett.)* **1975**, *34*, 451; P.-J. Bos, J. Pirs et al., *Mol. Cryst. Liq. Cryst.* **1977**, *40*, 59.
[86] L. Blum, A. J. Torruella, *J. Chem. Phys.* **1972**, *56*, 303.
[87] J. W. Goodby, G. W. Gray, D. G. McDonnell, *Mol. Cryst. Liq. Cryst. (Lett.)* **1977**, *34*, 183.
[88] W. H. de Jeu, *J. Phys. (France)* **1977**, *38*, 1263.
[89] G. Barbero, G. Durand, *Mol. Cryst. Liq. Cryst.* **1990**, *179*, 57.

3 Molecular Modelling

Mark R. Wilson

Over the past 10 years, rapid progress has been made in the field of molecular modelling. Advances have been led by two important factors: the increase in speed and reduction in cost of modern computers, and an accompanying improvement in the accuracy and ease of use of molecular modelling software. Modelling packages are now commonly available in many laboratories. Their ability accurately to predict molecular structures of simple organic molecules makes them a useful tool in the study of liquid crystals. In this article developments in molecular modelling are discussed in the context of liquid crystal systems. The article is divided into two main sections: Sec. 3.1 covers the main molecular modelling techniques that are currently available; whilst Sec. 3.2 covers specific applications of these techniques in the study of liquid crystal molecules.

3.1 Techniques of Molecular Modelling

3.1.1 Molecular Mechanics

Molecular mechanics is the simplest and most commonly used molecular modelling technique. It is concerned with the determination of molecular structure, and is of particular relevance to liquid crystal chemists concerned with the design of appropriate molecular structures, or to physicists interested in calculating molecular properties. The molecular mechanics approach has been reviewed in a number of places [1–4], and so here only the basics of the technique are described. The standard approximation employed in molecular mechanics is to consider a molecule as a collection of atoms held together by elastic restoring forces. These forces are described by simple functions that characterise the distortion of each structural feature within a molecule. Usually separate functions exist for each bond stretch, bond bend, and dihedral angle; as well as for each nonbonded interaction. Together these functions make up the molecular mechanics *force field* for a particular molecule. The steric energy E can then be defined with reference to the force field. E has no physical meaning in itself, but can be thought of as measuring how the energy of a particular molecular conformation varies from a hypothetical ideal geometry where all bonds, bond angles, etc. have their ideal (or natural) values.

$$E = \sum_{\substack{\text{bond} \\ \text{lengths}}} E_{\text{bond}} + \sum_{\substack{\text{bond} \\ \text{angles}}} E_{\text{angle}} \quad (1)$$

$$+ \sum_{\substack{\text{dihedral} \\ \text{angles}}} E_{\text{torsion}} + \sum_{i=1}^{N} \sum_{j<i}^{N} (E_{\text{nb}ij} + E_{\text{el}ij})$$

The symbols in Eq. (1) have the following meanings: E_{bond} is the energy of a bond that is stretched or compressed from its natural value, E_{angle} is the energy of a bond angle that is distorted from its natural value, E_{torsion} is the energy of a dihedral angle that is distorted from its natural value, $E_{\text{nb}ij}$ and $E_{\text{el}ij}$ are respectively the Lennard–Jones and electrostatic nonbonded interactions between the pair of atoms i and j, and N is the number of atoms in the system.

Bond stretches are usually characterised by a harmonic potential of the form

$$E_{\text{bond}} = \frac{1}{2} K_{\text{b}} (l - l_0)^2 \quad (2)$$

where l is the stretched or compressed bond length, l_0 is the natural bond length for an undistorted bond, and K_{b} is a force constant characterising the bond distortion. If a carbon–carbon bond is stretched from its ideal lowest energy value of $l_0 = 1.523 \text{Å}$, this results in a contribution to the steric energy in Eq. (1). Similarly, E_{angle} and E_{torsion} characterise bond angle and dihedral angle perturbations through a harmonic potential and a truncated Fourier series respectively:

$$E_{\text{angle}} = \frac{1}{2} K_\theta (\theta - \theta_0)^2,$$

$$E_{\text{torsion}} = \sum_m \frac{1}{2} K_{\phi_m} [1 + \cos(m\phi - \delta)] \quad (3)$$

where θ and θ_0 are distorted and natural bond angles, ϕ is a dihedral angle, δ is a phase angle, and K_θ and K_{ϕ_m} are force constants. Finally, 12–6 and Coulomb potentials are often used for nonbonded interactions:

$$E_{\text{nb}ij} = \frac{A_{ij}}{r_{ij}^{12}} - \frac{C_{ij}}{r_{ij}^{6}}, \quad E_{\text{el}ij} = \frac{q_i q_j}{r_{ij}} \quad (4)$$

Here, r_{ij} is the distance between atoms i and j, $A_{ij} = (A_{ii} A_{jj})^{1/2}$, $C_{ij} = (C_{ii} C_{jj})^{1/2}$, and q_i and q_j are the partial electronic charges on atoms i and j. A_{ii} and C_{ii} can be expressed in terms of the Lennard–Jones parameters ε and σ: $A_{ii} = 4 \varepsilon_{ii} \sigma_{ii}^{12}$ and $C_{ii} = 4 \varepsilon_{ii} \sigma_{ii}^{6}$.

Each molecular conformation has a different value of the steric energy. So, although E has no direct physical significance by itself, the differences between steric energies of any two conformations is equivalent to the energy difference between them. The terms in Eq. (1)–(4) are not unique, and the exact form of the potential functions differs from one force field to another. However, all force fields rely on the fact that a specific interaction is similar in every molecule (i.e. a pure C–C bond stretch in ethane is similar to a C–C stretch in decane or in a large liquid crystal molecule). In parameterising the force field, all the force constants are carefully optimised to predict the structures and relative conformational energies of a control set of small molecules.

Currently, a number of excellent force fields exist in the literature. For low molecular weight organic liquid crystals, the MM3 force field [5] (and its predecessors MM2 and MM1 [6, 7]) generally produce excellent structures and good conformational energies. Molecules containing mainly alkyl chains and saturated rings are described well by most force fields. However, calculations involving some functional groups commonly used in liquid crystal molecules (e.g. –N=N–, F, and CN will yield less accurate structures and energies. Parameters associated with these groups are often marked *preliminary* within force fields. This simply reflects the lack of molecules with these functional groups in the force field control set. The range of valence states adopted by metals, the lack of metals in force field control sets, and polarization effects associated with metal ions (which are not handled well by traditional force fields) combine to make structural predictions for metal-containing liquid crystal structures much less reliable than predictions for organic liquid crystals.

Once an appropriate force field has been chosen, the aim of a molecular mechanics study is to optimise molecular geometry by minimising E. In an *energy-minimised* conformation the strain associated with the steric energy will be spread throughout the molecule. In practice, molecular mechanics packages consist of a force field combined with energy minimisation routines, and often a graphical user interface (GUI) to provide an easy mechanism for carrying out molecular calculations. A typical molecular mechanics calculation consists of the following steps:

– build a trial molecular structure by providing coordinates from a crystal structure or generating *drawn* structures via a GUI;
– minimise the energy of the trial structure to provide a *minimum energy conformation*;
– undertake a search for other energy minima by adjusting dihedral angles within a molecule and re-minimising the steric energy for each conformation.

The final result of this process is a series of potential energy minima, one of which will be the global minimum and represent the lowest energy conformation of the molecule. For many liquid crystal molecules, many conformations exist that are similar in energy. This makes the tasks of finding the global energy minimum and characterising molecular structure difficult. New techniques for conformational searching have recently made this process easier [8–10]. However, the problem of energy minimisation on a multidimensional surface is still a difficult one, and for complicated molecules it is not always possible to guarantee that all relevant conformations have been found.

A single molecule with N_{conf} energy minima can be thought of as a N_{conf} state system with the probability P_j of the molecule being in state j given by the Boltzmann distribution

$$P_j = \frac{\exp(-\Delta E_j / k_B T)}{\sum_{i=1}^{N_{conf}} \exp(\Delta E_i / k_B T)} \quad (5)$$

where ΔE_i is the energy of conformation i relative to the lowest energy conformation and k_B is Boltzmann's constant. The use of Eq. (5) provides a simple weighting for each of the conformational states occupied at a particular temperature T.

3.1.2 Molecular Dynamics and Monte Carlo Simulation

One of the ways of circumventing the problem of finding multiple energy minima of complex molecules is to turn to more sophisticated techniques that are capable of sampling phase space efficiently without the need to home in on particular minimum energy conformations. The two most useful techniques are molecular dynamics (MD) and the Monte Carlo (MC) method. Both approaches make use of the same types of potential functions used in molecular mechanics, but are designed to sample conformation space such that a Boltzmann distribution of states is generated. MC and MD techniques for molecular systems have been widely reviewed [11–14], and only the basics of the two methods are described below.

In molecular dynamics Newton's equations of motion are solved for the system of atoms interacting via a potential such as that of Eq. (1). For each atom, the force $\mathbf{F}_i$ is given by

$$\mathbf{F}_i = -\nabla E_i \quad (6)$$

where E_i is the interaction energy of atom i in the force field. Typically, each atom is given a velocity sampled from a Maxwell–Boltzmann distribution, and the equations of motion are solved using finite difference techniques [13]. Simulations are broken down into a series of small time steps δt, and at each time step atomic forces are calculated and used to advance the velocities and atomic positions forward in time (see the schematic in Fig. 1). In the simplest form of MD the total energy of the system is conserved. However, it is usually more useful to employ a thermostat (such as the Nosé–Hoover thermostat [15]) in order to carry out MD calculations at constant temperature. When molecular mechanics force fields are used, the size of time step chosen depends on the fastest motion in the system, which is invariably a bond stretch. As a general rule, energy conservation improves dramatically as time steps are reduced, and δt should be at leat 25 times smaller than the period of the fastest motion in the system. For this reason, it is usual practice to constrain bond lengths in an atomic simulation using the SHAKE procedure [16]. This approximation works well because bond stretches are usually of sufficiently high frequency to be decoupled from bond bending motion and torsional angle rotations. With SHAKE, a typical MD time step is 2 fs; without bond length constraints, this must be reduced to at least 0.5 fs, with a consequential increase in computer time required for simulation.

Many molecular mechanics packages now include MD as an option. In a typical MD simulation of a single molecule, the molecule is slowly warmed from an energy-minimised (zero kelvin) structure to the required average temperature over a period of a few picoseconds. Simulations are then carried out for the desired length of time, with molecular conformations saved periodically for later analysis throughout the course of the simulation run. Later analysis of these conformations is then able to provide time-averaged information for a single molecule at the temperature of the simulation. This approach yields useful data on dihedral angle distributions, moment of inertia ellipsoids, average dipole moments, etc.

However, single molecule molecular dynamics for liquid crystal molecules can often be problematic. Many liquid crystal systems have torsional energy barriers in excess of 12 kJ mol^{-1} separating conformations of similar energy (see Sec. 3.2.1). Such barriers can be difficult to cross during the course of a short MD simulation, and this can result in molecules becoming periodically trapped in regions of phase space. This has led to the development of stochastic dynamics techniques where random noise added to the equations of motion is de-

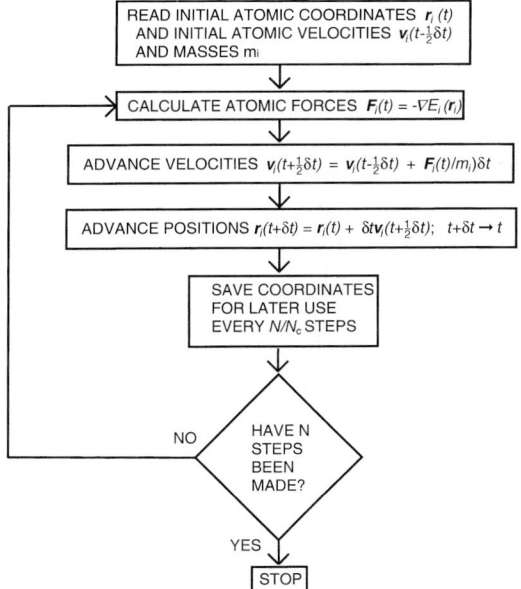

Figure 1. Schematic representation of a molecular dynamics simulation. The scheme for integrating the equations of motion is known as the leapfrog algorithm. The figure shows a flow diagram involving N_c samples of the coordinates for a simulation of N steps.

signed to mimic the effects of molecules colliding with a heat bath of surrounding molecules [17]. This process can lead to much faster barrier crossing rates than in standard MD. However, long simulations of 10–100 ns may still be needed if good dihedral angle distributions are required.

In Monte Carlo simulations the molecular mechanics interaction potential [Eq. (1)] can be used directly without the need to calculate atomic forces. Molecules must usually be represented by a set of internal coordinates (bond lengths, bond angles, and torsional angles), and the Metropolis approach is used to sample the configurational part of the partition function [11,12,14]. The Metropolis scheme involves making random changes to bond angles and torsional angles (bond lengths are usually held fixed) [11,18]. The energy of a new confirmation E_{new} is then calculated and compared with the previous energy E_{old}. If the trial energy is lower than E_{old}, the trial move is accepted. If the trial energy is higher than E_{old}, the move is accepted if $\exp[-(\Delta E)/k_B T]$ is greater than a random number between 0 and 1, where $\Delta E = E_{new} - E_{old}$. Consequently, over the course of a MC simulation, moves are accepted with a probability[1] of $\exp[-(\Delta E)/k_B T]$.

Single molecule MC simulations sample phase space much more efficiently than the corresponding MD calculations. Trial rotations about dihedral angles provide a mechanism to overcome the large energy barriers between liquid crystal conformers. The drawback of such calculations is complexity. They require a consistent representation of molecular structure, where Cartesian coordinates can be generated from a single reference point in terms of a set of specified

---

[1] When bond lengths are constrained, a small correction factor should be introduced in the configuration sampling to take account of the constraints [18].

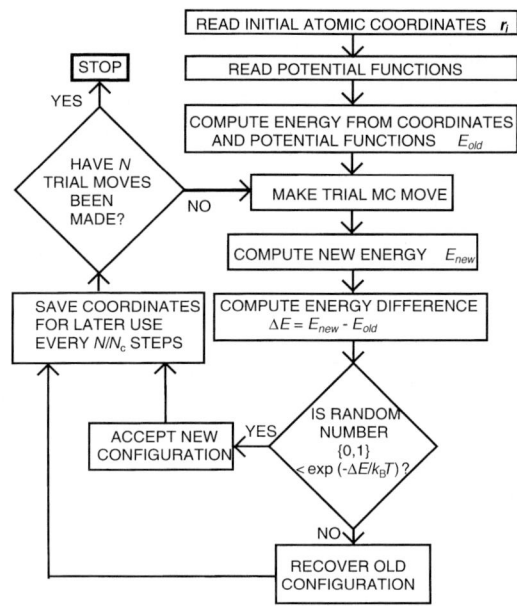

Figure 2. Schematic representation of a Metropolis Monte Carlo simulation. This scheme is suitable for most soft potentials, but constraints must be handled carefully (see footnote to text). N trial moves are attempted, with coordinate sampling carried out every N/N_c trial moves.

internal coordinates [18]; therefore it is harder to write general MC programs to handle any molecular structure. Despite this limitation, MC calculations are starting to become available in some modelling packages. Figure 2 shows a schematic diagram for the Metropolis method. The overall MC methodology is similar to that used in molecular dynamics, with conformations being periodically saved for later analysis after an initial equilibration period. A new technique has recently arisen that mixes stochastic dynamics with Monte Carlo sampling of dihedral angles [19]. This can greatly increase the barrier crossing rate for stochastic dynamics, and thereby reduce the length of runs required for efficient conformational sampling.

MD and MC are not restricted to single molecules, and in the last few years several simulations have appeared that have attempted to simulate mesogens in the liquid and liquid crystal phases [21–31]. These studies all make use of periodic boundary conditions, allowing a section of bulk fluid to be simulated without the need to worry about edge effects at the walls of the simulation box. However, care must always be taken to ensure that the simulation box is sufficiently large that atomic and centre of mass radial distribution functions are able to decay to a value of unity within the dimensions of the periodic box.

In contrast to single molecule calculations, in bulk simulations MD is preferable to Metropolis MC. A trial MC move involving a small rotation of a dihedral angle near the centre of a large rod-like molecule can lead to a large movement in the terminal parts of the molecule, resulting in collisions with neighbours in the bulk. This results in a large number of small rotations being rejected, and consequently a rather poor sampling of phase space occurs. Modern MC techniques such as configurational bias Monte Carlo are starting to tackle such problems successfully [32]; however, they have not as yet been applied to realistic simulations of liquid crystal systems. In contrast to MC methods, collisions with neighbours in the bulk fluid enable individual molecules within an MD simulation to sample phase space more efficiently than in the single molecule case.

The drawback with bulk simulations is their cost in terms of computer time. Computer time increases with the square of the number of atomic sites (truncation of short range interactions and the use of neighbour lists can improve this slightly), so a typical MD simulation of a few hundred molecules will involve the calculation of energies and forces for several thousand atomic sites at each time step. Because of this, it is usual in bulk simulations to employ the *united atom* approximation [21], in which carbons and attached hydrogens are replaced by single *extended* atomic sites. Use of this approximation requires a different force field to that employed in standard molecular mechanics studies. Internal molecular structures tend to be less accurately modelled by united atom force fields than by all-atom force fields. However, united-atom force fields have generally been designed with intermolecular forces in mind, and may well produce better intermolecular interaction energies than some all-atom force fields. In a series of papers [33–46], Jorgensen has carried out a large number of MC calculations for small molecules aimed at producing a set of transferable OPLS parameters (optimised parameters for liquid simulation) that model the thermodynamic properties of small molecules very well. These have recently been combined with the AMBER force field [47–50] to produce a combined AMBER/OPLS force field [51], which is ideal for the simulation of liquid crystal systems within the united-atom approximation [21]. Other united-atom force fields include the CHARMM force field [52, 53] and the AMBER force field itself.

In bulk MD simulations an initial equilibrium period of 200–300 ps is usually required to bring torsional angles into thermal equilibrium at the simulation temperature. However, molecular reorientation occurs on much longer timescales. Extrapolation from simpler models of liquid crystals suggest that the growth of a nematic liquid crystal from an isotropic liquid may require 1–10 ns of simulation time. This is currently at the limit of what can be achieved for atomic systems. However, the few bulk simulations that have appeared (see Sec. 3.2.4) suggest that this is a very exciting area of modelling that will develop strongly over the next few years.

3.1.3 Quantum Mechanical Techniques

The most natural way to determine molecular structure is by a direct quantum mechanical treatment of a molecule. This involves the solution of the Schrödinger equation for a given nuclear configuration, followed by the systematic adjustment of nuclear positions till the energy of a molecule is minimised. At the present time, quantum mechanical techniques are still extremely expensive, so that a full ab initio minimisation of molecular structure is only available for simple molecules such as methane [54]. Typically, computer time increases with the fourth power of the number of basis functions required in the calculation, and quantum energy minimisation is therefore extremely expensive for liquid crystal molecules. Despite this, quantum mechanical calculations are becoming useful in two guises. Firstly, accurate single point calculations can be carried out on energy-minimised structures produced by cheaper techniques such as molecular mechanics. Such calculations can provide reasonably accurate predictions for electric and magnetic properties of molecules [55]. However, it should be stressed that the molecular mechanics structure may not be the same as the molecular structure that would have been generated had the molecular geometry been allowed to relax in a quantum mechanical calculation. Secondly, semi-empirical quantum techniques have made rapid developments in the last few years [56]. These techniques are well suited to large liquid crystal molecules, and are starting to become useful in both the optimisation of molecular structures and in the determination of molecular properties [57].

3.2 Applications of Molecular Modelling

3.2.1 Determination of Molecular Structure

The determination of molecular structure is the simplest application of molecular modelling. In the first instance, a series of molecular mechanics calculations can produce a set of conformational energy minima for a liquid crystal that provide excellent information on molecular shape. For example, the mesogen 4-(trans-4-n-propylcyclohexyl]benzonitrile (PCH3) possesses a number of conformational energy minima corresponding to two possible chair conformations of the cyclohexane ring and rotations about torsional angles a, b, and c in Fig. 3. After an initial minimisation of a trial geometry, the next stage of conformational searching involves the driving of individual torsional angles [58]. The results of such calculations for PCH3 using the MM3 force field [5] are shown in Figs. 4a–c [59]. In its minimum-energy conformation the phenyl ring lies in the symmetry plane of the cyclohexyl ring, with the two torsional angles corresponding to the label c in Fig. 3 equal at 118° and –118°. However, the energy barrier corresponding to rotation about c is rather small, about 7.5 kJ mol^{-1} (Fig. 4c). The dihedral angle b involves the rotation of the propyl chain with respect to the cyclohexane ring. Figure 4b shows the standard

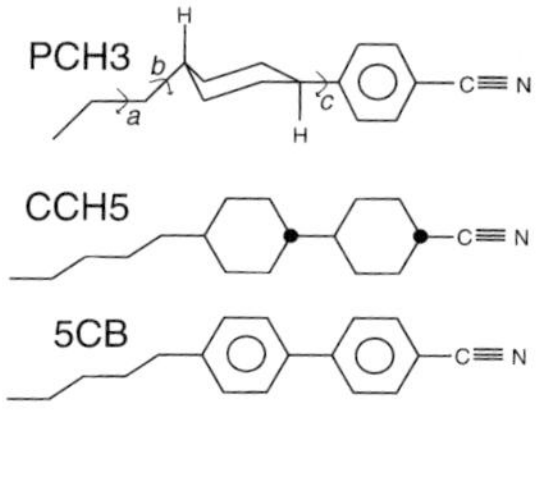

Figure 3. Structures of some common cyano-mesogens.

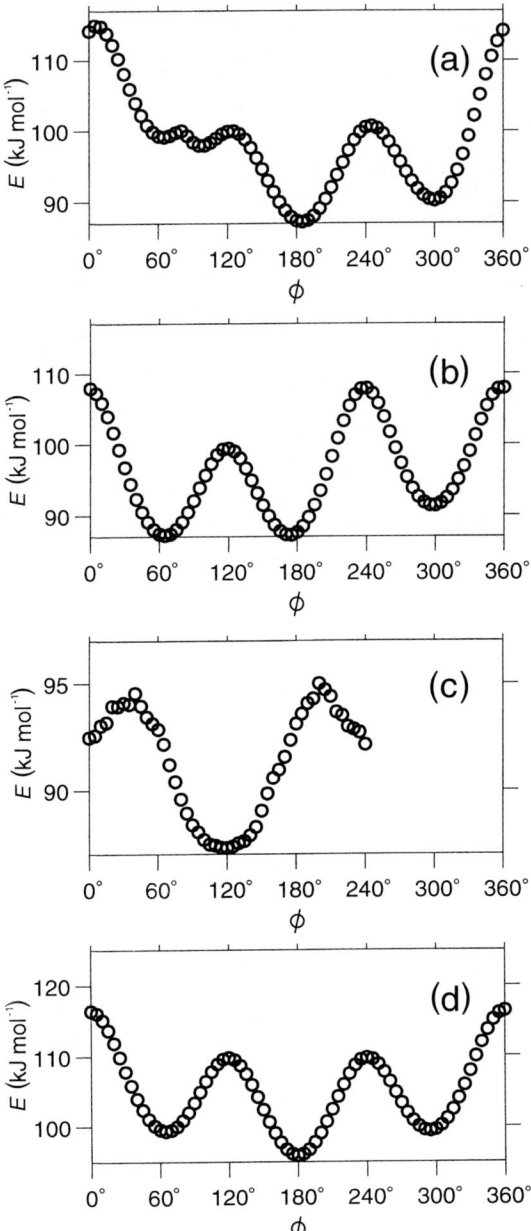

Figure 4. Typical torsional energy barriers for liquid crystal molecules. (a) Dihedral angle a for PCH3. (b) Dihedral angle b for PCH3. (c) Dihedral angle c for PCH3. (d) The terminal end-of-chain dihedral angle for PCH5. Torsional energy barriers are calculated using the MM3 force field [5] and the MacroModel molecular modelling package [60].

shape for this interaction potential, with two equal lowest energy conformations and a third energy minimum at 4.13 kJ mol^{-1} above the two low-energy states. The barriers to rotation occur at approximately 0° (20.69 kJ mol^{-1}), 120° (12.25 kJ mol^{-1}), and 240° (20.69 kJ mol^{-1}). Finally, the dihedral angle a is the end-of-chain dihedral for the propyl chain. Rotation about a produces two *gauche* conformations of unequal energy. The lowest energy *gauche* conformation is at 300° (3.05 kJ mol^{-1} above the ground state), whilst the second *gauche* conformation is split into two energy minima, the lowest of which occurs at 97.2° at an energy of 10.75 kJ mol^{-1} above the ground state. The energy barriers to rotation occur at 5° (27.8 kJ mol^{-1}), 80° (12.9 kJ mol^{-1}), 125° (12.8 kJ mol^{-1}), and 245° (13.5 kJ mol^{-1}). The assymmetry of the *gauche* conformations in Fig. 4a arises from strong repulsive interactions between the hydrogens on the terminal methyl group and axial cyclohexyl hydrogens as the alkyl chain rotates and collides with the core of the molecule. If the carbon chain of PCH3 is extended by two units (producing PCH5) then the terminal methyl group of the chain no longer interacts strongly with the cyclohexyl ring when rotation occurs about the end-of-chain dihedral angle (Fig. 4d). Consequently, the torsional angle energy profile of Fig. 4d is similar to that found in most liquid crystals with long alkyl chains: *gauche* conformations lie approximately 3.4 kJ mol^{-1} above the *trans* conformation, and the energy barriers to rotation are 20.5 kJ mol^{-1} and 13.9 kJ mol^{-1}.

From the dihedral angle energy profiles, a number of candidates for local potential energy minima can be identified and individually optimised. For PCH3, this results in the energy minima shown in Table 1, with the structure of the lowest energy conformation shown in Fig. 5. PCH3 exhibits a nematic phase between 36° and 46°C [61], and

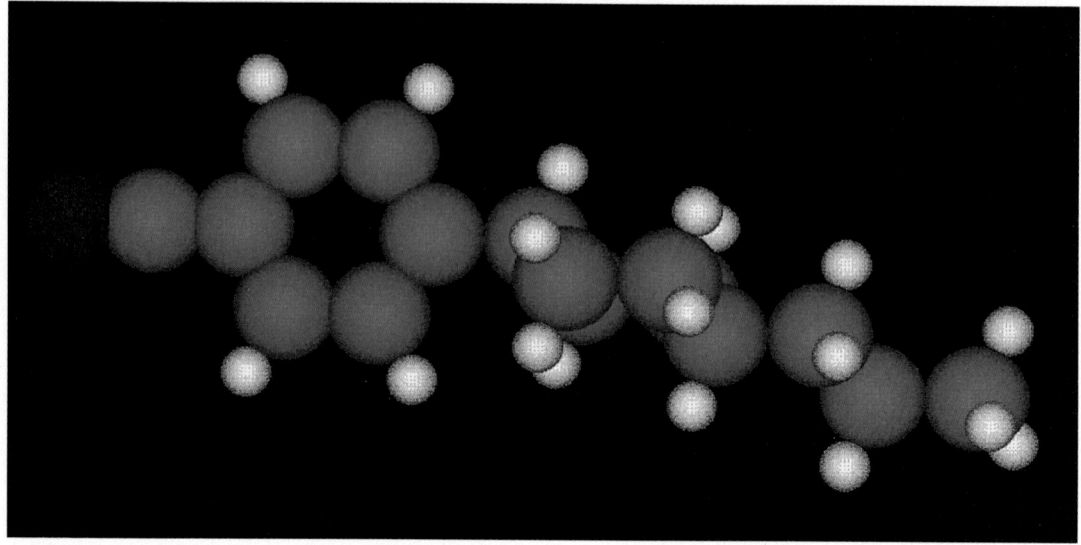

Figure 5. Minimum-energy conformation of PCH3.

Table 1. Local potential energy minima for PCH3 [59].

Energy (kJ mol^{-1})	Dihedral angle a	Dihedral angle b	Population at 46°C
87.34	−175.5°	173.2°	34.65
87.34	175.4°	64.9°	34.65
90.39	−60.6°	176.6°	10.98
90.39	60.6°	61.6°	10.98
91.47	180.0°	−62.2°	7.31
98.09	−97.2°	62.2°	0.60
98.09	97.2°	175.6°	0.60
102.41	−93.8°	−67.5°	0.12
102.41	93.9°	−57.1°	0.12

so application of Eq. (5) at $T = 46°C$ for the nine conformations in Table 1 yields appropriate weights for conformer populations, suggesting that at least five conformers have significant populations at 46°C. Concerted rotation about torsional angles in the cyclohexane ring produces a chair conformation with both the phenyl and propyl groups in axial positions. MM3 calculations show that the lowest energy conformation for this ring is 28.5 kJ mol^{-1} above the ground state, meaning that this conformation is unlikely to be significantly populated even at very high temperatures.

Recently, Dunmur et al. [57] have calculated the energy-minimised structures and rotational energy barriers for a number of chiral dopants using the MM2 force field and the techniques discussed above. In this work MM2 structures were further refined using the semi-empirical SCF quantum mechanical program MOPAC [62] (Sec. 3.1.3). Their results are shown in Fig. 6.

3.2.2 Determination of Molecular Properties

Energy-minimised structures generated by molecular mechanics or semi-empirical methods can be used as a starting point for the calculation of molecular properties. Dunmur et al. [57] have used MOPAC to calculate the dipole moments for the minimum-energy structures shown in Fig. 6. However, results have so far had mixed success. Calculated dipole moments can be compared with those measured in dilute so-

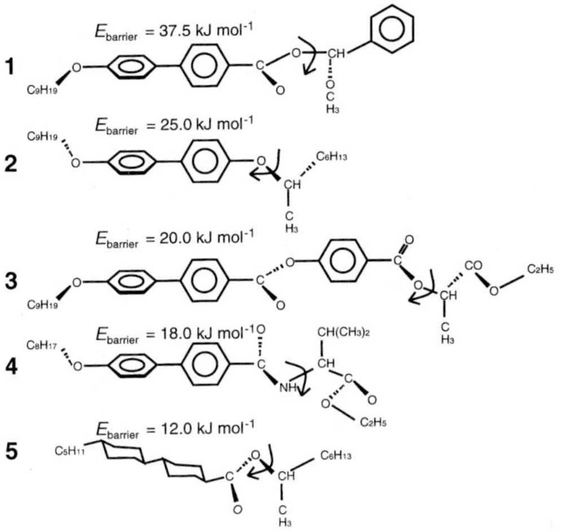

Figure 6. Energy-minimised structures for a series of chiral dopants adapted from the work of Dunmur et al. [57]. Structures were generated from MM2/MOPAC calculations, and $E_{barrier}$ is the (highest) rotational energy barrier calculated from the MM2 force field for the indicated bonds.

lutions of apolar solvents. For the materials in Fig. 6, good agreement was obtained with solution measurements for compounds **3** and **5**, but rather poor agreement was found for compounds **1**, **2**, and **4**. The errors have been attributed to the failure to take other molecular conformations into account in calculating the net dipole moment. In principle, this can be done using the analysis above (Sec. 3.2.1), but, in practice, the large number of possible conformations for compounds **1–5** makes this a rather lengthy process. Dunmur et al. [57] have also reported calculations of fragment dipoles for dipolar groups attached to the chiral centre in compounds **1–5**. They report reasonable correlation between transverse fragment dipole moments and measurements of spontaneous polarization (P_s) in SmC host solvents. In future, the single molecule Monte Carlo approach (described in Sec. 3.1.2) linked to a semi-empirical quantum method may provide a mechanism to generate useful conformationally averaged properties.

As yet, this combined technique has not been used for liquid crystal systems.

3.2.3 Determination of Intermolecular Potentials

Wilson and Dunmur have used molecular mechanics techniques to model the interaction energies of isolated pairs of liquid crystal molecules [58, 63, 64]. In this approach lowest energy conformations for liquid crystal molecules were first generated using MM2 [7], and these conformations were used to explore the potential energy surface for two molecules interacting via nonbonded terms of the form given in Eq. (4). For the mesogens 4-n-pentyl-4′-cyanobiphenyl (5CB) and 4-($trans$-4-n-pentylcyclohexyl)cyclohexylcarbonitrile (CCH5, Fig. 3), distinct lowest energy *dimers* were found, corresponding to parallel and antiparallel arrangements of molecular dipoles, with the antiparallel configuration energetically favoured [63]. The removal of partial charges from the calculations was found to have a negligible effect on the spatial configuration of molecular pairs, but largely removed the energy difference between parallel and antiparallel arrangements. Wilson and Dunmur concluded that dispersive forces provide the dominant factor in causing molecular association in liquid crystals, but that dipolar effects are important in determining the balance between parallel and antiparallel molecular pairs. These conclusions fit well with results from dielectric [65] and light scattering [66, 67] studies of molecular association in dilute solutions of mesogens.

Luckhurst and Simmonds [68] have used Lennard–Jones pair potentials to characterise the molecular interaction potential $U_{av}(\boldsymbol{u}_1, \boldsymbol{u}_2, \boldsymbol{R})$ for two liquid crystal molecules with molecular long axes in the directions u_1 and u_2, where R is the vector between the respective centres of mass. In order to do this successfully, molecular biaxiality must be projected out by taking Boltzmann-weighted averages of the interaction potential over rotations about both molecular long axes:

$$U_{av}(u_1, u_2, R)$$
$$= \frac{\int_0^{2\pi}\int_0^{2\pi} U_{LJ}\exp(-U_{LJ}/k_BT)\,d\alpha_1\,d\alpha_2}{\int_0^{2\pi}\int_0^{2\pi} \exp(-U_{LJ}/k_BT)\,d\alpha_1\,d\alpha_2} \quad (7)$$

where α_1 and α_2 are the rotation angles about the molecular long axes of molecules 1 and 2, and $U_{LJ} = U_{LJ}(u_1, u_2, R)$ is the interaction energy of two molecules for given u_1, u_2, and R. As above, U_{LJ} is equal to a sum of all nonbonded pair interactions between molecules 1 and 2. Luckhurst and Simmonds went on to show that it is possible to take $U_{av}(u_1, u_2, R)$ and use it to parameterise a version of the Gay–Berne potential [68]. This *simplified* single site potential can then be used to carry out bulk simulations of liquid crystal mesophases.

3.2.4 Large-Scale Simulation of Liquid Crystals

Bulk atomistic simulations of liquid crystal mesophases are extremely time-consuming and currently represent the limit of what can be achieved with today's computers. However, in the past few years a number of (mainly) *united-atom* models of small mesogens have started to appear in the literature. These simulations are summarised in Table 2, and snapshots of molecules taken from a MD simulation of CCH5 are shown in Fig. 7. Many of the studies in Table 2 suffer from common drawbacks, namely small numbers of molecules and rather short simulation

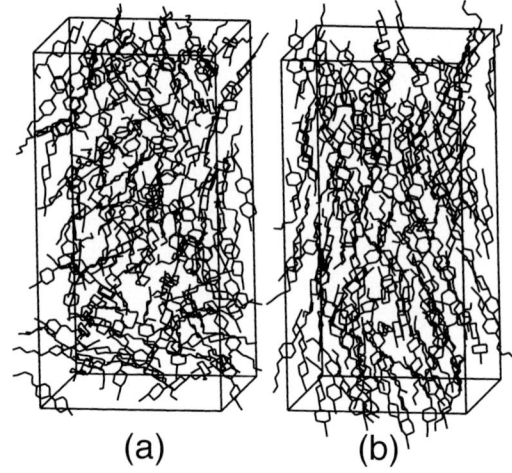

Figure 7. Snapshots showing the structure of CCH5 in liquid and liquid crystalline phases. (a) Isotropic phase at 390 K. (b) Nematic phase at 350 K, $S = 0.62$. (Molecular coordinates are from the simulations carried out by Mark Wilson and Mike Allen [23], courtesy of the authors.)

times. However, the very fact that such simulations can now be attempted bodes well for future developments in this field.

In atomistic simulations, molecular order can be characterised in a number of ways. For rigid rodlike molecules, orientational ordering may be defined by reference to the molecular long axis vector a_j:

$$Q_{\alpha\beta} = \frac{1}{N_{mol}} \sum_{j=1}^{N_{mol}} \frac{3}{2} a_{j\alpha} a_{j\beta} - \frac{1}{2}\delta_{\alpha\beta} \quad (8)$$

where N_{mol} is the number of molecules, and $\alpha, \beta = x, y, z$. The director n is the eigenvector associated with the largest eigenvalue (λ_+) of Q, and λ_+ can be equated to the uniaxial order parameter S:

$$S = \lambda_+ = \langle P_2(n \cdot a) \rangle = \langle P_2(\cos\theta) \rangle \quad (9)$$

where P_2 is a second-rank Legendre polynomial, θ is the angle between a molecule and the director, and the angular brackets denote an ensemble average. For the flexible molecules often used in atomistic simulations, a_j can be assigned to the eigenvec-

Table 2. Atomistic simulations of bulk liquid crystal systems.

Mesogen	Ref.	N_{mol}	Study	Notes
5CB	[20]	64	NpT MD bond constraints	60 ps simulation for charged and uncharged systems
EBBA[a]	[24]	60	NpT MC (all-atom)	312 K, 10^3 moves per molecule (nematic $S=0.8$)
pHB[b]	[23]	8, 16	Energy minimisation and NVT MD (all-atom model)	300 K, 70 ps (crystal) 500 K, 116 ps (nematic)
CCH5	[21] [22]	128	NpT MD bond constraints	Up to 720 ps 390 K (isotropic) 370 K (nematic $S=0.38$) 350 K (nematic $S=0.62$)
THE5[c]	[25]	54	NVT MD bond and angle constraints	380 K, 100 ps (discotic columnar)
nOCB[d] $n=5, 6, 7, 8$	[26]	64	NpT MD bond and angle constraints	Up to 180 ps 5OCB, 331 K 6OCB, 330 K, 339 K, 359 K 7OCB, 337 K; 8OCB, 359 K
2MBCB[e]	[30]	32	NpT MD	230 K, 10 ps with/without twisted periodic boundaries
HBA[f]	[29]	125	NpT MD bond constraints (all-atom model)	Up to 590 ps, 475 K (nematic $S=0.86$)
PCH5	[31]	50, 100	NpT MD bond constraints	Up to 360 ps, 333 K (nematic, isotropic)
5CB	[28]	80	NpT MD bond constraints	300 K, 1600 ps (nematic $S=0.6$)

[a] 4-Ethoxybenzylidene-4'-n-butylaniline;
[b] truncated *tetramer* segment of the liquid crystal polyester of 4-hydroxybenzoic acid [phenyl-4-(4-benzoyloxy-)benzoyloxy benzoate];
[c] hexakispentyloxytriphenylene;
[d] 4-alkoxy-4'-cyanobiphenyl;
[e] (+)-4-(2-methylbutyl)-4'-cyanobiphenyl;
[f] *tetramer* segment of the liquid crystal polyester of 4-hydroxybenzoic acid.

tor associated with the smallest eigenvalue of the inertia tensor,

$$I_{j\alpha\beta} = \sum_i m_i(r_i^2 \delta_{\alpha\beta} - r_{i\alpha}r_{i\beta}) \quad (10)$$

where m_i are atomic masses and the atomic distance vector r_i is measured relative to the molecular centre of mass. Alternatively, individual values of S can be calculated for different parts of the molecule. Wilson and Allen [22] (for CCH5) and Cross and Fung [28] (for 5CB) have considered the ordering of individual segments within a molecule. Their simulations demonstrate a classic odd–even effect in the ordering of individual bonds in the alkyl chain: odd bonds have much higher order parameters than even bonds. For CCH5 [22], the orientational distribution function $f(\cos\theta)$ for odd bonds is strongly peaked along the director. In contrast, $f(\cos\theta)$ exhibits a broad distribution for even bonds, and peaks at an angle to the director.

It is interesting to look at the dihedral angles for the alkyl chains in 5CB and CCH5. In the nematic phase at 350 K, Wilson and Allen found that almost 50% of molecules have the fully extended all-*trans* (ttt) conformation. However, dihedral angle distributions are temperature-dependent. The observed dihedral angle distribution $S(\phi)$ can be written in terms of an effective torsional potential $E_{\text{eff}}(\phi)$ (conformational free energy) [22, 28]:

$$S(\phi) = C \exp\left(-\frac{E_{\text{eff}}(\phi)}{k_B T}\right) \quad (11)$$

where C is a normalization factor,

$$E_{\text{eff}}(\phi) = E_{\text{ext}}(\phi) + E_{\text{torsion}}(\phi) + E_{\text{int}}(\phi) \quad (12)$$

where $E_{\text{torsion}}(\phi)$ is the torsional angle potential (see, e.g., Eq. (3)), $E_{\text{int}}(\phi)$ is due to internal nonbonded interactions (mainly 1–4 interactions), and $E_{\text{ext}}(\phi)$ depends on the local molecular environment of a molecule in a bulk fluid. $E_{\text{eff}}(\phi)$ can therefore be used to monitor the effect of the nematic field on molecular structure [28]. As the nematic phase is entered, the effect of $E_{\text{ext}}(\phi)$ is to favour conformations where bonds lie along the molecular long axis. In CCH5 this leads to a small change in shape, so that molecules become more elongated in the nematic phase.

For the five carbon chains of 5CB and CCH5, the three lowest energy chain conformations correspond to ttt, gtt, tgt, and ttg (Fig. 8). Whilst ttt is the lowest energy conformation for 5CB and CCH5, molecular mechanics predicts that the preferred ordering of *gauche* conformations is gtt, tgt, ttg owing to favourable chain ring nonbonded interactions, which reduce the value of $E_{\text{int}}(\phi)$ for *gauche* conformations close to the core [22]. However, in the nematic phase the effect of $E_{\text{ext}}(\phi)$ is such as to produce

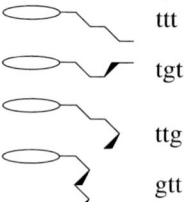

Figure 8. Chain conformations for the mesogen CCH5.

the preferred ordering tgt, ttg, gtt for *gauche* conformations in both 5CB and CCH5. This order arises because (unlike tgt) the *gauche* conformations ttg, gtt cause bonds to lie at an angle to the director, and so are strongly disfavoured by the local structure of the nematic fluid. The contributions of $E_{\text{ext}}(\phi)$ to $E_{\text{eff}}(\phi)$ significantly increase both the energy differences between *gauche* and *trans* conformers and the rotational energy barriers between conformers. This is something that is largely (and incorrectly) ignored in molecular mechanics studies of mesogens (Sec. 3.2.1).

A number of bulk simulations have attempted to study the dynamical properties of liquid crystals. Translational diffusion coefficients are available from the Einstein relation, which is valid for long times t:

$$D = \frac{1}{2t}\left\langle |\mathbf{r}_i(t) - \mathbf{r}_i(t_0)|^2 \right\rangle \quad (13)$$

where $\mathbf{r}_i$ is the centre of mass position of molecule i at time t and t_0 is an appropriate time origin. D may be resolved to monitor diffusion parallel ($D_\parallel$) and perpendicular ($D_\perp$) to the director separately. Table 3 lists values of $D_\parallel$ and $D_\perp$ from a variety of studies. In most cases a clear anisotropy in diffusion is seen, with diffusion along the nematic director favoured. The one exception in Table 3 is for the molecule THE5. In this case the simulations are for a discotic columnar phase, with diffusion favoured perpendicular to the direction of order. Both $D_\parallel$ and $D_\perp$ are very small for THE5.

The small number of molecules present in the simulations in Table 2 means that no

Table 3. Diffusion constants from atomistic simulations of liquid crystals.

Mesogen	Ref.	T (K)	$D_\parallel$ (10^{-9} m^2 s^{-1})	$D_\perp$ (10^{-9} m^2 s^{-1})	$D_\parallel/D_\perp$	S
5CB	[28]	300	1.12	0.30	3.73	0.60
CCH5	[21]	350	0.554	0.192	2.89	0.62
		370	1.076	0.517	2.08	0.38
PCH5	[31]	333	0.157	0.046	3.4	0.58
5OCB	[27]	331	0.36	0.188	1.91	0.53
6OCB		339	0.33	0.168	1.96	0.50
7OCB		337	0.316	0.176	1.80	0.50
8OCB		342	0.282	0.177	1.59	0.47
THE5	[26]	380	0.00984	0.0189	0.05	0.95

convincing atomistic simulations of smectic mesophases currently exist. However, a number of studies of bilayers and Langmuir–Blodgett films have been made in recent years [69–72], and these are starting to prove useful in studying molecular ordering in layered structures. One atomistic study has appeared that attempts to study a chiral system by molecular dynamics [30]. In this study twisted periodic boundary conditions were introduced to look at a (pseudo) chiral nematic phase. In this initial study only a very small system (32 2MBCB molecules) was used, so few definitive conclusions were available. However, this technique could, in principle, be extended to look at large chiral systems in the future.

Finally, it should be stressed that no one has yet proved the thermodynamic stability of atomistic model mesophases by growing a nematic phase directly from an isotropic liquid. This is a relatively easy process for single site models [68], but is extremely expensive for atom-based models, requiring long simulations (of the order of 10 ns) on systems of several hundred molecules. On account of this, data from the simulations in Table 2 should be treated as preliminary at this stage. However, the rapid increases in speed (and reduction in cost) of modern computers suggest that definitive atomistic simulations may be only a few years away.

3.3 References

[1] N. L. Allinger, *Rev. Phys. Org. Chem.* **1976**, *13*, 1–82.
[2] O. Burkert, N. L. Allinger, *Molecular Mechanics*, ACS Monograph 177, American Chemical Society, Washington, DC **1982**.
[3] D. B. Boyd, K. B. Lipkowitz, *J. Chem. Educ.* **1982**, *59*, 269–274.
[4] M. R. Wilson, Ph. D. Thesis, University of Sheffield **1988**, Chap. 1.
[5] N. L. Allinger, Y. H. Yuh, J. Lii, *J. Am. Chem. Soc.* **1989**, *111*, 8511–8582.
[6] N. L. Allinger, *J. Am. Chem. Soc.* **1977**, *99*, 8127–8134.
[7] N. L. Allinger, M. T. Tribble, M. A. Miller, D. H. Wertz, *J. Am. Chem. Soc.* **1971**, *93*, 1637–1647.
[8] A. J. Hopfinger, R. A. Pearlstein, *J. Comput. Chem.* **1984**, *5*, 486–499.
[9] D. M. Ferguson, D. J. Raber, *J. Am. Chem. Soc.* **1989**, *111*, 4371–4378.
[10] G. Chang, W. C. Guida, W. C. Still, *J. Am. Chem. Soc.* **1989**, *111*, 4379–4386.
[11] W. L. Jorgensen, *J. Phys. Chem.* **1983**, *87*, 5304–5314.
[12] D. Levesque, J. J. Weis, J. P. Hansen in *Applications of the Monte Carlo Method in Statistical Physics* (Ed.: K. Binder), Topics in Current Physics 36, Springer-Verlag, Berlin **1984**, Chap. 2.
[13] M. P. Allen, D. J. Tildesley, *Computer Simulation of Liquids*, Oxford University Press, Oxford **1987**, Chap. 3.
[14] M. P. Allen, D. J. Tildesley, *Computer Simulation of Liquids*, Oxford University Press, Oxford **1987**, Chap. 4.
[15] W. G. Hoover, *Phys. Rev.* **1985**, *A31*, 1695–1697.
[16] J. P. Ryckaert, G. Ciccotti, H. J. C. Berendsen, *J. Comput. Phys.* **1977**, *23*, 327–341.
[17] W. F. van Gunsteren, H. J. C. Berendsen, *Mol. Simul.* **1988**, *1*, 173–185.

[18] S. Leggetter, D. J. Tildesley, *Mol. Phys.* **1989**, *68*, 519–546.
[19] F. Guarnieri, W. C. Still, *J. Comput. Chem.* **1994**, *15*, 1302–1310.
[20] S. J. Picken, W. F. van Gunsteren, P. Th. van Duijnen, W. H. de Jeu, *Liq. Cryst.* **1989**, *6*, 357–371.
[21] M. R. Wilson, M. P. Allen, *Mol. Cryst. Liq. Cryst.* **1991**, *198*, 465–477.
[22] M. R. Wilson, M. P. Allen, *Liq. Cryst.* **1992**, *12*, 157–176.
[23] B. Jung, B. L. Schürmann, *Mol. Cryst. Liq. Cryst.* **1990**, *185*, 141–153.
[24] A. V. Komolkin, Yu. V. Molchanov, P. P. Yakutseni, *Liq. Cryst.* **1989**, *6*, 39–45.
[25] I. Ono, S. Kondo, *Mol. Cryst. Liq. Cryst.* **1991**, *8*, 69.
[26] I. Ono, S. Kondo, *Bull. Chem. Soc. Jpn.* **1992**, *65*, 1057–1061.
[27] I. Ono, S. Kondo, *Bull. Chem. Soc. Jpn.* **1993**, *66*, 633–638.
[28] C. W. Cross, B. M. Fung, *J. Chem. Phys.* **1994**, *101*, 6839–6848.
[29] J. Huth, T. Mosell, K. Nicklas, A. Sariban, J. Brickmann, *J. Phys. Chem.* **1994**, *98*, 7685–7691.
[30] M. Yoneya, H. J. C. Berendsen, *J. Phys. Soc. Jpn.* **1994**, *63*, 1025–1030.
[31] G. Krömer, D. Paschek, A. Geiger, *Ber. Bunsenges. Phys. Chem.* **1993**, *97*, 1188–1192.
[32] J. I. Siepmann, D. Frenkel, *Mol. Phys.* **1992**, *75*, 59–70.
[33] G. Kaminski, E. M. Duffy, T. Matsui, W. L. Jorgensen, *J. Phys. Chem.* **1994**, *98*, 13077–13082.
[34] E. M. Duffy, W. L. Jorgensen, *J. Am. Chem. Soc.* **1994**, *116*, 6337–6343.
[35] W. L. Jorgensen, E. R. Laird, T. B. Nguyen, J. Tiradorives, *J. Comput. Chem.* **1993**, *14*, 206–215.
[36] W. L. Jorgensen, T. B. Nguyen, *J. Comput. Chem.* **1993**, *14*, 195–205.
[37] J. M. Briggs, T. Matsui, W. L. Jorgensen, *J. Comput. Chem.* **1990**, *11*, 958–971.
[38] W. L. Jorgensen, J. M. Briggs, *Mol. Phys.* **1988**, *63*, 547–558.
[39] W. L. Jorgensen, J. Tiradorives, *J. Am. Chem. Soc.* **1988**, *110*, 1666–1671.
[40] W. L. Jorgensen, *J. Phys. Chem.* **1986**, *90*, 6379–6388.
[41] W. L. Jorgensen, J. Gao, *J. Phys. Chem.* **1986**, *90*, 2174–2182.
[42] W. L. Jorgensen, *J. Phys. Chem.* **1986**, *90*, 1276–1284.
[43] W. L. Jorgensen, J. D. Madura, *Mol. Phys.* **1985**, *56*, 1381–1392.
[44] W. L. Jorgensen, C. J. Swenson, *J. Am. Chem. Soc.* **1985**, *107*, 1489–1496.
[45] W. L. Jorgensen, C. J. Swenson, *J. Am. Chem. Soc.* **1985**, *107*, 569–578.
[46] W. L. Jorgensen, J. D. Madura, C. J. Swenson, *J. Am. Chem. Soc.* **1984**, *106*, 6638–6646.
[47] U. C. Singh, P. K. Weiner, J. Caldwell, P. A. Kollman, *AMBER 3.0*, University of California, San Francisco **1987**.
[48] P. K. Weiner, P. A. Kollmann, *J. Comput. Chem.* **1981**, *2*, 287–303.
[49] S. J. Weiner, P. A. Kollman, D. T. Nguyen, D. A. Case, *J. Comput. Chem.* **1986**, *7*, 230–252.
[50] S. J. Weiner, P. A. Kollman, D. A. Case, U. C. Singh, C. Ghio, G. Alagona, S. Profeta, P. Weiner, *J. Am. Chem. Soc.* **1984**, *106*, 765–784.
[51] W. L. Jorgensen, J. Tirado-Rives, *J. Am. Chem. Soc.* **1988**, *110*, 1666–1671.
[52] B. R. Gelin, M. Karplus, *Biochemistry* **1979**, *18*, 1256–1268.
[53] B. R. Brooks, R. E. Bruccoleri, B. D. Olafson, D. J. States, S. Swaminathan, M. Karplus, *J. Comput. Chem.* **1983**, *4*, 187–217.
[54] M. J. M. Pepper, I. Shavitt, P. V. Schleyer, M. N. Glukhovtsev, R. Janoschek, M. Quack, *J. Comput. Chem.* **1995**, *16*, 207–225.
[55] C. E. Dykstra, J. D. Augspurger, B. K. Kirtman, D. J. Malik in *Reviews in Computational Chemistry* (Eds.: K. B. Lipkowitz, D. B. Boyd), VCH, New York **1990**, Chap. 3.
[56] J. J. P. Stewart in *Reviews in Computational Chemistry* (Eds.: K. B. Lipkowitz, D. B. Boyd), VCH, New York **1990**, Chap. 2.
[57] D. A. Dunmur, M. Grayson, S. K. Roy, *Liq. Cryst.* **1994**, *16*, 95–104.
[58] M. R. Wilson, Ph. D. Thesis, University of Sheffield **1988**, Chap. 3.
[59] M. R. Wilson, unpublished work.
[60] *MacroModel V 3.5 X, Interactive Molecular Modelling System*, Department of Chemistry, Columbia University, New York **1992**.
[61] K. J. Toyne in *Thermotropic Liquid Crystals* (Ed.: G. W. Gray), Wiley, New York **1987**, Chap. 2.
[62] J. J. P. Stewart, *J. Comput. Aided Mol. Design* **1990**, *4*, 1–45.
[63] M. R. Wilson, D. A. Dunmur, *Liq. Cryst.* **1989**, *5*, 987–999.
[64] D. A. Dunmur, M. R. Wilson, *Mol. Simul.* **1989**, *4*, 37–59.
[65] K. Toriyama, D. A. Dunmur, *Mol. Phys.* **1985**, *56*, 479–484.
[66] K. Toriyama, D. A. Dunmur, *Mol. Cryst. Liq. Cryst.* **1986**, *139*, 123–142.
[67] K. Toriyama, D. A. Dunmur, *Liq. Cryst.* **1986**, *1*, 169–180.
[68] G. R. Luckhurst, P. S. J. Simmonds, *Mol. Phys.* **1993**, *80*, 233–252.
[69] P. van der Ploeg, H. J. C. Berendsen, *J. Chem. Phys.* **1982**, *76*, 3271–3276.
[70] P. van der Ploeg, H. J. C. Berendsen, *Mol. Phys.* **1983**, *49*, 233–248.
[71] A. Biswas, B. L. Schürmann, *J. Chem. Phys.* **1991**, *95*, 5377–5386.
[72] M. A. Moller, D. J. Tildesley, K. S. Kim, N. Quirke, *J. Chem. Phys.* **1991**, *94*, 8390–8401.

Chapter IV
General Synthetic Strategies

Thies Thiemann and Volkmar Vill

1 Introduction

This chapter provides a general overview of the methodology of liquid crystal (LC) synthesis. A certain degree of overlap with chapters on the synthesis of compounds showing specific mesophases (Vol. 2, Chap. III.1, IV.1, V.1, VI.1, and VII) is allowed for. Nevertheless, the reader is asked to refer to these chapters too for more detailed information of interest. Metallomesogens (see Vol. 2, Chap. XVI) are excluded from this chapter as are, for the most part, polymeric LCs (see Vol. 3).

Section 2 of this chapter presents a general guideline for planning the synthesis of a liquid-crystalline material. Brief comments on the historical development of LC synthesis show the change in synthetic strategy with time as reflected not only in the advance of new reaction methodology, but also in new target molecules. The access to more traditional systems, such as azoxy- and azo-compounds is also described. Section 4 of this chapter gives an overview of common reactions in LC synthesis used in more recent times. Section 5 lists typical building blocks, the methods for their connection with other building blocks, as well as their retrosynthetic disconnection. Section 6 shows methods of introducing centres of chirality into LC materials. Section 7 provides comments on using LC synthesis for educational purposes and a note on the purification of LC materials.

As much of the synthesis of LCs is in fact part of general synthetic organic chemistry, most of the synthetic methodology can be referred to in standard reference texts [1], compendia [2] and reaction data bases [3]. The list [1–3] provided is not exclusive, but gives the sources the authors of this chapter frequently consult.

2 General Guidelines

2.1 Thermotropic Calamitic Liquid Crystals

Liquid crystals derived from rod- or lath-shaped molecules, forming calamitic phases, have been the most comprehensively studied. Although no exact theory has been developed up to now, which relates the structure of molecules to their specific liquid-crystalline behavior, rules of thumb have been formulated based on the very large number of compounds [4] known, and the related physical data.

Calamitic mesogens usually follow the general structural formula R^1-A-[L]-B-[L]-C-[L]-D-R^2, where R^1, R^2 are terminal groups, A–D ring systems, and [L] linking units. At least two ring systems are usually required for the stabilization of a calamitic phase; compounds with one ring system and readily forming dimers are the only apparent exceptions to this rule. The sequence for mesophase stabilization by terminal groups R^1, R^2: $CN > OCH_3 > NO_2 > Cl > CH_3 > I > CF_3 > H$ is well known. In general, alkyl ter-

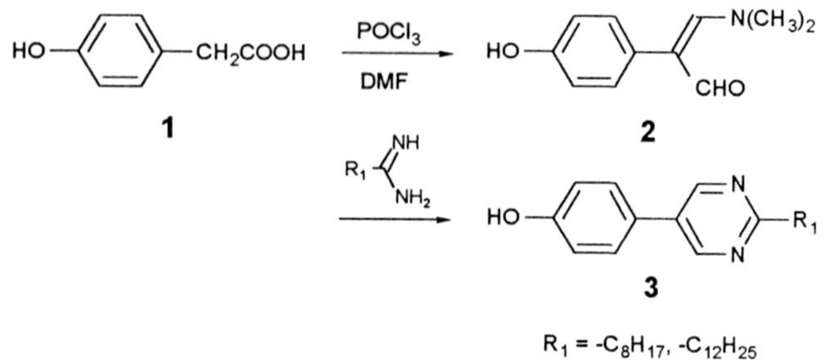

Scheme 1. Example of the construction of a heterocyclic unit upon linkage [14].

minal groups of odd carbon number give higher clearing points than those with even carbon numbers. Furthermore, the flexibility and conformation within a compound play a role. Thus the more flexible cyclohexane-moieties may be used beneficially, especially in lower temperature systems, while at higher temperatures the more rigid phenylene group may be preferred as a core-unit. Interactions among fragments are also important. While alkoxy-substituents give better results than alkyl-substituents with phenyl-groups, in cyclohexane core-units, just the opposite is true. Predictions about the characteristics of a substance on the basis of its structure have been ventured either by a statistical method using a large set of compounds [4] with known specific liquid-crystalline properties or by using neural nets [5]. Bearing all these factors in mind, further issues have to be considered in the choice of a target molecule, such as the desired chemical and physical properties. For example stability towards various influences (moisture, exposure to radiation) and switching times.

The synthesis of calamitic LCs often follows the R^1-A-[L]-B-[L]-C-[L]-D-R^2 scheme [6] in so far as a compound of such form is synthetically connected up at one of the points denoted by the separation lines (–). There are two main approaches.

(1) Two ring systems are synthetically constituted with simultaneous formation of the linking unit.

(2) Two ring systems are synthetically built up by attaching one ring system to the preformed linking unit on the other [8].

Most often in cases (1) and (2), the lateral groups (e.g. substituents on aromatic [ring] systems) are already in place. This circumvents later transformations of functional groups in complex systems. Moreover the pool of commercially available functionalized 'subunits' can be used more extensively.

In very few cases are ring systems themselves created upon linkage. Some *N*-heterocycles (e.g. pyrimidines (see Scheme 1), pyrazoles [9], oxazoles [10], thiadiazoles [11]) are, however, prepared in this way, as are acetals [12] and borinanes [13].

2.2 Columnar Liquid Crystals

Columnar liquid crystals are derived from disc- or plate-like molecules. For the most part, these compounds possess a core unit with pendant chains (usually 4, 6, or 8). Synthetically two main options of synthesis are pursued (see Scheme 2):

– construction of the core from fragments with preformed pendant chains; here pendant chains may be transformed after the synthesis of the core;
– attachment of the chains to a preformed (e.g. commercially available) core.

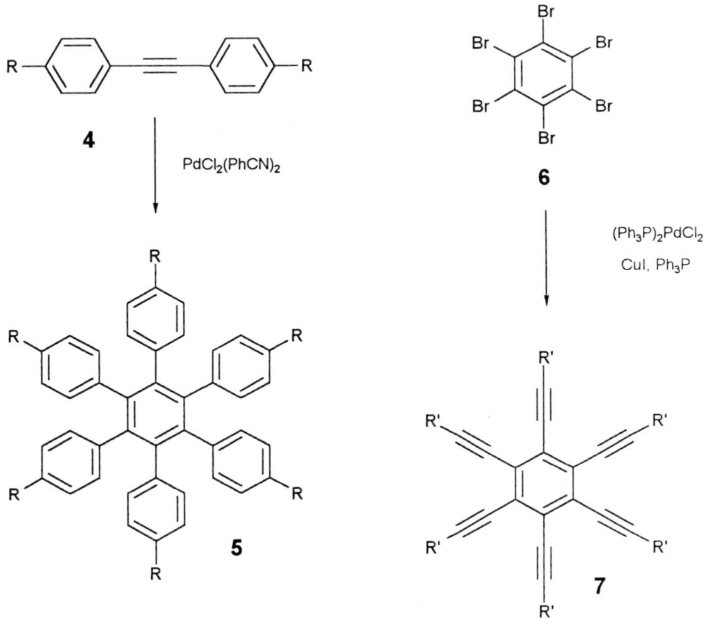

Scheme 2. Synthesis of discotic liquid crystals.
Left: Construction of core from fragments with preformed pendant chains [15a].
Right: Attachment of the chains to a preformed core [15b].

2.3 Concept of Leading Structures

In order to glean as much information as possible from novel compounds, it is important to compare them to as many other structures as possible. A number of leading structures have been developed over the years. Vorländer was the first to use one basic structure with a systematic variation of substructures. Figure 1 shows his main basic structure and the variations Vorländer performed on it. In recent times MBBA (4′-methoxybenzylidene-4-butylaniline) has been the most extensively studied compound, as well as being the compound with one of the largest number of substructural variations. Other structures include 5CB (4-cyano-4′-pentylbiphenyl) for nematics with a high dielectric anisotropy, and in more recent times, MHPOPC for antiferroelectrics.

3 Brief Survey of the History of Liquid Crystal Synthesis

The first materials studied [16] showing an unusual melting behaviour (*double melting point*), later classified as LCs, were isolated from nature. Thus in 1854 R. Virchow [17] described the lyotropic liquid-crystalline properties of myelin. In 1855, W. Heintz [18] reported on the stepwise melting of magnesium myristate, and the first unwitting use of liquid-crystalline substances probably dates back to the soaps of the Phoenicians, Sumerers, Egyptians, and Israelites [16c]. In 1910 D. Vorländer [19] published a detailed treatise on the liquid-crystalline properties of fatty acid salts. The prerequisite for these studies was the capability of isolating fatty acids in a pure, uniform state.

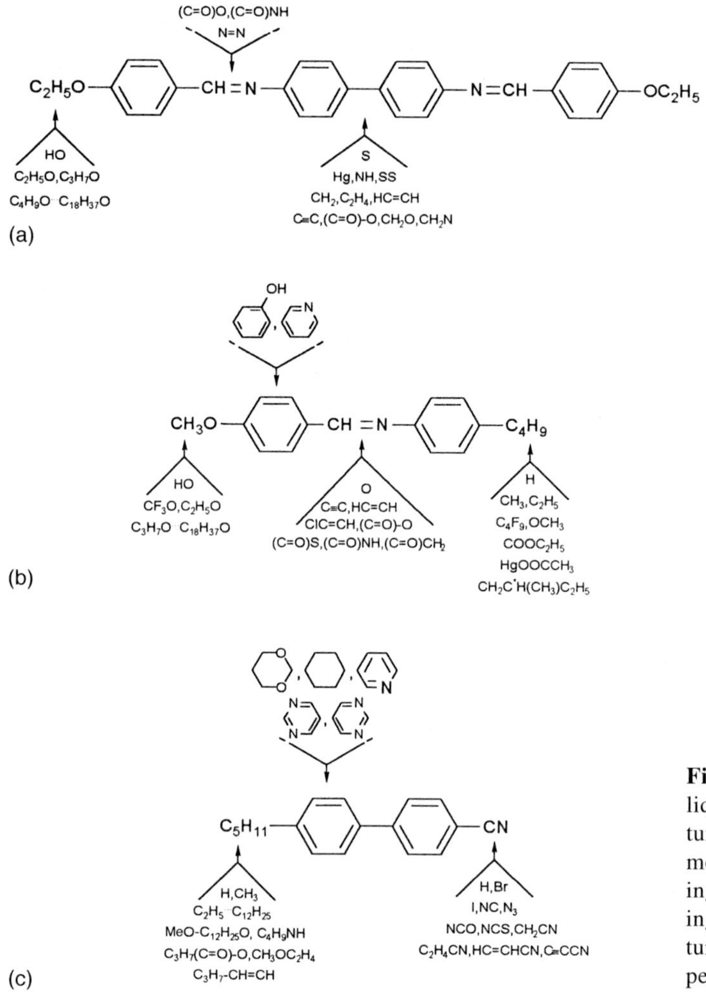

Figure 1. Concept of basic structures in liquid crystal synthesis: (a) basic structure developed by D. Vorländer and its modification; (b) possibilities of deriving liquid crystal structures by exchanging a single subunit in MBBA; (c) structural modifications of 5CB (4-cyano-4′-pentylbiphenyl).

Also the observations of P. Planer [20] (1861) on the cholesteric phase of cholesteryl chloride and by Reinitzer [21] (1888) on cholesteryl benzoate and acetate belong here, although in both cases chemical transformations had been undertaken on the natural products (chlorination; esterification with acetic and benzoic acid anhydride, respectively) [22]. Also today nature is a source of mesogenic structures (steroids [23a], triterpenes, carbohydrates [23b–d]) and of precursors of liquid-crystalline compounds (see sources of chirality).

In 1890 L. Gattermann and A. Ritschke [24] reported on the synthesis of 4,4′-dimethoxyazoxybenzene by the reduction of 4-methoxynitrobenzene in sodium methanolate, at that time already a known reaction [25]. 4,4′-dimethoxyazoxybenzene remained one of the focal points for the LC chemist over some years [26]. Other azoxy compounds showing liquid crystalline behaviour were later synthesized, such as 4,4′-azoxychalcones, azoxy benzoates, and azoxy cinnamates, for which the first homologous series (of alkyl esters) was inves-

Table 1. Typical substances with liquid crystalline behaviour and their preparation (from D. Vorländer et al., before 1910).

Liquid crystal target molecules	Precursors, starting material
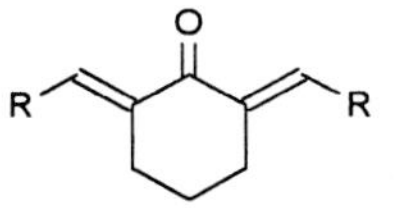 R–CH=C(C=O)C=CH–R (2,6-bis(arylidene)cyclohexanone)	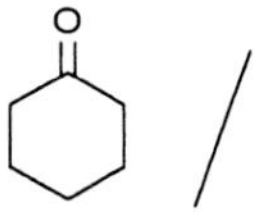 cyclohexanone / RCHO (Aldol-condensation) Ref.: [31,43e,f]
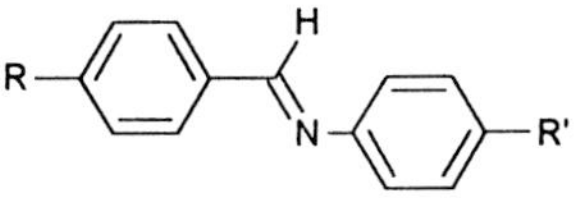 R–C₆H₄–CH=N–C₆H₄–R'	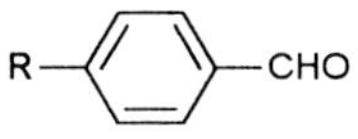 R–C₆H₄–CHO / H₂NPh-R' Ref.: [31a,d,32-34,42]
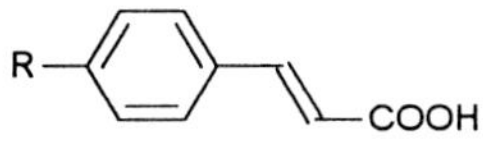 R–C₆H₄–CH=CH–COOH	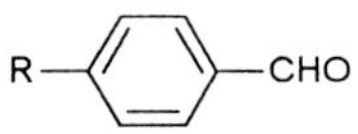 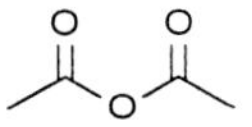  R–C₆H₄–CHO / (CH₃CO)₂O or HOOC–CH₂–COOH (Perkin) Ref.: [35,48]
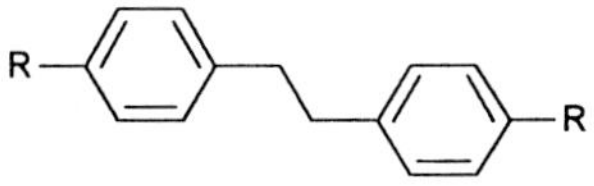 R–C₆H₄–CH₂–CH₂–C₆H₄–R	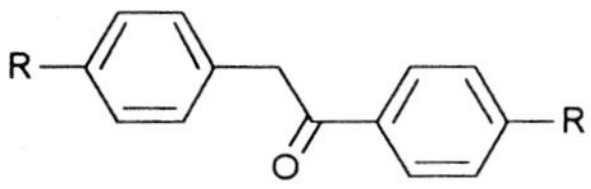 R–C₆H₄–CH₂–CO–C₆H₄–R (prepared by Friedel–Crafts-acylation) Ref.: [32d]
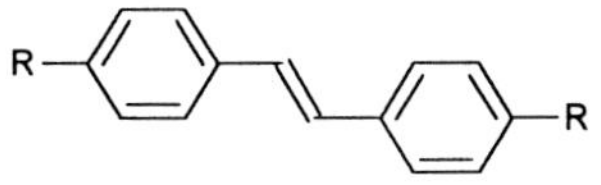 R–C₆H₄–CH=CH–C₆H₄–R	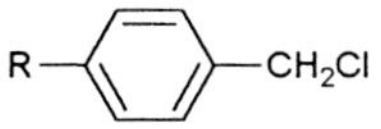 R–C₆H₄–CH₂Cl / NaOH Ref.: [36]
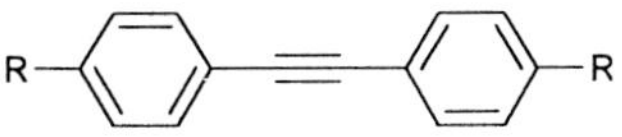 R–C₆H₄–C≡C–C₆H₄–R	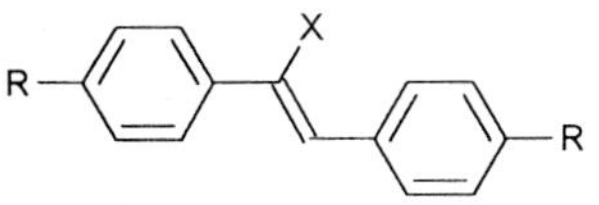 R–C₆H₄–CX=CH–C₆H₄–R / base commercially available tolanes were also used (1908) Ref.: [32d,35b]
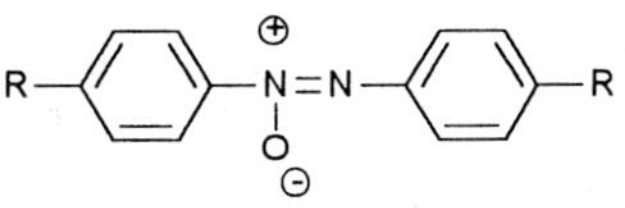 R–C₆H₄–N⁺(O⁻)=N–C₆H₄–R	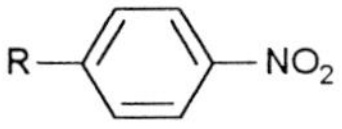 R–C₆H₄–NO₂ (Reductive coupling, using: NaOCH₃, or As₂O₃, OH⁻) Ref.: [24]

tigated [27]. With the turn of the century, D. Vorländer started his historic methodical studies on the synthesis of LC compounds, taking advantage of the 'lego-scheme' [28], that is, systematically exchanging one part (block) in the molecule for another. In an overview on liquid-crystalline substances by D. Vorländer [29] in 1908, a number of key chemical structures can already be noted, some of which are shown in Table 1 as are typical synthetic procedures of the time. By 1924 D. Vorländer [30] had synthesized over 1000 LCs and was able to draw general conclusions about their phase behaviour from their molecular structures.

Many structures included an aromatic subunit. At the time, synthetic aromatic chemistry was at its height and electrophilic aromatic substitution reactions (S_E-reactions) such as Friedel–Crafts alkylations [37] and acylations [38], nitrations [39] and diazotisations [40] were effectively used. Vorländer and others centred most procedures for the elongation of rod-like structures on carbonyl-reactions. It must be noted, however, that the scope of most of these reactions had not been realized until the late 1940s. Typical preparations of imines/anils (azines) [41] via aldehyde/ketone–amine/aniline (hydrazine) condensations have remained standard synthetic methods of today [42] (Scheme 3).

Olefinations (preparations of alkenes) by Knoevenagel and aldol condensations [43], and the Perkin reaction [44] have been complemented by Wittig [45] or Horner olefinations [46] and organometallic C–C coupling reactions (Heck reaction) [6, 47]. Complications resulting from a product mixture of E/Z isomers had been noted early on and it had been shown that only the E-isomers are mesomorphic [48]. Wittig reactions are known to give E- and Z-alkenes, while Knoevenagel condensations with aromatic aldehydes give exclusively trans- or (E)-products. Nevertheless, much work on the Wittig olefination has subsequently been devoted to the selective preparation of E- or Z-alkenes by the right choice of solvent, temperature and base system [49]. In many cases a chemical or photochemical isomerization of E/Z mixtures leads to a selective conversion to the desired (E)-isomer [50].

Although Vorländer included studies of molecules with cyclohexane core units [51], heterocyclic systems, and biaryls, it was not until after a much later date that those systems were explored more systematically. Thus in the 1920s, not many versatile reactions were known for the preparation of cyclohexanes [52]. It was only when the hydrogenation of arenes, with concurrent problems of cis-/trans-isomerization of 1,4-substituents in the hydrogenated products, was solved that more extensive research on LC compounds with cyclohexyl units developed in the 1970s. Heterocyclic compounds were investigated systematically by Schubert and Zaschke, who continued the city of Halle's tradition of LC research, in the 1950s and 1960s. Gray studied intensively

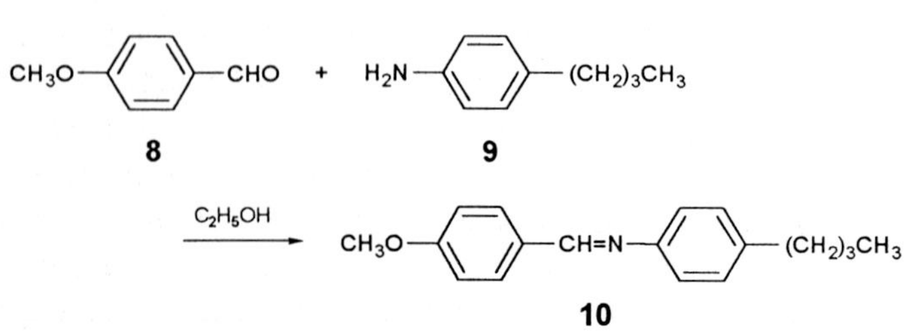

Scheme 3. MBBA (4'-methoxybenzylidene-4-butylaniline) [7].

Perkin-Reaction

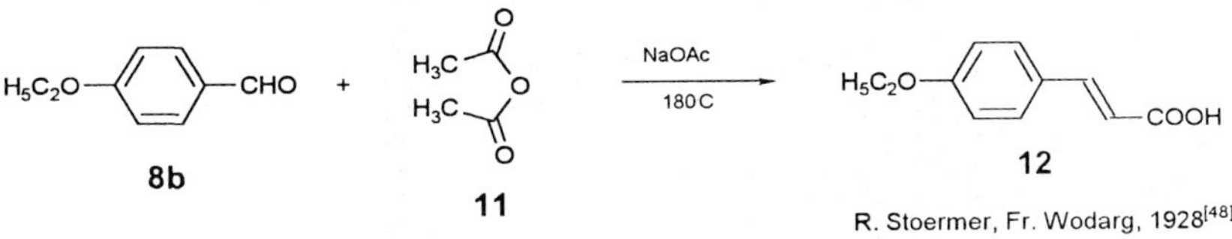

R. Stoermer, Fr. Wodarg, 1928[48]

Horner-Emmons-Olefination

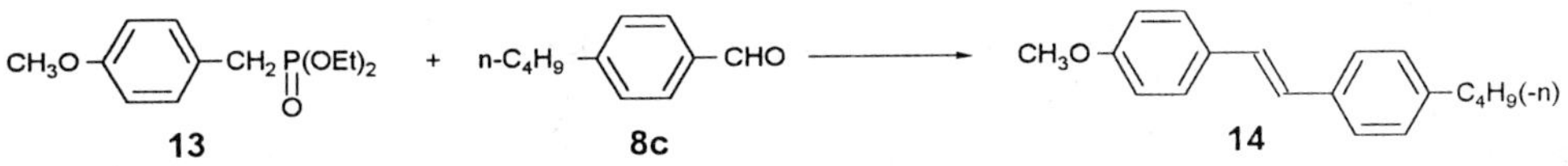

R. Steinsträßer, 1972[98d]

(only trans) (51%)

Heck-Coupling

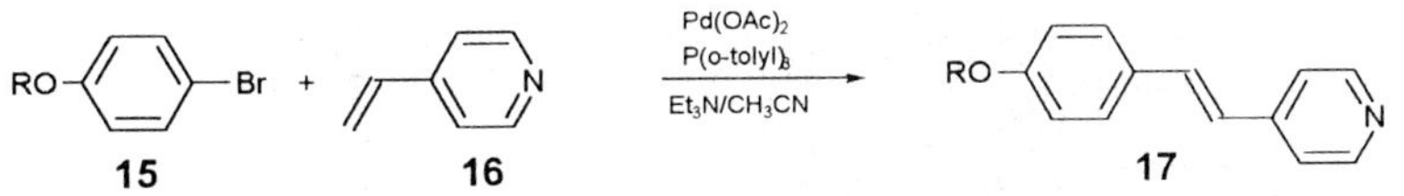

D.W.Bruce et al., 1988[47e]

Scheme 4. Typical methods of olefination.

a large variety of carbocycles including rationalization of the effects of lateral substituents. The preparation of these and other more recent compounds will be covered in the next chapter.

4 Common Reactions in Liquid Crystal Synthesis

The following paragraphs on aryl/aryl(heteroaryl)- and aryl/alkyl-bond formation centre on organometallic mediated C–C coupling reactions. For want of space, certain methods have had to be omitted and only those most commonly used have been included. Indeed, in certain cases combinations of organometallic reagents and metal catalysts that are not listed here may be preferable. For a more complete list of possible methods of organometallic C–C bond formation reactions, the reader is referred to references [1b] and [6].

4.1 Methods of Aryl–Aryl Bond Formation

Compounds with biaryl, triaryl and aryl/heteroaryl substructures account for one of the largest groups of LC materials. In former years, the parent compounds biphenyl, terphenyl and related arenes and heteroarenes were used as precursors and were selectively functionalized [51], for the most part by S_E-reactions, to give the desired liquid-crystalline materials. In more recent years, prefunctionalized aromatic compounds are joined by metal catalysed C–C coupling reactions [6, 53]. Historically, one of the first of such reactions was the Ullmann synthesis of biaryls [54a], in which two molecules of aryl halide react in the presence of finely divided copper [54]. Although reaction temperatures are quite high and yields are not always good, a number of functional groups (F [55], alkyl, alkoxy, NO_2, ROOC) are tolerated. Other groups, however, deactivate the process (NH_2, OH, COOH, bulky R). For the most part the synthesis has been employed for symmetrical biaryls, although unsymmetrical biaryls have also been prepared [56]. Nevertheless, the Ullmann synthesis has been used effectively in the preparation of LC material.

With the developments in organo-palladium and -nickel chemistry, a plethora of C–C coupling reactions have been developed for the synthesis of biaryls and homologues [6, 53]. Thus haloarenes can be coupled with aryl-magnesium [58], -zinc [58c, 58f], -lithium [6], -tin [59], -mercury [59] -copper [59], -zirconium [59], -titanium [6], and -fluorosilanes [60] in the presence of palladium [61a] [(PPh$_3$)$_2$Pd(Ph)I] [61b] or nickel complexes [Ni(acac)$_2$ [58a], NidppCl$_2$ [58b], Ni(PPh$_3$)$_4$ [58b], Ni(PPh$_3$)$_4$

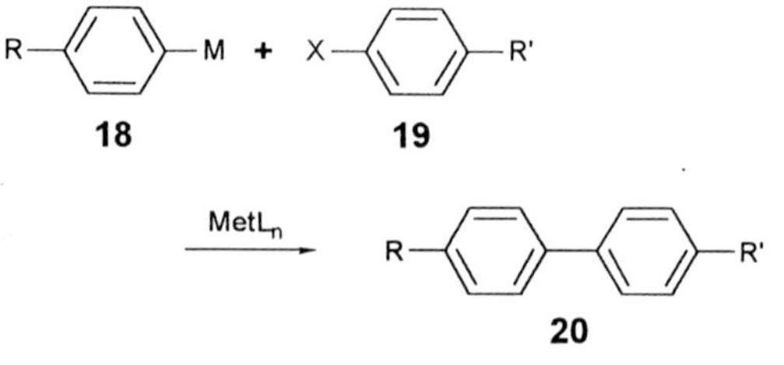

M = Mg,[58] Zn,[58c,f] Li,[6] Sn,[59] Hg,[59] Cu,[59] Zr,[59] Ti,[6] SiF$_2$[60]
Met = Ni, Pd, Pt, Rh
X = Hal, OSO$_2$CF$_3$, OCF$_3$

Scheme 5. Metal catalysed aryl-aryl coupling reactions.

[58c]]. These methods have also been used extensively for the coupling of heteroarenes [62].

Electro-reductive reactions in the presence of Ni, Pd, Rh, Pt are also known [63], but usually [63b] are only applicable to the synthesis of symmetric biaryls.

In many instances, the scope of the reactions listed above is limited as to the functional groups that are unaffected. Except for arylstannanes and arylfluorosilanes, carbonyl functionalities are often not compatible with the reactions. A number of cross-coupling reactions are known of aryl triflates [64a] and arylfluorosulfonates [64b] with aryl stannanes catalysed by Pd(0)-complexes (Stille reaction) [65], aryl zinc chlorides [64b], and aryl boron compounds (boronic acids and esters, Suzuki reaction [66]). Aryl stannates are relatively stable to air and moisture and many functional groups are tolerated. The coupling reaction using Ar–X/Ar–SnR$_3$ can be accelerated by Cu(I) [67] or Ag(I) [68] catalysis.

Because of its simplicity and versatility, due to the fact that most functional groups are not affected by the reaction, the Suzuki coupling has become one of the most frequently used C–C forming reactions for the preparation of biaryls, aryl/heteroaryls, biheteroaryls and homologes in liquid crystal science [45b, 53, 69]. The corresponding aryl boronic acids can easily be prepared from aryl bromides by transmetallation with boric acid trimethyl ester. Other advantages of this reaction lie in the easy work-up, as there are virtually no side-products from homo-coupling as seen in the other reactions mentioned. In some coupling reactions Ar–B(OH)$_2$/Ar–X, especially when Pd(PPh$_3$)$_4$ is used as catalyst, by-products Ar–Ph were found [70], Ph stemming from the ligand of the catalyst. In these cases, the use of another palladium catalyst with different ligands Pd[P(tolyl)$_3$]$_4$, Pd(AsPh$_3$)$_4$ often gives better results [70]. Toyne and

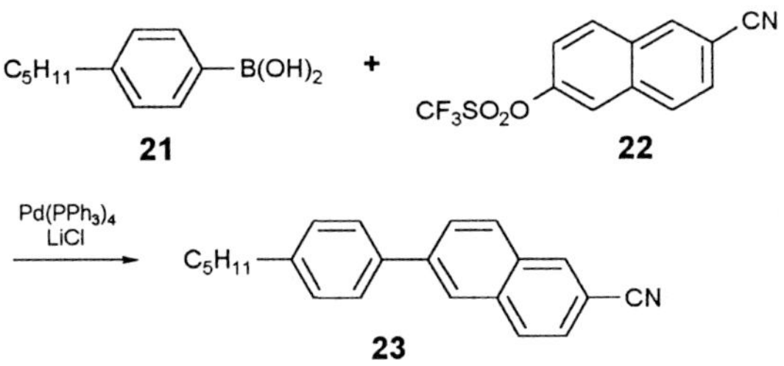

Scheme 6. After Hird et al. [53].

others have shown that it is possible to lithiate *ortho-* to a ring-fluorine [69b], thus enabling the transformation of fluoroaryl halides into the corresponding boronic acids and the subsequent C–C coupling. A most valuable selectivity of coupling has also been demonstrated using mixed dihalogenoarenes or halogenoaryl triflates. The Suzuki-coupling has also been used for the preparation of liquid-crystalline polymers, where either *p*-bromoaryl boronic acids are polymerized or aryl-1,4-diboronic acids and 1,4-dibromoarenes are copolymerized [71].

4.2 Alkyl-Functionalization of Arenes

The two most frequently used ways of introducing a *n*-alkyl chain into an aryl system are:

- Friedel–Crafts acylation with subsequent reduction of the keto-functionality;
- C–C cross coupling of arylhalides/triflates with alkylmetallates using palladium catalysis.

In contrast to Friedel–Crafts alkylations, which with longer chain *n*-alkyl halides usually give a substantial proportion of the corresponding secondary alkylarenes, Friedel–Crafts acylation also goes well with longer chain *n*-alkanoyl chlorides (and anhydrides). There are numerous methods for the reduction of the resulting arylketones. Most often used are reductions with $LiAlH_4$/$AlCl_3$ [72], Et_3SiH/CF_3COOH [58e, 73], Et_3SiH/CF_3SO_3H [74] (for reduction of hindered or electron-deficient arylketones), the classical Wolff–Kishner reduction [75] and its Huang–Minlon modification [76], although the high reaction temperatures required in the last of these are not compatible with all other functional groups. $BH_3 \cdot Py$/CF_3COOH [77], $NaBH_4$/$BF_3 \cdot Et_2O$ [78], and $NaBH_4$/CF_3COOH [79] have also been used as reducing agents.

Alkyl Grignard reagents prepared by standard methods from alkyl bromides undergo cross-coupling reactions with aryl halides and tosylates using palladium ($PdCl_2dppf$) catalysis [80]. Alkylboranes also undergo this reaction, where the reactivity gradation of the leaving group is $I > Br > OTf \gg Cl$ as shown in Scheme 7. This allows for an introduction of an alkyl chain to an aryl system with the retention of a second, less reactive leaving group which can be used in a subsequent coupling reaction with a different fragment [66a].

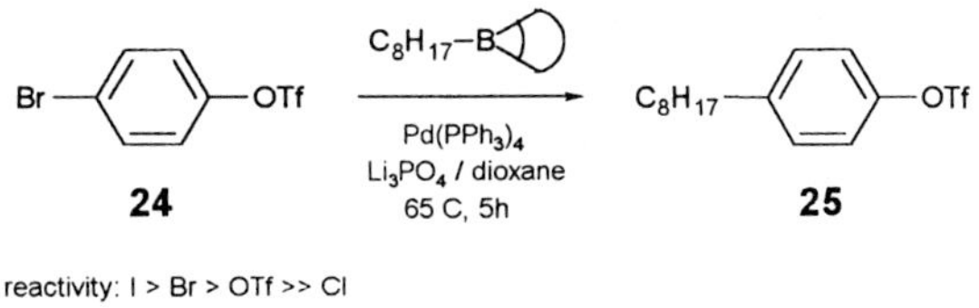

reactivity: I > Br > OTf >> Cl

Scheme 7. After Oh-e et al. [66a].

4.3 Aryl–Cycloalkyl Linkages

There are two major routes for the preparation of directly linked aryl–cycloalkyl substructures:

- the direct linkage of the two substructures;
- the construction of the alicycle from a precursor already linked to the arene.

4.3.1 4-E-Substituted Cyclohexylarenes

A synthesis of 1,4-E-disubstituted arylcyclohexanes via direct transition metal (e.g., Pd, Ni) catalysed cross-coupling [6] poses major problems due to β-elimination and lack of chemo- and stereo-selectivity. In principle, metal cycloalkyls can be reacted with aryl halides or tosylates. Nevertheless, for many sec-halides, even the preparation of the cycloalkyl metals leads to side-products stemming from β-elimination, especially in the case of direct metallation to cycloalkyl-zinc, -aluminum, and -tin. Cycloalkyl-magnesium and -lithium are more stable, albeit less reactive in the coupling reactions. Bis(cycloalkyl)zinc can be prepared by in situ transmetallation of the cycloalkyl-lithium intermediate [6], obtained by treating the cycloalkyl halide with lithium metal and zinc chloride using ultrasound [81]. Bis(cycloalkyl)zinc undergoes cross-coupling with various aryl-bromides [6]. The use of bidentate nickel and palladium catalysts with fixed cis-configuration, such as 1,1'-bis(diphenylphosphino)ferrocene-palladium(II)dichloride [$PdCl_2$(dppf)] and 1,2-bis(diphenylphosphino)ethane-nickel(II)dichloride [$NiCl_2$(dppe)] suppresses β-elimination [82]. An inverse addition of readily accessible arylmetals to cycloalkyl halides cannot be recommended in general, as the oxidative addition of the Pd(0) species to the cycloalkyl halide is slow [83] and β-elimination is more likely to occur.

A frequent method of coupling aryl fragments to cycloalkanes is the addition of aryllithium [84] or arylmagnesium compounds (Grignard reagents) [45b, 85] to cyclohexanones with subsequent dehydration of the ensuing tertiary alcohols to the corresponding alkenes, which are then hydrogenated. Reactive functionalities on both the aryl and the cycloalkyl substrates, such as formyl, carboalkoxy or alkanoyl, have to be protected before the coupling or have to be introduced at a later stage of the reaction sequence.

In the Nenitzescu reaction, an alkanoyl moiety and an aryl group are added simultaneously to a cyclohexyl ring at positions 1,4 [86]. The reaction is not well researched in its scope and is more widely used to prepare 1,3-disubstituted cyclopentanes (see below). Nevertheless in the synthesis of 4-alkylcyclohexylarenes it can be useful, as the substrates (arene, alkanoyl chloride and cyclohexene) are readily accessible.

As an example of a route of the second type (see Sect. 4.3.1), the Diels–Alder reaction is one of the classical ways of synthesizing six-membered ring systems. 2-Aryl-substituted dienes react with mono-activated alkenes (alkenes possessing one electron-withdrawing group such as cyano, alkanoyl, or alkoxycarbonyl) to form regioselectively 1,4-substituted arylcyclohexenes [87], which can be hydrogenated to the corresponding arylcyclohexanes.

4.3.2 3-Substituted Arylcyclopentanes

The Nenitzescu reaction [88] of alkanoyl chloride, cyclopentene, and an arene component directly leads to 3-alkanoyl substituted arylcyclopentanes. The scope of the reaction, as with the cyclohexene analogues (see above), has not been researched sufficiently. The alkanoyl functionality can be reduced to alkyl. The carbonyl group can also be used to construct further ring systems.

Grignard reaction of aryl-magnesium or -lithium on 3-substituted cyclopentanones leads to 3-substituted 1-arylcyclopentanols, which can be further transformed to arylcyclopentanes via the corresponding arylcyclopentenes, as described above for 3-substituted cyclopentanones. Substrates for the Grignard reaction, can often be furnished from cyclopentenone (Michael addition) or from 4-substituted cyclohexanones (via oxidative cleavage and ring contraction).

4.4 Linking Blocks via Simultaneous Construction of Ethenyl and Ethynyl Bridges: Tolanes and Ethenes

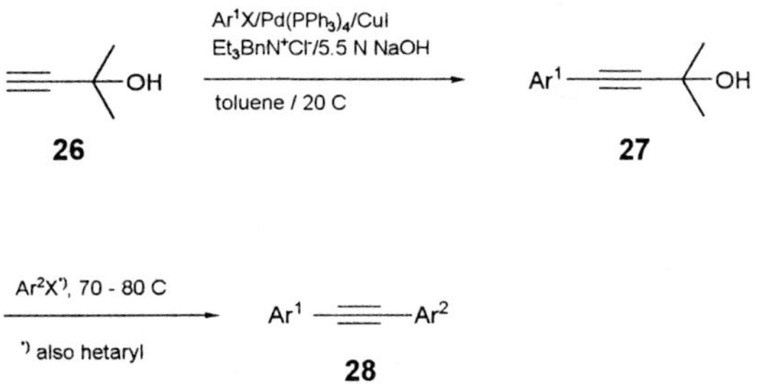

Scheme 8. After Ames et al. [92a], deprotection of 27 to give $Ar^1C \equiv CH$ occurs with refl. NaOH (aq) and after Carpita et al. [92b].

For many decades the standard method of preparation of diphenylacetylenes (tolanes) based upon the dehydrohalogenation of diphenylbromoethenes or diphenyldibromoethanes [89], which are easily available by hydrobromination and bromination, respectively, of the corresponding stilbenes [90]. This method of preparation can be used for diheteroarylacetylenes and phenylheteroarylacetylenes, usually with equal ease.

In recent years a multitude of metal-promoted C–C coupling reactions of arene-1-alkynes to haloarenes and haloheteroarenes [91] have been reported. A prerequisite for the use of these reactions is the ready availability of the corresponding arene-1-alkynes. Bromo- or iodo-arenes with trimethylsilylacetylenes in the presence of CuI and palladium/triphenylphosphine complex give trimethylsilylethynylarenes which are readily desilylated to the arene-1-acetylenes [69a]. Alternatively, aryl halides can be reacted with commercially available 1-methylbut-3-yn-2-ol with subsequent deprotection of **27** [92].

This second method, run as a one-pot reaction of 2-methylbut-3-yn-2-ol (**26**) with two different aryl components Ar^1X and Ar^2X, added at different temperatures, directly leads to diaryl/diheteroylacetylenes **28** [92] (see Scheme 8).

When coupled to arenes by palladium catalysis, arene-1-alkynes are often used as their metallates. For this purpose the alkynes are first lithiated and subsequently transmetallated. Although a wide range of alkynes has been reacted under these conditions, certain electrophilic functional groups undergo transformation and are not stable under these conditions.

Arene-1-alkynes can be coupled to aryl halides (iodides and bromides) in the presence of palladium phosphine complexes to yield tolanes directly [6, 93]. Reaction conditions usually are mild and many functional groups are tolerated.

Also, multiple coupling reactions are known to proceed in high yields and have been used in the preparation of discotic compounds [15, 94, 95].

The preparation of substituted diarylethenes, stilbenes, and analogues (e.g., stilbazoles) has been achieved by Wittig [45] or Horner–Emmons reactions [46] of substituted benzaldehydes and benzyltriphenylphosphonium halides or benzyldiethoxyphosphonates (see Scheme 4). 4-Methylpicolinium salts [96a,b] or 4-methylpicoline-N-oxides [96c], upon reaction with benzaldehydes under basic conditions lead directly to the corresponding stilbazolium salts or stilbazole-N-oxides, respectively. As of late, the metal-catalysed cross-coupling of vinylarenes, styrenes, and aryl halides, known as the Heck reaction [47], has been used frequently in the synthesis of styrenes. As with multiple acetylenations, multiple vinylations of aryl oligohalides are possible.

5 Building Blocks and Their Precursors

The building blocks for liquid crystals and their precursors are shown in Table 2 A–G.

Table 2 (A). Diarylethenes/Bisheteroarylethenes.

Target structure	Precursors of target	Precursors of precursor [a]
R—R' number of LCs with the given structure [b]	R-Z / R-Y starting materials for the given target structure; in most cases means R-Y + R-Z under the conditions found in the ref. given. R-Z ⇒ R-X R-X is precursor of R-Z R-X ⟶ R-Z R-X is reacted to give R-Z (R-X is precursor of R-Z)	R-L; R-N precursors of the compounds listed under heading 2 in the same line; the structure on the left (R-L) is the precursor of the left compound of heading 2, in this case of R-Z. R-Z ⇒ R-L R-L is precursor of R-Z R-L ⟶ R-Z R-L is reacted to give R-Z (R-L is precursor of R-Z)
 R'⎓R (996)	R'-CHO / Ph$_3$P$^+$CH$_2$R X$^-$ (Wittig-reaction) Ref.: [97]	R'CH$_2$X ⟶ R'-CHO where R'= Ar; Kroehnke-reaction R'-COOH, R'-CH$_2$OH R-CH$_2$X, where X = Br, Cl
	R'-CHO / (EtO)$_2$P(O)CH$_2$R (Horner-Emmons-reaction) Ref.: [98]	R'-COOH, R'-CH$_2$OH R-CH$_2$X, (Arbusov-reaction) Ref.: [99]
	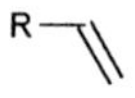 R⎓ R'-X; where R = Aryl, Heteroaryl and X = Br, I, OTf (Heck-reaction) Ref.: [100]	
	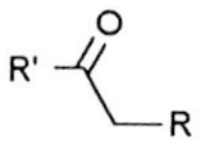 R'-C(O)-R (reduction to alcohol and subsequent elimination) Ref.: [101]	RCH$_2$COCl, where R' = Ar, Hetaryl (Friedel-Crafts-Acylation)

[a] Optional. [b] As taken from database Liqcryst [4].

Table 2 (B). Acetylenes, diacetylenes.

Target structure	Precursors of target	Precursors of precursor [a]
R—R' number of LCs with the given structure [b]	R-Z / R-Y starting materials for the given target structure; in most cases means R-Y + R-Z under the conditions found in the ref. given. R-Z ⇒ R-X R-X is precursor of R-Z R-X → R-Z R-X is reacted to give R-Z (R-X is precursor of R-Z)	R-L; R-N precursors of the compounds listed under heading 2 in the same line; the structure on the left (R-L) is the precursor of the left compound of heading 2, in this case of R-Z. R-Z ⇒ R-L R-L is precursor of R-Z R-L → R-Z R-L is reacted to give R-Z (R-L is precursor of R-Z)

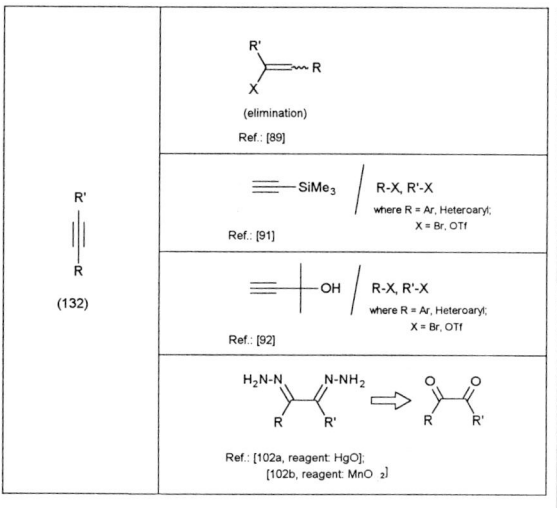

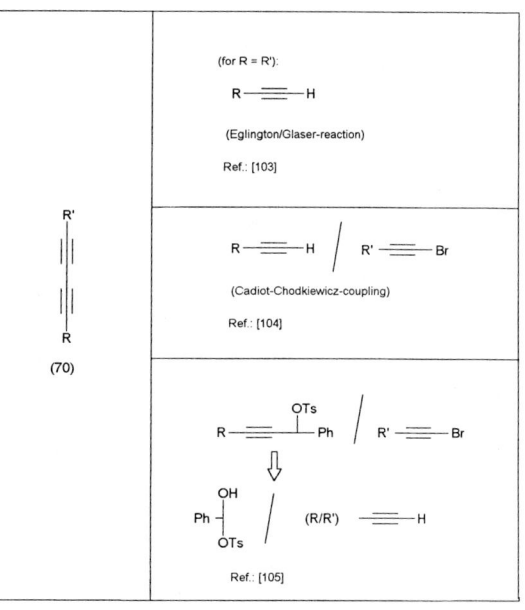

[a]) Optional.
[b]) As taken from database Liqcryst [4].

IV General Synthetic Strategies

Table 2 (C). Cyclohexanes.

Target structure	Precursors of target	Precursors of precursor [a]
R—R' number of LCs with the given structure [b]	R-Z / R-Y starting materials for the given target structure; in most cases means R-Y + R-Z under the conditions found in the ref. given. R-Z ⇨ R-X R-X is precursor of R-Z R-X ⟶ R-Z R-X is reacted to give R-Z (R-X is precursor of R-Z)	R-L; R-N precursors of the compounds listed under heading 2 in the same line; the structure on the left (R-L) is the precursor of the left compound of heading 2, in this case of R-Z. R-Z ⇨ R-L R-L is precursor of R-Z R-L ⟶ R-Z R-L is reacted to give R-Z (R-L is precursor of R-Z)
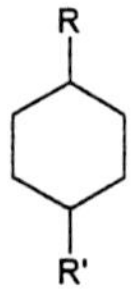 ![cyclohexane R/R'] (6967)	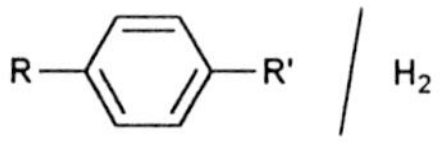 R—⟨phenyl⟩—R' / H$_2$ (reduction) Ref.: [106]	For the synthesis of 1,4-disubstituted phenylenes please consult chapters 4.1., 4.2., 4.3..
	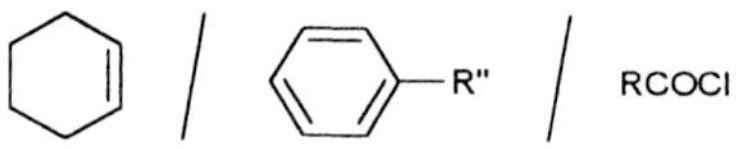 ⟨cyclohexene⟩ / ⟨phenyl⟩—R'' / RCOCl Nenitzescu-reaction Ref.: [86,88]	
	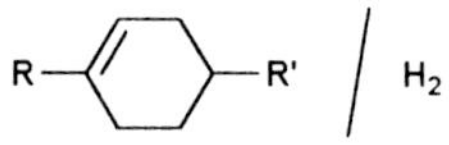 R—⟨cyclohexene⟩—R' / H$_2$ (reduction) Ref.: [87]	 R-diene / R'-ene [4+2]-cycloaddition for R'= COOEt, Ref.: [87]
	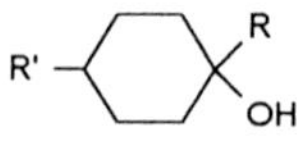 R'—⟨cyclohexane⟩—R,OH (dehydration/reduction) Ref.: [45b,85]	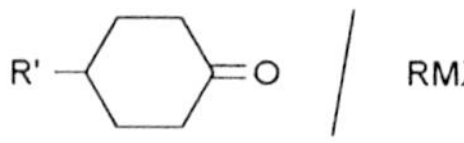 R'—⟨cyclohexanone⟩=O / RMX Grignard and related reactions Ref.: [45b,85]

[a]) Optional.
[b]) As taken from database Liqcryst [4].

5 Building Blocks and Their Precursors

Table 2 (D). Pyrimidines.

Target structure	Precursors of target	Precursors of precursor [a]
R—R' number of LCs with the given structure [b]	R-Z / R-Y starting materials for the given target structure; in most cases means R-Y + R-Z under the conditions found in the ref. given. R-Z ⇒ R-X R-X is precursor of R-Z R-X ⟶ R-Z R-X is reacted to give R-Z (R-X is precursor of R-Z)	R-L; R-N precursors of the compounds listed under heading 2 in the same line; the structure on the left (R-L) is the precursor of the left compound of heading 2, in this case of R-Z. R-Z ⇒ R-L R-L is precursor of R-Z R-L ⟶ R-Z R-L is reacted to give R-Z (R-L is precursor of R-Z)
Pyrimidine structure with R at 5-position, R' at 2-position (2384)	R'-C(NH·HCl)(NH₂) / OHC-CH(R)-CHOEt (CHNMe₂) Ref.: [107,108] R'-C(NH·HCl)(NH₂) / R-C=CH-N⁺(Me)=CH-NMe₂ ClO₄⁻ Ref.: [109] R'=R"S- can be modified subsequently to R'=-NC R'-C(NH·HCl)(NH₂) / R-CH(COOEt)₂ (subsequent reduction with POCl₃) Ref.: [110]	R-CH(CH(OEt)₂)₂ ; RCH₂COOH R-CH(CH(OEt)₂)₂ ; RCH₂COOH for R'=R"S- : R'Br + S=C(NH₂)₂ R-CH(CH(OEt)₂)₂ ; R-Hal + CH(COOEt)₂

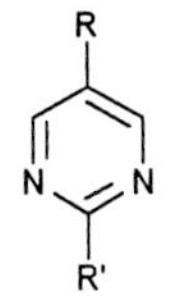

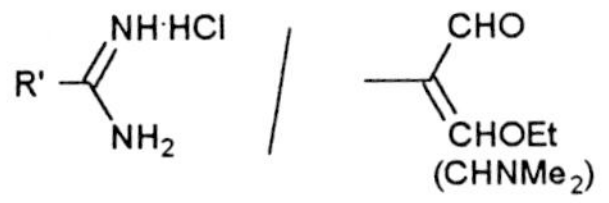

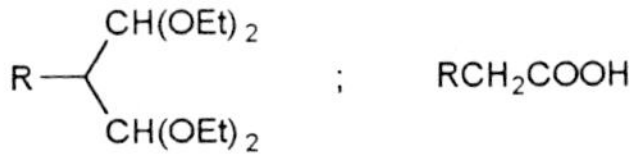

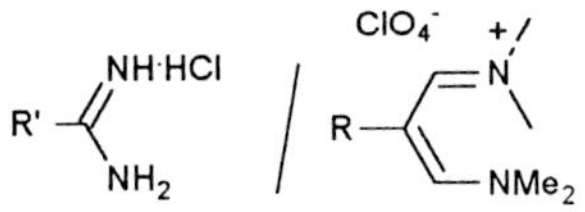

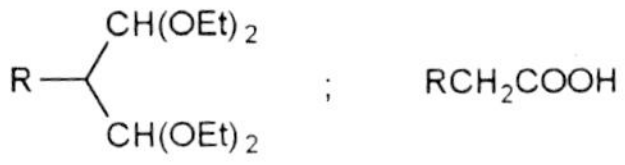

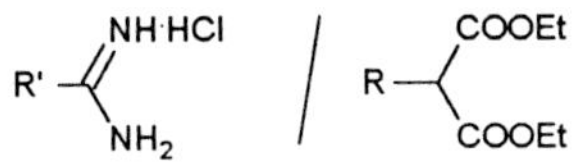

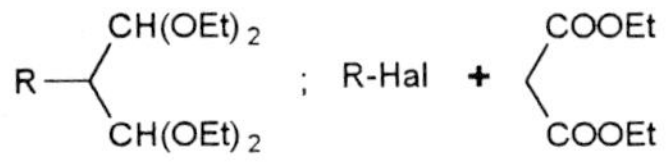

[a]) Optional.
[b]) As taken from database Liqcryst [4].

Table 2 (E). Triazines/pyrazines.

Target structure	Precursors of target	Precursors of precursor [a]
R—R' number of LCs with the given structure [b]	R-Z / R-Y starting materials for the given target structure; in most cases means R-Y + R-Z under the conditions found in the ref. given. R-Z ⇒ R-X R-X is precursor of R-Z R-X ⟶ R-Z R-X is reacted to give R-Z (R-X is precursor of R-Z)	R-L; R-N precursors of the compounds listed under heading 2 in the same line; the structure on the left (R-L) is the precursor of the left compound of heading 2, in this case of R-Z. R-Z ⇒ R-L R-L is precursor of R-Z R-L ⟶ R-Z R-L is reacted to give R-Z (R-L is precursor of R-Z)
Ar, R" triazine with R' (200)	(for R" = H) BrCH$_2$COAr / 2 × R'CONHNH$_2$ Ref.: [111] (for R" = H, Alkyl) [structure with Ar, R", N-O, R'] ⇒ [Ar–C(=NNH$_2$)–C(=NOH)–R'] + R"C(OEt)$_3$ Ref.: [112]	CH$_3$COAr / R'COOH [Ar–C(=O)–C(=NOH)–R'] ⟶ [Ar–C(=NNH$_2$)–C(=NOH)–R']
(Ar/R) pyrazine Ar' (251)	(for R = R'): Ref.: [113,(117)] 2 × ArCOCH$_2$NH$_2$ (not isolated, In situ reaction to pyrazine) ArCOCH$_2$NH$_3^+$Cl$^-$ Ref.: [114,115] ← ArCOCH$_2$N(phthalimide) → ArCOCH$_2$N$_3$ Ar'COCHO / H$_2$NCH$_2$CHRNH$_2$ Ref.: [(114),117] (only for R = OH:) Ar'COCHO / H$_2$NCOCH$_2$NH$_3^+$Cl$^-$ Ref.: [116]	NH$_3$ ← 2 × ArCOCH$_2$Hal Ref.: [115a] ArCOCH$_2$Hal Ref.: [115b]

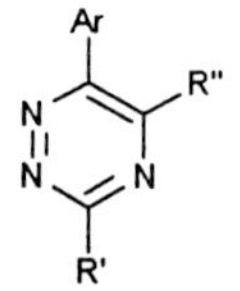

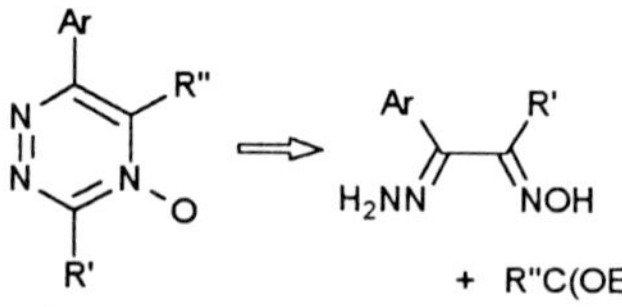

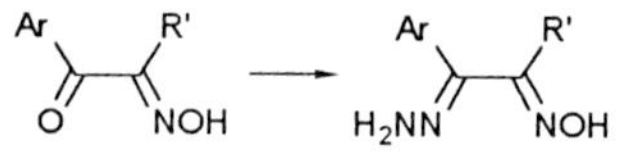

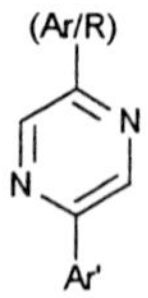

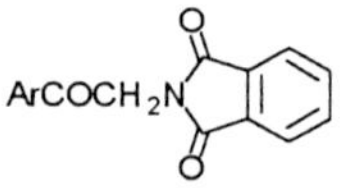

[a]) Optional.   [b]) As taken from database Liqcryst [4].

Table 2 (F). O/O-, S/O-, S/S-acetals, 1,3,2-dioxaborinanes, orthoesters.

Target structure R—R' number of LCs with the given structure [b]	Precursors of target R-Z / R-Y starting materials for the given target structure; in most cases means R-Y + R-Z under the conditions found in the ref. given. R-Z ⇒ R-X R-X is precursor of R-Z R-X ⟶ R-Z R-X is reacted to give R-Z (R-X is precursor of R-Z)	Precursors of precursor [a] R-L; R-N precursors of the compounds listed under heading 2 in the same line; the structure on the left (R-L) is the precursor of the left compound of heading 2, in this case of R-Z. R-Z ⇒ R-L R-L is precursor of R-Z R-L ⟶ R-Z R-L is reacted to give R-Z (R-L is precursor of R-Z)
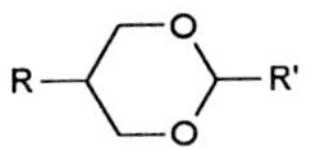 (809) 1,3-dioxane R—R'	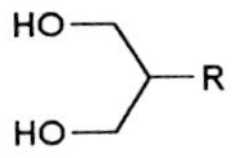 R'CHO / HO–⟨R⟩–OH Ref.: [118-120]	R'CH$_2$OH (Oxidation) / EtOOC–⟨R⟩–COOEt R'COOH (Reduction) 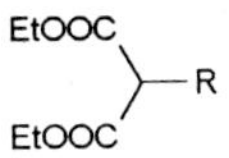 Ox.: ref. [1b, pp.607] Red.: ref. [1b, pp.619]
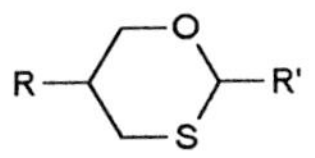 (70) 1,3-oxathiane R—R'	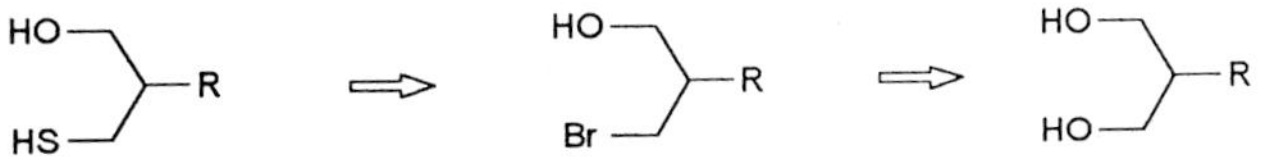 R'CHO / HO–⟨R⟩–SH Ref.: [120,121]	HO–⟨R⟩–Br ⇒ HO–⟨R⟩–OH
(74) 1,3-dithiane R—R'	R'CHO / HS–⟨R⟩–SH Ref.: [120-122]	Br–⟨R⟩–Br ⇒ HO–⟨R⟩–OH
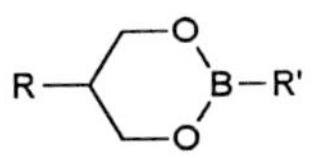 (212) 1,3,2-dioxaborinane R—B—R'	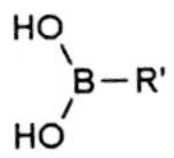 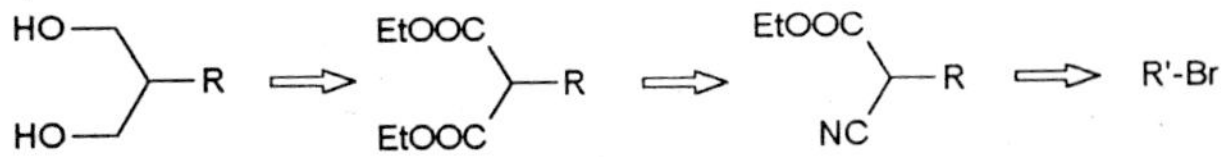 HO–B(OH)–R' / HO–⟨R⟩–OH Ref.: [123a-c]	EtOOC–⟨R⟩–COOEt ⇒ EtOOC–⟨R⟩(NC)–R ⇒ R'-Br Ref.: [123d]
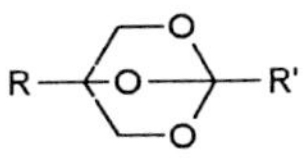 (29) orthoester R—⟨O,O,O⟩—R'	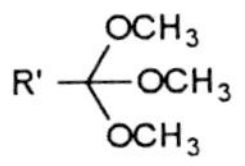 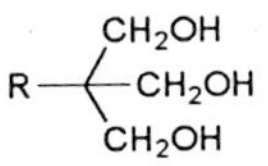 R'–C(OCH$_3$)$_3$ / R–C(CH$_2$OH)$_3$ Ref.: [124]	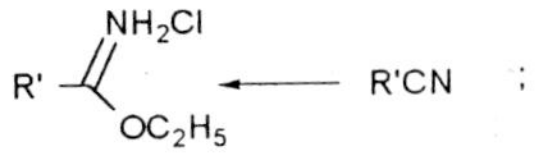 R'–C(=NH$_2$Cl)(OC$_2$H$_5$) ← R'CN ; R–C(COOEt)$_2$(COOEt) or: (CH$_3$O)$_3$CN + R'MgBr

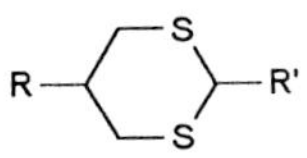

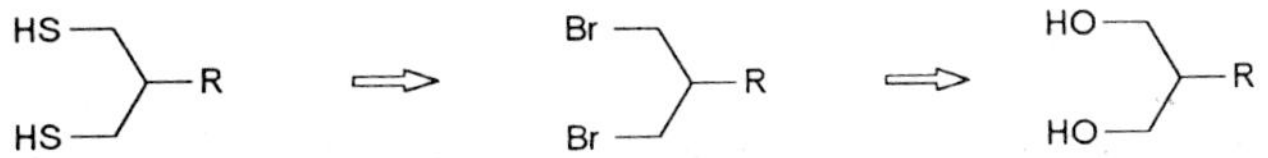

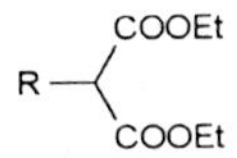

[a]) Optional. [b]) As taken from database Liqcryst [4].

Table 2 (G). Thiophenes, furans, pyrazines, thiadiazoles.

Target structure R—R' number of LCs with the given structure [b]	Precursors of target R-Z / R-Y starting materials for the given target structure; in most cases means R-Y + R-Z under the conditions found in the ref. given. R-Z ⇒ R-X R-X is precursor of R-Z R-X ⟶ R-Z R-X is reacted to give R-Z (R-X is precursor of R-Z)	Precursors of precursor [a] R-L; R-N precursors of the compounds listed under heading 2 in the same line; the structure on the left (R-L) is the precursor of the left compound of heading 2, in this case of R-Z. R-Z ⇒ R-L R-L is precursor of R-Z R-L ⟶ R-Z R-L is reacted to give R-Z (R-L is precursor of R-Z)
R—[thiophene-2,5-diyl]—R' (631)	P_4S_{10} or Lawesson's reagent [a] ref.:[125, 131]	diketone R-CO-CH_2-CO-R'
	R-Y / X—[thiophene]—R' Y = C≡CH [62a, 126a] for R = Ar, Hetaryl, see [62a]	
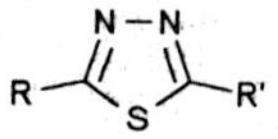 R—[1,3,4-thiadiazole]—R' (386)	S or Lawesson's reagent 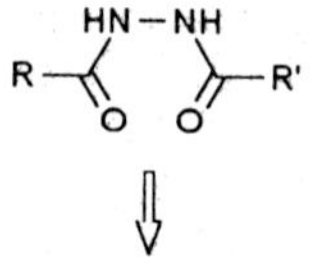 ref.:[11, 127, 128, 129] HN—NH / R-CO-NH-NH-CO-R' ⇓ R-COCl + R'CONHNH$_2$	
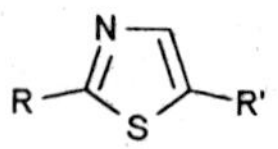 R—[thiazole]—R' (41)	P_4S_{10} or Lawesson's reagent 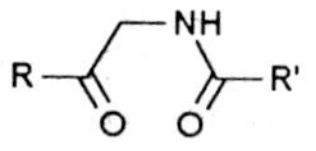 ref.:[130] R-CO-CH$_2$-NH-CO-R'	
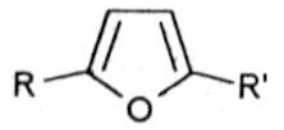 R—[furan-2,5-diyl]—R' (89)	$POCl_3$  ref.:[131] R-CO-CH$_2$-CO-R' 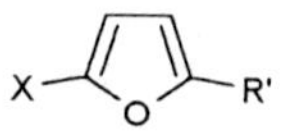 R-Y / X—[furan]—R' Y = C≡CH [132] for R = Aryl, see [62a]	
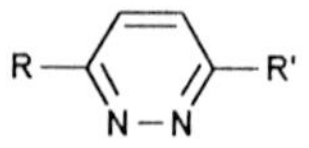 R—[pyridazine]—R' (N—N) (250)	$H_2NNH_2 \cdot H_2O$ 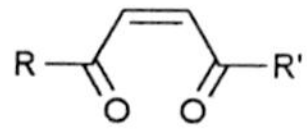 ref.:[133] R-CO-CH=CH-CO-R'	

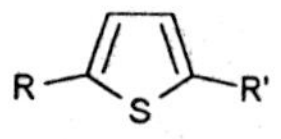

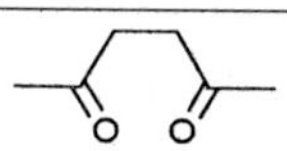

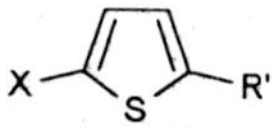

[a]) Optional.
[b]) As taken from database Liqcryst [4].

6 Chirality: the Preparation and Use of Enantiomers

With the advent of ferroelectric liquid crystals [134], the development of leading structures encompassing at least one centre of chirality has become of ever increasing importance [135]. In fact, for most of the modern technical applications (twisted nematics, ferroelectrics, and colour pigments) as well as for many of today's academic research interests in liquid-crystalline phases (TGBA phases, blue phases, helical inversions) chiral compounds are needed. Of specific interest is the creation of centres of chirality with a substituent of high polarity, the ensuing dipole moment often leading to high spontaneous polarization (Ps) if the compound has a SmC phase. Steric fixation of groups on the stereocentre is often wanted and leads to short helical pitches and also often to high spontaneous polarization. In the following text, typical chiral substructures for liquid crystal materials and methods for their synthesis are summarized.

6.1 Synthesis

For the synthesis of chiral-liquid crystalline material virtually all known techniques have been used.

(1) Many companies offer a good selection of chiral compounds, both from the natural chiral pool [137] and those that are industrially made, non-natural products (see also Tables 3 and 4). Why make,

Table 3. Commercially available compounds used directly as chiral substructures in liquid crystal synthesis.

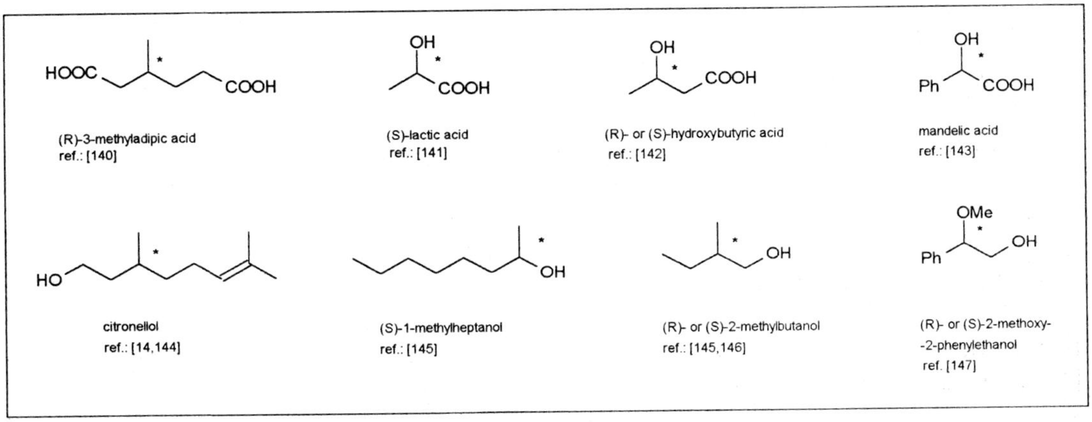

Table 4. Commercially available chiral starting materials for chiral substructures in liquid crystal synthesis.

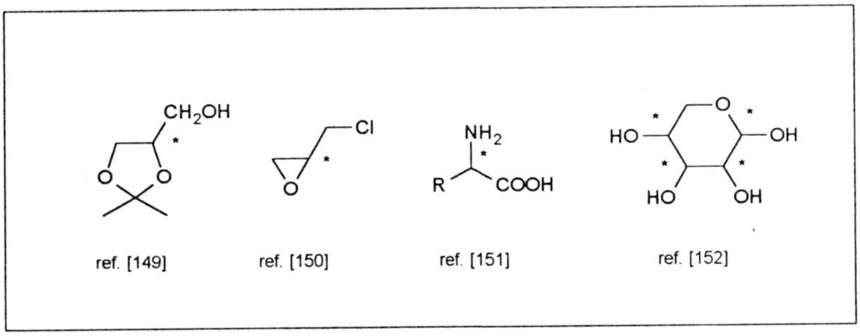

what you can buy at a reasonable price? There are several overviews on available chiral carbon fragments and their use in organic synthesis [136].
(2) Resolution of racemates by separation of intermediate diastereoisomers [137].
(3) Enantioselective synthesis [138]; this includes preparations using chiral auxiliaries, chiral synthetic catalysts [139] and enzymatic transformations.

6.2 Esterification: Chiral Alkanoic Acids and Chiral Alkanols

In many cases the wing group carrying the chiral centre has been linked to the mesogenic core by esterification of either a chiral alkanoic acid or an alkanol. Often, commercially available chiral material is used (Table 3). For alkanoic acids, techniques for the resolution of synthetically prepared racemates are well established [137] and have been used in the preparation of chiral substructures in liquid crystalline material [148]. α-Haloalkanoic acids have found frequent application [153]. An easy access to these structures is given by the one-step diazotation/halogenation of natural α-amino acids. The reaction proceeds with retention of configuration. Fluorination with pyridinium polyhydrogen fluoride (Py/HF 30:70) [154a] induces rearrangements in some substrates (Val, Iso-leu, Phe-Ala, Tyr, Thr) [154b]. These anchimerically assisted rearrangements can be suppressed by using a lower HF concentration (Py/HF 52:48) [154c]. Another synthetic method uses α-hydroxy esters as precursors. Transformation into the α-bromoester and subsequent treatment with Amberlite IRA 900 [F$^-$] and Amberlyst-A 26 [F$^-$] leads by double inversion to the chiral α-fluoro ester with retention of the chiral centre. Alternatively, mesylation of the hydroxy ester with subsequent reaction with fluoride ions on anion exchange resins leads to the chiral α-fluoro esters with an inversion of configuration [155]. Non-activated alkanols can be fluorinated with DAST with inversion of configuration. Fluorinations of this sort have also been achieved with Bu$_4$NF.

The α-haloalkanoic acids have been reduced to the corresponding 2-haloalkanols and these have been esterified to yield various mesogenic cores with chiral terminal groups [146, 156]. Another rapid access to enantiomerically pure 2-chloro- and 2-fluoro-alkanols uses the stereoselective addition of Hal-equivalents (SiF$_4$, ^{i}Pr$_2$NH) to chiral epoxides [157]. While there are possibilities of preparing chiral epoxides directly from alk-1-enes [158], a wealth of enantiomerically pure epoxides is commercially available. Epoxides have also been reacted with C-nucleophiles to give chiral alcohols with concomitant C–C linkage [159].

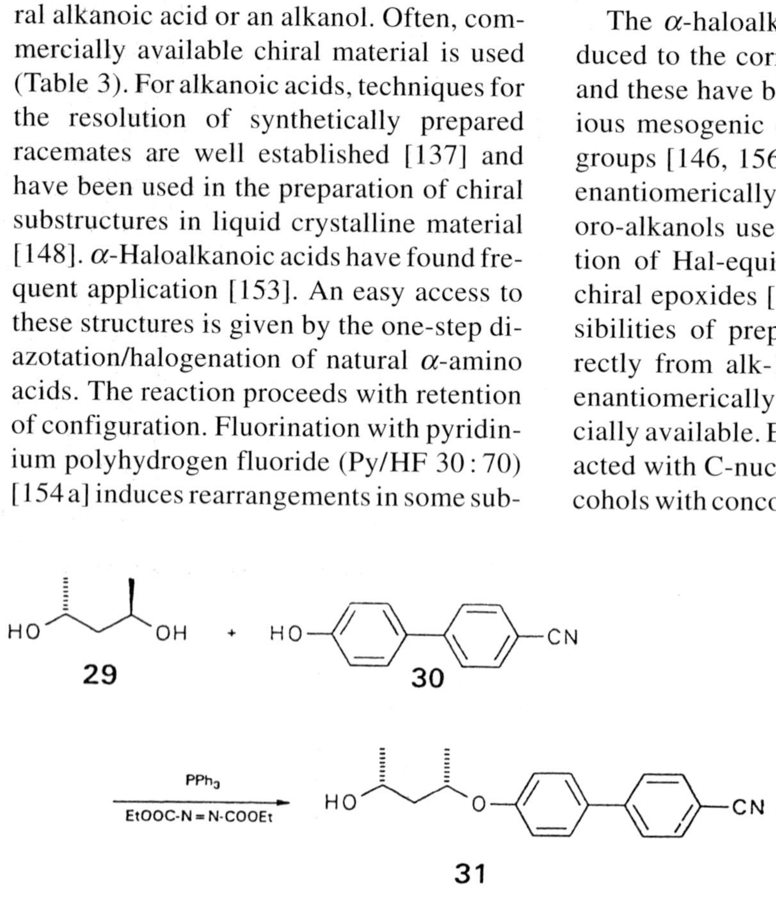

Scheme 9. Etherification with inversion of configuration at the reaction centre (after Shibata et al. [160b]).

6.3 Etherification

Chiral alkanols have also been linked as ethers. Of special interest for etherification is the Mitsunobu reaction [160], which can be used for inversion of configuration at the reaction centre of the chiral alcohol component (Scheme 9).

7 Liquid Crystal Synthesis in Education and a Note on the Purification of Liquid Crystals

The study of LCs is valuable and suitable for educational purposes. Often the synthesis of structures of great educational interest exhibiting liquid-crystalline behaviour is uncomplicated from the synthetic point of view and can readily be handled by students. Articles have been devoted specifically to the use of LCs for educational purposes and detailed procedures [161] for classical compounds have been included (e.g. cholesteryl nonanoate [161b], cholesteryl chloride [161b], and MBBA [161a]). Moreover, novel compounds can be reached by exchanging the building blocks (e.g. homologation) without alteration of the general method and procedure for their synthesis. The synthesis can be complimented by the study of structure–property relationships, the effects of solvents and optical and other properties.

Purification of LC compounds should be done with utmost care [162]. Thus, the phase transition temperatures are much more dependent on impurities than are melting points. For many compounds a simple purification by column chromatography is not sufficient and subsequent recrystallizations are recommended, until constant transition temperatures are reached. Usually samples can easily be checked for inhomogeneity, and for example, the behaviour of isolated droplets of a substance should be identical. Also, commercial products not specifically designated for use as LCs may have to be recrystallized before use. For amphiphilic LCs the purification of any synthetic precursor is recommended, followed by a simple final step towards the LC. This avoids the need for purification of the final amphiphilic products with which micelle formation generally prohibits normal purification procedures.

8 References

[1] (a) J. Fuhrhop, G. Penzlin, *Organic Synthesis: Concepts, Methods, Starting Materials*, VCH, Weinheim, 2nd ed., **1994**; (b) R. C. Larock, *Comprehensive Organic Transformations – A Guide to Functional Group Preparations*, VCH, NY **1989**.
[2] (a) Houben-Weyl, *Methoden der organischen Chemie*, Thieme, Stuttgart; (b) S. Patai (Ed.) *The Chemistry of Functional Groups*, Interscience Publishers, New York.
[3] (a) REACCS (Reaction Access System): Molecular Design Ltd., 2132 Farallon Drive, San Leandro, CA 94577; Molecular Design MDA AG, Mühlebachweg 9, CH-4123 Allschwil 2; (b) SYNLIB (Synthetic Library): Distributed Chemical Graphics.
[4] *LiqCryst 2.0 – Database of Liquid Crystalline Compounds for Personal Computers*, V. Vill, Fujitsu Kyushu System (FQS) Ltd, Fukuoka **1996**; LCI Publisher, Hamburg **1996**.
[5] R. Schroeder, H. Kränz, V. Vill, B. Meyer, *J. Chem. Soc., Perkin Trans.* **1996**, *2*, 1685.
[6] E. Poetsch, *Kontakte (Darmstadt)* **1988**, *2*, 15–28.
[7] H. Kelker, B. Scheurle, *Angew. Chem.* **1969**, *81*, 903–904; *Angew. Chem., Int. Ed. Engl.* **1969**, *8*, 884–885.
[8] Y. Zhang, G. B. Schuster, *J. Org. Chem.* **1994**, *59*, 1855–1862.
[9] K. Ohta, A. Ishii, H. Muroki, I. Yamamoto, K. Matsuzaki, *Mol. Cryst. Liq. Cryst.* **1985**, *116*, 299–307.

[10] J. Bartulin, R. Martinez, H. J. Müller, Z. X. Fan, W. Haase, *Mol. Cryst. Liq. Cryst. Sci. Technol. Sect. A* **1992**, *220*, 67–75.
[11] C. Tschierske, R. Girdziunaite, *J. Prakt. Chem.* **1991**, *333*, 135–137.
[12] L. Yu, J. M. Wu, *Mol. Cryst. Liq. Cryst.* **1991**, *204*, 37–42.
[13] K. Seto, S. Takahashi, T. Tahara, *J. Chem. Soc., Chem. Commun.* **1985**, 122–123.
[14] S.-I. Sugita, H. Takeno, T. Teraji, *Mol. Cryst. Liq. Cryst.* **1991**, *206*, 139–146.
[15] (a) C. Pugh, V. Percec, *J. Mater. Chem.* **1991**, *1*, 765–773; (b) B. Kohne, K. Praefcke, *Chimia* **1987**, *41*, 196–198.
[16] Excellent reviews on the history of liquid crystals are given by H. Kelker: (a) H. Kelker, *Mol. Cryst. Liq. Cryst.* **1988**, *165*, 1–43; (b) H. Kelker, *Mol. Cryst. Liq. Cryst.* **1973**, *21*, 1–48; see also (c) V. Vill, *Mol. Cryst. Liq. Cryst.* **1992**, *213*, 67–71.
[17] R. Virchow, *Virchows Arch.* **1854**, *6*, 571.
[18] W. Heintz, *J. Prakt. Chem.* **1855**, *66*, 1–81.
[19] (a) D. Vorländer, *Ber. Dtsch. Chem. Ges.* **1910**, *43*, 3120; (b) D. Vorländer, W. Selke, *Z. Phys. Chem.* **1928**, *129*, 435.
[20] P. Planer, *Liebigs Ann. Chem.* **1861**, *118*, 25–27.
[21] (a) F. Reinitzer, *Monatsh. Chem.* **1888**, *9*, 421–441; (b) H. Falk, P. Laggner, *Ö. Chem. Zeitg.* **1988**, *9*, 251–258.
[22] Both compounds had been prepared before: cholesteryl acetate (a) W. Löbisch, *Ber. Dtsch. Chem. Ges.* **1872**, *5*, 510–514; (b) O. Berthelot, *Ann. Chim. Phys.* **1874**, *56*, 54; (c) E. Schulze, *J. Prakt. Chem.* **1873**, *7*, 163–178.
[23] (a) T. Thiemann, V. Vill, *J. Phys. Chem. Ref. Data* **1997**, *26*, 291; (b) G. A. Jeffrey, L. M. Wingert, *Liq. Cryst.* **1992**, *12*, 179–202; (c) H. Prade, R. Mietchen, V. Vill, *J. Prakt. Chem.* **1995**, *337*, 427–448; (d) V. Vill, H.-W. Tunger, *Liebigs Ann.* **1995**, 1055–1059.
[24] (a) L. Gattermann, A. Ritschke, *Ber. Dtsch. Chem. Ges.* **1890**, *23*, 1738–1750; (b) H. Loesner, *J. Prakt. Chem.* **1894**, *50*, 563–567.
[25] (a) H. Klinger, *Ber. Dtsch. Chem. Ges.* **1882**, *15*, 865–867; (b) H. Klinger, R. Pitschke, *Ber. Dtsch. Chem. Ges.* **1885**, *18*, 2551–2556.
[26] (a) Th. Rotarski, *Ber. Dtsch. Chem. Ges.* **1903**, *36*, 3158–3163; (b) F. Meyer, K. Dahlem, *Liebigs Ann. Chem.* **1903**, *326*, 331–346.
[27] D. Vorländer, *Ber. Dtsch. Chem. Ges.* **1906**, *39*, 803–810.
[28] Lego is a trade-mark of LEGO System A/S, DK 71000 Billund, Denmark.
[29] D. Vorländer, *Kristallinisch-flüssige Substanzen*, Sammlung chemischer und chemisch-technischer Vorträge **1908**, *12*, 321–402.
[30] D. Vorländer, *Chemische Kristallographie der Flüssigkeiten*, Akademische Verlagsgemeinschaft, Leipzig **1924**.
[31] (a) V. Bertleff, Ph. D. Thesis, Halle **1908**; (b) D. Vorländer, *Ber. Dtsch. Chem. Ges.* **1907**, *40*, 1415–1432; (c) E. Wolferts, Ph. D. Thesis, Halle **1909**; (d) M. E. Huth, Ph. D. Thesis, Halle **1909**; (e) D. Vorländer, *Ber. Dtsch. Chem. Ges.* **1921**, *54*, 2261–2264; (f) D. Vorländer, *Naturwiss.* **1936**, *24*, 113–117.
[32] (a) P. Hansen, Ph. D. Thesis, Halle **1907**; (b) O. Meye, Ph. D. Thesis, Halle **1909**; (c) C. Reichardt, Ph. D. Thesis, Halle **1909**; (d) F. Janecke, Ph. D. Thesis, Halle **1910**.
[33] G. Bredig, G. v. Schukowsky, *Ber. Dtsch. Chem. Ges.* **1904**, *37*, 3419–3425.
[34] L. Gattermann, *Liebigs Ann.* **1906**, *347*, 351.
[35] (a) P. van Romburgh, *Proc. Kon. Nederl. Akad. Wetenschapen* **1901**, *9*, 9; (b) J. E. Hulme, Ph. D. Thesis, Halle **1907**.
[36] A. Gahren, Ph. D. Thesis, Halle **1908**.
[37] (a) F. Asinger, H. H. Vogel, *Houben-Weyl* **1970**, *5/1a*, 501–539; (b) G. A. Olah, *Friedel–Crafts and Related Reactions*, Vol. 2, Interscience, NY **1964**; (c) G. A. Olah, *Friedel–Crafts Chemistry*, Wiley, NY **1979**; (d) C. C. Price, *Org. React.* **1946**, *3*, 1–82; (e) R. M. Roberts, A. A. Khalif, *Friedel–Crafts Alkylation Chemistry*, Dekker, NY **1984**.
[38] (a) E. Berliner, *Org. React.* **1949**, *5*, 229–289; (b) P. H. Gore, *Chem. Rev.* **1955**, *55*, 229–281; (c) G. A. Olah, *Friedel–Crafts and Related Reactions*, Vol. 3, Interscience, NY **1965**; (d) C. W. Schellhammer, *Houben-Weyl* **1973**, *7/2a*, 15–378; (e) C. Friedel, J. M. Crafts, *C. R. Hebd. Séances Acad. Sci.* **1877**, *84*, 1392, 1450.
[39] (a) K. Schofield, *Aromatic Nitration*, Cambridge University Press, Cambridge **1980**; (b) W. Seidenfaden, D. Pawellek, *Houben-Weyl* **1979**, *10/1*, 479–818.
[40] (a) D. Vorländer, F. Meyer, *Liebigs Ann. Chem.* **1901**, *320*, 122–144; (b) F. Meyer, Ph. D. Thesis, Halle **1900**.
[41] R. W. Layer, *Chem. Rev.* **1963**, *63*, 489–510.
[42] (a) D. Vorländer, *Trans. Faraday Soc.* **1933**, *29*, 907–910; (b) C. Wiegand, *Z. Naturforschg.* **1954**, *B9*, 516–518; (c) G. W. Gray, J. B. Hartley, A. Ibbotson, B. Jones, *J. Chem. Soc.* **1955**, 4359–4368; (d) H. Kelker, B. Scheurle, R. Hatz, W. Bartsch, *Angew. Chem.* **1970**, *82*, 984–985; *Angew. Chem., Int. Ed. Engl.* **1970**, *9*, 962–963; (e) S. Sakagami, M. Nakamizo, *Bull. Chem. Soc. Jpn.* **1977**, *50*, 1009–1010; (f) M. H. Wann, G. S. Harbison, *J. Am. Chem. Soc.* **1989**, *111*, 7273–7274; (g) J. Malthête, *Adv. Mater.* **1990**, *2*, 150–151; (h) N. Hoshino, R. Hayakawa, T. Shibuya, Y. Matsunaga, *Inorg. Chem.* **1990**, *29*, 5129–5131.
[43] (a) O. Bayer, *Houben-Weyl* **1954**, *7/1*, 76–92; (b) T. Mukaiyama, *Org. React.* **1982**, *28*, 203–332; (c) A. T. Nielsen, W. J. Houlihan, *Org. React.* **1968**, *16*, 1–438. (d) E. Knoevenagel,

Ber. Dtsch. Chem. Ges. **1896**, *29*, 172–174; (e) K. Hanemann, Ph. D. Thesis, Halle **1972**; (f) K. Hanemann, H. Schubert, D. Demus, G. Pelzl, DD 115.283 (1975); *Chem. Abstr.* **1977**, *87*, 76418j; (g) Y. Matsunaga, S. Miyamoto, *Mol. Cryst. Liq. Cryst.* **1993**, *237*, 311–317.

[44] (a) H. Henecka, E. Ott, *Houben-Weyl* **1952**, *8*, 442–450; (b) J. R. Johnson, *Org. React.* **1942**, *1*, 210–265; (c) W. H. Perkin, *J. Chem. Soc.* **1868**, *21*, 53, 181.

[45] (a) H. J. Bestmann, *Pure Appl. Chem.* **1979**, *51*, 515–533 (review); (b) S. M. Kelly, A. Germann, R. Buchecker, M. Schadt, *Liq. Cryst.* **1994**, *16*, 67–93.

[46] W. S. Wadsworth, *Org. React.* **1977**, *25*, 73–253 (review).

[47] (a) R. F. Heck, *Org. React.* **1982**, *27*, 345–398 (review); (b) S. M. Kelly, J. Fünfschilling, A. Villiger, *Liq. Cryst.* **1994**, *16*, 813–829; (c) T. Richardson, A. Topacli, W. H. A. Mujid, M. B. Greenwood, D. W. Bruce, A. Thornton, J. R. Marsden, *Adv. Mat. f. Optics and Electronics* **1994**, *4*, 243–251; (d) H. Adams, N. A. Bailey, D. W. Bruce, S. A. Hudson, J. R. Marsden, *Liq. Cryst.* **1994**, *16*, 643–653; (e) D. W. Bruce, D. A. Dunmur, E. Lalinde, P. M. Maitlis, P. Styring, *Liq. Cryst.* **1988**, *3*, 385–395.

[48] R. Stoermer, Fr. Wodarg, *Ber. Dt. Chem. Ges.* **1928**, *61*, 2323–2330.

[49] M. Schlosser, *Topics Stereochem.* **1970**, *5*, 1–30.

[50] (a) P. E. Sonnet, *Tetrahedron* **1980**, *36*, 557–604 (review); (b) G. M. Wyman, *Chem. Rev.* **1955**, *55*, 625–657.

[51] D. Vorländer, *Ber. Dtsch. Chem. Ges.* **1910**, *43*, 3120–3135.

[52] At that time the most important precursor of cyclohexyl-units, cyclohexanone, was prepared by hydrogenation of phenol.

[53] M. Hird, G. W. Gray, K. J. Toyne, *Mol. Cryst. Liq. Cryst.* **1991**, *206*, 187–204.

[54] (a) F. Ullmann, J. Bielicki, *Ber. Dtsch. Chem. Ges.* **1901**, *34*, 2174; (b) P. E. Fanta, *Synthesis* **1974**, 9–21; (c) P. E. Fanta, *Chem. Rev.* **1964**, *64*, 613–632; (d) P. E. Fanta, *Chem. Rev.* **1946**, *38*, 139–196.

[55] D. E. Fenton, A. J. Park, D. Shaw, A. G. Massey, *Tetrahedron Lett.* **1964**, 949–950.

[56] P. E. Fanta, *Synthesis* **1974**, 13–14.

[57] (a) H. Dehne, R. Zahnow, H. G. Steinhagen, *Z. Chem.* **1971**, *11*, 305–306; (b) C. Vauchier, F. Vinet, N. Maiser, *Liq. Cryst.* **1989**, *5*, 141–151; (c) C. Vauchier, F. Vinet, EP 0.236.215 (1987); *Chem. Abstr.* **1988**, *108*, 104094y.

[58] (a) R. J. P. Corriu, J. P. Masse, *J. Chem. Soc. Chem. Commun.* **1972**, 144; (b) K. Tamao, A. Minato, M. Miyake, T. Matsuda, Y. Kiso, M. Kumada, *Chem. Lett.* **1975**, 133–136; (c) E. Negishi, A. O. King, N. Okukado, *J. Org. Chem.* **1977**, *42*, 1821–1823; (d) U. Lauk, P. Skrabal, H. Zollinger, *Helv. Chim. Acta* **1981**, *64*, 1847–1848; (e) U. Lauk, P. Skrabal, H. Zollinger, *Helv. Chim. Acta* **1983**, *66*, 1574–1575; (f) E. Poetsch, V. Meyer, H. Böttcher, DBP3 736 489 (1987) as cit. in [6]; *Chem. Abstr.* **1990**, *112*, 88951a; (g) H.-J. Altenbach in *Organic Synthesis Highlights* (Eds.: J. Mulzer, H.-J. Altenbach, M. Braun, K. Krohn, H.-U. Reissig), VCH, Weinheim **1991**, 181–185.

[59] I. P. Beletskaya, *J. Organomet. Chem.* **1983**, *250*, 551–564.

[60] Y. Hatanaka, S. Fukushima, T. Hiyama, *Chem. Lett.* **1989**, 1711–1714.

[61] (a) D. A. Widdowson, Y.-Z. Zhang, *Tetrahedron* **1986**, *42*, 2111–2116; (b) A. Sekiya, N. Ishikawa, *J. Organomet. Chem.* **1976**, *118*, 349–354.

[62] (a) V. N. Kalin, *Synthesis* **1992**, 413–432; (b) J. Malm, P. Björk, S. Gronowitz, A.-B. Hörnfeldt, *Tetrahedron Lett.* **1992**, *33*, 2199–2202.

[63] (a) S. Torii, *Synthesis* **1986**, 873–886; (b) S. Torii, H. Tanaka, K. Morisaki, *Tetrahedron Lett.* **1985**, *26*, 1655–1658.

[64] (a) K. Ritter, *Synthesis* **1993**, 735–762; (b) G. P. Roth, C. E. Fuller, *J. Org. Chem.* **1991**, *56*, 3493–3496.

[65] A. M. Echavarren, J. K. Stille, *J. Am. Chem. Soc.* **1987**, *109*, 5478–5486.

[66] (a) T. Oh-e, N. Miyaura, A. Suzuki, *Synlett* **1990**, 221–223; (b) T. Watanabe, N. Miyaura, A. Suzuki, *Synlett* **1992**, 207–210.

[67] V. Farina, S. Kapadia, B. Krishnan, C. Wang, L. S. Liebeskind, *J. Org. Chem.* **1994**, *59*, 5905–5911.

[68] (a) A. Huth, I. Beetz, I. Schumann, *Tetrahedron* **1989**, 6679–6682; (b) J. M. Fu, V. Snieckus, *Tetrahedron Lett.* **1990**, *31*, 1665–1668.

[69] G. W. Gray, M. Hird, D. Lacey, K. J. Toyne, *J. Chem. Soc., Perkin Trans. 2* **1989**, 2041–2053.

[70] (a) D. F. O-Keefe, M. C. Dannock, S. M. Marcuccio, *Tetrahedron Lett.* **1992**, *33*, 6679–6680; (b) D. K. Morita, J. K. Stille, J. R. Norton, *J. Am. Chem. Soc.* **1995**, *117*, 8576–8581.

[71] (a) H. Witteler, G. Lieser, G. Wegner, M. Schulze, *Makromol. Chem. Rapid Commun.* **1993**, *14*, 471–480; (b) M. Rehahn, A.-D. Schlüter, G. Wegner, W. J. Feast, *Polymer* **1993**, *30*, 1060–1062; (c) M. Rehahn, A.-D. Schlüter, G. Wegner, *Makromol. Chem.* **1990**, *191*, 1991–2003.

[72] (a) R. F. Nystrom, C. R. A. Berger, *J. Am. Chem. Soc.* **1958**, *80*, 2896–2898; (b) V. Percec, M. Zuber, S. Z. D. Cheng, A.-Q. Zhang, *J. Mater. Chem.* **1992**, *2*, 407–414.

[73] (a) Ch. T. West, St. J. Donnolly, D. A Kooistra, M. P. Doyle, *J. Org. Chem.* **1973**, *38*, 2675–

2681; (b) J. L. Butcher, D. J. Byron, S. N. R. Shirazi, A. R. Tajbakhsh, R. C. Wilson, *Mol. Cryst. Liq. Cryst.* **1991**, *199*, 327–343.

[74] G. A. Olah, M. Arvanaghi, L. Ohannesian, *Synthesis* **1986**, 770–772.

[75] (a) D. Todd, *Org. React.* **1948**, *4*, 378–422; (b) A. L. Bailey, G. S. Bates, *Mol. Cryst. Liq. Cryst.* **1991**, *198*, 417–428.

[76] (a) Huang-Minlon, *J. Am. Chem. Soc.* **1946**, *68*, 2487–2488; **1949**, *71*, 3301–3303; (b) J. Malthête, J. Canceill, J. Gabard, J. Jacques, *Tetrahedron* **1981**, *37*, 2815–2821; (c) M. Chambers, R. Clemitson, D. Coates, S. Greenfield, J. A. Jenner, I. C. Sage, *Liq. Cryst.* **1989**, *5*, 153–158.

[77] Y. Kikugawa, Y. Ogawa, *Chem. Pharm. Bull.* **1979**, *27*, 2405–2410.

[78] G. R. Pettit, B. Green, P. Hofer, D. C. Ayres, P. J. S. Pauwels, *Proc. Chem. Soc.* **1962**, 357.

[79] D. M. Ketcha, G. W. Gribble, *J. Org. Chem.* **1985**, *50*, 5451–5457.

[80] (a) M. A. Tius, X. Gu, T. Truesdell, S. Savariar, P. P. Crooker, *Synthesis* **1988**, 36–40; (b) A. L. Bailey, G. S. Bates, *Mol. Cryst. Liq. Cryst.* **1992**, *298*, 417–428.

[81] (a) C. Petrier, J. L. Luche, C. Dupuy, *Tetrahedron Lett.* **1984**, *25*, 3463–3466; (b) C. Petrier, J. C. de Souza Barbosa, C. Dupuy, J. L. Luche, *J. Org. Chem.* **1985**, *50*, 5761–5765.

[82] (a) T. Hayashi, M. Konishi, Y. Kobori, M. Kumada, T. Higuchi, K. Hirotsu, *J. Am. Chem. Soc.* **1984**, *106*, 158–163; (b) K. Tamao, K. Sumitani, M. Kumada, *J. Am. Chem. Soc.* **1972**, *94*, 4374–4376; (c) K. Tamao, Y. Kiso, K. Sumitani, M. Kumada, *J. Am. Chem. Soc.* **1972**, *94*, 9268–9269.

[83] P. Fitton, M. P. Johnson, J. E. McKeon, *J. Chem. Soc., Chem. Commun.* **1968**, 6–7.

[84] (a) S. Sugimori, T. Kojima, Y. Goto, T. Isoyama, K. Nigorikawa, EP 01 19 756 (1984); *Chem. Abstr.* **1985**, *102*, 24260k; (b) M. Osman, EP 00 19 665 B2 (1987) as cited in [6].

[85] (a) S. M. Kelly, A. Germann, M. Schadt, *Liq. Cryst.* **1994**, *16*, 491–507; (b) V. S. Bezborodov et al., *Zh. Org. Khim.* **1983**, *19*, 1669–1674; *Chem. Abstr.* **1984**, *100*, 5948h.

[86] H. Zhao, X. Pei, L. Wang, *Quinghua Duaxue Xuebao* **1986**, *26*, 99–105; *Chem. Abstr.* **1987**, *107*, 87024k.

[87] V. S. Bezborodov, VPC 547 595 2.

[88] C. D. Nenitzescu, J. G. Gavat, *Liebigs Ann.* **1935**, *519*, 266.

[89] (a) V. Jäger in *Methoden der Organischen Chemie (Houben-Weyl)*, Vol. V/2a (Ed.: E. Müller), 4th edn, Thieme, Stuttgart **1977**, 53–168; (b) Th. Zincke, S. Münch, *Liebigs Ann. Chem.* **1905**, *335*, 157–191; (c) G. W. Gray, A. Mosley, *Mol. Cryst. Liq. Cryst.* **1976**, *37*, 213–231.

[90] (a) A. Roedig in *Methoden der Organischen Chemie (Houben-Weyl)*, Vol. V/4 (Ed.: E. Müller), 4th edn, Thieme, Stuttgart **1960**, 38–152; (b) G. R. Newkome, D. L. Koppersmith, *J. Org. Chem.* **1973**, *38*, 4461–4463 (for heteroanaloga).

[91] (a) T. Sakamoto, M. Shiraiwa, Y. Kondo, H. Yamanaka, *Synthesis* **1983**, 312–314; (b) S. Takahashi, Y. Kuroyama, K. Sonogashira, N. Hagihara, *Synthesis* **1980**, 627–630; (c) E. C. Taylor, P. S. Ray, *J. Org. Chem.* **1987**, *52*, 3997–4000.

[92] (a) D. E. Ames, D. Bull, C. Takundwa, *Synthesis* **1981**, 364–365; (b) A. Carpita, A. Lessi, R. Rossi, *Synthesis* **1984**, 571–572.

[93] Y. Xu, P. Fan, Q. Chen, J. Wen, *J. Chem. Res. (S)* **1994**, 240–241.

[94] R. Diercks, J. C. Armstrong, R. Boese, K. P. C. Vollhardt, *Angew. Chem.* **1986**, *98*, 270; *Angew. Chem., Int. Ed. Engl.* **1986**, *25*, 268.

[95] (a) K. Praefcke, D. Singer, B. Gündogan, K. Gutbier, M. Langner, *Ber. Bunsenges. Phys. Chem.* **1993**, *97*, 1358–1361; (b) K. Praefcke, B. Kohne, D. Singer, *Angew. Chem.* **1990**, *102*, 200–202; *Angew. Chem., Int. Ed. Engl.* **1990**, *29*, 177–179; (c) K. Praefcke, B. Kohne, K. Gutbier, N. Johnen, D. Singer, *Liq. Cryst.* **1989**, *5*, 233–249; (d) M. Ebert, D. A. Jungbauer, R. Kleppinger, J. H. Wendorff, B. Kohne, K. Praefcke, *Liq. Cryst.* **1989**, *4*, 53–67.

[96] (a) D. W. Bruce, R. G. Denning, M. Grayson, R. Legadec, K. K. Lai, B. T. Pickup, A. Thornton, *Adv. Mater. Opt. Electron.* **1994**, *4*, 293–301; (b) S. R. Marder, J. W. Perry, B. G. Tiemann, R. E. Marsh, W. P. Schaefer, *Chem. Mater.* **1990**, *2*, 685–690; (c) F. Tournilhac, J. F. Nicoud, J. Simon, P. Weber, P. Guillon, A. Skoulios, *Liq. Cryst.* **1987**, *2*, 55–61.

[97] (a) A. W. Johnson, *Ylid Chemistry*, Academic Press, New York **1979**; (b) H. J. Bestmann, O. Vostrowsky, *Top. Curr. Chem.* **1983**, *109*, 85–164; (c) for use of catalytic amounts of trialkylarsine for exclusive preparation of *E*-isomer: L. Shi, W. Wang, Y. Wang, Y.-Z. Huang, *J. Org. Chem.* **1989**, *54*, 2027–2028.

[98] (a) W. S. Wadsworth, *Org. React.* **1977**, *25*, 73–253; (b) W. J. Stec, *Acc. Chem. Res.* **1983**, *16*, 411–417; (c) J. Boutagy, R. Thomas, *Chem. Rev.* **1974**, *74*, 87–99; (d) R. Steinsträsser, *Z. Naturforschg.* **1972**, *27B*, 774–779.

[99] B. A. Arbuzov, *Pure Appl. Chem.* **1964**, *9*, 307–335.

[100] (a) R. F. Heck, *Acc. Chem. Res.* **1979**, *12*, 146–151; (b) S. M. Kelly, J. Fünfschilling, A. Villiger, *Liq. Cryst.* **1994**, *16*, 813–829; (c) J. E. Beecher, T. Durst, J. M. J. Frechet, A. Godt, A. Pangborn, D. R. Robello, C. S. Willand, D. J. Williams, *Adv. Mater.* **1993**, *5*, 632–634.

[101] (a) W. R. Young, A. Aviram, R. J. Cox, *Angew. Chem.* **1971**, *83*, 399–400; *Angew. Chem., Int. Ed. Engl.* **1971**, *10*, 410; (b) W. R. Young, A. Aviram, R. J. Cox, *J. Am. Chem. Soc.* **1972**, *94*, 3976–3981.

[102] (a) T. Curtius, *Chem. Ber.* **1889**, *22*, 2161 (reagent HgO); (b) V. Jäger in *Methoden der Organischen Chemie (Houben-Weyl)*, Vol. V/2a (Ed.: E. Müller), 4th edn, Thieme, Stuttgart **1974**, 187–193; (c) F. Wöhrle, H. Kropf, unpublished results (reagent MnO_2).

[103] (a) L. I. Simandi in *The Chemistry of Functional Groups, Supplement C* (Eds.: S. Patai, Z. Rappoport), pt. 1, Wiley, New York **1983**, 529–534; (b) P. Cadiot, W. Chodkiewicz in *Acetylenes* (Ed.: H. G. Viehe), Dekker, New York **1969**, 597–647; (c) G. Eglington, W. McCrae, *Adv. Org. Chem.* **1963**, *4*, 225–328.

[104] (a) W. Chodkiewicz, *Ann. Chim. (Paris)* **1957**, *2*, 819; (b) A. Sevin, W. Chodkiewicz, P. Cadiot, *Bull. Soc. Chim. Fr.* **1974**, 913–917; (c) T. M. Juang, Y. N. Chen, S. H. Lung, Y. H. Lu, C. S. Hsu, S. T. Wu, *Liq. Cryst.* **1993**, *15*, 529–540; (d) Y. H. Lu, C. S. Hsu, *Mol. Cryst. Liq. Cryst. Sci. Technol. Sect. A* **1993**, *225*, 1–14.

[105] T. Kitamura, C. H. Lee, H. Taniguchi, M. Matsumoto, Y. Sano, *J. Org. Chem.* **1994**, *59*, 8053–8057.

[106] (a) R. Krieg, H.-J. Deutscher, U. Baumeister, H. Hartung, M. Jaskolski, *Mol. Cryst. Liq. Cryst.* **1989**, *166*, 109–122; (b) W. Schäfer, K. Altmann, H. Zaschke, H.-J. Deutscher, *Mol. Cryst. Liq. Cryst.* **1983**, *95*, 63–70; (c) H. Zaschke, W. Schäfer, H.-J. Deutscher, D. Demus, G. Pelzl, DDR WPC 07e/224 160 (1980).

[107] A. Boller, M. Cereghetti, Hp. Scherrer, *Z. Naturforschg.* **1978**, *33b*, 433–438.

[108] (a) H. Zaschke, S. Arndt, V. Wagner, H. Schubert, *Z. Chem.* **1977**, *17*, 293–294; (b) G. Kraus, H. Zaschke, *J. Prakt. Chem.* **1981**, *323*, 199–206; (c) H. Zaschke, *Z. Chem.* **1977**, *17*, 63–64; (d) G. G. Urquhart, J. W. Gates, R. Connor, *Org. Synth.* **1941**, *21*, 36; *Org. Synth. Coll. Vol. III* **1955**, 363–365.

[109] (a) H. Schubert, H. Zaschke, *J. Prakt. Chem.* **1975**, *312*, 494–506; (b) X. Yu, G. Li, *J. East China Inst. Chem. Technol.* **1991**, *17*, 663.

[110] (a) Z. Arnold, F. Sorm, *Coll. Czech. Chem. Commun.* **1958**, *23*, 452–461; (b) Z. Arnold, *Coll. Czech. Chem. Commun.* **1973**, *38*, 1168–1172.

[111] (a) T. V. Saraswathi, V. R. Srinivasan, *Tetrahedron Lett.* **1971**, 2315–2316; (b) H. Zaschke, K. Nitsche, H. Schubert, *J. Prakt. Chem.* **1977**, *319*, 475–484; (c) S.-i. Sugita, S. Toda, T. Yoshiyasu, T. Teraji, A. Murayama, M. Ishikawa, *Mol. Cryst. Liq. Cryst.* **1993**, *237*, 319–328.

[112] H. Neunhoeffer, V. Böhnisch, *Tetrahedron Lett.* **1973**, 1429–1432.

[113] Fr. Tutin, *J. Chem. Soc. (London)* **1910**, *97*, 2495–2524.

[114] G. B. Barlin, *Heterocyclic Compounds, The Pyrazines*, Vol. 41 (Eds.: A. Weissberger, E. C. Taylor), 11–17, 28–35.

[115] (a) H. Schubert, R. Hacker, K. Kindermann, *J. Prakt. Chem.* **1968**, *37*, 12–20; (b) H. Bretschneider, H. Hörmann, *Monatsh. Chem.* **1953**, *84*, 1021–1032.

[116] M. Matsumoto, Y. Sano, T. Naguishi, S. Yoshinaga, K. Isomura, H. Taniguchi, *Senryo* **1992**, *37*, 12–25 (only for R = OH).

[117] W. Staedel, L. Rügheimer, *Ber. Dtsch. Chem. Ges.* **1876**, *9*, 563–564.

[118] (a) H. M. Vorbrodt, S. Deresch, H. Kresse, A. Wiegeleben, D. Demus, H. Zaschke, *J. Prakt. Chem.* **1981**, *323*, 902–913; (b) H. Zaschke, H. M. Vorbrodt, D. Demus, W. Weissflog, DD-WPat. 139 852 (1980); *Chem. Abstr.* **1980**, *93*, 213415w; (c) H. Mineta, T. Yui, Y. Gocho, T. Matsumoto, EP 559.088 (1993); *Chem. Abstr.* **1994**, *120*, 91030v.

[119] (a) A. Villiger, F. Leenhouts, *Mol. Cryst. Liq. Cryst.* **1991**, *209*, 297–307; (b) J. Szulc, K. Czuprynski, M. Dabrowski, J. Przedmojski, *Liq. Cryst.* **1993**, *14*, 1377–1387.

[120] (a) Y. Haramoto, H. Kamagawa, *Chem. Lett.* **1987**, 755–758; (b) Y. Haramoto, K. Kawashima, H. Kamagawa, *Bull. Chem. Soc. Jpn.* **1988**, *61*, 431–434; (c) Y. Haramoto, H. Kamagawa, *Bull. Chem. Soc. Jpn.* **1990**, *63*, 156–158; (d) Y. Haramoto, H. Kamogawa, *Bull. Chem. Soc. Jpn.* **1990**, *63*, 3063–3065.

[121] (a) Y. Haramoto, H. Kamogawa, *J. Chem. Soc., Chem. Commun.* **1983**, 75–76; (b) Y. Haramoto, M. Sano, H. Kamogawa, *Bull. Chem. Soc. Jpn.* **1986**, *59*, 1337–1340.

[122] (a) Y. Haramoto, Y. Tomita, H. Kamogawa, *Bull. Chem. Soc. Jpn.* **1986**, *59*, 3877–3880; (b) Y. Haramoto, A. Nobe, H. Kamogawa, *Bull. Chem. Soc. Jpn.* **1984**, *57*, 1966–1969; (c) Y. Haramoto, K. Akazawa, H. Kamogawa, *Bull. Chem. Soc. Jpn.* **1984**, *57*, 3173–3176; (d) Y. Haramoto, H. Kamogawa, *Bull. Chem. Soc. Jpn.* **1985**, *58*, 477–480.

[123] (a) K. Seto, S. Takahashi, T. Tahara, *J. Chem. Soc., Chem. Commun.* **1985**, 122–123; (b) V. S. Bezborodov, *Zh. Org. Khim.* **1989**, *25*, 2168–2170; *Chem. Abstr.* **1990**, *112*, 189424m; (c) M. Matsubara, K. Seto, T. Tahara, S. Takahashi, *Bull. Chem. Soc. Jpn.* **1989**, *62*, 3896–3901; (d) M. Uno, K. Seto, S. Takahashi, *J. Chem. Soc., Chem. Commun.* **1984**, 932–933.

[124] R. Paschke, H. Zaschke, A. Hauser, D. Demus, *Liq. Cryst.* **1989**, *6*, 397–407.

[125] (a) W. D. Rudorf in *Houben-Weyl XX*, E6a, 196–207; (b) E. Campaigne, W. O. Foye, *J. Org. Chem.* **1952**, *17*, 1405–1412.
[126] (a) H. A. Dieck, F. R. Heck, *J. Organomet. Chem.* **1975**, *93*, 259; (b) M. Altmann, V. Enkelmann, G. Lieser, U. Bunz, *Adv. Mater.* **1995**, *7*, 726–728 (polymer).
[127] (a) K. J. Reubke in *Houben-Weyl XX*, E8d, 189–304; (b) S.-i. Sugita, S. Toda, T. Teraji, *Mol. Cryst. Liq. Cryst.* **1993**, *237*, 33–38; (c) J. Hausschild, H. Kresse, H. Schubert, D. Demus, Ger. (East) 117.014 (1975); *Chem. Abstr.* **1976**, *84*, 187561.
[128] (a) D. Demus, H. Kresse, H. Zaschke, W. Schäfer, U. Rosenfeld, H. Stettin, DDP. 247 221 (1987); *Chem. Abstr.* **1988**, *109*, 14939s; (b) W. Schäfer, U. Rosenfeld, H. Zaschke, H. Stettin, H. Kresse, *J. Prakt. Chem.* **1989**, *331*, 631.
[129] (a) G. Gattow, S. Lotz, *Z. Anorg. Allg. Chem.* **1986**, *533*, 109; (b) G. W. Gray, R. M. Scrowston, K. J. Toyne, D. Lacey, A. Jackson, J. Krause, E. Poetsch, T. Geelhaar, G. W. L. Weber, A. E. F. Waechtler, DEP 3 712 995 (1988); *Chem. Abstr.* **1989**, *110*, 85748n.
[130] K. Dölling, H. Zaschke, H. Schubert, *J. Prakt. Chem.* **1979**, *321*, 643–654.
[131] H. Schubert, I. Sagitdinov, J. V. Svetkin, *Z. Chem.* **1975**, *15*, 222–223.
[132] (a) A. Pelter, M. Rowlands, G. Clements, *Synthesis* **1987**, 51–53; (b) A. Arcadi, A. Burini, S. Cacchi, M. Delmastro, F. Marinelli, B. Pietroni, *Synlett* **1990**, 47–48.
[133] (a) C. Paal, H. Schulze, *Ber. Dtsch. Chem. Ges.* **1900**, *33*, 3800; (b) C. Weygand, W. Lanzendorf, *J. Prakt. Chem.* **1938**, *151*, 221–226; (c) H. Schubert, R. Koch, Ch. Weinbrecher, *Z. Chem.* **1966**, *6*, 467.
[134] (a) D. M. Walba, *Adv. Synth. React. Solids* **1991**, *1*, 173–235 (review); (b) K. Skarp, M. A. Handschy, *Mol. Cryst. Liq. Cryst.* **1989**, *165*, 439–509 (review).
[135] J. W. Goodby, *J. Mater. Chem.* **1991**, *1*, 307–318.
[136] On available chiral carbon fragments: (a) J. W. Scott in *Asymmetric Synthesis* (Eds.: J. D. Morrison, J. W. Scott), Academic Press, Orlando **1984**, Vol. 4, pp. 1–226 (review); on chiral intermediates from the chiral pool: (b) W. Graf in *Biocatalysts in Organic Syntheses* (Eds.: J. Tramper, H. C. van der Plas, P. Linko), Elsevier Science Publishers, Amsterdam **1985**, 41–58.
[137] J. Jacques, A. Collet, S. H. Wilen, *Enantiomers, Racemates and Resolutions*, Wiley Interscience, New York **1981**.
[138] A short overview is given by D. Seebach: D. Seebach, *Angew. Chem.* **1990**, *102*, 1363–1409; *Angew. Chem., Int. Ed. Engl.* **1990**, *29*, 1320.

[139] I. Ojima, Ed., *Catalytic asymmetric synthesis*, VCH, Weinheim **1993**; H. Brunner, W. Zettlmeier, *Handbook of Enantioselective Catalysis with Transition Metal Compounds*, 2 Vols., VCH, Weinheim **1993**.
[140] (a) J. Barbera, A. Omenat, J. L. Serrano, *Mol. Cryst. Liq. Cryst.* **1989**, *166*, 167–171; (b) M. Marcos, A. Omenat, J. L. Serrano, T. Sierra, *Chem. Mater.* **1992**, *4*, 331–338; (c) I. Nishiyama, H. Ishizuka, A. Yoshizawa, *Ferroelectrics* **1993**, *147*, 193–204.
[141] (a) H. Taniguchi, M. Ozaki, K. Yoshino, N. Yamasaki, K. Satoh, *J. Jpn. Appl. Phys.* **1987**, *26*, 4558–4560; (b) C. J. Booth, G. W. Gray, K. J. Toyne, J. Hardy, *Mol. Cryst. Liq. Cryst.* **1992**, *210*, 31–57.
[142] J. Nakauchi, K. Sakashita, Y. Kageyama, S. Hayashi, K. Mori, *Bull. Chem. Soc. Jpn.* **1989**, *62*, 1011–1016.
[143] G. W. Gray, D. Lacey, K. J. Toyne, R. M. Scrowston, A. Jackson, Eur. Patent 0334845B1, **1992**.
[144] M. Shimazaki, K. Minami, Jpn. Kokai Tokkyo Koho JP 05 201 995 (1993); *Chem. Abstr.* **1994**, *120*, 91037c.
[145] (a) E. Chiellini, G. Galli, F. Cioni, E. Dossi, B. Gallot, *J. Mater. Chem.* **1993**, *3*, 1065–1073; (b) S.-T. Nishiyama, Y. Ouchi, H. Takezoe, A. Fukuda, *Jpn. J. Appl. Phys.* **1987**, *26*, L1787–L1789; (c) K. Kühnpast, J. Springer, G. Scherowsky, F. Giesselmann, P. Zugenmaier, *Liq. Cryst.* **1993**, *14*, 861–869.
[146] (a) S. Takehara, T. Shoji, M. Osawa, N. Fujisawa, H. Ogawa, JP 01 131 134 (1989); *Chem. Abstr.* **1990**, *112*, 88952b; (b) J. Wen, M. Tian, Q. Chen, *Liq. Cryst.* **1994**, *16*, 445–451.
[147] S. B. Evans, J. I. Weinschenk, III, A. B. Padias, H. K. Hall, T. M. Leslie, Jr., *Mol. Cryst. Liq. Cryst.* **1990**, *183*, 361–364.
[148] (a) A. Sakaigawa, Y. Tashiro, Y. Aoki, H. Nohira, *Mol. Cryst. Liq. Cryst.* **1991**, *206*, 147–157; (b) H. Gerlach, *Helv. Chim. Acta* **1966**, *49*, 1291; (c) H. Poths, R. Zentel, S. U. Vallerien, F. Kremer, *Mol. Cryst. Liq. Cryst.* **1991**, *203*, 101–111.
[149] E. Chiellini, R. Po, S. Carrozzino, G. Galli, B. Gallot, *Mol. Cryst. Liq. Cryst.* **1990**, *179*, 405–418.
[150] (a) H. Nohira, *Yuki Gosei Kagaku Kyokaishi* **1991**, *49*, 467–474; (b) H. Nohira, S. Nakamura, M. Kamei, *Mol. Cryst. Liq. Cryst.* **1990**, *180B*, 379–388; (c) M. Koden, T. Kuratete, F. Funada, K. Awane, K. Sakaguchi, Y. Shiomi, T. Kitamura, *Jpn. J. Appl. Phys.* **1990**, *29*, L981–L983.
[151] (a) P. J. Alonso, M. Marcos, J. I. Martinez, J. L. Serrano, T. Sierra, *Adv. Mater.* **1996**, *6*, 667–670; (b) J. Slaney, M. Watson, J. W. Goodby, *J. Mater. Chem.* **1995**, *5*, 2145–2152.

[152] V. Vill, F. Bachmann, J. Thiem, I. F. Pelyvás, P. Pudlo, *Mol. Cryst. Liq. Cryst.* **1992**, *213*, 57–65.

[153] (a) T. Sakurai, N. Mikami, R. Higuchi, M. Honma, M. Ozaki, K. Yoshino, *J. Chem. Soc., Chem. Commun.* **1986**, 978–979; (b) Chr. Bahr, G. Heppke, *Mol. Cryst. Liq. Cryst.* **1986**, *4*, 31; (c) J. Bömelburg, Chr. Hänsel, G. Heppke, J. Hollidt, D. Lötzsch, K.-D. Scherf, K. Wuthe, H. Zaschke, *Mol. Cryst. Liq. Cryst.* **1990**, *192*, 335–343; (d) J. Bömelburg, G. Heppke, A. Ranft, *Z. Naturforschg.* **1989**, *44b*, 1127–1131; (e) M. D. Wand, R. Vohra, D. M. Walba, N. A. Clark, R. Shao, *Mol. Cryst. Liq. Cryst.* **1991**, *202*, 183–192.

[154] (a) G. A. Olah, J. Welch, *Synthesis* **1974**, 652; (b) J. Barber, R. Keck, J. Retey, *Tetrahedron Lett.* **1982**, *23*, 1549–1552; (c) F. Faustini, S. de Munari, A. Panzeri, V. Villa, C. A. Gandolfi, *Tetrahedron Lett.* **1981**, *22*, 4533–4536; (d) G. A. Olah, G. K. S. Prakash, Y. Li Chao, *Helv. Chim. Acta* **1981**, *64*, 2528–2530.

[155] (a) S. Colonna, A. Re, G. Gelbard, E. Cesarotti, *J. Chem. Soc., Perkin I* **1979**, 2248–2252; (b) S. Arakawa, K. Nito, J. Seto, *Mol. Cryst. Liq. Cryst.* **1991**, *204*, 15–25.

[156] (a) S. K. Prasad, S. M. Khened, S. Chandrasekhar, B. S. Shivkumar, B. K. Sadashiva, *Mol. Cryst. Liq. Cryst.* **1990**, *182B*, 313–323; (b) Chr. Bahr, G. Heppke, *Mol. Cryst. Liq. Cryst.* **1987**, *148*, 29–43; (c) A. J. Slaney, I. Nishiyama, P. Styring, J. W. Goodby, *J. Mater. Chem.* **1992**, *2*, 805–810.

[157] N. Shiratori, I. Nishiyama, A. Yoshizawa, T. Hirai, *Jpn. J. Appl. Phys.* **1990**, *29*, L2086–L2088.

[158] N. Shiratori, A. Yoshizawa, I. Nishiyama, M. Fukumasa, A. Yokoyama, T. Hirai, M. Yamane, *Mol. Cryst. Liq. Cryst.* **1991**, *199*, 129–140.

[159] T. Kusumoto, A. Nakayama, K.-i. Sato, T. Hiyama, S. Takehara, M. Osawa, K. Nakamura, *J. Mater. Chem.* **1991**, *1*, 707–708.

[160] (a) O. Mitsunobu, *Synthesis* **1981**, 1–28; (b) T. Shibata, M. Kimura, S. Takano, K. Ogasawara, *Mol. Cryst. Liq. Cryst.* **1993**, *237*, 483–485.

[161] (a) L. Verbit, *J. Chem. Educ.* **1972**, *49*, 36–39; (b) G. Patch, G. A. Hope, *J. Chem. Educ.* **1985**, *62*, 454–455.

[162] L. T. Creagh, *Proc. IEEE* **1973**, *61*, 814–822; *Chem. Abstr.* **1973**, *79*, 98008n.

Chapter V
Symmetry and Chirality in Liquid Crystals

John W. Goodby

1 Symmetry Operations in Liquid Crystals

Chirality has become arguably the most important topic of research in liquid crystals today. The reduced symmetry in these organized phases leads to a variety of novel phase structures, properties, and applications. Molecular asymmetry imparts form chirality to liquid crystal phases, which is manifested in the formation of helical ordering of the constituent molecules of the phase. Similarly, molecular asymmetry imposes a reduction in the space symmetry, which leads to some phases having unusual nonlinear properties, such as ferroelectricity and pyroelectricity.

However, there are a number of different symmetry concepts utilized in liquid crystals which often generate considerable confusion. This is because chemists and physicists use different symmetry arguments and different labelling and language in describing various systems. In the following sections, the various symmetry arguments and definitions that are more commonly used to describe the structures of chiral liquid crystals will be discussed.

2 Molecular Asymmetry

Molecular symmetry is used to describe the spatial configuration of a single molecular structure, inasmuch as it describes the geometric, conformational, and configurational properties of the material [1]. Optical activity, on the other hand, is simply the property of the molecules of an enantiomer to rotate the polarization plane as plane-polarized light interacts with them. Such molecules (and the compounds they constitute) are said to be optically active. The words optically active and chiral are frequently used interchangeably, even though a chiral molecule shows optical activity only when exposed to plane-polarized light.

Generally, in liquid crystals, it is only phases that contain optically active materials that exhibit chiral liquid crystal modifications. Thus at least one substance in the liquid crystal system must be a stereoisomer that contains at least one asymmetrically substituted atom, and which is present in a greater concentration than its opposite enantiomer. It is the configurational isomers in the system that give rise to chiral properties. Included in configurational isomers are two distinct classes of stereoisomers: *enantiomers and diastereoisomers* [2]. Enantiomers are two molecules that are related to one another as object and nonsuperimposable mirror image, as shown in the upper part of Fig. 1. Diastereoisomers usually contain more than one asymmetric atom, and pairs of diastereoisomers do not share a superimposable mirror image, as shown in the lower part of Fig. 1 [1, 2].

In Fig. 1 the spatial configuration about each asymmetric carbon atom, which is marked by an asterisk, is denoted by an absolute configuration label as either *R* (rec-

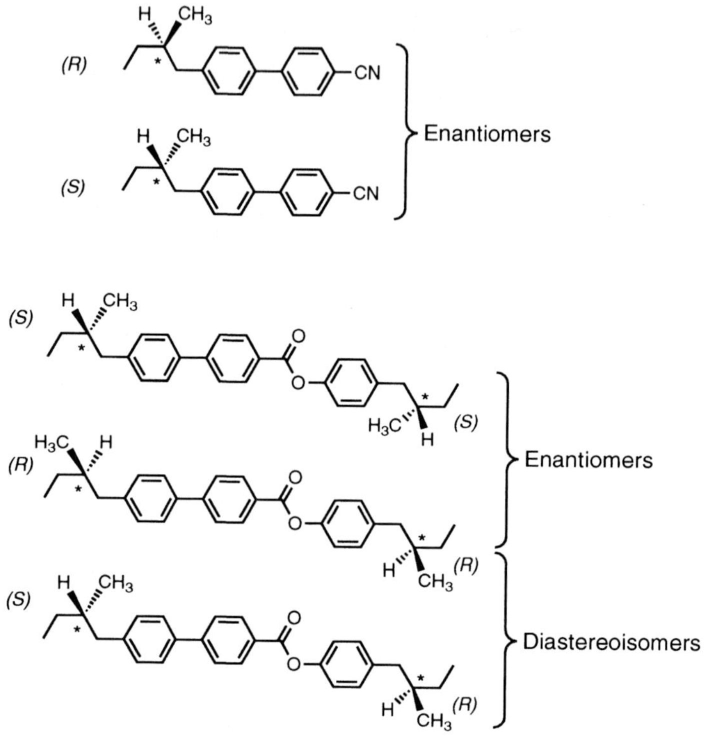

116   V Symmetry and Chirality in Liquid Crystals

Figure 1. Structures of enantiomers and diastereoisomers.

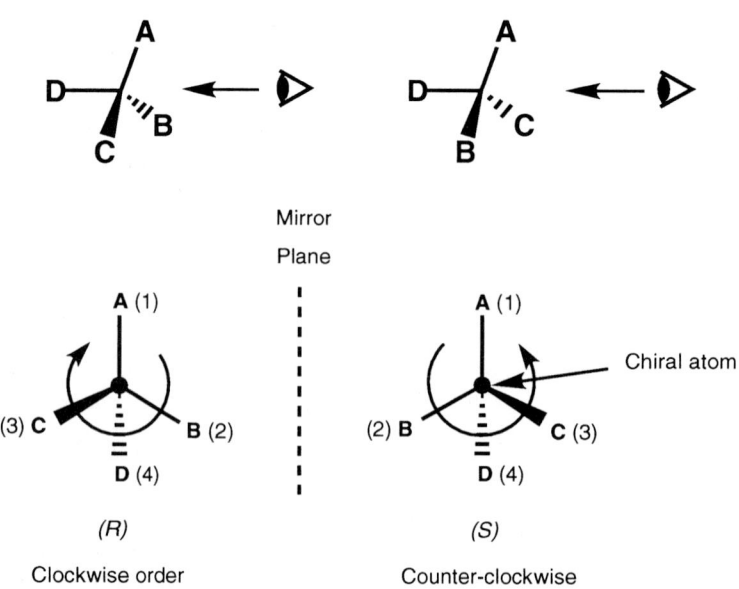

Figure 2. Determination of absolute spatial configuration. The groups A, B, C, & D are arranged in ascending atomic number (1 to 4) with the group of the lowest value pointing back into the plane of the page.

tus, right) or *S* (*sinister*, left). This system of nomenclature was first introduced by Cahn and co-workers in the 1950s [3, 4]. The configuration designation of an asymmetric carbon atom is obtained by first determining the relationship of the group priorities about the asymmetric atom based on the atomic number. The group with the lowest priority is positioned, see Fig. 2, according to the *conversion rule*, to the rear of the tetrahedral asymmetric carbon atom. The other three groups are viewed from the opposite side of the asymmetric center to the group of lowest priority. If the remaining groups are arranged in descending order of priority relative to a clockwise direction about the asymmetric center, the spatial configuration is designated *R*. For the reverse situation, when the priority order descends in a counterclockwise direction, the spatial configuration is given the absolute configuration label *S*.

2.1 Group Priority

This involves an ordering of functional groups based on their atomic number [2] (heaviest isotope first). A partial list (highest priority first) has $I > Br > Cl > SO_2R > SOR > SR > SH > F > OCOR > OR > OH > NO_2 > NO > NHCOR > NR_2 > NHR > NH_2 > CX_3$ (X = halogen) $> COX > CO_2R > CO_2H > CONH_2 > COR > CHO > CR_2OH > CH(OH)R > CH_2OH > C\equiv CR > C\equiv CH > C(R)=CR_2 > C_6H_5 > CH_2 > CR_3 > CHR_2 > CH_2R > CH_3 > D > H >$ electron pair.

2.2 Diastereoisomers

When more than one asymmetric center is present in a molecular structure, then the material can be diastereoisomeric under certain circumstances. For example, a molecule that has two asymmetric centers with (*R*) and (*S*) configurations will be the diastereoisomer of the one with an (*RR*) configuration. Similarly, the (*SS*) configuration is the diastereoisomer of the (*RS*) molecule. However, the (*RS*) compound is the enantiomer of the (*SR*) variation, and the (*RR*) compound is the enantiomer of the (*SS*) form. Enantiomers are expected to have the same physical properties as one another, but diastereoisomers can have properties that are different from one another.

2.3 Dissymmetry

So far we have only discussed chirality in molecules that contain an asymmetric atom; however, some molecules are optically active even when they do not possess an asymmetric atom, e.g., substituted allenes, spirocyclobutanes, etc. For molecules with a dissymmetric structural grouping, the (*R*) or (*S*) configuration is found by assigning priority 1 to the higher priority group in front, 2 to the lower priority group in front, 3 to the higher priority group in the back, etc., then examining the path $1 \rightarrow 2 \rightarrow 3 \rightarrow 4$. For example, the absolute spatial configurational assignment for 1,3-dimethylallene enantiomers is shown in Fig. 3.

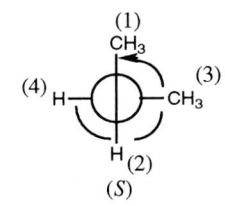

Figure 3. Determination of spatial configuration for the dissymmetric material 1,3-dimethylallene.

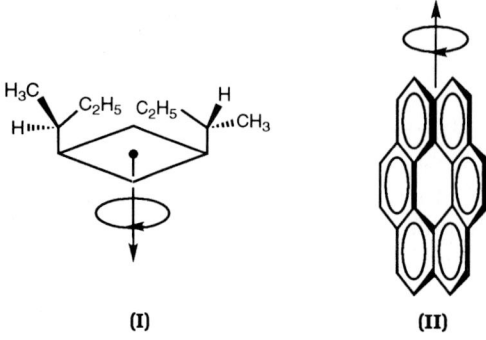

Figure 4. Rotational symmetry operations in chiral materials.

Dissymmetry can also occur for materials that possess rotational symmetry elements. Typically dissymmetric materials that have two- or threefold rotational axes of symmetry can also exhibit optical activity. For example, consider the following materials: S,S-1,3-(2-methylbutyl)cyclobutane and helicene (I and II respectively in Fig. 4). Compound I contains two asymmetric carbon atoms, does not possess a superimposable mirror image, and is also optically active. However, this compound is of a higher symmetry class than asymmetric compounds because it has a simple twofold axis of rotation passing through the center of, and normal to, the cyclobutane ring system. Helicene, compound II, does not possess any chiral carbon atoms as they are all sp^2 hybridized, and therefore each contains a mirror plane. However, helicene is still optically active because it has a helical structure that does not have a superimposable mirror image. This molecule also possesses a twofold axis of rotation across the hexagonal structure of the rings.

Thus, dissymmetric molecules commonly have a simple axis of symmetry, and in asymmetric molecules this axis is absent; however, both species are usually optically active. In liquid crystal systems both types of material are capable of exhibiting chiral properties. Table 1 summarizes the relationships between optical activity, molecular structure, and rotational symmetry operations [1].

3 Space Symmetry in Liquid Crystals

There are a variety of ways in which the space or environmental symmetry and asymmetry can be expressed in liquid crystals, with the most commonly discussed system being that of the chiral smectic C* phase. Thus, for the purposes of describing space symmetry in liquid crystals, the structure and symmetry properties of the smectic C phase will be described in the following sections [5, 6].

Consider initially the environmental symmetry for a smectic C phase that is composed of nonchiral molecules arranged in disordered layers with their long axes tilted in the same direction with respect to the layer planes (note there are as many molecules pointing up as there are pointing down). The space or environmental symme-

Table 1. Relationships between optical activity, molecular structure, and rotational symmetry operations (after Eliel [1]).

Term	Alternating axis	Simple axis	Optical activity
Symmetric	present	may or may not be present	inactive
Dissymmetric	absent	may or may not be present	usually active
Asymmetric	absent	absent	usually active

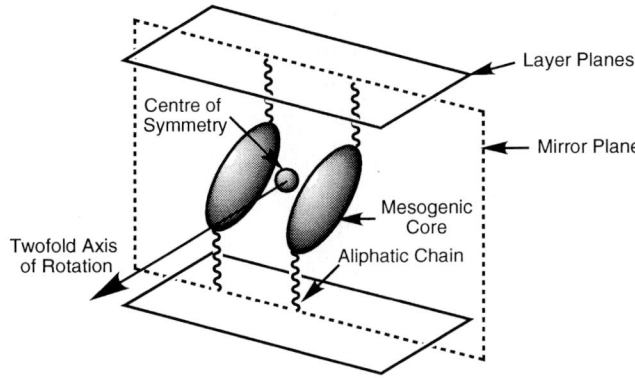

C$_{2h}$ Symmetry in the Achiral Smectic C Phase

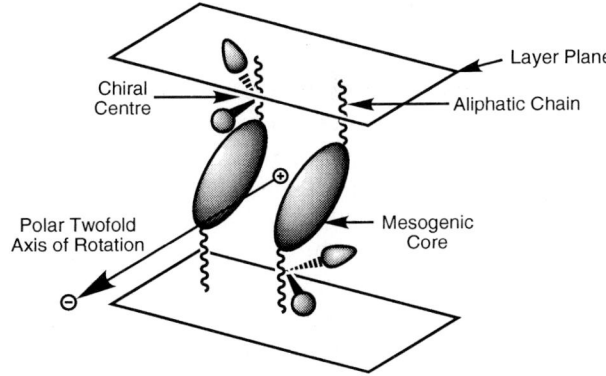

C$_2$ Symmetry in the Chiral Smectic C* Phase

Figure 5. Symmetry operations in the smectic C and chiral smectic C* phases.

try elements for the molecules in this phase are: a mirror plane, a twofold axis of rotation, and a center of symmetry. Thus the symmetry is classed as C_{2h}. However, when this phase contains optically active material, these symmetry elements are reduced to a single polar twofold axis parallel to the layer planes and normal to the vertical planes that contain the long axes of the molecules; consequently the phase has a reduced C_2 symmetry, as shown in Fig. 5.

The result of the packing of the dipolar regions of the molecules in these phases requires that a *spontaneous polarization* (P_S) acts along the C_2 twofold axis normal to the tilt direction, as predicted by Meyer and coworkers [5, 7] in the theoretical and experimental verification of this phenomenon in liquid crystals (Fig. 6). As this smectic phase does not have a well-organized structure, the molecules can be reoriented by applying a field of known polarity. These qualitites give rise to the term *ferroelectric* to describe such titled phases.

The presence of a spontaneous polarization in smectic liquid crystals is assumed to be due to a time-dependent coupling of the lateral components of the dipoles of the individual molecules with the chiral environment. Consequently, only the time-averaged projections of the dipole moments along the polar C_2 axis are effective in producing the macroscopic spontaneous polarization.

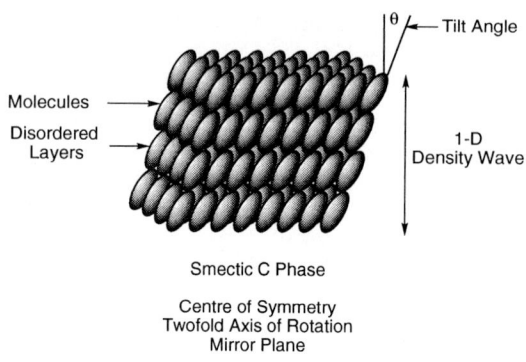

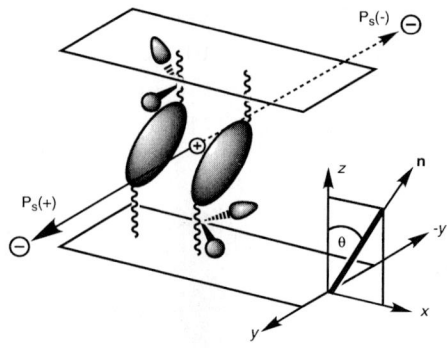

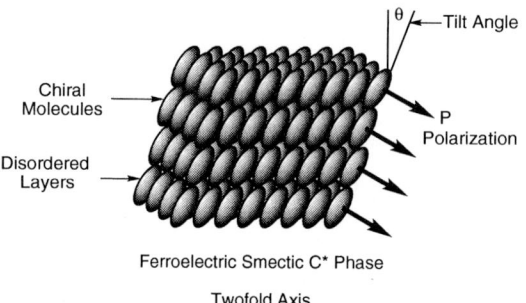

Figure 6. Comparison of the local structures of the smectic C and the ferroelectric smectic C* phases.

Figure 7. Positive and negative polarization directions in chiral smectic C* phases.

One result that stems from the above symmetry arguments is that the spontaneous polarization must have directionality [5, 6, 8, 9]. As the polarization acts along the normal to the tilt direction of the molecules and parallel to the layers, two directions for the spontaneous polarization become possible, as shown in Fig. 7. If the dipoles that produce the spontaneous polarization act along the positive y direction, when an object molecule is tilted to the right in the plane of the page, this results in positive polarization [$P_S(+)$]; the reverse situation is denoted as negative polarization [$P_S(-)$].

Here again there is some confusion between chemists and physicists, because they use different symbols to depict dipole moments. For chemists, the arrow showing the direction of a dipole points towards the negative pole, whereas for physicists it points towards the positive pole [10]. The diagram shown in Fig. 7 depicts the dipole moments as might be drawn by a chemist. However, depictions by chemists or physicists would nevertheless yield the same nomenclature for the polarization direction.

For the ferroelectric smectic C* phase, molecular chirality gives rise to a reduced space symmetry ($C_{2h} \rightarrow C_2$); however, the effects of a reduced molecular symmetry and space symmetry can also come into conflict causing frustrated structures to be formed. For example, some smectic C* materials can show strong asymmetric intermolecular interactions, which in turn can affect the space symmetry. If the polar coupling along the C_2 axis in a smectic C* phase is strong, then its effects can be reduced by having the direction of the polarization alternate between layers, thereby producing a phase with no spontaneous polarization. In order for this to occur, the tilt directions of the molecules must alternate from one layer to the next, thereby producing an antiferroelectric structure [11, 12] as shown in Fig. 8. The symmetry of the antiferroelectric phase is now higher than that of the ferroelectric state. Intermediate ferrielectric phases can be formed between the ferroelectric and antiferroelectric extremes. In ferrielectric phases there is a pattern of alternating tilts, e.g., two layers tilting to the left for every three tilting to the right, etc. Thus, as we move from the ferroelectric state to the antiferroelectric phase via intermediate ferrielectric phases, the polarization will fall in an uneven fashion, i.e., as in a devil's staircase.

The symmetry arguments for achiral or chiral smectic C* phases can also be applied in similar ways to the other smectic and soft crystal smectic phases. For example, all of the tilted phases (smectics I* and F*, and crystal phases J*, G*, H*, and K*) would have broken symmetries leading to polar noncentrosymmetric structures resulting in ferroelectric properties. Even orthogonal phases, such as the smectic A phase would have different symmetries for the chiral versus the achiral forms. For instance, the smectic A phase has $D_{\infty h}$ symmetry, whereas the chiral smectic A* phase containing chiral molecules has D_∞ symmetry [13].

In an alternative way to how it is broken by chiral molecules, space symmetry can also be broken by certain arrangements of achiral molecules. For example, achiral bent or banana-shaped or bowl-shaped molecules can pack together in certain ways to give noncentrosymmetric structures [14]. Bent molecules, as represented by structure III in Fig. 9, can pack together with their bent shapes snugly fitting together one curve inside another. Similarly bowl-shaped compounds, e.g., structure IV, can pack so that one bowl fits snugly inside another. Both of these arrangements will generate a

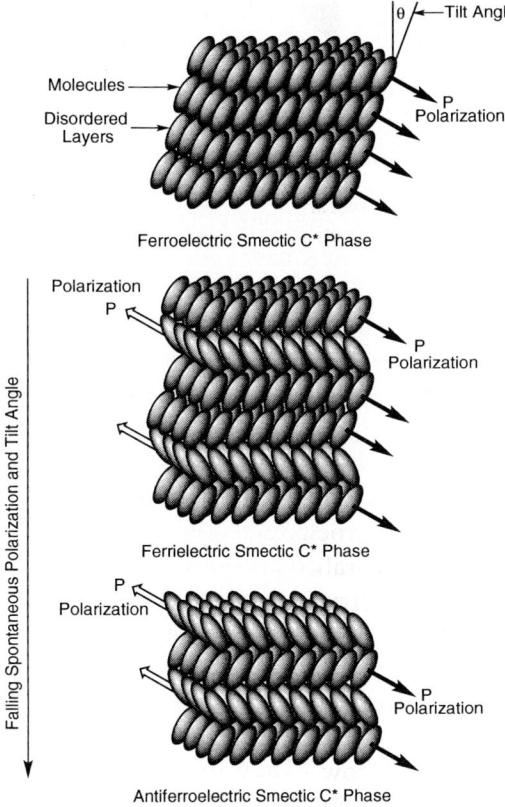

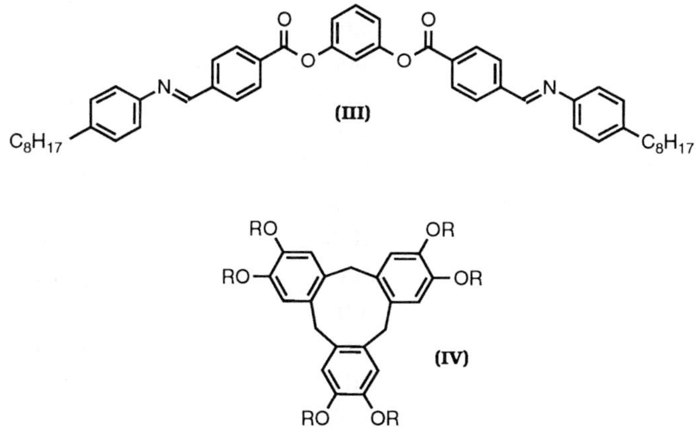

Figure 8. Comparison of the local structures of the ferroelectric, ferrielectric, and antiferroelectric smectic C* phases.

Figure 9. Bent and bowl-shaped molecules that are capable of exhibiting mesophases with reduced symmetries.

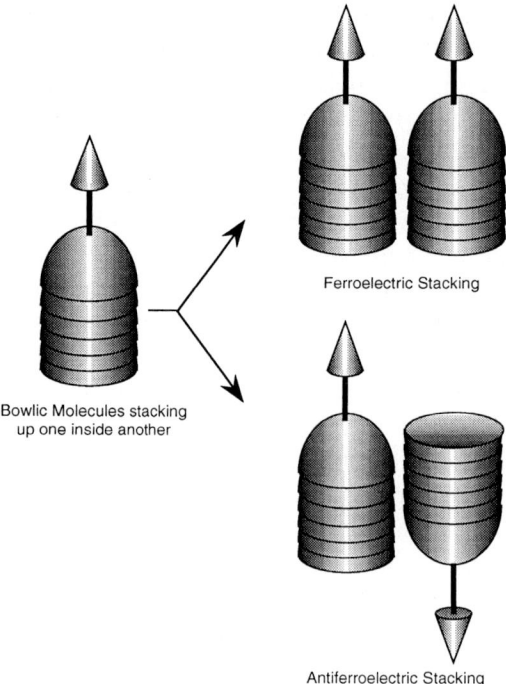

Figure 10. Parallel and antiparallel packing of columns of bowl-shaped molecules.

polar axis pointing in a direction perpendicular to the curve in the banana or bowl (Fig. 10). If the bananas or bowls preferentially stack so that all of the sheets or columns of bananas or bowls point in the same direction, then the phase will have a non-centrosymmetric, and hence ferroelectric, structure [14]. Alternatively, the sheets or columns can have an up–down arrangement leading to an antiparallel, and hence antiferroelectric, structure (Fig. 10).

4 Mesophase Phase Symmetry

When chiral molecules pack together to give a condensed liquid crystal phase, under certain conditions they can do so to give a phase with a helical macrostructure. For calamitic and discotic systems, the calamitic chiral nematic, discotic chiral nematic, and the smectic C*, I*, and F* phases are all capable of possessing helical structures, whereas the orthogonal calamitic smectic phases A* and B* (hexatic), and the crystal phases B*, E*, J*, G*, H*, and K* do not apparently exhibit helicity. In addition, some columnar phases are thought to have a helical disposition of the molecules along the axes of the columns [15].

Helical phases have similar symmetry properties to those of a molecule of helicene in that they are dissymmetric because they possess a twofold (infinite) axis of rotation normal to the helical axis through the midpoint of the helical structure. As helical phases are dissymmetric and do not have superimposable mirror images, like helicene, they are also optically active. Thus a beam of plane polarized light traversing such an arrangement of molecules will be rotated by the helical ordering [16]. However, the specific rotation (α) is usually much greater for helical phases than for the sum of the specific rotation(s) of the individual molecules. Thus, in this case, as the optical activity is due to the helical macrostructure of the phase, rather than due to the sum of the optical activities of the individual molecules, it is termed *form optical activity*.

The helical ordering can be defined as left- or right-handed by determining the direction of the rotation, to the right or the left, on moving away from an observation point along the helical axis, as described by Cahn et al. [3] (see the motor car wheel in Fig. 11, the direction of rotation with respect to the direction of travel determines the handedness of the helix). A rotation to the left is denoted as a left-hand helix, and rotation to the right results in a right-hand helix. Plane polarized light traversing a spiraling struc-

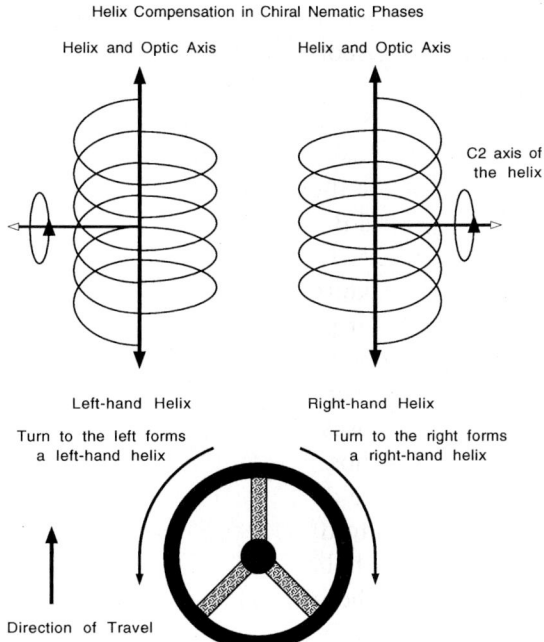

Figure 11. Definitions of helical twist direction in chiral nematic phases.

ible light, the helical phase will scatter or reflect iridescent light [15]. Moreover, the pitch of the helix is temperature-sensitive, because it depends on the orientations and interactions of the molecules, which are also temperature-dependent. Therefore the color of the scattered light also depends on the ambient temperature, consequently being capable of producing a *thermochromic* effect when the pitch of the helix is comparable to the wavelength of light. Furthermore, helical structures can be unwound by the application of an electric field which couples to the polarity or dielectric anisotropy of the material. Thus if a helix is scattering visible light it can be prevented from doing so by applying an electric field, thereby producing an electro-optic effect.

4.1 The Chiral Nematic Phase

The structure of the chiral nematic or cholesteric phase is one where the local molecular ordering is identical to that of the nematic phase. In the direction normal to the director the molecules pack to form a helical macrostructure (see Fig. 12), i.e., in the calamitic variation, the director is rotated in a direction perpendicular to the long axes of the molecules. As in the nematic phase, the molecules have no long range positional order, and no layering exists. The pitch of the helix can vary from about 0.1×10^{-6} m to almost infinity. The optic axis of the phase is along the helical axis so the phase has a negative birefringence and is optically uniaxial. Disc-like molecules can also exhibit cholesteric phases when they possess chiral molecular structures [18], as shown in Fig. 13.

ture will have its plane of polarization rotated in the same direction as the helix, i.e., the designation of helical handedness can be made from viewing the light beam moving away from the observer. However, in polarimetry experiments on optically active liquids or solutions, the observation point for classification of a system as dextro- or laevo-rotatory is made from the viewpoint of looking into the oncoming beam of light [2, 17]. Thus, by definition, rotation of the incoming plane of plane polarized light to the right is dextrorotatory and rotation to the left is laevorotatory. Consequently, in a polarimeter-style experiment a left-handed helix will produce a dextrorotation, and a right-hand helix will produce a laevorotation (see Fig. 11 for the cholesteric phase).

When the pitch of the helix formed by the orientational ordering of the molecules is comparable to the wavelength range of vis-

As noted earlier, when plane-polarized light traverses the helical structure of a chiral nematic phase, its plane is rotated in the

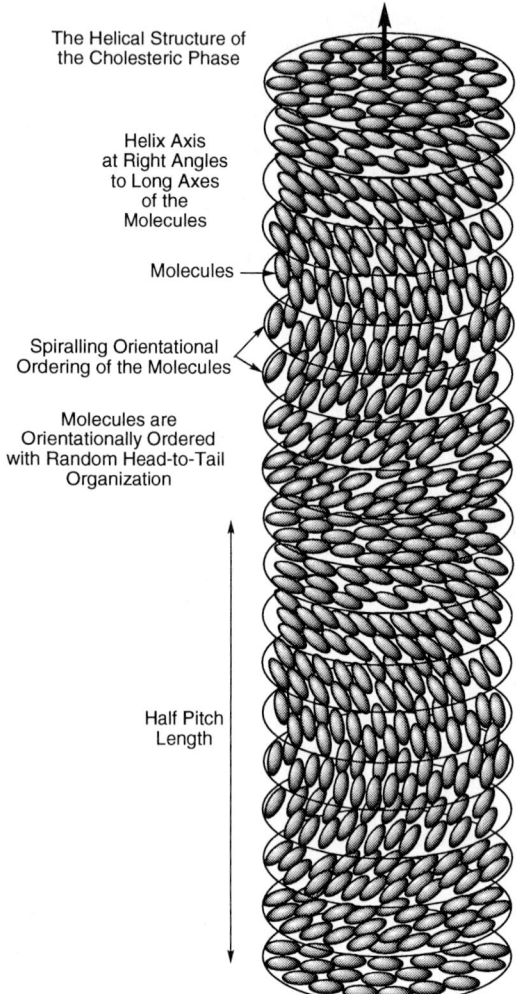

Figure 12. Helical structure of the chiral nematic phase.

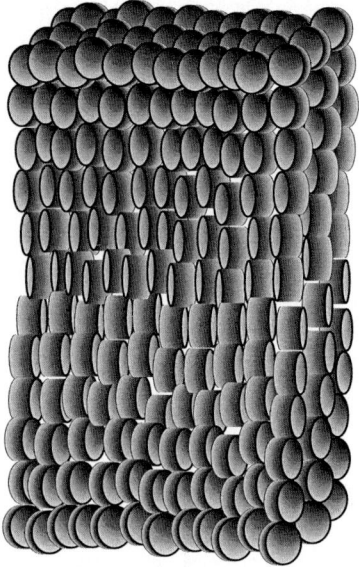

Figure 13. Helical structure of the chiral nematic discotic phase.

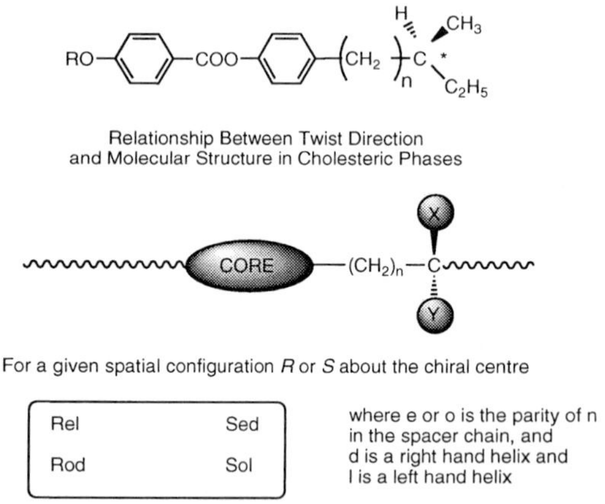

Relationship Between Twist Direction and Molecular Structure in Cholesteric Phases

For a given spatial configuration R or S about the chiral centre

| Rel | Sed |
| Rod | Sol |

where e or o is the parity of n in the spacer chain, and d is a right hand helix and l is a left hand helix

Figure 14. Relationship between helical twist direction and molecular structure in chiral nematic phases.

same direction as the helix by the helical ordering. Thus a cholestrogen can be defined in terms of its absolute spatial configuration (R) and (S) and its helical orientational ordering properties, i.e., dextrorotatory (d) and laevorotatory (l).

If we make a more detailed examination of cholesterogens that have molecular structures that contain single chiral carbon atoms, as exemplified by the typical homologous series shown in the top part of Fig. 14, then it is found that the number of atoms (n) that the chiral center is removed from the rigid central core determines the handedness of the helical structure of the chiral nematic phase. As the atom count by which the chiral center is removed from the core (n) switches from odd (o) to even (e) (parity), so the handedness of the helix alternates from left to right or *vice versa*. Sim-

Helical Twist - Spatial Configuration - Parity Rules

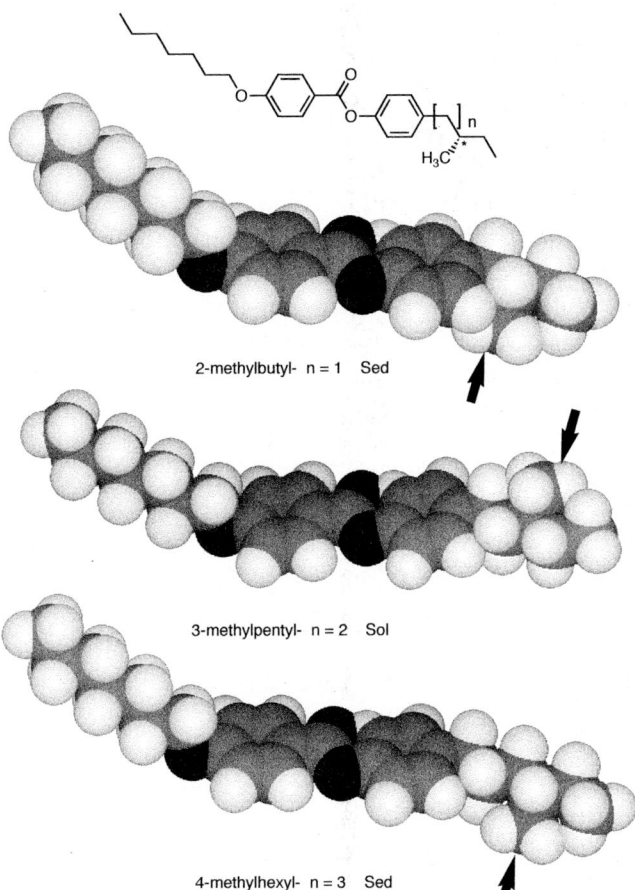

Figure 15. Alternation in position of a methyl group located at the chiral center as the chiral center is moved sequentially down a terminal aliphatic chain.

ilarly, if the absolute spatial configuration of the chiral center is inverted, say from (R) to (S), so the handedness of the helix also reverses. Thus the twist sense of the helix is dependent on the absolute spatial configuration and the position of the asymmetric center within the structure of the material relative to the rigid central core region.

Gray and McDonnell [19] suggested that the spatial configuration of the chiral center of the molecular structure is related to the screw direction of helical structures in cholesteric phases in the following way:

Sol Sed

Rod Rel

where (S) or (R) is the absolute spatial configuration of the chiral center, e or o is the odd or even parity for the atom count that the chiral center is removed from the central rigid core, and d or l refers to the handedness of the helical structure (see Fig. 14). Figure 15 shows, with the aid of space-filling molecular structures, how the location of the chiral center, and its related substituents, oscillates back and forth with the parity for a simple homologous series. The alternation in twist sense with change in parity suggests that the helical twist sense is related to the conformational structure and the steric packing of the molecules. This relationship works very well for simple compounds that possess only a single chiral center and which exhibit the cholesteric phase, but for more complex molecular structures (e.g., those containing more than one chiral center, or having strong polar groups attached to the chiral center, or large bulky groups attched to the asymmetric center) and smectic phases the relationship is not as useful.

In helical phases, it is generally found that when the chiral center is brought closer to the core the pitch of the helix becomes shorter, and therefore the chirality increases [20]. This is thought to be caused by the increased steric hindrance to the rotation of the asymmetric center about the long axis of the molecule. Hence the degree of chirality of the mesophase, determined by the pitch of the helix, can be predicted to some degree from the molecular structure of the material in question.

The twist sense of a chiral nematic phase can be determined, relatively easily, by observing contact preparations in the microscope [19]. A standard material of known twist sense is allowed to make contact with a chiral nematic of unknown twist direction and then the area of contact between the two compounds is observed in the microscope. If the two materials have the same twist sense, then the contact region will exhibit a chiral nematic (cholesteric) phase, but if they have opposite twist senses then a nematic phase will separate the two chiral nematic regions, as shown in Fig. 16.

4.2 Helical Smectic Phases

The smectic C*, I*, and F* phases have similar optical activity properties to the cholesteric phase, and also possess a helical distribution of their molecules. In the chiral smectic C* phase, the constituent molecules are arranged in diffuse layers where the molecules are tilted at a temperature-dependent angle with respect to the layer planes. The molecules within the layers are locally hexagonally close-packed with respect to the director of the phase; however, this ordering is only very short range, extending over distances of approximately 1.5–2.5 nm. Over large distances, therefore, the molecules are randomly packed, and in any one domain the molecules are tilted roughly in the same direction. Thus the tilt orientational ordering between successive layers is preserved over long distances.

A helical macrostructure is generated by a precession of the tilt about an axis normal to the layers, as shown in Fig. 17. The tilt direction of the molecules in a layer above or below an object layer is rotated through an azimuthal angle relative to the object layer. This rotation always occurs in the same direction for a particular material, thus forming a helix. The helix can be either right-handed or left-handed depending on the chirality of the constituent molecules, i.e., in a similar way to chiral nematic phases. The pitch of the helix for most C* phases is commonly greater than one micrometer in length, indicating that a full twist of the helix is made up of many thousands of layers. Thus the azimuthal angle is relatively small and is usually of the order of one-tenth to one-hundredth of a degree.

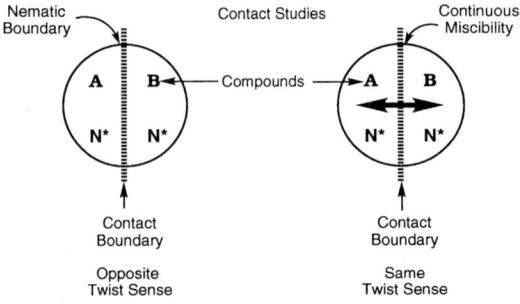

Figure 16. Contact method for determining the twist direction in chiral nematic phases.

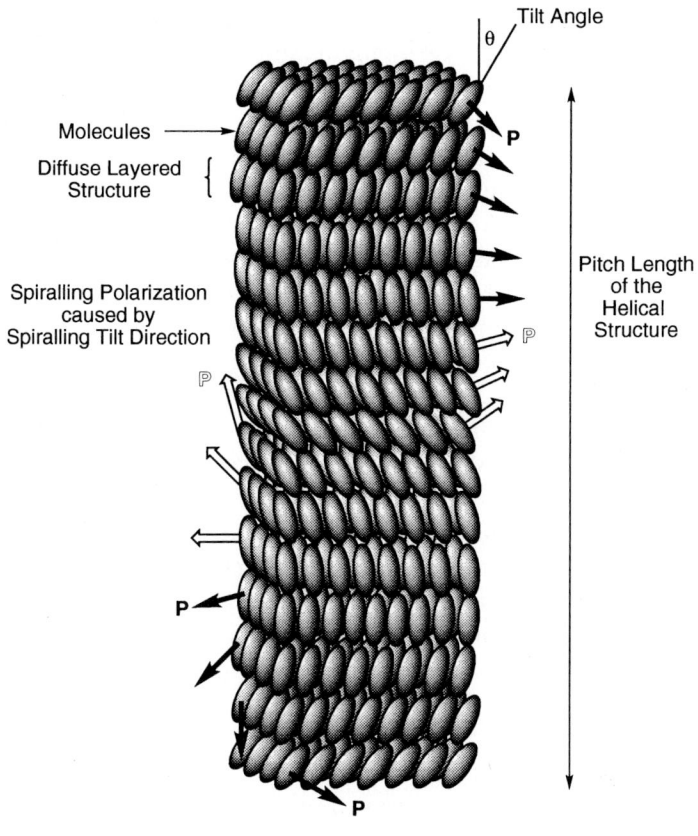

Figure 17. Helical macrostructure of the chiral smectic C* phase.

In the chiral smectic I* phase the molecules are arranged in a similar fashion to the way they are organized in the C* phase. In the smectic I phase, however, the in-plane ordering is much more extensive, with the molecules being hexagonally close packed with respect to the director of the phase. The positional ordering of the molecules extends over distances of 15–60 nm within the layeres and is therefore short range in nature. The phase, however, possesses long-range bond orientational order in that the hexagonal packing of the molecules remains in the same orientation over long distances in three dimensions, even though the positional order is only short range. The tilt orientation between layers, as in the smectic C* phase, is preserved. Another feature associated with the tilt in the smectic I/I* phase is that it is directed towards an apex of the hexagonal packing net – a structural parameter that distinguishes the phase from smectic F [21, 22].

As with the smectic C* phase, there is a spiralling of the tilt orientational order normal to the layer planes forming a macroscopic helix [23]. Again the twist of the tilt direction between successive layers can be either to the right or to the left, depending on the structure(s) of the constituent molecules. The pitch of the helix, although temperature-dependent, is usually greater for smectic I* phases than it is in preceding smectic C* phases in the same material. The azimuthal angle is consequently smaller for the smectic I* phase in comparison to the smectic C* phase for materials that exhibit both phases. As the phase has spiralling tilt orientational order, the hexagonal packing net is also rotated about an axis normal to the layers on passing successively from one layer to the next. Thus the bond orientational ordering presumably rotates in a direction normal to the layers.

The structure of the chiral smectic F* phase is similar to that of the chiral smectic I* phase, except that the tilt direction is towards the side of the local hexagonal packing net, rather than towards the apex.

The more ordered crystalline smectic-like phases J*, G*, H*, and K* have not been definitely shown to exhibit a helical structure. Certainly, X-ray diffraction and optical studies have failed to detect this ordering over many hundreds of layers, indicating that if a helix exists in any of these phases it must have an extremely long pitch length [24]. It is more likely, however, that this ordering is suppressed by the crystal forces of the phase. Consequently, unlike the chiral nematic and smectic C*, I*, and F* phases, which have two levels of optical activity (molecular and form), the crystalline phases J*, G*, H*, and K* probably only have a single level of molecular optical activity.

Another consequence of the formation of a helix in the smectic C*, I*, and F* phases is that the spontaneous polarization direction is tied to the tilt orientational ordering, which is itself spiralling in a direction normal to the layers. Hence the sponta-

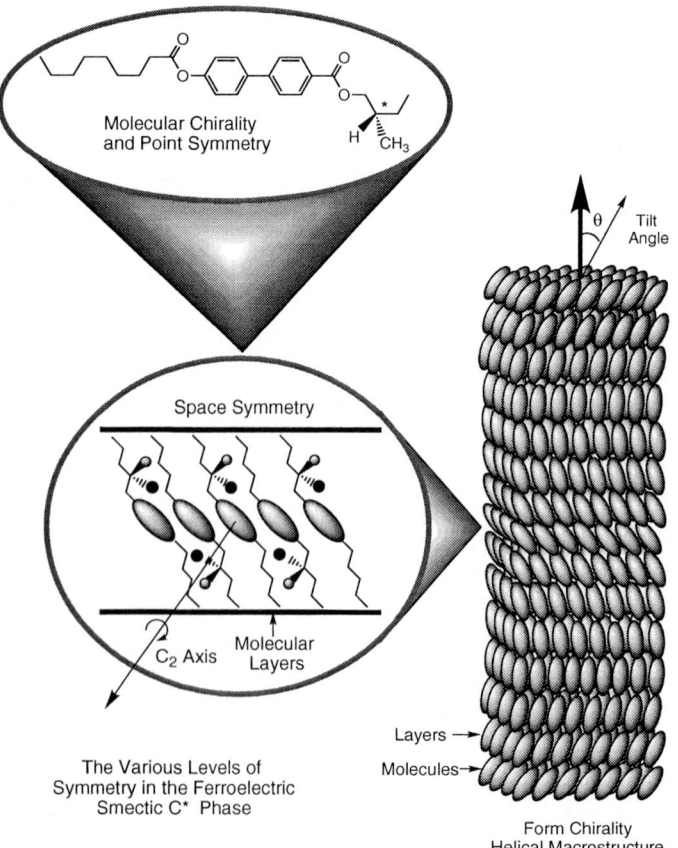

Figure 18. Relationship between molecular, environmental, and form chirality for the chiral smectic C* phase.

neous polarization is also rotated in a spiralling manner in a direction normal to the layers. Thus when the helix makes one full 360° turn, the polarization is averaged to zero. Therefore in a bulk undisturbed phase which is not influenced by external forces the spontaneous polarization will be averaged to zero. However, if the helix is unwound by external forces, such as surface interactions, or by compensating the helix in a mixture so that its pitch length is infinite, the phase will again become ferroelectric. It is for this reason that Brand et al. have described optically active helical smectic C* phases as *helielectric* rather than *ferroelectric* [25].

Overall, for the smectic C* phase we can relate the three symmetry elements to one another in a simple diagram, as shown in Fig. 18. At the microscopic level we have molecular chirality; locally for molecular clusters we must consider the space or environmental chirality, and for the bulk phase we have to include form chirality.

5 Frustrated Phases

5.1 Double Twist and Blue Phase Helical Structures

Consider first a uniaxial phase that is composed of chiral rod-like molecules. In the simplest situation, a helix can form in a direction perpendicular to the long axis of an object molecule. This example is analogous to the structure of the chiral nematic phase. In the direction parallel to the long axes of the molecules no twist can be effected. Now consider a similar situation, but this time the twist in the orientational order can occur in more than one direction in the plane perpendicular to the long axis of an object molecule. This structure is called a double twist cylinder, and is shown in Fig. 19.

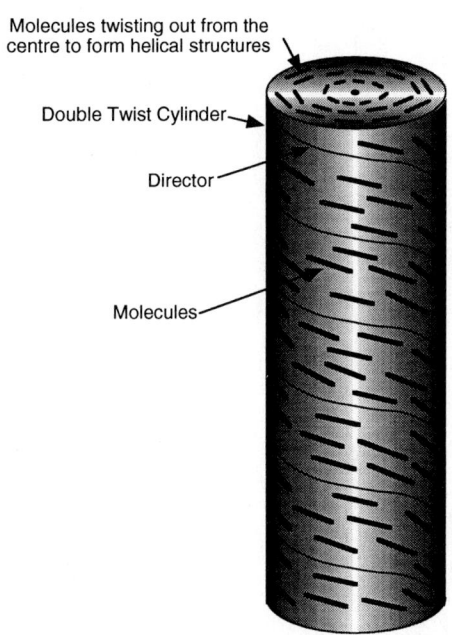

Figure 19. Double twist cylinder found in blue phases.

In the simplest form of the double twist cylinder, two helices are formed with their axes perpendicular to one another in the plane at right angles to the direction of the long axis of the molecule [26]. Expanding this structure in two dimensions, the two helices can intersect to form a 2-D lattice. However, the helices cannot fill space uniformly and completely, and hence defects are formed. As helices are periodic structures, the locations of the defects created by their inability to fill space uniformly are also periodic. Thus a 2-D lattice of defects is created. This inability to pack molecules uniformly can be extended to three dimensions to give various cubic arrays of defects [27]. The different lattices of defects provide the structural network required for the formation of a range of novel liquid crystal phases that are called blue phases. In prin-

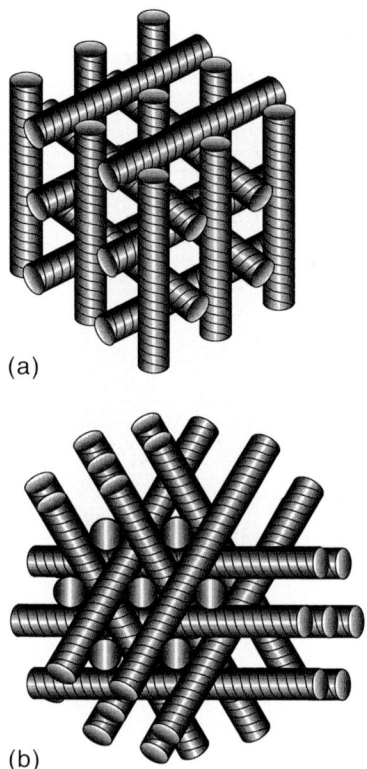

Figure 20. Two possible cubic defect structures for blue phases: (a) a simple cubic structure, (b) a body-centered cubic structure.

ciple these phases are frustrated structures where the molecules would like to fill space with a double twist structure but are prevented from doing so, and the result is the formation of defects. Thus the formation of defects stabilizes the structure of the phase. Two possible cubic structures of defects for the blue phase are shown in Fig. 20.

These frustrated phases were called blue phases because when they were first observed microscopically by Coates and Gray [28] they appeared blue. Their strong blue color is due to the selective reflection of light. Other materials were later discovered which exhibited blue phases where the selective reflection was in the red or green region of the spectrum. Experimentally it was found, however, that the helical pitch length must be of a similar length to the wavelength of visible light for a material to exhibit a blue phase, so the lattice period for the defects is of the order of 5000 Å (500 nm).

5.2 Twist Grain Boundary Phases

Blue phases are not the only frustrated structures that can be formed in liquid crystals. Twist grain boundary phases were predicted, by de Gennes [29] and later by Renn and Lubensky [30], to be a resultant of the competition between bend or twist deformations in smectic phases and the desire for the mesophase to form a layered structure. Typically, therefore, twist grain boundary (TGB) phases are usually detected at the phase transitions from the liquid or chiral nematic states to the smectic state [21]. So far stable twist grain boundary phases have been found to mediate the chiral nematic to smectic A*, isotropic liquid to smectic A* [31], and chiral nematic to smectic C* transitions. This gives rise to corresponding TGBA* and TGBC* phases which generally exist over a temperature range of a few degrees.

At a normal chiral nematic to smectic A* transition, the helical ordering of the chiral nematic phase collapses to give the layered structure of the smectic A* phase. However, for a transition mediated by a TGB phase, there is a competition between the need for the molecules to form a helical structure due to their chiral packing requirements and the need for the phase to form a layered structure. Consequently, the molecules relieve this frustration by trying to form a helical structure, where the axis of the helix is perpendicular to the long axes of the molecules (as in the chiral nematic phase), yet at the same time they also try to form a lamellar structure, as shown in Fig. 21. These two

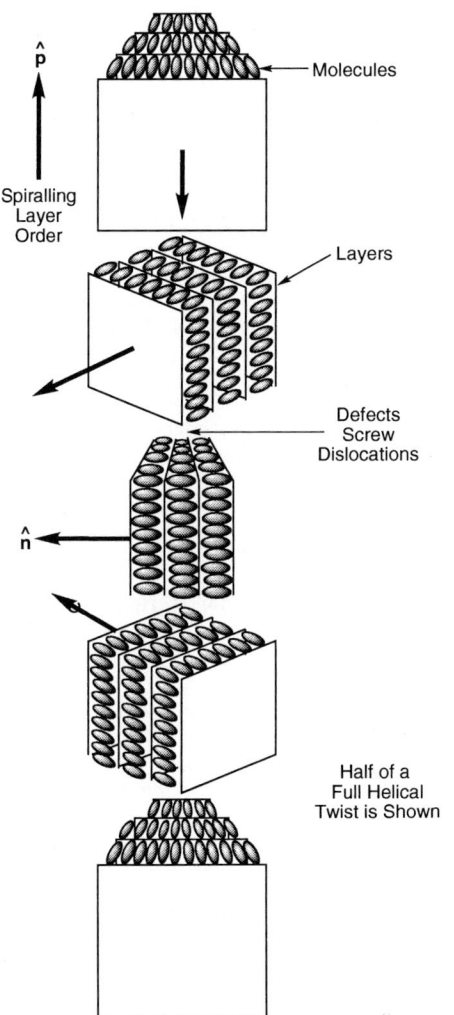

Figure 21. Helical structure of the TGBA* phase.

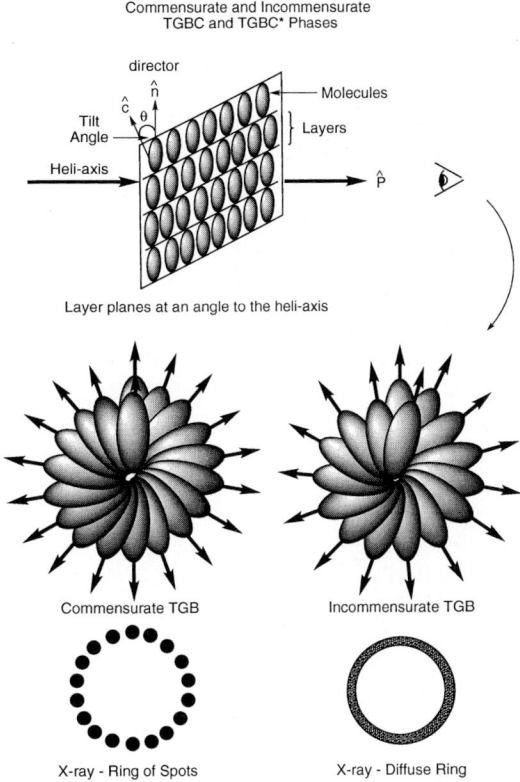

Figure 22. Commensurate and incommensurate structures in the TGBC phase.

structures are incompatible with one another and cannot co-exist and still fill space uniformly without forming defects. The matter is resolved by the formation of a periodic ordering of screw dislocations which enables a quasi-helical structure to co-exist with a layered structure. This is achieved by having small blocks of molecules, which have a local smectic A structure, being rotated with respect to one another by the screw dislocations [32], thereby forming a helical structure. As the macroscopic helix is formed with the aid of scew dislocations, the dislocations themselves must be periodic. It is predicted that rows of screw dislocations in the lattice will form grain boundaries in the phase, and hence this phase was called the twist grain boundary (TGB) phase.

In this analysis it must be emphasized that the TGB phase is not simply a layered cholesteric phase, and should not be confused with this concept. A layered cholesteric phase simply cannot exist on a macroscopic scale, and it is a requirement that defects must be formed.

Subsequently, the tilted analog of the TGBA* phase, the TGBC* phase, was detected at the chiral nematic to smectic C* transition. Two forms of this phase are predicted; one where the molecules are simply inclined to the layer planes with no interlayer twist within a block, and another where they are allowed to form helical structures normal to the layer planes in addition to the helices formed by the screw dislocations.

In both TGB phases the helical structure can be discrete or indiscrete, i.e., the number of blocks required to form a 360° rotation of the helix can be either a whole number, in which case the phase is said to be commensurate, or else it is not formed by a whole number, in which case the phase is said to be incommensurate, as shown in Fig. 22.

References

[1] E. L. Eliel, *Stereochemicstry of Carbon Compounds*, McGraw-Hill, New York **1962**.
[2] M. Orchin, F. Kaplan, R. S. Macomber, R. M. Wilson, H. Zimmer, *The Vocabulary of Organic Chemistry*, Wiley, New York **1980**.
[3] R. S. Cahn, C. K. Ingold, V. Prelog, *Angew. Chem., Int. Ed.* **1966**, *5*, 385.
[4] R. S. Cahn, C. K. Ingold, *J. Chem. Soc.* **1951**, 612.
[5] R. B. Meyer, *Mol. Cryst. Liq. Cryst.* **1976**, *40*, 74.
[6] J. W. Goodby, *J. Mater. Chem.* **1991**, *1*, 307.
[7] R. B. Meyer, L. Liebert, L. Strezlecki, P. Keller, *J. Phys. Lett. (Paris)* **1975**, *36*, L69.
[8] M. Handschy, N. A. Clark, *Ferroelectrics* **1984**, *59*, 69.
[9] N. A. Clark, S. T. Lagerwall, *Ferroelectrics* **1984**, *59*, 25.
[10] J. W. Goodby in *Ferroelectric Liquid Crystals; Principles, Properties and Applications* (Eds.: J. W. Goodby, R. Blinc, N. A. Clark, S. T. Lagerwall, M. A. Osipov, S. A. Pikin, T. Sakurai, K. Yoshino, B. Zeks), Gordon and Breach, Philadelphia, PA **1991**.
[11] M. Johno, A. D. L. Chandani, J. Lee, Y. Ouchi, H. Takezoe, A. Fukuda, K. Itoh, T. Kitazume, Proc. Jpn. Display **1989**, 22.
[12] A. D. L. Chandani, Y. Ouchi, H. Takezoe, A. Fukuda, K. Terashima, K. Furukawa, A. Kishi, *Jpn. J. Appl. Phys.* **1989**, *28*, L1261; E. Gorecka, A. D. L. Chandani, Y. Ouchi, M. Takezoe, A. Fukuda, *Jpn. J. Appl. Phys.* **1990**, *29*, 131; S. Inui, S. Kawano, M. Saito, H. Iwane, Y. Takanishi, K. Hiraoka, Y. Ouchi, H. Takezoe, A. Fukuda, *Jpn. J. Appl. Phys.* **1990**, *29*, L987.
[13] S. A. Pikin, M. A. Osipov in *Ferroelectric Liquid Crystals; Principles, Properties and Applications* (Eds.: J. W. Goodby, R. Blinc, N. A. Clark, S. T. Lagerwall, M. A. Osipov, S. A. Pikin, T. Sakurai, K. Yoshino, B. Zeks), Gordon and Breach, Philidelphia, PA **1991**.
[14] T. Nori, T. Sekine, J. Watanabe, T. Furukawa, H. Takezoe, *J. Mater. Chem.* **1996**, *6*, 1231.
[15] P. A. Heiney, E. Fontel, W. H. de Jeu, A. Riera, P. Carroll, A. B. Smith, *J. Phys. (Paris)* **1989**, *50*, 461.
[16] J. G. Grabmaier in *Applications of Liquid Crystals* (Eds.: G. Meier, E. Sackmann, J. G. Grabmaier), Springer-Verlag, Berlin **1975**; J. W. Goodby, *Science* **1986**, *231*, 350.
[17] J. W. Goodby, E. Chin, T. M. Leslie, J. M. Geary, J. S. Patel, *J. Am. Chem. Soc.* **1986**, *108*, 4729; J. W. Goodby, E. Chin, *J. Am. Chem. Soc.* **1986**, *108*, 4736; J. S. Patel, J. W. Goodby, *Opt. Eng.* **1987**, *26*, 373.
[18] J. Malthete, C. Destrade, N. H. Tinh, J. Jacques, *Mol. Cryst. Liq. Cryst. Lett.* **1981**, *64*, 233.
[19] G. W. Gray, D. G. McDonnell, *Mol. Cryst. Liq. Cryst.* **1977**, *34*, 211.
[20] G. W. Gray, D. G. McDonnell, *Mol. Cryst. Liq. Cryst.* **1978**, *48*, 37.
[21] A. J. Leadbetter, J. P. Gaughan, B. A. Kelly, G. W. Gray, J. W. Goodby, *J. Phys. (Paris)* C3 **1979**, *40*, 178.
[22] P. A. C. Gane, A. J. Leadbetter, P. G. Wrighton, *Mol. Cryst. Liq. Cryst.* **1981**, *66*, 247.
[23] J. Doucet, P. Keller, A.-M. Levelut, P. Porquet, *J. Phys. (Paris) Lett.* **1975**, *36*, 69.
[24] J. Budai, R. Pindak, S. C. Davey, J. W. Goodby, *Phys. Rev. Lett.* **1981**, *46*, 1135.
[25] H. R. Brand, P. E. Cladis, P. L. Finn, *Phys. Rev. A* **1985**, *31*, 361.
[26] P. P. Crooker, *Liq. Cryst.* **1989**, *5*, 751.
[27] D. W. Berremann in *Liquid Crystals and Ordered Fluids*, Vol. 4 (Eds.: A. C. Griffin, J. F. Johnson), Plenum, New York **1984**, pp. 925–943.
[28] D. Coates, G. W. Gray, *Phys. Lett.* **1973**, *45A*, 115.
[29] P. G. de Gennes, *Sol. State Commun.* **1972**, *10*, 753.
[30] S. R. Renn, T. C. Lubensky, *Phys. Rev. A* **1988**, *38*, 2132.
[31] J. W. Goodby, M. A. Waugh, S. M. Stein, E. Chin, R. Pindak, J. S. Patel, *J. Am. Chem. Soc.* **1989**, *111*, 8119.
[32] G. Srajer, R. Pindak, M. A. Waugh, J. W. Goodby, J. S. Patel, *Phys. Rev. Lett.* **1990**, *64*, 1545; K. J. Ihn, J. A. N. Zasadzinski, R. Pindak, A. J. Slaney, J. W. Goodby, *Science* **1992**, *258*, 275.

Chapter VI
Chemical Structure and Mesogenic Properties

Dietrich Demus

1 Introduction

In the early years of liquid crystal research several compounds with liquid crystalline properties were synthesized, sometimes by chance, without any knowledge of the relation between the molecular structure and the mesogenic properties [1, 2]. In the case of cholesterol derivatives, even the chemical structure of the basic moiety was not known in that time. Vorländer and coworkers also found their first mesogenic compounds by chance [3]; however, they very quickly changed to a systematic investigation of the relation between molecular structure and mesogeneity [4, 5]. As early as 1908, Vorländer was able to establish his rule regarding the most elongated molecular structure of the molecules of liquid crystals [4]. This rule is still valid for calamitic compounds. Table 1 shows some early examples from Vorländer, demonstrating the rule.

Vorländer also had the idea to look for mesogenic properties in the cases of star-like or cross-like molecules, however, his coworker did not have any success in this area [5, 6]. About 1977, Chandrasekhar et al. [7] and Billard et al. [8] were, independently, able to prove that disk-like molecules can in fact form mesophases, called columnar phases. Since then, several hundred discotic compounds have been synthesized [9–14].

During his systematic investigations, Vorländer had already found liquid crystalline compounds, with a molecular structure deviating strongly from a simple rod-like shape [15–18]. Such compounds having an 'unconventional' molecular structure have been a topic of research for the last 15 years, which has led to a variety of different dimers, trimers, branched and laterally substituted, polycatenar compounds and others [1, 2, 19–21]. Many of these compounds exhibit conventional nematic and smectic phases, while others show deviations from the classical layer thickness (double layers, interpenetrating layers) or form cubic or columnar phases.

Table 1. Transition temperatures of the compounds synthesized by Vorländer [4, 5]*.

1. 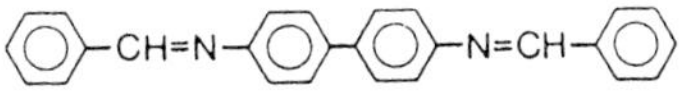 ⬡–CH=N–⬡–⬡–N=CH–⬡
   Cr 234 N 260 I

2. 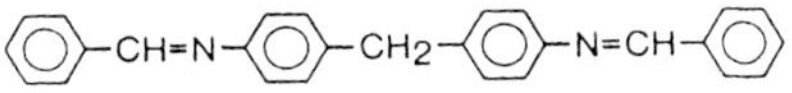 ⬡–CH=N–⬡–CH₂–⬡–N=CH–⬡
   Not liquid crystalline

3. 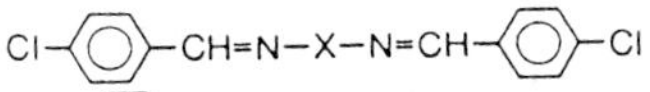 Cl–⬡–CH=N–X–N=CH–⬡–Cl
   X = –⬡–   Cr 180 N 288 I 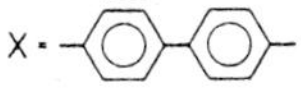
   X = –⬡–⬡– Cr 265 N 318 I

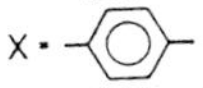

4. H₅C₂O–⬡–CH=N–⬡–CH=C–COOC₂H₅
                                |
                                R

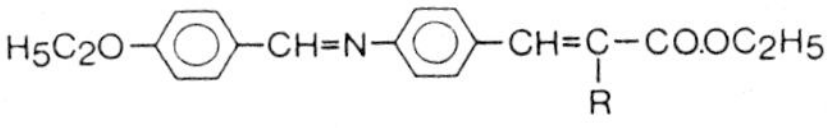

R	Cr		Sm		N		I
H	• 81	•	157	•	160		I
CH₃	• 95	(•	77)	•	123	•	
C₂H₅	• 73	–	–	(•	62)	•	
C₆H₅	• 104	–	–	–	–	•	

* From [2], by permission of the publishers.

Vorländer also investigated the influence of the length of the molecules on their mesomorphic properties, and during these investigations he made the first observation of oligomeric or polymeric liquid crystals [17]. Stimulated by the aim of producing superstrong fibres, polymer liquid crystals have been synthesized systematically since about 1967 [22], and at present many different varieties are known.

The clear dependence of the mesomorphic properties on the geometrical shape of the molecule has allowed the derivation of simple models for describing the systems theoretically.

Calamitic and discotic compounds can be described theoretically using the same basic model, i.e. the cylinder, by elongation or compression, respectively, of the central axis of the cylinder [23]. Of course, the cylinder model is a oversimplification of real mesogenic molecules. As most of the compounds contain ring systems, which are somewhat flat, it could be proposed that parallelepipeds provide better models. In fact, such models have been used for molecular-statistical theories, and these have allowed the description of, in addition to calamitic and discotic molecules, flat elongated (board-like, sanidic) molecules [24].

In order to synthesize liquid crystals with predicted properties, one of the basic questions for the chemist is: Which intermolecular interactions are responsible for the stabilization of liquid crystalline phases? The known dependence of mesogenity on the molecular geometry prompted the derivation of molecular-statistical theories, based on the shape-induced anisotropy of the repulsion of the molecules [25]. Despite the remarkable success of such theories, it is clear that real condensed phases always contain strong attractive interactions, which are completely neglected by this procedure. Therefore, attempts have been made to prove the stability of liquid crystalline phases by using only attractive interactions [25]. From the formal standpoint, by fitting certain constants of the theory, this treatment also works; however, in many cases it leads to wrong conclusions regarding the chemistry of the systems [26–28].

The combination of repulsive and attractive interactions of the van der Waals type in theories is the most satisfactory treatment of the problem [25, 29]. Because of the dense packing of the molecules in liquid crystalline phases, the anisotropy of the repulsion is the dominant quantity for the stabilization of the mesophase. The major part of the attraction, which is also isotropic in mesophases, couples with the shape anisotropy, producing the necessary density of the stabilization of the mesophase. However, the anisotropic part of the attraction, which is used as the essential basis of certain molecular-statistical theories, is, in most cases, only a minor correction of the total attraction [30]. Restricting the consideration to steric repulsive and dispersion attractive forces, the van der Waals theories predict a dependence of the clearing temperature of nematics on the length-to-breadth ratio X of the molecules, the average polarizability α, and the anisotropy of the polarizability $\Delta\alpha$. The examples in Figure 1 show that in fact only X has a clear functional connection with the clearing temperatures; in other words, X is the most important factor of the three. In compounds containing strongly polar groups (e.g. –CN, –NO$_2$) the interaction of permanent and induced electrical dipoles plays a remarkable role [28, 31, 32].

Molecular-statistical theories are available for several different smectic phase types [25]. In addition to the ingredients of the nematic phases, the theories incorporate parameters responsible for the formation of layers. The clearing temperatures, however,

2 Rod-Like (Calamitic) Liquid Crystalline Compounds

2.1 General

There are several published compilations of liquid crystalline compounds and their transition temperatures [33–36], showing the respective state of the art. The number of known compounds was 1412 in 1960 [33], 5059 in 1974 [34], 12876 in 1984 [35] and according to recent estimations more than 72000 [35a]. There are many reviews of the chemistry of liquid crystals. We deal here mainly with the newer reviews [37–44]; with regard to the older reviews, we refer simply to the near-complete citation of the older literature given in [45].

Most of the rod-like liquid crystalline compounds consist of two or more rings, which are bonded directly to one another or connected by linking groups (L), and may have terminal (R) and lateral (Z) substituents. The chemical structure of many mesogens can be represented by the general formula 1. The rings, represented in 1 by circles, and the linking groups L form the core of the compound. The core is usually a relatively stiff unit, compared to the terminal substituents, which in most cases are flexible moieties such as alkyl groups. The lateral substituents Z are, in most cases, small units such as halogens, methyl groups and $-CN$ groups.

The major anisotropy of the molecules, which is necessary for their mesogenity, results from the cores, which are also the cause

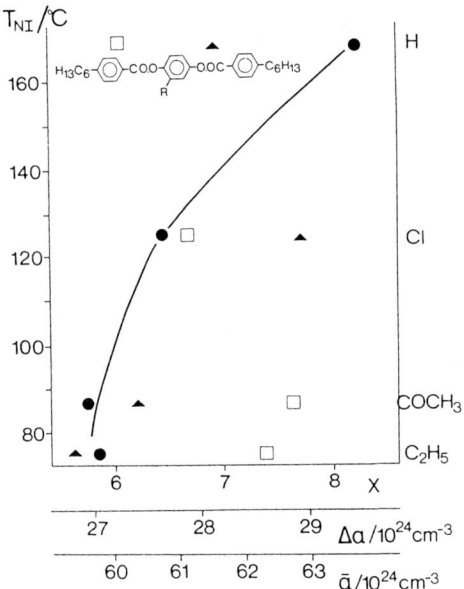

Figure 1. The nematic–isotropic transition temperatures, T_{N-I} of some liquid crystals plotted against the length-to-breadth ratio, X (●), the mean polarizability, α (□, and the anisotropy of the polarizability $\Delta\alpha$ (▲). (From Demus [2]).

are determined by the same properties which determine the T_{N-I} of nematics. The existence of columnar and discotic nematic phases can be explained by theories that are similar to those of nematics, but the model is extended from the rod-like elongated cylinder to disk-like flat cylinders [23].

In accordance with the philosophy explained above, in order to obtain pronounced mesogenic properties, the chemist has to produce molecules with strong shape anisotropy (rods, disks) and strong attraction (moieties with large polarizability, strong dipoles, hydrogen bonds, electron donor acceptor (EDA) interaction). In the following sections this will be proved using experimental data.

of the relatively high melting temperatures. The competition between the clearing temperature and the melting temperature of a compound controls the possibility of observing the mesogenic properties. In principle, anisotropic molecules of all shapes are mesogenic; however, in many cases, because of too high melting temperatures the mesogenic properties cannot be detected. In many cases the isotropic melts can be supercooled, which offers the possibility of observing monotropic liquid crystalline phases. The systematic decrease in the melting temperatures by the attachment of flexible terminal groups to the cores is therefore very important. Lateral substituents can also have the same effect.

In the following sections the effects of the different parts of the molecule on its mesogenic properties will be discussed.

2.2 Ring Systems

The core consists of rings that are connected to one another either directly or by linking groups. Any rings that allow a stretched configuration of the molecule can be used. In practice, mainly six- and five-membered rings are used, and some more complex ring systems such as cholesterol.

2.2.1 Six-Membered Rings

The classical ring, which in first decades of liquid crystal chemistry was used almost exclusively, is benzene. Table 2 presents the transition temperatures of its *para* oligomers. The melting temperatures increase rapidly in this series. The increase in melting temperature with the increasing number of rings is a general phenomenon, irrespective of the nature of the rings. It can also be seen

Table 2. Transition temperatures of *p*-oligophenyls.

Structure	T_m	T_{SmN}	T_{NI}	Ref.
Ph	5.5	–	–	[46]
Ph–Ph	70.8	–	–	[46]
Ph–Ph–Ph	213.8	–	–	[46]
Ph–Ph–Ph–Ph	320	–	–	[46]
Ph–Ph–(Ph)$_2$–Ph	388	–	418	[47]
Ph–Ph–(Ph)$_3$–Ph	440	475	565	[47]
Ph–Ph–(Ph)$_4$–Ph	545	?	?	[48]

from Table 2 that the mesogenity of the compounds increases with the number of linearly connected rings. *p*-Quinquiphenyl has pronounced nematogenity, and sexi- and heptiphenyl possess, in addition, smectic phases, probably smectic A phases. These oligomers are ideal linear molecules with high stiffness, and are the prototypes of rod-like molecules. Due to the large conjugated aromatic systems, the intermolecular attractions of the molecules are very large, and this explains the high melting temperatures.

Suitable terminal group substitution produces mesomorphic materials with biphenyl, terphenyl and quaterphenyl as cores. Table 3 presents some examples. It is interest-

Table 3. Oligophenyl derivatives.

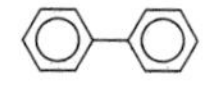

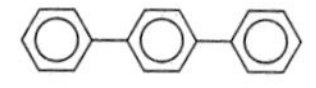

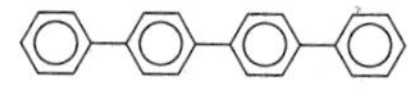

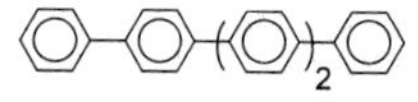

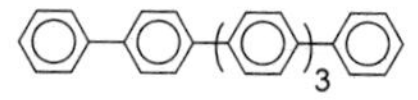

1. 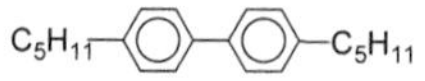 C_5H_{11}–Ph–Ph–C_5H_{11}
   Cr 12 SmE 47 SmB 52 I   [49]

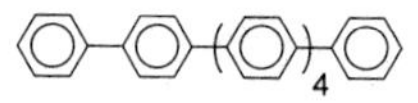

2.  C_5H_{11}–Ph–Ph–Ph–C_5H_{11}
   Cr 192 Sm 213 I   [50]

3. 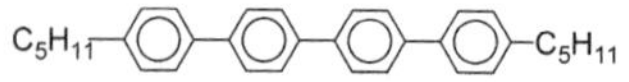 C_5H_{11}–Ph–Ph–Ph–Ph–C_5H_{11}
   Cr 297 SmA 352 I   [51]

ing to note that not only are the melting and clearing temperatures influenced by such substitution, but so too is the polymorphism.

The most important six membered rings introduced in liquid crystals are compiled in Table 4. These include aromatic rings and partially or completely saturated rings.

There are many liquid crystals, derived from nitrogen containing aromatic ring systems. Table 5 gives some examples. If a CH group in a ring is replaced by a nitrogen atom the shape of the ring is only slightly altered, but the electronic properties are changed and quite large electric dipoles are induced, which change the intermolecular attraction. The influence on the mesogenity is shown in Scheme 1.

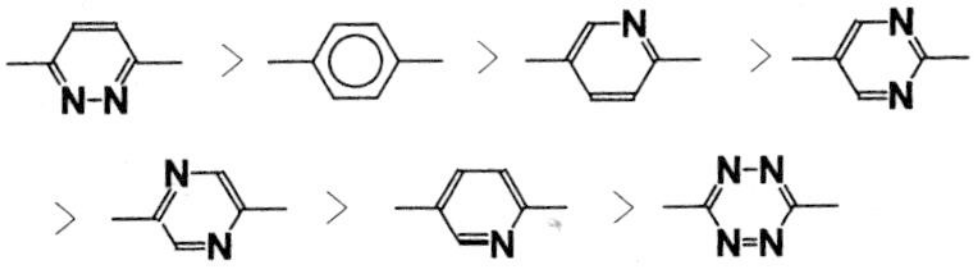

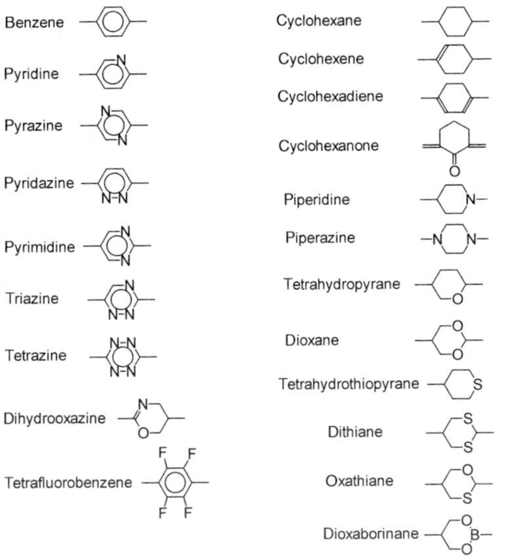

Table 5. Comparison of nitrogen heterocyclic six-membered ring systems (adapted from [52]).

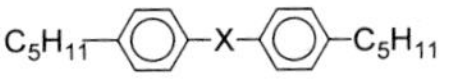

C_5H_{11}–◯–X–◯–C_5H_{11}

X	Cr		SmC		SmA		N		I
–◯–	•	192	–		•	213	–		•
pyridine	•	99	–		•	205	–		•
pyridazine	•	194	–		•	226	•	227	•
pyrimidine	•	106	•	195	–		–		•
pyrazine	•	143	•	174	•	182	•	191	•
N–N pyridazine isomer	•	150	•	169	–		•	185	•
tetrazine	•	163	–		–		–		•	172.5	•

Scheme 1. The influence of nitrogen substituents on the mesogeneity.

Table 4. Ring systems with six atoms.

Benzene	–◯–	Cyclohexane	–◯–
Pyridine	–◯N–	Cyclohexene	–◯–
Pyrazine	–◯–	Cyclohexadiene	–◯–
Pyridazine	–◯–	Cyclohexanone	–◯–
Pyrimidine	–◯–	Piperidine	–◯N–
Triazine	–◯–	Piperazine	–N◯N–
Tetrazine	–◯–	Tetrahydropyran	–◯O–
Dihydrooxazine	–◯–	Dioxane	–◯–
Tetrafluorobenzene	–◯–	Tetrahydrothiopyrane	–◯S–
		Dithiane	–◯–
		Oxathiane	–◯–
		Dioxaborinane	–◯B–

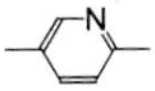

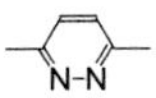

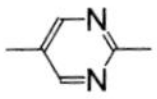

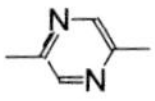

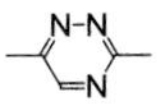

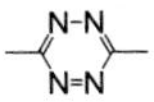

The cyclohexane ring is one of the most important moieties. It differs from benzene by being more bulky in shape, having some flexibility and being non-aromatic in character. The latter causes a strong decrease of the intermolecular attraction, leading to materials with a much lower packing fraction [27, 28]. Nevertheless, their clearing temperatures are, in many cases, much higher, due to higher length-to-breadth ratio. Since their anisotropy of polarizability is significantly lower than that of their aromatic analogues, it is clear that this property does not control the clearing temperature of liquid crystals [27, 28]. Cyclohexane derivatives belong to the most important class of substances for the application of liquid crystals in displays.

Table 6 gives some examples of two-ring cyclohexane derivatives, and shows the increase in the clearing temperature that occurs by exchanging a benzene ring by a cyclohexane ring. The second compound in

Table 6. Comparison of two-ring cyclohexane derivatives.

1. C₅H₁₁–⌬–⌬–C₅H₁₁
   Cr 25.1 SmE 46.1 SmE′ 47.1 SmB 52.2 I   [49]

2. C₅H₁₁–⌬–⌬–C₅H₁₁
   Cr –0.8 (Sm –8 N –5) I   [53]

3. C₅H₁₁–⌬–⌬–C₅H₁₁
   Cr 40 Sm 110.4 I   [54]

4. C₅H₁₁–⌬–COO–⌬–C₃H₇
   Cr 18.7 (N 2.7) I   [55]

5. C₅H₁₁–⌬–COO–⌬–C₃H₇
   Cr 44 (N 43) I   [56]

6. C₅H₁₁–⌬–COO–⌬–C₃H₇
   Cr 24 SmB 37.5 N 52.5 I   [57]

Table 6 does not fit the pattern of increasing clearing temperature, the reason for this being the high amount of *cis* isomer present.

Table 7 lists some three-ring compounds. In the first three examples the effect of cyclohexane substitution is not so clear; however, the last two compounds provide impressive proof of the strong mesogenity of cyclohexane derivatives. The first two examples in Table 8 follow the rule. In the third compound the central ring has two polar-group substituents, which produce lower clearing temperatures. The last three examples in Table 8 seem to show that the flexibility of the central ring has some negative influence on the mesogenity.

The cyclohexane derivatives, as well as other saturated ring systems, can exist in several conformations [67]. In order to obtain the most elongated molecular structure, the ring should exist in the chair form, and both the bonds linking the ring should be equatorial (*ee*). Substituents in the axial position (*aa*) produce a strongly bent molecular shape. It is well known that the chair form of the *trans*-1,4-disubstituted cyclohexane exhibits a thermodynamic equilibrium between the *aa* and *ee* conformers, which has been discussed in some detail by Deutscher et al. [43]. The differences in the Gibbs free energy between the two conformers ranges from about 2 to 20 kJ mol^{-1}. Depending on the nature of the substituents, the Boltzmann distribution between *aa* and *ee* conformers

Table 7. Comparison of three-ring cyclohexane derivatives.

1. C₅H₁₁–⌬–⌬–⌬–C₅H₁₁
   Cr 192 Sm 213 I   [50]

2. C₅H₁₁–⌬–⌬–⌬–C₅H₁₁
   Cr 50 Sm 196 I   [58]

3. C₅H₁₁–⌬–⌬–⌬–C₅H₁₁
   Cr 13 Sm 164 N 166   [59]

4. C₇H₁₅–⌬–⌬–⌬–C₇H₁₅
   Cr 181 Sm 205 I   [50]

5. C₇H₁₅–⌬–⌬–⌬–C₇H₁₅
   Cr 74 SmB 245 I   [52, 60]

Table 8. Comparison of three-ring cyclohexane esters.

1. C₅H₁₁–⌬–COO–⌬–OOC–⌬–C₅H₁₁
   Cr 125 N 188 I   [61]

2. C₅H₁₁–⌬–COO–⌬–OOC–⌬–C₅H₁₁
   Cr 123 N 216 I   [62]

3. C₅H₁₁–⌬–COO–⌬–OOC–⌬–C₅H₁₁
   Cr 101.7 Sm 176.1   [63]

4. C₅H₁₁–⌬–OOC–⌬–COO–⌬–C₅H₁₁
   Cr 152 N 178 I   [64]

5. C₅H₁₁–⌬–OOC–⌬–COO–⌬–C₅H₁₁
   Cr 101.5 Sm 135 N 161 I   [65]

6. C₅H₁₁–⌬–OOC–⌬–COO–⌬–C₅H₁₁
   Cr 125 Sm 159 N 167 I   [66]

can be very different, and it is temperature dependent. With substituents of low polarity, the *ee* conformer is more stable, while with polar substituents the *aa* form seems to be preferred and produces lower mesogenity [68].

Some non-aromatic ring systems are compared in Table 9. In the first and the third examples, the cyclohexene probably exists in the half-chair conformation [77], which does not provide good linearity of the molecules. The relatively high clearing temperatures of example 3 may be due to a conjugation effect with the –CN group. Comparing the cyclohexane and bicyclooctane derivatives, the latter have much stronger nematogenity. This has been discussed in terms of the different flexibility; however, bicyclooctane in connection with benzene gives an exactly linear core. The last four examples in Table 9 possess remarkable electric dipoles, which can influence T_{N-I}. Here the high melting point of the dithiane compound prevents the realization of the nematic phase but, as it is known from many other examples, dithiane derivatives have higher clearing temperatures than do the analogous dioxane derivatives.

2.2.2 Ring Systems with More than Six Atoms

Table 10 presents a selection of the large number of ring systems with more than six atoms that have been used for the synthesis of liquid crystals. Despite the fact that many of these rings produce liquid crystals with high clearing temperatures, these rings give rise to relatively large viscosities. For this reason, the application of such rings in displays has not been successful. In Table 11 some compounds containing bicyclic rings are compared with the analogous benzene and cyclohexane derivatives. Compared to the second cyclohexyl derivative, the more linear and less flexible third bicyclooctane compound has a higher T_c. The T_c of the fourth and fifth compounds is even higher, due to the presence of quite strong dipoles. Compared to the benzene compound, the compounds 2–5 in Table 11 are more bulky and do not have a conjugated core. This explains the differences in the T_c values.

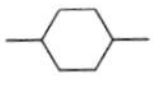

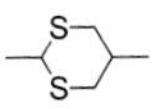

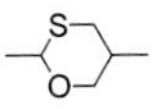

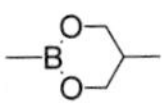

Table 12 gives some examples of compounds containing rings of more than six carbon atoms. All these compounds have a larger breadth than oligophenyl compounds, which reduces their mesogenity in comparison to the latter. The first three compounds are derived from partially or completely hydrogenated naphthalene. The high T_{N-I} of the completely hydrogenated compound is remarkable. Replacing CH_2 groups by heteroatoms does not remove the liquid crystalline properties; however, comparing compound numbers 3 and 4 T_{N-I} is strongly

Table 9. Comparison of two ring non-aromatic compounds.

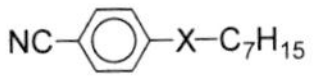

NC–⟨O⟩–X–C$_7$H$_{15}$

X	Cr		N		I	Ref.
–⟨⟩–	•	35	(•	5)	•	[69]
–⟨⟩–	•	30	•	59	•	[70]
–⟨⟩–	•	47.5	•	61	•	[71]
–⟨⟩–	•	61	•	95	•	[72]
–⟨O,O⟩–	•	54	(•	53)	•	[73]
–⟨S,S⟩–	•	98	?		•	[74]
–⟨S,O⟩–	•	78	(•	30)	•	[75]
–⟨O,B,O⟩–	•	53	?		•	[76]

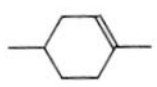

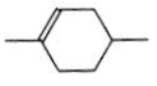

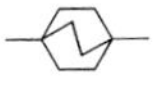

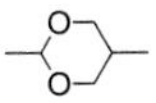

Table 10. Ring systems containing more than six atoms.

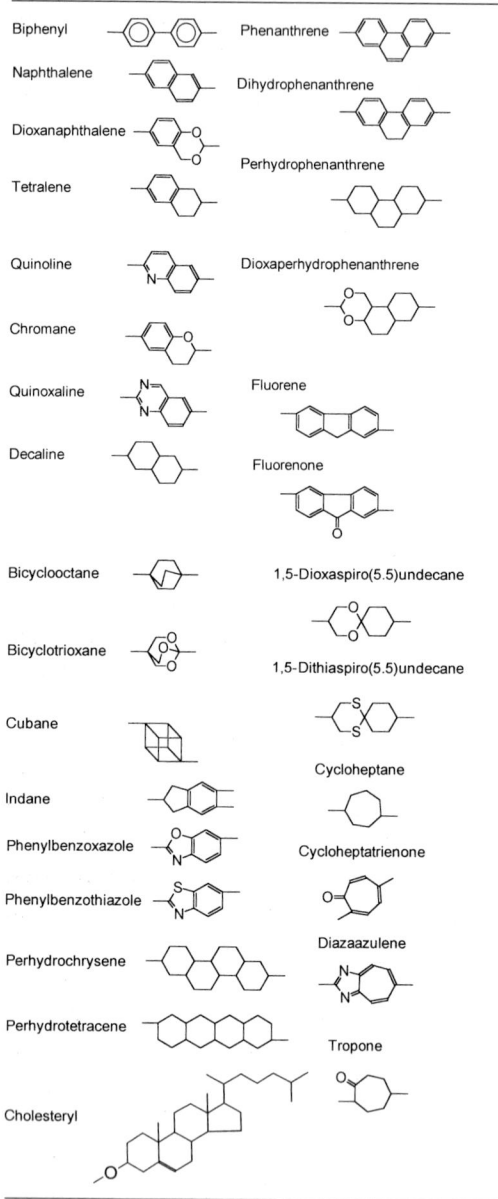

decreased. The last three compounds are derived from phenanthrene; the last compound, a perhydrogenated compound, shows inferior mesogenic properties. In these cases the degree of planarity of the ring, and the size of the conjugated part of the molecule should control the T_{N-I}.

Table 11. Bicyclic rings.

1. C_5H_{11}—◯—◯—C_7H_{15}
   Cr ? SmE 36 SmB 63 I [78]

2. C_7H_{15}—◯—◯—C_5H_{11}
   Cr 16 (Sm 31) I [53]

3. C_5H_{11}—◯—◯—C_7H_{15}
   Cr 52 (SmB 44) I [79]

4. C_5H_{11}—◯—◯—C_7H_{15}
   Cr 46 SmB 74 I [80]

5. C_5H_{11}—◯—◯—C_7H_{15}
   Cr 40 SmB 87 I [81]

Table 12. Large non-polar rings, with no linking groups.

1. C_6H_{13}—◯—◯—C_4H_9
   Cr 30.9 (SmA 28.5) Sm 56.7 N 60.5 I [82]

2. C_5H_{11}—◯—◯—C_5H_{11}
   Cr 39.9 N 59.7 I [83]

3. C_5H_{11}—◯—◯—C_5H_{11}
   Cr 70 SmA 90.5 N 140.5 I [83]

4. C_5H_{11}—◯—◯—C_5H_{11}
   Cr 65 SmA 79.5 N 95.8 I [84]

5. C_5H_{11}—◯—◯—C_5H_{11}
   Cr 146.5 (SmE 145.5) SmA 163.5 N 171.5 I [85]

6. C_9H_{19}—◯—COC_4H_9
   Cr 101.5 SmA 111.5 I [86]

7. C_9H_{19}—◯—COC_4H_9
   Cr 71 SmA 118 I [87]

8. C_9H_{19}—◯—COC_4H_9
   Cr 37 SmA 38 N 40 I [87]

9. C_7H_{15}—◯—COC_6H_{13}
   Cr 84 N 89 I [88]

Polar groups, especially the CN group, play an important role in the application of liquid crystals. Table 13 shows same examples of polar compounds derived from naphthalene. Due to the CN group, the molecules exhibit partial dimerization and, according to general experience, the density of the materials is higher than that of the analogous non-polar compounds. This explains the higher T_{N-I} (e.g. compare the first compound in Table 12 and the third compound in Table 13).

In the last 10 years a very large number of different ring systems have been introduced into liquid crystal chemistry [35a, 36]. Many of them are not very effective in terms of their mesogenity, and may be considered merely as curiosities. Table 14 presents some examples. Spiro ring systems such as compound numbers 1 and 9 in Table 14 are not planar; rather, the rings are perpendicular one to another. This causes an unfavourable bent molecular shape. In seven-membered rings (compound number 6) the directions of the substituent bonds are not parallel, which also leads to a bent molecular shape. Compound number 7 is derived from

Table 13. Large polar rings, without no linking groups.

1.	NC—[naphthalene-phenyl]—C_6H_{13}	
	Cr 59 N 117 I	[89]
2.	NC—[phenyl-naphthalene]—C_6H_{13}	
	Cr 49.5 N 79.7 I	[89]
3.	NC—[phenyl-naphthalene]—C_6H_{13}	
	Cr 75.5 N 102.4 I	[89]
4.	NC—[phenyl-pyrazine]—C_6H_{13}	
	Cr 90 mesophase 108 I	[90]
5.	NC—[phenyl-dioxane]—C_6H_{13}	
	Cr 87 (N 80) I	[84]

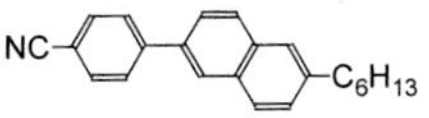

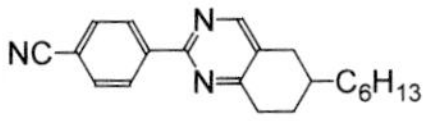

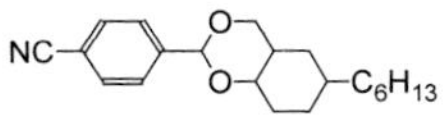

Table 14. Less common rings.

1.	$C_6H_{13}O$—[phenyl-dithiane]—C_5H_{11}	
	Cr ? Sm 46 I	[91]
2.	$C_6H_{13}O$—[phenyl-oxazole]—CH_3	
	Cr 94.8 (N 49) I	[92]
3.	$C_6H_{13}O$—[phenyl-thiazole]—OCH_3	
	Cr 105.5 N 119.2 I	[92]
4.	C_4H_9O—[phenyl-tetrathiafulvalene-phenyl]—OC_4H_9	
	Cr 171 SmG 186 N 210 I	[93]
5.	$C_5H_{11}O$—[phenyl-thiadiazole-phenyl]—OC_5H_{11}	
	Cr 145 SmE 180 SmC 228 N I	[94]
6.	NC—[phenyl-phenyl-cycloheptyl]—C_3H_7	
	Cr 67 N I	[95]
7.	C_7H_{15}—[phenyl]—COO—[tropone]—N=N—[phenyl]—C_7H_{15}	
	Cr (SmI 80 SmC 86) N 142 I	[96]
8.	[phenyl-phenyl-benzimidazole-cycloheptyl]—OC_4H_9	
	Cr 211 (SmA 207) N 240 I	[97]
9.	C_4H_9—[cyclohexyl-spirodioxane-cyclohexyl]—C_4H_9	
	Cr 188 SmB 235 I	[98]
10.	[phenyl]—COO—[cholesteryl]	
	Cr 150.5 N* 182.6 I	[99]
11.	[phenyl]—[cholesteryl]	
	Cr 113 N* 126 I	[100]

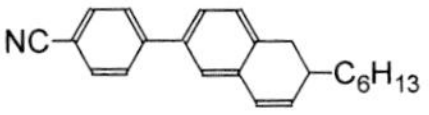

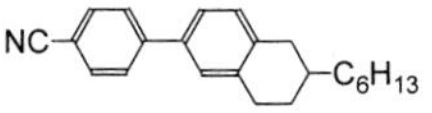

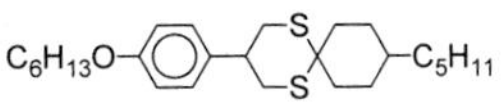

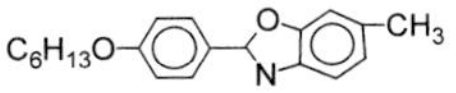

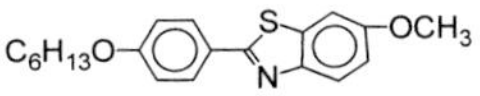

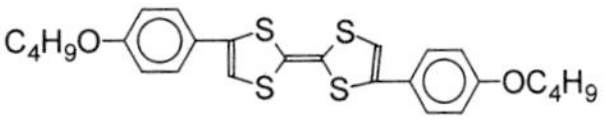

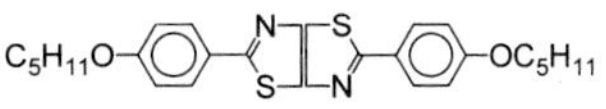

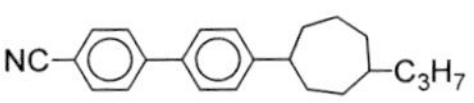

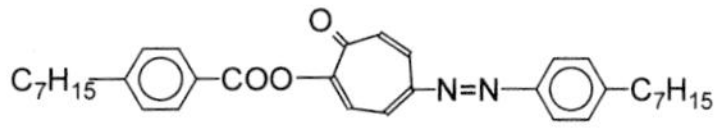

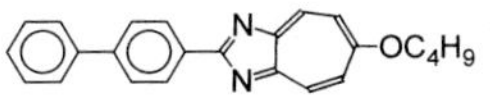

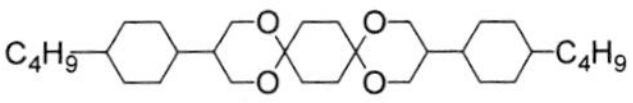

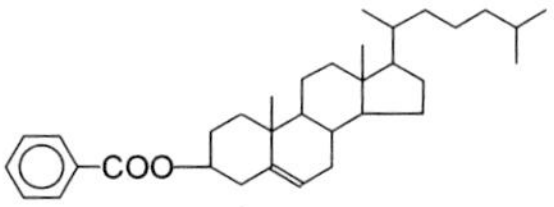

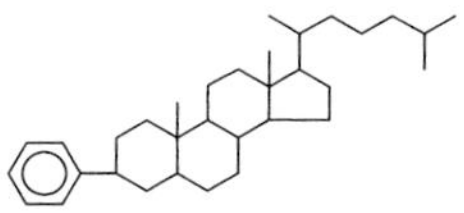

the seven-membered tropone ring. As is already known from investigations into isotropic solutions, 2-acyloxytropone derivatives show an intramolecular migration of the acyl substituents between the two oxygen atoms at C1 and C2, involving a sigmatropic rearrangement [101]. The existence of mesophases of compound number 7 and related tropone derivatives is discussed in terms of the mean rod-like shape of the molecules due to the temperature-dependent sigmatropic rearrangement [96, 102]. Therefore, such materials are also called 'sigmatropic liquid crystals'. The cholesterol compound (number 10) is among the first known liquid crystals, reported by Reinitzer in his famous work [99]. Although cholesterol does not have very good mesogenic potential, because it is a cheap natural product it has been used relatively often in the synthesis of chiral materials. Related steroids such as cholestane (number 11) have also been used. A more detailed discussion of mesogenic sterol derivatives can be found in Gray [38], and a short survey of the different ring systems is given in Deutscher et al. [43].

2.2.3 Rings with Three to Five Atoms

Several rings containing three to five atoms have been used in the design of liquid crystals. Table 15 shows some selected examples. In terms of their influence on the mesogenic properties, the small three and four membered rings are more comparable to bulky terminal substituents than to five- or six-membered rings [156]. We will discuss three-membered rings in Sec. 6.2.4, which deals with terminal substituents.

Table 16 shows some analogous compounds. In comparison to the benzene derivative (compound 1), all the other compounds have inferior mesogenic properties. It is well known that cyclobutane and cyclopentane are rather flexible, and the directions of the linking bonds are not parallel. On the other hand, the *cis* and *trans* isomers, as they are formed during synthesis, have quite different properties (e.g. compound numbers 2 and 3). The spiro rings (compound numbers 5 and 6) are linear, but rather flexible. The examples given in Table 17 further prove the superior mesogeneity of bicyclooctane derivatives. However, the cubane and bicyclopentane rings have a relatively poor mesogenic potential.

Table 15. Ring systems containing three to five atoms.

Cyclopropane	Furane
Oxirane	Pyrrole
Thiirane	Thiophene
Cyclobutane	Thiazole
Cyclopentane	Thiadiazole
Cyclopentanone	Selenophene
Tetrahydrofurane	Tellurophene
Dioxolane-2-one	

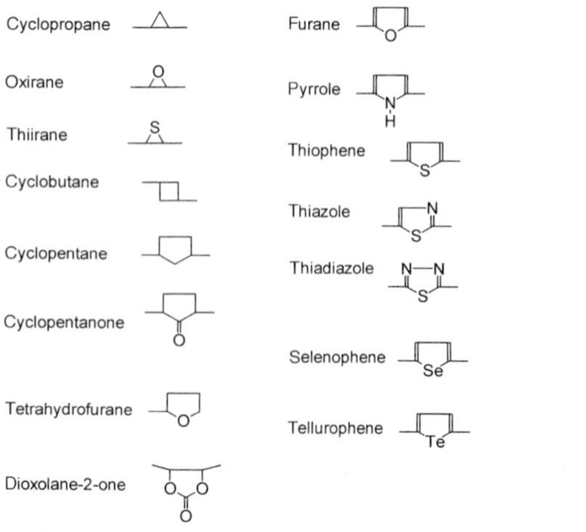

Table 16. Compounds containing small rings.

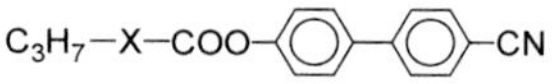

C_3H_7–X–COO–⟨ring⟩–⟨ring⟩–CN

	X	Cr	N	I	Ref.
1.	 phenyl	• 114	• 259	•	[104]
2.	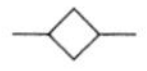 *trans*-Cyclobutane	• 47.5	• 141.6	•	[105]
3.	 *cis*-Cyclobutane	• 55.5	• 63	•	[105]
4.	 *cis/trans*-Cyclopentane	• 52.2	• 66.5	•	[106]
5.	 Spiro[3.3]heptane	• 82	• 154.7	•	[105]
6.	 Dispiro[3.1.3]decane	• 74	• 161.1	•	[105]

Table 17. Compounds of bicyclic rings (adapted from [107]).

CH₃O—◯—OOC—X—COO—◯—OCH₃

	X	Cr		N		I
1.	—◯—	•	211.0	•	281	•
2.	Bicyclo[2.2.2]octane	•	152.0	•	269.0	•
3.	Cubane	•	175.1	•	179.7	•
4.	Bicyclo[1.1.1]pentane	•	143.0	•	145.5	•

Table 18. Comparison of five-membered rings [108].

X-COO—◯—◯—OC₇H₁₅

	X	T_m	T_{N-I}
1.	◯	127.4	134.3
2.	(S)	132	124
3.	(S)	121.1	112.9
4.	(O)	127.5	130.3
5.	(O)	108.2	106.6
6.	(NH)	153.5	152.8
7.	(N-CH₃)	100.5	–

In Table 18 end-standing five-membered rings are compared with the mesogeneity of benzene. Apart from the quite polar compound 6, which probably shows association effects, all the rings show inferior mesogenity to that of benzene. This is probably mainly due to differences in the bond directions, as Zaschke [52] and Iglesias et al. [108a] investigated in detail (see Figure 2), and to differences in polarity. The differences in the bond angles have a greater effect in compounds with five-membered rings in the central core. Some examples are given in Table 19. In particular, the 1,3,4-thiadiazole ring bond in positions 2 and 5 (compound numbers 2 and 4) has good mesogenic potential, and seems to be useful in order to obtain compounds with smectic C or C* phases. The isomeric ring (compound number 5) is much less useful. Figure 2 shows that the thiazole rings provide more linear bond directions than do the other listed rings. This property is reflected in the relatively high clearing temperature of compound number 3.

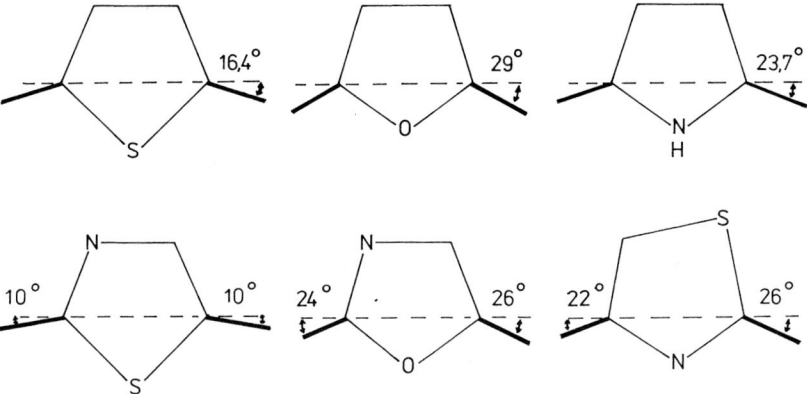

Figure 2. Comparison of the bond angles in five-membered heterocyclic rings. (By courtesy of Zaschke [52]).

Table 19. Compounds containing five-membered heterocyclic rings.

1. C_7H_{15}—⬡—⬡—⬡—C_7H_{15}
   Cr 181 Sm 205 I [50]

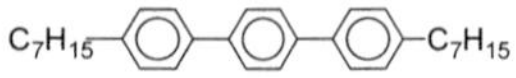

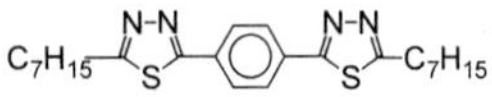

2. C_7H_{15}—[N-N/S]—⬡—[N-N/S]—C_7H_{15}
   Cr 165 SmC 195 I [109]

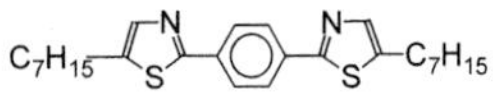

3. C_7H_{15}—[N/S]—⬡—[N/S]—C_7H_{15}
   Cr 125 (Sm 122) I [110]

4. C_7H_{15}—⬡—[N-N/S]—⬡—C_7H_{15}
   Cr 80 SmC 149 N 159 I [111]

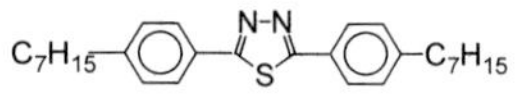

5. C_7H_{15}—⬡—[N-S/N]—⬡—C_7H_{15}
   Cr 49 I [109]

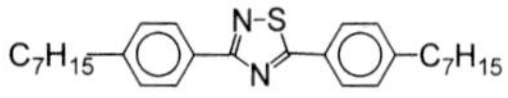

2.3 Linking Groups

Small chemical groups between the rings of a liquid crystal molecule can increase the length of the molecule, while preserving the linear shape. In other cases, however, linking groups produce a bent molecular shape and thus diminish the mesogenic potential. This has been clearly demonstrated by Vorländer [4, 5], and one example is given in Table 20. An even number of carbon atoms in the central linking group allows a linear molecular shape, while an uneven number of atoms induces strong bending,

Table 20. Influence of the length of the central linking group (data from Krücke and Zaschke [112]).

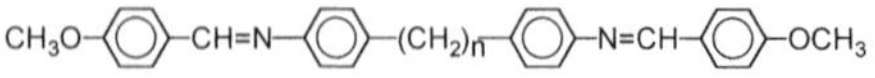

CH_3O—⬡—$CH=N$—⬡—$(CH_2)_n$—⬡—$N=CH$—⬡—OCH_3

n	Cr		N		I
0	•	266	•	390	•
1	•	161	–	–	•
2	•	171	•	312	•
3	•	134	–	–	•
4	•	156	•	270	•

in which case there are no longer nematic phases. Compounds of this type are called 'twins'.

Table 21 shows some of the common linking groups. The discussion of the effect of linking groups on liquid crystalline properties in general cannot be restricted to the geometry of the molecules, because there are the additional effects of conjugative interaction with aromatic groups, effects due to the polarity of the linking groups and the influence on the *cis/trans* isomer stability in compounds containing saturated rings. Table 22 shows some compounds that contain unsaturated linking groups. Without

Table 21. Linking groups.

-CH_2-CH_2-	-COS-
-CH_2-CH_2-CH_2-CH_2-	-OOC-$(CH_2)_n$-COO-
-CH=CH-	-N=N-
-C≡C-	-N=N- (O)
-C≡C-C≡C-	
-CH_2O-	-CH=N-
-COO-	-CH=N-OOC-
	-CH=N-N=HC-

Table 22. Compounds containing unsaturated linking groups (data from Goto et al. [113]).

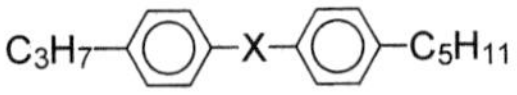

C_3H_7—⬡—X—⬡—C_5H_{11}

	X	T_m	T_{N-I}
1.	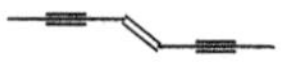 ⟋⟍	64.1	146.0
2.	 =	62.0	115.3
3.	 ⟍	62.5	112.3
4.	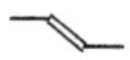 ⟍	107.8	80–90[a]
5.	—	–18	47.8[b]
6.	 =	39.0	35.8
7.	 ⟍	58.3	–[c]

[a] Estimated from the behaviour of homologues.
[b] Smectic B–isotropic transition.
[c] Isotropic melt supercooled to 15°C; not mesomorphic.

doubt the ranking of the linking groups with respect to their nematogenic potential is due to conjugation with the aromatic rings. As would be expected, the linking groups containing double bonds are more effective than are those containing triple bonds. The high T_{N-I} of compound 1 in Table 22 is typical for a three-ring compound. The lack of a nematic phase for compound 7 indicates a bent molecular shape.

Some compounds containing a methylenoxy group are compared in Table 23. If the oxygen atom has the chance to come into conjugative interaction with the benzene ring (compound 1), the hindrance of the rotation about the bonds to the oxygen atom is increased and thus the nematogenity is much higher than in the reverse case (compound 2). Comparison of compounds 3 and 4 shows that, probably because of the higher flexibility (lower rotational hindrance) of the methylenoxy group, these compounds exhibit poorer mesomorphic potential than do the compounds containing an ethylene group.

Table 24 is compiled from the literature, and includes a large number of different linking groups. Again the influence of conjugative effects can be seen; in addition, the effect of polarity of linking groups is illustrated, e.g. by comparison of the azomethine group with the nitrone groups, or the azo group with the azoxy group.

Table 23. Comparison of linking groups (data from Carr and Gray [114]).

C_5H_{11}—⬡—X—⬡—R_1

X	R_1	Cr		N		I
1. $-CH_2O-$	$-CH_3$	•	69	(•	33.5)	•
2. $-OCH_2-$	$-CH_3$	•	63	(•	-210^{a})	•
3. $-CH_2-CH_2-$	$-C_3H_7$	•	60	•	62	•
4. $-CH_2O-$	$-C_3H_7$	•	56	(•	30)	•

[a] Extrapolated from data for mixtures.

Table 24. Transition temperatures of compounds containing different linking groups (from Praefcke et al. [115]).

CH_3O—⬡—M—⬡—C_4H_9

M	K		N		I
$-CH=CH-$	•	116	•	121	•
$-CH=CCl-$	•	40	•	38	•
$-C\equiv C-$	•	49	•	37	•
$-CO-$ (=O)	•	40	•	24,5	•
$-CH=N-$	•	22	•	47	•
$-N=CH-\downarrow O$	•	108	•	70	•
$-CH=N-\downarrow O$	•	113	•	53	•
$-N=N-$	•	32	•	47	•
$-N=N-\downarrow O$	•	41	•	69	•
$-N=N-\downarrow O$	•	43	•	71	•

The effects of linking groups can be quite different in aromatic and non-aromatic compounds, because in the latter there are no conjugative effects. It can be seen from Table 25 that the ethylene (compound 3) and ethinylene (compound 4) groups, which in

Table 25. Linking groups in non-aromatic compounds.

1. C_5H_{11}—⬡—⬡—C_5H_{11}
   Cr 40 Sm 110.4 I [54]

2. C_5H_{11}—⬡—CH_2–CH_2—⬡—C_5H_{11}
   Cr 46 Sm 109 I [101]

3. C_5H_{11}—⬡—$HC=CH$—⬡—C_5H_{11}
   Cr 53 Sm 95 I [101]

4. C_5H_{11}—⬡—$C\equiv C$—⬡—C_5H_{11}
   Cr 52 (N 50) I [101]

aromatic compounds possess a high mesogenic potential, decrease the clearing temperatures in comparison to a single bond (compound 1) or an ethylene (compound 2) group.

The clearing temperatures of some homologous series are displayed in Figure 3. The influence of the linking groups is clearly visible in the different levels of the clearing temperatures; however, there is also a decisive influence of the terminal substituents, which will be discussed later.

It is worth noting that Deutscher et al. [43] synthesized compounds with bent moieties, in which the bending was compensated for by 'crooked' spacers. Several liquid crystalline compounds have been obtained in this way by using a CH_2 group as a 'crooked' linking group between two cyclohexane rings.

2.4 Terminal Substituents

Some of the more common terminal substituents are listed in Table 26. The most common ones are the alkyl and alkyloxy groups. The behaviour within the homologous series illustrated in Figure 3 shows that, in general, there is an alternation of T_{N-I}. This can be explained by the alternation of the length-to-breadth ratio. Figure 4 shows a typical six-membered ring, with an attached alkyl group. It is easily seen that the attachment of an odd-numbered carbon atom substituent increases the length-to-breadth ratio more than does the attachment of an even-numbered carbon atom substituent, i.e. the value of the length-to-breadth ratio will show an alternation. In the same sense, T_{N-I} alternates.

It is well known that alkyl and related groups are flexible, due to the relatively low energy barrier of about 3.4 kJ mol^{-1} neces-

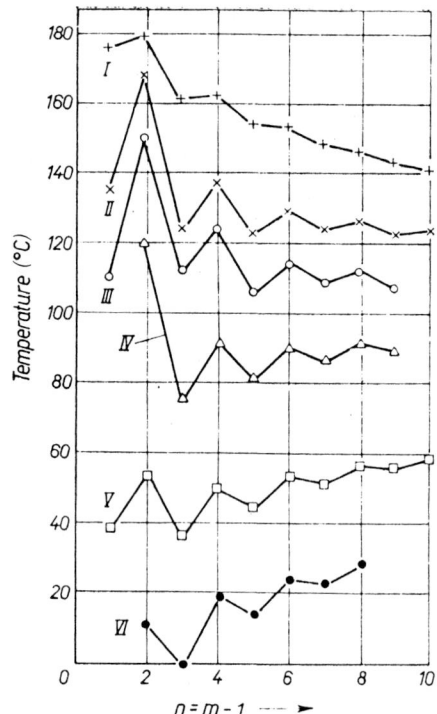

Figure 3. Nematic clearing temperatures in some homologous series. (From Demus [116]).

sary for the change from the *trans* to the *gauche* conformation. Alkyl chains in the liquid crystalline state never possess the ideal all-*trans* conformation, there being a mixture of all possible conformations in

Table 26. Terminal substituents.

Alkyl	$-C_nH_{2n+1}$
Deuterated alkyl	$-C_nD_{2n+1}$
Alkenyl	$-(CH_2)_m-CH=CH-C_nH_{2n+1}$
Alkynyl	$-(CH_2)_m-C\equiv C-C_nH_{2n+1}$
Alkyloxy	$-OC_nH_{2n+1}$
Alkylmercapto	$-SC_nH_{2n+1}$
Alkylamino	$-NH-C_nH_{2n+1}$
Acyl	$-CO-C_nH_{2n+1}$
Acyloxy	$-OCO-C_nH_{2n+1}$
Alkylester	$-COO-C_nH_{2n+1}$
Alkylcarbonates	$-OCOO-C_nH_{2n+1}$
Halogeno	$-F, -Cl, -Br, -I$
Cyano	$-CN$
Isothiocyanato	$-NCS$
Nitro	$-NO_2$
Cyanoalkyl	$-(CH_2)_n-CN$
Cyanoethenyl	$-CH=CH-CN$
Dicyanoethenyl	$-CH=CH(CN)_2$
Fluorinated methyl	$-CH_2F, -CHF_2, -CF_3$
Fluorinated methoxy	$-OCH_2F, -OCHF_2, -OCF_3$
Perfluoroalkyl	$-C_nF_{2n+1}$

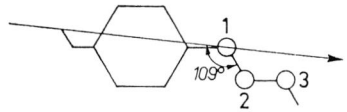

Figure 4. Fragment of a molecule containing a ring substituted at the 4-position. (From Demus [116]).

a temperature-dependent equilibrium. At higher temperatures the amount of *gauche* conformers will be higher than at lower temperatures. Of course, the number of possible conformers increases greatly with chain length. In homologous series with high clearing temperatures the T_{N-I} decreases, because of the reduced anisotropy of the alkyl chains, and the alternation seen at medium chain lengths disappears. However, in series with low T_{N-I}, because of the relatively strong anisotropy of the chains, the T_{N-I} increases with distinct alternation. The 'magic' temperature between the decrease and increase in T_{N-I} is about 70 °C, as can be derived from many homologous series. This principal behaviour seen in alkyl chains can also be found in other flexible chains.

There is another general rule in homologous series. Lower members often exhibit relatively high melting temperatures, T_m, the medium-sized members (about 4–8 carbon atoms) have the lowest T_m, and in higher members T_m again increases. This can be seen in the data for a typical series displayed in Figure 5. Despite this trend in the melting temperatures, they do not show a strictly regular behaviour like that of T_{N-I}. Because of the high T_m, in lower members the liquid crystalline phases are often monotropic, and only the higher members show enantiotropic behaviour (Figure 5).

In many homologous series the lower members are nematic, the medium members nematic and smectic, and the higher members are smectic only, eventually occurring in several modifications (Figure 6). Of course, there are also series where smectic phases occur in even the lowest member, and in a few series even in the highest members only nematic phases exist. Quite exceptionally, in a very few series the lower members are smectic and only at somewhat high-

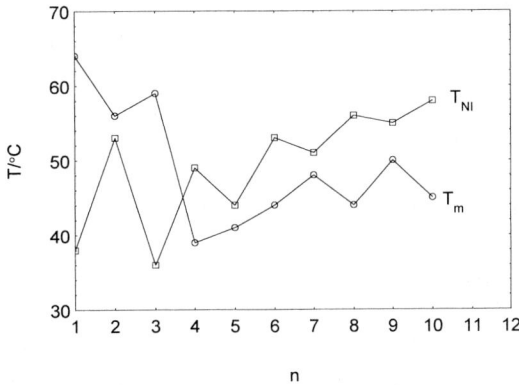

Figure 5. Transition temperatures in the homologous series of the 4-n-alkyloxyphenyl-4-n-hexylbenzoates. (Data from Schubert and Weissflog [117]).

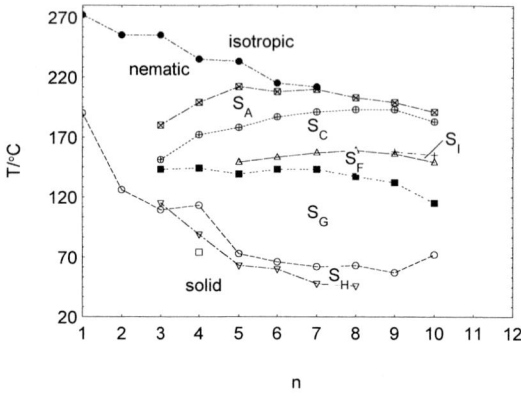

Figure 6. Transition temperatures in the homologous series of terephthalylidene-*bis*-*N*-(4-n-alkylanilines). (Data from Wiegeleben et al. [149]).

er chain lengths do nematic phases appear [118]. The above description shows that the prediction of smectic or nematic behaviour in mesogens is very difficult. The available theories [25, 119] start from the assumption that there must be some attractive forces for the formation of smectic layers to occur. The attraction should be maximal between the polarizable rings, compared to the attraction to the alkyl chains. In addition to the attraction, there needs to be a tendency to microphase separation in ring-containing and chain-containing microphases, the tendency being more pronounced at higher chain lengths, thus stabilizing the smectic layers. On the other hand, model calculations by Frenkel [120] showed that smectic phases can occur without any attraction, but merely by increasing the density of the system.

It is a well established rule that in completely non-aromatic compounds (e.g. dialkylbicyclohexanes) smectic B and E phases are the preferred phases [34–36]. A more detailed discussion of molecular structure and smectic properties is given in Gray [39b].

Partially or completely deuterated alkyl chains are prepared mainly for experiments such as neutron scattering or nuclear magnetic resonance (NMR). The transition temperatures are usually the same as or very similar to those of the analogous alkyl compounds [121]; however, exceptionally, there can be differences of about 10 K [96].

In work initiated by the research group of the Hoffmann La Roche company, many homologous series containing alkenyl groups have been developed (see [122–124, 124a] and references therein). There are remarkable alternating effects in several properties, depending on the position of the double bond. Table 27 presents the transition temperatures of some pyridine derivatives. The clearing temperatures show a distinct alternation. According to an explanation proposed by Schadt et al. [123], the direction of the double bonds in an odd-numbered position is more parallel to the director than that of double bonds in an even-numbered position, which is in contradiction to the usual presentation of the shape of alkenyl groups.

Compounds containing strongly polar groups (e.g. –CN, –NO$_2$) deserve special discussion. From investigations using X-rays and other method, it is well known that such compounds form double molecules that exist in equilibrium with single molecules [32, 125, 126]. Comparing the clearing temperatures of highly polar and low polar compounds, the former show much higher T_{N-I} (Table 28). Using simple arguments, this effect has been explained by the increase in the molecule length by dimerization. In fact, however, the length-to-breadth ratio controls T_{N-I}. Due to dimerization, the breadth increases by a factor of 2, and the length only by a factor of about 1.1–1.4, so that the effective length-to-breadth ratio should be reduced. It has been shown in many investigations that highly polar compounds have

Table 27. Transition temperatures of 5-n-heptyl- and 5-n-octyl-2-(4-n-octenyloxy-phenyl)pyridines (reproduced, with permission, from Kelly et al. [122]).

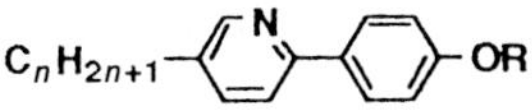

C_nH_{2n+1}—[pyridine]—[phenyl]—OR

n	R	C–Sm (°C)	SmG–Sm$_1$ (°C)	Sm$_1$–SmB or SmC (°C)	SmC or SmA–I (°C)
1. 7	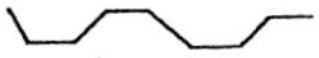	47	–	58	81
2. 7	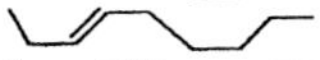	53	68	76	84
3. 7	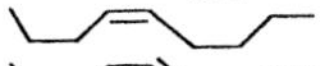	32	–	51	58
4. 7	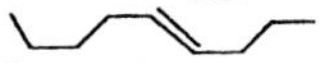	81	–	82	84
5. 7	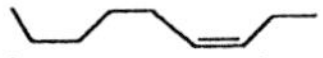	38	49	60	66
6. 7	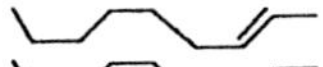	48	51	58	80
7. 7	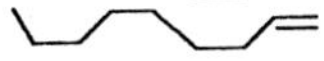	30	(29)	48	73
8. 8		38	–	62	82
9. 8	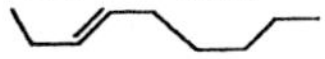	43	57	75	85
10. 8	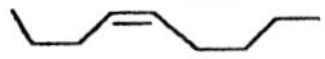	11	–	40	61
11. 8	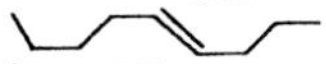	61	72	84	85
12. 8	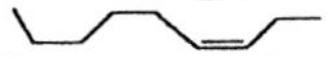	16	–	60	67
13. 8	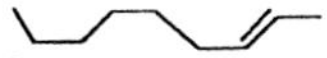	36	–	66	80
14. 8	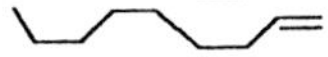	23	–	53	75

Table 28. Clearing temperatures of polar and non-polar liquid crystals (data from Demus and Hauser [28]).

	R	T_{N-I}
C_5H_{11}—[phenyl]—CH=N—[phenyl]—R	–CH$_3$	25.7
	–CN	75
$C_6H_{13}O$—[phenyl]—COO—[phenyl]—R	–CH$_3$	52
	–NO$_2$	58
	–CN	81.0

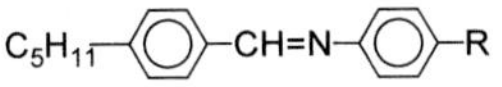

a much higher density (packing fraction) that do low polar compounds [27, 28]. This increase in density accounts for the increase in clearing temperature. This is an effect comparable to the increase in the transition temperature that occurs when the pressure is increased, because of density increase.

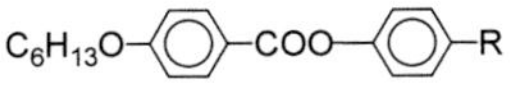

Table 29 contains some data on CN substituted biphenyl analogues. The highly polar compounds show association effects. The dimerization occurs by interaction of the CN group with a benzene ring (compounds 1 and 2), or interaction of two CN groups (compound 3). The sketches of the possible dimers elucidate the different effective length-to-breadth ratios, which explains the trend in the clearing temperatures.

Table 29. Association in polar biphenyl analogues (data from Ibrahim and Haase [127]).

		T_{N-I}
1.	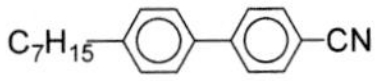 C_7H_{15}—[phenyl]—[phenyl]—CN	42.2
2.	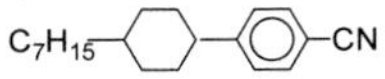 C_7H_{15}—[cyclohexyl]—[phenyl]—CN	56.8
3.	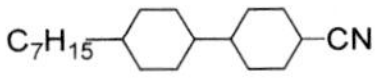 C_7H_{15}—[cyclohexyl]—[cyclohexyl]—CN	83.3

Possible associates

1. 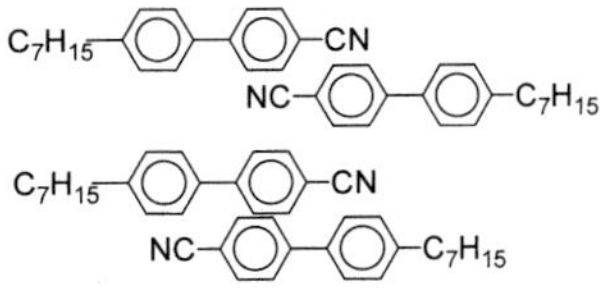 C_7H_{15}—[phenyl]—[phenyl]—CN
   NC—[phenyl]—[phenyl]—C_7H_{15}

   C_7H_{15}—[phenyl]—[phenyl]—CN
   NC—[phenyl]—[phenyl]—C_7H_{15}

2. 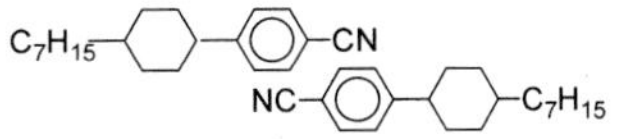 C_7H_{15}—[cyclohexyl]—[phenyl]—CN
   NC—[phenyl]—[cyclohexyl]—C_7H_{15}

3. 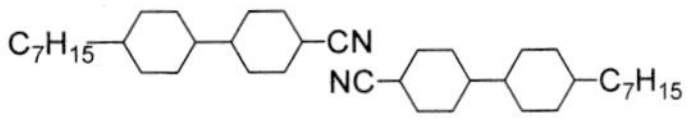 C_7H_{15}—[cyclohexyl]—[cyclohexyl]—CN
   NC—[cyclohexyl]—[cyclohexyl]—C_7H_{15}

The halogens and the isothiocyanato group introduce relatively large positive dielectric anisotropy into the molecules; however, there is no association [37].

Several of the substituents in Table 26 contain fluorine atoms. Because of the high energy of the C–F bond these substituents are very stable, and have become increasingly important in the application of nematics in TFT (Thin Film Technology) displays [128–131]. The perfluoroalkyl chains play a special role, for two reasons:

1. In a typical perfluorinated alkyl the energy difference between the *gauche* and *trans* conformations is relatively high (9.1 kJ mol^{-1}) compared to that in an alkyl (3.4 kJ mol^{-1}) [132, 133]. Therefore the perfluorinated chains are much more stretched and thus favourable for high clearing temperatures.
2. Perfluorinated moieties show a strong trend to segregation from alkyls and non-fluorinated cores [134]; perfluorinated benzene derivatives can be virtually immiscible with ordinary liquid crystal materials [135].

The trend to segregation has been used in the design of ferroelectric materials with a small positive or even negative temperature dependence of the layer thickness [136]. Considering this trend to segregation, Tournilhac et al. [137, 138] have designed 'polyphilic' liquid crystals, which consist of two fluorinated moieties, an ordinary alkyl part and the core. Compound **2** [137] is a typical

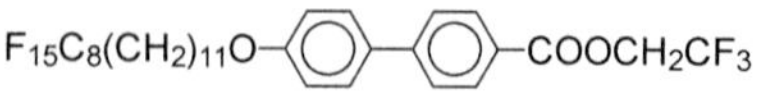

F$_{15}$C$_8$(CH$_2$)$_{11}$O—⬡—⬡—COOCH$_2$CF$_3$

2 Cr 95 (SmX 92) SmA 113 I

example. Despite the fact, that the polyphilic compounds are not chiral, they exhibit ferroelectric properties and have a small spontaneous polarization [137].

Table 30 contains transition data on some compounds with terminal small rings (compounds 2 and 3). Compared with the simple alkyloxy group (compound 1), the small rings have an effect comparable to branched alkyls, but unlike that of five- or six-membered rings. In order to obtain chiral materials, the introduction of an oxirane or thiirane ring in the terminal chains is useful [143].

There are many compounds with branched terminal substituents, particularly chiral materials. The effect of a branch depends substantially on its position in a chain [144–147]. This is demonstrated in Table 31. The reduction in the clearing temperature

Table 30. Compounds with terminal rings.

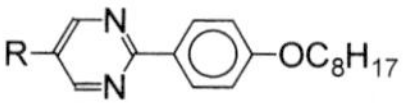

R—⬡(N,N pyrimidine)—⬡—OC$_8$H$_{17}$

	R		
1.	C$_{11}$H$_{23}$O—		
	Cr 56 SmC 84.9 SmA 97.1 I		[139]
2.	▷—(CH$_2$)$_9$O—		
	Cr 63.7 SmC 93.2 I		[140]
3.	◁—(CH$_2$)$_8$O—		
	Cr 62 SmC 83 I		[141, 142]

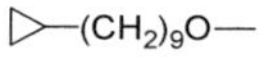

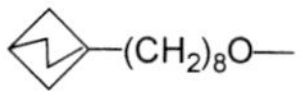

Table 31. Compounds containing branched pentyl groups (data from Gray and Harrison [144]).

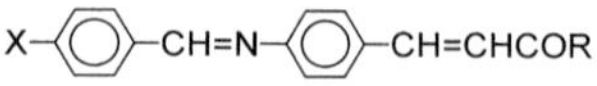

X—⬡—CH=N—⬡—CH=CHCOR

–R	X = phenyl T_{SmA-I}	X = NC– T_{N-I}
 ⋏⋏⋏	204	136.5
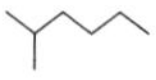 ⋎⋏⋏	180	<20
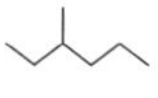 ⋏⋎⋏	190.5	112
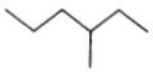 ⋏⋏⋎	196	103
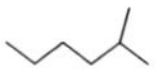 ⋏⋏⋀	199	119.5

is greater the nearer the branch is to the centre of the molecule. This can be understood by remembering the flexibility of alkyl chains, which increases from the centre towards the end of a molecule [148]. This causes an increase in the effective width of the chain towards its end, so that the effective shape of an alkyl is similar to a cone. An terminal branch can be easily packed into this cone without the need for additional space. A striking case is displayed in Figure 7. In this homologous series with a terminal phenyl group the members exhibit alternating nematic properties, i.e. only the even-number carbon atom members are nematic. Here the usual odd/even effect is reinforced by the bulky terminal phenyl moiety. In the terminal chains CH_2 groups may be replaced by an oxygen atom. This causes a substantial decrease in the clearing temperature [150, 151], the decrease being more pronounced the nearer the oxygen atom is to the core [151]. It seems that the major effect of the oxygen atom is to reduce the stiffness of the chain.

2.5 Lateral Substituents

Common lateral substituents are the halogens and methyl, ethyl, cyano and other small groups. There are some uncommon liquid crystals that contain large lateral substituents, even aromatic groups; these will be discussed later.

Every lateral substitution leads to an increase in the breadth of the molecule, and thus a reduction in the length-to-breadth ratio, X. In agreement with the van der Waals molecular–statistical theories, this usually reduces the clearing temperature. Figure 8 shows that there is a (not very strict) relation between X and T_{N-I} for laterally substituted compounds. Similar relations have been found between the width of the lateral substituents and the clearing temperature [153]. There are several reports in the literature that prove that small lateral substituents depress the clearing temperature less than do large ones [27, 28, 37–39, 153]. Exceptionally, in the case of highly polar lateral substituents T_{N-I} may be even enhanced (see Table 32). Of course, in these cases al-

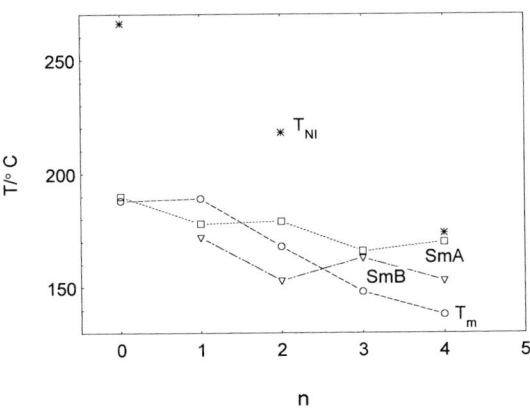

Figure 7. Transition temperatures in the homologous series of ω-phenylalkyl-4-[4-phenylbenzylideneamino)cinnamates. (Data from Gray [144]).

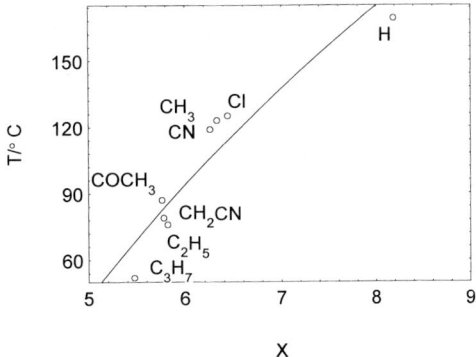

Figure 8. Clearing temperatures in dependence on the length-to-breadth ratio X in 2-substituted hydroquinone-*bis*-[4-n-hexylbenzoates]. (Data from Demus et al. [152]).

Table 32. Polar and non-polar lateral substituents (data from Demus and Hauser [28]).

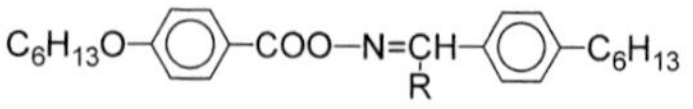

$C_6H_{13}O-\bigcirc-COO-N=CH-\bigcirc-C_6H_{13}$
                         $|$
                         R

	R	T_{N-I}
1.	–H	71
2.	–CN	83

$C_4H_9-\bigcirc-COO-\bigcirc-OOC-\bigcirc-C_4H_9$
                R_1 R_2

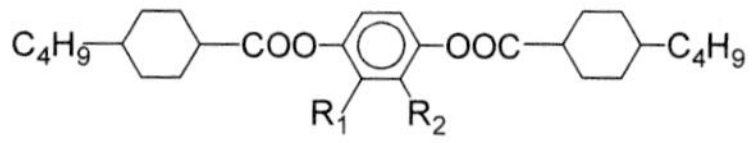

	R_1	R_2	T_{N-I}
3.	–CH$_3$	–H	180
4.	–CN	–CN	210

so the L/B breadth ratio will be decreased. However, because of the strong attraction in polar compounds the density is increased, and this may overcompensate the decrease in T_{N-I} resulting from molecular broadening.

To be useful for applications, compounds should have as low as possible melting temperatures. There are many examples in the literature (e.g. [154]) where the T_m of a compound (and thus its clearing temperature) has been decreased by means of lateral substitution (see also the examples given in Table 33). This effect is probably due to the reduction in the symmetry of the molecule, because in less symmetric parent molecules it seems that the effect of lateral substitution is suppressed.

The position of lateral substituents is quite important [154, 159–161], as can be seen from the examples given in Table 34. Because they do not reduce the conjugation of the rings, substituents in the positions 3 and 3' depress T_{N-I} far less than do those in positions 2 and 2'. On the other hand, the influence of the size of the substituent can be nicely observed. The influence of lateral substituents on the stability of the smectic phase by far exceeds that on the nematic phase. This is clearly demonstrated by compounds 4 in Table 1. Suppression of the

Table 33. Decrease in the melting and clearing temperatures by the addition of lateral substituents.

$C_5H_{11}-\bigcirc-COO-\bigcirc-OOC-\bigcirc-C_5H_{11}$
                         $|$
                         R

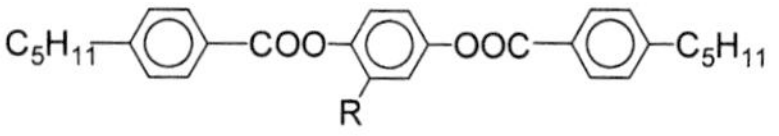

	R	T_m	T_{N-I}	Ref.
1.	–H	125	188	[155a]
2.	–Cl	79	145	[155b]
3.	–Br	76	134	[155b]
4.	–CH$_3$	78	146	[155b]
5.	–C$_2$H$_5$	63	100	[155b]

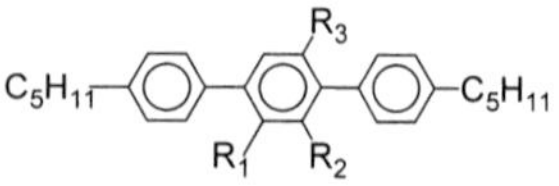

$C_5H_{11}-\bigcirc-\overset{R_3}{\bigcirc}-\bigcirc-C_5H_{11}$
             R_1 R_2

	R_1	R_2	R_3		Ref.
6.	H	H	H	Cr 192 SmA 213 I	[156]
7.	F	H	H	Cr 51.1 SmB 62 SmA 109.5 N 136.5 I	[157]
8.	CN	H	H	Cr 40 (SmA 35 N 38) I	[158]
9.	F	F	H	Cr 60 N 120 I	[156]
10.	F	H	F	Cr 63 N 85.5 I	[156]

Table 34. Lateral substituents in different positions (courtesy of Coates [37]).

3' 2'2 3
⟨○⟩-⟨○⟩-N=CH-⟨○⟩-OC₇H₁₅

Substituent	Depression of T_{N-I} (°C)
2 F	−53.5
2 Cl	−118.5
2 Br	−138.5
2 I	−160
2′ Cl	−111.5
2′ Br	−127.5
2′ I	−153
3 Cl	−73.5
3 Br	−81.5
3′ Cl	−27
3′ Br	−39

smectic phase by lateral substituents is systematically used in the design of liquid crystals for applications in displays.

There are compounds in which the lateral substituents are shielded, so that they are less effective in broadening the molecule. Table 35 shows some impressive examples of this kind. There are also compounds with axial CN groups in cyclohexane [162, 163] or dioxane [164] rings, in which partly the shielding effect and partly the increase in density due to the polar group may be responsible for the unexpectedly high clearing temperature.

Table 35. Shielded lateral substituents (courtesy of Coates [37]).

RO-⟨○⟩-(X)-C(=O···H-O)-O-H···O=C-⟨○⟩(X)-OR

X	T_{SmA-N} or T_{SmA-I} (°C)	T_{N-I} (°C)
H	147	181
Cl	186.5	192.5
Br	182.5	189.5
I	164.5	178.5
NO₂	166.5	–

Table 36. Lateral substituents with intramolecular association.

CH₃O-⟨○⟩-CH=N-⟨○⟩-C₄H₉
          R

R		
−H	Cr 22 N 47 I	[165]
−OH	Cr 43 N 63 I	[166]

C₁₀H₂₁O-⟨○⟩-CH=N-⟨○⟩-CH=CH-COO-C*H-C₂H₅
         R   CH₃

R		
−H	Cr 82 Sm 61 SmC 89 SmA 106 I	[167]
−OH	Cr 124 (SmC 118) SmA 135 I	[167]

Lateral substituents are may also give rise to shielding effects due to intramolecular association. Two examples involving the formation of intramolecular hydrogen bonds are given in Table 36. In both cases the laterally −OH substituted compounds have notably higher clearing temperatures. The intramolecular hydrogen bonds probably enhance the stiffness of the molecules, while at the same time dramatically improving the chemical stability of the highly reactive unsubstituted Schiff's bases.

3 Liquid Crystals with Unconventional Molecular Shapes

As discussed in Sec. 2.1, the prototype of the rod-like liquid crystal molecule (**1**) consists of a rigid core substituted with terminal flexible substituents, and eventually small lateral substituents. In the last few years in particular many mesomorphic compounds that do not correspond to this formula have been synthesized. Such materials may be called 'unconventional' liquid crystalline compounds [1, 2].

3.1 Acyclic Compounds

There are several reports of semifluorinated alkanes, proving the existence of smectic phases [168–171]. For example, perfluorodecyldecane $(F(CF_2)_{10}(CH_2)_{10}H)$ has the following transition scheme [169]:

tilted Sm 46.9 tilted SmG or SmJ 63.5 I

The investigated semifluorinated alkanes exhibit smectic phases, mainly of the hexagonal ordered type. However, a smectic A phase in an iodine derivative has also been reported [170].

The n-alkanes with chain lengths above 20 carbon atoms show phases similar to smectic B; however, there is no final proof of the phase type [172]. Mesophases have also been observed in some unsaturated acids and derivatives of acids and aldehydes [34, 35].

3.2 Flexible Cyclic Compounds and Cyclophanes

Cycloalkanes possess quite high flexibility in their rings. Möller et al. [173, 174] found that cycloalkanes with 12–96 methylene groups are able to form mesomorphic phases. The flexible rings are folded and build lamellar structures that show hexagonal order.

Esters of cyclic alkane carboxylic acids (7–11 carbon atoms in the rings) are nematic; their T_{N-I} decrease with increasing ring size [175].

Several crown ether derivatives with ring sizes of 15 and 18 atoms are mesomorphic [176–178]; an example is compound 3 [176].

The cyclophanes, in which the long flexible chains are interrupted by typical mesogenic rigid cores, are related to cycloal-

3 Cr 135 (N 133) I

kanes. Ashton et al. [179, 179a] found smectic phases in several cyclophanes containing alkyl or polyether linking groups

4 Cr 193 SmE 196 SmA 209 I

(e.g. **4** [179]). In the solid state the molecule is folded, with the biphenyl rings parallel one to another. This folded conformation would be a reasonable explanation for the existence of classical smectic phases in the cyclophanes. By comparison with the clearing temperatures of dialkylbiphenyls, there is a strong stabilizing effect of the smectic layers by the flexible linking groups.

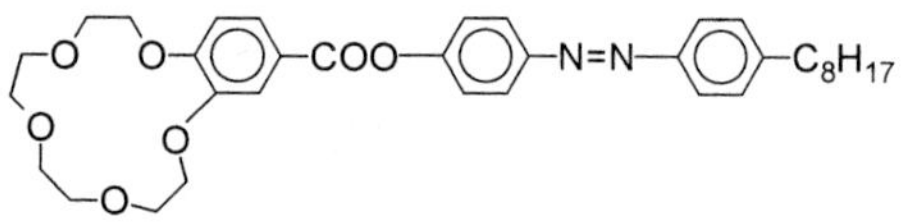

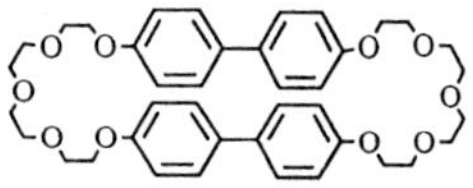

3.3 Compounds with Large Lateral Substituents

3.3.1 Acyclic Lateral Substituents

Until 1983 it was generally accepted that lateral substituents diminish the mesogeneity of a compound, the extent of the effect depending on their size. As found by Weissflog and Demus [180,181], surprisingly, compounds with large flexible lateral substituents exhibit liquid crystalline phases. Figure 9 shows a typical example, and illustrates the dramatic decrease in the clearing temperature with increasing length of the lateral chain and the final convergence with long lateral chains. A similar trend can be

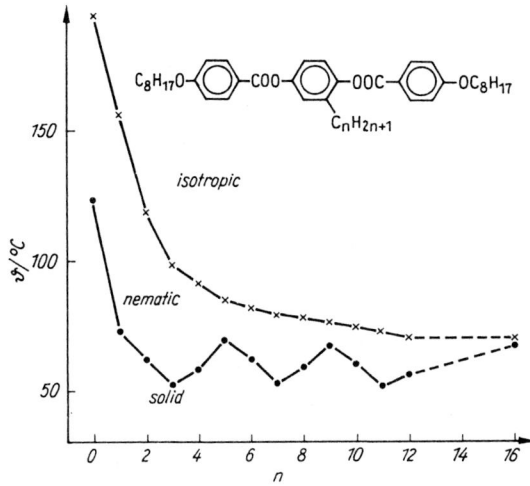

Figure 9. Transition temperatures in the homologous series of the 1,4-*bis*-(4-n-octyloxybenzoyloxy)-2-n-alkylbenzenes. (By courtesy of Weissflog and Demus [180]).

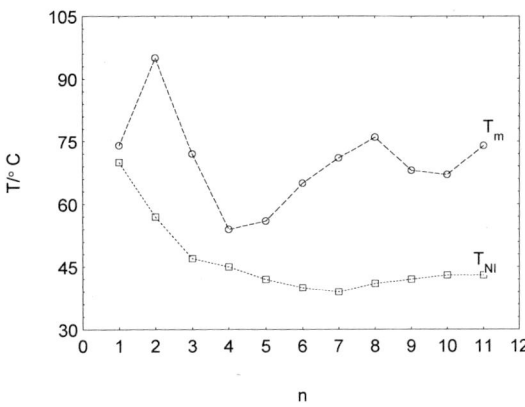

Figure 10. The transition temperatures of the di-n-alkylketoximino 2,5-*bis*-[4-n-octyloxybenzoyloxy]-benzoates. (Data from Weissflog and Demus [184]).

found in many homologous series. The investigation of several physical properties of such compounds delivered evidence favouring the conformation where that part of the flexible lateral chain exceeding five carbon atoms is more or less parallel to the basic molecule [182]. Te compound with $n=9$ shown in Figure 9 has been investigated in the solid state by X-ray diffraction [183]. Surprisingly, the lateral alkyl chain is in the all-*trans* conformation, and all three alkyl chains are nearly parallel. This conformation deviates strongly from the rod-like shape, and it seems improbable that it is maintained in the nematic state.

Figure 10 adds another argument to this discussion. The compounds with swallow-tailed lateral substituents show nematic properties. The important point in this series is that, for higher members, T_{N-I} increases with elongation of the lateral chains. This can only be explained if the length-to-breadth ratios increase at the same time.

There are compounds containing two long chain lateral substituents that are nematic [185]. If the basic molecule is large enough (e.g. **5** [185]) the clearing temperatures that can be achieved are quite high.

5   Cr 145 N 160 I

3.3.2 Lateral Ring-Containing Substituents

Whereas compounds with large flexible lateral substituents, assuming that the lateral chains are oriented nearly parallel to the basic molecule, may be considered as variants of the classical rod-like molecules, compounds with lateral ring-containing substituents lead to completely new concepts of mesogens.

The first examples of mesogens with lateral aromatic substituents were synthesized already in Vorländer's group [18, 186].

Mauerhoff prepared the nematic compound **6** [186]. This field of liquid crystal chemistry, which was for a long time forgotten, has been re-activated by Cox et al. [188], Gallardo and Müller [189] and Weissflog et al. [190–192].

CH₃O–⟨O⟩–CH=N–⟨O⟩–⟨O⟩–N=CH–⟨O⟩–OCH₃
            N=CH–⟨O⟩–OCH₃

6   Cr 159 N 218 I

In the series represented by compound **7** [188] which have monotropic nematic properties, the lateral phenyl group is attached without a spacer. It is not easy to understand, why compounds with very bulky substituents are mesogenic. Hoffmann et al. [193] investigated compound **8** in the solid state by X-ray analysis. As was already

$C_6H_{13}O$–⟨O⟩–COO–⟨O⟩–OOC–⟨O⟩–OC_6H_{13}
                      ⟨O⟩

7   Cr 65 N 70 I

C_2H_5O–⟨O⟩–COO–⟨O⟩–OOC–⟨O⟩–OC_2H_5
O_2N–⟨O⟩–CH_2–OOC

8   Cr 185 (N 166) I

known for related compounds, the phenylene *bis*(benzoate) three-ring skeleton was found to have a non-planar, but greatly extended shape. The lateral substituent is largely aligned parallel to the long axis of the basic molecule. Assuming that a similar conformation also exists in the nematic state (as it has been found by Perez et al. [193a] in similar case), this bulky but, in total, rodlike molecular shape can explain the mesogenic properties. Similar results have been obtained by Weissflog et al. [194] by X-ray analysis of an aromatic carboxylic acid with a bulky lateral substituent.

In more recent papers, compounds have been described in which the lateral substituents are attached via spacers [191–192]. As the spacer length is increased, the clearing temperatures are seen to alternate distinctly (Figure 11).

Lateral substituents can be in different positions of the basic molecule [195], and lateral aromatic substituents can have several substituents themselves [191, 192] or they can be alicyclic [191]. In the so-called λ-shaped mesogens [196], which occur in nematic and smectic A phases, the aromatic lateral substituents, as in older examples, are bound by carboxylic groups. Matsunaga et al. [20] synthesized 1,2-benzene derivatives and 2,3-naphthalene derivatives, which may be considered as compounds with ring-containing lateral substituents, attached at a terminal benzene ring. Fig. 12 presents the transition temperatures of a homologous series of this type. The first compounds of this U-shaped type were already synthesized by Vorländer and Apel [16]. Recent investigations by Attard et al. [187] in U-shaped compounds proved the existence of bilayers in the different smectic phases.

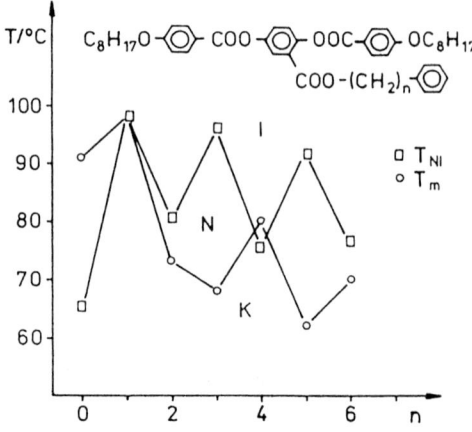

Figure 11. Transition temperatures of ω-alkylphenyl-2,5-*bis*-(4-n-octyloxybenzoyloxy)benzoates. (By courtesy of Weissflog et al. [191]).

ring-containing substituents (see **9** and Table 37). The compounds exhibit classical smectic B and smectic A phases. Because of the excessive steric crowding at the central benzene ring, the three substituents probably cannot be parallel to one another, which may explain the relatively low transition temperatures compared to similar benzene derivatives with two substituents. It should be mentioned here that benzene derivatives with three large substituents in positions 1, 3 and 4 exhibit columnar phases [198].

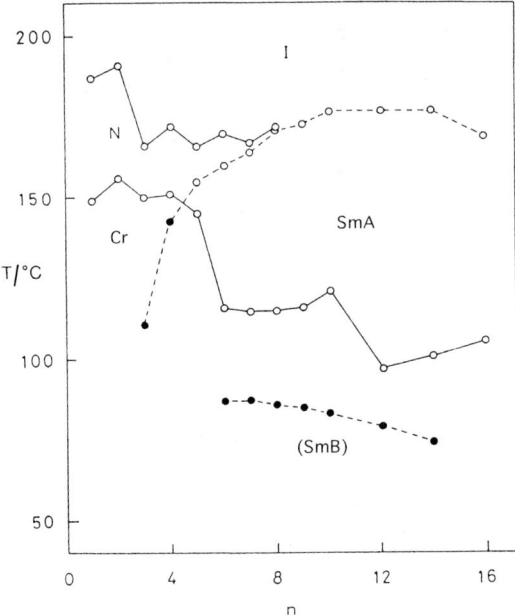

9 Cr 97.7 Sm 2 156.4 SmC$_A$ 161.4 I   [197b]

Figure 12. Transition temperatures of 1,2-phenylene-*bis*-[4-(4-n-alkyloxyphenyliminomethyl)benzoates]. (By courtesy of Matsuzaki and Matsunaga [20]).

Table 37. 1,2,3-*tris*[4-(4-n-Alkyloxybenzylideneamino)benzoyloxy]benzenes (data from Matsuzaki and Matsunaga [20]).

In 1,3-benzene derivatives like compound **9**, first synthesized by Matsunaga et al. [197], recently ferroelectric properties have been claimed [197a, 197b]. The reinvestigation and new synthesis of several such "banana-shaped" molecules by different groups [197c, 197d] proved antiferroelectric behaviour, which is striking, because the molecules are achiral.

3.3.3 Lateral Two-ring-Containing Substituents

Matsunaga and coworkers [20] also synthesized compounds of benzene substituted at positions 1, 2 and 3, which may also be considered as basic molecules with two large

$R = C_nH_{2n+1}$

n	Cr		SmB		SmA		I
6	•	89	(•	87)	•	119	•
7	•	96	(•	87)	•	119	•
8	•	96	(•	88)	•	122	•
9	•	101	(•	85)	•	121	•
10	•	100	(•	83)	•	122	•
12	•	104	–	–	•	119	•
14	•	106	–	–	•	118	•

Göring et al. [199] in three-fold substituted benzenes found calamitic phases (e.g. **10** [199]). The molecules are described as having a 'tuning-fork-like' shape. Due to the long spacers between the central ring and the substituents, the latter are decoupled, and therefore layer structures can be formed. In **10**, there are two smectic C*

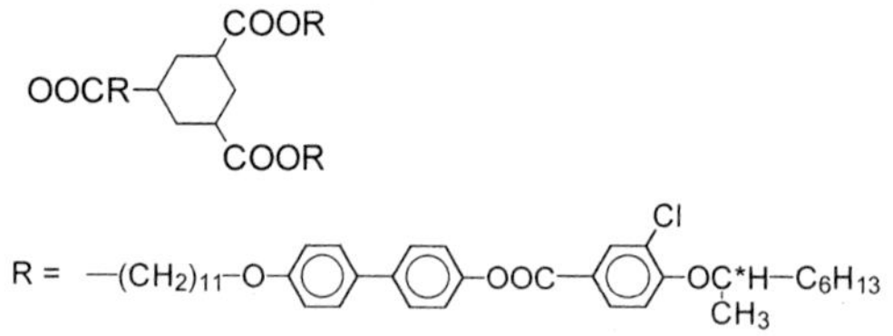

R = —(CH$_2$)$_{11}$—O—⬡—⬡—OOC—⬡—OC*H—C$_6$H$_{13}$
with Cl on ring and CH$_3$ on chiral center

10 Cr 82 SmC$_1^*$ 92 SmC$_2^*$ 112 I

phases of different structures. The smectic C$_1^*$ phase has a layer thickness related to the half-molecule length, and the smectic C$_2^*$ phase has a layer thickness related to the full molecule length. The authors compare the situation of their sterically polar molecules with the situation in molecules with strong electric dipoles, in which polymorphism of smectic A and C phases is well known. With somewhat different substituents in the given example, the compound exhibits two smectic A phases, one of which has an undulating structure [199].

Berg et al. [200] have reported on the liquid crystalline properties of four-fold substituted benzene derivatives, the structure of which may be considered as a basic molecule with two ring-containing lateral substituents (**11** [200]). According to X-ray studies in the solid state, compounds of this type are cross shaped, with the ring planes

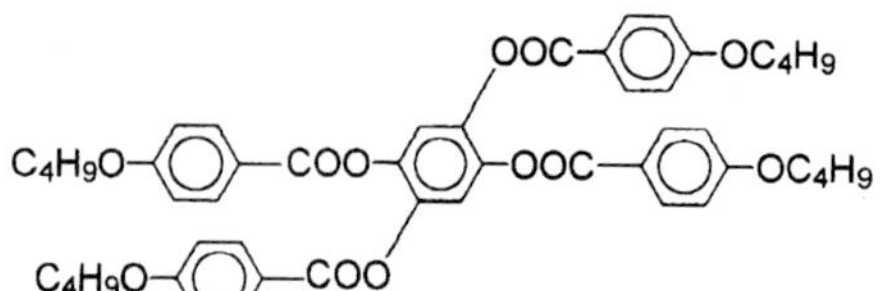

11 Cr 123 (N 121) I

of the laterally attached substituents lying perpendicular to those of the basic moiety. From this evidence the compounds would be expected to be discotic nematic, but the authors claim some arguments for classical nematic behaviour.

3.4 Swallow-Tailed Compounds

The swallow-tailed compounds are a special case of compounds with branched terminal substituents, with unusually large branches.

Weissflog et al. [201] were able to show that the clearing temperatures with increasing length of the two terminal chains of the swallow tail, after passing a minimum, increase (Figure 13). This may be considered to indicate a somewhat stretched and parallel orientation of the two alkyl chains.

Malthete et al. [87, 202] reported on some series with branching of the flexible chain at a greater distance to the central fluorene core (e.g. **12** [87]). In the original paper these compounds were called 'biforked'.

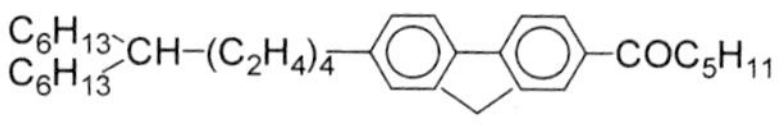

12 Cr 36 SmA 52 I

There are also compounds with swallow tails on both ends of the molecule ('bi-swallow-tailed' compounds) (e.g. **13** [201]).

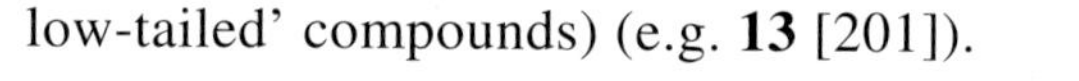

13 Cr 94 (84 SmC) N 89 I

Because of the very bulky swallow tails, the packing bi-swallow-tailed compounds in layers of smectic A phases would lead to large gaps between the cores. Therefore in smectic C layers the packing in which the molecules are shifted somewhat in the lon-

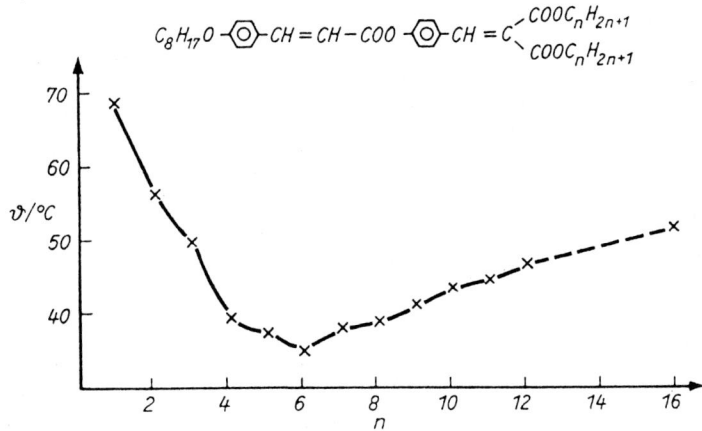

Figure 13. Transition temperatures of di-n-alkyl 4-(4-n-octyloxycinnamoyloxy)benzylidene malonates. (By courtesy of Weissflog et al. [201].

gitudinal direction is favoured. Smectic A structures are stabilized by filling the gaps with small molecules; these are called 'filled phases' [203, 204].

Recently, bi-swallow-tailed compounds that show a very unusual polymorphism have been described. Weissflog and coworkers [205] have presented a series (Table 38) of compounds the lower members of which show classical smectic C and nematic phases, while discotic oblique phases begin to appear at chain lengths of 12 carbon atoms. The re-entrant smectic C phase in the member with $n = 12$ is unique. This series shows a position between calamitic and discotic mesomorphism, similar to that shown by the tetracatenar compounds (discussed in Sec. 3.5.3).

Weissflog et al. [206] in another series of bi-swallow-tailed compounds, found very complex polymorphism (see **14**). In the

14 $n=8$ Cr 126 SmC 195 Cub 243 I_{re} 250 N 312 I
      $n=9$ Cr 125 Col_{ob} 158 SmC 195 Cub 238 I

Table 38. Bi-swallow-tailed compounds (data from Weissflog et al. [205]).

n	Cr		SmC_{re}		Col_{ob}		SmC		N		I
8	•	89	–		–		•	191	•	276	•
9	•	91	–		–		•	197	•	245	•
10	•	104	–		–		•	194	•	229	•
11	•	106	–		–		•	189	•	219	•
12	•	95	•	117	•	154	•	187	•	209	•
13	•	102	–		•	175	•	184	–		•
14	•	101	–		•	180[a]	–		–		•
16	•	104	–		•	178	–		–		•

[a] On cooling the isotropic liquid, the smectic C phase appears, which transforms to Col_{ob} at 175 °C.

member with $n = 8$ the very rare phenomenon of a re-entrant isotropic phase occurs. All phase transitions have been proven by calorimetry. This series also shows a behaviour comparable to that of polycatenar compounds, with some compounds being able to form smectic and nematic as well as columnar and cubic phases. A comparing investigation deals with the behaviour of swallow-tailed compounds with linear respectively cyclic siloxane substituents [206a].

3.5 Polycatenar Compounds

Polycatenar compounds possess two to six flexible chains attached at the terminal rings of large rod-like cores, and are called bi-, tri-, tetra-, penta- and hexacatenar compounds, respectively. Malthete et al. [21] have published a good review that explains many details about this substance class.

3.5.1 Bicatenar Compounds

The bicatenar compounds are, of course, in most cases classical rod-like molecules, which have been dealt with already. However, there are some examples in which the two substituents are not in the *para* position, and these deserve special attention.

Compound **15** [207,208] compares a classical rod-like molecule with the analogous bicatenar compound, which has an addi-

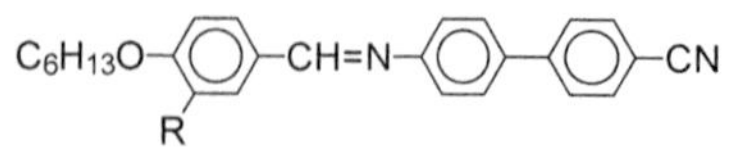

15 R = H– Cr 92 SmB 111 N 280 I
     R = $C_6H_{13}O$– Cr 131 SmA$_d$ 141 I

tional long flexible chain in the *meta* position. The substitution in the *meta* position causes a substantial decrease in the clearing temperature. Bicatenar compounds of similar structure have also been synthesized by Nguyen et al. [209].

3.5.2 Tricatenar Compounds

There are several examples of tricatenar compounds with three-ring cores, which exhibit the classical nematic and smectic phases [21]. Compared to related compounds that are unsubstituted in *meta* position, there is a remarkable decrease in the clearing temperatures in these compounds. The phase behaviour of the tricatenar compound **16** [210] is quite exceptional.

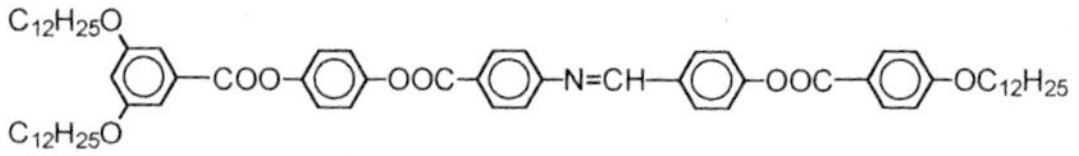

16 Cr 83 Cub 172 I

3.5.3 Tetracatenar Compounds

Considering the different positions (*ortho*, *meta* and *para*) of substituents, there are several possibilities, as discussed by Malthete et al. [21]. We present here just one homologous series, which shows the intermediate character of the tetracatenar compounds between that of calamitic and that of discotic compounds. The transition temperatures of this series are given in Table 39 [21]. The lower members exhibit classical smectic phases, the middle members exhibit, in addition, cubic phases, and the higher members exist in columnar phases, which according to the proposal of the Bordeaux group [21] are designated as Φ (from 'phasmidic') phases.

The change from lamellar to columnar phases has been observed in several tetracatenar series [21]. Obviously, the predominant influence of the cores in the lower members stabilizes lamellar phases, while the increasing influence of the flexible chains in the higher members leads to the dominance of columnar phases in these compounds. Cyclohexane rings have also been introduced into tetracatenar com-

Table 39. Transition temperatures of tetracatenar compounds (by courtesy of Malthete et al. [21]).

R–⟨⟩–CH=N–⟨⟩–OOC–⟨⟩–COO–⟨⟩–N=CH–⟨⟩–R

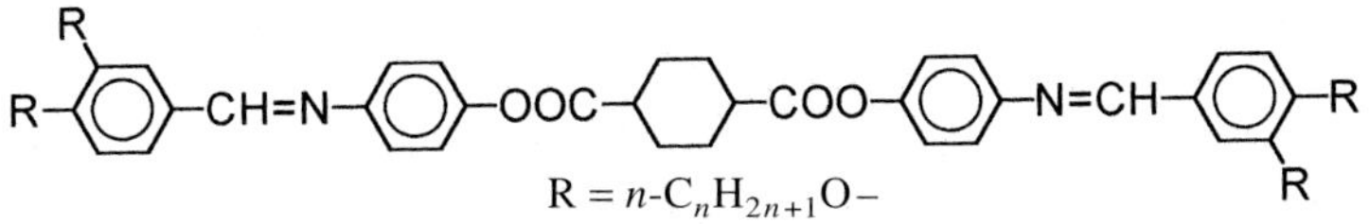

R = $n\text{-}C_nH_{2n+1}O-$

n	Cr		SmC		Cub		Φ_h		I
7	•	152	•	183	–		–		•
8	•	148	•	176	–		–		•
9	•	146.5	•	168.5	–		–		•
10[a]	•	144	•	156	•	165	–		•
11[b]	•	144	•	146	•	163	–		•
12[c]	•	142	–		•	162	–		•
13	•	141	–		–		•	163	•
14	•	140	–		–		•	163	•

Φ, Hexagonal columnar (phasmidic) phase.
[a]) On cooling: I 157 SmC 138 Cr.
[b]) On cooling: I 158 Φ_h 147 Cub 140 SmC 135 Cr.
[c]) On cooling: I 160 Φ_h 138 Cr (the hexagonal columnar phase has a large lattice constant).

pounds [21], and ester groups have been exchanged for thioester groups [212]. An X-ray investigation of the solid state of a thioester compound [212] has shown the nearly parallel orientation of the long-chain substituents, and the smectic C like lamellar structure, which is typical of a segregation of an aromatic core and aliphatic flexible substituents.

3.5.4 Pentacatenar and Hexacatenar (Phasmidic) Compounds

The few pentacatenar compounds that have been described in the literature all exhibit columnar (phasmidic) phases [21]. Typical examples of hexacatenar materials are presented in Table 40 [21]. Despite their similarity of having five benzene rings in the

Table 40. Transition temperatures of phasmidic compounds (by courtesy of Malthete et al. [21]).

R–⟨⟩–COO–⟨⟩–OCO–⟨⟩–COO–⟨⟩–OCO–⟨⟩–R

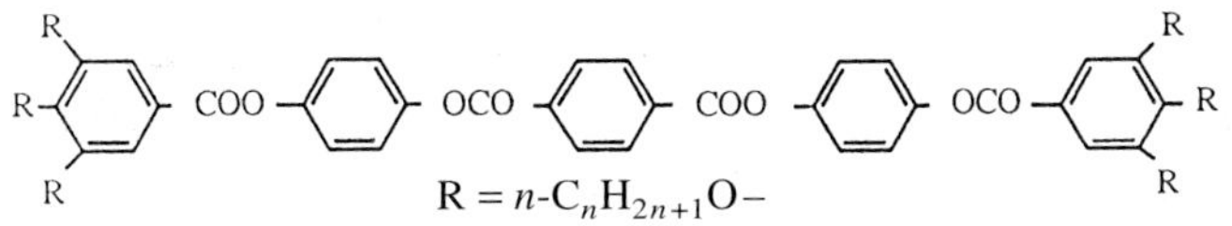

R = $n\text{-}C_nH_{2n+1}O-$

n	Cr		Φ_{ob}		Φ_h		I
7	•	82	•	87	–		•
8	•	69	•	87	–		•
9	•	73	•	87	–		•
10	•	60	•	90	–		•
11	•	83	–		•	92.5	•
12	•	88	–		•	94	•
13	•	90	–		•	94	•
14	•	85	–		•	94	•

Φ_h, Hexagonal columnar (phasmidic) phase; Φ_{ob}, Oblique columnar (phasmidic) phase.

Table 41. Transition temperatures of 2,3,4-trialkyloxycinnamic acids (data from Praefcke et al. [213]).

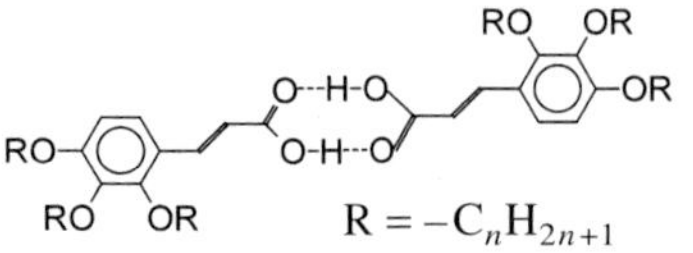

$R = -C_nH_{2n+1}$

n	Cr		N_b		I
4	•	75.6	(•	67.9)	•
6	•	51.2	•	59.6	•
8	•	54.2	(•	50.9)	•

core, the melting and clearing temperatures of these compounds are very low compared to those of classical five-ring calamitic compounds. This fact and the occurrence of the columnar (phasmidic) phase are clearly due to the high concentration of aliphatic chains in the molecules.

Hexacatenar compounds derived from cinnamic acid exist as dimeric molecules (Table 41) [213]. The compounds represent the seldom occurring biaxial nematic phase. Some compounds with four rings in the core are also nematic; however, their biaxiality has not been proven [213].

3.5.5 Summary of the Phase Behaviour of Compounds with More than Two Flexible Chains

The phase behaviour of compounds with more than two flexible chains is summarized in Table 42.

3.6 Twins and Oligomers

Twins consist of two entities typical for liquid crystals that are connected either directly or indirectly by a spacer, which can be rigid or flexible. Vorländer's group have synthesized compounds which, according to the modern nomenclature, are called 'twins' (e.g. [214]; see compounds 3767–3769 in [34]). In 'Siamese' twins [215] two exactly equal moieties are attached to one another. The bond between the entities can occur in different positions and be composed of different linking groups. A survey of the twin types known up until 1988, is given in [2]. Some twins and oligomers are shown schematically in Table 43.

Table 42. Phase behaviour of lateral long-chain substituted, swallow-tailed and polycatenar compounds.

Compound type	No. of chains in		Phase		
	meta-position	*para*-position	N or Sm	Cubic	Columnar
Lateral long chain	1	2	•	–	–
Swallow-tailed	0	2–3	•	–	–
Bi-swallow-tailed	0	4	•	•	•
Hexacatenar	4	2	•[a]	–	•
Pentacatenar	4	1	–	–	•
	3	2	–	–	•
Tetracatenar	4	0	–	•	•
	3	1	–	•	•
	2	2	•	•	•
Tricatenar	2	1	•	•	–
	1	2	•	–	–

[a] Nematic biaxial.

Table 43. Types of twins and oligomers.

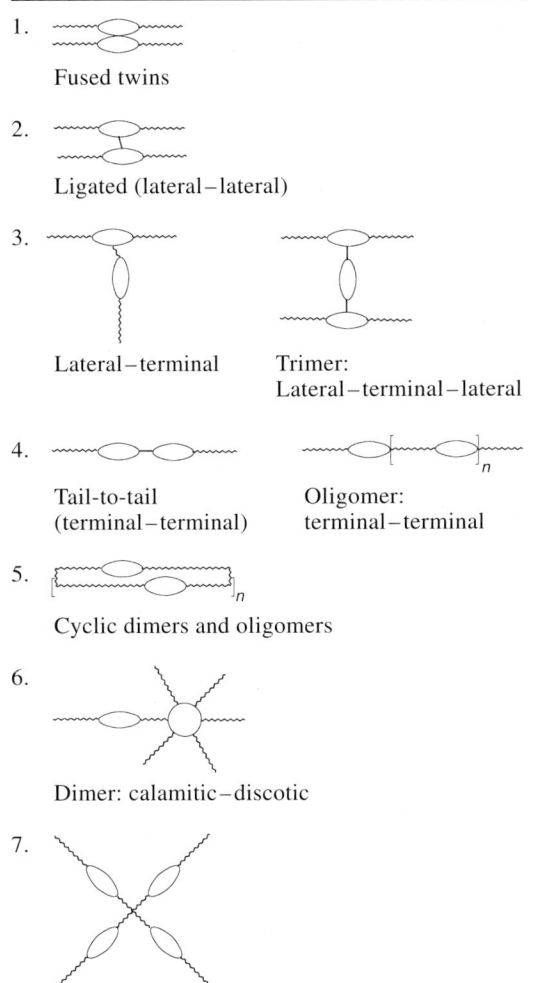

1. Fused twins
2. Ligated (lateral–lateral)
3. Lateral–terminal   Trimer: Lateral–terminal–lateral
4. Tail-to-tail (terminal–terminal)   Oligomer: terminal–terminal
5. Cyclic dimers and oligomers
6. Dimer: calamitic–discotic
7. Star-like oligomer

3.6.1 Fused Twins

In fused twins the two mesogenic moieties are linked rigidly. Four-fold substituted ring systems (e.g. **17** [216]) can belong to this class. Induced in the research in this area are many metal complexes (see Sec. 3.9.3 of this chapter) that, in principle, belong to the class of fused twins.

17 Cr 218.5 N 287 I

3.6.2 Ligated Twins

In ligated twins the mesogenic units are connected in a central position by a rigid or flexible spacers. The first examples of such compounds with a rigid spacer (**18**) were synthesized by Griffin et al. [217, 218].

18 Cr 119.1 (SmC 103 N 112) I

Twins of this type have much higher clearing temperatures than do the 'single' mesogens; however, because of the increased melting temperatures, the mesogenity in some cases appears less pronounced [217–219]. Weissflog et al. [220, 221] have reported twins with long flexible linking groups. With regard to the dependence of the clearing temperature on the length of the linking group, there is a pronounced alternation [220] (Figure 14). This alternation in the T_{N-I} is caused by the alternating molecular length-to-breadth ratios.

There are also twins in which the linking groups contain ring systems [220, 221]. Some of these may be considered as trimers (e.g. **19** [220]).

19 Cr 113 N 178 I

The behaviour of oligomers with such high clearing temperatures cannot be discussed simply on the basis of their length-to-breadth ratios. Obviously the parallel

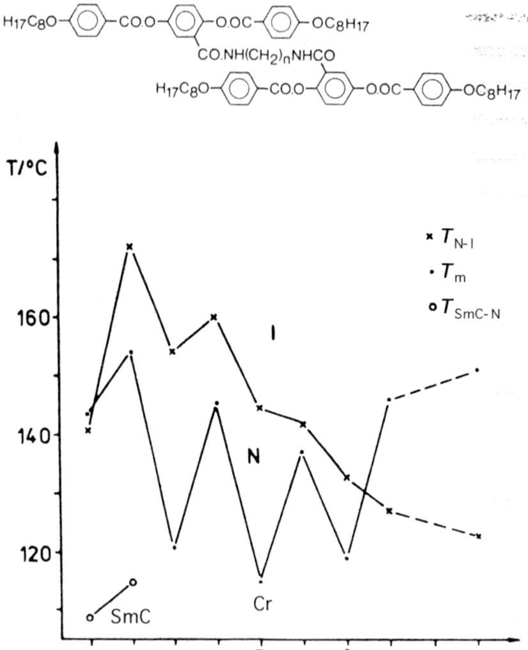

Figure 14. Transition temperatures of the α,ω-*bis*[2,5-*bis*(4-n-octyloxybenzoyloxy)benzamido] alkanes. (By courtesy of Weissflog et al. [220]).

Table 44. From laterally substituted compounds to twins (data from Weissflog et al. [221]).

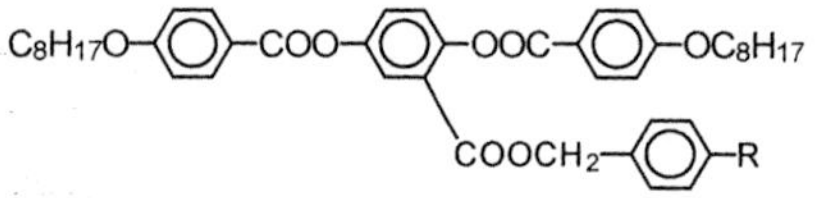

R	Cr		SmC		N		I
-H	•	98	–		•	98	•
$-OC_8H_{17}$	•	89	(•	67)	•	104	•
$-C_6H_4-OC_8H_{17}$	•	99	(•	88.5)	•	151.5	•
$OOC-C_6H_4-OOC-C_6H_4-OC_8H_{17}$	•	130	–		•	197	•

orientation of the mesogenic units, which is necessary for the stabilization of the nematic state, is already present in the oligomer molecules.

3.6.3 Twins with Lateral–Terminal and Lateral–Lateral Linking

If in the above discussed compounds with substituents containing large lateral rings (see Sec. 3.3.2 of this Chapter) the latter are large enough, they may themselves be considered as mesogenic units and the compounds that contain them then represent a new class of twins. By means of systematic elongation of the lateral substituents, Weissflog et al. [220, 221] have produced examples of this kind of compound (Table 44).

Tschierske et al. [222, 222a–c, 241] have synthesized liquid crystalline trimers and tetramers with lateral linking. Some of them are compared with related compounds in Table 45.

3.6.4 Twins with Tail-to-Tail (Terminal–Terminal) Linking

Terminal bond twins have been synthesized by Vorländer [223] in his studies on the influence of central linking groups on the mesomorphic properties of molecules. Figure 15 shows the pronounced alternation in the transition temperature in a series of compounds. This is due to the alternation of the molecules between the most elongated and the bent shape.

There are several reports of twins with flexible tail-to-tail linking. In most cases two equal units are linked; twins of this kind have been called 'symmetric' [224, 225, 225a]. Recently a series has been reported in which the lower members are smectic, and only at spacer lengths above six carbon atoms do the compounds appear nematic [225]. In other cases two unequal units are connected, producing a 'non-symmetric' dimer [226–228, 228a–b]. There have also been detailed investigations of the flexibility of the linking groups [228, 229], and

Table 45. Laterally linked trimers (by courtesy of Andersch and Tschierske [222]).

1. 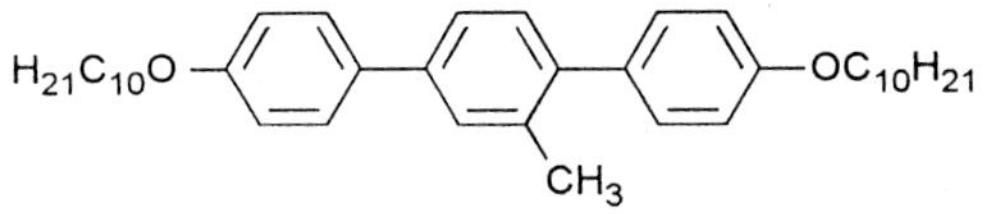 [structure: H$_{21}$C$_{10}$O–[biphenyl]–[phenyl with CH$_3$]–[phenyl]–OC$_{10}$H$_{21}$]

   Cr 78 (SmC 74) SmA 109 N 113 I

2. 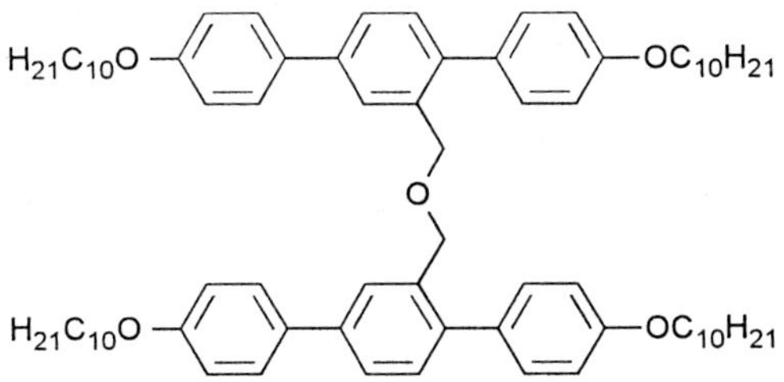 [structure: two arms H$_{21}$C$_{10}$O–...–OC$_{10}$H$_{21}$ connected via –CH$_2$–O–CH$_2$– lateral linker]

   Cr 93 SmC 120 SmA 164 I

3. 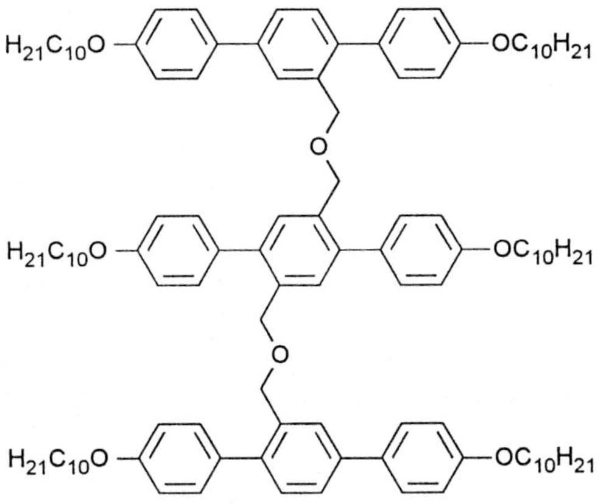 [structure: three arms H$_{21}$C$_{10}$O–...–OC$_{10}$H$_{21}$ laterally linked via –CH$_2$–O–CH$_2$– bridges on central phenyl]

   Cr ? SmA 159 I

4. 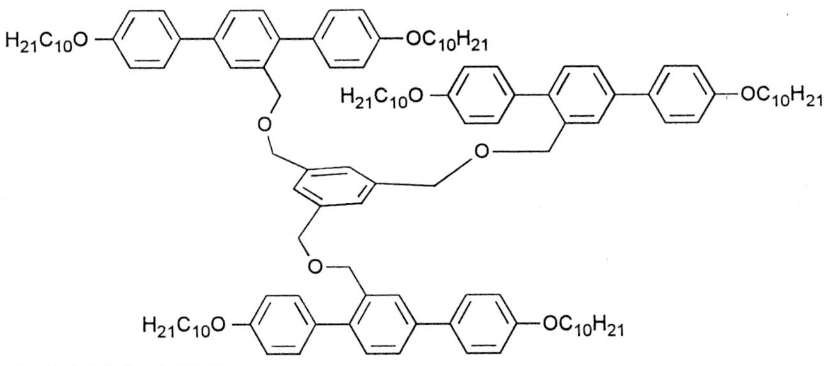 [structure: four arms H$_{21}$C$_{10}$O–...–OC$_{10}$H$_{21}$ connected via –CH$_2$–O–CH$_2$– linkers to a central benzene ring]

   Cr 73 (g 8) SmA 124 I

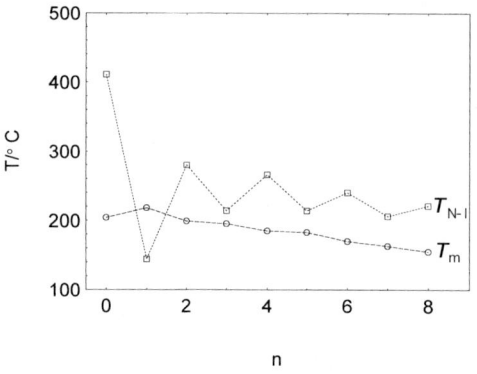

Figure 15. Transition temperatures of tail-to-tail twins. (Data from Vorländer [223]).

C$_2$H$_5$O-⟨O⟩-N=N-⟨O⟩-OOC-(CH$_2$)$_n$-OOC-⟨O⟩-N=N-⟨O⟩-OC$_2$H$_5$

dimers involving siloxane containing linking groups have been synthesized [220, 231, 231a]. Attard and Imrie [232] have synthesized dimers with terminal linking groups that contain rings bearing long-chain lateral substituents. Because of the ring-containing centre, these molecules are called 'trimers' (see **20** [232]).

NC-⟨O⟩-⟨O⟩-O-C$_4$H$_8$-O-⟨O⟩-O-C$_4$H$_8$-O-⟨O⟩-⟨O⟩-CN
                        |
                    COO(CH$_2$)$_5$CH$_3$

20   Cr 130 N 141 I

Symmetrically substituted linear trimers also exhibit the odd–even effect [232a]. Twins with phasmidic-like molecular structure occur in columnar phases [232b].

There are a number of mesogenic salts of group II metals and of mesogenic metal complexes, which have the typical molecular architecture of twins. Such materials are dealt with in Sec. 3.9 of this chapter.

3.6.5 Cyclic Dimers and Oligomers

Percec and Kawasumi [233, 234] have synthesized cyclic oligomers using polyether chains to link biphenylphenylbutane derivatives. Beginning with the trimer, the oligomers are mesomorphic (Table 46). The compounds show a pronounced trend to the glassy state. The already mentioned cyclophanes [179, 179a] also can be considered as cyclic dimers.

Table 46. Cyclic oligomers (data from Percec and Kawasumi [233]).

$X + Y$	
2	Cr 180 I
3	g 52 N 81 I
4	g 52 Cr 64 N 115 I
5	g 47 N 108 I

3.6.6 Calamitic–Discotic Dimers

In order to investigate the gap between discotic and calamitic liquid crystals and in an attempt to obtain biaxial nematics, Fletcher and Luckhurst [235] synthesized hybrids of rod-like and disk-like mesogens, by linking two units to dimers. Among the several compounds of this type is one example that shows nematic behaviour (**21**); however, the

21   Cr 125 (N 179) I

detailed nature of this phase could not be investigated. In mixtures with 2,4,7-trinitro-9-fluorenone compounds of this type give rise to EDA complexes, which are responsible for the occurrence of nematic phases consisting of columns of these complexes [235]. Novel twin compounds in which disk-like electron donor and acceptor groups are covalently bond, form columnar phases [235a].

Budig et al. [236, 237] combined a pyramidal core with six rod-like units by synthesizing tribenzocyclononene derivatives. With regard to the lengths of the spacers between the core and the rods, with short spacers they obtained columnar phases, and with long spacers they obtained smectic (mainly A and C) and nematic phases (Table 47). Obviously the long spacers decouple the core and the rod-like units. However, some effect of stabilizing the mesophase remains, as can be seen by comparing the clearing temperatures of the analogous monomeric compounds with those of the oligomers listed in Table 47.

Kreuder et al. [238] have synthesized a trimer consisting of two discotic units linked by a rod-like moiety. The mesophase of this compound has not been clearly classified.

Table 47. Transition temperatures of oligomers derived from tribenzocyclononene with rod-like units (by courtesy of Budig et al. [237]).

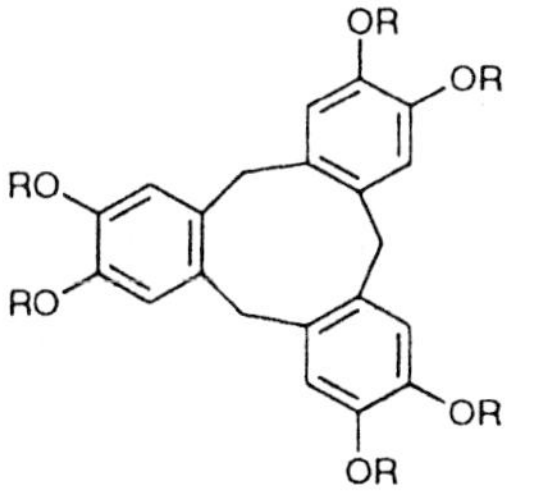

R	
$C_{12}H_{25}$—CO—	Cr 67 Col 139 I
 C_8H_{17}—⬡—CO—	Cr 60 Col 278 I
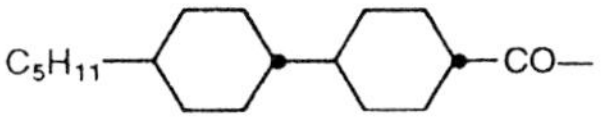 C_5H_{11}—⬡—⬡—CO—	Cr 118 Col 355 I_{decomp}
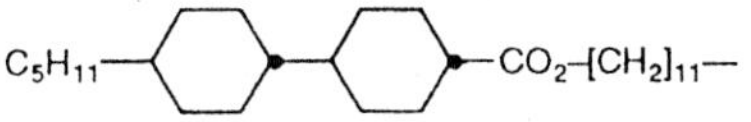 C_5H_{11}—⬡—⬡—CO_2—$[CH_2]_{11}$—	Cr 114–115 I
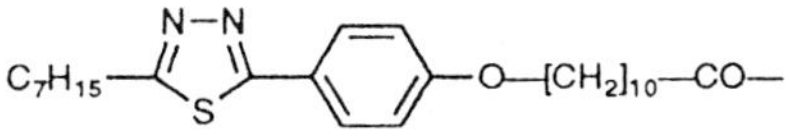 C_7H_{15}—(N—N thiadiazole)—⌬—O—$[CH_2]_{10}$—CO—	Cr 140–142 I
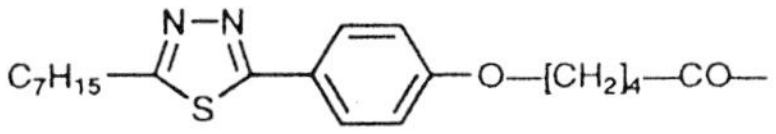 C_7H_{15}—(N—N thiadiazole)—⌬—O—$[CH_2]_4$—CO—	Cr 117 SmA 159 I
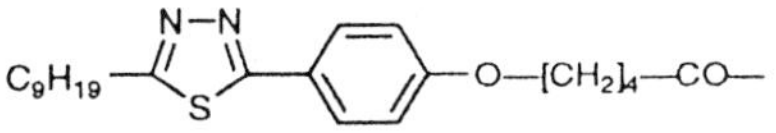 C_9H_{19}—(N—N thiadiazole)—⌬—O—$[CH_2]_4$—CO—	Cr 144/152 SmA 160 I
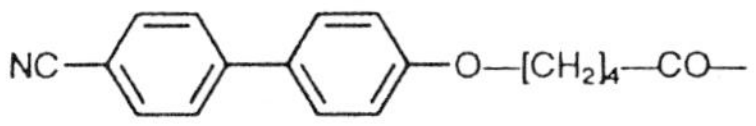 NC—⌬—⌬—O—$[CH_2]_4$—CO—	Cr 165 (SmA 133) I
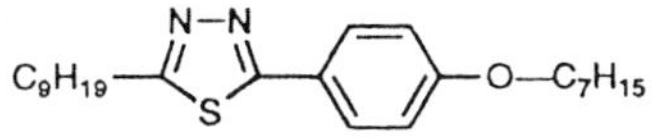 C_9H_{19}—(N—N thiadiazole)—⌬—O—C_7H_{15}	Cr 77 SmC 89 I
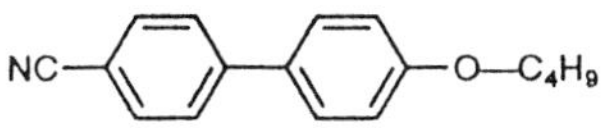 NC—⌬—⌬—O—C_4H_9	Cr 78 (N 75.5) I

3.6.7 Star-Like Compounds

In contrast to older models of the classical molecular shape of mesogens, some mesogens are star-like oligomers with flexible central units.

Eidenschink et al. [239] have synthesized derivatives of pentaerythritol (e.g. **22**).

$$\begin{array}{c} CH_2OOCR \\ RCOOH_2C-C-CH_2OOCR \\ CH_2OOCR \end{array}$$

R = —(CH$_2$)$_3$—◯—◯—◯—C$_5$H$_{11}$

22   SmX 152 SmB 229 SmA 275 I

Compounds with shorter spacers showed poor or no liquid crystalline properties. From this and the results of X-ray investigations it is concluded that the long mesogenic substituents are bent at the spacers, yielding a more elongated molecular shape despite the tetrahedral symmetry of the central carbon atom.

Using nitromethane–trispropanol, pentaerythritol or dipentaerythritol as the central unit and typical rod-like units with no spacers, Wilson [240] produced three-, four- and six-armed star-like molecules that showed smectic (mostly type A) and nematic phases. In order to explain the existence of the mesophases, Wilson compared the molecules to small sections of side-group polymers, where the flexibility of the central unit is emphasized.

Zab et al. [241] linked glycerol, pentaerythritol and 1,1,1-*tris*(hydroxymethyl)-ethane via spacers of different lengths to 2-phenyl-1,3,4-thiadiazole as rod-like units. Several of the trimers and tetramers obtained yielded smectic phases, usually of type C.

3.7 Epitaxygens

Norvez and Simon [242] have reported triptycene derivatives. The mesophases occurring in these materials are unique, and the compounds have therefore been called epitaxygens. Figure 16a shows the chemical formula of a five-fold substituted triptycene, which in its mesomorphic state forms layers containing the hard cores in a hexagonal arrangement and sublayers containing the flexible alkyl chains (Figure 16b).

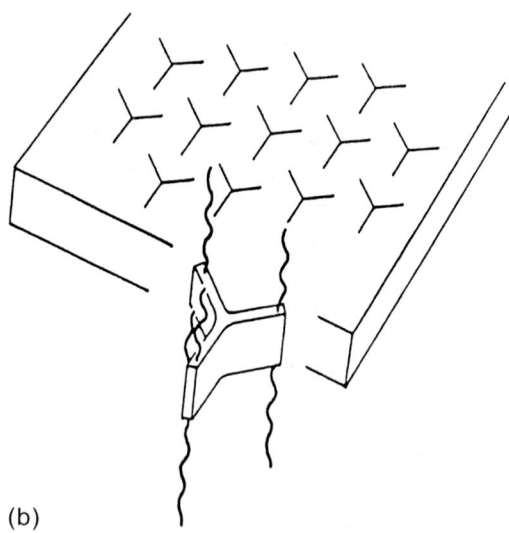

Figure 16. (a) A triptycene derivative substituted with five chains of equal length. (b) The lamellar structure of trypticene derivatives. (By courtesy of Norvez and Simon [242]).

3.8 Associated Liquid Crystals

Since Vorländer's work in the 1920s, it has been known that carboxylic acids can be liquid crystals, but that their esters are either not mesogenic or are less so [5]. Weygand [151] concluded that the acids occur as dimers, the carboxylic groups forming an additional ring. In their investigations on carboxylic acids Bennett and Jones [243] found that alkyloxybenzoic acids are nematic. A solid-state X-ray investigation of anisic acid [244] produced structure 23 for this ring.

23

At present several hundred mesogenic carboxylic acids are known. The dimerization occurs due to hydrogen bonding. Exchanging the hydrogen atom in the hydroxyl group for deuterium leads to a slight decrease in the clearing temperatures, because deuterium bridges are weaker than hydrogen bridges [245] (see 24).

24 X=H: Cr 100.8 SmC 107.8 N 147.8 I
     X=D: Cr 99.5 SmC 105.0 N 144.5 I

There are few examples of liquid crystalline primary amides of carboxylic acids (e.g. 25 [246], further [247]). The unusually high transition temperatures of these compounds (two ring Schiff bases usually have a phase transition temperature below 100 °C) are due to the formation of dimers.

25 Cr 238 Sm 268 I

Compounds containing polar groups such as –CN or –NO_2 show a strong tendency to dimer formation. The effect on the mesogenic properties has been discussed in Sec. 2.4 of this chapter.

Derivatives of pyridine [248], derivatives of pyrimidine [249], pyrazine [250] and other substance classes contain lateral hydroxyl groups, which give rise to strong association effects. The formation of mesophases in binary systems of pyridine derivatives and carboxylic acids is a specific field of research [250a–253, 253a, 253b]. In most cases complex formation leads to an increase in the length-to-breadth ratio, and different liquid crystalline phases can be induced. This is illustrated by the example shown in Figure 17 [250].

Using a tetrapyridyl compound and a dicarboxylic acid, Wilson [254] has been able to construct a highly ordered polymeric liquid crystalline network.

Liquid crystalline diols may be considered as having a structure intermediate between that of non-amphiphilic and amphiphilic liquid crystals, i.e. like soaps, tensides, phospholipids and sugars. Compound 26 [255] exhibits phases that are similar in

26 Cr 85.9 SmB* 113.3 SmA* 116.0 I

structure to the classical smectic A and B phases, but because of strong association it is not miscible with such phases. Even simple n-alkane-1,2-diols with sufficiently long alkyl chains form thermotropic and lyotropic mesophases [256]. Staufer et al. [257] have produced cis,cis-(3,5-dihydroxycyclohexyl)-3,4-bis(alkyloxy)benzoates that, depending on the length of the alkyl chains exhibit smectic, cubic or hexagonal columnar phases.

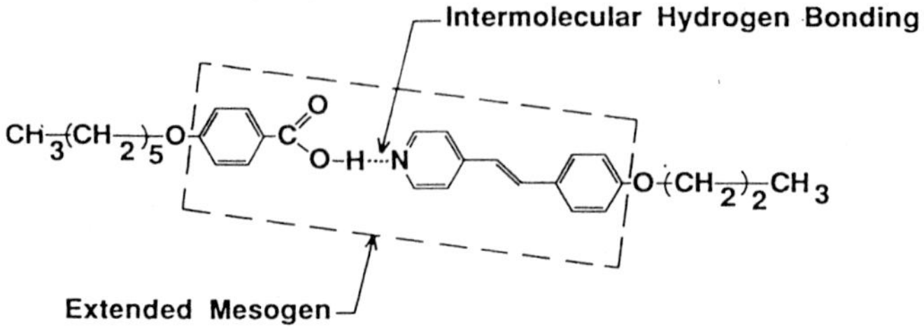

Figure 17. The formation of a mesogenic complex from hexyloxybenzoic acid (HBA) and a pyridine derivative (PYR). (By courtesy of Kato et al. [250]).

(HBA)$_2$ CR 106 N 153 I
HBA–PYR   CR 102 SmA 130 N 155 I
PYR   Cr 110 I

In recent years many different diols (e.g. [258–261]), tetraols [262–265] and polyols (mainly derived from sugars [266–270, 270a]) have been synthesized. Compounds of these classes are amphiphilic. They are mentioned briefly in Sec. 4.5 of this chapter and are dealt with in detail in Volume 3 of this book.

3.9 Salt-Like Compounds and Metal Complexes

Salts of carboxylic acids were among the earliest known liquid crystalline materials [271] and were investigated systematically by Vorländer [272]. There are mesomorphic salts of aliphatic and aromatic carboxylic acids, and in recent years several additional metals have been introduced. Mesophases have also been found in 'inverse' salts, i.e. salts consisting of organic cations and inorganic anions. A specific area of investigation is liquid crystalline metallorganic compounds, because such materials can have special physical properties (colour, magnetic properties, labels for Mössbauer spectroscopy, and electrical conductivity). More details can be found in the literature [273–275].

3.9.1 Salts

Most mesogenic salts derived from aliphatic acids, (R–COO)$_n$M (R = alkyl (normal and branched) or alkenyl; M = Li, Na, K, Rb, Cs, NH_4^+, Tl, Pb or other metal) form layered structures (lamellar phases, neat phases) that are similar to smectic A phases. However, mesogenic salts form double layers and are not miscible with smectic A phases [276]. Some of the materials show very complicated polymorphism with a large number of mesophases [276–279]. In general, the transition temperatures of the salts are quite high compared with those of nonpolar liquid crystals. Most of the salts can also form lyotropic liquid crystals.

Mesogenic salts can also be derived from ring-containing carboxylic acids. They include salts of Na, K, Rb, Cs or Tl with substituted benzoic acids, substituted cinnamic acids, mandelic acid or cyclohexyl carboxylic acid [277]. The materials exhibit lamellar phases.

3.9.2 Inverse Salts

Compound 27 [280] may be used as an example to demonstrate this substance class, the members of which form smectic-A-like layer structures and are usually also lyotropic liquid crystalline.

[C₄H₉O—⟨⟩—NH₃⁺] Cl⁻

27   Cr 118 SmA 211.5 I

Arkas et al. [281] have prepared cyano-propylalkyldimethylammonium bromides that show smectic A phases. There are also materials that show smectic A and C phases [282]. The smectic A phases of substituted pyrimidinium salts are completely miscible with non-polar pyrimidine derivatives [283].

The lipids, in which salts are formed by groups of different polarity within the same molecules, are typical amphitrophic liquid crystals [42].

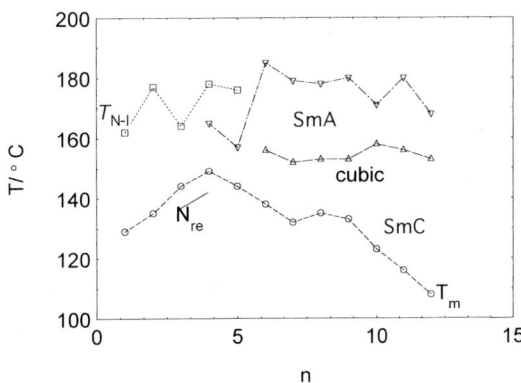

Figure 18. Transition temperatures of silver complexes. (Data from Bruce et al. [289].

3.9.3 Metallomesogens

The metal atoms can be bond by coordination bonds, either in a slat-like fashion or, more rarely, by σ-bonds [284]. Irrespective of the nature of the bonds the compounds may be called metallomesogens [274]. Depending on the geometry of the molecules, metallomesogens can be calamitic or discotic. Rod-like mercury containing compounds were prepared by Vorländer's group [5, 285].

Compound **28** [285] and many other metallomesogens can be considered as twins. In many cases (e.g. **29** [286]) the ligands are themselves liquid crystals.

⟨⟩—CH=N—⟨⟩—Hg—⟨⟩—N=HC—⟨⟩

28   Cr 180 N 184 I

C₈H₁₇—⟨⟩—⟨⟩—CN—Pt(Cl)(Cl)—CN—⟨⟩—⟨⟩—C₈H₁₇

29   Cr 166 N 176 I

There are metallomesogens that contain Ni, Pd, Pt, Cu, Ag, Au, V, Fe, Rh, Ir, Zn, Cd, Hg, Pb, rare earths [274], Mn [287], Rh [287] or Co [288]. Conventional nematic and smectic A, C and G phases have been found.

The silver complexes reported by Bruce et al. [289] show quite unusual polymorphism. In addition to conventional nematic and smectic phases, in several members of the series cubic phases have been observed (Figure 18). These cubic phases, in contrast to the case in other thermotrophic homologous series, exist even in members having relatively short alkyl chains.

It is worth noting that many of metallomesogens are intensely coloured, and thus their use as guest–host effect dyestuffs has been tried.

4 Discotics

The first papers on discotic liquid crystals were those by Chandrasekhar et al. [7] and Billard et al. [8]. There are several reviews [9–14, 290] of this topic, which now comprises more than 1000 compounds of quite different substance classes. In the following, with the aid of selected examples, some of these substance classes are described.

4.1 Derivatives of Benzene and Cyclohexane

Most of the discotic compounds consist of a central core to which are attached 3–12 substituents, which may contain ring systems and/or flexible chains. The most simple central core is the benzene ring, which was used for the synthesis of the first discotic compounds [7].

Some of these compounds have three flexible substituents at positions 1, 3 and 4 (**30** [291]). The molecular shape does not

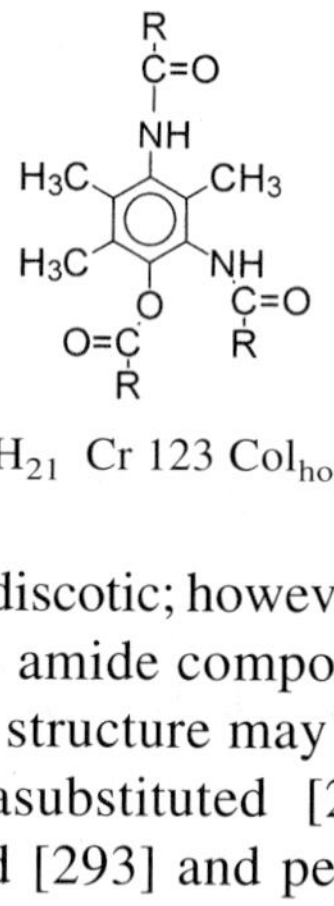

30 R = $-C_{10}H_{21}$ Cr 123 Col$_{ho}$ 152 N$_D$ 192 I

seem typically discotic; however, due to associations in this amide compound the effective molecular structure may be different.

1,2,3,4-Tetrasubstituted [292], 1,2,3,5-tetrasubstituted [293] and pentasubstituted [293] benzene derivatives can also be discotic. The most typical structure is the sixfold substituted benzene, which was used in the first discotic materials [7]. The substituents can be different in nature (**31** [7]).

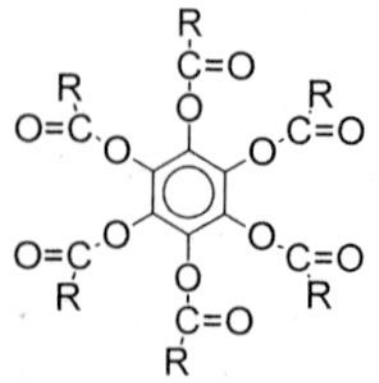

31 R = $-C_7H_{15}$ Cr 79.8 Col$_{rd}$ 83.4 I

Instead of the flat, stiff benzene ring, the more flexible and chair-like cyclohexane ring (a derivative of scylloinositol, a natural sugar) can be the central moiety (**32** [294]).

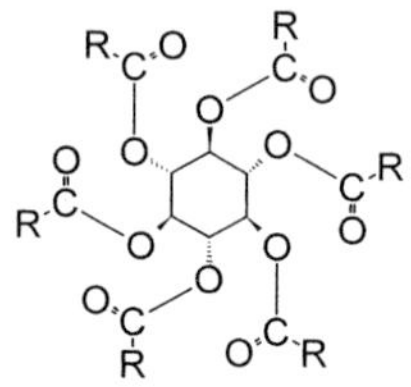

32 R = C_nH_{2n+1}

4.2 Large Ring Systems

Some large ring systems used as central cores in discotic compounds are listed in Table 48. In particular, triphenylene has been used frequently. Most of the rings are flat and stiff. The substituents are flexible, diminish the melting temperatures and have been considered as essential for the formation of discotic structures. However, starting from scylloinositol (a cyclohexane derivative), Kohne et al. [305] prepared the hexaacetyloxy derivative, which proved to be columnar. Variants have been synthesized from hexakis(phenyl-ethynyl)benzene in which one of the phenylethynyl groups is exchanged by alkyloxy or a flexible group linking two of the ring systems into a dimer; both yield biaxial nematic discotic phases [306].

Keinan et al. [307] have prepared discotic tricycloquinazoline derivatives.

The triphenylene derivatives **33**, which have been synthesized by Ringsdorf et al. [308], can be considered as heptamers. Like many of the compounds containing large molecules as their cores, the heptamers have a strong tendency to glass formation, and solid crystals could not be obtained. The clearing temperatures indicate that the

Table 48. Large discotic cores.

	Core compound	R	Ref.
1.	Rufigallol	Hexa-n-octanoate	[295]
2.	Triphenylene	$C_nH_{2n+1}COO-$	[296]
		$C_nH_{2n+1}O-$	[8, 296]
		$C_nH_{2n+1}O-\langle\bigcirc\rangle-COO-$	[297]
		$C_nH_{2n+1}-\langle\bigcirc\rangle-COO-$	[297]
		$C_4H_9-\langle\bigcirc\rangle-COO-$	[298]
		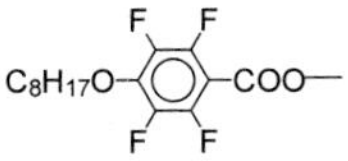 $C_8H_{17}O-\langle\bigcirc\rangle-COO-$ (with F,F,F,F)	[298]
		$C_nF_{2n+1}-(CH_2)_2-OOC-CH_2-O-$	[299]
		Unequal substituents	[9, 300]
3.	Truxene	$C_nH_{2n+1}COO-$	[301]
4.	Bipyran-4-ylidene	$C_nH_{2n+1}-$	[302]
5.	Dibenzopyrene		[300, 303]
6.	Hexakis(phenyl-ethynyl)benzene	$C_nH_{2n+1}-$	[304]

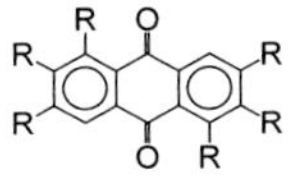

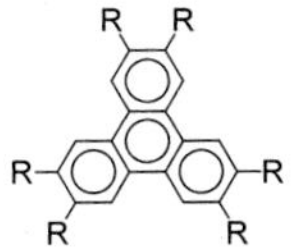

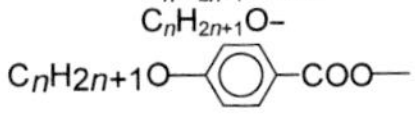

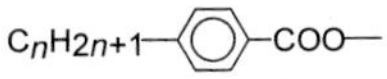

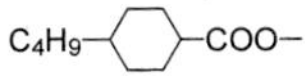

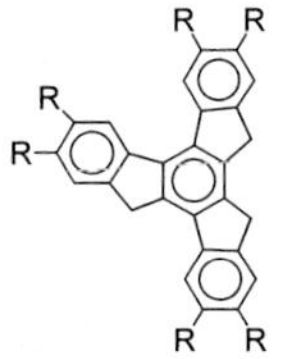

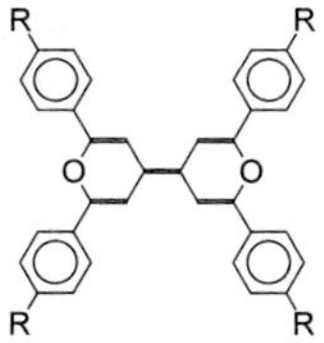

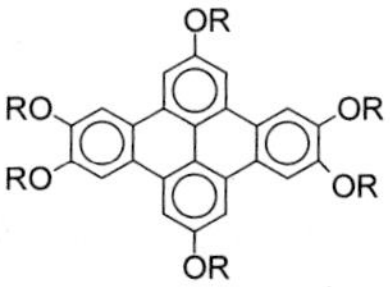

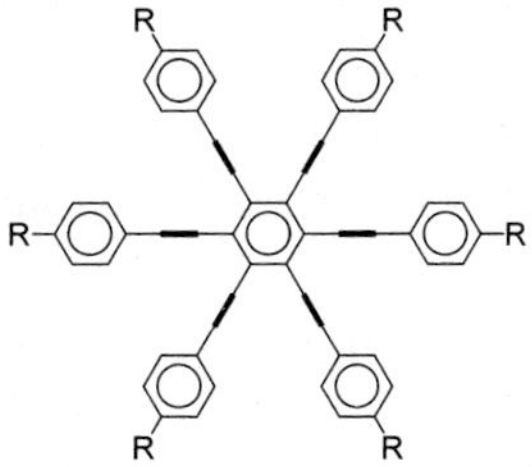

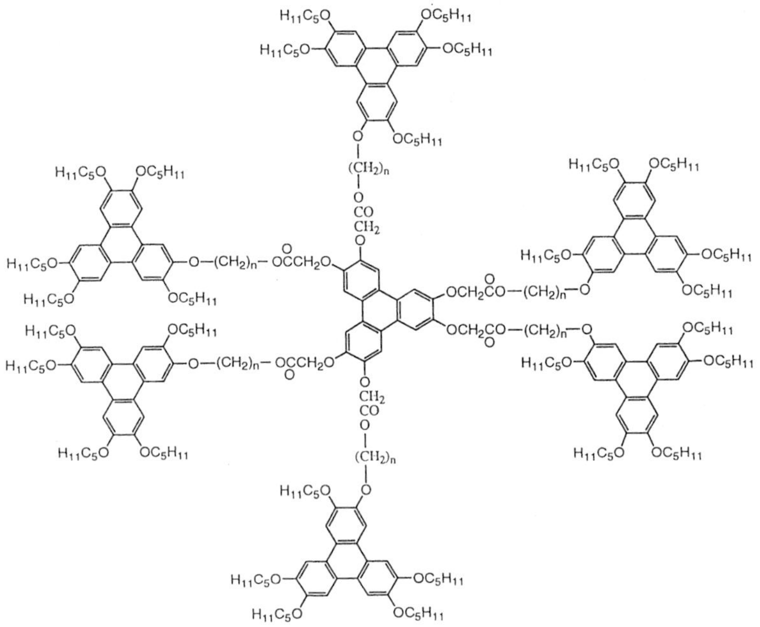

33

$n = 6$ g ... Col$_h$ 137 I
$n = 9$ g ... Col$_h$ 152 I
$n = 11$ g ... Col$_h$ 132 I

lengths of the spacers control the mesogenity. A chain length of $n=9$ seems to be the most appropriate, probably because of the optimal space filling.

4.3 Complex-Forming Salts and Related Compounds

Some salt-like compounds with wedge-shaped molecules (e.g. **34** [309] and **35** [310]) show columnar phases over wide temperature ranges.

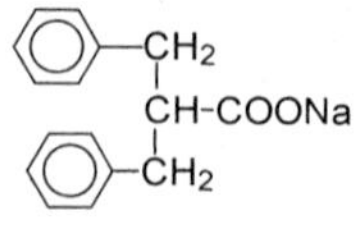

34   Cr 266 Col$_h$ 311 I

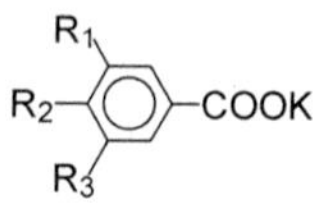

35

In order to form a disk-shaped unit, two or three of the molecules should be associated, similar to hydroxy compounds (discussed in Sec. 4.5 of this chapter). It is interesting that the corresponding free acid is not mesogenic; however, the related compound **36** [311] does exhibit a columnar phase.

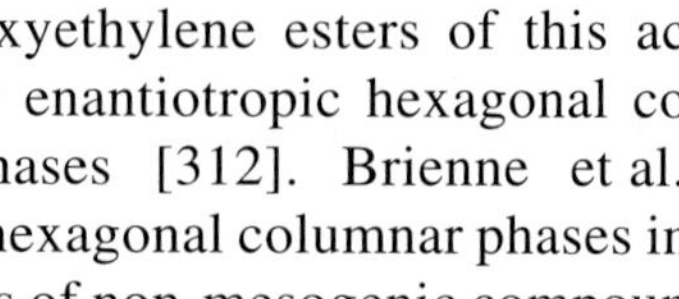

36   Cr 73 Col 136 I

Oligooxyethylene esters of this acid also display enantiotropic hexagonal columnar mesophases [312]. Brienne et al. [313] found hexagonal columnar phases in binary systems of non-mesogenic compounds, due to the formation of hydrogen bonded complexes.

Compounds 1 and 2 in Table 49 are examples of discotic metal complexes. They consist of two extended units, linked by the

Table 49. Discotic metallomesogens (by courtesy of Chandrasekhar [14]).

1. *bis*(4-n-Decylbenzoyl)methanato copper(II) [314]

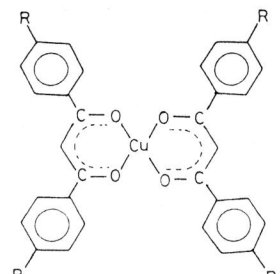

2. *bis*-(3,4-Dinonyloxybenzoyl)methanato copper(II) [315]

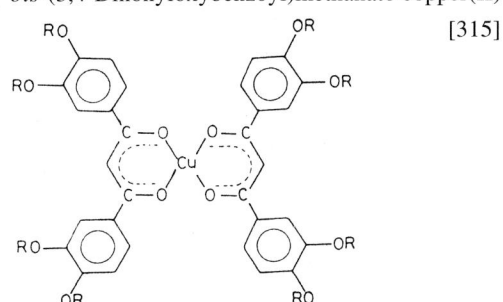

3. Octa-substituted metallophthalocyanine [316]

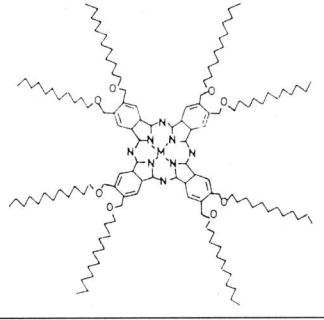

4. Benzo-15-crown-5-substituted phthalocyanine [317]

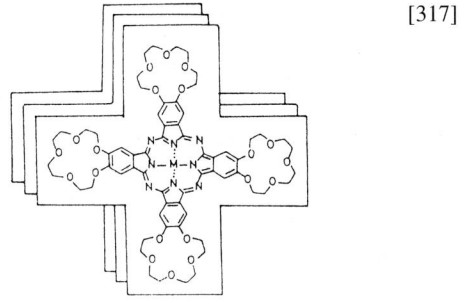

5. Tetrakis(alkyldithiolato)dinickel(II) [318]

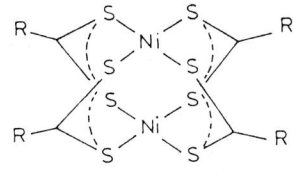

6. Binuclear copper carboxylate [319]

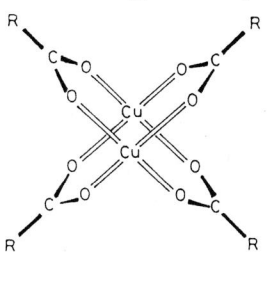

tetracoordinated metal atom. Examples 5 and 6 are also of this type. After complex formation, the phase behaviour depends on the resulting shape of the molecule. As discussed in Sec. 3.9, complexes with more rod-like shapes form nematic and smectic phases, while complexes with board- or disk-like shapes tend to discotic behaviour.

Several metallomesogens with nonpolar solvents like alkanes are able to form columnar nematic phases [319a–319f], in which the building units are columns consisting of the discotic molecules, separated by the nonpolar solvent. Columnar nematic phases can be formed also in mixtures of columnar charge transfer complexes and nonpolar solvents [319a, 319b, 319d, 319g].

Large disk-shaped rings like substituted tetraphenylporphyrins [320] or substituted phthalocyanines [321] can exhibit columnar phases. However, the mesogenity is more pronounced in the metal complexes of such materials [316, 317, 320–324]. Two exampales of phthalocyanine complexes, one with a flexible long-chain (compound 3) and one with flexible cyclic substituents (compound 4) are given in Table 49.

Cyclic oligoamides such as **37** [324] are mesomorphic. The open chain analogues (e.g. **38** [324]) show columnar phases [324–326].

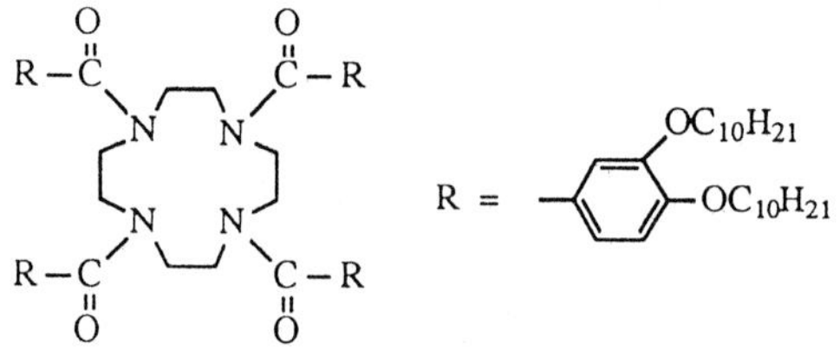

37 Cr 108 Col$_h$ 154 I

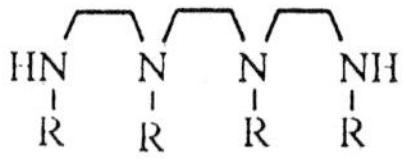

Cr 99 (Col$_h$ 93) I 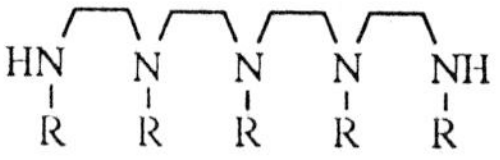   Cr 117 (Col$_h$ 105) I

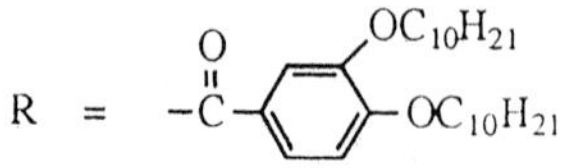

38

The formation of charge transfer (EDA) complexes between suitable components may induce columnar [327–329] or nematic discotic phases [329–331]. Also the formation of intramolecular EDA complexes in twin molecules can influence the mesomorphic properties substantially [310, 332].

4.4 Pyramidal (Bowlic), Tubular and Related Compounds

The cores of many discotic compounds are relatively flat. In contrast to this, there are derivatives of macrocycles that deviate substantially from the flat shape, being rather cone-shaped [333, 334]. These compounds are called 'pyramidic' or 'bowlic' compounds.

Zimmermann et al. [333] have prepared the substituted tribenzocyclononatriene (cyclotriveratrylene) **39**. The related substituted orthocyclophanes (cyclotetraveratrylenes) also form discotic phases (e.g. **40** [335, 336]).

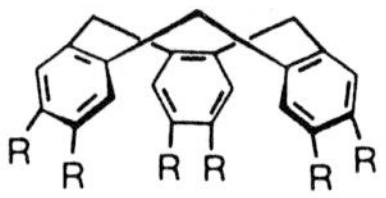

R = C$_9$H$_{19}$COO–

39 Cr 23.9 Columnar 152.6 I

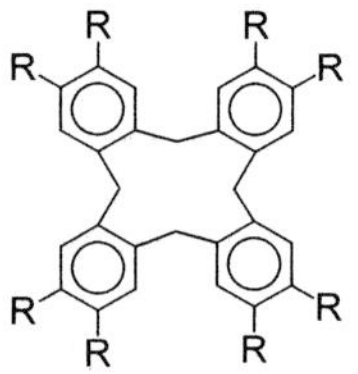

40 R = C$_n$H$_{2n+1}$–, C$_n$H$_{2n+1}$COO–

Pyramidic phases have been found also in cone-shaped calix[4]arene derivatives [337]. Lehn et al. [338] synthesized macrocyclic compounds with a hollow core and the ability to form complexes (e.g. **41**).

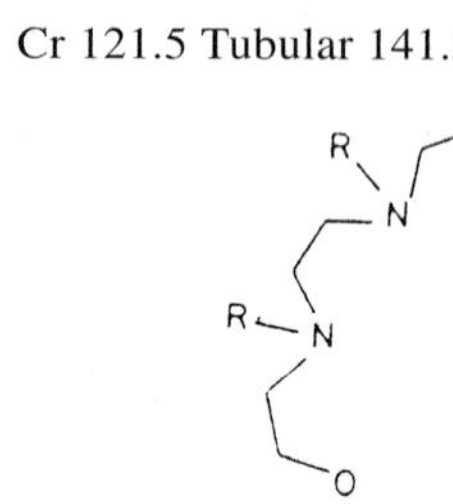

Cr 121.5 Tubular 141.5 I 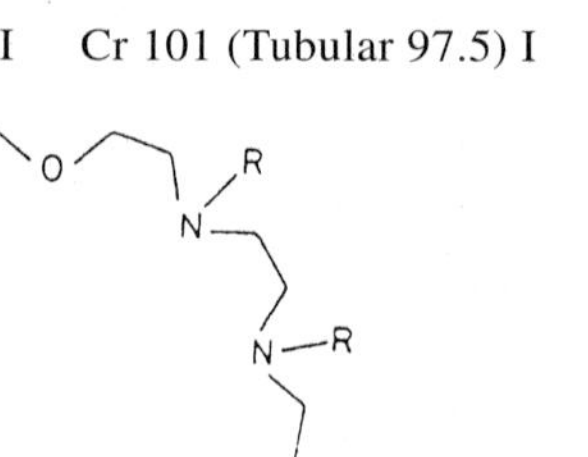   Cr 101 (Tubular 97.5) I

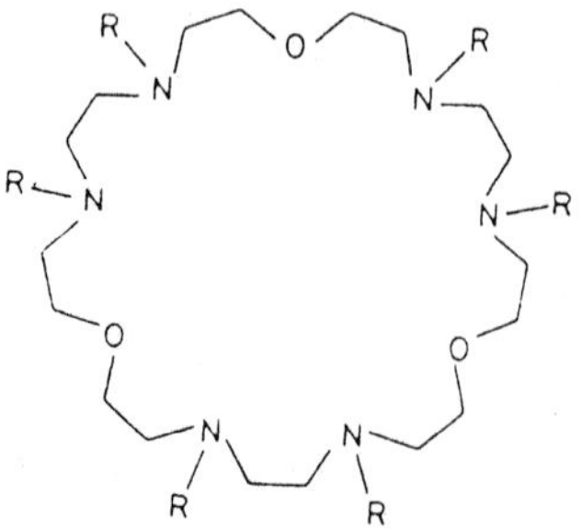

Cr 85 (Tubular 80) I

41   R = C$_{12}$H$_{25}$O–⟨⟩–CO–

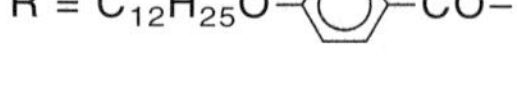

By stacking the macrocycles, hollow columns are formed. These phases are called 'tubular' and are expected to be useful in the development of ion-conducting channels [339]. Tubular phases have also been found by Idziak et al. [340] in hexacyclene derivatives and by Johansson et al. [341] in complexes of crown ethers.

4.5 Substituted Sugars

There are several polyols that exhibit discotic mesophases. Diisobutylsilanediol, which has been known since 1955 [342] but was not classified as hexagonal columnar until 1980 [343], associates in dimers [343] or oligomers [344] in order to form disk-like units (**42**).

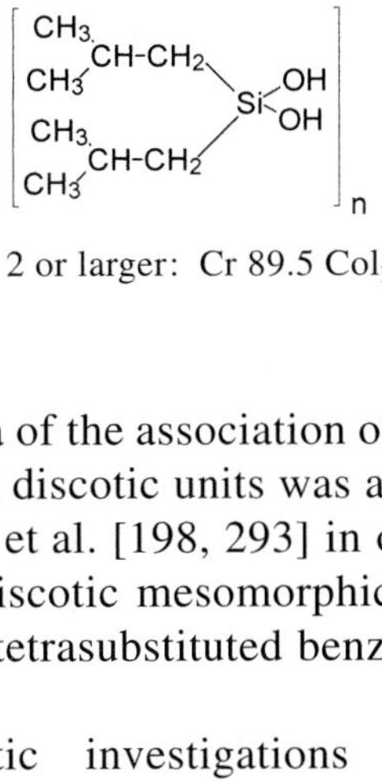

$$\left[\begin{array}{c} CH_3 \\ CH_3 \end{array} CH-CH_2 \begin{array}{c} \\ \\ \end{array} Si \begin{array}{c} OH \\ OH \end{array} \right]_n$$
$$CH_3 \begin{array}{c} \\ CH-CH_2 \end{array}$$

42 $n = 2$ or larger: Cr 89.5 Col$_h$ 101.5 I

This idea of the association of molecules resulting in discotic units was also used by Matsunaga et al. [198, 293] in order to explain the discotic mesomorphic properties of tri- and tetrasubstituted benzene derivatives.

Systematic investigations of sugars [266–268] have provided additional proof of the presence of associated building units in discotic structures. Several isomers, differing in the steric positions of the hydroxyl groups, of inositol are known. Because inositols contain six hydroxyl groups, many derivatives differing in the number and positions of the substituents are possible. By suitable substitution the derivatives can be mesomorphic. The kind of mesophase (calamitic or discotic) depends in a very delicate manner on the number and position of the substituents. Figure 19 gives an impression of this complicated situation.

Some additional examples may help clarity the confusing phase behaviour of inositol derivatives [268]. The monododecyl ethers of myoinositol exhibit smectic A phases, as do monosubstituted glucoside [346] and nojirimycin (an amino sugar [347]) derivatives. In diethers of myo-inositol, however, the phase behaviour depends on the position of the substituents: 4,5-diethers are hexagonal columnar and 3,6-diethers are smectic A. But the 4,5-diethers of chiroinositol show smectic A phases, as do the triethers of myoinositol.

The hypothesis of Praefcke et al. [345], that geminally branched amphiphiles form, by association, disk-like units, has been proved also for the case of galactopyranose derivatives [348, 349].

5 Conclusion

The discussion of the dependence of mesomorphic properties on molecular structure shows that the mesogenity of rod-like compounds is quite well understood, on the basis of both experimental material and theoretical explanations. The general supposition of the necessary shape anisotropy has allowed the derivation of simple procedures for predicting the clearing temperatures of rod-like compounds, e.g. the procedure using additive increments by Knaak and Rosenberg [116, 350] and the computer-aided predictions as elaborated by Vill [351]. In addition, the mesogeny of clearly disk-like compounds is well understood on the basis of experimental and theoretical results.

There are, however, difficulties with a lot of unconventional liquid crystalline materi-

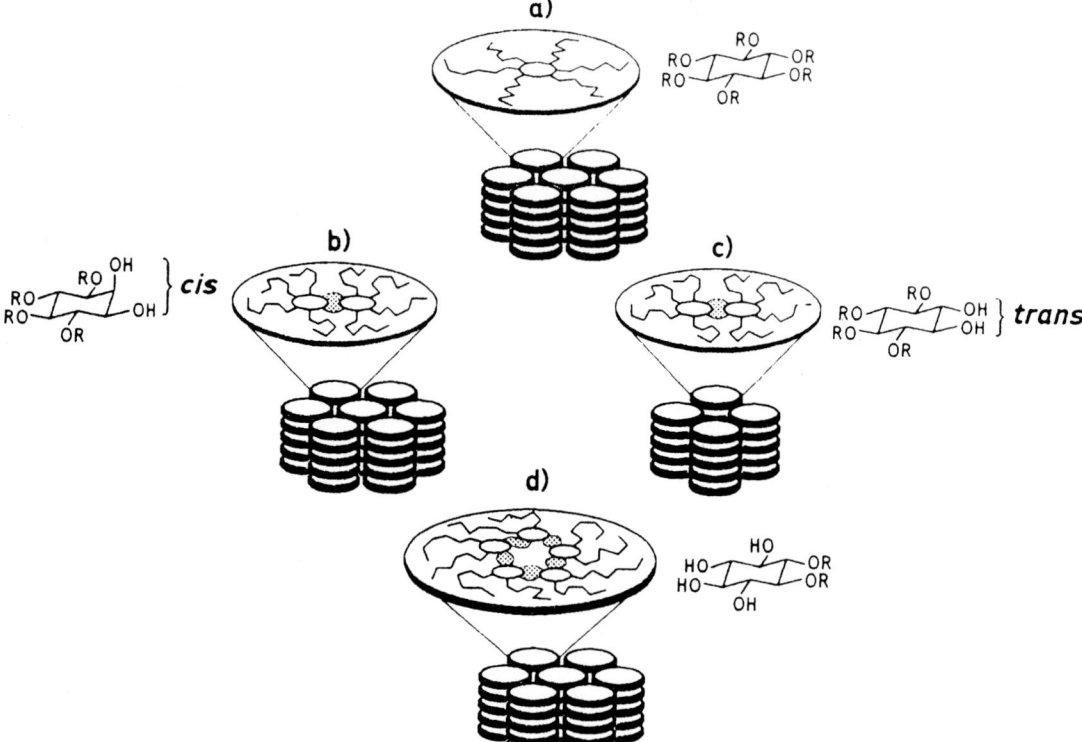

Figure 19. Models of the mesomorphic phases of some inositol derivatives. By courtesy of Praefcke et al. [345]. The enlarged parts represent slices of single columns and show the differences in their compositions. The hydrogen bonding regions of the molecules are shaded, and the zig-zags symbolize alkyloxy groups. The phase behaviour of the compounds with R = $-C_6H_{13}$ is as follows: (a) monomeric scylloinositol hexaether, Cr 18.4 Col_{ho} 90.8 I; (b) hydrogen-bridged vicinal diol dimer of myoinositol tetraether, Cr 27.7 Col_h 35.8 I; (c) hydrogen-bridged vicinal diol dimer of scylloinositol tetraether, Cr 48.9 Col_m 104.4 I; (d) hydrogen-bridged pentamer of scylloinositol diether, an all-*trans*-tetrol, Cr 111.5 Col_h 167.1 I. (Col_h, Columnar hexagonal phase; Col_m, columnar monoclinic phase.)

als. In many cases the effective molecular structure is not clear because of unknown conformations or association phenomena. In some cases the key to understanding their phase behaviour lies in the flexibility of the alkyl chains used as lateral or terminal substituents or, in particular, as spacers. The decoupling of molecular units by flexible spacers explains both the properties of several classes of mesogenic polymers as well as those of unconventional liquid crystalline compounds such as materials with cyclic lateral substituents, dimers and oligomers, because this flexibility allows a molecular shape that is effectively rod-like. 'Unconventional' molecules can assume effectively 'classical' (rod-like, discotic) molecular shapes by means of association phenomena, which explains the observed mesogenic properties.

Many unconventional and discotic compounds, especially those with large molecules, show a pronounced tendency towards the glassy state [196, 319f, 319g, 352–354, 358–365]. Glassy states are well known and common in polymers, and liquid crystalline oligomers show phase behaviours that are intermediate between those of low molecu-

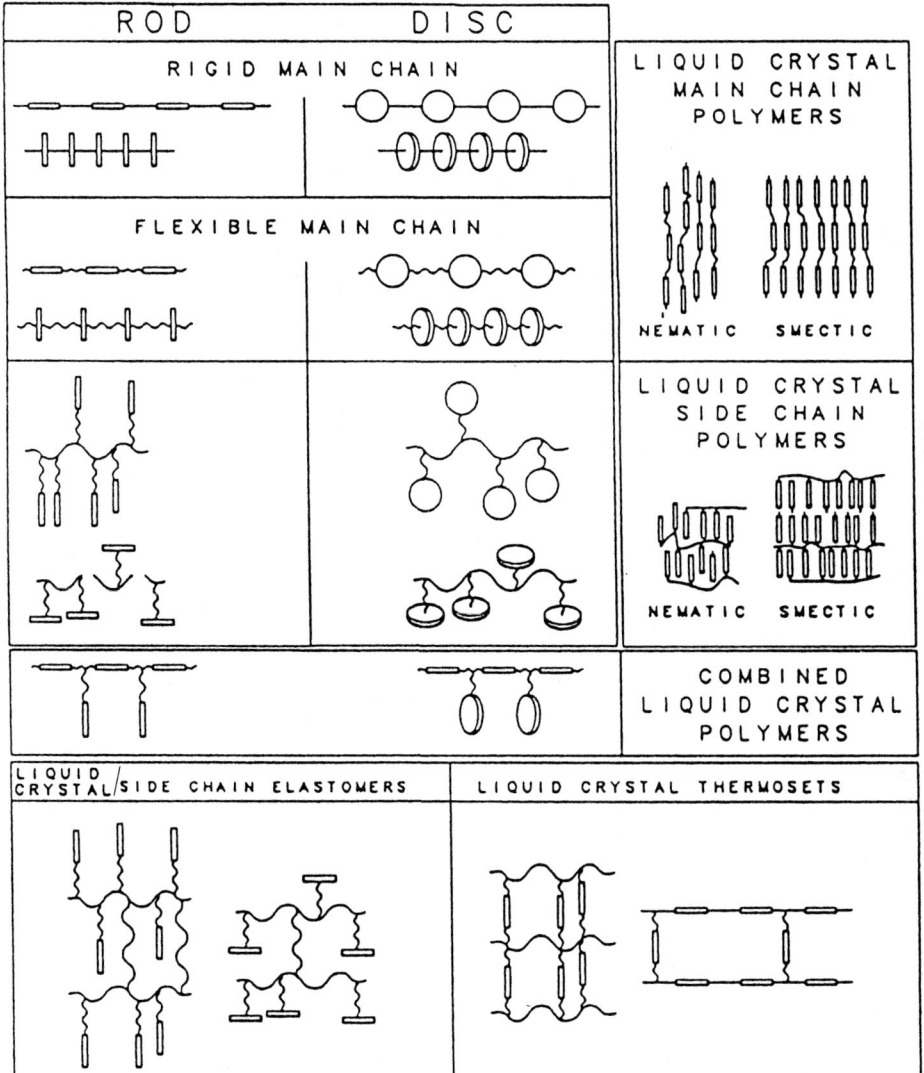

Figure 20. Structures of molecules of some liquid crystal polymers. The rod-like and disk-like mesogenic groups can be included as main groups in rigid or flexible polymeric chains, or can be attached as side groups of flexible chains. Cross-linking gives the liquid crystal elastomers or thermosets. (By courtesy of Adamczyk [355]).

lar and polymeric liquid crystals. By using the typical rod-like or discotic units in combination with polymer backbones and flexible spacers, the different possible means of linking these moieties give rise to many possibilities for the formation of mesogenic polymers [22, 272, 355–357, 366]. Some of these possibilities are shown schematically in Figure 20. A detailed overview of polymeric liquid crystals can be found in Volume 3 of this book. The author is indebted to Chisso Corporation, Tokyo, for continuous support and to Prof. W. Weissflog, Halle, for valuable comments.

6 References

[1] D. Demus, *Mol. Cryst. Liq. Cryst.* **1988**, *165*, 45.
[2] D. Demus, *Liq. Cryst.* **1989**, *5*, 75.
[3] F. Meyer, K. Dahlem, *Liebigs Ann. Chem.* **1903**, *326*, 331.
[4] D. Vorländer, *Kristallinisch-flüssige Substanzen*, Enke Verlag, Stuttgart, **1908**.
[5] D. Vorländer, *Chemische Kristallographie der Flüssigkeiten*, Akademische Verlagsgesellschaft, Leipzig, **1924**.
[6] H. Voigt, *Besitzen kreuzförmig aufgebaute Moleküle flüssig-kristalline Eigenschaften?* PhD Dissertation, Halle, **1924**.
[7] S. Chandrasekhar, B. K. Sadashiva, K. A. Suresh, *Pramana* **1977**, *9*, 471.
[8] J. Billard, J. C. Dubois, H. T. Nguyen, A. Zann, *Nouv. J. Chim.* **1978**, *2*, 535.
[9] S. Chandrasekhar, in *Advances in Liquid Crystals* (Ed.: G. H. Brown), Academic Press, New York, **1982**, *5*, 47.
[10] S. Chandrasekhar, *Phil. Trans. R. Soc., London, Ser. A* **1983**, *309*, 93.
[11] A. M. Levelut, *J. Chim. Phys.* **1983**, *88*, 149.
[12] C. Destrade, P. Foucher, H. Gasparoux, H. T. Nguyen, A. M. Levelut, J. Malthete, *Mol. Cryst. Liq. Cryst.* **1984**, *106*, 121.
[13] S. Chandrasekhar, G. S. Ranganath, *Rep. Prog. Phys.* **1990**, *53*, 57.
[14] S. Chandrasekhar, *Liq. Cryst.* **1993**, *14*, 3.
[15] D. Vorländer, *Ber. Deutsch. Chem. Ges.* **1929**, *62*, 2831.
[16] D. Vorländer, A. Apel *Chem. Ber.* **1932**, *65*, 1101.
[17] D. Vorländer, *Z. Phys. Chem.* **1923**, *105*, 211.
[18] C. Kuhrmann, Ph.D. Dissertation, Halle, **1926**.
[19] D. Demus in *Jahrbuch der Deutschen Akademie der Naturforscher Leopoldina*, Reihe 3, Halle, **1992**, *37*, p. 147.
[20] H. Matsuzaki, Y. Matsunaga, *Liq. Cryst.* **1993**, *14*, 105.
[21] J. Malthete, H. T. Nguyen, C. Destrade, *Liq. Cryst.* **1993**, *13*, 171.
[22] E. T. Samulski, *Faraday Discuss. Chem. Soc.* **1985**, *79*, 7.
[23] K. L. Savithramma, N. V. Madhusudana, *Mol. Cryst. Liq. Cryst.* **1981**, *74*, 243.
[24] N. P. Tumanyan, E. P. Sokolova, *Z. Fiz. Chim.* **1984**, *58*, 2444.
[25] N. V. Madhusudana in *Liquid Crystals. Applications and Uses*, Vol. 1 (Ed.: B. Bahadur), World Scientific, Singapore, **1990**, p. 37.
[26] A. Wulf, *J. Chem. Phys.* **1976**, *64*, 104.
[27] D. Demus, *Z. Chem.* **1986**, *26*, 6.
[28] D. Demus, A. Hauser in *Selected Topics in Liquid Crystal Research* (Ed.: H.-D. Koswig), Akademie-Verlag, Berlin, **1990**, p. 19.
[29] M. A. Cotter, *Phil. Trans. R. Soc., London, Ser. A* **1983**, *309*, 127.
[30] M. Warner, *J. Chem. Phys.* **1980**, *73*, 5874.
[31] L. Longa, W. H. de Jeu, *Phys. Rev., Ser. A* **1982**, *26*, 1632.
[32] W. H. de Jeu, *Phil. Trans. R. Soc., London, Ser. A* **1983**, *309*, 217.
[33] W. Kast in *Landolt–Börnstein: Zahlenwerte und Funktionen aus Physik-Chemie-Astronomie-Geophysik-Technik*, Vol. 2, Part 2, Springer-Verlag, Berlin, **1960**, p. 266.
[34] D. Demus, H. Demus, H. Zaschke, *Flüssige Kristalle in Tabellen*, VEB Deutscher Verlag f. Grundstoffindustrie, Leipzig, **1974**.
[35] D. Demus, H. Zaschke, *Flüssige Kristalle in Tabellen II*, VEB Deutscher Verlag f. Grundstoffindustrie, Leipzig, **1984**.
[35a] V. Vill, *LiqCryst 3.0, Database of Liquid Crystalline Compounds for Personal Computers*, LCI Publisher GmbH, Hamburg 1997.
[36] V. Vill in *Landolt–Börnstein: Numerical Data and Functional Relationships in Science and Technology, New Series* (Ed.: O. Madelung), Group IV: Macroscopic and Technical Properties of Matter, Vol. 7, Liquid Crystals, Springer-Verlag, Berlin, **1992–1994**; (a) *Subvol. a: One-Ring Systems and Two-Ring Systems without Bridging Groups*; (b) *Subvol. b: Two-Ring Systems with Bridging Group*; (c) *Subvol. c: Three-Ring Systems without Bridging Groups*; (d) *Subvol. d: Three-Ring Systems with One Bridging Group*; (e) *Subvol. d: Three-Ring Systems with Two Bridging Groups*; (f) *Four Ring Systems, Five Ring Systems, and more than Five Rings*. Additional volumes in preparation.
[37] D. Coates in *Liquid Crystals. Applications and Uses*, Vol. 1 (Ed.: B. Bahadur), World Scientific, Singapore, **1990**, Chap. 3.
[38] G. W. Gray in *Advances in Liquid Crystals*, Vol. 2 (Ed.: G. H. Brown), Academic Press, New York, **1976**, p. 1.
[39] G. W. Gray in *The Molecular Physics of Liquid Crystals* (Eds.: G. R. Luckhurst, G. W. Gray), Academic Press, London, **1979**; (a) Chap. 1; (b) Chap. 12.
[40] G. W. Gray, *Phil. Trans. R. Soc., London, Ser. A* **1983**, *309*, 77.
[41] G. W. Gray, *Proc. R. Soc. London, Ser. A* **1985**, *402*, 1.
[42] H. Ringsdorf, B. Schlarb, J. Venzmer, *Angew. Chem.* **1988**, *100*, 117.
[43] H. J. Deutscher, R. Frach, C. Tschierske, H. Zaschke in *Selected Topics in Liquid Crystal Research* (Ed.: H.-D. Koswig), Akademie-Verlag, Berlin, **1990**, p. 1.
[44] D. Demus in *Topics in Physical Chemistry* (Eds.; H. Baumgärtel, E. U. Franck, W. Grünbein), Vol. 3, *Liquid Crystals* (Ed.: H. Stegemeyer), Steinkopff, Darmstadt, **1994**, Chap. 1.
[45] H. Kelker, R. Hatz, *Handbook of Liquid Crystals*, Verlag Chemie, Weinheim, **1980**, Chap. 2.

[46] G. W. Smith, *Mol. Cryst. Liq. Cryst. Lett.* **1979**, *49*, 207.
[47] P. A. Irvine, D. C. Wu, P. J. Flory, *J. Chem. Soc., Faraday Trans. 1* **1984**, *80*, 1795.
[48] M. Busch, W. Weber, *J. Prakt. Chem.* **1936**, *146*, 1.
[49] K. Czuprynski, *Liq. Cryst.* **1994**, *16*, 399.
[50] H. Schubert, H. J. Lorenz, R. Hoffmann, F. Franke, *Z. Chem.* **1966**, *6*, 337.
[51] (a) H. Schubert, G. Hänse, unpublished. (b) G. Hänse, Dipl. Arbeit Halle, **1967**. (c) See also [34], compound no. 3665.
[52] H. Zaschke, *Wiss. Z. Univ. Halle XXIX'* **80** M, H. 3, **1980**, p. 35.
[53] M. A. Osman, *Z. Naturforsch., Teil A* **1983**, *38*, 693.
[54] M. A. Osman, T. Huynh-Ba, *Mol. Cryst. Liq. Cryst.* **1984**, *116*, 141.
[55] M. E. Neubert, L. T. Carlino, D. L. Fishel, R. M. D'Sidocky, *Mol. Cryst. Liq. Cryst.* **1980**, *59*, 253.
[56] H.-J. Deutscher, B. Laaser, W. Dölling, H. Schubert, *J. Prakt. Chem.* **1978**, *320*, 191.
[57] H.-J. Deutscher, M. Körber, H. Schubert, *Advances in Liquid Crystal Research and Application* (Ed.: L. Bata), Pergamon Press, Oxford, **1980**, p. 1075.
[58] See [36c], p. 176.
[59] R. Eidenschink, J. Krause, L. Pohl, J. Eichler in *Liquid Crystals, Proc. Int. Conf.* (Ed.: S. Chandrasekhar), Heyden, London, **1980**, p. 515.
[60] J. Billard, L. Mamlok, *Mol. Cryst. Liq. Cryst.* **1978**, *41*, 217.
[61] H. Schubert, W. Weissflog, unpublished. See [34], No. 975, p. 81.
[62] H.-J. Deutscher, M. Körber, H. Altmann, H. Schubert, *J. Prakt. Chem.* **1979**, *321*, 969.
[63] K. Praefcke, D. Schmidt, *Chem.-Ztg.* **1981**, *105*, 61.
[64] H.-J. Deutscher, H. Altmann, unpublished. See [35], No. 8344, p. 188.
[65] L. Verbit, R. L. Tuggey, A. R. Pinhas, *Mol. Cryst. Liq. Cryst.* **1975**, *30*, 201.
[66] H.-J. Deutscher, H. Altmann, unpublished. See [35], No. 5546, p. 49.
[67] E. L. Eliel, S. H. Wilen, L. N. Mander, *Stereochemistry of Carbon Compounds*, Wiley, New York, **1994**.
[68] H.-J. Deutscher, H.-M. Vorbrodt, H. Zaschke, *Z. Chem.* **1981**, *21*, 9.
[69] M. A. Osman, L. Revesz, *Mol. Cryst. Liq. Cryst.* **1982**, *82*, 41.
[70] L. Pohl, R. Eidenschink, G. Krause, D. Erdmann, *Phys. Lett.* **1977**, *A60*, 421.
[71] T. Inukai, H. Inoue, H. Sato, *US 4211666, EP 2136, JP 77-139, JP 78-67938*, **1980**.
[72] G. W. Gray, S. M. Kelly, *Angew. Chem.* **1981**, *93*, 412.
[73] H.-M. Vorbrodt, S. Deresch, H. Kresse, A. Wiegeleben, D. Demus, H. Zaschke, *J. Prakt. Chem.* **1981**, *323*, 902.

[74] H. Zaschke, A. Isenberg, H.-M. Vorbrodt in *Liquid Crystals. Ordered Fluids*, Vol. 4 (Eds.: A. Griffin, J. F. Johnson), Plenum Press, New York, **1984**, p. 75.
[75] Y. Haramoto, H. Kamogawa, *Chem. Lett.* **1985**, 79.
[76] V. S. Bezborodov, *Zh. Org. Khim.* **1989**, *25*, 2168.
[77] P. W. Rabideau (Ed.), *The Conformational Analysis of Cyclohexanes, Cyclohexadienes, and Related Hydroaromatic Compounds*, VCH, Weinheim, **1989**.
[78] R. Dabrowski, E. Zytynski, *Mol. Cryst. Liq. Cryst.* **1982**, *87*, 109.
[79] N. Carr, G. W. Gray, S. M. Kelly, *Mol. Cryst. Liq. Cryst.* **1985**, *129*, 301.
[80] R. Paschke, H. Zaschke, A. Hauser, D. Demus, *Liq. Cryst.* **1989**, *6*, 397.
[81] R. Paschke, Ph.D. Thesis, Halle, **1983**.
[82] M. Cereghetti, R. Marbet, K. Schleich, *Helv. Chim. Acta* **1982**, *65*, 1318.
[83] M. Petrzilka, K. Schleich, *Helv. Chim. Acta* **1982**, *65*, 1242.
[84] C. Tschierske, H. Zaschke, *J. Prakt. Chem.* **1988**, *330*, 1.
[85] U. H. Lauk, P. Skrabal, H. Zollinger, *Helv. Chim. Acta* **1985**, *68*, 1406.
[86] J. Canceill, C. Gros, J. Billard, *Pramana Suppl.* **1975**, *1*, 397.
[87] J. Malthete, J. Canceill, J. Gabard, J. Jacques, *Tetrahedron* **1981**, *37*, 2815.
[88] D. Varech, L. Lacombe, J. Jacques, *Nouv. J. Chim.* **1984**, *8*, 445.
[89] G. W. Gray, D. Lacey, *Mol. Cryst. Liq. Cryst.* **1983**, *99*, 123.
[90] J. Krause, L. Pohl, *DE-OS 2 951 099*, **1979**.
[91] D. Demus, H.-J. Deutscher, C. Tschierske, A. Wiegeleben, H. Zaschke, R. Frach, *DE-OS 3 806 716*, **1988**.
[92] A. I. Pavluchenko, N. I. Smirnova, V. V. Titov, E. I. Kovshev, K. M. Djumaev, *Mol. Cryst. Liq. Cryst.* **1976**, *37*, 35.
[93] J. Bartulin, C. Zuniga, J. Rematal, *Bol. Soc. Chil. Quim.* **1982**, *27*, 144.
[94] W. Weissflog, M. Süsse, *Z. Chem.* **1979**, *19*, 24.
[95] L. A. Karamysheva, I. F. Agafonova, R. K. Geivandov, V. F. Petrov, *Liq. Cryst.* **1991**, *10*, 875.
[96] M. Cavazza, C. Forte, G. Galli, M. Geppi, F. Pietra, C. A. Veracini, *Liq. Cryst.* **1993**, *15*, 275.
[97] B. Krieg, T. Kuhn, *Mol. Cryst. Liq. Cryst.* **1987**, *145*, 59.
[98] E. Bartmann, D. Dorsch, U. Finkenzeller, *Mol. Cryst. Liq. Cryst.* **1991**, *204*, 77.
[99] F. Reinitzer, *Monatsh. Chem.* **1888**, *9*, 421.
[100] T. Kawasaki, *J. Pharm. Soc. Jpn.* **1937**, *57*, 713.
[101] S. Masamune, A. V. Kemp-Jones, J. Green, D. L. Rabenstein, M. Yasunami, K. Takase, T. Nozoe, *Chem. Commun.* **1973**, 283.

[102] A. Mori, H. Takeshita, K. Kida, M. Uchida, *J. Am. Chem. Soc.* **1990**, *112*, 8635.
[103] L. K. M. Chan, G. W. Gray, D. Lacey, K. J. Toyne, *Mol. Cryst. Liq. Cryst.* **1988**, *158B*, 209.
[104] D. M. Gavrilovic, *US 3 951 846*, **1976**.
[105] L. K. M. Chan, G. W. Gray, D. Lacey, T. Srithanratana, K. J. Toyne, *Mol. Cryst. Liq. Cryst.* **1987**, *150B*, 335.
[106] L. A. Karamysheva, T. A. Geyvandova, I. F. Agafonova, K. V. Roitman, S. I. Torgova, R. K. Geyvandov, V. F. Petrov, A. Z. Rabinovich, M. F. Grebyonkin, *Mol. Cryst. Liq. Cryst.* **1990**, *191*, 237.
[107] P. Kaszynski, A. C. Friedli, J. Michl, *Mol. Cryst. Liq. Cryst. Lett.* **1988**, *6*, 27.
[108] J. W. Brown, D. J. Byron, D. J. Harwood, R. C. Wilson, A. R. Tajbakhsh, *Mol. Cryst. Liq. Cryst.* **1989**, *173*, 121.
[108a] R. Iglesias, J. L. Serrano, T. Sierra, *Liq. Cryst.* **1997**, *22*, 37.
[109] K. Dimitrova, J. Hauschild, H. Zaschke, H. Schubert, *J. Prakt. Chem.* **1980**, *322*, 933.
[110] K. Dölling, H. Zaschke, H. Schubert, *J. Prakt. Chem.* **1979**, *321*, 643.
[111] J. Hauschild, Ph.D. Thesis, Halle, **1974**.
[112] B. Krücke, H. Zaschke, *Wiss. Z. Univ. Halle XXXXI'92* M, H. 3, **1992**, p. 65.
[113] Y. Goto, T. Inukai, A. Fujita, D. Demus, *Mol. Cryst. Liq. Cryst.* **1995**, *260*, 23.
[114] N. Carr, G. W. Gray, *Mol. Cryst. Liq. Cryst.* **1985**, *124*, 27.
[115] K. Praefcke, D. Schmidt, G. Heppke, *Chem. Ztg.* **1980**, *104*, 269.
[116] D. Demus, *Z. Chem.* **1975**, *15*, 1.
[117] H. Schubert, W. Weissflog, unpublished. Data from [34], p. 64.
[118] H. Zaschke, R. Stolle, *Z. Chem.* **1975**, *15*, 441.
[119] W. L. McMillan, *Phys. Rev., Ser. A* **1971**, *4*, 1238.
[120] D. Frenkel in *Phase Transitions in Liquid Crystals* (Eds.: S. Martellucci, A. N. Chester), Plenum Press, New York, **1992**, p. 67.
[121] H. Zimmermann, *Liq. Cryst.* **1989**, *4*, 591.
[122] S. M. Kelly, J. Fünfschilling, A. Villiger, *Liq. Cryst.* **1993**, *14*, 1169.
[123] S. M. Kelly, *Liq. Cryst.* **1993**, *14*, 675.
[124] M. Schadt, R. Buchecker, K. Müller, *Liq. Cryst.* **1989**, *5*, 293.
[124a] S. M. Kelly, J. Fünfschilling, *Liq. Cryst.* **1996**, *20*, 77.
[125] D. Demus, S. Diele, S. Grande, H. Sackmann, in *Advances in Liquid Crystals* (Ed.: G. H. Brown), Academic Press, New York, **1983**, 6, 1.
[126] A. J. Leadbetter, R. M. Richardson, C. N. Colling, *J. Phys. Coll.* **1975**, *C1*, C1.
[127] I. H. Ibrahim, W. Haase, *Mol. Cryst. Liq. Cryst.* **1981**, *66*, 189.
[128] D. Demus, Y. Goto, S. Sawada, E. Nakagawa, H. Saito, R. Tarao, *Mol. Cryst. Liq. Cryst.* **1995**, *260*, 1.

[129] G. Weber, U. Finkenzeller, T. Geelhaar, B. Rieger, L. Pohl, *Liq. Cryst.* **1989**, *5*, 1381.
[130] S. Matsumoto, H. Hatoh, A. Murayama, *Liq. Cryst.* **1989**, *5*, 1345.
[131] E. Bartmann, J. Krause, K. Tarumi, Proc. 23. *Freiburger Arbeitstagung Flüssigkristalle*, **1994**, lecture No. 27.
[132] T. Doi, Y. Sakurai, A. Tamatani, S. Takenaka, S. Kusabayashi, Y. Nishihata, H. Terauchi, *J. Mater. Chem.* **1991**, *1*, 169.
[133] M. Koden, K. Nakagawa, Y. Ishii, F. Funada, M. Matsuura, K. Awane, *Mol. Cryst. Liq. Cryst. Lett.* **1989**, *6*, 185.
[134] E. P. Janulis, J. C. Novack, G. A. Papapolymerou, M. Tristani-Kendra, W. A. Huffman, *Ferroelectrics* **1988**, *85*, 375.
[135] V. M. Polosin, G. I. Sitnikova, O. V. Zvolinsky, A. N. Levov, A. V. Ivashchenko, V. F. Grebyonkin, presented at *14th International Liquid Crystal Conference*, Pisa, **1992**, Poster A-P 48.
[136] K. A. Epstein, M. D. Radcliffe, M. L. Brostrom, A. G. Rappaport, B. N. Thomas, N. A. Clark, presented at *4th International Conference on Ferroelectric Liquid Crystals*, Tokyo, **1993**, Poster P-46.
[137] F. G. Tournilhac, L. Bosio, J. Simon, L. M. Blinov, S. V. Yablonsky, *Liq. Cryst.* **1993**, *14*, 405.
[138] F. Tournilhac, J. Simon, *Ferroelectrics* **1991**, *114*, 283.
[139] K. Ohno, M. Ushioda, S. Saito, K. Miyazawa, *EP 313 991*, **1988**; *JP 87-269 923*, **1989**.
[140] A. de Meijere, H.-R. Dübal, C. Escher, W. Hemmerling, I. Müller, D. Ohlendorf, R. Wingen, *EP 318 423*, **1988**; *DE-OS 3 739 884*, **1987**.
[141] A. de Meijere, M. Messner, *Mol. Cryst. Liq. Cryst.* **1994**, *207*, 161.
[142] M. Messner, Ph.D. Thesis, Hamburg, **1992**.
[143] G. Scherowsky, A. Lotz, *Liq. Cryst.* **1993**, *14*, 1295.
[144] G. W. Gray, K. J. Harrison, *Symp. Chem. Soc. Faraday Div.* **1971**, *5*, 54.
[145] H. Matsuzaki, Y. Matsunaga, *Bull. Chem. Soc. Jpn.* **1990**, *63*, 2300.
[146] Y. Matsunaga, H. Matsuzaki, N. Miyajima, *Bull. Chem. Soc. Jpn.* **1990**, *63*, 886.
[147] Y. Matsunaga, N. Miyajima, *Mol. Cryst. Liq. Cryst.* **1990**, *178*, 157.
[148] A. Kloczkowski, G. R. Luckhurst, R. W. Phippen, *Liq. Cryst.* **1988**, *3*, 185.
[149] A. Wiegeleben, L. Richter, J. Deresch, D. Demus, *Mol. Cryst. Liq. Cryst.* **1980**, *59*, 329.
[150] I. Haller, R. J. Cox, *Liquid Crystals and Ordered Fluids* (Eds.: J. F. Johnson, R. S. Porter), Plenum Publ. New York, **1970**, p. 393.
[151] C. Weygand, *Chemische Morphologie der Flüssigkeiten und Kristalle, Hand- und Jahrbuch der Chemischen Physik*, Vol. 2, Akademische Verlagsges., Leipzig, **1941**, p. 33.

[152] D. Demus, A. Hauser, Ch. Selbmann, W. Weissflog, *Cryst. Res. Technol.* **1984**, *19*, 271.
[153] G. W. Gray, *Mol. Cryst.* **1966**, *1*, 333.
[154] M. Hird, K. J. Toyne, G. W. Gray, D. G. McDonnell, I. C. Sage, *Liq. Cryst.* **1995**, *18*, 1.
[155] H. Schubert, W. Weissflog, unpublished. (a) [34], p. 81; (b) [34], p. 84.
[156] G. W. Gray, M. Hird, K. J. Toyne, *Mol. Cryst. Liq. Cryst.* **1991**, *204*, 43.
[157] L. K. M. Chan, G. W. Gray, D. Lacey, *Mol. Cryst. Liq. Cryst.* **1985**, *123*, 185.
[158] G. W. Gray, D. Lacey, M. Hird, *GB 87-24.458*, **1987**; *PTC-WO 89/03.821*, **1988**.
[159] H. Sugiura, Y. Sakurai, Y. Masuda, H. Takeda, S. Kusabayashi, S. Takenaka, *Liq. Cryst.* **1991**, *9*, 441.
[160] Y. Masuda, Y. Sakurai, H. Sugiura, S. Miyake, S. Takenaka, S. Kusabayashi, *Liq. Cryst.* **1991**, *10*, 623.
[161] R. Dabrowski, V. S. Bezborodov, V. J. Lapanik, J. Dziaduszek, K. Czuprinski, *Liq. Cryst.* **1995**, *18*, 213.
[162] R. Eidenschink, *Mol. Cryst. Liq. Cryst.* **1985**, *123*, 57.
[163] R. Eidenschink, R. Haas, G. Römer, B. S. Scheuble, *Angew. Chem., Int. Ed. Engl.* **1984**, *23*, 151.
[164] C. Tschierske, H.-M. Vorbrodt, H. Kresse, H. Zaschke, *Mol. Cryst. Liq. Cryst.* **1989**, *117*, 113.
[165] H. Kelker, B. Scheurle, R. Hatz, W. Bartsch, *Angew. Chem.* **1970**, *82*, 984.
[166] S. Mori, N. Matsumura, *Oyo Butsuri* **1973**, *42*, 60.
[167] K. Yoshino, M. Ozaki, *Ferroelectrics* **1984**, *58*, 21.
[168] W. Mahler, D. Guillon, A. Skoulios, *Mol. Cryst. Liq. Cryst. Lett.* **1985**, *2*, 111.
[169] C. Viney, T. P. Russell, L. E. Depero, R. J. Twieg, *Mol. Cryst. Liq. Cryst.* **1989**, *168*, 63.
[170] C. Viney, R. J. Twieg, T. P. Russell, *Mol. Cryst. Liq. Cryst.* **1990**, *182B*, 291.
[171] J. Höpken, M. Möller, *Macromulecules* **1992**, *25*, 2482.
[172] G. W. Gray, P. A. Winsor, *Liquid Crystals and Plastic Crystals*, Halstead, New York, Vols. 1, 2, **1974**.
[173] G. Kögler, H. Drottloff, M. Möller, *Mol. Cryst. Liq. Cryst.* **1987**, *153*, 179.
[174] H. Drottloff, H. Rotter, D. Emeis, M. Möller, *J. Am. Chem. Soc.* **1987**, *109*, 7797.
[175] C. L. Hillemann, G. R. van Hecke, *J. Phys. Chem.* **1976**, *80*, 944.
[176] K. Kimura, M. Hirao, M. Yokoyama, *J. Mater. Chem.* **1991**, *1*, 293.
[177] J. Qin, M. G. Xie, Z.-L. Hu, H.-M. Zhao, *Syn. Commun.* **1992**, *22*, 2253.
[178] G. Zerban, H. Meier, *Z. Naturforsch., Teil B* **1993**, *48*, 171.
[179] P. R. Ashton, D. Joachimi, N. Spencer, J. F. Stoddart, C. Tschierske, A. J. P. White, D. J. Williams, K. Zab, *Angew. Chem.* **1994**, *106*, 1563.
[179a] D. Joachimi, P. R. Ashton, C. Sauer, N. Spencer, C. Tschierske, K. Zab, *Liq. Cryst.* **1996**, *20*, 337
[180] W. Weissflog, D. Demus, *Cryst. Res. Technol.* **1983**, *18*, K21.
[181] W. Weissflog, D. Demus, *Cryst. Res. Technol.* **1984**, *19*, 55.
[182] D. Demus, S. Diele, A. Hauser, I. Latif, Ch. Selbmann, W. Weissflog, *Cryst. Res. Technol.* **1985**, *20*, 1547.
[183] F. Hoffmann, H. Hartung, W. Weissflog, P. G. Jones, A. Chrapkowski, *Mol. Cryst. Liq. Cryst.* **1995**, *258*, 61.
[184] W. Weissflog, D. Demus, *Mol. Cryst. Liq. Cryst.* **1985**, *129*, 235.
[185] W. Weissflog, D. Demus, *Mater. Chem. Phys.* **1985**, *12*, 461.
[186] E. Mauerhoff, Ph.D. Thesis, Halle, **1922**.
[187] G. S. Attard, A. G. Douglass, *Liq. Cryst.* **1997**, *22*, 349.
[188] R. Cox, W. Volksen, B. L. Dawson in *Liquid Crystals and Ordered Fluids*, Vol. 4 (Eds.: A. C. Griffin, J. F. Johnson), Plenum, New York, **1984**, p. 33.
[189] V. Gallardo, H. J. Müller, *Mol. Cryst. Liq. Cryst.* **1984**, *102*, 13. There is strong doubt about the results of the synthesis: the compounds are suspicious to be simple Schiff's bases without the claimed lateral aromatic branches, see [191].
[190] S. Diele, W. Weissflog, G. Pelzl, H. Manke, D. Demus, *Liq. Cryst.* **1986**, *1*, 101.
[191] W. Weissflog, D. Demus, *Liq. Cryst.* **1988**, *3*, 275.
[192] W. Weissflog, D. Demus, S. Diele, *Mol. Cryst. Liq. Cryst.* **1990**, *191*, 9.
[193] F. Hoffmann, H. Hartung, W. Weissflog, P. G. Jones, A. Chrapkowski, *Mol. Cryst. Liq. Cryst.* **1996**, *281*, 205.
[193a] F. Perez, P. Judeinstein, J.-P. Bayle, F. Roussel, B. M. Fung, *Liq. Cryst.* **1997**, *22*, 711.
[194] W. Weissflog, E. Dietzmann, C. Stützer, M. Drewello, F. Hoffmann, H. Hartung, *Mol. Cryst. Liq. Cryst.* **1996**, *275*, 75.
[195] P. Berdague, J. P. Bayle, M. S. Ho, B. M. Fung, *Liq. Cryst.* **1993**, *14*, 667.
[196] D. Braun, M. Reubold, L. Schneider, M. Wegmann, J. H. Wendorff, *Liq. Cryst.* **1994**, *16*, 429.
[197] T. Akutagawa, Y. Matsunaga, K. Yasuhara, *Liq. Cryst.* **1994**, *17*, 659.
[197a] T. Niori, T. Sekine, J. Watanabe, T. Furukawa, H. Takezoe, *J. Mater. Chem.* **1996**, *6*, 1231.
[197b] G. Heppke, A. Jakli, D. Krüerke, C. Löhning, D. Lötzsch, S. Paus, S. Rauch, N. K. Sharma, *Eur. Conf. Liq. Cryst. Zakopane*, **1997**, Abstract no. O-8.

[197c] W. Weissflog, Ch. Lischka, I. Benne, T. Scharf, G. Pelzl, S. Diele, H. Kruth, *Proc. 26. Freiburger Arbeitstagung Flüssigkristalle*, **1997**, Contribution No. P 84.

[197d] *6th Int. Conf. Ferroelectric Liq. Cryst.*, Brest, July 20–24, **1997**, Poster by G. Heppke, S. Rauch, N. K. Sharma; Poster by D. M. Walba, J. E. Maclennan, M. A. Glaser, N. A. Clark.

[198] H. Kawada, Y. Matsunaga, *Bull. Chem. Soc. Jpn.* **1990**, *63*, 1691.

[199] P. Göring, G. Pelzl, S. Diele, P. Delavier, K. Siemensmeyer, K. H. Etzbach, *Liq. Cryst.* **1995**, *19*, 629.

[200] S. Berg, V. Krone, H. Ringsdorf, U. Quotschalla, H. Paulus, *Liq. Cryst.* **1991**, *9*, 151.

[201] W. Weissflog, A. Wiegeleben, S. Diele, D. Demus, *Cryst. Res. Technol.* **1984**, *19*, 583.

[202] J. Malthete, J. Billard, J. Canceill, J. Gabard, J. Jacques, *J. Phys. Coll. Paris* **1976**, *37*, C3.

[203] G. Pelzl, A. Humke, S. Diele, D. Demus, W. Weissflog, *Liq. Cryst.* **1990**, *7*, 115.

[204] G. Pelzl, S. Diele, K. Ziebarth, W. Weissflog, D. Demus, *Liq. Cryst.* **1990**, *8*, 765.

[205] W. Weissflog, M. Rogunova, I. Letko, S. Diele, G. Pelzl, *Liq. Cryst.* **1996**, *21*, 13.

[206] W. Weissflog, G. Pelzl, I. Letko, S. Diele, *Mol. Cryst. Liq. Cryst.* **1995**, *260*, 157.

[206a] B. Schiewe, H. Kresse, *Liq. Cryst.* **1995**, *19*, 659.

[207] D. Demus, G. Pelzl, A. Wiegeleben, W. Weissflog, *Mol. Cryst. Liq. Cryst.* **1980**, *56*, 289.

[208] W. Weissflog, S. Diele, D. Demus, *Mater. Chem. Phys.* **1986**, *15*, 475.

[209] H. T. Nguyen, J. Malthete, C. Destrade, *Mol. Cryst. Liq. Cryst. Lett.* **1985**, *2*, 133.

[210] A. M. Levelut, Y. Fang, *J. Phys. Paris, Coll. C7* **1990**, *51*, 229.

[211] H. T. Nguyen, C. Destrade, J. Malthete, *Liq. Cryst.* **1990**, *8*, 797.

[212] H. T. Nguyen, C. Destrade, H. Allouchi, J. P. Bideau, M. Cotrait, D. Guillon, P. Weber, J. Malthete, *Liq. Cryst.* **1993**, *15*, 435.

[213] K. Praefcke, B. Kohne, B. Gündogan, D. Singer, D. Demus, S. Diele, G. Pelzl, U. Bakowsky, *Mol. Cryst. Liq. Cryst.* **1991**, *198*, 393.

[214] K. Leister, Ph.D. Thesis, Halle, **1922**.

[215] J. Malthete, J. Billard, J. Jacques, *Comp. Rend. Hebd. Sci. Acad. Sci. Paris* **1975**, *C281*, 333.

[216] H. Kelker, B. Scheurle, *Mol. Cryst.* **1969**, *7*, 381.

[217] A. C. Griffin, S. F. Thames, M. S. Bonner, *Mol. Cryst. Liq. Cryst. Lett.* **1977**, *43*, 135.

[218] A. C. Griffin, N. W. Buckley, D. L. Hughes, D. L. Wertz, *Mol. Cryst. Liq. Cryst. Lett.* **1981**, *64*, 139.

[219] H. Dehne, A. Roger, D. Demus, S. Diele, H. Kresse, G. Pelzl, W. Wedler, W. Weissflog, *Liq. Cryst.* **1989**, *6*, 47.

[220] W. Weissflog, D. Demus, S. Diele, P. Nitschke, W. Wedler, *Liq. Cryst.* **1989**, *5*, 111.

[221] W. Weissflog, D. Demus, S. Diele, *Mol. Cryst. Liq. Cryst.* **1990**, *191*, 9.

[222] J. Andersch, C. Tschierske presented at *24 Freiburger Arbeitstagung Flüssigkristalle*, April **1995**, paper No. 29.

[222a] K. Zab, D. Joachimi, E. Novotna, S. Diele, C. Tschierske, *Liq. Cryst.* **1995**, *18*, 631.

[222b] J. Andersch, C. Tschierske, *Liq. Cryst.* **1996**, *21*, 51.

[222c] J. Andersch, S. Diele, D. Lose, C. Tschierske, *Liq. Cryst.* **1996**, *21*, 103.

[223] D. Vorländer, *Z. Phys. Chem. A* **1927**, *126*, 449.

[224] J. W. Emsley, G. R. Luckhurst, N. G. Shilstone, I. Sage, *Mol. Cryst. Liq. Cryst.* **1984**, *102*, 223.

[225] C. Aguilera, S. Ahmad, J. Bartulin, H. J. Müller, *Mol. Cryst. Liq. Cryst.* **1988**, *162B*, 277.

[225a] I. L. Rozhanskii, I. Tomita, T. Endo, *Liq. Cryst.* **1996**, *21*, 631.

[226] J. L. Hogan, C. T. Imrie, G. R. Luckhurst, *Liq. Cryst.* **1988**, *3*, 645.

[227] C. T. Imrie, *Liq. Cryst.* **1989**, *6*, 391.

[228] G. S. Attard, R. W. Date, C. T. Imrie, G. R. Luckhurst, S. J. Roskilly, J. M. Seddon, L. Taylor, *Liq. Cryst.* **1994**, *16*, 529.

[228a] I. D. Fletcher, G. R. Luckhurst, *Liq. Cryst.* **1995**, *18*, 175.

[228b] A. E. Blatch, I. D. Fletcher, G. R. Luckhurst, *Liq. Cryst.* **1995**, *18*, 801.

[229] D. J. Photinos, E. T. Samulski, H. Toriumi, *J. Chem. Soc., Faraday Trans.* **1992**, *88*, 1875.

[230] D. Creed, J. R. D. Gross, S. L. Sullivan, A. C. Griffin, C. E. Hoyle, *Mol. Cryst. Liq. Cryst.* **1987**, *149*, 185.

[231] F. Hardouin, H. Richard, M. F. Achard, *Liq. Cryst.* **1993**, *14*, 971.

[231a] A. Hohmuth, B. Schiewe, S. Heinemann, H. Kresse, *Liq. Cryst.* **1997**, *22*, 211.

[232] G. S. Attard, C. T. Imrie, *Liq. Cryst.* **1989**, *6*, 387.

[232a] N. V. Tsvetkov, V. V. Zuev, V. N. Tsvetkov, *Liq. Cryst.* **1997**, *22*, 245.

[232b] G. Ungar, D. Abramic, V. Percec, J. A. Heck, *Liq. Cryst.* **1996**, *21*, 73.

[233] V. Percec, M. Kawasumi, *Liq. Cryst.* **1993**, *13*, 83.

[234] V. Percec, M. Kawasumi, *J. Chem. Soc., Perkin Trans. 1* **1993**, 1319.

[235] I. D. Fletcher, G. R. Luckhurst, *Liq. Cryst.* **1995**, *18*, 175.

[235a] P. Busch, C. Schmidt, A. Stracke, J. H. Wendorff, D. Janietz, S. Mahlstedt, *Proc. 26. Freiburger Arbeitstagung Flüssigkristalle*, **1997**, Contribution P 61.

[236] H. Budig, S. Diele, P. Göring, R. Paschke, C. Sauer, C. Tschierske, *J. Chem. Soc., Chem. Commun.* **1994**, 2359.

[237] H. Budig, S. Diele, P. Göring, R. Paschke, C. Sauer, C. Tschierske, *J. Chem. Soc., Perkin Trans.* **1995**, *2*, 767.

[238] W. Kreuder, H. Ringsdorf, O. Herrmann-Schönherr, J. H. Wendorff, *Angew. Chem.* **1987**, *99*, 1300; *Angew. Chem., Int. Ed. Engl.* **1987**, *26*, 1249.
[239] R. Eidenschink, F.-H. Kreuzer, W. H. de Jeu, *Liq. Cryst.* **1990**, *8*, 879.
[240] L. M. Wilson, *Liq. Cryst.* **1994**, *16*, 1005.
[241] K. Zab, D. Joachimi, O. Agert, B. Neumann, C. Tschierske, *Liq. Cryst.* **1995**, *18*, 489.
[242] S. Norvez, J. Simon, *Liq. Cryst.* **1993**, *14*, 1389.
[243] M. G. Bennett, B. Jones, *J. Chem. Soc., London* **1935**, 1874; **1929**, 2660.
[244] R. F. Bryan, *J. Chem. Soc., Ser. B* **1967**, 1311.
[245] A. Kolbe, D. Demus, *Z. Naturforsch., Teil A* **1968**, *23*, 1237.
[246] J. Goldmacher, L. A. Barton, *J. Org. Chem.* **1967**, *32*, 476.
[247] D. Pucci, M. Veber, J. Malthete, *Liq. Cryst.* **1996**, *21*, 153.
[248] H. Schubert, K.-D. Münzner, not published; see compounds no. 4016–4030 in [34].
[249] H. Schubert, H. Zaschke, *J. Prakt. Chem.* **1970**, *312*, 494.
[250] H. Schubert, I. Eissfeldt, R. Lange, F. Trefflich, *J. Prakt. Chem.* **1966**, *33*, 265.
[250a] L. J. Yu, J. M. Wu, S. L. Wu, *Mol. Cryst. Liq. Cryst.* **1991**, *198*, 407.
[250b] L. J. Yu, J. S. Pan, *Liq. Cryst.* **1993**, *14*, 829.
[250c] L. J. Yu, *Liq. Cryst.* **1993**, *14*, 1303.
[250d] T. Kato, T. Uryu, F. Kaneuchi, C. Jin, J. M. Frechet, *Liq. Cryst.* **1993**, *14*, 1311.
[250e] T. Kato, M. Fukumasa, J. M. J. Frechet, *Chem. Mater.* **1995**, *7*, 368.
[251] H. Kresse, C. Dorscheid, I. Szulzewski, R. Frank, R. Paschke, *Ber. Bunsenges. Phys. Chem.* **1993**, *97*, 1345.
[252] H. Kresse, I. Szulzewski, S. Diele, R. Paschke, *Mol. Cryst. Liq. Cryst.* **1994**, *238*, 13.
[253] A. Treybig, C. Dorscheid, W. Weissflog, H. Kresse, *Mol. Cryst. Liq. Cryst.* **1995**, *260*, 369.
[253a] H. Bernhardt, W. Weissflog, H. Kresse, *Chem. Lett.* **1997**, 151.
[253b] A. Treybig, H. Bernhardt, S. Diele, W. Weissflog, H. Kresse, *Mol. Cryst. Liq. Cryst.* **1997**, *293*, 7.
[254] L. M. Wilson, *Liq. Cryst.* **1995**, *18*, 381.
[255] S. Diele, A. Mädicke, E. Geissler, K. Meinel, D. Demus, H. Sackmann, *Mol. Cryst. Liq. Cryst.* **1989**, *166*, 131.
[256] C. Tschierske, G. Brecesinski, F. Kuschel, H. Zaschke, *Mol. Cryst. Liq. Cryst. Lett.* **1989**, *6*, 139.
[257] G. Staufer, M. Schellhorn, G. Lattermann, *Liq. Cryst.* **1995**, *18*, 519.
[258] C. Tschierske, F. Hentrich, D. Joachimi, O. Agert, H. Zaschke, *Liq. Cryst.* **1991**, *9*, 571.
[259] G. Lattermann, G. Staufer, *Mol. Cryst. Liq. Cryst.* **1990**, *191*, 199.

[260] H. A. van Doren, R. van der Geest, R. M. Kellog, H. Wynberg, *Recl. Trav. Chim. Pays-Bas* **1990**, *109*, 197.
[261] D. Joachimi, C. Tschierske, A. Öhlmann, W. Rettig. *J. Mater. Chem.* **1994**, *4*, 1021.
[262] C. Tschierske, H. Zaschke, *J. Chem. Soc., Chem. Commun.* **1990**, 1013.
[263] F. Hentrich, C. Tschierske, H. Zaschke, *Angew. Chem., Int. Ed. Engl.* **1991**, *30*, 440.
[264] F. Hentrich, S. Diele, C. Tschierske, *Liq. Cryst.* **1994**, *17*, 827.
[265] F. Hentrich, C. Tschierske, S. Diele, C. Sauer, *J. Mater. Chem.* **1994**, *4*, 1547.
[266] G. A. Jeffrey, *Acc. Chem. Res.* **1986**, *19*, 168.
[267] G. A. Jeffrey, L. M. Wingert, *Liq. Cryst.* **1992**, *12*, 179.
[268] K. Praefcke, D. Blunk, J. Hempel, *Mol. Cryst. Liq. Cryst.* **1994**, *243*, 323.
[269] C. Tschierske, A. Lunow, H. Zaschke, *Liq. Cryst.* **1990**, *8*, 885.
[270] D. Blunk, K. Praefcke, G. Legler, *Liq. Cryst.* **1995**, *18*, 149.
[270a] K. Borisch, S. Diele, P. Göring, H. Müller, C. Tschierske, *Liq. Cryst.* **1997**, *22*, 427.
[271] V. Vill, *Mol. Cryst. Liq. Cryst.* **1992**, *213*, 67.
[272] D. Vorländer. *Ber. Deutsch. Chem. Ges.* **1910**, *43*, 3120.
[273] A.-M. Giroud-Godquin, P. M. Maitlis, *Angew. Chem.* **1991**, *103*, 370; *Angew. Chem., Int. Ed. Engl.* **1991**, *30*, 375.
[274] S. A. Hudson, P. M. Maitlis, *Chem. Rev.* **1993**, *93*, 861.
[275] S. Takahashi, T. Kaharu, *Mem. Inst. Sci. Ind. Res., Osaka* **1992**, *49*, 47.
[276] E. Baum, D. Demus, H. Sackmann, *Wiss. Z. Univ. Halle XIX'70 M*, H. 5, **1970**, p. 37.
[276a] B. Gallot, A. Skoulios, *Kolloid-Z. Z. Polym.* **1966**, *213*, 143.
[277] See: [34], Chap. 12; [35], Chap. 16.
[278] B. Gallot, A. Skoulios, *Mol. Cryst.* **1966**, *1*, 263.
[279] A. Sanchez Arenas, M. V. Garcia, M. I. Redondo, J. A. R. Cheda, M. V. Roux, C. Turrison, *Liq. Cryst.* **1995**, *18*, 431.
[280] D. C. Schroeder, J. P. Schroeder, *Mol. Cryst. Liq. Cryst.* **1976**, *34*, 43.
[281] M. Arkas, K. Yannakopoulou, C. M. Paleos, P. Weber, A. Skoulios, *Liq. Cryst.* **1995**, *18*, 563.
[282] M. Tabrizian, A. Soldera, M. Couturier, C. G. Bazuin, *Liq. Cryst.* **1995**, *18*, 475.
[283] G. Pelzl, S. Diele, A. Humke, D. Demus, H. Zaschke, *Mol. Cryst. Liq. Cryst.* **1990**, *191*, 307.
[284] T. Kaharu, H. Matsubara, S. Takahashi, *J. Mater. Chem.* **1992**, *2*, 43.
[285] R. Urban, Ph. D. Thesis, Halle, **1923**.
[286] H. Adams, N. A. Bailey, D. W. Bruce, D. A. Dunmur, E. Lalinde, M. Marcos, C. Ridgeway, A. J. Smith, P. Styring, P. M. Maitlis, *Liq. Cryst.* **1987**, *2*, 381.

[287] D. W. Bruce, X.-H. Liu, *Liq. Cryst.* **1995**, *18*, 165.
[288] R. Paschke, S. Diele, I. Letko, A. Wiegeleben, G. Pelzl, K. Griesar, M. Athanassopoulou, W. Haase, *Liq. Cryst.* **1995**, *18*, 451.
[289] D. W. Bruce, D. A. Dunmur, S. A. Hudson, E. Lalinde, P. M. Maitlis, M. P. McDonald, R. Orr, P. Styring, A. S. Cherodian, R. M. Richardson, J. L. Feijoo, G. Ungar, *Mol. Cryst. Liq. Cryst.* **1991**, *206*, 79.
[290] J. C. Dubois, J. Billard in *Liquid Crystals and Ordered Fluids*, Vol. 4 (Eds.: A. C. Griffin, J. F. Johnson), Plenum, New York, **1984**, p. 1043.
[291] H. Kawada, Y. Matsunaga, *Bull. Chem. Soc. Jpn.* **1990**, *63*, 1691.
[292] H. Kawada, Y. Matsunaga, *Bull. Chem. Soc. Jpn.* **1988**, *61*, 3083.
[293] J. Kawamata, Y. Matsunaga, *Mol. Cryst. Liq. Cryst.* **1993**, *231*, 79.
[294] B. Kohne, K. Praefcke, *Angew. Chem.* **1984**, *96*, 70.
[295] J. Billard, J. C. Dubois, C. Vaucher, A. M. Levelut, *Mol. Cryst. Liq. Cryst.* **1981**, *66*, 115.
[296] C. Destrade, M. C. Mondon, J. Malthete, *J. Phys. Paris* **1979**, *40*, C3-17.
[297] H. T. Nguyen, H. Gasparoux, C. Destrade, *Mol. Cryst. Liq. Cryst.* **1981**, *68*, 101.
[298] C. Vauchier, A. Zann, P. Le Barny, J. C. Dubois, J. Billard, *Mol. Cryst. Liq. Cryst.* **1981**, *66*, 103.
[299] H. Ringsdorf, U. Dahn, J. H. Wendorff, R. Festag, P. A. Heiney, N. Maliszewskyi, presented at *23 Freiburger Arbeitstagung Flüssigkristalle*, **1994**, poster P50.
[300] P. Henderson, S. Kumar, J. A. Rego. H. Ringsdorf, P. Schuhmacher, presented at *24 Freiburger Arbeitstagung Flüssigkristalle*, **1995**, Paper 06.
[301] C. Destrade, H. Gasparoux, A. Babeau, H. T. Nguyen, J. Malthete, *Mol. Cryst. Liq. Cryst.* **1981**, *67*, 37.
[302] R. Fugnitto, H. Strzelecka, A. Zann, J. C. Dubois, J. Billard, *J. Chem. Soc., Chem. Commun.* **1980**, 271.
[303] H. Bock, W. Helfrich, *Liq. Cryst.* **1995**, *18*, 387.
[304] K. Praefcke, B. Kohne, K. Gutbier, N. Johnen, D. Singer, *Liq. Cryst.* **1989**, *5*, 233.
[305] B. Kohne, K. Praefcke, J. Billard, *Z. Naturforsch., Teil B* **1986**, *41*, 1036.
[306] K. Praefcke, B. Kohne, D. Singer, D. Demus, G. Pelzl, S. Diele, *Liq. Cryst.* **1990**, *7*, 589.
[307] E. Keinan, S. Kumar, S. P. Singh, R. Ghirlando, E. J. Wachtel, *Liq. Cryst.* **1992**, *11*, 157.
[308] T. Plesnivy, H. Ringsdorf, P. Schuhmacher, U. Nütz, S. Diele, *Liq. Cryst.* **1995**, *18*, 185.
[309] D. Demus, H. Sackmann, K. Seibert, *Wiss. Z. Univ. Halle, XIX'70 M*, H. 5, **1970**, p. 47.

[310] D. Janietz, D. Daute, R. Festag, J. H. Wendorff, presented at *24 Freiburger Arbeitstagung Flüssigkristalle*, April **1995**, poster P44.
[311] V. Percec, J. Heck, *J. Polym. Sci. Polym. Chem.* **1991**, *29*, 591.
[312] D. Tomazos, G. Out, J. A. Heck, G. Johansson, V. Percec, M. Möller, *Liq. Cryst.* **1994**, *16*, 509.
[313] M. J. Brienne, J. Gabard, J. M. Lehn, I. Stibor, *J. Chem. Soc., Chem. Commun.* **1989**, 1868.
[314] A. M. Giroud-Godquin, J. Billard, *Mol. Cryst. Liq. Cryst.* **1981**, *66*, 147.
[315] A. M. Giroud-Godquin, M. M.Gauthier, G. Sigaud, F. Hardouin, M. F. Achard, *Mol. Cryst. Liq. Cryst.* **1986**, *132*, 35.
[316] C. Piechocki, J. Simon, A. Skoulios, D. Guillon, P. Weber, *J. Am. Chem. Soc.* **1982**, *104*, 5245.
[317] C. Sirlin, L. Bosio, J. Simon, V. Ahsen, E. Yilmazer, O. Bekaroglu, *Chem. Phys. Lett.* **1987**, *139*, 362.
[318] K. Ohta, Y. Morizumi, H. Ema, T. Fujimoto, I. Yamamoto, *Mol. Cryst. Liq. Cryst.* **1991**, *208*, 55.
[319] A. M. Giroud-Godquin, J. C. Marchon, D. Guillon, A. Skoulios, *J. Phys. Lett., Paris* **1984**, *45*, L681.
[319a] N. Usoltseva, K. Praefcke, D. Singer, B. Gündogan, *Mol. Mater.* **1994**, *4*, 253.
[319b] K. Praefcke, B. Bilgin, N. Usoltseva, B. Heinrich, D. Guillon, *J. Mater. Chem.* **1995**, *5*, 2257.
[319c] N. Usoltseva, G. Hauck, H. D. Koswig, K. Praefcke, B. Heinrich, *Liq. Cryst.* **1996**, *6*, 731.
[319d] K. Praefcke, J. D. Holbrey, N. Usoltseva, D. Blunk, *Mol. Cryst. Liq. Cryst.* **1997**, *292*, 123.
[319e] R. Segrouchni, A. Skoulios, *J. Phys. II France* **1995**, *5*, 1385.
[319f] H. Eichhorn, D. Wöhrle, D. Pressner, *Liq. Cryst.* **1997**, *22*, 643.
[319g] S. Zamir, D. Singer, N. Spielberg, E. J. Wachtel, H. Zimmermann, R. Pupko, Z. Luz, *Liq. Cryst.* **1996**, *21*, 39.
[320] Y. Shimizu, M. Miya, A. Nagata, K. Ohta, I. Yamamoto, S. Kusabayashi, *Liq. Cryst.* **1993**, *14*, 795.
[321] A. N. Cammidge, M. J. Cook, S. D. Haslam, R. M. Richardson, K. J. Harrison, *Liq. Cryst.* **1993**, *14*, 1847.
[322] M. K. Engel, P. Bassoul, L. Bosio, H. Lehmann, M. Hanck, J. Simon, *Liq. Cryst.* **1993**, *15*, 709.
[323] H. Eichhorn, D. Pressner, D. Wöhrle, H. W. Spiess, presented at *24 Freiburger Arbeitstagung Flüssigkristalle*, April **1995**, poster P49.
[324] H. Fischer, S. Kobayashi, T. Plesnivy, H. Ringsdorf, M. Seitz, presented at *24 Freiburger Arbeitstagung Flüssigkristalle*, April **1995**, paper 27.
[325] U. Stebani, G. Lattermann, M. Wittenberg, R. Festag, J. H. Wendorff, *Adv. Mater.* **1994**, *6*, 572.

[326] H. Fischer, S. S. Ghosh, P. A. Heiney, N. C. Maliszewskyj, T. Plesnivy, H. Ringsdorf, M. Seitz, *Angew. Chem.* **1995**, *107*, 879.
[327] H. Bengs, M. Ebert, O. Karthaus, B. Kohne, K. Praefcke, H. Ringsdorf, J. H. Wendorff, R. Wüstefeld, *Adv. Mater.* **1990**, *2*, 141.
[328] M. Ebert, G. Frick, Ch. Baehr, J. H. Wendorff, R. Wüstefeld, H. Ringsdorf, *Liq. Cryst.* **1992**, *11*, 293.
[329] D. Singer, A. Liebmann, K. Praefcke, J. H. Wendorff, *Liq. Cryst.* **1993**, *14*, 785.
[330] K. Praefcke, D. Singer, B. Kohne, M. Ebert, A. Liebmann, J. H. Wendorff, *Liq. Cryst.* **1991**, *10*, 147.
[331] H. Bengs, O. Karthaus, H. Ringsdorf, Ch. Baehr, M. Ebert, J. H. Wendorff, *Liq. Cryst.* **1991**, *10*, 161.
[332] M. Möller, V. Tsukruk, J. H. Wendorf, H. Bengs, H. Ringsdorf, *Liq. Cryst.* **1992**, *12*, 17.
[333] H. Zimmermann, R. Poupko, Z. Luz, J. Billard, *Z. Naturforsch., Teil A* **1985**, *40*, 149.
[334] J. Malthete, A. Collet, *Nouv. J. Chim.* **1985**, *9*, 151.
[335] N. Spielberg, M. Sarkar, Z. Luz, R. Poupko, J. Billard, H. Zimmermann, *Liq. Cryst.* **1993**, *15*, 311.
[336] S. C. Kuebler, C. Boeffel, H. W. Spiess, *Liq. Cryst.* **1995**, *18*, 309.
[337] T. Komori, S. Shinkai, *Chem. Lett.* **1993**, 1455.
[338] J. M. Lehn, J. Malthete, A.-M. Levelut, *J. Chem. Soc., Chem. Commun.* **1985**, 1794.
[339] J. M. Lehn, *Angew. Chem.* **1988**, 89.
[340] S. H. J. Idziak, N. C. Maliszewskyj, G. B. M. Vaughan, P. A. Heiney, C. Mertesdorf, H. Ringsdorf, J. P. McCauley, A. B. Smith III, *J. Chem. Soc., Chem. Commun.* **1992**, 98.
[341] G. Johansson, V. Percec, G. Ungar, D. Abramic, *J. Chem. Soc., Perkin Trans. 1* **1994**, 447.
[342] C. Eaborn, N. H. Hartshorne, *J. Chem. Soc., London* **1955**, 549.
[343] J. D. Bunning, J. W. Goodby, G. W. Gray, J. E. Lydon in *Liquid Crystals of One- and Two-Dimensional Order* (Eds.: W. Helfrich, G. Heppke), Springer-Verlag, Berlin, **1980**, p. 397.
[344] J. D. Bunning, J. E. Lydon, C. Eaborn, P. M. Jackson, J. W. Goodby, G. W. Gray, *J. Chem. Soc., Faraday Trans. 1* **1982**, 713.
[345] K. Praefcke, P. Marquardt, B. Kohne, W. Stephan, A.-M. Levelut, E. Wachtel, *Mol. Cryst. Liq. Cryst.* **1991**, *203*, 149.

[346] D. Blunk, K. Praefcke, G. Legler, *Liq. Cryst.* **1995**, *18*, 149.
[347] K. Praefcke, D. Blunk, A. Eckert, G. Legler, M. Langner, G. Pohl, presented at *24 Freiburger Arbeitstagung Flüssigkristalle*, April **1995**, poster P 46.
[348] R. Miethchen, M. Schwarze, J. Holz, *Liq. Cryst.* **1993**, *15*, 185.
[349] V. Vill, B. Sauerbrei, H. Fischer, J. Thiem, *Liq. Cryst.* **1992**, *11*, 949.
[350] L. E. Knaak, H. M. Rosenberg, *Mol. Cryst. Liq. Cryst.* **1972**, *17*, 171.
[351] T. Thiemann, V. Vill, *Liq. Cryst.* **1997**, *22*, 519.
[352] W. Wedler, D. Demus, H. Zaschke, K. Mohr, W. Schäfer, W. Weissflog, *J. Mater. Chem.* **1991**, *1*, 347.
[353] D. Demus in *Modern Topics in Liquid Crystals* (Ed.: A. Buka), World Scientific, Singapore, **1993**, p. 99.
[354] D. Demus, A. Hauser, M. Keil, W. Wedler, *Mol. Cryst. Liq. Cryst.* **1990**, *191*, 153.
[355] A. Adamczyk, *Mol. Cryst. Liq. Cryst.* **1994**, *249*, 75.
[356] R. Zentel in *Liquid Crystals, Topics in Physical Chemistry*, Vol. 3 (Ed.: H. Stegemeyer), Steinkopff, Darmstadt, **1994**, Chap. 3, p. 103.
[357] A. Ciferri (Ed.), *Liquid Crystallinity in Polymers: Principles and Fundamental Properties*, VCH, Weinheim, **1991**.
[358] E. A. Corsellis, H. J. Coles, *Mol. Cryst. Liq. Cryst.* **1995**, *261*, 71.
[359] D. W. Bruce, B. Donnio, D. Guillon, B. Heinrich, M. Ibn-Elhaj, *Liq. Cryst.* **1995**, *19*, 537.
[360] H. Shi, S. H. Chen, *Liq. Cryst.* **1995**, *18*, 733.
[361] H. Shi, S. H. Chen, *Liq. Cryst.* **1995**, *19*, 785.
[362] H. Shi, S. H. Chen, *Liq. Cryst.* **1995**, *19*, 849.
[363] S. H. Chen, J. C. Mastrangelo, T. N. Blanton, A. Bashir-Hashemi, K. L. Marshall, *Liq. Cryst.* **1996**, *21*, 683.
[364] H. Shi, S. H. Chen, M. E. De Rosa, T. J. Bunning, W. W. Adams, *Liq. Cryst.* **1996**, *20*, 277.
[365] W. Otowski, Y. Gonzalez, B. Placios, M. R. De La Fuente, M. A. Perez Jubindo, *Ferroelectrics* **1996**, *180*, 93.
[366] P. Magagnini in *Thermotropic Liquid Crystal Polymer Blends*, F. P. La Mantia (Ed.), Technomic Publ. Co. Inc. Lancaster, USA, **1994**, Pages 1–42.

Chapter VII
Physical Properties of Liquid Crystals

1 Tensor Properties of Anisotropic Materials

David Dunmur and Kazuhisa Toriyama

Liquid crystals are anisotropic, so like non-cubic crystals some of their properties depend on the direction along which they are measured. Such properties are known as tensor properties, and in order to provide a formal basis for the description of orientation-dependent physical properties of liquid crystals, we will give a brief introduction to tensors. An authoritative account of the tensor properties of crystals has been written by Nye [1], but liquid crystals are not explicitly dealt with. A convenient way of categorizing tensor properties is through their behavior on changing the orientation of a defining axis system. A scalar or zero rank tensor property is independent of direction, and examples are density, volume, energy or any orientationally averaged property such as the mean polarizability or mean electric permittivity (dielectric constant). The orientation dependence of a vector property such as dipole moment μ can be understood by considering how the components of the dipole moment change as the axis system is rotated. In Fig. 1 μ_β ($\beta = x, y, z$) are the components in the original coordinate frame, while μ'_α ($\alpha = X, Y, Z$) are the components in the new axis system. If the quantities $a_{\alpha\beta}$ are the nine direction cosines between the axes of the two coordinate frames, the transformation law for the vector property becomes:

$$\mu'_\alpha = a_{\alpha x} \mu_x + a_{\alpha y} \mu_y + a_{\alpha z} \mu_z = a_{\alpha\beta} \mu_\beta \quad (1)$$

where the repeated suffix β implies summation over all values of $\beta = x, y, z$. A consequence of the transformation law Eq. (1), is that under inversion all the components change sign, $\mu'_\alpha = -\mu_\alpha$: this is a polar vector. There are some quantities (axial or pseudo-

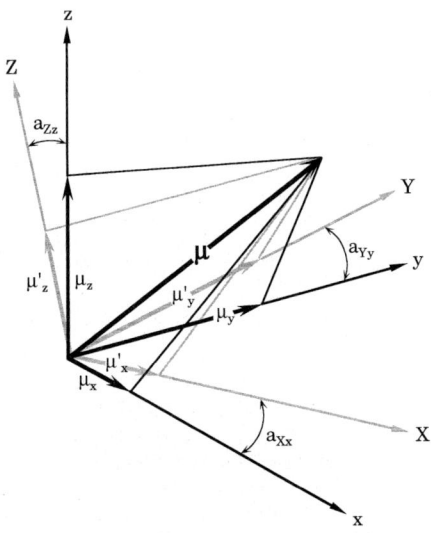

Figure 1. Change in vector components on rotating axes.

vectors) which transform according to:

$$\mu'_\alpha = -a_{\alpha\beta}\mu_\beta \qquad (2)$$

These are usually associated with magnetic phenomena (angular momentum, magnetic moments and magnetic fields are axial vectors) and related properties. The nine direction cosines can conveniently be represented as a 3×3 matrix, but the components of this matrix ($a_{\alpha\beta}$) are not independent, since for orthogonal axes the sums of squares of components in columns or rows are unity, while the sums of products of components in adjacent rows or columns are zero. It is clear that one set of orthogonal axes can be rigidly related to another by just three angles, and Euler angles (θ, ϕ, ψ) provide a consistent definition of three such angles which are frequently used. The Euler angles are defined as follows. Assuming that the axes (x, y, z) and (X, Y, Z) are initially coincident, rotation around $Z=z$, by an angle ψ gives $X \to x'$, $Y \to y'$, rotation about the new axis y' by θ gives $Z \to z$, $x' \to x''$, and finally rotation about z by an amount ϕ gives $x'' \to x$ and $y' \to y$, see Fig. 2 [2]. The direction cosine matrix represented by $a_{\alpha\beta}$, can now be expressed in terms of Euler angles by:

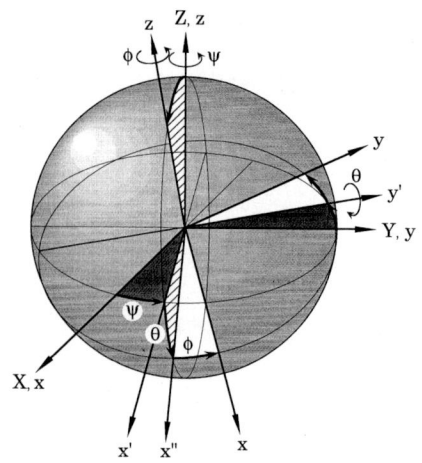

Figure 2. Euler angles.

sor quantity depends on another vector or tensor quantity. A simple example of importance to liquid crystals is the second rank tensor property electric susceptibility χ, which relates electric polarization (vector) to an applied electric field (vector). Neglecting nonlinear effects, the electric polarization P_α is proportional to the magnitude of the applied electric field E, but for anisotropic materials it may not be in the same direction. Consider a linear (one-dimensional) array of charges aligned at an angle

$$a_{\alpha\beta} = \mathbf{A} = \begin{bmatrix} a_{Xx} & a_{Xy} & a_{Xz} \\ a_{Yx} & a_{Yy} & a_{Yz} \\ a_{Zx} & a_{Yz} & a_{Zz} \end{bmatrix}$$

$$= \begin{bmatrix} \cos\theta\cos\phi\cos\psi - \sin\phi\sin\psi & -\cos\theta\sin\phi\cos\psi - \cos\phi\sin\psi & \sin\theta\cos\phi \\ \cos\theta\cos\phi\sin\psi + \sin\phi\cos\psi & -\cos\theta\sin\phi\sin\psi + \cos\phi\cos\psi & \sin\theta\sin\phi \\ -\sin\theta\cos\psi & \sin\theta\sin\psi & \cos\theta \end{bmatrix} \qquad (3)$$

Having established the transformation law Eq. (1), first rank tensor quantities (i.e. vectors) may be defined as those properties which transform according to Eq. (1). Higher order tensors can be defined in terms of different transformation laws, and they arise in a general sense when one vector or tensor quantity depends on another vector or to an applied electric field (see Fig. 3). A separation of charge will be induced by the field resulting in a polarization necessarily along the direction of the array of charges: the electric field has induced a component of the polarization in a direction orthogonal

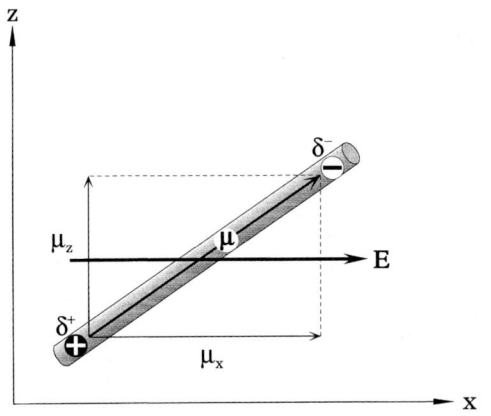

Figure 3. The polarization of a linear array of charges by a field.

to the field. Such a result cannot be described in terms of a simple proportionality between the polarization and the field ($P_\alpha \neq \chi E_\alpha$), because a field along the x-axis has induced a polarization along the y-axis. Obviously the susceptibility also depends on direction, and for the example given, the relationship between polarization and electric field can be written as (for the configuration depicted in Fig. 3, $E_y = 0$):

$$P_x = \chi_{xx} E_x + \chi_{xy} E_y$$
$$P_y = \chi_{yx} E_x + \chi_{yy} E_y \qquad (4)$$

This is readily extended to three dimensions, and using the notation introduced above:

$$P_\alpha = \chi_{\alpha\beta} E_\beta \qquad (5)$$

where $\chi_{\alpha\beta}$ represents nine coefficients or a 3×3 matrix of nine quantities. The electric susceptibility of an anisotropic material can be represented by a matrix of nine components, and we can now consider how such a second rank tensor property changes as a result of rotation with respect to an orthogonal axis system. The transformation rule for second rank tensor properties is:

$$\chi'_{\alpha\beta} = a_{\alpha\gamma} a_{\beta\delta} \chi_{\gamma\delta} \qquad (6)$$

where, as before, the repeated suffixes γ and δ indicate a summation over all possible values (x, y, z). Thus the expression for each component of the transformed property, for example χ_{xy}, contains nine terms, although in practice symmetry usually reduces this number. It is possible to use an alternative transformation rule for second rank tensor properties, equivalent to Eq. (6), but which can be expressed in terms of the matrix of direction cosines, such that:

$$\chi' = \mathbf{A} \chi \mathbf{A}^t \qquad (7)$$

where $\mathbf{A}^t$ is the transpose of the direction cosine matrix.

Although the simplest direction-dependent property is a vector, most physical properties of liquid crystals are higher order tensors. All tensor properties can be categorized by their transformation properties under rotation of the coordinate frame, and the transformation law for a third rank tensor such as the piezoelectric tensor $P_{\alpha\beta\gamma}$ is:

$$P'_{\alpha\beta\gamma} = a_{\alpha\delta} a_{\beta\varepsilon} a_{\gamma\phi} P_{\delta\varepsilon\phi} \qquad (8)$$

The tensor rank of a property is also established by the number of components it has: first rank (vector) 3, second rank 9, third rank 27 and so on. This definition is correct for three dimensions, but in two dimensions the number of components for a tensor property of rank n becomes 2^n. It must be emphasized that not all the components are necessarily independent, and symmetry can provide relationships between them, thereby reducing the number that have to be separately measured.

Many physical properties can be described in terms of the response of a system to an external force or perturbation. This force might be an electric field or a magnetic field, a mechanical force (stress), a torque or a combination of these. The effect of an external perturbation may be described in terms of a new tensor property, or a modification to an existing one; polarization by an electric field results in an induced polarization which adds to any permanent polarization already present. To lowest order, the induced polarization is linear in the applied field, but nonlinear terms can be important for strong fields, as in nonlinear optics. The nonlinear contributions represented in Eq. (9) below are described in terms of third $\chi_{\alpha\beta\gamma}$ and fourth rank tensors $\chi_{\alpha\beta\gamma\delta}$:

$$p_\alpha = p_\alpha^{(0)} + \sum_{\beta=x,y,z} \chi_{\alpha\beta} E_\beta$$
$$+ \sum_{\beta,\gamma=x,y,z} \chi_{\alpha\beta\gamma} E_\beta E_\gamma$$
$$+ \sum_{\beta,\gamma,\delta=x,y,z} \chi_{\alpha\beta\gamma\delta} E_\beta E_\gamma E_\delta \quad (9)$$

This equation includes a number of pairs of repeated suffixes, and each pair indicates a summation over all possible values. Having understood this convention (the Einstein summation convention) it is no longer necessary to write the summation signs explicitly in an equation. It is sometimes useful to represent the physical properties in terms of their contribution to the internal energy, since all types of energies are scalar quantities. Thus the energy of a polarized body in an electric field becomes:

$$u = -\int \mathbf{p} \cdot d\mathbf{E} = -\int p_\alpha \, dE_\alpha$$
$$= p_\alpha^{(0)} E_\alpha + \frac{1}{2} \chi_{\alpha\beta} E_\beta E_\alpha$$
$$+ \frac{1}{3} \chi_{\alpha\beta\gamma} E_\beta E_\gamma E_\alpha$$
$$+ \frac{1}{4} \chi_{\alpha\beta\gamma\delta} E_\beta E_\gamma E_\delta E_\alpha \quad (10)$$

This manipulation is a simple application of tensor calculus, and illustrates the simplification of the summation convention (summation signs have been omitted). The intrinsic symmetry of a property can reduce the number of independent components; thus neglecting complications that may arise if electric fields of more than one frequency are present, the order of the field components in Eq. (10) is immaterial. Hence the properties $\chi_{\alpha\beta}$, $\chi_{\alpha\beta\gamma}$ and $\chi_{\alpha\beta\gamma\delta}$ must be symmetric with respect to interchange of suffixes, and this immediately reduces the number of independent components of the tensor properties to six for $\chi_{\alpha\beta}$, ten for $\chi_{\alpha\beta\gamma}$ and 15 for $\chi_{\alpha\beta\gamma\delta}$.

Another example of a second rank tensor property is electrical conductivity $\sigma_{\alpha\beta}$ which relates the current flow j_α in a particular direction to the electric field:

$$j_\alpha = \sigma_{\alpha\beta} E_\beta \quad (11)$$

This may also be written in terms of resistivity $\rho_{\alpha\beta}$ as:

$$E_\alpha = \rho_{\alpha\beta} j_\beta \quad (12)$$

and the Joule heating which a current generates in a sample is a scalar given by:

$$u = j \cdot E = j_\alpha E_\alpha = \rho_{\alpha\beta} j_\alpha j_\beta \quad (13)$$

Each tensor property has an intrinsic symmetry, which relates to the interchangeability of suffixes. However, the number of independent tensor components for a property also depends on the symmetry of the system it is describing. Thus the properties of an isotropic liquid, which has full rotational symmetry, can be defined in terms of a single independent coefficient. The number of independent components for a particular tensor property depends on the point group symmetry of the phase to which it refers. This is expressed by Neumann's principle which states that the symmetry elements of any physical property of a crystal must include the symmetry elements of the point group of the crystal.

Many properties of interest for liquid crystals are second rank tensors, and these have some special properties, since they can be represented as a 3×3 matrix. Those second rank tensor properties which have an intrinsic symmetry with respect to interchange of suffixes are called symmetric and only have six independent components. A symmetric 3×3 matrix can always be diagonalized to give three principal components, which is equivalent to finding an axis system for which the off-diagonal components are zero. The principal axis system requires three angles to define it with respect to an arbitrary axis frame, so there is no loss of variables: three principal components and three angles being equivalent to the six independent components. However, if the material being described (i.e. the liquid crystal) has some symmetry, then the principal axes will be defined, and so it is possible to reduce the number of independent components of a second rank tensor to three (the principal values). For uniaxial liquid crystal phases of symmetry $D_{\infty h}$ such as N and SmA, as well as SmB and Col$_{ho}$ and Col$_{hd}$ phases, a unique symmetry axis can be defined, parallel to the director, and there are just two independent components of any second rank tensor property $\chi_{\alpha\beta}$: $\chi_\parallel$ (parallel) and $\chi_\perp$ (perpendicular) to the symmetry axis:

$$\chi_{\alpha\beta} = \begin{matrix} x \\ y \\ z \end{matrix} \begin{bmatrix} \chi_\perp & 0 & 0 \\ 0 & \chi_\perp & 0 \\ 0 & 0 & \chi_\parallel \end{bmatrix} \quad (14)$$

where the z-axis is defined as the symmetry axis. The values of the components in any other axis system can be obtained from the transformation law Eqs. (6) or (7). Two useful quantities defined in terms of these independent components are the anisotropy $\Delta\chi$ and the mean $\bar{\chi}$:

$$\Delta\chi = \chi_\parallel - \chi_\perp$$

$$\bar{\chi} = \frac{1}{3}(\chi_\parallel + 2\chi_\perp) \quad (15)$$

Another special feature of second rank tensor properties in three dimensions is that they can be represented by a property ellipsoid, such that the value of the property in a particular direction is represented by the length of the corresponding radius vector of the property ellipsoid. A three dimensional surface representing an ellipsoid can be defined by

$$C_{xx} x^2 + C_{yy} y^2 + C_{zz} z^2$$
$$+ 2 C_{xy} xy + 2 C_{xz} xz + 2 C_{yz} yz = 1 \quad (16)$$

where the coefficients $C_{\alpha\beta}$ behave as the components of a second rank tensor. If the ellipsoid is expressed in terms of principal axes, then the off diagonal terms in Eq. (16) are zero, and the equation of the ellipsoid becomes:

$$C_{xx} x^2 + C_{yy} y^2 + C_{zz} z^2 = 1 \quad (17)$$

where the lengths of the semi-axes are given by $[C_{ii}]^{-1/2}$. The value of this property in any direction (l) defined by direction cosines a_{lx}, a_{ly}, a_{lz} will be:

$$C = C_{xx} a_{lx}^2 + C_{yy} a_{ly}^2 + C_{zz} a_{lz}^2$$
$$= (x^2 + y^2 + z^2)^{-1} \quad (18)$$

so the length of any radius of the property ellipsoid (Fig. 4) is equal to the reciprocal of the square root of the magnitude of the property in that direction.

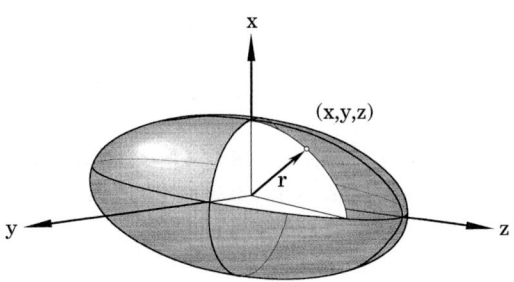

Figure 4. Property ellipsoid.

1.1 Macroscopic and Microscopic Properties

Liquid crystals as anisotropic fluids exhibit a wide range of complex physical phenomena that can only be understood if the appropriate macroscopic tensor properties are fully characterized. This involves a determination of the number of independent components of the property tensor, and their measurement. Thus a knowledge of refractive indices, electric permittivity, electrical conductivity, magnetic susceptibilities, elastic and viscosity tensors are necessary to describe the switching of liquid crystal films by electric and magnetic fields. Development of new and improved materials relies on the design of liquid crystals having particular macroscopic tensor properties, and the optimum performance of liquid crystal devices is often only possible for materials with carefully specified optical and electrical properties.

The anisotropy of liquid crystals stems from the orientational order of the constituent molecules, but the macroscopic anisotropy can only be determined through measurement of tensor properties, and macroscopic tensor order parameters can be defined in terms of various physical properties. The anisotropic part of a second rank tensor property can be obtained by subtracting the mean value of its principal components:

$$\chi_{\alpha\beta}^{(a)} = \chi_{\alpha\beta} - \bar{\chi}\delta_{\alpha\beta} \qquad (19)$$

where the Kronecker delta is equivalent to a unit second rank tensor or unit matrix, such that for $\alpha=\beta$, $\delta_{xx}=\delta_{yy}=\delta_{zz}=1$, otherwise all other components are zero. Note that $\delta_{\alpha\alpha}=3$, because of the repeated suffix convention. The advantage of using $\chi_{\alpha\beta}^{(a)}$ is that it is traceless, and the sum of the diagonal elements is zero. For an isotropic material $\chi_{\alpha\beta}=\chi\delta_{\alpha\beta}$, and so $\chi_{\alpha\beta}^{(a)}=0$; thus the quantity $\chi_{\alpha\beta}^{(a)}$ can be used as an order parameter. Alternatively a dimensionless order parameter tensor can be defined as

$$Q_{\alpha\beta} = \text{constant}\left(\chi_{\alpha\beta} - \bar{\chi}\delta_{\alpha\beta}\right) \qquad (20)$$

and without loss of generality the maximum value of a principal component (say Q_{zz}) of $Q_{\alpha\beta}$ can be set equal to one, so that:

$$Q_{zz} = 1$$
$$= \text{constant} \times \frac{2}{3}\left[\chi_{zz} - \frac{1}{2}\left(\chi_{xx} + \chi_{yy}\right)\right]$$
$$= \text{constant} \times \frac{2}{3}[\Delta\chi]_{\max} \qquad (21)$$

This provides the definition of the constant, and the macroscopic second rank tensor parameter becomes:

$$Q_{\alpha\beta} = \frac{3}{2}[\Delta\chi]_{\max}^{-1}\left(\chi_{\alpha\beta} - \frac{1}{3}\chi_{\gamma\gamma}\delta_{\alpha\beta}\right) \qquad (22)$$

The quantity $[\Delta\chi]_{\max}$ refers to the anisotropy of the tensor for a fully aligned state for which the order parameter is one. For a biaxial phase (i.e. a phase which has different properties along each of the three principal axes), the macroscopic order parameter in principal axes can be written as:

$$Q_{\alpha\beta} = \begin{bmatrix} -\frac{1}{2}(Q-P) & 0 & 0 \\ 0 & -\frac{1}{2}(Q+P) & 0 \\ 0 & 0 & Q \end{bmatrix} \qquad (23)$$

defining $Q = [\Delta\chi_{\max}]^{-1}\left(\chi_{zz} - \frac{1}{2}(\chi_{xx} + \chi_{yy})\right)$ and $P = \frac{3}{2}[\Delta\chi_{\max}]^{-1}\left(\chi_{xx} - \chi_{yy}\right)$. Other definitions of $Q_{\alpha\beta}$ are possible, but the chosen one maintains the correspondence between the macroscopic order parameter and the microscopic one to be introduced next. One problem with the macroscopic order parameter as defined through Eq. (23) is that it has been assumed that the principal axes of the property $\chi_{\alpha\beta}$ and the order parameter tensor coincide. While this is necessarily true for uniaxial materials, it is not true for biaxial materials, the principal axes for which will be different for different properties.

The relationship between macroscopic properties and molecular properties is a major area of interest, since it is through manipulation of the molecular structure of mesogens, that the macroscopic liquid crystal properties can be adjusted towards paricular values which optimize performance in applications. The theoretical connection between the tensor properties of molecules and the macroscopic tensor properties of liquid crystal phases provides a considerable challenge to statistical mechanics. A key factor is of course the molecular orientational order, but interactions between molecules are also important especially for elastic and viscoelastic properties. It is possible to divide properties into two categories, those for which molecular contributions are approximately additive (i.e. they are proportional to the number density), and those properties such as elasticity, viscosity, thermal conductivity etc. for which intermolecular forces are responsible, and so have a much more complex dependence on number density. For the former it is possible to develop a fairly simple theory using single particle orientational order parameters.

In the context of liquid crystals, single particle angular distribution functions are of major interest. They give the probability of a single molecule having a particular orientation with respect to some defined axis system, but they contain no information on pair correlations between molecules. A familiar example of single particle angular distribution functions is the hydrogenic s, p, d-orbitals, which are used via the square of the wave function to determine the probability distribution for a single electronic charge in an atom. For a many-electron atom the wave function for the electrons is often written as the sum of component orbital contributions. In an analogous fashion, the distribution of molecular orientations in a liquid crystal can be represented as a sum of contributions of particular symmetries. Restricting attention to axially symmetric molecules having an axially symmetric distribution of orientations with respect to the director (z-axis), it is convenient to use Legendre functions $P_L(\cos\theta)$ to describe the angular distribution function $f(\theta)$:

$$f(\theta) = \frac{1}{2}\begin{bmatrix} 1 + a_1 P_1(\cos\theta) + a_2 P_2(\cos\theta) \\ + a_3 P_3(\cos\theta) + a_4 P_4(\cos\theta) + \ldots \end{bmatrix} \quad (24)$$

The expansion in Legendre polynomials or more generally spherical harmonics is chosen because they are orthogonal functions. The coefficients a_L in the expansion can be obtained by multiplying both sides of Eq. (24) by $P_L(\cos\theta)$ and integrating over θ, with the result:

$$a_L = (2L+1)\frac{\int_0^\pi P_L(\cos\theta) f(\theta) \sin\theta \, d\theta}{\int_0^\pi f(\theta) \sin\theta \, d\theta} \quad (25)$$

The ratio of integrals in Eq. (25) is the definition of the average value of $P_L(\cos\theta)$ over the single particle angular distribution function, so the single particle distribution function can now be written as:

$$f(\theta) = \frac{1}{2}\begin{bmatrix} 1 + 3\langle P_1(\cos\theta)\rangle P_1(\cos\theta) \\ + 5\langle P_2(\cos\theta)\rangle P_2(\cos\theta) \\ + 7\langle P_3(\cos\theta)\rangle P_3(\cos\theta) \\ + 9\langle P_4(\cos\theta)\rangle P_4(\cos\theta) \\ + \ldots \end{bmatrix} \quad (26)$$

The amplitudes of the coefficients in this expansion now have a special significance:

they are the order parameters for the distribution function. A knowledge of all order parameters will provide a complete description of the single particle angular distribution function, and the magnitude of each order parameter gives the contribution of a particular symmetry to the distribution function for the disordered structure. If we wish to consider the distribution function for a molecule requiring three angles to specify its orientation, $f(\Omega = \theta, \phi, \psi)$ must be expanded in terms of a set of orthogonal functions which span the orientation space of the Euler angles. A convenient set of such functions are the Wigner rotation matrices, denoted $\mathbf{D}^L_{m,n}(\theta, \phi, \psi)$, so the single particle angular distribution function becomes (see [3] for further details):

$$f(\Omega) = \frac{1}{8\pi^2} \sum_L \sum_{m,n=-L}^{L} (2L+1) a_{L,m,n} D^L_{m,n}(\theta,\phi,\psi) \quad (27)$$

where as before the coefficients $a_{L,m,n}$ can be obtained from the orthogonality condition to give:

$$a_{L,m,n} = \langle D^L_{m,n}(\theta,\phi,\psi) \rangle \quad (28)$$

and the quantities $\langle D^L_{m,n}(\theta, \phi, \psi)\rangle$ are generalized orientational order parameters, the indices L, m, n relating to the angular variables θ, ϕ, ψ.

The symmetry of the constituent molecules and the symmetry of the liquid crystal phase provide some constraints on the terms which contribute to the distribution function. For example if a molecule has inversion symmetry, then only terms even in L will contribute; similarly for molecules with a C_2 rotation axis along the z-direction, only terms even in n will survive. A full set of symmetry operations and non-vanishing order parameters is given by Zannoni [4]. The number of independent order parameters necessary to specify the angular distribution function will also be reduced by the symmetry of the liquid crystal phase, see for example [5]. Returning to the simple case of uniaxial molecules having inversion symmetry in a uniaxial phase, the distribution function can be written as:

$$f(\theta) = \frac{1}{2}\begin{bmatrix} 1 + 5\langle P_2(\cos\theta)\rangle P_2(\cos\theta) \\ + 9\langle P_4(\cos\theta)\rangle P_4(\cos\theta) + \ldots \end{bmatrix} \quad (29)$$

and the two leading order parameters are $\langle P_2(\cos\theta)\rangle$ and $\langle P_4(\cos\theta)\rangle$: the former is often referred to as S. One point to notice is that order parameters are multivalued in the sense that different distributions may give the same value for an order parameter, hence the value of information on more than one order parameter. For example, a distribution in which the molecular axes were at an average angle of about 55° to the z-axis would give $\langle P_2(\cos\theta)\rangle$ close to zero with a negative $\langle P_4(\cos\theta)\rangle$. An isotropic distribution of molecules is indicated by all order parameters being zero. Alternative but equivalent definitions of order parameters as tensor quantities are sometimes more convenient, particularly in relation to physical properties. These definitions are only usefully compact for systems in which either the molecules or the phase are uniaxial. For a uniaxial liquid crystal phase of biaxial molecules, the single particle angular distribution function can be written as [3]:

$$f(\theta,\phi) = \frac{1}{4\pi}\begin{bmatrix} 1 + 3 S_\alpha l_\alpha + 5 S_{\alpha\beta} l_\alpha l_\beta \\ + 7 S_{\alpha\beta\gamma} l_\alpha l_\beta l_\gamma \\ + 9 S_{\alpha\beta\gamma\delta} l_\alpha l_\beta l_\gamma l_\delta + \ldots \end{bmatrix} \quad (30)$$

where $l_\alpha = l_x$, l_y, l_z are direction cosines of the director with respect to the molecular axes, and the quantities S_α, $S_{\alpha\beta}$, $S_{\alpha\beta\gamma}$ and $S_{\alpha\beta\gamma\delta}$ are ordering tensors of ranks 1, 2, 3, and 4, respectively, and the summation convention has been adopted. For phases with inversion

Table 1. Relationship between different definitions of order parameters.

	Uniaxial phase		Biaxial phase	
	Uniaxial molecule	Biaxial molecule	Uniaxial molecule	Biaxial molecule
Saupe ordering matrix $S_{\alpha\beta} = \left\langle \frac{1}{2}(3l_\alpha l_\beta - \delta_{\alpha\beta}) \right\rangle$	$S = \left\langle \frac{1}{2}(3l_z^2 - 1) \right\rangle$	$S_{zz} = \left\langle \frac{1}{2}(3l_z^2 - 1) \right\rangle$ $D = S_{xx} - S_{yy} = \left\langle \frac{3}{2}(l_x^2 - l_y^2) \right\rangle$	$S_{zz}^Z = \left\langle \frac{1}{2}(3l_{Z,z}^2 - 1) \right\rangle$ $P = S_{zz}^X - S_{zz}^Y = \left\langle \frac{3}{2}(l_{X,z}^2 - l_{Y,z}^2) \right\rangle$	$S_{zz}^Z = \left\langle \frac{1}{2}(3l_{Z,z}^2 - 1) \right\rangle$ $D = S_{xx}^Z - S_{yy}^Z = \left\langle \frac{3}{2}(l_{Z,x}^2 - l_{Z,y}^2) \right\rangle$ $P = S_{zz}^X - S_{zz}^Y = \left\langle \frac{3}{2}(l_{X,z}^2 - l_{Y,z}^2) \right\rangle$ $C = (S_{xx}^X - S_{yy}^X) - (S_{xx}^Y - S_{yy}^Y)$ $= \left\langle \frac{3}{2}\left[(l_{X,x}^2 - l_{X,y}^2) - (l_{Y,x}^2 - l_{Y,y}^2)\right] \right\rangle$
Legendre functions $\langle P_L(\cos\theta) \rangle$ and spherical harmonics $\langle Y_{Lm}(\theta,\psi) \rangle$	$S = \langle P_2(\cos\theta) \rangle$ $= \left\langle \frac{1}{2}(3\cos^2\theta - 1) \right\rangle$	$S = \left(\frac{5}{4\pi}\right)^{1/2} \langle Y_{20}(\theta,\psi) \rangle$ $D = \sqrt{6}\left(\frac{5}{4\pi}\right)^{1/2} \langle Y_{22}(\theta,\psi) \rangle$ $= \frac{3}{2}\langle \sin^2\theta \cos 2\psi \rangle$		
Wigner rotation matrices $\langle D_{mn}^L(\theta,\phi,\psi) \rangle$	$S = \langle D_{0,0}^2 \rangle$	$S = \langle D_{0,0}^2 \rangle$ $D = \sqrt{\frac{3}{2}}\langle D_{0,2}^2 + D_{0,-2}^2 \rangle$	$S = \langle D_{0,0}^2 \rangle$ $P = \sqrt{\frac{3}{2}}\langle D_{2,0}^2 + D_{-2,0}^2 \rangle$ $= \frac{3}{2}\langle \sin^2\theta \cos 2\phi \rangle$	$S = \langle D_{0,0}^2 \rangle$ $D = \sqrt{\frac{3}{2}}\langle D_{0,2}^2 + D_{0,-2}^2 \rangle$ $P = \sqrt{\frac{3}{2}}\langle D_{2,0}^2 + D_{-2,0}^2 \rangle$ $C = \frac{3}{2}\langle D_{2,2}^2 + D_{-2,-2}^2 + D_{2,-2}^2 + D_{-2,2}^2 \rangle$ $= \frac{3}{2}\left\langle (1+\cos^2\theta)\cos 2\phi \cos 2\psi - 2\cos\theta \sin 2\phi \sin 2\psi \right\rangle$

symmetry all ordering tensors of odd rank are zero, and the first nonvanishing tensor order parameter $S_{\alpha\beta}$ is sometimes known as the Saupe ordering matrix. Definitions of the ordering tensors as averages over direction cosines are:

$$S_\alpha = \langle l_\alpha \rangle \tag{31}$$

$$S_{\alpha\beta} = \left\langle \frac{1}{2}\left(3 l_\alpha l_\beta - \delta_{\alpha\beta}\right)\right\rangle$$

$$S_{\alpha\beta\gamma} = \left\langle \frac{1}{2}\begin{pmatrix} 5 l_\alpha l_\beta l_\gamma \\ -\left(l_\alpha \delta_{\beta\gamma} + l_\beta \delta_{\gamma\alpha} + l_\gamma \delta_{\alpha\beta}\right)\end{pmatrix}\right\rangle$$

$$S_{\alpha\beta\gamma\delta} = \left\langle \frac{1}{8}\begin{pmatrix} 35 l_\alpha l_\beta l_\gamma l_\delta \\ -5\begin{pmatrix} l_\alpha l_\beta \delta_{\gamma\delta} + l_\alpha l_\gamma \delta_{\beta\delta} \\ + l_\alpha l_\delta \delta_{\beta\gamma} + l_\beta l_\gamma \delta_{\alpha\delta} \\ + l_\beta l_\delta \delta_{\alpha\gamma} + l_\gamma l_\delta \delta_{\alpha\beta}\end{pmatrix}\end{pmatrix} \right. $$
$$\left. + \frac{1}{8}\begin{pmatrix} \delta_{\alpha\beta}\delta_{\gamma\delta} + \delta_{\alpha\gamma}\delta_{\beta\delta} \\ + \delta_{\beta\gamma}\delta_{\alpha\delta}\end{pmatrix}\right\rangle$$

These tensors are defined to be zero for an isotropic phase. For uniaxial molecules there is only one independent component for each of the tensors:

$$S_z = \langle \cos\theta \rangle = \langle P_1(\cos\theta)\rangle$$

$$S_{zz} = \left\langle \frac{1}{2}\left(3\cos^2\theta - 1\right)\right\rangle = \langle P_2(\cos\theta)\rangle$$

$$S_{zzz} = \left\langle \frac{1}{2}\left(5\cos^3\theta - 3\cos\theta\right)\right\rangle = \langle P_3(\cos\theta)\rangle$$

$$S_{zzzz} = \left\langle \frac{1}{8}\left(35\cos^4\theta - 30\cos^2\theta + 3\right)\right\rangle$$
$$= \langle P_4(\cos\theta)\rangle$$

One advantage of this representation of order parameters is that it readily describes nonuniaxial molecular order in a macroscopically uniaxial liquid crystal phase. If the reference axis frame has been chosen to diagonalize the ordering tensor, then its principal components are the order parameters of the different molecular axes with respect to a uniaxial director. If the distributions of the two shorter molecular axes (say x and y) are not identical, then $S_{xx} \neq S_{yy}$, which represents a local biaxial ordering of molecular axes. This may be expressed in terms of a new order parameter $D = S_{xx} - S_{yy}$, and the Saupe ordering matrix can be written as:

$$S_{\alpha\beta} = \begin{bmatrix} -\frac{1}{2}(S-D) & 0 & 0 \\ 0 & -\frac{1}{2}(S+D) & 0 \\ 0 & 0 & S \end{bmatrix} \tag{33}$$

If the Saupe ordering matrix is written in terms of the laboratory axis frame, but assuming now that the molecules are uniaxial, then phase biaxiality can be described in terms of the order parameter P, which is nonzero for tilted smectic phases and other intrinsically biaxial phases. For example the diagonal ordering matrix for the molecular long axis z can be written as:

$$S_{ij}^{(z)} = \left\langle \frac{1}{2}\left(3 l_{i,z} l_{j,z} - \delta_{ij}\right)\right\rangle$$

$$= \begin{bmatrix} -\frac{1}{2}(S-P) & 0 & 0 \\ 0 & -\frac{1}{2}(S+P) & 0 \\ 0 & 0 & S \end{bmatrix} \tag{34}$$

and the magnitude of P is a measure of the different probabilities of finding the z-molecular axis along the X and Y directions of the laboratory or phase reference frame.

The Cartesian tensor representation can be extended to describe the orientational ordering of biaxial molecules in biaxial phases by introducing [6] a fourth rank ordering tensor:

$$S_{\alpha\beta,ij} = \left\langle \frac{1}{2}\left(3 l_{i,\alpha} l_{j,\beta} - \delta_{ij}\delta_{\alpha\beta}\right)\right\rangle \tag{35}$$

where $l_{i,\alpha}$ is the cosine of the angle between the molecular axis α and the laboratory or phase axis i. The tensor $S_{\alpha\beta,ij}$ still describes second rank orientational ordering, and should not be confused with the ordering tensor $S_{\alpha\beta\gamma\delta}$ which refers to fourth rank orientational order. For a suitable choice of both sets of axes, the 81 components of $S_{\alpha\beta,ij}$ can be reduced to nine such that $i=j$ and $\alpha=\beta$. This is equivalent to defining three diagonal Saupe ordering matrices, one for each of the three axes, $i=X, Y, Z$:

$$S^{(i)}_{\alpha\beta} = \left\langle \frac{1}{2}\left(3l_{i,\alpha}l_{i,\beta} - \delta_{\alpha\beta}\right)\right\rangle \quad (36)$$

Taking the diagonal components of these three matrices allows the construction of a 3×3 matrix:

$$S^{(i)}_{\alpha\beta} = \begin{bmatrix} S^X_{xx} & S^X_{yy} & S^X_{zz} \\ S^Y_{xx} & S^Y_{yy} & S^Y_{zz} \\ S^Z_{xx} & S^Z_{yy} & S^Z_{zz} \end{bmatrix} \quad (37)$$

and the generalized biaxial order parameters can be defined as follows. The long axis ordering is described by $S=S^Z_{zz}$, while the phase biaxiality for a uniaxial molecule is given by $P=S^X_{zz}-S^Y_{zz}$. For biaxial molecules in a uniaxial phase the biaxial order parameter is $D=S^Z_{xx}-S^Z_{yy}$, but it would be equally possible to define a biaxial order parameter with respect to the X-axis, such that $D'=S^X_{xx}-S^X_{yy}$, and this would equal $D''=S^Y_{xx}-S^Y_{yy}$ for uniaxial phases. However, if the phase is biaxial, then the biaxiality defined with respect to phase axes X and Y will be different, and this new form of biaxiality is described in terms of a new biaxial order parameter:

$$C = D' - D''$$
$$= \left(S^X_{xx} - S^X_{yy}\right) - \left(S^Y_{xx} - S^Y_{yy}\right) \quad (37a)$$

In terms of averages over Euler angles, these may be defined as:

$$S = \left\langle \frac{1}{2}\left(3\cos^2\theta - 1\right)\right\rangle$$

$$D = \left\langle \frac{3}{2}\left(\sin^2\theta \cos 2\psi\right)\right\rangle$$

$$P = \left\langle \frac{3}{2}\left(\sin^2\theta \cos 2\phi\right)\right\rangle$$

$$C = \left\langle \frac{3}{2}\left[\left(1+\cos^2\theta\right)\cos 2\phi \cos 2\psi \right.\right.$$
$$\left.\left. - 2\cos\theta \sin 2\phi \sin 2\psi\right]\right\rangle \quad (38)$$

Slightly different definitions have been adopted by some other authors with different numerical factors. The advantage of the definitions in Eq. (38) is that these order parameters are simply related to the components of the Saupe ordering matrices Eq. (34), as indicated in Table 1.

The order parameters introduced in the preceding paragraphs are sufficient to describe the orientational order/disorder of rigid molecules in liquid crystal phases. They will be used to relate molecular properties to macroscopic physical properties, but there are additional sources of order/disorder which may affect physical properties. For flexible molecules certain physical properties or responses may be sensitive to a particular group or bond within the molecule, and under these circumstances it is the order parameter of that moiety which determines the measured anisotropy. As an example, the degree of order of flexible alkyl chains attached to a rigid molecular core is reduced by internal rotation of the chain segments. Using selectively deuterated mesogens, deuterium magnetic resonance is able to measure the order parameters of different segments of a flexible chain, and it has been shown that as expected the orientational order decreases along the chain away from the rigid core of the molecule [7].

Local biaxial ordering of molecules in uniaxial liquid crystal phases can be detected by spectroscopic techniques such as linear dichroism and NMR. It is often more convenient to probe the biaxial order of solutes in liquid crystal hosts, for example a detailed analysis of the ^{2}D NMR of fully deuteriated anthracene d_{10} in various liquid crystal solvents yields both S and D order parameters [8]. These are illustrated in Fig. 5, where they are compared to mean field calculations with the ratio (λ) of uniaxial and biaxial energies as an adjustable parameter. It is reasonable to assume that the orientational order parameters introduced will also be appropriate for smectic and columnar phases. However, there are additional contributions to the order/disorder, which can contribute to the measured anisotropy in physical properties. The characteristic structural feature of smectic and columnar phases is the presence of some translational order, and so the radial distribution function will have long range periodicity in certain directions, the amplitude of which will be determined by a suitable order parameter. Furthermore, there is the likelihood of coupled orientational and translational order: for example in a smectic A phase molecules will be more likely to be aligned parallel to the layer normal (the director) when their centres of mass coincide with the average layer position. For disordered smectics there is one dimensional positional order, and assuming uniaxial molecules in a uniaxial smectic phase, the corresponding single particle distribution function can be written as [9]:

$$f(z,\theta) = \frac{1}{2d} \left[\begin{array}{l} 1 + 2\left\langle \cos\left(\frac{2\pi z}{d}\right) \right\rangle \cos\left(\frac{2\pi z}{d}\right) \\ + 5\left\langle P_2(\cos\theta) \right\rangle P_2(\cos\theta) \\ + 10\left\langle P_2(\cos\theta)\cos\left(\frac{2\pi z}{d}\right) \right\rangle \\ \cdot P_2(\cos\theta)\cos\left(\frac{2\pi z}{d}\right) + \ldots \end{array} \right]$$

(39)

where translational (τ) and translational–rotational (σ) order parameters can be defined as:

$$\tau = \left\langle \cos\left(\frac{2\pi z}{d}\right) \right\rangle$$

$$\sigma = \left\langle P_2(\cos\theta)\cos\left(\frac{2\pi z}{d}\right) \right\rangle \qquad (40)$$

Columnar phases have two degrees of translational order, and the corresponding order parameters now require averages over periodic functions in two spatial dimensions; for completeness we include the distribution function for a uniaxial columnar phase consisting of uniaxial disc-like molecules:

$$f(x,y,\theta) = \frac{1}{2d_x d_y} \qquad (41)$$

$$\left[\begin{array}{l} 1 + 2\left\langle \cos\left(\frac{2\pi x}{d_x}\right) \cos\left(\frac{2\pi y}{d_y}\right) \right\rangle \\ \cdot \cos\left(\frac{2\pi x}{d_x}\right) \cos\left(\frac{2\pi y}{d_y}\right) \\ + 5\left\langle P_2(\cos\theta) \right\rangle P_2(\cos\theta) \\ + 20\left\langle P_2(\cos\theta)\cos\left(\frac{2\pi x}{d_x}\right) \cos\left(\frac{2\pi y}{d_y}\right) \right\rangle \\ \cdot P_2(\cos\theta)\cos\left(\frac{2\pi x}{d_x}\right) \cos\left(\frac{2\pi y}{d_y}\right) + \ldots \end{array} \right]$$

The contribution of translational order parameters to the anisotropy of physical properties of liquid crystals has not been studied in detail. Evidence suggests that there is a very small influence of translational ordering on the optical properties, but effects of translational order can be detected in the measurement of dielectric properties. There are strong effects in both elastic properties and viscosity, but the statistical theories of these properties have not been extended to include explicitly the effects of translational order.

1.1 Macroscopic and Microscopic Properties

the molecules:

$$\chi_{\alpha\beta} = N\langle\kappa_{\alpha\beta}\rangle = N\langle a_{\alpha\gamma} a_{\beta\delta}\rangle \kappa_{\gamma\delta}^{(m)} \quad (43)$$

The average over the products of direction cosine matrices contains the orientational order parameters, and in terms of the principal components of $\kappa_{\alpha\beta}^{(m)}$ the anisotropic part of the macroscopic tensor property becomes:

$$\chi_{\alpha\beta}^{(a)} = \frac{N}{3}\left\{\begin{array}{l}\langle 3a_{\alpha z} a_{\beta z} - \delta_{\alpha\beta}\rangle \\ \cdot\left[\kappa_{zz}^{(m)} - \frac{1}{2}\left(\kappa_{xx}^{(m)} + \kappa_{yy}^{(m)}\right)\right] \\ + \frac{1}{2}\langle 3a_{\alpha x} a_{\beta x} - 3a_{\alpha y} a_{\beta y}\rangle \\ \cdot\left(\kappa_{xx}^{(m)} - \kappa_{yy}^{(m)}\right)\end{array}\right\} \quad (44)$$

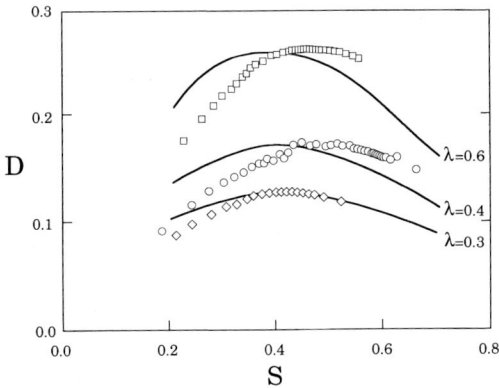

Figure 5. Uniaxial and biaxial order parameters for anthracene d_{10} in different liquid crystal solvents: (□) – ZLI1167 (best fit $\lambda=0.6$); (○) – E9 ($\lambda=0.4$); (◇) – Phase 5 ($\lambda=0.3$), reproduced from [8].

Having defined orientational order parameters, it is now possible to develop a general statistical theory which relates the physical properties of molecules to the macroscopic properties of liquid crystal phases. There are however many simplifying approximations which have to be introduced to give usable results. An important factor is that the nature of the property will determine the order parameters that will be included: in particular a property of tensor rank n will in general require order parameters up to tensor rank n to specify it.

If $\kappa_{\alpha\beta}$ is a molecular second rank tensor property, the principal components of which are $\kappa_{xx}^{(m)}$, $\kappa_{yy}^{(m)}$ and $\kappa_{zz}^{(m)}$ defined in a molecular axis system, then using the transformation rule for second rank tensors, the property in a laboratory frame is:

$$\kappa_{\alpha\beta} = a_{\alpha\gamma} a_{\beta\delta} \kappa_{\gamma\delta}^{(m)} \quad (42)$$

Ignoring the effect of molecular interactions, which is a gross assumption, the macroscopic response $\chi_{\alpha\beta}$ measured in a laboratory axis frame will be the molecular property multiplied by the number density, averaged over all possible orientations of

Those terms in Eq. (44) which have nonzero averages depend on the symmetry of $\chi_{\alpha\beta}^{(a)}$, which by Neumann's principle must contain the symmetry of the phase to which it relates. If the phase is uniaxial, the principal components of $\chi_{\alpha\beta}^{(m)}$ become:

$$\chi_{\parallel}^{(a)} = \frac{2N}{3}\left\{\begin{array}{l}S\left[\kappa_{zz}^{(m)} - \frac{1}{2}\left(\kappa_{xx}^{(m)} + \kappa_{yy}^{(m)}\right)\right] \\ + \frac{D}{2}\left(\kappa_{xx}^{(m)} - \kappa_{yy}^{(m)}\right)\end{array}\right\}$$

$$\chi_{\perp}^{(a)} = -\frac{N}{3}\left\{\begin{array}{l}S\left[\kappa_{zz}^{(m)} - \frac{1}{2}\left(\kappa_{xx}^{(m)} + \kappa_{yy}^{(m)}\right)\right] \\ + \frac{D}{2}\left(\kappa_{xx}^{(m)} - \kappa_{yy}^{(m)}\right)\end{array}\right\} \quad (45)$$

where the order parameters S and D are defined by Eq. (38). Comparison of Eq. (45) with the definition of the macroscopic order parameter Eq. (22) shows that:

$$Q_{zz} = \frac{N}{[\Delta\chi]_{max}}$$

$$\cdot\left\{\begin{array}{l}S\left[\kappa_{zz}^{(m)} - \frac{1}{2}\left(\kappa_{xx}^{(m)} + \kappa_{yy}^{(m)}\right)\right] \\ + \frac{D}{2}\left(\kappa_{xx}^{(m)} - \kappa_{yy}^{(m)}\right)\end{array}\right\} \quad (46)$$

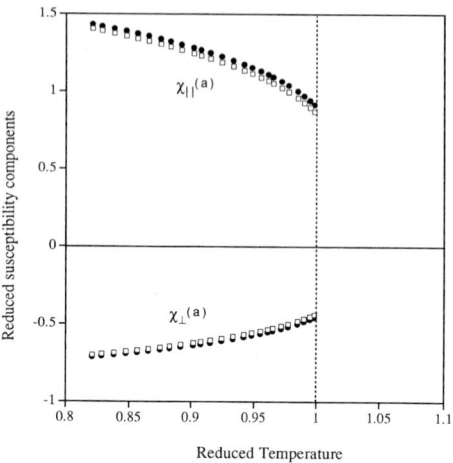

Figure 6. Reduced values $\chi_\parallel^{(a)}$ and $\chi_\perp^{(a)}$ as a function of reduced temperature plotted according to Eq. (45) using order parameters calculated using mean field theory. Open squares (□) assume no molecular biaxiality, so $D = 0$; full circles (●) are for an assumed molecular biaxiality of 0.3 and a λ value of 0.3.

For perfect alignment $Q_{zz} = S = 1$, and the macroscopic anisotropy $(\Delta \chi)_{max} = N[k_{zz}^{(m)} - 1/2\, k_{xx}^{(m)} + k_{yy}^{(m)}]$, which is simply N times the molecular anisotropy. Both order parameters S and D contribute to the anisotropy of second rank tensor properties, but they cannot be separated from a single measurement of the macroscopic anisotropy. The behavior of the macroscopic property components as a function of temperature is illustrated in Fig. 6 for a material of positive molecular anisotropy, for various values of the order parameters S and D calculated from mean field theory; for a negative anisotropy material the signs of the parallel and antiparallel components are interchanged.

If the liquid crystal phase is biaxial, then any second rank tensor property has three independent principal components. These are the diagonal elements of the anisotropic tensor $\chi_{\alpha\beta}^{(a)}$, and can be expressed in terms of the order parameters introduced for bi-axial phases:

$$\chi_\parallel^{(a)} = \frac{2N}{3} \left\{ \begin{array}{l} S\left[\kappa_{zz}^{(m)} - \frac{1}{2}\left(\kappa_{xx}^{(m)} + \kappa_{yy}^{(m)}\right)\right] \\ + \frac{D}{2}\left(\kappa_{xx}^{(m)} - \kappa_{yy}^{(m)}\right) \end{array} \right\}$$

$$\chi_{\perp'}^{(a)} = -\frac{N}{3} \left\{ \begin{array}{l} (S+P)\left[\kappa_{zz}^{(m)} - \frac{1}{2}\left(\kappa_{xx}^{(m)} + \kappa_{yy}^{(m)}\right)\right] \\ + \frac{D+C}{2}\left(\kappa_{xx}^{(m)} - \kappa_{yy}^{(m)}\right) \end{array} \right\}$$

$$\chi_\perp^{(a)} = -\frac{N}{3} \left\{ \begin{array}{l} (S-P)\left[\kappa_{zz}^{(m)} - \frac{1}{2}\left(\kappa_{xx}^{(m)} + \kappa_{yy}^{(m)}\right)\right] \\ + \frac{D-C}{2}\left(\kappa_{xx}^{(m)} - \kappa_{yy}^{(m)}\right) \end{array} \right\}$$

(47)

It has been assumed that molecular properties contribute additively to the macroscopic tensor components, which are consequently proportional to the number density. If intermolecular interactions contribute to the physical property, then deviations from a linear dependence of the property on density are expected. Also the contribution of orientational order will be more complex, since the properties will depend on the degree of order of interacting molecules. Effects of molecular interactions contribute to the dielectric properties of polar mesogens, and are particularly important for elastic and viscoelastic properties. Molecular mean field theories of elastic properties predict that elastic constants should be proportional to the square of the order parameter; this result highlights the significance of pairwise interactions.

1.2 References

[1] J. F. Nye, *Physical Properties of Crystals*, Oxford University Press, Oxford, **1985**.
[2] S. Altmann, *Rotations, Quaternions and Double Groups*, Oxford University Press, Oxford, **1986**, p. 65.

[3] C. Zannoni in *Nuclear Magnetic Resonance of Liquid Crystals* (Ed.: J. W. Emsley), D. Reidel, Dordrecht, Netherlands, **1983**, p. 1.

[4] C. Zannoni in *Molecular Physics of Liquid Crystals* (Eds.: G. W. Gray, G. R. Luckhurst), Academic Press, London, **1979**, p. 51.

[5] Z. Luz, D. Goldfarb, H. Zimmermann in *Nuclear Magnetic Resonance of Liquid Crystals* (Ed.: J. W. Emsley), D. Reidel, Dordrecht, Netherlands, **1983**, p. 351.

[6] D. W. Allender, M. A. Lee, N. Hafiz, *Mol. Cryst. Liq. Cryst.* **1985**, *124*, 45–52.

[7] J. W. Emsley, G. R. Luckhurst, C. P. Stockley, *Proc. R. Soc. Lond.* **1982**, *A381*, 117.

[8] J. W. Emsley in *Nuclear Magnetic Resonance of Liquid Crystals* (Ed.: J. W. Emsley), D. Reidel, Dordrecht, Netherlands, **1983**, p. 379.

[9] C. Zannoni in *The Molecular Dynamics of Liquid Crystals* (Eds.: G. R. Luckhurst, C. A. Veracini), Kluwer, Dordrecht, Netherlands, **1994**, p. 34.

2 Magnetic Properties of Liquid Crystals

David Dunmur and Kazuhisa Toriyama

In Sec. 1 of this chapter it is shown that macroscopic anisotropy in liquid crystals can be related to molecular properties through appropriate microscopic orientational order parameters, as in Eq. (47) of that section. This relationship assumes that the macroscopic response of a liquid crystal is simply the sum of the individual molecular responses averaged over an orientational distribution function (i.e. interactions between molecules are ignored, except to the extent that they determine the orientational order). For most physical properties such an approximation is very crude; however, magnetic properties are only very weakly influenced by intermolecular interactions, and so it can be assumed that the magnetic response of liquid crystals is simply the aggregated molecular response. The weak interaction between molecules and magnetic fields is shown by the magnetic permeability relative to that of free space for nonferromagnetic materials, which is close to unity. The magnetic response of materials depends on their electronic structure, and the susceptibility may be negative, characteristic of diamagnetic compounds, positive denoting a paramagnetic response, or ferromagnetic which indicates a permanent magnetization resulting from coupling between electrons on constituent atoms, ions or molecules; ferromagnetism is largely restricted to the solid state. Both diamagnetic and paramagnetic liquid crystals are known, and ferromagnetic liquid crystals have been prepared from colloidal suspensions of ferromagnetic materials in a liquid crystal host.

2.1 Magnetic Anisotropy

Like other tensor properties of liquid crystals, the magnetic susceptibility is anisotropic, and so magnetic fields can be used to control the alignment of liquid crystal samples. This is perhaps the single most useful application of the magnetic properties of liquid crystals, and the combination of magnetic field alignment with some other measurement forms the basis of many experimental investigations.

Macroscopically the magnetic susceptibility relates the induced magnetization M to the strength of the magnetic field, but assuming that local field effects are ignored, the susceptibility is usually defined in terms of magnetic induction B. For this chapter the magnetic induction will be referred to as the magnatic field, and the magnetization is given by:

$$M_\alpha = \mu_0^{-1} \chi_{\alpha\beta}^{\mathrm{mag}} B_\beta \qquad (1)$$

where μ_0 is the permeability of free space. The magnetic contribution to the free energy density becomes:

$$\begin{aligned} g_{\mathrm{mag}} &= -\int B_\alpha \, dM_\alpha \\ &= -\mu_0^{-1} \int B_\alpha \chi_{\alpha\beta} \, dB_\beta \\ &= g_0 - \frac{1}{2} \mu_0^{-1} \chi_{\alpha\beta} B_\alpha B_\beta \end{aligned} \qquad (2)$$

$\chi_{\alpha\beta}$ is a volume susceptibility, but a molar susceptibility $\chi_{\alpha\beta}^{\mathrm{mol}}$ may be defined as

$$\chi_{\alpha\beta}^{\mathrm{mol}} = \chi_{\alpha\beta} V^{\mathrm{mol}} \qquad (3)$$

where V^{mol} is the molar volume. The susceptibility has the symmetry of the material, so expressing this in terms of the principal axes of χ gives for the free energy density:

$$\begin{aligned} g_{\mathrm{mag}} &= g_0 - \frac{1}{2} \mu_0^{-1} \\ &\quad \cdot \left(\chi_\| B_\|^2 + \chi_{\perp'} B_{\perp'}^2 + \chi_\perp B_\perp^2 \right) \\ &= g_0 - \frac{1}{2} \mu_0^{-1} B^2 \\ &\quad \cdot \left(\chi_\| \cos^2\theta + \chi_{\perp'} \sin^2\theta \sin^2\phi \right. \\ &\quad \left. + \chi_\perp \sin^2\theta \cos^2\phi \right) \end{aligned} \qquad (4)$$

where θ and ϕ are polar angles defining the orientation of B with respect to the principal axes of the susceptibility. For a uniaxial material $\chi_{\perp'} = \chi_\perp$ and:

$$\begin{aligned} g_{\mathrm{mag}} &= g_0 - \frac{1}{2} \mu_0^{-1} B^2 \left(\chi_\perp + \Delta\chi \cos^2\theta \right) \\ &= g_0 - \frac{1}{2} \mu_0^{-1} B^2 \chi_\perp \\ &\quad - \frac{1}{2} \mu_0^{-1} \Delta\chi (\boldsymbol{B} \cdot \boldsymbol{n})^2 \end{aligned} \qquad (5)$$

where $\Delta\chi = \chi_\| - \chi_\perp$, and $\boldsymbol{n}$ is the director, which also defines the principal axis of the susceptibility for uniaxial materials. From Eq. (5) it is clear that the sign of the anisotropy $\Delta\chi$ will determine the orientation of the director with respect to a magnetic field. In order to minimize the free energy, the director will align parallel to the magnetic field for a material having positive $\Delta\chi$, while for negative $\Delta\chi$ the director will be perpendicular to $\boldsymbol{B}$; both situations can occur in practice.

As explained earlier any anisotropic property can be used to define a macroscopic order parameter, and because it is largely unaffected by molecular interactions, the magnetic susceptibility is a particularly useful measure: definitions are given as Eqs. (21) and (22) of Sec. 1 of this chapter. The value of $\Delta\chi_{\mathrm{max}}$ corresponding to perfect alignment can in principal be obtained from measurements on single crystals or from molecular susceptibilities. Eqs. (45) and (47) of Sec. 1 of this chapter relate a macroscopic susceptibility to a microscopic molecular property $\kappa_{\alpha\beta}$, and introduce appropriate order parameters. The molecular susceptibilities κ_{xx}, κ_{yy}, κ_{zz} are defined for the principal axes of the molecular susceptibility, which may not coincide with the axes that define the local orientational order; however, it is usually assumed that any differences can be neglected. Measurements of the magnetic susceptibility can provide a useful route to the order parameters of liquid crystals [1, 2], but require a knowledge of the molecular susceptibilities. These are not usually available for mesogens, but they can be obtained from single crystal measurements, provided full details are available for the crystal structure; the method for deriving molecular susceptibilities from crystal susceptibilities is explained in detail in [3].

2.2 Types of Magnetic Polarization

2.2.1 Diamagnetism

A diamagnetic response is the induction of a magnetic moment in opposition to an applied magnetic field, which thereby raises the free energy. Thus a diamagnetic material will be expelled from a magnetic field, or will adjust itself to minimize the diamagnetic interaction. Most liquid crystals are diamagnetic and this diamagnetism originates from the dispersed electron distribution associated with the molecular electronic structure. The diamagnetic susceptibility is a second rank tensor, and its principal components can be expressed as:

$$\kappa_{ll}^{\text{dia}} = -\frac{e^2 \mu_0}{4 m_e} \langle m^2 + n^2 \rangle$$

$$\kappa_{mm}^{\text{dia}} = -\frac{e^2 \mu_0}{4 m_e} \langle l^2 + n^2 \rangle$$

$$\kappa_{nn}^{\text{dia}} = -\frac{e^2 \mu_0}{4 m_e} \langle m^2 + l^2 \rangle \quad (6)$$

where e is the electronic charge, m_e is the mass of an electron, l, m, n are the molecular axes, and the quantities $\langle m^2 + n^2 \rangle$ are averages over the electron distribution for a plane perpendicular to the component axis (l in this case see Fig. 1).

The induced diamagnetic moment depends on the extent of the electron distribution in a plane perpendicular to an applied magnetic field. In a molecule, delocalized charge makes a major contribution to κ^{dia}, and in particular the ring currents associated with aromatic units give a large negative component of diamagnetic susceptibility for directions perpendicular to the plane of the aromatic unit. It is for this reason that the diamagnetic anisotropy of most calamitic

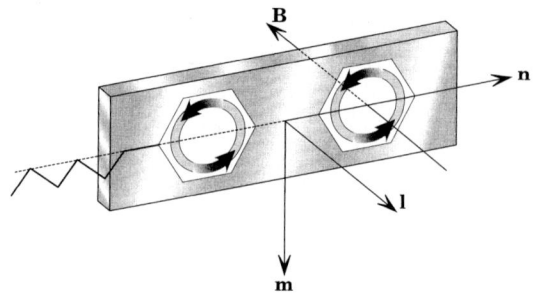

Figure 1. Diagram of molecular axes with representation of the perpendicular plane.

mesogens is positive since both components are negative, but $|\chi_\perp| > |\chi_\parallel|$.

2.2.2 Paramagnetism

Molecular paramagnetism is mostly connected with unpaired electron spins, which have associated magnetic moments. For paramagnetic mesogens the electron spin is introduced by metal centres, and one of the motivations for research into metal-containing mesogens is the desire to prepare paramagnetic liquid crystals. Orientation by an external magnetic field of the magnetic moment derived from an electron spin will induce a magnetization along the field direction, and so provides a positive contribution to the magnetic susceptibility:

$$\bar{\kappa}^{\text{para}} = \frac{\mu_0 g_e^2 \mu_B^2 s(s+1)}{3 k_B T} \quad (7)$$

where g_e is the electronic g-value, s is the total electron spin quantum number and μ_B is the Bohr magneton. κ^{para} is the isotropic molecular paramagnetic susceptibility, and the coefficient of 1/3 arises from the isotropic average of $\langle \cos^2 \theta \rangle$ over the spin orientation in a magnetic field. The temperature dependence of κ^{para} follows a simple Curie Law, and as written there is no anisotropy in κ^{para}, since in most molecules the unpaired electron spin is decoupled from the molecular structure and will align with an external magnetic field independently of the orientation of the molecule. Anisotropy in κ^{para} can be introduced if the g-value becomes anisotropic. The g-value is in fact a tensor quantity which describes the modification to the magnetic field experienced by the electron spin arising from the electron distribution in the molecule. It is analogous to the nuclear shielding in nuclear magnetic resonance, and it contributes to the magnetic internal energy as:

$$u_{\text{mag}} = -g^{\text{e}}_{\alpha\beta}\, \gamma_{\text{e}}\, s_\alpha\, B_\alpha \tag{8}$$

where γ_{e} is the electronic magnetogyric ratio and s_α is the vector component of the electronic spin. Interaction between an electron spin and its local electronic environment changes g_{e} from its free-electron value of 2.0023. These interactions are termed spin–orbit interactions, since they arise from a coupling between the electron spin and the angular momentum of the molecular orbitals. The principal components of the molecular paramagnetic susceptibility can be written as:

$$\kappa^{\text{para}}_{ii} = \frac{\mu_0 \left(g^{\text{e}}_{ii}\right)^2 \mu_{\text{B}}^2 s(s+1)}{3 k_{\text{B}} T} \tag{9}$$

Hence using Eq. (45) of Sec. 1 of this chapter for a uniaxial liquid crystal, and neglecting any biaxial local order gives for the anisotropic part of the paramagnetic susceptibility:

$$\chi^{(a)\text{para}}_{\parallel} = \frac{2N \mu_0 \mu_{\text{B}}^2 s(s+1) S}{3 k_{\text{B}} T} \\ \cdot \left[\left(g^{\text{e}}_{nn}\right)^2 - \frac{1}{2}\left\{\left(g^{\text{e}}_{ll}\right)^2 + \left(g^{\text{e}}_{mm}\right)^2\right\}\right] \tag{10}$$

$$\chi^{(a)\text{para}}_{\perp} = -\frac{N \mu_0 \mu_{\text{B}}^2 s(s+1) S}{3 k_{\text{B}} T} \\ \cdot \left[\left(g^{\text{e}}_{nn}\right)^2 - \frac{1}{2}\left\{\left(g^{\text{e}}_{ll}\right)^2 + \left(g^{\text{e}}_{mm}\right)^2\right\}\right]$$

Thus for a uniaxial $\mathbf{g}_{\text{e}}$ tensor, the paramagnetic susceptibility anisotropy and mean susceptibility can be written as:

$$\Delta\chi^{(a)\text{para}} = \frac{N \mu_0 \mu_{\text{B}}^2 s(s+1) S}{k_{\text{B}} T} \\ \cdot \left[\left(g^{\text{e}}_{nn}\right)^2 - \left(g^{\text{e}}_{ll}\right)^2\right] \tag{11}$$

$$\bar{\chi}^{\text{para}} = \frac{N \mu_0 \mu_{\text{B}}^2 s(s+1)}{9 k_{\text{B}} T} \\ \cdot \left[\left(g^{\text{e}}_{nn}\right)^2 + 2\left(g^{\text{e}}_{ll}\right)^2\right] \tag{12}$$

Care is necessary in defining the principal axes of the $\mathbf{g}_{\text{e}}$ tensor, since they are determined by the local symmetry of the free-electron spin, and therefore usually differ from the molecular axes that define the orientational order of the mesogen. A consequence of Eq. (11) is that there can be competition between paramagnetic and diamagnetic contributions to the macroscopic anisotropy, so the alignment of a paramagnetic mesogen in a magnetic field will be determined by the larger of $\Delta\chi^{\text{dia}}$ and $\Delta\chi^{\text{para}}$.

There is a further contribution to the molecular paramagnetic susceptibility from magnetic field induced distortion of the orbital angular momentum: this is known as temperature independent paramagnetism (TIP), and is the precise magnetic analogue of the electronic polarizability that determines the response of a molecule to a high frequency electric field. The importance of $\kappa^{\text{TIP}}_{\alpha\beta}$ for mesogens is yet to be established: it is only likely to be significant for molecules with low-lying excited electronic states that are connected to the ground state by magnetically allowed dipole transitions; such states are also important in the circular dichroism spectra of molecules. Taking account of all contributions to the molecular susceptibility, the components of the macroscopic susceptibility for a uniaxial liquid crystal composed of uniaxial molecules can be written as:

$$\chi_\parallel = \bar{\chi} + \frac{2N\Delta\kappa^{\text{dia}}S}{3} + \frac{2N\Delta\kappa^{\text{TIP}}S}{3}$$
$$+ \frac{2N\mu_0\mu_B^2 s(s+1)S}{9k_BT}\left[\left(g_{nn}^e\right)^2 - \left(g_{ll}^e\right)^2\right]$$

$$\chi_\perp = \bar{\chi} - \frac{N\Delta\kappa^{\text{dia}}S}{3} - \frac{N\Delta\kappa^{\text{TIP}}S}{3}$$
$$- \frac{N\mu_0\mu_B^2 s(s+1)S}{9k_BT}\left[\left(g_{nn}^e\right)^2 - \left(g_{ll}^e\right)^2\right]$$
(13)

2.2.3 Ferromagnetism

At sufficiently low reduced temperatures, and/or strong spin–spin interactions, spin magnetic moments can become ordered in a parallel array to give ferromagnetic materials, or ordered in an antiparallel fashion to give an antiferromagnetic structure. This behavior is rare in organic materials, for which ferromagnetic organization only occurs at low temperatures. There is a requirement for unpaired electron spins, and so in the context of liquid crystals, possible ferromagnetic materials will almost certainly require metal-containing mesogens. The contribution of ferromagnetic or antiferromagnetic coupling to the susceptibility can be described in terms of a modified Curie Law, known as the Curie–Weiss Law, and the isotropic susceptibility can be written as:

$$\bar{\chi}^{\text{para}} = \frac{N\mu_0 g_e^2 \mu_B^2 s(s-1)}{3k_B(T-\Theta)} \quad (14)$$

where the characteristic temperature Θ, the Curie temperature, is a measure of the ferromagnetic coupling, and marks the onset of permanent magnetization. If Θ is negative, then the local magnetic interactions are antiferromagnetic and Θ is called the Néel temperature. Although no ferromagnetic liquid crystals have been discovered so far, ferromagnetic liquid crystals can be prepared by the dispersion of ferromagnetic particles in a liquid crystal host. In preparing such systems it is desirable to make the ferromagnetic particles very small, so that each particle has a permanent magnetic moment. Normally in the absence of an external magnetic field or special conditioning, the permanent magnetization characteristic of ferromagnetic materials forms in domains of opposing magnetic moments, so that the total magnetisation is cancelled. However for sufficiently small particle sizes, the domain wall energies become relatively too high to sustain, and single domain particles are preferred. This effect is known as superparamagnetism or collective paramagnetism, and dispersed particles satisfying the requirements for single magnetic domains act in fluids as micromagnets: such systems are known as ferrofluids. The dispersion of single domain ferromagnetic particles in a liquid crystal host can form anisotropic ferrofluids or ferromagnetic liquid crystals.

2.3 Diamagnetic Liquid Crystals

Most liquid crystals are diamagnetic and their magnetic anisotropy arises from the electronic structure of the mesogens. Delocalisation of electronic charge will enhance the diamagnetic susceptibility and aromatic groups in particular make a large contribution to the diamagnetic susceptibility. In Table 1 are listed molecular susceptibility components for a number of molecules to indicate the likely contributions of various groups to mesogenic structures. These values have been obtained from susceptibility measurements on single crystals.

Table 1. Molecular susceptibility components from single crystal measurements.

Compound and molecular axes	Molecular susceptibility components			Ref.
	$N\kappa_l/10^{-9}$ m^3 mol^{-1}	$N\kappa_m/10^{-9}$ m^3 mol^{-1}	$N\kappa_n/10^{-9}$ m^3 mol^{-1}	
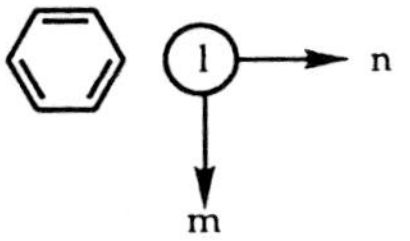 benzene (with axes l, m, n)	−1.19	−0.44	−0.44	[3]
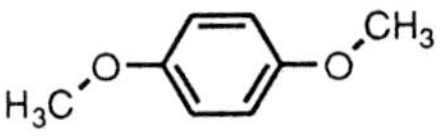 H$_3$C−O−C$_6$H$_4$−O−CH$_3$	−1.46	−0.99	−0.82	[3]
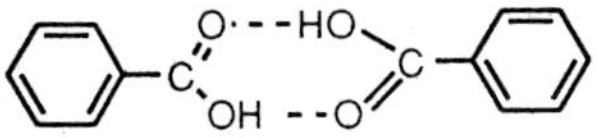 benzoic acid dimer	−1.38	−0.59	−0.69	[3]
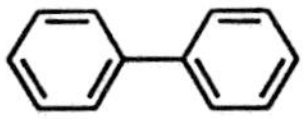 biphenyl	−2.31	−0.78	−0.85	[3]
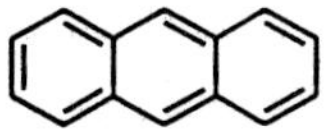 anthracene	−3.13	−0.96	−0.97	[3]
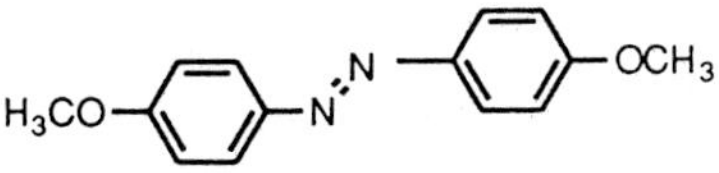 H$_3$CO−C$_6$H$_4$−N=N−C$_6$H$_4$−OCH$_3$	−3.08	−1.13	−1.33	[23]

Table 2. Diamagnetic susceptibilities for liquid crystals.

Compound and acronym	Diamagnetic susceptibilities		Ref.
	$\Delta\chi/10^{-9}$ m^3 kg^{-1}	$\chi/10^{-9}$ m^3 kg^{-1}	
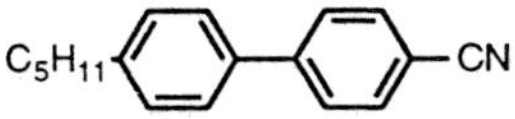 C$_5$H$_{11}$−biphenyl−CN	1.51	8.43	[2]
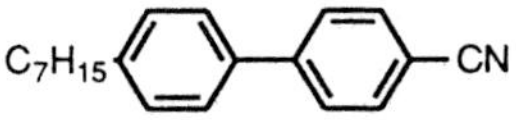 C$_7$H$_{15}$−biphenyl−CN	1.37	8.66	[2]
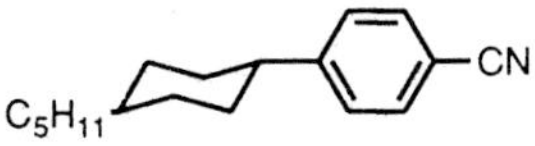 C$_5$H$_{11}$−cyclohexyl−phenyl−CN	0.46		[4]
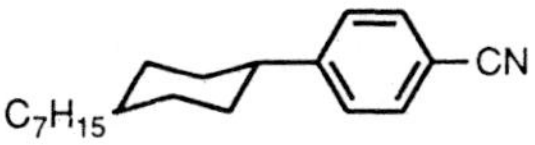 C$_7$H$_{15}$−cyclohexyl−phenyl−CN	0.42	9.32	[4]
 C$_7$H$_{15}$−cyclohexyl−cyclohexyl−CN	−0.38	8.87	[2]

Most thermotropic mesogens contain aromatic groups, and since the component of the diamagnetic susceptibility perpendicular to a benzene ring is greater than the in-plane component, liquid crystals composed of calamitic mesogens will have a positive diamagnetic anisotropy, while liquid crystals of disc-like molecules will have a negative diamagnetic anisotropy. Thus calamitic nematics and smectics will tend to align with their directors along the direction of an external magnetic field, while discotic liquid crystals will align with the director perpendicular to the field. Replacement of aromatic rings by saturated groups such as cyclohexyl, bicyclo-octyl or alkyl chains will reduce the anisotropy of the molecular core, so that liquid crystals based on the *trans-trans*-cyclohexylcyclohexyl core have a negative anisotropy due to the attached terminal groups. Some results for the magnetic susceptibilities of liquid crystals are given in Table 2.

2.4 Paramagnetic Liquid Crystals

Known paramagnetic liquid crystals are based on metal-containing mesogens, which have a variety of metal centres and coordination geometries [5, 6]. A requirement for paramagnetism is an unpaired spin, but to have an influence on the magnetic anisotropy, there must also be a significant **g**-tensor anisotropy. The effect of competition between diamagnetic and paramagnetic contributions to the susceptibility is illustrated by the behavior of salicylaldimine complexes of copper [7, 8]. These are formed from copper(II) having a d^9 electron configuration, which results in a square planar geometry around the metal centre. The **g**-tensor anisotropy is such that $(g_{nn}^e)^2 < \frac{1}{2}\left((g_{ll}^e)^2 + (g_{mm}^e)^2\right)$ so that the paramagnetic contribution to the anisotropy is negative. For complexes with four benzene rings in the structure (Table 3, compounds 1 and 2), the paramagnetic term is larger than the diamagnetic term in the anisotropy, and so the complexes align with the major axis (**n**) perpendicular to the field direction: free rotation about the molecular long axis is assumed. Increasing the number of benzene rings to six (Table 3, compounds 3 and 4) causes the diamagnetic anisotropy to dominate, and the director aligns parallel to a magnetic field (see Fig. 2).

Electron paramagnetic resonance measurements on these liquid crystals give **g**-values of $g_{nn}^e = 2.053$ and $1/2(g_{ll}^e + g_{mm}^e) = 2.082$. By contrast the corresponding

Table 3. Structures, susceptibility anisotropies and alignment of salicylaldimine complexes of copper [7, 8].

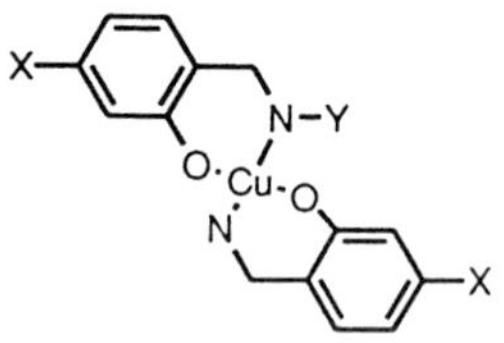

X	Y	$\Delta\chi^{para}/10^{-9}$ m^3 mol^{-1}	$\Delta\chi^{dia}/10^{-9}$ m^3 mol^{-1}	Orientation to magnetic field
C$_7$H$_{15}$O—	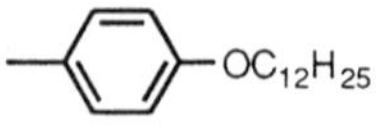 —⌬—OC$_{12}$H$_{25}$	−84.4	72.4	⊥
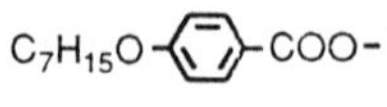 C$_7$H$_{15}$O—⌬—COO—	—C$_{12}$H$_{25}$	−91.5	79.5	⊥
C$_7$H$_{15}$O—	 —⌬—COO—⌬—OC$_{12}$H$_{25}$	−74.7	117	∥
C$_7$H$_{15}$O—	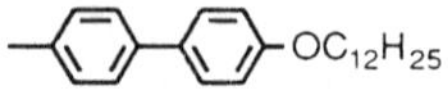 —⌬—⌬—OC$_{12}$H$_{25}$	−66.8	120	∥

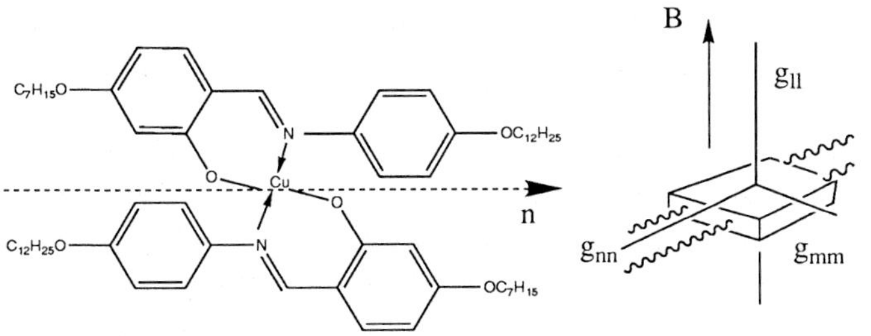

Δχ -ve, **n** perpendicular to field

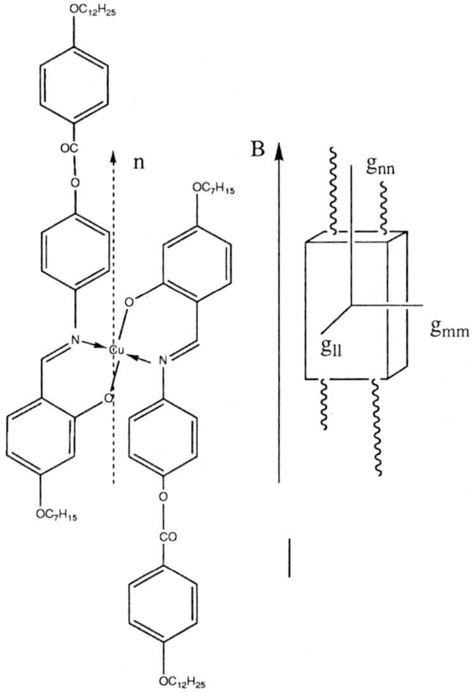

Δχ +ve **n** parallel to B

Figure 2. Alignment of salicyldimine complexes in a magnetic field.

vanadyl (VO) d^1 complexes have a reversed **g**-tensor anisotropy $g_{nn}^e = 1.987$ and $1/2(g_{ll}^e + g_{mm}^e) = 1.966$, and so these complexes always align with their molecular long axes along the magnetic field direction. Mesogenic paramagnetic salicylaldimine complexes of a number of rare earths have been reported showing SmA phases [9], and found to have large magnetic anisotropies. A similar result has been obtained [10] for a β-enaminoketone complex of dysprosium, but the corresponding gadolinium complex had a very small paramagnetic anisotropy.

2.5 Ferromagnetic Liquid Crystals

The possibility for a mesogenic material exhibiting ferromagnetism is at the present time remote. Organic ferromagnets have been prepared [11, 12], but they mostly have very low Curie temperatures, well below the melting points of the compounds. Since the origin of ferromagnetism is long range spin–spin interactions, it is unlikely that these will persist in a fluid liquid crystalline state, although there may be more chance of preparing metal-containing liquid crystal polymers having a potential for magnetic ordering [13]. A different approach to the preparation of ferromagnetic liquid crystals was proposed by Brochard and de Gennes [14] based on the dispersion of ferromagnetic particles in a liquid crystal host. As explained above, ferrofluids can be formed from dispersions of ferromagnetic materials that will form essentially single domain particles. Examples are colloidal suspensions of ferrite γ-Fe_2O_3, magnetite Fe_3O_4 or cobalt metal in either hydrocarbon or water-based solvents. Ferromagnetic particles may be imagined to couple with a liquid crystal host through interaction between the magnetic moment of the particle and the magnetic anisotropy of the surrounding liquid crystal [14]. Depending on the anisotropy of the liquid crystal the magnetisation of the particles should align parallel (positive $\Delta\chi$) or perpendicular (negative $\Delta\chi$) to the director.

Another mechanism for coupling the orientation of a magnetic particle to a liquid crystal is through elastic interactions. If the magnetic particles are anisotropic, then defining the director orientation at the surface of the particle with a suitable surfactant will cause a preferred alignment of the particle in a liquid crystal host. Chen and Amer [15] succeeded in stabilizing a suspension of particles of length 0.35 μm an 0.04 μm diameter in MBBA. The particles were coated with a surfactant which defined the director orientation at the particle surface as perpendicular to the particle axis, and changes in the observed optical anisotropy in the presence of a magnetic field were consistent with the reorientation of the liquid crystal director perpendicular to the field.

Lyotropic liquid crystals doped with ferromagnetic particles have also been studied [16, 17], and the magnetic particles can be stabilised in either hydrophobic or hydrophilic regions. Changes in birefringence with magnetic fields have been observed, suggesting that the optical anisotropy of the liquid crystal has been coupled to the magnetic anisotropy of the dispersed particles. It is possible that such magnetic field effects in anisotropic ferrofluids may find application in the future.

2.6 Applications of Magnetic Properties

Since intermolecular forces scarcely affect the magnetic susceptibility, measurements of the magnetic anisotropy can provide a direct measure of the orientational order. Using Eq. (13), the anisotropy of susceptibility for a uniaxial liquid crystal phase formed from uniaxial mesogens can be written as:

$$\Delta\chi = N\Delta\kappa S \qquad (15)$$

where $\Delta\kappa$ contains contributions from diamagnetic, paramagnetic and temperature independent paramagnetic terms:

$$\Delta\kappa = \Delta\kappa^{dia} + \Delta\kappa^{TIP} \qquad (16)$$
$$+ \frac{N\mu_0\mu_B^2 s(s+1)}{3k_B T}\left[\left(g_{nn}^e\right)^2 - \left(g_{ll}^e\right)^2\right]$$

$\Delta\kappa$ can be obtained from measurements on single crystals, and provided that accurate values are available for the density, the order parameter S can be obtained directly. An alternative way to determine S from measurements of the temperature dependence of the susceptibility is to fit values to a functional form for the variation of S with temperature. The simplest procedure known as the Haller extrapolation is described in the context of birefringence measurements in Sec. 3.2 of this chapter. The effects of local biaxial ordering on the measured susceptibility for cyanobiphenyls has been considered by Bunning, Crellin and Faber [2] using crystal data for biphenyl.

Magnetic properties have an importance in the NMR of liquid crystals [18,19], but the moments of the nuclear spins responsible for the NMR signal are far too small to make any contribution to magnetic susceptibilities. However, bulk susceptibility corrections to the NMR chemical shift of a standard immersed in the sample can be used to determine diamagnetic susceptibilities. The chemical shift of the standard is shifted to lower fields in a cylindrical sample due to the bulk magnetization, according to:

$$\sigma_{observed} = \sigma_{standard} - \frac{2\pi}{3}\chi_{ii} \quad (17)$$

where χ_{ii} is the susceptibility component in the direction of the external magnetic field. Diamagnetic liquid crystals will align such that the smallest component of χ is along the magnetic field direction, and this causes a splitting in the NMR lines, which can be related to the order parameter. This technique is extremely useful for obtaining detailed information on the ordering of different segments of flexible molecules [18,19] and can also yield values for the local biaxial order parameters of molecules [18]. The method has been successfully applied to both pure liquid crystals and to dopant molecules dissolved in liquid crystal hosts, which serve to orient the solute molecules. For these experiments the direction of alignment of the director with respect to the magnetic field is important, and since most liquid crystals have a positive susceptibility anisotropy, the director will align parallel to the magnetic field.

The standard method for measuring magnetic susceptibilities is to use a Faraday balance, which involves the measurement of the force on a sample in an inhomogeneous magnetic field [20]; other methods use a Gouy balance [3], or a SQUID magnetometer [21, 22]. All methods measure a single component of the susceptibility – the largest for paramagnetic samples, and the smallest for diamagnetic samples, assuming that the alignment of the sample liquid crystal is not constrained by other forces. In order to obtain the anisotropy, a second measurement is required, and this is usually taken as the mean susceptibility measured in the isotropic phase. Diamagnetic susceptibilities are independent of temperature, and it is reasonable to assume that the mean susceptibility in the liquid crystal phase is the same as in the isotropic phase: hence for positive materials, the susceptibility anisotropy is given by:

$$\Delta\chi = \frac{3}{2}(\chi_\| - \bar{\chi}) \quad (18)$$

For liquid crystals having a negative susceptibility, the anisotropy is:

$$\Delta\chi = 3(\bar{\chi} - \chi_\perp) \quad (19)$$

It is not possible to determine the sign of $\Delta\chi$ from magnetic measurements alone.

The ability of magnetic fields to control the alignment of liquid crystals is widely used, for example in X-ray structural studies of liquid crystals, or for optical measurements on aligned liquid crystal films.

An advantage of magnetic fields over electric fields for controlling alignment is that complications due to electrical conduction or electrohydrodynamic effects are not present. Competition between aligning fields has been used to obtain direct measurements of susceptibility anisotropies. The basis of the method can be understood from Eq. (5). A similar equation can be written for the free energy density of a liquid crystal in an electric field, such that:

$$\mathbf{g}^e = g_0 - \frac{1}{2}\varepsilon_0 E^2 \left(\varepsilon_\perp + \Delta\varepsilon \cos^2\theta\right) \quad (20)$$

For balancing torques of the electric and magnetic fields on a liquid crystal:

$$\mu_0^{-1}\Delta\chi B^2 = \varepsilon_0 \Delta\varepsilon E^2 \quad (21)$$

Thus by measuring the corresponding fields, and knowing the permittivity anisotropy, it is possible to determine $\Delta\chi$ directly [4].

2.7 References

[1] P. L. Sherrell, D. A. Crellin, *J. de Phys.* **1979**, *40*, C3-211.
[2] J. D. Bunning, D. A. Crellin, T. E. Faber, *Liq. Cryst.* **1986**, *1*, 37.
[3] J. W. Rohleder, R. W. Munn, *Magnetism and Optics of Molecular Crystals*, John Wiley, Chichester, **1992**, p. 27.
[4] H. Schad, G. Baur, G. Meier, *J. Chem. Phys.* **1979**, *70*, 2770.
[5] D. W. Bruce, *J. Chem. Soc. Dalton Trans.* **1993**, 2983.
[6] S. A. Hudson, P. M. Maitlis, *Chem. Rev.* **1993**, *93*, 861.
[7] J.-L. Serrano, P. Romero, M. Marcos, P. J. Alonso, *J. Chem. Soc., Chem. Commun.* **1990**, 859.
[8] I. Bikchantaev, Yu. Galyametdinov, A. Prosvirin, K. Griesar, E. A. Soto-Bustamente, W. Haase, *Liq. Cryst.* **1995**, *18*, 231.
[9] Yu. Galyametdinov, M. A. Athanassopoulou, K. Griesar, O. Kharitonova, E. A. Soto-Bustamente, L. Tinchurina, I. Ovchinnikov, W. Haase, *Chem. Mat.* **1996**, *8*, 922.
[10] I. Bikchantaev, Yu. Galyametdinov, O. Kharitonova, I. Ovchinnikov, D. W. Bruce, D. A. Dunmur, D. Guillon, B. Heinrich, *Liq. Cryst.* **1996**, *20*, 831.
[11] O. Kahn, Yu. Pie, Y. Journaux in *Inorganic Materials* (Ed.: D. W. Bruce, D. O'Hare), John Wiley, Chichester, **1992**, p. 61.
[12] D. Gatteschi, *Europhys. News* **1994**, *25*, 50.
[13] S. Takahashi, Y. Takai, H. Morimoto, K. Sonogashira, *J. Chem. Soc., Chem. Commun.* **1984**, 3.
[14] F. Brochard, P.-G. de Gennes, *J. de Phys.* **1970**, *31*, 691.
[15] S. H. Chen, N. M. Amer, *Phys. Rev. Lett.* **1983**, *51*, 2298.
[16] P. Fabre, C. Casagrande, M. Veyssie, V. Cabuil, R. Massart, *Phys. Rev. Lett.* **1990**, *64*, 539.
[17] C. Y. Matuo, F. A. Tourinho, A. M. Figueiredo-Neto, *J. Mag. Mag. Mat.* **1993**, *122*, 53.
[18] J. W. Emsley (Ed.), *Nuclear Magnetic Resonance of Liquid Crystals*, D. Reidel, Dordrecht, Netherlands, **1983**.
[19] R. Dong, *Nuclear Magnetic Resonance of Liquid Crystals*, Springer-Verlag, Berlin, **1994**.
[20] D. Jiles, *Introduction to Magnetism and Magnetic Materials*, Chapman & Hall, **1991**.
[21] J. S. Philo, W. M. Fairbank, *Rev. Sci. Instrum.* **1977**, *48*, 1529.
[22] C. Butzlaff, A. X. Trantwein, H. Winkler, *Meth. Enzym.* **1993**, *227*, 412.
[23] W. H. de Jeu, *Physical Properties of Liquid Crystalline Materials,* Gordon and Breach, New York, **1980**, p 31.

3 Optical Properties

David Dunmur and Kazuhisa Toriyama

The optical properties of liquid crystals determine their response to high frequency electromagnetic radiation, and encompass the properties of reflection, refraction, optical absorption, optical activity, nonlinear response (harmonic generation), optical waveguiding, and light scattering [1]. Most applications of thermotropic liquid crystals rely on their optical properties and how they respond to changes of the electric field, temperature or pressure. The optical properties can be described in terms of refractive indices, and anisotropic materials have up to three independent principal refractive indices defined by a refractive index ellipsoid.

Solution of Maxwell's equations for the propagation of a wave through an anisotropic medium gives three principal wave velocities for directions $i = 1, 2, 3$ as:

$$v_i^2 = \left(\mu_{ii}^m \varepsilon_{ii}\right)^{-1} \qquad (1)$$

where ε_{ii} and μ_{ii}^m are the principal components of the electric permittivity and magnetic permeability tensors, which are assumed to be diagonal in the same frame of axes. For other than ferromagnetic materials μ_{ii}^m is very close to unity, and comparing the velocities of the wave in a vacuum with the velocities of the wave in an anisotropic medium gives the principal refractive indices as:

$$n_i = (\varepsilon_{ii}/\varepsilon_0)^{1/2} \qquad (2)$$

In fact, two waves of different velocity (having the same wave normal but orthogonal polarizations) can propagate through an optically anisotropic medium along two different directions. This results in the appearance of a double image of an object viewed through anisotropic crystals, and is termed double refraction. These two rays have different refractive indices: the *ordinary* ray propagates along the wave normal, and its direction obeys the normal Snell's law of refraction so that $n_o = \sin i / \sin r$, while for the other *extraordinary* ray, the ray direction and wave normal are not parallel. The two refractive indices for a particular wave normal can be obtained from the refractive index indicatrix [2]. This ellipsoid is defined by the equation:

$$\frac{x^2}{n_1^2} + \frac{y^2}{n_2^2} + \frac{z^2}{n_3^2} = 1 \qquad (3)$$

where n_1, n_2, and n_3 are termed the three principal refractive indices, and the directions x, y, z are the principal axes of the electric permittivity tensor (Fig. 1).

For any direction in a crystal (OP) in Fig. 1, the refractive indices of the two wave

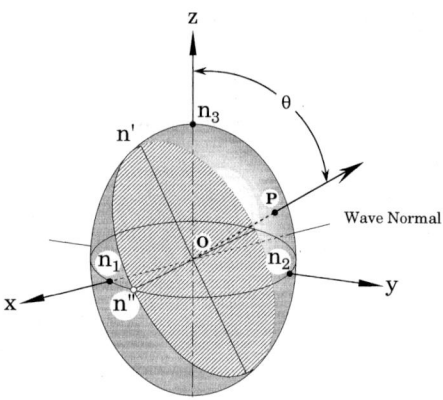

Figure 1. The optical indicatrix, where the principal refractive indices are labelled as n_1, n_2 and n_3. Refractive indices corresponding to the wave front normal OP are shown as n' and n''.

indices. For any direction, one ray has a refractive index of n_o, but the refractive index of the extraordinary ray depends on direction, such that:

$$n_e(\theta)^2 = \left(\frac{\cos^2 \theta}{n_1^2} + \frac{\sin^2 \theta}{n_3^2} \right)^{-1} \quad (4)$$

where the direction makes an angle θ with the z-axis. If $n_1 \neq n_2 \neq n_3$, there are two optic axes for which the perpendicular cross-section of the indicatrix is circular, such materials are biaxial, and there are two directions along which the material appears to be optically isotropic.

Many of the interesting properties of liquid crystals are a result of chirality or handedness, which is manifest in optical properties by optical activity. For isotropic materials or anisotropic materials viewed along their optic axes, optical activity causes the plane of polarization of propagating light to be rotated by an angle ϕ. This can be expressed in terms of a difference between refractive indices for left (n_l) and right (n_r) circularly polarized light:

$$\phi = \frac{\pi d}{\lambda} (n_l - n_r) \quad (5)$$

where d is the path length and λ is the vacuum wavelength of the light. For anisotropic materials including liquid crystals, the optical activity interacts with the linear birefringence, and the two propagating waves which correspond to a particular wave normal are elliptically polarized, the axes of the ellipses being perpendicular.

For an absorbing medium the refractive index can be represented as a complex quantity consisting of a real part (n_i') and an imaginary part (k_i):

$$n_i = n_i' - i k_i \quad (6)$$

and using the relation with the permittivity, we have:

fronts normal to OP that can propagate are given by the semi-major and semi-minor axes of the ellipse perpendicular to OP. In the case of an indicatrix of revolution, which would represent a uniaxial liquid crystal, $n_1 = n_2$ which is the ordinary refractive index n_o, and for light propagating along the z-direction (i.e. $\theta = 0$), which is the symmetry axis of the indicatrix, the ordinary and extraordinary rays are coincident; the largest refractive index (n_3 in the figure) is called n_e. Materials having two equal principal refractive indices are referred to as uniaxial, and the unique direction z is the optic axis. The special feature of the optic axis is that light of any polarization travels along this axis without any change in its polarization (i.e. the material responds as an optically isotropic medium. The difference between the two independent principal refractive indices $\Delta n = n_3 - n_1$ is termed the birefringence. For light propagating along a direction which is not the optic axis, the ordinary and extraordinary rays are not coincident and they travel with different velocities corresponding to different refractive

$$n_i^2 = \varepsilon_o^{-1}\left[\varepsilon_{ii}^{(\text{real})} - i\,\varepsilon_{ii}^{(\text{imaginary})}\right]$$

$$n_i'^2 - k_i^2 = \varepsilon_o^{-1}\varepsilon_{ii}^{(\text{real})}$$

$$2n_i' k_i = \varepsilon_o^{-1}\varepsilon_{ii}^{(\text{imaginary})} \qquad (7)$$

The real part of the refractive index determines the speed of light through the medium, while the imaginary part measures the attenuation of its intensity, so k_{ii} is defined as the absorption coefficient. If the absorption coefficients for light plane-polarized along different directions are different, then the material is said to exhibit linear dichroism. Nonchiral materials can only be linearly dichroic, but chiral materials can also show circular dichroism, which arises from a difference between the imaginary parts (absorption coefficients) of the refractive indices for left and right circularly polarized light.

3.1 Symmetry of Liquid Crystal Phases and the Optical Indicatrix

The symmetry of liquid crystalline phases can be categorized in terms of their orientational and translational degrees of freedom. Thus nematic, smectic and columnar phase types have respectively three, two, and one degrees of translational freedom, and within each type there can be different phases depending on the orientational or point group symmetry. Their optics (uniaxial, biaxial, optically active) are determined by the point group symmetries, which are listed in Table 1 for common liquid crystal phases; the optical symmetries of variants of these phases can usually be established directly from their structures.

It will be seen that most nematics and orthogonal smectic and columnar phases are uniaxial, with two equal principal refractive indices, while the tilted smectic and columnar phases are biaxial. The optical symmetries of liquid crystal phases can be determined by conoscopic observation of aligned thin films [3] but the technique is difficult, and made more so by the small biaxiality of tilted liquid crystal phases. Principal refractive indices of liquid crystals range from 1.4 to 1.9, and uniaxial birefringences $\Delta n = n_e - n_o$, can be between 0.02 and 0.4; negative birefringences are associated with discotic versions of liquid crystal phases (e.g. discotic nematic or columnar phases). For biaxial liquid crystals, all three principal refractive indices are different, but usually one (n_3) is significantly greater (or less) than the other two, in which case the uniaxial birefringence can be defined as $\Delta n = n_3 - \frac{1}{2}(n_2 + n_1)$ and the biaxiality is $\partial n = n_2 - n_1$. The biaxiality of liquid crystals is small (≈ 0.01) [4, 5], which is a consequence of the small degree of structural biaxiality of these phases.

3.2 Molecular Theory of Refractive Indices

The characteristics of optical and electro-optical liquid crystal devices are determined by the refractive indices of the materials, thus an understanding of the relationship between refractive indices and molecular properties is necessary for the design of improved liquid crystal materials. In developing a molecular theory for any electrical or optical property, the problem of the internal or local electric field has to be addressed. This arises because the field experienced by a molecule in a condensed phase differs from that applied across the macroscopic sample. The internal field has

3 Optical Properties

Table 1. Symmetries of common liquid crystal phase types: only those phases with well-established phase structures are included. Crystal smectic phases, including the cubic D phase have been omitted, as have recently discovered twist grain boundary phases, and structurally modulated variants of smectic phases.

Liquid crystal phase	Point group and translational degrees of freedom – $T(n)$	Optical symmetry; uniaxial – $u(+)$ $n_3 > n_2 = n_1$ $u(-)$ $n_3 < n_2 = n_1$; biaxial (b); helicoidal (h)
Achiral calamitic, micellar nematic N or N_u	$D_{\infty h} \times T(3)$	$u(+)$
Achiral nematic discotic (N_D), columnar nematic (N_{Col})	$D_{\infty h} \times T(3)$	$u(-)$
Chiral nematic (cholesteric) N^*	$D_{\infty} \times T(3)$	b, h, locally biaxial but globally $u(-)$
Biaxial nematic (N_b) (only micellar confirmed)	$D_{2h} \times T(3)$	b
Achiral calamitic orthogonal smectic or lamellar phases (SmA)	$D_{\infty h} \times T(2)$	$u(+)$
Achiral tilted smectic phase (SmC)	$C_{2h} \times T(2)$	b
Chiral tilted smectic phase (SmC^*)	$C_2 \times T(2)$	b, h
Orthogonal and lamellar hexatic phase (SmB)	$D_{6h} \times T(1)$ locally, $D_{6h} \times T(2)$ globally	$u(+)$
Tilted and lamellar hexatic phases (SmF and SmI)	$C_{2h} \times T$ (1 or 2 see above)	b
Chiral tilted and lamellar hexatic phases (SmF^* and SmI^*)	$C_2 \times T$ (1 or 2 see above)	b, h
Discotic columnar: hexagonal order of columns, ordered or disordered within columns (Col_{ho} or $_{hd}$)	$D_{6h} \times T(1)$	$u(-)$
Rectangular array of columns (Col_{ro} or $_{rd}$)	$D_{2h} \times T(1)$	b
Molecules tilted within columns (Col_{to} or $_{td}$)	$C_{2h} \times T(1)$	b

a special significance for anisotropic materials such as liquid crystals. For isotropic media the Lorentz local field is used $\left(E_{loc} = \frac{(\varepsilon+2)}{3} E\right)$ where ε is the mean permittivity, and this results in the Lorenz–Lorentz expression relating the refractive index to the mean molecular polarizability:

$$\frac{n^2-1}{n^2+2} = \frac{N\alpha}{3\varepsilon_0} \quad (8)$$

where N is the number density and α is the mean polarizability. For low density gases, this will be the molecular polarizability, but in condensed fluids it is an effective or dressed property, which takes account of short range intermolecular interactions. In anisotropic liquid crystals it is reasonable to adopt the isotropic model for the internal field, and their principal refractive indices can be written as:

$$\frac{n_i^2-1}{n^2+2} = \frac{N\langle\alpha_{ii}\rangle}{3\varepsilon_0} \quad (9)$$

The polarizability component $\langle\alpha_{ii}\rangle$ is the average value along the direction of the principal refractive index n_i, and $n^2 = \frac{1}{3}(n_1^2 + n_2^2 + n_3^2)$ is a mean refractive index. Using the general results for the transformation of second rank tensor properties, these polarizabilities can be expressed in terms of molecular components and orientational order parameters. For liquid crystal phases of uniaxial symmetry the optic axis coincides with the average alignment direction of the molecules, termed the director, and from Eq. (45) of Sec. 1 of this chapter we obtain for components parallel ($\langle\alpha_\parallel\rangle$) and perpendicular ($\langle\alpha_\perp\rangle$) to the director:

$$\langle\alpha_\parallel\rangle = \alpha + \frac{2}{3}\left\{\begin{array}{l} S\left[\alpha_{nn} - \frac{1}{2}(\alpha_{ll} + \alpha_{mm})\right] + \\ \frac{1}{2}D(\alpha_{ll} - \alpha_{mm}) \end{array}\right\}$$

$$\langle\alpha_\perp\rangle = \alpha + \frac{1}{3}\left\{\begin{array}{l} S\left[\alpha_{nn} - \frac{1}{2}(\alpha_{ll} + \alpha_{mm})\right] + \\ \frac{1}{2}D(\alpha_{ll} - \alpha_{mm}) \end{array}\right\} \quad (10)$$

The principal axes of the molecular polarizability tensor are labelled l, m, n, as shown in Fig. 2. Thus the importance of order parameters in determining the anisotropy of optical properties is clearly demonstrated. Both order parameters S and D contribute to the anisotropy of second rank tensor properties even in uniaxial liquid crystals, but they cannot be separated from a single measurement of the birefringence.

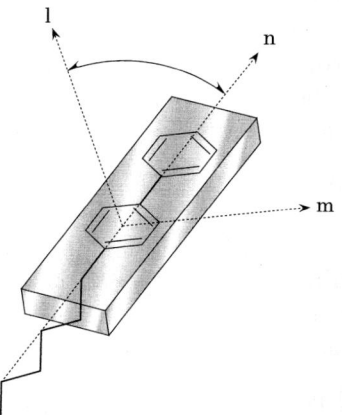

Figure 2. The principal axes of a molecular polarizability tensor.

If the liquid crystal phase is biaxial, as with SmC phases, then any second rank tensor property has three independent principal components and the average polarizabilities corresponding to the three refractive indices can be expressed in terms of the orientational order parameters introduced for biaxial phases:

$$\langle\alpha_{33}\rangle = \alpha + \frac{2}{3}\left\{\begin{array}{l} S\left[\alpha_{nn} - \frac{1}{2}(\alpha_{ll} + \alpha_{mm})\right] + \\ \frac{1}{2}D(\alpha_{ll} - \alpha_{mm}) \end{array}\right\}$$

$$\langle\alpha_{22}\rangle = \alpha - \frac{1}{3}\left\{\begin{array}{l} (S+P)\left[\alpha_{nn} - \frac{1}{2}(\alpha_{ll} + \alpha_{mm})\right] + \\ \frac{1}{2}(D+C)(\alpha_{ll} - \alpha_{mm}) \end{array}\right\}$$

$$\langle\alpha_{11}\rangle = \alpha - \frac{1}{3}\left\{\begin{array}{l} (S-P)\left[\alpha_{nn} - \frac{1}{2}(\alpha_{ll} + \alpha_{mm})\right] + \\ \frac{1}{2}(D-C)(\alpha_{ll} - \alpha_{mm}) \end{array}\right\} \quad (11)$$

Refractive indices of liquid crystals may be measured by a variety of methods, but all require a well-aligned thin film. The simplest method is to use a refractometer with a suitably coated prism surface to give an aligned sample, and use of a polarizer permits the separation of the two refracted rays. This works well for N and SmA phases, but it is not usually possible to align both directors for SmC phases; there are also limitations of temperature. Wedge cells have also been used to obtain refractive indices of liquid crystals, relying on external magnetic fields for alignment [6, 7]. Guided-mode methods can be used to obtain refractive indices [8, 9], while the Z-scan method [10, 11] can be used to obtain information on the field dependence of refractives indices. It is easier to measure birefringence changes of liquid crystals directly using interferometric methods, and high precision can be achieved [12–14]. The wavelength dependence of refractive indices can be obtained from so-called channelled spectra, which are interference bands observed from thin films in spectrophotometers [15–17].

Not only are refractive indices needed to optimise device and materials design, but also they provide a simple route to order parameters. Extrapolation methods can be used to determine the uniaxial order parameter S, if contributions from molecular biaxiality (D) are ignored [18–20]. For molecules having axial symmetry ($\alpha_{ll} = \alpha_{mm}$) Eq. (9) can be written as:

$$\frac{n_\parallel^2 - n_\perp^2}{n^2 - 1} = \frac{S \Delta \alpha}{\alpha} \quad (12)$$

where $\Delta\alpha = (\alpha_{nn} - \alpha_{ll})$. If it is assumed that the order parameter can be written in a simple form as:

$$S = \left(1 - \frac{T}{T_{N-I}}\right)^b \quad (13)$$

where T/T_{N-I} is a reduced temperature then a plot of $\log\left(\frac{n_\parallel^2 - n_\perp^2}{n^2 - 1}\right)$ against $\log\left(1 - \frac{T}{T_{N-I}}\right)$ will give a straight line of intercept $\frac{\Delta\alpha}{\alpha}$, using which the order parameters S can be calculated from Eq. (12). It is also possible to

Table 2. Molecular polarizabilities and refractive indices for a selection of mesogens.

Mesogen	Polarizability		Refractive index; reduced $T_R = 0.95$		λ (nm)	Ref.
	$\Delta\alpha/10^{-40}$ $C^2 J^{-1} m^2$	$\alpha/10^{-40}$ $C^2 J^{-1} m^2$	Δn	$\bar{n}$		
C_5H_{11}—⬡—⬡—CN	19.4	37.5	0.194 0.194	1.589 1.595	633 589	(a, f)
C_5H_{11}—⬡—⬡—CN	16.0	36.2	0.125	1.533	589	(b, f)
C_7H_{15}—⬡—⬡—CN	11.1	40.4	0.045[a]	1.471[a]	589	(b, c)
H_3CO—⬡—C=N—⬡—C_4H_9	31.2	41.1	0.21 0.22	1.61 1.62	633 589	(g, h)
C_8H_{17}—⬡—C=C—⬡—OC_2H_5	33.4	52.8	0.172 0.179	1.523 1.528	633 589	(d)
C_5H_{11}—⬡—C-O—⬡—OC_6H_{13}	26.4	43.9	0.121	1.530	589	(e)
C_5H_{11}—⬡—C-S—⬡—CN	25.3	37.0	0.197	1.592	589	(e)

[a]) $T_R = 0.965$.
(a) R. G. Horn, *J. de Phys.* **1978**, *39*, 105; or Ref. [19].
(b) M. M. M. Abdoh, S. N. C. Shivaprakash, J. S. Prasad, *J. Chem. Phys.* **1982**, *77*, 2570.
(c) I. H. Ibrahim, W. Haase, *Mol. Cryst. Liq. Cryst.* **1981**, *66*, 189.
(d) Reference [18].
(e) I. H. Ibrahim, W. Haase, *J. de Phys.* **1979**, *40*, 191.
(f) D. A. Dunmur, A. E. Tomes, *Mol. Cryst. Liq. Cryst.* **1983**, *97*, 241.
(g) M. Mitra, B. Majumdar, R. Paul, S. Paul, *Mol. Cryst. Liq. Cryst.* **1990**, *180B*, 187.
(h) I. Haller, H. A. Huggins, M. J. Freiser, *Mol. Cryst. Liq. Cryst.* **1972**, *16*, 53.

fit refractive index data to a mean field expression for the order parameter using a numerical procedure [6]. The refractive indices also provide a possible route to determining molecular polarizabilities, and design of molecules with specifically high or low polarizability is important for particular applications.

3.3 Optical Absorption and Linear Dichroism

The attenuation of the intensity of a beam of light on passing through an absorbing medium can be measured by the absorption coefficient, which is the imaginary part of the refractive index. A more usual measure is the optical absorbance (A) or molar extinction coefficient ε, which is defined in terms of the Beer–Lambert law as:

$$\log_{10}(I_0/I) = \varepsilon C l \tag{14}$$

where C is the concentration in moles per m^3, l is the optical path length in the sample, and I_0/I is the ratio of the incident intensity I_0 to the transmitted intensity I. The extinction coefficient is a function of the frequency of the light, and integrating over the absorption band $i(\omega_i \pm \Gamma_{on}/2)$, centered on the frequency of maximum absorption ω_i gives the optical absorbance as:

$$A = \frac{\log_e 10}{2\pi} \int_{-\Gamma_{on}/2}^{\Gamma_{on}/2} \varepsilon(\omega) \, d\omega \tag{15}$$

Finally for a narrow absorption, the absorption coefficient, k, can be related to the molar extinction coefficient by:

$$k(\omega) = \frac{6c\varepsilon(\omega) C \log_e 10}{2\pi n \omega_i} \tag{16}$$

where c is the velocity of light and n is the real part of the refractive index. The real and imaginary parts of the refractive index are related through a Kramers–Kronig relation, such that:

$$k(\omega) = \frac{2}{\pi} \int_0^\infty \frac{n(\omega)-1}{\omega - \omega_i} \, d\omega \tag{17}$$

Equations (14)–(17) apply to isotropic media. In an orientationally ordered material the extinction coefficient becomes dependent on the angle between the alignment axis and the polarization direction of the incident light, and has the characteristics of a second rank tensor. At a microscopic level, the optical absorption depends on the angle between the molecular transition dipole moment μ_i for the particular absorption band, and the electric field of the light wave. Restricting attention to uniaxial systems, an effective order parameter (S_{op}) for optical absorption can be defined as:

$$S = \frac{A_\parallel - A_\perp}{A_\parallel + 2A_\perp} \tag{18}$$

in which $A_\parallel$ and $A_\perp$ are the extinction coefficients for light polarized parallel or perpendicular to the director; the difference $A_\parallel - A_\perp$ is known as the linear dichroism. These extinction coefficients can be related to the transition dipole moments using the general result Eq. (45) of Sec. 1 of this chapter, such that:

$$A_\parallel = A_0 + B\left\{\frac{2}{3} S \left[(\mu_i)_n^2 - \frac{1}{2}\left((\mu_i)_l^2 + (\mu_i)_m^2\right)\right] \right.$$
$$\left. + \frac{1}{3} D \left((\mu_i)_l^2 - (\mu_i)_m^2\right)\right\} \tag{19}$$

$$A_\perp = A_0 - B\left\{\frac{1}{3} S \left[(\mu_i)_n^2 - \frac{1}{2}\left((\mu_i)_l^2 + (\mu_i)_m^2\right)\right] \right.$$
$$\left. + \frac{1}{6} D \left((\mu_i)_l^2 - (\mu_i)_m^2\right)\right\} \tag{20}$$

where $A_0 = \frac{1}{3}(A_\parallel + 2A_\perp)$ is the mean optical absorbance or the optical absorbance of the isotropic fluid. The factor B contains various fundamental constants, and μ_{il}, μ_{im} and μ_{in} are components of the transition moment along the principal axes of the molecule. Using the angles defined in Fig. 3 a simple manipulation of Eqs. (18)–(20) leads to the result [21]:

$$S_{op} = \frac{1}{2}(3\cos^2\beta - 1)S_\mu$$
$$- \frac{1}{2}(\sin^2\beta \cos 2\alpha) D_\mu \qquad (21)$$

where S_μ and D_μ are order parameters for the transition moment. A more complex result can be derived for biaxial samples using Eq. (47) of Sec. 1 of this chapter. S_{op} defined above is the order parameter for an optically absorbing mesogen; often the chromophore in a liquid crystal is not the mesogen but a solute dye molecule, which may or may not be mesogenic itself. There can still be dichroism because the liquid crystal host orders the dye molecule, but the order parameter of the dye may be substantially different from that of the host. In a mixture, the order parameters for the chromophore can in principle be related to the order parameters of the host material using the mean field theory of mixtures [23, 24].

It is sometimes useful to express the dichroism in terms of the dichroic ratio

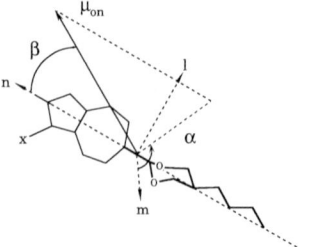

Figure 3. Orientation of the transition moment with respect to molecular axes for a dichroic mesogen [22].

$R = A_\parallel/A_\perp$. Assuming that there is no local biaxial order, the order parameter can be written in terms of R and the angle between the transition dipole and the ordered axis of the absorbing molecule:

$$R = \frac{1 + S(3\cos^2\beta - 1)}{1 - \frac{1}{2}S(3\cos^2\beta - 1)} \qquad (22)$$

For the special cases where the transition dipole is parallel or perpendicular to the molecular axis, the relationships between the dichroic ratio and the order parameter reduce to simple forms:

$$R(\beta = 0°) = \frac{1 + 2S}{1 - S}$$

$$R(\beta = 90°) = \frac{1 - S}{1 + \frac{1}{2}S} \qquad (23)$$

A particularly useful aspect of dichroic measurements is the chance to probe orientational order using more than one electronic transition in a molecule. Thus optical order parameters can be determined for different absorption bands, and if the transition moment directions are known, it is possible to determine both order parameters S and D. If it is assumed that there is a relationship between the uniaxial and biaxial order parameters, as given in Fig. 5 of Sec. 1 in this chapter, then it is possible to obtain both order parameters from the polarized spectra from a single absorption band [25]. This method has been applied [26] to the determination of order parameters of rigid aromatic probes, such as azulene, phenanthrene, and anthracene and related compounds. Dichroism measurements on impurity molecules in liquid crystal solvents have also been used [27, 28] to study intermolecular interactions, and their influence on electronic absorption bands. Polarization effects of the type described above for simple optical absorption spectroscopy can be observed and interpreted in a similar manner for many other types of spectroscopy such as Raman scattering [29] and resonance Raman scattering [30].

The Kramers–Kronig relationship between the real and imaginary parts of the refractive index shows that materials having a strong electronic absorption will tend to have a high refractive index. Conjugated mesogens or polarizable mesogens will therefore have relatively large refractive indices, and the birefringence will be determined by the polarization of the electronic absorptions. Some typical values for refractive indices of a range of different liquid crystals are given in Table 2. Materials with electronic absorptions in the UV at wavelengths less than 200 nm such as substituted bicyclohexanes will have small refractive indices and usually small birefringences. Changes in refractive indices with wavelength are also determined by the electronic absorptions for particular mesogens, and a polarized UV/visible spectrum for a standard liquid crystal is illustrated in Fig. 4. The dispersion in the corresponding refractive indices can be readily obtained from the following equations based on Drude's theory of optical dispersion:

$$n_e = 1 + n_{0e} + b_{1e}\left(\frac{1}{\lambda_1^2} - \frac{1}{\lambda^2}\right)^{-1} +$$

$$b_{2e}\left(\frac{1}{\lambda_2^2} - \frac{1}{\lambda^2}\right)^{-1} + \ldots$$

$$n_o = 1 + n_{0o} + b_{1o}\left(\frac{1}{\lambda_1^2} - \frac{1}{\lambda^2}\right)^{-1} +$$

$$b_{2o}\left(\frac{1}{\lambda_2^2} - \frac{1}{\lambda^2}\right)^{-1} + \ldots \quad (24)$$

Measured refractive indices for 5CB fitted to these equations are given in Fig. 5 [31].

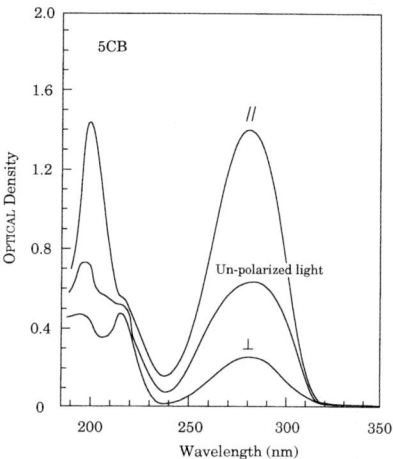

Figure 4. UV/visible spectrum of a mesogen (5CB).

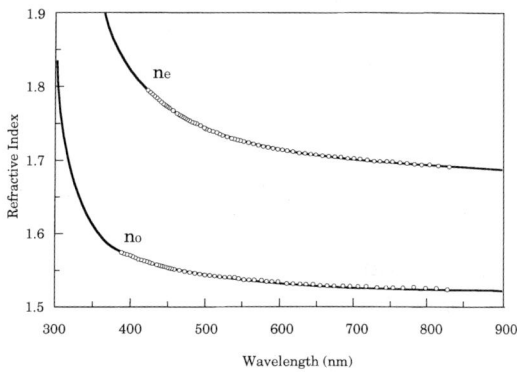

Figure 5. Fitted dispersion of refractive indices for 5CB.

3.4 Refractive Indices and Liquid Crystal Phase Structure

In Eqs. (9)–(11) the electronic polarizability is independent of temperature, so the temperature dependence of refractive indices is determined primarily by the order parameter, and to a lesser extent by changes in the density: the latter may be important at phase transitions. The variation of refractive indices of n-pentyloxyphenyl trans-4-noctylcyclohexanoate with temperature is given in Fig. 6, where the effect of phase changes is clearly seen. The changes mostly reflect changes in the order parameter at the transitions.

Thin films of oriented liquid crystals can act as optical wave guides, and examination of the eigenmodes of thin liquid crystal films can be used to obtain values for the real and imaginary parts of the refractive indices, as well as giving information on the director configuration in thin films [32]. The liquid crystal film is contained between two metallized (silver) reflecting surfaces, and the reflectivity is measured as a function of the angle and polarization of incident monochromatic light. The intensities of the reflected beam for light polarized in the plane of incidence (p) and perpendicular to the plane of incidence (s) are measured, and can be fitted to a model for the refractive indices, film thickness and director configuration. An example of experimental and fitted results is shown in Fig. 7 for the SmA phase of a commercial liquid crystal mixture Merck (UK) SCE3 [33].

A significant advance provided by guided mode experiments is the ability to meas-

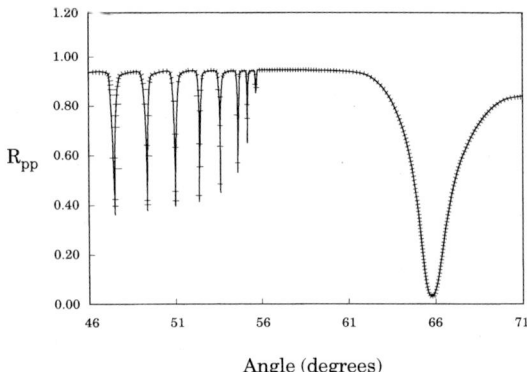

Figure 7. Reflectivity as a function of angle with fitted function for SCE3 at 72.8 °C (SmA); fitting parameters $\varepsilon_\parallel = 2.760$ and $\varepsilon_\perp = 2.208$.

ure not only the optical parameters of a liquid crystal film, but also the director configuration in complex geometries, such as the chevron structure in a SmC phase (see Fig. 8). The analysis of the optical response of complex liquid crystal structures is most conveniently achieved using the methods of matrix optics. There are a number of variants which can be used, depending on the particular problem, but the basic method is to represent an optical element as a matrix, which acts on the incident light, represented as a column vector, to give a resultant vector characteristic of the transmitted light. The approach is particularly suited to the geometries encountered in liquid crystal systems, since a complex optical structure can be split up into a series of elements each having its own characteristic matrix. Within each element it is assumed that the director is uniform, so the optical properties can be simply described in terms of principal refractive indices, the absorption coefficients for absorbing materials and the orientation of the optical indicatrix. The resultant response of the complex structure is then given by successive multiplication of the matrices for each element. The simplest method is that due to Jones, where the light wave

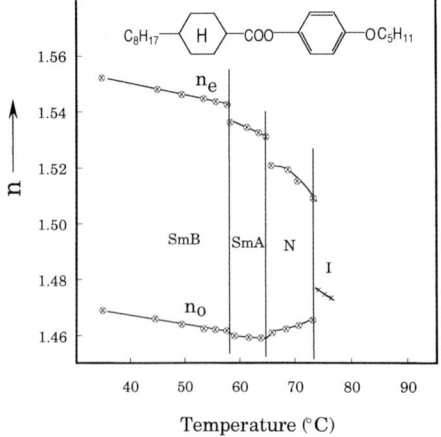

Figure 6. Refractive indices against temperature, including phase transitions [31]

is represented as a 2×1 column vector consisting of the electric field components E_x and E_y in a plane perpendicular to the propagation direction z. Thus the transmitted wave (E_{tx}, E_{ty}) through a birefringent element of thickness d and principal refractive indices n_e and n_o making an angle of α with the x axis can be written as:

$$\begin{bmatrix} E_{tx} \\ E_{ty} \end{bmatrix} = \begin{bmatrix} \cos\alpha & -\sin\alpha \\ \sin\alpha & \cos\alpha \end{bmatrix}$$
$$\begin{bmatrix} e^{-2\pi i n_e d/\lambda} & 0 \\ 0 & e^{-2\pi i n_o d/\lambda} \end{bmatrix}$$
$$\begin{bmatrix} \cos\alpha & \sin\alpha \\ -\sin\alpha & \cos\alpha \end{bmatrix} \begin{bmatrix} E_{ix} \\ E_{iy} \end{bmatrix}$$

or

$$\mathbf{E}_t = \mathbf{R}^{\text{trans}} \mathbf{B} \mathbf{R} \mathbf{E}_i \tag{25}$$

where the incident wave is represented by E_{ix}, E_{iy}, $\mathbf{R}$ is the rotation matrix and the matrix $\mathbf{B}$ represents the birefringent element. Extra elements such as polarizers and other birefringent elements can be included as appropriate matrix multipliers, and the transmitted light intensity is given by $I = E_{tx}^2 + E_{ty}^2$.

As an example of the application of the Jones matrices, the director configuration in a thin liquid crystal film of a SmC material has been investigated [34] by measuring the transmission of polarized light as a function of wavelength. Knowing the refractive indices of the material, the experimental results can be fitted to a model for the director configuration, as illustrated in the Fig. 8.

Another analytical method which had been extensively applied to liquid crystal structures was developed by Berreman [35]. It is based on a 4×4 matrix representation of optical elements [36], where the incident and transmitted light are described by a 4×1 column vector consisting of components of both the electric and magnetic field associated with the electromagnetic wave.

A widely used technique to study the properties of thin films is ellipsometry, and it has been used to investigate the structure of free-standing films of smectic liquid crystals consisting of only a few layers [37]. The method involves measuring the polarization characteristics of a transmitted or reflected beam of monochromatic light for different angles of incidence. Writing the phase difference between the s- and p-polarizations of the transmitted beam as $\Delta = \phi_p - \phi_s$, the value of Δ will depend on the integrated optical path difference across the smectic layers for light of the two polarizations. The technique has been used to probe the structure of ferroelectric and antiferro-

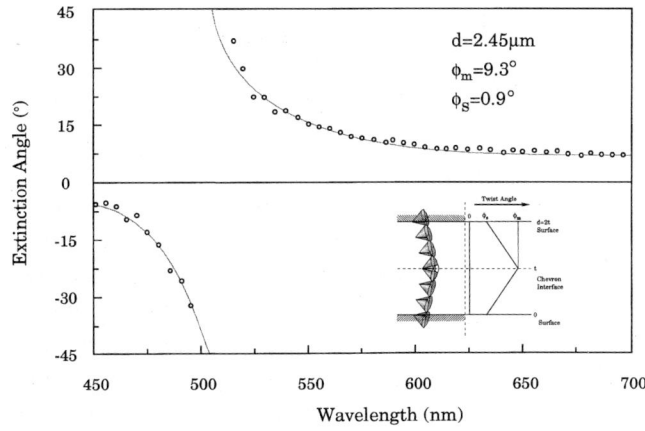

Figure 8. Fitted transmission against wavelength for a chevron structure [34].

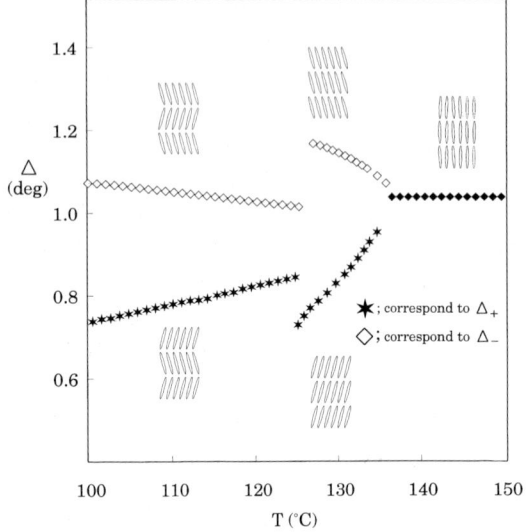

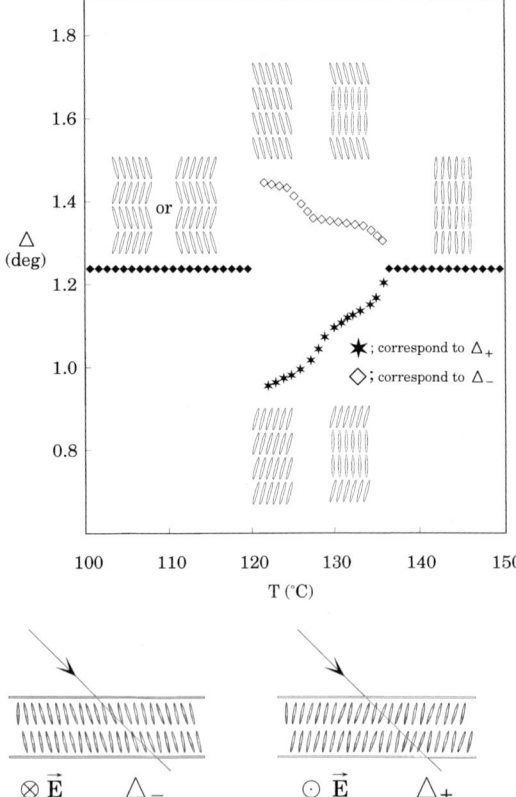

Figure 9. Measured optical path difference for three and four layer smectic structures of antiferroelectric, ferrielectric and ferroelectric SmC phases [37]. Bottom figure indicates angle of incidence for Δ_+ and Δ_- measurements.

electric smectic liquid crystals. For smectic films with alternating tilt directions, as in antiferroelectric or ferrielectric chiral SmC phases, the optical path difference will differ for odd or even numbers of layers. This is illustrated in Fig. 9 for a 3 and 4 layer film of MHPOBC, and the observed values for Δ measured in the antiferroelectric phase for 2 angles of incidence are consistent with the proposed structure for antiferroelectric phases.

3.5 Optics of Helicoidal Liquid Crystal Structures

In simple fluids of chiral molecules (molecules having structures such that mirror images are not superposable) the molecular chirality can be identified through the associated optical activity (i.e. the rotation of the plane of incident plane polarized light). If the light is of a frequency that corresponds with an electronic or vibrational absorption, then the fluid can exhibit differential absorption for left and right circularly polarized light, and the transmitted light is elliptically polarized. Optical activity can also be observed in differential scattering of left or right circularly polarized light. Chiral liquid crystals as well as exhibiting optical activity have a number of characteristic chiral properties, the most important of which is a tendency for the phases to develop helicoidal structures. Thus optical effects due to chirality in liquid crystals can be due to molecular chirality or a consequence of the helicoidal structure. This may seem to be an unnecessary distinction, since the helicoidal structures of chiral liquid crystal phases are a result of chiral interactions between molecules. However there is a difference when it comes to calculating the optical response of chiral liquid crystals, which can usually be modelled by twisted layers of linearly birefringent material; the intrinsic molecular optical activity of a liquid crystal is normally neglected in modelling the optical properties of chiral liquid crystals. For example the specific rotation measured in isotropic solution of a typical liquid crystal material such as CE6 ($C_{10}H_{21}OC_6H_4COOC_6H_4CH_2CH(CH_3)C_2H_5$) is $3 \cdot \text{cm}^{-1}$, while the optical rotation of an aligned film of the chiral nematic phase of such a material, which arises primarily from the helicoidal structure, is around $10^4 \cdot \text{cm}^{-1}$.

One optical feature of helicoidal structures is the ability to rotate the plane of incident polarized light. Since most of the characteristic optical properties of chiral liquid crystals result from the helicoidal structure, it is necessary to understand the origin of the chiral interactions responsible for the twisted structures. The continuum theory of liquid crystals is based on the Frank–Oseen approach to curvature elasticity in anisotropic fluids. It is assumed that the free energy is a quadratic function of curvature elastic strain, and for positive elastic constants the equilibrium state in the absence of surface or external forces is one of zero deformation with a uniform, parallel director. If a term linear in the twist strain is permitted, then spontaneously twisted structures can result, characterized by a pitch p, or wave-vector $\mathbf{q} = 2\pi p^{-1} \mathbf{i}$, where $\mathbf{i}$ is the axis of the helicoidal structure. For the simplest case of a nematic, the twist elastic free energy density can be written as:

$$g = -k_2(\mathbf{n} \cdot \nabla \times \mathbf{n}) + k_{22}(\mathbf{n} \cdot \nabla \times \mathbf{n})^2 \quad (26)$$

and the pitch of the corresponding chiral nematic is given by $p^{-1} = k_2/2\pi k_{22}$. The optical properties of such twisted structures have been determined using a model of twisted layers of linearly birefringent material, which predicts a variety of optical response depending on the value of $p\Delta n$ in comparison with the wavelength λ. For long pitch materials such that $p\Delta n \gg 1$, the structure behaves as a rotator, and linear polarized modes rotate with the helix: this is the mode utilized in twisted nematic displays. For shorter pitches the optical response is more complicated, but analytical results can be obtained for normal incidence, leading to the de Vries equation [38] for the optical rotation per unit length $\rho = \phi/d$:

$$\rho = -\frac{\pi}{4p}\left(\frac{n_e^2 - n_o^2}{n_e^2 + n_o^2}\right)^2 \left(\frac{\lambda_o}{\lambda}\right)^2 \left(1 - \left(\frac{\lambda}{\lambda_o}\right)^2\right)^{-1} \quad (27)$$

This equation predicts that the optical rotation diverges at a critical wavelength $\lambda_o = np$. The variation of ρ with wavelength is illustrated schematically in Fig. 10.

However the de Vries equation is not valid in the region of λ_o, which corresponds to a total reflection of circularly polarized light having the same sense as the helical pitch. This is often referred to as Bragg reflection, by analogy with X-ray diffraction, but only first order reflections are allowed for nor-

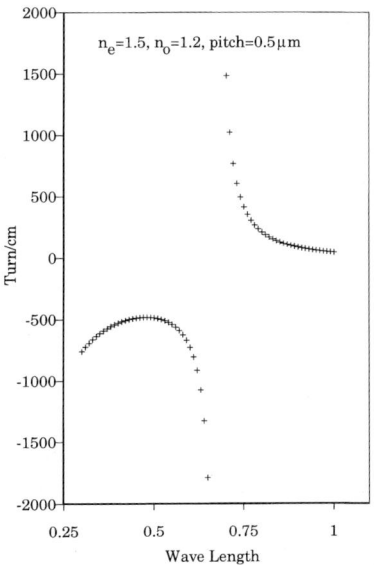

Figure 10. Variation of ρ with $\lambda/\mu\text{m}$ according to Eqn. 27.

mal incidence. The band width for total reflection centered on a wavelength of λ_o is $\Delta\lambda = p\Delta n$, and the sign of the optical rotation reverses on crossing the band at λ_o. It is observed that the wavelength of reflected light varies as a function of the angle of incidence, but an analytical solution of the optic response for this situation is not possible. Away from normal incidence, all orders of reflection are permissible [39], but an approximate result for the angular dependence of the wavelength of reflected light is

$$\lambda(\theta) = \lambda(0)\cos\left(\sin^{-1}\frac{\theta}{n}\right) \quad (28)$$

From the above it is clear that the optical response of helicoidal structures of liquid crystal molecules depends on the pitch, and in order to relate these optical properties to molecular structures, the dependence of pitch on molecular structure must be considered.

Chiral liquid crystal phases readily form in mixtures of chiral and nonchiral materials, and for mixtures of a chiral dopant and a nonchiral host it is convenient to define a twisting power b, which is a measure of the pitch induced per unit concentration of chiral dopant:

$$b = \frac{dp^{-1}}{dc} \quad (29)$$

The twist induced by different molecular species can be qualitatively related to the molecular structure of the chiral species [40]. The twist induced in a nonchiral liquid crystal solvent by a chiral dopant also depends on the nature of the solvent, and it has been proposed that chiral dopants can preferentially promote chiral conformations of the solvent molecules [41]. This effect has also been observed in isotropic solutions, where an enhanced optical rotation in solutions of a chiral biaryl in a cyanobiphenyl solvent was attributed [42] to an induction of chirality via preferential interactions between solute and solvent conformations of the same chirality. Attempts have been made to relate the twisting power to some molecular measure of chirality based on a variety of geometric indices [43], but with only limited success.

Chiral interactions in the isotropic phase of liquid crystals are clearly seen in the pretransitional increase in optical activity observed at the isotropic to chiral nematic or chiral smectic phase transition. This was first observed by Cheng and Meyer [44] and explained by them in terms of fluctuations in the off-diagonal elements of the correlation function for fluctuations in the ordering matrix [45]. There are five fluctuation modes which can contribute to the pretransitional optical activity, and the experimentally observed behaviour depends on the relative amplitudes of these modes and coupling between them [46]. For some systems of high chirality there is a reversal in the sign of the pretransitional optical activity which has been attributed to mode-coupling; similar results have been obtained in the pretransitional region of smectic phases [47]. Scattering of circularly polarized light has also been used [46] to probe different chiral fluctuation modes, and by selection of combinations of incident right or left and scattered left or right circularly polarized light it is possible to identify scattering from three of the five individual modes.

The imaginary part of the complex refractive indices for left and right circularly polarized light relates to circular dichroism, that is differential absorption for light of different circular polarizations. It is treated in a similar manner to linear dichroism, except that the definition of principal components follows a different convention. For linear birefringence and dichroism the principal values of the complex refractive index relate to the electric field polarization direction which is transverse to the propagation direction (TE); principal components of the circular dichroism tensor are defined for the propagation direction. Thus components of the transition moments will contribute differently to corresponding principal values of the linear dichroism and circular dichroism tensors [48].

The circular dichroism (CD) ΔA_{CD} is defined as the difference in the optical absorption for left and right circularly polarized light, and for small values of the dichroism this gives rise to a corresponding ellipticity ψ for transmitted plane polarized light:

$$\Delta A_{CD} = k_l - k_r \qquad (30)$$

$$\psi = \frac{(k_l - k_r)\omega d}{2c} \qquad (31)$$

where ω is the frequency of the light in radians/sec, d is the pathlength and c the velocity of light.

The use of CD as a probe of liquid crystalline properties has been rather limited. A helicoidal structure will induce circular dichroism at an absorption band of a nonchiral chromophore, and the magnitude of the induced CD absorption depends on the pitch of the helix, the sign of the CD changing if pitch inversion occurs. This technique has been used to investigate phase transitions between ferrielectric, ferroelectric and antiferroelectric smectic C* states of MHPOBC ($C_8H_{17}OC_6H_4COOC_6H_4COOCH(CH_3)C_6H_{13}$) [49]. Spectra were recorded in the visible for a dissolved nonchiral dye molecule, and a small induced CD was detected in the UV originating in the intrinsic absorption of MHPOBC. A change in sign of the CD of the dye molecule was observed at the phase transition between ferroelectric and ferrielectric states. Measurements were made on 100 μm cells homeotropically aligned along the helix axis, to avoid complications from linear birefringence or dichroism effects.

3.6 References

[1] I.-C. Khoo, S.-T. Wu, *Optics and Nonlinear Optics of Liquid Crystals*, World Scientific, Singapore, 1993.
[2] J. F. Nye, *Physical Properties of Crystals*, Oxford University Press, Oxford, **1985**.
[3] N. H. Hartshorne, *Microscopy of Liquid Crystals*, Microscope Publications, London, **1974**.
[4] T. E. Lockhart, D. W. Allender, E. Gelerinter, D. C. Johnson, *Phys. Rev.* **1979**, *A20*, 1655.
[5] F. Yang, J. R. Sambles, *Liq. Cryst.* **1993**, *13*, 1.
[6] D. A. Dunmur, D. A. Hitchen, Xi-Yun Hong, *Mol. Cryst. Liq. Cryst.* **1986**, *140*, 303.
[7] V. P. Arora, S. A. Prakash, V. K. Garwal, B. Bahadur, *Ind. J. Appl. Phys.* **1992**, *30*, 406.
[8] K. R. Welford, J. R. Sambles, *Appl. Phys. Lett.* **1987**, *50*, 871.
[9] C. R. Lavers, *Liq. Cryst.* **1992**, *11*, 819.
[10] M. Sheik-Bahae, A. A. Said, T.-H. Wei, D. J. Hagan, E. W. Van Stryland, *IEEE J. Quant. Electron.* **1990**, *26*, 760.
[11] P. Palffy-Muhoray, H. J. Yuan, L. Li, *Mol. Cryst. Liq. Cryst.* **1991**, *199*, 223.
[12] P. Palffy-Muhoray, D. A. Balzarini, *Can. J. Phys.* **1981**, *59*, 515.
[13] K. C. Lim, J. T. Ho, *Mol. Cryst. Liq. Cryst.* **1978**, *47*, 173.
[14] D. W. Bruce, D. A. Dunmur, P. M. Maitlis, M. R. Manterfield, R. Orr, *J. Mater. Chem.* **1991**, *1*, 288.
[15] M. Warenghem, C. P. Grover, *Rev. Phys. Appl.* **1988**, *23*, 1169.
[16] M. Warenghem, G. Joly, *Mol. Cryst. Liq. Cryst.* **1991**, *207*, 205.
[17] M. Boschmans, *Mol. Cryst. Liq. Cryst.* **1991**, *199*, 267.
[18] I. Haller, H. A. Huggins, H. R. Lillenthal, T. R. McGuire, *J. Phys. Chem.* **1973**, *77*, 950.
[19] R. G. Horn, *J. Phys.* **1978**, *39*, 167.
[20] R. G. Horn, T. E. Faber, *Proc. Roy. Soc. Lond.* **1979**, *A368*, 199.
[21] E. H. Korte, *Mol. Cryst. Liq. Cryst.* **1983**, *100*, 127.
[22] R. Brettle, D. A. Dunmur, S. Estdale, C. M. Mason, *J. Mater. Chem.* **1993**, *3*, 327.
[23] P. Palffy-Muhoray, D. A. Dunmur, D. A. Balzarini, *Ordered Fluids and Liquid Crystals* (Ed.: A. C. Griffin), Plenum Press, New York, **1984**, *4*, p. 615.
[24] E. M. Averyanov, *Nuovo Cim* **1990**, *12*, 1281.
[25] H.-G. Kuball, R. Memmer, A. Straus, M. Junge, G. Scherowsky, A. Schonhofer, *Liq. Cryst.* **1989**, *5*, 969.
[26] H.-G. Kuball, M. Junge, B. Schulteis, A. Scholhofer, *Ber. Bunsenges. Phys. Chem.* **1991**, *95*, 1219.
[27] E. M. Averyanov, V. M. Muratov, V. G. Rumyantsev, *Sov. Phys. JETP* **1985**, *61*, 476.

[28] E. M. Averyanov, V. M. Muratov, V. G. Rumyantsev, V. A. Churkina, *Sov. Phys. JETP* **1986**, *63*, 57.
[29] A. A. Minko, V. S. Rachevich, S. Ye. Yakovenko, *Liq. Cryst.* **1989**, *4*, 1.
[30] S. Yakovenko, R. Ignatovich, *Mol. Cryst. Liq. Cryst.* **1990**, *179*, 93.
[31] G. Pelzl in *Liquid Crystals* (Ed.: H. Stegemeyer), Sternkopff, Darmstadt, Germany, **1994**, p. 51.
[32] E. L. Wood, J. R. Sambles, S. J. Elston, *J. Mod. Opt.* **1991**, *38*, 1385.
[33] S. J. Elston, J. R. Sambles, M. G. Clark, *J. Mod. Opt.* **1989**, *36*, 1019.
[34] M. H. Anderson, J. C. Jones, E. P. Raynes, M. J. Tower, *J. Phys. D.: Appl. Phys.* **1991**, *24*, 338.
[35] D. W. Berremen, *J. Opt. Soc. Am.* **1973**, *63*, 1374.
[36] R. M. A. Azzam, N. M. Bashara, *J. Opt. Soc. Am.* **1972**, *62*, 1252.
[37] Ch. Bahr, D. Flieger, *Phys. Rev. E* **1993**, *70*, 1842.
[38] G. Vertogen, W. H. de Jeu, *Thermotropic Liquid Crystals, Fundamentals 1988*, Springer-Verlag, Berlin, Heidelberg, p. 133.
[39] R. Dreher, G. Meier, *Phys. Rev.* **1973**, *A8*, 1616.
[40] G. Solladie, R. G. Zimmerman, *Angew. Chem., Int. Ed. Engl.* **1984**, *23*, 348.
[41] G. Gottarelli, *Liq. Cryst.* **1986**, *1*, 29.
[42] G. Gottarelli, M. A. Osipov, G. P. Spada, *J. Phys. Chem.* **1991**, *95*, 3879.
[43] L. A. Kutulya, V. E. Kuzmin, I. B. Stelmakh, T. V. Handrimailova, P. P. Shtifanyuk, *J. Phys. Org. Chem.* **1992**, *5*, 308.
[44] J. Cheng, R. B. Meyer, *Phys. Rev. Lett.* **1972**, *29*, 1240.
[45] J. Cheng, R. B. Meyer, *Phys. Rev. A* **1974**, 2744.
[46] P. R. Battle, J. D. Miller, P. J. Collings, *Phys. Rev.* **1987**, *A36*, 369.
[47] K. C. Frame, J. L. Walker, P. J. Collings, *Mol. Cryst. Liq. Cryst.* **1991**, *198*, 91.
[48] H.-G. Kuball, S. Neubrech, A. Schonhofer, *J. Spectroscopy* **1989**, *10*, 91.
[49] J. Watanabe, *J. Phys. Condens. Matter* **1990**, *2*, SA271–SA274.

4 Dielectric Properties

David Dunmur and Kazuhisa Toriyama

The purpose of this Chapter is to describe the dielectric properties of liquid crystals, and relate them to the relevant molecular properties. In order to do this, account must be taken of the orientational order of liquid crystal molecules, their number density and any interactions between molecules which influence molecular properties. Dielectric properties measure the response of a charge-free system to an applied electric field, and are a probe of molecular polarizability and dipole moment. Interactions between dipoles are of long range, and cannot be discounted in the molecular interpretation of the dielectric properties of condensed fluids, and so the theories for these properties are more complicated than for magnetic or optical properties. The dielectric behavior of liquid crystals reflects the collective response of mesogens as well as their molecular properties, and there is a coupling between the macroscopic polarization and the molecular response through the internal electric field. Consequently, the molecular description of the dielectric properties of liquid crystals phases requires the specification of the internal electric field in anisotropic media which is difficult.

4.1 Dielectric Response of Isotropic Fluids

The various factors which influence the dielectric properties of a liquid crystal can be identified by recalling the results for isotropic fluids. The Debye equation (1) for an isotropic fluid relates the permittivity to the mean polarizability ($\bar{\alpha}$) and molecular dipole moment (μ):

$$(\varepsilon - 1) = \frac{NFh}{\varepsilon_0}\left[\bar{\alpha} + \frac{\mu^2 F}{3k_B T}\right] \quad (1)$$

where F and h are reaction field and cavity field factors which account for the field dependent interaction of a molecule with its environment, and N is the number density. Molecular contributions to the permittivity are approximately additive, since $(\varepsilon - 1)$ is proportional to N, but the internal field factors for the reaction field (F) and cavity field (h) are also density dependent. For isotropic fluids, the internal field factors are given by:

$$F = (1 - \alpha f)^{-1} \text{ and } f = \frac{2(\varepsilon - 1)}{4\pi\varepsilon_0 a^3(2\varepsilon + 1)}$$

$$h = \frac{3\varepsilon}{(2\varepsilon + 1)} \quad (2)$$

where a is the radius of the spherical cavity which accommodates the molecule. Using the Lorenz–Lorentz equation for the isotropic polarizability, Eq. (8) of Sec. 3 of this chapter gives for the reaction field factor:

$$F = \frac{(2\varepsilon+1)(n^2+2)}{3(2\varepsilon+n^2)} \quad (3)$$

Specific pairwise dipole–dipole interactions can be accounted for by introducing the Kirkwood correlation factor g_1, such that the mean square dipole moment is replaced by an effective mean square moment defined by:

$$\mu_{\text{effective}}^2 = g_1 \mu^2 \quad (4)$$

and this correlation factor g_1 can be related to the spatial dipole correlation function $G(\mathbf{r})$ in the fluid:

$$g_1 = 1 + V^{-1} \int G_1(\mathbf{r}) d\mathbf{r} \text{ and}$$

$$G_1(\mathbf{r}) = \frac{\langle \mu(0)\mu(\mathbf{r}) \rangle}{\langle \mu(0)\mu(0) \rangle} \quad (5)$$

The Kirkwood–Frohlich equation incorporates this factor, and enables the mean square effective dipole moment to be deduced from measurements of the electric permittivity, refractive index and number density of a fluid:

$$\frac{(\varepsilon-n^2)(2\varepsilon+n^2)}{\varepsilon(n^2+2)^2} = \frac{N g_1 \mu^2}{9\varepsilon_0 k_B T} \quad (6)$$

The electric permittivity determines the polarization (dipole moment per unit volume) induced in a material by an electric field. If the applied field varies with time, then the frequency dependence of the permittivity is an additional property of the material. A complication with any time-dependent response is that it may not be in-phase with the applied field. Thus to describe the frequency-dependent dielectric response of a material, the amplitude and phase of the induced polarization must be measured. A convenient way of representing phase and amplitude is through complex notation, so that ε' (real) measures the in-phase response, and ε'' (imaginary) measures the 90° out-of-phase response:

$$\varepsilon^*(\omega) = \varepsilon'(\omega) - i\varepsilon''(\omega) \quad (7)$$

and the phase angle is $\tan^{-1}(\varepsilon''/\varepsilon')$. The effective response of a molecule to an alternating field of frequency ω, assuming a single molecular dipole relaxation, can be described through the complex permittivity as:

$$\varepsilon^*(\omega) - 1 = (1 + i\omega\tau)^{-1} \frac{N\mu^2}{3\varepsilon_0 k_B T} \quad (8)$$

which gives:

$$\varepsilon'(\omega) - 1 = \left[1 + \omega^2\tau^2\right]^{-1} \frac{N\mu^2}{3\varepsilon_0 k_B T};$$

$$\varepsilon''(\omega) = \omega\tau\left[1 + \omega^2\tau^2\right]^{-1} \frac{N\mu^2}{3\varepsilon_0 k_B T} \quad (9)$$

The time τ is the relaxation time for dipole reorientation in an electric field of frequency ω (radians s^{-1}). For real systems there may be a number of contributions to the electric permittivity, each relaxing at a different frequency, for example due to internal dipole motion in flexible molecules or collective dipole motion. If these contributions to the electric permittivity are at sufficiently different frequencies, they can be separated in the dielectric spectrum, and it is possible to apply Eq. (9) to each relaxation process. At low frequencies ($\omega \to 0$), the orientation polarization contribution to the permittivity is $\frac{N\mu^2}{3\varepsilon_0 k_B T}$, neglecting any internal field effects, while at high frequencies ($\omega \to \infty$), the molecular dipoles do not rotate fast enough to contribute to the dielectric response. More generally the real and imaginary parts of the permittivity can be expressed as:

$$\varepsilon^*(\omega) - \varepsilon'(\infty) = \frac{[\varepsilon'(0) - \varepsilon'(\infty)]}{[1 + i\omega\tau]}$$

$$\varepsilon'(\omega) - \varepsilon'(\infty) = \frac{[\varepsilon'(0) - \varepsilon'(\infty)]}{[1 + \omega^2\tau^2]};$$

$$\varepsilon''(\omega) = \frac{\omega\tau[\varepsilon'(0) - \varepsilon'(\infty)]}{[1 + \omega^2\tau^2]} \quad (10)$$

where $\dfrac{N\mu^2}{3\varepsilon_0 k_B T} = [\varepsilon'(0) - \varepsilon'(\infty)]$.

These equations, due to Debye, can be used to describe any relaxation process in a material, but in such cases the frequencies $\omega=0$ and $\omega=\infty$ refer to frequencies below and above the relaxation frequency $\omega_0 = \tau^{-1}$, as illustrated in Fig. 1.

If the variable $\omega\tau$ is eliminated from Eq. (10), we obtain:

$$\varepsilon''(\omega)^2 + \left\{\varepsilon'(\omega) - \frac{1}{2}[\varepsilon'(0) + \varepsilon'(\infty)]\right\}^2$$

$$= \frac{1}{4}[\varepsilon'(0) - \varepsilon'(\infty)]^2 \quad (11)$$

which is the equation of a circle of radius $\frac{1}{2}[\varepsilon(0) + \varepsilon(\infty)]$ centered on the point $\varepsilon''(\omega) = 0$, $\varepsilon'(\omega) = \frac{1}{2}[\varepsilon(0) + \varepsilon(\infty)]$. This represen-

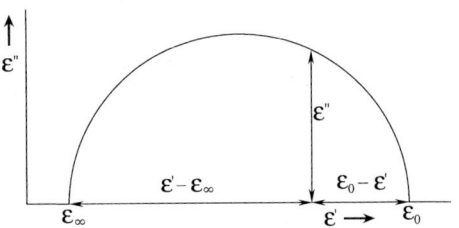

Figure 2. Cole–Cole plot of the real and imaginary parts of the permittivity.

tation of the real and imaginary parts of the electric permittivity is known as a Cole–Cole plot, and is illustrated in Fig. 2.

In plotting experimental data it is sometimes more convenient to use the Debye–Pellat equations which are obtained by rearranging Eq. (10) thus:

$$\varepsilon'(\omega) = \varepsilon'(\infty) + \frac{\varepsilon''(\omega)}{\omega\tau}$$

$$\varepsilon'(\omega) = \varepsilon'(0) - \omega\tau\varepsilon''(\omega) \quad (12)$$

since these provide a simple way to determine the low $\varepsilon'(0)$ and high $\varepsilon'(\infty)$ frequency contributions to the real part of the permittivity.

The measured frequency dependences of $\varepsilon''(\omega)$ and $\varepsilon'(\omega)$ in real fluids do not always fit the Debye–Pellat equations, and many methods [1] have been proposed to analyse skewed or displaced Cole–Cole plots. Debye's theory of dipole relaxation assumes that rotational motion can be described in terms of a single relaxation time. In a real system, fluctuations in the local structure of a molecule or its environment may result in a distribution of relaxation times about the Debye value, and such a situation can be described by a modification to Eq. (10)

$$\varepsilon^*(\omega) - \varepsilon'(\infty) = \frac{[\varepsilon'(0) - \varepsilon'(\infty)]}{[1 + (i\omega\tau)^{1-\alpha}]} \quad (13)$$

where α is a parameter introduced by Cole and Cole [2]. The effect of α is to produce

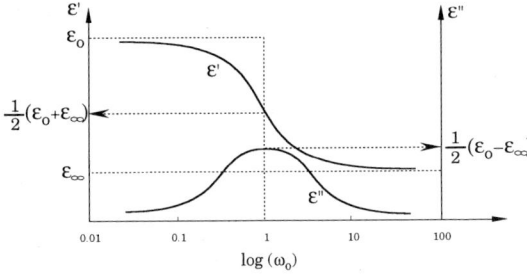

Figure 1. Schematic plot of the real and imaginary parts of the complex permittivity as a function of relative frequency $\omega_0 = \omega\tau$.

a semicircular Cole–Cole plot, the center of which is depressed below the abscissa. Asymmetric or skewed plots of $\varepsilon''(\omega)$ against $\varepsilon'(\omega)$ can sometimes be fitted by the Cole–Davidson equation:

$$\varepsilon^*(\omega) - \varepsilon'(\infty) = \frac{[\varepsilon'(0) - \varepsilon'(\infty)]}{[1 + (i\omega\tau)]^\beta} \quad (14)$$

where β is another parameter which is a measure of an asymmetric distribution of relaxation times. Another empirical functional form which is sometimes used to describe a non-Debye like relaxation is due to Fuoss and Kirkwood. This function, Eq. (15), gives a symmetric Cole–Cole plot, but depressed with respect to the result for a single relaxation: (α here is not the same as the α parameter in Eq. (13))

$$\varepsilon''(\omega) = \frac{1}{2}\alpha(\varepsilon(0) - \varepsilon(\infty))\operatorname{sech}\alpha(\ln\omega\tau) \quad (15)$$

In some isotropic liquids, and quite generally in polar liquid crystals, the plot of $\varepsilon''(\omega)$ or $\varepsilon'(\omega)$ against frequency shows evidence of two or more separate relaxation times. In some circumstances these may be well separated in frequency giving distinct Cole–Cole arcs, but more usually they overlap to give a composite Cole–Cole arc. The simplest analysis of such measurements is to assume that the relaxation processes contribute additively to the permittivity so that the complex permittivity can be written:

$$\varepsilon^*(\omega) - \varepsilon'(\infty) = \sum_j \frac{x_j}{[1 + i\omega\tau_j]} \quad (16)$$

where x_j is a weighting factor for each relaxation centred at frequency $\omega_j = \tau_j^{-1}$. Using this, nonsemi-circular Cole–Cole plots can be analysed in terms of a sum of contributions (see Fig. 3).

In deriving the macroscopic equations from the microscopic result, Eq. (8) the effect of the environment on a rotating molec-

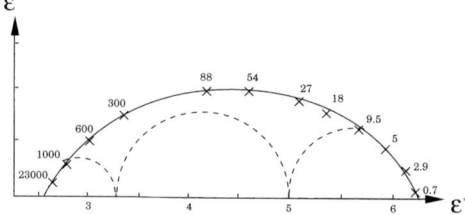

Figure 3. Cole–Cole plot of the perpendicular component of the permittivity for the nematic phase of 4-heptyl-4-cyanobiphenyl [3].

ular dipole has been neglected. This does not invalidate the Debye–Pellat equations or Cole–Cole plot, but requires a different interpretation of the relaxation time. Thus $\omega_j = \tau_j^{-1}$ is no longer the angular velocity of a rotating dipole, but is a macroscopic relaxation frequency. For an isotropic fluid an approximate relationship between the macroscopic and microscopic relaxation times is [4]:

$$\tau_{\text{macroscopic}} = \frac{3\varepsilon(0)}{2\varepsilon(0) + \varepsilon(\infty)}\tau_{\text{microscopic}} \quad (17)$$

but this result is not applicable to liquid crystals.

If the rotating molecule is assumed to be rigid and axially symmetric, then the dipole will lie along the symmetry axis, and the reorientation of the dipole in an electric field will be governed by the rotational motion of the molecule about an axis perpendicular to the symmetry axis. This in turn can be related to a rotational diffusion constant $D_\perp$:

$$\tau_{\text{molecular}} = (2D_\perp)^{-1} = I(2k_B T \tau_J)^{-1} \quad (18)$$

where τ_J is the relaxation time for the angular momentum about the short axis, and I is the moment of inertia. Assuming an isotropic distribution of angular momenta, Debye showed that τ_J could be simply related to the moment of inertia, an effective molecular radius (a) and a microscopic viscosity (η), giving:

$$\tau_{\text{molecular}} = \frac{4\pi \eta a^3}{k_B T} \quad (19)$$

Various extensions of this to other molecular shapes have been reported [5].

In the above discussion of the frequency dependent permittivity, the analysis has been based on either the single particle rotational diffusion model of Debye, or empirical extensions of this model. A more general approach can be developed in terms of time correlation functions [6], which in turn have to be interpreted in terms of a suitable molecular model. While using the correlation function approach does not simplify the analysis, it is useful, since experimental correlation functions can be compared with those deduced from approximate theories, and perhaps more usefully with the results of molecular dynamics simulations. Since the use of correlation functions will be mentioned in the context of liquid crystals, they will be briefly introduced here. The dipole–dipole time correlation function $C(t)$ is related to the frequency dependent permittivity through a Laplace transform such that:

$$\frac{\varepsilon^*(\omega) - n^2}{\varepsilon(0) - n^2} \left[\frac{\left(2\varepsilon'(\omega) + n^2\right)\varepsilon'(0)}{\left(2\varepsilon'(0) + n^2\right)\varepsilon'(\omega)} \right]$$
$$= 1 - i\omega \int_0^\infty C(t) e^{i\omega t} \, dt \quad (20)$$

where $C(t)$ is the dipole time correlation function defined by:

$$C(t) = \frac{\langle \mu(0)\mu(t) \rangle}{\langle \mu(0)\mu(0) \rangle} \quad (21)$$

4.2 Dielectric Properties of Anisotropic Fluids

In an orientationally ordered fluid the electric permittivity becomes a second rank tensor $\varepsilon_{\alpha\beta}$, and an electric susceptibility can be defined which relates the induced polarization P to the applied electric field E:

$$P_\alpha = \varepsilon_0 (\varepsilon_{\alpha\beta} - \delta_{\alpha\beta}) E_\beta$$
$$= \varepsilon_0 \chi_{\alpha\beta}^{\text{electric}} E_\beta \quad (22)$$

Thus the number of independent components of the permittivity tensor will depend on the symmetry of the liquid crystal phase. The frequency dependence of the permittivity is described in terms of real and imaginary parts, and these also will be tensor quantities. Apart from complications of anisotropic internal fields, the static or low frequency part of the permittivity tensor can be related to the molecular polarizability and dipole moment averaged over the appropriate orientational distribution functions.

4.2.1 The Electric Permittivity at Low Frequencies: The Static Case

4.2.1.1 Nematic Phase

In this section we wish to consider all the possible contributions to the electric permittivity of liquid crystals, regardless of the time-scale of the observation. Conventionally this permittivity is the static dielectric constant (i.e. it measures the response of a system to a d.c. electric field) in practice experiments are usually conducted with low frequency a.c. fields to avoid conduction and space charge effects. For isotropic dipolar fluids of small molecules, the permittivity is effectively independent of frequency below 100 MHz, but for liquid crystals it may be necessary to go below 1 kHz to measure the static permittivity; polymer liquid crystals can have relaxation processes at very low frequencies.

The electric susceptibility for anisotropic materials is a second rank tensor, and we can use the general results of Sec. 1 of this chapter. Firstly we will only consider the anisotropic part of the permittivity:

$$\chi^{(a)}_{\alpha\beta} = \chi_{\alpha\beta} - \frac{1}{3}\chi_{\gamma\gamma}\delta_{\alpha\beta} \qquad (23)$$

The principal components of this can be evaluated in terms of order parameters and molecular properties as given in Eq. (47) of Sec. 1 of this chapter. All that remains is to evaluate the components of the molecular susceptibility tensor $\kappa^{(\text{electric})}_{\alpha\beta}$. This contains contributions from (1) the dipole induced by the internal electric field $\boldsymbol{E}^{(\text{int})}$, and (2) the orientation polarization arising from the partial alignment of dipoles by the directing field $\boldsymbol{E}^{(\text{dir})}$. Thus in molecular axes the principal components of the microscopic electric susceptibility tensor will be obtained from:

$$\kappa^{(\text{electric})}_{ii} E_i = \alpha_{ii} E_i^{(\text{int})} + \mu_i \qquad (24)$$

If a molecule is freely rotating then the average dipole moment along any axis will be zero. The effect of an electric field is to break the $\pm$ symmetry of the axis, and the value of μ_i is the average over the $+i$ and $-i$ directions weighted by the Boltzmann energy associated with the directing electric field, so that:

$$\bar{\mu}_i = \frac{\mu_{+i}\exp-\left(\dfrac{u_0+\mu_i E_i^{(\text{dir})}}{k_B T}\right) + \mu_{-i}\exp-\left(\dfrac{u_0-\mu_i E_i^{(\text{dir})}}{k_B T}\right)}{\exp-\left(\dfrac{u_0+\mu_i E_i^{(\text{dir})}}{k_B T}\right) + \exp-\left(\dfrac{u_0+\mu_i E_i^{(\text{dir})}}{k_B T}\right)} \qquad (25)$$

where $\mu_{-i} = -\mu_{+i}$. Assuming that $\mu_i E_i^{(\text{dir})} \ll k_B T$, expanding the exponentials gives:

$$\bar{\mu}_i = \frac{\mu_i^2 E_i^{(\text{dir})}}{k_B T} \qquad (26)$$

and using this in Eq. (24) along with the results for $E^{(\text{int})}$ and $E^{(\text{dir})}$ obtained for isotropic liquids gives:

$$\kappa^{(\text{electric})}_{ii} = Fh\left[\alpha_{ii} + \frac{\mu_i^2 F}{k_B T}\right] \qquad (27)$$

This can be substituted into Eq. (47) of Sec. 1 to obtain the anisotropic part of the electric susceptibility tensor. The isotropic part of $\chi^{(\text{electric})}_{\alpha\beta}$ is the result obtained for isotropic fluids:

$$\frac{1}{3}\chi^{(\text{electric})}_{\gamma\gamma} = \frac{NFh}{\varepsilon_0}\left[\bar{\alpha} + \frac{\mu^2 F}{3k_B T}\right] \qquad (28)$$

so the final result for the principal components of the permittivity tensor can be obtained from Eqs. (27) of this section and (47) of Sec. 1 to give:

$$\chi^{(\text{electric})}_{\parallel} = \varepsilon_{\parallel} - 1 = \varepsilon_0^{-1} NFh \bigg\{ \bar{\alpha} + \frac{2}{3}\alpha_1 S + \frac{1}{3}\alpha_2 D$$

$$+ \frac{Fg_1^{(\parallel)}}{3k_B T}[\mu_z^2(1+2S) + \mu_y^2(1-S-D) + \mu_x^2(1-S+D)] \bigg\} \qquad (29\text{a})$$

$$\chi_{\perp'}^{(\text{electric})} = \varepsilon_{\perp'} - 1 = \varepsilon_0^{-1} NFh \left\{ \bar{\alpha} - \frac{1}{3} \alpha_1 (S+P) - \frac{1}{6} \alpha_2 (D+C) \right. \tag{29b}$$
$$\left. + \frac{Fg_1^{(\perp')}}{6k_B T} [2\mu_z^2 (1-S-P) + \mu_y^2 (2+S+P+D+C) + \mu_x^2 (2+S+P-D-C)] \right\}$$

$$\chi_{\perp}^{(\text{electric})} = \varepsilon_{\perp} - 1 = \varepsilon_0^{-1} NFh \left\{ \bar{\alpha} - \frac{1}{3} \alpha_1 (S-P) - \frac{1}{6} \alpha_2 (D-C) \right. \tag{29c}$$
$$\left. + \frac{Fg_1^{(\perp)}}{6k_B T} [2\mu_z^2 (1-S+P) + \mu_y^2 (2+S-P+D-C) + \mu_x^2 (2+S-P-D+C)] \right\}$$

These are the Maier and Meier equations [7] for the low frequency components of the permittivity extended [8] to include all orientational order parameters. They predict that the mean value of the permittivity $\bar{\varepsilon} = \frac{1}{3}(\varepsilon_{\parallel} + \varepsilon_{\perp'} + \varepsilon_{\perp})$ should be independent of the orientational order, and apart from changes in density it is expected to be continuous through all liquid crystal phase changes. This prediction is not always confirmed by experiment, and for polar mesogens there are often detectable changes in $\bar{\varepsilon}$ at phase transitions, including those to the isotropic liquid.

A simplified form of Eq. (29) may be used to write down the permittivity anisotropy of a uniaxial liquid crystal consisting of uniaxial molecules having an off-axis dipole which makes an angle of β with the principal molecular axis (see Fig. 4):

$$\Delta\varepsilon = \frac{NhFS}{\varepsilon_0} \left[\Delta\alpha + \frac{\mu^2 F}{2k_B T} \left(3\cos^2\beta - 1 \right) \right] \tag{30}$$

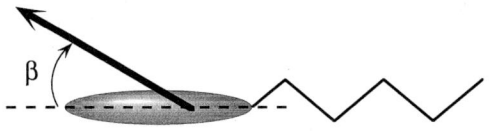

Figure 4. Schematic of a mesogen with an off-axis dipole moment making an angle of β with the molecular long axis.

For values of β less than 54.7°, the dipolar term is positive, while for angles greater than this it becomes negative, and may result in an overall negative dielectric anisotropy. For a particular combination of molecular properties, the polarisability anisotropy and the dipolar terms in Eq. (30) may cancel at a particular temperature. This has been observed in certain fluorinated cyclohexyl ethanyl biphenyls [9], which change the sign of their anisotropy from negative at low temperatures to positive at high temperatures. There is, however, a basic inconsistency with the model described by Eq. (30), since any molecule with an off-axis dipole is necessarily biaxial; rotation about the molecular long axis may result in the biaxial order parameter D averaging to zero.

Dielectric measurements on liquid crystal phases probe the dipole organization of molecules, and changes in the permittivity components as a liquid crystal undergoes transitions from nematic to various smectic phases will primarily reflect changes in orientational order and symmetry changes. Different degrees of translational order will only influence the permittivity components indirectly through macroscopic internal field corrections and through short range dipole–dipole interactions. The effect of macroscopic anisotropy on the dielectric properties of a material has been calculated for the model of a polarized sphere immersed in a dielectric continuum, with the result: [10]

$$\frac{(\varepsilon_i - n_i^2)[\varepsilon_i + \Omega_i(1-\varepsilon_i)]^2}{\varepsilon_i[\varepsilon_i + (n_i^2 - \varepsilon_i)\Omega_i]} = \frac{\langle M_i^2 \rangle}{\varepsilon_0 k_B T} \quad (31)$$

where $\langle M_i^2 \rangle$ is the mean square dipole moment of the sphere in the i-direction, and the factor Ω_i accounts for the depolarization field associated with a sphere in an anisotropic medium. Ω_i is defined in terms of the components of the permittivity tensor by:

$$\Omega_i = \frac{1}{2}\varepsilon_i \int_0^\infty \frac{(s\varepsilon_{ii}+1)^{-1} ds}{\sqrt{(s\varepsilon_{xx}+1)(s\varepsilon_{yy}+1)(s\varepsilon_{zz}+1)}} \quad (32)$$

and equals 1/3 for an isotropic dielectric. To proceed beyond Eq. (32) it is necessary to model both the anisotropic local field acting on molecules and the short range interactions between molecular dipoles in the sphere. The former depends on the long range anisotropy in the radial distribution function, while dipole–dipole correlations can be described in terms of anisotropic Kirkwood g-factors defined for different directions in the sample. These are most usefully defined in terms of the appropriate dipole correlation functions as:

$$g_1^i = 1 + V^{-1} \int G_1^i(\mathbf{r}) d\mathbf{r}$$

$$G_1^i(\mathbf{r}) = \frac{\langle \mu_i(0) \mu_i(\mathbf{r}) \rangle}{\langle \mu_i(0) \mu_i(0) \rangle} \quad (33)$$

where (i) refers to the parallel and perpendicular directions. Thus $g_1^\parallel$ and $g_1^\perp$ measure the extent to which the projections of molecular dipole components are correlated along the principal axes of an anisotropic fluid, assumed to be uniaxial in this case. If the rotational motion of molecules was isotropic, then clearly the correlation factors along the axes would be the same. This anisotropic dipole correlation is illustrated in

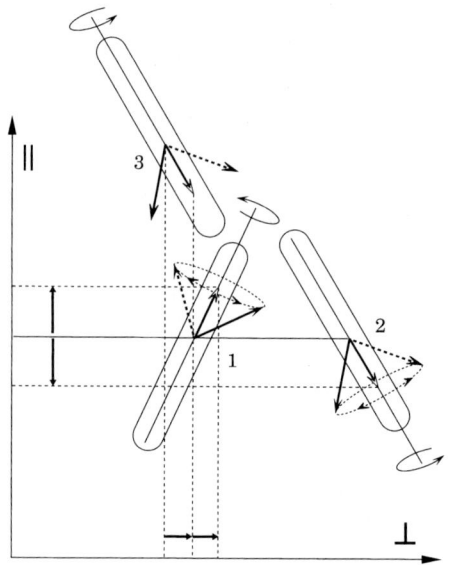

Figure 5. Anisotropic correlation of molecular dipole components.

Fig. 5, where the projected parallel components of the dipoles on molecules (1) and (2) are opposed ($g_1^\parallel < 1$), while the corresponding dipoles projected on to the perpendicular axis, molecules (1) and (3) are reinforcing ($g_1^\perp$).

Evaluating the mean square moment of a sphere (Eq. (31)), and using the definition of the correlation factor above gives the anisotropic version of the Kirkwood–Frohlich equation:

$$\frac{(\varepsilon_i - n_i^2)[\varepsilon_i + \Omega_i(n_i^2 - \varepsilon_i)]}{\varepsilon_i[n_i^2 + 2]^2} = \frac{N g_1^i \langle \mu_i^2 \rangle}{9\varepsilon_0 k_B T} \quad (34)$$

This equation can be written as:

$$(\varepsilon_i - n_i^2) = \frac{N h_i F_i^2 g_1^i \langle \mu_i^2 \rangle}{\varepsilon_0 k_B T} \quad (35)$$

where $h_i = \varepsilon_i [\varepsilon_i + \Omega_i(n_i^2 - \varepsilon_i)]^{-1}$ is the anisotropic cavity field factor, and $F_i = \frac{1}{3}(n_i^2 + 2)$

is an approximation to the anisotropic reaction field factor. While some progress has been made in describing local dipole correlations in partially ordered phases, the formulation of the internal electric field for a macroscopically anisotropic dipolar fluid remains a formidable problem. Various approximate models have been described [11–13], but it is difficult to assess the relative merits of such approaches. In liquid crystals, it has been proposed [14] that the long range anisotropy in the radial distribution function can be modelled by assuming an ellipsoidal cavity of dimensions a, b and c which depend on molecular shape. However comparison of local field tensors calculated for a continuum model with the results of dipole–dipole lattice sums have led to the conclusion [15] the contribution of shape anisotropy to the local field in anisotropic fluids can be neglected, and the assumption of isotropic internal field factors is justified. Under these circumstances, Eq. (35) can be written as more familiar Maier and Meier equations in terms of these effective mean square dipole moments:

$$\varepsilon_\parallel - 1 = \varepsilon_0^{-1} N L F$$
$$\cdot \left\{ \bar{\alpha} + \frac{2}{3} \Delta \alpha S + \frac{F}{k_B T} \left[\mu_{\text{eff}}^\parallel \right]^2 \right\} \quad (36)$$

$$\varepsilon_\perp - 1 = \varepsilon_0^{-1} N L F$$
$$\cdot \left\{ \bar{\alpha} - \frac{1}{3} \Delta \alpha S + \frac{F}{k_B T} \left[\mu_{\text{eff}}^\perp \right]^2 \right\} \quad (37)$$

where the effective mean square dipole moments can be written in terms of the order parameter and the longitudinal (μ_l) and transverse (μ_t) components of the molecular dipole as:

$$\left[\mu_{\text{eff}}^\parallel \right]^2 = \frac{g_1^\parallel}{3} \left[\mu_l^2 (1 + 2S) + \mu_t^2 (1 - S) \right] \quad (38)$$

$$\left[\mu_{\text{eff}}^\perp \right]^2 = \frac{g_1^\perp}{3} \left[\mu_l^2 (1 - S) + \mu_t^2 \left(1 + \frac{1}{2} S \right) \right] \quad (39)$$

A number of studies of dipole association in liquid crystalline systems have been reported [16–18], and it is clear that the orientation of the molecular dipole with respect to the molecular axis has a large influence on the local dipole organisation. A mean field theory of short range dipole–dipole correlation between interacting hard ellipsoids with embedded dipoles has been developed [19], and this predicts that prolate ellipsoidal molecules (rod-like) with longitudinal dipoles will exhibit local antiferroelectric order in ordered fluids, while oblate ellipsoids with dipoles along the shortest axis will order ferroelectrically. These studies can aid the development of new materials, for which carefully tailored dielectric properties are required [20], but are also of relevance in the research on anisotropic fluids having long range dipole organization.

As was pointed out in Sec. 1 of this chapter the symmetry of the phases will determine the number of independent components of the second rank electric permittivity; furthermore the point group symmetry of the phase and the constituent molecules will fix the orientational order parameters that contribute to a microscopic expression for the permittivity. In order to complete the description of the low frequency or static electric permittivity of liquid crystals, it is necessary to consider the additional effects of chirality, and the translational order associated with smectic phases.

The chiral nematic phase is characterized by a helical structure, and so the electric permittivity is biaxial, with three independent components along the principal axes, which are the local director axis, the helix and a third orthogonal axis. Since the pitches of chiral nematics are usually many molecular diameters, chiral nematics are locally uniaxial, and the pitch does not affect the symmetry or the magnitude of the permittivity.

4.2.1.2 The Smectic Phases

Experimentally there are changes in the components of the permittivity at nematic/smectic and smectic/smectic phase transitions, as illustrated in Figs. 6 and 7.

These changes reflect the molecular reorganization that takes place at transitions between different liquid crystal phases. The interpretation of the dielectric properties of smectic phases can be carried out using Eq. (34). Differences in orientational order in smectic phases are accounted for through the appropriate orientational order parameters given in Eq. (29), while other influences of the translational smectic order will affect the internal field factors and short range dipole–dipole interactions. For strongly polar mesogens dipole–dipole association is an important contributor to the physical properties of liquid crystal phases, and this is particularly important in smectic phases, where translational order can affect the dipole–dipole correlation factors. Dielectric studies have been made on materials exhibiting a number of different SmA phases (SmA$_1$, SmA$_2$, SmÃ), which are characterized by different degrees of dipole–dipole organization [23, 24]. The effect of different local interactions can be measured through the dipole correlation factors $g_1^{(i)}$, but evaluation of the anisotropic Kirkwood correlation factors either requires a detailed microscopic model for the liquid crystal, or it can be calculated from computer simulation. One model approach [25, 26] is to assume perfect orientational order (i.e. $S=1$) so that the molecules are constrained to be parallel or antiparallel to the director axis. The problem is then reduced to a two-state model, and if the dipole moment is assumed to be along the molecular axis, (i.e. parallel or antiparallel to the layer normal (z-axis), the net correlation is given by the relative probabilities for parallel or antiparallel dipole organization. Thus the dipole–dipole correlation factor can be written as:

$$g_1^{(\parallel)} = 1 + \frac{n\langle \mu_{1z}\mu_{2z}\rangle}{\langle \mu_{1z}^2\rangle} = 1 + n\langle p_+ - p_-\rangle \quad (40)$$

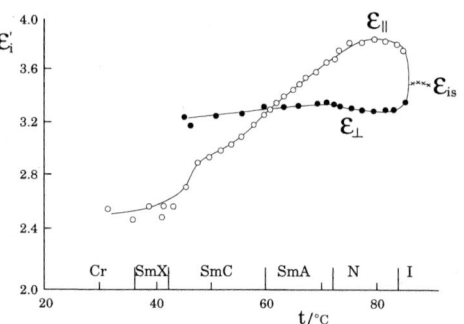

Figure 6. Dielectric permittivities for 95S showing effect of smectic phase transitions [21].

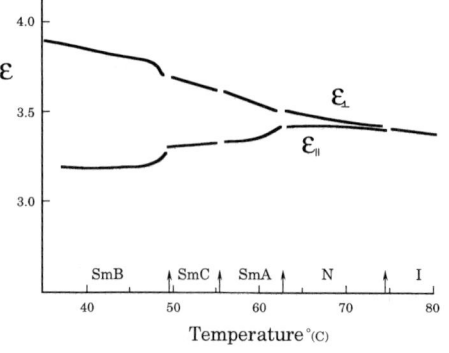

Figure 7. Dielectric permittivities of 4-pentyloxyl-benzylidene-4'-heptylaniline [22].

where n is the number of neighbors. The probabilities for parallel (+) or antiparallel (−) dipole orientation are determined by the dipole–dipole energy $u(\mu_1, \mu_2)$ (Eq. 41), and the average in Eq. (40) must be evaluated over the microscopic structure of the liquid crystal:

$$u(\mu_1, \mu_2) = -\frac{\mu^2}{4\pi r^3 \varepsilon_0} \begin{pmatrix} 2\cos\theta_1 \cos\theta_2 - \\ \sin\theta_1 \sin\theta_2 \cos\phi \end{pmatrix} \quad (41)$$

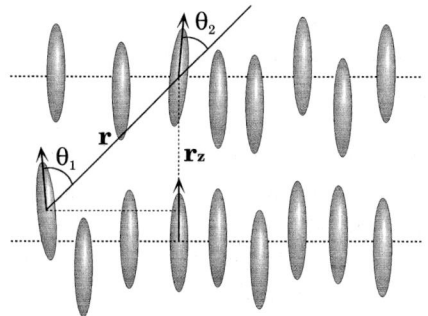

Figure 8. Dipole correlation in smectic phases.

The angles θ_1 and θ_2 are defined in Fig. 8 where the dihedral angle ϕ has been set equal to 0. Writing the probabilities in terms of the dipole–dipole energy:

$$g_1^{(\parallel)} = 1 + \quad (42)$$

$$n \frac{\left\langle \exp\left(-\frac{u_{\text{parallel}}}{k_B T}\right) - \exp\left(-\frac{u_{\text{antiparallel}}}{k_B T}\right) \right\rangle}{\left\langle \exp\left(-\frac{u_{\text{parallel}}}{k_B T}\right) + \exp\left(-\frac{u_{\text{antiparallel}}}{k_B T}\right) \right\rangle}$$

expanding the exponentials, and assuming that $u(\mu_1, \mu_2) \ll k_B T$, gives:

$$g_1^{(\parallel)} = 1 + n \frac{\langle u_{\text{antiparallel}} - u_{\text{parallel}} \rangle}{k_B T} \quad (43)$$

and for the model depicted in Fig. 8, this can be evaluated to give:

$$g_1^{(\parallel)} = 1 - \frac{n\mu^2 \left\langle 3\left(\frac{r_z}{r}\right)^2 - 1 \right\rangle}{4\pi \varepsilon_0 r^3 k_B T} \quad (44)$$

In a smectic phase, the average separation perpendicular to the layers (r_z) is likely to be greater than the in-plane separation, and this results in a $g_1^{(\parallel)} < 1$, while $g_1^{(\perp)}$ would be unity because the perpendicular component of the dipole is zero. This simple model can be extended to molecules with a molecular dipole inclined at an angle β to the molecu-lar alignment axis, in which case:

$$g_1^{(\parallel)} = 1 - \frac{n\mu^2 \cos^2 \beta \left\langle 3\left(\frac{r_z}{r}\right)^2 - 1 \right\rangle}{4\pi \varepsilon_0 r^3 k_B T} \quad (45)$$

$$g_1^{(\perp)} = 1 - \frac{n\mu^2 \sin^2 \beta \left\langle 3\left(\frac{r_x}{r}\right)^2 - 1 \right\rangle}{8\pi \varepsilon_0 r^3 k_B T} \quad (46)$$

If $\langle r_x^2 \rangle$ is less than $\langle r^2/3 \rangle$, the perpendicular dipole correlation factor will be greater than one, indicating a preferred parallel alignment of dipoles in the smectic layer. This model has been used [26] to explain the change in sign of the dielectric anisotropy from positive to negative on passing from the nematic to the smectic A phase of p-heptylphenylazoxy-p'-heptylbenzene.

Low frequency dielectric studies on smectic C, F and I phases are complicated by the intrinsic biaxiality of these phases. It is possible to use dielectric measurements to determine the tilt angle in SmC materials [27], but of more interest is the direct determination of the three principal components of the dielectric tensor, since such measurements can give additional information on the local molecular organization from Eq. (29). The orientation of the principal axes for tilted smectic phases is not determined by symmetry, except that one principal axis coincides with the C_2 rotation axis perpendicular to the tilt-plane. It is assumed that a further principal axis lies along the tilt direction, and this appears to be justified by experiment; the orientation of the axes are indicated in Fig. 9.

Most recent studies [28, 29] of the dielectric properties of the SmC phase have focussed on the ferroelectric chiral smectic C phase, because of its importance in applications. The molecular interpretation of the principal permittivities is contained in Eqs. (27), with appropriate correlation fac-

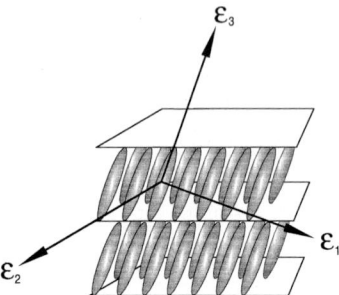

Figure 9. Definition of principal axes for SmC phases.

Specimen results for SCE13(R), a commercial racemic host, are given in Fig. 11 [28].

The reduced symmetry of chiral phases results in additional contributions to the low frequency permittivity. Tilted chiral phases such as smectic C*, F* and I* lack a centre of symmetry, and it is possible for these materials to be ferroelectric. The resulting spontaneous polarization P_S is directed along the C_2 symmetry axis, and is perpendicular to the tilt plane; it also depends di-

tors to account for the tilted layered structure [25], but the experimental problem is to find suitable alignment geometries that allow the independent measurement of the principal components of the permittivity, denoted as ε_1, ε_2, and ε_3 in Fig. 9. One approach [28], which is equally applicable to nonchiral and chiral tilted phases is to measure the permittivity of a homeotropically aligned sample in which the measurement direction is along the smectic layer normal. This gives a result for $\varepsilon_{\text{homo}}$, defined by:

$$\varepsilon_{\text{homo}} = \varepsilon_1 + (\varepsilon_3 - \varepsilon_1)\cos^2\theta \tag{47}$$

where $(\varepsilon_3 - \varepsilon_1)$ is defined as the dielectric anisotropy (ε_3 is equivalent to $\varepsilon_\parallel$ and ε_1 is equivalent to $\varepsilon_\perp$). A second permittivity can be measured for the so-called planar state, for which a chevron structure is assumed (Fig. 10) with a layer tilt angle of δ: the corresponding permittivity is:

$$\varepsilon_{\text{planar}} = \varepsilon_2 - \partial\varepsilon \frac{\sin^2\delta}{\sin^2\theta} \tag{48}$$

the quantity $\partial\varepsilon = (\varepsilon_2 - \varepsilon_1)$ is defined as the dielectric biaxiality.

In order to obtain the three components, it is assumed that the mean permittivity extrapolated from higher smectic and nematic phases may be used, such that:

$$\bar{\varepsilon} = \frac{1}{3}(\varepsilon_1 + \varepsilon_2 + \varepsilon_3) \tag{49}$$

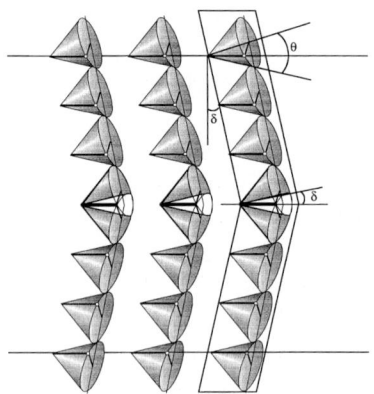

Figure 10. Chevron structure on a SmC phase.

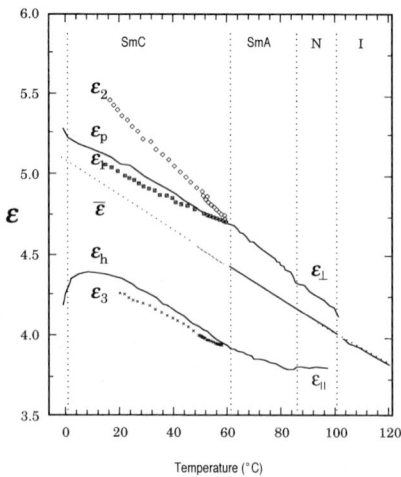

Figure 11. Dielectric results for SCE13(R) taken from [28].

rectly on the tilt angle θ:

$$P_S = P_0 \sin\theta \qquad (50)$$

In the absence of any constraints, the direction of P_S rotates from one smectic layer to the next, with a period equal to the smectic C* pitch, and so the average polarization for a sample would be zero. However, surface treatment or application of a field can cause the helix to untwist, resulting in a permanently polarized sample. The spontaneous polarization arises from a preferred alignment of molecular dipole components which are perpendicular to the molecular long axis, but it behaves differently from the ferroelectric and ferromagnetic polarization characterised for crystals. The liquid crystalline ferroelectric phases identified so far are *improper* ferroelectrics, since the spontaneous polarization results from a symmetry constraint, whereas in *proper* ferroelectrics the polarization results from dipole–dipole interactions. The Curie–Weiss law for proper ferroelectrics predicts a second order phase transition at the Curie temperature from the high temperature paraelectric state to a permanently polarized ferroelectric state:

$$P_S = \frac{N\mu^2 E}{3k_B(T-T_c)} \qquad (51)$$

However in chiral tilted liquid crystal smectic phases, the polarization is driven by the tilt angle, and the phase transition will not necessarily be of second order.

The helical structure which can develop in thick cells of chiral smectic C phases having planar surface alignment conditions can be used to obtain measurements of the components of the dielectric permittivity tensor [29], but the technique is restricted to chiral smectic phases. Measurements are made (see Fig. 9) of the homeotropic state, as above, and additionally the helical state (Fig. 12), and the uniformly-tilted state ob-

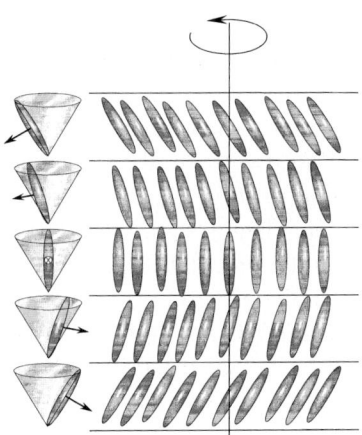

Figure 12. Alignment states for the helical state of chiral SmC phase.

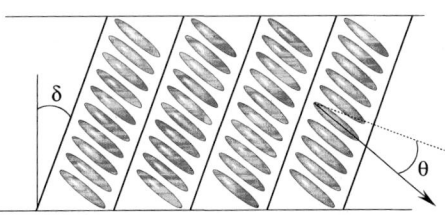

Figure 13. Alignment of the unwound chiral SmC phase.

tained by applying a field to unwind the helix (Fig. 13):

$$\varepsilon_{\text{helix}} = \frac{1}{2}\cos^2\delta\left(\varepsilon_1\cos^2\theta + \varepsilon_3\sin^2\theta + \varepsilon_2\right) + \sin^2\delta\left(\varepsilon_1\sin^2\theta + \varepsilon_3\cos^2\theta\right)$$

$$\varepsilon_{\text{unwound}} = \varepsilon_2\cos^2\delta + \sin^2\delta\left(\varepsilon_1\sin^2\theta + \varepsilon_3\cos^2\theta\right) \qquad (52)$$

The use of Eqs. (47) and (52) then allows the evaluation of three principal permittivity components.

The electric susceptibilities of chiral smectic A and chiral smectic C, I and F contain terms related to the permanent dipole polarization which can develop in tilted smectic phases (ferroelectricity), or which can be induced in orthogonal smectic phases through an induced tilt (electroclinism) by electric fields in the plane of the layers. The broken symmetry associated with these phases is along an axis perpendicular to the director, and the layer normal for smectics, and so the extra dielectric contributions are to a single perpendicular component of the permittivity. The origin of the polarization contributions to the electric permittivity of chiral smectic A and smectic C phases is illustrated in Fig. 14 [30].

Orientation of the transverse molecular dipole moments become biased along the y-axis (corresponding to ε_2 above) when a tilt develops either induced by an electric field along the y-axis (orthogonal smectics) or spontaneously in tilted smectics. This is known as the soft mode contribution to the electric susceptibility, and as well as contributing to all smectic phases there is a detectable effect in chiral nematic phases, close to phase transitions. Tilted chiral smectic phases can develop helicoidal structures, the helix axis being perpendicular to the layers. In the unperturbed state the polarization associated with the spontaneously aligned transverse dipole components rotates with the helix, however application of an electric field perpendicular to the helix axis gives a contribution to the electric polarization, and hence to the electric permittivity component: this is known as the Goldstone mode. In terms of Fig. 14 the Goldstone mode describes polarization resulting from changes in the azimuthal angle (ϕ) of the director, while the soft mode is polarization from changes in the tilt angle θ. A Goldstone mode contribution will only be measured if there is a helicoidal structure, and so will be absent in surface-stabilized chiral smectic C structures, such as the planar and unwound states described above.

The theory of the dielectric properties of chiral smectic liquid crystals is far from complete, particularly with respect to a molecular statistical approach. Simple Landau theory [31] gives expressions for the contributions of soft modes (χ_S) and Goldstone modes (χ_G) to the low frequency permittivity as:

$$\chi_S = \frac{\varepsilon_{\infty y}^2 \mu_p^2}{kq_0^2 + 2a(T_c - T)}$$

and

$$\chi_G = \frac{\varepsilon_{\infty y}^2 \mu_p^2}{2kq_0^2} \quad (53)$$

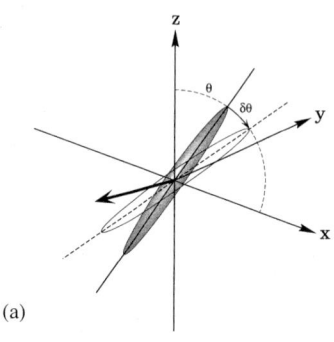

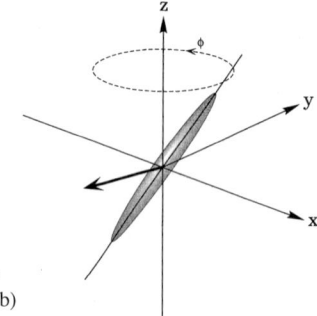

Figure 14. Contributions of soft mode (a) and Goldstone mode (b) deformations to the electric susceptibility.

where μ_p is the piezoelectric coefficient in the Landau free energy, which measures the coupling between the director and the polarization, and $\varepsilon_{\infty y}$ is the high frequency part

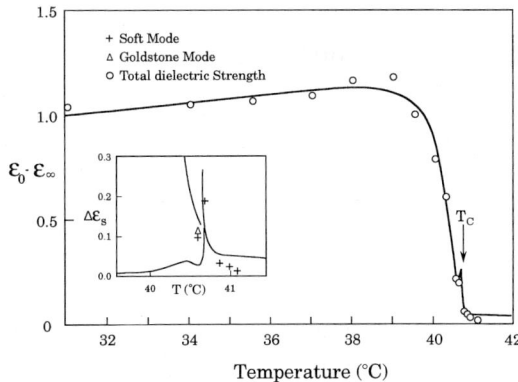

Figure 15. Soft mode and Goldstone mode contributions to the electric permittivity for the mixture BAHABAC [32]; the full lines are theoretical results from a generalized Landau theory.

of the permittivity component along the y-axis, k is an effective elastic constant for the helicoidal distortion, $q_0 = 2\pi/p$ is the wave vector of the helical pitch p and a is the coefficient in the Landau free energy of the term quadratic in the primary order parameter. The temperature dependence of the soft mode and Goldstone mode contributions to the electric permittivity is illustrated schematically in Fig. 15.

4.2.2 Frequency Dependence of the Electric Permittivity: Dielectric Relaxation

A full theory of the frequency dependence of the electric permittivity of liquid crystals cannot yet be given. It is a formidable problem since it requires proper account to be taken of the influence of orientational order on molecular motion, as well as the effects of macroscopic dielectric anisotropy. The dielectric response of a rigid dipolar molecule rotating in an isotropic fluid can be described in terms of a single relaxation time or rotational diffusion constant. In a liquid crystal phase, the rotational motion is no longer isotropic, and for a rigid rod-like or disc-like mesogen in a uniaxial phase two rotational diffusion constants can be defined, parallel to or perpendicular to the unique inertial axis. In general a dipole moment will not be along the inertial axis of a molecule, and there is likely to be a complicated relationship between the motion of the dipole, which determines the frequency dependence of the permittivity, and the rotational motion of the molecule. For rigid molecules, an anisotropic rotational diffusion model in which the isotropic motion of the molecule is modified by the orienting potential of the phase is sufficient to describe the dielectric response of nematic. For smectic phases translational order can affect the reorientation of molecules, and hence the dielectric relaxation, although this seems to have a minor influence for SmA and SmC phases. In tilted smectic phases the molecular tilt causes biaxiality, and provides an additional environmental constraint on molecular rotation; such effects are sometimes detectable in the dielectric properties. The increased in-plane order associated with hexatic liquid crystals and more ordered crystal smectic phases can be detected through measurements of the permittivity components. For chiral liquid crystal phases such as chiral SmA* SmC*, SmF* and SmI*, there are new contributions to the permittivity which arise from the alignment and collective motion of molecular dipoles, and these collective relaxation modes contribute to the dielectric behaviour of these phases.

The origin of a frequency dependent permittivity is molecular motion associated with a dipole moment. In an oriented fluid, induced or permanent dipole moments contribute differently to the components of the permittivity tensor; similarly the effects of molecular motion as reflected by the frequency dependence of the permittivity will also be different for different components.

Both real and imaginary parts of the permittivity tensor are frequency dependent, and there is a relationship between them known as a Kramers–Kronig relation:

$$\varepsilon_i(\omega)' = \varepsilon_i(\infty)' + \frac{2}{\pi} \int_0^\infty \frac{\omega \varepsilon_i(\omega)''}{(\omega^2 - \omega_0^2)} d\omega$$

$$\varepsilon_i(\omega)'' = \frac{2\omega_0}{\pi} \int_0^\infty \frac{\varepsilon_i(\omega)'}{(\omega^2 - \omega_0^2)} d\omega \qquad (54)$$

where the integral excludes the singular point at $\omega = \omega_0$. It is normal to measure and analyse both the real and imaginary parts of $\varepsilon(\omega)^*$, but any model which violates Eq. (54) must be open to doubt.

An idealized picture of the frequency dependence of $\varepsilon_\parallel$ and $\varepsilon_\perp$ may be constructed by examining Eqs. (36) and (37) for a uniaxial liquid crystal having no local biaxial order. The effects of induced moments can be removed by subtracting the high frequency part of the permittivity or square of the refractive index, giving for the real parts of the permittivity:

$$\varepsilon_\parallel(\omega) - n_\parallel^2 = \frac{NLF^2 g_1^\parallel}{3\varepsilon_0 k_B T}$$
$$[\mu_l^2(1+2S) + \mu_t^2(1-S)]$$

$$\varepsilon_\perp(\omega) - n_\perp^2 = \frac{NLF^2 g_1^\perp}{3\varepsilon_0 k_B T}$$
$$\left[\mu_l^2(1-S) + \mu_t^2\left(1+\frac{1}{2}S\right)\right] \qquad (55)$$

Each component of $\varepsilon(\omega)'$ contains two contributions from the molecular dipole moment, and each can have different relaxation times or frequencies. Thus frequency scans of $\varepsilon_\parallel(\omega)^*$ and $\varepsilon_\perp(\omega)^*$ are each expected to show two relaxation regions. The characteristic frequencies or relaxation times will be related to the rotational motion of a molecule in an anisotropic environment, and for a uniaxial molecule with two independent moments of inertia, the dynamics can be approximately described in terms of three rotational modes. There is some arbitrariness in choosing these modes, but for illustration we assert that the rotation of a molecule can be broken down into contributions from end-over-end rotation (ω_1), precessional motion (ω_2) about the director and rotation (ω_3) about its own long molecular axis. These motions are illustrated in Fig. 16.

A dipole component will contribute to a principal permittivity if there is a mechanism for that component to follow an electric field applied along the particular principal direction of the permittivity. The manner in which the rotational modes allow different dipole components to reverse in particular directions is seen in Fig. 16. In a nematic environment, it is expected that the magnitudes of the characteristic frequencies for the rotational modes will be in the order $\omega_1 \ll \omega_2 < \omega_3$. Contributions to the permittivity from different dipole components will be lost above frequencies corresponding to ω_1, ω_2 and ω_3, and the variation of $\varepsilon_\parallel(\omega)^*$ and $\varepsilon_\perp(\omega)^*$ with frequency is shown in Fig. 17.

This over simplified model matches the experimental measurements obtained for a

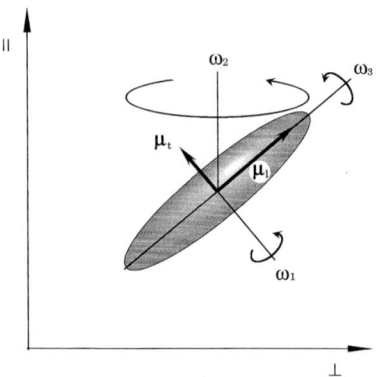

Figure 16. Molecular rotational modes that contribute to the dielectric relaxation.

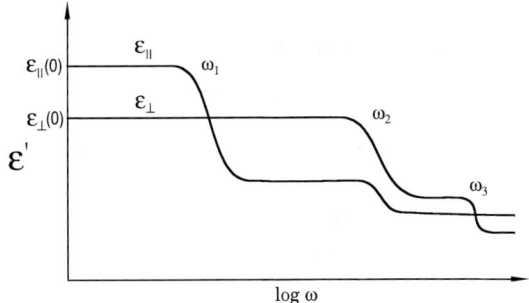

Figure 17. Schematic plot of $\varepsilon'_\|$ and $\varepsilon'_\perp$ against frequency.

number of nematic materials [33], except that the relaxations associated with ω_2 and ω_3 are not separated in the frequency spectrum. The low frequency absorption measured in most liquid crystals can usually be fitted very accurately by a semicircular Cole–Cole plot, but higher frequency relaxations tend to be broader, indicating a range of relaxation times. If other dipolar contributions are included in the permittivity expression, for example those arising from local biaxial order, then these would be expected to relax at different frequencies, and the corresponding dielectric spectrum would be more complicated.

For rigid molecules the frequency dependence of the orientational polarization in isotropic liquids can be calculated using Debye's model for rotational diffusion. This may be modified to describe rotational diffusion in a liquid crystal potential of appropriate symmetry, but the resulting equation is no longer soluble in closed form. Martin, Meier and Saupe [34] obtained numerical solutions for a nematic pseudopotential of the form:

$$u = -bS \cos^2 \theta \tag{56}$$

where θ is the angle between the long axis of a uniaxial molecule and the director. They assumed that the nematic potential had no effect on molecular rotation about the long axis of the molecule, so that relaxation processes involving μ_t are not influenced by orientational order.

The solutions to the anisotropic diffusion equation can be written as a series expansion, each term of which can be associated with a particular relaxation time. For a harmonic perturbation of the rotational distribution function, as occurs in a dielectric relaxation experiment with an ac electric field, it was found that a single relaxation time was sufficient to describe the relaxation of μ_l, and this could be expressed in terms of the relaxation time (τ_0) for μ_l in the absence of a nematic potential by:

$$\tau_i = j_i \tau_0 \tag{57}$$

The subscript $i = \|$ or $\perp$ identified the component of the permittivity, and the quantity j_i is a retardation factor calculated numerically from the model, which depends on the coefficient b of the pseudopotential, (j_i is often written as g_i, which we have avoided because of confusion with the Kirkwood dipole correlation factor $g_1^{(i)}$). An approximate result for the retardation factor was calculated [35] such that:

$$j_\| = \frac{k_B T}{bS} \left[\exp\left(\frac{bS}{k_B T}\right) - 1 \right] \tag{58}$$

where bS is identified as the height of the potential barrier to end-over-end rotation of a molecule.

The full solution of the rotational diffusion equation including a general single particle potential of $D_{\infty h}$ symmetry has been investigated [36], and it is found that the dipole correlation function, which can be related to the permittivity as a function of frequency, is a sum over many exponential terms each characterised by a different relaxation time. Extending Eq. (20) for an anisotropic fluid gives:

$$\frac{\varepsilon_i(\omega)^* - n_i^2}{\varepsilon_i(0) - n_i^2} \left[\frac{(2\varepsilon_i(\omega) + n_i^2)\varepsilon_i(0)}{(2\varepsilon_i(0) + n_i^2)\varepsilon_i(\omega)} \right]$$

$$= 1 - i\omega \int_0^\infty C_i(t) e^{i\omega t} dt \quad (59)$$

where $C_i(t)$ is the dipole time correlation function defined by:

$$C_i(t) = \frac{\langle \mu_i(0)\mu_i(t) \rangle}{\langle \mu_i(0)\mu_i(0) \rangle} \quad (60)$$

It was observed that in many cases $C_i(t)$ could be approximated by the first term, and this gives rise to four relaxation times corresponding to the four dipole contributions appearing in Eq. (55). Resolving the dipole into its longitudinal and transverse components, as above, enables the correlation functions for directions parallel and perpendicular to the director to be written as [36, 37]:

$$C_\parallel(t) = \frac{1}{3\mu_0^2} \left[\begin{array}{c} \mu_l^2(1+2S)\Phi_{00}(t) + \\ \mu_t^2(1-S)\Phi_{01}(t) \end{array} \right] \quad (61)$$

$$C_\perp(t) = \frac{1}{3\mu_0^2} \left[\begin{array}{c} \mu_l^2(1-S)\Phi_{10}(t) + \\ \mu_t^2\left(1+\frac{1}{2}S\right)\Phi_{11}(t) \end{array} \right] \quad (62)$$

where $\Phi_{kl}(t)$ describe the time dependence of different angular functions representing different relaxation modes for the molecular dipole in an anisotropic environment [38]. In the rotational diffusion model each $\Phi_{kl}(t)$ can be written in terms of a single relaxation time related to a particular rotational mode. For example the low frequency end-over-end rotational is given by Φ_{00}, and can be accurately represented by a single exponential, so that:

$$\Phi_{00}(t) = e^{-\frac{t}{\tau_{00}}} = \langle \cos\theta(0)\cos\theta(t) \rangle \quad (63)$$

The relaxation times τ_{ij} depend on the parameters of the assumed nematic pseudopotential and the anisotropy in the rotational diffusion constants $D_\parallel$ and $D_\perp$, which are related to the molecular shape and the local viscosity. If $D_\parallel$ and $D_\perp$ are assumed to be equal, then $\Phi_{01}(t) = \Phi_{10}(t)$, and the rotational modes can be represented as shown in Fig. 16 and the effect of an ordering potential on the relaxation times is shown schematically in Fig. 18.

These results indicate that the effect of the liquid crystal ordering potential is to decrease the relaxation frequency for end-over-end rotation (τ_{00}^{-1}) and increase it for rotation about the molecular long axis. If anisotropy in the rotational diffusion constants is included, then the relaxation time τ_{00} is further retarded.

Experimental measurements of dielectric relaxation confirm qualitatively the predictions of the rotational diffusion model, in that $\varepsilon_\parallel$ has a low and high frequency relaxation, while $\varepsilon_\perp$ only shows relaxations at higher frequencies. Unfortunately there are few liquid crystal systems that have been studied over wide frequency ranges, and measurements at high frequencies >50 MHz on aligned samples are difficult. Some typical results are shown in Fig. 19.

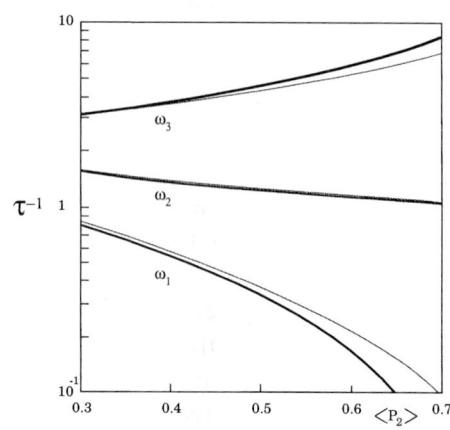

Figure 18. Calculated relaxation frequencies in reduced units plotted as a function of order parameter for para-azoxyanisole [36].

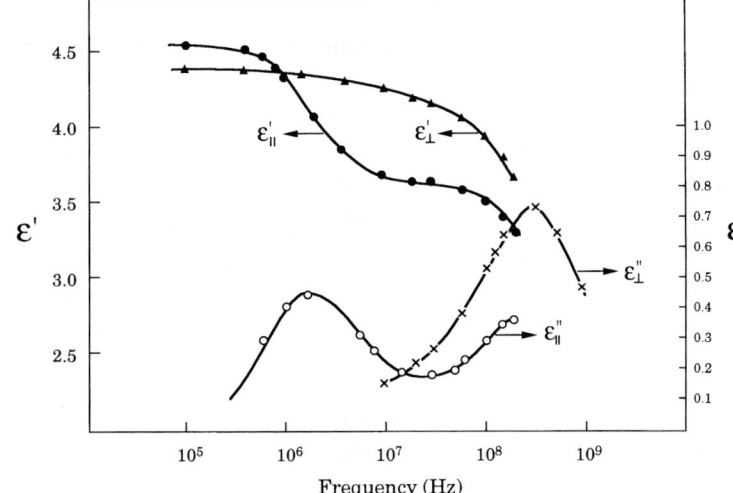

Figure 19. Complex permittivity components for 4-pentylphenyl 4-propylbenzoate [39].

In the study of dielectric relaxation, temperature is an important variable, and it is observed that relaxation times decrease as the temperature increases. In Debye's model for the rotational diffusion of dipoles, the temperature dependence of the relaxation is determined by the diffusion constant or microscopic viscosity. For liquid crystals the nematic ordering potential contributes to rotational relaxation, and the temperature dependence of the order parameter influences the retardation factors. If rotational diffusion is an activated process, then it is appropriate to use an Arrhenius equation for the relaxation times:

$$\tau = A \exp \frac{E_a}{k_B T} \quad (64)$$

where E_a is the activation energy or barrier to dipole reorientation, and A is another parameter of the model. This would seem to be a useful way to describe end-over-end rotation in liquid crystals [40].

It is observed experimentally that the activation energies for dipole reorientation often change at liquid crystal phase transitions. Changes occur due to differences in the degree of order and the local viscosity, and it is expected that the activation energies will be higher in smectic phases. In practice is often observed that the end-over-end rotation has a lower activation energy in the smectic A phase than in the higher temperature nematic phase, and this casts some doubt on the role of the order parameter and viscosity, which are both higher in smectic phases than the nematic phase. Benguigui [41] has satisfactorily explained the results for τ_{00} measured for nematic and SmA phases in terms of a free-volume model due to Vogel and Fulcher. Their result for for the relaxation time is:

$$\tau = B \exp \frac{E_a}{k_B (T - T_0)} \quad (65)$$

where T_0 is a hypothetical glass transition temperature. At phase transitions from disordered smectics (A and C) to those with hexatic order (B, F, and I) there is an order of magnitude increase in relaxation times, but activation energies for end-over-end dipole reorientation are similar in all smectic phases. This is illustrated in Fig. 20 using results for 4-pentylphenyl 4'-heptylbiphenyl-4-carboxylate [42]. The relaxation frequencies decrease by nearly a decade on go-

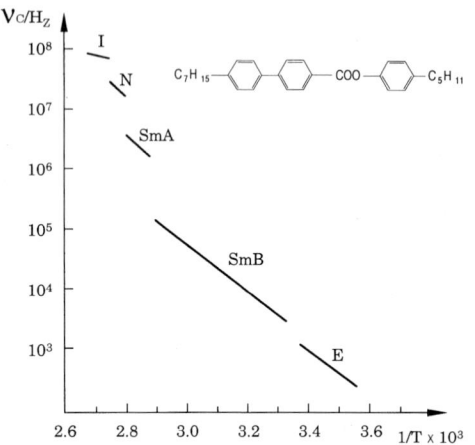

Figure 20. Relaxation frequencies plotted on a logarithmic scale as a function of inverse temperature (K) for 4-pentylphenyl 4'-heptylbiphenyl-4'-carboxylate.

ing from the nematic phase or the smectic A phase, and from the smectic A phase to the smectic B phase, but the activation energies, given by the slope of the lines in Fig. 20 are approximately the same.

Models for the interpretation of the low frequency relaxation in liquid crystals are often based on a single particle relaxation process, but spectroscopic probes of molecular motion such as magnetic resonance, neutron scattering and time-resolved fluorescence depolariastion, suggest that reorientation times for mesogens are of the order of 10^{-9} s to 10^{-10} s in isotropic, nematic and disordered smectic phases. Thus interpretation of dielectric relaxation processes at MHz or even kHz frequencies in terms of single molecule rotation is not likely to be correct. The low frequency relaxations observed in liquid crystals are the result of collective molecular motion, although the models outlined above are useful in analysing results and comparing materials.

It has been demonstrated that the different dipolar contributions to the permittivity components in Eq. (55) cease to contribute at different frequencies, and if these frequencies are well-separated the Debye equations can be applied separately to each term. The dielectric anisotropy changes as a function of frequency, and for particular relative values of longitudinal and transverse dipole moments, it is possible for $\Delta\varepsilon$ to change sign from positive at low frequencies to negative at high frequencies [39]. Liquid crystals exhibiting this behaviour are known as dual-frequency or two-frequency materials, because of potential applications in fast-switched liquid crystal devices, which are addressed by both low frequency ($\Delta\varepsilon_{lf}$ + ve) and high frequency ($\Delta\varepsilon_{hf}$ – ve) signals. Application of the Debye Equation (10) for $\varepsilon_\parallel$ gives the frequency at which $\Delta\varepsilon(\omega)$ changes sign:

$$\omega(\Delta\varepsilon = 0) = \tau_\parallel^{-1}\sqrt{\frac{\Delta\varepsilon_{lf}}{-\Delta\varepsilon_{hf}}} \qquad (66)$$

The frequency dependence of the permittivity component of chiral smectic phases along the director is similar to that observed for nonchiral materials, but for dielectric measurements perpendicular to the layer normal, ferroelectric polarization results in additional contributions to the electric susceptibility. Both the soft mode and Goldstone mode contributions are frequency dependent, and the latter gives rise to a low frequency relaxation for helicoidal smectic structures corresponding to the rotation of the polarization about the helix axis. This relaxation frequency is approximately independent of temperature, indicating that rotation of the polarization about the helical axis has zero activation energy, and so is identified as a Goldstone process. There are in principle four relaxation processes that can contribute to $\varepsilon_\perp$ in chiral smectic phases: two at low frequency are associated with the motion of the director, but there are also two high frequency modes which relate to changes in the polarization for a fixed director orientation. The former low frequency process exists for the orthogonal smectic phases. Typical relaxation frequencies for these modes are $\omega_S \approx 1 \rightarrow 10^4$ kHz and $\omega_G \approx 10 \rightarrow 200$ Hz. The high frequency polarization modes (ω_{PS}, ω_{PG}) are essentially the same as the dipole relaxation modes for $\varepsilon_\perp$ already discussed in the context of non-chiral phases, except they relate to a linear transverse dipolar contribution to the permittivity; they are degenerate in orthogonal smectic phases, and have relaxation frequencies in the region of 500 MHz. The temperature dependence of these relaxation processes is illustrated schematically in Fig. 21.

Contributions to the permittivity from fluctuations in the amplitude of the tilt angle are expected to be small away from phase transitions, but this process is strongly temperature dependent, and the relaxation frequency tends to zero at a phase transition: this relaxation is known as the soft mode in common with similar behaviour in crystals. Both the Goldstone mode and the soft mode relaxation processes are a result of the cooperative motion of molecular dipoles, and there is no proper molecular theory for them. The real and imaginary parts of the permittivity fit semicircular Cole–Cole plots, and so each mode can be characterized by a single relaxation frequency, although the dielectric absorptions for the two processes may overlap in the frequency spectrum.

The dielectric response of smectic C* liquid crystals can be derived from a Landau model [43] with the result:

$$\varepsilon_\perp(\omega)^* - \varepsilon_\perp(\infty)'$$
$$= \frac{\chi_G}{1+i\omega\tau_G} + \frac{\chi_S}{1+i\omega\tau_S}$$
$$+ \frac{\chi_{PG}}{1+i\omega\tau_{PG}} + \frac{\chi_{PS}}{1+i\omega\tau_{PS}} \quad (67)$$

where χ_i are the increments in the electric susceptibility associated with the Goldstone and soft modes for the director and the polarization having relaxation frequencies (τ_i). The dielectric increment for the Goldstone and soft modes is given in Eq. (53). χ_G can also be expressed in terms of the spontaneous polarization as:

$$\varepsilon_0 \chi_G = \frac{1}{2k}\left(\frac{P_S}{q\sin\theta}\right)^2 \quad (68)$$

As explained earlier, both the Goldstone and soft modes contribute to the perpendicular permittivity component in the smectic C* phase, although away from T_c the Goldstone mode dominates in twisted structures.

4.3 References

[1] C. J. F. Bottcher, P. Bordewijk, *Theory of Electric Polarization*, **1978**, Vol. 2, Elsevier, Amsterdam, Netherlands, p. 61.
[2] K. S. Cole, R. H. Cole, *J. Chem. Phys.* **1941**, 9, 341.
[3] D. Lippens, J. P. Parneix, A. Chapoton, *J. de Phys.* **1977**, 38, 1465.
[4] J. M. Deutch, *Faraday Symposium of the Chemical Society* **1977**, p. 26.
[5] See Ref. [1], p. 206.
[6] D. D. Klug, D. E. Kranbuehl, W. E. Vaughan, *J. Chem. Phys.* **1969**, 50, 3904.
[7] W. Maier, G. Meier, *Z Naturforsch.* **1961**, 16a, 262.
[8] K. Toriyama, D. A. Dunmur, S. E. Hunt, *Liquid Crystals* **1988**, 5, 1001.

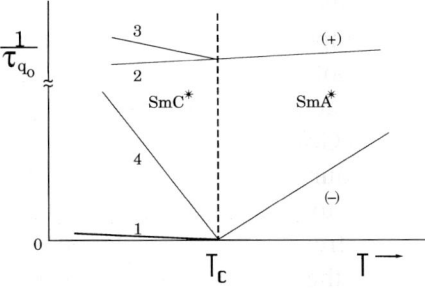

Figure 21. Schematic variation of the frequency of relaxation modes for chiral SmA and SmC phases [30].

[9] D. A. Dunmur, D. A. Hitchen, Xi-Jun Hong, *Mol. Cryst. Liq. Cryst.* **1986**, *140*, 303.
[10] See Ref. [1], p. 444.
[11] A. A. Minko, V. S. Rachevich, S. Ye Yakovenko, *Liq. Cryst.* **1989**, *4*, 1.
[12] E. M. Averyanov, M. A. Osipov, *Sov. Phys. Usp.* **1990**, *33*, 365.
[13] P. Palffy-Muhoray, D. A. Balzarini, D. A. Dunmur, *Mol. Cryst. Liq. Cryst.* **1984**, *110*, 315.
[14] P. Bordewijk, W. H. de Jeu, *J. Chem. Phys.* **1978**, *68*, 116.
[15] D. A. Dunmur, R. W. Munn, *Chem. Phys.* **1983**, *76*, 249.
[16] K. Toriyama, D. A. Dunmur, *Mol. Phys.* **1985**, *56*, 478.
[17] D. A. Dunmur, K. Toriyama, *Mol. Cryst. Liq. Cryst.* **1995**, *264*, 131.
[18] K. Toriyama, S. Sugimari, K. Moriya, D. A. Dunmur, R. Hanson, *J. Phys. Chem.* **1996**, *100*, 307.
[19] D. A. Dunmur, P. Palffy-Muhoray, *Mol. Phys.* **1992**, *76*, 1015.
[20] D. A. Dunmur, R. H. H. Okada, H. Onnagawa, S. Sugimori, K. Toriyama, *Proceedings of 15th Int. Display Res. Conference*, Hamamatsu, Japan, **1995**, p. 563.
[21] L. Bata, A. Buka, *Mol. Cryst. Liq. Cryst.* **1981**, *63*, 307.
[22] L. Benguigui, *J. de Phys.* **1980**, *41*, 341.
[23] C. Druon, J. M. Wacrenier, F. Hardouin, N. H. Tinh, H. Gasparoux, *J. de Phys.* **1983**, *44*, 1195.
[24] B. R. Ratna, C. Nagabhushana, V. N. Raja, R. Shashidhar, S. Chandrasekhar, G. Heppke, *Mol. Cryst. Liq. Cryst.* **1986**, *138*, 245.
[25] L. Benguigui, *J. de Phys.* **1979**, *40*, 705.
[26] W. H. de Jeu, W. J. A. Goosens, P. Bordewijk, *J. Chem. Phys.* **1974**, *61*, 1985.
[27] L. Benguigui, D. Cahib, *Phys. Stat. Sol.* **1978**, *a47*, 71.
[28] C. Jones, E. P. Raynes, *Liq. Cryst.* **1992**, *11*, 199.
[29] F. Gouda, W. Kuczynski, S. T. Lagerwall, M. Matuszczyk, K. Sharp, *Phys. Rev. A* **1992**, *46*, 951.
[30] B. Zeks, R. Blinc, *Ferroelectric Liquid Crystals* (Ed.: G. W. Taylor), Gordon and Breach, Philadelphia, USA **1991**, p. 395.
[31] See Ref. [30], p. 372.
[32] T. Carlsson, B. Zeks, C. Filipic, A. Levstik, *Ferroelectrics* **1988**, *84*, 223.
[33] C. Druon, J. M. Wacrenier, *Mol. Cryst. Liq. Cryst.* **1982**, *88*, 99.
[34] A. J. Martin, G. Meier, A. Saupe, *Faraday Society Symposium No 5*, **1971**, 119.
[35] G. Meier, A. Saupe, *Mol. Cryst. Liq. Cryst.* **1996**, *1*, 515.
[36] P. L. Nordio, G. Rigatti, U. Segre, *Mol. Phys.* **1973**, *25*, 129.
[37] W. Otowski, W. Demol, W. van Dael, *Mol. Cryst. Liq. Cryst.* **1993**, *226*, 103.
[38] G. S. Attard, *Mol. Phys.* **1986**, *58*, 1087.
[39] M. F. Bone, A. H. Price, M. G. Clark, D. G. McDonnell, *Liquid Crystals and Ordered Fluids*, Vol. 4 (Eds.: A. C. Griffin, J. P. Johnson), Plenum, New York **1984**, p. 799.
[40] W. H. de Jeu, Th. W. Latouwers, *Mol. Cryst. Liq. Cryst.* **1974**, *26*, 225.
[41] L. Benguigui, *Mol. Cryst. Liq. Cryst.* **1984**, *114*, 51.
[42] A. Buka, L. Bata, *Cryst. Res. Tech.* **1981**, *16*, 1439.
[43] T. Carlsson, A. Levstik, B. Zeks, R. Blinc, F. Gouda, S. T. Lagerwall, K. Skarp, C. Filipic, *Phys. Rev.* **1988**, *A38*, 5833.

5 Elastic Properties

David Dunmur and Kazuhisa Toriyama

Elasticity is a macroscopic property of matter defined as the ratio of an applied static stress (force per unit area) to the strain or deformation produced in the material; the dynamic response of a material to stress is determined by its viscosity. In this section we give a simplified formulation of the theory of torsional elasticity and how it applies to liquid crystals. The elastic properties of liquid crystals are perhaps their most characteristic feature, since the response to torsional stress is directly related to the orientational anisotropy of the material. An important aspect of elastic properties is that they depend on intermolecular interactions, and for liquid crystals the elastic constants depend on the two fundamental structural features of these mesophases: anisotropy and orientational order. The dependence of torsional elastic constants on intermolecular interactions is explained, and some models which enable elastic constants to be related to molecular properties are described. The important area of field-induced elastic deformations is introduced, since these are the basis for most electro-optic liquid crystal display devices.

5.1 Introduction to Torsional Elasticity

An important aspect of the macroscopic structure of liquid crystals is their mechanical stability, which is described in terms of elastic properties. In the absence of flow, ordinary liquids cannot support a shear stress, while solids will support compressional, shear and torsional stresses. As might be expected the elastic properties of liquid crystals are intermediate between those of liquids and solids, and depend on the symmetry and phase type. Thus smectic phases with translational order in one direction will have elastic properties similar to those of a solid along that direction, and as the translational order of mesophases increases, so their mechanical properties become more solid-like. The development of the so-called continuum theory for nematic liquid crystals is recorded in a number of publications by Oseen [1], Frank [2], de Gennes and Prost [3] and Vertogen and de Jeu [4]; extensions of the theory to smectic [5] and columnar phases [6] have also been developed. In this section it is intended to give an introduction to elasticity that we hope will make more detailed accounts accessible: the importance of elastic properties in determining the behavior of the mesophases will also be described.

The starting point for a discussion of elastic properties is Hooke's Law, which states that the relative extension of a wire (strain) is proportional to the force per unit area (stress) applied to the wire, and the constant of proportionality is the elastic constant:

$$\frac{F}{A} = k\frac{x}{l} \qquad (1)$$

The corresponding elastic energy can be obtained by integrating with respect to strain to give:

$$U = \int F\,dx \quad \text{or} \quad \frac{U}{V} = \frac{1}{2}ke^2 \qquad (2)$$

where $e = x/l$ is the strain, and U/V is the energy density. To describe the relationship between stress and strain in three dimensions requires a tensorial representation of the elements of stress and strain. If a body is subjected to a general stress, then the relative positions of points within the body will change.

The change in positions of particles at points r_1 and r_2 (see Fig. 1) due to some applied stress can be written as:

$$r'_1 = r_1 + u$$
$$r'_2 = r_2 + v \qquad (3)$$

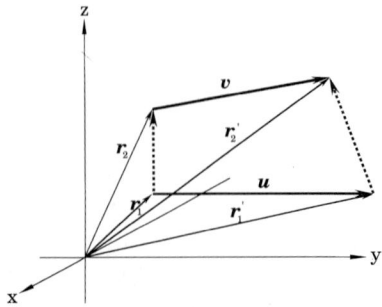

Figure 1. Strain in a body.

for small strains we can write using the tensor suffix notation:

$$v_\alpha = u_\alpha + \left(\frac{du_\beta}{dr_\alpha}\right)r_\beta \qquad (4)$$

where $r_\alpha = r_{2\alpha} - r_{1\alpha}$, so that the change in relative positions of r_1 and r_2 becomes:

$$(r'_{2\alpha} - r'_{1\alpha}) - (r_{2\alpha} - r_{1\alpha}) = \left(\frac{du_\beta}{dr_\alpha}\right)r_\beta \qquad (5)$$

The tensor $e_{\alpha\beta} = \left(\frac{du_\beta}{dr_\alpha}\right)$ is known as the strain tensor, the diagonal elements of which measure extensional strain along the x, y, z axes, while off-diagonal elements are a measure of the shear strains (i.e. the change in the angle between the position vectors r_1 and r_2 in the strained state). For small strains, the change in this angle ($\theta_{1,2}$ defined in radians) projected onto the xy, yz and zx planes is the sum of appropriate off-diagonal elements:

$$\gamma_{xy} = e_{xy} + e_{yx}$$
$$\gamma_{yz} = e_{yz} + e_{zy} \qquad (6)$$
$$\gamma_{zx} = e_{zx} + e_{xz}$$

Clearly if the strain tensor is antisymmetric such that $e_{\alpha\beta} = -e_{\beta\alpha}$, then the shear angles are zero, which corresponds to a whole body rotation without distortion. It is usual, therefore, to redefine the strain tensor in terms of its symmetric and antisymmetric parts, so that:

(7)
$$e_{\alpha\beta} = \frac{1}{2}(e_{\alpha\beta} + e_{\beta\alpha})_{\text{sym}} + \frac{1}{2}(e_{\alpha\beta} - e_{\beta\alpha})_{\text{antisym}}$$

The symmetric part of the strain tensor can be associated with changes in the relative positions of particles within a strained sample. For incompressible materials this is zero, and such an assumption is normally applied to nematic liquid crystals. However, in smectic and columnar phases the translational order results in some nonzero components of the symmetric part of the strain tensor.

We have defined the stress applied to a body as the force per unit area: the force may be perpendicular to the unit area, as with normal stress or pressure, or it can be in the plane of the unit area when it is known as shear stress. For any particular direction defining the normal to an element of area A, there will be a single component of normal stress and two shear stress components. Thus a system of forces acting on a body can be described in terms of the nine components of the stress tensor $\sigma_{\alpha\beta}$, defined as:

$$\sigma_{\alpha\beta} = \frac{n_\alpha F_\beta}{A} \tag{8}$$

where F_α is a force acting on an element of area A, the direction of which is defined by its normal n_α. Diagonal components of the stress are pressure, while off-diagonal elements refer to shear stress; for a stressed body to be in mechanical equilibrium, the normal forces on opposite faces must be equal, and the turning moment represented by off-diagonal elements must be zero (see Fig. 2).

Clearly the strain is a consequence of stress, and for small strains there is a linear relationship between them. Since both stress and strain are second rank tensors, the material property that links them must in general be a fourth rank tensor, so that:

$$\sigma_{\alpha\beta} = \sum_{\gamma,\delta=x,y,z} c_{\alpha\beta,\gamma\delta} e_{\gamma\delta} \tag{9}$$

The intrinsic symmetry of $\sigma_{\alpha\beta}$ and $e_{\gamma\delta}$ reduces the number of components of $c_{\alpha\beta,\gamma\delta}$ to 36: the elasticity tensor $c_{\alpha\beta,\gamma\delta}$ is also symmetric with respect to interchange of pairs of suffixes $\alpha\beta$ and $\gamma\delta$ which further limits the number of independent components to 21. Phase symmetry also lowers the numbers of elasticity components that have to be independently measured.

For small strains, the elastic energy density is second order in the strain so that:

$$u = \frac{1}{2} c_{\alpha\beta,\gamma\delta} e_{\alpha\beta} e_{\gamma\delta} \tag{10}$$

and for isothermal strains this is a direct contribution to the free energy density of the system; in Eq. (10) summation over all suffixes is assumed, representing 81 terms. There is an alternative notation widely used in elasticity theory, which enables the elasticity to be expressed in a more compact way. For homogeneous strain, and in the absence of a turning moment, both $\sigma_{\alpha\beta}$ and $e_{\alpha\beta}$ are symmetric tensors having six independent components. New elastic constants can be defined by:

$$\sigma_i = \sum_{j=1,6} c_{ij} e_j \tag{11}$$

but care must be exercised in manipulating the components with reduced indices, since they no longer transform as tensors.

Nematic liquid crystals having no translational order will not support extensional or shear strains, but they will support torsional strain which results from application of a torque. The torsional strain is conveniently represented in terms of the angular

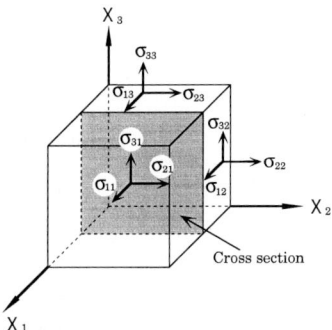

Figure 2. Elements of stress tensor.

displacement of a unit vector (the director) at position r from its equilibrium orientation at the origin, and by analogy with Eq. (4) the torsional strain tensor $e_{\alpha\beta}$ can be defined:

$$n'_\alpha = n_\alpha + \left(\frac{dn_\beta}{dr_\alpha}\right) r_\beta$$

or

$$n'_\alpha = n_\alpha + e_{\alpha\beta} r_\beta \tag{12}$$

As before the energy density can be written as:

$$g = \frac{1}{2} k_{\alpha\beta,\gamma\delta} e_{\alpha\beta} e_{\gamma\delta} \tag{13}$$

where we have now used free energy for an isothermal strain, and the tensor $k_{\alpha\beta,\gamma\delta}$ is the torsional elasticity. If $\boldsymbol{n}$ is chosen to be along the z-axis at the origin, for small strains, the components $e_{\beta z}$ can be neglected, and the nonvanishing elements of the strain tensor can be identified as follows with splay, twist and bend deformations of the director:

$$\text{splay}: \quad e_{xx} = \frac{dn_x}{dx}; \quad e_{yy} = \frac{dn_y}{dy}$$

$$\text{twist}: \quad e_{xy} = \frac{dn_y}{dx}; \quad e_{yx} = \frac{dn_x}{dy} \tag{14}$$

$$\text{bend}: \quad e_{zx} = \frac{dn_x}{dz}; \quad e_{zy} = \frac{dn_y}{dz}$$

these are illustrated in Fig. 3.

The condensed notation for the elements of the torsional elasticity tensor is normally used, and the torsional strain elements are written as a column vector with the components:

$$e_i (i = 1-6) = e_{xx}, e_{yx}, e_{zx}, e_{xy}, e_{yy}, e_{zy} \tag{15}$$

so the free energy density can now be written as:

$$g = \frac{1}{2} \sum_{i,j=1,6} k_{ij} e_i e_j \tag{16}$$

The strain tensor must conform to the symmetry of the liquid crystal phase, and as a result, for nonpolar, nonchiral uniaxial phases there are ten nonzero components of k_{ij}, of which four are independent (k_{11}, k_{22}, k_{33} and k_{24}). These material constants are known as torsional elastic constants for splay (k_{11}), twist (k_{22}), bend (k_{33}) and saddle-splay (k_{24}); terms in k_{24} do not contribute to the free energy for configurations in which the director is constant within a plane, or parallel to a plane. The simplest torsional strains considered for liquid crystals are one dimensional, and so neglect of k_{24} is reasonable, but for more complex director configurations and at surfaces, k_{24} can contribute to the free energy [7]. In particular k_{24} is important for curved interfaces of liquid crystals, and so must be included in the description of lyotropic and membrane liquid crystals [8]. Evaluation of Eq. (16) making the stated assumptions, leads to [9]:

$$g = \frac{1}{2} k_{11} \left(e_{xx} + e_{yy}\right)^2 + \frac{1}{2} k_{22} \left(e_{xy} - e_{yx}\right)^2$$
$$+ \frac{1}{2} k_{33} \left(e_{zx} + e_{zy}\right)^2$$
$$- \left(k_{22} - k_{24}\right)\left(e_{xx} e_{yy} - e_{xy} e_{yx}\right) \tag{17}$$

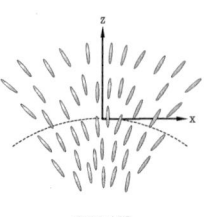
SPLAY

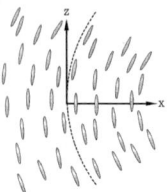
BEND

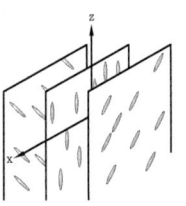
TWIST

Figure 3. Principal torsional elastic deformations.

This expression is often written in a more compact vector notation as:

$$g = \frac{1}{2}\left[k_{11}(\nabla \cdot n)^2 + k_{22}(n \cdot \nabla \times n)^2 \right.$$
$$+ k_{33}(n \times \nabla \times n)^2 - (k_{22} - k_{24})$$
$$\left. \cdot (\nabla \cdot \{n\nabla \cdot n + n \times \nabla \times n\})\right] \quad (18)$$

In the above we have assumed that the lowest energy state is one of uniform parallel alignment of the director; however, it is possible to modify these expressions for situations where the lowest energy state might be one of uniform splay, bend or twist. Of these, the last is important because it describes the helical liquid crystal phases that result from chiral molecules. Such states arise if there are terms in the free energy that are linear in the strain, and for a chiral nematic the free energy density becomes:

$$g = -k_2(e_{xy} - e_{yx}) + \frac{1}{2}k_{11}(e_{xx} + e_{yy})^2 \quad (19)$$
$$+ \frac{1}{2}k_{22}(e_{xy} - e_{yx})^2 + \frac{1}{2}k_{33}(e_{zx} + e_{zy})^2$$

Defining the torsional twist strain as $t = (e_{xy} - e_{yx})$, and minimising Eq. (19) with respect to t results in a stabilized helical structure having a finite twist strain $t_0 = k_2/k_{22}$, and the free energy density can be written as:

$$g = \frac{1}{2}k_{11}(e_{xx} + e_{yy})^2$$
$$+ \frac{1}{2}k_{22}(e_{xy} - e_{yx} - t_0)^2$$
$$+ \frac{1}{2}k_{33}(e_{zx} + e_{zy})^2 - \frac{1}{2}k_{22}t_0^2 \quad (20)$$

The lowest energy state for this structure is where the director describes a helix along one of the axes. Thus assuming that there is no bend or splay strain energy, a chiral nematic has a director configuration such that:

$$n_x = \cos(2\pi z/p); \; n_y = \sin(2\pi z/p); \; n_z = 0$$

where p is the pitch of the helix, as illustrated in Fig. 4.

Using this director distribution, the non-vanishing terms in the free energy expression Eq. (20) are:

$$g = \frac{1}{2}k_{22}\left[n_y\left(\frac{\partial n_x}{\partial n_z}\right) - n_x\left(\frac{\partial n_y}{\partial n_z}\right) - t_0\right]^2$$
$$- \frac{1}{2}k_{22}t_0^2 \quad (21)$$

which is a minimum when $t_0 = 2\pi/p$, i.e. the lowest energy state is one of uniform pitch.

It is also possible to envisage minimum energy structures with non-zero splay or bend strain, when other terms linear in strain contribute to the free energy. However, a uniformly splayed or bent structure will no longer have the reversal symmetry $+n = -n$ and so will be associated with permanently polarized structures; a uniformly bent structure would have to be biaxial with a polar structure perpendicular to the major axis. These structures, illustrated in Fig. 5, are associated with the phenomenon known as flexoelectricity, where polarization is coupled to elastic strain.

It is not essential that the molecular asymmetry is linked to an electric polarization, although symmetry suggests that the two are

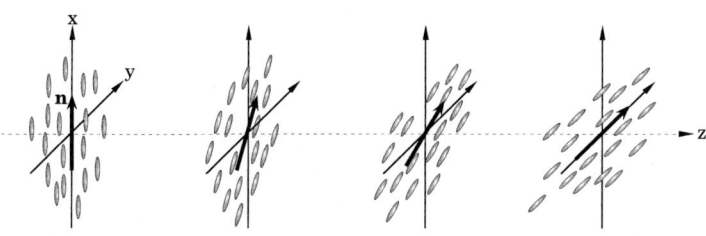

Figure 4. Director helix in a chiral nematic.

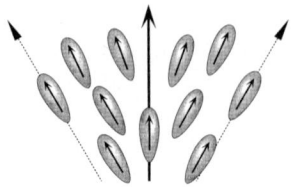

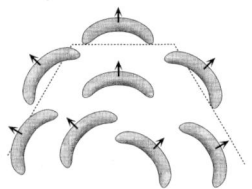

Figure 5. Flexo-electric structures.

coupled. Equivalent structures to those illustrated may be obtained on the basis of shape asymmetry, the so called steric dipole.

So far only torsional contributions to the elastic strain energy have been considered, but smectic and columnar liquid crystals having translational order may support extensional and shear strain like solids along certain directions. For a uniaxial SmA phase with one degree of translational freedom, the only homogeneous strain supported is an extension or compression along the axis of translational order (i.e. perpendicular to the layers). Bend and twist strains involve changes in layer spacing, and so are likely to be of very high energy; they will couple with layer compression but will be high order contributions to the free energy, and are neglected for small strains. Thus the elastic free energy of a SmA phase reduces to:

$$g = \frac{1}{2} B \left(\frac{du_z}{dz} \right)^2 + \frac{1}{2} k_{11} (\nabla \cdot \mathbf{n})^2 \qquad (22)$$

where B is a compression elastic constant for the smectic layers; (du_z/dz) is the layer strain along the layer normal, and k_{11} is the splay elastic constant.

The description of the elastic properties of columnar phases, biaxial smectics, chiral smectics and more ordered liquid crystal phases is being developed [10]. Since columnar phases are two dimensional solids, it is expected that compression elasticity in the plane perpendicular to the columns will be important, and the high strain energy associated with in-plane deformations prevents splay or twist torsional distortion [11]. It is however possible to bend the columns, while maintaining a constant in-plane separation of the columns, so bend distortions can be expected in columnar liquid crystals: the corresponding free energy density for a uniaxial columnar phase becomes:

$$g = \frac{1}{2} B \left(\frac{du_x}{dx} + \frac{du_y}{dy} \right)^2$$
$$+ \frac{1}{2} C \left[\left(\frac{du_x}{dx} - \frac{du_y}{dy} \right)^2 + \left(\frac{du_x}{dy} + \frac{du_y}{dx} \right)^2 \right]$$
$$+ \frac{1}{2} k_{33} (\mathbf{n} \times \nabla \times \mathbf{n})^2 \qquad (23)$$

Results for biaxial smectics and columnar phases have additional compressional terms, but tilted smectic phases can support additional torsional distortions. Such phases are conveniently described in terms of two directors, one along the tilt direction, corresponding to the nematic director $\mathbf{n}$, and the projection of $\mathbf{n}$ on the layer plane, known as the $\mathbf{c}$-director (see Fig. 6).

In practice it is mathematically more convenient to define a director (**a**) normal to the

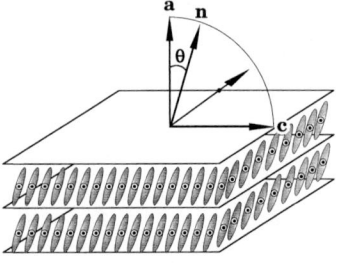

Figure 6. Definition of **a** and **c** directors.

layer such that:

$$\mathbf{n} = \mathbf{a}\cos\theta + \mathbf{c}\sin\theta \quad (24)$$

Torsional distortions can now be written in terms of derivatives of $\mathbf{a}$ and $\mathbf{c}$, and it is found [10] that nine torsional elastic constants are required for the smectic C phase. Mention should be made of the biaxial smectic C* phase, which has a twist axis along the normal to the smectic layers. This helix is associated with a twist in the **c**-director, and so elastic strain energy associated with this can be described by terms similar to those evaluated for the chiral nematic phase.

5.2 Director Distribution, Defects and Fluctuations

The development of Eq. (18) was based on the idea that torsional strain resulted from a director that is a function of position. Provided that the first nonvanishing term in the free energy is quadratic in torsional strain, the minimum energy configuration of a liquid crystal will be that in which the torsional strain is zero and the director is everywhere parallel to a symmetry axis. Linear contributions to the strain energy result in equilibrium structures in which the director is not uniformly parallel, but has some particular spatial dependence, as with the chiral nematic phase. External influences will also affect the director distribution in space, and surface interaction, external fields and thermal fluctuations give rise to structures in which the director is a function of position. Neglecting fluctuations, the director distribution will be that which minimizes the free energy given by:

$$g = \frac{1}{2}\int \left[k_{11}(\boldsymbol{\nabla}\cdot\mathbf{n})^2 + k_{22}(\mathbf{n}\cdot\boldsymbol{\nabla}\times\mathbf{n})^2 + k_{33}(\mathbf{n}\times\boldsymbol{\nabla}\times\mathbf{n})^2 + g_{\text{ext}}\right]d\mathbf{r} \quad (25)$$

here g_{ext} contains contributions to the free energy from external forces.

Experimental observations indicate that it is possible to define the director orientation by surface treatment or external electric or magnetic fields, and a simple example of the effect of surfaces on the director distribution in a nematic is illustrated in Fig. 7.

Two rubbed plates are held at a distance l apart, such that the alignment directions include an angle θ. The director orientation of a nematic between these plates will be defined by the rubbing directions at the plate surfaces, but in the bulk the director distribution will be that which minimizes Eq. (25). De Gennes [12] has shown that for this simple twist deformation, the director orientation varies linearly with position as indicated in the figure. Under these circumstances, the director can be written in terms of:

$$n_x = \sin\left(\frac{\theta y}{l}\right); \quad n_y = 0; \quad n_z = \cos\left(\frac{\theta y}{l}\right) \quad (26)$$

and the corresponding free energy density per unit wall area becomes:

$$g = \frac{1}{2}k_{22}\frac{\theta^2}{l} \quad (27)$$

The torque t is given by:

$$t = -\frac{dg}{d\theta} = -\frac{k_{22}\theta}{l} \quad (28)$$

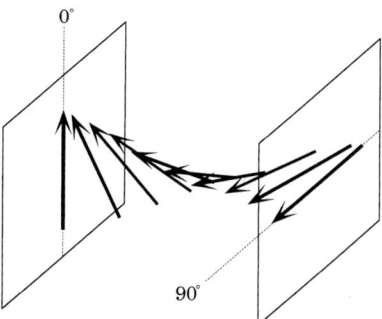

Figure 7. Director distribution between twisted plates for $\theta = 90°$.

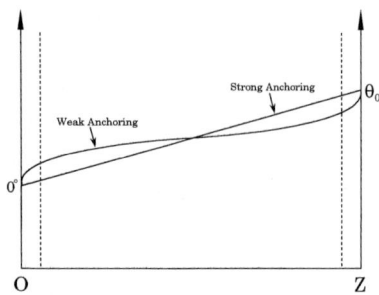

Figure 8. Plot of director orientation for strong and weak anchoring.

and is constant through the sample. Thus plate 1 exerts a torque of $k_{22}\theta/l$ on plate 2, while plate 2 exerts a torque of $-k_{22}\theta/l$ on plate 1.

This calculation assumes that the torques of opposing plates are insufficient to disturb the orientation of the director at the surface: such a condition is known as strong anchoring. In practice the strength of surface interactions may have to be considered, and in the example given, weak anchoring would cause the director orientation close to the boundary surfaces to depart from a linear dependence on position, as illustrated in Fig. 8.

For weak anchoring there is a competition between the torque in the bulk due to one plate, and the torque resulting from the other surface; this is usually confined to a boundary region as indicated in the figure. External electric and magnetic fields will also affect the equilibrium director distribution, and this is the basis of many liquid crystal applications; these effects will be considered later.

5.2.1 Defects in Liquid Crystals

It has been explained how nonuniform director distributions can arise owing to surface interactions or external fields, and these can be detected optically or by other physical methods. In many liquid crystal samples, large inhomogeneities of director orientation associated with structural defects can be observed by polarized light microscopy, and these clearly involve elastic strain energy. In this section we will outline the elastic theory of these defects: reviews of this topic have been given by Chandrasekhar [13] and Kleman [14]. As an illustration we will consider the simplest type of fluid defect observed in nematics named disclination lines by Frank.

In two dimensions and setting all the torsional elastic constants equal, the free energy density expression Eq. (17) can be written as:

$$g = \frac{1}{2}k\left[\left(e_{xx} + e_{yy}\right)^2 + \left(e_{xy} - e_{yx}\right)^2\right] \quad (29)$$

If the director is confined to a plane, then its components may be represented by:

$$n_x = \cos\theta(r) \quad \text{and} \quad n_y = \sin\theta(r)$$

where $\theta(r)$ is a function of x and y in the plane. The free energy density then becomes:

$$g = \frac{1}{2}k\left[\left(\frac{d\theta}{dx}\right)^2 + \left(\frac{d\theta}{dy}\right)^2\right] \quad (30)$$

This is minimized, corresponding to an equilibrium structure when:

$$\frac{d^2\theta}{dx^2} + \frac{d^2\theta}{dy^2} = 0 \quad (31)$$

The defect-free solution to Eq. (31) is obviously when θ is independent of position, but other solutions corresponding to disclination lines are given by:

$$\theta = s\tan^{-1}\left(\frac{y}{x}\right) + \theta_0 \quad (32)$$

where s is the strength of the disclination.

5.2 Director Distribution, Defects and Fluctuations

For a fixed distance from the origin, the director orientation as given by Eq. (32) changes by $s\pi$ as y/x varies from $+\infty$ to $-\infty$; contours of solutions for different s are given in Fig. 9. These director distributions give rise to characteristic images of disclination lines, which are readily observed for thin films in a polarizing microscope.

The elastic strain energy associated with a disclination line can now be calculated from Eq. (30) as:

$$g/\text{unit length} = s^2 \pi k \ln\left(\frac{r_{max}}{r_{min}}\right) \quad (33)$$

r_{max} is the distance from the centre of the disclination at which the strain energy becomes effectively zero, while r_{min} defines the size of the core of the disclination, the energy of which cannot be calculated from the above equations, which apply to small strains. The quadratic dependence of the strain energy on s explains why usually only disclinations of low s are observed in low molecular weight liquid crystals: higher strength disclinations have been observed in polymer liquid crystals [14], since these can support higher strain energies.

In real samples there will be a number of disclinations corresponding to different magnitudes and signs of s. The resultant effect of a number of disclinations on the director orientation at some point in the fluid can be obtained by combining the corresponding director angles. The corresponding free energy shows that disclinations of similar sign will repel each other, while those of opposite sign will be mutually attracted.

This simple treatment of liquid crystalline defects is only applicable to nematics, and the detailed appearance of disclination lines will differ from the simple structures described above because of differences between the elastic constants for splay, twist and bend. In smectic phases, defects associated with positional disorder of layers will also be important, and some smectic phase defects such as edge dislocations have topologies similar to those described for crystals. The defect structures of liquid crystals contribute to the characteristic optical tex-

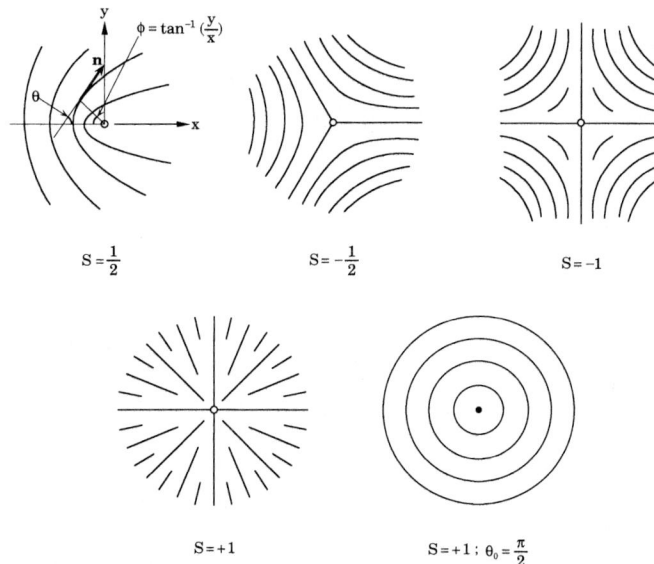

Figure 9. Disclinations obtained by solution of Eq. (32). The lines represent contours of the director.

tures, which are used to identify different liquid crystal phases, and it is apparent that there is a wealth of information on elastic properties contained in the observed optical textures.

5.2.2 Fluctuations

Neglecting static distortions which may result from interactions with surfaces or external fields, the lowest energy director configuration is predicted to be one of uniform director orientation. However, even in the absence of external influences, locally the director fluctuates about its lowest energy configuration, and this local director orientation disorder can contribute to liquid crystal properties. For example the characteristic turbidity of liquid crystalline phases is due to fluctuations in the refractive indices on the scale of the order of the wavelength of light. The origin of this nonuniform alignment is thermal excitation of librational modes associated with the director. The energy is assumed to be a quadratic function of the torsional strain; but since the torsional elastic constants are very small, little thermal energy is required to excite a torsional libration and so disturb the uniform director configuration. In reality there are many torsional modes that can be simultaneously excited, and the long-range orientational structure of the liquid crystal can be extensively disordered. This picture of director disorder assumes that the molecular organization is unaffected by the long-range macroscopic disorder, although it must be remembered that the director has no independent existence, and is itself defined by the molecular orientations averaged over an appropriate volume.

The director orientational disorder can be formally represented in exactly the same way as the molecular disorder by defining a director order parameter as:

$$Q_{\alpha\beta} = \frac{1}{2}\langle 3n_\alpha(r)n_\beta(r) - \delta_{\alpha\beta}\rangle \quad (34)$$

where $n_\alpha(r)$ are the components of the director as a function of position.

The spatial variation of the director can be expressed in terms of Fourier components of wave vector q, such that:

$$n_\alpha(q) = V^{-1}\int n_\alpha(r)\exp(i q \cdot r)d^3r \quad (35)$$
$$n_\alpha(r) = V(2\pi)^{-3}\int n_\alpha(q)\exp(-i q \cdot r)d^3q$$

Small distortions from a director axis z can be expressed in terms of $n_x(q)$ and $n_y(q)$; however these are not the normal coordinates for a mode of wavelength $\lambda = 2\pi/q$, and using the following transformation, the free energy can be written as the sum of quadratic contributions from the normal modes:

$$\begin{bmatrix} n_1(q) \\ n_2(q) \end{bmatrix} = \frac{1}{\sqrt{(q_x^2 + q_y^2)}} \begin{bmatrix} q_x & q_y \\ -q_y & q_x \end{bmatrix} \begin{bmatrix} n_x \\ n_y \end{bmatrix} \quad (36)$$

For each q in a uniaxial phase there are two normal modes corresponding to a splay–bend distortion $n_1(q)$ and a twist–bend distortion $n_2(q)$: biaxial liquid crystal phases have five normal modes for each value of q. The free energy density can be written in terms of the normal coordinates for torsional displacement in a uniaxial nematic as:

$$g = \frac{1}{2}\sum_q \left[\lambda_+(q)n_1(q)^2 + \lambda_-(q)n_2(q)^2\right] \quad (37)$$

The coefficients λ_+ and λ_- are given by:

$$\lambda_+ = k_{11}(q_x^2 + q_y^2) + k_{33}q_z^2$$
$$\lambda_- = k_{22}(q_x^2 + q_y^2) + k_{33}q_z^2 \quad (38)$$

Because the torsional elastic constants are small, the terms in the expression for the free energy density are treated classically so that according to the equipartition theorem the average contribution of each mode is supposed to be $k_B T/2$, giving:

$$\langle n_1(\mathbf{q})^2 \rangle = \frac{k_B T}{V \lambda_+(\mathbf{q})}$$

$$\langle n_2(\mathbf{q})^2 \rangle = \frac{k_B T}{V \lambda_-(\mathbf{q})} \quad (39)$$

It is now possible to determine the mean square fluctuation of the director from its aligned state along the z-axis, such that:

$$\langle n_x(\mathbf{r})^2 + n_y(\mathbf{r})^2 \rangle$$

$$= \frac{V}{(2\pi)^3} \int_{q_{min}}^{q_{max}} \langle n_1(\mathbf{q})^2 + n_2(\mathbf{q})^2 \rangle d^3 \mathbf{q} \quad (40)$$

There is one problem remaining which is to fix the limits for the integration over q [15]. The lower limit for q is determined by the volume of the sample, since it is unrealistic to have distortion modes of wavelength longer than the maximum dimensions of the sample container: for an infinite sample $q_{min} = 0$. The upper limit for q corresponds to the fluctuation mode of shortest wavelength, which is likely to be of the order of molecular dimensions.

If the elastic constants for splay twist and bend are assumed to be independent of q, and for simplicity are set equal to k, then evaluation of Eq. (40) enables the director order parameter to be determined as follows:

$$\langle Q_{zz}(\mathbf{r}) \rangle = \frac{1}{2} \langle 3 n_z(\mathbf{r})^2 - 1 \rangle$$

$$= 1 - \frac{3}{2} \langle n_x(\mathbf{r})^2 + n_y(\mathbf{r})^2 \rangle$$

$$= 1 - \frac{3 k_B T q_{max}}{2 \pi^2 k} \quad (41)$$

Fluctuations in the director orientation will modulate the anisotropy of physical properties such as the electric permittivity tensor, which can be written as:

$$\varepsilon_{\alpha\beta} = \bar{\varepsilon} + \frac{\Delta\varepsilon}{3} \left(3 n_\alpha(\mathbf{r}) n_\beta(\mathbf{r}) - \delta_{\alpha\beta} \right) \quad (42)$$

where $\bar{\varepsilon}$ is the average permittivity and $\Delta\varepsilon = \varepsilon_{\parallel} - \varepsilon_{\perp}$ is the permittivity anisotropy. Light scattering arises because of spatial and time-dependent fluctuations of the local dielectric tensor:

$$\delta\varepsilon_{\alpha\beta}(\mathbf{r}, t) = \varepsilon_{\alpha\beta}(\mathbf{r}, t) - \varepsilon_{\alpha\beta}(0) \quad (43)$$

and if these fluctuations are expressed in terms of Fourier components of wave-vector $\mathbf{q}$, the intensity of scattered light can be derived as [16]:

$$I_{i,f}(\mathbf{q}, \omega) = \frac{\text{const}}{2\pi}$$

$$\cdot \int_{-\infty}^{+\infty} \langle \delta\varepsilon_{if}(\mathbf{q},0) \delta\varepsilon_{if}(\mathbf{q},t) \rangle \exp - i\omega t \, dt \quad (44)$$

The subscripts i, f refer to the polarization directions of the incident and scattered light, $\mathbf{q}$ is the wave-vector of the scattered light, and the constant is given by:

$$\text{const} = \frac{\pi^2 I_0}{\lambda^4 R^2 \varepsilon_0^2} \quad (45)$$

where I_0 is the incident light intensity, wavelength λ, and R is the distance from the scattering volume to the detector. Choice of suitable scattering geometries [17] and measurement of the intensity as a function of scattering angle allow the determination of the elastic constant ratios k_{33}/k_{11} and k_{22}/k_{11} [18].

5.3 Curvature Elasticity of Liquid Crystals in Three Dimensions

As pointed out for Eq. (17) there is a contribution to the elastic energy from saddle-splay distortions, which can be neglected in some circumstances. However for distortions such that the director is not constant in one plane, the term in k_{24} must be included in the free energy. Liquid crystal structures in which there is curvature in two dimensions can support saddle-splay distortion, and such curvature is often associated with fluid interfaces in liquid crystals. The main area of importance for interfacial contributions to the elastic energy is lyotropic liquid crystals formed by molecular association and segregation in materials having a minimum of two chemical constituents. A typical lyotropic system consisting of water and an amphiphile exhibits lamellar (smectic), hexagonal (columnar) and occasionally nematic liquid crystal phases, and in the phase diagram these are often separated by or bounded by narrow regions of cubic phases. The molecular organization in lyotropics represents an extreme example of amphiphilic liquid crystal phases in which the spatial separation of molecular subunits is stabilized by a second component, usually water, to produce a partitioning of space into polar and nonpolar regions. Inverse lyotropic phase structures can also be prepared where the spatial separation of alkyl chains is stabilized by a nonpolar solvent such as hexane. Spatial organization of molecular subunits can also occur in nonlyotropic liquid crystals resulting in a variety of modulated phase structures, and the formation of curved interfaces is of especial importance for biological membranes.

The macroscopic topology of lyotropic or liquid crystal phases involving segregation is determined by the curvature of the interface; a lamellar structure has zero curvature, while micellar phases or hexagonal phases exhibit interfacial curvature. An interface is defined by the segregation of different molecules or molecular subunits. Deformation of this interface may occur in a variety of ways: tangential stress will cause a planar stretch, while molecular tilting will generate a stress normal to the interface. Both have a high associated elastic energy, but deformation through interfacial curvature involves much lower elastic energies, and so is the preferred mechanism through which topological changes occur. The formation of spherical micelles requires curvature strain in three dimensions, and a director defined along the normal to the interface is no longer uniform within a plane or parallel to a plane. Bearing this in mind the curvature elastic energy per unit area (w) can be written as (c.f. Eq. (17)):

$$w = \frac{1}{2} k_{11}^{(a)} \left(e_{xx} + e_{yy} - c_0\right)^2 + k_{24}^{(a)} \left(e_{xx} e_{yy} - e_{xy} e_{yx}\right) \tag{46}$$

where c_0 allows for equilibrium structures having a non-zero splay. Since the deformation is confined to the interface and the director n remains everywhere normal to the interface, there are no bend or twist contributions to the elastic energy, and $e_{yx} = e_{xy} = 0$. The first term in Eq. (46) is clearly a splay elastic energy, while the second term was designated by Frank as a saddle-splay energy. From the definition of curvature strain elements $e_{\alpha\beta}$ (Eq. 14), it can be seen that for small strains:

$$e_{xx} = \frac{dn_x}{dx} = \frac{1}{R_1}$$

and

$$e_{yy} = \frac{dn_y}{dy} = \frac{1}{R_2} \tag{47}$$

where R_1 and R_2 are the radii of curvature of the surface in the zx and zy planes respectively (see Fig. 10), and c_0 is the spontaneous curvature of the surface in the absence of strain. Thus the elastic free energy per unit area can now be written in terms of the

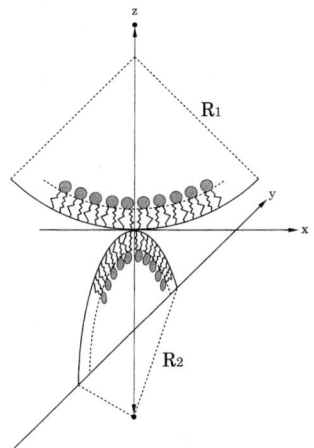

Figure 10. Principal radii of curvature in a saddle-splay distortion.

principal radii of curvature as:

$$w = \frac{1}{2} k_{\mathrm{m}}^{(a)} \left(\frac{1}{R_1} + \frac{1}{R_2} - c_0 \right)^2 + k_{\mathrm{g}}^{(a)} \frac{1}{R_1 R_2} \quad (48)$$

with $k_{\mathrm{m}}^{(a)}$ and $k_{\mathrm{g}}^{(a)}$ defined as elastic constants for the mean curvature $\frac{1}{2}(R_1^{-1} + R_2^{-1})$ and gaussian curvature $(R_1 R_2)^{-1}$ respectively.

Different types of surface can be categorized in terms of their mean and gaussian curvatures. For example, a sphere of radius R has a mean curvature of R^{-1} and a gaussian curvature of R^{-2}, while a cylinder has one principal radius of curvature equal to infinity, so the gaussian curvature is zero. Surfaces with zero mean curvature such that $R_1^{-1} = -R_2^{-1}$ are known as minimal surfaces and have been proposed as structures for some cubic phases [19].

5.4 Electric and Magnetic Field-induced Deformations

Competition between two competing elastic torques results in a director distribution that minimises the free energy. The origin of the torque may be mechanical as represented in Fig. 7, or it may arise through a coupling between an external field and the corresponding susceptibility anisotropy. For a uniaxial material the internal energy density in the presence of a field (F) can be written as:

$$u = -\frac{F^2}{2} \left[\bar{\chi} + \frac{\Delta \chi}{3} (3 \sin^2 \theta - 1) \right] \quad (49)$$

where $(90 - \theta°)$ is the angle between the field and the director (see Fig. 11), and so the torque is given by:

$$t = -\frac{\mathrm{d}u}{\mathrm{d}\theta} = -\Delta \chi F^2 \cos\theta \sin\theta$$
$$= -\Delta \chi (\boldsymbol{F} \cdot \boldsymbol{n})(\boldsymbol{F} \times \boldsymbol{n}) \quad (50)$$

The torque is zero when the field is parallel or perpendicular to the director, but depending on the sign of $\Delta \chi$, one of these states is in stable equilibrium, while the other is in unstable equilibrium. For positive $\Delta \chi$, increasing a field perpendicular to n will raise the energy and hence destabilize this state: eventually the increase in energy due to the field exceeds the elastic energy, and so the liquid crystal adopts a new director configuration. Reorientation of the liquid crystal does not necessarily occur as soon as the field is applied (because the torque is zero), and can be a threshold phenomenon occur-

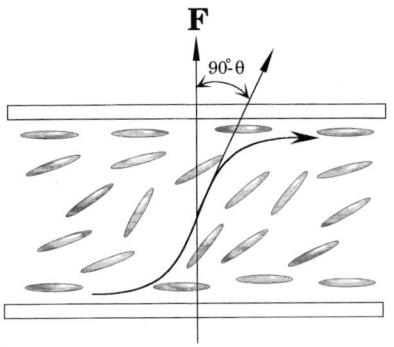

Figure 11. Director reorientation by a field.

ring above a critical field strength. The magnitude of the threshold field can be estimated as follows; if the director is deformed over a distance l, the strain is of order l^{-1}, and the elastic free energy density is k/l^2. The angle-dependent contribution to the free energy density from the field is $\approx \Delta\chi F^2$, and the threshold field for director reorientation is when these energies become equal:

$$\Delta\chi F^2 \approx \frac{k}{l^2}$$

i.e. $F_{\text{threshold}} \approx \frac{1}{l}\sqrt{\frac{k}{\Delta\chi}}$ (51)

the length l is typically of order of the sample thickness. This reorientation of the director by a field is known as a Fréedericksz transition [20], and forms the basis of most display applications of nematic liquid crystals. If the director is not perpendicular (or parallel) to the field because of misalignment or surface tilt of the director, then there is a torque for vanishingly small field strengths, and deformation occurs without a threshold.

5.4.1 Director Distribution in Magnetic Fields

Before outlining the effect of fields on the orientation of the director, it must be emphasized that surface interactions and boundary conditions are usually of importance. The models applied here assume that there is a well-defined director distribution in the absence of any external field, and in practice this can only be provided by suitable treatment of boundary surfaces. The strength of surface interactions must also be considered as this will influence the equilibrium director configuration in the presence of a field. For the simplest description of field effects in liquid crystals it is usual to assume an infinite anchoring energy for the director at the surface: this is the strong anchoring limit, but it is possible to include finite surface coupling energies [21–24]. It is found that there can still be a threshold response to external fields, but the critical field strength is reduced in comparison with the strong anchoring limit.

It is easier to model the deformation induced by magnetic fields than electric fields, because for nonferromagnetic liquid crystals the relative volume magnetic susceptibility anisotropy is very small ($\approx 10^{-6}$) in comparison with the corresponding dielectric anisotropy (≈ 10). This means that the induced magnetization is always parallel to the field direction, in contrast with the electric polarization, which in general makes some angle with the electric field. The magnetic contribution to the free energy of a liquid crystal can be written as (see Eq. 49):

$$g_{\text{mag}} = -\frac{1}{2\mu_0}\int \left(\chi_\perp + \Delta\chi \sin^2\theta\right) B^2 \, dV \quad (52)$$

where $90-\theta$ is the angle between the director and the magnetic field. Thus the total free energy including any elastic deformation becomes:

$$g = \frac{1}{2}\int k_{11}(\nabla \cdot \boldsymbol{n})^2 + k_{22}(\boldsymbol{n} \cdot \nabla \times \boldsymbol{n})^2$$
$$+ k_{33}(\boldsymbol{n} \times \nabla \times \boldsymbol{n})^2$$
$$- \mu_0^{-1}\left(\chi_\perp B^2 + \Delta\chi(\boldsymbol{n} \cdot \boldsymbol{B})^2\right) dr^3 \quad (53)$$

and the equilibrium director distribution $\boldsymbol{n}(r)$ is that which minimizes g. To proceed further, it is necessary to specify the geometry of the system more closely, and it is usual to consider three standard configurations (Fig. 12) where the applied field is perpendicular to the director of a uniformly aligned sample of positive susceptibility anisotropy.

These configurations define Fréedericksz transitions from undeformed to deformed states for which the threshold fields are sep-

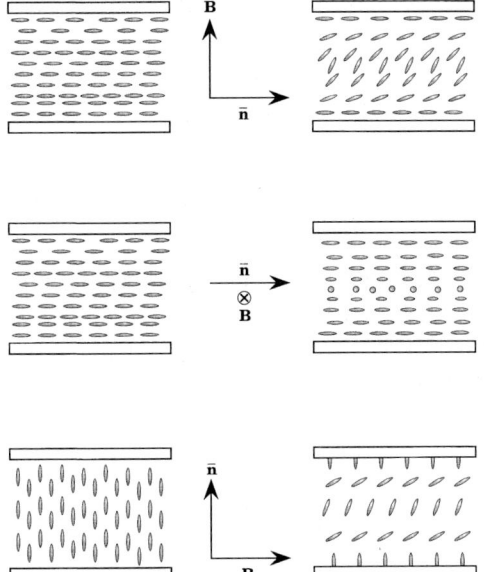

Figure 12. Field induced splay, twist and bend deformations.

arately related to the elastic constants for splay twist and bend. Only deformations for which the director is uniform in a plane are considered (so k_{24} is unimportant), and applying the Euler–Lagrange equations [25] to minimize g (Eq. (53)) gives the following results for the three cases in Fig. 12.

(1) Splay:

$$\frac{d}{dz}\left[\{k_{11}\cos^2\theta + k_{33}\sin^2\theta\}\left(\frac{d\theta}{dz}\right)^2 + \mu_0^{-1}\Delta\chi B^2 \sin^2\theta\right] = 0$$

(2) Twist:

$$\frac{d}{dz}\left[k_{22}\left(\frac{d\phi}{dz}\right)^2 + \mu_0^{-1}\Delta\chi B^2 \sin^2\phi\right] = 0$$

(3) Bend:

$$\frac{d}{dz}\left[\{k_{33}\cos^2\theta + k_{11}\sin^2\theta\}\left(\frac{d\theta}{dz}\right)^2 + \mu_0^{-1}\Delta\chi B^2 \sin^2\theta\right] = 0 \quad (54)$$

where the deformation plane is defined by the directions $z-\mathbf{n}$. The case (1) will be considered in a little more detail [26], and results will be quoted for other boundary conditions.

Eq. (54) implies that:

$$[k_{11}\cos^2\theta + k_{33}\sin^2\theta]\left(\frac{d\theta}{dz}\right)^2 + \mu_0^{-1}\Delta\chi B^2 \sin^2\theta = \text{constant } C \quad (55)$$

and this constant may be evaluated by recognizing that for the selected geometry, θ will be a maximum at the mid-point of the cell such that $z=d/2$. Setting this condition in Eq. (55) gives that:

$$C = \mu_0^{-1}\Delta\chi B_2 \sin^2\theta_m \quad (56)$$

Substitution in Eq. (55) and rearranging now gives a differential equation for the equilibrium director distribution in terms of the angle θ, the field strength and the position in the cell:

$$\frac{d\theta}{dz} = \left(\mu_0^{-1}\Delta\chi B^2\right)^{1/2} \cdot \left[\frac{\sin^2\theta_m - \sin^2\theta}{k_{11}\cos^2\theta + k_{33}\sin^2\theta}\right]^{1/2} \quad (57)$$

or in integral form:

$$\int_0^{\theta_m}\left[\frac{1+k'\sin^2\theta}{\sin^2\theta_m - \sin^2\theta}\right]^{1/2} d\theta$$

$$= \int_0^{d/2}\left(\frac{\mu_0^{-1}\Delta\chi}{k_{11}}\right)^{1/2} B \, dz \quad (58)$$

where a reduced elastic constant $k' = (k_{33}-k_{11})/k_{11}$ has been introduced. It is now convenient to change the variable θ to ψ using:

$$\sin\theta = \sin\theta_m \sin\psi, \text{ and}$$

$$\frac{d\theta}{d\psi} = \left[\frac{\sin^2\theta_m - \sin^2\theta}{1-\sin^2\theta}\right]^{1/2}$$

so that when $\theta=\theta_m$, $\psi=\pi/2$ and $z=d/2$, hence Eq. (58) becomes:

$$\int_0^{\pi/2}\left[\frac{1+k'\sin^2\theta_m\sin^2\psi}{1-\sin^2\theta_m\sin^2\psi}\right]^{1/2}d\psi$$
$$=\int_0^{d/2}\left(\frac{\mu_0^{-1}\Delta\chi}{k_{11}}\right)^{1/2}B\,dz \quad (59)$$

The right hand side of Eq. (59) is simply $\frac{Bd}{2}\left(\frac{\mu_0^{-1}\Delta\chi}{k_{11}}\right)^{1/2}$, but the left hand side is an elliptic integral of the third kind and must be evaluated numerically. At the threshold $\theta_m=0$, the left hand side reduces to $\pi/2$ giving the result for the Fréedericksz threshold magnetic field (B_c) as:

$$B_c=\frac{\pi}{d}\left(\frac{k_{11}}{\mu_0^{-1}\Delta\chi}\right)^{1/2} \quad (60)$$

for fields above threshold:

$$\frac{B}{B_c}=\frac{2}{\pi}\int_0^{\pi/2}\left[\frac{1+k'\sin^2\theta_m\sin^2\psi}{1-\sin^2\theta_m\sin^2\psi}\right]^{1/2}d\psi \quad (61)$$

The director distribution through the sample may be obtained for a particular value of B/B_c by finding θ_m from Eq. (61), and the value of $\theta(z)$ at position z within the sample is obtained (Eq. 58) from:

$$\int_0^{\theta(z)}\left[\frac{1+k'\sin^2\theta}{\sin^2\theta_m-\sin^2\theta}\right]^{1/2}d\theta$$
$$=zB\left(\frac{\mu_0^{-1}\Delta\chi}{k_{11}}\right)^{1/2}=\frac{z\pi B}{dB_c} \quad (62)$$

Results for $\theta(z)$ at various values of $\frac{B}{B_c}$ are plotted in Fig. 13 for an assumed value of $k'=1$, corresponding to $k_{33}=2k_{11}$.

Similar derivations can be applied to the other geometries illustrated in Fig. 12 with the results:

(2) Twist:

$$\frac{B}{B_c}=\frac{2}{\pi}\int_0^{\pi/2}\left[\frac{1}{1-\sin^2\phi_m\sin^2\psi}\right]^{1/2}d\psi$$

and $B_c=\frac{\pi}{d}\left(\frac{k_{22}}{\mu_0^{-1}\Delta\chi}\right)^{1/2}$ (63)

where the right hand side is an elliptic integral of the first kind.

The result for a bend deformation is the same as that given for splay, except the elastic constants k_{11} and k_{33} are interchanged; thus the threshold field is given by:

$$B_c=\frac{\pi}{d}\left(\frac{k_{33}}{\mu_0^{-1}\Delta\chi}\right)^{1/2} \quad (64)$$

In the above treatment, splay and bend deformations can be regarded as limiting cases of the effect of a field on a uniformly tilted structure with a zero-field tilt $\theta_0=0°$ (splay) and $\theta_0=90°$ (bend). For cell configurations with a uniform tilt between 0° and 90°, there will still be a threshold response provided that the field is perpendicular or parallel (depending on the susceptibility an-

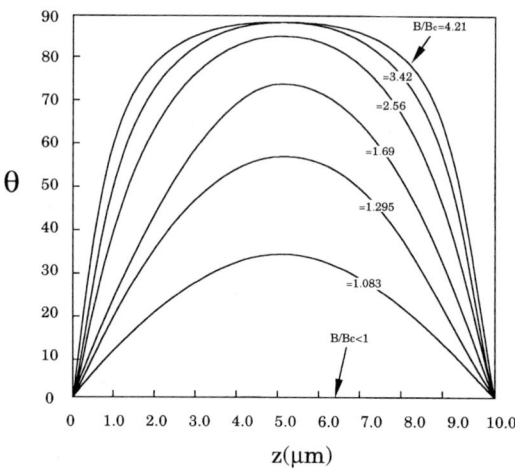

Figure 13. Director distribution above threshold.

isotropy) to the initial alignment direction; the corresponding threshold field is [24]:

$$B_c = \frac{\pi}{d}\left\{\frac{k_{11}}{\mu_0^{-1}\Delta\chi}\left[1+k'\sin^2\theta_0\right]\right\}^{1/2} \quad (65)$$

For the twisted (TN) geometry widely encountered in displays, the director distribution in three dimensions has to be evaluated in terms of both $\theta(z)$ and $\phi(z)$. This has been done [27] and the corresponding threshold field is:

$$B_c = \frac{\pi}{d} \quad (66)$$
$$\cdot\left\{\frac{1}{\mu_0^{-1}\Delta\chi}\left[k_{11}+(k_{33}-2k_{22})\left(\frac{\phi_0}{\pi}\right)^2\right]\right\}^{1/2}$$

where ϕ_0 is the zero-field twist angle. In these simple derivations the condition of strong anchoring has been assumed; however effects of weak anchoring where a surface interaction term is included have been examined [24].

5.4.2 Director Distribution in Electric Fields

The complication associated with electric fields is due to the large anisotropy of the electric permittivity, which means that above threshold the induced electric polarization is no longer parallel to the applied field. In a deformed sample the director orientation is inhomogeneous through the cell, and as a consequence the electric field is also nonuniform. An additional problem can arise with conducting samples, for which there is a contribution to the electric torque from the conductivity anisotropy. Neglecting this, the expressions for threshold electric fields are similar to those obtained for magnetic fields:

(1) Splay

$$E_c = \frac{\pi}{d}\left(\frac{k_{11}}{\varepsilon_0\Delta\varepsilon}\right)^{1/2}$$

(2) Twist

$$E_c = \frac{\pi}{d}\left(\frac{k_{22}}{\varepsilon_0\Delta\varepsilon}\right)^{1/2}$$

(3) Bend

$$E_c = \frac{\pi}{d}\left(\frac{k_{33}}{\varepsilon_0\Delta\varepsilon}\right)^{1/2} \quad (67)$$

and the corresponding threshold voltages are independent of sample thickness.

In order to calculate the electric response above threshold, it is necessary to start from the free energy:

$$g = \frac{1}{2}\int k_{11}(\nabla\cdot\mathbf{n})^2 + k_{22}(\mathbf{n}\cdot\nabla\times\mathbf{n})^2 \quad (68)$$
$$+ k_{33}(\mathbf{n}\times\nabla\times\mathbf{n})^2 - \varepsilon_0\mathbf{D}\cdot\mathbf{E}\,d\mathbf{r}^3$$
$$= \frac{1}{2}\int k_{11}(\nabla\cdot\mathbf{n})^2 + k_{22}(\mathbf{n}\cdot\nabla\times\mathbf{n})^2$$
$$+ k_{33}(\mathbf{n}\times\nabla\times\mathbf{n})^2$$
$$- \varepsilon_0\left(\varepsilon_\perp E^2 + \Delta\varepsilon(\mathbf{n}\cdot\mathbf{E})^2\right)d\mathbf{r}^3$$

In the absence of charge $\nabla\cdot\mathbf{D}=0$, so that D_z is constant; by symmetry E will only have a component along the z-direction, but its magnitude will be a function of z. Considering for a moment just the electric field contribution to the energy per unit area, this can be written as:

$$g_{elec}/\text{area} = -\frac{1}{2}D_z\int E_z\,dz$$
$$= -\frac{1}{2}D_z V \quad (69)$$

where the voltage V is just the integral of the field. The electric displacement $D_z=\varepsilon_0\varepsilon_{zz}E_z$, so from Eq. (69), D_z can be written as

$$D_z = V/\int(\varepsilon_0\varepsilon_{zz})^{-1}dz$$

For an initially planar cell (case (1) above), $\varepsilon_{zz} = \varepsilon_{\parallel} \sin^2\theta + \varepsilon_{\perp} \cos^2\theta$, so the free energy per unit area becomes:

$$g_{elec}/area \qquad (70)$$
$$= \frac{1}{2}\int \left(k_{11}\cos^2\theta + k_{33}\sin^2\theta\right)\left(\frac{d\theta}{dz}\right)^2 dz$$
$$- \frac{1}{2}\varepsilon_0 V^2 \left[\int \left(\varepsilon_{\parallel}\sin^2\theta + \varepsilon_{\perp}\cos^2\theta\right)^{-1} dz\right]^{-1}$$

Applying variational calculus to this expression for the free energy gives an integral equation (c.f. Eq. (58) for the director distribution:

$$\frac{V}{V_c} = \frac{2}{\pi}\left(1 + \gamma\sin^2\theta_m\right)^{1/2} \qquad (71)$$
$$\cdot \int_0^{\theta_m} \left[\frac{1 + k'\sin^2\theta}{(1+\gamma\sin^2\theta)(\sin^2\theta_m - \sin^2\theta)}\right]^{1/2} d\theta$$

where $\gamma = \Delta\varepsilon/\varepsilon_{\perp}$, with a similar expression for the bend deformation (case (3) above), except that $\varepsilon_{\parallel}$ and $\varepsilon_{\perp}$ and k_{11} and k_{33} are interchanged: the case (2) for a twisted deformation is more complicated [26]. This treatment has neglected any coupling between elastic deformation and electric polarization, but such flexoelectricity can contribute to Fréedericksz transitions [28].

5.4.3 Fréedericksz Transitions as a Method for Measuring Elastic Constants

As well as providing the basis for electro-optic displays, the Fréedericksz transitions can be used to determine the elastic constants of liquid crystals. Any physical technique that is sensitive to a change in the director distribution can be used to obtain elastic constants, but the most common methods rely on measurement of capacitance or birefringence changes during a Fréedericksz transition. As before the simplest configuration to consider is the planar to homeotropic transition observed in materials having a positive electric or magnetic susceptibility anisotropy.

5.4.3.1 Capacitance Method

For a particular applied field, the measured capacitance of a sample will be an integral over the permittivity component ε_{zz} across the cell. If the cell is imagined as a series of thin slices each of which acts as a parallel plate capacitor, the addition theorem for series capacitors gives the cell capacitance C as:

$$C^{-1} = \int \frac{dz}{\varepsilon_0 A \varepsilon_{zz}(z)}$$
$$= \frac{2}{A\varepsilon_0\varepsilon_{\perp}} \int_0^{d/2} \left(1 + \gamma\sin^2\theta\right)^{-1} dz \qquad (72)$$

It is convenient to change the variable z in Eq. (72) to one involving the director deformation θ, and for a magnetic field-induced Fréedericksz transition, Eq. (57) can be used for $\left(\frac{d\theta}{dz}\right)$ to give:

$$C^{-1} = \frac{2\left(\mu_0^{-1}\Delta\chi B^2\right)^{-1/2}}{A\varepsilon_0\varepsilon_{\perp}} \int_0^{\theta_m} \left[\frac{k_{11}\cos^2\theta + k_{33}\sin^2\theta}{\sin^2\theta_m - \sin^2\theta}\right]^{1/2} \left(1 + \gamma\sin^2\theta\right)^{-1} d\theta$$
$$= \frac{2B_c}{\pi C_0 B} \int_0^{\theta_m} \left[\frac{1 + k'\sin^2\theta}{\sin^2\theta_m - \sin^2\theta}\right]^{1/2} \left(1 + \gamma\sin^2\theta\right)^{-1} d\theta \qquad (73)$$

where $C_0 = \dfrac{A\varepsilon_0\varepsilon_\perp}{d}$ is the zero-field capacitance. For excitation by an electric field the result is:

$$\frac{C_0}{C} = \frac{\displaystyle\int_0^{\theta_m}\left[\frac{1+k'\sin^2\theta}{(1+\gamma\sin^2\theta)(\sin^2\theta_m-\sin^2\theta)}\right]^{1/2}d\theta}{\displaystyle\int_0^{\theta_m}\left[\frac{(1+\gamma\sin^2\theta)(1+k'\sin^2\theta)}{\sin^2\theta_m-\sin^2\theta}\right]^{1/2}d\theta} \qquad (74)$$

5.4.3.2 Birefringence Method

This technique usually measures the change in birefringence as a cell is switched from planar to homeotropic with an electric or magnetic field. Initially the birefringence is $(n_e - n_0)$, which at infinite field becomes zero. The change (Δ) in the birefringence can be written as:

$$(n_e - n_0) - (n_{\text{eff}} - n_0) = (n_e - n_{\text{eff}}) = \Delta \qquad (75)$$

where n_{eff} is the effective refractive index along the field direction, and is given by:

$$n_{\text{eff}} = \frac{1}{d}\int_0^d n(z)\,dz \qquad (76)$$

$n(z)$ depends on the director orientation, and can be obtained from Eq. (4) of Sec. 3 of this chapter, but note that the angle θ used in the earlier equation is $(90° - \theta)$ used here. Thus $n(z) = n_e n_0 (n_e^2 \sin^2\theta + n_0^2 \cos^2\theta)^{-1/2} = n_e(1+v\sin^2\theta)^{-1/2}$, and $v = (n_e^2 - n_0^2)/n_0^2$, and the birefringence change Δ can be obtained by transforming Eq. (76) to an integral over θ using Eq. (57), so that:

$$\Delta = n_e - \frac{2n_e\left(\mu_0^{-1}\Delta\chi B^2\right)^{-1/2}}{d}$$

$$\cdot \int_0^{\theta_m}\left[\frac{k_{11}\cos^2\theta + k_{33}\sin^2\theta}{(1+v\sin^2\theta)(\sin^2\theta_m-\sin^2\theta)}\right]^{1/2}d\theta \qquad (77)$$

and using the result for the threshold field, this becomes:

$$\frac{\Delta}{n_e} = 1 - \frac{2B_c}{\pi B} \qquad (78)$$

$$\cdot \int_0^{\theta_m}\left[\frac{1+k'\sin^2\theta}{(1+v\sin^2\theta)(\sin^2\theta_m-\sin^2\theta)}\right]^{1/2}d\theta$$

It will be useful to express this in terms of the new angular variable ψ introduced earlier, to give:

$$\frac{\Delta}{n_e} = 1 - \frac{2B_c}{\pi B}$$

$$\cdot \int_0^{\pi/2}\left[\frac{1+k'\sin^2\theta_m\sin^2\psi}{1+v\sin^2\theta_m\sin^2\theta}\right]^{1/2}$$

$$\cdot \left(1-\sin^2\theta_m\sin^2\psi\right)^{-1}d\theta \qquad (79)$$

The usual way to observe birefringence changes in a nematic slab undergoing a Fréedericksz transition is to illuminate the sample with polarized light, the plane of which makes an angle of 45° with the director axis. Increasing the field above threshold and measuring the intensity transmitted through a crossed polariser gives rise to a series of maxima and minima (fringes). These can be related to the expression (79) by the result:

$$\frac{I}{I_0} = \frac{1}{4}\left(1-\cos\left(\frac{2\pi\,d\Delta}{\lambda}\right)\right) \qquad (80)$$

where the ratio I/I_0 refers to changes in the intensity of unpolarized incident light (I_0).

The result for electric field excitation is more complex for the reasons given earlier, but has a similar form to Eq. (79)

$$\frac{\Delta}{n_e} = 1 - \frac{\int_0^{\pi/2} \left[\frac{\left(1 + k' \sin^2 \theta_m \sin^2 \psi\right)\left(1 + \gamma \sin^2 \theta_m \sin^2 \psi\right)}{\left(1 + \nu \sin^2 \theta_m \sin^2 \theta\right)\left(1 - \sin^2 \theta_m \sin^2 \theta\right)} \right]^{1/2} d\theta}{\int_0^{\pi/2} \left[\frac{\left(1 + k' \sin^2 \theta_m \sin^2 \psi\right)\left(1 + \gamma \sin^2 \theta_m \sin^2 \psi\right)}{1 - \sin^2 \theta_m \sin^2 \theta} \right]^{1/2} d\theta} \quad (81)$$

Numerical methods have been developed [29–31] to fit experimental values of $\frac{C}{C_0}$ or $\frac{\Delta}{n_e}$ to the theoretical expressions with B_c, k', ν, V_c and γ as adjustable parameters appropriate to the particular experiment being considered. Another way of using these expressions is to develop low-field or high-field expansions, so that material parameters can be obtained directly from a linear or polynomial expression. For example the low and high field expansions for electric field excitation have been derived [32, 33] as:

$$\left(\frac{C - C_0}{C_0}\right)_{E \to 0} = 2\gamma(1 + k' + \gamma)^{-1}\left(\frac{V - V_c}{V_c}\right)$$

$$\left(\frac{C - C_0}{C_0}\right)_{E \to \infty} = \gamma - 2\gamma \, \pi^{-1}(1 + \gamma)^{1/2} \frac{V_c}{V}$$

(82)

Care is required in using these expressions, since they are usually only valid close to the limiting field values where measurements are difficult and liable to error.

In this section the effects of magnetic and electric fields have been considered for a few standard geometries for materials having positive susceptibility anisotropies, but there are many possible variations with negative anisotropies or simultaneous excitation with both electric and magnetic fields of materials having both positive, both negative or different signs of electric and magnetic susceptibility anisotropies. Another variation which may be introduced is to have different boundary conditions for the containing surfaces of the liquid crystal film [34], so-called hybrid-aligned cells. The theoretical treatment outlined here has excluded the possibility of defect formation, although this can in principle be described by the elasticity theory already developed. Defect structures can be formed as a result of deformation by electric or magnetic fields. In some situations they may arise from natural degeneracies in the sample: for example in a twist cell, states of opposite twist will be of equal energy, and so may form distinct regions separated by a disclination. In real device cells the structures are designed to avoid the formation of defects.

5.4.4 Fréedericksz Transition for Chiral Nematics

Field effects on chiral nematics can be interpreted by adding a pitch term to the free energy, so it might be expected that the Fréedericksz transitions observed for chiral nematics will be similar to those described above for achiral nematics. In reality this is not the case because the helical structure in chiral phases prevents the formation of uniformly aligned films, and so defects and defect-modulated structures are unavoidable in many field-induced orientational changes. The effects of external fields on chiral nematic films have been described [35, 36], and some of the associated phenomena form the basis for display devices. In most cases a theoretical description of the induced deformation is only possible by numerical solution of the Euler–Lagrange equations, but one simple effect that has an analytical solution is the so-called field-induced cholesteric to nematic phase transition. An external field applied perpendicular to the helix axis of a material having a positive susceptibility anisotropy will cause the helix to unwind and the pitch to increase. A treatment similar to that given for the twist Fréedericksz transition shows that the critical field for divergence of the pitch to infinity is:

$$F_c = \frac{\pi q_0}{2} \left(\frac{k_{22}}{\Delta \chi \mu_0^{-1}} \right)^{1/2} \quad (83)$$

This results contrasts with the threshold field for a nematic Fréedericksz transition, which is thickness dependent: also F_c marks the end of the deformation rather than the beginning which defines the normal threshold fields.

predicts a threshold field which is inversely proportional to the square-root of the sample thickness [37] e.g. for a magnetic field:

$$B_c = \left(\frac{2\pi k_{11}}{\Delta \chi \mu_0^{-1} \lambda d} \right)^{1/2} \quad (84)$$

where d is the cell thickness and $\lambda = \sqrt{k_{11}/B}$ (see Fig. 14).

External field distortions in SmC and chiral SmC phases have been investigated [38], but the large number of elastic terms in the free-energy, and the coupling between the permanent polarization and electric fields for chiral phases considerably complicates the description. In the chiral smectic C phase a simple helix unwinding Fréedericksz transition can be detected for the c director. This is similar to the chiral nematic–nematic transition described by Eq. (83), and the result is identical for the SmC* phase. Indeed it appears that at least in interactions with magnetic fields in the plane of the layers, SmC and SmC* phases behave as two dimensional nematics [39].

5.4.5 Fréedericksz Transitions for Smectic Phases

The simplest elasticity theory for SmA phases includes two elastic contants, one for splay and one for layer compressibility. It might therefore be expected that a Fréedericksz transition for splay deformation should be observed corresponding to an initial deformation of layers in a planar to homeotropic transition. This is not observed, and field induced deformations in smectic A phases are accompanied by defect formation. The Helfrich–Hurault mechanism for the homeotropic to planar transition via the formation of undulations

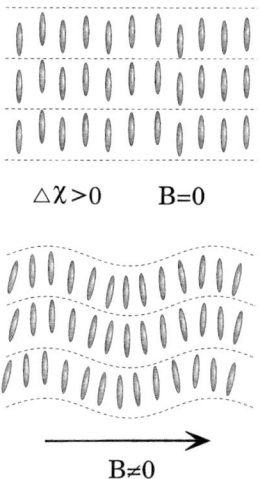

Figure 14. Helfrich–Hurault mechanism for field deformation of smectic layers.

5.5 Molecular Aspects of Torsional Elasticity

5.5.1 van der Waals Theory

Torsional elasticity is of special interest from a microscopic viewpoint since it is a property characteristic of liquid crystals, which distinguishes them from ordinary liquids. The elastic properties contribute to many physical phenomena observed for liquid crystals, and a molecular theory of torsional elasticity should enable the identification of particular molecular properties responsible for many aspects of liquid crystalline behavior.

The principal elastic constants for a nematic liquid crystal have already been defined in Sec. 5.1 as splay (k_{11}), twist (k_{22}) and bend (k_{33}). In this section we shall outline the statistical theory of elastic constants, and show how they depend on molecular properties. The approach follows that of the generalised van der Waals theory developed by Gelbart and Ben-Shaul [40], which itself embraces a number of earlier models for the elasticity of nematic liquid crystals. Corresponding theories for smectic, columnar and biaxial phases have yet to be developed.

Elastic constants are defined in terms of the deformation free energy of a liquid crystal subjected to torsional strain. Statistical models for liquid crystals result in equations for the free energy in an undistorted state: thus to calculate elastic constants it is necessary to obtain a statistical expression for the free energy of a strained liquid crystal. In developing a statistical theory it is easier to use the Helmholtz free energy to calculate elastic constants, although Frank originally defined them in terms of the Gibbs free energy, corresponding to strain at constant external pressure. We shall be considering torsional strain at constant volume, for which changes in both the internal energy and entropy of a liquid crystal will contribute to the elastic constants.

The most widely used statistical model for fluids is that due to van der Waals, which includes a mean attractive potential with a hard particle excluded volume. For such a model the Helmholtz free energy can be written as [41, 42]:

$$A\left[f(\Omega_1, \Omega_2, R_1, R_2)\right] \quad (85)$$
$$= A\left[f(\Omega_1, \Omega_2, R_1, R_2)\right]_{\mathrm{hp}}$$
$$- \frac{1}{2}\rho^2 \int f(\Omega_1, R_1) f(\Omega_2, R_2)$$
$$\cdot u_{12}(R_{12}, \Omega_1, \Omega_2)$$
$$\cdot g_{\mathrm{hp}}(R_{12}, \Omega_1, \Omega_2) \mathrm{d}\Omega_1, \mathrm{d}\Omega_2, \mathrm{d}R_1, \mathrm{d}R_2$$

where $f(\Omega_1, R_1)$ is a single particle distribution function for molecule 1, $u_{12}(R_{12}, \Omega_1, \Omega_2)$ is the attractive part of the pair potential and $g_{\mathrm{hp}}(R_{12}, \Omega_1, \Omega_2)$ is the pair distribution function for the hard particle interactions in the isotropic state.

To proceed we need to know how the functional $A[f(\Omega_1, \Omega_2, R_1, R_2)]$ varies when the equilibrium state of the liquid crystal is elastically distorted. A macroscopic strain will not influence u_{12} or g_{hp}, since these are dependent only on molecular parameters of the model: the free energy changes because the single particle distribution functions change. We assume that for the small distortions described by the Frank elastic constants, the single particle orientational distribution function, defined with respect to a local director axis, is also independent of strain i.e. elastic torques do not change the molecular order parameters. The product of distribution functions $f(\Omega_1, R_1) f(\Omega_2, R_2)$ will change with strain because the director orientations at R_1 and R_2 will differ, and the evaluation of the strain dependence of the

Helmholtz free energy has reduced to determining the strain derivatives of the single particle distribution functions. Both the terms in Eq. (85) will contribute to the elastic free energy, but the first term, which describes hard particle contributions to the Helmholtz free energy will only appear in the entropy of the system, since the internal energy of a hard particle fluid is zero. Furthermore it can be shown [40] that the rotational hard particle entropy is independent of strain, provided that the single particle orientational distribution defined with respect to the local director does not depend on strain. Thus the hard particle repulsion contributes to the elastic strain energy in two ways: firstly through the orientation-dependent excluded volume, which affects the transitional entropy, and secondly because the integration over the pair attractive potential energy is convoluted with the hard particle distribution function. A particularly convenient form for the hard particle translational entropy is provided by the 'y-expansion' [41], and using this the Helmholtz free energy becomes:

$$A[f(\boldsymbol{\Omega}_1,\boldsymbol{\Omega}_2,\boldsymbol{R}_1,\boldsymbol{R}_2)] \qquad (86)$$

$$= T\mathbf{S}_{hp}^{rot} - \frac{kT\rho^2}{2(1-v_0\rho)}$$

$$\cdot \int f(\boldsymbol{\Omega}_1,\boldsymbol{R}_1) f(\boldsymbol{\Omega}_2,\boldsymbol{R}_2)$$

$$\cdot \left\{ g_{hp}^{(0)} - 1 \right\} d\boldsymbol{\Omega}_1, d\boldsymbol{\Omega}_2, d\boldsymbol{R}_1, d\boldsymbol{R}_2$$

$$- \frac{1}{2}\rho^2 \int f(\boldsymbol{\Omega}_1,\boldsymbol{R}_1) f(\boldsymbol{\Omega}_2,\boldsymbol{R}_2)$$

$$\cdot u_{12}(\boldsymbol{R}_{12},\boldsymbol{\Omega}_1,\boldsymbol{\Omega}_2) g_{hp}^{(0)} d\boldsymbol{\Omega}_1, d\boldsymbol{\Omega}_2, d\boldsymbol{R}_1, d\boldsymbol{R}_2$$

$$= -T\mathbf{S}_{hp}^{rot} - \frac{1}{2}\rho^2$$

$$\cdot \int f(\boldsymbol{\Omega}_1,\boldsymbol{R}_1) f(\boldsymbol{\Omega}_2,\boldsymbol{R}_1+\boldsymbol{R}_{12})$$

$$\cdot \left\{ u_{12}(\boldsymbol{R}_{12},\boldsymbol{\Omega}_1,\boldsymbol{\Omega}_2) g_{hp}^{(0)} - \frac{kT}{(1-v_0\rho)} \left(g_{hp}^{(0)} - 1 \right) \right\}$$

$$\cdot d\boldsymbol{\Omega}_1, d\boldsymbol{\Omega}_2, d\boldsymbol{R}_1, d\boldsymbol{R}_{12}$$

where $g_{hp}^{(0)} = \exp - u_{hp}/k_B T$ is the pair correlation function for the hard particles, v_0 is the particle volume, u_{hp} is the hard particle potential, and $\mathbf{S}_{hp}^{rot}$ is the hard particle rotational entropy. In order to obtain the elastic-distortion-free energy from Eq. (86), we assume that molecule 1 is located at some arbitary origin in the fluid. The orientational distribution function for molecule 2 at position $\boldsymbol{R}_{12}$ only depends on the orientation of the director at $\boldsymbol{R}_{12}$ with respect to the director at the origin. Thus in the undeformed state the director at $\boldsymbol{R}_{12}$ is parallel to that at the origin (at least for non-chiral liquid crystals), but in the deformed state the director at $\boldsymbol{R}_{12}$ makes an angle of $\theta(\boldsymbol{R}_{12})$ to that at the origin; see Fig. 15.

The distortion free energy density is then the difference in free energies given by Eq. (86) between the distorted and undistorted states, that is:

$$\frac{A_{\text{distorted}}}{V} = \frac{1}{2}\rho^2 \int f(\boldsymbol{\Omega}_1,0) \qquad (87)$$

$$\cdot \left\{ f(\boldsymbol{\Omega}_2, \theta(\boldsymbol{R}_{12})) - f(\boldsymbol{\Omega}_2, 0) \right\}$$

$$\cdot \left\{ u_{12}(\boldsymbol{R}_{12},\boldsymbol{\Omega}_1,\boldsymbol{\Omega}_2) g_{hp}^{(0)} - \frac{kT}{(1-v_0\rho)} \left(g_{hp}^{(0)} - 1 \right) \right\}$$

$$\cdot d\boldsymbol{\Omega}_1, d\boldsymbol{\Omega}_2, d\boldsymbol{R}_{12}$$

The change in the single particle orientational distribution function for small defor-

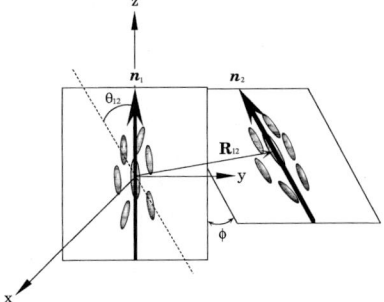

Figure 15. Molecular distributions in a deformed nematic liquid crystal.

mations can be obtained from a Taylor expansion, such that:

$$f(\Omega,\theta(R)) - f(\Omega,0)$$
$$= \left(\frac{df(\Omega,\theta(R))}{d\theta}\right)_0 \theta$$
$$+ \frac{1}{2}\left(\frac{d^2 f(\Omega,\theta(R))}{d\theta^2}\right)_0 \theta^2 + \ldots \quad (88)$$

The way in which $\theta(R)$ varies with position depends on the form of the torsional deformation applied to the liquid crystal, and in order to calculate the principal elastic constants, it makes sense to calculate the free energy density for 'normal mode' deformations, i.e. those which correspond to splay, twist and bend. These can be achieved easily by confining the director to a plane, and assuming the undisturbed director at the origin to be along the z-axis. q is the wavevector of the deformation, and for q constrained to the x, z plane, the components of the director as a function of position become:

$$n_x = \sin q \cdot R; \quad n_y = 0; \quad n_z = \cos q \cdot R$$

Director configurations corresponding to pure splay, twist and bend are illustrated in Fig. 16 and to the lowest order of approximation can be described in terms of long wavelength deformations parallel to the x-axis for splay, parallel to the y-axis for twist and parallel to the z-axis for bend. Under these circumstances, the angle between the director at the origin and that at position R becomes:

$$\theta(R) = \begin{cases} qx & \text{splay} \\ qy & \text{twist} \\ qz & \text{bend} \end{cases} \quad (89)$$

Gelbart and Ben Shaul have shown that a more consistent director distribution giving rise to normal mode elastic deformations is:

$$\tan\theta(R) = \begin{cases} qx(1+qz)^{-1} & \text{splay} \\ qy & \text{twist} \\ qz(1-qx)^{-1} & \text{bend} \end{cases}$$

or $$\theta(R) = \begin{cases} qx(1-qz) & \text{splay} \\ qy & \text{twist} \\ qz(1+qx) & \text{bend} \end{cases} \quad (90)$$

Substituting these results into Eq. (88) gives:

$$f(\Omega,\theta(R)) - f(\Omega,0)$$
$$= \left(\frac{df(\Omega,\theta(R))}{d\theta}\right)_0 \begin{cases} (qx - q^2 xz) \\ qy \\ (qz + q^2 zx) \end{cases}$$
$$+ \frac{1}{2}\left(\frac{d^2 f(\Omega,\theta(R))}{d\theta^2}\right)_0 \begin{cases} q^2 x^2 \\ q^2 y^2 \\ q^2 z^2 \end{cases} \quad (91)$$

The macroscopic expression for the elastic free energy density (Eq. 25) is:

$$g - g_0 = \frac{1}{2V}\int k_{11}(\nabla \cdot n)^2 + k_{22}(n \cdot \nabla \times n)^2$$
$$+ k_{33}((n \cdot \nabla)n)^2 \, dR \quad (92)$$

Using the expressions for $\theta(R)$ corresponding to pure splay, twist and bend deformations gives the result:

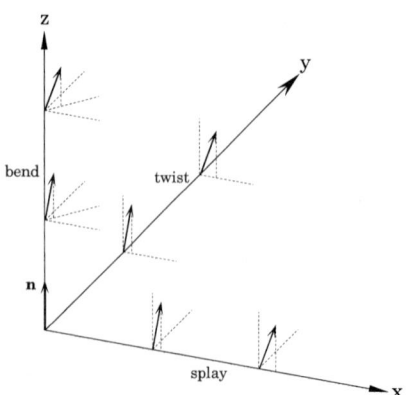

Figure 16. Deformations of director for splay, twist, bend.

$$g - g_0 = \begin{cases} \frac{1}{2} k_{11} q^2 & \text{splay} \\ \frac{1}{2} k_{22} q^2 & \text{twist} \\ \frac{1}{2} k_{33} q^2 & \text{bend} \end{cases} \quad (93)$$

Combining Eqs. (91) and (87) and equating the result for the free energy density to the macroscopic expression Eq. (93) gives:

$$\begin{Bmatrix} k_{11} \\ k_{22} \\ k_{33} \end{Bmatrix} = \rho^2 \int f(\Omega_1, 0)$$

$$\cdot \left\{ \left(\frac{\mathrm{d}f(\Omega_1, \theta(R))}{\mathrm{d}\theta} \right)_0 \begin{Bmatrix} -xz \\ 0 \\ +zx \end{Bmatrix} \right.$$

$$+ \frac{1}{2} \left(\frac{\mathrm{d}^2 f(\Omega_1, \theta(R))}{\mathrm{d}\theta^2} \right)_0 \left. \begin{Bmatrix} x^2 \\ y^2 \\ z^2 \end{Bmatrix} \right\}$$

$$\cdot \left[u_{12}(R_{12}, \Omega_1, \Omega_2) g_{hp}^0 - \frac{kT}{(1 - v_0 \rho)} (g_{hp}^0 - 1) \right]$$

$$\cdot \mathrm{d}\Omega_1, \mathrm{d}\Omega_2, \mathrm{d}R_{12} \quad (94)$$

The two terms in the square brackets of Eq. (94) can be identified as a temperature independent internal energy term, and a temperature dependent entropy term resulting from the hard particle pair distribution function. From this equation it can be seen that the calculation of the principal elastic constants of a nematic liquid crystal depends on the first and second derivatives with respect to the angle θ of the single particle orientational distribution function. Any appropriate angular function may be used for $f(\Omega_1, \theta(R))$, but the usual approach is to use an expansion in terms of spherical harmonics. The necessary mathematical manipulations are complicated, but give relatively compact results. Thus the ingredients of a molecular calculation of torsional elastic constants within the van der Waals theory, are a single particle angular distribution function, an attractive intermolecular potential and a hard particle pair distribution function. An immediate result of the above theory is that since the elastic constants depend on the product of single particle orientational distribution functions, they will depend on the product of order parameters.

5.5.2 Results from Lattice Models

For the simplest distribution function, only the term involving the second derivative in Eq. (94) is nonzero, and the torsional elastic constants are given by an average over the square of the intermolecular distances x, y and z. Since macroscopic uniaxiality is assumed, the averages over x and y, perpendicular to the undisturbed director, will be equal, with the result [43]:

$$k_{11} = k_{22} = \text{constant } \rho^2 S^2 b \langle x^2 \rangle \quad \text{and}$$
$$k_{33} = \text{constant } \rho^2 S^2 b \langle z^2 \rangle \quad (95)$$

where b is an energy parameter. If a lattice model is assumed, then the averages over x and z will relate to the unit cell dimensions, or the dimensions of an 'interaction volume'. The result, Eq. (95), fails to account for the observed difference between splay and twist elastic constants, and it fails to provide a useful basis for investigating the effects of molecular structure on elastic properties.

5.5.3 Mean Field and Hard Particle Theories

The first attempt at a molecular theory of elastic constants due to Saupe and Nehring [44] assumed an attractive pair potential of the form $u_{12} = -b r^{-6} P_2(\cos \theta_{12})$, and set

$g_{(\text{hp})}^{(0)}$ equal to that for hard spheres of diameter σ, with the result that:

$$\left. \begin{array}{c} k_{11} \\ k_{22} \\ k_{33} \end{array} \right\} = \frac{12\pi\rho^2 S^2 b}{5\sigma} \begin{pmatrix} 5 \\ 11 \\ 5 \end{pmatrix} \quad (96)$$

This suggests that the elastic constants for splay and bend should be equal. The difference between k_{11} and k_{33} has been identified as due to a term in $\langle P_2(\cos\theta)\rangle \langle P_4(\cos\theta)\rangle$, and while this usually leads to $k_{33} > k_{11}$, particular choices of intermolecular potential can result in $k_{33} < k_{11}$ or indeed a change in sign of the quantity $\left(\frac{k_{33}}{k_{11}} - 1\right)$ as a function of temperature [45].

Various authors [46, 47] have reported calculations of torsional elastic constants for hard spherocylinders. These neglect any attractive interactions, and so only give entropic contributions to the elastic free energy; their results can be summarized as:

$$\left. \begin{array}{c} \frac{k_{11}-k}{k} \\ \frac{k_{22}-k}{k} \\ \frac{k_{33}-k}{k} \end{array} \right\} = \left\{ \begin{array}{c} \Delta - 3\Delta' \frac{\langle P_2(\cos\theta)\rangle}{\langle P_4(\cos\theta)\rangle} \\ -2\Delta - \Delta' \frac{\langle P_2(\cos\theta)\rangle}{\langle P_4(\cos\theta)\rangle} \\ \Delta + 4\Delta' \frac{\langle P_2(\cos\theta)\rangle}{\langle P_4(\cos\theta)\rangle} \end{array} \right. \quad (97)$$

where $k = \frac{1}{3}(k_{11}+k_{22}+k_{33})$, and Δ and Δ' depend on the details of the potential. For hard spherocylinders, the parameters Δ and Δ' are:

$$\Delta = \frac{2R^2 - 2}{7R^2 + 20} \quad \text{and} \quad \Delta' = \frac{9(3R^2 - 8)}{16(7R^2 + 20)} \quad (98)$$

where $R + 1 = (l/w)$; l is the length and w is the width of the spherocylinder. Calculations of the principal elastic constants for various molecular models, with both attractive and repulsive interactions have been reported by various authors [48–52].

5.5.4 Computer Simulations

Some progress has been made in computer simulation of torsional elastic properties [53, 54]. A variety of methods is available for modelling elastic distortions, but the most used technique is based on the direct calculation of the amplitude of director fluctuations. In Sec. 5.2.2 of this Chapter the background to long range director fluctuations was outlined, and it was found that the mean square amplitude of the director component along z could be expressed in terms of the splay and bend elastic constants and the wave vector for a particular deformation mode (it is assumed that for the undistorted liquid crystal, the director lies along the z axis). In order to allow for the influence of local molecular disorder, the fluctuations in the director order parameter are calculated, assuming that for long wavelength fluctuations, the molecular order parameter simply scales the director order parameter. Thus the Fourier components of the order parameter are written as:

$$Q_{\alpha\beta}(\mathbf{q})$$
$$= \langle P_2(\cos\theta)\rangle \left(\frac{1}{2}(3n_\alpha(\mathbf{q})n_\beta(\mathbf{q}) - \delta_{\alpha\beta})\right) \quad (99)$$

Using the results Eqs. (38) and (39) give expressions for the fluctuation in the order parameter as [53]:

$$\langle Q_{xz}(\mathbf{q})Q_{xz}(-\mathbf{q})\rangle = \frac{9}{4}\langle P_2(\cos\theta)\rangle^2 \frac{k_B T}{V(k_{11}q_\perp^2 + k_{33}q_\parallel^2)}$$

$$\langle Q_{yz}(\mathbf{q})Q_{yz}(-\mathbf{q})\rangle = \frac{9}{4}\langle P_2(\cos\theta)\rangle^2 \frac{k_B T}{V(k_{22}q_\perp^2 + k_{33}q_\parallel^2)} \quad (100)$$

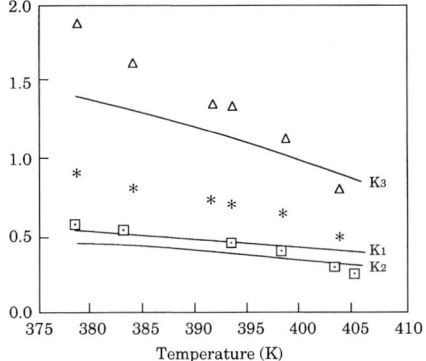

Figure 17. Comparison of calculated and theoretical elastic constants for *p*-azoxyanisole [53] (units are 10^{-11} N).

where $q_\parallel$ and $q_\perp$ are the components of the distortion mode wave vector parallel and perpendicular to the undisturbed director. Comparisons between theory and experiment are encouraging (see Fig. 17), although sometimes at the expense of assuming rather high values for the local molecular order parameters; this is attributed to a weakness in the theory which neglects long-range correlations [53].

5.6 Experimental Aspects of Elastic Constants and Comparison with Theoretical Predictions

The statistical theories of elasticity have shown that the principal elastic constants depend on the single particle distribution functions and the intermolecular forces. The former can be accounted for in terms of order parameters, but intermolecular parameters are more difficult to interpret in terms of molecular properties. Results for hard particle potentials relate the elastic constants to particle dimensions, but the dependence of elastic properties on details of molecular structure is more obscure. Measured values for the torsional elastic constants of nematics are of the order of 10^{-12} N, corresponding to $kT(4 \times 10^{-21}$ J$)/l(10^{-9}$ m$)$, where $T \approx 300$ K and l is a molecular dimension. The effect of elastic distortions on molecular interactions is very small, so that the assumption of strain-independent distribution functions is justified. For example the elastic energy in a 1 µm liquid crystal film in a $\pi/2$ twisted nematic cell is only $\approx 10^{-27}$ J/molecule, which compares with an intermolecular energy of about 10^{-20} J/molecule. The angular displacement per molecule in such a cell is 10^{-3} rad, and the small energies associated with such torsional strain mean that direct mechanical measurement of elastic constants in liquid crystals is difficult.

Indirect experimental measurements of torsional elastic constants indicate that for many materials $k_{33} > k_{11} > k_{22}$. It is usual to consider ratios of elastic constants, and most measured values lie in the ranges $0.5 < k_{33}/k_{11} < 3.0$ and $0.5 < k_{22}/k_{11} < 0.8$. Simple mean field theory predicted that $k_{33} = k_{11} < k_{22}$, which is clearly in error. For rigid molecules increasing the molecular length increases k_{33}/k_{11}, while increasing the width decreases k_{33}/k_{11}, in qualitative agreement with the results of hard particle theories. However within a homologous series, increasing the length of an alkyl chain causes the ratio k_{33}/k_{11} to decrease, which is likely to be a result of increased flexibility rather than any change in effective shape. For SmA phases the torsional elastic constants for twist and bend are expected to be very large, and it is observed that k_{33} and sometimes k_{22} measured for a nematic phase increase dramatically with decreasing temperature as an underlying smectic phase is approached. Short-range smectic-like ordering will influence the measured elastic constants; long alkyl chains promote smectic phase formation, and as a result k_{33}/k_{11} may increase in a homologous series with increasing chain length.

The main temperature dependence of elastic constants is due to the order parameter. Following the predictions of mean field theory Eq. (96), reduced elastic constants c_i have been introduced, which should be independent of temperature, defined by:

$$c_i = \frac{k_{ii}}{S^2 \rho^{7/3}} \qquad (101)$$

Experimentally the c_is do change with temperature, and this can be attributed to the influence of repulsive interactions on the elastic properties.

5.7 References

[1] C. W. Oseen, *Trans. Faraday Soc.* **1933**, *29*, 883.
[2] F. C. Frank, *Discuss. Faraday Soc.* **1958**, *25*, 19.
[3] P. G. de Gennes, J. Prost, *The Physics of Liquid Crystals*, Oxford University Press, Oxford, **1993**, p. 98.
[4] G. Vertogen, W. H. de Jeu, *Thermotropic Liquid Crystals, Fundamentals* **1988**, p. 74.
[5] Ref. [3], p. 341.
[6] Ref. [3], p. 349.
[7] R. D. Polak, G. P. Crawford, B. C. Kostival, J. W. Doane, S. Zumer, *Phys. Rev. E* **1994**, *49*, R978.
[8] J. Charvolin, J. F. Sadoc, *J. Phys. Chem.* **1988**, *92*, 5787.
[9] J. Nehring, A. Saupe, *J. Chem. Phys.* **1971**, *54*, 337.
[10] F. M. Leslie, I. W. Stewart, M. Nakagawa, *Mol. Cryst. Liq. Cryst.* **1991**, *198*, 443.
[11] Ref. [3], p. 351.
[12] Ref. [3], p. 74.
[13] S. Chandrasekhar, G. S. Ranganath, *Adv. Phys.* **1986**, *35*, 507.
[14] M. Kleman, *Liq. Cryst.* **1989**, *5*, 399.
[15] T. E. Faber, *Liq. Cryst.* **1991**, *9*, 95.
[16] B. J. Berne, R. Pecora, *Dynamic Light Scattering*, John Wiley, New York, **1976**, p. 27.
[17] S. Chandrasekhar, *Liquid Crystals*, Cambridge University Press, Cambridge, **1992**, p. 167.
[18] J. P. van der Meulen, R. J. J. Zijlstra, *J. de Phys.* **1984**, *45*, 1627.
[19] J. H. Seddon, R. H. Templer, *Phil. Trans. R. Soc. Lond. A* **1993**, *334*, 377.

[20] V. Fréedericksz, V. Zolina, *Trans. Faraday Soc.* **1933**, *29*, 919.
[21] J. Nehring, A. Saupe, *J. Chem. Phys.* **1971**, *54*, 337.
[22] A. J. Derzhanski, A. G. Petrov, M. D. Mitov, *J. de Phys.* **1978**, *39*, 272.
[23] T. Motooka, A. Fukuhara, K. Suzuki, *Appl. Phys. Lett.* **1979**, *34*, 305.
[24] T. Motooka, A. Fukuhara, *J. Appl. Phys.* **1979**, *50*, 6907.
[25] Ref. [4], p. 94.
[26] H. J. Deuling, *Sol. State Phys. Suppl.* 14 (Liquid Crystals), Ed.: L. Liebert, Academic Press, London 1978, p. 77.
[27] F. M. Leslie, *Mol. Cryst. Liq. Cryst.* **1970**, *12*, 57.
[28] H. J. Deuling, *Mol. Cryst. Liq. Cryst.* **1974**, *26*, 281.
[29] C. Maze, *Mol. Cryst. Liq. Cryst.* **1978**, *48*, 273.
[30] M. J. Bradshaw, E. P. Raynes, *Mol. Cryst. Liq. Cryst.* **1983**, *91*, 145.
[31] H. J. Deuling, *Sol. Stat. Commun.* **1974**, *14*, 1073.
[32] M. G. Clark, E. P. Raynes, R. A. Smith, R. J. A. Tough, *J. Phys. D.* **1980**, *13*, 2151.
[33] T. Uchida, Y. Takahashi, *Mol. Cryst. Liq. Cryst.* **1981**, *72*, 133.
[34] G. Barbero, *Z. Naturforsch.* **1984**, *39a*, 575.
[35] L. M. Blinov, *Electro-optical and Magneto-optical Properties of Liquid Crystals*, J. Wiley, Chichester, **1983**, p. 240.
[36] V. G. Chigrinov, *Kristallografiya* **1988**, *33*, 260.
[37] Ref. [3], p. 364.
[38] Ref. [3], pp. 347, 376.
[39] S. P. A. Gill, F. M. Leslie, *Liq. Cryst.* **1993**, *14*, 1901.
[40] W. M. Gelbart, A. Ben-Shaul, *J. Chem. Phys.* **1982**, *77*, 916.
[41] W. M. Gelbart, *J. Chem. Phys.* **1979**, *71*, 3053.
[42] M. A. Cotter, *J. Chem. Phys.* **1977**, *66*, 4710.
[43] H. Gruler, *Z Naturforsch.* **1975**, *30a*, 230.
[44] J. Nehring, A. Saupe, *J. Chem. Phys.* **1972**, *56*, 5527.
[45] W. M. Gelbart, J. Stecki, *Mol. Phys.* **1981**, *41*, 1451.
[46] R. Priest, *Phys. Rev.* **1973**, *A7*, 720.
[47] J. P. Straley, *Phys. Rev.* **1973**, *A8*, 2181.
[48] A. Ponerwierski, J. Stecki, *Mol. Phys.* **1979**, *38*, 1931.
[49] H. Kimura, M. Hoshino, H. Nakano, *Mol. Cryst. Liq. Cryst.* **1981**, *74*, 55.
[50] B. W. van der Meer, F. Postma, A. J. Dekker, W. H. de Jeu, *Mol. Phys.* **1982**, *45*, 1227.
[51] Y. Singh, S. Singh, K. Rajesh, *Phys. Rev.* **1992**, *A44*, 974.
[52] M. A. Osipov, S. Hess, *Mol. Phys.* **1993**, *78*, 1191.
[53] B. Tjipto-Margo, G. T. Evans, M. P. Allen, D. Frenkel, *J. Phys. Chem.* **1992**, *96*, 3942.
[54] M. A. Osipov, S. Hess, *Liq. Cryst.* **1994**, *16*, 845.

6 Phase Transitions

6.1 Phase Transition Theories

Philippe Barois

6.1.1 Introduction

The understanding of continuous phase transitions and critical phenomena has been one of the important breakthrough in condensed matter physics in the early seventies. The concepts of scaling behavior and universality introduced by Kadanoff and Widom and the calculation of non-gaussian exponents by Wilson and Fisher are undeniably brilliant successes of statistical physics in the study of low temperature phase transitions (normal to superconductor, normal to superfluid helium) and liquid–gas critical points.

But no other field in condensed matter physics has shown such a rich variety of continuous or weakly first-order phase transitions than liquid crystals: order parameters of various symmetrics, anisotropic scaling behaviors, coupled order parameters, multicritical points, wide critical domains, defect mediated transitions, spaces of low dimensionality, multiply reentrant topologies are currently found in liquid crystals, at easily accessible temperatures. Beside their famous technical applications in optics and electronic displays, liquid crystals can certainly be regarded as a paradise of the physics of phase transitions.

The main features of the most general phase transitions encountered in liquid crystals are presented in this chapter.

6.1.2 The Isotropic–Nematic Transition

6.1.2.1 Mean Field Approach (Landau–de Gennes)

The nematic phase being the liquid crystal of highest symmetry, its condensation from the isotropic liquid should be the simplest to describe. Indeed, molecular theories convincingly explain the natural onset of nematic ordering in a population of anisotropic molecules with excluded volume interaction (Onsager) or in mean field theory (Maier–Saupe). Regarding the effect of symmetry on the isotropic to nematic (I–N) phase transition, the phenomenological approach is useful too.

It was explained in Chapter III of this volume that the order parameter of the nematic phase is a symmetric traceless tensor Q_{ij} that can be constructed from any macroscopic tensor property such as the magnetic susceptibility χ_{ij}. For a uniaxial nematic for instance, the order parameter can be defined as [1]:

$$Q_{ij} = G\left(\chi_{ij} - \frac{1}{3}(\chi_\| + 2\chi_\perp)\delta_{ij}\right) \quad (1)$$

in which $\chi_\|$ and $\chi_\perp$ are susceptibilities parallel and perpendicular to the director respectively and δ_{ij} is the Kronecker unity tensor. The normalization constant G is generally chosen to ensure $Q_{zz} = 1$ for a perfectly ordered nematic. Such normalization suggests that Q_{zz} is similar to the scalar order parameter S defined in microscopic theories from the statistical distribution of molecular axes (see Chap. III, Sec. 1 of this volume). This is not exactly true, however, since macroscopic quantities χ_{ij} may not be a simple sum of uncorrelated microscopic susceptibilities (think of dielectric polarizability). Consequently, macroscopic nematic order parameters Q_{zz} defined from different macroscopic tensor properties (such as magnetic susceptibilities and refractive indices) may not show the same temperature behaviour. Experimental differences are, however, very weak [2].

In the eigenframe, the Q_{ij} matrix reads [1]:

$$Q = \begin{pmatrix} -\frac{q+\eta}{2} & 0 & 0 \\ 0 & -\frac{q-\eta}{2} & 0 \\ 0 & 0 & q \end{pmatrix} \quad (2)$$

The phenomenological scalar order parameter $q = Q_{zz}$ is non-zero in nematic phases and vanishes in the isotropic phase. η is non-zero in biaxial nematics only and referred to as the biaxial order parameter. The symmetric invariants of Q_{ij} may be written as [1, 3]:

$$\sigma_1 = Q_{ii} = 0$$
$$\sigma_2 = \frac{2}{3} Q_{ij} Q_{ij} = q^2 + \frac{\eta^2}{3}$$
$$\sigma_3 = 4 Q_{ij} Q_{jk} Q_{ki} = q(q^2 - \eta^2) \quad (3)$$

The second and third order invariants can be regarded as independent in the biaxial phase with the constraint $\sigma_3^2 < \sigma_2^3$. Equality $\sigma_3^2 = \sigma_2^3$ is reached in the uniaxial nematic.

A Landau–de Gennes free energy may now be expanded in powers of the two invariants σ_2 and σ_3 [1, 3–5]:

$$\Delta F = F_N - F_{Iso} \quad (4)$$
$$= a\sigma_2 + b\sigma_3 + \frac{1}{2}c\sigma_2^2 + d\sigma_2\sigma_3 + \frac{1}{2}e\sigma_3^2$$

Uniaxial Nematic

The most common case of uniaxial nematics can be treated first: the free energy expansion reduces then to [1]:

$$\Delta F_{Uniaxial} = \frac{1}{2}A(T)q^2 + \frac{1}{3}Bq^3$$
$$+ \frac{1}{4}Cq^4 + O(q^5) \quad (5)$$

As usual with Landau theories, the phase transition is governed by the coefficient of the quadratic term $A(T) = a(T - T^*)$. Other coefficients B and C are supposed to have weak temperature dependance. The presence of a third order term, imposed by the symmetry, drives the transition first order [1, 6] at a Landau temperature T_{N-I} determined by $2B^2 = 9aC(T_{N-I} - T^*)$. It must be emphasized that such third order term reflects the physical relevance of the sign of q: think about a nematic arrangement of long rods with director $\mathbf{n}$ parallel to reference axis $\hat{z}$, $q > 0$ corresponds to long axes lining up along $\hat{z}$ whereas $q < 0$ corresponds to rods parallel to the (x, y) plane with no angular order in this plane. If the first case is, of course, physical for rods, the second is more appropriate for discs. Corresponding uniaxial nematic phases are often called cylindric and discotic respectively [4, 5].

Below T_{N-I}, the order parameter varies as:

$$q = -\frac{B}{2c}\left(1 + \frac{4aC}{B^2}(T^{**} - T)^{1/2}\right) \quad (6)$$

in which T^{**} is the absolute limit of superheating of the nematic phase. The order parameter at the transition is:

$$q_c = -\frac{3a}{B}(T_{N-I} - T^*) \quad (7)$$

Experiments do confirm the first order character of the I–N transition: discontinuities are observed but are weak enough to justify the relevance of the Landau expansion.

Biaxial Nematic

The stability of biaxial nematics is calculated by minimizing the free energy (Eq. 4). The stability of the model requires $c > 0$, $e > 0$ and $ce > d^2$ (higher order terms have to be considered in the expansion otherwise). The absolute minimum of ΔF is straightforwardly obtained at [1, 3, 5]:

$$\tilde{\sigma}_2 = \frac{bd - ae}{ce - d^2} \quad \text{and} \quad \tilde{\sigma}_3 = \frac{ad - bc}{ce - d^2} \quad (8)$$

The phase diagram is calculated in the (a, b) plane, as usual with Landau theories. The various possible transition lines are obtained as follows:

- The first order isotropic–uniaxial nematic corresponds to the set of constraints $\Delta F = 0$ and $\sigma_2^{1/2} = \sigma_3^{1/3} = q$. At first order in b, its equation is:

$$a_{I-N} = \frac{b^2}{2c}\left(1 + \frac{8bd}{c^2} + O(b^2)\right) \quad (9)$$

- The isotropic–biaxial nematic line is determined by $\Delta F = 0$ and independent σ_2 and σ_3 which leads to the trivial equation:

$$a_{I-N_b} = \pm b\sqrt{\frac{c}{e}} \quad (10)$$

- Finally, the transition lines from uniaxial nematic to biaxial are determined Eq. (8) with the constraint $\tilde{\sigma}_2^{1/2} = \tilde{\sigma}_3^{1/3}$. At lowest order in b, the slope of the two lines at their meeting point $a = b = 0$ is given by:

$$a_{N_\pm - N_b} = b\frac{cd}{2ce - d^2} + O_\pm(b^2) \quad (11)$$

The complete phase diagram is obtained with the selection rule $\tilde{\sigma}_2^3 \geq \tilde{\sigma}_3^2$ and reproduced in Fig. 1 [1]. Note that a direct transition I–N_b occurs at a single point $a = b = 0$ on the line (Eq. 10) [7]. The biaxial nematic region separates two uniaxial nematics N_+ and N_- of opposite sign. $N_\pm$–N_b transitions are second order since the condition $\tilde{\sigma}_2^{1/2} = \tilde{\sigma}_3^{1/3}$ can be approached continuously from the biaxial phase.

Experiments on lyotropic systems [4, 5] do confirm the existence of a biaxial nematic phase separating two uniaxial nematics of opposite birefringence. Experimental features are in good agreement with the Landau–de Gennes mean field approach.

If the condition $ce > d^2$ is not satisfied, the minimum energy is clearly obtained with σ_2 as large as possible within the limit

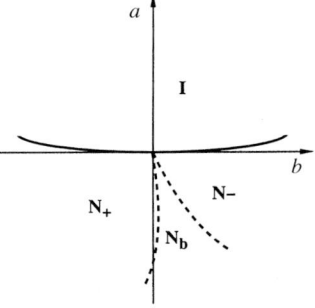

Figure 1. Phase diagram in the a, b plane resulting from minimization of Eq. (4) with biaxiality in the case $ce < d^2$. N_+ and N_- are uniaxial nematics with positive and negative order parameter respectively. N_b is a biaxial nematic. Solid lines are first-order transformations, dashed lines are second-order transformations.

$\tilde{\sigma}_2^3 \geq \tilde{\sigma}_3^2 \cdot \tilde{\sigma}_2 = \tilde{\sigma}_3$ is then the physical limit in this case and the biaxial domain is replaced by a direct first order transition from a positive to negative nematic [1].

6.1.2.2 Fluctuations

Like for any weakly first order phase transition, the mean field approach is expected to fail close to the temperature of transition T_{N-I} when fluctuations of the order parameter reach correlation lengths much larger than molecular sizes. Spatial variations of the order parameter are classically accounted for by adding a gradient term $½ L (\nabla q)^2$ to the expansion (Eq. 5) and taking the new expression as a local Hamiltonian (L is a positive Landau coefficient). Up to second order in q, the model is exactly soluble (Gaussian model) but has no condensed phase (it becomes unstable for $A(T)<0$). In the isotropic phase, the Gaussian correlation length is then $\xi_G = (a/L(T-T^*))^{-1/2}$ for $T > T_{N-I} > T^*$.

The first step beyond mean field and Gaussian approximations is to expand the partition function in powers of the interaction (i.e. terms of order higher than 2 in q) and integrate over Gaussian fluctuations. It is a standard result that such perturbation expansion presents infrared divergences (i.e. for wavenumber $k \to 0$) in three dimensions [8]. It can be used, however, to estimate the critical domain $T_{Ginz}-T^*$, T_{Ginz} being the Ginzburg temperature at which the lowest order contribution of the interaction is of order the mean field value [1, 9, 10]:

$$T_{Ginz} - T^* \approx \frac{L}{a}\left(\frac{7B^2 k_B T}{4\pi^2 L^3}\right)^{2/3} \quad (12)$$

Experimental estimates of B lead to a noticeable critical domain ($T_{Ginz}-T^*$ of order 1 to 10 K [1]). Significant deviations from mean field predictions are indeed commonly observed [11]. For instance, the difference ($T^{**}-T_{N-I}$) is smaller than its mean field value $B^2/36aC$ which shows that fluctuations decrease the effective value of B. Note that this leads to underestimate the critical domain by setting the experimental value of B in Eq. (12).

The divergences of the perturbation expansion can be overcome by length scale transformations of the renormalization group method [12]. Each step of the process consists in integrating out fluctuations of large wavenumber k close to the upper cut-off Λ, say $\Lambda/b < k < \Lambda$ with $0 < b < 1$ and rescaling all lengths ($k \to bk$) to restore original momentum space. High k integrations are safe with respect to infrared divergences whereas length rescaling decreases the correlation length and thus brings the system away from the critical point towards 'safer' (i.e. less diverging) regions. Iterations of this process lead to recursion relations which define trajectories of the Landau–Ginzburg coefficients $B(b)$ and $C(b)$. Stable fixed points of these trajectories are linked to second order phase transitions whereas unstable flows denote first order transitions. Given that the Landau rule predicts a first order N–I transition, the renormalization flow is expected to hit such unstable domain for large enough correlation length. It is interesting to notice that besides this first order behavior, the existence of an unexpected stable fixed point has been reported [13]. This fixed point should be physically accessible for small values of the ratio B^2/C, i.e. close to the biaxial nematic. Although conjectural though, this result suggests the possibility of a second order isotropic to nematic transition, despite the third order invariant!

6.1.2.3 Isotropic–Nematic Transition in Restricted Geometries

The size or shape of the liquid crystal sample has not been specified so far: phase transition theories implicitely deal with infinite samples, that is samples of macroscopic size D much larger than any relevant physical length such as the correlation length ξ of the order parameter. Liquid crystals, however, are commonly used in restricted geometries such as thin films or dispersions in a polymer matrix (PDLC) or a solid labyrinth (aerogels). There are cases where the restriction of the available space plays a role on the phase transition. We will discuss, as an example, the influence of a solid interface that develops different microscopic interactions with molecules in the isotropic and the nematic phase. A macroscopic way of considering this difference is to assume different wetting of the solid by the two phases. This problem was addressed by Poniewiersky and Sluckin [14] who showed that the isotropic to nematic transition is shifted towards lower temperatures if the isotropic phase wets the solid surface more than the nematic phase. The slope of the coexistence curve is given by a generalized Clausius-Clapeyron equation:

$$\frac{dT_{N-I}}{dD} = \frac{2}{D^2}\frac{(\gamma_N - \gamma_I)}{L} T_{N-I}(D) \quad (13)$$

in which D is the thickness of the liquid crystal film, γ_N and γ_I the surface tension of the nematic and isotropic phase relative to the substrate and L the latent heat of the N–I transition.

It could happen, on the contrary, that a solid substrate promotes local nematic ordering in the isotropic phase which amounts to setting $(\gamma_N - \gamma_I) < 0$ in Eq. (13) (this was experimentally observed for mica surfaces in lyotropic nematics [15]). The nematic phase may then form above T_{N-I} for thin enough liquid crystal films. This is the so-called capillary condensation.

Geometrical constraints may affect the N–I transition in different ways: strong anchoring conditions at the boundaries of limited volumes (PDLC or aerogels) produce topological defects in the nematic phase. The N–I transition may then be decreased to compensate for the extra energetical cost of the defects.

6.1.3 The Nematic–Smectic A Transition

6.1.3.1 Definition of an Order Parameter

After the isotropic to nematic transition, the next step towards more ordered mesophases is the condensation of SmA order when the continuous translational symmetry is broken along the director. The theoretical description of the N–SmA transition begins with the identification of an order parameter. Following de Gennes and McMillan [1, 16], we notice that the layered structures of a SmA phase is characterized by a periodic modulation of all the microscopic properties along the direction $\hat{z}$ perpendicular to the layers. The electron density for instance, commonly detected by X-ray scattering can be expanded in Fourier series:

$$\rho(\mathbf{r}) = \rho_0 + \sum_{n=1}^{\infty} \Psi_n e^{iq_{Sm}\cdot\mathbf{r}} + \Psi_n^* e^{-iq_{Sm}\cdot\mathbf{r}} \quad (14)$$

in which q_{Sm} is the smectic wavenumber of modules $2\pi/d$. The smectic period d (commonly referred to as layer thickness) is of the order of a molecular length in most thermotropic smectics and varies essentially as the inverse fraction Φ^{-1} of the material constituting the lamellae in lyotropic smectics. The average electron density ρ_0 is about the same in the nematic and smectic A phases. The fundamental term Ψ_1 of the Fourier expansion is obviously zero in the nematic phase. It is a natural choice for the N–SmA order parameter. As a complex, it has two independent components:

$$\Psi_1(\mathbf{r}) = \rho_1(\mathbf{r})e^{i\Phi_1(\mathbf{r})} \quad (15)$$

The modulus $\rho_1(\mathbf{r})$ is a measure of the strength of the local SmA ordering: the more the molecules are segregated into well defined layers, the higher ρ_1. Note that unlike the nematic scalar order parameter S, ρ_1 cannot be defined on an absolute scale that would assign 1 to perfect SmA ordering for instance. The amplitude of the electron density wave depends on each particular system. Variations of the phase $\Phi_1(\mathbf{r})$ are related to the local displacements of the layers commonly denoted as $u(\mathbf{r})$ with:

$$\Phi_1(\mathbf{r}) = -iq_{sm} u(\mathbf{r}) \qquad (16)$$

The basic elastic distortions of the layers, namely compression and curvature correspond to $\nabla_\parallel u$ and $\Delta_\perp u$ respectively (subscripts $\parallel$ and $\perp$ denote directions parallel and perpendicular to the director respectively).

6.1.3.2 Mean Field Description (McMillan, de Gennes)

Mean field approximation neglects all spatial inhomogeneities. The order parameter is thus considered as constant all over the sample. In the vicinity of the N–SmA phase transition, a Landau free energy density may be expanded in invariant combinations of the order parameter Ψ (the subscript 1 referring to the first term of the Fourier expansion will be omitted in the following). Translational invariance requires that the phase Φ does not enter the free energy. The Landau expansion thus reduces to:

$$\Delta F_1 = \frac{r}{2}|\Psi|^2 + \frac{u_0}{4}|\Psi|^4 + \ldots \qquad (17)$$

With $r = a(T - T_{N-A})$ and $u_0 > 0$, the N–SmA transition is second order at the mean field temperature T_{N-A}. The modulus $|\Psi|$ grows as $(r/u_0)^{1/2}$ below T_{N-A}. The critical exponent β of the order parameter is thus 1/2 as usual with mean field theories.

This picture is slightly more complicated if the coupling with the nematic order parameter $S_0(T) = 1/2 \langle 3\cos^2\theta_i - 1\rangle$ is included (θ_i denotes the angle between the long axis of the ith molecule and the director $\mathbf{n} = \hat{z}$ and the brackets $\langle\ldots\rangle$ represent an average over the volume of the sample).

If δS measures the deviation of the microscopic alignment from its equilibrium value S_0 in the nematic phase (calculated from the Maier–Saupe theory [17] for instance), the free energy density reads:

$$\Delta F_2 = \frac{r}{2}|\Psi|^2 + \frac{u_0}{4}|\Psi|^4 + \frac{1}{2\chi}\delta S^2 - C|\Psi|^2 \delta S \qquad (18)$$

$\chi(T)$ is a response function (susceptibility) and C is a generally positive constant (the onset of the smectic order usually increases the average attraction between the molecules and hence reinforces the alignment).

Minimization with respect to δS yields $\delta S = \chi C |\Psi|^2 \; (>0)$ and the new coefficient of the fourth order term is now:

$$u = u_0 - 2C^2\chi \qquad (19)$$

If the nematic susceptibility χ is low (i.e. far enough from T_{N-I}, temperature of the isotropic–nematic transition), u is positive and the N–SmA transition is second order at T_{N-A} again.

If χ is larger (i.e. T_{N-A} close to T_{N-I}, u is negative. A sixth order term $v/6|\Psi|^6$ must be added to Eq. (18) and the N–SmA transition is first order at a temperature $T_{N-A} + 3\,u^2/16\,a\,v > T_{N-A}$.

$\chi = u_0/2\,C^2$ (i.e. $u = 0$) defines a tricritical point on the N–SmA line.

McMillan estimated that the tricritical point corresponds to a value 0.87 of the ratio T_{N-I}/T_{N-A} [18] which agrees reasonably well with many experimental data.

6.1.3.3 Analogy with Superconductors

Because the smectic order is one dimensional, the fluctuations of the layers described by the local displacement field $u(\mathbf{r})$ are known to play an important role, even far away from any phase transition: the square amplitude $\langle u^2(\mathbf{r}) \rangle$ diverges like the logarithm of the size of the sample (this is the well known Landau–Peierls instability [19, 20]) hence killing a true long range order.

Close to the N–SmA transition, the vanishing of the elastic constant of compression of the layers amplifies the fluctuations of the phase whereas critical fluctuations of the amplitude $|\Psi(\mathbf{r})|$ are expected to be important too.

Fluctuations are accounted for in a Landau–Ginzburg expansion of a local Hamiltonian. Once again even powers of $|\psi|$ only are permitted. Including gradient terms and fluctuations of the nematic director $\delta\mathbf{n}_\perp = \mathbf{n}(\mathbf{r}) - \mathbf{n}_0 \, (\mathbf{n}_0 = \hat{z})$ yields the following Landau–Ginzburg functional:

$$F_S = \frac{1}{2} \int d^3r \Big\{ r|\psi|^2 + \frac{u}{2}|\psi|^4$$
$$+ C_\| |\nabla_z \psi|^2 + C_\perp |(\nabla_\perp - i q_{sm} \delta\mathbf{n}_\perp)\psi|^2$$
$$+ K_1 (\mathrm{div}\,\delta\mathbf{n}_\perp)^2 + K_2 (\hat{z} \cdot \mathbf{curl}\,\delta\mathbf{n}_\perp)^2$$
$$+ K_3 (\nabla_z \delta\mathbf{n}_\perp)^2 \Big\} \quad (20)$$

The first two terms are the Landau part, Eq. (17). Because of the nematic anisotropy, the gradient terms exhibit anisotropic coefficients ($C_\| \neq C_\perp$) along directions parallel and perpendicular to the director $\mathbf{n}$. With the notation $\nabla_z = \partial/\partial z$ and $\nabla_\perp = (\partial/\partial x, \partial/\partial y)$ and at lowest relevant order in ($\delta\mathbf{n}_\perp$), these gradients have the form Eq. (20). The last three terms are the usual Frank–Oseen elastic energy of the nematic [21].

If one forgets about the fluctuations of the director (i.e. set $\delta\mathbf{n}_\perp = 0$ in (20)) the N–SmA problem becomes equivalent to the condensation of superfluid helium (XY model, $d = 3$) since the smectic order parameter has two independent components (the anisotropy of the elastic coefficients $C_\|$ and $C_\perp$ can be removed by a simple anisotropic rescaling).

With non-zero $\delta\mathbf{n}_\perp$, expression (Eq. (20)) is very similar to the Landau–Ginzburg functional describing the normal–superconductor transition [1, 22, 23]:

$$F_{SC} = \frac{1}{2}\int d^3r \Big\{ a|\psi|^2 + \frac{u}{2}|\psi|^4$$
$$+ \frac{1}{4m}\Big|(\hbar\nabla - i\frac{2e}{c}\mathbf{A})\psi\Big|^2$$
$$+ \frac{1}{8\pi\mu}(\mathbf{curl}\,\mathbf{A})^2 \Big\} \quad (21)$$

Here, ψ is the superconductor gap order parameter. It corresponds to the wave function of the superconducting pair in BCS theory and has the XY symmetry of the smectic order parameter. The magnetic vector potential $\mathbf{A}$ comes analogous to the director $\mathbf{n}$ (m and e are the mass and charge of a single electron, $\hbar$ Planck's constant, c the velocity of light and μ the magnetic permittivity).

In liquid crystals, $\mathbf{curl}\,\mathbf{n}$ is then analogous to the magnetic field $\mathbf{B} = \mathbf{curl}\,\mathbf{A}$. Twist ($\mathbf{n} \cdot \mathbf{curl}\,\mathbf{n}$) and bend ($\mathbf{n} \times \mathbf{curl}\,\mathbf{n}$) are components parallel and perpendicular to the director respectively (they correspond to components of $\mathbf{B}$ parallel and perpendicular to $\mathbf{A}$ in superconductors). The anisotropy of the field $\mathbf{curl}\,\mathbf{n}$ follows from $K_2 \neq K_3$.

Interesting behaviors of the smectic state can be deduced from this analogy.

Just as superconductors expell magnetic fields $\mathbf{curl}\,\mathbf{A}$ (Meisner effect [24]) smectics expell bend and twist. The bend and twist moduli K_2 and K_3 should therefore diverge upon approaching the smectic state from the nematic.

Two important lengths characterize the superconductors: the order parameter coherence length $\xi = (m/|r|)^{1/2}$ over which ψ can vary and the London penetration depth of a magnetic field $\lambda = (mc^2/2\mu e^2|\psi|^2)^{1/2}$.

For smectics, two order parameter coherence lengths

$$\xi_{\parallel,\perp} = \left(\frac{C_{\parallel,\perp}}{|r|}\right)^{\frac{1}{2}} \quad (22)$$

and four penetration lengths associated with twist and bend:

$$\lambda_{\parallel,\perp}^{2,3} = \left(\frac{K_{2,3} u}{2 C_{\parallel,\perp} q_{Sm}^2 |r|}\right)^{\frac{1}{2}} \quad (23)$$

can be identified, which precludes a simple classification.

The superconductor analogy suggests, however, two distinct behaviors:

- Type I ($\lambda\sqrt{2} > \xi$). The superconducting state is observed with perfect Meisner effect below a critical field H_c.
- Type II ($\lambda\sqrt{2} < \xi$). Two critical values of the field are found. Vortices bearing a quantum flux $\phi_0 = h/q$ penetrate the system for $H < H_{c1}$ whereas the normal–type I superconductor transition occurs at $H_{c2} > H_{c1}$.

The analog of the magnetic intensity **H** (produced by external currents) would be an external field coupled to the two components of **curl n**. A local microscopic source of bend [23] is not easy to imagine. On the other hand, chirality is naturally coupled to twist: chiral mesogens develop a spontaneous twist in the nematic phase. The cholesteric is thus analogous to a normal metal in a magnetic field.

The smectic analog of a vortex is a dislocation. Type II smectics would therefore develop dislocations when submitted to a bending or twisting stress whereas type I would resist until a critical stress eventually induces the nematic state.

The observation of edge dislocation arrays in a wedge (i.e. bending stress) [25] and of screw dislocation arrays in chiral compounds [26] (the so-called twist grain boundary (TGB) phase anologous to the Abrikosov flux phase in superconductors) suggests that type II smectics do exist, but type I behaviors have also been reported [27]. The N–SmA transition in chiral smectics and the ability of TGB phases will be discussed in Sec. 6.1.5 of this chapter.

Finally, the bare smectic coherence length at 0 temperature $\xi_0 = (C_\perp / a T_{N-A})^{1/2}$ with $r = a(T - T_{N-A})$ is significantly shorter than its superconductor equivalent (10–20 Å instead of 5000 Å) because of the higher value of T_{N-A}. An interesting consequence is that the critical domain is expected to be much larger (i.e. more easily accessible) in the smectic case.

Important differences exist, however, about gauge invariance and the absence of true smectic long range order. How severe these differences are is not fully understood yet.

The superconductor Hamiltonian (Eq. 21) is invariant under the following gauge transformation:

$$\begin{cases} \mathbf{A}' = \mathbf{A} - \nabla L \\ \psi' = \psi' \exp\left(-\frac{q}{c} L\right) \end{cases} \quad (24)$$

so that any gauge choice is physically acceptable. The Coulomb gauge $\text{div} \mathbf{A} = 0$ is generally used.

In the smectic case, a gauge transformation reads:

$$\begin{cases} \delta \mathbf{n}' = \delta \mathbf{n} - \nabla L \\ \psi' = \psi \exp(-i q_{Sm} L) \end{cases} \quad (25)$$

but the Frank–Oseen energy is not gauge invariant because of the splay term K_1. The lack of gauge invariance reflects the fact that the director field **n** is a physical observable in liquid crystals (unlike its magnetic analog **A**). There is in fact only one physical gauge, namely $\delta n_z = 0$ (remember that the condition $n^2 = 1$ implies that $\delta \mathbf{n}$ lies in the

x, y plane). In the unphysical Coulomb gauge (div $\delta \mathbf{n} = 0$), the splay term disappears and it can be shown that true long range smectic order is not killed by the layer fluctuations [28]. If the physical $\delta \mathbf{n}$ is expressed in the Coulomb gauge via Eq. (25), the splay term reads:

$$\frac{K_1}{2} \int (\Delta L)^2 d^3x \qquad (26)$$

and with the condition $\delta n_z = 0$ (i.e. $A_z(q) - iq_z L(q) = 0$):

$$\frac{K_1}{2} \int \frac{q^4}{q_z^2} |A_z(q)|^2 \frac{d^3q}{(2\pi)^3} \qquad (27)$$

The splay term is non-analytic, which implies that K_1 cannot be renormalized by the fluctuations of ψ at any order in perturbation theory. In a renormalization flow, K_1 will simply evolve according to simple power counting.

6.1.3.4 Critical Exponents and Scaling

Experiments suggest the existence of anisotropic scaling laws: $\xi_\parallel \propto t^{-\nu_\parallel}$ and $\xi_\perp \propto t^{-\nu_\perp}$ with $\nu_\parallel \ne \nu_\perp$ for instance for the correlation lengths parallel and perpendicular to the director $\mathbf{n}$ ($t = (T - T_{N-A})/T_{N-A}$ is the reduced temperature).

Anisotropic homogeneous functions can indeed be defined to describe the scaling behavior of the correlation functions of the order parameters:

$$\langle \psi(q)\psi^*(q) \rangle = G(q_\parallel, q_\perp, t) \qquad (28)$$
$$= t^{-(2-\eta_\perp)\nu_\perp} G(t^{-\nu_\parallel} q_\parallel, t^{-\nu_\perp} q_\perp)$$

$$\langle \delta n_\alpha(q) \delta n_\beta^*(q) \rangle = G_{\alpha\beta}(q_\parallel, q_\perp, t) \qquad (29)$$
$$= t^{-(2-\eta_n)\nu_\perp} G_{\alpha\beta}(t^{-\nu_\parallel} q_\parallel, t^{-\nu_\perp} q_\perp)$$

G and $G_{\alpha\beta}$ are the two point correlation functions of the smectic order parameter ψ and of the director $\delta \mathbf{n}_\perp$ respectively (indices α, β can be x or y).

The exponent γ of the susceptibility ($G(q=0) \approx t^{-\gamma}$) comes immediately from Eq. (28):

$$\gamma = (2 - \eta_\perp)\nu_\perp = (2 - \eta_\parallel)\nu_\parallel \qquad (30)$$

The free energy density $f(t)$ classically scales like $k_B T$ over the volume of a 'block' of condensed phase ξ^3. In the smectic case, this yields:

$$f(t) \sim \frac{k_B T}{\xi_\parallel \xi_\perp^2} \sim t^{2\nu_\perp + \nu_\parallel} \sim t^{2-\alpha} \qquad (31)$$

in which α is the critical exponent of the singular part of the specific heat δC_P.

The critical behavior of the twist and bend elastic constants on the nematic phase follows from the equipartition theorem

$$\begin{cases} \langle \delta n_1(q) \delta n_1(-q) \rangle = \dfrac{k_B T}{K_1 q_\perp^2 + K_3 q_\parallel^2} \\ \langle \delta n_2(q) \delta n_2(-q) \rangle = \dfrac{k_B T}{K_2 q_\perp^2 + K_3 q_\parallel^2} \end{cases} \qquad (32)$$

Axes 1 and 2 in Eq. (32) are the eigendirections of the Frank–Oseen elastic energy in the Fourier plane (q_x, q_y). The corresponding unit vectors $\mathbf{e}_1$ and $\mathbf{e}_2$ are such that $\mathbf{e}_2$ is normal to $\mathbf{q}$ and $\mathbf{e}_1$ is normal to $\mathbf{e}_2$ [1]. With the structure (Eq. 29), one gets:

$$K_2 \sim t^{-\eta_n \nu_\perp}, \quad K_3 \sim t^{2(\nu_\perp - \nu_\parallel) - \eta_n \nu_\perp} \qquad (33)$$

On the smectic side:

$$\langle \delta n_\perp(q) \delta n_\perp(-q) \rangle = \frac{k_B T q_\perp^2}{B q_z^2 + K_2 q_\perp^4} \qquad (34)$$

Equations (29) and (34) yield:

$$B \approx t^{(4-\eta_n)\nu_\perp - 2\nu_\parallel} \qquad (35)$$

The final scaling laws are then:

$$\begin{cases} \text{(a)} \;\; \xi_\parallel = \xi_0 \, t^{-\nu_\parallel} \qquad \xi_\perp = \xi_0 \, t^{-\nu_\perp} \\ \text{(b)} \;\; \delta K_2 \propto \xi_\perp^2 / \xi_\parallel \qquad \delta K_3 \propto \xi_\parallel \\ \text{(c)} \;\; B \propto \xi_\parallel / \xi_\perp^2 \\ \text{(d)} \;\; 2 - \alpha = \nu_\parallel + 2\nu_\perp \quad \gamma = (2-\eta_\perp)\nu_\perp \end{cases} \qquad (36)$$

Equations (36b) and (36d) are valid above and below T_{N-A}.

6.1.3.5 Renormalization Group Procedures

According to the classical Wilson–Fisher renormalization process [12], the parameters of the Hamiltonian (Eq. 20) will evolve under an anisotropic rescaling of the wavevectors of the form ($q_\perp \to e^l q_\perp$ and $q_z \to e^{v_\parallel/v_\perp l} q_z$). The recursion relations defining the renormalization flow of the physical parameters are controlled by simple dimensional analysis (power counting) and Gaussian integration of the fluctuations of large wavevectors (perturbation theory).

The splay constant K_1 plays a particular role since its flow depends on power counting only (see Sec. 6.1.3.3 of this volume) which yields:

$$\frac{dK_1}{dl} = -\eta_n K_1(l) = -\frac{(2v_\perp - v_\parallel)}{v_\perp} K_1(l) \quad (37)$$

There are therefore only three possibilities for the fixed point:

(1) $K_1^* = 0$: for $2v_\perp > v_\parallel$, the K_1 eigendirection is stable. The problem reduces exactly to the normal–superconductor transition [29] which implies $v_\perp = v_\parallel$. There is no such stable fixed point, however, in the smectic case ($n=2$, $d=3$) [30].
(2) $K_1^* = \infty$: the fixed point can be stable if $2v_\perp < v_\parallel$ only, but none is found [29].
(3) K_1^* finite and $2v_\perp - v_\parallel = 0$: this anisotropy is too large to fit any experimental behavior.

Classical theories fail to find a stable fixed point (not even isotropic) and consequently cannot corroborate the experimental evidence that the transition can be second order. Two possible situations can be imagined upon moving away from the mean field tricritical point towards negative values of the Landau coefficient, u: either the transition is still first order but the discontinuities are too small to be detected (this would agree with the lack of stable fixed point of the renormalization flow) or beyond some original multicritical point the transition becomes truly second order. The theory of a transition induced by dislocation loops provides an example that fits in this last situation.

6.1.3.6 Dislocation Loops Theory

Dislocations are the most elementar defects of a SmA phase. Since they cannot end in the smectic material, they must form closed loops (at least in the thermodynamic limit of an infinite sample). The elastic energy of a dislocation line is proportional to its length [31]. It is then characterized by a line tension γ_0. The total free energy per unit length is:

$$\gamma = \gamma_0 - \frac{k_B T}{x} \quad (38)$$

k_B/x measures the entropy contribution that favors spontaneous nucleation of dislocation loops (x has the dimension of a length). γ becomes negative above some temperature so that the density of dislocations becomes finite. Helfrich [32] and Nelson and Toner [33] have shown that such proliferation of dislocations destroys smectic order: the elastic response of a SmA with a finite density of unbound dislocations is equivalent to that of a nematic [34] (unbound dislocation loops are those thermally excited defects that can move around freely under the effect of thermal fluctuations). Nelson and Toner [33], Dasgupta and Halperin [35] and Toner [34] have formalized this problem: the main result is the existence of a stable fixed point which for the first time gave the theoretical possibility of a true second order N–SmA transition.

This fixed point is isotropic and belongs to the so-called inverted XY universality class. Inverted refers to the inversion of the high and low temperature sides of the transition since the melting is described by a 'disorder' parameter ψ (density of dislocations). $|\psi| \neq 0$ corresponds to the nematic phase. The universal ratio of the amplitudes A^+ and A^- of the heat capacity singularity for instance ($\delta C_p = A^{\pm} t^{-\alpha}$) is the inverse of the superfluid helium value.

6.1.4 The Smectic A–Smectic C Transition

6.1.4.1 The Superfluid Helium Analogy

The SmC state differs from the SmA by a tilt θ of the director **n** with respect to the direction $\hat{z}$ normal to the layers. The director **n** is totally specified by θ and the azimuthal angle φ. The SmC order can thus be described by two real angles θ and φ or equivalently by the complex order parameter:

$$\psi(\mathbf{r}) = \theta(\mathbf{r}) e^{i\varphi(\mathbf{r})} \qquad (39)$$

A modulus $\theta = 0$ corresponds to the SmA state. A Landau–Ginzburg functional similar to Eq. (20) with $\delta \mathbf{n} = 0$, and therefore to the superfluid–normal helium problem, can be constructed to describe the SmA–SmC transition.

The straightforward consequence of this analogy is that the SmA–SmC transition may be continuous at a temperature $T_{SmC-SmA}$ with XY critical exponents. Below $T_{SmC-SmA}$ the tilt angle θ for instance should vary as $\theta = \theta_0 |t|^\beta$ with $\beta = 0.35$. Above $T_{SmC-SmA}$ an external magnetic field can induce a tilt θ proportional to the susceptibility $\chi \approx t^{-\gamma}$ with $\gamma = 1.33$.

Experiments (heat capacity measurements in particular [36]) rather show a mean field behavior which may be due to the narrowness of the critical domain or to the influence of a close by tricritical point.

The width of the XY critical domain can be estimated from the usual Ginzburg criterion: equating the mean field heat capacity discontinuity and the contribution of the fluctuations in the Gaussian approximation leads to a Ginzburg temperature T_{Ginz} [1, 37]:

$$T_{Ginz} - T_{SmC-SmA} \approx T_{SmC-SmA}/64\pi^2 \qquad (40)$$

so that the critical regime extends over a fraction of a degree. It may easily switch from observable to non-observable with a slight change of the roughly estimated numerical constant that lead to Eq. (40). Furthermore, the crossover regime may be different from one observable to another. Recent estimates of the Ginzburg criterion, for instance, indicate that elastic constants measurements are one hundred times more sensitive to fluctuations than heat capacity ones [37]. Experiments tend to confirm this point: a mean field behavior of the heat capacity has been reported [36] whereas fluctuations seem to be important in some tilt, susceptibility or bulk modulus measurements [38, 39].

Finally, a first order SmA–SmC transition is always possible.

6.1.4.2 The N–SmA–SmC Point

The existence of a N–SmA–SmC multicritical point (i.e. a point where the N–SmA, SmA–SmC and N–SmC lines meet) was demonstrated in the late seventies [40, 41]. Various theories have been proposed to describe the N–SmA–SmC diagram [42–45]. The phenomenological model of Chen and Lubensky [46] (referred to as the N–SmA–SmC model) captures most of the experimental features. The starting point is the observation that the X-ray scattering in the nematic phase in the vicinity of the N–SmA transition shows strong peaks at wavenumber $\mathbf{q}_A = \pm q_0 \mathbf{n}$. Near the N–SmC transition, these two peaks spread out into two rings at $\mathbf{q}_C = (\pm g_\parallel, q_\perp \cos \varphi, q_\perp \sin \varphi)$.

Once again, the order parameter is defined from the mass density wave $\rho(\mathbf{r})$. In the smectic phases, $\rho(\mathbf{r})$ becomes periodic with fundamental wave number $\mathbf{q}_0 = (\pm q_\parallel, \mathbf{q}_\perp)$ ($q_\perp = 0$ in the SmA phase). Chen and Lubensky defined a covariant Landau–Ginzburg Hamiltonian including fluctuations of the director through the Frank free energy. A mean field description of the N–SmA–SmC model is given here. The director is chosen along z: $\mathbf{n} = \mathbf{n}_0\,(0,0,1)$. Like in the SmA phase, the order parameter Ψ is chosen as the part of ρ with wave numbers in the vicinity of $\mathbf{q}_0$:

$$\Psi(\mathbf{r}) = \psi \exp i(\mathbf{q}_0 \cdot \mathbf{r}) = \psi \exp i(q_\parallel z + \mathbf{q}_\perp \cdot \mathbf{x}_\perp) \quad (41)$$

$\mathbf{x}_\perp$ and $z_\perp$ are two dimensional vectors in the plane (x, y) and (q_x, q_y) respectively. The mean field free energy density reads:

$$\Delta F_{\text{N–SmA–SmC}}$$
$$= \frac{1}{2}\left\{ a|\psi|^2 + D_\parallel |\nabla_z^2 \psi|^2 \right.$$
$$- C_\parallel |\nabla_z \psi|^2 + \frac{C_\parallel}{4 D_\parallel}|\psi|^2$$
$$\left. - C_\perp |\nabla_\perp \psi|^2 + D_\parallel |\nabla_\perp^2 \psi|^2 \right\} + u|\psi|^2 \quad (42)$$

with $C_\parallel$, $D_\parallel$, $D_\perp$, and $u > 0$ and $a = a'\left(\dfrac{T - T_{\text{N–SmA}}}{T_{\text{N–SmA}}}\right)$.

In Fourier space, $\Delta F_{\text{N–SmA–SmC}}$ reads:

$$\Delta F_{\text{N–SmA–SmC}}$$
$$= \frac{1}{2}\left\{ a + D_\parallel \left(q_z^2 - Q_\parallel^2\right)^2 + C_\perp q_\perp^2 + D_\perp q_\perp^4 \right\}$$
$$\times \psi^2 + u\psi^4 \quad (43)$$

where $Q_\parallel^2 = C_\parallel / 2 D_\parallel$.

The free energy (Eq. 43) is minimized the usual way:

– For $C_\perp > 0$, the minimum is reached for $q_\perp = 0$ and $q_z 0 \pm q_\parallel$. A second order N–SmA transition occurs at the mean field temperature $T = T_{\text{N–SmA}}$.

– For $C_\perp < 0$, $\Delta F_{\text{N–SmA–SmC}}$ can be rewritten as:

$$\Delta F_{\text{N–SmA–SmC}}$$
$$= \frac{1}{2}\left\{ \tilde{a} + D_\parallel \left(q_z^2 - Q_\parallel^2\right)^2 + D_\perp \left(q_\perp^2 - Q_\perp^2\right)^2 \right\}$$
$$\times \psi^2 + u\psi^4 \quad (44)$$

with $Q_\perp^2 = |C_\perp|/2 D_\perp$ and
$\tilde{a} = a'(T - T_{\text{N–SmC}})/T_{\text{N–SmA}}$,
$T_{\text{N–SmC}} = T_{\text{N–SmA}} + C_\perp^2 T_{\text{N–SmA}} / 4 D_\perp a'$.

The minimum corresponds to $q_\perp = \pm Q_\perp$, $q_z = \pm Q_\parallel$ and a N–SmC transition occurs at the mean field temperature $T_{\text{N–SmC}}$. The mean field N–SmA–SmC diagram is shown in Fig. 2.

The scattered X-ray intensity $I(q)$ can be estimated in the nematic phase from Gaussian fluctuations about the mean field solution $\psi = 0$:

$$I(q) \approx \langle \rho(q) \rho(-q) \rangle \quad (45)$$
$$\approx \frac{1}{a + D_\parallel\left(q_\parallel^2 - Q_\parallel^2\right) + C_\perp q_\perp^2 + D_\perp q_\perp^4}$$

For $C_\perp > 0$, $I(q)$ has peaks at $q_\parallel = \pm Q_\parallel$ corresponding to fluctuations into the SmA

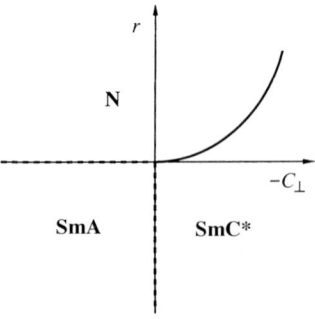

Figure 2. N–SmA–SmC phase diagram from Chen and Lubensky [46]. All transitions are second order in mean field but fluctuations lead to a first order N–SmC transition [47]. The line $C_\perp = 0$ separates two regions in the nematic phase with diffuse X-ray scattering centered about $q_\perp = 0$ ($C_\perp > 0$) and $q_\perp \neq 0$ ($C_\perp < 0$). The point $C_\perp = 0$, $r = 0$ is a Lifshitz point.

phase. When $C_\perp$ is negative (see Eq. 44) $I(q)$ is maximum on two rings ($\pm q_\parallel$, $q_\perp \cos\varphi$, $q_\perp \sin\varphi$) as observed.

The N–SmA–SmC problem bears strong similarities with the transition from paramagnetic to helimagnetic states [47]. Fluctuations are maximum at a finite m-dimensional vector $q_\perp$ in a d-dimensional space. Mean field N–SmA–SmC theory predicts a second oder N–SmC transition as shown. Fluctuations, however, are believed to lead to a first-order transition when $m = d$ or $d-1$. The N–SmC transition with $d = 3$ and $m = 2$ is therefore expected to be first order.

Finally, we notice that the N–SmA–SmC point is an example of Lifschitz point defined as the place where some of the coefficients of the gradient terms ($C_\perp$ here) vanish [48, 49].

Complications arise from the vanishing of the N–SmC latent heat at the N–SmA–SmC point and from the difficulties connected to the smectic state (Landau–Peierls instability) the N–SmA transition (lack of gauge invariance) and the SmA–SmC transition (proximity of a tricritical point).

The description of the Chen–Lubensky model is reasonably well borne out by experiment [50, 51]: the universal topology of the phase diagram and the existence of a $C_\perp = 0$ line in particular are well established. An interesting possibility pointed out by Grinstein and Toner [52] with a model of dislocation unbinding is the existence of a biaxial nematic phase: if the N–SmA and SmA–SmC lines are second order with XY critical exponents, the N–SmA–SmC point should be tetracritical [53] and a mixed phase (i.e. a biaxial nematic) should show up. It has however not been observed so far.

6.1.5 The Smectic A–Hexatic Smectic B Transition

The *crystalline* smectic B differs from the SmA phase by the existence of a regular hexagonal packing of the molecules in the plane of the layers. The B–SmA transition corresponds to the melting of a two-dimensional crystalline order in a three-dimensional material whereas the layered structure along the third direction switches from true to quasi-long range order. In two dimensions (i.e. in a hypothetic isolated single layer) a Kosterlitz–Thouless [54] mechanism of dislocations unbinding may happen to make the transition continuous with no singularities. In three dimensions, a direct transition B–SmA is expected to be first order.

An intermediate layered structure SmB, however, exists in which the two-dimensional ordering of the molecules within a layer is organized as follows: the correlations of the position of the center of mass of the molecules decay exponentially (like in a liquid) but the direction of the bonds linking two adjacent molecules exhibit a long range order with six-fold symmetry. Halperin, Nelson and Young [55, 56] have pointed out that dislocations can create such a two dimensional hexatic order. The existence of this hexatic order in smectics is a beautiful success of the Halperin–Nelson–Young concept of bond ordering.

Accounting for the six-fold symmetry of the B and SmB phases, the modulation of the mass density within the (x, y) plane of the layers can be expanded in a Fourier series:

$$\rho(\mathbf{r}) - \bar{\rho} = \rho_0(r) + \rho_6(r) \cos 6(\theta - \phi) + \text{higher harmonics} \quad (46)$$

$\mathbf{r}$ is a 2-d vector of cartesian coordinates $x = r\cos\theta$ and $y = r\sin\theta$. The correlation function of the radial distribution $\rho_0(r)$ falls off exponentially in the SmB and SmA phases. The amplitude $\rho_6(r)$ describing the six-fold order vanishes in the SmA phase only. A possible order parameter of the SmA–SmB transition is therefore:

$$\psi_6 = \rho_6\, e^{6i\phi} \qquad (47)$$

The phase ϕ accounts for spatial fluctuations of the orientation of the local six-fold axes.

The order parameter ψ_6 has two components and the SmA–SmB transition is expected to belong to the XY universality class.

Experiments do confirm the second order nature of the transition but high resolution a.c. calorimetry gives values of the specific heat exponent α about 0.6 [57] inconsistent wih the XY class ($\alpha = -0.06$). High sensitivity heat capacity measurements on freely suspended thin films (down to four layer) show a crossover from bulk ($\alpha = 0.59$) to a two dimensional cusp-like behavior with $\alpha = -0.26$ [58].

6.1.6 Phase Transitions in Chiral Liquid Crystals

6.1.6.1 Chirality in Nematic and Smectic Liquid Crystals

Experiments show that the introduction of chirality (chiral mesogen or chiral dopant) in a nematic phase generates a spontaneous twist of the director, $\mathbf{n} = (\cos 2\pi z/P, \sin 2\pi z/P, 0)$ for instance for a helical twist of pitch P along the direction z. The higher the chirality (i.e. fraction of chiral dopant for instance), the higher the twist (i.e. the shorter the pitch). It is thus clear that twist is a structural response coupled to the microscopic constraint chirality. How can these two physical quantities be linked in a quantitative way? The covariant expression of the local twist, commonly used in the Frank energy, is $\mathbf{n}\cdot \mathrm{curl}\,\mathbf{n}$ ($= -2\pi/P$ in the example of a simple cholesteric helix). A natural extension of the Frank energy density to chiral nematics is [1, 59]:

$$\Delta F_{\text{Frank}}^{\text{Chiral}} = \frac{K_1}{2}(\nabla\cdot\mathbf{n})^2 + \frac{K_2}{2}(\mathbf{n}\cdot\mathrm{curl}\,\mathbf{n})^2$$
$$+ \frac{K_3}{2}(\mathbf{n}\times\mathrm{curl}\,\mathbf{n})^2 - h\,\mathbf{n}\cdot\mathrm{curl}\,\mathbf{n} \qquad (48)$$

in which the field h coupled with the twist is a measure of the chirality. In the superconductor analogy, h is analogous to the magnetic field $\mathbf{H}$ imposed by external currents (or more exactly, to the component of $\mathbf{H}$ along $\mathbf{A}$; other components are coupled to bend).

In cholesterics, no other energy depends on twist and Eq. (48) is minimum ($= -h^2/2K_2$) for the helical solution of pitch $P = 2\pi K_2/h = 2\pi/k_0$: $\mathbf{n} = (\sin 2\pi z/P, \cos 2\pi z/P, 0)$ which satisfies $\nabla\cdot\mathbf{n} = 0$, $\mathbf{n}\times\mathrm{curl}\,\mathbf{n} = 0$ and $\mathbf{n}\cdot\mathrm{curl}\,\mathbf{n} = h/K_2$. Note that the pitch then gives a measure of the chirality. The sign of h reflects the handedness of the helix.

In defect-free smectic phases of constant layer thickness d, twisted arrangement of the layers are forbidden as can be shown easily: let A and B be two points in a smectic phase and $\mathbf{N}(\mathbf{r})$ the (oriented) layer normal at position $\mathbf{r}$. The (algebraic) number of layers crossed along a path going from A to B is $\int_A^B \mathbf{dl}\cdot\left(\dfrac{\mathbf{N}}{d}\right)$ independent of the path AB. This property implies that there exists a potential ϕ such as $\mathbf{N}/d = \nabla\phi$ which in turn leads to $\mathrm{curl}\,(\mathbf{N}/d) = 0$.

In a SmA phase, the layer normal $\mathbf{N}$ identifies with the director field $\mathbf{n}$ so that $\mathrm{curl}\,\mathbf{n} = 0$: twist and bend cannot develop on a macroscopic scale. The penetration of twist requires either non-uniform layer thickness $d(\mathbf{r})$ (which can be achieved on a limited scale: the twist penetration depth λ_2, Eq. (23)) or defects such as dislocations which make path-dependent the integral $\int_A^B \mathbf{dl} \cdot \left(\dfrac{\mathbf{N}}{d}\right)$. In mean field, $\mathbf{n} \cdot \mathrm{curl}\,\mathbf{n} = 0$ everywhere in the SmA phase with no defect and Eq. (48) implies that the free energy density of the cholesteric phase is decreased by the twist term $-h^2/2K_2$ with respect to the untwisted SmA. The cholesteric to SmA transition becomes first order at a temperature $T_{\mathrm{N}^*-\mathrm{A}} = T_{\mathrm{N-A}} - \sqrt{2u_0/K_2}\,|h|/a$ lower than $T_{\mathrm{N-A}}$ (u_0 and a are defined from Eq. (17)).

In a SmC phase, the layer normal $\mathbf{N}$ no longer coincides with the director field $\mathbf{n}$. Non-zero twist and bend (i.e. $\mathrm{curl}\,\mathbf{n} \neq 0$) are permitted. The well known structure of the director helix is $\mathbf{n}(\mathbf{r}) = (\sin\theta \cos 2\pi z/P, \sin\theta \sin 2\pi z/P, \cos\theta)$ for a layer normal along z and a SmC tilt θ ($\mathbf{N} \cdot \mathbf{n} = \cos\theta$). The twist and bend terms of the Frank energy are:

$$|\mathbf{n} \cdot \mathrm{curl}\,\mathbf{n}| = \frac{2\pi}{P}\sin^2\theta$$

$$|\mathbf{n} \times \mathrm{curl}\,\mathbf{n}| = \frac{2\pi}{P}\sin\theta\cos\theta \qquad (49)$$

Note that the gain in twist energy associated with the SmC* helix is coupled to a loss in bend energy. Phase transitions between chiral phases will be described below in the frame of the mean field approximation.

6.1.6.2 Mean Field Chiral N–SmA–SmC model

The chiral version of the N–SmA–SmC model of Chen and Lubensky was investigated by Lubensky and Renn [59]. The mean field free energy density is:
$$\Delta F_{\mathrm{N-SmA-SmC^*}} = \Delta F_{\mathrm{N-SmA-SmC}} + \Delta F_{\mathrm{Frank}}^{\mathrm{Chiral}} \qquad (50)$$

in which $\Delta F_{\mathrm{N-SmA-SmC}}$ is the smectic part of the N–SmA–SmC model and $\Delta F_{\mathrm{Frank}}^{\mathrm{Chiral}}$ is given by Eq. (48). In the simplified version of the mean field model (Eq. 41) of the smectic order parameter, $\Delta F_{\mathrm{N-SmA-SmC}}$ reduces to Eq. (42). It is more convenient to define the z axis along the layer normal in the smectic phases:

$$\Psi(\mathbf{r}) = \psi_0 e^{i\mathbf{k}\cdot\mathbf{r}} = \psi_0 e^{ikz} \quad \text{with} \quad \mathbf{k} = (0,0,k) \qquad (51)$$

and the director $\mathbf{n}$ as:

$$\mathbf{n} = (\sin\theta\cos\phi, \sin\theta\sin\phi, \cos\theta) \qquad (52)$$

ψ_0, k, θ, and ϕ are then treated as variational parameters. The important parameter is $\cos\theta = \mathbf{n}\cdot\mathbf{k}/k$ independent of the choice of the coordinate system. The three phases of interest are thus characterized by:

$$\psi_0 = 0, \quad \theta = \frac{\pi}{2}, \quad \nabla_z \phi = k_0$$

for the N* phase

$$\psi_0 \neq 0, \quad \theta = 0, \quad k = q_0$$

for the SmA phase

$$\psi_0 \neq 0, \quad \theta \neq 0, \frac{\pi}{2}, \quad k \neq q_0$$

for the SmC* phase $\qquad (53)$

where $k_0 = h/K_2 = 2\pi/P$ and q_0 is the wavenumber of the SmA phase. Using Eqs. (51) and (52) in the free energy density, Renn and Lubensky obtain:

$$\Delta F_{\mathrm{N^*-SmA-SmC^*}} \qquad (54)$$
$$= a|\psi_0|^2 + \frac{1}{2}u|\psi_0|^4$$
$$+ C_\parallel(k\cos\theta - q_0)^2|\psi_0|^2$$
$$+ (C_\perp k^2\sin^2\theta + D_\perp k^4\sin^4\theta)|\psi_0|^2$$
$$+ \frac{1}{2}K(\theta)\sin^2\theta(\nabla_z\phi)^2 + h\sin^2\theta(\nabla_z\phi)$$

where $a = a'(T - T_{\mathrm{N-SmA}})$ and

$$K(\theta) = K_3\cos^2\theta + K_2\sin^2\theta \qquad (55)$$

Equation (54) forms the basis for the derivation of the N*–SmA–SmC* phase diagram. Rather than using k and θ as independent parameters, it is convenient to use $k_\parallel = k\cos\theta$ and $k_\perp = k\sin\theta$. Straightforward minimization with respect to k and $\nabla_z\phi$ yields:

$$k_\parallel = q_0 \quad \text{and} \quad \nabla_z \phi = -\frac{h}{K(\theta)} \quad (56)$$

and the free energy density is finally a function of only two variational parameters $\tan\theta$ and ψ_0:

$$\Delta F_{N^*-SmA-SmC^*} = a|\psi_0|^2 + \frac{1}{2}u|\psi_0|^4$$
$$+ (C_\perp q_0^2 \tan^2\theta + D_\perp q_0^4 \tan^4\theta)|\psi_0|^2$$
$$- \frac{1}{2}\frac{h^2}{K_3}\frac{\tan^2\theta}{1+(K_2/K_3)\tan^2\theta} \quad (57)$$

The final minimization is somewhat tedious but presents no difficulties [59]. The equations for the phase boundaries are

$$a_{N^*-SmA} = -\left(\frac{u}{K_2}\right)^{1/2} h \quad (58\,a)$$

$$a_{N^*-SmC^*} = -\left(\frac{u}{K_2}\right)^{1/2} h$$
$$- \frac{1}{4D_\perp q_0^4}\left[-C_\perp q_0^2 + \frac{1}{2}\left(\frac{K_2 u}{K_3^2}\right)^2\right]^2 \quad (58\,b)$$

$$a_{SmA-SmC^*} = -\frac{u}{2K_3 C_\perp q_0^2} h^2 \quad (58\,c)$$

The N*–SmA–SmC* phase diagram is shown in Fig. 3. The N*–SmA and N*–SmC* lines are first order whereas the

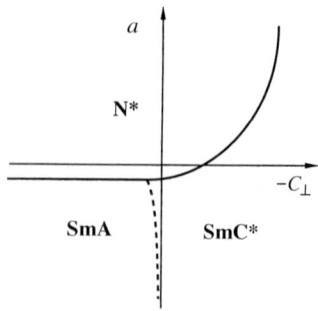

Figure 3. Chiral N*–SmA–SmC* phase diagram from Lubensky and Renn [59]. Solid lines are first order and dashed lines second order.

SmA–CmC* line is second order: the director tilt θ grows continuously at finite twist $\nabla_z \phi = 2\pi/P = -h/K_3$ across the SmA–SmC* transition (Eq. 56 with $\theta = 0$).

It is interesting to calculate the value of the Ginzburg parameter $\kappa = \lambda_2/\xi$ that controls the transition from type I to type II behavior. With the coherence length ξ and the twist penetration depth λ_2 given by Eqs. (22) and (23), one gets:

$$\kappa = \frac{\lambda_2}{\xi} = \frac{1}{C_\perp q_0}\left(\frac{uK_2}{2}\right)^2 \quad (59)$$

κ diverges in the vicinity of the N–SmA–SmC point where $C_\perp$ vanishes. The type II condition is thus expected to be fulfilled close to the N–SmA–SmC point. In presence of chirality, this is precisely the place where a liquid crystal analog of the Abrikosov flux phase should show up. The structure of this new liquid crystalline state will be described in subsequent section.

6.1.6.3 Twist Grain Boundary Phases

Structural Properties

The most spectacular outcome of the analogy with superconductors is undoubtedly the identification of twist grain boundary smectic phases (TGB for short) as liquid crystal analogs of the Abrikosov flux phase in type II superconductors. The existence and structural properties of the TGB phases were first predicted theoretically by Renn and Lubensky [60] (RL) in 1988 and discovered and characterized experimentally by Goodby and coworkers shortly after [26, 61].

The highly dislocated structure of a TGB phase is shown in Fig. 4. Slabs of SmA materials of thickness l_b are regularly stacked in a helical fashion along an axis $\hat{\mathbf{x}}$ parallel to the smectic layers. Adjacent slabs are continuously connected via a grain boundary constituted of a grid of parallel equi-

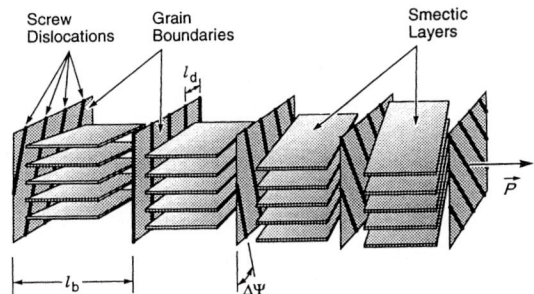

Figure 4. Structure of the TGB phase (from [60, 62]).

spaced screw dislocation lines analogous to magnetic vortices. The finite twist angle of each grain boundary is $\Delta\theta = 2\sin^{-1}(d/2 l_d)$ where d is the smectic period and l_d the distance between parallel screw dislocations within a grain boundary. Twist penetrates the smectic structure just as magnetic induction penetrates the type II superconducting phase via the Abrikosov lattice of magnetic vortices. The difference lies in the lattice structure: two dimensional hexagonal packing of parallel lines for superconductors, twisted array of rows in TGBs.

RL pointed out that the crystallographic nature of a TGB stack depends on the ratio of the slab thickness to the helical pitch l_b/P (or equivalently $\Delta\theta/2\pi$). If the ratio l_b/P is rational ($=p/q$ with p, q mutually prime integers) the structure is periodic of period $pP = q l_b$, with q-fold symmetry about the $\hat{\mathbf{x}}$ axis. Such TGB stack is referred to as commensurate. Reciprocal space is formed of q equispaced Bragg spots distributed on rings of radius $2\pi/d$ in planes q_y, q_z. If q differs from crystallographic values 2, 3, 4 or 6, the commensurate TGB structure is quasi-crystalline rather than crystalline. If on the other hand l_b/P is irrational, the TGB stack is non-periodic (or incommensurate). The scattering is dense on a ring of radius $2\pi/d$ and axis q_x. Reciprocal spaces of the commensurate and incommensurate TGB phases are consequently qualitatively different.

On the experimental point of view, a TGB structure is characterized by five parameters (d, P, l_b, l_d and $\Delta\theta$) linked by two structural relations:

$$\Delta\theta = 2\sin^{-1}(d/2 l_d) \quad \text{and} \quad \frac{l_p}{P} = \Delta\theta/2\pi \tag{60}$$

Full experimental characterization requires measurement of three independent parameters. The layer spacing d (of the order of 3 to 4 nm) and the helical pitch P (a few µm) are easily deduced from X-ray and optical measurements. The third parameter can be the twist boundary angle $\Delta\theta$ which can be measured in the case of commensurate TGB only. For incommensurate TGB [61], other parameters are inferred from the RL theoretical estimate $l_b \approx l_d$ which yields l_b and l_d of the order of a few tens of nm and $\Delta\theta$ about 15°.

Critical Fields h_{c1} and h_{c2}

The thermodynamic critical field h_c above with the SmA phase becomes globally unstable to the N* phase at some temperature T_{N^*-SmA} lower than T_0 follows from Eq. (58a):

$$h_c = -a_{N^*-SmA}\left(\frac{K_2}{u}\right)^{1/2}$$

$$= a'(T_0 - T_{N^*-SmA})\left(\frac{K_2}{u}\right)^{1/2} \tag{61}$$

The critical fields h_{c1} and h_{c2} of the SmA–TGB and TGB–N* transitions were calculated by RL from the superconductor model.

The lower critical field h_{c1} is reached when the SmA phase becomes unstable to the formation of TGB, that is when the gain in twist energy compensates the cost in the formation of grain boundaries. On the other hand, the upper critical field h_{c2} corresponds to the stable growth of a non-zero smectic order parameter in the N* phase, that is the inverse susceptibility of the smectic order parameter changes sign.

The result is [60]:

$$h_{c1} = \frac{h_c}{\kappa\sqrt{2}} \ln\kappa$$
$$h_{c2} = \sqrt{2}\,\kappa h_c \left(1 + O(k_0/q_0)\right) \quad (62)$$

For large enough κ (i.e. type II condition $\kappa > 1/\sqrt{2}$), $h_{c1} < h_c < h_{c2}$ and a stable TGB domain opens up in between the SmA and N* phases.

Experimental Situation

Most features of the RL model were confirmed by experiment. The expected Gaussian shape of the scattering along the pitch direction was confirmed [61]. Networks of screw dislocations were observed by freeze fracture experiments [62]. The link between the existence of TGB phases and the proximity of the N*–SmA–SmC* region was demonstrated by Nguyen and coworkers [63] with the discovery of a TGB phase with local SmC order (so called TGBc also predicted by RL). Quasi-crystalline (i.e. commensurate) TGB structures are commonly observed in TGBc systems [64]. Furthermore, a supercritical transformation within the cholesteric domain above the TGB phase was detected by high resolution calorimetry [65]. The low temperature cholesteric may be identified with the vortex liquid phase appearing in high T_c superconductors [66].

Although a number of experimental question remain, like the origin of commensurability of TGBc, the analogy of smectic liquid crystals with type II superconductors is now clearly established.

6.1.7 Frustrated Smectics

6.1.7.1 Polar Smectics

Liquid crystal molecules with long aromatic cores and strongly polar head groups like –CN or NO_2 exhibit a rich SmA (and SmC) polymorphism. Since the discovery of the first SmA–SmA transition by Sigaud, Hardouin and Achard in 1979 [67], seven different smectic phases have been identified in pure compounds or in binary mixtures of polar molecule.

Extensive experimental studies have been carried out about the structures, phase diagrams and physical properties of these thermotropic liquid crystals [68, 69].

The variety of structures arises from the asymmetry of the molecules: in addition to the classical N–SmA–SmC polymorphism, the long range organization of the position of the polar heads generates new phases. An antiferroelectric stacking of polarized layers for instance generates the bilayer SmA called SmA_2. If the dipoles are randomly oriented, the asymmetry can be forgotten and a monolayer SmA_1 phase is obtained. X-ray diffraction patterns clearly show the doubling of the lattice spacing at the SmA_1–SmA_2 transition.

Another experimental characteristic of polar mesogens is the intrinsic incommensurability of their structures. Nematic phases of polar compounds often exhibit diffuse X-ray scattering corresponding to a short range smectic order. Two sets of diffuse spots centered around incommensurate wavevectors $\pm q_1$ and $\pm q_2$ with $q_1 < q_2 < 2q_1$ are usually found. The wavevector q_2 associated with the classical monolayer order is clearly of order $2\pi/l$ where l is the length of a molecule in its most extended configuration. The wavevector q_1 associated with the head to tail association of the polar molecules reveals the existence of another natural length l' such as $l < l' < 2l$. l' is commonly identified with the length of a pair of antiparallel partially overlapping molecules, although microscopic approaches suggest that the emergence of l' involves more than just two molecules [70, 71]. Condensed smectic phases also exhibit incommensurate behaviours: the so-called partially bilayer smectic phase SmA_d with a lattice period $d \cdot l$ ($l < d < 2$) is commonly observed.

More rarely encountered, but definitely revealing of the incommensurate nature of polar smectics, are the incommensurate SmA phases SmA_{inc} in which the phase of the bilayer modulation shifts periodically (with a period Z) with respect to the monolayer order [72, 73]. The structure is truly incommensurate if the ratio Z over the period of the underlying smectic lattice (Z/l) is irrational. If it is rational ($Z/l = m/n$), the structure is rather modulated and has a period nZ.

The four structures described above are uniaxial since all their modulations have collinear wavevectors but polar mesogens can form biaxial structures too.

The smectic antiphase $Sm\tilde{A}$, first discovered by Sigaud et al. [74] exhibits a periodic modulation of the antiferroelectric order along a direction x parallel to the plane of the layers. The incommensurate wavevectors tilt over to lock in in two dimensions (Fig. 5).

The tilted antiphase (or ribbon phase) $Sm\tilde{C}$ [75] arises from an asymmetric 2-d lock in of the wavevectors. The denomination C emphasizes the fact that both the layers and the antiferroelectric modulations are tilted with respect to the director $\mathbf{n}$.

At last, careful studies of the SmA_2–$Sm\tilde{A}$ and SmA_2–$Sm\tilde{C}$ transitions have revealed the existence of the so-called crenelated SmA_{cre} phase over a very narrow range of temperature [76, 77]. SmA_{cre} exhibits the basic transverse modulation of $Sm\tilde{A}$ but with non-equal up and down domains in the plane of the layers.

The experimental phase diagrams of polar compounds are usually represented in axes temperature–pressure or temperature–concentration in binary mixtures. Although the whole set of structures described above is not found in one single system, most phase diagrams fit in a common topology: N, SmA_1, SmA_2 and SmA_d form the generic phase diagram of polar systems [67, 78] (Fig. 6). If biaxial phases ($Sm\tilde{A}$ and $Sm\tilde{C}$) are present, their domain opens up between SmA_1 and SmA_d [77, 79]. Tricritical points are observed on the N–SmA_1 and N–SmA_2 lines [78] whereas re-entrant behavior is often associated with the triple (or multicritical) point N–SmA_1–SmA_2 [80]. The SmA_2–SmA_d line may end up on a critical point beyond which no transition is detected [81].

6.1.7.2 The Model of Frustrated Smectics (Prost)

Prost showed that the properties and structures of frustrated smectics can be described by two order parameters [72, 82]. The first $\rho(\mathbf{r})$ measures mass density modulation familiar in SmA phases [1]. The second $P_z(\mathbf{r})$, often referred to as a polarization wave, describes long range head-to-tail correlations of asymmetric molecules along the z axis

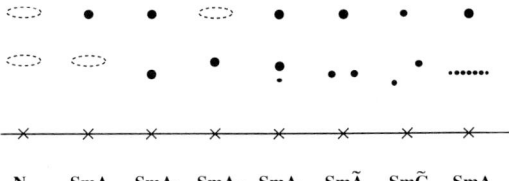

Figure 5. Sketches of the X-ray scattering patterns for various phases appearing in polar smectics. Axes q_x and q_z are horizontal and vertical respectively. Note incommensurate diffuse spots in the nematic phase at natural wavenumbers q_1 and q_2.

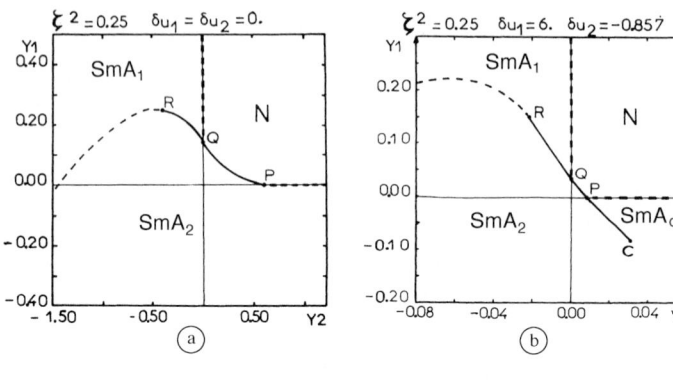

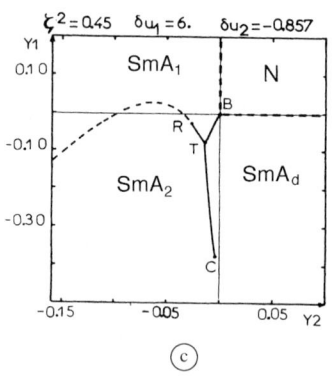

Figure 6. Mean field phase diagram obtained from the model of frustrated smectics for different values of the incommensurability parameter ζ. (a) Very weak incommensurability: N, SmA$_1$ and SmA$_2$ form the generic phase diagram of frustrated smectics. First calculated by Prost [82] this diagram is similar to the experimental one observed on the mixture DB5-TBBA [67]. (b) For slightly larger incomensurability, a SmA$_d$ domain appears in between the N and SmA$_2$ phases. The SmA$_2$–SmA$_d$ line terminates at an isolated critical point C. This diagram reproduces well the behavior of DB6–TBBA mixtures [78] including the order of the transition and the tricritical point R, but not C. (c) For large incommensurability, a new SmA$_1$–SmA$_d$ line and a mean field bicritical point B appear. The critical point C was discovered in this topology by Hardouin et al. [93].

(normal to the smectic layers). Although emphasized by the notation P_z, the antiferroelectric nature of this modulation is not a necessary input of the model.

In the absence of coupling between ρ and P_z, ρ would develop spatial modulation along the z axis at wavevector $q_2 = 2\pi/l$ where l is of the order of a molecular length whereas P_z would develop modulations at wavevector $q_1 = 2\pi/l'$ where l' is identified with the length of a pair of antiparallel partially overlapping molecules as discussed in the previous section.

To describe the appearance of modulated order, two complex fields ψ_1 and ψ_2 are introduced:

$P_z(\mathbf{r}) = \text{Re}(\psi_1(\mathbf{r}))$
and $(\psi_1(\mathbf{r}) = |\psi_1| e^{i(\mathbf{Q}_1 \cdot \mathbf{r} + \varphi_1)}$

$\rho(\mathbf{r}) = \text{Re}(\psi_2(\mathbf{r}))$
and $(\psi_2(\mathbf{r}) = |\psi_2| e^{i(\mathbf{Q}_2 \cdot \mathbf{r} + \varphi_2)}$ (63)

In terms of these fields, the Landau free energy of the model reads:

$$\Delta F = \frac{1}{V} \int_V dr \left\{ \frac{A_1}{2} |\psi_1|^2 + \frac{D_1}{2} |(\Delta + q_1^2)\psi_1|^2 \right.$$
$$+ \frac{C_1}{2} |\nabla_\perp \psi_1|^2 + \frac{U_1}{2} |\psi_1|^4$$
$$+ \frac{A_2}{2} |\psi_2|^2 + \frac{D_2}{2} |(\Delta + q_2^2)\psi_2|^2$$
$$+ \frac{C_2}{2} |\nabla_\perp \psi_2|^2 + \frac{U_2}{2} |\psi_2|^4$$
$$\left. + \frac{U_{12}}{2} |\psi_1|^2 |\psi_2|^2 - w \,\text{Re}(\psi_1^2 \psi_2^*) \right\}$$
(64)

where V is the volume, $A_1 = a_1(T - T_{c1})$ and $A_2 = a_2(T - T_{c2})$ measure the temperatures from the non-interacting mean field transition temperatures T_{c1} and T_{c2} of the fields ψ_1 and ψ_2. $\nabla_\perp$ is a derivative in the plane perpendicular to the director $\mathbf{n}$. The terms in D_1 and D_2 favor $Q_1^2 = q_1^2$ and $Q_2^2 = q_2^2$

respectively whereas the coupling term $w\,\mathrm{Re}(\psi_1^2\psi_2^*)$ favors lock-in conditions $2\mathbf{Q}_1=\mathbf{Q}_2$ in the case of a weak overlap (l' close to $2l$). A linear coupling term $w'\,\mathrm{Re}(\psi_1\psi_2^*)$ would be more appropriate in the strong overlapping limit $l' \approx l$ [72].

Frustration arises from the impossibility to satisfy simultaneously all these tendencies.

Note that the fluctuations of the director and the coupling with the nematic order are not included.

6.1.7.3 The Mean Field Model

To study the different structures and the phase diagrams in mean field, the free energy (Eq. 64) has to be minimized with respect to the smectic amplitudes $|\psi_1|$ and $|\psi_2|$ and the wavevectors $\mathbf{Q}_1$ and $\mathbf{Q}_2$. The following phases are expected:

(1) The nematic phase (N) with $|\psi_1|=|\psi_2|=0$. The director $\mathbf{n}$ defines the z axis.
(2) Uniaxial layered structures with $\mathbf{Q}_1$ and $\mathbf{Q}_2$ along z. $|\psi_1|=0$, $|\psi_2|\neq 0$ and $Q_2=q_2$ defines the monolayer smectic phase SmA$_1$. $|\psi_1|\neq 0$, $|\psi_2|\neq 0$ and $Q_2=2Q_1$ defines the bilayer antiferroelectric smectic phase SmA$_2$.
(3) Biaxial layered structures: $|\psi_1|\neq 0$, $|\psi_2|\neq 0$ and at least $\mathbf{Q}_1$ non collinear with $\mathbf{Q}_2$.
(4) Uniaxial modulated structures: $|\psi_1|\neq 0$, $|\psi_2|\neq 0$ and modulated phases $\varphi_1(z)$ and $\varphi_2(z)$.

The number of Landau parameters in Eq. (64) can be reduced by an apropriate straightforward rescaling of variables:

with complex order parameters:

$$\begin{cases} \theta_1 = x_1 \exp[i(k_0 z + k_x x + \alpha_1(z))] \\ \theta_2 = x_2 \exp[i(2k_0 z + k'_x x + \alpha_2(z))] \end{cases} \quad (66)$$

The phase diagrams are classically calculated in the plane y_1, y_2.

The difference $\zeta = k_1^2 - k_2^2/4$ of the reduced wavevectors k_1 and k_2 turns out to be an important parameter. The physical significance of ζ is clear with the original parameters $\zeta \propto (q_1^2 - q_2^2/4)w$ measures incommensurability over coupling strength. The frustration is thus essentially controlled by ζ, referred to as the incommensurability parameter. Incommensurate structures for instance are expected at high values of ζ.

The elastic coefficients of tilt γ_1 and γ_2 control the appearance of the biaxial phases SmÃ and SmC̃.

The last two coefficients δu_1 and δu_2 account for anisotropic fourth order terms. Although they can have a significant effect on the shape of the phases diagrams in the plane y_1, y_2 they do not change the qualitative features of their topology.

The most significant phase diagrams calculated from this model are given in Figs. 6 and 7.

Uniaxial Structures

For a small incommensurability parameter ζ and symmetric fourth order coefficients ($\delta u_1 = \delta u_2$) the phase diagram shown in Fig. 6a is similar to the very first diagram calculated by Prost in which incommensurability was not considered [82]. A second order N–SmA$_1$ line terminates at a mean

$$\left[\frac{2w^2}{u_{12}D_1}\right]^{3/4} \frac{u_{12}^3}{64w^4}\frac{D_1}{D_2}\frac{\Delta F}{V} = \Delta F[\theta_1,\theta_2]$$

$$= \frac{1}{V}\int_V d^3x\{y_1|\theta_1|^2 + |(\Delta+k_1^2)\theta_1|^2 + \gamma_1|\nabla_\perp\theta_1|^2 + \delta u_1|\theta_1|^4 + y_2|\theta_2|^2 + |(\Delta+k_2^2)\theta_2|^2$$

$$+\gamma_2|\nabla_\perp\theta_2|^2 + \delta u_2|\theta_2|^4 + (|\theta_1|^2+|\theta_2|^2)^2 - (\theta_1^2\theta_2^* + \theta_1^{*2}\theta_2)\} \quad (65)$$

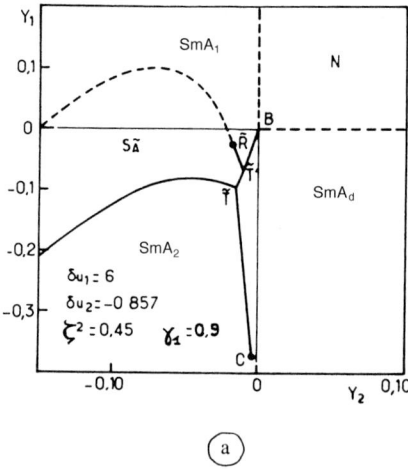
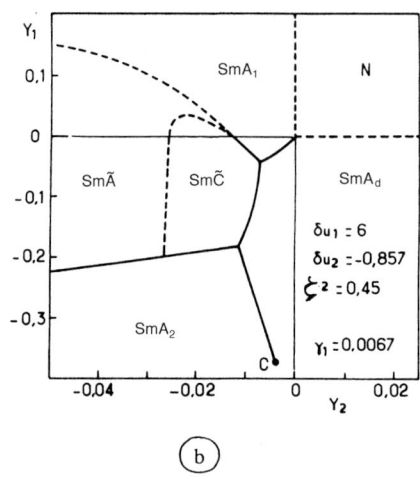

Figure 7. Mean field phase diagrams with biaxial phases. (a) For weak incommensurability, the antiphase SmÃ shows up in between the SmA$_1$ and SmA$_2$ phases. The SmA$_1$–SmÃ transition is second order in mean field but first order with fluctuations [47]. (b) For larger incommensurability, the tilted ribbon phase SmC̃ is stable instead of SmÃ close to SmA$_d$. These topologies again compare well with experimental situations [74, 75, 77, 93].

field critical end point Q where the N, SmA$_1$ and SmA$_2$ phases meet. A second order N–SmA$_2$ line terminates at a tricritical point P. The N–SmA$_2$ line QP is first order and continues into the smectic region as a first order SmA$_1$–SmA$_2$ line. Beyond a tricritical point R, the SmA$_1$–SmA$_2$ line is second order. Because of the coupling term w, the phase of the bilayer order parameter ψ_1 is locked in the monolayer smectic SmA$_1$ so that the amplitude x_1 only is critical at the SmA$_1$–SmA$_2$ transition. It is therefore expected to be in the Ising universality class [82, 83].

For higher incommensurability parameter ζ (and/or asymmetric fourth order terms $\delta u_1 > \delta u_2$) [84] a new phase boundary separating two SmA$_2$ phases appears (Fig. 6b). The two SmA$_2$ phases are distinguished by different values of the amplitudes of the order parameters x_1 and x_2 and therefore of the wavevector k_0. For $x_1 > x_2$, k_0 is of order k_1 (i.e. the smectic period is close to $l' = d \times l < 2l$) and the modulus x_1 is much larger than x_2. The SmA$_2$ phase is identified with the partially bilayer SmA$_d$ phase.

The new SmA$_2$–SmA$_d$ phase boundary terminates at a critical point C where the jump in wavevector goes to zero. Although the fluctuating parameter is a scalar (layer thickness) it will be shown in this chapter, Sec. 6.1.7.4 that the new critical point C is not expected to belong to the Ising universality class [85].

When the incommensurability parameter ζ is further increased, a new SmA$_1$–SmA$_d$ line appears (Fig. 6c) terminating at a mean field bicritical point B where the N, SmA$_1$ and SmA$_d$ phases meet, and a triple point T where the SmA$_1$, SmA$_2$, and SmA$_d$ phases coexist.

Biaxial Structures

Two different structures will be investigated:

$$\begin{cases} \theta_1 = x_1 \exp[i(k_0 z + k_x x)] \\ \quad\quad + x_1 \exp[i(k_0 z - k_x x)] \\ \theta_2 = x_2 \exp[2 i k_0 z] \end{cases} \quad (67)$$

describes the smectic antiphase SmÃ [86] and

$$\begin{cases} \theta_1 = x_1 \exp[\mathrm{i}((k_0 + \delta k_0)z + k_x x)] \\ \quad + x_1' \exp[\mathrm{i}((k_0 - \delta k_0)z - (k_x' - k_x)x)] \\ \theta_2 = x_2 \exp[\mathrm{i}(2k_0 z - k_x' x)] \end{cases} \quad (68)$$

the tilted antiphase SmC̃ [87].

Note that the name smectic is improperly but commonly used to denote these two dimensionally ordered phases.

Figure 7a shows the phase diagram in the case of weak/medium incommensurability parameter. A biaxial domains opens up for $\gamma_1 < 2\zeta$. The SmA$_2$–SmÃ transition is always first order since the free energies cannot link up continuously when k_x goes to zero. The SmA$_1$–SmÃ line is found second or first order in mean field with a tricritical point R̃. It meets the first order SmA$_1$–SmA$_2$ line at a triple point T̃. If variations of the elastic coefficient γ_1 with temperature are allowed for, the second order SmA$_1$–SmÃ line may meet the second order SmA$_1$–SmA$_2$ line at a SmA$_1$–SmA$_2$–SmÃ Lifshitz point [88]. Because of the continuous degeneracy of the fluctuations of the SmÃ order parameter in reciprocal space, the SmA$_1$–SmÃ transition (like N–SmC) is, however, expected to be always first order [47].

At higher incommensurability parameter, the biaxial domain reaches the SmA$_d$ region. The tilted antiphase SmC̃ is stable in between SmA$_d$ and SmÃ so that two new lines are found: a first order SmC̃–SmA$_d$ and a second order SmÃ–SmC̃ (Fig. 7b).

Uniaxial Modulated Structures

Incommensurate smectics with modulated phases $\alpha_1(z)$ and $\alpha_2(z)$ were among the first to be predicted by Prost [72]. An exact functional minimization of Eq. (65) with respect to the phases $\alpha_1(z)$ and $\alpha_2(z)$ is possible in terms of elliptic integrals [72, 89, 90]. The problem is in fact isomorphous to the cholesteric–nematic transition induced by a magnetic field [91]. In the modulated smectic, the phase difference $\alpha_1(z) - \alpha_2(z)$ undergoes π jumps (discommensurations or solitons) with a period Z. Such modulated structures (or SmA$_{\text{inc}}$) are found to be stable in the vicinity of the SmA$_1$–SmA$_2$–SmA$_d$ point [90, 92]. The SmA$_2$–SmA$_{\text{inc}}$ and SmA$_d$–SmA$_{\text{inc}}$ lines are first order, whereas the SmA$_1$–SmA$_{\text{inc}}$ line is second order. The period Z of the discommensurations does not diverge except at the SmA$_1$–SmA$_2$–SmA$_{\text{inc}}$ mean field bicritical point. Further away from this point, Z is not much larger than the layer thickness so that the phase difference $\alpha_1(z) - \alpha_2(z)$ is close to a linear function of z (weak coupling limit) [90]. An unfortunate consequence of this remark is that X-ray diffraction experiments may not easily distinguish SmA$_{\text{inc}}$ from a simple coexistence of SmA$_2$ and SmA$_d$.

Conclusion

Experimental phase diagrams turn out to compare remarkably well to the calculated ones. Actually most of the experimental topologies exhibiting uniaxial phases only fall in one of the theoretical schemes of Figs. 6 and 7.

Binary mixtures of TBBA (terephthal bis butyl aniline) and homologous compounds of the DBn (alkylphenyl cyanobenzoyloxy benzoate) series for instance illustrate well cases in Fig. 6a and b. The DB6–TBBA system in particular [78] exhibits the SmA$_d$ phase, the tricritical SmA$_1$–SmA$_2$ point and the order of all the phase transitions agrees well with Fig. 6b.

The DBn–C$_5$ stilbene binary mixtures [77] and the nOPCBOB (alkoxyphenyl cyanobenzyloxy benzoate) series [93] reproduce well the theoretical diagram of Fig. 7a and b respectively.

The SmA$_2$–SmA$_d$ critical point was first unambiguously identified by Shashidar et al. [81].

The existence of coexisting incommensurate smectic modulations was reported by Brownsey and Leadbetter [94] and demonstrated by Mang et al. [73].

The critical properties of the various transitions were also investigated by high resolution X-ray diffraction or a.c. calorimetry. 3-d XY critical exponents with non-inverted temperatures are measured for the N–SmA$_1$ transition [95].

The Ising nature of the SmA$_1$–SmA$_2$ transition is confirmed [78, 96, 97] with Fisher-renormalized exponents [98].

The SmA$_1$–SmÃ transition is found to be weakly first order with unusually large pretransitional fluctuations [99].

The stability of the crenelated phase in the general theoretical diagram has not been discussed yet. SmA$_{cre}$ can be roughly described by the following order parameter:

$$\begin{cases} \theta_1 = \left(x_0 + x_1 \exp[ik_x x] + x_1 \exp[-ik_x x]\right) \\ \qquad \times \exp[ik_0 z] \\ \theta_2 = x_2 \exp[2ik_0 z] \end{cases} \quad (69)$$

If the structure of Eq. (69) is inserted in the free energy density Eq. (65), SmA$_{cre}$ is found to be always unstable with respect to both SmA$_2$ and SmÃ. If a more realistic profile of the polarization wave along the layers is considered (i.e. include higher harmonics or compute the profile of lowest energy) SmA$_{cre}$ is found to be only marginally stable at the SmA$_2$–SmÃ transition: it has exactly the same free energy as SmA$_2$ and SmÃ right on the SmA$_2$–SmÃ line [100]. Whether the experimental stability is due to higher order terms in the free energy density [101] or to fluctuations is still an open question. The mean field result suggests anyway that the SmA$_{cre}$ domain should be narrow, as indeed observed [76, 77].

The mean field analysis of the Prost's model of frustrated smectics describes most of the reported experimental observations on polar smectics. Situations where fluctuations are important are however not correctly described: the understanding of multiply re-entrant behavior, the appearance of nematic bubbles and the scaling properties of the SmA$_2$–SmA$_d$ critical point for instance require more elaborate analysis. Some of these points will be discussed in the following sections.

6.1.7.4 Critical Properties of the Isolated SmA–SmA Critical Point

The behavior of the smectic wavevector q_0 in the vicinity of the SmA$_2$–SmA$_d$ critical point C [61, 81] clearly suggests a close similarity with a liquid-gas critical point.

In order to describe the fluctuations around C, it is convenient to express the smectic order parameter denoted $m(\mathbf{r})$ as follows:

$$m(\mathbf{r}) = \left[S_{01} + a_1 S(\mathbf{r})\right] e^{i\theta} e^{iq_0[z+u(\mathbf{r})]} \\ + \left[S_{02} - a_2 S(\mathbf{r})\right] e^{2i\theta} e^{2iq_0[z+u(\mathbf{r})]} \quad (70)$$

where S_{01} and S_{02} are the amplitudes of ψ_1 and ψ_2 at the critical point. The equilibrium wavevector is $q_0 = q_{0c}(1 + \langle \nabla_z u \rangle)$. $S(\mathbf{r})$ measures the deviation of the amplitudes from the critical values.

The part $H_{Am}[S(\mathbf{r'})]$ of the Landau–Ginzburg Hamiltonian depending on the amplitude of order parameter $S(\mathbf{r})$ has the liquid–gas form:

$$H_{Am}[S] = \int d^d r \left\{ h_s S + \frac{1}{2}[rS^2 + c(\nabla S)^2] \\ + \frac{w_s}{3!} S^3 + \frac{u_s}{4!} S^4 \right\} \quad (71)$$

where $r = a_s(T - T_s)$. In absence of coupling, H_{Am} is critical when r vanishes.

The energy of elastic deformations in SmA phases $H_{Sm}[u(\mathbf{r})]$ depends on phase $u(\mathbf{r})$ only. It can be developed likewise with the constraint that it must be invariant with respect to both rigid translations and rigid rotations. The former invariance is assured if H_{Sm} is a function only of the gradients of $u(\mathbf{r})$, the latter by requiring that the compressional term appears only in the rotationally invariant combination $E[u(\mathbf{r})] = \nabla_z u + 1/2 (\nabla u)^2$ [102]. The elastic Hamiltonian is thus:

$$H_{Sm}[u] = \int d^d r \left\{ h_0 E(u) + B_0 E^2(u) \right.$$
$$+ \frac{w_0}{3!} E^3(u) + \frac{v_0}{4!} E^4(u)$$
$$+ \frac{1}{2} \left[K_1^0 (\nabla_\perp^2 u)^2 + K_2^0 (\nabla_z^2 u)^2 \right.$$
$$\left. \left. + K_3^0 (\nabla_z \nabla_\perp u)^2 \right] \right\} \quad (72)$$

The classical physics of SmA phases is recovered by expanding Eq. (72) in powers of the gradients of $u(r)$. The quadratic terms in $(\nabla_z u)^2$ and $(\nabla_\perp^2 u)^2$ are responsible for the Landau–Peierls instability [1, 6]. The non-harmonic terms in $(\nabla_z u)(\nabla_\perp u)^2$ and $(\nabla_\perp u)^4$ lead to a break-down of conventional elasticity: $B(q)$ and $K_1(q)$ respectively vanishes and diverges as powers of $\ln(q)$ at small wavevectors q [102]. Other anharmonic terms are irrelevant in the renormalization group sense in the smectic phase.

The Hamiltonian coupling S to u is:

$$H_{compl.} = \int d^d r \left\{ \lambda_{11} SE + \lambda_{12} S^2 E \right.$$
$$\left. + \lambda_{21} SE^2 + \lambda_{22} S^2 E^2 \right\} \quad (73)$$

The final Hamiltonian is $H[S, u] = H_{Am} + H_{Sm} + H_{compl.}$. The physics of the critical point can be understood from the Gaussian truncation of H in reciprocal space:

$$H_G[S, u] = \frac{1}{2} \int \frac{d^d q}{(2\pi)^d} \left\{ \left(B_1 q_z^2 + K_1 q_\perp^4 \right. \right.$$
$$\left. + K_2 q_z^4 + K_{12} q_z^2 q_\perp^2 \right) |u(q)|^2$$
$$+ (r + Cq^2) |S(q)|^2$$
$$- \lambda_{11} (iq_z u(q) S(-q)$$
$$\left. + \text{conj. complex} \right\} \quad (74)$$

with $B_1 = B_0 + h_0$.

The linear coupling of S and u in Eq. (74) can be removed via the transformation:

$$S(q) = \sigma(q) + \frac{i\lambda_{11} q_z u(q)}{r + Cq^2} \quad (75)$$

The Gaussian correlation functions of the new variables are:

$$\begin{cases} G_{uu}^{(2)}(q, r) \quad (76) \\ = \frac{1}{B_1 \frac{(r - r_c)}{r} q_z^2 + K_1 q_\perp^2 + K_2' q_z^4 + 2K_{12}' q_\perp^4} \\ G_{\sigma\sigma}^{(2)}(q, r) = \frac{1}{r + Cq^2} \end{cases}$$

with $r_c = \lambda_{11}^2 > 0$. The layer compression is controlled by an effective elastic constant $B_{eff}(q)$

$$B_{eff}(q = 0) = B_1 \frac{(r - r_c)}{r} \quad (77)$$

which goes to zero for a positive value of r. $G_{\sigma\sigma}^{(2)}$ is therefore not critical at C and a Gaussian integration of the fluctuations of σ is acceptable. $u(r)$ is the critical variable and the critical point is reached when B_{eff} vanishes. This behavior is to be compared to the divergence of the compressibility in the liquid–vapor problem: $1/B$ is analogous to a smectic compressibility along the direction z.

The critical Landau–Ginzburg Hamiltonian is then:

$$H = \int d^d r \left\{ h \nabla_z u + \frac{B_1}{2} (\nabla_z u)^2 + \frac{B_2}{2} (\nabla_\perp u)^2 \right.$$
$$+ \frac{1}{2} \left[K_1 (\nabla_z u)^2 + K_2 (\nabla_\perp u)^2 \right.$$
$$\left. + 2 K_{12} (\nabla_z \nabla_\perp u)^2 \right]$$
$$+ \frac{w_1}{3!} (\nabla_z u)^3 + \frac{w_2}{3!} (\nabla_z u)(\nabla_\perp u)^2$$
$$+ \frac{v_1}{4!} (\nabla_z u)^4 + \frac{v_1}{4!} (\nabla_\perp u)^4$$
$$\left. + \frac{v_{12}}{12} (\nabla_z u)^2 (\nabla_\perp u)^2 \right\} \quad (78)$$

In the absence of external field, $h = B_2 = 0$. In mean field theory, $\langle \nabla_\perp u \rangle$ is zero, the order parameter is $\langle \nabla_z u \rangle$ and the critical point is found for $B_1 = w_1 = 0$.

When fluctuations are included, $\langle \nabla_\perp u \rangle$ cannot be ignored. In the liquid–gas Hamiltonian, there is only one third-order invariant ϕ^3 which can be removed by shifting ϕ. This implies that the liquid–gas transition belongs to the same universality as the Ising model with additional higher order irrelevant potentials. In the SmA–SmA Hamiltonian (Eq. 78) there are two distinct third order potentials $(\nabla_z u)^2$ and $\nabla_z u (\nabla_\perp u)^2$. They cannot be removed by shifting the order parameter $\langle \nabla_z u \rangle$. A renormalized theory should therefore take them into account explicitly. The general ϕ^3-field theory suggests that the critical dimension is $d_c = 6$.

The complete perturbation theory of the Hamiltonian (Eq. 78) has been developed by Park et al. [85]. The inverse susceptibility $\tilde{B}$ is found to be related to the coefficients of Eq. (78) as follows:

$$B_1 - B_{1c} = \tilde{B}\left(1 - w_{2c}^2 \tilde{B}^{d/2-3} + v \tilde{B}^{d/2-2} + \ldots\right) \quad (79)$$

since $w_{2c} \neq 0$, the expansion (Eq. 79) breaks down for spatial dimensions d lower than 6.

For $4 < d < 6$, fourth order terms are irrelevant in Eq. (78). An anisotropic ϕ^3-renormalized theory was constructed [85]. The Gaussian and Ising fixed point are unstable, as expected. A stable non-trivial fixed point exists and anisotropic critical exponents were calculated in $6 - \varepsilon$ dimensions.

Although a number of theoretical questions remain (unknown equation of state, non analytic corrections to the third order vertex for $6 < d < 8$) the main result is that the SmA–SmA critical point is expected to belong to a new universality class. Experiments confirm this point: a high resolution calorimetric study on a binary mixture gives a critical exponent γ/Δ of 0.6 ± 0.2 [103] whereas elastic constant measurements show a vanishing of B with a value 0.4 ± 0.2 of the same exponent γ/Δ [104]. These numbers are only marginally consistent with one another but definitely different from the Ising value 0.79. On the other hand, a mean-field behavior ($\gamma/\Delta = 0.67$) would not explain the asymmetry of the calorimetric and X-ray data.

The mean field theory of frustrated smectics in the limit of a strong overlap ($l' \approx l$) also predicts several critical points [105]. All of them (including the present SmA_2–SmA_d point) are in fact of SmA_d–SmA_d type (with $1 \leq d \leq 2$) and belong to the same universality class. It is worth emphasizing the unusual possibility of going from SmA_1 to SmA_2 via a second-order transition, a first-order jump or without any phase transition at all by a continuous growth of the smectic wavevector since all SmA phases have the same symmetry. Second order transitions may only occur with exact doubling (or tripling) of the period by continuous condensation of a subharmonic modulation.

6.1.7.5 The Re-entrant Phenomenon

Although not specific of liquid crystals, the re-entrant phenomenon (i.e. re-appearance of the phase of higher symmetry upon cooling) is often observed in frustrated smectics [80, 106]. Theoretical analyses suggest that there is no universal explanation for re-entrance in the N–SmA problem.

Single re-entrance may occur in mean field: the coupling of two order parameters may generate a curved N–SmA phase boundary and thus produce a re-entrant phenomenon if the physical temperature axis is a suitable combination of the two Landau control parameters.

Experiments suggest that double re-entrance is associated with the bi- or tetracritical N–SmA$_1$–SmA$_d$ point. Prost and Lubensky [107, 108] have argued that the correlation function of the smectic order parameter in both SmA$_1$ and SmA$_d$ obeys a scaling relation such as:

$$G(t,p) = \tilde{t}^{-\gamma} f\left(\frac{\tilde{p}}{\tilde{t}^x}\right) \quad (80)$$

where t, p are the reduced temperature and pressure in the vicinity of the critical point (defined by $t=p=0$) $\tilde{t}$ and $\tilde{p}$ are linear combinations of t and p. Equation (80) implies that the phase boundaries obey a relation

$$\tilde{p} = w \pm \tilde{t}^x \quad (81)$$

which gives a doubly re-entrant behavior with suitable rotation of the physical axes p, t with respect to $\tilde{p}$ and $\tilde{t}$.

Beside these phenomenological approaches that say nothing about the microscopic origin of re-entrance, molecular theories have been proposed. The frustrated spin glass model of Indekeu and Berker [70, 71] has been particularly successful.

Re-entrance may also occur as closed nematic domains (or nematic 'bubbles') deep in the smectic region. Such bubbles seem to be closely related to the existence of a SmA$_2$–SmA$_d$ or SmA$_1$–SmA$_d$ critical point [105, 109].

As explained in Sec. 6.1.7.4 of this chapter, the compressional elastic constant B vanishes at the SmA–SmA critical point so that the system is very close to a nematic. An interesting consequence is that the energy of dislocations becomes very weak and their proliferation may lead to a destruction of smectic order. Prost and Toner [109] have shown that depending on bare parameters of a particular system, either the nematic bubble or the critical point could be observed.

Beside this fluctuation corrected mean field theory, an exact model has been developed by Pommier and Prost [105] in the limit of an infinite number of components n of the order parameter (recall $n=2$ for the smectic case). With the same physics of frustration, exact phase diagrams have been calculated that reproduce fairly well the continuous inclusion of the re-entrant nematic phase that leads to a closed island.

6.1.8 Conclusions

The study of phase transitions in thermotropic liquid crystals has made considerable progress since 1980. Strong theoretical efforts coupled with high resolution and high sensitivity experimental techniques have provided a good understanding of the physics of the phase transformations that occur in mesophases. Clear experimental universal behaviors have been identified and most exceptions have received a reasonable theoretical explanation (Ginzburg criterion for the SmA–SmC transition, crossover towards tricritical behavior, analogy with superconductors).

A number of open questions remain: the N–SmA transition for instance is almost understood but not quite which suggests more efforts have to be done.

Transitions involving disk-like molecules, mesogenic polymers or lyotropic micelles have certainly not received such a great deal of attention and still constitute an active field of research.

Acknowledgments

I wish to thank the editors and particularly Professor Demus for infinite patience and constant encouragement for writing this chapter. I am also indebted to C. Coulon, C. W. Garland, F. Hardouin, A.-M. Levelut, T. C. Lubensky, J. P. Marcerou, F. Nallet, L. Navailles, H. T. Nguyen, Y. Park, R. Pindak, J. Pommier, J. Prost, R. Shashidar and G. Sigaud for helpful discussions about the work presented in this chapter.

6.1.9 References

[1] P. G. de Gennes, J. Prost, *The Physics of Liquid Crystals*, 2nd ed., Oxford Science Publications Clarendon Press, Oxford, **1993**.
[2] B. Deloche, B. Cabane, D. Jérôme, *Mol. Cryst. Liq. Cryst.* **1971**, *15*, 197.
[3] M. J. Freiser, *Phys. Rev. Lett.* **1970**, *24*, 1041; C. S. Shih, R. Alben, *J. Chem. Phys.* **1972**, *57*, 3055; Y. Rabin, W. E. Mullen, W. M. Gelbart, *Mol. Cryst. Liq. Cryst.* **1982**, *89*, 67.
[4] A. Saupe, P. Boonbrahm, L. J. Yu, *J. Chim. Phys.* **1983**, *80*, 7.
[5] Y. Galerne, J. P. Marcerou, *Phys. Rev. Lett.* **1983**, *51*, 2109.
[6] L. D. Landau in *Collected Papers* (Ed.: D. Ter Haar), Gordon and Breach, New York, **1965**.
[7] R. Alben, *J. Chem. Phys.* **1973**, *59*, 4299.
[8] D. J. Wallace in *Phase Transitions and Critical Phenomena*, Vol. 6 (Eds.: C. Domb, M. S. Green), Academic Press, London, **1976**.
[9] E. F. Gramsbergen, L. Longa, W. H. De Jeu, *Phys. Rep.* **1986**, *135*, 195.
[10] C. P. Fan, M. J. Stephen, *Phys. Rev. Lett.* **1970**, *25*, 500.
[11] J. Thoen, H. Marynissen, W. Van Dael, *Phys. Rev. A* **1982**, *A26*, 2886.
[12] K. G. Wilson, J. Kogut, *Phys. Rep.* **1974**, *12*, 77.
[13] A. L. Korzhenevskii, B. N. Shalaev, *Sov. Phys. JETP* **1979**, *49*, 1094.
[14] A. Poniewiersky, T. J. Sluckin, *Liq. Cryst.* **1987**, *2*, 281.
[15] P. Richetti, P. Moreau, P. Barois, P. Kékicheff, *Phys. Rev. E* **1996**, *54*, 1749.
[16] W. L. McMillan, *Phys. Rev.* **1971**, *A4*, 1238.
[17] W. Maier, A. Saupe, *Z. Naturforsch.* **1959**, *14A*, 882; **1960**, *15A*, 287.
[18] W. L. McMillan, *Phys. Rev. A* **1972**, *A6*, 936; *Phys. Rev.* **1973**, *A7*, 1673.
[19] L. D. Landau, *Phys. Z. Sowj. Un.* **1937**, *2*, 26.
[20] R. E. Peierls, *Annls Inst. H. Poincaré* **1935**, *5*, 177.
[21] C. W. Oseen, *Trans. Faraday Soc.* **1933**, *29*, 883; F. C. Frank, *Disc. Faraday Soc.* **1958**, *25*, 19.
[22] V. L. Ginzburg, L. D. Landau, *J. Exptl. Theoret. Phys. (USSR)* **1950**, *20*, 1064.
[23] P. G. de Gennes, *Solid State Commun.* **1972**, *10*, 753.
[24] W. Meisner, R. Ochsenfeld, *Naturwiss.* **1933**, *21*, 787.
[25] S. T. Lagerwall, R. B. Meyer, B. Stebler, *Ann. Phys. (France)* **1978**, *3*, 249.
[26] J. Goodby, M. Waugh, S. Stein, R. Pindak, J. Patel, *Nature* **1988**, *337*, 449.
[27] P. E. Cladis, S. Torza, *J. Applied Phys. (USA)* **1975**, *46*, 584.
[28] F. Jähnig, F. Brochard, *J. Phys. (France)* **1974**, *35*, 301.
[29] T. C. Lubensky, Jing-Huei Chen, *Phys. Rev. A* **1978**, *17*, 366.
[30] B. Halperin, T. C. Lubensky, Shang-Keng Ma, *Phys. Rev. Lett.* **1974**, *32*, 292.
[31] C. E. Williams, M. Kléman, *J. Phys. Colloq.* **1975**, *36*, C1–315; M. Kléman, *Points, Lignes, Parois,* Les Editions de Physique, Orsay, France, **1977**.
[32] W. Helfrich, *J. Phys. (Paris)* **1978**, *39*, 1199.
[33] D. R. Nelson, J. Toner, *Phys. Rev. B* **1981**, *B24*, 363.
[34] J. Toner, *Phys. Rev. B* **1982**, *B26*, 462.
[35] C. Dasgupta, B. I. Halperin, *Phys. Rev. Lett.* **1981**, *47*, 1556.
[36] M. A. Anisimov, V. P. Voronov, A. O. Kulkov, F. Kholmurodov, *J. Phys. (France)* **1985**, *46*, 2157.
[37] L. Benguigui, P. Martinoty, *Phys. Rev. Lett.* **1989**, *63*, 774.
[38] Y. Galerne, *J. Phys. (France)* **1985**, *46*, 733.
[39] D. Collin, J. L. Gallani, P. Martinoty, *Phys. Rev. Lett.* **1988**, *61*, 102.
[40] G. Sigaud, F. Hardouin, M. F. Achard, *Solid State Commun.* **1977**, *23*, 35.
[41] D. Johnson, D. Allender, D. Dehoff, C. Maze, E. Oppenheim, R. Reynolds, *Phys. Rev. B* **1977**, *B16*, 470.
[42] P. G. de Gennes, *Mol. Cryst. Liq. Cryst.* **1973**, *21*, 49.
[43] W. L. McMillan, *Phys. Rev. A* **1973**, *8*, 1921; R. B. Meyer, W. McMillan, *Phys. Rev. A* **1974**, *9*, 899.
[44] A. Wulf, *Phys. Rev. A* **1975**, *A11*, 365.
[45] R. J. Priest, *J. Phys. (Paris)* **1975**, *36*, 437.
[46] Jing-Huei Chen, T. C. Lubensky, *Phys. Rev. A* **1976**, *A14*, 1202.
[47] S. A. Brazowskii, *Soviet Phys. JETP* **1975**, *41*, 85.
[48] L. D. Landau, E. M. Lifschitz, *Statistical Physics,* Pergamon, New York, **1968**.
[49] R. M. Hornreich, M. Luban, S. Shtrikman, *Phys. Rev. Lett.* **1975**, *35*, 1678.
[50] C. R. Safinya, L. J. Martinez-Miranda, M. Kaplan, J. D. Litster, R. J. Birgeneau, *Phys. Rev. Lett.* **1983**, *50*, 56.
[51] L. J. Martinez-Miranda, A. R. Kortan, R. J. Birgeneau, *Phys. Rev. Lett.* **1986**, *56*, 2264; *Phys. Rev. A* **1987**, *A36*, 2372.
[52] G. Grinstein, J. Toner, *Phys. Rev. Lett.* **1983**, *51*, 2386.
[53] M. E. Fisher, D. R. Nelson, *Phys. Rev. Lett.* **1974**, *32*, 1350.
[54] J. M. Kosterlitz, D. J. Thouless, *J. Phys. (France)* **1973**, *C6*, 1181.
[55] B. I. Halperin, D. R. Nelson, *Phys. Rev. Lett.* **1978**, *41*, 121; D. R. Nelson, B. I. Halperin, *Phys. Rev. B* **1979**, *B19*, 2457.
[56] A. P. Young, *Phys. Rev. B* **1979**, *B19*, 1855.
[57] C. C. Huang, J. M. Viner, R. Pindak, J. W. Goodby, *Phys. Rev. Lett.* **1981**, *46*, 1289.

[58] R. Geer, C. C. Huang, R. Pindak, J. W. Goodby, *Phys. Rev. Lett.* **1989**, *63*, 540.
[59] T. C. Lubensky, S. R. Renn, *Phys. Rev. A* **1990**, *41*, 4392.
[60] S. R. Renn, T. C. Lubensky, *Phys. Rev. A* **1988**, *A38*, 2132.
[61] G. Srajer, R. Pindak, M. A. Waugh, J. W. Goodby, J. S. Patel, *Phys. Rev. Lett.* **1990**, *64*, 1545.
[62] K. J. Ihn, J. A. N. Zasadzinski, R. Pindak, A. J. Slaney, J. W. Goodby, *Science* **1992**, *258*, 275.
[63] H. T. Nguyen, A. Bouchta, L. Navailles, P. Barois, N. Isaert, R. J. Twieg, A. Maaroufi, C. Destrade, *J. Phys. (France)* **1992**, *2*, 1889.
[64] L. Navailles, P. Barois, H. T. Nguyen, *Phys. Rev. Lett.* **1993**, *71*, 545.
[65] T. Chan, C. W. Garland, H. T. Nguyen, *Phys. Rev. E* **1995**, *52*, 5000.
[66] R. D. Kamien, T. C. Lubensky, *J. Phys. I (France)* **1994**, *3*, 2123.
[67] G. Sigaud, F. Hardouin, M. F. Achard, *Phys. Lett.* **1979**, *72A*, 24.
[68] F. Hardouin, A. M. Levelut, M. F. Achard, G. Sigaud, *J. Chim. Phys.* **1983**, *80*, 53.
[69] R. Shashidar, B. R. Ratna, *Liq. Cryst.* **1989**, *5*, 421.
[70] J. O. Indekeu, A. N. Berker, *J. Phys. (France)* **1988**, *49*, 353.
[71] A. N. Berker, J. S. Walker, *Phys. Rev. Lett.* **1981**, *47*, 1469.
[72] J. Prost, *Proceedings of the Conf. on Liq. Cryst. of One and Two Dimensional Order, Garmisch Partenkirchen,* Springer Verlag, Berlin, **1980**.
[73] J. T. Mang, B. Cull, Y. Shi, P. Patel, S. Kumar, *Phys. Rev. Lett.* **1995**, *74*, 21.
[74] G. Sigaud, F. Hardouin, M. F. Achard, A. M. Levelut, *J. Phys. (France)* **1981**, *42*, 107.
[75] F. Hardouin, H. T. Nguyen, M. F. Achard, A. M. Levelut, *J. Phys. Lett. (France)* **1982**, *43*, L-327.
[76] A. M. Leelut, *J. Phys. Lett. (France)* **1984**, *45*, L-603.
[77] G. Sigaud, M. F. Achard, F. Hardouin, *J. Phys. Lett. (France)* **1985**, *46*, L-825.
[78] K. K. Chan, P. S. Pershan, L. B. Sorensen, F. Hardouin, *Phys. Rev. Lett.* **1985**, *54*, 1694; *Phys. Rev. A* **1986**, *A34*, 1420.
[79] H. T. Nguyen, C. Destrade, *Mol. Cryst. Liq. Cryst. Lett.* **1984**, *92*, 257.
[80] P. E. Cladis, *Phys. Rev. Lett.* **1975**, *35*, 48.
[81] R. Shashidar, B. R. Ratna, S. Krishna, S. Somasekhar, G. Heppke, *Phys. Rev. Lett.* **1987**, *59*, 1209.
[82] J. Prost, *J. Phys. (France)* **1979**, *40*, 581.
[83] Wang Jiang, T. C. Lubensky, *Phys. Rev. A* **1984**, *A29*, 2210.
[84] P. Barois, J. Prost, T. C. Lubensky, *J. Phys. (France)* **1985**, *46*, 391.
[85] Y. Park, T. C. Lubensky, P. Barois, J. Prost, *Phys. Rev. A* **1988**, *A37*, 2197.
[86] P. Barois, C. Coulon, J. Prost, *J. Phys. Lett. (France)* **1981**, *42*, L-107.
[87] J. Prost, P. Barois, *J. Chim. Phys. (France)* **1983**, *80*, 66.
[88] L. G. Benguigui, *J. Phys. (France)* **1983**, *44*, 273.
[89] P. Barois, J. Prost, *Ferroelectrics* **1984**, *58*, 193.
[90] P. Barois, J. Pommier, J. Prost, *Solitons in Liquid Crystals* (Ed.: Lui Lam, J. Prost), Springer Verlag, Berlin **1991**, Chapter 6.
[91] P. G. de Gennes, *Solid State Commun.* **1968**, *6*, 163.
[92] P. Barois, *Phys. Rev. A* **1986**, *A33*, 3632.
[93] F. Hardouin, M. F. Achard, C. Destrade, H. T. Nguyen, *J. Phys. (France)* **1984**, *45*, 765.
[94] G. J. Brownsey, A. J. Leadbetter, *Phys. Rev. Lett.* **1980**, *44*, 1608.
[95] C. W. Garland, G. Nounesis, K. J. Stine, G. Heppke, *J. Phys. (France)* **1989**, *50*, 2291; C. W. Garland, G. Nounesis, K. J. Stine, *Phys. Rev. A* **1989**, *A39*, 4919.
[96] C. Chiang, C. W. Garland, *Mol. Cryst. Liq. Cryst.* **1985**, *122*, 25.
[97] C. W. Garland, C. Chiang, F. Hardouin, *Liq. Cryst.* **1986**, *1*, 81.
[98] D. A. Huse, *Phys. Rev. Lett.* **1985**, *55*, 2228; M. A. Anisimov, A. V. Voronel, Gorodetskii, *Sov. Phys. JETP* **1971**, *33*, 605.
[99] K. Ema, C. W. Garland, G. Sigaud, H. T. Nguyen, *Phys. Rev. A* **1989**, *A39*, 1369.
[100] J. Pommier, P. Barois, unpublished.
[101] L. G. Benguigui, *Phys. Rev. A* **1986**, *A33*, 1429.
[102] G. Grinstein, R. A. Pelcovits, *Phys. Rev. A* **1982**, *A26*, 915.
[103] Y. H. Jeong, G. Nounesis, C. W. Garland, R. Shashidar, *Phys. Rev. A* **1989**, *A40*, 4022.
[104] J. Prost, J. Pommier, J. C. Rouillon, J. P. Marcerou, P. Barois, M. Benzekri, A. Babeau, H. T. Nguyen, *Phys. Rev. B* **1990**, *B42*, 2521.
[105] J. Pommier, Thèse de l'Université de Bordeaux 1, n° 274, **1989**.
[106] H. T. Nguyen, F. Hardouin, C. Destrade, *J. Phys. (France)* **1982**, *43*, 1127.
[107] J. Prost, *Adv. Phys.* **1984**, *33*, 1.
[108] T. C. Lubensky, unpublished.
[109] J. Prost, J. Toner, *Phys. Rev. A* **1987**, *A36*, 5008.

6.2 Experimental Methods and Typical Results

6.2.1 Thermal Methods

Jan Thoen

6.2.1.1 Introduction

Liquid crystals exhibit a rich variety of phases and phase transitions. The phase transitions can either be first-order or second-order and critical fluctuations often play an important role. In order to elucidate the nature of these phase transitions the application of high-resolution measuring techniques are essential.

Calorimetric studies have played a significant role in providing information on energy effects near many liquid crystal phase transitions and complimented structural information from X-ray investigations. The vast majority of thermal information concerns calorimetric data on the static thermal quantities enthalpy, H, or specific heat capacity, C_p. However, recently high-resolution results for thermal transport properties have also been obtained.

The rest of this section is divided in subsections: Section 6.2.1.2 gives some general thermodynamic aspects of phase transitions as well as on the crucial role played by critical fluctuations. Section 6.2.1.3 describes several high-resolution techniques and the way in which they allow to extract information on static and/or dynamic thermal quantities. Section 6.2.1.4 deals with calorimetric results obtained for a variety of phase transitions with the purpose of illustrating the possibilities of several of the measuring techniques in arriving at information on static thermal quantities. In Section 6.2.1.5 the capability of photoacoustic and photopyroelectric methods to obtain static and dynamic thermal quantities is illustrated.

6.2.1.2 Theoretical Background

Thermal Characteristics of Phase Transition

Phase transitions can be first-order or second-order (or continuous), and critical energy fluctuations quite often have a significant impact on thermal parameters. First-order transitions are characterized by discontinuous jumps in the first derivatives of the free energy, resulting in finite density ρ and enthalpy H differences between two distinct coexisting phases at the transition temperature T_{tr}. For a second-order transition there are no discontinuities in the density or the enthalpy but the specific heat capacity C_p will exhibit either a discontinuous-jump (for mean-field regime) or a critical anomaly (critical fluctuation regime). In the left hand part of Fig. 1 characteristic behavior for H as a function of temperature near T_{tr} is shown schematically for first-order as well as for second-order transitions. The right hand part of Fig. 1 gives the corresponding temperature behavior of the specific heat capacity $C_p = (\partial H/\partial T)_p$.

Figure 1(a) gives the temperature dependence of H near a strongly first-order transition with a large latent ΔH_L at T_{tr}. The variation of H with T is nearly linear above and below T_{tr} resulting in almost temperature independent C_p values in the low and high temperature phases. Figure 1(b) represents the case of a weakly first-order transition with only a small latent heat $\Delta H_L \neq 0$, but H shows substantial pretransitional temperature variation, which show up as anomalous pretransitional increases in the corresponding C_p behavior. The total enthalpy change associated with the phase transition can be written as the sum of two terms:

$$\Delta H_L + \delta H = \Delta H_L + \int \Delta C_p dT \quad (1)$$

with $\Delta C_p = C_p - C_p^b$ the excess specific heat capacity (above the background C_p^b) associated with changes in ordering.

Figure 1(c–e) schematically represent three commonly encountered cases for second-order phase transitions. At a second-order phase transition the latent heat $\Delta H_L = 0$ and the specific heat capacity C_p (the temperature derivative of H) shows singular behavior at the critical point (CP). Figure 1(c) is the case of a mean-field second-order transition with a normal linear behavior above the critical temperature $T_c = T_{tr}$ and a rapid variation of H below T_c due to changes in long-range order with temperature. This is reflected in a rapid change of C_p below T_c on approaching T_c and a discontinuous jump at T_c. Both cases given in Fig. 1(d and e) are critical fluctuations dominated second-order phase transitions with pretran-

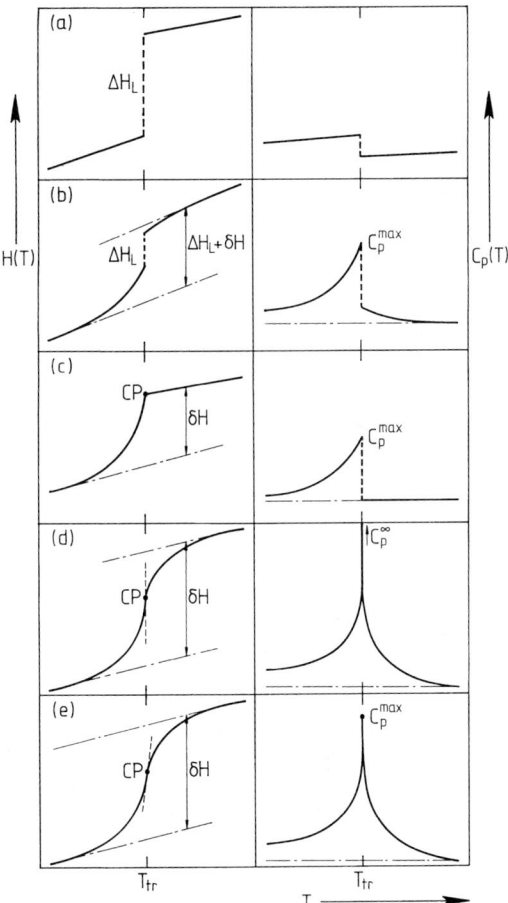

Figure 1. Schematic variation of the enthalpy H and specific heat capacity C_p with temperature T for transitions (at T_{tr}) that are: (a) strongly first-order, (b) weakly first-order with pretransitional fluctuation behavior, (c) mean-field second-order (CP indicates the critical point on the enthalpy curve for the Landau second-order transition temperature $T_c = T_{tr}$), (d) and (e) are critical fluctuation dominated second-order transitions with a diverging (d) or large but finite (e) specific heat capacity at the critical temperature $T_c = T_{tr}$. For the first-order transitions the latent heats ΔH_L correspond with the steps in $H(T)$ at $T = T_{tr}$. δH represent the fluctuation induced enthalpy change associated with the phase transition.

sitional enthalpy variations and anomalies in the specific heat capacity above and below T_c. The main difference between the two cases in the difference is the slope $(\partial H/\partial T)_p$

at the critical point, resulting in a divergence of the specific heat capacity C_p to infinity in Fig. 1(d), or in a large finite value at T_c (Fig. 1e). In all three of these second-order cases the total enthalpy change associated with the transition contains the δH contribution, which is indicated in the figures.

All these types of phase transitions have been encountered in the large amount of high-quality calorimetric studies during the last two decades. In Section 6.2.1.4 a series of typical cases will be considered in detail. More extensive overviews can be found in reviews that have appeared recently [1–6].

The calorimetric measurements provide information only on the static thermal quantities H and C_p. The dynamic thermal quantities such as the thermal diffusivity D and thermal conductivity κ, which are linked to the dynamics of the fluctuations near phase transitions, are substantially more difficult to measure. Only recently have high-resolution techniques with sufficiently small thermal gradients been used for the study of liquid crystal phase transitions.

Fluctuation Effects

In order to be able to make, even qualitatively, a determination that a transition is first-order or second-order one has to apply a measuring technique probing the temperature dependence of the enthalpy H. Very important is also the magnitude of the latent heat ΔH_L relative to the pretransitional enthalpy change δH (see Fig. 1). Quite often it is also important to know how this ratio changes with some physical parameter like pressure or concentration (for mixtures). As a function of these parameters the order of the transition may, indeed, change at a tricritical point or reach an isolated critical point (liquid–gas type transition in simple fluids).

Accurate determination of the temperature dependence of C_p above and below the transition temperature is also very important because the direct link with critical energy fluctuations. This allows one to find out whether mean-field or critical fluctuation theories are relevant for the transition under investigation [7–9].

If the intermolecular interactions responsible for ordering in the system are sufficiently long-range, the system can be described by a Landau mean-field model. Mean-field models are approximations in which only the long-range aspects of enthalpy fluctuations are taken into account and the short-range fluctuations in the order are ignored. The Landau model assumes the Gibbs free energy, G, to be analytic and a function of a long-range order parameter θ, given by

$$G = G_0 + a\varepsilon\theta^2 + b\theta^4 + c\theta^6 \quad (2)$$

where G_0 is the free energy of the disordered phase and $\varepsilon = (T_c - T)/T_c$ the reduced temperature with T_c the critical temperature. The coefficients a and c are positive, but b can be positive (second-order), zero (tricritical), or negative (first-order). In Eq. (2) odd powers of θ are omitted under the symmetry assumption $G(\theta) = G(-\theta)$. This assumption is, however, not always valid and a third-order term has to be included in some cases (see further the nematic-isotropic transition). The presence of a θ^3 term will always cause the transition to be (weakly) first-order [7, 9].

On the basis of Eq. (2) one arrives at

$$C_p = C_p^0, \quad \text{for} \quad T > T_c \quad (3)$$

$$C_p = C_p^0 + A\frac{T}{T_c}\left(\frac{T_m - T_c}{T_m - T}\right)^{1/2}$$
$$\text{for} \quad T < T_c \quad \text{and} \quad b > 0 \quad (4)$$

$$C_p = C_p^0 + 2A\frac{T}{T_c}\left(\frac{T_m - T_1}{T_m - T}\right)^{1/2}$$
$$\text{for} \quad T < T_c \quad \text{and} \quad b < 0 \quad (5)$$

where $A = |a^2/2bT_c|$, $T_m = T_c + (b^2 T_c/3ac)$ and C_p^0 the background heat capacity arising from the regular part G_0 of the free energy. When $b < 0$, one has a first-order transition at T_1 with a specific heat capacity jump given by $2A$. For a continuous transition, one must have $b \geq 0$ and $a, c > 0$. One has an ordinary second-order transition for $b > 0$ with a specific heat capacity jump (see Fig. 1c) given by A. If $b = 0$, a tricritical point is observed at $T_m = T_c$ and below T_m a divergent C_p is predicted with a critical exponent of 1/2.

In liquid crystals intermolecular interactions are short-range and C_p is dominantly related to short-range order fluctuations. Since fluctuations in local order occur above and below the transition, one expects, in contrast to the mean-field situation, anomalous C_p behavior above as well as below T_c. Unfortunately models taking short-range order parameter fluctuations into account are much more difficult to handle and analytic solutions are only available for some special cases. However, sophisticated renormalization group (RG) analyses have been carried out for several types of n vector order parameter models of which the three-dimensional cases are of particular relevance for liquid crystals [7, 10, 11].

In the renormalization group theory one obtains for the anomaly in the specific heat capacity an expression of the following form:

$$C_p^\pm = A^\pm |\varepsilon|^{-\alpha} \left(1 + D_1^\pm |\varepsilon|^{\Delta_1} + D_2^\pm |\varepsilon|^{\Delta_2}\right) + B \quad (6)$$

The superscripts ± indicate the quantities above and below the transition temperature. A^+ and A^- are the critical amplitudes. The constant term B contains a regular as well as a critical background contribution. $D_1^\pm$ and $D_2^\pm$ are the coefficients of the first- and second-order correction-to-scaling terms, while Δ_1 and Δ_2 are the correction exponents (typically $\Delta_1 \approx 0.5$ and $\Delta_2 \approx 2\Delta_1$). The critical exponent α has a unique theoretical value for a given universality class (e.g. $\alpha = 0.11$ for the three-dimensional $n = 1$ Ising model). Although the amplitudes in Eq. (6) are not universal, the ratios A^-/A^+ and D_1^-/D_1^+ have fixed values for each universality class.

The fact that many liquid crystalline compounds exhibit several phase transitions of different types in a limited temperature range makes them very suitable for testing phase transition theories. However, in several cases the closeness of two phase transitions may cause coupling between different order parameters and result in complex crossover between different kinds of critical behavior. An important and complicated example of crossover is the N–SmA transition, where order parameter coupling results in an evolution from second-order to first-order via a tricritical point by varying pressure or the composition of a binary mixture. In many experimental systems an intermediate crossover behavior with effective critical exponents which differ from the ones theoretically expected for the uncoupled cases, is often encountered. Crossover from mean-field behavior at large reduced temperatures to critical fluctuation behavior close to T_c is another important case. There are also situations where critical fluctuation results are expected but mean-field behavior is observed in the experimentally accessible reduced temperature range. This can be understood on the basis of the Ginzburg criterion [12] relating the size of the critical region to the range of correlated fluctuations.

In many cases one studies binary mixtures of liquid crystals in order to change or avoid crossover effects or follow a given phase transition as it evolves from one type to another. In principle, however, one should then apply Eq. (6) to $C_{p\phi}$ as a function of $\varepsilon_\phi = |(T - T_c)/T_c|$, where ϕ designates a path of constant chemical potential difference, and not to C_{px} as measured along a line of constant concentration. The distinction between measurements along paths of constant x or ϕ is not very important as long as dT_c/dx is not too large. Fisher [13] found that approaching a critical point via a constant-composition path results in exponent renormalization. For sufficiently large dT_c/dx values one may experimentally observe this Fisher renormalization with, for example, a renormalized critical exponent $\alpha_R = -\alpha/(1-\alpha)$, which results in a non-divergent specific heat capacity for $\alpha > 0$.

6.2.1.3 Experimental Methods

Differential Scanning Calorimetry

Differential scanning calorimetry (DSC) is by far the most commonly used thermal technique for studying liquid crystals. It is a very useful and sensitive survey technique for discovering new phase transitions and for determining the qualitative magnitude of thermal features. However, DSC is not well suited for making detailed quantitative measurements near liquid crystal phase transitions.

The reason for this can be understood by considering Fig. 2. In Fig. 2(a) representative enthalpy curves for a first-order and second-order transition are given. In part (b) the corresponding DSC responses (for heating runs) are compared. In a DSC measurement, a constant heating (or cooling) rate is imposed on a reference sample, which results in a constant and rapid temperature ramp (see Fig. 2). A servosystem forces the sample temperature to follow that of the reference by varying the power input to the sample. What is recorded then is the differential power dH/dt (between reference and sample), and the integral $\int (dH/dt) \, dt$ for a DSC peak approximates the enthalpy change associated with the corresponding

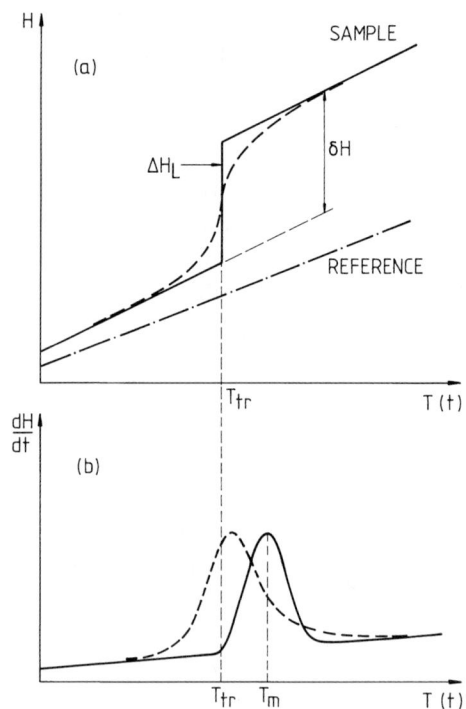

Figure 2. (a) Schematic representation of the temperature dependence of the enthalpy H of a liquid crystal sample near a first-order (solid curve) or second-order (dashed curve) phase transition at T_{tr}. The dashed-dotted line gives the enthalpy of a (DSC) reference material with a nearly constant heat capacity and without a phase transition. (b) Corresponding DSC responses (from heating runs) for the first-order (solid curve) and second-order (dashed curve) cases of part (a).

transition. However, as can be seen in Fig. 2(b) the DSC response is very similar for a first-order transition with latent heat ΔH_L and a second-order transition with a comparable integrated enthalpy change δH. The broadening of the DSC signal for the first-order case is due to the fact that the latent heat cannot be delivered to the sample instantaneously. The DSC response will be at a maximum for a temperature T_m well above T_{tr} for a heating run and well below T_{tr} for a cooling run. The difference between T_m and T_{tr} depends on the scan rate that is chosen. The DSC response for a second-order transition, given by the dashed curve in Fig. 2(b) is qualitatively the same. It is thus very difficult to distinguish between first- and second-order transitions and almost impossible to detect the difference between the latter ones and a weak first-order transition with significant pretransitional specific heat capacity variation. Furthermore, there is additionally a problem of thermodynamic equilibrium with DSC instruments. Although slow rates of the order of 0.1 K min^{-1} are possible, the best operational conditions of DSC machines are realized for fairly rapid scan rate (above 1 K min^{-1}). However, many liquid crystal phase transitions require scan rates which are typically two to three orders of magnitude slower than slow DSC scans.

Traditional Adiabatic Calorimetry

The classical method for measuring the specific heat capacity of a sample as a function of temperature is called adiabatic calorimetry and carried out by the stepwise addition of accurately measured small increments of electrically supplied heat [14, 15]. The heat capacity (at constant pressure p), C_p, of a sample can be obtained from the measured total heat capacity

$$C_1 = C_p + C_h = \frac{\Delta Q}{\Delta T} = \frac{\Delta H}{\Delta T} \qquad (7)$$

provided the heat capacity of the sample holder, C_h, has been independently determined in a calibration experiment. In the experiment a known amount of heat ΔQ is thus applied and the corresponding temperature rise ΔT measured. This ΔT step has of course to be sufficiently small compared to the curvature of the $H(T)$ curve at a given (average) temperature T, especially near a transition, which makes the method somewhat tedious and slow. Anisimov and coworkers [2,16] have mainly used this method for the investigations of liquid crystals.

Adiabatic Scanning Calorimetry

Much of the intrinsic difficulties with DSC measurements and the tedious data collecting process of traditional adiabatic calorimetry can be avoided by adiabatic scanning calorimetry. In this technique a measured heating power is continuously applied to (or extracted from) the sample and sample holder. It was used in the 1970s for the study of liquid–gas [17] and liquid–liquid critical points [18] and first applied to liquid crystals by us [19] and later also by Anisimov et al. [20]. In the dynamic modes the total heat capacity C_1 is now given by:

$$C_1 = C_p + C_h = \frac{P_1}{\dot{T}_1} = \frac{P_1^e + P_1^l}{\dot{T}_1} \qquad (8)$$

In Eq. (8) the total heating power P_1 has been divided in two parts: P_1^e the power applied electrically to a heater, heating the sample and sample holder, and P_1^l representing leaks with an (adiabatic) shield surrounding the sample holder. For cooling runs P_1^l has to be given a controlled negative value. By keeping P_1 or $\dot{T}_1$ constant, combined with increasing or decreasing the sample temperature, four practical modes of operation are obtained [3, 21]. These modes require different settings for the servo-systems controlling the temperature and maintaining adiabatic conditions, or a controlled heat transfer between sample holder and shielding. The most interesting operating conditions are the ones with constant heating or cooling power (P_1 constant). As will be pointed out, this has distinct advantages at first-order transitions. On the other hand, in modes in which $\dot{T}_1$ is kept constant, one runs basically into the same kind of problems as in case of the DSC technique, even if one uses substantially lower scanning rates. It is sufficient to consider here only the constant heating mode (the constant cooling mode is analogous). In this case $P_1 =$

P_1^e and P_1^l is kept negligibly small. In order to obtain the heat capacity of the sample one has to measure P_1^e, $\dot{T}_1$ and C_h. P_1^e is easily obtainable from a measurement of the d.c. current through and the voltage drop across the heating resistor on the sample holder. The rate $\dot{T}_1$ has to be obtained by numerical differentiation of the carefully measured time dependence of T with the sensor in close thermal contact with the sample. Values of $C_h(T)$ can be derived from calibration runs without the sample or with calibration fluids in the sample holder. In fact, the numerical differentiation is not necessary because the time dependence of temperature can be easily converted into a more basic result, namely, the enthalpy versus temperature. Indeed, after subtracting the contribution due to the sample holder, one gets $H(T) - H(T_s) = P(t - t_s)$, the index s refers to the starting conditions of the run. $P = P_1 - P_h$, with P_h the power needed to heat the sample holder. Here, T is the sample temperature at time t. In this manner, the latent heat can also be obtained. If a first-order transition occurs at a certain temperature T_{tr}, a step change will occur in $H(T)$ in the interval $\Delta T = t_f - t_i$ until the necessary (latent) heat ΔH_L has been supplied to cross the transition. One then can write:

$$H(T) - H(T_s) = P(t_i - t_s) + P(t_f - t_i)$$
$$+ P(t - t_f) = \int_{T_s}^{T_{tr}} C_p dT + \Delta H_L(T_{tr}) + \int_{T_{tr}}^{T} C_p dt \quad (9)$$

From the above considerations, it should be clear that running an adiabatic scanning calorimeter in the constant heating (or cooling) modes makes it possible to determine latent heats when present and distinguish between first-order and second-order phase transitions. On the basis of $C_p = P/\dot{T}$, it is also possible to obtain information on the pretransitional heat capacity behavior, provided one is able to collect sufficiently detailed and accurate information on the temperature evolution $T(t)$ during a scan [19, 21]. Details of the mechanical construction and alternative operating modes of previously used adiabatic scanning calorimeters, can be found elsewhere [3, 21]. Figure 3 gives a schematic representation of a recently constructed wide temperature range version.

A. C. Calorimetry

The a.c. calorimetric technique is very well suited to measure pretransitional specific heat capacities. Indeed it measures C_p directly and not the enthalpy H. This method has been extensively used for the investigation of liquid crystals by Johnson [22], Huang [23–25], Garland [4, 26, 27] and their coworkers.

For a.c. calorimetry an oscillating heating power $P_{ac} = P_0(1 + \cos \omega t)$ is supplied to a sample (in a sample holder), usually loosely coupled to a heat bath at a given temperature T_0, and from the amplitude ΔT_{ac} of the resulting temperature oscillation (see Fig. 4) the specific heat capacity of the sample can be derived. From Fig. 4, it should also be clear that for a first-order transition one can not determine the latent heat ΔH_L in this way.

In a basic one-dimensional heat flow model [28, 25] for a planar sample with thermal conductivity κ_s, the sample is assumed to be thermally coupled to the bath at T_0 with a finite thermal conductance K_b, but the coupling medium (gas) is assumed to have zero heat capacity. The following results are obtained:

$$\Delta T_{ac} \quad (10)$$
$$= \frac{P_0}{\omega C_p} \left[1 + \frac{1}{(\omega \tau_1)^2} + (\omega \tau_2)^2 + \frac{2K_b}{3K_s} \right]^{-1/2}$$

$$T_a = T_0 + \Delta T_{dc} = T_0 + \frac{P_0}{K_b} \quad (11)$$

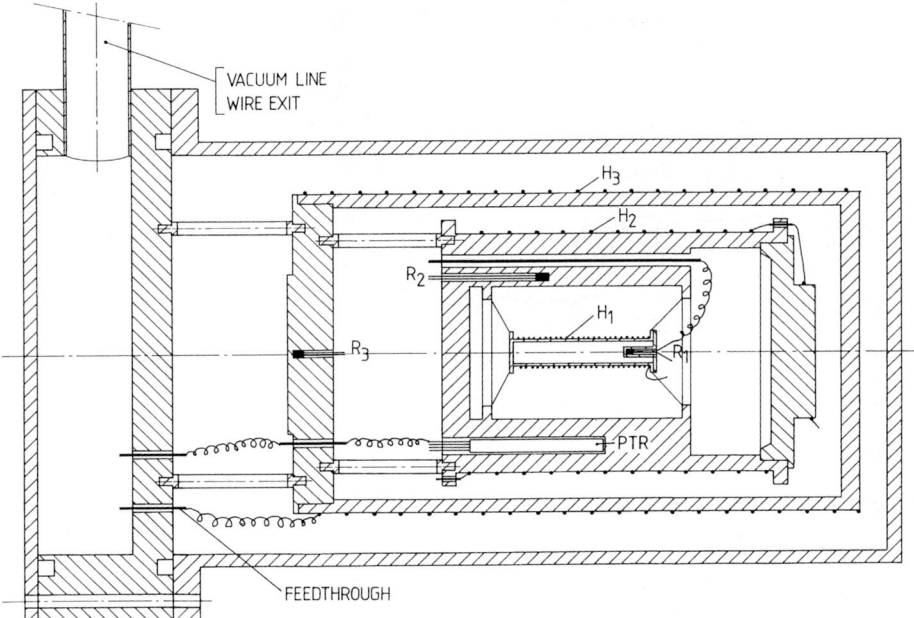

Figure 3. Schematic diagram of an adiabatic scanning calorimeter. Electric heaters and thermistors are denoted by H and R. PTR is a platinum resistance thermometer. The whole calorimeter is placed in a hot air temperature controlled oven. Details on a sample holder with stirring capabilities have been given elsewhere [21].

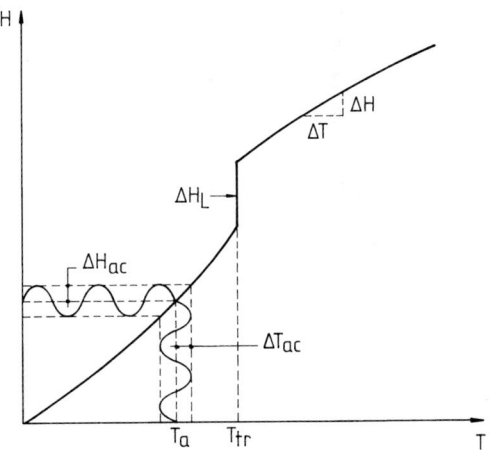

Figure 4. Enthalpy H curve for a first-order transition at T_{tr} and schematic representation of the a.c. calorimetric and standard adiabatic methods for determining the heat capacity C_p. ΔH_{ac} and ΔT_{ac} are, respectively the amplitude of the enthalpy variation and the corresponding temperature change. T_a is given by Eq. (11). In the standard adiabatic method, $C_p = (\Delta H/\Delta T)_p$ is obtained from the temperature increase ΔT resulting from the heat input ΔH.

In Eq. (11) T_a is the average sample temperature (see Fig. 4) and ΔT_{dc} is the offset between bath and sample temperature. In Eq. (10) K_s is the thermal conductance for heat flow perpendicular to the surface of the planar sample. τ_1 is the sample bath relaxation time and τ_2 is a combined (internal) relaxation time for the sample and addenda (heater, temperature sensor, and sample holder when present). τ_1 and τ_2 result, respectively, in a low and high frequency cut-off. In the normal a.c. calorimetric mode one chooses a frequency ω so that $\omega \tau_2 \ll 1$ in order to avoid temperature gradients in the sample. In Eq. (10) the third term can then be neglected. If one further assume the thermal conductance of the sample K_s to be much larger than K_b also the last term can be omitted. In addition to ΔT_{ac} one may also observe a phase shift $(\varphi - \pi/2)$ between $T(t)$ and $P(t)$, with φ given by [6]:

$$\tan \varphi = (\omega \tau_1)^{-1} = K_b(\omega C_p)^{-1} \qquad (12)$$

Eliminating τ_1 between Eqs. (12) and (10) (without the last two terms) results in:

$$C_p = \frac{P_0}{\omega \Delta T_{ac}} \cos\varphi \qquad (13)$$

When ω is well above the low frequency cutoff ($\omega\tau_1 \gg 1$), which corresponds to the normal operating conditions, one can set $\cos\varphi = 1$ as a good approximation in many cases. If necessary, one can correct for small non-zero φ values [6].

Satisfying the above normal operating conditions ($\omega\tau_1 \gg 1 \gg \omega\tau_2$) can be achieved by limiting the thermal coupling between the sample and bath and by using flat thin sample holders. Several designs for a.c. calorimeters have been described in the literature [22–27]. Figure 5 shows the mechanical and thermal design of a calorimeter capable of both a.c.-mode and relaxation-mode (see further) operation [6].

In Huang's group a special purpose a.c. calorimeter was built with the aim of studying free-standing liquid crystal films [25]. This calorimeter contains a special constant-temperature oven that allows the manipulations required for spreading smectic films, and a thermocouple detector placed very near (≈ 10 μm) but not touching the film. Periodic heating of the sample is achieved by utilizing a chopped laser beam. Thick smectic films (≈ 100 layers) as well as films as thin as two smectic layers have been studied [25, 29, 30].

Relaxation Calorimetry

In conventional relaxation calorimetry the bath temperature is held constant at a value T_0 and a step function d.c. power is supplied to the sample cell [31–33]. For heating runs, P is switched from 0 at time $t=0$ to a con-

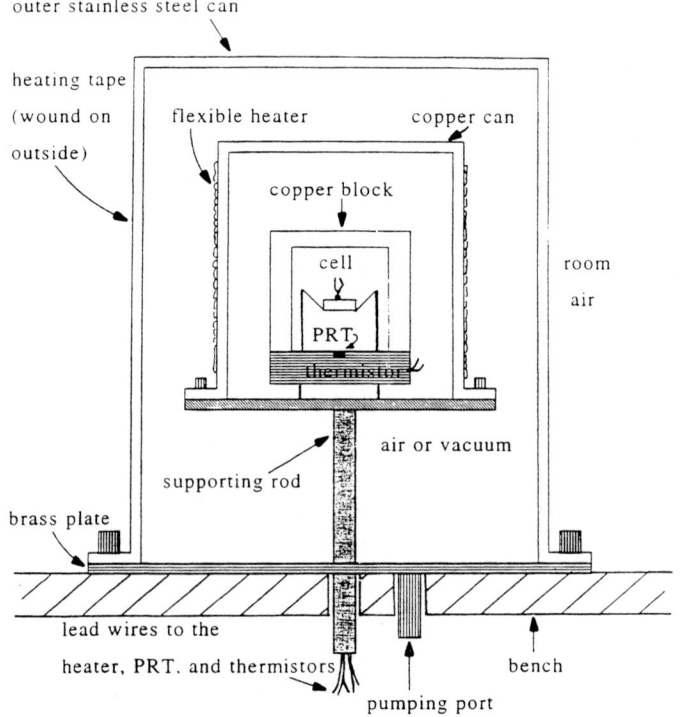

Figure 5. Mechanical/thermal design of a computer-controlled calorimeter capable of operating in the a.c. mode or in relaxation modes including non-adiabatic scanning [34, 35].

stant value P_0, while for cooling runs the power is switched from P_0 to 0. In the absence of latent heat effects the cell temperature $T(t)$ relaxes exponentially from $T(0)$ to $T(\infty)$. For the heating regime one has:

$$T(t) = T_0 + \Delta T_\infty [1 - \exp(-t/\tau_1)] \qquad (14)$$

For the cooling case one has:

$$T(t) = T_0 + \Delta T_\infty \exp(-t/\tau_1) \qquad (15)$$

In both equations $\tau_1 = C/K_b$ and $\Delta T_\infty = P_0/K_b$, with C the heat capacity of the cell and K_b the thermal conductance between cell and bath. In deriving Eqs. (14) and (15) it is assumed that K_b and C are constant over the (narrow) range from $T(0)$ to $T(\infty)$. From a fit of the $T(t)$ data with Eq. (14) or Eq. (15) both τ_1 and ΔT_∞ are obtained. Provided P_0 is known, this results in $C = \tau_1 K_b$.

If a first-order phase transition occurs between $T(0)$ and $T(\infty)$, there will be a non-exponential $T(t)$ variation due to latent heat effects. This situation can be handled [32, 33] by defining a time-dependent heat capacity $C(t) = \dot{P}/\dot{T}$, where $\dot{P} = P_0 - (T-T_0)K_b$. In this procedure it is possible to obtain the latent heat using:

$$\Delta H_L = \int_0^\infty [C(t) - C(0)] \dot{T}(t) dt \qquad (16)$$

Nonadiabatic Scanning Calorimetry

Recently a new type of relaxation calorimetry has been developed [34] in which the heater power is linearly ramped. This new method has been called nonadiabatic scanning calorimetry [6, 34]. For a heating run, $P=0$ for $t<0$, $P = \dot{P}t$ for $0 \le t \le t_1$ where $\dot{P} = dP/dt$ is a constant, and $P = P_0 = \dot{P}t_1$ for $t > t_1$. The initial ($t \le 0$) sample temperature is equal to the bath temperature T_0 and the final ($t \gg t_1$) sample temperature is $T(\infty) = T_0 + P_0 K_b^{-1}$. For a cooling run a reversed power profile is used: $P = P_0$ for $t<0$, $P = P_0 + \dot{P}t$ with $\dot{P}$ negative and constant for $0 \le t \le t_1$, and $P = 0$ for $t > t_1$. Now the initial temperature is $T(\infty)$ and the final one T_0. For the heating run the time dependence of the temperature of the cell over the time regime $0 \le t \le t_1$ is given by [6, 34]:

$$T(t) = T_0 + K_b^{-1} \dot{P}(t - \tau_1) + \tau_1 K_b^{-1} \dot{P} \exp(-t/\tau_1) \qquad (17)$$

For a cooling run ($\dot{P}$ negative) a similar expression is obtained:

$$T(t) = T(\infty) + K_b^{-1} \dot{P}(t - \tau_1) + \tau_1 K_b^{-1} \dot{P} \exp(-t/\tau_1) \qquad (18)$$

The thermal conductance can be derived from:

$$K_b = P_0 (T_\infty - T_0)^{-1} \qquad (19)$$

and the heat capacity in both cases is given by:

$$C(T) = \frac{dH}{dT} = \frac{P - (T - T_0) K_b}{dT/dt} \qquad (20)$$

Equation (20) is identical to Eq. (8) if one identifies P_1^l with $K_b(T_0 - T)$. In normal adiabatic scanning calorimetry one imposes $P_1^l = 0$ (or $T = T_0$) for heating runs or P_1^l constant for a cooling run [3, 21]. Here, however, $(T - T_0)$ and P are time dependent, because $P(t')$ is the power at time t' corresponding to the cell temperature $T(t')$ lying in the interval T_0 to $T(\infty)$. dT/dt is obtained by fitting $T(t)$ data over a short time interval centered at t'. In comparison with the conventional transient method, this method has the advantage of avoiding large transient disturbances (by step increases or decreases of P) and also an optimal nearly linear behavior of dT/dt, except for regions with very rapid H variation with T.

This new method also allows a determination of the latent heat of a first-order transition with a two-phase region [35]. In the two-phase region Eq. (20) represents an effective heat capacity C_e. Outside the two-phase region C_e is identical to $C(T)$. One identifies a first-order transition by observing anomalous behavior of C_e and the occurrence of hysteresis. The latent heat can be derived from the following expression [35]:

$$\Delta H_L = \int_{t_1}^{t_2}[P - K_b(T-T_0)]dt - \int_{T_1}^{T_2}[C_h + C_c]dt \quad (21)$$

where two-phase coexistence exists between $T_1(t_1)$ and $T_2(t_2)$, C_h is the heat capacity of the sample holder, and $C_c = X_a C_a + X_b C_b$ (with X the mole fraction) the heat capacity of the two coexisting phases that would be observed in the absence of phase conversion.

By integrating Eq. (20) between $T_1(t_1)$ and $T(t)$, and assuming $\dot{P}$, K_b and T_0 constant one also arrives at the following expression for the temperature dependence of the enthalpy:

$$H(t) - H(t_1) = \frac{\dot{P}}{2}(t^2 - t_1^2) \quad (22)$$
$$- K_b \int_{t_1}^{t} T(t)dt + K_b T_0(t - t_1)$$

The calorimeter developed at MIT [6, 35] and shown in Fig. 5, can also be operated in the new relaxation mode.

The Photoacoustic Method

Photoacoustics is well established [36] for the investigation of optical and thermal properties of condensed matter, but it has only rather recently been applied to liquid crystals [37–41]. The photoacoustic technique is based on the periodic heating of a sample, induced by the absorption of modulated or chopped (electromagnetic) radiation. In the gas microphone detection configuration the sample is contained in a gas-tight cell (Fig. 6). The thermal wave produced in the sample by the absorbed radiation couples back to the gas above the sample and periodically changes the tempera-

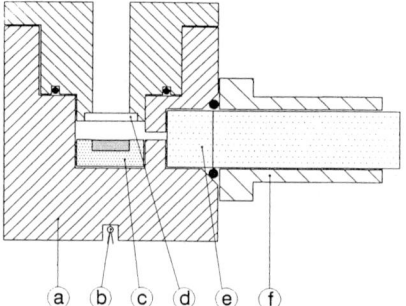

Figure 6. Schematic representation of a photoacoustic measuring cell with microphone detection: (a) copper cell body, (b) thermistor, (c) gold plated sample holder with sample, (d) quartz window, (e) condenser microphone, (f) Teflon microphone holder, solid dots are O-rings; heating elements for temperature control are not shown.

ture of a thin gas layer (with a thickness determined by the thermal diffusion length μ_g) above the sample surface. This will result in a periodic pressure change in the gas cell which can be detected by a microphone. A general one-dimensional theoretical model has been developed by Rosencwaig and Gersho [42].

Assuming a sinusoidal variation, part $I = I_0/2 \exp(i\omega t)$ of the radiation intensity will give rise to a time-dependent harmonic surface temperature variation $\theta = \theta_0 \exp(i\omega t)$. Starting with this temperature variation one arrives at the following result for the corresponding pressure variation in the gas [42]:

$$\delta p = \left[\gamma_g P_0 \mu_g / (\sqrt{2} T_0 l_g)\right] \theta e^{-i\pi/4}$$
$$\equiv Q e^{i(\omega t - \pi/4)} \quad (23)$$

where γ_g is the ratio of the specific heat capacities at constant pressure and constant volume of the gas, P_0 and T_0 the static pressure and temperature, μ_g the thermal diffusion length of the gas and l_g the gas length between the sample and the light window in the planar geometric configuration. Rosencwaig and Gersho [42] derived the following explicit relation for $\theta_0(\omega)$ for a homogeneous sample on top of a backing material:

$$\theta_0 = \frac{I_0 \beta}{2\kappa(\beta^2 - \sigma^2)} \tag{24}$$

$$\cdot \left[\frac{(r-1)(b+1)e^{\sigma l} - (r+1)(b-1)e^{-\sigma l} + 2(b-r)e^{\beta l}}{(g+1)(b+1)e^{\sigma l} - (g-1)(b-1)e^{-\sigma l}} \right]$$

In this equation β is the optical absorption coefficient of the sample, l is the sample thickness and $\sigma = (1+i)a$ with $a = 1/\mu$ the thermal diffusion coefficient and μ the thermal diffusion length of the sample. One further has: $b = \kappa_b a_b \mu \kappa^{-1}$, $g = \kappa_g a_g \mu \kappa^{-1}$, $r = (1-i)\beta\mu/2$. Here, and also in Eq. (23), quantities without subscript refer to the sample and the subscripts g and b refer to the cell gas and the sample backing material.

For measurements in liquid crystals, one usually works in the optically and thermally thick regime ($e^{\sigma l} \gg e^{-\sigma l}$, $e^{-\beta l}$), resulting in substantially simplified expressions. This means that β^{-1} and μ of the sample have to be much smaller than the sample thickness, l. The contribution of the backing material then disappears from the expressions. One then obtains for the photoacoustic microphone signal $Q = q e^{-i\psi}$, with amplitude q and phase ψ (with respect to the radiation modulation) [38]:

$$q = \frac{\gamma_g P_0 I_0 t (2t^2 + 2t + 1)^{-1/2}}{2\sqrt{2} T_0 l_g (1+s) \kappa_g a_g^2} \tag{25}$$

$$\tan \psi = 1 + 1/t \tag{26}$$

with $t = \mu \beta / 2$ and $s = a \kappa (\kappa_g a_g)^{-1}$.

The Eqs. (25) and (26) allow the simultaneous determination of the heat capacity per unit volume (ρC_p) and the thermal conductivity κ of the sample by solving for t and s. This gives

$$\rho C_p = \beta \kappa_g a_g s t^{-1} \omega^{-1} \tag{27}$$

$$\kappa = 2 \kappa_g a_g s t \beta^{-1} \tag{28}$$

However, one should arrange the measuring conditions in such a way that $1/t = 2/(\mu\beta)$ is not too small compared to l in Eq. (26). This can to some extent be done by choosing the proper modulation frequency (changing μ) and/or changing β by choosing an appropriate wavelength λ for the modulated light source. In any case one also has to measure (in a separate experiment) $\beta(\lambda)$. Moreover, Eqs. (25) and (26) for q and ψ contain several other quantities related to the cell characteristics, the cell gas and the light source, which have to be known in order to arrive at quantitative results for ρC_p and κ. This problem is usually solved by simultaneous or separate calibration runs with a sample of known optical and thermal parameters [39, 40].

In photoacoustics, as in a.c. calorimetry, relatively small periodic temperature variations (in the mK range) are used, and only small amounts of samples are needed. The small temperature variations allow measurements very close to phase transitions but latent heats of first-order phase transitions can also not be obtained in this way.

One of the important advantages of photoacoustical techniques is the possibility of simultaneously measuring the heat capacity and the thermal conductivity in small samples. These methods also offer the possibility of investigating anisotropic thermal transport properties and to carry out depth profiling in thermally inhomogeneous samples [40]. Studying the thermal conductivity anisotropy requires homogeneous samples with known director orientation. This can be achieved by placing the photoacoustic cell and/or sample in an external magnetic field.

It should also be pointed out that with the extension of the standard a.c. calorimetric technique by Huang et al. [24], one can also measure simultaneously the heat capacity and the thermal conductivity. In this case a.c. calorimetric measurements have to be extended to high frequencies into the regime where the thermal diffusion length $\mu = [2\kappa/(\rho C_p \omega)]^{1/2}$ of the sample becomes comparable to the sample thickness in which case the second term in Eq. (10) can no longer be neglected.

The Photopyroelectric Method

If we consider a pyroelectric transducer, with thickness l_p and surface area A, in a one-dimensional configuration, a change of the temperature distribution $\theta(x, t)$ relative to an initial reference situation $\theta(x, t_0)$ will cause a change of polarization. This in turn induces an electric charge given by:

$$q(t) = \frac{pA}{l_p} \int_0^{l_p} [\theta(x,t) - \theta(x,t_0)] dx = \frac{pA}{l_p} \theta(t) \quad (29)$$

with p the pyroelectric coefficient of the transducer. Consequently also a current is produced:

$$i_p = \frac{pA}{l_p} \frac{d\theta(t)}{dt} \quad (30)$$

Usually the electric signal from the pyroelectric transducer is detected by a lock-in amplifier. Thus only the a.c. component of the temperature variation gives rise to the detected signal. The signal depends, of course, on the impedance of the transducer and of the detecting electronics. The pyroelectric element can be represented by an ideal current source with a parallel leakage resistance and capacitance, while the detection electronics can be described by an input capacitance and a parallel load resistance [43]. Circuit analysis results then in a general expression for the signal $V(\omega)$ [44, 45].

Under properly chosen experimental conditions [44], however, it is possible to arrive at a simultaneous determination of the specific heat capacity and the thermal conductivity of a sample in thermal contact with the pyroelectric detector. A suitable setup for liquid crystal samples is given in Fig. 7. The wavelength of the modulated light, the modulation frequency, sample and detector thicknesses l_s and l_p are chosen in such a way that the sample and transducer are optically opaque, the detector thermally very thick ($\mu_p \ll l_p$) and the sample quasithermally thick ($\mu_s \leq l_s$). One then obtains for the signal amplitude and phase [46]:

$$|V(\omega)| = \frac{I_0 \eta_s A R p e_p}{l_p [1+(\omega\tau)^2]^{1/2} \rho_p C_p}$$

$$\cdot \frac{\exp[-(\omega/2\alpha_s)^{1/2} l_s]}{e_s(e_m/e_s + 1)(e_p/e_s + 1)} \quad (31)$$

$$\phi(\omega) = \tan^{-1}(\omega\tau) - (\omega/2\alpha_s)^{-1/2} l_s \quad (32)$$

where subscripts s, p and m, respectively, refer to the sample, the pyroelectric transducer and to the medium in contact with the sample front surface. I_0 is the nonreflected light source intensity, η_s is the nonradiative conversion efficiency, ρ is the density and $e = (\rho C \kappa)^{1/2}$ is the thermal effusivity. R and τ are a circuit equivalent resistance and time constant. From Eq. (32) it is possible to determine the sample thermal diffusivity α_s. Inserting α_s in Eq. (31) yields the sample effusivity. The thermal conductivity and specific heat capacity are then given by:

$$C_s = e_s \rho_s^{-1} \alpha_s^{-1/2} \quad \text{and} \quad \kappa_s = e_s \alpha_s^{-1/2} \quad (33)$$

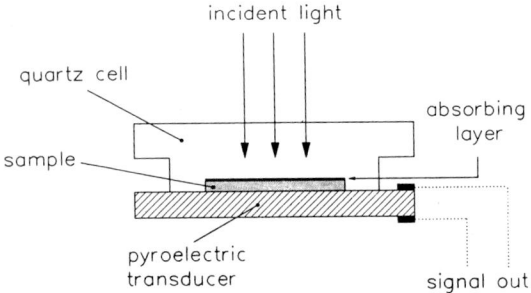

Figure 7. Schematic diagram of a photopyroelectric cell for liquid crystal samples [46].

6.2.1.4 Calorimetric Results

High resolution calorimetry has been extensively used to study the many important and often puzzling phase transitions in thermotropic liquid crystals of rodlike molecules. Discotic liquid crystals seem to be, from the calorimetric point of view, less exciting because most phase transitions are first order with little pretransitional fluctuation effects [47]. Although lyotropic liquid crystal systems show complex phase behavior, they have not been studied calorimetrically in great detail, because calorimetric signatures of phase transitions tend to turn out quite small [48]. No attempt will be made to give an exhaustive treatment of any of the transitions discussed further. Thermal results will be presented to show what can be accomplished for typical cases with high-resolution calorimetric experiments, careful analysis and theoretical interpretation. More extensive overviews can be found elsewhere [1–6]. A detailed account of calorimetric studies of free standing liquid crystal films can be found in a review article by Huang [49].

The I–N Transition

The only difference between the nematic phase and the isotropic phase is the orientational order. A proper description of this orientational order requires the introduction of a tensor of the second rank [7, 8]. This tensor can be diagonalized and for anisotropic liquids with uniaxial symmetry, the nematic phase can be described by only one scalar order parameter. The thermodynamic behavior in the vicinity of the N–I transition is usually described in terms of the mean-field Landau–de Gennes theory [7]. For the uniaxial nematic phase one can obtain the expansion of the free energy G in terms of the modulus of an order parameter Q.

$$G - G_0 = \frac{1}{2} A Q^2 - \frac{1}{3} B Q^3 + \frac{1}{4} C Q^4 + \frac{1}{6} D Q^6 + \ldots \quad (34)$$

In the isotropic phase $Q=0$ and in the nematic phase $Q \neq 0$. In Eq. (34) one has $A = a(T-T^*)/T_{N-I}$ and $B > 0$. The presence of the cubic term, which does not disappear at T_{N-I}, leads to a first-order transition with a finite discontinuity in the order parameter ($Q_{N-I} = 2B/3C$). T^* is the stability limit of the isotropic phase. For $B=0$, a normal second-order transition at T_{N-I} is expected. The excess heat capacity in the nematic phase is given by [16]:

$$C_p = -aQ \left(\frac{\partial Q}{\partial T} \right)_p \quad (35)$$

$$= \frac{a^2}{2CT_{N-I}} \left[1 + \frac{B}{2(aC)^{1/2}} \left(\frac{T^{**} - T}{T_{N-I}} \right)^{-1/2} \right]$$

with T^{**} the stability limit of the nematic phase. One, thus, has an anomalous contribution with an exponent $\alpha = 1/2$ in the nematic phase, resulting in a jump in C_p at T_{N-I} equal to $\Delta C_p = 2a^2/CT_{N-I}$. In the case of a mean field second-order ($B=0$ and $C>0$) transition the singular contribution (see Fig. 1c) follows from the mean-field behavior (with the critical exponent $\beta = 1/2$) of the order parameter [16].

For the enthalpy discontinuity at T_{N-I} one obtains $H_I - H_N = 2aB^2/9C^2$. When $B=0$, there is no enthalpy jump or latent heat and one has a critical point (Landau point) on an otherwise first-order line. Because of the presence of this (small) cubic term in Eq. (34), the N–I transition should (normally) be weakly first order. This is in agreement with experimental observations. In Fig. 8, part of the enthalpy curve near T_{N-I} is shown for hexylcyanobiphenyl (6CB), a compound of the alkylcyanobiphenyl (nCB) homologous series [5]. The nearly vertical part of the enthalpy curve corresponds to the

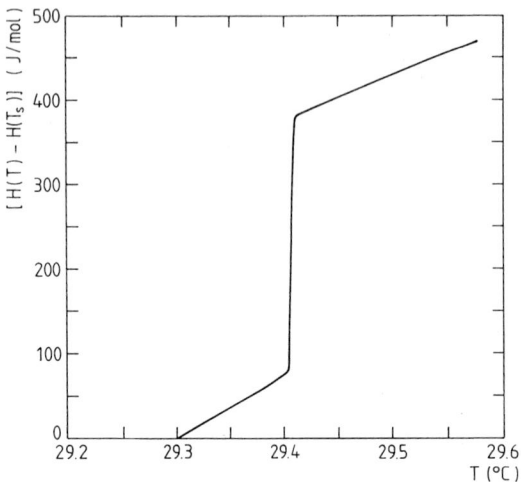

Figure 8. Temperature dependence of the enthalpy near the nematic to isotropic transition for hexylcyanobiphenyl (6CB) [5].

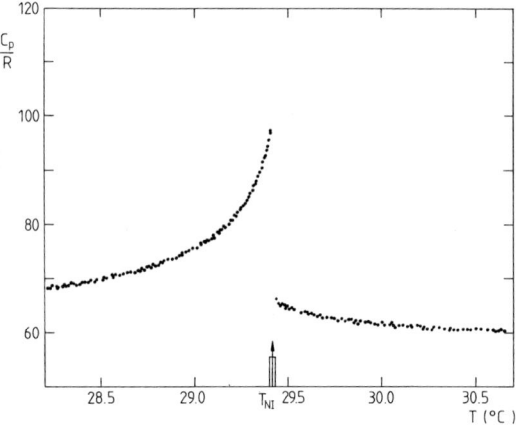

Figure 9. Temperature dependence of the reduced heat capacity per mole (R is the gas constant) near the nematic to isotropic (N–I) transition for hexylcyanobiphenyl (6CB). The width of the arrow on the temperature axis represents the two-phase region [5].

latent heat of the N–I transition. It is not perfectly vertical because of the small ($\approx$20 mK wide) impurity-induced two-phase region. ΔH_L for the transitions in 6CB is (293 ± 3) J/mol. This latent heat is indeed quite small, about two orders of magnitude smaller than the latent heat at the melting transitions in these cyanobiphenyl compounds [19, 50].

Because the latent heat at the N–I transition is very small and since the orientational interactions between liquid crystal molecules are short range, one might expect many physical properties to display critical-like behavior, described by power laws with the appropriate critical exponents. These fluctuation effects are, indeed, observed in many properties including the specific heat capacity. Figure 9 gives the temperature dependence of C_p near the N–I transition for the same substance as in Fig. 8. No data points are displayed in the coexistence region, whose width is indicated by the narrow box around the arrow marking T_{N-I}.

Since the Landau–de Gennes theory is a mean-field theory, one can only expect a qualitative description of the specific heat capacity anomaly [2, 16]. Attempts to determine a critical exponent α have been largely unsuccessful due to the first-order character of the N–I transition. Fits with Eq. (6) must be made separately for data above and below T_{N-I} with different effective critical temperatures (T^* and T^{**}), which both are different from the first-order transition temperature. For $T < T_{N-I}$, typical α_{eff} values lie in the range 0.3–0.4 but depend strongly on the fitting range. For $T > T_{N-I}$, an even wider range of α_{eff} values between 0.1 and 0.5 is observed. Furthermore $(T^* - T_{N-I})$ and $(T^{**} - T_{N-I})$ values obtained from C_p fits are consistently about one-tenth the magnitude of those from several other properties [2, 16]. The experimental behavior of the N–I transition is quite well characterized, but theoretical understanding is still rather poor.

Blue Phase Transitions

In the case of optically active molecules one arrives at a special situation for the nematic phase. In addition to the long range orientational order there is a spatial variation of the director leading to a helical structure. The local preferred direction of alignment of the molecules is slightly rotated in adjacent planes perpendicular to the pitch axis. This phase is usually called chiral nematic (N*) or cholesteric phase. The helical pitch in the N* phase is substantially larger than molecular dimensions and varies with the type of molecules. For a long pitch one has a direct first-order transition between the isotropic phase and the chiral nematic phase with characteristics very similar to the normal N–I transition in nonchiral compounds. For chiral nematics with short pitch (typically less than 0.5 μm), a set of intermediate blue phases (BP) are observed between the isotropic and the N* phase [51–53]. In order of increasing temperature these phases are denoted BP_I, BP_{II} and BP_{III}. The first two blue phases have three-dimensional cubic defect structures, whereas BP_{III}, which is also called the fog phase appears to be amorphous [51–53].

Different phase transitions involving blue phases are expected theoretically to be first-order [54] except the BP_{III} to isotropic transition which could become second order [54] in an isolated critical point at the termination of a first-order line [6, 55]. Figure 10 shows C_p results of cholesteryl nonnanoate (CN) with adiabatic scanning calorimetry [56]. From Fig. 10 it is clear that there are substantial pretransitional heat capacity effects associated with the BP_{III}–I transition, which means that a large amount of energy is going into changing the local nematic order. The other transitions appear as small, narrow features on the BP_{III}–I transition peak. From the inspection of the en-

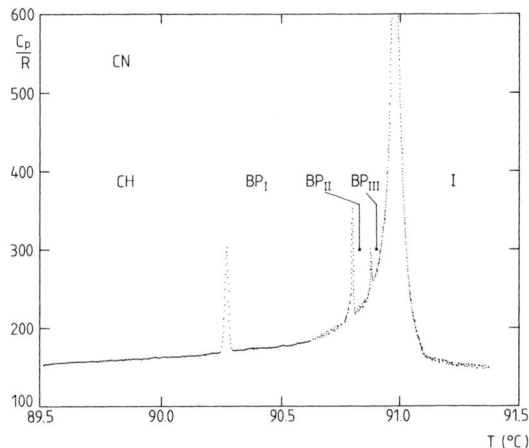

Figure 10. General overview for the reduced heat capacity per mole C_p/R for a temperature range covering all phase transitions involving the blue phases in cholesteryl nonanoate (CN) [56].

thalpy behavior, it was found that these small features correspond to first-order transitions with very small latent heats [56]. In CN the BP_{III}–I transition was also observed to be first order. This can be seen in Fig. 11 where the enthalpy versus temperature is given for a T-range of 0.2 K below and 0.2 K above the transition.

Further calorimetric studies by Voets and Van Dael [57, 58] have shown that the latent heat for BP_{III}–I decreases as the chirality of the molecule increases and that its magnitude varies considerably as a function of composition in binary mixtures of R- and S-enantiomers. Recently an a.c. and non-adiabatic scanning calorimetry investigation of the chiral compound S,S-(+)-4″-(2-methylbutylphenyl)-4′-(2-methylbutyl)-4-biphenyl-4-carboxylate (S,S-MBBPC), which has very short pitch, yielded the results given in Fig. 12 [55]. By combining these two calorimetric techniques one can show that there are sharp first-order transitions with small latent heats at the N*–BP_I and the BP_I–BP_{III} transitions (the BP_{II} phase is absent in this compound). However, no thermodynamic

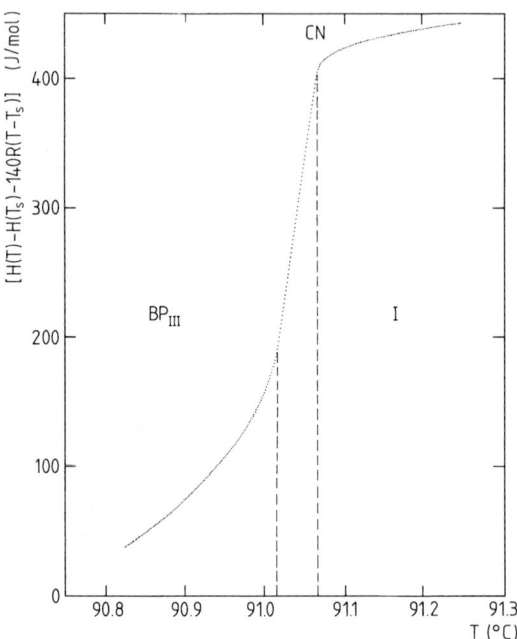

Figure 11. Detailed plot of the enthalpy near the phase transition from the blue phase III to the isotropic phase in cholesteryl nonanoate (CN). Note that for clarity a large linear background $140 R (T-T_s)$ with $T_s = 90.73\,°C$, has been subtracted from the direct data. The two vertical dashed lines indicate the width of the two-phase region [56].

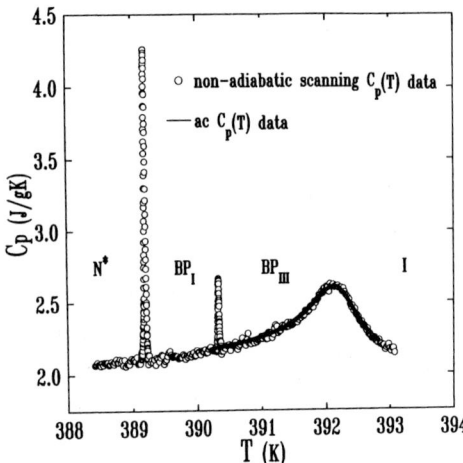

Figure 12. The specific heat capacity variation in the N^*–BP_I–BP_{III}–I transition region for S,S-MBBPC. The sharp peaks from non-adiabatic scanning through the N^*–BP_I and BP_I–BP_{III} regions are due to latent heat effects at the first-order transitions. No such effects are seen for the BP_{III}–I transition, because this is only a continuous supercritical evolution not a true phase transition in this material [55].

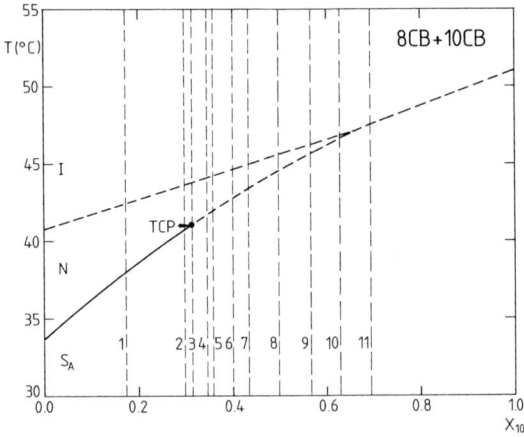

Figure 13. Phase diagram of 8CB+10CB mixtures. Heavy solid and dashed lines are, respectively, second-order and first-order transitions. TCP is the N–A tricritical point. Vertical dashed lines indicate the measured mixtures [64].

BP_{III}–I transition occurs in this S,S-enantiomer of MBBPC. One only observes a supercritical evolution. As pointed out by Kutnyak et al. [55] an investigation of S,S- and R,R-enantiomers of MBBPC should allow one to find the BP_{III}–I critical point and determine its critical exponent. The observed continuous supercritical evolution implies that the BP_{III} and I phases must have the same macroscopic symmetry allowing, as in the case of the liquid–gas transition of simple fluids, the first-order line to terminate in an isolated critical point.

The N–SmA Transition

This transition has been the most extensively studied of all liquid crystal phase transitions. In spite of vigorous experimental and theoretical efforts, many aspects of the critical behavior of this simple kind of one-dimensional freezing, are not yet understood, making this transition an intriguing and challenging problem in the statistical mechanics of condensed matter.

The SmA liquid crystalline phase results from the development of a one-dimensional density wave in the orientationally ordered nematic phase. The smectic wave vector q is parallel to the nematic director (along the z-axis) and the SmA order parameter $\psi = |\psi|e^{i\phi}$ is introduced by $\rho(r) = \rho_0[1 + \mathrm{Re}\,\psi e^{iqz}]$. Thus the order parameter has a magnitude and a phase. This led de Gennes to point out the analogy with superfluid helium and the normal-superconductor transition in metals [7, 59]. This would than place the N–SmA transition in the three-dimensional XY universality class. However, there are two important sources of deviations from isotropic 3D-XY behavior. The first one is crossover from second-order to first-order behavior via a tricritical point due to coupling between the smectic order parameter ψ and the nematic order parameter Q. The second source of deviation from isotropic 3D-XY behavior arises from the coupling between director fluctuations and the smectic order parameter, which is intrinsically anisotropic [60–62].

The first-order to second-order crossover can be qualitatively understood on the basis of a mean-field approximation for the excess smectic free energy (above the nematic one) as formulated by de Gennes [7]:

$$\Delta G = a|\psi|^2 + b_0|\psi|^4 + (\delta Q)^2/2\chi - C|\psi|^2\delta Q \tag{36}$$

where $a = a_0(T - T_{\mathrm{N-A}})/T_{\mathrm{N-A}} = a_0\varepsilon$; b_0, $C > 0$; and δQ the change in the nematic order induced by the formation of the smectic layers, and χ the temperature dependent nematic susceptibility [7] whose value at $T_{\mathrm{N-A}}$ depends on the width of the nematic range. Minimization of ΔG with respect to δQ yields:

$$\Delta G = a|\psi|^2 + b|\psi|^4 + c|\psi|^6 \tag{37}$$

with $b = b_0 - C^2\chi/2$. For narrow nematic ranges $\chi(T_{\mathrm{N-A}})$ is large and $b < 0$, resulting in a first-order transition ($c > 0$ for stability reasons). For wide nematic ranges $\chi(T_{\mathrm{N-A}})$ is small and $b > 0$, the transition is then second-order. In this mean-field approach the C_p behavior would then be given in Fig. 1(c). A tricritical point occurs for $b = 0$. The specific heat capacity in the low temperature side then diverges with a critical exponent $\alpha = 1/2$. In Fig. 1(c) one then has $C_p^{\max} = \infty$.

The crossover from first- to second-order behavior and the disappearance of measurable latent heats with increased widths of the nematic ranges has been shown on the basis of adiabatic scanning calorimetry of the 4′-n-alkyl-4-cyano-biphenyl (nCB) binary mixtures 9B + 10CB [63] and 8CB + 10CB [64]. The investigation of 8CB + 10CB mixtures with nematic ranges from 7 K to zero (see the phase diagram in Fig. 13) resulted in a partly first- and partly second-order SmA line with a tricritical point at $X_{10\mathrm{CB}} = 0.314$. In Fig. 14 the enthalpy curves near $T_{\mathrm{N-A}}$ are given for several mixtures of 8CB + 10CB with first-order transitions.

Along the second-order part of the transition line one would expect 3D-XY critical behavior at least in the limit of the critical point. In real experiments, one observes crossover between XY critical and tricritical behavior, resulting in α_{eff} values where $\alpha_{\mathrm{XY}} < \alpha_{\mathrm{eff}} < \alpha_{\mathrm{TC}}$ because the temperature range for most experimental data is limited to $10^{-5} < \varepsilon < 10^{-2}$. In addition to that the coupling between director fluctuations and the smectic order parameter intervenes as a second source of deviation from isotropic behavior and influences the behavior of the smectic susceptibility and the correlation lengths (parallel and perpendicular to the director) much more than the specific heat. Here also a broad crossover should be ob-

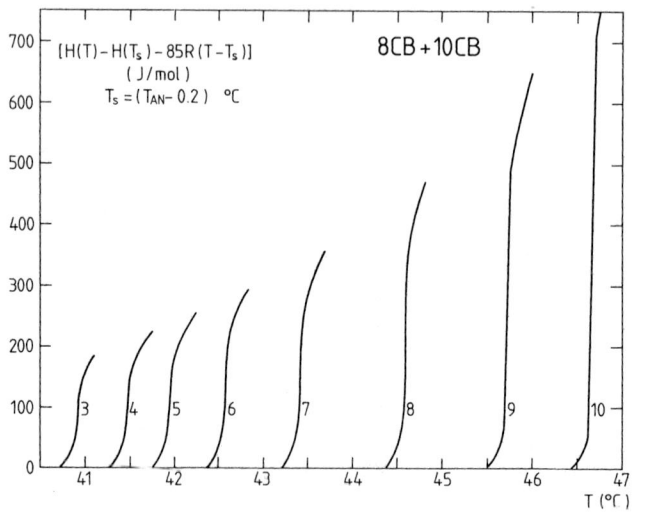

Figure 14. Enthalpy versus temperature near T_{N-A} for the tricritical mixture (no. 3) and several other 8CB+10CB mixtures with a first-order N–A-transition. A regular part 85 R $(T-T_s)$ has been subtracted for display reasons [64].

served from isotropic XY to a weakly anisotropic regime to the strong coupling limit with highly anisotropic correlation behavior. Narrow nematic ranges would result in highly anisotropic behavior and (very) wide nematic ranges should show isotropic or weakly anisotropic behavior. Figure 15

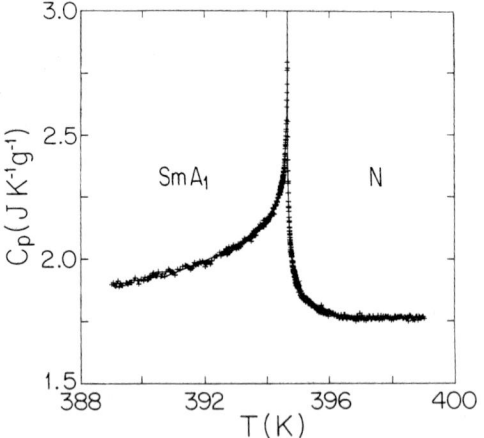

Figure 15. Heat capacity of 8OPCBOB near the nematic to smectic A_1 phase transition. The smooth curve represents a fit to the data with Eq. (6) based on critical parameters in agreement with the three-dimensional XY model. The index 1 in SmA refers to the monolayer structure of this frustrated smectic compound [65].

shows a typical example of 3D-XY specific heat capacity behavior for the compound 4-n-octyloxyphenyl-4-(4-cyanobenzyloxy)-benzoate (8OPCBOB) with a large nematic range [65]. The evolution from second-order to tricritical to first-order is illustrated in Fig. 16 by the behavior of mixtures of 4O.8 and 6O.8 (butyl- and hexyloxybenzylidene octylaniline) [66]. Pure 4O.8 has a nematic range of 14.72 K and an $\alpha_{\text{eff}} = 0.13$ is obtained. Pure 6O.8 has a nematic range of only 0.87 K and the transition is strongly first-order with a latent heat $\Delta H_L = 3700$ J mol^{-1}. Fits to the data in Fig. 16 yield the following α_{eff} values: 0.13, 0.22, 0.30, 0.45, 0.50 for X_1 to X_5 respectively. The mixtures with X_6 and X_7 are weakly first-order.

More extensive general discussions of the N–SmA problem can be found elsewhere [3, 5, 7, 9, 67] and detailed theoretical treatments are also available [60–62, 68].

The SmA–SmC Transition

The SmC phase differs from the SmA phase by a tilt angle θ of the director with respect to the layer normal. However, in order to fully specify the director one also needs the az-

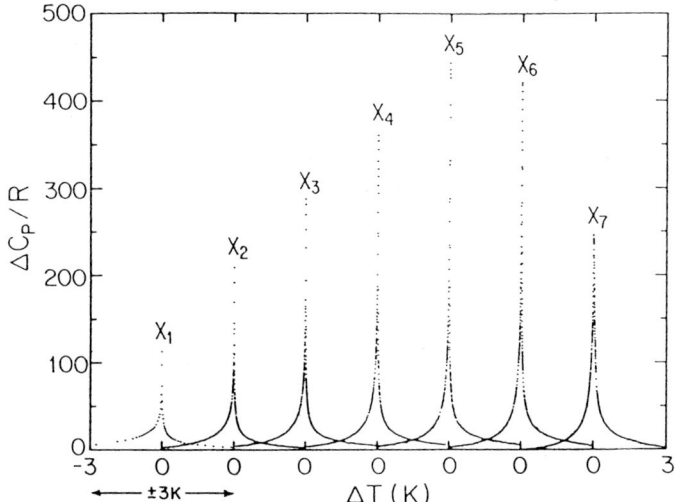

Figure 16. Reduced excess molar heat capacity $\Delta C_p/R$ for the N–A transition in mixtures of 4O.8 + 6O.8. In each case data are shown over the range $-3\,\text{K} < T - T_c < 3\,\text{K}$. The values of the mole fraction X of 6O.8 are $X_1 = 0$, $X_2 = 0.10$, $X_3 = 0.20$, $X_4 = 0.35$, $X_5 = 0.35$, $X_6 = 0.40$, and $X_7 = 0.50$ [66].

imuthal angle φ. This results in an order parameter $\varphi = \theta\,e^{i\varphi}$ with two components, and one would expect that the second-order AC transition should belong to the XY universality class. However, AC transitions in nonpolar nonchiral materials are second-order transitions that are very well described by the Landau theory with a large sixth-order coefficient c in Eq. (2). An example of this kind of behavior in N-(4-n-heptyloxy-benzylidene)-4'-n-heptylaniline (7O.7) is shown in Fig. 17 [69]. The explanation for this mean-field behavior is the fact that the Ginsburg [12] criterion indicates that the critical region is extremely small ($\varepsilon_{\text{crit}} < 10^{-5}$). It should, however, also be noted that the magnitude and sharpness of the Landau heat capacity peak for the SmA–SmC transition varies greatly from one material to another. In contrast to the monolayer SmA–SmC transitions (as e.g. in 7O.7) bilayer SmA_2–SmC_2 transitions show a steplike C_p Landau behavior with $c \approx 0$ [70].

If the constituent molecules are optically active, the chiral SmC* phase will be observed instead of the normal SmC phase. There will be a helical precession of the di-

rection of the director tilt with respect to the layer normal. Chiral compounds can exhibit strongly first-order SmA–SmC* transitions. By varying the composition of a bi-

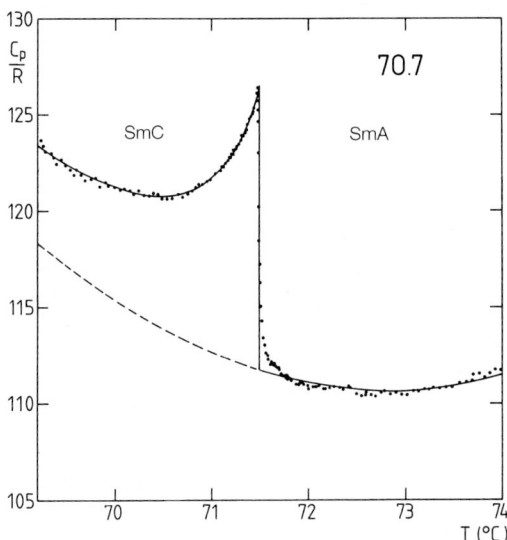

Figure 17. The temperature dependence of the reduced heat capacity per mole for 7O.7 near the SmA–SmC phase transition. The solid line represents a fit with Eq. (4) combined with the background behavior given by the dashed line. The curvature of the background is caused by the nearby SmC–B transition [69].

nary mixture the first-order can crossover via a tricritical point to a second-order transition [71, 72].

The N–SmA–SmC Multicritical Point

In some mixtures of liquid crystals exhibiting N, SmA and SmC phases, the N–SmA, SmC–SmC, and N–SmC transition lines meet at the N–SmA–SmC multicritical point. Fig. 18 gives the phase diagram with an N–SmA–SmC point for mixtures of $\overline{5}O.\overline{8} + \overline{6}O.\overline{8}$ (4-n-alkyloxyphenyl-4′-n-alkyloxybenzoate) compounds [73]. The N–SmA–SmC point has been the subject of extensive theoretical and experimental studies during the past decade. The nature of the point is, however, still not clearly established.

Since the first discovery [74, 75] of multicritical points in 1977, several systems have been investigated experimentally, also calorimetrically [5]. The calorimetric data show essentially a similar behavior near the multicritical point for all materials investigated in detail. The N–SmC transitions are first-order with a latent heat becoming zero at or very near the multicritical point (see e.g. Fig. 19). SmA–SmC/N–SmA transitions are found to be second-order. The heat capacity anomalies along the SmA–SmC line as well as along the N–SmC line become larger and sharper on approaching the multicritical point, thus suggesting a mean-field tricritical character of the point. On the other hand, high resolution X-ray scattering results [76] are in general agreement with the point being a Lifshitz point [77]. The multicritical point seems to simultaneously exhibit the characteristics of a Lifshitz point and of a tricritical point.

When chiral compounds are involved the phase diagram near the intersection of the N*–SmA, SmA–SmC and N*–(SmC*) transition lines is significantly more complicated. Several kinds of TGB phases have

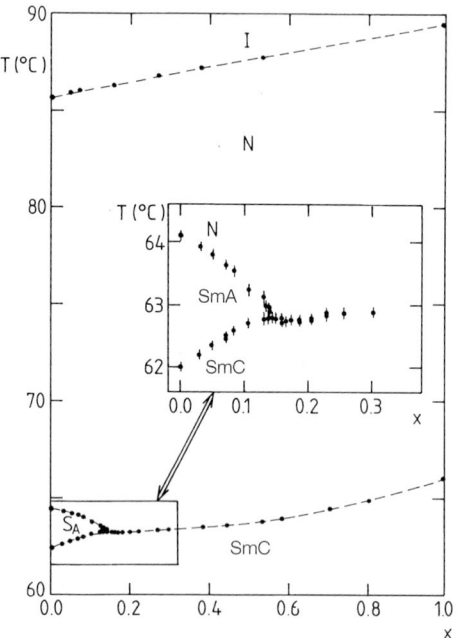

Figure 18. Phase diagram for the binary liquid crystal system $\overline{5}O.\overline{8}$ (left)–$\overline{6}O.\overline{8}$ (right), with X the weight fraction of $\overline{6}O.\overline{8}$ in the mixture [73].

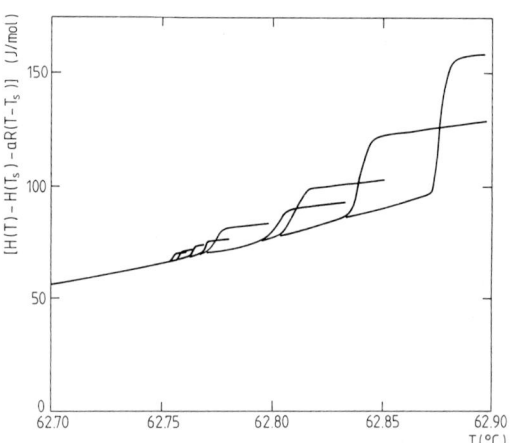

Figure 19. Temperature dependence of the enthalpy for several mixtures of $\overline{5}O.\overline{8} + \overline{6}O.\overline{8}$ with direct N–SmC transitions. For clarity reasons a large linear background $C_p^b = aR$ has been subtracted from the direct experimental data. The curve with the smallest steplike increase is closest to the N–SmA–SmC point, and has a latent heat of (1.7 ± 0.3) J/mol [73].

been theoretically predicted [78, 79] and experimentally observed [35, 80].

6.2.1.5 Photoacoustic and Photopyroelectric Results

As pointed out above the photoacoustic and photopyroelectric techniques permit the simultaneous measurement of static as well as transport thermal properties. In comparison with the vast amount of high-quality calorimetric information on the static thermal quantities specific heat capacity and enthalpy, there is rather limited information on thermal transport properties of liquid crystals in general and near phase transitions in particular. In early studies of the 1970s, and later on conventional steady-state gradient and transient techniques have been used to detect the expected anisotropy (in oriented samples) of the thermal diffusivity or the thermal conductivity [81, 82]. However, the techniques usually applied were not very well suited for measurements very close to phase transitions, because large samples and sizable temperature gradients are usually needed. Although in the photoacoustic and photothermal method, one works with very small samples and small temperature gradients, only rather recently these methods have been applied for liquid crystal studies. In fact, high-resolution investigations of thermal transport near liquid crystal phase transitions were pioneered about ten years ago by Huang and coworkers [24] by a high-frequency extension of the standard a.c. calorimetric technique (see Section for a.c. calorimetry). Instead of using Eq. (10) for ΔT_{ac} in a ω regime such that $\omega \tau_1 \gg 1$ and $\omega \tau_2 \ll 1$, one measures ΔT_{ac} as a function of ω in the region $\omega \tau_1 \gg 1$ including the high frequency cutoff $\omega_H \tau_2 = 1$. From this quantity the sample thermal diffusivity can be deduced. By also simultaneously measuring the heat capacity in the normal low ω, a.c. calorimetric regime, one can derive also the thermal conductivity κ. A number of phase transitions have been studied in this way by Huang and coworkers. An overview and discussion of the major results can be found elsewhere [5].

The first high-resolution photoacoustic investigations of phase transitions have been carried out by Zammit et al. [41] for a series of samples with N–SmA transitions. These investigations were carried out on samples of 9CB, 8CB and a mixture of 8CB + 7CB with a mole fraction $X = 0.76$ of 8CB. No effort was made to prepare homogeneously aligned samples. It is, however, likely that at least a substantial part of the sample near the free surface is homeotropically aligned, because these cyanobiphenyls strongly prefer homeotropic alignment at free surfaces [40]. The results showed a critical anomaly both in the specific heat capacity C_p and the thermal conductivity κ. The results for 9CB were subsequently also confirmed by measurements (also for nonaligned samples) in a photopyroelectric setup [46]. The values of the heat capacity critical exponents from fits to power laws of the type of Eq. (6), are fully consistent with previous results obtained by Thoen et al. [19, 63] by means of adiabatic scanning calorimetry (see above). For the critical exponent values which have been obtained from power law fits to the thermal conductivity, no agreement with theory or a systematic trend could be obtained [41]. However, in our photoacoustic measurements on homeotropically aligned 8CB samples [83] we did not see any evidence for a critical increase of the thermal conductivity at the N–SmA transition.

Very recently M. Marinelli et al. [84] carefully remeasured photopyroelectrically properly homeotropically as well as planar aligned 8CB samples. In Fig. 20 the specific heat capacity for both configurations is

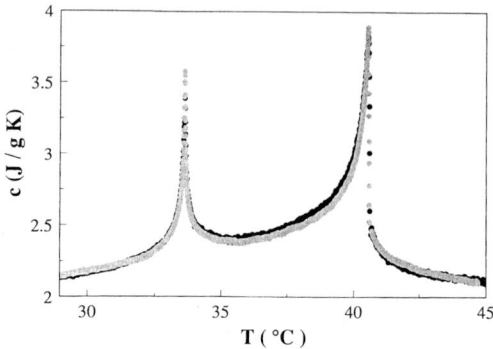

Figure 20. The specific heat capacity of 8CB as a function of temperature for planar (gray dots) and homeotropic (black dots) aligned samples [84].

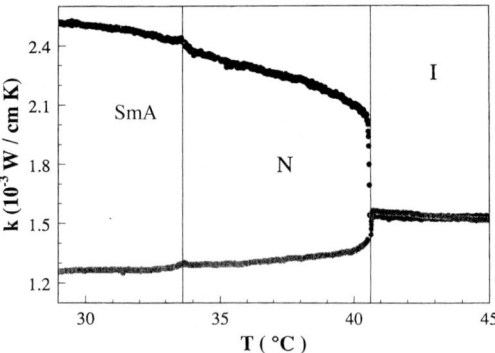

Figure 21. The thermal conductivity of 8CB as a function of temperature for planar (gray dots, lower curve) and homeotropic (black dots, upper curve) aligned samples [84].

given. Corresponding data for the thermal conductivity are given in Fig. 21. Contrary to their previous results [41, 46] a critical anomaly in κ is virtually absent. Whether this difference has to do with the fact that in the new results aligned samples have been used is not entirely clear, because also recently no evidence for a thermal conductivity anomaly at the N–SmA transition was found for nonaligned samples of octylcyanothiolbenzoate by the same group [85]. A possible explanation of the unusual κ behavior of the older results could be the presence of thermal gradients [84, 85].

6.2.1.6 Acknowledgements

The author wishes to thank C. W. Garland for many helpful discussions. He also thanks him, M. Marinelli and U. Zammit for the permission to reproduce figures from their publications.

6.2.1.7 References

[1] C. W. Garland in *Geometry and Thermodynamics* (Ed.: J.-C. Tolédano), NATO ASI Ser. B 229, Plenum, New York **1990**, pp. 221–254.
[2] M. A. Anisimov, *Critical Phenomena in Liquids and Liquid Crystals*, Gordon and Breach, Philadelphia **1990**, Chap. 10.
[3] J. Thoen in *Phase Transitions in Liquid Crystals* (Eds.: S. Martellucci, A. N. Chester), NATO ASI Ser. B 290, Plenum, New York **1992**, Chap. 10.
[4] C. W. Garland in *Phase Transitions in Liquid Crystals* (Eds.: S. Martellucci, A. N. Chester), NATO ASI Ser. B 290, Plenum, New York **1992**, Chap. 11.
[5] J. Thoen, *Int. J. Mod. Phys. B* **1995**, *9*, 2157–2218.
[6] C. W. Garland in *Liquid Crystals: Physical Properties and Phase Transitions* (Ed.: S. Kumar), Oxford University Press **1997**, Chap. 10 (in preparation).
[7] P. G. de Gennes and J. Prost, *The Physics of Liquid Crystals*, 2nd edn, Clarendon Press, Oxford **1993**.
[8] G. Vertogen, W. de Jeu, *Thermotropic Liquid Crystals, Fundamentals*, Springer-Verlag, Berlin **1988**.
[9] P. M. Charkin, T. C. Lubensky, *Principles of Condensed Matter Physics*, Cambridge University Press, **1995**.
[10] C. Bagnuls, C. Bervillier, *Phys. Rev. B* **1985**, *32*, 7209–7231.
[11] C. Bagnuls, C. Bervillier, D. I. Meiron, B. G. Nickel, *Phys. Rev. B* **1987**, *35*, 3585–3607.
[12] V. I. Ginzburg, *Sov. Phys. Solid State* **1961**, *2*, 1824–1829.
[13] M. E. Fisher, *Phys. Rev.* **1968**, *176*, 257–272.
[14] M. A. Anisimov, A. V. Voronel, T. M. Ovodova, *Sov. Phys. JETP* **1972**, *35*, 536–539.
[15] M. Sorai, S. Asahina, C. Destrade, N. H. Tinh, *Liq. Cryst.* **1990**, *7*, 163–180.
[16] M. A. Anisimov, *Mol. Cryst. Liq. Cryst.* **1988**, *162A*, 1–96.
[17] J. A. Lipa, C. Edwards, M. J. Buckingham, *Phys. Rev. Lett.* **1970**, *25*, 1086–1090; *Phys. Rev. A* **1977**, *15*, 778–789.
[18] E. Bloemen, J. Thoen, W. van Dael, *J. Chem. Phys.* **1980**, *73*, 4628–4635.
[19] J. Thoen, H. Marijnissen, W. Van Dael, *Phys. Rev. A* **1982**, *26*, 2886–2905.
[20] M. A. Anisimov, V. P. Voronov, A. O. Kulkov, F. Kholmurodov, *J. Phys. (Paris)* **1985**, *46*, 2137–2143.

[21] J. Thoen, E. Bloemen, H. Marijnissen, W. Van Dael, *Proceedings of the 8th Symposium on Thermophysical Properties*, Nat. Bur. Stand. Maryland, 1981, Am. Soc. Mech. Eng., New York **1982**, 422–428.
[22] C. A. Schantz, D. L. Johnson, *Phys. Rev. A* **1978**, *17*, 1504–1512.
[23] J. M. Viner, D. Lamey, C. C. Huang, R. Pindak, J. W. Goodby, *Phys. Rev. A* **1983**, *28*, 2433–2441.
[24] C. C. Huang, J. M. Viner, J. C. Novak, *Rev. Sci. Instrum.* **1985**, *56*, 1390–1393.
[25] R. Geer, T. Stoebe, T. Pitchford, C. C. Huang, *Rev. Sci. Instrum.* **1991**, *62*, 415–421.
[26] G. B. Kasting, K. J. Lushington, C. W. Garland, *Phys. Rev. B* **1980**, *22*, 321–331.
[27] C. W. Garland, *Thermochim. Acta* **1985**, *88*, 127–142.
[28] P. Sullivan, G. Seidel, *Phys. Rev.* **1968**, *173*, 679–685.
[29] A. J. Jin, T. Stoebe, C. C. Huang, *Phys. Rev. E* **1994**, *49*, R4791–R4791.
[30] T. Stoebe, C. C. Huang, *Phys. Rev. E* **1994**, *50*, R32–R35.
[31] D. Djurek, J. Baturic-Rubcic, K. Franulovic, *Phys. Rev. Lett.* **1974**, *33*, 1126–1129.
[32] K. Ema, T. Uematsu, A. Sugata, H. Yao, *Jpn. J. Appl. Phys.* **1993**, *32*, 1846–1850.
[33] K. Ema, H. Yao, I. Kawamura, T. Chan, C. W. Garland, *Phys. Rev. E* **1993**, *47*, 1203–1211.
[34] H. Yao, T. Chan, C. W. Garland, *Phys. Rev. E* **1995**, *51*, 4585–4597.
[35] T. Chan, PhD Thesis in Physics, Massachusetts Institute of Technology **1995**.
[36] A. Rosencwaig, *Photoacoustics and Photothermal Spectroscopy*, Wiley & Sons, New York **1980**.
[37] G. Louis, P. Peretti, G. Billard, *Mol. Cryst. Liq. Cryst.* **1985**, *122*, 261–267.
[38] M. Marinelli, U. Zammit, F. Scudieri, S. Martellucci, J. Quartieri, F. Bloisi, L. Vicari, *Nuovo Cimento D* **1987**, *9*, 557–563.
[39] C. Glorieux, E. Schoubs, J. Thoen, *Mat. Sci. and Eng. A* **1989**, *122*, 87–91.
[40] J. Thoen, C. Glorieux, E. Schoubs, W. Lauriks, *Mol. Cryst. Liq. Cryst.* **1990**, *191*, 29–36.
[41] U. Zammit, M. Marinelli, R. Pizzoferrato, F. Scudieri, S. Martellucci, *Phys. Rev. A* **1990**, *41*, 1153–1155.
[42] A. Rosencwaig, A. Gersho, *J. Appl. Phys.* **1976**, *47*, 64–69.
[43] H. J. Coufal, R. K. Grygier, D. E. Horne, J. E. Fromm, *J. Vac. Technol. A* **1987**, *5*, 2875–2889.
[44] A. Mandelis, M. M. Zver, *J. Appl. Phys.* **1985**, *57*, 4421–4429.
[45] C. Christofides, *Crit. Rev. Sol. State Mater. Sci.* **1993**, *18*, 113–174.
[46] M. Marinelli, U. Zammit, F. Mercuri, R. Pizzoferrato, *J. Appl. Phys.* **1992**, *72*, 1096–1100.

[47] S. Shandrasekhar, *Liquid Crystals*, 2nd edn, Cambridge University Press **1992**.
[48] S. T. Shin, S. Kumar, D. Finotello, S. S. Keast, M. E. Neubert, *Phys. Rev. A* **1992**, *45*, 8683–8692.
[49] T. Stoebe, C. C. Huang, *Int. J. Mod. B* **1995**, *9*, 147–181.
[50] H. Marijnissen, J. Thoen, W. Van Dael, *Mol. Cryst. Liq. Cryst.* **1983**, *97*, 149–161.
[51] P. P. Crooker, *Liq. Cryst.* **1989**, *5*, 751–775.
[52] D. J. Wright, N. D. Mermin, *Rev. Mod. Phys.* **1989**, *61*, 385–432.
[53] T. Seideman, *Rep. Prog. Phys.* **1990**, *53*, 659–705.
[54] H. Grebel, R. M. Hornreich, S. Shtrikman, *Phys. Rev. A* **1984**, *30*, 3264–3278.
[55] Z. Kutnyak, C. W. Garland, J. L. Passmore, P. J. Collings, *Phys. Rev. Lett.* **1995**, *74*, 4859–4862.
[56] J. Thoen, *Phys. Rev. A* **1988**, *37*, 1754–1759.
[57] G. Voets, W. Van Dael, *Liq. Cryst.* **1993**, *14*, 617–627.
[58] G. Voets, PhD Thesis, Katholieke Univ. Leuven, Belgium **1993**.
[59] P. G. de Gennes, *Sol. St. Commun.* **1972**, *10*, 753–756.
[60] T. C. Lubensky, *J. Chim. Phys.* **1983**, *80*, 31–43.
[61] B. R. Patton, B. S. Andereck, *Phys. Rev. Lett.* **1992**, *69*, 1556–1559.
[62] B. S. Andereck, B. R. Patton, *Phys. Rev. E* **1994**, *49*, 1393–1403.
[63] J. Thoen, H. Marijnissen, W. Van Dael, *Phys. Rev. Lett.* **1984**, *52*, 204–207.
[64] H. Marijnissen, J. Thoen, W. Van Dael, *Mol. Cryst. Liq. Cryst.* **1985**, *124*, 195–203.
[65] C. W. Garland, G. Nounesis, K. J. Stine, G. Heppke, *J. Phys. (Paris)* **1989**, *50*, 2291–2301.
[66] K. J. Stine, C. W. Garland, *Phys. Rev. A* **1989**, *39*, 3148–3156.
[67] C. W. Garland, G. Nounesis, *Phys. Rev. E* **1994**, *49*, 2964–2971.
[68] J. Prost, *Adv. Phys.* **1984**, *33*, 1–46.
[69] J. Thoen, G. Seynaeve, *Mol. Cryst. Liq. Cryst.* **1985**, *127*, 229–256.
[70] Y. H. Jeong, K. J. Stine, C. W. Garland, N. H. Tinh, *Phys. Rev. A* **1988**, *37*, 3465–3468.
[71] T. Chan, Ch. Bahr, G. Heppke, C. W. Garland, *Liq. Cryst.* **1993**, *13*, 667–675.
[72] H. Y. Liu, C. C. Huang, Ch. Bahr, G. Heppke, *Phys. Rev. Lett.* **1988**, *61*, 345–348.
[73] J. Thoen, R. Parret, *Liq. Cryst.* **1989**, *5*, 479–488.
[74] G. Sigaud, F. Hardouin, M. F. Achard, *Sol. St. Commun.* **1977**, *23*, 35–36.
[75] D. L. Johnson, D. Allender, R. DeHoff, C. Maze, E. Oppenheim, R. Reynolds, *Phys. Rev. B* **1977**, *16*, 470–475.
[76] L. J. Martinez-Miranda, A. R. Kortan, R. J. Birgeneau, *Phys. Rev. Lett.* **1986**, *56*, 2264–2267; *Phys. Rev. A* **1987**, *36*, 2372–2383.
[77] J. Chen, T. C. Lubensky, *Phys. Rev. A* **1976**, *14*, 1202–1207.

[78] S. R. Renn, T. C. Lubensky, *Phys. Rev. A* **1988**, *38*, 2132–2147; *Phys. Rev. A* **1990**, *41*, 4392–4401.
[79] S. R. Renn, *Phys. Rev. A* **1992**, *45*, 953–976.
[80] T. Chan, C. W. Garland, H. T. Nguyen, *Phys. Rev. E* **1995**, *52*, 5000–5003.
[81] R. Vilanove, E. Guyon, G. Mitescu, P. Pieranski, *J. Phys. (Paris)* **1974**, *35*, 153–162.
[82] T. Akahane, M. Kondoh, K. Hashimoto, M. Nagakawo, *Jpn. J. Appl. Phys.* **1987**, *26*, L1000–1005.
[83] J. Thoen, E. Schoubs, V. Fagard in *Physical Acoustics: Fundamentals and Applications* (Eds. O. Leroy, M. Breazeale), Plenum Press, New York **1992**, p. 179–187.
[84] M. Marinelli, F. Mercuri, S. Foglietta, U. Zammit, F. Scudieri, *Phys. Rev. E* **1996**, *54*, 1604–1609.
[85] M. Marinelli, F. Mercuri, U. Zammit, F. Scudieri, *Phys. Rev. E* **1996**, *53*, 701–705.

6.2.2 Density

Wolfgang Wedler

The publication of density and specific volume studies on mesogens started soon after the discovery of liquid crystalline phases. In 1898/99 and 1905 Schenck [1, 2] reported specific volume data for PAA and *p*-methoxycinnamic acid. Until 1958/1960, with the publication of the Maier–Saupe theory [3], specific volume and density data on mesophases were the topics of only a few publications, and were always combined with results of other measurements [4–15]. A brief historical review of this aspect of liquid crystal research has been given by Bahadur [16].

6.2.2.1 Instrumentation

Density and specific volume of mesophases are measured under atmospheric pressure with pycnometers [17–22], dilatometers [16, 23–25] (e.g., differential scanning dilatometers, DSD [26–31]), with a buoyancy method [32], and with a vibrating device, known as a digital precision density meter system, or densitometer (Anton Paar K.G.) [33–38]. Devices for high-pressure measurements of specific volumes have been described by Kuss [39, 40] and Dörrer et al. [41], but were also mentioned earlier [42]. More recently, a device for simultaneous scans of heat capacities and in-plane densities of smectic free-standing films, using reflectivity measurements, was reported by Stoebe et al. [43].

6.2.2.2 General Conclusions from Density Studies on Liquid Crystals

Specific volume and density are, as enthalpy changes, among the important parameters that indicate the order of a phase transition [44–47]. In fact, parallel development of enthalpy and relative volume changes, although not exactly proportional, is frequently observed [48–51]. The volume effects at mesophase–mesophase or mesophase–isotropic transitions are considerably smaller than those at melting into the mesophases, which have rarely been reported. For example, for the transition Cr/N in PAA, Bahadur [45] found fractional volume changes of 7.74% and a smaller thermal expansion coefficient in the crystalline phase (4.11×10^{-4} K^{-1}) than in the nematic phase (9.26×10^{-4} K^{-1}). Similar results have been obtained for chiral substances [52–56]. Despite the small magnitude of the observed discontinuities, first-order (e.g., Is/A [57], N/A [46, 47, 58, 301], A/B [57], and B/G [47]) weakly first-order (e.g., N/A [57, 59],

A/B [60], and B/F [61]), and second-order (e.g., N/A [62–65], A/C [47, 66–68], I/F [69, 70]) phase transitions, as well as singularities in the critical behavior [71], have been inferred. Lists of selected values of density and specific volume changes at different phase transitions are given in Tables 1 and 2. Collections of data and reviews have been given by Bahadur [16], Beguin et al. [72], Pisipati et al. [73], and Tsykalo [74].

From the temperature dependence of volume and density, the thermal expansion coefficient α and the thermal expansivity β [34] can easily be obtained. These quantities make phase transitions more apparent than does the density [33, 48]. Figures 1 and 2 demonstrate this with two examples. Densities, molar volumes, and thermal expansion coefficients show systematic behavior in homologous series. It has been shown that the magnitude of density changes at N/Is transitions [51, 57, 75–80, 301] and N/A transitions [75, 77–79, 301], and the absolute values of densities [79–81], are subject to even–odd effects with respect to the length of longitudinally attached alkyl

Table 1. Fractional volume changes (in %, upper right from diagonal, maximum and minimum reported values), and related information (lower left from diagonal) for phase transitions in mesophases.

Phase I / Phase II	Is	N	A	C
Is	–	0.51 [301] (7O.7) 0.0711 [112] (NPOOB)	2.00 [117] Diethyl (4,4'-azoxy-benzoate) 0.22 [73] (4O.10)	1.94 [123] (OOAB) 0.55 [69] (TBDA)
N	First-order, $\Delta V = 0.95 \pm 0.15$ cm^3 mol^{-1} (8CB) [146]	–	0.65 [301] (7O.2) 0–0.08 [37]	0.15 [299] (HOAB) 0.01–0.023 [37]
A	First-order (Two-phase coexistence [117])	First-order [138] (7O.4) Weakly first-order (CBOOA) [44] Second-order (CBOOA) [31] $\Delta V = 0.14 \pm 0.04$ cm^3 mol^{-1} (8CB) [146]	–	0.146 [48] (DOBACA) 0–0.005 [37]
C	First-order	First-order [299]	Second-order	–
B	No data available	First-order	First-order Weakly first-order [60]	First-order (Two phase coexistence [116])
I	No data available	No data available	No data available	First-order [69, 70]
F	First-order	No data available	First-order	First-order
G	No data available	First-order	No data available	First-order
H	No data available	No data available	No data available	First-order $\Delta V/V = 5.37\%$ (DOBAMBC) [162]

Table 1. (continued)

Phase II \ Phase I	B		I		F		G	
Is	No data available		No data available		2.14 (10O.14)	[302]	No data available	
N	1.50 (HBT) 1.36 (PMMA)	[223] [285]	No data available		No data available		1.16 (6O.2) 0.38 (5O.2)	[139] [37]
A	1.59 (7O.2) 0.05 (7O.1)	[73] [60]	No data available		1.44 (9O.4) 1.42 (9O.4)	[308] [283]	No data available	
C	1.23 (OOAB) 0.41 (4O.7)	[123] [79]	1.00 (TBAA9) 0.29 (TBAA12)	[309] [70]	1.07 (7O.6) 0.39 (5O.5)	[73] [65]	>0.7 0.14 (7O.4)	[73] [101]
B	–		No data available		0.03 (5O.6) 0.02 (5O.6)	[65] [61]	0.06 (8O.4)	[115]
I	No data available		–		0.00	[69,70]	No data available	
F	Weakly first-order Second-order		Second-order [69, 70]		–		0.04 (10O.14) 0.03 (5O.6)	[302] [61]
G	First-order		No data available		First-order (5O.6) [120] Weakly first-order (10O.14) [302] Second-order (TBOA) [118]			
H	First-order, measured, no data reported [46]		No data available		No data available		No data available	

chains in homologous series. The discontinuities in the absolute volume, like those in the enthalpy, increase with the length of the alkyl chains [50, 51, 82, 301], as can be seen for the two homologous series of 4O.m and 5O.m compounds in Table 3. This also was found for the Cr/N* transition [83]. The discontinuities are directly proportional to the molar mass, as can be seen when different substances are compared [48]. Absolute densities tend to decrease, and specific and molar volumes to increase, with chain length in a series of crystalline, smectic, and nematic phases [51, 84–88]; see the example in Fig. 3. A predicted inverse relation between molecular flexibility (due mainly to aliphatic chains) and the magnitude of the fractional volume change at the N/I transition [89] was verified by experiments in homologous series [23, 90–93] for longitudinal and lateral chain attachment. Furthermore, an increase in the thermal expansion coefficient with alkyl chain length has been found for N, A, and B phases [50, 94].

Table 2. Fractional volume changes (in %, upper right) and related information (lower left) for phase transitions in mesophases of chiral compounds.

Phase II \ Phase I	Is	Blue phase	N*	A*
Is	–	No data reported	0.17 [56] (cholesteryl stearate) 0.10 [27] (cholesteryl nonanoate)	No data reported
Blue phase	$\Delta V = 7.5 \times 10^{-4}$ cm^3 g^{-1} (cholesteryl oleate) [28]	–	0.004 [27] (cholesteryl nonanoate)	No data reported
N*		First-order [28] $\Delta V = 4 \times 10^{-5}$ cm^3 g^{-1} (cholesteryl nonanoate) [28]	–	0.14 [53] (cholesteryl myristate) 0.00 (cholesteryl linolenate) [107]
A*			$\Delta V = 154 \times 10^{-5}$ cm^3 g^{-1} (cholesteryl myristate) [28]	–

Based on density measurements, molar volume increments for methylene chain units, separated from the contribution of the aromatic part [95, 96], have been determined. The literature mentions mostly increments for longitudinal chain position [50, 51, 57, 60, 75, 79, 85, 86, 90, 94–97]. They agree well with those found for lateral chain position [91] and columnar mesophases [87]; see Table 4. On the basis of

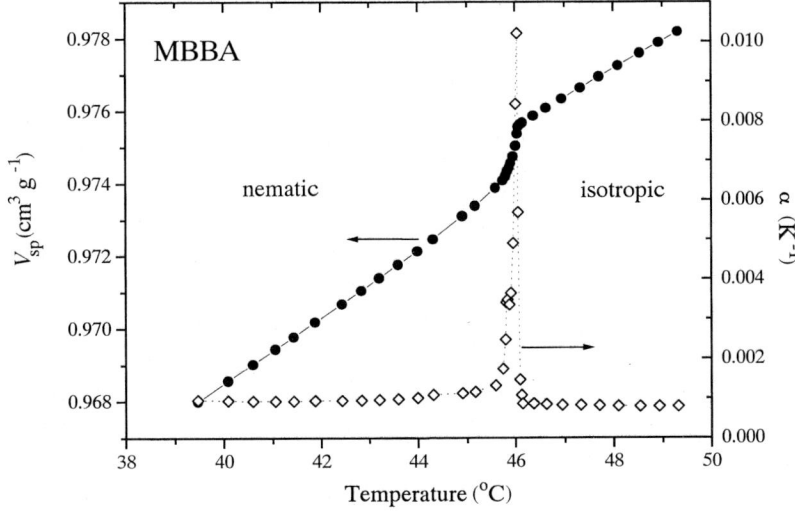

Figure 1. Temperature dependence of specific volume V_{sp}, and thermal expansion coefficient α for MBBA. Adapted from data set 2 of Gulari and Chu [33].

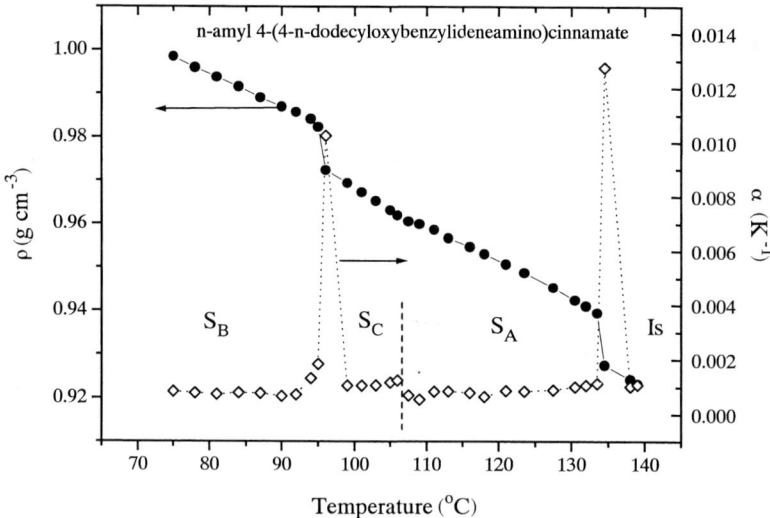

Figure 2. Temperature dependence of density, and thermal expansion coefficient for n-amyl 4-(4-n-dodecyloxybenzylidene-amino)cinnamate. Adapted from Demus and Rurainski [48].

these increments, conclusions have been drawn about the packing of the alkyl groups, locating the packing density in the nematic phase closely to that in the isotropic phase, with slightly higher values in the nematic phase [60]. An investigation of the influence of aromatic ring number on the packing density in homologous series can be found elsewhere [51, 91].

Table 3. Fractional volume and density changes (in %) in the homologous series nO.m: Is/N and N/A phase transitions.

$C_n H_{2n+1} O - \bigcirc - CH=N - \bigcirc - C_m H_{2m+1}$

m	$n=4$		$n=5$	
	$\frac{\Delta V}{V}$ at Is/N	$\frac{\Delta \rho}{\rho}$ at N/A	$\frac{\Delta \rho}{\rho}$ at Is/N	$\frac{\Delta \rho}{\rho}$ at N/A
4	0.28 [79]			
5	0.298 [79]		0.22 [88]	
6	0.22 [79]		0.34 [88]	
7	0.35 [79]		0.30 [88]	0.10 [88]
8	0.31 [79]		0.33 [88]	
9	0.39 [79]		0.25 [88]	0.08 [88]
10	0.21 [144]	0.02 [88]	0.46 [88]	
12	0.39 [88]	0.02 [88]	0.32 [88]	0.06 [88]

6.2.2.3 Studies of Calamitic Compounds

Density and Order of Phase Transitions

A systematic development of density and thermal expansion coefficients in calamitic compounds is frequently observed when neighboring mesophases are compared. At the N/Is transition, for example, the thermal expansion coefficient was found on average to be larger in the nematic than in the isotropic phase [45–47, 55, 57, 98–104]. This was also confirmed for the N*/Is transition [53, 105] (with a few exceptions [106]). For example in PAA, Price and Wendorff [55] reported 9.4×10^{-4} K^{-1} in the nematic and 8.4×10^{-4} K^{-1} in the isotropic phase. Similar data were given by Bahadur [45] and by Rao and co-workers. These observations were attributed to the rapid development of high order in the nematic phase. It was further suggested that the discontinuities at the N*/Is transition are slightly lower than those at the N/Is transition [107, 108] (compare Tables 1 and 2), and that the thermal expansion coefficient in N* phases is smaller than in N phases [109]. The temperature dependence of the thermal expansion coefficient

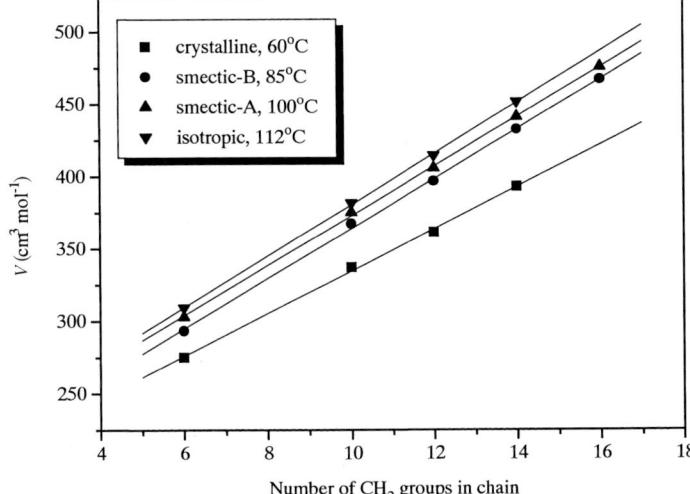

Figure 3. Molar volume development in the crystalline, smectic B, smectic A, and isotropic phases within the homologous series of 4-bromo-N-(4-n-alkyloxybenzylidene)anilines. Adapted from Seurin et al. [94].

Table 4. Selected molar volume increments for one methylene unit at different molecular positions in the isotropic and nematic phases.

	ΔV_{CH_2} (cm^3 mol^{-1})	Substance(s)	Reference
(a)	Rod-shaped molecules, longitudinal position		
	14.6–18.66 (isotropic phase)	4O.m	[79]
	17.14 ± 0.44 (isotropic phase, $T_{NI}+3$ K)	nO.1	[60]
	16.75 ± 0.35 (nematic phase, $T_{NI}-3$ K)	nO.1	[60]
(b)	Rod-shaped molecules, lateral position		
	18.7–22.3	1.4-bis[4-n-hexylbenzyloxy]-2-substituted benzenes	[91]
(c)	Columnar phases		
	16.9 ± 0.3	Copper(II) alkanoates	[87]

in the isotropic state was found to be linear [84, 100, 110], and has been treated by linear regression [19, 92]. In mesophases like the nematic phase, mostly nonlinear behavior is observed [111]. Consequently, some authors [19, 92, 112] have used polynomials to describe the temperature dependence in the nematic and isotropic state outside the pretransitional zone. Regarding other phase transitions, N phases [101, 113] and one blue phase [114] were shown to have higher thermal expansion coefficients than the A phases, or the respective N* phase, into which they transform at lower temperatures. However, contrary reports are also found [57, 66]. In a similar way, different results have been found for the A/Is transition. For example, OBBA [115] or 7O.8 and 8O.8 [116] were shown to have higher thermal expansion coefficients in the A phase, whereas the opposite was shown for diethyl 4,4'-azoxybenzoate [117]. Further, it should be mentioned that the C phase reportedly has higher thermal expansion coefficients than the adjacent A [66, 68, 79] and B phases, which was attributed to higher flexibility and rotational freedom [101]. However, there are also reports of the opposite case

[118]. Furthermore, in TBDA and TBAA12, it was found that the C phase has higher values than the adjacent I phase, whereas the I phase in turn has higher values than the low-temperature F phase [69, 70] with respect to the thermal expansion coefficient. Also, the F phase was found to have higher values than an adjacent A phase [308]. This was explained by chain stiffening and increasingly closer packing in the smectic layers during cooling. It is noteworthy here that the difficulty of density data acquisition by dilatometric/pycnometric methods due to high viscosity in the smectic phases with higher order in the layer structure is frequently mentioned [79].

Dilatometric investigations of materials with more complex polymorphisms revealed a second-order character for the F/I [69], C/A [68], C/N [37], and N*/blue phase [114] transitions; the last result was confirmed by dilatometric and ultrasonic experiments [119]. Uncertainties about the G/F transition [61, 73, 118, 120, 129] are noted in Table 1. Further results on polymorphous systems have been reported [48, 121–124]. A measurable density change at the cybotactic nematic/ordinary nematic transition [125, 126], occurring at 59.6°C in BBBA is doubtful [46]. Also, densitometric scanning in two mixtures showed a continuous change at the A_1/A_2 transition, and a slight increase in density (less than 0.02%) at the A_d/A_1 transition [127]. Only a slight change in slope of specific volume versus temperature was observed in an experiment scanning the hexatic B/crystal B transition [128].

The order of the A/N transition was the subject of an extended theoretical discussion during the early 1970s [130–135], which has been surveyed with respect to density measurements by Rao et al. [79, 103] and Potukuchi et al. [136]. Extensive density measurements on $nO.m$ compounds [57, 58, 60, 61, 65, 73, 79, 88, 101–104, 113, 115, 136–145, 301] were employed, along with other methods, to test these theories, and to find a predicted A/N transition order change within homologous series. The A/N transition occurs over a remarkable temperature interval, and the possible existence of a tricritical point was the subject of density studies, also on other materials, for example 8CB [146], CBOOA [31, 44], NPOOB [112], nCBs [34], HOPDOP [124], TBBA [66], side-group polysiloxanes [147], H_xBPAA [148], 8OCB, 8S5 [64], and 9CB/10CB blends ([149], successfully using a method described previously [150–153]). Based on the evidence of density measurements and other data, influences of molecular architecture (e.g., the dominant influence of the alkyloxy chains in $nO.m$ compounds [88]) and polarity, formation of cybotactic groups, and mono/bilayer organization in the smectic phase on the A/N transition order were found [60, 65, 66, 73, 79, 146, 154]. Several dilatometric studies have indicated a change from first to second order of the N*/A transition in a mixture of 62 mol% cholesteryl oleyl carbonate and 38 mol% cholesteryl chloride [155, 156], and blends of cholesteryl myristate with cholesteryl propionate [119]. The opposite influence of cholesterol was also studied [157]. Dilatometric testing in cholesteryl myristate revealed an untwisting of the cholesteric helix when the volume discontinuity at the N*/A transition vanishes [36], after high-pressure studies [158].

Further specific volume measurements with chiral mesogens, cholesteric, and blue phases have been made [26–28, 159–161]. The influence of chiral dopants on C/A and C/N phase transitions has been studied [37, 162]. Earlier studies, particularly on cholesteric mesophases and their mixtures, have been published, for example, for cholesteryl acetate [163], cholesteryl myristate [53], cholesteryl nonanoate [54], cholesteryl stearate [56], and a variety of other compounds [27, 106, 164–166]. In particular, Price et al. have published Avrami exponents from kinetic studies of density changes on two compounds, yielding values of 2 for the cholesteric/smectic and isotropic/cholesteric transitions [53, 56].

Pretransition Behavior

Precise measurements have shown that the density is subject to pronounced pretransition behavior for a variety of phase transitions. This may depend upon purity, and also upon the cooling/heating rate [60]. The N/Is and N*/Is transitions exhibit such effects only on the low-temperature side [53–55], which was theoretically explained and compared with enthalpy changes by Chandrasekhar and Sashidar [167, 168]. In contrast to mesophase/mesophase transitions, on the high-temperature side post-transitional behavior was detectable only with high-sensitivity testing [28, 67]. Several workers pointed out that post-transitional effects vanish in accordance with the predictions of the Maier–Saupe theory [3] when the samples are carefully purified [51, 53, 112]. The density jump at this specific phase transition was shown to occur in a temperature interval of 0.1–0.5 K, depending upon the purity of the sample [19, 92]. In the special case of ultra-high purity materials, Dunmur and Miller [34] and Demus et al. [112] found interval widths of less than 0.1 and 0.02 K, respectively. Evidence of the influence of domain size on the discontinuity at the transition was found [44]. Pretransitional effects pose a main problem for the exact evaluation of changes in molar volume, specific volume, and density at phase transitions. A precise method was devised for dilatometric and calorimetric studies on PAA by Klement and Cohen [169]. It was further developed by Dunmur and Miller [34] for nCBs, and used in other work [57, 170]. The method uses plots of $1/\rho$ versus $|T-T_t|^{0.5}$, and linearly extrapolates to $T=T_t$ (where T_t is the phase transition temperature) for the low- and high-temperature sides. Other functional forms for the modelling of density in the nematic/isotropic pretransitional region have been given [31, 79, 171–173]. Based on the Landau–de Gennes theory [174], the expansion $\Delta V = B\Delta T + E[(\Delta T + D)^{1-\alpha} - D^{1-\alpha}]$, in terms of the critical exponent α, was applied for CBOOA [31], MBBA and PAA [29], and several cholesterol derivatives [28], and used to describe pretransition behavior for other transitions (e.g., N/C in HOAOB [30]). The influence of the order parameter was taken into account by an expansion given by Chang [172]. Bendler [175, 176], using high-precision data on MBBA [33], attempted to model expansion coefficients in isotropic phases very close to the N/Is transition in terms of order parameter fluctuations within the framework of the de Gennes theory. This was extended to the pretransitional behavior on the nematic side, and applied to nematic mixtures [177] with good coincidence between theory and experiment. Note finally that the equation given by Chang and Gyspers [178], $\alpha_N = [e + f(T_{NI}-T)^h] \times 10^{-3}$, enjoys widespread popularity for data evaluation in the nematic state.

Molecular Interaction, Order Parameter, and Density

The influence of attractive and repulsive interactions between molecules on volumetric behavior, packing density, and the resulting phase stability have been studied by several authors [19, 20, 35, 51, 91–93, 111]. In mixtures, strong attractive forces due to dipole–dipole interactions or EDA complex formation [179–183] result in the formation of induced phases [184–186]. Sadowska et al. [35] reported results of density measurements for two blends, indicating a measurable shrinking effect. EDA complex formation was investigated by Demus et al. [20], who obtained complex stability constants, complex formation enthalpies, and complex concentrations as temperature functions by dilatometry and calorimetry. An interpretation was given using the theories of Cotter and Gelbart [187, 188]. Further density investigations on blends with nematic reentrant phases [189, 190] and induced smectic phases [191], were reported, and related to anomalies in pressure–temperature diagrams [192–194]. Aleksandriiskii and coworkers [111] assessed the influence of hydrogen bonds on phase transition characteristics. Enhanced values of fractional volume changes at the N/Is transition, compared with those for other polar compounds, were found, which is consistent with observations by Bahadur [16]. Obviously, hydrogen bond formation and also polar interactions [195] favor the formation of cybotactic groups, and lead to the nonlinear behavior that is also observed. However, differing results were found in a systematic study of longitudinally polar and nonpolar substituted liquid crystal blend systems [196], using a method described by Press and Arrott [32]. The dominant role of steric repulsion in the stabilization of the nematic phase was suggested on the basis of high-pressure experiments on PAA, BMAB, and EBBA by Stishov et al. [197] and Kachinskii et al. [198], and demonstrated by density studies [19, 91–93] under atmospheric pressure on laterally branched compounds [199]. The results are in qualitative agreement with molecular statistical theories based on hard rod models [187, 200, 201] and with van der Waals theories [202]. There has been discussion of repulsive forces [203–205].

Density studies help to assess the influence of molecular interaction, cybotactic groups, and the order parameter in nematic phases, in combination with refractive index measurements and other methods for pure substances [51, 76, 206–212] and binary blends [213–216]. This has been demonstrated, using different theoretical concepts for the inner field [217–222], also for smectic phases [59, 223, 224]. There have been additional density studies of binary systems [81, 170, 225–228].

Relation between Density and Other Properties in Calamitic Compounds

Combination with calorimetric data, and use of the Clausius–Clapeyron equation or the Ehrenfest equations (e.g., for the A/C transition [67]), yields slopes in the p–T phase diagram. These can be verified by high-pressure measurements of transition temperatures (e.g. [229], compared with density data [70]) or metabolemetric scans [230] (see Chap. VII, Sec. 6.2.3). On the other hand, direct measurements of densities or volumes at high pressures were also obtained [39–42, 158, 198, 231–240], yielding isothermal compressibilities and permitting conclusions about phase transition mechanisms. For example, using temperature as a parameter, Dörrer et al. [41] combined data gained from two blends with high-pressure data of the rotational viscosity γ_1, and constructed master curves, introducing reduced density coordinates.

After initial tests with N-(4-n-nonyloxybenzylidene)-4-toluidine [241], it was demonstrated in a variety of work on azomethine compounds of the nO.m type [58, 98, 100, 103, 104, 116, 137, 140–142, 148, 242–244] and other compounds [119, 148, 160, 177] that, with the ultrasonic velocity v, temperature T, and molar weight M, parameters like adiabatic compressibility $\beta_{ad} = \frac{V_{sp}}{v^2}$, Rao number (molar sound velocity) [245]

$R_n = M V_{sp} v^{1/3}$, and Wada constant (molar compressibility) [246] $B = \dfrac{M V_{sp}}{\beta_{ad}^{1/7}}$ can be calculated from the specific volume V_{sp} and compared with theoretical concepts [3, 132, 247, 248].

6.2.2.4 Discotics

Few density measurements on discotic phases have been reported. Data have been given for the Cr/D_{ho} and D_{ho}/Is transitions [249] and the D_{hd}/Is [18], D_{rd}/Is [21], and Cr/columnar [87] transitions. The last reference mentions data in older publications [250, 251]. Similar fractional volume changes and pretransitional effects, but higher transition enthalpies and entropies, compared with the nematic/isotropic transition, were found [18] – in some disagreement with theory [252, 253]. The transition from the high-viscosity D_{rd} phase into the isotropic phase was reported to show an increase in density [21]. The density within homologous series decreases with increasing alkyl chain length [87]. Further data are available [254].

6.2.2.5 Lyotropics and Anhydrous Soaps

Density behavior at phase transitions in lyotropic or micellar mesophases [110, 255–257] are found to be less pronounced than in thermotropic mesophases (for example, there is no discontinuity at the N/Is transition in $DACl/NH_4Cl/H_2O$ [257], and there is a 0.012% volume change at the N/Is transition in $CsPFO/H_2O$ [110]). Several publications [38, 110, 258–260] have discussed the system cesium pentadecafluorooctanoate (CsPFO)/water, where decreases in volume were found at the lamellar/nematic and nematic/isotropic transitions [258, 259]. This was interpreted in terms of significant changes in the micellar structure during the transitions. Several explanations for this observation were attempted [38, 260]. Effects of additional electrolytes on the density in these systems have been discussed [110]. The older literature also refers to density studies in anhydrous surfactants [261–270], for example, in alkali metal stearates, laurates, and palmitates. Finally, we mention reports of density measurements on mesophasic systems involving liposomes and membranes [271, 272].

6.2.2.6 Polymeric Liquid Crystals

Density measurements in liquid crystalline polymeric systems have rarely been reported. Density scans of A/N transitions in liquid crystalline side-chain polysiloxanes [147] and of N/Is transitions in side-chain polysiloxanes and polymethacrylates [273] have been published. Increased densities and lower expansion coefficients compared with corresponding low molecular weight model systems were found, and a van der Waals analysis has been given [273]. One main-chain polybenzoate was tested in solution [274].

6.2.2.7 Further Studies

Density and its changes at phase transitions have been the subject of molecular theories [275–280]. It is also an important parameter in molecular dynamics (MD) studies. Two examples are an MD study of the density dependence of diffusion coefficients in nematic phases [281] and simulations of the density dependence of orientational order in nematics using a Gay–Berne potential [282]. Further explanations and references are given in Chap. III.

Investigations of the density and expansion coefficients in mesogens, in connection with data from other experiments, and especially in the transition regions have been described [61, 82, 108, 129, 234, 235, 283–309].

6.2.2.8 References

[1] F. R. Schenck, (a) *Z. Phys. Chem.* **1898**, *25*, 337–352; (b) *Z. Phys. Chem.* **1898**, *27*, 167–171; (c) *Z. Phys. Chem.* **1899**, *28*, 280–288; Royal Society of London, Catalogue of Scientific Papers **1923**, *18*, 503.
[2] F. R. Schenck, *Kristallinische Flüssigkeiten und Flüssige Kristalle*, Engelmann, Leipzig **1905**.
[3] W. Maier, A. Saupe, (a) *Z. Naturforsch.* **1958**, *13a*, 564–566; (b) *Z. Naturforsch.* **1959**, *14a*, 882–889; (c) *Z. Naturforsch.* **1960**, *15a*, 287–292.
[4] E. Eichwald, Untersuchungen über flüssige Kristalle, Dissertation, Marburg **1905**.
[5] F. Dickenschied, Untersuchungen über Dichte, Reibung, Kapillarität kristalliner Flüssigkeiten, Dissertation, Halle (Saale) **1908**.
[6] F. Conrat, *Phys. Z.* **1909**, *10*, 202 [*Chem. Abstr.* **1909**, *3*, 1609].
[7] E. Schäfer, Dichte, Reibung und Kapillarität kristallinischer Flüssigkeiten, Dissertation, Halle (Saale) **1911**.
[8] F. M. Jaeger, *Z. Anorg. Chem.* **1917**, *101*, 152 [*Chem. Abstr.* **1918**, *12*, 875].
[9] D. Vorländer, *Phys. Z.* **1930**, *31*, 428–435 [*Chem. Abstr.* **1930**, *24*, 4433].
[10] E. Bauer, J. Bernamont, *J. Phys. Radium* **1936**, *7*, 19–22 [*Chem. Abstr.* **1936**, *30*, 5088[1]].
[11] G. Becherer, W. Kast, *Ann. Physik (5. Folge)* **1942**, *41*, 355–374 [*Chem. Abstr.* **1942**, *36*, 6390[5]].
[12] F. Linsert, Thesis, Halle (Saale) **1945**.
[13] I. Gabrielli, L. Verdini, *Nuovo Cim. (Sr. 10)* **1955**, *2*, 526–541 [*Chem. Abstr.* **1956**, *50*, 6122b].
[14] W. A. Hoyer, A. W. Nolle, *J. Chem. Phys.* **1956**, *24*, 803–811.
[15] W. R. Runyan, A. W. Nolle, *J. Chem. Phys.* **1957**, *27*, 1081–1087.
[16] B. Bahadur, *J. Chim. Phys., Phys.-Chim. Biol.* **1976**, *73*, 255–267 [*Chem. Abstr.* **1976**, *85*, 54670s].
[17] H. Sackmann, F. Sauerwald, *Z. Phys. Chem.* **1950**, *195*, 295–312 [*Chem. Abstr.* **1951**, *45*, 8314d].
[18] T. H. Smith, G. R. van Hecke, Proceedings of the 8th International Liquid Crystal Conference, Kyoto, Part D, *Mol. Cryst. Liq. Cryst.* **1981**, *68*, 23–28.
[19] A. Hauser, R. Rettig, C. Selbmann, W. Weissflog, J. Wulf, D. Demus, *Cryst. Res. Technol.* **1984**, *19*, 261–270.
[20] D. Demus, A. Hauser, G. Pelzl, U. Böttger, S. Schönburg, *Cryst. Res. Technol.* **1985**, *20*, 381–390.
[21] K. A. Lawler, G. R. Van Hecke, *Liq. Cryst.* **1991**, *10*, 341–346.
[22] ASTM Standard D 1481-93, Standard Test Method for Density and Relative Density (Specific Gravity) of Viscous Materials by Lipkin Bicapillary Pycnometer, *Annual Book of ASTM Standards* **1993**, Vol. 05.01, pp. 521–525.
[23] D. Guillon, A. Skoulios, *Mol. Cryst. Liq. Cryst.* **1977**, *39*, 139–157.
[24] D. Grasso, *J. Calorim., Anal. Therm. Thermodyn. Chim.* **1986**, *17*, 318–320 [*Chem. Abstr.* **1987**, *107*, 49926y].
[25] D. Grasso, *Liq. Cryst.* **1987**, *2*, 557–560.
[26] D. Armitage, F. P. Price, *Bull. Am. Phys. Soc.* **1975**, *20*, 886.
[27] D. Armitage, F. P. Price, *J. Appl. Phys.* **1976**, *47*, 2735–2739.
[28] D. Armitage, F. P. Price, *J. Chem. Phys.* **1977**, *66*, 3414–3417.
[29] D. Armitage, F. P. Price, *Phys. Rev. A* **1977**, *15*, 2496–2500.
[30] D. Armitage, F. P. Price, *Phys. Rev. A* **1977**, *15*, 2069–2071.
[31] D. Armitage, F. P. Price, *Mol. Cryst. Liq. Cryst.* **1977**, *38*, 229–237.
[32] M. J. Press, A. S. Arrott, *Phys. Rev. A* **1973**, *8*, 1459–1465.
[33] E. Gulari, B. Chu, *J. Chem. Phys.* **1975**, *62*, 795–797.
[34] D. A. Dunmur, W. H. Miller, *J. Phys. (Paris) Colloque C3* **1979**, *40*, 141–146.
[35] K. W. Sadowska, A. Zywocinski, J. Stecki, R. Dabrowski, *J. Phys.* **1982**, *43*, 1673–1678.
[36] P. Pollmann, K. Schulte, *Ber. Bunsenges. Phys. Chem.* **1985**, *89*, 780–786.
[37] R. Kiefer, G. Baur, *Liq. Cryst.* **1990**, *7*, 815–837.
[38] N. Boden, K. W. Jolley, *Phys. Rev. A* **1992**, *45*, 8751–8758.
[39] E. Kuss, *Mol. Cryst. Liq. Cryst.* **1978**, *47*, 71–83.
[40] E. Kuss, *Mol. Cryst. Liq. Cryst.* **1981**, *76*, 199–210.
[41] H. L. Dörrer, H. Kneppe, F. Schneider, *Liq. Cryst.* **1992**, *11*, 905–915.
[42] Y. B. Kim, K. Ogino, *Phys. Lett. A* **1977**, *61*, 40–42.
[43] T. Stoebe, C. C. Huang, J. W. Goodby, *Phys. Rev. Lett.* **1992**, *68*, 2944–2947.
[44] S. Torza, P. E. Cladis, *Phys. Rev. Lett.* **1974**, *32*, 1406–1409.
[45] B. Bahadur, *Mol. Cryst. Liq. Cryst.* **1976**, *35*, 83–89.
[46] J. V. Rao, N. V. S. Rao, V. G. K. M. Pisipati, C. R. K. Murty, *Ber. Bunsenges. Phys. Chem.* **1980**, *84*, 1157–1160.
[47] J. V. Rao, K. R. K. Rao, L. V. Choudary, P. Venkatacharyulu, *Cryst. Res. Technol.* **1986**, *21*, 1245–1249.
[48] D. Demus, R. Rurainski, *Z. Phys. Chem. (Leipzig)* **1973**, *253*, 53–67.
[49] D. Demus, S. Diele, S. Grande, H. Sackmann in *Advances in Liquid Crystals* Vol. 6 (Ed.: G. H. Brown), Academic Press, New York **1983**, pp. 1–107.

[50] G. Poeti, E. Fanelli, G. Torquati, D. Guillon, *Nuovo Cim.* D **1983**, *2*, 1335–1346.
[51] A. Hauser, C. Selbmann, R. Rettig, D. Demus, *Cryst. Res. Technol.* **1986**, *21*, 685–695.
[52] E. McLaughlin, M. A. Shakespeare, A. R. Ubbelohde, *Trans. Faraday Soc.* **1964**, *60*, 25–32.
[53] F. P. Price, J. H. Wendorff, *J. Phys. Chem.* **1971**, *75*, 2839–2849.
[54] F. P. Price, J. H. Wendorff, *J. Phys. Chem.* **1972**, *76*, 276–280.
[55] F. P. Price, J. H. Wendorff, *J. Phys. Chem.* **1972**, *76*, 2605–2608.
[56] F. P. Price, J. H. Wendorff, *J. Phys. Chem.* **1973**, *77*, 2342–2346.
[57] V. G. K. M. Pisipati, N. V. S. Rao, M. V. V. N. Reddy, C. G. Rama Rao, G. Padmavathi, *Cryst. Res. Technol.* **1991**, *26*, 709–716.
[58] V. G. K. M. Pisipati, N. V. S. Rao, M. K. Rao, D. M. Potukuchi, P. R. Alapati, Proceedings of the 11th International Liquid Crystal Conference, Berkeley, CA 1986, Part C, *Mol. Cryst. Liq. Cryst.* **1987**, *146*, 89–96.
[59] R. K. Sarna, V. G. Bhide, B. Bahadur, *Mol. Cryst. Liq. Cryst.* **1982**, *88*, 65–79.
[60] N. V. S. Rao, P. V. Datta Prasad, V. G. K. M. Pisipati, *Mol. Cryst. Liq. Cryst.* **1985**, *126*, 175–186.
[61] M. Takahashi, S. Kondo, *Tokyo Kogyo Koto Senmon Gakko Kenkyu Hokokusho (Research Report of Tokyo Technical College)* **1987**, *19*, 19–23 [*Chem. Abstr.* **1988**, *109*, 220081m].
[62] L. Longa, *J. Chem. Phys.* **1986**, *85*, 2974–2985.
[63] L. Longa, *Z. Phys.* B **1986**, *64*, 357–361.
[64] A. Zywocinski, S. A. Wieczorek, J. Stecki, *Phys. Rev.* A **1987**, *36*, 1901–1907.
[65] P. R. Alapati, D. M. Potukuchi, N. V. S. Rao, V. G. K. M. Pisipati, A. S. Paranjpe, U. R. K. Rao, *Liq. Cryst.* **1988**, *3*, 1461–1479.
[66] N. V. S. Rao, V. G. K. M. Pisipati, *Mol. Cryst. Liq. Cryst.* **1984**, *104*, 301–306.
[67] A. Zywocinski, S. A. Wieczorek, *Phys. Rev.* A **1985**, *31*, 479–482.
[68] A. Zywocinski, S. A. Wieczorek, *Mol. Cryst. Liq. Cryst.* **1987**, *151*, 399–410.
[69] P. R. Alapati, D. M. Potukuchi, N. V. S. Rao, V. G. K. M. Pisipati, D. Saran, Proceedings of the 11th International Liquid Crystal Conference, Berkeley, CA, 1986, Part C, *Mol. Cryst. Liq. Cryst.* **1987**, *146*, 111–119.
[70] N. V. S. Rao, V. G. K. M. Pisipati, P. R. Alapati, D. M. Potukuchi, *Mol. Cryst. Liq. Cryst.* **1988**, *162B*, 119–125.
[71] J. Stecki, A. Zywocinski, S. A. Wieczorek, *Phys. Rev.* A **1983**, *28*, 434–439.
[72] A. Beguin, J. Billard, F. Bonamy, J. M. Buisine, P. Cuvelier, J. C. Dubois, P. Le Barny, Sources of Thermodynamic Data on Mesogens, *Mol. Cryst. Liq. Cryst.* **1984**, *115*, 1–326.
[73] V. G. K. M. Pisipati, N. V. S. Rao, A. Alapati, *Cryst. Res. Technol.* **1989**, *24*, 1285–1290.
[74] A. L. Tsykalo, *Thermophysical Properties of Liquid Crystals*, Gordon and Breach, New York **1991**.
[75] P. Adomenas, V. A. Grozhik, *Vestsi Akad. Navuk BSSR, Ser. Khim. Navuk* **1977**, *2*, 39–42 [*Chem. Abstr.* **1977**, *87*, 29178w].
[76] W. H. de Jeu, W. A. P. Claasen, *J. Chem. Phys.* **1978**, *68*, 102–108.
[77] M. Takahashi, S. Mita, S. Kondo, *Mol. Cryst. Liq. Cryst.* **1987**, *147*, 99–105.
[78] M. Takahashi, S. Mita, S. Kondo, *Phase Transitions* **1987**, *9*, 1–9.
[79] N. V. S. Rao, D. M. Potukuchi, V. G. K. M. Pisipati, *Mol. Cryst. Liq. Cryst.* **1991**, *196*, 71–87.
[80] O. N. Puchkov, V. A. Molochko, *Zh. Prikl. Khim. (S.-Peterburg)* **1992**, *65*, 825–829 [*Chem. Abstr.* **1993**, *118*, 212851h].
[81] W. Labno, J. Jadzyn, *Pr. Kom. Mat.-Przyr., Poznan. Tow. Przyj. Nauk, Fiz. Dielektr. Radiospektrosk.* **1981**, *12*, 75–84 [*Chem. Abstr.* **1983**, *98*, 99290c].
[82] R. A. Orwoll, V. J. Sullivan, G. C. Campbell, *Mol. Cryst. Liq. Cryst.* **1987**, *149*, 121–140.
[83] N. A. Nedostup, V. V. Gal'tsev, (a) *Russ. J. Phys. Chem.* **1977**, *51*, 121–122; (b) *Zh. Fiz. Khim.* **1977**, *51*, 214–216.
[84] R. Dabrowski, K. Kenig, Z. Raszewski, J. Kedzierski, K. Sadowska, *Mol. Cryst. Liq. Cryst.* **1980**, *61*, 61–78.
[85] G. Albertini, E. Fanelli, D. Guillon, S. Melone, G. Poeti, F. Rustichelli, G. Torquati, *J. Chem. Phys.* **1983**, *78*, 2013–2016.
[86] G. Albertini, E. Fanelli, D. Guillon, S. Melone, G. Poeti, F. Rustichelli, G. Torquati, *J. Phys.* **1984**, *45*, 341–346.
[87] H. Abied, D. Guillon, A. Skoulios, A. M. Giroud-Godquin, P. Maldivi, J. C. Marchon, *Colloid Polym. Sci.* **1988**, *266*, 579–582.
[88] V. G. K. M. Pisipati, N. V. S. Rao, D. M. Potukuchi, P. R. Alapati, P. B. Rao, *Mol. Cryst. Liq. Cryst.* **1989**, *167*, 167–171.
[89] A. Wulf, A. G. De Rocco, *J. Chem. Phys.* **1971**, *55*, 1–27.
[90] I. Haller, H. A. Huggins, H. R. Lilienthal, T. R. McGuire, *J. Phys. Chem.* **1973**, *77*, 950–954.
[91] D. Demus, A. Hauser, C. Selbmann, W. Weissflog, *Cryst. Res. Technol.* **1984**, *19*, 271–283.
[92] D. Demus, A. Hauser, A. Isenberg, M. Pohl, C. Selbmann, W. Weissflog, S. Wieczorek, *Cryst. Res. Technol.* **1985**, *20*, 1413–1421.
[93] D. Demus, S. Diele, A. Hauser, I. Latif, C. Selbmann, W. Weissflog, *Cryst. Res. Technol.* **1985**, *20*, 1547–1558.
[94] P. Seurin, D. Guillon, A. Skoulios, *Mol. Cryst. Liq. Cryst.* **1981**, *65*, 85–110.
[95] D. Guillon, A. Skoulios, *J. Phys. (Paris)* **1976**, *37*, 797–800.
[96] D. Guillon, A. Skoulios, *J. Phys. (Paris)* C3 **1976**, *37*, 83–84.

[97] J. H. Ibrahim, W. Haase, *Z. Naturforsch. A* **1976**, *31*, 1644–1650.
[98] B. Bahadur, S. Chandra, *J. Phys. C* **1976**, *9*, 5–9.
[99] M. N. Rao, V. G. K. M. Pisipati, N. V. S. Rao, J. V. Rao, C. R. K. Murty, *Phase Transitions* **1981**, *2*, 231–238.
[100] K. R. K. Rao, J. V. Rao, L. V. Choudary, C. R. K. Murty, *Acta Phys. Pol. A* **1983**, *63*, 419–424 [*Chem. Abstr.* **1983**, *98*, 189509k].
[101] N. V. S. Rao, V. G. K. M. Pisipati, *Phase Transitions* **1983**, *3*, 317–327.
[102] N. V. S. Rao, V. G. K. M. Pisipati, P. V. Datta Prasad, P. R. Alapati, *Phase Transitions* **1985**, *5*, 187–195.
[103] K. R. K. Rao, J. V. Rao, P. Venkatacharyulu, *Acta Phys. Pol. A* **1986**, *69*, 261–265 [*Chem. Abstr.* **1984**, *104*, 159999v].
[104] J. V. Rao, L. V. Choudary, K. R. K. Rao, P. Venkatacharyulu, *Acta Phys. Pol. A* **1987**, *72*, 517–522 [*Chem. Abstr.* **1988**, *108*, 66511x].
[105] N. M. Sakevich, (a) *Zh. Fiz. Khim.* **1968**, *42*, 2930; (b) *Russ. J. Phys. Chem.* **1968**, *42*, 1555–1557.
[106] L. E. Hajdo, A. C. Eringen, J. Giancola, A. E. Lord Jr., *Lett. Appl. Eng. Sci.* **1975**, *3*, 61 [*Chem. Abstr.* **1975**, *83*, 69402p].
[107] A. E. Lord Jr., F. E. Wargocki, L. E. Hajdo, A. C. Eringen, *Lett. Appl. Eng. Sci.* **1975**, *3*, 125–132 [*Chem. Abstr.* **1975**, *83*, 88976v].
[108] R. S. Porter, J. F. Johnson, *J. Appl. Phys.* **1963**, *34*, 55–59.
[109] T. Matsumoto, *Kogakuin Daigaku Kenkyu Hokoku (Research Reports of the Kogakuin University)* **1974**, *36*, 1–8 [*Chem. Abstr.* **1975**, *83*, 171116k].
[110] S. Plumley, M. R. Kuzma, *Mol. Cryst. Liq. Cryst.* **1991**, *200*, 33–41.
[111] V. V. Aleksandriiski, V. A. Burmistrov, O. I. Koifman, (a) *Zh. Fiz. Khim.* **1993**, *67*, 1623–1625; (b) *Russ. J. Phys. Chem.* **1993**, *67*, 1456–1458.
[112] D. Demus, H.-J. Deutscher, S. König, H. Kresse, F. Kuschel, G. Pelzl, H. Schubert, C. Selbmann, W. Weissflog, A. Wiegeleben, J. Wulf, *Wiss. Beitr. Martin-Luther-Univ. Halle-Wittenberg* **1978**, *21*, 9–20 [*Chem. Abstr.* **1978**, *89*, 189209b].
[113] N. V. S. Rao, V. G. K. M. Pisipati, *J. Phys. Chem.* **1983**, *87*, 899–902.
[114] D. Demus, H. G. Hahn, F. Kuschel, *Mol. Cryst. Liq. Cryst.* **1978**, *44*, 61–70.
[115] V. G. K. M. Pisipati, N. V. S. Rao, *Phase Transitions* **1983**, *3*, 169–175.
[116] V. G. K. M. Pisipati, N. V. S. Rao, Y. Gouri Sankar, J. S. R. Murty, *Acustica* **1986**, *60*, 163–168.
[117] L. E. Hajdo, A. C. Eringen, A. E. Lord Jr., *Lett. Appl. Eng. Sci.* **1974**, *2*, 367–371 [*Chem. Abstr.* **1975**, *83*, 19359j].
[118] N. V. S. Rao, V. G. K. M. Pisipati, Y. Gouri Sankar, *Mol. Cryst. Liq. Cryst.* **1985**, *131*, 237–243.
[119] A. K. George, A. R. K. L. Padmini, *Mol. Cryst. Liq. Cryst.* **1981**, *65*, 217–226.
[120] Y. Thiriet, J. A. Schulz, P. Martinoty, D. Guillon, *J. Phys.* **1984**, *45*, 323–329.
[121] D. Demus, S. Diele, M. Klapperstück, V. Link, H. Zaschke, *Mol. Cryst. Liq. Cryst.* **1971**, *15*, 161–174.
[122] D. Demus, R. Rurainski, *Mol. Cryst. Liq. Cryst.* **1972**, *16*, 171–174.
[123] D. Demus, H. König, D. Marzotko, R. Rurainski, *Mol. Cryst. Liq. Cryst.* **1973**, *23*, 207–214.
[124] D. Demus, M. Pohl, S. Schönberg, L. Weber, A. Wiegeleben, W. Weissflog, *Wiss. Beitr. Martin-Luther-Univ. Halle-Wittenberg* **1983**, *41* (Forsch. Flüss. Krist.), 18–28 [*Chem. Abstr.* **1984**, *101*, 178680p].
[125] A. De Vries, *Mol. Cryst. Liq. Cryst.* **1970**, *10*, 31–37.
[126] A. De Vries, *Liquid Crystals: Proceedings of International Conference 1973, Pramana Suppl.* **1975**, *1*, 93–113 [*Chem. Abstr.* **1976**, *84*, 114666m].
[127] B. R. Ratna, C. Nagabhushana, V. N. Raja, R. Shashidar, G. Heppke, *Mol. Cryst. Liq. Cryst.* **1986**, *138*, 245–257.
[128] G. Poeti, E. Fanelli, D. Guillon, *Mol. Cryst. Liq. Cryst. Lett.* **1982**, *82*, 107–114.
[129] P. Bhaskara Rao, N. V. S. Rao, V. G. K. M. Pisipati, D. Saran, *Cryst. Res. Technol.* **1989**, *24*, 723–731.
[130] K. Kobayashi, *Phys. Lett. A* **1970**, *31*, 125–126.
[131] K. K. Kobayashi, *J. Phys. Soc. Jpn.* **1970**, *29*, 101–105 [*Phys. Abstr.* **1970**, *73*, 55591].
[132] W. L. McMillan, (a) *Phys. Rev. A* **1971**, *4*, 1238–1246; (b) *Phys. Rev. A* **1972**, *6*, 936–947; (c) *Phys. Rev. A* **1973**, *7*, 1419–1422.
[133] P. G. de Gennes, *Solid State Commun.* **1972**, *10*, 753–756.
[134] R. Alben, *Solid State Commun.* **1972**, *13*, 1783–1785.
[135] B. I. Halperin, T. C. Lubensky, S. K. Ma, *Phys. Rev. Lett.* **1974**, *32*, 292–295.
[136] D. M. Potukuchi, K. Prabhakar, N. V. S. Rao, V. G. K. M. Pisipati, D. Saran, *Mol. Cryst. Liq. Cryst.* **1989**, *167*, 181–189.
[137] L. V. Choudary, J. V. Rao, P. N. Murty, C. R. K. Murty, *Z. Naturforsch. A* **1983**, *38*, 762–764.
[138] N. V. S. Rao, S. M. Rao, V. G. K. M. Pisipati, *Phase Transitions* **1983**, *3*, 159–167.
[139] V. G. K. M. Pisipati, N. V. S. Rao, *Phase Transitions* **1984**, *4*, 91–96.
[140] V. G. K. M. Pisipati, N. V. S. Rao, P. V. Datta Prasad, P. R. Alapati, *Z. Naturforsch. A* **1985**, *40*, 472–475.
[141] A. K. Jaiswal, G. L. Patel, *Acta Phys. Pol. A* **1986**, *69*, 723–726 [*Chem. Abstr.* **1986**, *104*, 234839y].

[142] K. R. K. Rao, J. V. Rao, P. Venkatacharyulu, V. Baliah, *Acta Phys. Pol. A* **1986**, *70*, 541–547 [*Chem. Abstr.* **1987**, *106*, 59321r].

[143] N. V. S. Rao, V. G. K. M. Pisipati, Y. Gouri Sankar, D. M. Potukuchi, *Phase Transitions* **1986**, *7*, 49–57.

[144] N. V. S. Rao, D. M. Potukuchi, P. B. Rao, V. G. K. M. Pisipati, *Cryst. Res. Technol.* **1989**, *24*, 219–225.

[145] S. Lakshminarayana, C. R. Prabhu, D. M. Potukuchi, N. V. S. Rao, V. G. K. M. Pisipati, *Liq. Cryst.* **1993**, *15*, 909–914.

[146] A. J. Leadbetter, J. L. A. Durrant, M. Ruyman, *Mol. Cryst. Liq. Cryst. Lett.* **1977**, *34*, 231–235.

[147] M. F. Achard, F. Hardouin, G. Sigaud, M. Mauzac, *Liq. Cryst.* **1986**, *1*, 203–207.

[148] K. R. K. Rao, J. V. Rao, P. Venkatacharyulu, V. Baliah, *Mol. Cryst. Liq. Cryst.* **1986**, *136*, 307–316.

[149] V. N. Raja, S. Krishna Prasad, D. S. Shankar Rao, S. Chandrasekhar, *Liq. Cryst.* **1992**, *12*, 239–243.

[150] C. W. Garland, G. B. Kasting, K. J. Lushington, *Phys. Rev. Lett.* **1979**, *43*, 1420–1423.

[151] C. Rosenblatt, J. T. Ho, *Phys. Rev. A* **1982**, *26*, 2293–2296.

[152] H. Marynissen, J. Thoen, W. Van Dael, *Mol. Cryst. Liq. Cryst.* **1983**, *97*, 149–161.

[153] T. Pitchford, G. Nounesis, S. Dumrongrattana, J. M. Viner, C. C. Huang, J. W. Goodby, *Phys. Rev. A* **1985**, *32*, 1938–1940.

[154] V. G. K. M. Pisipati, S. B. Rananavare, *Liq. Cryst.* **1993**, *13*, 757–764.

[155] W. U. Müller, H. Stegemeyer, *Chem. Phys. Lett.* **1974**, *27*, 130–132.

[156] W. U. Müller, H. Stegemeyer, *Ber. Bunsenges. Phys. Chem.* **1974**, *78*, 880–883.

[157] H. Stegemeyer, W. U. Müller, *Naturwissenschaften* **1976**, *63*, 388.

[158] V. K. Semenchenko, V. M. Byankin, V. Yu. Baskakov, (a) *Sov. Phys. Crystallogr.* **1975**, *20*, 111–113; (b) *Kristallografija* **1975**, *20*, 187–191.

[159] A. W. Neumann, L. J. Klementowski, R. W. Springer, *J. Colloid Interface Sci.* **1972**, *41*, 538–541.

[160] J. R. Otia, A. R. K. L. Padmini, *Mol. Cryst. Liq. Cryst.* **1976**, *36*, 25–39.

[161] M. Nakahara, Y. Yoshimura, J. Osugi, *Nippon Kogakukai (Bull. Chem. Soc. Jpn.)* **1981**, *54*, 99–102 [*Chem. Abstr.* **1981**, *94*, 74987q].

[162] D. M. Potukuchi, N. V. S. Rao, V. G. K. M. Pisipati, *Ferroelectrics* **1993**, *141*, 287–296.

[163] F. P. Price, J. H. Wendorff, *J. Phys. Chem.* **1971**, *75*, 2849–2853.

[164] S. N. Mochalin, P. P. Pugachevich, *Uch. Zap., Ivanov. Gos. Pedagog. Inst.* **1972**, *99*, 200–207. From: Ref. Zh., Fiz., E. **1972**, Abstr. No. 10E138 [*Chem. Abstr.* **1973**, *78*, 152444w].

[165] S. N. Mochalin, P. P. Pugachevich in *Sb. Dokladov II. nauchnoi konferentsii po zhidkim kristallam (Collection of Contributions to the 2nd Conference on Liquid Crystal Science)*, Ivanovo **1972**, p. 200.

[166] S. N. Mochalin, *Uch. zap. Ivanov., un-t* **1974**, *128*, 86–89. *Ref. Zh., Khim.* **1975**, Abstr. No. 413795, only title translated [*Chem. Abstr.* **1975**, *83*, 152689h].

[167] S. Chandrasekhar, R. Shashidar, N. Tara, *Mol. Cryst. Liq. Cryst.* **1971**, *12*, 245–250.

[168] S. Chandrasekhar, R. Sashidar, *Mol. Cryst. Liq. Cryst.* **1972**, *16*, 21–32.

[169] W. Klement, L. H. Cohen, *Mol. Cryst. Liq. Cryst.* **1974**, *27*, 359–373.

[170] G. A. Oweimreen, A. K. Shihab, K. Halhouli, S. F. Sikander, *Mol. Cryst. Liq. Cryst.* **1986**, *138*, 327–338.

[171] H. Imura, K. Okano, *Chem. Phys. Lett.* **1972**, *17*, 111–113.

[172] R. Chang, *Solid State Commun.* **1974**, *14*, 403–406.

[173] G. R. Van Hecke, J. Stecki, *Phys. Rev. A* **1982**, *25*, 1123–1126.

[174] P. G. de Gennes, *The Physics of Liquid Crystals*, Clarendon, Oxford **1974**.

[175] J. T. Bendler, Theory of pretransitional effects in the i phase of liquid crystals, Dissertation Yale University **1974**.

[176] J. Bendler, *Mol. Cryst. Liq. Cryst.* **1977**, *38*, 19–30.

[177] A. K. George, R. A. Vora, A. R. K. L. Padmini, *Mol. Cryst. Liq. Cryst.* **1980**, *60*, 297–310.

[178] R. Chang, J. C. Gyspers, *J. Phys. (Paris) C1* **1975**, *36*, 147–149.

[179] G. Pelzl, D. Demus, H. Sackmann, *Z. Phys. Chem.* **1968**, *238*, 22–32.

[180] J. W. Park, C. S. Bak, M. M. Labes, *J. Am. Chem. Soc.* **1975**, *97*, 4398–4400.

[181] J. W. Park, M. M. Labes, (a) *J. Appl. Phys.* **1977**, *48*, 22–24; (b) *Mol. Cryst. Liq. Cryst. Lett.* **1977**, *34*, 147–152.

[182] A. C. Griffin, T. R. Britt, N. W. Buckley, R. F. Fisher, S. J. Havens, D. W. Goodman in *Liquid Crystals and Ordered Fluids*, Vol. 3 (Eds.: J. F. Johnson, R. S. Porter), Plenum Press, New York **1978**, pp. 61–73.

[183] N. K. Sharma, G. Pelzl, D. Demus, W. Weissflog, *Z. Phys. Chem.* **1980**, *261*, 579–584.

[184] L. Longa, W. H. De Jeu, *Phys. Rev. A* **1982**, *26*, 1632–1647.

[185] M. Domon, J. Billard, *J. Phys. Colloq.* **1979**, *40*, 413–418.

[186] F. Schneider, N. K. Sharma, *Z. Naturforsch.* **1981**, *36a*, 62–67.

[187] M. Cotter in *The Molecular Physics of Liquid Crystals* (Eds.: G. R. Luckhurst, G. W. Gray), Academic Press, London **1979**, p. 181.

[188] W. M. Gelbart, *J. Phys. Chem.* **1982**, *86*, 4298–4307.

[189] Y. Guichard, G. Sigaud, F. Hardouin, *Mol. Cryst. Liq. Cryst. Lett.* **1984**, *102*, 325–330.

[190] F. R. Bouchet, P. E. Cladis, *Mol. Cryst. Liq. Cryst.* **1980**, *64*, 81–87.

[191] V. V. Belyaev, T. P. Antonyan, L. N. Lisetski, M. F. Grebenkin, G. G. Salshchova, V. F. Petrov, *Mol. Cryst. Liq. Cryst.* **1985**, *129*, 221–233.

[192] R. Shashidar, H. D. Kleinhans, G. M. Schneider, *Mol. Cryst. Liq. Cryst. Lett.* **1981**, *72*, 119–126.

[193] H. D. Kleinhans, G. M. Schneider, R. Shashidar, *Mol. Cryst. Liq. Cryst.* **1982**, *82*, 19–24.

[194] H. D. Kleinhans, G. M. Schneider, R. Shashidar, *Mol. Cryst. Liq. Cryst.* **1983**, *103*, 255–259.

[195] A. E. White, P. E. Cladis, S. Torza, *Mol. Cryst. Liq. Cryst.* **1977**, *43*, 13–31.

[196] R. Kiefer, G. Baur, *Mol. Cryst. Liq. Cryst.* **1990**, *188*, 13–24.

[197] S. M. Stishov, V. A. Ivanov, V. N. Kachinskii, (a) *Pis'ma Zh. Eksp. Teor. Fiz.* **1976**, *24 (6)*, 329–332; (b) *JETP Lett.* **1976**, *24 (6)*, 297–300.

[198] V. N. Kachinski, V. A. Ivanov, A. N. Zisman, S. M. Stishov, (a) *Sov. Phys. JETP* **1978**, *48*, 273–277; (b) *Zh. Eksp. Teor. Fiz.* **1978**, *75*, 545–553.

[199] W. Weissflog, D. Demus, Proceedings of the 10th International Liquid Crystal Conference, York, U.K. 1984, Part E, *Mol. Cryst. Liq. Cryst.* **1985**, *129*, 235–243.

[200] M. Cotter, (a) *Phys. Rev. A* **1974**, *10*, 625–636; (b) *Mol. Cryst. Liq. Cryst.* **1976**, *35*, 33–70.

[201] R. Pynn, *J. Chem. Phys.* **1974**, *60*, 4579–4581.

[202] B. A. Baron, W. M. Gelbart, *J. Chem. Phys.* **1978**, *67*, 5795–5801.

[203] D. Demus, *Mol. Cryst. Liq. Cryst.* **1988**, *165*, 45–84.

[204] D. Demus, A. Hauser in *Selected topics in Liquid Crystal Research* (Ed.: H.-D. Koswig), Akademie-Verlag, Berlin **1990**, Chap. 2 [*Chem. Abstr.* **1991**, *115*, 61027d].

[205] D. Demus, A. Hauser, M. Keil, W. Wedler, *Mol. Cryst. Liq. Cryst.* **1990**, *191*, 153–161.

[206] A. P. Kovshik, Yu. I. Denite, E. I. Ryumtsev, V. N. Tsvetkov, (a) *Kristallografiya* **1975**, *20*, 861–864; (b) *Sov. Phys. Crystallogr.* **1975**, *20*, 532–534.

[207] H. S. Subramhanyan, J. Shashidara Prasad, *Mol. Cryst. Liq. Cryst.* **1976**, *37*, 23–27.

[208] J. Shashidara Prasad, H. S. Subramhanyam, *Mol. Cryst. Liq. Cryst.* **1976**, *33*, 77–82.

[209] W. H. de Jeu, P. Bordewijk, *J. Chem. Phys.* **1978**, *68*, 109–115.

[210] F. Leenhouts, W. H. De Jeu, A. J. Dekker, *J. Phys. (Paris)* **1979**, *40*, 989–995.

[211] N. C. Shivaprakash, M. M. M. Abdoh, S. and J. Shashidara Prasad, *Mol. Cryst. Liq. Cryst.* **1982**, *80*, 179–193.

[212] A. K. Garg, G. K. Gupta, V. P. Arora, V. K. Agarwal, B. Bahadur, Proceedings of Nuclear Physics and Solid State Physics Symposium 1981, **1982**, *24C*, 343–344 [*Chem. Abstr.* **1982**, *97*, 102189p].

[213] C. Cabos, J. Sicard, *C. R. Acad. Sci. Paris, Sér. B* **1975**, *281*, 109–111.

[214] S. Denprayoonwong, P. Limcharoen, O. Phaovibul, I. M. Tang, *Mol. Cryst. Liq. Cryst.* **1981**, *69*, 313–326.

[215] O. Phaovibul, K. Chantanasmit, I. M. Tang, *Mol. Cryst. Liq. Cryst.* **1981**, *71*, 233–247.

[216] O. Phaovibul, S. Denprayoonwong, I. M. Tang, *Mol. Cryst. Liq. Cryst.* **1981**, *73*, 71–79.

[217] M. F. Vuks, *Opt. Spectrosc.* **1966**, *20*, 361–364 [*Chem. Abstr.* **1966**, *65*, 3174e].

[218] H. E. J. Neugebauer, (a) *Phys. Rev.* **1952**, *88*, 1210 [*Chem. Abstr.* **1953**, *47*, 3076b]; (b) *Can. J. Phys.* **1954**, *32*, 1–8 [*Phys. Abstr.* **1954**, *57*, 3442].

[219] E. M. Aver'yanov, V. F. Shabanov, (a) *Kristallografiya* **1979**, *24*, 184–186; (b) *Sov. Phys. Cryst.* **1979**, *24*, 107–109.

[220] E. M. Aver'yanov, V. F. Shabanov, (a) *Kristallografiya* **1979**, *24*, 992–997; (b) *Sov. Phys. Cryst.* **1979**, *24*, 567–570.

[221] P. Palffy-Muhoray, D. A. Balzarini, *Can. J. Phys.* **1981**, *59*, 375–377.

[222] P. Palffy-Muhoray, D. A. Balzarini, D. A. Dunmur, *Mol. Cryst. Liq. Cryst.* **1984**, *110*, 315–330.

[223] R. K. Sarna, B. Bahadur, V. G. Bhide, *Mol. Cryst. Liq. Cryst.* **1979**, *51*, 117–136.

[224] B. Bahadur, R. K. Sarna, V. G. Bhide, *Mol. Cryst. Liq. Cryst. Lett.* **1982**, *72*, 139–145.

[225] A. I. Pirogov, I. V. Novikov, *Zh. Prikl. Khim. (Leningrad)* **1988**, *61*, 6, 1382–1384 [*Chem. Abstr.* **1988**, *109*, 139708z].

[226] O. Phaovibul, K. Pongthana-Ananta, I. Ming Tang, *Mol. Cryst. Liq. Cryst.* **1980**, *62*, 25–32.

[227] G. A. Oweinreen, M. Hasan, *Mol. Cryst. Liq. Cryst.* **1983**, *100*, 357–371.

[228] J. Jadzyn, W. Labno, *Chem. Phys. Lett.* **1980**, *73*, 307–310.

[229] A. Bartelt, H. Reisig, J. Herrmann, G. M. Schneider, *Mol. Cryst. Liq. Cryst. Lett.* **1984**, *102*, 133–138.

[230] J. M. Buisine, R. Cayuela, C. Destrade, N. H. Tinh, Proceedings of the 11th International Liquid Crystal Conference, Berkeley, CA, 1986, Part B, *Mol. Cryst. Liq. Cryst.* **1987**, *144*, 137–160.

[231] V. K. Semenchenko, N. A. Nedostup, V. Yu. Baskakov, 1(a) *Russ. J. Phys. Chem.* **1975**, *49*, 909–912; 1(b) *Zh. Fiz. Khim.* **1975**, *49*, 1543–1547; 2(a) *Russ. J. Phys. Chem.* **1975**, *49*, 912–914; 2(b) *Zh. Fiz. Khim.* **1975**, *49*, 1547–1550.

[232] A. C. Zawisza, J. Stecki, *Solid State Commun.* **1976**, *19*, 1173–1175.

[233] N. A. Nedostup, V. K. Semenchenko, (a) *Russ. J. Phys. Chem.* **1977**, *51*, 958–959; (b) *Zh. Fiz. Khim.* **1977**, *51*, 1628–1631.

[234] A. P. Kapustin, *Eksperimental'nye issledovaniya zhidkikh kristallov (Experimental Study of Liquid Crystals)*, Nauka, Moscow **1978**, p. 368 [*Chem. Abstr.* **1979**, *90*, B 95971n].

[235] E. A. S. Lewis, H. M. Strong, G. H. Brown, *Mol. Cryst. Liq. Cryst.* **1979**, *53*, 89–99.

[236] T. Shirakawa, T. Inoue, T. Tokuda, *J. Phys. Chem.* **1982**, *86*, 1700–1702.

[237] T. Shirakawa, M. Arai, T. Tokuda, *Mol. Cryst. Liq. Cryst.* **1984**, *104*, 131–139.

[238] C. S. Johnson, P. J. Collings, *J. Chem. Phys.* **1983**, *79*, 4056–4061.

[239] S. N. Nefedov, A. N. Zisman, S. M. Stishov, *Zh. Eksp. Teor. Fiz.* **1984**, *86*, 125–132 [*Chem. Abstr.* **1984**, *100*, 94933q].

[240] T. Shirakawa, Y. Kikuchi, T. Seimiya, *Thermochim. Acta* **1992**, *197*, 399–405.

[241] A. P. Kapustin, G. E. Zvereva, (a) *Kristallografiya* **1965**, *10*, 723–726; (b) *Sov. Phys. Cryst.* **1966**, *10*, 603–606.

[242] B. Bahadur, J. Prakash, K. Tripathi, S. Chandra, *Acustica* **1975**, *33*, 217–219.

[243] B. Bahadur, *Z. Naturforsch. A* **1975**, *30*, 1093–1096.

[244] V. G. K. M. Pisipati, N. V. S. Rao, *Z. Naturforsch. A* **1984**, *39*, 696–699.

[245] M. R. Rao, *J. Chem. Phys.* **1941**, *9*, 682–685.

[246] Y. Wada, *J. Phys. Soc. Jpn.* **1949**, *4*, 280–283 [*Chem. Abstr.* **1950**, *44*, 6703c].

[247] J. Frenkel, *Kinetic Theory of Liquids*, Dover, New York **1955**.

[248] F. T. Lee, H. T. Tan, Y. M. Shin, C. N. Woo, *Phys. Rev. Lett.* **1973**, *31*, 1117–1120.

[249] H. Gasparoux, M. F. Achard, F. Hardouin, G. Sigaud, *C. R. Acad Sci. Paris II* **1981**, *293*, 1029–1032 [*Chem. Abstr.* **1982**, *97*, 47743n].

[250] P. A. Spegt, A. E. Skoulios, *Acta Crystallogr.* **1963**, *16*, 301–306.

[251] P. A. Spegt, A. E. Skoulios, *Acta Crystallogr.* **1964**, *17*, 198–207.

[252] W. M. Gelbart, B. Barboy, *Mol. Cryst. Liq. Cryst.* **1979**, *55*, 209–226.

[253] W. M. Gelbart, B. Barboy, *Acc. Chem. Res.* **1980**, *13*, 290–296.

[254] V. N. Raja, R. Shashidar, S. Chandrasekhar, R. E. Boehm, D. E. Martire, *Pramana* **1985**, *25*, L119–L122 [*Chem. Abstr.* **1985**, *103*, 113798e].

[255] J. S. Clunie, J. M. Corkill, J. F. Goodman, *Proc. R. Soc. Lond. Ser. A* **1965**, *285*, 520–533.

[256] S. Yano, K. Tadano, K. Aoki, *Mol. Cryst. Liq. Cryst. Lett.* **1983**, *92*, 99–104.

[257] M. Stefanov, A. Saupe, *Mol. Cryst. Liq. Cryst.* **1984**, *108*, 309–316.

[258] P. Photinos, A. Saupe, *J. Chem. Phys.* **1989**, *90*, 5011–5015.

[259] P. Photinos, A. Saupe, *Phys. Rev. A* **1990**, *41*, 954–959.

[260] A. A. Barbosa, A. V. A. Pinto, *J. Chem. Phys.* **1993**, *98*, 8345–8346.

[261] A. S. C. Lawrence, *Trans. Faraday Soc.* **1938**, *34*, 660–677 [*Chem. Abstr.* **1938**, *32*, 4033[1]].

[262] R. D. Vold, M. J. Vold, *J. Am. Chem. Soc.* **1939**, *61*, 808–816 [*Chem. Abstr.* **1939**, *33*, 9108[7]].

[263] R. D. Vold, F. B. Rosevear, R. H. Ferguson, *Oil Soap* **1939**, *16*, 48–51.

[264] M. J. Vold, M. Macomber, R. D. Vold, *J. Am. Chem. Soc.* **1941**, *63*, 168–175 [*Chem. Abstr.* **1941**, *35*, 1658[2]].

[265] W. Gallay, I. E. Puddington, *Can. J. Res. B* **1943**, *21*, 202–210.

[266] R. D. Vold, M. J. Vold, *J. Phys. Chem.* **1945**, *49*, 32–42 [*Chem. Abstr.* **1945**, *39*, 1588[6]].

[267] F. W. Southam, I. E. Puddington, *Can. J. Res. B* **1947**, *25*, 121–124.

[268] G. Stainsby, R. Farnand, I. E. Puddington, *Can. J. Chem.* **1951**, *29*, 838–842 [*Chem. Abstr.* **1952**, *46*, 3300i].

[269] D. P. Benton, P. G. Howe, J. R. Farnand, I. E. Puddington, *Can. J. Chem.* **1955**, *33*, 1798–1805 [*Chem. Abstr.* **1956**, *50*, 11688i].

[270] K. U. Ingold, I. E. Puddington, *J. Inst. Petrol.* **1958**, *44*, 41–44 [*Chem. Abstr.* **1958**, *52*, 6774i].

[271] A. G. MacDonald, *Biochim. Biophys. Acta* **1978**, *507*, 26–37.

[272] K. Ohki, K. Tamura, I. Hatta, *Biochim. Biophys. Acta* **1990**, *1028*, 215–222.

[273] M. Wolf, J. H. Wendorff, Proceedings of the 5th European Winter Liquid Crystal Conference, Borovets, Bulgaria, 1987, *Mol. Cryst. Liq. Cryst.* **1987**, *149*, 141–162.

[274] M.-J. Gonzalez-Tejera, J. M. Perena, A. Bello, I. Hernandez-Fuentes, *Polym. Bull. (Berlin)* **1993**, *31*, 111–115.

[275] S. K. Ghosh, S. Amadesi, *Phys. Lett. A* **1976**, *59*, 282–284.

[276] J. G. Ypma, G. Vertogen, *Phys. Lett. A* **1977**, *61*, 45–47.

[277] L. Feijoo, V. J. Rajan, Chia-Wei Woo, *Phys. Rev. A* **1979**, *19*, 1263–1271.

[278] K. Singh, S. Singh, *Mol. Cryst. Liq. Cryst.* **1984**, *108*, 133–148.

[279] A. V. Belik, V. A. Potemkin, Yu. N. Grevtseva, *Dokl. Akad. Nauk* **1994**, *336*, 361–364 [*Chem. Abstr.* **1994**, *121*, 242457p].

[280] A. V. Belik, V. A. Potemkin, Yu. N. Grevtseva, (a) *Zh. Fiz. Khim.* **1995**, *69*, 101–105; (b) *Russ. J. Phys. Chem.* **1995**, *69*, 91–94.

[281] M. P. Allen, *Phys. Rev. Lett.* **1990**, *65*, 2881–2884.

[282] J. W. Emsley, G. R. Luckhurst, W. E. Palke, D. J. Tildesley, *Liq. Cryst.* **1992**, *11*, 519–530.

[283] P. R. Alapati, D. M. Potukuchi, P. Bhaskara Rao, N. V. S. Rao, V. G. K. M. Pisipati, A. S. Paranjpe, *Liq. Cryst.* **1989**, *5*, 545–551.

[284] V. V. Aleksandriiski, V. A. Burmistrov, O. I. Koifman, *Izv. Vyssh. Uchebn. Zaved., Khim.*

Khim. Tekhnol. **1988**, *31*, 111–114 [*Chem. Abstr.* **1988**, *109*, 30485y].
[285] D. Demus, M. Klapperstück, R. Rurainski, D. Marzotko, *Z. Phys. Chem.* **1971**, *246*, 385–395.
[286] D. Guillon, A. Skoulios, *C. R. Acad. Sci., Sér. C* **1974**, *278*, 389–391.
[287] D. Guillon, A. Skoulios, *J. Phys. (Paris)* **1977**, *38*, 79–83.
[288] D. Guillon, A. Skoulios, *Mol. Cryst. Liq. Cryst.* **1979**, *51*, 149–160.
[289] S. D. Lotke, S. B. Desai, R. N. Patil, Proceedings of the 9th International Liquid Crystal Conference, Bangalore, 1982, Part C, *Mol. Cryst. Liq. Cryst.* **1983**, *99*, 267–277.
[290] A. P. Kapustin, *Elektroopticheskije i akusticheskie svoistva zhidkikh kristallov (Electro-Optic and Acoustic Properties of Liquid Crystals)*, Nauka, Moscow **1973**, p. 232.
[291] R. Kiefer, G. Baur, Proceedings of the 17th Freiburger Arbeitstagung Flüssigkristalle, Freiburg, 1987.
[292] T. Matsumoto, *Kogakuin Daigaku Kenkyu Hokoku (Research Reports of the Kogakuin University)* **1973**, *34*, 1–6 [*Chem. Abstr.* **1974**, *80*, 149676j].
[293] R. Paul, B. Jha, D. A. Dunmur, *Liq. Cryst.* **1993**, *13*, 629–636.
[294] A. I. Pirogov, I. V. Novikov, *Izv. Vyssh. Uchebn. Zaved. Khim. Khim. Tekhnol.* **1987**, *30*, 3, 63–68 [*Chem. Abstr.* **1987**, *107*, 145296p].
[295] A. I. Pirogov, I. V. Novikov, *Izv. Vyssh. Uchebn. Zaved. Khim. Khim. Tekhnol.* **1987**, *30*, 10, 63–67 [*Chem. Abstr.* **1988**, *108*, 66525e].
[296] A. I. Pirogov, N. Kodabakas, *Zh. Fiz. Khim.* **1987**, *61*, 1754–1760 [*Chem. Abstr.* **1987**, *107*, 209494m].
[297] A. I. Pirogov, N. Kodabakas, *Zh. Fiz. Khim.* **1989**, *63*, 368–372 [*Chem. Abstr.* **1989**, *110*, 203502c].
[298] R. S. Porter, J. F. Johnson, *J. Appl. Phys.* **1963**, *34*, 51–54.
[299] N. V. S. Rao, V. G. K. M. Pisipati, D. Saran, *Phase Transitions* **1984**, *4*, 275–279.
[300] N. V. S. Rao, D. M. Potukuchi, P. V. Sankar Rao, V. G. K. M. Pisipati, *Liq. Cryst.* **1992**, *12*, 127–135.
[301] P. Bhaskara Rao, D. M. Potukuchi, J. S. R. Murty, N. V. S. Rao, V. G. K. M. Pisipati, *Cryst. Res. Technol.* **1992**, *27*, 839–849.
[302] N. V. S. Rao, G. Padmaja Rani, D. M. Potukuchi, V. G. K. M. Pisipati, *Z. Naturforsch.* **1994**, *49a*, 559–562.
[303] P. I. Rose, *Mol. Cryst. Liq. Cryst.* **1974**, *26*, 75–85.
[304] P. Seurin, D. Guillon, A. Skoulios, *Mol. Cryst. Liq. Cryst.* **1980**, *61*, 185–190.
[305] A. G. Shashkov, I. P. Zhuk, V. A. Karolik, *High Temp.–High Press.* **1979**, *11*, 485–490 [*Chem. Abstr.* **1980**, *93*, 58566a].
[306] R. Somashekar, M. S. Madhava, *Mol. Cryst. Liq. Cryst.* **1987**, *147*, 79–84.
[307] C. I. Venkatamana Shastry, J. Shashidara Prasad, Proceedings of the 11th International Liquid Crystal Conference, Berkeley, CA 1986, Part A, *Mol. Cryst. Liq. Cryst.* **1986**, *141*, 191–200.
[308] D. M. Potukuchi, P. B. Rao, N. V. S. Rao, V. G. K. M. Pisipati, *Z. Naturforsch. A* **1989**, *44*, 23–25.
[309] N. V. S. Rao, V. G. K. M. Pisipati, J. S. R. Murthy, P. Bhaskara Rao, P. R. Alapati, *Liq. Cryst.* **1989**, *5*, 539–544.

6.2.3 Metabolemeter

Wolfgang Wedler

In 1983, Buisine and coworkers [1] described a device which, in a small cell with constant volume, permits the detection of pressure differences in condensed, especially liquid-crystalline, phases as a function of the temperature. This device was designed to gain quickly comprehensive information about the phase behaviour with very small quantities of material. It produces plots of pressure difference–temperature which were named thermobarograms. Clearly, it relates two intensive properties. The authors were able to show that the only occurrence of intensive properties conveniently permits a significant miniaturization of the apparatus. Furthermore, a theoretical estimate of the pressure changes in the chamber when heated or cooled beyond first-order phase transitions was carried out. This was combined with first experimental data from scans of MBBA, EBBA and Octylcyanobiphenyl, which indicated that the sensitiv-

6.2.3 Metabolemeter

ity of the method was sufficiently high to detect phase transitions which proceeded with only 0.06% fractional volume change. The authors suggested to name this apparatus metabolemeter, from μεταβολη (transformation) and μετρον (to measure).

Several publications showed its usefulness by describing phase transitions in mesogenic materials with a rich polymorphism [2–4]. Liquid-crystalline polymorphism of chiral compounds, having multicritical points [5, 6], as well as mesogens with discotic or pyramidic molecular geometry, and re-entrant behaviour [7–11] were subject to thermobarometric analysis (TBA). Furthermore, the metabolemeter proved its usefulness in studies of binary mixtures of mesogens [12], and in polymer research [13, 14].

An overview of the construction is given in Fig. 1. Additionally, Fig. 2 shows a schematic representation of the sample cell. The following description bases on the informations given by the authors [1].

To verify simultaneous recording of sample pressure and temperature, a rigid but dilatable cell is used. This cell can be heated and contains a pressure transducer and a thermometer. The pressure transducer used is a HEM 375-20000-Kulite International. It has a flushing sensible metallic membrane, a working temperature range between −55°C and 260°C, and a maximum sustainable pressure of 1700 bar. The cell is composed of a crucible (3) in which is machined a cavity (4) giving a sample volume of 5.97 mm^3. The pressure transducer (1) covers the upper part of the measurement cell. For successful measurements, sensor and crucible have to be aligned parallel to each other in order to tightly close the chamber for measurement. The hemispheric shape of the crucible together with its placement on

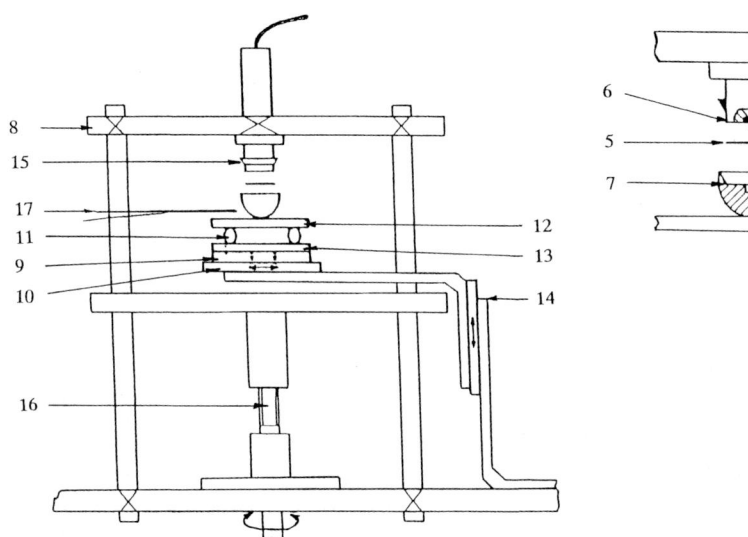

Figure 1. Schematic representation of the metabolemeter (initial design). The following parts are indicated by numbers: (1) pressure transducer (Kulite International, HEM 375 20000), (2) flushing sensible membrane, (3) crucible (17–4–PH stainless steel), (4) chamber (6 mm^3), (5) tin joint, (6) crowned insensible surface of the head transducer, (7) plane surface of the crucible, (8) pressure sensor support, (9, 10) horizontal translation movements, (11) steel balls, (12, 13) aluminium plates, (14) vertical translation movement, (15) centring cone, (16) screw, (17) temperature sensor. (Reproduced from [1] by permission of Gordon and Breach Science Publishers, Inc).

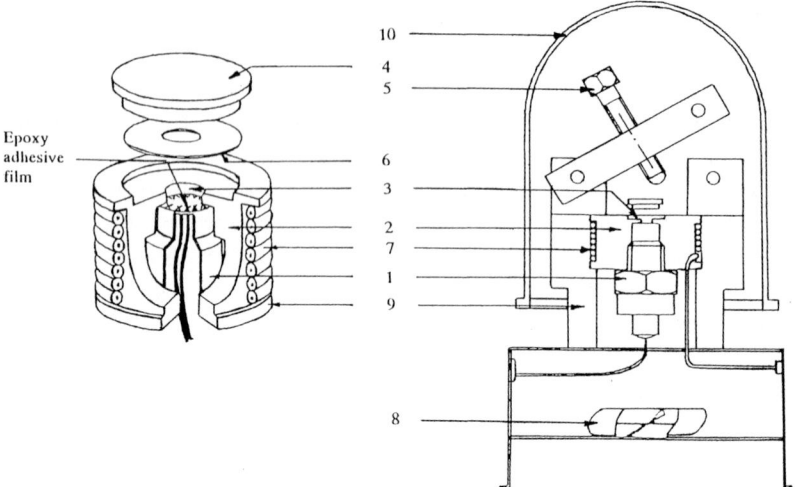

Figure 2. Schematic representation of the scanning numerical metabolemeter metallic sample cell. (1) pressure transducer, (2) steel crucible, (3) cavity, (4) steel cover, (5) set screw, (6) annular joint, (7) heater (8) fan, (9) steel stand, (10) glass housing. (Reproduced from [10] by permission of Société Française Physique).

ball bearings (11) and aluminium plates (12) and (13), is intended to solve this problem. When the lower part is moved towards the stationary upper part, the centring cone (15) moves the crucible into the right position, without sliding. Another essential point is to guarantee a homogenous deformation of the chamber during measurements. Therefore, the cell is constructed of the same material as the pressure sensor (17–4–PH stainless steel). A tin joint (5) insures tightness when exposed to pressures. This material also introduces an upper limit to the available temperature, being at 230°C, beyond which the tin melts. Later studies extended the temperature range to 270°C, marking the maximum working temperature of the pressure transducer, by using a gold joint [3]. Also, attempts were made to obtain tightness of the cell up to 2500 bar [13]. The apparatus is hermetically closed by vertical motion (14) of the lower part by means of a screw (16), located at the base of the apparatus. The cell then can be heated with an oven, and the temperature is measured with a platinum resistance thermometer (17). Configured in the described way, heating rates of 5 K min^{-1}, and cooling rates of 1 K min^{-1} were achieved [3]. The signals which come from the thermometer and the pressure transducer, are respectively transmitted to the X- and Y-inputs of a recorder, or can be fed into a computer, giving the thermobarograms.

The apparatus has been further developed into a scanning numerical metabolemeter [11, 15, 16].

6.2.3.1 Thermobarograms

Plots which are produced by TBA, so-called thermobarograms, are similar and related to the pressure–temperature phase diagrams which are commonly used to express equilibrium between phases. However, the first and main difference between such a phase diagram and a thermobarogram is that the latter also shows out-of-equilibrium states. An estimate [1] from literature data leads to the result, that in between phase transitions, the slope is often less than 13 bar K^{-1}, whereas it is higher than 26 bar K^{-1} at first-order phase transitions for most cases. Hence, these phase transitions express themselves in a thermobarogram by noticable slope changes. The second difference between a phase diagram and a thermobarogram is the choice of the pressure variable. Phase diagrams depict the absolute pressure at equilibrium as a function of phase transition temperature. In contrast, thermobarograms show pressure differences evolving with the temperature, which is not necessarily the phase transition temperature at equilibrium. Therefore, experiments can be started at different temperatures: close to, or far from phase transitions. The zero pressure difference then always starts with ambient pressure, but the trajectory in the phase diagram is different for every case. When crossing phase transitions in this way, one finally is able to piece together the phase diagram from several different thermobarograms. Figure 3 illustrates this procedure for the example of HBPD [3].

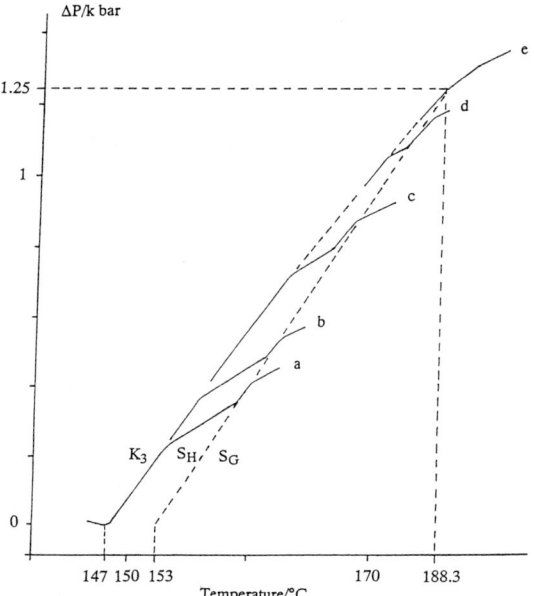

Figure 3. Generation of a pressure–temperature phase diagram from thermobarograms of HBPD. Reproduced from [3] by permission of Gordon and Breach Science Publishers, Inc. Sequence of four thermobarograms (a–d) showing the K_3–S_H, and S_H–S_G phase transitions. The fifth curve indicates the triple point K_2–K_3–S_G (HBPD was shown to have three different solid phases)

It was shown [1] that an assessment of the isothermal pressure increment $(\Delta P)_T$, occurring at phase transitions in the thermobarograms, is possible. Assuming fractional volume changes $\Delta V/V_m$ of 6%, 0.6%, and 0.06%, which are realistic (see Sect. 6.2.2 of this chapter), together with reasonable assumptions about the other parameters give pressure increments of 1600 bar maximum, and 12 bar minimum, which are well in the range of the sensitivity of the pressure transducer. Furthermore, the influence of respiration (incurvation variation of the membrane) as a source of error has been discussed [1]. Respiration increases the cavity volume v_c, and decreases the observed pressure increment, setting a lower limit to sample cell minimization. It was found that for 6% relative volume change, and a cavity volume of 6 mm^3, independence of $(\Delta P)_T$ from v_c occured. A drop in cavity size to 0.4 mm^3 leaves the pressure increment still higher than 90% of the maximum value for large cavity volumes. During data evaluation, the expression for the pressure increment provides the possibility to calculate fractional volume changes and phase transition enthalpies [2].

In blends, incompatibility between two mesogenic compounds A and B in the crystalline state, decrease of the melting temperature and eutectic point formation cause more complexity in the thermobarograms. For the case of the perfect solution Buisine [17], and Buisine and Billard [12] derived mathematical relations which allow calculation and understanding of thermobarograms of blends and estimation of the detectability of phase transition effects. Using these relations, pressure increments of 238 bar and 768 bar were estimated, which are well in the range of detectability of the metabolemeter, for the eutectic melting and the dissolution of the majority component, respectively. According to [12], the eutectic melting is reflected in the thermobarogram by a slope of approximately 40 bar K^{-1} which is similar to the melting of the pure components. The subsequent dissolution starts with a drop of slope to 15 bar K^{-1}, and proceeds by increasing up to around 33 bar K^{-1}. Further equations were given [12] which model pressure increments and slopes for fluid – fluid phase transitions in blends. It could be shown that, for example, the slope always has to be positive. For the simple case of horizontal curves of coexistence in the isobaric temperature–concentration phase diagram, the slope becomes definetely bigger than the out-of-transition slope. Also, its value stays always smaller than that of the corresponding transitions in the pure compounds. Deviations from the horizontal shape of the coexistence curve in the phase diagram, and a consequent widening of the spindle, decrease the slope value. Estimates for the pressure increments at several different fluid–fluid transitions are compiled from [12], and summarized in Table 1. In some cases, new phases were discovered and identified. Also, phase transitions of weakly first order [3] or second order [6], and glass transitions [13] were detected. A general introduction into theory and practice of TBA has been given by Buisine [17, 18].

As an example we mention the TBA results for the binary system 4-methoxy-4'-nonyltolane (MNT)/4-methoxy-4'-ethyltolane (MET). This system has only one nematic phase, and has been described previously [19]. Figure 4 shows the thermobarogram, as obtained on heating from a blend, containing 11.9 wt% MET. The melting at 32°C under atmospheric pressure, and T_{N-I} at 63.5°C under 500 bar, are visible. The dissolution process of excess MNT crystals starts with a slope of 6.2 bar K^{-1}, and continuously develops to a slope of 13.6 bar K^{-1}. The nematic dilatation begins with a faint change of slope at 60°C, under 450 bar.

Finally, we note that the further development of TBA was not restricted exclusively to the phase characterization of liquid crystals. Since 1988, a more exended use in

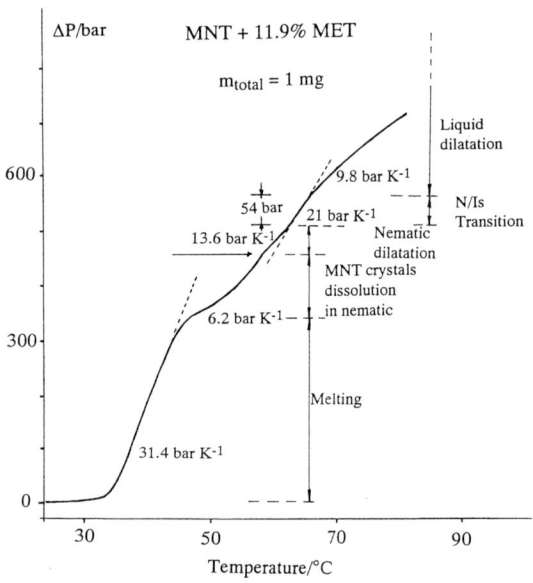

Figure 4. Thermobarogram on heating for the mixture MET/MNT with 11.9 wt% MET. Sample weight was 1 mg. Sequence of phase transitions: eutectic melting, crystal dissolution of excess MNT, nematic/isotropic phase transition. Reproduced from [12] by permission of Gordon and Breach Science Publishers, Inc. and OPA Ltd

Table 1. Estimated typical values for the thermobarometric pressure increments at different fluid–fluid phase transitions (compiled from [12]).

Transition	ΔP_I (bar K^{-1})
Smectic-A/nematic	20
Nematic/isotropic, or smectic-C/nematic	30
Highly organized mesophase/ smectic-A (C), or highly organized mesophase/nematic	40
Highly organized mesophase/ highly organized mesophase	150

polymer science, paraffin and bitumen research, and in the characterization of reacting systems has been reported [20], underlining the versatility of the apparatus.

6.2.3.2 References

[1] J. M. Buisine, B. Soulestin, J. Billard, *Mol. Cryst. Liq. Cryst.* **1983**, *91*, 115–127.
[2] J. M. Buisine, B. Soulestin, J. Billard, *Proceedings of the 9th International Liquid Crystal Conference*, Bangalore, India, Part D, **1982**, *Mol. Cryst. Liq. Cryst.* **1983**, *97*, 397–406.
[3] J. M. Buisine, *Mol. Cryst. Liq. Cryst.* **1984**, *109*, 143–157.
[4] J. M. Buisine, *C. R. Acad. Sci. Paris, Sér. II*, **1983**, *297*, 323–326.
[5] C. Legrand, N. Isaert, J. Hmine, J. M. Buisine, J. P. Parneix, H. T. Nguyen, C. Destrade,
    (a) *Ferroelectrics* **1991**, *121*, 21–31.
    (b) *J. Phys. II* **1992**, *2*, 1545–1562.
[6] A. Anakkar, A. Daoudi, J. M. Buisine, N. Isaert, T. Delattre, H. T. Nguyen, C. Destrade, *J. Therm. Anal.* **1994**, *41*, 1501–1513.
[7] J. M. Buisine, R. Cayuela, C. Destrade, N. H. Tinh, *Proceedings of the 11th International Liquid Crystal Conference*, Berkeley, Part B, 1986. *Mol. Cryst. Liq. Cryst.* **1987**, *144*, 137–160.
[8] J. M. Buisine, J. Malthête, C. Destrade, N. H. Tinh, *Physica B* **1986**, *139/140*, 631–635.
[9] J. M. Buisine, M. Domon, *C. R. Acad. Sc. Paris, Sér. II*, **1986**, *303*, 1769–1772.
[10] J. M. Buisine, B. Soulestin, *Rev. Phys. Appl.* **1987**, *22*, 1211–1214.
[11] J. M. Buisine, H. Zimmermann, P. Poupko, Z. Luz, J. Billard, *Mol. Cryst. Liq. Cyst.* **1987**, *151*, 391–398.
[12] J. M. Buisine, J. Billard, *Proceedings of the 10th International Liquid Crystal Conference*, York, United Kingdom, Part D, 1984. *Mol. Cryst. Liq. Cryst.* **1985**, *127*, 353–379.
[13] J. M. Buisine, P. Le Barny, J. C. Dubois, *J. Polym. Sci.: Polym. Lett.* **1984**, *22*, 149–152.
[14] C. Lahmamssi, X. Coqueret, J. M. Buisine, C. Gors, *Calorim. Anal. Therm.* **1992**, *23*, 359–366.
[15] The scanning numerical metabolemeter was manufactured and distributed by Micro Technique Métropole LEADER, Moulin 1, 2 Rue de la Créativité 59650 Villeneuve d'Ascq, France, under the reference MAB 02 (Information from Ref. [11]).
[16] J. M. Buisine, J. L. Bigotte, M. T. M. Leader, *Calorim. Anal. Therm.* **1987**, *18*, 387–391.
[17] J. M. Buisine, *Thèse*, Lille, France, **1984**.
[18] J. M. Buisine, *Calorim. Anal. Therm.* **1988**, *19*, C24.1-C24.8.
[19] J. Malthête, M. Leclercq, M. Dvolaitzky, J. Gabard, J. Billard, V. Pontikis, J. Jacques, *Mol. Cryst. Liq. Cryst.* **1973**, *23*, 233–260.
[20] The metabolemeter was also used in experiments with different, non-liquid-crystalline systems. Publications regarding these studies are listed below. (a) Bitumen, Paraffins: J. M. Buisine, C. Such, A. Eiadlani, *Calorim. Anal. Therm.* **1988**, *19*, P 21.1-P 21.8. J. M. Buisine, C. Such, A. Eiadlani, *Prepr.-Am. Chem Soc., Div. Pet. Chem.* **1990**, *35*, 320–329. J. M. Buisine, C. Such, A. Eiadlani, *Fuel Sci. Technol. Int.* **1992**, *10*, 835–853. G. Joly, F. Farcas, A. Eiadlani, C. Such, J. M. Buisine, *Calorim. Anal. Therm.* **1992**, *23*, 343–349. D. Lourdin, A. H. Roux, J. P. E. Grolier, J. M. Buisine, *Thermochim. Acta* **1992**, *204*, 99–110. J. Li. Tamarit, B. Legendre, J. M. Buisine, *Mol. Cryst. Liq. Cryst. Sci. Technol., Sect. A* **1994**, *250*, 347–358.
(b) Polymers, Polymeric Blends, Filled Systems: J. M. Buisine, P. Cuvelier, B. Addadi, N. Elbounia, *Calorim. Anal. Therm.* **1991**, *22*, 185–192. D. Lourdin, J. R. Quint, A. H. Roux, J. P. E. Grolier, *Calorim. Anal. Therm.* **1992**, *23*, 225–232. P. Cuvelier, B. Haddadi, J. M. Buisine, N. Elbounia, *Thermochim. Acta* **1992**, *204*, 123–135.
(c) Reacting Systems: A. Squalli, L. Montagne, P. Vast, G. Palavit, J. M. Buisine, *J. Therm. Anal.* **1991**, *37*, 1673–1678. A. Squalli, J. M. Buisine, *Calorim. Anal. Therm.* **1992**, *23*, 401–408.

6.2.4 High Pressure Investigations

P. Pollmann

6.2.4.1 Introduction

By application of hydrostatic pressure the liquid crystalline range of existence can be varied: it can be increased, decreased (in the extreme case suppressed) or induced at all (Fig. 1). Liquid crystalline phases can disappear with increasing pressure and appear again at still higher pressures. Thus there exists a valuable tool for influencing the phase behavior of liquid crystals.

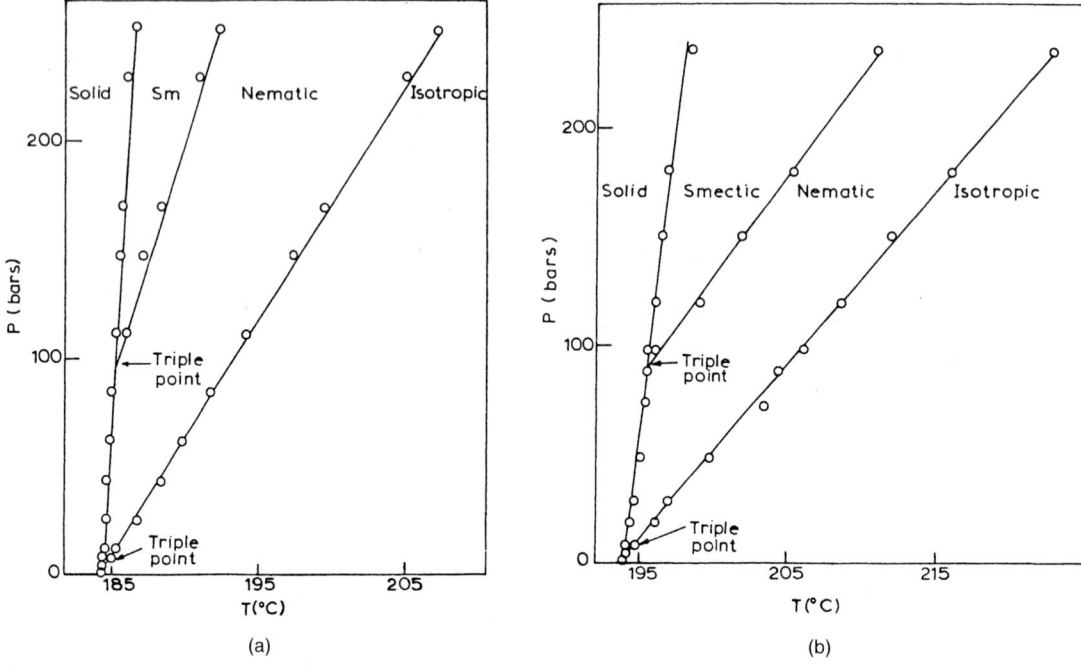

Figure 1. Experimental pressure–temperature phase diagrams for 4-methoxybenzoic acid (a) and 4-ethoxybenzoic acid (b) showing the solid–nematic–isotropic and solid–smectic–nematic triple points. (From [19], reproduced by permission of Indian Academy of Sciences.)

Pressure plays an important role in the field of critical phenomena in liquid crystal systems. Contrary to investigations at atmospheric pressure where, for instance, the composition of a mixture must be varied, single compounds can be observed only by increasing the pressure.

The measurement of the pressure dependence of a physical property of a liquid crystalline phase at constant temperatures offers the advantage of studying only the influence of density without temperature effects.

6.2.4.2 Phase transitions at High Pressures

Phase transitions involving liquid crystalline phases have been the subject of experimental work for many years. By 1899 Hulett [1] had already investigated 4,4'-bis-methoxy-azoxybenzene, 4,4'-bis-ethoxy-azoxybenzene and cholesteryl benzoate up to 30 MPa. After a long period of inactivity in this field, in 1926 Puschin and Grebenschtschikow [2] studied 4,4'-bis-methoxy-azoxybenzene up to 250 MPa, and in 1938 Robberecht [3] studied cholesteryl pentanoate and cholesteryl hexanoate up to 76 MPa.

For an indication of a discontinuous phase transition each physical quantity can be used in principle which is changed sensitively enough by the phase transition also at higher pressures. The detection methods for a continuous phase transition are naturally restricted.

Experimental Techniques

Differential Thermal Analysis (DTA)/ Differential Scanning Calorimetry (DSC)

DTA and DSC are often utilized for detecting liquid crystalline phase transitions, which are associated with low transition enthalpies.

After the first measurements of Garn [4] up to 1 MPa, Reshamwala and Shashidhar [5] designed a DTA device with a coaxial cell working up to 750 MPa. The sample was enclosed in Teflon and thus totally separated from the pressure-transmitting medium.

Herrmann and coworkers [6] developed a DTA apparatus with a diamond anvil cell for measurements up to 300 MPa. For measuring the temperature difference between sample and reference cell the thermocouples are soldered into two blind plugs. The samples are encapsulated in lead, indium or nickel cells.

Sandrock, Bartelt and Schneider [7, 8] later on used a microcomputer-assisted apparatus with a pressure range up to 1000 MPa. It consisted of a twin autoclave. The sample is encapsulated in a lead or indium cup.

Investigations of phase transitions with DSC [9] are advantageous when, aside from the phase boundary lines, enthalpy changes of the transitions shall be determined. Using the transition enthalpies and the slopes of the transition curves the volume changes accompanying the transitions can additionally be calculated by means of the Clausius–Clapeyron equation.

Garland [10] developed an a.c. calorimeter technique for the study of second-order phase transitions. The device works up to 300 MPa and requires only a small sample (50–100 mg). In [10] two versions of this calorimeter are described: a manually operated calorimeter with a computerized data-acquisition system, and a fully computerized calorimeter which can work in a scanning mode.

Optical Methods

A simple method for detecting phase transitions is the optical transmission technique [11]. The transition is indicated by a discontinuity in the intensity of transmitted light or a change in its pressure dependence. The measuring substance is contained in an optical high pressure cell which is usually sealed by two sapphire windows. In many cases the sample is filled into a special cuvette which in the pressure chamber is surrounded by the pressure transmitting not light absorbing fluid [12, 13]. The pressure then is transmitted to the contents of the cuvette by a variety of devices (e.g. bellows, shrinking hose, thin-walled tube).

A diamond anvil cell can also serve as an optical high pressure cell [14]. Transitions between liquid crystalline phases which exhibit optical activity can be indicated by jumps in the angle of optical rotation [15].

The characteristic behavior of the selective light reflection (e.g. of a cholesteric phase near a smectic A phase) can be used to observe the transition of phases. In some cases a distinction between a discontinuous and a continuous smectic A–cholesteric phase transition is possible [16].

Often a phase transition between two liquid crystalline phases is associated with a change in texture, which is characteristic for a special phase. The observation of the sample in the high pressure optical cell by a polarizing microscope then offers the advantage not only of detecting the phase transition but also of identifying the type of the involved liquid crystalline phases [13, 17].

The apparatus described in [18] allows not only the observation of the sample with a polarizing microscope but also includes the application of optical spectroscopy electrooptic and electrical measurements.

A multipurpose and a constant sample thickness cell can be used alternatively as specimen cells. In the first case the pressure transmitting medium is argon. Samples which are affected by argon then are contained between two sheets of glass sealed with PTFE. The latter arrangement, however, has the disadvantage of a layer thickness which varies with pressure. If the sample is not affected by argon, it can be placed in an open cell which consists of two panes of glass with 10 µm thick stainless steel spacers or even it can be observed as a free-standing film.

The pressure transmitting medium in the case of the constant sample thickness cell is oil. The sample is contained in a small reservoir, the deformable bottom (copper membrane) of which transmits the pressure to the specimen. The observation is performed in such a way that the layer thickness is not influenced by the deformation of the copper membrane.

Very special methods of detecting phase transitions include the following.

Normal Behavior (including pressure-induced and suppressed behavior)

For facilitating the survey of the numerous publications concerning phase transitions the latter are classified by the method of examination. The letter after the reference gives the types of the liquid crystalline phases which are involved in the transitions investigated. The different smectic phases are all listed under the symbol S; if only transitions between smectic phases were observed see letter **k**. In the cases of twisted-grain-boundary phases (TGB), cholesteric phases (N*) and blue phases (BP) their symbols are positioned beside the reference number of the literature and the referring letter. The letters mean the following combinations of phase transitions: (a) N–I; (b) S–I; (c) S–N–I; (d) Cr–N–I; (e) Cr–S–I; (f) Cr–S–N–I; (g) S–N; (h) Cr–N; (i) Cr–S–N; (j) Cr–S; (k) S–S.

Differential Thermal Analysis (DTA)

[19] – **c, f (N*)**: Whereas the nematic range of 4,4′-bis-n-(heptyloxy)-azoxybenzene is extended by increasing pressure, that of the SmC one disappears at a pressure of about 668 MPa.

The cholesteric range of cholesteryl oleyl carbonate increases with pressure. There is evidence of a tricritical point. In the case of methoxy- and ethoxybenzoic acid a pressure-induced nematic and a smectic phase are observed.

[20] – **f**: The DTA measurements of 4-n-hexyloxyphenyl-4-n-decyloxybenzoate are combined with microscopic observations of the sample in a diamond anvil cell. A SmA–SmA phase transition is supposed to occur. Monotropic SmB and SmE phases are observed.

[21] – **e, f**: The SmC phases of the three compounds di-(4′-n-octyloxyphenyl)-trans-cyclohexane-1,4′-dicarboxylate), 4-n-nonyloxybenzoic acid and n-pentyl-4(4′-n-decyloxybenzylideneamino) cinnamate exhibit a different temperature dependence of the tilt angle in all three cases but the same pressure behavior: the SmC phases are suppressed at higher pressures.

[22] – **d, f**: The experimental values for the slopes of the nematic–isotropic phase boundary lines dT/dp of the first six homologs of the 4,4′-di-n-alkoxyazoxybenzene are in good agreement with those calculated from the Clausius–Clapeyron equation. The fifth and sixth homologs show in contrast to the other ones additionally a smectic phase with an unusual pressure behavior.

[23] – **c (N*), d, f (N*)**: The monotropic nematic phase of 4-cyanophenyl-4-n-propyloxy-α-methyl cinnamate becomes enantiotropic at 58 MPa and 86 °C. The cholester-

ic phase of cholesteryl nonanoate is suppressed at 285 MPa and 196 °C. All transitions of cholesteryl tetradecanoate remain enantiotropic up to 160 MPa. 4,4′-bis-n-butyl-azobenzene and 4-biphenyl-4-ethylbenzoate exhibit a pressure-induced mesophase.

[24] – d, f: The pressure dependence of the phase transition temperatures of the pentyl to octyl homologs of the 4′-n-alkyl-4-cyanobiphenyls has been determined. The slopes of the nematic-isotropic phase boundary lines dT/dp exhibit an alternation with the number of carbon atoms in the alkyl chain. This is not the case for the Cr–N (or Cr–S)transition. The SmA–N transition of 4′-n-octyl-4-cyanobiphenyl becomes tricritical at 268 MPa and 92.5 °C.

[25] – d, e, f: 4-cyanophenyl-trans-4-n-butylcyclohexane-1-carboxylate, 4-n-pentylphenyl-trans-4-n-pentylcyclohexane-1-carboxylate, trans-4,4′-di-n-propyl-1,1-bicyclohexyl-cis-4-carbonitrile, trans-4-methoxy-4′-propyl-1,1′-bicyclohexane, and trans-4-methoxy-4′-n-butyl-1,1′-bicyclohexane were investigated in the temperature range 300–500 K up to 800 MPa. Some transitions of these compounds undergo a change from monotropic to enantiotropic behavior at higher pressures. An extended Simon equation reflects the influence of the chain length on the phase behavior of the homologous series.

[26] – c, d: The phase transitions of N-(4-methoxybenzylidene)-4-n-butylaniline, N-(4-ethoxybenzylidene)-4-n-butylaniline, N-(4-pentoxybenzylidene)-4-n-butylaniline, 4-ethoxybenzylidene-4-aminobenzonitrile, 4,4′-dimethoxyazoxybenzene, and 4,4′-diethoxyazoxybenzene were followed to 300 MPa in the temperature range 273–550 K. From the experimentally determined transition entropies at atmospheric pressure and the slopes of the obtained transition curves the transition volumes were evaluated by means of the Clausius–Clapeyron equation.

[27] – d, f: The influence of saturation with helium, nitrogen, argon or argon/carbondioxid mixtures on the phase behavior of 4-n-hexyloxy- and 4-n-decyloxy-4′-cyanobiphenyl was studied up to 270 MPa. The latter compound exhibits a pressure-induced nematic phase above 35 MPa.

The nematic and the smectic phases are destabilized by the saturation with the gases mentioned above. The pressure-induced nematic phase does not appear in the presence of the gases with the exception of helium.

[28] – e, f: The phase transitions of terephthaldiylidene-bis-(4-n-octyl)aniline and terephthaldiylidene-bis-(4-n-dodecyl)aniline were observed in a diamond anvil cell up to 300 MPa in the temperature range 300–600 K. For identifying the different smectic phases (A, C, F, G, I) and detecting phase transitions of higher order a polarizing microscope was used. For the SmA–SmC transition line of the octyl member there seems to be a tricritical point at 110 MPa and 512 K. At 23 MPa and 488 K a pressure-induced nematic phase appears. The dodecyl member shows a pressure-induced phase at 80.5 MPa and 492 K, which is probably a SmA phase. For the butyl and pentyl member of this homologous series see [29] – f.

[6] – d, f: The phase behavior of eight members of the homologous series of the 4,4′-bis-(n-alkoxybenzylidene)-1,4-phenylenediamine (butoxy-, pentyloxy-, hexyloxy, heptyloxy, octyloxy, dodecyloxy-, tridecyloxy-, tetradecyloxy-) are observed up to 300 MPa. The nematic phase of all members is stabilized by increasing pressure.

This is confirmed of a pressure-induced nematic phase in the case of the tri- and tetradecyloxy members. The pressure behavior of the particular smectic ranges (C, G, H, I) is very complex. Some smectic phases vanish at higher pressures. Two compounds investigated additionally: 4-n-octyloxy-4'-cyanobiphenyl and 4-cyanobenzylidene-4'-n-octyloxy aniline and their mixtures show pressure-induced nematic re-entrant behavior.

[30] – d: 4,4'-bis-methoxy-azoxybenzene was studied up to 200 MPa. The slopes of the phase boundaries obtained are compared with those calculated from the Clausius–Clapeyron equation.

[31] – d: The clearing temperatures of numerous nematic 2-substituted hydroquinone-bis-(4-substituted benzoates) could be measured up to 500 MPa. The N–I phase transition curves are strongly nonlinear and can be described by a modified Simon–Glatzel equation. A preferred orientation of the lateral substituents parallel to the molecular long axis was expected to be strengthened by increasing pressure. The latter effect should manifest itself in enhanced values of the slopes of the transition lines. However, despite the strong deviation of the compounds from the ideal rod-like shape, the data are similar to those of classical nematics.

[32] – d: The DTA measurements were combined with a method which allows recording of dielectric losses. The pressure dependence of the clearing and melting temperatures of 4,4'-bis-methoxy-azoxybenzene and N-(4-methoxybenzylidene)-4-n-butylaniline could be determined up to 700 MPa.

[33]: Pressure-temperature phase diagrams of the disc-like mesogens benzene-hexa-n-hexanoate and benzene-hexa-n-octanoate are presented up to 200 MPa and 110 °C. Both compounds exhibit one mesophase. The mesophase of the latter compound is bounded about 140 MPa.

[34] – d: The phase diagram of 4-(trans-4-pentyl-cyclohexyl)-benzonitrile is presented up to 260 MPa and 97 °C. Dielectric studies were performed in the nematic phase in the pressure range 0.1–140 MPa, the frequency range 1 kHz–13 MHz and the temperature range 38–77 °C (in the isotropic phase up to 45 MPa).

Differential Scanning Calorimetry (DSC)

[35] – f (N*): In order to examine if there is a tricritical point on the SmA–N* transition line of cholesteryl tetradecanoate the corresponding phase transition enthalpy was measured up to 300 MPa in the temperature range 300–600 K. The pressure–temperature phase diagram was obtained by DTA.

[36] – d: Numerical data for the transition temperatures, enthalpies, entropies and volumes are presented for the N–I and two Cr–N phase transitions of N-(4'-ethoxy-benzylidene)-4-n-butylaniline. The measurements were performed up to 250 MPa.

[37]: Simultaneous measurements of the rate of heat evolution and volume changes by phase transitions were carried out with a pressure–volume–temperature controlled scanning calorimeter up to 175 MPa. No pressure–temperature phase diagram for the compound under test 4-n-pentylphenyl-4-n-decyloxythiobenzoate, which has a nematic and three smectic phases, is given.

Optical Transmission Technique

[38] – i: The effect of pressure on the phase transitions in the seventh to tenth homologs

of 4-alkoxy-4-(4'-nitrobenzoyloxy)-benzoates (DB.kO.NO$_2$) was investigated up to 260 MPa and 200°C. The pressure does not cause any peculiarities for the $k=7$ and 8 homologs, which have a SmA and SmC phase. Only both SmC phases are suppressed at higher pressures. The $k=9$ and 10 homologs, however, exhibit a very complex re-entrant phase behavior.

[39] – i, f: The phase transitions of four pure compounds with strongly polar terminal groups (CN or NO$_2$) and two binary mixtures of components with these groups are observed up to 400 MPa and 190°C. In all cases the partially bilayer SmA$_d$ is bounded, while the nematic phase is stabilized with higher pressures. Furthermore some compounds show re-entrant nematic behavior.

[40] – f, g, i: The pressure–temperature phase diagrams of 4-n-pentylphenyl-4-n-heptyloxy-thiobenzoate and 4-n-pentylphenyl-4-n-octyloxy-thiobenzoate and their mixtures are given up to 350 MPa and 130°C. While in all cases the stability of the SmA phase is increased by increasing pressure, the SmC phase of the pure compounds gets bounded at higher pressures.

[41] – b, e, k: The phase behavior of three ferroelectric compounds, the eighth and tenth homologs of 2-methylbutyl-N-(4-n-alkoxybenzylidene)-4-amino-cinnamate and 2-chloro-propyl-N-(4-n-hexyloxybenzylidene)-4-amino-cinnamate was studied up to 400 MPa and 175°C. All the three compounds possess three smectic phases: SmA, SmC* and SmI*. The chiral C phases of both homologs are suppressed at higher pressures. The pressures and temperatures where these phases disappear are 170 MPa, 115°C and 380 MPa, 162°C, respectively. They are quite different, although their transition temperatures at atmospheric pressure are nearly the same. A suppression of the SmC* phase of the third compound is supposed only about 800 MPa.

[11] – a (N*), c, d: The pressure dependence of the melting and clearing temperatures of numerous 4.4' disubstituted azobenzenes, azoxybenzenes and benzylideneanilines are determined up to 250 MPa and 150°C. Only the clearing temperatures are given for cholesteryl oleyl carbonate and cholesteryl geranyl carbonate up to 300 MPa and 150°C. For the mathematical description of the pressure dependence of the phase transition temperatures the melting point equation of Simon–Glatzel in the modified form of Kraut and Kennedy is successfully applied.

[42] – c (N*): The effect of pressure on the phase transition temperatures of cholesteryl oleate, cholesteryl linoleate and cholesteryl linolenate was studied up to 100 MPa and 80°C. The transition volume of the Sm–N* transition increases with the number of the double bonds in the molecule, while that of the N*–I transition shows little effect.

[20] – f (see DTA)

[43] – a: Turbidity measurements in the isotropic phases of N-(4-methoxybenzylidene)-4-n-butylaniline, N-(4-ethoxybenzylidene)-4-n-butylaniline, and N-(4-cyanobenzylidene)-4-n-nonylaniline were performed up to 120 MPa and 110°C. N–I transition parameters for the three compounds as clearing pressure, temperature, and order parameter are presented in tabular form.

[44] – d (N*): The N*–I and Cr–I transitions of cholesteryl stearate are observed up to 80 MPa and 103°C. The transitions are monotropic with respect to the solid phase.

[45] – a (N*, BP): The BP_{III}–I, BP_I–BP_{III} and N*–BP_I transitions of 4-(2-methylbutylphenyl)-4′-(2-methylbutyl)-4-biphenylcarboxylate were recorded up to 120 MPa and 170 °C. The phase boundaries in the pressure–temperature diagram are linear.

Optical Microscopy

[20] – f: The optical microscopy studies were combined with DTA measurements. Compounds and pressure–temperature range are therefore described in the DTA subsection.

An opposed diamond anvil cell was used as a high pressure optical cell. For getting good textures a 0.1 mm thick gasket made of hardened steel was used. A Sm–N transition is identified by a change from the Schlieren to the focal conic texture with ellipses. The monotropic SmB phase shows a mosaic texture.

[14] – f: A diamond anvil cell was used as a high pressure optical cell. The pressure–temperature phase diagram of trans-4-n-propyl-1-(4-n-butoxyphenyl)cyclohexane is given up to 1600 MPa and 277 °C. The compound exhibits a pressure-induced N and SmB phase. The phases were identified by their textures.

[46] – e, f: Pressure–temperature phase diagrams of N-(4-n-butyloxybenzylidene)-4-n-octylaniline and n-hexyl-4-n-pentyloxybiphenyl-4′-carboxylate are presented up to 300 MPa and 150 °C. Both compounds possess in addition to a SmA phase a SmB phase. One B phase is liquid (hexatic B) the other one crystalline in nature. The neighboring SmA phase of the latter one is bounded above 194 MPa, while the other A phase remains stable with increasing pressure. To explain this differing behavior a model is suggested.

[47] – c: The phase transitions of a racemic composition of 4-(2′-methylbutyl)phenyl-4′-n-nonyloxybiphenyl-4-carboxylate were observed up to 220 MPa and 200 °C. The phase boundaries are extremely linear. The compound under test has five smectic phases: A, C, I, G and H.

[48] – f: The phase diagrams of 4,4′-bis-n-decyloxytolane and 3β-n-tetradecyl-(5α)-cholestane are presented up to 400 MPa and 160°C. While in the case of the former compound the SmA and nematic phase disappear at higher pressures, these phases are stabilized by increasing pressure in the case of the latter compound. This contrary effect of pressure is related to the different dipole moments of both compounds.

[49] – f: The phase transitions of 4-n-octyl-4-cyanobiphenyl and 4-n-octyloxy-4-cyanobiphenyl are observed up to 500 MPa and 140 °C. The SmA phase of the latter compound is a bilayer SmA and is destabilized by increasing pressure.

[50] – d: The pressure–temperature phase diagrams of the 4-n-butyl-, 4-n-heptyl- and 4-n-nonyl-4-methoxy-tolanes are presented up to 400 MPa and 140 °C. The compounds exhibit no peculiarities.

[18] – a, b (TGB): For testing their apparatus the pressure–temperature phase diagrams of 4′-n-pentyl-4-cyanobiphenyl and 1-methylheptyl-4′-(4-n-tetradecyloxyphenylpropioloyloxy)biphenyl-4-carboxylate are determined up to 60 MPa and 370 K. The latter compound exhibits a TGB_A phase which is suppressed at 25 MPa. The TGB_A–I phase boundary line shows a negative gradient. The measurements could be confirmed on the free-standing film.

Barometric Method

This method allows study of first- and second-order phase transitions of pure compounds. The mesogenic sample is enclosed in a rigid but dilatable metallic cell. The sample pressure and temperature are simultaneously recorded. The thermobarograms obtained exhibit a clear change of slope at the phase transition. The measuring apparatus is called a metabolmeter and requires only a very small amount of mesogen.

[51] – d, e: For testing the barometric method the pressure dependence of the phase transition temperatures of 4-n-octyl-4′-cyanobiphenyl and methoxy- and N-(4-ethoxybenzylidene)-4-n-butylaniline are determined.

[52] – f: For further testing the barometric method the SmA–N transition of 4-n-octyl-4′-cyanobiphenyl is additionally observed. This transition, associated with a very small transition volume, is detected as well.

[53] – e, j: The pressure–temperature phase diagrams of bis-(4-n-heptyloxybenzylidene)-1,4-phenylenediamine and terephthalylidene-bis-(4-n-decylaniline) are presented up to 120 MPa and 200 °C. In the case of the former compound the SmI–SmG and SmI–SmC transitions can be individually observed. The SmG–SmF, SmI–SmF, SmC–SmA and SmA–isotropic transitions of the latter compound are all detectable.

[54] – d, f: The method is extended to phase transitions of binary mixtures of 4-methoxy-4′-ethyltolane and 4-methoxy-4′-n-nonyltolane as well as of terephthaldiylidene-N,N′-bis-(1-methyl-heptyl-4-aminocinnamate) and N,N′-bis-(4-n-heptyloxybenzylidene)-phenylene-1,4-diamine.

[55] – f (TGB, N*): The phase transitions of two homologs ($n = 16$ and $n = 18$) of the 4-(3-fluoro-4(R)- or (S)-methylheptyloxy)-4-(4-fluorobenzoyloxy)-tolanes were studied up to 140 MPa and 125 °C. Both compounds have a TGB_A phase which is stabilized by increasing pressure. The $n = 18$ homolog exhibits a pressure-induced cholesteric phase resulting in a TGB_A–N*–I triple point.

[56] – discotic: The pressure–temperature phase diagrams of two disc-like mesogens are presented up to 90 MPa and 170 °C. The compounds investigated are (–)-2.3.6.7.10.11-hexa-[S-(3-methyl)-n-nonanoyloxy]-triphenylene and 2.3.6.7.10.11-hexa-(n-dodecanoyloxy) triphenylene. The transitions between both columnar mesophases of the first compound and between the D_0, hexagonal and rectangular columnar mesophases of the second one can be detected.

Selective Light Reflection

Chiral mesogens can reflect light within a narrow region of wavelength. The reflection is represented by the wavelength of maximum light reflection, which shows a characteristic pressure and temperature behavior near and at a phase transition point. The light reflection causes a quasiabsorption and in the most cases therefore is measured by an absorption spectrophotometer.

[57] – a (N*, BP): The measurements were performed with cholesteryl nonanoate up to 115 MPa and 137 °C. The BP_{II}–BP_{III} transition is detected by a disappearance of the reflection band, the BP_I–BP_{II} and N*–BP_I transitions by jumps in the wavelength of maximum reflection.

[58] – g (N*): The SmA–N* phase transitions of four cholesteryl n-alkanoates, the cholesteryl decanoate, tridecanoate, pentadecanoate and heptadecanoate are observed

up to 260 MPa and 130 °C. Since the wavelength of maximum light reflection is directly proportional to the pitch of the cholesteric phase, the pressure behavior of the reflection wavelength corresponds to that of the pitch. If the *n*-alkyl chain is not too short, the cholesteric phase transforms into the SmA phase with a finite pitch. With increasing transition pressure this transition pitch is shifted to higher values till at a definite transition pressure the pitch is infinite. For the tri-, penta- and heptadecanoate this pressure is about 100 MPa.

[59] – d (N*): The pressure–temperature phase diagrams of cholesteryl *n*-pentanoate and *n*-hexanoate are presented up to 190 MPa and 150 °C. The pressure–temperature behavior of the wavelength of maximum light reflection of both compounds reveals no indication of a pressure-induced SmA phase, which was supposed for the pentanoate by other authors. For both compounds separate crystallization and melting curves were found.

[60] – g (N*): The SmC*–N* phase transition of 4-*n*-hexyloxyphenyl-4'-(2-methylbutyl)biphenyl-4-carboxylate was studied up to 200 MPa and 115 °C. For the SmC* phase also light reflection measurements were performed. Probably because of the high viscosity of this phase the reproducibility was low. No crossing over of the phase transition from first- to second-order was observed at higher pressures.

X-ray

[61] – j: The high pressure X-ray investigations were performed with a high pressure X-ray camera equipped with two cone shaped X-ray windows made of boron single crystals. The apparatus works up to 500 MPa and 120 °C. The pressure dependence of the crystalline-SmG* melting temperature of 4-(2'-methylbutyl)phenyl-4'-*n*-octylbiphenyl-4-carboxylate is given (see also [62]).

[63] – k: The high pressure X-ray measurements were carried out with an X-ray diffraction system equipped with a rotating anode and a bent quartz crystal. The patterns are registered with a stable position-sensitive X-ray detector. The sample is contained in a quartz capillary, which is located in a beryllium cylinder. The pressure is transmitted to the sample by a flexible Teflon bellows, which is connected to the upper part of the cylinder (for details see [64]). The thickness of the smectic layers (SmA, SmC*, SmI*) is studied as a function of pressure at different temperatures. While the layer spacing of a SmC phase with a nematic phase as neighbor is independent of pressure, that of a chiral C phase with a SmA phase as neighbor changes strongly as the pressure increases from the SmC*–SmA transition pressure. The compounds investigated are 4,4'-bis(heptyloxy)azoxybenzene (no SmA phase) and 2-methylbutyl-N-(4-*n*-decyloxybenzylidene)-4-aminocinnamate (with SmA phase).

[65] – f (TGB, N*): The phase behavior of *N*-[4-((6-cholesteryloxycarbonyl)pentyloxy)benzylidene]-4-*n*-butylaniline was investigated by a wide-angle X-ray scattering apparatus up to 100 MPa and 200 °C. The compound exhibits a complex phase sequence with a TGB phase at atmospheric pressure. Only a few phase transitions (without TGB phase) can be observed at higher pressures. For details of the apparatus see [66].

Nuclear Magnetic Resonance (NMR)

[67] – d: The high pressure vessel was machined of nonmagnetic Cu–Be and

Cu–Ni alloys. The equipment works up to 700 MPa. The pressure transmitting medium was purified helium gas. The NMR coil and the sample are contained inside a heating tube. The N–I and the Cr–N transitions of 4,4′-bis-methoxy-azoxybenzene were observed up to 400 MPa and 210 °C. The order parameter of the nematic phase could be determined from a characteristic doublet of the NMR CW spectrum. The order parameter at the N–I transition is found to be constant up to 300 MPa.

Optical Activity

[15] – a (N*, BP): The optical rotation was determined by a half-shade polarimeter. The high pressure optical cell is similar to that of the optical transmission technique.

All phase transitions involving blue phases (BP$_I$, BP$_{II}$, BP$_{III}$) of S-(+)-4′-(2-methylbutyl)phenyl-4-n-decyloxy and -dodecyloxy benzoate could be observed up to 280 MPa and 103 °C. The decyloxy homolog exhibiting all three blue phases at atmospheric pressure looses BP$_{II}$ at 120 MPa. The dodecyloxy homolog only with BP$_I$ shows a pressure-induced BP$_{II}$ already at lower pressures. A correlation between the pretransitional behavior of the optical activity in the isotropic liquid phase and the phase behavior of the blue phases at high pressures is found.

Re-entrant Behavior

In 1975 Cladis discovered the sequence of phases nematic, smectic, and again nematic at atmospheric pressure. The lower-temperature nematic phase was designated as the 're-entrant nematic' phase (N$_{re}$). By 1977 Cladis [68] was successful in giving evidence of a pressure-induced 're-entrant nematic' phase (Fig. 2). The investigated compounds were cyano Schiff bases and cyanobiphenyls with terminal n-alkyl or n-alkoxy chains, which all were known (or suspected) to exhibit a bilayer SmA phase.

The transitions were observed with a high pressure optical microscope stage working up to 1000 MPa.

The most important results are the following (Fig. 2):
1. The re-entrant phase exists only in the supercooled region of the liquid.
2. There is a maximum pressure p_m above which the bilayers are destabilized.

It turned out that p_m increases with increasing transition enthalpy of the SmA–N transition determined at atmospheric pressure. p_m decreases with decreasing number of methylene groups interacting within a layer.

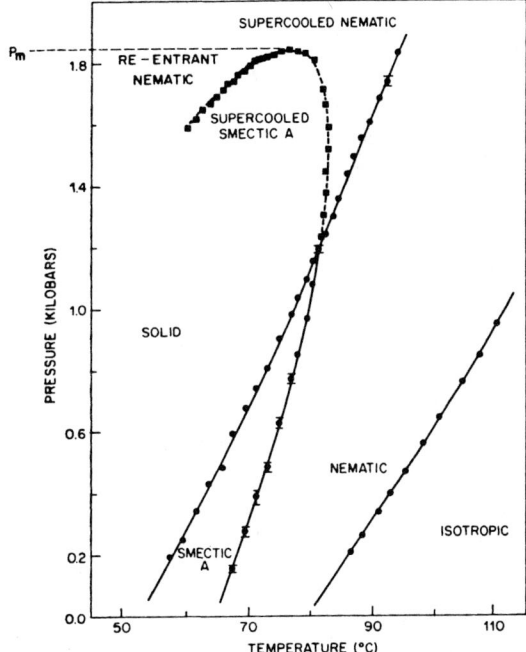

Figure 2. Pressure–temperature phase diagram for 4-cyano-4′-octyloxybiphenyl. Data taken on the re-entrant nematic–smectic A transition in the supercooled liquid are shown as squares. P_m is the maximum pressure at which the smectic exists for this compound. (From [68], reproduced by permission of American Physical Society.)

The studied compounds all have a terminal cyano group which means a very strong dipole. Cladis et al. [69] proposed a structural model of the bilayer smectic A phase for this kind of molecules (Fig. 3). The molecules are assumed to be associated in antiparallel pairs, which results in a weak interacting between the different polar parts of the pairs and a less dense packing of the molecules. Thus a transition of such bilayer SmA phase to a re-entrant nematic phase seems to be evident, because in this phase empty spaces of the structure are filled up more efficiently. Probably for similar reasons Pollmann et al. [70–72] found a pressure-induced re-entrant cholesteric phase behavior for ternary mixtures of cholesteryl n-alkanoates which, however, are terminally nonpolar (see Fig. 4). The mixtures of cholesteryl tetradecanoate (Ch-14), nonanoate (Ch-9) and propionate (Ch-3) possess one component which has a considerably shorter n-alkylcarboxy chain than the other components. This probably leads to 'holes' in the smectic structure. With increasing pressure the molecules are increasingly forced into these 'holes' until the layer structure of the SmA phase finally is destroyed and the re-entrant cholesteric phase appears.

Guillon et al. [73] investigated the pressure dependence of the SmA phase layer spacing of N-(4-cyanobenzylidene)-4-n-octyloxyaniline by X-ray-diffraction measurements. While the layer spacing decreases, when the SmA phase is pressurized towards the solid phase, it remains constant when this smectic phase is pressurized towards the re-entrant nematic phase.

Shashidhar and Rao [74] performed high pressure X-ray studies on liquid crystals with re-entrant behavior with an opposed diamond anvil cell. They found that the layer spacing of the SmA phase of 4-n-octyloxy-4'-cyanobiphenyl first decreases more or less linearly with increasing pressure up to 140 MPa, then increases at still higher pressures. Since this compound shows re-entrant nematic behavior at high pressures, this result confirms the prediction of Cladis et al. that the occurrence of a re-entrant nematic phase is associated with an expansion of the SmA phase layer spacing.

Shashidhar et al. [75] studied the influence of pressure on the SmA-(re-entrant) nematic and N–I phase boundaries of mixtures of 4-n-hexyloxy- and 4-n-octyloxy-4'-cyanobiphenyl. The maximum pressure where the SmA and re-entrant nematic phase, respectively, still exist, decreases with increasing mole fraction, x, of the hexyloxy homolog till at $x \approx 0.30$ the SmA phase disappears. Just in this mole fraction region the slope of the N–I transition

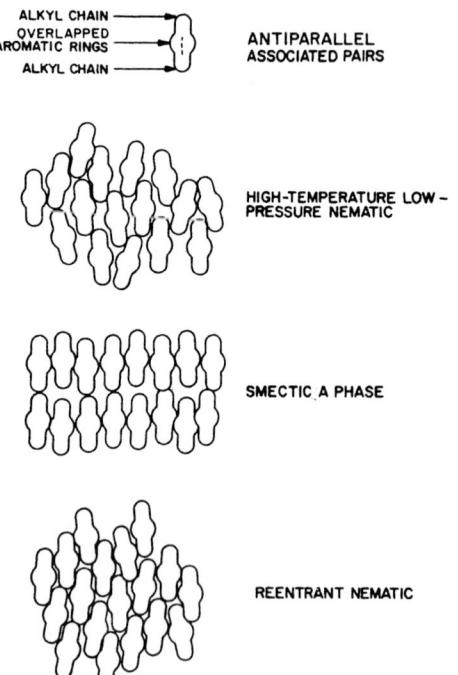

Figure 3. Schematic arrangement of antiparallel associated pairs in the nematic, smectic A, and re-entrant nematic phase. (From [64], reproduced by permission of American Physical Society.)

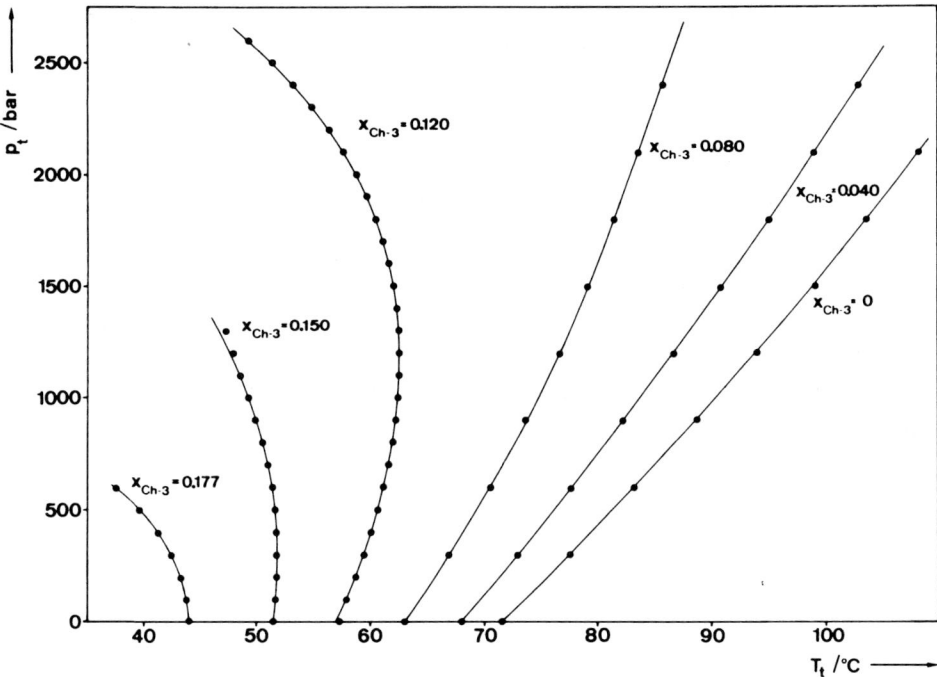

Figure 4. Pressure–temperature smectic A–cholesteric phase boundary lines of the ternary system Ch-9/Ch-14/Ch-3; $x_{Ch-9}/x_{Ch-14} = 2.11$ (x = mole fraction). (From [71], reproduced by permission of Taylor & Francis Ltd.)

(at 0.1 MPa) exhibits a marked anomaly. Corresponding investigations of mixtures of 4-n-octyloxy-4′-cyanobiphenyl and N-(4-n-cyanobenzylidene)-4′-octyloxyaniline were carried out by Herrmann [76].

Prasad et al. [77] studied the phase transition behavior of 4-n-octyloxy- and 4-n-nonyloxybenzoyloxy-4′-cyanoazobenzene. While the pressure–temperature diagram of the eighth homolog shows no peculiarities, the ninth one shows already at atmospheric pressure a SmA (partially bilayer)–N, N_{re}–SmA (partially bilayer) and a re-entrant SmA (monolayer)–N_{re} phase transition. The partially bilayer SmA–(re-entrant) N phase boundary shows an elliptic shape and the partially bilayer SmA is suppressed at higher pressures. The re-entrant monolayer SmA–re-entrant N phase boundary is a straight line at all pressures.

Similar investigations of eight pure compounds were performed by Kalkura et al. [78] and of further five pure compounds and some mixtures by Prasad et al. [79]. While pressure-induced re-entrant cholesteric phase behavior of mixtures of terminally nonpolar components is already known [70], Ratna et al. [80] found, that a mixture of the terminally nonpolar components n-dodecyl-N-4-(4-ethoxybenzylideneamino)-α-methylcinnamate and 4-n-heptyloxyphenyl-4-(4-ethylcyclohexanoyloxybenzoate) not only shows a re-entrant N phase behavior at atmospheric pressure but also at higher pressures.

Illian et al. [81] observed a closed loop re-entrant N–SmA phase boundary in a temperature–mole fraction diagram at atmospheric pressure. At higher pressures the SmA region of existence decreases until at a critical pressure of 145 MPa this region is reduced to a point. The measurements were carried out with mixtures of 4-n-heptyloxy-4′-cyanobiphenyl and a mixture of N-4-cyanobenzylidene-4′-n-octyloxyaniline and 4-n-pentylphenyl-4-n-octyloxybenzoate.

The pressure-induced re-entrant phase behavior is not restricted to one re-entrant N phase and not only to N phases. Raja et al. [38] studied the effect of high pressure on the phase transitions in the seventh to tenth homologs of 4-alkoxyphenyl-4′-nitrobenzoyloxybenzoates. The decyloxy member, and only this one, exhibits a conspicuous phase diagram: besides two pressure-induced SmA phases and a bounded SmC phase a 'quadruple re-entrance' is observed: SmA–N–SmA$_d$–N–SmA$_d$–N.

Illian et al. [82] studied the pressure effect on the phase transition behavior of binary mixtures of terminally polar and nonpolar components which exhibit induced SmA phases. The re-entrant N phase is stabilized by increasing pressure and at about 101 MPa the SmA–re-entrant N phase boundary meets the N–SmA one. At higher pressures a nematic gap appears and finally the SmA phase on the polar side of the temperature–mole fraction diagram (175 MPa) disappears.

Daoudi et al. [83] applied the thermodynamic approach 'equal Gibbs energy analysis' to the pressure–composition phase diagrams of binary mixtures exhibiting re-entrant phase behavior. Three different solution models are tested. The experimental data for the 4-n-hexyloxy- and 4-n-octyloxy-4′-cyanobiphenyl system are successfully described by the regular solution hypothesis.

6.2.4.3 Critical Phase Transitions Under Pressure

The different types of liquid crystals are closely related to one another in a thermodynamic sense so that a great potential for multicritical phenomena can be supposed.

Tricritical Behavior

Following the terminology of Landau and Lifshitz [84] order parameters and quantities of state, respectively, for instance the enthalpy, entropy and volume, change discontinuously at a phase transition of the first kind, and continuously at a phase transition of the second kind. More commonly the terms 'first order' and 'second order' are used.

When a line of first-order phase transitions changes to a line of second-order phase transitions the corresponding point in the phase diagram is called a critical point of the second kind (Tisza [85]) or a tricritical point (Griffiths [86]). In the following the latter expression will be taken. In 1973 Keyes et al. [87] could find such a tricritical point (TCP) for the SmA–N* phase transition of cholesteryl oleylcarbonate at 266 MPa and 60.3 °C. The TCP was indicated by a sharp discontinuity in the transmitted light. Shashidhar et al. [19] could confirm this discovery by DTA measurements (p(TCP) = 267 MPa) and later in Shichijyo et al. [88] by volume and DTA investigation (p(TCP) = 300 MPa) (see also [89]). The TCP p–T coordinates obtained in this way are probably essentially too high since the entire disappearance of the DTA peak is taken as indication of the TCP. In the case of cholesterogens following de Gennes [90, 91] Pollmann et al. [92] determined by selective light reflection measurements that transition pressure as the tricritical pressure, where the pitch of the cholesteric helix turns to infinity according to the divergence of the characteristic coherence length. At this tricritical pressure clear transition volume [92] and enthalpy effects [35] remain, which can be understood as pretransitional effects. The latter disappear only at still higher pressures. The role of the pretransitional enthalpy of the SmA–N* transition can be seen in [16], where the tricritical pressures of nine cholesteryl n-alkanoates (nonanoate until heptadecanoate) were determined. For 4'-n-octyl-4-cyanobiphenyl, Shashidhar et al. [24] found a TCP of this transition at 268 MPa and 92.5 °C.

Garland [10] studied second-order SmA–N and SmC–SmA phase transitions by very precise heat capacity measurements up to 300 MPa. Similar measurements of the critical heat capacity near the SmA–N transition were performed by Kasting et al. [93, 94]. McKee et al. [95] carried out orientational order determinations near a possible SmA–N TCP ($p=289$ MPa, 140 °C) by NMR. From McMillan's theory [96] in the case of a TCP of the SmA–N (N*) transition at atmospheric pressure a value of 0.866 (model parameter $\delta=0$) for $T(\text{SmA–N})/T(\text{N–I})$ follows. Thus a rough test by the corresponding transition temperatures at higher tricritical pressures is possible.

Bartelt et al. [28] suppose a TCP also for the SmC–SmA transition. The first-order SmC–SmA transition of terephthaldiylidene-bis-(4-n-octylaniline) probably changes to second-order at 110 MPa and 239 °C.

Multicritical Behavior

Re-entrant N–SmC–SmA and N–SmA–SmC Multicritical Behavior

The re-entrant N–SmC–SmA multicritical point and the N–SmA–SmC multicritical point are, by definition, points in the temperature–concentration or pressure–temperature plane at which three second order phase boundaries meet. At this point all involved three phases become indistinguishable. After no N–SmA–SmC point for a single component liquid crystal system could be detected by studying the pressure–temperature phase behavior of N-(4-n-pentyloxybenzylidene)-4'-n-hexylaniline up to 800 MPa [97], Shashidhar et al. [98, 99] were successful in finding another kind of multicritical point at 52 MPa and 86.2 °C: a N_{re}–SmC–SmA point for N-4-(4-n-decyloxybenzoyloxy)-benzylidene-4'-cyanoaniline.

Shortly after finding this point Shashidhar et al. also observed a N–SmA–SmC point under high pressures [100]. The N–SmA–SmC point of 4-n-heptacylphenyl-4'-(4''-cyanobenzoyloxybenzoate) occurs at 30.4 MPa and 149.9 °C. The most remarkable feature of the obtained p–T phase diagram is its topology near the N–SmA–SmC point which is nearly the same as obtained for binary mixtures in a temperature-concentration diagram at atmospheric pressure (Fig. 5a and b). A quantitative description of the three phase boundaries of the pure compound near the N–SmA–SmC point agrees so closely with that of the binary mixtures that the universal behavior of the N–SmA–SmC point of the latter at atmospheric pressure can be extended to such a point at higher pressures for single component systems. This is not valid for the behavior of the N_{re}–SmC–SmA point under high pressure of the above mentioned compound which is different from that at atmospheric pressure in the case of binary mixtures. The N_{re}–SmC–SmA point at atmospheric pressure, however, can on certain conditions exhibit the same universal behavior as the N–SmA–SmC point [101]. There is no satisfying theoretical explanation for the uniqueness in the topology of the phase transition lines near the N–SmA–SmC and N_{re}–SmC–SmA multicritical points.

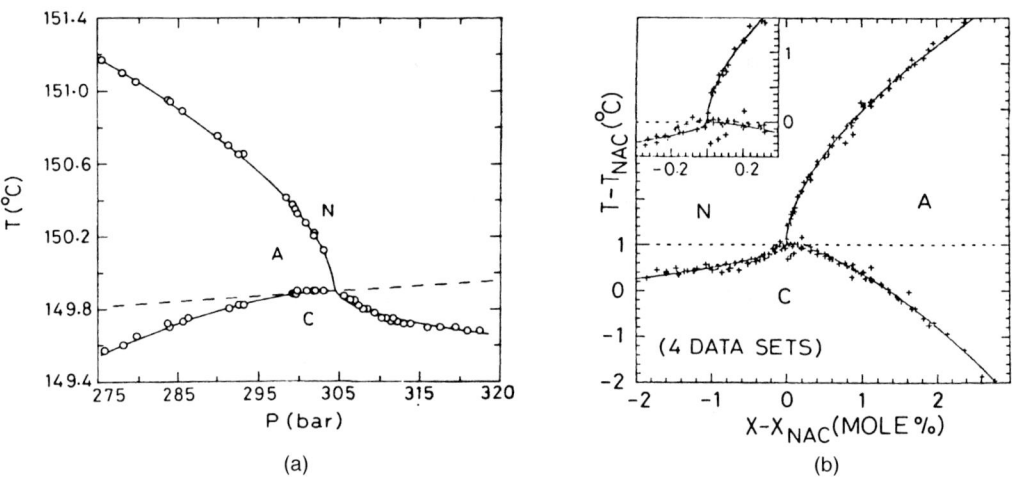

Figure 5. (a) High resolution temperature–pressure diagram in the vicinity of the N–A–C multicritical point in 4-n-heptacylphenyl-4'-(4"-cyanobenzoyloxybenzoate). The solid lines are computer fits of data evaluated with equations representing the NA, NC, and AC phase boundaries, respectively. (b) The universal temperature–concentration plot showing the data for four binary liquid crystal systems. (From [100], reproduced by permission of American Physical Society.)

Legrand et al. [102] report an electroclinic effect in the cholesteric phase near a N*–SmA–SmC* multicritical point of the $n = 8$ homolog of a biphenyl benzoate series. This 'chiral' analogon to the N–SmA–SmC point is found at 15 MPa and 144 °C.

Smectic C–Smectic A-TGB$_A$ and Smectic A-TGB$_A$–Cholesteric Multicritical Behavior*

Experimental evidence under high pressures is given by Anakkar et al. [103] for two new kinds of multicritical points involving Twist Grain Boundary (TGB) phases: SmC*–SmA-TGB$_A$ and SmA-TGB$_A$-N* points for single component systems. The investigated compound was the $n = 12$ homolog of the series 3-fluoro-4((R) or (S)-methylheptyloxy)-4'-(4-alkyloxy-3-fluorobenzoyloxy)tolanes. The p–T coordinates of the multicritical points are: SmC*–SmA-TGB$_A$: 67.5 MPa, 91.5 °C; SmA-TGB$_A$-N*: 106 MPa, 113 °C. The result is in good agreement with the Renn–Lubensky theory.

Anakkar et al. [104] also investigated the $n = 10$ and $n = 11$ members of the above tolane series which both exhibit the TGB$_A$ phase. For $n = 11$ also experimental evidence is given for a SmC*–SmA-TGB$_A$ multicritical point ($p = 13$ MPa, $T = 367$ K).

Bicritical Behavior

A bicritical point is a point in the temperature–concentration or pressure–temperature plane at which two second-order phase boundaries and one first-order phase boundary meet. In order to find such a bicritical point in the pressure–temperature plane Bahr et al. [105] studied the pressure–temperature phase behavior of binary mixtures of 4-(4-n-butyloxybenzoyloxy)-4'-nitroazobenzene and 4-(4-n-nonyloxybenzoyloxy)-4'-cyanoazobenzene. From the p–T phase diagrams of the pure components which are also presented a bicritical behavior of their mixtures can be expected. Two mixtures with molar fractions of about 0.80 for the cyano component show an abrupt change of the slope of the SmA–N phase boundary at 30 MPa. Here the transition of the partially bilayer SmA (A_d) phase to the nematic phase changes to a transition of the monolayer SmA phase (A_1) to the nematic phase. Just at this crossover point the SmA_1–SmA_d phase boundary, which, however, could not be detected by the used optical transmission technique, should meet the SmA–N phase boundary.

Since the SmA_1–N and SmA_d–N transitions are expected to be second order (low T_{A-N}/T_{N-I}) values [97]) and the SmA_1–SmA_d transition first-order (symmetry arguments), the meeting point should be a bicritical point. Moreover the topology of the p–T phase diagram obtained resembles that of a diagram exhibiting known bicritical points [106].

Prasad et al. [107] studied the critical character of the N*–TGB–SmA and TGB–SmA–SmC* meeting points in the pressure–temperature phase diagram of mixtures of 4-(2-methylbutyl)phenyl-4'-n-octyl-biphenyl-4-carboxylate and 4-n-dodecyloxy-biphenyl-4'-yl-4-(2'-methylbutyl)-benzoate. For a weight fraction, $x = 0.25$, of the benzoate there is a pressure-induced TGB phase between the N* and the SmA phase. Analysis of the topology of the pressure–temperature diagram in the vicinity of the resulting N*–TGB–SmA meeting point more points to a critical than to a bicritical point. At $x = 0.64$ there appears a pressure-induced SmA phase between the TGB and the SmC* phase so that a TGB–SmA–SmC* meeting point is observed. The topology of this phase diagram suggests that the latter point is a bicritical point.

6.2.4.4 Pressure-Volume-Temperature Behavior

The physical properties of a substance depend on the intermolecular distance and since the change of the specific volume or density with pressure and temperature means a change of the mean distance between the molecules, p–V–T data are required to understand the pressure–temperature behavior of physical properties.

Experimental Techniques

(a) Tungsten Tip Surface Scanning by Screw Spindle [108]

The apparatus allows the measurement of p–V–T data of liquid crystal compounds up to 400 MPa.

The sample is enclosed in a glass piezometer and separated from the pressure transmitting fluid by mercury. As the sample volume decreases with increasing pressure the level of the mercury, which is contained in a calibrated glass tube, sinks and thus opens the electric contact to a tungsten tip. By means of a screw spindle the contact of the tungsten tip with the mercury level is made again. The number of revolutions needed for it is a measure for the decrease in volume of the sample and can be measured by an electronic altimeter outside the autoclave. The reproducibility of the density is ± 0.0003 g cm^{-3}.

(b) The Bridgman Flexible Bellows with Slide Wire [109]

In this experimental set up the liquid crystal is contained in a stainless-steel bellows which is placed inside a pressure-vessel-double-oven arrangement. The oven temperature is regulated by a proportional servomechanism using a sensor thermocouple. The Bridgman bellows with slide-wire technique [110] is used for the measurement of the sample volume. As the sample is compressed the flexible bellows contracts and moves a slide-wire under a contact. The fixed end does not move with respect to the contact.

The position of the slide-wire from which the change in volume of the liquid crystal sample can be determined is measured by means of an a.c. bridge arrangement. The accuracy in determination of the specific volume is ± 0.00015 cm^3 g^{-1}.

(c) Piston Displacement [111]

A steel cylinder is used to contain the sample under investigation. The sample is compressed by a steel piston, the displacement of which is measured by means of a differential transformer. From the displacement of the piston the volume decrease of the liquid crystal sample can be evaluated.

The pressure is determined from the pressure dependence of the electrical resistance of a manganin wire which is placed inside the pressure chamber with the sample. The accuracy of the measurements of the volume changes is within $\pm 0.3\%$.

Experimental Results and Theoretical Description

In the following only those publications which contain important theoretical descriptions are specified. The p–V–T investigated compounds of the publications not mentioned here, are however, in 'the list of compounds' with the referring citation.

The letter (a–c) after the following citations means the experimental method which has been used (see 'experimental techniques').

[112] (a): p–V–T and viscosity data of the methoxy (MBBA) and ethoxy (EBBA) homologs of the N-(4-alkyloxybenzylidene)-4'-n-butylaniline series are presented up to 200 MPa.

The p–V–T behavior of both compounds can be well described by the semiempirical Tait equation:

$$\kappa = -\frac{1}{V_0}\left(\frac{\partial V}{\partial p}\right)_T = \frac{C}{B+p} \quad (1)$$

here used in the form

$$\frac{V_0 - V}{V_0} = C \cdot \log \frac{B+p}{(B+p_0)} \quad (2)$$

The constants B and C were evaluated from the experimental data and are given together with the compressibility coefficient κ at 1 bar for MBBA in Table 1.

C can be assumed to be independent of temperature. B is correlated with the molecular structure of the substance.

[108] (a): The p–V–T data for three trans-4-n-alkyl-1-(4-cyanophenyl)-cyclohexanes (propyl, pentyl, heptyl) and one eutectic mixture were measured up to 200 MPa.

For isotherms in the nematic phase beginning at high N–I transition pressures and temperatures (p_1, V_1), the V_0 of the Tait equation (see [112]) is not known with suf-

Table 1. The compressibility coefficient and the coefficients of the Tait equation in both phases of MBBA.

| | $\kappa \cdot 10^6$ [bar^{-1}] | | | | B [bar] | |
| | 1 bar | Transition point | | C | | |
°C	isotropic	isotropic	nematic	for all temperatures	isotropic	nematic
60	54.55	45.88	43.96	isotropic	2116.2	2087.4
70	55.94	41.77	40.13	0.2658	2063.7	2034.5
80	57.35	38.92	36.65	nematic	2012.5	1977.9
90	59.10	36.60	34.38	0.2619	1947.0	1908.7

ficient accuracy. The equation is therefore modified to:

$$\frac{V_1 - V}{V_1} = \frac{C \cdot V_0}{V_1} \cdot \log\left(\frac{B+p}{B+p_1}\right)$$

$$= C^* \cdot \log\left(\frac{B+p}{B+p_1}\right) \qquad (3)$$

The constants B, C, and C^* (not independent of temperature) are listed in Tables 2 and 3 for the isotropic and nematic phase, respectively. An influence of the length of the alkyl chain is obvious.

[109] (b): p–V–T data were obtained for the nematic range of the first six members of the homologous series of 4,4′-bis-alkoxyazoxybenzenes up to 123 MPa.

Measurements performed at constant volume showed that the temperature range of the nematic phase is considerably larger at constant volume than at constant pressure.

Measurements at constant temperature established a decrease in N–I transition volume and fractional volume of a substance as one moves up to higher transition pressures and temperatures. The volume at the N–I transition point also decreases. The results in this work, especially the influence of the length of the flexible tail of the molecules on the data, are compared with the predictions of numerous theories.

[113] (b): The p–V–T behavior of the non-re-entrant nematic (N) 4,4′-bis-n-heptyloxy-azoxybenzene (7OAB) and the re-entrant nematic (N_{re}) 4-n-octyloxy-4′-cyanobiphenyl (8OCB) were compared with one another. The N–I transition volumes of these two compounds are similar at low transition temperatures and pressures but exhibit quite a different transition temperature dependence.

The SmC–N transition volumes are an order of magnitude larger than those of the N_{re}–SmA transition. For theoretical considerations the isothermal compressibility coefficient and the parameter

$$\gamma = -\frac{\partial(\ln T/T_0)}{\partial(\ln V/V_0)} \qquad (4)$$

are used, where T_0 and V_0 represent a point on the phase equilibrium curve. The latter parameter is sensitive to the relative

Table 2. Tait parameter for the isotropic range, B [bar].

	Temperature (°C)			
	60	70	80	90
Propyl (30%)	1867	1812	1756	1701
Pentyl (40%)	1847	1793	1738	1683
Heptyl (30%)	1822	1767	1713	1657
Eutectic mixture	1834	1779	1724	1670

$C = 0.2850$ for all compounds and temperatures.

Table 3. Tait parameter for the nematic range, B [bar].

	$C^* \times 10^4$ Temperature (°C)				B [bar] Temperature (°C)			
	60	70	80	90	60	70	80	90
Propyl	1834	1862	1889	1917	890	814	715	622
Pentyl	1818	1845	1872	1899	969	865	761	657
Heptyl	1807	1834	1861	1889	984	879	774	669
Eutectic mixture	1827	1854	1882	1907	959	855	751	647

$C = 0.1800$ for all compounds and temperatures.

[114] (b): p–V–T measurements were used to investigate short-range smectic fluctuations in the N–I transitions of the seventh to ninth members of the homologous series of 4,4′-bis-n-alkylazoxybenzenes.

The experimental results confirm the theoretical prediction that the N–I transition volume increases when the N–I transition gets closer to a SmA–N transition. This is due to short-range smectic order. This influence of the nematic range on the N–I transition volume can be very obviously seen for the ninth member of the series which exhibits a SmA–N–I triple point at 92 MPa and 103 °C.

[111] (c): Pressure–volume isotherms of the isotropic and nematic phase are presented in the temperature range 57–83 °C up to 90 MPa. Whereas the N–I transition volume remains constant with increasing transition pressure that of the Cr–N transition clearly decreases.

According to an equation of Hiwatari and Matsuda [115] the p–V–T data for the nematic state yields an $1/r^{10.8}$ dependence of the repulsive potential energy on the intermolecular distance r or $\rho^{3.6}$, when the density ρ is used. McColl [116] had found $\rho^{3.7}$ for 4,4′-bis-methoxy-azoxybenzene.

[117] (c): The pressure–volume isotherms of 4-(trans-4-pentylcyclohexyl)benzonitrile were measured near its clearing transition. For the parameter $\gamma = -\partial \ln T_C / \partial \ln V_C$ a value of +5.24 is determined and related to the intermolecular potential energy. The value of γ' depends on the number of carbons in the alkyl chain.

[118] (b): p–V–T measurements were carried out on N-(4-methoxybenzylidene)-4-n-butylaniline (MBBA), N-(4-cyanobenzylidene)-4-n-octyloxyaniline (CBOOA), 4,4′-bis-n-heptyloxy-azoxybenzene (HAB) and cholesteryl nonanoate (CN).

The variations in the N–I transition volumes and the isothermal compressibility coefficients of the different compounds can be attributed to a large extent to the effects of end chain flexibility. The N (N*)–I transition volumes of CBOOA and CN decrease more rapidly with increasing transition temperature than that of the relatively inflexible MBBA.

The transition densities of all four compounds increase with increasing transition temperatures, for MBBA the slope is the smallest, for HAB the largest. The experimental result for MBBA is compared with the predictions of three theoretical models. The nematic ordering seems to be almost entireley a result of the repulsive forces.

[119] (b): The p–V–T equation of state of 4-n-octyloxy-4′-cyanobiphenyl was measured to test the hypothesis of Pershan and Prost [120] that the re-entrant nematic phase occurs because the SmA phase can exist only near an optimum density. The experimental results are in substantial disagreement with the hypothesis.

[121–124] (c): The p–V–T behavior of 4-n-octyl-4′-cyanobiphenyl was observed near the N–I transition. While the N–I transition volume was almost constant with increasing transition temperature (T_{tr}) the volume of the nematic phase at the transition temperature V_{tr} decreases.

The plot of $\ln T_{tr}$ vs. $\ln V_{tr}$ yields a straight line with a slope of -4.3. This value agrees with the prediction of the Pople–Karasz theory not with that of Maier–Saupe. An important role of repulsive forces in the N–I transition is suggested.

Analogous measurements with the pentyl, hexyl, and heptyl members of the series result −7.62, −6.10, and −5.15 for the above plot, hence a decrease with the length of the alkyl chain [122]. The larger values of these compounds compared to that of the octyl member make it obvious that in this case the volume-dependent part of the intermolecular potential for the nematic state is harder than the repulsive part of the Lennard–Jones potential but softer than the hard-rod potential.

In a $p–V–T$ study on 4-fluorophenyl-trans-4-n-heptylcyclohexanecarboxylate $d\ln T_{tr}/d\ln V_{tr} = -2.98$ was obtained. This value is noticeably lower than those of the former compounds and attributed to a larger intramolecular potential at the phase transition point (see also 4′-mono-4-propyl-cyanophenylcyclohexane with a value of −8.29 [124]).

The results are discussed on the basis of the molecular structure, particularly of the molecular rigidity.

6.2.4.5 Appendix

List of Compounds

4-n-alkyl-4′-cyanobiphenyl (pentyl to heptyl) [122]
trans-4-alkyl-(4-cyanophenyl)-cyclohexane (propyl, pentyl, heptyl) [108]
4-(trans-4-n-butylcyclohexyl)benzonitrile [127]
cholesteryl butyrate [133]
cholesteryl hexadecanoate (palmitate) [130]
cholesteryl hexanoate (capronate) [129]
cholesteryl nonanoate (pelargonate) [118, 131]
cholesteryl octadecanoate (stearate) [137]
cholesteryl oleyl carbonate [88]
cholesteryl pentanoate (valerate) [128]
cholesteryl propionate [134]
cholesteryl tetradecanoate (myristate) [131]
α,ω-bis[4-cyanobiphenyl-4′-yl]alkane (nonane, decane) [126]
4,4′-bis-alkyloxy-azoxybenzene (methyl to hexyl) [109]
4,4′-bis-alkyl-azoxybenzene (heptyl to nonyl) [114]
4,4′-bis-methoxy-azoxybenzene [109, 135, 136]
4,4′-bis-n-octyloxy-azoxybenzene [132]
N-(4-ethoxybenzylidene)-4-n-butylaniline [112]
4-fluorophenyl-trans-4-heptylcyclohexane-carboxylate [123]
4,4′-bis-n-heptyloxy-azoxybenzene [113, 118, 131]
N-(4-methoxybenzylidene)-4-n-butylaniline [112, 118, 125]
4-n-octyl-4′-cyanobiphenyl [121]
4-n-octyloxy-4′-cyanobiphenyl [113, 119]
4-(trans-4-n-pentylcyclohexyl)benzonitrile [108, 117]
N-(4-n-pentyloxybenzylidene)-4-n-butylaniline [111]
4′-mono-4-propyl-cyanophenyl-cyclohexane [108, 124]

Physical Properties under Pressure

Property	Refs.
Viscosity	[112, 138, 139]
Dielectricity	[140–145, 146–148, 152, 153]
Ferroelectricity	[150]
Refraction	[149, 151]
Elasticity (elastic constants)	[151]
Ultrasound	[88, 154]
Electrooptic (elastic constant/ dielectric anisotropy ratio)	[155]
Acoustic	[156, 157]
Acoustooptic	[158]

6.2.4.6 References

[1] G. A. Hulett, *Z. Phys. Chem.* **1899**, *28*, 629–672.
[2] N. A. Puschin, I. W. Grebenschtschikow, *Z. Phys. Chem.* Stöchiometrie und Verwandtschaftslehre (Leipzig) **1926**, *124*, 270–276.
[3] J. Robberecht, *Bull. Soc. Chim. Belg.* **1938**, *47*, 597–639.
[4] P. D. Garn, R. J. Richardson, *Thermal Analysis* **1971**, *3*, 123–130.
[5] A. S. Reshamwala, R. Shashidhar, *J. Phys. E* **1977**, *10*, 180–183.
[6] J. Herrmann, H. D. Kleinhans, G. M. Schneider, *J. de Chimie Physique* **1983**, *80*, 111–117.
[7] R. Sandrock, Doctoral Thesis, **1982**, Ruhr-Universität Bochum, Germany.
[8] A. Bartelt, G. M. Schneider, *Rev. Sci. Instrum.* **1989**, *60*, 926–929.
[9] M. Kamphausen, *Rev. Sci. Instrum.* **1975**, *46*, 668–669.
[10] C. W. Garland, *Thermochimica Acta* **1985**, *88*, 127–142.
[11] M. Feyz, E. Kuss, *Ber. Bunsenges. Phys. Chem.* **1974**, *78*, 834–842.
[12] P. H. Keyes, H. T. Weston, W. B. Daniels, *Phys. Rev. (Lett.)* **1973**, *31*, 628–630.
[13] P. Pollmann, *J. Phys. E* **1974**, *7*, 490–492.
[14] A. Rothert, G. M. Schneider, *High Press. Res.* **1990**, *3*, 285–287.
[15] P. Pollmann, E. Voß, *Liq. Cryst.* **1997**, *23*, 299–307.
[16] P. Pollmann, B. Wiege, *Liq. Cryst.* **1989**, *6*, 657–666.
[17] G. Illian, H. Kneppe, F. Schneider, *Liq. Cryst.* **1989**, *4*, 643–652.
[18] C. Carboni, H. F. Gleeson, J. W. Goodby, A. J. Slaney, *Liq. Cryst.* **1993**, *14*, 1991–2000.
[19] R. Shashidhar, S. Chandrasekhar, *J. de Physique Col. C 1* **1975**, *36*, 49–51.
[20] R. Shashidhar, H. Herrmann, H. D. Kleinhans, *Mol. Cryst. Liq. Cryst.* **1982**, *72*, 177–182.
[21] A. N. Kalkura, R. Shashidhar, G. Venkatesh, M. E. Neubert, J. P. Ferrato, *Mol. Cryst. Liq. Cryst.* **1983**, *99*, 177–183.
[22] G. Venkatesh, R. Shashidhar, D. S. Parmar, *Proceedings of Int. Liq. Crystals Conference* **1979**, Bangalore.
[23] R. Shashidhar, *Mol. Cryst. Liq. Cryst.* **1977**, *43*, 71–81.
[24] R. Shashidhar, G. Venkatesh, *J. de Physique Col. C3* **1979**, *40*, 396–400.
[25] J. Rübesamen, G. M. Schneider, *Liq. Cryst.* **1993**, *13*, 711–719.
[26] W. Spratte, G. M. Schneider, *Ber. Bunsenges. Phys. Chem.* **1976**, *80*, 886–891.
[27] R. Krombach, G. M. Schneider, *Thermochimica Acta* **1994**, *231*, 169–175.
[28] A. Bartelt, H. Reisig, J. Herrmann, G. M. Schneider, *Mol. Cryst. Liq. Cryst. (Lett.)* **1984**, *102*, 133–138.

[29] J. Herrmann, A. Bartelt, H. D. Kleinhans, H. Reisig, G. M. Schneider, *Mol. Cryst. Liq. Cryst. (Lett.)* **1983**, *92*, 225–230.
[30] W. Klement, L. H. Cohen, *Mol. Cryst. Liq. Cryst.* **1974**, *27*, 359–373.
[31] C. Rein, D. Demus, *Liq. Cryst.* **1993**, *15*, 193–202.
[32] N. A. Tikhomirova, L. K. Vishin, V. N. Nosow, *Sov. Physics – Crystallography* **1973**, *17*, 878–880.
[33] S. Chandrasekhar, B. K. Sadashiva, K. A. Suresh, N. V. Madhusudana, S. Kumar, R. Shashidhar, G. Venkatesh, *J. de Physique C3* **1979**, *40*, 120–124.
[34] T. Brückert, T. Büsing, A. Würflinger, S. Urban, *Mol. Cryst. Liq. Cryst. Sci. Technol. Sect. A* **1995**, *262*, 1497–1508.
[35] J. Herrmann, R. Sandrock, W. Spratte, G. M. Schneider, *Mol. Cryst. Liq. Cryst. (Lett.)* **1980**, *56*, 183–188.
[36] R. Sandrock, M. Kamphausen, G. M. Schneider, *Mol. Cryst. Liq. Cryst.* **1978**, *45*, 257–265.
[37] S. L. Randzio, *J. Thermal. Anal.* **1992**, *38*, 1989–1993.
[38] V. N. Raja, B. R. Ratna, R. Shashidhar, G. Heppke, C. Bahr, J. F. Marko, J. O. Indeken, A. N. Berker, *Phys. Rev. A* **1989**, *39*, 4341–4344.
[39] S. K. Prasad, S. Pfeiffer, G. Heppke, R. Shashidhar, *Z. Naturforsch.* **1985**, *40a*, 632–635.
[40] A. N. Kalkura, S. K. Prasad, R. Shashidhar, *Mol. Cryst. Liq. Cryst.* **1983**, *99*, 193–202.
[41] S. K. Prasad, B. R. Ratna, R. Shashidhar, V. Surendranath, *Ferroelectrics* **1984**, *58*, 101–105.
[42] M. Nakahara, K. Maeda, J. Osugi, *Bull. Chem. Soc. Jpn.* **1980**, *53*, 2499–2501.
[43] W. J. Lin, P. H. Keyes, *J. Physique* **1980**, *41*, 633–638.
[44] K. Maeda, M. Nakahara, K. Hara, J. Osugi, *Rev. Phys. Chem. Jpn.* **1979**, *49*, 85–90.
[45] J. Hollmann, P. Pollmann, P. J. Collings, *Liq. Cryst.* **1993**, *15*, 651–658.
[46] P. E. Cladis, J. W. Goodby, *Mol. Cryst. Liq. Cryst. (Lett.)* **1982**, *72*, 307–312.
[47] P. E. Cladis, J. W. Goodby, *Mol. Cryst. Liq. Cryst. (Lett.)* **1982**, *72*, 313–317.
[48] L. Liebert, W. B. Daniels, J. Billard, J. Malthete, *C. R. Acad. Sci. Paris* **1977**, *285*, Série C 451–453.
[49] L. Liebert, W. B. Daniels, *J. de Physique (Lett.)* **1977**, *38*, L 333–335.
[50] L. Liebert, W. B. Daniels, J. Billard, *Mol. Cryst. Liq. Cryst.* **1977**, *41*, 57–62.
[51] J. M. Buisine, B. Soulestin, J. Billard, *Mol. Cryst. Liq. Cryst.* **1983**, *91*, 115–127.
[52] J. M. Buisine, B. Soulestin, J. Billard, *Mol. Cryst. Liq. Cryst.* **1983**, *97*, 397–406.
[53] J. M. Buisine, *Mol. Cryst. Liq. Cryst.* **1984**, *109*, 143–157.
[54] J. M. Buisine, J. Billard, *Mol. Cryst. Liq. Cryst.* **1985**, *127*, 353–379.

[55] A. Daoudi, A. Anakkar, J. M. Buisine, F. Bougrioua, N. Isaert, H. T. Nguyen, *J. Thermal. Anal.* **1996**, *46*, 337–345.
[56] J. M. Buisine, J. Malthete, C. Destrade, N. H. Tinh, *Physica (Amsterdam)* **1986**, *139/140 B*, 631–635.
[57] P. Pollmann, G. Scherer, *Z. Naturforsch.* **1979**, *34a*, 255–256.
[58] F. Pakusch, P. Pollmann, *Mol. Cryst. Liq. Cryst.* **1982**, *88*, 255–271.
[59] P. Pollmann, B. Wiege, *Mol. Cryst. Liq. Cryst. (Lett.)* **1982**, *72*, 271–276.
[60] P. Pollmann, K. Schulte, *Liq. Cryst.* **1987**, *2*, 701–706.
[61] J. Przedmojski, S. Gierlotka, R. Wisniewski, B. Pura, W. Zajac, *Ferroelectrics* **1989**, *92*, 345–348.
[62] J. Przedmojski, S. Gierlotka, B. Pura, R. Wisniewski, *Cryst. Res. Technol.* **1988**, *23*, K72–K73.
[63] D. Guillon, J. Stamatoff, P. E. Cladis, *J. Chem. Phys.* **1982**, *76*, 2056–2063.
[64] D. Guillon, P. E. Cladis, D. Aadsen, W. B. Daniels, *Phys. Rev. A* **1980**, *21*, 658–665.
[65] Y. Maeda, Y. Yun, J. Jin, *Mol. Cryst. Liq. Cryst.* **1996**, *280*, 85–90.
[66] Y. Maeda, N. Tanigaki, A. Blumstein, *Mol. Cryst. Liq. Cryst.* **1993**, *237*, 407–418.
[67] B. Deloche, B. Cabane, D. Jerome, *Mol. Cryst. Liq. Cryst.* **1971**, *15*, 197–209.
[68] P. E. Cladis, R. K. Bogardus, W. B. Daniels, G. N. Taylor, *Phys. Rev. (Lett.)* **1977**, *39*, 720–723.
[69] P. E. Cladis, R. K. Bogardus, D. Aadsen, *Phys. Rev. A* **1978**, *18*, 2292–2306.
[70] P. Pollmann, B. Wiege, *Mol. Cryst. Liq. Cryst. (Lett.)* **1984**, *102*, 119–124.
[71] P. Pollmann, B. Wiege, A. Rothert, *Liq. Cryst.* **1988**, *3*, 225–233.
[72] P. Pollmann, B. Wiege, *Liq. Cryst.* **1988**, *3*, 1203–1213.
[73] D. Guillon, P. E. Cladis, D. Aadsen, W. B. Daniels, *Phys. Rev. A* **1980**, *21*, 658–665.
[74] R. Shashidhar, K. V. Rao, *Proceedings of Int. Liq. Crystals Conference* **1979**, Bangalore.
[75] R. Shashidhar, H. D. Kleinhans, G. M. Schneider, *Mol. Cryst. Liq. Cryst. (Lett.)* **1981**, *72*, 119–126.
[76] J. Herrmann, *Mol. Cryst. Liq. Cryst.* **1982**, *72*, 219–224.
[77] S. K. Prasad, R. Shashidhar, K. A. Suresh, A. N. Kalkura, G. Heppke, R. Hopf, *Mol. Cryst. Liq. Cryst.* **1983**, *99*, 185–191.
[78] A. N. Kalkura, R. Shashidhar, N. Subramanya Raj Urs, *J. Physique* **1983**, *44*, 51–55.
[79] S. K. Prasad, S. Pfeiffer, G. Heppke, R. Shashidhar, *Z. Naturforsch.* **1985**, *40a*, 632–635.
[80] B. R. Ratna, R. Shashidhar, V. N. Raja, C. Nagabhushan, *Mol. Cryst. Liq. Cryst.* **1989**, *167*, 233–237.
[81] G. Illian, H. Kneppe, F. Schneider, *Ber. Bunsenges. Phys. Chem.* **1988**, *92*, 776–780.
[82] G. Illian, H. Kneppe, F. Schneider, *Liq. Cryst.* **1989**, *4*, 643–652.
[83] A. Daoudi, A. Anakkar, J. Buisine, *Thermochimica Acta* **1994**, *245*, 219–229.
[84] L. D. Landau, E. M. Lifschitz in *Lehrbuch der theoretischen Physik V*, Akademie Verlag, Berlin **1971**, p. 642.
[85] L. Tisza, *Ann. Phys.* **1961**, *13*, 1–92.
[86] R. B. Griffiths, *Phys. Rev. (Lett.)* **1970**, *24*, 715–717.
[87] P. H. Keyes, H. T. Weston, W. B. Daniels, *Phys. Rev. (Lett.)* **1973**, *31*, 628–630.
[88] S. Shichijyo, T. Okamoto, T. Takemura, *Jpn. J. Appl. Phys.* **1982**, *21*, 1260–1267.
[89] K. J. Lushington, G. B. Kasting, C. W. Garland, *Phys. Lett.* **1979**, *74A*, 143–145.
[90] P. G. de Gennes, *Solid State Commun.* **1972**, *10*, 753–756.
[91] P. G. de Gennes in *The Physics of Liquid Crystals*, Clarendon Press, Oxford, **1975**, p. 331.
[92] P. Pollmann, K. Schulte, *Ber. Bunsenges. Phys. Chem.* **1985**, *89*, 780–786.
[93] G. B. Kasting, C. W. Garland, K. J. Lushington, *J. Physique* **1980**, *41*, 879–884.
[94] G. B. Kasting, K. J. Lushington, C. W. Garland, *Phys. Rev. B* **1980**, *22*, 321–331.
[95] T. J. McKee, J. R. McColl, *Phys. Rev. (Lett.)* **1975**, *34*, 1076–1080.
[96] W. L. McMillian, *Phys. Rev. A* **1971**, *4*, 1238–1246.
[97] R. Shashidhar, A. N. Kalkura, S. Chandrasekhar, *Mol. Cryst. Liq. Cryst. (Lett.)* **1980**, *64*, 101–107.
[98] R. Shashidhar, A. N. Kalkura, S. Chandrasekhar, *Mol. Cryst. Liq. Cryst. (Lett.)* **1982**, *82*, 311–316.
[99] R. Shashidhar, S. K. Prasad, S. Chandrasekhar, *Mol. Cryst. Liq. Cryst.* **1983**, *103*, 137–142.
[100] R. Shashidhar, B. R. Ratna, S. K. Prasad, *Phys. Rev. (Lett.)* **1984**, *53*, 2141–2144.
[101] S. Somasekhara, R. Shashidhar, B. R. Ratna, *Phys. Rev. A* **1986**, *34*, 2561–2563.
[102] C. Legrand, N. Isaert, J. Hmine, J. M. Buisine, J. P. Parneix, H. T. Nguyen, C. Destrade, *Ferroelectrics* **1991**, *121*, 21–31.
[103] A. Anakkar, A. Daoudi, J. M. Buisine, N. Isaert, F. Bougrioua, H. T. Nguyen, *Liq. Cryst.* **1996**, *20*, 411–415.
[104] A. Anakkar, A. Daoudi, J. M. Buisine, N. Isaert, T. Delattre, H. T. Nguyen, C. Destrade, *J. Thermal Anal.* **1994**, *41*, 1501–1513.
[105] Ch. Bahr, G. Heppke, R. Shashidhar, *Z. Naturforsch.* **1985**, *40a*, 1311–1315.
[106] J. Prost, *Adv. Physics* **1984**, *33*, 1–46.
[107] S. K. Prasad, G. G. Nair, S. Chandrasekhar, J. W. Goodby, *Mol. Cryst. Liq. Cryst.* **1995**, *260*, 387–394.

[108] E. Kuss, *Mol. Cryst. Liq. Cryst.* **1981**, *76*, 199–210.
[109] R. V. Transfield, P. J. Collings, *Phys. Rev. A* **1982**, *25*, 2744–2749.
[110] W. A. Steele, W. Webb in *High Pressure Physics and Chemistry* (Ed. R. S. Bradley), Academic Press, New York, **1963**, Vol. 1, p. 145.
[111] C. Hanawa, T. Shirakawa, T. Tokuda, *Chem. Lett. Chem. Soc. Jpn.* **1977**, 1223–1226.
[112] E. Kuss, *Mol. Cryst. Liq. Cryst.* **1978**, *47*, 71–83.
[113] C. S. Johnson, P. J. Collings, *J. Chem. Phys.* **1983**, *79*, 4056–4061.
[114] M. W. Lampe, P. J. Collings, *Phys. Rev. A* **1986**, *34*, 524–528.
[115] Y. Hiwatari, H. Matsuda, *Progr. Theor. Phys.* **1972**, *47*, 741–764.
[116] J. R. McColl, *Physics Lett.* **1972**, *38A*, 55–57.
[117] H. Ichimura, T. Shirakawa, T. Tokuda, T. Seimiya, *Bull. Chem. Soc. Jpn.* **1983**, *56*, 2238–2240.
[118] P. H. Keyes, W. B. Daniels, *J. Physique C3* **1979**, *40*, 380–383.
[119] R. Shashidhar, P. H. Keyes, W. B. Daniels, *Mol. Cryst. Liq. Cryst.* **1986**, *3*, 169–175.
[120] P. S. Pershan, J. Prost, *J. Physique (Lett.)* **1979**, *40*, L-27–30.
[121] T. Shirakawa, T. Inoue, T. Tokuda, *J. Phys. Chem.* **1982**, 1700–1702.
[122] T. Shirakawa, T. Hayakawa, T. Tokuda, *J. Phys. Chem.* **1983**, *87*, 1406–1408.
[123] T. Shirakawa, H. Eura, H. Ichimura, T. Ito, K. Toi, T. Seimija, *Thermochim. Acta* **1986**, *105*, 251–256.
[124] T. Shirakawa, M. Arai, T. Tokuda, *Mol. Cryst. Liq. Cryst.* **1984**, *104*, 131–139.
[125] E. A. S. Lewis, H. M. Strong, G. H. Brown, *Mol. Cryst. Liq. Cryst.* **1979**, *53*, 89–99.
[126] A. Abe, S. Y. Nam, *Macromol.* **1995**, *28*, 90–95.
[127] T. Shirakawa, H. Ichimura, I. Ikemoto, *Mol. Cryst. Liq. Cryst.* **1987**, *142*, 101–106.
[128] V. Ya. Baskakov, V. K. Semenchenko, *ZhETF Pis. Red.* **1973**, *17*, 580–583; *JETP Lett.* **1973**, *17*, 414.
[129] V. M. Byankin, V. K. Semenchenko, V. Ya. Baskakov, *Zh. Fiz. Khim.* **1974**, *48*, 1250–1253.
[130] N. A. Nedostup, V. K. Semenchenko, *Zh. Fiz. Khim.* **1977**, *51*, 1628–1631.
[131] V. K. Semenchenko, V. M. Byankin, V. Ya. Baskakov, *Sov. Phys. Crystallogr.* **1975**, *20*, 187–191.
[132] V. Ya. Baskakov, V. K. Semenchenko, V. M. Byankin, *Zh. Fiz. Khim.* **1976**, *50*, 200–202.

[133] V. K. Semenchenko, N. A. Nedostup, V. Ya. Baskakov, *Zh. Fiz. Khim.* **1975**, *49*, 1547–1550.
[134] V. K. Semenchenko, V. M. Byankin, V. Ya. Baskakov, *Zh. Fiz. Khim.* **1974**, *48*, 2353–2355.
[135] V. Ya. Baskakov, V. K. Semenchenko, N. A. Nedostup, *Kristallogr.* **1974**, *19*, 185–187.
[136] V. Ya. Baskakov, V. K. Semenchenko, V. M. Byankin, *Zh. Eksp. Teor. Fiz.* **1974**, *66*, 792–797; *Sov. Phys.-JETP* **1974**, *39*, 383–385.
[137] V. K. Semenchenko, N. A. Nedostup, Y. Ya. Baskakov, *Zh. Fiz. Khim.* **1975**, *49*, 1543.
[138] A. C. Diogo, *Solid State Commun.* **1984**, *50*, 895–897.
[139] H. Doerrer, H. Kneppe, F. Schneider, *Liq. Cryst.* **1992**, *11*, 905–915.
[140] M. Ozaki, N. Yasuda, K. Yoshino, *Jpn. J. Appl. Phys.* **1987**, *26*, L 1927–L 1929.
[141] T. Brückert, A. Würflinger, *Ber. Bunsenges. Phys. Chem.* **1993**, *97*, 1209–1213.
[142] H.-G. Kreul, S. Urban, A. Würflinger, *Phys. Rev. A* **1992**, *45*, 8624–8631.
[143] A. Würflinger, *Internat. Rev. Phys. Chem.* **1993**, *12*, 89–121.
[144] T. Brückert, T. Büsing, A. Würflinger, S. Urban, *Mol. Cryst. Liq. Cryst. Sci. Technol. Sect. A* **1995**, *262*, 1497–1508.
[145] T. Brückert, S. Urban, A. Würflinger, *Ber. Bunsenges. Phys. Chem.* **1996**, *100*, 1133–1137.
[146] S. Urban, B. Gestblom, T. Brückert, A. Würflinger, *Z. Naturforsch. A: Phys.-Sci.* **1995**, *50*, 984–990.
[147] S. Urban, *Z. Naturforsch. A: Phys.-Sci.* **1995**, *50*, 826–830.
[148] N. Yasuda, S. Fujimoto, S. Funado, *J. Phys. D: Appl. Phys.* **1985**, *18*, 521–530.
[149] R. G. Horn, *J. Physique* **1978**, *39*, 167–172.
[150] S. M. Khened, S. K. Prasad, V. N. Raja, S. Chandrasekhar, B. Shivkumar, *Ferroelectrics* **1991**, *121*, 307–318.
[151] P. L. Sherrell, J. D. Bünning, T. E. Faber, *Liq. Cryst.* **1987**, *2*, 3–19.
[152] T. Uemoto, K. Yoshino, Y. Inuishi, *Mol. Cryst. Liq. Cryst.* **1981**, *67*, 137–152.
[153] T. Uemoto, K. Yoshino, Y. Inuishi, *Jpn. J. Appl. Phys.* **1980**, *19*, 1467–1472.
[154] D. Ergashev, *Mezhvuz. Sb. Nauch. Tr. Vses. Zaoch. Mashinostr. Int.* **1982**, *34*, 129–133.
[155] V. I. Kireev, S. V. Pasechnik, V. A. Balandin, *Pis'ma Zh. Tekh. Fiz.* **1989**, *15*, 88–90.
[156] A. S. Lagunov, A. N. Larionov, *Akust. Zh.* **1984**, *30*, 344–351.
[157] S. V. Pasechnik et al., *J. Phys. (Les Ulis. Fr.)* **1984**, *45*, 441–449.
[158] S. V. Pasechnik, V. A. Balandin, V. I. Kireev, *Pis'ma Zh. Tekh. Fiz.* **1988**, *14*, 1756–1760.

6.3 Fluctuations and Liquid Crystal Phase Transitions

P. E. Cladis

6.3.1 Introduction

Many liquid crystal phase transitions involve broken continuous symmetries in real space and their interactions on a molecular scale are short range [1]. As a result, fluctuations have long been known to be an important feature of liquid crystal phase transitions: even weakly first order (discontinuous) ones. Compared to major advances in our understanding of fluctuation controlled second-order (continuous) phase transitions, relatively little is known about fluctuation phenomena (critical phenomena) at first-order phase transitions such as the nematic–isotropic transition.

The central concept of critical phenomena is "universality". Simply put, this means that fluctuation dominated continuous phase transitions are controlled by a unique length, $\xi \equiv \xi_o/\varepsilon^\nu$, where ξ is a measure of the length over which order parameter fluctuations are correlated. $\varepsilon \equiv |T-T_c|/T_c$, with T_c the second-order phase transition temperature. ν is the correlation length exponent that is completely defined by the symmetry of the system (i.e. the number of components of its order parameter) and the dimensionality, d, of the space in which the material is embedded [2]. ξ_o, the 'bare' correlation length, is a measure of what fluctuations have to beat to become critical (i.e. to take control of the phase transition). In low-temperature superconductors, $\xi_o \simeq 200$ nm while in liquid crystals, $\xi_o \approx 0.5$–1 nm. Because ξ_o is so large, low temperature superconductors have only been studied in a 'mean field' limit. In liquid crystals, a mean field limit with a cross-over to a critical regime as $T \to T_c$ can also be observed.

A surprising result is that the mean field limit is exact for spatial dimension, d, greater than or equal to a critical dimension, d_c. For phase transitions far from a tricritical point, $d_c = 4$. For phase transitions in the vicinity of a tricritical point, $d_c = 3$: exponents in the vicinity of a tricritical point are mean field [2].

In the vicinity of fluctuation dominated phase transitions, the temperature dependence of thermodynamic parameters such as the specific heat at constant pressure, $C_p \approx \varepsilon^\alpha$, and the order parameter, $\psi \approx \varepsilon^\beta$, are all related to ξ through a free energy density giving rise to scaling relations. For example: $\alpha = 2 - \nu d$ and $\beta = (d-2)\nu/2$ [2]. Despite the variety of their continuous broken symmetries, most liquid crystal phase transitions are expected to fall in the 3D-XY (helium) universality class with, $\alpha \approx -0.01$, $\nu \approx 0.67$ and $\beta \approx 0.33$.

Here we give an overview of fluctuation dominated thermotropic liquid crystal phase transitions with a few hints of emerging aspects. From this perspective, the situation may be fairly summarized by noting that while analogies to phase transition models in spin-space (e.g. XY model with two components for the order parameter or Ising model with one) or momentum space (superconductivity) are a powerful tool to predict qualitative behavior for fluctuation dominated, real space, high temperature liquid crystal phase transitions, there is a significant gap between several quantitative (and qualitative) expectations and experimental measurements.

One reason may be that liquid crystals have first-order phase transitions that are so weakly discontinuous, their first-order nature escapes detection by traditional methods such as adiabatic calorimetry [3]. Recently, a macroscopic qualitative test of phase transition order [4] revealed that even immeasurably small discontinuities at first-order phase transitions [5] using static tests, have a distinct dynamic signature in interface (front) propagation compared to second-order or continuous phase transitions [4]. It is important to know the order of a phase transition because for universality to apply at all levels of its hierarchy [2], ξ must approach infinity continuously: there can be no discontinuities at T_c. If there are, 'all bets are off' [6].

A salient feature of molecular materials, including liquid crystals, is that they are liquid well above $T(K) \equiv 0$. As a result they have generic long range correlations even far from critical points or hydrodynamic instabilities [7] that could make it difficult to access critical regimes before being finessed by a first-order phase transition. At these high temperatures, externally supplied noise [8] (e.g. expressed by random thermal fluctuations) can *suppress* the onset of macroscopic instabilities such as spatial turbulence far from any phase transition [9]. Dynamic correlation functions decay with long time tails supporting nonequilibrium steady states and contribute to divergences in transport coefficients that cannot be accounted for by a 'static' theory of phase transitions [7]. Liquid crystal contributions to dynamic critical phenomena [10] are beginning to emerge [11].

The richness implicit in what we have learned and can learn from fluctuations in liquid crystals seems endless. Here we give only the merest hint of the enormous volume of information contained in a vast, and still growing, body of research.

6.3.2 The Nematic–Isotropic Transition

Liquid crystal materials in the isotropic liquid state are transparent. As temperature decreases towards the nematic liquid state, the material becomes more and more turbid (i.e. scatters more and more light). The picture is that nematic droplets are forming in the isotropic liquid to scatter the light. As the temperature is lowered, more and more droplets of the ordered state appear until the whole system is nematic. In liquid crystals, such droplets, called 'cybotatic groups', were an early precursor of the notion of 'fluctuations'. The idea of droplets of a disordered state appearing in an ordered state, and vice versa, as phase transitions are approached is still a useful picture for explaining pretransitional changes in light scattering data at first-order phase transitions, for example. While similar to classical explanations of critical opalescence as the liquid–gas critical point is approached, the difference is that in liquid crystals, droplets in the nematic state have a different symmetry from the disordered state, for example isotropic liquid [12]. Because there is no surface energy between the ordered and disordered states at continuous or second-order phase transitions, the picture of compact structures driven by surface tension is unsatisfactory. At second-order phase transitions, it has turned out that a more useful concept is to think in terms of statistical objects termed fluctuations with a characteristic length over which ordering is correlated [1].

When there are no external magnetic or electric fields, the nematic and isotropic liquids do not have the same symmetry. The effect of an applied field is to induce orientational order in the isotropic phase that grows with increasing field intensity. Eventually, with increasing external fields, the jump at the transition vanishes at a field induced critical point. The increase in orientational order in the isotropic liquid results in an enhancement of the nematic–isotropic transition temperature, δT_{N-I}. In analogy to paramagnetism, the isotropic liquid with field induced orientational order is called paranematic. In low molecular weight liquid crystals, $\delta T_{N-I} \approx 1$ mK even in the most intense fields currently available. However, as can be seen in another chapter of this volume [13], in liquid crystal elastomers, $\delta T_{N-I} \approx 10$ K (i.e. is large for even quite modest mechanical fields). As a result, it is now possible to pass through the field induced critical point and observe the state beyond where nematic and paranematic states are indistinguishable.

Fluctuation effects are large in polymeric liquid crystals even far from phase transition temperatures. For example, in a novel liquid crystalline elastomer with a SmA–I transition, under an external mechanical stress, it was found in a mean field limit that, well in the isotropic phase, nematic fluctuations dominate with a cross-over temperature closer to the transition where SmA fluctuations become more important [14].

6.3.3 The Uniaxial–Biaxial Nematic Transition

Both uniaxial positive (rod-like) and uniaxial negative (disc-like) nematics exist. The nematic order parameter, $Q_{\alpha\beta}$, distinguishes between the two possibilities [15]. If $Q_{\alpha\beta}$ refers to uniaxial positive nematics (rod-like), then, $-Q_{\alpha\beta}$ describes uniaxial negative nematics (disc-like). As a result, the N–I transition is necessarily first order for geometric reasons [15].

While both side-on side-chain liquid crystal polymers [16] and low molecular weight liquid crystals [17] have been reported to have biaxial nematic phases, only in a lyotropic liquid crystal system [18], has the uniaxial–biaxial nematic transition been studied in detail from the point of view of critical phenomena. This transition is found to be second order. So far, it is the one liquid crystal system where earlier theoretical expectations [19] of fluctuation dominated phase transitions and later experimental results are most fully in agreement with respect to both static [20] and dynamic [21] aspects of critical phenomena. In particular, static critical phenomena predicts 3D-XY exponents which have been observed (with irrelevant corrections to scaling) by Saupe et al. [19]. Transport parameters were not expected [19] to show any singularities at the transition [19] as later verified by Roy et al. [21] because the dynamics of the biaxial order parameter is nonconserved (Model A) [19].

6.3.4 Type I Smectic A's: Nematic–Smectic A Transition

Following McMillan's prediction of the conditions under which the N–SmA transition could be second order [22] and de Gennes' analogy between the N–SmA transition and the normal-metal–superconducting transition [23], many experiments followed to test new scaling ideas at phase transitions. This was because de Gennes also pointed out that ξ_0 was small in liquid crystals, so fluctuations did not have to work so hard to take over the transition (i.e. liquid crystal phase transitions could have observable critical regimes). At the normal-metal–superconducting transition in Type I superconductors, magnetic field lines ($\mathbf{H} = \text{curl}\mathbf{A}$, where $\mathbf{A}$ is the magnetic vector potential) are expelled (Meissner effect). Type I superconductors are perfect diamagnets. In a similar way, at the transition to an ordered Sm A phase from the nematic phase, twist and bend deformations contained in curl$\mathbf{n}$, with $\mathbf{n}$ the liquid crystal director, are expelled [23].

The most recent extensive overview of the experimental situation for the N–SmA transition has been given by Garland and Nounesis [24]. The overall picture is that exponents for the correlation length parallel ($v_\parallel$) and perpendicular ($v_\perp$) to the director are non-universal and different. Hyperscaling (substituting $v = (v_\parallel + 2v_\perp)/3$ in scaling relations e.g. $\alpha = 2 - vd$) approximately works. McMillan's number, $M = T_{N-A}/T_{N-I}$, emerges as a robust measure of a relatively sharp cross-over in the specific heat exponent form the 3D-XY value when $M < 0.93$ with a systematic transition to the tricritical value ($\alpha = 0.5$) as $M \to 1$. As the authors emphasize, the 3D-XY model cannot account for the observation $v_\parallel \neq v_\perp$.

Many of the compounds in the large list of Garland and Nounesis have N–SmA phase transitions determined to be continuous by calorimetry and X-ray diffraction and discontinuous using the more powerful dynamic test of phase transition order [4]. In particular, the compounds known as 8CB and 9CB, with second order N–SmA phase transitions by the standard tests [25], were found to exhibit dynamic behavior consistent with a first-order phase transition. This test further enabled a determination of the small cubic term (the HLM effect) [26] predicted to be a feature of the N–SmA transition in the vicinity of a tricritical point, an appropriate limit for this result [26] to apply [5]. The conclusion is that while the N–SmA transition is intrinsically first order, the discontinuity is 'small'. Indeed, precise measurements of changes in sound speed and the elastic constants deduced from these measurements find evidence for 3D-XY fluctuations even far from the transition temperature [27].

The theoretical picture for the N–SmA transition has been succinctly summarized by Lubensky. [28] He explains that there are several important differences between the normal-metal–superconducting and the N–SmA transitions where K_1, the splay elastic constant, emerges as a 'dangerous irrelevant parameter' [28].

First, unlike superconductors, SmA order is not long range (Peierls' argument). This introduces an additional length, l. As a result, correlations in director fluctuations and SmA order parameter fluctuations have *different* lengths. Where director orientational fluctuations are correlated on a length, ξ_{thermo}, SmA order parameter fluctuations are correlated on a length ξ_{eff} where $1/\xi_{\text{eff}} = 1/\xi_{\text{thermo}} + 1/l$. The temperature dependence (i.e. critical exponents) of l depends on splay fluctuations which in turn depend on K_1.

Second, in the N–SmA case, 'nature has so arranged it that measurable quantities are in a gauge where fluctuations are most violent' [28]. Measurable quantities in superconductors are gauge invariant, (i.e. independent of div$\mathbf{A}$). In the N–SmA case, this corresponds to $K_1 \equiv 0$. The Coulomb gauge, div$\mathbf{A} \equiv 0$ in superconductors, corresponds to div$\mathbf{n} \equiv 0$ or $K_1 \to \infty$ [29]. In both these cases $v_\| \equiv v_\perp$. However, in the $K_1 \to \infty$ limit, gauge transformation theory predicts that the 3D-XY fixed point is unstable leaving only the stable fixed point at $K_1 \equiv 0$.

Dislocation-loop melting theory [30], taking into account entropic effects, introduced two new fixed points on the K_1 axis, one stable and the other unstable, where $v_\| = 2v_\perp$ is 'built-in' [28]. As the value of the unstable fixed point is at smaller K_1 than the stable one, K_1 has to be larger than its unstable value for fluctuations to converge to dislocation-loop theory's fixed point [31]. Recent self-consistent one loop calculations by Andereck and Patton [32] and the persistent anisotropy in the critical behavior of ξ have renewed interest in this still mysterious aspect of the N–SmA transition [24]. However, Lubensky emphasizes that the only way theoretically to obtain the larger $K_1 \neq 0$ stable fixed point is by ignoring l [28]. If one includes l, the only stable fixed point is at $K_1 = 0$ where $v_\| = v_\perp$.

It may be that mixtures can be found [33] to tune K_1 sufficiently well to locate the stable fixed point where $K_1 \neq 0$. For example, some early experiments [34] measuring the elastic constants of K_2 and K_3 in mixtures of CBOOA (cyanooctyloxybiphenyl) and ortho-MBBA (o-methoxybenzilidene butylaniline), found at sufficiently large concentration of o-MBBA that $K_3 \approx \xi_\|$ diverged with an exponent $v_\| \approx 1$. In these same mixtures, $K_2 \approx \xi_\perp^2/\xi_\|$ showed no divergence [34] consistent with $v_\| = 2v_\perp$. In view of significant advances in materials and measuring techniques since these early days, redoing these experiments, including precise measurements of K_1 [35], in a system where this 'dangerous irrelevant parameter' can be varied through a large range of values, may be a way to find the $K_1 \neq 0$ stable fixed point.

A first indication that long range dynamic correlations [7] existed in liquid crystals appeared with the discovery of divergences in the Ericksen–Leslie viscous coefficients, α_1, α_3 and α_6 [36]. Brochard's result was found using Kawasaki's mode coupling theory [37]. One consequence of divergences in these viscous coefficients is that 'flow alignment' breaks down in the nematic phase on approaching the N–SmA transition [38]. While this result has inspired, and is continuing to inspire, the study of macroscopic instabilities in liquid crystals under well-controlled and well-defined conditions [39], more traditional features of dynamic critical phenomena are beginning to emerge, especially at other liquid crystal phase transitions [40]. In particular, it has been shown that dynamic behavior in the vicinity of the N–SmA transition depends on material parameters specific to a compound. The implication is that the universal behavior expected for dynamic critical phenomena will only be observed in certain compounds (e.g. those with relatively stiff and incompressible layers in the case of SmA) and not others [11].

6.3.5 Type II Smectic A's: Cholesteric–Smectic A Transition

The parameter distinguishing between Type I and Type II superconducting behavior in a magnetic field [41] is the Ginzburg–Landau parameter, κ. When $\kappa < 1/\sqrt{2}$, the superconductor is Type I (Meissner effect) and when $\kappa > 1/\sqrt{2}$, it is Type II (flux lines penetrate but the penetration is not complete). The analogue of studying superconductors in a magnetic field (curl$\mathbf{A} \neq 0$) is to turn on spontaneous twist deformations ($\mathbf{n}$curl$\mathbf{n}$, chirality, a property of cholesteric liquid crystals) in the liquid crystal case [23]. The transition is analogous to a Type I superconductor and the N*–SmA transition, in a high enough chirality limit, is analogous to Type II superconductors exhibiting the analogue of a vortex lattice in what are now known as twist grain boundary phases (TGB) [42].

Renn and Lubensky proposed a model for the analogue of the vortex lattice [43] for the N*–SmA transition [42]. Simultaneously and independently such a phase was discovered by direct observation, supported by X-ray analysis as well as freeze-fracture, by Goodby et al. [44] between the isotropic liquid and a SmC* phase. The first TGBA phase found to exist between N* and SmA was studied in a dynamic experiment [45]. In the Renn-Lubensky model, uniform sheets of SmA of extent $\xi_\perp$, separated by parallel planes of screw dislocations, twist relative to each other [46].

When Dasgupta and Halperin included fluctuations in the HLM effect [47], they found that the second order nature was restored in the high κ limit of superconductors. Recently in a dynamic experiment [45], there was no interface between cholesteric and TGBA in a material for which κ was estimated to be ≈ 3 times larger than $1/\sqrt{2}$. In addition, the N phase region did not propagate (i.e. either advance or retreat): TGBA grew as SmA melted and was squeezed out by an advancing SmA phase. These observations [45] support a second order TGBA–N* phase transition [4].

High resolution specific heat measurements raise the possibility of another interesting scenario [48]. These measurements show a large broad feature well above the TGB–N* transition and only a tiny (if any) latent heat at the N*–TGB transition. A similar behavior was observed at the first TGBA–N* transition [45] in a different compound making it a possible generic feature of TGB–N* transitions. In this case [45], the size of the heat signature at the TGBA–N* transition was time dependent. Chan et al. [48] suggest that the broad pretransitional heat feature was consistent with the formation of a liquid of screw dislocations condensing into either a glassy or an ordered TGB phase [49]. They point out [48] the resemblance between their phase diagram and a theoretical one for a Type II superconductor with strong thermal fluctuations where, with decreasing temperature, a vortex liquid state condenses into either a glassy or an ordered state [50]. Whether the second order nature of the N*-TGBA transition [45] is a result of fluctuations obliterating the small HLM singularity or a proposed [50] (but not yet experimentally verified) [51] liquid–glass transition calls for more work. Materials with TGB phases with a large temperature range [52] may help clarify this question relevant not only to liquid crystal and high T_c superconducting materials, but also other complex materials such as polymers, electrorheological fluids and ferrofluids [42].

6.3.6 Transitions between Tilted and Orthogonal Smectic Phases

In the SmC phase, the director tilts relative to the layer normal at an angle, $\theta_T \neq 0$, breaking the continuous rotational symmetry in the plane of SmA layers. The order parameter for this transition is θ_T, the angle between the layer normal and the director. Guillon and Skoulios [53] suggested that a direct measure of θ_T is given by cos $\theta_T = d_C/d_A$ where d_C is the layer spacing measured in the C phase and d_A is its (maximum) value at $T_{SmA-SmC}$. According to theory [54], this transition is in the 3D-XY universality class. Although measurements by the MIT group found mean field exponents and unobservably small critical regimes ($\xi_0 \approx 700$ nm) [55], a compound studied by Delaye [56] in light scattering, as well as optical measurements in other compounds by Ostrovskii et al. [57] and Galerne [58], and

X-ray layer spacing measurements of Keller et al. [59], observable critical regimes were found with 3D-XY exponents as expected by theory [54].

Recently Ema et al. [60] found that, with 'significant corrections-to-scaling', their high resolution specific heat measurements agreed with theoretical expectations [54] at the SmC$_\alpha^*$–SmA transition. The SmC$_\alpha^*$ phase is composed of alternating layers of equal and opposite tilt [61]. Ema et al. [60] give a nice summary of these transitions in the context of critical phenomena as well as a sense of the detailed scrutiny high quality data are subjected to in such studies.

The controversy over the nature of the SmA–SmC transition was essentially resolved by Benguigui and Martinoty [62]. These authors addressed the question of why fluctuation effects can be observed on long lengths scales with ultra-sound propagation but not by specific heat or X-ray measurements. In particular in $\overline{8}$S5, where specific heat, X-ray and dilatometric measurements found mean-field behavior, ultrasound damping measurements showed significant fluctuation effects [63]. Most recently [64], they further elucidated the question of how certain compounds do show mean-field behavior while others show critical behavior at long wavelengths. Their explanation rests on the Andereck–Swift theory of the SmA–SmC transition [65] that couples the SmC order parameter to gradients in the density and gradients in the layer spacing.

6.3.7 B–SmB–SmA Transitions

The SmB phase is another example of a phase where entropic effects are important. SmA is a two dimensional isotropic liquid in the plane of its layers. B is a three dimensional hexagonal crystal. SmB in-plane ordering is between these two with long range hexagonal bond orientational order and short range translational order that is not correlated from layer to layer [66]. A simple picture is that in the SmB phase, molecules are delocalized on a hexagonal grid [59]. X-ray diffraction patterns of B are resolution limited while those of SmB are not [67]. Although a first order SmB–SmC* transition has been observed [59], most of the work has concentrated on the expected fluctuation dominated SmB–SmA transition [68].

So far, none of the exponents observed at this transition fit a recognizable universality class. Nounesis et al. measured the divergence in the thermal conductivity and found dynamic critical exponents consistent with tricritical values [69]. As they had already established that $\alpha \approx 0.6$, very different from expectations of static critical phenomena (i.e. 3D-XY) they suggested that the SmB–SmA transition, at least in the compound they were studying (65OBC) [70], was driven towards a tricritical point because of a coupling between hexatic order and a short range ordering field such as a 'herringbone' structure from a nearby, lower temperature crystal E phase [1]. However, later extensive work on different compounds along a SmB–SmA transition line, with SmB temperature ranges between 0.8 K and 22 K, found $\alpha \approx 0.6$ along the entire line [71]. These authors point out that only a tricritical point (*not* a line) is compatible with a coupling between bond-orientational order and short range positional order.

Gorecka et al. [72] studied a compound in thick freely suspended films exhibiting a B–SmB–SmA phase sequence. They found no significant difference in hexatic properties in this system and those with SmB–E transitions [71]. The exponent for the hexatic order parameter was found to be $\beta = 0.15$

±0.03 which the authors point out is consistent with the three state Potts model in three dimensions [68]. They could not exclude the possibility that the SmB–A transition in their material was first order. The suggestion was that the hexatic order parameter did not fully cross-over from mean field to 3D-XY. It was also found that while there is a sharp discontinuity in the in-plane positional correlation length at the SmB–B transition, suggesting a strongly first-order transition, there was no measurable enthalpy change.

Fluctuation effects at the more fluid of liquid crystal phase transitions have been sensitively probed using techniques exploiting long length scales. Indeed, ultrasonic wavelengths are highly sensitive to hexatic fluctuations in SmA. Gallani et al. [73] found evidence of hexatic fluctuations at the SmA–SmB transition resulting in a strong damping and velocity anomaly in the A phase of an ultrasonic wave. Their data suggests that the in-plane hexatic ordering couples to SmA layer undulations which in turn have a nonlinear coupling to the sound velocity field.

6.3.8 Fluctuations at Critical Points

A line of first order phase transitions between two phases with the same symmetry may end at a critical point where fluctuations are expected to dominate. The classic example of this is the liquid–gas critical point.

6.3.8.1 BP$_{III}$–Isotropic

The analogue of the pressure–temperature plane in a liquid–gas system is the temperature–chirality plane for blue phase transitions. Along the lines of the classic liquid–gas example, Keyes [74] suggested that as the BP$_{III}$ and isotropic phases have the same symmetry, the line of first order phase transitions could end at a critical point. Voigts and van Dael found, [75] later verified by others, that, with increasing fraction of a chiral component, the latent heat at the BP$_{III}$–I transition decreased. More recently, Kutjnak et al. [76] located a BP$_{III}$-I critical point and characterized its behavior with high resolution calorimetry and measurements of optical activity. They report that both measurements are consistent with mean field behavior instead of the theoretically expected [77] Ising fluctuation behavior. They suggest that mean field behavior is observed because ξ_o is large. However, measured values [74] of $\xi_o \approx 1.5$ nm (i.e. a molecular length scale).

6.3.8.2 SmA$_d$ and SmA$_I$

These two smectic phases have the same symmetry but their layer spacings are different and incommensurate. In SmA$_I$, the layer spacing is about the molecular length while in SmA$_d$ it is 1.2–1.3 times the molecular length. As the critical point is approached, fluctuations become so large the critical point explodes into a bubble of a disordered state–cholesteric if the compounds are chiral and nematic if they are not–embedded in SmA. [78] On the other hand, a line of first order phase transitions between SmA$_d$ and SmA$_2$, for which the layer spacing is about twice the molecular length, ends at a critical point [79]. Prost and Toner [80] developed a dislocation melting theory for a line of first-order SmA–SmA phase transitions ending at a N–SmA$_d$–SmA, triple point. The first-order nature of this triple point for a nematic bubble has been verified by Wu et al. [81]. Furthermore, it was predicted that this line could continue linearly into the nematic region to end at a N–N critical point [80]. At the triple point on a nematic bubble [78], such a line would have to bend nearly 90°. But, at a nematic 'estuary' connected to a nematic 'ocean' (i.e. not surrounded by SmA), evidence has been found for a critical point terminating a first-order line of N–N transitions. [82]

6.3.8.3 NAC Multicritical Point

A point where three fluctuation dominated phase transition lines meet in a 2-dimensional parameter space is also expected to exhibit universal features. An extensively studied liquid crystal candidate was the N–SmA–SmC point in mixtures [83], in a pure compound under pressure [84] and at the re-entrant N–SmA–SmC multicritical point [85]. The situation may be summarized as follows. The systems studied showed qualitative and quantitative similarities. However, the exponents exhibited were not in the expected universality class for three second order phase transition lines meeting at a point. This is likely because, in the N–SmA–SmC case, the N–SmC transition line is first order [86] as is the N–SmA transition line, [26] leaving only the SmA–SmC second-order phase transition line.

6.3.9 Conclusion

Given the rich variety to their broken symmetries, enormous strides in perfecting low molecular weight organic liquid crystal materials for applications in industry, as well as important advances in measurement technologies of material properties under well-controlled conditions, liquid crystals emerge as useful materials to explore the intricate and beautiful interplay in nature between symmetry and spatial dimensionality at phase transitions.

6.3.10 References and Notes

[1] P. G. de Gennes, J. Prost, *The Physics of Liquid Crystals*, Clarendon Press, Oxford **1993**, Chapters I and II.
[2] P. Pfeuty, G. Toulouse, *Introduction to the Renormalization Group and to Critical Phenomena*, John Wiley and Sons, New York **1977**.
[3] J. Thoen, in *Phase Transitions in Liquid Crystals*, (Eds.: S. Martellucci, A. N. Chester) Plenum, New York **1992**, pp. 155–174; C. W. Garland, *ibid*, pp. 175–187; J. Thoen, *Int. J. Mod. Phys. B*, **1995**, *9*, 2157; in *Liquid Crystals in the Nineties and Beyond*, (Ed.: S. Kumar) World Scientific Publ. Co, **1995**, pp. 19–80.
[4] P. E. Cladis, W. van Saarloos, D. A. Huse, J. S. Patel, J. W. Goodby, P. L. Finn, *Phy. Rev. Lett.* **1989**, *62*, 1764; G. Dee, J. S. Langer, *Phys. Rev. Lett.* **1983**, *50*, 383.
[5] M. A. Anisimov, P. E. Cladis, E. E. Gorodetskii, D. A. Huse, V. E. Podneks, V. G. Taratuta, W. van Saarloos, V. P. Voronov, *Phys. Rev.* **1990**, *A41*, 6749, See also: P. E. Cladis, *J. Stat. Phys.* **1991**, *62*, 899.
[6] G. Grinstein, private communication, **1980**, Pfeuty and Toulouse [2].
[7] A recent review is: J. R. Dorfman, T. R. Kirkpatrick, J. V. Sengers, *Annu. Phys. Chem.* **1994**, *45*, 213.
[8] A. Schenzle, H. R. Brand, *Phys. Rev.* **1979**, *A20*, 1628.
[9] H. R. Brand, S. Kai, S. Wakabayashi, *Phys. Rev. Lett.* **1985**, *54*, 555.
[10] A recent overview of this aspect of critical phenomena is: B. Schmittmann, R. K. P. Zia, *Statistical Mechanics of Driven Difusive Systems*, Academic Press, New York **1995**.
[11] See for example: L. Benguigui, D. Collin, P. Martinoty, *J. Phys. I (Paris)* **1996**, *6*, 1469. In this paper, the authors show that the dynamic exponent, y, associated with high frequency damping at a second order phase transition is frequency dependent so that simple scaling laws obtained from mode coupling calculations should not apply. However, in some smectics, where the regular part of the specific heat behavior is large compared to the fluctuation amplitude, deviations from these laws may be too small to measure.
[12] For a recent perspective on fluctuation effects in nematic liquid crystals see: Z. H. Wang, P. H. Keyes, *Phys. Rev. E*, **1996**, *54*, 5249.
[13] H. R. Brand, H. Finkelmann, *Liquid Crystal Elastomers*, Vol. III and references therein.
[14] M. Olbrich, H. R. Brand, H. Finkelmann, K. Kawasaki, *Europhys. Lett.* **1995**, *31*, 281.
[15] P. G. de Gennes, *Mol. Cryst. Liq. Cryst.* **1971**, *12*, 193. See also ref: [1]
[16] F. Hessel, H. Finkelmann, *Polym. Bull.* **1985**, *14*, 375; H. Leube, H. Finkelmann, *Makro-*

mol. Chem. **1990**, *191*, 2707, ibid **1991**, *192*, 1317.

[17] See for example: W. Wedler, P. Hartmann, U. Bakowsky, S. Diele, D. Demus, *J. Mater. Chem.* **1993**, 2, 1195.

[18] L. J. Yu, A. Saupe, *Phys. Rev. Lett.* **1980**, *45*, 1000. The system is mixtures of potassium laurate, 1-decanol and D_2O. This system also exhibits a reentrant nematic transition (cf. the Chapter in this Handbook on Reentrant Phase Transitions). The higher temperature nematic is rod-like while the lower temperature nematic is disc-like. In addition, the isotropic liquid is also re-entrant in this system.

[19] H. Brand, J. Swift, *J. Phys. Lett.* **1983**, *44*, L-333. C. Cajas, J. B. Swift, H. R. Brand, *Phys. Rev.* **1983**, *A28*, 505 (statics); ibid **1984**, *A30*, 1579 (dynamics).

[20] A. Saupe, P. Boonbrahm, L. J. Yu, *J. Chem. Phys.* **1984**, *81*, 2076.

[21] M. Roy, J. P. McClymer, P. H. Keyes, *Mol. Cryst. Liq. Cryst. Lett.* **1985**, *1*, 25.

[22] W. McMillan, *Phys. Rev.* **1971**, *A4*, 1238; ibid. **1972**, *A6*, 936; ibid **1974**, *A9*, 1720.; K. K. Kobayashi, *Phys. Lett.* **1970**, *A31*, 125.

[23] P. G. de Gennes, *Solid State Commun.* **1972**, *10*, 753; P. G. de Gennes, J. Prost, *The Physics of Liquid Crystals*, Clarendon, Oxford **1993**, Chapter X.

[24] C. W. Garland, G. Nounesis, *Phys. Rev. E.* **1994**, *49*, 2964 and references therein.

[25] cyano-octylbiphenyl and cyano-nonylbiphenyl respectively. The small discontinuities in 8CB and 9CB escaped the fine net of adiabatic calorimetry (J. Thoen, H. Marynissen, W. van Dael, *Phys. Rev. Lett.* **1984**, *52*, 204; *Phys. Rev.* **1982**, *A26*, 2888; H. Marynissen, J. Thoen, W. van Dael, *Mol. Cryst. Liq. Cryst.* **1985**, *124*, 195) and high resolution X-ray diffraction (B. M. Ocko, R. J. Birgeneau, J. D. Litster, *Z. Phys.* **1986**, *B62*, 487) but were caught by the front propagation test [5].

[26] B. I. Halperin, T. C. Lubensky, S. K. Ma, *Phys. Rev. Lett.* **1974**, *32*, 292 for superconductors and B. I. Halperin, T. C. Lubensky, *Solid State. Comm.* **1974**, *14*, 997 for the N–SmA transition.

[27] P. Sonntag, PhD Thesis, University of Louis Pasteur de Strasbourg I (1996) and to be published (1997).

[28] T. C. Lubensky, *J. Chim. Phys.* **1983**, *80*, 31.

[29] T. C. Lubensky, G. Grinstein, R. Pelcovits, *Phys. Rev.* **1982**, *B25*, 6022.

[30] D. R. Nelson, J. Toner, *Phys. Rev.* **1981**, *B24*, 363 and J. Toner, *Phys. Rev.* **1982**, *B26*, 462. A. J. McKane, T. C. Lubensky, *J. Phys. (Paris) Lett.* **1982**, *43*, L217.

[31] A. R. Day, A. J. McKane, T. C. Lubensky, *Phys. Rev.* **1983**, *A27*, 1461.

[32] B. R. Patton, B. S. Andereck, *Phys. Rev. Lett*, **1992**, *69*, 1556; B. S. Andereck, B. R. Patton, *Phys. Rev.* **1994**, *E49*, 1393.

[33] See for example [27].

[34] P. E. Cladis, *Phys Lett.* **1974**, *48A*, 179 and unpublished. This is a qualitative result. While it may not be possible to make elastic constant measurements for K_2 and K_3 deep enough into the N–SmA critical regime to establish quantitative certainty about critical exponents, there should be no problem for K_1 which does not diverge at this transition.

[35] Unlike para-MBBA, the ortho-MBBA isomer is not liquid crystalline. It was synthesized by Gary Taylor at Bell Labs (ca. 1970) to 'fine tune' liquid crystal electro-optic response for devive applications. While the shape of *p*-MBBA is rod-like, that of *o*-MBBA resembles a bent rod.

[36] F. Brochard, *J. Phys. (Paris)* **1973**, *34*, 411. F. Jähnig and F. Brochard, *J. Phys. (Paris*, **1974**, *35*, 301. See also: K. A. Hossain, J. Swift, J.-H. Chen and T. C. Lubensky, *Phys. Rev.* **1979**, *B19*, 432.

[37] K. Kawasaki, *Phys. Rev.* **1966**, *150*, 291; *Ann Phys.* **1970**, *61*, 1. See also review article [7].

[38] Ch. Gähwiller, *Phys. Rev. Lett.* **1972**, *28*, 1554; *Mol. Cryst. Liq. Cryst.* **1973**, *20*, 301. P. Pieranski, E. Guyon, *Phys. Rev. Lett.* **1974**, *32*, 924.

[39] See for example: P. E. Cladis in *Nematics: Mathematical and Physical Aspects*, (Eds.: J.-M. Coron, J.-M. Ghidaglia, F. Helein) Kluwer Academic Publishers, Boston **1990**; and H. R. Brand, C. Fradin, P. L. Finn, P. E. Cladis, *Phys. Lett. A* **1997**.

[40] P. Martinoty et al. (*Phys. Rev. E* **1997**) at the first-order SmC–SmF transition where nevertheless, fluctuations are observed by anomalies in ultrasonic measurements and H. Yao, T. Chan, C. W. Garland, *Phys. Rev.* **1995**, *E51*, 4584 at the first-order SmC–SmI critical point. See also [11].

[41] Type II superconductors in a magnetic field (and in the appropriate geometry) are characterized by the following properties. P. G. de Gennes, *Superconductivity of Metals and Alloys*, W. A. Benjamin, Inc. (pub.), New York, **1966**. The Meissner effect is not observed except in weak fields, $H<H_{c1}$. For $H>H_{c1}$, magnetic field lines penetrate but the penetration is not complete. de Gennes points out that this region was first experimentally discovered by Schubnikov in 1937 but is known as the vortex state from the microscopic theory derived by A. Abrikosov, *Zh. Eksp. Teor. Fiz.* **1957**, *32*, 1442 [*Soviet Physics JETP* **1957**, *5*, 1174]. For $H>H_{c2}$, no expulsion of flux is observed however superconductivity is not completely destroyed. Experimentally the transition at H_{c2} is second order. In the interval $H_{c2}<H<H_{c3}\approx 1.69H_{c2}$, there is a super-

conducting surface sheath. P. B. Vigman, V. M. Filev, *Zh. Eksp. Teor. Fiz.* **1975**, *69*, 1466 [*Soviet Physics JETP*, **1975**, *42*, 747] suggested looking for the analogue of H_{c3} in liquid crystals to support de Gennes' analogy [23].
[42] T. C. Lubensky, *Physica A*, **1995**, *220*, 99.
[43] S. R. Renn, T. C. Lubensky, *Phys. Rev.* **1988**, *A38*, 2132.
[44] J. W. Goodby, M. A. Waugh, S. M. Stein, R. Pindak, J. S. Patel, *Nature*, **1988**, *337*, 449; *J. Am. Chem. Soc.* **1989**, *111*, 8119. SmC* is the chiral analogue of SmC with a macroscopic helix axis parallel to the layer normal.
[45] P. E. Cladis, A. J. Slaney, J. W. Goodby, H. R. Brand, *Phys. Rev. Lett.* **1994**, *72*, 226; *Il Nuovo Cimento*, **1994**, *16D*, 765 and unpublished. This Slaney–Goodby compound is (S)-2-chloro-4-methylpentyl 4′-(4-n-dodecyloxypropioloyloxy)-4-biphenylcarboxylate (12O2Cl4M5T).
[46] On a microscopic scale, the geometry of this model may be more reminiscent of Type I intermediate states rather than a vortex state [41]. The analogue of the vortices are the screw dislocations. A analogue of the intermediate state at the N-SmA transition is a static stripe pattern; P. E. Cladis, S. Torza, *J. Appl. Phys.* **1975**, *46*, 584.
[47] C. Dasgupta, B. I. Halperin, *Phys. Rev. Lett.* **1981**, *47*, 1556.
[48] T. Chan, C. W. Garland, H. T. Nguyen, *Phys. Rev.* **1995**, *E52*, 5000; L. Navailles, C. W. Garland, H. T. Nguyen, *J. de Phys. II* **1996**, *6*, 1243.
[49] R. D. Kamien, T. C. Lubensky, *J. Phys. (Paris) I* **1994**, *3*, 2123.
[50] D. S. Fisher, M. P. A. Fisher, D. A. Huse, *Phys. Rev.* **1991**, *B43*, 130; D. A. Huse, M. P. A. Fisher, D. S. Fisher, *Nature* **1992**, *358*, 553.
[51] The time dependence [45] of the heat singularity at TGBA–N* in the Slaney–Goodby compound is suggestive of a liquid–glass transition in this material.
[52] A. C. Ribeiro, A. Dreyer, L. Oswald, J. F. Nicoud, A. Soldera, D. Guillon, Y. Galerne, *J. de Phys. II* **1994**, *4*, 407 and unpublished; V. Vill, H.-W. Tunger, D. Peters, *Liq. Cryst.* **1996**, *20*, 547. J. W. Goodby, private communication, Capri (1996).
[53] D. Guillon, A. Skoulios, *J. Phys.* **1977**, *38*, 79.
[54] P. G. de Gennes, *Mol. Cryst. Liq. Cryst.* **1973**, *21*, 49.
[55] C. R. Safinya, M. Kaplan, J. Als-Nielsen, R. J. Birgeneau, D. Davidov, J. D. Litster, D. L. Johnson, M. E. Neubert, *Phys. Rev.* **1980**, *21*, 4149. These experiments were made on the sulfur compound, 4-n-pentyl-phenylthiol-4′-n-octyloxybenzoate ($\bar{8}$S5).
[56] M. Delaye, *J. Phys. Colloq.* **1979**, *40*, C3-350, 4-nonyloxybenzoate 4′-butyloxyphenyl (9E4).

[57] B. I. Ostrovskii, A. Z. Rabinovich, A. S. Sonin, E. L. Sorkin, B. A. Strukov, S. A. Taraskin, *Ferroelectrics*, **1980**, *24*, 309.
[58] Y. Galerne, *Phys. Rev.* **1981**, *A24*, 2284.
[59] P. Keller, P. E. Cladis, P. L. Finn, H. R. Brand, *J. Phys.* **1985**, *46*, 2203.
[60] K. Ema, J. Watanabe, A. Takagi, H. Yao, *Phys. Rev.* **1995**, *E52*, 1216 and references therein.
[61] A. Fukuda, Y. Takanishi, T. Isozaki, K. Ishikawa, H. Takezoe, *J. Mater. Chem.* **1994**, *4*, 997.
[62] L. Benguigui, P. Martinoty, *Phys. Rev. Lett.* **1989**, *63*, 774.
[63] D. Collin, S. Moyses, M. E. Neubert, P. Martinoty, *Phys. Rev. Lett.* **1994**, *73*, 983.
[64] L. Benguigui, P. Martinoty, *J. de Phys. II.*, **1997**, *7*, 225. and references therein.
[65] B. S. Andereck, J. Swift, *Phys. Rev.* **1982**, *A5*, 1084.
[66] B. I. Halperin, D. R. Nelson, *Phys. Rev. Lett.* **1978**, *41*, 121; D. R. Nelson, B. I. Halperin, *Phys. Rev.* **1979**, *19*, 2457. R. J. Birgeneau, J. D. Litster, *J. Phys. Lett. (Paris)*, **1978**, *39*, L399.
[67] A. J. Leadbetter, J. C. Frost, M. A. Mazid, *J. Phys. Lett.* **1979**, *40*, 325; R. Pindak, D. E. Moncton, S. D. Davey, J. W. Goodby, *Phys. Rev. Lett.* **1981**, *46*, 1135.
[68] C. C. Huang, *Adv Phys.* **1993**, *43*, 343.
[69] G. Nounesis, C. C. Huang, J. W. Goodby, *Phys. Rev. Lett.*, **1986**, *56*, 1712. In these experiments, the temperature oscillations were at 1 Hz.
[70] hexylalkyl-4′-pentylalkoxybiphenyl-4-carboxylate
[71] G. Nounesis, R. Geer, H. Y. Liu, C. C. Huang, J. W. Goodby, *Phys. Rev.* **1989**, *A40*, 5468.
[72] E. Gorecka, L. Chen, W. Pyzuk, A. Króczyński, S. Kumar, *Phys. Rev.* **1994**, *E50*, 2863.
[73] J. L. Gallani, P. Martinoty, D. Guillon, G. Poeti, *Phys. Rev.* **1988**, *A37* (Rap. Comm.), 3638; *Phys. Rev. E* **1997**.
[74] P. H. Keyes, *Phys. Rev. Lett.* **1987**, *59*, 83; E. P. Koistinen and P. H. Keyes, *Phys. Rev. Lett*, **1995**, *74*, 4460 and unpublished.
[75] G. Voigts, W. van Dael, *Liq. Cryst.* **1993**, *14*, 617.
[76] Z. Kutjnak, C. W. Garland, C. G. Schatz, P. J. Collings, C. J. Booth, J. W. Goodby, *Phys. Rev.* **1996**, *E53*, 4955 and to be published.
[77] T. C. Lubensky, H. Stark, *Phys. Rev.* **1996**, *E53*, 714. These authors introduce a pseudo-scalar order parameter for this critical point.
[78] P. E. Cladis, H. R. Brand, *Phys. Rev. Lett.* **1984**, *52*, 2210.
[79] F. Hardouin, M. F. Achard, H. T. Nguyen, G. Sigaud, *J. Phys. (Paris) Lett.*, **1984**, *46*, L123. An extensive review of smectic A polymorphism is contained in F. Hardouin, A. M. Levelut, M. F. Achard, G. Sigaud, *J. Chim. Phys.* **1983**, *80*, 53.

[80] J. Prost, J. Toner, *Phys. Rev.* **1987**, *A36*, 5008.
[81] L. Wu, C. W. Garland, S. Pfeiffer, *Phys. Rev.* **1992**, *A46*, 973, ibid **1992**, *A46*, 6761.
[82] G. Nounesis, S. Kumar, S. Pfeiffer, R. Shashidhar, C. W. Garland, *Phys. Rev. Lett.* **1994**, *73*, 565.
[83] D. L. Johnson, *J. Chim. Phys*, **1983**, *80*, 45.
[84] R. Shashidhar, B. R. Ratna, S. K. Prasad, *Phys. Rev. Lett.* **1984**, *53*, 2141.
[85] S. Somasekhara, R. Shashidhar, B. R. Ratna, *Phys. Rev.* (Rap. Comm.) **1986**, *A34*, 2561.
[86] J. B. Swift, *Phys. Rev.* **1976**, *A14*, 2274. Swift explains that the N–SmC transition is driven first order by fluctuations in a system with a characteristic length, $2\pi/k$. In such a system, fluctuations are confined to a shell around $k \neq 0$ rather than a sphere around $k = 0$. The effect is to drive the second order N–SmC transition temperature to lower temperatures and so is pre-empted by a first-order phase transition.

6.4 Re-entrant Phase Transitions in Liquid Crystals

P. E. Cladis

6.4.1 Introduction

In a re-entrant phase transition, a higher temperature thermodynamic phase with higher symmetry (and most dramatically with greater fluidity) reappears at temperatures below a stable thermodynamic phase with lower symmetry (and less fluidity). Molecules have many more internal degrees of freedom than atoms. As a result, liquid crystals have more than 30 different thermodynamic phases, many involving broken continuous symmetries. It is not surprising then that there is not a unique mechanism to account for re-entrant behavior in liquid crystals. In addition, the characterization of liquid crystal structure (i.e. microscopic structure) versus property (i.e. macroscopic expression) relations requires measurements on many different length and time scales: thermal measurements; k-space measurements (e.g. X-ray and light scattering); direct observations with microscopes, particularly the polarizing light microscope; acoustic measurements; etc. While a phase diagram in, for example, the temperature–concentration plane or temperature–electric field plane gives a macroscopic picture, that these various techniques generate, it does not distinguish between different microscopic mechanisms for re-entrant behavior. In the absence of a theoretical framework covering all known cases of re-entrance in liquid crystals, we have sorted them into three broad classes. However, as is typical of liquid crystal phenomena, and indeed of complex natural phenomena in general, the boundaries between the different classes are not sharp.

The sequence of phase transitions, with decreasing temperature, N SmA N_{re} was discovered in 1975 in cyano compounds with two benzene rings [1] (Fig. 1, [2]). Since 1975, liquid crystals have been found to

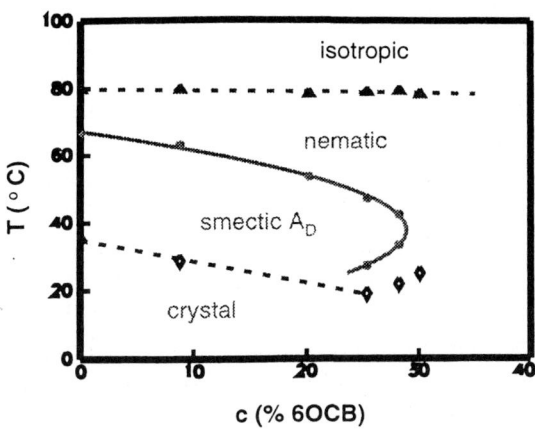

Figure 1. The N_{re} phase diagram for 8OCB/6OCB mixtures [2].

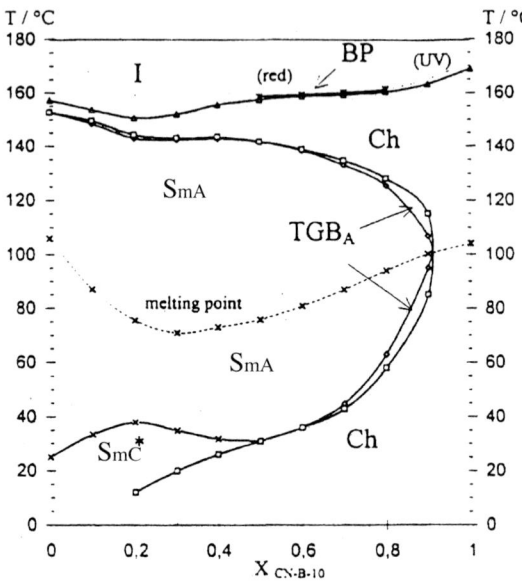

Figure 2. The N_{re}^* and $TGBA_{re}$ phases [6].

show many different examples of re-entrant behavior arising from several different mechanisms, most of which have no obvious solid-state analog. For example, the first re-entrant transition [1] is now understood to result when 'short-range effects surprisingly do the long-range ones in!' [3]. We call this category R1, re-entrance from frustration. A comprehensive review of R1 re-entrance appeared in 1988 [4] to mark the 100th anniversary of the discovery of liquid crystals. As R1 is by far the most novel example of re-entrance in liquid crystals, and robust materials exhibiting R1 re-entrance are widely available [5], it has generated the most amount of work [4]. Therefore, R1 is the transition that is currently best understood. The most recent R1 example is the phase sequence, with decreasing temperature N*, TGBA, SmA, re-entrant TGBA, N_{re}^* which is shown in mixtures involving a cyano compound with a carbohydrate link [6] (Fig. 2). These materials [6] may help develop an understanding of biologically relevant materials where more-fluid states are known to occur at lower temperatures than less-fluid ones [7].

Re-entrance in liquid crystals can also involve temperature-driven change in short-range steric forces competing for long-range order. Molecular shape is a crucial factor in liquid crystal structure–property relations. Rod-like nematics (uniaxial positive) cannot continuously transform to disc-like nematics (uniaxial negative) without an intervening biaxial nematic phase or an isotropic liquid state in the case of discontinuous transitions. The phase sequence with decreasing temperature – isotropic liquid, uniaxial positive nematic, biaxial nematic, uniaxial negative nematic, re-entrant isotropic liquid (Fig. 3, [8]) – is an example from lyotropic liquid crystals of both a re-entrant nematic (N_{re}) and the first re-entrant isotropic liquid.

Some thermotropic liquid crystal structures are so complex both rod-like and disc-like liquid crystal phases can be stabilized at different temperatures and for different members in a homologous series [9]. For example, in double-swallow-tailed compounds with aliphatic chain lengths less than 10, 'rod-like' liquid crystal phases are observed, and in longer chains, discotic liquid crystal phases are observed [10]. We call the above re-entrant phenomenon, 're-entrance triggered by complex geometrical factors', R2. Other re-entrant phase sequences we list in R2 are the phase sequences SmA SmC_{re} SmA [11] and SmC SmO_{re} SmC [12] found in a symmetrical, relatively long compound [13] the homologous series of which has an alternating stability for SmO and SmC phase. As materials exhibiting R2 re-entrance tend to be in the 'exotic' limit, apart from temperature–concentration phase diagrams for given molecular structures little is known about R2 re-entrance.

The term 're-entrant' was first used to describe the reappearance of a lower temper-

6.4.2 R1: Re-entrance from Frustration

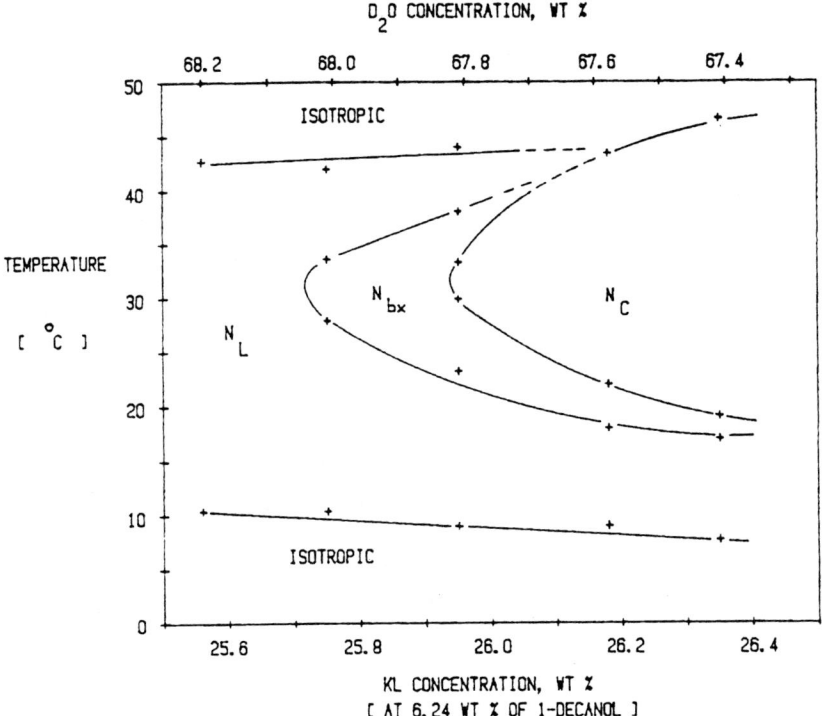

Figure 3. The first I_{re} phase [8].

ature normal metal phase in superconductors (see, for example, Cladis [4] and references therein). Re-entrance in this case, R3, results from the competition between long-range forces, e.g. a new order parameter grows [14] or there is a temperature-dependent coupling between competing order parameters [15]. The phase diagram used to discuss superconducting re-entrance is the magnetic field–temperature plane. R3 re-entrance has been known longer than has re-entrance from frustration, R1. Only the terminology is new for R3. In R3 re-entrance, fluctuations often play an important role, leading to universal features such as the magnificant spiral at the N_{re}–SmC–SmA multicritical point (Fig. 4) [16]; the N–SmA–NSmC step [17] (Fig. 5) showing the dramatic suppression in the N–SmC transition temperature resulting from Brazovskii fluctuations at the N–SmC transition [18] compared to the N–SmA transition; and the nematic bubble when fluctuations from two SmA phases with nearly similar but different layer spacings compete [19]. As it has been eloquently argued [20] that re-entrance in nonpolar compounds [21] depends sensitively on universal features of the NSmA SmC multicritical point, we include it in R3 re-entrance.

6.4.2 R1: Re-entrance from Frustration

In the re-entrant nematic transition, a liquid phase without translational order (nematic) occurs at a lower temperature or higher pressure than one with one-dimensional translational order (SmA). With decreasing tem-

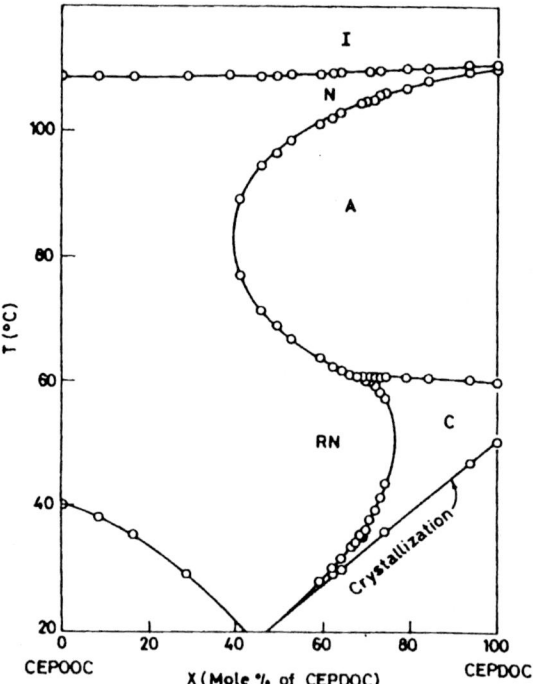

Figure 4. The N_{re}–SmA–SmC multicritical point [16].

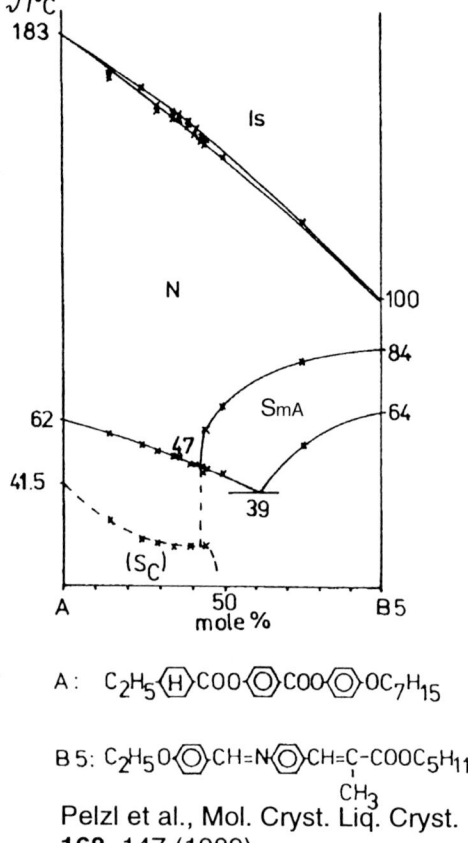

A: C_2H_5‒⟨H⟩‒COO‒⟨O⟩‒COO‒⟨O⟩‒OC_7H_{15}

B5: C_2H_5O‒⟨O⟩‒CH=N‒⟨O⟩‒CH=C‒$COOC_5H_{11}$
$\quad\quad\quad\quad\quad\quad\quad\quad\quad\quad\quad |$
$\quad\quad\quad\quad\quad\quad\quad\quad\quad\quad CH_3$

Pelzl et al., Mol. Cryst. Liq. Cryst. **168**, 147 (1989)

Figure 5. Step down to lower temperatures of the N–SmC transition line from the N–SmA transition line [17].

perature (see Fig. 1) or increasing pressure (Fig. 6), the phase sequence is N–SmA–N. The lower temperature nematic phase is called 're-entrant' [22]. This particular phase sequence was first discovered in 1975 in mixtures of cyano-Schiff base compounds (cyanooctyloxy aniline and heptyloxybenzilidene aniline) at 1 atm [1], and later in pure materials, including cyanobiphenyloctyloxy aniline (8OCB) under pressure [23]. These materials have two benzene rings.

As it was known that SmA phases of cyanobiphenyl compounds exhibit layer spacings that are larger and incommensurate with their molecular length [23], the suggestion was made [22] that the N_{re} phase in these materials was the result of dimer-type associations on a molecular level and that it could transform back to a SmA phase with layer spacing comparable to its molecular length, even when there was an intervening SmC phase (Fig. 7, [22]).

Goodby and coworkers [24] elegantly showed that re-entrance in cyano compounds resulted from a sensitive balance between dipolar and steric factors. They synthesized two benzene ring materials with an ester link that either reinforced (resulting in exclusively re-entrant nematic behavior) or opposed (resulting in the appearance of a SmC phase below the SmA) the cyanophenyl 'mesomeric relay' [24]. The conclusion is that, when dipolar forces domi-

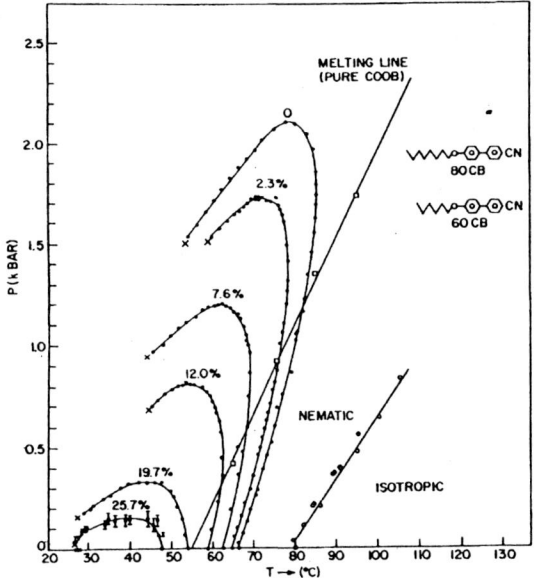

Figure 6. The N_{re} phase as a function of temperature and pressure for 8OCB/6OCB mixtures [4].

nate we have re-entrance from frustration, and when steric forces dominate the door opens to R2 and R3 re-entrance.

The incommensurate SmA phase was called SmA_d [25], where d denotes 'dimer' for the pairwise overlapped associations of aromatic cores fromed by these compounds [26], and the lowest temperature SmA phase was called SmA_I [19], where I stands for 'interdigitated', as the aliphatic chains from one layer formed of dimers are interdigitated with those of neighboring layers [22]. Because the layer spacing is about the same as the molecular length, these kinds of SmA phases are also known as SmA_1. However, stimulating the formation of monomolecular SmA phases in cyano compounds leads to the enhancement of the SmA phase at the expense of a nematic phase (Fig. 8), introducing the need to distinguish between

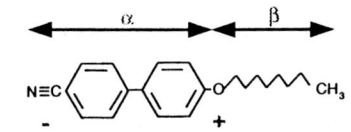

	cyano aromatic moiety: α	aliphatic chain length: β	total length: α + β	model length: α + 2β	Measured length
CBOOA	16.0 Å	10.4 Å	26.4 Å	36.8 Å	36.3 Å
HBAB	16.0	8.2	24.2	-	-
8OCB	13.9	9.2	23.1	32.3	30.8

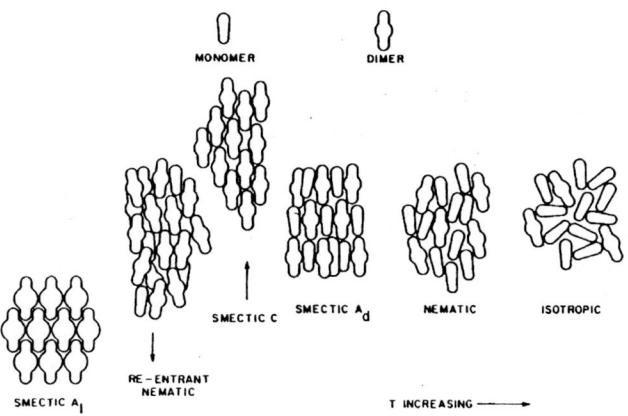

Figure 7. The re-entrant scenario for cyano compounds [22] with the model for SmA_d and SmA_I.

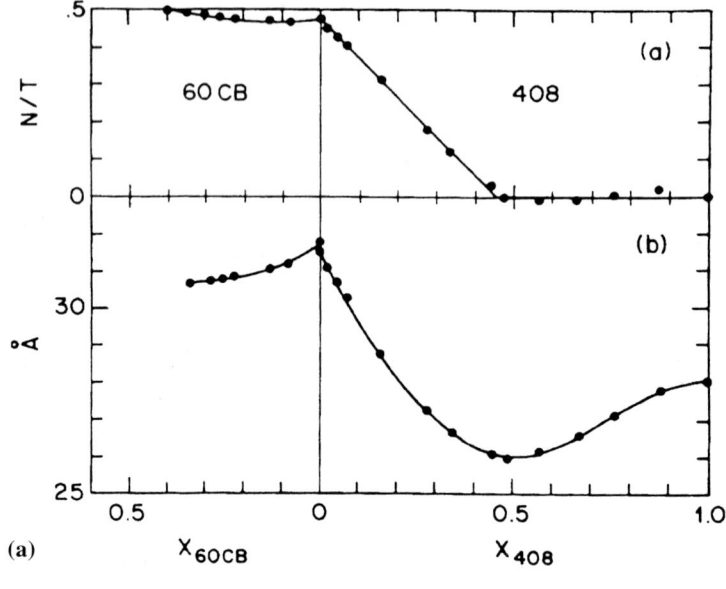

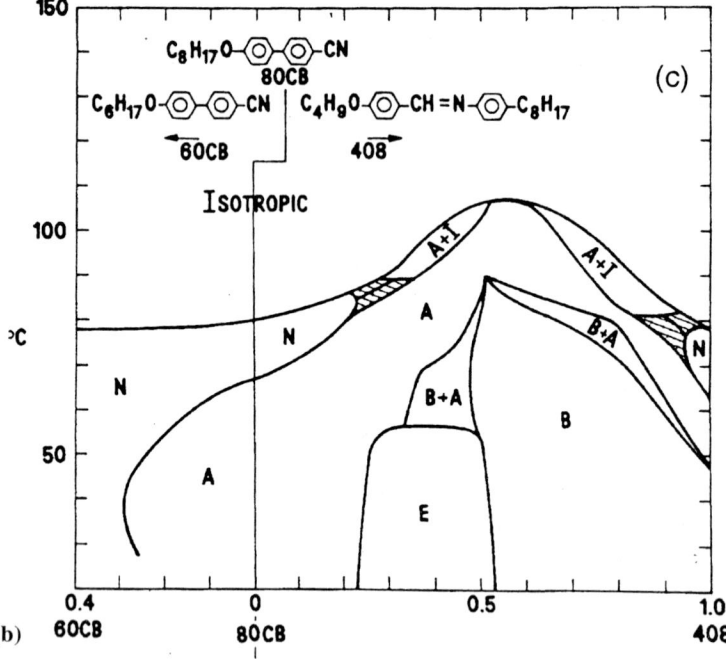

Figure 8. The number of dimer pairs (a) deduced from the observed SmA layer spacing (b) in 6OCB/8OCB and 4O8/8OCB mixtures. (c) At complete pair saturation, the number of pairs (N) compared to the total number of molecules (T) is 0.5. In a mixture of monomers, $N=0$ (c), as $N/T \rightarrow 0.5$, the SmA$_d$ phase disappears, and when $N/T \rightarrow 0$, the nematic phase is squeezed out by the enhanced SmA$_1$ phase [27].

SmA$_1$ and SmA$_I$. Thus the reappearance of a SmA phase that has a layer spacing close to the molecular length may be an indication that its layer structure is more than a simple packing of single molecules (see Fig. 7, [27]).

While the N$_{re}$ phase was also found in nitro compounds with two benzene rings [28], re-entrant phenomena in liquid crystals made a giant leapforward in 1979, when the stable N$_{re}$–SmA transition [29] and multiple re-entrance [30] was discovered at 1 atm

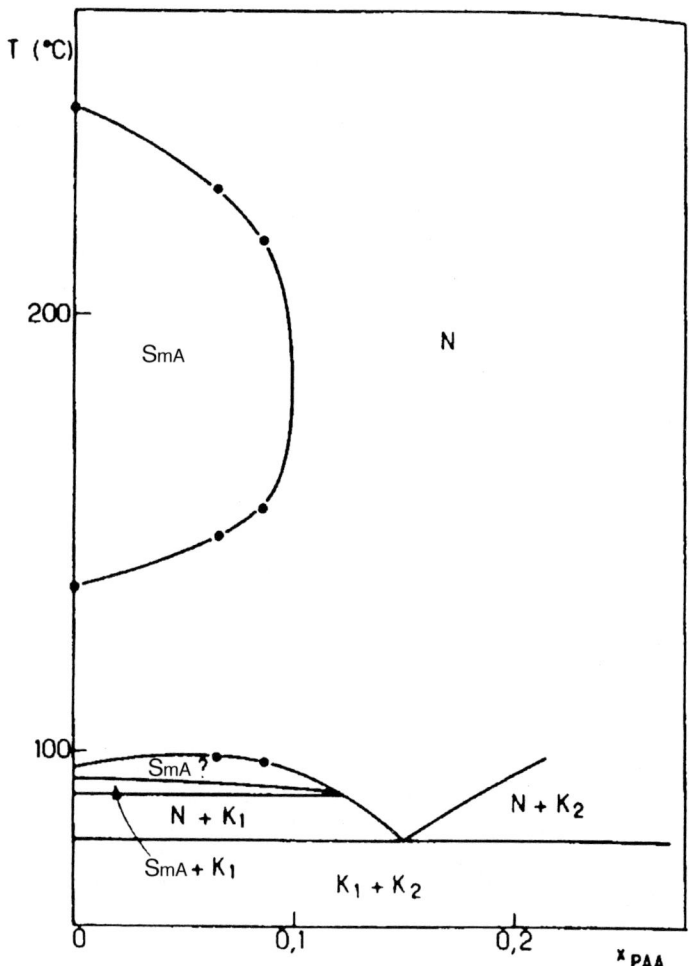

Figure 9. One of the first pure compounds to show a stable N_{re} phase (RN) with two RNA transitions [29].

by the Bordeaux group in several different cyano compounds containing three benzene rings (Fig. 9). A multiple re-entrant phase sequence with decreasing temperature is, for example, N SmA N_{re} SmA N_{re} SmA. These discoveries had a major impact on both the chemical and physical frontiers of liquid crystal research in the polymorphism of SmA and re-entrant behavior [31]. It should also be noted that the authors pointed out that the material in which they first identified a N_{re} phase [29] had previously been identified as a more ordered smectic phase [32].

One of the first temperature–concentration re-entrant phase diagrams from the Bordeaux group (Fig. 9) [29] bears an uncanny resemblance to the magnetic field–temperature phase diagram observed in a complex superconducting material at very low temperatures and very high magnetic fields (Fig. 10) [32]. This superconducting re-entrance is the closest solid-state analog to R1 re-entrance. In these systems, the internal magnetic field first shields the superconducting state from the applied magnetic field, so that the superconducting state re-enters at higher fields, and then, eventually,

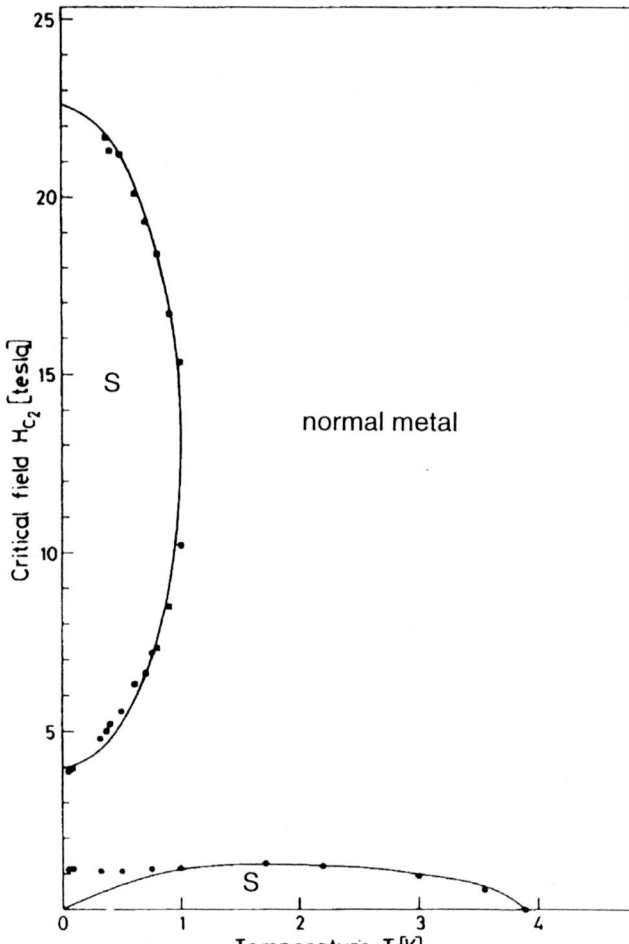

Figure 10. An example of the re-entrant phenomenon in complex superconductors [33] having a phase-diagram topology similar to that in Figure 9. At low fields the normal metal state is not re-entrant, as predicted from theory [34]. However, at extremely low temperatures and with increasing applied field, the superconducting state (S) is first restored and then destroyed again above 20 T.

when its shielding powers are exhausted, superconductivity is again destroyed [34]. However, unlike the case of magnetic superconductors [14] the normal metal state both above and below the superconducting 'nose' in Fig. 10 are similar.

An internal/external field competition [33] may be at work in the SmC$^*_{re}$ phase observed in an applied electric field [35] when the helix structure is suppressed by an applied electric field. However, while our understanding of the re-entrant mechanism behind the behavior illustrated by Fig. 10 is relatively complete [34], nonlinear theories of smectic phase transitions [36] have not yet accounted for the observed helielec-

tric–ferroelectric–helielectric phase sequence with decreasing temperature in an applied electric field [35] or the C* O* C$^*_{re}$ [13] sequence with decreasing temperature (discussed in Sec. 4).

Following the Bordeaux discovery by Hardouin et al. [29], many three-ring cyano compounds were discovered, particularly by the Halle group [37], that showed stable N$_{re}$ phases in cyano compounds with three benzene rings. An overview of the chemical structure and re-entrant behavior of three benzene ring compounds has been given by Weissflog et al. [38].

The theoretical work by Indekeu and Berker [39] showed the new physics to emerge from R1 re-entrance. Simply put, they showed that treating molecular associations from the perspective of statistical physics, the static dimer concept [22] was consistent with an ensemble of triplets correlated on a microscopic scale. They considered a two-particle dipolar potential that had both parallel ('up–up') and antiparallel ('up–down') exchange interactions. When dipolar forces between two members of the triplet cancel, the third member experiences no force and is free to permeate from layer to layer, 'frustrating' smectic order. By allowing displacements of three dipoles relative to each other, a population of triplets is generated that has a net short-range dipole interacting with each other via short-range dipolar forces stabilizing the layered structure. The energy fluctuations associated with a triplet ensemble is evaluated using an Ising criterion. If satisfied, a layer structure can be stabilized; if not, the nematic re-enters.

The N_{re} phase shares frustration properties with a 'spin-glass'. Roughly speaking, a spin-glass results from the presence of a wide variety of competing interactions arising from random interactions in a manybody system. In a spin-glass, different regions of phase space become irretrievably separated by energy barriers. The system is called 'nonergodic', as each small piece of the system is in a different region of phase space from all others and only 'knows' its own local conditions. The Liouville theorem breaks down and the system is called 'frustrated' [40]. As the temperature is lowered, these isolated states proliferate.

In this sense, SmA_d and the N_{re} phases are frustrated. They escape being nonergodic because they are liquids in the usual sense. To underscore the uniqueness of this particular example of frustration, Berker and Indekeu [39] refer to their theory as a 'spin-gas' theory, rather than a spin-glass theory.

By combining a cyanoethyl compound with a cholesteric liquid crystal with a sufficiently high twisting power, Pelzl et al. [41] obtained re-entrant cholesteric phases. Near the N*–SmA phase transition they found the opposite color sequence resulting from the untwisting of the helix structure by SmA fluctuations from the higher temperature N*–SmA transition. They pointed out that binary mixtures exhibiting both the usual N*–SmA and the N^*_{re}–SmA transition could, in principle, be used to make temperature-sensing devices in two temperature ranges.

Vill and coworkers [6] give a summary of N^*_{re} phases, and have shown that quite different phase diagrams can result from subtle changes in molecular structure. In particular, when mixing a cyanophenyl carbohydrate based compound containing a boron link (CNB10), which shows I–BP–N* transitions with decreasing temperature, with a slightly longer compound, in which boron is replaced by carbon (CNC12) and which has a large SmA temperature range, they observed the novel phase diagram (Fig. 2) exhibiting a re-entrant TGBA phase. Contact preparations of CNC12 with (1) a relatively nonpolar smaller molecule results in a cholesteric mountain surrounded by SmA and SmC phases and with (2) a N* compound, a lower temperature N* phase ending in the middle of the SmA phase where the phase boundaries (presumably) were lost in the contact preparation [42]. In both contact preparations, the high-temperature N* phases of the two components are immiscible. The high-temperature N* phase of CNB10 is also immiscible when mixed with the same two compounds; however, in the contact mixture with the N* compound there is an almost vertical N*–SmA phase boundary rather similar to the ones observed in the presence of competing SmA and SmC phase fluctuations (see, for example,

Fig. 5), with the TGBA squeezed in between. By suitably choosing components in binary mixtures, carbohydrate based cyanophenyl compounds can exhibit features associated with both R1 and R3 re-entrance.

6.4.3 Re-entrance from Geometric Complexity

Molecular shape plays a significant role in determining the stability of liquid crystal phases. Even the number of carbon atoms in the aliphatic chain can lead to an odd–even variation in transition temperatures in a homologous series and an odd–even stability of liquid crystal phases. A dramatic example of the latter is the racemic 1-methylterephthalidene-bis(aminocinamates) (MnTAC) [13]. These compounds contain three benzene rings and are chemically identical on both sides of the center of the molecular long axis. In this homologous series, MnTAC exhibits only a SmO phase [12] when n is odd and ≥ 5, and only a SmC phase when n is even and >4. M4TAC exhibits a stable SmC_{re} transition; that is, with decreasing temperature, the phase sequence is I–SmC–SmO–SmC_{re}. The chiral analogs with $4 \leq n \leq 7$ exhibit the SmO* phase independent of the parity of n. By mixing racemic M4TAC with materials showing only either SmO* or a SmC* phase, Heppke et al. [13] obtained phase diagrams with showing SmC^*_{re}.

In lyotropic liquid crystals, uniaxial negative nematics (N_L, i.e. disk-like phase) were found in aqueous solutions of potassium laurate (KL), 1-decanol, and potassium chloride [43]. Within a very narrow temperature and concentration range of a heavy water solution of this ternary mixture, Yu and Saupe [8] found a novel phase diagram (see Fig. 3). In this phase diagram, N_C is uniaxial positive (rod-like). Between the N_L and N_C phases is the first stable biaxial N phase, which Yu and Saupe [8] denoted by N_{bx}, and the first I_{re} liquid state (see Fig. 3).

Weissflogg et al. [10] give an overview of the literature on I_{re} phases in thermotropic liquid crystals. They point out that mesogens consisting of a rod-like core ending in two half-disk shaped moieties (polycatenar mesogens) have steric features between those of rod-lilke and disk-like mesogens, and can exhibit nematic, lamellar, columnar, and cubic mesophases [44]. Increasing the molecular complexity to six-ring mesogens, they found that slight variations in molecular structure can give rise to large changes in liquid crystal properties [10] and stable I_{re} phases. A fascinating example is the transition sequence with decreasing temperature found in six-ringe double-swallow-tailed compounds: I N I Cub SmC [10]. In molecular structures where conformational and corresponding entropic factors play an important role, it is possible that entropic elasticity similar to that which has been proven useful for polymers [45] (in addition to complex steric factors) should be considered.

Dowell [46] has considered the case of, for example, discotic (and lamellar) structures in which rigid aromatic moieties stack in a column (or layers), which can be stabilized by the collective dynamics of disordered alkyl chains (the 'floppy tails') surrounding the columns. The conclusion is that, as the temperature decreases, the alkyl chains become more ordered (and less dynamic), thus destabilizing the columnar stacks. Conversely, as the temperature increases the chains become more active and perhaps even entangled, leading to a more stable columnar structure, and thus there is no higher temperature N phase as is found in some discotic materials [47]. However, in other discotic materials, both the high- and

6.4.3 Re-entrance from Geometric Complexity

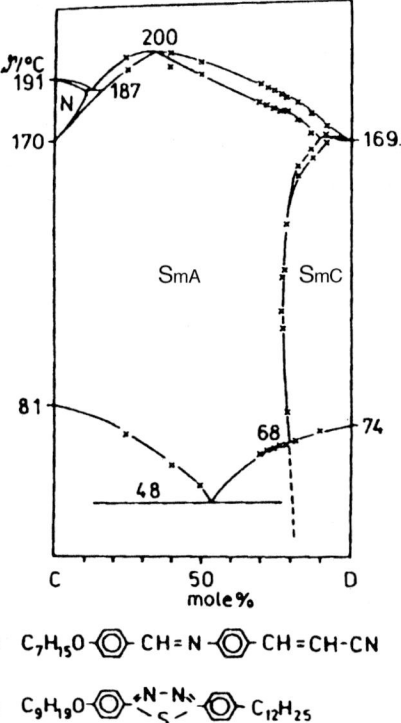

Figure 11. An example of the SmA–SmC–SmA$_{re}$ transition. Here the high-temperature SmA phase has a layer spacing somewhat smaller than that in a mixture of monomers, while the SmA$_{re}$ phase has a layer spacing comparable to that of a mixture of fully extended monomers [11].

low-temperature N$_{re}$ phases as well as re-entrant hexagonal columnar phases have been observed [50]. Thus, while observation of temperature-driven changes in the conformation and dynamics of the aliphatic chains help us to develop on understanding of the subtle features that suppress re-entrance in some cyano compounds [49], it seems clear that interactions of the aromatic moieties in disk-like compounds cannot be ignored, and may account for the wide variety of re-entrant behaviors observed in discotic liquid crystals.

Perhaps a somewhat simpler situation is presented by the SmA–SmC–SmA$_{re}$ transitions [11], which may be triggered by significant temperature-dependent conformational changes in the long alkyl chains (Fig. 11). A vertical SmA–SmC phase boundary is associated the SmA–SmC–SmA$_{re}$ transition [11], where a polar compound (C in Fig. 11) with an SmA$_d$ phase is mixed with a nonpolar one (D in Fig. 11) with two relatively long aliphatic chains and only a SmC phase. X-ray studies of the layer spacings with increasing concentration of D (x_D) show three distinct regions. When $x_D < 0.5$, the layer spacing is larger than the average molecular length, indicating that in this range of concentrations the SmA phase consists of monomers and dimers. When $0.6 < x_D < 0.7$ the layer spacing corresponds to the monomer average of C and D. In the third regime, the layer spacing is smaller than the average of that of fully extended monomers as the SmC phase is approached. However, the layer spacing in the SmA$_{re}$ phase is somewhat larger than that of the high-temperature SmA phase, being approximately the same as that of the monomer average when the alkyl chains are in an all-*trans* configuration [11].

As thermotropic liquid crystal materials become more complex more sophisticated numerical computations will be needed in order to develop a corresponding hierarchy of structure–property relationships, as is currently in progress for polymers [50]. For example, the SmC to oblique columnar phase transitions [51] in six-ring, double-swallow-tailed compounds represents a worthy challenge for such computations. Furthermore, in order that the use of data on increasingly complex liquid crystal materials can be optimized in terms of their contributions to the development of new fundamental knowledge potentially relevant to polymeric and biological materials as well as new liquid crystal technologies and applications, a broader range of experimental data correlating to, for example, mechanical, magnetic, electrical, rheological, and optical properties with chemical structure is needed.

6.4.4 R3: Re-entrance from Competing Fluctuations

To describe a line of phase transitions between two SmA phases (SmA_d and SmA_I, for example), Barois et al. [52] have used two order parameters. As the symmetries are the same, an SmA_d–SmA_I line of first-order phase transitions should end at a critical point. However, the observation is that, as the critical point is approached and fluctuations increase on the two sides of the transition line, where the symmetry is the same but the layer spacings are different, a nematic bubble spontaneously appears embedded in SmA [19]. Mean field theories [52] cannot account for this observation.

At a second-order SmA–SmC phase transition, the symmetries are different but the layer spacing is the same. Fluctuations can drive a line of second-order SmA–SmC phase transitions to an N–SmA–SmC multicritical point (see Fig. 4) [53]. Competing N–SmA and N–SmC fluctuations pull the N phase under the SmA phase in the temperature–concentration phase plane, leading to the N_{re}–SmA–SmC multicritical point [16]. High-resolution studies, as a function of both concentration and pressure, resolve the fluctuation-driven N–SmA/N–SmC step (see, e.g. Fig. 5) into a universal spiral (Fig. 12) [16] around the N–SmA–SmC and the N_{re}–SmA–SmC multicritical points. Loosely speaking, the N–SmA transition line is dominated by N–SmA fluctuations, and the N–SmC transition line is dominated by Brazovskii fluctuations [54] that drive the N–SmC transition to lower temperatures compared to the N–SmA transition [18].

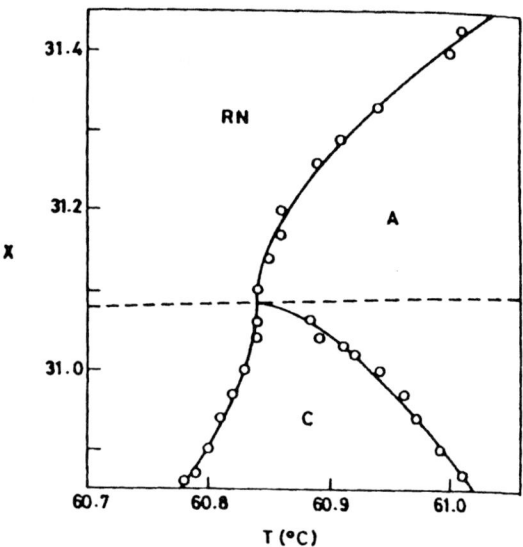

Figure 12. Detail of the N_{re}–SmA–SmC multicritical point showing the universal spiral topology [16] of the N–SmA–SmC multicritical point [54].

The N_{re} phase has also been observed in binary mixtures of 'nonpolar' compounds [55]. Shashidhar [20] has pointed out that in these two-component systems only one component is weakly polar or nonpolar. The second component has a dipole moment that is pronounced but not as strong as that of, for example, cyanobiphenyls [21]. The Bangalore group found that practically all liquid crystals possessing a net molecular dipole moment also show near-neighbor antiparallel dipolar correlations, as evidenced by the difference in the average and isotropic values of the dielectric constant. Such correlations are absent only when the molecular structure is such that one half of the molecule is an exact mirror image of the other half. Neither of the two 'nonpolar' components showing N_{re} phase behavior have this property. Nonpolar re-entrance is observed over a very small temperature range and an even narrower concentration range [21].

On further investigation, Shashidar [20] explains, a remarkable situation was found. In every case, one of the materials had an inherent SmC phase, while the other did not. As there was no observed miscibility gap, in every case there had to be a N–SmA–SMC point at very low temperatures. They also knew from their studies at high pressure [54] as well as their detailed studies on mixtures exhibiting the N–SmA–SmC and the N_{re}–SmC–SmA point (see Figs. 4 and 12) [16], that the phase diagram has the universal spiral topology where phase boundaries curve as the N–SmA–SmC point is approached. Shashidhar [20] concludes that, as re-entrance in nonpolar compounds results from the universal curvature of the phase boundaries as the N–SmA–SMC point is approached, its origin is the fluctuations associated with the N–SmA–SmC multicritical point.

6.4.5 Conclusions

In this overview of re-entrant phase transitions in liquid crystals we have identified three main causes of re-entrance: R1, re-entrance from frustration; R2, re-entrance from complex steric factors; and R3, re-entrance from competing fluctuations (e.g. 'a new order parameter grows'). Apart from (possibly) the re-entrant hexagonal transition in discotic liquid crystals, most re-entrant transitions in liquid crystals have no obvious solid-state analog. To optimize the impact of research into liquid crystal re-entrant phase transitions on fundamental knowledge and applications in general, and polymer and biological physics in particular, a braod variety of experimental data and sophisticated theories of structure–property relationships is required. So far, such a depth of investigation has been accomplished only for R1 re-entrance, as robust compounds exhibiting R1 re-entrance are widely available. Thus, while we have learned new physics from liquid crystals by studying R1 re-entrance, the new physics to emerge from the less widely studied, more 'exotic' materials exhibiting R2 and R3 re-entrance is still to come.

6.4.6 References

[1] P. E. Cladis, *Phys. Rev. Lett.* **1975**, *35*, 48.
[2] D. Guillon, P. E. Cladis, J. B. Stamatoff, *Phys. Rev. Lett.* **1978**, *41*, 1598.
[3] N. Berker, personal communication (1996). As SmA translational order is not 'really' long range, it may not be totally irrelevant.
[4] P. E. Cladis, *Mol. Cryst. Liq. Cryst.* **1988**, *165*, 85.
[5] See e.g. P. E. Cladis, *Liq. Cryst.* (in press).
[6] V. Vill, H.-W. Tunger, *J. Chem. Soc., Chem. Commun.* **1995**, 1047. The authors found that the 11 chain homolog (1S,6R,8R)-8-(4′-undecoxy-pheny)-3-(4′-cyanophenyl)-2,4,7-trioxa-3-borabicyclo[4.4.0]decane, CNB11, exhibits TGBA$_{re}$. V. Vill, H.-W. Tunger, D. Peters, *Liq. Cryst.* **1996**, *20*, 547. Binary mixtures of CNB10 and CNC12 (Figure 2) were studied. In CNC12, boron is replaced by carbon.
[7] For example haemoglobin becomes more fluid as it is cooled towards 0°C.
[8] L. J. Yu, A. Saupe, *Phys. Rev. Lett.* **1980**, *45*, 1000.
[9] N. H. Tinh, J. Malthête, C. Destrade, *J. Phys.* **1981**, *42*, L417.
[10] W. Weissflogg, I. Letki, G. Pelzl, S. Diele, *Liq. Cryst.* **1995**, *18*, 867; *Mol. Cryst. Liq. Cryst.* **1995**, *260*, 157, and references therein.
[11] S. Diele, G. Pelzl, A. Humke, S. Wünsch, W. Schäfer, H. Zaschke, D. Demus, *Mol. Cryst. Liq. Cryst.* **1989**, *173*, 113, and references therein.
[12] The SmO and SmC phases have layered structures. In the SmC phase the molecular tilt in a layer is inclined at an angle to the layer normal and is the same in all layers. In SmO, the molecular tilt alternates in a herringbone pattern with equal and opposite tilts in neighboring layers. In the chiral case, the SmC* and SmO* phases have, in addition, a helix structure, and so they are both helielectric, although SmO* has a double-helix structure because of its herringbone structure. For a recent review of the SmO phase see: A. Fukuda, Y. Takanishi, T. Isozaki, K. Ishikawa, H. Takezoe, *J. Mater. Chem.* **1994**, *4*, 997.
[13] G. Heppke, P. Kleinberg, D. Lötzsch, *Liq. Cryst.* **1993**, *14*, 67. A. Fukuda, personal communication (1996).

[14] W. A. Fertig, D. C. Johnson, L. E. DeLong, R. W. McCallum, M. B. Maple, B. T. Matthias, *Phys. Rev. Lett.* **1977**, *38*, 987.
[15] Perhaps the earliest example of this is the temperature–concentration phase diagram where FeC alloys show a fcc phase both above and below a bcc phase. J. Billard, personal communication.
[16] S. Somasekhara, R. Shashidhar, B. R. Ratna, *Phys. Rev. A* **1986**, *34*, 2561.
[17] G. Pelzl, C. Scholz, D. Demus, H. Sackmann, *Mol. Cryst. Liq. Cryst.* **1989**, *168*, 147.
[18] J. B. Swift, *Phys. Rev. A* **1976**, *14*, 2274.
[19] P. E. Cladis, H. R. Brand, *Phys. Rev. Lett.* **1984**, *52*, 2261. F. Hardouin, M. F. Achard, N. H. Tinh, G. Sigaud, *Mol. Cryst. Liq. Cryst. Lett.* **1986**, *7*.
[20] R. Shashidhar, personal communication.
[21] B. R. Ratna, R. Shashidhar, V. N. Raja, C. Nagabhusan, S. Chandrasekhar, G. Pelzl, S. Diele, I. Latif, D. Demus, *Mol. Cryst. Liq. Cryst.* **1989**, *167*, 233.
[22] P. E. Cladis, R. K. Bogardus, W. B. Daniels, G. N. Taylor, *Phys. Rev. Lett.* **1977**, *39*, 720. P. E. Cladis, R. K. Bogardus, D. Aadsen, *Phys. Rev. A* **1978**, *18*, 2292.
[23] G. W. Gray, J. E. Lydon, *Nature* **1974**, *252*, 221. J. E. Lydon, C. J. Coakley, *J. Phys. Coll.* **1975**, *36*, C1–45.
[24] P. E. Cladis, P. L. Finn, J. W. Goodby in *Liquid Crystals and Ordered Fluids*, Vol. 4 (Eds.: A. C. Griffin, J. F. Johnson), Plenum, New York, **1984**, p. 203.
[25] D. Guillon, P. E. Cladis, D. Aadsen, W. B. Daniels, *Phys. Rev. A* **1979**, *21*, 659.
[26] A. J. Leadbetter, J. L. A. Durrant, M. Rugman, *Mol. Cryst. Liq. Cryst. Lett.* **1977**, *34*, 231; unpublished data.
[27] P. E. Cladis, *Mol. Cryst. Liq. Cryst.* **1981**, *67*, 177.
[28] W. Weissflogg, N. K. Sharma, G. Pelzl, D. Demus, *Krist. Techn.* **1980**, *15*, K35. G. Pelzl, U. Böttger, S. Kallweit, D. Demus, H. Sackmann, *Cryst. Res. Technol.* **1981**, *16*, K119.
[29] Nguyen Tinh, H. Gasparoux, *Mol. Cryst. Liq. Cryst.* **1979**, *49*, 287. F. Hardouin, G. Sigaud, M. F. Achard, H. Gasparoux, *Phys. Lett.* **1979**, *71A*, 347; *Solid. State Commun.* **1979**, *30*, 265.
[30] F. Hardouin, A. M. Levelut, N. H. Tinh, G. Sigaud, *Mol. Cryst. Liq. Cryst. Lett.* **1979**, *56*, 35.
[31] For reviews of this group's work see: F. Hardouin, *Physica* **1986**, *140A*, 359. F. Hardouin, A. M. Levelut, M. F. Achard, G. Sigaud, *J. Chim. Phys.* **1983**, *80*, 53.
[32] J. C. Dubois, A. Zann, *J. Phys.* **1976**, *C3-37*, C3-35.
[33] H. W. Meul, C. Rossel, M. Decroux, O. Fischer, G. Remenyi, A. Briggs, *Phys. Rev. Lett.* **1984**, *53*, 497. In the pseudo-ternary Eu–Sn molybdenum chalcogenides for different Eu concentrations.
[34] The authors [33] call this the Jaccarino–Peter compensation effect from V. Jaccarino and M. Peter, *Phys. Rev. Lett.* **1962**, *9*, 920.
[35] K. Kondo, H. Takezoe, A. Fukuda, E. Kuze, *Jpn. J. Appl. Phys.* **1983**, *22*, L43. H. Takezoe, K. Kondo, K. Miyasato, S. Abe, T. Tsuchiya, A. Fukuda, E. Kuze, *Ferroelectrics* **1984**, *58*, 55, and references therein.
[36] See, for example, M. Yamashita in *Solitons in Liquid Crystals* (Eds.: L. Lam, J. Prost), Springer Verlag, New York **1991**.
[37] See, for example, G. Pelzl, D. Demus, *Z. Chem.* **1952**, *21*, 152. W. Weissflogg, N. K. Sharma, G. Pelzl, D. Demus, *Krist. Techn.* **1980**, *15*, K35.
[38] W. Weissflogg, G. Pelzl, D. Demus, *Mol. Cryst. Liq. Cryst.* **1981**, *76*, 261.
[39] J. O. Indekeu, A. N. Berker, *Physics* **1986**, *140A*, 368; *Phys. Rev. A* **1986**, *33*, 1158; *J. Phys.* **1988**, *49*, 353. A. Nihat Berker, J. O. Indekeu, in *Commensurate Crystals, Liquid Crystals and Quasicrystals* (Eds.: J. F. Scott, N. A. Clark), Plenum, New York **1987**, p. 205.
[40] This term was coined by G. Toulouse after a remark by P. W. Anderson. See, for example, P. W. Anderson, in *Future Trends in Materials Science* (Ed.: J. Keller), World Scientific, New Jersey **1988**.
[41] G. Pelzl, B. Oertel, D. Demus, *Cryst. Res. Technol.* **1983**, *18*, K18. See also Vill and coworkers [6].
[42] As SmA and N* (and N) phases do not have the same symmetry, a critical point is excluded for this transition.
[43] R. C. Long, Jr., J. H. Goldstein in *Liquid Crystals and Ordered Fluids*, Vol. 2 (Eds.: J. F. Johnson, R. S. Porter), Plenum, New York **1974**, p. 147.
[44] J. Malthête, N.-H. Tinh, C. Destrade, *Liq. Cryst.* **1993**, *13*, 171.
[45] For a description of the basic aspects of entropic elasticity see P. G. de Gennes, *Scaling Concepts in Polymer Physics*, Cornell University Press, Ithaca, NY **1979**. For details of how it has been applied to polymers, see E. R. Duering, K. Kremer, G. S. Prest, *Phys. Rev. Lett.* **1991**, *67*, 3531; *J. Chem. Phys.* **1994**, *101*, 8169, and references therein. R. Everaers, K. Kremer, *Phys. Rev. E* **1996**, *53*, R37; *J. Mol. Model.* **1996**, *2*, 293.
[46] F. Dowell, *Phys. Rev. A* **1983**, *28*, 3526; **1987**, *36*, 5046.
[47] C. Destrade, H. Gasparoux, A. Babeau, N. H. Tinh, J. Malthête, *Mol. Cryst. Liq. Cryst.* **1981**, *67*, 37.
[48] N. H. Tinh, P. Foucher, C. Destrade, A. M. Levelut, J. Malthête, *Mol. Cryst. Liq. Cryst.* **1984**, *111*, 277.

[49] A. Nayeem, J. H. Freed, *J. Phys. Chem.* **1989**, *93*, 6539.
[50] See, for example, F. Müller-Plathe, *Chem. Phys. Lett.* **1996**, *252*, 419. K. Binder, C. Cicotto (Eds), *Monte Carlo and Molecular Dynamics of Condensed Matter Systems*, SIF, Bologna **1996**.
[51] W. Weissflogg, M. Rogunova, I. Letko, S.-Diele, G. Pelzl, *Liq. Cryst.* **1995**, *19*, 541.
[52] For a review of SmA polymorphism in the context of two order parameter mean field theories see, for example, P. Barois, J. Pommier, J. Prost, *Solitons in Liquid Crystals* (Eds: L. Lam, J. Prost), Springer Verlag, New York **1991**.
[53] G. Pelzl, D. Demus, *Z. Chem.* **1981**, *21 (4)*, 151. G. Pelzl, U. Böttger, D. Demus, *Cryst. Res. Technol.* **1981**, *16*, 5; *Mol. Cryst. Liq. Cryst. Lett.* **1981**, *64*, 283.
[54] R. Shashidhar, B. R. Ratna, S. Krishna Prasad, *Phys. Rev. Lett.* **1984**, *53*, 2141.
[55] G. Pelzl, S. Diele, I. Latif, D. Demus, *Cryst. Res. Technol.* **1982**, *17*, 78.

7 Defects and Textures

Y. Bouligand

7.1 Definitions, Conventions and Methods

7.1.1 Local Molecular Alignment

Ordered media are never perfect, but present deformations and even defects. In liquid crystals, the order is defined by several parameters, mainly the *local mean direction* of approximately parallel molecules, usually represented by a unit vector *n* chosen parallel either to the long axis if the molecule is elongated, or normal to the molecules if the molecule is discoidal. A second local variable is the *order parameter*, which corresponds to a more or less accurate alignment of molecules. The orientation of *n* is often chosen arbitratily (+*n* equivalent to −*n*), since there are no polarities in the distribution of molecules, even if the chemical formula is 'arrowed', as in the classical example of 4-methyloxy-4′-*n*-butylbenzylidene aniline (MBBA) molecules. Both the parallel and the antiparallel alignment occur in equal proportions. In this case, *n* is called a *director*, with only the direction of the molecules being defined, with no preferred orientation.

Several conventions are used to represent directors or molecules in figures. Rod-like or disk-like molecules can be drawn as circular cylinders, either elongated or flat [1]. Another representation is given by a point for a director normal to the figure plane P, a segment of constant length for a director parallel to P, and a nail with length proportional to $\cos\alpha$ for directors lying at an angle α from P, the sharp end of the nail corresponding to the director extremity pointing towards the observer [2, 3]. The opposite orientation for nails is also adopted [4, 5].

7.1.2 Microscopic Preparations of Liquid Crystals

When a liquid crystal is introduced between two parallel glass plates (a slide and a coverslip), without any particular care, and is examined between crossed polarizers under a polarizing microscope, within a thermostated stage if necessary, the general chromatic polarization is observed and multiple patterns appear, showing that alignment is only local, with the director varying continuously over large distances. Discontinuities in the optical image suggest the presence of discontinuities in director distribution, in the form of singular points, lines and walls. Liquid crystals oppose an elastic energy to these deformations and defects. As these media are liquid, the singularities move easily until they reach an equilibrium position, this often resulting in a regular arrangement of defects and domains, called *texture*.

Descriptions of defects and textures in liquid crystals are based on concepts of *differential geometry* (see Sec. 7.4 of this Chapter). An intuitive approach is proposed in the books by Hilbert and Cohn-Vossen [6] and Coxeter [7], both of which contain chapters devoted to topology, in a style of reasoning close to that used in the study of liquid crystals. For details on polarizing microscopy see Hartshorne and Stuart [8], and for an illustrated presentation of defects in true crystals see Amelinckx [9]. Numerous micrographs and interpretative drawings can be found in early works such as those by Lehmann [10] and Friedel [11, 12] (see Chap. I of this volume). More recent books cover the essential knowledge about mesogenic molecules, phase transitions, liquid crystalline structures, and defects and their arrangement around domains, with the whole illustrated by numerous figures and micrographs, colour plates showing the main textures seen in polarizing microscopy [13, 14]. The structure and energetics of defects in liquid crystals have been reviewed by Chandrasekhar and Ranganath [15], while Kléman [16] has covered not only liquid crystals but also other ordered media and, in particular, magnetic systems.

7.1.2.1 Thermotropic Textures

Examples of textures are present in rodlets, spherulites and more or less expanded germs of a mesophase, floating within the isotropic liquid or attached to one of the two glasses, at the transition. A long and brilliant SmA rodlet or *bâtonnet* is shown in Fig. 1, contrasted against the black background due to the isotropic liquid, as observed between crossed polarizers. Smaller germs are also present and, on the left of the micrograph, the smectic A phase shows continuous shade variations and lines of discontinuity, in the form of ellipses, parabolae and hyperbolae. These conics correspond to lines of discontinuity extending either from an interface or in the bulk of the mesophase. As the liquid crystal structure is elastically deformed all around the defect, each of these lines lies at the origin of a part of the elastic energy, and when defects attach along the interface with an isotropic phase a strong minimization is obtained. Numerous defects adhere to the isotropic interface of the smectic rodlet in Fig. 1, the layers of which lie almost normal to the long axis, but are invisible in photon microscopy, their thickness being approximately equal to the molecular length. The surface defects in this bâtonnet form regular patterns with discrete rotation symmetry, and some irregularities.

Spherical germs of a twisted nematic phase (a mixture of a nematic liquid and an asymmetric compound) of different sizes and orientations, as observed under natural light are shown in Fig. 2. The molecules align according to the cholesteric model, but the helicoidal pitch is larger than in pure cholesteric phases. The existence of a periodicity creates a series of contrasted stripes. The general aspect is lamellate, with 'layers' normal to the diameter of the spherulite, but curved at the periphery, and lying perpendicular to the isotropic interface. The parallel arrangement of 'layers' can be disturbed by the coalescence of several spherulites, leading to the formation of defect lines, which often annihilate at the isotropic interface. Also due to the helical struc-

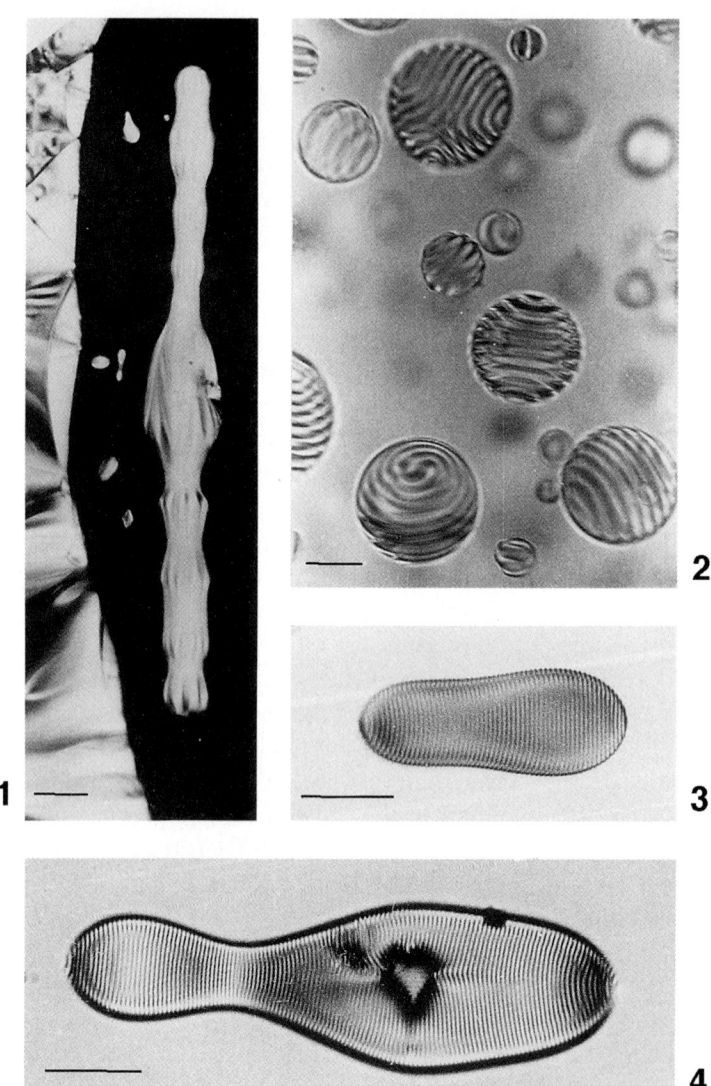

Figure 1. A smectic A phase extended on the left-hand side, and showing defect lines in the form of conics; a smectic rodlet and small germs are floating in the isotropic phase. The mesogenic product, 4-cyano-4'-n-octylbiphenyl (8CB), was added together with a small amount of Canada balsam, and the mixture was observed between crossed polarizers at room temperature. Canada balsam is an isotropic but optically active and fluid resin, extracted from the conifer *Abies balsamea*, which facilitates the production of regular smectic textures in several thermotropic liquids. Collophony, or rosin, is a stabilized pine tree resin, used to rub violin bows; Friedel used this substance not only for his violin, but also to obtain remarkable smectic textures. Scale bar: 20 µm.

Figure 2. Spherical cholesteric germs of MBBA, plus a small amount of cholesterol benzoate. There is a disclination similar to the τ^- pattern shown in Fig. 25 and a spiral decoration in another spherulite. One polarizer only. Scale bar: 20 µm.

Figure 3. Elongated germs of MBBA cholesterized by a small amount of Canada balsam as observed between parallel polarizer and analyser. Layers lie mainly perpendicular to the long axis, without defects, and orientate normally to the isotropic interface. This germ does not float within the bulk, but is slightly sandwiched between the slide and the coverslip. Scale bar: 20 µm.

Figure 4. A more developed cholesteric rodlet in the same preparation as in Fig. 3, showing a set of internal defects. Scale bar: 20 µm.

ture, spiral patterns generally appear at two opposite poles of the spherulite.

When the helical pitch is much smaller, instead of being spherical, germs grow along an axis normal to 'layers' (Fig. 3), as in smectic phases. The shape of the smectic and cholesteric germs reveals the *anisotropy of surface tension*. Molecules often tend to lie parallel to the isotropic interface in smectic and nematic phases, whereas they prefer to orientate normally to the air interface. There are strong similarities between the textures of smectic A and cholesteric phases, but the three-dimensional arrangement of cholesteric layers is often resolved in optical microscopy, whereas that of smectics is not. However, many textural patterns are similar in smectic A and cholesteric phases. Comparisons between the textures of these two phases concern mainly domains extending over distances that are very long compared to the helical pitch. For instance, when several tens of layers are present elongated cholesteric drops are observed. It should be noted that the germs shown in Figs. 3 and 4 have grown to dimensions larger than the distance from the slide to the coverslip, and are 'flattened', but a thin isotropic film is still present, separating them from direct contact with the glass plates.

7.1.2.2 Lyotropic Textures

Amphiphilic molecules have two parts, one hydrophilic and one hydrophobic, and form liquid crystals the textures of which are of great interest, since they are very close in morphology to biological materials and, in particular, cell membranes [17–19], showing a polymorphism related to that known in water–lipid systems (see Chaps. XV to XVII of Vol. 2 of this Handbook). The lamellar structure displays some usual defects and textures (Fig. 5). The bilayers are more or less separated by water (Fig. 5 a, b); this

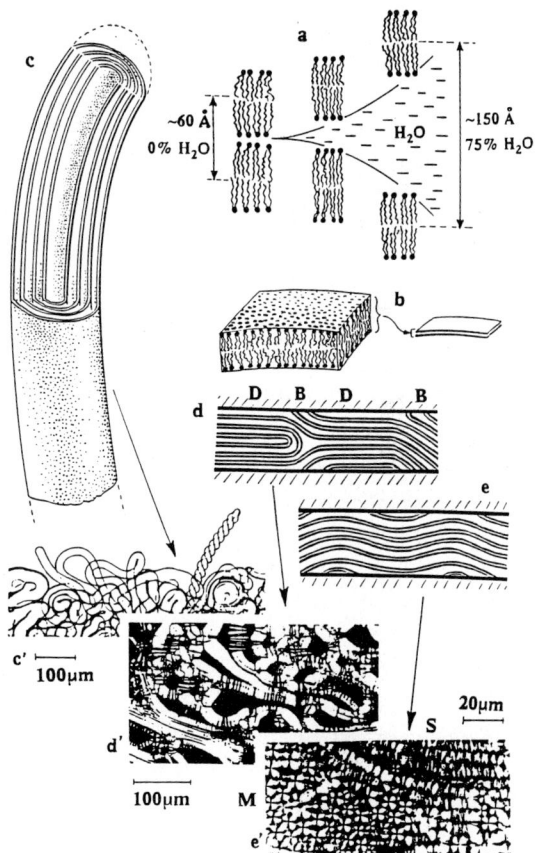

Figure 5. Schematic representations of lyotropic lamellar systems. (a) Bilayers of phospholipids (one polar end and two paraffinic chains) are more or less separated by water (after Schmitt and Palmer [18]). (b) A bilayer (sketched on a different scale) with two parallel surfaces (or two parallel lines in cross-section). (c) Myelinic finger, with cylindrically nested bilayers, the number of which is strongly reduced (possibly by a factor of 10^3), compared to the lecithin fingers in (c'). (d) The bilayers parallel to the preparation plane are said to be 'horizontal' and form a dark background (D) between crossed polarizer, whereas oblique or vertical bilayers appear bright (B), mainly if they extend at 45° from the orientation of the polarizers, as shown in (d') for lecithin. (e) Bilayers that were initially horizontal often corrugate in one or two directions to form domes and basins, each appearing as a 'Maltese cross' (M) when observed between crossed polarizers (e').

swelling is important in the fluid *myelin textures*, a very common texture in lyotropic systems, with concentrically nested layers (Fig. 5c). Some cylindrical shapes, either straight or curved, or forming simple or double helices, are shown in Fig. 5c'.

When a myelinic system is progressively dehydrated between the slide and the coverslip, many bilayers arrange horizontally (Fig. 5d). These appear black between crossed polarizers, whereas those that remain vertical or oblique form brilliant stripes called 'oily streaks' (Fig. 5d'). Further dehydration leads to the formation of air bubbles between the slide and coverslip; the bubbles extend progressively, and lateral compressions result in wrinkles that superimpose in several directions. The horizontal bilayers often transform into nested domes or nested basins, each being characterized by a 'Maltese cross', when examined between crossed polarizers (Fig. 5e). The 'Maltese crosses' arrange in a more or less regular fashion and form polygonal mosaics, which are frequent in many liquid crystals. The oily streaks also corrugate through dehydration, giving them a cross-striated aspect in polarizing microscopy.

7.1.2.3 Liquid Monocrystals

Instead of defects and complex textures, some experimental situations allow one to obtain a very good alignment and thus are useful for comparisons with local aspects of complex preparations. Large monocrystalline domains are mainly obtained with thermotropic liquid crystals and, much less often, in lyotropic systems. Such alignments can be produced by external fields (see Sec. 9 of this Chapter) or by treating the slide and coverslip in order to create a regular *anchoring* of molecules in a uniform direction (*homeotropic* if the direction is normal to the glass, and *horizontal* if parallel to the glass), with a preferred or *easy direction* within the interface plane (see Sec. 10 of this Chapter). In polarizing microscopy, homeotropic preparations appear uniformly dark, with more or less intense light scattering due to director fluctuations. Liquid monocrystals with horizontally aligned molecules show an extinction in four positions of the rotating stage in the polarizing microscope, each extinction being separated by 90°.

7.1.3 Images of Liquid Crystals in Polarizing Microscopy

One of the difficulties in studying liquid crystals is to attain skill in making microscopic preparations and a general knowledge about images and their interpretation. In general, defects and textures are observed within preparations at equilibrium, but heating of or pressure exerted on the coverslip produce motions and streams, which cease as a new equilibrium is progressively reached; this generally takes some minutes, but there are textures with remarkable regularities which only appear after days.

Consider a liquid crystalline slab sandwiched between a slide and a coverslip, that is sufficiently thin (of the order of micrometers) and has no strong anchoring conditions. The directors will align along practically straight lines, lying normal, oblique or parallel to the glasses, since the elastic constants prevent the occurrence of high curvatures (however, those exist in the vicinity of defects). This means that, in large domains, the length of curvature radii is notably larger than the liquid crystal thickness. Projection of the local director onto the preparation plane then corresponds to one of the two local directions of extinction between crossed polarizers, and the director is determined with the help of an additive plate, generally a quartz first-order retardation plate, a $\lambda/4$, or a compensator, as in conventional crystallography [8]. The use of quasi-monochromatic waves (e.g., sodium light) often facilitates these investigations.

When the liquid crystal forms a thick layer (20–50 µm) between the slide and the coverslip, the interpretation is more difficult. Examination under natural light or between parallel polarizers can be useful. It is remarkable that even with such thick preparations, which show an extreme complexity of textures and optics, it is still possible to obtain clear images, focused either at the upper level of the liquid crystal in contact with the coverslip or at the bottom level, at the slide interface. This is illustrated in Fig. 6 for a cholesteric texture [20]. The picture quality is better at the coverslip level in Fig. 6a, but both views are useful.

Some good pictures can be obtained using intermediate thickness optical sections, at horizontal levels between the slide and the coverslip. The quality of the images facilitates the preparation of stereo-pairs, in order to observe textures in three dimensions. The two examples shown in Fig. 7 were obtained simply by tilting the preparation slightly differently. The stereoviews of Fig. 7a and b show hyperbolae branches

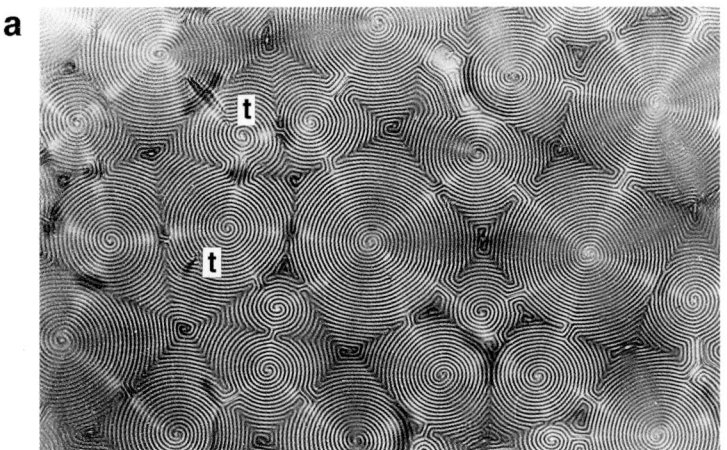

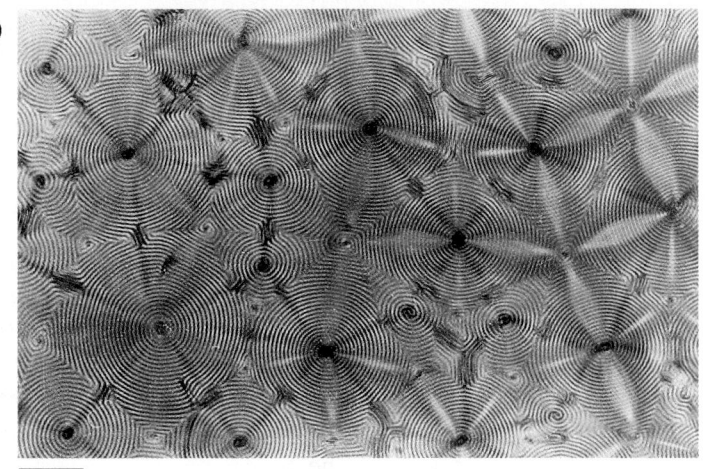

Figure 6. Polygonal texture in a cholesteric liquid (MBBA and Canada balsam) observed under natural light: (a) coverslip level, (b) lower glass level. t, Translation defects (there are other unmarked translation defects in the micrographs). Scale bar: 20 µm.

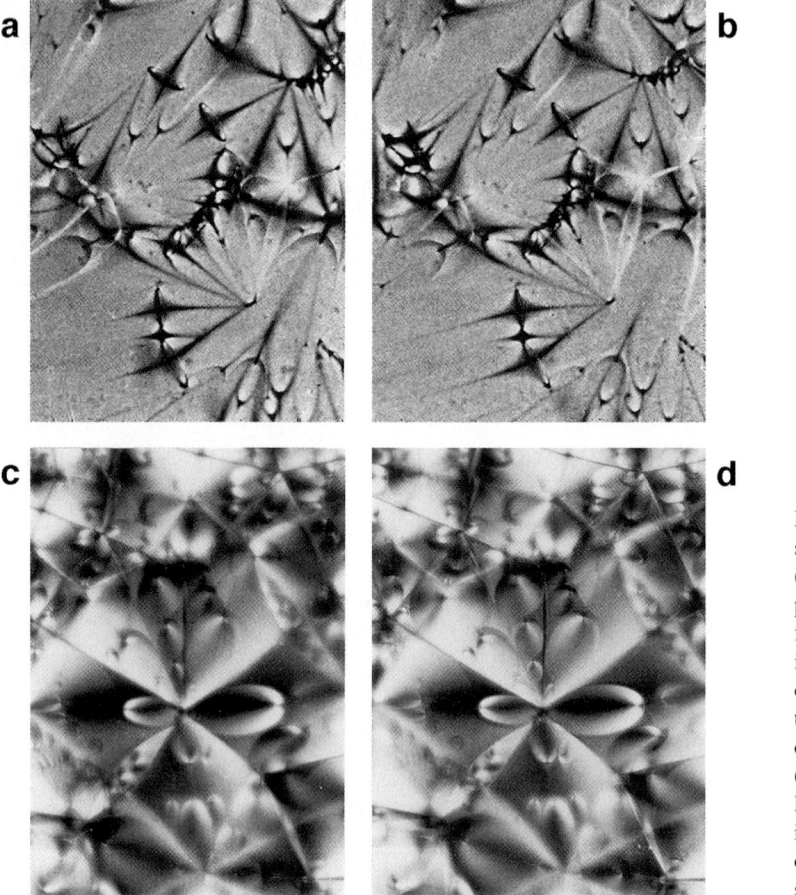

Figure 7. Stereoviews of smectic A textures of 8CB (see Fig. 1). (a, b) A stereopair in natural light, showing focal conics, each ellipse being penetrated by a branch of a hyperbola passing through one of the two foci of the corresponding ellipse. (c, d) A stereopair of a polygonal texture, with ellipses in the two separated planes of the slide and the coverslip. Scale bar: 100 µm.

passing through ellipses, a classical pair of defects in smectics, and named focal conics, that it is worth seeing in three dimensions. Figures 7c and d show a polygonal field in a smectic A phase; there are lattices of ellipses, and the slide and the coverslip levels can be distinguished. All vertices of the upper polygons superimpose vertically to central points in polygons of the bottom latice, and vice versa as in the cholesteric polygonal fields shown in Fig. 6. These textures were mainly studied by Friedel and Grandjean [11, 21].

There is another method that can be used to explore textures in three dimensions. In some favorable cases, mesogenic molecules can polymerize within the mesophase and form a resin, without any modification of the initial distribution of directors and with a texture that remains stable. The whole transformation can be observed in the hot stage in polarizing microscopy [22]. The slide and coverslip can be unstuck and the resin slab polished on its two opposing faces, to suppress the thin layers in contact with the glasses. In this way the director distributions at the glass interface and in the bulk can be compared. Semi-thin sections of the resin slab can also be prepared by ultramicrotomy, using a glass or diamond knife, and observed using polarizing microscopy.

7.1.4 Other Microscopic Methods

Techniques introduced by biologists to study cells and tissues [23, 24] have been extremely useful for the examination of liquid crystals, mainly lyotropic crystals. Microtomy methods are applied to embedded biological materials, after stabilization with small amounts of OsO_4 in water, introducing numerous cross-links into the structure. The material must be progressively dehydrated in ethyl alcohol before it is embedded in an epon resin. By using an electron-dense contrast agent (e.g. phosphotungstic acid, uranyl acetate) one can obtain excellent images, and artefacts are generally well controlled. The lamellar and middle phases of soaps or other water–lipid systems have been studied using these techniques, and excellent views of defects and textures have been obtained [25].

An alternative method is the freeze-etching technique [24, 25], which consists of producing platinum shadowed replicas of fractures created in rapidly frozen biological systems or lyotropic liquid crystals. The fracture orientation is somewhat haphazard, but occurs preferentially within the paraffinic level of the bilayers. These methods offer the possibility of viewing bilayers directly and preparing stereoviews. Beautiful pictures of liquid crystalline DNA, both cholesteric and hexagonal, have been obtained, but individual molecules are not easily resolved. However, the director distribution can be deduced from the images [26].

Microtomy and freeze-etching techniques have revealed many similarities between liquid crystals and a large series of biological materials, which show the same symmetries as nematic, smectic and cholesteric phases, but without being fluid [27–34]. The biological materials are stabilized analogues, and it has been verified that most of these systems can be obtained in a true mesomorphic state [35, 36]. Defects and textures also are present in these biological counterparts of liquid crystals, and are associated with different shapes in tissues and organs; this aspect represents a new axis of research in biological morphogenesis [37].

7.2 Strong Deformations Resulting in Defects

7.2.1 Singular Points

Simple experimental situations lead to the formation of singular points, lines or walls. For instance, glass capillary tubes (diameter ~0.5 mm) can be treated with a surfactant to ensure radial anchoring at the inside wall, which corresponds to homeotropy [38]. When filled with a nematic liquid, the director lines are radial at the periphery and lie parallel to the capillary along its axis (Fig. 8a). The main curvatures are of the splay (mainly along the axis) and bend types. Twist is not excluded, because minimization of the curvature energy requires contributions from the three terms. As there are two possible splay orientations along the tube axis, punctual singularities occur here and there, the patterns of which are shown in meridian section in Fig. 8a [39–41].

The director lines distributed around one of these point singularities form a pattern reminiscent of the radial electric field produced by a charged particle, whereas the other singular pattern recalls the electric field about the zero point, lying in the middle of a segment the extremities of which are occupied by two identical electric charges. These singular points are very common in nematics and cholesterics. The two con-

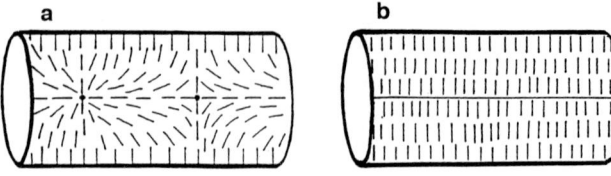

Figure 8. Textures produced by homeotropic boundary conditions in a capillary tube filled with a nematic liquid (a) or a smectic liquid (b).

figurations shown in Fig. 8a are said to be complementary, since they can fuse and disappear, or can be created by pairs (e.g. due to strong disturbances in fluid streams).

7.2.2 Singular Lines

When a nematic–smectic A transition occurs in a capillary tube (Fig. 8a), smectic layers nucleate at disclination points and a singular line forms along the axis, with cylindrical layers (as shown in Fig. 8b), but the presence of beads along the axial defect shows that the situation is less schematic [42].

Another situation occurs in a nematic liquid between two parallel glasses with horizontal anchorage, the two easy directions lying at right angles to one another [43]. The nematic liquid can 'choose' between two orientations for the twist: a right- or left-handed helix (Fig. 9). In Figs. 9a–c, a narrow domain, limited by an arc or a complete circle, is represented at the boundary between regions of left- and right-handed twist. Within this domain, the directors diverge by angles of 0–90°; this discontinuity is either attached to the coverslip (C) or the slide (S), or lies in the bulk. These narrow discontinuities, present in each figure plane, correspond to the successive sections

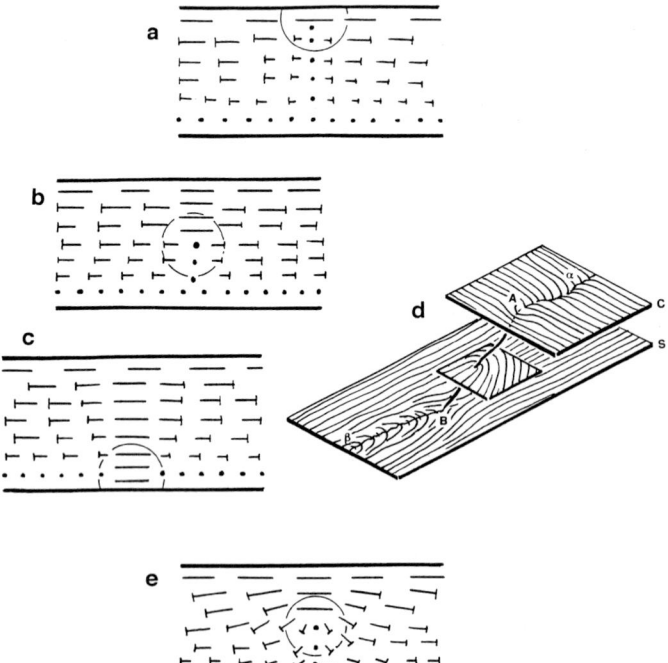

Figure 9. Defect line produced in a nematic liquid by two parallel glasses so treated that they have planar anchoring conditions, with two mutually perpendicular easy directions. The discontinuity line is either attached to the upper glass (a) or to the lower one (c), or lies in the bulk (b). (d) The three localizations of the defect thread along a path $\alpha AB\beta$, with a planar configuration in the bulk. (e) A nonplanar distribution of directors is more likely.

of a thread, attached to the coverslip along αA, penetrating the bulk from A to B, and attached to the glass slide along Bβ (Fig. 9d). The thread observed in the bulk forms a smooth curve, but when attached to glass it often follows a complex path, with a somewhat fractal aspect, as indicated in Fig. 9d. The directors are not necessarily horizontal in this texture (Fig. 9e).

Now, let us consider the distribution of directors along a circuit closed around such a thread of discontinuity in the bulk of the liquid. Following this circuit, it appears that the chosen orientation for the director n is reversed after one turn. The ribbon generated by a short segment parallel to n, and centred on a point M describing this circuit, is a Möbius strip and is, therefore, a non-orientable surface [44]. If the length of the circuit around the thread is progressively reduced, it appears that a discontinuous distribution of n is present at each point of the thread. More generally, the director field is said to form a non-orientable manifold, and physically this means that the director is not 'arrowed', there being no preferred orientation of the molecules (either parallel or antiparallel to n), as indicated above. This type of thread comes from a topological obstruction, a discontinuity due to the non-orientable distribution of directors at the periphery.

These two simple experimental systems show the presence in liquid crystals of two types of defect: lines and point singularities. Liquid crystals contain a large variety of lines with well-defined geometries or topologies. There are also lines that have a continuous core (for example, in the capillary tube); the axial zone corresponds to a maximum of splay and is generally considered to be a defect line, although no discontinuities apart from the singular points are present. This situation is also encountered in the third type of defect – walls.

7.2.3 Walls

Discontinuity walls do not exist in principle in usual liquid crystals, since there are no apparent limits to certain curvatures, as far as this can be observed in the vicinity of singular points and discontinuity lines. Several types of wall are usually considered:

- Twins possibly occur in liquid crystals that have a structure close to that of solid crystals, with more or less extended three-dimensional order, possibly in a smectic E phase (see Sec. 7.4.4.6).
- Defect lines arrange along narrow zones separating domains of different orientations in the liquid crystal, like grain boundaries in true crystals.
- When the anchoring conditions are sufficiently strong, at the slide and coverslip levels for instance, a reversal in the orientation of the directors can be confined in a narrow band [45]. Such zones of rapid rotation of directors are analogous to those observed in magnetic materials and, according to the relative values of the elastic constants, there are 'Bloch walls' showing mainly a twist curvature and 'Néel walls' associating bend and splay [46].

7.2.4 Interface Defects (Points and Lines)

A frequent situation in nematic phases close to the isotropic transition is the presence of a thin film of an isotropic phase separating the liquid crystal from the preparation glasses. The molecules lie either horizontally at this interface, or at an angle other than 90°, with no preferred direction in the horizontal plane. Defects can be numerous, as shown in Figs. 10 and 11.

The two vertical threads L in Fig. 10a and b correspond to singular lines surrounded by director lines following approximately the planar solutions of the Laplace equation $\nabla^2 \Phi = 0$, as shown by Oseen and later reviewed by Frank [54]. Φ is the azimuth of n supposed to be horizontal in a plane xy parallel to the glass plates. The solutions are of the type $= \Phi_0 + (N\varphi/2)$, where N is a positive or negative integer or zero (with $\tan\varphi = y/x$ and $\tan\Phi = n_y/n_x$), the horizontal Cartesian coordinate system xy being centred on the disclination trace. In Fig. 10a $N=1$, and in Fig. 10b $N=-1$. In the recent literature, N is replaced by $s = N/2$ a multiple of $+1/2$. The fact that, in general, the three elastic constants are not equal, and that spontaneous curvatures can be present, the twist, for example, modifies the shape of the director lines and even the symmetries, but not the general aspect of these disclinations in cross-section. The expected and observed aspects between crossed polarizer in the case $N=+1$ are pairs of dark 'brushes' attached to a central point (Fig. 10c and d). A slight shift of the coverslip relative to the slide separates the two brushes, and a thin thread joins them, as shown in Fig. 9d.

The preparations also show sets of four dark brushes attached to a common center, and a weak shear generally separates these patterns into pairs of dark brushes linked by a thread that is much thicker and less sharply contrasted than the threads described above. This texture was studied in resins produced by polymerized nematics, after abrasion and polishing of the upper and lower faces of the nematic analogue slab [22]. Different patterns were obtained, some of which are shown in Fig. 11 a–c, the types (a) and (c) being the most common. The corresponding aspects in polarizing microscopy are shown in Fig. 11 a'–c'. Examination using a compensator showed strong obliquities or even verticality of the directors in the core region, and continuous variations in their orientation. Splay, twist and bend are present, and there are no discontinuities along the thick threads in the bulk.

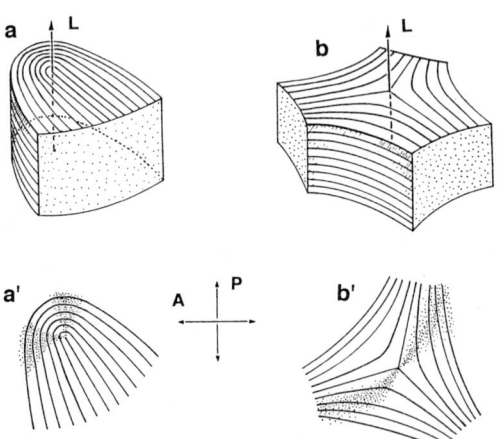

Figure 10. Two common defects in nematics, between parallel interfaces, with horizontal anchoring and no preferred orientation. (a, b) Three-dimensional views; the director lines, rather than separate segments as in Fig. 15, are represented and lie parallel or normal to the faces of the curved polyhedra. Bend and splay are concentrated in alternating sectors, the twist not being excluded from the bulk. (c, d) Corresponding aspects between crossed polarizers; the director lines are not visible under the microscope.

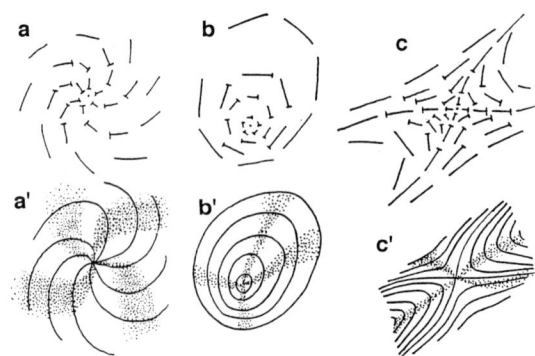

Figure 11. (a, b, c) Continuous distribution of directors within the median horizontal plane of three different types of nematic nucleus with four brushes (a', b', c'). Corresponding aspects between crossed polarizers; the curves correspond to the alignment of nails, but are not visible under the microscope.

Each of these patterns was given a name by Lehmann [10] – *halber Kernpunkt* ($s=1/2$), *halber Konvergenz Punkt* ($s=-1/2$), *ganzer Kernpunkt* ($s=1$) and *ganzer Konvergenz Punkt* ($s=-1$) – and were called *noyaux* by Friedel and Grandjean [21], or *nuclei* in English works. The assemblies of nuclei were named *plages à noyaux* [10, 11], or *nuclei textures* or *Schlieren textures* [45].

In a nematic preparation, between a horizontal slide and coverslip, the locus of vertical directors is made up of one or several lines, corresponding to intersections of surfaces $n_x(x, y, z)=0$ and $n_y(x, y, z)=0$ (within a Cartesian coordinate system x, y, z attached to the preparation, z being normal to the preparation plane). Such lines, cut the interfaces at isolated points, where the anchoring conditions can make them singular, and their structure is represented in Fig. 12, the twist excepted [44, 47]. Associations of these interfacial singular points are shown in Fig. 13. Similarly, the locus of horizontal directors corresponds to surfaces $n_z(x, y, z)=0$, cutting the limiting interfaces along curves, also made singular by the anchoring conditions, in particular, if horizontal directors are forbidden along an isotropic interface such as the one due to the thin isotropic films that are often observed along the glasses, near the isotropic transition. This is shown in Fig. 14 for nuclei with $N=1$

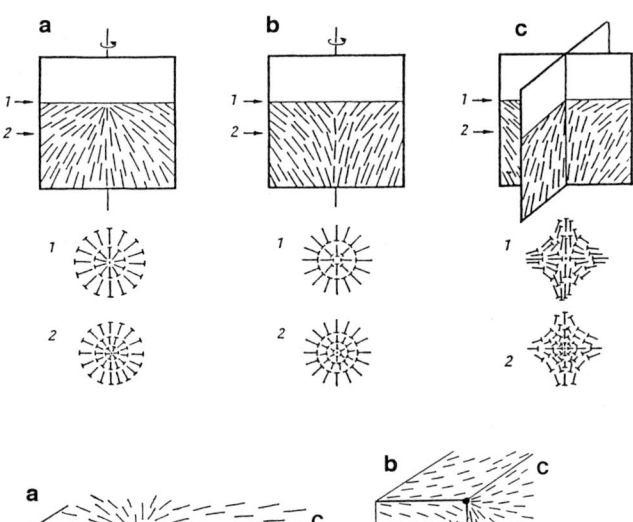

Figure 12. Singular points of the director distribution at an interface presenting a 45° anchoring angle, and the corresponding patterns in top view, at the interface (1) and just below in the bulk (2). The introduction of a twist allows one to pass continuously from the radial structure of point (a) to that of (b), with a constant revolution symmetry. Point (c) does not present this symmetry.

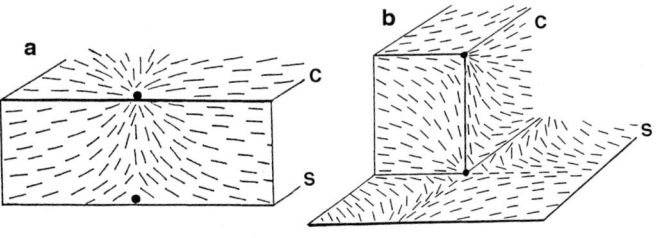

Figure 13. Vertical pairs of singular points at the slide (S) and coverslip (C) levels: (a) association of two radial points; (b) association of two non-radial points.

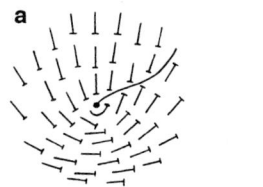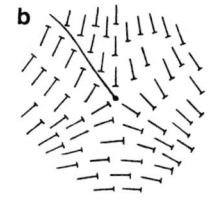

Figure 14. Distribution of directors at an interface presenting a constant anchoring angle (as in Fig. 12) and around the extremity of a thin thread, with patterns (as in Fig. 10). A discontinuity line appears unavoidably at the interface and is attached to the core of the defect. In the bulk, these lines create Néel or Bloch walls, or hybrid textures.

and $N=-1$, at the interface level, when the anchoring angle is constant and different from zero. The presence of such discontinuity lines originating from the core extremity of the nucleus is unavoidable for odd values of N.

7.2.5 The Nature of Thick and Thin Threads

Since a horizontal shift of the coverslip relative to the slide transforms a nuclei texture into a set of threads, this indicates that nuclei correspond to vertical threads with minimized length and elastic energy, this being confirmed by the generally observed stability of this texture. However, nuclei often fuse, and either disappear if their indices N are opposite, or form a new nucleus, the index N of which is the algebraic sum of the corresponding indices of the two parent nuclei [10, 11]. Similarly, threads annihilate or recombine, following the same type of laws [48, 49]. Threads were considered therefore as disclinations, the cross-sectional structure of which is that shown in Figs. 10 and 11 [16]. However, the situation is not that simple since, at any point M of a director field n (M), the director distribution within a plane P normal to n (M) generally forms patterns very similar to those shown in Fig. 11 a–c [44]. Each four-brush nucleus is continuous in the bulk and presents two singular points at its extremities, whereas a two-brush nucleus corresponds to a vertical line of discontinuity of the Möbius type. Conversely, a simple circuit closed around a thick thread joining the two singular extremities of a four-brush nucleus is not of the Möbius type.

7.3 Topological Structure of Defects

7.3.1 The Volterra Process

Defects can be defined by the geometrical operations necessary to pass from the perfect crystal to the disturbed structure. This point of view was developed in works by Volterra and Love [50, 51], and was later applied to crystals [52]. Consider an ordered material and suppose that it can be cut along surfaces and then deformed and restuck such that the two facing materials show parallel crystallographic orientations. Such operations are topological since they may change the connectivity of the material and the deformations are equivalent to symmetry operations. Such a procedure is called the 'Volterra process' and leads, in general, to a theoretical structure close to that of singular lines. This method was first introduced into the study of liquid crystals by Kléman and Friedel [2, 3].

7.3.2 The Volterra Process in Nematic, Smectic A and Cholesteric Phases

The local symmetries of nematics are all the translations, all the rotations about the director axis and all the $\pm\pi$ rotations about any axis normal to the director. The operations combining these translations and rotations also correspond to local symmetries of the nematic structure.

Topological rehandlings of nematics, after cutting, deforming and resticking, do not necessarily result in defect lines. For instance, if the two lips S_1 and S_2 of a cut surface S, limited by a line L, are separated by a translation and restuck after an eventual addition or subtraction of nematic matter, the line L would simply disappear by viscoelastic relaxation. However, if we separate the two lips S_1 and S_2 by an angle π, through rotation about an axis normal to $\boldsymbol{n}$, and add nematic matter as indicated in Fig. 15 a–c, the material relaxes into a differently connected system, with a defect line as represented in (c) and called a disclination. Figure 15 d–f represent the situation when a piece of matter is subtracted and the two lips S_1 and S_2 are stuck after another π rotation about an axis normal to $\boldsymbol{n}$ [2, 3]. The peripheral distribution of directors corresponds nearly to that observed in many preparations, and it was concluded, more or less explicitly, that the director lines lie in parallel planes as in Fig. 10. However, all three types of deformation (splay, twist and bend) are likely to be present.

In smectic A phases the symmetries associate: all translations normal to the director and all those parallel to the director, the length of which is an integer multiple of the layer thickness; all the rotations about an axis parallel to the director; and all $\pm \pi$ rotations about axes normal to the director, either at the limit between successive layers or at the half-thickness of a layer.

Several examples of the Volterra process in smectic A phases are indicated in Fig. 16 [2, 3]. The symmetry involved can be a translation normal to the layers, in which case the defect is said to be a *dislocation*, the translation vector $\boldsymbol{b}$ or Burgers vector being either parallel to the line L as in a *screw dislocation* (Fig. 16 a and a′), or normal to L as in an *edge dislocation* (Fig. 16 b and b′). As all rotations about any axis L normal to the layers superimpose a smectic A phase onto itself, any sector centred on L can be subtracted, and one obtains, be resticking, a set of nested conic layers, which remains stable for certain boundary conditions (Fig. 16 c and c′) [53]. This will be considered further in the study of focal conics (see Sec. 7.4.2). When the symmetry involved is a rotation, the defect line is called a *disclination* and is mainly characterized by the corresponding angle. The production of disclinations in smectic A phases is similar to that presented for nematics in Fig. 15. The addition of smectic layers is shown in Fig. 16 d′. The resulting viscoelastic relaxation often leads to a symmetrical disclination (Fig. 16″), L being then a three-fold axis, as in nematics (Fig. 15 c).

Disclinations are rotation defects, and are rare or absent in three-dimensional crystals, owing to their prohibitive energies, but are in general compatible with liquid crystalline structures. They were initially called 'disinclinations' [2, 3, 54], but the term was later simplified to 'disclination'.

Cholesterics are helically twisted nematics, the local symmetry of which is slightly biaxial, but close to that of nematics. If the

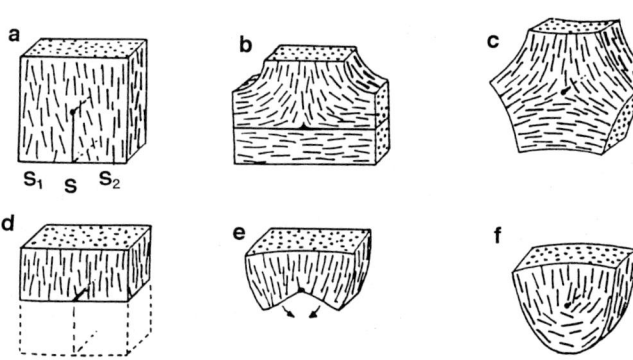

Figure 15. The Volterra process applied to a nematic liquid. A planar section limited by a line normal to the director $\boldsymbol{n}$ allows the two lips S_1 and S_2 to be separated by an angle π (b), and nematic material to be added to obtain the disclination structure (c). An initial matter subtraction creates two lips S_1 and S_2 (d), both rotated by an angle $\pi/2$ (e) and restuck to obtain another disclination (f).

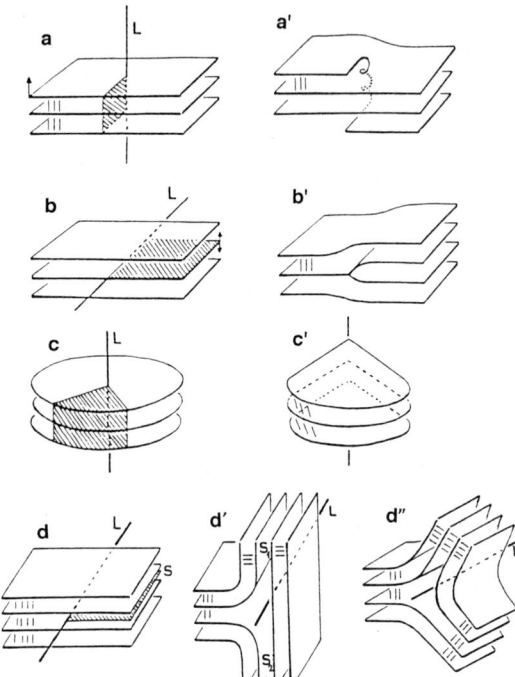

Figure 16. Creation of defects in a smectic A phase. (a, a') Creation of a right-handed screw dislocation. (b, b') Creation of an edge dislocation. (c, c') Creation of a stack of nested conic layers, as observed along focal conics. (d, d', d'') Creation of a disclination from a planar cut surface limited by a line L; a $+\pi$ separation of lips S_1 and S_2 is followed by the addition of matter and relaxation.

spontaneous twist is included in the structure definition, the symmetry group is changed and associates: all translations normal to the twist axis and all translations parallel to the twist axis that are integer multiples of the half-helical pitch; and all $+\pi$ rotations about any axis parallel either to the director or the twist axis, or normal to both [2, 3].

Examples of defect lines created in cholesterics according to the Volterra process are shown in Fig. 17. The core structure will be considered below this being replaced by a narrow cylinder. Figures 17a and b represent translation dislocations: in both cases, the Burgers vector lies parallel to the cholesteric axis, but is either normal to the line L, as an edge dislocation, or parallel to it, as a screw dislocation, as shown in Fig. 16 for smectic A phases. No additional material is necessary to obtain the screw dislocation in the Volterra process. Figure 17c–g show the two main disclinations created (see Sec. 7.3.4).

7.3.3 A Different Version of the Volterra Process

De Gennes and Friedel [55] proposed a simple interpretation of thin threads forming loops in nematic liquids (Fig. 18a) and produced a modified version of the Volterra process by taking into account the liquid character of the mesophase. For instance, consider a perfect nematic liquid aligned along a common easy direction, defined at the surfcae of the slide and the coverslip. The medium is cut along S, a horizontal disk in the bulk, limited by a circular loop L, and two lips S_1 and S_2 are created. Each director close to S is rotated about a vertical axis by an angle $+\pi/2$ in S_1 and $-\pi/2$ in S_2, the structure being restuck along S. Such topological rehandlings are only possible in liquid ordered media, and not in true crystals. After relaxation, the director shows a $+\pi$ twist through L, between the slide and the coverslip, whereas this global twist is absent from the remainder of the preparation. The Volterra process adapted for nematic liquids is represented locally in the vicinity of L (Fig. 18b–d), in a purely planar model.

At the interface limiting a mesophase, the anchoring conditions often lead to the creation of singular lines lying either at the interface or in the bulk, when the anchoring conditions are very strong. The case of such lines at an interface in a cholesteric liquid is

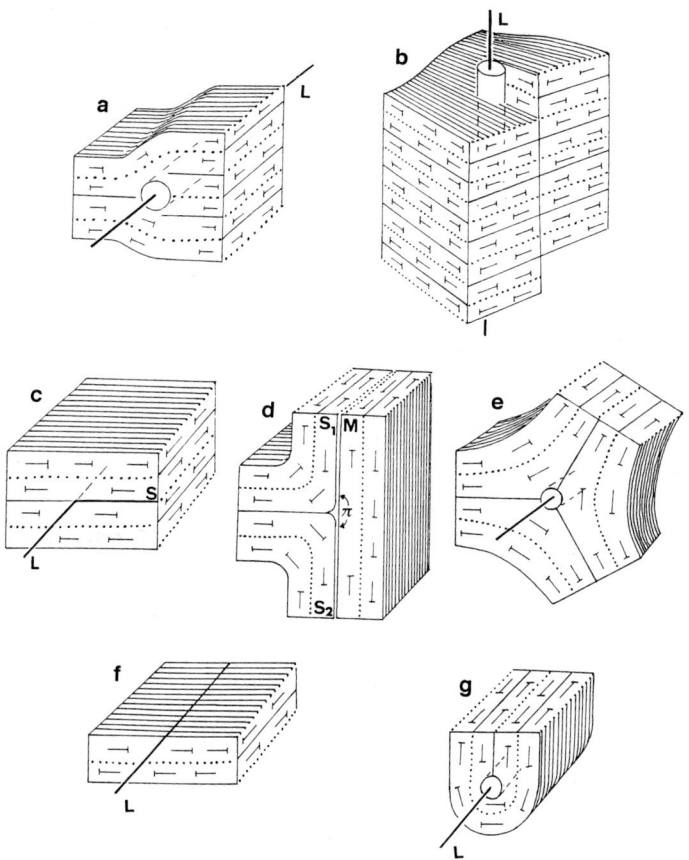

Figure 17. The Volterra process applied to cholesteric phases. The core structure is masked by a cylinder along the line L. (a, b) Edge and screw dislocation. (c–e) A section S limited by L, normal to the cholesteric axis, allows one to build either the edge dislocation (a) or a disclination (d), as in smectics (Fig. 16 d and d″). (f, g) Construction of the opposite disclination. (Drawing made in collaboration with F. Livolant).

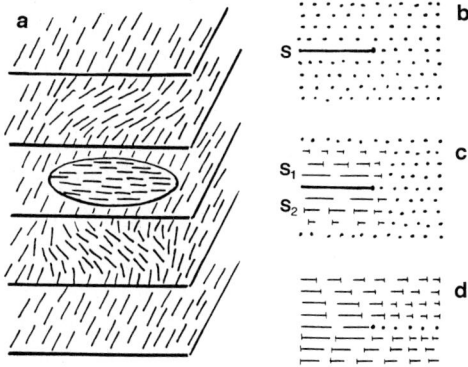

Figure 18. Planar model of a very thin thread in a nematic liquid and the modified Volterra principle used to produce it. (a) The distribution of molecules in five successive planes. (b) Directors are initially normal to the plane of the page. (c) A local twist about an angle $\pi/2$ is introduced in each lip S_1 and S_2, after a section S. (d) Relaxation occurs after resticking of the directors, which are supposed to remain horizontal.

illustrated in Fig. 19 [56]. Here, purely local rotations of directors near the interface are sufficient for the defect structure to be attained, whereas surface cuts would be necessary for defect lines to appear in the bulk. Different aspects of director distributions are shown in nematic liquids in the vicinity of a singular line at an interface, which forbids horizontal anchoring (Fig. 20). Starting from a uniform nematic alignment, a cut and a local reorientation of directors may be necessary to obtain such defect structures at an interface. (Such lines are also considered in Fig. 14).

Another example of topologically stable lines is observed in smectic C liquid crystals [57]. When layers lie horizontally between the slide and the coverslip, four-

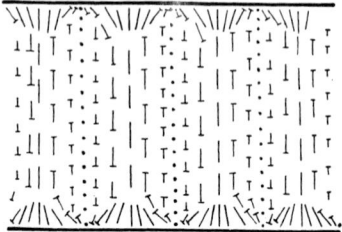

Figure 19. Distortions of a left-handed cholesteric liquid, introduced between the slide and the coverslip, with forbidden horizontal anchoring; the twist axis is horizontal between the plates.

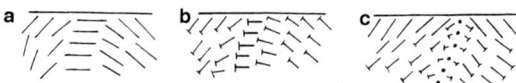

Figure 20. Distribution of nematic directors at an interface, when horizontal anchoring is forbidden. Interface line defects appear, with three possible cross-sections (a–c).

branch nuclei resembling those shown in Fig. 11 for nematics, are frequently seen under the polarizing microscope. These arise from the oblique orientation of the molecules relative to the layers and the existence of stable arrangements defined along closed circuits within the layers (Fig. 21 a and b). These patterns are transmitted from layer to layer along nearly vertical singular lines. There are also horizontal lines of the same type, the characteristic circuits of which are shown in Fig. 21 c and d [57]. Such disclinations are produced by the Volterra process applied to a perfect smectic C liquid and, therefore, according to the following operations: an appropriate cut, a rotation of molecules within the layers of one lip, resticking and viscoelastic relaxation. Brunet and Williams [58] have described examples of such lines running parallel to layers in chiral samples of smectic C phases.

The helicoidal periodicity of smectic C* phases, due to the twisted distribution of the molecular tilt is not given by the half-pitch as in cholesterics, but by the whole pitch p. Edge dislocations of this periodicity are easily recognizable in preparations, as they have Burgers vectors the lengths of which are a multiple of p. In principle, these defects of the helical periodicity can form without defects of the lamellar structure (Fig. 21 e). The equidistant stripes resulting from the helicoidal periodicity do not lie strictly parallel to the smectic layers. These dislocations need the presence in their core of a disclination line of the type shown in Fig. 21 d (or 21 c), as shown in Fig. 21 e.

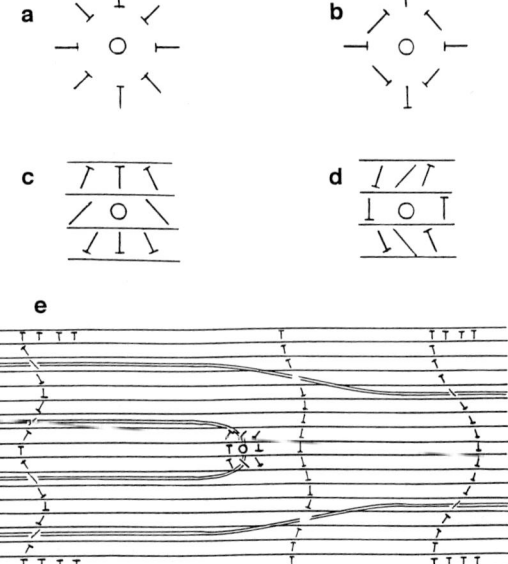

Figure 21. Schematic representation of singular lines in a smectic C phase. (a, b) The distribution of the directors in four-branch nuclei, the layers being seen from the top; open circles indicate the singularity position. (c, d) Cross-sections of singular lines parallel to layers. (e) Superimposed layers of a smectic C* phase can be devoid of defects, whereas an edge dislocation is present for the rotation periodicity (left-handed helix). The tilted molecules show a -4π rotation on the left-hand side of the diagam and only a -2π one on the right-hand side. This situation involves the presence of a line similar to (d) in the defect core. The double lines indicate the locus of directors parallel to the plane of the page.

7.3.4 Continuous and Discontinuous Defects in Cholesteric Phases

Thin threads in nematics are due to the presence of a discontinuity line, whereas thick threads correspond to the locus of vertical directors. The situation is similar in cholesterics, which are spontaneously twisted nematics, the threads in which are recognizable in large-pitch samples ($p > 10$ mm) [48, 49]. This situation is appropriate for studying the core structure of dislocations and disclinations. Let us start from a perfect cholesteric liquid (assumed to be left-handed) and isolate a virtual cylinder, the axis of which coincides with the twist axis (Fig. 22a). The arrangement of the directors at the lateral surface of this cylinder is represented in Fig. 22a′, a rectangle obtained by unfolding the limiting cylindrical surface, after cutting along a generator AA′ (or BB′). The positions of the directors lying within the drawing plane form a series of oblique stripes, corresponding to a double helix at the cylinder periphery, and this is also the case for positions of directors normal to this plane. Directors at $+45°$ form a quadruple helix and are represented by a series of nails. A cross-section of this cholesteric cylinder, at the level of line nm is shown in Fig. 21 a″ to recall the constant director orientation within a thin 'cholesteric layer', and it can be verified that the nail orientations along line nm in Fig. 22a′ correspond well to the expected obliquities of the directors relative to the interface in the successive sectors of the cylinder in Fig. 22a″.

If we now apply the Volterra process, there are two possible ways to obtain a screw dislocation, as indicated in Fig. 22b and c, when the length of the Burgers vector is $p/2$, the half-helicoidal pitch. In a simple model it is assumed that the directors have a planar distribution, normal to the screw axis. The arrangement of directors remains that of parallel stripes in rectangles AA′BB′ transformed into parallelograms in Fig. 22b′ and c′. However, due to the Volterra process, the double helices in Fig. 22a and a′ are replaced by triple helices in Figures 22b and 22b′ and simple helices in Fig. 22c and c′, which also are left-handed. The lateral distribution of directors along the circumference nn′ can be prolonged towards the core, the director orientation being constant along any radius. One then refinds the two classical patterns of disclinations in nematics (Fig. 22b″ and c″), considered above in Fig. 10.

The patterns of various screw dislocations are represented in Fig. 23, either in meridian or cross-sectional views. The directors are coplanar and lie normal to the screw axis, as in Fig. 22, but have various Burgers vector lengths (multiples of $p/2$). The orientations of vectors b and p (Burgers vector and helical pitch vector) are identical if the spontaneous twist and the screw dislocation follow the same handedness. By convention, we say that the Burgers vector length b and the helical pitch p are positive when the corresponding distortions (and the spontaneous twist) are right-handed. These two lengths are negative if they both correspond to left-handed orientations. Each screw dislocation is characterized by an integer Z, which can be positive or negative, such that $b = Zp/2$; the cholesteric liquid is perfect when $Z = 0$.

The locus of the horizontal directors lying at a constant angle to the cylinder forms helices at the surface of the limiting cylinder, or a set of equidistant circles for $Z = -2$. A $180°$ arc of these helices for radial directors is represented in each of the drawings in Fig. 23, except for the case when $Z = -2$. All meridian sections of these screw dislocations are equal and superimpose through

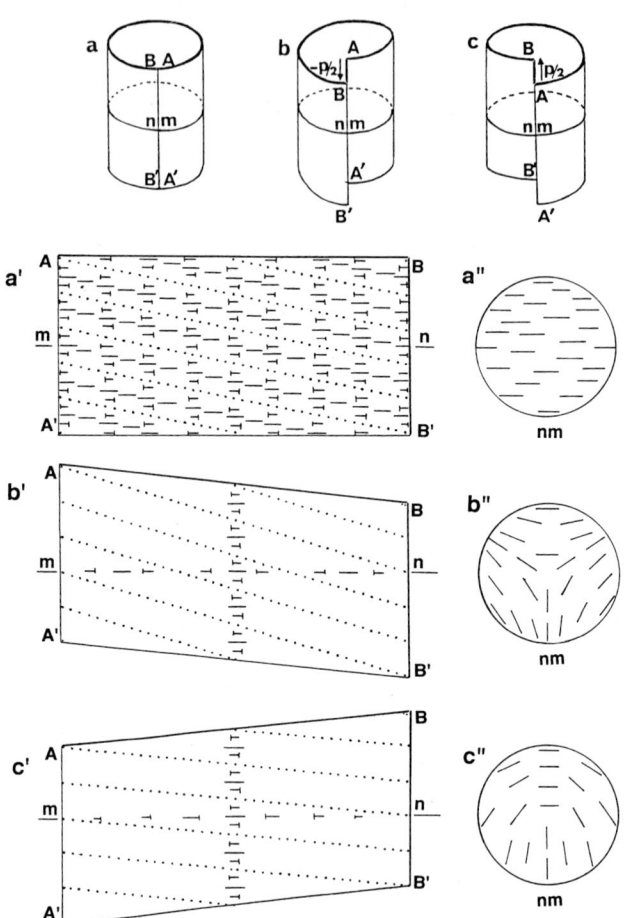

Figure 22. The distribution of directors along circular cylinders, within a left-handed cholesteric phase; the cylinder axis corresponds in (b) and (c) to the core of a screw dislocation. (a) The cholesteric order is perfect and the cylinder is cut along a generator, to be developed into a planar rectangle ABA′B′ represented in (a′), the drawing corresponding to the external view of the cylinder. Along a circle nm, directors show a definite distribution that is coherent, with a uniform alignment within the horizontal plane (a″). When the Volterra process is applied along the cylinder axis to create screw dislocations (b or c), the rectangles ABA′B′ are transformed into parallelograms (b′ or c′) and the distributions of the directors are changed along a transverse circle nm. This leads to two different patterns in the core (b″ or c″) when the molecules are assumed to remain horizontal. (Drawn in collaboration with F. Livolant.)

the corresponding helicoidal displacement. It can be verified that the pitch of this helix is $h=(Z+2)p/2$. The hatched part of the meridian section allows one to generate the whole screw dislocation by means of the corresponding helical motion. This hatched region is absent from the top figure, since the helix is replaced by a series of coaxial circles and a pure rotation suffices to generate the whole structure, with different patterns according to the section level (concentric circles, spirals, diverging radii).

The cross-sections of the various screw dislocations in cholesterics show disclination patterns similar to those encountered in nematic liquid crystals and correspond to solutions of $\nabla^2 \Phi = 0$ in the plane [54, 59]. The core patterns corresponding to odd values of Z cannot be made continuous, since the ribbon of directors centred on a simple circuit closed about this defect line is a Möbius strip. Conversely, the disclination structures obtained for even values of Z (2, −2, −4, −6) can be made continuous, as indicated in Fig. 24 a–f, the expected meridian structure corresponding to Fig. 24 g [59–61]. The director that was normal to L in Fig. 23 is now made oblique by the presence of a component n_z which grows continuously from 0 to 1, from the periphery of the core (ρ) to its axis (ω) in Fig. 24 g. The two dotted areas correspond to zones of in-

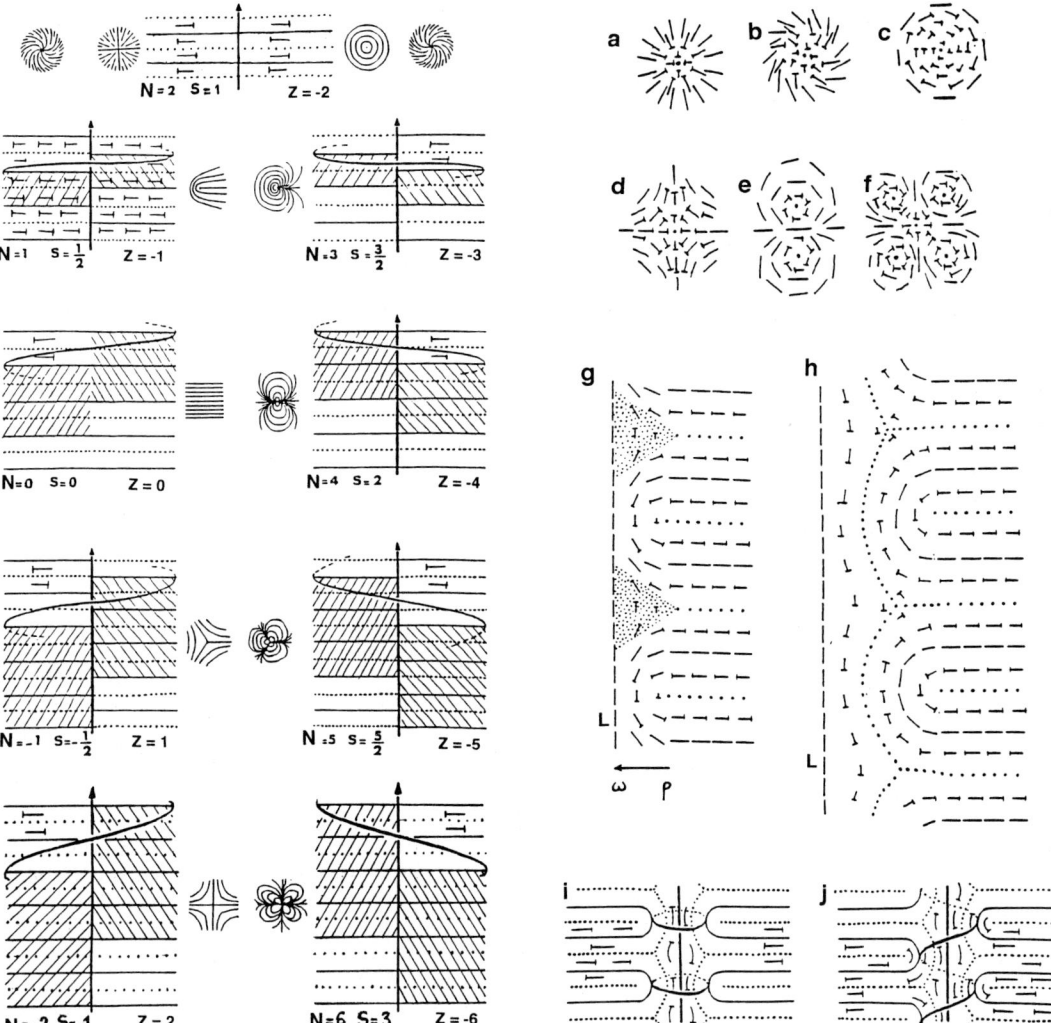

Figure 23. Meridian views of a right-handed cholesteric liquid, after the application of the Volterra process creating screw dislocations, and the corresponding patterns in cross-section. The parameters indicated below each pattern are N, $s = N/2$ and $Z = 2b/p$, the ratio of the Burgers vector to the half-helical pitch, both of which have positive length if the twist and the screw distortion are right-handed. The helical arcs indicate how one passes by helical displacement from the pattern on the left (hatched rectangle) to the corresponding one on the right (also hatched). The pitch of these helices is $h = p + b$. The cross-section patterns verify that $\Phi = \Phi_0 + N\varphi/2$, and one has $N = -Z$.

Figure 24. Transverse and meridian patterns of continuous cores in even-indexed (or non-Möbian) screw dislocations in cholesterics. (a–f) Transformations of transverse patterns for $Z=-2$ (a–c), 2 (d), –4 (e) and –6 (f), by 'escape in the third dimension'. (g) Meridian section expected from (a–d), L representing the axis of the screw line. (h) Another type of meridian section, also compatible with the production of screw dislocations along the axis L. (i, j) Two patterns deduced from (h), where the locus of the vertical directors consists of L and a set of equidistant rings with $Z=-2$ (i) or a helix with $Z=0$ (j), the observation axis being either parallel to L or close to it.

verted twist. The meridian section in Fig. 24e is quite different: this screw dislocation was observed in cholesteric spherulites, the cholesteric layers of which are nested concentrically and, therefore, lie parallel to the external interface [62]. This defect, often called a 'radius of disclination', presents a continuous structure and has been modelled in both cross-sections and in meridian sections [63]. A planar twist zone is differentiated along the screw axis, with an orientation opposite to that of the spontaneous twist, with a pitch slightly larger than the spontaneous one. Alternative meridian structures are possible, and avoid twist inversion by the introduction of point singularities at regular intervals along the screw axis [59] (as has been supposed to occur along the disclination diameter of some cholesteric spherulites [63]). The continuous meridian structure shown in Fig. 24h also avoids twist inversions but these have been shown to occur in the two structures shown in Fig. 24i and j [49, 60].

The core structure of cholesteric disclinations was interpreted by Kléman and Friedel [2, 3]. The rotation vector considered in the Volterra process is normal to the cholesteric axis and is either parallel to the molecules or normal to them, this resulting in a core structure that is either continuous, with a longitudinal nematic alignment of directors in the core (λ disclinations), or discontinuous (τ disclinations), with a singular line of the type encountered in non-twisted nematic liquids.

Now, consider around any of these disclination lines a closed circuit C and the ribbon generated by the set of directors centred along C [44, 53]. Figure 26 shows a short straight segment of such a circuit, at a place where the director n is nearly constant, and the corresponding ribbon. One can replace a segment C_1 of this circuit by a curved segment C_1', which adds one helical turn and

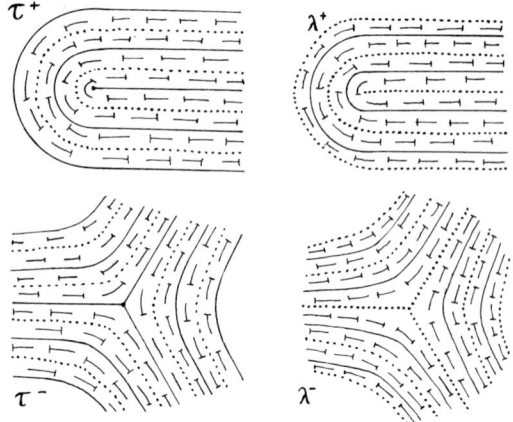

Figure 25. Local addition of a helical turn to a ribbon of directors (see text).

joins C tangentially at the two extremities of C_1. The addition of a helical turn modifies by one unit the 'linking number' of the two closed lines formed by the ribbon edges, but does not change the status of the ribbon, which is either the separation of the two faces, or their communication in the case of a Möbius strip. Note that the edges of a closed strip form either two closed curves, in a non-Möbius strip, or a single closed curve, in the case of a Möbius strip. In the study of such strips, there is also the problem of points the director of which is locally tangential to C, but in this case a slight modification of the circuit suffices to avoid such points [44].

The main result in this study is that simple circuits closed about λ disclinations give non-Möbius strips with a vector n, whereas those closed around τ disclinations are of the Möbius type. The core needs, to be continuous in order to be surrounded by non-Möbius strips. Similar ribbons built with the twist vector C centred along such circuits are of the Möbius type for the four disclinations shown in Fig. 25.

7.3 Topological Structure of Defects

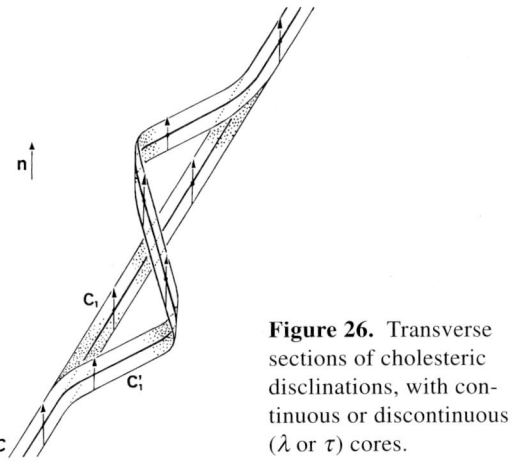

Figure 26. Transverse sections of cholesteric disclinations, with continuous or discontinuous (λ or τ) cores.

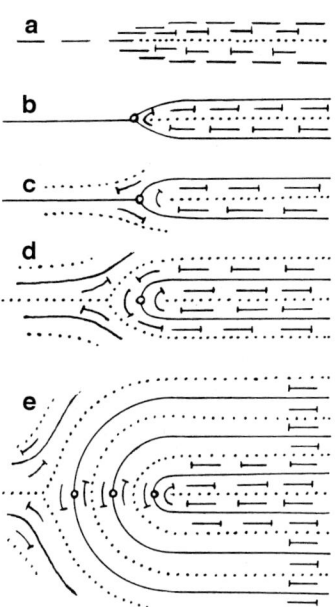

Figure 28. Progressive disjunction of edge dislocations in cholesterics (left-handed twist). The open circles show the locus of vertical directors. (a) Planar model of a 'very thin thread'. (b) First step of the disjunction into a λ^- and a τ^+ dislocation. (c) Cross-section of a 'thin thread', with complete disjunction of the two disclinations λ^- and τ^+, corresponding to a Burgers vector of length $b = p/2$. (d) Edge-dislocation of the type $\lambda^- \lambda^+$, with $b = p$. (e) A $\lambda^- \lambda^+$ pair with $b = 3p$.

7.3.5 Core Disjunction into Pairs of Disclinations

The main variations of the core structure in edge dislocations are illustrated in Fig. 27 and 28. Consider a large-pitch cholesteric liquid introduced between two parallel glasses, with a unique direction of easy alignment lying in the plane of the page, and assume that the distance separating the two glasses is $3p/4$, intermediate between $p/2$ and p. Two types of domain are expected: those corresponding to a $+\pi$ rotation and those to a $+2\pi$ rotation from the slide to the coverslip, and these are separated by a discontinuity line analogous to that found in a very close situation, when a nematic liquid is twisted either by an angle of $+\pi/2$ or $-\pi/2$, between two parallel glasses rubbed at right angles (see Fig. 9).

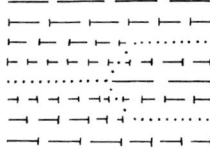

Figure 27. Planar model of a discontinuous thread separating a 2π twist zone (right) from a π one (left) (right-handed twist).

When a large pitch cholesteric liquid is introduced within a wedge of two rubbed glasses, the 'Grandjean–Cano wedge' [64], edge dislocations form and arrange in parallel, those with large Burgers vectors lying in the thickest regions of the wedge. Planar models such as the one shown in Fig. 27 apply in the thinnest zones, whereas in thicker areas the directors present vertical components in the vicinity of the defect and this transforms the dislocation lines into pairs of disclinations (Fig. 28a–c), with a τ^- on the left and a λ^+ on the right [48, 65]. Edge dislocations having larger Burgers vectors transform into dislocation pairs of the type $\lambda^+ \lambda^-$, with a fully continuous distribution of directors (Fig. 28d and e).

These structures are well evidenced when cholesteric layers extend normally to the preparation plane, with weak anchoring conditions, resulting in so-called 'fingerprint textures'. Disclinations are well observed in these conditions between crossed polarizers, if the pitch is large enough, and the presence of a central black zone is demonstrative of a λ core; this is not the case with a τ disclination. The $\lambda^+\lambda^-$ associations are the most frequent. It can be verified that the set of directors centred on a circuit forming a simple loop around such disclination pairs is not a Möbius strip when the core is continuous, but is of the Möbius type when one of the two disclinations of the pair is a τ disclination. Conversely, the ribbon obtained with the cholesteric axis C along such circuits is not of the Möbius type, whatever the nature of the two disclinations forming the dislocation.

Disjunctions into disclination pairs are general in liquid crystals when the Burgers vector is large enough, and this holds for both edge and screw dislocations. This was first considered for edge dislocations in cholesterics [2, 3], but also applies to screw dislocations (see Fig. 24 i and j).

7.3.6 Optical Contrast

All the translation defects shown in Fig. 28 are easily prepared in a wedge obtained, for instance, when a coverslip lies tangentially along a generator of a cylindrical lens, both the lens and the coverslip having been treated to obtain a planar anchorage with a preferential direction. A drop of nematic liquid added together with a small amount of a twisting agent allows one to obtain a pitch of 20–30 µm. After this cholesteric mixture has been introduced between the two glasses, edge dislocations stabilize on both sides of the contact generator and form a series of parallel lines that is easily observed under natural light.

The use of a single polarizer or analyser is sufficient to vary the defect contrast considerably. As the local ellipsoid of indices is positively uniaxial and the twist is weak, the electric vector of any penetrating wave rotates as the director orientation (Mauguin's condition: the wavelength being very small compared to the cholesteric pitch) [66, 67].

If the electric vector of the incident beam is normal to the anchoring direction, the wave propagates as an ordinary ray, whereas if it is parallel the resulting extraordinary ray deviates and the image of the threads is much more shadowy and fuzzy [11, 68]. The use of a polarizer the main axis of which is perpendicular to the common anchoring direction of the two glasses gives clear-cut ordinary images of threads, much thinner than the half cholesteric pitch $p/2$, and by analysis of the whole system several types of thread can be distinguished.

Very thin threads in the vicinity of the contact generator between the cylindrical lens and the coverslip are poorly contrasted, and correspond to the planar structure shown in Fig. 28a.

The thin threads lie in slightly thicker regions of the preparation and correspond to the model shown in Fig. 28c. Horizontal streams decrease the contrast of these threads, which then resemble 'very thin threads', this probably being due to a horizontal alignment of directors [48, 65].

The thick threads observed at the periphery correspond to the model shown in Fig. 28d, but often form groups of parallel lines, separated by a distance that is the half-helical pitch, as shown in Fig. 28e.

The positions of thin and thick threads are indicated in Fig. 28 by open circles; many observations have led to the conclusion that these correspond to the locus of directors parallel to the optical axis of the microscope, i.e. normal to the preparation plane, which is usually called the vertical [48]. Most papers dealing with thick and thin threads in nematics and large pitch cholesterics generally consider that these threads represent disclination cores. This is true for thin threads, but generally not so for thick ones. The confusion arises mainly from the fact that the core of a defect line corresponds to a strong maximum of curvature and it is highly likely that vertical directors will be found in its vicinity. A good approximation is to say that thick threads represent the locus of points where $n_x = n_y = 0$, and $n_z = 1$, but the optics are complex, and this simply remains a first approximation.

7.3.7 Classification of Defects

As shown in every part of this section, liquid crystals are states of matter some of which are most appropriate for illustrating remarkable situations in combinatorial topology. The fluidity of mesophases is compatible with different types of curvature resulting in large series of defects, particularly disclinations, and the best tools for their definition and classification are the 'homotopy groups', the applications of which to liquid crystals are numerous [69–78], but their use is somewhat arduous; they can be understood as a mathematical generalization of the Volterra process [72]. All the 'topologically plausible' points, lines and walls and their rehandlings are predicted without exception by this theory. The Volterra process has been used to define singular lines, but not defects as points, walls and certain structures with a continuous core. The theory also considers the rehandlings of defects [48, 53]. However, it is worth remembering that among the defects that are 'topologically plausible', many of them do not exist, since they involve considerable energies; the simplest example is that of disclinations which are topologically plausible in three-dimensional crystals – these have drawn [79] but are excluded in true crystals for energy reasons. This is also true for most liquid crystals. Geometrical and mechanical constraints are forgotten in purely topological considerations, and these are presented in the next section.

Translation and rotation vectors are still the best tools in crystallography and for defining and classifying defects in liquid crystals. We have distinguished above between edge and screw dislocations, which are translation defects characterized by a Burgers vector, either normal or parallel to the line, but these defects often lie oblique to this vector as in three-dimensional crystals, and the structure of jogs and kinks has mainly been studied in cholesterics [48]. Disclinations (or rotation dislocations) introduce a rotation vector v that is either normal to L, a *twist disclination*, encountered mainly in planar models (see Figs. 9, 18 and 27), or parallel to L, a *wedge disclination*, more frequent in smectics and cholesterics. In the latter, we deal with four vectors: the director n, the cholesteric axis C, the rotation axis v and the line axis L. Let us recall that, in λ disclinations, v is parallel to n, but normal to C. In τ disclinations, v is normal to n and C (see Fig. 25); in χ disclinations, v is normal to n, but parallel to C (see Fig. 23). Note also that these disclinations, when they run parallel to the twist axis, reproduce the structure of screw dislocations in cholesterics. The symbols λ, τ and χ are often followed by an exponent $+\pi$, $-\pi$, $+2\pi$, -2π, etc. corresponding to the rotation angle of the director after one 2π turn along a small circle surrounding the core line of the defect. These exponents are often reduced to +, −, 2+, 2−. Figure 25 shows a $\lambda^{+\pi}$, a $\lambda^{-\pi}$, a $\tau^{+\pi}$ and a $\tau^{-\pi}$ (often abbreviated as λ^+, λ^-, τ^+,

τ^-), whereas Fig. 23 shows a series of χ disclinations from $\chi^{+2\pi}$ to $\chi^{-6\pi}$.

The purely orientational order of nematics leads to discontinuities that are discrete structures in the form of points and thin threads. Continuous defects also exist, but some of these cannot be deformed continuously into the monocrystal and therefore should be classified as genuine defects. The presence of a spontaneous twist in a nematic liquid facilitates such situations, the main example being that of thick threads forming interlocked rings [48, 49, 53, 73].

7.4 Geometrical Constraints and Textures

7.4.1 Parallel Surfaces and Caustics in Liquid Crystals

Many liquid crystals show a lamellar aspect, due to layers of uniform thickness arranged along parallel surfaces, and this is well observed in polarizing microscopy for some cholesterics, when the layer outlines are apparent (see Fig. 6). In these two micrographs, the layers change direction abruptly along straight or curvilinear segments belonging to the hyperbola branches, since they correspond to successive intersections of nearly circular profiles centred on two different points lying in the same optical section. The layers in fact form spirals, owing to the helical structure of cholesterics, rather than circular rings, and their thickness varies due either to real changes in helical pitch, or to apparent variations when layers are oblique.

When the cholesteric pitch is less than 0.5 µm, the parallel profiles of the cholesteric layers cannot be resolved in light microscopy, but the straight or bent segments due to abrupt changes in orientation remain well contrasted. These segments belong to hyperbola branches or ellipses, or parabolae, and form well-structured arrangements that are found not only in short pitch cholesteric but also in smectic phases [11].

Parallel contours are also evident in smectic phases viewed under a polarizing microscope. In the case of smectic C* phases, for example, these contours are due to the helical periodicity, while in smectic E and lyotropic lamellar phases they occur when defects accumulate along certain lamellae or groups of lamellae.

Normals to parallel layers form straight lines, as do the light rays with respect to waves in an isotropic medium, and they envelop two surfaces, called *focal surfaces* or *caustics* [19, 80]. However, their presence in the mesophase produces a discontinuity wall, which generally disjoins into alternate $+\pi$ and $-\pi$ disclinations [20, 81]. Caustics are often absent from liquid crystals, since layers can be poorly curved, with 'virtual caustics' located outside the liquid crystal. For instance, layers are nearly horizontal in *stepped drops* of smectics, when the conditions are homeotropic along the glass and at the air interface. They appear to interrupt along oblique 'cliffs', arranged as more or less concentric level lines, and distortions are observed in their vicinity [82]. There are also examples of parallel spherical layers, with caustics reduced to a single point. The most extreme case of degeneracy of caustics occurs when each of the two surfaces reduces to a unique singular line and not to a series of lines as indicated above.

7.4.2 Dupin's Cyclides and Focal Conics

Surfaces Σ that are envelopes of spheres centred along a curve L have their normals passing through L; conversely, all surfaces with normals converging along a curve L are envelopes of such families of spheres (Fig. 29 a). The contact curves between the spheres S and the envelope Σ are circles γ. The normals along γ converge towards A, the centre of the sphere S, and form a *revolution cone*, the axis of which is tangent to L in A. Cylindrical bilayers, which are straight, bent, or helical in myelin figures, are an example of such surfaces, which are envelopes of spheres of equal radius centred on a straight, bent or helical line L.

Each caustic surface can reduce to lines L and L' and a simple example is presented in Fig. 29 b; the parallel surfaces are the nested and parallel tori, forming a $+2\pi$ disclination L, the whole system presenting a revolution symmetry about a vertical axis, which is itself a second defect line L'. These tori are surfaces generated by (at least) two families of circles, their meridians and parallels, which also are their lines of curvature.

The general problem of surfaces the normals of which pass through two curves L and L' was solved by Dupin [83] and the first applications were considered by Maxwell; the general theory was developed in the 19th century [83] and in more recent articles [84].

Such surfaces Σ are envelopes of spheres in two different ways – either a family of spheres S centred on L, or a family of spheres S' centred on L' – and there are two spheres S and S' tangent at any point M of Σ. The revolution cone associated with a contact circle γ of S contains L', and vice versa. If one knows one sphere S and three spheres S' tangent to S, the family S can be defined (and the S' family also). The three spheres S' cut each other at two points, O and Ω, either real or imaginary. Let O be an inversion centre with a coefficient $k = \mathbf{OM} \cdot \mathbf{Om} = O\Omega^2$, where M and m are two homologous points aligned with O. The three

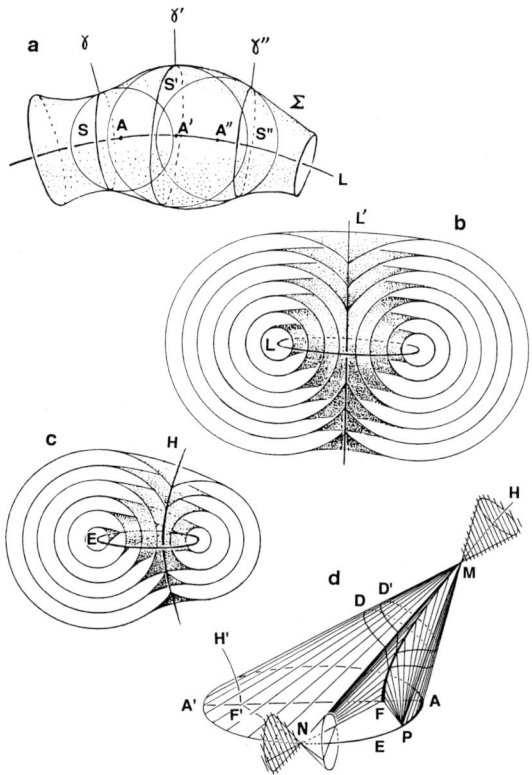

Figure 29. Cyclides are surfaces the lines of curvature of which are circles. (a) An envelope Σ of a family of spheres S, centred in A describing L, and contact circles γ. (b) Nested tori, the normals of which pass through the circle L and the axis L', one sheet being suppressed to avoid intersection of the surfaces. (c) General aspect of Dupin's cyclides, the normals of which pass through an ellipse E and a hyperbola H. (d) A focal domain built on the arc FM of a hyperbola and on the arc PAA'N of the ellipse E. AA' is the major axis with the two foci FF'. H' is the virtual hyperbola branch. D and D' are circular intersections of Dupin's cyclides, the revolution cones being represented by their generators or director lines.

spheres S' transform into three planes p' and the spheres S transform into spheres s tangent to the planes p', their envelope being a revolution cone c, and thus the inverse of the surface Σ. The tangent planes to this cone are the inverse of spheres S', which also pass through O and Ω, and the line L' lies in the mediator plane of OΩ. L' is planar and belongs to all revolution cones of the normals converging on L, and is therefore a conic, an ellipse or a hyperbola or a parabola; this also holds for L. These two lines are conics, one being the locus of the vertices of the revolution cones passing through the other one. This situation is that of *focal conics*, which lie in two normal planes intersecting along their major axes, the vertices of one conic coinciding with the foci of the other one. In general, an ellipse is associated with a hyperbola of inverse eccentricity, and this leads to a set of nested toroidal surfaces, the *Dupin cyclides* (Fig. 29c), a system that is less symmetrical than the one shown in Fig. 29b. In this case, instead of a revolution symmetry and a mirror plane (Fig. 29b), there are two mirror planes at right angles to one another, which are the planes of the two focal curves.

Smectic A phases often contain pairs of focal conics (see Fig. 7a and b) that are well contrasted in the liquid, even in the absence of polarizers. A more common situation is the presence of arcs of conics associated in pairs (Fig. 29d). The parallel Dupin cyclides and the associated revolution cones form three mutually orthogonal systems of surfaces [7], and therefore the lines of curvature of any surface in one system are its intersections with the surfaces of the other two systems. The toroidal shape of Dupin's

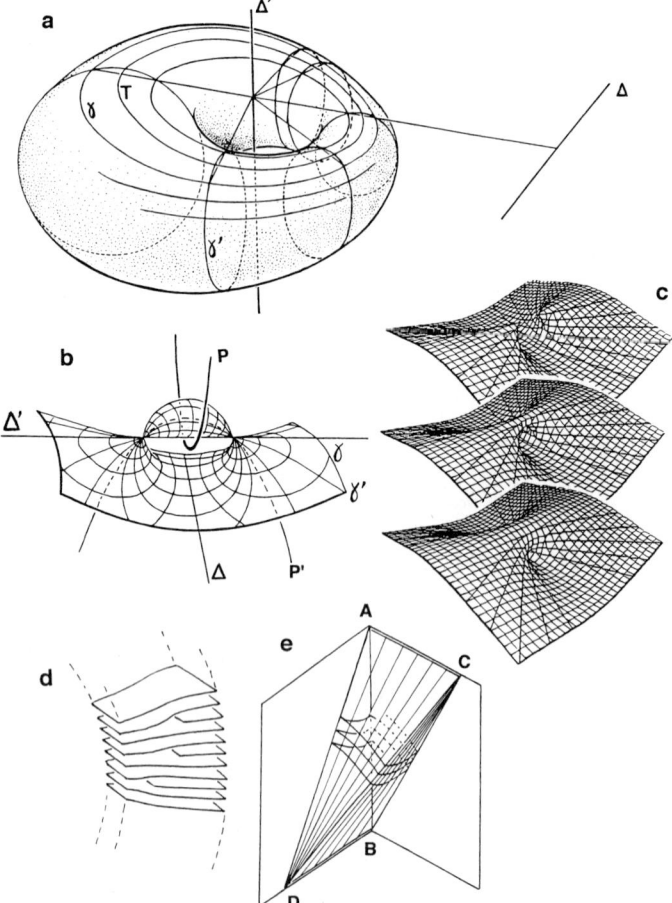

Figure 30. Dupin's cyclides and their deformations. (a) A toroidal Dupin cyclide and its circular lines of curvature γ and γ', the planes of which intersect along the axes Δ (joining O and Ω) and Δ', respectively. Two tangential planes along two symmetrical circles T intersect along Δ. (b) A parabolic Dupin's cyclide, with its two focal parabolae P and P', its circular curvature lines γ and γ', and its axes Δ and Δ'. (c) Three parallel, but separated, parabolic Dupin's cyclides. The lines represent sections by planes $x=0$, $y=0$ and $z=0$; at the top of the figure, two conical points are present, as in (b), but the spindle-shaped sheet has been removed (Rosenblatt et al. [85]). (d, e) Edge and screw dislocations 'compensate' the bend and the twist of directors in a set of equidistant surfaces normal to the director lines.

cyclide is indicated in Fig. 30a. Conical points of Dupin's cyclides are present all along the two conics, but one of the two sheets is absent, since interpenetration of layers is impossible. The forbidden conical sheets are shown explicitly in Fig. 29d. The case of confocal parabolae is particularly frequent in the lamellar phase of lyotropic systems [85] and the corresponding Dupin's cyclides (Fig. 30b and c).

Micrographs of focal conics were presented in the pioneer work by Lehmann [10], and, some years later, in that by Friedel [11] who, together with Grandjean, observed the alignment of the positively uniaxial ellipsoids of indices along the straight segments joining points on the two associated conics [11]. The transition to the solid crystal obtained after cooling often shows the formation of thin crystalline lamellae lying normal to these molecular alignments between the two conics (ethylazoxycinnamate or its mixture with, for example, ethylazoxybenzoate). The paired conics remain generally recognizable in this solid texture, which is called a pseudomorphosis [11] or paramorphosis [13, 14]. By reheating the preparations one can revert to the initial liquid crystalline texture, with the focal conics located in the same places. This suggested the presence in these liquids of fluid layers forming sets of parallel Dupin's cyclides, with molecules lying normal to them [86]. Very similar textures and defects were found in soaps and other lyotropic systems, which justifies the term 'smectic' chosen by Friedel for these lamellar liquid crystals [11], the structures of which were rapidly confirmed by the first X-ray diffraction studies [87]. The persistence of various textural aspects and domains through many phase transitions is common, and several examples have been described [13, 14].

7.4.3 Slight Deformations of Dupin's Cyclides

In a perfect system of parallel Dupin cyclides, the unit vectors normal to surfaces align along straight segments joining two focal conics. A deformation resulting in a local bend modifies the layer thickness or involves edge dislocations (Fig. 30d). Similarly, screw dislocations compensate a twist, when for instance focal conics are replaced by different curves. This is indicated in Fig. 30e, in which the layers are assumed to extend normally to straight segments, the extremities of which belong to two rectilinear segments AC and BD instead of to two arcs of focal conics. Suppose, for example, that AB, AC and BD are parallel to the axes of a tri-rectangle trihedron. The intersections of layers with the four faces of the tetrahedron ABCD form concentric circular arcs, the concavities of which alternate, showing that layers are saddle-shaped. Moreover, successive arcs joined at their extremities do not form a closed loop, but a helical path. The geometry of this tetrahedral domain requires the presence of right-handed screw dislocations, whereas the mirror image of ABCD would lead to left-handed defects. The local density of these defects within a lamellar structure (total Burgers vector in the unit volume) is $|\boldsymbol{n} \times \operatorname{curl} \boldsymbol{n}|$ for pure edge dislocations and $|\boldsymbol{n} \cdot \operatorname{curl} \boldsymbol{n}|$ for pure screw dislocations [88].

Polygonal textures such as those shown in Fig. 6 associate not only horizontal segments (curvi- or rectilinear), but also vertical ones, joining centres to the vertices of polygons at the upper and lower faces of the mesomorphic slab, between the slide and the coverslip. The positions of these segments are shown schematically in Fig. 31a for the cholesteric phase occupying the vertical prism AA'CC'BB'DD'. The cholesteric axes, normal to layers, join points belong-

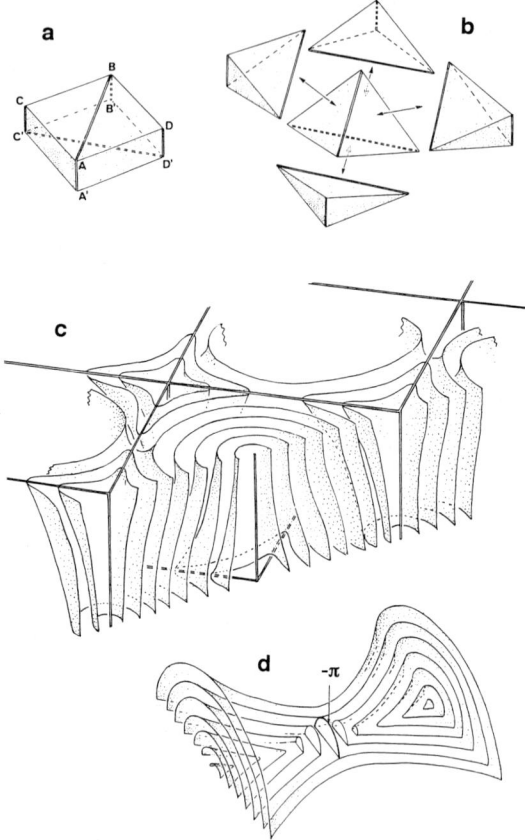

Figure 31. (a) Prismatic subunit of a polygonal field, in cholesteric and smectic A phases. Double lines represent focal segments. (b) The prismatic unit shown in (a) decomposes into five tetrahedral domains. (c) General aspect of the layers in a square polygonal field; double lines represent the array of focal segments. (d) Very acute conical points along focal segments are not only rounded, but are also rehandled by a $-\pi$ disclination. This actually introduces two conical walls, but the layers also are slightly rounded and more focal curves are involved.

ing to the two contours A'ABB' and CC'D'D indicated by a double line. Therefore, the prism decomposes into five tetrahedra (Fig. 31b) filled with saddle-shaped cholesteric layers. In each tetrahedron two opposite edges, called focal curves or focal segments, play the role of focal conics. These segments often are conics, but not in a focal position.

The whole texture is made of such prisms, juxtaposed in continuity. These divergences from the Dupin cyclides model can result in the presence of either translation dislocations or layer thickness variations, or in layers that are slightly oblique to segments joining the conjugated contours. The cholesteric layers extend continuously across the triangular faces joining the tetrahedra, except for the focal segments A'ABB' and CC'D'D. There are no other discontinuities across the vertical faces joining all the prisms into a texture such as the one shown in Fig. 10a and b. Note, however, that the two radii of curvature show a discontinuity along the oblique edges of the tetrahedral domains, which is probably attenuated by slight deformations.

Polygonal textures often form square lattices in cholesteric and in smectic A phases. The global arrangement of nested parallel layers, between the horizontal slide and coverslip is shown in Fig. 31c. The layers lie vertically at the upper and lower faces of the mesophase and are oblique in between, but they are not perfectly perpendicular to straight lines joining the conjugated segments, and thus show important thickness variations, particularly in the vicinity of focal segments. This is often corrected by a rehandling such as the one illustrated in Fig. 31d.

7.4.4 Textures Produced by Parallel Fluid Layers

The main patterns present in these textures are shown in Fig. 5 for lyotropic lamellar phases and in Fig. 32 for thermotropic liquid crystals.

7.4.4.1 Planar Textures

Layers lie parallel to the slide and the coverslip, a more or less frequent situation in

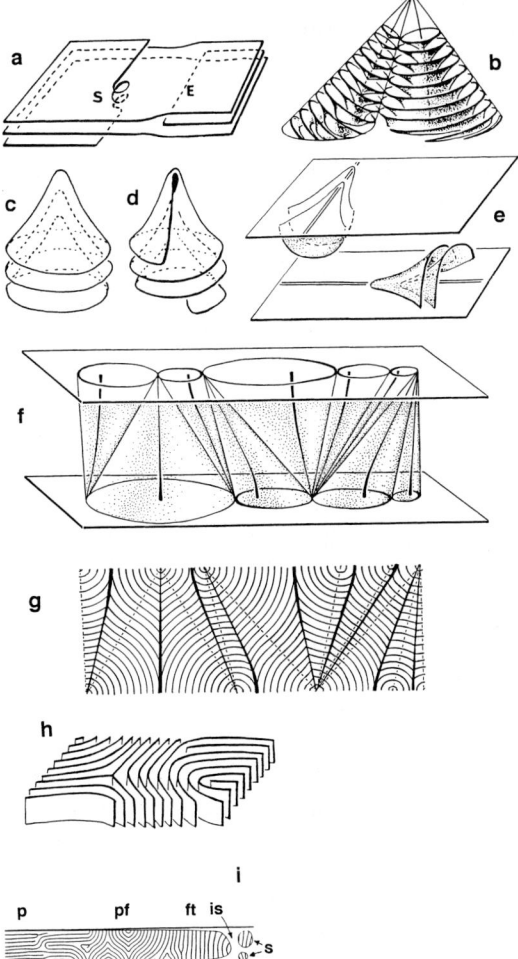

smectic A and cholesteric phases, when the glasses have not been treated for molecular anchoring. These textures are produced regularly in smectic A phases, where strong homeotropic conditions prevail, while a planar anchoring with an easy horizontal direction is required for cholesterics. In general, edge and screw dislocations are present within these planar textures (Fig. 32 a) and are often seen by light microscopy of cholesteric phases of sufficiently large pitch [20], whereas these defects are detected only by electron microscopy in cryofractured, mainly lyotropic, smectic phases [89]. In some cases, edge dislocations intercalating a unique smectic layer can be visualized by means of light microscopy; an example of this has been described in smectic C phases, together with all the technical details of the optics [90]. Some edge-dislocations, however, present large Burgers vectors and disjoin into $+\pi$ and $-\pi$ disclinations; layers intercalated between these disclinations lie vertically and form brilliant stripes in polarizing microscopy or *oily streaks* (see Fig. 5 d), a common textural pattern in smectic A and cholesteric phases [11].

7.4.4.2 Focal Conics and Polygonal Textures

We have already considered the polygonal networks of cholesteric liquids, which assemble saddle-shaped layers close to Dupin's cyclides, and the focal segments of which differ from those of focal conics (see Sec. 7.4.3). Friedel and Grandjean [21] described in smectic A phases a texture much closer to Dupin's cyclides, with genuine focal conics. As for cholesterics, this texture is obtained when a thin film of the isotropic phase of the mesogenic compound separates the mesophase from the two glasses.

The mesophase itself is divided into focal domains, each one being defined by its

Figure 32. The main types of texture in layered liquid crystals. (a) Planar texture with edge (E) and screw (S) dislocations. (b) Two tangent focal domains, and their tangent ellipses and the converging arcs of hyperbola. (c) Nested 'hats' along a focal line. (d) Screw superimposed on a focal line. (e) Focal lines attached to the upper and lower interfaces. (f) Focal domains tangent along a series of coplanar generators; contacts between domains are parallel or antiparallel. (g) Section of (d) along the symmetry plane, showing the discontinuities in the curvature at the antiparallel contacts. (h) General patterns in fans, associating $+\pi$ and $-\pi$ disclinations and edge dislocations (and also focal curves, not shown). (i) Schematic distribution of textures in the vicinity of the isotropic phase in lamellar thermotropic liquids; ft, fan texture; is, isotropic phase of the mesogen; p, planar texture; pf, polygonal fields; s, spherulites.

two conjugated arcs of conics (see Fig. 29 d), which determines uniquely the corresponding family of parallel Dupin cyclides. Each domain is limited by fragments of revolution cones joining along edges, which are either generators or arcs of these conics. The fragments of Dupin cyclides filling these domains are essentially saddle-shaped (see Fig. 32 b), but are hat-shaped along the focal curves, with a strong maximum of curvature and layer thickness (see Fig. 32 c). This can be viewed in cholesterics by using light microscopy (see Fig. 6) and in lyotropic smectic phases by using electron microscopy [89]. Screw dislocations sometimes superimpose on focal curves, as shown in Fig. 32 d [54]. When focal lines run horizontally, attached to the upper and lower interfaces limiting the mesophase, the layers form half-cones, which also present a rounded apex (Fig. 32 e).

Focal domains in smectic A phases are tangent along generators of their limiting cones, the contact being either parallel or antiparallel in Fig. 32 f, where the major axes of ellipses are aligned, and this allows one to obtain a view of layers in the mediator plane (Fig. 32 g). The dashed lines in Fig. 32 g represent the contact generators between the tangent focal domains, either parallel and without discontinuity of the radii of curvature, or antiparallel and with a discontinuity. It is worth remembering that, as for cholesterics, these discontinuities of curvature present in the model are attenuated or smoothed out by a rapid but continuous change in the sign of the curvature. Such tangent domains are illustrated in Fig. 33, which shows a lattice of ellipses, just below the coverglass. The ellipses are separated into different sectors, which end in a polygonal field at the top left-hand side of the micrograph. The sectors are separated by lines that are particular polygon edges, and in fact each sector belongs to a polygon.

The ellipses are tangent to other ellipses and to the polygon edges. Their focal hyperbolae converge to a point on the opposite interface, which is a vertex of the conjugated polygonal lattice. This means that, within the polygon plane, the major axes of the ellipses converge to a point that is the vertical projection of a vertex common to several polygons of the conjugated lattice. When three ellipses are tangent, the free space left between them is often occupied by another ellipse at the base of a narrower focal domain, tangential to the surrounding domains, and the interstices can in turn be filled by narrower cones. It has been proposed that this iterative process persists down to a few molecular lengths [91], and therefore below the resolving power of the light microscope. This view is justified in models that keep all layers normal to the horizontal interfaces, thus limiting the polygonal field. It is also justified by energy estimates and by cryofracture images, which often show very local conical deformation involving a few bilayers only [89]. However, in thermotropic smectic phases forming polygonal networks, very small ellipses seem to be absent, their size rarely being less than the 1/20 of the polygon diameter. On the contrary, the interstices are assumed to form spherical domains, the layers being spherical and centred on the polygon vertex, where the hyperbolae of the surrounding focal domains converge [92].

7.4.4.3 Fan Textures

Layers that are mainly vertical and the presence of disclinations (see Fig. 32 h) are the two very common characteristics of fan textures in smectic and cholesteric phases. The name 'fan texture' comes from the circular contours that are often visible around disclinations in, for example, cholesterics, and from the frequent radial decorations (see

Figure 33. Polygonal field (top left) and elongated polygons, with their long edges converging at a $+\pi$ disclination. 4-Cyano-4'-n-octylbiphenyl plus Canada balsam; crossed polarizer. Scale bar: 50 µm.

Fig. 33) seen in smectic A phases. Focal curves and translation dislocations are not excluded from fan textures. The fan patterns themselves can be absent from certain arrangements of disclinations, which however belong to the family of fan textures. This is illustrated by the rhombus in Fig. 34a and b, where two $+\pi$ and two $-\pi$ disclinations are associated with several focal curves at the top and bottom levels of the mesophase.

7.4.4.4 Texture Distribution in Lamellar Mesophases

When smectic A or cholesteric phases are in equilibrium with their isotropic phase, the mesophase is present in the preparation in the form of bâtonnets or droplets, and extends here and there over larger domains, with a definite distribution of textures. Fans lie in the vicinity of the isotropic interface, whereas planar textures are rarely in contact with it, and polygons are observed along zones separating fans from planes.

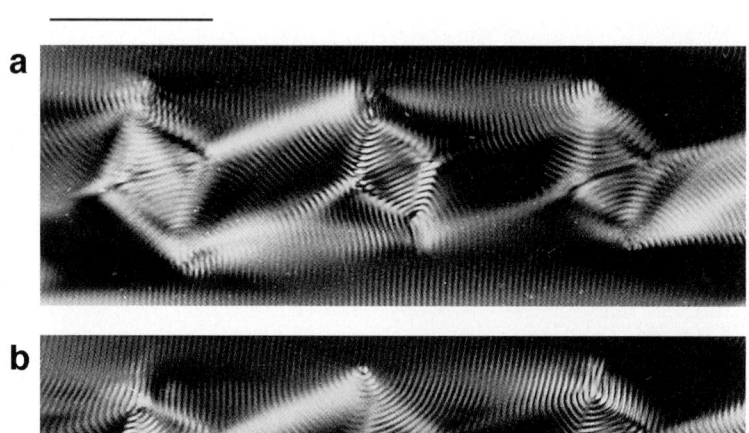
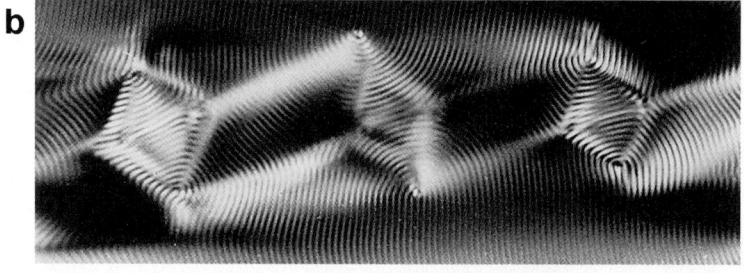
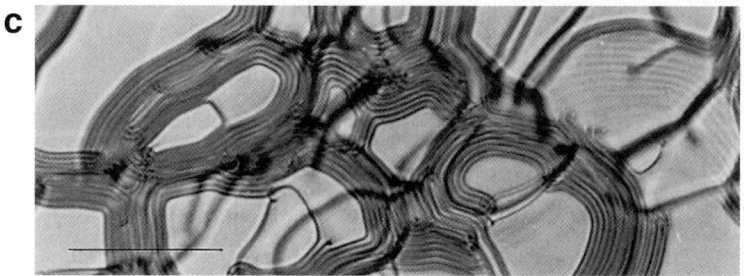
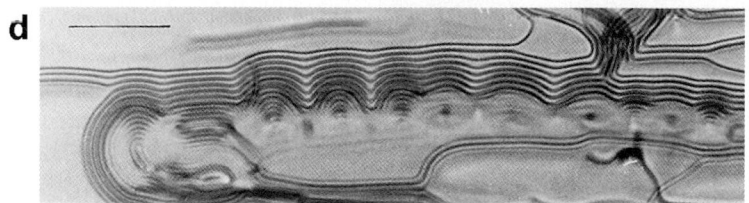

Figure 34. Cholesteric textures: (a–c) MBBA plus Canada balsam; (d) MBBA plus cholesterol benzoate. (a, b) Two views at the coverslip and slide levels of a fan texture, with lozenges associating two $+\pi$ and two $-\pi$ disclinations, linked by pairs of focal lines. (c) Planar domains separated by walls of nearly vertical layers. (d) One of the walls is associated with horizontal and vertical focal lines, appearing as intercalated black spots. (a, b) Crossed polarizers; (c, d) natural light. Scale bars: (a, b) 20 μm; (c, d) 50 μm.

As indicated above and illustrated in Fig. 32i, layers lie horizontally in planes and show some translation dislocations. They are oblique in polygonal fields and generally cross the focal lines at points where they are strongly hat-shaped, whereas elsewhere they are saddle-shaped and dislocations are present, often superimposed on the focal lines. In fans, the layers are vertical and all types of defect (disclinations, focal curves, dislocations) are present.

This texture distribution is common, but corresponds to a rather schematic model, indicating that high-energy defects, such as disclinations, are found mainly in the vicinity of the isotropic transition. Other situations are observed in thick preparations of cholesterics, for example, where planar domains can be interrupted by walls of vertical layers (Fig. 34c and d), due to edge dislocations disjoining into disclination pairs (see Fig. 5d). Despite these particular examples, the presence of disclinations remains a general character of fan textures.

The textures described above belong to the same phase and are not separated by sharp interfaces. The frontiers are fuzzy and are often the sites of hybrid, but interesting, textures. Among these, at the limit between polygons and fans, one finds occasional chevrons, the organization of which is shown in Fig. 35 [93].

Chevrons and related textures are common in smectic phases but, at the horizontal interfaces, the directors must be adjusted to the anchoring conditions [88, 93]. They are regularly produced when opposite sides of polygons are extremely elongated in a given direction [20]. This leads to arrangements as those represented in Fig. 35. When vertical anchoring of layers is preferred, ellipses or parabolae can form at the interfaces, with the corresponding focal domains, but the layer anchoring remains oblique in the interstices between domains [93]. This situation is also realized in Fig. 33 with a radiating series of extremely elongated polygons, centered around a $+\pi$ disclination.

7.4.4.5 Illusory Conics

The analyis of focal conic systems in liquid crystals is made difficult by the presence in pictures of lines, which often resemble focal conics or focal curves but actually are not. It was indicated above that conics such as polygon edges are not in a focal position, and that many lines differ from conics but play a focal role. Figure 36 shows an example of a cholesteric texture simulating the presence of a series of parabolae, but this aspect is due to complex behavior of the polarized light, associated with a frequent moiré pattern. This texture is simply interpreted in Fig. 37 on the basis of an examination of the orientations of the layer and the focal lines, which are straight or slightly curved in this texture, but not parabolic.

7.4.4.6 Walls, Pseudowalls and Broken Aspects

True walls are defined by a director discontinuity extending over a surface within a liquid crystal, but these defects are excluded from nematic, cholesteric and smectic A phases (see Sec. 7.2.3). Grain boundaries comparable to those of true crystals are associated with translation dislocations, which lie parallel within the wall or form

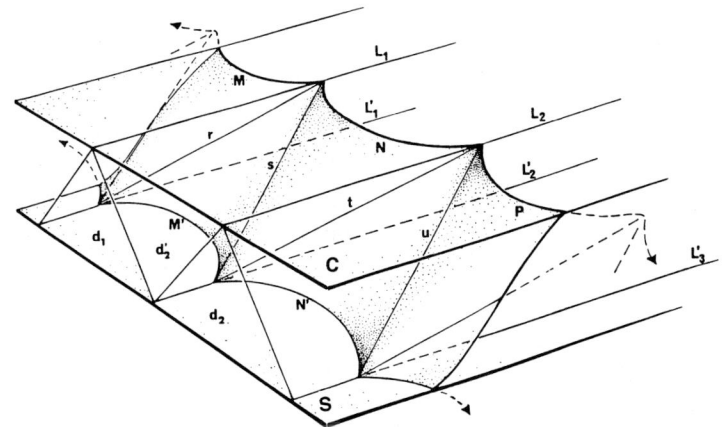

Figure 35. A model of chevron textures (the chevrons are viewed from the top) showing alternating angular and rounded contours, at both the slide and the coverslip level. Prismatic domains d and d' are filled with antiparallel conical sheets, in continuity along the generators r, s, t and u. Parabolic contours M, N and P join and form angles along the lines L at the coverslip level C, and similarly arcs M', N' and P' join and form angles along the lines L' at the slide level S.

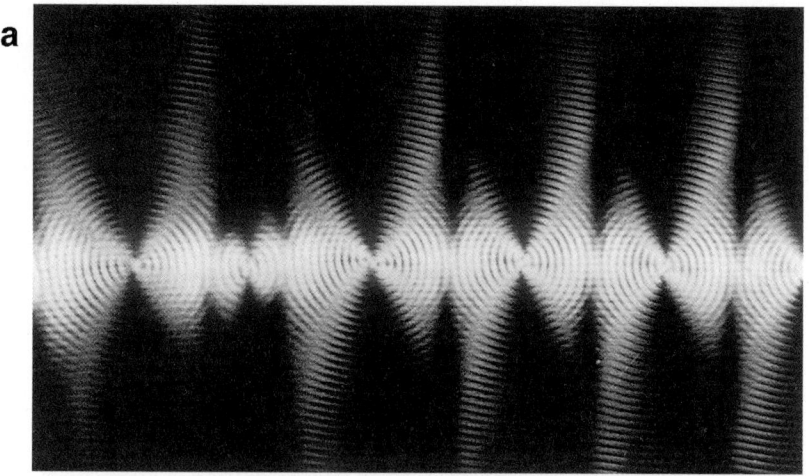

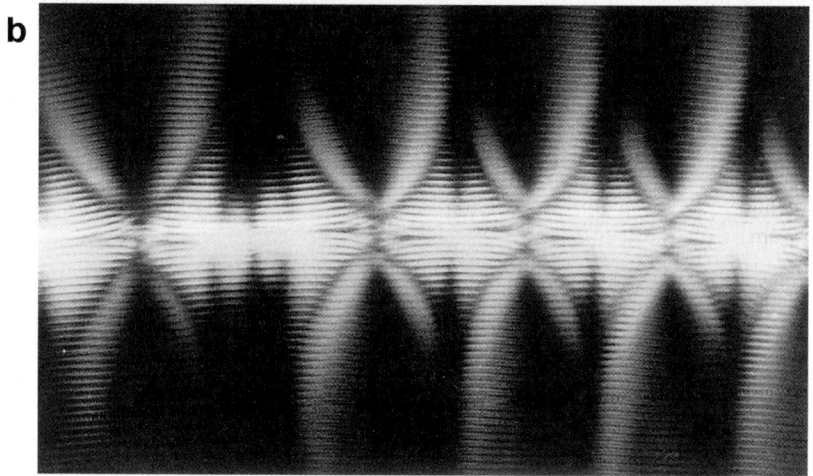

Figure 36. A cholesteric texture simulating the presence of focal parabolae. (a, b) Optical sections at the coverslip and slide levels. MBBA plus cholesterol benzoate, crossed polarizers. Scale bar: 20 μm.

diverse networks [9]. The presence of such defects is probable in smectic phases having two-dimensional order within layers, and in certain columnar phases. This also holds for smectic systems, when the hexagonal or rectangular arrangements are coherent over several lamellae (smectics B, E, G, H, etc.) the order being three-dimensional.

When smectic layers are mainly horizontal, the existence of walls often leads to *mosaic textures*, each domain extending over the entire thickness of the preparation. A *platelet texture* is observed when the domains of the horizontal layers are much thinner than the mesomorphic slab, and superimpose here and there, with their distinct orientations [13, 14]. When vertical layers bend concentrically about $+\pi$ disclinations, circular bands of different colour or shade appear, indicating changes in the crystallographic orientations along grain boundaries. These banded textures are common in orthorhombic smectic E phases.

A *planar wall* has been described in smectic C* phases, giving rise to a particular chevron texture. This has been shown

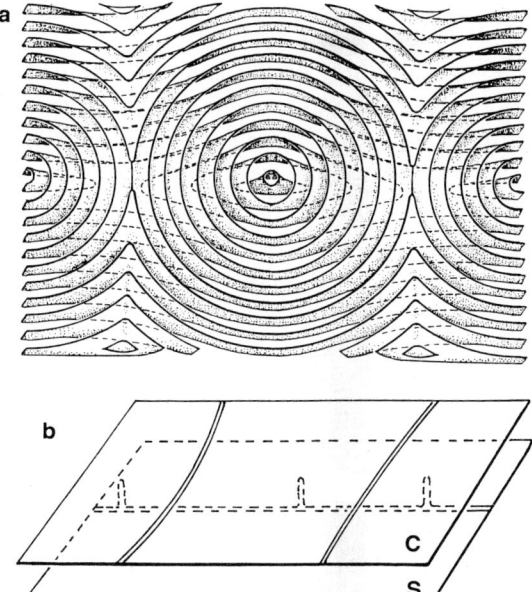

Figure 37. Model of the arrangement of layers and focal lines in the texture shown in Fig. 36. (a) Distribution of layers. (b) Vertical and horizontal focal lines at the top (coverslip C) and bottom (slide S) levels of the mesophase.

by high-resolution X-ray scattering in surface stabilized ferroelectric samples (see Chap. VI, Sec. 2 of Vol. 2 of this Handbook) obtained by cooling from the smectic A phase [94]. At the smectic C transition, the layers progressively tilt symmetrically in two opposite directions, by an angle of about 10°, whereas molecules are tilted horizontally within layers at about 20° from the normal to the layers. The long axis of molecules show a quasi uniform orientation within this wall and the two domains, but the ferroelectric polarization changes abruptly across the wall and breaks its apparent symmetry. This type of chevrons leads to complex textures [95].

In smectic A phases, such a change in the orientation of layers along a wall involves a discontinuity of molecular alignment. If the wall is replaced by a narrow zone of strongly curved layers, this involves an equal splay of molecules, and the structure is generally handled by dividing it into a set of focal curves or even disclinations. The case for smectic C phases is different, as such walls often are compatible with a uniform alignment of molecules, facilitated by a dilatation modulus of the layer thickness that is much smaller than that for smectic A phases. This leads to noticeable undulation instabilities [96]. Molecules rotate within layers and, conversely, layers show different possible orientations with respect to a molecular alignment. In chevrons such as those shown in Fig. 35, the layers are at a constant angle to the lines L and L', an angle which can be chosen to be that of the tilt [57]. The discontinuity in the layer orientation in the ferroelectric display can be smoothed by focal curves and other defects, which are weakly contrasted because the molecular orientation is not really altered by their presence.

Singular lines often attach to interfaces. Among the numerous examples given here is the case that occurs in nematic phases at the slide or coverslip level (see Fig. 9) and that of large-pitch cholesteric phases at various interfaces (see Figs. 19 and 20). Such lines are due to the presence of horizontal directors that are forbidden at a horizontal interface. The *locus of horizontal directors* in the bulk form a surface, which shows a higher mean refractive index and appears contrasted under natural light or in phase-contrast microscopy. The clear-cut interface line seems to extend into a wall within the bulk, but actually this is a mere optical illusion, which is the origin of the so-called Grandjean planes in cholesteric phases [44].

Broken aspects are classical in focal conics or fan textures of tilted smectics, and also originate from singular lines attached to an interface. This is observed in smectic C phases when disclinations like those shown in Fig. 21a–d adhere at an interface and separate two different molecular orientations. The smectic layers lie almost vertically (i.e. perpendicular to the interface), whereas the molecules are horizontal within the layers at the interface contact. However, the molecular orientation remains continuous in the bulk. The lines limiting these interfacial domains are either parallel or normal to layers and this creates a contour composed of straight segments alternating with arcs at right angles and with some interspersed irregularities [97]. These interface textures superimpose broken patterns onto smectic focal domains and fans, not only in smectic C phases but also in more ordered types of tilted smectic phases such as SmF, SmG and SmH [13, 14].

7.4.5 Origin of Spirals in Chiral Liquid Crystals

Cholesteric layers show spiral contours (Fig. 6a and b) around polygon centres and vertices (see Fig. 6a and b). Analogous patterns decorate cholesteric droplets (see Fig. 2). Such spirals have also been found in thin sections of stabilized analogues of cholesterics, i.e. in biological materials assembling long biopolymers such as chitin, in various arthropod carapaces (crabs, insects) [27–33]. To explain these patterns, let us start from cholesteric layers arranged as nested toroidal surfaces (see Fig. 29b). This situation is shown schematically along a meridian plane in Fig. 38a; the left-handed twist and three section planes (1–3) are illustrated in Fig. 38b–d. The hatched zone in Fig. 38a indicates the part of the drawing to be suppressed in order to obtain a situation similar to that of layers in contact with an isotropic phase of the mesogen at level 1 (Fig. 38b). Going up Fig. 38a one passes continuously from a uniform alignment of directors (d) to a double-spiral pattern (c) and a pair of $\lambda^{+\pi}$ disclinations (b). These three steps are assembled into a unique pattern in Fig. 38e, which indicates how a continuous deformation resembling a whorl produces a disclination pair and a spiral the orientation of which is linked to that of the twist.

Figure 31d shows how the presence of a $-\pi$ disclination eliminates very acute conical shapes of nested layers. This arrangement of layers creates a rhombic and conical domain in the midpart of most of the polygon edges in Fig. 6a. Careful examination of the layers under natural light at a level close to the coverslip generally shows a structure like the one shown in Fig. 38f. The $\lambda^{-\pi}$ disclination forms a helical half-loop [20] due to the cholesteric twist, a situation very similar to that described in other helical cholesteric patterns (see Sec. 7.3.4).

Spirals and concentric circles are common in the focal domains of smectic C* phases [97]. They are produced by lines such as those shown in Fig. 21c and d. These disclination lines have often been observed in preparations of vertical or slightly tilted layers. The molecules lie almost parallel to the glass plates at their contact, with uniform alignment, whereas the helical arrangement is present in the bulk. Equidistant and parallel disclinations of the type indicated shown in Fig. 21c and d separate the chiral and non-chiral domains; these have been called 'unwinding lines' [58]. They are present at two levels: in the neighbourhood of the upper and lower plates; and adjoining the thinner regions of the preparation, where they form loops or simple helical lines [97].

When focal domains are differentiated in smectic C* phases, with ellipses attached to a glass, the molecules also lie parallel to the glass, and this prevents the formation of a helical structure, which does, however, occur at a distance in the bulk. Unwinding

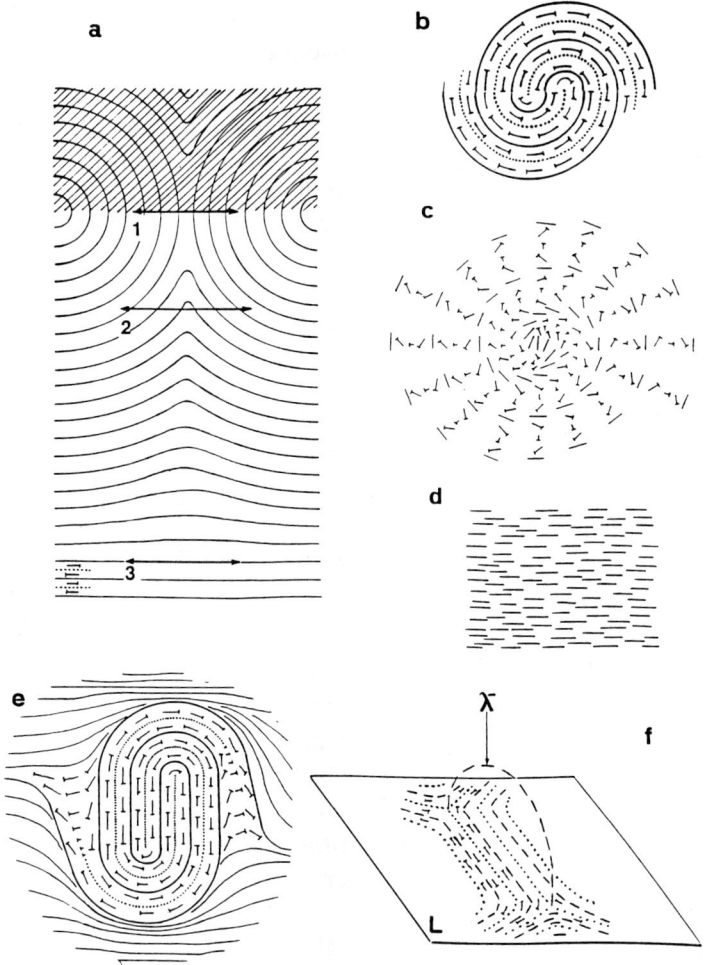

Figure 38. Continuous passage from a focal line to disclination pairs in a cholesteric liquid. (a) Meridian section of the texture corresponding to a three-dimensional representation of the structure shown in Fig. 29b; the nail convention indicates the presence of a left-handed twist. (b–d) Distribution of directors at levels 1–3 in (a), presented with the nail convention. (e) An imagined pattern to show in one picture the continuous passage from a uniform alignment at the periphery to a whorl-like texture, with two $\lambda^{+\pi}$ disclinations in the core. In (b) and (e), the separated segments corresponding to horizontal directors are represented by continuous lines. (f) Helical shape of the $\lambda^{-\pi}$ disclination in Fig. 31d and the lozenges of polygonal edges shown in Fig. 6.

lines therefore appear, and produce patterns the principle of which is closely related to that described above for cholesterics. The main differences are that in smectic C* phases, the distance between two successive lines is p, the helical pitch, whereas in cholesterics the periodicity is $p/2$, and the double spiral in cholesterics is reduced to a single one in smectic C phases. One can also observe concentric circular lines, the successive radii of which differ by p, and this is explained by the presence of a screw line superimposed on the vertical focal line, as in cholesterics (see Fig. 32d).

7.4.6 Defects and Mesophase Growth

Defects, particularly screw dislocations, play an important role in the growth of true crystals [9]. Liquid crystal germs present the defects that were described in detail by Friedel and Grandjean [21] for smectic A phases. The germs elongate perpendicular to the mean direction of the layers, the surface tension being anisotropic, and the focal domains present in the bâtonnet are arranged such that the layers lie normal to the isotropic interface. A focal line is often present along the axis of the bâtonnet, and screw dislocations can be superimposed [54], which could accelerate the transition and facilitate the elongation.

Cholesteric rodlets and spherulites having parallel layers can be devoid of inner defects, but this does not prevent their growth. However, surface points or surface lines are present. Some $-\pi$ disclinations, resulting from germ coalescence, are frequent, but disappear by confluence with the isotropic interface, as for nematic droplets.

Spherulites showing concentric layers present a disclination radius or diameter, but this structure is due to a topological constraint and does not seem to be linked to liquid crystal growth. Very rapid growth of cholesteric phases often generates screw dislocations of the two types shown in Fig. 24i and j, and this has been filmed by Rault in p-azoxyanisol added to cholesterol benzoate [98, 99]. Slow growth does not result in the production of these defects.

Experiments have also been done at phase transitions from nematic to smectic A phases, starting from a twisted situation between two glasses with a strong planar anchoring, the two easy directions being at right angles (see Fig. 9). Thin threads are present at the junction of twisted zones (left-handed and right-handed). These transform at the smectic transition into a broken helical line, composed of a series of triplets comprising (1) a focal curve, (2) a segment of a $+\pi$ disclination and (3) a focal curve, the two focal segments (1 and 3) being conjugated [100]. Considerations such as those developed for Fig. 30e show that numerous screw dislocations attach to the $+\pi$ line and form a radiating pattern that is visible under the light microscope, which suggests the presence of important Burgers vectors. Lyotropic lamellar systems often show undulated or helical arrangements of focal domains [101], possibly similar to those of thermotropic phases [100].

7.4.7 Defect Energies and Texture Transformations

Translation dislocations do not strongly modify the orientation of the directors or layers at long distances, as disclinations in liquid crystals do, and focal conics occupy an intermediate position (see Fig. 32i). This indicates a gradation of energies, which will be considered below using some examples.

7.4.7.1 Disclination Points

Let us begin with simple defects, such as those that occur in capillary tubes (see Fig. 8). In the left-hand part of Fig. 8a there is a radial point in the vicinity of which the directors are supposed to be aligned radially: $\boldsymbol{n}(M) = \boldsymbol{r}/r$, with $\boldsymbol{r} = OM$ and r being the positive distance separating M from O, the core of the defect. Then div $\boldsymbol{n} = 2/r$; the twist and bend, whether spontaneous or not, are absent, and the density of the elastic energy is given by

$$\frac{dF}{dv} = \frac{2k_{11}}{r^2} \quad \text{or} \quad dF = 8\pi k_{11} dr$$

Integrating the elastic energy over a sphere centred on O, of radius R, yields

$$F = 8\pi k_{11} R$$

Such a purely radial arrangement of directors is more plausible when twist and bend are forbidden, particularly in smectic drops immersed in an isotropic fluid creating homeotropic conditions. This is realized for some spherical 4-cyano-4′-n-octylbiphenyl (8CB) droplets immersed in water or glycerol (Fig. 39a), but for most droplets in this domain the spherical symmetry is reduced to revolution symmetry by a focal domain attached at the centre, with smectic layers normal to the interface (Fig. 39b). This means that interfacial tension depends on the angle of directors at the interface, with

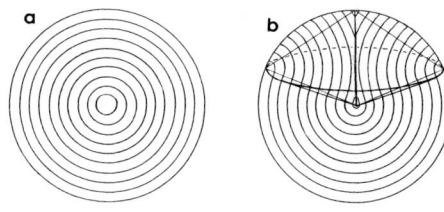

Figure 39. Spherical smectic droplets of 8CB in water or glycerol. (a) Presence of a radial singularity. (b) A focal domain associating a focal radius and a circle. The distribution of the layers is drawn in a meridian plane.

a strong minimum for molecules lying normally and a less marked one for those parallel to this interface with water or glycerol. In nematics, director lines form diverse patterns around singular points, associating splay, twist and bend in different proportions [102]. In a spherical nematic droplet floating in an isotropic medium, with the directors parallel to the interface, two singularities appear at two diametrically opposite poles, with radial arrangements similar to the one shown in Fig. 12a. For a spherulite of radius R, with $k_{11} = k_{33} = k$, and assuming that the twist is absent, $F = 5\pi k R$. This means that the volume energy due to a radial point at the centre of the spherulite is appreciably higher than that of two surface points. This confirms that defects 'prefer' to join an outer interface in order to minimize their energy, and this is reinforced by the fact that nematic molecules generally align with the isotropic interface, rather than lying normally.

7.4.7.2 Disclination Lines

As shown in Fig. 8b, lateral homeotropy in a capillary tube filled with a smectic A phase results in pure splay that is distributed homogeneously about a disclination line, together with cylindrical symmetry. Note, however, that core rehandlings are observed [42]. If r represents the transverse component of OM (O being an arbitrary point of the line), then

$$\operatorname{div} \mathbf{n} = \frac{1}{r} \text{ and } \frac{dF}{dv} = \frac{k_{11}}{2r^2}$$

or $dF = \pi L \left(\dfrac{k_{11}}{r} \right) dr$

where L is a given length of the singular line. The energy cannot be integrated from 0 to R, as for the radial point, and a cut-off is required. A radius core R_c that is close to molecular dimensions is introduced; E_c is the core energy per unit length. Hence the line energy density E in the capillary or in a myelin form is

$$E = E_c + \pi k_{11} \ln\left(\frac{R}{R_c}\right)$$

The disclination patterns usually considered in energy calculations in nematics are the solutions of a Laplace equation $\nabla^2 \Phi = 0$ of the type

$$\Phi = \Phi_0 + \frac{N\varphi}{2} = \Phi_0 + s\varphi$$

(see Sec. 7.2.4; in energy calculations, the topological parameter $s = N/2$ is now generally used rather than N). In this case $k_{11} = k_{22} = k_{33} = k$, and

$$\operatorname{div} \mathbf{n} = \cos(\Phi - \varphi) \frac{d\Phi}{r d\varphi}$$

$$(\mathbf{n} \times \operatorname{curl} \mathbf{n})^2 = \frac{d\Phi^2}{dr^2 + r^2 d\varphi^2}$$

from which we deduce

$$2\frac{dF}{dv} = \frac{ks^2}{r^2} \text{ or } dF = \frac{k\pi L s^2 \, dr}{r}$$

$$E = E_c + k\pi s^2 \ln\left(\frac{R}{R_c}\right)$$

For $s = 1$, we now re-find the energy of the radial disclination of a smectic liquid in a capillary tube. At the nematic transition, this disclination relaxes, with an 'escape in the third dimension' (see Fig. 8a), the splay lying mainly along the axis and the bend at the periphery. In the absence of twist, one has: $r/R = \tan\theta/2$ (where R is the inner radius of the capillary, r is the distance from point M to the axis, and θ is the angle separating the director at point M from the axis). This means that the director lines have a meridian profile of constant shape and, to a first approximation, the textures in two different tubes of radii R_1 and R_2 differ only by a pure dilation of ratio R_2/R_1. The total deformation remains constant and the energy does not depend on the inner radius R. The energy per unit length is $E = 3k\pi$, in disclinations $s = +1$, with a continuous core [39–41]. For $s = -1$, $E = k\pi$. The passage from a planar structure of a disclination to a continuous core suppresses the factor $\ln(R/R_c)$, and represents a non-negligible energy saving.

Vertical disclination lines normal to horizontal layers in smectic C phases also form nuclei. Polarizing microscopy shows that these nuclei have four branches, and when examined in projection onto the layer plane the observed patterns correspond to $\Phi = \Phi_0 + s\varphi$, with $s = \pm 1$. It has been demonstrated that the tilt angle of molecules with respect to the normal to layers is variable, but decreases to zero in the vicinity of the disclination core [103]. This also resembles an 'escape in the third dimension', and is mainly due to the low value of the dilatation modulus B.

Interactions of disclinations have been studied in *nematic* and *smectic C schlieren textures*, which associate nuclei interpreted as vertical disclinations. Assuming a purely horizontal distribution of directors and an 'elastic isotropy', within any horizontal plane and in the vicinity of disclination D_i, centred on O_i, we have $\Phi = \Phi_i + s_i \varphi_i$. At any point M in the texture, the director is defined as $\Phi = \Phi_0 + \Sigma s_i \varphi_i$, but this superposition principle requires that $R_c \ll r_{12} \ll r$ [104]. The energy of a disclination pair is given by

$$E = k\pi(s_1 + s_2)^2 \ln\left(\frac{R}{R_c}\right)$$
$$- 2k\pi s_1 s_2 \ln\left(\frac{r_{12}}{2R_c}\right)$$

The first term disappears when $s_1 = -s_2$, a common situation in schlieren textures of nematic and smectic C phases. In general, $\Sigma s_i/m \sim 0$, m being the total number of nuclei in the texture [11, 45]. The second term in the energy represents the interaction: disclinations of opposite sign ($s_1 s_2 < 0$) attract because variations in E and r_{12} occur in parallel [45], whereas they repel when they are of like signs ($s_1 s_2 > 0$).

7.4.7.3 Focal Curves

The energy of a focal domain defined by the ellipse E and the corresponding hyperbola branch, extending from one focus of E to infinity, is:

$$E = k_{11} \pi L (1 - e^2) \ln(a/R_c)$$

L being the ellipse perimeter, e its eccentricity and a its half major axis. The energy of focal domains was also calculated for pairs of parabolae [105].

It was indicated in Fig. 32c that layers along the conics are not exactly conical but present a rounded apex, with a local thickness maximum. Assuming curvature radii very large relative to the layer thickness, θ being the angle between the normal to layer and the focal curve, whose maximum θ_∞ remains small, the energy associatings play and layer dilatation, is per unit length:

$$E = \pi k_{11}(\theta_\infty)^2 \ln(1.46 R/r)$$

with $r \sim 2\lambda/\theta_\infty$, λ being the penetration length $(k_{11}/B)^{1/2}$ [106].

7.4.7.4 Translation Dislocations

All the expressions given above for disclination and focal curve energies are similar to those for screw and edge dislocations in true crystals [9, 15], but the situation is different for translation dislocations in smectics. The presence of a screw dislocation of weak Burgers vector does not introduce any splay and does not really modify the layer thickness even at small distances [16]. The energy is reduced to a core term and therefore interaction terms are absent in the frame of the linear theory. This is not the case for edge dislocations in this approximation [106], and one has the expression:

$$E = E_c + k_{11} b^2 / 2\lambda R_c$$

with $\lambda^2 = k_{11}/B$, b being the Burgers vector length. Actually, edge dislocations in smectics generally split into parallel disclinations $+\pi$ and $-\pi$, distant of $b/2$ [16] and this introduces a pure splay in domains of nested cylindrical layers, whose energy is: $\pi k_{11} \ln(b/2R_c)/2$. A third term τ_c is to be added and corresponds to the core energy of the $-\pi$, what leads to a more complex expression:

$$E = k_{11} b^2 / 2\lambda R_c + \pi k_{11} \ln(b/2R_c)/2 + \tau_c$$

When parallel edge dislocations lie in the same plane, their interaction is negligible; but if they lie at different levels, dislocations attract or repel according to their relative positions and orientations [107].

7.4.7.5 Simulations of Defects and Textures

Computer models have been used to simulate the distribution of directors in nematic disclinations and textures, the partial differential equations being replaced by finite difference equations [108, 109]. For instance, in a square lattice, the cell i, the director of which is specified by an angle Φ_i, has four neighbouring cells i'. If the elasticity is isotropic $\nabla^2 \Phi = 0$, the interaction energy is $E_i = \Sigma \sin^2(\Phi_{i'} - \Phi_i)$. Differentiating with respect to Φ_i gives the best orientation in cell i (generally a minimum of E_i, not a maximum) if the $\Phi_{i'}$ only differ by a few degrees:

$$\frac{\sin 2\Phi_i}{\cos 2\Phi_i} = \frac{\Sigma_i \sin 2\Phi'_{i'}}{\Sigma_i \cos 2\Phi_{i'}}$$

The total energy over the model is then:

$$E = \sum_{i,i'} \sin^2(\Phi_{i'} - \Phi_i)$$

Note that a knowledge of Φ along the cells at the periphery is generally sufficient to deduce the entire pattern, since it leads to a linear system with an equal number of equations and unknowns, plus a determinant different from zero, as indicated in early work on Dirichlet's problem [110]. It is also possible to start from a given distribution of Φ, with various boundary conditions such as those known to produce disclinations, and to recalculate the different Φ_i averaged over the four neighbouring cells. This leads first to a decrease in E, the process being repeated iteratively towards a complete relaxation. One can verify that a planar $+2\pi$ disclination disjoins into a pair of $+\pi$ disclinations, due to the presence of the s^2 term in the energy: The total energy of two disclinations $s = 1/2$ represents the half-energy of a single disclination $s = 1$. Obviously, there are textures where integer indices are preferred, for many reasons: escape in the third dimension, point singularities, certain anchoring conditions, and topological constraints (such as in smectic C phases) all forbid half-integer values of the s indices.

These methods have been refined to give a constant splay/bend ratio in purely planar systems, and they allow one to reproduce the well-known aspects of disclinations in pure bend or pure splay situations. These models have also been extended to three dimensions by taking the twist into account, and one can re-find the geometries of director lines in capillary tubes or other cases of 'escapes in the third dimension'.

7.4.7.6 Defect Nucleation

Many defects originate from a non-parallel coalescence of germs (Fig. 40). In addition, mechanical constraints generate edge and screw dislocations, or another pattern called 'elementary pinch' (Fig. 41), which is also found in cholesterics and probably at the origin of focal domains in smectics. This pattern seems, therefore, to be essential in texture transformations.

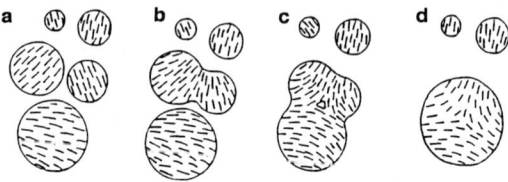

Figure 40. Schematic representation of defects created by the coalescence of nematic germs.

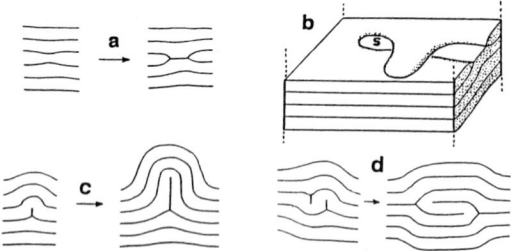

Figure 41. Creation of defects in a lamellar liquid crystal. (a) Edge dislocation due to a layer rupture. (b) Deformation of an edge dislocation into the helical structure of a screw dislocation. (c) Differentiation of an elementary pinch. (d) A pair of elementary pinches transforming into a pair of edge dislocations.

Elementary pinches differentiate from local fluctuations that are able to diverge and produce large defects, if certain constraints are present. In cholesterics, the formation of elementary pinches is continuous, since it involves the nucleation of a $\lambda^- \lambda^+$ pair [48]. These pinches are produced in the same way in nematics submitted to electric fields (see Fig. 5a in Stieb et al. [111]). In large-pitch cholesterics, the transition from a constrained aligned state to the relaxed twisted state also involves the nucleation of these elementary pinches. The constrained state is due, for example, to strong homeotropic conditions, and is relaxed by a progressive increase in the distance separating the two horizontal plates [112], or by a transverse electric field, if the dielectric anisotropy is negatively uniaxial [113, 114].

Many textures arise from a field or a dynamics created in a previously aligned liquid crystal [115, 116]. Such evolutions are also observed in cholesteric mixtures in equilibrium with the isotropic phase, with fans along the interface, and with polygons and planar textures generally at a greater distance from the isotropic transition. Gentle heating modified the positions of the textures by means of local transformations, but hydrodynamic streams are absent, or almost so. Cholesteric layers, which are initially planar, arrange progressively into domes and basins, and the transformation is continuous, the explanation being derived from Fig. 38 a–e. This also arises from local fluctuations forming directly elementary pinches, which easily transform into spiral patterns.

7.4.7.7 Textures and Defect Associations

Defects form complex networks, but between these singular structures the liquid crystal is regularly curved. The geometry is often well defined, and one may think that a knowledge of the network would be sufficient to rebuild the complete director distribution. Actually, this is a difficult problem, the solution of which depends on its accurate formulation. For instance, two parallel disclinations λ^+ and λ^- can be associated to give either an edge dislocation (as in Fig. 28 d and e) or an elementary pinch (see Fig. 41 b), or very different textures. On the contrary, the presence of two conjugated focal curves defines unambiguously the director distribution within the corresponding domain, with the exception of the exact layer position.

Three types of defect association have been encountered in this discussion: *superposition* of defects, *concurrence* of lines at singular points and *conjugation* between two lines or two points.

Among superpositions, the first case is simply that of lines of identical nature, for instance translation dislocations that join or disjoin, with additive laws for the Burgers vectors. As indicated above, the energy of these dislocations increases as the square of the Burgers vector b^2, and in principle they split into series of elementary dislocations. However, in practice, they are seen to assemble or separate in, for example, cholesterics. This fact underlines the importance of boundary conditions and the presence of other defects in their vicinity. Another common association is that of two $+\pi$ disclinations, which add to form a 2π disclination. The reverse transformation occurs during the course of water evaporation in a myelin form which is squashed between a slide and a coverslip. The dissociation of one line into several lines often occurs at definite points, or in restricted volumes, and these singular points have their own topologies and energies that are rarely considered in theories. Focal curves can branch into pairs and ramify (see Fig. 27 in Friedel and Grandjean [21]). Trees of focal conics have even been photographed in the bulk, near an air–smectic A interface, with several ramification steps, as in fractals [117]. Lines of different nature can also be superimposed. Screw dislocations often run along focal lines (see Fig. 32 d), and disclination lines can play the role of focal curves (see Fig. 33 and other examples in Bouligand [20]).

The main systems of concurrent lines are those of converging hyperbolae at the vertices of polygons in smectic phases and tangent conics (ellipses and hyperbolic edges of polygons), which are often replaced by systems of branching focal curves. Focal curves can attach at definite intervals along disclinations. All the points considered where such sets of lines converge, also have their own topologies and energies, depending on the textural environment.

The vertices of two focal conics are singular, since they correspond to a reverse in the orientation of the materialized conic sheets of Dupin's cyclides. These points often disjoin into pairs of lower energy (see Fig. 31 d and Fig. 11 in Bouligand [53]). At a transition from a smectic A to a smectic C phase disclination lines appear in focal domains, linking the two conics, and are of the type shown in Fig. 21 a and b [118]. Their origin is easily understood from Figs. 14 and 25 in Bouligand and Kléman [57].

Much more work has been concerned with systems of conjugated lines, mainly pairs of focal conics and pairs of disclination lines of opposite indices, forming typical patterns of dislocations or elementary pinches. Such associations also exist between singular points showing complementary topological properties, as occur in the capillary tube (see Figs. 8 a, 13 a and b). Note that certain associations are strictly obligatory. For example, a singular point in the bulk of a nematic liquid, or in a large-pitch cholesteric, always lies along a thick thread, since these threads correspond to the locus of vertical directors (i.e. parallel to the microscope axis), and all possible directions (vertical in particular) are represented in the immediate vicinity of such points. This thick thread is itself vertical in the case in Fig. 8a for an observer looking from above the capillary tube, and the pattern seen under a polarizing microscope is that of nuclei (see Fig. 4.16 in Meyer [47]).

7.4.7.8 Crystals of Defects

The distribution of defects in mesophases is often regular, owing to their fluidity, and this introduces pattern repeats. For instance, square polygonal fields are frequent in smectics and cholesteric liquids. Such repeats occur on different scales – at the level of structural units or even at the molecular level. Several types of amphiphilic mesophase can be considered as 'made of defects'. In many examples the defect enters the architecture of a unit cell in a three-dimensional array and the mesophase forms a 'crystal of defects' [119]. Such a situation is found in certain cubic phases in water–lipid systems [120] and in blue phases [121] (see Chap. XII of Vol. 2 of this Handbook). Several blue phases have been modeled as being cubic centred lattices of disclinations in a 'cholesteric matrix'. Möbius disclinations are assumed to join in groups of 4×4 or 8×8, but in nematics or in large-pitch cholesterics such junctions between thin threads are unstable and correspond to brief steps in recombinations. An isotropic droplet or a Ginsburg decrease to zero of the order parameter probably stabilizes these junctions in blue phases.

Similarly, the recently discovered smectic A* phases [122, 123] are layered like ordinary smectic A phases, but show a regular twist and the director follows the distribution that is classical in cholesteric liquids, whereas bend and twist are forbidden in principle (curl n =0). Layers are disrupted by nearly equidistant twist grain boundaries, each of which is composed of quasi-equidistant and parallel screw dislocations, normal to the helical axis, and well evidenced by freeze–fracture studies [124].

In columnar liquid crystals also, twist and splay are excluded in principle, but twist boundaries are plausible when these systems are strongly chiral [125]. The existence of such cholesteric columnar phases awaits confirmation.

7.5 References

[1] M. J. Press, A. S. Arrott, *J. Phys. Paris* **1975**, *36*, C1-177.
[2] M. Kléman, J. Friedel. *J. Phys. Paris* **1969**, *30*, C4-43.
[3] J. Friedel, M. Kléman in *Fundamental Aspects of Dislocation Theory* (Eds.: J. A. Simmons, R. de Wit, R. Bullough), National Bureau of Standards, Washington, DC, Special Publication 317, No. I, **1970**, pp. 607–636.
[4] Y. Ouchi, K. Ishikawa, H. Takezoe, A. Fukuda, K. Kondo, S. Era, A. Mukoh, *Jpn. J. Appl. Phys.* **1985**, *24* (Supp. 2), 899.
[5] N. A. Clark, T. P. Rieker, *Phys. Rev. A* **1988**, *37*, 1053.
[6] D. Hilbert, S. Cohn-Vossen, *Anschauliche Geometrie (Geometry and the Imagination)*, Chelsea, New York **1990**.
[7] H. S. M. Coxeter, *Introduction to Geometry*, 2nd ed., Wiley, New York **1989**.
[8] N. H. Hartshorne, A. Stuart, *Crystals and the Polarizing Microscope*, 4th edn, Arnold, London **1970**.
[9] S. Amelinckx, *Solid State Phys.* **1964**, Suppl. 6, 1.
[10] O. Lehmann, *Flüssige Kristalle*, Verlag W. Engelmann, Leipzig **1904**.
[11] G. Friedel, *Ann. Phys.* **1922**, *18*, 273.
[12] G. Friedel, *Colloid Chemistry*, Vol. 1, Chemical Catalog Co., **1926**, pp. 102–136.
[13] D. Demus, L. Richter, *Textures of Liquid Crystals*, 2nd edn, VEB Deutscher Verlag, Leipzig **1980**.
[14] G. W. Gray, J. W. G. Goodby, *Smectic Liquid Crystals, Textures and Structures*, Leonard Hill, Philadelphia **1984**.
[15] S. Chandrasekhar, G. S. Ranganath, *Adv. Phys.* **1986**, *35*, 507. S. Chandrasekhar, *Liquid Crystals*, Cambridge University Press, London **1977**.

[16] M. Kléman, *Points, Lines and Walls in Liquid Crystals, Magnetic Systems and Various Ordered Media*, Wiley, Chichester, NY **1983**. M. Kléman, *Rep. Prog. Phys.* **1989**, *52*, 555; *Liq. Cryst.* **1989**, *5*, 399.

[17] D. Chapman (Ed.), *The Structure of Lipids*, Methuen, London **1965**.

[18] J. Nageotte, *Morphologie des Gels Lipoïdes, Myéline, Cristaux Liquides, Vacuoles, Actualités (Sci. Ind., 431/434)*, Hermann, Paris **1936**. F. O. Schmitt, K. J. Palmer, *Cold Spring Harbor Symp. Quant. Biol.* **1940**, 94. D. M. Small, M. C. Bourgès, D. G. Dervichian, *Biophys. Biochim. Acta* **1966**, *125*, 563. F. B. Rosevear, *J. Am. Oil Chem. Soc.* **1954**, *31*, 628.

[19] Y. Bouligand in *Geometry in Condensed Matter Physics*, World Scientific, Singapore **1990**, Chap. 4; *J. Phys. Paris* **1990**, *51*, C7-35.

[20] Y. Bouligand, *J. Phys.* **1972**, *33*, 715. Y. Bouligand in *Dislocations in Solids* (Ed.: F. R. N. Nabarro), North Holland, Amsterdam **1979**, Chap. 23.

[21] G. Friedel, F. Grandjean, *Bull. Soc. Mineral.* **1910**, *33*, 409.

[22] Y. Bouligand, P. E. Cladis, L. Liébert, L. Strzelecki, *Mol. Cryst. Liq. Cryst.* **1974**, *25*, 233.

[23] W. A. Jensen, R. B. Park, *Cell Ultrastructure*, Wadworth, Belmont, CA **1967**. D. W. Fawcett, *The Cell*, 2nd edn, W. B. Saunders, Philadelphia **1981**.

[24] H. Moor, K. Mühlenthaler, *J. Cell. Biol.* **1963**, *17*, 609.

[25] R. R. Balmbra, J. S. Clunie, J. F. Goodman, *Proc. R. Soc. London, Ser. A* **1965**, *285*, 534. J. F. Goodman, J. S. Clunie in *Liquid Crystals & Plastic Crystals*, Vol. 2 (Eds.: G. Gray, P. A. Winsor), Ellis Horwood, New York **1974**, pp. 1–23.

[26] F. Livolant, *Physica* **1991**, *A176*, 117; *J. Mol. Biol.* **1991**, *218*, 165.

[27] Y. Bouligand, *Solid State Phys.* **1978**, Supp. 14, 259.

[28] Y. Bouligand in *Liquid Crystalline Order in Polymers* (Ed.: A. Blumstein), Academic Press, New York **1978**, Chap. 8.

[29] Y. Bouligand, *J. Phys. Paris* **1969**, *30*, C4-90.

[30] S. Caveney, *Proc. R. Soc. London, Ser. B* **1971**, *178*, 205.

[31] M.-M. Giraud-Guille, H. Chanzy, R. Vuong, *J. Struct. Biol.* **1990**, *103*, 232.

[32] M.-M. Giraud-Guille, *Int. Rev. Cytol.* **1996**, *166*, 59.

[33] Y. Bouligand, *Tissue Cell* **1972**, *4*, 189.

[34] Y. Bouligand, *7th Int. Congr. Electron Microscopy*, Grenoble **1970**, *3*, 105.

[35] M.-M. Giraud-Guille, *Calcif. Tissue Int.* **1988**, *42*, 167.

[36] M.-M. Giraud-Guille, *J. Mol. Biol.* **1992**, *224*, 861. L. Besseau, M.-M. Giraud-Guille, *J. Mol. Biol.* **1995**, *251*, 197.

[37] Y. Bouligand in *Physics of Defects* (Eds.: R. Ballian, M. Kléman, J.-P. Poirier), North Holland, Amsterdam **1980**, pp. 780–811.

[38] J. E. Proust, L. Ter Minassian, E. Guyon, *Solid State Commun.* **1972**, *11*, 1227.

[39] C. Williams, P. Pieranski, P. E. Cladis, *Phys. Rev. Lett.* **1972**, *29*, 90.

[40] R. B. Meyer, *Phil Mag.* **1973**, *27*, 405.

[41] A. Saupe, *Mol. Cryst. Liq. Cryst.* **1973**, *21*, 211.

[42] P. E. Cladis, *Phil. Mag.* **1974**, *29*, 641. P. E. Cladis, A. E. White, W. F. Brinkman, *J. Phys. Paris* **1979**, *40*, 325.

[43] A. M. J. Spruijt, *Solid. State Commun.* **1973**, *13*, 1919.

[44] Y. Bouligand, *J. Phys. Paris* **1974**, *35*, 215.

[45] J. Nehring, A. Saupe, *J. Chem. Soc., Faraday Trans. II* **1972**, *68*, 1.

[46] M. Kléman in *Dislocations in Solids* (Ed.: F. R. N. Nabarro), North Holland, Amsterdam **1980**, pp. 5, 245–297, 351–402. G. Ryschenkov, M. Kléman, *J. Chem. Phys.* **1976**, *64*, 413.

[47] R. B. Meyer, *Mol. Cryst. Liq. Cryst.* **1972**, *16*, 335.

[48] Y. Bouligand, *J. Phys.* **1974**, *35*, 959.

[49] Y. Bouligand in *Dislocations in Solids*, Vol. 5 (Ed.: F. R. N. Nabarro), North Holland, Amsterdam **1980**, pp. 301–347.

[50] V. Volterra, *Ann. Ecole Normale Sup.* **1907**, *24*, 401.

[51] A. E. H. Love, *A Treatise on the Mathematical Theory of Elasticity*, Dover, New York 1944.

[52] J. Friedel, *Dislocations, International Series Monographs on Solid State Physics*, No. 3, Pergamon, Oxford **1964**.

[53] Y. Bouligand in *Physics of Defects* (Eds.: R. Ballian, M. Kléman, J.-P. Poirier), North Holland, Amsterdam **1980**, pp. 668–711.

[54] F. C. Frank, *Discuss. Faraday Soc.* **1958**, *25*, 19.

[55] P. G. de Gennes, J. Friedel, *C. R. Heb. Acad. Sci., Paris B* **1965**, *268*, 257.

[56] P. E. Cladis, M. Kléman, *Mol. Cryst. Liq. Cryst.* **1972**, *16*, 1.

[57] Y. Bouligand, M. Kléman, *J. Phys. Paris* **1979**, *40*, 79.

[58] M. Brunet, C. Williams, *Ann. Phys. Paris* **1978**, *3*, 237.

[59] Y. Bouligand, M. Kléman, *J. Phys. Paris* **1970**, *31*, 1041.

[60] J. Rault, *Phil. Mag. A* **1973**, *28*, 11.

[61] M. Kléman in *Dislocations in Solids*, Vol. 5 (Ed.: F. R. N. Nabarro), North Holland, Amsterdam **1980**, pp. 245–297.

[62] C. Robinson, J. C. Ward, R. B. Beevers, *Faraday Soc. Discuss.* **1958**, *25*, 29.

[63] Y. Bouligand, F. Livolant, *J. Phys. Paris* **1984**, *45*, 1899.

[64] R. Cano, *Bull. Soc. Fr. Minér. Cristallogr.* **1967**, *90*, 333; **1968**, *91*, 20.

[65] G. Malet, J. Marignan, O. Parodi, *J. Phys.* **1978**, *9*, 863. G. Malet, J. C. Martin, *J. Phys. Paris* **1979**, *40*, 355.
[66] C. Mauguin, *Bull. Soc. Minéral.* **1911**, *34*, 6, 71.
[67] P. G. de Gennes, *The Physics of Liquid Crystals*, Oxford University Press, Oxford **1974**.
[68] F. Grandjean, *Bull. Soc. Fr. Minéral.* **1919**, *42*, 42.
[69] G. Toulouse, M. Kléman, *J. Phys. Paris* **1976**, *37*, L-149.
[70] V. Poenaru, G. Toulouse, *J. Phys. Paris* **1977**, *38*, 887.
[71] M. Kléman, L. Michel, G. Toulouse, *J. Phys. Paris* **1977**, *38*, L-195.
[72] M. Kléman, *J. Phys. Paris* **1977**, *38*, L-199.
[73] Y. Bouligand, B. Derrida, V. Poenaru, Y. Pomeau, G. Toulouse, *J. Phys. Paris* **1978**, *39*, 863.
[74] M. Kléman, L. Michel, *J. Phys. Paris* **1978**, *39*, L-29; *Phys. Rev. Lett.* **1978**, *40*, 1387.
[75] V. Poenaru in *Ill-Condensed Matter* (Eds.: R. Ballian et al.), *Les Houches XXI Summer School Proc.*, 1978, p. 265.
[76] M. D. Mermin, *Rev. Mod. Phys.* **1979**, *51*, 581.
[77] H. R. Trebin, *Adv. Phys.* **1982**, *31*, 195.
[78] M. V. Kurik, O. D. Lavrentovich, *Sov. Phys. Usp.* **1988**, *31*, 196.
[79] W. F. Harris, *Sci. Am.* **1977**, *237 (6)*, 130.
[80] M. Born, E. Wolf, *Principles of Optics*, 6th edn, Pergamon, Oxford **1980**.
[81] Y. Bouligand, *J. Phys. Paris* **1973**, *35*, 603.
[82] F. Grandjean, *Bull. Soc. Fr. Minéral.* **1916**, *39*, 164.
[83] C. Dupin, *Applications de Géométrie et de Mécanique*, Bachelier, Paris **1822**. J. C. Maxwell, *Q. J. Pure Appl. Maths.* **1867**, 34; *On the Cyclide*, Scientific Papers, Cambridge University Press, Cambridge **1890**, 2, 144–159. C. M. Jessop, *Quartic Surfaces*, Cambridge University Press, Cambridge **1916**. G. Darboux, *Géométrie Analytique*, Gauthier-Villars, Paris **1917**. H. Bouasse, *Optique Cristalline, Polarisation Rotatoire, Etats Mésomorphes*, Delagrave, Paris **1925**.
[84] W. Bragg, *Trans. Faraday Soc.* **1933**, *29*, 1056. J. P. Meunier, J. Billard, *Mol. Cryst. Liq. Cryst.* **1969**, *7*, 421. M. Kléman, *J. Phys. Paris* **1977**, *38*, 1511.
[85] C. S. Rosenblatt, R. Pindak, N. A. Clark, R. Meyer, *J. Phys. Paris* **1977**, *38*, 1105.
[86] G. Friedel, G. Grandjean, *C. R. Acad. Sci., Paris* **1911**, *152*, 322.
[87] M. de Broglie, E. Friedel, *C. R. Acad. Sci., Paris* **1923**, *176*, 738. E. Friedel, *C. R. Acad. Sci., Paris* **1925**, *180*, 269.
[88] Y. Bouligand, *J. Phys. Paris* **1972**, *33*, 525.
[89] M. Kléman, C. Williams, C. E. Costello, T. Gulik-Krzywicki, *Phil. Mag.* **1977**, *35*, 33.
[90] S. T. Lagerwall, S. Stebler, *Les Houches Summer School Proc.* **1981**, *XXXV*, 757.

[91] R. Bidaux, N. Boccara, G. Sarma, P. G. de Gennes, O. Parodi, *J. Phys. Paris* **1973**, *34*, 661.
[92] J. P. Sethna, M. Kléman, *Phys. Rev. A* **1982**, *26*, 3037.
[93] Y. Bouligand, *J. Microsc. France* **1973**, *17*, 145.
[94] T. P. Rieker, N. A. Clark, G. S. Smith, D. S. Parmar, E. B. Sirota, C. R. Safinya, *Phys. Rev. Lett.* **1987**, *59*, 2658.
[95] N. A. Clark, T. P. Rieker, *Phys. Rev. A* **1988**, *37*, 1053.
[96] D. Johnson, A. Saupe, *Phys. Rev. A* **1977**, *15*, 2079.
[97] L. Bourdon, J. Sommeria, M. Kléman, *J. Phys. Paris* **1982**, *43*, 77.
[98] J. Rault, *Solid State Com.* **1971**, *9*, 1965.
[99] J. Rault, *Phil. Mag.* **1973**, *28*, 11.
[100] C. E. Williams, *Phil. Mag.* **1975**, *32*, 313.
[101] M. B. Schneider, W. W. Webb, *J. Phys. Paris* **1984**, *45*, 273.
[102] F. R. N. Nabarro, *J. Phys. Paris* **1972**, *33*, 1089.
[103] S. T. Lagerwall, S. Stebler, *Ordering in Strongly Fluctuating Condensed Matter* (Ed.: T. Riste), Plenum, New York **1972**, p. 383.
[104] C. M. Dafermos, *Q. J. Mech. Appl. Math.* **1970**, *23*, 49.
[105] M. Kléman, *J. Phys. Paris* **1977**, *38*, 1511.
[106] P. G. de Gennes, *C. R. Hebd. Séanc. Acad. Sci., Paris* **1972**, *275B*, 549.
[107] M. Kléman, C. Williams, *J. Phys. Paris, Lettres* **1974**, *35*, L49.
[108] S. E. Bedford, T. M. Nicholson, A. H. Windle, *Liq. Cryst.* **1991**, *10*, 63. S. E. Bedford, A. H. Windle, *Liq. Cryst.* **1993**, *15*, 31.
[109] T. Kimura, D. G. Gray, *Liq. Cryst.* **1993**, *13*, 2330; *Macromolecules* **1993**, *26*, 3455.
[110] H. B. Philips, N. Wiener, *J. Math. Phys. MIT 2nd Ser.*, March 1923.
[111] A. Stieb, G. Baur, G. Meier, *J. Phys. Paris* **1975**, *36-C1*, 185. A. E. Stieb, *J. Phys. Paris* **1980**, *41*, 961.
[112] F. Lequeux, P. Oswald, J. Bechhoefer, *Phys. Rev. A* **1989**, *40*, 3974.
[113] P. Ribière, S. Pierkl, P. Oswald, *Phys. Rev. A* **1991**, *44*, 8198.
[114] F. Lequeux, M. Kléman, *C. R. Hebd. Acad. Sci., Paris* **1986**, *303*, 765; *J. Phys. Paris* **1988**, *49*, 845.
[115] F. Rondelez, J. P. Hulin, *Solid State Commun.* **1972**, *10*, 1009. F. Rondelez, H. Arnould, *C. R. Hebd. Acad. Sci., Paris B* **1971**, *273*, 549.
[116] P. Oswald, J. Behar, M. Kléman, *Phil. Mag. A* **1982**, *46*, 899.
[117] J. B. Fournier, G. Durand, *J. Phys. II France* **1991**, *1*, 845.
[118] A. Perez, M. Brunet, O. Parodi, *J. Phys. Paris* **1978**, *39*, L-353.

[119] W. Helfrich in *Physics of Defects* (Eds.: R. Ballian, M. Kléman, J.-P. Poirier), North Holland, Amsterdam **1980**, Course 12, pp. 716–755.
[120] H. Delacroix, A. Gulik, T. Gulik-Krzywicki, V. Luzzati, P. Mariani, R. Vargas, *J. Mol. Biol.* **1988**, *204*, 165; **1993**, *229*, 526.
[121] S. Meiboom, S. Sammon, D. W. Berreman, *Phys. Rev. A* **1983**, *28*, 3553.
[122] J. W. Goodby, M. A. Waugh, S. M. Stein, E. Chin, R. Pindak, J. S. Patel, *Nature* **1989**, *337*, 449.
[123] G. Strajer, R. Pindak, M. A. Waugh, J. W. Goodby, *Phys. Rev. Lett.* **1990**, *64*, 1545.
[124] K. J. Ihn, J. A. N. Zasadzinski, R. Pindak, *Science* **1992**, *258*, 275.
[125] R. D. Kamien, D. R. Nelson, *Phys. Rev. e, Statistical Physics* **1996**, *53*, 650.

8 Flow Phenomena and Viscosity

F. Schneider and H. Kneppe

A complete assessment of the literature on the hydrodynamics of liquid crystals is beyond the scope of this handbook. Therefore, only the most important fundamentals for the description of flow phenomena will be discussed. Further details can be found in the references cited at the end of each section and in the review articles and books of Porter and Johnson [1], de Gennes [2], Stephen and Straley [3], Jenkins [4], de Jeu [5], Vertogen and de Jeu [6], and de Gennes and Prost [7]. Two review articles of Leslie [8, 9] are exclusively devoted to this subject. Experimental results for uniaxial nematic liquid crystals are described in Chap. III, 2.6 of Vol. 2A of this Handbook.

The hydrodynamic theory for uniaxial nematic liquid crystals was developed around 1968 by Leslie [10, 11] and Ericksen [12, 13] (Leslie–Ericksen theory, LE theory). An introduction into this theory is presented by F. M. Leslie (see Chap. III, Sec. 1 of this Volume). In 1970 Parodi [14] showed that there are only five independent coefficients among the six coefficients of the original LE theory. This LEP theory has been tested in numerous experiments and has been proved to be valid between the same limits as the Navier–Stokes theory. An alternative derivation of the stress tensor was given by Vertogen [15].

In 1971 the Harvard group [16] presented a different form of the constitutive equations which gives very compact expressions for systems with nearly uniform director orientation. The results of both theories agree. Nevertheless, the LEP presentation is preferred in most of the experimental and theoretical studies in this field.

The constitutive hydrodynamic equations for uniaxial nematic calamitic and nematic discotic liquid crystals are identical. In comparison to nematic phases the hydrodynamic theory of smectic phases and its experimental verification is by far less elaborated. Martin et al. [17] have developed a hydrodynamic theory (MPP theory) covering all smectic phases but only for small deformations of the director and the smectic layers, respectively. The theories of Schiller [18] and Leslie et al. [19, 20] for SmC-phases are direct continuations of the theory of Leslie and Ericksen for nematic phases. The Leslie theory is still valid in the case of deformations of the smectic layers and the director alignment whereas the theory of Schiller assumes undeformed layers. The discussion of smectic phases will be restricted to some flow phenomena observed in SmA, SmC, and SmC* phases.

Although the constitutive hydrodynamic equations for nematic and polymeric liquid crystals are identical the following discussion is only devoted to monomeric systems. Polymeric liquid crystals exhibit a great variety of special effects which restrict the application of standard hydrodynamic theory. Reviews on this topic are given by Wissbrun [21] and Marrucci and Greco [22].

8.1 Nematic Liquid Crystals

Most of the hydrodynamic effects observed in nematic liquid crystals can be explained by the LEP theory. Its generality is sufficient for the following discussion. According to this theory the viscous part of the stress tensor σ_{ij} for an incompressible uniaxial nematic liquid crystal is

$$\sigma_{ij} = \alpha_1 n_i n_j V_{kp} n_k n_p + \alpha_2 n_j N_i + \alpha_3 n_i N_j + \alpha_4 V_{ij} + \alpha_5 n_j V_{ik} n_k + \alpha_6 n_i V_{jk} n_k \quad (1)$$

using the notation of Clark and Leslie [23] and usual summation convention. **V** is the symmetric part of the velocity gradient tensor

$$V_{ij} = \frac{1}{2}(v_{i,j} + v_{j,i}) \quad (2)$$

and **N** is the rotation of the director **n** relative to the fluid

$$N_i = \dot{n}_i - \frac{1}{2}(v_{i,j} - v_{j,i})n_j$$

or $\quad \mathbf{N} = \dot{\mathbf{n}} - \frac{1}{2}(\text{curl}\,\mathbf{v}) \times \mathbf{n} \quad (3)$

A superposed dot denotes the material time derivative. The Leslie coefficients α_i are assumed to be independent on velocity gradient and time. They depend on temperature and pressure. Four of the coefficients are related one with another via the Parodi equation [14, 24]

$$\alpha_2 + \alpha_3 = \alpha_6 - \alpha_5 \quad (4)$$

which is a consequence of the Onsager relations. Thus, the viscous properties of an incompressible nematic liquid crystal can be described by five independent coefficients.

In the following, we assume a steady and laminar flow of the liquid crystal.

8.1.1 Shear Viscosity Coefficients η_1, η_2, η_3, and η_{12}

Figure 1 shows the system of coordinates which is used in the following discussion. The liquid crystal flows parallel to the x-axis, and the velocity gradient is parallel to the y-axis. The orientation of the director **n** is described by the angles Φ and Θ

$$\mathbf{n} = (\sin\Theta \cos\Phi, \sin\Theta \sin\Phi, \cos\Theta) \quad (5)$$

Assuming that the velocity is only a function of y, the only non-vanishing components of the velocity gradient tensor are

$$V_{xy} = V_{yx} = \frac{1}{2} v_{x,y} \quad (6)$$

If the orientation of the director is fixed, for example by application of an external magnetic or electric field, Eq. (3) reduces to

$$\mathbf{N} = \frac{1}{2}(-\sin\Theta \sin\Phi, \sin\Theta \cos\Phi, 0) v_{x,y} \quad (7)$$

If a liquid crystal is sheared between two infinite plates parallel to the xz-plane, a constant velocity gradient parallel to the y-axis results. Assuming stationary flow the stress component σ_{xy} can be calculated from

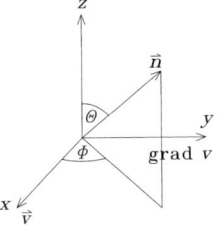

Figure 1. Director orientation with respect to the flow velocity and the velocity gradient.

Eqs. (1) and (5–7).

$$\sigma_{xy} = \Big(\alpha_1 \sin^4\Theta \sin^2\Phi \cos^2\Phi \\
- \frac{\alpha_2}{2}\sin^2\Theta\sin^2\Phi + \frac{\alpha_3}{2}\sin^2\Theta\cos^2\Phi \\
+ \frac{\alpha_4}{2} + \frac{\alpha_5}{2}\sin^2\Theta\sin^2\Phi \\
+ \frac{\alpha_6}{2}\sin^2\Theta\cos^2\Phi\Big) v_{x,y} \quad (8)$$

Collecting terms with the same dependence on the director orientation and calculation of the viscosity $\eta = \sigma_{xy}/v_{x,y}$ gives

$$\eta(\Theta,\Phi) = \frac{1}{2}(\alpha_4 + \alpha_5 - \alpha_2)\sin^2\Theta\sin^2\Phi \\
+ \frac{1}{2}(\alpha_4 + \alpha_6 + \alpha_3)\sin^2\Theta\cos^2\Phi \\
+ \frac{\alpha_4}{2}\cos^2\Theta \\
+ \alpha_1 \sin^4\Theta\cos^2\Phi\sin^2\Phi \quad (9)$$

Therefore, it makes sense to introduce a new set of viscosity coefficients (see Fig. 2)

$$\eta_1 = \frac{1}{2}(\alpha_4 + \alpha_5 - \alpha_2)$$

$$\eta_2 = \frac{1}{2}(\alpha_4 + \alpha_6 + \alpha_3)$$

$$\eta_3 = \frac{1}{2}\alpha_4$$

$$\eta_{12} = \alpha_1 \quad (10)$$

Rewriting Eq. (9) gives

$$\eta(\Theta,\Phi) = \eta_1 \sin^2\Theta\sin^2\Phi + \eta_2 \sin^2\Theta\cos^2\Phi \\
+ \eta_3 \cos^2\Theta \\
+ \eta_{12}\sin^4\Theta\cos^2\Phi\sin^2\Phi \quad (11)$$

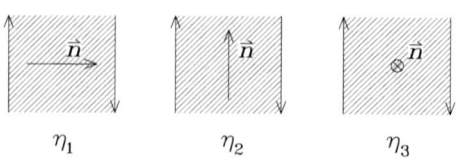

Figure 2. Definition of the shear viscosity coefficients η_1, η_2, and η_3. The outer pair of arrows symbolizes the shear flow.

Several notations for the shear viscosity coefficients are used. The notation used here stems from Helfrich [25]. A different notation was proposed by Miesowicz [26].

The shear viscosity coefficients η_1, η_2, and η_3 can separately be determined in shear flow experiments with adequate director orientations [27].

η_1: $\mathbf{n} \| \text{grad } v$

η_2: $\mathbf{n} \| v$

η_3: $\mathbf{n} \perp v$, $\mathbf{n} \perp \text{grad } v$ \quad (12)

For the determination of η_{12} at least three experiments are necessary. Two for the determination of η_1 and η_2 and one experiment with a director orientation where the influence of η_{12} on the shear viscosity is maximal ($\Theta = 90°$, $\Phi = 45°$). Then, η_{12} can be calculated according to

$$\eta_{12} = 4\eta(90°, 45°) - 2(\eta_1 + \eta_2) \quad (13)$$

Methods for the determination of the shear viscosity coefficients will be described in Chap. III, Sect. 2.6 of Vol. 2A of this Handbook.

8.1.2 Rotational Viscosity

Viscous torques are exerted on the director of a liquid crystal during a rotation of the director and by a shear flow with a fixed director orientation. The density Γ_i of the viscous torque is obtained by application of the Levi–Civita tensor

$$\varepsilon_{ijk} = \begin{cases} 1 & i,j,k \text{ cyclic} \\ -1 & i,j,k \text{ anticyclic} \\ 0 & \text{otherwise} \end{cases} \quad (14)$$

on the stress tensor. Only the skew symmetric part

$$\sigma_{ij}^{ss} = \frac{\alpha_2 - \alpha_3}{2}(n_j N_i - n_i N_j) \\
+ \frac{\alpha_5 - \alpha_6}{2}(n_j V_{ik} n_k - n_i V_{jk} n_k) \quad (15)$$

of the stress tensor (Eq. 1) gives a contribution

$$\Gamma_i = \varepsilon_{ijk}\sigma_{kj}^{ss} = \varepsilon_{ijk}\{(\alpha_2 - \alpha_3)n_j N_k + (\alpha_5 - \alpha_6)n_j V_{kp} n_p\} \quad (16)$$

or

$$\Gamma = -n \times (\gamma_1 N + \gamma_2 Vn) \quad (17)$$

with

$$\gamma_1 = \alpha_3 - \alpha_2; \quad \gamma_2 = \alpha_6 - \alpha_5 \quad (18)$$

γ_1 is called rotational viscosity or rotational viscosity coefficient.

At first the viscous torque on a rotating director is calculated. If the liquid crystal is at rest the last term in Eq. (17) vanishes. A director rotation in the xy-plane gives

$$\Gamma_z = -\gamma_1 \dot{\Phi} \quad (19)$$

Multiplying with the volume of the liquid crystal we obtain the viscous torque

$$M_z = -\gamma_1 \dot{\Phi} V \quad (20)$$

The torque which has to be applied on the director for this rotation amounts to $-M_z$.

If a liquid crystal at rest is surrounded by a vessel the torque exerted by a rotating magnetic or electric field on the director must be transferred to the vessel.

This torque will be calculated for a completely filled vessel with quadratic cross section. The liquid crystal and the vessel are assumed to be at rest and the director is rotating in the xy-plane (see Fig. 3).

As in this case neither the director and its rotation N nor the velocity gradient tensor V contains components depending on z, the only non-vanishing stress tensor components are σ_{xy} and σ_{yx} and in these components only terms with the coefficients α_2 and α_3 are non-vanishing. We obtain from Eqs. (3) and (1)

$$N = (-\sin\Phi, \cos\Phi, 0)\dot{\Phi} \quad (21)$$

and

$$\sigma_{xy} = (-\alpha_2 \sin^2\Phi + \alpha_3 \cos^2\Phi)\dot{\Phi}$$
$$\sigma_{yx} = (\alpha_2 \cos^2\Phi - \alpha_3 \sin^2\Phi)\dot{\Phi} \quad (22)$$

Multiplying by the areas a_y and a_x of the vessel gives the forces f_x and f_y

$$f_x = -\sigma_{xy} a_y = (\alpha_2 \sin^2\Phi - \alpha_3 \cos^2\Phi)\dot{\Phi} a_y$$
$$f_y = -\sigma_{yx} a_x = (-\alpha_2 \cos^2\Phi + \alpha_3 \sin^2\Phi)\dot{\Phi} a_x \quad (23)$$

There are no forces on the base and the top plate. The total torque is, therefore,

$$M_z = (\alpha_3 - \alpha_2)\dot{\Phi} V \quad (24)$$

which is in agreement with the torque exerted on the director (Eq. 20). In principle, Eq. (23) allows determination of the Leslie coefficients α_2 and α_3 separately. α_2 corresponds to the force on the surface if the director is perpendicular to it, α_3 corresponds to the force for the parallel case.

The magnitude of the torque does not depend on the shape of the container. It is only a function of the volume of the liquid crystal, that is the number of rotating molecules. The torque on a body with volume V suspended in a liquid crystal has the same magnitude but different sign.

The rotation of the director can be achieved by application of rotating magnetic or electric fields [5, 28]. For larger samples magnetic fields can be applied with less experimental problems. In the following we will therefore consider a liquid crystal in a rotating magnetic field with constant field strength and angular velocity. The results are identical for a rotating sample in a fixed

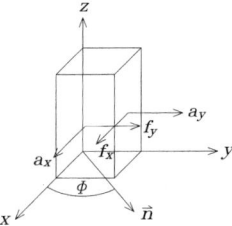

Figure 3. Surface forces on a vessel due to a rotating director.

field if the calculation is carried out with respect to the rotating sample.

The torque exerted by the magnetic field on the director amounts to

$$M = \frac{1}{2}\mu_0 \chi_a H^2 V \sin 2(\Psi - \Phi) \quad (25)$$

where

$$\chi_a = \chi_\parallel - \chi_\perp \quad (26)$$

is the anisotropy of the magnetic susceptibility which is assumed to be positive and H is the magnetic field strength. Under stationary conditions and small angular velocities of the field rotation the director follows the magnetic field direction with a constant phase lag $\Psi - \Phi$, see Fig. 4. The viscous and the magnetic torque become equal

$$\gamma_1 \dot{\Phi} = \frac{1}{2}\mu_0 \chi_a H^2 \sin 2(\Psi - \Phi) \quad (27)$$

Increasing the angular velocity of the rotating field increases the phase lag between field and director until the maximum magnetic torque is exerted at a phase lag of $\pi/4$. The corresponding angular velocity is the critical angular velocity $\dot{\Phi}_c$

$$\dot{\Phi}_c = \frac{\mu_0 \chi_a H^2}{2\gamma_1} \quad (28)$$

A further increase of the angular velocity leads to a non-stationary behaviour. The solutions of the corresponding differential equation

$$\dot{\Phi} = \dot{\Phi}_c \sin 2(\dot{\Psi} t - \Phi) \quad (29)$$

are known [29]. Their characteristic feature is a periodicity in time. Because of disturbances by the surrounding surfaces the predicted oscillations are usually not observable and an inhomogeneous director rotation results [30].

Measuring the torque on a sample of a nematic liquid crystal in a magnetic field rotating with an angular velocity smaller than the critical one represents a relatively simple method for the determination of the rotational viscosity coefficient. Below the critical angular velocity Eq. (24) is valid with $\dot{\Phi} = \dot{\Psi}$. Neither the phase lag $\Psi - \Phi$ nor the anisotropy of the magnetic susceptibility have to be known. This method will be thoroughly discussed in Chap. III, Sect. 2.6 of Vol. 2A of this Handbook.

8.1.3 Flow Alignment

If a nematic liquid crystal with a fixed director orientation is sheared a torque according to Eq. (17) is exerted on the director

$$\boldsymbol{\Gamma} = \boldsymbol{n} \times \left(\frac{1}{2}\gamma_1 (\text{curl}\,\boldsymbol{v}) \times \boldsymbol{n} - \gamma_2 \mathbf{V}\boldsymbol{n} \right) \quad (30)$$

If the direction of flow and shear gradient are the same as in Fig. 1, the torque on the director becomes

$$\Gamma = v_{x,y} (\alpha_3 \sin\Theta \cos\Theta \cos\Phi, \\ - \alpha_2 \sin\Theta \cos\Theta \sin\Phi, \\ \sin^2\Theta (\alpha_2 \sin^2\Phi - \alpha_3 \cos^2\Phi)) \quad (31)$$

The following discussion is simpler in a coordinate system which refers to the director (see Fig. 5). The Θ-axis is parallel to $z \times n$ and the Φ-axis is parallel to the $n \times \Theta$ vector.

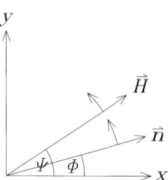

Figure 4. Director alignment in a rotating magnetic field.

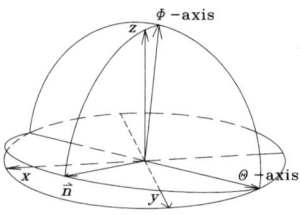

Figure 5. System of coordinates for the director orientation.

The torques with respect to this system of coordinates are

$$\Gamma_\Theta = -v_{x,y} (\alpha_2 + \alpha_3) \sin\Theta \cos\Theta \sin\Phi \cos\Phi \qquad (32)$$

$$\Gamma_\Phi = v_{x,y} \sin\Theta (\alpha_2 \sin^2\Phi - \alpha_3 \cos^2\Phi) \qquad (33)$$

If the director is free to rotate there will be either a rotation to a stable orientation (flow alignment, see Fig. 6) or a continuous rotation (tumbling) under the influence of the shear gradient. Which case is observed depends on the signs of α_2 and α_3. Because of thermodynamical arguments (see Sect. 8.1.9) the rotational viscosity coefficient must be positive.

$$\gamma_1 = \alpha_3 - \alpha_2 > 0 \qquad (34)$$

That is, $\alpha_2 > 0$ and $\alpha_3 < 0$ is not allowed. All nematic liquid crystals investigated up to now exhibit negative α_2 values. Therefore we will discuss in the following only different signs for α_3.

Some nematic liquid crystals show positive α_3 values in the neighbourhood of a nematic/smectic transition. For positive α_3 the sign of the torque component Γ_Φ does not depend on Φ. This leads to a continuous rotation if the director is orientated in the shear plane ($\Theta = 90°$). The sign of the torque component Γ_Θ depends on Φ, that is, the director is stabilized in the shear plane for two quarters of a revolution and destabilized for the other two quarters. Because of the additional influence of surface alignment and elastic torques the real movement of the director is difficult to predict [31].

Negative α_2 and α_3 values are the most frequent combination for nematic liquid crystals. Under these circumstances no torque is exerted on the director in three orientations

(a) $\Theta = 0°$
(b) $\Theta = 90°$, $\Phi = \Phi_0$
(c) $\Theta = 90°$, $\Phi = -\Phi_0$

where

$$\tan\Phi_0 = +\sqrt{\alpha_3/\alpha_2} \qquad (35)$$

Equations (32) and (33) show that only the orientation (b) is stable against deviations from this orientation. Experiments have shown that $|\alpha_2| \gg |\alpha_3|$ is usually valid. Microscopic theories [32, 33] in which the molecules are assumed to be ellipsoids of revolution predict

$$\frac{\alpha_3}{\alpha_2} \approx \left(\frac{b}{a}\right)^2 \qquad (36)$$

where a and b are the length of the ellipsoid parallel and perpendicular to the symmetry axis and a perfect parallel alignment of the molecules ($S = 1$) is assumed. Thus, the flow alignment angle is usually small. For a flow aligned liquid crystal the viscosity is given by

$$\begin{aligned}\eta_0 &= \eta_1 \sin^2\Phi_0 + \eta_2 \cos^2\Phi_0 \\ &= \eta_2 + (\eta_1 - \eta_2) \sin^2\Phi_0 \\ &= \eta_2 - (\alpha_2 + \alpha_3) \sin^2\Phi_0\end{aligned} \qquad (37)$$

neglecting a small term with the coefficient α_1. Inserting Eq. (35) yields

$$\eta_0 = \eta_2 - \alpha_3 \qquad (38)$$

As α_3 is negative and usually small as compared with η_2, the viscosity under flow alignment is somewhat larger than η_2. Mostly the difference can be neglected.

Equation (38) is only valid for a uniform director orientation. In a capillary with circular cross section there are regions along the surface and the capillary axis where the orientation of the director disagrees with the

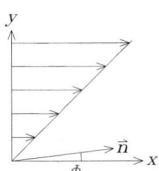

Figure 6. Flow alignment.

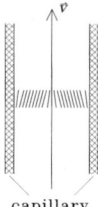

Figure 7. Flow alignment in a capillary.

flow alignment angle (see Fig. 7 and Sec. 8.1.4). By choosing proper values for radius and flow velocity and by a surface treatment the influence of these regions can be minimized.

Theoretical studies on the flow alignment can be found in the papers of Manneville [34], Carlsson [35], and Zuniga and Leslie [31].

8.1.4 Viscous Flow under the Influence of Elastic Torques

Except for the short discussion at the end of the last section, a uniform director alignment has been assumed up to now. Surface alignment and inhomogeneous fields can lead to an inhomogeneous alignment and the occurrence of elastic torques. For a complete description of hydrodynamics of nematic liquid crystals these elastic torques have to be included.

The equilibrium condition for the elastic torques corresponds to the minimum of the elastic energy under the constraint of constant director length. It is convenient to introduce a molecular field $\boldsymbol{h}$ [36]

$$\boldsymbol{h} = \boldsymbol{h}_S + \boldsymbol{h}_T + \boldsymbol{h}_B \tag{39}$$

where the three terms describe splay, twist and bend distortion of the liquid crystal.

$$\boldsymbol{h}_S = k_{11} \operatorname{grad} \operatorname{div} \boldsymbol{n};$$
$$\boldsymbol{h}_T = -k_{22}\{(\boldsymbol{n}\operatorname{curl}\boldsymbol{n})\operatorname{curl}\boldsymbol{n} + \operatorname{curl}[(\boldsymbol{n}\operatorname{curl}\boldsymbol{n})\boldsymbol{n}]\}$$
$$\boldsymbol{h}_B = k_{33}\{(\boldsymbol{n}\times\operatorname{curl}\boldsymbol{n})\times\operatorname{curl}\boldsymbol{n}$$
$$\quad + \operatorname{curl}[\boldsymbol{n}\times(\boldsymbol{n}\times\operatorname{curl}\boldsymbol{n})]\} \tag{40}$$

In hydrostatic equilibrium without electric and magnetic fields the molecular field is parallel to the director and the elastic torque

$$\boldsymbol{\Gamma}^e = \boldsymbol{n}\times\boldsymbol{h} \tag{41}$$

vanishes.

As the values of the three elastic constants are comparable and the expression for the molecular field is rather complicated, the one constant approximation

$$k_{11} = k_{22} = k_{33} = k \tag{42}$$

is often preferred and the molecular field becomes

$$\boldsymbol{h} = k\nabla^2\boldsymbol{n} \tag{43}$$

Elastic terms have to be taken into account in the equations for the balance of linear and angular momentum. If the body forces are assumed to be conservative their scalar potential can be added to the pressure in the equation for the linear momentum. Nevertheless, as director rotation and shear stress are coupled both equations have to be combined for a solution of a given problem.

At low shear rates viscous torques on the director can be neglected. Then the equations for linear and angular momentum are uncoupled and the orientation of the director in the bulk can be calculated for any given surface orientation by solving the equation for angular momentum. Using the obtained director profile the equation for linear momentum can then be integrated. The apparent viscosity will be a function of the choosen surface orientation and the elastic constants.

At high shear rates the director orientation in the bulk is dominated by flow alignment. The influence of elastic torques is restricted to small boundary layers at the surfaces and regions where the velocity gradient changes sign. Therefore the apparent viscosity is close to the value for flow alignment and does not depend on the surface orientation or elastic constants.

At medium shear rates director orientation and velocity profile can only be obtained by numerical calculations. In the following the flow between parallel plates with a surface alignment parallel to the pressure gradient is calculated [37]. The orientation of the capillary is the same as in Fig. 10. The equation for linear momentum is:

$$v_{x,y} = p_{,x} \frac{y}{g(\Phi)} \tag{44}$$

where $g(\Phi)$ is the viscosity

$$g(\Phi) = \eta_1 \sin^2\Phi + \eta_2 \cos^2\Phi \tag{45}$$

The small α_1-term has been neglected. The equation for angular momentum is:

$$k\Phi'' + v_{x,y}(g(\Phi_0) - g(\Phi)) = 0 \tag{46}$$

or with the aid of Eq. (44)

$$k\Phi'' + p_{,x}\left(\frac{g(\Phi_0)}{g(\Phi)} - 1\right) = 0 \tag{47}$$

where Φ_0 is the flow alignment angle. This equation is solved numerically. The alignment at the surface is assumed to be $\Phi=0$. Figure 8 shows Φ as a function of y/T for $(\eta_1-\eta_2)/\eta_2=8$, $\Phi_0=0.1$, and different values of the non-dimensional parameter

$$D = \frac{T^3 p_{,x}}{k} \tag{48}$$

where $2T$ is the distance of the plates. The different slopes of the curves for $y=0$ and T are due to the different velocity gradients in these areas.

Integration of the velocity with the aid of Eqs. (44) and (46) gives the apparent viscosity

$$\eta = \frac{g(\Phi_0)}{1 - \frac{3k\Phi'(T)}{T^2 p_{,x}}} \tag{49}$$

$\Phi'(T)$ (that is, Φ' at the plates) is taken from the numerical solution. Figure 9 shows the viscosity ratio η_2/η as a function of $1/D = k/T^3 p_{,x}$ for $(\eta_1-\eta_2)/\eta_2=8$ and different flow alignment angles. $1/\eta$ gives an asymptotically linear dependence on $1/D$ for large D-values.

Some general predictions can be made with the aid of the scaling properties [19] of the Leslie–Ericksen equations. Neglecting the molecular inertia, the substitution

$$r^* = h\,r; \quad t^* = h^2\,t; \quad p^* = p/h^2;$$
$$\sigma^*_{ij} = \sigma_{ij}/h^2 \tag{50}$$

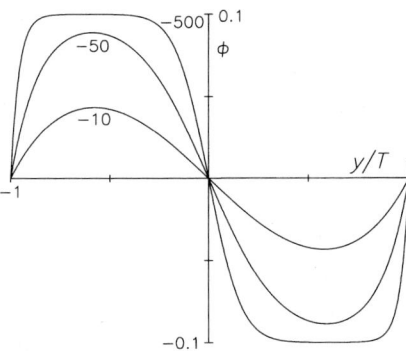

Figure 8. Alignment angle Φ as a function of y/T for different values of the non-dimensional parameter D.

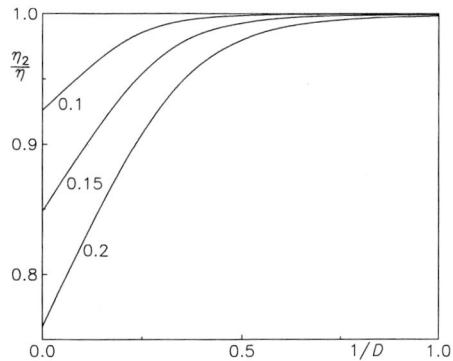

Figure 9. Viscosity ratio η_2/η as a function of $1/D = k/T^3 p_{,x}$ for different flow alignment angles.

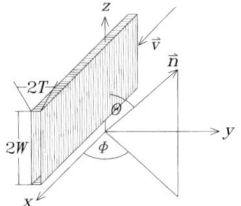

Figure 10. Transverse pressure in a capillary with rectangular cross section.

leaves the form of the differential equations unchanged. The scaling parameter h can be the gap width. Currie [37] has shown that the apparent viscosity is mainly a universal function of a non-dimensional parameter depending on the specific geometry of the flow and a parameter which depends on the surface orientation and the flow alignment angle.

Flow phenomena in the presence of elastic torques have been studied in the Couette [38, 39] and the Poiseuille geometry [40].

8.1.5 Transverse Pressure

Due to the tensor properties of the viscosity a nematic liquid crystal will generally not flow along the pressure gradient. If flow direction and director are neither parallel nor perpendicular a pressure gradient is observed transverse to the flow direction. The skew director orientation can be accomplished by application of a magnetic field to the liquid crystal flowing in a capillary. As an example the flow in a long capillary with rectangular cross section will be discussed (see Fig. 10). The width $2W$ of the capillary is assumed to be considerably larger than its thickness $2T$. Therefore the velocity gradient across the capillary width can be neglected.

The largest transverse pressure difference is observed for a director orientation in the xz-plane ($\Phi=0$). The transverse pressure can be derived from the equation for the conservation of linear momentum

$$\rho \dot{v}_i = (-p\delta_{ij} + \sigma_{ij})_{,j} \tag{51}$$

Stationary flow and $i=z$ gives

$$-p_{,z} + \sigma_{zx,x} + \sigma_{zy,y} + \sigma_{zz,z} = 0 \tag{52}$$

The second term vanishes because of the translational symmetry in flow direction and the last term vanishes for an infinite width of the capillary. The remaining stress tensor element amounts to

$$\sigma_{zy} = (\eta_2 - \eta_3)\sin\Theta\cos\Theta\, v_{x,y} \tag{53}$$

and therefore

$$p_{,z} = (\eta_2 - \eta_3)\sin\Theta\cos\Theta\, v_{x,yy} \tag{54}$$

The x-component of Eq. (51) becomes

$$-p_{,z} + \sigma_{xx,x} + \sigma_{xy,y} + \sigma_{xz,z} = 0 \tag{55}$$

For the case studied this leads to

$$p_{,x} = (\eta_2 \sin^2\Theta - \eta_3 \cos^2\Theta)\, v_{x,yy} \tag{56}$$

leading to the final result

$$\frac{p_{,z}}{p_{,x}} = \frac{(\eta_2 - \eta_3)\sin\Theta\cos\Theta}{\eta_2\sin^2\Theta + \eta_3\cos^2\Theta} \tag{57}$$

The transverse pressure gradient passes through a maximum at approximately $\Theta=45°$. A transverse pressure for this case and an angle dependence according to Eq. (57) has been experimentally confirmed [41]. In principle this experiment can be used for the determination of viscosity coefficient ratios. Because of experimental difficulties it should only be used to demonstrate the tensor property of the viscosity of nematic liquid crystals.

If the director is orientated in the shear plane ($\Theta=90°$) a pressure gradient is also generated along the velocity gradient. The conservation of linear momentum gives for $i=y$

$$-p_{,y} + \sigma_{yx,x} + \sigma_{yy,y} + \sigma_{yz,z} = 0 \tag{58}$$

The second term vanishes for symmetry reasons. The last term is the symmetric counterpart to the third term in Eq. (52), but vanishes under our assumptions (infinite width). Neglecting a small term with the coefficient α_1 gives

$$p_{,y} = \alpha_6 \sin\Phi \cos\Phi \, v_{x,yy} \tag{59}$$

and with Eq. (55)

$$p_{,x} = (\eta_1 \sin^2\Phi + \eta_2 \cos^2\Phi) \, v_{x,yy} \tag{60}$$

the final result becomes

$$\frac{p_{,y}}{p_{,x}} = \frac{\alpha_6 \sin\Phi \cos\Phi}{\eta_1 \sin^2\Phi + \eta_2 \cos^2\Phi} \tag{61}$$

As the thickness of the capillary is smaller than its width the pressure difference across the thickness is considerably smaller than the difference across the width.

8.1.6 Backflow

An externally applied torque on the director can only be transmitted to the surfaces of a vessel without shear, if the director rotation is homogeneous throughout the sample as assumed in Sect. 8.1.2 for the rotational viscosity. Otherwise this transmission occurs partially by shear stresses. The resulting shear flow is called backflow. As there is usually a fixed director orientation at the surfaces of the sample container, a director rotation in the bulk of the sample by application of a roating magnetic field leads to an inhomogeneous rotation of the director and to a backflow [42].

We study a nematic layer (Fig. 11) confined between two parallel plates of infinite dimension. The director is assumed to lie in the x,y plane and the pressure is constant. The inertial terms in Eq. (51) can be neglected under usual conditions and the x component gives

$$\sigma_{xx,x} + \sigma_{xy,y} + \sigma_{xz,z} = 0 \tag{62}$$

The first and third term in this equation vanish due to the translational symmetry. Neglecting a term with the usually small coefficient α_1 gives

$$\frac{\partial}{\partial y}(\dot{\Phi}(\alpha_3 \cos^2\Phi - \alpha_2 \sin^2\Phi)$$
$$+ v_{x,y}(\eta_1 \sin^2\Phi + \eta_2 \cos^2\Phi)) = 0 \tag{63}$$

and integration with respect to y leads to

$$\dot{\Phi}(\alpha_3 \cos^2\Phi - \alpha_2 \sin^2\Phi)$$
$$+ v_{x,y}(\eta_1 \sin^2\Phi + \eta_2 \cos^2\Phi) = \text{const.} \tag{64}$$

The integration constant may still depend on time and can be determined from the boundary condition at the solid surface or in the middle of the cell. Equation (63) shows that every inhomogeneous director rotation is coupled with a velocity gradient.

We assume a homeotropic orientation of the director at the solid surfaces and apply a magnetic field in the x,y plane with an orientation nearly parallel to the y axis. The magnetic anisotropy is assumed to be positive and the strength of the magnetic field to be small. Then the deformation will be small and one gets from Eq. (64)

$$-\alpha_2 \dot{\Phi} + \eta_1 v_{x,y} = -\alpha_2 \dot{\Phi}_{\max} \tag{65}$$

For a common liquid crystal with negative α_2 and positive η_1 a positive angular velocity $\dot{\Phi}$ of the director will be coupled with positive shear gradients at the solid surfaces as the angular velocity vanishes there. This leads to a positive rotation of the bulk and an increase of the angular velocity of the director with respect to a fixed system of coordinates. This effect is usually described by introduction of an effective rotational viscosity coefficient γ_1^* which is smaller than γ_1.

The exact solution for this and other cases [42, 43] shows that the effective rotation-

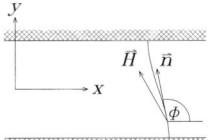

Figure 11. Director orientation under the influence of a magnetic field.

al viscosity coefficient can be as small as 25% of γ_1. Backflow effects are not observed for a planar orientation of the director and a rotation of the magnetic field in the surface plane (twist geometry).

8.1.7 Discotic Liquid Crystals

Nematic discotic liquid crystals consist of disc-like molecules with a preferred parallel orientation of their short molecular (symmetry) axes. There is no translational order. The symmetry of the liquid crystal agrees with that of usual calamitic nematic liquid crystals consisting of rod-like molecules. Accordingly, the stress tensor for calamitic and discotic nematic liquid crystals are identical. Nevertheless, there are some differences in the flow phenomena due to differences in flow alignment.

The flow alignment angle depends on the ratio of the axes of the molecules according to Eq. (36)

$$\tan^2 \Phi_0 \approx \left(\frac{b}{a}\right)^2 \quad (66)$$

The flow alignment angle for discotics should therefore be approximately 90°. A stability analysis shows that the angle above 90° is the stable one [44]. Thus for both types of nematic liquid crystals the configuration with the large dimension nearly parallel to the flow direction is the stable alignment. Figure 12 demonstrates this phenomenon.

8.1.8 Influence of Temperature and Order Parameter on the Viscosity Coefficients

The theory of Leslie and Ericksen is a macroscopic theory. Predictions for the temperature dependence of the viscosity coefficients can only be derived from microscopic theories. Because of several simplifications whose validity is not obvious the results of these microscopic theories are in some cases contradictory and at best qualitatively correct. A general feature is an exponential temperature dependence as known for isotropic liquids with comparable activation energies for all viscosity coefficients far away from the clearing point. This temperature dependence is superposed by an individual dependence on the order parameter whose influence dominates in the neighbourhood of the clearing point. Up to first order all authors predict a linear dependence of the Leslie coefficients α_2, α_3, α_5 as well as α_6 and a quadratic dependence of α_1 [12, 33, 45, 46]. The viscosity coefficient α_4 should be independent of the degree of order. The dependences of the shear viscosities on the order parameter are more complicated. Different results were obtained for the rotational viscosity. However, there are strong arguments that the rotational viscosity vanishes at the clearing point with S^2.

According to most theories the dependence on the order parameter and the gen-

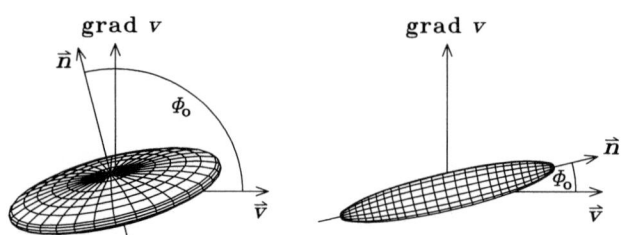

Figure 12. Flow alignment of nematic discotic and a nematic calamitic liquid crystals. The director alignment is shown schematically by the alignment of one molecule.

eral temperature dependence can be separated

$$\alpha_i = f_i[S(T)] \cdot g_i(T) \quad (67)$$

where $g_i(T)$ is related to a relaxation process in the nematic liquid crystal. If the activation energies for the various relaxation processes are equal the ratio between two Leslie or shear viscosity coefficient should be constant far away from the clearing point which has been observed experimentally.

Kuzuu and Doi [45] obtained the following expressions for the Leslie coefficients

$$\alpha_1 = -2f \frac{p^2-1}{p^2+1} S_4 \quad (68)$$

$$\alpha_2 = -f\left(1+\frac{1}{\lambda}\right)S_2; \quad \alpha_3 = -f\left(1-\frac{1}{\lambda}\right)S_2 \quad (69)$$

$$\alpha_4 = f \frac{2}{35} \frac{p^2-1}{p^2+1}(7 - 5S_2 - 2S_4) \quad (70)$$

$$\alpha_5 = f\left[\frac{1}{7}\frac{p^2-1}{p^2+1}(3S_2 + 4S_4) + S_2\right];$$

$$\alpha_6 = f\left[\frac{1}{7}\frac{p^2-1}{p^2+1}(3S_2 + 4S_4) - S_2\right] \quad (71)$$

where the common factor f is

$$f = \frac{ckT}{2D_r}\frac{p^2-1}{p^2+1} \quad (72)$$

c is the number density of molecules, $p=a/b$ the aspect ratio and D_r is the effective rotational diffusion coefficient. The parameter λ is connected to the flow alignment angle

$$\cos 2\Phi_0 = \frac{1}{\lambda} \quad (73)$$

and can be calculated from the theory. The expression for λ contains the order parameter S_2. S_2 and S_4 are the averages over the Legendre polynomials of second and forth order of the equilibrium distribution function. S_2 corresponds to the usual order parameter. As S_4 is usually not known Ehrentraut and Hess [33] suggest the ansatz

$$S_4 = S_2(1 - (1-S_2)^\nu) \quad (74)$$

with $\nu=0.6$.

This and other mesoscopic theories assume that the molecules consist of rigid rods or ellipsoids, that a special form of the interaction potential is valid and that there are no pretransitional phenomena due to a smectic phase. As real liquid crystal molecules do not show such properties these theories only allow a very rough description of the experimental observations. A comparison between some theories has been presented by Kröger and Sellers [47].

8.1.9 Concluding Remarks

The six Leslie coefficients can not be measured directly. They can only be determined with the aid of several experimental methods which ususally lead to combinations of these coefficients. Taking into account the Parodi equation, the six coefficients can be obtained from five linear independent viscosity coefficients. Thus, the four viscosity coefficients η_1, η_2, η_3, and η_{12} and the rotational viscosity coefficient γ_1 give

$$\alpha_1 = \eta_{12}$$
$$\alpha_2 = \frac{1}{2}(\eta_2 - \eta_1 - \gamma_1)$$
$$\alpha_3 = \frac{1}{2}(\eta_2 - \eta_1 + \gamma_1)$$
$$\alpha_4 = 2\eta_3$$
$$\alpha_5 = \frac{1}{2}(3\eta_1 + \eta_2 - 4\eta_3 - \gamma_1)$$
$$\alpha_6 = \frac{1}{2}(\eta_1 + 3\eta_2 - 4\eta_3 - \gamma_1) \quad (75)$$

Furthermore, Eqs (10), (18) and (35) lead to

$$(\eta_1 - \eta_2)\cos 2\Phi_0 = \gamma_1 \quad (76)$$

which can be used to introduce the flow alignment angle Φ_0 into Eq. (75) instead of η_1, η_2 or γ_1. Owing to thermodynamical arguments [8] the values of the Leslie coefficients are subject to the conditions

$\alpha_4 > 0;\quad 2\alpha_1 + 3\alpha_4 + 2\alpha_5 + 2\alpha_6 > 0;$
$\gamma_1 > 0;\quad 2\alpha_4 + \alpha_5 + \alpha_6 > 0;\quad\quad (77)$
$4\gamma_1(2\alpha_4 + \alpha_5 + \alpha_6) > (\alpha_2 + \alpha_3 + \alpha_6 - \alpha_5)^2$

or

$\eta_3 > 0;\quad \eta_{12} + 2(\eta_1 + \eta_2) > \eta_3 + \gamma_1$
$\gamma_1 > 0;\quad \gamma_1\{2(\eta_1 + \eta_2) - \gamma_1\} > (\eta_2 - \eta_1)^2$
(78)

The fourth condition in Eq. (77) has been omitted in Eq. (78) as it follows from the third and the fifth. Whereas η_1, η_2, η_3, and γ_1 must be positive η_{12} can have either sign.

8.2 Cholesteric Liquid Crystals

Under isothermal conditions the constitutive equations for the description of flow phenomena in nematic and cholesteric liquid crystals are identical [48]. Nevertheless, a series of novel effects are caused by the helical structure of cholesteric phases. They arise firstly because of the inhomogeneous director orientation in the undistorted helix and secondly because of the winding or unwinding of the helix due to viscous torques.

Regarding the orientation of the helical axis with respect to velocity and velocity gradient there are three main cases (see Fig. 13) which are labelled according to the indices of the shear viscosity coefficients of nematic liquid crystals. The helical axis $\boldsymbol{h}$ is assumed to be parallel to the z-axis for the following discussion. Φ describes the angle between local director $\boldsymbol{n}$ and x-axis (see Fig. 14).

8.2.1 Helical Axis Parallel to the Shear Gradient (Case I)

The cholesteric liquid crystal is sheared between two parallel walls at $z=\pm T$. Leslie [49] and Kini [50] have studied this case with different boundary conditions. Leslie assumes that the pitch of the helix remains constant at the walls whereas Kini assumes a fixed orientation of the director at the wall with an orientation which corresponds to that of the undistorted helix. Both boundary conditions are difficult to realize in an experiment. In spite of the different orientations at the walls the results of the calculations are qualitatively the same. In the bulk the director orientation is determined by elastic and viscous torques. Transverse flow in z-direction is excluded.

The differential equations can analytically be solved for small helix distortions. If the pitch P_0 is comparable to the cell thickness $2T$ the volume velocity and the appar-

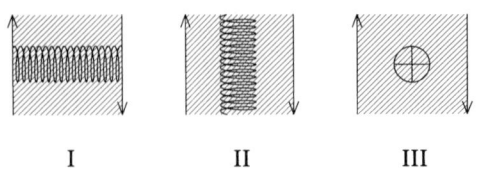

Figure 13. Orientation of the helix with respect to the shear gradient. The pair of arrows symbolizes the shear flow.

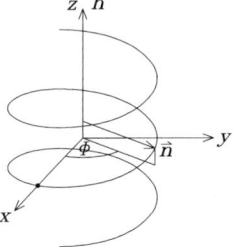

Figure 14. Orientation of the helical axis and the director with respect to the coordinate system.

ent viscosity respectively are oscillating functions of the pitch for constant cell thickness. The reason for this effect is the oscillating mean value of the director orientation. With decreasing pitch these oscillations die out and the apparent viscosity approaches the limiting value

$$\eta(P_0 = 0) = 2\eta_2\eta_3/(\eta_2 + \eta_3) \tag{79}$$

For larger shear rates the helix will be distorted. The corresponding differential equation can only numerically be solved. The calculations of Leslie and Kini show, that for a given pitch and cell thickness the apparent viscosity starts with a plateau for small shear rates and decreases with increasing shear rates to the viscosity which is observed for flow alignment. The necessary shear rates for this transition depend on the pitch. The smaller the pitch the larger are the necessary shear rates as the resistance against a distortion increases with decreasing pitch.

The influence of static shear deformations has been studied in several papers [51, 52].

8.2.2 Helical Axis Parallel to the Flow Direction (Case II)

The flow of a cholesteric liquid crystal through a narrow capillary with circular cross section according to Fig. 15 is the most simple example for this case. The orientation at the capillary wall should be consistent with the helical structure and the pressure gradient along the capillary axis should be small. Helfrich [53] assumes that the helical structure remains undistorted and a permanent rotation of the director in the flowing liquid crystal results. The director rotation is coupled with a specific energy dissipation of

$$\frac{\dot{E}}{V} = \gamma_1 \dot{\Phi}^2 = \gamma_1 \left(\frac{2\pi v}{P}\right)^2 \tag{80}$$

For small pitches and large velocities this term exceeds the energy dissipation by Poiseuille flow and the resulting flow velocity is nearly constant across the cross section (plug flow). The dissipated energy is taken from the pressure gradient along the capillary. This leads to

$$\dot{V} = \frac{R^2 P^2}{4\pi\gamma_1} p_{,z}; \quad \eta_{\text{app}} = \frac{\pi^2}{2}\frac{\gamma_1 R^2}{P^2} \tag{81}$$

In contrast to Poiseuille flow the volume flow depends only on the square of the capillary radius. For $P \ll R$ the apparent viscosity exceeds γ_1 by far. According to Helfrich this type of flow is called permeation. Kini et al. [54] have studied the same problem using the equations of continuum theory. They could show that the velocity profile is almost flat except a small layer near the capillary wall. Neglecting the boundary layer the equations presented above were obtained.

8.2.3 Helical Axis Normal to v and grad v (Case III)

The cholesteric liquid crystal is sheared between two parallel walls. It is assumed that the walls do not influence the director orientation (weak anchoring) and that the flow velocity is parallel to the movement of

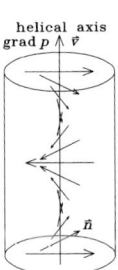

Figure 15. Flow of a cholesteric liquid crystal parallel to the helical axis.

the walls. Under these assumptions the continuum theory gives the differential equation

$$\Phi_{,\tau} = \Phi_{,\xi\xi} + k\left(1 - \frac{\cos^2 \Phi}{\cos^2 \Phi_0}\right) \quad (82)$$

where

$$\tau = \frac{k_{22}}{\gamma_1 P_0^2} t; \quad \xi = \frac{z}{P_0}; \quad k = \frac{\alpha_2 v_{x,y} P_0^2}{k_{22}} \quad (83)$$

are a non-dimensional time τ, position ξ, and a non-dimensional parameter k of the system. k_{22} is the twist elastic constant and P_0 is the pitch of the undistorted helix.

For small k values the helical structure remains undistorted and its pitch constant. Then Eq. (82) allows a simple calculation of the director rotation:

$$\Phi_{,\tau} = k\left(1 - \frac{\cos^2 \Phi}{\cos^2 \Phi_0}\right) = k\left(1 - \frac{1}{2\cos^2 \Phi_0}\right) \quad (84)$$

which gives

$$\Phi_{,t} = -\frac{v_{x,y}}{2} \quad (85)$$

Thus a permanent director rotation results.

The apparent viscosity amounts to

$$\eta_{app} = \frac{1}{2}(\eta_1 + \eta_2) \quad (86)$$

which agrees with the value for a fixed helix, as there is no rotation of the director with respect to the fluid in both cases.

For larger k values the helix becomes distorted and its pitch increases with k. Finally, the helix unwinds and a flow aligned structure appears.

8.2.4 Torque Generation Under Flow

In the case of geometry II the director of the cholesteric liquid crystal rotates during the flow through a round capillary. According to the section on the rotational viscosity this director rotation is coupled with a torque on the liquid crystal. The torque generation must be compensated by elastic deformations or additional circular flow.

For a flow of a cholesteric liquid crystal through the gap between two coaxial cylinders the torque is partly transferred to the walls of the cylinders. Fischer et al. [55] used a fixed outer cylinder and determined the torque on the inner cylinder.

Equation (20) gives for the torque on the walls

$$M = \gamma_1 \dot{\Phi} V = \pm \gamma_1 \frac{2\pi}{P} v V \quad (87)$$

for a sample with volume V.

The experiments could only demonstrate the existence of the effect, as the flow in the experiment was dominated by shear flow and only a minor part (some %) resulted from permeation.

8.3 Biaxial Nematic Liquid Crystals

Shortly after the first observation of a biaxial nematic phase by Yu and Saupe [56] the hydrodynamic theory of the uniaxial nematic phase was extended to the biaxial case in several papers [57–60]. The following description is similar to that given by Saupe [59] as well as Govers and Vertogen [60], but the notation is different.

A biaxial nematic phase is characterized by two orthogonal directors a and b of length 1. Introduction of a third orthogonal vector c of length 1 simplifies the notation of the following equations. According to Saupe, Govers and Vertogen the viscous part of the stress tensor of an incompressible biaxial nematic phase is the sum of the symmetric part

$$\sigma_{ij}^s = \mu_1(a_i a_j V^{bb} + b_i b_j V^{aa})$$
$$+ \mu_2(b_i b_j V^{cc} + c_i c_j V^{bb})$$
$$+ \mu_3(c_i c_j V^{aa} + a_i a_j V^{cc})$$
$$+ \mu_4(a_i b_j + a_j b_i)V^{ab}$$
$$+ \mu_5(b_i c_j + b_j c_i)V^{bc}$$
$$+ \mu_6(c_i a_j + c_j a_i)V^{ca}$$
$$+ \mu_7(a_i b_j + a_j b_i)A^b$$
$$+ \mu_8(b_i c_j + b_j c_i)B^c$$
$$+ \mu_9(c_i a_j + c_j a_i)C^a \quad (88)$$

and the skew symmetric part

$$\sigma_{ij}^{ss} = \mu_7(a_i b_j - a_j b_i)V^{ab}$$
$$+ \mu_8(b_i c_j - b_j c_i)V^{bc}$$
$$+ \mu_9(c_i a_j - c_j a_i)V^{ca}$$
$$+ \mu_{10}(a_i b_j - a_j b_i)A^b$$
$$+ \mu_{11}(b_i c_j - b_j c_i)B^c$$
$$+ \mu_{12}(c_i a_j - c_j a_i)C^a \quad (89)$$

where

$$V_{ij} = \frac{1}{2}(v_{i,j} + v_{j,i}); \quad W_{ij} = \frac{1}{2}(v_{i,j} - v_{j,i}) \quad (90)$$

are the symmetric and the skew symmetric part of the velocity gradient tensor and

$$A_i = \dot{a}_i - W_{ij}a_j; \quad B_i = \dot{b}_i - W_{ij}b_j;$$
$$C_i = \dot{c}_i - W_{ij}c_j \quad (91)$$

are the director rotations relative to the fluid. Furthermore, the following notation is used

$$X^u = X_p u_p; \quad Y_p^u = Y_{pq}u_q; \quad Y^{uv} = Y_{pq}u_q v_p \quad (92)$$

The viscous torque is obtained according to the procedure in Sect. 8.1.2 on the rotational viscosity.

$$\Gamma_i = \varepsilon_{ijk}\sigma_{kj}^{ss} = 2\varepsilon_{ijk}(\mu_7 a_k b_j V^{ab} + \mu_8 b_k c_j V^{bc} + \mu_9 c_k a_j V^{ca} + \mu_{10} a_k b_j A^b + \mu_{11} b_k c_j B^c + \mu_{12} c_k a_j C^a) \quad (93)$$

Equations (88) and (89) include the case of the stress tensor for the uniaxial nematic phase [59].

8.3.1 Shear Viscosity Coefficients

For a fixed director orientation and a flow according to Fig. 16 the velocity gradients and the director rotations relative to the fluid are

$$V_{xy} = V_{yx} = W_{xy} = -W_{yx} = \frac{1}{2}v_{x,y} \quad (94)$$

$$A_i = -W_{ij}a_j; \quad B_i = -W_{ij}b_j; \quad C_i = -W_{ij}c_j \quad (95)$$

and the viscosity $\eta = \sigma_{xy}/v_{x,y}$ becomes

$$\eta = \frac{1}{2}\{a_x^2 b_y^2(\mu_4 + 2\mu_7 + \mu_{10})$$
$$+ b_x^2 c_y^2(\mu_5 + 2\mu_8 + \mu_{11})$$
$$+ c_x^2 a_y^2(\mu_6 + 2\mu_9 + \mu_{12})$$
$$+ a_y^2 b_x^2(\mu_4 - 2\mu_7 + \mu_{10})$$
$$+ b_y^2 c_x^2(\mu_5 - 2\mu_8 + \mu_{11})$$
$$+ c_y^2 a_x^2(\mu_6 - 2\mu_9 + \mu_{12})$$
$$+ 2a_x a_y b_x b_y(2\mu_1 + \mu_4 - \mu_{10})$$
$$+ 2b_x b_y c_x c_y(2\mu_2 + \mu_5 - \mu_{11})$$
$$+ 2c_x c_y a_x a_y(2\mu_3 + \mu_6 - \mu_{12})\} \quad (96)$$

In principle, the director components can be expressed in terms of Eulerian angles. However, the resulting expression is rather lengthy and the cyclic structure of Eq. (96) is lost.

Director orientations with all directors parallel to the coordinate axes are described by the first six terms, for example, the orientation $a \| x$ and $b \| y$ corresponds to the

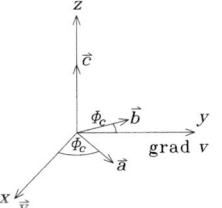

Figure 16. Director orientation in a biaxial nematic phase with respect to the flow velocity and the velocity gradient. The case $c \| z$ is shown whereas the equations in the text correspond to the general case.

viscosity coefficient $\frac{1}{2}(\mu_4 - 2\mu_7 + \mu_{10})$. The last three terms give a maximum contribution if two directors lie under 45° in the xy plane, i.e. they correspond to the α_1 term of uniaxial nematics. Thus, nine coefficient combinations can be determined with the aid of flow experiments.

8.3.2 Rotational Viscosity Coefficients

The last three terms in the torque Eq. (93) describe the torques due to director rotations. The three main cases are rotations around a, around b, and around c. The latter corresponds to $\dot{\Phi}_c \neq 0$ in Fig. 16. The viscous torques for these cases are

$$\Gamma_a = -2\mu_{11}\dot{\Phi}_a; \quad \Gamma_b = -2\mu_{12}\dot{\Phi}_b;$$
$$\Gamma_c = -2\mu_{10}\dot{\Phi}_c \tag{97}$$

where Φ_i describes the rotation around the i axis.

The three rotational and the nine shear viscosity coefficients according to Eq. (96) represent a complete set of coefficients.

8.3.3 Flow Alignment

It is assumed that the c director is parallel to the z axis, e.g. by application of a magnetic field. A shear flow $v_{x,y}$ exhibits a torque on the directors around the z axis. Calculation with Eq. (93) gives

$$\Gamma_c = -2\mu_{10}\dot{\Phi}_c - v_{x,y}(\mu_7 \cos 2\Phi_c + \mu_{10}) \tag{98}$$

$\Gamma_c = 0$ and $\dot{\Phi}_c = 0$ give

$$\cos 2\Phi_c = -\frac{\mu_{10}}{\mu_7} \tag{99}$$

This equation only has solutions for

$$\left|\frac{\mu_{10}}{\mu_7}\right| < 1 \tag{100}$$

Correspondingly, the flow alignment angles around all axes are

$$\cos 2\Phi_a = -\frac{\mu_{11}}{\mu_8}; \quad \cos 2\Phi_b = -\frac{\mu_{12}}{\mu_9};$$
$$\cos 2\Phi_c = -\frac{\mu_{10}}{\mu_7} \tag{101}$$

The conditions for the first two viscosity coefficient ratios correspond to Eq. (100). A discussion of the stability of the various solutions is presented in the paper of Saupe [59]. Brand and Pleiner [61] as well as Leslie [62] discuss the flow alignment without the restriction that one director is perpendicular to the shear plane.

8.4 SmC Phase

The hydrodynamic theory for SmC phases was developed by Schiller [18] and Leslie et al. [19]. Schiller as well as Leslie et al. assume a constant layer thickness, a constant tilt angle and do not include the permeation, that is, the dissipation of energy by the penetration of molecules through the layers. Furthermore, Schiller neglects the distorsion of the layers in contrast to Leslie et al. The Leslie theory is, therefore, somewhat more general. The orientation of the SmC phase is described by two directors. The director a is normal to the layers and c is parallel to the layers in the direction of the tilt. The a director, the preferred direc-

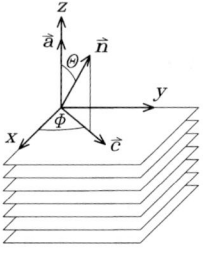

Figure 17. Definition of the directors a and c for an SmC phase.

tion of the long molecular axes, n, and c lie, therefore, in a plane (Fig. 17). According to the theory of Leslie et al. [19] the viscous part of the stress tensor is the sum of the symmetric part

$$\sigma_{ij}^s = \mu_0 V_{ij} + \mu_1 V^{aa} a_i a_j + \mu_2 (V_i^a a_j + V_j^a a_i)$$
$$+ \mu_3 V^{cc} c_i c_j + \mu_4 (V_i^c c_j + V_j^c c_i)$$
$$+ \mu_5 V^{ac} (a_i c_j + a_j c_i) + \lambda_1 (A_i a_j + A_j a_i)$$
$$+ \lambda_2 (C_i c_j + C_j c_i) + \lambda_3 A^c (a_i c_j + a_j c_i)$$
$$+ \kappa_1 (V_i^a c_j + V_j^a c_i + V_i^c a_j + V_j^c a_i)$$
$$+ \kappa_2 (V^{aa}(a_i c_j + a_j c_i) + 2 V^{ac} a_i a_j)$$
$$+ \kappa_3 (V^{cc}(a_i c_j + a_j c_i) + 2 V^{ac} c_i c_j)$$
$$+ \tau_1 (C_i a_j + C_j a_i) + \tau_2 (A_i c_j + A_j c_i)$$
$$+ 2\tau_3 A^c a_i a_j + 2\tau_4 A^c c_i c_j \quad (102)$$

and the skew symmetric part

$$\sigma_{ij}^{ss} = \lambda_1 (V_j^a a_i - V_i^a a_j) + \lambda_2 (V_j^c c_i - V_i^c c_j)$$
$$+ \lambda_3 V^{ac}(a_i c_j - a_j c_i) + \lambda_4 (A_j a_i - A_i a_j)$$
$$+ \lambda_5 (C_j c_i - C_i c_j) + \lambda_6 A^c (a_i c_j - a_j c_i)$$
$$+ \tau_1 (V_j^a c_i - V_i^a c_j) + \tau_2 (V_j^c a_i - V_i^c a_j)$$
$$+ \tau_3 V^{aa}(a_i c_j - a_j c_i)$$
$$+ \tau_4 V^{cc}(a_i c_j - a_j c_i)$$
$$+ \tau_5 (A_j c_i - A_i c_j + C_j a_i - C_i a_j) \quad (103)$$

The notation is the same as described in Sect. 8.3. The stress tensor contains 20 independent coefficients. It fulfills the Onsager relations.

The viscous torque is obtained according to the procedure used for nematic liquid crystals

$$\Gamma_i = \varepsilon_{ijk} \sigma_{kj}^{ss} = -2\varepsilon_{ijk} \{ a_j (\lambda_1 V_k^a + \lambda_3 c_k V^{ac}$$
$$+ \lambda_4 A_k + \lambda_6 c_k A^c + \tau_2 V_k^c$$
$$+ \tau_3 c_k V^{aa} + \tau_4 c_k V^{cc} + \tau_5 C_k)$$
$$+ c_j (\lambda_2 V_k^c + \lambda_5 C_k + \tau_1 V_k^a$$
$$+ \tau_5 A_k) \} \quad (104)$$

8.4.1 Shear Flow with a Fixed Director Orientation

The xy plane of the system of coordinates (Fig. 18) used for the following calculation is assumed to be parallel to the smectic

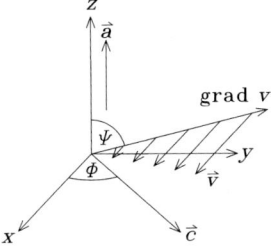

Figure 18. Definition of angles for a flow experiment in the SmC phase.

layers. The a director and the z axis are, therefore, parallel to one another. As a flow perpendicular to the layers is largely suppressed due to the permeation effect, the flow has to be parallel to the xy plane. Furthermore, it can be assumed without loss of generality for this calculation that the flow is parallel to the x axis. The c director lies in the xy plane and its direction is described by the angle Φ. The velocity gradient lies in the yz plane and is described by the angle Ψ.

Under these conditions the effective viscosity calculated from the stress tensor Eqs. (102) and (103) becomes

$$\eta = \left\{ \frac{1}{2}(\mu_0 + \mu_4 - 2\lambda_2 + \lambda_5) \sin^2 \Phi \right.$$
$$+ \frac{1}{2}(\mu_0 + \mu_4 + 2\lambda_2 + \lambda_5) \cos^2 \Phi$$
$$\left. + \mu_3 \sin^2 \Phi \cos^2 \Phi \right\} \sin^2 \Psi$$
$$+ \left\{ \frac{1}{2}(\mu_0 + \mu_2 - 2\lambda_1 + \lambda_4) \sin^2 \Phi \right.$$
$$+ \frac{1}{2}(\mu_0 + \mu_2 + \mu_4 + \mu_5 - 2\lambda_1 + 2\lambda_2$$
$$\left. -2\lambda_3 + \lambda_4 + \lambda_5 + \lambda_6) \cos^2 \Phi \right\} \cos^2 \Psi$$
$$+ \left\{ \kappa_1 - \tau_1 - \tau_2 + \tau_5 + 2(\tau_3 - \tau_4) \cos^2 \Phi \right\}$$
$$\cdot \sin \Phi \sin \Psi \cos \Psi \quad (105)$$

We define, therefore, a new set of viscosity coefficients

$$\eta_1 = \frac{1}{2}(\mu_0 + \mu_4 - 2\lambda_2 + \lambda_5)$$
$$\eta_2 = \frac{1}{2}(\mu_0 + \mu_4 + 2\lambda_2 + \lambda_5)$$
$$\eta_3 = \mu_3$$

$$\eta_4 = \frac{1}{2}(\mu_0 + \mu_2 - 2\lambda_1 + \lambda_4)$$

$$\eta_5 = \frac{1}{2}(\mu_0 + \mu_2 + \mu_4 + \mu_5$$
$$- 2\lambda_1 + 2\lambda_2 - 2\lambda_3 + \lambda_4 + \lambda_5 + \lambda_6)$$

$$\eta_6 = \kappa_1 - \tau_1 - \tau_2 + \tau_5$$

$$\eta_7 = 2(\tau_3 - \tau_4) \tag{106}$$

and get finally

$$\eta = (\eta_1 \sin^2 \Phi + \eta_2 \cos^2 \Phi$$
$$+ \eta_3 \sin^2 \Phi \cos^2 \Phi) \sin^2 \Psi \tag{107}$$
$$+ (\eta_4 \sin^2 \Phi + \eta_5 \cos^2 \Phi) \cos^2 \Psi$$
$$+ (\eta_6 + \eta_7 \cos^2 \Phi) \sin \Phi \sin \Psi \cos \Psi$$

The flow properties of a SmC phase with fixed director orientation and a flow parallel to the layers can, therefore, be described by seven independent viscosity coefficients. The experimental determination of these coefficients should be connected with a series of problems. If the coefficients η_4 or η_5 are determined in a capillary with a rectangular cross section with $T \ll W$ and the layer parallel to one of the plates as in Fig. 20 I. The thickness has to be constant over the whole sample with an accuracy that is not easy achieved [63,64]. There are similar problems in the measurement of the other coefficients. Minor difficulties should occur in a shear experiment with a small lateral movement of one of the plates.

8.4.2 Rotational Viscosity

Equation (104) allows to calculate the viscous torque which is exerted on the c director during a rotation around the a director with a constant tilt angle Θ. It is assumed that the liquid crystal is at rest. The only non-vanishing term in Eq. (104) is that with the coefficient λ_5.

$$\Gamma_z = -2\lambda_5 \varepsilon_{zjk} c_j \dot{C}_k = -2\lambda_5 (c_x \dot{C}_y - c_y \dot{C}_x)$$
$$= -2\lambda_5 \dot{\Phi} \tag{108}$$

The viscous dissipation requires

$$\lambda_5 > 0 \tag{109}$$

It is often interesting to compare the rotational viscosity coefficient γ_1 in the nematic phase and

$$\gamma_1^c = 2\lambda_5 \tag{110}$$

in the SmC phase. Then it has to be taken into account that $\dot{n}$ and $\dot{c}$ are different and that the torque on c is only the projection of the torque on n. A rotational viscosity coefficient suitable for a comparison with the coefficient γ_1 in the nematic phase is, therefore

$$\gamma_1^{c*} = \gamma_1^c / \sin^2 \Phi \tag{111}$$

A magnetic field rotating in the xy plane leads to a rotation of the c director. The calculation is similar to that in the nematic phase. However, it has to be taken into account that the interaction with the magnetic field is different for the two phases. Principally, the magnetic susceptibility of an SmC phase is biaxial with the main susceptibilities parallel to n (χ_1), perpendicular to n in the layer plane (χ_2) and perpendicular to these two directions (χ_3). In a good approximation $\chi_2 = \chi_3$ can be assumed and the magnetic torque on the c director becomes

$$\Gamma = \frac{1}{2} \mu_0 \chi_a H^2 \sin^2 \Theta \sin 2(\gamma - \Phi) \tag{112}$$

where γ is the angle between field direction and x axis and

$$\chi_a = \chi_1 - \chi_2 \tag{113}$$

In the biaxial case $\chi_1 \sin^2 \Theta + \chi_3 \cos^2 \Theta - \chi_2$ has to be used instead of $\chi_a \sin^2 \Theta$. As in the nematic phase there is a critical velocity

$$\dot{\Phi}_c = \frac{\mu_0 \chi_a H^2 \sin^2 \Theta}{4\lambda_5} \tag{114}$$

up to which the director follows the field rotation with a phase lag. λ_5 can be determined from the torque on the sample

$$M_z = 2\lambda_5 \dot{\gamma} V \tag{115}$$

8.4.3 Flow Alignment

The SmC phase is sheared between two parallel plates of infinite dimension. The distance between the plates is so large that the influence of the surface alignment at the plates can be neglected in the bulk. The orientation of the directors, the velocity and the velocity gradient is the same as already discussed for the shear flow experiment (Fig. 18). Furthermore, a rotation of the director c is now allowed. Calculation of the torque according to Eq. (104) and $\Gamma_z = 0$ gives

$$2\lambda_5 \dot{\Phi} + \{(\lambda_5 + \lambda_2 \cos 2\Phi) \sin \Psi \\ + (\tau_5 - \tau_1) \sin \Phi \cos \Psi\} v_{,q} = 0 \tag{116}$$

where $v_{,q}$ is the velocity gradient. For $\psi = 0$ this equation gives

$$2\lambda_5 \dot{\Phi} + (\tau_5 - \tau_1) \sin \Phi \, v_{x,z} = 0 \tag{117}$$

Time independent solutions of this equation are

$$\Phi = 0; \quad \Phi = \pi \tag{118}$$

Assuming $v_{x,z} > 0$, a stable flow alignment is obtained at $\Phi = 0$ for $\tau_5 > \tau_1$ and at $\Phi = \pi$ for $\tau_5 < \tau_1$. For $\psi = 90°$ Eq. (116) gives

$$2\lambda_5 \dot{\Phi} + (\lambda_5 + \lambda_2 \cos 2\Phi) v_{x,y} = 0 \tag{119}$$

λ_5 is always positive (see Eq. 109). For $\lambda_5 > |\lambda_2|$ there is no flow alignment. For $\lambda_5 < |\lambda_2|$ a time independent solution is

$$\cos 2\Phi_0 = -\frac{\lambda_5}{\lambda_2} \tag{120}$$

which gives four solutions for Φ_0. The stable solutions are found by an expansion of Eq. (119) around the flow alignment angle ($\Phi = \Phi_0 + \Delta \Phi$) for small deviations. This leads to

$$\lambda_2 v_{x,y} \sin 2\Phi_0 < 0 \tag{121}$$

Assuming $v_{x,y} > 0$ and $\lambda_2 > 0$ gives a stable flow alignment for $\sin 2\Phi_0 < 0$ and vice versa.

An extensive discussion of the flow alignment in SmC phases can be found in a paper of Carlsson et al. [65].

The viscous properties of a SmC phase with a fixed orientation of the layers can be described by nine independent coefficients. Seven coefficients can be determined with the aid of flow experiments (see Eq. 107). The investigation of the flow alignment including the relaxation time gives the three combinations λ_5, λ_2, and $\tau_5 - \tau_1$. λ_2 can, however, be determined from

$$\eta_2 - \eta_1 = 2\lambda_2 \tag{122}$$

The nine coefficients can, therefore, be determined from the seven viscosity coefficients, the rotational viscosity coefficient $2\lambda_5$ and the relaxation time for the flow alignment for $\psi = 0$. Instead of the two last determinations it is also possible to use two flow alignment angles at different ψ values.

A mesoscopic theory for the SmC phase has been presented by Osipov and Terentjev [66].

8.4.4 SmC* Phase

The viscous part of the stress tensor for the SmC and the ferroelectric chiral smectic (SmC*) phase agree with one another. The flow phenomena with a fixed director orientation discussed in the foregoing section can not be observed due to the inhomogeneous director orientation in the SmC* phase. However, there is a large interest in rotational movements of the director in ferroelectric displays.

In surface stabilized ferroelectric liquid crystal cells the liquid crystal is sandwiched

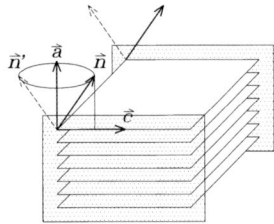

Figure 19. Director alignment in a ferroelectric liquid crystal cell.

between two glass plates which are treated to give the book shelf alignment shown in Fig. 19 with two stable director orientations at the glass plates. Due to the alignment of the director at the glass plates and due to the short distance of the glass plates the helical structure of the SmC* phase is surpressed and the orientation of the c director is nearly homogeneous within the cell. The Leslie theory assumes a rotational movement on the cone during the switching from n to n'. Other trajectories have also been discussed. Carlsson and Žekš [67] found out that the viscosities for a movement of the n director on the cone and perpendicular to it are identical. Experiments [68] have shown that the movement occurs on the cone apart from a small deviation due to the electroclinic effect.

8.5 SmA Phase

The stress tensor for the SmA phase can be obtained from the two director theory for the SmC phase by elimination of all terms containing the c director. The sum of the symmetric and the skew symmetric part becomes

$$\sigma_{ij} = \mu_0 V_{ij} + \mu_1 V_p^{aa} a_i a_j + \mu_2(V_i^a a_j + V_j^a a_i) + \lambda_1(A_i a_j + A_j a_i + V_j^a a_i - V_i^a a_j) + \lambda_4(A_j a_i - A_i a_j) \quad (123)$$

and the viscous torque

$$\Gamma_i = -2\varepsilon_{ijk} a_j (\lambda_1 V_k^a + \lambda_4 A_k) \quad (124)$$

A rearrangement gives

$$\sigma_{ij} = \mu_1 a_i a_j V_{kp} a_k a_p + (\lambda_1 - \lambda_4) a_j A_i + (\lambda_1 + \lambda_4) a_i A_j + \mu_0 V_{ij} + (\mu_2 - \lambda_1) a_j V_{ik} a_k + (\mu_2 + \lambda_1) a_i V_{jk} a_k \quad (125)$$

and

$$\Gamma = -2a \times (\lambda_4 A + \lambda_1 Va) \quad (126)$$

As the symmetry of the SmA and the nematic phase agree, the structure of the stress tensors is identical and the connection between the coefficients is

$$\begin{array}{lll} \alpha_1 \cong \mu_1 & \alpha_4 \cong \mu_0 & \gamma_1 \cong 2\lambda_4 \\ \alpha_2 \cong \lambda_1 - \lambda_4 & \alpha_5 \cong \mu_2 - \lambda_1 & \gamma_2 \cong 2\lambda_1 \\ \alpha_3 \cong \lambda_1 + \lambda_4 & \alpha_6 \cong \mu_2 + \lambda_1 & \end{array} \quad (127)$$

Up to now the influence of the permeation effect has been neglected and the Eqs. (123–126) give only meaningful results if the flow velocity component perpendicular to the layers can be neglected.

In the other case the permeation effect, that is, an energy dissipation for a flow perpendicular to the layers, has to be taken into account. For this purpose the layer normal is assumed to be parallel to the z axis (Fig. 20) and the stress tensor becomes

$$\sigma_{ij} = \mu_1 V_{zz} \delta_{iz} \delta_{jz} + (\lambda_1 - \lambda_4) A_i \delta_{jz} + (\lambda_1 + \lambda_4) A_j \delta_{iz} + \mu_0 V_{ij} + (\mu_2 - \lambda_1) V_{iz} \delta_{jz} + (\mu_2 + \lambda_1) V_{jz} \delta_{iz} \quad (128)$$

According to Helfrich [53], de Gennes [69], and Martin et al. [17] the permeation can be described with the additional term

$$p_{,z} = \frac{1}{\zeta}(\dot{z} - v_z) \quad (129)$$

in the equation of motion, $(v_z - \dot{z})$ is the flow velocity with respect to the layer structure.

In the following the flow of a SmA phase between two infinite plates will be considered. In contrast to the discussion of nematic phases these plates will be assumed to be at rest as moving plates are not always compatible with the layered structure. The three main cases are outlined in Fig. 20.

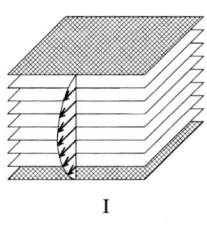

I

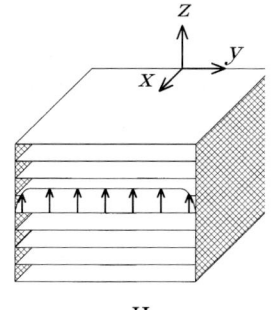

II

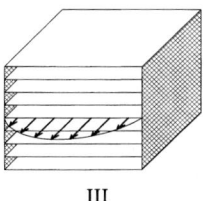

III

Figure 20. Flow of an SmA phase between two parallel plates (hatched). The layer normal is parallel to the z-axis.

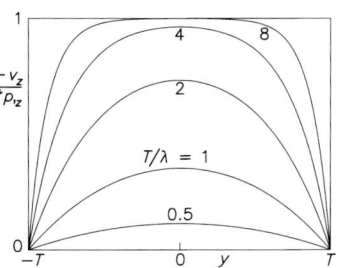

Figure 21. Velocity profile for a flow with permeation (case II of Fig. 20).

For the cases I and III parabolic velocity profiles as known for isotropic liquids result. The viscosity coefficients are

case I: $\eta_1 = \frac{1}{2}(\mu_0 + \mu_2 - 2\lambda_1 + \lambda_4)$

case III: $\eta_3 = \frac{1}{2}\mu_0$ \hfill (130)

For case II the permeation has to be taken into account.

$$-p_{,z} + \eta_2 v_{z,yy} - \frac{1}{\zeta}v_z = 0 \quad (131)$$

with

$$\eta_2 = \frac{1}{2}(\mu_0 + \mu_2 + 2\lambda_1 + \lambda_4) \quad (132)$$

Integration gives the velocity profile

$$v_z = -\zeta p_{,z}\left(1 - \frac{\cosh y/\lambda}{\cosh T/\lambda}\right) \quad (133)$$

where

$$\lambda = \sqrt{\eta_2 \zeta} \quad (134)$$

is the thickness of the shear dominated part and $2T$ is the distance of the plates. Thus $T/\lambda \gg 1$ gives a plug flow and $T/\lambda \ll 1$ gives a shear dominated velocity profile (Fig. 21).

The viscosity coefficient μ_1 can only be determined in a flow experiment where the flow is neither parallel nor perpendicular to the layer normal.

The elastic behaviour of the smectic layers, the permeation effect and the viscous flow lead to a great variety of phenomena in SmA phases: effects due to the compression of the layers [70, 71], flow behind obstacles [69, 72], and instabilities [73, 74].

8.6 References

[1] R. S. Porter, J. F. Johnson in *Rheology, Theory and Applications* (Ed.: F. R. Eirich), Academic Press, New York **1967**.
[2] P. G. de Gennes, *The Physics of Liquid Crystals*, Clarendon Press, Oxford **1975**.
[3] M. J. Stephen, J. P. Straley, *Rev. Mod. Phys.* **1974**, *46*, 617–704.
[4] J. G. Jenkins, *Ann. Rev. Fluid. Mech.* **1978**, *10*, 197–219.
[5] W. H. de Jeu, *Physical Properties of Liquid Crystalline Materials*, Gordon and Breach, New York **1980**.
[6] G. Vertogen, W. H. de Jeu, *Thermotropic Liquid Crystals, Fundamentals*, Springer Series in Chemical Physics 45, Springer, Berlin **1988**.
[7] P. G. de Gennes, J. Prost, *The Physics of Liquid Crystals*, 2nd edn, Clarendon Press, Oxford **1995**.
[8] F. M. Leslie in *Advances in Liquid Crystals* (Ed.: G. H. Brown), Academic Press, New York **1979**.
[9] F. M. Leslie, *Mol. Cryst. Liq. Cryst.* **1981**, *63*, 111–128.
[10] F. M. Leslie, *Quart. J. Mech. Appl. Math.* **1966**, *19*, 357–370.
[11] F. M. Leslie, *Arch. Ratl. Mech. Anal.* **1968**, *28*, 265–283.
[12] J. L. Ericksen, *Arch. Ratl. Mech. Anal.* **1966**, *23*, 266–275.

[13] J. L. Ericksen, *Mol. Cryst. Liq. Cryst.* **1969**, *7*, 153–164.
[14] O. Parodi, *J. Physique* **1970**, *31*, 581–584.
[15] G. Vertogen, *Z. Naturforsch.* **1983**, *38a*, 1273–1275.
[16] D. Forster, T. C. Lubensky, P. C. Martin, J. Swift, P. S. Pershan, *Phys. Rev. Lett.* **1971**, *26*, 1016–1019.
[17] P. C. Martin, O. Parodi, P. S. Pershan, *Phys. Rev. A* **1972**, *6*, 2401–2420.
[18] P. Schiller, *Wiss. Z. Univ. Halle* **1985**, *34*, 61–78.
[19] F. M. Leslie, I. W. Stewart, M. Nakagawa, *Mol. Cryst. Liq. Cryst.* **1991**, *198*, 443–454.
[20] S. P. A. Gill, F. M. Leslie, *Liq. Cryst.* **1993**, *14*, 1905–1923.
[21] K. F. Wissbrun, *J. Rheol.* **1981**, *25*, 619–662.
[22] G. Marrucci, F. Greco, *Adv. Chem. Phys.* **1993**, *86*, 331–404.
[23] M. G. Clark, F. M. Leslie, *Proc. R. Soc. Lond. A* **1978**, *361*, 463–485.
[24] P. K. Currie, *Mol. Cryst. Liq. Cryst.* **1974**, *28*, 335–338.
[25] W. Helfrich, *J. Chem. Phys.* **1969**, *51*, 4092–4105.
[26] M. Miesowicz, *Bull. Acad. Polon. Sci. Lett., Ser. A, Sci. Math.* **1936**, 228–247.
[27] Ch. Gähwiller, *Phys. Lett.* **1971**, *36A*, 311–312.
[28] V. Zwetkoff, *Acta Physicochimica URSS* **1939**, *10*, 555–578.
[29] F. M. Leslie, G. R. Luckhurst, H. J. Smith, *Chem. Phys. Lett.* **1972**, *13*, 368–371.
[30] K. B. Migler, R. B. Meyer, *Physica D* **1994**, *71*, 412–420.
[31] I. Zuniga, F. M. Leslie, *Europhys. Lett.* **1989**, *9*, 689–693.
[32] W. Helfrich, *J. Chem. Phys.* **1969**, *50*, 100–106.
[33] H. Ehrentraut, S. Hess, *Phys. Rev. E* **1995**, *51*, 2203–2212.
[34] P. Manneville, *Mol. Cryst. Liq. Cryst.* **1981**, *70*, 223–250.
[35] T. Carlsson, *Mol. Cryst. Liq. Cryst.* **1984**, *104*, 307–334.
[36] P. G. de Gennes, *The Physics of Liquid Crystals*, Clarendon Press, Oxford **1975**, p. 68.
[37] P. K. Currie, *J. Physique* **1979**, *40*, 501–505.
[38] R. J. Atkin, F. M. Leslie, *Quart. J. Mech. Appl. Math.* **1970**, *23*, S3–S24.
[39] P. K. Currie, *Arch. Ratl. Mech. Anal.* **1970**, *37*, 222–242.
[40] R. J. Atkin, *Arch. Ratl. Mech. Anal.* **1970**, *38*, 224–240.
[41] P. Pieranski, E. Guyon, *Phys. Lett.* **1974**, *49A*, 3, 237–238.
[42] P. Pieranski, F. Brochard, E. Guyon, *J. Physique* **1973**, *34*, 35–48.
[43] F. Brochard, *Mol. Cryst. Liq. Cryst.* **1973**, *23*, 51–58.
[44] T. Carlsson, *J. Physique* **1983**, *44*, 909–911.
[45] N. Kuzuu, M. Doi, *J. Phys. Soc. Jpn.* **1983**, *52*, 3486–3494.

[46] M. A. Osipov, E. M. Terentjev, *Z. Naturforsch.* **1989**, *44a*, 785–792.
[47] M. Kröger, H. S. Sellers, *J. Chem. Phys.* **1995**, *103*, 807–817.
[48] F. M. Leslie, *Proc. R. Soc. A* **1968**, *307*, 359–372.
[49] F. M. Leslie, *Mol. Cryst. Liq. Cryst.* **1969**, *7*, 407–420.
[50] U. D. Kini, *J. Physique Coll. C3* **1979**, *40*, 62–66.
[51] N. Scaramuzza, R. Barberi, F. Simoni, F. Xu, G. Barbero, R. Bartolino, *Phys. Rev. A* **1985**, *32*, 1134–1143.
[52] V. G. Kamenskii, E. I. Kats, *Sov. Phys. JETP* **1988**, *66*, 1007–1012.
[53] W. Helfrich, *Phys. Rev. Lett.* **1969**, *23*, 372–374.
[54] U. D. Kini, G. S. Ranganath, S. Chandrsekhar, *Pramana* **1975**, *5*, 101–106.
[55] F. Fischer, J. Grupp, *J. Physique Lett.* **1984**, *45*, L1091–L1095.
[56] L. J. Yu, A. Saupe, *Phys. Rev. Lett.* **1980**, *45*, 1000–1003.
[57] H. Brand, H. Pleiner, *Phys. Rev. A* **1981**, *24*, 2777–2787.
[58] M. Liu, *Phys. Rev. A* **1981**, *24*, 2720–2726.
[59] A. Saupe, *J. Chem. Phys.* **1981**, *75*, 5118–5124.
[60] E. Govers, G. Vertogen, *Physica* **1985**, *133A*, 337–344.
[61] H. Brand, H. Pleiner, *J. Physique* **1982**, *43*, 853–858.
[62] F. M. Leslie, *J. Non-Newtonian Fluid Mech.* **1994**, *54*, 241–250.
[63] L. Léger, A. Martinet, *J. Physique* **1976**, *37*, C3-89–97.
[64] P. G. de Gennes, J. Prost, *The Physics of Liquid Crystals*, 2nd edn, Clarendon Press, Oxford **1995**, p. 432.
[65] T. Carlsson, F. M. Leslie, N. A. Clark, *Phys. Rev. E* **1995**, *51*, 4509–4525.
[66] M. Osipov, E. M. Terentjev, *Mol. Cryst. Liq. Cryst.* **1991**, *198*, 429–435.
[67] T. Carlsson, B. Žekš, *Liq. Cryst.* **1989**, *5*, 359–365.
[68] I. Dierking, F. Giesselmann, P. Zugenmaier, G. Pelzl, P. Schiller, *Cryst. Res. Technol.* **1992**, *27*, 727–739.
[69] P. G. de Gennes, *Phys. Fluids* **1974**, *17*, 1645–1654.
[70] R. Bartolino, G. Durand, *J. Physique* **1981**, *42*, 1445–1451.
[71] R. Bartolino, G. Durand, *J. Physique Lett.* **1983**, *44*, L79–L83.
[72] P. Oswald, *J. Physique Lett.* **1983**, *44*, L303–L309.
[73] P. Oswald, S. I. Ben-Abraham, *J. Physique* **1982**, *43*, 1193–1197.
[74] J. Marignan, O. Parodi, *J. Physique* **1983**, *44*, 263–271.

9 Behavior of Liquid Crystals in Electric and Magnetic Fields

Lev M. Blinov

9.1 Introduction

As is well known, liquid crystals are very sensitive to electric and magnetic fields [1] and it is this property which allows their application in display and other optical devices technology. The field effects may be divided into three groups. The first includes the electric- or magnetic-field-induced changes in the microscopic structure of a liquid crystal phase. This implies that the field is directed in such a way with respect to the director that, under the action of the field, the orientational state of the mesophase is unchanged (the field stabilizes the director distribution). The second group includes changes in the macroscopic structure, due to a destabilization and reorientation of the director under direct action of the electric or magnetic field. The backflow effects accompanying the distortion in the transient regime are driven by the director reorientation (primary process) and are of secondary importance. The fluid is considered to be nonconducting. The effects caused by the electric field and related to the conductivity of a liquid crystal are included in the third group (electrohydrodynamic instabilities). All these effects are dissipative in nature and have no magnetic analogs.

Due to the strong anisotropy of liquid crystals, field effects are accompanied by dramatic changes in optical properties. The electro-optical properties of liquid crystals have been studied very actively since the beginning of the century, and many important phenomena have been discovered. The earlier theoretical and experimental results on the physical properties of liquid crystals, including certain electric field effects, have been reviewed in a variety of books [1–5], as have the details of electro-optical effects [6–9], and many results relating to the properties of liquid crystalline materials and their application in devices have been reported [10–12].

The present chapter is devoted to a discussion of electric field effects in various liquid crystal phases, with an emphasis on the physical aspects of the phenomena. The discussion is based on classical results, although the most important recent achievements in the field are also mentioned. As ferroelectric liquid crystals are covered in detail in other chapters in this book, they are discussed only briefly here for the sake of completeness. Electro-optical properties of polymer liquid crystals have recently been discussed in detail [13]; here, only specific features of polymer mesophases relevant to their field behavior are mentioned.

9.2 Direct Influence of an Electric or Magnetic Field on Liquid Crystal Structure

9.2.1 Shift of Phase Transition Temperatures

In general, when an external field is applied to a nonpolar or polar liquid crystal, a quadratic- or linear-in-field term, respectively, must be added to the expression for the free energy density of the medium. The quadratic-in-field energy terms describe the interaction of the electric or magnetic field with the dielectric χ_a or diamagnetic γ_a anisotropy of susceptibility:

$$g_E = \frac{1}{2} \chi_a E^2 \quad (1\,a)$$

$$g_H = \frac{1}{2} \gamma_a H^2 \quad (1\,b)$$

The linear terms describe the interaction of the fields with media, polarized or magnetized either spontaneously or by external factors other than electric field itself. In ferroelectrics and ferromagnetics the magnitude of the spontaneous polarization P_S or magnetization M_S is finite and the energy term is

$$g_E = P_S E \quad (2\,a)$$

$$g_H = M_S H \quad (2\,b)$$

As ferromagnetic mesophases are still to be discovered the linear term (Eq. (2a) or (2b)) is important only when considering ferroelectric liquid crystals. More generally, the PE term is also used to describe other polarized systems; for example, in the discussion of the flexoelectric effects in nonpolar phases as nematics or cholesterics where the polarization P is induced by a mechanical distortion.

The field terms may stabilize or destabilize the intrinsic thermodynamic order of a mesophase, and hence increase or decrease the temperature of its transition to a less ordered phase. The shift in the transition temperature is calculated by comparing the electric (magnetic) energy with the transition enthalpy or some other competing thermodynamic quantity (e.g. elastic energy). Below we consider examles.

9.2.1.1 Second-Order Transitions

Smectic A – Smectic C Transition

In this case we must compare a gain in the electric (or magnetic) energy with a loss of the soft-mode elastic energy characterized by a change in tilt angle θ. Thus, for the electric field case, the shift in the A–C transition temperature is [7]:

$$\Delta T_{C\text{-}A} = \frac{\varepsilon_a (E \cdot n)^2}{8 \pi \alpha} \quad (3)$$

where n is the director, $\varepsilon_a = 1 + 4\pi \chi_a$ is the dielectric anisotropy and α is a parameter of the Landau expansion for the A–C transition:

$$g = \frac{1}{2} \alpha (T - T_{C\text{-}A}) \theta^2 + \frac{1}{4} b \theta^4 \quad (4)$$

and g_E is taken in the form of Eq. (1a). The sign of the shift depends on the field direction. If E is parallel to the normal h to the smectic layers and $\varepsilon_a > 0$, the field stabilizes the smectic A phase and $T_{C\text{-}A}$ decreases. An oblique field ($\varepsilon_a > 0$) induces a tilt, and thus stabilizes the smectic C phase and $T_{C\text{-}A}$ increases.

Smectic A–Smectic C* Transition

For the chiral smectic C* phase, which is ferroelectric, in addition to the quadratic term (Eq. (1a) or (1b)) the linear term proportional to polarization P must be taken into consideration. As a rule, the linear term exceeds the quadratic one and the free energy may be taken in form:

$$g = \frac{1}{2}a_0\theta^2 + \frac{1}{4}b\theta^4 + \frac{P^2}{2\chi_\perp} - CP\theta - PE \quad (5)$$

where $a_0 = \alpha(T - T_{C-A})$ describes the elasticity for $\delta\theta$ changes, P is the polarization, C is the tilt-polarization coupling constant (or piezocoefficient), and $\chi_\perp = (\varepsilon_\perp - 1)/4\pi$ is the background dielectric susceptibility, which may be taken from the isotropic phase. In Eq. (5) the smectic C* is assumed to be unwound and all chiral terms important for a helical ferroelectric are omitted.

Strictly speaking, even for an infinitely small electric field the second-order A–C* transition disappears. However, the soft-mode dielectric susceptibility maximum characteristic of that transition is still observed at the "apparent" transition temperature T_m. With increasing field T_m increases according to expression [14]:

$$T_m - T_{C-A} = 3\left(\frac{bC^2}{16\alpha^3}\right)^{\frac{1}{3}} E^{\frac{2}{3}} \quad (6)$$

Such a shift governed by the second term in the Landau expansion and strongly dependent on piezocoefficient C has been observed experimentally [14, 15]. The experimental points for a multicomponent ferroelectric mixture [14] and the fit using Eq. (6) is shown in Fig. 1.

9.2.1.2 Strong First-Order Transitions

Isotropic Liquid–Nematic Transition

In the case of positive dielectric anisotropy of a nematic, even a weak field makes the isotropic phase uniaxial and the N–I phase transition disappears (see Pikin [7], Chap. 4). However, the apparent N–I phase transition temperature may change with the electric or magnetic field. For $\varepsilon_a > 0$ and $\mathbf{E} \| \mathbf{n}$, the quadratic-in-field energy terms (Eq. (1a) and (1b)) reduce the free energy and stabilize the anisotropic phase. In the

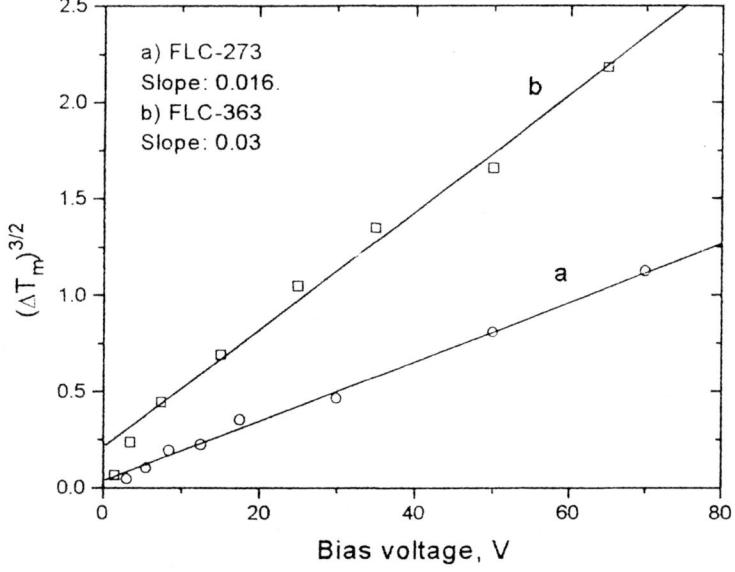

Figure 1. The voltage dependence of the shift in the soft-mode susceptibility maximum temperature [14]: (a) mixture 1; (b) mixture 2. Dotted lines correspond to Eq. (6). Cell thickness 10 μm.

case of an electric field, the transition temperature increases by the value [16]

$$\Delta T_{N-I} = \frac{(\varepsilon_{\parallel} - \varepsilon_{iso})\,E^2}{8\pi \cdot \Delta H_t} \quad (7)$$

where ΔH_t is the enthalpy of the N–I phase transition, $\varepsilon_{\parallel}$ is the dielectric permittivity of the nematic phase parallel to the director, and ε_{iso} is the dielectric permittivity of the isotropic phase. Experimentally, for rather high positive dielectric anisotropy of the nematic phase when $\varepsilon_{\parallel} - \varepsilon_{iso} \approx 10$, and a typical value of $\Delta H_t \approx 5 \times 10^4$ erg cm^{-3}, a shift of the order of 0.8 K is anticipated for a field strength of 10^5 V cm^{-1} (about 300 CGS units).

The shift of the first-order N–I transition is described theoretically with a Landau type expansion over two variables, orientational order parameter S, and polarization P [17]. In the case of the electric field directed along the nematic optical axis the expansion is as follows:

$$g = \frac{1}{2}\alpha(T - T_c^*)S^2 - \frac{1}{3}bS^3 + \frac{1}{4}cS^4$$
$$+ \frac{P^2}{2\chi_0} + \sigma P^2 S + \frac{1}{2}\kappa P^2 S - PE \quad (8)$$

where χ_0 is the susceptibility of the isotropic phase, while parameters σ and κ represent the anisotropy of the susceptibility of the nematic phase. With this form of the free energy, as expected the shift of the transition is proportional to the square of the field, the coefficient α playing the role of the transition heat in Eq. (7):

$$\Delta T_c^* = \chi_0^2 \alpha^{-1} (4\sigma^2 \chi_0 - \kappa)\,E^2$$

The experimental investigation of the field-induced shift of the N–I transition was carried out using a pulse technique [17], which allowed rather strong electric fields to be applied to samples. The data [17] are shown in Fig. 2 for two nematic cyanobiphenyls, 5-CB and 6-CB, that differ in

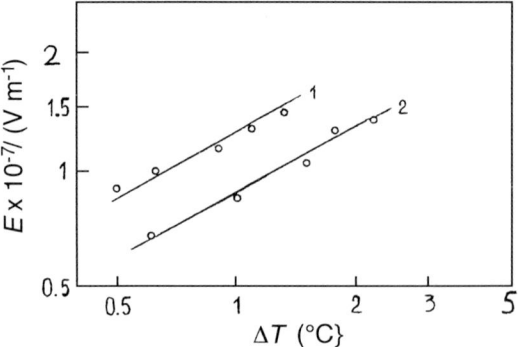

Figure 2. Change in the N–I transition temperature with the electric field in (1) 5-CB and (2) 6-CB [17]. A double logarithmic scale is used.

their latent heats of the N–I transition (1.5×10^4 erg cm^{-3} and 2.5×10^4 erg cm^{-3} for 5-CB and 6-CB, respectively). The results are consistent with both the phenomenological approach (Eq. (8)) and microscopic theory [17], operating with the number density, and the magnitude, and direction of the molecular dipoles.

At extremely high fields the isotropic phase should be indistinguishable from the nematic one, even well above the zero field transition temperature, since the uniaxial order induced by a magnetic or ac electric field in the isotropic phase will be comparable with the nematic orientational order. However, such fields are hardly accessible, even with the pulse technique. Much stronger changes in order parameter may be achieved with ferroelectric transitions (see below).

For negative dielectric anisotropy of the nematic phase, $\varepsilon_a < 0$, the external field may induce a biaxial nematic phase. With increasing field one may reach the tricritical point [18] where the first-order N–I transition becomes a second-order one.

A–C*, A–G*, C*–G*, and N–C* Ferroelectric Transitions

For these transitions the electric-field-induced shifts of transition temperatures are linear functions of the field. The A–C* transition may be either second or first order, while the A–G* and C*–G* transitions are first order [19, 20]. Thus, to calculate the transition shift, we should compare the transition enthalpy with the linear-in-field free energy term (Eq. (2)). The Clausius–Clapeyron equation for solid ferroelectrics is well known:

$$\frac{\Delta T}{E} = \frac{\Delta P_s}{\Delta H_t} \qquad (9)$$

Here, ΔP_S is the difference in spontaneous polarization between the two phases under consideration. For the first-order transition the coefficient b in the Landau expansion must be negative and a term of the sixth order in the tilt angle has to be included:

$$g = \frac{1}{2} a_0 \theta^2 + \frac{1}{4} b \theta^4 + \frac{1}{6} c \theta^6 \\ + \frac{P^2}{2\chi_\perp} - CP\theta - PE \qquad (10)$$

The coefficient c enters the expression for the shift in the transition temperature [19, 20]:

$$\Delta T = \frac{4\chi_\perp}{a} \left(-\frac{c}{3b}\right)^{\frac{1}{2}} E$$

Experimentally, a field strength of the order of 10^5 V cm^{-1} induces a ΔT value of the order of 1 K for a ΔH_t of about 3×10^5 erg cm^{-3} K^{-1} and $\Delta P_S \approx 300$ nC cm^{-2}. An example of the $\Delta T(E)$ experimental dependence is shown in Fig. 3.

Field dependence was also observed for the first-order N–C* transition temperature. For a field strength of about 2×10^5 V cm^{-1} the phase transition point increased by 0.6 K [21]. For a compound with a much higher spontaneous polarization in the C* phase, approximately the same shift is induced by a field one order of magnitude lower [22].

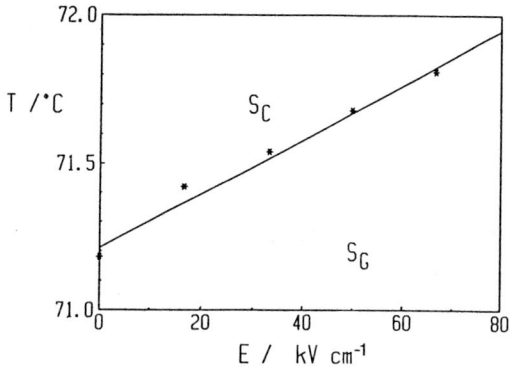

Figure 3. Electric field strength dependence for the C*–G* phase transition (compound A7, [19]).

9.2.1.3 Weak First-Order Transitions

N–A Transition

The latent heat of this transition is usually small and may even vanish if the width of the nematic temperature range is sufficiently large [23]. Thus, the transition can be either of first or second order. For the second-order transition the discontinuity in the orientational order parameter S and, hence, the dielectric (or diamagnetic) susceptibility, disappears and the field influence on both phases is the same. Thus we do not anticipate any field-induced shift in the N–A transition temperature. For the weak first-order transition there is a small discontinuity in both S and n_e dielectric (and magnetic) susceptibilities, and the shift depends on the competition between two small quantities: the difference in susceptibilities for the nematic and smectic A phases on the one hand and transition enthalpy on the other. In particular, the field may induce a change in the phase transition order, from first to the second order, as shown in Fig. 4 [24].

Smectic A–C Ferroelectric Transition*

This case has been analyzed in detail both theoretically and experimentally [25]. The

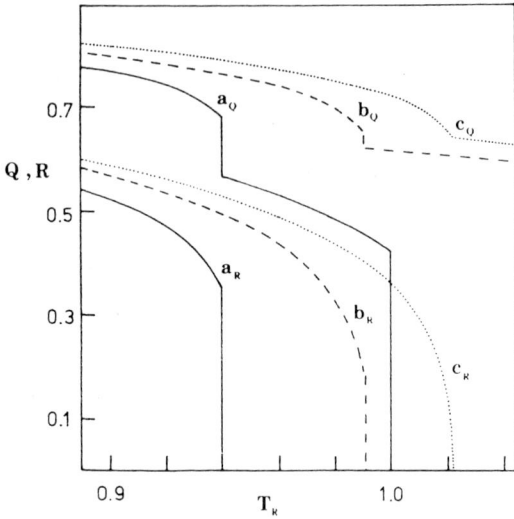

Figure 4. Field-induced changes in order parameters for a smectic A liquid crystal. Q, orientational order parameter; curves a_Q, b_Q, and c_Q show the temperature dependence of Q at various fields. R, order parameter describing the coupling of translational and orientational order, curves a_R, b_R, and c_R show the temperature dependence of R. The field increases from curves a to curves c and results in a change in the N–C transition from first to second order [24].

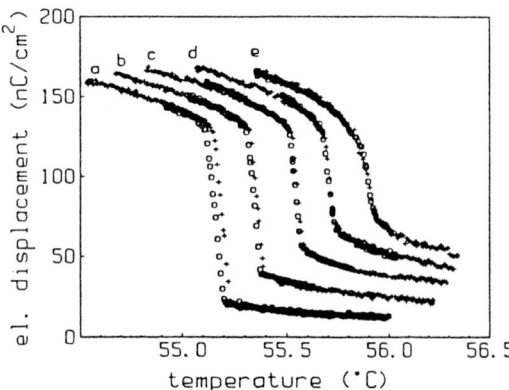

Figure 5. Temperature dependence of the electric displacement of C7 (*I*) at various field strengths. *E* (in kV cm^{-1}): (a) 10, (b) 20, (c) 30, (d) 40, and (e) 50. (+) Increasing and (○) decreasing temperature runs [25].

temperature dependence of the electric displacement D at various bias fields for the chiral compound C7 (**I**) is shown in Fig. 5.

$$C_2H_5O-(CH_3)C^*H-C^*H(Cl)-$$

$$-COO-\bigcirc-\bigcirc-C_7H_{15}$$

I

The change in the A–C* transition from first to second order can be clearly seen in Fig. 5. The tricritical point corresponds to the following sets of parameters: $T_C \approx 55.8\,°C$ (i.e. 0.8 K above the zero-field point), $E_C \approx 5 \times 10^4$ V cm^{-1} and $D_C \approx 100$ nC cm^{-2}.

A theoretical description of the experiment can be obtained using the same approach; that is, using the Landau expansion (Eq. (10)). Equation (10) has been solved numerically and the results of the calculations are shown in Fig. 6, where the molecular tilt angle, which is the order parameter of the A–C* transition, is given as a function of the external electric field for various temperatures. The parameters used to fit the experimental (Fig. 5) and theoretical (Fig. 6) curves were (in CGS units): $\alpha \approx 8.9 \times 10^5$ erg cm^{-3} K^{-1}, $b \approx -1.3 \times 10^7$ erg cm^{-3}, $c \approx 6.6 \times 10^7$ erg cm^{-3}, $T_0 = 51.4\,°C$, $\chi_\perp = 0.33$, $C \approx 2.9 \times 10^3$ CGS (in [25] these parameters are given in SI units).

9.2.2 Influence of the Field on Order Parameters

As the field-induced reorientation of the director occurs at rather weak fields, the geometries worth discussing here for nonpolar mesophases are: (i) the field parallel to the director and $\varepsilon_a > 0$; and (ii) the field perpendicular to the director and $\varepsilon_a < 0$. In the first case, uniaxial symmetry is conserved and the field stabilizes thermal fluctuations and increases the orientational order parameter in both the nematic and smectic A phase [26]. A qualitative picture is

9.2.3 Changes in Symmetry

The field-induced tilt in the achiral smectic A phase (see above) results in a change in symmetry from $D_{\infty h}$ to C_{2h}. The induced tilt angle is proportional to the field squared and the factor $(T_{CA}/T-T_{C-A})\gamma$, where $\gamma=1.3$ [1]. Due to the tilt, optical biaxiality can be observed in the vicinity of the transition.

9.2.3.1 Induced Biaxiality in Nematics

Another example of a symmetry change is the field-induced biaxiality in nematic liquid crystals with negative dielectric anisotropy [29]. In this case the field is applied perpendicular to the director. The latter is parallel to the z axis. The orientational order at a point r in a nematic liquid crystal is defined by the ordering matrix

$$S_{\alpha\beta}(r) = \frac{1}{2} S_0 [3 n_\alpha(r) n_\beta(r) - \delta_{\alpha\beta}] \quad (11)$$

where S_0 is the order parameter defined with respect to a local director $n(r)$. Averaging over time and spatial fluctuations of $n(r)$ gives the ordering matrix in a principal axis system as:

$$\langle S_{\alpha\beta}\rangle \begin{bmatrix} -\frac{1}{2}(Q-B) & 0 & 0 \\ 0 & -\frac{1}{2}(Q+B) & 0 \\ 0 & 0 & Q \end{bmatrix} \quad (12)$$

where Q and B are uniaxial and biaxial order parameters

$$Q = S_0 \left[1 - \frac{1}{2} \langle n_x^2 + n_y^2 \rangle \right] \quad (13)$$

$$B = \frac{3}{2} S_0 \langle n_x^2 + n_y^2 \rangle \quad (14)$$

Now Q and B are presented in terms of the director fluctuations which are not assumed to be isotropic and the x, y plane. For $n_x^2 = n_y^2$ the biaxial parameter vanishes and the uni-

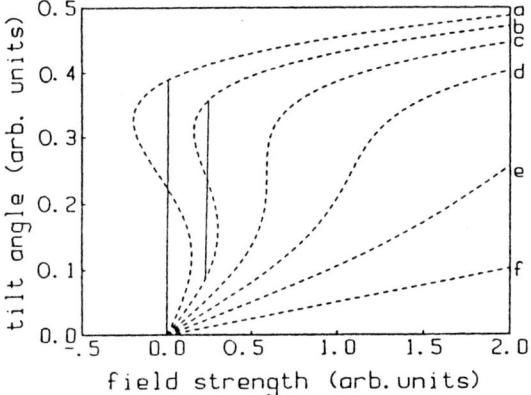

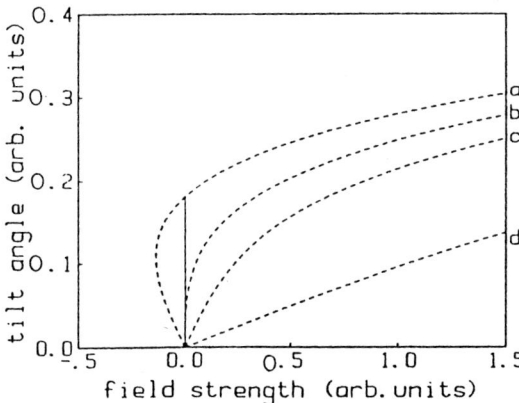

Figure 6. Theoretical tilt angle versus field strength (dashed lines): vertical solid lines indicate discontinuous transitions [25].
First-order transition, $b<0$: (a) $T=T_c(0)$; (b) $T_c(0)<T<T_c$; (c) $T=T_c$; (d) $T>T_c$; (e,f) $T\gg T_c$. $T_c(0)$ and T_c are the zero-field first-order transition temperature and the critical point temperature, respectively. (Top)
Second-order transition, $b>0$: (a) $T<T_c$; (b) $T=T_c$; (c) $T>T_c$; (d) $T\gg T_c$. T_c denotes the second-order transition temperature. (Bottom)

shown in Fig. 4 [24]. With increasing field the orientational order parameter S increases and T_{NA} increases, as discussed above. The positional order parameter, being coupled with S also increases.

The additional optical anisotropy induced by the field is proportional to either E or $|E|$ for a weak and strong field, respectively [27, 28].

axial one becomes equal to the well-known parameter:

$$S = \frac{1}{2} \langle 3\cos^2\theta - 1 \rangle \quad (15)$$

where θ is an angle between the longitudinal axis of a molecule and the director of the nematic mesophase.

When a field is applied, for example, along the x direction, and $\varepsilon_{xx} = \varepsilon_{yy} = \varepsilon_\perp > \varepsilon_{zz} = \varepsilon_\parallel$ the n_x fluctuations are quenched to some extent and B increases. The theory [30] considers the quenching of fluctuations with various wavevectors and results in the following field dependence of the biaxial order parameter:

$$B = \frac{3 S_0 k_B T}{8\pi K} \left(\frac{\varepsilon_a E^2}{4\pi K} \right)^{\frac{1}{2}} \quad (16)$$

where K is an average elastic constant and k_B is the Boltzmann constant.

The field-induced biaxiality has been observed in a planar-oriented layer of compound **II**,

C$_5$H$_{11}$—⟨H⟩—⟨H⟩—C$_5$H$_{11}$
               |
               CN

II

with $\varepsilon_a = -5.4$, by measuring the ellipticity of laser light transmitted through a sandwich cell ($E \parallel$ light wavevector) [29].

9.2.3.2 The Kerr Effect

The Kerr effect in the isotropic phase may also be discussed in terms of the field-induced symmetry change of the medium. A birefringence is induced in an ordinary liquid if a uniform electric field is applied at right-angles to the direction of a beam passing through the cell; its value is related to the field strength E:

$$\Delta n(E) = \lambda B' E^2 \quad (17)$$

where B' is the Kerr constant and λ is the wavelength of the light.

At temperatures significantly above the transition point from the nematic to the isotropic liquid, materials forming liquid crystals behave like normal liquids. In an electric field they display the Kerr effect, with an order of magnitude equal to or less than that of nitrobenzene ($+4.1 \times 10^{-12}$ m V^{-2}). However, as the temperature decreases toward T_{N-I} the Kerr constant of a nematogen changes considerably. For some substances it diverges when $T \to T_{N-I}$, being either positive or negative. In general, there is a correlation between the sign of the Kerr effect in the isotropic phase and the dielectric anisotropy of the nematic phase.

The growth of the Kerr constant is accounted for by considerable contribution of fluctuations of the orientational order parameter to dielectric properties of the isotropic phase. This contribution can be calculated within the framework of the Landau theory [31]. Field E induces the orientational order [32]

$$Q(E) = \frac{\rho \varepsilon_a E^2}{12\pi\alpha(T - T^*)} \quad (18)$$

where ρ is the number density of molecules and α is the parameter of the Landau expansion for the N–I transition.

The field-induced order can be related to the Kerr constant

$$B = \frac{\Delta n^0 Q(E)}{\lambda E^2} \quad (19)$$

where Δn^0 is the birefringence of the perfectly aligned liquid crystal [33]. Experiments carried out on 5-CB showed that the law $B \propto (T - T_{N-I})^{-1}$ is fulfilled in the $T - T_{N-I}$ range from several kelvin down to 0.5 K (the Kerr constant is equal to 6×10^{-11} m V^{-2} at $T - T_{N-I} = 1$ K and $\lambda = 633$ nm). Closer to T_{N-I} the law is violated and the discrepancy can be accounted for by higher order terms in the Landau expansion [33]. The microscopic theory of the Kerr effect is described by Rjumtsev et al. [17].

The Kerr effect in the pretransition temperature region has been investigated very actively [17, 34–39]. Tsvetkov et al. [39] determined the values for $B' > 0$ (phenylbenzoates) and $B' < 0$ (cyanostilbenes) close to T_{N-I}. These values exceed the absolute value of the Kerr constant for nitrobenzene by 200 and 6.5 times, respectively. Such high values of B enable the control voltage of the Kerr cell to be reduced several fold and give a decrease of an order of magnitude or more in power consumption [40].

Relaxation times τ_1 associated with order-parameter fluctuations increase when approaching T_{N-I}, but still remain short enough for light modulation (within the range 0.1–1 µS for 5-CB at $T - T_{N-I} = 5$ to 0.5 K [41]). The other time (τ_2) related to molecular relaxation is independent of temperature.

9.2.4 Specific Features of Twisted Phases and Polymers

In Section 9.2.1 of this Chapter we discussed field-induced changes in the microstructure of liquid crystals. However, field-induced unwinding of the cholesteric (macroscopic) helix (see Section 9.3.2.3 of this Chapter) shows that the transition from a twisted to a uniform nematic may also be considered as a phase transition. In the latter case the field energy term competes with a rather small elastic energy proportional to nematic-like elastic moduli and the squared wave vector of the helical superstructure Kq^2. As the pitch of the helix $p = 2\pi/q$ is large, the field threshold for the transition is very low. On the other hand, between the two extreme cases (a microstructure with a molecular characteristic dimension and a large-pitch macrostructure) there are cases where the pitch might be smaller and the elastic moduli higher than those in cholesterics. Examples are blue phases in cholesterics and twist grain boundary (TGB) phases in smectics A and C* (TGBA and TGBC). The structures of these phases are discussed in other chapters in this book. What is essential for their field behavior is a periodic, solid-like defect structure which has its own wavevector and elasticity.

9.2.4.1 Blue Phases

There are three known thermodynamically stable blue phases: BP_I and BP_{II}, and BP_{III} (or the foggy phase). The structure of the first two is already established: BP_I is a body-centered cubic phase (symmetry group O^8 or $I4_132$) and BP_{II} is simple cubic (symmetry group O^2 or $P4_232$) [42]. The foggy phase, BP_{III}, can probably be described using one of the quasicrystal models [43].

The three-dimensional crystalline structure of blue phases with lattice periods comparable to the wavelength of visible light results in optical diffraction, which is dependent on the orientation of the light wavevector with respect to the crystalline planes. The most pronounced features of the optics of BP_I and BP_{II} are [44]: (i) selective reflection in the visible range which gives rise to the blue color of the phases (in contrast to conventional cholesterics, blue phases manifest several reflection orders); (ii) only one circular polarization, as in cholesterics, is back-scattered; (iii) blue phases are optically active and the sign of the light polarization rotation changes at the wavelengths of selective scattering maxima; and (iv) the linear birefringence is absent (i.e. blue phases are optically isotropic), but multiple scattering in perfect samples can result in a small apparent optical anisotropy. The optical properties of blue phases are completely defined by the spatially periodic tensor of the high-frequency dielectric permittivity [44]:

$$\varepsilon(\mathbf{r}) = \varepsilon_0 + \varepsilon_a^{ik}(x, y, z) \quad (20)$$

where ε_0 is the average dielectric permittivity and ε_a^{ik} is its three-dimensional periodic part.

An external electric field interacts with the local dielectric anisotropy of a blue phase and contributes $\varepsilon_a E^2/4\pi$ to the energy of the liquid crystal [45]. The field distorts the cubic lattice and results in a change in the angular (or spectral) positions of Bragg's reflections. Moreover, field-induced phase transitions to novel phases have been observed [42, 46, 47]. The field can also induce birefringence parallel to the field direction, due to the optical biaxiality of the distorted cubic lattice [48].

In a weak field no phase transition occurs and a cubic lattice is distorted according to the sign of the local dielectric anisotropy of the medium. The effect is described using the electrostriction tensor γ_{iknp} [41], which relates the electric-field components to the strain tensor:

$$e_{ik} = \gamma_{iknp} E_n E_p \quad (21)$$

The distortion of the lattice is accompanied by a change in the Fourier harmonics of the dielectric tensor (Eq. (20)) and corresponding changes in the optical properties. For example, the wavelengths of the Bragg peaks for a substance with $\varepsilon_a > 0$ may increase or decrease with increasing field depending on the field direction with respect to the crystallographic axes (Fig. 7) [49]. When the dielectric anisotropy is negative, all the signs of components of the electrostriction tensor are inverted and all the field-induced red shifts in the Bragg maxima are replaced by blue shifts, and vice versa [50]. Red and blue shifts may reach values of 26 and

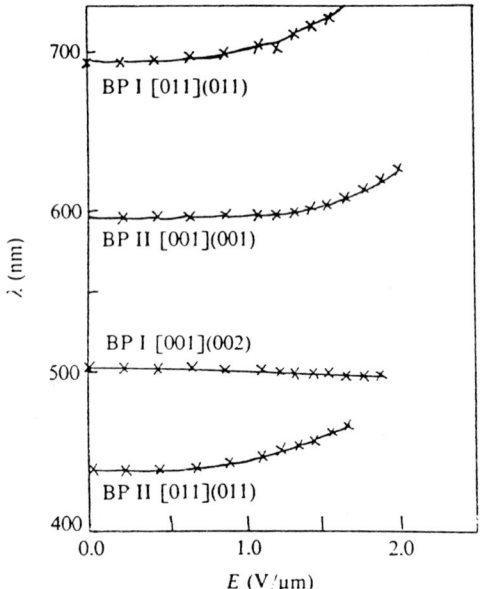

Figure 7. The wavelengths of the Bragg peaks corresponding to the planes (hkm) versus the applied field strength for BP_I (34.1 °C) and BP_{II} (34.3 °C) oriented with either a four-fold [001] or two-fold [011] axis parallel to the field [49].

35 nm, respectively, in a rather weak field of about 0.2–0.4 V mm^{-1} and a negative ε_a [51].

With increasing external field a series of field-induced phase transitions is observed: $BP_I \rightarrow$ cholesteric, $BP_{II} \rightarrow$ cholesteric, and then cholesteric $\rightarrow$ nematic [52]. This is illustrated in Fig. 8 [42] which shows a voltage–temperature phase diagram for a mixture (47–53 mol%) of chiral 15-CB with 4-n-hexyloxycyanobiphenyl (60-CB). BP_I loses its stability, first transforming into the cholesteric phase, because the transition enthalpy ΔH_t is extremely small (~ 50 J mol^{-1}) for the BP_I–Ch transition. This enthalpy, normalized to a unit volume (0.2 J cm^{-3}) and compared with the difference in electrostatic energy density between the two phases $\delta\varepsilon \cdot E^2$ ($\delta\varepsilon = \varepsilon_{BP_I} - \varepsilon_{Ch} \sim 0.2$) explains the observed results. For other materials, novel phases induced by an external

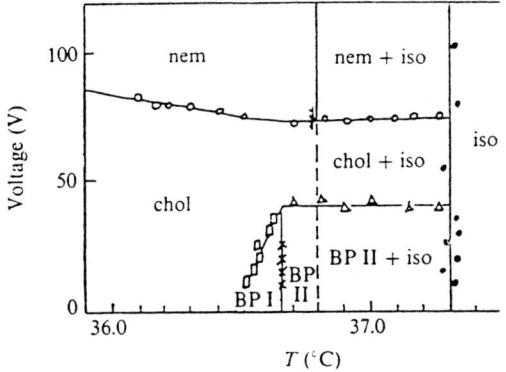

Figure 8. Voltage–temperature phase diagram for a 47–53 mol% mixture of 15-CB and 6-OCB [42].

field have been observed. Among them a tetragonal (BPX) and two hexagonal (a three-dimensional BPH^{3d}, and a two-dimensional BPH^{2d}) phases were distinguished [46, 47]. Chilaya and Petriashvili [51] observed the field-induced transition from BP$_{II}$ to the isotropic phase in a mixture with $\varepsilon_a > 0$.

When an electric field is applied to the BP$_{III}$ (foggy) phase with $\varepsilon_a > 0$, the broad selective reflection peak typical of this less ordered phase decreases in intensity and, at some threshold voltage, is replaced by a sharp peak at longer wavelength [43]. Thus, the transition to a new phase occurs. The symmetry of the new phase has not yet been established. For a system with $\varepsilon_a < 0$ [53], the increase in the field results in a considerable increase in the selective reflection peak. An explanation of such behavior and many other examples of field effects in blue phases can be found in a comprehensive review by Kitzerow [54].

9.2.4.2 Twist Grain Boundary Phases

TGB phases are observed only in chiral substances. The director is twisted in the presence of layering, the layer twist being mediated by the twist grain boundary walls with screw dislocations. In the TGBA phase the layering is of the smectic A type, while in TGBC* it is of the smectic C* (ferroelectric) type.

TGBA – Smectic A Transition

The electric field may push the defect walls out of the sample and induce the transition from TGBA to conventional smectic A. The field threshold of such a transition decreases dramatically on approaching the chiral nematic phase and may be of the order of 2×10^4 V cm^{-1} [55]. With increasing fields up to 1×10^5 V cm^{-1}, the apparent transition temperature decreases dramatically (by 18 K). Since in both the TGBA and SmA phases the spontaneous polarization is zero, the transition should be driven by dielectric anisotropy, which appears to be almost independent of temperature in both phases. Thus the elastic properties of the layered TGBA structure competing for the field action must play the principal role in the temperature dependence of the transition parameters.

TGBA–Smectic C Transition*

The results of an investigation of the electric-field-induced transition between these two phases is shown in Fig. 9 [56]. At any given temperature the short pitch helix of TGBA is unwound by the field and either a uniform smectic C* structure (I) or a modulated one in the form of stripes (II) or parquet (III) appears. The slope of the boundary between the TGBA and SmC* phases can be explained by the Clausius–Clapeyron equation (Eq. (9)) where, in this case, $\Delta P = P_S(C^*)$ as $P_S(TGBA) = 0$. Note that the transition temperature from TGBA to the isotropic phase is, in fact, field independent due to the much higher enthalpy of that transition.

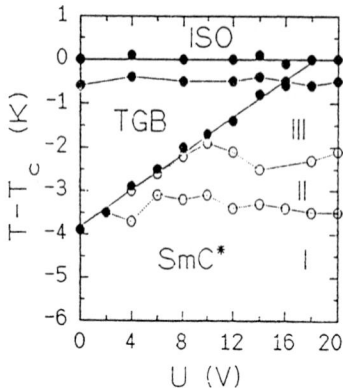

Figure 9. Voltage–temperature phase diagram obtained on heating 14P1M7 [56]. Cell thickness 2 μm.

9.2.4.3 Comment on Polymer Liquid Crystals

In addition to the effects observed in low-molecular-weight liquid crystals, an electric field causes some specific transformations in the structure of polymeric mesophases. As pointed out by Shibayev [57] the correlation length of the short-range smectic order in the nematic phase of comb-like polymers is changed considerably under the action of an electric field. The conformation and the orientational order parameter of flexible spacers separating mesogenic units from the backbone are also changed (more *gauche* isomers than *trans* isomers appear, as indicated by infrared spectroscopy [58]). The anisotropy in the orientation of polymeric backbones is induced by the electric field due to a torque exerted by the mesogenic units. In some cases the field induces smectic phases in thermodynamically stable nematic ones [59].

9.3 Distortions due to Direct Interaction of a Field with the Director

We now turn to the changes that occur in the macroscopic structure of a liquid crystal due to a destabilization and reorientation of the director under direct action of an electric or magnetic field. The external field might be coupled either to the dielectric (diamagnetic) anisotropy (magnetically or electrically driven uniform Frederiks transition and periodic pattern formation) or to the macroscopic polarization (flexoelectric effect and ferroelectric switching) of the substance. The fluid is considered to be nonconductive.

9.3.1 Nematics

9.3.1.1 Classical Frederiks Transition

The process of the field-induced reorientation of the director, called the Frederiks transition [60, 61], is usually discussed for the three typical geometries shown in Fig. 10.

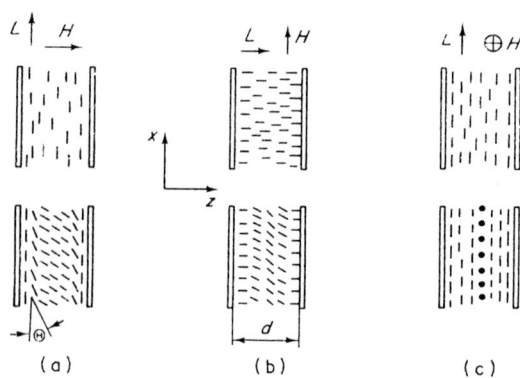

Figure 10. Three typical geometries for observing the Frederiks transition: the splay (a), bend (b), and twist (c) distortions induced by a magnetic field H. (Top) Initially, the director n is along the L axis, but when the field exceeds a certain threshold (bottom) distortion occurs.

The geometries correspond to the splay (with subsequent development of the bend), bend (combined with a splay) and pure twist distortions, which we may call the S-, B-, and T-distortions.

In the simplest model the director n is reoriented by the magnetic field under the action of the diamagnetic torque, which is proportional to the diamagnetic anisotropy γ_a. The corresponding contribution to the density of the nematic free energy is

$$g_H = \gamma_a (Hn)^2 \qquad (22)$$

that is, the director n tends to align itself along the field ($n \parallel H$) if $\gamma_a > 0$ and perpendicular to it ($n \perp H$) if $\gamma_a < 0$.

The elastic torque supports the initial director orientation, fixed by the boundary conditions on the surface. In the case of strong (infinite) anchoring, the condition $n_S = n_0$ holds during the reorientation process and, as a result, a compromised director profile pertains, which satisfies the condition of the minimum free energy:

$$F_V = \int_V (g_k + g_H) \, d\tau \qquad (23)$$

where g_k is the Frank energy including splay, twist and bend terms, respectively:

$$g_k = \frac{1}{2} [K_{11}(\text{div}\,n)^2 + K_{22}(n \cdot \text{curl}\,n)^2 + K_{33}(n \times \text{curl}\,n)^2] \qquad (24)$$

In the more general case of a finite director anchoring at the boundaries, the total energy of a liquid crystal is written as the sum of the bulk and the surface terms:

$$F = F_V + F_S \qquad (25)$$

Les us consider the splay Frederiks transition (all the expressions are also valid for the bend effect if K_{11} and $\gamma_\parallel$ are replaced by K_{33} and $\gamma_\perp$, respectively). In the case of the uniform orientation of the director in the plane of the sample (the x,y-plane, wavevectors of distortion $q_x, q_y = \infty$) the director profile $n(z)$ takes a two-dimensional form:

$$n(z) = [\cos \theta(z), \sin \theta(z)] \qquad (26)$$

which can be described by elliptic functions [8, 9].

The distortion starts at a well-defined threshold field inversely proportional to the cell thickness d:

$$H_S = \frac{\pi}{d} \left(\frac{K_{11}}{\gamma_a} \right)^{\frac{1}{2}} \qquad (27)$$

since, with increasing thickness, a weaker field is necessary to overcome the elastic energy $K_{11} q^2$ due to a smaller wavevector of the distortion $q = \pi/d$.

For a weak surface anchoring (surface energy $W^S < \infty$) the thickness dependence of the threshold field

$$H(W) = \frac{H_S}{d+b} \qquad (28)$$

includes a so-called 'surface extrapolation length'

$$b = \frac{K_{11}}{W^S} \qquad (29)$$

Equation (28) holds only for the zero pretilt angle of the director at the limiting boundaries. Any finite pretilt angle results in a loss of the threshold character of the effect (a continuous distortion instead of the Frederiks transition).

The dynamics of the splay and bend distortions inevitably involve the flow processes coupled with the director rotation. Such a backflow effect usually renormalizes the viscosity coefficients. Only a pure twist distortion is not accompanied by the flow. In the latter case, and for the infinite anchoring energy, the equation of motion of the director ϕ (angle variation) expresses the balance between the torques due to the elastic and viscous forces and the external field (and does not contain the fluid velocity) [62, 63]:

$$K_{22} \frac{\partial^2 \phi}{\partial z^2} + \gamma_a H^2 \sin\phi \cdot \cos\phi = \gamma_1 \frac{\partial \phi}{\partial t} \quad (30)$$

This equation describes director rotation in the magnetic field with the inertial term $I\partial^2\phi/\partial t^2$ being disregarded, $\gamma_1 = \alpha_3 - \alpha_2$ is the rotational viscosity, α_i are Leslie's coefficients. In the limit of small ϕ angles, $\phi \ll 1$, Eq. (30) reduces to a linear form:

$$K_{22} \frac{\partial^2 \phi}{\partial z^2} + \gamma_a H^2 \phi = \gamma_1 \frac{\partial \phi}{\partial t} \quad (31)$$

with the solution:

$$\phi = \phi_m \exp\left(\frac{t}{\tau_r}\right) \cdot \sin\frac{\pi z}{d} \quad (32)$$

where

$$\tau_r = \frac{\gamma_1}{\gamma_a H^2 - K_{22}\pi^2/d^2} \quad (33)$$

is the reaction or switching-on time.

The corresponding relaxation or decay time is found from Eq. (31) for $H = 0$ in a similar way:

$$\tau_d = \frac{\gamma_1}{K_{22} q^2} = \frac{\gamma_1 d^2}{K_{22} \pi^2} \quad (34)$$

Backflow effects may accompany the transient process of the director reorientation [64, 65]. The process is opposite to the flow orientation of the director known from rheological experiments. Disregarding the backflow, we can use the same equations for the splay (with K_{11}) and bend (K_{33}) small-angle distortions. The backflow effects renormalize the rotational viscosity of a nematic:

$$\gamma_1(S) = \frac{\gamma_1 - 2(\alpha_3)^2}{(\alpha_3 + \alpha_4 + \alpha_6)(\alpha_3 - \alpha_2)} \quad (35\,\text{a})$$

$$\gamma_1(B) = \frac{\gamma_1 - 2(\alpha_2)^2}{(\alpha_4 + \alpha_5 - \alpha_2)(\alpha_3 - \alpha_2)} \quad (35\,\text{b})$$

for the splay and bend distortion, respectively. The effective viscosity for the splay distortion changes by less than 1% and the bend distortion viscosity is smaller than γ_1 by 10–20%; thus the corresponding transient process including the backflow effect is faster.

Weak boundary anchoring increases both τ_r and τ_d. This is easily understood, if we remember that a finite anchoring energy W^S results in an increase in the apparent thickness of the cell according (Eq. (28)).

The classical results just discussed have been applied successfully to the determination of the elastic moduli K_{ii} and Leslie's viscosity coefficients α_i of conventional nematics. The experimental data may be found in books [3, 8–10]. The measurement of the director profile throughout the cell, including the surface layers, allows the anchoring energy to be calculated. The profile of the director distortion above the threshold of the Frederiks transition is shown in Fig. 11. The difference between curves 1 ($W^S = \infty$) and 2 ($0 < W^S < \infty$) is a measure of the anchoring energy. The simplest estimate comes from the comparison of the magnetic field coherence length at the threshold field H_c [66–68]:

$$\xi_H^c = \frac{1}{H_c}\left(\frac{K_{ii}}{\gamma_a}\right)^{1/2} \quad (36)$$

with the sum $d + 2b$ (Eq. (28)). When the applied field is much higher than the threshold of the Frederiks transition, the field coherence length becomes comparable with the surface extrapolation length itself $\xi_H \sim b$. This condition corresponds to the second threshold of the complete reorientation of a liquid crystal, including the surface layers (curve 3, Fig. 11). Thus, the second threshold field also allows b (and W^S) to be calculated.

The direction of the field-induced rotation of the director is, in principle, degenerate. The clockwise and anticlockwise directions are equally probable. Thus, domain

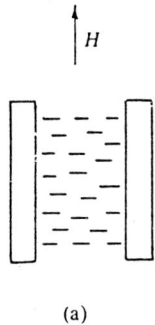 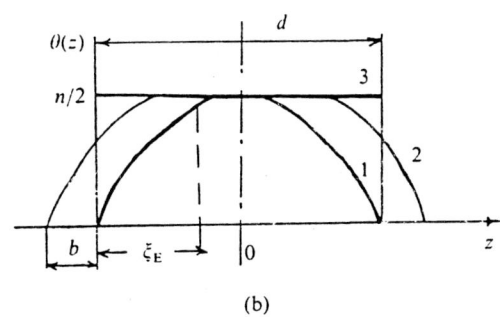

Figure 11. The geometry of the bend Frederiks transition (a) and the profile of the director deviation, induced by an external field (b).
(1) $H > H_c$, $W^S = \infty$;
(2) $H^* > H > H_c$, $0 < W^S < \infty$;
(3) $H > H^*$, $0 < W^S < \infty$.

walls appear, separating the two areas (domains) with different director orientations [69]. The wall movement influences the dynamics of the transition; the corresponding theory is presented in [70].

The Frederiks transition induced by an electric field has certain specific features not observed in the magnetic field equivalent. First, as the dielectric anisotropy is usually much higher than the magnetic one, the distortion leads to a spatial dependence of the dielectric susceptibility and a certain inhomogeneity in the electric field. Secondly, the character of the distortion may be affected by the electric conductivity and the influence of the flexoelectric effect. A discussion of these effects may be found in literature (e.g. [8, 9]).

More interesting features of the Frederiks transition are observed in the case of the simultaneous action of an electric and magnetic field. If a magnetic field is applied parallel to the nematic director $n \parallel H$ and $\gamma_a > 0$, it increases the apparent elastic coefficient by an additional quantity of $\gamma_a H^2 d^2/\pi^2$, which results in higher values of the corresponding threshold voltages. However, this is not the only result of applying a magnetic field [71]. If a stabilizing magnetic field is applied along the homogeneously or homeotropically oriented director and a destabilizing electric field is applied perpendicular or parallel to the substrate plane, first-order Frederiks transitions with a discontinuous jump of the director orientation at the threshold voltage may take place for sufficiently high H values.

When $E \perp H \perp n$ and both E and H are destabilizing, several distortion regions result [72]. For sufficiently high magnetic fields and small electric fields, twist distortion takes place in the direction of H. If the magnetic field is low and the electric field is strong, a splay–bend distortion occurs. In the region, where the actions of the electric and magnetic fields are comparable, a mixed splay–bend–twist deformation arises. Both second- and first-order phase transitions occur between the differently distorted regions; first-order transitions are accompanied by bistability and hysteresis phenomena [73].

The Frederiks transition in linear-chain and comb-like polymers may be treated in an analogous way as low-molecular-weight compounds, with some precautions [13], because, in general, the electric and viscoelastic properties of liquid crystal polymers are field dependent and the response of the materials to an external field is essentially nonlinear. Unfortunately, in the major part of electro-optical experiments this nonlinearity is not taken into account and results are interpreted in terms of conventional nematodynamics and constant material parameters. In addition, in only a few papers (e.g.

[74]) is a certain preliminary orientation of a polymer specified, and only then can one speak of the true Frederiks transition with a well-defined threshold voltage.

9.3.1.2 Field-Induced Periodic Structures

Under certain conditions the field-induced reorientation caused by the dielectric torque may result not in a uniform distribution of the director in the plane of a liquid crystal layer but in a spatially periodic distortion, either steady-state or transient.

Steady-State Patterns

At present, at least three types of steady-state dielectrically driven pattern are known for nematics. The electric-field-induced periodic bend distortion in the form of parallel stripes has been observed in a homeotropically oriented layer of 5-CB ($\varepsilon_a = 13$) in the presence of a stabilizing magnetic field H_z [75, 76]. The stripes with a wavevector q were parallel to the electric field E and stationary at low fields. It was shown that a stable periodic pattern of the director minimizes the free energy of the cell when the elastic moduli K_{11} and K_{33} are similar to each other. In these experiments the Frederiks transition is of first order, the nondeformed and deformed areas coexist at a given voltage, and the front between them may propagate along the direction y perpendicular to both fields [77].

The other stationary periodic pattern has been observed in polymer lyotropic liquid crystals [78]. The smallness of the ratio K_{22}/K_{11} of a poly-g-benzylglutamate solution results in a structure in which the period of the longitudinal domains is of the same order of magnitude as the cell thickness. Due to the high viscosity of the material ($\gamma_1 = 34\,P$) the time for domain formation exceeds 2 h. If the director was initially oriented along the x-axis (n_x) and the field (magnetic in [68]) acts along the cell normal z, the structure of the distortion may be searched for in the form

$$n_z = f(z)\cos qy;\quad n_y = g(z)\sin qy;\quad n_x = 1 \tag{37}$$

(the uniform Frederiks distortion corresponds to $q = 0$).

The threshold voltage may be plotted as a function of K_{22}/K_{11} [79] (Fig. 12). The curves shown in the figure were calculated numerically for the zero anchoring energy ($W^S = 0$) at both boundaries. It can be seen that for $K_{22}/K_{11} < 0.303$ the periodic splay–twist distortion is more favorable than a uniform splay.

The same result has been confirmed [80, 81] for a finite anchoring energy. The critical ratio K_{22}/K_{11} increases with increasing W^S, and reaches 0.5 for infinitely strong anchoring. The periodic distortion has been observed in a magnetic field, but an analogous case must exist in an electric field. Different geometries for studying the effect have been reviewed by Kini [82].

The steady-state modulated structure is also observed in homogeneously oriented nematic layers at frequencies of the applied

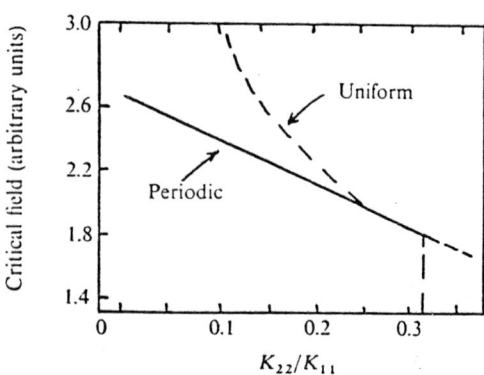

Figure 12. Calculated values of the critical field (in units of $(\pi/d)(K_{22}/\gamma_a)^{\frac{1}{2}}$) for the periodic splay–twist and uniform splay distortions [78].

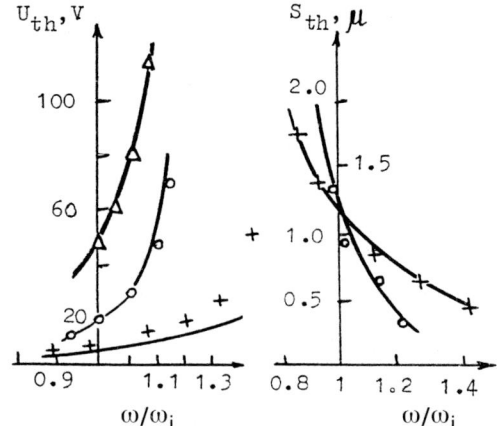

Figure 13. Frequency dependence of the threshold voltage U_{th} and domain period S_{th} for Frederiks domains near the dielectric anisotropy sign inversion frequency [83, 84]. (—) Calculated values. Experimental values: (+) $\varepsilon_0 - \varepsilon_\perp = +4.75$; (○) $\varepsilon_0 - \varepsilon_\perp = +0.35$; (△) $\varepsilon_0 - \varepsilon_\perp = +0.05$. ε_0 is $\varepsilon_\parallel$ for $\omega \to 0$.

ac voltage close to the frequency of the sign inversion ω_i of the dielectric anisotropy [83, 84]. The calculations show that the free energy is minimum when the periodic splay distortion and periodic potential distribution in the plane of a layer occur. The frequency region of the modulated Frederiks structure is very narrow near $\omega = \omega_i$. For $\omega < \omega_i$ the uniform Frederiks transition becomes more favorable, while for $\omega > \omega_i$ the threshold diverges rapidly (Fig. 13). If an initial director orientation is homeotropic the curves for the threshold voltage U_{th} and the period of the pattern at the threshold S_{th} reverse their slope with respect to the axis $\omega = \omega_i$ [84].

Transient Patterns

The formation of transient domain patterns aligned perpendicular to the initial director during the relaxation process of the magnetically driven Frederiks transition has been known since the earliest observations [85, 86]. The parallel stripes observed are, in fact, walls separating splayed regions with the clockwise and anticlockwise rotations of the director, arising due to coupling of the director orientation with a flow of liquid (the backflow effect [64] discussed above). The rate of the pattern relaxation is governed by the dynamics of the domain walls [87]. The same mechanism is responsible for the appearance of oblique structures [88].

In some cases, magnetically induced transient twist distortions have been observed in both thermotropic (MBBA [89]) and lyotropic (PBG [90]) systems. In this case, backflow effects are allowed only in a nonlinear regime, for strong distortions. The physical origin of this phenomenon could be the faster response times of modulated structures, as compared with uniform ones. When the equilibrium director distribution is approached, i.e. a relaxation process is over, the transient structures disappear. The emergence and subsequent evolution of the spatial periodicity of the transient structures have been considered theoretically [89, 90]. In addition, the pattern kinetics have been studied in detail experimentally [91] on a mixture of a polymer compound with a low-molecular-mass matrix. The polymer considerably increases the rotational viscosity of the substance and reduces the threshold for pattern formation. This indicates the possibility of recording the pattern using a video camera. A typical transient pattern is shown in Fig. 14 [91].

Some transient patterns overlapping the Frederiks transition (e.g. as observed by Buka et al. [92]) may be electrohydrodynamic in nature; these are discussed in Section 9.4 of this Chapter.

9.3.1.3 Flexo-Electric Phenomena

When liquid crystals possess electric polarization P (either spontaneous or induced by external factors), in addition to the qua-

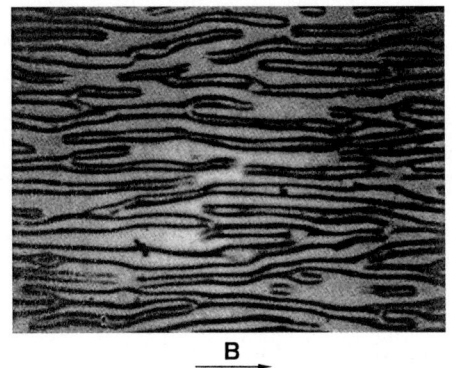
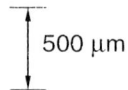

Figure 14. Transient pattern observed in 5-CB doped with a polymer. The magnetic induction B is in the plane of the liquid crystal layer (thickness 100 μm) and perpendicular to the initial director (twist distortion) [91].

dratic-in-field coupling of an external electric field to the dielectrically anisotropic medium (Eq. (1)) a new, linear-in-field interaction appears (free energy density, PE). One of the sources of the electric polarization is an orientational distortion of the liquid crystal.

Classical Results

The macroscopic dipole moment of a unit volume (i.e. the electric polarization) may result from both the nonuniform director orientation [93]

$$P_f = e_1 n \, \text{div} \, n - e_3 (n \times \text{curl} \, n) \qquad (38)$$

and the spatially nonuniform orientational order parameter [94]

$$P_0 = r_1 (n \cdot \text{grad} \, S) n + r_2 \, \text{grad} \, S \qquad (39)$$

The flexoelectric polarization P_f depends on the curvature of the director field (div n and curl n) at constant modulus of the order parameter S. For uniform director orientation (n = constant), $P_f = 0$. A simplified molecular picture of the phenomenon is shown in Fig. 15a, b. Equation (39) defines the so-called ordoelectric polarization P_0, which depends on the gradient of the nematic orientational order parameter S (quadrupolar in nature) and does not vanish for a uniform director distribution. For instance, near a solid wall the nematic order parameter is a function of the z coordinate normal to the surface. The ordoelectric coefficients r_1 and r_2 are of the order of the flexoelectric coefficients e_1 and e_3, which are approximately 10^{-4} CGS. We will not discuss ordoelectricity here as it is mostly related to

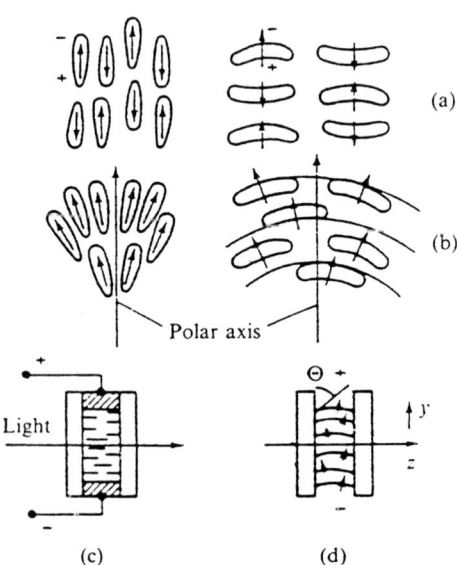

Figure 15. Flexoelectric effect: (a) structure of an undeformed nematic liquid crystal with pear- and banana-shaped molecules; (b) the same nematic liquid crystal subjected to splay and bend deformations, respectively; (c) experimental configuration; (d) distribution of the director along the z axis due to the flexoelectric effect.

the surface properties reviewed in other chapters in this book.

The flexoelectric effect results in a distortion of a dielectrically stable configuration (homeotropic nematic with $\varepsilon_a < 0$), provided that the director anchoring at the boundaries is weak (Fig. 15c). For small distortions the components of the director corresponding to the experimental geometry may be written as [95, 96]:

$$n_z = \cos \theta(z) \approx 1$$

and

$$n_y = \sin \theta(z) \approx \theta(z)$$

Then Eq. (38) reduces to

$$P_y = e_3 \frac{\partial \theta}{\partial z} \qquad (40)$$

For small dielectric anisotropy ($\varepsilon_a \approx 0$) the free energy density includes only two terms: the flexoelectric and the elastic terms derived from the Frank energy (Eq. (24))

$$g = P_y E + \frac{K_{33}}{2} \cdot \frac{\partial^2 \theta}{\partial z^2} \qquad (41)$$

By minimizing Eq. (41) with substitution of Eq. (40), one obtains the linear dependence of angle θ on the z coordinate:

$$\frac{\partial \theta}{\partial z} = \frac{e_3 E}{K_{33}} \qquad (42a)$$

$$\theta = \frac{e_3 E}{K_{33}} z \qquad (42b)$$

where $z=0$ corresponds to the center of a layer.

The resulting distortion is shown in Fig. 15d. It differs from the distortion in a Frederiks transition (see e.g. Fig. 11): the maximum angle of deviation of the director due to the flexoeffect occurs at the restricting surfaces, and the angle of deviation in the middle of the cell is zero. This is due to a particular geometry of the distortion: the polarization under a field occurs uniformly throughout the volume and, as a result, the torque capable of rotating the director through angle θ occurs only at the surfaces. Because of this, weak anchoring of the nematic liquid crystal at the surface is a necessary requirement for the occurrence of such a flexoelectric distortion. The case of finite dielectric anisotropy and other geometries of the flexoelectric distortion have also been considered [96].

The dynamics of the flexoelectric effect have been studied [97, 98]. For a nematic layer with a geometry similar to that shown in Fig. 12c and finite dielectric anisotropy, the torque balance equation for the bulk of a sample, written in the linearized form ($\sin \theta \approx \theta$, $\cos \theta \approx 1$)

$$\gamma_1 \frac{\partial \theta}{\partial z} = K_{33} \frac{\partial^2 \theta}{\partial z^2} - \frac{\varepsilon_a E^2 \theta}{4\pi} \qquad (43)$$

does not contain a destabilizing flexoelectric torque, as it is linear in the derivatives of the director. However, the flexoelectric term appears in the boundary conditions:

$$K_{33} \frac{\partial \theta}{\partial z}\bigg|_{z=\pm d/2} \pm W_i^s \theta|_{z=\pm d/2} - e_3 E = 0 \qquad (44)$$

Here W_i^S corresponds to different anchoring energies at the two opposite surfaces ($i=1$ or 2), which have coordinates of $\pm d/2$ (where d is the thickness of the layer). The solution of the linear equations (Eqs. (43) and (44)) is given for various values of ε_a and W_i^S. For an ac electric field of frequency ω, two spatially periodic viscoelastic waves with wavevectors depending on the field frequency are predicted. For sufficiently high frequencies the waves decay in the vicinity of the glass plates limiting the cell. At low frequency, the waves interfere with each other in the bulk, and for $\omega=0$ the steady state distortion shown in Fig. 15d occurs.

The bulk flexoelectric response is limited by a narrow frequency region below the director relaxation frequency determined by Eq. (34). The viscoelastic wave of the director curvature decaying into the bulk of the substance has been probed experimentally by using an evanescent light wave and the modulation ellipsometry technique [99]. The wavevector of the curvature oscillations increases with increasing field frequency, $q_\omega = (\omega \gamma_1 / 2 K)^{\frac{1}{2}}$. In such a case the linear-in-field response was observed up to frequencies of the order of 100 kHz. The characteristic frequency $f_c = K/\pi \gamma_1 \lambda^2$ is determined by a crossover of two distances: the spatial period of the curvature wave $2\pi/q_\omega$ and the penetration length of the optical evanescent wave (a few tenths of the light wavelengths λ).

Flexoelectric Domains

The flexoelectric term δF in nematic free energy

$$\delta F_f = -\int_V [e_1 \, \mathbf{E}\mathbf{n} \, \mathrm{div}\,\mathbf{n} - e_3 \, \mathbf{E}(\mathbf{n} \times \mathrm{curl}\,\mathbf{n})]\,\mathrm{d}\tau \qquad (45)$$

may result in a two-dimensional spatially periodic structure in the bulk. The application of an external dc electric field to a homogeneously oriented nonconductive nematic layer gives rise to a pattern of the domains parallel to the initial director orientation [100]. The period of the stripes w_{th} and the threshold voltage U_{th} of their appearance has been found [101–103] by minimizing the free energy of the nematic in an electric field, taking into account the term given by Eq. (45). A detailed experimental study of the dependence of the threshold and period of the longitudinal domains on the dielectric anisotropy and other parameters of nematic liquid crystals [104] has shown very good agreement between theory and experiment. The flexoelectric modulated structure in nematic liquid crystals is also known as the "variable grating mode" [105–108], since for $U > U_{th}$ the period of domains w varies as $w \sim U^{-1}$ (Fig. 16).

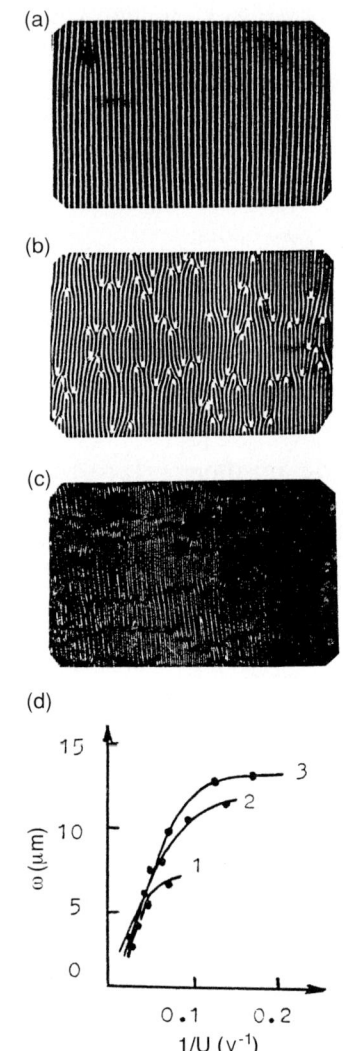

Figure 16. Flexoelectric domains in the nematic liquid crystal 4-butyl-4′-methoxyazoxybenzene [104]: (a) $U = 16$ V, (b) $U = 25$ V, (c) $U = 50$ V ($\varepsilon_a = -0.25$, $d = 11.7$ μm). (d) Variation in the domain period for different ε_a values: (1) $\varepsilon_a = -0.25$; (2) 0; (3) +0.15.

9.3.2 Twisted Nematics and Cholesterics

9.3.2.1 Twist and Supertwist Structures

Twist Effect

If the directions x and y of the planar orientation of nematic liquid crystal molecules on opposite electrodes are perpendicular to each other and the material has a positive dielectric anisotropy, then when an electric field is applied along the z axis (Fig. 17) a director reorientation occurs which is a combination of the splay, bend, and pure torsional distortions [109]. In the absence of the field (Fig. 17a), the light polarization vector follows the director and, consequently, the structure rotates the light polarization vector through an angle $\pi/2$ [110]. This specific waveguide regime (Mauguin's regime) occurs when the phase delay satisfies the condition $\Delta n \, d/\lambda \gg 1$.

As in the case of the Frederiks transition (discussed in Section 9.3.1.1 of this Chapter), the theoretical interpretation of the twist effect is based on the minimization of the free energy of the system. In this case, however, the problem is two dimensional, since both the azimuthal angle $\phi(z)$ and the tilt angle $\theta(z)$ are considered to be dependent on the z coordinate. In the case of the infinitely strong anchoring, the threshold for the distortion includes all three elastic moduli of a nematic liquid crystal [111]:

$$U_{\text{tw}} = \pi \left[\frac{\pi}{\varepsilon_a} (4K_{11} + K_{33} - 2K_{22}) \right]^{\frac{1}{2}} \quad (46)$$

When the applied voltage exceeds this threshold, the director deviates from the initial orientation so that the linear z dependence of the azimuthal angle disappears and the tilt angle becomes nonzero (Fig. 17b). The qualitative character of the functions $\phi(z)$ and $\theta(z)$ for different voltages is shown

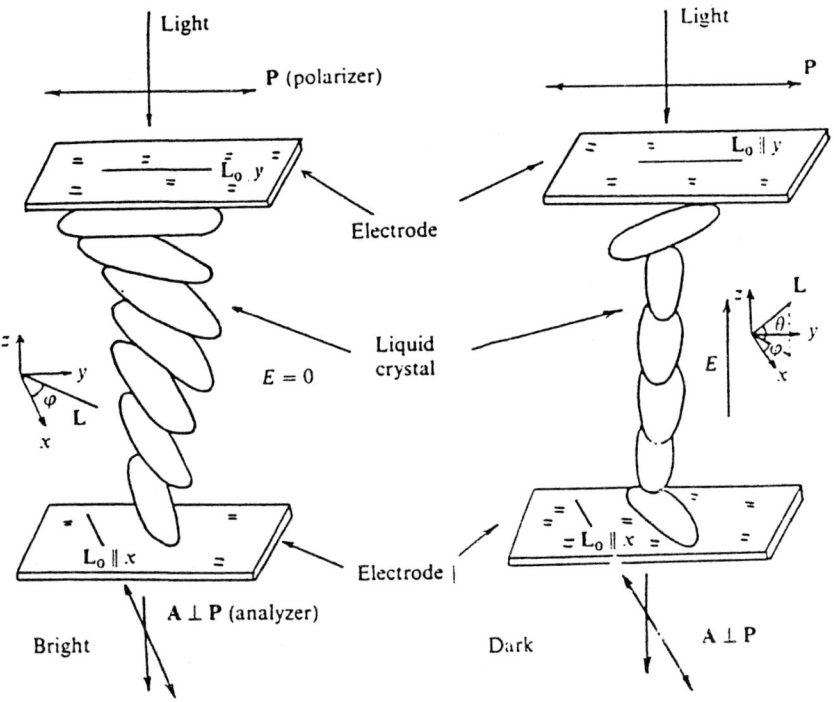

Figure 17. Twist-effect. Left: below threshold; Right above threshold

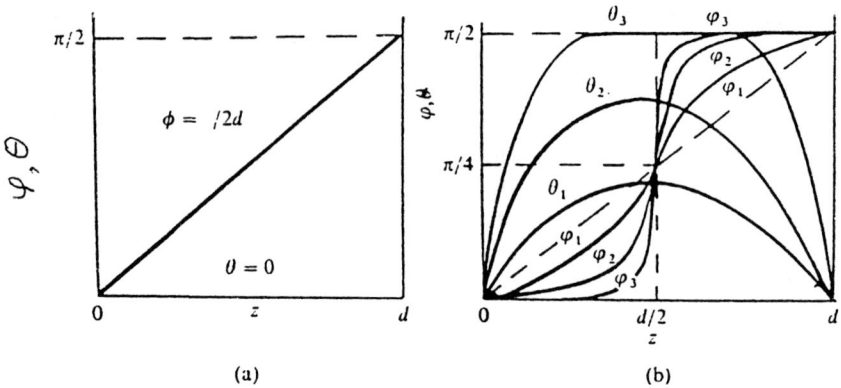

Figure 18. Dependence of director angles $\phi(z)$ and $\theta(z)$ in the twist effect for different voltages: (a) $U < U_{tw}$, (b) $U_{tw} < U_1 < U_2 < U_3$.

in Fig. 18. As the director tends to orient perpendicular to substrates the average value of $\langle \Delta n \rangle$ decreases, and for a certain voltage (the optical threshold for the twist effect) the waveguide regime vanishes.

Supertwist Effect

Supertwisted (i.e. twisted through an angle exceeding 90°) liquid crystal cells [112] are prepared from nematics doped with a small amount of an optically active material. Thus a cholesteric (or chiral nematics) with a large pitch P is created so that the helical pitch could be fitted to the boundary conditions for the directors at the substrates. For a cell with a supertwist angle ϕ_m, strong anchoring, and the zero director pretilt at the boundaries, the director distortion appears at a certain threshold voltage [113–115]:

$$U_{th}(\phi_m) = U_S$$
$$\cdot \left\{ 1 + \frac{\phi_m}{\pi} \left[\frac{\phi_m}{\pi} \left(\frac{K_{33}}{K_{11}} - \frac{2K_{22}}{K_{11}} \right) + \frac{4K_{22}d}{K_{11}P} \right] \right\}^{\frac{1}{2}} \quad (47)$$

where $U_S = \pi (4\pi K_{11}/\varepsilon_a)^{1/2}$ is the threshold for the splay Frederiks distortion in a planar nematic cell.

For a finite director pretilt at the boundaries θ_S, a sharp threshold disappears; however, it is possible to find a voltage at which the director angle at the center of the layer becomes equal to the boundary pretilt angle θ_S [116]. This voltage is very close to the optical threshold for the supertwist transition.

In supertwisted structures, bistable states are often observed [117]; for example, in the Grandjean texture with a tilted director orientation at the boundaries [118, 119]. In the absence of the tilt, the free (elastic) energy is a minimum at the thickness/pitch ratio $d/P_0 = n/2$ ($n = 1, 2, 3$, etc.), corresponding to n half-turns of the helix, located along the layer normal. If the directors at the boundaries are tilted then the corresponding free energy minima are no longer equidistant with thickness. At a certain thickness/pitch ratio, an electric field applied to a cell (the substance has $\varepsilon_a > 0$) switches the twisted structure from one state to another, differing in the number of half-turns. The relaxation of a new state might be very slow, and such a state may be considered stable. Thus, a field-induced bistability (or even multistability) effect occurs.

9.3.2.2 Instability of the Planar Cholesteric Texture

Texture Transitions

When an electric or magnetic field is applied to a liquid crystal cell, a texture transition occurs to minimize the free energy of the system. These texture changes in cholesteric liquid crystals are physically similar to the Frederiks transition in a nematic liquid crystal and result in a significant change in the optical properties of the layer. Texture transitions have been reviewed previously [8, 9] with allowance made for the sign of the dielectric or diamagnetic anisotropy, the initial texture, and the direction of the applied field. Here, we consider only the instability of the planar cholesteric texture, which has been widely discussed in recent literature.

Field Behavior of the Planar Texture

Let us discuss the case of when both the helical axis and the electric field are parallel to the normal z of a liquid crystal layer and the substance has positive dielectric anisotropy. The physical reason for the purely dielectric instability is rather well understood. When an electric field is not high enough to untwist a helix with a rather short pitch P_0, the dielectric torque tends to reorient the local directors into an arrangement similar to a homeotropically oriented nematic. In contrast, the elastic forces tend to preserve the distribution of the cholesteric layers (Fig. 19a). With not too large fields a pattern which undulates in the x and y directions appears as a compromise distribution of the director (Fig. 19b). The wavevector of this distortion along the z axis is approximately π/d. Related to the two-dimensional director pattern is a two-dimensional periodicity in the distribution of the refractive index, and hence a two-dimensional optical grating is formed. This grating is similar to the grid pattern formed due to the electrohydrodynamic instability (see Fig. 31c). With increasing voltage, the angle of deviation of the director increases, tending towards the limit $\theta = \pi/2$ for the whole layer (helix untwisting).

The threshold field for the formation of the periodic distortion has been calculated [120, 121] based on the expression for the

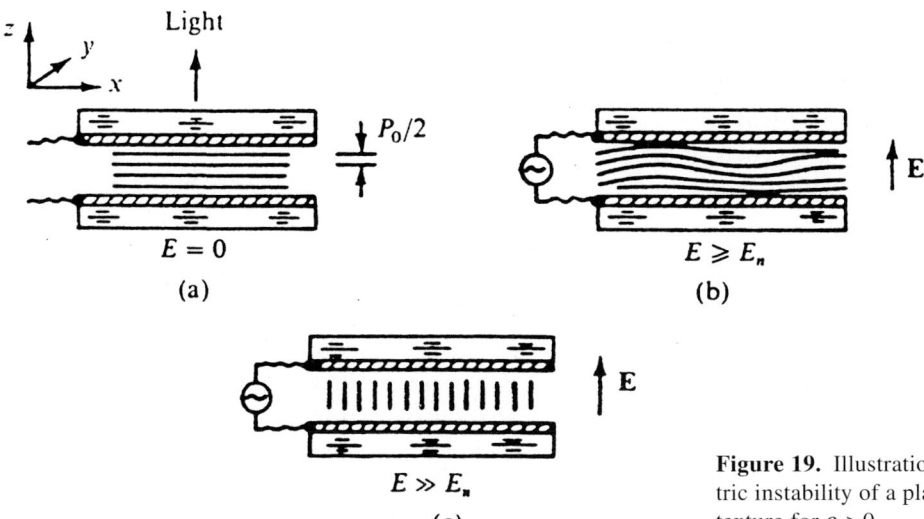

Figure 19. Illustration of the dielectric instability of a planar cholesteric texture for $\varepsilon_a > 0$.

free energy of a cholesteric liquid crystal and assuming $\varepsilon_a \ll \varepsilon$ and $d \ll P_0$:

$$g = \frac{1}{16} \cdot \left[3K_{33}\left(\frac{\pi}{w}\right)^2 + 8K_{22}\left(\frac{wq_0}{d}\right)^2 - \varepsilon_a \boldsymbol{E}^2 \right] \theta_m^2 \quad (48)$$

In addition, it was assumed that there were only small deviations of the axis of the cholesteric helix from the normal to the glass surfaces or, equivalently, small deviations of the director from the xy planes (i.e. $\sin\theta \simeq \theta$ and $\cos\theta \simeq 1$). Futhermore, the deformation was assumed to be sinusoidal along both the x axis (the half-period equals w) and the z axis (the half-period equals d) and is described, in contrast to the nematic case, by two variables. One is the angle of deviation of the helical axis from the normal to the surfaces

$$\theta = -\theta_m \sin\left(\frac{\pi z}{d}\right) \times \sin\left(\frac{\pi x}{w}\right) \quad (49)$$

and the other is the difference in wavevectors for the distorted and the equilibrium helix

$$\Delta q = q - q_0 \, (q_0 = 2\pi/P_0) \quad (50a)$$

$$\Delta q = \Delta q_m \cos\left(\frac{\pi z}{d}\right) \times \sin\left(\frac{\pi x}{w}\right) \quad (50b)$$

The maximum compression Δq_m of the cholesteric planes and the maximum angle of deviation θ_m of the helical axis from the normal can be related to each other using purely geometric considerations:

$$\Delta q_m = \theta_m \left(\frac{q_0 w}{d}\right) \quad (51)$$

This value is substituted into the term with the K_{22} coefficient in Eq. (48). The numerical factors in Eq. (48) appear as a result of averaging the energy over the period of deformation [121].

The minimization of the elastic part of the free energy (Eq. (48)) (without the term containing the field) by choosing the proper value of w gives the period of the distortion

$$w = \frac{1}{2}\left(\frac{3K_{33}}{2K_{22}}\right)^{\frac{1}{4}} (P_0 \, d)^{\frac{1}{2}} \quad (52)$$

and, using the condition $g=0$ with allowance for the field term, we find the threshold voltage for the distortion

$$U_{\text{th}} = 2\pi^{\frac{3}{2}} (24 K_{22} K_{33})^{\frac{1}{4}} \left(\frac{d}{\varepsilon_a P_0}\right)^{\frac{1}{2}} \quad (53)$$

The threshold field is independent of the frequency up to the dispersion region ω_D of the dielectric permittivity.

The action of the field on a planar texture of a cholesteric liquid crystal for cell thicknesses comparable to the pitch and with a rigid anchoring of the directors to the surfaces does not cause two-dimensional deformations. In this case one-dimensional periodic patterns are observed [8, 9], with the orientation of the domains depending on the number of half-turns of the helix contained within the cell thickness. In fact, the domains are always perpendicular to the director in the middle of the layer.

A theoretical consideration of the case of a pitch that is comparable to the layer thickness for a purely dielectric destabilization of a planar texture in a field $\boldsymbol{E} \parallel \boldsymbol{h}$ has been given both numerically [122] and analytically [123, 124]. In the latter case the perturbation theory was used to search for the structure of the director field just above the threshold of the instability. Two variables, the polar angle θ and the azimuthal angle ϕ were considered, with orientation of the director at opposite walls differing by a twist angle α (pretilt angles at boundaries were also taken into account). It has been shown that two types of instability can be observed depending on the elastic moduli of the material: a total twist of the structure between two boundaries α, with a ratio of the cell thickness d to cholesteric pitch P_0; and dielectric anisotropy. The first type is a homogeneous distortion analogous to the Frederiks transition in nematic liquid crystals. The second type is a periodic distortion (unidimensional in the layer plane) with a wavevector q directed at certain angles with respect to the director at boundaries.

With increasing voltage we enter one of the regimes (uniform or periodic distortion) separated by the Lifshits point where the wavevector of distortion is zero. The corresponding phase diagram has already been discussed for nematics (see Fig. 12). The periodic instability is favored when the original twist and the normalized cell thickness (d/P_0) increase and the dielectric anisotropy decreases. The principal predictions of the theory, namely, the disappearance of the periodic instability in thin cells and for small twist angles, agree with experiments performed on the Grandjean planar texture with varying d/P_0 [122].

Qualitatively similar results were obtained by Cohen and Hornreich [125]. In addition, it was claimed that the presence of a pretilt in the planar cholesteric texture decreases the tendency of the system to undergo transition to a periodically modulated (ripple) phase.

9.3.2.3 Field Untwisting of the Cholesteric Helix

Untwisting of a helix is observed only with positive dielectric or diamagnetic anisotropy and only in a field perpendicular to the helical axis h. Thus, in some cases, the director reorientation (or a texture transition) has to occur before untwisting is possible.

The Static Case

The model investigated by Meyer [126] and de Gennes [127] is illustrated in Fig. 20. Let

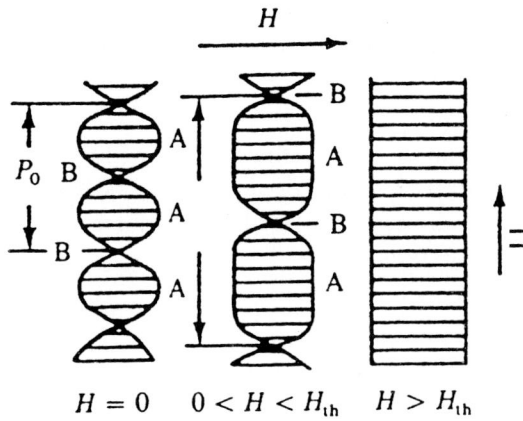

Figure 20. Schematic representation of untwisting of the cholesteric helix by a magnetic field [1].

us assume that we have a helix with pitch P_0 in the absence of a field, and that the thickness of the sample is sufficiently large (relative to P_0); thus the boundary conditions can be neglected. If $H \perp h$, then, in the initial state, in certain parts of the helix (A) the molecules are arranged favorably relative to the field, but in other parts (B) they are arranged unfavorably and tend to reorient themselves with their axes aligned along the field. As a result, the A regions will increase in size not so much due to a sharp decrease in the dimensions of the B regions (this would require too large an elastic energy $K_{22}(\partial\phi/\partial z)$), as due to a decrease in their number. Therefore, the helical pitch increases and the helix itself becomes nonideal; that is, the sinusoidal dependence of the azimuth of the director ϕ upon the z coordinate is destroyed. When the last B region has been eliminated from the sample, the liquid crystal becomes a nematic, oriented uniformly along the field. Consequently, a field-induced cholesteric to nematic phase transition has occurred in the field.

The free energy of cholesterics in a magnetic field is

$$F = \frac{1}{2}\int\left[K_{22}\left(\frac{\partial\phi}{\partial z}-q_0\right)^2 - \gamma_a H^2 \sin^2\phi\right]dz \quad (54)$$

The function $\phi(z)$ should satisfy the condition for minimum F, and the corresponding Euler equation has a solution in terms of the elliptic function

$$\sin\phi(z) = \mathrm{sn}\left[\left(\frac{z}{\xi k}\right), k\right] \quad (55)$$

The values $u = z/\xi k$ and k are the argument and modulus, respectively, of the elliptic function, and

$$\xi = \frac{1}{H}\left(\frac{K_{22}}{\gamma_a}\right)^{\frac{1}{2}} \quad (56)$$

is the magnetic coherence length, see also the more general form given by Eq. (36).

The condition for the minimum of free energy (Eq. (54)) is

$$q_0\,\xi = \frac{2E(k)}{\pi k} \quad (57)$$

and the new pitch of the helix satisfying the condition of a minimum in F is

$$P(H) = P_0\left(\frac{2}{\pi}\right)F(k)\,E(k) = 4\,\xi\,k\,F(k) \quad (58)$$

Here $F(k)$ and $E(k)$ are complete elliptic integrals of the first and second kind. When $k \to 1$ the elliptic integral $F(k)$ diverges logarithmically, i.e. the helix pitch $P(H) \to \infty$. Simultaneously, when $k \to 1$, the integral $E(k) = 1$ and Eq. (57) give the critical field for the untwisting of the helix:

$$H_c = \frac{\pi^2}{P_0}\left(\frac{K_{22}}{\gamma_a}\right)^{\frac{1}{2}} \quad (59)$$

The pitch of the helix increases gradually, beginning with infinitely small fields. Expansion of Eq. (58) gives:

$$P(H) = P_0\left[1 + \frac{\gamma_a^2 P_0^4 E^4}{2^9\,\pi^4\,K_{22}^2} + \ldots\right] \quad (60)$$

The increase in the pitch with increasing electric (or magnetic) field has been confirmed in numerous experiments carried out on rather thick samples [8, 9].

Field untwisting of a helix for thin cells with planar boundary conditions occurs differently from the case of an infinite cholesteric medium. For a field perpendicular to the axis of a cholesteric helix in the planar texture, a stepwise change in the pitch with increase in the field is predicted [128]. The size of the step increases with a decrease in the ratio $2d/P_0$. Stepwise untwisting of the helix by an electric field perpendicular to the helix axis has been observed for the case of planar cholesteric texture with $\varepsilon_a > 0$ and with strong anchoring of the molecules to the limiting surfaces [129]. Under these conditions the relaxation of the field-induced (i.e. untwisted) state is accompanied by the formation of spatially modulated structures in the form of strips or grids.

The role of the boundary conditions has been extensively discussed in [130–133], and the influence of the flexoelectric effect has been analyzed [134].

Dynamics of the Untwisting

The dynamics of the helix unwinding is described by

$$\gamma_1\frac{\partial\phi}{\partial t} = K_{22}\frac{\partial^2\phi}{\partial t^2} + \gamma_a\,H^2\sin\phi\cos\phi \quad (61)$$

The approximate solution of Eq. (61) results in the following expression for the response τ_r and decay τ_d times for the field-induced helix distortion [135]:

$$\tau_r = \frac{2\gamma_1}{2K_{22}\,q^2 + \gamma_a\,H^2} \quad (62)$$

and

$$\tau_d = \frac{\gamma_1}{K_{22} q^2} \quad (63)$$

Here γ_1 and K_{22} are the rotational viscosity and twist elastic constant, q is a wavevector of the distortion mode under investigation. The plus and minus signs in Eq. (62) are related to the cases $q > q_0$ and $q < q_0$, respectively, where $q_0 = 2\pi/P_0$ is the equilibrium pitch.

In general, the dynamics of the untwisting of a helix have not been investigated sufficiently, and only the following fundamental facts have so far been established. The relaxation time is determined by the helical pitch P_0 when $d \gg P_0$, and by the thickness of the cell when $d \ll P_0$. The state of complete untwisting of the helix is metastable, and relaxation begins at defects in the structure. The response times in the field are inversely proportional to E^2.

Dynamics of Blue Phases

The field-induced distortion of the orientational state of blue phases was discussed in Section 9.2.4.3. These distortions decay rather rapidly. In the geometry of a Kerr cell (the field is perpendicular to the light wavevector) the field induces birefringence δn which results in electro-optical modulation ΔI of the transmitted light. Modulation characteristics have been studied over a wide temperature range of the blue phase with negative dielectric anisotropy $\varepsilon_a = -0.75$ [136]. The time constant for relaxation of the field-induced birefringence is well approximated by the simplest expression, Eq. (63) with $q = \pi/P_0$, where P_0 is the lattice constant and K is an averaged Frank modulus.

The dynamics of the BP$_{III}$ (foggy) phase have some features related to the field-induced changes in the size of the domains that this phase forms.

9.3.2.4 Flexoelectric Effects

As discussed for nematics, flexoelectric effects are caused by the linear coupling of an external electric field with the flexoelectric polarization. In cholesterics, these effects are manifested in the three specific phenomena described below.

Fast Linear-in-Field Rotation of the Cholesteric Helix

This effect is observed in a geometry where the cholesteric axis h is homogeneously oriented in the plane of the cell (along x) and an electric field is applied to the electrodes of a sandwich cell along the z axis [137,138]. In this case, the helical structure, even the ideal one, is incompatible with the planar boundary conditions, and splayed and bended regions form near the boundaries. Thus, according to Eq. (38), the flexoelectric polarization arises in those regions which can interact with the electric field. The distortion is very similar to that observed in the ferroelectric smectic C* phase (see Fig. 24) for a so-called 'deformed helix ferroelectric' effect [139].

Such a distortion results in a deviation of the optical axis in the plane of a cell. The sign and the magnitude of the deviation angle depend on the polarity and strength, respectively, of the applied field. In the field-off state the helix is undistorted and the cell behaves as a uniaxial optical plate, with the optical axis coincident with h. When the field is applied molecules leave the yz plane due to the flexoelectric deformation coming from the surface regions where the flexoelectric torque $M_f = e_f E$ is developed ($e_f = e_1 = e_3$ is assumed). Now the optical axis does not coincide with the initial orientation of the helix but forms the angle ψ with respect to it and proportional to E. The field-induced distortion of the helix is shown in Fig. 21 [137].

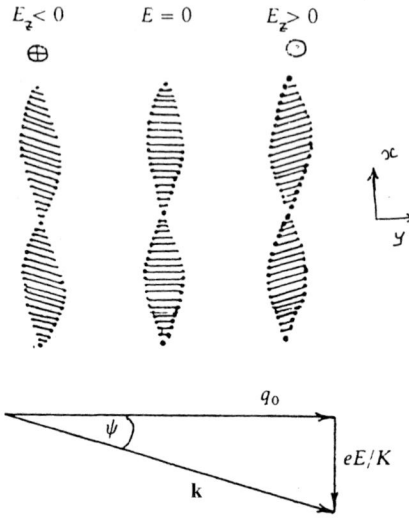

Figure 21. (Top) The pattern of the director rotation induced by an electric field applied perpendicular to the plane of the drawing [137]. (Bottom) Components of the wavevector k of the distorted helix.

In the steady-state regime, for director components parallel to the yz plane, $n_z = \cos\phi$, $n_y = \sin\phi$ (no conical distortions), the free energy density of the system in an electric field includes both the flexoelectric and dielectric terms:

$$g = \frac{1}{2} K \left(\frac{\partial\phi}{\partial y}\right)^2 + \frac{1}{2} K_{22}\left(q_0 - \frac{\partial\phi}{\partial x}\right)^2$$
$$- e_f E \frac{\partial\phi}{\partial y} + \frac{\varepsilon_a}{8\pi} E^2 \sin^2\phi \quad (64)$$

where $K = K_{11} = K_{33}$.

For negligible dielectric anisotropy ($\varepsilon_a \to 0$), minimization of Eq. (64) results in $\partial\phi/\partial x = q_0$ and $\partial\phi/\partial y = e_f E/K$ (the anchoring energy is considered to be small). If the rotated axis of the helix is represented by a wavevector k then $|k|\cos\psi = q_0$ and $|k|\sin\psi = e_f E/K$ (Fig. 21) and the rotation angle is given by

$$\tan\psi = \frac{e_f E}{q_0 K} \quad (65)$$

which is linear in E for small rotations.

For a finite value of dielectric anisotropy, with increasing field the linear response is affected by the helix unwinding [140] due to the quadratic-in-field interaction (last term in Eq. (64)). In order to increase the field-induced linear deviation of the optical axis, shot-pitch materials with small ε_a must be used [141].

Experiments show that the effect of the reorientation of the helical axis is fairly fast (in the range 10–100 µs) and the speed of the response is independent of field strength. This can be accounted for under the assumption that the pitch of the helix does not change in the field-on state, $|k| = q_0$. Then, the free energy becomes only a function of the angle ψ [137]. For small ψ, the rate of the response is determined by the dynamics of the field-free helix (Eq. (63)) with an effective viscosity coefficient instead of γ_1.

Electroclinic Response

When the helical structure of the chiral nematic phase is unwound by the influence of the limiting walls, one can observe a linear-in-field light modulation which is caused by a small molecular tilt [142]. The effect is analogous to the electroclinic effect observed in the smectic A phase as a pretransitional phenomenon in the vicinity of the A–C* transition. It is particularly strong in the vicinity of the N–A–C* multicritical point [143].

Flexoelectric Domains in Cholesterics

In Section 9.3.1.3 we discussed the longitudinal flexoelectric domains in nematics, the period of which depends on the inducing field, $w \propto E^{-1}$. When changing the sign of the field, the phases of the distortion angles $\theta(y)$ and $\phi(y)$ are changed for the opposite ones, but the direction of the wavevector of the instability (perpendicular to the initial director orientation n_0) remains unchanged. In cholesterics the mirror symmetry is broken and the same instability has specific features [144, 145].

First, the orientation of the domains depends on the number of half-turns of the helix between the cell boundaries. In the Grandjean texture the direction of the domains is different in neighboring Grandjean zones [146] for the same polarity of the field. When the polarity of the applied voltage is changed, the directions of the domain lines also change. With an oscillating field (at very low frequencies, of the order of 1 Hz), the direction of domains oscillates between two positions, characterized by an angle $\pm\alpha$ with respect to the director in the middle of the cell (in the first two Grandjean zones or in a twist nematic cell). The threshold voltage for the domain formation, and the domain period and angle α have been calculated [145].

For the zero zone with the directors at opposite boundaries parallel to each other (no twist) and zero dielectric anisotropy, both the threshold voltage and the wavevector can be calculated analytically as a function of the cholesteric pitch, despite the completely unwound state of the helix. The sign and amplitude of the angle of the field-induced domain rotation (at threshold) depends on the handedness and the pitch of the helix and the sign of the difference in the flexoelectric coefficients $e^* = e_{11} - e_{33}$:

$$\alpha_{th} = \frac{2 d K_{22}}{P_0 K_{33}} \text{sign}(e^* E) \qquad (66)$$

With increasing frequency of a pulsed electric field, $\omega > \tau_d$ where τ_d is the relaxation time of the director (Eq. (34)); the deviation angle decreases and the threshold voltage diverges [146].

The other flexoelectric structure can arise in a very thin planar cholesteric layer with a certain set of material parameters. The instability occurs in the form of spiral domains [147] the handedness of which depends on the sign of the electric field. While the threshold voltage for the linear domains discussed above is almost independent of cell thickness, the threshold for spiral domains is proportional to the thickness, and the number of turns of the helix seen under a microscope in the plane of the layer increases with increasing field. Spiral domains have been observed experimentally [148].

9.3.3 Smectics and Discotics

9.3.3.1 Field Behavior of Achiral Smectics

Frederiks Transition in a Smectic A Phase

In the smectic A phase the director is always perpendicular to the plane of the smectic layers. Thus, only the splay distortion leaves the interlayer distance unchanged and only the elastic modulus K_{11} is finite, while K_{22} an K_{33} diverge when approaching the smectic A phase from the nematic one. However, the compressibility of the layered structure and the corresponding elastic modulus B should be taken into account when discussing the elastic properties of smectic phases. The free energy density for the smectic A phase subjected to the action of an external magnetic field is [1]:

$$g = \left[B \left(\frac{\partial u}{\partial z} \right)^2 + K_{11} (\text{div} \, n)^2 - \gamma_a (n \, H)^2 \right] \qquad (67)$$

where $u(r)$ is the displacement of the smectic layers from their equilibrium position, and the z direction coincides with the normal to the layers. Thus, it makes sense to investigate the Frederiks transition in the geometry corresponding to the splay distortion only. Let us assume a so-called 'bookshelf geometry', in which the smectic layers are perpendicular to the substrates, forming a sandwich-type cell of thickness d, and the director is oriented in the plane of the cell (e.g. along the x axis). If the magnetic anisotropy is positive ($\gamma_a > 0$) and a magnetic field is applied, the Frederiks transition should take place. The threshold for the transition coincides formally with that for a nematic (Eq. 27)); however, the amplitude of the distortion θ_m for fields exceeding the threshold is very small, being of the order of λ_S/d [149], where a smectic characteristic length

$$\lambda_S = \left(\frac{K_{11}}{B}\right)^{\frac{1}{2}} \qquad (68)$$

is of the order of few layer thicknesses. Thus, even in this, favorable case of the allowed splay distortion, the Frederiks transition is, in fact, unobservable (ghost transition). Instead, one observes a texture transition accompanied by the appearance of a number of defects.

Dielectrically Induced Texture Transitions

Texture transitions are particularly pronounced when an electric field is applied to materials having a large dielectric anisotropy. A planar texture undergoes transition to a quasihomeotropic optically transparent texture via intermediate structural defects [150, 151]. The threshold voltage observed experimentally for a transition from a planar to a homeotropic texture depends on the layer thickness according to $U \propto d^{\frac{1}{2}}$ (for the Frederiks transition the threshold voltage is independent of thickness). A model that accounts for the experimental data (at least partly) has been developed by Parodi [152] who assumed the formation of transition layers between the surface and the bulk of a sample. A discrepancy between the calculated and observed periods of the texture instability may be due to a nonuniform 'chevron' structure of the original 'bookshelf' geometry. Chevron structures with smectic layers broken by a certain angle are often observed in both smectic C and smectic A layers [153].

The kinetics of the relaxation of the homeotropic field-induced texture to a focal conic one depends on the surface treatment [154–156]. A rough surface which provides a strong anchorage facilitates relaxation [156].

Frederiks Transition in a Smectic C Phase

Depending on the arrangement of the smectic layers and the director of the liquid crystal relative to the limiting surfaces, four different configurations can be identified. For each of them there are at least three alternatives for the direction of the field. Thus, 12 variations in all can be obtained and these have been studied theoretically by Rapini [149]. In most of the configurations the Frederiks transition is a ghost one, since it requires the collapse of the smectic layers. However, in three instances [157] the field only induces rotation of the director around the normal to the layers, leaving the layer structure unchanged.

Let us identify the three principal values of the dielectric permittivity as ε_i ($i=1, 2, 3$), where ε_3 corresponds to the direction along the director, ε_1 corresponds to the direction perpendicular to the plane of the tilt, and ε_2 corresponds to the direction perpendicular to the preceding two [8, 9]. When the field is oriented along the director, a Frederiks transition is possible when $\varepsilon_2 > \varepsilon_3$. The director should rotate around the normal to smectic layers, not changing its angle Ω. The corresponding threshold field is proportional to $(\varepsilon_2 - \varepsilon_3)^{-\frac{1}{2}}$. A field perpendicular to the director and lying in the plane of the tilt induces the same distortion when $\varepsilon_2 > \varepsilon_1$. Now the threshold for reorientation of the director is proportional to $(\varepsilon_2-\varepsilon_1)^{-\frac{1}{2}}$. Finally, when the field is perpendicular to the tilt plane the same rotation takes place at threshold proportional to $(\varepsilon_1\cos^2\Omega+\varepsilon_3\sin^2\Omega-\varepsilon_2)^{-\frac{1}{2}}$.

Measurements have been made of the Frederiks transition in various smectic C liquid crystals [158–161], using the conventional sandwich cells and optically transparent electrodes. The director was oriented uniaxially by rubbing the electrodes, and the smectic layers were tilted with respect to the cell plane (a 'bookshelf' geometry). Due to positive dielectric anisotropy, an electric field applied along the cell normal (z direction) (Fig. 22) induces director rotation around the normal to the smectic layers.

The threshold field for the director deviation can be calculated neglecting the distortion of the smectic layers [158] and assuming $\varepsilon_2 \approx \varepsilon_3$. The threshold depends on the angle μ between the normal and xy plane. For $\mu=0$, when the xy plane cuts half of the director cone (with an angle Ω at the apex) the threshold field is

$$E_c = \frac{\pi}{d\sin\Omega}\cdot\left(\frac{4\pi K^*}{\varepsilon_a}\right)^{\frac{1}{2}} \quad (69)$$

where K^* is an effective elastic constant. When the cone just touches the xz plane

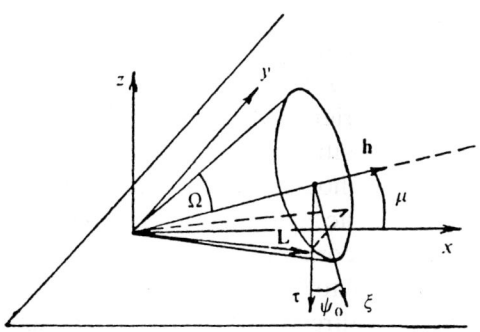

Figure 22. Geometry of the Frederiks transition for a planar-oriented smectic C liquid crystal.

($\mu=\Omega$) (Fig. 22), the threshold field increases proportionally to $\cos\Omega$. Both such cases have been studied using various substances [159–161]. For a certain set of parameters, the theory predicts a bistable behavior of the smectic C phase in a strong field. Such bistability has been observed in the form of the field-induced motion of domain walls, accompanied by a hysteresis in the process of the director reorientation. The walls appear due to the equally probable field-induced rotation of the director clockwise and anticlockwise, and separate the two areas of different director orientation.

In the absence of walls the dynamic behavior may be described, as in nematics, by adding a viscous torque to the elastic and electric ones. For the field-off state, the distortion decays with the 'nematic' time constant (see Eq. (34), where K^* should be substituted for K_{22}). The time of the response to an electric field can only be calculated for the simplest case when $\mu=0$, $\tau_E=\gamma_1/\varepsilon_a\sin^2\Omega(E^2-E_c^2)$, where E_c is defined by Eq. (69). Thus, the response of the smectic C phase is $\sin^2\Omega$ times slower than that of the nematic phase. However, in experiments the same substance often responds faster in the smectic C phase than in the nematic one [159–161]. This may be due to the smaller value of γ_1 when the motion of the director is confined by the cone surface. The same phenomenon has been observed for the ferroelectric smectic C phase [162]. The domain-wall motion makes the dynamics of switching more complicated; the field-induced wall velocity has been calculated by Schiller et al. [70].

Frederiks Transition in the Hexagonal Smectic I Phase

Frederiks transition in the hexagonal smectic I phase (and also in the smectic C phase with a weak hexagonal order [163]) has some specific features. In the smectic I phase the director n forms a fixed angle θ with respect to the smectic layer normal. The lines between the centers of gravity of neighbouring molecules within a smectic layer form a hexagonal structure and the director is oriented toward one of the six vertices of the hexagon. Thus a field-induced reorientation of the director is accompanied by a reorientation of the bond direction. The latter requires some additional energy, which must be added to the elastic term in the free energy expansion. The calculations of the threshold electric voltage for the Frederiks transition from a 'bookshelf' geometry [164] show that it may be presented in the form

$$U_I^2 = U_C^2 \left[1 + f(K, L, G, h, d)\right] \qquad (70)$$

where U_C is the 'smectic C threshold voltage' from Eq. (69); K, L, and G are elastic moduli related to the director, bond reorientation, and the coupling between the two; h is a constant related to the constraint on the director to be along one of the six preferred directions due to the hexagonal order; and d is the cell thickness. Thus, the threshold voltage for a smectic I phase, is in contrast to nematic and smectic C phases, thickness dependent.

Polar Achiral Polyphilics

Polar crystalline and mesomorphic phases are divided into two groups according to their symmetry. One group is composed of chiral molecules having point group symmetries ranging from C_1 to C_∞. Examples of this type are liquid crystalline chiral phases (smectic C*, discussed in Section 9.3.3.2 of this Chapter, F*, I*, etc.) differing in the character of the in-plane packing of their molecules within a smectic layer. These ferroelectric phases manifest reversible switching of their spontaneous polarization.

The other group of polar materials comprises mirror-symmetric (achiral) phases of point symmetry groups from C_{2v} to $C_{\infty v}$. Examples of these are many classical solid crystalline ferroelectrics, such as BaTiO, triglycinsulfate, and thiourea. Recently polar properties (probably antiferroelectric) have also been observed in a noncrystalline (mesomorphic) achiral molecular material [165,166]. The basic idea was to use the polyphilic effect to form 'building elements' of a polar phase. According to this concept, chemically different moieties of a molecule tend to segregate to form polar aggregates or lamellas which result in the formation of polar phases. The direction of the spontaneous polarization is allowed to be in the plane of the lamellas or layers (transverse ferro- or antiferroelectric) or perpendicular to the layers (longitudinal ferro- or antiferroelectric [167]). In the particular case by Tournilhac et al. [165, 166], field reversible piezoelectric and pyroelectric effects were observed in a tilted, highly disordered metastable phase [168] of a polyphilic compound.

9.3.3.2 Chiral Ferroelectrics and Antiferroelectrics

In 1975, the first ferroelectric liquid crystal (FLC), called DOBAMBC (**III**),

$$C_{10}H_{21}-O--\text{N}--\text{COO}-CH_2-\overset{*}{\underset{|}{CH}}-C_2H_5$$
$$\overset{|}{CH_3}$$

III

was synthesized [169]. Since then FLCs have been the object of intensive investigation. A contemporary version of the general phenomenological theory, including the dynamic behavior, can be found in various books [7, 170]. Many experimental results and the discussion of FLC parameters may be found in other publications [8, 9, 171, 172] and in the other chapters of this book.

A comprehensive review devoted to antiferroelectric liquid crystals has also been published [173]. Thus, for the sake of completeness, only the field effects of fundamental importance are discussed here.

Smectic C* Phase Ferroelectrics

The symmetry of the ferroelectric smectic C* phase corresponds to the polar point group C_2. When proceeding along the z coordinate perpendicular to the smectic layers, the director n and the polarization vector P directed along the C_2 axis rotate; that is, a helix of pitch h is formed.

In the presence of an external field the total free energy includes the elastic energy of director deformations F_d, the surface energy F_S, and energy F_E of the interaction of a ferroelectric phase with the field E:

$$F = F_d + F_s + F_E = \int_V g_d \, d\tau + \int_S W_s \, d\sigma$$

$$+ \int_V \left(-PE - \frac{DE}{8\pi} \right) d\tau \quad (71)$$

In general, the elastic term includes the elasticity related to the reorientation of the director (nematic moduli K_{ii}), and to a change in the tilt angle (coefficients a, b, c, etc., of the Landau expansion; see e.g. Eq. (10)). The corresponding expression for g_d is very complex and may be found in Pikin [7]. The surface energy includes both the dispersion and the polar terms of the anchoring energy W_S. Below we consider a few rather simple cases.

The Clark–Lagerwall Effect. This effect is observed in thin surface-stabilized FLC (SSFLC) cells where the smectic layers are perpendicular to the substrates, the thickness is less than the helical pitch ($d \leq h$), and the helix is unwound by the walls [174–176] (Fig. 23). The electric field of opposite polarity switches the direction of the spontaneous polarization between the UP and

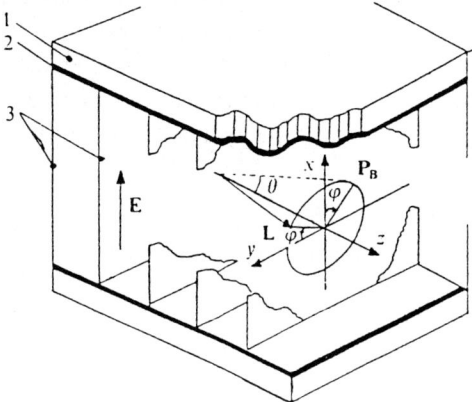

Figure 23. The Clark–Lagerwall effect. In a thin SSFLC cell the electric field of opposite polarity switches the spontaneous polarization between the UP and DOWN positions, which correspond to the LEFT and RIGHT positions of the director.

DOWN positions which correspond to the LEFT and RIGHT positions for the director that moves along the surface of a cone, the axis of which is normal to the layers and parallel to the cell substrates. In the LEFT and RIGHT positions the director remains parallel to the substrates and the SSFLC cell behaves as a uniaxial phase plate. The total angle of switching equals the double tilt angle θ.

The variation in the azimuthal director angle ϕ in the Clark–Lagerwall effect is described by the equation for the torque equilibrium, which follows from the minimization of the free energy (Eq. 71):

$$\gamma_\phi \frac{\partial \phi}{\partial t} = K \frac{\partial^2 \phi}{\partial x^2} + P_s E \sin \phi$$

$$+ \frac{\varepsilon_a}{4\pi} E^2 \sin \phi \cos \phi \quad (72)$$

where the FLC is assumed to be uniaxial and γ_ϕ is the viscosity related to the motion of the director along the cone surface mentioned. When the helix is unwound by the cell walls, the first term on the left-hand side vanishes.

The polar W_p and dispersion W_d surface terms are included in the boundary conditions:

$$K \frac{\partial \phi}{\partial t} + W_p \sin \phi \pm W_d \sin 2\phi|_{x=0,d} \quad (73)$$

For polarizations P greater than 10 nC/cm^{-2}, driving fields E less than 10 V μm^{-1}, and dielectric anisotropies $|\varepsilon_a|$ less than 1 we have

$$|\varepsilon_a| E/4\pi \ll P \quad (74)$$

and, consequently, the third term on the right-hand side of Eq. (72) may be omitted. For $|\varepsilon_a| \approx 0$ and infinitely strong anchoring there is a threshold for the switching of a SSFLC cell [177]

$$E_c = \frac{K\pi^2}{P_s d^2} \quad (75)$$

Here the elastic modulus K is defined in such a way that it includes θ^2 (see e.g. [7]). In the absence of backflow and at not very high voltages, the response times in the Clark–Lagerwall effect are determined by

$$\tau_\phi = \frac{\gamma_\phi}{P_s E} \quad (76)$$

The backflow effect may decrease the response time several-fold [177]. If the inequality given by Eq. (74) is invalid, as occurs for sufficiently high fields, the response times of the Clark–Lagerwall effect increase sharply for positive ε_a values. Conversely, for negative values of ε_a, the corresponding switching times become shorter [178, 179]. For $|\varepsilon_a| E/4\pi \gg P$ the FLC switching times are approximately governed by the field squared, as in the Frederiks effect in nematic phases.

For weak anchoring conditions, two regimes of switching exist in SSFLC cells, separated by threshold field E_{th} [180]:

$$E_{th} = \frac{4W_d}{P_s d} \quad (77)$$

For $E < E_{th}$ one observes the motion of domain walls, separating the regions of differently oriented polarization P and $-P$. The switching time is defined by the domain-wall motion. If $E > E_{th}$ (the Clark–Lagerwall regime), the switching time is determined by Eq. (76). At certain pretilt angles at the boundaries and high applied voltages, the switching may have the character of solitons or kinks in the function $\phi(z)$ spreading from one electrode to the other [181].

Deformed Helix Ferroelectric Effect. The deformed helix ferroelectric (DHF) effect observed in short-pitch FLCs [139, 182, 183] is a particular case of a more general phenomenon of the field-induced helix distortion observed in the very first investigations of FLCs [169, 184]. A theoretical explanation of the effect is given by Ostrovskii and coworkers [185–187]. The geometry of a FLC cell with the DHF effect is presented in Fig. 24. The polarizer at the first substrate makes an angle β with the helix axis and the analyzer is crossed with the polarizer. The FLC layers are perpendicular to the substrates and the layer thickness d is much larger than the helix pitch h_0. A light beam passes through an aperture $a \gg h_0$, parallel to the FLC layers, through the FLC cell.

In an electric field $\pm E$ the FLC helical structure becomes deformed, so that the corresponding dependence of the director distribution $\cos \phi$ as a function of coordinate $2\pi z/h_0$ oscillates between the two plots shown. These oscillations result in a variation in the effective refractive index (i.e. an electrically controlled birefringence appears). The effect takes place up to strengths sufficient to unwind the FLC helix

$$E_u = \frac{\pi^2 K q_0^2}{16 P_s} \quad (78)$$

Here the elastic modulus K is defined as for Eq. (75).

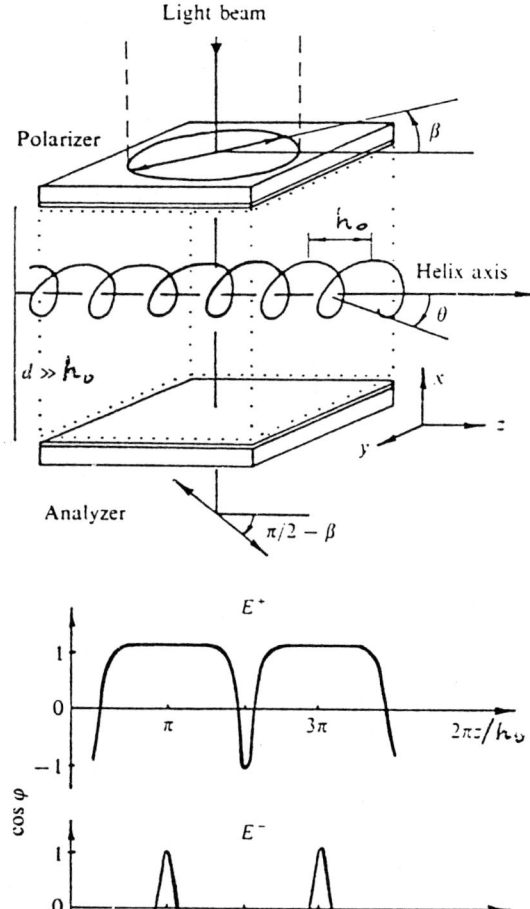

Figure 24. The deformed helix ferroelectric effect.

This means that for $E \approx E_u$ it is possible to establish the memory state. If the helix is weakly distorted, fast and reversible switching could be obtained in the DHF mode [188].

Electroclinic Effect Near the Smectic A–Smectic C* Phase Transition. The electric field may induce a tilt of the director in the orthogonal smectic A phase near the smectic A–smectic C* phase transition (electroclinic effect [189, 191]). The electroclinic effect may be understood within framework of the Landau phase transition theory [7]. If the dielectric anisotropy is negligible, the free energy density of the smectic A phase may be expanded over the field-induced tilt angle θ:

$$g = \frac{1}{2}a\theta^2 + \frac{1}{4}b\theta^4 + \frac{P^2}{2\chi_\perp} - \mu_p P\theta - PE \quad (80)$$

where $a = \alpha(T - T_c)$, $b > 0$, $\chi_\perp$ is the background (far from the transition) contribution to the dielectric susceptibility, and μ_p is the tilt–polarization coupling constant (piezocoefficient). Under the assumption that $b = 0$, the minimization of g with respect to θ and the polarization P results in the field-dependent tilt angle:

$$\theta = \frac{\mu_p \chi_\perp E}{\alpha(T - T_c)} = e_c E \quad (81)$$

The coefficient e_c is known as the electroclinic coefficient. According to Eq. (81), the electroclinic coefficient becomes infinite at $T = T_c$ (in fact, it is limited by the term $b\theta^4$ [14]).

The dynamics of the electroclinic effect are, in fact, the same as the dynamics of the ferroelectric soft mode. The switching time may be derived from the equation for the balance of the viscous, elastic, and electric torques:

$$\gamma_\theta \frac{\partial \theta}{\partial t} = -\alpha(T - T_c)\theta + \mu_p \chi_\perp E \quad (82)$$

The characteristic response times of the effect at low fields $E/E_u \ll 1$ are independent of FLC polarization P_S and field E and are defined only by the rotational viscosity γ_ϕ and the helix pitch h_0:

$$\tau_c = \frac{\gamma_\phi}{K q_0^2} = \frac{\gamma_\phi h_0^2}{\pi^2 K} \quad (79)$$

For E close to the unwinding field E_u, the pitch h increases sharply to $h \gg h_0$. Consequently, the helix relaxation times τ_d to the initial state also increase: $\tau_d/\tau_c \propto h^2/h_0^2$.

It follows from Eq. (82) that the switching time of the electroclinic effect is independent of the electric field, and is defined only by the tilt rotational viscosity γ_θ and the elastic modulus α:

$$\tau_\theta = \frac{\gamma_\theta}{\alpha(T-T_c)} \quad (83)$$

The experimental data confirm the linear increase in the tilt angle with increasing E for the electroclinic effect and independence τ_θ of E [190] (Fig. 25). The switching time τ_θ may be calculated more precisely if the fourth-order term in $b\theta^4$ is taken into account. In this case, a decrease in τ_θ for larger electric fields is observed [192]. At present, the electroclinic effect is the fastest of the known electro-optical effects in liquid crystals.

Domain Mode. A remarkable domain structure has recently been observed in FLCs possessing high spontaneous polarization (>100 nC cm^{-1}) [193–195]. A very stable spatially periodic optical pattern arises in layers of FLCs with their molecules oriented parallel to the substrates after a dc field treatment of the cell. The stripes are oriented perpendicular to the director orientation and their period is inversely proportional to the polarization magnitude squared, and independent of cell thickness.

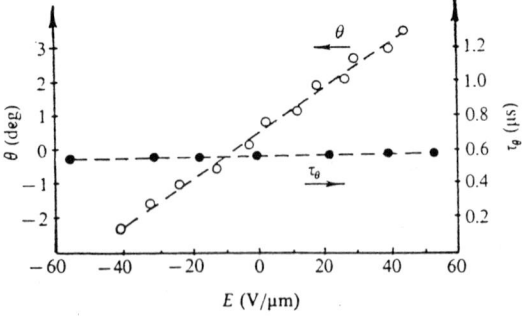

Figure 25. Dependence of the tilt angle θ and the response time τ_θ on the electroclinic effect in a FLC cell ($T = 25\,°C$) [190].

The optical pattern may be observed for several months after switching off the dc field. The theoretical model is based on the consideration of a stabilizing role of the structural defects (dislocations) interacting with free charges in a FLC layer. The tilted layer structure responsible for such defects has been observed by direct X-ray investigation. The field-off domain structure has also been observed after the application of the electric field to the smectic A phase of a ferroelectric substance with very high spontaneous polarization [196]. The phenomenon is assumed to arise due to a break in the smectic layers (very similar to that just mentioned) induced by a strong electroclinic effect.

Ferroelectric Liquid Crystal Polymers. The symmetry requirements necessary for ferroelectricity in low-molecular-mass compounds are also valid for polymer mesophases. If a tilted chiral smectic phase is stable after a polymerization process, it must be ferroelectric. Following this idea the first polymeric liquid crystalline ferroelectric was synthesized by Shibayev et al. [197]. The substance is a comb-like homopolymer (*IV*)

IV

with a polymethacrylate chain as a backbone and flexible spacers, $COO(CH_2)_{10}-COO-$, separating the phenylbenzoate mesogenic units from the backbone. Transverse (with respect to mesogenic units) dipole moments provided by several $COO-$ groups are weakly coupled to the chiral terminal fragment, and thus the spontaneous polarization of the substance is small. Later, a variety of comblike and main chain polymers manifesting chiral smectic C*, F*, and other phases were synthesized (for reviews see [13, 198]). The field behavior of polymer ferroelectric liquid crystals is very similar to that typical of low-molecular-mass compounds. In general, all processes are very slow due to the high viscosity of polymer materials. An interesting feature of the polymers is their ability to form, on cooling, a glassy state with a frozen spontaneous polarization. Such solid materials have pyroelectric and piezoelectric properties.

Antiferroelectrics

A possibility of the antiferroelectric behavior of low temperature, highly ordered, chiral phases and the herringbone packing of smectic layers with an alternating tilt of the director was first discussed in connection with their very strong field-induced pyroelectric response [199]. In such a structure the direction of the in-plane polarization also alternates and the whole structure is antiferroelectric. The characteristic switching time in such phases is very long (seconds). The slowly switchable smectic X phase [200] appears to belong to the same family of antiferroelectrics. A rather fast tristable switching typical of solid antiferroelectrics has recently been reported [201–204] for chiral liquid crystalline phases of MHPOBC (**V**).

$$H_{17}C_8-O-\phenyl-\phenyl-CO_2-\phenyl-$$
$$-CO_2-\overset{*}{C}H-(CH_3)-C_6H_{13}$$
V

A third, field-off, state, in addition to the two stable states known for the Clark–Lagerwall effect, has also been shown to be stable, and a layered structure with alternating tilt has been established by optical [205, 206] and scanning tunneling microscopy [207] techniques.

In fact, MHPOBC shows a very rich polymorphism [208]:

$$SmI_A^* - 64°C - SmC_A^* - 118.4°C$$
$$-SmC_\gamma^* - 119.2°C - SmC* - 120.9°C$$
$$-SmC_\alpha^* - 122°C - SmA - 148°C - I$$

where SmC_A^* and SmC_α^* are antiferroelectric phases, SmC^* is a typical ferroelectric phase (discussed in the previous section), and SmC_γ^* is a ferrielectric phase. On a macroscopic scale the phases are twisted, with a characteristic pitch depending on temperature. Antiferroelectric phases have been observed in many other compounds, including polymers [209], and have been investigated in detail (for a review see e.g. [173]).

The application of an electric field has a considerable effect on the phase transition points in MHPOBC [210]; the phase diagram (electric field versus temperature) is shown in Fig. 26. A very narrow low-temperature ferrielectric phase is seen (FIL) between the SmC_A^* and SmC_γ^* phases mentioned earlier. In addition, even a weak field strongly influences the boundary between the SmC_α^* and SmC^* phases. Further inves-

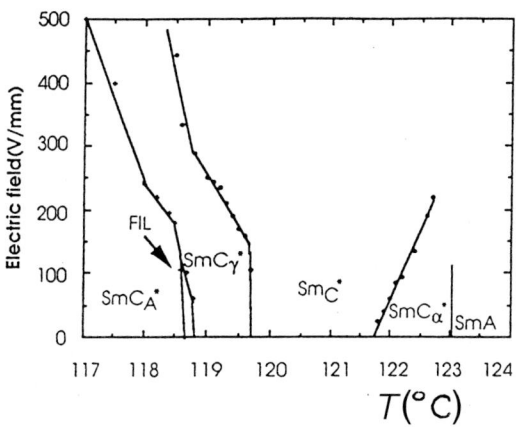

Figure 26. Electric field versus temperature phase diagram of MHPOBC [210].

tigations have shown [211] that the field-induced transition between the two phases in MHPBOC involves the formation of a variety of ferrielectric configurations. An even richer field-induced polymorphism has been observed in binary mixtures of MHPOBC and MHPOCBC [210].

A typical double hysteresis loop for electro-optical switching of the antiferroelectric SmC$_A^*$ phase is shown in Fig. 27. There are two field-induced uniform states and a zero-field third state which may be either uniform or twisted [212]. The transition from the antiferroelectric to the ferroelectric state proceeds through intermediate ferrielectric states. With increasing frequency in the range 10–100 Hz, the double hysteresis loop gradually transforms into the conventional ferroelectric loop [213]. The characteristic time corresponds to the time of the formation of domains of the antiferroelectric phase. The low-frequency switching (0.2 Hz) of the SmC$_\gamma^*$ phase reveals a triple hysteresis loop with two intermediate (+ and −) ferrielectric states and two final ferroelectric states. At higher frequency, conventional ferroelectric switching is observed.

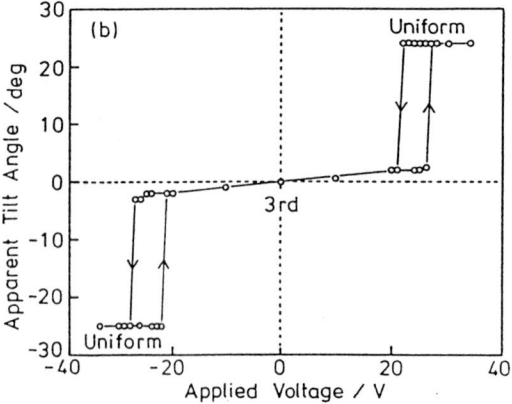

Figure 27. A double hysteresis loop typical of the electro-optical switching of the antiferroelectric phase [212].

Chiral Lyotropics and Discotics

Lyotropic Systems. Water solutions of certain amiphiphilic molecules (e.g. lipids) form lamellar lyotropic phases with a tilt of molecules to the layer normal. When the same molecules are chiral or other chiral molecules (guests) are introduced into the lamellas (host), the lamellar phase satisfies all the symmetry requirements of a ferroelectic phase (in formal analogy to SmC*). One of the best chiral additives to be introduced in lyotropic phases is cholesterol, which induces rather high spontaneous polarization in thermotropic SmC matrices [214]. In the L' phase, where matrix molecules are tilted, the chiral mixture (ethylene glycol doped with cholesterol) manifests a one order of magnitude higher piezo response than the undoped mixture. An even more striking difference between the undoped and doped samples (two orders of magnitude) is observed when a dc poling field is applied [215]. With increasing concentration of the dopant the piezo response grows. A dramatic difference between the piezo response of the undoped and the cholesterol doped samples has also been observed for water solutions [216]. Certainly the electric behavior of chiral analogs of the lyotropic phases is much more pronounced than that of their achiral counterparts and, therefore, cholesterol induces ferroelectricity in both systems studied (based on ethylene glycol and water).

Discotic Ferroelectrics. Columnar mesophases with chiral disk-shaped molecules whose normals are tilted with respect to the column axis also manifest ferroelectric switching [217]. In such phases the direction of the spontaneous polarization is perpendicular to both the column axis and to the normal to the molecular disks. The tilt of the disk forms a helix with its axis oriented along the columns. In an amphiphilic pyrene derivative, the positive and negative voltages rotate the direction of the minimum refractive index (which is perpendicular to the column axis) through a certain angle (about ±20°) clockwise or anticlockwise, respectively. The other homolog from the same series shows a field-induced change in the structure of the phase with an increase in the spontaneous polarization from 60 to 180 nC cm^{-2} [218]. The two phases are thought to have different sublattices of elliptic and polar columns. The switching times vary over a very broad range (from 1 ms to 100 s), being strongly dependent on the field (as $\tau \propto E^4$) and temperature. The low-field phase is switched five times faster than its high-field counterpart. The columnar phases have some advantages over ferroelectric smectics (e.g. they are shock resistant) [218].

9.4 Electrohydrodynamic Instabilities

In this section we discuss electrohydrodynamic (EHD) instabilities, that is electric-field-induced phenomena that are caused by the flow of a liquid crystal (see also [8, 219]). The reason for the flow is electrical conductivity, which has been disregarded in previous sections. The flow may arise either independently of the anisotropic properties of substance, as in isotropic liquids (isotropic modes of the electrohydrodynamic instability), or may be driven by the conductivity anisotropy, as in liquid crystals (anisotropic modes). The threshold for EHD instabilities depends on many parameters, such as the electrical and viscoelastic properties of substance, the temperature, and the applied field frequency. Due to flow distortion of the director alignment, the instability is usually accompanied by a characteristic optical pattern that depends on optical anisotropy of substance. This pattern formation is a special branch of physics dealing with the nonlinear response of dissipative media to external fields, and liquid crystals are well accepted model objects for the investigation of such phenomena [220]. To date, the most important results on EHD instabilities have been obtained for nematic liquid crystals (for a review see e.g. [221]).

9.4.1 Nematics

9.4.1.1 Classification of Instabilities

Electrohydrodynamic instabilities in nematics may be classified according to their threshold voltage (or field) dependence on the physical parameters of the liquid crystal (cell geometry, field frequency, etc.). The most important case for which all the typical instabilities may be observed is a planar cell with homogeneously oriented director. Instabilities in cells with homeotropic or tilted molecular orientations have some specific features, but the general mechanisms are the same. A qualitative phase diagram (the threshold voltage versus the frequency of the applied ac field) for different electrohydrodynamic instabilities in homogeneously oriented liquid crystals is shown in Fig. 28.

Among the instabilities shown there are two isotropic modes. One of these is well known from experiments on isotropic liquids and occurs at very low frequencies due to some injection processes at electrodes (injection mode). The other is also observable in ordinary liquids (such as silicon oil), but is seen particularly clearly in nematics, due to their optical anisotropy. The latter mode occurs over a wide frequency range due to ion drift to the electrodes as in electrolysis (electrolytic mode). Both isotropic

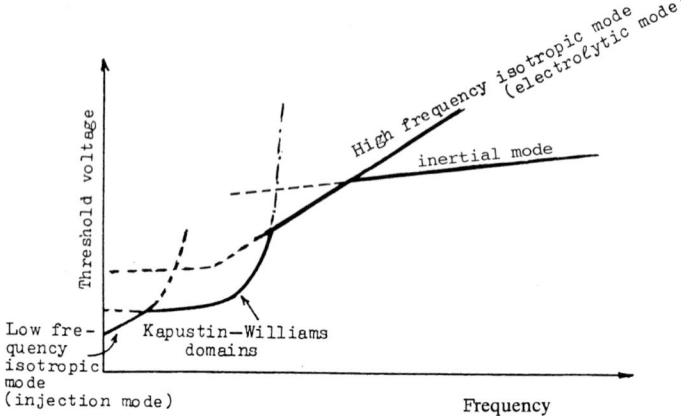

Figure 28. Threshold voltage as a function of frequency for different electrohydrodynamic instabilities in homogeneously oriented nematics.

EHD instabilities are driven by the electric-current-induced flow, even at zero conductivity anisotropy.

The other two instabilities shown in Fig. 28 may be observed only in liquid crystals (nematic, cholesteric, and smectic C). The first is the Carr–Helfrich instability, which is caused by a low-frequency electric field and occurs in the form of elongated vortices with their axis perpendicular to the original director alignment. The vortices cause a distortion of the director orientation, which is observed optically as a one-dimensional periodic pattern (Kapustin-Williams domains). The other anisotropic mode is observed only in highly conductive liquid crystals. For its interpretation the inertial term $\partial v/\partial t$ for the fluid velocity must be taken into account, which is why this mode may be called inertial mode.

The different modes may interfere with each other, even near their thresholds, and a variety of patterns may be observed. For example, 'chevrons' occur due to interference between the electrolytic and inertial modes, and a transient periodic pattern due to backflow effects accompanies the Frederiks transition. Well above the threshold, when the induced distortion is nonlinear with the applied field, many other patterns are observed, including travelling waves or periodically pulsing modes [221].

The most important EHD instabilities are discussed below.

9.4.1.2 Isotropic Modes

Nematic liquid crystals are weak electrolytes containing a certain amount of ions. The ions may be intrinsic or created by an external electric field, which favors the dissociation of neutral molecules (chemical degradation). They may also appear as a result of electrochemical reactions at electrodes due to the injection of additional charge carriers. In some cases, the appearance of EHD instabilities in the nematic phase does not correlate with such inherent liquid crystalline properties as dielectric ε_a or conductivity σ_a anisotropy. The only physical reason for these instabilities might be a nonuniform ion distribution along the direction z parallel to electric field.

Standard Equations

To describe the field-induced motion of a liquid, a set of fundamental equations should be written which may be solved under certain assumptions. The first is the equation of motion in the Euler form:

$$\frac{\partial \boldsymbol{v}}{\partial t} + (\boldsymbol{v}\nabla)\boldsymbol{v} = \frac{\boldsymbol{f}}{\rho} \quad (84)$$

where ρ is the density of the liquid, $\boldsymbol{v}$ is its velocity, t is time, and $\boldsymbol{f}$ is the force per unit volume, which includes the pressure gradient (grad p), the gravity force $\rho \boldsymbol{g}$, the electric force $\boldsymbol{f}_E$, and forces due to viscous friction $\boldsymbol{f}_{\text{visc}}$ that depend on the fluid velocity $\boldsymbol{v}$:

$$\boldsymbol{f} = -\text{grad}\, p + \rho \boldsymbol{g} + \boldsymbol{f}_E + \boldsymbol{f}_{\text{visc}} \quad (85)$$

The second equation to be taken into consideration is the equation for the conservation of mass (or the continuity of liquid):

$$\frac{\partial \rho}{\partial t} + \text{div}\,\rho \boldsymbol{v} = 0 \quad (86)$$

The force exerted on the liquid by an external field is due, in the simplest case, to the space charge of density Q:

$$\boldsymbol{f}_E = Q\boldsymbol{E} \quad (87)$$

In the case of a space and time dependent force $\boldsymbol{f}$, Eqs (84)–(87) have no general solution and the problem must be simplified. For example, we may ignore the gravity force and the pressure gradient and, in addition, consider the fluid to be incompressible (i.e. ρ=constant). Then the friction force is written simply as [222]:

$$\boldsymbol{f}_{\text{visc}} = \eta \nabla^2 \boldsymbol{v} \quad (88)$$

where η is the viscosity coefficient, and Eqs (84), (86), and (87) take the form of the flow continuity equation

$$\text{div}\,\boldsymbol{v} = 0 \quad (89)$$

and the Navier–Stokes equation

$$\frac{\partial \boldsymbol{v}}{\partial t} + (\boldsymbol{v}\nabla)\boldsymbol{v} = \eta \frac{\nabla^2 \boldsymbol{v}}{\rho} + \frac{\boldsymbol{f}_E}{\rho} \quad (90)$$

As before, Eq. (90) is non-linear and has a simple solution only for low fluid velocities when the Reynolds number is small (Re $= \rho v l/\eta \ll 1$, where l is a characteristic length of an experiment). Then, the second term on the left-hand side of the Eq. (90) vanishes and we get a new form:

$$\frac{\partial \boldsymbol{v}}{\partial t} - \eta \frac{\nabla^2 \boldsymbol{v}}{\rho} - \frac{Q\boldsymbol{E}}{\rho} = 0 \quad (91)$$

Even now there remain serious problems to be solved. In fact the charge density $Q(\boldsymbol{r},t)$ in Eqs (87) and (91) depends not only on the external field but also on the velocity of charged fluid. Thus, Eqs (89) and (91) have to be complemented by an equation for the conservation of charge (continuity of the electric current):

$$\frac{\partial Q}{\partial t} + \text{div}\,\boldsymbol{J} = 0 \quad (92)$$

where $\boldsymbol{J}$ is the current density, which includes the ohmic term, the convective term due to the charge transfer by the moving liquid, and the diffusion term due to the gradient of the charge carrier density:

$$\boldsymbol{J} = \sigma \boldsymbol{E} + \boldsymbol{J}_{\text{conv}} + \boldsymbol{J}_{\text{diff}} \quad (93)$$

Finally, there is a coupling between the charge and the electric field strength given by Maxwell's equation:

$$\text{div}\,\boldsymbol{E} = \frac{4\pi Q}{\varepsilon} \quad (94)$$

Equations (89)–(94) govern the behavior of an isotropic liquid in an electric field. For very weak fields the liquid is conducting but immobile. With increasing field a non-uniform spatial distribution of Q and $\boldsymbol{E}$ arises due to an ion drift to electrodes or/and charge injection, and the liquid starts moving in order to satisfy the minimum of entropy production. The corresponding critical field E_{crit} is considered to be the threshold field for the appearance of an EHD instability. Below we discuss the two particular cases mentioned earlier, namely the injection mode and the electrolytic mode.

Injection Mode

Let us consider a plane sandwich cell into which the cathode injects electrons that negatively charge neutral molecules. Due to an excess negative charge density $Q(z)$ near the cathode, a force QE is directed to the anode and tries to shift the charged layer of a liquid to the right (Fig. 29a). Since the cell is sealed and the liquid is incompressible, a circular convective flow occurs in order to reduce the internal pressure (Fig. 29b, c). This case resembles the well-known Benard problem in the thermoconvection of liquids.

The threshold of the instability may be estimated from Eqs (89)–(94). For simplicity, consider the one-dimensional case when the z component of the velocity is harmonically dependent on the transverse x coordinate:

$$v_z(x) = v_z \cos\frac{\pi}{d} x \qquad (95)$$

The half-period of the cosine wave is assumed to be equal to the vortex diameter, that is to the cell thickness d. Differentiating Eq. (95) twice and substituting the result into Eq. (91) we get

$$\frac{\partial v_z}{\partial t} = \frac{v_z}{\tau_v} - \frac{QE_z}{\rho} = 0 \qquad (96)$$

where

$$\tau_v = \frac{\rho d^2}{\eta \pi^2} \qquad (97)$$

is a characteristic time for the relaxation of a vortex with diameter d.

In the continuity equation (Eq. (92)) we leave only the ohmic and convective terms:

$$J = \sigma E + Q v \qquad (98)$$

When calculating div J it should be borne in mind that electrical conductivity results mainly in the relaxation of the space charge inhomogeneities along the x coordinate and E_z may be considered to be uniform. Thus,

$$\text{div } J_{\text{ohm}} = \sigma \frac{\partial E_x}{\partial x} = \sigma \frac{4\pi Q}{\varepsilon} = \frac{Q}{\tau_Q} \qquad (99)$$

where

$$\tau_Q = \frac{4\pi\sigma}{\varepsilon} \qquad (100)$$

is the Maxwell time for the space charge relaxation.

The convective term in the current obeys the equation:

$$\text{div } J_{\text{conv}} = v_z \frac{\partial Q}{\partial z} + Q_z \frac{\partial v_z}{\partial z} = \alpha \cdot v_z \qquad (101)$$

where v_z is assumed to be constant and α is the gradient of the injected charge density. Therefore, the current continuity equation

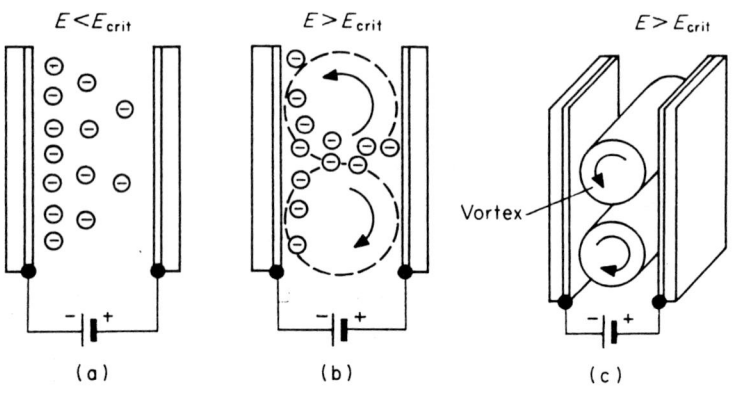

Figure 29. EHD instability caused by the injection mechanism. (a) Distribution of the injected charge at voltages below the threshold for convection. (b) Change in the charge distribution due to convection. (c) Cylindrical convective vortices.

(Eq. (92)) becomes

$$\frac{\partial Q}{\partial t} + \frac{Q}{\tau_Q} + \alpha v_z = 0 \qquad (102)$$

and we have a pair of equations (Eqs (96) and (102)) completely analogous to the set describing the thermoconvective instability [223–225].

To calculate the critical field for the injection instability it is assumed that, *precisely at the threshold*, the time dependences $v_z(t)$ and $Q(t)$ are described by an exponential growth function $\exp(st)$ with $s=0$, which means that at the threshold the velocity and charge density neither increase nor decay. This is a general principle and is called the "exchange of stability" (see e.g. [226]). Calculating the determinant for the set of linear equations (Eqs (96) and (102)) we obtain the critical field for any given injection strength α:

$$E_{\text{crit}} = \frac{\rho}{\alpha \tau_v \tau_Q} = \frac{\pi \eta \sigma}{\alpha \varepsilon d^2} \qquad (103)$$

Finally, assuming a linear space charge distribution along the normal to the cell, $\alpha = \partial Q / \partial z \approx Q_0 / d$, and introducing the ion mobility $\mu = \sigma / Q_0$, we arrive at the threshold voltage independent of the cell thickness in the form

$$U_{\text{crit}} = \pi \frac{\eta \mu}{\varepsilon} \qquad (104)$$

This differs from the more rigorous theoretical expressions only by the use of the numerical coefficient T (instead of π). For typical parameters of a dielectric liquid ($\eta \approx 1$ P, $\mu \approx 10^{-5}$ cm^{-2} V^{-1} s^{-1} and $\varepsilon \approx 10$), $U_{\text{crit}} \approx 1$ V. However, in some cases (e.g. for conductive liquids) T may be as high as 100, or even more [227, 228]. Experimentally, in thin nematic liquid crystal cells, the critical voltage is of the order of few volts. The instability appears at a dc field or at very low frequencies (Fig. 28). The optical pattern depends on the geometry of the experiment; in the isotropic phase the fluid motion may be observed using a strioscopic technique [229], or by following the motion of foreign particles. In nematics the areas of strong velocity gradients are easily visualized under a polarizing microscope due to the director distortion by shear. For example, in planarly oriented nematics an optical pattern in the form of beans is observed [230, 231].

Electrolytic Mode

The other important 'isotropic' instability is the so-called electrolytic mode [232, 233]. It appears in any fluid phase (isotropic, nematic, cholesteric, smectic A) within a wide frequency range in the form of small vortices near cell electrodes. The reason for the instability is a non-uniform space charge distribution (Fig. 30) due to an electrolytic process in the bulk of the substance (positive and negative ionic charges are separated during the period of the applied external ac field and move to opposite electrodes). The vortices are seen under a microscope in the form of small linear pre-chevron domains (in homogeneously oriented nematics), Maltese crosses (homeotropic orientation) and more complicated structures (cholesterics and nematics with a short-range smectic order). However, when there is no coupling between the flow and the di-

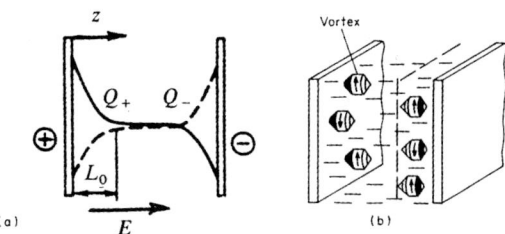

Figure 30. Electrolytic EHD mode. (a) Distribution of the density of positive (Q^+) and negative (Q^-) charges during one period of the applied ac field. (b) Vortex motion of the fluid near the electrodes.

rector (the isotropic and smectic A phases) the vortices may be observed only by using foreign floating particles. The instability has a well-defined field (not voltage) threshold that is dependent on the frequency of the applied field according to $E \propto \omega^{\frac{1}{2}}$.

To calculate the threshold it is assumed [7] that the charge density gradient along the layer thickness obeys the law:

$$\frac{\partial Q}{\partial z} = -\chi E(z) \qquad (105)$$

where χ is the electrokinetic coefficient, which may be evaluated using the Maxwell equation:

$$\frac{\partial Q}{\partial z} = -\frac{\varepsilon}{4\pi} \cdot \frac{\partial^2 E}{\partial z^2} \approx -\frac{\varepsilon E}{4\pi L_D^2};$$

$$\chi = \frac{\varepsilon}{4\pi L_D^2} \qquad (106)$$

Here L_D is a characteristic (Debye) length where the space charge is concentrated:

$$L_D = \left(\frac{\varepsilon D}{8\pi\sigma}\right)^{\frac{1}{2}} \qquad (107)$$

where D is an ion diffusion coefficient. When the frequency ω of the applied field is increased, the ions do not have time to overcome the gap between the electrodes and concentrate in more extended regions of thickness

$$L_D(\omega) \approx L_D(0) \left[1 + \left(\frac{\varepsilon\omega}{8\pi\sigma}\right)^2\right]^{-\frac{1}{4}} \qquad (108)$$

Thus, the electrohydrodynamic instability appears in regions near the electrodes and vortices have a characteristic dimension L_D, which is smaller than the gap between electrodes.

Using the same current continuity equation (Eq. (98)) we arrive at the same set of equations (Eqs (96) and (102)) for small amplitudes of velocity v and space charge Q.

The only difference is in the characteristic relaxation time for a hydrodynamic vortex:

$$\tau_v \approx \frac{\rho}{\eta} \cdot \frac{q_x^2}{\left(q_x^2 + q_z^2\right)^2} \qquad (109)$$

Now the vortex has dimensions $w_x = \pi/q_x$ and $w_z = L_D = \pi/q_z$ along the x and z axes, respectively. The latter depends explicitly on the field frequency (see Eq. (108)).

The threshold field E_{crit} may be found using Eqs (96) and (102) and the same procedure described as earlier. For a square wave excitation with frequency f and amplitude E the different expressions are obtained in two frequency ranges: At low frequencies, $\omega \ll 4\pi\sigma/\varepsilon$,

$$E_{\text{crit}} = A \frac{(\sigma\eta)^{\frac{1}{2}}}{\varepsilon} \qquad (110)$$

and at high frequencies, $\omega \gg 4\pi\sigma/\varepsilon$,

$$E_{\text{crit}} = B \left(\frac{\rho D \sigma \omega}{\varepsilon}\right)^{\frac{1}{2}} \qquad (111)$$

where A and B are numerical constants (given in [233, 234]). The characteristic size of the vortices at the threshold w_{crit} has also been found.

The thresholds for the formation of vortices in silicone oil, carbon tetrachloride, acetone, and in the isotropic phase of MBBA have been measured [232, 233] and it was shown that $U_{\text{crit}} \propto \omega^{\frac{1}{2}}$ and that the threshold decreases with decreasing viscosity. More detailed measurements of the dependences $U_{\text{crit}}(\omega)$ in pure and doped MBBA at various temperatures (for sandwich cells) have also been made. The threshold for the vortex motion was taken as the onset of the circulation of the solid particles in the electrode plane. The shape of the $U_{\text{crit}}(\omega)$ curves depends on the electrical conductivity. For high electrical conductivity the curves have a plateau proportional to σ in the low-frequency region and a characteristic dependence $U_{\text{crit}} \propto \omega^{\frac{1}{2}}$ independent of σ at frequencies above the critical one, which is defined by inverse time τ_Q (Eq. (100)). At the phase transition to the nematic phase, the threshold voltage of the instability has no peculiarity.

In homeotropically oriented layers the instability appears in the form of Maltese crosses (Fig. 31). The crosses are caused by small vortices (Fig. 30b) (see in [235]). In the homogeneously oriented layers, *the same* instability has a form of linear, short-pitch domains which are precursors of the well-known *chevrons*. Thus, the high-frequency chevron mode, observed in various materials, is an isotropic mode and cannot be related to the high-frequency dielectric regime of the Carr–Helfrich instability where the threshold must diverge for zero anisotropy of conductivity (see next section). Up to now, all attempts to use this model to interpret experimental data at high frequencies have failed. Of course, in this frequency range, at relatively high threshold fields the director of a nematic has a short relaxation time and oscillates with the field frequency, as predicted by the Carr–Helfrich–Orsay approach, but this effect is irrelevant to the onset of instability.

9.4.1.3 Anisotropic Modes

We now explicitly include in the discussion the anisotropy of the electrical conductivity σ_a of a liquid crystal. This anisotropy itself turns out to be a reason for electrohydrodynamic destabilization. First, we discuss the Carr–Helfrich mode of the instability [236, 237], which arises in a homogeneously oriented liquid crystal layer in a sandwich cell between transparent electrodes.

Carr–Helfrich Mode

When a low frequency ac electric field ($\omega \ll 1/\tau_Q$) is applied to homogeneously oriented, fairly conductive nematics, a very regular vortex motion (normal rolls) is often observed, which is accompanied by a strip domain optical pattern (Kapustin–Williams domains) [238, 239] (Fig. 31b). The reason for this instability is the space charge accumulated in the bulk due to the anisotropy of conductivity. It appears in thin cells ($d = 10$–100 μm) and has a well-defined voltage threshold that is independent of the thickness. The threshold can be easily calculated for the simplest steady-state, one-dimensional model shown in Fig. 32.

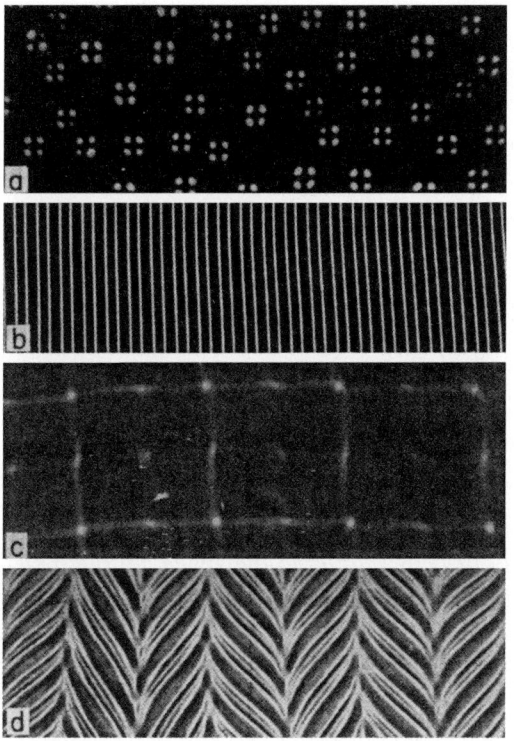

Figure 31. Optical patterns accompanying different EHD processes. (a) Electrolytic mode for the homeotropic orientation of a nematic liquid crystal. (b) Kapustin–Williams domains (KWD) in homogeneously oriented nematic. (c) Anisotropic EHD mode for the planar texture of a cholesteric. (d) A chevron structure due to interference of two instabilities (KWD and inertial mode).

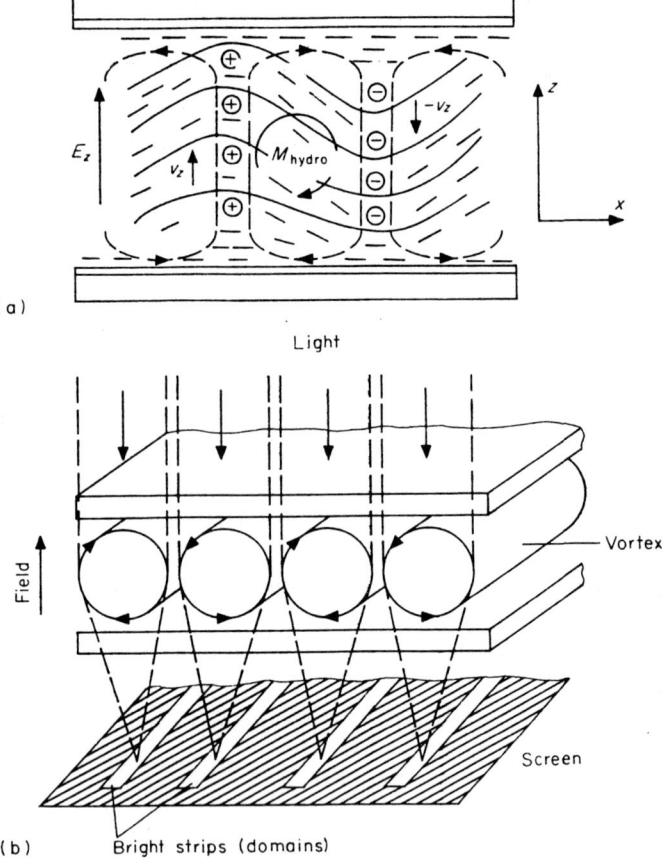

Figure 32. Carr–Helfrich EHD instability in nematic liquid crystals: (a) onset of the instability; (b) vortex motion of a liquid crystal and the pattern of black and white stripes in the screen plane.

The simplest model. The physical mechanism of the instability could be described as follows [236]. A homogeneously oriented nematic liquid crystal is stabilized by the elastic torque (due to surface anchoring). The dielectric torque is considered to be negligible ($\varepsilon_a=0$). Now, let us imagine a small incident director fluctuation with a period w_x of the order of cell thickness d:

$$\theta(x) = \theta_m \cos \frac{\pi x}{w_x} \quad (112)$$

With the field applied, this fluctuation results in a slight deviation of the electric current lines in the x direction proportional to the anisotropy of conductivity $\sigma_a = \sigma_\parallel - \sigma_\perp > 0$, and, in turn, in an accumulation of the space charge around extremum points ($\theta_m = 0$) of the director distribution (Fig. 32a). The x component of the current is

$$J_x = \sigma_\parallel E_x + \Delta\sigma E \theta \quad (113)$$

where the new component of the electric field E_x is related to the space charge distribution $Q(x)$ and the director angle $\theta(x)$ by Maxwell's relationships (div $\boldsymbol{D}=0$, div $\boldsymbol{J}=0$):

$$\frac{\partial E_x}{\partial x} = \frac{4\pi Q(x)}{\varepsilon} \quad (114)$$

and

$$\frac{\partial J_x}{\partial x} = \sigma_\parallel \frac{\partial E_x}{\partial x} + \Delta\sigma E \frac{\partial \theta}{\partial x} = 0 \quad (115)$$

Substituting θ from Eq. (112) into Eqs (114) and (115) we have for the space charge distribution:

$$Q(x) = \frac{\Delta\sigma \, \varepsilon \, E \, \theta_m}{4\pi \, \sigma_\| \, w_x} \sin \frac{\pi x}{w_x} \qquad (116)$$

At a certain critical voltage the destabilizing viscous torque M_V, which comes from the interaction of the space charge with an external field, becomes sufficiently large and is not compensated for by the stabilizing elastic M_k torque. The nematic liquid begins to move with a velocity v under the action of a pushing force proportional to $-Q(x)E$ in accordance with the Navier–Stokes equation (Eq. (91)) (written for the steady state regime):

$$\eta \frac{\partial^2 v_z}{\partial x^2} = -Q(x) \, E \qquad (117)$$

where $\eta = (\frac{1}{2})(\alpha_4 + \alpha_5 - \alpha_2)$ is a combination of Leslie's viscosity coefficients α_i.

The velocity v, in turn, supports the director fluctuation $\theta(x)$, in accordance with the torque balance equation:

$$K_{33} \frac{\partial^2 \theta}{\partial x^2} = \alpha_2 \frac{\partial v_z}{\partial x} \qquad (118)$$

Combining Eqs (116)–(118) we derive the following formula for the threshold voltage:

$$U_{\text{crit}} = E_{\text{crit}} \, d$$
$$= \left[\frac{4\pi^3 K_{33} \sigma_\| \eta}{(-\alpha_2) \Delta\sigma \varepsilon} \right] \cdot \frac{d}{w_x} \qquad (119)$$

above which a periodic pattern of vortices forms with a period $w_x \approx d$ along the x axis. The whole process is governed by the anisotropy of conductivity. The threshold is diverged when σ_a vanishes (e.g. in the case of nematics with short-range smectic order). The threshold is proportional to the ratio η/α_2 of the two viscosities which, roughly speaking, are proportional to each other. Thus the threshold seems to be independent of viscosity and the instability may be easily observed in very viscous materials (e.g. polymer liquid crystals). The reason for this is the compensation for two effects: (a) in very viscous media the velocity of vortex motion is lower (η coefficient); and (b) the coupling between the flow and the director (α_2 coefficient) is much stronger. A more precise expression for the threshold voltage, derived by Helfrich [236], includes also a nonvanishing value of the dielectric anisotropy. All the dependences predicted by the simplest theory have been confirmed qualitatively by experiment [8, 9]. In order to interpret the experimental data more precisely, some additional approaches have been developed.

Generalization of the Simplest Model. In the one-dimensional case, the diameter of the vortex is assumed to be equal to the cell thickness. In a more precise two-dimensional approach [240–242] this assumption is not used and the stability of the systems is studied using the variation of the wavevector $q_x = \pi/w_x$ along the x axis. The threshold voltage found is in excellent agreement with experiment [243]. Later, a three-dimensional version of the theory [244], which allows for the arbitrary anchoring energy, explained more complex domain patterns, such as oblique rolls [245–247]. Such rolls are seen for materials having a small negative dielectric anisotropy and large σ_a. If the flexoelectric term is included in the field interaction energy [248–250] the symmetry of the problem is changed; this may also account for two-dimensional vortex patterns (e.g. oblique rolls). A recent review [251] contains many interesting details concerning this problem.

In order to discuss the behavior of the instability threshold as a function of the applied field frequency, the time dependent terms $\gamma_1(\partial\theta/\partial t)$ and $\partial Q/\partial t$ must be added to the equations for the director and charge continuity. Estimates show that the inertial term for the velocity $\partial v/\partial t$ in the Navier–Stokes equation may be disregarded even in the transient regime (except when discussing the inertial mode in strongly conducting nematics; see below). Thus we have a set of two coupled linear equations for the space charge $Q(x)$ and the curvature $\psi = \partial\theta/\partial x$ [225, 252]:

$$\frac{\partial \psi}{\partial t} + \frac{\psi}{\tau_r(E)} + \frac{QE}{\eta} = 0 \quad (120)$$

and

$$\frac{\partial Q}{\partial t} + \frac{Q}{\tau_Q} + \sigma_{\text{eff}} \psi E = 0 \quad (121)$$

This pair of equations has the same form as the set (Eqs (96) and (102)) for the injection instability. However, the fluid velocity is not included in the new set explicitly, and the destabilization is due to the coupling terms QE/η and $\sigma_{\text{eff}} \psi E$, where σ_{eff} is an effective constant including the anisotropy of conductivity that is responsible for the instability. The field-dependent time constant for the reaction of the director to the field was discussed earlier (Eqs (33) and (62)). Near the threshold $\tau_r \gg \tau_Q$.

The solution of set (Eqs (120) and (121)) represents a threshold voltage that is independent of frequency in the range $\omega \ll 1/\tau_Q$ (*conductance regime*) and increases critically when $\omega \to \omega_c \approx 1/\tau_Q$. Such a dependence is shown qualitatively in Fig. 28.

For frequencies above ω_c another regime is predicted by the same theory where the director *driven by flow* is oscillating clockwise and anticlockwise in the ac field. The threshold for this '*dielectric regime*' should also diverge with $\sigma_a \to 0$. In experiments, however, the modes observed at $\omega > \omega_c$ survive even at zero conductive anisotropy. Evidently, these instabilities (electrolytic and inertia modes) have a lower threshold than that predicted for the dielectric regime of the Carr–Helfrich mode.

If a substance has a small positive ε_a, the homeotropically oriented samples are stable against pure dielectric perturbation; however, the convection may be observed due to EHD destabilization. The vortices have a short period and result in an optical pattern in the form of a short pitch grid. The period of the pattern is $(\alpha_2/\alpha_3)^{\frac{1}{2}}$ times less than d (10 times in typical nematics) [1, 8, 9, 251]. For higher ε_a the direct transition to the chaotic state may be observed [253].

Behavior Above the Threshold. At voltages higher than the threshold of the Carr–Helfrich instability, the normal rolls transform in more complex hydrodynamic patterns. One can distinguish zig-zag and fluctuating domain patterns which, in turn, are substituted by the turbulent motion of a liquid crystal accompanied by a strong (dynamic) light scattering. The corresponding 'stability diagram' [226] is shown in Fig. 33 (see the discussion of the dielectric regime above). To calculate the wavevectors and amplitude of the distortions, a set of nonlinear equations must be solved. More generally, the problem of describing the transition from a regular electrohydrodynamic vortex motion to turbulence is a part of the classical problem, concerning the transition from laminar to turbulent flow of a liquid. Some progress has been achieved

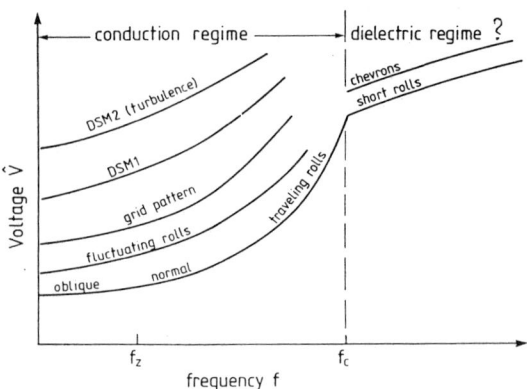

Figure 33. Voltage versus frequency phase diagram for electrohydrodynamic (convection) instabilities in the low frequency regime [226].

recently in understanding the nonlinear behavior of nematics in terms of bifurcation, phase transition, and dynamic chaos theories (for reviews see [226, 251, 254].

In the splay Frederiks transition geometry (homogeneous orientation, large dielectric anisotropy $\varepsilon_a \gg 0$) an interesting transient pattern in the form of domain stripes oriented *parallel to the initial director* has been observed [92]. The periodic pattern arises because the response of a liquid crystal placed suddenly far from equilibrium is faster to a spatially periodic director distribution than to a homogeneous orientation. The nonlinear theory of the transient patterns takes into account both the dielectric and the conductivity anisotropies [255, 256].

Dynamic Scattering of Light. As is known from the general theory of dissipative dynamic systems, after a finite number of bifurcations a system reaches dynamic chaos. This scenario is also observed in the case of EHD convective motion. With increasing voltage the velocity of the vortices increases rapidly and the periodic flow of the liquid transforms to turbulence. Turbulent motion in nematic liquid crystals results in a nonuniform distribution of the director accompanied by dynamic scattering of light [257, 258]. As the field is increased further, dynamic texturing or secondary dynamic light scattering is observed. This is related to the appearance of disclination loops or/and walls separating regions with different orientations of the director.

The light dynamic scattering effect was initially proposed for use in the manufacture of field-controllable shutters and displays, and was the starting point of modern studies of liquid crystals [259, 260].

Inertial Mode

This anisotropic mode is observed at rather high frequencies (10^3–10^6 Hz) in the case of highly conductive homogeneously oriented nematics with $\varepsilon_a < 0$. It appears in the form of long, periodic vortices of a liquid in the plane of a cell perpendicular to the initial director (Fig. 34). Such a flow results in an optical pattern of wide domains [261]. The threshold voltage of the instability diverges for vanishing conductivity anisotropy and also increases with decreasing mean conductivity, being proportional to $\sigma^{-\frac{1}{3}}$.

The physical mechanism of the instability involves the inertial term $\partial v/\partial t$ for the fluid velocity (for this reason the mode is called 'inertial'). As already mentioned, when constructing a theory for the Carr–Helfrich instability, the inertia term $\rho(\partial v_z/\partial t)$ in the Navier–Stokes equation was disregarded and two modes were predicted theoretically for different frequency ranges (the conductance and the dielectric regimes). Inclusion of the inertial term [262] led to the prediction of another high-frequency mode involving steady-state motion of the liquid and a stationary distribution of the director (as in Kapustin–Williams domains). In this case the space charge oscillates in counterphase with the external field. The approximate expression for the

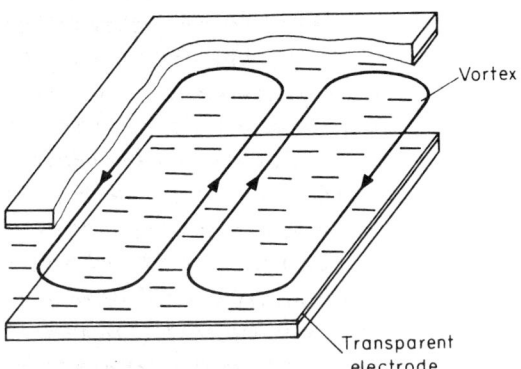

Figure 34. Convective motion of a strongly conductive nematic liquid crystal characteristic of the inertial mode.

threshold voltage of the inertial mode is

$$U_{\text{crit}} = A\omega \left(\frac{\varepsilon K}{\sigma \Delta \sigma}\right)^{\frac{1}{2}} \qquad (122)$$

where ε, K, and σ are average values of the dielectric, elastic, and conductivity constants. This expression agrees with most experimental observations.

The investigations of wide domains have identified the cause of the chevron (herringbone) structures. They result from interference between two modes with neighboring thresholds: the linear pre-chevron domains (deformation in the xz plane) and the wide domains (deformation in the xy plane). It has also been shown that such a herringbone structure can result from interference between the wide domains and the Kapustin–Williams domains (Fig. 31 d).

9.4.2 Cholesterics and Smectics

9.4.2.1 Cholesterics

The EHD behavior of cholesteric liquid crystals is very similar to that of nematics. When the anisotropy of the electrical conductivity is positive ($\sigma_a > 0$), the planar texture of a cholesteric liquid crystal in a field parallel to the helical axis is unstable for any sign of ε_a [263, 264]. The instability is caused by the torque induced by the electrical conductivity acting against the elastic torque of the cholesteric and, although the cause is different from the purely dielectric case (see Sec. 9.3.2.2 of this Chapter), the result obtained is the same; that is, the appearance of a two-dimensional periodic pattern for the distribution of the director.

Investigation of the EHD instability ($\varepsilon_a < 0$) in a planar Grandjean texture [265] shows that the directions of the domains alternate with a transition from one Grandjean zone to another, while the domains always form perpendicular to the director of a cholesteric in the middle of a layer. With increasing thickness, one-dimensional deformations transform to a two-dimensional grid (Fig. 31 c).

The theory for the threshold of the instability in cells with thicknesses considerably exceeding the equilibrium pitch ($d \gg P_0$) has been considered by analogy with the case of dielectric instability [121, 266], but with allowance being made for the additional, destabilizing term in the free energy which is caused by the space charge. The frequency dependence of the threshold field for $\varepsilon_a < 0$ has been shown to be similar to that calculated for nematics. For a cholesteric liquid crystal with $\varepsilon_a > 0$ the presence of electrical conductivity is revealed by a lowering of the threshold of the instability at low frequencies.

For voltages above the threshold, as in nematics the regular pattern of periodic deformations changes to a turbulent motion of the liquid and dynamic scattering of light. However, in contrast to nematics, the cholesteric cell remains rather turbid even after removal of the voltage (dynamic scattering storage mode [267]). The residual turbidity of the cell is accounted for by the scattering of light by the focal-conic texture obtained through the influence of the flow [268]. The residual scattering texture can be erased by a high-frequency field, using a purely dielectric transition from the focal-conic to the planar texture.

If the cell walls are treated to produce a homeotropic orientation, in a cholesteric liquid crystal with $\varepsilon_a < 0$ for a particular relationship between the thickness and the helical pitch ($1 < P_0/d < 2$), a new texture (bubble domains [269]) appears under the influence of the electric field. In its initial state the liquid crystal has a homeotropic nematic structure, since the helix is untwisted by the walls. The application of a low-frequency field induces an electrohydrodynamic instability. After the removal of the field a new stable texture appears in the form of cholesteric 'bubbles' dispersed through the homeotropic nematic phase. Thus a memory is created, while erasure of the information can be achieved either by applying a high-frequency field or by mechanical displacement of glass plates.

At frequencies higher than the critical one $\omega \gg \omega_c = 4\pi\sigma/\varepsilon$ defined by inverse time τ_Q (Eq. (100)), a characteristic dependence $E_{crit} \propto \omega^{\frac{1}{2}}$ is observed for the threshold field of the convective distortion. This EHD instability has been shown [270] to be the same electrolytic mode as the one discussed above for nematic liquid crystals.

EHD instabilities have also been observed in blue phases of cholesterics [271]; the threshold voltage increases with frequency [272].

9.4.2.2 Smectics A

Anisotropic Modes

Theoretically, the Helfrich approach to anisotropic EHD instabilities in nematics may be applied to the smectic A phase, using its specific viscoelastic properties and the sign of the conductivity anisotropy. It makes sense to consider only the onset of a splay deformation in a homeotropic structure for smectic A, since $K_{33} \to \infty$. This approach is developed in Geurst and Goossens [273]. For the field directed along the normal to the smectic layers, the threshold field of an instability is given by

$$E_{crit} = q_x \left[\frac{K_{11}\varepsilon_\perp}{\varepsilon_\parallel(\sigma_\perp - \sigma_\parallel)\tau_\perp} \cdot \frac{1+\omega^2\tau_\perp^2}{1-\omega^2\tau_\perp\tau_\Delta} \right]^{\frac{1}{2}} \quad (123)$$

where $\tau_\perp = \varepsilon_\perp/4\pi\sigma_\perp$, $\tau_\Delta = \varepsilon_a/4\pi(-\sigma_a)$, $\sigma_a = \sigma_\parallel - \sigma_\perp$, and q_x is the wavevector of the instability in the direction perpendicular to the field. If it is assumed that the half-period of the instability equals the thickness of the cell, as was assumed in the Helfrich one-dimensional model for nematics, $q_x = \pi/d$, then Eq. (123) is analogous to Eq. (119), but with the condition that the viscosity coefficient η diverges in the smectic A phase so that the term including the ratio of the viscosities in the expression for the threshold field vanishes. In addition, the anisotropy of the electrical conductivity is negative in smectic A phases.

No quantitative experimental data on EHD instability in the smectic A phase are available. EHD processes are often observed in the form of the motion of a liquid, which may be detected using solid foreign particles mixed with a liquid crystal [274, 275]. In a weak field, such a motion cannot destabilize the director and the initial orientation (e.g. homeotropic) is stable. It is possible to observe the optical patterns induced by EHD processes in strongly conducting samples of smectic A phases [276–278]. An EHD instability is observed in homeotropically oriented samples of material with $\varepsilon_a > 0$ and $\sigma_a < 0$ and, in general, the situation corresponds to the model by Guerst and Goossens [273]. After switching the low-frequency field off, the optical pattern is maintained for a long period, but it can be erased by the application of a high-frequency field, which reorients the director via the dielectric torque (the controllable memory effect).

Isotropic Modes

In the pretransition temperature region above the smectic A–nematic phase transition, where the nematic phase has a certain degree of the short-range smectic A order, it is easy to detect both the motion of a liquid and the formation of domain patterns. The latter show marked differences from the domain patterns seen in pure nematic phases. For instance, in the temperature region where the ratio of the electrical conductivity $\sigma_\parallel/\sigma_\perp$ becomes less than 1, Kapustin–Williams domains are not observed with an initial planar orientation. Instead, different domains are formed at low frequencies ($\omega<\omega_c$) of the applied field. They are arranged parallel to the initial orientation of the director and have a well-defined field strength threshold, $U_{\text{crit}} \propto d$ [275]. The threshold voltage for the Carr–Helfrich instability diverges at the temperature corresponding to the condition $\sigma_\parallel/\sigma_\perp=1$. The disappearance of the Kapustin–Williams domains allows a new EHD mode to be seen (longitudinal domain mode). For the appearance of this new mode the sign of σ_a is irrelevant. Without doubt, the high-frequency EHD mode is caused by the isotropic mechanism of destabilization, since for $\sigma_\parallel/\sigma_\perp<1$ the Carr–Helfrich model does not hold. Analysis shows [275] that the longitudinal domain mode is caused by the isotropic mechanism (electrolytic mode; as discussed above for nematics).

Planar oriented smectic A layers ($\sigma_a<0$, $\varepsilon_a<0$) also show domain instability that is EHD in nature [279]. This is probably caused by a destabilization of thin layers adjacent to the electrodes (the same isotropic mechanism). However, in this case, the conductivity anisotropy can influence the visual appearance of the effect.

9.4.2.3 Smectics C

In general, it can be anticipated that the EHD processes will result in a change in the director orientation in the smectic C phase, even when there is no change in the arrangement of the smectic layers, since the director has an extra degree of freedom – the possibility of rotation around the normal to the layers. In fact, the EHD instability is observed only in smectic C phases with a large tilt angle almost independent of temperature (C_1 phase). In C_2 smectics with a small, temperature-dependent tilt the Frederiks transition is usually observed (discussed in Section 9.3.3.1 of this Chapter). The reason for this seems to be different textures formed by the two phases [280].

The theory of the Carr–Helfrich instability in the smectic C phase has been arrived at [280] by analogy with the one presented above for nematic liquid crystals, but making allowance for the biaxiality of the former. The same set of equations for the director curvature and space charge (Eqs (120) and (121)) are valid for the C phase; however, material constants such as elastic moduli, viscosity coefficients, relaxation times for the director τ_r, and the space charge τ_Q and conductivity anisotropy σ_{eff} are more complicated. The frequency-dependent threshold voltage for the instability has been calculated for a wide range of dielectric anisotropy and the crossover of the Frederiks transition and EHD instability has been studied in detail. In particular, in the low-frequency range, both the ascending and descending types of threshold behavior with increasing frequency have been predicted.

EHD instabilities in the smectic C phase have been repeatedly observed experimentally. The formation of domains in a smectic C phase under the influence of an electric voltage was first observed in [281], then it has been studied in detail [281–283]. The results of these studies can be summarized as follows. At low frequencies, $\omega<\omega_c$, where the critical frequency ω_c is now defined taking into account the biaxiality of the smectic C phase, an EHD instability is observed (when $\sigma_a>0$) with a threshold voltage that is independent of the cell thickness (so-called 'initial domains'). In a certain sense, this instability is similar to the

Carr–Helfrich one in nematics and may be studied quantitatively theoretically [280]. At high frequencies, an instability is observed with a characteristic frequency dependence of the threshold field $E \propto \omega^{\frac{1}{2}}$ that is independent of the layer thickness (fundamental domains). This has been regarded as an analog of the dielectric regime [282], but it can also be interpreted as the electrolytic mode [283] with some specific features. In some special cases a new domain mode is observed [284], which has been referred to the inertial (anisotropic) mode (discussed in Section 9.4.1.3 of this Chapter).

terparts. However, the inverse is not true, because liquid crystals possess strong dielectric (not magnetic) anisotropy and finite spontaneous polarization (but no spontaneous magnetization has been found in liquid crystals). Besides, a steady-state magnetic field cannot induce hydrodynamic and other dissipative processes. To a certain extent the situation may change in the near future, as mesophase have recently been synthesized that have extremely high magnetic susceptibilities [285]. The discovery of ferromagnetic mesophases cannot be excluded either.

9.5 Conclusion

In conclusion, electric field effects in liquid crystals is a well-developed branch of condensed matter physics. The field behavior of nematic liquid crystals in the bulk is well understood. To a certain extent the same is true for the cholesteric mesophase, although the discovery of bistability phenomena and field effects in blue phases opened up new fundamental problems to be solved. Ferroelectric and antiferroelectric mesophases in chiral compounds are a subject of current study. The other ferroelectric substances, such as discotic and lyotropic chiral systems and some achiral (like polyphilic) mesogenes, should attract more attention in the near future*. The same is true for a variety of polymer ferroelectric substances, including elastomers.

Until now, all the magnetic field effects have been found to have electric field coun-

---

* In fact, two new representatives of polar achiral systems have been discovered quite recently: antiferroelectric polymer-monomer mixtures [286] and ferroelectric biaxial smectic A phases composed of banana-like molecules [287].

9.6 References

[1] P. G. de Gennes, *The Physics of Liquid Crystals*, Clarendon Press, Oxford **1974**.
[2] S. Chandrasekhar, *Liquid Crystals*, Cambridge University Press, Oxford **1974**.
[3] W. H. de Jeu, *Physical Properties of Liquid Crystalline Materials*, Gordon & Breach, New York **1980**.
[4] V. A. Belyakov, A. S. Sonin, *Optika Kholestericheskikh Zhidkikh Kristallov (Optics of Cholesteric Liquid Crystals)*, Nauka, Moscow **1982** [in Russian].
[5] G. Vertogen, W. H. de Jeu, *Thermotropic Liquid Crystals, Fundamentals*, Springer-Verlag, Berlin **1988**.
[6] A. P. Kapustin, *Elektroopticheskiye i akusticheskiye svoistva zhidkikh kristallov (Electro-Optical and Acoustical Properties of Liquid Crystals)*, Nauka, Moscow **1973** [in Russian].
[7] S. A. Pikin, *Structural Transformations in Liquid Crystals*, Gordon & Breach, New York **1991**.
[8] L. M. Blinov, *Electro-Optical and Magneto-Optical Properties of Liquid Crystals*, Wiley, Chichester **1983**.
[9] L. M. Blinov, V. G. Chigrinov, *Electrooptic Effects in Liquid Crystal Materials*, Springer-Verlag, New York **1993**.
[10] M. F. Grebenkin, A. V. Ivashchenko, *Zhidkokristallicheskiye Materialy (Liquid Crystalline Materials)*, Khimiya, Moscow **1989** [in Russian].
[11] M. V. Loseva, E. P. Pozhidayev, A. Z. Rabinovich, N. I. Chernova, A. V. Ivashchenko, *Itogi Nauk. Tekh. (Fiz. Khim.)* **1990**, *3*, 3.

[12] B. Bahadur, *Mol. Cryst. Liq. Cryst.* **1984**, *109*, 1.
[13] L. M. Blinov, W. Haase, *Mol. Mater.* **1993**, *2*, 145.
[14] L. M. Blinov, L. A. Beresnev, W. Haase, *Ferroelectrics*, **1996**, *181*, 187.
[15] K. Yoshino, K. Nakano, H. Taniguchi, M. Ozaki, *Jpn. J. Appl. Phys.* **1987**, *26* (Suppl. 26-2), 97.
[16] W. Helfrich, *Phys. Rev. Lett.* **1970**, *24*, 201.
[17] E. I. Rjumtsev, M. A. Osipov, T. A. Rotinyan, N. P. Yevlampieva, *Liq. Cryst.* **1995**, *18*, 87.
[18] M. J. Stephen, C. P. Fan, *Phys. Rev. Lett.* **1970**, *25*, 500.
[19] Ch. Bahr, G. Heppke, *Mol. Cryst. Liq. Cryst.* **1987**, *150B*, 313.
[20] Ch. Bahr, G. Heppke, B. Subaschus, *Liq. Cryst.* **1992**, *11*, 41.
[21] G. Andersson, K. Flatischler, L. Komitov, S. T. Lagerwall, K. Skarp, B. Stebler, *Ferroelectrics* **1991**, *113*, 361.
[22] L. M. Blinov, L. A. Beresnev, W. Haase, *Ferroelectrics*, **1996**, *181*, 211.
[23] M. A. Anisimov, *Mol. Cryst. Liq. Cryst.* **1988**, *162A*, 1.
[24] H. Hama, *J. Phys. Soc. Jpn.* **1985**, *54*, 2204.
[25] Ch. Bahr, G. Heppke, *Phys. Rev. A* **1990**, *41*, 4335.
[26] D. A. Dunmur, P. Palffy-Muhoray, *J. Phys. Chem.* **1988**, *92*, 1406.
[27] D. A. Dunmur, T. F. Waterworth, P. Palffy-Muhoray, *Mol. Cryst. Liq. Cryst.* **1985**, *124*, 73.
[28] D. A. Dunmur, T. F. Waterworth, preprint.
[29] D. A. Dunmur, K. Szumilin, T. F. Waterworth, *Mol. Cryst. Liq. Cryst.* **1987**, *149*, 385.
[30] B. Malraison, Y. Poggi, E. Guyon, *Phys. Rev. A* **1980**, *21*, 1012.
[31] N. V. Madhusudana, S. Chandrasekhar, *Liquid Crystals and Ordered Fluids*, Vol. 2 (Eds. J. F. Johnson, R. S. Porter), Plenum Press, New York **1974**, p. 657.
[32] D. Dunmur, P. Palffy-Muhoray, *J. Phys. Chem.* **1988**, *92*, 1406.
[33] D. A. Dunmur, A. E. Tomes, *Mol. Cryst. Liq. Cryst.* **1981**, *76*, 231.
[34] M. Schadt, W. Helfrich, *Mol. Cryst. Liq. Cryst.* **1972**, *17*, 355.
[35] H. J. Coles, R. R. Jennings, *Mol. Phys.* **1976**, *31*, 571.
[36] J. C. Filippini, Y. Poggi, *J. Phys. D* **1975**, *8*, 201.
[37] B. R. Ratna, M. S. Vijaya, R. Shashidhar, B. K. J. Sadashiva, *Pramana* **1975**, *1* (Suppl.), 69.
[38] R. Yamamoto, S. Ishihara, K. Moritomo, *Phys. Lett.* **1978**, *69A*, 276.
[39] V. N. Tsvetkov, E. I. Ryumtsev, A. P. Kovshik, G. I. Denis, Yu. Daugvila, *Dokl. Akad. Nauk SSSR* **1974**, *216*, 1105.
[40] J. C. Filippini, *J. Phys. D* **1975**, *8*, 201.
[41] V. N. Tsvetkov, E. I. Ryumtsev, *Mol. Cryst. Liq. Cryst.* **1986**, *133*, 125.
[42] H. Stegemeyer, Th. Bluemel, K. Hiltrop, H. Onusseit, F. Porsch, *Liq. Cryst.* **1986**, *1*, 3.
[43] D. K. Yang, P. P. Crooker, *Phys. Rev. A* **1988**, *37*, 4001.
[44] V. A. Belyakov, *Difraktsionnaya optika periodicheskikh sred slozhnoi struktury (Diffraction Optics of Periodical Media with Complex Structure)*, Nauka, Moscow **1988**, p. 255 [in Russian].
[45] V. E. Dmitrienko, *Liq. Cryst.* **1989**, *5*, 847.
[46] B. Jerome, P. Pieranski, *Liq. Cryst.* **1989**, *5*, 799.
[47] G. Heppke, B. Jerome, H. S. Kitzerow, P. Pieranski, *Liq. Cryst.* **1989**, *5*, 813.
[48] F. Porsch, H. Stegemeyer, K. Hiltrop, *Z. Naturforsch. A* **1984**, *39A*, 475.
[49] G. Heppke, B. Jerôme, H. S. Kitzerow, P. Pieranski, *J. Phys. (Paris)* **1991**, *52*, 2991.
[50] G. Heppke, B. Jerôme, H. S. Kitzerow, P. Pieranski, *J. Phys. (Paris)* **1989**, *50*, 549.
[51] G. Chilaya, G. Petriashvili, *Mol. Mater.* **1993**, *2*, 239.
[52] D. Armitage, R. J. Cox, *Mol. Cryst. Liq. Cryst.* **1980**, *64*, 41.
[53] H.-S. Kitzerow, P. P. Crooker, S. L. Kwok, G. Heppke, *J. Phys. France* **1990**, *51*, 1303.
[54] H.-S. Kitzerow, *Mol. Cryst. Liq. Cryst.* **1991**, *202*, 51.
[55] R. Shao, J. Pang, N. A. Clark, J. A. Rego, D. M. Walba, *Ferroelectrics* **1993**, *147*, 255.
[56] H.-S. Kitzerow, A. J. Slaney, J. W. Goodby, *Ferroelectrics* **1996**, *179*, 61.
[57] V. Shibayev, *Mol. Cryst. Liq. Cryst.* **1988**, *155*, 189.
[58] A. L. Smolyansky, Z. A. Roganova, R. V. Tal'rose, L. A. Kazarin, S. G. Kostromin, *Kristallografiya* **1987**, *32*, 265.
[59] H. Finkelmann, D. Naegele, H. Ringsdorf, *Macromol. Chem.* **1979**, *180*, 803.
[60] V. Frederiks, V. Zolina, *Zh. Russ. Fiz. Khim. Obshch.* **1930**, *62*, 457.
[61] V. Frederiks, V. Zolina, *Trans. Faraday Soc.* **1933**, *29*, 919.
[62] B. Brochard, P. Pieranski, E. Guyon, *Phys. Rev. Lett.* **1972**, *26*, 1681.
[63] P. Pieranski, E. Brochard, E. Guyon, *J. Phys. France* **1973**, *34*, 35.
[64] C. Z. van Doorn, *Phys. Lett. A* **1973**, *42A*, 537.
[65] M. Schiekel, K. Fahrenschon, H. Gruler, *Appl. Phys.* **1975**, *7*, 99.
[66] L. M. Blinov, E. I. Kats, A. A. Sonin, *Usp. Fiz. Nauk.* **1987**, *152*, 449.
[67] H. Yokoyama, *Mol. Cryst. Liq. Cryst.* **1988**, *165*, 265.
[68] B. Jerome, *Rep. Prog. Phys.* **1991**, *54*, 391.
[69] L. Leger, *Mol. Cryst. Liq. Cryst.* **1973**, *24*, 33.
[70] P. Schiller, G. Pelzl, D. Demus, *Liq. Cryst.* **1989**, *6*, 417.

[71] B. J. Frisken, P. Palffy-Muhoray, *Phys. Rev. A* **1989**, *40*, 6099.
[72] G. Barbero, E. Miraldi, C. Oldano, P. Taverna Valabrega, *Z. Naturforsch. A* **1988**, *43*, 547.
[73] G. Barbero, E. Miraldi, C. Oldano, *Phys. Rev. A* **1988**, *38*, 3027.
[74] V. N. Tsvetkov, L. N. Andreeva, A. P. Filippov, *Mol. Cryst. Liq. Cryst.* **1987**, *153*, 217.
[75] B. J. Frisken, P. Palffy-Muhoray, *Phys. Rev. A* **1989**, *39*, 1513.
[76] D. Allender, B. J. Frisken, P. Palffy-Muhoray, *Liq. Cryst.* **1989**, *5*, 735.
[77] P. Palffy-Muhoray, H. J. Yuan, B. J. Frisken, W. van Saarloos, *NATO ASI, Ser. B* **1990**, *225*, 313.
[78] F. Longberg, R. B. Meyer, *Phys. Rev. Lett.* **1985**, *55*, 718.
[79] V. G. Taratuta, A. J. Hurd, R. B. Meyer, *Phys. Rev. Lett.* **1985**, *55*, 246.
[80] C. Oldano, *Phys. Rev. Lett.* **1986**, *56*, 1098.
[81] W. Zimmermann, L. Kramer, *Phys. Rev. Lett.* **1986**, *56*, 2655.
[82] V. D. Kini, *Mol. Cryst. Liq. Cryst.* **1987**, *153*, 1.
[83] M. I. Barnik, L. M. Blinov, A. N. Trufanov, V. G. Chigrinov, T. V. Korkishko, *Zh. Eksp. Teor. Fiz.* **1984**, *87*, 196.
[84] V. G. Chigrinov, T. V. Korkishko, M. I. Barnik, A. N. Trufanov, *Mol. Cryst. Liq. Cryst.* **1985**, *129*, 283.
[85] E. F. Carr, *Mol. Cryst. Liq. Cryst.* **1977**, *34*, L-159.
[86] E. Guyon, R. Meyer, J. Salan, *Mol. Cryst. Liq. Cryst.* **1979**, *54*, 261.
[87] A. J. Hurd, S. Fraden, F. Longberg, R. B. Meyer, *J. Phys. France* **1985**, *46*, 905.
[88] F. Sagues, M. San Miguel, *Phys. Rev. A* **1989**, *39*, 6567.
[89] F. Longberg, S. Fraden, A. J. Hurd, R. B. Meyer, *Phys. Rev. Lett.* **1984**, *52*, 1903.
[90] A. D. Rey, M. M. Denn, *Liq. Cryst.* **1989**, *4*, 409.
[91] E. E. Pashkovsky, W. Stille, G. Strobl, *J. Phys. II, France* **1995**, *5*, 397.
[92] A. Buka, M. Juarez, L. Kramer, I. Rehberg, *Phys. Rev. A* **1989**, *40*, 7427.
[93] R. B. Meyer, *Phys. Rev. Lett.* **1969**, *22*, 319.
[94] G. Barbero, I. Dozov, I. Palierne, G. Durand, *Phys. Rev. Lett.* **1986**, *56*, 2056.
[95] D. Schmidt, M. Schadt, W. Helfrich, *Z. Naturforsch. A* **1972**, *27*, 277.
[96] A. I. Derzhanski, A. G. Petrov, M. D. Mitov, *J. Phys. (Paris)* **1978**, *39*, 273.
[97] J. Prost, P. S. Pershan, *J. Appl. Phys.* **1976**, *47*, 2298.
[98] A. I. Derzhanski, A. G. Petrov, *Acta Phys. Pol.* **1979**, *A55*, 747.
[99] L. M. Blinov, G. Durand, S. V. Yablonsky, *J. Phys. II, France* **1992**, *2*, 1287.
[100] L. K. Vistin, *Dokl. Akad. Nauk SSSR* **1970**, *194*, 1318.
[101] Yu. P. Bobylev, S. A. Pikin, *Zh. Eksp. Teor. Fiz.* **1977**, *72*, 369.
[102] Yu. P. Bobylev, V. G. Chigrinov, S. A. Pikin, *J. Phys. France* **1979**, *40*, C3-331.
[103] P. Schiller, G. Pelzl, D. Demus, *Cryst. Res. Technol.* **1990**, *25*, 111.
[104] M. I. Barnik, L. M. Blinov, A. N. Trufanov, B. A. Umansky, *Zh. Eksp. Teor. Fiz.* **1977**, *73*, 1936.
[105] Kh. Khinov, L. K. Vistin, *J. Phys. France* **1979**, *40*, 269.
[106] Kh. Khinov. *Z. Naturforsch. A* **1982**, *37*, 334.
[107] B. H. Soffer, J. D. Margerum, A. M. Lackner, D. Boswell, A. R. Tanguay, T. S. Strand, A. Sawchuk, P. Chavel, *Mol. Cryst. Liq. Cryst.* **1981**, *70*, 145.
[108] Ho-In Jeon, A. A. Sawchuk, *Appl. Opt.* **1987**, *26*, 261.
[109] M. Schadt, W. Helfrich, *Appl. Phys. Lett.* **1971**, *18*, 127.
[110] C. Mauguin, *Bull Soc. Fr. Miner.* **1911**, *34*, 71.
[111] F. M. Leslie, *Mol. Cryst.* **1970**, *12*, 57.
[112] M. Becker, J. Nehring, T. Scheffer, *J. Appl. Phys.* **1985**, *57*, 4539.
[113] D. W. Berreman, *J. Opt. Soc. Am.* **1973**, *63*, 1374.
[114] D. W. Berreman, *Phil. Trans. R. Soc., London, Ser. A* **1983**, *309*, 203.
[115] E. P. Raynes, *Mol. Cryst. Liq. Cryst. Lett.* **1986**, *4*, 1.
[116] P. A. Breddels, H. A. van Sprang, *J. Appl. Phys.* **1985**, *58*, 2162.
[117] R. Thurston, J. Cheng, G. Boid, *IEEE Trans. Electr. Dev.* **1980**, *ED-27*, 2969.
[118] W. R. Heffner, D. W. Berreman, *J. Appl. Phys.* **1982**, *53*, 8599.
[119] D. W. Berreman, *J. Appl. Phys.* **1984**, *55*, 806.
[120] W. Helfrich, *Appl. Phys. Lett.* **1970**, *17*, 531.
[121] J. P. Hurault, *J. Chem. Phys.* **1973**, *59*, 2068.
[122] V. G. Chigrinov, V. V. Belyayev, S. V. Belyayev, M. F. Grebenkin, *Zh. Eksp. Teor. Fiz.* **1979**, *77*, 2081.
[123] P. Schiller, K. Schiller, *Liq. Cryst.* **1990**, *8*, 553.
[124] P. Schiller, *Phase Trans.* **1990**, *29*, 59.
[125] G. Cohen, R. M. Hornreich, *Phys. Rev. A* **1990**, *41*, 4402.
[126] R. Meyer, *Appl. Phys. Lett.* **1968**, *12*, 281.
[127] P. G. de Gennes, *Sol. State Commun.* **1968**, *6*, 163.
[128] R. Dreher, *Solid State Commun.* **1973**, *13*, 1571.
[129] S. V. Belyayev, L. M. Blinov, *Pis'ma Zh. Eksp. Teor. Fiz.* **1979**, *30*, 111.
[130] W. J. A. Goossens, *J. Phys. France* **1982**, *43*, 1469.
[131] J. Brokx, G. Vertogen, E. W. C. van Groesen, *Z. Naturforsch. A* **1983**, *38*, 1.

[132] A. Mochizuki, S. Kobayashi, *Mol. Cryst. Liq. Cryst.* **1993**, *225*, 89.
[133] G. Hauk, H. D. Koswig, *Cryst. Res. Technol.* **1987**, *22*, 1333.
[134] J. S. Patel, R. B. Meyer, *Phys. Rev. Lett.* **1987**, *58*, 1538.
[135] J. D. Parson, C. F. Hayes, *Phys. Rev. A* **1974**, *9*, 2652.
[136] P. R. Gerber, *Mol. Cryst. Liq. Cryst.* **1985**, *116*, 197.
[137] J. S. Patel, S. D. Lee, *J. Appl. Phys.* **1989**, *66*, 1879.
[138] S. D. Lee, J. S. Patel, R. B. Meyer, *J. Appl. Phys.* **1990**, *67*, 1293.
[139] L. A. Beresnev, L. M. Blinov, D. I. Dergachev, S. B. Kondrat'yev, *Pis'ma Zh. Eksp. Teor. Fiz.* **1987**, *46*, 328.
[140] R. Rundquist, L. Komitov, S. T. Lagerwall, *Phys. Rev. E* **1994**, *50*, 4735.
[141] R. Rundquist, M. Buivydas, L. Komitov, S. T. Lagerwall, *J. Appl. Phys.* **1994**, *76*, 7778.
[142] L. Komitov, S. T. Lagerwall, B. Stebler, G. Andersson, K. Flatischer, *Ferroelectrics* **1991**, *114*, 167.
[143] G. Legrand, N. Isaert, J. Hmine, J. M. Buisine, J. P. Parneix, N. T. Nguyen, C. Destrade, *Ferroelectrics* **1991**, *121*, 21.
[144] B. A. Umansky, L. M. Blinov, M. I. Barnik, *Pis'ma Zh. Tekh. Fiz.* **1980**, *6*, 200.
[145] B. A. Umansky, V. G. Chigrinov, L. M. Blinov, Yu. B. Pod'yachev, *Zh. Eksp. Teor. Fiz.* **1981**, *81*, 1307.
[146] G. Chigrinov, *Kristallografiya* **1983**, *28*, 825.
[147] E. M. Terent'yev, S. A. Pikin, *Kristallografiya* **1985**, *30*, 227.
[148] L. Bourdon, M. Sommeria, M. Kleman, *J. Phys. France* **1982**, *43*, 77.
[149] A. Rapini, *J. Phys. France* **1972**, *33*, 237.
[150] M. Hareng, S. Le Berre, J. J. Metzger, *Appl. Phys. Lett.* **1975**, *27*, 575.
[151] G. Pelzl, H. J. Deutscher, D. Demus, *Cryst. Res. Technol.* **1981**, *16*, 603.
[152] O. Parodi, *Solid State Commun.* **1972**, *11*, 1503.
[153] Y. Takanishi, Y. Ouchii, H. Takezoe, A. Fukuda, *Jpn. J. Appl. Phys.* **1989**, *28*, L487.
[154] H. P. Hinov, *Liq. Cryst.* **1988**, *3*, 1481.
[155] H. P. Hinov, K. Avramova, *Liq. Cryst.* **1988**, *3*, 1505.
[156] A. Jakli, A. Saupe, *Mol. Cryst. Liq. Cryst.* **1992**, *222*, 101.
[157] M. J. Stephen, J. P. Straley, *Rev. Mod. Phys.* **1974**, *46*, 617.
[158] G. Pelzl, P. Kolbe, V. Preukschas, S. Diele, D. Demus, *Mol. Cryst. Liq. Cryst.* **1979**, *53*, 167.
[159] G. Pelzl, P. Schiller, D. Demus, *Liq. Cryst.* **1987**, *2*, 131.
[160] G. Pelzl, P. Schiller, D. Demus, *Cryst. Res. Technol.* **1990**, *25*, 215.
[161] P. Schiller, G. Pelzl, D. Demus, *Liq. Cryst.* **1987**, *2*, 21.
[162] E. P. Pozhidayev, M. A. Osipov, V. G. Chigrinov, V. A. Baikalov, L. A. Beresnev, L. M. Blinov, *Zh. Eksp. Teor. Fiz.* **1988**, *94*, 125.
[163] R. Bruinsma, D. R. Nelson, *Phys. Rev. B* **1981**, *21*, 5312.
[164] P. Schiller, G. Pelzl, C. Camara, *Liq. Cryst.*, submitted.
[165] F. Tournilhac, L. M. Blinov, J. Simon, S. V. Yablonsky, *Nature* **1992**, *359*, 621.
[166] F. Tournilhac, L. M. Blinov, J. Simon, D. Subachius, S. V. Yablonsky, *Synth. Met.* **1993**, *54*, 253.
[167] J. Prost, F. Barois, *J. Chim. Phys.* **1983**, *80*, 65.
[168] L. M. Blinov, T. A. Lobko, B. I. Ostrovsky, S. N. Sulianov, F. Tournilhac, *J. Phys. II, France* **1993**, *3*, 1121.
[169] R. B. Meyer, L. Liebert, L. Strzelecki, P. Keller, *J. Phys. France Lett.* **1975**, *36*, L-69.
[170] J. W. Goodby, R. Blinc, N. A. Clark, S. T. Lagerwall, M. A. Osipov, S. A. Pikin, T. Sakurai, K. Yoshino, B. Zeks, *Ferroelectric Liquid Crystals. Principles, Properties and Applications*, Gordon & Breach, Philadelphia **1991**.
[171] L. A. Beresnev, L. M. Blinov, M. A. Osipov, S. A. Pikin, *Mol. Cryst. Liq. Cryst.* **1988**, *158A*, 1–150.
[172] T. P. Rieker, N. A. Clark, in *Phase Transitions in Liquid Crystals* (Eds S. Martellucci, A. N. Chester), Plenum Press, New York **1992**.
[173] A. Fukuda, Y. Takanishi, T. Isozaki, K. Ishikawa, H. Takezoe, *J. Mater. Chem.* **1994**, *4*, 997.
[174] N. A. Clark, S. T. Lagerwall, *Appl. Phys. Lett.* **1980**, *36*, 899.
[175] M. A. Handschy, N. A. Clark, S. T. Lagerwall, *Phys. Rev. Lett.* **1983**, *51*, 471.
[176] N. A. Clark, S. T. Lagerwall, *Ferroelectrics* **1984**, *59*, 25.
[177] T. Carlson, N. A. Clark, Z. Zou, *Liq. Cryst.* **1993**, *15*, 461.
[178] J. Z. Xue, M. A. Handschy, N. A. Clark, *Ferroelectrics* **1987**, *73*, 305.
[179] H. Orihara, K. Nakamura, Y. Ishibashi, Y. Yamada, N. Yamamoto, M. Yamanaki, *Jpn. J. Appl. Phys.* **1986**, *25*, L-839.
[180] P. Schiller, *Cryst. Res. Technol.* **1986**, *21*, 167, 301.
[181] I. Abdulhalim, G. Moddel, N. A. Clark, *J. Appl. Phys.* **1994**, *76*, 820.
[182] L. A. Beresnev, L. M. Blinov, D. I. Dergachev, *Ferroelectrics* **1988**, *85*, 173.
[183] L. A. Beresnev, V. G. Chigrinov, D. I. Dergachev, E. P. Pozhidayev, J. Fuenfschilling, M. Schadt, *Liq. Cryst.* **1989**, *5*, 1171.
[184] B. I. Ostrovskii, A. Z. Rabinovich, A. S. Sonin, B. A. Strukov, *Zh. Eksp. Teor. Fiz.* **1978**, *74*, 1748.

[185] B. I. Ostrovskii, V. G. Chigrinov, *Kristallografiya* **1980**, *25*, 560.
[186] B. I. Ostrovskii, A. Z. Rabinovich, V. G. Chigrinov, in *Advances in Liquid Crystal Research and Applications* (Ed. L. Bata), Pergamon Press, Oxford **1980**, p. 469.
[187] B. I. Ostrovskii, S. A. Pikin, V. G. Chigrinov, *Zh. Eksp. Teor. Fiz.* **1979**, *77*, 1631.
[188] J. Funfschilling, M. Schadt, *J. Appl. Phys.* **1989**, *66*, 3877.
[189] S. Garoff, R. Meyer, *Phys. Rev. A* **1979**, *19*, 338.
[190] G. Andersson, I. Dahl, L. Komitov, S. T. Lagerwall, K. Sharp, B. Stebler, *J. Appl. Phys.* **1989**, *66*, 4983.
[191] L. M. Blinov, L. A. Beresnev, W. Haase, *Ferroelectrics* **1995**, *174*, 221.
[192] S. Lee, J. Patel, *Appl. Phys. Lett.* **1989**, *55*, 122.
[193] L. A. Beresnev, M. V. Loseva, N. I. Chernova, S. G. Kononov, P. V. Adomenas, E. P. Pozhidaev, *Pis'ma Zh. Tehn. Fiz.* **1990**, *51*, 457.
[194] S. A. Pikin, L. A. Beresnev, S. Hiller, M. Pfeiffer, W. Haase, *Mol. Mater.* **1993**, *3*, 1.
[195] L. A. Beresnev, E. Schumacher, S. A. Pikin, X. Fan, B. I. Ostrovski, S. Hiller, A. P. Onikhov, W. Haase, *Jpn. J. Appl. Phys.* **1995**, *34*, 2404.
[196] K. Skarp, G. Andersson, T. Hirai, A. Yoshizawa, K. Hiraoka, H. Takezoe, A. Fukuda, *Jpn. J. Appl. Phys.* **1992**, *31*, 1409.
[197] V. P. Shibayev, M. V. Kozlovsky, L. A. Beresnev, L. M. Blinov, N. A. Plate, *Polymer Bull.* **1984**, *12*, 299.
[198] M. V. Kozlovsky, L. A. Beresnev, *Phase Transitions* **1992**, *40*, 129.
[199] L. A. Beresnev, L. M. Blinov, V. A. Baikalov, E. P. Pozhidaev, G. V. Purvanetskas, A. I. Pavlyuchenko, *Mol. Cryst. Liq. Cryst.* **1982**, *89*, 327.
[200] H. R. Brand, P. E. Cladis, *J. Phys. France Lett.* **1984**, *45*, L-217.
[201] N. Hiji, A. D. L. Chandani, S. Nishiyama, Y. Ouchi, H. Takezoe, A. Fukuda, *Ferroelectrics* **1988**, *85*, 99.
[202] A. Chandani, T. Hagiwara, Y. Suzuki, Y. Ouchi, H. Takezoe, A. Fukuda, *Jpn. J. Appl. Phys.* **1988**, *27*, L-729.
[203] A. Chandani, Y. Ouchi, H. Takezoe, A. Fukuda, K. Terashima, K. Furukawa, A. Kishi, *Jpn. J. Appl. Phys.* **1989**, *28*, L-1265.
[204] A. Chandani, E. Gorecka, Y. Ouchi, H. Takezoe, A. Fukuda, *Jpn. J. Appl. Phys.* **1989**, *28*, L-1265.
[205] Y. Galerne, L. Liebert, *Phys. Rev. Lett.* **1990**, *64*, 906; **1991**, *66*, 2891.
[206] Ch. Bahr, D. Fliegner, *Phys. Rev. Lett.* **1993**, *70*, 1842.
[207] M. Hara, T. Umemoto, H. Takezoe, A. F. Garito, H. Sasabe, *Jpn. J. Appl. Phys.* **1991**, *30*, L2052.

[208] Ji Li, H. Takezoe, A. Fukuda, *Jpn. J. Appl. Phys.* **1991**, *30*, 532.
[209] K. Skarp, G. Andersson, F. Gouda, S. T. Lagerwall, H. Poths, R. Zentel, *Polym. Adv. Technol.* **1992**, *3*, 241.
[210] T. Isozaki, T. Fujikawa, H. Takezoe, A. Fukuda, T. Hagiwara, Y. Suzuki, I. Kawamura, *Jpn. J. Appl. Phys.* **1992**, *31*, L1435.
[211] K. Hiraoka, Y. Takanishi, K. Skarp, H. Takezoe, A. Fukuda, *Jpn. J. Appl. Phys.* **1991**, *30*, L1819.
[212] M. Johno, A. Chandani, J. Lee, Y. Ouchi, H. Takezoe, A. Fukuda, K. Itoh, *Proc. SID* **1990**, *31*, 129.
[213] J. Lee, A. Chandani, K. Itoh, Y. Ouchi, H. Takezoe, A. Fukuda, *Jpn. J. Appl. Phys.* **1990**, *29*, 1122.
[214] L. A. Beresnev, L. M. Blinov, E. I. Kovshev, *Dokl. Akad. Nauk SSSR* **1982**, *265*, 210.
[215] L. M. Blinov, S. A. Davidyan, A. G. Petrov, A. T. Todorov, S. V. Yablonsky, *Pis'ma Zh. Eksp. Teor. Fiz.* **1988**, *48*, 259.
[216] A. G. Petrov, A. T. Todorov, P. Bonev, L. M. Blinov, S. V. Yablonsky, D. B. Subachius, N. Tvetkova, *Ferroelectrics* **1991**, *114*, 415.
[217] H. Bock, W. Helfrich, *Liq. Cryst.* **1992**, *12*, 697.
[218] H. Bock, W. Helfrich, *Liq. Cryst.* **1994**, *14*, 345.
[219] L. M. Blinov, *Sci. Prog., Oxford* **1986**, *70*, 263.
[220] S. Kai (Ed.), *Pattern Formation in Complex Dissipative Systems and Global Dynamics*, World Scientific, Singapore **1992**.
[221] L. Kramer, W. Pesch, in *Pattern Formation in Liquid Crystals* (Eds A. Buka, L. Kramer), Springer-Verlag, New York **1995**.
[222] R. D. Feynman, R. B. Leighton, M. Sands, *The Feynman Lectures on Physics*, Vol. 2, Addison-Wesley, Reading, MA **1964**, Ch. 41.
[223] N. J. Felici, *J. Phys. France Colloq. C1* **1976**, *37*, C1–17.
[224] J. J. Felici, *Rev. Gen. Electr.* **1969**, *78*, 717.
[225] E. Guyon, P. Pieranski, *Physica* **1974**, *73*, 184.
[226] W. Zimmermann, *MRS Bull.* **1991**, *24*, 46.
[227] P. Atten, R. Moreau, *C. R. Acad. Sci.* **1970**, *270A*, 415.
[228] J. M. Schneider, P. K. Watson, *Phys. Fluids* **1970**, *13*, 1948.
[229] J. C. Filippini, J. P. Gosse, J. C. Lacroix, R. Tobazeon, *C. R. Acad. Sci.* **1969**, *69B*, 16, 736.
[230] H. Koelmans, A. M. van Boxtel, *Phys. Lett.* **1970**, *32A*, 32.
[231] H. Koelmans, A. M. van Boxtel, *Mol. Cryst. Liq. Cryst.* **1971**, *12*, 185.
[232] M. I. Barnik, L. M. Blinov, M. F. Grebenkin, A. N. Trufanov, *Mol. Cryst. Liq. Cryst.* **1976**, *37*, 47.
[233] M. I. Barnik, L. M. Blinov, S. A. Pikin, A. N. Trufanov, *Zh. Eksp. Teor. Fiz.* **1977**, *72*, 756.

[234] V. G. Chigrinov, S. A. Pikin, *Kristallografiya* **1978**, *23*, 333.
[235] A. N. Trufanov, M. I. Barnik, L. M. Blinov, V. G. Chigrinov, *Zh. Eksp. Teor. Fiz.* **1980**, *80*, 704.
[236] W. Helfrich, *J. Chem. Phys.* **1969**, *51*, 4092.
[237] E. F. Carr, *Mol. Cryst. Liq. Cryst.* **1977**, *34*, L-159.
[238] G. E. Zvereva, A. P. Kapustin in *Primeneniye ultraakustiki kissledovaniyu veshchestva (Application of Ultraacustics to Investigation of Substances)*, Moscow **1961**, Vol. 15, p. 69 [in Russian].
[239] R. Williams, *J. Chem. Phys.* **1963**, *39*, 384.
[240] S. A. Pikin, *Zh. Eksp. Teor. Fiz.* **1971**, *60*, 1185.
[241] S. A. Pikin, A. A. Shtol'berg, *Kristallografiya* **1973**, *18*, 445.
[242] P. A. Penz, G. W. Ford, *Phys. Rev. A* **1972**, *6*, 414, 1676.
[243] M. I. Barnik, L. M. Blinov, M. F. Grebenkin, S. A. Pikin, V. G. Chigrinov, *Zh. Eksp. Teor. Fiz.* **1975**, *69*, 1080.
[244] E. Bodenschatz, W. Zimmerman, L. Kramer, *J. Phys. France* **1988**, *49*, 1875.
[245] S. Kai, K. Hirakawa, *Prog. Theor. Phys.* **1978**, *64* (Suppl.), 212.
[246] M. Kohno, *Phys. Rev. A* **1989**, *40*, 6554.
[247] A. Joets, R. Ribotta, *J. Phys. France* **1986**, *47*, 595.
[248] N. V. Madhusudana, V. A. Raghunathan, K. R. Sumathy, *Pramana J. Phys.* **1987**, *28*, L311.
[249] V. A. Raghunathan, N. V. Madhusudana, *Pramana J. Phys.* **1988**, *31*, L163.
[250] W. Thom, W. Zimmerman, L. Kramer, *Liq. Cryst.* **1989**, *4*, 309.
[251] L. Kramer, W. Pesch in *Pattern Formation in Liquid Crystals* (Eds A. Buka, L. Kramer), Springer Verlag, New York **1995**.
[252] E. Dubois-Violette, P. G. de Gennes, O. Parodi, *J. Phys. France* **1971**, *32*, 305.
[253] A. Hertich, W. Decker, W. Pesch, L. Kramer, *J. Phys. France II* **1992**, *2*, 1915.
[254] I. Rehberg, B. L. Winkler, M. Torre Juarez, S. Raenat, W. Schoepf, *Festkoerperprobleme* **1989**, *29*, 35.
[255] B. L. Winkler, H. Richter, I. Rehberg, W. Zimmermann, L. Kramer, A. Buka, *Phys. Rev. A* **1991**, *43*, 1940.
[256] A. Buka, L. Kramer, *Phys. Rev. A* **1992**, *45*, 5624.
[257] V. Frederiks, V. Tsvetkov, *Dokl. Akad. Nauk SSSR* **1935**, *4*, 123.
[258] V. Frederiks, V. Tsvetkov, *Acta Physicochim. URSS* **1935**, *3*, 879.
[259] G. H. Heilmeier, L. A. Zanoni, L. A. Barton, *Proc. IEEE* **1968**, *56*, 1162.
[260] G. H. Heilmeier, L. A. Zanoni, L. A. Barton, *IEEE Trans. Elect. Dev.* **1990**, *ED-17*, 22.

[261] A. N. Trufanov, L. M. Blinov, M. I. Barnik, *Zh. Eksp. Teor. Fiz.* **1980**, *78*, 622.
[262] S. A. Pikin, V. G. Chigrinov, *Zh. Eksp. Teor. Fiz.* **1980**, *78*, 246.
[263] F. Rondelez, H. Arnould, *C. R. Acad. Sci.* **1971**, *273B*, 549.
[264] A. Arnould-Nettilard, F. Rondelez, *Mol. Cryst. Liq. Cryst.* **1974**, *24*, 11.
[265] S. V. Belyayev, L. M. Blinov, *Zh. Eksp. Teor. Fiz.* **1976**, *70*, 184.
[266] W. Helfrich, *J. Chem. Phys.* **1971**, *55*, 839.
[267] G. H. Heilmeier, J. E. Goldmacher, *Appl. Phys. Lett.* **1968**, *12*, 132.
[268] W. Haas, J. Adam, C. Dir, *Chem. Phys. Lett.* **1972**, *14*, 95.
[269] T. Akahane, T. Tako, *Jpn. J. Appl. Phys.* **1976**, *15*, 1559.
[270] S. V. Belyayev, *Zh. Eksp. Teor. Fiz.* **1978**, *75*, 663.
[271] P. L. Finn, P. E. Cladis, *Mol. Cryst. Liq. Cryst.* **1982**, *84*, 159.
[272] H. Gleeson, R. Simon, H. J. Coles, *Mol. Cryst. Liq. Cryst.* **1985**, *129*, 37.
[273] J. A. Guerst, W. J. A. Goossens, *Phys. Lett.* **1972**, *41*, 369.
[274] M. Gosciansky, *Philips Res. Rep.* **1975**, *30*, 37.
[275] L. M. Blinov, M. I. Barnik, V. T. Lasareva, A. N. Trufanov, *J. Phys. France Colloq. C3* **1979**, *40*, C3-263.
[276] V. N. Chirkov, D. F. Aliyev, A. Kh. Zeinally, *Pis'ma Zh. Eksp. Teor. Fiz.* **1977**, *3*, 1016.
[277] V. N. Chirkov, D. F. Aliyev, A. Kh. Zeinally, *Zh. Eksp. Teor. Fiz.* **1978**, *74*, 1822.
[278] D. F. Aliyev, H. F. Abbasov, *Liq. Cryst.* **1989**, *4*, 293.
[279] N. A. Tikhomirova, A. V. Ginzberg, E. A. Kirsanov, Yu. P. Bobylev, S. A. Pikin, P. V. Adomenas, *Pis'ma Zh. Eksp. Teor. Fiz.* **1976**, *24*, 301.
[280] M. P. Petrov, A. G. Petrov, G. Pelzl, *Liq. Cryst.* **1992**, *11*, 865.
[281] L. K. Vistin, A. P. Kapustin, *Kristallografiya* **1968**, *13*, 349.
[282] B. Petroff, M. Petrov, P. Simova, A. Angelov, *Ann. Phys.* **1978**, *3*, 331.
[283] D. F. Aliyev, A. Kh. Zeinally, N. A. Guseinov, *Kristallografiya* **1981**, *26*, 867.
[284] D. F. Aliyev, *Kristallografiya* **1983**, *28*, 358.
[285] Yu. Galyametdinov, M. A. Athanassopoulou, W. Haase, I. Ovchinnikov, *Koord. Khim.* **1995**, *21*, 9.
[286] E. A. Soto Bustamante, S. V. Yablonsky, B. I. Ostrovskii, L. A. Beresnev, L. M. Blinov, W. Haase, *Chem. Phys. Lett.* **1996**, *260*, 447.
[287] T. Niori, T. Sekine, J. Watanabe, T. Furukawa, H. Takezoe, *J. Mater. Chem.* **1996**, *6*, 1231.

10 Surface Alignment

Blandine Jérôme

10.1 Introduction

The phenomenon of orientation of liquid crystals by surfaces has been known nearly as long as have liquid crystals themselves [1]. The phenomenon has mainly been studied in low-molecular-weight nematic liquid crystals, both because of the simplicity of their structure and because of the use of this type of liquid crystal in displays. Most of the present chapter is therefore be dedicated to this type of liquid crystal.

When a nematic liquid crystal is placed in contact with another phase (solid or liquid), a surface bounding the liquid crystal is created. The presence of this surface induces a perturbation of the nematic order close to it (Fig. 1a). The anisotropic interactions between the molecules located right at the surface – in the surface layer – and the other phase favors certain orientations of the surface molecules. This leads to an orientational distribution of the liquid crystal molecules in the surface layer that is generally different from the bulk nematic order. The orientational order evolves from the one induced by the surface to the one in the bulk in an interfacial region of thickness ξ_i, which is of the order of the nematic coherence length. Just outside the interfacial region, the nematic director has a preferred orientation *a*. This macroscopic orientation of a liquid crystal by a surface is called anchoring.

The macroscopic anchoring of low-molecular-weight nematic liquid crystals is dis-

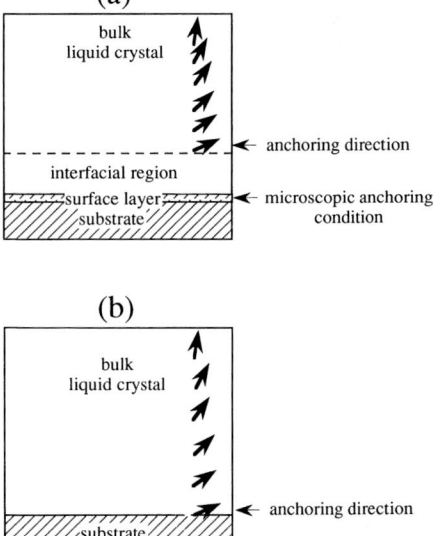

Figure 1. Schematic drawing of a liquid crystal in contact with a substrate: (a) taking into account the presence of the surface layer and interfacial region; (b) according to the macroscopic approach. The arrows indicate the director field in the bulk, represented here in a distorted configuration.

cussed in Section 10.2 and the microscopic aspects behind it in Section 10.3. The case of other liquid crystals is considered in Section 10.4. For more detailed reviews on the surface effects in liquid crystals see, for example, Sluckin and Poniewierski [2], Sonin [3], and Jérôme [4].

10.2 Macroscopic Alignment of Nematic Liquid Crystals

Until quite recently, anchoring has been described ignoring the existence of the surface layer and interfacial region, assuming that the bulk structure extends right up to the surface (Fig. 1b). The configuration of the liquid crystal is then described at each point by the director. At the surface, this director has a preferred orientation, which is the boundary condition for the field of the director in the rest of the liquid crystal. This macroscopic description works well when the director at the surface is not driven too far from its preferred orientation (by another orienting force) and when the origin of this preferred orientation is irrelevant for the problem under consideration.

10.2.1 Definitions

Both discussing anchoring in more details, it is necessary to define a terminology to describe it (for a more detailed lexicon, see Jérôme [4]).

The main concept is the anchoring induced by an interface. The energy γ_s of this interface – also called the anchoring energy – depends, among other things, on the orientation of the director at the surface. This energy has a certain number of minima obtained for orientations of the director a_α, which are the anchoring directions of the liquid crystal at the surface. The set of possible anchoring directions $\{a_\alpha\}$ characterizes the anchoring induced by an interface. This anchoring can be monostable, multistable, or degenerate, depending on whether the number of elements in the set $\{a_\alpha\}$ is one, a finite number greater than one, or infinite. It can also be planar, tilted or homeotropic, depending on whether the anchoring directions are parallel, tilted or perpendicular to the plane of the interface (Fig. 2).

In the macroscopic approach, which ignores the detailed structure of this interface, the expression of the interfacial energy γ_s has to be found following thermodynamic and symmetry considerations. γ_s is a periodic function of the azimuthal angle φ and the tilt angle θ (with respect to the surface normal) defining the orientation of the director at the surface. γ_s can thus be developed in a Fourier series [2, 10]:

$$\gamma_s(\theta,\varphi) = \sum A_{lm} Y_l^m(\theta,\varphi) \quad (1)$$

where $Y_l^m(\theta,\varphi)$ are spherical harmonics and A_{lm} are coefficients which are non-zero only if the corresponding Y_l^m is compatible with the symmetry of the nematic phase (even l) and that of the surface.

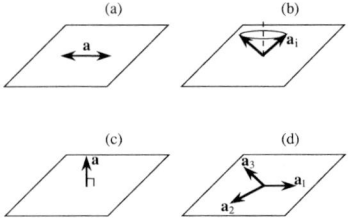

Figure 2. Some experimentally observed anchorings: (a) monostable planar (on grooved surfaces [5]); (b) degenerate tilted, also called conical (at the nematic/isotropic interface [6, 7]); (c) homeotropic (on surfactant-coated glass [8]); (d) tristable planar (on phlogopite mica [9]).

The term 'surface' means here the surface of the other phase in contact with the nematic liquid crystal. The symmetry of this surface (and of γ_s) is independent of the orientation taken by the nematic phase at the surface. In contrast, the symmetry of the interface depends on this orientation: it is the subgroup of the surface symmetry group containing the symmetry elements which leave invariant the anchoring direction effectively taken by the liquid crystal. If the other phase is a solid or liquid substrate, the surface is simply the surface of this substrate. In the case when the other phase is the gas or isotropic phase, the 'surface' is not a physical entity. However, one can still, in principle, distinguish this isotropic surface (C_∞ symmetry) from the interface with the nematic phase, the symmetry of which is C_∞, C_{2v}, and C_{1v} for homeotropic, planar, and tilted anchoring, respectively.

The interfaces at which anchoring has been mostly studied are those with crystal surfaces and treated-glass substrates. Glass treatments can be mechanical (rubbing) or chemical (deposition of a layer of surfactant, polymer, inorganic substance, etc.) or a combination of the two (for a review see Cognard [8]). The most commonly obtained anchorings are homeotropic, degenerate, and monostable planar (Fig. 2). Monostable tilted anchorings are less frequent, and multistable anchorings are seldom obtained.

The occurrence of multistable or degenerate anchorings raises the issue of the selection of anchoring directions; indeed, at each point of the surface, the director can take only one of the possible anchoring directions characterizing such anchorings. Generally speaking, the selection is made by the history of the sample, for instance cooling from the isotropic phase under a magnetic field [11, 12] or spreading [13]. In general, once the selection has been made, the anchoring direction is preserved by the surface, even in the case of degenerate anchorings [11, 14]. This appears to come from the adsorption of the surface molecules onto the substrate [12]. Switching between the anchoring directions of a multistable anchoring is, however, possible [15, 16].

10.2.2 Anchoring Directions

By minimizing the interfacial energy γ_s, one can find the anchoring diagram giving the different anchorings induced by a surface depending on the values of the coefficients A_{lm} appearing in Eq. (1) (Fig. 3). These diagrams give a general description of anchoring, independent of the nature of the system. However, the limitation of this macroscopic approach lies in the fact that it is, in general, not possible to establish a relationship between the coefficients A_{lm} and the parameters governing the structure of the interface.

Anchoring diagrams show the different ways in which anchoring can change when some parameters are varied. These anchoring transitions follow the same symmetry rules as phase transitions: transitions which do not involve a change of symmetry of the interface must be first order, while when a symmetry change is involved the transition can be first or second order [10]. Anchoring transitions also appear the same way as phase transitions: nucleation and growth of domains having a new orientation for first-order transitions [19], and director fluctuations for second-order transitions [18]. These fluctuations correspond to a divergence of the susceptibility of the system for a force driving the director away from the anchoring directions: indeed, the anchoring strength (see Section 10.2.3) does go to zero at second-order anchoring transitions [20–22]. One characteristic of anchoring transitions is that this susceptibility can be

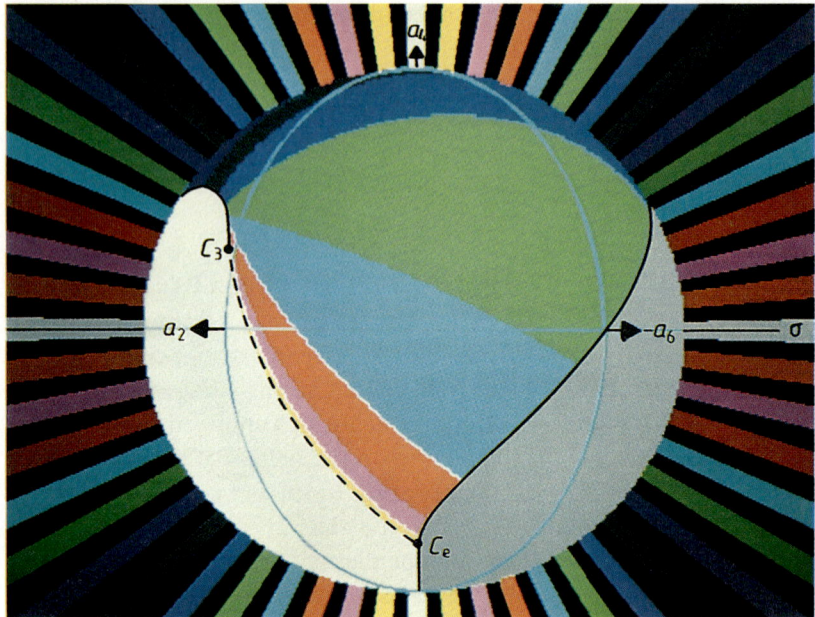

Figure 3. Anchoring diagram corresponding to a surface with C_{1v} symmetry (with mirror plane σ) inducing planar anchorings: $\gamma_s(\varphi) = a_2 \cos(2\varphi) + a_4 \cos(4\varphi) + a_6 \cos(6\varphi)$ [17, 18]. γ_s is normalized by the condition $a_2^2 + a_4^2 + a_6^2 = 1$, so that the anchoring diagram can be represented on a sphere. At each point on the sphere, the color indicates the predicted anchoring direction. The code is given by the background, the surface of which corresponds to the substrate surface. (———) First-order transition; (-----) second-order transition. C_e, Critical endpoint; C_3, tricritical point.

anisotropic (i.e. the response to an applied force can depend on the force direction) [23].

There are several examples of experimentally observed anchoring transitions (see Jérôme [4] and Pieranski and Jérôme [10], and references therein). Since anchoring transitions occurring at the surface of a liquid crystal cell induce a change of director configuration in the cell, these transitions can be used to switch a liquid crystal device between different states [24, 25].

10.2.3 Anchoring Energy

The director at an interface might depart from the anchoring directions induced by this interface under the effect of another orienting field. The way in which this happens is determined the dependence of γ_s on the orientation of the director $\boldsymbol{n}$ at the surface. Knowledge of this dependence is necessary in order to establish the surface boundary conditions on the director field [26]. Several methods have been developed in order to measure experimentally some characteristics of $\gamma_s(\boldsymbol{n})$ [27, 28]. Generally speaking, they consist in observing the change in orientation induced at a surface under the action of an external field.

Thermodynamically, γ_s can be defined by the equation [28]:

$$f = f_0 + \gamma_s[\boldsymbol{n}(z_d), z_d] + \int_{z_d}^{\infty} f_d\left[\boldsymbol{n}(z), \frac{d\boldsymbol{n}}{dz}(z)\right] dz \quad (2)$$

where f is the total free energy of the system per unit area, f_0 is the free energy in absence of surface and deformation in the bulk, f_d is the elastic energy density due to deformations of the director field $\boldsymbol{n}(z)$, and z_d is the position of the surface above which the nematic phase is considered to have bulk-like behavior. From this definition, it is clear that when the director field is not uniform, which is always the case in anchoring energy measurements, γ_s depends on z_d. Unfortunately, the value of z_d depends both on the technique used to measure the anchoring energy and on the structure of the interfacial region [28]. Therefore, the results of the anchoring energy measurements interpreted using this approach (as has generally been done until now) are not comparable with each other.

The dependence of the interfacial energy γ_s with the surface director was first considered by Rapini and Papoular [29] who assumed that the dependence of γ_s on the tilt angle θ was proportional to $\sin^2(\theta - \theta_s)$ where θ_s is the tilt of the anchoring direction. The coefficient of proportionality measures the ease with which the surface director can deviate from the anchoring direction and is called the anchoring strength or anchoring energy coefficient. Generally speaking, one defines two anchoring strengths, the polar and the azimuthal one, corresponding to deviations from the anchoring direction (θ_s, φ_s) at constant azimuthal angle φ and tilt angle θ, respectively:

$$W_p = \frac{1}{2}\left(\frac{\partial^2 \gamma_s}{\partial \theta^2}\right)_\varphi (\theta_s, \varphi_s) \quad (3a)$$

$$W_a = \frac{1}{2}\left(\frac{\partial^2 \gamma_s}{\partial \varphi^2}\right)_\theta (\theta_s, \varphi_s) \quad (3b)$$

All measurements of anchoring strengths give values of W_p ranging between 10^{-7} and 10^{-3} J m^{-2} (10^{-5}–1 erg cm^{-2}) and values of W_a one or two orders of magnitude smaller. This inequality might be intrinsic to the anchoring mechanisms, as predicted by some theoretical calculations of $\gamma_s(\theta, \varphi)$ [30–32]. It might also come from the order-parameter gradient existing at the interface, giving rise to an order-electric polarization [33] and to a contribution to γ_s [34]: owing to symmetry considerations, this contribution is zero for W_a while it is finite for W_p. It has also been suggested that the upper limit of the measured anchoring strengths could be due to the fact that, for strong anchorings, the interfacial region would prefer to melt instead of undergoing too strong a distortion [35].

The anchoring strengths (Eq. (3)) are useful only when the director remains close to the anchoring directions. For large deviations of the surface director, the whole dependence of γ_s should be known. With few exceptions [23], measurements of γ_s are performed assuming that the θ and φ variations of γ_s are independent. Concerning the θ dependence, neither experiments [36–39] nor theoretical calculations [31, 40] (see also Jérôme [4] and references therein) have come to an agreement on whether the expression of $\gamma_s(\theta)$ should contain one or more Fourier components. Part of the discrepancies in the anchoring energy measurements could be due to bulk effects such as ferroelectricity [39] or the presence of ions [27, 41, 42].

In addition, in none of these studies was the contribution to the anchoring energy of the adsorption of liquid crystalline molecules onto the substrate [12] taken into account. In fact, γ_s contains two terms [4]:

$$\gamma_s(\boldsymbol{n}) = \gamma_s^{\text{surf}}(\boldsymbol{n}) + \gamma_s^{\text{ad}}(\boldsymbol{n}; \boldsymbol{a}_{\text{ad}}) \quad (4)$$

γ_s^{surf} is the interfacial energy in the absence of adsorption; γ_s^{ad} arises from the adsorption of molecules onto the substrate and has two opposite minima corresponding to the anchoring direction $\boldsymbol{a}_{\text{ad}} \equiv -\boldsymbol{a}_{\text{ad}}$ initially taken by the liquid crystal. The orienting field felt by the surface director therefore has uniaxial symmetry, regardless of the symmetry of γ_s^{surf}. However, γ_s^{ad} is not an intrinsic property of the surface and can be subject to changes, depending on the history of the sample [43].

From the above discussion, it is clear that the macroscopic approach is inadequate for describing situations in which the director is forced to depart from anchoring directions. A proper description of these situations should take into account the response of the interfacial region to such distortions. There have been some attempts to describe the interfacial region in a semi-macroscopic way, assuming that the nematic order is retained in the interfacial region but with a varying order parameter [22, 34, 44, 45]. These semi-macroscopic theories define a surface anchoring energy giving the energy of the surface layer as a function of its orientational state defined by the tensorial nematic order parameter. The spatial evolution of this order parameter away from the surface is calculated using the Landau–de Gennes [46] theory under the appropriate conditions corresponding to the way, in which the orientation of the director is imposed in the bulk.

The main defect in this kind of theory is the assumption that the orientational order is nematic like in the interfacial region, which is generally not true (see Section 10.3.1). Another problem is linked to the fact that the surface anchoring energy is unknown. This surface anchoring energy is difficult to measure directly and the result of any indirect measurement depends on how the response of the system to a disorientation is modeled. One way out is to use the orientational distribution of the molecules in the surface layer as a boundary condition of the nematic order [47] (see also Section 10.3.2].

10.3 Microscopic Surface Order of Nematic Liquid Crystals

10.3.1 Surface Orientational Order

An experimental technique has recently been developed to investigate the surface orientational order of liquid crystals: optical second-harmonic generation [48]. Because of its high sensitivity to surface polar ordering and its surface specificity, this method is able to probe the orientational distribution of polar monolayers of liquid crystal molecules located at the surface of a substrate, whether they are covered by a bulk liquid crystal or not.

So far, such measurements have only been performed on cyanobiphenyl molecules (n-CB and n-OCB). These molecules exhibit polar ordering (with the cyano groups pointing towards the surface) on hydrophilic substrates such as water, glass, and certain polymer films [49–51], but not on hydrophobic layers [52]. On the latter, the molecules orient with their aliphatic chains in contact with the substrate, and arrange following their natural tendency to form apolar 'dimers' (two molecules oriented head to head). In all cases where a surface dipolar ordering exists, it is lost after the first molecular layer [49].

The tilt of the hard core of cyanobiphenyl molecules is found to be always approximately the same, namely 70° from the surface normal, resulting from a balance between the dipole–dipole interaction of the polar heads and the steric interaction of the biphenyl core with the substrate [49, 53]. By means of infrared sum-frequency generation, it has also been shown that the aliphatic chains are relatively straight and point away from the surface [54].

The distribution of azimuthal orientation strongly depends on the substrate [52, 55, 56]. On glass and surfactant- and polymer-coated glass this distribution is isotropic, as would be expected from the isotropy of these surfaces. When polymer films are rubbed by translation in one direction on a piece of cloth, the polymer chains at the surface of the film are oriented along the rubbing direction [57, 58]. The liquid-crystal molecules then orient preferentially along the polymer chains (Fig. 4a). As on crystal surfaces, this anisotropic orientational distribution is the result of the direct interaction between the molecules and the substrate. Conversely, monolayers at the surface of rubbed glass, certain rubbed surfactant layers, and evaporated SiO films exhibit an isotropic azimuthal distribution [51, 52]; despite their anisotropy, these substrates have no short-range azimuthal orienting action on nematic molecules.

The azimuthal orientational distribution of surface liquid-crystal molecules can, in principle, also be obtained by imaging the molecules. However, sufficiently high resolution can only be obtained with scanning tunneling microscopy, which limits the use of imaging techniques to special conducting substrates, such as cleaved pyrolytic graphite [59, 60] and MoS crystals [61]. A feature common to all the images obtained is the high orientational and positional order of the molecules in the surface monolayer.

The techniques described above allow for the determination of the orientational order in the surface layer. A complete description of the liquid crystalline ordering close to a surface should also include the evolution of the orientational order in the interfacial region from the one at the surface to the one in bulk. There are, however, no experimental techniques available to measure this evolution directly. Some experimental techniques allow one to measure some integrated quantities over the whole interfacial region, which depend on the evolution of the orientational order in this region (see e.g. [62–65]). To obtain the order profile from such measurements, one needs to assume the general shape of this evolution and fit the experimental data to obtain the profile parameters. Knowledge of the surface orientational order can, however, guide the choice of model profile.

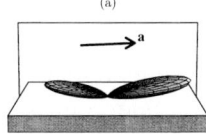
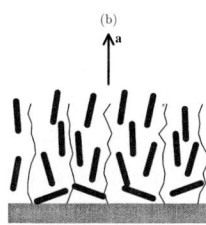
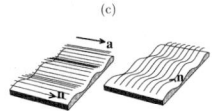

Figure 4. The three main substrate classes: (a) smooth surfaces on which surface molecules have a definite orientational distribution (represented: surface obtained on a rubbed polyimide film [52]); (b) interpenetrable surfaces of dangling chains; (c) topographies (represented: grooved surface) with a favorable (left) and unfavorable director field *n*. In all cases, *a* is the macroscopic anchoring direction.

10.3.2 Microscopic Anchoring Mechanisms at Solid Substrates

With the presently known information on the orientational order in surface layers and interfacial regions, it is possible to deduce some of the microscopic origins of the macroscopic anchoring of liquid crystals. One can distinguish three main classes of substrate, giving rise to three main classes of anchoring mechanism: smooth surfaces, interpenetrable layers, and topographies (Fig. 4). It should, however, be emphasized that the macroscopic anchoring is often due to a combination of effects, some of which are specific to certain substrates.

Smooth substrates include crystal and glass surfaces, and polymer films. The direct influence of such a substrate on the liquid crystal is limited to the molecules in the surface layer and is due to the short-range interaction of these molecules with the substrate (Fig. 4a). Once these microscopic anchoring conditions are known, the configuration of the rest of the liquid crystal (in particular the orientation of the director outside the interfacial region) can be determined from the properties of the liquid crystal regarding propagation of orientational order.

The microscopic anchoring conditions can be modeled by the energy of the surface layer as a function of its orientational order. The evolution of the orientational order away from the surface is then calculated by minimizing the sum of this surface energy and of the (orientational order dependent) bulk energy. Since the dependence of the energy of the surface layer on its orientational order is generally unknown, the orientational distribution of the surface molecules, if experimentally available (see Section 10.3.1), can be used as a boundary condition for the order in the rest of the liquid crystal [52, 55]. This surface orientational distribution does indeed seem to be strongly enforced by the surface; it is essentially independent of whether this monolayer is covered by a bulk liquid crystal or not, and whether this bulk is in the nematic or isotropic phase [52].

This type of calculation has been performed successfully, both by using a general form of the surface energy [22] and experimentally determined surface orientational distributions [66, 67], in order to explain the anchoring directions and anchoring transitions observed on different substrates. These calculations were made within the framework of the Landau–de Gennes theory [46], the liquid crystalline order at each point in space (including the interfacial region and surface layer) being described by the tensorial order parameter. The evolution of the director and the degree of molecular order in the interfacial region is coupled. This order-parameter/director coupling comes essentially from the L_2 term in the Landau–de Gennes energy; this term should therefore not be neglected (as is often the case). This implies that, since the degree of order is generally different at the surface and in the bulk, the anchoring direction is different from the mean orientation of the molecules at the surface [66–68]. This also means that the degree of surface ordering is a determinant factor in the tilt of the anchoring directions [43, 47, 55].

The Landau–de Gennes theory is, however, not applicable to all surfaces; in particular it is unable to predict the azimuthal orientation of the anchoring directions in the case of multistable anchorings, which occur on substrates with sufficiently symmetry. Indeed, reducing the information on the surface orientational distribution to a nematic tensorial order parameter suppresses the information on the symmetry of the surface. Order parameters of higher order reflecting the surface symmetry should then be included in the description of the orientational order, together with the corresponding terms in the energy of the system.

A second class of substrates corresponds to surfaces which are smooth but on which molecules with a long tail (such as surfactants) are grafted, forming a layer that the liquid crystal can penetrate (Fig. 4b). The influence of the substrate on the liquid crystal at a microscopic level can then be separated into two parts: the effect of the surface underlying the surfactant layer and the effect of the chains forming the layer. It seems that the latter always dominates; the liquid crystal adopts the orientation of the chains in the layer, which leads to a homeotropic or conical anchoring depending on the orientation of the chains [54, 69, 70].

It should be emphasized that not all substrates covered with aliphatic chains belong to this class of substrates. If the chains are closely packed, the liquid crystal molecules cannot penetrate the layer [54] and the substrate can be seen as a smooth surface belonging to the first class described above.

The third class of substrates corresponds to those having a surface of a certain topography (Fig. 4c). This topography is generally obtained by rubbing the surface with a hard material (rubbed glass or surfactant-coated glass) or by anisotropic vapor deposition (evaporated SiO films). Each point of such a surface induces a given orientational distribution of liquid crystal molecules. On all the substrates mentioned above, this local orientational distribution is isotropic [51, 52]. However, since the orientation and the height of the surface is not uniform, the microscopic anchoring conditions are not uniform over the surface, introducing distortions along the plane of the surface. The anchoring direction chosen by the liquid crystal is then the one that minimizes these distortions. For instance, grooved surfaces induce an anchoring direction parallel to the grooves. This anchoring mechanism can be modeled by minimizing the Franck–Oseen elastic free energy associated with director distortions [5].

All the anchoring mechanisms described above are based on the knowledge of the microscopic anchoring conditions. However, a complete understanding of anchoring requires knowledge of the origins of this microscopic anchoring, which should be found in the interactions between the liquid crystal molecules and the substrates. These interactions are basically unstudied.

10.3.3 The Nematic/Isotropic and Nematic/Vapor Interfaces

The case of the interface with an isotropic or vapor phase is relatively simple in that no interaction with a substrate is involved. The molecular ordering and the orientation of the nematic director at these interfaces arise mainly from the change in density and type of order taking place across the interface.

Studies of the orientational order at this type of interface concern exclusively nematic/vapor interfaces. No direct measurement of this order has been performed so far, but indirect observations have shown that the order at the free surface can be higher or lower than in the bulk [71–73]. Excess order occurs when the intermolecular interaction depends on the relative molecular positions (i.e. when the translational and rotational degrees of freedom are coupled) [74, 75]. This coupling is also essential for anchoring to occur at the surface (Fig. 5).

Since the surface breaks the bulk inversion symmetry, it creates a polarity which, in principle, should give rise to polar ordering near the surface. This ordering has been investigated in asymmetric molecules with a polar head. These molecules prefer to orient with their polar heads pointing in the direction of the more polar medium. This tends to create a polar surface layer with all the heads pointing away from the surface. However, in order to minimize the polar intermolecular interactions, the molecules tend to arrange antiparallel to each other with a partial or total overlap. This effect compensates for any surface-induced polar ordering, creating either a non-polar [76] or an anti-ferroelectric [49, 77, 78] surface ordering.

As far as macroscopic anchoring is concerned, it has been found experimentally that, at the nematic/isotropic interface, the director is tilted (in all the compounds so far studied, including n-CB and MBBA), with

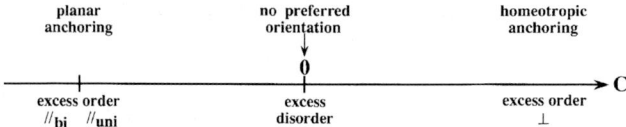

Figure 5. Anchoring (top) and surface ordering (bottom) at the free surface of a nematic liquid crystal as a function of the strength C of the coupling between translational and rotational degrees of freedom [75]. $\perp$, Homeotropic ordering; $\|_{uni}$ and $\|_{bi}$, uniaxial and biaxial ordering parallel to the surface.

an angle θ_s from the surface normal in the range 50–80° [6,7]. At the free surface, there are more differences from one liquid crystal to the other: $\theta_s = 0°$ in n-CB and PCHn [79, 80] (because the polar head of these molecules prefers to point away from the surface), $\theta_s = 90°$ in PAA [81], and θ_s decreases to 0° as the temperature increases in MBBA and EBBA [82].

The first attempt to account for these results was made by de Gennes using the Landau–de Gennes theory [46]. This theory predicts that the tilt angle can only be 0° or 90°, depending on the elastic properties of the liquid crystals. In order to account for tilted anchorings in this framework, different effects can be added in the model, such as order electricity arising from order-parameter gradients [33].

Another series of theoretical models is based on molecular interactions, either anisotropic hard-core interactions [40, 83–85] or combining different kinds of intermolecular interactions [86, 87] (see also Jérôme [4] and references therein). The results depend greatly on the details of the models, the validity of which still needs to be checked.

10.4 Orientation of Other Liquid Crystals

The study of surface orientation of non-nematic liquid crystals is still incomplete. Here, separate sections are dedicated to polymer liquid crystals and lyotropic liquid crystals, the latter being peculiar in that they are inhomogeneous phases.

10.4.1 Smectic and Chiral Liquid Crystals

In smectic A phases where the smectic layers are perpendicular to the molecules, the orientation of the whole structure is, in principle, fixed once the orientation of the molecules is defined by the interface. The surface orientation of achiral smectic A phases is then the same as that of the nematic phase [88, 89]. However, since splay deformations of smectic layers (director bend deformations) are forbidden and layer bend deformations (director splay deformations) require a lot of energy, smectic phases tend to adopt uniform configurations, even between two walls inducing two different orientations. In the latter case, the surface orientation of the smectic phase differs from that of the nematic phase, and depends on the layer configuration in the bulk [90, 91].

In the case of tilted smectic phases (for instance the smectic C phase), there is a degeneracy in the orientation of the director with respect to that of the layers. If a surface inducing a degenerate planar anchoring is placed perpendicular to the layers, there are two possible orientations of the director that satisfy both the surface and the liquid crystal structural constraints (Fig. 6).

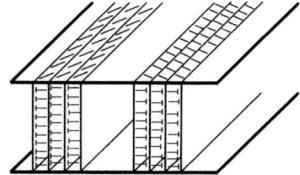

Figure 6. Two equivalent orientational states in a smectic C liquid crystal between two walls inducing a degenerate planar anchoring.

This surface bistability is at the basis of chiral smectic C surface stabilized ferroelectric liquid crystal (SSFLC) devices [92]. As their name indicates, these devices are made of thin cells in which the walls, imposing the orientation of the molecules at the surfaces, unwind the spontaneous smectic C* helix and stabilize two uniform configurations of the director in the cell. Switching between these two states can be done by applying an electric field.

The preparation of such cells requires both that the director is oriented parallel to the surface and that the layers are oriented perpendicular to the surfaces. Because the walls only have an influence on the orientation of the director, and because of the director/layer-normal degeneracy, the alignment of the layers must be achieved in the smectic A* phase, before cooling the system in the smectic C* phase [93]. Due to the positional anchoring of the layers at the surfaces [94], the location of the layers is pinned at the walls. Since the layer thickness d_A in the smectic A phase is larger than the thickness d_C in the smectic C phase, the layers must tilt in the bulk by an angle δ_C such that $d_C = d_A \cos \delta_C$ [95].

The combination of tilted orientation of the director at the surface (breaking the in-plane isotropy of the surface) and of the surface-induced up–down asymmetry [96] gives rise to new terms in the energy of the surface layer which favor splay configurations of the director [97, 98]. Such configurations have been observed at the free surface of achiral tilted smectic phases. The details of the patterns obtained depend on the type of smectic phase considered: in a smectic C phase, the director can rotate freely on a cone, while in a smectic phase presenting a bond-orientational order (smectic I, F, or L) the coupling existing between this order and the director imposes discrete rotations of the director through walls [99, 100]. The surface pattern also depends on whether or not there is a mechanism imposing the sign of the surface-induced polarization, and consequently that of the director rotation [100–103].

In a chiral compound, the liquid crystalline structure itself possesses a polarity which interacts with the surface. At the surface of a smectic C* liquid crystal, the ferroelectric polarization P points preferably either towards or away from the surface, depending on the material [104, 105]. Conversely, if the surface director of a chiral nematic liquid crystal is tilted, a polarization (dependent on the tilt angle) is created perpendicular to the tilt plane [106]. In principle, the presence of this polarization makes a contribution to the anchoring energy; this chiral contribution is, however, too small to be measured [107].

The combination of liquid crystalline chirality and surface polarization is also the origin of the so-called surface electroclinic effect in chiral smectic A phases: if the smectic layers are oriented perpendicular to the surface, the surface electric field tilts the layer normal away from the surface molecular orientation [108–110] (Fig. 7).

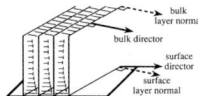

Figure 7. The surface electroclinic effect rotating the smectic layer normal away from the surface molecular orientation in smectic A* liquid crystals.

10.4.2 Polymer Liquid Crystals

The macroscopic orientation of polymer liquid crystals at surfaces can formally be described in the same way as that of low-molecular-weight liquid crystals. However, the polymer character of these materials has an influence on the surface behavior at a microscopic level. One can distinguish different aspects: loss of freedom of the mesogenic groups, change in polymer chain entropy due to the steric restriction near the surface, increase of longitudinal persistence length (along the director), and attraction/repulsion of the polymer chain at the surface.

When mesogenic groups are part of a polymer chain (in main-chain polymers) or attached by one of their ends to a main chain (in side-group polymers), they lose the ability to orient independently of each other. In particular, they might not be able to adopt the surface orientation which they would have if they were free. This effect can lead to a change in anchoring direction with respect to that of the free mesogens, as observed at the free surface of side-chain polymer liquid crystals the side-groups of which are terminated by a polar head (such as cyanobiphenyl groups) [111, 112]. As mentioned in Section 10.3.3, low-molecular-weight liquid crystalline compounds carrying a polar head orient perpendicularly to their free surface because the head prefers to point away from the surface. In a side-group polymer liquid crystal, satisfying this constraint would force the surface polymer molecules to have their main chain confined at the surface and all their side-groups pointing in the same direction. This is sterically impossible. The system solves this configurational problem by orienting the side-groups (and the surface director) parallel to the surface.

As in any polymer, the presence of a surface limits the number of configurations that the polymer chains are allowed to take in the vicinity of this surface. Since these chains and the mesogenic groups are mechanically coupled, this loss of chain entropy implies a limitation of the orientations accessible to the mesogenic groups. Moreover, the presence of a polymer chain linking the mesogenic groups increases the longitudinal persistence length along the director. These two effects can make the anchoring strength of polymer liquid crystals considerably larger than that of low-molecular-weight liquid crystals [113].

Finally, the polymer chain (the main chain in side-chain polymers or the intermesogen chain in main-chain polymers) interacts directly with the surface and can be either attracted to or expelled from the surface. For instance, side-groups terminated by a polar head and attached to a chain also carrying polar groups orient parallel to hydrophilic surfaces, instead of being tilted as the corresponding low-molecular-weight liquid-crystal molecules (see Section 10.3.1).

10.4.3 Lyotropic Liquid Crystals

The anisotropic micelles forming lyotropic liquid crystals are also oriented by surfaces. Both prolate and oblate micelles orient parallel to flat surfaces [114, 115] probably due to hard-core interactions [116]. Prolate micelles can also be azimuthally oriented by grooved surfaces, or homeotropically oriented by two-dimensional topographies [117].

However, one peculiarity of these systems is that the surfactant molecules forming the micelles are surface active and generally adsorb at the surface; this leads a modification of the surface [115]. The presence of anisotropic defects in this adsorbed layer can then induce an azimuthal anchoring of the bulk micelles. The slow reorientation of these surface defects under a magnetic field induces a slow in-plane reorientation – or gliding – of the anchoring directions [118].

10.5 References

[1] C. Mauguin, *C.R.A.S.* **1911**, *156*, 1246.
[2] T. J. Sluckin, A. Poniewierski, in *Fluid Interfacial Phenomena* (Ed. C. A. Croxton), Wiley, New York **1986**, Chap. 5, pp. 215–253.
[3] A. A. Sonin, *The Surface Physics of Liquid Crystals*, Gordon and Breach, Amsterdam **1995**.
[4] B. Jérôme, *Rep. Prog. Phys.* **1991**, *54*, 391.
[5] D. W. Berreman, *Phys. Rev. Lett.* **1972**, *28*, 1683.
[6] R. Vilanove, E. Guyon, C. Mitescu, P. Pieranski, *J. Phys. France* **1974**, *35*, 153.
[7] H. Yokoyama, S. Kobayashi, H. Kamei, *Mol. Cryst. Liq. Cryst.* **1984**, *107*, 311.
[8] J. Cognard, *Mol. Cryst. Liq. Cryst.* **1982**, *1* (Suppl.), 1.
[9] P. Pieranski, B. Jérôme, M. Gabay, *Mol. Cryst. Liq. Cryst.* **1990**, *179*, 285.
[10] P. Pieranski, B. Jérôme, *Mol. Cryst. Liq. Cryst.* **1991**, *199*, 167.
[11] J. Cheng, G. D. Boyd, *Appl. Phys. Lett.* **1979**, *35*, 444.
[12] Y. Ouchi, M. B. Feller, T. Moses, Y. R. Shen, *Phys. Rev. Lett.* **1992**, *68*, 3040.
[13] B. Jérôme, P. Pieranski, *J. Phys. France* **1988**, *49*, 1601.
[14] N. Koshida, S. Kibui, *Appl. Phys. Lett.* **1982**, *40*, 541.
[15] R. Barberi, M. Boix, G. Durand, *Appl. Phys. Lett.* **1989**, *55*, 2506.
[16] R. Barberi, G. Durand, *Liq. Cryst.* **1991**, *10*, 289.
[17] H. S. Kitzerow, B. Jérôme, P. Pieranski, *Physica A* **1991**, *174*, 163.
[18] J. Bechhoefer, J. L. Duvail, L. Masson, B. Jérôme, R. M. Hornreich, P. Pieranski, *Phys. Rev. Lett.* **1990**, *64*, 1911.
[19] P. Pieranski, B. Jérôme, *Phys. Rev. A* **1989**, *40*, 317.
[20] P. Chiarelli, S. Faetti, L. Fronzoni, *Phys. Lett.* **1984**, *101A*, 31.
[21] G. A. DiLisi, C. Rosenblatt, A. C. Griffin, U. Hari, *Liq. Cryst.* **1990**, *7*, 353.
[22] P. I. C. Teixeira, T. J. Sluckin, D. E. Sullivan, *Liq. Cryst.* **1993**, *14*, 1243.
[23] M. Nobili, G. Durand, *Europhys. Lett.* **1994**, *25*, 527.
[24] K. Ichimura, Y. Suzuki, T. Seki, A. Hosoki, K. Aoki, *Langmuir* **1988**, *4*, 1214.
[25] W. Gibbons, P. Shanon, S. T. Sun, B. Svetlin, *Nature* **1991**, *351*, 49.
[26] P. G. de Gennes, J. Prost, *The Physics of Liquid Crystals*, 2nd edn, Clarendon Press, Oxford **1993**.
[27] L. M. Blinov, A. Y. Kabayenkov, A. A. Sonin, *Liq. Cryst.* **1989**, *5*, 645.
[28] H. Yokoyama, *Mol. Cryst. Liq. Cryst.* **1988**, *165*, 265.
[29] A. Rapini, M. Papoular, *J. Phys. Coll.* **1969**, *30*, C4 54.
[30] J. Bernasconi, S. Strässler, H. R. Zeller, *Phys. Rev. A* **1980**, *22*, 276.
[31] K. Okano, N. Matsuura, S. Kobayashi, *Jpn. J. Appl. Phys.* **1982**, *21*, L109.
[32] S. Faetti, *Phys. Rev. A* **1987**, *36*, 408.
[33] G. Barbero, I. Dozov, J. F. Palierne, G. Durand, *Phys. Rev. Lett.* **1986**, *56*, 2056.
[34] H. Yokoyama, S. Kobayashi, H. Kamei, *J. Appl. Phys.* **1987**, *61*, 4501.
[35] G. Barbero, G. Durand, *Mol. Cryst. Liq. Cryst.* **1991**, *203*, 33.
[36] K. H. Yang, C. Rosenblatt, *Appl. Phys. Lett.* **1983**, *43*, 62.
[37] G. Barbero, N. V. Madhusudana, G. Durand, *Z. Naturforsch., Teil a* **1984**, *39*, 1066.
[38] H. Yokoyama, S. Kobayashi, H. Kamei, *Mol. Cryst. Liq. Cryst.* **1985**, *129*, 109.
[39] G. Barbero, G. Durand, *J. Phys. France* **1986**, *47*, 2129.
[40] J. D. Parsons, *Mol. Phys.* **1980**, *42*, 951.
[41] G. Barbero, G. Durand, *J. Phys. France* **1990**, *51*, 281.
[42] B. Valenti, M. Grillo, G. Barbero, P. Taverna Valabrega, *Europhys. Lett.* **1990**, *12*, 407.
[43] B. O. Myrvold, *Liq. Cryst.* **1995**, *18*, 287.
[44] G. Barbero, G. Durand, *J. Appl. Phys.* **1991**, *69*, 6968.
[45] M. Nobili, G. Durand, *Phys. Rev. A* **1992**, *46*, 6174.
[46] P. G. de Gennes, *Mol. Cryst. Liq. Cryst.* **1971**, *12*, 193.
[47] Y. Zhuang, L. Marrucci, Y. R. Shen, *Phys. Rev. Lett.* **1994**, *73*, 1513.
[48] Y. R. Shen, *Nature* **1989**, *337*, 519.
[49] P. Guyot-Sionnest, H. Hsiung, Y. R. Shen, *Phys. Rev. Lett.* **1986**, *57*, 2963.
[50] C. S. Mullin, P. Guyot-Sionnest, Y. R. Shen, *Phys. Rev. A* **1989**, *39*, 3745.
[51] W. Chen, M. Feller, Y. R. Shen, *Phys. Rev. Lett.* **1989**, *63*, 2665.
[52] M. B. Feller, W. Chen, Y. R. Shen, *Phys. Rev. A* **1991**, *43*, 6778.
[53] N. A. J. M. van Aerle, *Liq. Cryst.* **1994**, *17*, 585.
[54] J. Y. Huang, R. Superfine, Y. R. Shen, *Phys. Rev. A* **1990**, *42*, 3660.
[55] M. Barmentlo, R. W. J. Hollering, N. A. J. M. van Aerle, *Phys. Rev. A* **1982**, *46*, 4490.
[56] B. Jérôme, Y. R. Shen, *Phys. Rev. E* **1993**, *48*, 4556.
[57] J. M. Geary, J. M. Goodby, A. R. Kmetz, J. S. Patel, *J. Appl. Phys.* **1987**, *62*, 4100.

[58] M. F. Toney, T. P. Russel, J. A. Logan, H. Kikuchi, J. M. Sands, S. K. Kumar, *Nature* **1995**, *374*, 709.
[59] J. S. Foster, J. E. Frommer, *Nature* **1988**, *333*, 542.
[60] D. P. E. Smith, J. K. H. Hörber, G. Binnig, H. Nejoh, *Nature* **1990**, *344*, 641.
[61] M. Hara, Y. Iwakabe, K. Tochigi, H. Sasabe, A. F. Garito, A. Yamada, *Nature* **1990**, *344*, 228.
[62] K. Miyano, *Phys. Rev. Lett.* **1979**, *43*, 51.
[63] J. P. Nicholson, *J. Phys. France* **1987**, *48*, 131.
[64] W. Chen, L. J. Martinez-Miranda, H. Hsiung, Y. R. Shen, *Phys. Rev. Lett.* **1989**, *62*, 1860.
[65] G. P. Crawford, R. Stannarius, J. W. Doane, *Phys. Rev. A* **1991**, *44*, 2558.
[66] D. Johannsmann, H. Zhou, P. Sonderkaer, H. Wierenga, B. O. Myrvold, Y. R. Shen, *Phys. Rev. E* **1993**, *48*, 1889.
[67] B. Jérôme, *J. Phys.: Condens. Matter* **1994**, *6*, A269.
[68] B. Jérôme, J. O'Brien, Y. Ouchi, C. Stanners, Y. R. Shen, *Phys. Rev. Lett.* **1993**, *48*, 4556.
[69] G. Porte, *J. Phys. France* **1976**, *37*, 1245.
[70] K. Hiltrop, H. Stegemeyer, *Ber. Bunsenges. Phys. Chem.* **1981**, *85*, 582.
[71] A. W. Neumann, R. W. Springer, R. T. Bruce, *Mol. Cryst. Liq. Cryst.* **1974**, *27*, 23.
[72] D. Beaglehole, *Mol. Cryst. Liq. Cryst.* **1982**, *89*, 319.
[73] D. Langevin, *J. Phys. France* **1972**, *33*, 249.
[74] J. H. Thurtell, M. M. Telo da Gama, K. K. Gubbins, *Mol. Phys.* **1985**, *54*, 321.
[75] B. Tjipto-Margo, A. K. Sen, L. Mederos, D. E. Sullivan, *Mol. Phys.* **1989**, *67*, 601.
[76] E. F. Gramsbergen, W. H. de Jeu, *J. Phys. France* **1988**, *49*, 363.
[77] B. M. Ocko, P. S. Pershan, C. R. Safinya, L. Y. Chiang, *Phys. Rev. A* **1987**, *35*, 1868.
[78] E. F. Gramsbergen, J. Als Nielsen, W. H. de Jeu, *Phys. Rev. A* **1988**, *37*, 1335.
[79] M. G. J. Gannon, T. E. Faber, *Phil. Mag. A* **1978**, *37*, 117.
[80] S. Immerschitt, T. Kohl, W. Stille, G. Strobl, *J. Chem. Phys.* **1992**, *21*, 173.
[81] M. A. Bouchiat, D. Langevin-Cruchon, *Phys. Lett. A* **1971**, *34*, 331.
[82] S. Faetti, L. Fronzoni, *Solid State Commun.* **1978**, *25*, 1087.
[83] R. Holyst, A. Poniewierski, *Phys. Rev. A* **1988**, *38*, 1527.
[84] B. G. Moore, W. E. McMullen, *Phys. Rev. A* **1990**, *42*, 6042.
[85] M. A. Osipov, S. Hess, *J. Chem. Phys.* **1993**, *99*, 4181.
[86] J. D. Parsons, *Phys. Rev. Lett.* **1978**, *41*, 877.
[87] B. Thipto-Margo, D. E. Sullivan, *J. Chem. Phys.* **1988**, *88*, 6620.
[88] W. Urbach, M. Boix, E. Guyon, *Appl. Phys. Lett.* **1974**, *25*, 479.
[89] J. E. Proust, L. Ter-Minassian-Saraga, *J. Phys. Colloq.* **1979**, *40*, C3 490.
[90] S. J. Elston, *Liq. Cryst.* **1994**, *16*, 151.
[91] J. J. Bonvent, J. A. M. M. van Haaren, G. Cnossen, A. G. H. Verhulst, P. van der Sluis, *Liq. Cryst.* **1995**, *18*, 723.
[92] N. A. Clark, S. T. Lagerwall, *Appl. Phys. Lett.* **1980**, *36*, 899.
[93] K. Kondo, F. Kobayashi, A. Fukuda, E. Kuze, *Jpn. J. Appl. Phys.* **1981**, *20*, 1773.
[94] M. Cagnon, G. Durand, *Phys. Rev. Lett.* **1993**, *70*, 2742.
[95] M. A. Handschy, N. A. Clark, *Ferroelectrics* **1984**, *59*, 69.
[96] Y. Galerne, L. Liebert, *Phys. Rev. Lett.* **1990**, *64*, 906.
[97] R. B. Meyer, P. S. Pershan, *Solid State Commun.* **1973**, *13*, 989.
[98] J. V. Selinger, Z. G. Wang, R. F. Bruinsma, C. M. Knobler, *Phys. Rev. Lett.* **1988**, *70*, 1139.
[99] S. B. Dierker, R. Pindak, R. B. Meyer, *Phys. Rev. Lett.* **1986**, *56*, 1819.
[100] J. Maclennan, M. Seul, *Phys. Rev. Lett.* **1992**, *69*, 2082.
[101] T. J. Scheffer, H. Gruler, G. Meier, *Solid State Commun.* **1972**, *11*, 253.
[102] E. I. Demikhov, *Phys. Rev. E* **1995**, *51*, 12.
[103] J. Pang, N. A. Clark, *Phys. Rev. Lett.* **1994**, *73*, 2332.
[104] M. A. Handschy, N. A. Clark, S. T. Lagerwall, *Phys. Rev. Lett.* **1983**, *51*, 471.
[105] J. Xue, N. A. Clark, M. R. Meadows, *Appl. Phys. Lett.* **1988**, *53*, 2397.
[106] S. Tripathi, M. H. Lu, E. M. Terentjev, R. G. Petschek, C. Rosenblatt, *Phys. Rev. Lett.* **1991**, *67*, 3400.
[107] K. A. Crandall, C. Rosenblatt, R. M. Hornreich, *Liq. Cryst.* **1995**, *18*, 251.
[108] K. Nakagawa, T. Shinomiya, M. Koden, K. Tsubota, T. Kuratate, Y. Ishii, F. Funada, M. Matsuura, K. Awane, *Ferroelectrics* **1988**, *85*, 427.
[109] J. Xue, N. A. Clark, *Phys. Rev. Lett.* **1990**, *64*, 307.
[110] W. Chen, Y. Ouchi, T. Moses, Y. R. Shen, K. H. Yang, *Phys. Rev. Lett.* **1992**, *68*, 1547.
[111] S. Immerschitt, W. Stille, G. Strobl, *Macromolecules* **1992**, *25*, 3227.
[112] G. Decher, J. Reibel, M. Honig, I. G. Voigt-Martin, A. Dittrich, H. Ringsdorf, H. Poths, R. Zentel, *Ber. Bunsenges. Phys. Chem.* **1993**, *97*, 1386.
[113] E. M. Terentjev, *J. Phys. France* **1995**, *5*, 159.
[114] M. C. Holmes, N. Boden, K. Radley, *Mol. Cryst. Liq. Cryst.* **1983**, *100*, 93.
[115] U. Kaeder, K. Hiltrop, *Mol. Cryst. Liq. Cryst. Lett.* **1991**, *7*, 173.
[116] A. Poniewierski, R. Holyst, *Phys. Rev. A* **1988**, *38*, 3721.
[117] T. Yoshino, M. Suzuki, *J. Phys. Chem.* **1987**, *91*, 2009.
[118] E. A. Oliveira, A. M. Figueiredo Neto, G. Durand, *Phys. Rev. A* **1991**, *44*, 825.

11 Ultrasonic Properties

Olga A. Kapustina

Liquid crystal acoustics is an extremely diverse science insofar as the range of the problems that is covers is concerned. In recent years many experimental and applied investigations have been performed which have given rise to new branches of this science. We present here the current state of liquid crystal acoustics and discuss the most important advances in and the future potential of the field. The classical aspects of the subject are outlined briefly, with reference to associated chapters in this book.

11.1 Structural Transformation in Liquid Crystals

Since the very beginning of the research on mesomorphism, the influence of ultrasonic fields has been of great importance; only the goals of study have changed with time. The early research on which the later developments were based was done by Lehman [1], Zolina [2], Zvereva and Kapustin [3], and Fergason [4]. Discussion about the nature of the phenomena observed simulated much work in the pioneering days [5–13]. The various principles used to interpret the influence of ultrasonic fields on the mesophase were of exceptional theoretical importance. These included interpreting the effects of the field on the molecular arrangement of the liquid crystalline state on the basis of the Leslie–Erickson hydrodynamic theory or, as Kozhevnikov later demonstrated, using models adapted from nonequilibrium hydrodynamics. This variety of approaches to the problem was a result of the complex nature of the phenomena. The effect of ultrasound on the mesophase is, as a rule, associated mainly with the onset of flows induced by nonlinear phenomena. The flow process is characterized by an essentially stationary velocity distribution in the layer. Calculation and experimental determination of the flow velocity present major difficulties. Flow-induced disturbances in the molecular arrangement can be of several kinds; in particular, they can be both of the threshold or nonthreshold type and modulated or unmodulated. Modulated disturbances always show threshold behavior. Most of the data available at present refer to the nematics, the characteristic properties of which are manifested to certain degree in the smectics and in cholesterics. Ultrasonic field effects are covered in several reviews and monographes [13–19].

11.1.1 Orientation Phenomena in Nematics

All research on the changes in the macrostructure of nematics when in an ultrasonic field is done on samples 10–360 μm thick and using generally accepted optical methods (see Ch. VIII of this Volume). Any changes are observed either as a result of the depolarization of light transmitted through the sample or because of light scattering by the inhomogeneous structure [20]. Such experiments have shown three kinds of layer structure disturbance: homogeneous, spatially periodic, and inhomogeneous. Naturally, intermediate states between these three extreme phenomena are possible.

11.1.1.1 Homogeneous Distortion Stage

This stage involves longitudinal waves [5, 10, 21, 34] and surface acoustic waves (SAWs) [9, 35–40], with wavevectors $k = \omega/c$ and $k_R = \omega/c_R$, respectively, where c and c_R are phase velocities and ω is the angular frequency of the wave. Typical geometries of the equipment used for experiments on homeotropically aligned nematics are shown in Fig. 1. The wavevector is parallel or perpendicular to the director n and makes an angle θ with n. The measurement of ultrasound-induced birefringence offers an excellent tool for studying orientational phenomena and checking models.

The characteristic properties of the effect that can be elucidated experimentally are [5, 9, 10, 21–40]:

- Ultrasonic longitudinal and surface waves applied to a nematic sample change the birefringence properties of the fluid in the reflective and transmission modes when the ultrasound intensity J exceeds a certain minimum value [9, 10, 22, 26, 27].

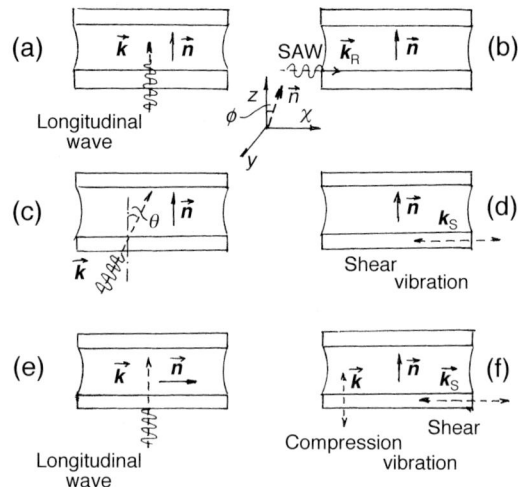

Figure 1. Equipment geometries used for studying ultrasonically induced structural transformations in liquid crystals: (a) longitudinal waves, $k \parallel n$; (b) SAW, $k_R \perp n$; (c) longitudinal waves, $k \parallel n$, $\theta \neq 0$; (d) shear vibrations, $k_S \perp n$; (e) longitudinal waves, $k \perp n$; (f) shear and compression vibrations, $k_S \perp n, k \parallel n$.

- The conditional clearing threshold of the ultrasound intensity J^{th} for birefringence is dependent on the layer thickness as d^{-2} [22, 25, 27].
- The frequency dependence of the threshold J^{th} is given by $J^{th} \simeq f^{-1/2}$ [17, 22, 27].
- The value of the threshold can be reduced by 'biasing' the sample with an electric field [10].
- Above the threshold: the relation between the optical transmission m of the nematic layer and the ultrasound intensity J is nonlinear [10, 22, 25, 27]; the transmission depends on the ultrasound intensity as $m \simeq \sin^2(\text{const.} J^2)$ and shows interference maxima and minima between ordinary and extraordinary light rays for monochromatic light [21, 25]; and for small values of the intensity $m \simeq J^4$ [21, 38].
- The optical transmission properties of nematic samples depend on the acoustic boundary conditions (cells with fixed

[22] or free [21] edges) and on the molecular orientation at the boundaries (planar or homeotropic configurations [27]).

- The variation in the orientation of the molecules in the ultrasonic field is observed as a system of alternating light and dark bands, the width and contrast of which depend on the ultrasound intensity. The distance between the centers of the light bands is of the order of the ultrasound wavelength [9, 12, 13, 21, 22, 27, 29, 35, 38–40]. The band configurations depend on the cell structure, the acoustic boundary conditions and the mutual orientation of the wavevector and director; these clearing patterns may be distorted by nonuniformity of the wave field inside the ultrasonic beam.

Several theories aimed at explaining the phenomena have been proposed, each of which is founded on completely different concepts. Sripaipan et al. [21] proposed a nematic layer with free ends, in which the interaction between the longitudinal oscillations (induced by the motion of the free ends of the layer in compression) and the traverse oscillations establishes steady flow of the liquid and, as a result, rotation of the molecules. However, these authors used incorrect dispersion relations and their calculations are not consistent with observed layer compression patterns. Nagai and coworkers [26, 27] hypothesized that with normal incidence of an ultrasound beam on the layer the rotation of molecules is attributable to radiation fluxes. Radiation fluxes are the steady acoustic flows caused by radiation forces in a traveling acoustic wave, the only provision being that the width of the ultrasound beam is smaller then the dimensions of the cell. In reality, radiation fluxes can only occur near the boundaries of the beam and produce a compression effect that is smaller than the one that is actually observed [22, 25]. Radiation fluxes are significant when an ultrasound beam is focused so that it is obliquely incident on the layer [28]. Helfrich [33] and Chaban [34] treated molecular rotation as a threshold effect associated with the fact that the equations of motion for nematics contain nonlinear stresses proportional to the angle of molecular rotation and the particle velocity. However, the theoretical values obtained by these authors for the threshold ultrasound intensity are 2–3 orders of magnitude greater than the intensities usually observed experimentally. Dion and coworkers [41–43] hypothesized that torques created by the sound absorption anisotropy are responsible for rotating the molecules, which tend in turn in such a way as to diminish the losses in the ultrasound wave. This effect is theoretically possible, but is two or three orders of magnitude smaller than what is observed experimentally.

Kozheynikov and coworkers [44–46] have developed the most general theory, which is based on Leslie–Ericksen hydrodynamics. These workers attributed the orientation effects to the steady inhomogeneous acoustic flows that result from the interaction between the periodic compression of the layer in the ultrasound field and the periodic motion of the liquid along the plates confining the layer. Unlike previous theories [27, 33, 34, 41–43], Kozhevnikov's model predicts the effect observed for normal incidence of the ultrasound on the layer. In the case of $\mathbf{k} \| \mathbf{n}$, according to the theory [44, 45], two physical reasons are responsible for the periodic liquid motion that occurs in the cell: vibrations of the cell plates in the layer having free ends [21, 44] and pressure gradients near the border of the beam for an infinitely wide layer [45]. In all instances, the acoustic flow velocity component along the x axis, V_x that initiates the director rotation, satisfies a well-known acoustics equation [47]. The equation for the small steady-state angle is [48]

$$\alpha_2 \frac{\partial V_x}{\partial z} = K_3 \frac{\partial^2 \psi}{\partial z^2} \qquad (1)$$

Information on the velocity flows, configuration, scale, velocity distribution and the suitable distortion profiles of the director field of the layer under the action of an ultrasonic field for the qualitatively different situations presented above can be found in the literature [44, 45].

The distortion of the layer structure is usually studied by estimating the optical phase differences Δ_o of the light that is polarized in the plane of the sandwich-cell plates, or the optical transmission m of the layer. The latter quantity can be calculated from the intensity I of the light transmitted through the layer parallel to the z axis and under crossed polarizers, the initial intensity I_0, and Δ_o according to $m = I/I_0 \sin^2 \Delta_o/2$ [20]. It is useful to introduce the effective clearing threshold J^{th} as the ultrasound intensity (or displacement amplitude) at which $m = 0.01$ [17]. Sometimes, in order to compare theoretical prediction with experimental results it is also possible to estimate the ultrasound intensity at which $\Delta_o = \pi$ (i.e. the depth of light modulation is 100%). Many experimental data are presented in some detail and are discussed quite throughly by Kapustin and Kapustina [17, 18]. Kozhevnikov's theory is borne out by the facts. For example, according to Gus'kov and Kozhevnikov [45], in a cell with fixed ends, $J^{th} \simeq d^{-2.5} f^{-1}$. The experimental values of Jth obtained by Kapustina and Lupanov [22] and Hatakeyama and Kagawa [25] also follow this law (Fig. 2). Recent data on the qualitative behavior of Jth for some equipment geometries are summarized in Table 1 [44–46].

It has recently been shown that the flow structure depends on both the inhomogene-

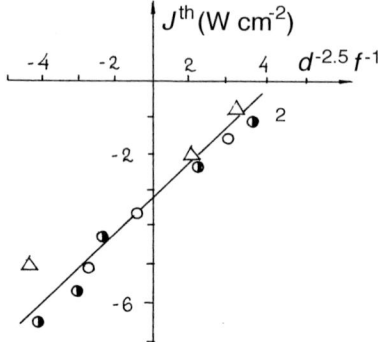

Figure 2. Effective clearing thresholds J^{th} for a 10–126 μm sample of MBBA in the ultrasound frequency range: (●) 0.4, (○) 1, (△) 3.2 MHz.

ity of the wave field in the cell and d/λ, where λ is the length of the elastic wave (see in [18]). In the ultrasound frequency range $(d/\lambda \ll 1)$ the efficiency of the flow generation is high if the typical value of the wave inhomogeneity length κ is in the vicinity of λ. In this case $J^{100\%} = (4\lambda/\kappa \delta d) J_0^{100\%}$, where $J_0^{100\%}$ is the value of $J^{100\%}$ for the wave field with $\kappa \neq \lambda$; $\delta = (\rho \omega/2\eta)^{1/2}$ is the wave number in a viscous wave. At hypersonic frequencies $(d/\lambda \gg 1)$, when the sound absorption in nematics is sufficiently high, $J^{100\%} = (4\alpha \lambda/\pi \kappa \delta d) J_0^{100\%}$. In nematics the sound absorption coefficient α is usually about 20 dB cm^{-1}. The preceding discussion is summarized in Table 2 using the terminology of Russian physists for the comparison of the flow efficiency [18].

Information concerning the ultrasound-induced birefringence in homeotropic nematics when the **k** makes an angle with **n** (see Fig. 1c) can be found in the literature [19, 28, 30]. The theoretical aspects of the problem cannot be detailed here and the reader is referred to the paper by Zhukovskaya et al. [46], wherein the flow mechanism of the aligning effect of an ultrasound wave or beam on a sample layer is discussed. In this case the flow is induced by nonline-

Table 1. Effective clearing thresholds.

Type of wave	$\theta\,(°)$	Boundary conditions	Equipment geometry	Effective threshold	Remarks
Compression wave	0	Acoustically rigid	Cell of radius R	$J^{th} \simeq d^{-2.5}f^{-1}$	–
Compression wave	0	Acoustically soft	Infinitely wide cell; ultrasound beam of radius R	$J^{th} \simeq d^{-2.5}f^{-0.5}$	thick layer
				$J^{th} \simeq d^{-2.5}f^{-0.5} \times \exp[-\delta_0(R-r)]$	thin layer
Compression wave	0	Acoustically soft	Infinitely wide cell; beam in the form of a strip	$J^{th} \simeq d^{-2.5}f^{-0.5}$	–
Compression wave	$\theta \neq 0$	Acoustically soft	Infinitely wide cell; beam of radius R	$J^{th} \simeq d^{-1.25}$ $J^{th} \simeq d^{-0.7}$	first maximum second maximum
SAW	90	Acoustically soft	–	$\xi^{th}_{0R} \simeq (fd)^{-0.75}$	two cell eigenmodes

Here value of R is the radius of a ultrasonically irradiated circular zone: an ultrasonic beam of radius R is incident on the layer of infinite width or an ultrasonic wave is incident on the layer of radius R pressed between a transparent plate and a thin wafer or elastic film; the value of r is a component of a cylindrical coordinate system (r, φ, z) with z-axis directed along the normal to the layer plane, while the boundary of the layer have the coordinates z = ±d/2; the value of $\delta_0 = k/2\,\delta d$ is the coefficient of attenuation of longitudinal waves in the layer; the source of these waves in the layer is nonuniform compression near the boundaries of the ultrasonic beam or it is the nonuniformity of compression of the layer as a result of bending of the thin wafer or film at the edges. So that the value of $\delta_0(R-r)$ describes the attenuation of longitudinal waves during propagation along the width of the liquid crystal layer in the radial direction both into and outside of the ultrasonically irradiated region; the value of ξ^{th}_{0R} is the threshold of displacement amplitude of SAW.

Table 2. Values of $J^{100\%}$ versus the frequency and order of the inhomogeneity of the field.

Layer thickness (µm)	Frequency (Hz)	Order of inhomogeneity, κ	$J^{100\%}$ (mW cm^{-2})
10	10^6	λ	10
10	10^6	2λ	20
10	10^9	–	1

ar boundary forces due to the interaction between the longitudinal and viscous waves. According to this theory, the maximum of the effect corresponds to the maximum of the acoustic transparency of the cell, while the relationship between J^{th} and d depends on the angle of incidence of the ultrasound wave or beam (see Table 1). This theory correlates with the experimental results reported by Hareng and coworkers [19, 30].

There have been unanimously verified reports of steady-state distortion of a homotropically aligned nematics in a SAW field [9, 35–40]. In this case the wave field in the layer is caused by the mixing of several cell eigenmodes that have different phase velocities excited in the layer when a SAW propagates in one of the bounding plates. Thus the number of modes depends on d/λ. Earlier work concentrated more on the phenomena themselves rather than their interpretation [9, 39]. Calculated values obtained using theories based on viscoelastic effects and the parametric instability [34–37] show considerable discrepancy with experimental data. Miyano and Shen [38] have investigated flows in a nematic layer with the SAW propagation along the substrate (see Sec. 11.2 of this chapter).

11.1.1.2 Spatially Periodic Distortion Stage

Domains of different nature in homeotropically and planar oriented layers may appear under the action of longitudinal or shear waves only above a certain threshold. The directions of the domains, their width, and the threshold value depend on the type of the wave, the wave frequency, and the layer thickness.

Russian physists were the first to observe ultrasonically induced domains in non-oriented samples [5]. Later, Italian and Japanese physists investigated the appearance of domains in a homeotropic layer subjected to shear vibrations, created by the ac motion of one of the cell plates in its plane (see Fig. 1 d) [49–52].

The domain lines always aligned perpendicular to the shear direction. A theoretical attempt to describe the domains [53] was concerned with a periodic shear strain, but this model does not fit what is observed experimentally: in particular, the threshold values of the shear amplitude to not fit the experimental values in terms of magnitude, do not exhibit the experimentally observed dependence on the frequency and layer thickness, and fail to give the domain dimensions. Kozhevnikov [54] has analyzed the effect on the basis of the equations of the hydrodynamics of nematics, retaining the quadratic terms proportional to the product of the angle of rotation of the molecules and the velocity of the liquid. The model leads to the following picture of the effect. A periodic shear, in the case of a random slowly varying and periodic (along the layer) deviation of the molecules from the normal, creates eddies oscillating at the shear frequency. The interaction of the eddies with the initial shear field gives rise to an average moment which increases the deflection of the molecules. The threshold amplitude and domain size are given by

$$\xi_0^{th} = \left(\frac{2K_3}{\rho}\right)^{1/2} \left(\frac{\pi}{d^2\omega}\right) F(q,p); \quad L \simeq \frac{\pi}{q_0} = 2d \quad (2)$$

where $p = \pi/d$, $q = \pi/L$, and q_0 is the wavenumber at the threshold of the effect; for information concerning the form of function $F(q, p)$ see Kozhevnikov [54]. Figure 3 shows the theoretical and experimental dependences of the threshold shear amplitude on the frequency for various values of d [49, 52, 55]. It is evident that in the range of validity of the calculations, Kozhevnikov's theory and the experimental data agree well. According to Kozhevnikov [54], the theory considers the frequencies f satisfying the inequalities $\pi K_3/2\gamma_1 d^2 \ll f \ll \rho c^2/2\pi\eta$ and $f < \eta/2\pi\rho d^2$.

Russian physists [56, 57] have performed systematic investigations of the condition

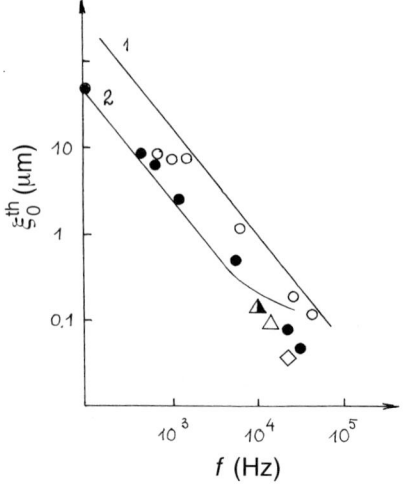

Figure 3. Calculated (lines 1 and 2) and measured (symbols) threshold displacements ξ_0^{th} versus frequency for various layer thicknesses: (1) 20, (2) 100, (○) 20, (□) 66, (●) 100, (△) 105, (▲) 200 μm.

under which domains appear in the planar layer provided that $k \perp n$ (see Fig. 1e). In this situation the domain lines are orthogonal to the director n in the undisturbed state. The results were interpreted on the basis of the model proposed by Kozhevnikov [58] who developed a new approach for describing the structural transformations. The model is based on nonequilibrium hydrodynamics and considers both the ultrasonic relaxation processes and the anisotropy of the viscoelastic properties of nematics (the dynamic modulus of elasticity and the bulk viscosity). A schematic drawing of a cell which illustrates the model is shown in Fig. 4. The mechanism of the effect lies in the amplification of the random and nonuniform distortion of the planar structure when oscillating ultrasonically produced vortex flows interact with the director field. The anisotropic shear stresses in the layer that generate these flows are given by

$$\sigma_{xz} = \psi_0 (\mu_3 \vartheta_{zz} + U_{zz} \Delta E)$$

where U_{zz} and ϑ_{zz} are the compression and the rate of compression of the medium in the ultrasound field; μ_3 and ΔE are the real and imaginary parts of the modulus of elasticity (both depend on the relaxation time of the order parameter τ_1 and the transformations of finite chains of molecules τ_2). The spatial harmonic of the form $\psi \simeq \psi_0 \sin q_x \cdot \sin pz$ is amplified most. The amplitudes of the threshold particle velocity and domain size are given by

$$\vartheta_0^{th} = \frac{1}{2\pi c} \left[\frac{K_3 c \eta F(S_{min})}{(\alpha_5 - \alpha_6) \Delta E} \right]^{1/2} d^{-1},$$

$$L = \frac{d}{S_{min}^{1/2}} \quad (3)$$

where S_{min} is the value of the parameter $S = (q^2/p^2)$ that minimizes the function $F(S)$. The expression for this function can be found in Kozhevnikov [58]. At low and high frequencies, which are determined by limiting values of the parameter $B = \pi^2 \eta \mu_3/\rho d^2 \Delta E$, minimum values of $F(S)$ correspond to $S_{min} = 3.7$ and 4.8. This gives for the spatial period L the values 0.52 d and 0.48 d, which are in agreement with the experimental data (Fig. 5a). In the frequency range satisfying the condition $\omega \tau_1 > 1$, the amplitude of the threshold particle velocity is virtually independent of temperature and follows the law: $v_0^{th} \simeq \Delta T^{-1/3}$ [58]. This agrees with the experimentally observed data (Fig. 5b). The foregoing discussion illustrates that the new approach proposed by Russian physists is valid.

11.1.1.3 Inhomogeneous Distortion Stage

This stage involves the phenomena corresponding to higher excitation: at a definite threshold the nematic layer exhibits orientational turbulence, corresponding to observable light scattering. Kapustin and Dmitriev [5] were the first to observe this effect on a nonoriented sample in a field of longitudinal waves. Later, Kessler and Sawyer [7] found that the intensity of this scattering depends nonlinearly on the excitation level. By analogy with a similar phe-

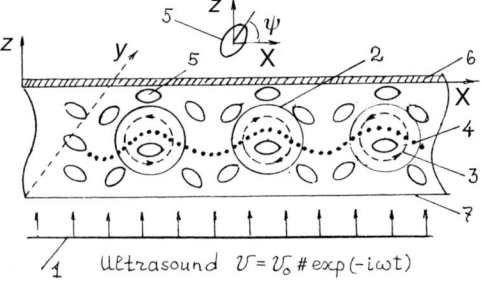

Figure 4. Diagram of a cell with domains in cross-section: (1) Ultrasound, (2) system of oscillatory vortex flow with the velocity $v_z' \simeq \cos qz$, (3) stationary flows, (4) profile of distortion of the nematic orientation, (5) nematics molecule, (6) acoustically rigid boundary, (7) acoustically soft boundary.

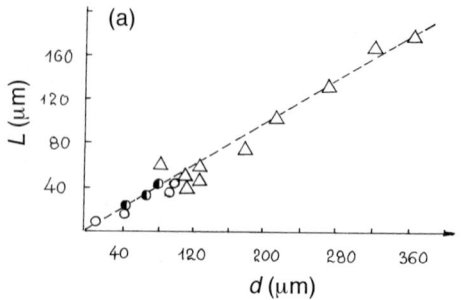

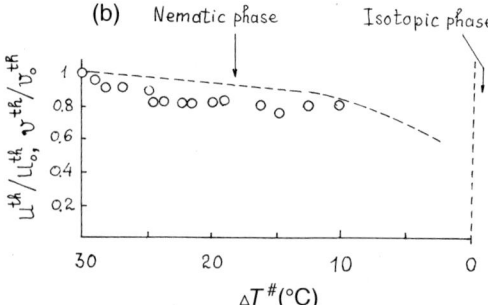

Figure 5. Ultrasonically induced orientational instability in planar nematics. (a) Spatial period of domains versus layer thickness, $f=3.2$ MHz, $T=30.2\,°C$. (b) Calculated and measured threshold parameters characterizing the wave field in the layer versus temperature of the nematics; the running values of Uth (tension) and v^{th} are normalized to the values of U_0^{th} and v_0^{th}, corresponding to the lower limit of the test temperature interval ($T=22.8\,°C$), $\Delta T = T_{N-I} - T$.

nomenon in an electric field, this effect is called the acoustic dynamic scattering mode (DSM).

The characteristic properties of this effect are as follows [22, 56, 59, 60]:

- The intensity and angular distribution of the scattered light are independent of the wavelength and polarization of the incident light, and the scattering is directed mainly forward.
- The intensity and frequency of the ultrasound play an important role. Instability begins just above a threshold value J_s^{th}. Sometimes domains appear first, and then gradually switch over to the DSM with increasing intensity of the ultrasound. The effect reaches maximum scattering (saturation) at an excitation level of about 6–8 times J_s^{th}. The threshold value is virtually independent of the substance, and hardly varies from the thickness of the layer ($J_s^{th} \simeq 1/d^2$) [22].
- The initial orientation of the molecules in relation to k and to the boundary surfaces does not play a critical role, but it does have importance in technical application [22, 56, 59].
- The rise and decay times are influenced by many parameters (viscosity, sample thickness, intensity, etc.) [17, 59].

This problem has not yet been studied theoretically. However, the progress made to date in the area of experimental research clearly foreshadows rapid advances in the future. The flows generated by ultrasound create optical inhomogeneities in the layer that act as individual scattering centers. Bertolotti and coworkers [50, 60] have measured the coherence length of scattered light on the hometropic samples. The transition to the disordered state is accompanied by a drop in the coherence length to a rather small value, and the sample behaves as a thermal source containing a large number of statistically independent scattering centers. The sample thickness and observation angle do not play a critical role. Many features of the acoustic DSM have been studied with regard to technical applications [59]; this applies in particular to rise and decay times [17, 59]. Several technical studies have been concerned with contrast ratio [17, 59]. Kapustina [61] has recently reported on the acoustic DSM induced by ultrasound (1.5 MHz) in a polymer-dispersed nematic film.

11.1.2 Cholesterics in an Ultrasonic Field

The behavior of cholesterics in an ultrasonic field has generally been far less well researched than that of nematics. Only a few aspects of the problem have been examined since the systematic investigations done by Russian and Italian physists [17, 58, 62–65]. It may therefore seem to be somewhat stretching a point to draw an analogy with the phenomena in nematics, but this is a valuable way to obtain an overview of the following rather complex phenomena:

- periodic two-dimensional distortion, analogous to nematic domains (k parallel to the helical axis);
- excitation of the storage mode via turbulent flow, analogous to acoustic DSM in nematics (k parallel or perpendicular to the helical axis);
- erasing of the storage mode (k parallel to the helical axis);
- focal–conic texture transformation into a planar one (k parallel to the normal to the layer);
- bubble domains texture formation (k perpendicular or parallel to the helical axis);
- fingerprint texture transformations (k parallel to the normal to layer).

11.1.2.1 Periodic Distortion

An ultrasonic field applied parallel to the helical axis h ($k \parallel h$) can cause a square grid-like pattern deformation of the planar texture [17, 58, 62]. This can be observed in nematic/cholesteric mixtures with a helix of large pitch. According to experimental data, the spatial period of distortion follows the equation $L \simeq (p_o d)^{1/2}$ and tends to decrease slightly with frequency. The threshold particle velocity in the wave ϑ_0^{th} is practically independent of frequency. By analogy with nematics, Kozhevnikov [62] described this phenomenon within the framework of an approach based on nonequilibrium hydrodynamics, but introduced the added assumption of an initial deformation of the layered systems in the crystal axis direction. The threshold amplitude ϑ_0^{th} and the size of the domains L in the equilibrium layers or the layers extended along the crystal axis are determined. As with nematics, two frequency bands (limiting value of $B = 38 \eta \pi^2 / \rho \omega^2 \tau d^2$) appear here also to be due to dissimilar behaviors of the threshold and domain size relative to d, p_o, f, etc. In particular, for $\omega \tau \gg 1$ ($B \ll 1$, high frequency) the value of $\vartheta_0^{th} \simeq (p_o d)^{-1/2} (1-l)^{1/2}$, whereas in the case of $\omega \tau \ll 1$ ($B \gg 1$, low frequency) the value of ϑ_0^{th} is also independent of frequency and follows the law: $\vartheta_0^{th} \simeq (d/p_o)^{1/2} (1-l)^{1/2}$. Here $l = \Delta / \Delta_c$ is a relative value of the layer extension (Δ_c is the critical extension of the layered system when the domains appear under static deformation [48, 66] and Δ is the extension of the layer). According to the theory [58], the domain size L is approximately $(p_o d)^{1/2}$ over all the frequency range. The correlation between the experimental data and this theory is satisfactory only if the initial deformation of the layered system is taken into account; the relative value of the layer extension is then found to be close to 0.8–0.9.

11.1.2.2 Storage Mode

The interest in this area is due more to the possibilities of technical application than theoretical considerations. Gurova, Kapustina, and Lupanov were the first to transform the uniform planar texture of a nematic/cholesteric mixture sample into a focal–conic one by means of an ultrasound field [17, 59]. The helical structure is still present in this state, having the same pitch on a microscopic level. Macroscopically, the sample is broken up into a myriad of randomly oriented domains with the size of a few micrometers, and consequently it strongly scatters visible light of all wavelengths. Such a texture is said to be stable for up to several days, during which time it gradually reverts to the planar one. The initial texture can be restored at any time by applying an ac pulse (20 kHz) [59, 64]. Russian physists have described the ultrasonically initiated storage effect [64]. Their data for the contrast ratio and for the rise and decay times have not as yet been improved upon. As mentioned above, the stored information slowly decays, but it persists in certain mixtures for several days after ultrasound excitation has been removed. Thickness, pitch and boundary conditions were found to play a crucial role in the memory properties of the planar to focal–conic texture transformations [17, 59].

11.1.2.3 Focal–Conic to Planar Texture Transition

Ultrasound waves may also create an oriented structure. This effect was observed in a sample with a focal–conic texture subjected to ultrasound (0.4 and 3.2 MHz [17, 67]). According to Hiroshima and Shimizu [67], the transmission of a sample (a nematic/cholesteric mixture) increases with increasing ultrasound intensity tending towards a value characteristic of a planar texture.

11.1.2.4 Bubble Domain Texture

Gurova and Kapustina were the first to describe ultrasonically induced bubble domains in large pitch nematic/cholesteric mixtures [17, 63]. It was found that relaxation of an ultrasound-induced perturbation at a critical layer thickness, which is commensurate with the pitch of the helix, is accompanied by the formation of a bubble domain, which is a strong light scatterer. These data and the mechanism of the formation and stabilization of bubble domains have been discussed within the framework of the Akahane and Tako model [68]. This model attributes the formation and high stability of the bubble domains to the presence of defects (disclinations) in the sample, which pin the domains. The model does not take into account the interaction of the domains, and is thus valid in the case of low density domain packing. Test bubble domains have provided the first possibility for a quantitative analysis of the model, since they have a rather low packing density. The correlation of the theory with the experimental data is fully satisfactory (Fig. 6).

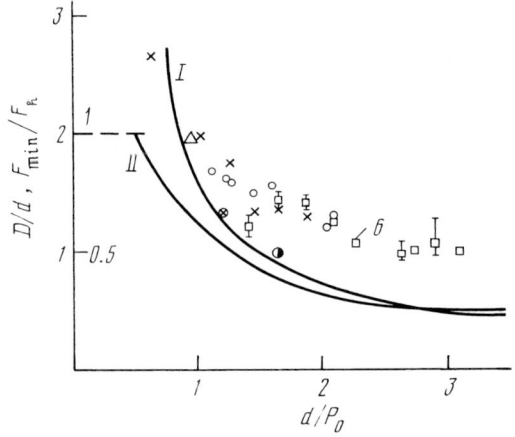

Figure 6. Comparison of Akahane and Tako's model and the experimental data. The calculated bubble domain diameter D and the free energy F_{min}, normalized to the layer thickness d and the free energy of the homeotropic structure F_h, respectively, versus the ratio $d/\bar{P}_0$ (curves I and II). Pitches (μm): (●) 3, (⊗) 4.2, (△) 10.5, (×) 48 ($f=0.2$ MHz). Frequency (MHz): (○) 0.5, (□) 3 ($P=48$ μm).

11.1.2.5 Fingerprint Texture Transition

Bertolotti et al. [65] have studied the behavior of a well-known fingerprint texture [69] formed in a sample (a nematic/cholesteric mixture) due to periodic compression in the frequency range 0.1 Hz to 130 kHz (free-edge cell). All the data obtained were analysed within the framework of Press–Arrot's model [70].

11.1.3 Smectic Phase in an Ultrasound Field

There has been no systematic study of this area because at present no potential technological applications of smectics in conjunction with ultrasound are known. Thus we have only a fragmentary picture from the few isolated investigations that have been made of smectic phases in ultrasound fields [5, 59, 71, 72].

Visual observations have shown that the optical transparency of a nonoriented sample of a smectic A phase under the action of longitudinal waves is diminished as a result of light scattering due to fluctuations in the refractive index [73]; transparency is not restored to its initial value when the disturbance is removed [5]. Italian physists have performed experiments on homeotropic smectic A layers and observed an ultrasonically induced spatially periodic structure distortion, including the formation of domains with relaxation times of up to several hours [71]. Russian physists have acknowledged the ability of smectic A phases to preserve remanent strain [72]. They investigated the behaviour of a homeotropic sample under the action of longitudinal waves. According to these data the smectic A phase exhibits an ultrasonically induced storage mode, where the remanent optical transparency barely differs from the value observed at the onset of the action of the ultrasound waves. It this case the storage effect is highly pronounced. The thickness of the sample and the method of preparation of the plate cell surfaces were found to affect both the contrast ratio and the efficiency of the transition from the transparent to the opaque structure. As in cholesterics, recovery of the initial orientation in smectic phases is realized by the application of an electric pulse (20 kHz, 200 V) [72]. All the above-described studies were concentrated more on the phenomena themselves than on their interpretation.

11.2 Wave Interactions in Nematics

All meaningful contributions in this area are relatively recent because, without the support of Vuzhva's theory [40], earlier observations have remained partly inexplicable [35, 36, 42, 74]. Since the systematic studies done by Miyano and Shen [38], only a few aspects of the problem have been examined. Three simple cases can be differentiated:

– the interaction of two nematic cell eigenmodes with different phase velocities [40];
– the interaction of the nematic cell eigenmode and the viscous wave [75];
– the interaction of the elastic and viscous waves [76].

The mathematical models proposed by Kozhevnikov and coworkers [44–46] paved the way for the development of a general wave interaction theory based on the unified concept: steady-state inhomogeneous acoustic flows. The validity of this theory [40, 75] is borne out by the following observations related to the traditional geometry. A homeotropically oriented nematic, several tenths of a micrometer thick, was placed between the substrate (1) and a glass plate (2) and observed under a microscope (Fig. 7).

In a first scenario (Fig. 7a), an interdigital transducer (3) generated the SAW at the substrate (1). Within the range of frequencies that satisfy the inequalities $d \ll \lambda$ and $d \ll \lambda_R$, the SAW excites only two nematic cell eigenmodes with phase velocities c_1 and $c_2 \simeq c/2$. The velocity components of the modes obey the following conditions: $\vartheta_{1x} \ll \vartheta_{1z}$, $\vartheta_{2z} \ll \vartheta_{2x}$ [40]. If the frequency ω is above 10^{-6} s^{-1}, the wave numbers δ and k_R are such that $\delta \gg k_R$ and $\delta d \gg 1$. According to Vuzhva's model, the velocity distribution of acoustic flows resulting from the mode interactions in the xz plane and the associated director tilt angle ψ distribution are given by

$$V \sim \vartheta_{01} \vartheta_{02} \omega^{-1} \sin(k_R - k) x F(\delta, z, d),$$
$$\psi \sim \vartheta_{01} \vartheta_{02} \omega^{-1} \sin(k_R - k) x \left(\frac{d - z^2}{d}\right) \quad (4)$$

The acoustic flow patterns and the director field distortion profile are shown in Fig. 8. (For the form of the function $F(\delta, z, d)$, see Anikeev et al. [40].)

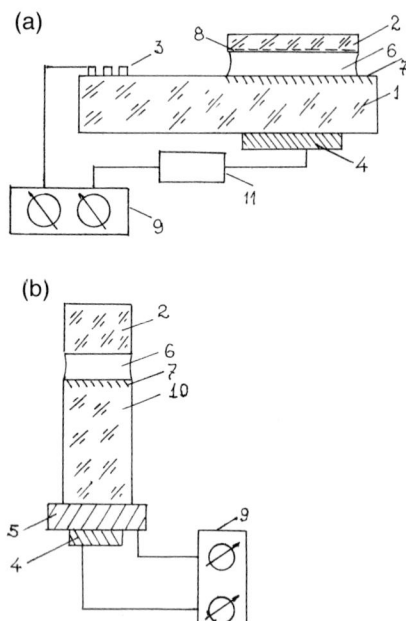

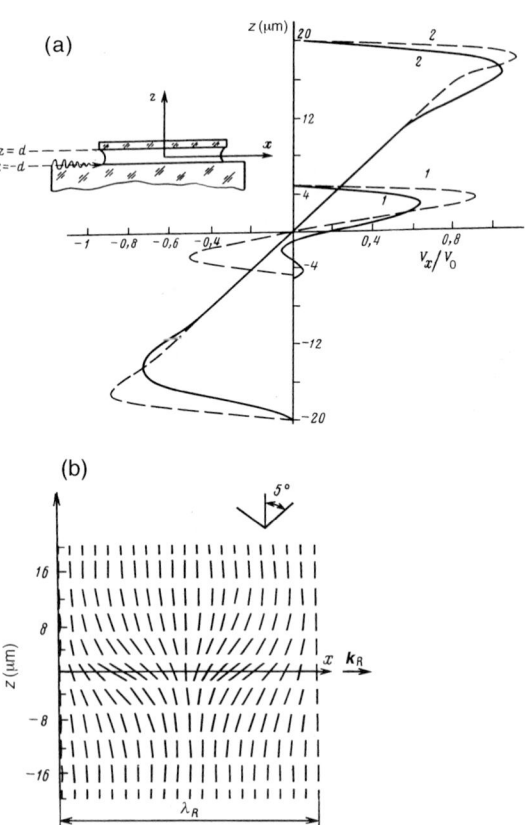

Figure 7. Equipment geometries for studying the wave and acousto-electrical interactions in nematics: (1) substrate (y cut, x oriented quartz), (2) glass plate, (3) interdigital transducer, (4) shear transducer (y cut quartz), (5) compression transducer (x cut quartz), (6) nematics, (7) mirror coating, (8) optically transparent electrode, (9) generator, (10) waveguide (substrate), (11) phase meter.

Figure 8. (a) Acoustic flow patterns for 10 and 40 μm MBBA samples (curves 1 and 2) at frequencies of (——) 6.47 and (---) 28.6 MHz for the flow velocity amplitude $V_0 = 10^{-4}$ cm·s^{-1} and $\sin(k_R - k)x = 1$. (b) Director field distortion profile in the xz plane for a 40 μm MBBA sample at a frequency of 6.47 MHz.

In a second scenario, the nematics also filled a flat capillary formed by the substrate (1) and the glass plate (2). A second transducer (4) in addition to the interdigital transducer (3) was placed under the substrate (1) (Fig. 7a). This second transducer was used to generate the viscous wave in the nematic layer. One of two nematic cell eigenmodes with a phase velocity c_2 quickly decays along the layer, and the viscous wave was able to interact only with the faster eigenmode. For information concerning the conformation, the scale and velocities of the flow resulting from the interaction the components of $\vartheta_x(\delta, \omega, z, t)$ and $\vartheta_z(\omega, z, x, t, \mathbf{k})$, and the tilt angle of director caused by this flow, the reader is referred to Anikeev et al. [75] and Bocharov [76].

It is of interest to compare the change in the optical responses of nematics due to the distortion of the director field caused by the flows in the first and second cases. The optical transmission m of the nematic layer between crossed polarizers when disturbed by a SAW with and without a viscous wave is shown in Fig. 9a. Under the action of the SAW only, $m \simeq \xi_{0R}^8$; this is in agreement with other reported results [38, 40]. With a viscous wave, $m \simeq \xi_{0R}^4$ [75]. The correlation between the calculated data and the measured values of m is evident [75]. It should be emphasized that in the case of the combined two-wave action the effective threshold is an order of magnitude lower.

Finally, in a third scenario the nematics is placed between the substrate (10) in the form of a bar and a thick glass plate (2) (Fig. 7b). One transducer, the plate (5) is bonded to the bar and creates the elastic wave of compression. The other transducer, the plate (4) is placed under the plate (5) and used to generate the viscous wave in the nematics. The data obtained using this geometry ($f = 15$ MHz) were reported by Bocharov [76]. It should be noted that only under

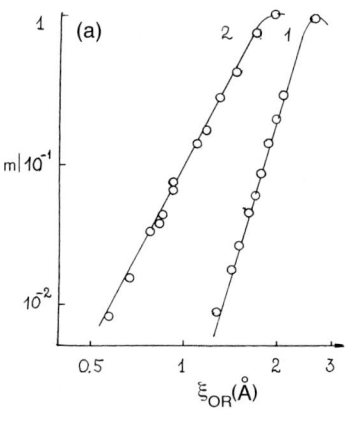

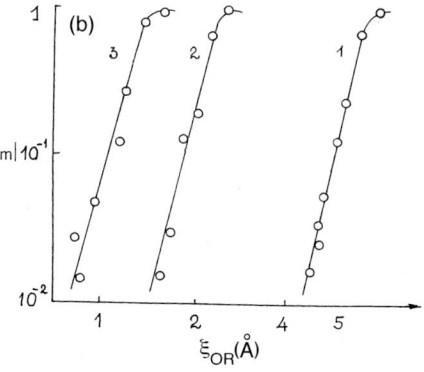

Figure 9. Optical transmission of the layer versus the amplitude of the SAW. (a) Viscous wave amplitude 0 (curve 1) and 2 Å (curve 2), $d = 40$ μm, $f = 30$ MHz, $\lambda_0 = 0.633$. (b) Optical transmission versus the amplitude of the SAW in the electric field: no voltage (curve 1), 4.5 and 4.6 V (curves 2 and 3), $d = 40$ μm, $f = 6$ MHz. (——) calculated data; (○) experimental data.

conditions of efficient viscous and elastic wave interactions is the optical response of the nematic layer subjected to the same regularities that result from the unified acoustic flow model. For example, the phase difference is $\Delta_0 \simeq \xi_{0x}^2 \xi_{0z}^2$ (where ξ_{0x} and ξ_{0z} are the viscous and elastic wave amplitude, respectively). This result corresponds well with Vuzhva's theory [75] which is based on the concept of acoustic flows. Bocharov [76] proposed a design for a new reliable acoustic holography system based on the coherent wave interaction effects considered above. Such a system, which uses the viscous wave as the reference wave, is more elaborate than earlier convertors [22, 41, 56, 77], and has the advantages of providing an order of magnitude higher sensitivity and a doubled dynamic range.

11.3 Acousto-electrical Interactions in Nematics

The interest in this area is due more to the possibilities of technical applications than to theoretical considerations. Today, three combinations of equipment geometry and the value of the dielectric anisotropy $\Delta\varepsilon$ can be differentiated:

– a crossed electric field (E directed along the z-axis; see Fig. 1) and $\Delta\varepsilon > 0$;
– a crossed electric field and $\Delta\varepsilon < 0$;
– a longitudinal electric field (E directed along the x axis) and $\Delta\varepsilon < 0$.

With the exception of the early work by Kapustin [14], who concentrated on the combined acousto-electric orientational effect in polycrystalline nematics, all basic studies of this phenomena have been done in recent years and have considered mainly homeotropically aligned nematics. The subject has been summarized in some reviews [14–18], which also outline potential applications.

Most studies have dealt with case (1) above. This equipment geometry can be used either to suppress an orientational effect [39, 78, 79] or to erase the ultrasonic images on the convertor surface [26, 59]. Nagai and Iizuka [26] have shown that under an applied electric pulse the erasure time is reduced by one or two orders of magnitude. Of the results that have reported, the influence of the anchoring energy on the acousto-electrical interaction mechanism is of particular interest [79].

Studies of acousto-electric interactions that fall into case (2), above, have been reported by various authors [30, 32, 75, 76, 80–84].

According to Kagawa et al. [32], the crossed electric field results in an increase in the sensitivity of nematics to ultrasound action; however, the rise and decay times are impaired [80]. Perbet et al. [30] controlled successfully all the characteristics mentioned above by switching the frequency of the electric field and by using a nematics with a frequency-dependent sign reversal of the dielectric anisotropy. For the ZLI-518 this frequency was affected in the range 200 Hz ($\Delta\varepsilon = +0.8$) to 20 kHz ($\Delta\varepsilon = -0.4$) and the critical frequency was 10 kHz. Ezhov and coworkers [80, 81] have tried to treat the acousto-electrical interaction problem for the peculiar geometry where $n \perp k$, by introducing a too simple assumption of the field's additivity, which is not supported by the results obtained. Hayes [83] was the first to interpret a mechanism of acousto-electric interactions for the case where $n \| k$ within the framework of the acoustic flows model. However, this concept is not completely borne out by the observed results. Several other authors have also tried to build a bridge to other models to account for the critical behavior of the threshold for the acousto-optic effect near the Frederiks transition. Akopyan et al. [74] performed an analysis on the basis of the unfaithful Helfrich's model [33] and drew mistaken conclusions. Strigazzi and Barbero [84] discussed the improvement obtained in the contrast ratio of an ultrasonic image by means of a crossed electric field. However, according to their calculations in the vicinity of the Fredericks transition, increasing the voltage applied to the electrodes gives rise to a small change of an average director tilt angle only. This is inconsistent with the reported experimental data [75, 85].

The most extensive theoretical and experimental studies of this problem have been done by Bocharov [76], who used an apparatus in which a SAW was excited at the substrate (1) (Fig. 7a). The nematics was sandwiched between the glass plate (2) and the substrate (1) bearing transparent electrodes, to which an ac voltage ($f \approx 50$ Hz) was applied. The dependence of the layer optical transmission m on the amplitude of the SAW with and without an applied electric field is shown in Fig. 9b. The threshold value of the Frederiks transition was 4.65 V. It is evident that in the vicinity of this transition the sensitivity of the nematics to the action of the SAW increases by more than an order of magnitude. According to Vuzhva's theory, near the threshold of the Frederiks transition ($E \rightarrow E_0$)

$$\frac{m(E)}{m(0)} = \frac{1}{\pi^2} \left(\frac{E_0 - E}{E_0} \right)^{-4} \quad (5)$$

which is in a good agreement with experimental results [75, 85]. These data have contributed to a coherent unified physical picture of the phenomena based on the general acoustic flows model.

Case 3 has been particularly well investigated by Belova [82] with the result that the transmission of a layer in the SAW and the longitudinal electric fields can be smaller than $m = 0.01$. However, in these experiments the optical clearing pattern was found to have high spatial homogeneity along the layer, which is inconsistent with many of the other data reported [35, 38–40]. It is likely that the SAW-created distortion in the nematics is masked by the flexo-electric effect one [68].

It should be noted also that all the results so far reported indicate great promise for practical applications [18].

11.4 Ultrasound Studies of Liquid Crystals

11.4.1 Ultrasonic Spectroscopy

In the last 25 years, the investigation of the propagation of ultrasound in liquid crystals has made a considerable contribution to the study of the following aspects of these systems:

– the phase transitions between two ordered states or between a mesophase and the isotropic state (see Chap. VII, Sec. 6 of this volume);
– the relaxation phenomena characteristic of the mesophase;
– the anisotropy of the acoustic properties in the ordered phase;
– the behaviour of acoustic parameters in variable (rotating and pulsating) magnetic fields.

Modern experimental techniques used for such studies include the fixed- and variable-path pulse-position methods as well as microwave measurement techniques. In most cases the ultrasound and microwave measurements complement each other nicely.

Since 1970 the main anomalies in the propagation of ultrasonic waves in non-oriented samples of liquid crystals have been discovered: these are the phase velocity dispersion and the nonclassical ($\alpha/f^2 =$ constant) behavior of the ultrasound absorption coefficient due to the relaxation phenomenon. The anomalies are manifested in the megahertz frequency range and are most pronounced in the vicinity of phase transitions. Kapustin [14] was the first to begin extensive investigations on the anomalies, and look at all types of liquid crystals. Most of the work done since 1980 has involved the use of a dc magnetic or electric field to measure the anisotropy of the properties of ultrasound in ordered samples. It has been established that the anisotropy of the velocity and absorption depends on the angle between the direction of the wave propagation and the applied field. The behavior of these properties of ultrasound in oriented liquid crystals does not obey classical theory and depends on the nature of the relaxation process. Relaxation mechanisms have been widely discussed in the literature; the data have been summarized in some reviews [13, 15, 56] and monographs [14, 16, 18, 86–89]. The essential factors concerning the nature of relaxation processes have been derived. These results can be grouped as follows:

- processes related to the relaxation of the order parameter (Landau–Khalatnikov mechanism [73]) and the relaxation of the order parameter fluctuations [90];
- processes resulting from the rotation of the molecules about their long and short axes and the translation of molecules;
- processes due to molecular motions, when the finite chains of molecules move as a whole, or full conformational transformations.

Most of the modern 'structure' models [18] can give only a qualitative picture of the phenomena, taking into account scalar parameters (temperature, pressure, rate of change in volume) and analyzing the isotropic component of absorption, which depends on the direction of propagation of the wave. Aero [91] has proposed a new approach to the treatment of the problem. According to his model, the propagation of the ultrasound wave produces nonequilibrium changes in the long-range orientational and short-range translation order due not only to the oscillations in the temperature and pressure, but also to shearing of the medium. Thus his model takes into account three specific mechanisms of ultrasound absorption ('critical', 'normal', and classical). The numerical estimates, obtained with this model, of the velocity and absorption of ultrasound for nonoriented and ordered nematics are consistent with the experimental data over a wide frequency range (2–500 MHz) [91].

For details concerning the behavior of the parameters of ultrasound waves in cholesterics the reader is refered to the literature [14–16, 18, 86, 92, 93]. However, the existence of an absorption gap (~1.5 dB) in cholesterics should be noted; this is observed experimentally at a wavelength that is close to the helix pitch [92, 93].

The absorption of ultrasound in smectic phases is significantly more anisotropic than that in nematics, and even the velocity has a measurable anisotropy of about 5%. Details of the behaviour of SmA, SmB, SmC and SmE phases can be found in the literature [14–16, 18, 86, 94–97]. The usual approach to the analysis of smectic phases, based on the linear theory of elasticity and hydrodynamics, results in the relationship $\alpha \simeq f^2$, which does not agree with the experimental data. In the low-frequency range the coefficients α_1, α_2, α_4 and α_5 demonstrate singularity, induced by nonlinear effects, in the form of ω^{-1}. This results in a linear frequency dependence of the ultrasound absorption. The corrections for the coefficients of elasticity B and K_1, taking into account the nonlinear fluctuation effects in smectic phases, depend on the wavevector of the smectic phase layer structure q_s: $B \simeq (\ln q_s)^{-4.5}$, $K_1 \simeq (\ln q_s)^{2.5}$ [96, 97]. In these relationships the numberical value of coefficient is very small; the vector q_s is associated with the sample dimensions and does not vanish. Thus only a small correction to the results of the linear theory of elasticity is needed.

11.4.2 Photoacoustic Spectroscopy

The revival of photoacoustic (PA) spectroscopy in the early 1970s stimulated the development of various photothermal detection techniques. The common feature of these techniques is the detection in a test sample of the oscillations in temperature that result from exposure of the sample to a modulated light beam [98]. The sensitivity of the cell depends on its radius, the length of the air column, the coefficient of thermal diffusion in the air, the frequency modulation and the size of the light beam [98]. Rosencwaig [99] was the first to demonstrate the possibility of applying the PA method to the study of the phase state of liquid crystals. The data obtained using PA spectroscopy can also be used to analyze pretransitional effects (see Chap. VII, Sec. 6.3 of this volume) in liquid crystal. Martinelli et al. [100] have elucidated the mutual relationship between the thermodynamic properties of liquid crystals and the main characteristics (phase and amplitude) of the PA signal. A new method for determining each thermal parameter individually has been developed on the basis of these data [100]. For more detailed infomation on the PA effect, see Chap. VII, Sec. 6.2.1 of this volume.

11.4.3 Acoustic Emission

For a long time it was thought that acoustic emission is inherent only to solids. However, it is now known that such emission is also a characteristic of liquid crystals, which have many structural features (defects, monopoles, solitones, etc.) [73] (see Chap. VIII, Sec. 7). The nature of the emission is closely connected with the type of liquid crystal arrangement. Therefore, if a dynamic process developing in the test sample is enacted on the defect, the release of elastic energy leads to acoustic emission. One simple situation, which makes it possible to observe this phenomenon, consists of varying the temperature in the test sample which then passes through a number of phase transitions [101]. It was found that in 1 cm thick nonoriented samples of MBBA, COOB, and CBOOA acoustic activity was most highly pronounced at the nematic–isotropic transition [102]. The high sensitivity of the method allows its use as a means of monitoring the alignment stability of mesophase samples.

11.4.4 Monitoring Boundary Effects

An important parameter describing the interaction of a mesophase with a solid substrate is the anchoring energy W (see Chap. VII, Sec. 10). New methods have been described of estimating W by means of the optical response of the nematic cell to an acoustic action: pressure variations [101], shear vibration [103], and joint shear and longitudinal vibrations [104]. A qualitative relationship between the value of W, consistent with the test cell and its optical response to the shear vibration, was discovered by Belova [103], using the equipment geometry shown in Fig. 1 d. A method has been described [105] for quantitatively estimating W. The method is based on the nonlinear interactions of the oscillations of the director with the field of velocities of flow of the nematic material due to a joint effect of shear and compressive vibrations of the same frequency, but shifted in phase by $\pi/2$ [106]. These nonlinear interactions are particularly important at amplitudes and frequencies satisfying the inequality $0.1\, \gamma_1 \beta \omega \xi_{0x} d/K_3 \gg 1$. In this case the relationship between W and the amplitude of shear vibration ξ_{0x} is [105]

$$W \simeq \beta \gamma_1 (\xi_{0x}^{max})^2 \frac{(\Delta n k_0 d)^{1/2}}{4 d \omega} \quad (6)$$

where ξ_{0x}^{max} is the amplitude corresponding to the first maximum of the function $m = f(\xi_x)$, $\beta = \xi_{0z}/\xi_{0x}$ (see Fig. 1 f). Kapustina and Reshetov [104] have reduced Eq. (6) to a form more convenient for subsequent analysis and report the results of tests.

11.4.5 Acoustic Microscopy

There are several types of acoustic microscopes, which differ from one another mainly in the principle of transforming the acoustic image into a visible one. The lens scanning acoustic microscope is widely used [107]. The specification of a modern acoustic microscope is as follows: frequency range 50–3000 MHz, resolution 50–0.1 µm, magnification 50–5000, and depth of penetration 1 µ to 1 mm.

Acoustic microscopy is rather sensitive to inhomogeneities in micro-objects, including optically opaque ones, and can also be used to detect disturbances in adhesion, flaking, microcracks, pores, deviations from the prescribed thickness of a layer, and coatings. Possible applications of the method are investigations of the topography and morphology of smooth and textured surfaces. It may also be useful in the qualitative monitoring of multilayer structures and composite materials (e.g. polymer dispersed nematics or cholesterics) in which the components are very similar in terms of their optical characteristics but have different acoustic characteristics.

Studies of the action of hypersound on nematics have shown that a reliable acoustic microscope can be constructed from nematics and standard optical components without the need for the scanner and focuser that are usually included. According to rough estimates for an equipment geometry like the one shown in Fig. 1 a, the acoustic-optical conversion sensitivity over the hypersonic frequency range is given by

$$J^{100\%} = \frac{4\rho c^2 k_1 k_2 K_3}{(\alpha_2 d \delta^2)} \left[\frac{105 \lambda_0}{\Delta n d} \right]^{1/2} \quad (7)$$

here k_1 and k_2 are components of the elastic wavevector $k = k_1 + i k_2$. For a 10 µm nematic layer at 10^9 Hz the sensitivity is estimated to be $J^{100\%} = 1$ mW cm^{-2}.

11.5 References

[1] O. Lehman, *Das Kristallisationsmikroskop und die damit gemachten Entdeckungen insbesondere die flüssige Kristalle*, Vieweg Verlag, Braunschweig **1910**.
[2] V. V. Zolina, *Trudy Lomonosovskogo Inst., AN SSSR*, **1936**, 11.
[3] G. E. Zvereva, A. P. Kapustin, *Appl. Ultraacoust. Study Matter* **1961**, *15*, 69 [in Russian].
[4] J. L. Fergason, *Sci. Am.* **1964**, *211*, 77.
[5] A. P. Kapustin, L. M. Dmitriev, *Kristallografiya* **1962**, *7*, 332.
[6] A. P. Kapustin, *Kristallografiya*, **1969**, *14*, 943; *Isv. Vysshikh Uchebnykh Zavedenii, Fiz.* **1967**, *11*, 55 [in Russian].
[7] L. W. Kessler, S. P. Sawyer, *Appl. Phys. Lett.* **1970**, *17*, 440.
[8] A. P. Kapustin, Z. Kh. Kuvatov, A. I. Trofimov, *Izv. Vysshikh Uchebnykh Zavedenii, Fiz.* **1971**, *4*, 150; *Uchenye Zapiski Ivanovskogo Pedagogicheskogo Inst.* **1972**, *96*, 64 [in Russian].
[9] L. E. Davis, J. Chambers, *Electron. Lett.* **1971**, *7*, 287.
[10] H. Mailler, L. L. Likins, T. R. Teilor, J. L. Fergason, *Appl. Phys. Lett.* **1971**, *18*, 105.
[11] M. Bertolotti, S. Martellucci, F. Scudieri, D. Sette, *Appl. Phys. Lett.* **1972**, *21*, 74.
[12] A. P. Kapustin, O. A. Kapustina, *Krist. Techn.* **1973**, *8*, 237.
[13] O. A. Kapustina, *Sov. Phys. Acoust.* **1974**, *20*, 1.
[14] A. P. Kapustin, *Electrooptical and Acoustical Properties of Liquid Crystals*, Nauka, Moscow **1973**, p. 120 [in Russian].
[15] G. G. Natal, *J. Acoust. Soc. Am.* **1978**, *65*, 1265.
[16] A. P. Kapustin, *Experimental Studies of Liquid Crystals*, Nauka, Moscow **1978**, p. 130 [in Russian].

[17] O. A. Kapustina, *Acoustooptical Phenomena in Liquid Crystals* (Ed. M. M. Labes), Gordon & Breach, London **1984**, pp. 1–164.
[18] A. P. Kapustin, O. A. Kapustina, *Acoustics of Liquid Crystals,* Nauka, Moscow **1986**, pp. 1–247 [in Russian].
[19] J. N. Perbet, M. Hareng, S. Le Berre and B. Mourrey, *Rev. Tech. Thomson CSF* **1979**, *11*, 837.
[20] M. Born, E. Wolf, *Principles of Optics,* Pergamon Press, New York **1970**.
[21] Ch. Sripaipan, Ch. F. Hayes, G. T. Fang, *Phys. Rev. A* **1977**, *15*, 1297.
[22] O. A. Kapustina, V. N. Lupanov, *Sov. Phys. Acoust.* **1977**, *23*, 218.
[23] R. Bartolino, M. Bertolotti, F. Scudieri, D. Sette, A. Sliwinski, *J. Appl. Phys.* **1975**, *46*, 1928.
[24] H. Bruchmuller, *Acoustica* **1978**, *40*, 155.
[25] T. Hatakeyama, Y. Kagawa, *J. Acoustic. Soc. Jpn.* **1976**, *32*, 92; *J. Sound Vibr.* **1976**, *46*, 551.
[26] S. Nagai, K. Iizuka, *Jpn. J. Appl. Phys.* **1974**, *13*, 189; **1978**, *17*, 723; *Mol. Cryst.* **1978**, *45*, 83–101.
[27] S. Nagai, A. Peters, S. Candau, *Rev. Phys. Appl.* **1977**, *12*, 21.
[28] S. Candau, A. Ferre, A. Peters, G. Waton, P. Pieranski, *Mol. Cryst. Liq. Cryst.* **1980**, *61*, 7.
[29] I. Lebran, S. Candau, S. Letcher, *J. Acoust. Soc. Am.* **1978**, *63*, 55; *J. Phys. (Paris)* **1979**, *40* (Suppl.), 298.
[30] I. N. Perbet, M. Hareng, S. Le Berre, *Rev. Phys. Appl.* **1979**, *14*, 569.
[31] Ch. F. Hayes, *Mol. Cryst. Liq. Cryst.* **1980**, *59*, 317.
[32] Y. Kagawa, T. Hatakeyama, Y. Tanako, *J. Sound Vibr.* **1974**, *36*, 407.
[33] W. Helfrich, *Phys. Rev. Lett.* **1972**, *29*, 1583.
[34] I. A. Chaban, *Sov. Phys., Acoust.* **1979**, *25*, 67.
[35] O. A. Kapustina, A. A. Talashev, *Sov. Phys., Acoust.* **1974**, *19*, 397.
[36] O. A. Kapustina, Yu. G. Statnikov, *Sov. Phys., JETP* **1973**, *37*, 117; *Sov. Phys., Acoust.* **1974**, *20*, 154.
[37] E. N. Kozhevnikov, I. A. Chaban, *Sov. Phys., Acoust.* **1976**, *21*, 550.
[38] K. Miyano, Y. R. Shen, *Appl. Phys. Lett.* **1976**, *28*, 473, 699; *Phys. Rev. A* **1977**, *15*, 2471.
[39] S. Sato, H. Uedo, *Jpn. J. Appl. Phys.* **1981**, *20*, 511.
[40] D. I. Anikeev, Yu. V. Bocharov, O. A. Kapustina, A. D. Vuzhva, *Sov. Phys., Acoust.* **1989**, *35*, 563.
[41] J. L. Dion, A. D. Jacob, *Appl. Phys. Lett.* **1977**, *31*, 490; *IEEE Trans. Ultrason., Ferroelec. Freq. Contr.* **1987**, *34*, 550.
[42] J. L. Dion, *CR Hebd. Sean. Acad. Sci. B* **1977**, *284*, 219; **1978**, *286*, 383; *J. Appl. Phys.* **1979**, *50*, 2965.
[43] J. L. Dion, A. LeBlanc, A. D. Jacob, *Acoust. Imaging* **1982**, *10*, 151.

[44] E. N. Kozhevnikov, *Sov. Phys., Acoust.* **1981**, *27*, 297; *Sov. Phys. JETP* **1982**, *55*, 96.
[45] N. K. Gus'kov, E. N. Kozhevnikov, *Sov. Phys., Acoust.* **1983**, *29*, 21.
[46] E. I. Zhukovskaya, E. N. Kozhevnikov, V. M. Podol'skii, *Sov. Phys. JETP* **1982**, *56*, 113.
[47] L. K. Zarembo, V. A. Krasil'nikov, *Introduction to Non-linear Acoustics*, Nauka Moscow **1966** [in Russian].
[48] S. A. Pikin, *Structural Transformation in Liquid Crystals*, Nauka, Moscow **1981**, p. 336 [in Russian].
[49] F. Scoudieri, *Appl. Phys. Lett.* **1976**, *29*, 398; *Ann. Phys.* **1978**, 311.
[50] F. Scudieri, A. Ferrari, M. Bertolotti, D. Apostol, *Opt. Commun.* **1975**, *15*, 57.
[51] F. Scudieri, M. Bertolotti, S. Melone, G. Albertini, *J. Appl. Phys.* **1976**, *47*, 3781.
[52] Y. Kagawa, T. Hatakeyama, Y. Tanaka, *J. Sound. Vibr.* **1975**, *41*, 1; *J. Acoust. Soc. Jpn.* **1975**, *31*, 81.
[53] I. A. Chaban, *Sov. Phys. Acoust.* **1978**, *24*, 145; **1985**, *31*, 77.
[54] E. N. Kozhevnikov, *Sov. Phys., JETP* **1986**, *64*, 793.
[55] G. N. Belova, E. I. Remizova, *Sov. Phys., Acoust.* **1985**, *31*, 171.
[56] O. A. Kapustina, V. N. Lupanov, *Sov. Phys., JETP* **1976**, *44*, 1225; *Wiss. Z. Univ. Halle,* **1977**, *24*, 49.
[57] D. I. Anikeev, O. A. Kapustina, V. N. Lupanov, *Soc. Phys., JETP* **1991**, *73*, 109.
[58] E. N. Kozhevnikov, in *Proc. Sixth All-Union Conf.,* Chernigovski Pedagogicheski Inst., Chernigov, Vol. 1, **1988**, p. 121.
[59] I. N. Gurova, O. A. Kapustina, V. N. Lupanov, *Advances in Liquid Crystal Research and Applications* (Ed. L. Batal), Pergamon Press/Akademiai Kiado, Oxford/Budapest, **1980**.
[60] M. Bertolotti, F. Scudieri, E. Sturla, *J. Appl. Phys.* **1978**, *49*, 3922.
[61] O. A. Kapustina, *Sov. Phys., Acoust.* **1991**, *97*, 153.
[62] E. N. Kozhevnikov, *Sov. Phys., JETP* **1987**, *65*, 731.
[63] I. N. Gurova, O. A. Kapustina, *Liquid Crystals,* **1989**, *6*, 525; *Sov. Phys., Acoust.* **1989**, *35*, 262.
[64] O. A. Kapustina, V. N. Lupanov, G. S. Chilay, *Sov. Phys., Acoust.* **1978**, *24*, 76.
[65] M. Bertolotti, I. Sbrolli, F. Scudieri, *Advances in Liquid Crystal Research and Applications,* Pergamon Press/Akademiai Kiado, Oxford/Budapest **1988**, pp. 433–439; *J. Appl. Phys.* **1982**, *53*, 4750.
[66] W. Haas, J. Adams, *Appl. Phys. Lett.* **1974**, *25*, 535.
[67] K. Hiroshima, H. Shimizu, *Jpn. J. Appl. Phys.* **1977**, *16*, 1889.
[68] T. Akahane, T. Tako, *Mol. Cryst. Liq. Cryst.* **1977**, *38*, 251.

[69] P. E. Cladis, M. Kleman, *Mol. Cryst. Liq. Cryst.* **1972**, *16*, 1.
[70] M. J. Press, A. S. Arrot, *J. Phys. (Paris)* **1976**, *37*, 387.
[71] F. Scudieri, A. Ferrari, E. Gunduz, *J. Phys. (Paris)* **1979**, *40* (Suppl.), 90.
[72] O. A. Kapustina, V. N. Lupanov, V. M. Shoshin, *Sov. Phys. Acoust.* **1980**, *26*, 406.
[73] P. G. De Gennes, *The Physics of Liquid Crystals,* Oxford University Press, London **1974**.
[74] R. S. Akopyan, B. Ya. Zel'dovich, N. V. Tabiryan, *Sov. Phys. Acoust.* **1988**, *34*, 337.
[75] D. I. Anikeev, Yu. V. Bocharov, A. D. Vuzhva, *Zh. Tekh. Fiz.* **1988**, *58*, 1554; *Liq. Cryst.* **1989**, *6*, 593.
[76] Yu. V. Bocharov, Dissertation, Moscow **1989** [in Russian].
[77] P. Greguss, *Acoustics* **1973**, *29*, 52.
[78] T. F. North, W. G. B. Britton, R. W. B. Stephens, in *Proc. Ultrasonic Int. Conf.*, London **1975**, pp. 120–123.
[79] H. Hakemi, *J. Appl. Phys.* **1982**, *53*, 6137.
[80] S. Ezhov, S. Pasechnic, V. Balandin, *Pis'ma Zh. Tekh. Fiz.* **1984**, *10*, 482.
[81] E. V. Gevorkyan, S. Ezhov, *Trudy All-Union Correspondence Inst. Mech. Eng.* **1984**, *36*, 46 [in Russian].
[82] G. N. Belova, *Sov. Acoust.* **1988**, *34*, 13.
[83] C. F. Hayes, in *Liquid Crystals and Ordered Fluids,* New York **1978**, pp. 287–296.
[84] A. Strigazzi, G. Barbero, *Mol. Cryst. Liq. Cryst.* **1983**, *103*, 193; *Mol. Cryst. Liq. Cryst. Lett.* **1982**, *82*, 5.
[85] Yu. V. Bocharov, A. D. Vuzhva, *Pis'ma Zh. Tekh. Fiz.* **1989**, *15*, 84; **1988**, *14*, 1460.
[86] K. Miyano, J. B. Ketterson, *Phys. Rev. A* **1975**, *12*, 615.
[87] P. K. Khabibullaev, E. V. Gevorkyan, A. S. Lagunov, *Rheology of Liquid Crystals,* Allerton Press, New York, **1994**, Chs 3.2, 6.3, 6.4.
[88] H. Kelker, R. Hatz, *Handbook of Liquid Crystals,* **1980**, Ch. 3.
[89] S. Candau, S. V. Letcher, in Advances in Liquid Crystal, ed. G. H. Brown, **1978**, vol. 3, 168–235.
[90] V. G. Kamensky, E. I. Katz, *Sov. Phys., JETP* **1982**, *56*, 591.
[91] E. L. Aero, Ph. D. Dissertation, Leningrad, **1982**, p. 408 [in Russian]; *Fiz. Tverdogo Tela* **1974**, *16*, 1245.
[92] I. Muscutariu, S. Bhattacharya, J. B. Ketterson, *Phys. Rev. Lett.* **1975**, *35*, 1584.
[93] J. D. Parsons, C. F. Hayes, *Sov. State. Commun.* **1974**, *15*, 429.
[94] E. I. Katz, V. V. Lebedev, *Sov. Phys., JETP Lett.* **1983**, *37*, 709.
[95] D. Mazenko, S. Ramaswary, J. Toner, *Phys. Rev. A* **1983**, *28*, 1618.
[96] E. I. Katz, *Sov. Phys., JETP* **1982**, *56*, 791.
[97] G. Grinstein, R. A. Pelcovitz, *Phys. Rev. A* **1982**, *26*, 915.
[98] V. P. Zharov, V. S. Letokhov, Laser *Opticoacoustic Spectroscopy,* Nauka, Moscow **1984** [in Russian].
[99] A. Rosencwaig, in *Optoacoustic Spectroscopy and Detection* (Ed. Yoh-Han Pao), Academic Press, New York **1977**, Ch. 8; *Photoacoustics and Photoacoustic Spectroscopy,* Wiley, New York, **1980**.
[100] M. Martinelli, U. Zamit, F. Scudieri, M. Martellucci, J. Quartieri, *Nuov. Cim.* **1987**, *90*, 557.
[101] L. M. Blinov, S. A. Davidian, N. N. Reshetov, D. V. Subachyus, S. V. Yablonsky, *Soc. Phys., JETP* **1990**, *70*, 902.
[102] F. Scudieri, T. Papa, D. Sette, M. Bertolotti, *Ann. Phys.* **1978**, 263.
[103] G. N. Belova, *Kristallografia,* **1988**, *33*, 1320.
[104] O. A. Kapustina, V. N. Reshetov, *Sov. Phys., JETP* **1984**, *70*, 122.
[105] O. A. Kapustina, *Mol. Cryst. Liq. Cryst.* **1990**, *179*, 173.
[106] O. A. Kapustina, E. N. Kozhevnikov, G. N. Yakovenko, *Sov. Phys., JETP* **1984**, *80*, 483.
[107] B. Hadimioglu, J. S. Foster, *J. Appl. Phys.* **1984**, *56*, 1976.

12 Nonlinear Optical Properties of Liquid Crystals

P. Palffy-Muhoray

12.1 Introduction

Optical nonlinearity is manifested by changes in the optical properties of a medium due to light propagating in that medium. A large number of distinct nonlinear optical processes have been observed; these are not only of fundamental scientific interest, but are also of great importance for both existing and potential applications in technology and industry.

The field of nonlinear optics originated in the early 1960s with experimental work on optical second harmonic generation [1], and theoretical work by Bloembergen and coworkers [2] on optical wave mixing. Since that time, the field has undergone explosive growth due to the development of high intensity laser sources with short pulse duration, and the discovery of materials with large optical nonlinearities.

Liquid crystals are orientationally ordered fluids; they are 'soft' materials in the sense that their physical properties can be readily altered by even modest fields. They are particularly well suited for optical applications requiring switching. In addition to their usefulness in display technology, they are also becoming increasingly important candidate materials for applications in nonlinear optics. A number of reviews of optical nonlinearities of liquid crystals have appeared [3–7]. This chapter provides a brief overview of the nonlinear optical response of liquid crystals and possible areas of application.

12.2 Interaction between Electromagnetic Radiation and Liquid Crystals

Liquid crystals are anisotropic fluids with optical properties similar to those of birefringent crystals. They are often inhomogeneous and, as a consequence, the dielectric tensor is a function of position. The inhomogeneity may be a property of the phase, such as the helical structure of chiral phases, or it may be the result of deformations.

12.2.1 Maxwell's Equations

Maxwell's equations in a dielectric material with no free charges are (in SI units):

$$\boldsymbol{\nabla} \times \boldsymbol{E} = -\frac{\partial \boldsymbol{B}}{\partial t} \qquad (1)$$

$$\nabla \times H = \frac{\partial}{\partial t}(\varepsilon_0 E + P) \quad (2)$$

$$\nabla \cdot (\varepsilon_0 E + P) = 0 \quad (3)$$

$$\nabla \cdot B = 0 \quad (4)$$

where E and D are the electric field and displacement, B and H are the magnetic induction and intensity, ε_0 is the permittivity of free space, P is the electric polarization, and higher order multipoles have been ignored. If the magnetic permeability μ, defined by $B = \mu H$, is a constant, then the wave equation

$$c^2 \frac{\mu_0}{\mu} \nabla \times \nabla \times E + \frac{\partial^2 E}{\partial t^2} = -\frac{1}{\varepsilon_0} \frac{\partial^2 P}{\partial t^2} \quad (5)$$

where μ_0 is the permeability of free space and c is the speed of light, follows. The polarization $P(r,t)$ at position r at time t is a time-varying source term which is a function of $E(r,t)$. The optical response of the material is thus governed by the constitutive equation relating P and E.

12.2.2 Nonlinear Susceptibility and Hyperpolarizability

The polarization P is, in general, a nonlinear function of E. Expanding P in terms of E gives, for the case where E can be written as a sum of plane waves [8] $E_i = E_{i0} \exp[i(k_i \cdot r - \omega t)]$:

$$P = \varepsilon_0 \{\chi^{(1)} E + \chi^{(2)} EE + \chi^{(3)} EEE + \ldots\} \quad (6)$$

where $\chi^{(n)}$ is the nth order susceptibility. The susceptibilities are tensors which depend on the wavevectors k_i and frequencies ω_i of the fields; explicitly [9], the contribution to the polarization from nth order effects is

$$P_\alpha^{(n)}(\omega, k) = \varepsilon_0 \chi_{\alpha\beta\gamma\ldots}$$
$$(-\omega; \omega_1, \omega_2 \ldots, -k; k_1, k_2, \ldots)$$
$$E_\beta(\omega_1, k_1) E_\gamma(\omega_2, k_2) \ldots \quad (7)$$

and momentum and energy conservation give $k = k_1 + k_2 + \ldots$ and $\omega = \omega_1 + \omega_2 + \ldots$. Equations (5) and (6) indicate how the nonlinear susceptibilities give rise to nonlinear phenomena, from harmonic generation to intensity-dependent effects. It is worth noting that, since the susceptibilities depend on frequency, in experiments involving pulsed lasers the observed polarization depends not only on the optical frequency but also on the pulse duration. In the first approximation [7], it is reduced from the steady state value by the factor of $\tau_l / \sqrt{\tau_l^2 + \tau^2}$, where τ_l is the laser pulse duration and τ is the response time of the material.

The nonlinear polarization can arise through the nonlinear polarizabilities of the constituent molecules, or through the collective response of molecules the individual polarizability of which may be linear.

The induced electric dipole moment p of individual atoms and molecules can be written in terms of the local electric field F and the polarizabilities, similar to Eq. (6):

$$p = \alpha F + \beta FF + \gamma FFF + \ldots \quad (8)$$

where α is the linear polarizability, and β and γ are the second- and third-order hyperpolarizability tensors, respectively.

Calculations of atomic and molecular hyperpolarizabilities usually proceed via time-dependent perturbation theory for the perturbed atomic states. Even for molecules of modest size, the calculation of the complete set of unperturbed wavefunctions, and exact calculation of the hyperpolarizabilities, is prohibitively difficult. Liquid crystals typically consist of organic molecules with aromatic cores, and there is considerable experimental [10] and theoretical [11, 12] evidence to indicate that the dominant contribution to the polarizabilities originates from the delocalized π-electrons in conjugated regions of these molecules. Even considering only π-electrons the calculations rapidly become computationally intensive; for this reason, hyperpolarizabilities of liquid crystal cores are usually calculated within the framework of the free-electron Hückel model [11], or using the Pariser–Parr–Pople model [13], which includes electron–electron interactions. Other approaches include self-consistent field theory [14], which includes electron–electron interactions and singly and doubly excited configurations; and simulations of molecules with known nonlinear constituents [15].

The susceptibilities are related to the polarizabilities via the macroscopic polarization $P = \langle \rho p \rangle$, where ρ is the number density and $\langle \rangle$ denotes the ensemble average. In general, susceptibilities are not simply proportional to the orientationally averaged hyperpolarizability of the same order. Both the orientation and density of molecules depend on the field, and local field corrections contain nonlinear contributions. Consequently, high order hyperpolarizabilities contribute to lower order susceptibilities, and low order hyperpolarizabilities contribute to higher order susceptibilities [16]. The large changes in the orientation of liquid crystals in electric fields is one example of anisotropic linear polarizability giving rise to third-order susceptibility. The relation between second-order macroscopic susceptibilities and molecular hyperpolarizabilities is discussed by Singer et al. [17].

12.3 Nonlinearities Originating in Director Reorientation

Liquid crystals possess strongly anisotropic linear susceptibilities originating in the orientational order of non-spherical molecules. Due to this anisotropy, liquid crystals experience body torques in the presence of applied fields, which may give rise to director reorientation and a change in the dielectric tensor. This mechanism is responsible for extremely large nonlinear susceptibilities in liquid crystals. For reasons of symmetry, the dielectric tensor of nonpolar nematics is a function of the square of the electric field; hence reorientational effects contribute to the intensity-dependent susceptibility $\chi^{(3)}(-\omega; \omega, -\omega, \omega)$.

12.3.1 D.C. Kerr Effect

The most widely exploited effect in display applications is the influence of low frequency electric fields on the birefringence. Here, due to a distortion of the director, elements of the dielectric tensor change as a function of the applied field. From Eq. (6), the contribution of the third-order susceptibility $\chi^{(3)}(-\omega; 0, 0, \omega)$ to the dielectric tensor is given by $\varepsilon = \varepsilon_0 \{ I + \chi^{(1)} + \chi^{(3)} E_{dc} E_{dc} \}$. Although the response is non-local, it is possible to obtain a crude estimate of the average susceptibility; $\chi^{(3)} \simeq \Delta\varepsilon/(\varepsilon_0 E_{dc}^2)$, where $\Delta\varepsilon$ is the change in ε due to the field. If the threshold voltage for the Freedericksz transition [18] V_{th} is applied to a cell of thickness d, then

$$\chi^{(3)}(-\omega; 0, 0, \omega) \simeq \frac{d^2}{V_{th}^2}(n_e^2 - n_o^2) \qquad (9)$$

where n_o and n_e are the ordinary and extraordinary refractive indices, and, for 5CB in a 20 μm thick cell, $\chi^{(3)} \simeq 4 \times 10^{-10}$ (SI). The sample-size dependence of $\chi^{(3)}$ is an indication of the spatial nonlocality of the response. The response time for this process is just the turn-on time of the cell [18].

12.4 Optical Field-Induced Reorientation

Optical field-induced director reorientation is responsible for the largest nonlinear optical susceptibility observed in liquid crystals, the largest in any known material. Although the process is slow, the nonlinearity is about 10^9 times greater than that of CS_2. Because of its magnitude, the orientational nonlinearity of liquid crystals has been termed 'giant optical nonlinearity' (GON).

The prediction [19] that a low power optical field can induce appreciable director reorientation just above the dc field induced Freedericksz transition has been verified experimentally [20, 21] concurrently with experimental and theoretical work on optical reorientation [22–24]. Since then, it has become one of the most intensively studied nonlinear optical effects in liquid crystals [3]. The phenomenon originates from the tendency of the director to align parallel to the electric field of light due to the anisotropic molecular polarizability. The free energy density arising from the interaction of a plane electromagnetic wave and the liquid crystals is

$$F_{opt} = -\frac{1}{2} E \cdot D - \frac{1}{2} B \cdot H$$
$$= -\frac{(E \times H) \cdot k}{\omega} = -\frac{I n_p}{c} \quad (10)$$

where k is the wavevector and c/n_p is the phase velocity. For materials with $n_e > n_o$, this is minimized if the polarization is along the director, since then $n_p = n_e$ is a maximum.

Experimental results and theoretical curves are shown in Fig. 1 [24]. A 250 μm thick cell containing the liquid crystal 5CB with homeotropic alignment at the cell walls was irradiated with light from an Ar^+ laser propagating at an angle α to the cell normal. For normal incidence ($\alpha = 0$), the optical-field-induced torque causes the director field to distort above a threshold intensity I_{th}.

The results may be simply interpreted by noting that light-induced torques are opposed by elastic torques, and the free energy density has the form

$$F \simeq \frac{1}{2} K \left(\frac{\partial \theta}{\partial z} \right)^2 - \frac{I n_o}{2c} \left(\frac{n_e^2 - n_o^2}{n_e^2} \right) \sin^2 \theta \quad (11)$$

where K is the bend elastic constant and θ is the angle the director makes with

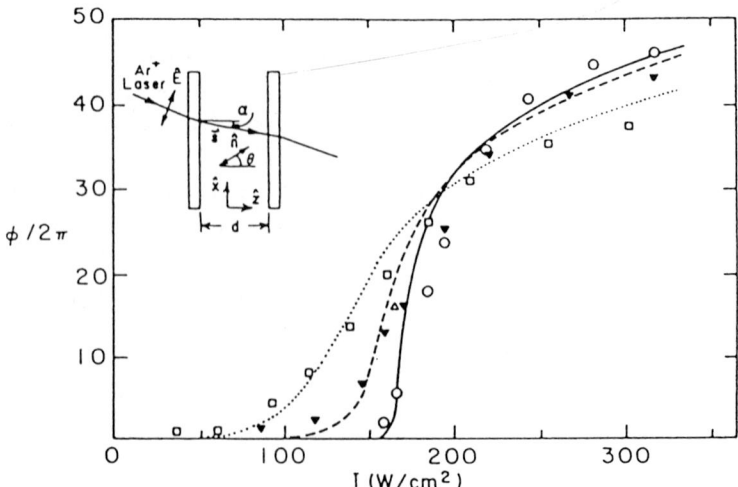

Figure 1. Experimentally observed laser-induced phase shift in 5CB.

the normal to the cell walls. If $\theta=0$ at the walls, minimizing F gives the threshold $I_{\text{th}} = \dfrac{\pi^2 K c n_e^2}{d^2 n_o (n_e^2 - n_o^2)}$. It is interesting to note that there exists a material constant $P_0 = \dfrac{Kc}{n_e^2 - n_o^2}$ with the units of power, which sets the scale for optical-field-induced effects. For 5CB, $P_0 \simeq 3$ mW. The nonlinear response is again nonlocal; the susceptibility may be estimated as in the case of the dc Kerr effect;

$$\chi^{(3)}(-\omega; \omega, -\omega, \omega) \simeq \frac{\varepsilon_0 c (n_e^2 - n_o^2)}{I_{\text{th}} n_e^2} \quad (12)$$

$$= \frac{d^2}{\pi^2 K} \frac{\varepsilon_0 n_o (n_e^2 - n_o^2)^2}{n_e^4}$$

and for 5CB in a 20 μm cell, $\chi^{(3)} \simeq 10^{-11}$ (SI). The response time again is the turn-on time of the cell:

$$\tau_{\text{on}} = \frac{\gamma d^2 I_{\text{th}}}{\pi^2 K (I - I_{\text{th}})} \quad (13)$$

$\gamma = 7 \times 10^{-2}$ Pa s and, for a 20 μm cell, for $I = 2 I_{\text{th}}$, $\tau_{\text{on}} \approx 300$ ms. For intense fields, fast response is possible. Director reorientation in 5CB due to 6 ns pulses at 1.06 m from a Q-switched Nd:YAG laser was observed [25] with response times of the order of 10–100 μs.

Anomalously low threshold intensities have been observed in nematics doped with dyes [27]. A theoretical model has been proposed [29], with the key feature that the interaction between dye and liquid crystal molecules is altered if the dye is optically excited. It has been found that the contribution of the dye to the nonlinear susceptibility may change sign as function of wavelength [29].

The optical-field-induced twist transition in planar samples has also been studied [30]. Here, because of the tendency of the polarization of the optical field to follow the principal axes of the dielectric tensor, the threshold intensity is found to be approximately four orders of magnitude greater than in the homeotropic case. In general, angular momentum from the radiation field is transferred to the liquid crystal [31], leading to precession of the director and a rich variety of nonlinear dynamic behavior [32, 33].

As a consequence of director reorientation, self-diffraction can occur in nematics [8, 34]. Laser-induced gratings can arise because two waves with ordinary and extraordinary polarizations propagate with different phase velocities. Wave mixing experiments have been performed in nematics, where two beams focused on the sample create an intensity grating. The phase grating which results from the nonlinear response of the material is then used to diffract light; the diffraction efficiency is proportional to the square of the third-order susceptibility, while the temporal profile of the diffracted beam gives information about the response time of the nonlinearity. Picosecond holographic gratings have been used to characterize the relaxation of orientation in liquid crystal films [36, 37]. Wave mixing has also been used to study optically generated hydrodynamic excitations in smectic A liquid crystals [38]. Cholesteric liquid crystals are important for laser applications. Optically induced reorientation has been studied [39]. Bistability due to light-induced pitch change of the cholesteric helix has been predicted [40] and retro-self-focusing caused by pitch dilation due to the optical field has been observed [41]. Optical-field-induced reorientation has also been studied in nonplanar geometries, such as PDLC films [42] and glass capillary arrays [43–45]. There is also evidence of photorefraction due to reorientation due to photoinduced charges [46].

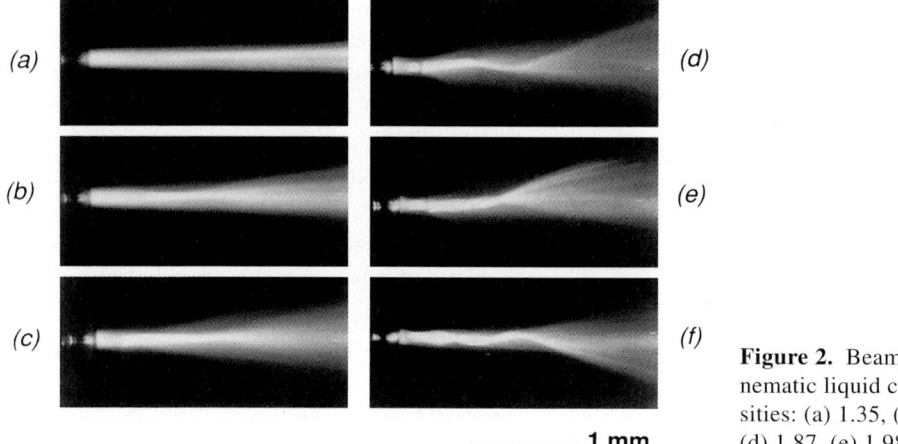

Figure 2. Beam undulation in the nematic liquid crystal E 209. Intensities: (a) 1.35, (b) 1.57, (c) 1.78, (d) 1.87, (e) 1.98, (f) 2.4 kW cm^{-2}.

Optical pattern formation in liquid crystals is of considerable interest; here, typically far-field patterns appear as result of optical field induced modulation of the director in directions transverse to the direction of propagation. Patterns have been observed in resonant cavities [47], systems with feedback mirrors [48], and in systems without feedback [49]. The propagation of a laser beam in a capillary filled with a nematic has been studied, showing self-focusing, undulation and filamentation [50]. Photographs, showing the beam profile as a function of intensity in a 1.5 mm capillary filled with the nematic mixture E209 are shown in Fig. 2.

Obtaining the full solution of Maxwell's equations together with the torque balance equations for the liquid crystal remains one of the most challenging problems in the nonlinear optics of liquid crystals [51].

12.5 Nonlinearities without Director Reorientation

Optical fields can affect the optical properties of liquid crystals without changing the orientation of the director. These nonlinearities originate from changes in the degree of orientational order, density, molecular conformation and electronic response.

12.5.1 Optical-Field-Induced Orientational Order

If a linearly polarized beam is propagating in the isotropic phase of a liquid crystal, the optical field will tend to align the molecules because of their anisotropic polarizability. Due to anisotropic interactions between the molecules, this alignment is enhanced, and the corresponding susceptibility diverges as the critical temperature associated with the phase transition is approached. The optical Kerr effect in the isotropic phase of nematics, as well as the effects of the optical field on the nematic–isotropic transition have been examined theoretically [52] by incorporating the orienting effects of the optical field into the Maier–Saupe theory [53]. Observations of optical-field-induced order in the isotropic phase of MBBA [54, 55] were shown to be in good agreement with the predictions of the Landau–de Gennes model.

Writing the free energy in terms of the scalar order parameter S gives

$$F = \frac{1}{2}a_0(T-T^*)S^2 + \ldots - \frac{In_p}{c} \quad (14)$$

Noting that, to lowest order, the refractive indices are linear in S, one obtains [7]

$$\chi^{(3)}(-\omega;\omega,-\omega,\omega) \simeq \frac{\varepsilon_0(n_e^2 - n_o^2)}{15k\rho(T-T^*)} \quad (15)$$

since $a_0 \simeq 5k\rho$ where ρ is the number density. For 5CB, this gives $\chi^{(3)} \simeq 8 \times 10^{-18}/(T-T^*)$ (in SI units). The response time is (see Eq. (13))

$$\tau = \frac{\gamma}{5k\rho(T-T^*)} \quad (16)$$

which has been verified by experiment [55, 56]. It is interesting to note that experimental measurements of $\chi^{(3)}$ with short laser pulses do not show this divergence [57] at $T=T^*$, since $\chi^{(3)}_{\text{meas.}} \simeq \chi^{(3)} \tau_l/\tau$ is independent of temperature [58].

In addition to affecting the orientation of the long axes of the liquid crystal molecules, evidence for reorientation of short axes by optical fields has been reported in smectic A materials [59].

12.5.2 Thermal Effects

Changes in temperature arising from the absorption of laser radiation can give rise to large changes in optical properties. The thermo-optic coefficients dn/dT in nematics are large; larger than in most other materials. The magnitude of the thermally induced nonlinearity can be calculated from the temperature rise due to absorption. In the adiabatic limit, $\Delta T \simeq (\alpha I \tau_l)/C$, where α is the linear absorption coefficient, τ_l is the laser pulse duration, and C is the specific heat per volume. For the nonadiabatic case, in the steady state limit, ΔT may be estimated by replacing τ_l by the thermal diffusion time $\tau = l^2/D$, where D is the thermal diffusivity and l the relevant diffusion length. The susceptibility due to laser heating is given, since $\chi^{(3)} \simeq \Delta \varepsilon n c/I$, by

$$\chi^{(3)}(-\omega;\omega,-\omega,\omega) \simeq \frac{2n^2 c \varepsilon_0 \alpha \tau}{C} \frac{dn}{dT} \quad (17)$$

in the steady state case. If $\tau_l \ll \tau$, the susceptibility is reduced by the factor τ_l/τ.

In general, both the linear absorption coefficient and the thermal diffusivity are anisotropic. Typical values for the thermal diffusivity are [60] $D_\parallel = 1.25 \times 10^{-7}$ m^2 s^{-1} and $D_\perp = 7.9 \times 10^{-8}$ m^2 s^{-1}. The absorption of light by nematic liquid crystals is typically very small in the visible region, $\alpha \simeq 10$ m^{-1} for the nematic mixture E7 and for MBBA in the isotropic phase [61]. From Eq. (17), for $l=10$ μm and $\tau \simeq 1$ ms, and estimating $C \simeq 2 \times 10^6$ J m^3 K^{-1} and $\alpha \simeq 6$ m^{-1}, for 20 ns pulses, $\chi^{(3)} \simeq 5 \times 10^{-18}$ (SI).

A wide variety of studies involving thermal effects in liquid crystals have been carried out. The dynamics of the generation of and diffraction from thermal gratings in nematic liquid crystals has been the subject of considerable interest, with good agreement between experiment and theory [62, 63].

12.5.3 Conformational Effects

In 1981, self-diffraction was observed in nematic MBBA [64] under circumstances where director reorientation could not occur. The measured nonlinear susceptibility was as large as the 'giant' nonlinearity associated with director reorientation; the mechanism proposed was photostimulated conformational changes of the liquid crystal molecules. Since the linear polarizability of the excited conformer differs from that of the molecule in the ground state, there is a direct contribution to the bulk susceptibility. The metastable conformers also act as impurities, and reduce the orientational order.

Photoinduced conformational changes in azo-dye-doped liquid crystals have been used for real-time holography [65]. The orientation of azo dyes by an optical field, the Weigert effect [66], has been used to induce orientation of liquid crystals both in the bulk and in alignment layers [67, 68].

12.5.4 Electronic Response

The fastest optical nonlinearity originates from the electronic response. To lowest order, $\chi^{(2)} = \langle \rho \beta \rangle / \varepsilon_0$ and $\chi^{(3)} = \langle \rho \gamma \rangle / \varepsilon_0$, where β and γ are molecular hyperpolarizabilities. As in the case of nonlinear susceptibilities, hyperpolarizabilities describe a variety of processes. Optical harmonic generation will be considered separately.

Contributions to the third-order susceptibility $\chi^{(3)}(-\omega; \omega, -\omega, \omega)$ from third-order hyperpolarizability have received considerable attention. Picosecond measurements have been carried out on liquid crystals in the isotropic phase [69, 70]; from the critical power for self-focusing, the nonlinear refractive index could be determined. More recently, the Z-scan method [71, 72] has been used to determined the susceptibility of a variety of liquid crystals [73]. It appears that in a variety of materials, the dominant contribution on the nanosecond timescale is excited state absorption, from a two-photon excited state [57, 74, 75]; thus this is a fifth-order process. This mechanism appears promising for optical power limiting applications [76].

Knowledge of the nonlinear susceptibilities in an aligned nematic phase allows, in principle, determination of the elements of the hyperpolarizability tensor of liquid crystal molecules if the pulse durations used in the measurements are short enough. Recent measurements of nonlinearities arising from optical-field-induced orientation processes in picosecond self-diffraction experiments [27] in the nematic and isotropic phases, suggest that pulses in the femtosecond range may be necessary to identify electronic contributions uniquely. A general review of organic materials for third-order nonlinear optics is given in Soileau et al. [77].

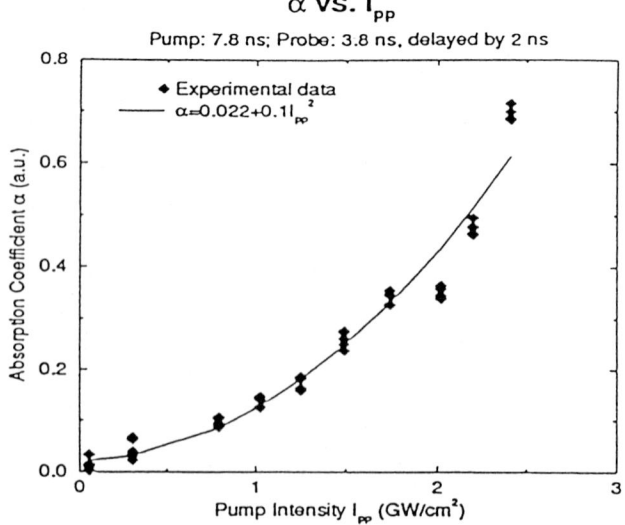

Figure 3. Intensity dependence of the absorption coefficient in the nematic liquid crystal 5CB.

12.6 Optical Harmonic Generation

Optical harmonic generation was one of the first areas of investigation in the field of nonlinear optics of liquid crystals; it has continued to receive a great deal of attention. The aim of the first studies was to observe second harmonic generation (SHG) from cholesterics; subsequent work has involved other phases, third harmonic generation (THG), and harmonic generation from surfaces as well as the bulk.

12.6.1 Bulk Second Harmonic Generation

The second-order susceptibility $\chi^{(2)}_{ijk}(-2\omega, \omega, \omega)$ is a third-rank tensor. For materials with inversion symmetry, in the electric dipole approximation $\chi^{(2)}_{ijk}=0$ for all i, j, k [78]. Nematic liquid crystals possess $D_{\infty h}$ symmetry; therefore, in the dipole approximation they are not be expected to give rise to SHG. However, SHG was observed in aligned samples of MBBA [79, 80]. It has been proposed that the origin of this is second harmonic polarization arising from the quadrupolar susceptibility and gradients of the optical field [81]. More recent work suggests that flexoelectric effects arising from thermal fluctuations [82] or, more generally, electric polarization due to spatial variation of the dielectric tensor [83] may be responsible.

Cholesteric liquid crystals are not centrosymmetric, and might be expected to give rise to SHG. They possess D_{∞} symmetry, which allows for one independent nonvanishing element of the second-order susceptibility. If the director is along the α direction, this is $\chi^{(2)}_{\gamma\alpha\beta}=-\chi^{(2)}_{\gamma\beta\alpha}$. For nonabsorbing materials Kleinman's rule holds, according to which elements of a nonlinear susceptibility tensor are invariant under any permutation of their indices [84]. As a consequence, $\chi^{(2)}$ vanishes, and SHG is disallowed. Contrary to these expectations, early experiments detected SHG in cholesteryl carbonate [85]. Subsequent work showed that the second harmonic signal originated from crystalline particles in the sample, and that the melting of these leads to disappearance of the second harmonic signal [86]. The blue phases of cholesterics may be studied by SGH [87].

If a dc electric field is present in addition to the optical field, the third-order nonlinearity can give rise to SHG. This results from contributions to the polarization of the form

$$P(2\omega) = \chi^{(3)}(-2\omega; \omega, \omega, 0)E_{\text{opt}}E_{\text{opt}}E_{\text{dc}} \tag{18}$$

Assuming that $\chi^{(3)}$ arises from the third-order hyperpolarizability, elements of $\chi^{(3)}$ give information about the degree of orientational order. Measurements of electric-field-induced second-harmonic generation (EFISH) have been carried out [88] on 5CB, showing the expected temperature dependence. In liquid crystals, E_{dc} may originate from order or flexoelectric effects, or other mechanisms resulting in electric polarization.

Ferroelectric chiral smectic C phases lack inversion symmetry, and are distinguished by spontaneous helicoidal electric polarization. The first experiment to measure SHG from ferroelectric liquid crystals was carried out [89] on unaligned samples under a dc electric field. Phase-matched SHG in ferroelectric liquid crystals has been carried out by using an electric field to unwind the helix [90]. Mechanical deformations in chiral smectic C elastomers have been shown to give rise to SHG [91]. A great deal of work has been carried out recently in studying SHG in ferroelectric liquid crystals [93], with particular emphasis on the relation between molecular structure and second-harmonic efficiency [94–96].

12.6.2 Surface Second Harmonic Generation

Second harmonic generation from surfaces has been shown to be a powerful probe of surface interactions [97], which can yield information about the structure and alignment of liquid crystal molecules on surfaces and at interfaces [98]. The role of the interface is similar to that of the dc electric field in electric-field-induced SHG, that is, to provide a symmetry-breaking field the direction of which is defined by the surface normal.

In general, liquid crystal molecules do not have the $D_{2\infty}$ symmetry of the uniaxial nematic phase. Since an interface acts as a field, its presence can provide polar order. Such surface polar ordering, confined to a single molecular layer, has been observed [99]. Surface SHG can also be used to probe the orientational distribution at the surface, and anchoring transitions [100].

12.6.3 Third Harmonic Generation

Third harmonic generation is characterized by the susceptibility $\chi^{(3)}(-3\omega; \omega, \omega, \omega)$ which is nonvanishing for all materials. Measurements of THG have been carried out on MBBA in the nematic and isotropic phase [101]. Considerable information about orientational order can be obtained from $\chi^{(3)}(-3\omega)$ [102].

In cholesterics, the situation regarding phase matching is more complex than in nematics. The normal modes are plane polarized waves in a coordinate system co-rotating with the director [87], and hence the wavevectors are functions of both the frequency and the cholesteric pitch. This pitch dependence can be used to achieve phase matching. In cholesterics, the momentum of the electromagnetic field is not necessarily conserved during harmonic generation; exchange of momentum between the field and the periodic structure of the cholesteric is possible in a 'coherent optical Umklapp process' [103].

12.7 Materials and Potential Applications

For a wide variety of proposed applications, material requirements are large and fast optical nonlinearities coupled with low losses. Low molecular weight liquid crystals have nonlinear susceptibilities which range from nine orders of magnitude greater than that of CS_2 on millisecond and slower time scales, to approximately the same as that of CS_2 on the picosecond scale. Novel materials, such as biaxial nematics and liquid crystal elastomers, offer the possibility of novel nonlinear behavior. Polymeric liquid crystals are particularly promising because of the possibility of a large nonlinear electronic response due to extended conjugated regions, as well as a fast reorientational response due to restricted reorientation of mesogenic units. They can also be used as host materials for nonlinear chromophores. Scattering losses, which can be considerable in low molecular weight materials, are reduced in polymer films.

The most widely utilized optical nonlinearity of liquid crystals is electric-field-induced director reorientation, which is used in display application. Although it is usually dealt with under the heading of 'electro-optics', it is a nonlinear optical process involving waves with frequencies 0 and ω.

Optically bistable devices are of considerable interest for information storage and processing applications. Bistable Fabry–Perot resonators [104] may be useful as dynamic optical memories as well as elements for information processing. Optically sensitive alignment layers [62], as well as a wide variety of laser addressable materials [105–107] promise the use of liquid crystals as materials for optical information storage.

Binary optical computing has been demonstrated [108] using the liquid crystal light valve made by Hughes [109].

Development of other spatial light modulators, using nematic and ferroelectric liquid crystals [110], and polymer dispersed liquid crystals [111] is in progress. Such liquid crystal light modulators can be used in binary optical computing, where high parallel processing density [112] compensates for the relatively slow device response time. A nonlinear liquid crystal film can act as an optical switching device [113] in signal processing applications.

The high optical damage threshold of liquid crystals makes them well suited for optical power limiting (OPL). This can occur for input energies as low as 0.3 µJ for 5 ns pulses at 532 nm [76]. Polymer dispersed liquid crystal films have been proposed for OPL applications both on the hybrid mode [114] and in the all optical mode using reorientational and thermal effects [115] to induce nonlinear scattering. Optic fibers with liquid crystalline cores have promise for both electro-optic and nonlinear optical applications; their potential for OPL has been investigated [26, 27, 34]. Liquid crystal can also be used as nonlinear couplers between optic fiber waveguides. Optical solitons, potentially useful in signal processing applications, have been theoretically considered in liquid crystals [116].

Ferroelectric liquid crystals are becoming increasingly promising for SHG devices and Pockels modulators [117]. The second harmonic intensity may be modulated [118]. A review of advanced liquid crystal polymers is given in Dubois [119]. Potential applications also include phase conjugation and real-time holography.

Nonlinear optical effects in liquid crystals encompass a remarkably wide range of phenomena, and offer broad potential for applications [120]. There can be little doubt that liquid crystals will play an increasingly important role in the nonlinear optical devices of the future.

12.8 References

[1] P. A. Franken, A. E. Hill, C. W. Peters, G. Weinreich, *Phys. Rev. Lett.* **1961**, *7*, 118.
[2] J. A. Armstrong, N. Bloembergen, J. Ducuing, P. S. Pershan, *Phys. Rev.* **1962**, *127*, 1918.
[3] N. V. Tabiryan, A. V. Sukhov, B. Ya. Zel'dovich, *Mol. Cryst. Liq. Cryst. Special Topics XIX* **1986**, *136*, 1.
[4] I. C. Khoo in *Progress in Optics*, Vol. 26 (Ed. E. Wolf), North Holland, New York **1988**.
[5] I. Janossy in *Perspectives in Condensed Matter Physics* (Ed. L. Miglio), Kluwer, Dordrecht **1990**.
[6] I. C. Khoo, *Liquid Crystals: Physical Properties and Nonlinear Optical Phenomena*, Wiley, New York **1995**.
[7] P. Palffy-Muhoray in *Liquid Crystals: Applications and Uses* (Ed. B. Bahadur), World Scientific, Singapore **1990**.
[8] Y. R. Shen, *The Principles of Nonlinear Optics*, Wiley, New York **1984**.
[9] D. C. Hannah, M. A. Yuratich, D. Cotter, *Nonlinear Optics of Free Atoms and Molecules*, Springer Series in Optical Sciences (Ed. D. L. MacAdam), Springer, New York **1979**.
[10] B. F. Levine, C. G. Bethea, *J. Chem. Phys.* **1975**, *63*, 2666.
[11] S. Risser, S. Klemm, D. W. Allender, M. A. Lee, *Mol. Cryst. Liq. Cryst.* **1987**, *150B*, 631.
[12] D. N. Beratan, J. N. Onuchic, J. W. Perry, *J. Chem. Phys.* **1987**, *91*, 2696.
[13] S. Risser, D. W. Allender, M. A. Lee, K. E. Schmidt, *Mol. Cryst. Liq. Cryst.* **1990**, *179*, 335.
[14] J. R. Heflin, K. Y. Wong, O. Zamani-Khamiri, A. F. Garito, *Mol. Cryst. Liq. Cryst.* **1988**, *160*, 37.

[15] R. Pachter, S. S. Patnaik, R. L. Crane, W. W. Adams, *SPIE Proc.* **1993**, *1916*, 2.
[16] G. R. Meredith in *Nonlinear Optics: Materials and Devices*, Springer Proceedings in Physics (Eds C. Flytzanis, J. L. Oudar), Springer, New York **1986**.
[17] K. D. Singer, M. G. Kuzyk, J. E. Sohn, *J. Opt. Soc. Am. B* **1987**, *4*, 968.
[18] L. M. Blinov, V. G. Chigrinov, *Electrooptic Effects in Liquid Crystal Material*, Springer, New York **1994**.
[19] R. M. Herman, R. J. Serinko, *Phys. Rev. A* **1979**, *19*, 1757.
[20] I. C. Khoo, S. L. Zhuang, *Appl. Phys. Lett.* **1980**, *37*, 3.
[21] I. C. Khoo, *Phys. Rev. A* **1981**, *23*, 2077.
[22] A. S. Zolotko, V. F. Kitaeva, N. Kroo, N. N. Sobolev, L. Chillag, *JETP Lett.* **1980**, *32*, 158.
[23] B. Ya. Zel'dovich, N. V. Tabiryan, *Sov. J. Quantum Electron.* **1990**, *10*, 440.
[24] S. D. Durbin, S. M. Arakelian, R. Y. Shen, *Opt. Lett.* **1981**, *6*, 411.
[25] P. S. Pershan, *Phys. Rev.* **1963**, *130*, 919.
[26] H. Hsiung, L. P. Shi, Y. R. Shen, *Phys. Rev. A* **1984**, *30*, 1453.
[27] I. Janossy, T. Kosa, *Opt. Lett.* **1992**, *17*, 1183.
[28] I. Janossy, *Phys. Rev. E* **1994**, *49*, 2957.
[29] T. Kosa, I. Janossy, *Opt. Lett.* **1995**, *20*, 1230.
[30] E. Santamato, G. Abbate, P. Maddalena, *Phys. Rev. A* **1987**, *36*, 2389.
[31] E. Santamato, B. Daino, R. Romagnoli, M. Settembre, Y. R. Shen, *Phys. Rev. Lett.* **1986**, *57*, 2423.
[32] E. Santamato, G. Abbate, P. Maddalena, L. Marrucci, *Phys. Rev. Lett.* **1990**, *64*, 1377.
[33] V. Carbone, G. Cipparrone, C. Versace, R. Bartolino, C. Umeton, *Mol. Cryst. Liq. Cryst.* **1994**, *251*, 167.
[34] S. M. Arakelian, Y. S. Chilingarian, *IEEE J. Quantum Electron.* **1986**, *QE-22*, 1276.
[35] G. Eyring, M. D. Fayer, *J. Chem. Phys.* **1984**, *81*, 4314.
[36] M. D. Fayer, *IEEE J. Quantum Electron.* **1986**, *QE-22*, 1437.
[37] H. J. Eichler, R. Macdonald, B. Trösken, *Mol. Cryst. Liq. Cryst.* **1993**, *231*, 1.
[38] G. F. Kventsel, B. I. Lembrikov, *Mol. Cryst. Liq. Cryst.* **1995**, *262*, 591.
[39] P. Maddalena, G. Arnone, G. Abbate, L. Marrucci, E. Santamato, *Mol. Cryst. Liq. Cryst.* **1995**, *261*, 113.
[40] H. G. Winful, *Phys. Rev. Lett.* **1982**, *49*, 1179.
[41] J.-C. Lee, S. D. Jacobs, A. Schmid, *Mol. Cryst. Liq. Cryst.* **1987**, *150b*, 617.
[42] P. Palffy-Muhoray, B. J. Frisken, J. Kelly, H. J. Yuan, *SPIE Proc.* **1989**, *1105*, 33.
[43] H. J. Yuan, H. Lin, P. Palffy-Muhoray, *Mol. Cryst. Liq. Cryst.* **1992**, *223*, 229.
[44] H. Lin, P. Palffy-Muhoray, *Opt. Lett.* **1994**, *19*, 436.
[45] I. C. Khoo, H. Li, *Appl. Phys. B* **1994**, *59*, 573.
[46] I. C. Khoo, H. Li, Y. Liang, *Opt. Lett.* **1994**, *19*, 1723.
[47] T. Vogeler, M. Kreuzer, T. Tschudi, N. V. Tabiryan, *Mol. Cryst. Liq. Cryst.* **1994**, *251*, 159.
[48] R. Macdonald, H. Danlewski, *Mol. Cryst. Liq. Cryst.* **1994**, *251*, 145.
[49] G. Hu. P. Palffy-Muhoray, N. V. Tabiryan in *Licht- und Teilchenoptik, Ann. Rep.*, Tech. Hochsch. Darmstadt **1992**.
[50] E. Braun, L. Faucheux, A. Libchaber, *Phys. Rev. A* **1993**, *48*, 611.
[51] E. Braun, L. Faucheux, A. Libchaber, D. W. McLaughlin, D. J. Muraki, M. J. Shelley, *Europhys. Lett.* **1993**, *23*, 4.
[52] J. Hanus, *Phys. Rev.* **1969**, *178*, 420.
[53] W. Maier, A. Saupe, *Z. Naturforsch., Teil A* **1958**, *13*, 564; **1959**, *14*, 882; **1960**, *15*, 287.
[54] J. Prost, J. R. Lalanne, *Phys. Rev. A* **1973**, *8*, 2090.
[55] G. K. Wong, Y. R. Shen, *Phys. Rev. Lett.* **1973**, *30*, 895.
[56] P. A. Madden, F. C. Saunders, A. M. Scott, *IEEE J. Quantum Electron.* **1986**, *QE-22*, 1287.
[57] P. Palffy-Muhoray, T. Wei, W. Zhao, *Mol. Cryst. Liq. Cryst.* **1994**, *251*, 19.
[58] C. W. Greef, *Mol. Cryst. Liq. Cryst.* **1994**, *238*, 179.
[59] J. R. Lalanne, J. Bouchert, C. Destrade, H. T. Nguyen, J. P. Marcerou, *Phys. Rev. Lett.* **1989**, *62*, 3046.
[60] W. Urbach, H. Hervet, F. Rondelez, *Mol. Cryst. Liq. Cryst.* **1978**, *46*, 209.
[61] S.-T. Wu, K.-C. Lim, *Appl. Opt.* **1987**, *26*, 1722.
[62] I. C. Khoo, R. Normandin, *IEEE J. Quantum Electron.* **1985**, *QE-21*, 329.
[63] I. C. Khoo, S. Shepard, *J. Appl. Phys.* **1983**, *54*, 5491.
[64] S. G. Odulov, Yu. A. Reznikov, M. S. Soskin, A. I. Khizhnyak, *Sov. Phys. JETP* **1982**, *55*, 854.
[65] A. G. Chen, D. J. Brady, *Opt. Lett.* **1992**, *17*, 441.
[66] F. Weigert, M. Nakashima, *Z. Phys. Chem.* **1929**, *34*, 258.
[67] W. M. Gibbons, P. J. Shannon, Shao-Tang Sun, B. J. Swetlin, *Nature* **1991**, *351*, 49.
[68] P. J. Shannon, W. M. Gibbons, S. T. Sun, *Nature* **1994**, *368*, 532.
[69] M. J. Soileau, S. Guha, W. E. Williams, E. W. Van Stryland, H. Vanherzeele, *Mol. Cryst. Liq. Cryst.* **1985**, *127*, 321.
[70] M. J. Soileau, E. W. Van Stryland, S. Guha, E. J. Sharp, G. L. Wood, J. L. W. Pohlmann, *Mol. Cryst. Liq. Cryst.* **1987**, *143*, 139.
[71] M. Sheik-bahae, A. A. Said, E. W. Van Stryland, *Opt. Lett.* **1989**, *14*, 955.
[72] M. Sheik-bahae, A. A. Said, T. H. Wei, D. J. Hagan, E. W. Van Stryland, *IEEE J. Quantum Electron.* **1990**, *26*, 760.
[73] L. Li, H. J. Yuan, G. Hu, P. Palffy-Muhoray, *Liq. Cryst.* **1994**, *16*, 703.

[74] K. J. McEwan, R. C. Hollins, *SPIE Proc.* **1994**, *2229*, 122.
[75] W. Zhao, J. H. Kim, P. Palffy-Muhoray, *SPIE Proc.* **1994**, *2229*, 131.
[76] A. Hochbaum, Y.-Y. Hsu, J. L. Fergason, *SPIE Proc.* **1994**, *2229*, 48.
[77] H. S. Nalwa, *Adv. Mater.* **1993**, *5*, 341.
[78] P. A. Franken, J. F. Ward, *Rev. Mod. Phys.* **1963**, *35*, 23.
[79] S. M. Arakelian, G. L. Grigorian, S. Ts. Nersiyan, M. A. Nshayan, Yu. S. Chilingaryan, *Sov. Phys. JETP Lett.* **1978**, *28*, 186.
[80] S. M. Arakelian, G. A. Lyakhov, Yu. Chilingaryan, *Sov. Phys. Usp.* **1980**, *23*, 245.
[81] Ou-Yang Zhong-can, Xie Yu-zhang, *Phys. Rev. A* **1985**, *32*, 1189.
[82] M. Copic, T. Ovsenik, *Europhys. Lett.* **1993**, *24*, 93.
[83] L. Li, H. J. Yuan, P. Palffy-Muhoray, *Mol. Cryst. Liq. Cryst.* **1991**, *198*, 239.
[84] P. G. Harper, G. S. Wherrett (Eds), *Nonlinear Optics*, Academic Press, London **1977**.
[85] I. Freund, P. M. Rentzepis, *Phys. Rev. Lett.* **1967**, *18*, 393.
[86] G. Durand, C. H. Lee, *Mol. Cryst.* **1968**, *5*, 171.
[87] V. A. Belyakov, *Diffraction Optics of Complex-Structured Periodic Media*, Springer, New York **1992**.
[88] S. K. Saha, G. K. Wong, *Appl. Phys. Lett.* **1978**, *34*, 423.
[89] A. N. Vtyurin, V. P. Yermakov, B. I. Ostrovsky, V. F. Shavanov, *Kristallografiya* **1981**, *26*, 546.
[90] M. I. Barnik, L. M. Blinov, N. M. Shtykov, *Sov. Phys. JETP* **1984**, *59*, 980.
[91] I. Benne, K. Semmler, H. Finkelmann, *Macromolecules* **1995**, *28*, 1854.
[92] A. Taguchi, Y. Ouchi, H. Takezoe, A. Fukuda, *Jpn. J. Appl. Phys.* **1989**, *28*, L997.
[93] D. M. Walba, M. B. Ros, N. A. Clark, R. Shao, K. M. Johnson, M. G. Robinson, J. Y. Liu, D. Doroski, *Mol. Cryst. Liq. Cryst.* **1991**, *189*, 51.
[94] D. M. Walba, D. A. Zummach, M. D. Wand, W. N. Thurmes, K. M. Moray, K. E. Arnett, *SPIE Proc.* **1993**, *1911*, 21.
[95] M. Loddoch, G. T. Marowsky, H. Schmidt, G. Heppke, *Appl. Phys. B* **1994**, *59*, 591.
[96] M. Ozaki, M. Sakuta, K. Yoshino, B. Helgee, M. Svensson, K. Skarp, *Appl. Phys. B* **1994**, *59*, 601.
[97] Y. R. Shen, *Nature* **1989**, *337*, 519.
[98] P. Guyot-Sionnest, H. Shiung, Y. R. Shen, *Phys. Rev. Lett.* **1986**, *57*, 2963.
[99] C. S. Mullin, P. Guyot-Sionnest, Y. R. Shen, *Phys. Rev. A* **1989**, *39*, 3745.
[100] B. Jerome, *Mol. Cryst. Liq. Cryst.* **1994**, *251*, 219.
[101] K. Y. Wong, A. F. Garito, *Phys. Rev. A* **1986**, *34*, 5051.
[102] H. Matoussi, G. C. Berry, *Mol. Cryst. Liq. Cryst.* **1992**, *223*, 41.
[103] J. W. Shelton, Y. R. Shen, *Phys. Rev. A* **1972**, *5*, 1867.
[104] M. Kreuzer, H. Gottschling, R. Neubecker, T. Tschudi, *Appl. Phys. B* **1994**, *59*, 581.
[105] G. Hu, T. Kosa, P. Palffy-Muhoray in *ECLE '95 Abstracts*, Ljubljana **1995**.
[106] R. S. Akopyan, N. V. Tabiryan, T. Schudi, *Phys. Rev. E* **1994**, *49*, 3143.
[107] M. Kreuzer, T. Tschudi, R. Eidenschink, *Mol. Cryst. Liq. Cryst.* **1992**, *223*, 219.
[108] S. A. Collins, *SPIE Proc.* **1989**, *1080*.
[109] W. P. Bleha, L. T. Lipton, E. Wiener-Avenar, J. Grienberg, P. G. Reif, D. Casasent, H. B. Brown, B. V. Markevitch, *Opt. Eng.* **1978**, *17*, 371.
[110] D. Armitage, J. I. Thackara, W. D. Eades, *Liq. Cryst.* **1989**, *5*, 1389.
[111] K. Takizawa, H. Kikuchi, H. Fujikake, K. Kodama, K. Kishi, *J. Appl. Phys.* **1994**, *75*, 3158.
[112] K. M. Johnson, M. A. Handschy, L. A. Pagano-Stauffer, *Opt. Eng.* **1987**, *26*, 385.
[113] I. C. Khoo, *Appl. Phys. Lett.* **1982**, *40*, 645.
[114] J. Y. Kim, P. Palffy-Muhoray, *J. Appl. Phys.* **1989**, *66*, 362.
[115] F. Simoni, G. Cipparrone, C. Umeton, G. Arabia, G. Chidichimo, *Appl. Phys. Lett.* **1989**, *54*, 896.
[116] Y. S. Yung, L. Lam in *Solitons and Chaos in Optical Systems* (Eds.: H. C. Morris, D. Heesernan), Plenum, New York **1990**.
[117] M. Schadt, K. Schmitt, *Appl. Phys. B* **1994**, *59*, 607.
[118] I. Drevensek, T. Renato, M. Copic, *Mol. Cryst. Liq. Cryst.* **1994**, *251*, 101.
[119] J. C. Dubois, *Polym. Adv. Technol.* **1995**, *6*, 10.
[120] N. V. Tabiryan, T. Vogeler, T. Tschudi, *SPIE Proc.* **1994**, *2175*, 191.

13 Diffusion in Liquid Crystals

F. Noack[†]

13.1 Introduction

Diffusion, that is molecular mass transport by, usually, thermally activated particle motions, is a property of liquid crystals which is relatively poorly understood both experimentally and theoretically, despite countless efforts. In general this process does not strongly affect the orientational order of the mesophase, and therefore it seems not to be of primary technical importance. Furthermore, the available experimental and theoretical results are often inconsistent because of great difficulties encountered in developing and applying adequate research procedures, and so most standard textbooks [1–3] do not treat the subject systematically. Nevertheless, diffusion constants can reveal directly details about the dynamics of liquid crystalline order and thus are involved in the understanding and tailoring of other material quantities, such as anisotropic viscosities and electrical conductivities, needed to optimize the performance of liquid crystal applications. This chapter covers some basic theoretical concepts and common experimental techniques used to study diffusion in liquid crystals, and summarizes some general results and references.

13.2 Theoretical Concepts

13.2.1 The Diffusion Tensor

The diffusion sensor or diffusivity **D** of a system of n identical or specially labeled molecules is defined by two phenomenological, related linear diffusion laws which, on the one hand, describe a particle current $J_r \equiv \partial n/\partial t$ within the sample in any selected direction r, perpendicular through an area A, by the concentration gradient $\partial c/\partial r$ of the particles in the direction of J_r (Fick's first law) and, on the other hand, express the time development of the concentration $\partial c/\partial t$ by the Laplace derivative $\Delta \equiv \partial^2 c/\partial r^2$ (Fick's second law), so [4, 5]

$$\left(\frac{\partial n}{\partial t}\right) = -\mathbf{D}\,A\left(\frac{\partial c}{\partial r}\right) \qquad (1a)$$

and

$$\left(\frac{\partial c}{\partial t}\right) = +\mathbf{D}\left(\frac{\partial^2 c}{\partial r^2}\right) \qquad (1b)$$

If the molecules under consideration are different from the host system, one must distinguish between *solute* or *binary* diffusion and *self*-diffusion. Due to the liquid crystalline order, described by the director field $n(r, t)$ and the order parameter S, diffusion in mesophases is usually anisotropic. In general, **D** will be a second-rank tensor, the symmetry of which is determined by the symmetry of the liquid crystal. In an (x, y, z) axis system with the first principal axis parallel to **n** and the other two principal axes perpendicular to **n**, the diffusion tensor can be diagonalized and thus involves, at most three independent components $(D_{xx} = D_{\perp 1}, D_{yy} = D_{\perp 2}, D_{zz} = D_{\parallel})$ [4], and for uniaxial symmetry, as for common nematic (N), smectic A (SmA), or smectic B (SmB) phases, one has $D_{\perp 1} = D_{\perp 2} = D_{\perp}$. Hence as a rule, the diffusion tensor of liquid crystals has the simpler form

$$\mathbf{D} = \begin{pmatrix} D_{\perp} & 0 & 0 \\ 0 & D_{\perp} & 0 \\ 0 & 0 & D_{\parallel} \end{pmatrix} \quad (2)$$

with two characteristic diffusion constants or coefficients of the order or 10^{-11} to 10^{-10} m^2 s^{-1}. The anisotropy ratio $\sigma = D_{\perp}/D_{\parallel}$ is a material parameter, which can be larger or smaller than 1; for nematic phases it is typically near ½. Above the transition to the isotropic liquid, where no macroscopic order persists, **D** discontinuously becomes a scalar D_{iso}. Often $\langle D \rangle \equiv (2 D_{\perp} + D_{\parallel})/3$ is defined as an average diffusion constant, which is expected to be similar to D_{iso}.

The most common, idealized solution of Eq. (1a) describes the concentration profiles $c(\mathbf{r}, t)$ obtained for an infinite medium by focalizing at the beginning ($t = 0$) some labeled particles at $\mathbf{r} = \mathbf{r}_0$, i.e. $c(\mathbf{r}, 0) = c_0 \delta(\mathbf{r} - \mathbf{r}_0)$. Using the symmetry of Eq. (2) and (x, y, z) coordinates aligned along the main tensor axes, one obtains a three-dimensional, ellipsoidal Gaussian concentration distribution [4, 5]:

$$c(\mathbf{r} - \mathbf{r}_0, t) = c_0 \left[(4\pi)^3 D_{\perp} D_{\perp} D_{\parallel} t^3 \right]^{-\frac{1}{2}} \quad (3)$$

$$\times \exp\left[-\frac{(x - x_0)^2}{(4 D_{\perp} t)} - \frac{(y - y_0)^2}{4 D_{\perp} t} - \frac{(z - z_0)^2}{(4 D_{\parallel} t)} \right]$$

which for convenience is sometimes interpreted (with $c_0 = 1$) as the conditional probability or propagator $P(\mathbf{r}_0 | \mathbf{r}, t)$ to find a molecule at time t in the volume element d$\mathbf{r}$ at $\mathbf{r}$, if it was initially at position $\mathbf{r}_0$. Equation (3) is directly or indirectly the basis for the experimental techniques used to study diffusivities. In particular, the mean square displacement at time t

$$\langle (\mathbf{r} - \mathbf{r}_0)^2 \rangle = (4 D_{\perp} + 2 D_{\parallel}) t \quad (4)$$

obtained by averaging $(\mathbf{r} - \mathbf{r}_0)^2$ of individual molecules with Eq. (3), provides a relatively easy and transparent means of estimating the diffusion constants. Furthermore, it should be noted that the propagator concept indicates the way to generalize Fick's laws by including additional terms [6], a behavior that has been found necessary in solid-state physics and is also discussed for liquid crystals.

13.2.2 Basic models

Assuming the validity of Fick's laws for liquid crystals, various theories [7–15] have been developed to describe the diffusion constants $D_{\parallel}$ and $D_{\perp}$, in particular their anisotropy ratio σ, in terms of the mesophase order and other adequate macroscopic and microscopic material parameters. However, in view of the unsatisfactory agreement with experimental data, so far this has only achieved limited success. By transforming the diffusion tensor from a local (cluster) to the laboratory director frame and by taking the orientational ensemble average, Blinc et al. [7] showed that the two tensor components of **D** for *thermotropic* N, S$_A$, and S$_C$ phases should be coupled to the order parameter S, the average diffusion constant $\langle D \rangle$, and the limiting values $D_{\parallel}^0$, $D_{\perp}^0$ of a perfectly ordered cluster ($S = 1$) in the form

$$D_\perp = \langle D \rangle (1 - S) + S D_\perp^0 \qquad (5a)$$

$$D_\parallel = \langle D \rangle (1 - S) + S D_\parallel^0 \qquad (5b)$$

This result was later confirmed and refined by Chu and Moroi [8, 9], who evaluated $\langle D \rangle$ explicitly by using a power series of the momentum autocorrelation function, and also suggested geometric estimations for the D^0 values in a perfectly ordered cluster region. With such extensions, Eq. (5) finally gives in the case of nematic order [8]

$$D_\perp = \langle D \rangle \left[1 - \frac{1-\gamma}{2\gamma+1} S \right] \qquad (6a)$$

$$D_\parallel = \langle D \rangle \left[1 + \frac{2(1-\gamma)}{2\gamma+1} S \right] \qquad (6b)$$

where in terms of the velocity $v(t)$, diameter d, and length L of rod-like molecules, one has

$$\langle D \rangle = \frac{1}{3} \int \langle v(t) v(0) \rangle \, dt \qquad (6c)$$

$$\gamma = D_\perp^0 / D_\parallel^0 \simeq \frac{\pi d}{4L} \qquad (6d)$$

Hence the predicted macroscopic anisotropy ratio σ is independent of $\langle D \rangle$, and for $S = 1$ is equal to $(\pi d)/(4L)$; for $S = 0.6$ with $d/L = 3$, Eq. (6) gives $\sigma = 0.5$. Relations equivalent to Eqs. (6a) and (6b) have been reported by Leadbetter et al. [10]. In smectic phases the final expression for γ becomes more complicated and lengthy, since a periodic potential barrier, hindering the diffusion process perpendicular to the smectic layers (i.e. $D_\parallel$) must be included [9–12]; this is often done by using an additional Boltzmann-factor for $D_\parallel$ in Eq. (6b), so that despite the fact that $d < L$, observed ratios σ and γ even considerably larger than unity can be explained by such a modification of the theoretical concept. Hess et al. [13] have more recently developed expressions rather different from Eqs. (5a, 5b) by an affine transformation of the isotropic diffusion law for hard spherical particals to an affine space with aligned, nematically ordered uniaxial ellipsoids (long axis a, short axis b), which leads to

$$D_\perp = D_0 \alpha \left[Q^{-2/3} + \frac{1}{3} Q^{-2/3} (Q^2 - 1)(1 - S) \right] \qquad (7a)$$

$$D_\parallel = D_0 \alpha \left[Q^{-4/3} - \frac{2}{3} Q^{-2/3} (Q^2 - 1)(1 - S) \right] \qquad (7b)$$

with

$$\alpha = \left[1 + \frac{2}{3}(Q^{-2} - 1)(1 - S) \right]^{-1/3}$$

$$\times \left[1 + \frac{1}{3}(Q^2 - 1)(1 - S) \right]^{-2/3} \qquad (7c)$$

$D_0 = D_{iso}$, and $Q = a/b$. Equations (7a) and (7b) imply that $\langle D \rangle = D_0$ for $S = 0$ and any value of Q; the main distinctions between Eqs. (6) and (7) arise from the dissimilar dependence on the geometrical parameters Q and γ.

A quite different model for nematic diffusivities, based on Oseen's hydrodynamic theory of isotropic liquids, was elaborated by Franklin [14]; it describes the diffusivity components in terms of the five Leslie viscosities α_1 to α_5, a scalar friction constant ξ, two geometric molecular quantities [14] μ and ϕ, and the order parameter S by the Franklin relations

$$D_\perp = kT \left[\frac{1}{\mu \xi} + \right. \qquad (8a)$$

$$\left. + \frac{5 + S}{12 \pi \mu^2 \phi (-2.75 \alpha_2 + 2.25 \alpha_3 + 6 \alpha_4 - 1.5 \alpha_5)} \right]$$

$$D_\parallel = kT \left[\frac{1}{\mu \xi} + \right. \qquad (8b)$$

$$\left. + \frac{2 - S}{12 \pi \mu^2 \phi (4 \alpha_1 + 1.5 \alpha_2 + 6.5 \alpha_3 + 6 \alpha_4 + 10 \alpha_5)} \right]$$

where k is Boltzmann's constant and T is the absolute temperature. Equation (8) obviously extends Einstein's diffusivity–viscosity law [4] of isotropic liquids.

13.2.3 Model Refinements

Because of the numerous, poorly known parameters these theories and reported modifications [6, 15] or extensions for *solute* molecules [6, 12, 16, 17] and *lyotropic* systems [18] are mainly qualified to calculate the anisotropy of the diffusion constants, whereas estimations of their absolute magnitudes and temperature dependences need material data that are usually not available. Therefore it is quite common to describe experimental results by simple or, with the free-volume concept, modified heuristic Arrhenius laws [19]:

$$D_i = D_i^0 \exp\left(\frac{-E_i}{kT}\right) \quad (9a)$$

or

$$D_i = D_i^0 \exp\left(\frac{-F_i}{T-T_g}\right)\exp\left(\frac{-E_i}{kT}\right) \quad (9b)$$

where i is either $\perp$ or $\parallel$; as usual D_i^0 denotes the pre-exponential factor, E_i is the activation energy, F_i is the free volume parameter, T is the absolute temperature, and T_g is the glass temperature of the thermally activated diffuse mass transport. It should be noted that either Eqs. (6), (7), or (8) exactly predicts this type of behavior, but over small mesophase ranges the deviations from Eqs. (9a) and (9b) are, as a rule, minor and hence difficult to detect experimentally. However, in view of the shortcomings of the model it is interesting to note that the more recent theoretical developments have shifted to a large degree to *molecular dynamics* (MD) and *Monte Carlo* (MC) simulations [13, 15, 20–22], where the geometry of the molecules (rods, disks, ellipsoids), and the shape of the interaction potentials (Lennard–Jones, Coulomb, Gay–Berne) are more easily controlled by means of powerful computers than in the model approaches.

13.3 Experimental Techniques

Measurements of the diffusion constants of liquid crystals involve difficulties not encountered for normal liquids or solids, as in liquid crystals there is both anisotropy and viscous flow. As a consequence, data obtained with different techniques often differ greatly, and so in recent years researchers have sometimes preferred to test the quality of the models using results from molecular dynamics simulations [13, 15]. The first investigation of solute diffusion in a nematic mesophase was reported by Svedberg [23] as early as 1917, and the first detailed review on thermotropic mesophases was presented in 1982 by Krüger [24]. A review on lyotropic systems by Lindblom and Orädd [25] followed in 1994. One distinguishes between *direct* methods, where mass transport is studied directly by observing the evolution of the diffusion profile according to Eq. (3), and, for more the practicable initial conditions of Eq. (1), *indirect* methods, where such profiles govern other, usually more accessible system quantities [4, 5, 24]. In order to illustrate the main problems, these methods are briefly described below.

13.3.1 Tracer Techniques

Tracer techniques directly measure mass transport by means of putting adequate labels on some molecules, which are initially positioned with well-defined geometry on the sample surface to allow the observation of the propagating diffusion profile. In liquid crystals this has been achieved by radioactive tracers (i.e. isotope labels) or optical tracers (i.e. color labels). Suitable radioactive tracers can be realized by replacing ^{1}H by ^{3}H or ^{12}C by ^{14}C. Optical tracers (dyes, impurities, solutes) make use of molecules of similar size as the host liquid crystal (e.g. *m*-nitrophenol in *p*-azoxyanisole) [23]. Both approaches involve problems, namely the synthesis of a suitable radiotracer and the availability of a dye which really measures self-diffusion and not solute diffusion. Therefore, few results based on radioactive labels have been reported [23, 26, 27], and optical studies [28–31] are often considered not to show self-diffusion correctly. If the radiotracer or dye concentration profiles are measured as a function of time, penetration depth, and orientation for specific tracer injections, the diffusion constants can be evaluated by using the pertinent solution of Fick's laws (e.g. Eq. (3)). Several refinements have been used to broaden the limits of such procedures with regard to the profile analysis, and to take into account isotope or mass effects [24].

13.3.2 Quasielastic Neutron Scattering

Quasielastic neutron scattering (QENS) is a rather indirect method with many limitations. It makes use of the small ('quasielastic') energy shift that neutrons experience in any scattering by a moving particle, say by the diffusive translations of protons on a molecule. Mathematically, the normalized scattered neutron intensity as a function of kinetic neutron energy E (or frequency $\omega \equiv 2\pi E/h$) is related to the time Fourier transform of the dynamic pair-distribution function $G(r, t)$ of the sample material [6, 32]. Hence in Fick's approximation the linewidth increase $\Delta\omega(q)$ of the outgoing scattered beam for a selectable momentum transfer $\Delta p \equiv h q/(2\pi)$ with wavevector q, compared with the ingoing monochromatic beam linewidth [32]

$$\Delta\omega(q) = 2[D_\parallel q_\parallel^2 + D_\perp q_\perp^2] \tag{10}$$

in principle allows one to determine the two diffusion constants. Provided the instrumental resolution is sufficient for both energy and momentum, the diffusional broadening by Eq. (10) can be separated reliably from effects due to the generally much faster molecular rotations [33], and the beam geometry does not hinder the necessary director alignment. Most early QENS measurements did not check these conditions critically, but later work gave better results [34, 35] which are almost consistent with more directly obtained data.

13.3.3 Magnetic Resonance

Magnetic resonance (NMR, ESR) with nuclear (NMR) or electronic spins (ESR) provide a broad spectrum of direct and indirect techniques for diffusion studies of pure liquid crystals and solute molecules, and thus provide the most successful means of investigating these materials. Indirect methods consider the strongly model-dependent effects of diffusion on spin-relaxation times such as T_1, $T_{1\rho}$, T_{1d}, T_{1q}, and T_2, or on lineshape parameters such as the line-width $\Delta\nu$ or the second moment $\equiv M_2$ of suitable spin signals [36, 37]. This implies similar difficulties as with QENS, since any molecular motion induces signal changes, and an unambiguous assignment to $D_\perp$ and $D_\parallel$ proves problematical. Therefore many early papers reported incorrect results, as a rule too small diffusivities. Most problems have now been overcome by special nuclear spin–echo methods, which make it possible to observe directly the magnetization decay due to diffusion by applying strong magnetic field gradients, either stationary or pulsed. This requires an analysis of Fick's diffusion term for the spin magnetization M_T transverse to the applied NMR Zeeman field B_0 according to the generalized Bloch equation [36, 38]:

$$\frac{\partial M_T}{\partial t} = -\frac{M_T}{T_2} + \mathbf{D}\frac{\partial^2 M_T}{\partial r^2} \quad (11a)$$

where T_2 is the transverse relaxation time of the spins. In order to get the components of $\mathbf{D}$, the field gradients ($G = \partial B_0/\partial r$ must be oriented parallel and perpendicular to the director n by suitable gradient coils and, to evaluate them reliably, the diffusion-damping term must dominate the relaxation damping. At present, it is still a big problem to realize such conditions satisfactorily [24, 25, 38]. Explicitly, the diffusion decay of a spin with a gyromagnetic ratio γ is detectable in a selected direction i if

$$D_i \gtrsim \frac{10}{\gamma^2 T_2^3 G_i^2} \quad (11b)$$

which means that, in view of the typically short T_2 times (<50 μs) of mesophases, very strong gradients G_i (>1 to 10 T cm^{-1}) are necessary. Such 'brute-force' methods have recently been tried on solids [39], but they have not yet been used for liquid crystals where they would perturb the director field in a rather uncontrollable way. The presently used alternatives make use of lengthening the effective T_2 by a more or less sophisticated averaging of the dipolar interactions; for example, by adding poorly ordered solute molecules [24], by exciting special solid-like echos [24], by multiple radiofrequency pulses which reduce the relaxation term [40], or by orienting the director under the magic angle (54.74°) with respect to the Zeeman field [24, 41–43]. Results obtained in this way are illustrated in the next section.

Nematic phases give rise to more difficulties than smectic phases [24]. For chiral nematic (cholesteric) systems, where $\mathbf{D}$ may depend on the pitch length p, field-gradient methods are particularly problematical because of the coupling between p and B_0 [1, 2]. Thus several groups [44, 45] have described alternative procedures based on deuteron or proton spin lineshape calculations without the need for field gradients, which allow one to measure the diffusivity along the helical axis (i.e. $D_\perp$). However, in addition to this geometric restriction, the technique is rather indirect.

13.4 Selected Results

Mainly by using optical tracer and NMR pulsed-field-gradient techniques, the diffusivities of thermotropic and lyotropic materials have been studied more or less extensively in most of the known familiar liquid crystalline mesophases [24, 25, 38], including: nematic, cholesteric, and smectic (A, B, C, C*, G, H) order; cubic, lamellar, and hexagonal phaes; rod-like, discotic, and polymer molecules; biological systems; and often also for the related isotropic liquids. A large part of the data obtained has been collected and commented on in the designated reviews. However, in view of the large deviations between different experimental methods, and the rather minor distinctions between alternative models, the value of such numerous works in the understanding of the underlying processes is generally highly unsatisfactory, since only results for isotropic phases [38] (which, due to the liquid-like long T_2 instrumental NMR requirements, can be achieved easily) are consistent or undisputed. Some general findings and problems are outlined in the final sections of this chapter.

13.4.1 The Experimental Dilemma

Figure 1 illustrates, using the results of selected studies on the familiar nematogens *p*-azoxyanisole (PAA) [23, 46–50] and methoxybenzylidenebutylaniline (MBBA) [50–54], that the diffusion constants and their temperature dependences obtained by independent research groups and techniques differ significantly, particularly for the nematic mesophase where the greatest problems occur. Even results obtained using the so-called 'direct' techniques disagree well outside the specified error limits, obviously due to methodical problems. This should be kept in mind when comparing such data with subtle details of theoretical models, and when discussing distinctions between unalike materials or phases.

13.4.2 Nematic Mesophase

The NMR pulsed-field-gradient measurements given in Fig. 1 [50] for PAA and MBBA, and similar results for countless other materials [24, 42, 43] such as *n*-alkyl- or *n*-alkoxycyanobiphenyls (*n*-CBs, *n*-OCBs [43, 55], Fig. 2) in the nematic phase always reveal, within the error limits, a simple Arrhenius-type temperature behavior of both $D_\parallel$ and $D_\perp$, with only slightly different activation energies (20–70 kJ mol^{-1}), and so one obtains nearly temperature independent anisotropy ratios of 0.5–0.7. The transition to the isotropic phase is discontinuous, but not in the way expected according to $\langle D \rangle$. Such findings are in remarkable disagreement with many results obtained by radioactive and optical tracer and NMR T_1 methods (e.g. with optical measurements of 5-CB [56])

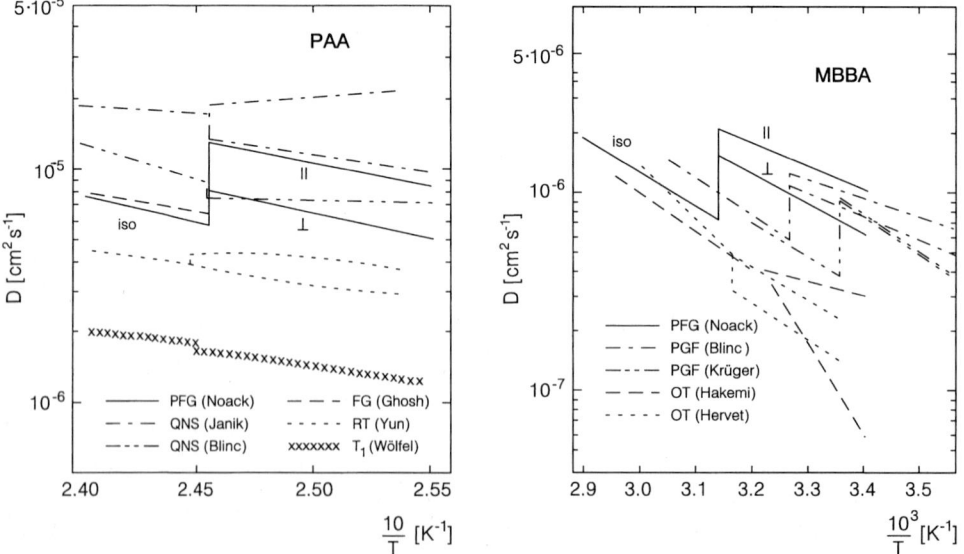

Figure 1. Comparison of selected diffusion measurements for PAA and MBBA from the literature to illustrate the scatter of experimental data. The methods are explained in the text; data are taken from the following studies: Yun and Fredrickson [26], Noack et al. [42, 50], Blinc et al. [46, 52], Janik et al. [47], Ghosh and Tettamanti [49], Wölfel [48], Krüger and Spiesecke [51], Hakemi and Labes [53], and Hervet et al. [54]. Note that even the isotropic-to-nematic transition temperatures of the samples deviate considerably.

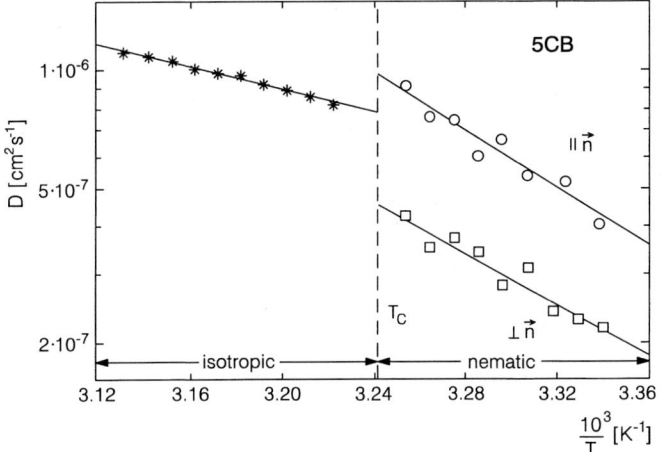

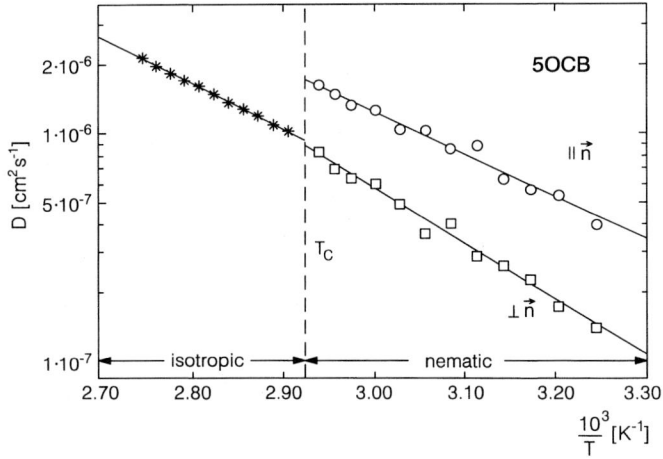

Figure 2. Arrhenius plots of pulsed-field-gradient diffusion constants [43, 55] of nematic and isotropic 5-CB and 5-OCB to illustrate the quality of the Arrhenius fits and the changes due to the slightly different side-groups of the two cyanobiphenyls. The activation energies for 5-CB are $E_{iso}=29.6$, $E_\perp=60.4$, and $E_\parallel=69.7$ kJ mol^{-1}, and for 5-OCB $E_{iso}=38.3$, $E_\perp=46.4$, and $E_\parallel=34.8$ kJ mol^{-1}. Neither the Arrhenius behavior nor the discontinuity at the phase transition can be described [43] satisfactorily by the available theoretical models [8, 13, 14]. Note that the major distinctions between the plots result from the different temperature ranges and clearing temperatures T_c.

and they also deviate from theoretical predictions [8, 13, 14] which would indicate a more visible influence of the non-Arrhenius variations upon the order parameter or the Leslie viscosities. In comparison with these strong inconsistencies, the distinctions between the various pulsed-field-gradient results, which must be ascribed to the different averaging procedures (by solute, multiple radiofrequency pulses, magic angle orientation, etc.), are relatively small. Some additional NMR lineshape studies indicate [45] a strong decrease in $D_\perp$ (i.e. of the transport along the helix axis) if the ordering of the phase is chiral.

13.4.3 Nematic Homologues

The literature contains very few systematic reports on the diffusion anisotropy of homologous molecules [35, 42, 43, 50] that allow a more critical analysis of the basic model parameters (diameter, length) than of chemically dissimilar systems. Figure 3 compares some results [42, 50] for members of the 4,4′-di-n-bialkoxyazoxybenzene (or PAA) series in both the nematic and the isotropic state, with the number $n'=n-1$ of –CH$_2$ side-groups varying from 0 to 6. It can be seen that, compared at constant temperature difference ΔT relative to the individu-

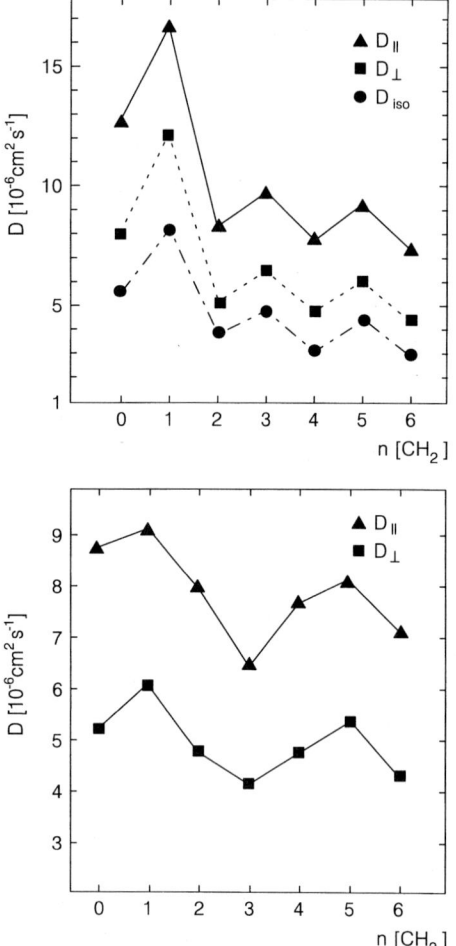

Figure 3. Diffusion constant alternation of seven homologous molecules in the *n*-bialkoxybenzene series ($n = 1$ to 7, $n' = 0$ to 6 –CH$_2$ groups) near the phase transition temperatures ($T = T_c \pm 0.5$°C) in the isotropic and nematic phases (top), and at constant temperature ($T = 119$°C) in the nematic phase [42, 50]. Note that the discontinuity at T_c ($D_\parallel > D_\perp > D_{iso}$) disagrees with theoretical predictions, and the alternation is opposite to that predicted by calculations using the Chu–Moroi model [8] (Eqs (6a) and (6b)).

al clearing points T_c, the diffusivity components show an odd–even alternation parallel to the clearing temperature, which changes to a $n' = 0$–3–6 alternation also exhibited by the nematic activation energies if compared at constant temperature T [50]. In the first case one has $D_\parallel > D_\perp$ and, surprisingly, $D_\perp > D_{iso}$, independent of the alternation. Such data, which were later confirmed by QNS studies [35], and analogous results for several *n*-OCB homologs [43] ($n = 5, 6, 7, 8$) cannot be described by the reported theories [8, 13, 14], although the cyanobiphenyls show the predicted order $D_\parallel > D_{iso} > D_\perp$ if a more sophisticated field-cycling magic angle averaging is applied [43].

13.4.4 Smectic Mesophases

The most striking distinction from nematic diffusivities is that NMR pulse-field-gradient measurements [24, 56] of smectic mesophases show Arrhenius-type behavior of $D_\perp$ and $D_\parallel$ with strongly different activation energies, namely $E_\perp < E_\parallel$. As a consequence, the two plots can cross [24], either within the smectic phase or outside by extrapolation, and with decreasing temperature the anisotropy ratio σ can change from <1, as for nematics, to $\gg 1$. Values up to 10 have been reported for the SmA phase of homologous alkanoylbenzylideneaminoazobenzenes (e.g. C$_{12}$-BAA [56]), whereas in the more familiar SmA phase of terephthalylidene-bis(4-*n*-butylaniline) (TBBA) one finds only $\sigma \approx 1$ due to the plot crossing [24]. Qualitatively, this has been explained using the Chu–Moroi concept [8, 9] assuming liquid-like diffusion within the layers ($D_\perp$) but solid-like jumps from layer to layer ($D_\parallel$). At phase transitions between different smectic order, the pre-exponential factors, activation barriers, and anisotropy ratio change discontinuously; no quantitative models are yet available.

13.4.5 Solute Diffusion

Despite some principal restrictions, it is generally easier to determine the diffusion coefficients of solute molecules [23, 24, 28–31] rather than true self-diffusion. However, the obvious distinctions have long been underestimated, and at present the two quantities cannot be related reliably. Solute diffusion constants do not follow the familiar relation of being reciprocally proportional to the square-root of the mass [5] (sometimes a strong concentration dependence is observed [57]) and theoretical estimations about the significance of the solute shape [16, 17] have not been examined sufficiently. Nevertheless, several findings are well established [24]. Due to the smaller order of the solute, the diffusion anisotropy σ in *nematic* systems turns out smaller than that of the host (i.e. near 1). On the other hand, for *smectic* phases, where the anisotropy depends primarily on the different activation potentials E_i for jumps within or perpendicular to the layers, σ becomes even larger than for self-diffusion. Values of σ up to 100 have been reported for TTE molecules (1,1,1-trichloro-2,2,2-trifluoroethane) in C_{12}-BAA [24, 56], which originate from the higher $E_\parallel/E_\perp$ ratio of the solute compared with the host. Usually, **D** values are smaller at phase transitions in the low-temperature phase [24], but the opposite jump is also observed [41].

13.4.6 Lyotropic Mesophases

Diffusivities of binary, ternary and multicomponent liquid crystalline mixtures, e.g. of soap (potassium laurate (PL), water [25, 58], and lipid (dipalmitoylphosphatidylcholine (DPPC) [25, 59] systems in lamellar, hexagonal, cubic, nematic and micellar mesophases [25, 60, 61] have been studied extensively by pulsed-field-gradient NMR [25] and optical techniques [62], partly because of their intimate relation to the structure and dynamical performance of biological membranes [18]. The main distinction from thermotropic phases is that for layered structures a noticeable diffusion occurs only *within* the layers (i.e. lateral, frequently written as D_L, but in our notation $D_\perp$), whereas it is negligibly small and difficult to detect *across* the layers [60–62] (transverse migration, for bilayers denoted by 'flip-flop'); so the mobility is essentially two dimensional, and the anisotropy ratio is so great that it is seldom specified explicit-

Table 1. Diffusion constants and anisotropy ratios of typical thermotropic and lyotropic liquid crystals considered in the text (Extensive data and references are collected in Krüger [24], Lindblom et al. [25], and Kärger et al. [38]).

Material	Temp. (°C)	Phase	$D_\perp \times 10^{-11}$ m^2 s^{-1}	σ	Refs.
PAA	119	N	55	0.62	42, 50
MBBA	28	N	2.9–7.1	0.57	42, 43, 50
5-CB	30	N	2.8	0.43	43, 55
5-OCB	35	N	1.4	0.35	43, 55
TBBA	175	SmA	25	~1.0	24
	135	SmC	8.5	2.5	24
C_{12}-BAA	120	SmA	5.0	4.8	24, 56
	115	SmB	0.25	~2.5	24, 56
PL (28% H$_2$O)	80	Lamellar	24		25, 58
DPPC (25% H$_2$O)	52	Lamellar (L$_\alpha$)	0.5–0.9	>20	25, 59, 60

PAA, 4,4'-di-*n*-bialkoxy-azoxybenzene; MBBA, 4-methoxybenzylidene-4'-*n*-butylaniline; 5-CB, 4'-*n*-pentyl-4-cyanobiphenyl; 5-OCB, 4'-*n*-pentoxy-4-cyanobiphenyl; TBBA, terephthalylidene-bis(4-*n*-butylaniline); C_{12}-BAA, C_{12}-benzylideneaminoazobenzene; PL, potassium laurate; DPPC, dipolmitoylphosphatidylcholine.

ly. In the easy direction diffusion constants are of similar order of magnitude as in thermotropics, namely typically about 10^{-11} to 10^{-10} m^2 s^{-1}.

13.4.7 Selected Diffusion Constants

Table 1 lists a few pulsed-field-gradient measurements of diffusion constants for some familiar mesophases considered in the text. Extensive data have been collected by Krüger [24] and by Lindblom and Orädd [25]; further more recent references have been summarized by Kärger et al. [38].

13.5 References

[1] G. Vertogen, W. H. de Jeu, *Thermotropic Liquid Crystals, Fundamentals*, Springer Verlag, Berlin **1988**, 201.
[2] P. de Gennes, J. Prost, *The Physics of Liquid Crystals*, Clarendon Press, Oxford **1993**, 250.
[3] L. M. Blinov, V. G. Chigrinov, *Electrooptic Effects in Liquid Crystal Materials*, Springer Verlag, Berlin **1994**, 89.
[4] W. Jost, K. Hauffe, *Diffusion, Methoden der Messung und Auswertung*, Dr. Dietrich Steinkopff Verlag, Darmstadt **1972**, 1.
[5] H. J. V. Tyrrel, K. R. Harris, *Diffusion in Liquids, a Theoretical and Experimental Study*, Butterworths, London **1984**, 56.
[6] A. Ferrarini, P. L. Nordio, G. J. Moro in *The Molecular Dynamics of Liquid Crystals* (Eds. G. R. Luckhurst, C. A. Veracini), Kluwer, Dordrecht **1994**, 41.
[7] R. Blinc, M. Burgar, M. Luzar, J. Piřš, L. Zupančič, S. Žumer, *Phys. Rev. Lett.* **1974**, *33*, 1192.
[8] K. S. Chu, D. S. Moroi, *J. Phys. Colloq.* **1975**, *36*, C1-99.
[9] K. S. Chu, D. S. Moroi, *Mol. Cryst. Liq. Cryst.* **1981**, *67*, 109.
[10] A. J. Leadbetter, F. P. Temme, A. Heidemann, W. S. Howells, *Chem. Phys. Lett.* **1975**, *34*, 363.
[11] F. Volino, A. J. Dianoux, A. Heidemann, *J. Phys. Lett.* **1979**, *40*, L-583.
[12] G. Moro, P. L. Nordio, U. Segre, *Mol. Cryst. Liq. Cryst.* **1984**, *114*, 113.

[13] S. Hess, D. Frenkel, M. Allen, *Mol. Phys.* **1991**, *74*, 765.
[14] W. Franklin, *Phys. Rev., Ser. A* **1975**, *11*, 2156.
[15] S. Tang, G. T. Evans, *J. Chem. Phys.* **1993**, *98*, 7281.
[16] M. E. Moseley, A. Loewenstein, *Mol. Cryst. Liq. Cryst.* **1982**, *90*, 117.
[17] A. Kozak, D. Sokolowska, J. K. Mosciki, *Proc. ECLC*, Bovec, Slovenia **1995**, 178.
[18] R. M Clegg, W. L. C. Vaz in *Progress in Protein–Lipid Interactions* (Eds. A. Watts, J. J. H. H. M. De Pont), Elsevier, Amsterdam **1985**, 173.
[19] M. H. Cohen, D. Turnbull, *J. Chem. Phys.* **1959**, *31*, 1164.
[20] M. P. Allen, D. J. Tildesley, *Computer Simulations of Liquids*, Clarendon Press, Oxford **1989**, 182.
[21] A. Alavi, D. Frenkel, *Phys. Rev.* **1992**, *45*, R5355.
[22] G. Krömer, D. Paschek, A. Geiger, *Ber. Bunsenges. Phys. Chem.* **1993**, *97*, 1188.
[23] T. Svedberg, *Kolloidzeitschrift* **1918**, *22*, 68.
[24] G. J. Krüger, *Phys. Rep.* **1982**, *82*, 229.
[25] G. Lindblom, G. Orädd, *Prog. NMR Spectrosc.* **1994**, *26*, 483.
[26] C. K. Yun, A. G. Fredrickson, *Mol. Cryst. Liq. Cryst.* **1970**, *12*, 73.
[27] A. V. Chadwick, M. Paykary, *Mol. Phys.* **1980**, *39*, 637.
[28] F. Rondelez, *Solid State Commun.* **1974**, *14*, 815.
[29] H. Hakemi, M. M. Labes, *J. Chem. Phys.* **1975**, *63*, 3708.
[30] M. Daoud, M. Gharbia, A. Gharbi, *J. Phys. II (France)* **1994**, *4*, 989.
[31] T. Moriyama, Y. Takanishi, K. Ishikawa, H. Takezoe, A. Fukuda, *Liq. Cryst.* **1995**, *18*, 639.
[32] P. A. Egelstaff, *An Introduction to the Liquid State*, Clarendon Press, Oxford, **1992**, 217.
[33] J. A. Janik, *Acta Phys. Pol.* **1978**, *A54*, 513.
[34] J. Töpler, B. Alefeld, T. Springer, *Mol. Cryst. Liq. Cryst.* **1973**, *26*, 297.
[35] M. Bée, A. J. Dianoux, J. A. Janik, J. M. Janik, R. Podsiadly, *Liq. Cryst.* **1991**, *10*, 199.
[36] A. Abragam, *The Principles of Nuclear Magnetism*, Clarendon Press, Oxford **1961**, 265.
[37] J. H. Freed, A. Nayeem, S. B. Rananavare in *The Molecular Dynamics of Liquid Crystals* (Eds. G. R. Luckhurst, C. A. Veracini), Kluwer, Dordrecht, **1994**, 71.
[38] J. Kärger, H. Pfeifer, W. Heink, *Adv. Magn. Reson.* **1988**, *12*, 1.
[39] I. Chang, F. Fujara, B. Geil, G. Hinze, H. Sillescu, A. Tölle, *J. Non-Cryst. Sol.* **1994**, *172*, 674.
[40] R. Blinc, J. Piřš, I. Zupančič, *Phys. Rev. Lett.* **1973**, *30*, 546.
[41] S. Miyajima, A. F. McDowell, R. M. Cotts, *Chem. Phys. Lett.* **1993**, *212*, 277.
[42] G. Rollmann, Thesis, University of Stuttgart **1984**.

[43] J. O. Mager, Thesis, University of Stuttgart **1993**.
[44] N. A. P. Vaz, G. Chidichimo, Z. Yaniv, J. V. Doane, *Phys. Rev.* **1982**, *A26*, 637.
[45] R. Stannarius, H. Schmiedel, *J. Magn. Reson.* **1989**, *81*, 339.
[46] R. Blinc, V. Dimic, *Phys. Lett.* **1970**, *31A*, 531.
[47] K. Otnes, R. Pynn, J. A. Janik, J. M. Janik, *Phys. Lett.* **1972**, *38A*, 335.
[48] W. Wölfel, Thesis, University of Stuttgart **1978**.
[49] S. Ghosh, T. Tettamanti, *Chem. Phys. Lett.* **1980**, *69*, 403.
[50] F. Noack, *Mol. Cryst. Liq. Cryst.* **1974**, *113*, 247.
[51] G. Krüger, H. Spiesecke, *Z. Naturforsch. Teil a* **1973**, *28*, 964.
[52] I. Zupančič, J. Pirš, M. Luzar, R. Blinc, *Solid State Commun.* **1974**, *15*, 227.
[53] H. Hakemi, M. M. Labes, *J. Chem. Phys.* **1974**, *61*, 4020.
[54] H. Hervet, W. Urbach, F. Rondelez, *J. Chem. Phys.* **1978**, *68*, 2725.
[55] F. Noack in *Proc. AMPERE Summer Institute on Magnetic Resonance* (Eds. R. Blinc, M. Vilfan), J. Stefan Institute, Ljubliana, Slovenia **1993**, 18.
[56] M. Hara, S. Ichikawa, H. Takezoe, A. Fukuda, *Jpn. J. Appl. Phys.* **1984**, *23*, 1420.
[56a] G. J. Krüger, H. Spiesecke, R. van Steenwinkel, F. Noack, *Mol. Cryst. Liq. Cryst.* **1977**, *40*, 103.
[57] H. Hakemi, *Mol. Cryst. Liq. Cryst.* **1983**, *95*, 309.
[58] R. T. Roberts, *Nature* **1973**, *242*, 348.
[59] A. L. Kuo, C. G. Wade, *Biochemistry* **1979**, *18*, 2300.
[60] P. Ukleja, J. W. Doane in *Ordering in Two Dimensions* (Ed. S. K. Sinha), Elsevier, Amsterdam **1980**, 427.
[61] G. Lahajnar, S. Žumer, M. Vilfan, R. Blinc, L. W. Reeves, *Mol. Cryst. Liq. Cryst.* **1984**, *113*, 8592.
[62] R. Homan, H. J. Pownall, *Biochim. Biophys. Acta* **1988**, *938*, 155.

Chapter VIII
Characterization Methods

1 Magnetic Resonance

Claudia Schmidt and Hans Wolfgang Spiess

1.1 Introduction

Magnetic resonance techniques are highly versatile in liquid crystal research. They can provide a wealth of information on the molecular level about both the solutes in liquid-crystalline phases and the mesogenic molecules or aggregates themselves, including thermotropic or lyotropic, monomeric or polymeric systems. The application of magnetic resonance spectroscopy to liquid crystals includes investigations of the chemical structure, mesomorphic polymorphism, phase transitions, molecular orientations and conformations, molecular and collective dynamics, self diffusion, spatial relationships between molecules, and viscoelastic properties. Nanostructures not observable by microscopy or any other techniques can be analyzed, providing information about phase separation on a nanometer scale. The arsenal of magnetic resonance provides tools to study mesomorphic compounds in any state, including isotropic liquids, mesophases of any type, and amorphous or crystalline solids.

The unique selectivity of magnetic resonance (in particular, when studying nuclei other than protons by NMR), its high versatility, and its potential for the further development of improved and new techniques make it an active, mature, yet expanding field of research. The number of publications per year on magnetic resonance of liquid crystals is still growing and reached a value well above 200 in 1996. This makes it impossible to give a full account of this field on a few pages. The purpose of this contribution is rather to convey the most basic concepts of magnetic resonance and to give a taste of its enormous capabilities by presenting a few examples, the selection of which is certainly a biased one. The references given are mainly of recent publications and reviews. The focus is on thermotropic rather than lyotropic, and on nuclear magnetic resonance (NMR) rather than electron spin resonance (ESR).

In-depth treatments of the topic are available in several books [1–6] and in a large number of review articles. The monograph by Dong [6], for example, focuses on aspects like the dynamics of nuclear spins, orientational order, molecular field theories, nuclear spin relaxation theory, director fluctuation and spin relaxation, rotational and translational dynamics, internal dynamics of flexible mesogens, and multiple-quantum and two-dimensional NMR, topics that will be touched upon very briefly here. Recent reviews on specific aspects of NMR on liquid crystals cover, for instance, molecules oriented in liquid crystals [7], the determination of order parameters [8], deuteron NMR spectroscopy of solids and liquid crystals [9], spin dynamics of exchanging quadrupolar nuclei in locally anisotropic systems [10], NMR of surfactant systems [11], microstructure and dynamics in lyotropic liquid crystals [12], studies of translational diffusion in lyotropic liquid crystals and lipid membranes [13], oriented phospolipid micelles for the study of membrane-associated molecules [14], local molecular dynamics in small molecules [15] and polymer liquid crystals [16], dynamic NMR in liquid crystal solvents [17], polymer liquid crystals [18], multidimensional NMR studies of the structure and dynamics of liquid crystal polymers [19], variable angle sample spinning [20], and noble gases dissolved in isotropic and anisotropic liquids [21]. For comprehensive annular listings of the literature on the NMR of liquid crystals, the series edited by Webb [22] may be consulted. ESR investigations on liquid crystals are discussed in several chapters of [5].

1.2 Basic Concepts of NMR (Nuclear Magnetic Resonance)

The fact that liquid-crystalline phases are intermediate between liquids and solids is reflected in the NMR of these materials, which combines aspects of both liquid and solid state NMR. While the instrumentation and techniques often show the features of liquid state NMR, the theoretical background of solid state NMR [23–26] is required to design experiments and to analyze spectra. The characteristic feature of this branch of NMR is the anisotropy of the interactions of spins with their surroundings. This anisotropy provides the basis for almost all NMR studies of liquid crystals.

1.2.1 Anisotropy of Spin Interactions

Spin interactions that can be exploited to obtain information on liquid crystals are chemical shifts, magnetic dipole–dipole interactions between spins, and, for nuclei with spins $I \geq 1$, quadrupole interactions between the electric quadrupole moment of the nucleus and the electric field gradient at the site of the nucleus. The anisotropy of all these spin interactions leads to an orientation dependence of the resonance frequency v (or the corresponding angular frequency $\omega = 2\pi v$), given by

$$v = v_L + \frac{1}{2} \Delta (3\cos^2\theta - 1 - \eta \sin^2\theta \cos 2\phi) \quad (1)$$

Here v_L is the Larmor frequency including isotropic chemical shifts, Δ describes the strength of the anisotropic coupling and η is the asymmetry parameter describing the deviation of the anisotropic coupling from axial symmetry ($0 \leq \eta \leq 1$). The angles θ and ϕ are the polar angles of the magnetic field B_0 in the principal axes system of the coupling tensor. Equation (1) describes the frequency position of a peak corresponding to the single transition of an isolated spin-1/2 system; in the case of dipolar couplings and quadrupole couplings of ^{2}H, each peak has a mirror image with respect to v_L, leading to a doublet of peaks at frequencies $v_\pm = v_L \pm \Delta/2 (3\cos^2\theta - 1 - \sin^2\theta \cos 2\phi)$.

The quadrupole coupling is a one-spin interaction, giving rise to simple spectra which are easy to analyze. Typical ^{2}H spectra are shown in Fig. 1. Dipolar couplings, in contrast, result from two-spin interactions. If more than two spins are coupled,

2 496 144, respectively [27]. Although the actual number of observed transitions is reduced by symmetry, either special spectroscopic techniques or isotopic substitution are required to tackle molecules containing more than a few spins.

The orientation dependence of NMR frequency shifts or splittings often provides information directly related to the geometry of the molecule, since the unique axis of the coupling coincides in many cases with a bond direction, for instance, the ^{13}C–^{1}H bond in the case of the heteronuclear dipole interaction between such a pair of $I=1/2$ nuclei or the C–^{2}H bond in the case of the quadrupole interaction of the $I=1$ nucleus ^{2}H. Thus the orientation dependence of the NMR signal is the basis for studies of molecular orientations and rotational motions.

In powder samples, for instance, polydomain liquid crystals, the individual signals from all orientations of the coupling tensors in the sample are superimposed to yield the powder line shape, as shown for the case of ^{2}H NMR in Fig. 1 (top and middle). If the sample is partially ordered, like a partially aligned liquid crystal polymer, the NMR line shape reflects the orientational distribution function of the coupling tensors [28, 29]. Figure 1 (bottom) depicts the special case of a completely aligned sample. Such a sample with macroscopically uniform alignment, for nematic phases usually induced by the magnetic field, gives rise to a simple doublet.

Not only the orientational distribution but also rotational motions affect the line shape, provided the reorientation rate is of the order of or larger than the static frequency shift or splitting. This phenomenon of motional narrowing is the subject of the following section.

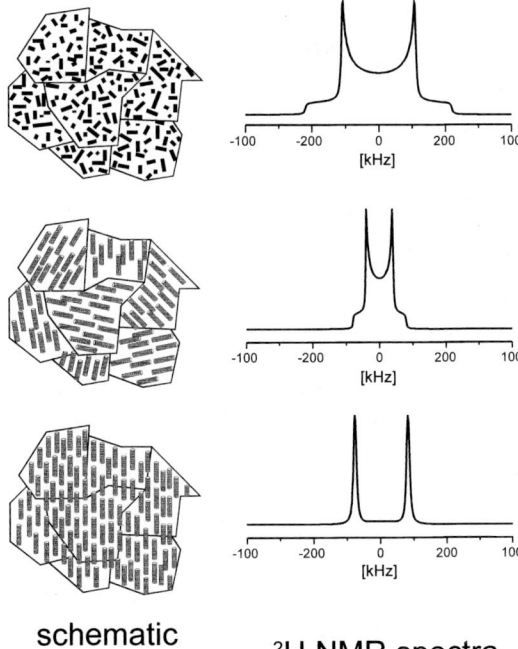

schematic phase structure ^{2}H NMR spectra

Figure 1. Deuteron NMR line shapes. *Top*: static powder pattern for a macroscopically disordered sample in the absence of molecular motions ($\eta=0$ case). *Middle*: motionally narrowed powder pattern for a macroscopically disordered mesophase with axially symmetric molecular motions ($\langle\eta\rangle=0$). The ratio of the peak splittings in the top and middle spectra defines the microscopic order parameter. *Bottom*: doublet spectrum of a macroscopically aligned mesophase. If the microscopic order parameter is known, the angle between the director and the magnetic field can be obtained from the splitting.

forming a network of n coupled spins, the spectrum rapidly becomes too complex to be analyzed, as the upper limit to the number of the normally observed one-quantum transitions, that is, transitions between energy levels differing in their total magnetic quantum number by $\Delta m=1$, increases as $\binom{2n}{n+1}$ ($I=1/2$ case). For $n=4$, $n=8$, and $n=12$, this number is 56, 11440, and

1.2.2 Exchange and Motional Narrowing

Molecular motions play an important role in liquid crystals. Fast motions, like conformational interconversions and tumbling motions of individual molecules in thermotropic liquid crystals or translational diffusion along the hydrophilic/hydrophobic interface in lyotropic systems, the rates of which are much higher than the static frequency shift or splitting (extreme narrowing limit), lead to an effective average of the coupling tensors, and the static values of Δ and η in Eq. (1) must be substituted by the motionally averaged parameters $\bar{\Delta}$ and $\bar{\eta}$ [24, 26].

Line shapes in the extreme narrowing limit contain information on the type of motion. Thus a continuous motion of a paraphenylene ring about its axis can be easily distinguished from 180° jumps between the two symmetry related orientations of the ring [26]. In macroscopically uniaxial phases, the local motions often also exhibit uniaxial symmetry. In this case, $\bar{\eta}=0$ and the NMR frequency is related to the angle between the director and the magnetic field, β, by the simple relationship

$$v = \bar{v}_L + \frac{1}{2} \bar{\Delta}(3\cos^2 \beta - 1) \qquad (2)$$

The spectrum in the middle of Fig. 1 shows this effect of motional narrowing.

If the averaged coupling can be assigned to a rigid segment of the molecule, as in the case of the quadrupole coupling of a ^2H nucleus in an X–^2H bond or the dipole–dipole coupling between a pair of spins on the same rigid moiety, e.g., a phenyl ring, the ratio between the motionally averaged coupling and the static coupling, $\bar{\Delta}/\Delta$, can be used to define an order parameter of the bond or rigid segment. This forms the basis for measurements of order parameters by NMR [8, 30, 31]. Obviously, the numerical value of an order parameter determined in such a way must be accompanied by a specification of the molecular segment and the assumptions on the geometry of that segment used in the calculation of the order parameter.

Slower motions with rates comparable to the static frequency shift or splitting, as often found in polymer liquid crystals, affect the spectrum, which is normally obtained from an echo pulse sequence, in a more complicated way [10, 32, 33]. A numerical line shape analysis in this regime of intermediate and slow motional exchange yields information about the geometry of the motional process that goes beyond the information content of the line shape in the extreme narrowing limit. Even slower motions with reciprocal rates in the millisecond or second range can be investigated by two-dimensional exchange spectroscopy (see Sec. 1.2.5).

1.2.3 Spin Relaxation

Relaxation measurements provide another way to study dynamical processes over a large dynamic range in both thermotropic and lyotropic liquid crystals (see Sec. 2.6 of Chap. III of Vol. 2A). The two basic relaxation times of a spin system are the spin–lattice or longitudinal relaxation time T_1 and the spin–spin or transverse relaxation time T_2. A detailed description, however, requires a more precise definition of the relaxation times. For spin $I=1$, for instance, two types of spin–lattice relaxation must be distinguished, related to the relaxation of Zeeman and quadrupolar order with rates T_{1Z}^{-1} and T_{1Q}^{-1}. The relaxation rates depend on spectral density functions which describe the spectrum of fluctuating fields due to molecular motions. A detailed discussion of spin relaxation is beyond the scope of this chapter; numerous articles on this subject are available [3, 6, 9, 12, 34–38]. Suffice it here to say that the maximum rate of spin–lattice relaxation occurs for motions with frequencies close to the Larmor frequency. Hence slow motions like director fluctuations can be investigated by NMR experiments in very low magnetic fields, which are accessible by field-cycling techniques [39–41] (see Sec. 1.2.4). Figure 2 illustrates the proton T_1 dispersion measured in a smectic C phase [41]. Regimes where the frequency dependence of T_1 is influenced by molecular motions, self-diffusion, and collective motions can be discriminated.

Another way to detect slow motions is by deuteron transverse spin relaxation experiments using a modified Carr–Purcell–Meiboom–Gill spin echo pulse sequence [42–44]. Information about the dynamics of liquid crystals can be retrieved from the frequency dispersion (see Fig. 2) and the anisotropy of relaxation rates.

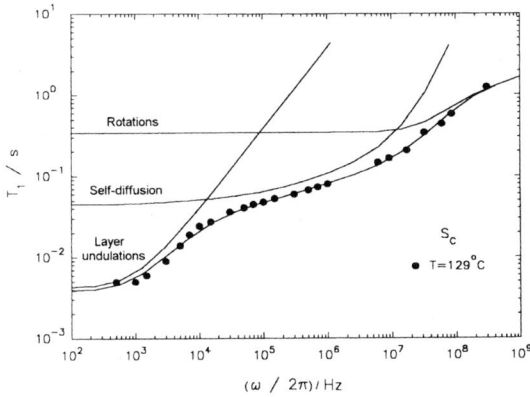

Figure 2. Proton spin–lattice relaxation time T_1 obtained by field cycling as a function of the Larmor frequency $\omega/2\pi$ for the smectic C phase of a biforked liquid crystal. The solid line is the best fit of a model taking into account collective movements, self diffusion, and molecular rotation. Reprinted with permission from [41].

1.2.4 Advanced Techniques

A large variety of techniques has been developed to simplify NMR spectra [23–26]. A comprehensive treatment is beyond the scope of this section; only the most important methods and some less frequently used techniques, which are referred to later in this section on magnetic resonance, will be introduced here.

Magic angle spinning (MAS) is used routinely in solid-state ^{13}C NMR to increase the resolution of spectra. By spinning the sample at a frequency of several kilohertz about an axis inclined at the magic angle of 54.7° to the magnetic field, the anisotropy of chemical shifts is removed, and liquid-like spectra with peaks at the positions of the isotropic chemical shift frequencies are obtained. If the sample cannot be spun with a frequency much higher than the width of the chemical shift, anisotropy sidebands appear at positions separated from the isotropic chemical shift line by integer multiples of the rotor frequency.

Other techniques used in ^{13}C NMR of solids are *cross polarization* (CP), which enhances the signal-to-noise ratio of ^{13}C spectra dramatically, and *proton decoupling* to eliminate line broadening due to dipolar couplings. Proton decoupling is also useful in ^{2}H NMR of liquid crystals to remove dipolar multiplets or broadening.

Another method to remove the anisotropy of spin interactions and to obtain high resolution spectra of solids is by *multiple-pulse sequences*. The combination of multiple-pulse sequences and MAS (CRAMPS: combined rotational and multiple-pulse spectroscopy) is used to obtain highly resolved proton spectra.

A technique particularly useful for liquid crystals, where intermolecular dipolar interactions are removed because of the tumbling motion of the molecules but intramolecular couplings remain due to the anisotropy of the motions, is *multiple-quantum* NMR. Multiple-quantum NMR simplifies the spectra of molecules containing a network of several coupled spins by reducing the number of observed transitions.

Field cycling is used to measure relaxation rates or line shapes in low or even zero fields. The spins are first polarized in a high magnetic field, then subjected to an evolution period at low or zero field, and finally detected again at high field. This field cycling is achieved either by switching of the magnetic field [45] or by shuttling the sample between a position in high field and a position in low field [27, 46].

1.2.5 Multidimensional Spectroscopy

Advanced NMR spectroscopy often requires multidimensional techniques. Therefore the basic ideas of two- and more-dimensional (2D and nD) Fourier spectroscopy will be presented in the following. Details on the enormous number of different experimental techniques can be found in the literature (for example, [25, 26]).

A 2D NMR spectrum is generated by recording a two-dimensional data set following pulsed irradiation as a function of two time variables, as shown schematically in Fig. 3a, and subsequent double Fourier transformation. The development of the nuclear spin system in the evolution period with step-wise incremented time t_1 at the beginning of the pulse sequence provides the basis for the first frequency dimension v_1. The NMR signal is recorded in the detection period as a function of time t_2, providing the basis for the second frequency dimension v_2, at the end of the pulse sequence. In many experiments, a variable mixing period of fixed duration t_m is inserted between evolution and detection. The concept of 2D spectroscopy is readily extended to three and higher dimensions by inserting additional evolution and mixing periods. During the evolution and detection periods, the spin

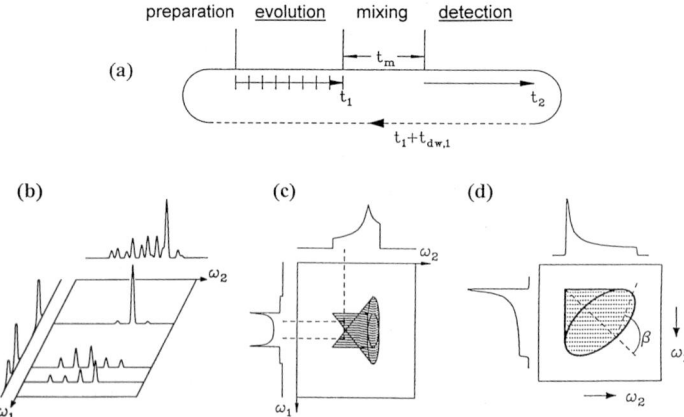

Figure 3. Principles of 2D-NMR spectroscopy: (a) Basic pulse sequence consisting of preparation, evolution, mixing, and detection periods. In the preparation period, relaxation filters can be applied to discriminate, e.g., against molecular motion. (b) Separation of anisotropic chemical shift side-band patterns (ω_2) according to the isotropic chemical shift values (ω_1). (c) Correlation of anisotropic chemical shift (ω_2) and dipolar interaction (ω_1). (d) 2D exchange spectrum revealing the reorientation of a molecular segment about the angle β through an elliptical exchange ridge.

system can be manipulated in many different ways in order to either remove or scale down some of the spin interactions.

The most important aspects of 2D NMR, from which additional information is gained, are exemplified in Fig. 3. First, 2D NMR is often used to increase the spectral resolution by separating different interactions. As an example, consider 1D MAS side-band spectra [23–26]. Spinning side-band spectra are often hampered by insufficient spectral resolution due to overlapping side-band patterns from different ^{13}C sites in the sample. This lack of resolution is circumvented when a new frequency dimension is introduced to separate the individual side-band patterns, as demonstrated in Fig. 3b.

Other 2D NMR techniques aim at obtaining new information by correlating different interactions. As a specific example, Fig. 3c displays the correlation of a ^{13}C chemical shift with a ^{13}C–^{1}H dipolar powder pattern. The correlation of different signals is often used for the assignment of peaks.

As far as applications to liquid-crystalline polymers with their slowed-down dynamics are concerned, *2D exchange* NMR proved to be particularly valuable [26]. First of all, by varying the mixing time t_m, slow dynamic processes in the range of milliseconds to seconds can be followed in real time. Moreover, 2D exchange spectra yield unique and model-independent information about the geometry of rotational motions. In fact, for axially symmetric coupling tensors, ubiquitous in ^{2}H NMR of liquid crystals, the 2D exchange spectrum yields elliptical ridge patterns from which the angle about which the molecules have rotated can be read off with a ruler (Fig. 3d).

Like many advanced NMR experiments, both multiple-quantum [25, 27] and field cycling [27, 45, 46] experiments are intrinsically two-dimensional, although most of the time only one dimension is presented. In the following, a few more examples from the large number of nD experiments for which all n frequency dimensions are essential will be given.

A 2D MAS experiment that provides information about orientational order based on spin-1/2 chemical shift anisotropies was introduced by Harbison and other workers [26, 47]. Compared to ^{2}H NMR spectroscopy of selectively deuterated liquid crystals, no isotopic labeling is required for this experiment. Spinning side-band spectra of oriented samples show variations in the phases and intensities of the spinning side-bands, which depend on the degree of order present and the position of the rotor at the start of acquisition of the NMR signal. These variations can be translated into a 2D side-band pattern by making the initial rotor position a function of the evolution time t_1 of a rotor-synchronized 2D NMR experiment. From the side-band intensities of the 2D spectrum, the moments of the orientational distribution function can be extracted.

This 2D experiment suffers from a lack of resolution if more than a few carbon sites are present. The problem of side-band overlap is overcome by a 3D extension of the experiment, in which the 2D side-band patterns of different sites are separated along the 3rd dimension according to their isotropic chemical shifts [26, 48]. An example for the application of this experiment will be given in Sec. 1.3.4.

The last example of 2D techniques presented here is the *wide-line separation* (WISE) experiment [26, 49]. In a WISE spectrum, high-resolution ^{13}C-CPMAS spectra are displayed on one axis, while conventional ^{1}H wide-line spectra dominated by the dipole–dipole coupling are displayed on the other. Thus the proton sites are separated according to the isotropic chemical shifts of the ^{13}C atoms to which they are attached. The widths of the ^{1}H lines reflect the mobility of the different sites: the higher the mobility, the narrower the line. Insertion of a mixing period into the pulse sequence of the WISE experiment allows for proton spin diffusion (transport of magnetization, no mass transport) to take place [26]. By this extension of the WISE experiment, the mobility at interfaces can be determined and domain sizes in the nanometer range can be estimated.

At this point, it must be mentioned that not only spin diffusion but also self diffusion of molecules can be studied by NMR [11, 13]. The available magnetic resonance techniques, relaxation and field-gradient methods, are discussed in Sec. 13.3.3 of Chap. VII of this Volume.

1.3 Applications of NMR

1.3.1 Phase Behavior

The perhaps most trivial but often fruitful application of NMR to liquid crystals is the detection and investigation of phase transitions. Any changes of the order parameter can easily be followed by NMR, since a change of the order parameter is connected with a change of the motionally averaged tensor components of the spin interactions. Most commonly, changes of the line splittings in ^{2}H NMR spectra or changes of the chemical shift in ^{13}C spectra are observed at a phase transition. In thermotropic systems the first-order nematic-to-isotropic phase transition is usually easy to detect. Transitions between different smectic or between smectic and nematic phases do not always give rise to a significant change in the NMR spectrum.

To study the phase diagrams of aqueous lyotropic liquid crystals, ^{2}H NMR on deuterated water, an easily accessible probe, is widely used, e.g. [50–53]. Cubic, hexagonal, and lamellar phases can easily be distinguished by the different averages of the quadrupole interaction, which result from the different curvature and symmetry of the hydrophobic–hydrophilic interface, along which the water molecules diffuse. Furthermore, different lyomesophases can also be discriminated based on their orientation in the magnetic field. Most surfactants with aliphatic tails have a negative anisotropy of the diamagnetic susceptibility. Therefore lamellar or disk-like aggregates tend to align with their normal perpendicular to the magnetic field, whereas cylindrical aggregates align with their symmetry axis parallel to the magnetic field. NMR can thus be used to determine the shape of aggregates and to discern N_D and N_C lyotropic nematic phases.

An advantage of NMR in the investigation of phase transitions is the fact that biphasic regions can be detected from the superposition of two different components in the spectrum. This facilitates the distinction between discontinuous (first-order) and continuous (higher-order) transitions (see Sec. 1.3.5). The tricritical point of the smectic A to nematic phase transition for binary liquid crystal mixtures, for example, has been determined by 2D ^{13}C NMR [54].

In polymers and elastomers where phase characterization by polarizing microscopy often fails because of poorly developed textures, NMR is a superior method because, as a molecular method, it is sensitive to the phase structure of very small domains.

Phase biaxiality in lyotropic system was detected by NMR a long time ago [55, 56]. For the ternary system, sodium dodecylsulphate/decanol/water, the transitions from the uniaxial nematic N_C and N_D phases to the biaxial nematic phase in-between were shown by ^{2}H NMR on D$_2$O to be first order [57]. For thermotropic nematic phases, however, NMR has failed to confirm the biaxiality found in several systems by other methods [58]. A possible reason could be a relatively large threshold value of $\bar{\eta}$, which might be as large as 0.2 [59] and is probably not reached in the thermotropic biaxial nematics discovered to date.

1.3.2 Molecular Orientation and Conformation

The investigation of orientational order and conformations is the most frequent application of NMR in liquid crystals. Magnetic resonance on molecules dissolved in oriented liquid crystals gains its power from the intermediate position of liquid crystals between isotropic liquids, on the one hand, and anisotropic solids, on the other. Because of the motional anisotropy of liquid crystals, a wealth of information not available from isotropic liquids becomes accessible. On the other hand, intermolecular interactions averaged to zero and intramolecular interactions scaled by motional narrowing lead to a better resolution and smaller overall width of liquid crystal spectra compared to the spectra of solids, thereby making requirements on instrumentation less demanding. In the past, these advantages were made use of mainly to investigate solute molecules in a liquid-crystalline matrix [1, 17, 60]. Some examples of recent studies on solutes are given in [61–70]. The goals of such solute studies have been twofold: in some cases, interest was exclusively in the determination of the structure of the solute molecule, while in other cases the solute was used as a spy to probe the molecular field of the liquid crystal. The boom of solute studies for structure elucidation appears to be over, at least for thermotropic liquid crystals; instead, the mesomorphic molecules themselves are now at the focus of interest, as the increasing number of studies on mesomorphic molecules, for example [71–77] indicates. For lyotropic liquid crystals, the situation may be somewhat different: The use of oriented lyotropic systems for the study of solvent molecules may only be at its beginning, since magnetically-oriented phospho-lipid bilayer micelles (bicelles) have been discovered as the matrix for structural studies of membrane-associated biomolecules [14].

To determine the orientation of a rigid molecule or a rigid segment of a flexible molecule in a macroscopically uniaxial phase completely, a segmental order matrix consisting of five independent elements is required [4]. Symmetry may reduce the number of independent matrix elements substantially. For flexible molecules the order matrix of each rigid subunit must be known and, in addition, information about the relative motions of the subunits, in other words, about the statistical weights of different conformations, is required. The latter is typically provided by long-range dipole–dipole couplings. Information of this type available to date is incomplete, but for the smallest molecules. However, even the limited data for larger molecules provides important constraints for testing inter- and intramolecular potentials of molecules in liquid crystals. Thus magnetic resonance is the most important source of experimental data against which models of orientational potentials and molecular dynamics simulations can be tested [8], as, for example, in [63, 64, 69, 78] (see also Chap. III of this Volume).

Information about the molecular order can be derived from any of the anisotropic spin interactions introduced in the previous section. Most commonly the quadrupole couplings of deuterated molecules or intramolecular dipole–dipole couplings between pairs of protons or pairs of different nuclei, like ^{13}C and ^{1}H, are used; chemical shift anisotropies have only been employed more recently [79, 80].

Proton studies of solutes and liquid crystals are difficult because the strongly coupled dipolar network of these many spin systems gives rise to spectra with an enourmous number of lines (see Sec. 1.2.1). Even for as few as six or seven coupled spins in a molecule, the spectrum is very complex. In general, for a spin system of a given size, the information content of the spectrum increases with its complexity. Thus the proton spectrum of a dipolarly coupled system is more complex than a proton-decoupled deuterium spectrum (in which most deuterium–deuterium couplings are too weak to be resolved). On the other hand, the dipole–dipole couplings, due to their larger number, provide more information about the system than quadrupole couplings alone. A spectrum with a large information content is of no use, however, if it is intractably complex. The crucial point here is that the same interactions that provide geometrical information are the cause of the complexity of the spectrum. Therefore a sufficient simplification of the spectrum combined with a minimum loss of information is required. Several approaches towards a solution of this problem have been taken.

One solution is the partial deuteration of the molecules. A recent example of this approach is a study on different partially deuterated phenyl-benzoate molecules [68]. The deuterium NMR spectra of partially deuterated molecules are easy to analyze because of the single-particle character of the quadrupole interaction. Also, proton dipolar spectra of partially deuterated molecules, measured under conditions of deuterium decoupling, are less difficult to analyze than the spectra of the fully protonated compounds. Therefore, in many studies, the considerable synthetic effort of deuteration is undertaken.

Other approaches use advanced spectroscopic techniques to simplify the spectrum. Examples from this category of solutions are multiple-quantum NMR [25, 27] to obtain ^{1}H–^{1}H couplings and separated local-field spectroscopy [23, 25, 26] with either ^{13}C (SLF) or proton detection (PDLF) to obtain ^{13}C–^{1}H couplings (or other heteronuclear interactions, such as ^{13}C–^{19}F couplings). Separated local-field techniques are usually combined with *variable angle spinning* (VAS or OMAS for off magic angle spinning) [20] to reduce the size of the couplings in order to further simplify the spectra (from second to first order) and to reduce the instrumental demands for decoupling power. In the following, the multiple-quantum and the separated local-field approach will be explained in more detail.

Multiple-quantum spectra have a greatly reduced number of lines because the number of different multiple-quantum transitions of order m (e.g., m spins flipping simultaneously) is reduced as the order increases. For a system consisting of n spins there is only one multiple-quantum transition of order $m=n$. Thus compared to the intractably complex 1-quantum spectrum of an n-spin system, the n-quantum spectrum of the same system consists of only one single line. In most investigations, the most simple and best resolved but nontrivial $n-1$-quantum spectra are analyzed. Figure 4 shows as an example the 1-quantum and the 5-quantum spectra of a 6-spin system, naphthaquinone in a nematic solvent [81].

Multiple-quantum spectra of a given order can be detected selectively in a two-dimensional experiment using conventional phase-cycling [25] or field-gradient selec-

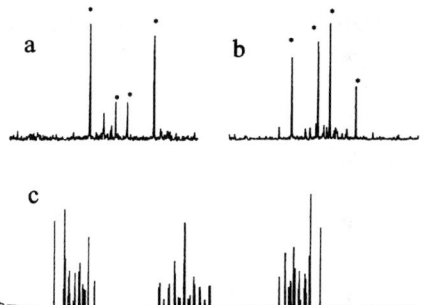

Figure 4. 300 MHz spectra of the protons in naphthaquinone dissolved in the nematic solvent ZLI 1132. (a) Five-quantum spectrum obtained from a pulse sequence without chemical-shift refocusation, (b) five-quantum spectrum with chemical-shift refocusation, (c) one-quantum spectrum. Real lines resulting from five-quantum transitions are marked with * to distinguish them from spurious peaks arising from pulse imperfections. Reprinted with permission from [81].

tion [82]. Some recent examples of solutes in liquid crystals, for which dipole–dipole couplings have been determined with the help of multiple-quantum NMR, are 3-chloroiodobenzene and 1-bromonaphthalene [83], biphenylene [84], and biphenyl [85]. From the spectra of highest orders ($m=n-1$ and $m=n-2$), start values for dipole–dipole coupling constants may be extracted, which can subsequently be used as input parameters for the numerical analysis of the 1-quantum spectrum. Such a procedure may even be automated, as has been demonstrated for the five-spin system of bromobenzene and the six-spin systems of ethynylbenzene and naphthaquinone [81].

In separated local field experiments, the ^{13}C sites are separated in the v_1 dimension of a 2D experiment according to their chemical shifts and correlated with their individual dipole–dipole couplings (their local fields) in the v_2 dimension. The original ^{13}C-detected local-field spectra are more complicated than the more recently introduced proton-detected local-field spectra [86]. When using ^{13}C detection, each ^{13}C atom is coupled to a multitude of protons on the same molecule, resulting in multiplet-of-multiplet type ^{13}C spectra in the v_2 dimension. With proton detection, however, each proton couples typically to only one ^{13}C atom on the same molecule (because of the low natural abundance of ^{13}C), and the dipolar-coupled ^{13}C spectra observed in the v_2 dimension are a simple superposition of doublets.

Figure 5 shows a PDLF spectrum of 4-n-pentyl-4'-cyanobiphenyl in the nematic phase [86]. Each ^{13}C cross section in the 2D spectrum shows several doublet splittings. The largest one results from the directly bonded ^{13}C–^{1}H pair, while smaller splittings originate from long-range ^{13}C–^{1}H couplings. For the VAS angle of 48.2° used to obtain the spectrum of Fig. 5, the signals from the carbon atoms 2 and 2' overlap. Owing to the scaling of the chemical shift anisotropy by sample spinning, the positions of the ^{13}C signals can be shifted by using a different spinning angle so that such accidental overlap can be resolved. Quantitative assignment of the long-range couplings can be achieved by combining the PDLF experiment with a heteronuclear chemical shift correlation (HETCOR) experiment in a reduced 3D experiment [87]. In a conventional 2D HETCOR experiment, a proton and a ^{13}C signal are correlated and show a crosspeak if there is a dipole–dipole coupling between the corresponding sites. In the reduced 3D experiment by Hong et al. [87], correlations between ^{13}C chemical shifts on the one hand (v_1 axis) and ^{1}H chemical shifts plus ^{13}C–^{1}H couplings on the other (v_2 axis) are detected and thus the nontrivial assignments of the long-range ^{13}C–^{1}H couplings to specific ^{13}C–^{1}H pairs become possible.

A different approach towards spectral simplification, yielding proton couplings, was demonstrated recently [88]. Individual

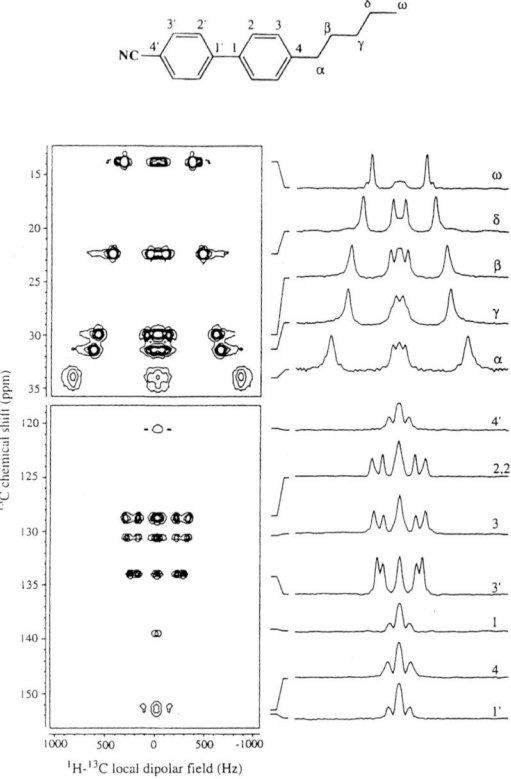

Figure 5. Contour plot of the OMAS-PDLF spectrum of 4-n-pentyl-4′-cyanobiphenyl recorded by spinning the sample at a rate of about 1.6 kHz at an angle of 48.2° with respect to the magnetic field. The signals from the different carbon atoms in the molecule are separated according to their ^{13}C chemical shifts along the v_2 axis (vertical) and correlated with their ^{13}C–^{1}H dipolar spectra along the v_1 axis (horizontal). Each dipolar spectrum shows several splittings due to directly bonded and long-range ^{13}C–^{1}H couplings. Reprinted with permission from [86] (Copyright 1996 American Chemical Society).

homonuclear ^{1}H–^{1}H couplings of a fully protonated liquid crystal molecule were obtained by a state-correlated two-dimensional experiment, which separates the proton sites according to their chemical shifts in the isotropic phase and correlates them with their local homonuclear dipolar fields in the liquid-crystalline phase. This was achieved by rapid switching of the temperature between the isotropic and the liquid-crystalline state during the mixing time of the pulse sequence.

The VAS technique to reduce the size of couplings can only be applied to samples in which the director aligns due to the sample spinning. Cholesteric or chiral smectic systems show an alignment not of the director but of the helicity axis. Therefore VAS does not show the desired effect in these systems [89]. A different technique to reduce the couplings, applicable to chiral liquid crystals, is coherent averaging in spin space by means of multiple-pulse techniques [23–26]. To retain the additional information gained from the anisotropy of liquid crystals, special multiple-pulse sequences have been developed which average the dipole–dipole couplings not to zero as conventional multiple-pulse sequences do, but scale them to a small value. Thus a complicated second-order spectrum can be turned into a simple first-order spectrum [89–91]. Such a multiple-pulse reduction of the dipole–dipole couplings was used for example in the v_2 dimension of a modified COSY experiment [89]. This is one example showing that powerful 2D techniques, which are standard today in the high-resolution NMR of isotropic liquids [25], are becoming more and more available for anisotropic ordered liquids as well.

Modern 2D techniques are applied, for example, in NMR on chiral solutes in chiral liquid crystalline solvents. NMR spectroscopy on such chiral host/chiral guest systems can be used for the spectral discrimination of enantiomeric solutes because the enantiomers show differential orientational ordering in the chiral environment. Based on this differential ordering effect (DOE), enantiomeric purity or excess can be quantitatively measured [92]. The DOE may also help to understand the intermolecular interactions involved in the chiral recognition mechanism. By the application of 1D and 2D NMR to measure and assign ^1H–^1H and ^1H–^{13}C couplings, the full order matrices of both enantiomers of β-(trichloromethyl)-β-propiolactone have been determined [93]. Two-dimensional correlation experiments have been shown to be powerful tools to discriminate and analyze the NMR spectra of the enantiomers [94]. An alternative approach is the derivatization of the enantiomers with a deuterated nonchiral agent, like acetyl-d_3 chloride, and measurement of the very simple deuteron NMR spectra [95].

1.3.3 Molecular Dynamics

Dynamic processes that can be investigated by NMR include both the motions of individual molecules, e.g., conformational dynamics and molecular rotations, and collective motions, e.g., director fluctuations in nematic systems, layer undulations in smectic systems, or density modulations in columnar phases of discotic systems. Self-diffusion can be measured by NMR relaxation or field gradient methods, as discussed in Sec. 13 of Chap. VII of this Volume. Table 1 gives an overview of the time scales accessible by the most common experimental techniques and examples of the type of motion that can be studied.

The application of relaxation studies for the investigation of liquid crystal dynamics was mentioned earlier in Sec. 1.2.3. Alternatively, motions can be studied by the analysis of 1D and 2D NMR line shapes. One-dimensional spectra are sensitive to motions with frequencies of the order of the static linewidth. Deuterium solid echo line shapes, for instance, are influenced by rotational motions with frequencies in the range of 10^4–10^8 Hz. ^{13}C spectra are sensitive to somewhat slower motions, since their static linewidth, due to chemical shift anisotropy, is smaller than in the case of deuterons. Slow motions with frequencies lower than the static (or preaveraged) linewidth, but higher than the spin–lattice relaxation rate can be investigated by 2D exchange NMR. The axial motion in columnar liquid crystals, for instance, was studied by this technique [96].

Often different motions with widely separated time scales are present. The fastest motions, for which the extreme narrowing limit is fulfilled, then lead to preaveraged line shapes, which are still sensitive to additional, now somewhat slower, motions with correlation times of the order of the inverse width of the preaveraged spectrum. A typical example, where such stepwise averaging is assumed, is the surfactant/water system. Local motions lead to strongly narrowed lines (by a factor of 10 for surfactant molecules and a factor of 100 for water), which can be narrowed further by diffusion along the hydrophobic–hydrophilic interface if this involves reorientations. This explains why the spectra of lamellar and hexagonal phases differ substantially in width, by a factor of approximately two: Diffusion

Table 1. Time scales of NMR methods.

Method	Correlation times	Examples of motional processes
T_1 (high field)	10^{-8}–10^{-12} s	Conformational exchange, molecular rotations in liquids
T_2	10^2–10^{-4} s	Director fluctuations, layer undulations
Line shape analysis (^2H)	10^{-4}–10^{-8} s	Rotations in viscous fluids and glasses
2D exchange	10^2–10^{-4} s	Rotations in molecular glasses and crystalline solids

along the flat interface of lamellar phases is not coupled to rotations like diffusion along the curved interface of hexagonal phases. This effect forms the basis for the easy recognition of phase transitions in lyotropic liquid crystals (see Sec. 1.3.1) and finds wide application for the study of phase diagrams of surfactant and membrane systems [50–53].

In low molar mass liquid crystals, conformational dynamics is typically too fast to be studied by 2D exchange NMR or line shape analysis. Dynamical studies of small molecules belong therefore almost exclusively to the realm of relaxation experiments. There are, however, cases of slow motions even in low molar mass systems, which can be investigated by 2D exchange spectroscopy. A recently investigated example is a discotic tetrabenzocyclophane derivative, which shows very slow motions in the columnar phase compared to other discotic compounds [97]. As illustrated in Fig. 6, deuterium 2D exchange NMR revealed the motional mechanism to be a combination of pseudorotation and proper rotation of the sofa conformer.

In polymers, motions are slowed down compared to when they occur in low molar mass liquid crystals. This often brings processes that are too fast in low molar mass liquid crystals into the kinetic window of line shape analysis or even 2D exchange experiments. Thus motional processes that cannot be studied in low molar mass liquid crystals become accessible in polymers [19]. Examples of NMR applications to liquid-crystalline polymers are given in the following section.

1.3.4 Liquid-Crystalline Polymers

A common feature of different types of liquid crystal polymers (LCPs), e.g., thermotropic side-chain or main-chain (either stiff or with flexible spacers) polymers, is their slowed-down dynamics compared to low molecular weight liquid crystals (LCs). Often polymers can be quenched to a glassy state in which the liquid-crystalline order is preserved but motions are completely frozen out. Such liquid-crystalline glasses provide a unique opportunity to determine, in principle, the full orientational distribution function, whereas only its second moment is available from motionally averaged NMR spectra. Thus LCP studies have made fundamental contributions to LC science.

The first studies of the orientational distribution function in liquid crystalline side chain polymers have employed ^{2}H NMR on selectively labeled polymers [31]. By revealing a gradient of decreasing order from mesogen via spacer to main chain, these studies have shown the basic validity of the spacer concept [98, 99], namely, the decou-

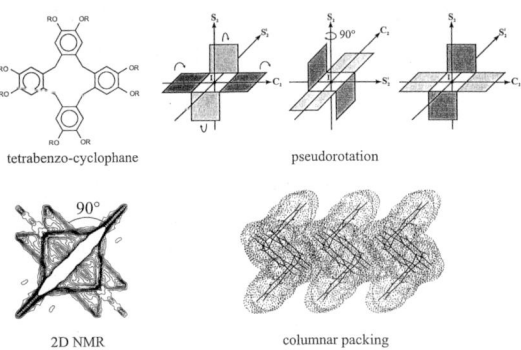

Figure 6. Analysis of molecular motions of the columnar phase of a tetrabenzocyclophane derivative by ^{2}H 2D exchange NMR. *Top:* molecular structure and schematic diagram of the motion. The first sofa conformer is converted by pseudorotation into a second sofa conformer, this first step being followed by a proper rotation of the molecule. *Bottom left:* deuteron 2D exchange spectrum of the phenyl-ring deuterated compound showing the ridge pattern characteristic of 90° jumps. *Bottom right:* stacking of the molecules in a column. For details see [97].

pling of the mesogens with their tendency to form ordered structures from the polymer chain with its tendency to form a random coil by inserting a flexible spacer (see Fig. 7a).

More recent studies of the order in side chain liquid crystal polymers used the rotor-synchronized 2D and 3D ^{13}C MAS experiments described in Sec. 1.2.5. Figure 7b shows an example [48]. The values of the order parameters given in the figure are for the ^{29}Si chemical shift tensor in the main chain, the CH$_2$ groups of the spacer, and the long axis of the mesogen (values averaged over several ^{13}C sites on the mesogen). The results obtained for the series of polysiloxanes with 4-hydroxy-4'-methoxyphenylene-benzoate as the mesogen and spacers containing 3, 4, 5, and 6 methylene groups, respectively, reveal significant differences for the polymers of this homologous series. The order parameter ($\bar{P}_2$) of the ^{29}Si chemical shift tensor of the main chain, for example, shows a pronounced odd–even behavior with $\bar{P}_2$ values of 0.21, <0.05, 0.38, and 0.09 for a spacer of 3, 4, 5, and 6 methylene groups, respectively. The NMR data obtained for this homologous series make it evident that the spacer does not simply decouple the ordered mesogen from the random coil main chain. Rather, both the spacer and the main chain must be considered as making an integral contribution to the overall structure.

Further examples of characteristic dynamical processes in liquid crystal polymers, discussed in detail elsewhere [19, 31], include the phenyl ring flip in side chain liquid crystal polymers [100] and the conformational dynamics of the spacer in a main chain liquid crystal polymer [101].

An example illustrating the ability of NMR to study the spatial proximity of different structural elements is shown in Fig. 8, which depicts 2D WISE spectra of stiff chain macromolecules with flexible C$_{16}$H$_{33}$ side chains [49]. Only that portion of the spectra arising from the side chains is shown. The ^{1}H wide-line spectra reflect the molecular dynamics of the different methylene groups. They are superpositions of broad (immobile) and narrow (mobile) components, even at the chain end (C$_{16}$). Overall, the side chains show a higher mobility in polyamide compared to polyimide. By invoking ^{1}H spin diffusion in the WISE experiments, the domain sizes of the regions with different side-chain mobility were estimated. From these experiments, the following picture of the packing in these systems results: The main chains are arranged in planar arrays with the side chains perpen-

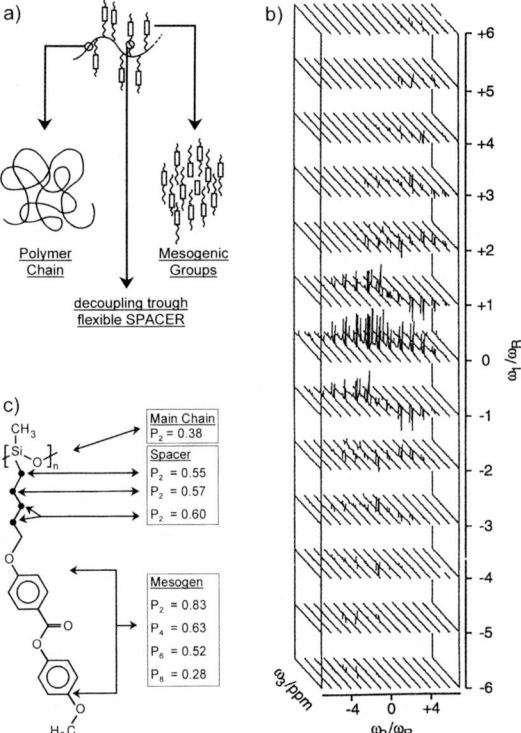

Figure 7. Molecular order in calamitic side-group polymers: (a) Spacer model, (b) 3D-MAS-NMR spectrum of ^{13}C of a polysiloxane liquid crystal polymer, (c) order parameters of different segments of the molecule. For details see [48].

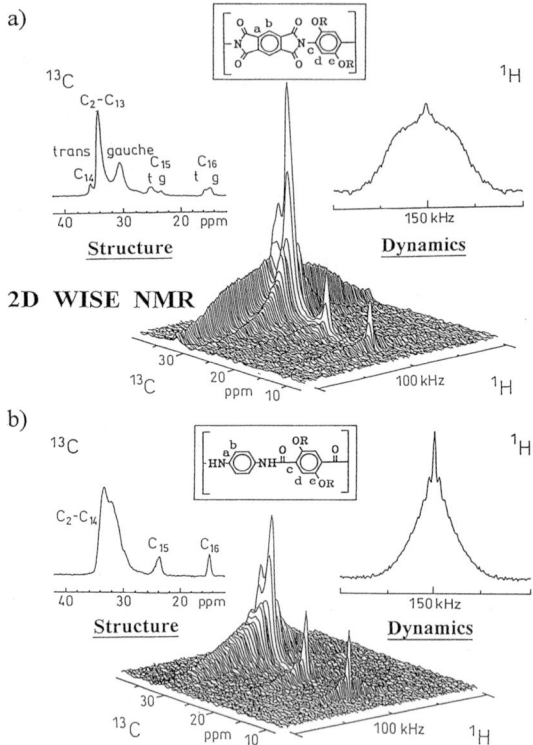

Figure 8. Two-dimensional WISE NMR spectra of stiff macromolecules relating structure and mobility of side chains: (a) Polyimide, (b) polyamide. For details see [49].

dicular to them. In the polyimide, the side chains exhibit substantial crystallinity with all-*trans* conformations in the crystalline regions. At the same time, amorphous regions with high side chain mobility are present. In the polyamide, regions with highly extended but somewhat mobile side chains and regions with disordered side chains are present. For both types of polymers, the size of the regions is comparable to the layer spacing.

1.3.5 Liquid Crystals in Microconfined Environments

Magnetic resonance can be used not only to study bulk liquid crystals, but also to investigate liquid crystals confined in a polymer matrix (polymer-dispersed liquid crystals, PDLCs) [18, 102–104] or in pores of various shapes [18, 105–113] (see Sec. 2.6.10 to 2.6.12 of Chap. III of Vol. 2a of this Handbook). The physical properties of these microconfined systems are modified by surface and finite size effects. Again, deuterium NMR has proven particularly useful for the study of such systems [103–105, 108–111], but ^{1}H relaxation [102] and ^{13}C spectroscopical measurements [112] have also been employed.

First, the phase transitions of liquid crystals in microcavities of submicrometer size are strongly affected by finite size effects. The nematic-to-isotropic phase transition, for instance, has been shown to become continuous [105, 106, 111]. This phenomenon can be explained by Landau-type models [105–107, 111, 114, 115]. The same effect of a continuous nematic-to-isotropic transition was also observed in bulk liquid single crystal elastomers [116, 117] (see Chap. V of Vol. 3 of this Handbook), whereas the corresponding linear polymer shows a discontinuity of the order parameter at the phase transition. For the elastomers, both a confinement due to the crosslinking and an internal mechanical field, resulting from a second crosslinking performed under mechanical stress, may explain the continuous character of the nematic-to-isotropic transition.

In addition, the orientational distribution of the liquid crystal is affected by surface effects. Deuterium NMR can be applied to study the director field in microcavities [103, 107–110, 112]. Figure 9 shows deuteron NMR spectra of selectively deuterated 4-*n*-pentyl-4′-cyanobiphenyl confined in parallel polyimide-modified cylindrical pores of a Nuclepore membrane. The suggested models for the director field in such cylindrical pores with parallel anchoring,

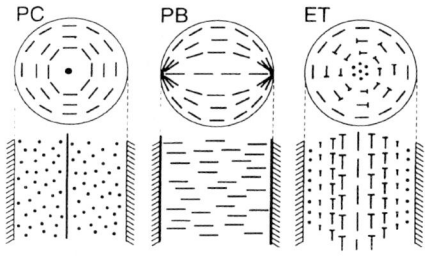

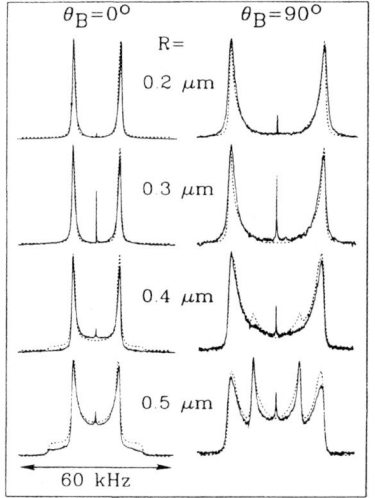

Figure 9. ^{2}H NMR spectroscopy on a microconfined nematic liquid crystal. *Top:* possible nematic director fields in cylindrical pores with parallel anchoring: the planar-concentric (PC), planar-bipolar (PB), and escaped-twisted (ET) structure, where the top of the nails represents the part of the director beneath the plane of the cross section. *Bottom:* ^{2}H NMR line shapes of 4-*n*-pentyl-4'-cyanobiphenyl-βd_2 confined in polyimide-modified Nuclepore membranes. Θ_B is the angle between the pore axis and the magnetic field and R is the pore radius. The dashed lines correspond to theoretical fits. The $R=0.2$ μm and $R=0.3$ μm theoretical spectra are composed of 100% PB, the $R=0.4$ μm spectrum of 15% ET and 85% PB, and the $R=0.5$ μm spectrum of 50% ET and 50% PB. Reprinted with permission from [109].

i.e., the planar-concentric, the planar-bipolar, and the escaped-twisted structure, are depicted at the top of Fig. 9. Fits of the line shapes (dashed lines in Fig. 9) indicate a transition from the escaped-twisted to the planar-bipolar structure with decreasing pore radius. The polar and azimuthal anchoring energies and elastic constants may be determined from such studies [109, 110].

Finally, due to the surface anchoring, the dynamics of confined liquid crystals differs from the bulk dynamics. As for bulk systems the dynamics can be investigated by relaxation studies [102, 104, 113] or line-shape analysis [112].

1.3.6 Viscoelastic Properties

As mentioned in the previous section on confined systems, viscoelastic properties of liquid crystals (see Sec. 1 of Chap. III and Sec. 8 of Chap. VII of this Volume) can be determined from experiments in which the nematic field is distorted by surface forces. Similarly, the influence of an external field, namely, the magnetic field, on the director orientation, is the basis of several magnetic resonance experiments which can be used to determine viscoelastic parameters of nematic phases.

The most common experiments of this type involve reorientation of the sample in the magnetic field [20, 118–125]. Either a sudden reorientation of the initially magnetically aligned sample by a fixed angle θ_0 about an axis perpendicular to the magnetic field, followed by observation of the development of the transient orientational distribution of the director, or continuous rotation of the sample can be used. Both types of experiments yield γ_1, the twist or rotational viscosity. In sudden reorientation experiments with $\theta_0 > \pi/4$, spatially periodic patterns of the director reorientation form and the director realignment becomes inhomogeneous. In this case, NMR spectra can yield four of the five independent viscosity constants and the ratio of two of the three elastic constants of the nematic phase [121].

Information on viscoelastic properties can also be obtained from relaxation measurements [39, 44, 126] and, in the case of polymers, from line shape analysis [127, 128].

An example of an experiment involving continuous sample rotation with synchronized data acquisition is shown in Fig. 10 [123]. A thin liquid crystal cell filled with a side chain liquid crystal polymer was continuously rotated about an axis perpendicular to the magnetic field. The director behavior was followed by deuteron NMR as well as by polarizing microscopy. The optical texture and the orientation-dependent NMR line shapes resulting from the convective director pattern formed after 2 h are shown in Fig. 10b and c.

Instead of rotating the sample, the direction of the applied magnetic field can be changed [129]. A sudden field switching can be performed within milliseconds so that low molecular weight samples, which realign fast because of their low viscosity, can be studied by this method. Four viscosity parameters and the average Frank elastic constant could be determined by such a field-cycling experiment [129].

An alternative to reorientation of the sample or the magnetic field is the application of shear during the NMR measurement [130–134]. For liquid-crystalline samples with high viscosity, such as liquid crystal polymers, the steady-state director orientation is governed by the competition between magnetic and hydrodynamic torques. Deuteron NMR can be used to measure the director orientation as a function of the applied shear rate and to determine two Leslie coefficients, α_2 and α_3, of nematic polymers [131, 134]. With this experiment, flow-aligning and tumbling nematics can be discriminated. Simultaneous measurement of the apparent shear viscosity as a function of the shear rate makes it possible to determine two more independent viscosity parameters [131, 134].

Figure 11 demonstrates the application of deuteron NMR spectroscopy on sheared samples with the simultaneous measurement of the viscosity to two different nematic polymer systems. The different behavior of these systems is apparent from the data: one system is flow-aligning, the other system is tumbling. The Leslie coefficients α_2 and α_3 obtained for the shear-aligning system are both negative, whereas $\alpha_2 < 0$ and $\alpha_3 > 0$ for the tumbling system. Deuteron NMR has also been employed to study the director orientation of lyotropic lamel-

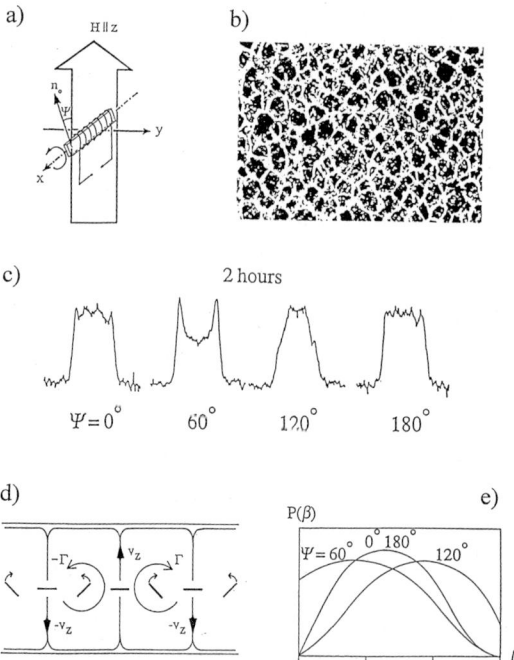

Figure 10. Convective director structures in a nematic side group polymer. (a) Schematic diagram of the experimental set-up for continuous rotation about an axis perpendicular to the magnetic field. (b) Optical texture after two hours of rotation. (c) ^{2}H NMR spectra obtained during one revolution after two hours. (d) Schematic diagram of convection rolls evolving due to nonlinear coupling between director rotation and viscous flow. (e) Director distribution $P(\beta)$ extracted from the ^{2}H NMR spectra in (c). For details see [123].

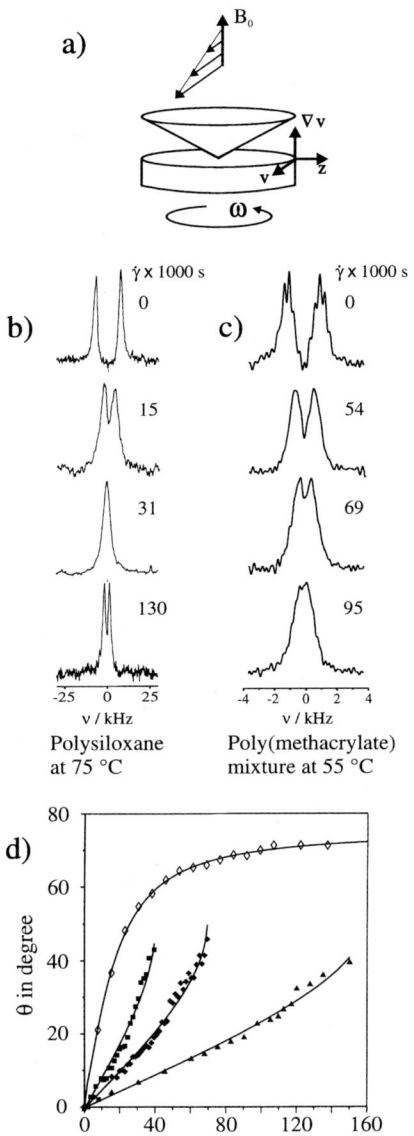

Figure 11. Flow-alignment behavior of side chain liquid crystal polymers. (a) Schematic diagram of the cone-and-plate shear geometry for NMR measurements (v: velocity parallel to tangential axis, ∇v: velocity gradient parallel to magnetic field axis). (b) ^{2}H NMR spectra of a flow-aligning polysiloxane at different shear rates. (c) ^{2}H NMR spectra of a nonflow-aligning nematic mixture at different shear rates. (d) Director orientation θ with respect to the magnetic field as a function of shear rate measured from the NMR splittings. Open symbols: polysiloxane; filled symbols: polymethacrylate mixture (squares: 52°C; diamonds: 55°C; triangles 60°C). For details see [134].

lar [132] and hexagonal phases [133] under shear flow.

1.4 ESR (Electron Spin Resonance) of Liquid Crystals

Electron spin resonance (ESR), often called electron paramagnetic resonance (EPR), requires the presence of unpaired electrons in the sample [5, 15, 16, 135]. Since most organic molecules do not carry free radicals, the most common method is to simply dissolve stable free radicals as spin probes in the sample. Alternatively, free radical groups can be chemically bonded to the molecule of interest providing a spin label. Probably the most frequently used spin probes are nitroxide free radicals like 2,2,5,5-tetramethyl-piperidinoxide (TEMPO).

As in NMR, the interactions that govern the shape of the resonance line are anisotropic. The spin Hamiltonian contains two relevant contributions

$$H = \beta S \mathbf{g} B + I \mathbf{A} S \qquad (3)$$

namely g-shift and hyperfine interaction. Here S is the spin operator of the unpaired electron (spin quantum number $S=1/2$), I the spin operator of the nucleus ($I=1$ for nitrogen), B the magnetic field, $\mathbf{g}$ the electron gyromagnetic tensor, $\mathbf{A}$ the hyperfine tensor, and β the Bohr magneton. While the basic principles of ESR and NMR are similar, their time scales are very different because of the different strengths of the interactions. The time window in which ESR spectra of nitroxide spin labels are sensitive to molecular motions is centered around 10^{-9} s, which is four orders of magnitude shorter than for ^{2}H NMR line shapes. Thus much faster rotational motions are accessible by ESR spectra. The line shapes are mostly analyzed in terms of the stochastic Liouville equation [136].

An example of typical ESR spectra, measured in the first derivative mode, is shown in Fig. 12. Just like NMR, ESR can be used to detect phase transitions and to study the orientation and dynamics of liquid crystals. The spectra shown in Fig. 12, for example, are from a study comparing the dynamics of the spin label at the end of the polymer chain and the freely dissolved spin probe in a liquid-crystalline polyether by continuous wave ESR (Fig. 12) and 2D Fourier transform ESR experiments [137]. The end label showed smaller ordering and larger reorientational rates than the dissolved spin probe. Furthermore, it was demonstrated that the advanced 2D FT ESR experiments (see below) on the end-labeled polymer chain could not be explained by the conventional Brownian model of reorientation, although this model could explain the 1D spectra. This led to the development of a new motional model of a slowly relaxing local structure, which enabled differentiation between the local internal modes experienced by the end label and the collective reorganization of the polymer molecules around the label. The latter was shown to be slower by two orders of magnitude.

The first 2D experiments in ESR were 2D electron spin echo experiments in which a Hahn echo sequence was applied [138]. The echo height was monitored while a slow sweep through the ESR spectrum was performed by sweeping the DC magnetic field. This experiment was repeated for several values of the pulse separation time τ. By Fourier transformation with respect to τ, a 2D spectrum resulted yielding the inhomogeneously broadened absorption-like echo spectrum in one dimension and the homogeneous line shape in the other dimension. Today, owing to improved hardware, full time-domain 2D FT experiments as in NMR can be performed in ESR as well.

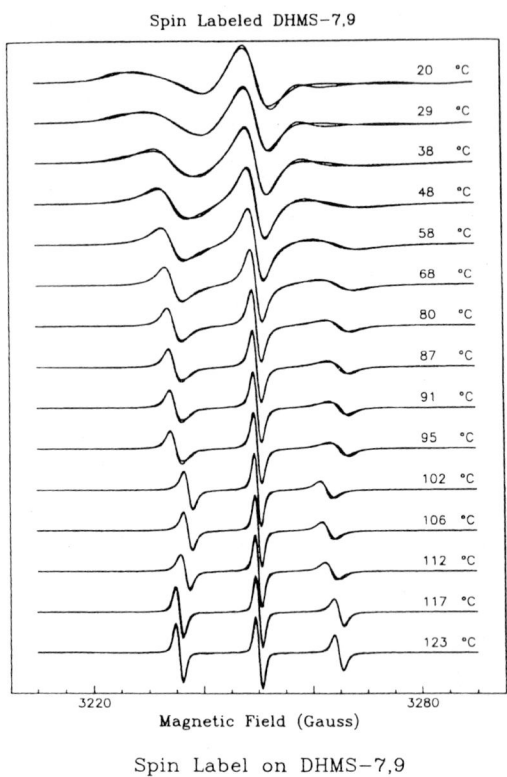

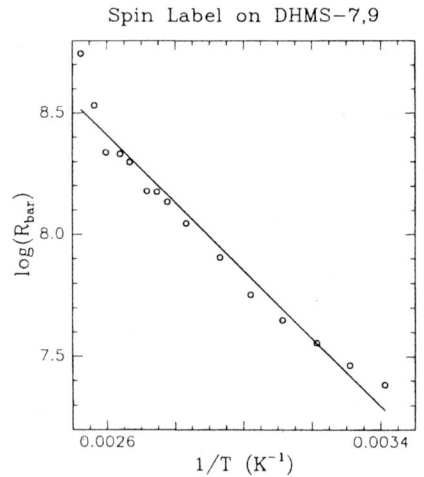

Figure 12. *Top:* ESR spectra of main chain liquid crystalline poly(ether) DHMS-7,9 (based on 4,4′-dihydroxy-α-methylstilbene and mixed aliphatic spacers of 7 and 9 methylene groups), spin-labeled with CSL (3β-doxyl-5α-cholestane). Solid lines: experimental; dashed lines: simulated. *Bottom:* average rotational diffusion coefficient, $\bar{R} = (R_\perp^2 R_\parallel)^{1/3}$, obtained from the fits to the ESR line shapes. The activation energy, E_{label}, is calculated to be $1.16 \pm 0.07 \times 10^4$ J/mol. Reprinted with permission from [137] (Copyright 1996 American Chemical Society).

Examples of time-domain 2D FT ESR spectra are the 2D FT electron–electron double resonance (ELDOR) spectra obtained for the rod-like cholestane spin probe in a smectic phase, which are shown in Fig. 13. The pulse sequence used in this type of ESR experiment is essentially the same as for the 2D exchange experiment in NMR. As in NMR, 2D experiments have greatly enhanced the ability to distinguish different types of rotational dynamics.

In polymeric liquid crystals, not only the mobility in the liquid-crystalline phase but also the highly restricted dynamics of the probe in the frozen state can be studied by 2D ELDOR spectroscopy [140]. In high performance materials based on main chain liquid crystalline polymers, stable radicals are created during production and processing. The nature and the dynamics of these defects has been elucidated by pulsed electron–nuclear double resonance (ENDOR) spectroscopy [141].

The faster time scale of ESR combined with its much higher sensitivity compared to NMR (resulting from the bigger magnetic moment of the electron) makes the analysis of relatively fast kinetics possible. One application of this type is the study of electron transfer processes after laser irradiation, which is of great interest for investigations on model systems of photosynthesis [142].

1.5 Summary

The few examples given here illustrate that magnetic resonance is a powerful tool for the study of liquid crystals. Although many aspects, like the investigation of chemical structure or magnetic resonance imaging, have been completely omitted, the versatility of magnetic resonance techniques in their application to liquid crystals should

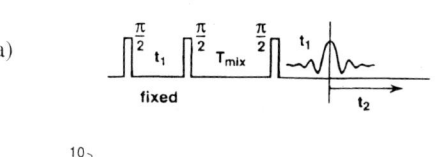
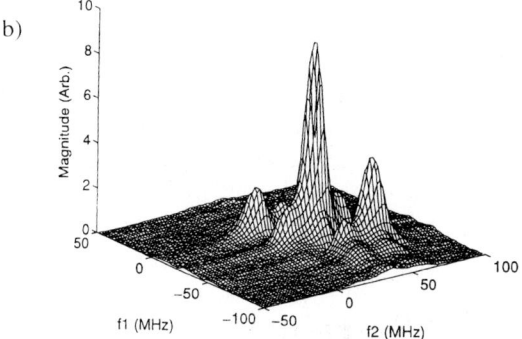
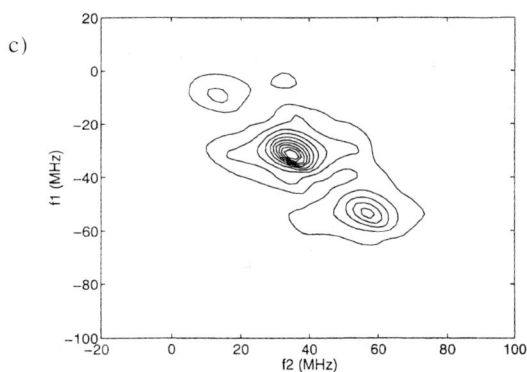
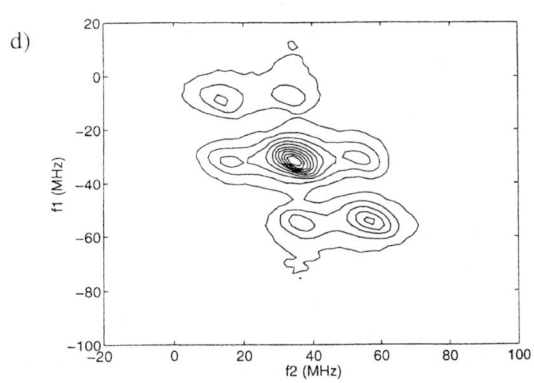

Figure 13. Two-dimensional ELDOR spectra of a CSL spin probe in the smectic A phase of N-(p-butoxybenzylidene)-p-n-octylaniline at 59°C. (a) Pulse sequence, reprinted in part with permission from [137] (Copyright 1996 American Chemical Society), (b) mixing time $T_{mix} = 110$ ns, (c) $T_{mix} = 90$ ns, (d) $T_{mix} = 250$ ns. The frequencies corresponding to the time parameters t_1 and t_2 are f_1 and f_2. (b)–(d) reprinted in part with permission from [139].

have become obvious. Being spectroscopic methods, NMR and ESR provide information on the molecular scale. This makes them complementary to other common methods of investigation, e.g., polarizing microscopy and scattering. The potential of magnetic resonance for further developments has by no means been exhausted, and many improved techniques and novel applications to liquid crystals will most certainly appear within the next few years.

1.6 References

[1] J. W. Emsley, J. C. Lindon (Eds.), *NMR Spectroscopy using Liquid Crystal Solvents*, Pergamon, Elmsford, NY **1975**.
[2] G. R. Luckhurst, G. W. Gray (Eds.), *The Molecular Physics of Liquid Crystals*, Academic, London **1979**.
[3] R. L. Vold, R. R. Vold, *Nuclear Magnetic Relaxation in Liquid Crystals*, Reidel, Dordrecht **1983**.
[4] J. W. Emsley (Ed.), *Nuclear Magnetic Resonance of Liquid Crystals*, Nato ASI Series C: Mathematical and Physical Sciences, Vol. 141, Reidel, Dordrecht **1985**.
[5] G. R. Luckhurst, C. A. Veracini (Eds.), *The Molecular Dynamics of Liquid Crystals*, Kluwer, Dordrecht **1994**.
[6] R. Y. Dong, *Nuclear Magnetic Resonance of Liquid Crystals*, Springer, New York **1994**.
[7] C. L. Khetrapal, *Int. J. Mod. Phys. B* **1995**, *9*, 2573.
[8] G. R. Luckhurst, *Ber. Bunsenges. Phys. Chem.* **1993**, *97*, 1169.
[9] G. L. Hoatson, R. L. Vold, in: *NMR – Basic Principles and Progress,* Vol. 32 (Eds.: P. Diehl, E. Fluck, H. Günther, R. Kosfeld, J. Seelig, B. Blümich), Springer, Berlin **1994**, p. 1.
[10] B. Halle, *Progr. NMR Spectrosc.* **1996**, *28*, 137.
[11] O. Söderman, P. Stilbs, *Progr. NMR Spectrosc.* **1994**, *26*, 445.
[12] B. Halle, P.-O. Quist, I. Furó, *Liq. Cryst.* **1993**, *14*, 227.
[13] G. Lindblom, G. Orädd, *Progr. NMR Spectrosc.* **1994**, *26*, 483.
[14] C. R. Sanders, II, B. J. Hare, K. P. Howard, J. H. Prestegard, *Progr. NMR Spectrosc.* **1994**, *26*, 421.
[15] L. Monnerie, F. Lauprêtre, C. Noël, *Liq. Cryst.* **1988**, *3*, 1.
[16] L. Monnerie, F. Lauprêtre, C. Noël, *Liq. Cryst.* **1988**, *3*, 1013.
[17] R. Poupko, Z. Luz in *Encyclopedia of Nuclear Magnetic Resonance*, Vol. 1 (Eds.: D. M. Grant, R. K. Harris), Wiley, Chichester **1996**, p. 1783.
[18] J. W. Doane, G. P. Crawford in *Encyclopedia of Nuclear Magnetic Resonance* (Eds.: D. M. Grant, R. K. Harris), Wiley, Chichester **1996**, p. 3663.
[19] H. W. Spiess, *Ber. Bunsenges. Phys. Chem.* **1993**, *97*, 1294.
[20] J. Courtieu, J. P. Bayle, B. M. Fung, *Progr. NMR Spectrosc.* **1994**, *26*, 141.
[21] J. Jokisaari, *Progr. NMR Spectrosc.* **1994**, *26*, 1.
[22] G. A. Webb (Ed.), *Nuclear Magnetic Resonance (Specialist Periodical Reports)*, Vol. 25, The Royal Society of London, London **1996**, and earlier volumes.
[23] C. A. Fyfe, *Solid State NMR for Chemists*, C. F. C. Press, Guelph, Ontario, Canada **1983**.
[24] M. Mehring, *High Resolution NMR in Solids*, Springer, Berlin **1983**.
[25] R. R. Ernst, G. Bodenhausen, A. Wokaun, *Principles of Nuclear Magnetic Resonance in One and Two Dimensions*, Oxford University Press, Oxford **1987**.
[26] K. Schmidt-Rohr, H. W. Spiess, *Multidimensional Solid-State NMR and Polymers*, Academic, London **1994**.
[27] A. Pines in *Proc. of the 100th School of Physics "Enrico Fermi"* (Ed.: B. Maraviglia), North-Holland, Amsterdam **1988**, p. 43.
[28] H. W. Spiess in *Developments in Oriented Polymers*, Vol. 1 (Ed.: I. M. Ward), Applied Science, Barking, Essex **1982**.
[29] H. W. Spiess in *Structure and Properties of Oriented Polymers* (Ed.: I. M. Ward), Chapman & Hall, London, 1997, Chap. 5, in press.
[30] P. G. de Gennes, J. Prost, *The Physics of Liquid Crystals*, 2nd ed., Oxford University Press, New York **1993**.
[31] C. Boeffel, H. W. Spiess in *Side Chain Liquid Crystal Polymers* (Ed.: C. B. McArdle), Blackie, Glasgow **1989**, Chap. 8, p. 224.
[32] H. W. Spiess, H. Sillescu, *J. Magn. Reson.* **1981**, *42*, 381.
[33] N. J. Heaton, *Chem. Phys. Lett.* **1996**, *252*, 77.
[34] F. Noack in *NMR – Basic Principles and Progress*, Vol. 3 (Eds.: P. Diehl, F. Fluck, R. Kosfeld), Springer, Berlin **1971**, p. 83.
[35] J. H. Freed, *J. Chem. Phys.* **1977**, *66*, 4183.
[36] R. R. Vold, R. L. Vold, *J. Chem. Phys.* **1977**, *66*, 4018.
[37] H. W. Spiess in *NMR – Basic Principles and Progress*, Vol. 15 (Eds.: P. Diehl, F. Fluck, R. Kosfeld, Springer, Berlin **1978**, p. 55.
[38] E. A. Joghems, G. van der Zwan, *J. Phys. II France* **1996**, *6*, 845.
[39] F. Noack, *Progr. NMR Spectrosc.* **1986**, *18*, 171.
[40] J. Struppe, F. Noack, *Liq. Cryst.* **1996**, *20*, 595.

[41] C. Cruz, J. L. Figueirinhas, P. J. Sebastião, A. C. Ribeiro, F. Noack, H. T. Nguyen, B. Heinrich, D. Guillon, *Z. Naturforsch.* **1996**, *51a*, 155.
[42] K. Müller, R. Poupko, Z. Luz, *J. Magn. Reson.* **1990**, *90*, 19.
[43] N. Heaton, D. Reimer, G. Kothe, *J. Non-Cryst. Solids* **1994**, *172–174*, 917.
[44] G. Kothe, N. Heaton in *Encyclopedia of Nuclear Magnetic Resonance* (Eds.: D. M. Grant, R. K. Harris), Wiley, Chichester **1996**, p. 4436.
[45] F. Noack, M. Notter, W. Weiss, *Liq. Cryst.* **1988**, *3*, 907.
[46] A. M. Thayer, A. Pines, *Acc. Chem. Res.* **1987**, *20*, 47.
[47] G. S. Harbison, V.-D. Vogt, H. W. Spiess, *J. Chem. Phys.* **1987**, *86*, 1206.
[48] J. J. Titman, D.-L. Tzou, S. Féaux de Lacroix, C. Boeffel, H. W. Spiess, *Acta Polym.* **1994**, *45*, 204.
[49] J. Clauss, K. Schmidt-Rohr, A. Adam, C. Boeffel, H. W. Spiess, *Macromolecules* **1992**, *25*, 5208.
[50] J. H. Davis, *Biochim. Biophys. Acta* **1983**, *737*, 117.
[51] N. Boden, G. R. Hedwig, M. C. Holmes, K. W. Jolley, D. Parker, *Liq. Cryst.* **1992**, *11*, 311.
[52] S. S. Funari, M. C. Holmes, G. J. T. Tiddy, *J. Phys. Chem.* **1994**, *98*, 3015.
[53] J. C. Blackburn, P. K. Kilpatrick, *Langmuir* **1992**, *8*, 1679.
[54] C. W. Cross, B. M. Fung, *Liq. Cryst.* **1995**, *19*, 863.
[55] L. J. Yu, A. Saupe, *Phys. Rev. Lett.* **1980**, *45*, 1000.
[56] G. Chidichimo, N. A. P. Vaz, Z. Yaniv, W. J. Doane, *Phys. Rev. Lett.* **1982**, *49*, 1950.
[57] P.-O. Quist, *Liq. Cryst.* **1995**, *18*, 623.
[58] S. M. Fan, I. D. Fletcher, B. Gündogan, N. J. Heaton, G. Kothe, G. R. Luckhurst, K. Praefcke, *Chem. Phys. Lett.* **1993**, *204*, 517.
[59] F. Biscarini, C. Chiccoli, P. Pasini, F. Semeria, C. Zannoni, *Phys. Rev. Lett.* **1995**, *75*, 1803.
[60] P. Diehl, C. L. Khetrapal in *NMR – Basic Principles and Progress* (Eds.: P. Diehl, E. Fluck, R. Kosfeld), Springer, Berlin **1969**, p. 1.
[61] R. Wasser, P. Diehl, *Struct. Chem.* **1990**, *1*, 259.
[62] E. J. Delikatny, M. A. Heldman, E. E. Burnell, *J. Magn. Reson.* **1991**, *95*, 32.
[63] D. J. Photinos, E. T. Samulski, H. Toriumi, *J. Chem. Soc. Faraday Trans.* **1992**, *88*, 1875.
[64] J. W. Emsley, I. D. Wallington, D. Catalano, C. A. Veracini, G. Celebre, M. Longeri, *J. Phys. Chem.* **1993**, *97*, 6518.
[65] P. J. Barnes, A. G. Douglass, S. K. Heeks, G. R. Luckhurst, *Liq. Cryst.* **1993**, *13*, 603.
[66] A. Abe. E. Iizumi, Y. Sasanuma, *Polym. J.* **1993**, *25*, 1087.
[67] G. A. Naganagowda, S. Vivekanandan, N. Suryaprakash, C. L. Khetrapal, *J. Magn. Reson., Ser. A* **1996**, *118*, 33.

[68] J. W. Emsley, M. I. C. Furby, G. de Luca, *Liq. Cryst.* **1996**, *21*, 877.
[69] M. Luzar, M. E. Rosen, S. Caldarelli, *J. Phys. Chem.* **1996**, *100*, 5098.
[70] A. F. Terzis, C.-D. Poon, E. T. Samulski, Z. Luz, R. Poupko, H. Zimmermann, K. Müller, H. Toriumi, D. J. Photinos, *J. Am. Chem. Soc.* **1996**, *118*, 2226.
[71] J. P. Bayle, B. M. Fung, *Liq. Cryst.* **1993**, *15*, 87.
[72] J. L. Figueirinhas, J. W. Doane, *Mol. Cryst. Liq. Cryst.* **1994**, *238*, 61.
[73] M. L. Magnuson, B. M. Fung, M. Schadt, *Liq. Cryst.* **1995**, *19*, 333.
[74] A. Yoshizawa, H. Kikuzaki, M. Fukumasa, *Liq. Cryst.* **1995**, *18*, 351.
[75] S. Miyajima, T. Hosokawa, *Phys. Rev. B* **1995**, *52*, 4060.
[76] A. Abe, H. Furuya, R. N. Shimizu, S. Y. Nam, *Macromolecules* **1995**, *28*, 96.
[77] J. W. Emsley, G. De Luca, G. Celebre, M. Longeri, *Liq. Cryst.* **1996**, *20*, 569.
[78] D. Sandström, A. V. Komolkin, A. Maliniak, *J. Chem. Phys.* **1996**, *104*, 9620.
[79] W. Guo, B. M. Fung, *J. Chem. Phys.* **1991**, *95*, 3917.
[80] T. H. Tong, B. M. Fung, J. P. Bayle, *Liq. Cryst.* **1997**, *22*, 165.
[81] G. Celebre, F. Castiglione, M. Longeri, J. W. Emsley, *J. Magn. Reson., Ser. A* **1996**, *121*, 139.
[82] G. K. Pierens, T. A. Carpenter, L. D. Colebrook, L. D. Field, L. D. Hall, *J. Magn. Reson.* **1992**, *99*, 398.
[83] L. D. Field, G. K. Pierens, K. J. Cross, M. L. Terry, *J. Magn. Reson.* **1992**, *97*, 451.
[84] J. M. Polson, E. E. Burnell, *J. Magn. Reson., Ser. A* **1994**, *106*, 223.
[85] T. Chandrakumar, J. M. Polson, E. E. Burnell, *J. Magn. Reson., Ser. A* **1996**, *118*, 264.
[86] S. Caldarelli, M. Hong, L. Emsley, A. Pines, *J. Phys. Chem.* **1996**, *100*, 18696.
[87] M. Hong, A. Pines, S. Caldarelli, *J. Phys. Chem.* **1996**, *100*, 14815.
[88] A. Naito, M. Imanari, K. Akasaka, *J. Chem. Phys.* **1996**, *105*, 4504.
[89] P. Lesot, F. Nielson, J. M. Ouvrard, J. Courtieu, *J. Phys. Chem.* **1994**, *98*, 12849.
[90] J.-M. Ouvrard, B. N. Ouvrard, J. Courtieu, C. L. Mayne, D. M. Grant, *J. Magn. Reson.* **1991**, *93*, 225.
[91] L. Frydman, P. C. Rossomando, B. Frydman, *J. Magn. Reson.* **1991**, *95*, 484.
[92] J.-L. Canet, A. Fadel, J. Salaün, I. Canet-Fresse, J. Courtieu, *Tetrahedron: Asymmetry* **1993**, *4*, 31.
[93] P. Lesot, Y. Gounelle, D. Merlet, A. Loewenstein, J. Courtieu, *J. Phys. Chem.* **1995**, *99*, 14871.
[94] P. Lesot, D. Merlet, J. Courtieu, J. W. Emsley, *Liq. Cryst.* **1996**, *21*, 427.

[95] J.-L. Canet, I. Canet, J. Courtieu, S. Da Silva, J. Gelas, Y. Troin, *J. Org. Chem.* **1996**, *61*, 9035.
[96] J. Leisen, M. Werth, C. Boeffel, H. W. Spiess, *J. Chem. Phys.* **1992**, *97*, 3749.
[97] S. C. Kuebler, C. Boeffel, H. W. Spiess, *Liq. Cryst.* **1995**, *18*, 309.
[98] H. Finkelmann, H. Ringsdorf, J. H. Wendorff, *Makromol. Chem.* **1978**, *179*, 273.
[99] V. Shibaev, N. Platé, Y. Freidzon, *Polym. Sci. Polym. Chem. Ed.* **1979**, *17*, 1655.
[100] J. Leisen, C. Boeffel, R. Y. Dong, H. W. Spiess, *Liq. Cryst.* **1993**, *14*, 215.
[101] J. Leisen, C. Boeffel, H. W. Spiess, D. Y. Yoon, M. H. Sherwood, M. Kawasumi, V. Percec, *Macromolecules* **1995**, *28*, 6937.
[102] D. Schwarze-Haller, F. Noack, M. Vilfan, G. P. Crawford, *J. Chem. Phys.* **1996**, *105*, 4823.
[103] G. S. Iannacchione, D. Finotello, L. V. Natarajan, R. L. Sutherland, V. P. Tondiglia, T. J. Bunning, W. W. Adams, *Europhys. Lett.* **1996**, *36*, 425.
[104] J. Dolinšek, O. Jarh, M. Vilfan, S. Žumer, R. Blinc, J. W. Doane, G. Crawford, *J. Chem. Phys.* **1991**, *95*, 2154.
[105] G. S. Iannacchione, G. P. Crawford, S. Žumer, J. W. Doane, D. Finotello, *Phys. Rev. Lett.* **1993**, *71*, 2595.
[106] G. S. Iannacchione, G. P. Crawford, S. Qian, J. W. Doane, D. Finotello, *Phys. Rev. E* **1996**, *53*, 2402.
[107] G. P. Crawford, R. Ondris-Crawford, S. Žumer, J. W. Doane, *Phys. Rev. Lett.* **1993**, *70*, 1838.
[108] G. P. Crawford, R. J. Ondris-Crawford, S. Žumer, S. Keast, M. Neubert, J. W. Doane, *Liq. Cryst.* **1993**, *14*, 1573.
[109] R. J. Ondris-Crawford, G. P. Crawford, S. Žumer, J. W. Doane, *Phys. Rev. Lett.* **1993**, *70*, 194.
[110] R. J. Ondris-Crawford, G. P. Crawford, J. W. Doane, S. Žumer, M. Vilfan, I. Vilfan, *Phys. Rev. E* **1993**, *48*, 1998.
[111] A. Zidanšek, S. Kralj, G. Lahajnar, R. Blinc, *Phys. Rev. E* **1995**, *51*, 3332.
[112] H. Schmiedel, R. Stannarius, C. Cramer, G. Feller, H.-E. Müller, *Mol. Cryst. Liq. Cryst.* **1995**, *262*, 167.
[113] F. Grinberg, R. Kimmich, *J. Chem. Phys.* **1996**, *105*, 3301.
[114] P. Sheng, *Phys. Rev. Lett.* **1976**, *37*, 1059.
[115] P. Sheng, *Phys. Rev. A* **1982**, *26*, 1610.
[116] S. Disch, C. Schmidt, H. Finkelmann, *Macromol. Rapid. Commun.* **1994**, *15*, 303.
[117] H. Finkelmann, S. Disch, C. Schmidt in *Polymeric Materials Encyclopedia*, Vol. 5 (Ed.: J. C. Salamone), CRC, Boca Raton, FL **1996**, p. 3794.

[118] F. M. Leslie, G. R. Luckhurst, H. J. Smith, *Chem. Phys. Lett.* **1972**, *13*, 368.
[119] J. W. Emsley, J. C. Lindon, G. R. Luckhurst, D. Shaw, *Chem. Phys. Lett.* **1973**, *19*, 345.
[120] S. K. Khoo, G. R. Luckhurst, *Liq. Cryst.* **1993**, *15*, 729.
[121] A. F. Martins, P. Esnault, F. Volino, *Phys. Rev. Lett.* **1986**, *57*, 1745.
[122] P. Esnault, J. P. Casquilho, F. Volino, A. F. Martins, A. Blumstein, *Liq. Cryst.* **1990**, *7*, 607.
[123] N. Schwenk, C. Boeffel, H. W. Spiess, *Liq. Cryst.* **1992**, *12*, 735.
[124] A. S. Sailaja, K. Venu, V. S. S. Sastry, *Mol. Cryst. Liq. Cryst.* **1994**, *250*, 177.
[125] E. Ciampi, J. W. Emsley, *Liq. Cryst.* **1997**, *22*, 543.
[126] N. Heaton, D. Reimer, G. Kothe, *Ber. Bunsenges. Phys. Chem.* **1993**, *97*, 1320.
[127] P. Esnault, J. P. Casquilho, F. Volino, *Liq. Cryst.* **1988**, *3*, 1425.
[128] J. B. Ferreira, H. Gerard, D. Galland, F. Volino, *Liq. Cryst.* **1993**, *13*, 645.
[129] H. Gotzig, S. Grunenberg-Hassanein, F. Noack, *Z. Naturforsch.* **1994**, *49a*, 1179.
[130] A. I. Nakatani, M. D. Poliks, E. T. Samulski, *Macromolecules* **1990**, *23*, 2686.
[131] D. A. Grabowski, C. Schmidt, *Macromolecules* **1994**, *27*, 2632.
[132] M. Lukaschek, S. Müller, A. Hasenhindl, D. A. Grabowski, C. Schmidt, *Colloid Polym. Sci.* **1996**, *274*, 1.
[133] S. Müller, P. Fischer, C. Schmidt, *J. Phys. II France* **1997**, *7*, 421.
[134] H. Siebert, D. A. Grabowski, C. Schmidt, *Rheol. Acta* **1997**, *36*, in press.
[135] L. J. Berliner (Ed.), *Spin Labeling Theory and Application*, Academic, New York **1976**.
[136] D. J. Schneider, J. H. Freed, *Adv. Chem. Phys.* **1993**, *73*, 387.
[137] D. Xu, R. H. Crepeau, C. K. Ober, J. H. Freed, *J. Phys. Chem.* **1996**, *100*, 15873.
[138] G. L. Millhauser, J. H. Freed, *J. Chem. Phys.* **1984**, *81*, 37.
[139] V. S. S. Sastry, A. Polimeno, R. H. Crepeau, J. H. Freed, *J. Chem. Phys.* **1996**, *105*, 5753.
[140] J. W. Saalmueller, H. W. Long, T. Volkmer, U. Wiesner, G. G. Maresch, H. W. Spiess, *J. Polym. Sci. B: Polym. Phys.* **1996**, *34*, 1093.
[141] A. Sammet, M. Hubrich, G. Maresch, H. W. Spiess, *Adv. Mater.* **1995**, *7*, 747.
[142] K. Hasharoni, H. Levanon, J. von Gersdorff, H. Kurreck, K. Möbius, *J. Chem. Phys.* **1993**, *98*, 2916.

2 X-Ray Characterization of Liquid Crystals: Instrumentation

Richard H. Templer

2.1 Origins of X-Rays

Electrons with kinetic energies in excess of a few thousand electron volts will produce X-radiation when subjected to large accelerations. Laboratory X-ray sources do this by bombarding a metal target with electrons boiled off from a hot filament, which is held at a high potential with respect to the target. The electrons lose energy to X-ray conversion either by direct collision with the metal atoms or by the acceleration they experience in the presence of the field near the atomic nucleus.

In the former case, inner shell electrons in the metal atom are ionized and the filling of the now vacant inner shell by an outer shell electron produces an X-ray of a characteristic wavelength. Of course there are a number of different routes by which the ionized atom may reach its lowest energy state, which leads to a series of distinct X-ray emission peaks. The emissions are typically characterized by the initial and final energy states of the atom. In Fig. 1 K_α and K_β transitions are illustrated.

Energy lost by an electron in the deceleration experienced during interaction with the atom's nuclear electric field leads to the emission of a continuous spectrum of X-ray wavelengths called bremsstrahlung. The spectral distribution is governed by the laws of quantum electrodynamics, which predict a spectrum whose form depends upon the incoming electron energy alone, with the low wavelength cut-off in the spectrum being equal to the maximum electron kinetic energy. The combination of atomic transitions and bremsstrahlung leads to typical

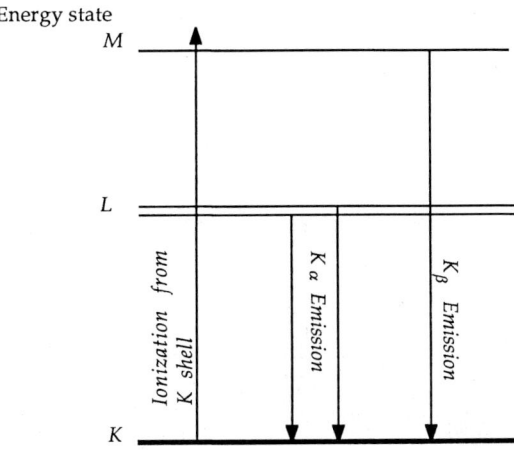

Figure 1. Energy level diagram showing a K shell electron being ionized and the transitions of electrons from the L and M shells which lead to X-radiation.

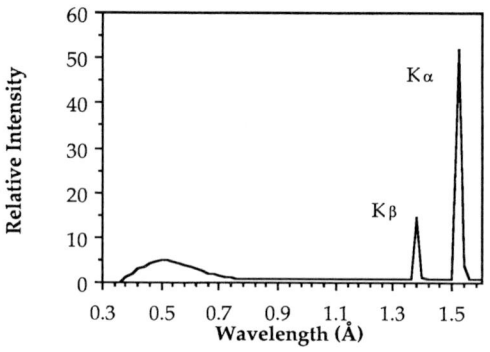

Figure 2. The X-ray spectrum of copper at 35 kV. The K_α peak is in fact a doublet with the most intense peak at 1.5405 Å.

X-ray emission spectra such as that shown in Fig. 2 for copper.

2.2 Generation of X-Rays in the Laboratory

An appropriate X-ray source for routine liquid crystal work is illustrated schematically in Fig. 3. Electrons are produced by electrically heating a tungsten wire filament to temperatures in excess of 2000 K. This heating produces a cloud of electrons around the wire. The potential gradient between the wire (cathode) and the target (anode) accelerates the thermalized electrons into the target where X-rays are produced. In order to ensure that the mean free path of the electron is greater than the anode to cathode distance, and in order to avoid high voltage breakdown, the apparatus is held under high vacuum, either in a sealed glass tube or in an evacuable chamber. Typically, a copper target is used, where the brightest X-ray emission is at 1.5405 Å.

Most X-ray sources also contain some electron focusing device intermediate between the cathode and the anode, with the most common configuration being one in which the focusing cup produces a rectangular focal spot on the target, with dimensions of approximately 1×10 mm^2. This means that by viewing the target end on to the rectangular focal spot and at a shallow take-off angle (6° or so) one gets a projected image of the spot of 1×1 mm^2. At 90° to this view the projected image is a narrow beam 100 μm $\times$ 10 mm. Many X-ray sources provide four ports at 90° intervals, giving two point sources and two line sources. Each port has a beryllium window, which transmits over 90% of the emitted X-rays.

The major objective in all X-ray tube design is to provide as much X-ray intensity

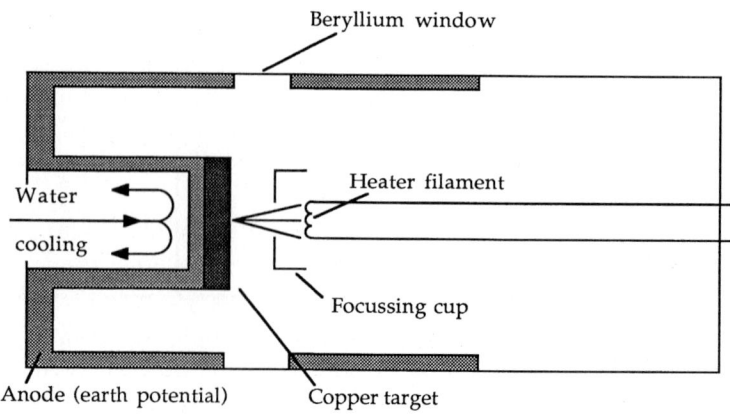

Figure 3. The hot-cathode X-ray tube.

as possible. This is particularly important when examining the structure of liquid crystals, where the scattering power of the material is, in general, relatively low. The greatest stumbling block to increased X-ray intensity is the high proportion of electron energy (98% or so), which generates heat in the target material. Increasing the electron beam current will eventually result in pitting or melting of the target metal. This is prevented in the type of X-ray source shown in Fig. 3 by drawing heat out of the target with a flow of coolant.

Using such cooling, tubes can be run at levels of up to 40 mA beam current at around 40 kV. Running flat out, the average source should last at least 2000 h before the filament burns out. Most X-ray sources of the type described above are made as a sealed glass X-ray tube to be fitted onto the manufacturers high voltage supply units. These tubes are discarded once the filament is burned out. Alternatively, sources are available where the filaments are replaceable, and hence the X-ray chamber must be continuously evacuated.

The standard X-ray sources described above are suitable for most of the measurements likely to be made routinely. However, where greater X-ray fluxes are required, increases in X-ray intensities of approximately 25× may be obtained by using a rotating anode generator. Here the target is both cooled and continuously rotated in the electron beam, thereby allowing higher beam current densities. Its greatest disadvantage over the standard X-ray tube is the need for more regular servicing and repair of the moving components in the vacuum system.

For even greater X-ray intensities, an X-ray synchrotron source is necessary. Here X-rays are produced by accelerating relativistic electrons (or positrons) around curved paths. The X-rays produced by these means are by their nature highly collimated (compare this to the systems described so-far, where X-rays are emitted isotropically) and appear as a continuum of energies. The total intensity at a typical sample at presently available sources is approximately four orders of magnitude greater than a rotating anode generator. At the moment such sources are only to be found at centralized national and international facilities.

Before passing on to a discussion of X-ray optics, mention should be made of microfocus diffraction tubes. These X-ray sources are in all respects similar to the standard, continuously evacuable X-ray sources, with the difference that the electron source is small. This means that smaller electron beam currents are maintained, but very small focal spots can be created on the target. This is extremely useful in high resolution work for small-angle X-ray scattering studies, a common requirement in liquid crystalline studies, where lattice spacings may exceed 10 nm.

2.3 X-Ray Cameras

For most X-ray diffraction techniques, a relatively monochromatic X-ray beam is needed. From Fig. 2, it is clear that the best part of the spectrum to select, from the point of view of intensity, is the K_α peak. In fact the K_α peak is composed of a closely spaced doublet, $K_{\alpha 1}$ and $K_{\alpha 2}$. So for the highest monochromaticity, and hence the best resolution, the $K_{\alpha 1}$ peak should be selected, being twice as bright as $K_{\alpha 2}$.

For liquid crystalline materials, where lattice parameters are large, a well collimated or focused beam with extremely small amounts of radiation scattered from sources external to the sample under investigation is additionally required. Such scattering is frequently termed "parasitic", and can arise from X-rays scattering from materials in their path, e. g., air and the beryllium vacuum window. The net effect of such parasitic scatter is to introduce a background signal at small angles, which can obscure the low angle X-ray peaks of materials with large lattices.

2.3.1 The Debye–Scherrer and Flat Film Cameras

The Debye–Scherrer camera is the most basic system with which measurements on a limited range of liquid crystalline materials can be made. The camera can only be used with "powder" samples, i. e., a sample with many microdomains covering all orientations with respect to the X-ray beam.

The Debye–Scherrer uses X-ray filtering to reduce the K_β and continuum components of the spectrum with respect to the K_α peak. This is a poor man's monochromator, and is best suited to low resolution work. In Fig. 4 the absorption characteristics of nickel have been plotted over the radiation spectrum for copper at 35 kV. The sharp rise in absorption with increasing photon energy is due to X-ray absorption in the K-shell of nickel. To install such a "monochromator" the correct thickness of filter has to be used in order to reduce K_β to negligible values without reducing K_α by too much. For the nickel filter a suitable thickness is 17 µm, giving 98.4% absorption of the K_β peak, while reducing K_α by 50%.

An elaboration of the above method is to introduce a second filter whose K absorption edge is at a wavelength slightly greater than that of the K_α peak. For copper, nickel and cobalt are used. This reduces the intensity of emission at shorter wavelengths by even more with respect to K_α. However, filter thicknesses have to be carefully bal-

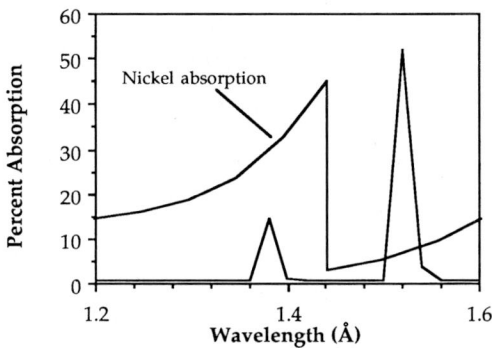

Figure 4. The X-ray absorption curve for a nickel filter approximately 8 µm thick. The copper K peaks are shown in order to indicate the selective filtration of K_β radiation.

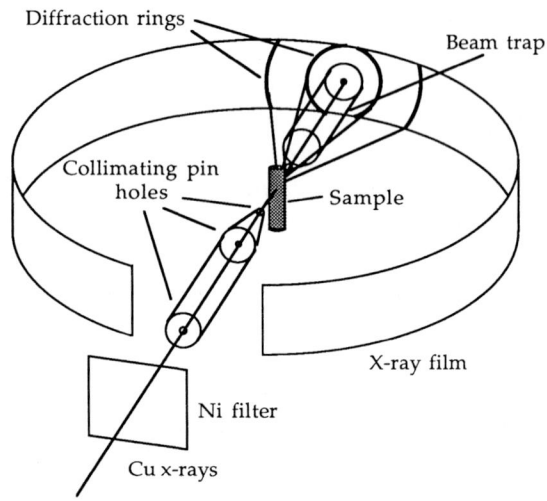

Figure 5. The Debye–Scherrer camera. The X-ray film is held in place inside the cylindrical camera housing, which has not been shown.

anced to give the optimum results. In this case the best values are 10 µm of nickel and 10.8 µm of cobalt, reducing K_β to 2.5% of K_α. The continuum radiation is also reduced to a very small fraction of K_α.

Radiation from the standard laboratory X-ray source is emitted in all directions. The Debye–Scherrer produces an X-ray beam simply by looking at a small, collimated fraction of the emitted radiation. The system is shown in Fig. 5 in schematic form.

Three pinholes at the input form a collimated beam with low parasitic scatter. The first two holes are of equal diameter and produce the collimation. The third hole is somewhat larger and positioned as close as possible to the sample. Its function is to shield any scattering off the edges of the second aperture, and by being placed close to the sample it minimizes air scatter. There is also an exit tube, whose function is to prevent air-scattered X-rays from contributing to the background. Diffraction from the powder sample is recorded on film along the chamber circumference.

Standard Debye–Scherrer cameras have diameters of 114.6 mm, making 1 mm on the circumference equal to 1° in angle. Because of the size of the exit tube needed to reduce parasitic scatter, the standard cameras miss out the first 6° of scattering. Using copper K_α X-rays, this relates to spacings of 1.5 nm, this is not of enormous value for most liquid crystal work. However, since liquid crystalline materials by their very nature will produce no diffraction at large angles [spacings greater than 0.2 nm or $2\theta = 45°$], the exit tube can be replaced by a small concave button in line with the direct beam. X-rays backscattered by the beam stop will for the most part fall outside the regions of interest. There will be increased scattering from gas, but this can be reduced, either by evacuating the chamber or flowing helium gas into it. The realistic limit on maximum spacings with such a system is approximately 4 nm, given a beam diameter of the order of 1 mm.

A variant of the Debye–Scherrer is the flat film camera shown in Fig. 6. Here the filtering and pinhole collimation of the X-ray beam are identical to those in a Debye–Scherrer camera, but the pattern is recorded on the flat film. This has the ad-

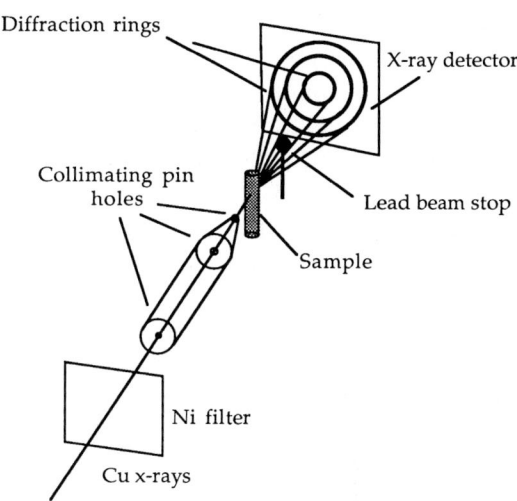

Figure 6. The flat film camera. Normally the entire camera is enclosed in a vacuum chamber or a chamber filled with a gas which causes minimal X-ray scattering, such as helium. Note that because of the smaller size of the beam, smaller reciprocal spacings can be observed.

vantage of the entire ring of diffraction being recorded, which means that oriented samples can be examined and crystal symmetries determined directly.

Clearly both of these systems will be of limited utility for liquid crystal work, because of the systems rather inadequate resolution at small angles. For better resolution, greater monochromaticity is required along with low levels of parasitic scatter.

2.3.2 The Guinier Camera

The Guinier camera (e.g. [1]) is altogether more sophisticated than the pinhole and filtration techniques used in Debye–Scherrer type cameras, and will enable the user to make accurate determinations of lattice spacings up to 15 nm with exposure times of the order of a few tens of minutes for an average sample on a standard X-ray generator.

The Guinier produces highly monochromatic radiation by using a crystal monochromator. A monochromator disperses the light from an X-ray source in accordance with Bragg's law, with longer wavelengths being diffracted through larger angles. By setting a narrow aperture at the angle corresponding to the desired wavelength, it is possible to separate the K_α doublet. Of course Bragg's equation at the given angle will also be satisfied by wavelengths of $\lambda/2$, $\lambda/3$, and so on. However, for a copper target they are in the white light part of the spectrum, which is of greatly reduced intensity, and, furthermore, by choosing a suitable material the crystal planes corresponding to the second harmonic wavelength can be suppressed or eliminated. From this point of view, large single crystals of diamond, silicon, or germanium are excellent materials for a monochromator.

The Guinier camera goes a step further by bringing the divergent X-ray beam to a focus. In this way, the usable beam intensity can be increased to levels close to those in a Debye–Scherrer camera. A focusing crystal monochromator is illustrated in Fig. 7. The curved crystal is shown as a full line which falls on a circle of radius R. A line source of X-rays is at P with the long axis coming out of the page. The crystal layer orientation has been indicated at two points along the curved crystal surface. The change in layer orientation across the crystal surface is created by carefully polishing a radiused curve on the crystal before it is bent into a cylinder. When in its final configuration, the polished curve falls on a radius of R and the bent crystal itself falls on a radius $2R$. This then brings monochromatic radiation to a focus at Q.

A quartz crystal can be polished and elastically bent to a radius in the range 200–400 mm, with the crystal being 2 mm thick. The actual value of R must be chosen to suit the wavelength range of interest and the size of the camera. The configuration of a Guinier camera is illustrated in Fig. 8. The sample is interposed in the beam some distance before its focal point, so that a reasonable volume can be irradiated. The diffraction pattern then falls on a circle whose di-

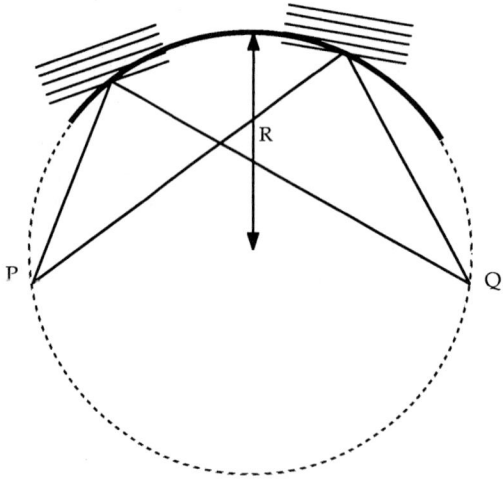

Figure 7. A curved crystal monochromator.

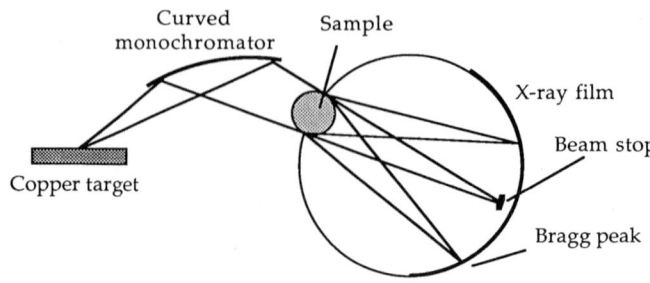

Figure 8. The Guinier camera. Sharp monochromatic Bragg peaks are brought to a focus on a cylindrically curved film strip.

ameter is given by the distance from the specimen to the focus. The size of the focused beam is equal to that of the projected area on the X-ray target, so a standard X-ray tube will give us a beam width at a focus of 100 µm. Finer lines are of course possible with finer focus X-ray tubes (widths down to a few tens of micrometers are achievable). A properly adjusted camera will give an extremely monochromatic and sharp beam profile at small scattering angles. The camera chamber is normally evacuated, so that air scatter is eliminated. The insertion of beam stops to block the direct beam and light scattered from parts of the monochromator reduces the parasitic scatter to levels where extremely high resolution is possible at low angles. At larger angles, the resolution is reduced somewhat as the beam becomes somewhat broadened. However, in the range 15–0.3 nm the Guinier is an excellent camera for liquid crystal work on powder samples.

2.3.3 The Franks Camera

The Franks camera allows the resolution of features in excess of 60 nm with a point focus [2]. However, with this camera lower levels of monochromaticity are achieved than with a Guinier. As shown in Fig. 9, the Franks camera produces a point source image by reflection off two curved surfaces.

At the extremely short wavelengths of X-radiation, total reflection only occurs at glancing incidence. Copper K_α radiation has a critical angle of incidence for total reflection of 11' for glass and 30' for a gold surface. Most of the shorter wavelength radiation passes through the reflector. However, it is prudent to use a nickel filter prior to the mirrors to reduce the K_β component.

By elastically bending the mirrors to rather large radii of curvature, it is possible to bring the beam to a point focus. The first mirror focuses in the horizontal direction, the second in the vertical direction. The distance from the X-ray source to focus can lie in the range 200–400 mm. The aperture is used to cut out parasitic scatter arising from scattering off the reflector surface, and this is placed just in front of the sample. The distance from sample to focal plane is normally around 20–40 mm. However, because the X-ray beam is only very slowly convergent, the diffraction pattern can be recorded quite a long way away from the focal plane with no significant loss of resolution in the Bragg peaks. This makes it possible to separate closely spaced diffraction peaks by simply recording the pattern further away from the sample.

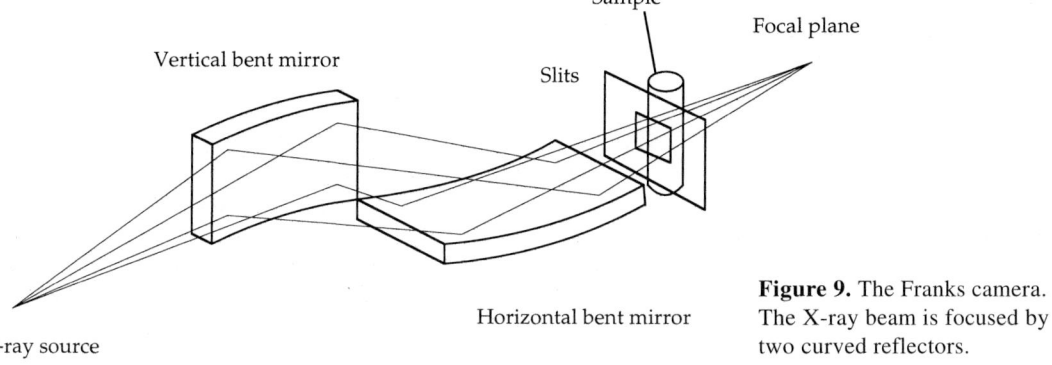

Figure 9. The Franks camera. The X-ray beam is focused by two curved reflectors.

Clearly the Franks camera cannot produce monochromatic light like the Guinier, but this is compensated for by the lower levels of parasitic scatter, which have enabled measurements up to 100 nm to be made. This low parasitic scatter arises primarily because the slow convergence of the X-ray beam allows the exclusion of scattered radiation rather effectively. Naturally enough this comes at the expense of lower beam intensity, so for most experimental configurations a high intensity X-ray source, such as a rotating anode generator, is required. In this configuration the Franks camera is an excellent device for making measurements on macroscopically oriented liquid crystal phases, with fairly large repeat distances.

2.3.4 The Huxley–Holmes Mirror–Monochromator Camera

This camera (see Fig. 10) uses a cross between the Franks and Guinier optics [3]. The monochromator crystal is used to select $K_{\alpha 1}$ and focuses the beam in the horizontal direction. The glancing incidence mirror focuses rays in the orthogonal direction. This camera can resolve repeat spacings beyond the 100 nm mark. This camera is not commercially available, although component parts can be purchased. For those with the ability to construct and adjust such a camera, there are many advantages. It produces a brighter and more monochromatic beam than the Franks, and with less parasitic scatter at low angles than the Guinier. To realize the full potential of the camera it should be used with a high intensity source and a sensitive X-ray detector.

2.3.5 The Elliott Toroid Camera and Others

Where a more intense spot focus of fairly monochromatic X-radiation is required than is available in either a Franks or a mirror–monochromator camera, an Elliott toroid camera can be used [4]. This camera uses a gold-coated mirror distributed on a toroidal surface, as shown in Fig. 11, to bring an anulus of X-radiation to a focus.

The toroidal mirror is created from an epoxy resin cast. The angle of incidence for point source X-rays on the gold coated surface is 29.5′, and additional nickel filtering reduces the K_β radiation to negligible levels. The first aperture cuts out any beam not incident on the toroid. Next, a beam stop is used to cut out the direct beam, leaving only a ring of light which will strike the toroid. The next two beam stops cut out some of the radiation parasitically scattered off the mirror surfaces. Indeed it is increased parasitic scatter that is the cost paid for the higher intensity available from the toroid camera. The scatter reduces the maximum resolvable repeat spacing to somewhat less than 8 nm, when working with the full intensity beam. However, for oriented liquid crystals with repeat spacings less than 8 nm, where reasonably short exposure times are required, the Elliott camera is a good solution.

A camera with similarities to the Franks and Guinier cameras is due to Kratky and

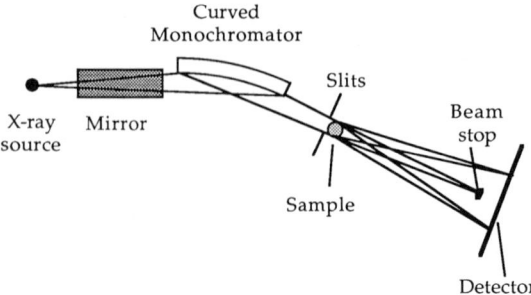

Figure 10. A top view of the Huxley–Holmes camera. A curved crystal monochromator replaces the first glass reflector of the Franks camera.

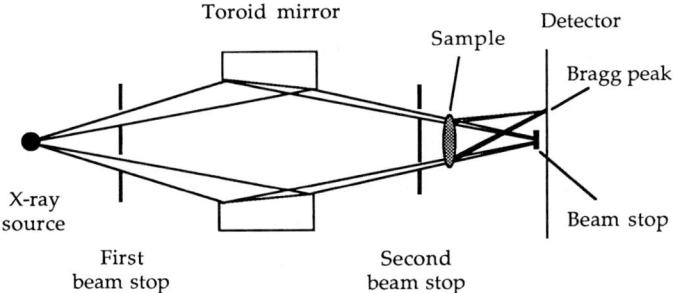

Figure 11. The Elliott toroid camera. Rays are brought to a focus by the toroidal mirror; in this figure the focal length has been considerably exaggerated, normally the radius of curvature of the mirror is 20 m.

Leopold [5]. Here, either a flat monochromator or a flat glass block is used to monochromate the X-ray beam. A second glass block is positioned so that its face just grazes the outgoing monochromated beam. The reflected beam is of course not focused, but even so the use of a glass surface to reflect away parasitic scatter enables spacings up to 100 nm to be resolved. Without focusing, the exposure times tend to be longer than for a Franks or a Guinier single mirror system.

2.3.6 What Camera Should be Used

The range of available cameras is rather large and confusing. The list given here has been severely shortened by personal experience and no doubt some bias as well. In that spirit the following comments are made over choosing a camera:

a) No single camera will cover all work.
b) Fortunately, most X-ray generators have several windows, so that different cameras can be put on the ports.
c) A good basic set of cameras should include one line focus and one point focus camera.
d) A good selection of cameras will include a Guinier, on which most basic work will probably be done.
e) The camera required is not always commercially available. Guinier, Franks, and Kratky cameras are readily available through commercial manufacturers, but home-made versions always appear to perform better.

2.4 The Recording of X-Ray Diffraction Patterns

There are a wide array of detectors available for X-ray imaging. We have limited this discussion to two-dimensional X-ray detectors: film, image plates, multi-wire proportional counters, and optoelectronic detectors. Other two-dimensional systems, such as directly bombarded, solid state X-ray detectors are also available, but are still in the early stages of development.

2.4.1 X-Ray Film

Each X-ray photon that is stopped by X-ray film produces a single developable silver halide grain in the film emulsion. A typical film thickness will stop approximately 60% of the photons. Film has an inherent background, which means that X-ray peaks only become visible once they exceed the noise level in this background. For a 50×50 μm^2 beam this corresponds to something in the

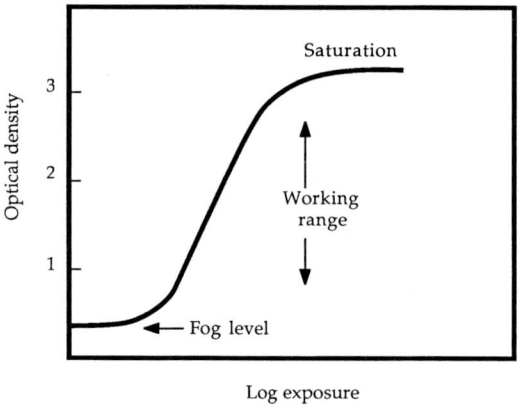

Figure 12. The response of X-ray film.

region of 10 000 photons in a Bragg spot before the peak becomes detectable at a signal-to-noise ratio of 1. Once above this level, it becomes possible to measure peak-to-peak spacings on the film, either under a traveling microscope or by microdensitometerization. These measurements can be used to work out the mesophase symmetries and lattice repeat spacings. When it is required to determine structures from the diffraction, the intensities of each peak must be determined. To do this with film is a relatively complex procedure, because of the nonlinear response of the film's optical density with exposure.

As can be seen in Fig. 12, there is only a limited region of linearity. Of course one solution to this problem might be to count each developed silver halide grain, but the more usual approach is to expose a stack of X-ray films of known X-ray absorption and then make measurements in the linear region of optical density. This means that the most intense peaks may be measured from the third film from the top of the stack, whilst the weakest reflections are measured from the top of the stack.

Although X-ray film suffers from nonlinearity and poor low dose response, it has many attractive features over other detectors. It is easy to use for semi-quantitative work; it is inexpensive; spatial resolution is limited by grain size, although in most cases it is the effective resolution of the densitometer (25 μm) that puts a limit on a measurement; it is available in all sorts of sizes and it is pliable; it has good absorption efficiency for X-rays, and at higher exposure values approaches an ideal quantum limited detector.

Every X-ray laboratory uses film, no matter what other detectors it may possess.

2.4.2 Image Plates

A direct and popular replacement for film has recently appeared on the market: the image plate [6]. Diffraction images are recorded by exposing a plastic sheet coated with a 150 μm layer of $BaFBr:Eu^{2+}$ powder in a white organic binder. This creates a latent image in the sheet in the form of electronic defects, which can subsequently be read out by scanning the sheet with a focused HeNe laser beam. The red light stimulates a blue (390 nm) emission wherever the screen was exposed to X-rays (see Fig. 13). Since the emission energy is greater than the stimulating photon's energy, it is clear that we are recording a stored signal. The blue light is detected with a filtered photomultiplier. The photomultiplier current is digitized and the resulting signal is recorded as a function of position on a computer. Unlike standard X-ray film, the phosphor screen may apparently be used repeatedly, by simply zeroing any remaining signal on the image plate by exposure to visible light.

The image plate is more sensitive than X-ray film, although it is not quite sensitive enough to pick up single X-rays. It has a linear response to X-rays over a wider dynamic range than film, something like a factor of 10^5 between the dimmest and the bright-

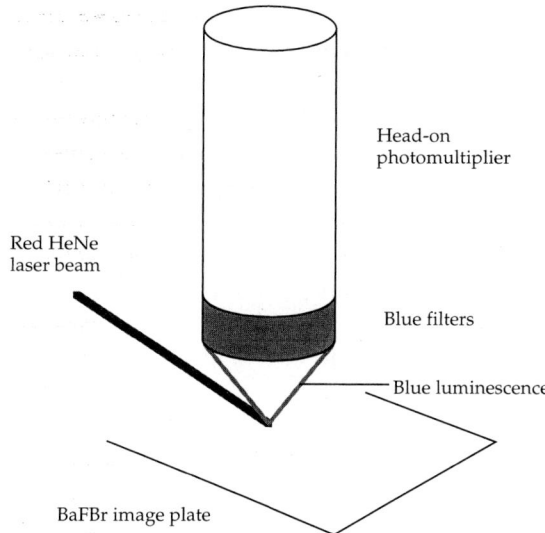

Figure 13. Optical scanning of an image plate. The red laser beam raster scans the plate, causing blue emission where there has been radiation ionization.

est signal. However, it has lower spatial resolution, at 100 μm, it is far more costly than film, and it has a slower read out and zeroing cycle than other modern detectors.

2.4.3 Multi-Wire Proportional Counters

Multi-wire proportional counters work by detecting the spatial position of X-ray induced gas ionization [7]. The location of the ionizing event is picked up on the set of perpendicular anode wires shown in Fig. 14.

The chamber is normally filled with an inert gas mixture consisting mainly of xenon, which has a high capture cross section for X-ray photons. For a xenon-filled chamber 12 mm thick at 1 atm (1.013×10^5 Nm^{-2}), copper radiation is absorbed at an efficiency of 80%. The upper and lower cathode are held somewhere in the region of a few kilovolts below the anode voltage. The ionization induced by the passage of an X-ray can be spatially localized by the amplitude of the electrical pulse picked up by the grid of parallel wires in the upper and lower cathode planes. The cathode wires are spaced at approximately 1 mm intervals, giving a resolution of 1 mm in 200–1000 mm. Resolution towards the edges of the detector is actually somewhat worse than this because of parallax effects on precise position location.

The multi-wire proportional counter has a number of advantages over film. It detects single quanta with very small noise levels, so this leads to a great improvement over film in terms of sensitivity. On average, exposures with a multi-wire proportional counter will take 1/50th of the time of exposures on film. Intensity measurements are also far more accurate; the reproducibility

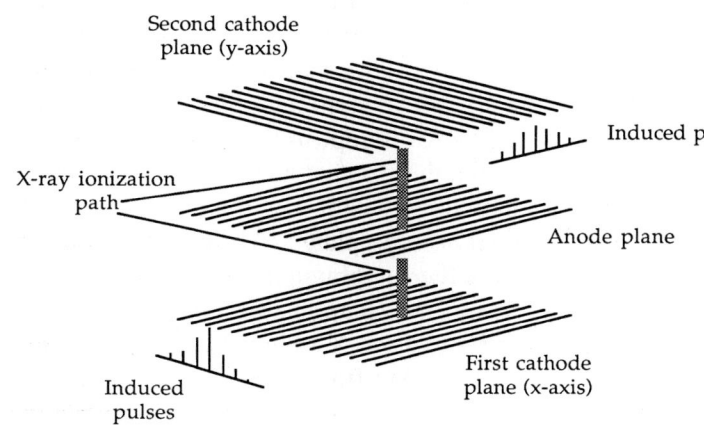

Figure 14. A multi-wire proportional counter. Electrical pulses induced by the X-ray ionization avalanche are picked up on the cathode wires and are used to locate the X-ray's position.

of measurements is simply limited by noise in the X-ray flux. The best that can be expected with X-ray film is 5% reproducibility. The resolution of the detector is better than film for low intensity peaks, because of the better signal-to-noise ratio. At higher intensity, the resolution for small-angle X-ray scattering is comparable to film, so long as the diffraction pattern is focused across the entire detector. This implies rather large distances from the X-ray sources, so Franks or Huxley–Holmes optics have to be used. The image is of course digital, making immediate quantitative analysis of the diffraction pattern possible. The multi-wire proportional counter, like the image plate, has greater dynamic range than film, for example, $10^4:1$ as compared to maybe $200:1$. Clearly it is possible to quantitatively measure a wider range of peak intensities simultaneously on a multi-wire proportional counter or image plate than on film.

The multi-wire proportional counter is an excellent device for accurate quantitative work. It has drawbacks, certainly in terms of expense and complexity, but commercial devices of a high standard are available. The multi-wire proportional counter is a detector for those who can afford the cost and need sensitive, quantitative X-ray detection.

2.4.4 Opto-Electronic X-Ray Imaging Devices

These devices give similar levels of performance for small-angle X-ray scattering as a multi-wire proportional counter, at somewhat lower levels of instrumental complexity [8]. As shown in Fig. 15, these detectors work by converting the X-ray pattern to an optical image using an X-ray phosphor. The optical image is then amplified with an image intensifier, which works somewhat like a photo-multiplier, but retains spatial information.

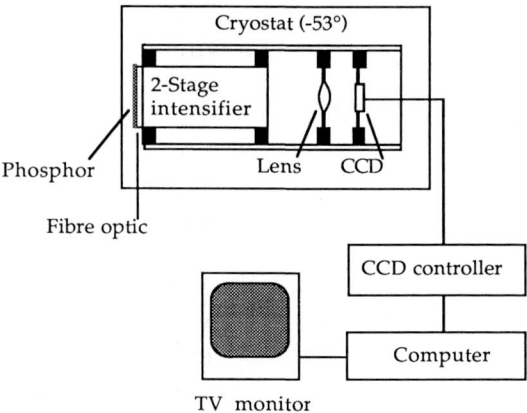

Figure 15. A CCD based optoelectronic X-ray imager.

The output image on the intensifier tube is then focused onto an electronic image device such as a TV tube, or more recently a CCD (charge coupled device), from where it can be read out. The image tube and the imaging device are cooled to low temperatures in order to reduce dark current and noise levels.

These devices are in general used as integrating imaging devices, which allow an image to be accumulated over some specified time without simultaneous read out. This leads to a great reduction of noise, quantum-limited detection of X-rays, and a dynamic range which for CCDs can rise to $10^5:1$. Over this dynamic range linearity is better than a fraction of a percent. Because the image is integrated there is no count-rate limitation. The input phosphor screen is normally between 40 and 80 mm in diameter, with a resolution in the region of 150 μm. Again, these detectors give better resolution than film at low X-ray intensity, but are less good where peaks are bright. In addition, since the images are already digitized, accurate and reproducible measurements of peak intensities can be made.

These detectors are suitable for use with Franks, Huxley–Holmes, or flat film cameras. The technology is slightly less complex than for the multi-wire proportional counter, and less expensive than both the counter and the image plate systems (although still a great deal more costly than film).

These detectors can be put together from commercial components, but recently a number of companies have sprung up selling such detectors. The new CCD based detectors have such low dark current and noise that very faint images can be integrated over many hours. So, although these detectors have the same sensitivity gain as a multi-wire proportional counter, they also extend exposure times with extremely low noise and background.

2.4.5 What Detector Should be Used

Again the range of detectors available is large and confusing. Using a multi-wire proportional counter or an optoelectronic detector in conjunction with a rotating anode generator and toroid or Huxley–Holmes optics, it is possible to obtain diffraction images from liquid crystal phases in less than 1 s. Using a standard X-ray generator with a Guinier camera and film, the same quality of image might taken an hour or two.

Clearly detectors and intense X-ray sources make it possible to do experiments that cannot be done with a standard system. The greatest advantage they have is, however, not just to do with the fact that they are so much faster (although this is an important element), but with the fact that data can be read out and analyzed in situ. This enables the experimenter to interact with an experiment in a way that film and image plate do not allow. In our experience, this has been the single greatest advance in our ability to probe and understand the structural and phase behavior of liquid crystalline systems.

The following recommendations stem from this experience:
a) It is advisable to have at least one beam line using film and one with a detector.
b) For liquid crystal research, optoelectronic detectors are probably the best choice. They are quite robust, cheaper than the alternatives, and the moderate format is entirely compatible with the diffraction angles obtained with liquid crystals.
c) The more complex X-ray camera should be used with the detector. This enhances the performance of the beam line, and the detector also makes it easier to make camera alignment.

2.5 Holding a Liquid Crystal in the X-Ray Beam

A number of methods are available for getting a liquid crystal sample into the path of the X-ray beam.

The most widely used method of mounting powder samples uses a thin wall capillary. Glass capillary tubes with overall diameters in the range 0.5–2.0 mm are available, and have been drawn to wall thicknesses of a few tens of micrometres. The scatter from the glass is insignificant. Capillaries in excess of 2.0 mm cause too much X-ray absorption to be of interest, and for most liquid crystalline materials anything below 1.0 mm will not produce sufficient diffraction. The exceptions to this latter case are where the material under investigation contains heavy atoms that scatter strongly.

Getting samples into capillaries can often be a frustratingly difficult process; every practitioner finds their own suite of preferred methods for removing air bubbles from thermotropics or making homogeneous mixtures with lyotropics. However, sometimes the difficulties are insuperable and alternative sample containment has to be used. A method used routinely in some laboratories is shown in Fig. 16. The X-ray "transparent" windows are made from thin sheets of mica cut to size. The sample is placed in the cavity produced by the Teflon spacer before the second mica window is used to seal the sample in place. Unlike glass, mica produces some sharp Bragg peaks which will appear in the diffraction pattern. These are easily identified and pose no problem as long as they do not mask sample diffraction. Alternative materials for X-ray windows include thin Mylar sheet, cling film and Kapton.

If there is any propensity for macroscopic alignment, then a powder pattern can still be produced by rotating the sample in the beam during exposure. More often than not, any sample alignment in the above sample containers is only partial.

Samples need not be confined in an X-ray transparent container in order to make measurements. Freely suspended films and pulled strands can be formed in situ for studies on aligned samples (see Figs. 17 and 18). The literature is littered with mentions of alternative methods of sample alignment, with surface-induced alignment, and magnetic alignment being among the more common [10]. In our laboratory we have a samarium/cobalt permanent magnet, giving a field of 1.6 T with a 1 mm pole gap. The magnet can be taken up to above 200 °C, which makes it ideal for in situ alignment of thermotropic liquid crystals in the nematic phase.

2.6 Controlling the Sample Environment

The major concern of X-ray structural work on liquid crystal systems is the need to make

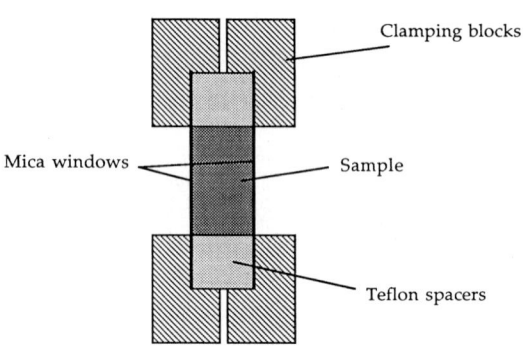

Figure 16. A simple X-ray sample holder.

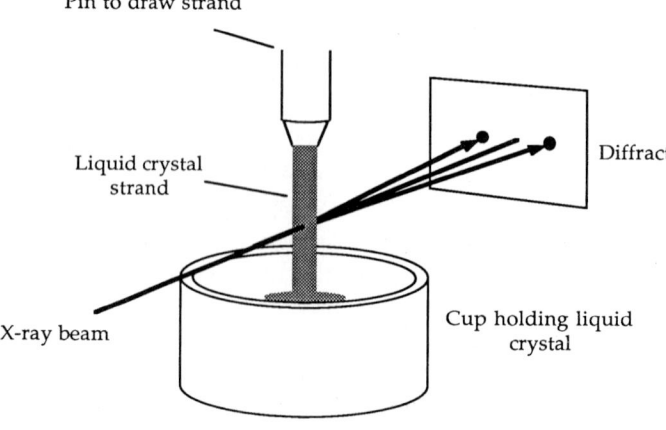

Figure 17. A liquid crystal strand is pulled out of a cup containing a pool of sample. Normally this will be in the smectic phase, with temperature being varied after an aligned strand has been formed.

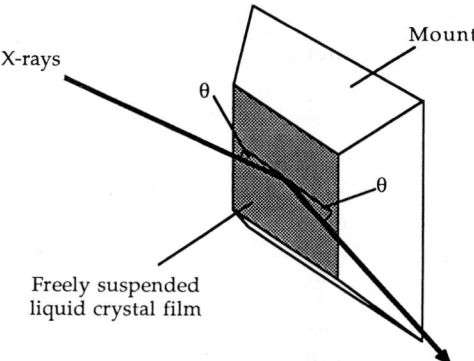

Figure 18. An aligned liquid crystal film is formed by wiping sample across an aperture, whilst the sample is in the smectic A or C phase [9]. With a protruding sample mount, X-ray scattering at small angles may be performed on aligned samples.

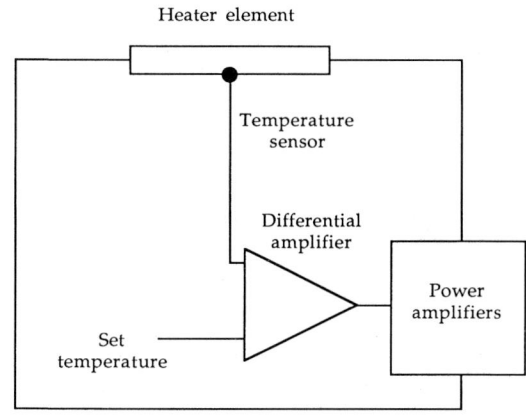

Figure 19. Schematic diagram of a servo controlled heater.

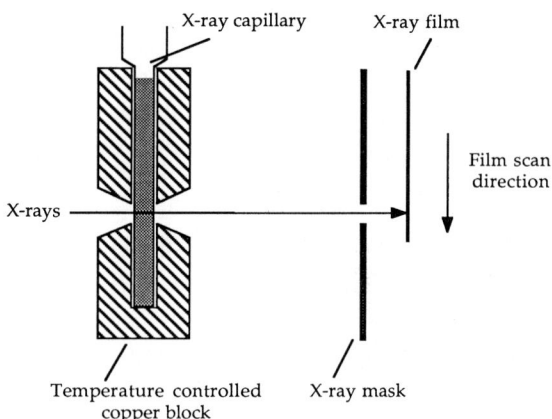

Figure 20. A continuous sequence of X-ray diffraction patterns from a powder sample can be recorded as a function of temperature by ramping the sample temperature whilst simultaneously scanning the X-ray film past a linear aperture in a metal X-ray mask.

X-ray measurements whilst thermodynamic parameters are varied. From the foregoing mention of sample alignment methods it should be clear that the same is true when creating aligned samples.

Temperature control is a universal requirement. Normally ±0.1 °C control or better will be required. This sort of precision requires servo control of the heater. A schematic servo controlled system is shown in Fig. 19.

The sample temperature is measured by a miniaturized sensor, such as a thermocouple or a platinum resistance thermometer. The sensor voltage is then compared to a voltage set by the experimenter and an error voltage is used to drive current through the heater. For the best precision, three term control is used in which power is reduced as the set temperature is approached and optimum heating is provided to settle on the set temperature.

By producing an electronic ramp, the temperature of the sample can be increased continuously. By simultaneously scanning a film across the diffraction pattern as it varies with temperature, an entire thermotropic phase sequence can be recorded in one measurement (e. g., [11]) (see Fig. 20).

Where cooling is required as well as heating, a closed loop refrigeration circuit, servoed in a similar manner, can be used. A more sophisticated temperature controller, which can both heat and cool, can be made from a Peltier thermoelectric device. In this device a voltage across a series of thermocouple junctions sets up a temperature difference across the couples. By suitably arranging the thermocouple junctions, one face of the device cools whilst the other heats. Reversing the polarity of the junction connections reverses the heating and cooling faces. Servo controlling a Peltier is more complex than separately controlling a heater and a refrigerator, but performance at ±0.03 °C is possible over a temperature range from −50 °C to 80 °C.

In lyotropic systems where single crystals are being grown, humidity control may be necessary. This can be done crudely by circulating humidified helium gas through the sample chamber. Various salt solutions can be mixed to give a number of humidity values. Care of course must be taken to ensure that condensation does not occur in the pipework to the chamber, and that in the chamber temperatures are controlled sufficiently to ensure a small enough variation in the relative humidity.

A humidity generator with servo control of the humidity is a far more satisfactory solution to the difficult problem of humidity control [12]. The atmosphere in a large temperature-controlled reservoir is regulated by pumping the gas through a drier or bubbling it through water. The humidity is sensed by an electronic dew point hygrometer. The drying or wetting end is turned on depending on whether the sensor is above or below the dewpoint value set by the user. With this set up it is possible to control the dewpoint to ±0.1 °C.

2.7 References

[1] A. Guinier, *Ann. Phys., Paris* **1939**, *12*, 161; A. Guinier, *X-Ray Crystallographic Technology*, Hilger and Watts, London **1952**; J. Witz, *Acta Cryst.* **1969**, *A25*, 30.
[2] A. Franks, *Brit. J. Appl. Phys.* **1958**, *9*, 349.
[3] M. K. Reedy, K. C. Holmes, R. T. Tregear, *Nature* **1965**, *207*, 1267; H. E. Huxley, W. Brown, *J. Mol. Biol.* **1967**, *30*, 383.
[4] A. Elliott, *J. Sci. Instrum.* **1965**, *42*, 312.
[5] O. Kratky, H. Leopold, *Makromol. Chem.* **1970**, *133*, 181.
[6] Y. Amemiya, J. Miyahara, *Nature* **1988**, *336*, 89.
[7] J. R. Helliwell, *Nucl. Inst. Methods Phys. Res.* **1982**, *201*, 153; U. W. Arndt, *J. Appl. Cryst.* **1986**, *19*, 145.
[8] S. M. Gruner, *Science* **1987**, *238*, 305; E. F. Eikenberry, S. M. Gruner, J. L. Lowrance, *IEEE Trans. Nucl. Sci.* **1986**, *33*, 542.
[9] D. E. Moncton, R. Pindak, *Phys. Rev. Lett.* **1979**, *43*, 701; W. Klämke, Z. X. Fan, W. Haase, H. J. Müller, H. Gallardo, *Ber. Bunsenges. Phys. Chem.* **1989**, *93*, 478.
[10] J. Als-Nielsen, P. S. Pershan, *Nucl. Inst. Methods Phys. Res.* **1983**, *208*, 545; R. Oldenbourg, W. C. Phillips, *Rev. Sci. Instrum.* **1986**, *57*, 2362.
[11] C. J. Oakley, J. E. Lydon, *J. Phys. E: Sci. Instrum.* **1977**, *10*, 296.
[12] S. M. Gruner, *Rev. Sci. Instrum.* **1981**, *52*, 134.

3 Structural Studies of Liquid Crystals by X-ray Diffraction

John M. Seddon

3.1 Introduction

This section is concerned with how structural information about liquid crystal phases may be obtained from X-ray diffraction studies. In addition to presenting a brief overview of some of the more quantitative aspects, we will also attempt to give a simple pictorial view of how diffraction patterns may be analyzed qualitatively in terms of the mesophase structures. For the background X-ray diffraction theory relevant to partially-ordered systems, a number of classical texts should be consulted [1–3].

Here we will focus on the bulk static structures of the phases formed by low molar mass calamitic (rod-like) mesogens. The principles involved in structural studies of columnar phases, and of polymeric and lyotropic systems, are very similar. The study of liquid crystal surfaces requires reflectivity and/or glancing angle diffraction techniques [4–6].

A number of reviews describe X-ray studies of liquid crystals up to about 1980 [7–10]. A monograph by P. S. Pershan [11] reviews X-ray work up to 1987, and reprints many of the most significant papers in the field. A number of more recent texts cover various structural aspects of liquid crystal phases [12–14]. Other reviews describe the structural classification of liquid crystals [15], the nature of the ordering within smectic phases [16–18], the structures of ferroelectric and chiral smectic phases [19], the structures of frustrated smectics [20–24], the structures of columnar discotic liquid crystals [25], and X-ray studies of side group liquid crystalline polymers [26].

Orientational order in liquid crystals is most effectively studied using 'molecular probe' techniques such as NMR spectroscopy. However, the study of translational order, that is structure, requires diffraction techniques, usually involving X-rays or neutrons. A disadvantage of diffraction is that it does not distinguish between static and dynamic disorder. However, this can be achieved using quasielastic neutron scattering.

Diffraction is a phenomenon which comes about when radiation is elastically scattered from atoms in a sample (primarily from electrons for X-ray diffraction). The various scattered wavelets from the different atomic sites combine, undergoing constructive or destructive interference, depending on the relative phases of the different wavelets. These phases depend upon the relative positions of all of the atomic sites, and hence the amplitude (and hence intensity) of the diffracted wave in a given direction depends upon the spatial distribution of the atoms in the sample. Thus a diffraction experiment probes the static structure, that is, the translational order within the sample. The total scattered intensity does not depend on the positional order, but only on the total number of scattering centers (and their scattering powers) in the sample; correlations in position between atoms or groups of atoms lead to a redistribution of the scattering, intense in some directions, weak in others. The degree of redistribution is related to the extent of positional order in the sample.

X-rays are a form of electromagnetic radiation, with typical wavelength, $\lambda = 0.154$ nm. The scattering power of atoms is simply proportional to the atomic number Z; it is weak for hydrogen, and becomes progressively stronger on moving down the periodic table. Unlike visible light, X-rays are practically undeviated by refraction on passing through matter, a typical value of refractive index is $n = (1 - 10^{-5})$. This means that they cannot be focused with lenses, and hence cannot form a direct image of the sample. This constitutes the essential problem of diffraction studies, that the 'image' must be deduced mathematically from an analysis of the diffraction pattern. However, unfortunately this is not directly possible, because all information on the relative phases of the scattered waves is lost, as only the modulus squared of the wave amplitude (the intensity) is recorded experimentally. This is the well-known phase problem of crystallography.

3.2 Bragg's Law and Powder Diffraction

Bragg visualized the scattering of X-rays by a crystal in terms of reflections from sets of lattice planes, as shown in Fig. 1 (note that there is normally a phase change of 180° upon scattering, for X-rays).

For one particular set of planes, constructive interference between rays reflected by successive planes will only occur when the path difference, $2d \sin \theta$, equals an integral number of wavelengths:

$$2d \sin \theta = n\lambda \tag{1}$$

where d is the separation of the planes, θ is the angle of incidence, n is an integer and λ is the wavelength. Note that the diffraction angle is 2θ.

To observe a particular diffraction peak, labelled 'n' in Fig. 2a, we must align the planes at an angle θ_n to the incident beam. The incident and diffracted beam directions are specified, as shown in Fig. 2b, by their wavevectors k_i and k_s, whose moduli are $|k_i| = |k_s| = \frac{2\pi}{\lambda}$. Note that since the scattering is elastic (no energy change), the magnitudes of the wavevectors are equal. The diffracted intensity is most conveniently plotted as a function of the scattering vector Q, where $Q = (k_s - k_i)$, that is, it represents the change in wavevector of the diffracted beam (Fig. 2). Its modulus is given by

$$|Q| \equiv Q = \frac{4\pi \sin \theta}{\lambda} \tag{2}$$

An equivalent statement of Bragg's law is then

$$Q_n = n\left(\frac{2\pi}{d}\right) \tag{3}$$

For a set of equally spaced planes, the scattered intensity $I(Q)$ is everywhere zero ex-

3.2 Pseudo-Bragg's Law and Powder Diffraction

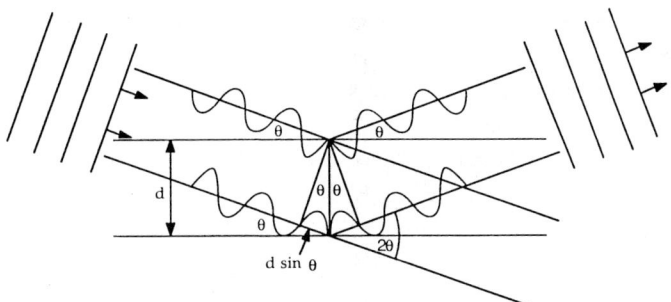

Figure 1. Bragg's law.

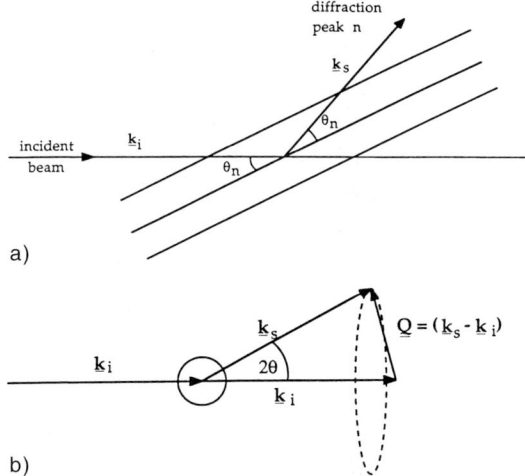

Figure 2. (a) Reflection from parallel planes, and (b) definition of the scattering vector Q.

cept where Bragg's law is satisfied, and the diffraction pattern consists of a set of equally-spaced Bragg peaks, $2\pi/d$ apart, along a direction normal to the planes.

The simplest way to observe the various diffraction peaks is to employ a powder sample, that is, a sample which is composed of very many microdomains, each with a random orientation (Fig. 3a). The Bragg condition will thus automatically be satisfied for all values of n, and all of the possible diffraction peaks will be simultaneously observed. Furthermore, each spot will be averaged into a ring around the incident beam, since the sample will contain many different domains aligned at a given angle θ_n, but with random angles of rotation around the beam.

The averaging is equivalent to rotating the scattering vector Q around the incident beam k_i, producing a ring of possible scattered beams k_s for each Bragg peak. Such powder patterns are hence intrinsically one-dimensional, being completely specified by the radii of the various rings (i.e. the diffraction angles 2θ). Thus the intensity measured at the detector varies only with the magnitude of Q, not its direction, and hence we need only plot $I(Q)$ versus scalar Q to describe the diffraction pattern, as illustrated in Fig. 3b.

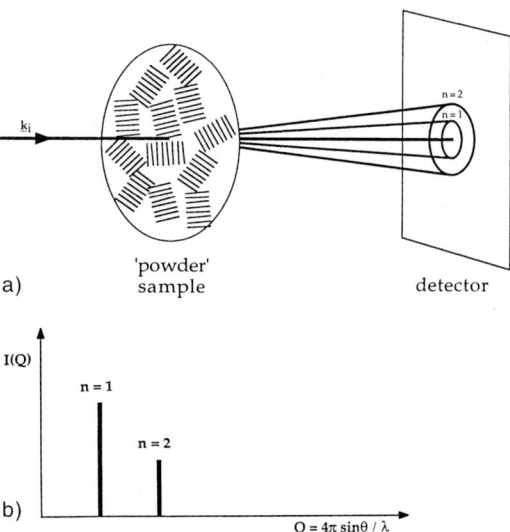

Figure 3. (a) Origin of powder patterns, and (b) radial plot of intensity.

3.3 Diffraction Patterns: Unaligned and Aligned Samples

3.3.1 Unaligned Samples

The positions in Q (angle) of the diffraction peaks are reciprocally related to the separations between molecules (or groups of molecules) within the liquid crystalline phase. The sharpness of the peaks is related to the extent to which the separations extend periodically over large distances. For phases with Bragg peaks, the ratio of the peak positions reveals the long-range organization of the phase (Fig. 4). In the case of the low-angle peaks, ratios of 1, 2, 3 ... indicate a smectic phase (1-D stack of layers); ratios of 1, $\sqrt{3}$, 2, $\sqrt{7}$, 3 ... indicate a hexagonal phase (2-D lattice of rods, e.g., columnar phase of discotics); ratios of 1, $\sqrt{2}$, $\sqrt{3}$, 2, $\sqrt{5}$, $\sqrt{6}$, $\sqrt{8}$, 3 ... indicate a cubic phase. The wide angle peaks relate to the translational order over short distances (or having short periodicities), for example the lateral (in-plane) order in the various smectic phases.

Probably the most useful information to be obtained from unaligned smectic phases is the temperature dependences of the layer spacings. Such data have, for example, been reported for terephthal-bis-(4-n-alkylanilines) [27, 28], cyano-, alkoxybenzylidene alkylanilines [29], three-ring esters [30], mesogens with siloxane tails [31], and bipyridine stilbene compounds [32].

For the nematic phase, which is an orientationally-ordered fluid, diffuse maxima are seen in both the low angle (low-Q) and the wide angle (high-Q) regions (Fig. 5).

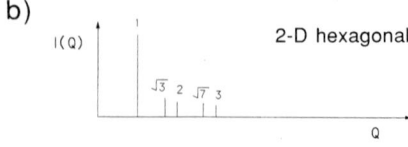

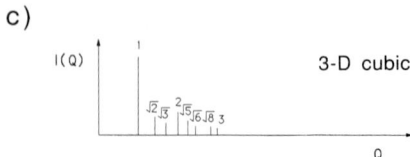

Figure 4. Low angle diffraction patterns characteristic of (a) 1-D layer (smectic) phases, (b) 2-D hexagonal phases, and (c) 3-D cubic phases.

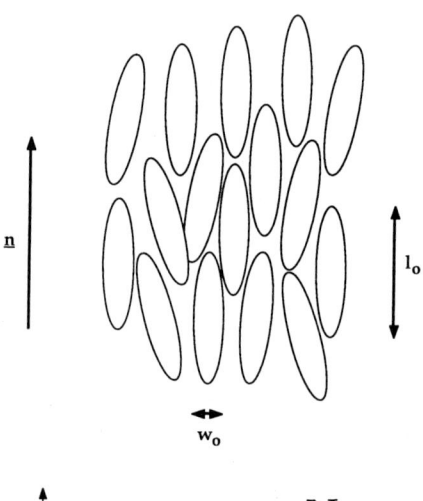

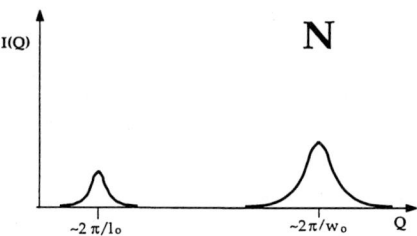

Figure 5. Local structure and X-ray powder pattern from the unaligned nematic phase of calamitic mesogens.

These peaks are normally centred close to $Q \approx 2\pi/l_0$ and $Q \approx 2\pi/w_0$ respectively, where l_0 and w_0 are the molecular length and width (typically, $l_0 \approx 2.5$ nm; $w_0 \approx 0.5$ nm). They arise from the average end-to-end and side-to-side separations of the close-packed molecules. The peaks are diffuse because these positional correlations only extend over short distances, typically, a few molecular diameters. The widths of the diffuse maxima are inversely proportional to these correlation lengths, ξ. In fact, such correlations usually also exist in the isotropic phase, so it may be difficult to distinguish the powder patterns of the nematic from that of the isotropic phase. Note that the chiral nematic (cholesteric), where the pitch is typically $\gg 0.1$ μm, will also give a very similar diffraction pattern.

For the simplest kind of SmA phase, denoted SmA$_1$, there is a layer structure with a periodicity d slightly less than the molecular length l_0. This effect is due to a combination of imperfect molecular orientational order ($\langle P_2 \rangle < 1$) and conformational disorder in the molecules. This leads to a number of equally-spaced low-angle (00l) pseudo-Bragg peaks at $Q_l = \left(\dfrac{2\pi l}{d}\right)$ (Fig. 6a). The tendency for layering is rather weak: for many smectic phases the density distribution along the layer normal is close to a cosine wave. This means that only the first one or two pseudo-Bragg peaks have observable intensity. The ratio I_{002}/I_{001} can be as low as 2×10^{-7}. The wide angle diffuse peak is similar to that observed from the nematic phase, since the lateral packing of the molecules remains liquid-like. Other types of SmA phase are possible, as will be described later, but these may be difficult to infer from the powder patterns. One indication for a more complex SmA structure is if the layer spacing d is very different from the molecular length l_0 (deduced from molecular modeling).

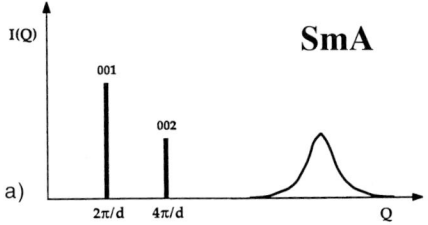

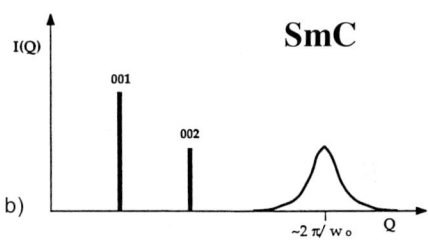

Figure 6. Powder pattern of (a) the SmA phase and (b) the SmC phase.

The powder diffraction pattern of the SmC phase (Fig. 6b), where the director is tilted with respect to the layer normal, is difficult to distinguish from that of the SmA phase. The only difference is that the layer spacing will be reduced due to the tilt. If the transition to the SmC phase occurs from a SmA phase, then it is possible to deduce the tilt angle θ_t from plots of layer spacing versus temperature, as shown in Fig. 7. By extrapolating the SmA spacing d' into the SmC phase, the tilt angle may be deduced approximately using the relation $\theta_t = \cos^{-1}(d/d')$. Such estimates agree extremely well with the directly-measured tilt angle, at least for the system 4O.7 [33].

The hexatic phase SmB, and the tilted versions SmF and SmI, are characterized by the onset of increased order within the layers, with the positional coupling between layers remaining weak. In general, one cannot distinguish these phases, nor deduce the tilt angles, from the powder patterns, shown for the SmB phase in Fig. 8. The lateral packing of the molecules is locally hexago-

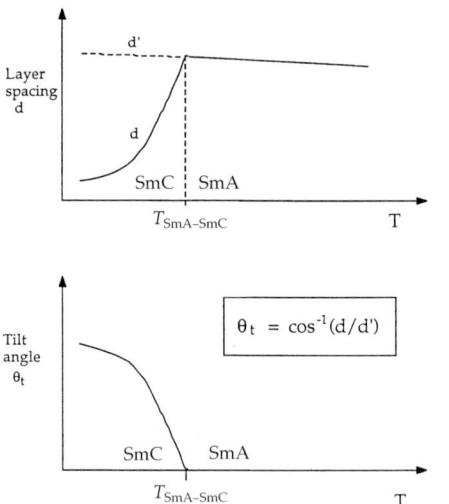

Figure 7. Tilt angle measurements for SmC phase, from powder patterns.

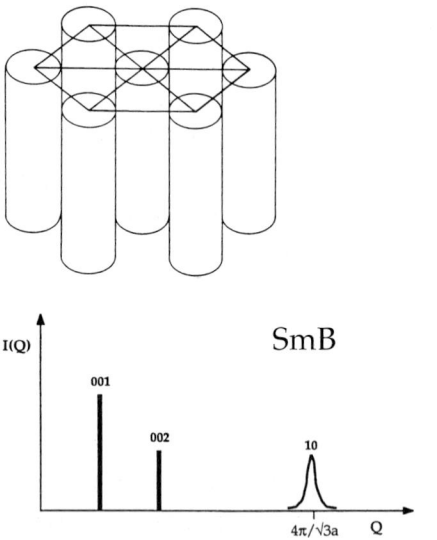

Figure 8. Powder pattern of the SmB phase.

nal (implying long axis rotation) but this positional order is not long range, and disappears over distances of the order of 10 nm. However, the hexatic phases display long range lattice orientational order. The low-angle patterns are usually similar to those from the SmA and SmC phases, but the wideangle peak becomes much sharper. The position of this peak, indexed on a hexagonal lattice as the (10) peak, is at a value of $Q_{10} = 4\pi/\sqrt{3}a$, where a is the lateral separation of the molecules. Note that one cannot quantify the extent of in-plane structure from the width of this peak, since other effects, such as the dimensionality and the tilt, contribute to its width. However, usually the widths are in the following order: SmB < SmI < SmF.

The signature for the ordered smectic mesophases B, G, J, and E, H, K is the apperance of a (small) number of sharp (hkl) reflections in the wide-angle region. This indicates the onset of coupling between the layers, and the development of long-ange order within the layer. For phases B, G and J, which derive from the hexatic phases SmB, SmF and SmI, there is six-fold rotational disorder about the molecular long axis, which leads to hexagonal or quasihexagonal packing of the molecules. The B phase is (3-D) hexagonal, whereas the tilted G and J phases are C-centered monoclinic (or triclinic). Quenching of the six-fold rotations leads to a herring-bone packing, which lowers the symmetry of the lattices, converting the untilted phase B to E (orthorhombic), and G and J to H and K (monoclinic). However, rapid axial rotations of 180° are still present. These crystal mesophases are distinct from normal crystals in that considerable disorder is present: this leads to only a few pseudo-Bragg peaks being observed. For the crystal B phase, different layer stacking sequences, such as AAAAAA, ABABAB, and ABCABC, have been observed; in addition, a ripple-like modulation of the layers has been detected, on approaching a transition to the tilted G phase [34, 35].

3.3.2 Thermotropic Cubic Phases

Early work showed that the optically isotropic phase once called smectic D gave low-angle X-ray powder patterns consistent with cubic symmetry [36], later suggested to be of spacegroup Ia3d [37]. A number of models were proposed for the structure of the cubic D phase [38, 39]. It should be noted that cubic liquid crystalline phases are very common in lyotropic systems [40]. Typically, thermotropic cubic phases have been found to occur either between the SmC and SmA phase, or between the crystal and a SmC phase. Their unit cell dimensions are in the range of 4.5–9.0 nm. Their diffraction patterns consist of sharp (h, k, l) Bragg peaks in the low-angle region, whose positions are in the ratios 1, $\sqrt{2}$, $\sqrt{3}$, 2, $\sqrt{5}$, $\sqrt{6}$, $\sqrt{8}$, 3..., characteristic of cubic lattices (these peaks may not all be observed). The wide-angle region shows diffuse scatter similar to that from the SmA phase, indicating that the local structure is liquid-like. One of these cubic phases was shown by monodomain diffraction studies to have a primitive cubic structure [41, 42]. Recently, many more examples of thermotropic cubic phases have been found, for example in metallomesogens [43–49], in silver thiolates [50], in ellagic acid derivatives [51], in carbohydrate liquid crystals [52–55], in polyhydroxy amphiphiles [56, 57], in biforked mesogens [58], phasmidic systems [59], tetrasubstituted tribenzosilatrane [60], dendrimers [61], substituted esters of benzopyranobenzopyran-dione [62], six-ring double-swallow-tail mesogens [63], and in mixtures of the previous two examples with suitable electron acceptor compounds [64, 65].

3.3.3 Aligned Samples

The use of aligned (monodomain) samples is essential for detailed structural studies of liquid crystals. The alignment may be obtained by electric or magnetic fields, by surfaces, or by the free-standing film method. The following additional features may then be investigated:

1. long-range, quasilong-range and short-range order;
2. cybotactic (smectic-like) clusters in the nematic phase, and critical behavior;
3. fitting of high resolution data to obtain translational correlation lengths;
4. direct measurement of tilt angles;
5. modulated phases and incommensurate smectic phases;
6. structural diffuse scatter: defects, stacking faults, etc.;
7. lattice orientational order;
8. molecular orientational order; and
9. twist grain-boundary (TGB) phases.

By studying free-standing films of liquid crystals [17, 66–84], the structure can be probed free from any substrate, and the effects of dimensionality can be determined by reducing the film thickness, to as few as two molecular layers.

3.4 Diffracted Intensity: Molecular Transforms and Reciprocal Lattices

Bragg's law predicts the direction of any diffracted ray and shows that the smaller the spacing, d, the larger is the diffraction angle 2θ. Thus there is a *reciprocal* relationship between the separations of the planes d in real space, and the diffraction angles $2\theta_n$. However, Bragg's law does not say anything about the intensities of the various peaks. Diffraction theory shows that the amplitude of the wave scattered by a crystal in a particular direction, $F(Q)$, is given by the Fourier transform of the 'structure', where for X-rays the 'structure' means the electron density distribution within the sample, $\rho(r)$:

$$F(Q) = \int \rho(r) \exp[i Q \cdot r] \, dV \qquad (4)$$

The observed intensity is then the modulus-squared of the amplitude:

$$I(Q) = |F(Q)|^2 \qquad (5)$$

If we are considering a crystal, where the mean atomic positions are fixed, then the Fourier integral in Eq. (4) can be replaced by a Fourier summation, involving the atomic scattering factors, over the atomic sites within the unit cell. The effects of (anisotropic) thermal vibrations are taken into account by use of Debye–Waller factors.

To gain a qualitative understanding of liquid crystal diffraction patterns, it is very useful to consider the form of the molecular transform. This may be calculated by carrying out a Fourier summation over the atomic sites in the molecule, in an assumed conformational state. However, for illustrative purposes it is adequate to model a calamitic molecule as a uniform cylinder of electron density ρ, length $2L$, radius r, and volume V. For such an isolated cylinder, embedded in a medium of electron density ρ_s, the scattered amplitude is given [85] by

$$F(Q) \equiv F(Q, \beta) = 2(\rho - \rho_s)$$
$$\cdot V \left(\frac{\sin(QL \cos \beta)}{QL \cos \beta} \right) \left(\frac{J_1(Qr \sin \beta)}{Qr \sin \beta} \right) \qquad (6)$$

where β is the angle between the cylinder axis and the scattering vector Q, and J_1 is the first-order Bessel function of the first kind.

This 'molecular' transform, as indicated in Fig. 9, has roughly the form in reciprocal space of a flat 'pancake', aligned with its symmetry axis parallel to the real space cylinder. In terms of the surface at which the amplitude first passes through zero, the diameter of the 'pancake' is roughly $(7.7/r)$, and its thickness is roughly $(2\pi/L)$, in reciprocal space (subsidiary nonzero regions exist beyond this surface). The transform $F(Q)$

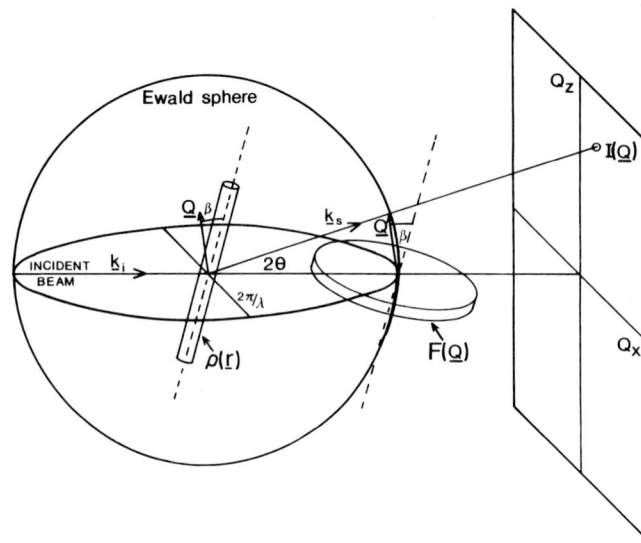

Figure 9. Transform of a cylindrical molecule, and geometry of diffraction. From [390].

3.4 Diffracted Intensity: Molecular Transforms and Reciprocal Lattices

is centered on the surface of the Ewald sphere, at the point where the undiffracted X-ray beam emerges. The orientation of the sample (defined by the electron density function $\rho(r)$), and its transform $F(Q)$, are coupled together: rotating the sample around any axis through its centre has the effect of rotating the transform about its centre by the same amount. It is the intersection of this transform with the Ewald sphere which yields the diffraction pattern of the cylinder (see Fig. 9).

For a periodic structure, $F(Q)$ is only non-zero for those discrete diffraction directions, denoted by the reciprocal lattice vectors Q_{hkl}, for which the Bragg condition is satisfied. The Q_{hkl} constitute a periodic set of δ-functions, known as the reciprocal lattice, and the intensity $I(Q)$ thus consists of a set of sharp (Bragg) peaks. Note that the set of integers (hkl) are the Miller indices of the various sets of lattice planes of the crystal. The real lattice and the reciprocal lattice are related by the standard expressions:

$$a^* = 2\pi \frac{(b \times c)}{V};$$
$$b^* = 2\pi \frac{(c \times a)}{V};$$
$$c^* = 2\pi \frac{(a \times b)}{V} \qquad (7)$$

where $V = a \cdot (b \times c)$.

Thus a^* is perpendicular to b and c, and similarly for b^* and c^*.

We can represent the allowed diffraction from any lattice $\rho(r)$ by constructing a map of the scattered amplitude $F(Q)$ in reciprocal ('Q') space, as shown in Fig. 10.

A periodic stack of infinitely-wide, thin flat layers, a distance a apart in real space, corresponds to a set of points $2\pi/a$ apart in reciprocal space, aligned along the layer normal. Such a structure gives a one-dimensional diffraction pattern, characterized

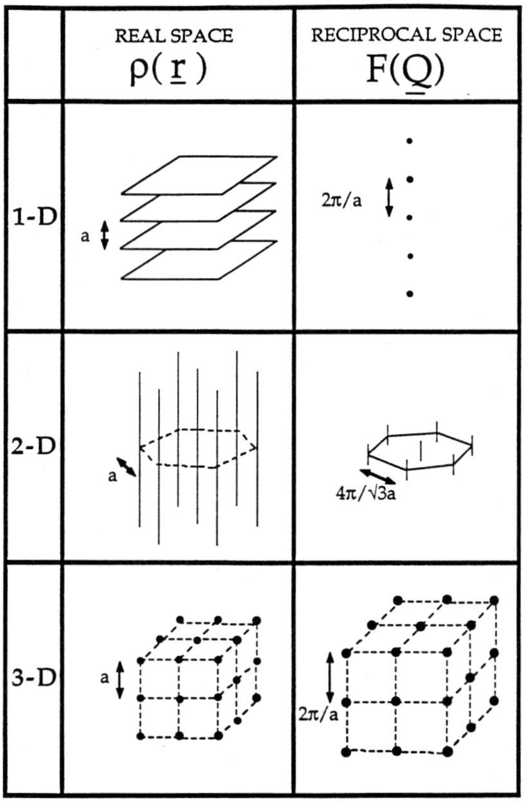

Figure 10. Real space structures and their Fourier transforms. From [391].

by $(00l)$ Bragg reflections (strictly speaking, the h and k labels are redundant in this case).

Because of the properties of Fourier transforms, the labels 'real' and 'reciprocal' are interchangeable, that is a one-dimensional row of points of separation a will give rise to periodic sheets of scattering extending normal to the row, of separation $2\pi/a$.

A two-dimensional lattice in real space corresponds to a two-dimensional lattice in reciprocal space. For example, a hexagonal lattice of infinite rods corresponds to another hexagonal lattice, but of points, and with the lattice rotated by 30°. The real space and reciprocal space lattice parameters a and $4\pi/\sqrt{3}a$) are again reciprocally related. When the rods are long but not infinite, the points in reciprocal space extend along the normal direction to the 2D-lattice. In other words, the points become short rods, whose length is inversely related to the length of the rods of the real space structure. Such a lattice gives a two-dimensional diffraction pattern, characterized by $(hk0)$ reflections (strictly speaking, the l label is redundant in this case).

Finally, a three-dimensional lattice in real space corresponds to a 3-D lattice of points in reciprocal space. For example, a cubic lattice corresponds to a cubic reciprocal lattice; the lattice parameters are a and $2\pi/a$ respectively. Such a lattice gives a three-dimensional diffraction pattern, characterized by (hkl) reflections.

To help us visualize and predict the form of the diffraction pattern, we can consider the structure of an ordered mesophase to consist of the convolution of a 'unit cell' structure with an appropriate lattice. This is useful because of the convolution theorem, which tells us that the scattered amplitude (and hence intensity) from such a mesophase will consist of the product of the transform of the unit cell with the transform of the lattice. For example, the (wide-angle) diffraction pattern of a layer of hexagonally-packed untilted cylindrical molecules (e.g., in the SmB phase) will consist of the product of the transform of a cylindrical molecule (see Fig. 9) with the transform of a hexagonal lattice of points (see Fig. 10). To a first approximation this will have the form of six, hexagonally arranged rods of scattering length $\approx 4\pi/l_0$, with the rods extending normal to the layer (see Fig. 22; note that the hexatic SmB phase does not, in fact, have long-range in-plane order).

3.5 Translational Order

For a partially disordered system, such as a liquid crystalline phase there is normally not a three-dimensional lattice, or in the case of the nematic phase, there may be no periodic structure whatsoever. In these cases $F(Q)$, and hence the intensity $I(Q)$, is typically a continuous function in reciprocal space, with maxima in certain angular regions, and minima in others. In addition, where periodicities do exist in the structure, such as a set of layers, a certain number of Bragg peaks will also be observed. We can define the degree of translational order [15] by a distribution function $D(r)$, which is a measure of the probability of finding the center of mass of a molecule at position r with respect to a test molecule at $r=0$.

3.5.1 Long-range Order

For true long-range order (a crystal), at a temperature of absolute zero, this function will be a periodic set of δ-functions in 3-D, repeating itself whenever r is a multiple of any combination of the lattice vectors a, b and c. In this case, the diffraction pattern consists of series of equally-spaced Bragg peaks (δ-functions), whose positions are reciprocally related to the lattice spacings. The intensity of the peaks can be expressed as:

$$I(Q) \alpha \delta(Q - Q_{hkl}) \qquad (8)$$

where the Q_{hkl} are reciprocal lattice vectors.

For simplicity, we will consider the situation for a smectic layered structure in 1-D, as shown in Fig. 11.

If the structure is smeared-out (by molecular motion: rotation, vibration and/or translation), but remains strictly periodic, then

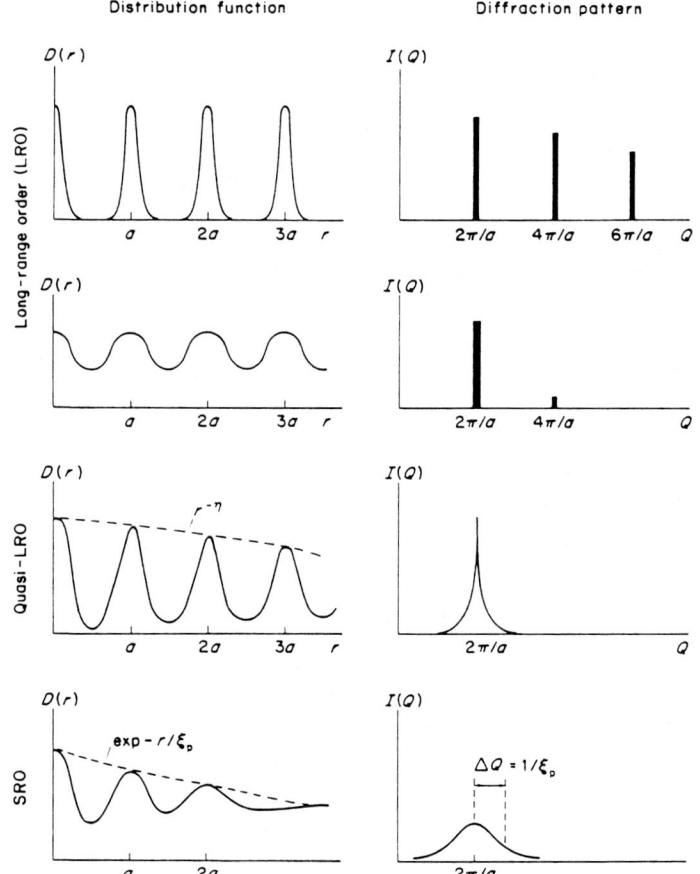

Figure 11. Molecular distribution functions and their diffraction patterns for different kinds of positional order. (From [15]).

the diffraction pattern still consists of sharp Bragg peaks, and in the same positions, but the peaks towards higher angles (higher Q) are progressively reduced in intensity. The more smoothed out the structure, the stronger this effect is. In the limit of a purely sinusoidal distribution of electron density, only the first Bragg peak is observed, at $Q_{001} = 2\pi/a$. Of course, for a completely uniform distribution, there are no observable peaks at all.

The intensities of the $(00l)$ X-ray peaks, suitably corrected for geometric and experimental effects, can be used to derive the electron density distribution along the layer normal. The square root of the intensities yields the moduli of the amplitudes, $\sqrt{I_{00l}} = |F_{00l}|$. Assuming (as is usually the case) that the layer structure is centrosymmetric, then the phases of the amplitudes can only take the values 0 or π, and so only the signs (± 1) of the amplitudes need to be deduced, to yield the structure factors F_{00l}, equivalent to the coefficients α_l in Eq. (9) below. A number of approaches can be employed in order to deduce these signs. Fourier transformation of the F_{00l} then gives the one-dimensional electron density profile (the structure). However, this profile will usually be of low-resolution, because only a few Bragg peaks are detected for thermotropic smectic phases.

3.5.2 Smectic Order Parameters

In such cases it is preferable to analyze the relative intensities of the (00l) Bragg peaks in terms of the smectic order parameters [7, 86–95]. We can represent the periodic modulation in electron density along the layer normal, taken as the z axis, as a Fourier series

$$\rho(z) = \sum_{l=0}^{\infty} \alpha_l \cos\left(\frac{2\pi l z}{a}\right) \qquad (9)$$

If $\alpha_l = 0$ for all $l > 0$, $\rho(z) = \alpha_0 = \frac{1}{a}$, that is uniform, and the structure is completely 'smeared out' (i.e., averaged) along z. For $l \geq 1$, the coefficients α_l are given by the cosine terms in Eq. (9) averaged over the normalized positional distribution function $D(z)$

$$\alpha_l = \frac{2}{a} \int_{-a/2}^{a/2} D(z) \cos\left(\frac{2\pi l z}{a}\right) dz = \frac{2}{a} \tau_l \qquad (10)$$

where the τ_l are defined as the smectic order parameters.

If $\alpha_l = 0$ for all $l > 1$, i.e. only the first harmonic τ_1 (often referred to as 'the smectic order parameter') is nonzero, then

$$\rho(z) = \alpha_0 + \alpha_1 \cos\left(\frac{2\pi z}{a}\right) \qquad (11)$$

and the structure exhibits a pure cosine modulation, of period $z = a$. Note that τ_1 gives the strength of the sinusoidal smectic ordering, that is, is directly proportional to the amplitude of the cosine modulation in electron density along z.

Nonzero higher order Fourier components ($l > 1$) give the deviation from a purely sinusoidal density distribution along z, that is, represent a sharpening of the layer structure, and the coefficients α_l are then direct measures of the higher smectic order parameters τ_l. In the limit of perfect smectic order (perfect crystalline order along z), all of the smectic order parameters τ_l are equal to unity, and $D(z)$ reduces to a set of δ-functions, a constant distance a apart.

Each term in the expansion for $\rho(z)$ in Eq. (9) corresponds to a pair of Bragg peaks in the diffraction pattern. For example, for $l = 2$, the term is a cosine wave of period $z = a/2$, which gives a pair of Bragg peaks at $Q_2 = \pm 4\pi/a = \pm 2 Q_1$. The intensity of this pair of Bragg peaks, relative to that which would be observed for a perfectly ordered (crystalline) smectic phase, is proportional to the square of the amplitude coefficient α_2, and the same is true for all the higher order peaks. We thus see that the intensities of the various Bragg peaks are direct measures of the smectic order parameters. To a good approximation [7] we may write

$$\tau_l^2 = \left(\frac{I_l}{I_l^P}\right) \qquad (12)$$

where I_l^P is the intensity which would be obtained from a perfectly ordered smectic phase, $I_l^P = |F_l|^2$, where the F_l are the molecular structure factors along the z direction.

The problem with applying Eq. (12) directly to experimental X-ray data is that absolute intensities are needed, and this is difficult to determine. When only two Bragg peaks are detected, as for many smectic phases, if a model structure can be assumed for the molecules, then the intensities which would be observed from a perfectly-ordered smectic phase, I_{00l}^P can be calculated. The ratio of the first two smectic order parameters is then given by:

$$\left(\frac{\tau_2}{\tau_1}\right)^2 = \left(\frac{I_{002}}{I_{001}}\right)\left(\frac{I_{001}^P}{I_{002}^P}\right) \qquad (13)$$

Note that in practice it may be necessary to correct the right hand side of Eq. (13) by a factor of 2 [33, 90] to take account of the Lorentz factor arising from mosaic spread in the sample. Furthermore, accurate measurement of I_{002} may be complicated by multiple scattering, which may be difficult to separate from the true second-order peak.

If a model is assumed for the distribution function $D(z)$, then both the smectic order parameter τ_l and the r.m.s fluctuation of the molecules along z can be determined. Typical results for the SmA phase [7] are $\tau_1 = 0.56$ and $\langle z^2 \rangle^{1/2} = 0.38$ nm.

For two different SmA phases, the ratio $\left(\dfrac{\tau_2}{\tau_1}\right)$ has been found to vary from approximately 0.05 to 0.5, the former value corresponding to a weak SmA phase [90], but in other systems, such as 4O.7 where I_{002}/I_{001} is less than 2×10^{-7}, the ratio can be less than 0.001 [33].

3.5.3 Quasi-long-range Order

For systems whose dimensionality is less than three, a situation can arise in which thermally-excited fluctuations destroy the long-range order (Landau–Peierls instability). As a result of this, the distribution function is no longer exactly periodic, but slowly decays over large distances (see Fig. 11). In this case there is only quasi-long-range order, that is the translational order is slowly lost on moving away from a given site. Typically, the positional correlations undergo an algebraic decay with distance [96, 97]:

$$D(r) \propto r^{-\eta} \quad (14)$$

Where η is a temperature-dependent exponent of the order of 0.1–0.4. For a smectic phase:

$$\eta = \frac{kTQ_0^2}{8\pi\sqrt{BK_1}} \quad (15)$$

where B is the compressibility modulus of the layers, K_1 is the splay elastic constant, and Q_0 is the Q-value corresponding to the layer spacing (we now use Q_0 instead of Q_{001} to emphasize that it is not a true Bragg peak). The effect on the diffraction pattern is to broaden the X-ray peak slightly, without shifting its position, converting it into a so-called 'power-law singularity', with the intensity varying along the direction normal to the layers through the peak as:

$$I(Q_\parallel, Q_\perp = 0) \propto |Q_\parallel - Q_0|^{-2+\eta} \quad (16)$$

and perpendicular to this direction through the peak as

$$I(Q_\perp, Q_\parallel = Q_0) \propto |Q_\perp|^{-4+2\eta} \quad (17)$$

This gives rise to peaks which are quite sharp, but with characteristic 'wings' on each side, which fall off with a different Q-dependence in directions parallel and perpendicular to the layer normal. Furthermore, the broadening is progressively greater for higher order peaks, if these are observed. High resolution diffraction equipment, employing synchrotron radiation or at least a rotating anode X-ray generator is required to study these effects in detail. In later sections, we will refer to these peaks as 'layer reflections' or even '00l Bragg peaks', although it must be remembered that this is formally incorrect, at high resolution.

In conjunction with light scattering measurements of the penetration depth $\lambda = \left(\dfrac{K_1}{B}\right)^{1/2}$, fits to the X-ray data yield values for K_1 and B [97]. Typical values for the SmA phase of 8-OCB within 0.5 °C of the SmA–N transition are $K_1 \approx 8 \times 10^{-12}$ N (8×10^{-7} dyne) and $B \approx 2 \times 10^6$ N m^{-2} (2×10^7 dyne cm^{-2}); note that B tends to zero as the transition is approached.

3.5.4 Short-range Order

When the positional order is only short range, corresponding to a liquid-like structure, $D(r)$ falls off very rapidly, reaching a uniform value within 2 or 3 molecular diameters. Typically the correlations fall off exponentially with distance:

$$D(r) \propto \exp\left[-\frac{r}{\xi}\right] \quad (18)$$

where the correlation length ξ is typically of the order of less than 1 nm (but for the hexatic phases it can be larger than 10 nm). This leads to a Lorentzian distribution of intensity:

$$I(Q) \propto \frac{1}{1+\xi^2|Q-Q_0|^2} \quad (19)$$

where Q_0, the position of the peak maximum, is reciprocally related to the average particle separation a.

In this case the diffraction pattern consists of a broad peak, with a HWHM (half-width at half-maximum) of $\frac{1}{\xi}$. For a two-dimensional system of close-packed cylinders, the maximum would occur at $Q_0 = 1.155(2\pi/a)$, whereas for randomly packed cylinders the maximum would occur at $Q_0 = 1.12(2\pi/a)$, the factor of 1.12 being the first maximum in the Bessel function $J_0(Qa)$ [7].

3.5.5 Lattice Orientational Order

In addition to positional order, any crystal lattice has associated with it a so-called 'lattice orientational order', as shown in Fig. 12a. What this means is that each set of lattice vectors (a, b, or c) all point in precisely the same direction throughout the crystal. For a normal crystal, both types of order disappear simultaneously upon melting. However, for two-dimensional systems, it is possible to destroy the long range positional order, without affecting the lattice orientational order, creating a so-called hexatic phase [16, 17, 98]. This can be achieved by introducing noninteger dislocations along one of the lattice vectors, as shown in Fig. 12b.

This leads to a peculiar situation where the symmetry of the original lattice is maintained, yet without long-range positional order being present. The hexatic liquid crystalline phases SmB, SmF and SmI are essentially the only examples where this effect (which is of fundamental importance in theories of melting) has been demonstrated experimentally.

An overview of the different types of order relevant to various smectic phases [11, 15] is shown in Table 1. The structural relationships between the common smectic phases are as shown in Table 2.

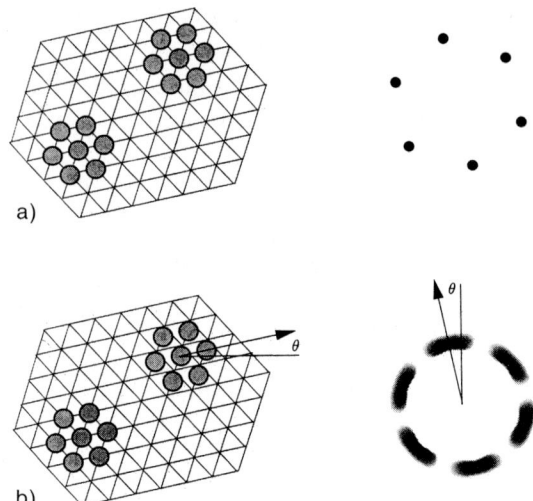

Figure 12. Long-range lattice orientational order with (a) long-range positional order, and (b) short-range positional order. The corresponding wide angle diffraction patterns are shown to the right (From [14]).

Table 1. The types of order relevant to smectic phases.

Structure	Positional order	Lattice orientational order
3-D crystal	LRO	LRO
2-D crystal	Quasi-LRO	LRO
Stacked 2-D crystal (smectic B, G, J, E, H, K)	LRO	LRO
Hexatic	SRO	Quasi-LRO
Stacked hexatic (SmB, SmF, SmI)	SRO	LRO
Liquid	SRO	SRO
Stacked liquid layers (SmA, SmC)	SRO	SRO[a]

LRO = long-range order; SRO = short-range order; [a] the SmC phase has quasi-LRO of the tilt direction.

Table 2. Structural relationships between smectic phases.

Nematic	Fluid smectic	Stacked hexatic	3-D ordered smectic (pseudo-hexagonal)	3-D ordered smectic (herringbone)	Tilt
N	SmA	SmB	B	E	Untilted
		SmF	G	H	Tilted to face of hexagon
	SmC	SmI	J	K	Tilted to apex of hexagon

3.5.6 Diffuse Scattering

Structured diffuse scattering arising from nonperiodic modulations in electron density arising from localized layering or modulations in the layers, layer undulations, defects, stacking faults, edge dislocations, localized molecular vacancies, or other types of disorder, is commonly observed in smectic phases [8, 21, 99, 100], columnar phases [101] and liquid crystalline polymers [26, 102–104]. An analysis of the form of the diffuse scattering may allow the nature of the disorder to be determined. For example, the SmC phase of hexa-aryloxycyclotriphosphazenes, which have an unusual 'cluster' molecular shape (like two three-legged stools stacked on top of each other) has been shown to contain local stacking fault lines [105]. Similarly, for a siloxane mesogen in the nematic phase, evidence for uncorrelated strings of molecules extending along the director has been obtained from an analysis of the diffuse meridional peaks [106].

3.6 Nematic Phase

A clear difference between the scattering from the isotropic and the nematic phases is revealed when a field is applied to produce a uniform alignment of the nematic director n. The pattern from the isotropic phase is unaltered, whereas that from the nematic phase is changed dramatically, the low-angle scattering becoming concentrated on the

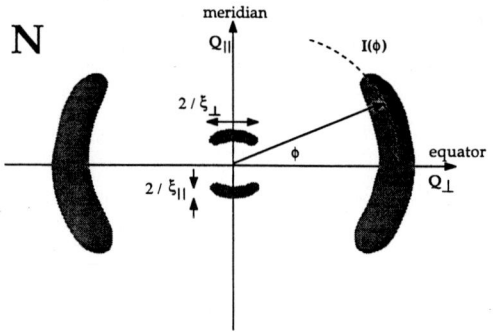

Figure 13. Diffraction pattern from an aligned nematic phase.

meridian (i.e. Q parallel to the director), and the wide-angle scattering on the equator (Q perpendicular to the director), as shown in Fig. 13.

The uniaxial symmetry of the nematic phase is reflected in the cylindrical symmetry of the diffraction pattern: rotating the sample around the z-axis has no effect on the diffraction pattern. Note also that the diffraction pattern is symmetrical above and below the equator: Friedel's law states that all diffraction patterns are centrosymmetric, irrespective of whether or not the sample is.

3.6.1 Biaxial Nematics

Mesogens combining molecular features of both rod and disc appear to favor the formation of biaxial nematic phases [107–109]. However, a subsequent NMR study of 2,3,4-tri-n-hexyloxycinnamic acid showed that any biaxiality of the nematic phase is very small, less than 0.1 [110]. The influence of molecular biaxiality on intramolecular X-ray scattering has been assessed [111], and the calculations show that inclusion of biaxial ordering makes a very small contribution to the single molecule scattering.

If an aligned nematic phase exhibits more than one wide angle equatorial diffuse peak, then this could be an indication of a biaxial nematic phase. Such a feature has been observed in a number of thermotropic systems, but, although indicating local biaxial molecular ordering, does not in itself prove that the phase is biaxial, and direct evidence from X-ray studies has not yet been obtained [112]. To provide such proof would require a true monodomain sample, with both the director n and the other axes macroscopically aligned. If such a sample were rotated around the director n, the equatorial peaks would exhibit maxima in intensity for different values of the rotation angle α (probably but not necessarily 90° apart).

3.6.2 Critical Behavior of the N–SmA Transition

The nematic phase tends to contain localized, fluctuating regions of layer-like structure, or cybotactic groups, as shown in Fig. 14a. As the nematic phase is cooled towards a SmA phase, these critical smectic fluctuations become much more pronounced, leading to a sharpening and increased intensity of the low-angle peak [86]. If the regions are SmA-like, then the meridional scattering peak becomes sharper, without shifting in position. The widths of this peak along the meridian and the equator are inversely proportional to the average thickness, $\xi_\parallel$ and width, $\xi_\perp$, of the smectic-like clusters, respectively. These correlation lengths, which have values in the nematic phase of the order of 30 nm and 3 nm respectively, diverge with decreasing temperature on approaching the transition to the smectic phase. If the cybotactic groups are SmC-like, then the scattering peak splits into two maxima displaced on either side of the meridian by the tilt angle of the cybotactic cluster, as shown in Fig. 14b.

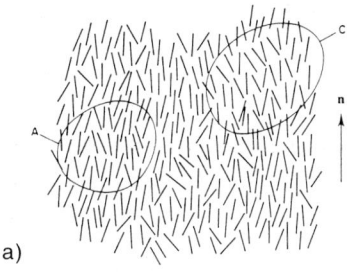

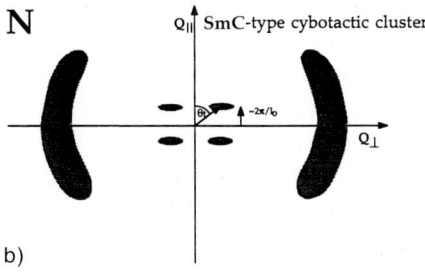

Figure 14. (a) Cybotactic groups in the nematic phase; (b) diffraction pattern from a nematic phase with SmC-type cybotactic clusters.

Upon cooling a nematic phase towards a second-order N–SmA transition, the susceptibility of smectic fluctuations χ diverges as $t^{-\gamma}$, where $t = \dfrac{(T - T_{N-A})}{T_{N-A}}$ and γ is a critical exponent with a value close to unity [113]. Experimental values of γ lie in the region of 1.10 to 1.53 [12]. Experimentally, the X-ray intensity above T_{N-A} has been found to vary along $Q_{\parallel}$ and $Q_{\perp}$ as

$$I(Q_{\parallel}, Q_{\perp}) \propto \frac{\chi}{1 + \xi_{\parallel}^2 (Q_{\parallel} - Q_0)^2 + \xi_{\perp}^2 Q_{\perp}^2 + c\xi_{\perp}^4 Q_{\perp}^4} \quad (20)$$

This equation has the expected theoretical form, but modified by an empirical term quartic in $Q_{\perp}$ which was required to fit the data [114–117]. The quartic term has been alternatively written as $\xi_s^4 Q_{\perp}^4$, where the correlation length ξ_s is related to the splay elastic constant K_1 [118].

Extensive high resolution X-ray studies of the N–SmA transition in a range of systems [33, 115, 118–123] have shown that the correlation lengths $Q_{\parallel}$ and $Q_{\perp}$ diverge with different exponents $v_{\parallel}$ and $v_{\perp}$ upon approaching the SmA phase. Normally, $v_{\parallel} > v_{\perp}$, which means that the smectic fluctuations grow more rapidly along, rather than perpendicular to, the director. Furthermore, these exponents are nonuniversal, with smaller values tending to be observed for systems where the ratio of T_{N-A}/T_{N-I} is larger (under these conditions the N–SmA transition tends towards becoming first-order). For example, for 8CB, where $T_{N-A}/T_{N-I} = 0.977$, $v_{\parallel} = 0.67$ and $v_{\perp} = 0.51$, whereas for 4O.7, where $T_{N-A}/T_{N-I} = 0.926$, $v_{\parallel} = 0.79$ and $v_{\perp} = 0.66$ (compilations of experimentally-determined values of critical exponents are available [11, 12]).

3.6.3 Orientational Order Parameters

The variation of intensity $I(\phi)$ around the wide-angle arc (see Fig. 13) is related to the degree of orientational order $f(\beta)$ of the molecules. The higher the degree of orientational order, the more the scattered intensity is concentrated on the equator. By making certain rather drastic assumptions [7, 87, 88], the following relationship between $I(\phi)$ and the molecular orientational distribution function $f(\beta)$ was derived:

$$I(\phi) = \int_{\beta=\phi}^{\pi/2} \frac{f(\beta)\sin(\beta)}{\cos^2\phi(\tan^2\beta - \tan^2\phi)^{1/2}} \, d\beta \quad (21)$$

From a suitable expansion of $f(\beta)$, one can then extract estimates for not only the second rank orientational order parameter $\langle P_2 \rangle$, but also one or even more of the higher rank terms $\langle P_4 \rangle$, $\langle P_6 \rangle$ etc. On the other hand, if a Maier–Saupe form can be assumed for $f(\beta)$, then a simplified form of Eq. (21) results [124], and the order parameter $\langle P_2 \rangle$ can be directly obtained from the width of the intensity distribution around ϕ. The validity and accuracy of these approaches has recently been tested on a range of calamitic and phasmidic mesogens, and main- chain and side- group liquid crystal polymers [125] and it has been found that in most cases the simplified form is adequate.

Another approach, utilizing the intensities of layer reflections from both X-ray and neutron diffraction experiments to obtain estimates for $\langle P_2 \rangle$, has been applied to the SmA, SmB and smectic E phases of IBP-BAC [126].

A totally different approach is to employ neutron scattering with protonated/perdeuteriated mixtures [127–129], which allows the scattering from effectively-isolated single molecules to be observed directly.

3.7 Smectic A and Smectic C Phases

On cooling an aligned nematic into the SmA phase, the low-angle diffuse peaks sharpen into a small number of equally-spaced (pseudo-)Bragg peaks at $Q_l = l\left(\dfrac{2\pi}{d}\right)$, as shown in Fig. 15. The peaks are quite sharp because the layer periodicity is maintained over many layers. However, using high-resolution diffraction equipment, it is found that they are slightly broadened into power law singularities due to the slow algebraic decay of the periodicity, caused by the Landau-Peierls instability. The correlation length ξ along the layer normal is greater than 3 µm for 4O.7 [33]. It should be

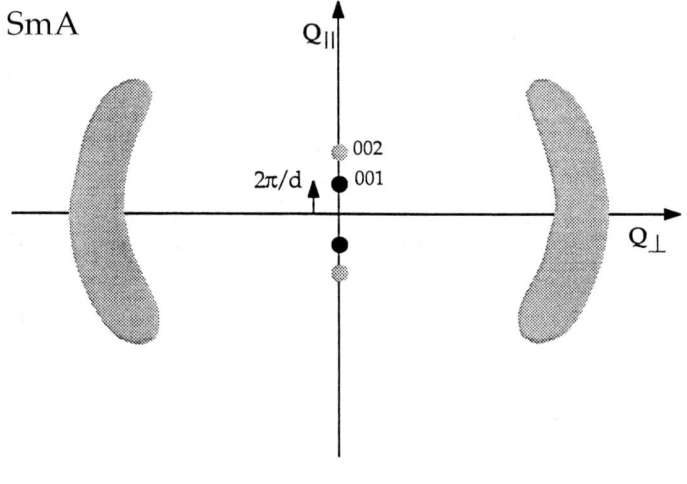

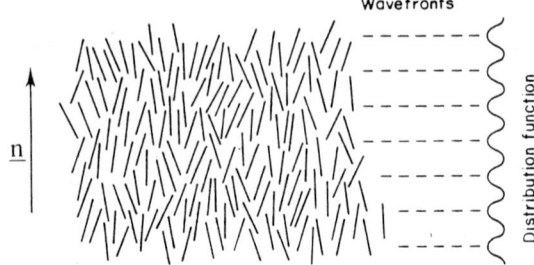

Figure 15. Diffraction pattern from an aligned SmA phase.

stressed that this effect cannot be detected using conventional X-ray equipment. It is not unusual to observe diffuse scattering extending horizontally about the Bragg peaks, arising from effects such as out-of-plane coupled layer deformations.

The wide-angle diffuse maximum on the equator stays essentially unchanged on passing from the nematic to the SmA or SmC phase, indicating that the short-range structure remains liquid-like. Typically, for 4O.7 [33], it is centered at $Q_\perp^0 = 14.3$ nm^{-1}. The width along the $Q_\perp$ direction (i.e., along the equator) of this peak is inversely proportional to the range of the lateral positional order within the layer, which is of the order of 1.5 nm.

3.7.1 Smectic A polymorphism

There are a number of different types of SmA phase [20, 130, 131]. For the simplest SmA phase, denoted SmA$_1$ (Fig. 16b), the layer spacing calculated from the pseudo-Bragg peaks corresponds to slightly less than the molecular length. This is found for systems in which there is little tendency for molecular association, and there is hence complete head-to-tail disorder. Polar mesogens (e.g., with terminal CN or NO$_2$ groups) have some tendency to form antiparallel dimers, giving an antiferroelectric ordering, which minimizes the dipolar energy. This can lead to formation of the partially-bilayer SmA$_d$ phase, with a layer spacing $d = l_0' \approx (1.4 - 1.8) \, l_0$ or to the bilayer SmA$_2$ phase, with $d \approx 2 l_0$. For dimeric mesogens, an intercalated SmA$_{1/2}$ phase (Fig. 16a), where the layer periodicity is close to half the molecular length, has been observed in terminal nonpolar, tuning-fork shaped mesogens [132], in nonsymmetric [133–135] and symmetric dimers [136, 137], and in main chain liquid crystal polymers having

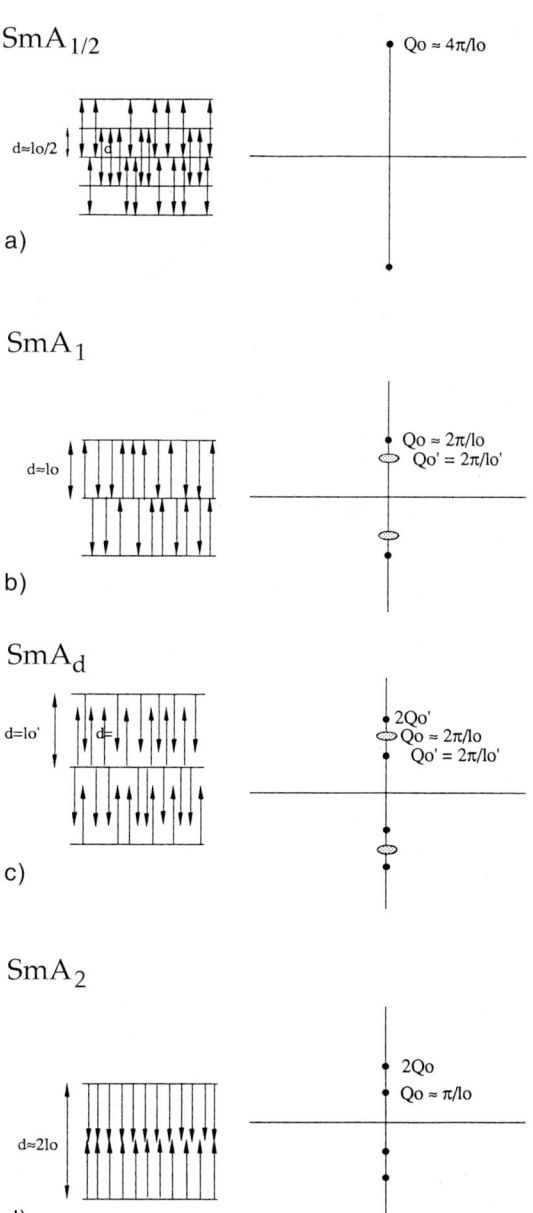

Figure 16. Structures of different types of SmA phase, and the schematic forms of the observed diffraction patterns: (a) intercalated SmA$_{1/2}$, (b) monolayer SmA$_1$, (c) partial bilayer SmA$_d$, (d) bilayer SmA$_2$. Note that the diffuse peaks shown in (b) and (c), arising from local domains of SmA$_d$-like and SmA$_1$-like ordering respectively, may not actually be observed.

two different spacers [138]. Such dimeric or polymeric systems with spacers show distinct odd–even effects in the structures of the smectic phases as the number of methylene groups in the spacer chains linking the mesogenic groups are varied.

The schematic low-angle diffraction patterns from the various types of SmA phase are also shown in Fig. 16. The $SmA_{1/2}$, SmA_1, SmA_d and SmA_2 phases are discriminated partly by the positions of the pseudo-Bragg peaks on the meridian, at $Q_0 \approx 4\pi/l_0$, $Q_0 \approx 2\pi/l_0$, $Q'_0 \approx 2\pi/l'_0$, and $Q_0 \approx \pi/l_0$. In addition, the SmA_1 and SmA_d phases may exhibit diffuse maxima at $Q'_0 \approx 2\pi/l'_0$ and $Q_0 \approx 2\pi/l_0$, corresponding to small domains of SmA_d-like and SmA_1-like ordering, respectively.

Transitions between these different SmA phases may be second-order or first-order, with critical points terminating the first-order transition line [139], and such a critical point has been observed experimentally [18, 140]. X-ray scattering measurements of critical SmA_2 fluctuations have been carried out along a line of second-order transitions between the SmA_1 and SmA_2 phases in mixtures of DB6 and TBBA [117, 141]. The critical exponents were measured to be $\gamma = 1.46$; $v_{\parallel} = v_{\perp} = 0.74$, and were constant along the second-order line, even close to the tricritical point where the transition became first-order.

3.7.2 Smectic C and C* Phases

As for the SmA phases, similar variants exist for SmC (i.e. SmC_1, SmC_2, SmC_d) and intercalated versions. The use of aligned samples permits direct measurement of the tilt angle in the SmC phase of nonchiral molecules. However, there is a complicating factor, in that on cooling an aligned sample from the SmA phase into the SmC phase, a number of different patterns may be observed, depending on the type of alignment of the SmC phase (Fig. 17).

a) If the various layer normals tilt so as to give a cylindrical symmetry around the director (chevron-like structure), then four first order pseudo-Bragg peaks will be observed, tilted at $\pm \theta_t$ to the meridian.

b) If the tilting occurs in one unique direction, then only two peaks will appear, either at $+\theta_t$ or at $-\theta_t$ to the meridian.

c) If the aligning field is weak, the layers may stay fixed with the same orientation as the SmA phase, the director tilting over to the angle θ_t. This corresponds to an antiferroelectric SmC_{alt} structure, if successive layers have the opposite tilt directions. In this case, the diffraction pattern will be difficult to distinguish from that of a SmA phase. However, for certain dimers [136] and liquid crystalline polymers [138] forming tilted smectic phases, the splitting of the wide-angle diffuse peak above and below the equator may be seen quite clearly.

For the chiral SmC* phase, the situation is complicated by the presence of the helical pitch along the layer normal, which will prevent the tilting occurring in a unique direction, unless the aligning field is strong enough to unwind the pitch.

Aligned samples are required to obtain the tilt angles in the SmC* phase, when the transition occurs directly from a nematic phase. In this case, the transition is usually first order, and the tilt angle jumps directly to some finite value at the transition point.

High resolution X-ray experiments have been carried out to study N–SmC and SmA–SmC transitions. For 4O.7, on crossing the SmA–SmC phase boundary the tilt angle θ_t was found to vary as $t^{-0.37}$, where $t = \dfrac{(T_{AC} - T)}{T_{AC}}$ [142]. This result could be

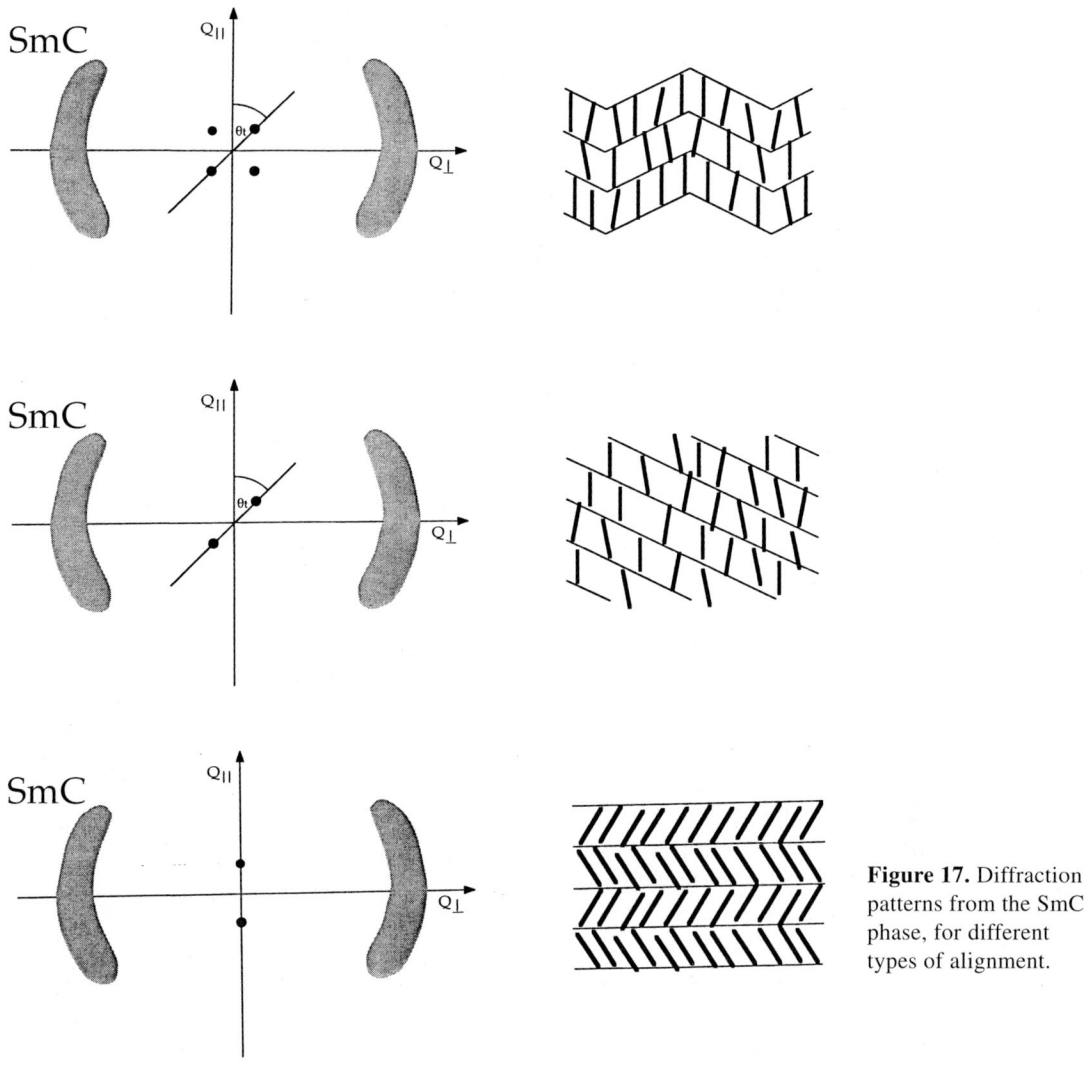

Figure 17. Diffraction patterns from the SmC phase, for different types of alignment.

consistent with either power-law or mean field behavior, although the latter is believed to be correct [113].

Scattering experiments have also been carried out close to the N–SmA–SmC multicritical point in binary mixtures of mesogens (see reference [143] and references therein). Above the first-order N–SmC transition, smectic fluctuations evolve from being SmA-like to SmC-like upon cooling.

Along the N–SmA phase boundary, second-order N–SmA transitions are observed. For mixtures of 8-OCB with $\overline{7}S5$ (pentylphenyl heptyloxythiol benzoate), multicritical behavior is observed very close to the N–SmA–SmC point [143].

3.7.3 Ferroelectric Smectic C Phases

In the ferroelectric SmC* phase the tilt direction precesses in a helical path from layer to layer, whereas in the antiferroelectric

SmC$_a^*$ phase the tilt points in opposite directions in adjacent layers, although the helical superstructure is still present [19, 144–147]. Competition between interactions tending to create ferroelectric and antiferroelectric ordering can lead to intermediate ferrielectric SmC$_\gamma^*$ phases [147]. On crossing between the SmC*, SmC$_\gamma^*$, and SmC$_a^*$ phases with varying temperature, the layer spacing changes only vary slightly [148, 149], making it difficult to distinguish these structures by their low-angle diffraction patterns [150].

The antiferroelectric SmC$_a$ structure (see Fig. 17) can also occur in racemates [94] or in nonchiral compounds such as symmetric dimers [136, 137], nonsymmetric dimers [133], and main chain liquid crystal polymers [138], where its formation is driven by steric and/or conformational effects. Antiferroelectric ordering has been shown to increase the smectic order parameters in ferroelectric liquid crystals [94, 95].

For thin samples of chiral SmC* phases between solid plates, the helical pitch is suppressed. In such a surface-stabilized ferroelectric SmC* phase, a bookshelf geometry with the layers normal to the plates, a tilted geometry with the layers uniformly tilted, or a chevron structure similar to the arrangement shown in Fig. 17 may be found. The detailed chevron structure has been determined by high resolution X-ray studies [151–153]. In a study of the SmA–SmC* transition of a ferroelectric system within a surface-stabilize cell, evidence has been obtained for an unusual tilted arrangement of the SmA phase [154]. The effect of an applied field on the SmC* chevron structure and orientation has been studied [155, 156], and it has been confirmed that electroclinic layer constriction, and consequent layer buckling, is due to field-induced molecular tilt [157]. Smectic layer reorientation during electric field switching of a ferroelectric liquid crystal device has been directly observed by *in-situ* X-ray diffraction [158]. Further studies have achieved microsecond time resolution, employing synchrotron diffraction [159]. It was found that during the switching, there is both a change of chevron angle, and a rotation of the layers by 1° about an orthogonal axis, both motions occurring on a timescale of 10 µs. A recent synchrotron study of a chiral ferroelectric system, employing an X-ray microbeam, has found evidence for a chevron structure with a spontaneous layer twist [160].

3.8 Modulated and Incommensurate Fluid Smectic Phases

A phenomenological model for frustrated smectics proposed that 2-D modulated smectic phases might arise as an escape from underlying incommensurability in the system [22, 161, 162]. The role of polar and steric molecular assymetry in smectic polymorphism has been comprehensively reviewed [23, 163]. The forms of the X-ray diffraction patterns from modulated and incommensurate smectic phases have been extensively discussed [20–23] and these papers should be consulted for earlier references and for details which will not be covered here.

3.8.1 Modulated Phases

Complex types of biaxial SmA phase have been discovered, such as the SmÃ phase (Fig. 18a), in which an antiferroelectric ordering is modulated to give a 'ribbon' structure, the ribbons being arranged on centered [164, 165] or primitive [166] rectangular lattices, the latter being denoted a 'crenellated' phase. These phases give rise to additional low angle Bragg peaks away from the meridian. Initially these 'antiphases' were observed with mesogens having strongly polar endgroups, but later studies showed that other types of molecular asymmetry, such as steric effects, can also induce the same structures to form. A corresponding modulated type of SmC antiphase, denoted SmC̃, was discovered in DB8NO$_2$ [167], and was shown to be based on a 2-D oblique lattice (Fig. 18b).

Note that within the SmA$_1$ phase above the SmÃ phase, the diffuse peak at $Q_0 \approx 2\pi/l'_0$ on the meridian splits into a pair of diffuse peaks on either side of the meridian, indicating the formation of small domains with a SmÃ-like transverse modulation, as the SmÃ phase is approached. A high resolution X-ray study of the N–SmA$_1$–SmÃ transition of DB7NO$_2$ [168, 169] found that two incommensurate smectic fluctuations present in the nematic phase cross over just above the SmA$_1$ phase to a region of coexisting incommensurate and antiphase fluctuations, the latter evolving from SmC̃-like to SmÃ-like with decreasing temperature before locking in to the SmÃ phase. Other high resolution studies have also found evidence for incommensurate fluctuations in SmA phases [170, 171], and have examined SmA$_1$–SmÃ–SmA$_2$ [172] and SmA$_1$–SmC̃ transitions [173]. Another X-ray study of binary mixtures of a polar mesogen exhibiting SmA$_1$–SmC̃, re-entrant N–SmC̃ and SmA$_d$–SmC̃ transitions has provided further support for the idea that antiphases occur as an escape from incommensurability [82].

Modulated phases have also been observed in biforked mesogens [58], tuning-fork shaped mesogens [132], nonpolar polymeric mesogens [102, 138, 174], mesogens with fluorinated terminal chains [163, 175], and symmetric [176] and nonsymmetric [177] dimeric mesogens. Evidence for modulations or distortions along the columns within discotic phases, arising from packing mismatch between the cores and the surrounding chains [178], has also been found [179].

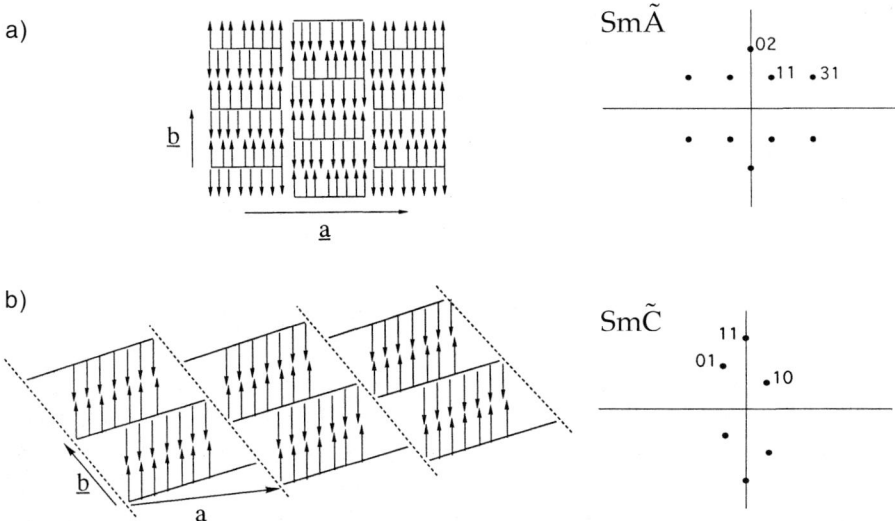

Figure 18. Structures of modulated (a) SmÃ and (b) SmC̃ antiphases. The schematic low angle diffraction patterns are shown to the right.

3.8.2 Incommensurate Phases

Some years ago, incommensurate smectic phases, in which SmA_d and either SmA_1 or SmA_2 periodic density waves coexist along the layer normal, were proposed [180]. The real space structures are quite difficult to describe [22]: in the weakly-coupled versions, the two interpenetrating incommensurate density waves are almost independent of each other, whereas in the strongly-coupled (soliton) version, regions of SmA_2 ordering are separated by thinner regions where the coexisting density waves are out of phase.

The weakly-coupled $SmA_{2,inc}$ phase, in which coexistence of SmA_d with a SmA_2 density wave occurs (Fig. 19a), should exhibit a diffraction pattern (Fig. 19b) with pseudo-Bragg peaks at Q_0, Q'_0 and $2Q_0$, corresponding to both coexisting periodicities. In the strongly-coupled $SmA_{inc,soliton}$ phase, in addition to a Q_1 peak close to $2\pi/lo$, lower-angle incommensurate peaks Q_2 and Q_3 ($=Q_1-Q_2$) should be observed (additional sattelite peaks should also occur).

The first evidence for incommensurate smectic modulations was obtained by Leadbetter and coworkers in the smectic E phase of n-alkyl cyanoterphenyls [181, 182]. Subsequent studies presented evidence for fluid $SmA_{1,inc}$, $SmA_{2,inc}$ and $SmA_{inc,soliton}$ phases in binary mixtures [183–185]. However, more recent high resolution X-ray studies have contradicted these latter findings and have found that the apparently incommensurate phases are actually coexistence regions between various smectic phases (e.g. $SmA_1 + SmA_d$; $SmA_2 + SmA_d$) [24, 186–189]. These authors have recently claimed to have found a true incommensurate ordered smectic phase in mixtures of TBBA and DB6 [190]. Evidence has also been presented for two soliton-type incommensurate SmA phases (one

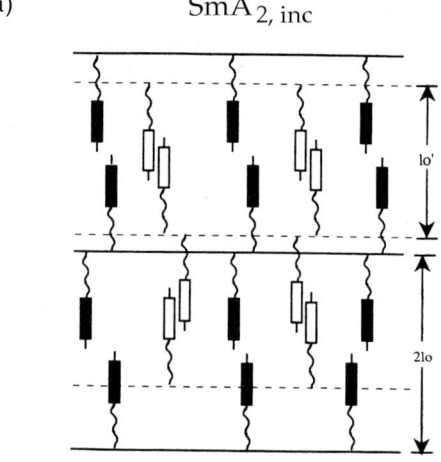

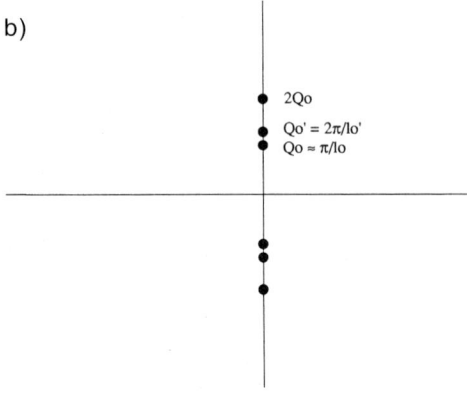

Figure 19. (a) Structure and (b) low angle diffraction pattern from the incommensurate $SmA_{2,inc}$ phase.

of them re-entrant) in a nonsymmetric dimeric system, KIn [191]. The incommensurate phases only appear for an intermediate spacer length of five methylene units ($n=5$), but not for shorter ($n=4$) or longer ($n=7$ or 10) spacers [192].

3.8.3 Re-entrant Phases

A phenomenon which is closely related to smectic polymorphism is that of re-entrant behavior, where unusual phase sequences, such as N–SmA–N–SmA, occur upon cooling. Such behavior is believed to be due to a tendency for the onset of two density waves within the sample, one connected with the molecular length, and one connected with associated pairs of molecules. If these two density waves are incommensurate (i.e., have periodicities which are not simple multiples of each other) then a competition between the two types of ordering occurs, and this can lead to a delicate balance being set up, which shifts backwards and forwards between a nematic, and different types of smectic ordering, as the temperature is varied.

A re-entrant nematic phase, occurring at lower temperature than a SmA phase was first observed in binary mixtures of two cyano compounds (6-OCB and 8-OCB) [193, 194]. However, reentrant behavior has also been observed in mixtures of mesogens which do not have strongly polar endgroups [195–201]. Re-entrance was also found for pure 8-OCB at high pressure [193, 202]. Multiple reentrance, with the phase sequence on cooling of: I–N–SmA_d–N–SmA_d–N–SmA_1–$Sm\tilde{C}$–SmA_2–SmC_2–Cr was subsequently observed in the pure compound 4-nonyloxyphenyl-4′-nitrobenzoyloxybenzoate [203, 204]. High resolution X-ray studies have shown that the low temperature reentrant nematic phases are highly ordered along the director, with the correlation lengths $\xi_\parallel$ in the three nematic phases increasing from 3nm to 100 nm to $\geq$330 nm, with decreasing temperature [170]. Another high resolution X-ray study of the critical behavior in the nematic phases of n-alkoxybenzyl-oxycyanostilbenes found that for the octyloxy compound, with the phase sequence on cooling of I–N–SmA_d–N–SmA_1, the SmA_d fluctuations within the re-entrant nematic phase switch over from being SmA-like to SmC-like with decreasing temperature [205].

A re-entrant isotropic phase between the nematic phase and a cubic phase has been observed in a pure double-swallow-tailed compound [206]. Furthermore, another such compound was found to exhibit a re-entrant SmC phase below an oblique columnar phase, with the sequence SmC–Col_{ob}–SmC–N–I upon heating [207].

3.8.4 Liquid Crystal Surfaces: X-ray Reflectivity

Studies of liquid crystal surfaces by X-ray reflectivity were initiated by Als-Nielsen and Pershan [208–210]. Quantized growth of 1 to 5 smectic layers was observed at the free surface of the isotropic phase of 12CB upon approaching the temperature of the bulk SmA phase [211]. In reflectivity studies on 8OCB and 4O.8 it was found that smectic ordering at the surface of the nematic phase decays exponentially into the bulk with a correlation length which is equal to that for bulk critical fluctuations in the nematic phase [212]. Incommensurate smectic order was observed at the surface of the nematic phase of DB7NO2 [213]. Antiferroelectric (SmA_2-like) ordering at the surface of the bulk SmA_1 phase of 8-OPCBOB has been observed at the surface of a liquid crystal [214]. However, for a slightly less polar mesogen, only SmA_1-like ordering was observed at the surface [215]. The feasibility of carrying out reflectivity on free-standing smectic films has been demonstrated [216].

3.9 Twist Grain Boundary Phases

Highly dislocated helical versions of the SmA and SmC phases, denoted twist grain

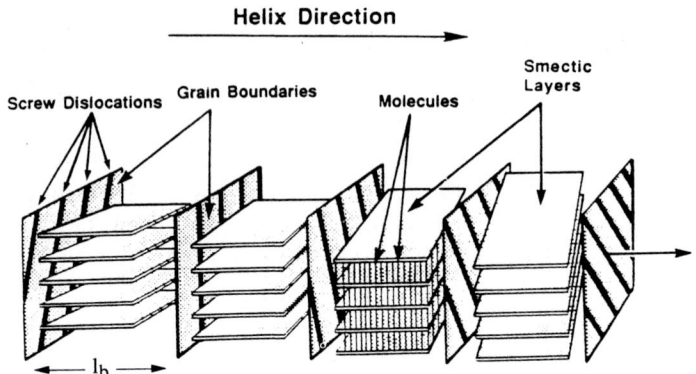

Figure 20. Structure of the twist grain boundary TGBA phase. (From [223]).

boundary (TGB) phases, were predicted theoretically by Renn and Lubensky [217–219]. In the TGBA phase (Fig. 20), smectic blocks of length l_b are rotated by an angle $\Delta\theta$ around an (x-) axis lying in the plane of the layers. The ratio $\alpha = \dfrac{\Delta\theta}{2\pi} = \dfrac{l_b}{\lambda_0}$, where λ_0 is the helix pitch, determines whether the TGB phase is incommensurate or commensurate. If the ratio is irrational then the X-ray pattern (Fig. 21a) has the form of a hollow cylinder of radius $Q_0 = \dfrac{2\pi}{d}$, extending along the Q_x direction for roughly a distance $\dfrac{2\pi}{l_b}$. On the other hand, if α is rational (i.e., $\alpha = p/q$ where p and q are integers: p seems usually to be equal to unity, and q is in the region of 20), the structure is periodic along the x-axis and a commensurate TGB phase results. The diffraction pattern then consists of a ring of q (or $2q$ if q is odd) equispaced pseudo-Bragg spots around the Q_x axis (strictly speaking, each spot is a set of very closely spaced peaks along Q_x, due to the helix pitch periodicity λ_0, which is of the order of one micron). Note that unless the number of spots around the Q_x axis is equal to 2, 4 or 6, the reciprocal lattice is quasicrystalline.

The first actual example of a TGBA phase, occurring between the isotropic liq-

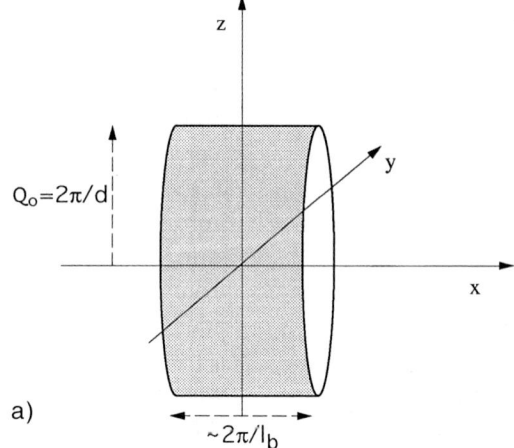

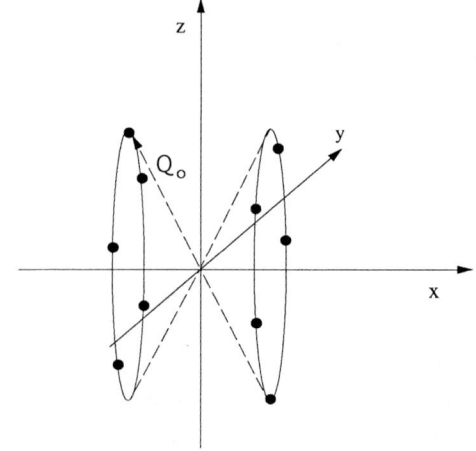

Figure 21. Schematic low angle diffraction patterns from (a) the incommensurate TGBA phase, and (b) the commensurate TGBC phase with $q=5$. (Adapted from [14]).

uid and a SmC* phase, was discovered by Goodby and coworkers in a series of optically-active 1-methylheptyl n-alkoxyphenyl propioloyl oxybiphenyl carboxylates [220, 221]. The layer spacing was found to be constant in the TGBA phase, with a value of 4.35 nm for the tetradecyl compound, equivalent to the fully-extended length of the molecule; optical studies found that the helix pitch decreased from approx 0.6 µm at 90 °C to approx 0.4 µm at 94 °C [222]. A freeze-fracture electron microscopy study confirmed the twist grain boundary structure of this phase, and showed that the rotation of individual smectic layer blocks occurred in discrete 17° steps, implying that the grain boundaries are approx 24 nm apart [223]. The fact that the layer rotation angle is not an exact multiple of 360° means that the TGBA phase is incommensurate, and this is confirmed by a detailed X-ray study which found no structure around the circumference of the Bragg cylinder [222].

A TGBC phase, with a tilt angle of the director of approx 18°, was subsequently also discovered [224], and has been shown to be the first example of a commensurate TGB phase (see Fig. 21b) [225, 226]. This was proved by the observation on a well-aligned sample that the continuous ring of scattering in the cholesteric phase split into 18 equispaced spots on cooling into the TGBC phase, corresponding to a width of the smectic slabs of about 72 nm. On cooling through the 1° range of the TGBC phase, the helical pitch increased from 1–2 µm, and the number of spots increased discontinuously from 16 to 18 to 20, showing that the TGBC phase evolves in a series of transitions between commensurate structures. A subsequent, more detailed X-ray study of systems exhibiting TGBA–TGBC transitions showed that the smectic layers in the TGBC phase are actually tilted with repect to the helix pitch axis [227]. For one compound, the layers were found to lie at an angle of approx 13° to the pitch axis (similar to the tilt angle of the director), and the number of spots increased from 18 to 19 to 20 upon cooling within the TGBC phase. However, for another compound the tilt angle of the layers was found to vary continuously (showing power law behavior) on cooling, with the structure varying between commensurate (26 to 25 to 24 spots) and incommensurate (continuous ring of scattering) over the range of the TGBC phase. The TGBC helix can be unwound by an external DC electric field applied perpendicular to the pitch direction [228]. A chiral tolane has been found to exhibit both the TGBC phase and an antiferroelectric SmC* phase [229]. TGB phases have also been found in non-symmetric dimeric mesogens [135, 191].

3.10 Hexatic Phases

In the hexatic phases, the local cross-sectional packing of the molecules is hexagonal (Fig. 22a). The diffraction pattern from an aligned sample of SmB (Fig. 22b) is quite similar to that from SmA, except that the diffuse wide-angle scattering on the equator becomes a much sharper peak, because of the increased in-plane positional order. The d-spacing of a hexagonal lattice is $d_{10}=(\sqrt{3}/2)a$, where a is the lattice parameter (center–center separation between molecules), and so the position of these peaks is $Q_{10}=2\pi/d_{10}=4\pi/\sqrt{3}a$. Note that an equivalent view of the packing is in terms of a centered rectangular lattice, with $b=\sqrt{3}a$; in this case the wide angle peak indexes as the (11) and (20) reflections (which coincide for this X-ray beam orientation). The intensity distribution along the $Q_\perp$ direction (i.e. along the equator) of these peaks has a Lorentzian form, which shows

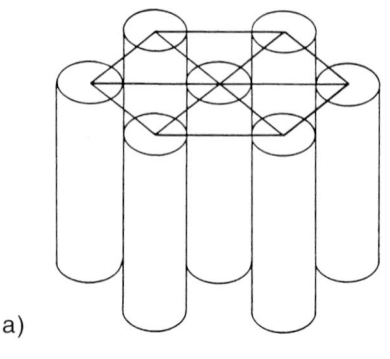

SmB

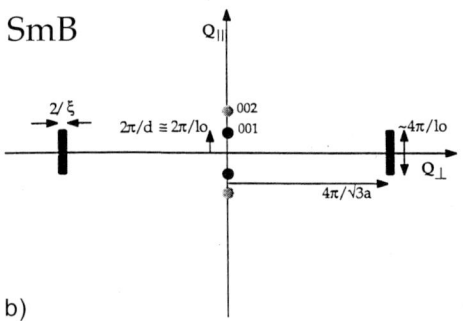

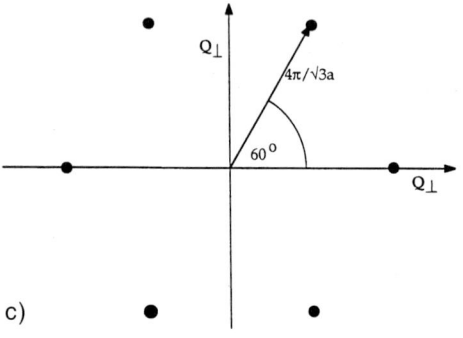

Figure 22. (a) Hexagonal molecular packing in the hexatic SmB phase, and the diffraction patterns from the SmB phase, with the incident beam either (b) perpendicular, or (c) parallel to the layer normal.

quite large. These peaks extend along the $Q_\parallel$ direction (i.e., along the layer normal) a distance of approximately $\pm 2\pi/l_0$, because the layer is a two-dimensional lattice of rods of length l_0 (see Fig. 9 and 10).

A striking feature of the monodomain SmB phase is that if the X-ray beam is directed along the layer normal, a hexagonal pattern of wide-angle peaks is observed (Fig. 22c). Strictly speaking, a 2D hexatic phase should exhibit only quasi-long range lattice orientational order, and for a large enough sample the six-fold modulation should be lost. However, the weak interlayer coupling in an actual liquid crystal hexatic phase converts it to true long range order. For this reason SmB is referred to as a stacked hexatic phase; note that the lattice orientational order extends over all of the layers.

The basic structural features of, and differences between, the SmB, SmF and SmI phases were established by X-ray studies primarily of the nO.ms, TBnAs, 8OSI, HOBACPC and PDOBAC [28, 35, 230–241]. Unusually, 5O.6 shows a hexatic SmF phase occurring at lower temperatures than a more ordered SmB phase [235]. Various types of hexatic phase have been identified in symmetric [242] and nonsymmetric [133] dimeric systems.

The SmF and SmI phases are tilted versions of the SmB phase, with similar structural features (Fig. 23). They may be viewed as having C-centered monoclinic unit cells, with axes $a > b$ and $b > a$, respectively, where b is the unique axis around which the tilting occurs (the c-axis is then at an angle of $(90° + \theta_t)$ to the a-axis). For the SmF phase, the tilt direction is thus towards a face of the quasihexagonal net, (i.e. halfway between nearest neighbors). For the SmI phase, on the other hand, the tilting occurs towards an apex (i.e. towards a neighboring molecule). In a lyotropic phospholipid system, a fur-

that the order is, formally speaking, short-range. However, the correlation length $\xi_\perp$, which is inversely proportional to the peak width, has a value in the region of 15 nm, showing that the range of the lateral positional order within the layers is actually

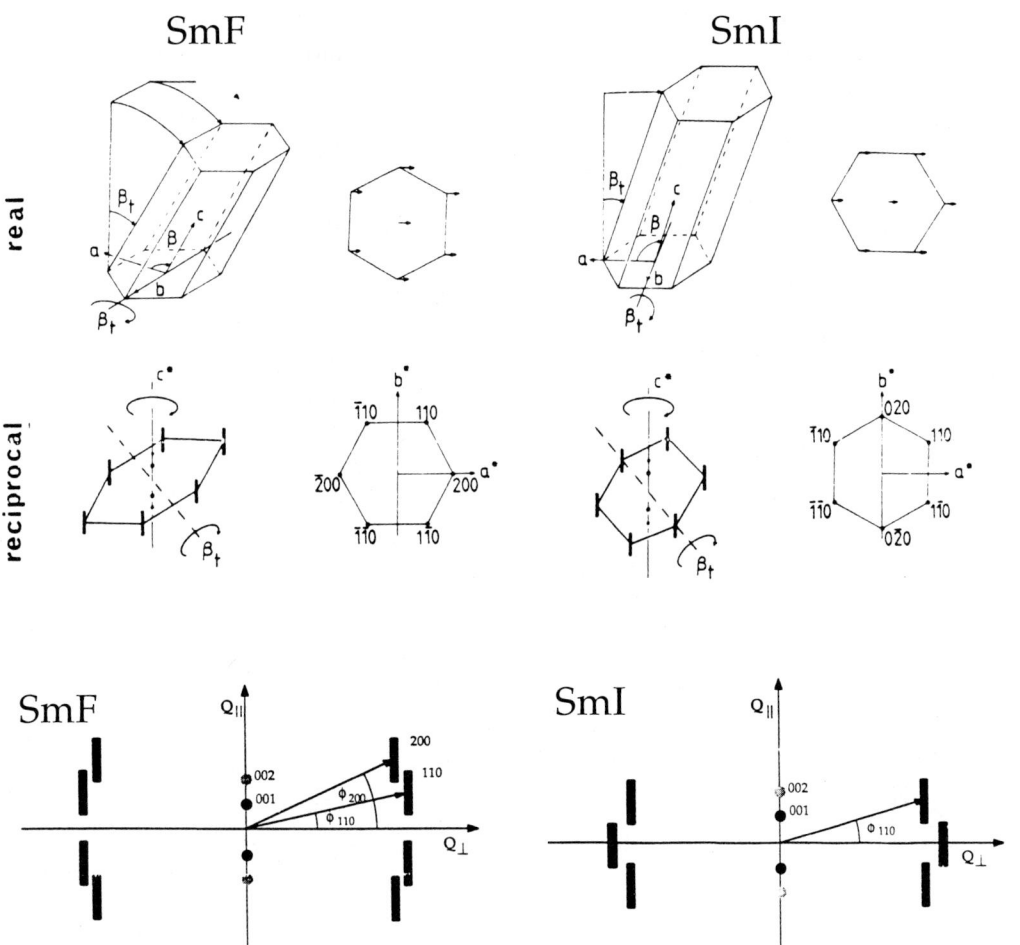

Figure 23. Structures of the tilted hexatic phases SmF and SmI, and their diffraction patterns with the incident beam perpendicular to the layer normals. (Adapted from [238]).

ther tilted hexatic phase has been found between a SmF and a SmI phase, with increasing hydration [243]. In this phase, denoted SmL, the tilt direction changes continuously between the directions in the SmF and SmI phases.

The effect of the tilt on the diffraction pattern depends on the type of alignment. If the director remained pinned, with the layers tilting, then the low angle Bragg peaks would become displaced from the meridian by the tilt angle, and the wide-angle pattern would remain as for the SmB. However, it is more usual for the layers to remain fixed and the director to tilt, which causes the wide angle peaks to tilt away from the equator, as shown in Fig. 23.

This leads to characteristic wide-angle patterns which are different for the SmF and SmI phases, and from which the tilt angle θ_t may be obtained. For the SmF phase, the (200) and (110) peaks are at angles of θ_t and $\sin^{-1}(\pm(1/2)\sin\theta_t)$ to the equator, respectively. For the SmI phase, the (020) and (110) peaks are at angles of zero and $\sin^{-1}(\pm(\sqrt{3}/2)\sin\theta_t)$ to the equator, respectively.

As in the SmB phase, the lattice orientational order is long range in the SmF and SmI phases. However, to observe the sixfold modulation of the wide-angle peaks, the director must point in a unique direction, so that the X-ray beam can be aligned along it.

High resolution X-ray studies employing free-standing films has allowed the nature of the hexatic ordering (reviewed in [16, 17]) to be probed in great detail [66, 76–78]. On cooling a well-aligned sample of 8OSI from the SmC phase into the SmI phase, weak lattice orientational order in the SmC phase was found to evolve continously into the strong hexatic ordering, characterized by nonzero higher harmonic (up to 8) lattice orientational order parameters, C_{6n} [76]. A surprising discovery was that a simple scaling relationship exists between the various harmonics. A subsequent study of the same system found that the effect of reducing the film thickness to 23 or to 4 molecular layers was to supress the growth of the higher harmonics of the hexatic ordering on cooling into the SmI phase [77]. A study of SmC–SmF transitions in TBnAs found that, unlike for the SmC–SmI transition of 8OSI, there was no detectable pretransitional lattice orientational ordering in the SmC phase [78]. The evolution of the higher harmonics C_{12} and C_{18} across the transition switched from discontinuous to continuous with increasing chainlength of the mesogens, indicating that the SmC–SmF transition was changing from first order to second order.

A 2D modulated hexatic phase has been observed in 4-(4′-cyanobenzyloxy) benzylidene-4″-n-alkoxyanilines [244], with an oblique structure analogous to a SmC̃ phase. A similar structure has been found in a series of symmetric $m.OnO.m$ dimers [176, 242]. In the latter case the modulated structure arises from a steric molecular asymmetry rather than from a dipolar asymmetry. Evidence for a modulated SmB̃ phase has also been reported in terminal nonpolar, tuning fork shaped mesogens [132].

3.11 Ordered Smectic Phases

The molecular packing in the sets of crystal smectic mesophases B, G, J and E, H, K, with director orientations similar to the hexatic SmB, SmF and SmI phases, respectively, is shown in Fig. 24. For two dimen-

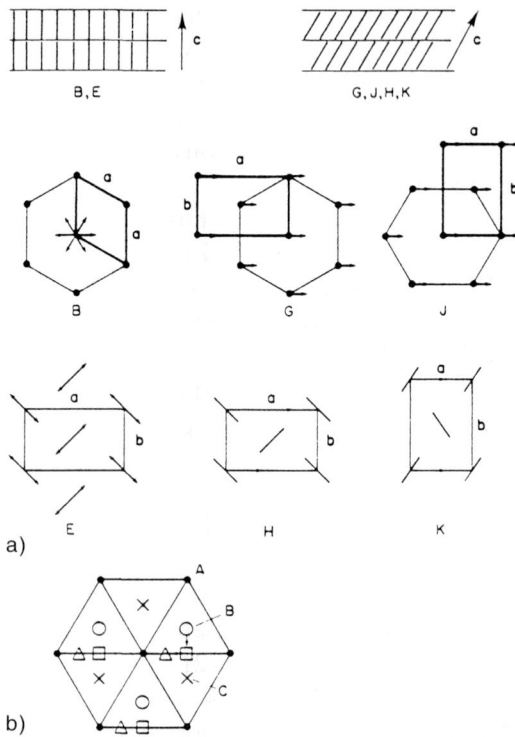

Figure 24. (a) Structures, and (b) layer stackings of the ordered smectic phases. A, B and C denote the highest symmetry positions of molecules in adjacent layers, whilst the open squares and triangles show the lower symmetry positions (ortho-F and mono-C) which have been observed in the smectic B phase. (From [15]).

sional crystals, the positional in-plane order should theoretically only be quasi-long range. However, the weak interlayer coupling within these ordered smectic phases converts this to true long-range order.

A series of X-ray studies, primarily on the nO.ms, established the nature of the packing and interlayer correlations within the ordered SmB phase [34, 35, 70, 72, 238, 245]. The in-plane molecular packing is hexagonal, and the director n is aligned along the layer normal (i.e., the phase is untilted). The wide-angle diffraction pattern of the ordered B phase differs from that of the hexatic SmB phase in that the diffuse rods of scattering are replaced by sets of Bragg peaks extending along the layer normal. The positions (and relative intensities) of these peaks depends on the type of stacking sequence: AAA (Fig. 25a) leads to wide-angle $(h0l)$ peaks with the same spacing in Q as the $(00l)$ layer reflections, whereas sequences ABABAB (Fig. 25b) and ABCABC give $(h0l)$ peaks whose spacings in Q are, respectively, 1/2 and 1/3 those of the $(00l)$ layer reflections.

These AAA, ABAB and ABC stacking sequences have been found in various systems. Restacking transitions may be observed on changing the temperature within the B phase. For example, the compound 7O.7 undergoes restacking transitions upon cooling, from hexagonal close packed (ABAB) – orthorhombic face-centered (ortho-F) – monoclinic C-centred (mono-C) – simple hexagonal (AAA) before transforming to the G phase [70, 72]. On reducing the film thickness below approximately 280 molecular layers, additional hexagonal-ABC and orthorhombic-a structures were observed [73, 75].

The C-centered monoclinic ordered phases G and J have structures analogous to the tilted hexatic SmF and SmI phases, with the rows of wide-angle spots being tilted above and below the equator in similar characteristic ways (Fig. 26). The C-centering causes those (hkl) reflections where the sum $(h+k)$ is odd to be systematically absent. It should be noted that a study of a monodomain free-standing film of TB6A found that the symmetry of the G phase for this compound is actually triclinic [78].

The structure of the J phase was first described for the compound in HOBACPC [230]. Subsequently, the phase sequence SmI–J–K upon cooling was observed in 8OSI [231], and HEPTOBPD [246]. The phase sequence G–J was observed in a binary mixture [247], and ordered J* and G* phases have been studied in chiral 2M4P8BC [248]. These phases, unlike the higher temperature SmI* phase, do not have a helical arrangement of the tilt director, but, unlike the racemic J and G counterparts, do exhibit ferroelectric behavior. Other X-ray studies of the G and J phases have been carried out [100, 247, 249]. A synchrotron high resolution study of the ordered smectic phases of 8OSI, which in bulk forms J and K phases, found that as the film thickness was reduced from 1000 to 5 molecular layers, the behavior changed from three- to two-dimensional [79].

The observation of satellite reflections in the equatorial plane close to various of the Bragg peaks provided evidence that the layers in certain B and G phases are corrugated, the in-plane modulation being sometimes one-dimensional and sometimes two-dimensional [34, 35, 70, 72, 250–252]. Furthermore, evidence for local antiphase modulations was obtained in the B phase of cyano bi-cyclohexanes [253]. The nature of the structured diffuse scattering observed in B and G phases has been analyzed in terms of defects [99, 100, 254].

The orthorhombic ordered E phase, derived from the B phase by quenching of the six-fold rotational disorder, shows a num-

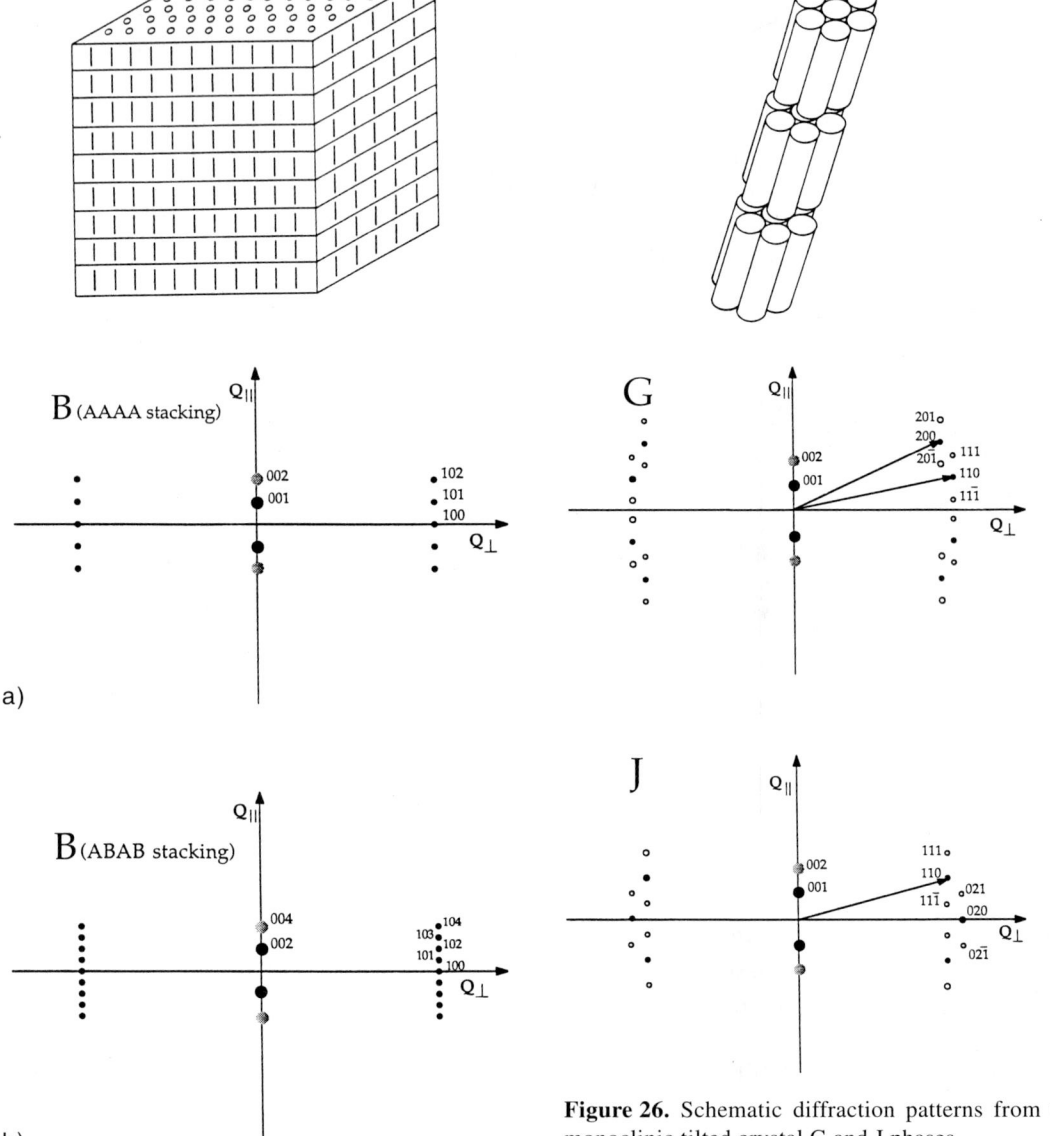

Figure 25. Schematic diffraction patterns from the ordered B phase, with stacking sequences (a) AAAA, and (b) ABAB.

Figure 26. Schematic diffraction patterns from the monoclinic tilted crystal G and J phases.

ber of rows of Bragg peaks in the wide-angle region, centered at different values of Q (Fig. 27). This is partly because the a, b lattice parameters distort from the value $b = \sqrt{3}a$ which they had in the hexagonal ordered phase B, causing the sets of reflections $(11l)$ and $(20l)$ to become resolved from each. But also, additional sets of reflections such as $(21l)$ become permitted by the reduction in symmetry due to the herringbone packing. X-ray studies of a series of mesogens with a biphenyl unit linked by a short spacer to a fluorinated alkyl chain

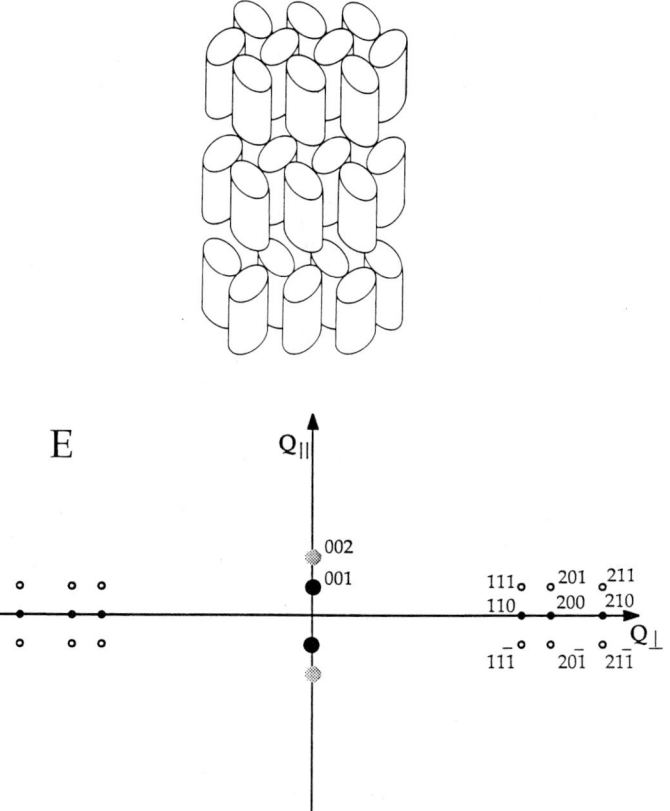

Figure 27. Schematic diffraction pattern from the orthorhombic untilted crystal E phase, with herringbone packing of the molecules.

have indicated that the perfluoroalkyl chains remain disordered in the E phase [225]. As previously discussed, a complex modulated structure has been found to exist along the layer normal in the E phase of the alkyl cyanoterphenyls, resulting from competing incommensurate density waves [182]. On forming the monoclinic ordered phases H and K, derived from phases G and J, similar modifications to the diffraction patterns as in the case of the E phase occur, due to the reduction in symmetry (X-ray patterns not shown). The system 4,4'-di-(2-methoxyethoxy)biphenyl has been found to be the first system where the K phase forms directly on cooling from the isotropic phase [256].

3.12 Crystalline Packing and Conformation of Mesogens

The relationship between the structures of the crystal and the E phase for IBPBAC [257], the crystal and the B phase for 4O.8 [258], and the crystal and the G phase for 8O.4 [259] have been examined, and in the latter two cases it was concluded that there is no close structural relationship between the crystal and the smectic phases. In the case of a biforked mesogen, however, it was concluded that there is a clear relationship between the crystalline and the SmC structures [260]. The crystal structures of semi-perfluorinated liquid crystals have been used to help understand the nature of the packing in the smectic phase [261]. Although studies of molecular conformation and packing in the crystalline state will thus not necessarily be directly relevant to understanding the mesomorphic behavior, such information nevertheless provides an extremely useful basis for modeling various features of liquid crystals, and for interpreting certain types of experiment. Full details of all crystal structures solved to date can be found in the Cambridge Crystallographic Database. A small selection of references is as follows:

4-cyano-4'-n-alkylbiphenyls: 2-CB [262], 3-CB [263], 4-CB [264] and 5-CB [265]; 4-cyano-4'-n-alkoxybiphenyls: 1-OCB, 2-OCB, 3-OCB and 4-OCB [266] and 5-OCB [267]; N-(4-n-alkoxybenzylidene)-4'-n-alkylanilines (nO.ms): 4O.2 [268], 1O.4 (MBBA) [269], 2O.4 [270], 8O.4 [259], 7O.6 [268], 4O.8 [258]; 4,4'-dimethoxyazoxybenzene (PAA) [271]; isobutyl 4-(4'-phenyl-benzylidene amino) cinnamate (IBPBAC): [257]; symmetric dimers: α,ω-bis(4-cyanobiphenyl-4'-oxy)heptane (BCBO7) [272] and m.OnO.m: α,ω-bis(4-n-alkylanilinebenzylidine-4'-oxy)alkanes (5.O4O.5 and 5.O5O.5) [273]; a ferroelectric liquid crystal [274]; metallomesogens [46, 275]. Recent single crystal studies of various discotic mesogens have been reviewed [276]. The relationship between the crystalline and the mesophase structures of aligned liquid crystal polymers has been explored [277–279].

3.13 Columnar Liquid Crystals

The principles involved in structural studies of discotic liquid crystals are identical to those in X-ray studies of calamitics, except that the natural molecular arrangement is in columns rather than layers. The packing of the columns then inevitably leads to phases which are ordered in at least two dimensions, with a range of different packing symmetries [25, 280]. Examples of the common symmetries observed are shown in Fig. 28.

Columnar phases of columnar mesogens formed by discogens were identified two decades ago by X-ray studies on hexa-alkanoyloxybenzenes [281], and in hexa-alkoxy and alkanoyloxy-triphenylenes [280, 282]. Columnar phases have also been found in copper mesogens [283–285], copper alkanoates [286–289], a copper alkadiynoate [290], nickel alkyldithiolate [291], tetrapalladium mesogens [292], conical or pyrimidical shaped mesogens [293, 294], in certain macrocyclic systems [295–297], phthalocyanine derivatives [179, 289, 298–305], phasmids [59, 306], polycatenar derivatives [307, 308], carbohydrate mesogens [309,

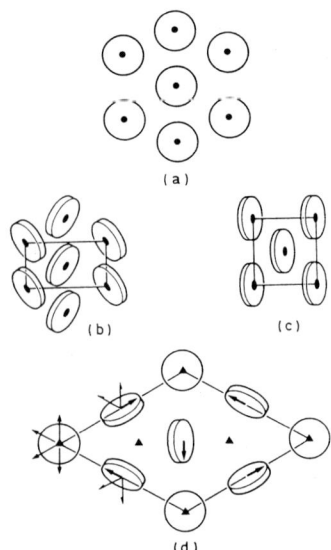

Figure 28. Structures and symmetries of various columnar phases of discotic mesogen: (a) hexagonal Col$_{hd}$; (b) rectangular Col$_{rd}$ (P2$_1$/a); (c) rectangular Col$_{rd}$ (C2/m); (d) the 4-sublattice triangular phase (P321). (From [392]).

310], triphenylene hexa-*n*-alkylcyclohexanoates [311], and star-like discotic mesogens [312]. Mesogens intermediate between calamitics and discotics, such as the phasmids or polycatenars (reviewed in [313]), may exhibit lamellocolumnar phases, intermediate in structure between conventional smectic and columnar phases [314, 315], and may undergo temperature-induced smectic–columnar transitions [59, 207, 316]. Smectic-like columnar phases have also been reported in certain metallomesogens [283, 285, 317], although the detailed structures are not completely clear.

The structure of the nematic phase of discogens is analogous to the nematic phase of calamitics. The normals to the average molecular planes of the disc-like molecules align preferentially along a director **n**, and the positional order is short range. At lower temperatures, the discs may become arranged into columns, which then pack into two-dimensional arrays. The positional order along the columns may then be short range, Col_h (disordered), or long range, Col_o (ordered). The 2-D symmetry of the packing of the columns may be hexagonal, rectangular, triangular or oblique [280, 318]. In many of these phases the cores are tilted with respect to the column axes. A synchrotron study of freely-suspended oriented discotic strands of triphenylene hexa-*n*-dodecanoate found that the cores are tilted in both the disordered rectangular Col_{rd} and hexagonal Col_{hd} phases, the transition Col_{rd}–Col_{hd} corresponding to a locking-in of the tilt directions into a herringbone arrangement [319, 320]. A high resolution study of oriented truxene hexa-*n*-tetradecanoate in the Col_{hd} phase showed that the correlation length of the 2-D hexagonal packing of the disordered columns was at least 400 nm [321]. A study of oriented hexa-hexylthiotriphenylene, which exhibits a hexagonal disordered–ordered Col_{hd}–Col_{ho} transition [322], found a helicoidal stacking of cores within each column in the ordered phase, as well as a three-column superlattice structure [323, 324].

3.14 Liquid Crystal Polymers

Liquid crystalline polymers will be dealt with here only briefly, and a number of reviews covering structural aspects should be consulted for further details [26, 325, 326]. The principles of X-ray studies described above for low molar mass systems can equally be applied to polymeric mesophases, allowing information about the long range organization and short range order of the backbones and the mesogenic cores to be deduced. For papers describing the packing in smectic phases (e.g., SmA, SmC, SmB, E, etc.) of side group liquid crystalline polymers, electron density profiles, correlation lengths, critical behavior, and so on, see references 4 to 61 in the review by Davidson and Levelut [26]. For a selection of more recent structural papers, see references [104, 106, 138, 297, 327–359].

In X-ray diffraction patterns from aligned liquid-crystalline polymers, compared with low molar mass systems, a number of differences are apparent. First, it is common to observe up to six layer reflections (00*l* Bragg peaks), indicating that the density wave along the layer normal deviates strongly from sinusoidal. This higher resolution can be usually exploited by calculating electron density profiles (see Eq. 9) from the observed structure factors, if their phases (signs) can be deduced. Second, strong structured diffuse scattering is apparent in the diffraction patterns, localised in different regions of reciprocal space (also invariably present in the patterns from low molar mass systems, but usually to a less marked extent).

SmA$_1$, SmA$_d$, and SmA$_2$ phases can be observed in various side-group liquid crystalline polymers, as can tilted SmC and SmC* phases. Opposite to the usual behavior of low molar mass systems, in a polymeric SmC phase, the tilt angle was found to decrease with decreasing temperature, apparently due to a stiffening of the backbone [360]. Evidence has been obtained of layer modulations in the SmA phase of polysiloxanes [361], and of undulations of SmÃ-like ordering in polymethacrylates [102]. A true SmÃ antiphase has been observed in a mesomorphic combined polymer (with mesogenic groups both in the main chain and in the side groups) [147]; a more complex 2-D modulated structure has been found in a main chain liquid crystalline polymer [138].

The polymeric SmB phase appears to be a stacked hexatic phase, although this cannot be confirmed until a true monodomain (rather than a fiber-averaged) sample becomes available for X-ray study [26]. Interestingly, in the polymeric smectic E phase, unlike that in molar mass systems, the adjacent smectic layers appear to be uncorrelated [362–364].

The schematic X-ray pattern from the aligned nematic phase of a laterally-attached side-group polysiloxane is shown in Fig. 29a. In addition to the usual diffuse peaks typical of a nematic phase of low molar mass mesogens, additional diffuse features are seen [342]. That on the equator labeled (b) is thought to arise from the siloxane backbones, preferentially aligned parallel to the director. The features labeled (c) on the meridian arise from uncorrelated locally periodic columns of mesogenic groups extending along the director. The low angle diffuse spots (d) can either arise from SmC-like cybotactic groups, or from local side-

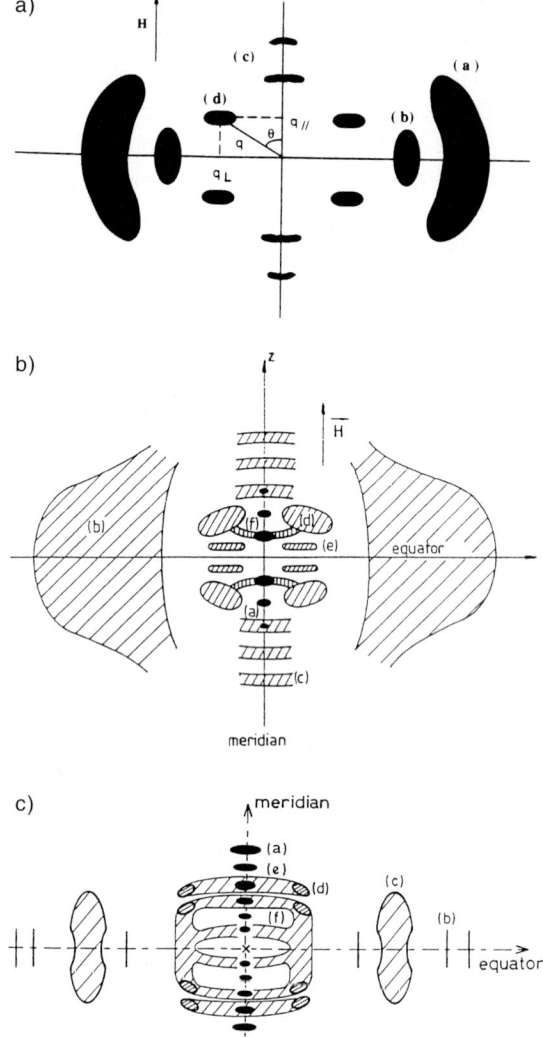

Figure 29. Schematic X-ray diffraction patterns from aligned liquid crystalline polymers in (a) the N phase of a laterally attached side group polysiloxane, (b) the SmA phase of a side-group polymethacrylate, and (c) the SmB phase of a side-group polysiloxane. (Taken from [342], part a; [26], parts b and c).

group blocks where the aliphatic tails are tilted with respect to the mesogenic cores.

The schematic diffraction pattern from the SmA phase of a side-group polymethacrylate is shown in Fig. 29b. In addition to the layer reflections on the meridian and the diffuse equatorial wide-angle scattering, diffuse features are seen both on and off the meridian. Diffuse streaks (c) are an indication of uncorrelated longitudinal disorder of rows of mesogenic cores, diffuse spots (d) arise from local layer undulations, spots (e) arise from local SmÃ-like antiphase regions, and the weak streaks ('moustaches') labeled (f) are an indication of layer crossing by the backbones. In side group polyacrylates, X-ray evidence has been obtained for the presence of edge dislocations in the SmA phase [103].

The schematic X-ray pattern from the SmB phase of a side-group polysiloxane is shown in Fig. 29 c. In addition to the meridional layer reflections and the (10), (11) and (20) equatorial rods of scattering from the in-plane hexagonal packing (which implies that it is a stacked hexatic phase), diffuse feature (c) indicates a local herringbone packing, slightly correlated between adjacent layers. Diffuse spots (d) along (1/2, 0, l) arise from local layer undulations, whose (local) period is locked to a value of twice the d_{10} spacing of the underlying in-plane hexagonal lattice.

X-ray studies of stretch-aligned fibers of main chain [365] or side chain [366] liquid crystalline polymers containing triphenylene units show that they both form hexagonal columnar phases, but that in the main chain polymer the columns are aligned perpendicular to the stretching direction, whereas in the side-group polymer they are aligned parallel to it. A hexagonal columnar phase has also been observed in a polymeric ethylenimine derivative [297].

Fairly rigid main chain liquid crystalline polymers with disc-shaped units in the chain have been reported to form a 'sanidic' (board-like) nematic phase, with a parallel alignment of the boards [367].

3.14.1 Block Copolymers

Block copolymers can adopt a wide range of 1-D, 2-D and 3-D ordered mesophases, such as for example lamellar, hexagonal and cubic, depending on the nature and lengths of the individual blocks [326, 368]. In fact their self-assembly behavior [369] is more akin to that of lyotropic (amphiphilic) systems, where a similar range of mesophases is found as the balance between the polar and the nonpolar regions is varied [40]. For recent X-ray diffraction studies of the bulk mesomorphic structures of block copolymers the following references should be consulted [370–383]. The use of X-ray reflectivity to study block copolymers has been reviewed [384, 385], and a selection of such recent surface structure studies, including off-specular scattering, may be found in references [386–389].

3.15 References

[1] A. Guinier, *X-Ray Diffraction in Crystals, Imperfect Crystals, and Amorphous Bodies*, W. H. Freeman, San Francisco, **1963**.
[2] B. E. Warren, *X-Ray Diffraction*, Addison-Wesley, Reading, **1963**.
[3] B. K. Vainshtein, *Diffraction of X-Rays by Chain Molecules*, Elsevier, Amsterdam, **1963**.
[4] P. S. Pershan, *Faraday Discuss. Chem. Soc.* **1990**, *89*, 231–245.
[5] R. H. Ottewill et al., *Faraday Discuss. Chem. Soc.* **1990**, *89*, 247–258.
[6] J. Daillant, K. Quinn, C. Gourier, F. Rieutord, *J. Chem. Soc. Faraday Trans.* **1996**, *92*, 505–513.
[7] A. J. Leadbetter in *The Molecular Physics of Liquid Crystals* (Eds.: G. R. Luckhurst, G. W. Gray), Academic Press, London, **1979**, 285–316.
[8] J. Doucet in *The Molecular Physics of Liquid Crystals* (Eds.: G. R. Luckhurst, G. W. Gray), Academic Press, London, **1979**, 317–341.
[9] H. Kelker, R. Hatz, *Handbook of Liquid Crystals*, Verlag Chemie, Weinheim, **1980**.
[10] L. V. Azaroff, *Mol. Cryst. Liq. Cryst.* **1980**, *60*, 73–98.
[11] P. S. Pershan, *Structure of Liquid Crystal Phases*, World Scientific, Singapore, **1988**.

[12] S. Chandrasekhar, *Liquid Crystals*, Cambridge University Press, Cambridge, **1992**.
[13] P. G. de Gennes, J. Prost, *The Physics of Liquid Crystals*, Clarendon Press, Oxford, **1993**.
[14] P. M. Chaikin, T. C. Lubensky, *Principles of Condensed Matter Physics*, Cambridge University Press, Cambridge, **1995**.
[15] A. J. Leadbetter in *Thermotropic Liquid Crystals* (Ed.: G. W. Gray), John Wiley and Sons, Chichester, **1987**, 1 – 27.
[16] J. D. Brock, R. J. Birgeneau, J. D. Litster, A. Aharony, *Phys. Today* **1989**, *42*, 52 – 59.
[17] J. D. Brock, R. J. Birgeneau, J. D. Litster, A. Aharony, *Contemp. Phys.* **1989**, *30*, 321 – 335.
[18] R. Shashidhar in *Phase Transitions in Liquid Crystals* (Eds.: S. Martellucci, A. N. Chester), Plenum Press, New York, **1992**, 227 – 237.
[19] J. W. Goodby et al. *Ferroelectric Liquid Crystals*, Gordon and Breach, Philadelphia, **1991**.
[20] F. Hardouin, *Physica A* **1986**, *140*, 359 – 367.
[21] A. M. Levelut in *Incommensurate Crystals, Liquid Crystals and Quasi-Crystals* (Eds.: J. F. Scott, N. A. Clark), Plenum Press, New York, **1987**, 283 – 296.
[22] P. Barois, J. Pommier, J. Prost in *Solitons in Liquid Crystals* (Eds.: L. Lam, J. Prost), Springer-Verlag, New York, **1992**, 191 – 234.
[23] B. I. Ostrovskii, *Liq. Cryst.* **1993**, *14*, 131 – 157.
[24] P. Patel, S. Kumar, P. Ukleja, *Liq. Cryst.* **1994**, *16*, 351 – 371.
[25] S. Chandrasekhar, *Liq. Cryst.* **1993**, *14*, 3 – 14.
[26] P. Davidson, A. M. Levelut, *Liq. Cryst.* **1992**, *11*, 469 – 517.
[27] S. Kumar, *Phys. Rev. A* **1981**, *23*, 3207 – 3214.
[28] D. Guillon, A. Skoulios, J. J. Benattar, *J. Phys.* **1986**, *47*, 133 – 138.
[29] F. Barbarin et al. *Liq. Cryst.* **1987**, *2*, 815 – 823.
[30] J. D. Bunning, J. L. Butcher, D. J. Byron, A. S. Matharu, R. C. Wilson, *Liq. Cryst.* **1995**, *19*, 693 – 698.
[31] C. A. Vieth, E. T. Samulski, N. S. Murthy, *Liq. Cryst.* **1995**, *19*, 557 – 563.
[32] L. Douce et al. *Liq. Cryst.* **1996**, *20*, 235 – 242.
[33] B. M. Ocko, A. R. Kortan, R. J. Birgeneau, J. W. Goodby, *J. Phys.* **1984**, *45*, 113 – 128.
[34] A. J. Leadbetter, M. A. Mazid, B. A. Kelly, J. W. Goodby, G. W. Gray, *Phys. Rev. (Lett.)* **1979**, *43*, 630 – 633.
[35] A. J. Leadbetter, M. A. Mazid, R. M. Richardson in *Liquid Crystals* (Ed.: S. Chandrasekhar), Heyden and Son, London, **1980**, 65 – 79.
[36] S. Diele, P. Brand, H. Sackmann, *Mol. Cryst. Cryst.* **1972**, *17*, 163 – 169.
[37] A. Tardieu, J. Billard, *J. Phys. Colloque France* **1976**, *37*, C3-79 – C3-81.
[38] J. E. Lydon, *Mol. Cryst. Liq. Cryst. Lett.* **1981**, *72*, 79 – 87.
[39] D. Guillon, A. Skoulios, *Europhys. (Lett.)* **1987**, *3*, 79 – 85.
[40] J. M. Seddon, R. H. Templer, *Phil. Trans. Roy. Soc. Lond. A – Phys. Sci. Eng.* **1993**, *344*, 377 – 401.
[41] G. Etherington, A. J. Langley, A. J. Leadbetter, X. J. Wang, *Liq. Cryst.* **1988**, *3*, 155 – 168.
[42] G. Etherington, A. J. Leadbetter, X. J. Wang, G. W. Gray, A. Tajbakhsh, *Liq. Cryst.* **1986**, *1*, 209 – 214.
[43] D. W. Bruce, D. A. Dunmur, E. Lalinde, P. M. Maitlis, P. Styring, *Nature* **1986**, *323*, 791 – 792.
[44] D. W. Bruce et al. *Mol. Cryst. Liq. Cryst.* **1991**, *206*, 79 – 92.
[45] D. W. Bruce et al. *Mol. Cryst. Liq. Cryst.* **1992**, *215*, 1 – 11.
[46] H. Adams, et al. *J. Mater. Chem.* **1992**, *2*, 395 – 400.
[47] D. W. Bruce, S. A. Hudson, *J. Mater. Chem.* **1994**, *4*, 479 – 486.
[48] D. W. Bruce, et al. *J. Phys. II* **1995**, *5*, 289 – 302.
[49] D. W. Bruce, B. Donnio, D. Guillon, B. Heinrich, M. Ibnelhaj, *Liq. Cryst.* **1995**, *19*, 537 – 539.
[50] M. J. Baena, P. Espinet, M. C. Lequerica, A. M. Levelut, *J. Am. Chem. Soc.* **1992**, *114*, 4182 – 4185.
[51] J. Billard, H. Zimmermann, R. Poupko, Z. Luz, *J. Phys.* **1989**, *50*, 539 – 547.
[52] B. Kohne, K. Praefcke, J. Billard, *Z. Naturforsch. B* **1986**, *41*, 1036 – 1044.
[53] K. Praefcke, B. Kohne, A. Eckert, J. Hempel, *Z. Naturforsch. B* **1990**, *45*, 1084 – 1090.
[54] S. Fischer et al. *Liq. Cryst.* **1994**, *17*, 855 – 861.
[55] K. Borisch, S. Diele, P. Goring, H. Muller, C. Tschierske, *Liq. Cryst.* **1997**, *22*, 427 – 443.
[56] G. Lattermann, G. Staufer, *Mol. Cryst. Liq. Cryst.* **1990**, *191*, 199 – 203.
[57] G. Staufer, M. Schellhorn, G. Lattermann, *Liq. Cryst.* **1995**, *18*, 519 – 527.
[58] N. H. Tinh, C. Destrade, A. M. Levelut, J. Malthete, *J. Phys.* **1986**, *47*, 553 – 557.
[59] A. M. Levelut, Y. Fang, *J. Phys.* **1990**, *51*, C7-229 – C7-236.
[60] C. Soulie, P. Bassoul, J. Simon, *J. Chem. Soc. Chem. Commun.* **1993**, 114 – 115.
[61] V. S. K. Balagurusamy, G. Ungar, V. Percec, G. Johansson, *J. Am. Chem. Soc.* **1997**, *119*, 1539 – 1555.
[62] H. Zimmermann et al., *Liq. Cryst.* **1992**, *12*, 245 – 262.
[63] W. Weissflog, G. Pelzl, I. Letko, S. Diele, *Mol. Cryst. Liq. Cryst. Sci. Technol. A: Mol. Cryst. Liq. Cryst.* **1995**, *260*, 157.
[64] I. Letko, S. Diele, G. Pelzl, W. Weissflog, *Mol. Cryst. Liq. Cryst. Sci. Technol. A: Mol. Cryst. Liq. Cryst.* **1995**, *260*, 171 – 183.
[65] I. Letko, S. Diele, G. Pelzl, W. Weissflog, *Liq. Cryst.* **1995**, *19*, 643 – 646.
[66] R. Pindak, D. E. Moncton, S. C. Davey, J. W. Goodby, *Phys. Rev. (Lett.)* **1981**, *46*, 1135 – 1138.

[67] R. Pindak, D. Moncton, *Phys. Today* **1982**, *35*, 57–62.
[68] D. E. Moncton, R. Pindak, S. C. Davey, J. W. Goodby, *Bull. Am. Phys. Soc.* **1981**, *26*, 274–275.
[69] D. E. Moncton, R. Pindak, S. C. Davey, G. S. Brown, *Phys. Rev. (Lett.)* **1982**, *49*, 1865–1868.
[70] J. Collett et al. *Phys. Rev. (Lett.)* **1982**, *49*, 553–556.
[71] J. Collett, P. S. Pershan, E. B. Sirota, L. B. Sorensen, *Phys. Rev. (Lett.)* **1984**, *52*, 356–359.
[72] J. Collett, L. B. Sorensen, P. S. Pershan, J. Als-Nielsen, *Phys. Rev. A: Gen. Phys.* **1985**, *32*, 1036–1043.
[73] E. B. Sirota, P. S. Pershan, L. B. Sorensen, J. Collett, *Phys. Rev. (Lett.)* **1985**, *55*, 2039–2042.
[74] E. B. Sirota, P. S. Pershan, S. Amador, L. B. Sorensen, *Phys. Rev. A: Gen. Phys.* **1987**, *35*, 2283–2287.
[75] E. B. Sirota, P. S. Pershan, L. B. Sorensen, J. Collett, *Phys. Rev. A: Gen Phys.* **1987**, *36*, 2890–2901.
[76] J. D. Brock et al. *Phys. Rev. (Lett.)* **1986**, *57*, 98–101.
[77] J. D. Brock et al. *Z. Phys. B: Cond. Matter* **1989**, *74*, 197–213.
[78] D. Y. Noh, J. D. Brock, J. D. Litster, R. J. Birgeneau, J. W. Goodby, *Phys. Rev. B: Cond. Matter* **1989**, *40*, 4920–4927.
[79] D. Y. Noh, et al. *Phys. Rev. B: Cond. Matter* **1991**, *43*, 842–850.
[80] A. S. Cherodian, R. M. Richardson, *Mol. Cryst. Liq. Cryst.* **1991**, *196*, 115–131.
[81] R. Geer et al., *Nature* **1992**, *355*, 152–154.
[82] T. A. Lobko, B. I. Ostrovskii, W. Haase, *J. Phys. II* **1992**, *2*, 1195–1213.
[83] K. Tajiri et al. *J. Phys. Soc. Jap.* **1995**, *64*, 3157–3159.
[84] J. D. Shindler, E. A. L. Mol, A. Shalaginov, W. H. de Jeu, *Phys. Rev. (Lett.)* **1995**, *74*, 722–725.
[85] A. Guinier, G. Fournet, *Small-Angle Scattering of X-Rays,* John Wiley and Sons, New York, **1955**.
[86] W. L. McMillan, *Phys. Rev. A: Gen. Phys.* **1972**, *6*, 936–947.
[87] A. J. Leadbetter, E. K. Norris, *Mol. Phys.* **1979**, *38*, 669–686.
[88] A. J. Leadbetter, P. G. Wrighton, *J. Phys. Colloque France* **1979**, *40*, C3-234–C3-242.
[89] P. E. Cladis et al. in *Advances in Liquid Crystal Research and Applications* (Ed.: L. Bata), Pergamon Press, Oxford, **1980**, 81–91.
[90] E. F. Gramsbergen, W. H. de Jeu, *Liq. Cryst.* **1989**, *4*, 449–455.
[91] Z. X. Fan, W. Haase, *J. Chem. Phys.* **1991**, *95*, 6066–6074.
[92] Z. X. Fan, C. G. Seguel, C. Aguilera, W. Haase, *Liq. Cryst.* **1992**, *11*, 401–409.
[93] Z. X. Fan, L. Y. Chiang, W. Haase, *Liq. Cryst.* **1995**, *18*, 13–19.
[94] Y. Takanishi, A. Ikeda, H. Takezoe, A. Fukuda, *SPIE Proc Ser.* **1993**, *1911*, 70–76.
[95] Y. Takanishi, A. Ikeda, H. Takezoe, A. Fukuda, *Phys. Rev. E.* **1995**, *51*, 400–406.
[96] A. Caillé, *C. R. Acad. Sci. Ser. B* **1972**, *274*, 891.
[97] J. Als-Nielsen et al. *Phys. Rev. (Lett)* **1980**, *22*, 312–320.
[98] D. R. Nelson, B. I. Halperin, *Phys. Rev. B* **1979**, *19*, 2457–2484.
[99] A. M. Levelut, *J. Phys.* **1990**, *51*, 1517–1526.
[100] P. Davidson, E. Dubois-Violette, A. M. Levelut, B. Pansu, *J. Phys. I* **1992**, *2*, 899–913.
[101] P. Davidson et al. *J. Phys. II* **1995**, *5*, 249–262.
[102] P. Davidson, P. Keller, A. M. Levelut, *J. Phys.* **1985**, *46*, 939–946.
[103] P. Davidson, B. Pansu, A. M. Levelut, L. Strzelecki, *J. Phys. II* **1991**, *1*, 61–74.
[104] R. E. Geer, R. Shashidhar, A. F. Thibodeaux, R. S. Duran, *Phys. Rev. (Lett.)* **1993**, *71*, 1391–1394.
[105] A. M. Levelut, K. Moriya, *Liq. Cryst.* **1996**, *20*, 119–124.
[106] S. G. McNamee et al. *Liq. Cryst.* **1995**, *18*, 787–794.
[107] S. Chandrasekhar, B. K. Sadashiva, B. R. Ratna, V. N. Raja, *Pramana* **1988**, *30*, L 491–L 494.
[108] S. Chandrasekhar, B. R. Ratna, B. K. Sadashiva, V. N. Raja, *Mol. Cryst. Liq. Cryst.* **1988**, *165*, 123–130.
[109] K. Praefcke et al. *Mol. Cryst. Liq. Cryst.* **1991**, *198*, 393–405.
[110] S. M. Fan et al., *Chem. Phys. (Lett.)* **1993**, *204*, 517–523.
[111] I. W. Hamley, G. R. Luckhurst, J. M. Seddon, *Mol. Phys.* **1991**, *74*, 1221–1231.
[112] S. Chandrasekhar, *Mol. Cryst. Liq. Cryst. Sci. Technol. A: Mol. Cryst. Liq. Cryst.* **1994**, *243*, 1–9.
[113] J. D. Litster, *Phil. Trans. Roy. Soc. Lond. A: Math. Phys. Sci.* **1983**, *309*, 145–153.
[114] J. Als-Nielsen, R. J. Birgeneau, M. Kaplan, J. D. Litster, C. R. Safinya, *Phys. Rev. Lett.* **1977**, *39*, 352–355.
[115] B. M. Ocko, R. J. Birgeneau, J. D. Litster, M. E. Neubert, *Phys. Rev. (Lett.)* **1984**, *52*, 208–211.
[116] B. M. Ocko, R. J. Birgeneau, J. D. Litster, *Z. Phys. B: Cond. Matter* **1986**, *62*, 487–497.
[117] K. K. Chan, P. S. Pershan, L. B. Sorensen, F. Hardouin, *Phys. Rev. A: Gen. Phys.* **1986**, *34*, 1420–1433.
[118] W. G. Bouwman, W. H. De Jeu, *Phys. Rev. (Lett.)* **1992**, *68*, 800–803.
[119] J. D. Litster et al., *J. Phys. (Paris) Colloque* **1979**, *40*, C3-339–C3-344.

[120] D. Davidov et al. *Phys. Rev. B* **1979**, *19*, 1657–1663.
[121] R. J. Birgeneau, C. W. Garland, G. B. Kasting, B. M. Ocko, *Phys. Rev. A: Gen. Phys.* **1981**, *24*, 2624–2634.
[122] C. R. Safinya, R. J. Birgeneau, J. D. Litster, M. E. Neubert, *Phys. Rev. (Lett.)* **1981**, *47*, 668–671.
[123] W. G. Bouwman, W. H. de Jeu, *J. Phys. II* **1994**, *4*, 787–804.
[124] V. K. Kelkar, A. S. Paranjpe, *Mol. Cryst. Liq. Cryst.* **1987**, *4*, 139–144.
[125] P. Davidson, D. Petermann, A. M. Levelut, *J. Phys. II* **1995**, *5*, 113–131.
[126] R. M. Richardson, A. J. Leadbetter, M. A. Mazid, P. A. Tucker, *Mol. Cryst. Liq. Cryst.* **1987**, *149*, 329–345.
[127] R. M. Richardson, J. M. Allman, G. J. McIntyre, *Liq. Cryst.* **1990**, *7*, 701–719.
[128] R. W. Date, I. W. Hamley, G. R. Luckhurst, J. M. Seddon, R. M. Richardson, *Mol. Phys.* **1992**, *76*, 951–977.
[129] I. W. Hamley et al. *J. Chem. Phys.* **1996**, *104*, 10046–10054.
[130] F. Hardouin, A. M. Levelut, M. F. Achard, G. Sigaud, *J. Chimie Phys. Physico-Chimie Biol.* **1983**, *80*, 53–64.
[131] F. Hardouin et al. *Liq. Cryst.* **1989**, *5*, 463–478.
[132] P. Goring et al. *Liq. Cryst.* **1995**, *19*, 629–635.
[133] G. S. Attard et al. *Liq. Cryst.* **1994**, *16*, 529–581.
[134] A. E. Blatch, I. D. Fletcher, G. R. Luckhurst, *Liq. Cryst.* **1995**, *18*, 801–809.
[135] V. Faye, P. Barois, H. T. Nguyen, V. Laux, N. Isaert, *New J. Chem.* **1996**, *20*, 283–286.
[136] J. Watanabe, H. Komura, T. Niori, *Liq. Cryst.* **1993**, *13*, 455–465.
[137] T. Niori, S. Adachi, J. Watanabe, *Liq. Cryst.* **1995**, *19*, 139–148.
[138] J. J. Watanabe, Y. Nakata, K. Simizu, *J. Phys. II* **1994**, *4*, 581–588.
[139] Y. Park, T. C. Lubensky, P. Barois, J. Prost, *Phys. Rev. A* **1988**, *37*, 2197–2213.
[140] R. Shashidhar, B. R. Ratna, S. K. Prasad, S. Somasekhara, G. Heppke, *Phys. Rev. (Lett.)* **1987**, *59*, 1209–1211.
[141] K. K. Chan, P. S. Pershan, L. B. Sorensen, F. Hardouin, *Phys. Rev. (Lett.)* **1985**, *54*, 1694–1697.
[142] R. J. Birgeneau et al. *Phys. Rev. A: Gen Phys.* **1983**, *27*, 1251–1254.
[143] L. J. Martinez-Miranda, A. R. Kortan, R. J. Birgeneau, *Phys. Rev. A: Gen. Phys.* **1987**, *36*, 2372–2382.
[144] A. D. L. Chandani et al. *Jap. J. Appl. Phys. 2 (Lett.)* **1988**, *27*, L729–L732.
[145] Y. Galerne, L. Liebert, *Phys. Rev. (Lett.)* **1990**, *64*, 906–909.
[146] Y. Galerne, L. Liebert, *Phys. Rev. (Lett.)* **1991**, *66*, 2891–2894.
[147] A. Fukuda, Y. Takanishi, T. Isozaki, K. Ishikawa, H. Takezoe, *J. Mater. Chem.* **1994**, *4*, 997–1016.
[148] A. Suzuki et al., *Jap. J. Appl. Phys. 2 (Lett.)* **1990**, *29*, L336–L338.
[149] M. Johno et al., *Jap. J. Appl. Phys. 2 (Lett.)* **1990**, *29*, L111–L114.
[150] V. L. Lorman, *Liq. Cryst.* **1996**, *20*, 267–276.
[151] T. P. Rieker et al. *Phys. Rev. (Lett.)* **1987**, *59*, 2658–2661.
[152] T. P. Rieker, N. A. Clark, G. S. Smith, C. R. Safinya, *Liq. Cryst.* **1989**, *6*, 565–576.
[153] T. P. Rieker, N. A. Clark, C. R. Safinya, *Ferroelectrics* **1991**, *113*, 245–256.
[154] L. Taylor, R. M. Richardson, J. Ebbutt, J. C. Jones. *Mol. Cryst. Liq. Cryst. Sci. Technol. A: Mol. Cryst. Liq. Cryst.* **1995**, *263*, 255–270.
[155] P. C. Willis, N. A. Clark, J. Z. Xue, C. R. Safinya, *Liq. Cryst.* **1992**, *12*, 891–904.
[156] P. C. Willis, N. A. Clark, C. R. Safinya, *Liq. Cryst.* **1992**, *11*, 581–592.
[157] A. G. Rappaport et al., *Appl. Phys. (Lett.)* **1995**, *67*, 362–364.
[158] R. A. M. Hikmet, *Liq. Cryst.* **1995**, *18*, 927–932.
[159] H. F. Gleeson, A. S. Morse, *Liq. Cryst.* **1996**, *21*, 755–758.
[160] Y. Takanishi, A. Iida, K. Ishikawa, H. Takezoe, A. Fukuda, *Jap. J. Appl. Phys. 1: Reg. Papers, Short Notes & Review Papers* **1996**, *35*, 683–687.
[161] P. Barois, C. Coulon, J. Prost, *J. Phys. (Lett.)* **1981**, *42*, L107–L110.
[162] P. Barois, *Phys. Rev. A: Gen. Phys.* **1986**, *33*, 3632–3634.
[163] T. A. Lobko, B. I. Ostrovskii, *Mol. Mater.* **1993**, *1*, 99–132.
[164] F. Hardouin, G. Sigaud, N. H. Tinh, M. F. Achard, *J. Phys. (Lett.)* **1981**, *42*, L63–L66.
[165] G. Sigaud, F. Hardouin, M. F. Achard, A. M. Levelut, *J. Phys.* **1981**, *42*, 107–111.
[166] A. M. Levelut, *J. Phys. (Lett.)* **1984**, *45*, L603–L611.
[167] F. Hardouin, N. H. Tinh, M. F. Achard, A. M. Levelut, *J. Phys. (Lett.)* **1982**, *43*, L327–L331.
[168] C. R. Safinya, W. A. Varady, L. Y. Chiang, P. Dimon, *Phys. Rev. (Lett.)* **1986**, *57*, 432–435.
[169] C. R. Safinya, L. Y. Chiang in *Incommensurate Crystals, Liquid Crystals, and Quasi-Crystals* (Eds.: J. F. Scott, N. A. Clark), Plenum Press, New York, **1987**, 271–281.
[170] E. Fontes, P. A. Heiney, J. L. Haseltine, A. B. Smith, *J. Phys.* **1986**, *47*, 1533–1539.
[171] E. Fontes, P. A. Heiney, P. Barois, A. M. Levelut, *Phys. Rev. (Lett.)* **1988**, *60*, 1138–1141.
[172] M. J. Young, W. Lei, G. Nounesis, C. W. Garland, R. J. Birgeneau, *Phys. Rev. E* **1994**, *50*, 368–376.

[173] Y. S. Shi, G. Nounesis, S. Kumar, *Phys. Rev. E* **1996**, *54*, 1570–1573.
[174] B. W. Endres, M. Ebert, J. H. Wendorff, B. Reck, H. Ringsdorf, *Liq. Cryst.* **1990**, *7*, 217–239.
[175] H. T. Nguyen et al., *Liq. Cryst.* **1991**, *10*, 389–396.
[176] R. W. Date, G. R. Luckhurst, M. Shuman, J. M. Seddon, *J. Phys. II* **1995**, *5*, 587–605.
[177] V. Faye et al. *Liq. Cryst.* **1996**, *21*, 485–503.
[178] P. G. de Gennes, *J. Phys. (Lett.)* **1983**, *44*, L657–L664.
[179] D. Guillon, A. Skoulios, C. Piechocki, J. Simon, P. Weber, *Mol. Cryst. Liq. Cryst.* **1983**, *100*, 275–284.
[180] J. Prost, P. Barois, *J. Chim. Phys. Physico-Chimie Biol.* **1983**, *80*, 65–81.
[181] G. J. Brownsey, A. J. Leadbetter, *Phys. Rev. (Lett.)* **1980**, *44*, 1608–1611.
[182] A. J. Leadbetter, P. A. Tucker, G. W. Gray, A. R. Tajbakhsh, *Liq. Cryst.* **1990**, *8*, 1–12.
[183] B. R. Ratna, R. Shashidhar, V. N. Raja in *Incommensurate Crystals, Liquid Crystals, and Quasi-Crystals* (Eds.: J. F. Scott, N. A. Clark), Plenum Press, New York, **1987**, 259–270.
[184] B. R. Ratna, R. Shashidhar, V. N. Raja, *Phys. Rev. (Lett.)* **1985**, *55*, 1476–1478.
[185] R. Shashidhar, B. R. Ratna, *Liq. Cryst.* **1989**, *5*, 421–442.
[186] S. Kumar, L. Chen, V. Surendranath, *Phys. Rev. (Lett.)* **1991**, *67*, 322–325.
[187] P. Patel, S. S. Keast, M. E. Neubert, S. Kumar, *Phys. Rev. (Lett.)* **1992**, *69*, 301–304.
[188] P. Patel, C. Li, S. Kumar, *Phys. Rev. E* **1993**, *47*, 2643–2653.
[189] P. Patel, S. Kumar, *Europhys. Lett.* **1993**, *23*, 135–140.
[190] J. T. Mang, B. Cull, Y. S. Shi, P. Patel, S. Kumar, *Phys. Rev. (Lett.)* **1995**, *74*, 4241–4244.
[191] F. Hardouin, M. F. Achard, J. L. Jin, J. W. Shin, Y. K. Yun, *J. Phys. II* **1994**, *4*, 627–643.
[192] F. Hardouin, M. F. Achard, J. I. Jin, Y. K. Yun, *J. Phys. II* **1995**, *5*, 927–935.
[193] P. E. Cladis, R. K. Bogardus, W. B. Daniels, G. N. Taylor, *Phys. Rev. (Lett.)* **1977**, *39*, 720–723.
[194] D. Guillon, P. E. Cladis, J. Stamatoff, *Phys. Rev. (Lett.)* **1978**, *41*, 1598–1601.
[195] S. Diele, G. Pelzl, I. Latif, D. Demus, *Mol. Cryst. Liq. Cryst.* **1983**, *92*, 27–33.
[196] G. Pelzl, S. Diele, I. Latif, W. Weissflog, D. Demus, *Cryst. Res. Technol.* **1982**, *17*, K78–K82.
[197] G. Pelzl et al., *Mol. Cryst. Liq. Cryst.* **1986**, *139*, 333–351.
[198] G. Pelzl, C. Scholz, H. Sackmann, *Cryst. Res. Technol.* **1988**, *23*, K47–K50.
[199] G. Pelzl et al., *Mol. Cryst. Liq. Cryst.* **1989**, *168*, 197–208.
[200] G. Pelzl, C. Scholz, D. Demus, H. Sackmann, *Mol. Cryst. Liq. Cryst.* **1989**, *168*, 147–154.
[201] G. Pelzl et al., *Cryst. Res. Technol.* **1990**, *25*, 223–233.
[202] P. E. Cladis, D. Guillon, F. R. Bouchet, P. L. Finn, *Phys. Rev. A: Gen. Phys.* **1981**, *23*, 2594–2601.
[203] N. H. Tinh, F. Hardouin, C. Destrade, *J. Phys.* **1982**, *43*, 1127–1131.
[204] R. Shashidhar et al., *J. Phys. (Lett.)* **1985**, *46*, L445–L450.
[205] K. W. Evans-Lutterodt et al., *Phys. Rev. A: Gen. Phys.* **1987**, *36*, 1387–1395.
[206] W. Weissflog, I. Letko, G. Pelzl, S. Diele, *Liq. Cryst.* **1995**, *18*, 867–870.
[207] W. Weissflog, M. Rogunova, I. Letko, S. Diele, G. Pelzl, *Liq. Cryst.* **1995**, *19*, 541–544.
[208] P. S. Pershan, J. Als-Nielsen, *Phys. Rev. (Lett.)* **1984**, *52*, 759–762.
[209] P. S. Pershan, *Proc. Nat. Acad. Sci. USA* **1987**, *84*, 4692–4693.
[210] P. S. Pershan, *J. Phys.* **1989**, *50*, C71–C720.
[211] B. M. Ocko, A. Braslau, P. S. Pershan, J. Als-Nielsen, M. Deutsch, *Phys. Rev. (Lett.)* **1986**, *57*, 94–97.
[212] P. S. Pershan, A. Braslau, A. H. Weiss, J. Als-Nielsen, *Phys. Rev. A: Gen. Phys.* **1987**, *35*, 4800–4813.
[213] B. M. Ocko, P. S. Pershan, C. R. Safinya, L. Y. Chiang, *Phys. Rev. A: Gen. Phys.* **1987**, *35*, 1868–1872.
[214] E. F. Gramsbergen, W. H. de Jeu, J. Als-Nielsen, *J. Phys.* **1986**, *47*, 711–718.
[215] E. F. Gramsbergen, W. H. de Jeu, *J. Phys.* **1988**, *49*, 363–367.
[216] S. Gierlotka, P. Lambooy, W. H. de Jeu, *Europhys. (Lett.)* **1990**, *12*, 341–345.
[217] S. R. Renn, T. C. Lubensky, *Phys. Rev. A* **1988**, *38*, 2132–2147.
[218] S. R. Renn, T. C. Lubensky, *Mol. Cryst. Liq. Cryst.* **1991**, *209*, 349–355.
[219] S. R. Renn, *Phys. Rev. A* **1992**, *45*, 953–973.
[220] J. W. Goodby et al., *Nature* **1989**, *337*, 449–452.
[221] J. W. Goodby et al., *J. Am. Chem. Soc.* **1989**, *111*, 8119–8125.
[222] G. Srajer, R. Pindak, M. A. Waugh, J. W. Goodby, J. S. Patel, *Phys. Rev. (Lett.)* **1990**, *64*, 1545–1548.
[223] K. J. Ihn, J. A. N. Zasadzinski, R. Pindak, A. J. Slaney, J. Goodby, *Science* **1992**, *258*, 275–278.
[224] H. T. Nguyen et al., *J. Phys. II* **1992**, *2*, 1889–1906.
[225] L. Navailles, P. Barois, H. T. Nguyen, *Phys. Rev. (Lett.)* **1993**, *71*, 545–548.
[226] A. Bouchta et al., *J. Mater. Chem.* **1995**, *5*, 2079–2092.
[227] L. Navailles, R. Pindak, P. Barois, H. T. Nguyen, *Phys. Rev. (Lett.)* **1995**, *74*, 5224–5227.
[228] M. Petit, P. Barois, H. T. Nguyen, *Europhysics (Lett.)* **1996**, *36*, 185–190.

[229] L. Navailles, H. T. Nguyen, P. Barois, N. Isaert, P. Delord, *Liq. Cryst.* **1996**, *20*, 653–664.
[230] J. Doucet, P. Keller, A. M. Levelut, P. Porquet, *J. Phys.* **1978**, *30*, 548–553.
[231] A. J. Leadbetter, J. P. Gaughan, B. Kelly, G. W. Gray, J. W. Goodby, *J. Phys. Colloque C3* **1979**, *40*, C3-178–C3-184.
[232] J. J. Benattar, J. Doucet, M. Lambert, A. M. Levelut, *Phys. Rev. A* **1979**, *20*, 2505–2509.
[233] J. J. Benattar, F. Moussa, M. Lambert, C. Germain, *C. R. Acad. Sci. Ser. B* **1980**, *291*, 283–286.
[234] J. J. Benattar, F. Moussa, M. Lambert, *J. Phys.* **1980**, *41*, 1371–1374.
[235] J. W. Goodby, G. W. Gray, A. J. Leadbetter, M. A. Mazid, *J. Phys.* **1980**, *41*, 591–595.
[236] S. Diele, D. Demus, H. Sackmann, *Mol. Cryst. Liq. Cryst. (Lett.)* **1980**, *56*, 217–224.
[237] P. A. C. Gane, A. J. Leadbetter, J. J. Benattar, F. Moussa, M. Lambert, *Phys. Rev. A: Gen. Phys.* **1981**, *24*, 2694–2700.
[238] P. A. C. Gane, A. J. Leadbetter, P. G. Wrighton, *Mol. Cryst. Liq. Cryst.* **1981**, *66*, 247–266.
[239] J. J. Benattar, F. Moussa, M. Lambert, C. Germain, *J. Phys. (Lett.)* **1981**, *42*, L67–L70.
[240] J. J. Benattar, F. Moussa, M. Lambert, *J. Chim. Phys. Physico-Chimie Biol.* **1983**, *80*, 99–110.
[241] F. Moussa, J. J. Benattar, C. Williams, *Mol. Cryst. Liq. Cryst.* **1983**, *99*, 145–154.
[242] R. W. Date, C. T. Imrie, G. R. Luckhurst, J. M. Seddon, *Liq. Cryst.* **1992**, *12*, 203–238.
[243] G. S. Smith, E. B. Sirota, C. R. Safinya, R. J. Plano, N. A. Clark, *J. Chem. Phys.* **1990**, *92*, 4519–4529.
[244] A. M. Levelut, N. H. Tinh, *J. Phys.* **1987**, *48*, 847–853.
[245] D. E. Moncton, R. Pindak, *Phys. Rev. (Lett.)* **1979**, *43*, 701–704.
[246] P. A. C. Gane et al. *Mol. Cryst. Liq. Cryst.* **1983**, *100*, 67–74.
[247] S. Diele, A. Madicke, E. M. Hoft, G. Wende, H. Sackmann, *Mol. Cryst. Liq. Cryst.* **1983**, *92*, 187–191.
[248] J. Budai, R. Pindak, S. C. Davey, J. W. Goodby, *J. Phys. (Lett.)* **1984**, *45*, 1053–1062.
[249] G. Pelzl et al., *Z. Phys. Chemie-Leipzig* **1981**, *262*, 815–819.
[250] P. A. C. Gane, A. J. Leadbetter, P. A. Tucker, G. W. Gray, A. R. Tajbakhsh, *J. Chem. Phys.* **1982**, *77*, 6215–6217.
[251] P. A. C. Gane, A. J. Leadbetter, *J. Phys. C: Solid State Phys.* **1983**, *16*, 2059–2067.
[252] E. B. Sirota, P. S. Pershan, M. Deutsch, *Phys. Rev. A: Gen. Phys.* **1987**, *36*, 2902–2913.
[253] G. J. Brownsey, A. J. Leadbetter, *J. Phys. (Lett.)* **1981**, *42*, L135–L139.
[254] E. Dubois-Violette, P. Pansu, P. Davidson, A. M. Levelut, *J. Phys. II* **1993**, *3*, 395–405.
[255] S. Diele et al., *Liq. Cryst.* **1996**, *21*, 603–608.
[256] C. J. Bowden, T. M. Herrington, A. M. Moseley, R. Richardson, *Liq. Cryst.* **1995**, *18*, 825–828.
[257] A. J. Leadbetter, M. A. Mazid, K. M. A. Malik, *Mol. Cryst. Liq. Cryst.* **1980**, *61*, 39–60.
[258] A. J. Leadbetter, M. A. Mazid, *Mol. Cryst. Liq. Cryst.* **1981**, *65*, 265–283.
[259] P. A. C. Gane, A. J. Leadbetter, *Mol. Cryst. Liq. Cryst.* **1981**, *78*, 183–200.
[260] H. Allouchi, M. Cotrait, D. Guillon, B. Heinrich, H. T. Nguyen, *Chem. Mater.* **1995**, *7*, 2252–2258.
[261] P. Kromm, M. Cotrait, J. C. Rouillon, P. Barois, H. T. Nguyen, *Liq. Cryst.* **1996**, *21*, 121–131.
[262] W. Haase, J. Loub, H. Paulus, *Z. Kristallogr.* **1992**, *202*, 7–16.
[263] W. Haase, H. Paulus, R. Pendzialek, *Mol. Cryst. Liq. Cryst.* **1983**, *100*, 211–221.
[264] G. V. Vani, *Mol. Cryst. Liq. Cryst.* **1983**, *99*, 21–27.
[265] T. Hanemann, W. Haase, I. Svoboda, H. Fuess, *Liq. Cryst.* **1995**, *19*, 699–702.
[266] L. Walz, H. Paulus, W. Haase, *Z. Kristallogr.* **1987**, *180*, 97–112.
[267] P. Mandal, S. Paul, *Mol. Cryst. Liq. Cryst.* **1985**, *131*, 223–235.
[268] W. Thyen, F. Heinemann, P. Zugenmaier, *Liq. Cryst.* **1994**, *16*, 993–1003.
[269] R. Boese, M. Y. Antipin, M. Nussbaumer, D. Blaser, *Liq. Cryst.* **1992**, *12*, 431–440.
[270] S. V. Sereda et al., *Kristallografiya* **1989**, *34*, 333–344.
[271] W. R. Krigbaum, Y. Chatani, P. G. Barber, *Acta Crystallogr. B* **1970**, *26*, 97–101.
[272] L. Malpezzi, S. Bruckner, E. Galbiati, G. R. Luckhurst, *Mol. Cryst. Liq. Cryst.* **1991**, *195*, 179–184.
[273] J. M. Seddon, D. J. Williams, *Liq. Cryst.* **1995**, *18*, 761–768.
[274] I. Zareba et al., *Liq. Cryst.* **1996**, *21*, 565–574.
[275] H. Adams et al., *J. Mater. Chem.* **1991**, *1*, 843–855.
[276] K. Usha, K. Vijayan, S. Chandrasekhar, *Liq. Cryst.* **1993**, *15*, 575–589.
[277] K. Hans, P. Zugenmaier, *Makromol. Chem.: Macromol. Chem. Phys.* **1988**, *189*, 1189–1197.
[278] P. Zugenmaier, J. Mugge, *Makromol. Chemie: Rapid Commun.* **1984**, *5*, 11–19.
[279] P. Zugenmaier, H. Menzel, *Makromol. Chem.: Macromol. Chem. Phys.* **1988**, *189*, 2647–2655.
[280] A. M. Levelut, *J. Chim. Phys. Physico-Chim. Biol* **1983**, *80*, 149–161.
[281] S. Chandrasekhar, B. K. Sadashiva, K. A. Suresh, *Pramana* **1977**, *7*, 471–480.
[282] A. M. Levelut, *J. Phys. (Lett.)* **1979**, *40*, L81–L84.
[283] A. M. Giroud, J. Billard, *Mol. Cryst. Liq. Cryst.* **1981**, *66*, 467–470.

[284] A. M. Giroud-Godquin, M. M. Gauthier, S. Sigaud, F. Hardouin, M. F. Achard, *Mol. Cryst. Liq. Cryst.* **1986**, *132*, 35–44.
[285] H. Sakashita et al., *Mol. Cryst. Liq. Cryst.* **1988**, *163*, 211–219.
[286] A. M. Godquin-Giroud, J. C. Marchon, D. Guillon, A. Skoulios, *J. Phys. (Lett.)* **1984**, *45*, L681–L684.
[287] H. Abied et al., *Liq. Cryst.* **1987**, *2*, 269–279.
[288] A. M. Giroud-Godquin et al., *J. Phys.* **1989**, *50*, 513–519.
[289] D. Guillon, A. Skoulios, *J. Phys. IV* **1996**, *6*, 41–48.
[290] G. S. Attard, R. H. Templer, *J. Mater. Chem.* **1993**, *3*, 207–213.
[291] K. Ohta, Y. Morizumi, H. Ema, T. Fujimoto, I. Yamamoto, *Mol. Cryst. Liq. Cryst.* **1991**, *208*, 55–63.
[292] K. Praefcke et al., *Liq. Cryst.* **1995**, *18*, 857–865.
[293] H. Zimmermann, R. Poupko, Z. Luz, J. Billard, *Z. Naturforsch. A: J. Phys. Sci.* **1985**, *40*, 149–160.
[294] A. M. Levelut, J. Malthete, A. Collet, *J. Phys.* **1986**, *47*, 351–357.
[295] J. M. Lehn, J. Malthete, A. M. Levelut, *J. Chem. Soc.: Chem. Commun.* **1985**, 1794–1796.
[296] S. H. J. Idziak et al., *J. Chem. Soc. Chem. Commun.* **1992**, 98–99.
[297] H. Fischer et al., *Angew. Chem.: Int. Edn Engl.* **1995**, *34*, 795–798.
[298] C. Piechocki, J. Simon, A. Skoulios, D. Guillon, P. Weber, *J. Am. Chem. Soc.* **1982**, *104*, 5245–5247.
[299] D. Guillon, P. Weber, A. Skoulios, C. Piechocki, J. Simon, *Mol. Cryst. Liq. Cryst.* **1985**, *130*, 223–229.
[300] P. Weber, D. Guillon, A. Skoulios, *J. Phys. Chem.* **1987**, *91*, 2242–2243.
[301] J. F. Vanderpol et al., *Liq. Cryst.* **1989**, *6*, 577–592.
[302] P. Weber, D. Guillon, A. Skoulios, *Liq. Cryst.* **1991**, *9*, 369–382.
[303] A. N. Cammidge, M. J. Cook, S. D. Haslam, R. M. Richardson, K. J. Harrison, *Liq. Cryst.* **1993**, *14*, 1847–1862.
[304] C. F. Vannostrum et al., *J. Chem. Soc.: Chem. Commun.* **1993**, 1120–1122.
[305] G. C. Bryant et al., *J. Mater. Chem.* **1994**, *4*, 209–216.
[306] A. M. Levelut, J. Malthete, C. Destrade, N. H. Tinh, *Liq. Cryst.* **1987**, *2*, 877–888.
[307] S. Diele, K. Ziebarth, G. Pelzl, D. Demus, W. Weissflog, *Liq. Cryst.* **1990**, *8*, 211–216.
[308] U. Nutz et al., *Liq. Cryst.* **1995**, *18*, 699–705.
[309] K. Praefcke, A. M. Levelut, B. Kohne, A. Eckert, *Liq. Cryst.* **1989**, *6*, 263–270.
[310] A. Takada et al., *Liq. Cryst.* **1995**, *19*, 441–448.
[311] D. R. Beattie, P. Hindmarsh, J. W. Goodby, S. D. Haslam, R. M. Richardson, *J. Mater. Chem.* **1992**, *2*, 1261–1266.
[312] T. Plesnivy, H. Ringsdorf, P. Schuhmacher, U. Nutz, S. Diele, *Liq. Cryst.* **1995**, *18*, 185–190.
[313] J. Malthete, H. T. Nguyen, C. Destrade, *Liq. Cryst.* **1993**, *13*, 171–187.
[314] P. Davidson, A. M. Levelut, H. Strzelecka, V. Gionis, *J. Phys. (Lett.)* **1983**, *44*, L823–L828.
[315] J. Malthete, A. M. Levelut, L. Liebert, *Adv. Mater.* **1992**, *4*, 37–41.
[316] W. Weissflog, A. Saupe, I. Letko, S. Diele, G. Pelzl, *Liq. Cryst.* **1996**, *20*, 483–487.
[317] K. Ohta et al., *Mol. Cryst. Liq. Cryst.* **1986**, *140*, 131–152.
[318] A. M. Levelut, P. Oswald, A. Ghanem, J. Malthete, *J. Phys.* **1984**, *45*, 745–754.
[319] C. R. Safinya, K. S. Liang, W. A. Varady, N. A. Clark, G. Andersson, *Phys. Rev. (Lett.)* **1984**, *53*, 1172–1175.
[320] C. R. Safinya, N. A. Clark, K. S. Liang, W. A. Varady, L. Y. Chiang, *Mol. Cryst. Liq. Cryst.* **1985**, *123*, 205–216.
[321] E. Fontes, P. A. Heiney, M. Ohba, J. N. Haseltine, A. B. Smith, *Phys. Rev. A* **1988**, *37*, 1329–1334.
[322] E. F. Gramsbergen, H. J. Hoving, W. H. de Jeu, K. Praefcke, B. Kohne, *Liq. Cryst.* **1986**, *1*, 397–400.
[323] E. Fontes, P. A. Heiney, W. H. de Jeu, *Phys. Rev. (Lett.)* **1988**, *61*, 1202–1205.
[324] P. A. Heiney et al., *J. Phys.* **1989**, *50*, 461–483.
[325] P. Davidson, *Prog. Poly. Sci.* **1996**, *21*, 893–950.
[326] A. Skoulios in *Advances in Liquid Crystals* (Ed: G. H. Brown) Academic Press, New York, **1975**, 169–188.
[327] E. Chiellini et al., *Mol. Cryst. Liq. Cryst.* **1992**, *215*, 279–286.
[328] D. Navarro-Rodriguez, D. Guillon, A. Skoulios, Y. Frere, P. Gramain, *Makromol. Chem.: Macromol. Chem. Phys.* **1992**, *193*, 3117–3128.
[329] P. J. Alonso et al. *Macromolecules* **1993**, *26*, 4304–4309.
[330] T. J. Bunning, H. E. Klei, E. T. Samulski, W. W. Adams, R. L. Crane, *Mol. Cryst. Liq. Cryst.* **1993**, *231*, 163–174.
[331] A. S. Cherodian, N. J. Hughes, R. M. Richardson, M. S. K. Lee, G. W. Gray, *Liq. Cryst.* **1993**, *14*, 1667–1682.
[332] P. Davidson, K. Kuhnpast, J. Springer, G. Scherowsky, *Liq. Cryst.* **1993**, *14*, 901–910.
[333] G. Decher et al., *Ber. Bunsenges. Phys. Chem.: Int. J. Phys. Chem.* **1993**, *97*, 1386–1394.
[334] O. Francescangeli et al., *Liq. Cryst.* **1993**, *13*, 353–363.
[335] O. Francescangeli et al., *Liq. Cryst.* **1993**, *14*, 981–990.

[336] P. Maldivi et al., *Adv. Mater.* **1993**, *5*, 909–912.
[337] R. Deschenaux, J. Santiago, D. Guillon, B. Heinrich, *J. Mater. Chem.* **1994**, *4*, 679–682.
[338] R. Deschenaux, I. Kosztics, U. Scholten, D. Guillon, M. Ibnelhaj, *J. Mater. Chem.* **1994**, *4*, 1351–1352.
[339] F. B. Dias, J. P. Voss, S. V. Batty, P. V. Wright, G. Ungar, *Macromol. Rapid Commun.* **1994**, *15*, 961–969.
[340] R. Geer, S. Qadri, R. Shashidhar, A. F. Thibodeaux, R. S. Duran, *Liq. Cryst.* **1994**, *16*, 869–875.
[341] S. G. McNamee, T. J. Bunning, C. M. McHugh, C. K. Ober, W. W. Adams, *Liq. Cryst.* **1994**, *17*, 179–190.
[342] M. F. Achard, S. Lecommandoux, F. Hardouin, *Liq. Cryst.* **1995**, *19*, 581–587.
[343] J. H. Chen, G. H. Hsiue, C. P. Hwang, J. L. Wu, *Liq. Cryst.* **1995**, *19*, 803–806.
[344] E. Chiellini, E. Dossi, G. Galli, R. Solaro, B. Gallot, *Macromol. Chem. Phys.* **1995**, *196*, 3859–3875.
[345] J. Ebbutt, R. M. Richardson, J. Blackmore, D. G. McDonnell, M. Verrall, *Mol. Cryst. Liq. Cryst. Sci. Technol. A: Mol. Cryst. Liq. Cryst.* **1995**, *261*, 549–566.
[346] O. Francescangeli et al., *J. Poly. Sci. B: Poly. Phys.* **1995**, *33*, 699–705.
[347] B. Gallot, G. Galli, E. Dossi, E. Chiellini, *Liq. Cryst.* **1995**, *18*, 463–473.
[348] R. E. Geer, S. B. Quadri, R. Shashidhar, A. F. Thibodeaux, R. S. Duran, *Phys. Rev. E* **1995**, *52*, 671–680.
[349] M. Kozlovsky, D. Subachius, T. Weyrauch, W. Haase, *Mol. Cryst. Liq. Cryst. Sci. Technol. A: Mol. Cryst. Liq. Cryst.* **1995**, *261*, 453–463.
[350] P. Masson, P. Gramain, D. Guillon, *Macromol. Chem. Phys.* **1995**, *196*, 3677–3686.
[351] T. J. Bunning, E. P. Socci, B. L. Farmer, A. L. Campbell, W. W. Adams, *Mol. Cryst. Liq. Cryst. Sci. Technol. A: Mol. Cryst. Liq. Cryst.* **1996**, *275*, 143–154.
[352] T. J. Bunning et al., *Macromolecules* **1996**, *29*, 8717–8725.
[353] E. A. S. Bustamante et al., *Liq. Cryst.* **1996**, *21*, 829–839.
[354] E. A. S. Bustamante et al., *Chem. Phys. (Lett.)* **1996**, *260*, 447–452.
[355] R. Deschenaux, V. Izvolenski, F. Turpin, D. Guillon, B. Heinrich, *Chem. Commun.* **1996**, 439–440.
[356] O. Francescangeli, D. Rinaldi, M. Laus, G. Galli, B. Gallot, *J. Phys. II* **1996**, *6*, 77–89.
[357] B. Gallot, F. Monnet, *Eur. Poly. J.* **1996**, *32*, 147–151.
[358] B. Gallot, M. Fafiotte, A. Fissi, O. Pieroni, *Macromol. Rapid. Commun.* **1996**, *17*, 493–501.
[359] B. I. Ostrovskii, E. A. S. Bustamante, S. N. Sulianov, Y. G. Galyametdinov, W. H. Haase, *Mol. Cryst. Liq. Cryst. Sci. Technol. C: Mol. Mater.* **1996**, *6*, 171–188.
[360] E. N. Keller, E. Nachaliel, D. Davidov, H. Zimmermann, *Phys. Rev. A* **1988**, *37*, 2251–2253.
[361] S. Diele, B. Hisgen, B. Reck, H. Ringsdorf, *Makromol. Chemie: Rapid Commun.* **1986**, *7*, 267–271.
[362] R. Duran, D. Guillon, P. Gramain, A. Skoulios, *Makromol. Chem.: Rapid Commun.* **1987**, *8*, 181–186.
[363] R. Duran, D. Guillon, P. Gramain, A. Skoulios, *J. Phys.* **1988**, *49*, 1455–1466.
[364] Y. Frere, F. Yang, P. Gramain, D. Guillon, A. Skoulios, *Makromol. Chem.: Macromol. Chem. Phys.* **1988**, *189*, 419–427.
[365] O. Herrmann-Schönherr, J. H. Wendorff, W. Kreuder, H. Ringsdorf, *Makromol. Chem.: Rapid Commun.* **1986**, *7*, 97–101.
[366] B. Hüser, T. Pakula, H. W. Spiess, *Macromolecules* **1989**, *22*, 1960–1963.
[367] M. Ebert, O. Herrmann-Schönherr, J. H. Wendorff, H. Ringsdorf, P. Tschirner, *Makromol. Chemie: Rapid Commun.* **1988**, *9*, 445–451.
[368] A. Skoulios, *Makromol. Chem.: Macromol. Symp.* **1992**, *62*, 201–211.
[369] M. W. Matsen, M. Schick, *Curr. Opin. Coll. Interface Sci* **1996**, *1*, 329–336.
[370] S. Forster et al., *Macromolecules* **1994**, *27*, 6922–6935.
[371] D. A. Hajduk et al., *Macromolecules* **1994**, *27*, 490–501.
[372] D. A. Hajduk et al., *Macromolecules* **1994**, *27*, 4063–4075.
[373] D. A. Hajduk et al., *Macromolecules* **1995**, *28*, 2570–2573.
[374] D. A. Hajduk et al., *Macromolecules* **1995**, *28*, 7148–7156.
[375] A. K. Khandpur et al., *Macromolecules* **1995**, *28*, 8796–8806.
[376] A. J. Ryan, I. W. Hamley, W. Bras, F. S. Bates, *Macromolecules* **1995**, *28*, 3860–3868.
[377] O. Francescangeli et al., *Poly. J.* **1996**, *28*, 193–197.
[378] D. A. Hajduk, S. M. Gruner, S. Erramilli, R. A. Register, L. J. Fetters, *Macromolecules* **1996**, *29*, 1473–1481.
[379] H. H. Lee, R. A. Register, D. A. Hajduk, S. M. Gruner, *Poly. Eng. Sci.* **1996**, *36*, 1414–1424.
[380] B. S. Pinheiro, D. A. Hajduk, S. M. Gruner, K. I. Winey, *Macromolecules* **1996**, *29*, 1482–1489.
[381] M. F. Schulz et al., *Macromolecules* **1996**, *29*, 2857–2867.
[382] R. J. Spontak et al., *Macromolecules* **1996**, *29*, 4494–4507.
[383] J. Zhao et al., *Macromolecules* **1996**, *29*, 1204–1215.
[384] M. D. Foster, *Crit. Rev. Anal. Chem.* **1993**, *24*, 179–241.

[385] T. P. Russell, *Phys. B* **1996**, *221*, 267–283.
[386] Z. H. Cai et al., *J. Chem. Phys.* **1993**, *98*, 2376–2386.
[387] G. Coulon, J. Daillant, B. Collin, J. J. Benattar, Y. Gallot, *Macromolecules* **1993**, *26*, 1582–1589.
[388] S. K. Sinha et al., *Phys A* **1996**, *231*, 99–110.
[389] G. Vignaud et al., *J. Physics Cond. Matter* **1997**, *9*, L125–L130.
[390] J. M. Seddon, *Biochim. Biophys. Acta* **1990**, *1031*, 1–69.
[391] G. Cevc, J. M. Seddon in *Phospholipids Handbook* (Ed.: G. Cevc), Marcel Dekker, New York, **1993**, 351–401.
[392] J.-P. Carton, E. Dubois-Violette, J. Prost, *Liq. Cryst* **1990**, *7*, 305–314.

4 Neutron Scattering

Robert M. Richardson

4.1 Introduction

Neutrons, like X-rays, are radiation that can be used to determine the structure of condensed matter such as liquid crystals. In principle they could be used to determine the period and range of any density waves, and hence perform the same type of structural determination as is usually made using X-rays as a probe. X-ray diffraction measurements have been used extensively to identify mesophases and investigate their translational order, and they were reviewed in the previous section. This type of investigation is rarely undertaken with neutrons because X-rays are much more appropriate for many reasons:

1. X-ray diffraction apparatus is often available in a liquid crystal laboratory whereas neutrons are only available at central facilities such as the Institut Laue Langevin (ILL), Grenoble, or the ISIS Neutron Source, Oxfordshire.
2. The intensity of X-rays is generally much greater than that of neutrons, and this leads to good scattering vector resolution and the use of small samples. The intensity now available from synchrotron sources is very high.
3. There is no need to label compounds isotopically for study by X-rays.

The theory of neutron scattering has been explained in several good text books [1–3]. The aim of this introduction is to emphasize the features that distinguish neutron scattering from X-ray diffraction, and allow information on the structure and dynamics of liquid crystals to be determined. Generally, such information is not available from other scattering or spectroscopic techniques.

Neutrons are scattered by the nuclei of atoms rather than their electron clouds, so most elements scatter them both coherently and incoherently. Coherent scattering is useful in diffraction experiments in that it is sensitive to structural features in the material. For instance, the coherent scattering would give rise to a Bragg peak from a periodic structure. Incoherent scattering is not sensitive to structure and provides a flat background in diffraction experiments. However, the energy distribution of this scattering is determined by the dynamics of the sample.

The following points indicate why neutron scattering provides information on the structure and dynamics of liquid crystals that is not available by other methods:

1. An extremely useful property of neutrons is that the coherent scattering lengths for hydrogen and deuterium atoms are completely different. Hydrogen has an (unusual) negative value of -3.74×10^{-15} m, while deuterium has a more typical value of 6.674×10^{-15} m. Since the nucleus has a negligible effect on the electrons, it is possible to substitute deuterium for hydrogen to highlight particular features in the scattering without influencing the intermolecular interactions. This isotopic labeling allows the scattering from single entities such as polymer chains to be measured, which is not possible with X-rays.

2. A neutron with wavelength comparable to intermolecular distances also has an energy comparable with the energy of the diffusive molecular motions in liquid crystals. A neutron may exchange a significant amount of energy with the sample during the scattering process, so it is feasible to investigate diffusive motions using quasi-elastic neutron scattering. In fact quasi-elastic neutron scattering can give simultaneous information on the structure and the dynamics of liquid crystals.

3. The incoherent scattering cross section of hydrogen is exceptionally large so incoherent quasi-elastic neutron scattering is dominated by the motion of hydrogen atoms, which simplifies the interpretation.

These features of the neutron scattering process have led to the method being widely used as a complement to X-ray diffraction. In Sec. 4.2, the principle of a neutron scattering experiment is outlined and the subsequent sections review applications of neutron scattering techniques to liquid crystals.

4.2 Neutron Scattering Experiments

In a typical neutron scattering experiment, an incident monochromatic beam is prepared. High energy neutrons (~1 MeV) are produced by fission in a nuclear reactor or by fission and spallation in a proton–synchrotron based source. Their energy distribution is shifted to values that are better suited to condensed matter studies by passing them into a moderator, so that their energy approaches the thermal energy of the material in the moderator. For research on soft condensed matter, such as liquid crystals, a cold (typically 20 K) moderator is used which produces neutrons of about 3.3 meV energy. The de Broglie relationship links the energy and the wavelength so, for neutrons, a 3.3 meV particle has a wavelength of 5 Å (0.5 nm) which is ideal for probing large distances such as those found in liquid crystals.

A single wavelength (λ_0) is then selected from this distribution, either by a mechanical velocity selector (since neutron wavelength is inversely proportional to their velocity) or by Bragg reflection from a crystal. (For a pulsed source, it is also possible to use a "white" beam and determine the wavelength by the time-of-flight.) The incident beam is brought onto the sample and the intensity of the scattered neutrons is measured as a function of the scattering angle (2θ) and the scattered neutron energy, as shown in Fig. 1.

The natural variables to describe the change in state of the neutron on scattering are the change in its energy, $\hbar\omega$

$$\hbar\omega = E - E_0 \qquad (1)$$

and the change in its momentum $\hbar \boldsymbol{Q}$

$$\hbar \boldsymbol{Q} = \hbar(\boldsymbol{k} - \boldsymbol{k}_0) \qquad (2)$$

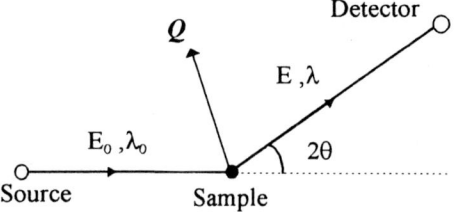

Figure 1. Schematic diagram of a neutron scattering experiment.

where k_0 and k are the incident and scattered wave vectors of the neutron wave.

Q is called the scattering vector, and for elastic and quasi-elastic scattering (where $\lambda \approx \lambda_0$), its magnitude is given by the formula

$$Q = \frac{4\pi \sin\theta}{\lambda} \qquad (3)$$

The direction of Q is defined by k and k_0, and for quasi-elastic scattering it bisects the incident and scattered beam, as shown in Fig. 1.

There are two broad classes of neutron scattering measurement. In diffraction measurements, the energy of the scattered radiation is not analyzed, so only the intensity versus Q is determined. Such measurements determine the structure of the sample. In a second class, the scattered energy is analyzed, so the intensity versus Q and $\hbar\omega$ is determined. This type of measurement is able to give structural and dynamic information about the sample, and is often called an inelastic measurement. In the special case where there are diffusive motions in the sample, which broaden the sharp incident spectrum, the term quasi-elastic is used. Neutron scattering instruments are optimized for one or the other of these types of measurement.

4.3 Neutron Diffraction from Isotopically Labeled Samples

4.3.1 Orientational Order in Low Molar Mass Materials

A long standing problem in liquid crystal research has been the difficulty in measuring the orientational distribution of the molecules in a mesophase. Nuclear magnetic resonance and many other techniques are only sensitive to second rank tensor properties of materials, and so are only able to measure $\overline{P}_2$, which is the mean value of the second Legendre polynomial, as defined in Eq. (5) below. Raman scattering and electron spin resonance are sensitive to fourth rank properties and so can measure up to $\overline{P}_4$. In principle, X-ray and neutron scattering are sensitive to all the order parameters and could determine the complete distribution function $f(\beta)$

$$f(\beta) = \sum_{L \text{ even}} \left(\frac{2L+1}{2}\right) \overline{P}_L P_L(\cos\beta) \qquad (4)$$

where the order parameters are defined as

$$\overline{P}_L = \int P_L(\cos\beta) f(\beta) \sin\beta \, d\beta \qquad (5)$$

The main features in a normal X-ray or neutron diffraction pattern arise from *inter*molecular interference, so it would be possible to determine the orientational order parameters for a distribution of local groups of molecules rather than the single molecule distribution. An early application of neutron diffraction was to find an *intra*molecular interference at high Q where intermolecular effects have died away. This is more practical with neutron diffraction, because the atomic form factor for X-rays also dies away at high Q. The angular distribution of such

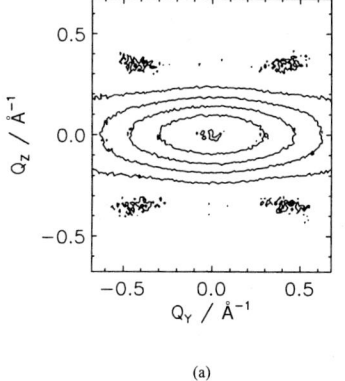
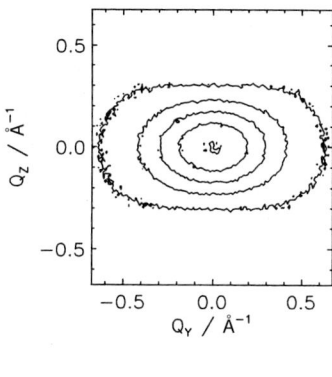

Figure 2. Calculated single molecule scattering from PAA "difference molecules" with (a) a sharp orientational distribution and (b) a Maier–Saupe distribution with $\bar{P}_2 = 0.429$.

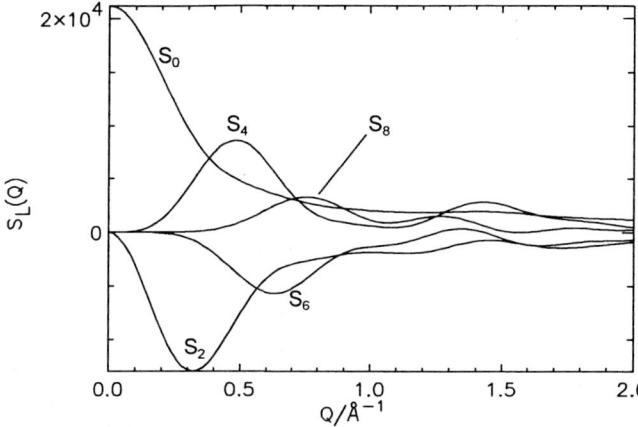

Figure 3. The structure factors for a PAA difference molecule.

an intramolecular peak about the director can be interpreted in terms of the distribution of molecules about the director, $f(\beta)$. In fact, difficulties in the absolute scaling of a high Q peak preclude the determination of individual order parameters, but the ratio of $\bar{P}_2/\bar{P}_4$ has been measured [4].

Single molecule scattering can be measured by utilizing the very different scattering lengths of H and D for neutrons. It has been shown that since a mixture of hydrogenous and deuteriated versions of the same molecule is random, the neutron scattering from such a mixture contains a new component, which is single molecule scattering from a "difference" molecule. The order parameters have been determined from the scattering either by assuming the molecules to be cylindrical [5] or (more rigorously) by assuming the molecular structure [6]. Figure 2 shows the scattering from a mixture [7] of hydrogenous and deuteriated para azoxyanisole (PAA). Perfectly aligned long rods would give a horizontal streak of scattering, but this is smeared out by the distribution of orientations in the nematic phase.

It has been shown [8] that the single molecule scattering depends on the order parameters and a series of "structure factors" which act as coefficients to the terms in the series

$$I(Q, \vartheta_Q) = \sum_{L \text{ even}} S_L(Q) \bar{P}_L P_L(\cos \vartheta_Q) \qquad (6)$$

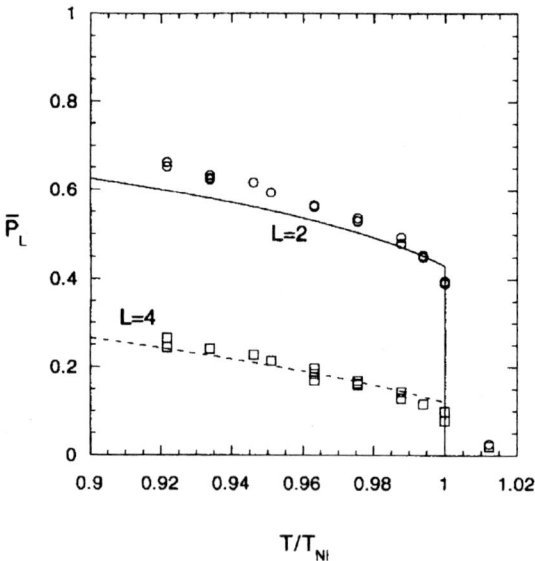

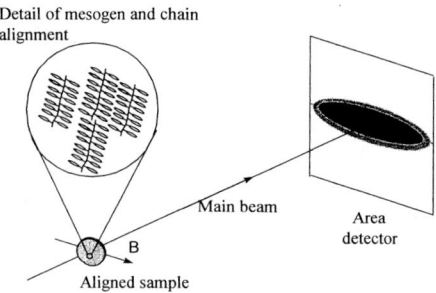

Figure 5. Schematic diagram of a SANS experiment on a side group liquid crystal polymer with the mesogenic units aligned by the magnetic field B. Some of the backbones are deuterium labeled so that the area detector measures an anisotropic scattering pattern which is used to determine the dimensions of the backbones.

Figure 4. The values of $\bar{P}_2$ and $\bar{P}_4$ determined by neutron scattering from a mixture of hydrogenous and deuteriated PAA (from [7], with permission).

where θ_Q is the angle between the scattering vector Q and the director, and the structure factors $S_L(Q)$ are determined by the molecular structure.

Figure 3 shows these structure factors for a PAA molecule. It seems that different regions in Q are influential in determining different order parameters, and in this case $\bar{P}_2$ and $\bar{P}_4$ were determined by fitting Eq. (6) to the data. $\bar{P}_6$ was indistinguishable from zero. The results are compared with the predictions of the mean field theory in Fig. 4.

4.3.2 The Background to Small Angle Neutron Scattering from Polymers

X-ray diffraction from liquid crystal polymers has been extensively used to determine the translational order of the mesogenic groups. For instance, in a smectic phase the layer spacing is easily determined from the layer reflection, and the lateral spacing of the mesogenic units can be inferred from the appropriate peak in the scattering. Since the materials are homogenous, there is no signature in the scattering from the polymeric backbone and so it is not possible to determine its conformation. Neutron diffraction from isotopically labeled samples again provides a technique to extend the structural information that can be determined. Polymer scientists [9] used mixtures of simple macromolecules with deuteriated and hydrogenous chains to determine the radius of gyration of single chains by small angle neutron scattering (SANS). Since the entropy of mixing is small for polymers, the free energy of mixing is sensitive to small differences in the interaction between H and D and care has to be taken to ensure that isotope effects are correctly treated, particularly for very high molecular weights [10]. For side group liquid crystal polymers, it is usual to label a few hydrogen sites as near as possible to the backbone, so that the perturbation due to deuteriation is expected to be negligible. For liquid crystal polymers the method is extended, because it is possible to make small-angle scattering measurements on macroscopically aligned materials. If an area detector is used, as shown in Fig. 5, the chain dimensions parallel and perpendicular to the director may be determined simultaneously using Guinier's approximation [11] for the scattering at low Q

$$I(Q_\parallel)^{-1} = I(Q_\parallel)^{-1}\left(1 + Q_\parallel^2 R_\parallel^2\right) \quad Q_\parallel R_\parallel \ll 1$$
$$I(Q_\perp)^{-1} = I(Q_\perp)^{-1}\left(1 + Q_\perp^2 R_\perp^2\right) \quad Q_\perp R_\perp \ll 1$$
(7)

where $R_\parallel$ and $R_\perp$ are the dimensions of the chain parallel and perpendicular to the director, which is generally imposed by magnetically aligning the sample.

4.3.3 Small Angle Neutron Scattering (SANS) Results from Side Chain Liquid Crystal Polymers

For a side group liquid crystal polymer with terminally attached mesogenic units, the backbone might be expected to be strongly elongated perpendicular to the director in a smectic phase, because it would be confined between the layers of mesogenic units. This was found to be the case in materials with polyacrylate and polymethacrylate backbones [12], but materials with siloxane backbones tended to show rather weaker anisotropy [13, 14]. This lack of perfect confinement between the layers has been predicted theoretically where a more sophisticated model takes account of the possibility of a backbone hopping from layer to layer, as indicated schematically in Fig. 6. The extent of such hopping will depend on the flexibility of the backbone and the spacer. The temperature dependence [15] was shown to follow the predictions of a theory [16] that suggested that the backbones hopped between layers more as the temperature was raised so as to increase their entropy.

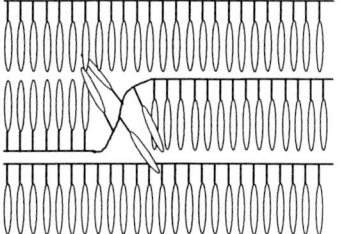

Figure 6. Highly schematic diagram of a side group liquid crystal polymer to illustrate a backbone hopping from one layer to another.

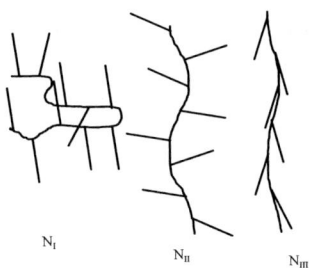

Figure 7. The three types of nematic side chain liquid crystal polymer.

The interplay between the backbone, the flexible spacer, and the mesogenic units is more subtle in a nematic phase. Three different nematic phases have been predicted from theoretical arguments concerning the strength of the ordering of the mesogenic groups, the backbone, and the flexibility of the spacer [17]. These are illustrated in Fig. 7. Most side group liquid crystal polymers show type I nematics because there is generally some smectic-like short-range order. An interesting exception to this is the re-entrant nematic phase (N_{Re}) which has been found in a polyacrylate with its mesogenic units terminated by a cyano group

—(CD$_2$—CD)—
   |
   CO$_2$—(CH$_2$)$_6$—O—⟨⟩—⟨⟩—CN

T_g 35 °C N_{Re} 83 °C Sm$_{Ad}$ 112 °C N 122 °C I

The backbone is a weakly oblate ellipsoid in the normal nematic phase, suggesting a type I arrangement. It becomes more strongly oblate on cooling into the smectic A phase, but dramatically reverses into a strongly prolate ellipsoid in the re-entrant nematic. Re-entrant nematics are quite common in low molar mass liquid crystals and arise when there is competition between two types of smectic ordering. In this case, it is suggested that the onset of smectic C_d-like order at low temperatures is responsible for the formation of the re-entrant nematic and reversal of the backbone anisotropy. X-ray results show that the orientation of the mesogenic units remains constant during this transition, so the structure must be a type III nematic where the director of the mesogenic units is parallel to the backbone [18].

Another striking example of the type III nematic is found when the mesogenic units are attached laterally to the backbone via a spacer, as shown in Fig. 8. The result is a mesogen jacket around the backbone, which causes the backbone to become highly extended. Such a structure can be inferred indirectly from X-ray and infrared dichroism measurements [19, 20], but is demonstrated unequivocally by small angle neutron scattering [21–23]. The strong extension of the backbone is rapidly lost if the mesogenic units on the backbone are diluted with small nonmesogenic groups. The oblate form of the backbone, which is found in smectic phases of side group liquid crystal polymers with terminally attached mesogenic groups, is much more stable to such a dilution [24].

Neutron diffraction from the layers is also able to supplement the X-ray results. This method is able to give information on the internal structure of the layers [14]. For X-rays, a Fourier transform of the layer intensities gives the electron density distribution in a direction perpendicular to the layers. It is difficult to interpret this because it contains contributions from all the constituent parts of the macromolecules. However, if the neutron diffraction intensities are measured from two polymers, differing only in that one has a deuteriated backbone, the difference in the intensities can be interpreted directly in terms of the distribution of the backbone along the layer normal. In one example, the amount of backbone in the sublayer of mesogenic units was estimated at 25% at room temperature [25] and increased as the temperature increased.

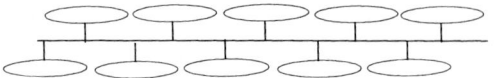

Figure 8. Laterally attatched side group liquid crystal polymer.

4.3.4 SANS Results from Main Chain Liquid Crystal Polymers

Similar measurements on main chain liquid crystal polymers have been used to determine the persistence length of the chains in the nematic phase. Again, there are theoretical predictions for the temperature behavior. It has been suggested that as the temperature increases the number of "hairpin" bends of the chain will increase and the conformation of the chain will become more isotropic [26]. This has been confirmed qualitatively, but since most of the materials have been polyesters and have had nematic phases at quite high temperatures the effects of transesterification have interfered [27–30]. However, similar results have been found in a polyether [31] where the chains were found to contain one hairpin defect on average.

SANS has also been applied to liquid crystal polymer solutions such as poly-γ (benzyl L-glutamate) and hydroxypropylcellulose [32–34]. The samples were aligned by shear flow in a Couette cell. These experiments have been used to find the shear rates at which changes in the alignement behavior take place. These have been correlated with relaxation times found by rheological methods. Similar macroscopic alignment has been induced in a liquid crystalline solution of DNA (deoxyribonucleic acid) fragments by a magnetic field [35].

4.4 Dynamics of Liquid Crystals

4.4.1 Time Resolved Diffraction

Neutron scattering has been used to investigate the "self-organizing critical state" that is produced when a nematic is contained in a large container (30×30×3 mm in this case). The fluctuations of the director are followed by time-resolving the intensity of the diffuse scattering at $Q = 1.8 \text{ Å}^{-1}$ (18 nm^{-1}) and comparing the results with theories for such processes [36, 37]. The time resolution of such experiments is typically 10 s, so the method is restricted to very large scale, slow director fluctuations. The dynamics of liquid crystals on a much faster time scale and at the molecular level may be studied using inelastic or quasi-elastic neutron scattering.

4.4.2 Coherent Inelastic Neutron scattering

Coherent neutron scattering is normally used to investigate collective excitations such as phonons in crystals. It has not been applied extensively to liquid crystals, but has been used to show that a transverse phonon can propagate perpendicular to the layers in a crystal B phase [38] thus demonstrating that the molecular positions in adjacent layers are well correlated and that the phase has three-dimensional long-range order.

4.4.3 Background to Incoherent Quasi-Elastic Neutron Scattering (IQENS)

Incoherent quasi-elastic neutron scattering (IQENS) is much more widely applied to the study of diffusive molecular motions in liquid crystals. The principles and applications of this method to a wide range of systems have been reviewed very thoroughly [39–41] and the application to liquid crystals has been reviewed from time to time [42–44]. Only aspects relevant to liquid crystals are covered here. Molecular vibrations and librations generally give an inelastic peak at an energy transfer that is well separated from the elastic peak in the scattered neutron spectra, whereas the slow diffuse motions such as translational or rotational diffusion change the shape of the elastic peak by broadening it. The principles behind the study of diffusive motion in low molar mass liquid crystals will be discussed in this section and some experimental results will be reviewed in those that follow. Many of the investigations on phases formed by simple calamitic molecules were done in the 1970s and early 1980s when the high resolution instruments at the ILL became available. Since then, interest in applying IQENS has broadened to include complex materials such as polymeric and lyotropic systems.

The incoherent scattering cross-section of hydrogen is very much greater than that of any other nucleus (e.g., $\sigma_H = 81.5$, $\sigma_D = 7.6$, $\sigma_C = 5.51$, and $\sigma_O = 4.42$ in units of

10^{-28} m^2), so the spectra are dominated by the motion of the hydrogen atoms. It is important to measure the spectra at Q values that are free from coherent scattering features such as Bragg peaks. Certain groups can be rendered "invisible" in the quasi-elastic or inelastic spectrum by deuteriating them so that their contribution to the incoherent scattering spectra is reduced. This is clearly illustrated in Fig. 9, which shows the neutron scattering spectra from PAA in its crystalline phase [45]. (The spectra are in the form of raw experimental results, so the abscissae are neutron time-of-flight rather than energy transfer.) Molecule PAA-CD3 has a hydrogenous core and shows an elastic peak and a weak inelastic feature from the lattice vibrations. There is not quasi-elastic broadening of the elastic peak because there is no diffusive motion of the aromatic core. Molecule PAA-ϕD4 shows a much more pronounced inelastic peak from the torsional motions of the methyl group and a broadened elastic peak from the random reorientation of the –OCH$_3$ group.

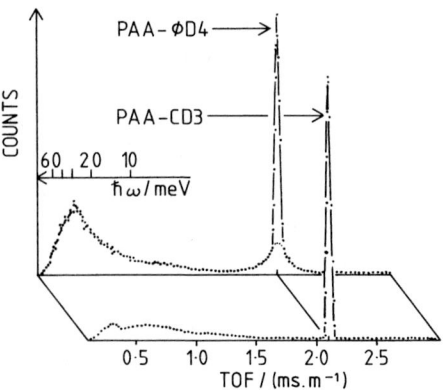

Figure 9. Time-of-flight (TOF) inelastic neutron scattering spectra from two differently deuteriated PAA molecules (from [45], with permission).

PAA-CD3

PAA-φD4

Many mesogenic molecules consist of a rigid elongated core which is terminated by flexible groups such as alkyl chains. These are usually deuteriated so that the motion of the core dominates the scattering.

4.4.4 Instruments for High Resolution IQENS

The range of time scales that are observable with quasi-elastic scattering is determined by the energy resolution of the spectrometer. If the width of the elastic peak in a spectrum is ΔE, then it will be difficult to observe motion on a time scale that is longer than τ

$$\tau = \frac{\hbar}{\Delta E} \tag{8}$$

because the change in width would be too small. The maximum range of energy transfers that is measurable will similarly determine the fastest motion that can be observed with a particular instrument. Considerable effort has been put into optimizing neutron spectrometers for different energy transfer ranges. The backscattering instruments IN10 and IN16 at the Institut Laue Langevin (ILL) give the highest energy resolution (~1 µeV), but cover a rather small energy transfer range (14 µeV). Multichopper instruments such as IN5 (also at ILL) cover the resolution range 10–100 µeV, but with a much larger energy transfer range [46], and can therefore measure diffusive molecular motions with correlation times up to 10^{-10} s. The IRIS instrument at the ISIS neutron source [47] and the MIBEMOL at the Laboratoire Leon Brillouin at Saclay, France, also have suitable energy resolution.

The Q range of a spectrometer is also important. Although there are no features in the incoherent scattering that result from interatomic or intermolecular interference, there is a Q dependence in the scattering that is determined by the geometry of the motion of the scattering atoms. The quasi-elastic scattering tends to be largest when the scalar product of Q and the amplitude, a, of the diffusive motion is large. A useful range of Q is defined by the formula

$$Q \cdot a \approx \pi \qquad (9)$$

The incoherent scattered intensity versus Q and $\hbar\omega$ is known as the incoherent scattering law, $S_{\text{inc}}(Q,\omega)$, of the sample, and this is directly related to the motion of the scattering atoms. The motion may be described by a self correlation function, $G_s(r,t)$. If the scattering atom is at an arbitrary origin at $t=0$, $G_s(r,t)$ is the probability of finding the same atom at a displacement r from the origin at a later time, t, averaged over all choices of origin. The incoherent scattering law is the spatial and temporal Fourier transform of this correlation function [48].

$$S_{\text{inc}}(Q,\omega)$$
$$= \frac{\pi}{2} \iint G_s(r,t) \exp[i(Q \cdot r - \omega t)] \, dr \, dt \qquad (10)$$

The usual way of interpreting the measured scattering law is to compare it with scattering laws that have been calculated for various models of molecular motion. There are several common themes in the scattering laws and these will be discussed below when considering some examples of calculated scattering laws and their application.

4.4.5 IQENS Measurements of Translational Diffusion

The scattering law for isotropic translational diffusion, such as that found in a normal liquid, is a single Lorentzian function $\mathcal{L}$ whose width increases with Q squared [49]

$$S_{\text{inc}}(Q,\omega) = \frac{1}{\pi} \frac{D_t Q^2}{\left(D_t Q^2\right)^2 + \omega^2} = \mathcal{L}\left(D_t Q^2\right) \qquad (11)$$

where D_t is the translational diffusion coefficient. Figure 10 shows what the incoherent scattering from such a system looks like. In practice the scattering would be measured as a series of energy transfer spectra at different values of the scattering vector, and the full width at half maximum would be given by the formula

$$\Delta \hbar \omega = 2 \hbar D_t Q^2 \qquad (12)$$

In liquid crystals, the anisotropic nature of the materials means that there are at least two different components to the diffusion tensor. Measurements of unaligned liquid crystal samples give a weighted average $(\bar{D})$ of the two principle values [49] that describe diffusion parallel and perpendicular to the director

$$\bar{D} \approx (D_\| + 2D\perp)/3 \qquad (13)$$

If a macroscopically aligned liquid crystal sample is used, these components can be measured simultaneously by arranging the scattering geometry such that for some detectors Q is perpendicular to the director and for others Q is nearly parallel to it [50, 51]. Translational diffusion measurements have been made at scattering vectors that are small compared to the molecular dimensions, so that the motion of the molecule is effectively followed for several molecular diameters. At low Q the broadening is small (because of the Q squared dependence) and

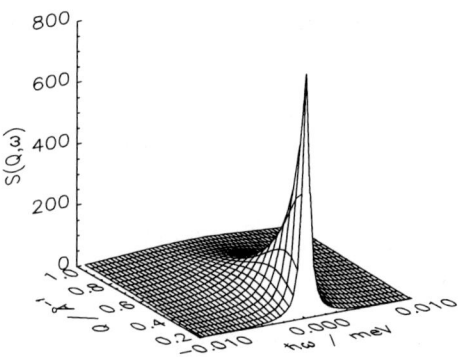

Figure 10. Incoherent quasi-elastic scattering law for a particle undergoing simple translational diffusion ($D_t = 1.5 \times 10^{-10}$ m^2 s^{-1}). This is normally measured as a series of constant Q spectra.

so backscattering instruments have generally been used to get adequate resolution [52]. High Q measurements are dominated by more rapid motions, such as molecular rotation [53] and do not give reliable diffusion coefficients. When the correct experimental conditions are applied, the values of diffusion coefficients perpendicular to the director are in good agreement with NMR (nuclear magnetic resonance) results [54], as shown in Fig. 11 for EABAC (ethyl-4-(4'-acetoxybenzylidene)aminocinnamate). For diffusion parallel to the director, there is a small discrepancy because of the weak periodic potential exerted by the layers. This has been used to estimate the height of the barrier [55, 56].

4.4.6 IQENS from the Localized Motion of Calamitic Molecules

The scattering from a localized motion such as molecular rotational diffusion is qualitatively different in that the energy spectra generally contain an unbroadened elastic component and at least one broadened component. Theoretical scattering laws have

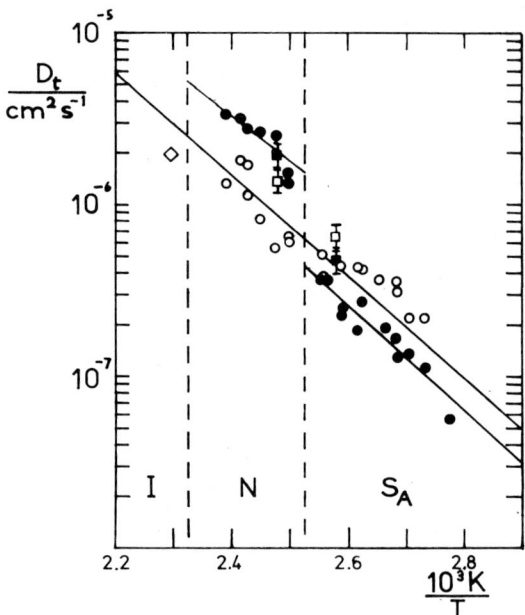

Figure 11. Translational diffusion coefficients of EABAC measured using NMR (circles) and IQENS (squares). The open points are for Q perpendicular to the director and the filled points are for Q parallel to it (from [56], with permission; S_A = smectic A, SmA).

been calculated for several different models that find application to liquid crystals. The simplest reasonable model is uniaxial rotational diffusion about the long molecular axes [57]. For a single hydrogen atom moving around a circle of radius a, this scattering law consists of a delta function $\delta(\omega)$ (the elastic scattering) and a series of Lorentzians

$$S_{\text{inc}}(\boldsymbol{Q}, \omega) = A_0(\boldsymbol{Q})\delta(\omega) + \sum_{l=1}^{\infty} A_l(\boldsymbol{Q}) \mathcal{L}(l^2 D_r) \quad (14)$$

where

$$A_0(\boldsymbol{Q}) = J_0(\boldsymbol{Q}a\sin\beta_Q) \text{ and}$$
$$A_l(\boldsymbol{Q}) = 2J_l(\boldsymbol{Q}a\sin\beta_Q) \quad (14a)$$

and J_l is a cylindrical Bessel function of order l, β_Q is the angle between $\boldsymbol{Q}$ and the axis of the circle, and D_r is the rotational dif-

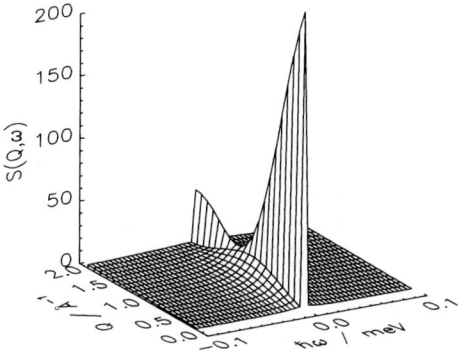

Figure 12. Incoherent quasi-elastic scattering law for a particle undergoing uniaxial rotational diffusion ($D_r = 1.5 \times 10^{10}$ rad^{2}s^{-1}) on a circle of radius 2 Å (0.2 nm) with Q perpendicular to the axis of rotation. The elastic part of the scattering is represented by a triangular peak with the correct area rather than a delta function.

fusion coefficient. To represent the scattering from a real molecule, this expression would be averaged over the different radii for the different hydrogen atoms. The scattering law is illustrated in Fig. 12.

The figure shows that the proportion of elastic scattering decreases as Q increases from zero. In fact the ratio of elastic to total (elastic plus quasi-elastic) scattering intensity provides a very useful quantity which is known as the elastic incoherent structure factor (EISF). It is often experimentally measurable and can be calculated easily from a theoretical model of the molecular motion

$$\text{EISF} = \frac{I_e}{I_e + I_q} \quad (15)$$

where I_e and I_q represent the intensity of the elastic and quasi-elastic components at a particular value of the scattering vector, Q. It can be seen that, since the total intensity in the theoretical scattering law is unity, the coefficient of the delta function [$A_0(Q)$ in Eq. 14 above] is the theoretical value of the EISF for that model (uniaxial rotational diffusion). In fact, the EISF depends only on the geometry and not on the time-scale of the molecular motion. It can be calculated from the infinite time spatial distribution of the scattering atom, $P(r)$

$$A_0(Q) = \left| \int P(r) \exp(iQ \cdot r) dr \right|^2 \quad (16)$$

since it is the modulus squared of the Fourier transform of this distribution. For calculation of the EISF of a rotating molecule, the EISF would be averaged over the values for each scattering (hydrogen) atom. A "powder" average would also be needed for unaligned samples. For the EISF to be a measurable quantity, it is essential for the molecular motion to be rapid enough for the quasi-elastic scattering to be clearly distinct from the elastic peak. If this is the case, the EISF may be determined by fitting a reasonable model with an adjustable amount of delta function.

We have considered the simple model of uniaxial rotational diffusion in the above section for heuristic reasons. Since the relevant molecules have irregular shapes, which often resemble laths rather than rods, it is likely that in some of the more ordered smectic phases the molecular rotation becomes restricted so that one or two orientations are preferred. In the disordered smectic phases and the nematic phase there is likely to be rotation around the short molecular axes, which will also be restricted by neighboring molecules. Such a motion has been described as a fluctuation of the long axis direction. Long range translational diffusion will take place on some time scale, but there is also likely to be some more rapid local translational motion of the molecules in the "cage" formed by their neighbors. This may have strong or weak coupling to the rotational motions. Finally, the possibility of internal rotation should be considered. Some, or all, of these motions may contribute to the quasi-elastic scattering from liquid crystal systems, and it is important to measure over a wide range of Q and ω to be able to separate their contributions to the scattering. For instance, translational diffusion is the main contribution to the broadening at low Q (i.e., $Q \ll 2\pi$/molecular dimension). At higher Q, the local motions tend to dominate and translational diffusion only contributes a small additional broadening.

Restriction of the rotation about the long molecular axes has been extensively studied for the compound TBBA [terephthalylidene-bis-(4-n-bytylaniline)] in its tilted phases

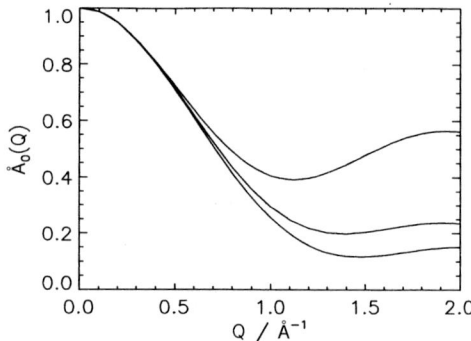

Figure 13. EISF of a particle undergoing uniaxial rotational diffusion on a circle of radius 2 Å (0.2 nm) in a powder sample. The lower line is for no preferred orientation, so that $\langle \cos 2\varphi \rangle = 0$. The upper line is for two sites separated by π, strongly preferred, so that $\langle \cos 2\varphi \rangle = 1$, and the intermediate line is for $\langle \cos 2\varphi \rangle = 0.5$, which occurs when the restricting potential is roughly comparable with $k_B T$.

The full scattering law for orientationally restricted rotation has been calculated numerically [58], although the EISF may be calculated from a simple formula. Figure 13 shows the EISF for a powder sample of a "model" molecule undergoing uniaxial rotational diffusion, and the same molecule where its rotation is severely restricted so that it only samples two orientations (separated by π).

The orientational order for such a model is characterized by the mean value of $\cos 2\varphi$, where φ is the angle of rotation about the long molecular axis. The two extremes shown in the figure correspond to $\langle \cos 2\varphi \rangle = 0$ and $\langle \cos 2\varphi \rangle = 1$. Any intermediate values of the ordering would give an EISF between these two extremes, and that for $\langle \cos 2\varphi \rangle = 0.5$ is shown as an example. If the rotation is restricted with one site preferred, the EISF approaches unity as the preference becomes strong and motion ceases. It can be seen that by comparing measured and calculated values of the EISF, the orientational order about the long axes

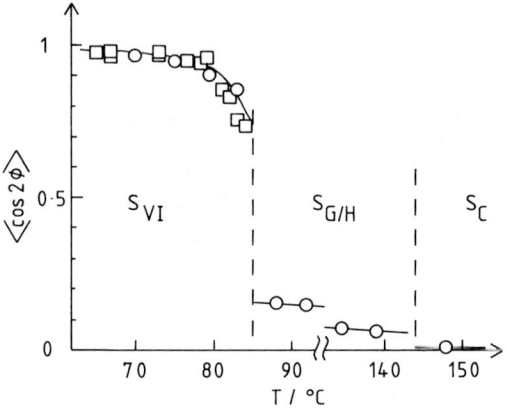

Figure 14. The values of the orientational order parameters $\langle \cos 2\varphi \rangle$ derived from the analysis of IQENS and magnetic resonance results from TBBA in its SmVI, SmH (now classified as crystalG), and SmC phases (from [60], with permission).

can be deduced. In a study of the tilted smectic phases of TBBA, the authors found the order parameters shown in Fig. 14 by assuming that two sites were preferred and fitting the full scattering law to the measured quasi-elastic scattering [59]. This gave excellent agreement with magnetic resonance results [60].

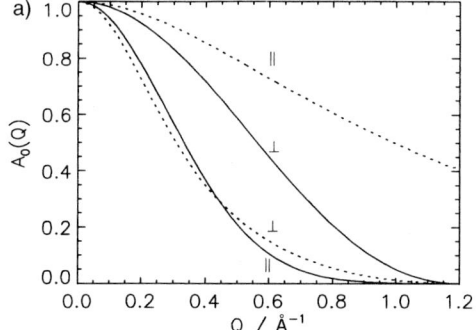

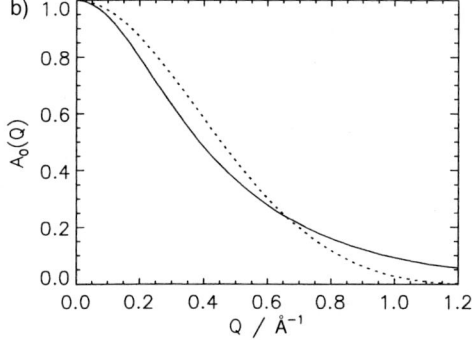

Figure 15. The EISF calculated for a model calamitic molecule that is 30 Å (3 nm) long with hydrogen atoms that rotate on a 2 Å (0.2 nm) radius. The dashed lines represent model (1) in the text where the molecules undergo uniaxial rotational diffusion and long axis fluctuations with $\langle P_2 \rangle = 0.73$. The solid lines represent model (2) where the molecules undergo uniaxial rotational diffusion and a local translational motion parallel to the long axes with an amplitude of 2.5 Å (0.25 nm). The EISF values observed parallel and perpendicular to the director in a sample with perfect macroscopic alignment are shown in (a). The EISF values for a powder sample are shown in (b).

In the smectic phases that exhibit disordered layers and the nematic phase, the molecular motions become more free, and many of the motions outlined above can contribute to the scattering. It is essential to measure the scattering from aligned samples with Q parallel and perpendicular to the director if some of the motions are to be distinguished. Figure 15 shows the EISF calculated for a "model" molecule that is 30 Å (3 nm) long and has protons that can rotate on a 2 Å (0.2 nm) radius about the long axes. Two models are shown:

1. It is assumed that the molecule rotates about its long axis and the direction of the long axis fluctuates about a mean rapidly enough to broaden a typical resolution function, while complete rotation by about π is much slower and so effectively contributes elastic scattering and does not reduce the EISF [61, 62]. In this model there is effectively a unipolar distribution of long axis directions, $f(\beta)$, where β is the angle between a long axis and the local director

$$f(\beta) = \frac{\delta}{4\pi \sinh \delta} \exp(\delta \cos \beta) \quad (17)$$

where δ is a parameter that controls the sharpness of the distribution. The amplitude of the fluctuations can be characterized by the value of $\langle P_2 \rangle$ for the distribution. A more rigorous approach [63] shows that this assumption is justified when $\langle P_2 \rangle$ is sufficiently large.

2. The molecule is undergoing uniaxial rotational diffusion together with a local translational motion parallel to the long axis of the molecules. This could result from some coupling with the rotation, so it would be on a similar time scale to the rotation. It is characterized by a root-mean-square amplitude, u, and it will tend to reduce the EISF for Q parallel to the director

$$A_0(Q) = \exp\left[-(Q \cdot u)^2\right] \quad (18)$$

Figure 15 shows EISF for the two models (a) from a well-aligned sample with Q parallel and perpendicular to the director, and (b) from a powder sample. It can be seen that the two models would be more difficult to distinguish if only data from powder samples were available.

Measurements on magnetically aligned samples such as IBPBAC [isobutyl-4(4'-phenylbenzylidene)aminocinnamate][64]

Cr 86 °C, CrE 114 °C, SmB 162 °C, SmA 206 °C, N 214 °C

have shown that the amplitude of motion parallel to the long axis is generally greater that that due to the rotation about the long axes, so the EISF for Q parallel to the director is lower. Figure 16 shows an example of IBPBAC in its smectic B phase. The EISF values at $Q=0.83$ Å^{-1} (8.3 nm^{-1}) are 0.17 for $Q \| n$ and 0.33 for $Q \perp n$ where n is the director.

The macroscopic alignment of the samples in these examples was not perfect, so a detailed analysis was done by fitting models that had been averaged over the static distribution of the layers. At 150 °C in the SmB phase, the scattering was consistent with uniaxial rotational diffusion about the long axes (possibly with six sites preferred) plus a localized translational motion parallel to the director. At this temperature in the SmB, there was also evidence of some long axis direction fluctuations, but this was not seen at a lower temperature, possibly because they became too slow to broaden the resolution. On cooling into the E phase rotation about the long axes becomes restricted. Table 1 summarizes the results from this material by giving the parameters to define the geometry and the correlation times of all of the motions. It shows that the motion parallel to the long axis direction persists in all the smectic phases. The similarity in the widths of the quasi-elastic scattering for Q parallel and perpendicular suggests there is some coupling between the translational motion and the rotation of the molecules. In the SmA phase coupling is weaker and there are two components to the parallel motion. The more rapid one results from coupling with the rotation and a slower one results

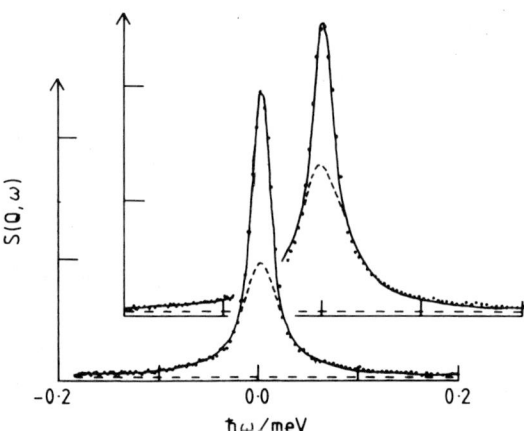

Figure 16. IQENS spectra at $Q = 0.83$ Å^{-1} (8.3 nm^{-1}) measured from IBPBAC in its SmB phase at 150 °C. The spectrum for $Q \perp n$ is superimposed on that for $Q \| n$. The dashed line separates the elastic and quasi-elastic components, and the solid line is a model fit (from [64], shown with permission).

Table 1. The parameters for the localized molecular motion of IBPBAC.

T (°C)	Phase	Rotation about the long axis		Long axis fluctuations		Local translation $\| n$	
		D_r (10^{10} rad^2 s^{-1})	$\langle \cos 2\varphi \rangle$	τ_f (10^{-11} s)	$\langle P_2 \rangle$	$\tau_{\|}$ (10^{-11} s)	u (Å)
100	E	0.32±0.03	0.5±0.2	>300	–	~4	0.9±0.2
108	E	0.3±0.2	0.3±0.1	–	–	~4	0.9±0.1
118	SmB	0.6±0.1	<0.15	–	–	~4	1.4±0.2
150	SmB	2.3±0.3	<0.10	36±10	0.85+0.5 or −0.1	3	1.8±0.2
172	SmA	4.7±0.3	0.0	2±1	0.6±0.1	2	0.8±0.3

from the restriction to translational diffusion by the smectic layers. Similar results were found for 5CB [65] and EABAC [56, 66] and have been treated theoretically [67].

4.4.7 IQENS from Other Types of Mesophase

The same incoherent quasi-elastic neutron scattering methods have been used to probe the molecular dynamics of more complex systems. Fatty acid salts of copper form a disk-like molecule comprising a binuclear "core" with four chains extending from it, and these can form a columnar mesophase where the disks are stacked. Copper palmitate has been examined in its crystalline phase and in a columnar mesophase. The quasi-elastic scattering was interpreted in terms of the motion of the alkyl chains. Each methylene group explored a spherical volume and the radius of the sphere increased with distance from the copper core [68]. A later experiment on oriented fibers of copper laurate [69] showed that this motion was anisotropic with a greater amplitude in the plane perpendicular to the columns.

The side group liquid crystal polymer with a re-entrant nematic phase that was discussed in Sec. 4.3.3 has also been studied by IQENS. It was found that the experimental EISF jumped at the I–N and the $SmA–N_{re}$ transitions but not at the N–SmA [70]. The changes in the dynamics of the mesogenic units mirror the changes in the backbone conformation and have been interpreted in terms of restricted rotation of the mesogenic group about its long axis and transverse motion of the spacer group [71].

It is possible to focus on the motion of one component in lyotropic liquid crystals by using D_2O instead of ordinary water. The high incoherent cross section of hydrogen then ensures that the results are sensitive to the motion of the lipid. The chain dynamics of a phospholipid have been shown to change with hydration in the L_α phase and on cooling into the gel phase. Only kink motion in the chains was observed in the gel phase, but whole molecule translation was invoked in the liquid crystalline phase [72]. Similar changes in the proton mobility have been observed at the gel point of poly(γ-benzyl L-glutamate) in deuterated benzyl alcohol [73]. The translational diffusion of the lipid has also been measured in the L_α phase [74].

4.4.8 Medium Resolution IQENS Studies

The very detailed picture of the diffuse translational and rotational molecular motions described above was revealed because of the high energy transfer resolution (typically less than 20 µeV) that is available with the instruments at ILL. However, useful results have also been obtained from medium resolution (~100 µeV) quasi-elastic scattering instruments by skilful choice of samples and by fitting different models. It is not generally possible to measure experimentally the EISF for the whole molecule rotational diffusion, because this motion is too slow to give quasi-elastic scattering that is clearly distinct from the elastic peak.

Quasi-elastic scattering from a homologous series of molecules based on PAA (*para*-azoxyanisole) in the nematic phase has been analyzed by fitting a rotational diffusion model with an extra parameter to adjust the proportion of delta function in the model. Values close to zero for this excess of elasticity were taken to indicate a good model for the molecular motion. Initially, uniaxial rotation of the whole molecule about its principle axis was assumed, but this gave large negative values of the excess

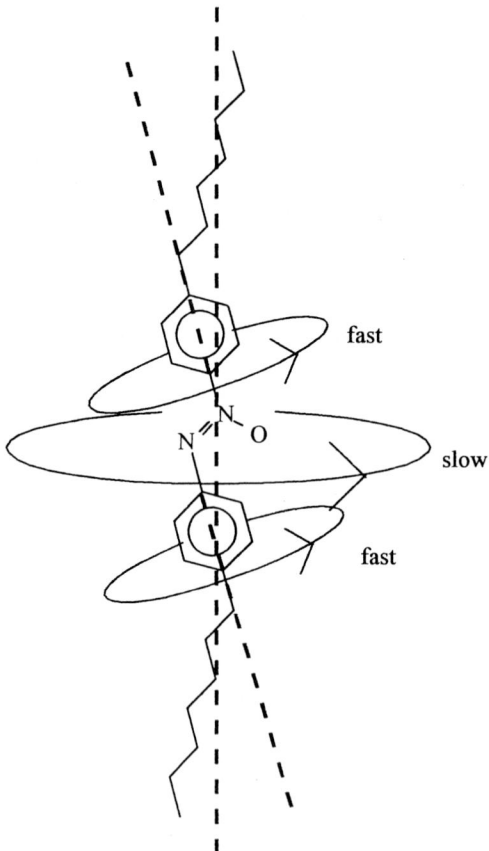

Figure 17. Schematic diagram of the model used to explain medium resolution IQENS results from PAA homologues. The "fast" (~10^{-11} s) motion is observed with medium resolution and implies an internal rotation. The "slow" (~10^{-10} s) motion is whole molecule rotation and requires high resolution to be observed.

elasticity which showed an odd-even effect along the homologous series. It was found that rotation of the two separate moieties about the C-N bonds in the aromatic core (as shown in Fig. 17) gave excess elasticity nearer to zero and so was a much better model for the molecular motion [75]. The implication of this result is that the whole molecule rotation is too slow to be seen with medium resolution, but that there are more rapid internal rotations which are observed in this experiment and by dielectric relaxation. This internal rotation model has been successfully applied to other materials [76, 77] where it gives good agreement between the correlation times observed by medium resolution quasi-elastic neutron scattering and dielectric relaxation (i.e., ~10^{-11} s).

Changes in the dielectric relaxation frequency at the SmA-SmB transition have also been reported to correlate with the appearance of a librational peak, suggesting a hardening of the potential [78].

Medium resolution measurements have also demonstrated the onset of stochastic motion of a flexible molecule in the crystal phase below the nematic. In TCDCBPh [di-(4-n-butyloxyphenyl) *trans*-cyclohexane-1,4-dicarboxylate] the cyclohexane ring starts large angle internal deformations of the molecule 30°C below the melting point [79]. Closer to the melting point, a more mobile intermediate crystal phase was discovered [80]. Studies of mixtures have shown that the correlation times of a cyanobiphenyl mixed with an azoxyanisole are not additive and this is attributed to the breaking up of the antiparallel correlations in the cyanobiphenyl [81].

4.5 Conclusions

Neutron scattering is unlikely to replace techniques such as X-ray diffraction and NMR as a method for characterizing liquid crystal phases. However, it is capable of giving unique information that is not accessible by other techniques. In neutron diffraction experiments, this relies on the ability to use isotopic labeling to highlight certain aspects of the structure, such as the backbone conformation. The simultaneous probing of geometry and the time scale of molecular dynamics is the great strength of the IQENS method and isotopic labeling is also a very useful part of this method.

4.6 References

[1] S. W. Lovesey, *Theory of Neutron Scattering from Condensed Matter: Nuclear Scattering*, Clarendon, Oxford, **1984**.

[2] G. L. Squires, *Introduction to Theory of Neutron Scattering*, Cambridge University Press, Cambridge **1978**.

[3] S. W. Lovesey, T. Springer, *Dynamics of Solids and Liquids by Neutron Scattering*, Springer, Berlin **1977**.

[4] M. Kohli, K. Otnes, R. Pynn, T. Riste, *Z. Phys.* **1976**, 147.

[5] R. M. Richardson, J. M. Allman, G. J. Mcintyre, *Liq. Cryst.* **1990**, 8, 701–719.

[6] R. W. Date, I. W. Hamley, G. R. Luckhurst, J. M. Seddon, R. M. Richardson, *Mol. Phys.* **1992**, 951–977.

[7] I. W. Hamley, S. Garnett, G. R. Luckhurst, S. J. Roskilly, J. S. Pedersen, R. M. Richardson, J. M. Seddon, *J. Chem. Phys.* **1996**, 10046–10054.

[8] P. G. de Gennes, *C. R. Acad. Sci. (Paris)* **1972**, 142.

[9] J. P. Cotton, D. Decker, H. Benoit, J. Farnoux, J. Higgins, G. Jannink, J. des Cloizeaux, R. Ober, C. Picot, *Macromolecules* **1974**, 863.

[10] J. Higgins, H. Benoit, *Polymers and Neutron Scattering*, Oxford Science Publications, Oxford **1994**.

[11] A. Guinier, G. Fournet, *Small Angle X-ray Scattering*, Wiley, New York **1955**.

[12] F. Moussa, J. P. Cotton, F. Hardouin, P. Keller, M. Lambert, G. Pepy, M. Mauzac, H. Richard, *J. Physique* **1987**, 1079–1083.

[13] F. Hardouin, L. Noirez, P. Keller, M. Lambert, F. Moussa, G. Pepy, H. Richard, *Mol. Cryst. Liq. Cryst.* **1988**, 389–397.

[14] R. M. Richardson, G. W. Gray, A. R. Tajbakhsh, *Liq. Cryst.* **1993**, 14, 871–879.

[15] L. Noirez, G. Pepy, P. Keller, L. Benguigui, *J. Physique II* **1991**, 821–830.

[16] W. Renz, M. Warner, *Phys. Rev. Lett.* **1986**, 1268–1271.

[17] W. Renz, M. Warner, *Proc. R. Soc. London, Ser. A – Math. Phys. Sci.* **1988**, 213–233.

[18] L. Noirez, P. Keller, P. Davidson, F. Hardouin, J. P. Cotton, *J. Physique* **1988**, 1993.

[19] A. S. Cherodian, N. J. Hughes, R. M. Richardson, M. S. K. Lee, G. W. Gray, *Liq. Cryst.* **1993**, 14, 1667–1682.

[20] G. R. Mitchell, F. J. Davis, W. Guo, R. Cywinski, *Polymer* **1991**, 1347–1353.

[21] F. Hardouin, N. Leroux, S. Mery, L. Noirez, *J. Physique II* **1992**, 271–278.

[22] N. Leroux, P. Keller, M. F. Achard, L. Noirez, F. Hardouin, *J. Physique II* **1993**, 1289–1296.

[23] F. Hardouin, N. Leroux, L. Noirez, P. Keller, M. Mauzac, M. F. Achard, *Mol. Cryst. Liq. Cryst. Sci. Technol. A – Mol. Cryst. Liq. Cryst.* **1994**, 267–282.

[24] S. Lecommandoux, L. Noirez, M. Mauzac, F. Hardouin, *J. Physique II* **1994**, 2249–2255.

[25] L. Noirez, P. Davidson, W. Schwarz, G. Pepy, *Liq. Cryst.* **1994**, 16, 1081–1092.

[26] X. J. Wang, M. Warner, *J. Phys. A – Math. General* **1986**, 2215–2227.

[27] V. Arrighi, J. S. Higgins, L. Abis, A. L. Cimecioglu, R. A. Weiss, *J. Macromol. Sci. – Pure Appl. Chem.* **1994**, 1101–1108.

[28] M. H. Li, A. Brulet, P. Davidson, P. Keller, J. P. Cotton, *Phys. Rev. Lett.* **1993**, 2297–2300.

[29] V. Arrighi, J. S. Higgins, R. A. Weiss, A. L. Cimecioglu, *Macromolecules* **1992**, 5297–5305.

[30] J. F. Dallest, P. Sixou, A. Blumstein, R. B. Blumstein, J. Teixeira, L. Noirez, *Mol. Cryst. Liq. Cryst.* **1988**, 581–592.

[31] F. Hardouin, G. Sigaud, M. F. Achard, A. Brulet, J. P. Cotton, D. Y. Yoon, V. Percec, M. Kawasumi, *Macromolecules* **1995**, 5427–5433.

[32] M. D. Dadmun, C. C. Hun, *Macromolecules* **1994**, 7522–7532.

[33] M. D. Dadmun, *ACS Symp. Ser.* **1995**, 320–334.

[34] K. Hongladarom, V. M. Ugaz, D. K. Cinader, W. R. Burghardt, J. P. Quintana, B. S. Hsiao, M. D. Damun, W. A. Hamilton, P. D. Butler, *Macromolecules* **1996**, 5346–5355.

[35] L. C. A. Groot, M. E. Kuil, J. C. Leyte, J. R. C. Vandermaarel, R. K. Heenan, S. M. King, *Liq. Cryst.* **1994**, 16, 263–276.

[36] T. Riste, K. Otnes, *Liq. Cryst.* **1993**, 14, 581–589.

[37] G. Pepy, J. F. Ravoux, *Mol. Cryst. Liq. Cryst. Sci. Technol. A – Mol. Cryst. Liq. Cryst.* **1995**, 303–309.

[38] P. Porquet, Thesis Universite Paris-Sud **1977**.

[39] M. Bee, *Quasielastic Neutron Scattering*, Adam Hilger, Bristol, U. K. **1988**.

[40] T. Springer, *Quasielastic Neutron Scattering for the Investigation of Diffuse Motions in Solids and Liquids*, Springer, Berlin **1972**.

[41] W. Press, *Single Particle Rotations in Molecular Crystals*, Springer, Berlin **1981**.

[42] A. J. Leadbetter, R. M. Richardson, in: G. R. Luckhurst, G. W. Gray (Eds.), *The Molecular Physics of Liquid Crystals*, Academic, London **1979**, p. 451–483.

[43] J. A. Janik, *Adv. Liq. Cryst.* **1982**, 215–233.

[44] R. M. Richardson, *NATO ASI Ser.* **1994**, 451–480.

[45] H. Hervet, A. J. Dianoux, R. E. Lechner, F. Volino, *J. Physique* **1976**, 587–594.

[46] K. Ibel (Ed.), *Guide to Neutron Research Facilities at the ILL*. Institut Laue Langevin; Grenoble **1994**.

[47] B. Boland, S. Whapham (Eds.), *User Guide to Experimental Facilities at ISIS*, Rutherford Appleton Laboratory, Oxfordshire **1992**; RAL 92-041.

[48] L. Van Hove, *Phys. Rev.* **1954**, 249–262.
[49] K. Rosciszewski, *Acta Phys. Pol. A* **1972**, 549–561.
[50] A. J. Leadbetter, F. P. Temme, A. Heidemann, W. S. Howells, *Chem. Phys. Lett.* **1975**, 363–368.
[51] A. J. Leadbetter, R. M. Richardson, B. A. Dasannacharya, W. S. Howells, *Chem. Phys. Lett.* **1976**, 501–504.
[52] J. Toepler, B. Alefeld, T. Springer, *Mol. Cryst. Liq. Cryst.* **1974**, 297–303.
[53] V. Dimic, L. Barbic, R. Blinc, *Phys. Status Solidi B* **1972**, 121–124.
[54] F. Volino, A. J. Dianoux, A. Heidemann, *J. Phys.* **1979**, 583–586.
[55] F. Volino, A. J. Dianoux, *Mol. Phys.* **1978**, 389–399.
[56] R. M. Richardson, A. J. Leadbetter, D. H. Bonsor, G. J. Kruger, *Mol. Phys.* **1980**, 741–757.
[57] L. D. Favro, R. E. Burgess (Eds.), *Fluctuation Phenomena in Solids*, Academic, New York **1965**, p. 79.
[58] A. J. Dianoux, F. Volino, *Mol. Phys.* **1977**, 1263–1277.
[59] A. J. Dianoux, H. Hervet, F. Volino, *J. Phys. (Paris)* **1977**, 809–816.
[60] A. J. Dianoux, F. Volino, *J. Physique* **1978**, 181.
[61] F. Volino, A. J. Dianoux, H. Hervet, *Mol. Cryst. Liq. Cryst.* **1977**, 125–140.
[62] F. Volino, A. J. Dianoux, H. Hervet, *J. Phys. (Paris)* **1976**, 55–64.
[63] P. L. Nordio, U. Segre, *Mol. Cryst. Liq. Cryst.* **1978**, 185–192.
[64] R. M. Richardson, A. J. Leadbetter, J. C. Frost, *Mol. Phys.* **1982** 1163–1191.
[65] D. H. Bonsor, A. J. Leadbetter, F. P. Temme, *Mol. Phys.* **1978**, 1805–1823.
[66] A. J. Leadbetter, R. M. Richardson, *Mol. Phys.* **1978**, 1191–1200.
[67] I. Dozov, N. Kirov, S. Temkin, *J. Mol. Liq.* **1992**, 1–10.
[68] L. Carpentier, M. Bee, A. M. Giroud-Godquin, P. Maldivi, J. C. Marchon, *Mol. Phys.* **1989**, 1367–1378.
[69] M. Bee, A. M. Giroud-Godquin, P. Maldivi, J. Williams, *Mol. Phys.* **1994**, 57–68.
[70] L. Benguigui, L. Noirez, R. Kahn, P. Keller, M. Lambert, E. Cohen de Lara, *J. Phys. II* **1991**, 451–458.
[71] B. Cvikl, J. Peternelj, *J. Phys. II* **1992**, 87–92.
[72] S. Konig, T. M. Bayerl, G. Coddens, D. Richter, E. Sackmann, *Biophys. J.* **1995**, 1871–1880.
[73] M. D. Dadmun, M. Muthukumar, R. Hempelmann, D. Schwahn, T. Springer, *J. Polym. Sci. Part B – Polym. Phys.* **1996**, 649–656.
[74] J. Tabony, B. Perly, *Biochim. Biophys. Acta* **1991**, 67–72.
[75] J. A. Janik, J. M. Janik, K. Otnes, T. Stanek, *Liq. Cryst.* **1989**, 5, 1045–1051.
[76] R. Podsiadly, J. A. Janik, *Mol. Cryst. Liq. Cryst. Sci. Technol.* **1995**, 1573–1580.
[77] K. Chledowska, D. Chrusciel, J. A. Janik, J. M. Janik, H. Kresse, H. Stettin, K. Otnes, T. Stanek, S. Urban, S. Wrobel, *Liq. Cryst.* **1988**, 4, 1339–1354.
[78] L. Bata, G. Pepy, L. Rosta, *Liq. Cryst.* **1988**, 4, 893–900.
[79] R. Podsiadly, J. A. Janik, J. M. Janik, K. Otnes, *Liq. Cryst.* **1993**, 14, 1519–1528.
[80] R. Podsiadly, J. Mayer, W. Witko, *Mol. Cryst. Liq. Cryst. Sci. Technol. A – Mol. Cryst. Liq. Cryst.* **1995**, 423.
[81] R. Dabrowski, J. A. Janik, J. M. Janik, K. Otnes, *Liq. Cryst.* **1988**, 4, 443–452.

5 Light Scattering from Liquid Crystals

Helen F. Gleeson

One of the most notable features of liquid crystals is that they scatter light strongly, a property that was employed in the earliest liquid crystal display devices. Light scattering phenomena may be analysed quantitatively to derive fundamental liquid crystal parameters, and here quasi-elastic (Rayleigh) light scattering is considered. It is useful to differentiate between *static* and *dynamic* experiments. In the former case, no hydrodynamic effects are considered, and as a consequence, it is not possible to deduce information on viscosity from the results of static experiments. On the other hand, dynamic experiments determine the frequency spectrum of the quasi-elastically scattered light and information may be deduced regarding the viscosities of the liquid-crystalline systems. In the following sections experimental light scattering studies of nematic, chiral nematic and blue phases, smectic materials and pretransitional systems are reviewed, together with their theoretical basis.

5.1 Light Scattering from Nematic Liquid Crystals

5.1.1 Static Light Scattering from Nematic Liquid Crystals

Thermally excited angular fluctuations in the director of a nematic liquid crystal give rise to strongly depolarized light scattering, with high angular asymmetry. The history of light scattering from nematic materials effectively began with the studies of Chatelain [1–3], who studied the depolarized scattering from nematic materials. Chatelain's work studied the intensity dependence of polarized scattering for all scattering geometries; he interpreted his data with respect to a system consisting of swarms of material ~0.2 µm in diameter [4–7] and introduced an order parameter expression to describe their relative orientation [8]. The swarm theory considered nematic materials as groups of ordered fluid with a certain coherence length, separated by isotropic fluid, and clearly had severe limitations in its basis. Chatelain's data were re-analysed by de Gennes [9] using a long range model of the nematic phase and the elastic constants defined by Frank [10] and Oseen [11].

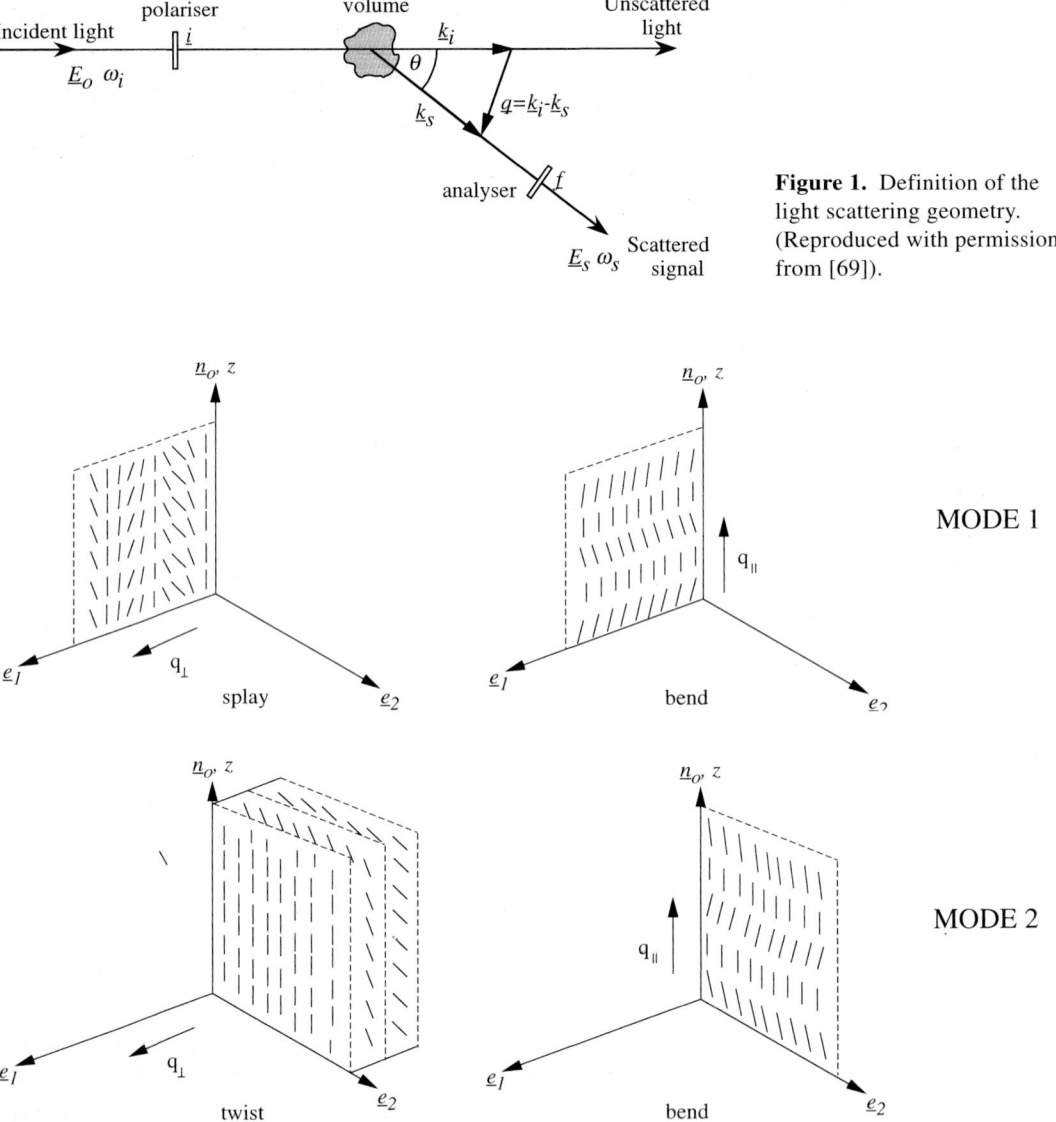

Figure 1. Definition of the light scattering geometry. (Reproduced with permission from [69]).

Figure 2. The common cartesian coordinate system used in light scattering, and the two modes of director fluctuation in the scattered light.

The intensity of scattered radiation depends on the geometry of the experiment (including the orientation of the director in the sample) and the angle of detection. These are shown in Fig. 1. The scattering plane contains the scattering vector q, scattering angle, θ and the wave vectors k_i and k_s of the incident and scattered light. The polarization of the light is defined with respect to the scattering plane such that it is parallel or perpendicular to it. The Cartesian coordinate system shown in Fig. 2 is commonly used in light scattering from nematic liquid crystals. Here, the z-axis is parallel to the nematic director and the scattering vector q lies in the e_1–z plane (the scattering plane). e_1 and e_2

are unit vectors that describe directions such that e_2 is perpendicular to n_0 and q, while e_1 is perpendicular to n_0 and e_2.

De Gennes suggested that by analysing the scattering cross-section of a nematic material, the ratios of the three Frank elastic constants, k_{11}/k_{22} and k_{33}/k_{22} may be determined from the angular dependence of the scattered light [12]. According to de Gennes, the differential cross-section per unit scattering volume ($d\sigma/d\Omega$), is:

$$\frac{d\sigma}{d\Omega} = \left(\frac{\pi \Delta \varepsilon}{\lambda^2}\right) k_B T \sum_{\alpha=1,2} \frac{(i_\alpha f_z + i_z f_\alpha)^2}{(k_{33} q_\parallel^2 + k_{\alpha\alpha} q_\perp^2)} \quad (1)$$

where $\Delta \varepsilon$ is the dielectric anisotropy, λ the wavelength of light used, k_B the Boltzmann constant and T the temperature. The parameters $q_\perp$ and $q_\parallel$ are the perpendicular and parallel components of the scattering vector with respect to the director, n_0. The unit vectors i and f specify the polarization of the incident and scattered beams. De Gennes assumed that the incident and scattered waves propagate in an isotropic medium. If the incident light is normal to a nematic sample and polarized either parallel or perpendicular to the director, the average scattered light intensity within the scattering medium is:

$$\langle I \rangle = \frac{\omega_0^4 E_0^2}{2c^4} \frac{(\Delta \varepsilon V)^2 k_B T}{(4\pi R)^2}$$
$$\cdot \sum_{\alpha=1,2} \frac{(i_\alpha f_z + i_z f_\alpha)^2}{V(k_{33} q_\parallel^2 + k_{\alpha\alpha} q_\perp^2)} \quad (2)$$

Here, E_0 is the amplitude of the incident optical field, ω_0 the frequency of the incident light, V the scattering volume and R the distance between the scattering volume and the detector. The scattered light consists of two modes, splay/bend ($\alpha=1$) and twist/bend ($\alpha=2$), as shown in Fig. 2. It can be seen from Fig. 3 (a) that the mode 1 fluctuations can contain only contributions from the bend and splay distortions. Mode 2 fluctuations are in the perpendicular plane (Fig. 3b and c) and can contain only contributions from twist and bend distortions.

Because scattered light is detected outside the sample, the birefringence of the nematic material must in general be accounted for. A few special situations occur when the scattered light is not split into ordinary and extraordinary rays. Static light scattering experiments are further complicated as absolute values of scattered light intensity are required. Relative intensity measurements may be used provided the angular range is restricted to those where the contributions from both modes are approximately equal. Further, suitable choices of polarization and geometry make it possible to deconvolute the modes and distortions, allowing values to be obtained for the various elastic constants $k_{\alpha\alpha}$ [13–15].

Langevin and Bouchiat [16] calculated the total cross section of light scattering by a nematic by integrating de Gennes' equation over all angles and including the influence of the optical anisotropy. In three particular geometries, (a) $k \parallel n_0$, (b) $k \perp n_0$, $i \perp n_0$, and (c) $k \perp n_0$, $i \parallel n_0$, absorption coefficients yield the Frank elastic constants. Langevin [17] and others [18–20] used the technique to measure the elastic constants of MBBA (see Fig. 4). Intensity contours of Rayleigh scattering from MBBA provided tilt bias angles in nematic cells [21]. Other measurements of the elastic ratios should be referred to [22–24].

It is also possible to use turbidity measurements to evaluate elastic constants (Fig. 5, though elastic constants measured in this way for alkylcyanobiphenyls (nCBs) [25] and mixtures [26] are consistently higher than those determined by Frederiks transition studies. Hakemi [27] measured k_{11}, k_{22} and k_{33} for 8CB (octylcyano biphenyl) close to the nematic–smectic A phase transition and deduced a value of the critical exponent from the divergence of k_{22} and k_{33}.

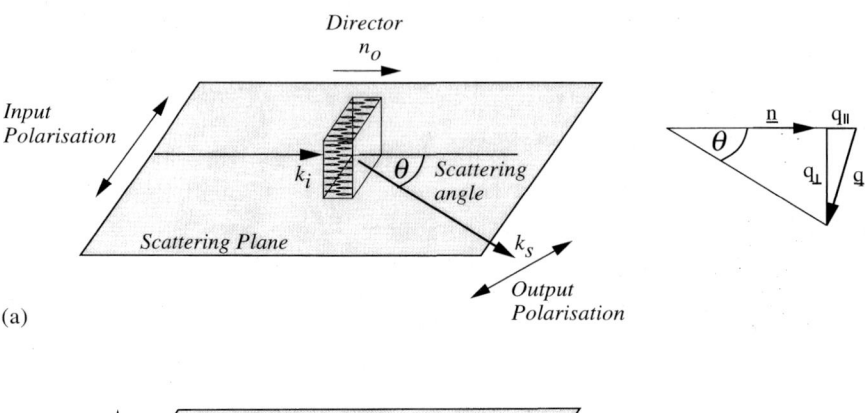

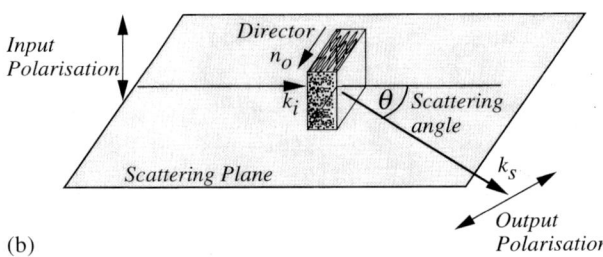

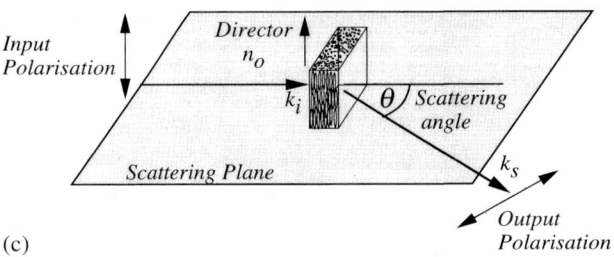

Figure 3. (a) Mode 1 geometry: the director is parallel to the scattering plane. (b) Mode 2 geometry: the scattering plane and input light polarization are perpendicular to the director. (c) Mode 2 geometry: the scattering plane perpendicular to the director, input light polarization parallel to the director. (Reproduced with permission from [69])

5.1.2 Dynamic Light Scattering from Nematic Liquid Crystals

The Orsay Liquid Crystal Group [28] determined a model of director modes of fluctuation based on the hydrodynamic description of nematic phases by Leslie [29] and Ericksen [30]. This approach developed the subject of dynamic light scattering in liquid crystals, and allows viscoelastic ratios to be deduced from light scattering data. For singly scattered light from a material, the time variation of the scattered field $E_s(\bm{R}, t)$ at a position $\bm{R}$ depends only on fluctuations in the dielectric tensor $\delta\varepsilon(\bm{r}, t)$, that is

$$E_s(\bm{R}, t) = \frac{k_s^2 E_0 V e^{i(k_s \cdot \bm{R} - \omega_i t)}}{4\pi\varepsilon_0 R} \delta\varepsilon_{if}(\bm{q}, t) \quad (3)$$

where $\delta\varepsilon_{if}(\bm{q}, t) = \hat{\bm{f}} \cdot \delta\varepsilon(\bm{q}, t) \cdot \hat{\bm{i}}$, $\bm{f}$ and $\bm{i}$ represent the initial and final polarization directions of the light, and $\delta\varepsilon(\bm{q}, t)$ is the spatial Fourier transform of the dielectric tensor.

The time dependence of the scattered light may be related to allowed deforma-

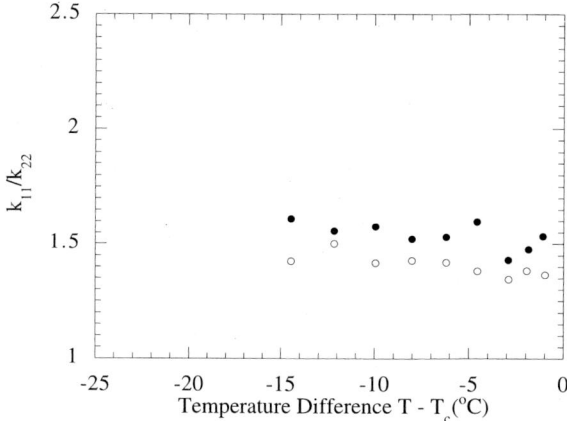

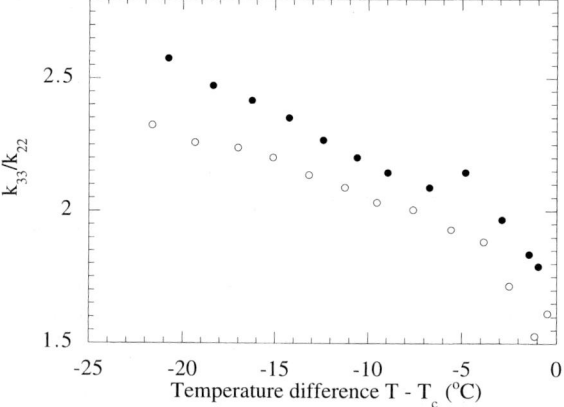

Figure 4. Elastic constant ratios k_{33}/k_{22} and k_{11}/k_{22} of MBBA as a function of temperature, determined by static light scattering. The open circles refer to $i \parallel (k_i, k_f)$ whilst the closed circles correspond to $i \perp (k_i, k_f)$. (Redrawn from [18]).

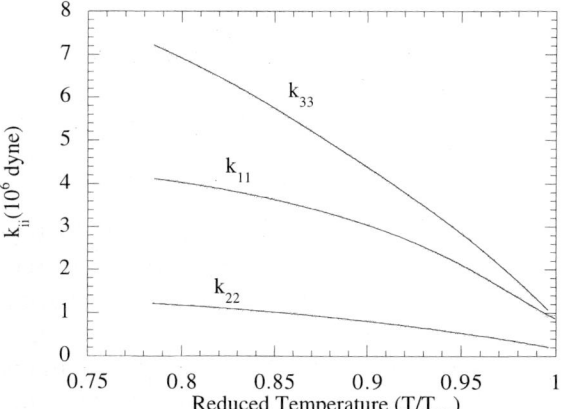

Figure 5. The elastic constants k_{ii} of the mixture E7 (Merck Ltd) as a function of reduced temperature determined via the turbidity method. (Redrawn from [26]).

tions in a nematic material via the 1st order field and 2nd order intensity correlation functions, $g_1(\tau)$ and $g_2(\tau)$. In *homodyne* detection experiments, E-field contributions from the scattering volume interfere at the detector and, assuming that the scattered field obeys Gaussian statistics [31], $g_2(\tau) = 1 + [g_1(\tau)]^2$. In *heterodyne* detection, the scattered field is allowed to beat with a portion of unscattered light derived from the same laser source. Provided that the fluctuations in the unscattered light are negligible, the scattered and unscattered fields are statistically independent, and the intensity of the heterodyne beam is far greater than that scattered by the sample, then,

$$g_2^{\text{het}}(\tau) \approx 1 + \frac{2\langle I_s(t)\rangle\langle I_u(t)\rangle}{\langle I_s(t) + I_u(t)\rangle^2} g_1(\tau) \quad (4)$$

The subscripts s and u refer to the scattered and unscattered fields respectively. Heterodyne experiments are technically more difficult than homodyne, and require longer detection times, but can give more accurate results as no assumptions are made about the scattered light. Further, heterodyne techniques avoid inaccuracies that may be incurred in homodyne experiments that unwittingly detect partially heterodyned signals.

Again, the geometries of the light scattering experiments are important in defining which of the viscoelastic constants are probed in a particular experiment. The time dependence of the scattered light may be related to viscoelastic parameters through the director fluctuations and allowed distortions of the local dielectric tensor. A suitable choice of coordinate system and geometry allows the fluctuations to be split into two modes:

Mode 1: splay/bend; director confined to the x–z (e_1–n_0) plane and
Mode 2: twist/bend; director confined to the y–z (e_2–n_0) plane.

In nematic systems the two uncoupled fluctuation modes are purely relaxational with Lorenzian spectral densities. In general, the autocorrelation function of the scattered field includes two exponentials corresponding to decay times of the two fluctuation modes,

$$\Gamma_\alpha = \frac{k_{\alpha\alpha} q_\perp^2 + k_{33} q_{11}^2}{\eta_\alpha(q)} \quad (5)$$

where $\alpha = 1, 2$. The viscosities, η_α are related to the Leslie coefficients in nematic materials [32] and hence the Miesowicz viscosities [33] that describe the flow of an anisotropic fluid. The Orsay Group used photon correlation techniques (light beating spectroscopy) to observe the fluctuation modes in nematic 4-azoxyanisole (PAA) that arose from director fluctuations [34, 35].

In the absence of any other parameters, that is for thermal light scattering, the ratios $k_{\alpha\alpha}/\eta_\alpha$ may be measured. Adding a further parameter that is independent of $k_{\alpha\alpha}$ (e.g. a magnetic or electric field term) allows the viscoelastic ratios to be decoupled, as observed by Martinand and Durand [36] who considered systems with negative dielectric anisotropy. The use of electric fields was further considered by Leslie and Waters [37] who showed that flexoelectric and conduction effects within the sample were negligible. In the presence of an electric field the fluctuation modes are described by,

Mode 1:
$$\Gamma_{\text{splay}} = \frac{k_{11} q_\perp^2 + k_{33} q_{11}^2}{\eta_{\text{splay}}} + \frac{\varepsilon_0 \Delta\varepsilon \varepsilon_{11} E^2}{\varepsilon(\phi) \eta_{\text{splay}}} \quad (6a)$$

Mode 2:
$$\Gamma_{\text{twist}} = \frac{k_{22} q_\perp^2 + k_{33} q_{11}^2}{\eta_{\text{twist}}} + \frac{\varepsilon_0 \Delta\varepsilon E^2}{\eta_{\text{twist}}} \quad (6b)$$

where $\varepsilon(\phi) = \varepsilon_\perp + \delta\varepsilon \sin^2\phi$, and ϕ is the angle between the director and the scattering vector, $\boldsymbol{q}$.

The linewidth is thus proportional to the square of the applied damping field and the gradient gives the viscosity coefficient. The intercept, corresponding to zero field still gives the ratio $k_{\alpha\alpha}/\eta_\alpha$.

A detailed study of light scattering geometries has been made by van Eck [38], this is summarized below.

Homeotropic Sample Cell
(Director $\boldsymbol{n}_0$ Parallel to $\boldsymbol{k}_i$)

Mode 1 (splay/bend): input and output polarization in the scattering plane.

Mode 2 (twist/bend): input polarization perpendicular to the scattering plane and output polarization parallel to the scattering plane.

The angular range over which measurements are made is constrained to <15° to minimize the contribution from k_{33}.

Planar Sample Cells

(a) The director $\boldsymbol{n}_0$ perpendicular to the scattering plane gives a specific angle within the liquid crystal, θ_{lc}, where $\theta_{lc} = \cos^{-1}(\boldsymbol{n}_0/\boldsymbol{n}_e)$ that allows pure splay to be studied as:

$$\Gamma_{splay} = \frac{k_{11}}{\eta_{splay}} q_\perp^2 + \frac{\varepsilon_0 \Delta\varepsilon \varepsilon_{11} E^2}{\varepsilon(\phi)\eta_{splay}} \quad (7)$$

(b) If the director is parallel to the scattering plane, again a specific angle $\theta_{lc} = \cos^{-1}(\boldsymbol{n}_0/\boldsymbol{n}(\theta))$ may be defined such that only bend contributions are observed:

$$\Gamma_{bend} = \frac{k_{33}}{\eta_{bend}} q_{11}^2 + \frac{\varepsilon_0 \Delta\varepsilon E^2}{\eta_{bend}} \quad (8)$$

Here, $n(\theta) = \dfrac{n_0 n_e}{\sqrt{\left(n_e^2 \sin^2\theta + n_0^2 \cos^2\theta\right)}}$

The pioneering work of the Orsay Group in the area of dynamic light scattering [39, 40] led to many accurate measurements of the Frank elastic constants and Leslie viscosity coefficients. Haller and Litster [41, 42] investigated the temperature dependence of the normalised scattering intensity and the scattering linewidth as a function of temperature, finding that the latter was far more temperature sensitive. Many materials have been studied using this technique: MBBA [43–45], di-butylazoxybenzene and p-methoxy-p'-n-butylazoxybenzene [46–48], Schiffs bases [49–51], and the cyanobiphenyls [52, 53]. Typical data for the cyanobiphenyls are shown in Fig. 6.

Extensive studies of nematic liquid crystals doped with various concentrations of side chain polymer liquid crystals [54–63], reveal that the viscosities increase in proportion to the polymer concentration for little change in the elastic constants (Fig. 7 [64]). The bend elastic modulus of straight and bent dimeric liquid crystals (with odd and even numbers of methylene units in the spacer) was determined by Dilisi et al. [65], finding that k_{22}/k_{11} was the same for both species, but k_{33}/k_{11} was far smaller for the bent dimer. Comparisons of monomers and dimers have also been made [66]. The use of electric field dynamic light scattering to determine k_{22} is widespread, with relatively few measurements of k_{11} [67, 68]. The use of electric field dynamic light scattering to determine k_{33} in materials of positive dielectric anisotropy (5CB and 8CB, pentyl and octylcyanobiphenyl respectively) has been reported recently [69] using planar cells with transverse electrodes, and thus applying the electric field in the plane of the cell. The bend elastic constants of both materials are shown in Fig. 8, and the pretransitional behaviour of k_{33} at the nematic to smectic A phase transition is clearly observed.

The viscoelastic constants of the lyotropic nematic polymer poly-γ-benzyl glutamate was measured by Taratuta et al. [70]. Santos and Durand [71] measured order and mi-

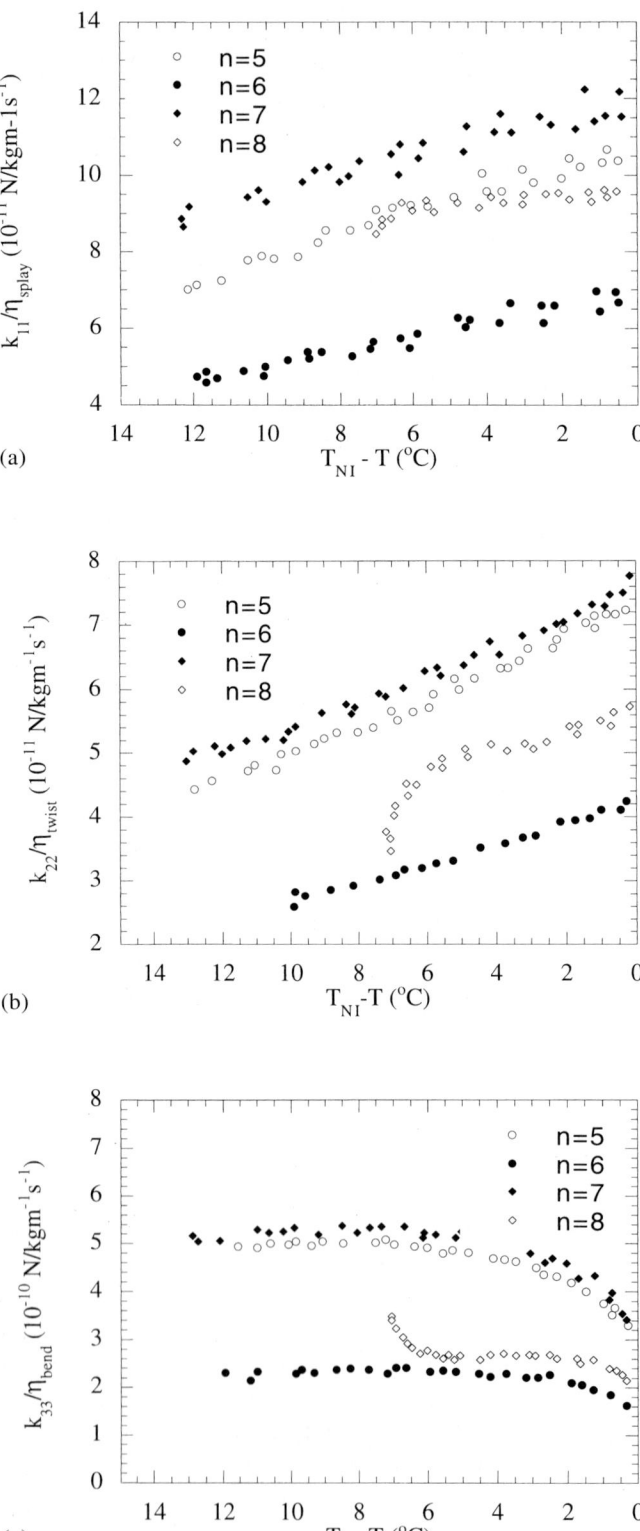

Figure 6. The viscoelastic ratios of some of the cyanobiphenyls as a function of reduced temperature (a) k_{11}/η_{splay}, (b) k_{22}/η_{twist}, and (c) k_{33}/η_{bend}. (Redrawn from [52]).

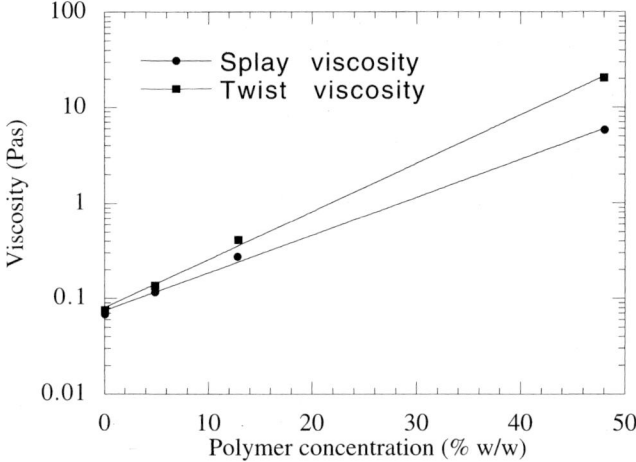

Figure 7. Viscosity versus polymer concentration for the side chain liquid crystal polymer PG296 [64] in E7 at a reduced temperature $T_C - T = 20\,°C$. (Redrawn from [62])

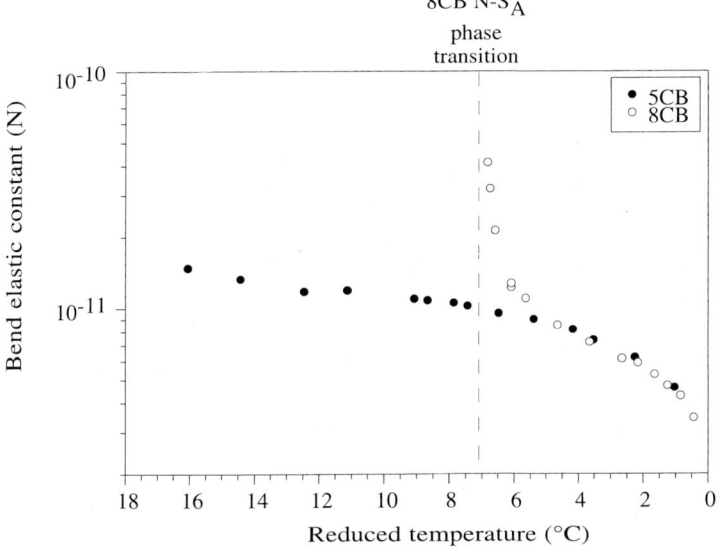

Figure 8. The bend elastic constant of 5CB and 8CB measured using electric field dynamic light scattering. (Reproduced with permission from [69]).

cellar density in the biaxial, uniaxial and isotropic phases of a lyotropic nematic material. A more detailed analysis of light scattering from a calamitic lyotropic nematic differentiates between signals associated with micellar fluctuations and the usual orientational fluctuations [72]. The micellar fluctuations are temperature independent and of a much lower relaxational frequency than those found previously in the same system exhibiting a biaxial nematic phase. The viscoelastic properties of metallomesogenic compounds for which the magnetic properties were unknown have also been measured by light scattering [73]. Anchoring energies at a nematic-rubbed nylon interface were determined using evanescent wave photon correlation spectroscopy, probing only the surface layer of the system [74].

5.1.3 Forced Rayleigh Scattering

Binary mass diffusion constants in nematic liquid crystals have been measured using forced Rayleigh scattering, in which two coherent laser beams interfere at the sample to produce a periodic light intensity within the sample [75]. The periodic light intensity influences the sample, which may contain photochromic species, to form a diffraction grating. If the lasers are pulsed, the grating becomes transient and smears due to diffusion of the excited species forming it. An analysis of the transient diffraction pattern produced from the sample by the probe beam allows the determination of diffusion constants. Experimental work includes the study of nematic, smectic A and smectic B phases [76], methyl red dissolved in members of a homologous series of alkylcyanobiphenyls [77], and self diffusion in MBBA [78]. The localization of electromagnetic modes has been studied via coherent back scattering in a nematic liquid crystal [79].

5.2 Cholesteric and Blue Phases

Both cholesteric and blue phases (BPs) exhibit spectacular light scattering effects due to their periodic helical structures (selective and Bragg-like reflection). However, Rayleigh light scattering from cholesteric phases can be considered in analogy to nematic phases using de Gennes phenomenological theory [80], with no special features. Two scattering modes should be observable and have been predicted [81]:

Mode 1: twisting and untwisting of the helical structure
Mode 2: a combination viscous splay mode.

Most light scattering studies on cholesteric liquid crystals have considered either the pretransitional region (see Chapter VIII, Section 5.5) or lyotropic polymers. Quasielastic light scattering from PBLG (poly-γ-benzyl-L-glutamate) solutions reveal two purely dissipative modes probably those predicted above [82–85]. Beevers et al. [86] undertook polarized and depolarized light scattering studies in the isotropic and blue phases of cholesteryl oleyl carbonate, and across the phase transition. Their correlation functions could not be fitted by single exponentials. Hornreich and Shtrikman [87] analysed light scattering in BPI and BPII (blue phase I and blue phase II) to yield structural information regarding the phases. Their theory described the experimental results of Flack et al. [88] and Meiboom and Sammon [89]. The anomalous scattering of blue phase III, BPIII, was also considered and found to be probably not attributable to harmonic fluctuations in the order parameter at the transition as experimentally observed scattering intensities are several orders of magnitude greater than would be predicted by this model. Marcus [90] discusses the twinkling of blue phases with respect to lattice fluctuations, and deduces that the shear modulus in BPII must be much smaller than the bulk modulus. Such twinkling is not observed in the cholesteric phase where no soft shear mode occurs (only rotations and a hard pitch change mode), but is similar to fluctuations observed in nematic systems. Studies of the lattice fluctuations via quasielastic light scattering [91] show a fluctuating quantity which is neither a simple relaxation, nor a propagating wave. Dmitrienko [92] used Landau theory of blue phases to calculate the elastic moduli and show that they could be determined by measuring light scattering of fluctuations in the order parameter. The kinematic theory of light scattering from blue phases was considered by Belyakov and Dmitrienko [93], and has been used recently to derive expressions to deduce order parameters in blue phases from Kossel diagrams [94].

5.3 Light Scattering from the Smectic A Phase

In nematic liquid crystals, there are two modes of angular fluctuation of the director that give rise to strongly depolarized Rayleigh scattering. In SmA systems, coupling between molecular orientation and the layer thickness is such that in principle thermally excited fluctuations should give rise to Rayleigh scattering no greater than in a normal fluid, except for undulations in the layers [95, 96]. The thermal average of angular fluctuations in a SmA liquid crystal may be written [97]:

$$\langle \theta \rangle = \frac{k_B T}{K\left(q_\perp^2 + l^2 q_c^4 / q_\perp^2\right)} \approx I \quad (9)$$

where K is the splay elastic constant (twist and bend are not allowed in the phase), q is the wave vector of the undulations (q_c is a critical value), and l is an integer greater than 1. A maximum occurs at $q_\perp = q_c$ for the fundamental mode. Delaye et al. [98] observed $q_c^2 \propto d^{-1}$, predicted by $q_c^2 \lambda d = \pi$ where λ is the de Gennes penetration depth. They initially found $\lambda = (2.2 \pm 0.3)$ nm, though later Ribotta et al. [99] reported $\lambda = 1.4$ nm.

In undertaking light scattering experiments from SmA samples, it is important to exclude effects due to defects and to wall induced undulations which are particularly pronounced in homeotropic geometries. Consequently, much of the work has been done on planar samples, showing that only splay distortions occur [100]. Clark and Pershan [101] studied the light scattering from racemic p-butoxybenzal-p-(β-methylbutyl)aniline (BBMBA) in the nematic and SmA phases to determine the ratio K/B in the SmA phase.

Experiments on lyotropic SmA systems determined the layer compressibility modulus, B (intermembrane interactions) and the bending modulus, K (interpreted as membrane flexibility) [102], supporting the view of membranes with flexibility of the order of $k_B T$. The special case of Rayleigh scattering from columnar undulations in a columnar discotic system [103] showed a 3-D solid like elasticity attributed to column entanglement.

Forced Rayleigh scattering experiments on smectic liquid crystals measured the binary mass diffusion constants parallel and perpendicular to the director ($D_\parallel$ and $D_\perp$), finding that $D_\parallel$ is larger than $D_\perp$ in both the smectic and nematic phase of 8CB [104]. Measure-ments for the SmB, SmA and N phases of p-butoxybenzylidene-p-n-octylaniline show $D_\parallel$ to be much smaller than $D_\perp$ in the SmB phase (see Fig. 9) due to the different activation energies between the phases [105].

5.4 Light Scattering from Achiral and Chiral Smectic C Phases

In smectic C materials, the relative twist of the planes is uncoupled to the layer thickness, giving rise to strong, nematic-like scattering [106]. There are relatively few light scattering studies of either achiral or chiral SmC phases, despite the technological importance of the ferroelectric SmC* phase. This is in part due to the few discussions of the elastic theories of these phases, and in particular descriptions that include dynamic behaviour. Indeed it is only relatively recently that Leslie et al. [107] de-

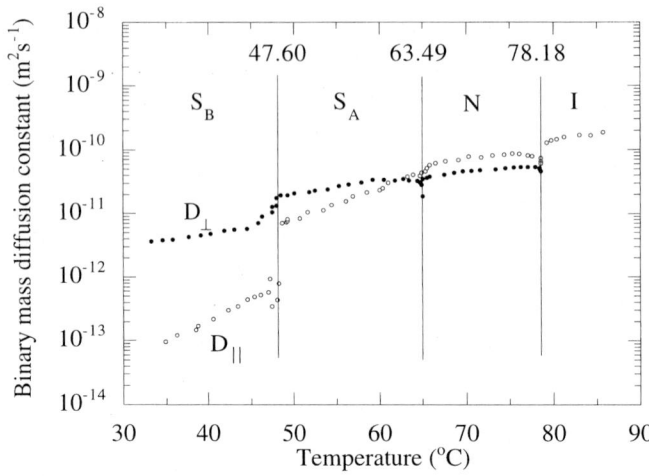

Figure 9. The temperature dependence of the binary mass diffusion constants of methyl red in 4O.8. (Redrawn from [105]).

duced a complete continuum theory of the SmC phase. In both the achiral and chiral SmC phases many more elastic constants are required to describe the systems than in either nematic or SmA systems (nine and fifteen in the achiral and chiral systems respectively). In achiral systems, where the symmetry of the phase is higher than that of the chiral analogue, the description is correspondingly simpler.

Light scattering from SmC liquid crystals was first described by the Orsay Group [108]; members of the group determined that the phase must be described by nine elastic constants, of which four could be determined via light scattering without significantly constraining the scattering vector, q. The other five elastic constants require that q is accurately parallel to the plane of the smectic layers. The free energy of the SmC phase was described with respect to a rotation vector $\Omega(r)$, where the direction defines the axis of rotation and the magnitude describes the rotation angle, allowing the free energy of the phase to be described in terms of gradients of $\Omega(r)$. An equivalent expression for the free energy was given by Rapini [109]. More recently, Carlsson et al. [110], described the free energy of the SmC phase in coordinate free notation, to deduce an expression almost identical to that of the Orsay Group apart from some differences in the terms, and neglecting surface energies. The expression for the free energy of the SmC phase was deduced by Carlsson et al. to be:

$$W = \frac{1}{2}A_{12}(\boldsymbol{b}\cdot\nabla\times\boldsymbol{c})^2 + \frac{1}{2}A_{21}(\boldsymbol{c}\cdot\nabla\times\boldsymbol{b})^2$$
$$+ A_{11}(\boldsymbol{b}\cdot\nabla\times\boldsymbol{c})(\boldsymbol{c}\cdot\nabla\times\boldsymbol{b})$$
$$+ \frac{1}{2}B_1(\nabla\cdot\boldsymbol{b})^2 + \frac{1}{2}B_2(\nabla\cdot\boldsymbol{c})^2$$
$$+ \frac{1}{2}B_3\left(\frac{1}{2}(\boldsymbol{b}\cdot\nabla\times\boldsymbol{b}+\boldsymbol{c}\cdot\nabla\times\boldsymbol{c})\right)^2$$
$$+ B_{13}(\nabla\cdot\boldsymbol{b})\left[\frac{1}{2}(\boldsymbol{b}\cdot\nabla\times\boldsymbol{b}+\boldsymbol{c}\cdot\nabla\times\boldsymbol{c})\right]^2$$
$$+ C_1(\nabla\cdot\boldsymbol{c})(\boldsymbol{b}\cdot\nabla\times\boldsymbol{c})$$
$$+ C_2(\nabla\cdot\boldsymbol{c})(\boldsymbol{c}\cdot\nabla\times\boldsymbol{b})$$

The vectors $\boldsymbol{a}$, $\boldsymbol{b}$ and $\boldsymbol{c}$ are defined with respect to a SmC layer, as shown in Fig. 10(a). The coefficients A describe curvature of the planes, B nonuniform rotations of the optic axis about z in a flat structure, and the C coefficients are coupling terms between A and B, and are shown schematically in Fig. 10(b). All of the temperature dependences of these coefficients may be included in the tilt angle.

In principle, the nine elastic coefficients of the SmC phase may be derived from the

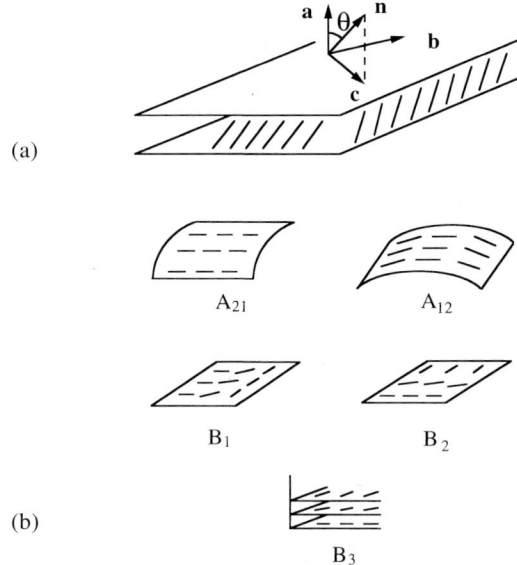

Figure 10. (a) Definitions of the vectors **a**, **b** and **c** in SmC layers. The director is shown by a unit vector, **n** which makes an angle θ with the layer normal, **a**. The c-director is a unit vector parallel to the projection of the director into the smectic planes, denoted by **c**. The unit vector **b** is defined by the relation $b = a \times c$. (Redrawn from [110]). (b) A schematic diagram of the elastic deformations of the SmC phase. (Redrawn from [110]).

intensity $I(q)$ of light scattered at wave vector q, as for nematics. The fluctuations of the dielectric tensor may be readily related to fluctuations in the rotation vector $\Omega(r)$. In fact, all such fluctuations are finite for all values of q, and it can be shown that:

$$\langle |\Omega_z(q)|^2 \rangle = \frac{kT}{B_1 q_x^2 + B_2 q_y^2 + B_3 q_z^2 + 2B_{13} q_x q_z} \tag{10}$$

$\Omega_x(q)$ and $\Omega_y(q)$ vanish for $q_z \neq 0$ because of the requirement for constant layer thickness. They contribute strongly for $q_z = 0$, though this condition is unphysical, requiring highly monodomain single crystals.

Galerne et al. [111] studied quasielastic light scattering in the nematic and SmC phases of di-(4-n-decyloxybenzal)-2-chloro-1,4-phenylenediamine (DOBCP), producing planar, monodomain alignment using a magnetic field. The experiment always detects heterodyne signals due to flare from the glass surfaces. The laser was normal to the glass plates and the smectic layers, with the scattering plane including the rubbing direction and light polarized either normal or parallel to the scattering plane. In the nematic phase, Mode 1 (**n** parallel to scattering plane) had a far greater intensity than Mode 2 (**n** perpendicular to scattering plane). Cooling to the smectic phase the Mode 1 signal reduced by two orders of magnitude, and was composed of two exponentials. This small signal was attributed to scattering off defects localized at the glass plates. The Mode 2 signals in the smectic phase are pure exponentials, and are due to the twist of the SmC planes. An analysis of the angular dependence of this Mode 2 in terms of $B(\theta)$, the elastic constant associated with layer twists of wave vector q along θ gave the ratio $B(\theta)_{max}/B(\theta)_{min} \sim 2.3$, analogous to $k_{33}/k_{22} \sim 2$ in the nematic phase. The elastic constants were typically 1.5 times larger than analogous constants in the nematic phase. An angular analysis of the damping time of this layer twist mode gave a value of viscosity typically an order of magnitude higher than in the nematic phase, though the ratios of viscosities were found to be comparable in both phases (twist/bend viscosity ~6 in nematics). Photon correlation experiments were also carried out by Bancroft [113] and Birtwistle [114] for the SmC phase of the racemic commercial mixture, SCE13R [112] and the viscoelastic ratios for this material were deduced.

Amongst the earliest light scattering experiments on ferroelectric liquid crystals were those of Young et al. [115], on thin free standing films of material. They deduced the parameters K_b/P_0^2, K_s/η_s, and K_b/η_b, where the subscripts b and s refer to splay and bend coefficients, and P_0 is the electric dipole moment per unit area of the material. The films were quantised in thickness and were oriented by a weak electric field. In the presence of a strong electric field [116], absolute measurements were made of the five parameters K_b, K_s, η_b, η_s and P_0 by coupling the molecular polarization with the strong electric field, damping the director fluctuations, and making measurements both with and without the field present. Thin free standing films of the ferroelectric liquid crystal DOBAMBC (n-decylorybenzylidene-N'-amine-2-methylbutylcinnimate) were also studied as a function of temperature by Rosenblatt et al. [117]. Light scattering experiments on a ferroelectric liquid crystal as a function of enantiomeric purity and electric field allowed the determination of the elastic constant B_1 and the associated viscosity of a ferroelectric chiral SmC material [118]. This work found that the elastic constant in fact consists of two parts, a small bare elasticity B_1^0, and a much larger term proportional to P_0^2/q^2 where P_0 is the magnitude of the polarization. The viscosity is independent of the polarization. The light scattering geometry examined only azimuthal modes. The free energy expression used in the work followed that of the Orsay Group and Saupe [119]:

$$F_b(r) = \frac{1}{2}\left\{B_1^0\left(\frac{\partial\varphi}{\partial x}\right)^2 + 2B_{13}\left(\frac{\partial\varphi}{\partial x}\frac{\partial\varphi}{\partial z}\right) + B_3\left(\frac{\partial\varphi}{\partial z}\right)^2\right\}$$
$$+ \frac{1}{2}P_0 E_\varphi^2 \qquad (11)$$

where the B terms are the elastic constants and φ is the azimuthal director rotation. Mode 1 corresponds to a bend-type distortion of the director projection in the smectic plane, and Mode 3 corresponds to a twist-type distortion. An additional component was included in the free energy due to space charge in the system, since for small azimuthal director fluctuations a space charge exists, to find an expression valid provided that no ionic screening occurred. A dynamic equation containing the viscosity allowed the viscoelastic ratio to be determined from the relaxation rate Γ of the correlation function [120]:

$$\Gamma = \frac{1}{\eta}\left[B_0^1 q_x^2 + \frac{(4\pi P_0^2)}{\varepsilon} + P_0 E\right] \qquad (12)$$

The slope of this equation (plotted with respect to E) clearly gives the viscosity η, provided that P_0 is independently known. The system exhibits an effective elasticity, and it's dependence on polarization may be probed using chiral–racemic mixtures of the commercial material, SCE12 [121]. The experimental geometry required the use of transverse electrodes in a cell and therefore a knowledge of the electric field in the strip between the electrodes. It is no longer possible to use the approximation $E = V/d$, where V is the applied voltage and d the device thickness, that is employed for homeotropic devices. In cells with transverse electrodes the electrode gap is typically much larger (~250 µm) than for homeotropic devices ($d \sim 50$ µm) and the electric field magnitude and direction is now a function of position in the electrode gap. However, the field was determined to a good approximation using conformal mapping techniques. Two relaxation processes were observed, one fast with a decay of $\sim 10^4$ Hz, and the second with a decay of between 10 and 100 Hz, attributed to ionic diffusion. The bare elastic constant B_1^0 was found to be $(2.2 \pm 0.5) \times 10^{-8}$ dyne, whereas the effective elastic constant is dominated by the polarization term. A value of 0.036 ± 0.009 P was measured for the viscosity.

Photon correlation spectroscopy has been used very recently [122] to examine the relaxation modes in the antiferroelectric liquid crystal (R)-MHPOBC [4-(1-methylheptyloxy-carbonyl)phenyl 4′-octyloxybiphenyl 4-carboxylate]. Viscoelastic ratios were obtained as a function of temperature from the dispersion relations of the Goldstone modes in the ferrielectric smectic C*, antiferroelectric smectic C* phases. The antiferroelectric smectic I* phase of this material was also examined.

5.5 Pretransitional Light Scattering Studies

5.5.1 The Isotropic to Nematic Phase Transition

Light scattering is a probe of molecular motions, and may be used effectively to study pretransitional phenomena. Stintson et al. [123] studied light scattered by order parameter fluctuations in MBBA. Their data follow the relationship:

$$I \propto (T - T_c^*) \tag{13}$$

as predicted by de Gennes [124] from Landau theory. Here, I is the intensity of the scattered light, T the temperature and T_c^* the temperature at which a second order phase transition would occur. A first order phase transition usually occurs at $T_K = T_c^* + 1$ K. The linear dependence of the above equation required that the order parameter fluctuations are not so large that the mean field approximation fails; the first order phase transition intervenes before this occurs [125]. Most of the work in the pretransitional regime includes estimates of the coherence lengths of ordered groups, as predicted by de Gennes phenomenological theory [126–131]. It is possible to obtain information on local rather than collective molecular motions by the laser frequency shifts in Rayleigh wing scattering [132, 133]. Analysis of the experimental results leads to a quantitative description of phenomenological parameters in de Gennes model. Local molecular motion may be studied in the time domain [134], possibly allowing local molecular associations such as antiferroelectric short range order to be probed.

5.5.2 The Isotropic to Cholesteric or Blue Phase Transition

Light scattering near the cholesteric phase [135] is expected to be very similar to the case of nematic systems (unsurprisingly since the two systems are so similar thermodynamically). However, there is little doubt that the existence of blue phases between the cholesteric and isotropic phase complicates the situation considerably. Early studies on cholesteryl oleyl carbonate [136, 137] are to some extent conflicting, probably due to impurities in the materials. The same material was studied by Bottger et al. [138]; using correlation techniques they showed that in the vicinity of the blue phase of the material the relaxation behavior was well described by stretched exponential decay rather than two exponentials, such as observed in the vicinity of a glass transition. Local orientational ordering similar to that within the blue phase was observed over a correlation length that grows rapidly as the I–BP transition is approached. This is in contrast to the nematic–isotropic phase transition where only the collective reorientation modes are affected. The light scattering properties of chiral systems have been worked out by Hornreich and Shtrikman [139], though most experimental studies have not made full use of these calculations. The results of various experiments [140, 141] gave information on fluctuations of some of the modes, but were not carried out in the most revealing geometry of back scatter. Light scattering measurements using circularly polarized back scattered light [142] in mixtures of CE2 and 7S5 showed that for mixtures of low chirality, fluctuations in the five structural modes expected in the isotropic phase of chiral mixtures are independent and, as predicted by theory the second order transition temperatures diverge from one another as the chirality increases. In highly chiral mixtures, fluctuations associated with the highest second order phase transition temperature deviate from the normal temperature dependence. It seems that the fluctuations in the structural modes are coupled in highly chiral systems, but no theoretical calculations of this phenomenon exist.

5.5.3 Smectic Phase Transitions

It is clear that the transition from a nematic (or cholesteric) to a smectic phase will result in the divergence of certain of the elastic constants. In particular, k_{22} and k_{33} will diverge at the N–SmA phase transition. The type of divergence observed will depend on the nature of the phase transition, which can be either first or second order [143, 144]. The transition is second order if the nematic phase is sufficiently wide, such that the nematic order parameter is saturated at the transition. Both de Gennes [145] and McMillan [146] developed theories of the SmA–N phase transition that have implications for light scattering. The form of the divergence of the twist and bend elastic constants can be written as:

$$k_{22} = k_{22}^0 + \left(\frac{k_B T}{24\pi}\right) q_0^2 \left(\frac{\xi_\perp^2}{\xi_\parallel}\right) = k_{22}^0 + \hat{k}_{22}$$

$$k_{33} = k_{33}^0 + \left(\frac{k_B T}{24\pi}\right) q_0^2 \xi_\parallel = k_{33}^0 + \hat{k}_{33} \quad (14)$$

The diverging part of the constant is included in the final term of the above expressions, containing the coherence lengths ($\xi_\parallel$ and $\xi_\perp$). The temperature dependence of the divergence depends on the model used. Mean field theory (McMillan) predicts that $\xi_\parallel$ and $\xi_\perp$ diverge as $(T-T_C)^{-1/2}$, while de Gennes predicts $(T-T_C)^{-\nu}$ and the critical exponent, $\nu = 0.66$ in the liquid helium analogy. On the smectic side of the transition, B and K diverge as $(T_C-T)^1$ (mean field) and $(T_C-T)^{0.66}$ (helium analogy). However, unlike the superfluid and superconducting problem, the correlation length exponents are weakly anisotropic for materials studied. The work of Hosain et al. [147, 148] recognizes the anisotropy of the de Gennes model and for k_{11} diverging to infinity predicts the exponent associated with splay to be half the twist coefficient exponent. The behaviour of the elastic constant above the nematic to SmC transition was predicted by Chen and Lubensky [149] to vary as ξ^2, in contrast to the $\xi^{3/2}$ for SmA systems.

Experimental studies of pretransitional phenomena in the nematic phase seem somewhat inconclusive. Static studies above the N–SmA transition found the twist coefficient exponent [150, 151] and other values [152] to be too small to agree with theory, though dynamic experiments found pure twist fluctuations to follow mean field theory [153]. For a second order N–SmA transition in CBOOA (p-cyanobenzilidene-p-octyloxyaniline) [154], the twist elastic constant was found to diverge with a critical exponent of 0.66, in agreement with de Gennes prediction. Pretransitional divergence in the splay, twist [155] and bend [156, 157] viscosity and elastic constants has also been observed in 8CB. Transitions to monolayer and bilayer smectics are found to be totally analogous, and it was found that the helium analogy holds only for the nematic phase [158], breaking down for $T < T_C$. It is clear that the origins of the dis-

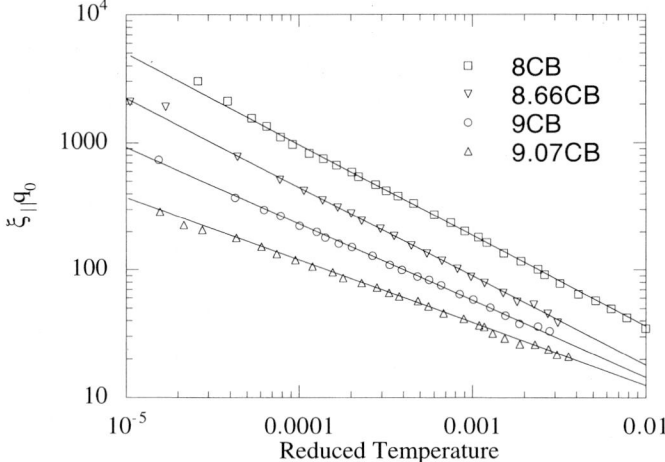

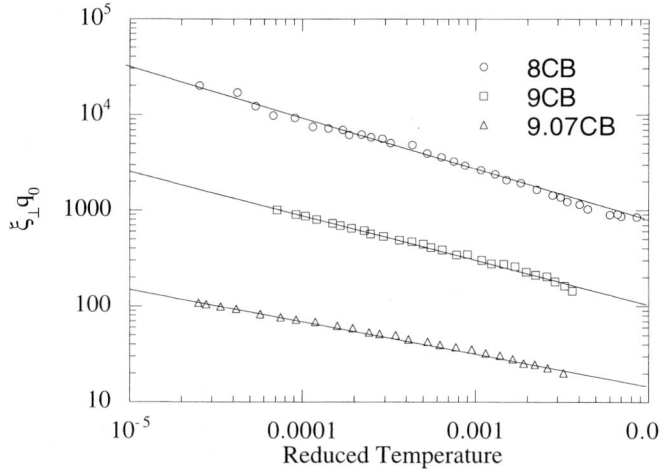

Figure 11. The parallel (a) and perpendicular (b) correlation lengths as a function of reduced temperature determined for 8CB, 9CB and mixtures thereof. In (b) the 8CB and 9CB data are multiplied by 100 and 10 respectively for clarity. (Redrawn from [166]). Non-integral chain lengths are based on mixture composition.

crepancies between theoretical predictions and experiment require further examination.

Experimental studies of the critical behaviour of the layer constant B [159–164] confirm the power law dependence, but again do not yield the expected exponents. In the case of mixtures of 6OCB/8OCB [165] the simple power law was found to apply only to pure 8OCB. A study of the nCB's by Litster et al. [166] found a linear decrease in the parallel correlation length as the tricritical point was approached, Fig. 11. The tricritical point was found to occur at an equivalent chain length of ~9.1 CB, determined from mixtures of the nCBs. An equivalent chain length of $n=9.1$ is a mixture of 90% 9CB plus 10% 10CB. The de Gennes scaling relation [167] between elastic constants and smectic correlation lengths holds as the effective critical exponents become smaller. In plane fluctuations at the SmA–N transition in 8CB and 8OCB have been studied by Mondain–Monval et al. [168], deducing experimental geometries to study the four components. Witanchchi et al. [169] found that the behaviour at the N–SmC transition followed the behaviour predicted by Chen and Lubensky, but near a N–SmA–SmC multicritical point, both layer and tilt fluctuations exist and none of the existing models adequately described the behaviour.

5.6 References

[1] P. Chatelain, *Compt. Rend.* **1944**, *218*, 652.
[2] P. Chatelain, *Compt. Rend.* **1946**, *222*, 229.
[3] P. Chatelain, *Acta Cryst.* **1948**, *1*, 315.
[4] P. Chatelain, *Compt. Rend.* **1947**, *224*, 130.
[5] P. Chatelain, *Compt. Rend.* **1948**, *227*, 136.
[6] P. Chatelain, *Acta Cryst.* **1951**, *4*, 453.
[7] P. Chatelain, *Bull. Soc. Franç. Mineral.* **1954**, *77*, 353.
[8] P. Chatelain, *Bull. Soc. Franç. Mineral.* **1955**, *78*, 262.
[9] P. G. de Gennes, *Compt. Rend.* **1968**, *266*, 15.
[10] F. C. Frank, *Discuss. Faraday Soc.* **1958**, *25*, 19.
[11] C. W. Oseen, *Trans. Faraday Soc.* **1933**, *29*, 883.
[12] P. G. de Gennes, *Mol. Cryst. Liq. Cryst.* **1969**, *7*, 325.
[13] H. Usui, H. Takazoe, A. Fukuda, E. Kuze, *Jpn. J. Appl. Phys.* **1979**, *18*, 1599.
[14] D. Langevin, *Solid State Commun.* **1974**, *14*, 435.
[15] J. P. van der Meulen, R. J. J. Zijlstra, *J. Phys. (Paris)* **1982**, *43*, 411.
[16] D. Langevin, M. A. Bouchiat, *J. Phys. (Paris)* **1975**, *36*, 197.
[17] D. Langevin, Thesis, University of Paris **1974**.
[18] H. Usui, H. Takezoe, A. Fukuda, E. Kuze, *Jpn. J. Appl. Phys.* **1979**, *18*, 1599.
[19] R. Akiyama, Y. Saito, A. Fukuda, E. Kuze, N. Goto, *Jpn. J. Appl. Phys.* **1980**, *19*, 1937.
[20] R. Akiyama, M. Hasegawa, A. Fukuda, E. Kuze, *Jpn. J. Appl. Phys.* **1981**, *20*, 2019.
[21] R. Akiyama, S. Abe, A. Fukuda, E. Kuze, *Jpn. J. Appl. Phys.* **1982**, *21*, L266.
[22] M. Hara, J. Hirakata, T. Toyooka, H. Takezoe, A. Fukuda, *Mol. Cryst. Liq. Cryst.* **1985**, *122*, 161.
[23] R. Akiyama, K. Tomida, A. Fukuda, E. Kuze, *Jpn. J. Appl. Phys.* **1986**, *25*, 769.
[24] M. P. Petrov, P. D. Simova, *J. Phys. D: Appl. Phys.* **1985**, *18*, 239.
[25] H. Hakemi, E. F. Jagodzinski, D. B. DuPre, *J. Chem. Phys.* **1983**, *78*, 1513.
[26] H. Hakemi, E. F. Jagodzinski, D. B. DuPre, *Mol. Cryst. Liq. Cryst.* **1983**, *91*, 129.
[27] H. Hakemi, *Liq. Cryst.* **1989**, *5*, 327.
[28] Orsay Liquid Crystal Group, *Phys. Rev. Lett.* **1969**, *22*, 1361, Errata **1969**, *23*, 208.
[29] F. M. Leslie, *Q. J. Mech. Appl. Math.* **1966**, *19*, 357.
[30] J. L. Ericksen, *Arch. Ration. Mech. Anal.* **1960**, *4*, 231 and **1962**, *9*, 371.
[31] B. Chu, *Laser Light Scattering*, Academic Press, New York, **1974**; P. N. Pusey, J. M. Vaughnan, *Specialist Periodical Reports: Dielectric and Related Molecular Processes, Vol. 2*, The Chemical Society, London, **1972**, 48–105.
[32] W. H. de Jeu, Chapter 7, in *Physical Properties of Liquid Crystalline Materials, Liquid Crystal Monographs, Vol. 1* (Ed. G. W. Gray) Gordon and Breach (New York, London, Paris) **1980**.
[33] M. Miesowicz, *Nature* **1946**, *158*, 27.
[34] G. Durand, L. Leger, F. Rondelez, M. Veyssie, *Phys. Rev. Lett.* **1969**, *22*, 1361.
[35] Orsay Liquid Crystal Group, *J. Phys. Paris*, C-4 **1969**, *30*, 71.
[36] J. L. Martinand, G. Durand, *Solid State Commun.* **1975**, *10*, 815.
[37] F. M. Leslie, C. M. Waters, *Mol. Cryst. Liq. Cryst.* **1975**, *123*, 101.
[38] D. C. van Eck, PhD Thesis, University of Utrecht, **1979**.
[39] Orsay Liquid Crystal Group, *Liquid Crystals and Ordered Fluids, Vol. 1* (Ed. J. F. Johnson, R. S. Porter) Plenum Press, **1970**, New York p. 447.
[40] Orsay Liquid Crystal Group, *Mol. Cryst. Liq. Cryst.* **1971**, *13*, 187.
[41] I. Haller, J. D. Litster, *Phys. Rev. Lett.* **1970**, *25*, 1550.
[42] I. Haller, J. D. Litster, *Mol. Cryst. Liq. Cryst.* **1971**, *12*, 27.
[43] H. Fellner, W. Franklin, S. Christensen, *Phys. Rev. A* **1975**, *11*, 1440.
[44] R. Akiyama, K. Tomida, A. Fukuda, E. Kuze, *Jpn. J. Appl. Phys.* **1986**, *25*, 769.
[45] S. Shaya, H. Yu, *J. Chem. Phys.* **1975**, *63*, 221.
[46] D. C. van Eck, W. Westera, *Mol. Cryst. Liq. Cryst.* **1977**, *38*, 319.
[47] D. C. van Eck, M. Perdeck, *Mol. Cryst. Liq. Cryst.* **1978**, *49*, 39.
[48] D. C. van Eck, R. J. J. Zijlstra, *J. Phys. (Paris)* **1980**, *41*, 351.
[49] J. P. van der Meulen, R. J. J. Zijlstra, *J. Phys. (Paris)* **1982**, *43*, 411.
[50] J. P. van der Meulen, R. J. J. Zijlstra, *J. Phys. (Paris)* **1984**, *45*, 1347.
[51] J. P. van der Meulen, R. J. J. Zijlstra, *J. Phys. (Paris)* **1984**, *45*, 1627.
[52] J. Hirakata, G. Chen, T. Toyooka, S. Kawamoto, H. Takezoe, A. Fukuda, *Jpn. J. Appl. Phys.* **1986**, *25*, L607.
[53] G. Chen, H. Takezoe, A. Fukuda, *Liq. Cryst.* **1989**, *5*, 341.

[54] M. S. Sefton, A. R. Bowdler, H. J. Coles, *Mol. Cryst. Liq. Cryst.* **1985**, *129*, 1.
[55] M. S. Sefton, H. J. Coles, *Polymer* **1985**, *26*, 1319.
[56] H. J. Coles, M. S. Sefton, *Mol. Cryst. Liq. Cryst.* **1988**, *4*, L123.
[57] R. G. Humphreys, H. A. Tarry, M. J. Bradshaw, *Liq. Cryst.* **1988**, *3*, 1039.
[58] D. Gu, A. M. Jamieson, C. Rosenblatt, D. Tomazos, M. Lee, V. Percec, *Macromolecules* **1991**, *24*, 2385.
[59] D. Gu, A. M. Jamieson, M. Kawasumi, M. S. Lee, V. Percec, *Macromolecules* **1992**, *25*, 2151.
[60] D. Gu, S. R. Smith, A. M. Jamieson, M. Lee, V. Percec, *J. de Physique II* **1993**, *3*, 937.
[61] F.-L. Chen, M. A. Jamieson, *Macromolecules* **1993**, *26*, 6576.
[62] H. J. Coles, M. S. Bancroft, *Mol. Cryst. Liq. Cryst.* **1993**, *237*, 97.
[63] F.-L. Chen, A. M. Jamieson, *Macromolecules* **1994**, *27*, 4691.
[64] P. A. Gemmell, G. W. Gray, D. Lacey, *Mol. Cryst. Liq. Cryst.* **1983**, *50*, 205.
[65] G. A. Dilisi, C. Rosenblatt, A. C. Griffin, *J. Phys. II* **1992**, *2*, 1065.
[66] G. A. Dilisi, C. Rosenblatt, A. C. Griffin, U. Hari, *Phys. Rev. A* **1992**, *45*, 5738.
[67] H. J. Coles, M. S. Sefton, *Mol. Cryst. Liq. Cryst.* **1987**, *4*, L123.
[68] H. J. Coles, M. S. Sefton, *Mol. Cryst. Liq. Cryst.* **1987**, *4*, L131.
[69] P. J. R. Birtwistle, PhD Thesis, University of Manchester, Manchester, UK **1995**.
[70] V. G. Taratuta, A. J. Hurd, R. B. Meyer, *Phys. Rev. Lett.* **1985**, *55*, L246.
[71] M. B. L. Santos, G. Durand, *J. Physique* **1986**, *47*, 529.
[72] M. B. L. Santos, W. B. Ferreira, M. A. Amato, *Liq. Cryst.* **1994**, *6*, 287.
[73] C. Versace, V. Formoso, D. Lucchetta, D. Pucci, C. Ferrero, M. Ghedini, R. Bartolino, *J. Chem. Phys.* **1993**, *98*, 8507.
[74] C. S. Perk, M. Copic, R. Mahmood, N. A. Clark, *Liq. Cryst.* **1994**, *16*, 135.
[75] M. Hara, S. Ichikawa, H. Takezoe, A. Fukuda, *Jpn. J. Appl. Phys.* **1984**, *23*, 1420.
[76] M. Hara, H. Tenmei, S. Ichikawa, H. Takezoe, A. Fukuda, *Jpn. J. Appl. Phys.* **1985**, *24*, L777; H. Takezoe, M. Hara, S. Ichikawa, A. Fukuda, *Mol. Cryst. Liq. Cryst.* **1985**, *122*, 169.
[77] M. Hara, H. Takezoe, A. Fukuda, *Jpn. J. Appl. Phys.* **1986**, *25*, 1756.
[78] W. Urbach, H. Hervet, F. Rondelez, *J. Chem. Phys.* **1985**, *83*, 1877.
[79] H. K. M. Vithana, L. Asvaw, D. L. Johnson, *Phys. Rev. Lett.* **1993**, *70*, 3561.
[80] P. A. Pincus, *C. R. Acad. Sci., Ser. B.* **1968**, *267*, 1290.
[81] C. Fan, L. Kramer, J. M. Stephen, *Phys. Rev. A* **1970**, *2*, 2482.

[82] R. W. Duke, Diss. Univ. Louisville, Louisville, Kentucky, **1974**.
[83] R. W. Duke, D. B. DuPre, *Mol. Cryst. Liq. Cryst.* **1977**, *43*, 33.
[84] D. B. DuPre, R. W. Duke, *Phys. Rev. Lett.* **1974**, *33*, 67.
[85] D. B. DuPre, R. W. Duke, *Fed. Proc.* **1974**, *33*, 1210.
[86] M. S. Beevers, D. A. Elliott, G. Williams, *Mol. Cryst. Liq. Cryst.* **1982**, *80*, 135.
[87] R. M. Hornreich, S. Shtrikman, *Phys. Rev. A* **1983**, *28*, 1781.
[88] J. H. Flack, P. P. Crooker, R. C. Svobada, *Phys. Rev. A* **1982**, *26*, 723.
[89] S. Meiboom, M. Sammon, *Phys. Rev. A* **1981**, *24*, 468.
[90] M. A. Marcus, *Phys. Rev. A* **1984**, *30*, 1109.
[91] M. A. Marcus, *Mol. Cryst. Liq. Cryst.* **1985**, *122*, 131.
[92] V. E. Dmitrienko, *Sov. Phys.: JEPT Lett.* **1986**, *43*, 419.
[93] V. A. Belyakov, V. E. Dmitrienko, *Sov. Phys. Usp.* **1985**, *28*, 535.
[94] R. J. Miller, H. F. Gleeson, *Phys. Rev. E* **1995**, *52*, 5011.
[95] R. Ribotta, G. Durand, J. D. Litster, *Solid State Commun.* **1973**, *12*, 27.
[96] R. Ribotta, *Proc. Int. Conf. Light Scattering Solids*, 3rd, M. Balanski, R. C. C. Leite, S. P. S. Porto (eds.), Pub. Flammarion Sci, Paris **1976**, 713.
[97] P. G. de Gennes, *J. Phys. Paris* **1969**, *30*, 65.
[98] M. Delaye, R. Ribotta, G. Durand, *Phys. Lett. A*, **1973**, *44*, 139.
[99] R. Ribotta, D. Salin, G. Durand, *Phys. Rev. Letts.*, **1974**, *32*, 6.
[100] F. Rondelez, H. Birecki, R. Schaetzing, J. D. Litster, *J. Phys. Paris* **1976**, *37*, 122.
[101] N. A. Clark, P. S. Pershan, *Phys. Rev. Lett.* **1973**, *30*, 3.
[102] F. Nallet, D. Roux, J. Prost, *J. Phys. (France)* **1989**, *50*, 3147.
[103] M. Gharbia, T. Othman, A. Gharbi, C. Destrade, G. Durand, *Phys. Lev. Lett.* **1992**, *68*, 2031.
[104] H. Takezoe, M. Hara, S. Ichikawa, A. Fukuda, *Mol. Cryst. Liq. Cryst.* **1985**, *122*, 169.
[105] M. Hara, H. Tenmei, S. Ichikawa, H. Takezoe, A. Fukuda, *Jpn. J. Appl. Phys.* **1985**, *24*, L777
[106] P. G. de Gennes, *J. Phys. Paris, Colloq.* **1969**, *30*, C-4, 65.
[107] F. M. Leslie, I. W. Stewart, M. Nakagawa, *Mol. Cryst. Liq., Cryst.* **1991**, *198*, 443.
[108] Orsay Liquid Crystal Group, *Solid State Commun.* **1971**, *9*, 653.
[109] A. Rapini, *J. Phys. (Paris)* **1972**, *33*, 237.
[110] T. Carlsson, I. W. Stewart, F. M. Leslie, *Liq. Cryst.* **1991**, *9*, 661.
[111] Y. Galerne, J. L. Martinand, G. Durand, M. Veyssie, *Phys. Rev. Lett.* **1972**, *29*, 562.

[112] Merck Ltd., Poole, Dorset, UK.
[113] M. Bancroft, PhD Thesis, Manchester University, UK **1989**.
[114] P. J. R. Birtwistle, PhD Thesis, Manchester University, UK **1995**.
[115] C. Y. Young, R. Pindak, N. A. Clark, R. B. Meyer, *Phys. Rev. Lett.* **1978**, *40*, 773.
[116] C. Rosenblatt, R. Pindak, N. A. Clark, R. B. Meyer, *Phys. Rev. Lett.* **1979**, *42*, 1220.
[117] C. Rosenblatt, R. B. Meyer, R. Pindak, N. A. Clark, *Phys. Rev. A* **1980**, *21*, 140.
[118] Min-Hua Lu, K. A. Crandall, C. Rosenblatt, *Phys. Rev. Lett.* **1992**, *68*, 3575.
[119] Orsay Liquid Crystal Group, *Solid State Commun.* **1971**, *9*, 653; A. Saupe, *Mol. Cryst. Liq. Cryst.* **1969**, *7*, 59.
[120] I. Drevensek, I. Musevic, M. Copic, *Mol. Cryst. Liq. Cryst.* **1991**, *207*, 199.
[121] Merck Ltd., Poole, Dorset, UK.
[122] H. Y. Sun, H. M. Orihara, H. Ishibashi, *J. Phys. Soc. Jpn.* **1993**, *62*, 2066.
[123] T. W. Stinson, J. D. Litster, N. A. Clark, *J. Phys. (Paris)* **1972**, *33*, 69.
[124] P. G. de Gennes, *Mol. Cryst. Liq. Cryst.* **1971**, *12*, 193.
[125] J. Als-Nielsen, R. J. Birgeneau, *Am. J. Phys.* **1977**, *45*, 554.
[126] B. Chu, C. S. Bak, F. L. Lin, *Phys. Rev. Lett.* **1972**, *28*, 1111.
[127] B. Chu, C. S. Bak, F. L. Lin, *J. Chem. Phys.* **1975**, *62*, 798.
[128] E. Gulari, B. Chu, *J. Chem. Phys.* **1975**, *62*, 798.
[129] E. Courtens, G. Koren, *Phys. Rev. Lett.* **1975**, *35*, 1711.
[130] E. Courtens, *J. Chem. Phys.* **1977**, *66*, 3995.
[131] S. M. Arakelian, L. E. Arushanian, Y. S. Chilingarian, Ahurnal Eksperimentalnoi i Teoreticheskoi Fiziki **1981**, *80*, 1186.
[132] N. Amer, Y. S. Lin, Y. R. Shen, *Solid State Commun.* **1975**, *16*, 1157.
[133] G. R. Alms, T. D. Gierke, G. D. Patterson, *J. Chem. Phys.* **1978**, *67*, 5779.
[134] J. R. Lalanne, B. Martin, B. Pouligny, S. Kielich, *Opt. Commun.* **1976**, *19*, 440.
[135] C. C. Yang, *Phys. Rev. Lett.* **1972**, *28*, 955.
[136] D. S. Mahler, P. H. Keyes, W. B. Daniels, *Phys. Rev. Lett.* **1976**, *36*, 491.
[137] T. Harada, P. P. Crooker, *Phys. Rev. Lett.* **1975**, *34*, 1259.
[138] A. Bottger, L. Niessen, J. Coelingh, D. Frenkel, R. J. J. Zijlstra, *Liq. Cryst.* **1988**, *3*, 337.
[139] R. M. Hornreich, S. Shtrikman, *Phys. Rev. A* **1983**, *28*, 1791.
[140] R. Parthasarathy, D. B. DuPre, *Liq. Cryst.* **1988**, *3*, 1231.
[141] H. Zink, W. Van Dael, *Liq. Cryst.* **1989**, *5*, 899.
[142] J. E. Wyse, P. J. Collings, *Phys. Rev. A* **1992**, *45*, 2449.
[143] W. L. McMillan, *Phys. Rev. A* **1971**, *4*, 1238.
[144] K. K. Kobayashi, *Phys. Lett. A* **1970**, *31*, 125; *J. Phys. Soc. Jpn.* **1970**, 29, 101; *Mol. Cryst. Liq. Cryst.* **1971**, *13*, 137.
[145] P. G. de Gennes, *Solid State Commun.* **1972**, *10*, 753.
[146] W. L. McMillan, *Phys. Rev. A* **1972**, *6*, 936.
[147] K. Hossain, J. Swift, *J. Phys.* **1979**, *40*, 1055.
[148] K. Hossain, J. Swift, J.-H. Chen, T. C. Lubansky, *Phys. Rev. B* **1979**, *19*, 432.
[149] J. H. Chen, T. C. Lubensky, *Phys. Rev. A.* **1976**, *14*, 1202.
[150] R. Mahmood, D. Brisbin, I. Khan, C. Gooden, A. Baldwin, D. L. Johnson, M. E. Neubert, *Phys. Rev. Lett.* **1985**, *54*, 1031.
[151] C. Gooden, R. Mahmood, D. Brisbin, A. Baldwin, D. L. Johnson, M. E. Neubert, *Phys. Rev. Lett.* **1985**, *54*, 1035.
[152] S. Sprunt, L. Solomon, J. D. Litster, *Phys. Rev. Lett.* **1984**, *53*, 1923.
[153] R. Mahmood, I. Khan, C. Gooden, A. Baldwin, D. L. Johnson, M. E. Neubert, *Phys. Rev. A* **1985**, *32*, 1286.
[154] M. Delaye, R. Robotta, G. Durand, *Phys. Rev. Lett.* **1973**, *31*, 443.
[155] H. J. Coles, M. S. Sefton, *Mol. Cryst. Liq. Cryst. Lett.* **1987**, *4*, 123.
[156] M. J. Bradshaw, E. P. Raynes, J. D. Bunning, T. E. Faber, *J. Phys. (Paris)* **1985**, *46*, 1513.
[157] P. J. R. Birtwistle, PhD Thesis, University of Manchester, Manchester, UK, **1995**.
[158] H. van Kanel, J. D. Litster, *Phys. Rev. A* **1981**, *23*, 3251.
[159] H. Birecki, R. Schaetzing, F. Rondolez, J. D. Litster, *Phys. Rev. Lett.* **1976**, *36*, 1376.
[160] D. Davidov, C. R. Safinya, M. Kaplan, S. S. Dana, R. Schaetzing, R. J. Birgeneau, J. D. Litster, *Phys. Rev. B* **1979**, *19*, 1657.
[161] J. D. Litster, J. Als-Nielsen, R. J. Birgeneau, S. S. Dana, D. Davidov, F. Garcia-Golding, M. Kaplan, C. R. Safinya, R. Schaetzing, *J. Phys. Colloque* **1979**, *40*, 339.
[162] J. D. Litster, C. W. Garland, K. Lushington, R. Schaetzing, *Mol. Cryst. Liq. Cryst.* **1981**, *63*, 145.
[163] J. D. Litster, *Phil. Trans. R. Soc. London* **1983**, A309, 145.
[164] H.-J. Fromm, *J. Phys.* **1987**, *48*, 647.
[165] M. R. Fisch, L. B. Sorensen, P. S. Pershan, *Phys. Rev. Lett.* **1982**, *48*, 943.
[166] L. Solomon, J. D. Litster, *J. Phys. B.-Condensed Matter* **1988**, *69*, 535.
[167] F. Jahnig, F. Brochard, *J. Phys. (Paris)* **1974**, *35*, 301.
[168] O. Mondain-Monval, H. J. Coles, J. R. Lalanne, J. P. Marcerou, *Phys. Rev. E* **1993**, *47*, 2201.
[169] S. Witanchchi, J. Huang, J. T. Ho, *Phys. Rev. Lett.* **1983**, *50*, 594.

6 Brillouin Scattering from Liquid Crystals

Helen F. Gleeson

With the exception of Raman scattering, light scattering probes matter on a length scale that is long compared with molecular dimensions, and therefore studies collective behaviour. It is a valuable tool to study the hydrodynamic modes of a system in which long lived modes occur with decay times proportional to some power of the wavelength. Behaviour such as this arises from either conservation laws or symmetry. In an isotropic fluid, conservation of momentum, energy and mass leads to five variables with hydrodynamic behaviour (density, temperature and three components of momentum). Whether such modes propagate or decay depends on the system and it should be noted that propagating modes must occur in pairs due to time reversal invariance. In an isotropic fluid, there are two propagating modes (longitudinal sound waves) and three diffusive ones (temperature and two transverse shear waves). In nematic liquid crystals, symmetry considerations show that there are long lived excitations that can be continuously broken, leading to two director modes that decay diffusively and which were discussed in detail in the preceding section of this chapter. In the case of the smectic A phase, there are six hydrodynamic variables; two always propagate (longi-

tudinal sound waves; 1st sound), two always decay diffusively (temperature and shear/transverse waves), and two are transverse waves that may either propagate (2nd sound) or decay diffusively, depending on the orientation of the mode wave vector. Static light scattering from smectic A systems, discussed previously, does not probe the hydrodynamic behaviour of the system. Indeed the first and second sound modes in smectic systems are probed by Brillouin scattering.

Brillouin scattering is inelastic scattering of light from alternate layers of compression and rarefaction produced by phonons in a material. The frequency shift of the Brillouin scattered light is very small (typically $<10^{11}$ Hz) and can therefore be difficult to separate from the Rayleigh peak. The frequency of Brillouin scattered light may be increased or decreased with respect to the central Rayleigh peak, as shown schematically in Fig. 1. Whereas the quasielastic Rayleigh scattering may be measured using techniques including photon correlation spectroscopy, the frequency range 10^8 to 10^{11} Hz is measured using interferometric techniques, usually a Fabry–Perot etalon.

Conservation of energy ($\hbar\omega_B$) and momentum ($\hbar k$) in the Brillouin scattering pro-

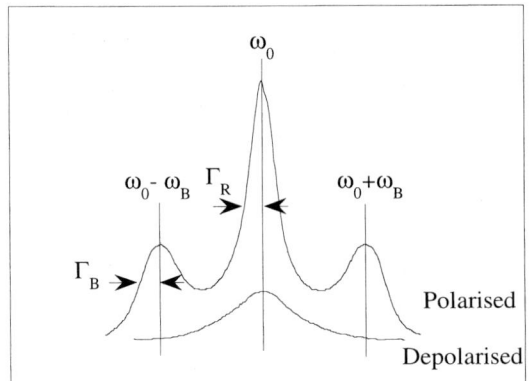

Figure 1. A schematic diagram showing the relative positions of Rayleigh and Brillouin scattering peaks. The incident light has frequency ω_0 and the Brillouin frequency shift is ω_B. The widths of the Brillouin and Rayleigh peaks (Γ_B and Γ_R) are also shown.

cess gives:

$$\hbar k_i - \hbar k_s = \hbar k$$
$$\hbar \omega_i - \hbar \omega_s = \omega_B \quad (1)$$

where k_i and k_s are the incident and scattered wave vectors and ω_i and ω_s the incident and scattered frequencies respectively. As the speed of sound is $v_s = \omega_s/k$, and since $\omega_i \sim \omega_s$, the Brillouin frequency shift is given by:

$$\omega_B = \pm \frac{4\pi n}{\lambda} v_s \sin\left(\frac{\theta}{2}\right) \quad (2)$$

Here, θ is the scattering angle and n the refractive index of the medium. The Brillouin doublet at $\omega_i \pm \omega_B$ is broadened due to acoustic damping of the propagating phonons, and in a non-relaxing medium the broadening is given approximately by:

$$\Gamma_B = \alpha \, v_s \quad (3)$$

The sound attenuation coefficient, α, is k^2 dependent. As already mentioned, light scattering is especially useful in the study of systems where long lived hydrodynamic modes occur with decay times proportional to some power of the wavelength. Here, the conservation of energy and momentum in a liquid decay imply times of the Brillouin scattering proportional to k^2.

6.1 Brillouin Scattering in the Isotropic and Nematic Phases

A number of studies of the velocity and absorption of second sound in liquid crystals have been carried out via Brillouin scattering, and much of the work is reviewed by Schaetzing and Litster [1]. Several studies have been made for the isotropic and pretransitional regime [2–13] of mesogens and for the nematic [14, 15] phase, where the results are essentially the same as for isotropic liquids. Adiabatic sound velocity and attentuation coefficients are determined from the width and shift of the Brillouin peaks. Vaughan [14] observed a discontinuity in sound velocity for the nematic to isotropic transition. Bradberry and Vaughan [16, 17] studied Brillouin scattering from 8CB (octylcyanobiphenyl) at the I–N and N–SmA transitions, as well as within the phases; the results within the smectic phase are discussed below. Hypersound studies of 8OCB (octyloxycyanobiphenyl) were made more recently by Gleed et al. [18], via Brillouin spectroscopy, again over the complete phase range, as shown in Fig. 2. In agreement with Vaughan, a pronounced discontinuity in the sound velocities measured at the I–N transition indicated a strongly first order phase transition.

The isotropic phases of the nCBs (alkylcyanobiphenyls) were also studied by Butler [19] over wide temperature ranges including both pretransitional and normal liquid regions of the fluid. No evidence of pretransitional effects was observed in the Brillouin shifts, even for 12CB which exhibits

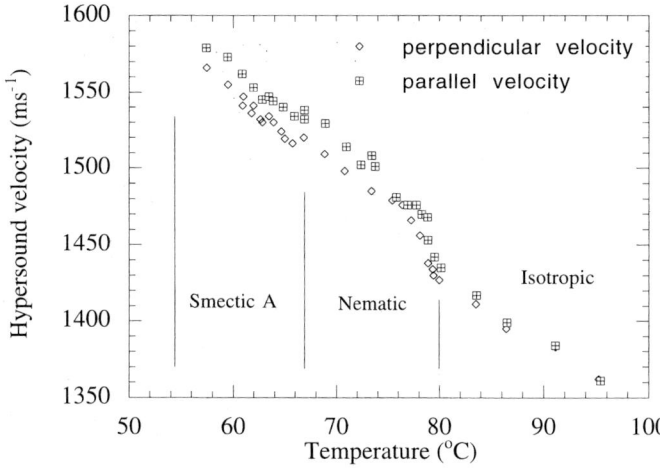

Figure 2. The temperature dependence of the hypersound velocity for different orientations parallel and perpendicular to the director in the SmA, N and isotropic phases of 8OCB. (Adapted from [18]).

a smectic A to isotropic phase transition. Similarly, there was no departure from the normal behaviour in the relaxations of the sound velocity, measured from the angular dependence of ω_B. Pure liquid like behaviour (shift is approximately linear with temperature) is observed at higher temperatures whereas at lower temperatures close to the phase transition, a dispersion is observed probably indicating the occurrence of some molecular association. The nematic to isotropic phase transition was examined in further detail as a function of pressure by Shichijo and Takemura [20], see Fig. 3. The pressure dependence of density and adiabatic compressibility were studied in the nematic materials MBBA [N-(p-(methoxybenzylidene)-p'-butylaniline] and EBBA [N-(p-(ethoxybenzylidene)-p'-butylaniline)]. No pretransitional effects were observed, even at high pressure, and they concluded that 2nd sound does not couple to molecular fluctuations.

Both polarized and depolarized light scattering from the isotropic phase of 6CB have been undertaken recently using a type

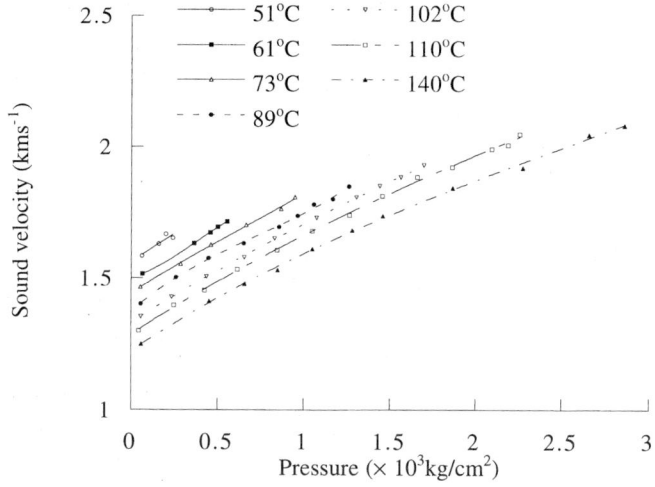

Figure 3. The pressure variation of the sound velocity for MBBA at various temperatures. (From [20]).

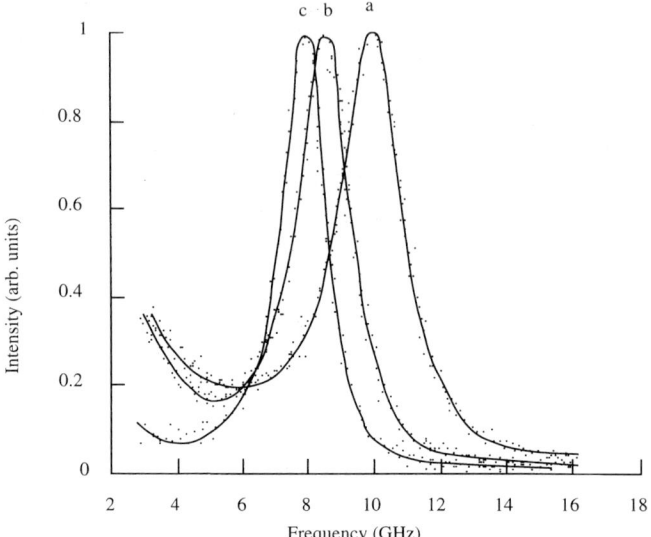

Figure 4. Typical Brillouin spectra showing the angular dependence at three different values of the scattering angle θ. (a) (θ = 7.0°); (b) θ = 48.6°; (c) θ = 83.0°. (Redrawn from [22]).

of optical beating spectroscopy, in which the usual Fabry–Perot etalon used in Brillouin scattering experiments was replaced with a spectrum analyser [21], and a resolution as high as 1 kHz over the range 10^4–10^9 Hz was achieved. The observed Brillouin and Raleigh spectra agree well with the theoretical predictions.

Brillouin scattering from the nematic (racemic) lyotropic polymer poly-γ-benzyl glutamate (PBG) shows a significant anisotropy in both the sound velocity (22%) and the Brillouin linewidth (40%) in the solutions [22], when measurements are made at different angles with respect to the nematic director. Figure 4 shows typical Brillouin spectra of PBG as a function of scattering angle. This work indicates the importance of elastic relaxations due to concentration fluctuations in these nematic polymeric materials.

6.2 Brillouin Scattering in the Cholesteric and Isotropic Phases

A theoretical treatment of Brillouin scattering from cholesteric liquid crystals is given by Parsons and Hayes [23, 24]. Brillouin scattering studies in cholesteric liquid crystals show the scattered light is polarised almost entirely in a plane perpendicular to the scattering plane. As the temperature is increased, the Brillouin peaks become larger and shift towards their related Rayleigh peak. They are shifted to either side of the exciting frequency ω_0 by the frequency of the hypersonic excitation given by:

$$\Omega = (2\omega_0 \, n \, v/c) \sin(\theta/2) \qquad (4)$$

Several studies of Brillouin [25–28] and stimulated Brillouin [29] scattering have been carried out for cholesteryl esters, though the reports are to some extent contradictory, probably due to impurities in the materials. The measurements of threshold power in stimulated Brillouin scattering all show a gain curve that is almost independent of temperature.

6.3 Brillouin Scattering in the Smectic A Phase

Martin et al. [30] showed theoretically that there are two acoustic modes that propagate in the smectic A phase with differing sound velocities. The propagating modes in a smectic A material are the two longitudinal modes that propagate for all orientations of the scattering vector, q (normal sound waves), and two transverse wave modes (second sound) due to changes in interlayer spacing at approximately constant density. The second sound modes propagate provided the scattering vector is neither along nor perpendicular to the director:

1. For q parallel to the director, n_0, the second sound mode becomes two diffusive modes (a shear mode and a permeation mode). The first sound mode is still exhibited in this case.
2. For q perpendicular to the director, n_0, the second sound mode again disappears, becoming a shear mode and a layer undulation mode. In this case the mode plays the same role in the SmA phase as the director Mode 1 does in nematic liquid crystals.

The speeds of the propagating waves are given by:

$$v_1^2 + v_2^2 = (1/\rho)\left\{C_{11} + (C_{33} - C_{11})\cos^2\phi\right\},$$
$$v_1^2 v_2^2 = (1/\rho^2)\left\{C_{11}C_{33} - C_{13}^2\right\}\cos^2\phi \sin^2\phi$$

v_1 and v_2 are the velocity of the 1st and 2nd sound respectively, ρ is the density and ϕ is the angle between the scattering vector and the director. The parameters C_{ij} are the elastic stiffness constants that relate stress and strain in elastic theory. Symmetry considerations in the smectic A phase together with the inability of smectic layers to support shear leave only the components of C_{ij} given in Equation (5). C_{ij} can be related to the elastic constants of de Gennes [31];

$$C_{11} = A_0,$$
$$C_{33} + C_{11} - 2C_{13} = B_0,$$
$$C_{11} - C_{13} = -C_0 \quad (6)$$

where A_0 and B_0 are elastic constants corresponding to changes in volume and interlayer spacing respectively, while C_0 represents the coupling between them. It should be noted that whilst the constant B_0 is analogous to the constant B discussed in Chapter VIII Section 5, it is not exactly the same; the elastic constants for Brillouin scattering are adiabatic rather than isothermal. The second sound velocity (given by $(B/\rho)^{1/2}$) [32] is typically $\sim 10^4$ cm s^{-1}, much less than the velocity of first sound. In ordinary smectics, propagation of 2nd sound is an overdamped mode with a relaxation rate given by:

$$\Gamma_q \approx \frac{k_{11} q_\perp^2}{\rho} \quad (7)$$

The first Brillouin scattering measurements on the smectic A phase were made by Liao et al. [33] who confirmed the predictions of the propagating mode structure in both smectic A and smectic B phases. The first and second sound velocities were measured for racemic β-methylbutyl p-[(p′-methoxybenzylidene)amino]cinnamate, and they showed that the first sound speed is almost isotropic with a minimum in the off-symmetry direction, with the speed of second sound being extremely anisotropic. The data are well described by theory. No 2nd sound measurements were made for $\phi > 45°$, indicating very anisotropic damping that was at-

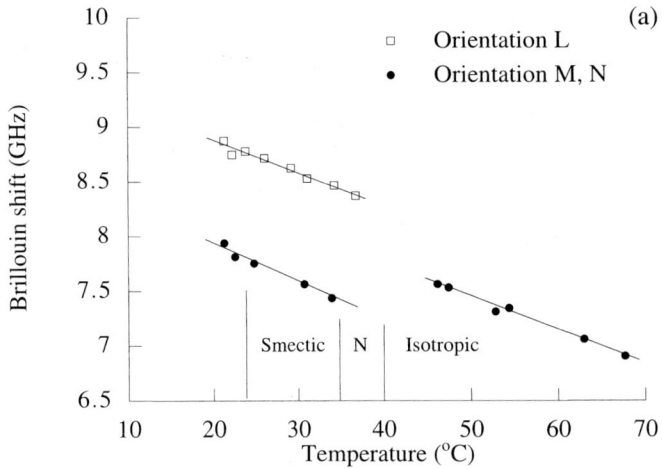

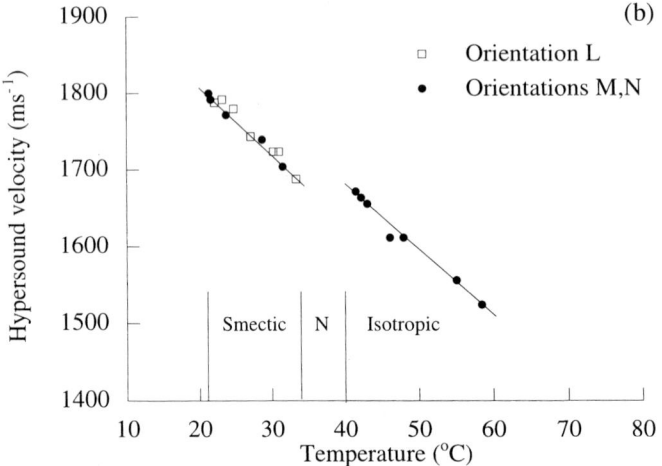

Figure 5. Hypersound velocities (a) and Brillouin frequency shifts (b) as a function of temperature for 8CB, redrawn from [16]. The notation of [16] is also used: L refers to the director perpendicular to k_i and k_s, M refers to the director parallel to k_i and perpendicular to k_s, and N has the director parallel to k_s, and perpendicular to k_i.

tributed [34] to a slow relaxation of the director dominant over the faster relaxation of the 2nd sound. An increase in the damping was predicted [35] for $\phi = 45°$, but this theory also predicts that the 2nd sound should be observed at higher angles. From the measurements of the velocities of 1st and 2nd sound, Liao et al. derived values for the adiabatic elastic constants as a function of temperature. A_0, associated with 1st sound, was continuous across the phase transitions, but B_0 and C_0 were very temperature dependent, and there is some evidence that they disappear at the SmA–N phase transition. Bradberry and Vaughan [16, 17] measured the speeds of 1st and 2nd sound very close to the SmA–N phase transition of 8CB, and whilst the results for A_0 agree with those of Liao et al. B_0 and C_0 were relatively independent of temperature and finite at the transition. The coefficient C_{11} was determined to be $(3.07 \pm 0.27) \times 10^9$ N m^{-2} at 25 °C, and $(2.50 \pm 0.23) \times 10^9$ N m^{-2} at 50 °C. Their data are shown in Fig. 5.

Bradberry and Clark [36] measured the hypersound velocity as a function of temperature in the homologues nCB, in both the nematic and smectic A phases. The measurements were made in the two principal directions, parallel and perpendicular to the director, and the velocities were observed to be asymmetric in all phases. Gleed et al. [18] considered the angular dependence of second sound in the smectic phase of 8OCB, finding it too to be asymmetrical. The discontinuity of the hypersound velocity at the nematic to smectic A phase transition led them to identify it as weakly first order. Brillouin scattering from lyotropic lamellar phases shows that at GHz frequencies the bending rigidity of surfactant layers in a swollen lyotropic lamellar liquid crystal is approximately 10^3 times larger than that found at low frequencies using other techniques [37]. Finally, mention should be made of a recent theoretical study of stimulated Brillouin scattering in smectic A liquid crystals [38]. Expressions are derived for the second sound wave amplitudes, and estimated gains are one to two orders of magnitude higher than those in isotropic liquids.

6.4 Brillouin Scattering in the Smectic C Phase

The number of dissipative coefficients in the smectic C phase is considerably more than for either of the uniaxial smectic A or nematic phases which have only five independent viscosity coefficients. Nonetheless, some similarities can be considered, particularly in the elastic constants of the smectic C and SmA phases. Both exhibit layered structures in which the layers can slide freely over one another. Thus the only uniform stresses that will produce strains are along the layer normal. To a first order, the elastic constants of the two phases are therefore similar, and it is these first order constants that determine the speeds of sound. Thus, the propagating mode structures of the smectic C and smectic A phases should be similar; the 1st sound should be relatively isotropic with an off-axis minimum, whereas the 2nd sound should be asymmetric and zero in the directions that the mode becomes diffusive. There are very few experimental studies of Brillouin scattering in the smectic C phase, though some ultrasonic measurements of 1st and 2nd sound have been made.

New ferroelectric smectic phases formed from achiral polar molecules (denoted SmAF phases) have been recently examined theoretically by Prost et al. [39]. Second sound in such phases is predicted to obey a more complex equation of motion than previously considered, and a new mode is shown to propagate at macroscopic wavelengths. So far, no experimental evaluation of the prediction exists.

6.5 References

[1] R. Schaetzing, J. D. Litster, Light scattering studies of liquid crystals, in *Advances in Liquid Crystals, Vol. 4*, Academic Press, London **1979**.
[2] T. Harada, P. P. Crooker, *Mol. Cryst. Liq. Cryst.* **1977**, *42*, 283.
[3] T. Steiger, J. D. Litster, in *Liquid Crystals and Ordered Fluids* (Ed. J. F. Johnson, R. S. Porter), Plenum, New York, **1973**, p. 671.
[4] N. A. Clark, Y. Liao, *J. Chem. Phys.* **1975**, *63*, 4133.
[5] Y. Liao, N. A. Clark, P. S. Perhan, *Phys. Rev. Lett.* **1973**, *30*, 639.
[6] A. Wergin, W. Krasser, M. Stillen, C. G. B. Frischkorn, *Phys. Rev. A* **1979**, *20*, 1120.
[7] S. Shihijo, T. Takemuru, *Mol. Cryst. Liq. Cryst.* **1974**, *78*, 251.
[8] I. M. Aref'ev, V. N. Biryukov, V. A. Gladkii, S. V. Krivokhizha, I. L. Fabelinskii, I. G. Chistyakov, *Zh. Eksp. Teor. Fiz.* **1972**, *63*, 1729.

[9] A. A. Adkhamov, I. M. Aref'ev, B. S. Umarov, *Dokl. Acad. Nauk Tadzh. SSR,* **1977**, *20*, 22.
[10] A. B. Aleinikov, G. P. Petrova, *Opt. Spektrosk.* **1977**, *43*, 267.
[11] M. Copic, B. B. Lavrencic, *J. Phys. (Paris)* **1975**, *36*, 89.
[12] G. P. Petrova, R. M. Bashirova, E. I. Koshel'nik, *Opt. Spektrosk.* **1975**, *39*, 399.
[13] T. R. Steger Jr., J. D. Litster, *Liq. Cryst. Ord. Fluids.* **1977**, *2*, 671.
[14] J. M. Vaughan, *Phys. Lett. A* **1976**, *58*, 325.
[15] G. P. Petrova, Y. M. Petrusevich, *Khimicheskaya Fizika* **1994**, *13*, 68.
[16] G. W. Bradberry, J. M. Vaughan, *J. Phys. C* **1976**, *9*, 3905.
[17] G. W. Bradberry, J. M. Vaughan, *Phys. Lett. A* **1977**, *62*, 225.
[18] D. G. Gleed, J. R. Sambles, G. W. Bradberry, *Phys. Lett. A* **1989**, *134*, 440.
[19] I. Butler, PhD Thesis, University of Manchester, Manchester, UK, **1984**.
[20] S. Shichijo, T. Takemura, *Mol. Cryst. Liq. Cryst.* **1981**, *78*, 251.
[21] T. Matsuoka, K. Sakai, K. Takagi, *Phys. Rev. Lett.* **1993**, *71*, 1510.
[22] Z. Hu, X. Wen, J. J. Wanderwal, X. Ao, D. Walton, R. B. Meyer, *J. Chem. Phys.* **1992**, *97*, 568.
[23] J. D. Parsons, C. F. Hayes, *Solid State Commun.* **1974**, *15*, 429.
[24] J. D. Parsons, C. F. Hayes, *Mol. Cryst. Liq. Cryst.* **1974**, *29*, 429.
[25] G. Durand, D. V. L. G. N. Rao, *Bull. Am. Phys. Soc.* **1967**, *12*, 1054.
[26] G. Durand, D. V. L. G. N. Rao, *Phys. Lett. A* **1968**, *27*, 455.
[27] W. A. Nordland, *J. Appl. Phys.* **1968**, *39*, 5033.
[28] H. Rosen, Y. R. Shen, *Mol. Cryst. Liq. Cryst.* **1972**, *18*, 285.
[29] D. V. G. L. N. Rao, *Phys. Lett. A* **1970**, *32*, 533.
[30] P. C. Martin, O. Parodi, P. S. Pershan, *Phys. Rev. A* **1972**, *6*, 2401.
[31] P. G. de Gennes, *J. Phys. (Paris)* **1969**, *30*, 65.
[32] L. Ricard, J. Prost, *J. Phys. (Paris)* **1979**, *40*, 83.
[33] Y. Liao, N. A. Clark, P. S. Pershan, *Phys. Rev. Lett.* **1973**, *30*, 639.
[34] F. Brochard, *Phys. Lett. A* **1974**, *49*, 315.
[35] F. Jahnig, *J. Phys. (Paris)* **1975**, *36*, 315.
[36] G. W. Bradberry, C. F. Clark, *Phys. Lett. A* **1983**, *95*, 305.
[37] S. Mangalampalli, N. A. Clark, J. F. N. A. Scott, *Phys. Rev. Lett.* **1991**, *67*, 2303.
[38] G. F. Kventsel, B. I. Lembrikov, *Liq. Cryst.* **1994**, *16*, 159.
[39] J. Prost, R. Bruinsma, F. Tournilhac, *J. Physique II* **1994**, *4*, 169.

7 Mössbauer Studies of Liquid Crystals

Helen F. Gleeson

Mössbauer investigations of materials consider the recoil free emission or resonant absorption of nuclear gamma rays. The Mössbauer effect [1] relies on the fact that some of the energies associated with nuclear events are relatively small, often no larger than chemical binding energies (~1–10 eV), or, more pertinent to liquid crystals, the energies characteristic of lattice vibrations (10^{-2}–10^{-1} eV). In particular, the energies associated with the recoil imparted to a nucleus by the emission of a low energy gamma ray falls into the appropriate regime.

Prior to the work of Mössbauer, the kinematics of gamma ray emission were usually studied with respect to free atoms. Consideration of the conservation of energy and momentum before and after emission of the gamma ray by an excited nucleus in a free atom shows that the recoil velocity of the atoms is independent of its initial velocity. Further, the fraction of available energy loss to the atom is extremely small, typically 1 part in 10^7, as the emitted gamma ray has most of the energy lost by the excited nucleus. However, the small recoil energy of the atom becomes significant when it is compared with the uncertainty in the energy of the gamma ray. The line width of the gamma radiation (defined by the half life of the excited state leading to its emission) is usually ~10^{-9} eV, that is very much smaller than the energy lost in nuclear recoil. Thus the absorption and emission lines do not overlap and nuclear resonance absorption is not observable. (This is in contrast to the case of atomic radiation where the energy emitted is typically 10^4 times smaller and the line widths are comparable to the nuclear case. As the recoil energy is smaller than the line width, resonant absorption of light is readily observable.) A radioactive source may be brought into the resonant scattering condition if it is rapidly spun. Then, the energy E of radiation emitted by the source moving towards an observer at a velocity V is Doppler increased by an amount EV/c, where c is the velocity of light. The velocity required to bring the system into the resonant scattering condition is just twice the recoil velocity, E/Mc, where M is the mass of the isotope emitting the gamma ray. Such velocities may be acquired in an ultracentrifuge and are thus experimentally attainable.

In considering the Mössbauer effect in the solid state or in condensed matter, three cases should be discussed.

(i) If the free atom recoil energy is large with respect to the binding energy of the atom in a solid, it will be dislodged from its lattice site. The minimum energy required to dislodge an atom is usually 15–30 eV, and in this case the free atom analysis is applicable.

(ii) If the free atom recoil energy is larger than the characteristic energy of the lattice vibrations (phonon energy), but less than the displacement energy, the atom will remain in its site and will dissipate the recoil energy thermally.

(iii) If the recoil energy is less than the phonon energy, the quantization of the lattice must be considered, and the energy transfer is thus quantized. The emission of the gamma ray is accompanied by the transfer of multiples of phonon energy, $(0, \pm \hbar\omega, \pm 2\hbar\omega, \ldots)$ to the lattice.

When the average is taken over many emission processes, the energy transferred per event is exactly the free atom recoil energy. The fraction of events f that take place without lattice excitation, provided the recoil energy E_R is much less than $\hbar\omega$ (so that two-quantum transitions can be neglected) is given by:

$$f = 1 - E_R/\hbar\omega \tag{1}$$

where f is often known as the Mössbauer coefficient since only these events give rise to the Mössbauer effect. The fraction of the elastic (zero phonon) processes is given by:

$$f = \exp\{-\kappa^2 \langle x^2 \rangle\} \tag{2}$$

where $\kappa = 2\pi/\lambda = E/\hbar c$ and $\langle x^2 \rangle$ is the component of the mean square vibrational amplitude of the emitting nucleus in the direction of the gamma ray. To obtain $f \sim 1$, $\kappa^2 \langle x^2 \rangle$ must be much less than 1 (i.e. the displacement of the nucleus must be small compared to the gamma ray wavelength). Information may be obtained from the recoil free gamma ray line since the proportion of the spectrum in the recoil free line depends on the ratio of $\langle x^2 \rangle$ to the wavelength. This will increase at low temperatures, and no recoil free component should occur in liquids where $\langle x^2 \rangle$ is unbounded as the atomic motion is associated with self diffusion. However, because $\langle x^2 \rangle$ must be averaged over a nuclear lifetime, it is possible that it remains sufficiently small in liquids to allow detection of a recoil free event. It is clearly more readily observable in glasses or viscous liquids. In an anisotropic system where the amplitude of motion is greatest along the axis of weakest binding, the recoil free fraction will be dependent on the direction of emission of the gamma ray.

Most of the Mössbauer investigations of liquid crystals have been in the higher order smectic phases or quenched glassy nematic or cholesteric states. Theoretical aspects of the intensity asymmetry of ^{57}Fe quadrupole split lines in smectic or frozen nematic solutions have been considered by Wilson and Uhrich [2] and Kaplan and Glasser [3]. Mössbauer studies can allow the evaluation of self diffusion coefficients, quadrupolar splitting, tilt angles and order parameters. Further, an oriented smectic liquid crystal in a magnetic field can serve as a matrix which cooperatively orients the solute, creating a monocrystal of the solute and allowing studies of the solute material in a specific geometry to be performed.

Mössbauer studies of glassy nematic [4–6], cholesteric [7, 8] and smectic [9–12] liquid crystals allow the determination of characteristic Debye temperatures, order parameters, intramolecular and lattice contributions to the nuclear vibrational anisotropy, and molecular orientation [13] in the liquid crystal states. For example, Bekeshev et al. have shown from $\ln f'$ versus temperature data at very low temperatures that MBBA [N-(p-methoxybenzylidene)-p-butylaniline] and some liquid-crystalline disubstituted derivatives of ferrocene have weak intermolecular bonds that result in the appearance of additional degrees of freedom above ~250 K. Measurements made of 1,1′-diacetylferrocene (DAF) in the supercooled smectic A glass state of 4-n-octyloxyphenyl 4-n-nonylbenzoate [14] allowed the determination of Debye temperatures that were lower in the smectic glass phase than in the crystalline state of the material. Similar measurements were made in the glassy smectic C phase of 7S5 [15]. Some Mössbauer studies have also been made on the nematic and smectic phases of materials. Mössbauer investigations were made on the smectic and nematic phases of 4,4′-diheptyloxyazobenzene at temperatures between 70 °C and 100 °C and used to deduce the angle of the molecules in the smectic layers to be $(40 \pm 15)°$ in a sample aligned in a magnetic field [16], in good agreement with ESR and X-ray measurements.

Many Mössbauer spectroscopy experiments have been carried out on the glassy smectic B phase [17–20]. For example ^{57}Fe was used to study a SmB phase using two different solute molecules [21]; ferrocenyl-4′-methoxyaniline (FMA) and ferrocenyl-4′-n-butylaniline (FBA). This work determined order parameters for the different probe molecules of $S = 0.28$ (FMA) and

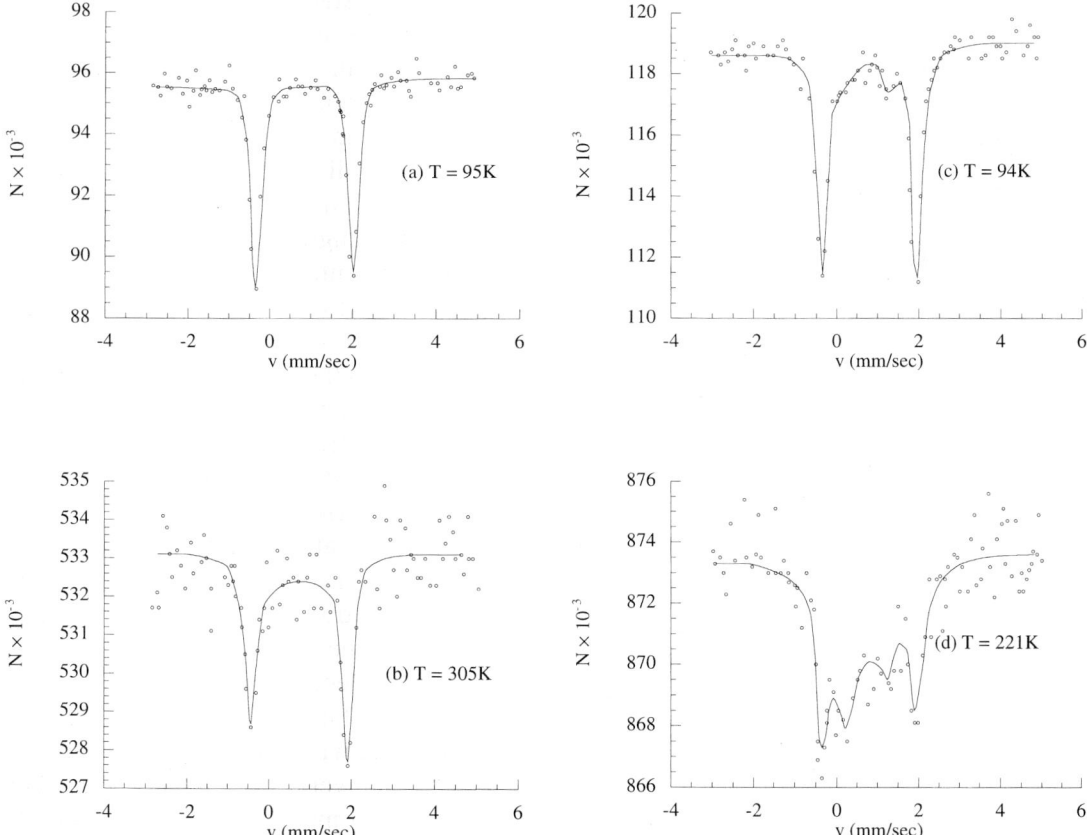

Figure 1. Mössbauer spectra of (a) CBO-II-PM+F @ 95 K, (b) CBO-II-PM+F @ 305 K, (c) CBO-II-PM+FA @ 94 K, and (d) CBO-II-PM+FA @ 221 K. Adapted from [25] using the notation of Kosova et al. (F and FA refer to the label molecules ferrocene and ferrocenealdehyde respectively, while the polymer CBO-II-PM is poly{1-[II-(4′-cyano-4-biphenyloxy)undecyloxycarbonyl]ethylene}).

$S = 0.24$ (FBA) at 110 K. Measurements of f versus temperature were made and Debye temperatures found to be ~23 K lower in the SmB glass than the crystalline phase of the material. A small anisotropy in the vibrational modes of the SmB phase was also reported from angular Mössbauer measurements. Most of the published work has used ^{57}Fe, though the Mössbauer effect of tin bearing solute molecules has also been studied [22].

The tilt angle and lattice contribution to the vibrational anisotropy in an H-phase of a mesogen was studied via the anisotropy of the recoil free fraction [23]. Other Mössbauer studies examined anisotropic diffusion at a glass transition involving an H phase to supercooled H phase transition temperature [24].

Studies of molecular mobility in various microregions of polymeric liquid crystals were made by Kosova et al. [25]. Estimates of characteristic Debye temperatures and mean square vibrational amplitudes of the label molecules ferrocene (F) and ferrocene aldehyde (FA) were used to determine the rigidity of regions in both the vitreous and liquid crystal phases of poly{1-[2-(4'-cyano-4-biphenyloxy)undecyloxycarbonyl]ethylene} (CBO-II-PM). Mössbauer spectra obtained in these studies are shown in Fig. 1.

References

[1] A good introduction to the Mössbauer effect may be found in *Mössbauer Effect: Principles and Applications* by G. K. Wertheim, Academic Press, New York, **1971**.
[2] J. M. Wilson, D. L. Uhrich, *Mol. Cryst. Liq. Cryst.* **1971**, *13*, 85.
[3] J. I. Kaplan, M. L. Glasser, *Mol. Cryst. Liq. Cryst.* **1970**, *11*, 103.
[4] M. J. Potasek, E. Muenck, J. L. Groves, P. G. Debrunner, *Chem. Phys. Lett.* **1972**, *15*, 55.
[5] V. G. Bekeshev, V. Y. Rochev, E. F. Makarov, *Hyperfine Interactions* **1991**, *67*, 661.
[6] M. C. Kandpal, V. G. Bhide, *Physica B & C*, **1982**, *112*, 57.
[7] V. I. Gol'danskii, O. P. Kevdin, N. K. Kivrina, E. F. Makarov, V. Ya, V. Y. Rochev, R. A. Stukan, I. G. Chistyakov, L. S. Shabyshev, *Dokl. Akad. Nauk. SSSR* **1973**, *209*, 1139.
[8] V. I. Gol'danskii, O. P. Kevdin, N. K. Kivrina, E. F. Makarov, V. Ya, V. Y. Rochev, R. A. Stukan, I. G. Chistyakov, *Int. Conf. Liq. Cryst., Bangalore,* Abstr. No. 11, **1973**.
[9] R. E. Detjen, Diss. Kent State Univ. Kent, Ohio, US, **1973**.
[10] V. I. Gol'danskii, O. P. Kevdin, N. K. Kivrina, E. M. Makarov, V. Y. Rochev, R. A. Stukan, *Sb. Dokl. Vses. Nauch. Konf. Zhidk. Krist. Simp. Ikh. Prakt. Primen,* 2nd, 63, **1972**.
[11] V. I. Gol'danskii, O. P. Kevdin, N. K. Kivrina, E. F. Makarov, V. Y. Rochev, V. Ya, R. A. Stukan, *Zh. Eksp. Teor. Fiz.* **1972**, *63*, 2323.
[12] V. G. Bekeshev, V. Y. Rochev, E. F. Makarov, *Khimicheskaya Fizika,* **1993**, *12*, 811.
[13] M. J. Potasek, P. G. Debrunner, G. DePasquali, *Phys. Rev. A* **1976**, *13*, 1605.
[14] D. L. Uhrich, R. P. Marande, *Mol. Cryst. Liq. Cryst.* **1986**, *133*, 97.
[15] D. L. Uhrich, R. P. Marande, *Mol. Cryst. Liq. Cryst.* **1984**, *111*, 171.
[16] V. I. Gol'danskii, O. P. Kevdin, N. K. Kivrina, V. Y. Rochev, R. A. Stukan, I. G. Chistyakov, L. S. Shabishev, *Mol Cryst. Liq. Cryst.* **1973**, *24*, 239.
[17] V. O. Aimiuwu, D. L. Uhrich, *Bull. Am. Phys. Soc.* **1975**, *20*, 887.
[18] V. G. Bekeshev, V. Y. Rochev, E. F. Makarov, *Khimicheskaya Fizika* **1990**, *9*, 11.
[19] R. P. Marande, *Phys. Rev. B* **1989**, *39*, 46.
[20] V. G. Bekeshev, V. Y. Rochev, E. F. Makarov, *Mol. Cryst. Liq. Cryst.* **1990**, *192*, 131.
[21] R. P. Marande, D. L. Uhrich, *Mol. Cryst. Liq. Cryst.* **1988**, *159*, 243.
[22] D. L. Uhrich, V. O. Aimiuwu, P. I. Ktorides, W. L. LaPrice, *Phys. Rev. A* **1975**, *12*, 211.
[23] D. L. Uhrich, J. Stroh, R. D'Sidocky, D. L. Fishel, *Chem. Phys. Lett.* **1974**, *24*, 539.
[24] R. E. Detjen, D. L. Uhrich, *Mössbauer Eff. Methodol* **1974**, *9*, 113.
[25] G. N. Kosova, V. Y. Rochev, S. G. Kostromin, V. P. Shibayev, *European Polymer Journal,* **1986**, *22*, 17.

Chapter IX
Applications

1 Displays

Ian C. Sage

1.1 Introduction

The development of electro-optic displays has provided a major impetus to all aspects of the study of calamitic thermotropic liquid crystals [1]. It is exclusively this class of materials that has been exploited in displays, although many of the electrical and optical phenomena exploited have direct counterparts in, for example, discotic systems [2]. In practical terms, at the time of writing, only three or four liquid crystal display effects are commercially important. Many more are of interest from the viewpoint of the historical development of the technology, as displays that may become of widespread importance in the future, as effects with some importance for special applications, or simply as illustrations of the range and generality of effects available from liquid crystal devices. An overview is given of the range of display effects that are known, bringing out the underlying relationships of the different devices. More detailed discussions of the important displays are presented in Chaps. III and VI of Vol. 2 of this Handbook.

First we consider the important classes of display which are based on light guiding, optical absorption or variable birefringence effects in nematic liquid crystal layers. These include the twisted nematic and supertwisted nematic devices that dominate the commercial production of displays, as well as dichroic dyed displays and other less well exploited electro-optic effects. The use of optical scattering effects in display devices is considered next, providing a diverse range of electro-optic devices. Finally, the use of ferroelectric liquid crystal phases in providing display effects is described, with the emphasis being placed on the use of ferroelectric layers as electrically rotatable retardation layers.

1.2 Display Construction

With a few exceptions, which are noted in later sections of this chapter, the construction of the different types of liquid crystal display cell is generally similar [3]. The liquid crystal material is contained between two sheets of glass, held parallel and spaced apart by a distance of a few micrometres. The glass is coated on the inner surfaces with a succession of thin optically transparent layers that serve specific purposes in the operation of the device. Those most commonly applied include, working successively away from the glass surface:

- A barrier layer, typically of silica, the purpose of which is to prevent contaminants leaching from the glass and changing the properties of either the alignment layer or the liquid crystal.
- Most colour displays have a subtractive filter array deposited on one sheet of glass at this point. The filter layer is typically fabricated from a dyed polymer layer or photographic emulsion and is photolithographically defined so that red, green and blue dots or stripes are in register with the pixel pattern of the panel.
- An electrically conductive layer, usually of indium tin oxide (ITO). This layer is patterned by photolithography to define the areas that can be activated on the finished display.
- An insulating layer is often applied above the electrodes in order to avoid short circuits across the cell caused by small conductive particles and to reduce the possibility of electrochemical degradation of the liquid crystal itself. A thin layer minimizes the AC voltage drop and maintains a low operating voltage in the display.
- An alignment material [4] forms the final layer on the glass and is in direct contact with the liquid crystal. This layer defines the direction along which the director of the liquid crystal in contact with it, lies. In most commercial displays, the alignment layer is composed of a polyimide, which is mechanically rubbed in a single direction with a cloth to impart a unique directionality to its surface. This type of layer will align a (nematic) liquid crystal with its director along the rubbing direction and almost parallel to the plane of the glass, but tilted out of this plane by a few degrees in the direction of the rubbing treatment. Many other alignment methods and materials are known, and some are noted as appropriate in the paragraphs below.

The glass plates are fastened and sealed round their edges with an adhesive; a gap left in the edge seal allows the liquid crystal to be introduced and is plugged afterwards. Small, uniformly sized spacer particles may be added to the edge seal material and/or dispersed over the area of the display to maintain a constant spacing.

The glass plates are offset at the edge to allow electrical contact to be made to the ITO electrodes, which exit there. Contact to simple displays is often made through elastomeric strips containing alternate layers of insulating and conductive polymer. Complex displays requiring many connections usually use an adhesive loaded with conductive particles to connect directly to a flexible printed circuit board. In order to achieve maximum economy in manufacture, most display cells are fabricated *en masse* on sheets of glass 14 inches square or larger. The individual cells are separated by cutting up the laminate before the liquid crystal is filled in.

After assembly of the cell, most displays require polarizing filters to be laminated onto each side to make the electro-optic effect visible. These filters are usually attached by pressure-sensitive adhesive, and one may incorporate a reflector or transflector according to the mode in which the device is to be used. In most cases, the liquid crystal panel is assembled into a module, which incorporates the drive and decoder electronics together with a backlight, before it is sold.

1.3 Quasi-Fréedericksz Effect Displays in Nematic Liquid Crystals

1.3.1 The Fréedericksz Transition

Consider first a nematic liquid crystal of positive dielectric anisotropy, contained between two parallel conductive glass plates. The glass plates are treated on their inner surfaces to impose a uniform planar alignment without pretilt, and in parallel directions on each plate. This alignment can be obtained by evaporation of various inorganic materials onto the glass from a slightly oblique angle of incidence. If a (AC or DC) voltage is applied between the plates, there is a tendency for the liquid crystal axis having the larger dielectric constant to align parallel to the electric field. However, because of the zero surface tilt, there is no torque on the director and no reorientation occurs at low applied voltages. As the voltage is increased, however, a critical point is reached at which the zero tilt state becomes unstable, and a finite tilt develops at the midplane of the cell [5]. Further increase of the voltage applied to the cell will bring about a corresponding increase in the midplane tilt, until at high voltages it approaches 90°. The voltage V_0 at which the first distortion of the director occurs is given by the expression:

$$V_0 = \pi \sqrt{\frac{k_{11}}{\varepsilon_0 \cdot \Delta \varepsilon}} \quad (1)$$

Note that, apart from the expected dependence on the dielectric anisotropy $\Delta\varepsilon$, the threshold voltage is independent of the cell thickness and depends only on the splay elastic constant k_{11}. This elastic constant dependence may easily be rationalized by the observation that the initial distortion in the Fréedericksz cell is a pure splay of the director (Fig. 1). At higher applied voltages, the director distortion also contains an important component of bend. Therefore, the slope of the tilt versus voltage curve depends on the elastic constant ratio k_{33}/k_{11} [6]. A small value of this ratio makes large magnitude distortions, which are mainly bend of the director, relatively easy compared to that at the threshold, and therefore makes the gradient steeper. The effect is shown in Fig. 2, which compares the tilt versus voltage profile of the commercial nematic liquid crystal mixture E7 [7] with that of a fictitious mixture LC A, having identical properties apart from a reduced k_{33}, sufficient to provide a k_{33}/k_{11} ratio of 0.8. Note that, in agreement with the discussion above, the threshold voltage is unchanged by the alteration in k_{33}.

A further influence on the form of the tilt versus voltage curve of a Fréedericksz cell is illustrated by the curve corresponding to LC B in Fig. 2. This fictitious mixture has properties similar to those of LC A, but the dielectric constant ratio $\Delta\varepsilon/\varepsilon_\perp$ has also been adjusted to a low value of 0.3. In contrast to the cases of E7 or LC A, in this material the effective dielectric constant is barely affected by the tilt angle of the director. This leads

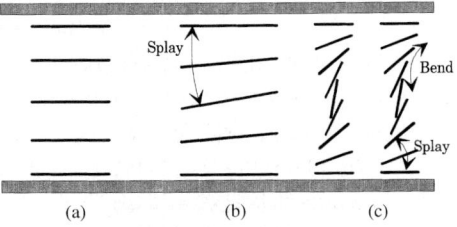

Figure 1. Director configuration and associated elastic distortion in a Fréedericksz cell below (a), slightly above (b), considerably above (c) the threshold voltage.

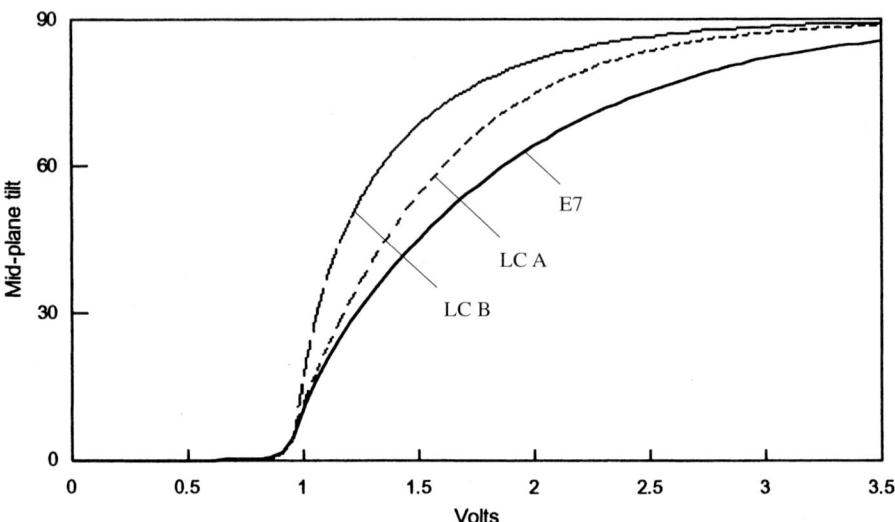

Figure 2. Tilt angle versus voltage profiles for Fréedericksz cells containing E7, and two fictitious liquid crystal (LC) materials. LC A has the k_{33}/k_{11} ratio reduced from 2.7 to 0.8; LC B also has the dielectric constant ratio $\delta\varepsilon/\varepsilon_\perp$ reduced to 0.3.

to a steeper threshold curve compared to the other materials, in which tilt of the director is accompanied by a change in the effective local dielectric constant and redistribution of the electric field in the cell, which is now strongest near to the aligning layers where the tilt angle is smallest.

At voltages above threshold, the director tilt is a maximum at the midplane of the cell, and is lower close to the cell walls. For most purposes, and using ordinary alignment methods, it can be assumed that the liquid crystal director in contact with the cell wall does not change its direction under an applied field. When the midplane tilt angle approaches 90°, further increase in the voltage applied to the cell results in the highly tilted region extending out to fill an increasing proportion of the cell thickness; the distance over which the change in orientation of the director occurs is in inverse proportion to the strength of the applied field [8]. The form of the tilt profile through the cell thickness at different applied voltages is illustrated in Fig. 3.

An important point about the Fréedericksz transition described above is that the director tilt can occur either to left or right, and the two deformations are degenerate in energy. The degeneracy can be lifted by imposing a small pretilt at the aligning surfaces; the transition then leads to tilt developing exclusively in one direction. A consequence of the finite pretilt in the cell is that a torque is experienced by the liquid crystal director even at the smallest applied voltages. This means that director deformation also begins at arbitrarily low voltages. In practical terms, the Fréedericksz threshold voltage remains important as the point about which the tilt begins to increase rapidly.

All of the above discussion has its direct counterpart in the bend Fréedericksz transition (Fig. 4). In this case, a nematic liquid crystal of negative dielectric anisotropy is confined between surfaces which impose homeotropic alignment, such as glass plates treated with an amine or silane surface active agent. The threshold voltage now de-

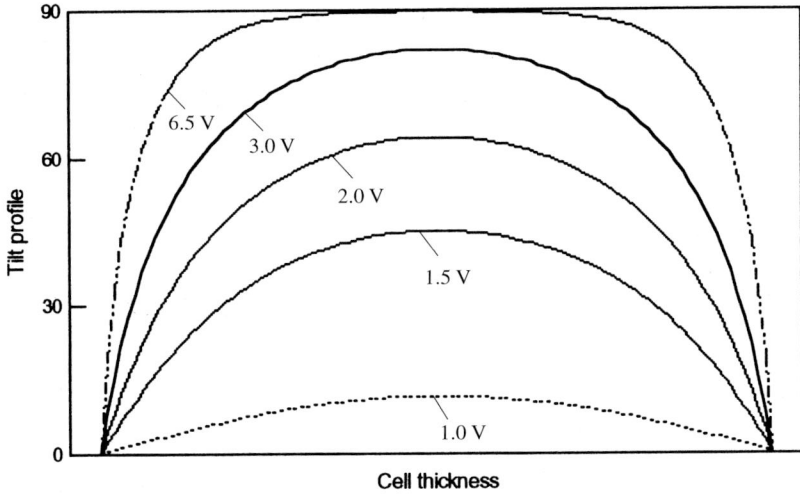

Figure 3. The director tilt in a Fréedericksz cell as a function of applied voltage.

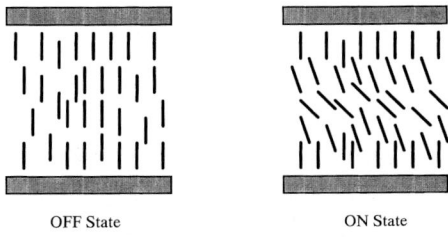

Figure 4. The bend Fréedericksz transition induced in a nematic fluid of negative dielectric anisotropy, with initially homeotropic alignment.

pends on k_{33}, and the steepness of the tilt versus voltage curve is maximized by small values of k_{11}/k_{33} and $|\Delta\varepsilon|/\varepsilon_\parallel$.

1.3.2 Transitions in Twisted Nematic Layers

If one containing surface of a splay Fréedericksz cell of the type described above is turned through an angle of up to 90° in the plane of the cell, the nematic director can be made to sustain a twisted structure; the director remains anchored to the aligning surfaces at their respective new orientations and forms a uniform helix in between. Attempts to generate a twist greater than 90° in this way lead to a relaxation to a lower twist, lower energy state through the generation of disclinations. The behaviour of the tilt versus voltage profile in such a twisted cell is very similar to that in a Fréedericksz device; but, as the applied voltage is increased, the twist profile changes in a way that at first appears unexpected. At zero and low applied voltage, the twist is uniform through the cell, but as the voltage rises above the threshold, it is increasingly concentrated into a narrower layer in the centre of the cell (Fig. 5). This effect is simply a consequence of the increasing tilt in this region of the cell; as the tilt increases less energy is required to sustain a simultaneous twist deformation. (In the limit where the tilt reaches 90° in any part of the cell, the twist angle at that point loses any meaning, and no torque can be sustained.) The importance of this redistribution of the total cell twist in determining the optical properties of the device above threshold is discussed below. Twisted nematic cells with a twist angle near 90° [9] form the mainstay of commercial production of liquid crystal displays for watches, calculators and similar low in-

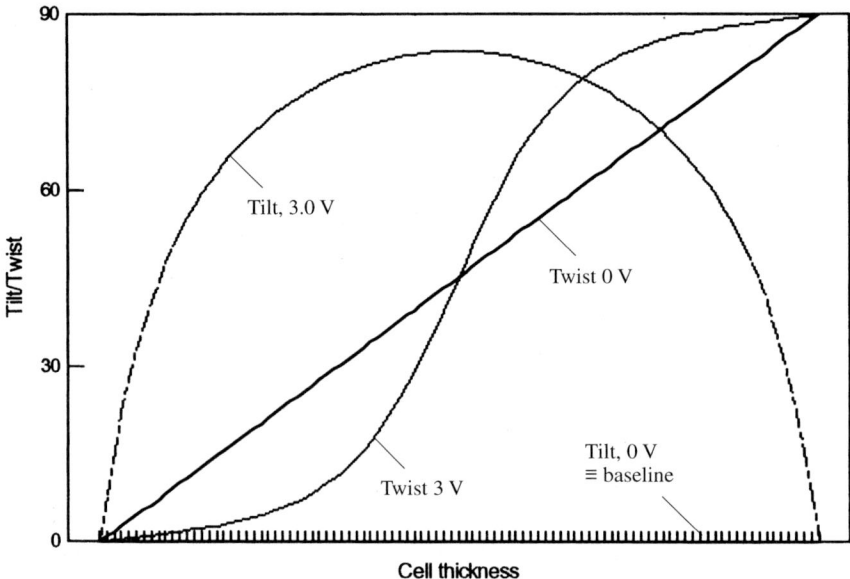

Figure 5. Through-cell tilt and twist profiles calculated for a 90° twisted nematic cell containing E7 in the 'off' and 'on' states.

formation content applications. They are also used extensively as the display elements above active matrix backplanes [10] for complex graphics panels. In twisted nematic displays, there is the possibility for the twist to form either a left- or right-handed helix. In practice, one sense of twist is selected by adding a small concentration of a chiral additive into the nematic liquid. The twist direction chosen [11] must be related to the direction of surface tilt at each cell wall in order to form a uniformly twisted and tilted structure throughout the cell.

The behaviour of the director tilt profile under an applied voltage is little changed by a cell twist up to 90°, but cells containing higher twist angles are of immense practical importance. A twist angle greater than 90° can only be sustained in a cell if the liquid crystal used is a cholesteric material of suitable pitch. In a cell with two conventional aligning layers, the liquid crystal will accommodate its natural pitch length and align with the molecular long axis in the rubbing directions on the surfaces. The range of twist angle available is then $360\,(d/P) \pm 90°$.

At twist angles above 90°, the twist and tilt of the chiral nematic director configuration begin to interact strongly. By far the most significant consequence of an increased twist angle [12] is that the gradient of the director tilt versus voltage curve becomes steeper. Ultimately, the gradient becomes vertical over some part of the switching curve, and at higher twist angles a bistable device results. This behaviour is shown in Fig. 6. Figure 7 shows the change in twist and tilt profiles resulting from a small change in voltage applied to a 270° twist cell. Comparison with Fig. 5 shows that, although the tilt changes much more rapidly than in the 90° TN cell, the redistribution of twist which occurs in twisted nematic cells is minimal at higher twist angles, partly due to the intrinsic cholesteric nature of the liquid crystal used. In order to interpret the effects that are observed it is worthwhile to consider the expression for the free

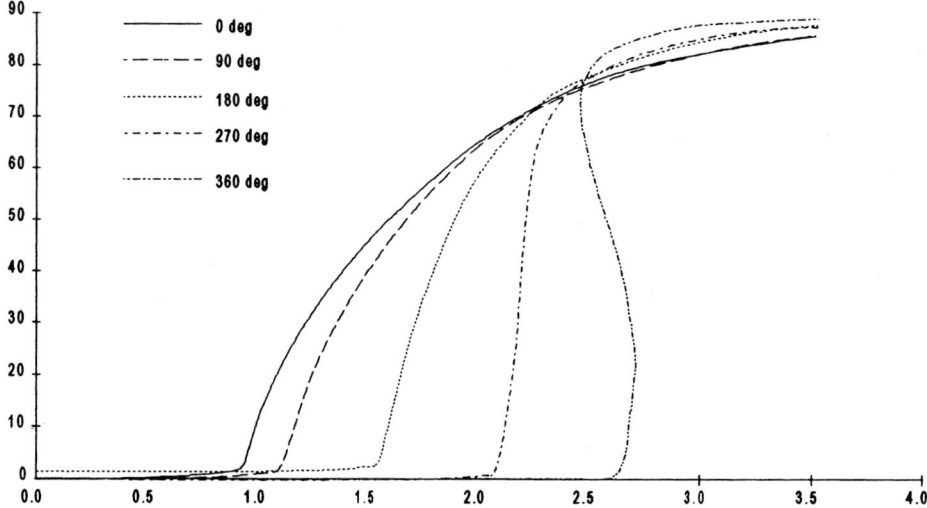

Figure 6. Mid-cell tilt angle versus voltage for a range of cells having different twist angles, showing the increase in slope and onset of bistability resulting from the supertwist effect. The liquid crystal is E7 in each case, with the pitch length matched to the twist angle.

energy density of the cell [13], which in polar coordinates and for a director configuration that is uniform within the plane of the cell can be written as

$$H = \frac{1}{2} k_{11} (\theta' \sin\theta)^2$$
$$+ \frac{1}{2} k_{22} \left(\phi' \sin^2\theta - \frac{2\pi}{P} \right)^2$$
$$+ \frac{1}{2} k_{33} (\theta' \cos\theta)^2$$
$$+ \frac{1}{2} k_{33} (\phi' \sin\theta \cos\theta)^2$$
$$+ \frac{1}{2} \frac{D^2}{\varepsilon_0 (\varepsilon_\parallel \cos^2\theta + \varepsilon_\perp \sin^2\theta)} \quad (2)$$

The Helmholtz free energy represented in this equation is extremal in configurations adopted by the liquid crystal director [14]. The symbols have their usual meaning: θ and ϕ are, respectively, the tilt (measured from the cell normal) and twist angles of the nematic director; P is the pitch length of the chiral nematic fluid; and D is the dielectric displacement, which is constant through the cell thickness and is related to the applied voltage by

$$V = \int_0^d \frac{D \cdot dz}{\varepsilon_0 \, \varepsilon(z)} \quad (3)$$

Five separate additive terms can be identified in the right-hand side of Eq. (2); these will be referred to as terms I to V in the discussion below. Terms I and III represent the energy due to the splay and bend deformations discussed earlier. As the magnitude of the director deformation increases, both of these terms grow monotonically, but term I rises more quickly immediately above the threshold. Term II represents the twist elastic energy. Both ϕ' and P are included in this term to allow for deviation of the rate of twist away from that due to the natural helical structure of a cholesteric liquid crystal. The final term V is the dielectric energy.

The key to the origin of the 'supertwist' effect, which leads to increased threshold steepness, is that at large twist angles where the cell becomes bistable, the energy of the

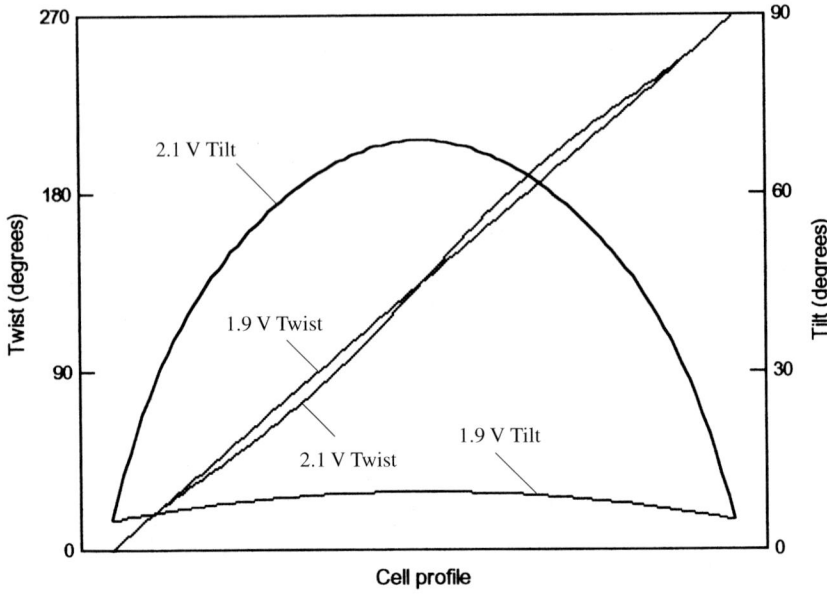

Figure 7. Through-cell tilt and twist profiles calculated for a 270° super-twisted nematic cell containing E7 at voltages corresponding to the 'off' and 'on' states under 100-way multiplex drive.

cell within the bistable range is lower at both high and low tilt angles than it is at intermediate values of tilt. In order to account for the bistability of highly twisted nematic devices, therefore, we require to find an energy term that is highest at intermediate values of director tilt angle. In this context, 'tilt angle' refers to some average of the tilt through the cell. Inspection of the energy terms in the equation above reveals that term IV has the required form, and moreover the factor ϕ'^2 means that this term will become increasingly important relative to the others as the twist angle increases. This energy term represents a coupling of twist and bend, and its effect is to resist deformation of the helical axis. Its effect is evident in Fig. 8, in which different terms of Eq. (2) are plotted separately as a function of dielectric displacement for a 270° twist cell, which just fails to be bistable. It can be seen that the increase in term IV at intermediate tilt angles does indeed make a critical contribution to the maintenance of a near-constant free energy over a range of tilt. Cells operating near this condition show a very steep slope of director tilt versus applied voltage, which can be used to provide the steep electro-optic threshold required for multiplex drive. The supertwist displays that dominate the production of alphanumeric and graphics panels using passive matrix technology use this type of device, with twist angles typically in the range 220–270°.

At still higher twist angles, the cell becomes fully bistable, and two configurations are possible within the cell over a finite range of voltage. In principle, this effect can be used to implement a storage display [15], but the switching time is relatively long and the bistability is imperfect in the presence of dust particles, pixel edges, etc. All these highly twisted devices are also subject to another constraint. It has been assumed in the above account that the deformation of the director is uniform in the plane of the cell, but this turns out not to be the case for some highly twisted devices. At twist angles above about 180° there is a tendency for the cell to form a two-dimensional striped structure [16, 17], which scatters light and inter-

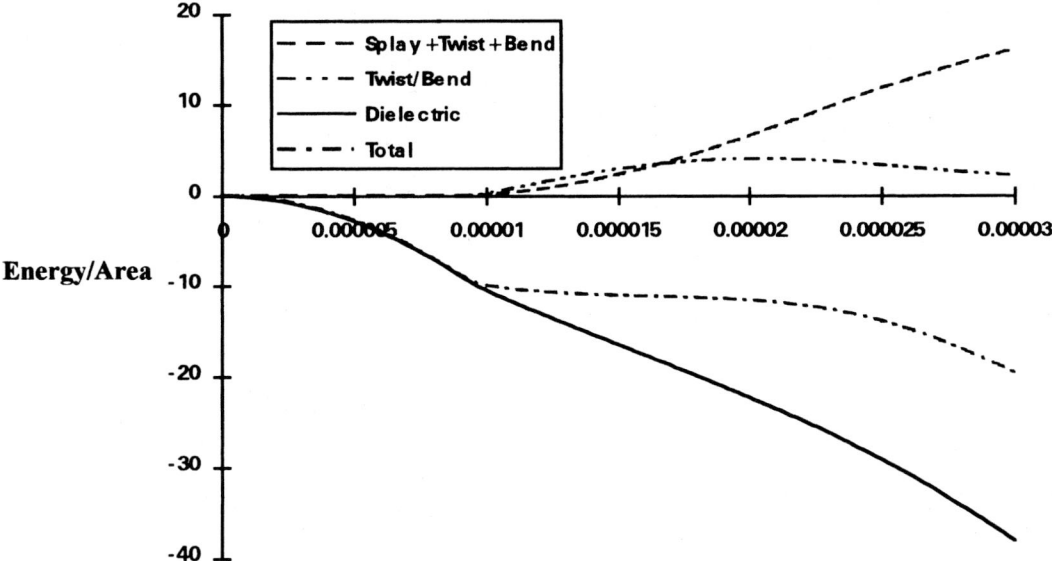

Figure 8. Elastic and dielectric energy terms in a 270° twist nematic cell as a function of dielectric displacement. The individual splay, twist and bend contributions to the energy have an approximately similar, monotonically increasing form, and together almost balance the change in dielectric energy. The twist/bend term then makes a decisive contribution towards providing a total energy, which is almost independent of the displacement; in the cell above the displacement range 1×10^{-5} to 2.2×10^{-5} the term corresponds to an increase in midplane tilt angle from 10° to 60° over the voltage range 2.1–2.26 V.

rupts the smooth switching processes described. For twist angles up to 360° or more, this scattering effect can be controlled by use of high pretilt aligning layers at each surface, and this is adequate for the manufacture of high-performance supertwist devices. Special high-tilt polymer aligning agents are available for this purpose.

When the pitch length of the liquid crystal is substantially shorter than the cell thickness, new effects become important. Typically, when a voltage is applied to such a cell, which contains several helical turns of the liquid crystal director within its thickness, the entire helical structure tilts over to give a scattering focal conic-like texture. Removal of the voltage at this stage allows a recovery to the initial state, which may take seconds or many days according to the nature of the surfaces and the liquid crystal used. If the voltage is raised further, however, the helical structure can be unwound above a critical voltage:

$$V_0 = \frac{\pi^2 d}{P} \sqrt{\frac{k_{22}}{\Delta\varepsilon \cdot \varepsilon_0}} \quad (4)$$

Such devices can be used in displays, when careful switching between the focal conic state and the homeotropic nematic state can provide a high information format [18, 19], albeit with slow writing speeds.

1.3.3 Optical Properties of Nematic Layers

The Fréedericksz transition and its variations described above can be used to implement electro-optic devices only if the change in director configuration is made to produce a corresponding change in optical transmission or reflection. Fortunately, there is no shortage of optical effects that can be exploited for this purpose. These rely either on the change of effective refractive index, or on the change in extinction coefficient of a pleochroic chromophore, which occurs when the director tilts. In general, both these effects occur together in any nematic device, but most practical displays are arranged so that one effect dominates the behaviour. Pragmatically, it is therefore reasonable to treat the effects separately, and this is the course followed below. A summary of the more common displays based on nematic liquid crystals is presented in Fig. 9.

The zero twist Fréedericksz cell has a particularly simple optical behavior. Consider a ray of light passing through the cell at normal incidence and polarized along the initial alignment direction in the plane of the cell. As the cell is switched from the zero voltage state to the fully switched condition, such a ray experiences an effective refractive index that changes progressively from n_e to n_o. The effective index is simply given by:

$$n_{\text{eff}} = \frac{1}{d}\int_0^d \left(n_e^2 \cos^2\theta + n_o^2 \sin^2\theta\right)^{\frac{1}{2}} dz \quad (5)$$

A ray of light with the orthogonal polarization experiences no change in refractive index; its electric vector remains perpendicular to the director as the cell switches, and it experiences a constant refractive index equal to n_o.

If a Fréedericksz cell is laminated between two polarizing films with their axes perpendicular to one another and at 45° to the cell alignment direction, it is easy to see that at high applied voltages (where the effective birefringence of the liquid crystal layer approaches zero) such a device blocks light. At lower applied voltages, the light may be resolved into components polarized, respectively, parallel and perpendicular to the liquid crystal alignment direction at the cell boundary. These emerge from the cell with a voltage-dependent phase difference

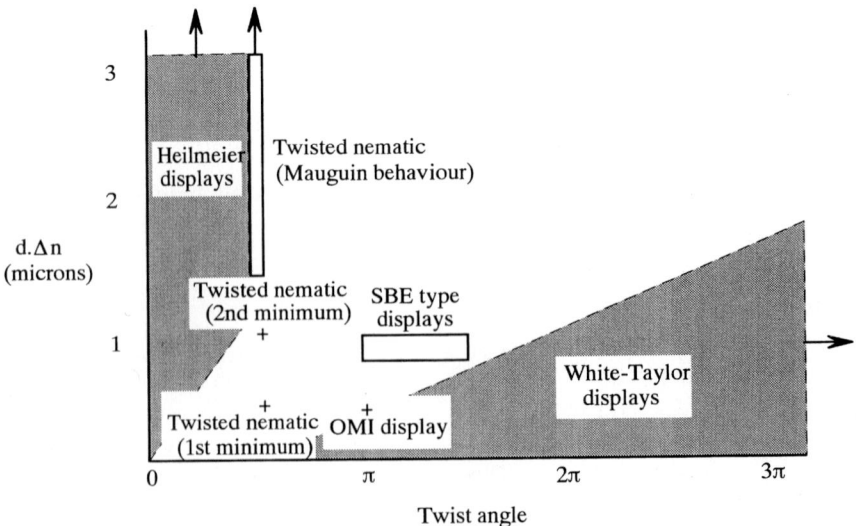

Figure 9. Operating regimes of some common display types based on Fréedericksz-like effects in nematic layers. Dichroic dyed devices are represented by the shaded areas, and birefringent and light guiding displays by '+' signs and clear boxes. The operating limits shown for each type of device are illustrative rather than decisive.

and recombine at the second polarizer. The transmitted intensity is then given by

$$\frac{I}{I_0} = \sin^2\left(\frac{\pi d \Delta n}{\lambda}\right) \quad (6)$$

If the condition holds that $d\Delta n \geq \lambda$, then as the cell is switched, the transmitted intensity of light passes through a series of maxima and minima. The change in transmission between the first maximum and minimum points occurs over a very small portion of the total switching event in such a cell, and therefore can occur over a small voltage interval. As will be discussed below, this is a desirable quality for a device intended for a complex information display. The transmitted intensity of the cell at zero applied voltage depends on the precise cell thickness and (temperature dependent) birefringence of the liquid crystal fluid, which is a serious disadvantage for a display device. The analogous bend Fréedericksz cell, however, does not suffer this defect as it is the zero field state in which the effective birefringence is zero (Fig. 10). In order to make a practical display [20], the surface alignment must impart a very small pretilt in order to define the direction of field-induced tilt relative to the polarizers. A less tractable problem in using a Fréedericksz cell for many purposes is that the transmitted intensity is wavelength dependent and highly dependent on the angle of view. Some approaches to reducing these difficulties are outlined below; in practice Fréedericksz cells with modest $d\Delta n$ have been exploited and achieved limited success in displays under various designations such as distortion of aligned phase (DAP), electrically controlled birefringence (ECB) and vertically aligned nematic (VAN) displays. Switchable colour filters have been demonstrated using the effect, but have not found widespread application.

In twisted cells in which the rate of twist is low and the product $d\Delta n$ is relatively large, there is an effective rotation of the plane of polarization of light passing through the cell equal to the twist angle. The optical behavior of such a twisted cell is equivalent to that of a stack of birefringent

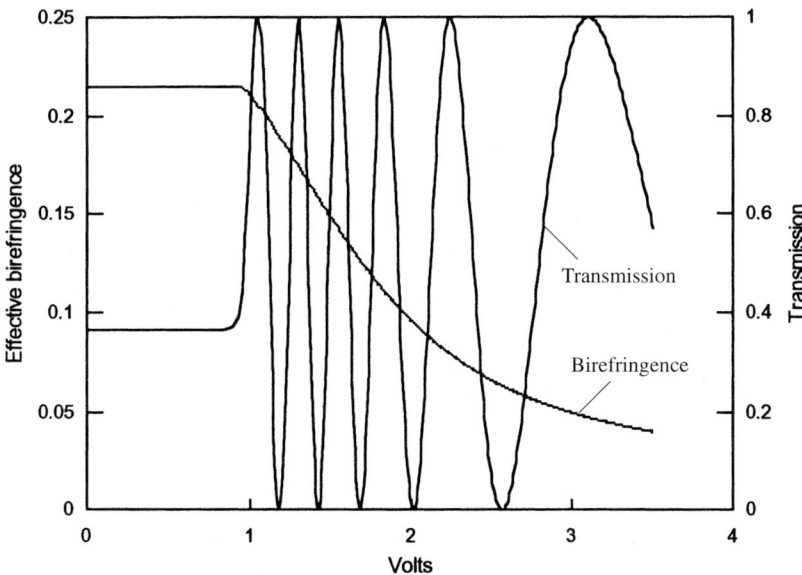

Figure 10. Effective birefringence and optical transmission calculated for a Fréedericksz cell operated in the tunable birefringence mode (cell thickness 10 µm, wavelength 633 nm, liquid crystal E7).

layers with a successive rotation of the optic axis of each layer. This arrangement is known as a Solc filter [21], and the analysis of its properties applied to a 90° twisted liquid crystal cell having the polarizing filters parallel to the adjacent surface alignment directions provides the Gooch–Tarry [22] curve shown in Fig. 11, the transmitted intensity being given by the expression

$$\frac{I}{I_0} = 1 - \frac{\sin^2[(\pi/2)\sqrt{1+u^2}]}{1+u^2} \quad (7)$$

$$u = \frac{2d\Delta n}{\lambda} \quad (8)$$

Such a 90° twist cell is of particular importance for electro-optic displays as it forms the basis of the twisted nematic display [23], which is the dominant technology both for panels of a low to medium information content and for use over an active backplane in complex displays. The important features of the rotation of light by the twisted layer may be summarized as follows:

- Ideal rotation of the plane of polarization is obtained for large values of $d\Delta n$, and for specific values of the parameter $u = 2d\Delta n/\lambda = \sqrt{3}, \sqrt{15}, \sqrt{35}$, etc.
- For appropriate values of $d\Delta n$, nearly ideal rotation of the plane of polarization can be obtained for a fairly wide range of wavelengths.
- At other values of $d\Delta n$, linearly polarized input light is changed to an elliptical polarization state. The major axis of the ellipse is rotated with respect to the input polarization.

In a twisted nematic display, the cell is constructed with a $d\Delta n$ product close to a point on the Gooch–Tarry curve that provides ideal rotation of light at visible wavelengths, i.e. with a $d\Delta n$ value usually close to 0.55 μm or 1.05 μm. Placed between two crossed polarizing films with their axes aligned with (or perpendicular to) the alignment directions at the cell surfaces, the resulting rotation of the optical plane of pola-

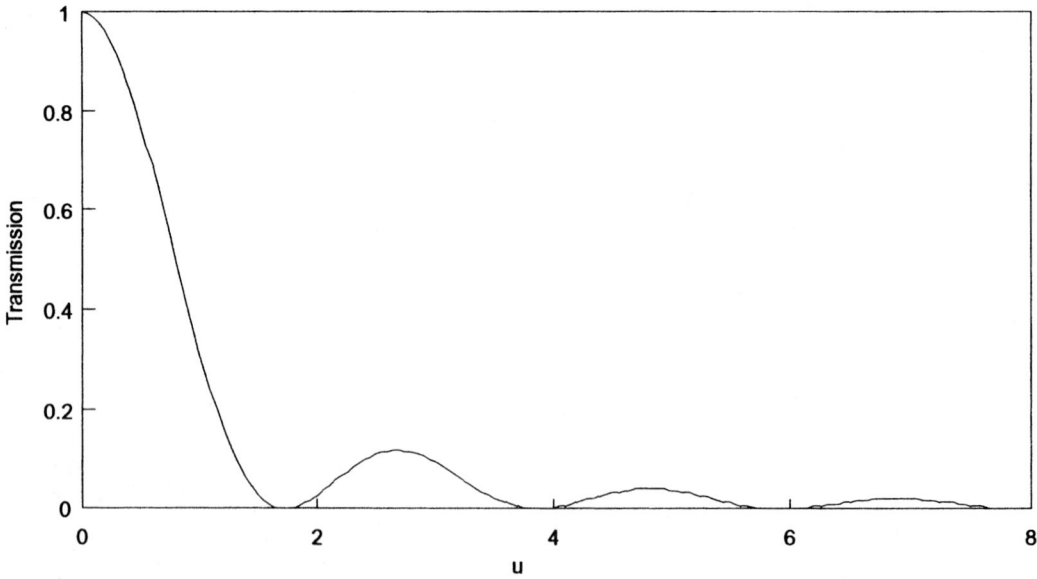

Figure 11. The Gooch-Tarry curve, showing off-state transmission for a 90° twisted nematic cell between parallel polarizers as a function of $u = 2d \Delta n/\lambda$.

rization results in a high transmission. When a voltage above the Fréedericksz threshold is applied to the cell two effects occur, as described above. The nematic director tilts, decreasing the effective Δn of the fluid, especially near the centre of the cell, and the twisted region is also compressed into the centre of the cell, thereby decreasing the effective d of the layer. The overall effect is therefore drastically to reduce the optical activity of the cell, and the display in this state shows a dark appearance representative of the two crossed polarizing films.

The optical behaviour of the twisted nematic cell at oblique angles of incidence [24] is complicated by the residual optical activity of the liquid crystal layers close to the cell walls. In these layers, the splay of the director can result in a change in the polarization state of obliquely incident light, similar to that induced in normally incident light by a director twist. The effect is highly directional, because the liquid crystal director in the centre of the cell is tilted in a unique direction (at 45° to each of the surface alignment directions). The overall result is that contours of constant contrast about the viewing cone of a twisted nematic cell have a pronounced cardoid or lobed shape. From certain viewing directions, the contrast of the cell can even be inverted.

As the rate of twist in a liquid crystal cell increases, the ability of the layer to rotate the plane of polarization of light is diminished. The light becomes elliptically rather than linearly polarized, and the major axis of the ellipse is rotated. This is the situation in the liquid crystal cells used for the supertwisted nematic display. Such displays are therefore normally operated in a variable birefringence mode [25]. In order to allow for the optical rotatory power of the liquid crystal layer, special choices of the optical thickness of the layer and the angles at which the polarizing filters are set allow efficient interference of two elliptical modes of light at the analyser and provide a high contrast supertwisted birefringence effect (SBE) display.

The first supertwisted nematic displays using the birefringence effect, like other birefringent displays of the time, suffered from objectional coloration effects. These occur because a birefringent slab provides a retardation that is almost constant as a function of wavelength. Only one wavelength of light matches this figure to undergo ideal constructive or destructive interference, and it follows that the transmitted intensity of light is highly wavelength dependent. In the case of supertwisted nematic displays, this led to a blue on yellow display. The effect can be minimized by using a small $d\Delta n$ product in the device. In the case of a supertwist device this results [26] in the optical mode interference (OMI) display, which is almost black and white, but has a brightness almost half that of SBE devices [27]. A more satisfactory solution to the coloration problem has been found in the use of external birefringent compensation layers on the display. Initially, a second (unpowered) liquid crystal cell was used for this purpose [28], but stretched polymer sheets [29–31] are now commonly used. Liquid crystal cell compensators are made with similar $d\Delta n$ to the display cell and identical twist angle with opposite sense of twist. Birefringent film compensators may be used singly or in pairs on one or on each side of the display cell. All these techniques are capable of providing, high contrast, high brightness black and white displays. Double cell devices are, however, little used today because of their greater weight and cost compared to those using plastic film.

The use of birefringent polymer layers has subsequently been applied to other types of display. These include compensated versions of the bend Fréedericksz cell operated in birefringence mode [32] and of the twisted nematic device [33]. In many cases, the main motivation has been improvement of the viewing angle properties of the display rather than removal of coloration.

1.3.4 Nematic Devices with Wider Viewing Angle

The widespread use of the twisted nematic device as the display mode above an active matrix array has highlighted its shortcomings. Such displays, which command a significant price premium in the marketplace, are subject to increasing scrutiny over the quality of the image. The narrow and non-uniform viewing cone of the normal twisted nematic display is increasingly viewed as an unacceptable aspect of performance, and various approaches have been described that seek to provide a better solution [34].

One important approach [35] is to divide each pixel in a twisted nematic display into separate domains, each of which provides its highest contrast when viewed from a different direction. When the display is in use, the eye averages the intensity of the domains and, by using a four-domain structure on each pixel [36], a nearly uniform angle of view can be achieved. Fabrication of devices using this technique is difficult; manipulation of the twisted nematic viewing cone relies on providing different alignment directions on the cell surfaces within each domain and/or achieving different senses of helical twist direction. Different alignment directions on a single pixel can be achieved by the process of depositing a polyimide layer, which is partly protected by a photoresist layer before rubbing. The resist is then stripped off, the rubbed area is protected by a second photoresist layer, and the substrate rubbed again along a different axis. This basic procedure has been elaborated in several ways. Notably, by use of two substrates each carrying patterned areas of high- and low-tilt polymer, it is possible to achieve a four-domain pixel having regions of opposing tilt and twist. The twist direction in this case is stabilized not by a chiral additive, but by the interaction of the pretilt directions at the two surfaces. Although these multi-domain devices achieve a very satisfactory uniform angle of view, this is at the expense of a reduction in overall contrast and a considerable increase in the complexity of the production process.

Standard variable birefringence effect devices have a poor angle of view because the effective through-cell birefringence of the device is a function both of the zenithal angle of incidence of light, and the azimuthal angle between the light incidence axis and the tilt direction of the liquid crystal director in the centre of the cell. The latter effect can be eliminated by exploiting the self-compensating capacity of a cell [37, 38] in which the liquid crystal director is near planar at the surfaces, but vertical at the centre of the device. This type of device has been termed a Pi-cell; the device has available liquid crystal configurations with a twist of Pi as well as that described above which has a director bend of near Pi. The electro-optic switching of the device involves a change in tilt of the liquid crystal layers close to the cell walls. Because the thickness of the effective switching layers is small, the speed of the device is significantly faster than would be expected from consideration of the full cell thickness, and is improved further by the fact that unlike most liquid crystal devices, the flow-related torque induced by switching does not oppose the director realignment. As a tunable birefringence shutter the device has a good uniform angle of view, as the tilt directions in the upper and lower halves of the cell oppose one another. A change in light incidence direction will, in general, lead to a decrease in effective birefringence in one half thickness of the cell, accompanied by a corresponding increase in effective birefringence in the other half. Unfortunately, the switching voltage is usually significantly higher than that of a twisted nematic cell.

A further device that shows some promise for high-quality displays is the in-plane twisted nematic cell [39, 40]. The switching of this device is achieved by the use of interdigitated electrodes on one cell wall, giving a transition from a planar untwisted structure to a planar twisted one. The twist does not develop uniformly through the cell as in a twisted nematic cell off state, but the maximum twist is developed near to the electrode-bearing surface. The optical behaviour of the structure is complicated; successful application of this device will rely on careful control of fringing field as well as device optimization and development of fabrication techniques for the electrode pattern. The device provides a wide viewing angle because of the absence of tilt in the structure.

1.3.5 Dichroic Dyed Displays

A liquid crystal layer containing a suitable dichroic dye (Fig. 12) absorbs polarized light efficiently only if the polarization axis is coincident with the director (assuming the dye is a normal dichroic one which has positive dichroism). The guest–host interaction between the rod-shaped dye molecule and the liquid crystal host serves to align the dye and its transition moment, even though the dye usually has no liquid crystal phase. This provides an alternative means to modulate light in a liquid crystal device, which can reduce or eliminate the need for external polarizing filters. In general, the effective order parameter of a dichroic dye is much

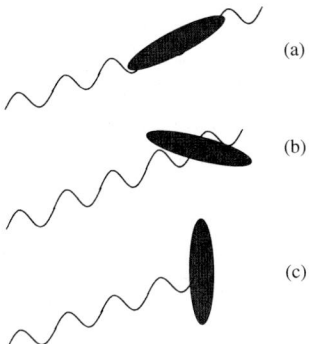

Figure 12. Absorption of light by a dichroic dye is weak in the geometries illustrated at (a) and (b), where the electric vector of the light is perpendicular to the chromophore axis. Strong absorption is obtained when the vector and axis are aligned as in geometry (c).

lower than that of a polarizing filter, and it is difficult to obtain a contrast ratio from a dichroic display as high as the one obtained from, for example, a twisted nematic device, but the brightness can be superior. The absorption of light is superimposed on the birefringence effects referred to above: for practical systems, refractive index anomalies arising from the absorption bands can generally be ignored in display applications.

A dichroic dyed Fréedericksz cell can absorb incident light only if the sense of polarization is parallel to the off-state alignment direction. The perpendicular mode must be removed by a polarizer if a high optical contrast is to be achieved. In the on state at high applied voltages, the dyed liquid crystal director is substantially perpendicular to the plane of polarization of incident light throughout the cell thickness, and light is not absorbed strongly. The resulting device [41] is known as a Heilmeier display. The optical contrast ratio in the Heilmeier display is determined by the order parameter of the dichroic dye used:

$$S = \frac{A_\| - A_\perp}{A_\| + 2A_\perp} \qquad (9)$$

Stable dyes in a variety of colours are available with order parameters up to around 0.8, and mixtures of dyes can provide black on white display effects. This order parameter implies a contrast of about 10:1 in absorbance ratio. In terms of transmittance or reflectance, high contrast can be achieved at the expense of brightness. The contrast ratio of dichroic displays is typically much lower than that apparent for polarizing filters, and it is difficult to achieve a balance of contrast and brightness that is competitive with, for example, that of twisted nematic devices. A further limitation arises from the fact that full contrast is obtained only at applied voltages that are much higher than the twisted nematic saturation voltage for a similar cell and material and that are sufficient to realign the nematic director throughout the cell, including the layers close to the cell walls. This general feature of dichroic displays makes it difficult to use them in conventional multiplexed applications with high information content.

Devices in which a dichroic nematic layer is sufficiently tightly twisted that light guiding does not occur can absorb incident light of arbitrary polarization. Practical displays are typically constructed with a cholesteric pitch sufficient to provide a few full turns of the cholesteric helix within the cell thickness. With a low birefringence liquid crystal, the optical absorption of such a cell can be almost independent of the polarization. A polarizer free display of this type [42], often referred to as a White–Taylor display, can therefore give good contrast, subject to similar considerations to those noted above for the Heilmeier device. Once again a high voltage is necessary to achieve full contrast, and the hysteresis effects noted above often lead to scattering from the cell in the off state.

It will be understood from the above discussion that dichroic dyed displays ideally operate in either the fully guiding, single polarizer Heilmeier regime, or in the non-guiding White–Taylor mode with no polarizer. In practice, however, twisted nematic cells are frequently operated in a condition that is intermediate between these conditions. The reasons are pragmatic: for example, host liquid crystals that combine low birefringence with a high dielectric anisotropy are required for low-voltage operation of a high-quality White–Taylor device, but are not easily available. This intermediate range of twist angle and birefringence also includes the important regime appropriate to supertwist displays, which were first demonstrated in a dichroic form [43]. Improvements that have been made to the birefringent effect, however, mean that dyed supertwist devices have been little used in practical applications.

1.3.6 Materials for Twisted and Supertwisted Nematic Displays

The paragraphs above have noted the basic relationships between the fundamental physical properties of liquid crystal materials, and some of the aspects of display performance that are of practical significance. Historically, the development of materials for these display effects (which together have dominated commercial production of liquid crystal devices since the mid-1970s) had followed the aspirations of the users for their display. The requirements have been:

- stable, colourless materials, initially for operation near room temperature
- low-voltage operation
- ability to multiplex drive at steadily increasing levels
- operation over a wider temperature range

- faster switching, both at low temperatures and at high multiplex rates
- especially for active matrix displays, very high and stable electrical resistivity

The issue of providing materials with room-temperature mesophases and with good stability to light and ordinary exposure to the atmosphere, which formed such a barrier to the exploitation of liquid crystal devices in the 1960s, was completely solved by the discovery of compound classes such as the cyanobiphenyls [44], the phenylpyrimidines [45] and the phenylcyclohexanes [46]. These compounds and their various derivatives rapidly displaced less stable systems such as Schiffs' base and azoxy derivatives from practical displays in the early to mid-1970s. The desire to operate displays from low supply voltages led at this time to the exploration of structure property relationships in these and other systems, since it was well understood from Eq. (1) and its analogues for twisted nematic devices that $\Delta\varepsilon$ and the magnitude of the elastic constants determined the threshold voltage. It rapidly became clear that there was a disparity between the observed dielectric properties of many liquid crystals and the magnitude of their dipole moments, and this was explained in terms of local antiparallel ordering of the molecules in the mesophase [47].

Once simple displays had been established in watches and calculators, there was pressure to provide liquid crystal materials that would permit multiplex drive more easily and at higher levels than the existing fluids. The motivation was the need to reduce the number of electrical interconnections between the display and the integrated circuit drivers, and hence to reduce costs. Multiplex drive of these displays is discussed below, but its consequence for the supplier of materials is that the voltage applied to both 'on' and 'off' pixels of the display cell is set by the drive electronics. Furthermore, these voltages move closer together as the multiplex level rises, implying the need for a rapid change in optical transmission with applied voltage once the threshold is exceeded. Additionally, the variation in threshold voltage with temperature in the display device should be minimized. For multiplexed displays, the steepness of the electro-optic switching curve becomes a key issue, and has been discussed above. For twisted nematic displays, the k_{33}/k_{11} ratio and the dielectric constant ratio are the main influences but, because of constraints on the switching voltage, the dielectric constants cannot be varied freely. It proved possible to improve many properties simultaneously by mixing components such as cyanobiphenyls with less polar liquid crystals such as esters, which have small dielectric anisotropy. In these mixtures, the local antiparallel order of the cyanobiphenyls is disrupted, with a beneficial effect on the elastic constant ratio and very little penalty in terms of threshold voltage at up to about 35% addition of the low $\Delta\varepsilon$ component. Addition of larger quantities of esters provides a means to increase the switching voltage to any desired figure. The same effect can be exploited in other liquid crystal families, with similar benefits. The best electro-optic threshold steepness is found in derivatives of aromatic and heterocyclic liquid crystals, while lower variation of threshold with temperature is found in alicyclic liquid crystal systems. Careful optimization of the mixtures is needed to avoid the intrusion of unwanted smectic phases, which arise on mixing nematic materials of high and low dielectric anisotropy.

The invention of ranges of liquid crystalline materials of very low polarity proved the key to formulating nematic mixtures with a very wide operating temperature range. Dialkyl and alkylalkoxy phenylcyclohexanes and cyclohexylcyclohexanes together with their three-ring counterparts and variations based on heterocyclic rings, lateral fluorination and the inclusion of non-polar linking groups such as dimethylene (CH_2CH_2) groups provide important examples. In many cases these compounds show only smectic phases, or have clearing temperatures well below room temperature. Mixtures are formulated by careful blending of two-, three- and four-ring non-polar mesogens with, mostly, two-ring polar compounds, which are needed to provide the necessary dielectric anisotropy for switching. The principal benefit of this approach to making the mixtures is that a very low viscosity can be achieved together with a small coefficient of viscosity against temperature. This means that a reasonable viscosity and switching time can be maintained down to working temperatures of $-20°$ to $-30\,°C$ or even lower while maintaining a clearing point near $100\,°C$.

The first commercial supertwist displays used nematic mixtures similar in character to those developed for wide temperature range operation. These mixtures provided the fast response and low coefficients of their physical properties with temperature needed for stable operation. Subsequently, mixtures with elastic properties specifically optimized for supertwist operation were developed. The invention and incorporation of materials carrying unsaturated alkylene links in their terminal groups provided an important advance [48, 49] in this field.

Development of mixtures for use on active matrix displays has revolved around the problem of providing materials with very high resistivity, which moreover retain this property after filling into the display cell and exposure to environmental heat and light. It is found experimentally that materials based on terminal cyano groups have great difficulty satisfying the stringent requirements placed on active matrix mixtures. Compounds based on terminal fluoro, trifluoromethyl and trifluoromethoxy groups have been widely exploited for this range of applications [50].

The subject of materials development has been reviewed rather frequently [51–53], but the depth of coverage has often been limited by issues of commercial confidentiality.

1.4 Scattering Mode Liquid Crystal Devices

Scattering of light occurs within media where the refractive index varies on a scale comparable with the wavelength of light. Common liquid crystal phases, especially the nematic phase, scatter light some 10^4 times more strongly than an isotropic liquid due to thermally driven director fluctuations. This scattering is nevertheless too weak to provide adequate contrast for a display from devices of practical thickness. If, however, the liquid crystal alignment is so strongly disturbed that alignment discontinuities are induced, the scattering intensity can rise to a practically useful level. This has been achieved in several ways and in different phases, but, as will be discussed, many of the effects are of historical rather than commercial importance. Finally, the enhancement of scattering by introduction of a separate solid material to the liquid crystal is described.

The dynamic scattering display [54] exploited scattering of light by a nematic liquid crystal induced by turbulence associated with electrical conductivity [55, 56] in the liquid crystal. The effect provided the first commercially exploited liquid crystal displays, using Schiff's base liquid crystals of negative dielectric anisotropy, which initially were conductive by virtue of impurities and degradation products that they contained. The same impurities commonly acted as surfactant agents to induce the necessary homeotropic alignment at the cell walls. These displays were marred by poor reliability and lifetime; later displays used highly purified materials with appropriate dopants to obtain conductivity and were more satisfactory.

On application of a low frequency AC or DC voltage, a liquid crystal of this type will show electrohydrodynamic effects. At low fields, stripe-like Williams domains [57] appear; dynamic scattering occurs at a higher applied voltage and results in a vigorous turbulent flow in the nematic layer. This turbulence is accompanied by a very strong scattering of light, which is suitable to provide a display effect, provided that the lighting and background reflectance are carefully arranged.

When the field is removed from a nematic dynamic scattering display, the liquid crystal relaxes to its quiescent alignment, resulting in a scattering on clear electro-optic effect, which responds to the applied root mean square (rms) voltage, in the same way as a field effect device. If dynamic scattering is induced in either an aligned (see Fig. 13a) cholesteric [58] or a homeotropic smectic A [59] liquid crystal, it will tend to relax to a scattering focal conic texture. This therefore can form the basis of a storage display device. The stability of the scattering state will depend on the nature of the cell walls, and, in the case of a cholesteric material, the pitch length. In the case of smectic A liquid crystals, the dynamic scattering effect can be obtained in a material of positive dielectric anisotropy. In the case of the cholesteric phase, the scattering state can be returned to a non-scattering Grandjean texture under the influence of a high frequency voltage acting on the negative dielectric anisotropy of the phase. Conversely, the smectic A phase can be switched back to the non-scattering homeotropic state by a high-frequency signal operating on the positive dielectric anisotropy. Each of these erase operations relies on the fact that the conductivity mediated dynamic scattering phenomenon does not occur above a critical frequency of the applied signal, usually of the order of a few kilohertz or less.

An alternative write/erase mechanism that may be exploited to provide a scattering display in smectic liquid crystals [60] relies on the electrothermal reorientation of a material of positive dielectric anisotropy (Fig. 14). The thermal input to the material can be provided either by ohmic heating of stripe electrodes [61] in the cell, or by scanning a visible or infrared laser spot across the device. A dye can with profit be incorporated into the liquid crystal [62] to improve the absorption of the incident light. Typically, the liquid crystal used will pos-

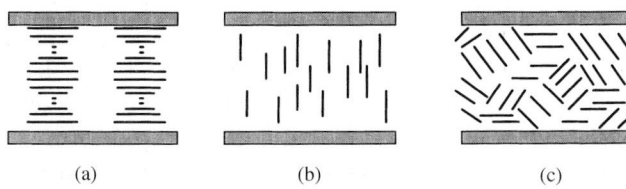

(a) (b) (c)

Figure 13. Three states of a cholesteric phase change device operated in light scattering mode. The grandjean texture (a) and homeotropic field-on state (b) are optically clear; the focal-conic state (c) represents an alternative field-off state, which is optically scattering.

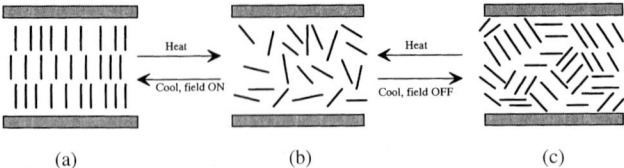

(a) (b) (c)

Figure 14. The read/write process in a smectic A thermoelectric scattering cell. Writing is accomplished by local heating of the cell to the isotropic state (b), which on cooling can be induced to form either a clear homeotropic state (a) if a field is simultaneously applied, or to a scattering focal-conic state (c) in the absence of a field. The field has no effect in the absence of simultaneous heating.

sess a narrow nematic phase range between the wide range smectic A and the isotropic phase. On heating into the isotropic phase and cooling rapidly to room temperature, the tendency of the liquid crystal is to form a focal conic texture that scatters light. If, however, a voltage is applied across the pixel simultaneously with the heating pulse, the material can realign in the field as it passes through the nematic phase and form a homeotropic texture directly.

Addressing of these displays is simplified by the fact that the voltage has no effect in the absence of a heating pulse; by scanning a laser, or sequentially heating row electrodes in the cell, a complex information format can be written.

A further means of achieving efficient light scattering in a liquid crystal device is to disperse a nematic material in a plastic matrix [63, 64]. The plastic is chosen so as to have a refractive index which is equal to n_0 of the liquid crystal. One fabrication route [63] involves forming an emulsion of liquid crystal in a water-borne polymer system, which is then coated onto a substrate and dried. The product is a plastic film containing many small, separate droplets of liquid crystal. Alternatively, the liquid crystal may be dissolved in a reactive monomer [65], which is polymerized in a thin layer. As the polymer forms, the liquid crystal is expelled from solution and forms a similar array of droplets. Another route uses a polymer in which the liquid crystal dissolves at high temperature. On cooling, the solute is excluded from the matrix as it solidifies, and forms droplets. Devices formed in any of these ways have roughly similar characteristics.

In the zero field state of these polymer dispersed droplet displays, the nematic phase in each droplet aligns in a manner that minimizes its own elastic energy. Commonly, the inside of the droplet allows a tangential disposition of the nematic director [66], and this is the alignment adopted, except at two points between which a line defect extends. This defect is necessary on purely topological grounds. If the droplet surfaces have a perpendicular boundary condition, then a radial director configuration with a point defect at the centre can result. In either case, the effective refractive index that the droplet presents to incident light is intermediate between n_e and n_0. This refractive index does not correspond with that of the polymer matrix, and light is scattered from the interface of each liquid crystal droplet with the polymer (Fig. 15). Under a moderate to high applied voltage, the liquid crystal in each droplet realigns so that the director is parallel to the field. The effective refractive index of the liquid crystal droplets in this alignment, for light normally incident on the device, is equal to n_0. This refractive index matches that of the polymer, and no light is scattered in this state.

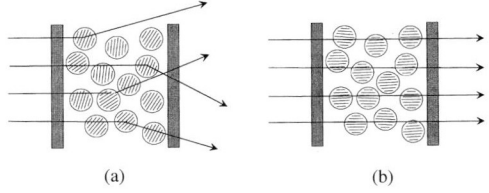

Figure 15. Liquid crystal dispersed droplet device in the scattering, field-off state (a) and the transparent field-on state (b).

There are numerous variations on the basic device structure described above. One of the most fundamental is the case where isotropic threads or particles are dispersed in a liquid crystal layer, with the latter forming the continuous phase. Such an architecture can be obtained, for example, using different kinds of prepolymer in which the liquid crystal is dissolved, or by absorbing a liquid crystal onto a porous material. These devices operate in a substantially similar way to dispersed droplet cells, but there are important differences in their behaviour. In particular, there is a tendency for substantial hysteresis to arise in the switching characteristic. The devices can, however, give low-voltage operation when suitably constructed.

A shortcoming in the performance of dispersed droplet devices arises from the fact that the apparent refractive index of the liquid crystal in the field-on state is sensitive to the angle between the applied field and the incidence direction of light. In practical terms this means that the refractive indices of the liquid crystal and the polymer matrix only match perfectly at one angle, normally close to normal incidence. At glancing incidence, there is a progressive mismatch in the refractive indices, and on-state optical scattering is observed. In principle, this off-axis scattering can be avoided by use of an aligned liquid crystalline polymer matrix. In practice, careful choice of the refractive indices of the liquid crystal and polymer minimizes the problem in practical applications.

The apparent optical contrast of scattering devices tends to be highly dependent on the lighting and viewing conditions used. Dispersed droplet devices have been proposed as switchable glazing components for privacy and comfort control. A further important possibility is their use in projection displays. The use of small aperture optics allows a very high contrast to be achieved, even from the small-angle, forward scattering typical of liquid crystal cells. These devices also work without any polarizer, increasing the optical efficiency of the system overall. In order to obtain a complex format display, the device must be operated over an active matrix; provision of a device compatible with the limited voltage available from standard backplanes requires special optimization.

1.5 Addressing Nematic Liquid Crystal Displays

The addressing of rms responding liquid crystal displays has historically revolved around the problems of how to provide a complex information format on the device, while minimizing the total number of electrical connections made to the panel and the number of integrated circuit drivers required. To realize these targets, liquid crystal displays from watches to computer screens are commonly driven by a multiplex scheme characterized by a multiplex level N. Such a scheme allows a display having P independently addressable pixels, to be driven through a total of $P/N+N$ connections. This reduction in the number of connections incurs a penalty; the maximum ratio of rms voltages [67] that can be applied

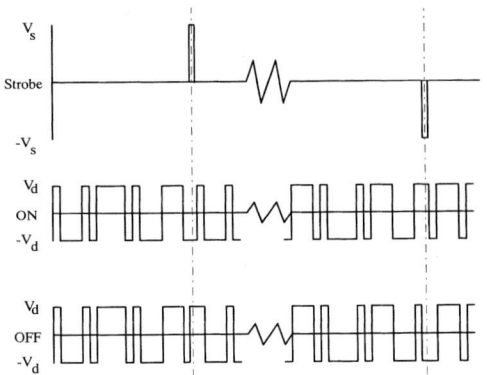

Figure 16. Waveforms for driving rms responding liquid crystals pixels into the 'on' and 'off' states (see text).

to the 'on' and 'off' pixels of the display is

$$\frac{V_{\text{ON}}}{V_{\text{OFF}}} = \sqrt{\frac{\sqrt{N}+1}{\sqrt{N}-1}} \qquad (10)$$

An implementation of such a multiplex scheme is shown in Fig. 16. Each pixel receives voltage pulses from both a row and a column driver circuit; the row supplies a 'strobe' signal, which is independent of the information shown on the pixel, and in this case comprises a single pulse occupying one time slot and is zero at all other times. Adjacent rows are driven with similar strobe signals, shifted one time slot earlier or later. The information displayed on the pixel depends only on the polarity of the data signal applied to the column at the time when the pixel in question receives its strobe pulse, which determines whether the strobe and data pulses effectively add together or subtract. In a second half of the addressing cycle, all the signals are repeated with their polarities reversed, in order to guarantee AC balance of the signal overall. The rms voltages for N multiplex driven rows are then:

$$V_{\text{ON}} = \sqrt{\frac{(N-1)V_{\text{d}}^2 + (V_{\text{s}}+V_{\text{d}})^2}{N}} \qquad (11)$$

$$V_{\text{OFF}} = \sqrt{\frac{(N-1)V_{\text{d}}^2 + (V_{\text{s}}-V_{\text{d}})^2}{N}} \qquad (12)$$

and Eq. (10) follows from selecting an optimal ratio of $V_{\text{s}}/V_{\text{d}}$.

The voltage ratio resulting from a given multiplex level leads directly to a requirement on the electro-optic response curve of the effect used. In order to realize the high multiplex level necessary for complex displays, a steep slope of the optical transmission versus voltage is needed, together with a small dependence of threshold voltage on angle of view and temperature. Electro-optic response curves of typical liquid crystal devices are shown in Fig. 17.

Dichroic displays rely on a high applied voltage to develop their full contrast, and are generally poorly suited to multiplex drive at high rates. (The dichroic supertwisted nematic display is an exception, but even in this case the contrast is reduced.) The twisted nematic device provides limited electro-optic steepness, and is limited to multiplex levels up to about 32 ways. The supertwisted nematic cell can be optimized to provide ideal steepness of the electro-optic response curve over a substantial range of contrast. The multiplex ratio that can be used is then limited by the critical slowing down of switching at high multiplex rates, and by the variation of switching voltage, which originates from manufacturing tolerances in the panel. Practically, multiplex rates up to 240 ways are common in supertwisted nematic displays, and levels up to more than 500 ways have been demonstrated with reasonably good appearance.

A consequence of the multiplex waveforms usually used for such high multiplex levels is that pure rms response is lost. The device tends to respond to periodic high peak voltages applied to the cell, and the result is a visible flicker or reduction in contrast of the display. The effect can be great-

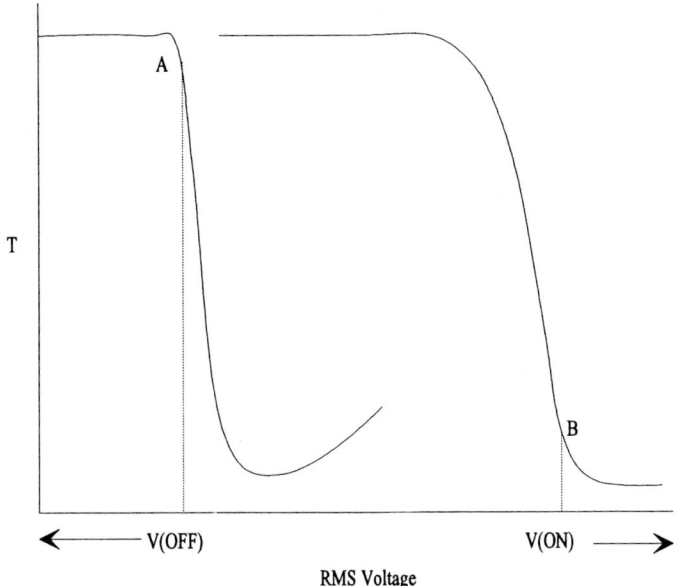

Figure 17. Electrooptic response curve of a typical liquid crystal device, showing the relationship of applied voltage to the threshold voltage at high temperature and oblique viewing angle (A) and the saturation voltage at low temperature and normal angle of view (B) required to achieve high contrast over the operating envelope of the display.

ly reduced by using novel drive schemes [68]. These use modified waveforms in which the high voltage pulse is reduced in magnitude and spread over a number of time slots to provide the same rms voltages at lower peak voltage, at the expense of a considerable overhead of increased complexity in the addressing circuitry.

The problems of rms addressing of liquid crystal devices can be circumvented by incorporating a semiconductor switch at each pixel of the display. The most usual architecture uses a field effect transistor as the switch [69], while other active components such as diode networks [70] and MIM (metal-insulator-metal) switches [71] have also been used (Fig. 18). The semiconductor material is usually amorphous silicon, which can be deposited and processed at temperatures compatible with a glass substrate. Polycrystalline silicon and other semiconductors, especially cadmium selenide, are also used in special applications such as those requiring very small pixel geometries and, by virtue of their higher carrier mobility, offer the possibility of fabricating drive circuits on the same substrate simultaneously with the manufacture of the display.

Use of an active backplane architecture for a liquid crystal device allows addressing of very large pixel arrays with very high performance. The column electrodes are supplied with voltages corresponding to 'on' and 'off' voltages for the liquid crystal device, and each row is selected in turn by applying a suitable signal to the TFT gate electrodes. The charge that activates the pixel must be

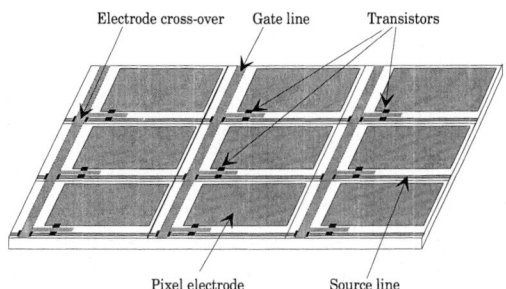

Figure 18. Diagrammatic structure of an active matrix structure using a simple field effect transistor array as the active element.

stored while the remainder of the display is addressed, and this is preferably accomplished without fabricating any additional storage capacitor at the pixel. The liquid crystal itself must then form the dielectric layer of the effective charge storage capacitor, which places stringent requirements on the resistivity and stability of the liquid crystal. These active matrix displays with a twisted nematic modulating layer are used extensively to provide high-quality panels with a complexity sufficient to provide video or VGA standard panels and higher. Colour is available from these displays through the incorporation [72] of a pixellated matrix of primary colour filters fabricated in registration with the pixels of the active matrix.

Nematic displays that exhibit bistability, or at least a decay time which is substantially slower than the write time, and that exhibit a pronounced voltage threshold, do not show rms response and are usually addressed by 'slow-scan' methods. In these drive schemes, each row of the display is written in turn by applying a voltage pulse. The information state written is determined by the polarity of pulses applied simultaneously to the columns. In favourable cases, where the bistability is strong, an indefinite number of lines can be written before the panel is refreshed. However, the total write time for the device is now dependent on the number of rows. For typical nematic effects with switching times in the range of tens of milliseconds, the addressing of complex displays in this way is incompatible with video applications, and problematic for most other uses of these devices. Slow-scan multiplexed nematic liquid crystal devices have always, therefore, been confined to a few niche applications. Recently reported developments that seek to provide bistable nematic display modes with much faster switching times [73, 74] have the potential to change this situation.

1.6 Ferroelectric Liquid Crystal Displays

The ferroelectric smectic C liquid crystal display has not, at the time of writing, achieved extensive commercial use. It nevertheless stands as an important device, both because of its potential application in complex displays, which will not require an active matrix, and because of its intrinsic scientific interest. In addition, ferroelectric liquid crystal displays show faster switching rates (of the order of microseconds) than conventional nematic-based displays.

The molecules in a smectic C phase are arranged in disordered layers, with their long axes tilted with respect to the layer normal. The tilt angle is typically of the order of 20°, and the direction of tilt is subject to long-range ordering, but can be distorted on a length scale of micrometres by weak imposed forces. In this respect the smectic C tilt shows analogies with a nematic phase director. As is illustrated in Fig. 19, the chiral smectic C phase has the correct local symmetry to show ferroelectric polarization

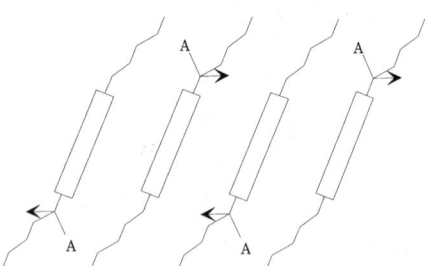

Figure 19. An illustrative justification of the source of spontaneous polarization in smectic C* phases. If the steric properties of substituent (A) at the chiral centre lead to an energetic preference for it to associate with the flexible end-chains in the tilted structure rather than the molecular cores, the dipole moments represented by arrows have a component perpendicular to the molecular tilt (out of the page) and independent of the head/tail symmetry of each layer.

[75], with the polarization vector in the plane of the smectic layer and perpendicular to the molecular tilt. By synthesis of smectic C materials having a strong lateral dipole moment close to the chiral centre, it is possible to obtain materials with a spontaneous polarization of 10^3 nC/cm^2 or more. The materials used in devices usually have a polarization 0.5–1.5 orders of magnitude lower than this.

On a macroscopic scale, the spontaneous polarization vector in the optically active phase spirals about an axis perpendicular to the smectic layers (Fig. 20), and sums to zero. This macroscopic cancellation of the polarization vectors can be avoided if the helical structure is unwound by surface forces, by an applied field, or by pitch compensation with an oppositely handed dopant. The surface stabilized ferroelectric liquid crystal display utilizes this structure and uses coupling between the electric field and the spontaneous polarization of the smectic C phase. The device uses a smectic C liquid crystal material in the so-called bookshelf structure shown in Fig. 21a. This device structure was fabricated by shearing thin (about 2 µm) layers of liquid crystal in the

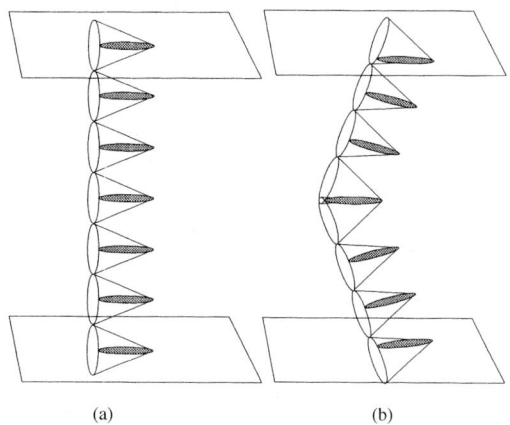

Figure 21. Illustrative diagram of a smectic C layer structure and the molecular configuration in the bookshelf (a) and chevron (b) states.

smectic C phase. The design of the device assumed a surface alignment in which the molecular long axes are constrained to lie in the plane of the cell, but with no preferred alignment direction within this plane.

The spontaneous polarization vector in a liquid crystal layer in the bookshelf structure lies perpendicular to the plane of the cell in either the 'up' or 'down' direction, and can reorientate in response to an applied DC voltage pulse. Each of these configurations has the same energy, and the switched states can be bistable indefinitely. The switching of polarization vectors is coupled (Fig. 22) to a rotation of the molecular long axes of the molecules in the smectic layers, and therefore to a rotation of the optic axis of the phase through twice the tilt angle. As is noted above, the molecular tilt angle is commonly of the order of 22.5°. Rotation of the optic axis through 45° can provide an efficient optical switch if the cell is placed between polarizers and the retardation of the liquid crystal layer is chosen to be $\lambda/2$, where λ is the wavelength of incident light. For a 2 µm cell and visible light, the latter condition can be satisfied by a liquid crys-

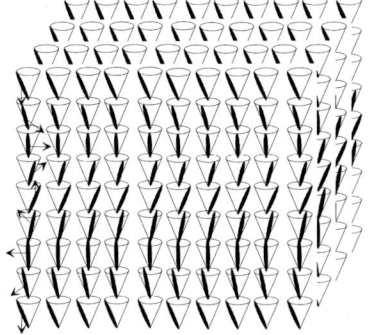

Figure 20. Schematic diagram of the arrangement of molecular tilt in a smectic C* phase showing the spiralling polarization vector (front left). The regular spatial arrangement of molecules is for clarity and is not present in the real phase.

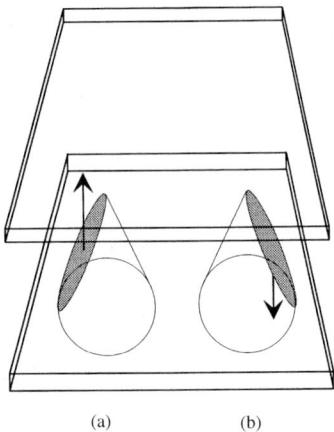

Figure 22. Mechanism of switching in smectic C* devices. Application of a pulse of one polarity to the cell can switch the polarization from the 'UP' state (a) to the 'DOWN' state (b), or vice versa, with consequent rotation of the molecular long axes and optic axis through twice the tilt angle.

tal with a convenient birefringence near 0.14.

The shear alignment used in the first ferroelectric cells was obviously unsatisfactory if such devices were to be fabricated in volume. Techniques were therefore developed [76–78] which allowed fabrication of ferroelectric devices using conventional rubbed polymer alignment layers on the cell walls in conjunction with a liquid crystal material having a I SmA* SmC* or I N* SmA* SmC* phase sequence. In the latter case, when the chiral nematic phase has a long pitch, excellent alignment of the smectic C* phase can be obtained. Devices made in this way show an electro-optic behavior that is inconsistent with the bookshelf structure, and contain visible alignment defects ('zig-zags'), which could not be explained in the framework of a bookshelf structure. X-ray diffraction studies [79] have also demonstrated that the smectic layers in such a cell are bimodally tilted at angles typically near 18–20° (a little less than the tilt angle of the molecules in the layers). The accepted configuration of the layers in these devices is the chevron structure shown in Fig. 21 b. The opposing direction of tilt in each half of the cell accounts for the X-ray data. Zig-zag defects are described further below.

The consequences of chevron formation in the ferroelectric cell are profound. The issue is not that the layer tilt angle is fairly small, but that it is of similar magnitude to the molecular tilt within the layers. The dual constraints of fixed tilt of the molecular long axes in the layers, and the small tilt at the cell walls away from the plane of the cell lead to the zero field configurations in the two switched states of the cell lying with the optic axes separated by a small angle, and with the spontaneous polarization vector nearly in the plane of the cell. At the centre of the cell, there is a defect in the liquid crystal structure corresponding to the interface of smectic layers tilted in opposite directions in each half of the chevron. The polarization in each half of the cell should lie in opposing directions also, and at present it is not clear how to reconcile the various forces acting on the liquid crystal at this point.

The probable, though tentative, explanation for chevron formation is a pinning of the layer spacing at the aligning layers, as the layers form in the smectic A* phase. On entering the smectic C* phase, the molecular tilt results in a reduction in the layer thickness. This reduction can be accommodated at the aligning surface without unpinning the periodicity of the anchoring by allowing the layers to tilt: tilt of equal and opposite magnitude at each surface leads to a chevron structure. This simple account would lead one to expect that the layer tilt angle δ should equal the molecular tilt angle ϕ. The discrepancy, usually of the order of 1–3° can be accounted for by an increase in order parameter as the temperature is reduced below the SmC*–SmA* transition.

The consequences of the chevron structure for the electro-optic behavior of a ferroelectric cell are multifaceted. Most evident is that the apparent rotation angle of the optic axis of the cell induced by voltage pulses relaxes to a value considerably less than that expected from the molecular tilt angle [80]. A switching angle close to the expected value can be recovered, however, by application of a high-frequency AC voltage. This phenomenon also stabilizes the two switched states in the cell (AC stabilization) by preventing reversion to an intermediate state of mixed alignment. These effects evidently originate from an interaction between the applied AC signal and the dielectric properties of the ferroelectric material. On a practical note, if a device is being addressed with a multiplex drive scheme, this will in itself serve to provide a high-frequency AC signal superimposed on the switching pulses and give AC stabilization.

Ferroelectric liquid crystal devices have suffered from a number of characteristic alignment defects which are absent from nematic liquid crystal devices. The most important of these, termed 'zig-zag' defects (from their characteristic shape) result from the chevron structure (Fig. 23). The chevron can form with the layer tilt toward either of the directions perpendicular to the layer planes; if there is also a surface pretilt, its direction and that of the layer tilt may be parallel or opposed, and the two chevron regions which result are denoted by C1 and C2, and have different optical characteristics. Where these regions meet, a zig-zag defect appears.

In contrast to that in nematic liquid crystal devices, the alignment in smectic cells is not readily or spontaneously re-established if it becomes damaged. This is a consequence of the very high effective viscosity of the phase, and the one-dimensional solid-like behaviour induced by the presence of the smectic layers. Damage to the alignment can be induced [81] by mechanical shock or vibration of the cell, or through electrically induced movement of the smectic layers. The latter effect can be used in a controlled manner [82] to increase the optical contrast and bistability of ferroelectric cells, apparently by changing the natural chevron structure to one which is closer to that of the bookshelf device. Under other conditions, however, application of a high electric field to a cell gives a poorly aligned layer that cannot provide a high optical contrast.

The switching behaviour of a ferroelectric liquid crystal layer in the chevron or bookshelf structure depends on the individual physical parameters of the material and the cell in a much more complex manner than is the case for nematic devices. The detailed elastic behaviour of the phases is not understood in sufficient detail to interpret the switching performance of devices, or even their static structures. It is, however, possible to give some interpretation of the key features of their behavior. The response of the ferroelectric liquid crystal layer at low applied voltages is dominated by the torque due to coupling between the electric field and the spontaneous polarization. This torque is proportional to the applied field. At higher applied voltages, coupling between the field and the dielec-

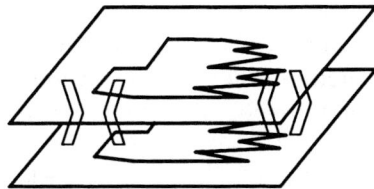

Figure 23. Diagrammatic structure of chevron layer disposition in the vicinity of a zig-zag defect.

tric constants becomes significant; this coupling rises as the square of the field, and can dominate the response to high voltages if the magnitude of the spontaneous polarization is modest. The details of the dielectric coupling can only be understood [83] by considering all three principal dielectric constants. This is demonstrated by considering the effect of an AC voltage on a chevron structure. In this case, the liquid crystal molecules lie almost in the plane of the cell in the zero field state, so that an imposed field is expected to have little effect if the material has the usual negative dielectric anisotropy. Experimentally, however, an AC voltage causes a reorientation of the liquid crystal alignment, which results in an increase in the apparent cone angle. This can only be explained by a coupling of the voltage to the (positive) biaxial dielectric anisotropy. This coupling stabilizes the two switched states of the device, but also provides a force which opposes the polarization vector mediated switching between them. Thus at high applied fields, the switching of such a cell becomes slower, in contrast to the low field behaviour, and ultimately under still higher fields the switching is suppressed altogether (Fig. 24).

The design of drive schemes for ferroelectric cells is much less well developed than are those for rms responding devices. It is complicated by the polarity sensitivity of the polarization vector switching process and the need to maintain a DC balance across each pixel, irrespective of the information displayed. Two distinct regimes have been identified and exploited for the multiplex drive of ferroelectric cells, using voltage pulses of a magnitude below and above those which give the minimum response time. Examples of such schemes are presented in Figs. 25 and 26, respectively. Consider first the scheme illustrated in Fig. 25, often called the 'Seiko scheme' [84], which operates in a region such as that denoted by the line XX in Fig. 24. In this region, switching occurs when an applied voltage pulse of a certain duration exceeds a critical value. As in the multiplex addressing of an rms responding display, strobe pulses that are independent of the information displayed are applied successively to the rows of the display matrix, while a succession of data pulses is applied to each column. The switching of each pixel is defined by the polarity of the data pulse applied simultaneously with a strobe pulse at that point in the display.

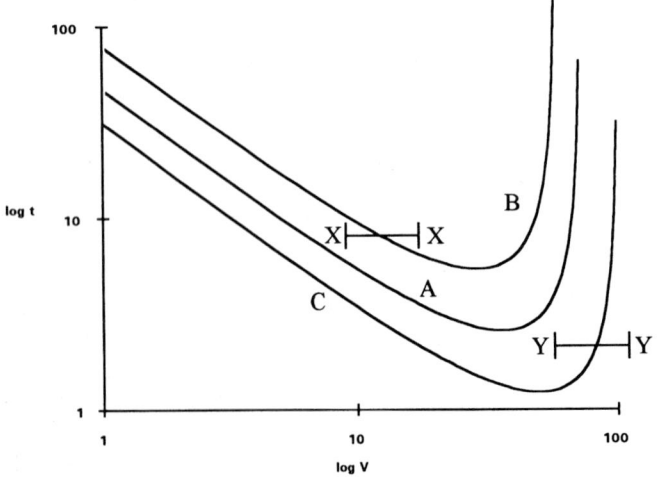

Figure 24. Variation of duration against the amplitude of the pulse required to achieve complete switching in a cell containing a ferroelectric liquid crystal of moderate polarization vector. The curves relate to the response to a single pulse (A), to a pulse when preceded by a similar pulse of opposite polarity (B), and to the same pulse preceded by a (smaller) data pulse of the same polarity (C). Lines XX and YY illustrate possible working regions for the multiplex schemes shown in Figures 25 and 26, respectively.

1.6 Ferroelectric Liquid Crystal Displays

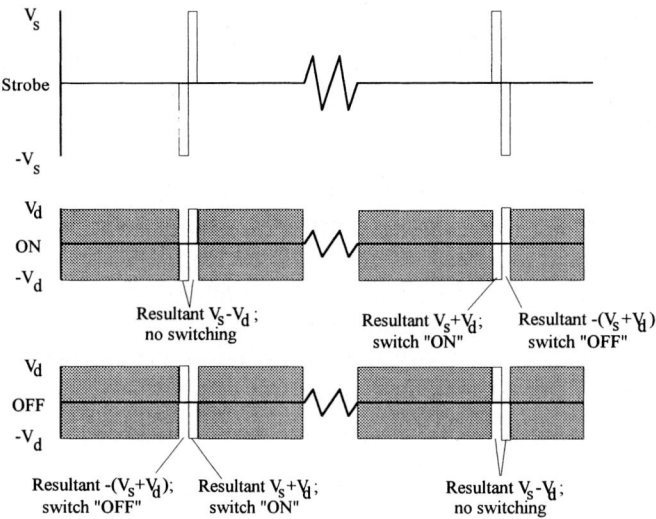

Figure 25. Example of a multiplex scheme for ferroelectric devices operated below the *V/t* minimum (see text).

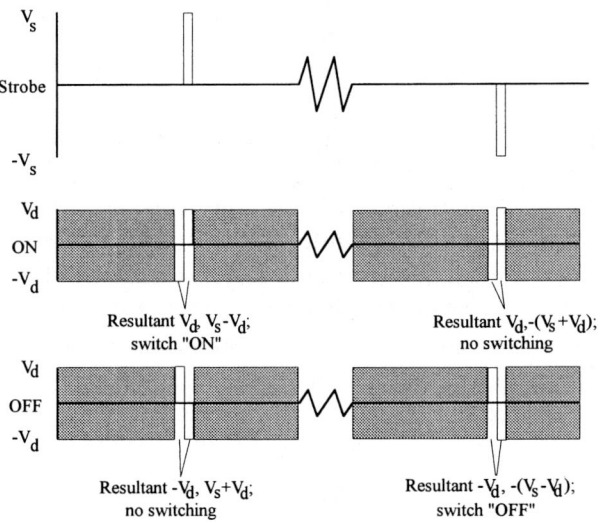

Figure 26. Example of a multiplex scheme for a ferroelectric device operated above the *V/t* minium (see text).

The magnitudes of the strobe and data signals are chosen such that the device will switch under a pulse of amplitude $V_s + V_d$, but not under pulses of magnitude V_d or $V_s - V_d$. Strict DC balance of the signal across each pixel is ensured by using a bipolar pulse for every data and strobe signal. Appropriate pixels can be switched to the 'off' state by a bipolar strobe signal with, say, a negative-going trailing pulse. In combination with an out of phase data signal, the pixel is switched 'on' and then immediately 'off'. As the addressing pulse occupies a very small fraction of the frame address time of the panel, the overall impression is that the pixel switches 'off' cleanly. An in-phase data signal has no effect on the pixel. Note that addressing the pixel in this way requires double the response time of the liquid crystal device to accommodate the bipolar pulse.

At this point, however, we have only switched 'off' pixels in the device. Switching 'on' of the remaining pixels is achieved by repeating the addressing of the whole panel using strobe pulses of the opposite polarity to that used before. In combination with the appropriate data signal, the target pixels are then switched first 'off' and then immediately 'on' to achieve the desired information state. In effect, the pixels switch to the state determined by the trailing pulse of the strobe signals, and the switching is determining by the polarity of the data signal at that instant. Addressing of the whole panel requires a period of four times the response time of the pixel, multiplied by the number of lines in the display.

The addressing scheme shown in Fig. 26 [85] is known as the JOERS–Alvey scheme. It uses a unipolar strobe pulse to switch a device in the regime above the V/t minimum, such as the region denoted by YY in Fig. 24. In this case, switching is accomplished by a voltage pulse of magnitude $V_s - V_d$, but not by pulses of magnitude V_d or $V_s + V_d$. In the scheme shown, the data signal is a bipolar pulse as before, but the strobe is a pulse of a single polarity. Pixels again switch according to the polarity of the strobe pulse, together with the trailing pulse of the corresponding data signal, but to achieve switching the polarity of the data pulse must be opposite to that in the first scheme described above. Each switching pulse is now preceded by a pulse on the pixel of magnitude V_d, which has the correct polarity to switch the device into its desired state. Although this leading pulse is too small in magnitude to cause actual switching, it conditions the pixel and reduces the time required to achieve switching under the influence of the following addressing pulse. Contrast this with the first scheme, in which each switching event is immediately preceded by a pulse that switches the pixel into the opposite state. The conditioning of the pixel by the leading pulses is the origin of the different curves on which lines XX and YY are drawn in Fig. 24, and is crucial in determining the relative merits of the two approaches. The latter scheme achieves not only an improvement in switching speed from this effect, but also an increase in the operating margin and optical contrast in the display. Typically, devices operated under the Seiko scheme are constructed so that the liquid crystal is in the C1 state. This maximizes the apparent switching angle, but requires a large motion of the director in the surface layers of the cell, which may slow the switching. The JOERS–Alvey scheme is usually used to address devices in the C2 state. A large apparent switching angle and high contrast are then achieved by coupling of the data signal (which is continuously present on every pixel of the device) to the dielectric anisotropy of the liquid crystal fluid. The switching occurs predominantly in the bulk layers of the cell rather than near the surfaces, giving a further switching speed advantage. The speed advantage is partly offset by the requirement to use lower spontaneous polarization liquid crystals, but the other gains remain decisive for many prospective applications. Still faster addressing schemes are known [86], and the development of optimized fluids for use in ferroelectric liquid crystal devices and further advances in addressing techniques are active areas for research.

1.7 References

[1] G. W. Gray (Ed.), *Thermotropic Liquid Crystals*, Wiley, Chichester **1987**.
[2] S. Chandrasekhar, *Liquid Crystals*, 2nd edn., Cambridge University Press, Cambridge **1992**.
[3] S. Morozumi in *Liquid Crystals Applications and Uses*, Vol. 1 (Ed.: B. Bahadur), World Scientific, Singapore **1990**, Chap. 7.

[4] T. Uchida, H. Seki in *Liquid Crystals Applications and Uses*, Vol. 3 (Ed.: B. Bahadur), World Scientific, Singapore **1992**, Chap. 1.
[5] V. Fréedericksz, V. Zolina, *Trans. Faraday Soc.* **1933**, *29*, 919.
[6] E. P. Raynes, R. J. A. Tough, K. A. Davies, *Mol. Cryst. Liq. Cryst.* **1979**, *56*, 63.
[7] Liquid Crystal Mixture E 7, Merck Ltd, Poole, Dorset, UK.
[8] P. G. de Gennes, *Mol. Cryst. Liq. Cryst.* **1971**, *12*, 193.
[9] M. Schadt, W. Helfrich, *Appl. Phys. Lett.* **1971**, *18*, 127.
[10] F. C. Luo in *Liquid Crystals Applications and Uses*, Vol. 1 (Ed.: B. Bahadur), World Scientific, Singapore **1990**, Chap. 15.
[11] E. P. Raynes, *Rev. Phys. Appl.* **1975**, *10*, 117.
[12] C. M. Waters, V. Brimmell, E. P. Raynes, *Proc. 3rd Disp. Res. Conf.* **1983**, 396.
[13] D. W. Berreman, *Phil. Trans. R. Soc. London, Ser. A* **1983**, *309*, 203.
[14] R. N. Thurston, D. W. Berreman, *J. Appl. Phys.* **1981**, *52*, 508.
[15] D. W. Berreman, W. R. Heffner, *J. Appl. Phys.* **1981**, *52*, 3032.
[16] V. G. Chigrinov, V. V. Belyaev, S. V. Belyaev, M. F. Grebenkin, *Sov. Phys. JETP* **1979**, *50*, 994.
[17] P. Schiller, K. Schiller, *Liq. Cryst.* **1990**, *8*, 553.
[18] M. Kawachi, K. Kato, O. Kogure, *Jpn. J. Appl. Phys.* **1977**, *16*, 1263.
[19] A. Mochizuki et al. *Proc. SID* **1985**, *26 (4)*, 243.
[20] M. F. Schiekel, K. Fahrenschon, *Appl. Phys. Lett.* **1971**, *19*, 393.
[21] P. Yeh, *Optical Waves in Layered Media*, Wiley, New York **1988**.
[22] C. H. Gooch, H. A. Tarry, *J. Phys. D: Appl. Phys.* **1975**, *8*, 1575.
[23] M. Schadt, W. Helfrich, *Appl. Phys. Lett.* **1971**, *18*, 127.
[24] D. Berreman, *J. Opt. Soc. Am.* **1973**, *63*, 1374.
[25] T. J. Scheffer, J. Nehring, *Appl. Phys. Lett.* **1984**, *45*, 1021.
[26] M. Schadt, F. Leenhouts, *Appl. Phys. Lett.* **1987**, *50*, 236.
[27] E. P. Raynes, *Mol. Cryst. Liq. Cryst.* **1987**, *4*, 159.
[28] K. Katoh et al., *Jpn. J. Appl. Phys.* **1987**, *26*, L1784.
[29] I. Fukuda, Y. Kotani, T. Uchida, *Proc. 8th Intl Disp. Res. Conf.* **1988**, 159.
[30] S. Matsumoto et al., *Proc. 8th Intl Disp. Res. Conf.* **1988**, 182.
[31] H. Odai et al., *Proc 8th Intl Disp. Res. Conf.* **1988**, 195.
[32] M. Yamauchi et al., *SID '89 Digest* **1989**, 378.
[33] S. N. Yamagishi, H. Watanabe, K. Yokoyama, *Abstracts, Japan Display '89* **1989**, 316.
[34] M. Ishikawa et al., *J. Soc. Inf. Disp.* **1995**, *3*, 237.
[35] K. H. Yang, *Proc. 11th Intl Disp. Res. Conf.* **1991**, 68.
[36] J. Chen et al., *SID '95 Digest* **1995**, 865.
[37] P. J. Bos, P. A. Johnson, K. R. Koehler, *SID Symp. Digest* **1983**, 30.
[38] T. Uchida, T. Miyashita, *Proc. 2nd Intl Disp. Workshops* **1995**, 39.
[39] R. Kiefer, B. Weber, F. Windscheid, G. Baur, *Proc. 12th Intl Disp. Res. Conf.* **1992**, 547.
[40] K. Kondo, N. Konoshi, K. Kinugawa, H. Kawakami, *Proc. 2nd Intl Disp. Workshops* **1995**, 42.
[41] G. Heilmeier, L. Zanoni, *Appl. Phys. Lett.* **1968**, *13*, 91.
[42] D. White, G. Taylor, *J. Appl. Phys.* **1974**, *45*, 4718.
[43] C. M. Waters, V. Brimmell, E. P. Raynes, *Proc. 3rd Disp. Res. Conf.* **1983**, 396.
[44] G. W. Gray, K. J. Harrison, J. A. Nash, *Electron. Lett.* **1973**, *9*, 130.
[45] A. Boller, M. Cereghetti, M. Schadt, H. Scherrer, *Mol. Cryst. Liq. Cryst.* **1977**, *42*, 1225.
[46] R. Eidenschink, D. Erdmann, J. Krause, L. Pohl, *Angew. Chem., Int. Ed. Engl.* **1977**, *16*, 100.
[47] W. H. de Jeu, *Phil. Trans. R. Soc. London, Ser. A* **1983**, *309*, 217.
[48] M. Schadt, M. Petrzilka, P. Gerber, A. Villiger, *Mol. Cryst. Liq. Cryst.* **1985**, *122*, 241.
[49] M. Schadt, R. Buchecker, K. Muller, *Liq. Cryst.* **1989**, *5*, 293.
[50] H. J. Plach, G. Weber, B. Rieger, *SID 1990 Digest* **1990**, 91.
[51] I. Sage in *Thermotropic Liquid Crystals* (Ed.: G. W. Gray), Wiley, Chichester **1987**.
[52] D. Coates in *Liquid Crystals Appications and Uses*, Vol. 1 (Ed.: B. Bahadur), World Scientific, Singapore **1990**, Chap. 3.
[53] L. Pohl, U. Finkenzeller in *Liquid Crystals Applications and Uses*, Vol. 1 (Ed.: B. Bahadur), World Scientific, Singapore **1990**, Chap. 4.
[54] G. H. Heilmeier, L. A. Zanoni, L. A. Barton, *Appl. Phys. Lett.* **1968**, *13*, 46.
[55] E. F. Carr, *Mol. Cryst. Liq. Cryst.* **1969**, *7*, 253.
[56] W. Helfrich, *J. Chem. Phys.* **1969**, *51*, 4092.
[57] R. Williams, *J. Chem. Phys.* **1963**, *39*, 384.
[58] G. H. Heilmeier, J. E. Goldmacher, *Proc. IEEE* **1969**, *57*, 34.
[59] D. Coates, W. A. Crossland, J. H. Morrissey, B. Needham, *J. Phys. D* **1978**, *11*, 2025.
[60] F. J. Kahn, *Appl. Phys. Lett.* **1973**, *22*, 111.
[61] S. LeBerre, M. Hareng, R. Hehlen, J. N. Perbet, *Displays* **1981**, 349.
[62] D. J. Armitage, *J. Appl. Phys.* **1981**, *52*, 4843.
[63] J. Fergason, *SID Digest* **1985**, *16*, 68.
[64] J. W. Doane, N. A. Vaz, B. G. Wu, S. Zumer, *Appl. Phys. Lett.* **1986**, *48*, 269.
[65] J. L. West, *Mol. Cryst. Liq. Cryst.* **1988**, *157*, 427.
[66] P. S. Drzaic, *Liq. Cryst.* **1988**, *3*, 1543.

[67] P. Alt, P. Pleshko, *IEEE Trans. Electron Devices* **1974**, *21*, 146.
[68] T. J. Scheffer, B. Clifton, D. Prince, A. R. Conner, *Displays* **1993**, *14*, 74.
[69] B. J. Lechner, *Proc. IEEE* **1971**, *59*, 1566.
[70] N. Szydia et al., *Proc. 1983 Japan Display* **1983**, 416.
[71] D. R. Baraff et al. *Proc. 1980 Bienn Display Research Conf.* **1980**, 109.
[72] Y. Hirai, H. Katoh, *Proc. 2nd Intl Display Workshops* **1995**, 49.
[73] R. Barberi, M. Boix, G. Durand, *Appl. Phys. Lett.* **1989**, *55*, 2506.
[74] R. Barberi, G. Durand, *Appl. Phys. Lett.* **1991**, *58*, 2709.
[75] R. B. Meyer, L. Liebert, L. Strzelecki, P. Keller, *J. Phys. Lett. (Paris)* **1975**, *36*, 69.
[76] J. S. Patel, T. M. Leslie, J. W. Goodby, *Ferroelectrics* **1984**, *61*, 137.
[77] M. J. Bradshaw, V. Brimmel, E. P. Raynes, *Proc. Japan Display 1986*, post-deadline paper 5, **1986**.
[78] M. J. Bradshaw, V. Brimmel, E. P. Raynes, British Patent Appl. 86:08 114, **1986**.
[79] N. A. Clark, T. P. Reiker, J. E. MacLennan, *Ferroelectrics* **1988**, *85*, 79.
[80] J. P. le Pesant et al., *J. Phys. (Paris)* **1984**, C-5, 217.
[81] N. Wakita et al., *Ferroelectrics* **1993**, *149*, 229.
[82] W. Hartmann, *Ferroelectrics* **1988**, *85*, 67.
[83] J. C. Jones, E. P. Raynes, *Liq. Cryst.* **1992**, *11*, 199.
[84] T. Harada et al., *Proc. SID Intl Symp.* **1985**, 131.
[85] P. W. H. Surguy et al., *Ferroelectrics* **1991**, *63*, 122.
[86] J. R. Hughes, E. P. Raynes, *Liq. Cryst.* **1993**, *13*, 597.

2 Nondisplay Applications of Liquid Crystals

William A. Crossland and Timothy D. Wilkinson

There are many nondisplay applications of liquid crystals, and it is not possible to do justice to all of them here in this section. Here we will outline some key examples, and summarize briefly a wider selection. Some of the most developed applications that are currently being researched are in optical correlators, optical interconnections, wavelength filters, and optoelectronic neural networks. Most of these systems use liquid crystal spatial light modulators (SLMs). In addition, we will describe how SLMs are used in autostereoscopic displays.

Display devices based on nematic liquid crystals, notably small liquid crystal televisions, are used as SLMs due to their increased availability [1]. However, in most of the above applications the speed of operation is important, and nematic liquid crystals are too slow, so the emphasis here is on SLMs that use the faster electrooptic effects to be found in chiral smectic liquid crystals. We will start by looking at how these interact with light.

2.1 Liquid Crystal Spatial Light Modulation

2.1.1 Polarized Light and Birefringence

Liquid crystals are birefringent and they therefore influence the state of polarization of light beams. This interaction is described by the Jones calculus [2], which is briefly outlined below, prior to using it to establish basic results in light modulation by ferroelectric liquid crystals.

Monochromatic light sources such as lasers can be represented in terms of an orthogonal set of propagating eigenwaves which are usually aligned to the x and y axes in a coordinate system with the direction or propagation along the z axis, as shown in Fig. 1. These eigenwaves can be used to describe the propagation of light through complex media. If the light is polarized in the direction of the y axis, then we have linearly polarized light in the y direction of amplitude V_y or vertically polarized light. This wave can be represented as a Jones matrix

$$V = \begin{pmatrix} 0 \\ V_y \end{pmatrix} \qquad (1)$$

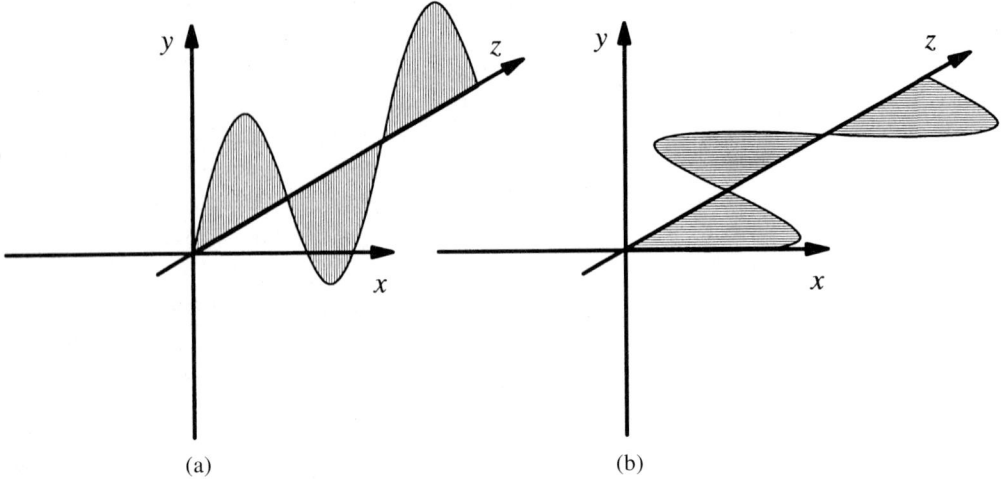

Figure 1. Eigenwaves of polarized light: (a) vertically polarized light, (b) horizontally polarized light.

If we have an electromagnetic wave propagating in the z direction along the x axis, then the light is classified as linearly polarized in the x direction or horizontally polarized.

$$V = \begin{pmatrix} V_x \\ 0 \end{pmatrix} \tag{2}$$

We can now combine these two eigenwaves to make any linear state of polarization that we require. If the polarization of the light were to bisect the x and y axes by 45°, we could represent this as $V_x = V_y$ and use the two states combined into a single Jones matrix

$$V = \begin{pmatrix} V_x \\ V_y \end{pmatrix} = V_x \begin{pmatrix} 1 \\ 1 \end{pmatrix} \tag{3}$$

We can also represent more complex states of polarization, such as circular states. So far we have assumed that the eigenwaves are in phase (i.e., they start at the same point). We can also introduce a phase difference ϕ between the two eigenwaves, which leads to circularly polarized light. In these examples, the phase difference ϕ is positive in the direction of the z axis and is always measured with reference to the vertically polarized eigenwave (parallel to the y axis), hence we can write the Jones matrix

$$V = \begin{pmatrix} V_x \\ V_y e^{j\phi} \end{pmatrix} \tag{4}$$

There are two states that give circularly polarized light. If ϕ is positive, then the horizontal component leads the vertical and the resultant director appears to rotate to the right around the z axis in a clockwise manner and gives right circularly polarized light. Conversely, if the horizontal lags behind the vertical, then the rotation is counter-clockwise and the light is left circularly polarized. In the case of pure circularly polarized light, $\phi = \pi/2$ for right circular and $\phi = -\pi/2$ for left circular

$$\text{Right circular } V = V_x \begin{pmatrix} 1 \\ j \end{pmatrix}$$

$$\text{Left circular } V = V_x \begin{pmatrix} 1 \\ -j \end{pmatrix} \tag{5}$$

Some crystalline materials such as sodium chloride have a cubic molecular structure.

When light passes through these structures it sees no preferred direction and is relatively unaffected. If the crystal has a structure such as hexagonal or trigonal, different directions of light will see very different crystalline structures. The effect of this is called birefringence, which is a property that is exploited in retarders; it is also the effect leading to the modulation of coherent light in liquid crystals. In a birefringent material, each eigenwave sees a different refractive index and will propagate at a different speed. This leads to a phase retardation between the two eigenwaves which is dependent on the thickness of the birefringent material and the wavelength of the light. The preferred directions of propagation within the crystal are defined as the *fast* (or extraordinary) axis and the *slow* (or ordinary axis). An eigenwave that passes in the same direction as the fast axis sees a refractive index n_f and the eigenwave that passes along the slow sees n_s. For light of wavelength λ passing through a birefringent crystal of thickness t, we define the retardation Γ as

$$\Gamma = \frac{2\pi t}{\lambda}(n_f - n_s) \quad (6)$$

The propagation of coherent light through such a birefringent material with this retardation can be expressed as a Jones matrix, assuming that the fast axis is parallel to the y axis

$$W_0 = \begin{pmatrix} e^{-j\Gamma/2} & 0 \\ 0 & e^{j\Gamma/2} \end{pmatrix} \quad (7)$$

It is more useful to be able to express the retardation from an arbitrary rotation of the fast axis by an angle ψ about the y axis. Rotation with Jones matrices can be done as with normal matrix rotation. If we define a counter-clockwise rotation of angle ψ about the y axis as positive, then the rotation matrix R is

$$R(\psi) = \begin{pmatrix} \cos\psi & \sin\psi \\ -\sin\psi & \cos\psi \end{pmatrix} \quad (8)$$

Hence the general form of the retardation plate is

$$W = R(-\psi)\, W_0\, R(\psi) \quad (9)$$

This can be expanded right to left (in normal matrix fashion) to give

$$W = \begin{pmatrix} e^{-j\Gamma/2}\cos^2\psi + e^{j\Gamma/2}\sin^2\psi & -j\sin\frac{\Gamma}{2}\sin(2\psi) \\ -j\sin\frac{\Gamma}{2}\sin(2\psi) & e^{j\Gamma/2}\cos^2\psi + e^{-j\Gamma/2}\sin^2\psi \end{pmatrix} \quad (10)$$

This matrix now allows us to propagate coherent light through an arbitrarily oriented birefringent material. We can also define useful retardation elements such as half and quarter waveplates, which can be used to make up functional optical systems. A combination of these elements can also be analyzed from right to left as a series of matrix multiplications.

2.1.1.1 The Half Wave plate

A half wave plate is a special example of the generalized retarder. In this case, the thickness of the plate has been chosen to give a phase retardation of exactly $\Gamma = \pi$. Hence the Jones matrix for a half wave plate at an angle ψ will be

$$j\begin{pmatrix} -\cos 2\psi & \sin 2\psi \\ \sin 2\psi & \cos 2\psi \end{pmatrix} \quad (11)$$

If the fast axis is aligned with the y axis, then $\psi = 0$, and the Jones matrix is

$$\begin{pmatrix} -j & 0 \\ 0 & j \end{pmatrix} \quad (11a)$$

A half wave plate can be used to rotate the direction of linearly polarized light from one linear state to another, which is a very useful property in an optical system.

2.1.1.2 The Quarter Wave plate

In a similar fashion, we can tailor the thickness to give a quarter wave retardation of $\Gamma = \pi/2$. Such a wave plate is useful for converting to and from circularly polarized light. For a quarter wave plate with its fast axis aligned to the y axis ($\psi = 0$), the Jones matrix will be

$$\frac{1}{\sqrt{2}} \begin{pmatrix} 1-j & 0 \\ 0 & 1+j \end{pmatrix} \qquad (11\,\text{b})$$

2.1.1.3 Linear Polarizers

An important function in an optical system is to be able to filter out unwanted polarization states whilst passing desired states, as shown in Fig. 2. This can be done using polarizers which pass a single linear state whilst blocking all others.

A polarizer oriented at an angle ψ from the y axis can be written as a Jones matrix

$$P = \begin{pmatrix} \sin^2 \psi & \sin 2\psi \\ \sin 2\psi & \cos^2 \psi \end{pmatrix} \qquad (12)$$

2.1.2 Electro-Optic Effects in Chiral Smectic C Phases

Molecular ordering and electrooptic switching in chiral smectic phases are discussed in detail elsewhere (see Chap. VII, Sec. 9 of this Volume). In the case of ferroelectric switching in chiral smectic C phases, the director **n** is free to move about a cone of angles which is centered on the horizontal axis. Each molecule has a ferroelectric dipole **P**, which is perpendicular to its length. This is depicted in Fig. 3.

When an electric field **E** is applied to the cell, there is an interaction between **E** and **P**, which forces the director to move around the cone to a point of equilibrium. If the field is changed, the director moves again. The phase adopts a structure in which the **n** director precesses helically around the cone from layer to layer to minimize the electrostatic interaction with the side dipoles, thus destroying any bulk spontaneous charge separation.

If the FLC (Ferroelectric Liquid Crystal) is constrained in a thin layer (usually 1–3 µm) between suitable aligning surfaces, it can adopt the structure shown in Fig. 4, i.e., the precessing of the **n** director can be suppressed and the material becomes ferroelectric [3]. In this case, applying an electric field will switch the **n** director (and there-

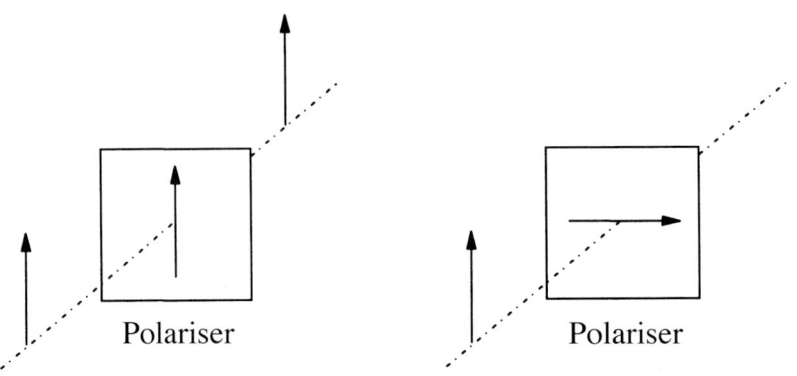

Figure 2. Action of polarizers with coherent illumination.

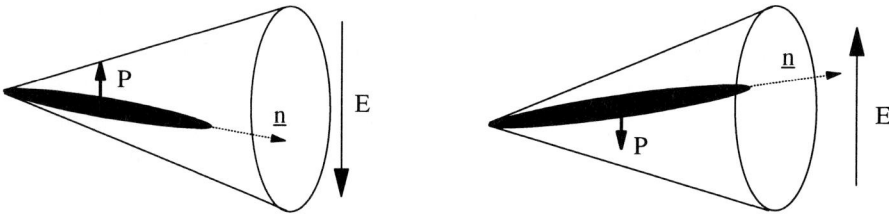

Figure 3. Molecular interaction with an applied electric field.

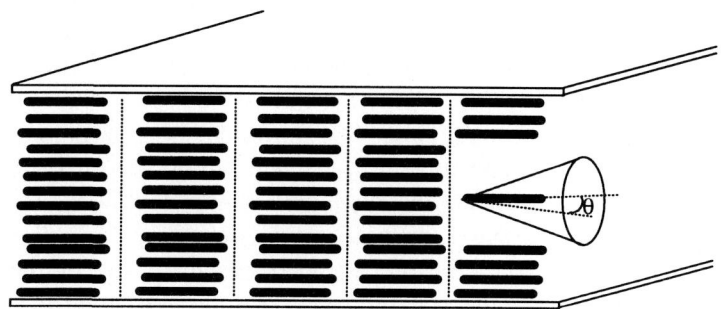

Figure 4. Layering in a surface stabilized ferroelectric liquid crystal (SSFLC) cell showing the degenerate cone for the tilt of the molecules.

fore the optical axis) from one side of the cone to the other. The angle between the two switched states is the switching angle θ, which for the so-called 'bookshelf' structure shown in Fig. 4 is twice the liquid crystal tilt angle [4]. Since only these two states are stable, the effect exhibits electrooptic bistability. Depending on the methods of device fabrication (surface alignment, pretreatment to the liquid crystal layer), the FLCs with nematic and smectic A phases above the SmC* phase may form the 'chevron' structure [5] (see Chap. X, Sec. 10 of this volume). In this case, the two equilibrium stable states will be separated by a switching angle that is less than twice the liquid crystal tilt angle.

The viscosity parameters associated with the electrooptic effect, and the increased torque that can be exerted on the liquid crystal by an applied electric field due to the spontaneous polarization, result in an electrooptic effect that can be several orders of magnitude faster than in nematics. Switching speeds below 10 μs can be achieved [6]. In such switching, the liquid crystal layer acts as an optically uniaxial medium with its optic axis in the plane of the layer. Hence it behaves like a switchable wave plate whose fast and slow axes can be in two possible states separated by the switching angle q, and whose retardation depends on the thickness and birefringence of the FLC. This is not only true for ferroelectric switching in the chiral smectic C phase, but also for electroclinic switching in chiral smectic A phases and for the deformed helix electrooptic effect in tight pitch helical smectic C structures. These latter two effects offer a proportional response to applied DC (direct current) voltages in contrast to the electrooptically bistable switching described above. The electrooptic symmetry of switching in chiral smectic phases is therefore fundamentally different from that of electrooptic effects in nematic liquid crystals that involve tilting of the optic axis out of the plane of the thin films.

2.1.3 The FLC Spatial Light Modulator

The FLC SLM is constructed as a thin layer (approx. 2 µm thick) of smectic C* FLC sandwiched between two transparent electrode arrays, as in Fig. 5a. The molecules are usually arranged in the bookshelf geometry and are all aligned to the glass electrodes by alignment layers which the FLC molecules follow.

The electric field is applied via a pixelated pattern of ITO (indium tin oxide) on the glass. One of the advantages of the binary modulation and bistability of the FLC is that pixels can be row/column addressed [7]. Hence very large matrices of pixels can be passively multiplexed one line at a time.

The rear glass wall can be replaced by a silicon VLSI die to make a silicon backplane SLM as in Fig. 5b. The SLM is now a reflective modulator with the aluminum from the second metal layer of the VLSI process acting as a mirror [8]. Circuitry on the backplane can be used to address the FLC pixels as either dynamic random access memory (DRAM) [9] or static RAM (SRAM) [10]. The silicon backplane SLM is very important to the development of nondisplay applications as it allows large arrays of small pixels to be built on a silicon wafer capable of high addressing speeds. Compared with line-at-a-time multiplexed SLMs, these silicon backplane SLMs allow a full frame of pixels to be addressed in one liquid crystal response period (as opposed to one row), hence they can be K times faster, where K is the number of rows. The pixel circuitry also isolates the liquid crystal from the electrical crosstalk inevitably occurring in passively multiplexed devices, so the effective contrast ratio will be much higher. However, metal layers other than standard VLSI metallization must be used to achieve high reflectances and contrast ratios (>20:1). A 320×240 array of 37 µm pixels capable of displaying 22 000 frames per second is shown in Fig. 6 [11]. Other silicon backplane devices include 256×256, 768×512, and 1280×1024 pixel arrays.

The interaction between the FLC molecules and the coherent polarized light can be modeled as a switchable wave plate with two possible axis positions. Hence we can use Jones matrices to model this interaction and polarizers to orient the light; then we can analyze the light after propagation through the FLC.

2.1.4 Binary Intensity Modulation

If the light is polarized so that it passes through an FLC pixel parallel to the fast axis in one state, then there is no change due

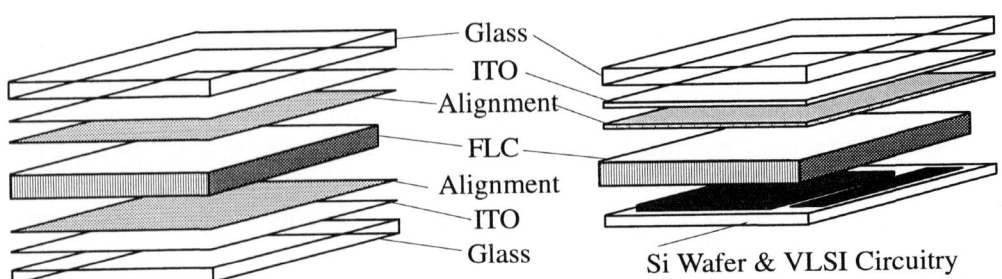

Figure 5. Construction of an SLM: (a) transmissive, (b) reflective.

Figure 6. A 320×240 pixel silicon backplane FLC SLM.

to the birefringence and the light will pass through a polarizer which is also parallel to the fast axis. This is demonstrated in Fig. 7. If the pixel is then switched into State 2, the fast axis is rotated by the FLC switching angle θ and the light now undergoes some birefringent interaction. We can use Jones matrices to represent the optical components

State 1

$$\begin{pmatrix} V_x \\ V_y \end{pmatrix} = \begin{pmatrix} 0 & 0 \\ 0 & 1 \end{pmatrix} \begin{pmatrix} e^{-j\Gamma/2} & 0 \\ 0 & e^{j\Gamma/2} \end{pmatrix} \begin{pmatrix} 0 \\ V_y \end{pmatrix}$$

$$= \begin{pmatrix} 0 \\ V_y e^{j\Gamma/2} \end{pmatrix} \quad (13)$$

State 2

$$\begin{pmatrix} V'_x \\ V'_y \end{pmatrix} = \begin{pmatrix} 0 & 0 \\ 0 & 1 \end{pmatrix} \quad (14)$$

$$\cdot \begin{pmatrix} e^{-j\Gamma/2}\cos^2\theta + e^{j\Gamma/2}\sin^2\theta & -j\sin\dfrac{\Gamma}{2}\sin(2\theta) \\ -j\sin\dfrac{\Gamma}{2}\sin(2\theta) & e^{j\Gamma/2}\cos^2\theta + e^{-j\Gamma/2}\sin^2\theta \end{pmatrix} \begin{pmatrix} 0 \\ V_y \end{pmatrix}$$

$$= \begin{pmatrix} 0 \\ V_y\left(e^{j\Gamma/2}\cos^2\theta + e^{-j\Gamma/2}\sin^2\theta\right) \end{pmatrix}$$

If the thickness of the FLC cell is set so that $\Gamma = \pi$ (as from Eq. 6), then the light in the

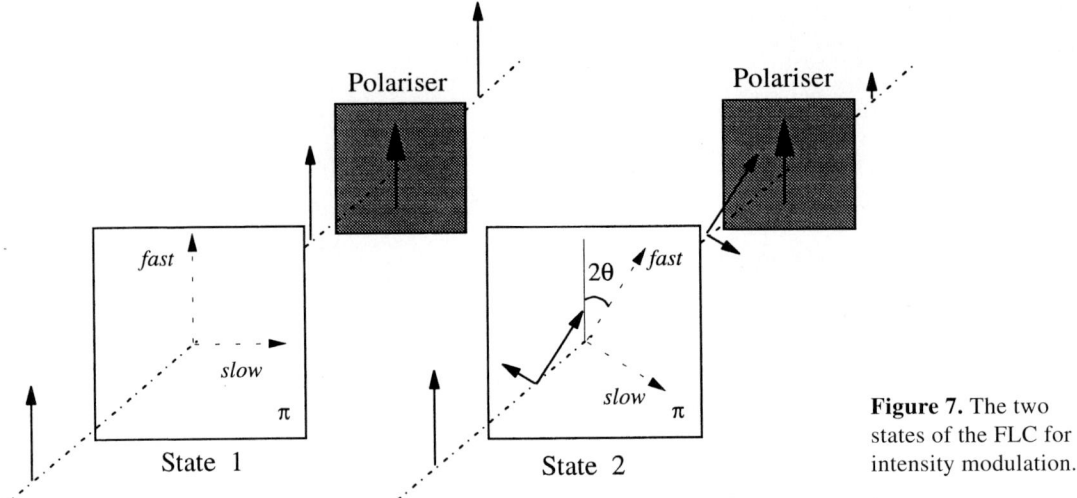

Figure 7. The two states of the FLC for intensity modulation.

direction of the slow axis will be rotated by 180°. This leads to a rotation of the polarization after the pixel, which is partially blocked by the following polarizer. A maximum contrast ratio will be achieved when State 2 is at 90° to the polarizer and the resulting horizontal polarization is blocked out. This will occur when

$$V_y \left(e^{j\Gamma/2} \cos^2\theta + e^{-j\Gamma/2} \sin^2\theta \right)$$
$$= V_y \left(j\cos^2\theta - j\sin^2\theta \right) = 0 \qquad (15)$$

Hence the optimum switching angle for an FLC performing intensity modulation is $\theta = 45°$. This is equivalent to a tilt angle of 22.5° and there are many fast FLC materials with this tilt available.

2.1.5 Binary Phase Modulation

If the light is polarized so that its direction bisects the switching angle and an analyzer (polarizer) is placed after the pixel at 90° to the input light, then phase modulation is possible [12], as shown in Fig. 8. If we start with vertically polarized light, then the FLC pixel fast axis positions must bisect the vertical axis and will be oriented at angles of

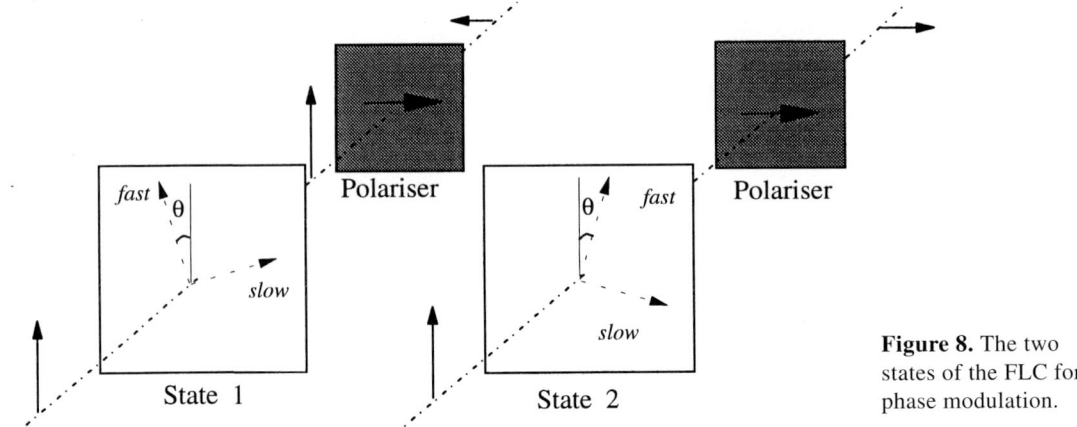

Figure 8. The two states of the FLC for phase modulation.

2.1 Liquid Crystal Spatial Light Modulation

$\theta/2$ and $-\theta/2$, respectively, for each state. Once again we can use Jones matrices to express the system

State 1

$$\begin{pmatrix} V'_x \\ V'_y \end{pmatrix} = \begin{pmatrix} 1 & 0 \\ 0 & 0 \end{pmatrix}$$

$$\cdot \begin{pmatrix} e^{-j\Gamma/2}\cos^2\frac{\theta}{2} + e^{j\Gamma/2}\sin^2\frac{\theta}{2} & -j\sin\frac{\Gamma}{2}\sin(\theta) \\ -j\sin\frac{\Gamma}{2}\sin(\theta) & e^{j\Gamma/2}\cos^2\frac{\theta}{2} + e^{-j\Gamma/2}\sin^2\frac{\theta}{2} \end{pmatrix} \begin{pmatrix} 0 \\ V_y \end{pmatrix}$$

$$= \begin{pmatrix} -V_y j\sin\frac{\Gamma}{2}\sin(\theta) \\ 0 \end{pmatrix} \quad (16)$$

State 2

$$\begin{pmatrix} V'_x \\ V'_y \end{pmatrix} = \begin{pmatrix} 1 & 0 \\ 0 & 0 \end{pmatrix}$$

$$\cdot \begin{pmatrix} e^{-j\Gamma/2}\cos^2\frac{\theta}{2} + e^{j\Gamma/2}\sin^2\frac{\theta}{2} & j\sin\frac{\Gamma}{2}\sin(\theta) \\ j\sin\frac{\Gamma}{2}\sin(\theta) & e^{j\Gamma/2}\cos^2\frac{\theta}{2} + e^{-j\Gamma/2}\sin^2\frac{\theta}{2} \end{pmatrix} \begin{pmatrix} 0 \\ V_y \end{pmatrix}$$

$$= \begin{pmatrix} -V_y j\sin\frac{\Gamma}{2}\sin(\theta) \\ 0 \end{pmatrix} \quad (17)$$

From these two expressions we can see that the difference between the two states is just the minus sign, which means that the light has been modulated by 180° (π phase modulation). Moreover, the phase modulation is independent of the switching angle θ and the retardation Γ. These parameters only affect the loss in transmission through the pixel, which can be gained by squaring the above expressions

$$T = V_y^2 \sin^2(\theta)\sin^2\left(\frac{\Gamma}{2}\right) \quad (18)$$

Hence maximum transmission (and therefore minimum loss) for phase modulation occurs when $\Gamma=\pi$ and $\theta=\pi/2$.

From this analysis it is possible to see why FLC SLMs form a vital part in optical systems. The liquid crystal modulation allows the display of information into coherent optical systems as both intensity and phase images. The nature of the FLC molecular interaction makes it fast and this ideally suits nondisplay applications. Even the restriction of binary modulation does not limit the usefulnes of FLC SLMs, although fast grey scale and multilevel phase modulation would be desirable. In fact, for some applications, binary modulation is a bonus and enhances performance.

2.1.6 The FLC Optically Addressed Spatial Light Modulator (OASLM)

The electric field that modulates the FLC molecules is externally supplied by interface circuitry, and hence the patterns displayed on the SLM must be loaded from a computer onto the pixels. An alternative method of addressing the FLC molecules is to use a photosensitive surface as part of the cell structure, which allows light intensity that is incident upon the OASLM to define the pattern displayed. The molecules are still switched by an electric field, but the field is now set up by the intensity pattern. The effect is much like a re-recordable photographic film.

An FLC OASLM is made up from a thin layer of FLC sandwiched between a conductive glass electrode and a photoconductive layer, as illustrated in Fig. 9. There is still an external electrical field applied, but this is common for the whole OASLM and is

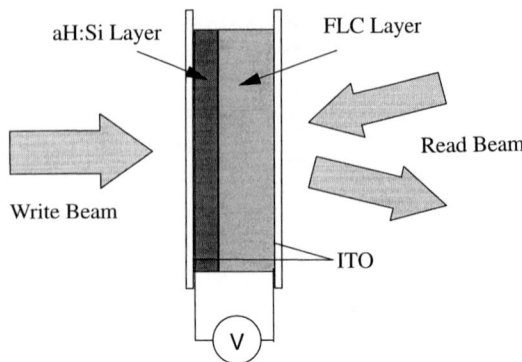

Figure 9. Construction of an FLC OASLM.

used to time the read and write cycles of the OASLM. The photoconductive layer is usually a thin film of amorphous silicon (aH:Si) which has the required photoactive properties. When the external electrical field is positive (write cycle), a voltage appears across the aH:Si and FLC layers. Figure 10 shows the addressing states of the OASLM. If the aH:Si is not illuminated (in a dark state), then its resistivity is very high, and most of the voltage appears across the aH:Si. This means that the voltage across the FLC is insufficient to switch the molecules. If the aH:Si is illuminated (bright state), then the resistivity of the aH:Si is low and the voltage appears across the FLC, causing it to switch.

A negative pulse must also be applied to achieve DC balancing and is used to erase the image on the OASLM. During the erase period, the OASLM must not be illuminated, and hence the write beam must be modulated in the same fashion as the electric field. The pattern that has been written onto the FLC can then be read by a read beam, which is either transmitted or reflected off the FLC layer to read the optically addressed pattern. The OASLM does not contain any pixels, and the resolution of the OASLM is only restricted by the structure and localized grouping of the aligned FLC. This is typically about 10–100 times higher than an electrically addressed SLM, and for a good OASLM values of 30 line pairs/mm have been measured, with quoted values reaching 100 line pairs/mm. The resolution of a photographic film is still in excess of this as it depends solely on the film grain size.

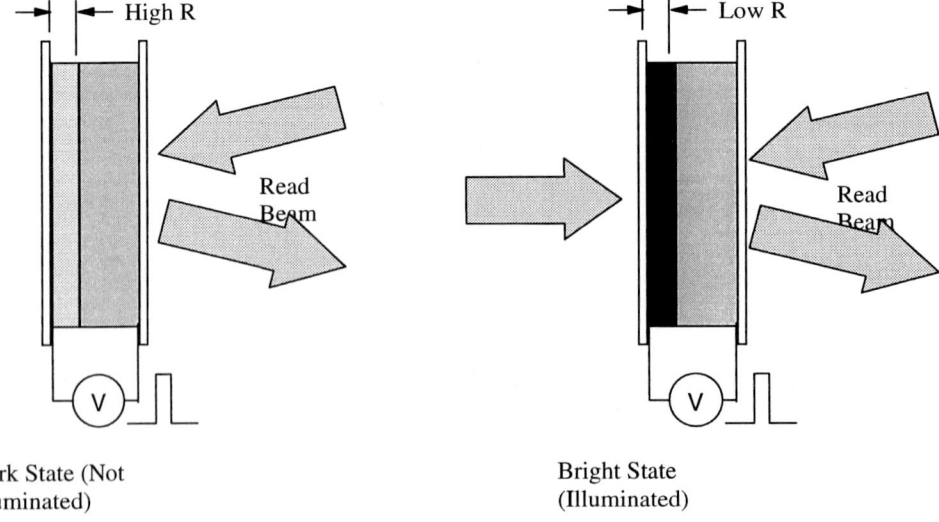

Figure 10. Modulation states for an FLC OASLM.

The devices shown in Figs. 9 and 10 often utilize the reflectivity of the aH:Si/FLC interface to reflect the read beam (~20%). The device is improved greatly if a good quality reflector is placed at the interface. To avoid short circuiting out the charge distribution defining the image, this mirror must be nonconducting. Dielectric mirrors can be used (as with nematic OASLMs), but for FLC devices (for which charge flow must occur on switching) this can make it difficult to remove charge from the interface. Metal mirrors (usually aluminum) can be used provided they are divided into a large number of small pixels, which will define the limiting resolution possible [13].

The response time (frame rate) of OASLMs is usually limited by the rate at which photogenerated carriers can be removed from the aH:Si [14]. This can be greatly speeded up if the aH:Si is doped and allowed to operate as a large area reverse biased photodiode [15]. This reduces the sensitivity compared with the purely photoconductive mode of operation and can result in difficulties in achieving full contrast. The photodiode structure can be achieved by depositing doped layers of amorphous silicon or by allowing the ITO layer adjacent to the aH:Si (see Fig. 9) to form a Schottky barrier [14].

2.2 Optical Correlation

2.2.1 The Positive Lens and the Fourier Transform

One of the most important elements in non-display applications of liquid crystals is the positive focal length lens and its relationship to the Fourier transform. It is a Fourier transform and its unique mathematical properties that allow us to perform many complex operation optically 'at the speed of light'. If we have an arbitrarily shaped aperture $A(x,y)$ with maximum dimensions x_{max} and y_{max} which is illuminated by plane wave coherent light of wavelength λ and we look at the electric field distribution $E(x,y)$ at a distance R [16], with

$$R \geqslant \frac{\pi \left(x_{max}^2 + y_{max}^2\right)}{\lambda} \qquad (19)$$

We find from diffraction theory that, $E(x,y)$ is in fact the analytical Fourier transform of $A(x,y)$. The pattern of $E(x,y)$ is called the far field diffraction pattern of the original aperture function. Hence we have the relationship between $E(x,y)$ and $A(x,y)$ linked by the Fourier transform

$$E(u,v) = \iint_A A(x,y)\, e^{2\pi j(ux+vy)}\, dx\, dy \qquad (20)$$

Where (u,v) are the co-ordinates in the domain of the Fourier transform and are defined as the measure of spatial frequency. Inversely we can calculate the aperture from the far field pattern by the inverse Fourier transform

$$A(x,y) = \iint_A E(u,y)\, e^{-2\pi j(ux+vy)}\, du\, dv \qquad (21)$$

This in itself is a very useful property for optical processing systems, but the restrictions set by the far field limit R make the fabrication of such systems difficult. Such a far field distance is difficult to achieve in practical terms, so a means of shortening the distance is needed. If a positive focal length lens is included directly after the aperture, as shown in Fig. 11, the far field pattern appears in the focal plane of the lens. A positive lens performs a Fourier transform of the aperture placed behind it.

If we consider the aperture $A(x,y)$ placed just before a positive lens of focal length f, then we can calculate the field after the lens

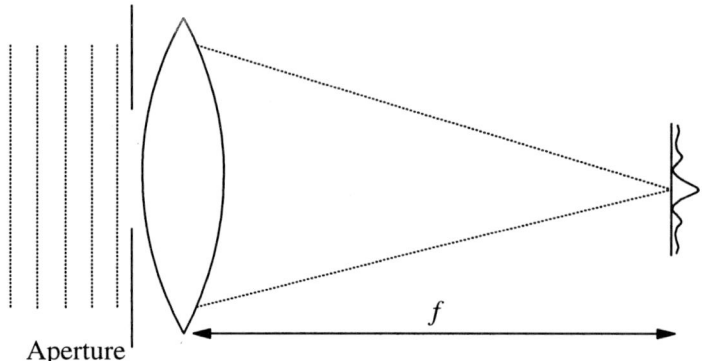

Figure 11. Diffraction through a positive focal length lens.

[17] by the paraxial approximation

$$A(x,y)' = A(x,y)e^{-j\frac{\pi}{\lambda f}(x^2+y^2)} \quad (22)$$

Application of Snells law at the spherical lens/air boundaries of the lens shows that the lens converts plane waves incident upon it into spherical waves convergent on the focal plane. For this reason, diffraction to the far field pattern now occurs in the focal plane of the lens. The effect of this is to shift the distant R to the single position specified by the local length f. The principles of diffraction can be applied to $A(x,y)'$ as if it were the aperture. The final result for the diffracted aperture $A(x,y)$ through the lens, in terms of absolute spatial co-ordinates, (α,β) is

$$E(\alpha,\beta) = e^{\frac{j\pi(\alpha^2+\beta^2)}{f\lambda}} \cdot \iint_A A(x,y) e^{\frac{j\pi(\alpha x+\beta y)}{f\lambda}} dx\,dy \quad (23)$$

We can translate this equation into spatial frequency coordinates, such that $u = k\alpha/2\pi f$ and $v = k\beta/2\pi f$, giving the Fourier transform relationship shown above for the far field region, with an added phase distortion due to the compression of R down to the focal plane f of the positive lens.

Far field region
= focal plane of a positive lens
= FT(aperture function)

It looks as though we are getting something for nothing, but this is not the case, as the lens introduces the quadratic phase distortion term in front of the transform. This can lead to a smearing of the Fourier transform and must be corrected for in compressed optical systems. There are several methods used to correct the phase, such as adding further lenses close to the focal plane or a compensating hologram to counter the phase distortion. If the aperture is placed a distance d behind the lens, then there will be a corresponding change in the phase distortion term of the Fourier transform

$$E(\alpha,\beta) = e^{\frac{j\pi}{f\lambda}\left(1-\frac{d}{f}\right)(\alpha^2+\beta^2)} \cdot \iint_A A(x,y) e^{\frac{j\pi}{f\lambda}(\alpha x+\beta y)} dx\,dy \quad (24)$$

From this equation we can see another way of removing the phase distortion. If the distance is set so that $d = f$, then the phase distortion is unity and we have the full Fourier transform scaled by the factor of the focal length, f. This is a very important feature used in the design of optical systems and is the principle behind the $4f$ system. In a $4f$ system, there are two identical lenses separated by a distance $2f$. This forms the basis of a low distortion optical system.

In both the examples of Fig. 12, the distortions are minimized. In the top system, the aperture is transformed in the focal plane of the first lens and then re-imaged by the inverse transform of the second lens. The image shown at the aperture (Give Way) appears at the output rotated by 180°. The reproduced image would be perfect if the two lenses were ideal, however, there are distortions within the lenses such as chromatic and spherical aberrations. Some of these distortions are reversible through the lenses and so cancel in the system, but some are not, leading to a slightly distorted image. The second $4f$ system is another configuration for preserving the wavefront from the input to the output, and is essential in optical systems such as holographic optical interconnects.

2.2.2 Correlation by Fourier Transform

If two functions are multiplied together and the Fourier transform (FT) taken, the result is the convolution of the FT of each function. In this notation we represent the FT by the change in case for the letter of the function, i.e., $F_T[g(x,y)] = G(u,v)$

$$F_T[g(x,y)h(x,y)] = G(u,v) \otimes H(u,v) \quad (25)$$

If we replace either of the functions $h(x,y)$ or $g(x,y)$ with its complex conjugate, then the result is the correlation of the two FTs

$$F_T[g(x,y)^* h(x,y)] = G(u,v)^* H(u,v) \quad (26)$$

Also, from Fourier theory we have this relationship for a complex conjugate

$$F_T[h(x,y)^*] = H(-u,-v) \quad (27)$$

If function $h(x,y)$ is real, then $h(x,y)^* = h(x,y)$, hence correlation and convolution

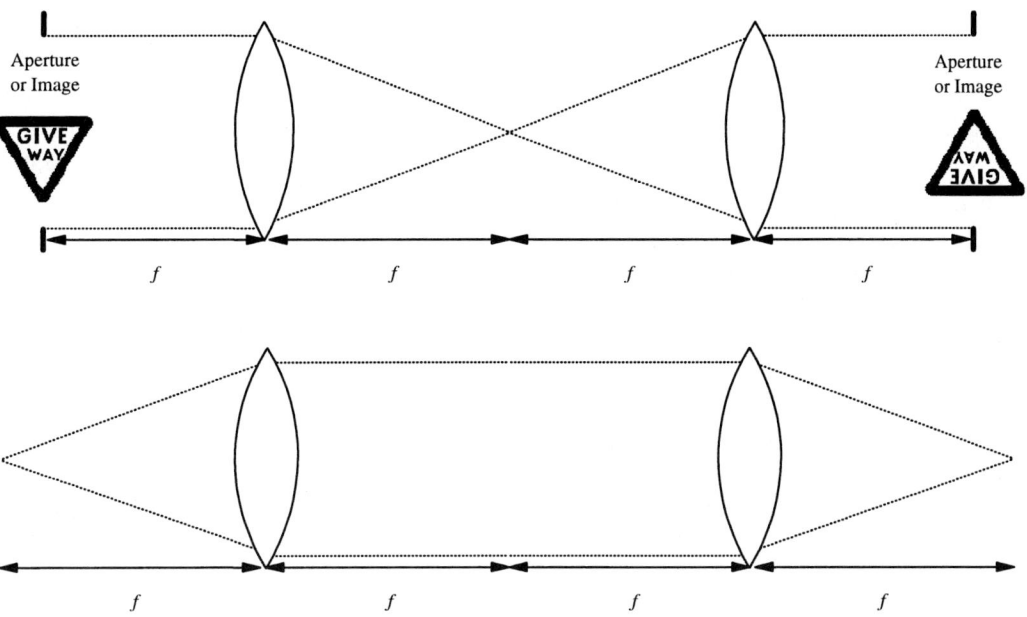

Figure 12. Two possible $4f$ optical systems.

are the same operation. Moreover, the correlation of a real object and the same object rotated by 180° is the same, making the differentiation between the two cases difficult.

Correlation gives us a means of comparing two functions. By correlating two functions with each other, we can judge how similar they are in contents and structure. If one of the functions is a reference image and we explicitly know its entire structure, then we are comparing an unknown image with a reference to see if the contents of the reference image are contained anywhere in the unknown image. This is demonstrated in the example of Fig. 13. We define the reference images $r(x,y)$ and the unknown input image as $s(x,y)$.

In order to perform a correlation, we need to multiply the FT of the reference image $r(x,y)$ and the unknown input image $s(x,y)$ in the Fourier domain before the final FT creates the correlation of the two original images

$$s(x,y) * r(x,y) = F_T\{F_T[s(x,y)] F_T[r(x,y)]\} \quad (28)$$

If the unknown input image $s(x,y)$ contains the same information as the reference $r(x,y)$, then an autocorrelation [maximum SNR (signal-to-noise ratio)] occurs. If $s(x,y)$ contains the same information as $r(x,y)$ but shifted to a different location in the image, then the autocorrelation will also be shifted by a proportional amount. If the unknown input $s(x,y)$ contains $r(x,y)$ shifted by (x_0, y_0), we can use the FT shift theorem, $F_T[g(x-x_0, y-y_0)] = G(u,v) e^{-j2\pi(x_0 u + y_0 v)}$ giving

$$s(x,y) = r(x-x_0, y-y_0) \quad (29)$$

$$F_T[s(x,y)] = S(u,v) = R(u,v) e^{-j2\pi(x_0 u + y_0 v)}$$

We then multiply this by the FT of the reference $r(x,y)$, which is centered on an origin or a known point of reference. We then take the FT of the product to give the correlation between the two images. The shifted reference contained within $s(x,y)$ leads to the exponential term above, which is now the shift of the correlation in the output plane. The correlation will be shifted by the same amount as the reference object was

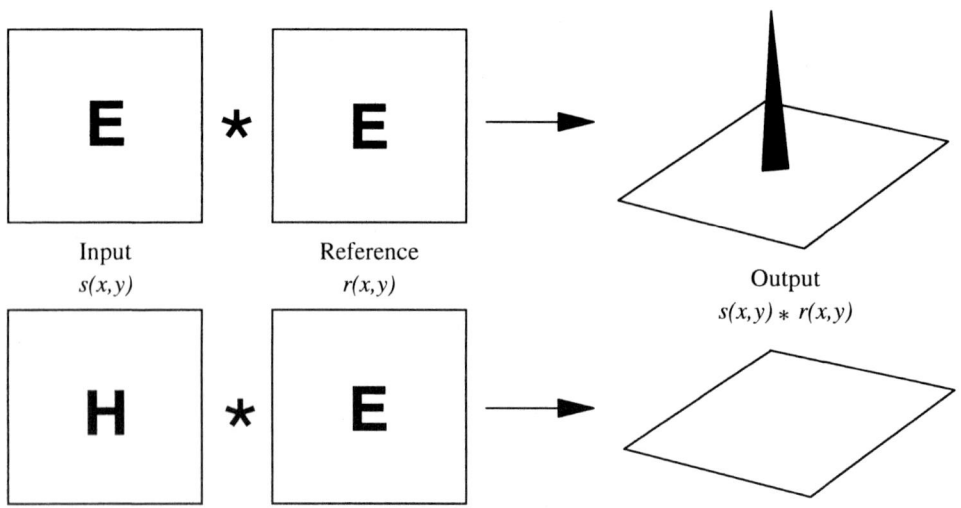

Figure 13. The principle of correlation.

shifted in the unknown input, (x_0, y_0), relative to the position of the object (zero as it was centered) in the reference image. With correlation we can not only detect if the object in $r(x,y)$ occurs in $s(x,y)$, but we can also state the position at which it occurs (if it occurs), as shown in Fig. 14. Correlation is shift invariant.

Shift invariance is an inherent property of correlators, however, other forms of invariance must be added by cunning techniques. The most common forms of invariance are rotation invariance [18] and scale (size) invariance [19].

The input image $s(x,y)$ may contain more than one reference object, in which case, both will correlate with $r(x,y)$ (Fig. 15). The result is two correlation peaks, each of which indicates if and where the reference object occurs in the input. For the sake of analysis of the correlation, we must consider each object in $s(x,y)$ as a separate function $s_1(x,y)$ and $s_2(x,y)$, with separate positions.

We can use the fact that a positive lens performs an FT in its focal plane to create architectures that can be used to perform optical correlation. There are two possible architectures which can easily be optically implemented.

2.2.3 The Matched Filter

The matched filter is the most logical method of performing optical correlation; this was first proposed by Vander Lugt in the 1960s [20]. The optical architecture is laid out in a linear fashion as a $4f$ system. The lower part of Fig. 16 depicts the transmission of light from left to right, from planes 1 to 4. The upper part is performed off-line by electronic processing and is then stored as a matched filter.

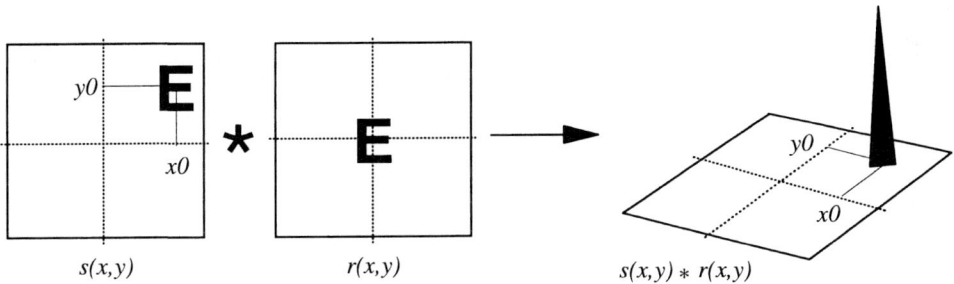

Figure 14. Shift invariance in a correlation.

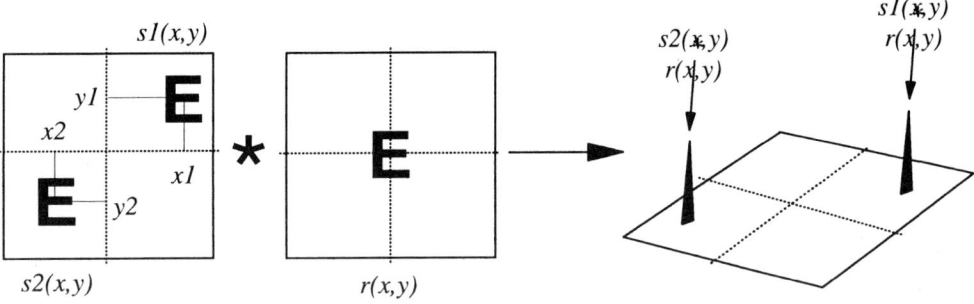

Figure 15. Multiple correlations.

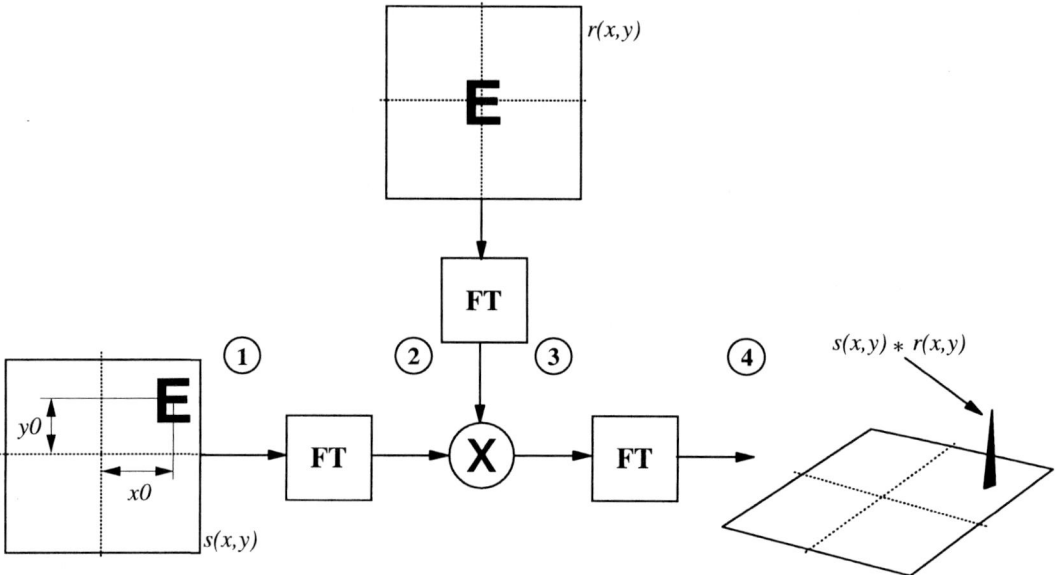

Figure 16. The matched filter.

The input image $s(x,y)$ is displayed in plane 1 before the FT into plane 2

$$S(u,v)e^{-j2\pi(x_0 u + y_0 v)} \qquad (30)$$

The FT of $s(x,y)$ is then multiplied by the FT of the reference $r(x,y)$

$$R(u,v)S(u,v)e^{-j2\pi(x_0 u + y_0 u)} \qquad (31)$$

The FT of the reference is done off line on a computer and is defined as the matched filter $R(u,v)$ for that particular reference $r(x,y)$. In fact, the generation of the filter may be more complicated (to include invariances), and it is advantageous to use only the phase information of the reference FT rather than the full complex amplitude and phase as it gives a more detectable narrow peak [21]. The product of the input FT and the filter then undergoes a further FT to give the correlation in plane 4. The object in the reference $r(x,y)$ is centered in the process of generating the filter $R(u,v)$, so that if a correlation peak occurs, its position is directly proportional to the object in the input image, with no need for any decoding. The best qualitative test for the matched filter is to perform an autocorrelation with the filter that has been generated. The reference image $r(x,y)$ is used as the input to the correlator to judge its performance. If the matched filter in Fig. 16 is used for the reference image of a letter E, then the autocorrelation will have optimum SNR.

Great improvements can be made to the usefulness of the correlation peak by using a phase only matched filter (POMF). The matched filter $R(u,v)$ is stripped of its phase information (i.e., the phase angle of the complex data at each pixel) and this is used as the filter in the correlator

$$R(u,v) = R_{\text{amp}}(u,v)e^{i\phi(u,v)} \qquad (32)$$

The autocorrelation for the POMF is much more desirable, even though there is a reduction in the SNR due to the increase in the background noise. The correlation peak is much narrower which is due to the information that is stored in the phase of the matched filter [21]. The impulse response (the inverse FT of the POMF) shows the structures that are recognized by the POMF, which tends to be the corners and points where intersections occur. The POMF is the most desirable filter to use as it has good narrow peaks but still remains selective of similar structured objects. The continuous phase structure of $\phi(u,v)$ means that it cannot easily be displayed in an optical system. Twisted nematic displays are capable of multilevel phase modulation [22], but the quality is poor and difficult to control, and they are slow. There is a possibility that new liquid crystals and phases may allow the implementation of four-level phase modulation in the near future.

The problems of displaying the filter can be alleviated by thresholding $\phi(u,v)$ to give a binary phase only matched filter (BROMF) [23]. The penalties associated with going to binary phase (listed below) are greatly outweighed by the advantages gained by being able to use FLC SLMs in the optical system:

1. The SNR is up to 6 dB worse than in the case of the POMF.
2. The filter cannot differentiate between an object and the same object rotated by 180° (due to the fact that the BPOMF is a real function).
3. The BPOMF is not as selective as the POMF due to the loss of information in the thresholding.

The binary phase is selected from the POMF by two thresholds δ_1 and δ_2. The thresholding is done such that

$$F_{\text{BPOMF}} = \begin{cases} 0 & \delta_1 \leq \phi(u,v) \leq \delta_2 \\ \pi & \text{Otherwise} \end{cases} \quad (33)$$

The selection of the two boundaries is made by exhaustive searching, as it depends on the shape and structure of the reference used to generate the filter. The benefits of this process are not high, and it is only likely to improve the SNR by a few percent. A safe threshold to get consistent results is $\delta_1 = -\pi/2$ and $\delta_2 = \pi/2$, as is the case in Fig. 17.

As can be seen from the impulse response of the BPOMF in Fig. 18, there is less information in the BPOMF. This can lead to erroneous results, especially where closely correlated objects (like E and F) are included in the reference library. More specialized techniques such as optimization and simulated annealing can be used to generate more robust filters under such conditions [19].

The use of SLMs in matched filters has been proposed on several occasions, but has always been limited by the availability to suitable SLM technologies. The first use of SLMs in a BPOMF was done with a magneto-optic SLM [24] and proved to be the first step in the development of the BPOMF. The choice of binary phase modulation allows us to use FLC SLMs to display the information in an all-optical system. The availability of high quality transmissive FLC SLMs [25, 26] has opened the door to the possibility of practically implementing

Figure 17. The BPOMF for the letter E, $\delta_1 = \pi/2$ and $\delta_2 = \pi/2$.

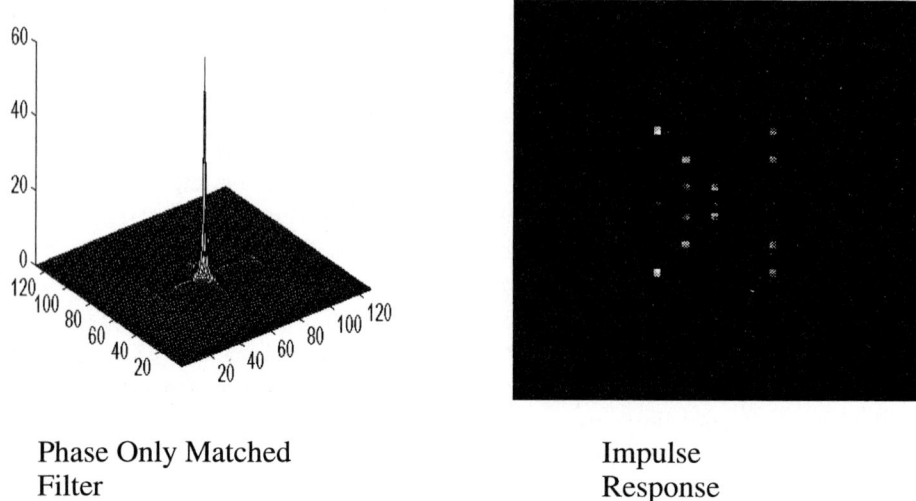

Phase Only Matched Filter

Impulse Response

Figure 18. Autocorrelation and filter impulse response for the BPOMF, letter E.

the BPOMF. Other FLC based devices have been used in the BPOMF, particularly the use of silicon backplane SLMs to miniaturize the overall correlator system [27].

The basic linear layout for an SLM based BPOMF is shown in Fig. 19. With the right choice of lenses, we can build a working optical BPOMF correlator.

The basic optical layout for a BPOMF follows directly from the theoretical expectations. The input light illuminates SLM1 which is used to display the input image $s(x,y)$. SLM1 is also an FLC SLM, but it is used in intensity mode (black and white), but it could equally well be a nematic grey scale device. The SLM is an $N_1 \times N_1$ array of square pixels, with a pitch of Δ_1, and we are assuming that there is no pixel dead space. The modulated light then passes through lens FT1 which performs the FT of the input image. The FT is formed in the focal plane of the lens and will have a finite resolution (or 'pixel' pitch) given by

$$\Delta_0 = \frac{f_0 \lambda}{N_1 \Delta_1} \quad (34)$$

There are N_1 'spatial frequency pixels' in the FT of the input image on SLM1, hence the total size of the FT will be $N_1 \Delta_0$. The BPOMF is displayed on SLM2 in binary phase mode. SLM2 is also an FLC device with $N_2 \times N_2$ pixels of pitch Δ_2. The size of

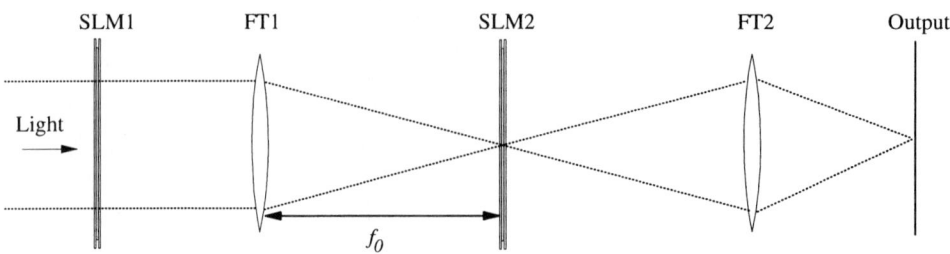

Figure 19. Basic layout of an SLM based BPOMF.

the FT of SLM1 must match pixel for pixel with the BPOMF on SLM2 in order for the correlation to occur. For this reason we must choose f_0 such that

$$f_0 = \frac{N_2 \Delta_2 \Delta_1}{\lambda} \qquad (35)$$

Once the FT of the input (matched in size to SLM2) has passed through SLM2, the product of the input FT and the BPOMF has been formed. This is then FT'ed again by the final lens and the output is imaged onto a CCD camera. An example is a BPOMF experiment where SLM1 and SLM2 are both 128×128 pixel FLC devices with a 220 µm pixel pitch at a wavelength of 633 nm. The required focal length to match the input Fourier transform to the BPOMF in this case is $f_0 = 9.787$ m, which is clearly impractical as an experimental system.

It is possible to shorten the actual length of the optical transform whilst still keeping the effective focal length that is desired by including further lenses in a combination lens. One technique is to combine a positive lens with a negative lens to make a two lens composite. This gives a length compression of around $f_0/5$ [28], which in the example above is still 2 m and impractical. Furthermore, the two lenses combine in aberrations, which leads to poor correlations due to poor optical quality. A second system is to use a three lens telescope which gives up to $f_0/15$ [19]. If the chosen lenses and SLMs fit the design equation, then the FT will be the same size as the BPOMF and the total length of the three lens system will be $z = f_1 + 2f_2 + f_3$. For the SLMs in the above example, a first lens $f_1 = 250$ mm was chosen to perform the initial FT. This means a magnification of 39.1 is required for the telescope. From the available catalog lenses, a combination of $f_2 = 10$ mm and $f_3 = 400$ mm was chosen to give a magnification of 40. The overall length of the system was $z = 670$ mm, which means that the BPOMF can now be constructed on an optical table.

Figure 20 shows the final layout of the experimental BPOMF described in the example [29]. The lenses were all high quality achromatic doublets and the laser was a collimated HeNe with a wavelength of 633 nm. There was sufficient tuning in the three lens combination to account for the slight mismatch between the desired magnification and the lens telescope. The three lens system had very low aberration as there is compensation between the three elements. Figure 21 shows a typical input image to the correlator as it would appear on SLM1 and the output from the correlator when the filter is based on a no-left turn roadsign of the same size.

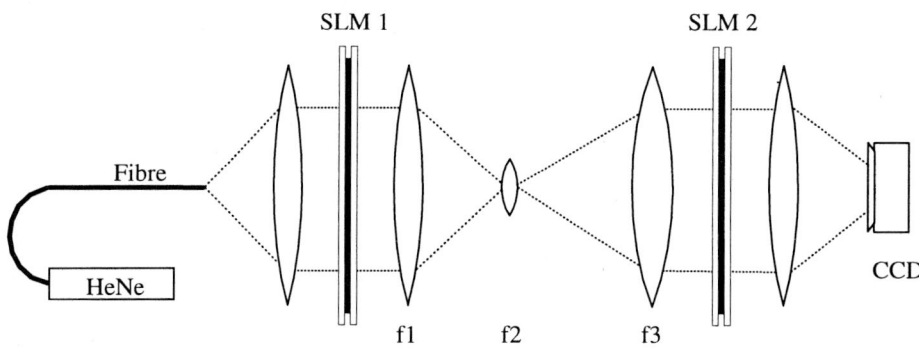

Figure 20. Layout of the experimental BPOMF.

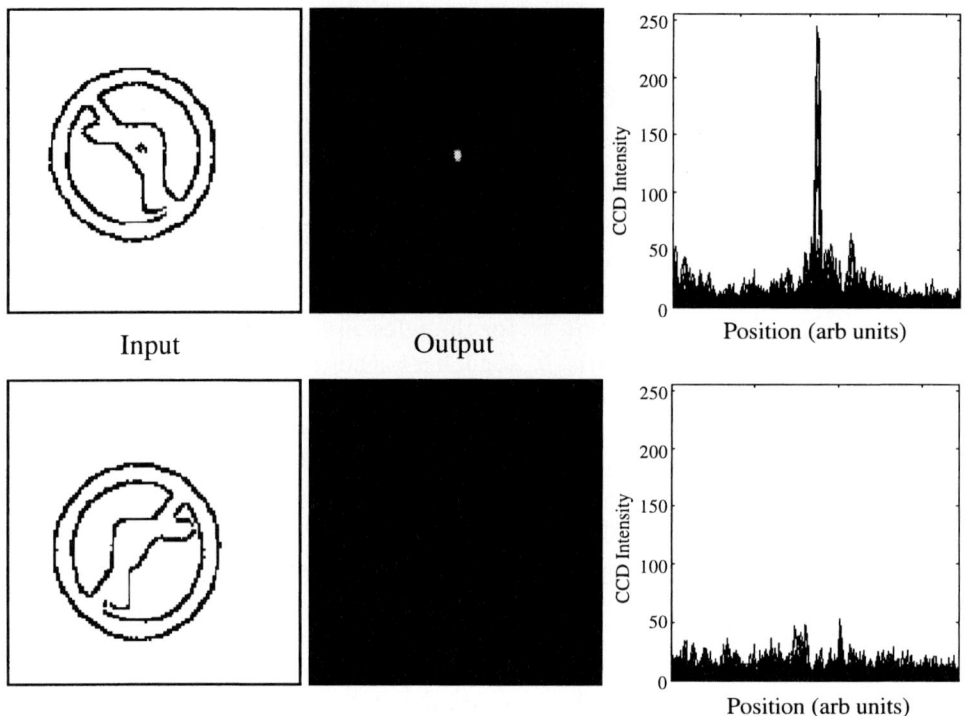

Figure 21. Experimental results from the BPOMF.

The overall size of the BPOMF was still sufficient to cover an 8×4′ (2.4×1.2 m) optical bench, making it very impractical for real world applications. The next step in developing the BPOMF is to reduce the size and pixel pitch of the SLMs. For this reason, FLC based silicon backplane VLSI SLMs are ideally suited to miniaturizing the BPOMF. If the SLMs are reduced to 256×256 pixel devices [10] with a pixel pitch of 40 μm, then the size of the BPOMF shrinks dramatically. The size of the devices makes the three lens compression system impractical, but we can still use the positive/negative lens system to shrink the BPOMF further. Figure 22 shows a miniature correlator designed for the 256×256 SLMs mounted on a magnetic optical baseplate. The total dimensions of this correlator are 180×150 mm, which is far more useful in real world applications [30].

As can be seen from Fig. 21, there are more practical applications for the BPOMF than telling an E from an F. In this application, it was envisaged that an optical correlator could be used in a car to act as a road-sign detector, providing driver assistance. One of the main results taken from this work was that the exact nature of the input scene must be carefully considered. The problem of scale invariance can be solved, but there are more complex distortions that can affect the performance of the BPOMF. For this reason, the roadsign application is perhaps in the future; however, there are several other applications that are ideally suited to the BPOMF. The majority of correlator research is performed in military institutions as the correlator has several qualities that appeal to military specifications. Other applications include fingerprint recognition for entry systems, credit card security, printed cir-

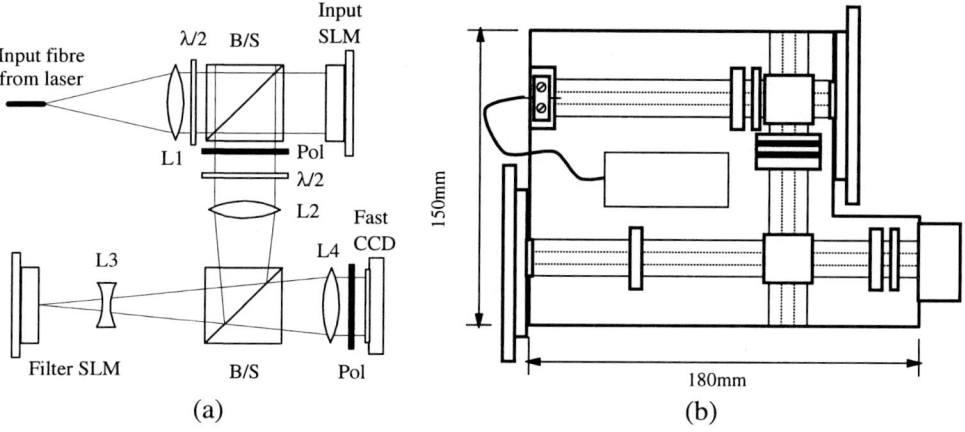

Figure 22. The miniature BPOMF: (a) optical layout, (b) baseplate layout.

cuit board inspection, and texture analysis. Perhaps the true place for the correlator lies as a morphological processor or image classifier. A correlation could easily be adapted to simplify or classify images to provide input to a trained classifier such as a neural network.

2.2.4 The Joint Transform Correlator

A second method for performing optical correlation is the joint transform correlator (JTC) [31]. In this equipment there is only one SLM and the input and reference images are displayed side by side upon it. It is also possible to incorporate an OASLM as the nonlinearity in the Fourier plane.

From Fig. 23 we can analyze the performance of the JTC; in plane 1, the input $s(x,y)$ and reference $r(x,y)$ are displayed side by side in an optical system and then Fourier transformed by a single lens into plane 2 giving the joint power spectrum (JPS)

$$\text{JPS} = S(u,v)e^{-j2\pi(x_1 u - y_1 v)} + R(u,v)e^{-j2\pi y_0 v} \tag{36}$$

The input image can appear anywhere within the lower half plane, whilst the reference

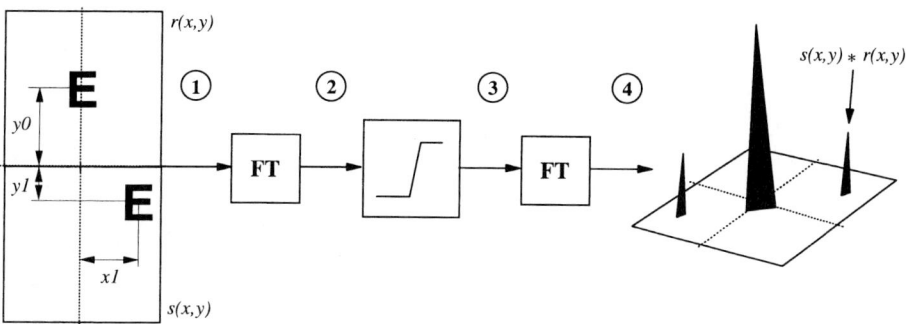

Figure 23. The joint transform correlator.

is centered in the upper half plane. The nonlinearity between planes 2 and 3 creates the term needed for correlation, and in its simplest form can be modeled by a square law detector such as a photodiode or CCD camera, which takes the magnitude squared of the light (JPS) falling upon it

$$S^2(u,v) + R^2(u,v)$$
$$+ S(u,v) R(u,v) e^{-j2\pi[x_1 u - (y_0 + y_1)v]}$$
$$+ S(u,v) R(u,v) e^{-j2\pi[-x_1 u + (y_0 + y_1)v]} \quad (37)$$

The output plane 4 is after the second FT. The output plane contains more information than just the correlation peaks, with the central DC terms proportional to $F(R^2 + S^2)$ and the two symmetrical correlation peaks spaced by $(x_0, y_1 + y_0)$ and $[-x_0, -(y_1 + y_0)]$, as shown in Fig. 24.

The central DC term is an unwanted source of noise which degrades the optical system. There is always a symmetrical pair of correlation peaks, so only half the output plane needs to be considered, but the position of the correlation peak has to be decoded to gain the actual position of the reference object in $r(x,y)$ if it correlates with $s(x,y)$.

The quality of the correlation can be improved by nonlinear processing of the JPS [32]. This can be done either by electronic processing or by an optically addressed component such as an OASLM [33]. Figure 25 shows a generic layout for an OASLM based nonlinear JTC. The performance of the JTC is very dependent on the quality of the OASLM. The first OASLM based JTC tests were done using the Hughes liquid crystal light valve (LCLV) [34]. This device takes the intensity of the JPS and uses it to modulate a nematic liquid crystal. The form of the nonlinearity is similar to a CCD camera, with the results being displayed in grey scale. The quality of the correlation peaks in this system was broad and fuzzy, with low SNR, partly due to the quality of the optical system and partly due to the nonlinear LCLV. The main drawback with the Hughes LCLV based systems was that the overall correlation rate was slow due to the writing of the intensity pattern and the addressing of the nematic liquid crystal.

Better JTC performance can now be achieved with the development of the FLC OASLM. This device offers an optically ad-

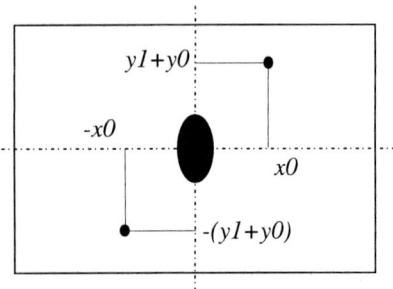

Figure 24. JTC output plane.

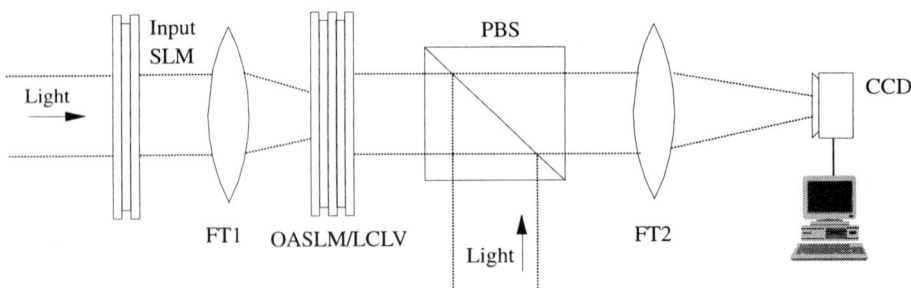

Figure 25. Optical layout of the OASLM based JTC.

dressed nonlinearity at speeds of 2 kHz or more, with high order nonlinearity. Because the FLC crystal modulation is binary, there must be a threshold in the OASLM. This is done automatically by the interaction between the amorphous silicon photodetector layer and the FLC. Such a nonlinearity is highly desirable in a JTC as it produces much sharper peaks with better SNR and reduces the power lost into the central DC term [35].

Figure 26 shows some experimental results taken from an FLC OASLM based JTC [29]. The optical setup was the same as in Fig. 25, on an in-car correlator 600×200 ×200 mm in size. The application was an in-car system capable of recognizing road signs. The left side image shows the input plane; the reference was designed to recognize the road sign over a range of different sizes. The input image to the JTC was supplied by a CCD camera mounted behind the rear view mirror. The car was driven past the road sign to collect the input data, and the output from the JTC was captured and recorded for later analysis. The JTC worked well with the car moving slowly (<20 km/h), but failed at higher speeds due to the limited resolution of the input SLM.

2.3 Optical Interconnects

Some of the most promising nondisplay applications for liquid crystals eligible for commercial exploitation are in the area of optical interconnects and switches. Potential applications include optically transparent fiber-to-fiber switches for use in the telecommunications transmission network,

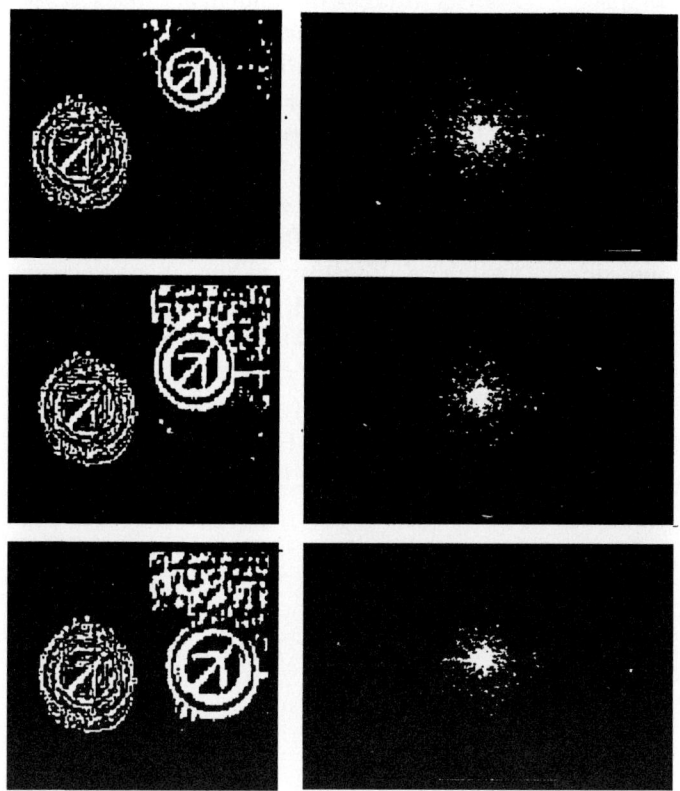

Figure 26. Experimental results from an FLC OASLM JTC (adapted from [29].

highly parallel computers, local area network, cable TV, etc. There are major economic advantages to retaining very high data rate optical signals in the optical domain while they are transported across cities, continents, and oceans. Expensive electronics can be avoided and the systems can be upgraded by only changing the terminal equipment. In telecomms transmissions, the erbium-doped fiber amplifier has made long distance 'light pulses' possible, and this has created the need for all optical switches. In computing, supercomputers are increasingly seen as arrays of RISC (Reduced Instruction Set Computer) processing connected optically, and high bandwidth fully connected operation may necessitate optical space switching.

Optical interconnections are also starting to be important for solving connectivity problems in large electronic systems [36]. Such interconnections may use free space optics rather than guided wave, and the interconnection patterns may be reconfigurable [37, 38]. Liquid crystal devices may have a role here [39].

Optical interconnects have progressed in both SLM technology and optical system design and fabrication to the point where semi-commercial systems are being designed and constructed. Ground breaking collaborations, as in the case of optically connected parallel machines (OCPMs) have not only furthered the theoretical concept of the system, but also the opto-mechanics which allow such systems to be more than just a bench-based demonstrator [40]. Reconfigurable optical interconnects and switches can be separated into two main areas: holographic switches, where the light is directed to the desired ports by reconfigurable holograms, and matrix or shutter based switches, where the routing mechanism is based on the blocking of light to unwanted ports.

2.3.1 Computer-Generated Holograms

An integral part of any optical switch or interconnect is the ability to direct the light throughout the system during the transition from input to output ports. For this reason, the computer-generated hologram (CGH) is a vital element in the optical system. As with this optical correlator, the CGH involves Fourier transforms and hence uses lenses to perform these transforms, as shown in Fig. 27. A CGH is essentially a computer-generated pattern or hologram which relates to a desired pattern or replay field via a Fourier transform.

The CGH can be either a fixed element as would be used for fan out in a shutter-based switch or a dynamic pattern displayed on a liquid crystal SLM. A simple example of a hologram is the 2-D grating or square wave shown in Fig. 28.

To find the replay field of this pattern we need to take its Fourier transform. The square grating can be represented by a sin-

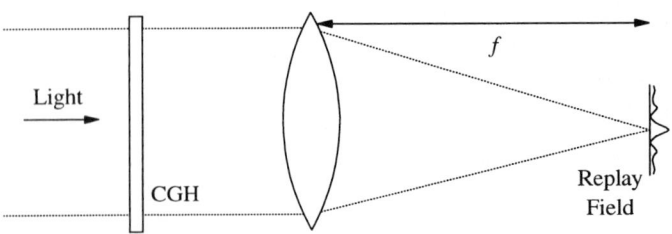

Figure 27. Fourier replay of a CGH.

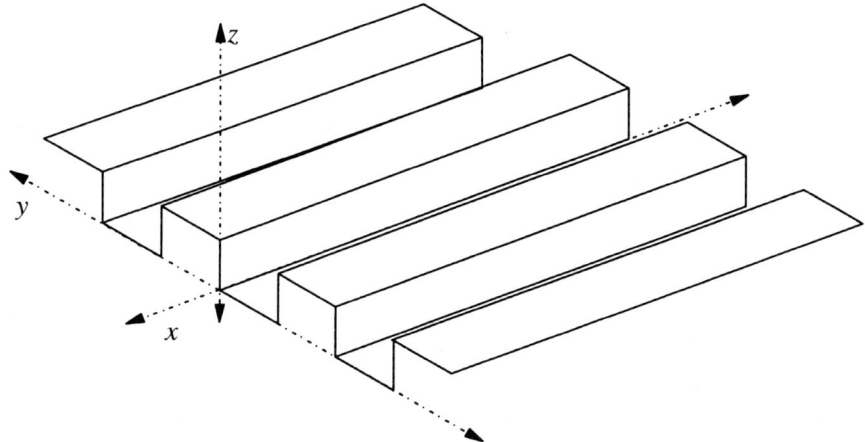

Figure 28. A two-dimensional grating.

gle square element (or pulse) convolved with an infinite train of delta functions. The result of the Fourier transform will be a sinc function $\left[\text{sinc}(x) = \frac{\sin(x)}{x}\right]$ multiplied by a different period train of delta functions. Hence, a train of delta functions has a sinc envelope and every second delta function is suppressed by the zeros of the sinc function. The 2-D far field region of the 2-D grating shown in Fig. 28 will be superimposed onto the y axis (at $x=0$) of the 2-D far field plane as in Fig. 29.

The next step is to look at a chequerboard pattern of pixels on an equally spaced grid (the pixel intensity has been restricted to 0 or 1 in the example of Fig. 30). The chequerboard can be generated by the XOR of the above 2-D grating with itself rotated by 180°. Hence the FT will be made from the convolution of the two gratings. This is the convolution of a sinc enveloped delta train in the x direction with the same in the y direction.

When an arbitrary pattern is generated by the FT of a hologram, it is contained within a sinc envelope based on the dimensions of the smaller or 'fundamental' pixel. For each lobe in the sinc there is an associated replication of the pattern. There is also a replication of the pattern at each zero in the sinc, even though the central value of the pattern is suppressed by the zero. With holograms, we are only interested in the central lobe of the sinc function. The other orders or lobes merely repeat the desired pattern in the replay field and waste the available intensity which can be placed into that desired pattern. The area of interest in the replay field must be limited to half of the area of the central lobe to prevent overlap of orders. As the pixel pitch decreases, the central lobe of the sinc envelope broadens, easing the restrictions that are placed on the replay field pattern.

There is a limitation if the intensity of the pixels is restricted to $T \in [0,1]$, which means

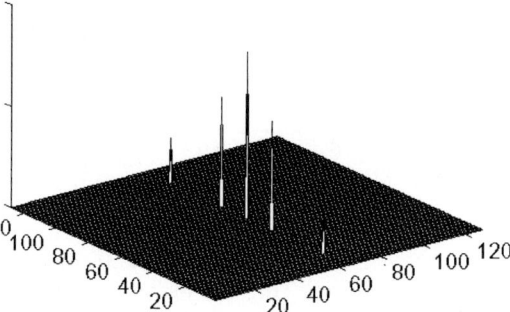

Figure 29. Two-dimensional FT of the grating in Fig. 28.

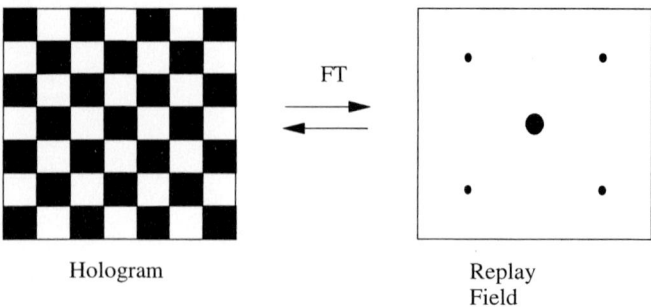

Figure 30. Hologram and replay field for a chequerboard.

that the central point at the origin (defined as the zero order) of the replay field can only be zero if all the pixel intensity values are zero. This is because the point at the origin of the replay field is proportional to the average of the pixels in the hologram. A better modulation scheme would be to have binary phase modulation i.e., $T \in [+1, -1]$. If there are the same number of pixels set to +1 as are set to −1, then the average will be zero and there will be no zero order. This can be demonstrated with the 1-D grating in Fig. 31.

With binary phase modulation ($T \in [+1, -1]$), the pixel in the center of the replay field can be defined by the structure of the hologram. A drawback of both these binary modulation schemes is that the hologram will always be a real function, which means that the FT of the hologram is the same as the FT of the hologram rotated about the origin by 180°. This symmetry restricts the useful area of the replay field to the upper half plane, as any pattern generated by the hologram will automatically appear as desired as well as rotated about the origin by 180°. Hence there will be a restriction on the maximum efficiency of the binary phase CGH, as half the power is wasted in the symmetry. If the desired pattern is asymmetric, then the total power which can be routed to that pattern will suffer at least a 3 dB penalty due to symmetry. There are advantages in going to more levels of phase than binary in order to increase the efficiency of the hologram and to break the symmetry [41]. This is theoretically possible, but the modulation technology is very limited. Nematic phase based devices are capable of up to 256 phase levels, but the optical quality is poor and the phase quantization is noisy, making them ineffective for CGH display. There is also a

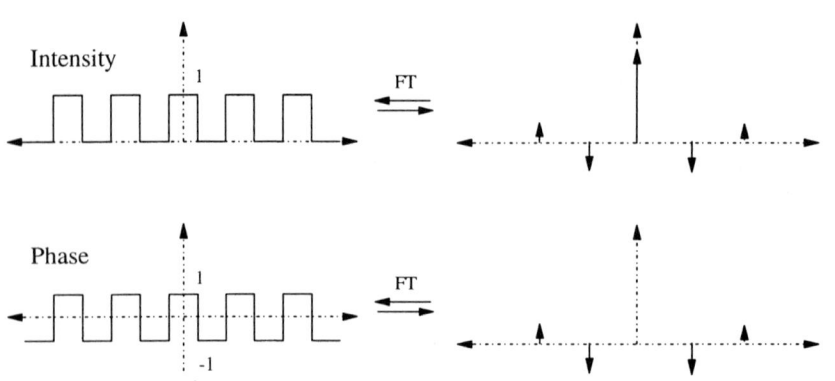

Figure 31. Binary intensity versus binary phase modulation.

penalty on modulation speed. FLC devices, as explained earlier, are capable of binary phase modulation, with the optimum performance being at a switching angle of $\theta=90°$. There are FLC mixtures that may deliver this switching angle, but they have yet to be demonstrated outside of the laboratory [42]. It is possible to perform four phase level modulation, but simulations have shown that to achieve this a switching angle of $\theta=270°$ must be achieved over all four states [43].

There is no simple way of generating a CGH except in simple cases such as gratings and chequerboards. In order to create a hologram that generates an arbitrary replay field, we need a more sophisticated algorithm. To achieve this we must use optimization techniques such as simulated annealing [44] or the genetic algorithm [45]. Say we want an array of 4×4 delta function in a square grid spaced as shown in Fig. 32, then the logical approach would be to take the FT of this replay field, take the phase and threshold it about $\pi/2$. The 'hologram' and replay field generated are not what we desire, as the correct threshold will vary over the FT of the replay field. Hence we need a better way of finding the optimum combination of pixels to give us the target replay field we desire. The simplest method of doing this is a direct binary search (DBS). In this technique we take a hologram of random pixel values and then calculate its replay field:

1. Define an ideal target replay field, T (desired pattern).
2. Start with a random array of binary phase pixels.
3. Calculate its replay field (FT), H_0.
4. Take the difference between T and H_0 and then sum up to make the first cost, C_0.
5. Flip a pixel state in a random position.
6. Calculate the new replay field, H_1.
7. Take the difference between T and H_1 and sum up to make the second cost, C_1.
8. If $C_0 < C_1$ then reject the pixel flip and flip it back (bad change).
9. If $C_0 > C_1$ then accept the pixel flip and update C_0 with the new cost C_1 (good change).
10. Repeat steps 4–9 until $|C_0 - C_1|$ reaches a minimum value.

This is not a fully optimum means of generating a hologram, but it gives a very good approximation, as shown in Fig. 33. More sophisticated techniques are required to fully exploit the possible combinations of pixel values in the hologram, as shown in Fig. 34. One such algorithm is simulated annealing, which uses DBS, but also includes a probabilistic evaluation of the cost function, which changes as the number of iterations increases. The idea is to allow the hologram to 'float' during the initial iteration, with good and bad pixel flips being accepted. This lets the optimization float into more global minima rather than getting stuck in local minima, as is the case with DBS.

The exact pattern of the replay field can be any arbitrary pattern of points that make up the target function. A good example of this is shown in Fig. 35, with the replay field of a hologram designed to generate the letters 'CRL'. The hologram was designed on a 64×64 grid and then replicated to fit the

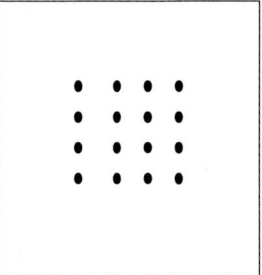

Figure 32. CGH desired replay field.

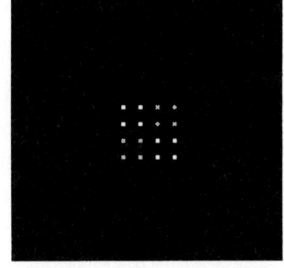

Figure 33. Hologram and replay field generated by DBS.

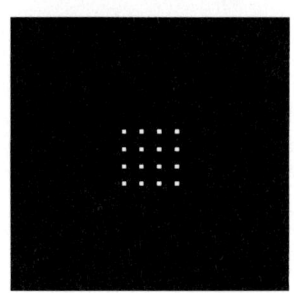

Figure 34. Hologram and replay field generated by simulated annealing.

Figure 35. Replay field of the 'CRL' hologram.

2.3.2 Polarization Insensitive Holographic Replay

In the optical system of Fig. 27, we have not stated any information about the polarization of the illumination source or the orientation of the SLM used to display the CGH. As stated earlier, the modulation of the CGH must be perfect binary phase to remove the zero order from the center of the plane. If the SLM were ideal, then there would be no need for polarizers at the input or output, making it truly polarization-insensitive. In the application of all-optical fiber to fiber switches, we cannot predict or easily detect the polarization state from the input fiber, making a polarization-insensitive switch essential. A recent discovery has led to the implementation of a polarization-insensitive switch based on an FLC SLM with no polarizers [46]. The penalty of this operation is that there will be a zero order due to the nonideal modulation of the SLM. However, the orders of the hologram remain po-

128 × 128 pixel resolution of the FLC SLM [25]. The replay field was transformed with a 250 mm focal length lens and imaged onto a CCD camera. The remains of the zero order can be seen in the center of the plane, mostly due to the electronic addressing of the SLM. The noise is mostly due to the limited resolution of the SLM.

larization independent. In order to understand this mechanism, we must look at the Jones matrix model of the modulation characteristics of the SLM. Given that the instantaneous input light has an arbitrary state of polarization, as stated earlier,

$$V_{inst} = \begin{pmatrix} V_x \\ V_y e^{j\phi} \end{pmatrix} \quad (38)$$

We can look at the propagation as before in Eq. (10) to find the modulated light through the SLM, as if it were a switchable wave plate with retardance Γ through the FLC switching angle θ. Hence we can rearrange Eq. (10) to a more convenient form for the two states

For pixel state 1

$$V_1 = \begin{pmatrix} c_1 + c_2 \\ c_3 e^{j\Gamma} + c_4 \end{pmatrix} \quad (39)$$

For pixel state 2

$$V_2 = \begin{pmatrix} c_1 + c_2 e^{j\Gamma} \\ c_3 + c_4 \end{pmatrix} \quad (40)$$

Where

$$\begin{array}{ll} c_1 = V_x e^{j\Gamma} \cos\theta & c_2 = V_y e^{j\phi} \sin\theta \\ c_3 = -V_x \sin\theta & c_4 = V_y e^{j\phi} \cos\theta \end{array} \quad (41)$$

From this we can see that each field component consists of a term (c_2 or c_3) that is modulated through a phase angle and a constant offset term (c_1 or c_4) that remains fixed as the pixels switch. This implies that the Fourier transform will contain some form of undiffracted DC term

$$\begin{aligned} F_T\left[a_{dc} + a_{ac} H_{ca}(x,y)\right] \\ = a_{dc}\,\delta(u,v) + a_{ac}\,r(u,v) \end{aligned} \quad (42)$$

where $\delta(u,v)$ is a Dirac delta function centered in the middle of the plane, $H_{CG}(x,y)$ is the binary phase hologram, $r(u,v)$ is the Fourier transform replay field of the CGH and a_{ac} and a_{dc} are the optical ratios diffracted into the replay field and the unwanted zero order, respectively, such that $a_{ac}^2 + a_{dc}^2 = 1$. To calculate the useful replay power efficiency given by a_{ac}^2, the polarization components of Eqs. (39) and (40) must be rewritten in terms of their mean DC value, superimposed onto a zero-mean AC (alternating current) phase modulation

$$V_1 = \begin{pmatrix} X_{dc} + X_{ac} \\ Y_{dc} - Y_{ac} \end{pmatrix} \text{ and } V_2 = \begin{pmatrix} X_{dc} - X_{ac} \\ Y_{dc} + Y_{ac} \end{pmatrix} \quad (43)$$

where

$$X_{dc} = c_1 + \frac{c_2}{2}\left(1 + e^{j\Gamma}\right) \quad X_{ac} = \frac{c_2}{2}\left(1 + e^{j\Gamma}\right)$$

$$Y_{dc} = c_4 + \frac{c_3}{2}\left(1 + e^{j\Gamma}\right) \quad Y_{ac} = \frac{c_3}{2}\left(1 + e^{j\Gamma}\right) \quad (44)$$

From Fourier theory we have that the power undiffracted into the zero order is the intensity of the mean complex field amplitude transmitted by the SLM. Hence a true binary phase hologram (with polarizers) will not form a zero order. These holograms have a mean field intensity of zero $a_{dc}^2 = \Sigma H_{CG}(x,y) = 0$, but suffer a transmission loss $a_{ac}^2 = T \leq 1$. In the absence of polarizers, the undiffracted zero order is derived from a mean of the pixel fields

$$a_{ac}^2 = \left|\frac{1}{2}(V_1 + V_2)\right|^2 \quad (45)$$

Since V_1 and V_2 have been normalized, we can assume that

$$a_{dc}^2 = 1 - a_{ac}^2 = \left|\frac{1}{2}(V_1 - V_2)\right|^2 \quad (46)$$

and as V_x and V_y are orthogonal and therefore independent, the proportion of power η_s that is replayed in the desired replay field $r(u,v)$ can be expressed as

$$\eta_x = \frac{a_{\mathrm{ac}}^2}{V_{\mathrm{inst}}} = \frac{\left(\left|\frac{1}{2}c_2\right|^2 + \left|\frac{1}{2}c_3\right|^2\right) \times \left|1 - e^{j\Gamma}\right|^2}{V_x^2 + V_y^2} \quad (47)$$

which simplifies into an expression that is completely independent of the original polarization variables, V_x, V_y, and ϕ

$$\eta_x = \sin^2\theta \sin^2\frac{\Gamma}{2} \quad (48)$$

This result tells us that the amount of power diffracted into the replay field $r(u,v)$ is constant and independent of the input state of polarization. Hence the light that is normally blocked by the input polarizer and analyzer is just directed to the central zero order. The benefit of this is twofold: with the absence of polarizers, there will be no fluctuations due to changes in the input polarization and there is an added bonus of 6 dB of extra power due to the 3 dB from each polarizer removed, due to the physical construction of each polarizer.

The results of this analysis have been verified by simulating the variation of input polarization states entering the FLC SLM and looking for a change in the intensity of the replay field, with no change observed. This result was also verified by an experimental test, the results of which can be seen in Fig. 36. A 128×128 pixel FLC SLM [25] was illuminated by a polarized coherent collimated light beam. The CGH displayed on the SLM was a replication of the pattern in Fig. 34, which was designed to give a replay field of 4×4 equally spaced dots. The CGH was Fourier transformed by a 250 mm focal length lens and imaged on to a CCD camera. The replay field in Fig. 36a shows the FLC SLM in binary phase mode with polarizers; there is no zero order visible, as expected. The replay field in Fig. 36b is the same CGH with the polarizers removed. The replay field is much brighter as expected due to the removal of the loss-producing polarizers, and there is a zero order as predicted. The important feature is that the diffracted replay field remained present and there was no variation in the peak intensities when the polarization state was varied.

2.3.3 Holographic Interconnects

The ability of holograms to route light to different positions in the replay field makes them ideal candidates for holographic switching [48]. Figure 37 demonstrates a one to sixteen switch; if the light illuminating the hologram comes from an input fiber and we place a four by four array of fibers in the replay field, then switching is

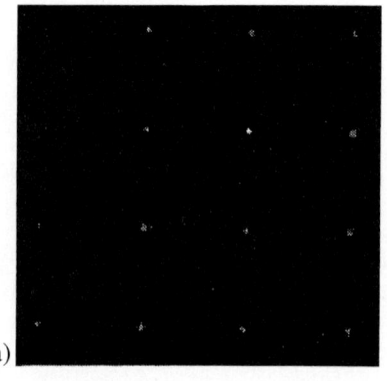
(a)

(b)

Figure 36. Replay field of a binary phase CGH: (a) with polarizers, (b) without polarizers (reproduced from [47]).

achieved. The ability to dynamically display the CGH in an optical system through the use of liquid crystal technology, such as FLC SLMs, means that we can reconfigure the switch at a rate set by the SLM. Such freedom allows us to perform optical interconnects through free space from fiber to fiber. Although SLM reconfiguration times are of the order of 100 µs there are significant advantages in making an optically transparent interconnect.

2.3.3.1 The One to *n* Holographic Switch

If we use the CGH in Fig. 34 as a binary phase image displayed on an FLC SLM, then it is possible to route light to several fibers in the replay field. We are limited however, by the binary phase modulation of the FLC SLM, which means that a symmetric copy of the desired replay field always appears rotated by 180°. Such a property is useful if the desired replay field is 180° rotationally symmetric, but this is not the case with a one to *n* interconnect, as this leads to asymmetric replay patterns. For this reason, we have to accept a 3 dB penalty in the power which can be routed from the input fiber to the output. In the analysis of this switch, we will assume that the replay field is limited to the upper half plane and that the symmetric order is repeated in the lower half plane. If polarization-insensitive modulation from the FLC SLM is to be used, then we must be careful to steer clear of the zero order, which will be fairly strong for a nonoptimized low switching angle FLC SLM. It is also important to note that the zero order appears with a sinc envelope and so will have side lobes extending out in all directions.

The operation of the 1 to *n* holographic switch (outlined in Fig. 37) is such that the

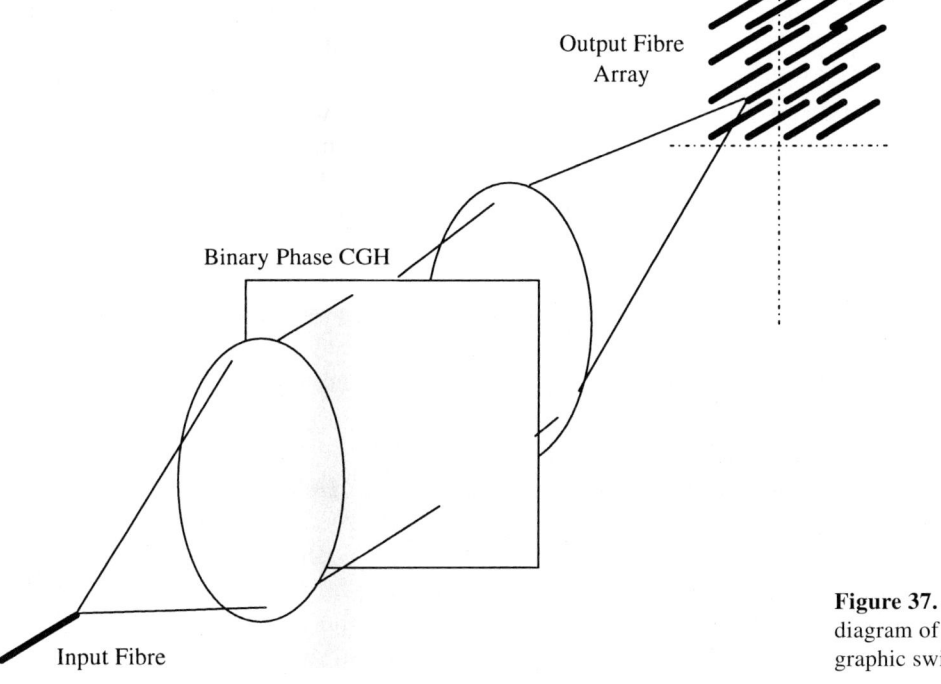

Figure 37. Schematic diagram of a 1 to *n* holographic switch.

CGH displayed on the SLM routes the light from the input fiber to a fiber in the output array of n fibers in the output plane. The desired replay pattern for one spot is asymmetric, hence the fiber array is played in the upper half plane and the 3 dB loss is tolerated. If we assume an ideal situation of lossless optics and perfect SLM and that the CGH has zero pixel pitch and an infinite number of pixels, then we can perform a simple analysis to gain an upper limit on the performance of the interconnect.

The total input power appears in the output plane as P_{in}. The total power that is routed into a spot by the CGH is P_{sp}, and the remaining power is dissipated into the whole plane as the background noise power P_{bk}

$$P_{in} = 2P_{sp} + P_{bk} \tag{49}$$

The factor of two is due to the symmetry of the pattern due to the binary phase. We can define the CGH efficiency η as the ratio between the power in the spot, P_{sp}, and the input power, P_{in}

$$\eta = \frac{P_{sp}}{P_{in}} \tag{50}$$

A typical value for this efficiency would be around 38% for a CGH generated by simulated annealing (the ideal maximum would be 50% due to the symmetry in the replay field). Higher efficiencies (up to 41%) are possible, depending on the position of the spot, the number of CGH pixels, and the method of generation. One of the most fundamental characteristics of an optical interconnect is the crosstalk. If the switch is configured to route light to the kth fiber in an array of n, then the crosstalk is the percentage of light launched down one of the other fibers to which we are not routed, compared with the power launched into the fiber to which we are routed. For n fibers in the output array of a 1 to n switch, the power into a single fiber will be ηP_{in}. If the CGH has $N \times N$ pixels, then the replay field can also be assumed to contain $N \times N$ 'spatial frequency pixels', and if we assume that the background power is uniformly distributed over the N^2 spatial frequency pixels in the replay field, then the crosstalk C will be given by

$$C = \frac{\eta}{1-2\eta} N^2 \tag{51}$$

The actual crosstalk is more complex than this as there are several factors that have been overlooked. Some of the more significant effects are:

1. Due to the binary phase modulation, the distribution of the background power is not uniform and there tend to be small peaks of intensity which may occur at fiber positions. This becomes less of a problem with large numbers of CGH pixels and careful CGH design.
2. The number of CGH pixels is finite and forms an overall aperture which leads to sinc or Bessel side lobes on the individual spots. There is also Gaussian illumination. These effects lose power into the side lobes and cause the spot to be broader than the original source fiber, leading to poor fiber launch efficiency.
3. The pixel pitch is finite, which leads to an overall sinc envelope that reduces the power into spots that occur further away from the center of the replay field. The sinc envelope also leads to power being lost in the outer orders, due to replication of the replay field.
4. The SLM used to display the CGH inevitably has dead space (optically inactive areas) between the pixels. This space limits the performance of the CGH as it alters the envelope of the replay field and increases the power replicated into the unwanted higher orders [43]. The effects

of dead space can be modeled as if the sinc envelope were due to the active area of the pixel, but the spacing in the replay field is related to the pitch of these active areas. This can be thought of, in the 1-D case, as a pulse train with zeros between the pulses. This means that the square shape of the pulse gives a sinc function, but the spacing of the orders no longer matches the zeros of the sinc function and gives rise to other orders that would normally be suppressed.

5. The physical alignment of the fibers in the output array is not perfect, so there are position errors in the spot locations which lead to poor interconnections. This can be corrected if N is large by slightly shifting the positions of the spots to match the fiber array.

6. So far we have assumed perfect optics with no limitations or distortions. In reality, it is optically more difficult to route light into the outer corners, leading to fan-in loss. This loss is usually modeled as a $1/n$. We can overcome this by placing the outputs close to the zero order and limiting n.

It is assumed that the light can be efficiently launched into a fiber without significant loss. For a 100% efficient launch, it is assumed that the distribution of the light focused into the fiber is exactly the same as when it was radiated from the input fiber. In the case of a single mode fiber, the distribution is very nearly Gaussian, which means that the FT will be Gaussian, aiding the launch efficiency. It is difficult to match this distribution when there are several distortions due to apertures, SLM imperfections, polarizers (if used), and optical aberrations. All these effects lead to distortions in the spot produced by the FT and greatly affect the loss of the interconnect due to launch down the fiber. This can be relieved by launching into multi-mode fiber, but a single mode to single mode fiber switch is far more desirable.

The biggest drawback with the holographic optical switch is the overall loss, which is a critical parameter along with crosstalk for long distance optical telecommunications. There is a growing need to improve on binary phase SLMs, even at the cost of overall switching speed. An added advantage of holographic switches is that they can be used to broadcast as well as route. Rather than address a single fiber, a simple change in the hologram pattern would allow us to send the information to any combination of the output fibers.

The Polarization Insensitive
1 to 16 Holographic Switch Experiment

A recent experimental system was set up to demonstrate a 1 to 16 polarization-insensitive holographic switch [49]. The optical system was identical to the one shown in Fig. 37, with a single mode input fiber being routed to a 4 by 4 array of single mode output fibers. The layout of the fiber array was slightly different to that in Fig. 37, to assist in the alignment of the optics. The output fiber array was placed slightly offset in the output plane to allow the zero order to be launched down the 10th fiber. This effectively makes a 1 to 15 fiber switch, but monitoring of the zero order helps the evaluation and alignment of the system.

The fiber array was fabricated to operate at a wavelength of 780 nm, so this was chosen as the routing wavelength. The SLM used to display the holograms was a 320×320 pixel FLC transmissive device. The results measured for the switch can be seen in Fig. 38. The plot shows the power measured in each fiber (including the zero order) and also the expected power from

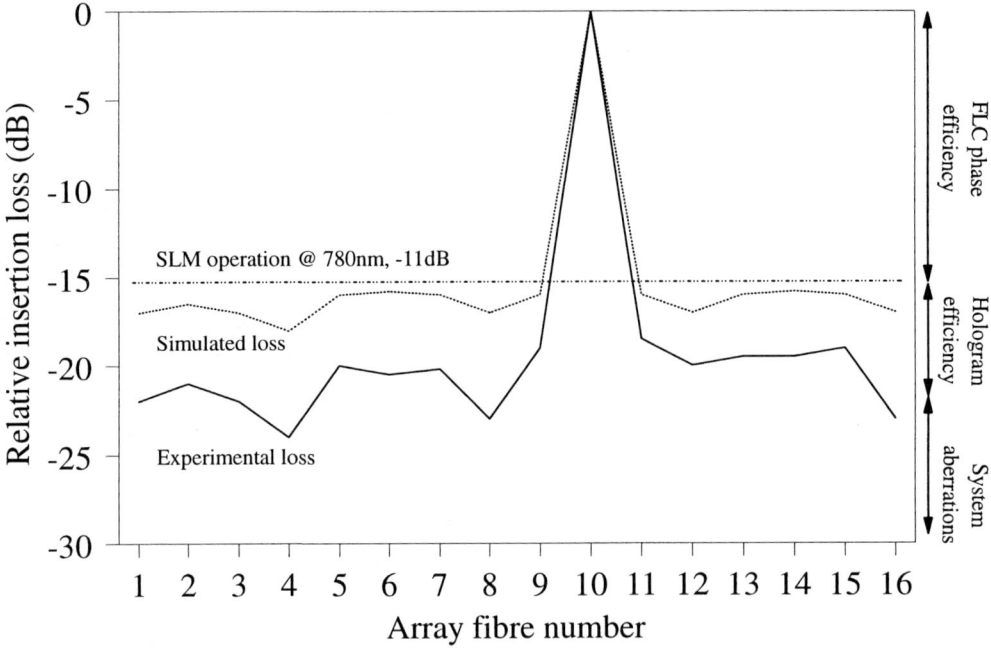

Figure 38. Experimental results for the 1 to 15 holographic switch (adapted from [47]).

simulations. One of the problems with this experiment was the irregular spacing of the fibers in the array, which led to miss-launch at the output. This was avoided by changing each hologram to scan around the area of the fiber to find the correct position for maximum launch. The experimental loss expected due to the nonideal switching angle of the FLC in the SLM was −11 dB, then there is the loss associated with the diffraction efficiency of the binary phase hologram. Finally, there were the losses due to the aberrations in the optical system. All of these effects are shown in Fig. 38. With an ideal phase modulation FLC and better optics and opto-mechanics, a theoretical loss closer to −6 dB could be expected. The crosstalk of the system was also measured, within the limits of the equipment, at over 30 dB isolation.

2.3.3.2 The n to n Holographic Switch

A further extension of the one to n holographic switch is to have multiple holograms and multiple input fibers in an array. This is in effect n one to n switches with overlapping outputs, as shown in Fig. 39.

In this case, each fiber in the input array is collimated by a lenslet (part of a lenslet array) and illuminates a portion of the SLM which contains the routing hologram for that input to a particular output fiber. The loss factors are similar to those of the one to n switch, but fan-in loss becomes more dominant. We can apply the same idealistic approach to the analysis as for the one to n switch, assuming uniform distribution of the nonrouted light in the background of the replay field. We also assume that the efficiency η of each CGH will be the same. The analysis for the crosstalk is the same except that we now have the background noise from

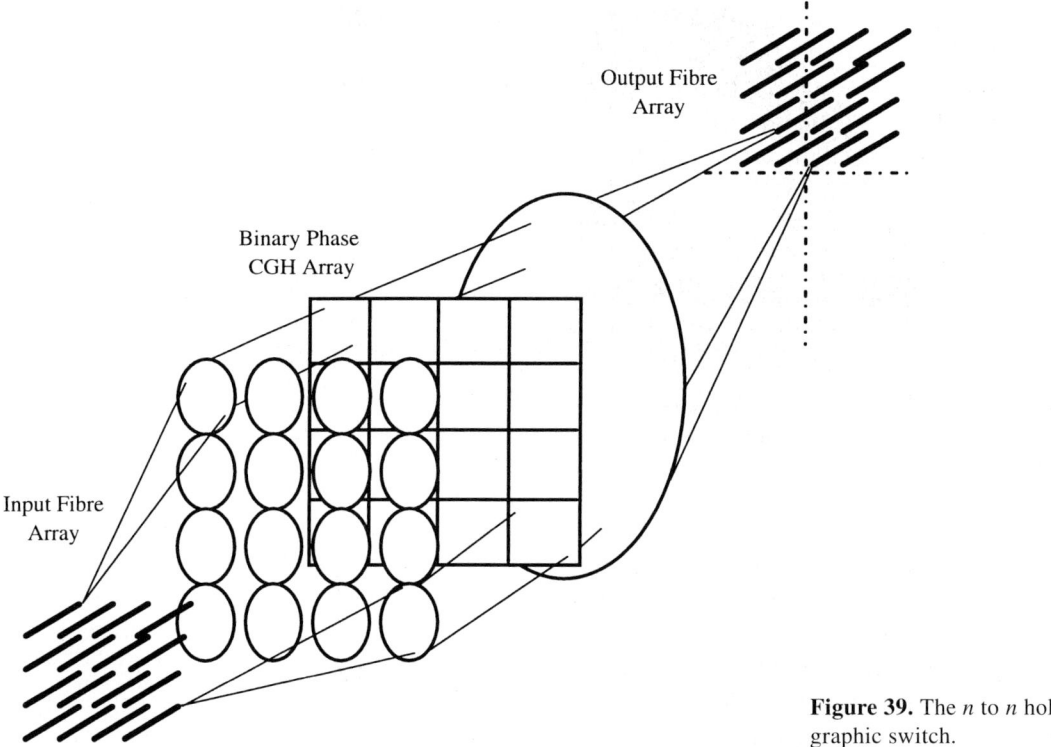

Figure 39. The n to n holographic switch.

each of the other $(n-1)$ input fibers appearing at the each output fiber along with the ηP_{in} from the routed input. Hence the crosstalk will be

$$C = \frac{\eta}{(n-1)(1-2\eta)} N^2 \qquad (52)$$

The optical design of the n to n holographic switch is more difficult as the alignment through the system is far more critical. Any errors in the alignment of the input will carry through to the output and affect the fiber launch efficiency.

One of the big problems with an n by n holographic switch is the loss due to fan-in of the light to the input fibers. This loss occurs because the outer corners of the n by n array have to be routed through large angles to reach the opposite corners of the output array. This angular restriction leads to distortions in the field as it reaches the desired output fiber, causing poor fiber launch efficiency due to mode mismatch. For crossbars operating between single mode fibers, at worst the optical power will suffer division by n, since (despite the beam steering fan-out) the structure is effectively attempting to fan-in n single mode fibers into one single mode fiber of the same numerical aperture. One of the most popular methods proposed to eliminate this loss is to correct for this angle with a second hologram at the output. This is an elegant solution to a difficult problem, but in itself creates further problems. The 3 dB loss due to symmetry from the binary phase CGH is now doubled to 6 dB, and the opto-mechanical complexity for laying out such a system has increased dramatically.

Optical interconnects have many applications, from telecommunications to aerospace, and there is a strong need for robust, compact, low loss and crosstalk all-optical switches. The holographic optical interconnect offers a promising solution to important applications in long haul telecommunications. Its crucial advantage being its potential to provide crossbar switches that operate between single mode fibers with an intrinsic loss that scales with n (the number of input and output fibers). Small VLSI backplane SLMs that contain either high speed and high switching angle liquid crystals or low speed multiphase modulation schemes could make compact low cost switches possible.

2.3.4 Shadow Routed Crossbars

Instead of providing each input channel with its own reconfigurable routing hologram, each input can be distributed to all the output ports via a fixed CGH and suitable optics, as shown in Fig. 40. This can be done by the illumination of an $n \times n$ spot CGH or Damman grating to replicate the source inputs. The use of the CGH to replicate the inputs comes from the CGH property that the spots in the replay field are the Fourier transform of input illumination. Since only one channel is likely to be required at each output, those not required can be blocked using liquid crystal shutters. Such switches are based on the Stanford vector matrix multiplier (SVMM) [50] related switching devices [51]. When implemented using a CGH to fan out, and with a 2D array of inputs (rather than the 1D arrays of the SVMM) to simplify the free space optics, these are called matrix-matrix switches [52]. This kind of structure is found in a range of optical processing architectures (see Sec. 2.4). For a symmetrical switch with n inputs and n outputs, an array of $n \times n$ shutters is required.

The fanned-out inputs are replicated as an n^2 array of the n inputs, all of which are incident on the shadow logic SLM or shutter. This device operates as an intensity shutter

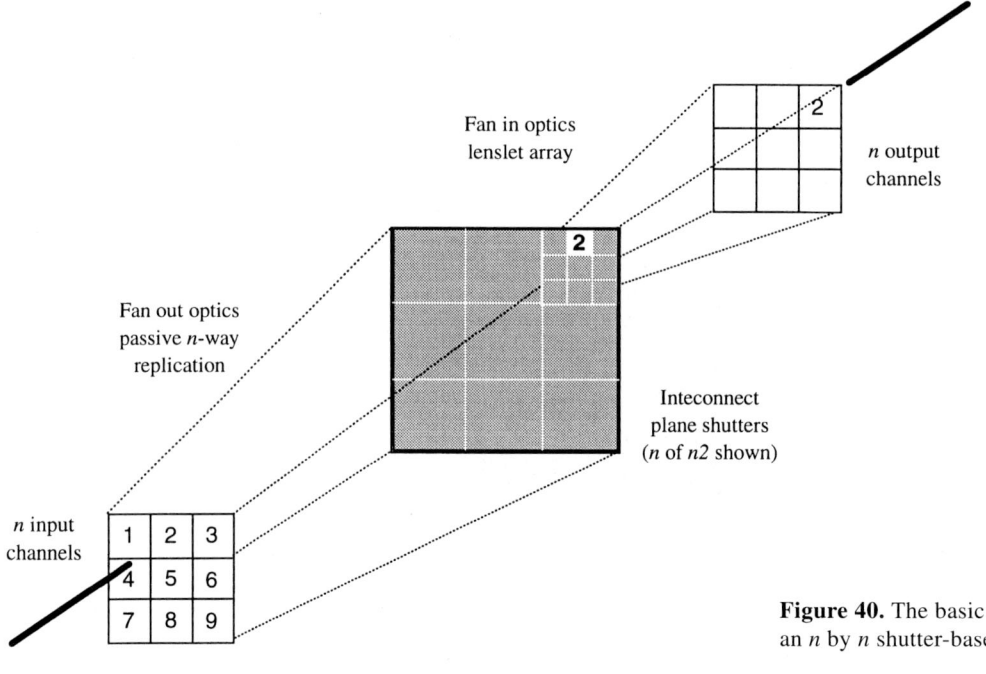

Figure 40. The basic format of an n by n shutter-based switch.

array to block or pass the desired light from the inputs. For each replicated input, there will be one open shutter which selects the input source to be routed to a particular output, hence there will be $n \times n$ shutters on the SLM. The position of the shutter selects the input source and the replication position selects the output for routing. The final stage is the most difficult as it fans in each replicated input array to a particular output position. The output plane is formed by the overlaying of all the replicated input sources.

2.3.4.1 The 1 to *n* Shadow Logic Switch

A simple example of a shadow logic switch is the 1 to *n* interconnect. As an example, a 1 to 16 (in a 4×4 array) switch is shown in Fig. 41.

The operation is such that the input fiber illuminates a 4×4 CGH or Damman grating. This is Fourier transformed by a lens to give a 4×4 replication of the input fiber. The replications then pass through a 4×4 shutter array (amplitude pixels on an SLM). The shutter array selects the input to be routed by opening a single pixel, the other 15 pixels remain closed to block the unwanted light. The selected input fiber is then fanned-in to the desired output fiber in the output array.

The problems with this type of switch arise when the characteristics of the interconnect are evaluated. It is obvious that the switch is loss-producing as it is based on the idea of blocking light. For the 1 to *n* switch (assuming ideal operation conditions), the output power P_{out} will be

$$P_{out} = \frac{P_{in}}{n} \quad (53)$$

The loss becomes intolerably high for a large value of *n*. The second parameter to consider is the crosstalk through the switch. If the SLM operates as an ideal shutter, there will be no crosstalk between the outputs. However, such SLMs do not exist, so we must consider the effects of a finite contrast ratio, *B*, on the possible crosstalk. If the output power P_{out} is as given above (the shutter has a transmission of one when open),

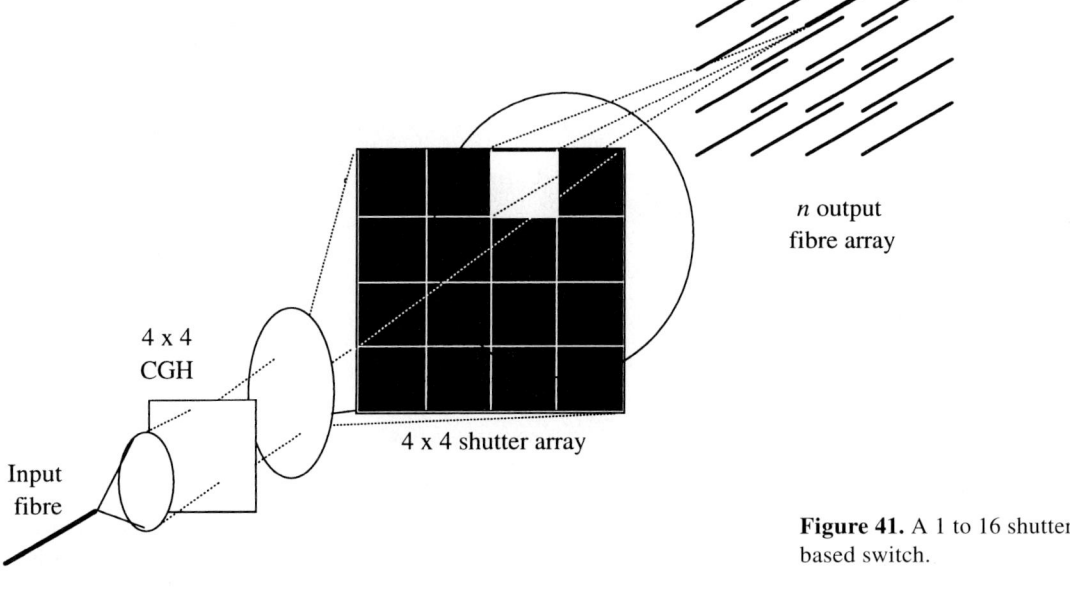

Figure 41. A 1 to 16 shutter-based switch.

then the closed shutters will each have a transmission of $1/B$. Hence the crosstalk will be

$$C = B \tag{54}$$

If we have an SLM with a contrast ratio of $B = n$ or less, then we cannot distinguish between a switched channel and all the others. A contrast of at least $2n$ would be required for reasonable operation. A typical SLM (the 128×128 FLC device) has a contrast ratio of 150, but some devices have contrast ratios as low as 10, which limits the maximum value of n.

2.3.4.2 The n by n Crossbar Switch

The 1 to n switch can be expanded for n inputs (in an array) to make an n by n optical crossbar. The requirements of the shutter SLM are now more critical. The input array of n fibers has been fanned out to an n^2 array of the n inputs. The shutter SLM requires a pixel for each element in the replicated input, so it must have a resolution of $n \times n$, as in Fig. 42.

The power appearing at the output will be the same for a given input fiber power (assuming all the inputs have the same power)

$$P_{out} = \frac{P_{in}}{n} \tag{55}$$

More importantly, the crosstalk at each output will increase as the crosstalk from each shuttered input not selected will add when the replicated images are fanned into the output

$$C = \frac{B}{n-1} \tag{56}$$

The expression of the crosstalk shows how the shadow routed crossbar can only operate for small values of n. If n is too large for a given B, then the crosstalk signal down the other channels will be larger than the actual routed signal, leading to incorrect signals. The fan-in to the output also becomes very critical, as each of the shuttered replications has to be overlapped onto a single output plane. This becomes an extremely difficult optical task when n is large.

2.3.4.3 The OCPM (Optically Connected Parallel Machine) Optical Switch

One of the best shadow logic crossbar switches was the one demonstrated on the

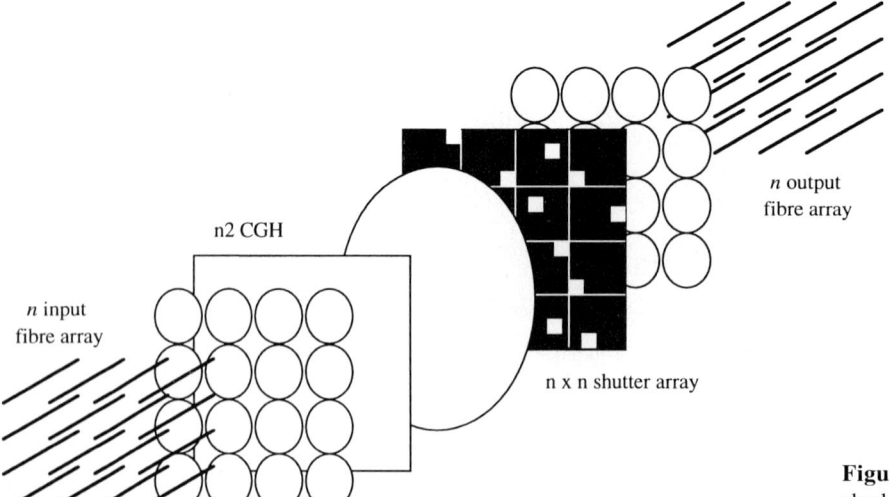

Figure 42. The n by n shadow routed crossbar.

optically connected parallel machine (OCPM) project [40]. This was a 1 to 16 (4×4 array) switch, with shutters provided by an FLC VLSI backplane SLM. One of the difficulties with this project was the design of the fan-in and fan-out optics, as the tolerances were extremely tight. The project routed single mode fibers to multimode in order to avoid the problem of fan-in loss. The switch was designed around an optical baseplate system based on a nickel-coated steel block with accurately machined mounting slots. The slots contain a small powerful magnet in the bottom. The optical components are centered in steel rings which rest on the edge of the slots and are held in place by the magnets. The system removes the need for x, y adjustment of the components, as they are all centered on the axis. The 16×16 switch is shown in Fig. 43.

Considerable effort was placed on the fabrication of custom multi-element lenses and diffractive optics to achieve the fan-out and fan-in. The SLM was a custom made 16×16 pixel FLC silicon VLSI backplane SLM with a typical contrast of 50:1, filled with a liquid crystal capable of a 15 µs address time. Considerable effort was also put into the accurate fabrication of the input and output fiber arrays. The 16×16 switch was run at a switching rate of 150 µs, with 1.5 MHz channels. The measured optical crosstalk was −17 dB and the total loss through the switch was 28 dB. A second 64×64 switch was also built and demonstrated at several conferences as a video signal switch [53].

A practical all-optical switch for long haul telecommunications must route single mode to single mode fibers. Matrix–matrix switches are not suitable for this as their loss in this case would scale with n^2. Locations where scaleable matrix–matrix switches might find applications operating between single and multimode fibers might include connection arrays in electronic computers (the original OCPM application) and other applications where optical switching is required between circuit boards, equipment racks, or computers. When switch structures fan-in into photodetectors (e. g., in chip-to-chip optical interconnections), then there is not intrinsic fan-in loss [as the effective NA (Numerical Aperture) of photodetectors is high] and therefore matrix–matrix switches are ideal.

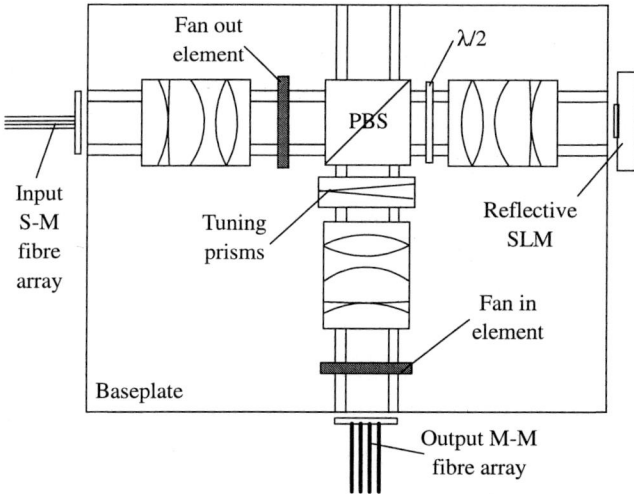

Figure 43. Baseplane layout of the OCPM 16×16 switch (adapted from [40]).

2.3.4.4 The ATM (Asynchronous Transfer Mode) Switch

A final extension of the n to n switch is to use the shadow logic to route packets of information [39]. A particular application is the asynchronous transfer mode (ATM) data, which is transmitted in finite size packets. For ATM chip-to-chip interconnects, it is desirable to switch packets or blocks of information rather than individual fibers, which can be done by shadow logic. In this case, the data are presented on an optically accessed RAM chip (opto-RAM) as a sequence of ATM packets in an n^2 array. The system layout is similar to the n by n crossbar in Fig. 42, except that each fiber is replaced with an ATM cell displayed on the opto-RAM SLM. The opto-RAM is an adaptation of an SRAM based silicon VLSI SLM capable of very high speed data modulation, with the pixel array broken up into ATM cells. The SLM consists of a 4×6 array of spatially separated ATM cells of 24×16 pixels each, on a silicon chip die with shift register electronics for pixel addressing. An ATM cell is shown in Fig. 44. The opto-RAM appears electronically to have the same functionality of a normal RAM.

The system for an ATM packed switch is shown in Fig. 45. The concept is the same as for the n by n crossbar, with the ATM cells being fanned out onto a shutter SLM before switching and then fanned-in to a similar structured chip containing photodetectors.

The opto-RAM pixels are illuminated with structured laser light generated by a photoresist CGH that matches the pixel pattern of the opto-RAM to avoid excessive loss due to the low fill factor of the SLM. The fan-out is performed by a second CGH

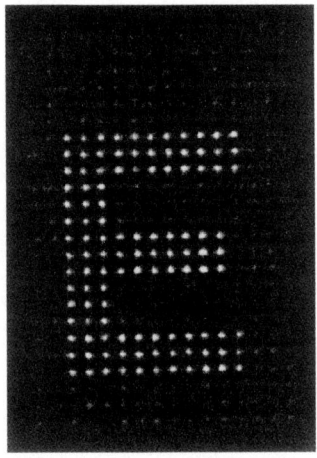

Figure 44. A single ATM cell on the opto-RAM chip.

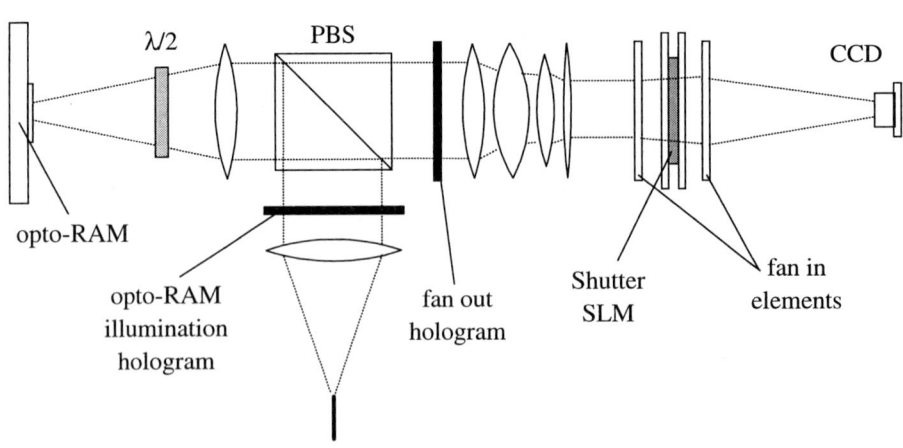

Figure 45. Optical layout of the ATM switch demonstrator.

and a custom designed four-element lens. The shutter SLM is a 128×128 FLC transmissive device with 220 µm pitch pixels and a contrast of over 130:1. At this stage, it is intended to implement the fan-in with diffractive Fresnel lens elements.

2.4 Wavelength Tuneable Filters and Lasers

A useful effect that can be exploited by liquid crystal modulators is wavelength dependence or filtering. This can be achieved by a variety of mechanisms, including dynamic grating display with spatial light modulators, multiple liquid crystal effects, and Fabry Perot based modulators. All of these devices and techniques are capable of controlling the wavelength; a very desirable property in both the displays field and the field of telecommunications. The exact mechanism used will dictate the wavelength operation: for displays, a broad RGB (Red Green Blue) wavelength control is required for color multiplexing, whereas for telecommunications applications, a narrow band response is desirable for wavelength division multiplexing (WDM).

2.4.1 The Digitally Tuneable Wavelength Filter

A simple grating can be used as a wavelength filter, as different wavelengths are diffracted at different angles as the light passes through the grating. In devices such as monochromators a rotating grating is used to pick out the desired wavelength for observation. The angle at which the light is diffracted through a grating is dependent on the grating pitch, so changing the pitch provides another mechanism for tuning the wavelength.

For the system shown in Fig. 46, illuminated with a wavelength of λ, we have

$$\sin\alpha - \sin\beta = \frac{m\lambda}{d} \qquad (57)$$

where d is the grating pitch, α and β are the angles shown in Fig. 46, and m is the integer order of the diffracted light. In general, we are only interested in the first diffracted order ($m = +1$). By changing the grating pitch d, we can vary the angle of the diffracted light β for the wavelength λ. This forms the basic tuning mechanism for the digital wavelength filter, as we use an SLM to effectively alter the grating pitch [54]. If we monitor a single position in the far field of the grating, then the position of the spot generated by the grating will sweep along with

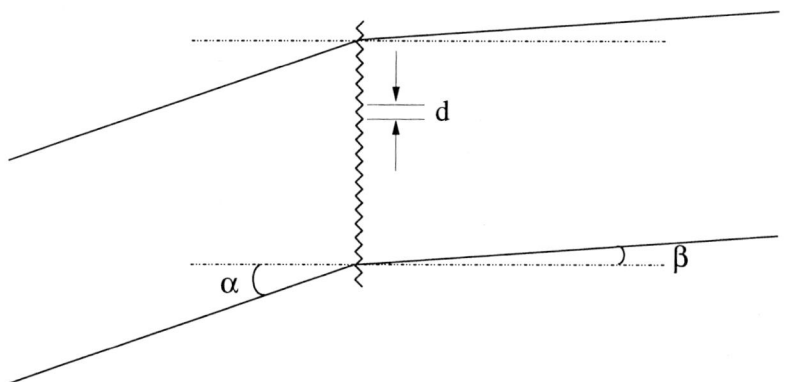

Figure 46. Illumination of a simple grating.

the change in pitch. If we have a multiple wavelength input source illuminating the grating and we monitor a fixed point in the replay field, then the wavelength will scan across that point with the changing pitch, making a wavelength tuneable filter.

If the grating pitch d was increased as an integer, we would have only a small number of fixed wavelengths that could be tuned. However, by using one-dimensional holograms, we can select any point in the output plane along the single axis, which means that we can select a range of angles and hence a range of wavelengths. The use of an FLC SLM is ideal for the high speed display of 1-D holograms to select the wavelength. Moreover, by using the phase modulation capabilities of the FLC, we can reduce the loss of the filter and prevent excess light being lost in the zero order of the hologram. As with other FLC phase applications, a switching angle of 90° would produce optimum results.

The most logical application of such a filter is in telecommunications WDM systems which require channels separated by 0.8 nm centered at a wavelength of 1550 nm. Such a filter would require a 1-D SLM with a pixel pitch of about 5 μm [55], which is unlikely to be built in the near future and may never work properly due to the properties of the FLC domains.

Figure 47 illustrates the system used to alter the tuning range of the large pitch grating by the addition of a finer pitch grating. From this system, we can define the two angles of diffraction such that

$$\sin\theta = \frac{\lambda}{D} \text{ and } \sin\phi = \frac{\lambda}{d} \tag{58}$$

where D and d are the respective grating pitches. Hence, for an SLM with pixel pitch D, we can choose the second grating pitch d based on the desired center wavelength and tuning range.

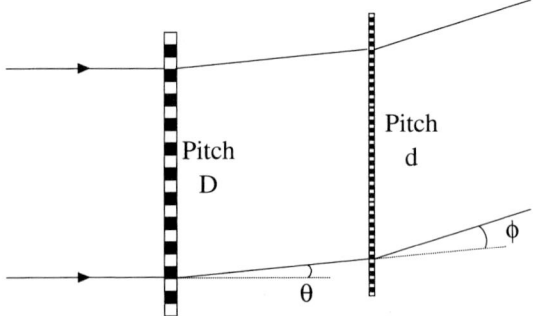

Figure 47. Two grating solutions to optimize tuning.

Figure 48 shows a practically laid out linear system for implementing a digitally tuneable filter based on an FLC SLM [55]. The SLM has N pixels at a pixel pitch D and the fixed grating has a pitch d. We also have a lens of focal length f to form the far field pattern of the filter and we are using a single mode fiber at a spatial position x to collect the tuned wavelength. From these parameters, we can estimate the wavelength tuned

$$\lambda \approx \frac{x}{f\left(\frac{n}{ND} + \frac{1}{d}\right)} \tag{59}$$

where n is an integer lying between 0 and $N/2$. In the practical example of Fig. 48, the parameters were set at $f = 96.1$ mm, $N = 128$, $D = 165$ μm, $d = 18$ μm, and $x = 8.3$ mm, which gives a tuning range of $l_{n=0} = 1592.1$ nm to $l_{n=64} = 1509.7$ nm in 64 steps of resolution 1.29 nm.

The filter was tested in an experimental bench setup and used to filter the amplified spontaneous emission of an erbium-doped fiber amplifier (EDFA). The SLM was a 128×128 2-D device with a cell thickness optimized for use at a wavelength of 1550 nm and was used with constant row data and columns set with the hologram data. The SLM was used as a binary phase modulator without polarizers to reduce the

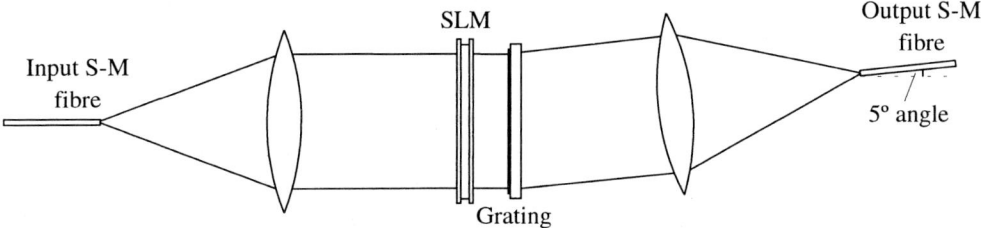

Figure 48. Experimental layout for the digitally tuneable wavelength filter.

loss through the system. The holograms were each calculated by simulated annealing and were designed to route light to one of 64 positions along the horizontal axis of the system, coinciding with the desired wavelength to be tuned. The fixed grating was fabricated as a binary photoresist relief pattern created by photolithography onto a $\lambda/10$ optical flat. The phase relief was designed for a 180° phase shift at 1550 nm, but the photoresist was too thin, leading to a phase step of only 168°.

Figure 49 shows the results of the experiment. The input source was the amplified spontaneous emission of an EDFA (Fig. 49 a) and the output was measured with a spectrum analyser. Figure 49 b shows the superimposed results for the filter with 11 different holograms. The measured loss through the filter was 22.8 dB (Table 1), which was high but could be accounted for in the power budget of the filter.

These losses could be reduced by further development of the SLM to give a switching angle of 90° and by using a blazed grating to replace the one in the photoresist. The diffraction efficiency of the hologram cannot be avoided as we are using binary phase modulation; perhaps a nematic based multilevel phase modulation could be used to reduce the loss, at the cost of overall tuning speed. Assuming an ideal FLC SLM and a high quality optical system, the loss could be reduced to as low as 6.5 dB.

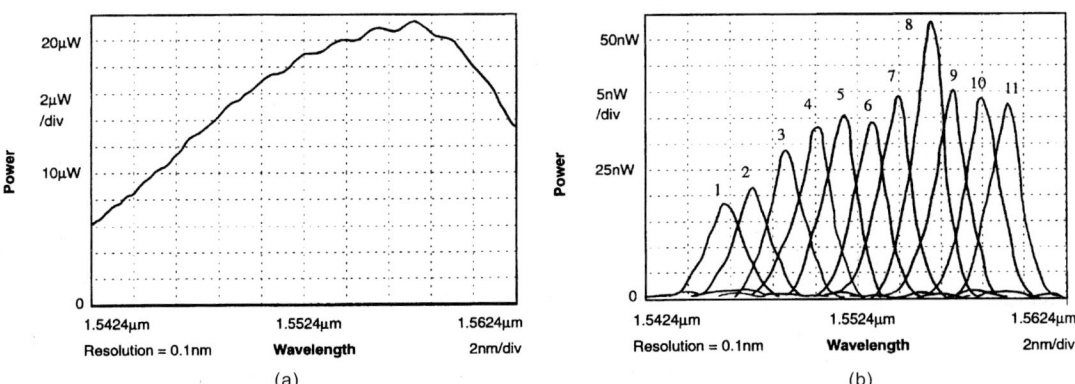

Figure 49. Experimental results of the wavelength filter: (a) EDFA spectrum, (b) filtered EDFA spectrum by 11 different holograms (reproduced from [55]).

Table 1. Filter losses.

SLM losses	dB
FLC switching angle $\theta = 28°$ (transmission = $\sin^2 2\theta$)	6.57
Diffraction efficiency ($\eta = 36.5\%$)	4.38
Aperture of SLM	0.79
Fixed grating losses	dB
Diffraction efficiency ($\eta = 36.5\%$)	4.38
Phase depth error (168°)	0.05
Other losses	dB
10 reflecting surfaces, 4% loss each	1.77
FC/PC patch cord losses	1.14
Fibre-to-lens coupling efficiency ($\approx 42\%$)	3.72
Total	22.8

2.4.2 Digitally Tuneable Fiber Laser

The next step from the wavelength tuneable filter is to use this filter as a tuning element to create a digitally tuneable laser [56]. The relatively long cavity length of fiber lasers (from meters to a few centimeters) result in very narrow linewidths and closely spaced longitudinal modes, enabling almost continuous tuning. Such attributes are very desirable in telecommunications WDM systems.

The wavelength filter of Fig. 48 remains unchanged when included in the tuneable laser. Figure 50 shows the layout of the tuneable laser. The output fiber is fed back to the input of the filter to tune the lasing of the fiber ring. Such a system could not lase as there is no net gain around the fiber ring and in fact there is a 22.8 dB loss. To counter this, an EDFA is included within the ring to counter the loss of the filter, with a gain of 25 dB across a typical bandwidth of 40 nm. An isolator was also included to cut any reflections propagating in the opposite direction of the fiber ring. Finally, a 3 dB coupler was added to allow the laser output to be monitored and coupled out of the laser.

Figure 51 shows the results measured from the digitally tuneable fiber laser.

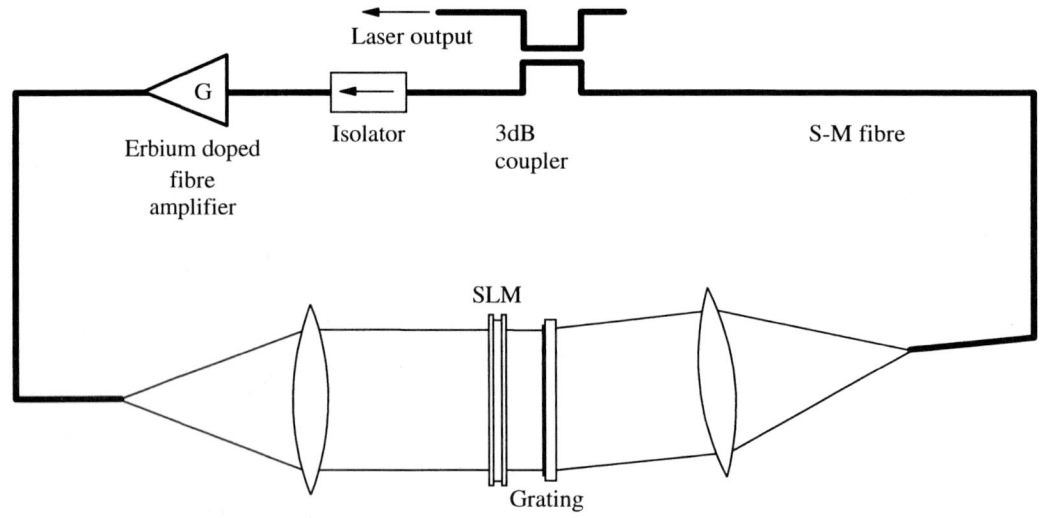

Figure 50. The digitally tuneable fiber laser.

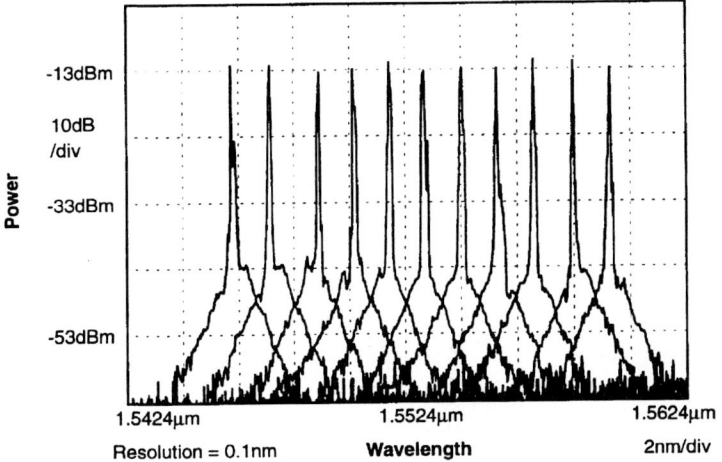

Figure 51. Eleven successively tuned wavelengths from the laser (reproduced from [55]).

Eleven holograms were used to tune the desired modes from the laser. Each mode had a power of −13 dBm with a side-mode suppression of over 30 dB; they were spaced 1.3 nm apart, as predicted by the performance of the tuneable filter calculations. The linewidth of the laser proved more difficult to measure, due to the fact that it was extremely narrow. A heterodyne self-beating technique was used with a delay line and a 100.6 MHz frequency shift. The use of a 14.6 km delay line yielded a linewidth of 1.7 kHz assuming a system resolution of 6.7 kHz. From this measurement we can conclude that the linewidth lies between 1.7 kHz and 6.7 kHz.

From these results, the digitally tuneable laser will prove a very useful tool in telecommunications WDM systems. With further reductions in the loss of the wavelength filter, the performance of the laser will improve and become more stable and reliable. A further development is the ability to equalize the output of the laser (i.e., give it constant output power over the tuning range) by altering the height of the peak generated by the hologram and therefore alter the loop gain of the ring laser and the output power [57].

2.4.3 Liquid Crystal Birefringent Wavelength Filters

The effects of birefringence have been exploited for phase and intensity modulation in FLC SLMs, but they can equally well be used to control the wavelength [58]. The birefringence Δn of a liquid crystal material of thickness d affects the retardance, Γ, such that

$$\Gamma(\lambda) = \frac{2\pi \Delta n d}{\lambda} \tag{60}$$

The effect of this modulation on the white polarized input light is to rotate the state of polarization according to the wavelength λ. These rotated polarization states vary from linear to elliptical, so an achromatic quarter wave plate is added to convert all the elliptically polarized light to linear states. The linearly polarized wavelength of the light can then be rotated further by a liquid crystal birefringent effect, and the resultant modulated wavelengths blocked by polarizers. The net effect of the wavelength filter is to convert the band of wavelengths to a polarization state that can be removed by a polarizer. By altering the magnitude of the liquid crystal modulation, it is possible to tune the band of wavelengths that is blocked. A system for implementing this filter is shown in Fig. 52.

The LC (liquid crystal) cell was filled with BDH764E electroclinic material and kept at 29 °C, 1 °C above the transition to the SmA* phase. The filter was tested by illumination with white light from a tungsten lamp and the output was monitored by a monochromator. The tuning range of the filter was limited by the switching angle of the liquid crystal and measured at 115 nm, centered on a wavelength of 540 nm with a maximum addressing voltage of ±30 V. The FWHM (full with at half maximum) of the filter characteristic was measured at 10 nm, which makes this sort of mechanism ideal for RGB color filters.

A better adaptation of this technique is to use FLC cells in the cascaded fashion shown in Fig. 53 [59]. Each cell has an associated birefringence and switching angle, by switching different combinations of cells it is possible to select five different filter states. The selection of the states is shown in Table 2.

The birefringence of the cells is governed by their thickness; for the system in Fig. 53,

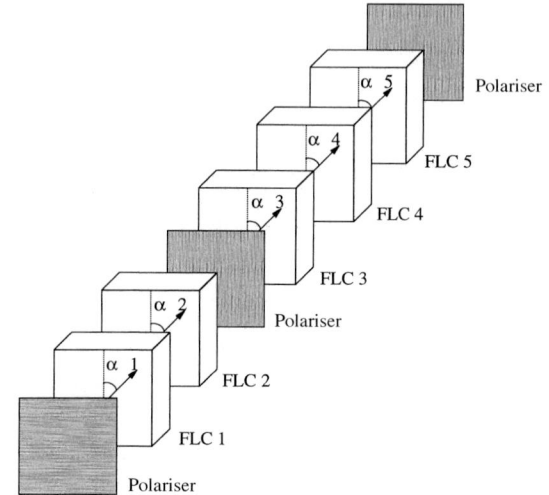

Figure 53. Cascaded FLC cell wavelength filter (adapted from [59]).

Table 2. FLC cell states for the RGB color filter.

Output	$\alpha 1$	$\alpha 2$	$\alpha 3$	$\alpha 4$	$\alpha 5$
White	$\pi/4$	0	0	0	0
Blue	$\pi/4$	$\pi/4$	0	0	0
Green	0	$\pi/4$	$\pi/4$	$\pi/4$	$\pi/4$
Red	$\pi/4$	$\pi/4$	0	$\pi/4$	$\pi/4$
Black	0	0			

the cells FLC 1–5 have thicknesses 1.8, 5.2, 2.6, 1.7, and 6.1 µm, respectively. In the blue state, the filter peak was centered at 465 nm and filtered out the red and green; the green state was centered at 530 nm, and the red state was centered at 653 nm. Each peak had an FWHM of around 70 nm. This characteristic is ideal for an RGB filter, and is suitable for applications such as camera filters and display systems. The RGB filter has since been commercialized by its inventor within the company Colorlink. It is a good example of liquid crystal technology making it into the market place in a nondisplay application.

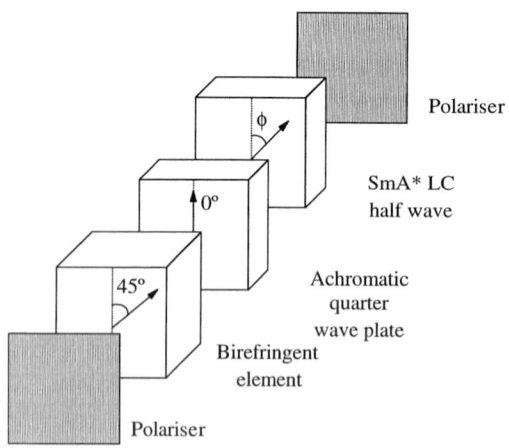

Figure 52. Smectic A* phase wavelength filter (adapted from [58]).

2.4.4 Fabry Perot Based Wavelength Filters

The wavelength properties of a Fabry Perot cavity have been known for a long time. A cavity can be made from two parallel, inward facing mirrors a small distance apart. Light incident on the cavity will be trapped within by multiple reflections. Those reflections that complete a whole round trip in a wavelength will not decay, whereas those that do not will be lost, leading to a resonant characteristic with peaks at integral multiples of whole wavelengths. The phase of this round trip δ can be expressed such that

$$\delta = \frac{4\pi n L}{\lambda} + \phi \qquad (61)$$

where L is the cavity thickness, n is the refractive index of the cavity, λ is the wavelength, and ϕ is the phase change on reflection from the mirror. A transmission peak will appear if the round trip phase $\delta = 2\pi$. The wavelength of the transmission peak can be varied by adding a substance to the cavity that has a variable refractive index. One such substance is an electroclinic liquid crystal with sufficient alignment [60]. The tuning of the cavity will now be

$$\Delta\lambda = \frac{\lambda}{n}\Delta n \qquad (62)$$

where n is now the average refractive index across the cavity and Δn is the variation in the refractive index.

The physical construction of a Fabry Perot filter is shown in Fig. 54. The liquid crystal modulation needs to be parallel to the direction of the light in order for a refractive index change to be achieved. For this reason, the liquid crystal molecules must be tilted off the surface of the mirrors to allow the modulation to occur. The electrical field which moves the molecules must be applied from the side of the cell rather than the front

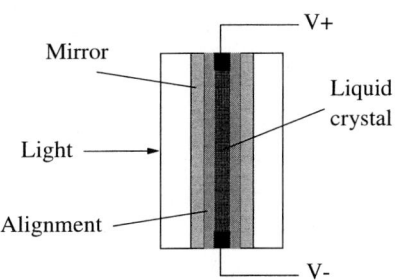

Figure 54. Construction of a Fabry Perot wavelength tuning element (adapted from [60]).

and back surfaces, which creates a difficult cell fabrication problem. The liquid crystal used in the cell of Fig. 54 was an electroclinic material, BDH764E [61]. This material has a birefringence of 0.15 and with a pre-tilt of 30° can give a refractive index change of 0.03–0.04. The alignment pre-tilt of 30° was achieved by evaporating a layer of SiOx at an angle of 10° and then adding a monolayer of homeotropic alignment agent, CTAB (cetyl trimethyl ammonium bromide). The pre-tilt greatly increases the refractive index modulation and hence the wavelength tuning. The filter was fabricated using photolithography to lay down the electrodes and the cavity thickness was measured as 4.8 μm. The filter was illuminated with white light and the output was monitored with a spectrometer. The width of the continuous tuning range was measured as 10.5 nm at the wavelength of 600 nm for a voltage swing of −8 to +8 V. The switching speed was measured at 9 μs.

A more compact version of the filter was then developed based on an optical fiber ferrule system. The basic structure was the same as in Fig. 53, but on a much smaller scale. The final size of the filter was about 40×5 mm and was built around two 1550 nm single mode fibers. The filter was designed as a telecommunications element and was optimized for a wavelength of 1550 nm. The tuning range for the device was

13 nm at about 1550 nm and the measured loss was 7 dB, but should be reducible to around 4 dB with good mirror finesse.

Another refractive index modulation mechanism based on distorted helix liquid crystals has also been enhanced by placing the filter into a Fabry Perot cavity [62]. Tests so far have been with silver and dielectric stack-based Fabry Perot cavities and have been encouraging, with a tuning range of 10 nm about a center wavelength of 630 nm being measured at a speed of 200 µs with the liquid crystal FLC6304. Several liquid crystals have been tested, including other distorted helix liquid crystals and several different electroclinics.

2.5 Optical Neural Networks and Smart Pixels

One of the first nondisplay applications which utilized liquid crystal devices was optical computing, especially where the problems were intrinsically highly parallel and very high speeds were not essential, such as in image processing. Large scale optical computeres have proved slow and cumbersome, and the modern role of optics is to combine with electronics to reduce the overall complexity of a system; rather than design an all-optical computer, the rationale today is to include optical systems as processing or interconnection units within an overall system. One field where considerable effort has been put into optics is neural networks, mainly because, as the complexity increases, the requirements on interconnects increase even more steeply, which creates an ideal candidate for the inherent parallelism for optical systems. In the field of neural networks, there are hundreds of different techniques, arrangements, and algorithms, each of which has specific applications and benefits. It would not be possible to cover every application where liquid crystal devices have been incorporated into optical and opto-electronic neural networks. For this reason, the intention of this section is to outline some of the issues and solutions, which most of these techniques have in common and which can be implemented with liquid crystal devices.

The basic format of a neural network is shown in Fig. 55. A neural network is a processor that maps a set of input conditions onto a set of output conditions to perform some function. Contained within the neural network are several layers of processing elements and interconnections between the processors. It is the setting of the interconnection pattern and their associated weights that creates the neural network, and the process of setting those weights is known as training the neural network.

The system in Fig. 55 contains all the vital elements needed to set up a neural network. There are two main components, the neurons and the synapses (or weights), both of which are modeled on observations made on the human brain's processing capabilities. The neuron acts as a unit which takes the sum of its inputs and then thresholds that sum before outputting to the next layer in the neural network. The synapses are the

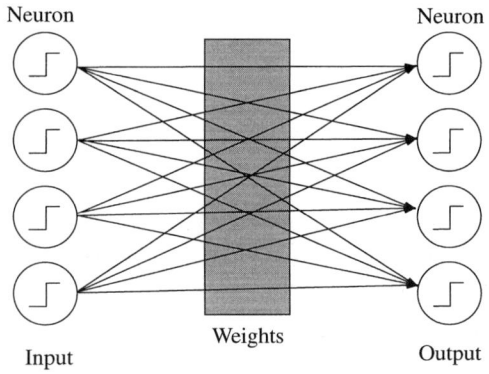

Figure 55. Basic format of a neural network.

weighted interconnections between the neurons which give each layer connectivity. In a neural network, the outputs from one layer are fanned out and weighted so as to connect to all of the inputs of the neurons in the next layer (as in Fig. 55). It is in the synapses that the potential limitations of optical neural networks lie. To fully implement a neural network and gain efficient training, bipolar grey scale weights are essential. Such weights are difficult to assign, especially the bipolar requirement. To get round this limitation of optics, several solutions have been suggested, including the use of polarization to represent the sign of the weight and the use of two light beams for each interconnect, one positive and one negative.

The most significant development, which led to the use of liquid crystal devices in neural networks, was the Stanford vector matrix multiplier (SVMM) by Goodman in 1978 [50, 63]. The basic structure of the system is shown in Fig. 56.

The input array (shown here as a 6×1 vector) is fanned out horizontally onto a weighting mask. The mask holds the weights for the layer of the neural network and the resultant multiplication is fanned-in vertically to the output. This system directly relates to the neural network in Fig. 55 and shows a possible optical arrangement for such a network. If a grey scale nematic SLM is used as the weighting mask, it is possible to implement and dynamically train the neural network [64]. This combination of elements forms the integral part in a whole variety of neural network arrangements and training structures in combination with a scheme for bipolar weights. More interesting is the fact that if the weighting mask is replaced with an optical shutter, then the SVMM becomes an optical switch.

2.5.1 The Optical Vector Processor

The expansion of the SVMM into a useful neural network with bipolar weights and multiple layers of neurons has been the topic of research for many years and has yielded several possible arrangements like that shown in Fig. 57 [65].

In this system, two FLC SLMs are used as binary weight masks and the combination of the two masks allows bipolar binary weights to be implemented in a single layer. The purpose of the diffuser in the middle is to remove any angular information between the two stages. This arrangement can now be combined with the Hopfield model of training binary weights [66] to simulate a full neural network without any need for grey scale modulators.

A practical demonstration of this system is shown in Fig. 58 (without polarizers to avoid confusion), which shows the optical construction of a 64 neuron network with five stored memories [67]. The FLC SLMs were early 64×64 pixel devices developed at STC [68]; two were used as the weight masks and the third was used to display the input data. The iteration time for the system was 50 ms and reconfiguration of the networks memory could be done in one cycle. The system was trained using the Hopfield model and used twelve vector sets each containing 10 memory vectors and two input

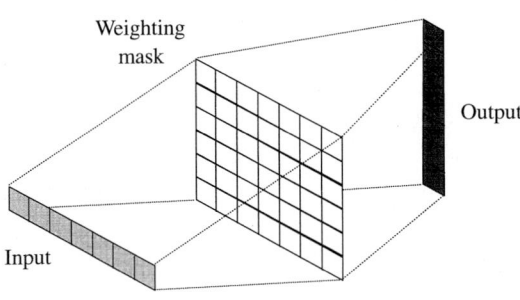

Figure 56. The Stanford vector matrix multiplier.

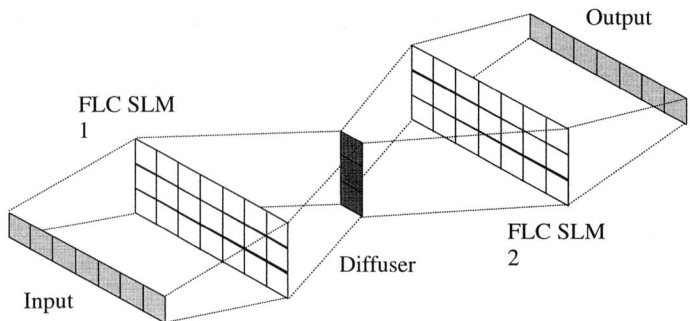

Figure 57. Optical vector neural network (adapted from [65]).

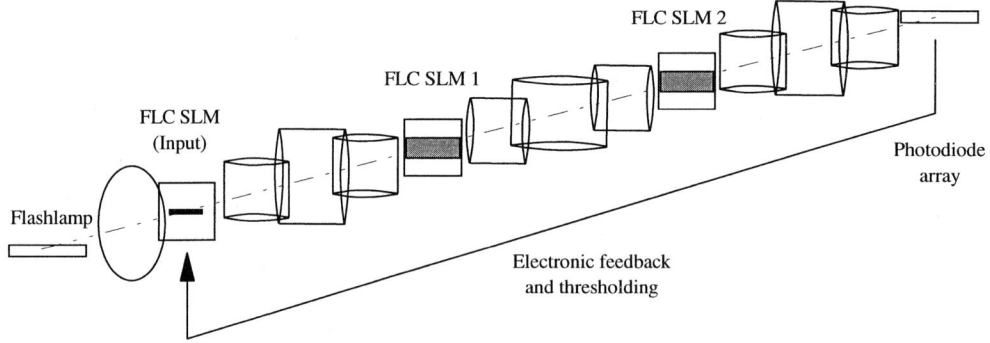

Figure 58. Practical neural network system based on FLC SLMs (adapted from [67]).

vectors. The system was designed to look for errors in the input vectors created by randomly generated corruptions. The results of the processing were compared with computer simulations and statistically analyzed over 200 tests per set. The test results showed that the network detected 80% of the vector errors compared to the computer simulation. Such a success rate is very good considering the number of optical elements in the alignment system and the limited capability of the FLC SLMs.

2.5.2 Computer-Generated Holograms as Synapses

The discovery of the CGH and its flexibility in generation has allowed further development in the field of neural networks. By controlling the direction of the light, CGHs can be used for fan-out and fan-in, as well as the setting of weights [69]. The CGHs discussed in the previous section were designed to generate maximum intensity in the replay field; however, it is possible to control this intensity within reason and form more versatile interconnects, as in Fig. 59.

The only drawback with this system is that we have to calculate the CGH for a given weight pattern, making the training process very long. If the calculations can be done off-line to generate the weights before the CGH is generated, then such a system is feasible. The CGH could be a fixed element, allowing the use of a high pixel density for each CGH and therefore generating accurate weights. However, reconfiguration of the network would require physically changing the hologram. The holographic

2.5.3 Hybrid Opto-Electronic Neural Networks

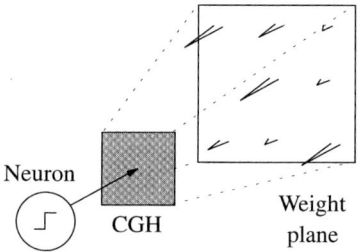

Figure 59. The CGH as weight control in a neural net.

weights can only be unipolar, so other techniques such as multiple beams or polarization modulation must be used to create bipolar weights.

The use of CGHs as interconnections has been proposed in several arrangements, such as the one in Fig. 60 [70]. In this example, the CGH is displayed on a refractive SLM. The input is presented at the CCD camera, weighted, and replicated by the appropriate CGH pattern, forming the appropriate synapses. The limitation on such a system is the optical quality and number of pixels available on the SLM, as this will affect the generation and quality of the CGH and therefore the quality of the weights. High resolution (>1000×1000) planarized silicon backplane SLMs in the near future will greatly enhance the capabilities of such neural networks.

Once again, there are a whole variety of neural network systems that come under this heading. Essentially, they are the neural networks that combine electronic processing with optical weights to draw some compromise between the two spheres. Almost all of these systems involve a weight mask that is optically imaged onto a photodetector of some sort. The weights could be fixed optical masks or intensity SLMs, and the photodetector is usually some form of CCD array. The summation of the weights and the thresholding of the neurons is performed by electronics.

The system in Fig. 61 is typical of a hybrid opto-electronic neural network [71]. It could equally well be implemented with optical fibers or other guided wave optics. A recent experimental neural network based on Fig. 61 was constructed with a custom FLC SLM. As the system uses one channel from each neuron, the number of pixels required for the weights is often small. In this case the SLM was an 8×8 array of 5 mm pixels. The photodetector was made from an array of amorphous silicon photoresistors, but these proved too nonuniform and so were replaced with an array of commercial-

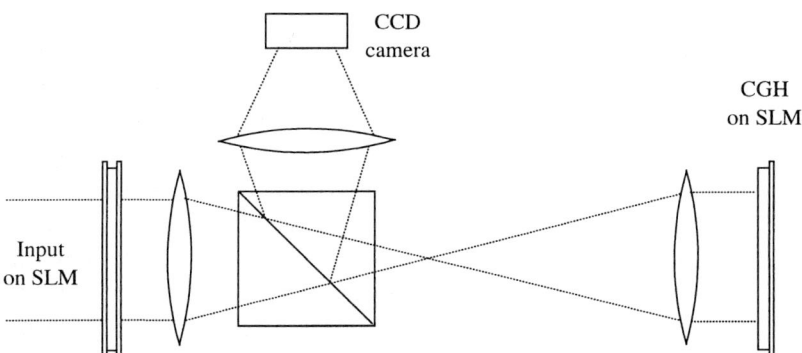

Figure 60. CGH based synaptic neural network (adapted from [70]).

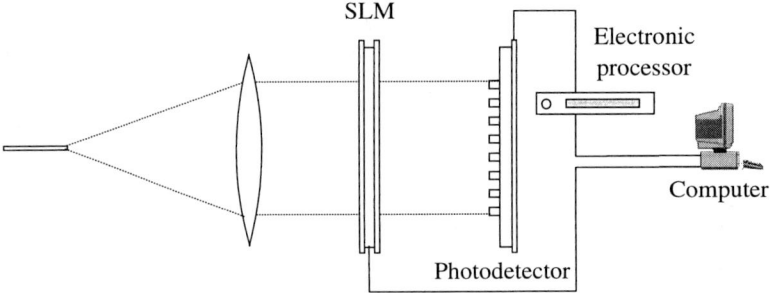

Figure 61. A hybrid opto-electronic neural network.

ly available photoresistors. The initial test of the system was as an exclusive-OR processor, made from three neurons in two layers (two in the first and one in the second). Such a processor only required 12 pixels from the SLM and used feedback to connect between the first and second layer neurons. The overall success of the system was limited by the modulation of the FLC SLM. It was origially envisaged to use pulse multiplexed data on the SLM pixels to obtain a limited grey scale. However, this attempted grey scale scheme proved unreliable and difficult to control, so pure binary weights had to be used.

2.5.4 Smart Pixels and SLMs

A logical advance in the design of an SLM is to combine the optical properties of the SLM with the processing ability of electronics [72]. This is especially important when VLSI silicon backplane SLMs are used, as the VLSI processes are identical to those used to create a standard silicon chip. For this reason, electronics are combined with the SLM to create smart pixels capable of processing information presented optically and then modulating light for further processing. Such pixels are very useful both for image processing and for optical neural networks.

In a neural network there are two distinct functions that can be performed by the optical components. The weighted interconnects can be obtained through a combination of CGHs and smart pixels, whilst the neuron threshold can be obtained by a smart pixel.

The smart pixel shown in Fig. 62 contains a photodiode connected via an electronically thresholded transimpedance amplifier to an SLM pixel [73]. The weighted interconnects are all incident on the photodiode, so a sum is formed and then converted to an electronic signal. This signal is thresholded about an external signal provided to each smart pixel, which means it can be varied to tune the pixel's performance or 'smartness'. The threshold then controls the SLM pixel, which interfaces optically to the next layer in the neural network.

A second variation on this theme was used in a processor based on the adaptive resonance theory [74]. This system was essentially an SVMM with a smart pixel array as the central weighting mask. The functionality of the smart pixel was basically the same as that for the pixel in Fig. 62. The only difference was that a pre-determined weight mask was loaded and stored on the smart pixel array. The array was then turned on and the photodetectors were illuminated with the input light. If the light was present at the photodetector and the corresponding pixel mask was selected, then the pixel modulator was turned off. This functionality provides a powerful technique for fast processing and learning in a neural network.

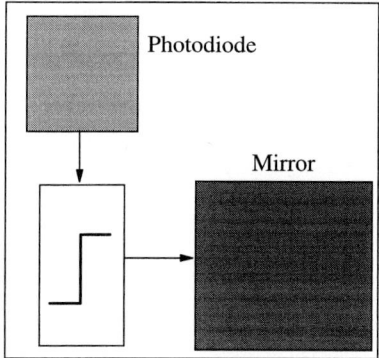

Figure 62. Basic structure of a smart pixel (SASLM 1).

A more advanced smart pixel is shown in Fig. 63, as it contains two photodetectors which are separately addressed by different optical sources [73]. If light is incident on photodiode 1, then the modulation of the pixel (i. e., the amount of light reflected) increases. If light is incident on photodiode 2, then the modulation of the pixel decreases. The pixel is capable of a 4 bit (16 levels) grey scale modulation (weight) based on an electroclinic or distorted helix FLC. It is this functionality that performs the interconnect weights in the neural network and allows them to be trained by the control of the intensity of the two beams addressing each smart pixel.

The two components (SASLM 1 and SASLM 2) can now be combined with CGHs to route (fan-out) the light through the optical neural network, as in Fig. 64. SASLM 1 is used as the neurons and SASLM 2 makes up the synaptic weights. The system is almost entirely optical and the advantages are inherent compared to electronic neural nets, as the interconnections are total and performed in free space. The problem is that for n SASLM 1 neurons, n^2 SASLM 2 weighting smart pixels are required, which limits the processing power of the neural net due to VLSI constraints [75].

A third type of smart pixel is shown in Fig. 65, which performs spatial image processing [76]. In this case, the pixel is fed by the four adjacent pixels to the left, right, above, and below. The photodiodes from the adjacent pixels are connected to the central pixel in the fashion shown. The functionality of the summation of these signals is set such that we have the left pixel subtracted from the right added to the top subtracted from the bottom. This performs the function of a Roberts Cross filter, which is a means of achieving edge enhancement. More complex processes such as Sobel filtering (a combination of all eight nearest neighbors) could also be implemented at the cost of VLSI facilities. Figure 66 shows the functionality of a smart pixel using the Roberts Cross filter to perform edge enhancement. Such a processor array would be very desirable as an image pre-processor for correlation systems like the binary phase only matched filter [29].

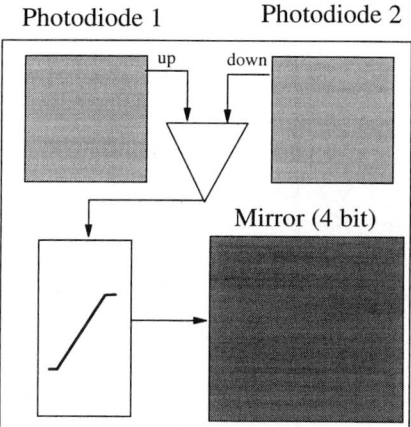

Figure 63. A more advanced smart pixel (SASLM 2).

816   2   Nondisplay Applications of Liquid Crystals

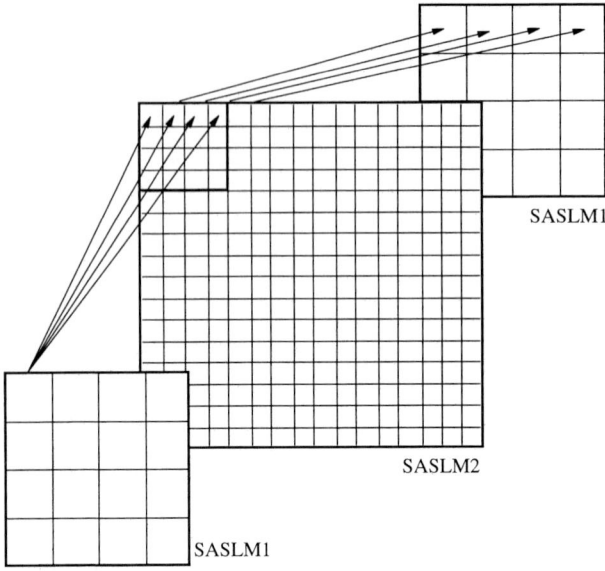

Figure 64. Smart pixel based neural network.

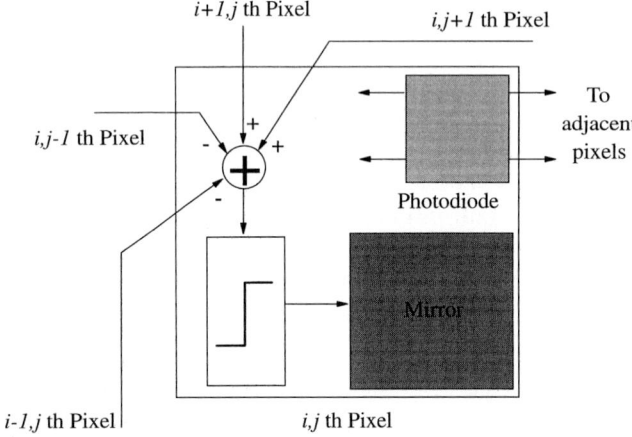

Figure 65. A spatial processing smart pixel.

Input

Output

Figure 66. Example of an image processed with the Roberts cross.

2.6 Other Applications

2.6.1 Three-Dimensional Video

This application of liquid crystal technology may not seem to fit into the category of nondisplay applications, but the actual use of the liquid crystal is not in the display of the images. Rather, it is used to perform the angular multiplexing which goes to make up the three-dimensional image [77]. The display itself is autostereoscopic, with no external optics, such as 3-D glasses, required. The viewer can see the 3-D image over a viewing angle of 40° with the naked eye.

The concept behind the operation of the 3-D display is the fact that a change in angle in one plane leads to a change in position after a Fourier transform (from a lens). The original 3-D image is broken up into a series of angular views from different angles in the horizontal plane (usually between 8 and 16 views). These views are then stored in angular sequence ready for display. The views are then displayed in sequence on a high speed and high brightness CRT (cathode ray tube), as shown in Fig. 67. Each separate view illuminates a lens that performs a Fourier transform. The transform plane now contains the information of the view, shifted along the horizontal axis according the original angle of the view. Also, in the plane is an FLC shutter, which contains a single pixel for each of the horizontal positions of the views. As the view is displayed, the corresponding shutter pixel is opened, allowing that view to pass through. The shuttered view then passes through another lens, where it is rotated in a spatial position according to the original angle of the view. The result is a sequence of images produced at discrete angular positions in the viewing plane. If the rate of display of the views is sufficiently fast, then the eye cannot detect the flicker as the views sequence through and it sees a continuous view. Moreover, as the eyes are spatially separated, they each see a separate view, which is then combined in the brain to form an autostereoscopic image.

The first generation of 3-D displays was eight-view monochromatic TVs. Further development has led to a 16-view display

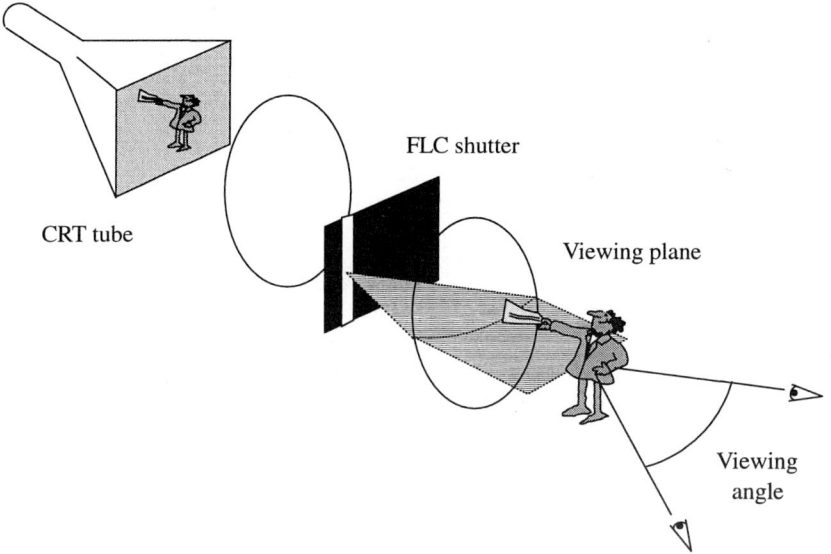

Figure 67. Basic principle of the 3-D display (adapted from [77]).

and a full color display, based on the inclusion of an RGB color filter to separate the different color components of each view. The launch of this display has created interest from both the medical surgical imaging field and the home games console companies.

2.6.2 α-Si/Novelty Filter

An interesting development beyond the FLC OASLM is the α-Si novelty filter [78]. This device uses an OASLM like structure to perform a simple motion detection to measure the difference between image frames. Rather than use the FLC material as a direct modulation medium, it uses the surface-stabilized characteristics of the FLC as a form of optic memory. The optical modulation of the FLC material is also converted from a polarization effect to an amplitude effect by the inclusion of a pleochroic dye with the FLC material.

The physical construction of the device is similar to that of an FLC OASLM, with the FLC sandwiched between an ITO coated glass cover, and a layer of α-Si. The inclusion of the dye in the FLC causes a change in the modulation characteristic of the FLC. When the first image is incident on the α-Si layer, the image is transferred to the FLC molecules as per the OASLM modulation. The dye molecules are also modulated, creating a stored image in the device.

When the second image is incident on the device, there is no addressing of the FLC, as the addressing pulse is removed. This means that the second image is filtered by the first and stored in the dye of the device. Any differences between the images will appear as photo-current in the α-Si surface, which can be read off using a structured pattern in the ITO. An example of this operation is shown in Fig. 68.

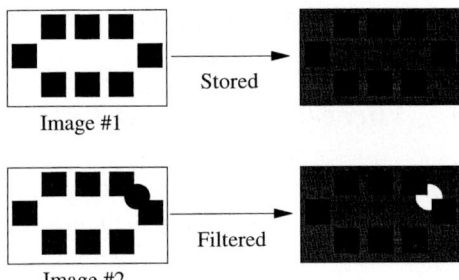

Figure 68. Operation of the α-Si novelty filter (adapted from [78]).

2.6.3 In-fiber Liquid Crystal Devices

LC materials have been used in a variety of fiber-based applications, such as LC waveguides and LC interactions with tapered fibers [79]. The inclusion of liquid crystals in hollow fiber tubes has created a new class of hybrid electro-optical devices with applications in telecommunications and fiber-based sensor systems. Initial work was done with nematic liquid crystals in hollow glass tubes and also with SmA materials. The alignment of the material was assumed to be homeotropic, and a transition in the liquid crystal from radial to axial alignment was achieved with an applied voltage. The liquid crystal waveguides were also demonstrated as being able to produce TE_{01} (transverse electric) and TM_{01} (transverse magnetic) modes in an optical fiber [80]. Since then a variety of waveguides has been produced containing FLC materials to improve the modulation and decrease loss. Several different alignment directions have been achieved, including radial (homeotropic), escaped radial (homeotropic/planar), and planar, by using surfactants like CTAB. Results so far have demonstrated in-fiber modulation of light at speeds of up to 100 kHz, with good modulation depth [81].

Another series of applications involve the interaction between liquid crystals and the modes of light propagating down the central core of the fiber. There are several ways of implementing this interaction, including the use of fiber tapers, D-section fibers, and fibers with partially etched cladding to expose the central core. Effects that can be generated include polarization state filtering, wavelength filtering, and intensity modulation. A recent demonstration of those principles was performed with a single mode fiber, side-etched to expose the core [82]. The etched portion was used as one side (including ITO contacts) of a nematic liquid crystal cell. The nematic liquid crystal was aligned to show a refractive index change with applied voltage. In one state, the refractive index was higher than that of the fiber core, which leaves the light within the fiber. When the liquid crystal is switched to match the fiber core refractive index, the light escapes from the core, causing an intensity modulation. Experiments so far have shown a modulator with an insertion loss of 0.5 dB and a modulation depth of 20 dB at a driving voltage of 5 V. The main limitation with the use of nematic liquid crystals is the relaxation time giving low speed modulation. The use of FLC and electroclinic materials should greatly increase the speed of modulation.

2.6.4 Liquid Crystal Lenses

The ability to control the refractive index of a liquid crystal allows the implementation of liquid crystal lenses. In these devices, a gradient of refractive indices is set up in the liquid crystal by addressing patterned ITO electrodes [83]. The refractive index profile acts as a curved surface of glass and hence acts as a lens. By tuning the structure of the refractive index profile, it is possible to alter dynamically the performance of the lens. Experiments so far have demonstrated lenses with swept focal lengths from 100 to 500 mm. Shorter focal lengths have proven more difficult due to the limitations of the ITO pattern. The lenses produced so far have not been close to diffraction limited and have contained several aberrations due to fabrication errors in the ITO. Until these problems are properly solved, the uses of these lenses are limited.

A novel form of liquid crystal 'lens' is the switchable hologram [84]. The exact functionality of the device depends on the design of the pattern, as it could be a CGH or a Fresnel lens. The device is designed to be blank in one state and to switch to a fixed ITO pattern when addressed electronically, such a device is useful when a lens or CGH is needed to be added or removed from an optical system. The lens is fabricated as a normal FLC single pixel cell, with a complex pattern placed in the ITO. The pattern is designed with a simulated annealing algorithm based on chemical crystal growth. The purpose of this algorithm is to ensure that the desired replay is generated while maintaining connectivity between all the pixels, allowing them to be switched as a single pixel. Results so far have demonstrated the principle with the fabrication of a switchable CGH of 128×128 pixels generating a complicated replay field pattern.

2.6.5 Optical Aberration Correction

A useful feature of the CGH is its ability to correct for aberrations in an optical system. This ability is becoming increasingly desirable for atmospheric correction in optical telescopes and laser-based long range weaponry. If the aberration of an optical system can be characterized as some form of modulation transfer function (MTF), then the CGH can be designed to correct for it, within the limitations of the SLM technology used to display the CGH. The ideal SLM for this application would be a 256 phase level modulator such as a nematic-based device. However, the requirement for speed is stringent in some applications, as the aberration may be changing at a faster rate than the SLM. For this reason, experiments have been performed using binary phase FLC SLMs with some success. One such FLC SLM experiment was to correct for an optical aberration inserted into a multilevel phase system [85]. The modulation was achieved with three FLC SLMs cascaded together with wave plate optics to form an eight level phase modulator. A low quality glass slide was then added to the system to aberrate the modulator to the point where it was no longer acceptable. A correction algorithm was then run to counter the effects of the added glass slide. The corrected phase planes were displayed on the SLMs and the performance was measured as being better than the system without the glass slide. The improvement in performance was due to the fact that the algorithm had corrected for all the other imperfections in the system as well. Such results show that aberration correction is possible, but ideally fast multilevel phase SLMs are needed to realistically implement these systems.

2.6.6 Switchable Phase Delays for Phase Array Antennae

A new, promising application of SLMs is in the control of delay lines in phased array antennae [86]. In a phased-array antenna, a sequence of antennae is addressed with a signal that is out of phase with the previous antennae in the array. Electronic delay lines have been used in the past to implement these phase delays, but such systems are large and bulky at microwave frequencies and they are influenced by external electromagnetic interference. Recent developments have shown that phased-array antennae can be fed by an optical signal that is interference resistant and more compact, and that the delays needed to perform the phase steps can be made from bulk optical components. It is the efficient and all-optical control of these phase steps that has been demonstrated with liquid crystal components. The phase delay is made by an extended optical path or bypass from the main signal path. By switching in these delays, it is possible to alter the phase of these signals. The delay path is made from a combination of mirrors fed from a polarizing beamsplitter (PBS). When the polarization is horizontal, then the light passes through the PBS undelayed. If the light is vertically polarized, then the light is reflected by the PBS along the path of the delay line. The switching of the polarization of the light is done with an FLC halfwave thickness cell. By cascading delay lines of different lengths, it is possible to select any phase step desired. The requirements of the FLC cells are that they should have a contrast of above 10 000:1 and a switching speed approaching 1 ms. Such cells can be made with current FLC technologies, making the stepped phase array antennae possible with optical signal addressing.

2.7 References

[1] F. Mok, J. Diep, H. K. Liu, D. Psaltis, *Opt. Lett.* **1986**, *11*, 748–750.
[2] R. C. Jones, *J. Opt. Soc. Am.* **1942**, *32*, 486.
[3] R. B. Meyer, L. Liebert, L. Strzelecki, P. Keller, *J. Phys. Lett.* **1975**, *36*, L69–L71.
[4] N. A. Clark, S. T. Lagerwall, *Appl. Phys. Lett.* **1980**, *36*, 899–901.
[5] N. A. Clark, T. P. Rieker, *Phys. Rev. A* **1988**, *37*, 1053–1056.

[6] Low viscosity difluoroterphenyl liquid crystal mixture developed at DRA Malvern.
[7] P. W. H. Surguy, et al., *Ferroelectrics* **1991**, *122*, 63–79.
[8] N. Collings, W. A. Crossland, P. J. Ayliffe, D. G. Vass, I. Underwood, *Appl. Opt.* **1989**, *28*, 4740–4747.
[9] D. J. McKnight, D. G. Vass, R. M. Sillito, *Appl. Opt.* **1989**, *28*, 4757–4762.
[10] D. C. Burns, M. L. Begbie, I. Underwood, D. G. Vass, *Ferroelectrics* **1996**, *181*, 93–97.
[11] W. A. Crossland, T. D. Wilkinson, T. M. Coker, T. C. B. Yu, M. Stanley, in: *OSA TOPS* **1997**, *14*, 102–106.
[12] S. E. Broomfield, M. A. A. Neil, E. G. S. Paige, G. G. Yang, *Elec. Lett.* **1992**, *28*, 26–28.
[13] L. W. K. Yim, A. B. Davey, A. R. L. Travis, *Ferroelectrics* **1986**, *181*, 1085–1098.
[14] F. Pérennès, W. A. Crossland, Z. Y. Wu, *Appl. Opt.* **1997**.
[15] G. Moddel, K. M. Johnson, W. Li, R. A. Rice, *Appl. Phys. Lett.* **1989**, *55*, 537–538.
[16] R. G. Wilson, *Fourier Series and Optical Transform Techniques in Contemporary Optics*, Wiley, New York, **1996**.
[17] J. W. Goodman, *Introduction to Fourier Optics*, McGraw-Hill, New York, **1968**.
[18] Y. N. Hsu, H. H. Arsenault, G. April, *Appl. Opt.* **1982**, *21*, 4012–4015.
[19] T. D. Wilkinson, D. C. O'Brien, R. J. Mears, *Appl. Opt.* **1993**, *33*, 4452–4453.
[20] A. Vander Lugt, *IEEE Trans. Inf. Theory* **1964**, *IT-10*, 139–145.
[21] J. L. Horner, P. D. Gianino, *Appl. Opt.* **1985**, *24*, 2889–2893.
[22] Y. Sheng, G. Paul-Hus, *Appl. Opt.* **1993**, *32*, 5782–5785.
[23] J. L. Horner, J. Leger, *Appl. Opt.* **1985**, *24*, 609–611.
[24] D. Psaltis, E. G. Paek, S. S. Venkatesh, *Opt. Eng.* **1984**, *23*, 698–704.
[25] A. P. Sparks, R. C. Chittick, W. A. Crossland, J. R. Brocklehurst, *IEE Tech. Dig.* **1988**, 121.
[26] M. A. Handschy, K. M. Johnson, G. Moddel, L. A. Pagano-Stauffer, *Ferroelectrics* **1988**, *85*, 279.
[27] R. M. Turner, D. A. Jared, G. D. Sharp, K. M. Johnson, *Appl. Opt.* **1993**, *32*, 3094.
[28] J. A. Davis, M. A. Waring, G. W. Bach, R. A. Lilly, D. M. Cottrell, *Appl. Opt.* **1989**, *28*, 10.
[29] T. D. Wilkinson, Y. Petillot, R. J. Mears, J. L. de Bougrenet, *Appl. Opt.* **1995**, *34*, 1885–1890.
[30] T. D. Wilkinson, R. J. Mears, *Ferroelectrics* **1996**, *181*, 47–52.
[31] C. S. Weaver, J. W. Goodman, *Appl. Opt.* **1966**, *5*, 1248–1249.
[32] B. Javidi, *Appl. Opt.* **1989**, *28*, 2358–2367.
[33] D. A. Jared, K. M. Johnson, G. Moddel, *Opt. Commun.* **1990**, *76*, 97–102.
[34] Q. Tang, B. Javidi, *Appl. Opt.* **1992**, *31*, 4016–4024.
[35] A. Kohler, B. Fracasso, P. Ambs, J. L. de Bougrenet de la Tocnaye, *Proc. SPIE* **1991**, *1564*, 236–243.
[36] C. Tocci, H. J. Caulfield, *Optical Interconnect Foundations and Applications*, Artech House, London, **1994**.
[37] F. B. McCormick, et al., *Appl. Opt.* **1994**, *33*, 1601–1618.
[38] M. P. Y. Demulliez, et al., *Appl. Opt.* **1995**, *34*, 5077–5090.
[39] R. J. Mears, W. A. Crossland, M. P. Dames, J. R. Collington, M. C. Parker, S. T. Warr, T. D. Wilkinson, A. B. Davey, IEEE *J. Sel. Top. Quant. Elect.*, **1996**, *2*, 35–46.
[40] H. J. White, et al., *OSA Proc. V16 Photonics Switching*, **1993**, *16*, 129–132.
[41] T. D. Wilkinson, D. C. O'Brien, R. J. Mears, *Opt. Commun.* **1994**, *109*, 222–226.
[42] P. Kloess, J. McComb, H. J. Coles, *Ferroelectrics* **1996**, *180*, 233–243.
[43] K. L. Tan, CUED Internal Report **1997**.
[44] M. P. Dames, R. J. Dowling, P. McKee, D. Wood, *Appl. Opt.* **1991**, *30*, 2685–2691.
[45] U. Mahlab, J. Shamir, H. J. Caulfield, *Opt. Lett.* **1991**, *16*, 648–650.
[46] S. T. Warr, R. J. Mears, *Elec. Lett.* **1995**, *31*, 714–716.
[47] S. T. Warr, Ph. D. Thesis, Cambridge University, **1996**.
[48] D. C. O'Brien, R. J. Mears, T. D. Wilkinson, W. A. Crossland, *Appl. Opt.* **1994**, *33*, 2795–2803.
[49] S. T. Warr, R. J. Mears, *Ferroelectrics* **1996**, *181*, 53–59.
[50] J. W. Goodman, A. R. Dias, L. M. Woody, *Opt. Lett.* **1978**, *2*, 1–3.
[51] A. R. Dias, R. F. Kalman, J. W. Goodman, A. A. Sawchuk, *Opt. Eng.* **1988**, *22*, 955–960.
[52] A. G. Kirk, W. A. Crossland, T. J. Hall, *Proc. 3rd Int. Conf. Holographic Systems* **1991**, *342*, 137–141.
[53] H. J. White, et al., *OSA 1995 Tech. Dig. V12 Photonics Switching* **1995**, 12.
[54] S. T. Warr, M. C. Parker, R. J. Mears, *Elec. Lett.* **1995**, *31*, 129–130.
[55] M. C. Parker, Ph. D. Thesis, Cambridge University **1996**.
[56] M. C. Parker, R. J. Mears, *IEEE Photon. Tech. Lett.* **1996**, *8*, 1007–1008.
[57] M. C. Parker, A. C. Cohen, R. J. Mears, *Photon. Techn. Lett.* **1997**, 9.
[58] G. D. Sharp, K. M. Johnson, D. Doroski, *Opt. Lett.* **1990**, *15*, 523–525.
[59] G. D. Sharp, D. Doroski, K. M. Johnson, *Opt. Lett.* **1991**, *16*, 875–877.
[60] A. Sneh, K. M. Johnson, *J. Lightwave Tech.* **1996**, *14*, 1067–1080.

[61] BDH764E liquid crystal available from BDH/Merck.
[62] W. K. Choi, A. B. Davey, W. A. Crossland, *Ferroelectrics* **1996**, *181*, 11–19.
[63] D. Psaltis, D. Brady, X.-G. Gu, S. Lin, *Nature* **1990**, *343*, 325–330.
[64] D. Psaltis, N. Farhat, *Opt. Lett.* **1985**, *10*, 98–100.
[65] N. H. Farhat, D. Psaltis, A. Prata, E. Paek, *Appl. Opt.* **1985**, *24*, 1469–1475.
[66] J. J. Hopfield, D. W. Tank, *Science* **1986**, *233*, 625–633.
[67] M. A. A. Neil, Ph. D. Thesis, Cambridge University **1989**.
[68] M. F. Bone, et al., *Displays* **1989**, 115–118.
[69] H. J. Caulfield, *Appl. Opt.* **1987**, *26*, 4039–4040.
[70] H. I. Jeon, J. Shamir, R. B. Johnson, H. J. Caulfield, *Neural Nets Perception* **1992**, *2*, 282–309.
[71] M. D. Binns, Ph. D. Thesis, Cambridge University **1996**.
[72] T. J. Drabik, M. A. Handschy, *Appl. Opt.* **1990**, *29*, 5220–5223.
[73] T. C. B. Yu, R. J. Mears, *Intelligent Eng. Through Artificial Neural Nets* **1992**, *2*, 45–50.
[74] I. Bar-Tana, J. P. Sharpe, D. J. McKnight, K. M. Johnson, *Opt. Lett.* **1995**, *20*, 303–305.
[75] T. C. B. Yu, Ph. D. Thesis, Cambridge University **1994**.
[76] K. Wagner, T. M. Slagle, *Appl. Opt.* **1993**, *32*, 1408–1435.
[77] A. R. L. Travis, S. R. Lang, J. R. Moore, N. A. Dodgson, *J. SID* **1995**, 203–205.
[78] M. G. Robinson, C. Tombling, *Ferroelectrics* **1996**, *181*, 87–92.
[79] E. S. Goldburt, P. S. J. Russell, *Appl. Phys. Lett.* **1985**, *46*, 338–340.
[80] H. Lin, P. Palffu-Muhoray, M. A. Lee, *Mol. Cryst. Liq. Cryst.* **1991**, *201*, 137.
[81] C. S. I. Wong, J.-Y. Liu, K. M. Johnson, *Ferroelectrics* **1995**, *181*, 61–67.
[82] Z. K. Ioannidis, I. P. Giles, C. Bowry, *Appl. Opt.* **1991**, *30*, 328–333.
[83] N. A. Riza, M. C. DeJule, *Inst. Phys. Conf. Ser. No. 139, Pt. II* **1995**, *139*, 231–233.
[84] S. S. Samus, P. W. Owan, W. J. Hossack, *Opt. Commun.* **1994**, *104*, 266–270.
[85] S. Broomfield, M. A. A. Neil, E. G. S. Paige, *Appl. Opt.* **1996**, 56–63.
[86] N. A. Riza, N. Madamopoulos, *Proc. IEEE LEOS* **1996**, 52–53.

3 Thermography Using Liquid Crystals

Helen F. Gleeson

3.1 Introduction

The dramatic variation of liquid crystalline properties with respect to temperature has resulted in the widespread use of cholesteric (chiral nematic) liquid crystals for thermography. The property that has been exploited most in liquid crystal thermography is the critical temperature dependence of the selective reflection from cholesteric liquid crystals, though other temperature dependent properties of mesophases have been utilized (e.g. the birefringence of nematic systems and selective reflection from other chiral phases). The helicoidal structure of cholesteric materials results in the selective reflection of visible light within a band of wavelengths of width $\Delta\lambda$ centered at a wavelength λ_0, such that:

$$\lambda_0 = \bar{n}p$$
$$\Delta\lambda = \Delta n p \qquad (1)$$

Here, p is the pitch of the helical structure, Δn is the birefringence of the medium and $\bar{n}$ is the average refractive index, $(n_o + n_e)/2$. The temperature range over which the cholesteric material selectively reflects visible light is known as the *color play* of the material. The temperature dependence of the selective reflection from a cholesteric material is a consequence of the temperature variation of the pitch of the system, defined by the twist elastic constants. A critical divergence of the selective reflection occurs if a cholesteric material cools when an underlying SmA phase is approached. This happens because the twist deformation is not allowed within a SmA phase, and the twist elastic constants (and therefore pitch) show pretransitional divergence. The pitch of the system increases rapidly as the temperature reduces towards the SmA phase transition temperature, typically resulting in the reflection of red light at lower temperatures. In cholesteric materials without a SmA phase, the critical divergence of the selective reflection is less pronounced. Indeed, in some materials the selective reflection is at a constant wavelength over the entire temperature range. Thus materials that exhibit a low lying SmA phase possess the most useful thermal properties for most thermography applications, though temperature insensitive materials have found application as shear sensitive systems (see Sec. 3.3). At higher temperatures within the cholesteric phase (above any pretransitional region) the pitch may either increase or decrease with increasing temperature, due to two competing mechanisms. Thermal expansion tends to cause an increase in the cholesteric pitch, though this phenomenon is usually overcome by the tendency for the average angular separation of molecules along the helical axis to increase with increasing temperature, thus decreasing the pitch of the system, and causing blue wavelengths to be selectively reflected at higher temperatures.

Equation (1) describes the selective reflection from a cholesteric material for normal incidence only. As the illuminating radiation becomes off-axis, the optics of the system become highly complicated. However, it has been shown that the angular dependence of the selectively reflected wavelengths varies *approximately* as

$$\lambda = \frac{p\bar{n}}{m} \cos\left[\frac{1}{2}\sin^{-1}\left(\frac{\sin\theta_i}{\bar{n}}\right) + \frac{1}{2}\sin^{-1}\left(\frac{\sin\theta_r}{\bar{n}}\right)\right] \quad (2)$$

where θ_i and θ_r are the angles of incidence and reflection respectively, and m is an integer, usually equal to unity. Equation (2) first derived by Fergason [1], is accurate for small values of θ_i and θ_r.

It is clear that the optical properties of a cholesteric material used for thermography depend on the birefringence of the material (low birefringence materials have only a narrow selective reflection band and thus reflect less of the visible spectrum). The birefringence of chiral nematic materials can be varied over quite a large range; Fig. 1 shows a series of materials with birefringence ranging from ≈0.02 to ≈0.18 [2]. Although systems with higher birefringence will appear brighter, the wavelength of reflection is less well defined, so the effective resolution of the material is lower. The brightness of a particular material will be to some extent temperature dependent since the birefringence is temperature dependent. The average refractive index, $\bar{n}$, also has an influence on the angular dependence of the selective reflection. It can be seen from Eq. (2) that the higher the refractive index, the more pronounced is the angular effect. Oron et al. [3] found that including certain nematic materials in mixtures of cholesteryl esters substantially reduced the angular dependence of the selective reflection.

The use of cholesteric liquid crystals for thermography was reviewed some years ago by Elser and Ennulat [4], and more recently in the commercial literature of Hallcrest [5]. Amongst the substances with the largest temperature coefficient are cholesteryl oleyl carbonate [6] and S-cholesteryl esters

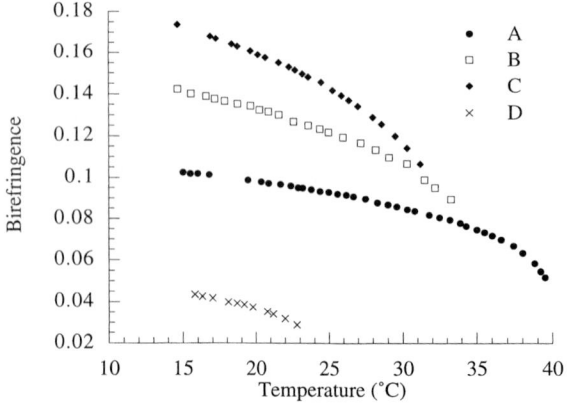

Figure 1. The temperature dependence of the birefringence of several commercial chiral nematic materials (redrawn from [2]). The symbols refer to mixtures of CB15 in different nematic hosts (all material names are as defined by Merck Ltd.): A is 37.7 wt% CB15 + 62.3 wt% ZLI1695 (a cyclohexylcyclohexane material based mixture); B is 39 wt% CB15 + 60.3 wt% ZLI1132 (a phenylcyclohexane material based mixture); C is 49.2 wt% CB15 in E63 (a biphenyl material based mixture); D is a mixture of two short chain Osman esters and a chiral Demus esters (see [2] for exact composition).

[7], which are capable of showing temperature variations of the order of a milliKelvin detectable to the human eye. Many hundreds of materials are now available for thermography applications, and mixture formulation for specific devices is common. The cholesteryl esters have been widely used in many devices, but are susceptible to impurities and degradation, and more stable chiral nematic materials were synthesized by Gray and McDonnell [8] following the discovery of the cyanobiphenyls (see Fig. 2). However, whilst the biphenyl based systems are more stable materials, their increased cost has hindered their general usage. Terminology frequently used in thermography refers to the phases of cholesteryl derivatives as *cholesteric* liquid crystals, whilst newer materials are known as *chiral nematics*.

The application of liquid crystals to thermography is widespread, and covers areas as diverse as infant bath and feed temperature indicators, medical screening and diagnostics, nondestructive testing, thermometers for fish tanks and refrigerators, battery testing, jewellery, advertising, aerodynamics research and art. Some specific applications areas are considered below, but it should be stressed that each section contains only representative examples, and by no means includes an exhaustive literature survey of the area.

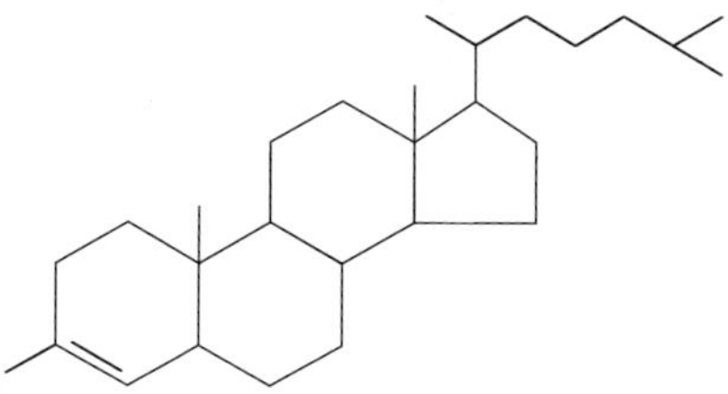

3.2 Device Structures

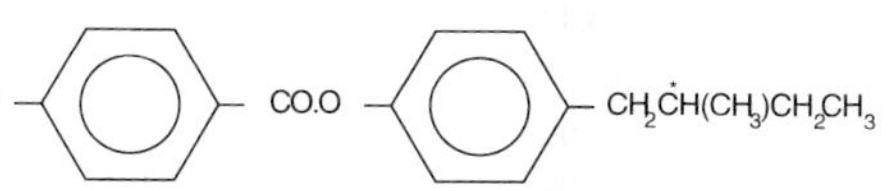

Liquid crystals are fluid systems, and whilst certain applications make use of their fluidity (e.g. shear flow measurements), by far the majority of applications require some form of encapsulation of the liquid crystal which serves to localize the material and to protect it from the environment. The iridescent colors of a cholesteric material are best observed in reflection and it is common to include a dark (preferably black) substrate in the device. In the simplest device, the material is enclosed between glass plates, as for a standard electro-optic device. Variations on this include spreading the material between plastic films, one of which may be black [9–15]. The production of plastic blisters filled with cholesteric material were particularly successful in the gimmick market. Textile fabrics coated with encapsulated cholesterics are known [16, 17], as are filled fibers [18, 19] and yarns wound with microencapsulated films [20]. The most common method of containing a cholesteric liquid crystal is microencapsulation followed by formulation into inks, sprays, pastes or slurries designed for a specific application. There are many procedures that may be followed to microencapsulate a liquid crystal [21–31]; these are summarized by Sliwka [32] and Vandegaer [33]. A typical encapsulation process includes the liquid crystal in gelatine–gum arabic capsules between 1 and 10 μm in diameter. The liquid crystal material is initially dispersed in an aqueous solution of gelatine and gum ar-

Figure 2. General structures of (a) cholesteric and (b) chiral nematic liquid crystals.

abic, and the size of the capsules is defined largely by the shear rate during the dispersion process. Air is removed from the dispersion during the stirring process, which must be sufficiently rapid to prevent the capsules from clumping. The capsule walls form through a mechanism known as complex coacervation which is shown schematically in Fig. 3. The walls are solidified by adjustment of the pH of the dispersion (often using acetic acid), and hardened using gluteraldehyde. Careful control of each stage of the microencapsulation process is necessary for the production of an even dispersion of capsules in the optimum size range (≈ 5 µm diameter).

Other techniques of encapsulating liquid crystalline materials exist, and find application in areas apart from thermography. In particular, the methods used in the fabrication of polymer dispersed liquid crystal N and Sm display devices can be readily applied to cholesteric materials. Polymer dispersed displays are fabricated by dissolving the liquid crystal in the monomeric form of the continuous medium of the final device.

On polymerizing the monomer, the liquid crystal comes out of solution, forming droplets dispersed in the polymeric matrix. The size of the droplets depends on polymerization rate, ratio of liquid crystal to monomer, and evaporation of solvent in solvated systems. This techique is considered in more detail in the chapter devoted to display devices in the discussion of polymer dispersed liquid crystal displays (PDLCs).

Microencapsulated cholesteric and chiral nematic liquid crystals can be incorporated into a variety of products suitable for a wide range of coating techniques. Materials are available [5] that have been optimized for coating using striping, doctor blade and knife pulling techniques, spraying, brushing, screen printing and gravure printing. The incorporation of cholesteric materials in microcapsules largely removes the angular dependence of the selective reflection. The capsules will invariably take up random orientations in the coating, though specific techniques have been suggested to retain directionality (application of shear or magnetic fields). Thus, the selectively reflected

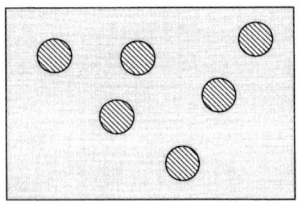

(a)
Dispersed droplets of liquid crystal in polymer solution.

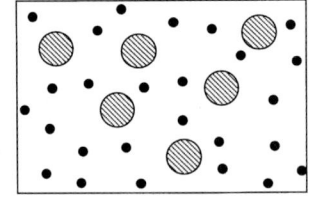

(b)
Separation of microdroplets of coacervate from solution

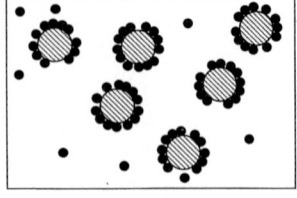

(c)
Coating of liquid crystal droplets by microdroplets of coacervate

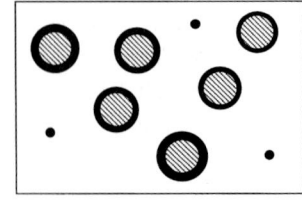

(d)
Coalescence of coacervate to form continuous droplet coating

Figure 3. A schematic representation of complex coacervation and microencapsulation.

light from a coating incorporating microcapsules is in general spectrally broad. It is not, however, generally brighter than a non-encapsulated film of comparable thickness; apart from less of the area actually containing liquid crystal, only a relatively small area of a droplet is in the optimum orientation to reflect light selectively. A top coating of a polymer of high refractive index can, however, reduce the range of incident angles and may increase the brightness of the device [34]. Another consequence of microencapsulating cholesteric materials can sometimes be observed if the cholesteric range of the system is not sufficiently broad. If the device is heated above the isotropic phase transition, it may become cloudy on cooling. This is due to the supercooling of the blue phases which almost always occur in the materials, and the subsequent formation of a nonplanar cholesteric structure. The phenomenon is difficult to avoid, and the only definite method of doing so is to arrange for the cholesteric phase range to extend beyond the temperature range over which the device will be used. Despite the drawbacks alluded to above, it is possible to use thermochromic liquid crystals to make highly accurate surface temperature measurements that are independent of the illumination spectrum, background lighting, angle of incidence and path [35]. The calibration range and uncertainty in temperature measurements that might be expected in typical careful use of a thermochromic system have also been detailed [36].

3.3 Engineering and Aerodynamic Research

The use of liquid crystals to observe shear stresses in wind tunnel models was first demonstrated by Klein [37] in 1968, who applied an unprotected liquid crystal solution to a test object in order to view the location of laminar and turbulent layers in aerodynamic models. The neat cholesteric material was applied to the surfaces and the boundaries between laminar and turbulent flow observed as differences in color. The regions of more turbulent flow conduct heat from the area more efficiently, resulting in a color difference between the regions. However, in this early work the materials used were not optimized for their shear properties, and were susceptible to contamination by their external environment, making quantitative measurements impossible. In later work, Klein and Margozzi [38, 39] employed the sensitivity of certain mixtures of cholesteric liquid crystals to stress to develop the principle of making quantitative measurements of stress visually. Whilst they formulated several mixtures with good sensitivity to shear, the visual information gained was difficult to interpret due to flow of the unencapsulated material. Since this early work, many mixtures have been developed commercially which have good shear response, whilst retaining low temperature, angle and pressure sensitivity.

Microencapsulation avoids many of the problems associated with the use of neat liquid crystal. McElderry [40] used microencapsulated material to study flat plates placed in a supersonic air stream, to produce a color display that had relatively low angular dependence. The encapsulation of material is particularly important if liquid crystal thermography is to be used in water tunnels. Here, it is useful to apply microencapsulated liquid crystal using an air brush, protect the capsules from contamination by solvents with an initial coating of acrylic polymer, and finally apply a waterproof polyurethane top coat, Fig. 4. Ogden and Hendricks [41] report the use of this technique for the study of turbulent water flow

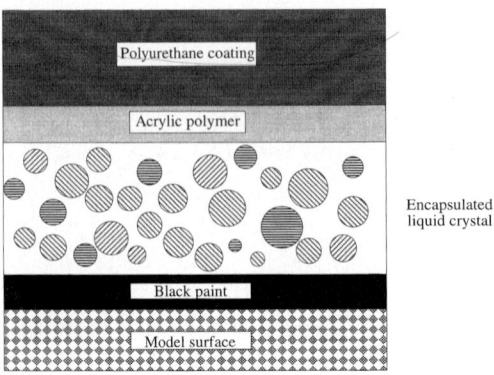

Figure 4. A schematic of the composition of liquid crystal films used in water tunnel studies.

over bodies. The areas of turbulent flow around the body result in more rapid cooling rates due to higher heat transfer rates. Similar materials were used in water tunnels by Maughan and Incropera [42], though shear sensitive unsealed mixtures have been used successfully to visualize hydrodynamic flow on surfaces with both steady and unsteady boundary layer separation and transition characteristics [43].

The use of thermochromic liquid crystals in qualitative studies of heat transfer is relatively common [44–48]; hot and cold regions are discriminated, but precise temperatures are not necessarily deduced from the results. Quantitative measurements were first made by Cooper et al. [49, 50], who evaluated the variation of the Nusselt number around a heated cylinder in cross-flow (the Nusselt number of a body, $Nu = hl/K\theta$, where h is heat loss per unit area, l is a typical dimension, θ the temperature difference between the body and surroundings and K is the thermal conductivity). Local heat transfer coefficients for plates in air jets have also been measured [51] and the temperature shift of the colour band due to the shearing action of the air flow determined [52]; this is particularly important in making quantitative measurements from exposed liquid crystals in wind tunnels. The use of thermochromic liquid crystals for flow visualization and heat flow is now relatively common. Local heat transfer coefficients of fins in wind tunnels are readily evaluated as described in [53]. Here, the air in the tunnel is held at a constant temperature, T_L, and the wind tunnel has two identical adjacent chambers, one of which is the test section. This test chamber is held at a temperature T_0. When steady flow conditions are established, the model, coated with a thin film of liquid crystal, is introduced to the test chamber. The subsequent transient heating of the test model changes the apparent colour of the liquid crystal, and a green line corresponding to a specific temperature T_G is readily observed and may be recorded as it moves across the fin. The local heat transfer coefficient α is computed by dividing the fin into elements of area ΔA. A transient energy balance relating the change in internal energy of an element to the net heat transfer from the air gives:

$$\rho c_p \delta \Delta A \frac{dT}{dt} = 2\alpha \Delta A (T_L - T) \qquad (3)$$

where ρ, δ, c_p and T are the density, thickness, heat capacity and temperature, respectively while T_L is the air temperature and t is time. Equation (3) assumes that the head conduction along the fin is negligible and the fin temperature is uniform across its thickness. Integration of this equation with the initial condition $T = T_0$ at $t = 0$ gives,

$$\alpha = \frac{\rho c_p \delta}{2 t_G} \ln\left[\frac{T_L - T_0}{T_L - T_G}\right] \qquad (4)$$

where t_G is the time necessary for the fin element in consideration to reach the temperature T_G.

There are many reports in the literature of similar work, in which either qualitative or quantitative evaluation of flow visualisation or heat transfer has been made using cholesteric liquid crystals (see for example references [54–77]. The vortex structures of triangular and rectangular longitudinal vortex generators were observed by Fiebig et al. [78], using liquid crystal thermography and the drag (a measure of flow losses), and local heat transfer coefficients were determined. The heat transfer enhancement per unit vortex generator area was found to be highest for delta wings followed by delta winglets and rectangular winglets. Further experiments have also employed liquid crystal thermography in studying the local heat transfer for various configurations of delta winglets in heat exchangers [79–82]. Steady state and transient studies of heat flow using liquid crystals, incorporating specially packaged heaters, have allowed the mapping and measurement of heat transfer coefficients as described by several authors (e.g. [83–86]). Such methods have relatively wide application, for example in the study of gas turbine components and recently the thermohydraulic behavior in a marine nuclear reactor [87]. Transient techniques using microencapsulated liquid crystals have been developed by Jones et al. [88–90] whereby heat transfer coefficients have been determined for specific geometries of gas turbine blades. In order to make quantitative measurements of transfer coefficients, Jones et al. evaluated the dynamic response of a rapidly heated microencapsulated liquid crystal film supported on a heated surface. The response times of the films were shown to be of the order of milliseconds. The influence of heat conduction on the determination of heat transfer coefficents as measured by liquid crystal thermography is considered by Valencia et al. [91] who show that neglect of tangential conduction can introduce significant errors, depending on the thermography experiment conducted.

Thermochromic liquid crystals have found extensive use in flow visualization measurements, both as coatings on models within water tunnels as discussed above, and as tracer particles in fluid flow. For example, capsules of thermochromic material have been doped at low concentration to visualize flow [92], and neat material dispersed into fluids to measure simultaneously the temperature and velocity distributions in the system [93]. Remote sensing of the temperatures of liquid drops in immiscible transparent liquids has been demonstrated by dispersing microcapsules of thermochromic liquid crystal in each drop [94]. Illuminating the system with a light sheet that cuts the drop at a meridian allows the spatial and time variation of temperature within drops in convective motion to be studied. Shear sensitive liquid crystal mixtures have been employed in the visualization of surface flow, and now play an important part in the study of laminar to turbulent flow transitions in wind tunnels and testing of aircraft wing and body surfaces. This follows the pioneering work of Klein et al. and NASA have developed several techniques for the study of such transitions using shear sensitive mixtures [95–98]. Nonencapsulated materials were used and applied to the test surface (often on aircraft wings) such that the shear stress induced a color change at the boundaries between high shear turbulent and low shear laminar flow. The work of Holmes et al. is considered to be pioneering in the use of shear sensitive liquid crystal mixtures for flow visualization. The major advantage over the transitionally employed oil flow or chemical sublimation techniques is the ability to use the materials over the entire subsonic range of aircraft flight. The use of liquid crystals for supersonic measurements has been considered by Reda [99], and Jones et al. [100] have suggested employing a shear induced texture change (from focal conic to planar) for flow visualization measurements.

3.4 Medical Thermography

Thermochromic liquid crystals have found extensive use in medical applications. In the simplest example, liquid crystals have been used for many years as forehead thermometers (also known as 'fever strips'). Usually, such devices are based on an indicator strip printed with several different thermochromic materials, each with a slightly different narrow color play around normal body temperature. The segment that appears bright green is that which displays the correct temperature. A different device structure is produced by Sharn Inc. [101], in which the liquid crystal is incorporated into the device in a single strip and treated such that the temperature is displayed as a narrow colour band that may move continuously across the device. The technology responsible for producing a continuous readout is not certain as the company have declined to patent it, but the resulting device offers a clear and unambiguous temperature read out.

Thermal mapping of various areas of the body has been used as a diagostic technique for a wide ranging group of medical conditions in which a temperature differential near the skin surface may be related to the disorder. Medical applications of liquid crystals are widespread; material seems to have been applied to almost every possible area of the body and the examples quoted here are considered to be representative rather than exhaustive. Device structures range from the application of neat material onto the skin [102], to including the liquid crystal in its microencapsulated form as a paint, or embedded in rigid or flexible substrates. The first example of the use of liquid crystals to indicate skin temperature was in 1964 [103], and many patents now exist in this area.

The use of liquid crystals in evaluating deep vein thrombosis is described by several authors [104–110]. Sandler and Martin, for example [111], consider liquid crystal thermography as a screening test for deep-vein thrombosis. The accuracy of liquid crystals compared with X-ray venography is considered for 80 patients, and out of 35 patients with confirmed deep vein thrombosis, 34 showed positive thermograms, giving a sensitivity measure of liquid crystal thermography of 97%: 17 false-positive thermograms gave a specificity of 62%, and the predictive value of a negative thermogram was 96.5%. Liquid crystal thermography is considered by Sandler and Martin to be a rapid and inexpensive screening test for patients suspected of having lower limb deep vein thrombosis. Other authors have, however, found the large numbers of false-positive results a problem in using liquid crystals as a screening technique. The location of veins [112–115] and the study of arterial disease [116, 117] using liquid crystal thermography is also common. The prediction of foot ulceration in diabetic patients through thermal mapping using liquid crystal technology is described by Benbow et al. [118]. Fifty patients were studied and liquid crystal thermography was found to be a cheap, simple and noninvasive method of identifying patients with increased risk of ulceration.

Liquid crystal thermography has been extensively evaluated in the area of oncology, especially breast cancer [119–128]. Subcutaneous and intracutaneous malignant tumours are typically 0.9–3.3 °C warmer than the surrounding tissue [129], making thermography an interesting candidate for cancer screening. Again, there is much discussion about its value as a screening technique, due to the relatively high number of reports of false-negative results [130], and the conclusions of most authors is that it is more useful as a screening tool than for diagnostics. Thermography has proved useful in the study of temporal variations in breast temperatures [131, 132]. The use of liquid crystal thermography has also been reported for the determination of the extent of basal cell carcinomas [133] and cancers in other areas of the body [134–138].

Other medical applications include the use of liquid crystal thermography in diagnosing acute paraproctitis [139]. A study of the use of liquid crystals in complex diagnostics in over one hundred patients with different forms of acute paraproctitis shows that the diagnostic value of the method increases with deepening and spreading of the inflammation focus and development of the scarry process in the pararectal fat. In 37.4% of the cases with deep forms of the disease liquid crystal thermography has a definite diagnostic value. Liquid crystal thermography has also proved of use in the localization of undescended testicles [140]. The position of the retained testicle is determined via a hot spot, though the intensity of the hot spot reflects only the amount of heat produced locally, and not the size of the undescended testicle. Thermography of the testicles has also been used in the diagnosis of infertility [141], and varicocele, the most common form of male infertility, proved to be easily detectable by this technique. One of the major advantages of liquid crystal thermography in this particular area is that it is one of the few available noninvasive diagnostic techniques: X-rays are unacceptable because of the radiosensitivity of the testicles. The diagnosis of scaphoid fractures using liquid crystal thermography has been suggested [142]. Flexible liquid crystal sheets have been used as a noninvasive investigative technique for the study of the nasal cycle [143]. Here, the air flow in healthy and diseased nasal cycles may be related to the area of colour change on a liquid crystal film which was found to be almost linear between flow rates of 1 and 3 $1 \text{min}^{-1} \text{cm}^{-2}$ (normal nasal air flow is $\approx 2.5\ 1\text{min}^{-1}\text{cm}^{-2}$). Indeed liquid crystal thermography has found use in areas as diverse as core body temperature measurements in anesthetic recovery [144], headache clinic setting [145], investigation of spinal root syndromes [146–150], lower back syndrome [151], chronic [152] and low [153] back pain, scrotal temperature in cases of spinal cord injury [154], knee joint stress [155], evaluation of the diabetic foot [156], the thyroid [157], lacrimal tract inflammation [158], as a diagnostic test in acute [159] and destructive [160] lactation mastitis, acute appendicitis [161], chronic liver disease [162], placental localization [163, 164], skin allergy [165–167] and carpal tunnel syndrome [168, 169]. Liquid crystal thermography in ophthalmologic practise is discussed by several authors [170–172].

It seems clear that while medical applications of liquid crystal thermography are extensive and imaginative, the technique is most useful for screening procedures, and there are only a few cases where it has proven diagnostic capability.

3.5 Thermal Mapping and Nondestructive Testing

The ease with which a film of cholesteric liquid crystal may be applied to large, uneven areas, makes it an ideal tool for thermal mapping and nondestructive testing. Indeed, the fact that specific mixtures can be formulated with cholesteric ranges extending over a wide range of temperatures (from well below zero to well above 100 °C), and with a great deal of flexibility in the color play range (from less than a degree to tens of degrees) allows for a great diversity in potential applications ranging from food processing [173] to the electronics and space applications described below.

In studying test objects, the surface temperature of the object may be varied by direct heating (e.g. electrically), or indirectly (e.g. by radiation). A flaw close to the surface is detectable through the temperature difference that might be caused by a decreased temperature conductivity in the structure. Tests may be made that are both static and dynamic; static approaches are somewhat limited due to lateral heat conduction which eventually obscures a discontinuity in temperature due to a small lesion. However, small defects are enhanced by transient measurements, because heat with short wavelengths propagates faster than heat with longer wavelengths, thus allowing fine detail to appear before the development of a steady state pattern. For example, liquid crystal thermography has been used to observe time dependent temperature patterns on an electrically heated surface during boiling [174]. Here, real time images were produced which revealed both spatial and temporal differences in the process by which boiling was initiated and developed in two different convection fields. Time dependent aspects of nucleate boiling have also been studied by Bergez [175] which showed that existing heat transfer models are inadequate for water under conditions of saturation, atmospheric pressure and low heat flux.

Most applications rely on the steady state method as it is cheaper and less complex than transient measurements. A bibliography of nondestructive testing using thermochromic liquid crystals was published in 1972 [176], and more recently by Hallcrest [5], so only a few examples are described here. Indeed, in most cases the approach is the same, and only the test object differs.

In composite materials, flaws, bonding faults and internal defects have all been examined using liquid crystal thermography, [177–184]. Surface and subsurface flaws have been detected, including regions of unstable plastic flow in aluminum alloys (Lueder lines) [185], faults in welded metals [186], and cracks, voids and leaks in pressure vessels [187–188]. Potential fracture sites have been determined in metals [189], together with defects in springs [190]. Other applications examine shrinkage cavities in metal castings [191], thermal isolation of aluminum rivets [192], and thermal gradients on the surface of heating coils [193].

Much attention has been paid to the use of thermochromic liquid crystals in the quality control of electronic components, and thermochromic 'paints' to examine overheating components on printed circuit boards are readily available. There are many examples of the use of cholesteric liquid crystals for thermal mapping of both ordinary [194–196] and multilayer [197] printed circuits boards. Shorts have been detected in field effect transistors [198] and toroidal transformers, and temperatures of other electrical components visualized [199–201]. Other applications include the study of deposits on thin film resistors [202], inhomogeneities in the sapphire windows of infrared detectors [203], switching phenomena in deposited metal films and junctions [204–206], and the effects of annealing and laser radiation on amorphous films of As, Te and Ge [207]. Liquid crystal thermography has found use in the verification of three dimensional thermal simulations of semiconductor chips, and in particular to AlGaAs/GaAs heterojunctions [208]. Breaks in electrical conductors embedded in windscreens [209], heat leaks in refrigerator doors [210], and heat patterns generated by piezoelectric and ultra high frequency devices [211, 212] have all also been studied via liquid crystal thermography. Finally, there are several examples of the use of cholesteric liquid crystals in the space programme aboard Apollo 14 [213, 214] and Apollo 17 [215], as temperature indicators in low gravity, and to estimate heat flow through textiles [216]. The areas in which liquid crystal thermography is of use in nondestructive testing continue to grow, particularly in the area of engineering research where the new chiral nematic materials offer improved performance over the cholesteryl esters used in early applications.

3.6 Radiation Detection

Films containing cholesteric liquid crystals are inexpensive and versatile tools for visualizing invisible radiation. Almost any absorbing medium that will convert incident radiation to heat may be backed by a liquid crystal, enabling almost any type of radiation to be visualized. In many cases (e.g., infrared radiation), the liquid crystal can itself act as the absorber, leading to a particularly simple device. Viscoelastic materials may be used to absorb ultrasound, carbon black, gold black and other materials for electromagnetic radiation and space matched metal films for microwave radiation. Technical details of the construction and operation of the detectors are readily available in the patent literature [217–221]. Membrane devices, including a thin layer of carbon black, are often used for infrared radiation detection and visualization [222, 223], and such detectors are particularly useful in the observation of the modes and emission parameters of infrared lasers [224–229]. A thermal imaging device that offers the possibility of night vision has also been described [230]. Other device constructions offer the possibility of thermally addressing and electrically erasing information on thermochromic liquid crystals using infrared laser beams, thus producing information storage devices, or even producing real time holograms using cholesteric materials [231]. Microwave detectors generally consist of thermochromic liquid crystal coupled onto a thin metal film [232]; the incident radiation induces currents in the metal film which in turn heat the thermochromic material. There are several examples of microwave detectors, and as with infrared radiation, thermochromic films have also been used to visualize microwave holograms [233, 234] and to allow internal examination of microwave transparent objects by interferometry [235, 236]. For the detection or visualization of ultrasound, thermochromic material is coated onto a viscoelastic substrate that becomes warmer in the areas of absorption. Acoustic images, and again, holograms, have been visualized [237–244]. Cholesteric liquid crystals are well known to have sensitivity to ultraviolet radiation, and the degradation of the materials that occurs on exposure changes the region of the color play band [245, 246] and this may be used as an indicator. The change in the materials' color play can be enhanced by the addition of radiation sensitive compounds (e.g., iodine containing materials). Information regarding the radiation stability of thermochromic materials is readily available in the literature. Indeed it has been shown that the shift in the selective reflection band is linear with respect to γ-ray dosage, and that the temperature response remained stable after exposure [247–250]. Finally, there have been suggestions that thermochromic liquid crystals may be used for the detection of elementary particles [251], heavy moving ions [252] and for monitoring the frequency of UHF electromagnetic fields [253].

3.7 References

[1] J. L. Fergason, *Liquid Crystals*, Proc. 2nd Kent Conf., Gordon and Breach, New York, **1966**.
[2] H. F. Gleeson, H. J. Coles, *Mol. Cryst. Liq. Cryst.* **1989**, *170*, 9.
[3] N. Oron, J. L. Yu, M. M. Labes, *Appl. Phys. Lett.* **1973**, *23*, 217.
[4] W. Elser, R. D. Ennulat, Selective Reflection of Cholesteric Liquid Crystals, in *Advances in Liquid Crystals, Vol. 2* (Ed. G. H. Brown) Academic Press, New York, **1976**, 73.
[5] M. Parsley, *The Hallcrest Handbook of Thermochromic Liquid Crystal Technology*. Hallcrest, Product and Application Information, 1820 Pickwick Lane, Glenview, IL 60025 USA, **1991**.
[6] R. D. Ennulat, *Mol. Cryst. Liq. Cryst.* **1971**, *13*, 337.
[7] R. D. Ennulat, L. E. Garn, J. D. White, *Mol. Cryst. Liq. Cryst.* **1974**, *26*, 245.
[8] G. W. Gray, D. G. McDonnell, *Electron. Lett.* **1975**, *11*, 556; *Mol. Cryst. Liq. Cryst.* **1976**, *34*, 189.
[9] D. E. Bishop, US Clearinghouse, Fed. Sci. Tech. Inform., PB Rep. PB-183838, **1968**.
[10] E. A. Kolenko, L. G. Lopatina, V. A. Borodulya, *Czech. Conf. Electron. Vac. Phys. Proc.*, 5th, IIa-17, **1972**.
[11] F. M. Mina, Fr Demande 2 038 104 (1 April) **1969**.
[12] N. A. Pichikyan, A. S. Sonin, N. B. Titova, Electron. **1976**, *3*, 1614.
[13] W. Woodmansee, US Patent 3 511 086, **1966**.
[14] R. D. Ennulat, J. L. Fergason, Mol. Cryst. Liq. Cryst. **1971**, *13*, 149.
[15] J. E. Adams, L. B. Leder, W. E. L. Haas, US Patent 3 679 290, **1972**.
[16] Y. Nisijima, K. Shimizu, *Japan. Kokai, 74*, **1972**, 784.
[17] M. Ono, T. Ito, T. Sawa, T. Kato, Japanese Patent 74 01 676, **1970**.
[18] N. Oguchi, I. Ikami, A. Hhe, Japanese Patent 74 35 114, **1970**.
[19] H. Arimoto, T. Kakishita, M. Takada, Japan. Kokai 73 44, 522, **1971**.
[20] Y. Matsuda, S. Aramoto, K. Iwata, Japan Kokai 74 66, 976, **1972**.
[21] H. Arimoto, T. Kakishita, M. Takada, Japan. Kokai 73 44, 177, **1971**.
[22] J. V. Cartmell, D. Churchill, US Patent 3 720 623, **1971**.
[23] D. Churchill, J. V. Cartmell, R. E. Miller, British Patent 1 138 590, **1966**.
[24] D. Churchill, J. V. Cartmell, R. E. Miller, US Patent 3 697 297, **1970**.
[25] D. Churchill, J. V. Cartmell, R. E. Miller, US Patent 3 732 119, **1968**.
[26] F. Hori, B. Kato, N. Arima, Japan. Kokai 74, 108, 187, **1973**.
[27] M. Kano, T. Ninomiya, Y. Nishimura, Japan. Kokai 73 71, 377, **1971**.
[28] S. Kubo, H. Arai, F. Hori, T. Kato, Japanese Patent 75 11, 344, **1970**.
[29] F. Kuhn-Weiss, *Materialprüfung* **1974**, *16*, 140.
[30] Z. Orido, Y. Uchida, T. Sawa, N. Tokuyama, T. Kato, Japan. 74 18, 914, **1970**.
[31] M. Ueda, F. Hori, B. Kato, N. Arima, Japan. Kokai 73 83, 903, **1972**.
[32] W. Sliwka, *Agnew. Chem.* **1975**, *87*, 556.
[33] J. E. Vandegaer, *Microencapsulation – Processes and Applications*, Plenum, New York, **1974**.
[34] L. D. Hodson, J. V. Cartmell, D. Churchill, J. W. Jones, US Patent 3, 585, 381 **1971**.
[35] D. J. Farina, J. M. Hacker, R. J. Moffat, J. K. Eaton, *Exp. Therm. Fluid Sci.* **1994**, *9*, 1.
[36] J. L. Hay, D. K. Hollingsworth, Experimental Thermal and Fluid Science, **1996**, *12*, 1.
[37] E. J. Klein, *Astronaut. Aeronaut.* **1968**, *6*, 70.
[38] E. J. Klein, A. P. Margozzi, NASA report TM-X-1774, **1969**.
[39] E. J. Klein, A. P. Margozzi, *Rev. Sci. Instrum.* **1970**, *41*, 238.
[40] E. D. Elderry, Air Force Flight Dynamics Lab., FDMG TM70-3 **1970**.
[41] T. R. Ogden, E. W. Hendricks, *Exp. Fluids* **1984**, *2*, 65.
[42] J. R. Maughan, F. P. Incropera, *Esp. Fluids*, **1987**, *5*, 334.
[43] P. K. Besch, T. B. Jones, J. P. Sikora, DTNSRDC-86/046, **1986**.
[44] T. E. Cooper, J. P. Groff, *J. Heat Transfer.* **1973**, *95*, 250.
[45] T. E. Cooper, W. K. Petrovic, *J. Heat. Transfer* **1974**, *96*, 415.
[46] G. M. Zharkova, A. P. Kapustin, Izv. Sib. Otd. Akad. Nauk SSSR, Ser. Tekh Nauk **1970**, *13*, 65 (Chem. Abst. **1971**, *74*, 80 809).
[47] A. Szymanski, *Post. Astronaut.* **1971**, *5*, 27.
[48] D. Vennemann, K. A. Buetefisch, *Deut. Luft-Raumfahrt, Rep.* DLR-FB 73-121 **1973**.
[49] T. E. Cooper, R. J. Field, J. F. Meyer, US Nat. Tech. Inform. Serv., Rep. Ad A-002 458, **1974**.
[50] T. E. Cooper, R. J. Field, J. F. Meyer, *J. Heat Trans.* **1975**, *97*, 442.
[51] C. den Ouden, *Delft Progr. Rep., Ser. A* **1973**/33.
[52] G. M. Zharkova, A. V. Lototko, Coll. Rep. 2nd All Union Sci. Conf. Liquid Cryst. Acad. Sci. USSR, **1973**, 271.
[53] M. Fiebig, P. Kallweit, N. Mitra, S. Tiggelbeck, *Exp. Therm. Fluid Sci.* **1991**, *4*, 103.
[54] M. M. Ardasheva, M. V. Ryzhkova, *Fluid Mech. Soc. Res.* **1977**, *6*, 128.
[55] C. J. Hoogendoorn, *Int. J. Heat. Mass Transfer.* **1977**, *20*, 1333.
[56] J. C. Simonich, R. J. Moffat, *Rev. Sci. Instrum.* **1982**, *53*, 678.
[57] C. den Ouden, C. J. Hoogendoorn, *Proc. Int. Heat Transfer Conf. 5th*, **1974**, 293.

[58] A. Brown, C. L. Saluja, *J. Phys. E.* **1978**, *11*, 1068.
[59] R. J. Goldstein, J. F. Timmers, *Int. J. Heat Mass Transfer* **1982**, *25*, 1857.
[60] V. M. Zharkova, *Fluid Mech. Sov. Res.* **1980**, *9*, 51.
[61] V. M. Zharkova, V. M. Khachaturyan, L. A. Vostokov, N. M. Alekseev, Study of Liquid Crystal Thermoindicators, in *Advances in Liquid Crystal Research and Applications* (Ed. L. Bata), Pergamon Press, Oxford, Akademiai, Kaido, Budapest, **1980**, p. 1221.
[62] R. J. Goldstein, M. E. Franchett, *J. Heat Transfer* **1988**, *110*, 1988.
[63] K. C. Cheng, T. Obata, R. R. Gilpin, *J. Heat Transfer.* **1988**, *110*, 596.
[64] J. Rojas, J. H. Whitelaw, M. Yianneskis, *J. Heat Transfer* **1987**, *109*, 866.
[65] N. Akino, T. Kunugi, K. Ichimiya, K. Mitsushiro, M. Ueda, *J. Heat Transfer* **1989**, *111*, 558.
[66] K. Ichimiya, N. Akino, T. Kunugi, K. Mitsushiro, *Int. J. Heat Mass Transfer* **1988**, *31*, 2215.
[67] G. J. van Fossen, R. J. Simoneau, *J. Heat Transfer* **1987**, *109*, 10.
[68] R. J. Moffat, *Exp. Therm. Fluid Sci* **1990**, *3*, 14.
[69] X. Zhang, J. Stasiek, M. W. Collins, *Exp. Therm. Fluid Sci* **1995**, *10(2)*, 229.
[70] S. Ashforthfrost, K. Jambunathan, *Int. Comm. in Heat and Mass Transfer* **1996**, *23(2)*, 155.
[71] S. J. Lee, J. H. Lee, D. H. Lee, *Int. Journal of Heat and Mass Transfer* **1994**, *37(6)*, 967.
[72] G. E. McCreery, *Heat Transfer Engineering* **1994**, *15(4)*, 33.
[73] Z. Wang, P. T. Ireland, T. V. Jones, *Journal of Turbomachinery: Trans ASME* **1995**, *117(2)*, 290.
[74] R. F. Matinezbotas, G. D. Lock, T. V. Jones, *Journal of Turbomachinery: Trans ASME* **1995**, *117(3)*, 425.
[75] J. Stasiek, M. W. Collins, M. Ciofalo, P. E. Chew, *Int. Journal of Heat and Mass Transfer* **1996**, *39(1)*, 149.
[76] N. Abauf, D. M. Kercher, *Journal of Turbomachinery: Trans ASME* **1994**, *116(1)*, 169.
[77] Y. W. Kim, J. P. Downs, F. O. Soechting, W. Abdelmesseh, G. D. Steuber, S. Tanrikut, Journal of Turbomachinery – Transactions of the ASME, **1995**, *117*, 1.
[78] M. Fiebig, P. Kallweit, N. Mitra, S. Tiggelbeck, *Exp. Therm. Fluid Sci.* **1991**, *4*, 103.
[79] S. Tiggelbeck, N. K. Mitra, M. Fiebig, *Int. J Heat Mass Trans* **1993**, *36(9)*, 2327.
[80] M. Fiebig, A. Valencia, N. K. Mitra, *Exp. Therm. Fluid Sci.* **1993**, *7(4)*, 287.
[81] M. Fiebig, A. Valencia, N. K. Mitra, *Exp. Therm. Fluid Sci.* **1994**, *8(1)*, 35.
[82] S. Tiggelbeck, N. K. Mitra, M. Fiebig, *J. Heat Transfer: Trans. ASME* **1994**, *116(4)*, 880.
[83] S. A. Hippensteele, L. M. Russell, F. S. Stepka, *J. Heat Transfer* **1983**, *105*, 184.
[84] S. A. Hippensteele, L. M. Russell, F. J. Torres, *J. Eng. For Gas Turbines and Power,* **1985**, *107*, 953.
[85] J. W. Baughn, M. A. Hoffman, D. B. Makel, *Rev. Sci. Instum.* **1986**, *57*, 650.
[86] J. Vonwofersdorf, R. Hoecker, T. Sattelmayer, *J. Heat Transfer: Trans. ASME* **1993**, *115(2)*, 319.
[87] A. Kurosawa, N. Akino, T. Otsuji, S. Kizu, K. Kobayashi, K. Iwahori, T. Takeda, Y. Ito, *J. Nucl. Sci. Technol.* **1993**, *30(2)*, 131.
[88] P. T. Ireland, T. V. Jones, *Proc. 8th Int. Heat Trans. Conf.,* San Francisco, **1986**, 975.
[89] T. V. Jones, P. T. Ireland, *J. Phys. E.* **1987**, *20*, 1195.
[90] K. W. Vantreuren, Z. Wang, P. T. Ireland, T. V. Jones, *J. Turbomachinery: Trans. ASME* **1994**, *116(3)*, 369.
[91] A. Valencia, M. Fiebig, N. K. Mitra, *Exp. Heat Trans.* **1995**, *8(4)*, 271.
[92] H. S. Rhee, J. R. Koseff, R. L. Street, *Exp. Fluids* **1984**, *2*, 57.
[93] W. J. Hiller, T. A. Kowalewski, *4th Int. Symposium on Flow Visualisation, Paris* **1986** (Ed. C. Veret), Hemisphere, Washington, **1987**, p. 617.
[94] T. Nozaki, T. Mochizuki, N. Kaji, Y. M. Mori, *Exp. Fluids* **1995**, *18(3)*, 137.
[95] B. J. Holmes, P. D. Gall, C. D. Croom, G. S. Manuel, W. C. Kelliher, *NASA Tech. Memo* 87666, **1986**.
[96] B. J. Holmes, P. D. Gall, *AIAA report 86-2592*, **1986**.
[97] B. J. Holmes, C. C. Croom, P. D. Gall, G. S. Manuel, D. L. Carraway, *AIAA report 86-9786*, **1986**.
[98] B. J. Holmes, C. J. Obara, *SAE Tech. Paper 871017*, **1987**.
[99] D. C. Reda, D. P. Aeschliman, *AIAA Paper 90-1513*, **1990**.
[100] P. Bonnet, T. V. Jones, D. G. McDonell, *Liq. Cryst.* **1989**, *6*, 271, and British Patent Application 2 218 215, **1989**.
[101] Sharn Inc. 4801 George Road, Tampa, Florida FL33634 USA, **1993**. See Designation E 1061-85, Annual book of ASTM standards, **1985**, *14.01*, 866.
[102] N. N. Goldberg, J. L. Fergason, US Patent, 3 533 399, **1970**.
[103] J. T. Crissey, E. Gordy, J. L. Fergason, R. B. Lyman, *J. Invest. Dermatol.* **1964**, *43*, 89.
[104] E. A. Thomas, M. J. D. Cobby, E. Rhys Davies, W. D. Jeans, J. T. Whicher, *BMS* **1989**, *299*, 951.
[105] R. Hoffman, *J. Vasc. Dis.* **1988**, *17*, 326.
[106] R. Pochaczevsky, G. Pillari, F. Feldman, *Am. J. Roentgenol.* **1982**, *138*, 717.
[107] E. D. Cooke, *Lancet* **1985**, *1*, 1109.
[108] D. A. Sandler, *J. Photo. Sci.* **1985**, *33*, 70.
[109] D. A. Sandler, J. F. Martin, *J. Photo. Sci.* **1985**, *33*, 231.

[110] C. Jensen, L. L. Knudsen, V. Hegedus, *Eur. J. Radiol.* **1983**, *3*, 99.
[111] D. A. Sandler, J. F. Martin, *Lancet,* **1985**, *1*, 665.
[112] F. Suzuki, T. W. Davison, US Patent 4015591, **1976**.
[113] F. Suzuki, T. W. Davison, US Patent 4175543, **1979**.
[114] E. R. Nosari, US Patent 3998210, **1976**.
[115] E. Kalodiki, L. Calahoras, G. Geroulakos, A. N. Nicolaides, *Phlebology* **1995**, *10*, 110.
[116] G. Belcaro, A. N. Domatos, R. Cotellese, E. Dinardo, A. Dellosa, A. Daulerio, L. Rosenkvist, A. Zukowski, F. Radocchia, M. W. Williams, F. Possati, *Acta Chirurgia Belgica* **1983**, *83*, 430.
[117] G. Belcaro, N. Dematos, M. Bull, M. Williams, A. N. Nicolaides, *Br. J. Surg.* **1983**, *70*, 692.
[118] S. J. Benbow, A. W. Chan, D. R. Bowsher, G. Williams, I. A. MacFarlane, *Diabetes Care* **1994**, *17*, 835.
[119] U. Moller, J. Bosjen, *Cancer Res.* **1975**, *35*, 3116.
[120] J. Tricoire, *J. Gynecol. Obstet. Biol. Reprod.* **1975**, *4*, 123.
[121] T. W. Davison, K. L. Ewing, J. L. Fergason, M. Chapman, A. Can, C. C. Voorhis, *Cancer* **1972**, *29*, 1123.
[122] M. Gautherie, *J. Phys.* **1969**, *30*, 122.
[123] M. Gautherie, Y. Quenneville, C. Rempp, C. Gros, *J. Radiol.* **1975**, *56*, 316.
[124] H. Kucera, E. Kubista, *Wein. Klin. Wochenschr.* **1976**, *88*, 737.
[125] W. W. Logan, *Inv. Radiol.* **1974**, *9*, 329.
[126] W. B. Hobbins, *Breast Disease Diagnosis and Treatment* (Ed. G. F. Schwartz and D. Marchant), Elsevier, North Holland, New York, **1981**, p. 87.
[127] M. Moskowitz, S. H. Fox, R. B. Delre, J. R. Milbrath, L. W. Bassett, R. H. Gold, K. A. Shaffer, *Radiology* **1981**, *140*, 659.
[128] E. E. Sterns, B. Zee, S. Sengupta, F. W. Saunders, *Cancer* **1996**, *77*, 1324.
[129] O. S. Selawry, H. S. Selawry, J. F. Holland, *Mol. Cryst.* **1966**, *1*, 495.
[130] G. Bothmann, U. van der Bussche, K. F. Kubli, G. Seybold, *Dtsch. Med. Wochenschr.* **1974**, *99*, 730.
[131] M. Borten, L. A. Dileo, E. A. Friedman, *Surg. Gynecol. Obstet.* **1983**, *156*, 785.
[132] M. Borten, B. J. Ransil, L. A. Dileo, E. A. Friedman, *J. Reprod. Med.* **1984**, *29*, 665.
[133] J. L. Ratz, P. L. Bailin, *J. Dermatol. Surg. Oncol.* **1981**, *7*, 27.
[134] C. Potanin, *Chest.* **1970**, *58*, 491.
[135] J. Tricoire, *Presse Med.* **1970**, *78*, 2481.
[136] O. S. Selawry, H. W. Neubauer, H. S. Selawry, F. S. Hoffmeister, *Am. J. Surg.* **1966**, *122*, 537.
[137] O. S. Selawry, J. F. Holland, *Proc. Am. Assoc. Cancer Res.* **1966**, *7*, 63.
[138] Several reports in *Coll. Rep., 2nd All Union Sci. Conf. Liquid Cryst., Adad. Sci. USSR,* **1972**, ed. G. N. Glazkov, A. M. Zhmud, G. G. Molokov, Yu K. Nepochatov; Yu M. Gerusov (page 304); Yu M. Gerusov, Yu P. Polosin, N. N. Gerusova (page 312); K. D. Panikrotov, Yu M. Gerusov, G. G. Maidachenko, V. A. Vorontsov (page 309).
[139] V. E. Smirnov, P. M. Lavreshin, *Vestnik Khirugii Imeni II Grekova* **1990**, *145*, 37.
[140] P. Goblyos, E. Suzule, *Eur. J. Radiol.* **1987**, *7*, 266.
[141] P. Goblyos, G. Vydra, I. Szabolcs, G. Irsy, M. Goth, G. Szilagyi, *Eur. J. Radiol.* **1982**, *2*, 246.
[142] K. B. Hosie, J. Wardrope, A. C. Crosby, D. G. Ferguson, *Arch. Emerg. Med.* **1987**, *4*, 117.
[143] R. J. Canter, *Clin. Otolaryngol.* **1986**, *11*, 329.
[144] M. R. Hecker, R. M. Brownfield, B. J. Rubal, *S. Med. J.* **1996**, *89*, 71.
[145] H. E. Burg, *S. Med. J.* **1986**, *79*, 31.
[146] G. H. Mills, G. K. Davies, C. J. M. Getty, J. Conway, *Spine* **1986**, *11*, 427.
[147] R. Pochaczevsky, *Orthop. Clin. N. Am.* **1983**, *14*, 271.
[148] R. Pochaczevsky, P. H. Meyers, C. H. Wexler, J. A. Epstein, *Arch. Phys. Med. Rehab.* **1982**, *63*, 517.
[149] P. H. Meyers, R. Pochaczevsky, C. Wexler, J. Epstein, J. A. Marc, *Headache,* **1982**, *22*, 155.
[150] R. Pochaczevsky, C. C. Wexler, P. H. Meyers, J. A. Epstein, J. A. Marc, *Neurology* **1982**, *32*, 75.
[151] P. S. Thomas, L. Chen, B. E. Frederickson, H. A. Yuan, I. C. Stehling, H. L. Zauder, *Adv. Pain Res. Ther.* **1985**, *9*, 403.
[152] R. I. Newman, J. L. Seres, E. B. Miller, *Pain,* **1984**, *20*, 293.
[153] B. J. Rubal, R. B. Traycoff, K. L. Ewing, *Phys. Ther.* **1982**, *62*, 1593.
[154] Y. H. Wang, T. S. Huang, M. C. Lin, C. S. Yeh, I. N. Lien, *Am. J. Phys. Med. Rehab.* **1993**, *72*, 6.
[155] D. J. Habes, A. Bhattacharya, M. Milliron, *Appl. Ergonom.* **1994**, *25*, 111.
[156] R. M. Stess, P. C. Sisney, K. M. Moss, P. M. Graf, K. S. Louie, G. A. W. Gooding, C. Grunfeld, *Diabetes Care* **1986**, *9*, 267.
[157] P. Goblyos, I. Szabolcs, I. Szilvasi, J. Gonczi, G. Irsy, M. Goth, G. Szilagyi, *Eur. J. Radiol.* **1985**, *5*, 291.
[158] P. Hinton, J. J. Hurwitz, P. L. Chart, *Can. J. Opthalmol* **1984**, *19*, 176.
[159] V. M. Udod, S. A. Seleznev, V. T. Storozhuk, S. I. Markelov, E. G. Karsten, *Vestink, Khirugii Imeni I I Gregkova* **1983**, *131*, 45.
[160] V. T. Storozhuk, S. I. Markelov, *Vestink, Khirugii Imeni I I Grekova* **1985**, *134*, 71.
[161] A. A. Lobenko, E. S. Chuzhina, A. K. Asmolov, N. V. Borschevskaja, *Vestnik Khirugii Imenii I I Grekova.* **1983**, *131*, 66.

[162] J. D. Steele, J. F. Dillon, J. N. Plevris, J. L. Hauer, I. A. D. Bouchier, P. C. Hayes, *J. Hepatol.* **1994**, *21*, 927.
[163] R. C. Margolis, L. S. Shaffer, *J. Am. Osteop. Assoc.* **1974**, *73*, 103.
[164] E. N. Peterson, G. D. Dixon, M. A. Levine, *Obstet. Gynecol.* **1971**, *37*, 468.
[165] R. M. Katz, A. S. Rohr, S. C. Siegel, G. S. Rachelefsky, S. L. Spector, *Ann. Aller.* **1985**, *54*, 359.
[166] J. K. Korotzer, *Ann. Aller.* **1972**, *30*, 473.
[167] J. Zuber, S. J. Klosowicz, J. Zmija, A. Makowski, J. Kruszewski, *Mol. Cryst. Liq. Crystl. Sci. Technol. Section A* **1994**, *249*, 171.
[168] R. Ysla, V. Ysla, H. Rosomoff, R. Rosomoff, *Arch. Phys. Med. Rehab.* **1986**, *67*, 662.
[169] S. Meyers, D. Cros, B. Sherry, P. Vermeire, *Neurology* **1989**, *39*, 1465.
[170] A. Y. Sverdlik, V. D. Belilovsky, *Vestnik Oftalmologii* **1985**, *4*, 54.
[171] J. B. Kinn, R. A. Tell, *IEEE Trans. Biomed. Eng.* **1973**, *20*, 387.
[172] P. Hinton, J. J. Hurwitz, P. L. Chart, *Can. J. Ophthalmol.* **1984**, *19*, 176.
[173] V. M. Bahasubramaniam, S. K. Sastry, *J. Food. Eng.* **1995**, *26*, 219.
[174] A. A. Watwe, D. K. Hollingsworth, *Exp. Ther. Fluid. Sci.* **1994**, *9(1)*, 22.
[175] W. Bergez, *J. Heat Mass Trans* **1995**, *38(10)*, 1799.
[176] M. A. Wall, *U. K. Atomic Energy Research Establishment, Bibliography* AERE-Bib 181, **1972**.
[177] M. Magne, P. Pinard, P. Thome, N. Chretien, *Colloq. Met.* **1968**, *12*, 241.
[178] M. Magne, P. Pinard, P. Thome, N. Chretien, *Bull. Inf. Sci. Techn. (Paris)* **1969**, *136*, 45.
[179] S. P. Brown, *Mater. Eval.* **1968**, *26*, 163; *Appl. Polym. Symp.* **1972**, *19*, 463.
[180] B. Boettcher, D. Gross, E. Mundry, *Materialprüfung*, **1969**, *11*, 156.
[181] A. J. Parisi, *Prod. Eng.* **1968**, *39*, 19.
[182] J. H. Williams, S. H. Mansouri, S. S. Lee, *U K J. Non-Destruct. Test.* **1980**, *22*, 184; J. H. Williams, S. S. Lee, *U K J. Non-Destruct. Test.* **1982**, *24*, 76.
[183] O. Altmann, *Kunststuffe* **1985**, *75*, 487; O. Altmann, L. R. Winter, *Kunststuffe* **1983**, *73*, 143.
[184] A. Kobayashi, H. Suemansu, *ASTM Spec. Tech. Publ.* (Recent Adv. Compos. US Japan), **1985**, *864*, 522.
[185] W. E. Woodmansee, *Appl. Opt.* **1968**, *7*, 1721.
[186] H. E. Steinicke, D D Patent Applications numbers 224 670A, **1984**, and 236 175A **1985**.
[187] J. C. Manarance, *J. Phys. D.* **1972**, *5*, 1120.
[188] W. E. Woodmansee, H. L. Southworth, *Mater. Eval.* **1968**, *26*, 149.
[189] L. J. Broutman, T. Kobayashi, D. Carrillo, *J. Comp. Mater.* **1969**, *3*, 702.
[190] B. Hampel, *Z. Werkstofftechn.* **1972**, *3*, 149.
[191] M. Pazdur, *Hutnik* **1970**, *37*, 205.
[192] T. A. Simcox, AIAA Report A69-25 293, **1969**.
[193] W. E. Kopp, *Prakt. Metallogr.* **1972**, *9*, 370.
[194] D. L. Uhls, *IBM Tech. Dis. Bull.* **1972**, *15*, 1670.
[195] Westinghouse Electric Corporation, US Patent 3 889 053, **1975**.
[196] I. P. Zhuk, V. A. Karolik, U. S. Mikhailov, V. P. Stepanov, *Vesti Akad. Navuk BSSR, Ser. Fiz-Energ. Navuk*, **1983**, 83.
[197] L. C. Mizell, *AIAA Rep.* A71–40738, **1971**.
[198] P. L. Garbarino, R. D. Sandison, *J. Electrochem. Soc.* **1973**, *120*, 834; *IBM Tech. Discl. Bull.* **1972**, *15*, 1738.
[199] G. V. Lukianoff, *Mol. Cryst. Liq. Cryst.* **1969**, *8*, 389.
[200] E. Grzejdziak, A. Rogowski, R. Szylhabel, A. Szymanski, J. Hejwowski, *Elektronika (Warsaw)* **1972**, *13*, 234.
[201] M. H. Perry, *NASA Report No.* 70-35867, **1970**.
[202] P. L. Gabarino, R. D. Sandison, *J. Electrochem. Soc.* **1973**, *120*, 834.
[203] R. S. Ziernicki, W. F. Leonard, *Rev. Sci. Instrum.* **1972**, *43*, 479.
[204] S. K. Key, G. D. Dick, *J. Vac. Sci. Technol.* **1974**, *11*, 97.
[205] C. Feldman, K. Moorjani, *Thin Solid Films* **1970**, *5*, R1.
[206] K. C. Park, S. Basavaiah, *J. Non-Cryst. Solids.* **1970**, *2*, 284.
[207] S. Kato, T. Uhida, H. Watanabe, M. Wada, *Oyo Butsuri*, **1975**, *44*, 156.
[208] M. Karner, U. Schaper, *Jap. J. Appl. Phys.* **1994**, *33(12A)*, 6501.
[209] H. E. Shaw, US Patent, 3 590 371, **1971**.
[210] E. Sprow, *Mach. Des.* **1969**, *41*, 37.
[211] P. M. Kendig, *Meet. Acoust. Soc. Amer. 84th*, **1972**; *Electr. Electron. Abst.* **1973**, *76*, 4973.
[212] G. M. Glazkov, A. M. Zhmud, G. G. Molokov, Yu K. Nepochatov, *Coll. Rep. All Union Sci. Conf. Liq. Cryst. Adac. Sci. USSR 1st* **1970**, 296.
[213] T. C. Bannister, *NASA Marshall Space Flt. Center, Summary Rep I and II,* Apollo 14, **1971**.
[214] P. G. Grodzka, T. C. Bannister, *Science* **1972**, *176*, 506.
[215] P. G. Grodzka, T. C. Bannister, *Science* **1975**, *187*, 165.
[216] T. L. Vigo, C. B. Hassenboehler, N. E. Wyatt, *Textile Res. J.* **1982**, 451.
[217] J. L. Fergason, P. Vogl, M. Garbuny, US Patent 3 114 836, **1960**.
[218] J. L. Fergason, A. E. Anderson, US Patent 3 401 262, **1965**.
[219] G. Jankowitz, US Patent 3 527 945, **1968**.
[220] P. U. Schulthess, Swiss Patent 520 939, **1970**.
[221] Westinghouse Electric Corp. British Patent, 1 309 558, **1969**.
[222] P. J. Allen, US Patent 3 604 930, **1970**.
[223] S. A. Hamdto, *Electr. Eng.* **1974**, *46*, 20.

[224] F. Keilmann, *Appl. Opt.* **1970**, *9*, 1319.
[225] F. Keilmann, K. F. Renk, *Appl. Phys. Lett.* **1971**, *18*, 452.
[226] J. P. Lesieur, M. C. Sexton, K. Veron, *J. Phys. D.* **1972**, *5*, 1212.
[227] M. Ohi, Y. Akimoto, T. Tako, *Oyo Butsuri* **1972**, *41*, 363.
[228] L. M. Klyukin, A. S. Sonin, B. M. Stepanov, I. N. Shibayev, *Kvantovaya Elektron.* **1975**, *2*, 61.
[229] A. S. Sonin, I. N. Shibayev, M. I. Epshtein, *Kvantovaya Elektron.* **1977**, *4*, 531.
[230] R. D. Ennulat, J. L. Fergason, *Mol. Cryst. Liq. Cryst.* **1971**, *13*, 149.
[231] W. A. Simpson, W. E. Deeds, *Appl. Opt.* **1970**, *9*, 499.
[232] C. F. Augustine, US Patent 3 693 084, **1972**.
[233] R. G. Pothier, US Patent 3 713 156, **1970**.
[234] K. Jizuka, *Electron. Lett.* **1969**, *5*, 26.
[235] C. F. Augustine, US Patent 3 693 084, **1972**.
[236] W. E. Kock, *Proc. IEEE*, *1972*, *60*, 1105.
[237] B. D. Cook, R. E. Werchan, *Ultrasonics* **1971**, *9*, 101.
[238] W. H. Sproat, S. E. Cohen, *Mater. Eval.* **1970**, *28*, 73.
[239] J. L. Dion, M. Bader, *Proc. Soc. Phot-Opt. Instrum. Eng.* **1973**, *38*, 42.
[240] J. F. Havlice, *Electron. Lett.* **1969**, *5*, 477.
[241] R. E. Werchan, B. D. Collk, *J. Acoust. Soc. Amer.* **1971**, *49*, 120.
[242] M. J. Intlekofer, D. C. Auth, *Appl. Phys. Lett.* **1972**, *20*, 151.
[243] K. Hiroshima, H. Shimizu, *Jap. J. Appl. Phys.* **1977**, *16*, 1889.
[244] M. J. Gamidov, F. F. Legusha, B. A. Finagin, *USSR Patent* 1 185 223, **1983**.
[245] W. E. L. Haas, J. E. Adams, J. J. Wysocki, *Mol. Cryst. Liq. Cryst.* **1969**, *7*, 371.
[246] J. E. Adams, W. E. L. Haas, *J. Electrochem Soc.* **1971**, *118*, 2026.
[247] W. E. L. Haas, J. E. Adams, J. H. Becker, J. J. Wysocki, US Patent 3 665 971, **1969**.
[248] K. F. Nelson, W. E. L. Haas, J. E. Adams, *J. Electrochem. Soc.* **1975**, *122*, 1564.
[249] Y. Yano, T. Takahashi, S. Harada, Japan Patent 74 16 806, **1969**.
[250] L. B. Leder, US Patent 3 789 225, **1971**.
[251] A. K. Jalaluddin, H. Husain, *Proc. Nucl. Phys. Solid. State Phys. Symp. 15th* **1971**, *2*, 527.
[252] D. E. Nagle, J. W. Doane, R. Madey, A. Saupe, *Mol. Cryst. Liq. Cryst.* **1974**, *26*, 71.
[253] I. A. Gromyko, O. S. Simoneko, V. E. Pustovarov, *USSR Patent* 1 191 840, **1984**.

4 Liquid Crystals as Solvents for Spectroscopic, Chemical Reaction, and Gas Chromatographic Applications

William J. Leigh and Mark S. Workentin

4.1 Introduction

The special combination of orientational order and mobility possessed by liquid crystals, the wide variation in these properties in different liquid crystalline types, and the fact that bulk orientation can be easily controlled by a number of different methods have led to numerous applications in which liquid crystals are employed as *anisotropic solvents* for the study of various physicochemical properties of molecules. This chapter deals with the use of liquid crystals as solvents or supports in spectroscopic applications, in gas chromatography, and for chemical reaction. In each case, the emphasis is on studies in which the *solute* is of primary interest, and in which calamitic liquid crystals are employed as solvents to control orientation or mobility. Spectroscopic studies in which some property of the mesophase is of primary interest, whether they involve a study of probe molecules or the liquid crystal itself, are treated in earlier chapters and will not be dealt with here. In many cases, studies of this type are intimately related, or provide information which is useful or essential to the complete interpretation of solute spectroscopic data.

All of the topics to be discussed in the following have been dealt with in detail in recent reviews; indeed, some are of such enormity that entire books have been written on the fundamental principles and recent applications in the area. We see little point in attempting to duplicate any of this material here. Thus, our coverage of each area will contain only an extremely cursory introduction, whose intent is simply to identify the particular physicochemical properties of molecules which are amenable to study or exploitation using liquid crystalline solvents, followed by a bibliography which directs the reader to the major resource materials in the area for details on theoretical background, experimental methods, data analysis techniques, etc. and the recent (1980–95) literature.

4.2 Liquid Crystals as Solvents in Spectroscopy

In principle, any molecular property which varies depending on the time-averaged orientation of the molecule with respect to the direction of applied electromagnetic radiation can be studied using liquid crystalline solvents. Liquid crystalline media – particularly nematics – provide the bulk molecular orientation necessary for observation of spectroscopic details analogous to those obtained in solid state experiments, yet allow sufficient mobility that linewidths similar to those obtained in fluid solution (due to averaging of intermolecular interactions) can be observed. Thus, in many respects, liquid crystalline solvents provide the spectroscopist with the means of carrying out 'single crystal' experiments in solution, and these media have received widespread attention as solvents in magnetic resonance (NMR and EPR) and optical spectroscopic studies of oriented molecules. The fundamental principles involved in the use of thermotropic liquid crystals for such applications, and the literature in these areas to 1977, were covered in detail in the 1st Edition of *The Handbook* [1]; a few general reviews have appeared in recent years [2, 3].

4.2.1 Nuclear Magnetic Resonance Spectroscopy

Following Khetrapal and Kunwar [4], the nuclear spin Hamiltonian ($\mathcal{H}$) for an oriented molecule is given by Eq. (1), where i and j are the interacting nuclei, σ_i and J_{ij} are the chemical shift and indirect spin–spin coupling constant as observed in the isotropic NMR spectrum, σ_i^{aniso} is the anisotropic component of the chemical shift, D_{ij} is the experimentally derived dipole–dipole coupling constant, B_i is the quadrupolar interaction (applicable for $I > 1/2$ spin systems), and L_{zi}, I_i^+, I_i^-, etc. are spin operators. In general, the dipole–dipole coupling D_{ij} is the sum of the direct dipolar coupling and the anisotropic component of the indirect spin–spin coupling ($D_{ij} = D_{ij}^{dir} + 1/2 J_{ij}^{aniso}$). The latter is usually negligible for a pair of protons, but becomes significant for heavier nuclei. In isotropic solvents, σ_i^{aniso}, D_{ij}, and B_i are averaged to zero, so that Eq. (1) transforms to the isotropic nuclear spin Hamiltonian. In the anisotropic medium afforded by liquid crystalline solvents, all three of these terms become non-zero and can be determined by analysis of the NMR spectrum.

The familiar expression for the dipolar coupling constant (D_{ij}) between two nuclei i and j is

$$D_{ij} = -\frac{h \gamma_i \gamma_j}{4\pi^2} \frac{1}{2} \left\langle \frac{3\cos^2 \theta_{ij} - 1}{r_{ij}^3} \right\rangle \quad (2)$$

where h is Planck's constant, γ_i and γ_j are the magnetogyric ratios, θ_{ij} is the angle between the internuclear axis and the magnetic field direction, and the angular brackets denote the average over all possible distances and orientations. Neglecting contributions from molecular vibrations, structural deformations due to solute–solvent interactions, and correlations between internuclear distances and ordering, this can be

$$\mathcal{H} = -\sum_i (1 - \sigma_i - \sigma_i^{aniso}) \nu_0 I_{zi} + \sum_{i<j} \sum (J_{ij} + 2D_{ij}) I_{zi} I_{zj}$$
$$+ (1/2) \sum_{i<j} \sum (J_{ij} - D_{ij})(I_i^+ I_j^- + I_i^- I_j^+) + \sum B_i I_{zi}^2 \quad (1)$$

rewritten as Eq. (3)

$$D_{ij} = -\frac{h\gamma_i\gamma_j}{4\pi^2}\frac{S_{ij}}{r_{ij}^3} \qquad (3)$$

where S_{ij} is the order parameter of the internuclear axis. Direct H–H dipolar couplings are typically of the order 10^3 Hz, and can be measured with a precision of ca. 0.1 Hz.

The NMR spectra of quadrupolar nuclei in liquid crystalline solvents are dominated by the quadrupolar interaction, which leads to splitting of magnitude $2B_i$ ($\approx 10^4 - 10^5$ Hz for ^{2}H) as given by Eq. (4),

$$2B_i = \frac{1}{2I_i(2I_i-1)}QCC_i[3S_{zz}+\eta(S_{xx}-S_{yy})] \qquad (4)$$

where QCC_i is the quadrupole coupling constant, η is the asymmetry parameter of the electric field gradient tensor, and S_{xx}, etc. are the diagonal elements of the order matrix for the nuclear quadrupole moment in molecule-fixed Cartesian coordinates.

The expressions given in Eqs. 1–4 are meant to illustrate, in a simple fashion, the rich information that can be gleaned from analysis of the NMR spectrum of a molecule oriented in a liquid crystalline solvent. The 'LCNMR' technique provides a direct method for the precise determination of (solute) bond lengths (r_{ij}) and bond angles *in solution* – in fact, the only one currently available. It also provides a useful method for the measurement of anisotropies in chemical shifts and in indirect spin–spin couplings, and for the determination of quadrupolar coupling constants. Since the magnitudes of dipole–dipole and quadrupolar splittings are dependent on the orientational ordering of the solute in the applied magnetic field, the two can be used for the measurement of order parameters and solvent diamagnetic anisotropies, or to monitor phase transitions. Finally, by analysis of dipolar or quadrupolar splittings due to exchangeable nuclei, it is possible to determine the dynamics of exchange processes which occur at rates too fast to be detected in isotropic solution at similar temperatures. This extends dramatically the timescale over which such processes (e.g. bond rotations, conformational changes, bond migrations, rearrangements, etc.) can be studied by NMR. In addition to these applications, relaxation time measurements afford information on the dynamics of translational and rotational motions of the solute; such experiments are of greater relevance to the properties of liquid crystals than to those of solutes, and thus will not be discussed here. The reader is directed elsewhere for reviews of this substantial branch of the field of NMR spectroscopy in liquid crystals [5–9].

4.2.1.1 Solute Structure Determination [4, 10–17]

The complete analysis of LCNMR spectra and extraction of the direct dipolar couplings is carried out by iterative computational procedures [4, 14], and usually requires input of indirect J-couplings. These are commonly taken from the isotropic solution spectrum of the solute, but heteronuclear couplings can be determined directly from the LCNMR spectrum when a mixture of liquid crystals of opposite diamagnetic anisotropies are used as solvent [18–20]. Knowledge of the relative signs of the indirect and direct couplings is desirable in such analyses. These can be determined by several methods which are reviewed briefly in a recent article by Tolman and Prestegard [21].

When the number of observable dipolar coupling constants exceeds the number of independent S_{ij}-values necessary for the description of the orientation, the ratios of internuclear distances between various nuclei in the molecule can be calculated from the observed dipolar splittings. With the choice of a suitable base internuclear distance, often taken from microwave, X-ray diffraction, or electron diffraction data, individual r_{ij}-values and bond angles can then be calculated. The ^{1}H NMR spectrum on its own affords information only on the proton internuclear distances and angles; more complete structural information can be obtained when ^{13}C-satellites can simultaneously be observed.

Two factors need to be accounted for in the calculation of the so-called 'r_α-structure' of a molecule from LCNMR data. The first is the effect of molecular harmonic vibrations on the observed dipolar splittings, which has been generally recognized since the late 1970s [22]; most LCNMR structures published since 1980 contain these corrections. The second factor is molecular deformations caused by interaction of the solute with the medium (due to correlations between molecular vibrations and solute reorientations), which as might be expected, are more important for some liquid crystals than others. These effects have been widely recognized since the early 1980s, and have been studied in detail (see for example [15, 23–26]). Procedures to correct for these effects have been published [27], and solvent systems which produce minimal structural distortions have been identified [12, 28, 29]. The problem and its solution have recently been re-emphasized by Diehl and coworkers [30].

Over 500 different molecular structures have been determined by LCNMR methods since the introduction of the technique in the 1960s [31]. Most have been derived from ^{1}H NMR spectra (and to a lesser extent, ^{2}H, ^{19}F or ^{31}P NMR), although numerous studies involving less abundant or less sensitive nuclei ($^nX = {}^{13}C$, ^{15}N, ^{111}Cd, ^{113}Cd, ^{117}Sn, ^{199}Hg) have been published as well. Most structures published since the late 1970s are derived from $^1H-{}^nX$ satellite spectra, which are generally more easily obtained than the heteronuclear spectrum, and contain much more complete structural detail than the simple proton spectrum can afford [14]. Most structures published since the early 1980s are corrected for harmonic vibrations, but only a dozen or so contain solvent-deformation corrections; the latter are true, solvent-independent structures which are in excellent agreement with those obtained by other methods [17]. Table 1 (see page 861) contains a list of compounds whose structures have been investigated using LCNMR techniques since the last compilation in 1978 [10].

A number of techniques have been developed for the simplification of LCNMR spectra of complex spin systems, which augment the time-honoured procedures of deuteration (selective, partial, or random) and double resonance methods. The most powerful of these include variable angle sample spinning [32], multiple quantum spectroscopy [33–35] and various other 2D or multiple pulse [36–41] techniques, and zero field NMR [42, 43].

4.2.1.2 Chemical Shift and Indirect Coupling Anisotropies [237]

Chemical shift anisotropies are determined by monitoring solute chemical shifts as the molecular orientation is varied [237]. This can be accomplished through changes in temperature, concentration, or phase [237], by use of mixed liquid crystals of opposite diamagnetic susceptibility anisotropies [12, 237, 238], or by use of the variable angle sample spinning technique [32, 239]. ^{1}H and ^{13}C chemical shift anisotropies are known to be solvent-dependent [237, 239, 240]. The problem is particularly significant for protons, where the changes in chemical shifts with orientation changes are small;

Burnell and de Lange have demonstrated that chemical shift anisotropies measured using mixtures of liquid crystals of opposite diamagnetic susceptibility anisotropies are prone to uncertainties as large as those for other methods [241], negating the widely-held view that such solvents afford 'solvent-independent' chemical shift anisotropies. The problem has recently been discussed in more detail, and the possible precisions in chemical shift anisotropy measurements for different nuclei are assessed [242].

Subsequent to the 1982 compilation of Lounila and Jokisaari [237], a large number of studies report chemical shift anisotropies for various nuclei in liquid crystalline solvents. Representative examples include $\Delta\sigma_H$ in dichloromethylphosphine [243], methylisothiocyanide [244], and norbornadiene [141]; $\Delta\sigma_C$ in chloro- and bromoform [245], bis(trimethylsilyl)diacetylene [246], and pyridine, pyrazine, pyridazine, and pyrimidine [247]; $\Delta\sigma_H$ and $\Delta\sigma_C$ in butyne [248], fluoro- [245, 249], bromo- [245], and iodomethane [244, 245, 250], ethylene [251], methylisocyanide [252], dimethylmercury [253], benzene [250, 254, 255], 1,3,5-trichloro- [99, 256], 1,3,5-tribromo- [244], and 1,3,5-trinitrobenzene [244]; $\Delta\sigma_D$ in iodomethane-d_3 [47]; $\Delta\sigma_F$ in fluoranil [257], hexafluorobenzene [91], and fluoromethane [249]; $\Delta\sigma_{Hg}$ in dimethylmercury [253]; $\Delta\sigma_{Se}$ in carbon diselenide [258]; $\Delta\sigma_{Te}$ in tellurophene [259]; and $\Delta\sigma_{Xe}$ [260].

In the case of J-anisotropies, the general coupling tensor (see Eq. 1) consists of contributions from both direct and indirect contributions, and it is the sum of the two types that are obtained from analysis of the LCNMR spectrum. As mentioned earlier, extraction of the direct couplings from the experimental data is normally accomplished by subtraction of the relevant J-couplings, which are either obtained directly or (more commonly) derived from isotropic solution spectra. There is no simple method for the additional correction that is required to account for J-anisotropy. It is well established that the J-anisotropy for non-directly bonded protons is very close to zero and can be safely ignored, those for C–H and N–H couplings are <0.1%, and those for couplings between heavier nuclei are significant [14, 237]. For this reason, it has been recommended that direct couplings between heavy (C–C, etc.) nuclei should not be used for structure determination, unless the J-anisotropy can be specifically evaluated [14]. Unfortunately, this is a complex and time-consuming task [237], but has continued to be of interest. The review by Lounila and Jokisaari contains a critical compilation of indirect coupling anisotropies from the literature to 1981 [237]. Numerous additional LCNMR studies have since been reported; for example, C–N [252], C–F [91, 110–112, 249], C–Hg [253], and F–F [112].

4.2.1.3 Quadrupolar Coupling Constants [10, 11, 261, 262]

Quadrupole coupling constants can be calculated from the splittings observed in the NMR spectra of quadrupolar nuclei, if the order parameter can be determined and the asymmetry parameter of the particular nucleus of interest is known (Eq. 4). The order parameter is normally calculated from dipolar splittings in the ^{1}H LCNMR spectrum of the protonated compound, or from those in the ^{13}C-satellite spectrum of the quadrupolar nucleus (if applicable). In the case of ^{2}H, the asymmetry parameter is usually neglected as it is known to be small, although it has been rigorously included on occasion by carrying out experiments in two different liquid crystals [263–266]. The calculation of order parameters from dipolar splittings requires corrections for harmonic vibrations and knowledge of the

r_α-structure. Corrections for solvent distortions are also required, as evidenced by the numerous reports of solvent dependent QCCs (see, for example [28, 46, 265, 267–269]).

4.2.1.4 Dynamics of Intramolecular Motions [8, 270–272]

The measurement of the rates and energetics of fast exchange processes is a well-known and powerful application of NMR spectroscopy [273]. In isotropic liquids, dynamic processes can be monitored through the chemical shifts or indirect spin-spin couplings of the exchanging nuclei, typically protons. The chemical shift is the most versatile since it has a larger range, but it still only varies between ≈10 and ≈10^3 Hz, limiting the range over which dynamic processes can be studied to $1-10^6$ s^{-1}. This necessitates the use of low temperatures for the study of many of the processes (e.g. conformational motions) which are of greatest interest.

In a nematic liquid crystalline solvent, where dipolar couplings in the range 10^2-10^3 Hz (for protons) and quadrupolar splittings in the range 10^3-10^5 Hz (for deuteria) can be observed, the dynamic range is extended to as high as 10^9 s^{-1}. This allows many processes, which require very low temperatures in isotropic solvents, to be studied at ambient temperatures with liquid crystalline solvents. Since the anisotropic forces responsible for the ordering in nematic liquid crystals are too weak to exert a distinct effect on the dynamics of fast conformational processes which involve only minor changes in solute shape (see Sec. 4.3 of this chapter), the results are generally directly comparable to those measured for isotropic solutions.

Studies of ring inversion or bond shifts using ^{1}H NMR have generally been restricted to fairly simple or highly symmetric compounds because of the difficulties in interpreting the more extensive dipolar couplings present in more complex systems. Examples include the ring inversion of s-trioxane [274, 275] and bond shift in cyclooctatetraene [275, 276]. For more complex systems, the use of ^{2}H NMR simplifies the situation considerably (and extends the dynamic range) because the dominant effect is the (much larger) quadrupolar interaction. Examples of systems which have been studied by ^{2}H NMR include the ring inversions of perdeuteriated cyclohexane [277–280], cis-decalin [281], and 1,1- and 1,4-dimethylcyclohexane [282], and the Cope rearrangement of bullvalene [278, 283]. Several of these systems have recently been investigated employing multiple quantum [284, 285] or other two-dimensional techniques [281, 282, 286], which are particularly useful for studies in the slow exchange range.

There has been considerably greater interest in the dynamics of single bond rotations, particularly in compounds which serve as models for common mesogens (see refs. [8, 272] and Chap. 4.3 of this volume for more complete discussions). These studies are most frequently carried out by ^{1}H NMR, since ^{1}H–^{1}H dipolar couplings are more sensitive to bond rotational processes than ^{2}H quadrupole splittings [170], and frequently employ selective deuteration to reduce spectral complexity or to provide additional information from the ^{2}H spectrum. The dynamics of single bond rotations have been studied by ^{1}H LC NMR for various substituted alkylbenzenes [147, 148, 150, 155, 167, 168, 175], anilines [129], anisoles [136, 142, 153, 154], thioanisoles [149], ethoxybenzenes [156–159, 170], biphenyls [163, 184, 192] and related molecules [181, 182, 287, 288], aromatic esters [152, 169] and aromatic ketones [151], to name but a few. Natural abundance ^{13}C NMR has also been employed successfully, for the study of C–N bond rotations in 4-dimethylaminopyrimidine [289] and 6-dialkylaminofulvenes [290].

4.2.1.5 Enantiomeric Purity of Optically Active Solutes

A relatively new development in LCNMR is the use of cholesteric liquid crystals as chiral NMR solvents for the determination of the enantiomeric purities of optically active solutes [291–293]. The initial reports of the method employed a cholesteric mixture of cholesteryl propionate in a commercial nematic solvent of negative diamagnetic anisotropy (Merck ZLI 2806), which was shown to give rise to impressively well-resolved ^{1}H NMR spectra of the two enantiomers of racemic 1,2-epoxy-3,3,3-trichloropropane [291]. The effect was shown to be due to differential ordering of the two enantiomers by the chiral solvent matrix, which results in the dipolar coupling taking on different values in the two enantiomers. Subsequent publications have employed ^{2}H NMR spectroscopy with selectively deuteriated solutes, and poly-γ-(benzyl-L-glutamate/dichloromethane mixtures as the (lyotropic) cholesteric solvents. The use of ^{2}H NMR, in spite of the fact that it requires deuteriation of the chiral solute, increases the power and general utility of the technique considerably because the spectra are so much simpler than ^{1}H spectra for more complex spin systems. Courtieu and his collaborators have demonstrated the power of this method for determining enantiomeric purities of benzylic alcohols [292, 294, 295], cyclic ketones [292, 296], carboxylic acids [292], and amino acids [297]. This is an important development in the field of organic synthesis, where there is an immense need for routine, accurate, and general methods for the determination of optical purities [298].

4.2.2 Electron Paramagnetic Resonance Spectroscopy [2, 3, 9, 299]

The EPR or ENDOR (electron nuclear double resonance) spectrum of a free radical, radical ion, triplet, or other paramagnetic species in a liquid crystalline solvent affords analogous information to that obtained by LCNMR of diamagnetic compounds: that is, information on various anisotropic interactions, from which knowledge regarding the molecular and electronic structure of the solute can be gleaned. Perhaps moreso than NMR, EPR is widely used (usually in conjunction with nitroxide spin probes) to study molecular mobility in liquid crystals, which affords information regarding the structure and physical properties of various mesophase types [9]. The principal advantage of EPR over NMR spectroscopy in such studies is the shortened timescale over which solute motional processes can be investigated ($10^{-8} – 10^{-9}$ s).

While the use of EPR spectroscopy for the study of liquid crystalline morphology continues to be an intensely active field [9], studies which deal specifically with the molecular or electronic structures of *solutes* are much less common. Representative examples include studies of semiquinone radical anions by EPR or ENDOR spectroscopy [300], Cr(V) [301] and Ag(II) [302] complexes by EPR, and spin polarization in porphyrin and chlorophyll triplet states by time-resolved EPR techniques [299, 303].

4.2.3 Polarized Optical Absorption and Emission Spectroscopy [1, 2, 304, 305]

The use of liquid crystals as orienting media for polarized optical spectroscopic studies was reviewed in the first edition of the *Handbook* [1], and has since been treated more comprehensively in books [304, 306, 307] and review articles [2, 308, 309]. Liquid crystals continue to provide an important technique for the partial alignment of solutes in polarized UV/VIS, IR, and fluorescence spectroscopic studies, complementing and extending that provided by the use of stretched polymer films (most commonly, polyethylene or poly(vinyl alcohol)) or other anisotropic media. From such studies, one can obtain information regarding the transition moments of various electronic and/or vibrational transitions of a solute, which in turn can be used to derive structural information on the solute. As usual, these techniques can alternatively be employed to investigate the structure and ordering of the mesophase, either by direct experiments on the mesophase itself, or by investigation of a suitable solute molecule oriented in the mesophase.

Liquid crystal linear dichroism (LCLD) examines the differential absorption of plane polarized (UV/VIS or IR) radiation by an uniaxially oriented solute in (usually) a nematic or compensated cholesteric liquid crystal. The observed dichroism ($LD(\bar{v})$) is given as the difference between the spectra measured with light polarized parallel and perpendicular to the optical axis of the absorbing species ($OD_{\parallel}(\bar{v})$ and $OD_{\perp}(\bar{v})$, respectively)

$$LD(\bar{v}) = OD_{\parallel}(\bar{v}) - OD_{\perp}(\bar{v}) \qquad (5)$$

Liquid crystal induced circular dichroism (LCICD) examines the differential absorption of circularly polarized radiation by an achiral solute oriented in a cholesteric liquid crystal. The circular dichroism results from an induced Cotton effect in the achiral solute due to the macroscopic chirality of its ordering in the helical solvent matrix. The effect has been shown theoretically to arise from the sum of dichroic absorbances from the solute in the individual nematic layers of the cholesteric phase (taken over the helical pitch length); it is thus related to the linear dichroism of the solute by

$$\begin{aligned} CD(\bar{v}) &= OD_L(\bar{v}) - OD_R(\bar{v}) \\ &= LD(\bar{v})\,[P\,\bar{v}^3 \Delta n / 2(\bar{v}^2 - \bar{v}_0^2)] \end{aligned} \qquad (6)$$

where P is the helical pitch length (positive for a right-handed cholesteric), Δn is the linear birefringence of an individual nematic layer, and $\bar{v}_0$ is the wavenumber of the cholesteric pitch band.

For a given absorption band, the magnitude of the observed dichroism depends on the polarization of the particular electronic or vibrational transition with respect to the solute molecular axes and the ordering matrix of the solute in the liquid crystal. Early work by Sackmann [310, 311] and Saeva [312, 313] and their co-workers demonstrated the utility of LCLD and LCICD in assigning transition polarizations in highly symmetrical (C_{2v} or D_{2h}) molecules, and in characterizing the ordering of various liquid crystalline phases. These measurements were carried out using the simplest experimental procedure, in which the dichroism is obtained by subtraction of two separately measured spectra corresponding to the two desired polarizations of the monitoring beam. This procedure is still evidently common [304], but modulation recording techniques afford much higher sensitivity [308] and are the norm in modern commercial spectrophotometers. For comprehensive discussions of the methods used for analysis of the data obtained in such experiments, the reader is directed to the books by Michl and Thulstrup [304] and Samori and Thulstrup [307].

The recent literature in this area has largely been concerned with the study of the orientational ordering of various pleochroic dyes (of potential use in display applications) [314, 315] or other aromatic molecules [316–319] in liquid crystalline hosts, rather than the specific study of solute spectroscopic properties.

The use of liquid crystals as orienting media in polarized infrared spectroscopy, though first demonstrated (as applied to solutes) in the late 1960s [310, 320], has received considerably less attention due to the fact that solvent absorptions severely limit the spectral range accessible to investigation, particularly with conventional dispersion instruments [1, 304]. Infrared linear and circular dichroism (IR–LD and IR–CD, respectively) of liquid crystalline materials themselves have, of course, been studied extensively [2]. With the advent of commercial FTIR spectrometers, whose enhanced sensitivity and superior spectral subtraction capabilities allow for far greater precision in IR spectral measurements than is possible with dispersion instruments, activity in this field has picked up considerably [314, 321–324] and has been reviewed recently [325].

Linearly polarized fluorescence (LPF) and *circularly polarized fluorescence (CPF)* provide complementary techniques to LCLD and LCICD for the assignment of solute electronic transition moments and for investigating solute orientational ordering and liquid crystalline properties [1, 2, 304, 326–328]. Additionally, these techniques can provide information on the electronic structure of excited state complexes (excimers and exciplexes) [329, 330]. Time-resolved luminescence depolarization experiments have been used by various workers to study the ordering and mobility of molecules in liquid crystalline phases [317–319, 331, 332]; the information obtainable from these studies in analogous to that obtained by NMR relaxation experiments. Since luminescence depolarization is the main result of probe molecular motions and is consequently very rapid, it leads to complications in the interpretation of steady state polarized luminescence experiments; for example, order parameters derived from LPF experiments are frequently different from those measured by LCLD spectroscopy [317].

For UV and fluorescence measurements, the most commonly used liquid crystals are mixtures of the nematic 4′-alkylbicyclohexyl-4-carbonitrile's (CCH) (e.g., ZLI 1167 and 1695), which are transparent down to ≈ 200 nm and exhibit nematic ranges between ≈ 30 and $80\,°C$ (see, for example, [315, 331, 333, 334]), and various cholesteric or compensated nematic phases of cholesteryl chloride/cholesteryl ester mixtures, which are transparent to ≈ 240 nm [310, 313, 326, 329]. Some use has also been made of 4′-(4-alkylcyclohexyl)benzonitriles (PCH-n), which are transparent to ≈ 290 nm [330, 335]. Several other mesophases, including thermotropic smectics, discotics, and lyotropic phases, have low absorption in the UV region and have been used from time to time as well. The most commonly used liquid crystals in FTIR studies are the CCH-mixtures ZLI 1167 and 1695 [314, 321, 336, 337]. The orientation of the liquid crystalline solution is most commonly achieved either by cell surface treatment or the application of an electric or magnetic field.

4.2.4 Enantiomeric Purity and Structure of Optically Active Solutes [338]

A final application that should be mentioned is the use of pitch band measurements in chiral-solute-induced twisted nematic phases to determine the enantiomeric purity and structures of chiral (solute) molecules [339, 340]. It has been shown that the ability of a chiral solute to twist a nematic phase (the so-called *twisting power*, β) can be defined in terms of the pitch length (p, in µm) and the concentration (c) and enantiomeric purity (r) of the solute according to

$$\beta = (p\,c\,r)^{-1} \qquad (7)$$

Positive values of β correspond to right-handed cholesterics, while negative values correspond to left-handed cholesterics; the magnitude varies between 1 and about 100 depending on solute structure and the nematic liquid crystal employed for the analysis [341]. The values of β are highest when the stuctures of the solvent and solute are similar, and the central cores of the nematic solvent molecules are conformationally mobile (see [341]) and references cited therein).

Empirical evaluation of the twisting powers of hundreds of chiral molecules of various structural types have revealed regular trends in the magnitude of β with solute structure within specific types; by interpolation, it is then possible to derive information on the absolute configuration and structure of new chiral molecules (i.e., that aspect of the structure that is responsible for the solute's chirality). Examples of this include calculations of the twist angle between aryl groups in biphenyls [338] and 1,1'-binaphthyls [340, 342, 343]. Alternatively, the technique can be used to determine solute enantiomeric purities; it requires very small quantities of solute and is particularly useful for compounds with extremely low optical rotations.

4.3 Liquid Crystals as Solvents in Chemical Reactions [344–348, 350, 440]

The number of studies of the use of thermotropic liquid crystals as solvents to alter the course or rates of uni- and bimolecular thermal and photochemical reactions has increased dramatically since the publication of the original *Handbook*. The impetus for these studies lies in the desire to be able to utilize the unique, anisotropic properties of liquid crystals to control the efficiency and specificity in microsyntheses, for fundamental studies of the elucidation of reaction mechanisms, and as models for organized environments involved in biological systems. To be useful for these purposes it is imperative to be able to predict accurately the specific effects that an anisotropic environment will have on a given reaction, and to be able to choose as a solvent a liquid crystal of the proper morphology to give the desired outcome. For these reasons the majority of studies over the past two decades have been directed at elucidating the factors that are important in defining the ability of liquid crystals to control solute reactivity; several critical reviews summarizing the progress in this area up to 1991 have appeared [344–348, 350].

4.3.1 Potential Effects of Anisotropic Solvents on Solute Chemical Reactivity

There are several ways in which liquid crystals may alter the chemical reactivity and dynamics of incorporated solutes compared to those in isotropic solvents. Most of these are based on the known ability of liquid crystals to control the orientation of solutes, impart constraints on their mobility, and remove the randomness of molecular motions and orientations. Thus, reactions that require a *specific relative orientation of reactants in the transition state,* undergo a *change in conformation (or shape) along the reaction coordinate,* or are sensitive to *anisotropic diffusion* are potentially alterable by liquid crystalline solvents. Depending on the precise details of the reactive system, a liquid crystalline solvent can act either to impede or favour a particular reaction pathway. These effects are expected to manifest themselves as a change in the kinetics or efficiency of a reaction or by changes in the stereochemistry and/or distribution of products, compared to that for the same reaction in typical isotropic solvents.

The ability of a liquid crystalline solvent to affect the reactivity of a solute will be determined by the degree and type of solvent order, and how effectively this is transmitted to the free energies of the reactant(s), product(s) and reaction transition state(s). These factors will depend on the structure of the reactive solute and the phase type and classification of the liquid crystalline solvent. Smectic phases are, in principle, the best candidates in which to observe effects on reactivity, due to their higher degree of orientational and translational ordering compared to nematic or cholesteric phases. While this has indeed been shown to be true in a large number of examples, nematic and cholesteric phases have also been shown to be capable of significantly altering solute chemical reactivity in selected cases.

For a particular phase type, the magnitude of the observed effects on reactivity correlates with the structural compatibility of the solute and solvent, to a first approximation. Since liquid crystals are expected to favour specific orientations or conformations of reactants, reactions involving solutes, transition states or products that are most closely related structurally to the liquid crystalline solvent often show the largest effect. However, examples also exist where the reactivity of small solutes that are not solvent-like, or of solutes whose reaction involves only minor changes in shape, is significantly altered in a liquid crystalline solvent. This is usually due to specific solvation effects. Likewise, solutes (or reaction pathways) that are seemingly well-suited based on the above ideas may not exhibit significant effects as a result of low solubility in the mesophase. This may result in a local environment which is considerably less ordered than the bulk phase; in the extreme, it can result in the formation of a biphasic mixture under the conditions of temperature and solute concentration employed. Results obtained under the latter conditions, though often intriguing, can be very difficult to interpret. They are especially common in cases employing smectic liquid crystalline solvents, where solubility problems have led to misinterpretation of results and under- or over-estimations of the true ability of the mesophase to affect the reactions of interest.

4.3.2 Solubility Factors and Phase Separation

In general, the more ordered the liquid crystal, the lower is its propensity to solubilize guest molecules. Thus, nematic phases can usually incorporate significantly higher concentrations of solute – an order of magnitude is not unusual – than smectic phases. This is unfortunate, because smectic phases have the greater potential to alter solute reactivity in the ways outlined above. It is well established that when the bulk solute concentration is higher than the solubility limit, the excess exists in a second, solute-rich phase. This second phase may be of a variety of distinct morphologies, such as an isotropic solution, a less-ordered mesophase (nematic or smectic), a binary crystalline solid, or the crystalline phase of the pure solute [351–358]. The relative proportion of the solute in the two components of the biphasic system, as well as the compositions of the two phases, will vary as a function of temperature and bulk solute concentration, as defined by the binary phase diagram for the solute/mesogen system. It is reasonable to assume that such phenomena are general, and not unique to the solute/liquid crystal systems that have been specifically examined in some detail. While such phenomena are generally considered to be a nuisance, the results of several studies prompt the intriguing suggestion that phase separation effects may ultimately prove to be useful [352–359].

Phase separation has important ramifications on bulk solute reactivity. Overall, the observed effect on reactivity in a biphasic system is the *weighted average* of the solute's reactivity in each of the two components. This often leads to complex variations in reactivity as the bulk solute concentration or the temperature is varied, particularly when the effects of liquid crystalline order on solute reactivity are pronounced. Generally, the presence of the less-ordered phase will result in a 'dilution' of the effect of the higher ordered phase. This has been illustrated succinctly in studies of the Norrish/Yang Type II reaction of aryl alkyl ketones in highly ordered smectic phases [352–355, 360], where the maximum effect of liquid crystalline order on product distribution (this is always gauged in relation to results obtained for the isotropic phase of the same mesogen, or a reasonable model solvent) was shown to occur at concentrations at or below the solubility limit of the solute. The magnitude of the effect decreases as the bulk solute concentration increases beyond the solubility limit in the mesophase; this is because an increasing *fraction* of solute resides (and reacts) in a coexisting, solute-rich isotropic or nematic phase. In the extreme, the results are indistinguishable from those obtained in the lower ordered phase.

In some cases, it has been demonstrated that the presence of the less-ordered phase in a two component system can magnify the apparent reactivity of the solute. For example, the increased yield (relative to that from irradiation in a model isotropic solvent) of dimerization products from irradiation of ethyl 4-methoxycinnamate in a smectic liquid crystal was attributed to reaction occurring mainly in a solute-enriched isotropic component of a biphasic system, and not to anisotropic ordering of the reactants in the smectic phase as might have been predicted [356]. Samori and co-workers have proposed that the (bimolecular) quaternization rearrangement reaction of allyl-*p*-dimethylaminobenzenesulphonate in smectic phases is *assisted* by phase separation phenomena, in the sense that a solute-rich isotropic phase acts as a reservoir for the reactant, which must diffuse into the smectic phase in order to react [357]. This has a clear analogy in phase transfer catalysis.

The variations in the composition of a biphasic mixture as a function of concentration and temperature have also revealed themselves in complex kinetic and (apparent) Arrhenius parameters. Examples of these effects are found in time-resolved studies of the dynamics of triplet excited state ketones [354, 355, 361, 362] and in kinetic studies of the allyl-*p*-dimethylaminobenzenesulphonate quaternization discussed above [357, 359, 363]. Prior to the development of a full understanding of the ramifications of phase separation phenomena, such complex behaviour was often misinterpreted and ascribed to the reaction of a homogeneously solubilized solute in the smectic phase [359, 361–363].

Examples of SmB–SmA biphasic reactions are also known [364]. To our knowledge, none involving smectic–solid biphases have yet been identified, although there is experimental evidence for the formation of such systems in some solute–liquid crystal mixtures [351, 352, 365, 366].

Phase separation is often difficult to detect by the more traditional thermal optical microscopy and differential scanning calorimetry techniques, especially when the bulk concentration of solute is less than a few times greater than the actual solubility limit. Deuterium NMR spectroscopy, using deuteriated solutes, is perhaps the most straightforward method of detecting phase separation phenomena and determining the solubility limits of solutes, provided that deuteriation is convenient. A number of studies outlining the detection of phase separation phenomena, the determination of solubility limits and simple approaches to the construction of solute–liquid crystal phase diagrams have appeared [351–355, 357]. Such experiments should be routine in any work related to evaluating the effectiveness of a smectic phase in controlling a reaction.

Finally, it is important to realize that the solubility properties of products are not necessarily the same as the reactants, and therefore, as a reaction proceeds, phase separation may be induced by the formation of less soluble products, resulting in a loss of 'control' as the reaction proceeds. This has not been studied in detail, but has well-known analogies in solid state chemistry [367].

4.3.3 Selection of a Liquid Crystal as a Solvent

There are a number of factors that must be considered in the selection of a suitable liquid crystalline solvent for a specific application. For photochemical reactions, the liquid crystal must be transparent to irradiation in the desired wavelength region (for most organic solutes this is typically at wavelengths above 250–270 nm), be an inefficient quencher of the excited state of the solute, and should exhibit the desired liquid crystalline phase at or near room temperature. For thermal reactions, the main requirements are that the solvent exhibits its liquid crystalline phases at the temperatures required for the reaction, and be thermally stable for extended periods of time under the required conditions.

For studies of solutes which are designed to probe the influence of liquid crystalline order on chemical reactivity, it is desirable to have a *series* of chemically similar mesogens with overlapping isotropic, nematic and smectic phases, so that meaningful conclusions can be drawn about the magnitude of the effect of solvent order on the reaction and about the generality of these effects. For the former, it is important to compare the reactivity of solutes in the liquid crystalline phases with the reactivity in an isotropic phase having similar electronic properties. Often the isotropic phase of the liquid crystal is used. However, for comparisons of reactivity at the same temperature one must employ either more typical isotropic organic solvents or, preferably, another member of the same family of mesogen that exists as an isotropic liquid over the same temperature range where the mesogen of interest is liquid crystalline. Clearly this latter choice provides the closest possible model phase in terms of structural, electronic and other physical properties. When a single compound cannot be found to serve as a model isotropic (or nematic) phase, a mixture of similar mesogens or a mixture of the liquid crystal and an organic solvent can be used.

While the solvent chosen for a desired application will be related to the type of reaction control desired, several commercial or readily synthesized materials are available that satisfy the aforementioned requirements. The structures of some of these liquid crystals and their common abbreviations are given in Table 2 (see page 872). More extensive tables of liquid crystals that are potentially suitable as reaction solvents are available throughout this *Handbook* and from other sources.

4.3.4 Selected Highlights: Reactions in Liquid Crystalline Solvents

The ability of liquid crystals to affect chemical reactions through orientational, conformational and diffusional control has been examined using a large variety of unimolecular and bimolecular reactions, and a number of experimental approaches. Liquid crystalline solvents can potentially alter the overall rate of a reaction of an excited state or reactive intermediate (which can be studied directly by time resolved spectroscopic methods), and ultimately, the product distribution. Most unimolecular reactions studied involve photochemical activation, since they have been found to be the most effective for probing the often weak effects of liquid crystals on conformational and other molecular motions. The outcome of a photochemical reaction is often completely, or at least partially, determined by the conformational dynamics of well-defined excited states or reactive intermediates generated from reaction of an excited state. Unimolecular photochemical/photophysical processes of this type that have been investigated include the Norrish/Yang type II reaction [352, 353, 360, 368–376] and intramolecular β-phenyl triplet quenching [351, 354, 355, 361, 362] of ketones, E–Z and other skeletal isomerizations [358, 377–382], and intramolecular fluorescence quenching or charge transfer reactions [383–390]. Photochemical fragmentation reactions such as the Norrish type I (α-cleavage) [391] and the photo-Fries [346, 392] reactions have been designed to probe mainly the effects of anisotropic diffusion. Fewer thermal unimolecular reactions have been studied, but examples involving equilibrium processes [393–395], isomerization [396–402], and sigmatropic rearrangements [403, 404] have been explored. Bimolecular reactions are most often designed to take advantage of the effects of orientational ordering of the solute on its reactivity. Examples investigated include photochemical cycloadditions [346, 349, 356, 399, 400, 405–408], excimer–exciplex chemistry [409–417], thermally activated hydrogen atom-abstraction (reduction) reactions [418–421], sulphonate ester rearrangements [357, 359, 363, 422–425], free radical reactions and dimerizations [394, 395, 418, 421, 426–431], and Diels–Alder and ene-addition reactions [364, 432–436]. As might be expected, the largest effects are observed when a reactant is incorporated into the structure of the liquid crystal itself, or with mesogenic reactant(s) in a liquid crystal of similar molecular structure [349, 358, 374, 375, 386, 392, 407, 408, 435, 436].

The most studied reaction in liquid crystalline solvents is the Norish/Yang type II reaction of alkyl and aryl alkyl ketones (Scheme 1). This reaction has been used extensively, mainly by Weiss and coworkers, to probe the properties of liquid crystals, since the distribution of fragmentation and cyclization products and the distribution of stereoisomeric *trans*- and *cis*-cyclobutanols are sensitive to the dynamics and conformational equilibria of the 1,4-biradical intermediate produced in the primary photochemical decay process of the ketone [352, 353, 360, 368–376]. Generally, in liquid crystals there is a preference for fragmentation over cyclization of the biradical, due to a preference for the biradical conformer that can *only* react by fragmentation. When cy-

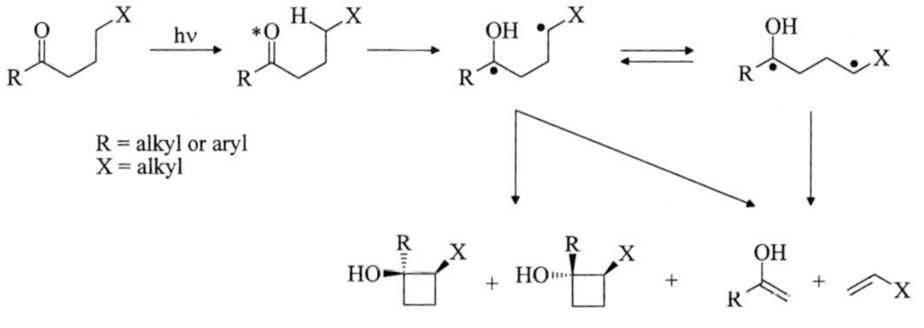

4.3 Liquid Crystals as Solvents in Chemical Reactions

Scheme 1

clization does occur, the less bulky *trans*-isomer of the cyclobutanol is often favoured over the *cis*-isomer.

For example, studies by Weiss and co-workers have shown that the liquid crystalline phases of *n*-butyl stearate (BS) can strongly influence the reaction of a number of ketones capable of Norrish/Yang type II reactivity [360, 368, 371, 372], including *sym*-alkanones (**1b**), 2-alkanones (**1a**) and *para*-alkyl-alkanophenones (**2a** and **2b**) [372, 373]. Increased selectivity for fragmentation products was found in the smectic phase over less ordered phases, and the selectivity was greatest for ketones whose length was nearly equal to that of fully extended BS; selectivity decreased as the solute was lengthened or shortened from the optimal length. The effect of solvent control is maximized when the solute and solvent are similar in length, since the former can be incorporated with minimal disruption to the ordered environment.

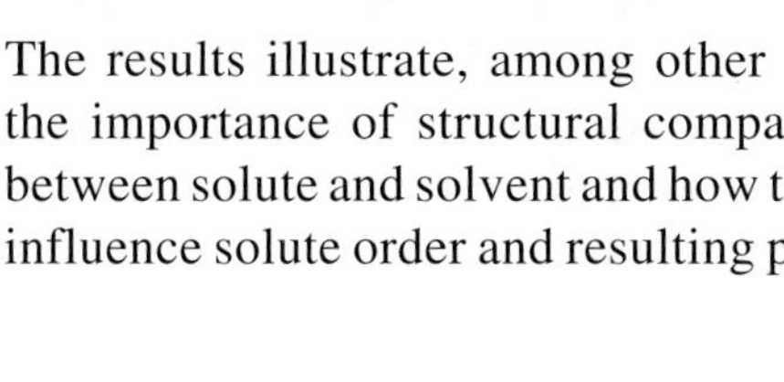

1a, **1b**, **2a**, **2b**

The results illustrate, among other things, the importance of structural compatibility between solute and solvent and how this can influence solute order and resulting product selectivity. However, results using ketones **1b** and ketones **2b** of constant (optimal) length and varying the length of either substituent (by varying m and n, respectively) showed that the relative position of the benzoyl group also effects solvent control of product selectivity. They rationalized the results by proposing that selectivity also depends on the location and orientation of the hydroxyl group of the intermediate biradical relative to the carbonyl groups of the solvent molecules as a result of their ability to interact by hydrogen bonding. Thus, factors beyond pure structural similarities are also involved in the solute–solvent interactions and the resulting reactivity. Complexities of the probe reaction and this particular system do not allow definitive conclusions as to what extent each of these factors, or others, contribute to the overall effect of the solvent on reactivity. However, it is clear that the ordered phases can influence reactivity by affecting the conformational mobility of a solute.

Similar effects on the conformational mobility of a solute were examined by Leigh and co-workers who used transient spectroscopic methods to probe the effect of liquid crystalline control on the rate of decay of excited state triplet ketones of the general structure **3** [354, 355]. In isotropic solution the rate of decay of the triplet excited state of these ketones is orders of magnitude greater than that of the analogous ketone without the β-phenyl ring, due to efficient intramolecular charge transfer quenching by the β-aryl ring. This interaction requires that the molecule adopt a *gauche* conformation, as illustrated below. In general, the decay rates of the probe solutes are decreased significantly in smectic phases compared to their values in nematic and isotropic phases.

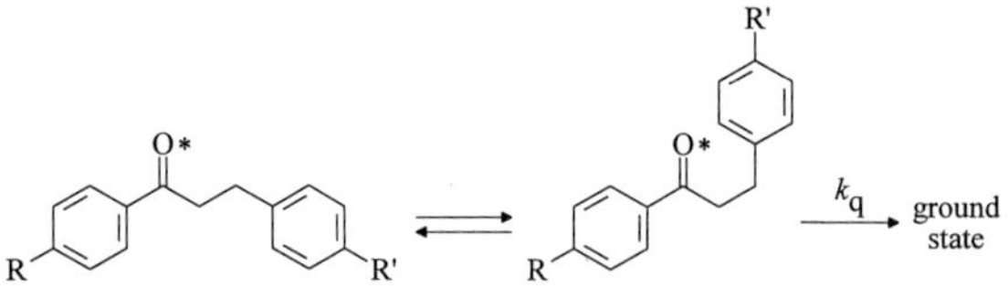

Larger differences were found for the more highly ordered smectic phases, and for ketones bearing substituents which mimic the structure of the mesogen. From an analysis of the results of a number of studies, it was concluded that the effects observed were the result of the solvent's ability to alter the equilibrium constant for *trans/gauche* interconversion in favour of the non-quenching *trans*-conformer. These studies included a careful determination of the solubility limits of each solute in the various phases by deuterium NMR spectroscopy and thus also provide several clear examples of the effects and complexities of biphasic solubilization on solute reactivity.

The largest reported effects of liquid crystalline solvents on thermal unimolecular reactivity involve *cis-trans*-olefin isomerizations. Examples include the *cis*- to *trans*-isomerization of N,N'-diacylindigos (**4**) [397] and the isomerization of merocyanines (**5**) to indolinospiropyrans (**6**) [378, 379].

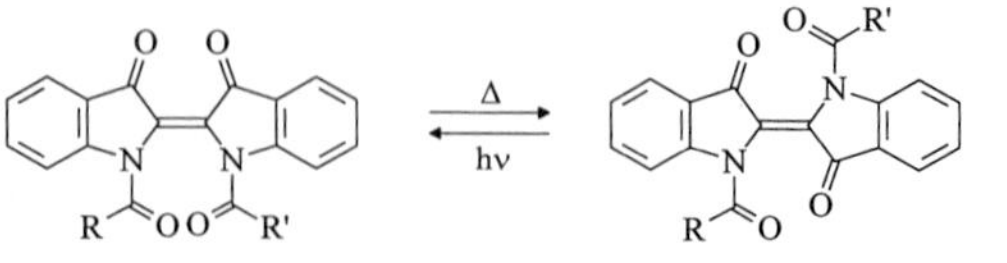

These isomerizations, which involve large shape changes in the transition state or intermediate leading to products, are inhibited in smectic phases compared to isotropic solution and exhibit substantial increases in activation enthalpy and entropy in the liquid crystalline phases. The influence on the reaction of compound **4** is largest when the two R-groups are long alkyl chains ($R=C_{17}H_{35}$), consistent with the solute being anchored in adjacent layers of the smectic phase. In both reactions one isomer is bulkier than the other (the *cis* product in the case of **4** and the spiropyran in the case of **5**), exhibits lower solubility in the mesophase, and gives significant reductions in phase transition temperatures. Under appropriate conditions of temperature and bulk concentration, the reaction can thus induce an isothermal phase transition as it proceeds. This, and the fact that both isomerizations are reversible photochemically, is of considerable potential value for device applications, as has been demonstrated in several more recent studies of these and analogous isomerization reactions in liquid crystalline solvents [358, 379, 398, 437].

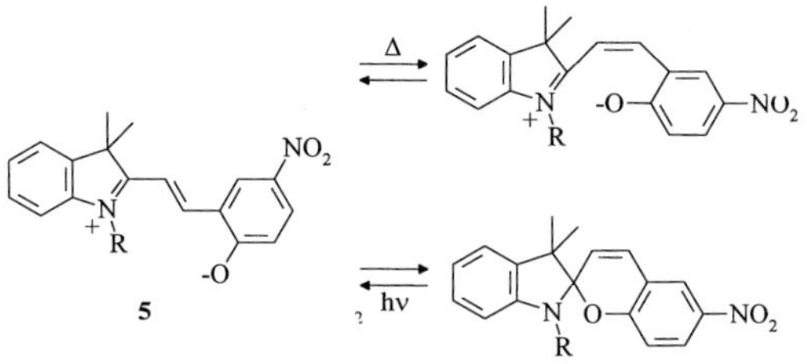

There are a number of noteworthy examples of thermal bimolecular reactions which exploit the ability of liquid crystals to control solute orientations. One of the earliest and most studied examples is the quaternization reaction of alkyl *p*-dimethylaminobenzenesulphonate esters (**7**) in cholesteric and smectic liquid crystalline phases [357, 359, 363, 422, 423, 425].

Me₂N―⟨⟩―SO₃R →Δ Me₂RN⁺―⟨⟩―SO₃⁻

7   R = CH₃ or CH₂CH=CH₂

This reaction requires a specific alignment of the migrating substituent and the amino group of an adjacent molecule and does not proceed in isotropic or nematic solvents. It does however, proceed in the neat crystalline phase when $R = CH_3$ (where the crystal structure of the lattice holds adjacent molecules in the proper orientation for reaction), and in the smectic phases of a large variety of liquid crystals. Initial studies of the kinetics of quaternization reported complex concentration and temperature dependence, including an inverse dependence of the rate constant on solute concentration and non-linear Arrhenius behaviour [359, 363]. Later studies have shown that these initial observations were the result of biphasic solubilization phenomena, with the reactant partitioned between the smectic phase and a nematic or isotropic phase [357]. The coexisting lower order phase (isotropic or nematic) was shown to participate in the reaction by providing a pool of reactant molecules which enter the smectic phase, react, and diffuse out as the zwitterionic product (see above).

Leigh and Mitchell have reported one of the largest liquid crystalline solvent effects on a thermal bimolecular reaction, and illustrated the effectiveness of these media in 'microsynthetic' applications [364, 433]. They studied the thermal ene and Diels–Alder addition reactions of the steroidal diene, 7-dehydrocholesteryl acetate (**8**), and a series of *N*-arylmaleimides (**9**) in cholesteric and smectic phases.

Of the four major products obtained in isotropic solvents, the relatively minor ene-adduct (**12**) is formed via a transition state in which the two reactants are required to assume a parallel relative orientation, the others require a perpendicular one. The net result in liquid crystalline solvents is an enhancement in the relative yield of the parallel product **12** at the expense of the others. The magnitude of the effect correlates with the length of the dienophile, increasing as the length increases, consistent with the idea that the increase in length magnifies the shape differences between the perpendicular and parallel transition states and thus, the solvents' ability to discriminate between them. The effect on reactivity was found to be greater in smectic phases than in cholesteric phases, although the difference was relatively small. This small difference was attributed partially to poor solubility of the reactants in the smectic phases and a dilution of the effect due to biphasic solubilization of the reactants in the smectogen employed.

856   4 Liquid Crystals as Solvents

In a more recent and more dramatic example, Kunieda and co-workers have studied the Diels-Alder cycloaddition reactions of 2,6-dialkoxyanthracenes (**14**) with a series of mesogenic (or mesogenic-like) dialkyl fumarates (**15**) [435, 436].

The cycloadditions proceed with a high degree of diastereoselectivity in liquid crystalline solvents or in the liquid crystalline phases of reactant **15**, with the ratio of the *syn*- to *anti*-products increasing dramatically in both cholesteric and smectic phases compared to that obtained in isotropic solution. While the actual differences in the steric requirements in the transition state leading to the two products are small (compared to the cycloaddition reactions reported by Leigh and Mitchell) the predominant formation of *syn*-adducts was rationalized as being due to preferential stabilization of the *syn*-transition state relative to that for formation of the *anti*-product. The *syn*-transition states have the molecular long axes of the two reactants oriented parallel to each other and close to the preferred dipole orientation. Further noted in this study was a marked decrease in diastereoselectivity as the concentration of reactants was increased beyond ≈4 mol%; this dilution of the effect of the mesogen was attributed to phase separation phenomena similar to that discussed previously.

This field has matured dramatically since publication of the first edition of *The Handbook*, and liquids crystals are now emerging as an important and valuable tool in synthetic and mechanistic chemistry. It is clear that so long as careful attention is paid to matching the structure of the reactant(s) to that of the mesogen (and hence optimizing solubilization in the liquid crystalline matrix), excellent control of solute reactivity can be obtained with liquid crystalline solvents – even with nematic and cholesteric phases. This applies to both uni- and bimolecular reactions, with either thermal or photochemical activation. The incorporation of reactive solutes in liquid crystals has been shown to be a viable tool in potential device applications which make use of chemically-induced isothermal phase transitions or optical pitch changes [438]. The field is ripe for further exploitation.

4.3.5 Reactions in Liquid Crystals: 1981–1996

Tables 3 and 4 (see pages 874 and 878) provide a bibliography of the wide variety of reactions that has been investigated using calamitic liquid crystals as solvents since publication of the first edition of *The Handbook*. In most studies, the reactions of interest were also investigated in model isotropic solvents for comparison. A number of the entries relate to reactions that exhibit little or no solvent control on the reactivity. In many cases, the lack of an effect may be attributed to the solvent's inability to impart its anisotropy to the substrate(s) or due to phase separation. In any case, these 'failures' are just as important as the 'successes' in contributing to the development of the field, and have been included in the tables. Those studies that include a careful and complete examination of the solubility properties of the solute/mesogen mixture are indicated by an asterisk (*) beside its reference in the tables. Absent from the table is the enormous amount of work directed towards the control of polymerization of solutes in liquid crystalline solvents; interested readers are referred instead to recent reviews of this area [350, 439].

4.4 Liquid Crystals in Gas Chromatographic Applications [444–449]

4.4.1 Liquid Crystals as Anisotropic Stationary Phases

Liquid crystals have found widespread use as stationary phases in gas chromatographic applications due to the benefits of coupling the usual analytical strengths of gas chromatography with the unique structure and shape selective properties of the liquid crystalline phase. Interaction of solutes with the orientational order provided by the anisotropy of the liquid crystal stationary phase allows for the effective and selective separation of positional and geometric isomers. This remarkable solute structural discrimination is especially important for the separation of isomers that have similar physical properties and thus cannot be conveniently separated on conventional capillary columns that mainly differentiate on the basis of boiling point/molecular weight or polarity differences. The mechanism of separation in liquid crystalline stationary phases is based on specific intermolecular interactions between a solute and the liquid crystal that are primarily related to the similarities between the solute's *structure* and the ordered geometry of the stationary phase (as opposed to polarity similarities); solutes most similar to the ordered environment will interact more strongly than those that are structurally incompatible.

A number of models have been developed to describe more quantitatively the enhancement in selectivity that is obtained from the anisotropic, orientational ordering of liquid crystalline stationary phases; most notable in this regard is the work by the groups of Janini [450, 451], Luffer [452], Martire [453 –456], and Wise [457, 458]. An understanding of the properties that are responsible for the selectivity is important to allow the prediction of retention indices [459] and to allow the choice of a proper liquid crystalline stationary phase for a particular analytical separation problem. However, given the complexities in accounting for all the factors that are involved in solute–liquid crystal interactions, it is not surprising that the use of these models for structure-retention type correlations still remains, with few exceptions [460], quite qualitative. Thus, a good deal of trial and error is still necessary in choosing a liquid crystalline phase for a particular application.

Early studies on the use of liquid crystals as stationary phases focused on packed columns; however as with other gas chromatographic applications, the technology has now almost exclusively shifted to capillary columns. Several reviews outlining the fundamentals and applications of liquid crystal stationary phases in both packed and capillary columns are available [444–449]. Stationary phases can be prepared from either monomeric or side-chain polymeric liquid crystals. Particularly relevant to the latter is the increased use of polysiloxane and polyarylate backbones for the attachment of mesogenic units, since these stationary phases are often more efficient and are able to operate at higher temperatures than most columns containing monomeric materials. To date, reported applications have relied on stationary phases based almost exclusively on calamitic liquid crystals, although a few examples using discotic stationary phases have appeared [461–463]. Nematic phases have found the most applicability since they are less viscous than smectics and usually have the widest mesophase temperature ranges; smectic stationary phases are, however, now also used extensively. Of the smectic phases that have been used, the less ordered phases are found to possess the better separation properties. This has been attributed to the fact that the more ordered smectic phases will be resistant to incorporation of the solute. In some cases, solubility properties may also play a role in the observed phase dependence; the more ordered smectic phases will have the lower solubility limit and thus, any added solute may act to disrupt the solvent order, so minimizing its ability to effect separation.

As with traditional stationary phases, liquid crystalline stationary phases must demonstrate high thermal stability and have low vapour pressures (to ensure column integrity and reproducibility). Furthermore, it is desirable that the liquid crystals exhibit their mesophases over as wide an operating temperature range as possible, so that the columns can be used for more generic separations and to permit their use with common temperature programming techniques.

In the preparation of any column, the two main considerations are *efficiency* and the *selectivity*. The efficiency of a liquid crystalline stationary phase is generally found to be lower than for traditional isotropic stationary phases due to their increased viscosity and decreased mass transfer properties.

This is particularly true for monomeric stationary phases; polymeric liquid crystalline stationary phases and phases containing a mixture of either a monomeric or polymeric liquid crystal blended with a traditional isotropic stationary phase are generally found to be more efficient [448, 464, 465]. The selectivity of a liquid crystalline stationary phase is determined by its ability to discriminate between two or more isomeric compounds that differ in shape, but whose physical properties are virtually identical. Selectivity is often expressed in terms of the relative retention time, α, with higher values indicating a better separation. The highest values of α are obtained within the mesophase temperature range at temperatures 2–5 °C above the onset of phase transition; α values are typically equal to 1 outside of the mesophase temperature region. The shape selectivity of high temperature stationary phases is often tested using a mixture of polynuclear aromatic hydrocarbons (anthracenes, phenanthrenes) (see, for example, refs. [461, 466, 467]) and for low temperature stationary phases, xylene isomers are typically used (see, for example, refs. [468, 469]). Recently, Krupcnik and coworkers have suggested that saturated cyclic compounds (methylcyclohexanes or decalins), being less polar than xylenes, are superior for testing the shape selectivity of a column [470].

Among the factors that affect column efficiency and selectivity are: the structure of the mesogen (in relation to the solutes of interest), the phase type(s) and degree of ordering, temperature, column loading and coverage, concentration (solubility) effects, and the surface structure and specialized treatment of the support. The column surface can dramatically affect the orientation of the liquid crystalline molecules (especially for monomeric mesogens) and thus alter separation efficiency. For polymeric liquid crystalline phases, additional factors such as the length and type of the polymer chain, length of the spacer between mesogenic unit and poylmer backbone, and the number and type of mesogenic units must also be considered. For maximum selectivity and separation efficiency, special attention should be paid to each of these factors. Numerous techniques, many of which are based on the technologies already used for the preparation of more traditional capillary columns [471], have been developed for the preparation of high resolution liquid crystal capillary columns for use over a wide variety of operating conditions; some of these techniques have been outlined in reviews [447, 448].

The number of liquid crystals that fulfil the necessary criteria for use as chromatographic stationary phases is large. In general, polymeric liquid crystals are now preferred due to their increased capacity for operating at higher column temperatures; however monomeric phases still find considerable application in the separation of mixtures of low boiling solutes. The ability of a liquid crystalline phase to exhibit high selectivity for separations based on solute size and structure has allowed for the analysis of many complex mixtures that are important in industrial, environmental, biochemical and geophysical applications. In these cases, the separation of structural or geometric isomers is often difficult or impossible using other techniques. The necessity for such separations is often tied to the fact that the physiochemical property of interest or the biological activity is intimately related to structure.

Common examples include the separation of positional isomers of substituted aryl- or naphthyl compounds, constitutional isomers of alkanes, geometric isomers of alkenes or alkynes, or isomers of polynuclear aromatic hydrocarbons. Other studies include the use of cholesteric and other chiral mesophases for the separation of enantiomers. Table 5 (see page 882) contains a survey of the types of molecules that have been effectively separated using capillary columns modified with monomeric and polymeric liquid crystalline stationary phases. The sheer number of liquid crystalline stationary phases employed preclude their summary here, but many are based on substituted biphenyl or terphenyl derivatives, isothiocyanates, Schiff bases, or azo- or azoxy-derivatives. A few liquid crystalline stationary phase columns are commercially available [472]. The readers are directed to the accompanying references in Table 5 and reviews [444–449] for lists of the specific mesogens used in preparation of the monomeric or polymeric stationary phase and for more specific details on column preparation and fundamental studies of column performance. As the number of anisotropic stationary phases continues to grow, so will the number of applications to the separation of new series of compounds. Table 5 is not complete, and readers are directed towards the bibliographic section of the *Journal of Chromatography* (published regularly) to find an up-to-date and continuing list of references to studies of the use of liquid crystalline stationary phases in gas chromatography, including references to the many valuable contributions made in the Russian and Chinese literature that have not been summarized comprehensively here.

4.4.2 Application of Liquid Crystal Stationary Phases for the Determination of Thermodynamic Data for Nonmesogenic Solutes

Gas chromatography is often the method of choice for the determination of reliable thermodynamic data for volatile solutes, under conditions approximating infinite dilution, due to the ease of the experiments and the large variety of solutes that can be studied. This methodology has been extended to the study of the thermodynamic properties of solutes in liquid crystals using mesogenic stationary phases. The thermodynamic properties obtained can provide information on the factors that are important in solute–liquid crystal interactions and how they vary as a function of mesophase structure and phase. This information is ultimately useful for further development of models of retention and selectivity that will permit the logical design and application of a particular liquid crystal stationary phase for a desired separation (see above).

Experimentally determined solute specific retention volumes, V_g^0, are used in the calculation of solute activity coefficients at infinite dilution, γ_f^∞, corrected for the non-ideality (fugacity) of the vapour phase, using Eq. (8).

$$\ln \gamma_f^\infty = \ln\left(\frac{273.2\,R}{M_1 P_2^g V_g^0}\right) - \left(\frac{B_{22} P_0^2}{RT}\right) \quad (8)$$

Here, R is the gas constant, M_1 is the solvent's molecular weight, and P_2^0 and B_{22} are the saturated vapour pressure and second virial coefficient of the pure solvent at temperature T, respectively. The required vapour pressures and virial coefficients can be calculated or estimated from known relationships [473–477]. Other treatments for the determination of γ_f^∞, which include a term to correct for the free volume contribution due to differences in the size of the solute and solvent, have also been used [478, 479]. The temperature dependence of the resulting solute activity coefficient is related to the infinite dilution partial molar excess Gibbs energy ($\Delta \overline{G}_e^\infty$) through Eq. (9).

$$\ln \gamma_f^\infty = \frac{\Delta \overline{G}_e^\infty}{RT} = \frac{\Delta \overline{H}_e^\infty}{RT} - \frac{\Delta \overline{S}_e^\infty}{R} \quad (9)$$

The $\Delta \overline{G}_e^\infty$ can then be separated into the respective enthalpic ($\Delta \overline{H}_e^\infty$) and entropic ($\Delta \overline{S}_e^\infty$) contributions to the activity coefficient, using the well known relationship that is also given in Eq. (9). The partial molar enthalpy ($\Delta \overline{H}^{sol}$) and entropy ($\Delta \overline{S}^{sol}$) of solution, with the reference state of the solute at infinite dilution in an ideal gas, can be calculated using Eqs. (10) and (11), respectively, where $\Delta \overline{H}^{vap}$ is the solute's molar heat of vapourization. The $\Delta \overline{H}^{vap}$ required can be determined from the Clausius–Clapeyron equation.

$$\Delta \overline{H}^{sol} = \Delta \overline{H}_e^\infty - \Delta \overline{H}^{vap} \quad (10)$$

$$\Delta \overline{S}^{sol} = \Delta \overline{S}_e^\infty - (\Delta \overline{H}^{vap}/T) \quad (11)$$

The thermodynamic parameters, measured at various temperatures for the isotropic, nematic, or smectic phase of a liquid crystalline stationary phase, are related to changes in the ability of a solute to interact with the solvent. Several theoretical treatments, including the refined infinite dilution solution model [453, 480, 481] and the lattice model [455, 456], have been postulated for the general interpretation of the thermodynamic parameters in terms of the retention and selectivity parameters in gas chromatographic separations. However, attempts to correlate the thermodynamic quantities with *specific* solute–solvent interactions remain qualitative. Nevertheless, a few general observations can be noted. Upon cooling from an isotropic phase into a liquid crystalline phase, the values of V_g^0 decrease and those of γ_f^∞ increase. Plots of V_g^0 or γ_f^∞ versus temperature show discontinuities at phase transition temperatures. The activity coefficient of any solute in an anisotropic environment (γ_A^∞) is found to be more positive (and typically, $\gamma_f^\infty > 1$) than in its isotropic phase (γ_I^∞). This trend in γ_f^∞ is a reflection of variations in $\Delta \overline{G}_e^\infty$, which result from contributions from both enthalpic and entropic factors. Partial molar enthalpic excesses are found to be more positive (i.e., more endothermic) in the anisotropic environment; this is a reflection of the greater amount of energy that is required to dissolve a solute in a more ordered environment. This large enthalpic term is compensated somewhat by the entropic term, which is also generally found to be positive. While more difficult to rationalize, the entropy term probably results from a combination of a decrease in the rotational and conformational entropies, due to the increase in solvent order, and an increase in the translational entropy as a result of poorer solute-solvent interactions. This combination of enthalpic and entropic factors usually results in activity coefficients which exhibit a positive deviation from Raoult's law ($\gamma_A^\infty/\gamma_I^\infty > 1$), indicating that liquid crystalline phases exhibit a reduced affinity for a solute compared with the isotropic phase; the lower the ordering, the better the dissolution of the solute. For a given series of solutes one expects the enthalpy and entropy terms to decrease for solvents that interact more strongly with the anisotropic phase (i.e. the ratio $\gamma_A^\infty/\gamma_I^\infty$ decreases as solute–solvent interactions increase). Values for $\Delta \overline{H}^{sol}$ and $\Delta \overline{S}^{sol}$ are negative in all phases and correlations between the two have been found for a homologous set of solutes in a number of liquid crystalline phases. However, there exists no unifying model that relates a solute's characteristics (size, shape, polarity) to the solution process in liquid crystalline solvents. Therefore, specific variations in values of γ_f^∞, $\Delta \overline{H}_e^\infty$, $\Delta \overline{S}_e^\infty$, $\Delta \overline{H}^{sol}$ and $\Delta \overline{S}^{sol}$ that are observed for a homologous series of solutes in the smectic, nematic and isotropic phases of a solvent or series of solvents cannot yet be predicted quantitatively.

Representative studies whose aims were to examine systematically the effects of solute and solvent structure on the solution process by determining the thermodynamic properties are summarized in Table 6 (see page 884). The reader is directed to the references for specific discussions of the thermodynamic values obtained for the series of solutes in relation to a particular solution model.

Table 1. Solute structural/conformational data by LCNMR.

Formula	Name	Structure	Techniques[a]	Notes[b]	Refs.
Organic compounds					
CH_3NO_2	nitromethane	CH_3NO_2	^{1}H, C-sat, N-sat, ^{2}H	1	[28, 44]
CH_3F	fluoromethane	CH_3F	^{1}H, labl	1	[45]
CH_3Br	bromomethane	CH_3Br	^{2}H, C-sat, labl	1	[28, 46]
CH_3I	iodomethane	CH_3I	^{2}H, labl	1	[28, 45, 47]
$C_2H_2F_2$	1,1-difluoroethylene	$F_2C=CH_2$	^{1}H, C-sat	1	[48]
C_2H_3N	acetonitrile	$CH_3C{\equiv}N$	^{1}H, N-sat, C-sat; ^{2}H, C-sat	1, 2	[28, 49, 50]
C_2H_3N	methyl isocyanide	$CH_3N{\equiv}C$	^{1}H, C-sat	1	[51]
C_2H_4	ethylene	$H_2C=CH_2$	^{1}H, C-sat	1	[48]
C_2H_4O	oxirane	△	^{1}H, C-sat	1	[52]
$C_2H_4O_2$	acetic acid	CH_3CO_2H	^{1}H, ^{13}C, ^{2}H, labl	1 a	[53]

Table 1. (continued)

Formula	Name	Structure	Techniques[a]	Notes[b]	Refs.
C_2H_5NO	methyl formamide	CH_3NHCHO	^1H, ^{13}C, labl	2	[54]
C_3H_3F	3-fluoroprop-1-yne	$HC\equiv CCH_2F$	^1H, ^{19}F, C-sat	1	[55]
C_3H_3Cl	3-chloroprop-1-yne	$HC\equiv CCH_2Cl$	^1H, ^{13}C	2	[56]
C_3H_3NO	oxazole		^1H, C-sat	1	[57]
C_3H_4	propyne	$HC\equiv CCH_3$	^1H, C-sat	1	[58]
C_3H_6O	oxetane		^1H, ^{13}C	1	[59]
C_3H_6S	thietane		^1H, ^{13}C	1	[60]
C_3H_6SO	thietane-1-oxide		^1H	2	[61]
$C_3H_6SO_2$	thietane-1,1-dioxide		^1H	2	[61]
C_4H_3Br	2-bromothiophene		^1H, C-sat; ^{13}C	1	[62]
C_4H_3Cl	2-chlorothiophene		^1H, C-sat; ^{13}C	1	[62]
$C_4H_4N_2$	pyridazine		^1H, C-sat	1	[63]
$C_4H_4N_2$	pyrazine		^1H, C-sat, N-sat	1	[64, 65]
$C_4H_4N_2$	pyrimidine		^1H, C-sat, N-sat	1	[64, 65]
$C_4H_4N_2$-I_2	pyrimidine-iodine complex		^1H, ^{13}C	1	[66]
$C_4H_4O^e$	furan		^1H, C-sat	1	[67–69]
C_4H_4S	thiophene		^1H, C-sat	1	[67, 70, 71]
C_4H_5N	pyrrole		^1H, C-sat	2	[29]
C_4H_5N	trans-crotonitrile		^1H	1	[72]
C_4H_6	but-2-yne	$H_3CC\equiv CCH_3$	^1H, C-sat	1	[73]
C_4H_6	methylenecyclopropane		^1H, ^{13}C	1	[74, 75]
$C_4H_6N_2$	N-methylimidazole		^1H	2	[76]
C_4H_8	cyclobutane		re-analysis	1	[77]
C_4H_8O	tetrahydrofuran		^1H	2	[78]
C_5H_{10}	cyclopentane		^1H, labl	3	[79]
C_5H_4FN	2-fluoropyridine		^1H	2	[80]

Table 1. (continued)

Formula	Name	Structure	Techniques[a]	Notes[b]	Refs.
$C_5H_4O_2$	4-pyrone	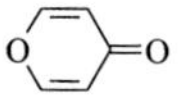	^{1}H, C-sat	1	[81]
$C_5H_4O_2S$	1,6-dioxa-6aλ^4-trithiapentalene	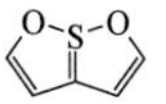	^{1}H, C-sat, ^{13}C	1	[82, 83]
$C_5H_4S_3$	1,6,6aλ^4-trithiapentalene	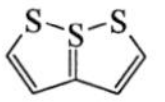	^{1}H, C-sat	1	[82, 83]
C_5H_5N-I_2	pyridine-iodine complex	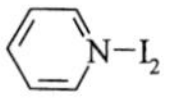	^{1}H, N-sat	1	[84]
C_5H_5NO	pyridine N-oxide	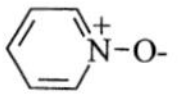	^{1}H	2	[85]
C_5H_6	vinylidenecylopropane	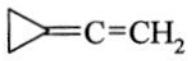	^{1}H	1a	[86]
C_5H_7N	N-methylpyrrole	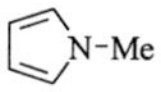	^{1}H	2	[87]
C_5H_8	spiro[2.2]pentane		^{1}H	1	[88]
C_5H_8	cyclopentene		^{1}H	3	[89, 90]
C_6F_6	hexafluorobenzene	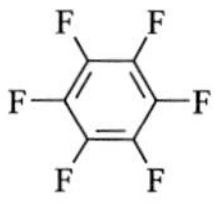	^{19}F, C-sat	1	[91]
$C_6H_5NO_2$	nitrobenzene	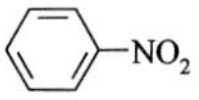	^{1}H, C-sat	1	[30]
$C_6H_{12}O_6$	myo-inositol	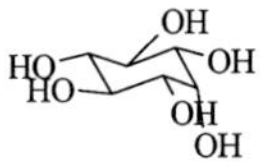	^{1}H	2, 4	[92]
C_6H_{14}	n-hexane	$CH_3(CH_2)_4CH_3$	^{1}H, 2D(MQ), labl	3	[93–95]
$C_6H_2Cl_4$	1,2,3,5-tetrachlorobenzene	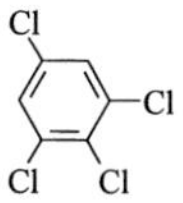	^{1}H, C-sat	1	[96, 97]
$C_6H_2Cl_4$	1,2,3,4-tetrachlorobenzene	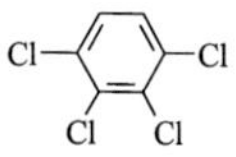	^{1}H, C-sat	1	[98]
$C_6H_2Cl_4$	1,2,4,5-tetrachlorobenzene	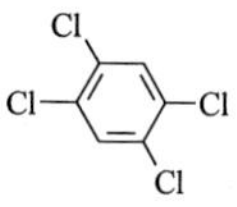	^{1}H, C-sat	1	[98]
$C_6H_3Cl_3$	1,3,5-trichlorobenzene	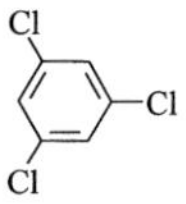	^{1}H, C-sat, ^{13}C	1	[99, 100]
$C_6H_3Cl_3$	1,2,3-trichlorobenzene	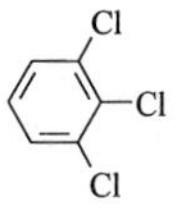	^{1}H, C-sat	1	[101, 102]
$C_6H_3Cl_3$	1,2,4-trichlorobenzene	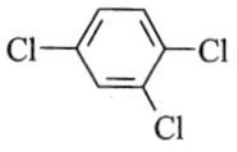	^{1}H, C-sat	1	[103]

Table 1. (continued)

Formula	Name	Structure	Techniques[a]	Notes[b]	Refs.
C_6H_4BrF	4-bromofluorobenzene	 F–⟨⟩–Br	^1H	2, 4	[104]
$C_6H_4Cl_2$	p-dichlorobenzene	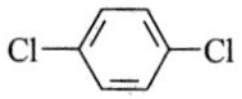 Cl–⟨⟩–Cl	^1H, C-sat, ^{13}C	1	[105]
$C_6H_4Cl_2$[e]	m-dichlorobenzene	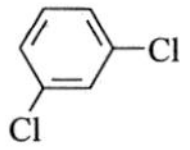	re-analysis	1	[106]
$C_6H_4Cl_2O$	2,6-dichlorophenol	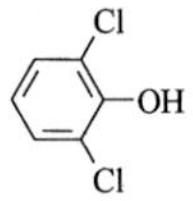	^1H, C-sat	1	[107]
C_6H_4ClF	3-chlorofluorobenzene	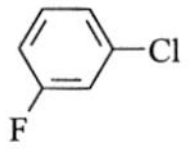	^1H	1 a	[108]
C_6H_4ClF	4-chlorofluorobenzene	 Cl–⟨⟩–F	^1H	2, 4	[104]
C_6H_4ClI	3-chloroiodobenzene	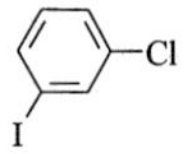	^1H, 2D(MQ)	2	[109]
$C_6H_4F_2$	o-difluorobenzene	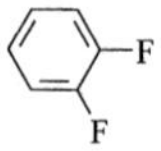	^1H, ^{19}F, C-sat	1	[110]
$C_6H_4F_2$	m-difluorobenzene	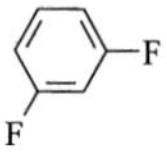	^1H, ^{19}F, C-sat	1	[111]
$C_6H_4F_2$	p-difluorobenzene	 F–⟨⟩–F	^1H, ^{19}F, C-sat	1	[112]
C_6H_4FI	m-fluoroiodobenzene	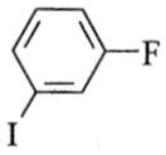	^1H	2	[108]
C_6H_4FI	p-fluoroiodobenzene	 I–⟨⟩–F	^1H	2, 4	[104]
$C_6H_4FNO_2$	p-fluoronitrobenzene	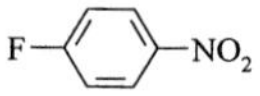 F–⟨⟩–NO$_2$	^1H, ^{19}F	2, 4	[104]
$C_6H_4S_2$	thieno[2,3-b]thiophene	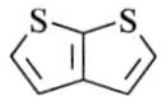	^1H, C-sat	1	[113]
C_6H_5Br	bromobenzene	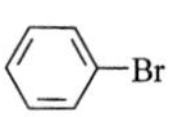 ⟨⟩–Br	^1H, C-sat	2, 4 1	[114] [115]
C_6H_5BrO	4-bromophenol	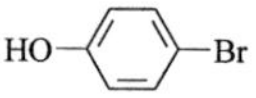 HO–⟨⟩–Br	^1H	2, 4	[116]
C_6H_5Cl	chlorobenzene	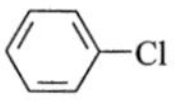 ⟨⟩–Cl	^1H, C-sat	1 2, 4	[117, 118] [114]
C_6H_5ClO	p-chlorophenol	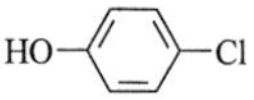 HO–⟨⟩–Cl	^1H	2, 4	[116]
C_6H_5ClS	p-chlorothiophenol	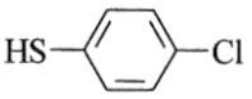 HS–⟨⟩–Cl	^1H	2, 3	[119]
C_6H_5F	fluorobenzene	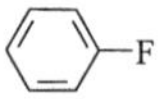 ⟨⟩–F	^1H, C-sat; ^{19}F, C-sat	1	[120–122]
C_6H_5FO	p-fluorophenol	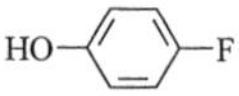 HO–⟨⟩–F	^1H	2, 4	[116]

Table 1. (continued)

Formula	Name	Structure	Techniques[a]	Notes[b]	Refs.
C_6H_5I	iodobenzene	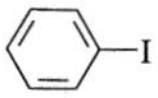	^{1}H; ^{1}H, C-sat	1, 4	[114, 123, 124]
$C_6H_5N_3$	1,2,4-triazanaphthalene	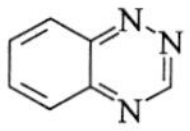	^{1}H	2	[125]
$C_6H_5NO_3$	4-nitrophenol	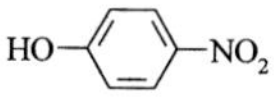	^{1}H	2, 4	[116]
C_6H_6	benzene		^{1}H, ^{2}H, labl, ^{3}H	1	[126–128]
C_6H_6BrN	4-bromoaniline	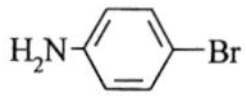	^{1}H	1	[129]
$C_6H_6Cl_6$	1a, 2a, 3e, 4e, 5e, 6e-hexachlorocyclohexane	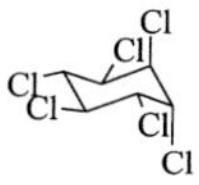	^{1}H	1	[130]
$C_6H_6Cl_6$	1e, 2e, 3e, 4e, 5e, 6e-hexachlorocyclohexane	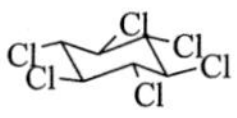	^{1}H	1	[130]
$C_6H_6Cl_6$	1a, 2a, 3e, 4e, 5e, 6a-hexachlorocyclohexane	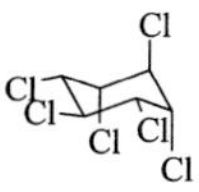	^{1}H	1	[131]
C_6H_6FN	p-fluoroaniline	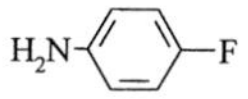	^{1}H	2, 4	[104]
$C_6H_6N_2O_2$	p-nitroaniline	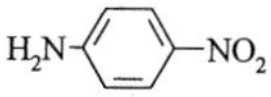	^{1}H	1	[129]
C_6H_6O	phenol	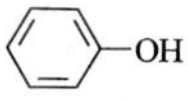	^{1}H, C-sat	1	[132]
C_6H_6S	thiophenol	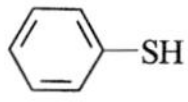	^{1}H	3	[119]
C_6H_8	cyclopropylidenecyclopropane	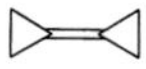	^{1}H	1	[133]
$C_6H_8N_2O_2$	dimethyluracil	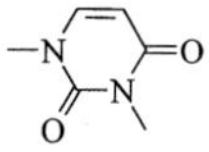	^{1}H	3	[134]
$C_7H_4ClF_3$	p-chlorobenzotrifluoride	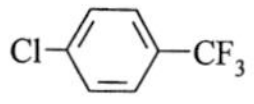	^{1}H, ^{19}F	2	[135]
$C_7H_4F_3IO$	p-iodotrifluoromethoxybenzene	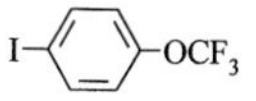	^{1}H, ^{19}F	3	[136]
$C_7H_4F_4O$	p-fluorotrifluoromethoxybenzene	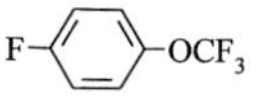	^{1}H, ^{19}F	3	[136]
C_7H_5N	benzonitrile	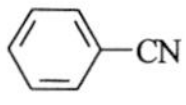	^{1}H, C-sat; ^{1}H, labl	1	[137]
C_7H_5NO	anthranil	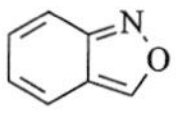	^{1}H	2	[138]
C_7H_5NS	phenyl isothiocyanate	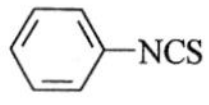	^{1}H, C-sat	1	[139]
C_7H_6	bicyclo[2.2.1]hepta-2,5-diene		^{1}H	2	[140, 141]

Table 1. (continued)

Formula	Name	Structure	Techniques[a]	Notes[b]	Refs.
$C_7H_6Cl_2O$	2,6-dichloroanisole	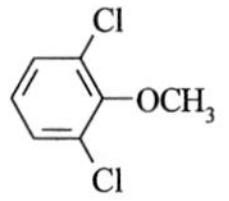	^{1}H, ^{2}H, labl	1	[142]
$C_7H_6Cl_2O$	3,5-dichloroanisole	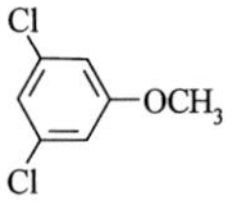	^{1}H, ^{2}H, labl	1	[142]
$C_7H_6N_2$	1,3-diazanaphthalene (quinazoline)	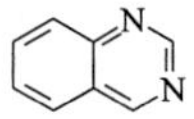	^{1}H	2	[125]
C_7H_6O	benzaldehyde	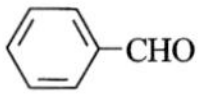	^{1}H, C-sat	1	[143]
$C_7H_6O_2$	4-hydroxybenzaldehye	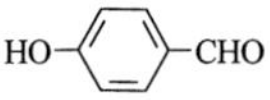	^{1}H	2, 4	[116]
$C_7H_6O_2$	1,3-benzodioxole	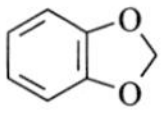	^{1}H	1	[144]
$C_7H_7BF_4$	tropylium tetrafluoroborate		^{1}H	2, 4	[145]
C_7H_7Cl	benzyl chloride	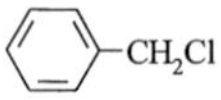	^{1}H	3	[146, 147]
C_7H_7Cl	4-chlorotoluene	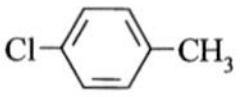	^{1}H, C-sat	1	[148]
C_7H_7ClS	4-chlorothioanisole	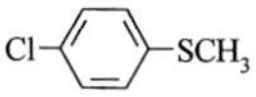	^{1}H, ^{2}H, ^{13}C, labl	3	[149]
C_7H_7F	benzyl fluoride	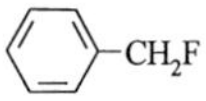	^{1}H, ^{19}F, labl	3	[150]
C_7H_7NO	4-acetylpyridine	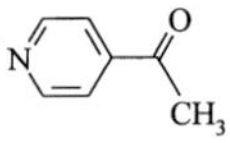	^{1}H, ^{2}H	1	[151]
$C_7H_7NO_2$	methyl 4-pyridinecarboxylic acid	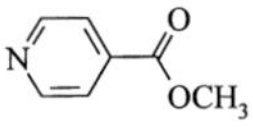	^{1}H	3	[152]
$C_7H_7NO_2$	methyl 3-pyridinecarboxylic acid	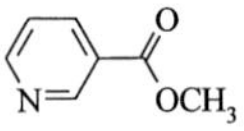	^{1}H	3	[152]
$C_7H_7NO_2S$	4-nitrothioanisole	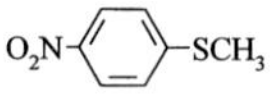	^{1}H, ^{13}C, labl	3	[149]
$C_7H_7NO_3$	4-nitroanisole	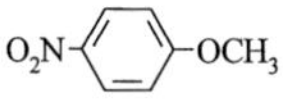	^{1}H, labl	3	[153]
C_7H_8O	anisole	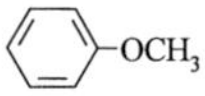	^{1}H, echo, labl; ^{13}C	3	[154]
C_7H_8S	thioanisole	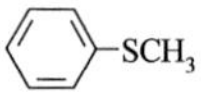	^{1}H, ^{13}C, labl	3	[149]
C_8H_{10}	ethylbenzene	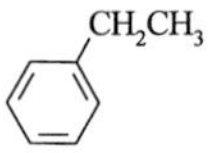	^{1}H, echo; ^{2}H, labl	3	[155]
$C_8H_{10}O$	ethoxybenzene	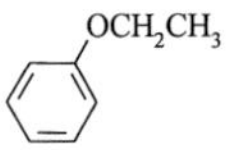	^{1}H, ^{2}H, labl	3	[156–159]

Table 1. (continued)

Formula	Name	Structure	Techniques[a]	Notes[b]	Refs.
$C_8H_{10}O_2$	1,4-dimethoxybenzene	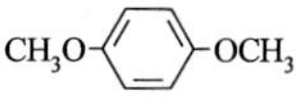	^{1}H, ^{2}D, 2D, echo, labl	3	[160]
$C_8H_4N_2$	phthalonitrile	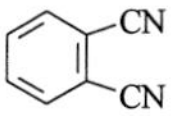	^{1}H, C-sat, N-sat	1	[161]
$C_8H_6N_2$-I_2	quinazoline-iodine complex	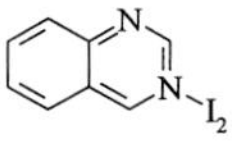	^{1}H	2	[162]
$C_8H_6O_2$	2,2'-bifuryl	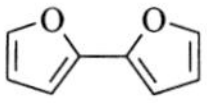	^{1}H	3	[163]
$C_8H_6O_2$	phthalide	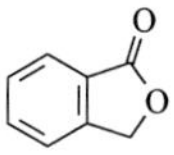	^{1}H	2	[164]
$C_8H_6O_2$	2-coumaranone	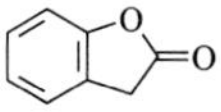	^{1}H	2	[164]
C_8H_6S	thialene	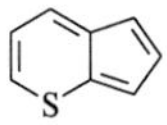	^{1}H	2	[165]
$C_8H_6S_2$	2,2'-dithiophene	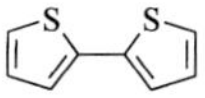	^{1}H	3	[166]
C_8H_7Br	4-bromostyrene	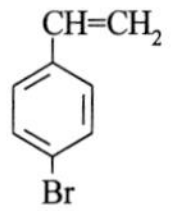	^{1}H	2	[167]
C_8H_8	styrene	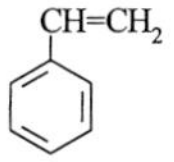	^{1}H, ^{2}H, labl	2	[167]
$C_8H_8Br_2$	3,5-dibromostyrene	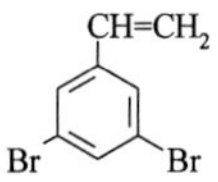	^{1}H, labl	3	[168]
$C_8H_8O_2$	phenyl acetate	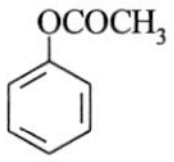	^{1}H, labl; ^{13}C; ^{2}H, labl	3	[169]
$C_8H_9ClO_2$	4-chloroethoxybenzene	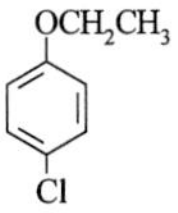	^{1}H, ^{2}H, labl	3	[157, 158, 170]
C_8H_9FO	4-fluoroethoxybenzene	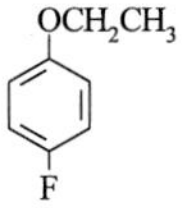	^{1}H, ^{2}H, labl	3	[156–158]
C_9H_{12}	1-propylbenzene	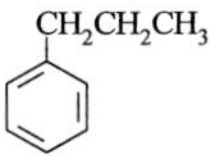	^{1}H, echo; ^{2}H, labl	3	[155]
$C_9H_{12}O_3$	1,3,5-trimethoxybenzene	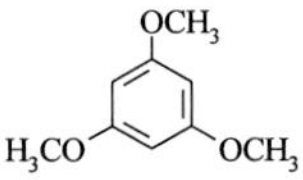	^{1}H, ^{2}H, labl	3	[171]

Table 1. (continued)

Formula	Name	Structure	Techniques[a]	Notes[b]	Refs.
$C_9H_5BrCl_2N_2$	2-(4-bromophenyl)-4,6-dichloropyrimidine	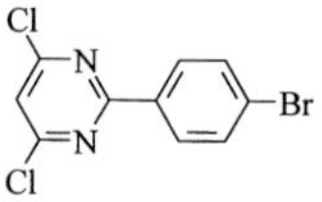	^1H	2	[172]
$C_9H_6O_2$	1,3-indandione	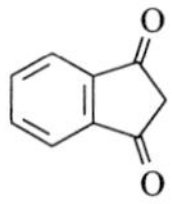	^1H	1	[144]
C_9H_6OS	thiocoumarin	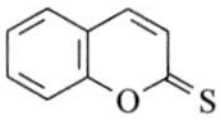	^1H	2	[173]
C_9H_7NO	3-phenylisoxazole	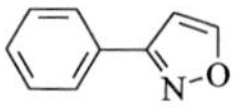	^1H	3	[174]
C_9H_7NO	4-phenylisoxazole	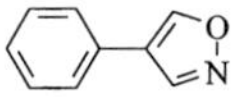	^1H	3	[174]
C_9H_7NO	5-phenylisoxazole	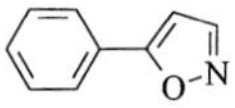	^1H	3	[174]
C_9H_8	1-phenylpropyne	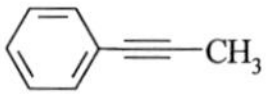	^1H	2	[175]
$C_9H_8O_2$	cubanecarboxylic acid	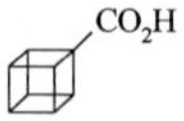	^1H	2, 4	[176]
$C_{10}H_{10}$	azulene	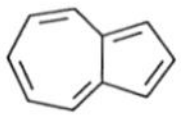	^1H	1	[177]
$C_{10}H_6N_2$	2,3-dicyanobicyclo[2.2.2]-octa-2,5,7-triene	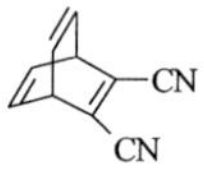	^1H	2	[178]
$C_{10}H_8$	naphthalene	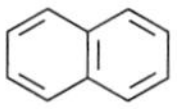	^1H	1	[179]
$C_{10}H_8N_2$	2,2'-bipyridyl	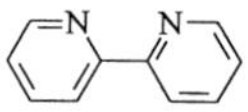	^1H, ^2H, labl	3 1	[180] [181]
$C_{10}H_8O_4$	cubane-1,4-dicarboxylic acid	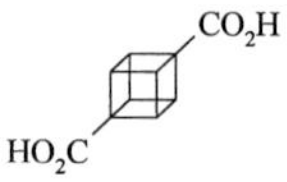	^1H	2, 4	[176]
$C_{10}H_8S$	3-phenylthiophene	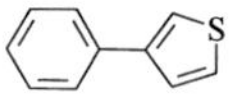	^1H	3	[182]
$C_{10}H_9Br$	1-bromonaphthalene	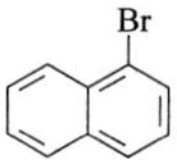	^1H-2D, MQ	2	[109]
$C_{12}H_6BrClF_2$	4'-bromo-4-chloro-2,6-difluorobiphenyl	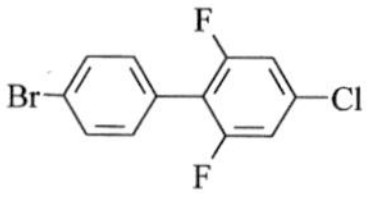	^1H	3	[183, 184]
$C_{12}H_6Cl_2S_2$	2,7-dichlorothianthrene	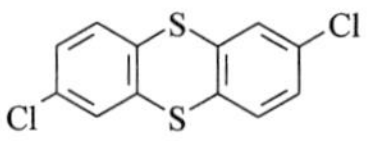	^1H	3	[185]
$C_{12}H_6Cl_4$	3,3',5,5'-tetrachlorobiphenyl	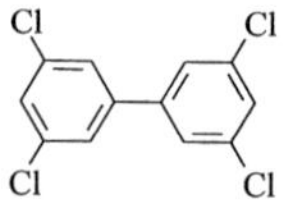	^1H	3	[163]

Table 1. (continued)

Formula	Name	Structure	Techniques[a]	Notes[b]	Refs.
$C_{12}H_7Br_2NS$	3,7-dibromo-10H-phenothiazene	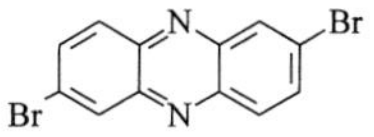	^1H	3	[186]
$C_{12}H_7Cl_2NS$	3,7-dichloro-10H-phenothiazene	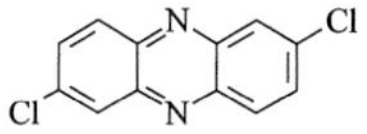	^1H	3	[187]
$C_{12}H_8Cl_2$	4,4′-dichlorobiphenyl	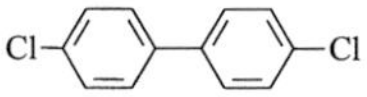	^1H	3	[184]
$C_{12}H_8N_2$	phenazine	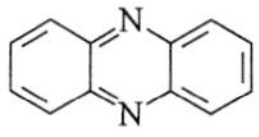	^1H	2	[188]
$C_{12}H_8O$	dibenzofuran	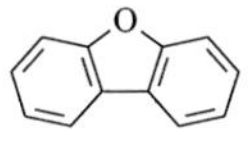	^1H	1a	[189]
$C_{12}H_8S$	dibenzothiophene	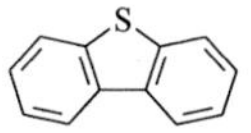	^1H	1a	[189]
$C_{12}H_8O_2$	dibenzo-*p*-dioxin	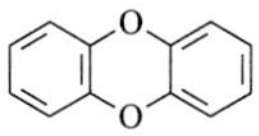	^1H	3	[190]
$C_{12}H_8S_2$	thianthrene	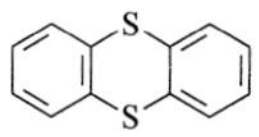	^1H	3	[185] [191]
$C_{13}H_9N$	4-cyanobiphenyl	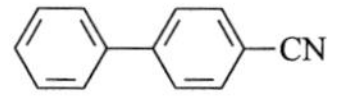	$^1H, ^2H$, labl	3	[192]
$C_{29}H_{50}O_2$	α-tocopherol	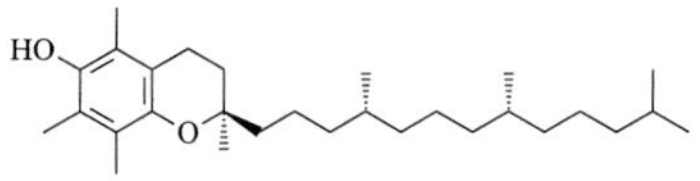	^2H, labl	2, 4	[193]
C_{60}	fullerene		^{13}C	2	[194]

Inorganic and Organometallic Compounds

P, Se, As, Sb

Formula	Name	Structure	Techniques[a]	Notes[b]	Refs.
$C_3H_9O_4P$	trimethyl phosphate	$O=P(OCH_3)_3$	$^1H, ^{31}P$	3	[20]
HF_2PS	difluorophosphine sulfide	$S=PF_2H$	$^1H, ^{19}F, ^{31}P$, 2D	1	[195]
$C_4F_{12}P_4$	tetrakis(trifluoromethyl)-cyclotetraphosphane	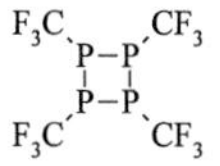	$^3P, ^{19}F$	2	[196]
C_5H_5P	phosphabenzene	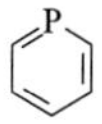	$^1H, ^{31}P$	1	[197]
$C_6H_{18}N_3P$	tris(dimethylamino)phosphine	$P(NMe_2)_3$	$^{15}N, ^{31}P$	2	[198]
$C_6H_4ClO_2P$	*o*-phenylene phosphorochloridite	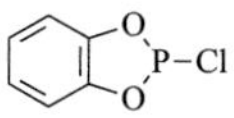	$^1H, ^{31}P$	2	[199]
$C_6H_4ClO_3P$	*o*-phenylene phosphorochloridate	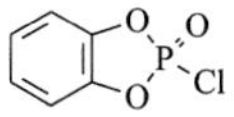	$^1H, ^{31}P$	2	[199]
$C_6H_5Cl_2O_2P$	phenyl dichlorophosphate		$^1H, ^{31}P$	2	[200]
$C_{21}H_{15}O_3P$	tribenzoyl phosphine	$P(COC_6H_5)_3$	$^{13}C, ^{31}P$, labl	2	[201]

Table 1. (continued)

Formula	Name	Structure	Techniques[a]	Notes[b]	Refs.
$C_8H_{20}P_4$	tetraethylcyclotetraphosphane	CH_3CH_2, CH_2CH_3 P–P / P–P CH_3CH_2 CH_2CH_3	^{31}P	2	[196]
C_5H_5As	arsabenzene	⬡–As	^1H	1	[197]
C_6H_7As	4-methylarsabenzene	Me–⬡–As	^1H	1	[202]
C_6H_7Sb	4-methylstibabenzene	Me–⬡–Sb	^1H	1	[202]
Se, Te					
HF_2PSe	difluorophosphine selenide	$Se=PF_2H$	^1H, ^{19}F	1	[203]
C_2H_6Se	dimethylselenide	$Se(CH_3)_2$	^1H, C-sat, Se-sat	1	[204]
C_3H_9PSe	trimethylphosphine selenide	$Se=P(CH_3)_3$	^{13}C, ^{31}P	1	[205]
$C_2H_2N_2Se$	1,2,5-selenadiazole	(N–Se–N ring)	^1H, C-sat, Se-sat	1	[206]
C_4H_4Se	selenophene	(Se ring)	^1H, C-sat, Se-sat	1	[207, 208]
$C_6H_4Se_4$	tetraselenafulvalene	(Se Se / Se Se)	^1H, Se-sat	1	[209]
C_6H_5ClSe	phenyl selenyl chloride	⬡–SeCl	^1H, C-sat	1	[18]
$C_6H_4N_2Se$	benzoselenadiazole	(benzo-N-Se-N)	^1H, C-sat, Se-sat; ^1H, 2D	1	[210] [211]
C_6H_5BrSe	phenyl selenyl bromide	⬡–SeBr	^1H, Se-sat	2	[19]
C_5H_4SeO	selenophene-2-carbaldehyde	(Se ring)–CHO	^1H, Se-sat	1	[212]
$C_8H_6Se_2$	2,2′-biselenophene	(Se–Se bi-ring)	^1H, Se-sat	3	[213]
C_7H_7ClSe	methyl(4-chlorophenyl)-selenide	Cl–⬡–SeCH$_3$	^1H, Se-sat, 2D, echo; ^2H, labl	1	[214]
C_2H_6Te	dimethyl telluride	$Te(CH_3)_2$	^1H, C-sat, Te-sat	1	[204]
C_4H_4Te	tellurophene	(Te ring)	^1H, C-sat, Te-sat	1	[215, 216]
Si, Sn					
C_2H_4Si[e]	silyl acetylene	$HC≡CSiH_3$	^1H, ^{13}C	1	[217]
C_6H_8Si	phenylsilane	⬡–SiH$_3$	^1H	3	[218]
$C_2H_6SnCl_2$	dichlorodimethylstannane	$Cl_2Sn(CH_3)_2$	^1H, ^{13}C, C-sat, Sn-sat	1	[219]
$C_1H_3Cl_3Sn$	trichloromethylstannane	Cl_3SnCH_3	^1H, ^{13}C, C-sat, Sn-sat	1	[219]
C_3H_9SnCl	chlorotrimethylstannane	$ClSn(CH_3)_3$	^1H, ^{13}C, C-sat, Sn-sat	1	[219]
$C_6H_5Cl_3Sn$	trichlorophenylstannane	⬡–SnCl$_3$	^1H, Sn-sat	1	[220]

Table 1. (continued)

Formula	Name	Structure	Techniques[a]	Notes[b]	Refs.
Transition Metals					
CH_3HgX	methylmercuric halide	X-HgCH$_3$	^{1}H, ^{13}C	1	[221]
CH_3HgNO_3	methylmercuric nitrate	O$_3$NOHgCH$_3$	^{1}H, ^{13}C	1	[222]
C_2H_6Hg	dimethylmercury	Hg(CH$_3$)$_2$	^{1}H, Hg-sat; ^{13}C, Hg-sat	1	[223, 224]
$C_2H_3HgF_3$	methyl-trifluoromethyl-mercury	F$_3$CHgCH$_3$	^{1}H, ^{13}C, ^{199}Hg	1	[225]
C_6H_5ClHg	phenylmercuric chloride		^{1}H, Hg-sat	2	[226]
$C_7H_4FeO_3$	cyclobutadieneiron tricarbonyl		^{1}H, C-sat, ^{2}H, labl	1	[227, 228]
$C_8H_5MnO_3$	(η^5-cyclopentadienyl) manganese tricarbonyl		^{1}H		[229]
$C_7H_4MnO_3P$	phosphacymantrene		^{1}H	2	[230]
$C_8H_6MnO_3P$	3-methyl-phosphacymantrene		^{1}H	2	[231]
$C_8H_5ReO_3$	(η^5-cyclopentadienyl)-rhenium tricarbonyl		^{1}H	2	[229]
$C_{10}H_4Fe_2O_6$	(μ-butatriene)diiron hexacarbonyl		^{1}H	2	[232]
$C_7H_9Cl_2NPt^{II}$	dichloro(ethylene)-pyridine platinum(II)		^{1}H, ^{2}H	1	[233]
$C_9H_6O_3Cr^{IV}$	(π-benzene)chromium tricarbonyl		^{1}H, C-sat, labl	2	[234, 235]
$C_8H_6O_3W$	(η^5-cyclopentadienyl)-hydrido tungsten tricarbonyl		^{1}H, C-sat, W-sat	1	[236]
$C_8H_6O_3Mo$	(η^5-cyclopentadienyl)-hydrido molybdenum tricarbonyl		^{1}H, C-sat	1	[236]
$C_{10}H_7O_3CrBF_4$	(π-tropylium)-chromium tricarbonyl tetrafluoroborate		^{13}C, labl	2	[235]

[a] ^nX = NMR nucleus; X-sat = satellite spectra; 2D = 2-dimensional technique; MQ = multiple quantum NMR spectroscopy; echo = spin-echo techniques; labl = isotopic labelling.
[b] 1 = r_α structure; 1a = partial r_α structure; 2 = vibrational corrections not applied; 3 = emphasis on conformational features, usually without vibrational corrections; 4 = lyotropic solvent employed.

Table 2 (a). Representative examples of thermotropic liquid crystals used as solvents for organic reactions.

Liquid crystal[a]	Abbreviation	Liquid crystal[a]	Abbreviation
Saturated systems		**Substituted biphenyls**	
H(CH$_2$)$_n$–⟨⟩–⟨⟩–CN	CCH-n[b]	C$_5$H$_{11}$–⟨⟩–⟨⟩–⟨⟩–(CH$_2$)$_3$H	S-1544[c]
CCH-3 + CCH-5 + CCH-7 (eutectic) 2:1 CCH-2 + CCH-4	CCH-357 CCH-24[e]	C$_5$H$_{11}$–⟨⟩–⟨⟩–C$_2$H$_5$	S-1409[c]
H(CH$_2$)$_n$–⟨⟩–⟨⟩–O(CH$_2$)$_m$H	CCH-nOm[b]	NC–⟨⟩–⟨⟩–O(CH$_2$)$_n$H	nOCB[b]
H(CH$_2$)$_n$–⟨⟩–O–CO–⟨⟩–(CH$_2$)$_m$H	OS-nm[b]	NC–⟨⟩–⟨⟩–(CH$_2$)$_n$H	nCB[b]
F$_{21}$C$_{10}$–C$_{10}$H$_{21}$	F10H10[b]	NC–⟨⟩–⟨⟩–⟨⟩–(CH$_2$)$_n$H	nCT[b]
H(CH$_2$)$_n$–⟨⟩–⟨⟩–COOH	CACH-n[b,c]	5CB + 7CB + 8OCB + 5CT (eutectic)	E7[b]
H(CH$_2$)$_{17}$–CO–OC$_4$H$_7$	BS[b]	**Other unsaturated systems**	
Substituted benzenes		H(H$_2$C)$_n$O–⟨⟩–CH=N–⟨⟩–C$_4$H$_9$	MBBA n=1[e] EBBA n=2
C$_3$H$_7$–⟨⟩–⟨⟩–O(CH$_2$)$_n$H	ZLI-1476 n=2 ZLI-1477 n=4	H$_3$CO–⟨⟩–N=N(O)–⟨⟩–C$_4$H$_9$	Phase IV[e]
H(CH$_2$)$_n$–⟨⟩–⟨⟩–CN	PCH-n		
C$_3$H$_7$–⟨⟩–⟨⟩–C$_2$H$_5$	S-1484[d]	H$_{11}$C$_5$COO–⟨⟩–CO–O–⟨⟩–OC$_4$H$_9$	CBPB[e]
H(CH$_2$)$_n$–⟨⟩–CO–O–⟨⟩–O(CH$_2$)$_m$H	nOm		
H(CH$_2$)$_n$–⟨⟩–CO–O–⟨⟩–(CH$_2$)$_m$H	nHm	H(CH$_2$)$_n$O–⟨⟩–CO–O–⟨⟩–C$_5$H$_{11}$	Phase 1052[e]
C$_4$H$_9$OC$_6$H$_4$O$_2$C–⟨⟩–CO$_2$C$_6$H$_4$OC$_4$H$_9$	BPCD[c]		
C$_4$H$_9$OC$_6$H$_4$CO$_2$–⟨⟩–O$_2$CC$_6$H$_4$OC$_4$H$_9$	CBPP[c]		

[a] General structure only. Abbreviations are those recommended by supplier or generally accepted for use in publications (in some cases, other abbreviations are known).
[b] Liquid crystals exhibit low temperature ($\approx 20-80$ °C) smectic and nematic phases.
[c] Liquid crystals exhibit high temperature smectic and nematic phases.
[d] Clearing point -70 °C.
[e] Nematic phase only; no smectic phase.

Table 2 (b). Representative cholesteric liquid crystals used as solvents for organic chemical reactions.

Liquid crystal[a]	R =	Abbreviation
Cholesteryl derivatives	$H_{17}C_8C=CHC_8H_{16}OCO_2-$	COC[b]
	p-Cl–C_6H_4–CO_2-	CBCl[c]
	$C_8H_{17}CO_2-$	CN[b]
	Cl	CCl[d]
	$H_{17}C_8CH=CHC_7H_{14}CO_2-$	CO[c]
	$C_{11}H_{23}CO_2-$	CL[b]
	3,5-Cl_2-$C_6H_4CO_2-$	CDCB[c]
	$C_6H_4(CH_2)_2CO_2-$	CHC[c]
35:65 CCl/CN		CCl/CN[c]
59.5:15.6:24.9: CO + CN + CBCl		CM[e]
60/26/14: CO + CN + 5α-cholestanyl-3β-yl acetate		CE[c]
Cholestanyl derivative	CH_3CO_2-	CHA
	$H_{17}C_8CO_2-$	CHN
	p-H_3C-$C_6H_4CO_2-$	CnT[c]
	C_6H_5-	CnB[c]
	CH_3CH_2-	CnP[c]
	$PhCH=CHCO_2-$	CHNN[c]
Cholestadienyl derivatives	o-$CH_3(C_6H_4)-$	ACD[f]
	m-$CH_3(C_6H_4)-$	
	p-$CH_3(C_6H_4)-$	
	Ph-; biphenyl-	
	p-CH_3O-C_6H_4-	

[a] General structure only. Abbreviations are those recommended by supplier or generally accepted for use in publications (in some cases, other abbreviations are known).
[b] Liquid crystals exhibit smectic and cholesteric phases.
[c] Liquid crystals exhibit cholesteric phases.
[d] Monotropic cholesteric
[e] Compensated nematic.
[f] Form cholesteric phases between 100–230 °C depending on substituent.

Table 3. Unimolecular chemical reactions of solutes in liquid crystalline solvents [a].

Substrate	Reaction	Solvent [b]	Phases [c]	Method [d]	Refs. [e]
Photophysical					
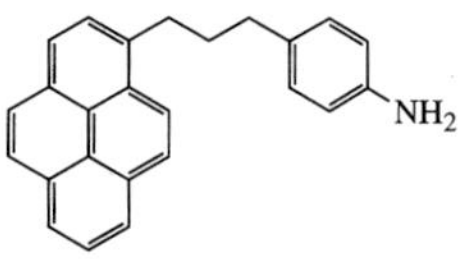 [pyrene-propyl-phenyl-NH$_2$ structure]	singlet excited state dynamics, intramolecular exciplex quenching	CM	Ch, I	fluorescence	[383] [440]
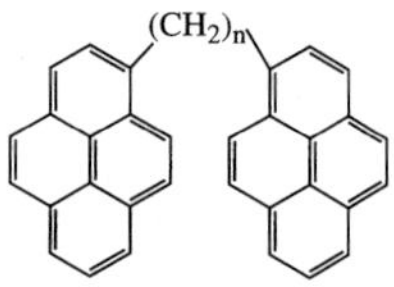 [bis-pyrene with (CH$_2$)$_m$ linker] n = 3,5,6,7,9-12, 13, 22	singlet excited state dynamics, intramolecular excimer quenching	CM CCH-4 CN/CCl	Ch, I CrSmB, N Ch	fluorescence SPC OM CD	[383] [384] [385] [440]
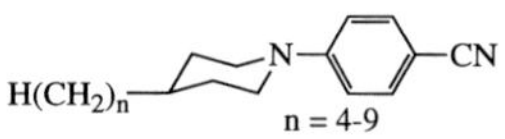 H(CH$_2$)$_n$–N(pyrrolidine)–C$_6$H$_4$–CN, n = 4-9	TICT fluorescence	neat (mixtures)	N, I	fluorescence SPC	[386]
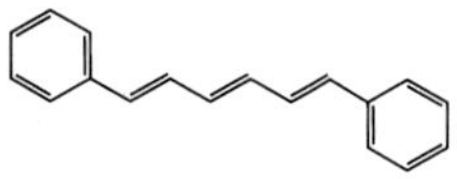 [diphenylhexatriene structure]	singlet excited state dynamics	CCH-357	N	fluorescence CPF	[387]
tetraphenylporphyrin (H$_2$TPP)	triplet excited state dynamics	E7 CCH-357	Cr, N, I, Cr, Sm, N, I	EPR	[388]
porphoryin–cyclohexylene–quinone	intramolecular electron transfer	E7 CCH-357	Cr, N, I, Cr, Sm, N, I	EPR	[389] [390]
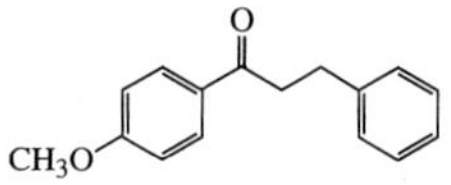 [CH$_3$O–C$_6$H$_4$–CO–CH$_2$–CH$_2$–C$_6$H$_5$]	intramolecular β-phenyl triplet quenching	CCH-4 CCH-3 CCH-2 CCH-24 OS-35 OS-53	CrSmB, N, I N, I N, I N, I SmB, N, I SmB, N, I	NLFP ^{2}H NMR	[354]*
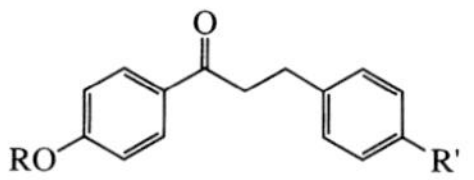 [RO–C$_6$H$_4$–CO–CH$_2$–CH$_2$–C$_6$H$_4$–R'] R = CH$_3$; C$_2$H$_5$; C$_3$H$_7$ R' = H; CH$_3$; (CH$_3$)$_2$CH; n-C$_6$H$_{11}$; c-C$_6$H$_{11}$	intramolecular β-phenyl triplet quenching	CCH-4 CCH-24 CCH-502 CCH-303 CCH-501	CrSmB, N, I N, I SmB, N, I Sm, N, I I	NLFP ^{2}H NMR	[361] [362] [351]* [355]*
Photochemical					
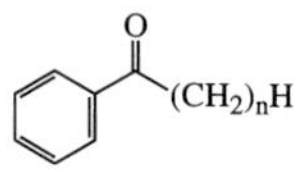 [C$_6$H$_5$–CO–(CH$_2$)$_n$H] n = 4, 10, 17, 19, 21	Norrish/Yang type II	BS	Cr, SmB, I	product study NLFP	[368]
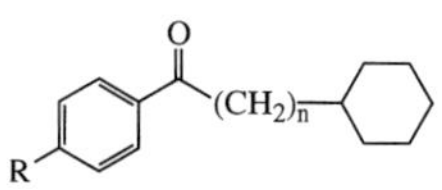 [R–C$_6$H$_4$–CO–(CH$_2$)$_n$–C$_6$H$_{11}$] R = Et; n = 1 R = H; n = 3 and 4	Norrish/Yang type II	CCH-4	Cr, CrSmB N, I	product study ^{2}H NMR	[369] [370]*

Table 3. (continued)

Substrate	Reaction	Solvent[b]	Phases[c]	Method[d]	Refs.[e]
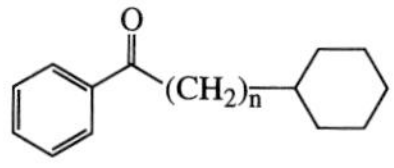 (structure, n = 3)	Norrish/Yang type II	CCH-4 CCH-24	CrSmB N, I	product study ^{2}H NMR DSC, OM	[352]*
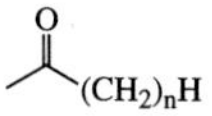 (structure, n = 11-31) 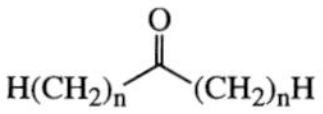 (structure, n = 11-31)	Norrish/Yang type II	BS F10H10	Cr, SmB, I Sm$_I$, Sm$_{II}$, I	product study ^{2}H NMR DSC, Om	[371] [372] [360]*
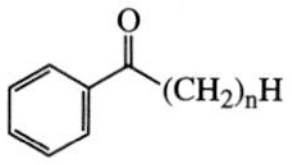 (structure, n = 3, 4, 5)	Norrish/Yang type II	CCH-4 EB; EC	Cr, SmB, I N, I	product study ^{2}H NMR	[353]*
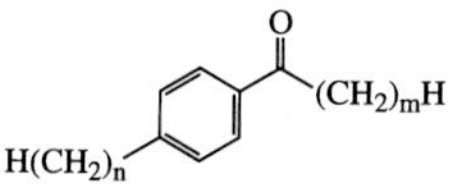 (structure, n = 0-15; m = 6-21)	Norrish/Yang type II	BS	Cr, SmB, I	product study DSC FTIR	[373]
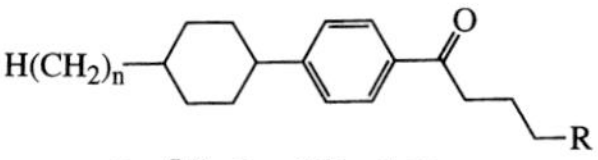 (structure, n = 5,7 R = CH$_3$; C$_3$H$_7$)	Norrish/Yang type II	neat or mixtures	Cr, SmB, N, I	product study ^{2}H NMR DSC, OM X-ray	[374] [375]
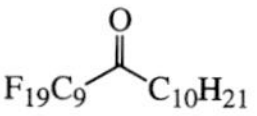 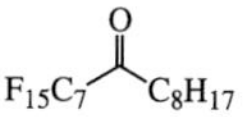 (F$_{19}$C$_9$-CO-C$_{10}$H$_{21}$) (F$_{15}$C$_7$-CO-C$_8$H$_{17}$)	Norrish/Yang type II	F10H10	Sm$_I$, Sm$_{II}$, I	product study ^{2}H NMR DSC OM	[376]* [360]*
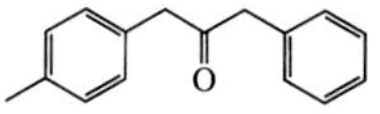 (structure)	Norrish I (α-cleavage)	CCl, CN CCH-4 BS	Ch Sm, N, I Sm, I	product study OM	[391]
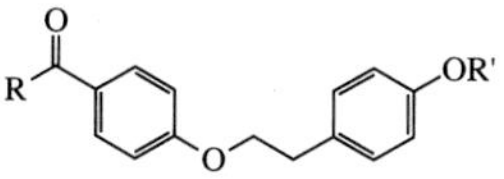 (structure) R = CH$_3$ R' = H, CH$_3$ R = C$_4$H$_9$ R' = H, CH$_3$	intramolecular phenolic quenching of triplet ketone R=C$_4$H$_9$ Norrish/Yang	CCH-4 CCH-2 CCH-24	CrSmB, N Sm N, I	NLFP product study OM	[441]
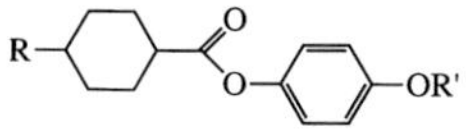 (structure)	photo-Fries	nOm CCl/CN	Cr, Sm, N, I Ch, I	product study UV-vis	[346]
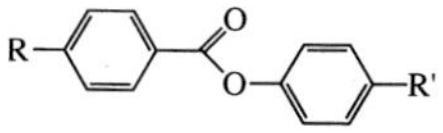 (structure)	photo-Fries	neat/ mixtures CCl/CN	Sm, N, Ch, I	product study UV-vis	[392]
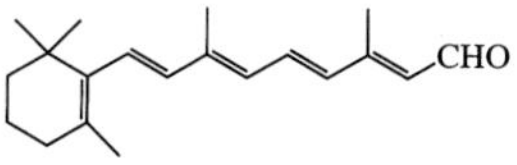 (retinal structure)	E-Z-isomerization	BS	Sm, I	UV-vis	[377]

Table 3. (continued)

Substrate	Reaction	Solvent[b]	Phases[c]	Method[d]	Refs.[e]
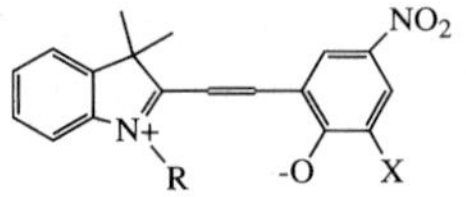 merocyanine	photochromic isomerization to indolinospiropyran	BS 5 CB nOm CN/CO/CP	Sm, I N, I N, I Ch, I	UV-vis	[378] [379]
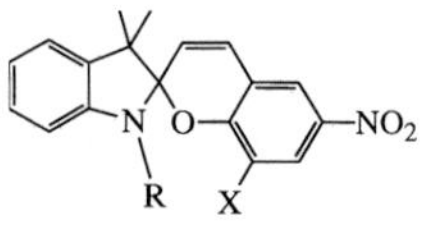 indolinospiropyran	photochromic isomerization to merocyanine	BS 5 CB nOm CN/CO/CP	Sm, I N, I N, I Ch, I	UV-vis	[378] [379]
R = H, -C$_4$H$_9$, -C$_8$H$_{17}$, -C$_{12}$H$_{25}$	E-Z-isomerization	BS	Sm, I, mI	UV-vis	[380]
	E-Z-isomerization	CCl/CN BS 5 CB, 6 CB 7 CB, 3–6 OCB	Ch, Sm, I Sm, I N, I	UV-vis DSC OM	[381] [382]
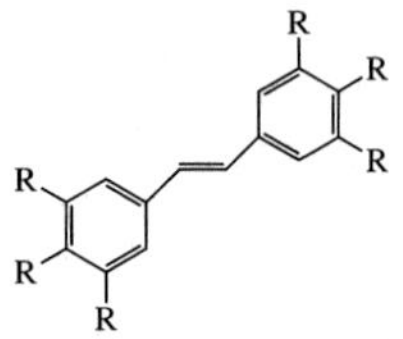 R = H, -OC$_6$H$_{13}$, -OC$_{12}$H$_{25}$	E-Z-isomerization	BS	Sm, N, I	product study DSC OM	[358]

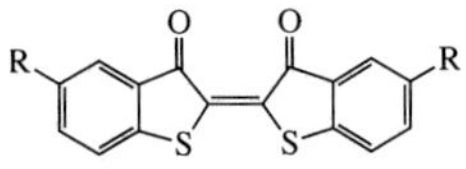

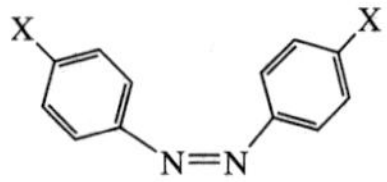

Thermal

Substrate	Reaction	Solvent[b]	Phases[c]	Method[d]	Refs.[e]
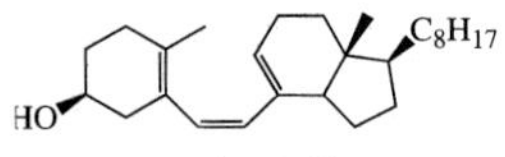 pre-vitamin D	1,7-sigmatropic rearrangement to Vitamin D	cholesteric CO/CN/CCl	Ch, I	UV-vis	[403]
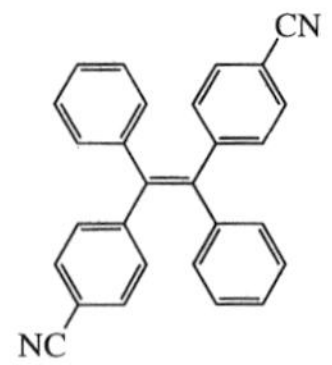	E-Z-isomerization	CB CBCl CB/CT/CBCl	I Ch Ch, I	product study	[396]
R = R' = CH$_3$ and C$_{17}$H$_{35}$ and Ph R = CH$_3$; R' = C$_{17}$H$_{35}$	E-Z-isomerization	BS CE 7 CB/8 OCB/ 5 CT mixture	Sm, I Ch, I Sm, N N	UV-vis UV-vis OM	[397] [398]

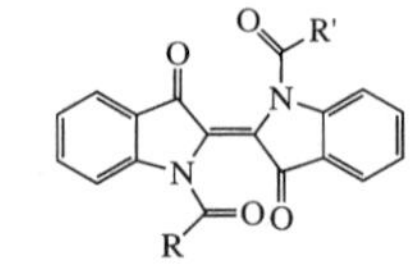

Table 3. (continued)

Substrate	Reaction	Solvent[b]	Phases[c]	Method[d]	Refs.[e]
(binaphthyl structure)	atropisomerization	MBBA	Cr, N, I	product study	[399]
		CCH-357	N, I	UV-vis	[400]
		phase IV	N, I	CD	[401]
		E7	Cr, N, I		[402]
		7 CB/8 OCB/	Sm, N, I		
		5 CT	Ch, I		
		CHN/CHA			
(terpene alcohol structure)	rearrangement	CACH-9	Cr, SmB, N, I	product study	[404]
		CACH-5	Cr, N, I	XRD	
		mixtures			
(cyclooctyl NMe$_3^+$OH$^-$ structure)	Hoffman elimination asymmetric induction	ACD	Ch	product study	[426]

[a] Studies where the substrate is added as a solute to a liquid crystalline solvent. Included are a few studies where the substrate is also the mesogen or is mesogenic.
[b] Structures of solvents are listed in Tables 2a and b.
[c] Cr = crystal; Sm = smectic; CrSmB = crystal smectic B; N = nematic; Ch = cholesteric; I = isotropic
[d] fluorescence = steady state fluorescence; SPC = time-resolved single photon counting; CPF = circularly polarized fluorescence; UV-vis = UV-visible absorption spectrophotometry; DSC = differential scanning calorimetry; OM = optical microscopy; XRD = X-ray diffraction; EPR = electron paramagnetic resonance; NMR = nuclear magnetic resonance.
[e] References with an asterisk. (*) indicate studies where phase separation has been noted and related to the reactivity.

Table 4. Bimolecular chemical reactions of solutes in liquid crystalline solvents [a].

Substrate 1	Substrate 2	Reaction	Solvent [b]	Phases [c]	Method [d]	Refs. [e]
Photophysical						
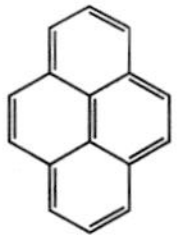 pyrene	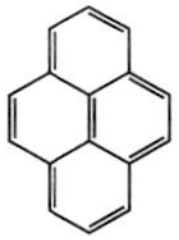 pyrene	singlet excited state dynamics excimer quenching	CM	Ch, I	SPC fluorescence	[409]
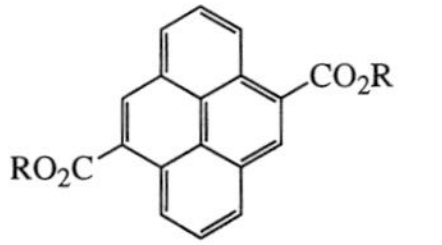 RO$_2$C-pyrene-CO$_2$R	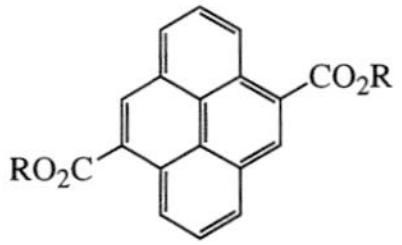 RO$_2$C-pyrene-CO$_2$R	singlet excited state dynamics excimer quenching	PCH-5	N, I	fluorescence CPF	[410]
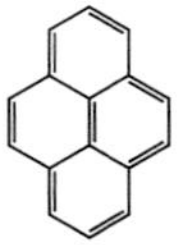 cholestane-Me$_2$N	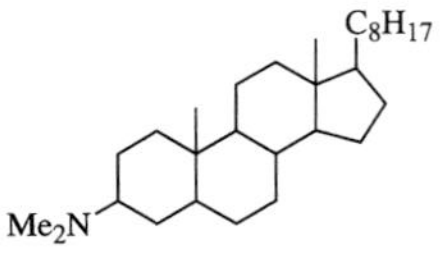 pyrene-C$_8$H$_{17}$	singlet excited state dynamics exciplex quenching	CM	Ch, I	SPC fluorescence	[411]
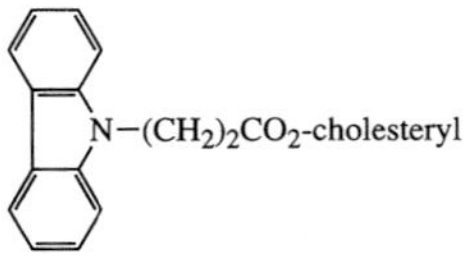 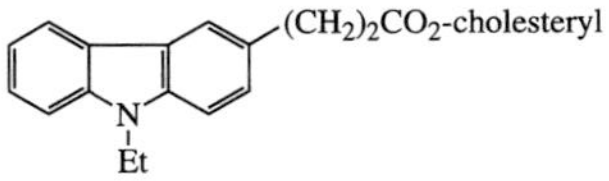 carbazole-N-(CH$_2$)$_2$CO$_2$-cholesteryl	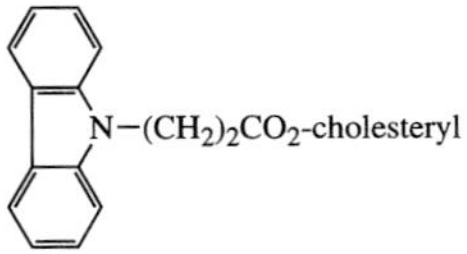 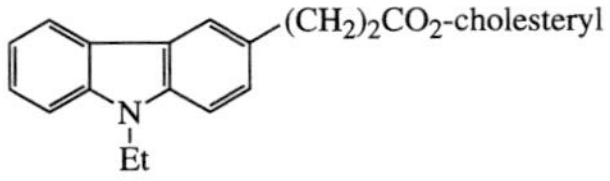 carbazole-N-(CH$_2$)$_2$CO$_2$-cholesteryl (N-Et)	singlet excited state dynamics excimer/exciplex quenching	cholesteryl derivatives	Ch, I	fluorescence CPF SPC CD	[412] [413] [414]
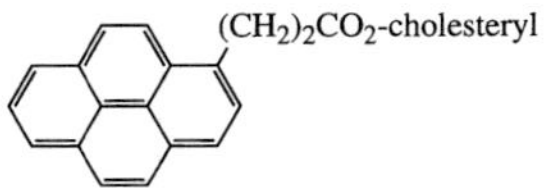 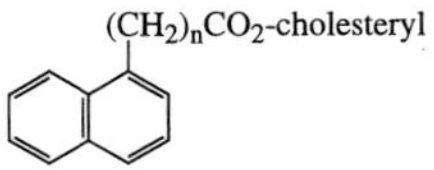 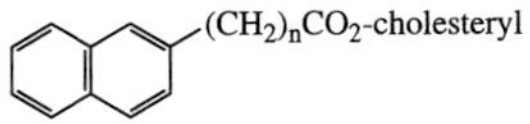 pyrene-(CH$_2$)$_2$CO$_2$-cholesteryl; naphthyl-(CH$_2$)$_n$CO$_2$-cholesteryl; naphthyl-(CH$_2$)$_n$CO$_2$-cholesteryl	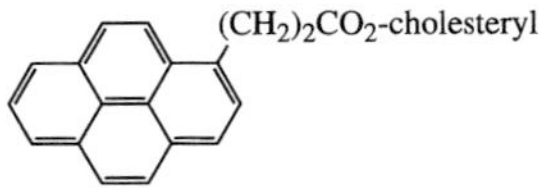 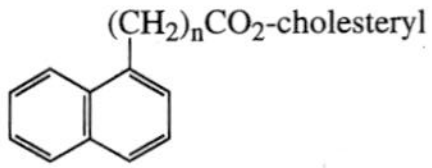 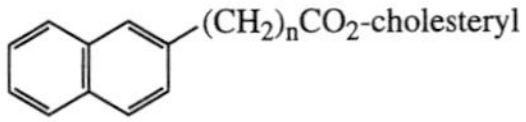 pyrene-(CH$_2$)$_2$CO$_2$-cholesteryl; naphthyl-(CH$_2$)$_n$CO$_2$-cholesteryl; naphthyl-(CH$_2$)$_n$CO$_2$-cholesteryl	singlet excited state dynamics	CCl/CN cholesteryl derivatives	Ch, I	fluorescence CPF SPC	[415] [416] [417]

4.4 Liquid Crystals in Gas Chromatographic Applications

Photochemical

Structure	Reaction	Solvent	Phases	Study	Ref.
acenaphthylene	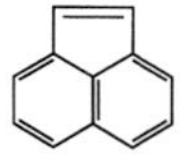 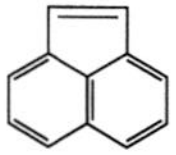 2+2 dimerization	BS / CHN/CHA	Sm, I / Ch, I	product study / quantum yield	[399] [400]
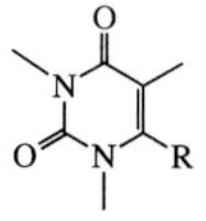 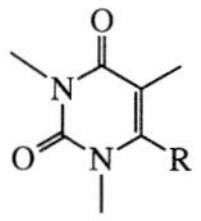 dimethyluracil (R = H, CH$_3$)	2+2 dimerization	BS / COC / CL	Sm, I / Cr, Sm, Ch, I / Cr, SM, I	product study	[405] [406]
cinnamate–Ocholestanyl	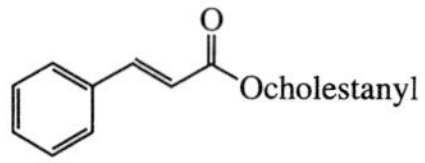 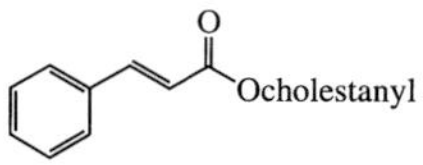 2+2 dimerization / E-Z-isomerization	itself	Cr, Ch, I	product study	[407]
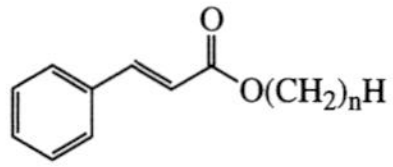 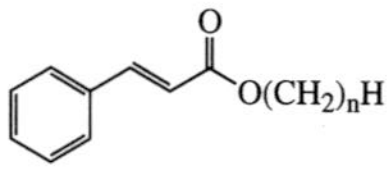 cinnamate–O(CH$_2$)$_n$H, n = 14, 16, 18	2+2 dimerization / E-Z-isomerization	BS	Cr, SmB, I	product study	[349]
ethyl cinnamate	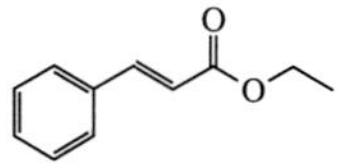 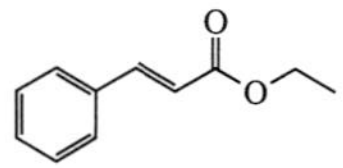 2+2 dimerization / E-Z-isomerization	CCH-4 / CCH-2 / BS / CCl/CN	CrSmB, N, I / SmB, N, I / SmB, I / Ch, I	product study / quantum yields / ^{2}HNMR / DSC / OM	[356]* [346]
anthracene diester	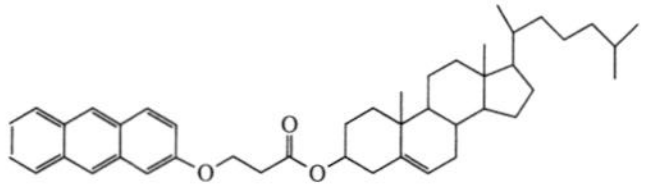 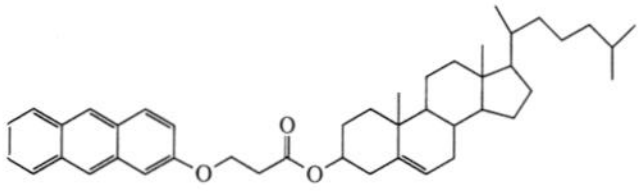 4+4 dimerization (anthracene units)	itself	Ch, I	product study / OM	[408]
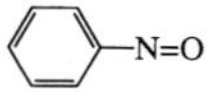 (CH$_3$)$_3$C• (from photolysis of (CH$_3$)$_3$C–N=O)	radical trapping spin probe	50, 60, 70 / 80 CB / 5 CB / 8 CB	Cr, Sm, N, I	EPR / DSC / OM	[427, 428] [429, 430] [418, 431] [421]
(CH$_3$)$_3$C–N=O / C$_6$H$_5$–N=O					

Table 4. (continued)

Substrate 1	Substrate 2	Reaction	Solvent[b]	Phases[c]	Method[d]	Refs.[e]
Thermal						
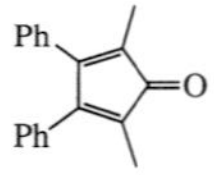 2,3-diphenyl-tetramethyl-cyclopentadienone (Ph, Ph substituted)	 cyclopentadiene; R = -(CH$_2$)$_3$-; -(CH$_2$)$_4$-; 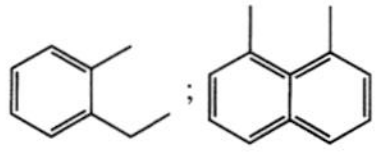 methylbenzyl	Diels–Alder cycloaddition	CN, S1409	Ch, I; SmB	product study OM	[432]
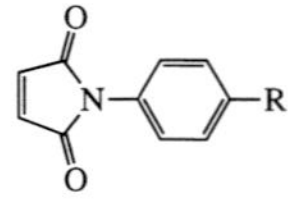 N-aryl maleimide (R = H, Ph, Ph-Ph)	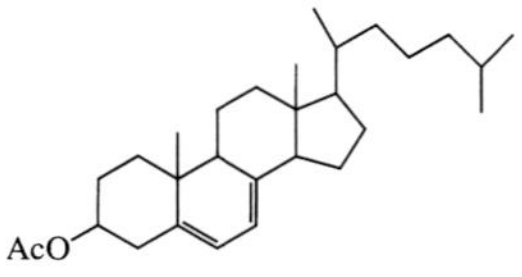 cholesteryl acetate derivative	Diels–Alder and ene-reactions	CnB, ChCB, CnT, S1409, S1544, CDCB	Ch, I; Ch; I; SmB, Ch; Ch	product study HPLC OM, DSC product study	[433] [364]*
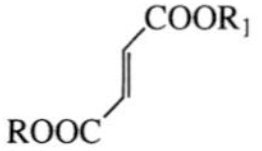 COOR$_1$ / ROOC alkene	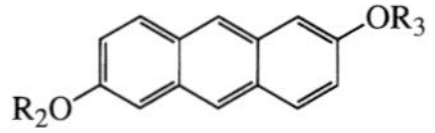 anthracene diether (OR$_3$, R$_2$O)	Diels–Alder cycloaddition	CHC, CBPB, BPCD	I; I; Sm	OM	[434] [435]*
	Cholesteryl fumarates			Ch, I	product study DSC, OM	[436]
 Me$_2$N—C$_6$H$_4$—SO$_3$R (R = CH$_3$, CH$_2$CH=CH$_2$)	 Me$_2$N—C$_6$H$_4$—SO$_3$R (R = CH$_3$, CH$_2$CH=CH$_2$)	sulphonate ester rearrangement (quaternization reaction)	S1484, S1409, S1544, CBPB, OS-35, OS-44, OS-53	I; SmB, N, I; Sm; N; Cr, SmB, N, I	product study UV-vis kinetics XRD product study UV-vis kinetics XRD, LC-LD ^{2}HNMR	[422] [423] [424] [363] [425] [359] [357]*
tetraphenylporphyrin (iodosobenzene)	 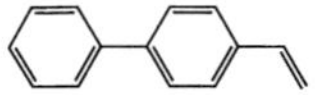 Ph—CH=CH—C$_6$H$_4$—Ph (stilbene)	epoxidation	6CB	N, I	product study EPR	[442]
acetylene	acetylene	polymerization	ZLI-1476, ZLI-1477	N	product study	[443]

4.4 Liquid Crystals in Gas Chromatographic Applications

	Reaction	Solvent	Phase	Method	Ref.
	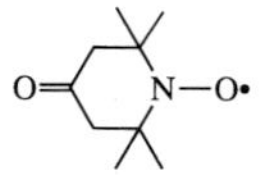 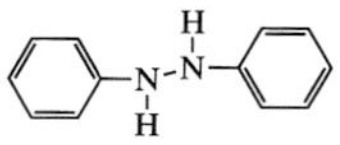 H-abstraction/reduction	MBBA, MBBA/EBBA	N, I	EPR	[418, 419] [420, 421]
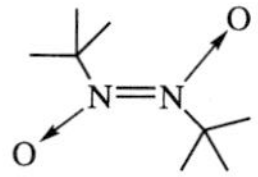	dimer/monomer equilibrium	MBBA	N	UV-vis	[394] [395]
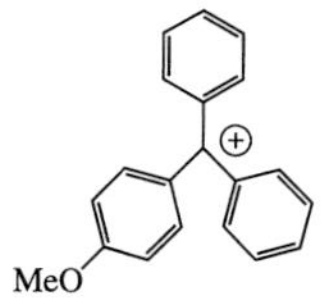	dimerization/equilibrium	Phase 1052	N, I	UV-vis, ^{19}FNMR	[393]

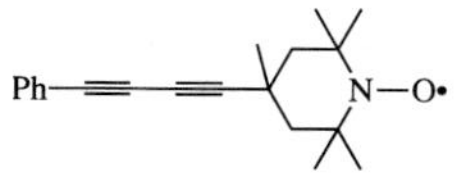

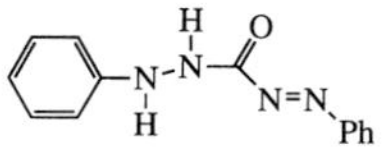

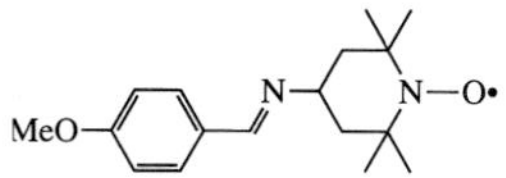

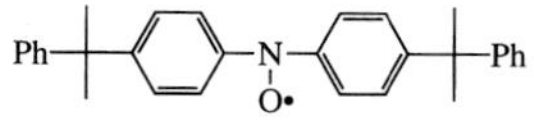

[a] Studies are those where the substrates are added as a solute to a liquid crystalline solvent. Included are a few studied where the substrate is also the mesogen or is mesogenic.
[b] Structures of solvents are listed in Tables 2 a and b.
[c] Cr = crystalline; Sm = smectic; CrSmB = smectic crystal B; N = nematic; Ch = cholesteric; I = isotropic.
[d] fluorescence = steady state fluorescence; SPC = time-resolved single photon counting; CPF = circularly polarized fluorescence; UV-vis = UV-visible absorption spectrophotometry; DSC = differential scanning calorimetry; OM = optical microscopy; XRD = X-ray diffraction; EPR = electron paramagnetic resonance; NMR = nulcear magnetic resonance.
[e] References with an asterisk (*) indicate studies where phase separation has been noted and related to reactivity.

Table 5. Examples of isomeric solutes separated using monomeric and polymeric liquid crystalline stationary phases.

Class of compound and isomers separated	Refs.
Polycyclic aromatic hydrocarbons (PAHs)	
complex mixtures (general separations) (mixtures including: anthracenes, fluoranthenes, phenanthrenes, pyrenes, perylenes, chrysenes, benzopyrenes etc.)	[461, 465–467, 482–495]
coal tar separation	[496, 497]
carbon black mixture	[498]
methylphenanthrenes	[487, 490, 497, 499]
ethylphenanthrenes	[499]
trimethylphenanthrenes	[499]
methylbenz(a)anthracenes	[465, 500]
methylbenzo(a)pyrenes	[465]
methylchrysenes	[483, 485, 487, 490, 500]
Aromatic halogen derivatives	
miscellaneous mixtures	[491, 492, 501]
polychlorinated benzenes	[484, 492]
polychlorinated biphenyls	[465, 484, 502, 503]
chlorophenols	[482, 484]
2,3,7,8-class congeners; polychlorodibenzodioxins	[484, 502–504]
2,3,7,8-class congeners; polychlorodibenzofurans	[504]
Polycyclic aromatic heterocycles	
sulfur heterocycles, coal tar	[485, 487, 490, 505]
methylcarbazoles	[497]
aminophenanthrenes	[487]
methyldibenzothiophenes; alkyldibenzothiophenes; dialkyldibenzothiophenes	[485, 506, 507]
hydroxydibenzothiophenes	[487]
methylnaphtho[1,2-b]thiophenes	[490]
methylnaphtho[2,1-b]thiophenes	[490]
mono- and di-methoxyquinones: (naphthaquinones; anthraquinones; phenanthraquinones; benzanthraquinones; dibenz(a,h)anthraquinones; benz(a,j)anthraquinones)	[508]
Naphthalenes	
miscellaneous mixtures	[509]
mono- and di-methylnaphthalenes	[510]
1- and 2-naphthols	[496, 509]
1- and 2-naphthyl alkanaotes and benzoates	[496]
alkoxynaphthalenes	[491]
Alkylbenzenes	
isomeric substituted benzenes; including mono- and di-substituted benzenes and xylenes	[461, 464, 468–470, 492, 493, 511]
trimethylbenzenes	[512]
isomeric alkylbenzenes	[513–515]
diastereomeric arylalkanes	[516]
alkylcyclohexylbenzenes	[517]
C_8 alkylbenzenes	[518]
C_{14}–C_{17} alkylbenzenes	[519]
C_{10}–C_{13} phenylalkanes	[519]

Table 5. (continued)

Class of compound and isomers separated	Refs.
Alkanes	
miscellaneous mixtures of alkanes	[463, 487]
C_8 linear and branched isomers	[520]
diastereomeric C_8-C_{10} isomers	[521]
C_9-C_{11} n-alkanes	[522]
cyclohexanes, decalins	[470]
n-nonadecanes	[523]
Alkenes	
C_8 linear and branched alkenes	[463]
C_8-C_9 alkenes and alkadienes	[524]
C_9-C_{11} n-alkenes	[522]
$C_{10}-C_{13}$ n-alkenes	[513–515, 525]
cis-,trans-tridecanes and tetradecanes	[526]
cis-,trans-$C_{10}-C_{14}$ n-alkenes	[518]
C_{15} isomers	[525]
cis-,trans-$C_{15}-C_{17}$ n-alkenes	[527]
$C_{17}-C_{18}$ n-alkenes	[528]
n-nonadecenes	[523]
Alkynes	
C_5-C_{13} branched alkynes	[529]
C_6-C_{14} n-alkynes	[530]
n-decyne, n-undecyne, n-tridecyne, n-tetradecyne	[530]
Acids and esters	
toluic acids	[509]
alkyl benzoates (methyl-hexyl)	[509]
alkyl hydroxy benzoates	[482]
cis- and trans-$C_{10}-C_{13}$ acetates (alkene and diene) (insect pheromones)	[531]
9,11-tetradecadienyl acetate isomers (insect pheromones)	[491]
cis- and trans-fatty acid methyl esters	[465]

Table 6. Examples of studies reporting the thermodynamic properties of non-mesogenic solutes in liquid crystalline stationary phases.

Solute (number of isomers)	Liquid crystal stationary phase; (phase types)	Refs.
C_5–C_9 n-alkanes (5) C_5–C_9 branched alkanes (5) cycloalkanes	p-n-hexyloxyphenyl p-methoxybenzoate; (N, I) p-n-pentyloxyphenyl p-ethoxyazoxybenzene; (N, I)	[532]
n-alkanes (5) hexanes (5) heptanes (9) nonanes (2) benzene, toluene	4-cyano-4'-pentylbiphenyl; (N, I) 4-cyano-4'-hexylbiphenyl; (N, I) 4-cyano-4'-heptylbiphenyl; (N, I) 4-cyano-4'-octylbiphenyl; (SmA, N, I)	[533]
chloroalkanes (7) and alkenes (2) C_6H_{12}, CH_3–C_6H_{11}, C_6H_{10} CH_3–C_5H_9, substituted benzenes (9)	cholesteryl myristate; (SmA, Ch, I)	[479]
chloroalkanes (7) and alkenes (2) C_6H_{12}, CH_3–C_6H_{11}, C_6H_{10} CH_3–C_5H_9 substituted benzenes (9)	4-4'-bis(heptyloxy)azoxybenzene; (Sm, N, I)	[478]
Et_4C Me_4Sn, Et_4Sn, nPr_4Sn, nBu_4Sn n-C_8H_{18}–n-$C_{16}H_{34}$ benzene, hexa-2,4-diyne biphenyl, terphenyl	4-cyano-4'-octylbiphenyl; (SmA, N, I)	[534]
decane, Et_4Sn, benzene	azoxy-4,4'-bis(undecyl-α-methyl cinnamate); (SmC, SmA, I) cholesteryl nononaote (Sm, Ch, I)	[535]
phenanthrene, anthracene	N,N'-bis(p-methoxybenzylidene)-α,α'-bi-p-toluidine; (N, I) N,N'-bis(p-methoxybenzylidene)-4,4'-diamino diphenylmethane; (I)	[536]
C_5–C_8 isomers (16) C_5–C_7 alkenes (8) hexadienes (5); benzene	4-cyano-4'-hexylbiphenyl; (N, I)	[537]
n-C_5–n-C_9 (5); C_6–C_8 isomers (10) C_5–C_7 alkenes (6); hexadienes (6) benzene, toluene, xylenes (3)	4-cyano-4'-heptylbiphenyl; (N, I)	[538]
decane, benzene, CCl_4	4,4'-bis(n-hexyloxy)azoxybenzene; (N, I) 4,4'-bis(n-heptyloxy)azoxybenzene; (SmC, N, I)	[539, 540]
Et_4C Me_4Sn, Et_4Sn, nPr_4Sn, nBu_4Sn n-C_8H_{18}–n-$C_{14}H_{30}$	N-(p-methoxybenzylidene)-p'-n-butylaniline; (N, I) 4-cyano-4'-pentylbiphenyl; (N, I) p-azoxyanisole; (N, I) 4,4'-bis(hexyloxy)azoxybenzene; (N, I)	[541]
xylenes (3) halotoluenes (9) substituted benzenes (4) C_8H_{18}–$C_{12}H_{26}$ (4) Cycloalkanes (3)	4,4'-bis(heptyloxy)azoxybenzene; (SmC, N, I) n-amyl 4-(4-dodecyloxybenzylidenamino)cinnamate; (SmB, SmC, SmA, N, I) 5-n-octyl-2-(4-n-hexyloxyphenyl)pyrimidine; (SmC, SmA, N, I)	[451, 542]
m-, p-xylenes	azoxybenzene-p-cyano-p'-alkylcarbonates (Sm, N, Ch, I, depending on alkyl group)	[543]
methylbenz(a)anthracenes (12)	N,N'-bis(p-alkoxyphenylbenzylidene)-α,α'-bi-p-toluidine; (N, I)	[544]

Table 6. (continued)

Solute (number of isomers)	Liquid crystal stationary phase; (phase types)	Refs.
m-,p-xylene; m-,p-dichlorobenene substituted toluenes (4)	4,4′-bis(n-hexyloxy)azoxybenzene; (N, I)	[453]
C_5-C_{10} n-alkanes (5) C_5-C_{10} branched alkanes (9) benzene, toluene, xylenes (3)	N-(p-methoxybenzylidene)-p'-n-butylaniline; MBBA (N, I) 4-cyano-4′-pentylbiphenyl; (N, I)	[545]
C_7-C_9 alkanes (3); benzene, toluene, ethylbenzene, xylenes 4-methylcyclohexyl alkanoates (10)	cholesteryl palmitate (Sm, Ch, I)	[546]
xylenes (3), halotoluenes (6), dihalobenzenes (3), substituted benzenes (4), C_8-C_{12} (4)	4-pentylacetophenone-(O-4-n-alkoxybenzyloximes)	[547]
C_7-C_9 alkanes (3); benzene, toluene, ethylbenzene, xylenes 4-methylcyclohexyl alkanoates (10)	(4-ethoxyphenylazo)phenyl heptanoate (N, I)	[548]
o-,m-,p-xylenes; n-C_6-C_9; C_6H_{12}	4,4′-bis(n-heptyloxy)azoxybenzene; (N, I) p-(p-(ethoxyphenylazo)phenyl undecylate; (N)	[549]
n-heptane	4,4′-bis(n-hexyloxy)azoxybenzene; (N, I)	[550]
octane(ene), nonane(ene), xylenes, chlorobenzene	cholesteryl myristate; (Sm, Ch, I) cholesteryl chloride; (Ch, I)	[551]
C_7 alkanes (12) C_7-C_{11} alkenes (11) haloakanes (6) xylenes, substituted benzenes (11)	p-azoxyanisole; (N, I) 4,4′-bis(n-hexyloxy)azoxybenzene; (N, I)	[480]
divinylbenzene, ethylvinylbenzene	4,4′-bis(n-hexyloxy)azoxybenzene; (N, I)	[552]

4.5 References

[1] H. Kelker, R. Hatz (Eds) *Handbook of Liquid Crystals* Verlag Chemie, Weinheim, **1980**.
[2] C. L. Khetrapal, R. G. Weiss, A. C. Kunwar, in *Liquid crystals – Applications and uses, Vol. 2* (Ed.: B. Bahadur) World Scientific, Singapore, **1991**, p. 225.
[3] L. Pohl, in *Topics in Physical Chemistry. Volume 3: Liquid Crystals* (Ed.: H. Stegemeyer) Springer, New York, **1994**, p. 173.
[4] C. L. Khetrapal, A. C. Kunwar, *Adv. Liq. Cryst.* **1983**, *6*, 173–242.
[5] G. J. Kruger, *Phys. Rep.* **1982**, *82*, 229–269.
[6] R. Y. Dong, *Isr. J. Chem.* **1983**, *23*, 370–379.
[7] L. Monnerie, F. Laupetre, C. Noel, *Liq. Cryst.* **1988**, *3*, 1–29.
[8] J. W. Emsley, *Isr. J. Chem.* **1988**, *28*, 297–308.
[9] *The Molecular Dynamics of Liquid Crystals.* NATO ASI Series C: Mathematical and Physical Sciences – Vol. 431 Kluwer Academic Publishers, Dordrecht, **1994**.
[10] C. Schumann, in *Handbook of Liquid Crystals,* (Eds.: H. Kelker, R. Hatz) Verlag Chemie, Weinheim, **1980**, p. 426.
[11] J. W. Emsley, J. C. Lindon, *NMR Spectroscopy using Liquid Crystalline Solvents,* Pergamon Press, New York, **1975**.
[12] C. L. Khetrapal, A. C. Kunwar, *Isr. J. Chem.* **1983**, *23*, 299–303.
[13] P. Diehl, in *Nuclear Magnetic Resonance of Liquid Crystals,* J. W. Emsley, Ed. D. Reidel, **1985**, p. 147.
[14] P. Diehl, J. Jokisaari, in *NMR in Stereochemical Analysis,* (Eds.: Y. Takeuchi, A. P. Marchand) VCH, Deerfield Beach, **1986**, p. 41.
[15] C. L. Khetrapal, E. D. Becker, *Magn. Res. Rev.* **1991**, *16*, 35–56.

[16] C. L. Khetrapal, K. V. Ramanathan, in *Nuclear Magnetic Resonance (Specialist Periodical Reports).* Volume 23 (Ed.: G. A. Webb) Royal Society of Chemistry, London, **1994**, p. 439.
[17] P. Diehl, in *Accurate Molecular Structures,* (Eds.: A. Domenicano, I. Hargittai) Oxford University Press, London, **1992**, p. 299.
[18] N. Suryaprakash, A. C. Kunwar, C. L. Khetrapal, *J. Mag. Reson.* **1983**, *43*, 502–506.
[19] N. Suryaprakash, A. C. Kunwar, C. L. Khetrapal, *J. Organomet. Chem.* **1983**, *252*, 301–304.
[20] C. L. Khetrapal, G. Govil, H. J. C. Yeh, *J. Mol. Struct.* **1984**, *116*, 303–311.
[21] J. R. Tolman, J. H. Prestegard, *J. Mag. Reson.* **1994**, *106A*, 203–208.
[22] S. Sykora, J. Vogt, H. Bösinger, P. Diehl, *J. Mag. Reson.* **1979**, *36*, 53–60.
[23] C. A. De Lange, J. G. Snijders, E. E. Burnell, in *Nuclear Magnetic Resonance of Liquid Crystals,* (Ed.: J. W. Emsley) D. Reidel, Boston, **1985**, p. 181.
[24] J. Lounila, *Mol. Phys.* **1986**, *58*, 897–918.
[25] P. Diehl, *Mag. Reson. Chem.* **1986**, *24*, 667–669.
[26] A. J. van der Est, E. E. Burnell, J. Lounila, *J. Chem. Soc., Faraday Trans. 2* **1988**, *84*, 1095–1108.
[27] R. Wasser, M. Kellerhals, P. Diehl, *Mag. Reson. Chem.* **1989**, *27*, 335–339.
[28] J. Jokisaari, P. Diehl, J. Amrein, E. Ijas, *J. Mag. Reson.* **1983**, *52*, 193–201.
[29] T. Väänänen, J. Jokisaari, A. Kääriäinen, J. Louinila, *J. Mol. Struct.* **1983**, *102*, 175–182.
[30] P. Streckeisen, O. Münster, R. Seydoux, P. Diehl, *J. Mol. Struct.* **1993**, *301*, 137–141.
[31] A. Saupe, G. Englert, *Phys. Rev. Lett.* **1963**, *11*, 462–464.
[32] J. Courtieu, J. P. Bayle, B. M. Fung, *Prog. NMR Spectrosc.* **1994**, *26*, 141–169.
[33] G. P. Drobny, *Ann. Rev. Phys. Chem.* **1985**, *36*, 451–489.
[34] D. N. Shykind, J. Baum, S.-B. Liu, A. Pines, A. N. Garroway, *J. Mag. Reson.* **1988**, *76*, 149–154.
[35] J. C. T. Rendell, *J. Mag. Reson.* **1995**, *A112*, 1–6.
[36] A. Kumar, *Curr. Sci.* **1988**, *57*, 109–120.
[37] R. Christy Rani Grace, A. Kumar, *J. Mag. Reson.* **1992**, *97*, 184–191.
[38] J.-M. Ouvrard, B. N. Ouvrard, J. Courtieu, C. L. Mayne, D. M. Grant, *J. Mag. Reson.* **1991**, *93*, 225–241.
[39] D. Nanz, M. Ernst, M. Hong, M. A. Ziegeweid, K. Schmidt-Rohr, A. Pines, *J. Mag. Reson.* **1995**, *A113*, 169–176.
[40] R. Christy Rani Grace, N. Suryaprakash, A. Kumar, C. L. Khetrapal, *J. Mag. Reson,* **1994**, *107A*, 79–82.
[41] P. Lesot, J. M. Ouvrard, B. N. Ouvrard, J. Courtieu, *J. Mag. Reson.* **1994**, *107A*, 141–150.

[42] A. M. Thayer, M. Luzar, A. Pines, *J. Phys. Chem.* **1987**, *91*, 2149–2197.
[43] A. M. Thayer, M. Luzar, A. Pines, *J. Mag. Reson.* **1987**, *72*, 567–573.
[44] P. Diehl, T. Bjorholm, J. Amrein, *Org. Mag. Reson.* **1981**, *15*, 104–105.
[45] J. Lounila, P. Diehl, Y. Hiltunen, J. Jokisaari, *J. Mag. Reson.* **1985**, *61*, 272–283.
[46] J. Jokisaari, Y. Hiltunen, *J. Mag. Reson.* **1986**, *67*, 319–327.
[47] J. Jokisaari, Y. Hiltunen, *J. Mag. Reson.* **1984**, *60*, 307–319.
[48] R. Wasser, P. Diehl, *Mag. Reson. Chem.* **1987**, *25*, 766–770.
[49] H. Bösiger, P. Diehl, *J. Mag. Reson.* **1980**, *38*, 361–364.
[50] C. L. Khetrapal, K. V. Ramanathan, *Chem. Phys. Lett.* **1985**, *118*, 589–590.
[51] Y. Hiltunen, J. Jokisaari, J. Lounila, A. Pulkkinen, *Chem. Phys. Lett.* **1988**, *148*, 353–357.
[52] W. E. Bechtold, J. H. Goldstein, *Chem. Phys. Lett.* **1981**, *83*, 401–404.
[53] J. Hansen, J. P. Jacobsen, *J. Mag. Reson.* **1980**, *41*, 381–388.
[54] C. L. Khetrapal, A. C. Kunwar, *J. Biochem. Biophys. Methods* **1981**, *4*, 185–190.
[55] E. Haloui, C. B. Nasr, D. Canet, *J. Mol. Struct.* **1982**, *95*, 231–235.
[56] E. E. Babcock, J. H. Goldstein, *J. Mag. Reson.* **1983**, *52*, 247–253.
[57] R. Ambrosetti, D. Catalano, C. A. Veracini, G. Chidichimo, M. Longeri, *J. Mol. Struct.* **1981**, *73*, 203–208.
[58] G. Dombi, P. Diehl, *Org. Mag. Reson.* **1984**, *22*, 187–189.
[59] A. L. Esteban, M. P. Galache, *Mol. Phys.* **1992**, *77*, 17–27.
[60] J. Jokisaari, J. Kuonanoja, A.-M. Häkkinen, *Org. Mag. Reson.* **1980**, *14*, 290–292.
[61] G. Fronza, R. Mondelli, S. Bradamante, *J. Mag. Reson.* **1979**, *36*, 343–350.
[62] J. Jokisaari, K. Räisänen, T. Väänänen, *J. Mag. Reson.* **1981**, *42*, 396–402.
[63] S. Cradock, C. Purves, D. W. H. Rankin, *J. Mol. Struct.* **1990**, *220*, 193–204.
[64] P. Diehl, T. Bjorholm, H. Bösiger, *J. Mag. Reson.* **1981**, *42*, 390–395.
[65] S. Cradock, P. B. Liescheski, D. W. H. Rankin, H. E. Robertson, *J. Am. Chem. Soc.* **1988**, *110*, 2758–2763.
[66] N. Suryaprakash, R. Ugolini, P. Diehl, *Mag. Reson. Chem.* **1991**, *29*, 1024–1027.
[67] P. Diehl, J. Jokisaari, S. Müller, T. Väänänen, *Org. Mag. Reson.* **1983**, *21*, 143–145.
[68] J. Jokisaari, Y. Hiltunen, H. Heikkilä, *Mag. Reson. Chem.* **1985**, *23*, 725–727.
[69] P. B. Liescheski, D. W. H. Rankin, *J. Mol. Struct.* **1989**, *196*, 1–19.

[70] J. Jokisaari, Y. Hiltunen, T. Väänänen, *Mol. Phys.* **1984**, *51*, 779–791.
[71] P. B. Liescheski, D. W. H. Rankin, *J. Mol. Struct.* **1988**, *178*, 227–241.
[72] K. H. M. E. Hashem, F. A. E. A. Ali, *Egypt. J. Chem.* **1987**, *30*, 97–108.
[73] P. Diehl, J. Amrein, *J. Mol. Struct.* **1982**, *82*, 245–249.
[74] J. Courtieu, Y. Gounelle, J. Jullien, G. Balavoine, M. Fan, *Org. Mag. Reson.* **1980**, *14*, 55–57.
[75] T. Turki, E. Haloui, *Mag. Reson. Chem.* **1990**, *28*, 571–572.
[76] C. L. Khetrapal, C. L. Chatterjee, H. J. Yeh, *Chem. Phys. Lett.* **1980**, *76*, 553–556.
[77] A. L. Esteban, M. P. Galache, *Mol. Phys.* **1991**, *74*, 283–292.
[78] A. L. Esteban, M. P. Galache, E. Diez, S. J. Fabian, J. Guilleme, *J. Mol. Struct.* **1986**, *142*, 375–378.
[79] R. Poupko, Z. Luz, H. Zimmermann, *J. Am. Chem. Soc.* **1982**, *104*, 5307–5314.
[80] N. Suryaprakash, A. C. Kunwar, C. L. Khetrapal, *Curr. Sci.* **1980**, *49*, 1–3.
[81] D. S. Williams, T. C. Wong, *J. Mol. Struct.* **1983**, *101*, 297–303.
[82] T. Bjorholm, J. P. Jacobsen, C. T. Pedersen, *J. Mol. Struct.* **1981**, *75*, 327–332.
[83] J. P. Jacobsen, J. Hansen, C. T. Pedersen, T. Pedersen, *J. Chem. Soc., Perkin Trans. II* **1979**, 1521–1524.
[84] D. Catalano, C. A. Veracini, P. L. Barili, M. Longeri, *J. Chem. Soc., Perkin Trans. II* **1983**, 171–174.
[85] N. Suryaprakash, S. Arumugam, A. C. Kunwar, C. L. Khetrapal, *J. Mag. Reson.* **1982**, *47*, 507–509.
[86] J. Courtieu, Y. Gounelle, J. Jullien, P. Gonord, S. K. Kan et al. *J. Mol. Struct.* **1981**, *72*, 257–260.
[87] N. Suryaprakash, A. C. Kunwar, C. L. Khetrapal, *Chem. Phys. Lett.* **1984**, *107*, 333–336.
[88] W. Bechtold, J. H. Goldstein, *J. Am. Chem. Soc.* **1981**, *103*, 4989–4991.
[89] D. S. Stephenson, G. Binsch, *Mol. Phys.* **1981**, *43*, 697–710.
[90] C. R. Counsell, J. W. Emsley, G. R. Luckhurst, *Mol. Phys.* **1981**, *43*, 711–715.
[91] D. Catalano, L. Marcolini, C. A. Veracini, *Chem. Phys. Lett.* **1982**, *88*, 342–346.
[92] H. Fujiwara, J. Hirai, Y. Sasaki, *J. Chem. Soc., Perkin Trans. II* **1988**, 791–794.
[93] G. Drobny, D. P. Weitekamp, A. Pines, *Chem. Phys.* **1986**, *108*, 179–185.
[94] M. Gochin, H. Zimmermann, A. Pines, *Chem. Phys. Lett.* **1987**, *137*, 51–56.
[95] M. E. Rosen, S. P. Rucker, C. Schmidt, A. Pines, *J. Phys. Chem.* **1993**, *97*, 3858–3866.
[96] G. Dombi, J. Amrein, P. Diehl, *Org. Mag. Reson.* **1980**, *13*, 224–227.
[97] R. Blom, D. W. H. Rankin, H. E. Robertson, *Acta Chem. Scand.* **1990**, *44*, 978–983.
[98] P. Diehl, G. Dombi, H. Boesiger, *Org. Mag. Reson.* **1980**, *14*, 280–284.
[99] Y. Hiltunen, J. Jokisaari, *J. Mag. Reson.* **1987**, *75*, 213–221.
[100] R. Blom, S. Cradock, S. L. Davidson, D. W. H. Rankin, *J. Mol. Struct.* **1991**, *245*, 369–377.
[101] P. Diehl, H. Bösiger, *J. Mag. Reson.* **1979**, *35*, 367–372.
[102] A. J. Blake, R. Blom, S. Cradock, S. L. Davidson, D. W. H. Rankin, *J. Mol. Struct.* **1990**, *240*, 95–110.
[103] P. Diehl, F. Moia, H. Bösiger, *J. Mag. Reson.* **1980**, *41*, 336–340.
[104] S. A. Spearman, J. H. Goldstein, *J. Mag. Reson.* **1978**, *30*, 59–63.
[105] S. Cradock, P. B. Liescheski, D. W. H. Rankin, *J. Mag. Reson.* **1991**, *91*, 316–325.
[106] D. G. Anderson, S. Cradock, P. B. Liescheski, D. W. H. Rankin, *J. Mol. Struct.* **1990**, *216*, 181–190.
[107] M. R. Lakshminarayana, A. Kumar, C. L. Khetrapal, *J. Mag. Reson.* **1984**, *60*, 184–189.
[108] N. Suryaprakash, A. C. Kunwar, C. L. Khetrapal, *J. Mol. Struct.* **1983**, *101*, 121–125.
[109] L. D. Field, G. K. Pierens, K. J. Cross, M. L. Terry, *J. Mag. Reson.* **1992**, *97*, 451–465.
[110] T. Väänänen, J. Jokisaari, J. Louinila, *J. Mag. Reson.* **1982**, *49*, 73–83.
[111] T. Väänänen, J. Jokisaari, J. Lounila, *J. Mag. Reson.* **1983**, *54*, 436–446.
[112] A. Pulkkinen, J. Jokisaari, T. Väänänen, *J. Mol. Struct.* **1986**, *144*, 359–369.
[113] J. P. Jacobsen, C. H. Hansen, *J. Mol. Struct.* **1984**, *118*, 345–350.
[114] R. C. Long, Jr., E. E. Babcock, J. H. Goldstein, *Mol. Phys.* **1980**, *39*, 1423–1436.
[115] J. Jokisaari, T. Väänänen, *J. Mag. Reson.* **1981**, *45*, 379–385.
[116] R. C. Long, Jr., J. H. Goldstein, *J. Mag. Reson.* **1979**, *34*, 361–375.
[117] P. Diehl, J. Jokisaari, *J. Mol. Struct.* **1979**, *53*, 55–60.
[118] S. Cradock, J. M. Muir, D. W. H. Rankin, *J. Mol. Struct.* **1990**, *220*, 205–215.
[119] L. Lunazzi, P. Bellomo, C. A. Veracini, A. Amanzi, *J. Chem. Soc., Perkin Trans. II* **1979**, 559–563.
[120] R. C. Long, Jr., E. E. Babcock, J. H. Goldstein, *J. Phys. Chem.* **1981**, *85*, 1165–1170.
[121] J. Jokisaari, J. Kuonanoja, A. Pulkkinen, T. Väänänen, *Mol. Phys.* **1981**, *44*, 197–208.
[122] J. Lounila, T. Väänänen, *Mol. Phys.* **1983**, *49*, 859–863.
[123] J. Jokisaari, T. Väänänen, J. Lounila, *Mol. Phys.* **1982**, *45*, 141–148.
[124] R. Ugolini, P. Diehl, *J. Mol. Struct.* **1990**, *216*, 325–332.

[125] S. Arumugam, A. C. Kunwar, C. L. Khetrapal, *Mol. Cryst. Liq. Cryst.* **1984**, *109*, 263–270.
[126] P. Diehl, M. Reinhold, *Mol. Phys.* **1978**, *36*, 143–149.
[127] P. Diehl, H. Bösinger, H. Zimmermann, *J. Mag. Reson.* **1979**, *33*, 113–126.
[128] T. C. Wong, L. J. Altman, *J. Mag. Reson.* **1980**, *37*, 285–291.
[129] G. Fronza, R. Mondelli, F. Lelj, E. W. Randall, C. A. Veracini, *J. Mag. Reson.* **1980**, *37*, 275–284.
[130] H. Fujiwara, N. Shimizu, T. Takagi, Y. Sasaki, *Bull. Chem. Soc. Jpn.* **1985**, *58*, 2285–2288.
[131] H. Fujiwara, N. Shimizu, T. Takagi, Y. Sasaki, *Mag. Reson. Chem.* **1986**, *24*, 71–74.
[132] T. Bjorholm, J. P. Jacobsen, *J. Mol. Struct.* **1983**, *96*, 315–324.
[133] J. Courtieu, Y. Gounelle, J. Jullien, J. P. Barnier, J. M. Conia, S. K. Kan, *Org. Mag. Reson.* **1982**, *19*, 63–65.
[134] C. L. Khetrapal, A. C. Kunwar, *J. Phys. Chem.* **1982**, *86*, 4815–4817.
[135] Y. Sugino, K. Takahashi, *Bull. Chem. Soc. Jpn.* **1978**, *51*, 2527–2529.
[136] M. Barnes, J. W. Emsley, T. J. Horne, G. M. Warnes, G. Celebre, M. Longeri, *J. Chem. Soc., Perkin Trans. II* **1989**, 1807–1810.
[137] P. Diehl, J. Amrein, C. A. Veracini, *Org. Mag. Reson.* **1982**, *20*, 276–278.
[138] S. Arumugam, A. C. Kunwar, C. L. Khetrapal, *J. Mol. Struct.* **1984**, *117*, 157–159.
[139] N. Suryaprakash, C. L. Khetrapal, *Mag. Reson. Chem.* **1986**, *24*, 247–250.
[140] K. C. Cole, D. F. R. Gilson, *J. Mol. Struct.* **1982**, *82*, 71–75.
[141] W. A. Heeschen, D. W. Alderman, D. M. Grant, *J. Phys. Chem.* **1988**, *92*, 6504–6511.
[142] J. W. Emsley, C. M. Exon, S. A. Slack, A. Giroud, *J. Chem. Soc., Perkin Trans. II* **1978**, 928–931.
[143] P. Diehl, J. Jokisaari, J. Amrein, *Org. Mag. Reson.* **1980**, *13*, 451–454.
[144] T. C. Wong, K. J. Koval, *J. Mol. Struct.* **1978**, *50*, 91–97.
[145] P. Diehl, J. Jokisaari, G. Dombi, *Org. Mag. Reson.* **1979**, *12*, 438–439.
[146] M. Longeri, G. Chidichimo, P. Bucci, *Org. Mag. Reson.* **1984**, *22*, 408–410.
[147] G. Celebre, M. Longeri, J. W. Emsley, *Mol. Phys.* **1988**, *64*, 715–723.
[148] P. Diehl, F. Moia, *Org. Mag. Reson.* **1981**, *15*, 326–328.
[149] J. W. Emsley, M. Longeri, C. A. Veracini, D. Catalano, G. F. Pedulli, *J. Chem. Soc., Perkin Trans. II* **1982**, 1289–1296.
[150] G. Celebre, G. De Luca, M. Longeri, J. W. Emsley, *Mol. Phys.* **1989**, *67*, 239–248.
[151] J. W. Emsley, M. Longeri, A. Liguori, *J. Chem. Soc., Perkin Trans. II* **1980**, 540–544.

[152] M. Kon, H. Kurokawa, H. Takeuchi, S. Konaka, *J. Mol. Struct.* **1992**, *268*, 155–167.
[153] D. Catalano, G. Celebre, M. Longeri, C. A. Veracini, *Gazz. Chim. Ital.* **1985**, *115*, 233–236.
[154] G. Celebre, G. De Luca, M. Longeri, J. W. Emsley, *J. Phys. Chem.* **1992**, *96*, 2466–2470.
[155] A. G. Avent, J. W. Emsley, S. Ng, S. M. Venables, *J. Chem. Soc., Perkin Trans. II* **1984**, 1855–1860.
[156] G. Celebre, M. Longeri, J. W. Emsley, *Liq. Cryst.* **1989**, *6*, 689–700.
[157] D. Gallard, F. Volino, *J. Phys. II* **1991**, *1*, 209–223.
[158] L. Di Bari, M. Perisco, C. A. Veracini, *J. Chem. Phys.* **1992**, *96(6)*, 4782–4791.
[159] D. Catalano, L. Di Bari, C. A. Veracini, *Mol. Cryst. Liq. Cryst.* **1992**, *212*, 155–162.
[160] J. W. Emsley, D. L. Turner, A. Giroud, M. Longeri, *J. Chem. Soc., Faraday Trans. 2* **1985**, *81*, 603–611.
[161] P. Diehl, J. Amrein, H. Bösinger, F. Moia, *Org. Mag. Reson.* **1982**, *18*, 20–21.
[162] N. Suryaprakash, C. L. Khetrapal, P. Diehl, *Mag. Reson. Chem.* **1992**, *30*, 940–943.
[163] R. Berardi, F. Spinozzi, C. Zannoni, *J. Chem. Soc., Faraday Trans. 2*, **1992**, *88(13)*, 1863–1873.
[164] T. C. Wong, K. J. Koval, *Org. Mag. Reson.* **1980**, *13*, 384–386.
[165] N. Suryaprakash, C. L. Khetrapal, R. F. X. Klein, V. Horak, *Mag. Reson. Chem.* **1985**, *23*, 1045–1047.
[166] L. C. Ter Beek, D. S. Zimmerman, E. E. Burnell, *Mol. Phys.* **1991**, *74*, 1027–1035.
[167] J. W. Emsley, M. Longeri, *Mol. Phys.* **1981**, *42*, 315–328.
[168] G. Celebre, M. Longeri, N. Russo, A. G. Avent, J. W. Emsley, V. N. Singleton, *Mol. Phys.* **1988**, *65*, 391–401.
[169] E. K. Foord, J. Cole, M. J. Crawford, J. W. Emsley, G. Celebre, M. Longeri, J. C. Lindon, *Liq. Cryst.* **1995**, *18*, 615–621.
[170] G. Celebre, M. Longeri, J. W. Emsley, *J. Chem. Soc., Faraday Trans. 2* **1988**, *84(8)*, 1041–1052.
[171] J. W. Emsley, S. Hadenfeldt, T. J. Horne, G. Celebre, M. Longeri, *J. Chem. Soc., Perkin Trans. II* **1992**, 419–423.
[172] C. L. Khetrapal, A. C. Kunwar, S. Arumugam, *Org. Mag. Reson.* **1984**, *22*, 707–709.
[173] S. Arumugam, C. L. Khetrapal, *Can. J. Chem.* **1986**, *64*, 714–716.
[174] C. A. Veracini, A. De Munno, G. Chidichimo, M. Longeri, *J. Chem. Soc., Perkin Trans. II* **1979**, 572–575.
[175] C. L. Khetrapal, R. Highet, *Org. Mag. Reson.* **1981**, *16*, 117–118.
[176] A. S. Tracey, *Can. J. Chem.* **1984**, *62*, 101–104.
[177] A. D. Hunter, E. E. Burnell, T. C. Wong, *J. Mol. Struct.* **1983**, *99*, 159–164.

[178] A. Tomishima, Y. Takakura, Y. Yokoyama, K. Saito, K. Takahashi, *Bull. Chem. Soc. Jpn.* **1992**, *65*, 906–908.
[179] W. Bechtold, R. C. Long, Jr., J. H. Goldstein, *J. Mag. Reson.* **1980**, *40*, 361–367.
[180] F. Lelj, N. Russo, G. Chidichimo, *Chem. Phys. Lett.* **1980**, *69*, 530–533.
[181] J. W. Emsley, J. G. Garnett, M. A. Long, L. Lunazzi, G. Spunta, C. A. Veracini, A. Zandanel, *J. Chem. Soc., Perkin Trans. II* **1979**, 853–856.
[182] G. Chidichimo, A. Liguori, M. Longeri, C. A. Veracini, *J. Mag. Reson.* **1983**, *51*, 438–445.
[183] L. D. Field, S. Sternhell, *J. Am. Chem. Soc.* **1981**, *103*, 738–741.
[184] D. Catalano, L. Di Bari, C. A. Veracini, G. N. Shilstone, C. Zannoni, *J. Chem. Phys.* **1991**, *94(5)*, 3928–3935.
[185] G. Fronza, E. Ragg, G. Ronsisvalle, *J. Chem. Soc., Perkin Trans. II* **1982**, 1209–1211.
[186] H. Fujiwara, M. Watanabe, I. Yamanaka, T. Takagi, Y. Sasaki, *Bull. Chem. Soc. Jpn.* **1988**, *61*, 369–373.
[187] H. Fujiwara, M. Watanabe, Y. Sasaki, *Bull. Chem. Soc. Jpn.* **1988**, *61*, 1091–1094.
[188] C. L. Khetrapal, *Org. Mag. Reson.* **1981**, *15*, 238–241.
[189] S. Arumugam, A. C. Kunwar, C. L. Khetrapal, *Org. Mag. Reson.* **1984**, *22*, 547–549.
[190] G. Fronza, E. Ragg, *J. Chem. Soc., Perkin Trans. II* **1982**, 291–293.
[191] H. Fujiwara, A. Kawamura, T. Takagi, Y. Sasaki, *J. Am. Chem. Soc.* **1983**, *105*, 125–126.
[192] J. W. Emsley, T. J. Horne, G. Celebre, G. De Luca, M. Longeri, *J. Chem. Soc., Faraday Trans. 2* **1992**, *88(12)*, 1679–1684.
[193] I. H. Ekiel, L. Hughes, G. W. Burton, J. Jovall, K. U. Ingold, I. C. P. Smith, *Biochemistry* **1988**, *27*, 1432–1440.
[194] C. L. Khetrapal, N. Suryaprakash, *Liq. Cryst.* **1993**, *14*, 1479–1484.
[195] P. D. Blair, S. Cradock, D. W. H. Rankin, *J. Chem. Soc. Dalton Trans.* **1985**, 755–759.
[196] J. P. Albrand, A. Cogne, J. B. Robert, *J. Am. Chem. Soc.* **1978**, *100:9*, 2600–2603.
[197] T. C. Wong, A. J. Ashe, III, *J. Mol. Struct.* **1978**, *48*, 219–225.
[198] A. Cogne, J.-B. Robert, L. Wiesenfeld, *Chem. Phys. Lett.* **1978**, *57*, 627–629.
[199] K. Saito, A. Yoshino, T. Yoshida, K. Takahashi, *Bull. Chem. Soc. Jpn.* **1992**, *65*, 1135–1137.
[200] J. Hansen, J. P. Jacobsen, *Org. Mag. Reson.* **1981**, *15*, 29–32.
[201] A. Cogne, L. Wiesenfeld, J. B. Robert, R. Tyka, *Org. Mag. Reson.* **1980**, *13*, 72–74.
[202] T. C. Wong, M. G. Ferguson, A. J. Ashe III, *J. Mol. Struct.* **1979**, *52*, 231–237.
[203] A. S. F. Boyd, G. S. Laurenson, D. W. H. Rankin, *J. Mol. Struct.* **1981**, *71*, 217–226.
[204] P. Diehl, A. C. Kunwar, H. Bösiger, *J. Organomet. Chem.* **1978**, *145*, 303–306.
[205] A. Cogne, A. Grand, J. Laugnier, J. B. Robert, L. Wiesenfeld, *J. Am. Chem. Soc.* **1980**, *102*, 2238–2242.
[206] G. Chidichimo, F. Lelj, P. L. Barili, C. A. Veracini, *Chem. Phys. Lett.* **1978**, *55*, 519–522.
[207] G. Chidichimo, F. Lelj, M. Longeri, N. Russo, *J. Mag. Reson.* **1980**, *41*, 35–41.
[208] W. Bechtold, B. G. Magruder, J. H. Goldstein, *Org. Mag. Reson.* **1982**, *19*, 173–175.
[209] T. C. Wong, E. M. Engler, *J. Mol. Struct.* **1980**, *67*, 279–284.
[210] C. L. Khetrapal, A. C. Kunwar, *J. Mol. Struct.* **1981**, *74*, 343–345.
[211] M. A. Thomas, K. V. Ramanathan, A. Kumar, *J. Mag. Reson.* **1983**, *55*, 386–396.
[212] P. Bucci, G. Chidichimo, F. Lelj, M. Longeri, N. Russo, *J. Chem. Soc., Perkin Trans. II* **1979**, 109–110.
[213] G. Chidichimo, F. Lelj, M. Longeri, N. Russo, C. A. Veracini, *Chem. Phys. Lett.* **1979**, *67*, 384–387.
[214] A. G. Avent, J. W. Emsley, M. Longeri, S. G. Murray, P. N. Nicholson, *J. Chem. Soc., Perkin Trans. II* **1986**, 487–493.
[215] G. Chidichimo, P. Bucci, F. Lelj, M. Longeri, *Mol. Phys.* **1981**, *43*, 877–886.
[216] P. Diehl, M. Kellerhals, J. Lounila, R. Wasser, Y. Hiltunen, et al. *Mag. Reson. Chem.* **1987**, *25*, 244–247.
[217] C. A. Brookman, S. Cradock, D. W. H. Rankin, N. Robertson, P. Vefghi, *J. Mol. Struct.* **1990**, *216*, 191–200.
[218] C. L. Khetrapal, E. D. Becker, *J. Mag. Reson.* **1981**, *43*, 8–14.
[219] H. Fujiwara, Y. Sasaki, *J. Phys. Chem.* **1987**, *91*, 481–486.
[220] P. Diehl, J. Amrein, *Org. Mag. Reson.* **1982**, *19*, 148–149.
[221] K. Räisänen, J. Kuonanoja, J. Jokisaari, *Mol. Phys.* **1979**, *38*, 1307–1310.
[222] J. Jokisaari, K. Räisänen, J. Kuonanoja, P. Pyykkö, L. Lajunen, *Mol. Phys.* **1980**, *39*, 715–723.
[223] J. Jokisaari, P. Diehl, *Mol. Phys.* **1979**, *37*, 981–983.
[224] J. Jokisaari, P. Diehl, *Org. Mag. Reson.* **1980**, *13*, 359–362.
[225] J. Jokisaari, J. Kuonanoja, *J. Mol. Struct.* **1980**, *68*, 173–182.
[226] C. L. Khetrapal, *Proc. Indian natn. Sci. Acad.* **1989**, *55*, 329–334.
[227] P. Diehl, F. Moia, H. Boesiger, J. Wirz, *J. Mol. Struct.* **1983**, *98*, 297–308.
[228] D. F. R. Gilson, *J. Mol. Struct.* **1985**, *127*, 121–126.
[229] D. F. R. Gilson, P. J. Fitzpatrick, I. S. Butler, *Organometallics* **1984**, *3*, 387–389.

[230] C. L. Khetrapal, A. C. Kunwar, F. Mathey, *J. Organomet. Chem.* **1979**, *181*, 349–353.
[231] N. Suryaprakash, A. C. Sunwar, C. L. Khetrapal, *J. Organomet. Chem.* **1984**, *275*, 53–56.
[232] S. Arumugam, A. C. Kunwar, C. L. Khetrapal, *J. Organomet. Chem.* **1984**, *265*, 73–76.
[233] J. W. Emsley, J. Evans, *J. Chem. Soc. Dalton Trans.* **1978**, 1355–1358.
[234] P. Diehl, A. C. Kunwar, *Org. Mag. Reson.* **1978**, *11*, 47–49.
[235] R. Aydin, H. Günther, J. Runsink, H. Schmickler, H. Seel, *Org. Mag. Reson.* **1980**, *13*, 210–215.
[236] S. Caldarelli, D. Catalano, L. Di Bari, M. Pasquali, C. A. Veracini, P. Leoni, *Gazz. Chim. Ital.* **1990**, *120*, 211–213.
[237] J. Lounila, J. Jokisaari, *Progr. NMR Spectrosc.* **1982**, *15*, 249–290.
[238] C. L. Khetrapal, A. C. Kunwar, *Chem. Phys. Lett.* **1981**, *82*, 170–171.
[239] T. Väänänen, J. Jokisaari, M. Seläntaus, *J. Mag. Reson.* **1987**, *72*, 414–421.
[240] P. Diehl, J. Jokisaari, F. Moia, *J. Mag. Reson.* **1982**, *49*, 498–506.
[241] E. E. Burnell, C. A. de Lange, *Chem. Phys. Lett.* **1987**, *136*, 87–92.
[242] P. Diehl, R. Seydoux, *J. Mag. Reson.* **1994**, *108A*, 215–218.
[243] C. L. Khetrapal, S. Raghothama, N. Suryaprakash, *J. Mag. Reson.* **1987**, *71*, 140–143.
[244] S. Raghothama, *J. Mag. Reson.* **1984**, *57*, 294–296.
[245] Y. Hiltunen, *Mol. Phys.* **1987**, *62*, 1187–1194.
[246] A. Ejchart, A. Gryff-Keller, P. Szczecinski, *J. Mag. Reson.* **1992**, *91*, 313–321.
[247] P. Parhami, B. M. Fung, *J. Am. Chem. Soc.* **1985**, *107*, 7304–7306.
[248] P. Diehl, F. Moia, *J. Mag. Reson.* **1983**, *54*, 312–313.
[249] J. Jokisaari, Y. Hiltunen, J. Lounila, *J. Chem. Phys.* **1986**, *85(6)*, 3198–3202.
[250] J. Jokisaari, Y. Hiltunen, *Mol. Phys.* **1983**, *50*, 1013–1023.
[251] P. Diehl, F. Moia, *Isr. J. Chem.* **1983**, *23*, 265–267.
[252] Y. Hiltunen, J. Jokisaari, J. Lounila, A. Pulkkinen, *J. Am. Chem. Soc.* **1989**, *111*, 3217–3220.
[253] A. Pulkkinen, Y. Hiltunen, J. Jokisaari, *Liq. Cryst.* **1988**, *3*, 737–744.
[254] B. B. Sharma, G. Chidichimo, A. Saupe, H. Chang, G. H. Brown, *J. Mag. Reson.* **1982**, *49*, 287–295.
[255] B. M. Fung, P. Parhami, *J. Mag. Reson.* **1985**, *63*, 168–173.
[256] J. Jokisaari, Y. Hiltunen, *Chem. Phys. Lett.* **1985**, *115*, 441–444.
[257] D. F. R. Gilson, S. Brownstein, *Can. J. Chem.* **1982**, *60*, 1078–1080.
[258] J. Jokisaari, P. Lazzeretti, P. Pyykkö, *Chem. Phys.* **1988**, *123*, 339–350.
[259] J. Jokisaari, T. Väänänen, *Mol. Phys.* **1986**, *58*, 959–964.
[260] O. Muenster, J. Jokisaari, P. Diehl, *Mol. Cryst. Liq. Cryst.* **1991**, *206*, 179–186.
[261] J. Voigt, T. Bjorholm, J. P. Jacobsen, *Isr. J. Chem.* **1983**, *23*, 291–294.
[262] A. Loewenstein, in *Advances in Nuclear Quadrupole Resonance.* Vol. 6. (Ed.: J. A. C. Smith) Wiley, New York, **1983**, p. 53.
[263] J. P. Jacobsen, E. J. Pederson, *J. Mag. Reson.* **1981**, *44*, 101–108.
[264] D. Catalano, C. Forte, C. A. Veracini, *J. Mag. Reson.* **1984**, *60*, 190–198.
[265] R. Ambrosetti, D. Catalano, C. Forte, C. A. Veracini, *Z. Naturforsch.* **1986**, *41a*, 431–435.
[266] A. Loewenstein, D. Igner, *J. Phys. Chem.* **1988**, *92*, 2124–2129.
[267] G. C. Lickfield, J. F. McKenna, A. L. Beyerlein, G. B. Savitsky, *J. Mag. Reson.* **1983**, *51*, 301–304.
[268] G. C. Lickfield, A. L. Beyerlein, G. B. Savitsky, L. E. Lewis, *J. Phys. Chem.* **1984**, *88*, 3566–3570.
[269] T. M. Barbara, *Mol. Phys.* **1985**, *54*, 651–656.
[270] Z. Luz, J. *Isr. J. Chem.* **1983**, *23*, 305–313.
[271] Z. Luz, in *Nuclear Magnetic Resonance of Liquid Crystals,* (Ed.: J. W. Emsley) D. Reidel, Boston, **1985**, p. 315.
[272] C. A. Veracini, M. Longeri, in *Nuclear Magnetic Resonance of Liquid Crystals*, (Ed.: J. W. Emsley) D. Reidel, Boston, **1985**, p. 123.
[273] *Dynamic Nuclear Magnetic Resonance Spectroscopy,* Academic Press, New York, **1979**.
[274] Z. Luz, R. Naor, E. Meirovitch, *J. Chem. Phys.* **1981**, *74(12)*, 6621–6630.
[275] D. Gamliel, Z. Luz, S. Vega, *J. Chem. Phys.* **1986**, *85*, 2516–2527.
[276] R. Naor, Z. Luz, *J. Chem. Phys.* **1982**, *76(12)*, 5662–5664.
[277] R. Poupko, Z. Luz, *J. Chem. Phys.* **1981**, *75(4)*, 1675–1681.
[278] R. Poupko, Z. Luz, A. J. Vega, H. Zimmermann, *J. Chem. Phys.* **1987**, *86*, 5358–5364.
[279] A. J. Vega, R. Poupka, Z. Luz, *J. Mag. Reson.* **1989**, *83*, 111–127.
[280] K. Müller, R. Poupko, Z. Luz, *J. Mag. Reson.* **1990**, *90*, 19–39.
[281] C. Boeffel, Z. Luz, R. Poupko, H. Zimmermann, *J. Am. Chem. Soc.* **1990**, *112*, 7158–7163.
[282] K. Müller, Z. Luz, R. Poupko, H. Zimmermann, *Liq. Cryst.* **1992**, *11*, 547–560.
[283] R. Poupko, H. Zimmermann, Z. Luz, *J. Am. Chem. Soc.* **1984**, *106*, 5391–5394.
[284] D. Gamliel, Z. Luz, S. Vega, *J. Chem. Phys.* **1988**, *88*, 25–42.

[285] D. Gamliel, Z. Luz, A. Maliniak, R. Poupko, A. J. Vega, *J. Chem. Phys.* **1990**, *93(8)*, 5379–5386.
[286] C. Boeffel, Z. Luz, R. Poupko, A. J. Vega, *Isr. J. Chem.* **1988**, *28*, 283–296.
[287] J. W. Emsley, C. A. Veracini, P. L. Barili, *J. Chem. Soc., Perkin Trans. II* **1980**, 816–818.
[288] L. Di Bari, C. Forte, C. A. Veracini, C. Zannoni, *Chem. Phys. Lett.* **1988**, *143*, 263–269.
[289] B. M. Fung, R. V. Singh, M. M. Alcock, *J. Am. Chem. Soc.* **1984**, *106*, 7301–7304.
[290] J. Afzal, B. M. Fung, *J. Chem. Phys.* **1986**, *84*, 6119–6122.
[291] E. Lafontaine, J. P. Bayle, J. Courtieu, *J. Am. Chem. Soc.* **1989**, *111*, 8294–8296.
[292] A. Meddour, I. Canet, A. Loewenstein, J. M. Pechine, J. Courtieu, *J. Am. Chem. Soc.* **1994**, *116*, 9652–9656.
[293] I. Canet, J. Lovschall, J. Courtieu, *Liq. Cryst.* **1994**, *16*, 405–412.
[294] E. Lafontaine, J. M. Pechine, J. Courtieu, C. L. Mayne, *Liq. Cryst.* **1990**, *7*, 293–298.
[295] J. P. Bayle, J. Courtieu, E. Gabetty, A. Loewenstein, J. M. Pechine, *New J. Chem.* **1992**, *16*, 837–838.
[296] J.-L. Canet, A. Fadel, J. Salaun, I. Canet-Fresse, J. Courtieu, *Tetrahedron: Asymmetry* **1993**, *4*, 31–34.
[297] I. Canet, A. Meddour, J. Courtieu, J.-L. Canet, J. Salaun, *J. Am. Chem. Soc.* **1994**, *116*, 2155–2156.
[298] D. Parker, *Chem. Rev.* **1991**, *91*, 1441–1457.
[299] H. Levanon, *Rev. Chem. Intermed.* **1987**, *8*, 287–320.
[300] B. Kirste, *Mag. Reson. Chem.* **1987**, *25*, 166–175.
[301] N. A. Novosadov, A. S. Mukhtarov, Z. I. Usmanov, *J. Struct. Chem. USSR* **1987**, *27*, 662–664.
[302] P. Chmielewski, A. Jezierski, *Polyhedron* **1988**, *7*, 25–28.
[303] O. Gonen, H. Levanon, *J. Chem. Phys.* **1986**, *84*, 4132–4141.
[304] J. Michl, E. W. Thulstrup, *Spectroscopy with Polarized Light*, VCH, Weinheim, **1986**.
[305] C. Zannoni, in *Polarized Spectroscopy of Ordered Systems*, (Eds.: B. Samori, E. W. Thulstrup) Kluwer Academic Publishers, Dordrecht, **1988**, p. 57.
[306] F. D. Saeva, in *Liquid Crystals. The Fourth State of Matter*, (Ed.: F. D. Saeva) Marcel Dekker, Inc. New York, **1979**, p. 249.
[307] B. Samori, E. W. Thulstrup (Eds.) *Polarized Spectroscopy of Ordered Systems*. Kluwer Academic Publishers, Dordrecht, **1988**.
[308] B. Samori, *Mol. Cryst. Liq. Cryst.* **1983**, *98*, 385–397.
[309] J. Schellman, H. P. Jensen, *Chem. Rev.* **1987**, *87*, 1359–1399.
[310] E. Sackmann, *J. Am. Chem. Soc.* **1968**, *90*, 3569.

[311] E. Sackmann, J. Voss, *Chem. Phys. Lett.* **1972**, *14*, 528.
[312] F. D. Saeva, J. J. Wysocki, *J. Am. Chem. Soc.* **1971**, *93*, 5928.
[313] F. D. Saeva, G. R. Olin, *J. Am. Chem. Soc.* **1973**, *95*, 7882.
[314] M. Belhakem, B. Jordanov, *J. Mol. Struct.* **1991**, *245*, 195–202.
[315] D. Baumann, H.-G. Kuball, *Chem. Phys.* **1993**, *176*, 221–231.
[316] M. Sisido, K. Takeuchi, Y. Imanishi, *Chem. Lett.* **1983**, 961–964.
[317] L. B.-A. Johansson, *Chem. Phys. Lett.* **1985**, *118*, 516–521.
[318] L. B.-A. Johansson, J. G. Molotkovsky, L. D. Bergelson, *J. Am. Chem. Soc.* **1987**, *109*, 7374–7381.
[319] B. Kalman, N. Clarke, L. B.-A. Johansson, *J. Phys. Chem.* **1989**, *93*, 4608–4615.
[320] G. P. Ceasar, R. A. Levenson, H. B. Gray, *J. Am. Chem. Soc.* **1969**, *91*, 772–774.
[321] F. Jede, A. W. Stratmann, B. Schrader, *Mol. Cryst. Liq. Cryst.* **1986**, *140*, 287–295.
[322] M. Belhakem, B. Jordanov, *J. Mol. Struct.* **1991**, *245*, 29–40.
[323] B. Jordanov, R. Nentchovska, B. Schrader, *J. Mol. Struct.* **1993**, *297*, 401–406.
[324] M. Rogojerov, D. Kolev, B. Jordanov, *J. Mol. Struct.* **1995**, *349*, 447–450.
[325] B. Jordanov, B. Schrader, *J. Mol. Struct.* **1995**, *347*, 389–398.
[326] E. Sackmann, D. Rehm, *Chem. Phys. Lett.* **1970**, *4*, 537–540.
[327] J. P. Riehl, F. S. Richardson, *Chem. Rev.* **1986**, *86*, 1–16.
[328] E. Wolarz, *Z. Naturforsch.* **1992**, *47A*, 807–812.
[329] H. Beens, H. Mohwald, D. Rehm, E. Sackmann, A. Weller, *Chem. Phys. Lett.* **1971**, *8*, 341–344.
[330] H. Stegemeyer, J. Haase, W. Laarhoven, *Chem. Phys. Lett.* **1987**, *137*, 516–520.
[331] A. Arcioni, F. Bertinelli, R. Tarroni, C. Zannoni, *Mol. Phys.* **1987**, *61*, 1161–1181.
[332] E. I. Demikhov, V. K. Dolganov, V. V. Korshunov, D. Demus, *Liq. Cryst.* **1988**, *3*, 1161–1164.
[333] B. O. Myrvold, P. Klaeboe, *Acta Chem. Scand. A* **1985**, *39*, 733–747.
[334] H.-G. Kuball, M. Junge, B. Schultheis, A. Schonhofer, *Ber. Bunsenges. Phys. Chem.* **1991**, *95*, 1219–1227.
[335] W. Haase, O. Trinquet, U. Quotschalla, J. K. Foitzik, *Mol. Cryst. Liq. Cryst.* **1987**, *148*, 15–27.
[336] F. Jede, A. W. Stratmann, B. Schrader, *Mol. Cryst. Liq. Cryst.* **1986**, *140*, 297–302.
[337] M. Belhakem, B. Jordanov, *J. Mol. Struct.* **1990**, *218*, 309–314.
[338] G. Solladie, G. Gottarelli, *Tetrahedron* **1987**, *43*, 1425–1437.

[339] G. Gottarelli, P. Mariani, G. P. Spada, B. Samori, G. Solladie, M. Hibert, *Tetrahedron* **1983**, *39*, 1337.
[340] C. Rosini, L. Franzini, P. Salvadorie, G. P. Spada, *J. Org. Chem.* **1992**, *57*, 6820–6824.
[341] G. Gottarelli, G. P. Spada, G. Solladie, *Nouv. J. Chim.* **1986**, *10*, 691.
[342] G. Gottarelli, M. Hibert, B. Samori, G. Solladie, G. P. Spada, R. Zimmermann, *J. Am. Chem. Soc.* **1983**, *105*, 7318–7321.
[343] G. Gottarelli, G. P. Spada, R. Bartsch, G. Solladie, R. Zimmermann, *J. Org. Chem.* **1986**, *51*, 589.
[344] W. J. Leigh, in *Photochemistry on Solid Surfaces* (Eds.: M. Anpo, T. Matsuura) Elsevier, Amsterdam, **1989**, p. 481.
[345] W. J. Leigh, in *Liquid Crystals. Applications and Uses* (Ed.: B. Bahadur) World Scientific, Singapore, **1991**, p. 357–396.
[346] D. Kreysig, J. Stumpe, in *Selected Topics in Liquid Crystal Research* (Ed.: H. D. Koswig) Akademie-Verlag, Berlin, **1990**, p. 69–114.
[347] R. G. Weiss, *Tetrahedron* **1988**, *44*, 3413–3475.
[348] R. G. Weiss, in *Photochemistry in Organized and Constrained Media*, (Ed.: V. Ramamurthy) VCH, Weinheim, **1991**, p. 603–689.
[349] V. Ramesh, R. G. Weiss, *J. Org. Chem.* **1986**, *51*, 2535–2539.
[350] V. Percec, H. Jonsson, D. Tomazos, in *Polym. Organ. Media,* (Ed.: C. Paleo) Gordon and Breach, Philadelphia, **1992**, p. 1–104.
[351] B. J. Fahie, D. S. Mitchell, W. J. Leigh, *Can. J. Chem.* **1988**, *67*, 148–159.
[352] B. J. Fahie, D. S. Mitchell, M. S. Workentin, W. J. Leigh, *J. Am. Chem. Soc.* **1989**, *111*, 2916–2929.
[353] M. S. Workentin, W. J. Leigh, K. R. Jeffrey, *J. Am. Chem. Soc.* **1990**, *112*, 7329–7336.
[354] M. S. Workentin, B. J. Fahie, W. J. Leigh, *Can. J. Chem.* **1991**, *69*, 1435–1444.
[355] M. S. Workentin, W. J. Leigh, *J. Phys. Chem.* **1992**, *96*, 9666–9673.
[356] J. Stumpe, S. Grande, S. Wolf, G. Hempel, *Liq. Cryst.* **1992**, *11*, 175–198.
[357] B. Samori, S. Masiero, G. R. Luckhurst, S. K. Heeks, B. A. Timimi, P. Mariani, *Liq. Cryst.* **1993**, *14*, 217–231.
[358] H. Meier, T. Lifka, K. Muller, *Recl. Trav. Chim. Pays-Bas* **1995**, *114*, 465.
[359] B. Samori, P. De Maria, F. Rustichelli, P. Zani, *Tetrahedron* **1987**, *43*, 1409–1424.
[360] P. M. Vilalta, G. S. Hammond, R. G. Weiss, *Langmuir* **1993**, *9*, 1910–1921.
[361] W. J. Leigh, *J. Am. Chem. Soc.* **1985**, *107*, 6114–6116.
[362] W. J. Leigh, *Can. J. Chem.* **1986**, *64*, 1130–1139.
[363] P. De Maria, B. Samori, A. Tampieri, P. Zani, *Bull. Chem. Soc. Jpn.* **1988**, *61*, 1773–1777.

[364] W. J. Leigh, D. S. Mitchell, *J. Am. Chem. Soc.* **1992**, *114*, 5005–5010.
[365] B. M. Fung, M. Gangoda, *J. Am. Chem. Soc.* **1985**, *107*, 3395.
[366] M. Gangoda, B. M. Fung, *Chem. Phys. Lett.* **1985**, *120*, 527.
[367] J. R. Scheffer, P. R. Pokkuluri, in *Photochemistry in Organized and Constrained Media*, (Ed.: V. Ramamurthy) VCH, Weinheim, **1991**, p. 185–246.
[368] D. A. Hrovat, J. H. Liu, N. J. Turro, R. G. Weiss, *J. Am. Chem. Soc.* **1984**, *106*, 7033–7037.
[369] R. G. Zimmerman, J. H. Liu, R. G. Weiss, *J. Am. Chem. Soc.* **1986**, *108*, 5264–5271.
[370] R. L. Treanor, R. G. Weiss, *J. Phys. Chem.* **1987**, *91*, 5552–5554.
[371] R. L. Treanor, R. G. Weiss, *J. Am. Chem. Soc.* **1986**, *108*, 3139–3140.
[372] R. L. Treanor, R. G. Weiss, *Tetrahedron* **1987**, *43*, 1371–1391.
[373] Z. He, R. G. Weiss, *J. Am. Chem. Soc.* **1990**, *112*, 5535–5541.
[374] I. Furman, R. G. Weiss, *J. Am. Chem. Soc.* **1992**, *114*, 1381–1388.
[375] I. Furman, R. J. Butcher, R. M. Catchings, R. G. Weiss, *J. Am. Chem. Soc.* **1992**, *114*, 6023–6030.
[376] P. M. Vilalta, G. S. Hammond, *Photochem. Photobiol.* **1991**, *54*, 563–569.
[377] P. Deval, A. K. Singh, *J. Photochem. Photobiol. A: Chem.* **1988**, *42*, 329–336.
[378] J. P. Otruba III, R. G. Weiss, *Mol. Cryst. Liq. Cryst.* **1982**, *80*, 165–178.
[379] S. Kurihara, T. Ikeda, S. Tazuke, J. Seto, *J. Chem. Soc., Faraday Trans.* **1991**, *87*, 3251–3254.
[380] K. Fukunishi, S. Kitamura, S. Kuwabara, H. Yamanaka, M. Nomura, *Chem. Lett.* **1989**, 587–590.
[381] J. P. Otruba III, R. G. Weiss, *J. Org. Chem.* **1983**, *48*, 3448–3453.
[382] S. Kurihara, T. Ikeda, S. Tazuke, *Mol. Cryst. Liq. Cryst.* **1990**, *178*, 117–132.
[383] V. C. Anderson, B. B. Craig, R. G. Weiss, *Mol. Cryst. Liq. Cryst.* **1983**, *97*, 351–363.
[384] V. C. Anderson, R. G. Weiss, *J. Am. Chem. Soc.* **1984**, *106*, 6628–6637.
[385] M. F. Sonnenschein, R. G. Weiss, *J. Phys. Chem.* **1988**, *92*, 6828–6835.
[386] B. M. Sheikh-Ali, R. G. Weiss, *J. Am. Chem. Soc.* **1994**, *116*, 6111–6120.
[387] A. Arcioni, F. Bertinelli, R. Tarroni, C. Zannoni, *Mol. Phys.* **1987**, *61*, 1161.
[388] A. Regev, T. Galili, H. Levanon, *J. Chem. Phys.* **1991**, *95*, 7907–7916.
[389] K. Hasharoni, H. Levanon, J. von Gersdorff, H. Kurreck, K. Möbius, *J. Chem. Phys.* **1993**, *98*, 2916–2926.

[390] K. Hasharoni, H. Levanon, *J. Phys. Chem.* **1995**, *99*, 4875–4878.
[391] D. A. Hrovat, J. H. Liu, N. J. Turro, R. G. Weiss, *J. Am. Chem. Soc.* **1984**, *106*, 5291–5295.
[392] J. Stumpe, C. Selbmann, D. Kreysig, *J. Photochem. Photobiol., A: Chem.* **1991**, *58*, 15–30.
[393] G. R. Bauer, F. Dickert, A. Hammerschmidt, *Angew. Chem. Int. Ed. Engl.* **1986**, *25*, 841–842.
[394] G. B. Sergeev, V. A. Batyuk, T. I. Shabatina, *Kinet. Katal.* **1983**, *24*, 538–541.
[395] V. A. Batyuk, T. N. Boronina, G. B. Sergeev, *Mol. Cryst. Liq. Cryst.* **1990**, *186*, 93–98.
[396] W. J. Leigh, D. T. Frendo, P. J. Klawunn, *Can. J. Chem.* **1985**, *63*, 2131–2138.
[397] S. Ganapathy, R. G. Zimmermann, R. G. Weiss, *J. Org. Chem.* **1986**, *51*, 2529–2535.
[398] A. Nuñez, T. Hollebeek, M. M. Labes, *J. Am. Chem. Soc.* **1992**, *114*, 4925–4926.
[399] J. M. Nerbonne, R. G. Weiss, *J. Am. Chem. Soc.* **1978**, *100*, 2571–2572.
[400] J. M. Nerbonne, R. G. Weiss, *J. Am. Chem. Soc.* **1979**, *101*, 402–407.
[401] J. Naciri, G. P. Spada, G. Gottarelli, R. G. Weiss, *J. Am. Chem. Soc.* **1987**, *109*, 4352–4357.
[402] M. M. Labes, V. Ramesh, A. Nuñez, W. Shang, X. Luo, *Mol. Cryst. Liq. Cryst.* **1994**, *240*, 25–31.
[403] E. G. Cassis Jr., R. G. Weiss, *Photochem. Photobiol.* **1982**, *35*, 439–444.
[404] S. Melone, V. Mosini, R. Nicoletti, B. Samori, G. Torquati, *Mol. Cryst. Liq. Cryst.* **1983**, *98*, 399–409.
[405] T. Kunieda, T. Takahashi, M. Hirobe, *Tet. Lett.* **1983**, *24*, 5107–5108.
[406] T. Nagamatsu, C. Kawano, Y. Orita, T. Kunieda, *Tet. Lett.* **1987**, *28*, 3263–3266.
[407] V. Ramesh, R. G. Weiss, *Mol. Cryst. Liq. Cryst.* **1986**, *135*, 13–22.
[408] Y. Lin, R. G. Weiss, *Liq. Cryst.* **1989**, *4*, 367–384.
[409] V. C. Anderson, B. B. Craig, R. G. Weiss, *J. Am. Chem. Soc.* **1981**, *103*, 7169–7176.
[410] H. Stegemeyer, J. Hasse, W. Laarhoven, *Chem. Phys. Lett.* **1987**, *137*, 516–520.
[411] V. C. Anderson, B. B. Craig, R. G. Weiss, *J. Am. Chem. Soc.* **1982**, *104*, 2972–2977.
[412] K. Kawaguchi, M. Sisido, Y. Imanishi, *J. Phys. Chem.* **1988**, *92*, 4806.
[413] M. Sisido, X. Wang, K. Kawaguchi, Y. Imanishi, *J. Phys. Chem.* **1988**, *92*, 4801–4806.
[414] K. Kawaguchi, M. Sisido, Y. Imanishi, M. Kiguchi, Y. Taniguchi, *Bull. Chem. Soc. Jpn.* **1989**, *62*, 2146.
[415] M. Sisido, K. Takeuchi, Y. Imanishi, *J. Phys. Chem.* **1984**, *88*, 2893.
[416] M. Sisido, K. Kawaguchi, K. Takeuchi, Y. Imanishi, *Mol. Cryst. Liq. Cryst.* **1988**, *162B*, 263.
[417] M. Sisido, K. Takeuchi, Y. Imanishi, *Chem. Lett.* **1983**, 961.
[418] V. A. Batyuk, T. I. Shabatina, Y. N. Morosov, G. B. Sergeev, *Mol. Cryst. Liq. Cryst. Nonlin. Opt.* **1988**, *161*, 109–117.
[419] V. A. Batyuk, T. N. Boronina, G. E. Chudinov, G. B. Sergeev, *Liq. Cryst.* **1989**, *4*, 181–196.
[420] V. A. Batyuk, T. N. Boronina, G. E. Chudinov, G. B. Sergeev, *Mol. Cryst. Liq. Cryst.* **1990**, 289–293.
[421] T. N. Boronina, G. E. Chudinov, V. A. Batyuk, G. B. Sergeev, *Mol. Mat.* **1992**, *2*, 47–50.
[422] B. Samori, L. Fiocco, *J. Am. Chem. Soc.* **1982**, *104*, 2634–2636.
[423] P. De Maria, A. Lodi, B. Samori, F. Rustichelli, G. Torquati, *J. Am. Chem. Soc.* **1984**, *106*, 653–656.
[424] G. Albertini, P. Mariani, F. Rustichelli, B. Samori, *Mol. Cryst. Liq. Cryst.* **1988**, *163*, 163.
[425] P. De Maria, P. Mariani, F. Rustichelli, B. Samori, *Mol. Cryst. Liq. Cryst.* **1984**, *116*, 155.
[426] P. Seuron, G. Solladie, *J. Org. Chem.* **1980**, *45*, 715–719.
[427] E. Sokolov, T. I. Shabatina, Y. N. Morozov, V. A. Batyuk, G. B. Sergeev, *Mol. Mat.* **1993**, *2*, 189–193.
[428] Y. N. Morozov, A. V. Reiter, T. I. Shabatina, V. A. Batyuk, *Mol. Mat.* **1992**, *2*, 35–40.
[429] V. A. Batyuk, T. I. Shabatina, Y. N. Morosov, G. B. Sergeev, *Mol. Cryst. Liq. Cryst.* **1990**, *186*, 87–92.
[430] T. I. Shabatina, Y. N. Morosov, V. A. Batyuk, G. B. Sergeev, *Mol. Cryst. Liq. Cryst.* **1990**, *191*, 277–281.
[431] T. I. Shabatina, Y. N. Morozov, A. I. Konstantinov, V. A. Batyuk, G. B. Sergeev, *Mol. Mat.* **1992**, *2*, 41–45.
[432] W. J. Leigh, *Can. J. Chem.* **1985**, *63*, 2736–2741.
[433] W. J. Leigh, D. S. Mitchell, *J. Am. Chem. Soc.* **1988**, *110*, 1311–1313.
[434] S. Hiraoka, T. Yoshida, T. Nagamutsu, T. Kunieda, *Tetrahedron* **1991**, *32*, 1729.
[435] H. Kansui, S. Hiraoka, T. Kunieda, *J. Am. Chem. Soc.* **1996**, *118*, 5346–5352.
[436] T. Yamaguchi, T. Yoshida, T. Nagamatsu, T. Kunieda, *Tet. Lett.* **1991**, *32*, 1729–1732.
[437] L. Dinescu, A. Mezo, G. Luyt, R. P. Lemieux, *Tet. Lett.* **1994**, *35*, 7549–7552.
[438] M. Zhang, G. B. Schuster, *J. Am. Chem. Soc.* **1994**, *116*, 4852–4857.
[439] C. M. Paleos, *Chem. Soc. Rev.* **1985**, *14*, 45.
[440] M. F. Sonnenschein, R. G. Weiss, in *Photochemistry of Solid Surfaces* (Eds.: M. Anpo, T. Matsuura) Elsevier, Amsterdam, **1989**, p. 526–550.
[441] W. J. Leigh, S. Jakobs, *Tetrahedron* **1987**, *43*, 1393–1408.
[442] R. Neumann, *Mol. Cryst. Liq. Cryst.* **1994**, *240*, 33–37.

[443] K. Araya, A. Mukoh, T. Narahara, H. Shirakawa, *Chem. Lett.* **1984**, 1141–1142.
[444] G. M. Janini, *Adv. Chromatogr.* **1979**, *17*, 231–277.
[445] Z. Witkiewicz, *J. Chromatogr.* **1982**, *251*, 311–337.
[446] M. S. Vigdergauz, N. F. Belyaev, M. S. Esin, *Fresenius Z. Anal. Chem.* **1989**, *335*, 70–76.
[447] Z. Witkiewicz, *J. Chromatogr.* **1989**, *466*, 37–87.
[448] Z. Witkiewicz, J. Mazur, *LC–GC* **1990**, *8*, 224–336.
[449] Z. Witckieicz, in *Liquid Crystals. Applications and Uses.* Vol. 2 (Ed.: B. Bahadur) World Scientific, Singapore, **1991**, p. 309–356
[450] G. M. Janini, K. Johnston, W. L. Zielinski Jr. *Anal. Chem.* **1975**, *47*, 670.
[451] W. L. Zielinski Jr., G. M. Janini, *J. Chromatogr.* **1979**, *186*, 237.
[452] D. R. Luffer, W. Ecking, M. Novotny, *J. Chromatogr.* **1990**, *505*, 79.
[453] D. E. Martire, A. Nikolic, K. L. Vasanth, *J. Chromatogr.* **1979**, *178*, 401.
[454] D. E. Martire, *J. Chromatogr.* **1987**, *406*, 27.
[455] C. Yan, D. E. Martire, *Anal. Chem.* **1992**, *64*, 1246–1253.
[456] C. Yan, D. E. Martire, *J. Phys. Chem.* **1992**, *96*, 3489–3504.
[457] S. A. Wise, W. J. Bonnett, F. R. Guenther, W. E. May, *J. Chromatogr. Sci.* **1981**, *19*, 457.
[458] S. A. Wise, L. C. Sander, *HRC–CC: J. High Resolut. Chromatogr. Chromatogr. Commun.* **1985**, *8*, 248.
[459] C. T. Peng, *J. Chromatogr.* **1994**, *678*, 189–200.
[460] R. Fu, P. Jing, J. Gu, Z. Huang, Y. Chen, *Anal. Chem.* **1993**, *65*, 2141–2144.
[461] A. Kraus, U. Schumann, G. Kraus, B. Kohne, K. Praefcke, *J. Chromatogr.* **1992**, *609*, 277–281.
[462] Z. Witkiewicz, J. Szulc, R. Dabrowski, *J. Chromatogr.* **1984**, *315*, 145–149.
[463] Z. Witkiewicz, B. Goca, *J. Chromatogr.* **1987**, *402*, 73–85.
[464] F. Hahne, G. Kraus, H. Zaschke, *J. Chromatogr.* **1990**, *520*, 85–90.
[465] G. M. Janini, G. M. Muschik, H. J. Issaq, R. J. Laub, *Anal. Chem.* **1988**, *60*, 1119–1124.
[466] K. Jinno, Y. Saito, R. Malhan nee Chopra, J. J. Pesek, J. C. Fetzer, W. B. Biggs, *J. Chromatogr.* **1991**, *557*, 459–468.
[467] H. Wang, T. Gen, R. Fu, J. Zhu, X. Wei, H. Li, J. Lui, *J. Chromatogr.* **1992**, *609*, 414–418.
[468] G. Kraus, J. M. Theirfelder, L. Sojak, *J. Chromatogr.* **1994**, *677*, 197–200.
[469] L. Sojak, G. Kraus, I. Ostrovsky, *J. Chromatogr.* **1985**, *323*, 414–417.
[470] J. Krupcuik, M. Valachovicova, G. Kraus, *J. Chromatogr.* **1994**, *665*, 111–116.
[471] B. Xu, N. P. E. Vermaulen, *J. Chromatogr.* **1988**, *445*, 1–28.

[472] J. A. Yancey, *J. Chromatogr. Sci.* **1994**, *32*, 403–413.
[473] R. R. Driesbach, *Adv. Chem. Ser.* **1955**, 15.
[474] R. R. Driesbach, *Adv. Chem. Ser.* **1959**, 22.
[475] M. L. McGlashan, D. J. B. Potter, *Proc. Roy. Soc. London, Ser. A* **1972**, *267*, 478.
[476] M. L. McGlashan, C. J. Wormald, *Trans. Faraday Soc.* **1964**, *60*, 646.
[477] E. A. Guggenheim, C. J. Wormald, *J. Chem. Phys.* **1965**, *74*, 3775.
[478] J. Coca, I. Medina, S. H. Langer, *Chromatographia* **1988**, *25*, 825–829.
[479] J. Coca, I. Medina, S. H. Langer, *Liq. Cryst.* **1989**, *4*, 175–180.
[480] L. C. Chow, D. E. Martire, *J. Phys. Chem.* **1971**, *75*, 2005.
[481] D. E. Martire, *Mol. Cryst. Liq. Cryst.* **1974**, *28*, 63.
[482] A. L. Jadhav, K. P. Naikwadi, S. Rokushika, H. Hatano, M. Oshima, *HRC–CC: J. High Resolut. Chromatogr. Chromatogr. Commun.* **1987**, *10*, 77–81.
[483] F. Janssen, *Chromatographia* **1983**, *17*, 477–480.
[484] A. Kaune, D. Lenoir, U. Nikolai, A. Kettrp, *Chemosphere* **1994**, *29*, 2083–2096.
[485] R. C. Kong, M. L. Lee, Y. Tominaga, R. Pratap, M. Iwao, R. N. Castle, *Anal. Chem.* **1982**, *54*, 1802–1806.
[486] T. Kreczmer, A. Gutorska, *Chem. Anal. (Warsaw)* **1985**, *30*, 419.
[487] K. E. Markides, M. Nishioka, B. J. Tarbet, J. S. Bradshaw, M. L. Lee, *Anal. Chem.* **1985**, *57*, 1296–1299.
[488] J. Mazur, Z. Witkiewicz, R. Dabrowski, *J. Chromatogr.* **1988**, *455*, 323–326.
[489] H. Moser, H. Arm, *HRC–CC: J. High Resolut. Chromatogr. Chromatogr. Commun.* **1984**, *7*, 637–638.
[490] M. Nishioka, B. A. Jones, B. J. Tarbet, J. S. Bradshaw, M. L. Lee, *J. Chromatogr.* **1986**, *357*, 79–91.
[491] K. P. Naikwadi, S. Rokushika, H. Hatano, M. Oshima, *J. Chromatogr.* **1985**, *331*, 69.
[492] K. P. Naikwadi, A. M. McGovern, F. W. Karasek, *Can. J. Chem.* **1987**, *65*, 970–975.
[493] S. Sakagami, *J. Chromatogr.* **1982**, *246*, 121.
[494] S. A. Wise, B. A. Brenner, G. D. Byrd, S. M. Chesler, R. E. Rebbert, M. M. Schantz, *Anal. Chem.* **1988**, *60*, 887–894.
[495] A. Ziolek, Z. Witkiewicz, R. Dabrowski, *J. Chromatogr.* **1984**, *299*, 159.
[496] G. Chiavari, L. Pastorelli, *J. Chromatogr.* **1983**, *262*, 175.
[497] K. E. Markides, H.-C. Chang, C. M. Schregenberger, B. J. Tarbet, J. S. Bradshaw, M. L. Lee, *HRC–CC: J. High Resolut. Chromatogr. Chromatogr. Commun.* **1985**, *8*, 516–520.
[498] L. Zoccolillo, L. Liberti, F. Coccioli, M. Ronchetti, *J. Chromatogr.* **1984**, *288*, 347.

[499] H. Budzinski, M. Radke, P. Garrigues, S. A. Wise, J. Bellocq, H. Willsch, *J. Chromatogr.* **1992**, *627*, 227–239.

[500] J. S. Bradshaw, C. M. Schregenberger, H.-C. Chang, K. E. Markides, M. L. Lee, *J. Chromatogr.* **1986**, *358*, 95–106.

[501] N. T. Karabanov, Z. P. Vetrova, L. A. Ivanova, N. M. Olefirenko, A. Y. Devjatjarov, *Zh. Anal. Khim.* **1984**, *39*, 749.

[502] K. P. Naikwadi, F. W. Karasek, *J. Chromatogr.* **1986**, *369*, 203–207.

[503] K. P. Naikwadi, F. W. Karasek, *Chemosphere* **1990**, *20*, 1379–1384.

[504] M. Swerev, K. Ballschmiter, *HRC–CC: J. High Resolut. Chromatogr. Chromatogr. Commun.* **1987**, *10*, 544–547.

[505] R. C. Kong, M. L. Lee, Y. Tominaga, R. Pratap, M. Iwao, R. N. Castle, S. A. Wise, *J. Chromatogr.* **1982**, *20*, 502–510.

[506] B. A. Jones, J. S. Bradshaw, M. Nishioka, M. L. Lee, *J. Org. Chem.* **1984**, *49*, 4947–4951.

[507] H. Budzinski, P. Garrigues, J. Bellocq, *J. Chromatogr.* **1992**, *590*, 297–303.

[508] G. M. Muschik, T. Kelly, W. B. Manning, *J. Chromatogr.* **1980**, *202*, 75–82.

[509] K. P. Naikwadi, A. L. Jadhav, S. Rokushika, H. Hatano, *Makromol. Chem.* **1986**, *187*, 1407–1414.

[510] B. B. Ghatge, N. V. Bhalerao, *J. Chromatogr.* **1991**, *549*, 423–428.

[511] J. Szulc, Z. Witkiewicz, A. Ziolek, *J. Chromatogr.* **1983**, *262*, 161.

[512] Z. P. Vetrova, N. T. Karabanov, T. N. Shuvalova, L. A. Ivanova, Y. I. Yashin, *Chromatographia* **1985**, *20, 41*.

[513] L. Sojak, G. Kraus, I. Ostrovsky, E. Kralovicova, J. Krupcik, *J. Chromatogr.* **1981**, *206*, 463–474.

[514] L. Sojak, G. Kraus, I. Ostrovsky, E. Kralovicova, P. Farkas, *J. Chromatogr.* **1981**, *219*, 225–234.

[515] L. Sojak, G. Kraus, I. Ostrovsky, E. Kralovicova, J. Krupcik, *J. Chromatogr.* **1981**, *206*, 475–483.

[516] P. E. Sonnet, R. R. Heath, *J. Chromatogr.* **1985**, *321*, 127.

[517] J. Mazur, Z. Witkiewicz, R. Dabrowski, *J. Chromatogr.* **1991**, *600*, 123–127.

[518] L. Sojak, G. Kraus, P. Farkas, I. Ostrovsky, *J. Chromatogr.* **1982**, *249*, 29–40.

[519] E. Smolkova-Keulemansova, L. Sojak in *Ordered Media in Chemical Separations* (Eds.: W. L. Hinze, D. W. Armstrong) The Society, Washington, **1987**, p. 247.

[520] J. Krupcik, D. Repka, T. Hevesi, J. Mocak, *J. Chromatogr.* **1986**, *355*, 99–105.

[521] L. Sojak, I. Ostrovsky, G. Kraus, T. G. Andronikashvili, *J. Chromatogr.* **1988**, *436*, 47–52.

[522] L. Sojak, I. Ostrovsky, P. Farkas, J. Janak, *J. Chromatogr.* **1986**, *356*, 105–114.

[523] L. Sojak, I. Ostrovsky, R. Kubinec, G. Kraus, A. Kraus, *J. Chromatogr.* **1992**, *609*, 283–288.

[524] L. Sojak, I. Ostrovsky, R. Kubinec, *J. Chromatogr.* **1990**, *509*, 93–99.

[525] L. Sojak, I. Ostrovsky, *J. Chromatogr.* **1988**, *446*, 339–346.

[526] L. Sojak, G. Kraus, P. Farkas, I. Ostrovsky, *J. Chromatogr.* **1982**, *238*, 51–57.

[527] L. Sojak, G. Kraus, P. Farkas, I. Ostrovsky, *J. Chromatogr.* **1984**, *294*, 155–165.

[528] L. Sojak, I. Ostrovsky, R. Kubinec, G. Kraus, A. Kraus, *J. Chromatogr.* **1990**, *520*, 75–83.

[529] L. Sojak, P. Farkasu, I. Ostrovsky, J. Janak, J. R. Chretien, *J. Chromatogr.* **1991**, *557*, 241–253.

[530] L. Sojak, P. Farkas, J. Janak, S. Rang, O. Eisen, *J. Chromatogr.* **1984**, *287*, 271–291.

[531] I. Nesterova, B. Rekhter, G. Roshka, Z. Witkiewicz, *J. Chromatogr.* **1991**, *537*, 482–488.

[532] A. E. Habboush, S. M. Farroha, A.-L. Kreishan, *J. Chromatogr.* **1994**, *664*, 71–76.

[533] S. Ghodbane, G. A. Oweimreen, D. E. Martire, *J. Chromatogr.* **1991**, *556*, 317–330.

[534] D. E. Martire, S. Ghodbane, *J. Phys. Chem.* **1987**, *91*, 6403.

[535] J. F. Bocquet, C. Pommier, *J. Chromatogr.* **1983**, *261*, 11.

[536] G. M. Janini, M. T. Ubeid, *J. Chromatogr.* **1982**, *236*, 329.

[537] G. A. Oweimreen, *J. Solution Chem.* **1982**, *11*, 105.

[538] G. A. Oweimreen, *Mol. Cryst. Liq. Cryst.* **1981**, *68*, 257.

[539] J. F. Bocquet, C. Pommier, *J. Chromatogr.* **1981**, *205*, 239.

[540] J. F. Bocquet, C. Pommier, *J. Chromatogr.* **1981**, *205*, 251.

[541] G. A. Oweimreen, D. E. Martire, *J. Chem. Phys.* **1980**, *72*, 2500.

[542] K. Seifert, G. Kraus, *J. Chromatogr.* **1980**, *192*, 97.

[543] Z. Witkiewicz, J. Szulc, R. Dabrowski, J. Sadowski, *J. Chromatogr.* **1980**, *200*, 65.

[544] G. M. Janini, R. I. Sato, G. M. Muschik, *Anal. Chem.* **1980**, *52*, 2417.

[545] G. A. Oweimreen, G. C. Lin, D. E. Martire, *J. Phys. Chem.* **1979**, *83*, 2111.

[546] A. A. Jeknavorian, P. Barrett, A. C. Watterson, E. F. Barry, *J. Chromatogr.* **1975**, *107*, 317.

[547] G. Kraus, K. Seifert, H. Schubert, *J. Chromatogr.* **1974**, *100*, 101.

[548] A. A. Jeknavorian, E. F. Barry, *J. Chromatogr.* **1974**, *101*, 299.

[549] D. G. Willey, G. H. Brown, *J. Phys. Chem.* **1972**, *76*, 99.

[550] H. T. Peterson, D.E. Martire, W. Linder, *J. Phys. Chem.* **1972**, *76*, 596.

[551] J. M. Schnur, D. E. Martire, *Anal. Chem.* **1971**, *43*, 1201.

[552] W. L. Zielinski Jr., D. H. Freeman, D. E. Martire, L. C. Chow, *Anal. Chem.* **1970**, *42*, 176.

Index

aberration, SLM 819
absorption, optical 221, 319
ac calorimetric techniques, phase transitions 314
acetals, synthesis 102
acetylenes, synthesis 99
achiral polyphilics, polar 508
achiral smectics 505, 709
acids, solvents 883
acoustics 373, 560 ff
acoustoelectrical interactions, nematics 562 f
acyclic compounds, mesogenic properties 154
acyl, terminal substituents 147
acylation 91, 98
addressing, displays 751
adiabatic calorimetry, phase transitions 313
aerodynamic research, thermography 827
Akahane–Tako model, ultrasonic properties 558
alignment
– local molecular 406
– nematics 32
– XRD samples 638 f
alignment defects, displays 756
alignment materials, displays 732
aliphatic chains 125, 395
alkanes, solvents 883
alkanoic acids, esterification 106
alkanols, esterification 106
alkenes, solvents 883
alkenyl, terminal substituents 147
alkyl functionalization, arenes 95
alkyl groups
– even-odd effect 333
– re-entrants 400
– terminal substituents 146
alkylation 92
alkylbenzenes 882
alkylcyanobiphenyls (nCBs) 701
alkyloxy groups, terminal substituents 146
alkynes, solvents 883
AMBER force field, molecular modelling 77
anchoring 12
– director-field interactions 500
– elastic properties 260
– external fields 492
– flow/viscosity 467
– instabilities 523

– nematics 30
– solid substrates 541
– surface alignment 536 f
– textures 410
Andereck–Swift theory 383
anhydrous soaps, density measurements 341
anisotropic fluids, dielectric properties 235 f
anisotropic interactions 44, 535
anisotropic materials, tensor properties 189–203
anisotropic modes 521, 527
anisotropic solvents 839
anisotropy
– external fields 483
– magnetic 204
– orientational 253
– ultrasonic properties 562
anisotropy ratios, diffusion 591
antiferroelectrics 120, 508, 513
antiphases, XRD 657
apertures, SLM 774 f
arenes, alkyl functionalization 95
argon saturation, phase transitions 357
Arrhenius law 249, 585
aryl–aryl bond formation 93 f
associations 169
– defects/textures 448
asymmetry 115 ff
asynchronous transfer mode switch (ATM) 802
atomistic simulations, bulk systems 83

B distortions, external fields 489
B–SmB–SmA transitions 383
backflow effect
– director-field interactions 490
– flow/viscosity 463
– smectics 510
background lightening, thermography 827
balance laws, nematics 35
band configuration, ultrasonic properties 551
barometric method, high-pressure experiments 361
Beer–Lambert law 221
bend deformation
– directors 544
– elastic properties 256, 267
bend Frederiks transition 735

benzene derivatives, discotics   172
benzenes
- re-entrants   391
- six-atom ring systems   137
- solvents   872
Bessel function, XRD   642
Bessel side lobes, SLM   794
biaxial nematics
- flow/viscosity   468
- transitions   281
- XRD   650
biaxial smectics, torsional elasticity   258
biaxial structures, mean field model   300
biaxiality   42, 56, 483
bicatenar compounds, mesogenic properties   160
bicritical behavior, phase transitions   368
bicyclic rings, mesogenic properties   140 f
bicyclooctane, ring systems   140
bilayers   85, 409
bimolecular chemical reactions, solvents   878
binary intensity modulation   768
binary phase modulations, CGH   788
binary phase only matched filter (BROMF)   779
biphasic systems, solvent applications   849
biphenyls   140, 872
bipyran-4-ylidene, discotic cores   173
birefringence
- displays   731
- ECB   741
- elastic constants   271
- external fields   484
- SLM   763
- thermography   823
- ultrasonic properties   550
birefringent wavelength filters   806
blisters, thermography   825
block copolymers, XRD   671
Block walls, defects   415
blue light detection, X rays   628
blue phases
- cholesterics   503, 577
- external fields   485
- helical structures   129
- light scattering   708
- transitions   323
body forces/moments, nematics   28
Bohr magneton   206
Boltzmann distribution, modelling   74
Boltzmann factor, diffusion   584
bond formations   93
bond ordering, phase transitions   291
bond shift, NMR   844
bond stretches, molecular modelling   73
bookshelf structures
- displays   755
- external fields   506
- SLM   767
boundary conditions
- clearing thresholds   553

- molecular modelling   77
- nematics   30
- twisted periodic   85
boundary effects, ultrasonic properties   565
bowlic compounds   176
BP_{III}-isotropic transitions, fluctuations   384
Bragg law   486, 636 f
Bragg peaks
- Guinier camera   624
- neutron scattering   680
bremsstrahlung, X rays   619
Bridgman flexible bellows with slide wire   369
Brillouin scattering   719–726
broken aspects, textures   439
bubble domains, ultrasonic fields   557
building blocks, synthesis   98
bulk fluids, molecular modelling   77
bulk SHG, optical properties   577
bulk systems, large-scale simulation   83
buoyancy method, density measurements   332
Burgers vectors, defects   425, 447

Cadiot–Chodkiewicz coupling   99
Cahn nomenclature   117
calamitic compounds   134 f
- density measurements   336, 340
calamitic mesogens   87
- magnetic properties   209
- XRD   635
calamitic molecules, IQENS   690
calamitic thermotropics, displays   731 f
calamitic-discotic dimers   166
calamitics   19
- flow/viscosity   454
calorimeter, phase transitions   313, 316
calorimetric data, phase transitions   308, 321 f
cameras, X rays   621 f, 627
Canada balsam   437
capacitance method, elastic constants   270
capillary, flow alignment   459
capillary colums, gas chromatography   857
capillary tubes, X rays   631
carbon black, radiation detection   833
carbon tetrachloride, vortex formation   520
Carnahan–Starling equation   55, 64
Carr–Helfrich–Orsay approach, instabilities
   516 f, 521
Carr–Purcell–Meiboom–Gill spin echo pulse
   sequence   599
cartesian representation, tensor properties   198
caustics, textures   430
central core   19
CGH synapses   812
chains, synthesis   88
charge transfer reactions, solvents   852
CHARMM force field, molecular modelling   77
Chatelain data, light scattering   699
Chen–Lubensky model, phase transitions   289

chevrons
- displays 755
- external fields 506
- instabilities 516, 521
- SLM 767
- textures 439
- XRD 654
chiral centers labelling 116
chiral ferroelectrics, external fields 508
chiral liquid crystals, surface alignment 544
chiral lyotropics/dicotics, external fields 514
chiral nematics 21, 123, 323
- Frederiks transitions 272
- thermography 825
chiral smectics 21
- light scattering 709
- XRD 635
chirality 18, 105 f, 115–132
- blue phase transitions 323
- phase transitions 292 f
cholestanyl derivatives 873
cholesteric flexoelectric domains 504
cholesteric helix, field untwisting 501
cholesteric textures 438, 442
cholesteric–smectic A transitions, fluctuations 381
cholesterics 323
- blue phases 485
- Brillouin scattering 722
- EHD behavior 526 f
- flow/viscosity 466 f
- light scattering 708
- NMR 845
- solvents 873
- thermography 823
- ultrasonic field 557
- Volterra process 418
cholesteryl, ring systems 140
cholesteryl derivatives, solvents 873
cholesteryl nonanoate (CN)
- blue phase transitions 323
- lateral substituents 135
cholestrogen 124
chromane 140
circular dichroism 229, 846
Clark–Lagerwall effect 509
classification, liquid crystals 17–23
Clausius–Clapeyron equation
- calamitic compounds 340
- external fields 481, 487
- gas chromatography 860
- high-pressure experiments 355
clearing parameters 23
clearing temperatures
- diffusion 590
- lateral substituents 151
- phase transitions 358
- terminal substituents 146 f
clearing threshold, ultrasonic properties 552
cleavage α, solvents 852

cling film, X rays 632
clusters, XRD 641, 651
CN groups, lateral substituents 135
coherent scattering 681, 687
Cole–Cole plots, dielectric properties 233
Cole–Davidson equation, dielectric properties 234
color, thermochromic effect 123
color, thermography 823, 827, 833
columnar phases
- physical properties 200, 253
- XRD 668
competing fluctuations, re-entrants 402
complex forming salts, discotics 174
compound separation, solvents 882
compounds list, high-pressure experiments 373
computer simulations
- elastic properties 278
- molecular theories 40
concurrence, singular points 449
condis crystals, definitions 23
conductivity
- Carr–Helferich mode 524
- light scattering 704
- phase transitions 314
- tensor properties 192
configuration labels, chirality 115
conformation
- mesogens 667
- molecular modelling 79
- NMR 603
- optical properties 575
- solvents 852
conical interfaces, surface alignment 536
conjugation, singular points 449
constitutive hydrodynamic equations 454 f
continous defects, cholesterics 423
continuum theory 25–39, 253
contrast ratio, displays 745
conversion rule 117
cooling runs, scanning calorimetry 313
Cope rearrangement, NMR 844
copolymers, XRD 671
copper, X ray spectrum 620
core disjunctions, defects 427
core units, synthesis 88
correlation, Fourier transform 775
correlation lengths
- fluctuations 377
- phase transitions 282
- XRD 641
correlation times, NMR 607
Cotter approach, nematics 55
Cotton effect 846
Coulomb gauge, phase transitions 287
Coulomb interaction, diffusion 585
Coulomb potentials, molecular modelling 73
coupling constants
- external fields 479
- NMR 843

cracks, thermography 832
crenellated phases, XRD 657
critical point, phase transitions 287, 309
critical point fluctuations 384
cross polarization, NMR 599
crossbar switch n by n* 800
crossover behavior, phase transitions 311, 325
crystal phases, definitions 18
crystalline packing, XRD 667 f
cubane, ring systems 140
cubic crystals, definitions 23
cubic phases, XRD 641
Curie law 206
Curie–Weiss law 208, 243
curvature elasticity, 3-dim 263 f
cyano, terminal substituents 147
cyano compounds, re-entrants 391
cyano mesogens, molecular modelling 78
cyanobiphenyls
– displays 747
– enthalpy curves 321
– external fields 480
– photoacoustic/photopyroelectric data 329 f
– re-entrants 396
– surface alignment 540
– textures/defects 444 f
– viscoelastic ratios 706
– 5CB 81, 85, 89
– – optical properties 223
– – phase shifts 572
 see also: alkylcyanobiphenyls (nCBs)
cybotactic groups
– phase transitions 340
– XRD 651
cyclic dimers/oligomers 166
cycloadditions 100
– solvents 852
cyclobutane 142
cycloheptane 140
cyclohexadiene 137
cyclohexanes 88, 100, 137
– discotics 172
cyclohexanone 137
cyclohexylarenes, 4-E-substituted 96
cyclopentane 142
cyclopentanone 142
cyclophanes, mesogenic properties 154
cyclopropane 142
cyobatic groups, fluctuations 378

Damman grating, SLM 798
dangling chains, anchoring 541
database Liqcryst 98 f
de Broglie relationship 681
de Gennes theory
– cholesteric phases 708
– fluctuations 379
– I-N transitions 321

– light scattering 701
– penetration depth 709
– phase transitions 284 f
de Vries equation, optical properties 227
Debye equations 233 f, 247
Debye temperature, Mössbauer studies 728
Debye–Pellat equation, dielectric properties 234
Debye–Scherrer camera 622
Debye–Waller factor, XRD 642
decaline 140
defect crystals 450
defects 406–453
– blue phases/TGB 130
– elastic properties 259 ff
– ultrasonic properties 558
– XRD 641
deformations
– Dupin cyclides 433
– elastic properties 253, 264
– external fields 491
– smectics 544
– solute–solvent interactions 840
– strong 413
– thermography 823
deformed helix ferroelectric effect 503, 510
dehydration, synthesis 100
delta winglets, heat exchangers 829
density, phase transitions 332 f
density functional approach 58
density wave vector, smectics 32
detectors, selection criteria 631
deuterium, coherent scattering 681
deuterium NMR 596 f, 610, 842
device structures, thermography 825 f
dextrorotary orientation 124
diacetylenes, synthesis 99
diamagnetic susceptibilities, nematics 28
diamagnetism 204 ff
diarylethenes/bisheteroarylethenes 98
diastereoisomers, chirality 115
diazaazulene 140
dibenzopyrene, discotic cores 173
dichroic displays 745, 752
dielectric anisotropy, ultrasonic properties 562
dielectric losses, DTA measurements 358
dielectric permittivities, nematics 28
dielectric properties 231–252, 373
dielectric regime, Carr–Helferich mode 524
Diels–Alder reactions, solvents 852, 880
differential ordering effect (DOE), NMR 606
differential scanning calorimetry (DSC) 312, 354
differential thermal analysis (DTA) 354
diffraction, SLM 773, 806
diffraction techniques 621 ff, 635 ff
diffuse scattering, XRD 649
diffusion 582–83
– anisotropic 848
– IQENS 689
diffusion constants 84, 591

diffusion control, solvents   852
diffusion tensor   582
diffusivity, phase transitions   320
digitally tuneable fiber laser   806
digitally tuneable wavelength filter   803
dihydrooxazine   137
dihydrophenantrene   140
dilatometers, density measurements   332
dilution, solvent applications   850
dimerization, solvents   852
dimers
– mesogens   81, 163
– re-entrants   396
– surface alignment   540
diols, mesogenic properties   169
dioxaborinanes   102, 137
dioxalane-2-one   142
dipolar couplings, NMR   596
dipole moments
– dielectric properties   231 f
– molecular modelling   44, 80
– tensor properties   189
dipole–dipole interactions
– NMR   601, 840
– phase transitions   339
– surface alignment   540
Dirac delta function   791
direct binary search (DBS), CGH   789
direct current Kerr effect   571
directions, easy   414
director defects   406
director deformations, smectics   544
director distribution   477
– elastic properties   259 f
director fields, surface alignment   535
director fluctuations, light scattering   700
director gradients, nematics   31
director helix, chiral nematics   257
director orientation, flow/viscosity   455, 471
director reorientation, optical properties   571, 574
director structures, NMR   612
director-field interactions, distortions   488
directors
– large-scale simulation   82
– molecular theories   43
– nematics   26
– torsional elasticity   258
Dirichlet problem, defects   447
disclination pairs, core disjunction   427
disclination points/lines   444
disclinations
– defects   414, 429
– elastic properties   260 f
– ultrasonic properties   558
– Volterra process   419
discontinous defects, cholesterics   423
discotic compounds   134 f
discotic cores, large   173
discotic ferroelectrics, external fields   514

discotic stationary phases, gas chromatography   858
discotics   22, 88, 133 f
– density measurements   341
– external fields   505 f, 514
– flow/viscosity   454, 464 f
– high-pressure experiments   361
– mesogenic properties   171 f
– re-entrants   400
– XRD   635, 657
dislocations
– loop theory   288, 381
– phase transitions   291
– TGB   131
– Volterra process   419
– XRD   660
dispersion, molecular theories   44
displays   731–762
dissipation, nematics   36
dissymmetry   117
distortion aligned phase (DAP)   741
distortions
– director-field interactions   488
– discotic phases   657
– elastic properties   262
– ultrasonic properties   550, 557
dithiane   137
divergence, pretransitional   823
DOBAMBC(III), external fields   508
domain mode, smectics   512
Doppler effect, Mössbauer studies   727
double twist helical structures   129
DPPC, diffusion   591
DSC, phase transitions   312, 358
DTA, phase transitions   354
Dupin cyclides, textures   430
dyes, displays   745
dynamic light scattering, nematics   702
dynamic random acces memory (DRAM), SLM   768
dynamic scattering displays   748
dynamic scattering mode (DSM)   556
dynamics   74 f
– nematics   35 f
– neutron scattering   687

easy directions, defects   414
EBBA
– anchoring   544
– atomistic simulations   83
– density measurements   340
– metabolemeter   348
– p–V–T data   371
– solvents   872
EDA complex formation, phase transitions   339
edge dislocation, Volterra process   419
effusity, phase transitions   320
Eglington–Glaser reaction, synthesis   99
eigenwaves, SLM   764
Einstein diffusivity–viscosity law   585

Einstein relation, atomistic simulations 84
Einstein summation convention 192
elastic constants
- external fields 510
- Frederiks transitions 270
- nematics 26
- thermography 823
- twist/splay/bend 274
  see also: Frank eleastic constants
elastic incoherent structure factor (EISF) 691
elastic properties 253–280, 373
elastic torques 460, 572
elastomers 20, 379
electric field induced deformations 265
electric field SHG (EFSHG) 577
electric fields 477–534
- ultrasonic properties 562
electric permittivity 215 f
electrically controlled birefringence (ECB) 741
electroclinic effect
- smectics 511
- surface alignment 545
electroclinic response, cholesteric helix 504
electrohydroynamic instabilities 515 f
electrokinetic coefficients, instabilities 520
electrolytic mode, instabilities 519
electromagnetic radiation, optical properties 569 f
electron nuclear double resonance (ENDOR) 845
electron paramagnetic resonance (EPR) 845
- solvents 874
electron spin resonance (ESR) 586, 595, 613, 682
electron–electron double resonance (ELDOR) 615
electron–nuclear double resonance (ENDOR) 615
electronic response, optical properties 576
electronic structures 204
electrooptic device, thermography 825
electrooptical effect, smectics 766
electrooptical properties 373, 477
electrostatic interaction, molecular theories 44
elementary pinch, defects 448
Elliott toroid camera 626
emission spectroscopy 845
emissions, X rays 619
enantiomer mixtures, blue phase transitions 323
enantiomeric purity, pitch band measurements 845 f
enantiomers, chirality 105 f, 115
energy profiles, molecular modelling 74, 79
engineering, thermography 827
enthalpy, phase transitions 308, 321, 332
epitaxygens, mesogenic properties 168
equilibrum equations
- nematics 26 f
- smectics 32 f
erbium doped fiber amplifier (EDFA) 804
Ericksen–Leslie viscous coefficients 381
esterification, chirality 106
esters, solvents 883
ethenes, ethynyl bridges 97
etherification 107

ethynyl bridges, linking blocks 97
Euler angles
- flow/viscosity 469
- tensor properties 190
Euler equations, instabilities 516
Euler–Lagrange equations
- elastic properties 273
- nematics 30
- smectics 35
eutectic mixture, Tait parameters 371
eutetic point formation, metabolemeter 351
Ewald sphere, XRD 643
excimer quenching, solvents 874
excimer–exciplex chemistry, solvents 852
expansion, thermography 824
expansivity, thermal 333
experimental techniques
- diffusion 585 f
- elastic properties 279
- phase transitions 308 ff
extinction coefficients, optical properties 221

Fabry–Perot based wavelength filters 809
Fabry–Perot etalon, Brillouin scattering 719
Fabry–Perot resonator 579
fan textures 436
fanned-out inputs, SLM 798
Faraday balance 213
faults, thermography 832
ferrielectric phases (FIP) 120
- external fields 513
ferroelectric compounds, phase transitions 359
ferroelectric displays 754
ferroelectric liquid crystals (FLC) 508
- SLM 766 f
ferroelectric smectics 120
- chiral 508
- XRD 635, 655
ferroelectric TGB, external fields 487
ferroelectric transitions 373, 480
ferromagnetism 204, 208, 212
fever strips, thermography 830
fiber arrays, SLM 795
Fick diffusion law 582
field cycling, NMR 600
field effect transistor, addressing 753
field untwisting, cholesteric helix 501
filled phases, shallow-tailed compounds 159
filter layers, displays 732
filter losses, SLM 806
filters, wavelength tuneable 803
fingerprint textures 428
- ultrasonic fields 559
first order transitions 479 f, 309
flat film camera 622
flexible cyclic compounds 154
flexoelectric effects
- cholesteric helix 503 f

- external fields 493 f
- light scattering 704
flexoelectric structure, elasticity 258
flow alignment
- nematics 458, 470
- SmC phase 473
flow distribution, ultrasonic properties 549
flows 454–476
- thermography 827
fluctuations
- elastic properties 259 ff
- phase transitions 282, 308, 377
fluid layers, textures 434
fluid smectic phases, XRD 656
fluid–fluid phase transition, metabolemeter 352
fluidity, re-entrants 391
fluorene/fluorenone 140
fluorescence
- polarized 847
- solvents 852, 874
fluorinated methyl 147
focal curves, textures/defects 446
focal surfaces, textures 430, 435 f
foggy phases
- cholesteric helix 503
- external fields 485
force field, molecular modelling 72
forced Rayleigh scattering 708
form optical activity 122
Fourier series, surface alignment 536
Fourier spectroscopy 600
Fourier transform
- SLM 773
- XRD 642
fractional volume changes, phase transitions 333
fragmentation reactions, solvents 852
frame rate, FCOASLM 773
Frank elastic constants
- cholesteric helix 503
- light scattering 701
- molecular theories 60
- NMR 612
  see also: elastic constants
Frank energy, field interactions 489, 495
Frank–Oseen approach, optical properties 227
Frank–Oseen energy
- anchoring 543
- nematics 26, 36
- phase transitions 285
Franklin relations, diffusion 584
Franks camera 625
Frederiks transitions
- displays 733 ff
- elastic properties 266 ff
- external fields 488, 497
- light scattering 701
- smectics 505 f
- ultrasonic properties 562
freedom degree, optical properties 218

freeze etching technique, textures 413
frequencies
- Brillouin scattering 719, 724
- electric permitivities 245
- ultrasonic properties 550
Friedel–Crafts acylation 91, 98
frustrated phases 129
frustrated smectics
- phase transitions 296
- XRD 635, 656
frustration re-entrants R1 392
fugacy, gas chromatography 860
functional groups 117
furans, synthesis 104
fused twins, mesogenic properties 163

γ rays
- Mössbauer studies 727
- thermography 833
gas chromatographic applications 839–895
gauche conformation
- molecular modelling 79
- solvents 854
- terminal substituents 146
Gaussian model, phase transitions 282, 303
Gay–Berne interaction
- diffusion 585
- phase transitions 341
gelatine-gum arabicum, encapsulation 825
Gelbhart–Barboy expansion, nematics 55
Gelbhart–Ben–Shaul theory 274
giant optical nonlinearity (GON) 572
Gibbs free energy 48
- elastic properties 274
- mesogenic properties 138
- phase transitions 310
Ginzburg criterion, phase transitions 311
Ginzburg temperature 282
Ginzburg–Landau parameter 381
glancing angle diffraction techniques 635
glass capillary tubes, X rays 631
gold black, radiation detection 833
Goldstone modes, dielectric properties 244
Gooch–Tarry curve, displays 742
Gouy balance 213
grain boundaries, textures 439
Grandjean textures
- cholesteric helix 505
- displays 749
- EHD behavior 526
- external fields 498
Grandjean–Cano wedge 427
grating, Damman 798
grating losses, SLM 806
Grignard reactions, synthesis 95, 100
group priority 117
groups, definitions 19
guest–host interactions, displays 745

Guinier camera 623
gyromagnetic ratio 587

half waveplate, SLM 765
half-width at half-maximum (HWHM), XRD 648
halogen derivatives, solvents 882
halogenes, substituents 135, 147
Halperin–Nelson–Young concept, bond ordering 291
Hamiltonians
- molecular theories 51
- NMR 840
- phase transitions 282 f
hard core repulsion, Maier function 56
hard particles theory, elastic properties 277
hard rod fluid, phase diagram 63
harmonic potentials, bond stretches 73
HBA, atomistic simulations 83
HBPD, thermobarogram 350
heat capacity, phase transitions 308
heat transfer, thermography 828
heating modes, scanning calorimetry 313
Heck reaction, synthesis 93, 98
Heilmeier displays 740, 745
Helfrich model, ultrasonic properties 562
Helfrich–Hurault mechanism 273
helical axis, cholesterics 466
helical smectic phases 126
helical structures, light scattering 708
helicity 18
helicoidal structures
- optical properties 226
- thermography 823
helium saturation, phase transitions 357
helix compensation, chiral nematics 123
Helmholtz energy
- displays 737
- elastic properties 274
heptyl, Tait parameters 371
heringbone structure, fluctuations 383
heterocycles 143, 882
heterocyclic unit, linkage 88
heterodane detection, light scattering 704
heteronuclear chemical shift correlation (HETCOR) 605
hexa-n-octanoate, discotic cores 173
hexagonal smectic phases, external fields 507
hexakis(phenyl-ethynyl)benzene 173
hexatic phases, XRD 639, 661 f
hexatic properties, fluctuations 383
high pressure experiments, phase transitions 353 f
high resolution IQENS 688
historical developments 1–16, 89 f
HOBA, EHD behavior 528
Hoffman elimination, solvents 877
holograms, computer-generated (CGH) 786
holographic interconnects, SLM 792
holographic replay, polarization insensitive 790
homeotropic nematics, external fields 495

homeotropic sample cell, light scattering 705
homodyne detection, light scattering 704
homogeneous distortions, ultrasonic properties 550
homologues, nematic 589
Hooke law, torsional elasticity 254
Horner-Emmons reactions 93, 97
hot cathode X ray tube 620
Hückel model, free-electron 571
Huxley–Holmes mirror-monochromator camera 626
hybrid aligned cells, elastic properties 272
hybrid optoelectronic neural networks 813
hydrocarbon mixtures, gas chromatography 858
hydrocarbons, solvents 882
hydrogen, coherent scattering 681
hydrogen bond formation 340
hydrophobic–hydrophilic interface, NMR 602
hyperpolarizability 570
hypersound velocities, Brillouin scattering 724

illumination spectrum, thermography 827
illuminations, SLM 772, 794
image plates, X ray techniques 628
images, real time 832
imidazole derivatives, mesogenic properties 169
in fiber devices, SLM 818
incidence angle, thermography 827
incoherent quasi-elastic neutron scattering (IQENS) 687
incommensurate smectic phases, XRD 656
incompressibility
- nematics 27
- smectics 34
indanes, ring systems 140
indicatrix, optical 216 f
indium tin oxide (ITO) 732, 768
inertia tensor, atomistic simulations 83
inertial mode, instabilities 525
inhomogeneous distortions, ultrasonic properties 555
injection modes, instabilities 518
inorganic compounds, NMR 869
inositol derivatives, mesogenic properties 178
instabilities 515 f
- electrohydrodynamic 477
- planar cholesteric textures 499
- XRD 647
instrumentation, density measurements 332
intensity
- SLM 768
- XRD 641
interaction types 44, 585
interconnects 775, 785 f
interdigitated smectic phases, re-entrants 395
interface defects 415
interface deformation, elastic properties 264
interfaces
- nematic 30, 543
- NMR 602
- surface alignment 535

interferometry, radiation detection   833
intermolecular interactions   40, 65, 253
– phase transitions   310
intermolecular interference   682
intermolecular motions, NMR   844
intermolecular potentials, molecular modelling   81
interpenetrable layers, anchoring   541
intramolecular association, lateral substituents   153
intramolecular interference   682
inverse salts, mesogenic properties   170
iridescent colors, cholesterics   825
isomer separation
– gas chromatography   857
– solvents   882
isomerization   92
– solvents   852, 875
isotropic fluids, dielectric properties   231 f
isotropic instabilities   516 f
isotropic interfaces, nematic   543
isotropic modes, EHD behavior   527
isotropic nematics, Brillouin scattering   720
isotropic–cholesteric transitions, light scatterin   713
isotropic–nematic transitions   279 f, 321 f
– scattering   713, 720
isotropically labeled samples, neutron diffraction   682
IUPAC   17

J couplings, NMR   843
JOERS–Alvey scheme, displays   760
jogs, defects   429
joint transform correlator JTC   783
Jones matrices, SLM   763 f
Joule heating, tensor properties   192

K transitions, X rays   619
Kapton, X rays   632
Kapustin–William domains (KPD)   516 f, 521
Kawasaki mode coupling, fluctuations   381
Kernpunkt, Lehmann convention   416
Kerr cell, cholesteric helix   503
Kerr effect
– direct current   571
– external fields   484 f
kinks, defects   429
Kirkwood correlation factor   232
Kirkwood–Frohlich equation   232
Kobayashi model   61
Konvergenzpunkt, Lehmann convention   416
Kosterlitz–Thoules mechanism, dislocations   291
Kozhevnikov theory, ultrasonic properties   551 f
Kramers–Krohnig relation   223, 245
Krat–Kennedy equation, phase transitions   359
Kratky camera   627
Kroehnke reaction, synthesis   98
Kronecker delta, tensor properties   194

laevorotary orientation   124
laminar flows, thermography   827
Landau theory   310
– blue phases   708
– external fields   478
– SmA–SmC transition   327
Landau–de Gennes theory   279 ff
– anchoring   542
– I–N transitions   321
– optical properties   574
Landau–Ginzburg coefficients   282
Landau–Ginzburg Hamiltonian, critical   303
Landau–Khalatnikov mechanism   564
Landau–Lifshitz theory, phase transitions   366
Landau–Peierls instabilities   285, 647
Langmuir–Blodgett film, atomistic simulations   85
Laplace equation   445
large-scale simulation, molecular modelling   82
Larmor frequencies, NMR   596
lasers, wavelength tuneable   803
lateral linked twins   164
lateral substituents, mesogenic properties   135, 151 ff
lattice models, elastic properties   277
lattice orientational order, XRD   648
lattice vibrations, Mössbauer studies   727
lattices, reciprocal   641
layer packing, smectics   126
layers
– displays   732
– textures   407 f, 434
leading structures, synthesis   89
leaks, scanning calorimetry   313
leapfrog algorithm, molecular modelling   75
Legendre functions
– flow/viscosity   465
– molecular theories   41
– tensor properties   195
Lehmann convention, defects   416
Lennard–Jones interaction, diffusion   585
lenses, SLM   819
Leslie viscosities
– diffusion   584
– external fields   490
– flows   455, 465
– light scattering   704
– NMR   612
Leslie–Erickson theory
– flow/viscosity   454 f
– nematics   702
– ultrasonic properties   549 f
lettering scheme   17
Levi–Civita tensor
– flow/viscosity   456
– molecular theories   43
Lifshitz point   328, 501
ligated twins, mesogenic properties   163
light band centers, ultrasonic properties   551
light scattering   699–718
– instabilities   525

light valves, SLM   784
line pairs, FCOASLM   772
line shapes, deuterium NMR   597
linear dichroism   846
– tensor properties   200
linear polarizers   766
linewidth
– Mössbauer studies   727
– spectroscopy   840
linking groups   19, 87
– ethynyl bridges   97
– mesogenic properties   135, 144 ff
Liouville theorem
– ESR   613
– re-entrants   399
liquid–gas type transitions   310
liquid–nematic transition, isotropic   479
London penetration depth, phase transitions   285 f
long-range order, XRD   644
longe-range translational diffusion   691
Lorentz factor, XRD   647
Lorenz spectral densities   704
Lorenz–Lorentz equation   218 f, 232 f
low-angle diffraction pattern   638
low-frequency permittivities   235
low-molecular mass materials, neutron scattering   682
Lubensky model, fluctuations   382
lyotropic mesophases, diffusion   591
lyotropic nematic polymers, light scattering   705
lyotropic nomenclature   18
lyotropic systems
– external fields   493
– XRD   635
lyotropic textures   409
lyotropics   11
– chiral   514
– density measurements   341
– elastic properties   264
– surface alignment   546

MacroModel, molecular structure   79
macroscopic structures   477
macroscopic surface alignment   536 f
macroscopic tensor properties   194
magic angle spinning (MAS)   599
magnetic applications   212, 602 ff
magnetic field induced deformations   265
magnetic fields   477–534
magnetic properties   28, 190, 204–215
magnetic resonance   595–618
– diffusion   586
magnetometers   213
Maier function, hard-core repulsion   56
Maier–Meier equations   236 f
Maier–Saupe theory   44 f, 279 f, 332
– optical properties   574
main chain liquid crystalline polymers (MCLCP)   20
– SANS   686

Maltese cross, textures   410
mapping, thermography   830
matched filters, SLM   777
matrix–matrix switches, SLM   798
Mauguin behavior
– defects   428
– displays   740
Maxwell equations
– instabilities   517
– optical properties   215, 569 f
MBBA   89
– anchoring   544
– bulk SHG   577
– clearing thresholds   552
– conformational effects   575
– defects   406
– diffusion   588
– metabolemeter   348
– polygonal textures   411
– Rayleigh scattering   701
– solvents   872
– sound velocity   721
– Tait parameters   370
– volume changes   335
– vortex formation   520
MBBA/Canada balsam, textures   438
2MBCB, atomistic simulations   83
McMillan theory   61, 284 f
– fluctuations   379
– light scattering   714
mean field theory   46 f
– elastic properties   277
– fluctuations   377
– light scattering   714
– N–SmA transition   60
– N–SmA–SmC   293
– phase transitions   299, 310
measuring cell, photoacoustic   318
mechanics, molecular modelling   72 f
medical thermography   830
medium resolution IQENS   695
Meissner effect   285 f
– fluctuations   380 f
melting point equation   359
melting temperatures, lateral substituents   151
mesogen conformation, XRD   667
mesogenic properties   133–186
mesogens   19, 78 f
– modulated phases   657
mesophase stabilization, terminal groups   88
mesophase structures, XRD   635
mesophase symmetry   122
mesophases   18, 133 f
– defects   443
– density measurements   332
– diffusion   588
– IQENS   695
– lamellar   437
MET(MNT) mixture, thermobarogram   352

metabolemeter   348 f
metal complexes, mesogenic properties   170
metal containing liquid crystals   74
metallomesogens   171 ff
methyl groups, lateral substituents   135
methyl red, mass diffusion   710
methylene, volume changes   337
Metropolis algorithm   76
microconfined environments, NMR   610
microencapsulation, thermography   825
microfocus diffraction tubes   621
microscopic anchoring mechanism, solid substrates   541
microscopic preparations, defects   406
microscopic structures   477
microscopic surface order, nematics   540 f
microscopic tensor properties   194
microtomy methods, textures   413
microwave detector, thermography   833
Miesowitz viscosities   704
Miesowitz notation   456
Miller indices   643
miscibility   17
mixtures
– displays   733
– molecular theories   41
– re-entrants   391, 394
Möbius strips, defects   426
model potentials, molecular theories   46
modelling defects, nematics   32
modelling sofware   72
modulated smectic phases, XRD   656
modulated structures, mean field model   301
molecular asymmetry   115 ff
molecular deformations   842
molecular dynamics, NMR   607
molecular interactions, phase transitions   339
molecular modelling   72–86
molecular motion, neutron scattering   691
molecular properties   80
molecular shapes
– re-entrants   400
– unconventional   153 f
molecular structures   78
– symmetry   118
molecular theories   40–71
– elastic properties   274 f
– refractive indices   217 f
– statistical   134 f
molecular transforms, XRD   641
monitoring boundary effects, ultrasonic properties   565
monocrystals, textures   410
Monte Carlo simulation, molecular modelling   74 f
MOPAC, molecular modelling   80
mosaic textures   440
Mössbauer studies   727 ff
motional narrowing, NMR   598
MPP theory, flow/viscosity   454

multicritical point
– N–SmA–SmC   655
– phase transitions   328, 367
– re-entrants   393
multidimensional spectroscopy, magnetic resonance   600
multiple correlations, SLM   777
multiple sequences, NMR   599
multiwire proportional counters   629
myelin textures   409
Mylar sheet, X rays   632

n to n* holographic switch   796
N–A transitions, external fields   481
N–I transitions   47, 59, 321
– density changes   333
– external fields   479
– fluctuations   378
– optical fields   574
N–Sm transitions
– fluctuations   379
– light scattering   714
N–SmA transitions   283 f, 324
– critical behavior   650
N–SmA–SmC point
– multicritical   328, 385
– phase transitions   289
N–SmC–SmA transition, re-entrant behavior   367
naphthalenes   140
– solvents   882
Navier–Stokes theory
– flow/viscosity   454
– instabilities   517, 523
Neel walls, defects   415
neighbor lists, molecular modelling   77
nematic biaxiality, induced   483
nematic mesophase, diffusion   588
nematic phases   17 ff, 47
– Brillouin scattering   720
– dielectric properties   235
– order parameters   41
– powder pattern   638
– XRD   649 f
nematics
– director-field interactions   488
– flow/viscosity   454 f
– instabilities   515
– light scattering   699
– surface alignment   535
– Volterra process   418
Nenitzescu reaction   100
Neumann principle, tensor properties   201
neural networks   810
neutron scattering   680–698
Newton equations, molecular modelling   74
nickel, X ray absorption curve   622
nitro, terminal substituents   147
nitrogen saturation, phase transitions   357

nitrogen, six-atom ring systems   137
nitroxide spin probes, EPR   845
nomenclature   17–23
nonadiabatic scanning calorimetry   317
nonaromatic compounds, ring systems   139
nondestructive testing, thermography   831
nondisplay applications   763–822
nonergodic systems, re-entrants   399
nonlinear optical properties   569–581
nonmesogenic solutes   859
– thermodynamic properties   884
nonpolar rings, mesogenic properties   140
normal-metal–superconducting transition   380
Norrish type I reactions, solvents   852
Norrish–Yang type II reactions   850
– solvents   874
Nose–Hoover thermostat, molecular modelling   75
nuclear magnetic resonance (NMR)   213, 595 ff, 840
– diffusion   586
– phase transitions   362
– solvents   874
– tensor properties   200
nucleation, defects   448
nuclei textures   417
Nusselt number   828

OBBA, volume/density changes   336
OCB
– atomistic simulations   83
– Brillouin scattering   721
– re-entrant phases   659
– metabolemeter   348
oily streaks, textures   410
olefin isomerization, solvents   854
olefinations, synthesis   92
oligomers, mesogenic properties   162 f
oligophenyls, transition temperatures   136
oncology, thermography   830
one to n* holographic switch   793, 795
one to n* shadow logic switch   799
Onsager theory   44 f
– flow/viscosity   455, 471
– nematics   37
– phase transitions   279 f
operating regimes, displays   740
optical absorption coefficient, phase transitions   319
optical activity   118
– high-pressure experiments   363
optical contrast, defects   428
optical correlation, SLM   773 f
optical harmonic generation   577
optical methods, high-pressure experiments   355, 360
optical neural networks   810
optical path, smectic structures   226
optical power limiting (OPL)   579
optical properties   215–230
– nematic layers   739
– nonlinear   569–581

– thermography   824
optical second harmonic generation, surfaces   540
optical systems   4 f, 775
optical transmission
– nematics   561
– ultrasonic properties   550
optical vector processor   811
optically addressed SLM (OASLM)   771
optically connected parallel machine switch   800
optimized parameters for liquid simulation   77
optoelectronic X ray imaging devices   630
order parameter-director coupling, anchoring   542
order parameters   41
– defects   406
– external fields   482
– large-scale simulation   82
– NMR   609
– phase transitions   280 f, 339
– tensor properties   194
– viscosity coefficients   464
order types
– phase transitions   336
– XRD   644 f, 649
ordered smectic phases, XRD   649, 664
ordoelectric polarization, external fields   494
organic compounds, NMR   861
organic liquid crystals, modelling   74
organo nickel/palladium chemistry   94
organometallic compounds, NMR   869
orientational anisotropy, elastic properties   253
orientational correlations, short-range   51
orientational order
– neutron scattering   682
– optical-field-induced   574
– XRD   641
orientations
– nematics   550, 651
– NMR   603
– solvents   839, 852
– surface alignment   535, 544
– transition states   848
Ornstein–Zernike equation   58
Orsay energy, smectics   32
orthoesters   102
orthogonal phases, definitions   22
Oseen theory, diffusion   584
oxathiane   137
oxirane   142

PAA
– density measurements   340
– diffusion   588
– IQENS   695
– structure factors   683
packing
– crystalline   667
– helical smectics   126
– thermotropic nematics   54

paints, overheating components 832
paramagnetism 210, 204 f
parasitic scattering, X rays 622
Pariser–Parr–Pople model, optical properties 571
Parodi equation, flow/viscosity 455, 465
Parsons approximation, molecular theories 60
Parsons–Lee approximation, thermotropic nematics 54
pattern formation, optical 574
PCH3
– energy minima 80
– molecular modelling 78
PCH5, atomistic simulations 83
Peierls argument, fluctuations 380
pendant chains, synthesis 88
penetration depth, phase transitions 285
pentyl, Tait parameters 371
Percus–Yevick approximation, nematics 55
perhydrochrysene 140
perhydrophenantrene 140
perhydrotetracene 140
periodic distortions, ultrasonic properties 554
periodic structures, external fields 492
Perkin reaction, synthesis 91
permittivity, dielectric 231 f, 506
pertubation theory
– tensor properties 191
– time-dependent 570
phase array antennae, SLM 820
phase behavior, NMR 602
phase delays, switcheable 820
phase diagrams
– hard-rod fluid 63
– N–SmA transition 324
– N–SmA–SmC multicritical point 328
– re-entrants 391
phase modulation, binary 770
phase only matched filter (POMF) 778
phase problem, XRD 636
phase separation, solvent applications 849
phase sequences, re-entrants 392 f
phase structures, optical properties 223
phase symmetry, optical properties 217 f
phase transitions 20, 279–376
– high-pressure experiments 356
– optical properties 224
– temperature shifts 478
– ultrasonic properties 563
– under pressure 366
phases 17 ff
– optical properties 218
– XRD 635
phasmidic compounds, mesogenic properties 161
pHB, atomistic simulations 83
phenanthrene 140
phenylbenzothiazole 140
phenylbenzoxazole 140
phenylcyclohexanes, displays 747
phenylpyrimidines, displays 747

phlogopite mica, interfaces 536
phospholipids, textures 409
photo-Fries, solvents 852, 875
photoacoustic methods, phase transitions 318
photoacoustic spectroscopy 565
photochemical reactions, solvents 851, 874
photomultiplier, X rays 628
photon correlation spectroscopy 712
photophysical reactions, solvents 874, 878
photopyroelectric method, phase transitions 320
physical properties
– orientation-dependent 189 f
– under pressure 373
see also individual types
Pi cell, displays 744
piezocoefficient, external fields 479
piezoelectric tensor 191
piperazine 137
piperidine 137
piston displacement, phase transitions 370
pitch length 846
pixel pitch, SLM 780 f
pixels
– CGH 787, 794
– displays 747
– smart 810
plages a noyaux, textures 417
planar sample cell, light scattering 705
planar textures 434
– external fields 499
– walls 440
Planck constant, NMR 840
plastic crystals, definitions 23
platelet textures 440
plug flow 467
Pockels modulator, optical properties 579
Poiselle flow 467
polar achiral polyphilics, external fields 508
polar interactions, phase transitions 340
polar smectics, phase transitions 32, 296
polarity, nematics 26
polarizabilities
– anisotropic 51
– dielectric properties 231 f
– mesogens 220
polarization 570
– chirality 105
– magnetic 206
– optical properties 216 f
– ordoelectric 494
– photopyroelectric method 320
– spontaneous 119
– tensor properties 190
polarized optical absorption 845
polarizers, SLM 763 f
polarizing microscopy, textures 410
polycatenar compounds, mesogenic properties 160 f
polygonal textures 435
polymer dispersed liquid crystals (PDLC) 12, 826

polymer liquid crystals   20
– external fields   488
– ferroelectric   512
– surface alignment   546
– XRD   669
polymeric systems, XRD   635
polymers
– density measurements   341
– external fields   485
– NMR   608
– SANS   684
polymorphism, smectic A phase   653
polyphilics, external fields   508
Poniwierski–Sluckin theory   68
positive lens, SLM   773
potentials, molecular modelling   73
Potts model, fluctuations   384
powder diffraction   636 f
powder samples, Debye–Scherrer camera   622
precursors, building blocks   98
precursors, synthesis   91
pressure, flow/viscosity   462
pressure transducer, metabolemeter   349
pressure–temperature phase diagram, calamitics   340
pressure–volume–temperature behavior   369
pretransition behavior, density measurements   339
pretransitional light scattering   713
pretransitional properties, thermography   823
propyl, Tait parameters   371
Prost model, phase transitions   297
proton decoupling, NMR   599, 604
pseudo-Bragg law, XRD   636
pseudowalls, textures   439
pycnometers, density measurements   332
pyramidal compounds   176
pyrazines   101, 104, 137
pyridazine   137
pyridine   137
pyrimide   137
pyrimidines   101
pyroelectric transducer, phase transitions   320

quadrupolar coupling, NMR   843
quadrupolar splitting, NMR   841
quadrupole interactions, NMR   596
quality control, thermography   832
quantum mechanical techniques, molecular modelling   78
quasielastic neutron scattering (QENS)   586, 681
quasilong-range order, XRD   647
quater waveplate, SLM   766
quaternization, solvents   855, 880
quenching, solvents   874
quinoaxaline   140
quinoline   140
quinquiphenyl   136

R1 re-entrants   393 f
R3 re-entrants   402
r_α structure, NMR   842
radiation detection, thermography   833 f
radiation fluxes, ultrasonic properties   551
Raman scattering   682, 719
rank, tensor properties   189 f
Raoult law   861
Rayleigh scattering   699 f, 708
reactants, solvent applications   849
reactivity, chemical   848
real space structures, Fourier transform   643
real time images, thermography   832
rearrangement, NMR   844
reciprocal lattices, XRD   641
recoil energy, Mössbauer studies   727
rectus, Cahn nomenclature   115
red shifts, external fields   486
redistribution, XRD   636
reduction, synthesis   100
re-entrant behavior
– high pressure experiments   363
– metabolemeter   349
re-entrant phase transitions   304, 391–405
re-entrant phases, XRD   658
re-entrants, definitions   23
reflection, thermography   823
reflectivity
– optical properties   224
– XRD   659
refraction
– phase transitions   373
– XRD   636
refractive indices
– fluctuations   262
– mesogens   220
– optical properties   215–230, 576
– thermography   824
– wavelength filters   809
relaxation
– dielectric properties   245 f
– NMR   598
– ultrasonic properties   563
relaxation rate, Brillouin scattering   723
relaxation times
– ac calorimetry   315
– dipole reorientation   232
– external fields   485
– NMR   841
Renn–Lubensky model, fluctuations   382
renormalization group   288, 311
reorientation
– directors   265
– optical-field-induced   572
replay field, CGH   788
resistivity, tensor properties   192
response time, FCOASLM   773
retardance, wavelength filters   806
retention, gas chromatography   857

Reynolds number 517
ring containing substituents, lateral 155 f
ring inversion, NMR 844
ring systems
– large 172 f
– mesogenic properties 136 ff
rod-like compounds, mesogenic properties 135 f
rotational symmetry operations 118
rotational viscosity 456, 470 f
rufigallol, discotic cores 173

S distortions, external fields 489
salicylaldimine complexes, magnetic properties 210
salt like compounds, mesogenic properties 170 f
salts, mesogenic properties 170
sample cells, light scattering 705
sample environment, X rays 632
saturated systems, solvents 872
Saupe ordering matrix, tensor properties 197
Saupe-Nehring theory, elastic properties 277
SAW, ultrasonic properties 550
scaling, phase transitions 287 f
scanning calorimetry, phase transitions 313
scattering experiments 621, 649, 681
scattering mode devices 748 f
Schiff base mixtures, re-entrants 394
Schiller theory, flow/viscosity 454
schlieren textures 417
– nematic 446
screw dislocations
– TGB 131
– Volterra process 419
– XRD 660
second harmonic generation (SHG) 577
second order transitions 309
– external fields 478
segregation, elastic properties 264
Seiko scheme, displays 758
selection criteria, solvents 851
selective reflection
– phase transitions 361
– temperature dependence 823
selectivity, gas chromatography 858
selfdiffraction, optical properties 573
separation efficiency, gas chromatography 858
shadow routed crossbars, SLM 798
SHAKE procedure, molecular modelling 75
shallow-tailed compounds 158 f
shear gradients, flow/viscosity 466
shear properties, thermography 827
shear stress 253
– wind tunnels 827
shear viscosities
– flow phenomenon 455
– nematics 469
shell electrons, X-rays 619
shift invariance, correlators 777
shifts, phase transitions 478

short-range dipole–dipole interactions 240
short-range order, XRD 648
shutter-based switch 1 to 16 799
side chain polymers, SANS 685
side groups liquid crystalline polymers (SGLCP) 20
sigmatropic rearrangements, solvents 852, 876
signal-to-noise ratio (SNR), SLM 776
silica layers, displays 732
α silicon noveltry filter 818
silicone oil, vortex formation 520
Simon equation, phase transitions 357
Simon–Glatzel equation, phase transitions 358
single crystal experiments 840
singular points/lines/walls, defects 413 f
sinister, Cahn nomenclature 116
six membered rings, mesogenic properties 136
SmA phases
– flow/viscosity 474
– interdigitated 395
SmA–C* transitions, ferroelectric 481
SmA–hexatic smectic B transition 291 f
SmA–SmC transition 65, 289 f, 326, 478
SmA-TGB$_A$–cholesteric, multicritical behavior 368
SmA–SmA critical point 302
SmA$_d$/SmA$_I$, fluctuations 384
SmC*–SmA-TGB$_A$, multicritical behavior 368
SmC/C* phases, flow/viscosity 470 f
small angle neutron scattering (SANS) 684
smart pixels 810, 814
smectic mesophase, diffusion 590
smectic order parameters, XRD 646
smectic phases 17, 43, 60
– Brillouin scattering 722
– dielectric properties 240
– EHD behavior 527
– light scattering 709 f
– order parameters 41, 664
– powder patterns 639
– solvents 849
– ultrasonic field 559
– XRD 635, 652
smectics 21
– EHD behavior 526
– external fields 505 f
– Frederiks transitions 273
– surface alignment 544
– torsional elasticity 258
– Volterra process 418
smooth surfaces, anchoring 541
Snells law, SLM 774
soft crystals, description 20
solid state, description 20
solubility factor, solvent applications 849
solute–solvent interactions 840
solutes
– chemical reactivity 848
– diffusion 590
– NMR 841
– nonmesogenic 859

solvents, chemical reactions   839–895
sound velocity, Brillouin scattering   721
space symmetry   118
spacer model, NMR   609
spatial configurations, symmetry   117
spatial frequency pixels   780
spatial light modulation (SLM)   763
specific volume, phase transitions   332
spectroscopic applications   839 57
spherical harmonics, tensor properties   197
spin, NMR   596 f
spin Hamiltonian, ESR   613
spin-glas, re-entrants   399
spin–spin coupling, NMR   840
spin–spin interactions   208
spirals origin, textures   442
spiro ring systems   141
splay deformation
– directors   544
– elastic properties   267
– torsional elasticity   256
splay Frederiks cell, displays   735
splay Frederiks distortion, external fields   498
splay-bend director, light scattering   704
splay–bend–twist deformations, external fields   491
splitting, NMR   841, 844
SQUID magnetometer   213
stability
– mechanical   253
– exchange   519
stacking, bowl-shaped molecules   122
stacking faults, XRD   641, 664
Stanford vector matrix multiplier   798, 811
star-like compounds, mesogenic properties   167
states, definitions   18
static light scattering, nematics   699
static random access memory (SRAM)   768
stationary phases, gas chromatography   857
statistical mechanics   40
statistical molecular theory   58
steady state patterns, external fields   492
stepped drops, textures   430
stereochemistry, definitions   23
stereoisomers, chirality   115
steric energy, molecular modelling   72
steric fixation, chirality   105
Stille reaction   94
stochastic dynamics, molecular modelling   75
storage model, ultrasonic fields   557
strain   253 f
strain tensor   37
stress   253 f
stress matrices   35
stress tensor   455
strong anchoring, elastic properties   260
structural formulas, mesogens   87
structural properties, TGB phases   294
structural studies, XRD   635–679
structural transformations   549 f

structure deformations, solute–solvent interaction   840
structure factors, neutron scattering   683
structures   78
– columnar phases   668
– magnetic/electric field influences   477
– ordered smectic phases   664
– periodic   492
– supertwist   497
– ultrasonic properties   564
substituted sugars, discotics   177
superconductor analogy, phase transitions   284 f
superconductors   377, 393
superfluid helium analogy, phase transitions   289
superposition, defects   449
supertwist structures, external fields   497
supertwisted birefringence effect (SBE)   743
supertwisted nematic displays   746
surface alignment   535–548
surface conditions, continuum theory   25
surface extrapolation length, distortions   489
surface orientational order, nematics   540
surface SHG, optical properties   578
surface-stabilized ferroelectrics (SSFLC)   509, 545
– flow/viscosity   473
surface temperature, thermography   827
surfaces
– textures   409, 430
– XRD   659
susceptibility
– diamagnetic   28, 209
– dielectric   236
– elastic   265
– external fields   479
– magnetic   204
– molecular   42, 209
– nonlinear   570
– phase transitions   280
– tensor properties   190
Suzuki reaction   94
switcheable phase delays, array antennae   820
switches, holographic   793
switching, displays   756
symmetry   115–132
– re-entrants   391
symmetry changes, external fields   483
synapses, neural networks   810
synthesis, strategies   87–113

T distortions, external fields   489
Tait equation, phase transitions   370
target molecules, synthesis   91
target structures, building blocks   98
TBBA, diffusion   590
TBDA, volume/density changes   336
temperature control, X ray experiments   633
temperature dependence
– density   333
– phase transitions   309

- refractive indices 224
- selective reflection 823
temperature effect, viscosity coefficients 464
temperature independent paramagnetism (TIP) 207
temperatures, Mössbauer studies 720
TEMPO, ESR 613
tensor properties 42, 189–103, 204
- torsional elasticity 254 f
terminal aliphatic chains 125
terminal groups 19
- synthesis 87
terminal linked twins 164
terminal substituents 135, 146 f
testicles, thermography 831
tetracatenar compounds, mesogenic properties 160
tetrafluorobenzene 137
tetrahydrofurane 142
tetrahydropyrane 137
tetralene 140
texture transformations 444 f
- ultrasonic fields 557
texture transitions, smectics 506
textures 406–453
- EHD behavior 526
  see also: Grandjean textures
thermal characteristics, phase transitions 308
thermal effects, optical properties 575
thermal expansion coefficients 332
thermal mapping 831
thermal methods, phase transitions 308 f
thermal reactions, solvents 876
thermistors, ac calorimetry 315
thermobarograms, phase transitions 348 f
thermobarometric analysis (TBA) 349
thermochromic effect 123
thermodynamic properties, nonmesognic solutes 884
thermography 823–838
thermometers 830
thermotropic calamitic liquid crystals 87 ff
thermotropic cubic phases, XRD 641
thermotropic mesogens, magnetic properties 209
thermotropic nematics, packing effects 54
thermotropic properties 18 ff
thermotropic systems, external fields 493
thermotropic textures 407 f
thermotropics, solvents 872
THES, atomistic simulations 83
thiadiazoles, synthesis 104
thilirane, ring systems 142
thiopenes, synthesis 104
third harmonic generation (THG) 577 f
threads, thin/thick 418
three dimensional video 817
three ring systems 138
tilt angles
- Frederiks cell 734
- SmC phase 640
- XRD 641

tilt-polarization coupling, external fields 479
tilted chiral phases 21
tilted hexatic phases, XRD 663
tilted-orthogonal smectic transitions, fluctuations 382
time-resolved diffraction 687
time scales, NMR 607
tolanes, ethynyl bridges 97
topographies, anchoring 541
torque
- elastic properties 259
- flow/viscosity 457
- generation under flow 468
torsional elasticity 253 f
torsional energy barriers, molecular modelling 75
tracer technique, diffusion 585
trans/gauche interconversions, solvents 854
transconformation
- molecular modelling 79
- terminal substituents 146
transient patterns, external fields 493
transition metals, NMR 871
transition temperatures 20
- linking groups 145 f
- mesogenic properties 133 ff
- polycatenar compounds 161 f
transitions
- external fields 478
- fluctuations 379
  see also: phase transitions
translation dislocations 447
translational diffusion, IQENS 689
translational order, XRD 644
transmission, optical properties 225
transmission techniques, high-pressure experiments 358
transparence, solvents 851
transparent windows, X rays 632
transport properties, phase transitions 308
transverse pressure, flow/viscosity 462
triazine 101, 137
tribenzocyclononene derivates, oligomers 167
tricatenar compounds, mesogenic properties 160
tricritical point
- anchoring diagram 538
- external fields 482
- fluctuations 377
- phase transitions 328, 366
trimers, mesogenic properties 163
triphenylene, discotic cores 173
triplet quenching, solvents 852, 874
triptycene, mesogenic properties 168
tropone 140
truxene, discotic cores 173
tubular compounds 176
tumbling, flow/viscosity 459
tungsten tip surface scanning by scew spindle 369
turbidity measurements, phase transitions 359
turbulent flows, thermography 827

twins, mesogenic properties   162 f
twist angles, displays   738
twist/bend director   704
twist–bend distortions   262
twist deformation
– elastic properties   267
– thermography   823
– torsional elasticity   256
twist disclination, defects   429
twist grain boundary (TGB)   22, 130
– external fields   485 f
– fluctuations   382
– high-pressure experiments   360
– phase transitions   286, 294 f
– XRD   641, 659 f
twist transitions, optical-field-induced   573
twisted nematic displays   746
twisted nematics/cholesterics   497
twisted phases, external fields   485
twisted transitions, displays   735
twisting power, nematics   847
two-dimensional approach, instabilities   523
two ring systems   138

Ullmann synthesis   94
ultrasonic properties   549–568
ultrasound
– absorption   564
– intensity   550
– phase transitions   373
– radiation detection   833
– studies   563
unaligned samples, diffraction patterns   638
uniaxial nematic transitions   280
uniaxial nematics, flow/viscosity   454
uniaxial structures, mean field model   299
uniaxial–biaxial nematic transitions   379
united atom model   77
– mesogens   82
untwisting dynamics, cholesteric helix   502

vacuum, X rays   620
van der Meer–Vertogen model   68
van der Waals interaction   134
– elastic properties   274 f
– molecular theories   44
– nematics   55
vapor interfaces, nematic   543
variable angle spinning (MAS)   604
variable grating mode, external fields   496
vector matrix multiplier   798
vector processor   811
vertically aligned nematic displays (VAN)   741
viewing angles, nematic devices   744
virtual work, nematics   27
viscoelastic properties, NMR   611

viscosity   253 ff, 454–476
– diffusion   584
– external fields   490
– light scattering   704
– transitions under pressure   373
viscous coefficients
– fluctuations   381
– nematics   36
Vogel–Fulcher model, dielectric properties   249
voids, thermography   832
voltage breakdown, X rays   620
voltage–temperature phase diagram, 14P1M7   488
Volterra process, defects   418
volume changes, phase transitions   333
volume effects, Onsager theory   52
vortex formation, instabilities   518 f
vortex generator, thermography   829
vortex patterns, two-dimensional   523
Vuzhva theory, ultrasonic properties   561

walls, textures   439
wave interactions, nematics   559 f
wave propagation, Brillouin scattering   719
wavelength
– thermography   823
– XRD   636
wavelength tuneable filters/lasers   803
waveplates, SLM   765
weak anchoring
– elastic properties   260
– flow/viscosity   467
wedge disclination, defects   429
White–Taylor displays   740, 746
wide line separation experiment (WISE)   601
Wigner rotation matrices   196
Williams domains, displays   749
winglets, heat exchangers   829
Wittig reaction, synthesis   98
write/erase mechanism, displays   749
Wulf model, smectic A–C transition   67

X ray characterization, instrumentation   619–634
X ray diffraction   621 f
– structural studies   635–679
X ray measurements, high-pressure experiments   362

y expansion
– elastic properties   275
– hard-core energy   55
Ypma–Vertogen approach, nematics   55

Z-scan method, optical properties   576